Highlights of the LIFE Companion Website www.thelifewire.c

Free Access, no password required!

The MAIN MENU gives you quick access to all the site's content, either by chapter or by category.

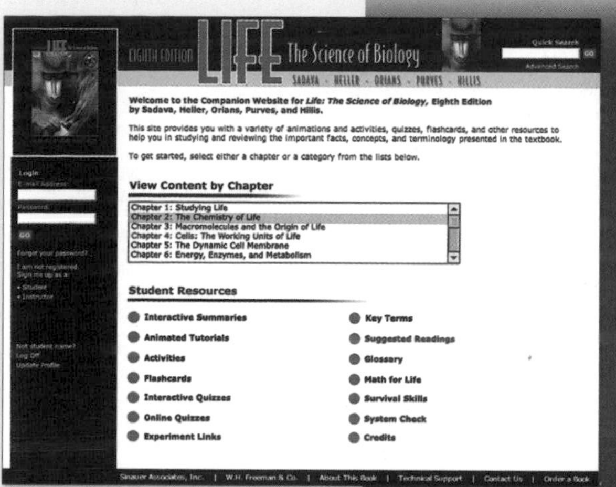

INTERACTIVE QUIZZES combine textbook figures, detailed answer feedback, and links to electronic versions of textbook pages for an integrated, self-directed review of important concepts in each chapter.

Subpopulations

7) This figure shows the distribution of subpopulations of the bay checkerspot butterfly population. This type of population structure has been maintained by

A. logistic growth.
B. density-independent factors.
C. density-dependent factors.
D. the rescue effect.
E. migration.

Well done! Choice D is correct.

The population as a whole consists of numerous subpopulations confined to suitable habitats and maintained by immigration of individuals between populations. This process is known as the rescue effect.

Textbook Review: Section 54.4 How Do Spatially Variable Environments Influence Population Dynamics? Figure 54.14

In-depth ANIMATED TUTORIALS present many of the dynamic processes and concepts of biology in an animated format. Each includes an Introduction, an Animation, a Conclusion, and a short Quiz.

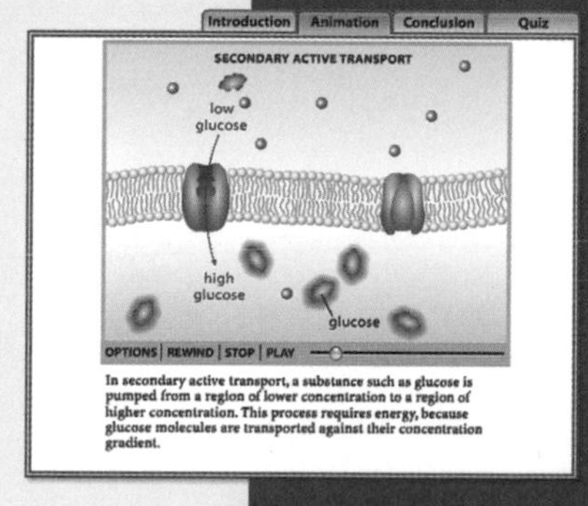

In secondary active transport, a substance such as glucose is pumped from a region of lower concentration to a region of higher concentration. This process requires energy, because glucose molecules are transported against their concentration gradient.

- A chapter-by-chapter list of these Tutorials and Activities starts on the next page.

A wide variety of interactive ACTIVITIES help you learn by doing: labeling parts of structures or steps in processes, identifying images, and completing diagrams.

Chapter 3: Macromolecules and the Origin of Life: Activity 3.4

Textbook Reference - Section 3.5 What Are the Chemical Structures and Functions of Nucleic Acids? Figure 3.24

Additional features of the LIFE Companion Website

- ONLINE QUIZZES test your knowledge of chapter content, and the results can be submitted electronically to your instructor.
- FLASHCARD ACTIVITIES help you master the large amount of new terminology introduced in the textbook. A full GLOSSARY WITH AUDIO pronunciations provides quick access to definitions of key terms.
- MATH FOR LIFE is a collection of handy mathematical shortcuts and references to help hone quantitative skills. STUDENT SURVIVAL SKILLS gives many practical tips on how to study biology most effectively.
- EXPERIMENT LINKS connect the experiments featured in the textbook to further research and applications.
- INTERACTIVE SUMMARIES provide a great way to review the entire chapter.
- The TREE OF LIFE lets you explore all the major groups of organisms and their relationships to each other.
- SUGGESTED READINGS for each chapter provide sources for further study.

Look for this icon throughout the book. It refers to the following Animated Tutorials and Activities on the Companion Website (www.thelifewire.com).

E indicates an Experiment Tutorial * indicates an activity that applies to an entire chapter

(Continued on inside back cover)

LIFE

The Science of Biology

EIGHTH EDITION

Sinauer Associates, Inc.

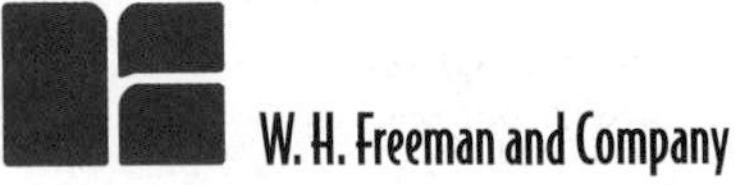
W. H. Freeman and Company

EIGHTH EDITION

LIFE The Science of Biology

DAVID SADAVA
The Claremont Colleges
Claremont, California

H. CRAIG HELLER
Stanford University
Stanford, California

GORDON H. ORIANS
Emeritus, University of Washington
Seattle, Washington

WILLIAM K. PURVES
Emeritus, Harvey Mudd College
Claremont, California

DAVID M. HILLIS
University of Texas
Austin, Texas

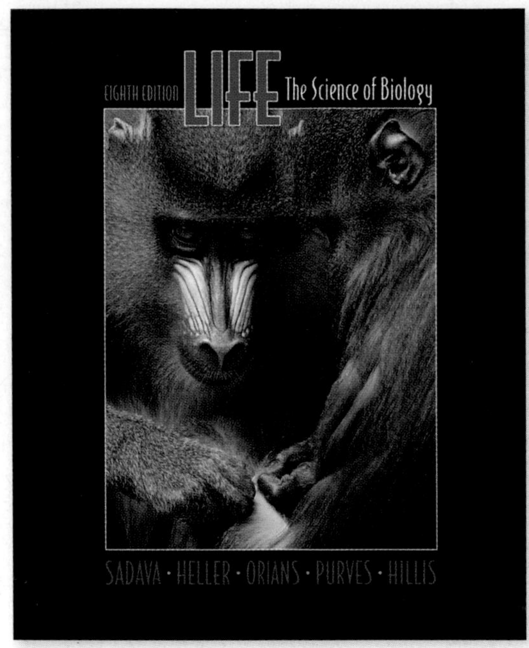

Cover Photograph
"Amor de Madre," photograph of mandrill (*Mandrillus sphinx*) mother and young.

Frontispiece
Grizzly bears (*Ursus arctos horribilis*) hunt salmon in an Alaskan river.

LIFE: The Science of Biology, Eighth Edition

Address editorial correspondence to:
Sinauer Associates Inc., 23 Plumtree Road, Sunderland, MA 01375 U.S.A.
www.sinauer.com
publish@sinauer.com

Address orders to:
VHPS/W.H. Freeman & Co., Order Dpt., 16365 James Madison Highway,
U.S. Route 15, Gordonsville, VA 22942 U.S.A.
www.whfreeman.com

Examination copy information: 1-800-446-8923
Orders: 1-888-330-8477

Library of Congress Cataloging-in-Publication Data
Life: the science of biology / David Sadava ... [et al.]. — 8th ed.
p. cm.
Includes index.
ISBN-13: 978-0-7167-7671-0 (hardcover) – ISBN 978-0-7167-7673-4 (Volume I) –
ISBN 978-0-7167-7674-1 (Volume 2) – ISBN 978-0-7167-7675-8 (Volume 3)
1. Biology. I. Sadava, David E.
QH308.2.L565 2007
570—dc22 2006031320

Printed in U.S.A.
First Printing December 2006
The Courier Companies, Inc.

To our students, especially the more than 30,000 we have collectively instructed in introductory biology over the years.

The Authors

Craig Heller Gordon Orians Bill Purves David Sadava David Hillis

David Sadava is the Pritzker Family Foundation Professor of Biology at the Keck Science Center of Claremont McKenna, Pitzer, and Scripps, three of The Claremont Colleges. Twice winner of the Huntoon Award for superior teaching, Dr. Sadava has taught courses on introductory biology, biotechnology, biochemistry, cell biology, molecular biology, plant biology, and cancer biology. He is a visiting scientist in medical oncology at the City of Hope Medical Center. He is the author or coauthor of five books on cell biology and on plants, genes, and crop biotechnology. His research has resulted in over 50 papers, many coauthored with undergraduates, on topics ranging from plant biochemistry to pharmacology of narcotic analgesics to human genetic diseases. For the past 15 years, he and his collaborators have investigated multi-drug resistance in human small-cell lung carcinoma cells with a view to understanding and overcoming this clinical challenge. Their current work focuses on new anti-cancer agents from plants.

Craig Heller is the Lorry I. Lokey/Business Wire Professor in Biological Sciences and Human Biology at Stanford University. He earned his Ph.D. from the Department of Biology at Yale University in 1970. Dr. Heller has taught in the core biology courses at Stanford since 1972 and served as Director of the Program in Human Biology, Chairman of the Biological Sciences Department, and Associate Dean of Research. Dr. Heller is a fellow of the American Association for the Advancement of Science and a recipient of the Walter J. Gores Award for excellence in teaching. His research is on the neurobiology of sleep and circadian rhythms, mammalian hibernation, the regulation of body temperature, and the physiology of human performance. Dr. Heller has done research on sleeping kangaroo rats, diving seals, hibernating bears, and exercising athletes. Some of his recent work on the effects of temperature on human performance is featured in the opener to Chapter 40.

Gordon Orians is Professor Emeritus of Biology at the University of Washington. He received his Ph.D. from the University of California, Berkeley in 1960 under Frank Pitelka. Dr. Orians has been elected to the National Academy of Sciences and the American Academy of Arts and Sciences, and is a Foreign Fellow of the Royal Netherlands Academy of Arts and Sciences. He was President of the Organization for Tropical Studies, 1988–1994, and President of the Ecological Society of America, 1995–1996. He is a recipient of the Distinguished Service Award of the American Institute of Biological Sciences. Dr. Orians is a leading authority in ecology, conservation biology, and evolution. His research on behavioral ecology, plant–herbivore interactions, community structure, and environmental policy has taken him to six continents. He now devotes full time to writing and to helping apply scientific information to environmental decision-making.

Bill Purves is Professor Emeritus of Biology as well as founder and former Chair of the Department of Biology at Harvey Mudd College in Claremont, California. He received his Ph.D. from Yale University in 1959 under Arthur Galston. A fellow of the American Association for the Advancement of Science, Dr. Purves has served as head of the Life Sciences Group at the University of Connecticut, Storrs, and as Chair of the Department of Biological Sciences, University of California, Santa Barbara, where he won the Harold J. Plous Award for teaching excellence. His research interests focused on the hormonal regulation of plant growth. Dr. Purves elected early retirement in 1995, after teaching introductory biology for 34 consecutive years, in order to concentrate entirely on research directed at learning and science education. He is currently participating in the development of a virtual technical high school, with responsibility for curriculum design in scientific reasoning and health science.

David Hillis is the Alfred W. Roark Centennial Professor in Integrative Biology and the Director of the Center for Computational Biology and Bioinformatics at the University of Texas at Austin, where he also has directed the School of Biological Sciences. Dr. Hillis has taught courses in introductory biology, genetics, evolution, systematics, and biodiversity. He has been elected into the membership of the American Academy of Arts and Sciences, awarded a John D. and Catherine T. MacArthur Fellowship, and has served as President of the Society for the Study of Evolution and of the Society of Systematic Biologists. His research interests span much of evolutionary biology, including experimental studies of evolving viruses, empirical studies of natural molecular evolution, applications of phylogenetics, analyses of biodiversity, and evolutionary modeling. He is particularly interested in teaching and research about the practical applications of evolutionary biology.

Preface

As active scientists working in a wide variety of both basic and applied biology, we are fortunate to be part of a field that is not only fascinating but also changes rapidly. It is apparent not just in the time span since we started our careers—we see it every day when we open a newspaper or a scientific journal. As educators of both introductory and advanced-level students, we desire to convey our excitement about biology's dynamic nature.

This new edition of *Life* looks, and is, quite different from its predecessors. In planning the Eighth Edition, we focused on three fundamental goals. The first was to maintain and enhance what has worked well in the past—an emphasis on not just what we know but how we came to know it; the incorporation of exciting new discoveries; an art program distinguished by its beauty and clarity; plus a unifying theme. As should be the case in any biology textbook, that theme is evolution by natural selection, a 150-year-old idea that more than ever ties together the living world. We have been greatly helped in this endeavor by the addition of a new author, David Hillis. His knowledge and insights have been invaluable in developing our chapters on evolution, phylogeny, and diversity, and they permeate the rest of the book as well.

Our second goal has been to make *Life* more pedagogically accessible. From the bold new design to the inclusion of numerous learning aids throughout each chapter (see New Pedagogical Features), we have worked to make our writing consistently easy to follow as well as engaging.

Third, between editions we asked seven distinguished ecologists—all of whom teach introductory biology—to provide detailed critiques of the Ecology unit. As a result of their extensive suggestions, Part Nine, Ecology, has a fresh organization (see The Nine Parts). And one of the seven, May Berenbaum, has agreed to join the *Life* author team for the Ninth Edition. The other six stalwarts are thanked in the "Reviewers of the Eighth Edition" section.

Enduring Features

As stated above, we are committed to a blending of a presentation of the core ideas of biology with an emphasis on introducing our readers to the process of scientific inquiry. Having pioneered the idea of depicting seminal experiments in specially designed figures, we continue to develop this here, with 96 **Experiment** figures (28 percent more than in the Seventh Edition). Each follows the structure: Hypothesis, Method, Result, and Conclusion. Many now include "Further Research," which asks students to conceive an experiment that explores a related question.

A related feature is the **Research Method** figures, depicting many laboratory and field methods used to do this research. All the Experiment and Research Method figures are listed in the endpapers at the back of the book.

Another much-praised feature—which we pioneered ten years ago in *Life*'s Fifth Edition—is the **balloon captions** used in our figures. We know that many students are visual learners. The balloon captions bring explanations of intricate, complex processes directly into the illustration, allowing the reader to integrate the information without repeatedly going back and forth between the figure and its legend.

Life is the only introductory biology book for science majors that begins each chapter with a story. These **Opening Stories**, most of which are new to this edition, are meant to intrigue students while helping them see how the chapter's biological subject relates to the world around them.

New Pedagogical Features

There are several new elements in the Eighth Edition chapters. Each has been designed as a study tool to aid the student in mastering the material. In the opening page spread, **In This Chapter** previews the chapter's content, and the **Chapter Outline** gives the major section headings, all numbered and all framed as questions to emphasize the inquiry basis of science.

Each main section of a chapter now ends with a **Recap**. This key element briefly summarizes the important concepts in the section, then provides two or three questions to stimulate immediate review. Each question includes reference to pertinent text or a figure or both.

The **Chapter Summary** boldfaces key terms introduced and defined in the chapter. We have kept the highlighted references to key figures and to the Web tutorials and activities that support a topic in the chapter.

Another new element, **BioBits**, is not strictly a learning aid, but offers intriguing (occasionally amusing) supplemental information. BioBits, like the opening stories, are intended to help students appreciate the interface between biology and other aspects of life.

Redesigned **Web Icons** alert the reader to the tutorials and activities on *Life*'s companion website (www.thelifewire.com). Each of these study and review resources, many of which are new for the Eighth Edition, has been created specifically for *Life*. A full list, by chapter, is found in the front endpapers of the book.

The Nine Parts

We have reorganized the book into nine parts. Part One sets the stage for the entire book, with the opening chapter on biology as an exciting science, starting with a student project, and how evolution unites the living world. This is followed by chapters on the basic chemical building blocks that underlie life. We have tried to tie this material together by relating it to theories on the origin of life, with new discoveries of water in our solar system as an impetus.

In Part Two, Cells and Energy, we present an integrated view of the structure and biochemical functions of cells. The discussions of biochemistry are often challenging for students; thus we have reworked both the text and illustrations for greater clarity. These discussions are presented in the context of the latest discoveries on the origin of life and evolution of cells.

Part Three, Heredity and the Genome, begins with continuity at the cellular level, and then outlines the principles of genetics and the identification of DNA as the genetic material. New examples, such as the genetics of coat color in dogs, enliven these chapters. This is followed by chapters on gene expression and on the prokaryotic and eukaryotic genomes. Many new discoveries have been made in this new field of genomics, ranging from tracking down the bird flu virus to the genomes of wild cats, such as the cheetah.

Part Four, Molecular Biology: The Genome in Action, reinforces the basic principles of classical and molecular genetics by applying them to such diverse topics as cell signaling, biotechnology, and medicine. We use many new experiments and examples from applied biology to illustrate these concepts. These include the latest information on the human genome and the emerging field of systems biology. The chapter on natural defenses now includes a discussion of allergy.

Part Five, The Patterns and Processes of Evolution, has been updated in several important ways. We have emphasized the importance of evolutionary biology as a basis for comparing and understanding all aspects of biology, and have described numerous practical applications of evolutionary biology that will be familiar and relevant to the everyday lives of most students.

Recent experimental studies of evolution are described and explained, to help students understand that evolution is an ongoing, observable process. The chapters on phylogenetics and molecular evolution have been completely rewritten to reflect recent advances in those fields. Other changes reflect our growing knowledge of the history of life on Earth and the mechanisms of evolution that have given rise to all of biodiversity.

Part Six, The Evolution of Diversity, reflects the latest views on phylogeny. It continues to emphasize groups united by evolutionary history over classically defined taxa. This emphasis is now supported by an appendix on the Tree of Life that clearly maps out and describes all groups discussed in the text, so that students can quickly look up unfamiliar names and see how they fit into the larger context of life. We now discuss aspects of phylogeny that are still under study or debate (among major groups of eukaryotes, plants, and animals, for instance).

In Part Seven, Flowering Plants: Form and Function, we report on several exciting new discoveries. These include the receptors for auxin, gibberellins, and brassinosteroids as well as great progress on the florigen problem. We have updated our treatment of signal transduction pathways and of circadian rhythms in plants. The already strong treatment of environmental challenges to plants has been augmented by new Experiment figures on plant defenses against herbivores, one confirming that nicotine does help tobacco plants resist certain insects.

Part Eight, Animals: Form and Function, is about how animals work. Although we give major attention to human physiology, we embed it in a background of comparative animal physiology. Our focus is systems physiology but we also introduce the underlying cellular and molecular mechanisms. For example, our explanations of nervous system phenomena—whether they be action potentials, sensation, learning, or sleep—are discussed in terms of the properties of ion channels. The actions of hormones are explained in terms of the molecular mechanisms. Maximum athletic performance is explained in terms of the underlying cellular energy systems. Throughout Part Eight we try to help the student make the connections across all levels of biology, from molecular to behavioral, and to see the relevance of physiology to issues of health and disease. Of central importance in each chapter is mechanisms of control and regulation.

Part Nine, Ecology, begins with a new chapter that describes the scope of ecological research and discusses recent advances in our understanding of the broad patterns in the distribution of life on Earth. The next chapter, also new, combines Behavior and Behavioral Ecology. It shows how the decisions that organisms make during their lives influence both their survival and reproductive success, and also the dynamics of populations and the structure of ecological communities. The chapter on Population Ecology has new material that explains how ecologists are able to mark and follow individual organisms in the wild to determine their survival and reproductive success. Following a chapter on Community Ecology, another new chapter, Ecosystems and Global Ecology, shows how ecologists are expanding the scope of their studies to encompass the functioning of the global ecosystem. This discussion leads naturally to the final chapter in the book, Conservation Biology, which describes how ecologists and conservation biologists work to reduce the rate at which species are becoming extinct as a result of human activities.

Full Books, Paperbacks, or Loose-Leaf

We again provide *Life* both as the full book and as a cluster of paperbacks. Thus, instructors who want to use less than the whole book, or who want their students to have more portable units, can choose from these split volumes:

Volume I, The Cell and Heredity, includes: Part One, The Science and Building Blocks of Life (Chapters 1–3); Part Two, Cells and Energy (Chapters 4–8); Part Three, Heredity and the Genome (Chapters 9–14); and Part Four, Molecular Biology: The Genome in Action (Chapters 15–20).

Volume II, Evolution, Diversity, and Ecology, includes: Chapter 1, Studying Life; Part Five, The Patterns and Processes of Evolution (Chapters 21–25); Part Six, The Evolution of Diversity (Chapters 26–33); and Part Nine, Ecology (Chapters 52–57).

Volume III, Plants and Animals, includes: Chapter 1, Studying Life; Part Seven, Flowering Plants: Form and Function (Chapters 34–39); and Part Eight, Animals: Form and Function (Chapters 40–51).

Note that each volume also includes the book's front matter, Appendixes, Glossary, and Index.

Life is also available in a loose-leaf version. This shrink-wrapped, unbound, 3-hole punched version is designed to fit into a 3-ring binder. Students take only what they need to class and can easily integrate any instructor handouts or other resources.

Media and Supplements for the Eighth Edition

The media and supplements for *Life,* Eighth Edition have been assembled with two main goals in mind: (1) to provide students with a collection of tools that helps them effectively master the vast amount of new information that is being presented to them in the introductory biology course; and (2) to provide instructors with the richest possible collection of resources to aid in teaching the course—preparing, presenting the lecture, providing course materials online, and assessing student comprehension.

All of the *Life* media and supplemental resources have been developed specifically for this textbook. This gives the student the greatest degree of consistency when studying across different media. For example, the animated tutorials and activities found on the Companion Website were built using textbook art, so that the manner in which structures are illustrated, the colors used to identify objects, and the terms and abbreviations used are all consistent.

The rich collection of visual resources in the Instructor's Media Library provides instructors with a wide range of options for enhancing lectures, course websites, and assignments. Highlights include: layered art PowerPoint® presentations that break down complex figures into detailed, step-by-step presentations; a collection of approximately 200 video segments that can help capture the attention and imagination of students; and the new set of PowerPoint® slides of textbook art with editable labels and leaders that allow easy customization of the figures.

For a detailed description of all the media and supplements available to accompany the Eighth Edition, please turn to "Life's Media and Supplements package" on page xiii.

Many People to Thank

One of the wisest pieces of advice ever given to a textbook author is to "be passionate about your subject, but don't put your ego on the page." Considering all the people who looked over our shoulders throughout the process of creating this book, this advice could not be more apt. We are indebted to many people, who gave invaluable help to make this book what it is. First and foremost are our colleagues, biologists from over 100 institutions. Some were users of the previous edition, who suggested many improvements. Others reviewed our chapter drafts in detail, including advice on how to improve the illustrations. Still others acted as accuracy reviewers when the book was almost completed. Our publishers created an advisory group of introductory course coordinators. They advised us on a variety of issues, ranging from book content and design to elements of the print and media supplements. All of these biologists are listed in the Reviewer credits.

We needed a fresh editorial eye for this edition, and we were fortunate to work with Carol Pritchard-Martinez as development editor. With a level head that comes from years of experience, she was a major presence as we wrote and revised. Elizabeth Morales, our artist, was on her second edition with us. This time, she extensively revised almost all of the prior art and translated our crude sketches into beautiful new art. We hope you agree that our art program remains superbly clear and elegant. Once again, we were lucky to have Norma Roche as the copy editor. Her firm hand and encyclopedic recall of our book's many chapters made our prose sharper and more accurate. For this edition, Norma was joined by the capable and affable Maggie Brown. Susan McGlew coordinated the hundreds of reviews that we described above. David McIntyre was a truly proactive photo editor. Not only did he find over 500 new photographs, including many new ones of his own, that enrich the book's content and visual statement, but he set up, performed, and photographed the experiment shown in Figure 36.1. The elegant new interior design is the creation of Jeff Johnson. He also coordinated the book's layout and designed the cover. Carol Wigg, for the eighth time in eight editions, oversaw the editorial process. Her influence pervades the entire book—she created many BioBits, shaped and improved the chapter-opening stories, interacted with David McIntyre in conceiving many photo subjects, and kept an eagle eye on every detail of text, art, and photographs.

W. H. Freeman continues to bring *Life* to a wider audience. Associate Director of Marketing Debbie Clare, the Regional Specialists, Regional Managers, and experienced sales force are effective ambassadors and skillful transmitters of the features and unique strengths of our book. We depend on their expertise and energy to keep us in touch with how *Life* is perceived by its users.

Finally, we are indebted to Andy Sinauer. Like ours, his name is on the cover of the book, and he truly cares deeply about what goes into it.

David Sadava

Craig Heller

Gordon Orians

Bill Purves

David Hillis

Reviewers for the Eighth Edition

Between-Edition Reviewers (Ecology and Animal Parts)

May Berenbaum, University of Illinois, Urbana-Champaign
Carol Boggs, Stanford University
Judie Bronstein, University of Arizona
F. Lynn Carpenter, University of California, Irvine
Dan Doak, University of California, Santa Cruz
Jessica Gurevitch, SUNY, Stony Brook
Margaret Palmer, University of Maryland
Marty Shankland, University of Texas, Austin

Advisory Board Members

Heather Addy, University of Calgary
Art Buikema, Virginia Polytechnic Institute and State University
Jung Choi, Georgia Technical University
Rolf Christoffersen, University of California, Santa Barbara
Alison Cleveland, Florida Southern University
Mark Decker, University of Minnesota
Ernie Dubrul, University of Toledo
Richard Hallick, University of Arizona
John Merrill, Michigan State University
Melissa Michael, University of Illinois
Deb Pires, University of California, Los Angeles
Sharon Rogers, University of Nevada, Las Vegas
Marty Shankland, University of Texas, Austin

Manuscript Reviewers

John Alcock, Arizona State University
Charles Baer, University of Florida
Amy Baird, University of Texas, Austin
Patrice Boily, University of New Orleans
Thomas Boyle, University of Massachusetts, Amherst
Mirjana Brockett, Georgia Institute of Technology
Arthur Buikema, Virginia Polytechnic Institute and State University
Hilary Callahan, Barnard College
David Champlin, University of Southern Maine
Chris Chanway, University of British Columbia
Mike Chao, California State University, San Bernardino
Rhonda Clark, University of Calgary
Elizabeth Connor, University of Massachusetts, Amherst
Deborah A. Cook, Clark Atlanta University
Elizabeth A. Cowles, Eastern Connecticut State University
Joseph R. Cowles, Virginia Polytechnic Institute and State University
William L. Crepet, Cornell University
Martin Crozier, Wayne State University
Donald Dearborn, Bucknell University
Mark Decker, University of Minnesota
Michael Denbow, Virginia Polytechnic Institute and State University
Jean DeSaix, University of North Carolina, Chapel Hill
William Eldred, Boston University
Andy Ellington, University of Texas, Austin
Gordon L. Fain, University of California, Los Angeles
Kevin M. Folta, University of Florida
Miriam Goldbert, College of the Canyons
Kenneth M. Halanych, Auburn University
Susan Han, University of Massachusetts, Amherst
Tracy Heath, University of Texas, Austin
Shannon Hedtke, University of Texas, Austin
Mark Hens, University of North Carolina, Greensboro
Albert Herrera, University of Southern California
Barbara Hetrich, University of Northern Iowa
Erec Hillis, University of California, Berkeley
Jonathan Hillis, Austin, Texas
Hopi Hoekstra, University of California, San Diego
Kelly Hogan, University of North Carolina, Chapel Hill
Carl Hopkins, Cornell University
Andrew Jarosz, Michigan State University
Norman Johnson, University of Massachusetts, Amherst
Walter Judd, University of Florida
David Julian, University of Florida
Laura Katz, Smith College
Melissa Kosinski-Collins, Massachusetts Institute of Technology
William Kroll, Loyola University of Chicago
Marc Kubasak, University of California, Los Angeles
Josephine Kurdziel, University of Michigan
John Latto, University of California, Berkeley
Brian Leander, University of British Columbia
Jennifer Leavey, Georgia Institute of Technology
Arne Lekven, Texas A&M University
Don Levin, University of Texas, Austin
Rachel Levin, Amherst College
Thomas Lonergan, University of New Orleans
Blase Maffia, University of Miami
Meredith Mahoney, University of Texas, Austin
Charles Mallery, University of Miami
Ron Markle, Northern Arizona University
Mike Meighan, University of California, Berkeley
Melissa Michael, University of Illinois, Urbana-Champaign
Jill Miller, Amherst College
Subhash Minocha, University of New Hampshire
Thomas W. Moon, University of Ottawa
Richard Moore, Miami University of Ohio
John Morrissey, Hofstra University
Leonie Moyle, University of Indiana
Mary Anne Nelson, University of New Mexico
Dennis O'Connor, University of Maryland, College Park
Robert Osuna, SUNY, Albany
Cynthia Paszkowski, University of Alberta
Diane Pataki, University of California, Irvine
Ron Patterson, Michigan State University
Craig Peebles, University of Pittsburgh
Debra Pires, University of California, Los Angeles
Greg Podgorski, Utah State University
Chuck Polson, Florida Institute of Technology

Donald Potts, University of California, Santa Cruz
Jill Raymond, Rock Valley College
Ken Robinson, Purdue University
Sharon L. Rogers, University of Nevada, Las Vegas
Laura Romano, Denison University
Pete Ruben, Utah State University, Logan
Albert Ruesink, Indiana University
Walter Sakai, Santa Monica College
Mary Alice Schaeffer, Virginia Polytechnic Institute and State University
Daniel Scheirer, Northeastern University
Stylianos Scordilis, Smith College
Kevin Scott, University of Calgary
Jim Shinkle, Trinity University
Denise Signorelli, Community College of Southern Nevada
Thomas Silva, Cornell University
Jeffrey Tamplin, University of Northern Iowa
Steve Theg, University of California, Davis
Sharon Thoma, University of Wisconsin, Madison
Jeff Thomas, University of California, Los Angeles
Christopher Todd, University of Saskatchewan
John True, SUNY, Stony Brook
Mary Tyler, University of Maine
Fred Wasserman, Boston University
John Weishampel, University of Central Florida
Elizabeth Willott, University of Arizona
David Wilson, University of Miami
Heather Wilson-Ashworth, Utah Valley State College

Accuracy Reviewers

John Alcock, Arizona State University
John Anderson, University of Minnesota
Brian Bagatto, University of Akron
Lisa Baird, University of San Diego
May Berenbaum, University of Illinois, Urbana-Champaign
Gerald Bergtrom, University of Wisconsin, Milwaukee
Stewart Berlocher, University of Illinois, Urbana-Champaign
Mary Bisson, SUNY, Buffalo
Arnold Bloom, University of California, Davis
Judie Bronstein, University of Arizona
Jorge Busciglio, University of California, Irvine
Steve Carr, Memorial University of Newfoundland
Thomas Chen, Santa Monica College
Randy Cohen, California State University, Northridge
Reid Compton, University of Maryland, College Park
James Courtright, Marquette University
Jerry Coyne, University of Chicago
Joel Cracraft, American Museum of Natural History
Joseph Crivello, University of Connecticut, Storrs
Gerrit De Boer, University of Kansas, Lawrence
Arturo DeLozanne, University of Texas, Austin
Stephen Devoto, Wesleyan University
Laura DiCaprio, Ohio University
John Dighton, Rutgers Pinelands Field Station
Jocelyne DiRuggiero, University of Maryland, College Park
W. Ford Doolittle, Dalhousie University
Emanuel Epstein, University of California, Davis
Gordon L. Fain, University of California, Los Angeles
Lewis J. Feldman, University of California, Berkeley
James Ferraro, Southern Illinois University
Cole Gilbert, Cornell University
Elizabeth Godrick, Boston University
Martha Groom, University of Washington
Kenneth M. Halanych, Auburn University
Mike Hasegawa, Purdue University
Mark Hens, University of North Carolina, Greensboro
Richard Hill, Michigan State University
Franz Hoffman, University of California, Irvine
Sara Hoot, University of Wisconsin, Milwaukee
Carl Hopkins, Cornell University
Alfredo Huerta, Miami University
Michael Ibba, The Ohio State University
Walter Judd, University of Florida
Laura Katz, Smith College
Manfred D. Laubichler, Arizona State University
Brian Leander, University of British Columbia
Mark V. Lomolino, SUNY College of Environmental Science and Forestry
Jim Lorenzen, University of Idaho
Denis Maxwell, University of Western Ontario
Brad Mehrtens, University of Illinois, Urbana-Champaign
John Merrill, Michigan State University
Allison Miller, Saint Louis University
Clara Moore, Franklin and Marshall College
Julie Noor, Duke University
Mohamed Noor, Duke University
Theresa O'Halloran, University of Texas, Austin
Norman R. Pace, University of Colorado
Randall Packer, George Washington University
Walt Ream, Oregon State University
Eric Richards, Washington University
Steve Rissing, The Ohio State University
R. Michael Roberts, University of Missouri, Columbia
Pete Ruben, Simon Fraser University
David A. Sanders, Purdue University
Mike Sanderson, University of California, Davis
Marty Shankland, University of Texas, Austin
Jeff Silberman, University of Arkansas
Margaret Silliker, DePaul University
Dee Silverthorn, University of Texas, Austin
M. Suzanne Simon-Westendorf, Ohio University
Alastair G.B. Simpson, Dalhousie University
John Skillman, California State University, San Bernardino
Frederick W. Spiegel, University of Arkansas
John J. Stachowicz, University of California, Davis
Heven Sze, University of Maryland
E.G. Robert Turgeon, Cornell University
Mary Tyler, University of Maine
Mike Wade, Indiana University
Leslie Winemiller, Texas A&M University
Mimi Zolan, Indiana University

Tree of Life Appendix Reviewers

John Abbott, University of Texas, Austin
Joseph Bischoff, National Center for Biotechnology Information
Ruth Buskirk, University of Texas, Austin
David Cannatella, University of Texas
Joel Cracraft, American Museum of Natural History
Scott Federhen, National Center for Biotechnology Information
Carol Hotton, National Center for Biotechnology Information
Robert Jansen, University of Texas, Austin
Brian Leander, University of British Columbia
Detlef Leipe, National Center for Biotechnology Information
Beryl Simpson, University of Texas, Austin
Richard Sternberg, National Center for Biotechnology Information
Edward Theriot, University of Texas
Sean Turner, National Center for Biotechnology Information

Supplements Authors

Dany Adams, The Forsyth Institute
Erica Bergquist, Holyoke Community College
Ian Craine, University of Toronto
Ernest Dubrul, University of Toledo
Edward Dzialowski, University of North Texas
Donna Francis, University of Massachusetts, Amherst
Jon Glase, Cornell University
Lindsay Goodloe, Cornell University
Celine Muis Griffin, Queen's University
Nancy Guild, University of Colorado at Boulder
Norman Johnson, University of Massachusetts, Amherst
James Knapp, Holyoke Community College
Jennifer Knight, University of Colorado, Boulder
David Kurjiaka, University of Arizona
Richard McCarty, Johns Hopkins University
Betty McGuire, Cornell University
Nancy Murray, Evergreen State College
Deb Pires, University of California, Los Angeles
Catherine Ueckert, Northern Arizona University
Jerry Waldvogel, Clemson University

LIFE's Media and Supplements Package

For the Student

Companion Website www.thelifewire.com

(Also available as a CD, optionally packaged with the book)

The *Life*, Eighth Edition Companion Website is available free of charge to all students (no access code required). The site features a variety of study and review resources designed to help students master the wide range of material presented in the introductory biology course. Features of the site include:

- *Interactive Summaries.* These summaries combine a review of important concepts with links to all the key figures from the chapter as well as all of the relevant animated tutorials and activities.
- *Animated Tutorials.* Over 100 in-depth animated tutorials present complex topics in a clear, easy-to-follow format that combines a detailed animation with an introduction, conclusion, and quiz.
- *Activities.* Over 120 interactive activities help the student learn important facts and concepts through a wide range of activities, such as labeling steps in processes or parts of structures, building diagrams, and identifying different types of organisms.
- *Flashcards.* For each chapter of the book, there is a set of flashcards that allows the student to review all the key terminology from the chapter. Students can review the terms in the study mode, and then quiz themselves on a list of terms.
- *New! Experiment Links.* New for the Eighth Edition, each experiment featured in the textbook has a corresponding treatment on the companion website that links to further information about the experiment, further research that followed, and applications derived from the research.
- *Interactive Quizzes.* Every question includes an image taken from the textbook, thorough feedback on both right and wrong answer choices, references to textbook pages, and links to electronic versions of book pages, where the related material is highlighted.
- *Online Quizzes.* These quizzes test the student's comprehension of the chapter material, and the results are stored in the online gradebook.
- *Key Terms.* The key terminology introduced in each chapter is listed, with definitions and audio pronunciations from the Glossary.
- *Suggested Readings.* For each chapter of the book, a list of suggested readings is provided as a resource for further study.
- *Glossary.* The language of biology is often difficult for students taking introductory biology, so we have created a full glossary with audio pronunciations.
- *Math for Life* (Dany Adams, *The Forsyth Institute*). *Math for Life* is a collection of mathematical shortcuts and references to help students with the quantitative skills they need in the laboratory.
- *Survival Skills* (Jerry Waldvogel, *Clemson University*). *Survival Skills* is a guide to more effective study habits. Topics include time management, note-taking, effective highlighting, and exam preparation.

Study Guide (ISBN 978-0-7167-7893-6)

Edward M. Dzialowski, *University of North Texas*; Jon Glase, *Cornell University*; Lindsay Goodloe, *Cornell University*; Nancy Guild, *University of Colorado*; and Betty McGuire, *Smith College*

For each chapter of the textbook, the *Life* Study Guide offers a variety of study and review tools. The contents of each chapter are broken down into both a detailed review of the Important Concepts covered and a boiled-down Big Picture snapshot. In addition, Common Problem Areas and Study Strategies are highlighted. A set of study questions (both multiple-choice and short-answer) allows students to test their comprehension. All questions include answers and explanations.

Lecture Notebook (ISBN 978-0-7167-7894-3)

This invaluable printed resource consists of all the artwork from the textbook (more than 1,000 images with labels) presented in the order in which they appear in the text, with ample space for note-taking. Because the Notebook has already done the drawing, students can focus more of their attention on the concepts. They will absorb the material more efficiently during class, and their notes will be clearer, more accurate, and more useful when they study from them later.

MCAT® Practice Test (ISBN 0-7167-5907-1)

A complete printed MCAT exam, with answers, allows students to test their knowledge of the full range of introductory biology content as they prepare for the medical school entrance exams.

CatchUp Math & Stats

Michael Harris, Gordon Taylor, and Jacquelyn Taylor

This primer will help your students quickly brush up on the quantitative skills they need to succeed in biology. Presented in brief, accessible units, the book covers topics such as working with powers, logarithms, using and understanding graphs, calculating standard deviation, preparing a dilution series, choosing the right statistical test, analyzing enzyme kinetics, and many more.

Student Handbook for Writing in Biology, Second Edition

Karen Knisely (ISBN 0-7167-6709-0)

This book provides practical advice to students who are learning to write according to the conventions in biology. Using the standards of journal publication as a model, the author provides, in a user-friendly format, specific instructions on: using biology databases to locate references; paraphrasing for improved comprehension; preparing lab reports, scientific papers, posters; preparing oral presentations in PowerPoint®, and more.

Bioethics and the New Embryology: Springboards for Debate

Scott F. Gilbert, Anna Tyler, and Emily Zackin (ISBN 0-7167-7345-7)

Our ability to alter the course of human development ranks among the most significant changes in modern science and has brought embryology into the public domain. The question that must be asked is: Even if we *can* do such things, *should* we do such things?

BioStats Basics: A Student Handbook

James L. Gould and Grant F. Gould (ISBN 0-7167-3416-8)

BioStats Basics provides introductory-level biology students with a practical, accessible introduction to statistical research. Engaging and informal, the book avoids excessive theoretical and mathematical detail to focus on how core statistical methods are put to work in biology.

Laboratory Manuals

W. H. Freeman publishes a range of high-quality biology lab texts, all of which are available for bundling with *Life,* Eighth Edition. Our laboratory texts are available as complete paperback texts, or as Freeman Laboratory Separates.

- *Biology in the Laboratory,* Third Edition
  Doris R. Helms, Carl W. Helms, Robert J. Kosinski, and John C. Cummings (ISBN 0-7167-3146-0)
- *Laboratory Outlines in Biology-VI*
  Peter Abramoff and Robert G. Thomson (ISBN 0-7167-2633-5)
- *Anatomy and Dissection of the Frog,* Second Edition
  Warren F. Walker, Jr. (ISBN 0-7167-2636-X)
- *Anatomy and Dissection of the Rat,* Third Edition
  Warren F. Walker, Jr. and Dominique Homberger (ISBN 0-7167-2635-1)
- *Anatomy and Dissection of the Fetal Pig,* Fifth Edition
  Warren F. Walker, Jr. and Dominique Homberger (ISBN 0-7167-2637-8)
- *Atlas and Dissection Guide for Comparative Anatomy,* Sixth Edition
  Saul Wischnitzer and Edith Wischnitzer (ISBN 0-7167-6959-X)

Custom Publishing for Laboratory Manuals

http://custompub.whfreeman.com

Instructors can build and order customized lab manuals in just minutes, choosing material from Freeman's biology laboratory manuals, as well as their own material.

For the Instructor

Instructor's Media Library

In order to give you the widest possible range of resources to help engage students and better communicate the material, we have assembled an unparalleled collection of media resources. The Eighth Edition of *Life* features an expanded Instructor's Media Library (available on a set of CDs and DVDs) that includes:

- *Textbook Figures and Tables.* Every image from the textbook is provided in both JPEG (high- and low-resolution) and PDF formats.
- *Unlabeled Figures.* Every figure in the textbook is provided in an unlabeled format. These are useful for student quizzing and custom presentation development.
- *Supplemental Photos.* The supplemental photograph collection contains over 1,500 photographs (all in addition to those in the text), forming a rich resource of visual imagery.
- *Animations.* A collection of over 100 in-depth animations, all of which were created from the textbook's art program, and which can be viewed in either narrated or step-through mode.
- *Videos.* This collection of approximately 200 video segments covering topics across the entire textbook helps demonstrate the complexity and beauty of life.
- *PowerPoint® Resources.* For each chapter of the textbook, we have created several different types of PowerPoint® presentations. These give instructors the flexibility to build a presentation in the manner that best suits their needs. Included are:
  - Figures and Tables
  - Lecture Presentation
  - New! Editable Labels
  - Layered Art
  - Supplemental Photos
  - Videos, Animations
- *Clicker Questions.* A set of questions written specifically to be used with classroom personal response systems ("clickers") is provided for each chapter. These questions are designed to reinforce concepts, gauge student comprehension, and provide an outlet for active participation.
- *Chapter Outlines, Lecture Notes,* and the complete *Test File* are all available in Microsoft Word® format for easy use in lecture and exam preparation.
- An intuitive *Browser Interface* provides a quick and easy way to preview all of the content on the Instructor's Media Library.
- *Computerized Test Bank.* The entire printed Test File, plus the textbook end-of-chapter Self-Quizzes, the Companion Website Online Quizzes, and the Study Guide questions are all included in Brownstone's easy-to-use Diploma® software.
- *Instructor's Website.* A wealth of instructor's media, as well as electronic versions of other instructor supplements, are available online for instant access anytime.
- *Online Quizzes.* The Companion Website includes an Online Quiz for each chapter of the textbook. Instructors can choose to use these quizzes as assignments, and can view the results in the online gradebook.

- *Course Management System Support.* As a service for adopters using WebCT, Blackboard, or ANGEL for their courses, full electronic course packs are available.

Instructor's Resource Kit

The *Life,* Eighth Edition Instructor's Resource Kit includes a wealth of information to help instructors in the planning and teaching of their course. The Kit includes:

- *Instructor's Manual,* featuring:
  - A "What's New" guide to the Eighth Edition
  - A brief chapter overview
  - A key terms section with all the boldface terms from the text
  - Chapter outlines
- *Lecture Notes*—detailed notes for each chapter that can serve as the basis for lectures, including references to figures and media resources.
- *Media Guide*—A visual guide to the extensive media resources available with the Eighth Edition of *Life*. The guide includes thumbnails and descriptions of every video, animation, PowerPoint®, and supplemental photo in the Media Library, all organized by chapter.
- *Lab manual* and custom lab manual information.

Overhead Transparencies

This set includes over 1,000 transparencies—all the four-color line art and all the tables from the text—along with convenient binders. Balloon captions have been removed and colors have been enhanced for clear projection in a wide range of conditions. Labels and images have been resized for improved readability.

Test File

Ernest Dubrul, *University of Toledo;* Jon Glase, *Cornell University;* Norman Johnson, *University of Massachusetts;* Catherine Ueckert, *Northern Arizona University*

The test file offers more than 5,000 questions, including fill-in-the-blank and multiple-choice test questions. The electronic version of the Test File also includes all of the textbook end-of-chapter Self-Quiz questions, all of the Student Website Online Quiz questions, and all of the Study Guide questions.

iclicker

Developed for educators by educators, iclicker is a hassle-free radio-frequency classroom response system that makes it easy for instructors to ask questions, record responses, take attendance, and direct students through lectures as active participants. For more information, visit www.iclicker.com.

The *Life* eBook

The *Life,* Eighth Edition eBook is a complete online version of the textbook that can be purchased online, or packaged with the printed textbook. This online version of *Life* is a substantially less expensive alternative that gives your students an efficient and rich learning experience by integrating all of the resources from the companion website directly into the eBook text. In addition, the eBook offers instructors unique opportunities to customize the text with their own content. Key features of the eBook include:

- *For Students:*
  - Integration of all website **activities** and **animated tutorials**
  - In-text **self quiz questions**
  - Interactive **summary exercises**
  - Custom text **highlighting**
  - A **notes** feature that allows students to annotate the text
  - Complete **glossary**, **index**, and **full-text search** features
- *For Instructors:*
  - A powerful notes feature that can incorporate text, Web links, and documents directly into the text
  - Easy integration of instructor media resources, including videos and supplemental photographs
  - Ability to link directly to any eBook page from other sites

BIOPORTAL

New for the Eighth Edition, BioPortal is the digital gateway to all of the teaching and learning resources that are available with *Life.* BioPortal integrates the *Life* textbook, all of the student and instructor media, extensive assessment resources, and course planning resources all into a powerful and easy-to-use learning management system. All of this means that your students get easy access to learning resources, presented in the proper context and at the proper time, and you get a complete learning management system, ready to use, without hours of prep work. Features of BioPortal include:

- *eBook*
  - Completely integrated with all media resources
  - Customizable with notes, sections, images, and more
- *Student Resources*
  - Animated Tutorials and Activities
  - Assessment: Online Quizzes and Interactive Quizzes
- *Instructor Resources*
  - Complete Test Bank
  - All quizzes
  - Media resources: Videos, PowerPoints, Supplemental Photos, and more
- *Assignments*
  - Quizzes and exams
  - Assignable textbook sections
  - Assignable animations and activities
  - Custom assignments
- *Easy-to-Use Course Management*
  - Complete course customization
  - Custom resources/Document posting
  - Announcements/Calendar/Course Email/Discussions
  - Robust Gradebook

Contents in Brief

Contents

Part One ■ The Science and Building Blocks of Life

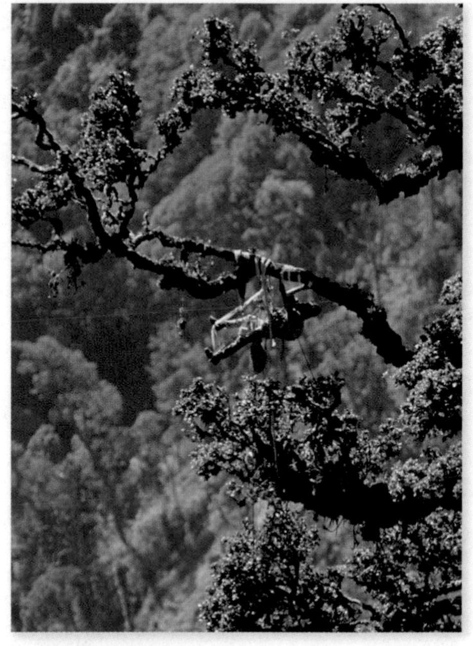

Part Two ■ Cells and Energy

Part Three ■ Heredity and the Genome

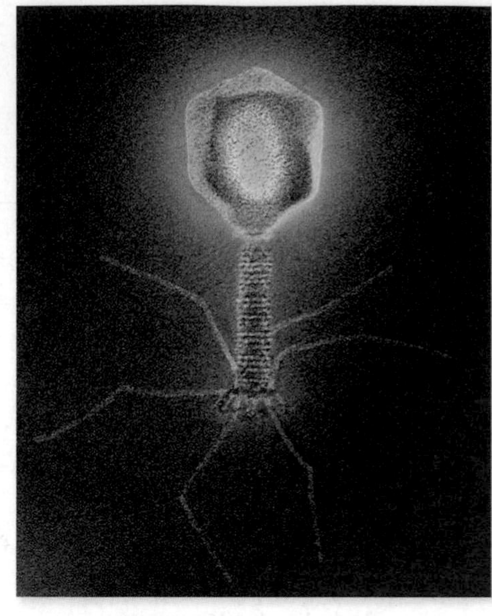

Part Four ■ Molecular Biology: The Genome in Action

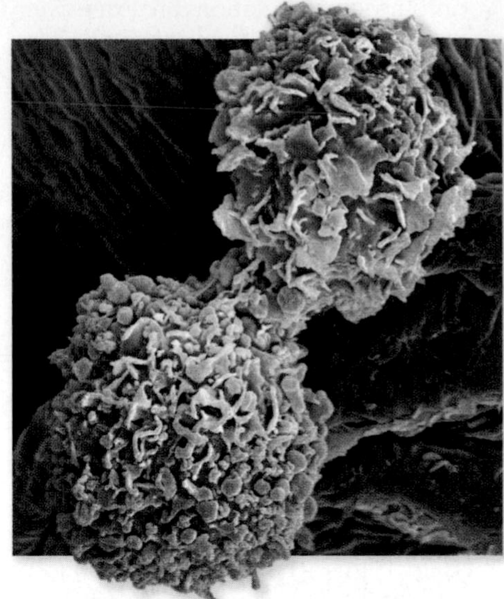

Part Five ▪ The Patterns and Processes of Evolution

Part Six ■ The Evolution of Diversity

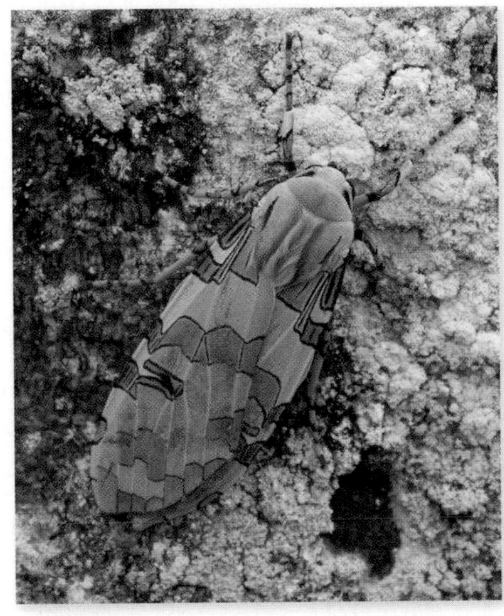

Part Seven ■ Flowering Plants: Form and Function

Part Eight ■ Animals: Form and Function

Part Nine ■ Ecology

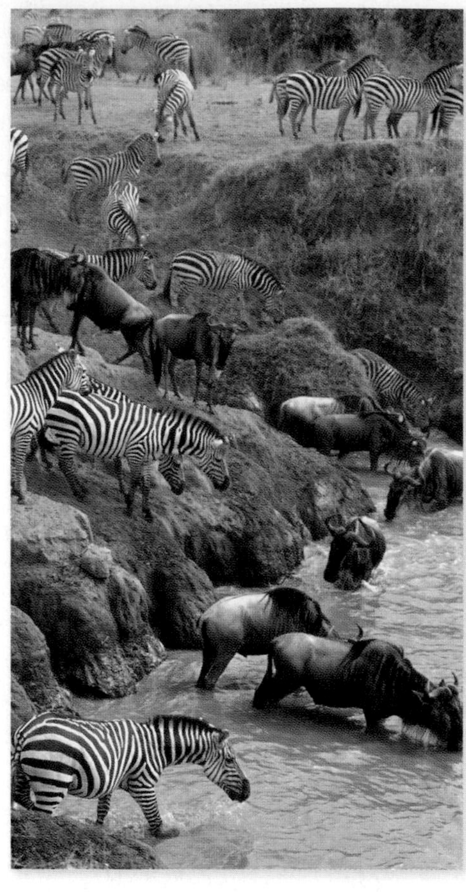

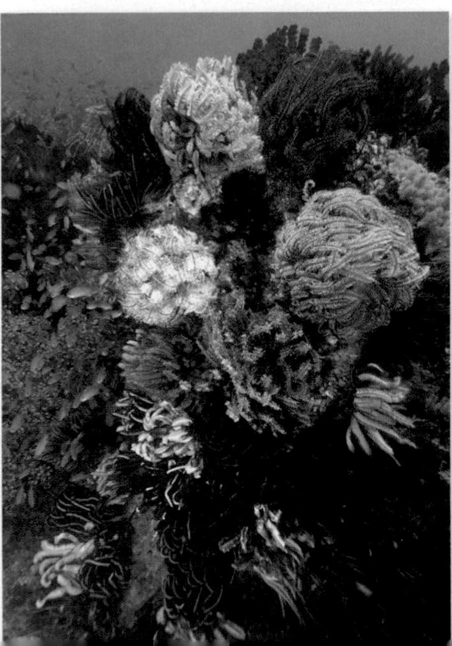

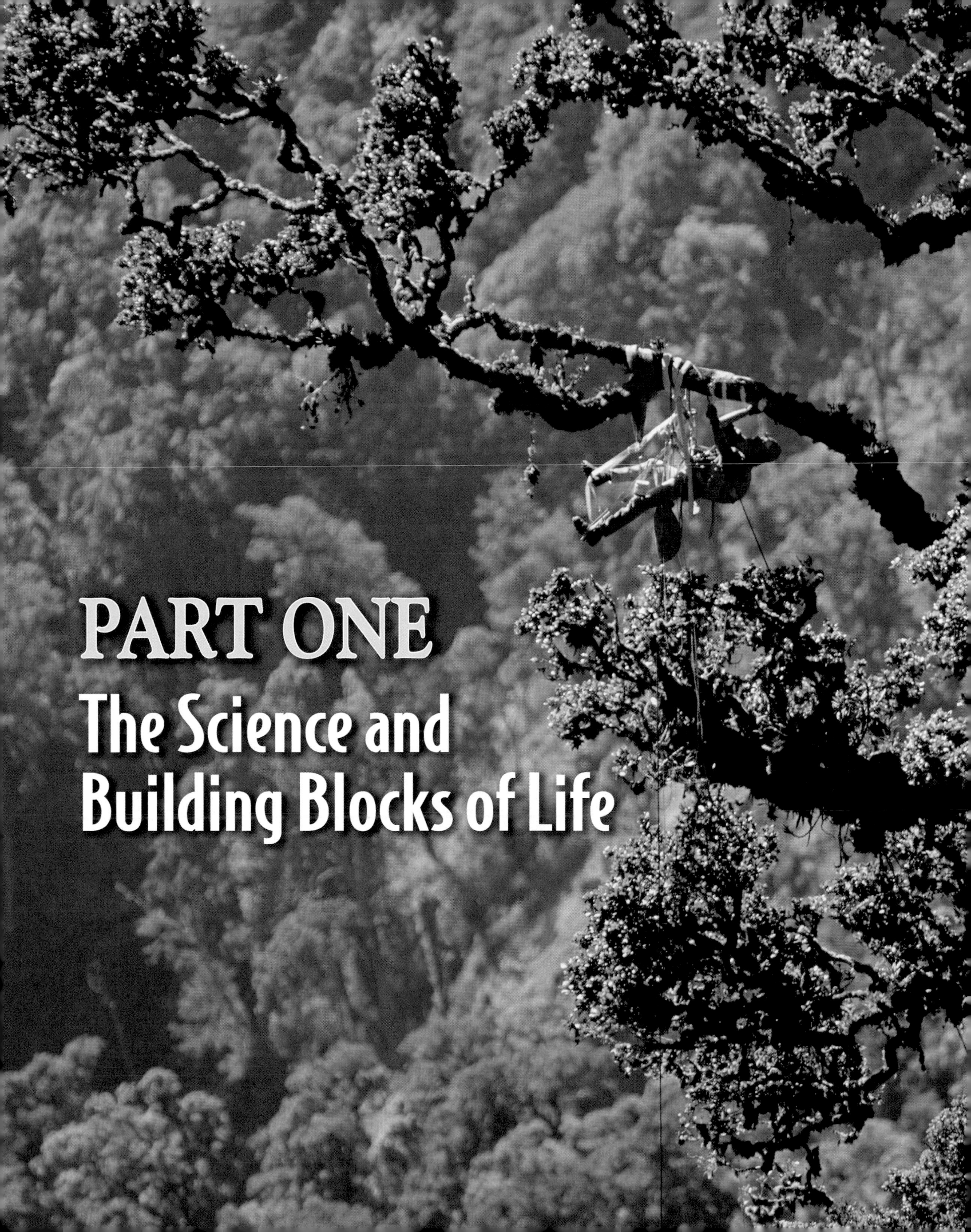

PART ONE
The Science and Building Blocks of Life

CHAPTER 1 Studying Life

Why are frogs croaking?

In August of 1995, a group of Minnesota middle school students on a field trip were hiking through some local wetlands when they discovered a horde of young frogs, most of them with deformed, missing, or extra legs. The students' find made the national news and focused public attention on amphibian population declines, an issue already being studied by many scientists.

There are a number of possible reasons for the problems amphibians are facing. Water pollution is an obvious possibility, since these animals breed and spend their early lives in ponds and streams. Acidic rain resulting from air pollution could also affect their watery homes. Could ultraviolet radiation be causing "mutant" frogs? Is global warming adversely affecting amphibians? Is some disease attacking them? Evidence exists to support each of these possibilities, and there is no a single answer. In one case, a college undergraduate came up with an answer, and in the process gave scientists a whole new perspective on the question.

In 1996, Stanford University sophomore Pieter Johnson was shown a collection of Pacific tree frogs with extra legs growing out of their bodies. He decided to focus his honors research project on finding out what caused these deformities. The frogs came from a pond in an agricultural region near abandoned mercury mines; thus two possible causes of the deformities were agricultural chemicals, and heavy metals from the old mines.

Pieter applied the *scientific method*. Based on what he knew and on his library research, he proposed a logical explanation for the monster frogs—environmental water pollution—and designed an experiment to test his idea. His experiment compared ponds where there were deformed frogs with ponds where the frogs were normal and tested for the presence or absence of pollutants. As frequently happens in science, his proposed explanation, or *hypothesis*, was disproved by his experiment. But his field work led to a new hypothesis: that the deformities are caused by a parasite. Pieter conducted laboratory experiments, the results of which supported the conclusion that a certain type of parasite is present in some ponds. These parasites burrow into newly hatched tadpoles and disrupt the development of the adult

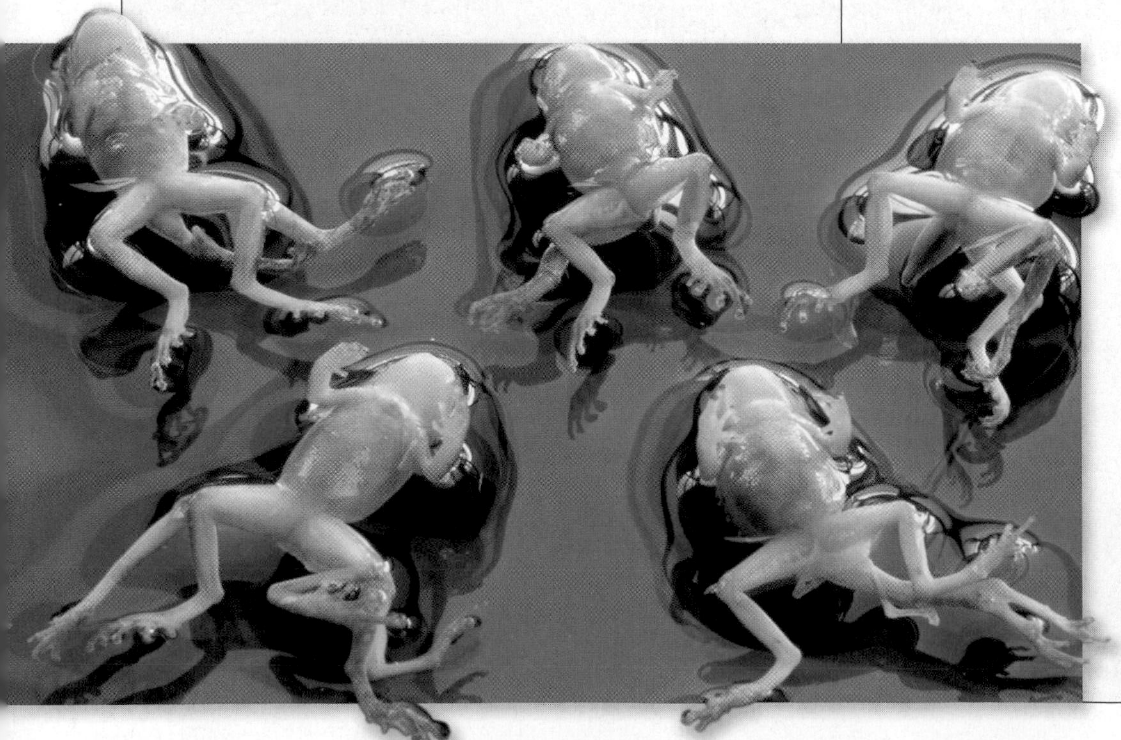

Frogs Are Having Serious Problems These preserved Pacific tree frogs (*Hyla regilla*) exhibit multiple deformities of the hind legs. Similar deformities have been found in frogs from different regions of the world.

A Biologist at Work As a college sophomore, Pieter Johnson studied numerous ponds that were home to Pacific tree frogs, trying to discover why some ponds had so many deformed frogs.

frog's hind legs. Pieter's research did not explain the global decline in amphibians, but it did illuminate one type of problem amphibians encounter. Science usually progresses in such small but solid steps.

Biologists use the scientific method to investigate the processes of life at all levels, from molecules to ecosystems. Some of these processes happen in millionths of seconds and others cover millions of years. The goals of biologists are to understand how organisms and groups of organisms function, and sometimes they use that knowledge in practical and beneficial ways.

IN THIS CHAPTER we examine the most common features of living organisms and put those features into the context of the major principles that underlie all biology. Next we offer a brief outline of how life evolved and how the different organisms on Earth are related. We then turn to the subjects of biological inquiry and the scientific method.

CHAPTER OUTLINE

1.1 What Is Biology?

Biology is the scientific study of living things. Biologists define "living things" as all the diverse organisms descended from a single-celled ancestor that appeared almost 4 billion years ago. Because of their common ancestry, living organisms share many characteristics that are not found in the nonliving world. Most living organisms:

- consist of one or more cells
- contain genetic information
- use genetic information to reproduce themselves
- are genetically related and have evolved
- can convert molecules obtained from their environment into new biological molecules
- can extract energy from the environment and use it to do biological work
- can regulate their internal environment

This list can serve as a rough guide to the major themes and unifying principles of biology that you will encounter in this book. A simple list, however, belies the incredible complexity and diversity of life. Some forms of life may not display all of these characteristics all of the time. For example, the seed of a desert plant may go for many years without extracting energy from the environment, converting molecules, regulating its internal environment, or reproducing; yet the seed is alive.

And what about viruses? Although they do not consist of cells, viruses probably evolved from cellular organisms, and many biologists consider them to be living organisms. Viruses cannot carry out physiological functions on their own, but must parasitize the machinery of host cells to do those jobs for them—including reproduction. Yet viruses contain genetic information, and they certainly evolve (as we know because evolving flu viruses require annual changes in the vaccines we create to combat them). Are viruses alive? What do you think?

This book explores the characteristics of life, how these characteristics vary among organisms, how they evolved, and how they work together to enable organisms to survive and reproduce. *Evolution* is a central theme in biology, and therefore in this book. Through differential survival and reproduction, living systems evolve and become adapted to Earth's many environments. The processes of evolution have generated the enormous diversity that we see today as life on Earth (**Figure 1.1**).

Living organisms consist of cells

Cells and the chemical processes within them are the topic of Part Two of this book. Some organisms are *unicellular,* consisting of a single cell that carries out all the functions of life, while others are *multicellular,* made up of a number of cells that are specialized for different functions.

The discovery of cells was made possible by the invention of the microscope in the 1590s by father and son Dutch spectacle makers Zaccharias and Hans Janssen. The first biologists to improve on their technology and use it to study living organisms were Antony van Leeuwenhoek (Dutch) and Robert Hooke (English) in the middle to late 1600s. It was van Leeuwenhoek who discovered that drops of pond water teemed with single-celled organisms, and he made many other discoveries as he progressively improved his microscopes over a long lifetime of research. Hooke carried out similar studies. From observations of plant tissues (cork, to be specific) he concluded that the tissues were made up of repeated units he called *cells* (**Figure 1.2**).

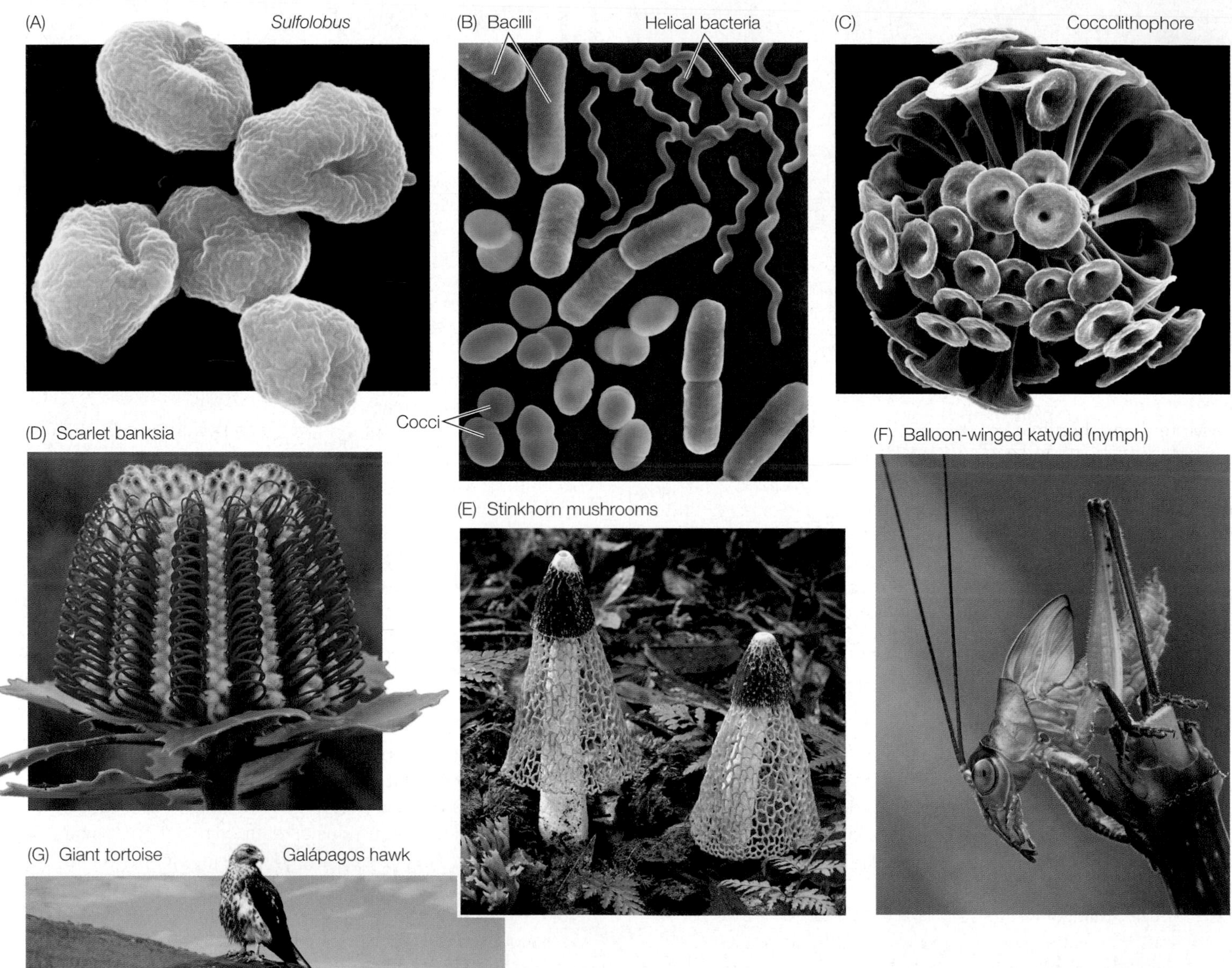

1.1 The Many Faces of Life The processes of evolution have led to the millions of diverse organisms living on Earth today. Archaea (A) and bacteria (B) are single-celled organisms. The three different bacteria in (B) represent the three shapes (rods, or bacilli; helices; and spheres, or cocci) often seen in these organisms, which are described in Chapter 26. (C) Organisms whose single cells display more complexity are commonly known as protists. Protists are the subjects of Chapter 27. (D) Chapters 28 and 29 cover the multicellular green plants. The other broad groups of multicellular organisms are (E) the fungi, discussed in Chapter 30, and (F,G) the animals, covered in Chapters 31–33.

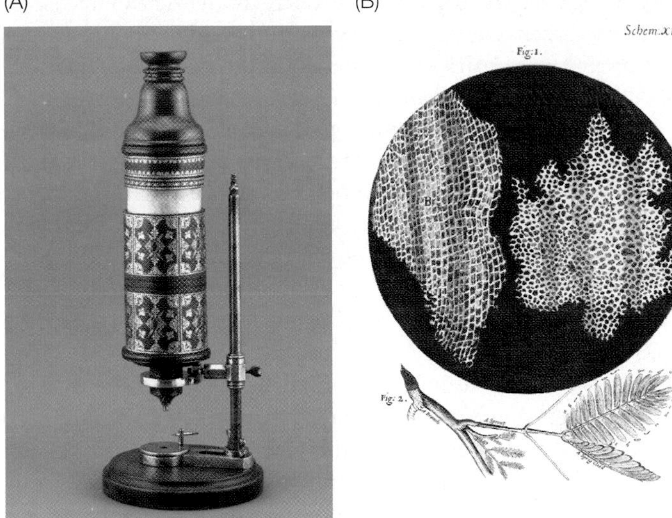

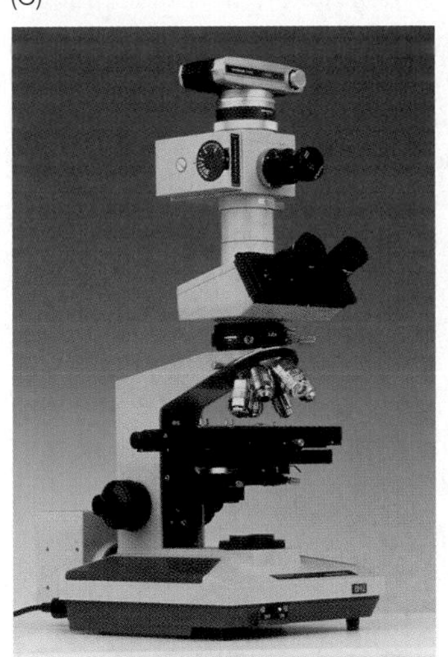

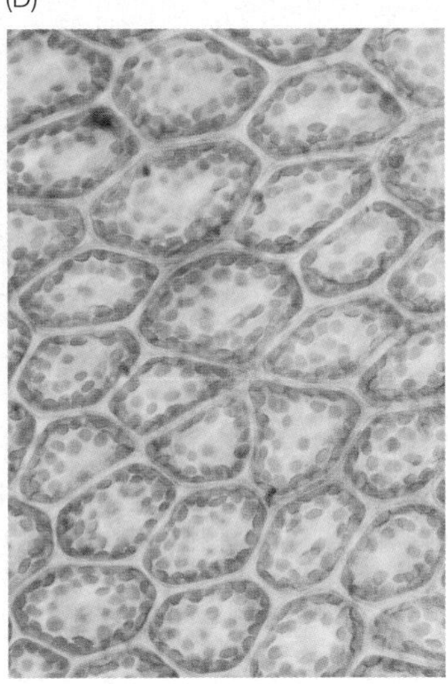

1.2 All Life Consists of Cells (A) The development of microscopes, such as this instrument of Robert Hooke's, revealed the microbial world to seventeenth-century scientists. (B) Hooke was the first to propose the concept of cells, based on his observations of thin slices of plant tissue (cork) under his microscope. (C) A modern version of the optical, or "light" microscope. (D) A modern light micrograph reveals the intricacies of cells in a leaf.

In 1676, Hooke wrote that van Leeuwenhoek had observed "a vast number of small animals in his Excrements which were most abounding when he was troubled with a Loosenesse and very few or none when he was well." This simple observation represents the discovery of bacteria.

More than a hundred years passed before studies of cells advanced significantly. In 1838, Matthias Schleiden, a German biologist, and Theodor Schwann, from Belgium, were having dinner together and discussing their work on plant and animal tissues, respectively. They were struck by the similarities in their observations and came to the conclusion that the structural elements of plants and animals were essentially the same. They formulated their conclusion as the **cell theory**, which states that:

- Cells are the basic structural and physiological units of all living organisms.
- Cells are both distinct entities and building blocks of more complex organisms.

But Schleiden and Schwann did not understand the origin of cells. They thought cells emerged by the self-assembly of nonliving materials, much as crystals form in a solution of salt. This conclusion was in accordance with the prevailing view of the day that life arises from nonlife by spontaneous generation—mice from dirty clothes, maggots from dead meat, insects from mixtures of straw and pond water. The debate over whether or not life could arise from nonlife continued until 1859, when the French Academy of Sciences sponsored a contest for the best experiment to prove or disprove spontaneous generation. The prize was won by the great French scientist Louis Pasteur; his experiment proving that life must be present in order for life to be generated is described in Figure 3.30. Pasteur's vision of microorganisms led him to propose the germ theory of disease and explain the role of single-celled organisms in the fermentation of beer and wine. He also designed a method of preserving milk by heating it to kill microorganisms, a process we now know as pasteurization.

Today we readily accept the fact that all cells come from preexisting cells. In addition, we understand that the functional properties of organisms derive from the properties of their cells. We also understand that cells of all kinds share many essential mechanisms because they share a common ancestry that goes back billions of years. We therefore add a few more elements to the cell theory:

- All cells come from preexisting cells.
- All cells are similar in chemical composition.
- Most of the chemical reactions of life occur within cells.
- Complete sets of genetic information are replicated and passed on during cell division.

At the same time Schleiden and Schwann were building the foundation for the cell theory, Charles Darwin was beginning to understand how organisms undergo evolutionary change.

The diversity of life is due to evolution by natural selection

Evolution by **natural selection**, as proposed by Charles Darwin, is perhaps the major unifying principle of biology and is the topic of Part Five of this book.

Darwin proposed that living organisms are descended from common ancestors and are therefore related to one another. He did not have the advantage of understanding the mechanisms of genetic inheritance that you will learn about in Part Three, but even so he surmised that such mechanisms existed because offspring resembled their parents in so many different ways. That simple fact is the basis for the concept of a **species**. Although the precise definition of a species is complicated, in its most widespread usage it refers to a group of organisms that look alike ("are morphologically similar") and can breed successfully with one another.

But offspring also differ from their parents. Any population of a plant or animal species displays variation, and if you select breeding pairs on the basis of some particular trait, that trait is more likely to be present in their offspring than in the general population. Darwin himself bred pigeons, and was well aware of how pigeon fanciers selected for unusual feather patterns, beak shapes, and body sizes. He realized that if humans could select for specific traits, the same process could operate in nature; hence the term *natural selection*.

How would selection function in nature? Darwin postulated that different probabilities of survival and reproductive success would do the job. He reasoned that the reproductive capacity of plants and animals, if unchecked, would result in unlimited growth of populations, but we do not observe such unlimited population growth in nature; therefore only a small percentage of offspring must survive to reproduce. So, any trait that confers even a small increase in the probability that its possessor will survive and reproduce would be strongly favored and would spread in the population. Darwin called this phenomenon *natural selection*.

Because organisms with certain traits survive and reproduce best under specific sets of conditions, natural selection leads to **adaptations**: structural, physiological, or behavioral traits that enhance an organism's chances of survival and reproduction in its environment (**Figure 1.3**). The many different environments and ecological communities organisms have adapted to over evolutionary history have led to a remarkable amount of diversity, which we will survey in Part Six of this book.

If all cells come from preexisting cells, and if all the diverse species of organisms on Earth are related by descent with modification from a common ancestor, then what is the source of information that is passed from parent to daughter cells and from parental organisms to their offspring?

Many leaves are wide and flat, a configuration that presents a maximum of photosynthetic surface to the sun. Some trees, such as this sugar maple, lose their leaves in response to cold or dry weather.

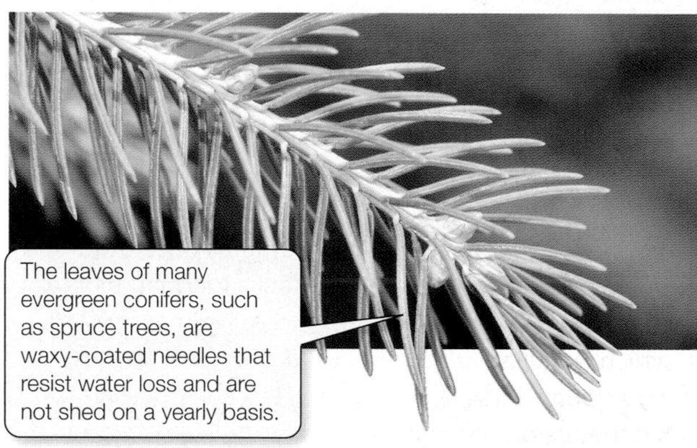

The leaves of many evergreen conifers, such as spruce trees, are waxy-coated needles that resist water loss and are not shed on a yearly basis.

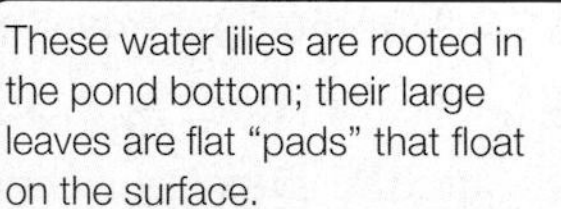

These water lilies are rooted in the pond bottom; their large leaves are flat "pads" that float on the surface.

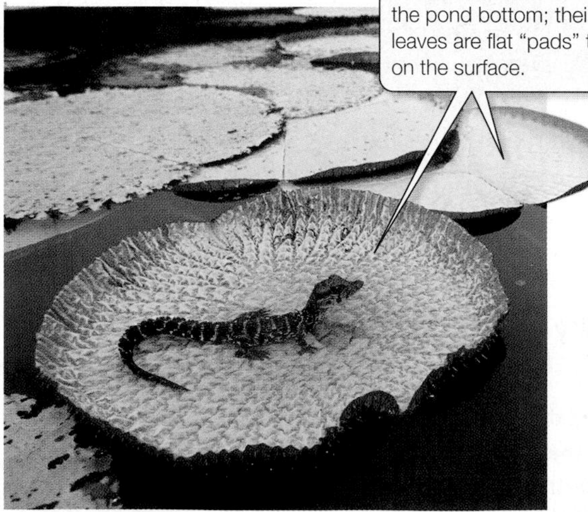

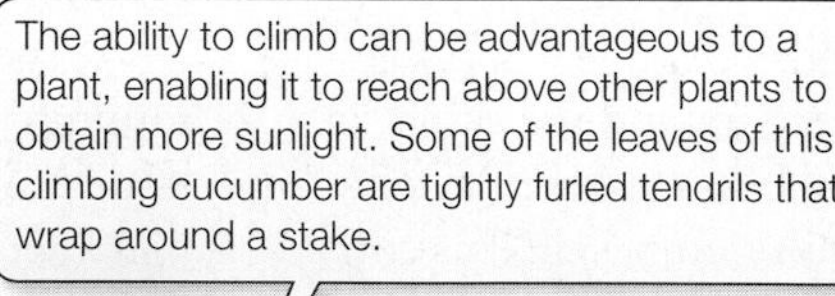

The ability to climb can be advantageous to a plant, enabling it to reach above other plants to obtain more sunlight. Some of the leaves of this climbing cucumber are tightly furled tendrils that wrap around a stake.

The leaves of pitcher plants form a vessel that holds water. The plant receives extra nutrients from the decomposing bodies of insects that drown in the pitcher.

1.3 Adaptations to the Environment The leaves of all plants are specialized for photosynthesis—the sunlight-powered transformation of water and carbon dioxide into larger structural molecules called carbohydrates. The leaves of different plants, however, display many different adaptations to their individual environments.

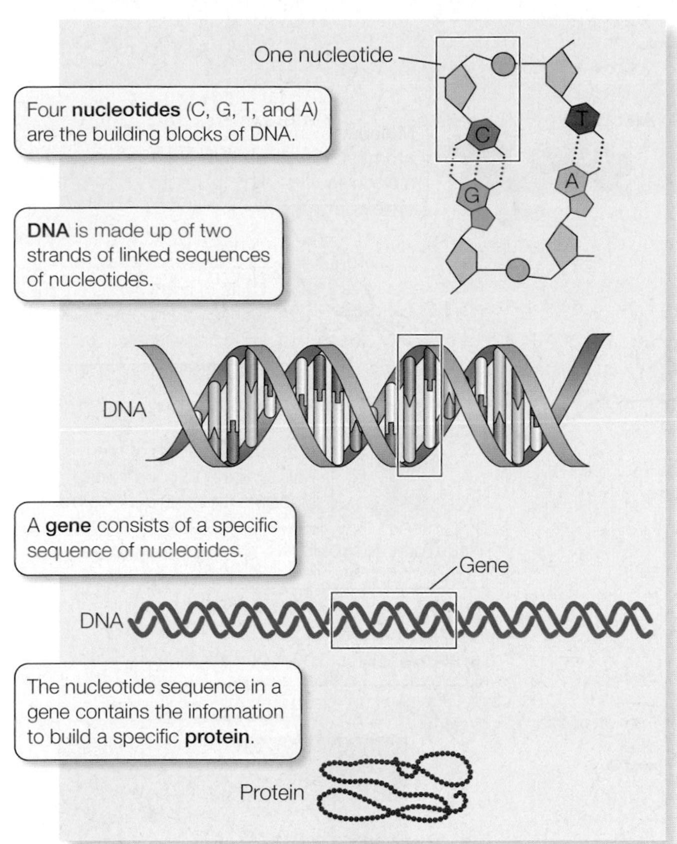

One nucleotide
Four **nucleotides** (C, G, T, and A) are the building blocks of DNA.
C
T
G
A
DNA is made up of two strands of linked sequences of nucleotides.
DNA
A **gene** consists of a specific sequence of nucleotides.
Gene
DNA
The nucleotide sequence in a gene contains the information to build a specific **protein**.
Protein

1.4 The Genetic Code Is Life's Blueprint The instructions for life are contained in the sequences of nucleotides in DNA molecules. Specific DNA sequences comprise genes, and the information in each gene provides the cell with the information it needs to manufacture a specific protein. The average length of a single human gene is 16,000 nucleotides.

Biological information is contained in a genetic language common to all organisms

A cell's instructions—or "blueprints" for existence—are contained in its **genome**, which is the sum total of all the DNA molecules in the cell. **DNA** (deoxyribonucleic acid) molecules are long sequences of four different subunits called **nucleotides**. The sequence of the nucleotides contains genetic information. Specific segments of DNA called **genes** contain the information the cell uses to make proteins (**Figure 1.4**). **Proteins** make up much of an organism's structure and are the molecules that govern the chemical reactions within cells. By analogy with a book, the nucleotides of DNA are like the letters of an alphabet. Protein molecules are the sentences they spell. Combinations of proteins that constitute structures and control biochemical processes are the paragraphs. The structures and processes that are organized into different systems with specific tasks (such as digestion or transport) are the chapters of the book, and the complete book is the organism. Natural selection is the author and editor of all the books in the library of life.

If you were to write out your own genome using four letters to represent the four nucleotides, you would have to write a total of more than 3 billion letters. If you used the same size type as you are reading in this book, your genome would fill about a thousand books the size of this one.

All the cells of a multicellular organism contain the same genome, yet different cells have different functions and form different structures. Therefore, different types of cells in an organism must express different parts of their genome. How the control of gene expression enables a complex organism to develop and function is a major focus of current biological research.

The genome of an organism consists of thousands of genes. If the nucleotide sequence of a gene is altered, it is likely that the protein that gene encodes will be altered. Alterations of genes are called *mutations*. Mutations occur spontaneously; they can also be induced by many outside factors, including chemicals and radiation. Most mutations are deleterious, but occasionally a change in the properties of a protein alters its function in a way that improves the functioning of the organism under the environmental conditions it encounters. Such beneficial mutations are the raw material of evolution.

Cells use nutrients to supply energy and to build new structures

Living organisms acquire substances called *nutrients* from the environment. Nutrients supply the organism with energy and raw materials for building biological structures. Cells take in nutrient molecules and break them down into smaller chemical units. In doing so, they can capture the energy contained in the chemical bonds of the nutrient molecules and use that energy to do different kinds of work. One kind of cellular work is the building, or *synthesis*, of new complex molecules and structures from smaller chemical units. For example, we are all familiar with the fact that carbohydrates eaten today may be deposited in the body as fat tomorrow (**Figure 1.5A**). Another kind of work cells do is mechanical work—for example, moving molecules from one cellular location to another, or even moving whole cells or tissues, as in the case of muscles (**Figure 1.5B**).

Living organisms control their internal environment

Life depends on thousands of biochemical reactions that occur inside cells. These reactions require materials to be moved into and out of cells in a controlled manner. Within the cell, those reactions are linked in that the products of one are the raw materials of the next. If this complex network of reactions is to be properly integrated, reaction rates within a cell must be precisely controlled. A large proportion of the activities of cells are directed toward the regulation of the multiple chemical reactions continuously in progress inside the cell.

Organisms that are made up of more than one cell have an *internal environment* that is not cellular. That is, their individual cells are bathed in extracellular fluids, from which they receive nutrients and into which they excrete wastes. The cells of multicellular organisms are specialized to contribute in some way to the maintenance of that internal environment. However, with the evolution of specialized

(B)
(A)

1.5 Energy from Nutrients Can Be Stored or Used Immediately (A) The cells of this Arctic ground squirrel have broken down the complex carbohydrates in plants and converted their molecules into fats, which are stored in the animal's body to provide an energy supply for the cold months. (B) The cells of this kangaroo are breaking down food molecules and using the energy in their chemical bonds to do mechanical work—in this case, to jump.

functions, these cells lost many of the functions carried out by single-celled organisms, and therefore depend on the internal environment for essential services. The interdependence of the different kinds of cells in a multicellular organism can by expressed in the famous motto of the Three Musketeers: "One for all, and all for one!"

To accomplish their specialized tasks, assemblages of similar cells are organized into *tissues.* For example, a single muscle cell cannot generate much force, but when many of these cells combine to form the tissue of a working muscle, considerable force and movement can be generated (see Figure 1.5B). Different tissue types are organized to form *organs* that accomplish specific functions. Familiar organs include the heart, brain, and stomach. Organs whose functions are interrelated can be grouped into *organ systems.* The functions of cells, tissues, organs, and organ systems are all integral to the multicellular *organism* (**Figure 1.6**). The biology of organisms is the subject of Parts Seven and Eight of this book.

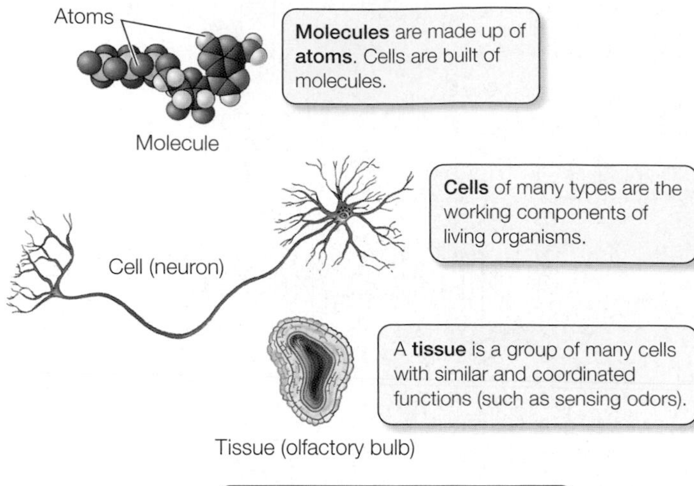
Atoms
Molecule
Molecules are made up of **atoms**. Cells are built of molecules.
Cell (neuron)
Cells of many types are the working components of living organisms.
Tissue (olfactory bulb)
A **tissue** is a group of many cells with similar and coordinated functions (such as sensing odors).
Organ (brain)
Organs combine several tissues that function together. Organs form **systems**, such as the nervous system.
Organism (fish)
An **organism** is a recognizable, self-contained individual. A multicellular organism is made up of organs and organ systems.
Population (school of fish)
A **population** is a group of many organisms of the same species.
Communities consist of populations of many different species.
Community (coral reef)
Biological communities in the same geographical location form **ecosystems**. Ecosystems exchange energy and create Earth's **biosphere**.
Biosphere

1.6 Biology Is Studied at Many Levels of Organization Life's properties emerge when DNA and other molecules are organized in cells. Energy flows through all the biological levels shown here.

(A)

(B)

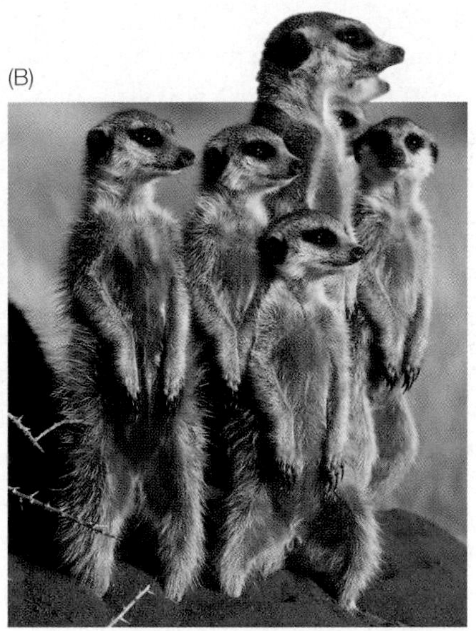

1.7 Conflict and Cooperation Organisms of the same species interact with one another in various ways. (A) Territorial elephant seal bulls defend stretches of beach from other males. The single male who controls a stretch of beach is able to mate with the many females (seen in the background) that live there. (B) Members of a meerkat colony are usually related to one another. Meerkats cooperate in many ways, such as watching out for predators and giving a warning bark if one appears.

As Figure 1.6 shows, individual organisms do not live in isolation, and the hierarchy of biology continues well beyond that level.

Living organisms interact with one another

Organisms interact with their external environments as well as the internal environment. Individual organisms are part of *populations* that interact among themselves and with populations of different organisms.

Organisms interact in many different ways. For example, some animals are *territorial* and will try to prevent other individuals of their species from exploiting the resource they are defending, whether it be food, nesting sites, or mates (**Figure 1.7A**). Animals may also *cooperate* with members of their species, forming social units such as a termite colony, a school of fish, or a meerkat colony (**Figure 1.7B**). Such interactions among individuals have resulted in the evolution of social behaviors such as communication.

The interaction of populations of many different species forms a *community*, and interactions between different species are a major evolutionary force. Adaptations that give an individual of one species an advantage in obtaining members of another species as food (and the converse, adaptations that lessen an individual's chances of becoming food) are paramount in evolutionary history. Organisms of different species may compete for the same resources, resulting in natural selection for specialized adaptations that allow some individuals to exploit those resources more efficiently than others can.

In any given geographic locality, the interacting communities form *ecosystems*. Organisms in the ecosystem can modify the environment in ways that affect other organisms. For example, in most terrestrial (land) environments, the dominant plants greatly modify the environmental conditions in which animals and other plants must live. The ways in which species interact with one another and with their environment is the subject of *ecology*, the topic of Part Nine of this book.

Discoveries in biology can be generalized

Because all life is related by descent from a common ancestor, shares a genetic code, and consists of similar building blocks—cells—knowledge gained from investigations of one type of organism can, with care, be generalized to other organisms. Therefore, biologists can use **model systems** for research, knowing that they can extend their findings to other organisms and to humans. For example, our basic understanding of the chemical reactions in cells came from research on bacteria, but is applicable to all cells, including those of humans. Similarly, the biochemistry of photosynthesis—the process by which plants use sunlight to produce biological molecules—was largely worked out from experiments on *Chlorella* (a type of pond scum; see Figure 8.12). We learned much of what we know about the genes that control plant development from work on a single species of plant (see Chapter 19). Knowledge about how animals develop has come from work on sea urchins, frogs, chickens, roundworms, and fruit flies. And recently, the discovery of a major gene controlling human skin color came from work on zebrafish. Being able to generalize from model systems is a powerful tool.

1.1 RECAP

Living organisms are made of cells, evolve by natural selection, contain genetic information, extract energy from their environment and use it to do biological work, control their internal environment, and interact with one another.

- Can you describe the relationship between evolution by natural selection and the genetic code? See pp. 5–7
- Do you understand why results of biological research on one species can be generalized to very different species? See p. 9

Now that we have an overview of the major features of life that will be explored in depth in this book, we can ask how and when life first emerged. In the next section we will describe the history of life from the earliest simple life forms to the complex and diverse organisms that inhabit our planet today.

1.2 How Is All Life on Earth Related?

What do biologists mean when they say that all organisms are *genetically related*? They mean that species on Earth share a *common ancestor*. If two species are similar, as dogs and wolves are, then they probably have a common ancestor in the fairly recent past. The common ancestor of two species that are more different—say, a dog and a deer—probably lived in the more distant past. And if two organisms are very different—such as a dog and a clam—then we must go back to the *very* distant past to find their common ancestor. How can we tell how far back in time the common ancestor of any two organisms lived? In other words, how do we discover the evolutionary relationships among organisms?

For many years, biologists investigated the history of life by studying the *fossil record*—the preserved remains of organisms that lived in the distant past (**Figure 1.8**). Geologists supplied knowledge about the ages of fossils and the nature of the environments in which they lived. Biologists then inferred the evolutionary relationships among living and fossil organisms by comparing their anatomical similarities and differences. The development of modern molecular methods for comparing genomes, described in Chapter 24, has enabled biologists to more accurately establish the degrees of relationship between living organisms and use that information to help us interpret the fossil record.

In general, the greater the differences between the genomes of two species, the more distant was their common ancestor. Using molecular techniques, biologists are exploring fundamental questions about the history of life on Earth. What were the earliest forms of life? How did those simple organisms give rise to the great diversity of organisms alive today? Can we reconstruct the family tree of all life?

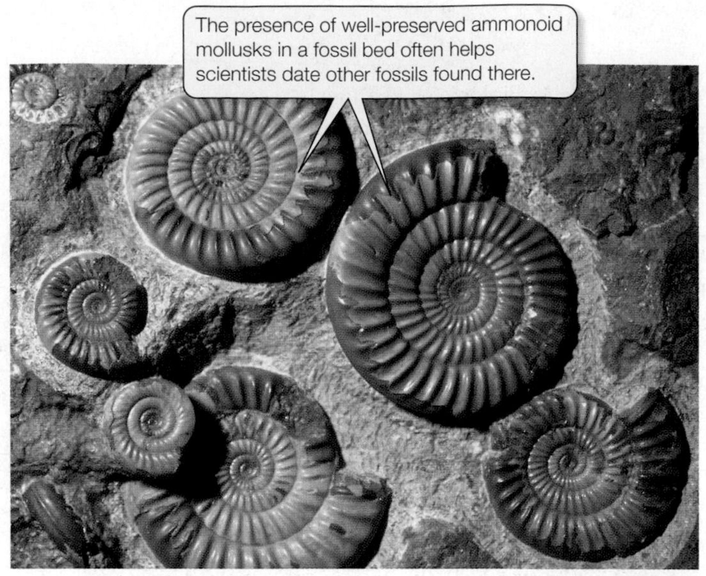

The presence of well-preserved ammonoid mollusks in a fossil bed often helps scientists date other fossils found there.

1.8 Fossils Give Us a View of Past Life The most prominent of the many fossilized organisms in this rock sample are ammonoids, an extinct mollusk group whose living relatives include squids and octopus. Ammonoids flourished between 200 million and 60 million years ago; this particular group of fossils is about 185 million years old.

Life arose from nonlife via chemical evolution

Geologists estimate that Earth is about 5 billion years old. For the first billion years, it was not a very hospitable place for life, and there was no life. If we picture the history of Earth as a 30-day calendar, life probably arose around the middle of the first week, or about 4 billion years ago (**Figure 1.9**).

When we consider how life might have arisen from nonliving matter, we must take into account the properties of the young Earth's atmosphere, oceans, and climate, all of which were very differ-

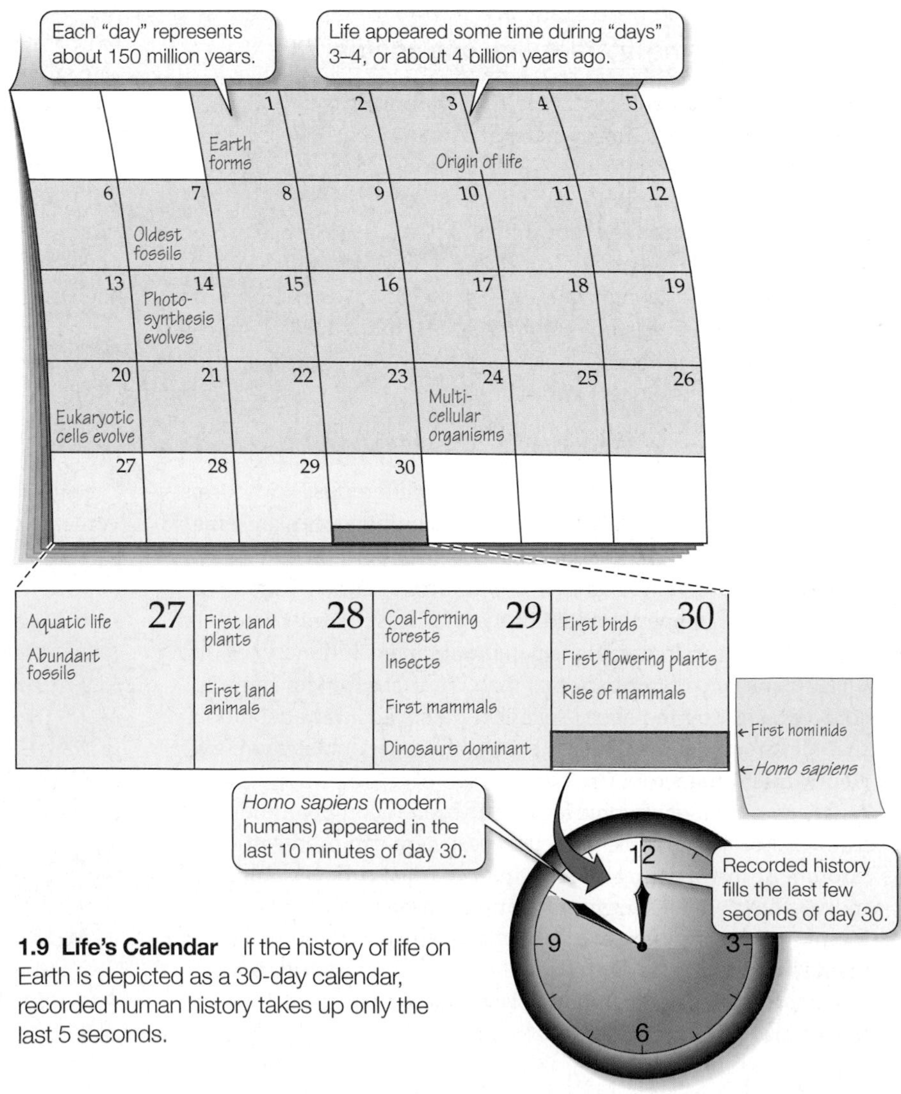

1.9 Life's Calendar If the history of life on Earth is depicted as a 30-day calendar, recorded human history takes up only the last 5 seconds.

ent from today's. Biologists postulate that complex biological molecules first arose through the random physical association of chemicals in that environment. Experiments that simulate the conditions on early Earth have confirmed that the generation of complex molecules under such conditions is possible, even probable. The critical step, however, for the evolution of life had to be the appearance of molecules that could reproduce themselves and also serve as templates for the synthesis of large molecules with complex but stable shapes. The variation of the shapes of these large, stable molecules (described in Chapter 3) enabled them to participate in increasing numbers and kinds of interactions with other molecules: chemical reactions.

Biological evolution began when cells formed

The second critical step in the origin of life was the enclosure of complex biological molecules in *membranes,* which kept them close together and increased the frequency with which they interacted. Fatlike molecules were the critical ingredient: because these molecules are not soluble in water, they form membrane-like films. These films tend to form spherical *vesicles,* which could have enveloped assemblages of other biological molecules. Scientists postulate that about 3.8 billion years ago, this natural process of membrane formation resulted in the first cells with the ability to replicate themselves—an event that marked the beginning of biological evolution.

For 2 billion years after cells originated, all organisms consisted of only one cell. These first unicellular organisms were (and are, as multitudes of their descendants exist in similar form today) **prokaryotes.** Prokaryotic cell structure consists of DNA and other biochemicals enclosed in a membrane.

These early prokaryotes were confined to the oceans, where there was an abundance of complex molecules they could use as raw materials and sources of energy. The ocean shielded them from the damaging effects of ultraviolet light, which was intense at that time because there was no oxygen in the atmosphere, and hence no protective ozone layer.

Photosynthesis changed the course of evolution

The sum total of all the chemical reactions that go on inside a cell constitutes the cell's **metabolism.** To fuel their metabolism, the earliest prokaryotes took in molecules directly from their environment, breaking these small molecules down to release the energy contained in their chemical bonds. Many modern species of prokaryotes still function this way, and very successfully.

An extremely important step that would change the nature of life on Earth occurred about 2.5 billion years ago with the evolution of **photosynthesis.** The chemical reactions of photosynthesis (which are explained in Chapter 8) transform the energy of sunlight into a form of energy that can power the synthesis of large biological molecules. These large molecules become the building blocks of cells; they can also be broken down to provide metabolic energy. Because its energy-capturing processes provide food for other organisms, photosynthesis is the basis of much of life on Earth today.

Early photosynthetic cells were probably similar to present-day prokaryotes called *cyanobacteria* (**Figure 1.10**). Over time, photosynthetic prokaryotes became so abundant that vast quantities of oxygen gas—O_2, which is a by-product of photosynthesis—began slowly to accumulate in the atmosphere. O_2 was poisonous to many of the prokaryotes that lived at that time. However, those organisms that tolerated O_2 were able to proliferate as the presence of oxygen opened up vast new avenues of evolution. Metabolism based on the use of O_2, called *aerobic metabolism,* is more efficient than the *anaerobic* (non-oxygen-using) *metabolism* that characterized earlier organisms. Aerobic metabolism allowed cells to grow larger, and today it is used by the majority of Earth's organisms.

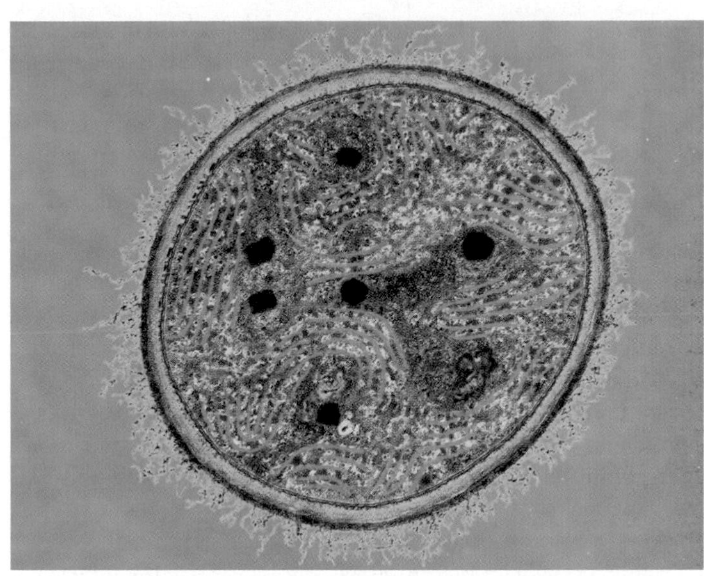

1.10 Photosynthetic Organisms Changed Earth's Atmosphere
This modern cyanobacterium may be very similar to the early photosynthetic prokaryotes that introduced oxygen into Earth's atmosphere.

Over millions of years, the vast quantities of oxygen released by photosynthesis formed a layer of ozone (O_3) in the upper atmosphere. As the ozone layer thickened, it intercepted more and more of the sun's deadly ultraviolet radiation. Only in the last 800 million years has the presence of a dense ozone layer allowed organisms to leave the protection of the ocean and live on land.

Eukaryotic cells evolved from prokaryotes

Another important step in the history of life was the evolution of cells with discrete intracellular compartments, called **organelles,** that were capable of taking on specialized cellular functions. This event happened about 3 weeks into our calendar of Earth's history (see Figure 1.9). One of these organelles, the *nucleus,* came to contain the cell's genetic information. The nucleus has the appearance of a dense kernel, giving these cells their name: **eukaryotes** (from the Greek *eu,*"true," and *karyon,*"kernel"), as distinguished from the cells of prokaryotes, which lack internal compartments (*pro,* "before").

Some organelles are hypothesized to have originated when cells ingested smaller cells (see Figure 4.26). For example, the organelle specialized to conduct photosynthesis, the *chloroplast,* could have originated as a photosynthetic prokaryote that was ingested by a larger eukaryote. If the larger cell failed to break down this intended food object, a partnership could have evolved in which the ingested prokaryote provided the products of photosynthesis and the host cell provided a good environment for its smaller partner.

Multicellularity arose and cells became specialized

Until slightly more than 1 billion years ago, all the organisms that existed—whether prokaryotic or eukaryotic—were unicellular. Yet another important evolutionary step occurred when some eukaryotes failed to separate after cell division, remaining attached to each other. The permanent association of cells made it possible for some cells to specialize in certain functions, such as reproduction, while other cells specialized in other functions, such as absorbing nutrients and distributing them to neighboring cells. This **cellular specialization** enabled multicellular eukaryotes to increase in size and become more efficient at gathering resources and adapting to specific environments.

Biologists can trace the evolutionary Tree of Life

If all the species of organisms on Earth today are the descendants of a single kind of unicellular organism that lived almost 4 billion years ago, how have they become so different? And why are there so many species?

As long as individuals within a population mate at random, structural and functional changes may evolve within that population, but the population will remain one species. However, if some event isolates some members of a population from the others, structural and functional differences between them may accumulate over time. In short, the evolutionary paths of the two groups may diverge to the point where their members can no longer reproduce with each other. They have evolved into different species. This evolutionary process, called *speciation*, is detailed in Chapters 22 and 23.

Biologists give each species a distinct scientific name formed from two Latinized names (a *binomial*). The first name identifies the species' *genus*—a group of species that share a recent common ancestor. The second is the name of the species. For example, the scientific name of the human species is *Homo sapiens*: *Homo* is our genus and *sapiens* is our species. Scientific names usually refer to some characteristic of the species. *Homo* is derived from the Latin word for "man," and *sapiens* is derived from the Latin word for "wise" or "rational."

As many as 30 million species of organisms may exist on Earth today. Many times that number lived in the past but are now extinct. Many millions of speciation events created this vast diversity, and the unfolding of these events can be diagrammed as an evolutionary "tree" showing the order in which populations split and eventually evolved into new species. An evolutionary tree traces the descendants of ancestors that lived at different times in the past. The organisms on any one branch share a common ancestor at the base of that branch. The most closely related groups are placed together on the same branch; more distantly related organisms are on different branches. In this book, we adopt the convention that time flows from left to right, so the tree in Figure 1.11 (and other trees in this book) lies on its side, with its root—the ancestor of all life—at the left. Although many details remain to be clarified, the broad outlines of the Tree of Life have been determined. Its branching patterns are based on a rich array of evidence from fossils, structures, metabolic processes, behavior, and molecular analyses of genomes.

No fossils exist to help us determine the earliest divisions in the lineage of life because those unicellular organisms had no parts that could be preserved as fossils. However, molecular evidence has been used to separate all living organisms into three major **domains**: Archaea, Bacteria, and Eukarya (**Figure 1.11**). The organisms of each domain have been evolving separately from organisms in the other domains for more than a billion years.

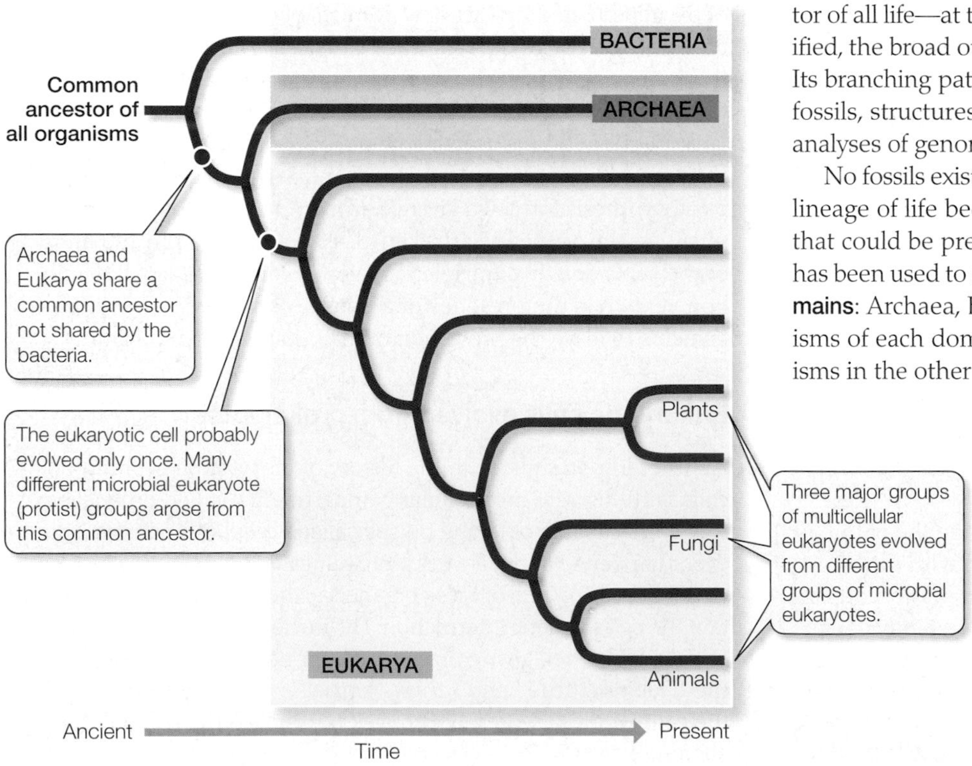

Organisms in the domains **Archaea** and **Bacteria** are all prokaryotes. Archaea and Bacteria differ so fundamentally from each other in their metabolic processes that they are believed to have separated into distinct evolutionary lineages very early.

Members of the third domain—**Eukarya**—have eukaryotic cells. Three major groups of multicellular eukaryotes—plants, fungi, and animals—all evolved from unicellular *microbial eukaryotes*, more generally referred to as *protists*. The photosynthetic protist that gave rise to plants was completely distinct from the protist that was ancestral to both animals and fungi, as can be seen from the branching pattern of Figure 1.11.

BACTERIA
ARCHAEA
Common ancestor of all organisms
Archaea and Eukarya share a common ancestor not shared by the bacteria.
The eukaryotic cell probably evolved only once. Many different microbial eukaryote (protist) groups arose from this common ancestor.
Plants
Fungi
Animals
Three major groups of multicellular eukaryotes evolved from different groups of microbial eukaryotes.
EUKARYA
Ancient
Present
Time

1.11 The Tree of Life The classification system used in this book divides Earth's organisms into three domains: Bacteria, Archaea, and Eukarya. The unlabeled blue branches within the Eukarya represent various groups of microbial eukaryotes, more commonly known as "protists."

Some bacteria, some archaea, some protists, and most plants are capable of photosynthesis. These organisms are called *autotrophs* ("self-feeders"). The biological molecules they produce are the primary food for nearly all other living organisms.

Fungi include molds, mushrooms, yeasts, and other similar organisms, all of which are *heterotrophs* ("other-feeders")—that is, they require a source of molecules synthesized by other organisms, which they then break down to obtain energy for their own metabolic processes. Fungi break down energy-rich food molecules in their environment and then absorb the breakdown products into their cells. Some fungi are important as decomposers of the waste products and dead bodies of other organisms.

Like fungi, animals are heterotrophs, but unlike fungi they ingest their food source, then break down the food in a digestive tract. Animals eat other forms of life, including plants, fungi, and other animals. Their cells absorb the breakdown products and obtain energy from them.

1.2 RECAP

The first cellular life on Earth was prokaryotic and arose about 4 billion years ago. The complexity of the organisms that exist today is the result of several important evolutionary events, including the evolution of photosynthesis, eukaryotic cells, and multicellularity. The genetic relationships of all organisms can be shown as a branching Tree of Life.

- Can you explain the evolutionary significance of photosynthesis? See p. 11
- What do the domains of life represent? What are the major groups of eukaryotes? See p. 12 and Figure 1.11

In February of 1676, Robert Hooke received a letter from the physicist Sir Isaac Newton. In this letter Newton famously remarked to Hooke, "If I have seen a little further, it is by standing on the shoulders of giants." We all stand on the shoulders of giants, building on the research of earlier scientists. By the end of this course, you will know more about evolution than Darwin ever could have, and you will know infinitely more about cells than Schleiden and Schwann did. Let's look at the methods biologists use to expand our knowledge of life.

1.3 How Do Biologists Investigate Life?

Biologists use many tools and methods in their research, but regardless of the methods they use, biologists take two basic approaches to their investigations of life: they observe and they conduct experiments.

Observation is an important skill

Biologists have always observed the world around them, but today their abilities to observe are greatly enhanced by many sophisticated technologies, such as electron microscopes, DNA chips, magnetic resonance imaging, and global positioning satellites. Advances in technology have been responsible for most major advances in biology. For example, not too long ago it was extremely difficult and time-consuming to decipher the nucleotide sequence that makes up a single gene. New technologies enabled biologists to sequence the entire human genome in only 13 years (1990–2003). Scientists now use these methods routinely, sequencing the genomes of organisms (including organisms that cause serious diseases) in only days. We will explore some of these technologies and what we have learned from them in Part Four of this book.

Our ability to observe the distributions of organisms, such as fish in the world's oceans, has also improved dramatically. A short time ago, researchers could put physical tags on fish and then only hope that someday a fisherman would catch one and send back the tag, which would at least reveal where the fish ended up. Today, electronic recording devices attached to fish can continuously record not only where the fish is, but also how deep it swims at different times of day and the temperature and salinity of the water around it (**Figure 1.12**). At set intervals, these tags download their information to a satellite, which relays it back to researchers. Suddenly we are acquiring a great deal of new knowledge about the distribution of life in the oceans.

The scientific method combines observation and logic

Observations lead to questions, and scientists make additional observations and do experiments to answer those questions. The conceptual approach that underlies the design and conduct of most modern scientific investigations is called the **scientific method**. This

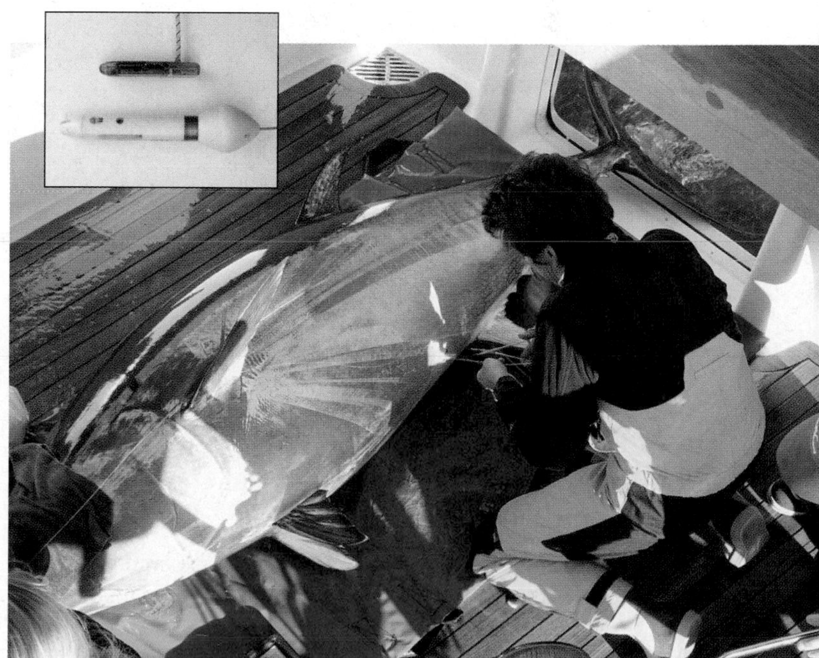

1.12 Tuna Tracking Marine biologist Barbara Block attaches a computerized data recording tag (inset) to a bluefin tuna. The use of such tags makes it possible to track individual tuna wherever they travel in the world's oceans.

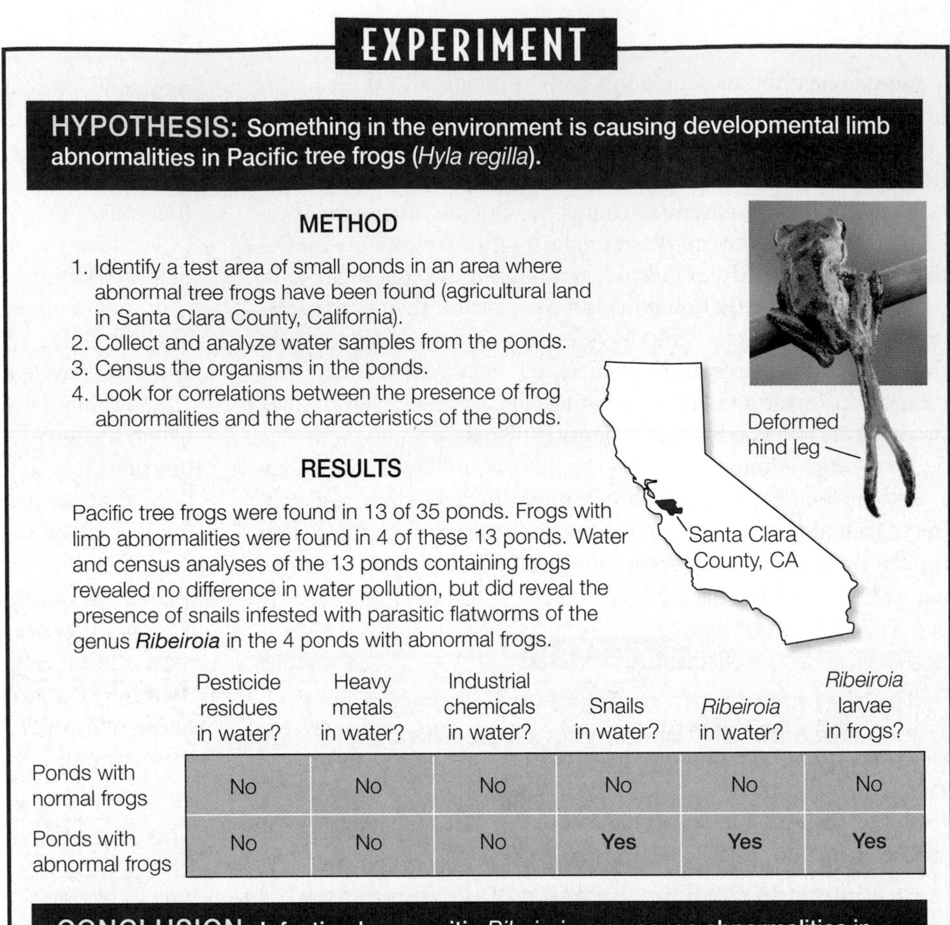

EXPERIMENT
HYPOTHESIS: Something in the environment is causing developmental limb abnormalities in Pacific tree frogs (*Hyla regilla*).
METHOD
1. Identify a test area of small ponds in an area where abnormal tree frogs have been found (agricultural land in Santa Clara County, California).
2. Collect and analyze water samples from the ponds.
3. Census the organisms in the ponds.
4. Look for correlations between the presence of frog abnormalities and the characteristics of the ponds.
Deformed hind leg
RESULTS
Pacific tree frogs were found in 13 of 35 ponds. Frogs with limb abnormalities were found in 4 of these 13 ponds. Water and census analyses of the 13 ponds containing frogs revealed no difference in water pollution, but did reveal the presence of snails infested with parasitic flatworms of the genus *Ribeiroia* in the 4 ponds with abnormal frogs.
Santa Clara County, CA
	Pesticide residues in water?	Heavy metals in water?	Industrial chemicals in water?	Snails in water?	*Ribeiroia* in water?	*Ribeiroia* larvae in frogs?
Ponds with normal frogs	No	No	No	No	No	No
Ponds with abnormal frogs	No	No	No	**Yes**	**Yes**	**Yes**
CONCLUSION: Infection by parasitic *Ribeiroia* may cause abnormalities in the limb development of Pacific tree frogs.

1.13 Comparative Experiments Look for Differences between Groups Pieter Johnson analyzed the differences between ponds in which deformed frogs were present versus nearby ponds in which there were no deformed frogs. Such comparisons can result in valuable insights.

powerful tool, also called the *hypothesis–prediction (H–P) method*, is a conceptual approach that provides a strong foundation for making advances in biological knowledge. The scientific method has five steps:

- Making *observations*
- Asking *questions*
- Forming *hypotheses*, or tentative answers to the questions
- Making *predictions* based on the hypotheses
- *Testing* the predictions by making additional observations or conducting experiments

Once a question has been posed, a scientist uses *inductive logic* to propose a tentative answer to the question. That tentative answer is called a **hypothesis**. For example, at the opening of this chapter, you learned that Pieter Johnson was shown abnormal frogs gathered in certain ponds. The first question stimulated by this observation was, is there something in these ponds that caused frogs to develop such extreme anatomical abnormalities?

In formulating a hypothesis, scientists put together the facts they already know to formulate one or more possible answers to the question. Pieter knew that there were likely to be contaminants in the ponds where deformed frogs were found because agricultural pesticides were used heavily in the region. In addition, mercury had once been mined nearby, and the abandoned mines could be a source of heavy metals in the water. He also knew that there were nearby ponds in which the frogs were normal. His first hypothesis, therefore, was that contaminants in the water caused mutations in the frog eggs.

The next step in the scientific method is to apply a different form of logic—*deductive logic*—to make predictions based on the hypothesis. Based on his hypothesis, Pieter predicted (1) that he would find contaminants in the ponds with the abnormal frogs, and (2) that eggs from those ponds would produce abnormal frogs when they were hatched in the laboratory.

Good experiments have the potential of falsifying hypotheses

Once predictions are made from a hypothesis, **experiments** can be designed to test those predictions. The most informative experiments are those that have the ability to show that the prediction is wrong. If the prediction is wrong, the hypothesis must be questioned, modified, or rejected.

Both of Pieter Johnson's initial predictions proved to be wrong. He counted frogs and other organisms in 35 ponds in the region where the deformed frogs had been found and measured chemicals in the water. Thirteen of the ponds were home to Pacific tree frogs, but he found deformed frogs in only four ponds. To Pieter's surprise, analysis of the water samples failed to reveal higher amounts of pesticides, industrial chemicals, or heavy metals in the ponds with deformed frogs. Also surprisingly, when he collected eggs from those ponds and hatched them in the laboratory, he always got normal frogs. The original hypothesis that contaminants caused mutations in the frog eggs had to be rejected. A new hypothesis had to be formulated and new experiments had to be conducted.

There are two general types of experiments and Pieter used both:

- In a **comparative experiment**, we predict that there will be a difference between samples or groups based on our hypothesis. We then test whether or not the predicted difference exists.
- In a **controlled experiment**, we also compare samples or groups, but in this case we start the experiment with groups that are as similar as possible. We predict on the basis of our hypothesis that some factor, or *variable*, plays a role in the phenomenon we are investigating. We then use some method to manipulate that variable in an "experimental" group while leaving the "control" group unaltered. We then

test to see if the manipulation created the predicted difference between the experimental and control groups.

COMPARATIVE EXPERIMENTS Comparative experiments are valuable when we do not know or cannot control the critical variables. Pieter Johnson performed a comparative experiment when he tested the water in the ponds (**Figure 1.13**). His challenge was to find some variable that differed between the ponds with normal and abnormal frogs. Finding no differences in the water chemistry of the two types of ponds, he had to reject his hypothesis that environmental contaminants were causing mutations in the frogs. So he compared the two types of ponds to see what variables *were* different between them.

Pieter found that a species of freshwater snail was present in the ponds with abnormal frogs, but absent from the ponds with normal ones. Freshwater snails are hosts for many parasites. His new hypothesis was that a parasite infecting the snail was in some way responsible for the frogs' deformities. To test that hypothesis, he performed controlled experiments.

CONTROLLED EXPERIMENTS In controlled experiments, one variable is manipulated while others are held constant. The variable that is manipulated is called the *independent variable* and the response that is measured is the *dependent variable*. A good controlled experiment is not easy to design because biological variables are so interrelated that it is difficult to alter just one.

Many parasites go through complex life cycles with several stages, each of which requires a specific host animal. Pieter focused on the possibility that some parasite that used freshwater snails as one of its hosts was infecting the frogs and causing their deformities. Pieter found a candidate parasite with this type of life cycle: a small flatworm called *Ribeiroia*, which was present in the ponds where the deformed frogs were found.

In Pieter's controlled experiment, the independent variable was the presence or absence of the snail and the parasite (**Figure 1.14**). He controlled all other variables by collecting frog eggs from ponds where there were no snails or flatworm parasites and hatching them in the laboratory. He divided the resulting tadpoles into two groups and placed them in separate tanks. He introduced snails and parasites into half of the tanks (the experimental group) and left the other tanks (the control group) free of snails and parasites. His dependent variable was the frequency of abnormalities in the frogs that developed under the different sets of conditions. He found that 85 percent of the frogs in the experimental tanks with *Ribeiroia* alone, but none of the frogs in the other tanks, developed abnormalities. Thus Pieter's results supported his hypothesis, and he could go on to investigate how the parasites caused abnormalities in the developing frogs.

Ribeiroia uses three hosts in California ponds: snails, frogs, and predatory birds such as herons. For the parasite to complete its life cycle and reproduce, it must be able to move from a frog to a bird. The limb deformities *Ribeiroia* causes may actually make infected frogs easier for predatory birds to capture and eat.

EXPERIMENT

HYPOTHESIS: Infection of Pacific tree frog tadpoles by the parasite *Ribeiroia* causes developmental limb abnormalities.

METHOD

1. Collect *Hyla regilla* eggs from a site with no record of abnormal frogs.
2. Allow eggs to hatch in laboratory aquaria. Randomly divide equal numbers of the resulting tadpoles into control and experimental groups.
3. Allow the control group to develop normally. Subject the experimental groups to infection with *Ribeiroia*, a different parasite (*Alaria*), and a combination of both parasites.
4. Follow tadpole development. Count and assess the resulting adult frogs.

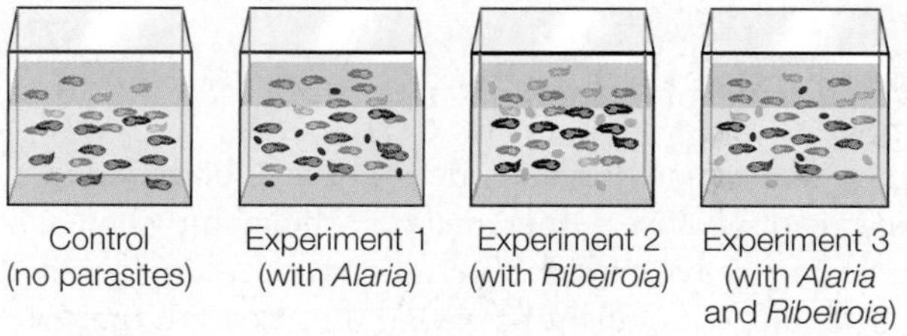

Control (no parasites) | Experiment 1 (with *Alaria*) | Experiment 2 (with *Ribeiroia*) | Experiment 3 (with *Alaria* and *Ribeiroia*)

RESULTS

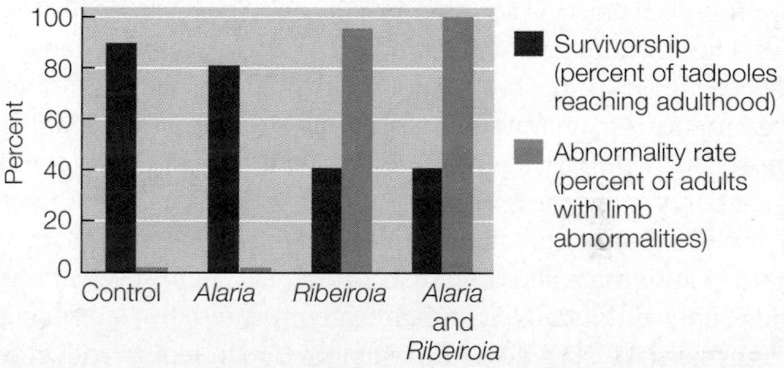

Percent: 0, 20, 40, 60, 80, 100

Control | *Alaria* | *Ribeiroia* | *Alaria* and *Ribeiroia*

Survivorship (percent of tadpoles reaching adulthood)

Abnormality rate (percent of adults with limb abnormalities)

CONCLUSION: *Ribeiroia* causes developmental limb abnormalities in Pacific tree frogs.

1.14 Controlled Experiments Manipulate a Variable The variable Johnson manipulated was the presence or absence of two species of parasitic flatworm. Other conditions of the experiment remained constant.

Statistical methods are essential scientific tools

Whether we are doing comparative or controlled experiments, at the end we have to decide whether there is a difference between the samples, individuals, groups, or populations in the study. How do we decide whether a measured difference is enough to support or falsify a hypothesis? In other words, how do we decide in an unbiased, objective way that the measured difference is significant?

Significance can be measured with statistical methods. Scientists use statistics because they recognize that variation is ubiquitous. Statistical tests analyze that variation and calculate the probability that the differences observed could be due to random variation. The results of statistical tests are therefore probabilities. A statistical test starts with a **null hypothesis**—the premise that no difference exists. When quantified observations, or **data**, are collected, statistical methods are applied to those data to calculate the likelihood that the null hypothesis is correct.

More specifically, statistical methods tell us the probability of obtaining the same results by chance even if the null hypothesis were true. Put another way, we need to eliminate insofar as possible the chance that any differences showing up in the data are merely the result of random variation in the samples tested. Scientists generally conclude that the differences they measure are significant if the statistical tests show that the *probability of error* (the probability that the results can be explained by chance) is 5 percent or lower. In particularly critical experiments, such as tests of the safety of a new drug, scientists require much lower probabilities of error, such as 1 percent or even 0.1 percent.

Not all forms of inquiry are scientific

Science is a unique human endeavor that is bounded by certain standards of practice. Other areas of scholarship share with science the practice of making observations and asking questions, but scientists are distinguished by what they do with their observations and how they answer their questions. Data, subjected to appropriate statistical analysis, are critical in the testing of hypotheses. The scientific method is the most powerful way humans have devised for learning about the world and how it works. Scientific explanations for natural processes are objective and reliable because the hypotheses proposed *must be testable* and *must have the potential of being rejected* by direct observations and experiments. Scientists clearly describe the methods they have used to test hypotheses so that other scientists can repeat their observations or experiments. Not all experiments are repeated, but surprising or controversial results are always subjected to independent verification. All scientists worldwide share this built-in process of testing and rejecting hypotheses, so they all contribute to a common body of scientific knowledge.

If you understand the methods of science, you can distinguish science from non-science. Art, music, and literature are activities that contribute to the quality of human life, but they are not science. They do not use the scientific method to establish what is fact. Religion is not science, although religions have historically purported to explain natural events ranging from unusual weather patterns to crop failures to human diseases and mental afflictions. Many such phenomena that at one time were mysterious are now explicable in terms of scientific principles.

The power of science derives from the uncompromising objectivity and absolute dependence on evidence that comes from *reproducible and quantifiable observations*. A religious or spiritual explanation of a natural phenomenon may be coherent and satisfying for the person or group holding that view, but it is not testable, and therefore it is not science. To invoke a supernatural explanation (such as an "intelligent designer" with no known bounds) is to depart from the world of science.

Science describes the facts about how the world works, not how it "ought to be." Many of the recent scientific advances that have contributed so much to human welfare also raise major ethical issues. Developments in genetics and developmental biology, for example, now enable us to select the sex of our children, to use stem cells to repair our bodies, and to modify the human genome. Although scientific knowledge allows us to do these things, science cannot tell us whether or not we should do them, or if we choose to do so, how we should regulate them.

Making wise decisions about such issues requires a clear understanding of the implications of available scientific information. Success in surgery depends on an accurate diagnosis. So does success in environmental management. However, to make wise decisions about public policy, we also need to employ the best possible ethical reasoning in deciding which outcomes we should strive for. For a bright future, society needs both good science and good ethics, as well as an educated public that understands the importance of both and the critical differences between them.

1.3 RECAP

The scientific method of inquiry starts with the formulation of hypotheses based on observations and data. Comparative and controlled experiments are carried out to test hypotheses.

- Can you explain the relationship between a hypothesis and an experiment? See p. 14
- What features characterize questions that can be answered only by using a comparative approach? See p. 14 and Figure 1.13
- What is controlled in a controlled experiment? See p. 15 and Figure 1.14
- Do you understand why arguments must be supported by quantifiable and reproducible data in order to be considered scientific? See p. 16

The vast amount of scientific knowledge accumulated over centuries of human civilization allows us to understand and manipulate aspects of the natural world in ways that no other species can. These abilities present us with challenges, opportunities, and, above all, responsibilities. Let's look at how knowledge of biology can affect the formulation of public policy.

1.4 How Does Biology Influence Public Policy?

The study of biology has long had major implications for human life. Agriculture and medicine are two important human activities that depend on biological knowledge. Our ancestors unknowingly applied the principles of evolutionary biology when they domesticated plants and animals. People have also been speculating about the causes of diseases and searching for methods of combating them since ancient times. Long before the causes of diseases were known, people recognized that diseases could be passed from one person to another. Isolation of infected persons has been practiced as long as written records have been available, but most so-called cures were not effective until scientists found out what caused diseases.

Today, thanks to the deciphering of genomes and the ability to manipulate them, vast new possibilities exist for improvements in the control of human diseases and agricultural productivity. At the same time, these capabilities have raised important ethical and policy issues. How much and in what ways should we tinker with the genetics of humans and other species? Does it matter whether our crops and domesticated animals are changed by traditional breeding experiments or by gene transfers? What rules should govern the release of genetically modified organisms into the environment? Science alone cannot provide answers to those questions, but wise policy decisions must be based on accurate scientific information.

Another reason for studying biology is to understand the effects of the vastly increased human population on its environment. Our use of natural resources is putting stress on the ability of Earth's ecosystems to continue to produce the goods and services on which our society depends. Human activities are changing global climates, causing the extinctions of a large number of species, and spreading new diseases while facilitating the resurgence of old ones. The rapid spread of the SARS and West Nile viruses, for example, was facilitated by modern modes of transportation, and the recent resurgence of tuberculosis is the result of the evolution of bacteria that are resistant to antibiotics. Biological knowledge is vital for determining the causes of these changes and for devising wise policies to deal with them. An understanding of biology also helps people appreciate the marvelous diversity of living organisms that provides goods and services for humankind and also enriches our lives aesthetically and spiritually.

Biologists are increasingly called on to advise government agencies concerning the laws, rules, and regulations by which society deals with the increasing number of problems and challenges that have at least a partial biological basis. As an example of the value of scientific knowledge for the assessment and formulation of public policy, let's return to the tracking study of bluefin tuna introduced in Section 1.3. Prior to this study, both scientists and fishermen knew that bluefins had a western Atlantic breeding ground in the Gulf of Mexico and an eastern Atlantic breeding ground in the Mediterranean Sea. Overfishing was endangering the western breeding population. Everyone assumed that the fish from the two breeding populations had geographically separate feeding grounds as well as separate breeding grounds, so an international commission drew a line down the middle of the Atlantic Ocean and established stricter fishing quotas on the western side of the line. The intent was to allow the western population to recover. However, new data revealed that in fact the eastern and western bluefin populations mix freely on the feeding (and hence fishing) grounds across the entire North Atlantic (**Figure 1.15**). Thus a fish caught on the eastern side of the line could be from the western breeding population, so the established policy was not appropriate for achieving its intended goal.

Throughout this book we will share with you the excitement of studying living things and illustrate the rich array of methods that biologists use to determine why the world of living things looks and functions as it does. The most important motivator of most biologists is curiosity. People are fascinated by the richness and diversity of life and want to learn more about organisms and how they interact with one another. The trait of human curiosity

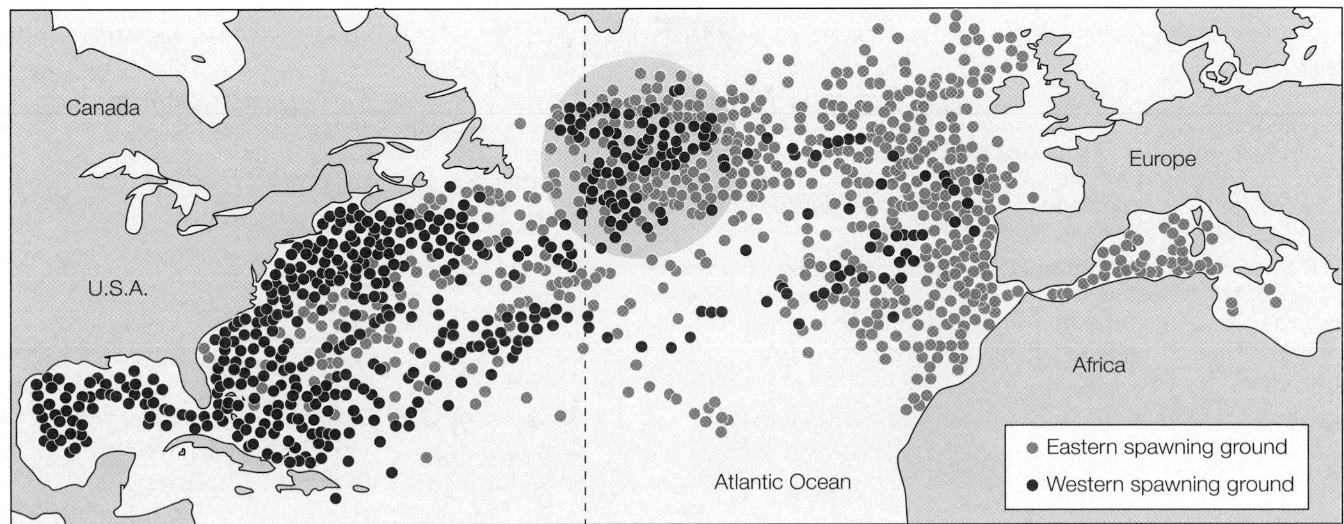

1.15 Bluefin Tuna Do Not Recognize the Lines Drawn on Maps by International Commissions Because it was assumed that western (red dots) and eastern (gold dots) breeding populations of bluefin tuna also fed on their respective sides of the Atlantic Ocean, separate fishing quotas were established to either side of 45°W longitude (dashed line). It was believed this would allow the endangered western population to recover. However, tracking data showed that the two populations mix freely, especially in the heavily fished waters of the northernmost Atlantic (blue circle); so in fact the established policy does not protect the western population.

might even be seen as adaptive, and could have been selected for if individuals who were motivated to learn about their surroundings were likely to have survived and reproduced better, on average, than their less curious relatives!

There are vast numbers of questions for which we do not yet have answers, and new discoveries usually engender questions no one thought to ask before. Perhaps you will eventually pose and answer one or more of those questions.

CHAPTER SUMMARY

1.1 What is biology?

Biology is the study of life at all levels of organization, ranging from molecules to the biosphere.

The **cell theory** states that all life consists of cells, and all cells come from preexisting cells.

All living organisms are related to one another through descent with modification. **Evolution** by **natural selection** is responsible for the diversity of **adaptations** found in living organisms.

The instructions for a cell are contained in its **genome**, which consists of DNA molecules made up of sequences of **nucleotides**. Specific segments of DNA called **genes** contain the information the cell uses to make **proteins**. Review Figure 1.4

Cells are the basic structural and physiological units of life. Most of the chemical reactions of life take place in cells. Living organisms control their internal environment. They also interact with other organisms of the same and different species. Biologists study life at all these levels of organization. Review Figure 1.6, Web/CD Activity 1.1

Biological knowledge obtained from a **model system** may be generalized to other species.

1.2 How is all life on Earth related?

Biologists use fossils, anatomical similarities and differences, and molecular comparisons of genomes to reconstruct the history of life. Review Figure 1.9

Life first arose by chemical evolution. Biological evolution began with the formation of cells.

Photosynthesis was an important evolutionary step because it changed Earth's atmosphere and provided a means of capturing energy from sunlight.

The earliest organisms were **prokaryotes**; organisms with more complex cells, called **eukaryotes**, arose later. Eukaryotic cells have discrete intracellular compartments, called **organelles**, including a **nucleus** that contains the cell's genetic material.

The genetic relationships of **species** can be represented as an evolutionary tree. Species are grouped into three **domains**: **Archaea**, **Bacteria**, and **Eukarya**. The domains Archaea and Bacteria consist of unicellular prokaryotes. The domain Eukarya contains the microbial eukaryotes (protists), plants, fungi, and animals. Review Figure 1.11, Web/CD Activity 1.2

1.3 How do biologists investigate life?

The **scientific method** used in most biological investigations involves five steps: making observations, asking questions, forming hypotheses, making predictions, and testing those predictions.

Hypotheses are tentative answers to questions. Predictions made on the basis of a hypothesis are tested with additional observations and two kinds of **experiments**: **comparative** and **controlled experiments**. Review Figures 1.13 and 1.14

Statistical methods are applied to **data** to establish whether or not the differences observed are significant or whether they could be expected by chance. These methods start with the **null hypothesis** that there are no differences.

Science can tell us how the world works, but it cannot tell us what we should or should not do.

1.4 How does biology influence public policy?

Wise public policy decisions must be based on accurate scientific information. Biologists are often called on to advise governmental agencies on the solution of important problems that have a biological component.

FOR DISCUSSION

1. Even if we knew the sequences of all of the genes of a single-celled organism and could cause those genes to be expressed in a test tube, we still could not create one of those organisms in the test tube. Why do you think this is so? In light of this fact, what do you think of the statement that the genome contains all of the information for a species?
2. If someone told you that giraffes developed long necks because they stretched their necks to reach leaves higher and higher on trees, how would you help that person think about giraffes more accurately, in terms of evolution by natural selection?
3. In a recent discovery of the genes that control skin color in zebrafish, why did the biologists assume that the same genes might be responsible for skin color in humans?
4. Why is it so important in science that we design and perform tests capable of falsifying a hypothesis?
5. What features characterize questions that can be answered only by using a comparative approach?

FOR INVESTIGATION

1. The abnormalities of frogs in Pieter Johnsons's study were associated with the presence of a parasite. How would you investigate how the parasites induced the formation of monster frogs? Hint: When tadpoles were exposed to the parasites *after* they began to develop legs, they did not show abnormalities.

2. Just as all cells come from preexisting cells, mitochondria—the cell organelles that convert energy in food to a form of energy that can do biological work—all come from preexisting mitochondria. Cells do not synthesize mitochondria from the genetic information in their nuclei. What investigations would you carry out to understand the nature of mitochondria?

CHAPTER 2 The Chemistry of Life

Where there is water, there can be life

On July 14, 2005, the spacecraft *Cassini*, launched 8 years earlier by NASA, flew 168 kilometers above the south pole of Enceladus, one of Saturn's moons. As it approached this small, frigid body, *Cassini*'s instruments relayed information on chemicals in the atmosphere while its camera snapped photos. Back on Earth, scientists looking at this data were in for a big surprise. Photographs showed an immense plume of water vapor, ice particles, and liquid water that spewed from the moon like the geysers of Earth's Yellowstone Park. Since the temperature on Enceladus is about −200°C, surface water would freeze solid; thus the source of the plume must have been a pool of liquid water below the surface, probably heated by molten rocks.

Meanwhile, a little closer to home, two robotic vehicles were looking for water on Mars. One of them landed on what photographs indicated was a huge dry lake bed. With instructions from scientists on Earth, the vehicle dug up and analyzed Martian soil samples, finding a mineral called dry hematite that is often a chemical signature of the presence of water. In addition, there was salt residue, suggesting that water had evaporated. These discoveries by geologists sparked the interest of biologists, because where there is water, there can be life.

Put another way, there is good reason to believe that life as we know it cannot exist without water. Animals and plants could not survive on Earth's land masses until adaptations evolved that allowed them to retain the water that makes up some 70 percent of their bodies. Aquatic organisms, of course, do not need such water-retention mechanisms, which leads biologists to conclude that life originated in a watery environment. That environment need not have been the lakes, rivers, and oceans with which we are familiar. Living organisms have been found in hot springs at temperatures above the usual boiling point of water, in a lake beneath the frozen Antarctic ice, in water trapped 2 miles below Earth's surface, in water 3 miles below the surface of the sea, in extremely acidic and extremely salty water, and even in the water that cools the interiors of nuclear reactors.

Biologists' attention focuses not only on the presence of water, but on what is dissolved in that water. A major discovery of biology is that living things are composed of the same chemical elements as the vast nonliving por-

Geysers on a Frozen Moon Images from the spacecraft *Cassini*, enhanced with computer-generated color, show huge plumes of water and water vapor being sprayed from the south polar surface of Saturn's moon Enceladus. Scientists believe that these jets are geysers erupting from pools of liquid water, heated by volcanic activity, that may lie just below the frozen surface of this moon.

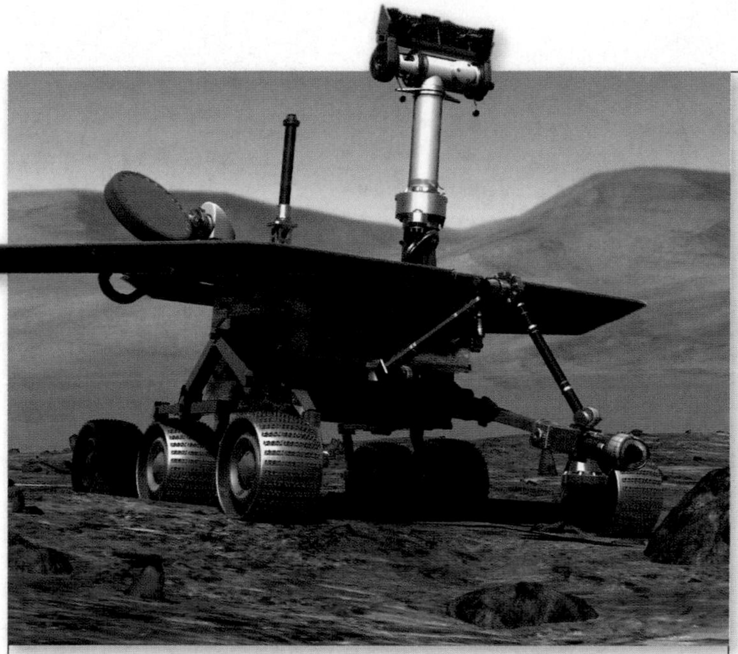

Looking for Water The robotic rover vehicle *Opportunity*, shown here in a NASA mock-up, was sent to Mars to look for evidence of water. The rover's instruments found salt residue and other evidence of evaporated water in the rock and sand of a region that appears to have once been at the bottom of a huge lake.

tion of the universe. This *mechanistic* view—that life is chemically based and obeys universal laws of chemistry and physics—is relatively new in human history. Until the nineteenth century, a "vital force" (from the Latin *vitalis*, "of life"), distinct from the mechanistic forces governing physics and chemistry, was presumed to be responsible for life. Many people still assume that a vital force exists, but the mechanistic view of life has led to great advances in biological science and is the cornerstone of modern medicine and agriculture. The search for life on Mars and the description of life here on Earth both begin with chemistry.

IN THIS CHAPTER we will introduce the constituents of matter: atoms. We will examine their variety, their properties, and their capacity to combine with other atoms. Then we will consider how matter changes. In addition to changes in state (solid to liquid to gas), substances undergo chemical reactions that transform both their composition and their characteristic properties. Finally, we will take a closer look at the structure and properties of water and its relationship to chemical acids and bases.

CHAPTER OUTLINE

2.1 What Are the Chemical Elements That Make Up Living Organisms?

All matter is composed of **atoms**. Atoms are tiny—more than a trillion (10^{12}) of them could fit on top of the period at the end of this sentence. Each atom consists of a dense, positively charged **nucleus**, around which one or more negatively charged **electrons** move (**Figure 2.1**). The nucleus contains one or more **protons** and may contain one or more **neutrons**. Atoms and their component particles have volume and mass, which are properties of all matter. *Mass* measures the quantity of matter present; the greater the mass, the greater the quantity of matter.

The mass of a proton serves as a standard unit of measure called the **atomic mass unit** (**amu**), or *dalton* (named after the English chemist John Dalton). A single proton or neutron has a mass of about 1 dalton (Da), which is 1.7×10^{-24} grams (0.0000000000000000000000017 g). The mass of an electron is 9×10^{-28} g (0.0005 Da). Because the mass of an electron is negligible compared with the mass of a proton or a neutron, the contribution of electrons to the mass of an atom can usually be ignored when measurements and calculations are made. It is electrons, however, that determine how atoms will interact in chemical reactions, and we will discuss them extensively later in this chapter.

Each proton has a positive electric charge, defined as +1 unit of charge. An electron has a negative charge equal and opposite to that of a proton; thus the charge of an electron is –1 unit. The neutron, as its name suggests, is electrically neutral, so its charge is 0. Charges that are not alike (+/–) attract each other, whereas charges that are alike (+/+, –1–1) repel each other. Atoms are electrically neutral because the number of electrons in an atom equals the number of protons.

An element consists of only one kind of atom

An **element** is a pure substance that contains only one kind of atom. The element hydrogen consists only of hydrogen atoms; the element iron consists only of iron atoms. The atoms of each element have certain characteristics or properties that distinguish them from the atoms of other elements. The more than 100 elements found in the universe are arranged in the *periodic table* (**Figure 2.2**). These elements are not found in equal amounts. Stars have abundant hydrogen and helium. Earth's crust, and the surfaces of the neighboring planets, are almost

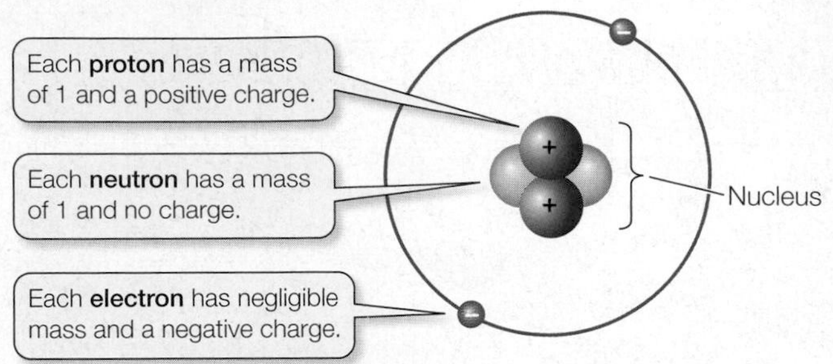
Each **proton** has a mass of 1 and a positive charge.
Each **neutron** has a mass of 1 and no charge.
Each **electron** has negligible mass and a negative charge.
Nucleus

2.1 The Helium Atom This representation of a helium atom is called a Bohr model. It exaggerates the space occupied by the nucleus. In reality, although the nucleus accounts for virtually all of the atomic mass, it occupies only about 1/10,000 of the atom's volume.

half oxygen, 28 percent silicon, 8 percent aluminum, and 2–5 percent each of sodium, magnesium, potassium, calcium, and iron; they contain much smaller amounts of the other elements.

About 98 percent of the mass of every living organism (bacterium, turnip, or human) is composed of just six elements: carbon, hydrogen, nitrogen, oxygen, phosphorus, and sulfur. The chemistry of these six elements will be our primary concern, but other elements found in living things are important as well. Sodium and potassium, for example, are essential for nerve function; calcium can act as a biological signal; iodine is a component of a vital hormone; and magnesium and molybdenum are essential to plants (magnesium as part of their chlorophyll pigment, and molybdenum for incorporating nitrogen into biologically useful substances).

Protons: Their number identifies an element

An element differs from other elements by the number of protons in each of its atoms. This **atomic number** is unique to each element and does not change. An atom of helium always has 2 protons, and an atom of oxygen always has 8 protons; the atomic numbers of helium and oxygen are thus 2 and 8, respectively.

2.2 The Periodic Table The periodic table groups the elements according to their physical and chemical properties. Elements 1–92 occur in nature; elements with atomic numbers above 92 were created in the laboratory.

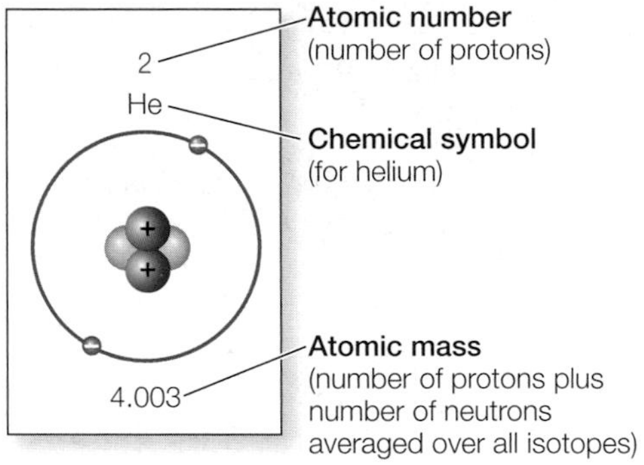

2 — **Atomic number** (number of protons)
He — **Chemical symbol** (for helium)
4.003 — **Atomic mass** (number of protons plus number of neutrons averaged over all isotopes)

The six elements highlighted in yellow make up 98% of the mass of most living organisms.

Vertical columns contain elements with similar properties.

Elements highlighted in orange are present in small amounts in many organisms.

1 H 1.0079																	2 He 4.003
3 Li 6.941	4 Be 9.012											5 B 10.81	6 C 12.011	7 N 14.007	8 O 15.999	9 F 18.998	10 Ne 20.179
11 Na 22.990	12 Mg 24.305											13 Al 26.982	14 Si 28.086	15 P 30.974	16 S 32.06	17 Cl 35.453	18 Ar 39.948
19 K 39.098	20 Ca 40.08	21 Sc 44.956	22 Ti 47.88	23 V 50.942	24 Cr 51.996	25 Mn 54.938	26 Fe 55.847	27 Co 58.933	28 Ni 58.69	29 Cu 63.546	30 Zn 65.38	31 Ga 69.72	32 Ge 72.59	33 As 74.922	34 Se 78.96	35 Br 79.909	36 Kr 83.80
37 Rb 85.4778	38 Sr 87.62	39 Y 88.906	40 Zr 91.22	41 Nb 92.906	42 Mo 95.94	43 Tc (99)	44 Ru 101.07	45 Rh 102.906	46 Pd 106.4	47 Ag 107.870	48 Cd 112.41	49 In 114.82	50 Sn 118.69	51 Sb 121.75	52 Te 127.60	53 I 126.904	54 Xe 131.30
55 Cs 132.905	56 Ba 137.34	71 Lu 174.97	72 Hf 178.49	73 Ta 180.948	74 W 183.85	75 Re 186.207	76 Os 190.2	77 Ir 192.2	78 Pt 195.08	79 Au 196.967	80 Hg 200.59	81 Tl 204.37	82 Pb 207.19	83 Bi 208.980	84 Po (209)	85 At (210)	86 Rn (222)
87 Fr (223)	88 Ra 226.025	103 Lr (260)	104 Rf (261)	105 Db (262)	106 Sg (266)	107 Bh (264)	108 Hs (269)	109 Mt (268)	110 (269)	111 (272)	112 (277)	113	114 (285)	115 (289)	116	117	118 (293)

Masses in parentheses indicate unstable elements that decay rapidly to form other elements.

Elements without a chemical symbol are as yet unnamed.

Lanthanide series	57 La 138.906	58 Ce 140.12	59 Pr 140.9077	60 Nd 144.24	61 Pm (145)	62 Sm 150.36	63 Eu 151.96	64 Gd 157.25	65 Tb 158.924	66 Dy 162.50	67 Ho 164.930	68 Er 167.26	69 Tm 168.934	70 Yb 173.04
Actinide series	89 Ac 227.028	90 Th 232.038	91 Pa 231.0359	92 U 238.02	93 Np 237.0482	94 Pu (244)	95 Am (243)	96 Cm (247)	97 Bk (247)	98 Cf (251)	99 Es (252)	100 Fm (257)	101 Md (258)	102 No (259)

Along with a definitive number of protons, every element except hydrogen has one or more neutrons in its nucleus. The **mass number** of an atom is the total number of protons and neutrons in its nucleus. The nucleus of a carbon atom contains 6 protons and 6 neutrons, and has a mass number of 12. Oxygen has 8 protons and 8 neutrons, and has a mass number of 16. The mass number is essentially the mass of the atom in daltons.

Each element has its own one- or two-letter chemical symbol. For example, H stands for hydrogen, C for carbon, and O for oxygen. Some symbols come from other languages: Fe (from the Latin, *ferrum*) stands for iron, Na (Latin, *natrium*) for sodium, and W (German, *wolfram*) for tungsten.

In text, immediately preceding the symbol for an element, the atomic number is written at the lower left and the mass number at the upper left. Thus hydrogen, carbon, and oxygen are written as ${}^{1}_{1}H$, ${}^{12}_{6}C$, and ${}^{16}_{8}O$, respectively.

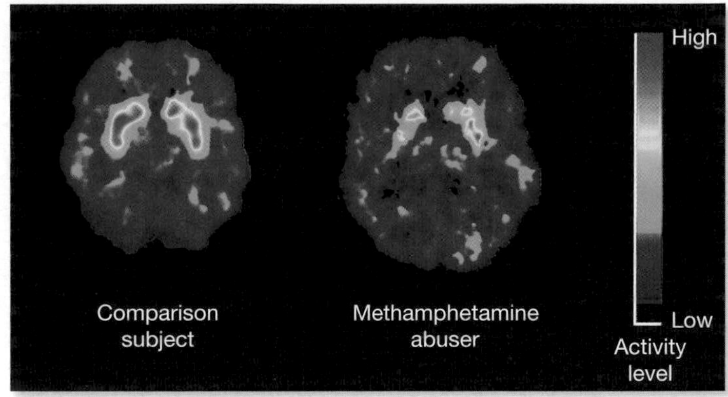

2.4 Tagging the Brain A radioactively labeled sugar detects differences between the brain activity of a healthy person and that of a person abusing methamphetamines. The more active a brain region is, the more sugar it takes up. The healthy brain (left) shows more activity in the region involved in memory (the red area) than the drug abuser's brain does.

Neutrons: Their number differs among isotopes

In some elements, the number of neutrons in the atomic nucleus is not constant. **Isotopes** of the same element all have the same, definitive, number of protons, but differ in their number of neutrons. Many elements have several isotopes. The isotopes of hydrogen shown in **Figure 2.3** have special names, but the isotopes of most elements do not have distinct names. The natural isotopes of carbon, for example, are ${}^{12}C$ (6 neutrons in the nucleus), ${}^{13}C$ (7 neutrons), and ${}^{14}C$ (8 neutrons). Note that all three (pronounced "carbon-12," "carbon-13," and "carbon-14") have 6 protons, so they are all carbon. Most carbon atoms are ${}^{12}C$, about 1.1 percent are ${}^{13}C$, and a tiny fraction are ${}^{14}C$. An element's atomic mass, or **atomic weight**, is the average of the mass numbers of a representative sample of atoms of the element, with all isotopes in their normally occurring proportions. The atomic weight of carbon, taking into account all of its isotopes and their abundances, is thus calculated to be 12.011.

Most isotopes are stable. But some, called **radioisotopes**, are unstable and spontaneously give off energy in the form of α (alpha), β (beta), or γ (gamma) radiation from the atomic nucleus. Known as *radioactive decay*, this release of energy transforms the original atom. These transformations can extend even to a change in the number of protons, so that the original atom is now a different element. The energy released by radioactive decay can interact with surrounding substances. Scientists can incorporate such radiation-sensitive substances into instruments that allow them to detect the presence of radioisotopes. For instance, if an earthworm is given food to which a radioisotope has been added, its path through the soil can be followed by a simple detector called a Geiger counter. If a radioisotope is incorporated into a molecule, it acts as a tag or label, allowing researchers and physicians to trace that molecule and to identify any changes the molecule undergoes inside the body (**Figure 2.4**). Radioisotopes are also used to date fossils, an application described in Section 21.1.

Although radioisotopes are useful in research and in medicine, even a low dose of the radiation they emit has the potential to damage molecules and cells. However, these damaging effects are sometimes used to our advantage; for example, the γ radiation from ${}^{60}Co$ (cobalt-60) is used in medical practice to kill cancer cells.

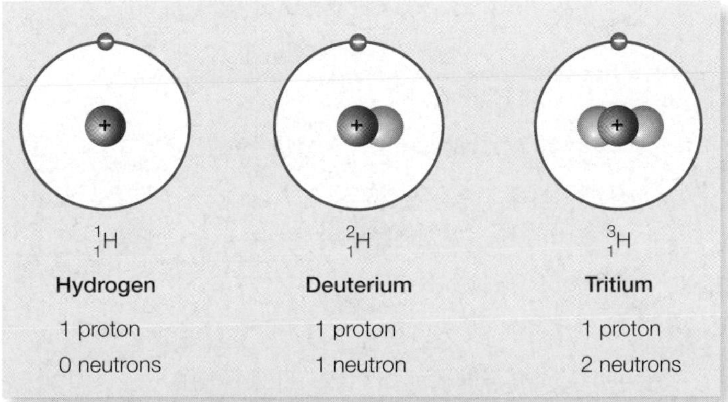

2.3 Isotopes Have Different Numbers of Neutrons The isotopes of hydrogen all have one proton in the nucleus, which defines them as hydrogen. Their differing mass numbers are due to different numbers of neutrons.

Electrons: Their behavior determines chemical bonding

The characteristic number of electrons in each atom of an element determines how its atoms will react with other atoms. Biologists are interested in how chemical changes take place in living cells. When considering atoms, they are concerned primarily with electrons because the behavior of electrons explains how chemical *reactions* occur. These reactions often cause the atomic composition of substances to be altered. Reactions usually involve changes in the distribution of electrons between atoms.

The location of a given electron in an atom at any given time is impossible to determine. We can only describe a volume of space within the atom where the electron is likely to be. The region of space where the electron is found at least 90 percent of the time is the electron's **orbital**. Orbitals have characteristic shapes and orientations, and a given orbital can be occupied by a maximum of two electrons. Thus any atom larger than helium (atomic number 2) must have electrons in two or more orbitals. The orbitals are filled in a specific sequence, in a series of what are known as **electron shells**, or *energy levels*, around the nucleus (**Figure 2.5**).

- *First shell*: The innermost electron shell consists of just one orbital, called an *s* orbital. A hydrogen atom has one electron in its first shell ($_1$H); helium has two ($_2$He). Atoms of all other elements have two or more shells to accommodate orbitals for additional electrons.
- *Second shell*: The second shell contains four orbitals (an *s* orbital and three *p* orbitals), and hence holds up to eight electrons.
- *Additional shells*: Elements with more than ten electrons have three or more electron shells. The farther a shell is from the nucleus, the higher the energy level is for an electron occupying that shell. (That is, the negatively charged electron needs to absorb a greater amount of energy to overcome the "pull" of the positively charged nucleus and remain in that shell.)

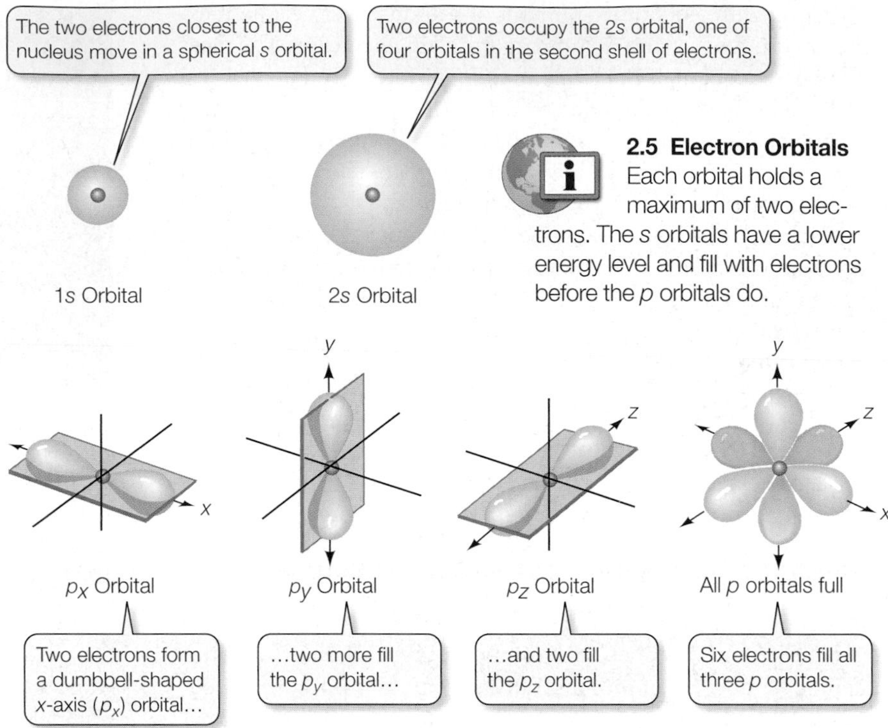
The two electrons closest to the nucleus move in a spherical *s* orbital.
Two electrons occupy the 2*s* orbital, one of four orbitals in the second shell of electrons.
1*s* Orbital
2*s* Orbital
y
z
x
p_x Orbital
p_y Orbital
p_z Orbital
All *p* orbitals full
Two electrons form a dumbbell-shaped *x*-axis (p_x) orbital…
…two more fill the p_y orbital…
…and two fill the p_z orbital.
Six electrons fill all three *p* orbitals.

2.5 Electron Orbitals Each orbital holds a maximum of two electrons. The *s* orbitals have a lower energy level and fill with electrons before the *p* orbitals do.

2.6 Electron Shells Determine the Reactivity of Atoms Each orbital holds a maximum of two electrons, and each shell can hold a specific maximum number of electrons. Each shell must be filled before electrons move into the next shell. The energy level of electrons is higher in shells farther from the nucleus. An atom with unpaired electrons in its outermost shell can react (bond) with other atoms.

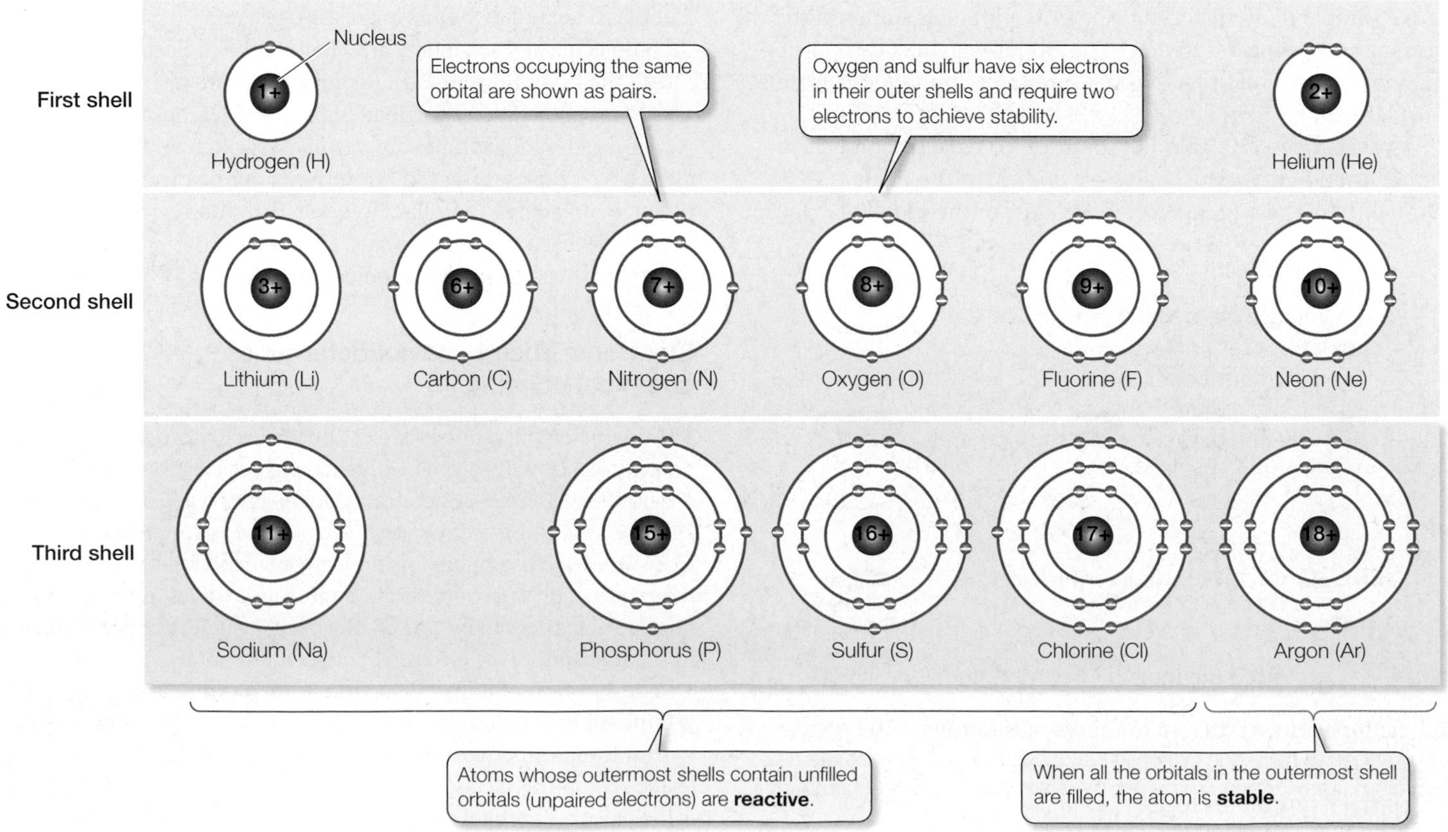

First shell
Nucleus
1+ Hydrogen (H)
Electrons occupying the same orbital are shown as pairs.
Oxygen and sulfur have six electrons in their outer shells and require two electrons to achieve stability.
2+ Helium (He)

Second shell
3+ Lithium (Li)
6+ Carbon (C)
7+ Nitrogen (N)
8+ Oxygen (O)
9+ Fluorine (F)
10+ Neon (Ne)

Third shell
11+ Sodium (Na)
15+ Phosphorus (P)
16+ Sulfur (S)
17+ Chlorine (Cl)
18+ Argon (Ar)

Atoms whose outermost shells contain unfilled orbitals (unpaired electrons) are **reactive**.
When all the orbitals in the outermost shell are filled, the atom is **stable**.

The *s* orbitals fill with electrons first, and their electrons have the lowest energy level. Subsequent shells have different numbers of orbitals, but the outermost shells usually hold only eight electrons. In any atom, the outermost electron shell (the *valence shell*) determines how the atom combines with other atoms—that is, how the atom behaves chemically. When a valence shell with four orbitals contains eight electrons, there are no unpaired electrons, and the atom is *stable*—it will not react with other atoms (**Figure 2.6**). Examples of chemically stable elements are helium, neon, and argon.

> Life on Earth is based on carbon chemistry. Because silicon shares many biochemical properties with carbon, scientists and science fiction writers alike have speculated about the possibility and probable nature of a silicon-based life form.

Non-stable, or *reactive*, atoms have unpaired electrons in their outermost shells. Reactive atoms can attain stability either by sharing electrons with other atoms or by losing or gaining one or more electrons. In either case, the atoms involved are *bonded* together into stable associations called **molecules**. The tendency of atoms in stable molecules to have eight electrons in their outermost shells is known as the *octet rule*. Many atoms in biologically important molecules—for example, carbon (C) and nitrogen (N)—follow this rule. An important exception is hydrogen (H), which attains stability when two electrons occupy its single shell (consisting of just one *s* orbital).

2.1 RECAP

The living world is composed of the same set of chemical elements as the rest of the universe. The structure of an atom—with its nucleus of protons and neutrons and its characteristic configuration of electrons in orbitals around the nucleus—determines its properties.

- Can you describe the arrangement of protons, neutrons, and electrons in an atom? See Figure 2.1
- Can you use the periodic table to identify some of the differences and similarities in atomic structure among different elements (for example, oxygen, carbon, and helium)? Do you understand how the configuration of the valence shell influences the placement of an element in the periodic table? See pp. 22–24 and Figures 2.2 and 2.6
- Do you understand how bonding can help a reactive atom achieve stability? See p. 25 and Figure 2.6

We have introduced the individual players on the biochemical stage—the atoms. We have shown how the energy levels of electrons drive an atomic "quest for stability." Next we will describe the different types of chemical bonds that can lead to stability, joining atoms together into molecular structures with hosts of different properties.

2.2 How Do Atoms Bond to Form Molecules?

A **chemical bond** is an attractive force that links two atoms together in a molecule. There are several kinds of chemical bonds (**Table 2.1**). In this section we will begin with *covalent bonds*, the strong bonds that result from the sharing of electrons. Next we will examine *ionic bonds*, which form when an atom gains or loses electrons to achieve stability. We will then consider other, weaker, kinds of interactions, including hydrogen bonds, that are enormously important to biology.

Covalent bonds consist of shared pairs of electrons

A **covalent bond** forms when two atoms attain stable electron numbers in their outermost shells by *sharing* one or more pairs of electrons. Consider two hydrogen atoms coming into close proximity, each with a single unpaired electron in its single shell (**Figure 2.7**). Each positively charged nucleus *attracts the other atom's* unpaired negatively charged electron, but this attraction is countered by each electron's *attraction to its own nucleus*. Thus the two unpaired electrons become shared, filling the shells of both atoms. The shared attraction links the two hydrogen atoms in a covalent bond, and a stable hydrogen gas molecule (H_2) is formed.

A **compound** is a molecule made up of atoms of two or more elements bonded together in a fixed ratio. Methane gas (CH_4), water (H_2O), and table sugar (sucrose, $C_{12}H_{22}O_{11}$) are examples of compounds. The chemical symbols identify the different elements in a compound, and the subscript numbers indicate how many atoms of each element are present. Every compound has a **molecular weight** (molecular mass) that is the sum of the atomic

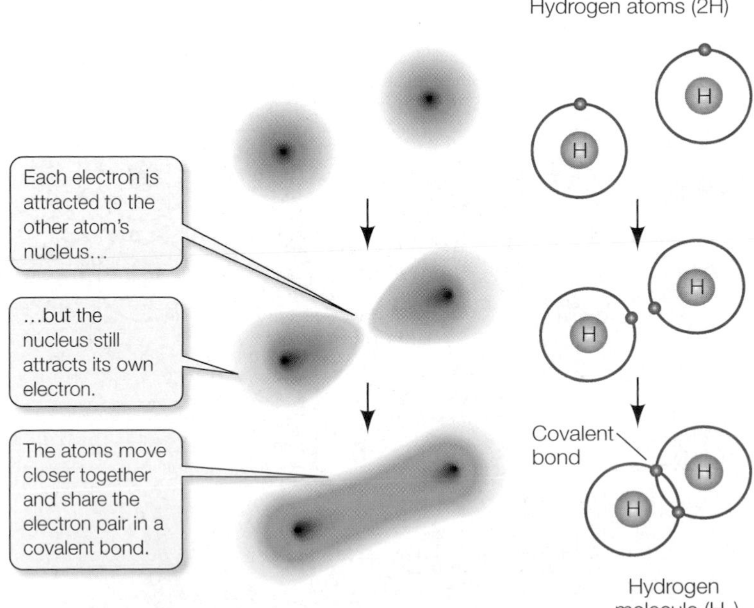
Hydrogen atoms (2H)
H
H
Each electron is attracted to the other atom's nucleus...
...but the nucleus still attracts its own electron.
H
H
The atoms move closer together and share the electron pair in a covalent bond.
Covalent bond
H
H
Hydrogen molecule (H_2)

2.7 Electrons Are Shared in Covalent Bonds Two hydrogen atoms can combine to form a hydrogen molecule. A covalent bond forms when the electron orbitals of the two atoms overlap.

TABLE 2.1

Chemical Bonds and Interactions

NAME	BASIS OF INTERACTION	STRUCTURE	BOND ENERGY[a] (KCAL/MOL)
Covalent bond	Sharing of electron pairs	H O; —N—C—	50–110
Ionic bond	Attraction of opposite changes	H, H; —N—H ⊕ ⊖ O—C—; O	3–7
Hydrogen bond	Sharing of H atom	H; δ+ δ−; —N—H•••••O=C—	3–7
Hydrophobic interaction	Interaction of nonpolar substances in the presence of polar substances (especially water)	H H H H; —C—C—H C—C; H H H H H	1–2
van der Waals interaction	Interaction of electrons of nonpolar substances	H—H · C (H, H, H, H)	1

[a]*Bond energy* is the amount of energy needed to separate two bonded or interacting atoms under physiological conditions.

weights of all atoms in the molecule. Looking at the periodic table in Figure 2.2, you can calculate the molecular weights of these three compounds to be 16, 18, and 342, respectively. Molecular weights are usually related to a molecule's size.

Consider the electrons involved in covalent bonding in methane gas. The carbon atom in this compound has six electrons: two electrons fill its inner shell, and four travel in its outer shell. Because its outer shell can hold up to eight electrons, carbon can share electrons with up to four other atoms—*it can form four covalent bonds* (**Figure 2.8A**). When an atom of carbon reacts with four hydrogen atoms, methane forms. Thanks to electron sharing, the outer shell of methane's carbon atom is now filled with eight electrons; the outer shell of each of the four hydrogen atoms is also filled. Four covalent bonds—four shared electron pairs—hold methane together.

Figure 2.8B shows several different ways to represent the molecular structure of methane. **Table 2.2** shows the covalent bonding capacities of some biologically significant elements.

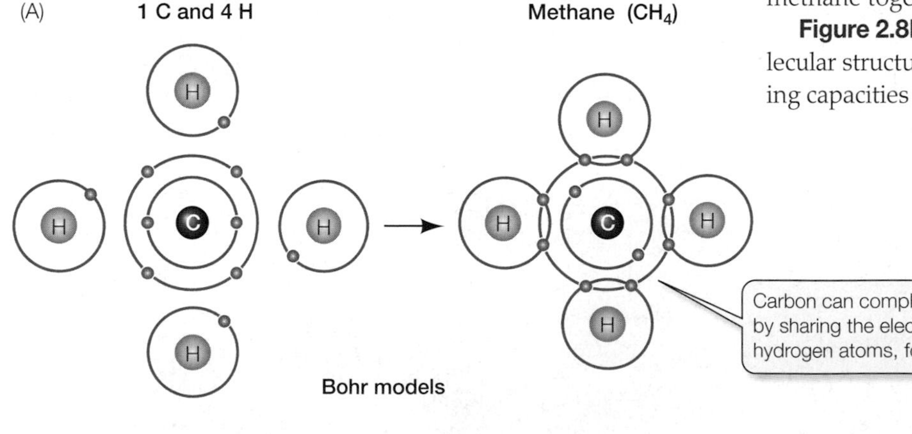

(B)

Each line or pair of dots represents a shared pair of electrons.

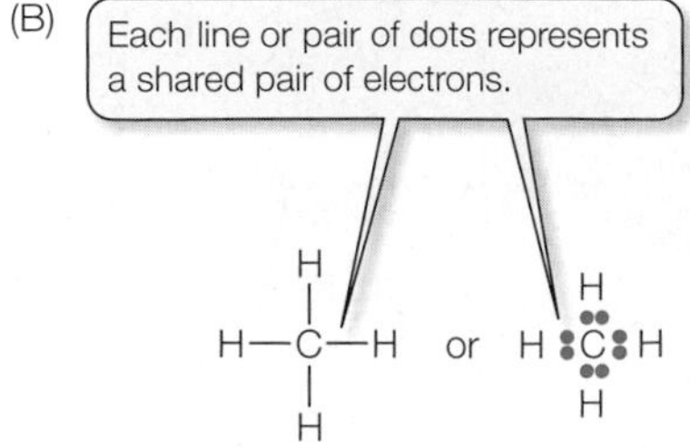

The hydrogen atoms form corners of a regular tetrahedron.

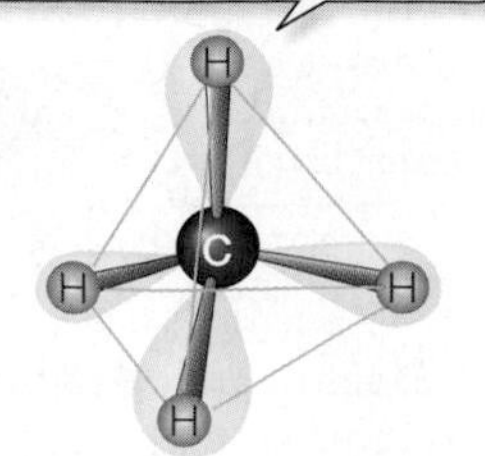

This space-filling model shows the shape methane presents to its environment.

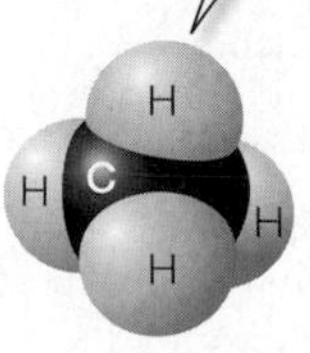

2.8 Covalent Bonding Can Form Compounds (A) Covalent bond formation in methane, whose molecular formula is CH_4. (B) Three different ways of representing the structure of methane. The ball-and-stick model and the space-filling model show the spatial orientation of the bonds.

TABLE 2.2

Covalent Bonding Capabilities of Some Biologically Important Elements

ELEMENT	USUAL NUMBER OF COVALENT BONDS
Hydrogen (H)	1
Oxygen (O)	2
Sulfur (S)	2
Nitrogen (N)	3
Carbon (C)	4
Phosphorus (P)	5

STRENGTH AND STABILITY Covalent bonds are very strong, meaning that it takes a lot of energy to break them. The thermal energy that biological molecules ordinarily have under physiological conditions (that is, the chemical and physical environment found in living tissues) is less than 1 percent of that needed to break covalent bonds. So biological molecules, most of which are put together with covalent bonds, are quite stable, as are their three-dimensional structures and the spaces they occupy. Most biochemical structures, ranging from the muscles you are using to move your eyes as you read this book to the paper it is printed on, are based on the stability of covalent bonds.

ORIENTATION For a given pair of elements—for example, carbon bonded to hydrogen—the length, angle, and direction of the covalent bonds are consistently the same, regardless of the larger molecule of which the particular bond is a part. The four filled orbitals around the carbon nucleus of methane, for example, are always distributed in space so that the bonded hydrogens define the corners of a regular tetrahedron, with carbon in the center (see Figure 2.8B). The three-dimensional structure formed by carbon and four hydrogens is the same in a large, complicated protein as it is in a simple methane molecule. This property of covalent bonds makes the prediction of biological structure possible. The shapes of molecules contribute to their biological functions, as we will see in Section 3.1.

Multiple covalent bonds

A covalent bond can be represented by a line between the chemical symbols for the linked atoms:

- A *single bond* involves the sharing of a single pair of electrons (for example, H—H or C—H).
- A *double bond* involves the sharing of four electrons (two pairs) (C═C).
- *Triple bonds*—six shared electrons—are rare, but there is one in nitrogen gas (N≡N), the major component of the air we breathe.

UNEQUAL SHARING OF ELECTRONS If two atoms of the same element are covalently bonded, there is an equal sharing of the pair(s) of electrons in the outermost shell. However, when the two atoms are of different elements, the sharing is not necessarily equal. One nucleus may exert a greater attractive force on the electron pair than the other nucleus, so that the pair tends to be closer to that atom.

The attractive force that an atomic nucleus exerts on electrons is its **electronegativity**. The electronegativity of a nucleus depends on how many positive charges it has (nuclei with more protons are more positive and thus more attractive to electrons) and on the distance between an electron and the nucleus (the closer the electron, the greater the pull of the electronegativity). The closer two atoms are in electronegativity, the more equal their sharing of electrons will be. **Table 2.3** shows the electronegativities of some elements important in biological systems.

If two atoms are close to each other in electronegativity, they will share electrons equally in what is called a *nonpolar covalent bond*. Two oxygen atoms, for example, both with electronegativity of 3.5, will share electrons equally. So will two hydrogen atoms (both with electronegativity of 2.1). But when hydrogen bonds with oxygen to form water, the electrons involved are unequally shared: they tend to be nearer to the oxygen nucleus because it is the more electronegative of the two. When electrons are drawn to one nucleus more than to the other, the result is a *polar covalent bond* (**Figure 2.9**).

Because of this unequal sharing of electrons, the oxygen end of the hydrogen–oxygen bond has a slightly negative charge (symbolized δ^- and spoken as "delta negative," meaning a partial unit of charge), and the hydrogen end has a slightly positive charge (δ^+). The bond is **polar** because these opposite charges are separated at the two ends, or poles, of the bond. The partial charges that result from polar covalent bonds produce polar molecules or polar regions of large molecules. Polar bonds greatly influence the interactions between molecules that contain them.

Ionic bonds form by electrical attraction

When one interacting atom is much more electronegative than the other, a complete transfer of one or more electrons may take place. Consider sodium (electronegativity 0.9) and chlorine (3.1). A sodium atom has only one electron in its outermost shell; this condition is unstable. A chlorine atom has seven electrons in its out-

TABLE 2.3

Some Electronegativities

ELEMENT	ELECTRONEGATIVITY
Oxygen (O)	3.5
Chlorine (Cl)	3.1
Nitrogen (N)	3.0
Carbon (C)	2.5
Phosphorus (P)	2.1
Hydrogen (H)	2.1
Sodium (Na)	0.9
Potassium (K)	0.8

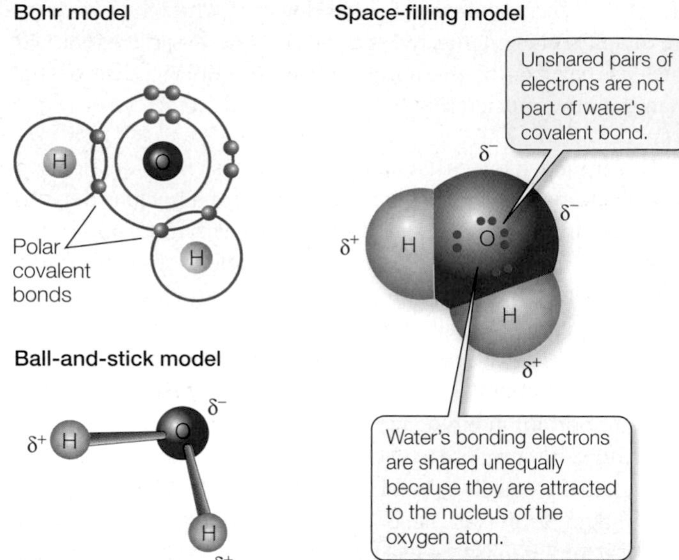

Bohr model
Space-filling model
Unshared pairs of electrons are not part of water's covalent bond.
Polar covalent bonds
Ball-and-stick model
Water's bonding electrons are shared unequally because they are attracted to the nucleus of the oxygen atom.

2.9 Water's Covalent Bonds are Polar These three representations all illustrate polar covalent bonding in water (H_2O). When atoms with different electronegativities, such as oxygen and hydrogen, form a covalent bond, the electrons are drawn to one nucleus more than to the other. A molecule held together by such a polar covalent bond has partial (δ) charges at the ends. In water, the shared electrons are displaced toward the oxygen atom's nucleus.

ermost shell—another unstable condition. Since the electronegativity of chlorine is so much greater than that of sodium, any electrons involved in bonding will tend to be much nearer to the chlorine nucleus—so near, in fact, that a complete transfer of the outermost electron from sodium to chlorine's outermost shell takes place (**Figure 2.10**). This reaction between sodium and chlorine makes the resulting atoms more stable. The result is two *ions*.

Ions are electrically charged particles that form when atoms gain or lose one or more electrons:

- The sodium ion (Na^+) in our example has a +1 unit of charge because it has one less electron than it has protons. The outermost electron shell of the sodium ion is full, with eight electrons, so the ion is stable. Positively charged ions are called **cations**.
- The chloride ion (Cl^-) has a –1 unit of charge because it has one more electron than it has protons. This additional electron gives Cl^- a stable outermost shell with eight electrons. Negatively charged ions are called **anions**.

Some elements can form ions with multiple charges by losing or gaining *more than one* electron. Examples are Ca^{2+} (calcium ion, a calcium atom that has lost two electrons) and Mg^{2+} (magnesium ion). Two biologically important elements can each yield more than one stable ion: iron yields Fe^{2+} (ferrous ion) and Fe^{3+} (ferric ion), and copper yields Cu^+ (cuprous ion) and Cu^{2+} (cupric ion). Groups of covalently bonded atoms that carry an electric charge are called *complex ions*; examples include NH_4^+ (ammonium ion), SO_4^{2-} (sulfate ion), and PO_4^{3-} (phosphate ion). The charge from an ion radiates in all directions. Once formed, ions are usually stable and no more electrons are lost or gained.

Ionic bonds are bonds formed by electrical attraction between ions bearing opposite charges. Ions can form bonds that result in stable solid compounds (referred to by the general term *salts*) such as sodium chloride (NaCl) and potassium phosphate (K_3PO_4). In sodium chloride—familiar to us as table salt—cations and anions are held together by ionic bonds. In solids, the ionic bonds are strong because the ions are close together. However, when ions are dispersed in water, the distance between them can be large; the strength of their attraction is thus greatly reduced. Under the conditions in living cells, an ionic attraction is usually less than one-tenth as strong as a nonpolar covalent bond (see Table 2.1).

Not surprisingly, ions can interact with polar molecules, since they both carry electric charges. Such an interaction results when a salt such as NaCl dissolves in water. Water molecules surround the individual ions, separating them (**Figure 2.11**). The negatively charged chloride ions attract the positive pole of the water molecules; the negative pole of the water molecules is, in contrast, oriented toward the positively charged sodium ions. The polarity of water gives it other special properties, as we will see.

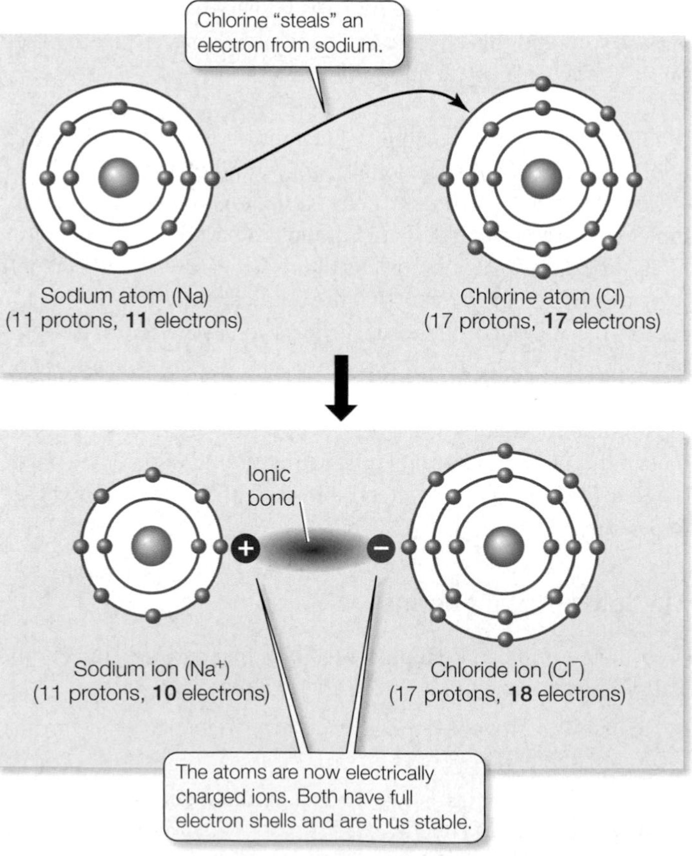

2.10 Formation of Sodium and Chloride Ions When a sodium atom reacts with a chlorine atom, the more electronegative chlorine fills its outermost shell by "stealing" an electron from the sodium. In so doing, the chlorine atom becomes a negatively charged chloride ion (Cl^-). The sodium atom, upon losing the electron, becomes a positively charged sodium ion (Na^+).

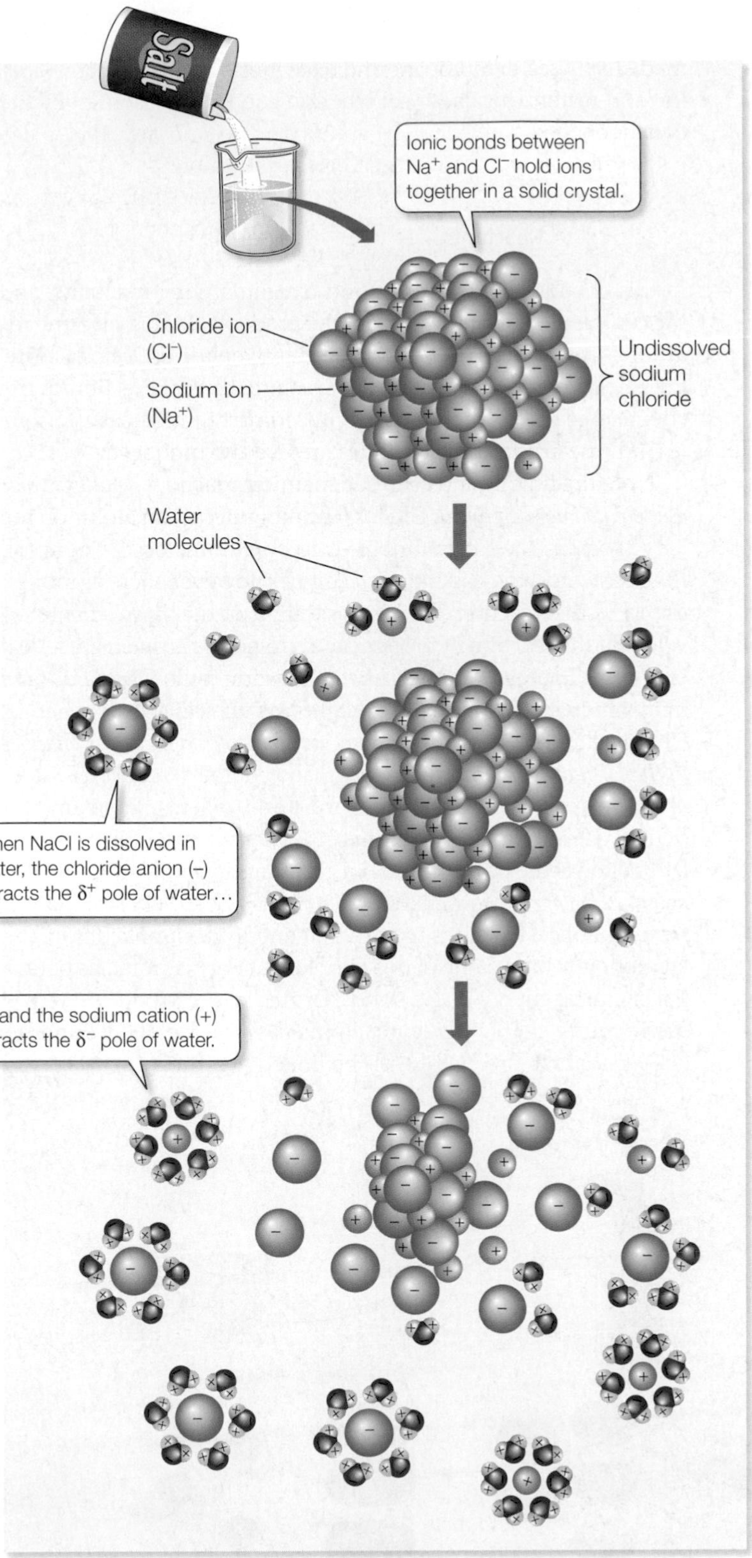

Salt
Ionic bonds between Na^+ and Cl^- hold ions together in a solid crystal.
Chloride ion (Cl^-)
Sodium ion (Na^+)
Undissolved sodium chloride
Water molecules
When NaCl is dissolved in water, the chloride anion (–) attracts the δ^+ pole of water...
... and the sodium cation (+) attracts the δ^- pole of water.

2.11 Water Molecules Surround Ions When an ionic solid dissolves in water, polar water molecules cluster around cations or anions, blocking their reassociation into a solid and forming a solution.

Hydrogen bonds may form within or between molecules with polar covalent bonds

In liquid water, the negatively charged oxygen (δ^-) atom of one water molecule is attracted to the positively charged hydrogen (δ^+) atoms of another water molecule (**Figure 2.12A**). The bond resulting from this attraction is called a **hydrogen bond**. Hydrogen bonds are not restricted to water molecules; they may also form between a strongly electronegative atom and a hydrogen atom covalently bonded to a different electronegative atom, as shown in **Figure 2.12B**. A hydrogen bond is weaker than most ionic bonds because it is formed by partial charges (δ^+ and δ^-). It has about one-tenth (10 percent) the strength of a covalent bond between a hydrogen atom and an oxygen atom (see Table 2.1). However, where many hydrogen bonds form, they have considerable strength and greatly influence the structure and properties of substances. Later in this chapter we'll see how hydrogen bonding between water molecules contributes to many of the properties that make water so significant for living systems. Hydrogen bonds also play important roles in determining and maintaining the three-dimensional shapes of giant molecules such as DNA and proteins (see Section 3.5).

Polar and nonpolar substances: Each interacts best with its own kind

Just as water molecules can interact with one another through polarity-induced hydrogen bonds, any molecule that is polar can interact with other polar molecules through the weak (δ^+ to δ^-) attractions of hydrogen bonds. If a polar molecule interacts with water in this way, it is called **hydrophilic** ("water-loving").

What about nonpolar molecules? Nonpolar molecules tend to interact with other nonpolar molecules. For example, carbon (electronegativity 2.5) forms nonpolar bonds with hydro-

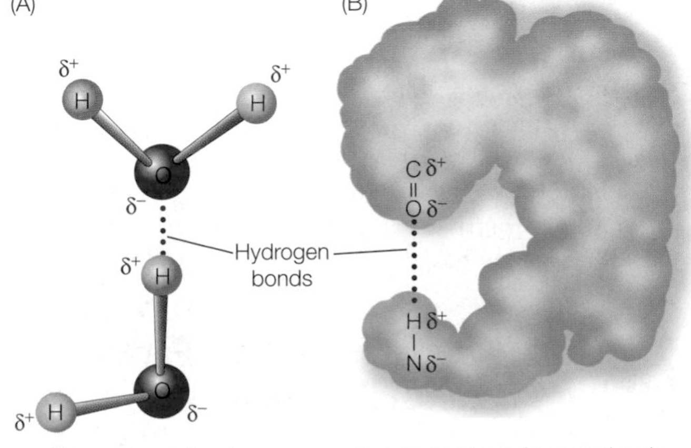

2.12 Hydrogen Bonds Can Form Between or Within Molecules (A) A hydrogen bond between two molecules is an attraction between negative charges on one molecule and positive charges on the hydrogen atoms of a second molecule. (B) Hydrogen bonds can form between large molecules, as well as between different parts of the same large molecule.

gen (electronegativity 2.1). The resulting *hydrocarbon molecule*—that is, a molecule containing only hydrogen and carbon atoms—is nonpolar, and in water it tends to aggregate with other nonpolar molecules rather than with polar water. Such nonpolar molecules are known as **hydrophobic** ("water-hating"), and the interactions between them are called *hydrophobic interactions*.

Of course, hydrophobic substances do not really "hate" water; they can form weak interactions with it, since the electronegativities of carbon and hydrogen are not exactly the same. But these interactions are far weaker than the hydrogen bonds between the water molecules, so the nonpolar substances keep to themselves.

Hydrophobic interactions between nonpolar substances are enhanced by **van der Waals forces**, which result when two atoms of a nonpolar molecule are in close proximity. These brief interactions are a result of random variations in the electron distribution in one molecule, which create an opposite charge distribution in the adjacent molecule. Although a single van der Waals interaction is brief and weak at any given site, the sum of many such interactions over the entire span of a large nonpolar molecule can produce substantial attraction.

2.2 RECAP

Chemical bonds link atoms together to form molecules. Strong covalent bonds are important in establishing biological structures, as are ionic bonds in salt formation. Polarity and weak hydrophobic interactions allow molecules to interact.

- How do two atoms' electronegativities result in the unequal sharing of electrons in polar molecules? See p. 27 and Figure 2.9
- Can you explain why a covalent bond is stronger than an ionic bond? See pp. 27–28 and Table 2.1
- What is a hydrogen bond and how is it important in biological systems? See p. 29 and Figure 2.12

The bonding of atoms into molecules is not necessarily a permanent affair. The dynamic of life involves constant change, even at the molecular level. Let's look at how molecules interact with one another—how they break up, how they find new partners, and what the consequences of those changes can be.

2.3 How Do Atoms Change Partners in Chemical Reactions?

A **chemical reaction** occurs when atoms combine or change their bonding partners. Consider the combustion reaction that takes place in the flame of a propane stove. When propane (C_3H_8) reacts with oxygen gas (O_2), the carbon atoms become bonded to oxygen atoms instead of hydrogen atoms, and the hydrogen atoms become bonded to oxygen instead of carbon (**Figure 2.13**). As the covalently bonded atoms change partners, the composition of the matter changes, and propane and oxygen gas become carbon dioxide and water. This chemical reaction can be represented by the equation

$$C_3H_8 + 5\,O_2 \rightarrow 3\,CO_2 + 4\,H_2O + \text{Energy}$$

$$\text{Reactants} \rightarrow \text{Products}$$

In this equation, the propane and oxygen are the **reactants**, and the carbon dioxide and water are the **products**. In this case, the reaction is *complete*: all the propane and oxygen are used up in forming the two products. The arrow symbolizes the direction of the chemical reaction. The numbers preceding the molecular formulas indicate how many molecules are used or produced.

Note that in this and all other chemical reactions, *matter is neither created nor destroyed*. The total number of carbon atoms on the left (3) equals the total number of carbon atoms on the right (3). In other words, the equation is *balanced*. However, there is another aspect of this reaction: the heat and light of the stove's flame reveal that the reaction of propane and oxygen released a great deal of energy. **Energy** is defined as the capacity to do work, but on a more intuitive level, it can be thought of as the capacity for change. Chemical reactions do not create or destroy energy, but *changes in the form of energy* usually accompany chemical reactions.

In the reaction between propane and oxygen, a large amount of heat energy was released. This energy was present in the covalent bonds of the propane and oxygen gases in another form, called *potential chemical energy*. Not all reactions release energy; indeed, many chemical reactions require that energy be supplied from the environment, and some of this supplied energy is stored as potential chemical energy in the bonds formed in the products. We will see in future chapters how reactions that release energy and reactions that require energy can be linked together.

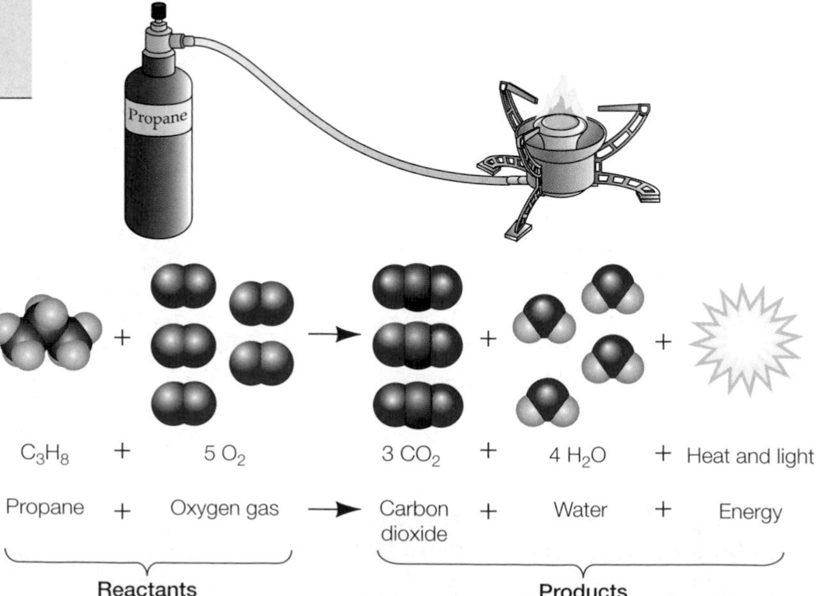

2.13 Bonding Partners and Energy May Change in a Chemical Reaction One molecule of propane reacts with five molecules of oxygen gas to give three molecules of carbon dioxide and four molecules of water. This reaction releases energy in the form of heat and light.

Many reactions take place in living cells, and some of these have much in common with the combustion of propane. The fuel is different—it is the sugar glucose, rather than propane—and the reactions proceed by many intermediate steps that permit the energy released from the glucose to be harvested and put to use by the cell. But the products are the same: carbon dioxide and water. These reactions, which we will study in detail in Chapter 7, are a key to the origin of life from simpler molecules.

2.3 RECAP

Chemical reactions involve the making and breaking of bonds between atoms. Many chemical reactions involve changes in the form of energy.

- What are the components of a chemical reaction? See Figure 2.13
- Do you understand how the form of energy can change during a chemical reaction?

We will present and discuss energy changes, oxidation–reduction reactions, and several other types of chemical reactions that are prevalent in living systems in Part Two of this book. First, however, we must understand the unique properties of the substance in which most biochemical reactions take place: water.

2.4 What Properties of Water Make It So Important in Biology?

Like most matter, water can exist in three states: solid (ice), liquid, or gas (vapor). As we mentioned at the start of this chapter, liquid water is probably the medium in which life originated, and it is in water that life on Earth evolved for its first billion years. In this section we will explore how the structure and interactions of water molecules make water essential to life.

Water has a unique structure and special properties

The molecule H_2O has unique chemical features. As we have already learned, water is a polar molecule that can form hydrogen bonds. In addition, the four pairs of electrons in the outer shell of the oxygen atom repel one another, giving the water molecule a tetrahedral shape:

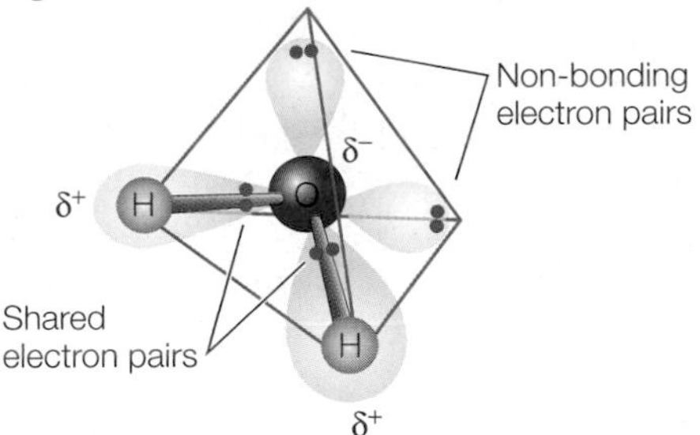

Non-bonding electron pairs
δ−
δ+
H
O
H
Shared electron pairs
δ+

These chemical features explain some of the interesting properties of water, such as the ability of ice to float, the melting and freezing temperatures of water, the ability of water to store heat, and the ability of water droplets to form.

ICE FLOATS In ice, water's solid state, individual water molecules are held in place by hydrogen bonds. Each water molecule is hydrogen-bonded to four other water molecules in a rigid, crystalline structure (**Figure 2.14**). Although the molecules are held firmly in place, they are not as tightly packed as they are in liquid water. In other words, *solid water is less dense than liquid water,* which is why ice floats.

Think of the biological consequences if ice were to sink in water. A pond would freeze from the bottom up, becoming a solid

Solid water (ice)
In ice, water molecules are held in a rigid state by hydrogen bonds.
Gaseous water (vapor)
In its gaseous state, water does not form hydrogen bonds.
Hydrogen bonds continually break and form as water molecules move.
Liquid water

2.14 Hydrogen Bonds Hold Water Molecules Together Hydrogen bonding exists between the molecules of water in both its liquid and solid states. Ice is more structured but less dense than liquid water, which is why ice floats. Water forms a gas when its hydrogen bonds are broken and the molecules move farther apart.

block of ice in winter and killing most of the organisms living there. Once the whole pond was frozen, its temperature could drop well below the freezing point of water. But in fact ice floats, forming a protective insulating layer on the top of the pond, reducing heat flow to the cold air above. Thus fish, plants, and other organisms in the pond are not subjected to temperatures lower than 0°C, the freezing point of pure water. The recent discovery by the *Global Surveyor* satellite of liquid water beneath the Martian polar ice has created speculation that life might exist there.

MELTING, FREEZING, AND HEAT CAPACITY Compared with many other substances of the same molecular size, ice requires a great deal of heat energy to melt, because hydrogen bonds must be broken in order for water to change from solid to liquid. In the opposite process—freezing—a great deal of energy is lost when water is transformed from liquid to solid.

Water contributes to the surprising constancy of the temperatures found in the oceans and other large bodies of water throughout the year. The temperature changes of coastal land masses are also moderated by large bodies of water. Indeed, water helps minimize variations in atmospheric temperature across the planet. This moderating ability is a result of the high *heat capacity* of liquid water, which is in turn a result of its high specific heat. The **specific heat** of a substance is the amount of heat energy required to raise the temperature of 1 gram of that substance by 1°C. Raising the temperature of liquid water takes a relatively large amount of heat because much of the heat energy is used to break the hydrogen bonds that hold the liquid together. Compared with other small molecules that are liquids, water has a high specific heat.

Water also has a high **heat of vaporization**, which means that a lot of heat is required to change water from its liquid to its gaseous state (the process of *evaporation*). Once again, much of the heat energy is used to break hydrogen bonds. This heat must be absorbed from the environment in contact with the water. Evaporation thus has a cooling effect on the environment—whether a leaf, a forest, or an entire land mass. This effect explains why sweating cools the human body: as sweat evaporates from the skin, it uses up some of the adjacent body heat.

COHESION AND SURFACE TENSION In liquid water, individual water molecules are free to move about. The hydrogen bonds between the molecules continually form and break. Chemists estimate that this occurs about a trillion times a minute in a single water molecule, making it a truly dynamic structure. At any given time, a water molecule will form an average of 3.4 hydrogen bonds with other water molecules. These hydrogen bonds explain the *cohesive strength* of liquid water. This cohesive strength, or **cohesion**, is defined as the capacity of water molecules to resist coming apart from one another when placed under tension. Water's cohesive strength permits narrow columns of liquid water to move from the roots to the leaves of trees more than 100 meters high. When water evaporates from the leaves, the entire column moves upward in response to the pull of the molecules at the top.

The surface of liquid water exposed to the air is difficult to puncture because the water molecules in this surface layer are hydrogen-bonded to other water molecules below them (**Figure 2.15**).

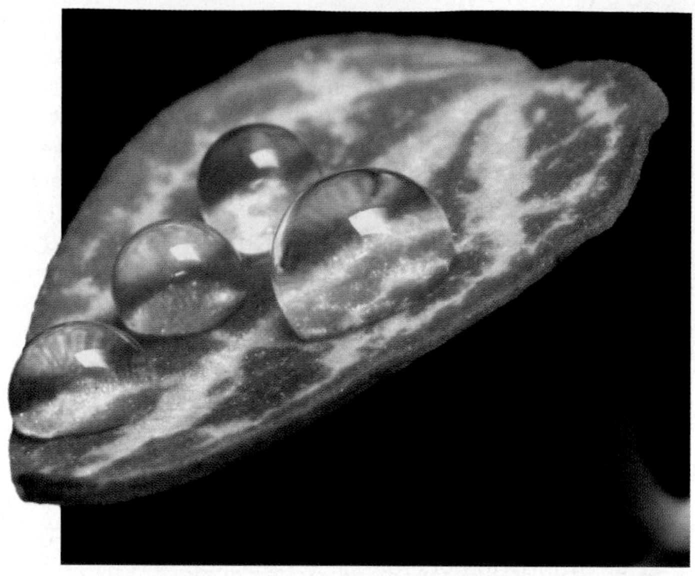

2.15 Surface Tension Water droplets form "beads" on the surface of a leaf because hydrogen bonds keep the water molecules together.

This *surface tension* of water permits a container to be filled slightly above its rim without overflowing, and it permits insects to walk on the surface of a pond.

There are at least 20 trillion galaxies in the universe and an average of 100 billion stars in each galaxy. Probability analyses of these numbers suggest that many undiscovered planets must exist, and that some of those planets probably have water—and thus the possibility of life.

Water is the solvent of life

A human body is over 70 percent water by weight, excluding the minerals contained in bones. Water is the dominant component of virtually all living organisms, and most biochemical reactions take place in this watery environment.

A **solution** is produced when a substance (the **solute**) is dissolved in a liquid (the **solvent**). If the solvent is water, then the solution is an *aqueous solution*. Many of the important molecules in biological systems are polar, and therefore soluble in water. Many important biochemical reactions occur in aqueous solutions. Biologists study what happens in these reactions, both in terms of the nature of the reactants and products and in terms of their amounts:

- *Qualitative analysis* deals with substances dissolved in water and the chemical reactions that occur there. For example, it would investigate the steps involved, and the products formed, during the combustion of glucose in living tissues. Qualitative analysis is the subject of much of the next few chapters.

- *Quantitative analysis* measures concentrations, or the amount of a substance in a given amount of solution. For example, it would seek to describe *how much* of a certain product is formed during the combustion of a given amount of glucose. What follows is a brief introduction to some of the quantitative chemical terms you will see in this book.

Fundamental to quantitative thinking in chemistry and biology is the mole concept. A **mole** is the amount of a substance (in grams) the mass of which is numerically equal to its molecular weight. So a mole of table sugar ($C_{12}H_{22}O_{11}$) weighs 342 grams; a mole of sodium ion (Na^+) weighs 23 grams; and a mole of hydrogen gas (H_2) weighs 2 grams.

Quantitative analysis does not yield direct counts of molecules. Chemists use a constant that relates the weight of any substance to the number of molecules of that substance. This constant is called **Avogadro's number**, which is 6.02×10^{23} molecules per mole. It allows chemists to work with moles of substances (which can be weighed out in the laboratory) instead of actual molecules (which are too numerous to be counted). Consider 34.2 grams (just over 1 ounce) of table sugar, $C_{12}H_{22}O_{11}$. This is one-tenth of a mole, or as Avogadro puts it, 6.02×10^{22} molecules.

If you have trouble grasping the idea of a mole, think of the mole concept the way you think of the concept of a dozen: we buy a dozen eggs or a dozen doughnuts, knowing that we will get 12 of whichever we buy, even though they don't weigh the same or take up the same amount of space.

When a physician injects a certain molar concentration of a drug into the bloodstream of a patient, a rough calculation can be made of the actual number of drug molecules that will interact with the patient's cells. In the same way, a chemist can dissolve a mole of sugar in water to make 1 liter of solution, knowing that the mole contains 6.02×10^{23} individual sugar molecules. This solution—1 mole of a substance dissolved in water to make 1 liter—is called a 1 molar (1 *M*) solution.

The many molecules dissolved in the water in living tissues are not present at anything close to 1 molar concentrations. Most are in the micromolar (millionths of a mole per liter of solution; μM) to millimolar (thousandths of a mole per liter; m*M*) range. Some, such as hormone molecules, are even less concentrated than that. While these molarities seem to indicate very low concentrations, remember that even a 1 μM solution has 6.02×10^{17} molecules of the solute per liter.

Aqueous solutions may be acidic or basic

When some substances dissolve in water, they release *hydrogen ions* (H^+), which are actually single, positively charged protons. Hydrogen ions can attach to other molecules and change their properties. For example, the protons in "acid rain" can damage plants, and you probably have experienced the excess of hydrogen ions we know as "acid indigestion."

Here we will examine the properties of **acids**, which release H^+, and **bases**, which accept H^+. We will distinguish between strong and weak acids and bases and provide a quantitative means for stating the concentration of H^+ in solutions: the pH scale.

ACIDS RELEASE H^+, BASES ACCEPT H^+ When hydrochloric acid (HCl) is added to water, it dissolves, releasing the ions H^+ and Cl^-:

$$HCl \rightarrow H^+ + Cl^-$$

Because its H^+ concentration has increased, such a solution is *acidic*. Just like the combustion reaction of propane and oxygen shown in Figure 2.13, the dissolution of HCl to form its ions is a *complete* reaction. HCl is therefore called a *strong acid*.

Acids are substances that *release* H^+ ions in solution. HCl is an acid, as is H_2SO_4 (sulfuric acid). One molecule of sulfuric acid may ionize to yield two H^+ and one SO_4^{2-}. Biological compounds that contain —COOH (the carboxyl group) are also acids because

$$\text{—COOH} \rightarrow \text{—COO}^- + H^+$$

However, not all acids ionize fully in water. For example, if acetic acid is added to water, at the end of the reaction, there are two ions, but some of the original acetic acid remains as well. Because the reaction is *not complete*, acetic acid is a *weak acid*.

Bases are substances that *accept* H^+ in solution. Just as with acids, there are strong and weak bases. If NaOH (sodium hydroxide) is added to water, it dissolves and ionizes, releasing OH^- and Na^+ ions:

$$NaOH \rightarrow Na^+ + OH^-$$

Because the concentration of OH^- increases and OH^- absorbs H^+ to form water, such a solution is *basic*. Because this reaction is complete, NaOH is a *strong base*.

Weak bases include the bicarbonate ion (HCO_3^-), which can accept a H^+ ion and become carbonic acid (H_2CO_3), and ammonia (NH_3), which can accept a H^+ and become an ammonium ion (NH_4^+). Biological compounds that contain —NH_2 (the amino group) are also bases because

$$\text{—}NH_2 + H^+ \rightarrow \text{—}NH_3^+$$

ACID–BASE REACTIONS MAY BE REVERSIBLE When acetic acid is dissolved in water, two reactions happen. First, the acetic acid forms its ions:

$$CH_3COOH \rightarrow CH_3COO^- + H^+$$

Then, once the ions are formed, they re-form acetic acid:

$$CH_3COO^- + H^+ \rightarrow CH_3COOH$$

This pair of reactions is reversible. A **reversible reaction** can proceed in either direction—left to right or right to left—depending on the relative starting concentrations of the reactants and products. The formula for a reversible reaction can be written using a double arrow:

$$CH_3COOH \rightleftharpoons CH_3COO^- + H^+$$

In terms of acids and bases, there are two types of reactions, depending on the extent of reversibility:

- The ionization of strong acids and bases is virtually irreversible.
- The ionization of weak acids and bases is somewhat reversible.

WATER IS A WEAK ACID The water molecule has a slight but significant tendency to ionize into a hydroxide ion (OH^-) and a hydrogen ion (H^+). Actually, two water molecules participate in this reaction. One of the two molecules "captures" a hydrogen ion from the other, forming a hydroxide ion and a hydronium ion:

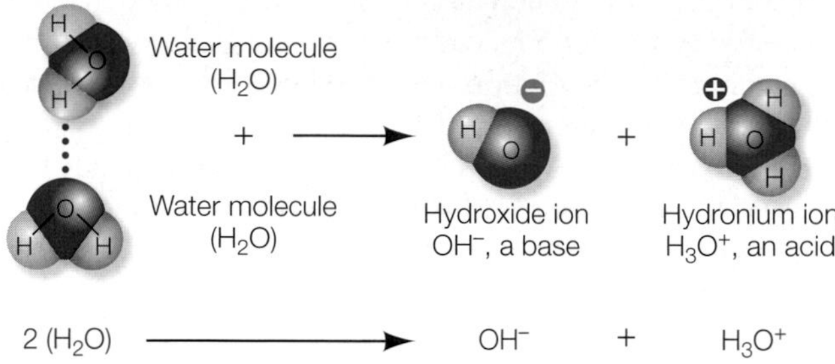

$$2\,(H_2O) \longrightarrow OH^- + H_3O^+$$

The hydronium ion is, in effect, a hydrogen ion bound to a water molecule. For simplicity, biochemists tend to use a modified representation of the ionization of water:

$$H_2O \rightarrow H^+ + OH^-$$

The ionization of water is important to all living creatures. This fact may seem surprising, since only about one water molecule in 500 million is ionized at any given time. But we will be less surprised if we focus on the abundance of water in living systems and the reactive nature of the H^+ produced by ionization.

pH is the measure of hydrogen ion concentration

Solutions are acidic or basic; compounds or ions can be acids or bases. We can measure how acidic or basic a solution is by measuring the concentration of H^+ in moles per liter, or *molarity* (see page 33). Here are some examples:

- Pure water has a H^+ concentration of 10^{-7} *M*.
- A 1 *M* HCl solution has a H^+ concentration of 1 *M* (recall that all the HCl dissolves into its ions).
- A 1 *M* NaOH solution has a H^+ concentration of 10^{-14} *M*.

This is a very wide range of numbers to work with (think about the decimals!). It is easier to work with the *logarithm* of the H^+ concentration, because logarithms compress this range: the $\log_{10}$ of 100, for example is 2, and the $\log_{10}$ of 0.01 is –2. Because most H^+ concentrations in living systems are less than 1, we convert these negative numbers into positive ones by using the *negative* of the logarithm of the H^+ molar concentration (designated by square brackets: $[H^+]$). This number is called the **pH** of the solution.

Since the H^+ concentration of pure water is 10^{-7} *M*, its pH is $-\log(10^{-7}) = -(-7)$, or 7. A smaller negative logarithm means a larger number. In practical terms, a lower pH means a higher H^+ concentration, or greater acidity. In 1 *M* HCl, the H^+ concentration is 1 *M*, so the pH is the negative logarithm of 1 ($-\log 10^0$), or 0. The pH of 1 *M* NaOH is the negative logarithm of 10^{-14}, or 14.

A solution with a pH of less than 7 is acidic—it contains more H^+ ions than OH^- ions. A solution with a pH of 7 is neutral, and a solution with a pH value greater than 7 is basic. **Figure 2.16** shows the pH values of some common substances.

Buffers minimize pH change

The maintenance of internal constancy—*homeostasis*— is a hallmark of all living things and extends to pH. As we mentioned earlier, when H^+ is added to many molecules, they change their properties, thus upsetting constancy. Maintaining internal constancy can be achieved by buffers: solutions that maintain a relatively constant pH even when substantial amounts of an acid or base are added to a system. How can this work?

A **buffer** is a solution of a weak acid and its corresponding base—for example, carbonic acid (H_2CO_3) and bicarbonate ions (HCO_3^-). If an acid is added to a solution containing this buffer, not all the H^+ ions from that acid stay in solution. Instead, many of them combine with the bicarbonate ions to produce more carbonic acid. This reaction uses up some of the H^+ ions in the solution and decreases the acidifying effect of the added acid:

$$HCO_3^- + H^+ \rightarrow H_2CO_3$$

If a base is added, the reaction essentially reverses. Some of the carbonic acid ionizes to produce bicarbonate ions and more H^+, which counteracts some of the added base. In this way, the buffer minimizes the effects of an added acid or base on pH. This is what happens in the blood, where this buffering system is important in preventing significant changes in pH that could disrupt the ability of the blood to carry vital oxygen to tissues. A given amount of acid or base causes a smaller change in pH in a buffered solution than in an unbuffered one (**Figure 2.17**).

How do you spell relief? The lining of the stomach constantly secretes hydrochloric acid, making the stomach contents acidic. Excessive stomach acid inhibits digestion and causes discomfort, but it can be relieved by ingesting a salt such as $NaHCO_3$ ("bicarbonate of soda") that acts as a buffer.

Buffers illustrate an important chemical principle of reversible reactions, called the *law of mass action*. Addition of a reactant on one side of a reversible system drives the reaction in the direction that uses up that compound. In the case of buffers, addition of an acid drives the reaction in one direction; addition of a base drives the reaction in the other direction.

Life's chemistry began in water

As we have emphasized throughout this chapter, the presence of water on a planet—Mars, Earth, or any other—is a necessary prerequisite for life as we know it. Astronomers believe our solar system began forming about 4.6 billion years ago, when a star exploded and collapsed to form the sun and 500 or so bodies, called planetesimals. These planetesimals collided with one another to form the inner planets, including Earth and Mars. The first chemical signatures indicating the presence of life on Earth now appear to be about 4 billion years old. So it took 600 million years, during a geological time frame called the Hadean, for the chemical conditions on Earth to become just right for life. Key among those conditions was the presence of water.

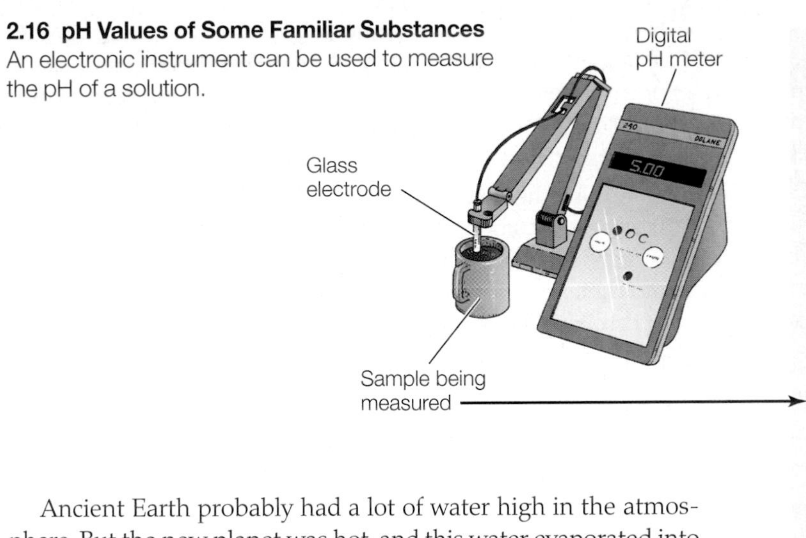

2.16 pH Values of Some Familiar Substances An electronic instrument can be used to measure the pH of a solution.

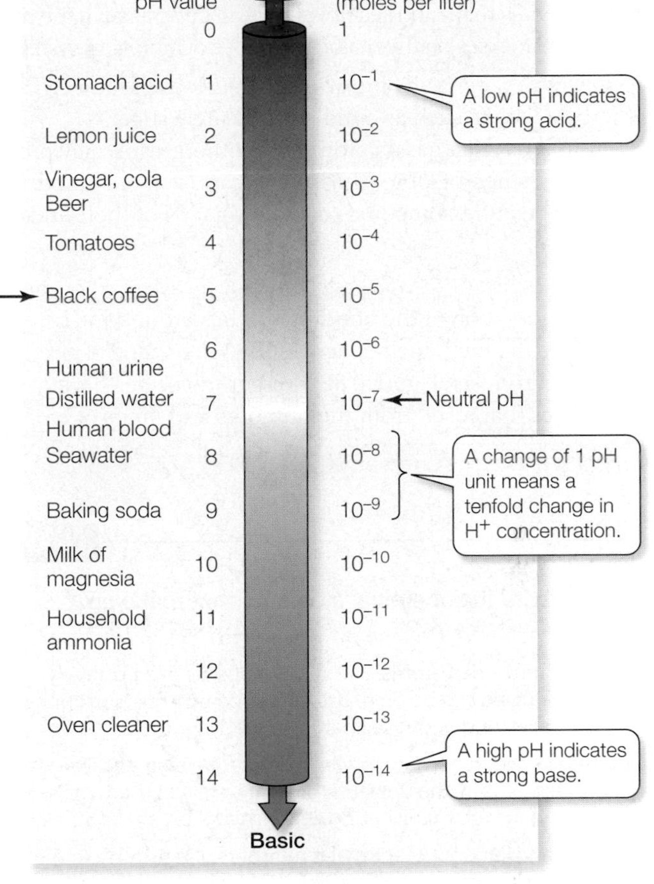

Ancient Earth probably had a lot of water high in the atmosphere. But the new planet was hot, and this water evaporated into space. As Earth cooled, it became possible for water to remain on its surface, but where did that water come from? One current view is that comets—loose agglomerations of dust and ice that have orbited the sun since the planets formed—struck Earth and Mars repeatedly and brought not only water but other chemical components of life, such as nitrogen. As the planets cooled, chemicals from their crusts dissolved in the water, and simple chemical reactions would have taken place. Some of these reactions could have led to life, but impacts by large comets and rocky meteorites would have released enough energy to heat the developing oceans almost to boiling, thus destroying any early life. On Earth, these large impacts eventually subsided, and life gained a foothold about 3.8 to 4 billion years ago. The prebiotic Hadean was over. The Archean had begun, and there has been life on Earth ever since.

In Section 3.6 we will return to the question of how the first life could have arisen from inanimate chemicals.

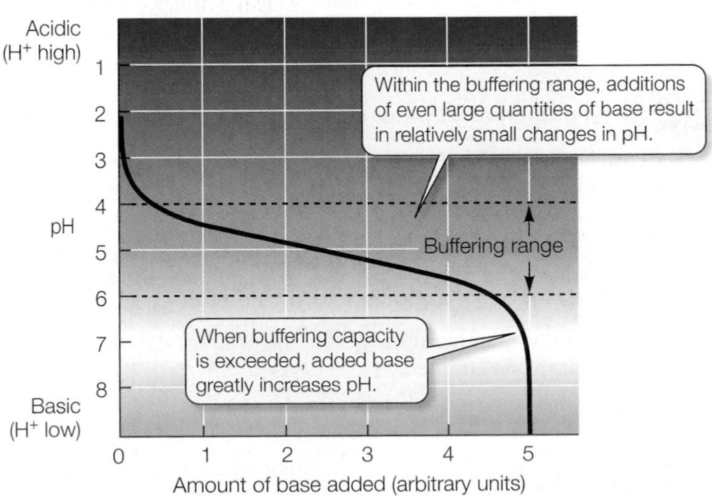

2.17 Buffers Minimize Changes in pH With increasing amounts of added base, the overall slope of a graph of pH is downward. Inside the buffering range, however, the slope is shallow. At very high and very low values of pH, where the buffer is ineffective, the slopes are much steeper.

2.4 RECAP

Most of the chemistry of life occurs in water, which has molecular properties that make it suitable for its important biochemical roles. A special property of water is its ability to ionize (release hydrogen ions). The presence of hydrogen ions in solution can change the properties of biological molecules.

- What are some biologically important properties of water arising from its molecular structure? See pp. 31–32 and Figure 2.14
- Do you understand what a solution is, and why we call water "the solvent of life"? See pp. 32–33
- Do you see the relationship between hydrogen ions, acids, and bases? Can you explain what the pH scale measures? See pp. 33–34 and Figure 2.16
- Can you explain how a buffer works, and why buffering is important to living systems? See p. 34 and Figure 2.17

An Overview and a Preview

Now that we have covered the major properties of atoms and molecules, let's review them and see how they will be applied in the next chapter, which focuses on the major molecules of biological systems.

- *Molecules vary in size.* Some are small, such as H_2 and CH_4. Others are larger, such as a molecule of table sugar ($C_{12}H_{22}O_{11}$), which has 45 atoms. Still others, especially proteins such as hemoglobin (the oxygen carrier in red blood cells), are gigantic, sometimes containing tens of thousands of atoms.
- *All molecules have a specific three-dimensional shape.* For example, the orientation of the bonding orbitals around the carbon atom gives the methane molecule (CH_4) the shape of a regular tetrahedron (see Figure 2.8B). Larger molecules have complex shapes that result from the numbers and kinds of atoms present and the ways in which they are linked together. Some large molecules, such as hemoglobin, have compact, ball-like shapes. Others, such as the protein called keratin that makes up your hair, have long, thin, ropelike structures. Their shapes relate to the roles these molecules play in living cells.
- *Molecules are characterized by certain chemical properties* that determine their biological roles. Chemists use the characteristics of composition, structure (three-dimensional shape), reactivity, and solubility to distinguish a pure sample of one molecule from a sample of a different molecule. The presence of certain groups of atoms can impart distinctive chemical properties to a molecule.

Between the small molecules discussed in this chapter and the world of the living cell stand the macromolecules. These huge molecules—proteins, lipids, carbohydrates, and nucleic acids—are the subject of the next chapter.

CHAPTER SUMMARY

2.1 What are the chemical elements that make up living organisms?

Matter is composed of atoms. Each **atom** consists of a positively charged **nucleus** made up of **protons** and **neutrons**, surrounded by **electrons** bearing negative charges. Review Figure 2.1

The number of protons in the nucleus defines an **element**. There are many elements in the universe, but only a few of them make up the bulk of living organisms. Review Figure 2.2

Isotopes of an element differ in their numbers of neutrons. **Radioisotopes** are radioactive, emitting radiation as they decay.

Electrons are distributed in **shells**, which are volumes of space defined by specific numbers of orbitals. Each **orbital** contains a maximum of two electrons. Review Figure 2.6, Web/CD Activity 2.1

In losing, gaining, or sharing electrons to become more stable, an atom can combine with other atoms to form **molecules**.

2.2 How do atoms bond to form molecules?

See Web/CD Tutorial 2.1

A **chemical bond** is an attractive force that links two atoms together in a molecule. Review Table 2.1

A **compound** is a molecule made up of atoms of two or more elements bonded together in a fixed ratio, such as water (H_2O) or table sugar ($C_6H_{12}O_6$).

Covalent bonds are strong bonds formed when two atoms share one or more pairs of electrons. Review Figure 2.7

When two atoms of unequal electronegativity bond with each other, a **polar** covalent bond is formed. The two ends, or poles, of the bond have partial charges (δ^+ or δ^-). Review Figure 2.9

Ions are electrically charged bodies that form when an atom gains or loses one or more electrons. **Anions** and **cations** are negatively and positively charged ions, respectively.

Ionic bonds are electrical attractions between oppositely charged ions. Ionic bonds are strong in solids (salts), but weaken when the ions are separated from one another in solution. Review Figure 2.10

Hydrogen bonds are weak electrical attractions that form between a δ^+ hydrogen atom in one molecule and a δ^- atom in another molecule (or in another part of a large molecule). Hydrogen bonds are abundant in water.

Nonpolar molecules interact very little with polar molecules, including water. Nonpolar molecules are attracted to one another by very weak bonds called **van der Waals forces**.

2.3 How do atoms change partners in chemical reactions?

In **chemical reactions**, atoms combine or change their bonding partners. **Reactants** are converted into **products**.

Some chemical reactions release **energy** as one of their products; other reactions can occur only if energy is provided to the reactants.

Neither matter nor energy is created or destroyed in a chemical reaction, but both change form. Review Figure 2.13

Some chemical reactions, especially in biology, are reversible. That is, the products formed may be converted back to the reactants.

In living cells, chemical reactions take place in multiple steps so that the released energy can be harvested for cellular activities.

2.4 What properties of water make it so important in biology?

Water's molecular structure and its capacity to form hydrogen bonds give it unique properties that are significant for life. Review Figure 2.14

The high **specific heat** of water means that water gains or loses a great deal of heat when it changes state. Water's high **heat of vaporization** ensures effective cooling when water evaporates.

The **cohesion** of water molecules refers to their capacity to resist coming apart from one another.

Solutions are produced when solid substances (**solutes**) dissolve in a liquid (the **solvent**). Water is the critically important solvent for life.

Acids are solutes that release hydrogen ions in aqueous solution. **Bases** accept hydrogen ions.

The **pH** of a solution is the negative logarithm of its hydrogen ion concentration. Values lower than pH 7 indicate a solution is acidic; values above pH 7 indicate a basic solution. Review Figure 2.16

Buffers are mixtures of weak acids and bases that limit the change in the pH of a solution when acids or bases are added.

SELF-QUIZ

1. The atomic number of an element
   a. equals the number of neutrons in an atom.
   b. equals the number of protons in an atom.
   c. equals the number of protons minus the number of neutrons.
   d. equals the number of neutrons plus the number of protons.
   e. depends on the isotope.
2. The atomic weight (atomic mass) of an element
   a. equals the number of neutrons in an atom.
   b. equals the number of protons in an atom.
   c. equals the number of electrons in an atom.
   d. equals the number of neutrons plus the number of protons.
   e. depends on the relative abundances of its electrons and neutrons.
3. Which of the following statements about the isotopes of an element is *not* true?
   a. They all have the same atomic number.
   b. They all have the same number of protons.
   c. They all have the same number of neutrons.
   d. They all have the same number of electrons.
   e. They all have identical chemical properties.
4. Which of the following statements about covalent bonds is *not* true?
   a. A covalent bond is stronger than a hydrogen bond.
   b. A covalent bond can form between atoms of the same element.
   c. Only a single covalent bond can form between two atoms.
   d. A covalent bond results from the sharing of electrons by two atoms.
   e. A covalent bond can form between atoms of different elements.
5. Hydrophobic interactions
   a. are stronger than hydrogen bonds.
   b. are stronger than covalent bonds.
   c. can hold two ions together.
   d. can hold two nonpolar molecules together.
   e. are responsible for the surface tension of water.
6. Which of the following statements about water is *not* true?
   a. It releases a large amount of heat when changing from liquid into vapor.
   b. Its solid form is less dense than its liquid form.
   c. It is the most effective solvent of polar molecules.
   d. It is typically the most abundant substance in a living organism.
   e. It takes part in some important chemical reactions.
7. The reaction $HCl \rightarrow H^+ + Cl^-$ in the human stomach is an example of the
   a. cleavage of a hydrophobic bond.
   b. formation of a hydrogen bond.
   c. elevation of the pH of the stomach.
   d. formation of ions by dissolving an acid.
   e. formation of polar covalent bonds.
8. The hydrogen bond between two water molecules arises because water is
   a. polar.
   b. nonpolar.
   c. a liquid.
   d. small.
   e. hydrophobic.
9. When table salt (NaCl) is added to water,
   a. a covalent bond is broken.
   b. an acidic solution is formed.
   c. Na^+ and Cl^- ions are separated.
   d. Na^+ is attracted to the hydrogen atoms of water.
   e. water molecules surround Na (but not Cl) atoms.
10. The three most abundant elements in a human skin cell are
    a. calcium, carbon, and oxygen.
    b. carbon, hydrogen, and oxygen.
    c. carbon, hydrogen, and sodium.
    d. carbon, nitrogen, and potassium.
    e. nitrogen, hydrogen, and argon.

FOR DISCUSSION

1. Using the information in the periodic table (Figure 2.2), draw a Bohr model (see Figure 2.8) of silicon dioxide showing electrons shared in covalent bonds.
2. Compare a covalent bond between two hydrogen atoms and a hydrogen bond between hydrogen and oxygen atoms with regard to the electrons involved, the role of polarity, and the strength of the bond.
3. Write an equation describing the combustion of glucose ($C_6H_{12}O_6$) to produce carbon dioxide and water.
4. The pH of the human stomach is about 2.0, while the pH of the small intestine is about 10.0. What are the hydrogen ion (H^+) concentrations inside these two organs?

FOR INVESTIGATION

Would you expect the elemental composition of Earth's crust to be the same as that of the human body? How could you find out?

CHAPTER 3

Macromolecules and the Origin of Life

How sweet it is

Sometimes a scientist gets lucky. In 1989, a British sugar company was looking for nontraditional uses for its product, and wondered if sugar might be chemically modified to make other useful substances. As often happens in industry, the company sought the expertise of university scientists. A professor at the University of London knew how to modify table sugar (sucrose, $C_{12}H_{22}O_{11}$) by replacing hydrogen and oxygen atoms with other elements. The professor assigned a graduate student the job of making sugar with chlorine atoms in place of some of the OH (hydroxyl) groups.

The professor's idea was to test each new molecule for possible use as a starting material for other compounds. But the graduate student, Shahikant Phadnis, responded to his professor's request for "testing" one molecule by "tasting." To his astonishment, the new molecule—$C_{12}H_{19}O_8Cl_3$—was sweet. Indeed, it was many times sweeter than sugar. But unlike sugar, it could not be used by the body as food energy. Phadnis had discovered a new artificial sweetener, now known as sucralose.

The commercial demand for artificial sweeteners is huge. About one-third of all people in Europe, the Americas, and Australia consume low-calorie foods and beverages. Some people, such as diabetics, must restrict their sugar intake or suffer serious health consequences. People fighting obesity want to reduce their intake of sugar, which is a major calorie source. Still others are prone to tooth decay from bacteria in the mouth that feed on sugar. With sucralose, all these people can have their cake and eat it too. It is used in hundreds of products, and is sold by itself as a sugar substitute, packaged in yellow under the brand name "Splenda."

Ordinary sugar has two important chemical properties that affect people who eat it. First, it binds noncovalently to certain proteins found in taste buds on the tongue. Sucralose binds to the taste buds even better than sugar does, hence its intense sweet taste. Second, sugar molecules have a three-dimensional structure that allows them to bind to digestive molecules in the body, which break sugar down for energy and conversion to other substances (including fat). It is in this second chemical property that sucralose differs strikingly from ordinary sugar. Sucralose does not bind to digestive molecules, but passes through the body largely unchanged.

Sucrose and its artificial cousin sucralose are carbohydrates, one of the four major kinds of large molecules that characterize living systems. These *macromolecules,*

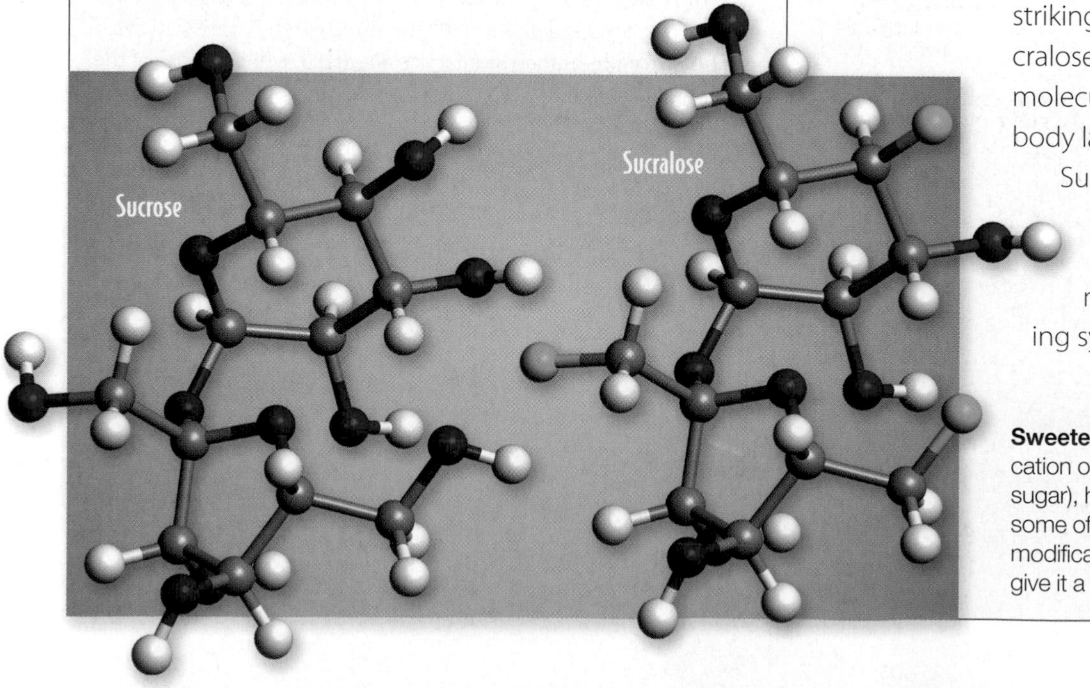

Sucrose

Sucralose

Sweeter Than Sugar Sucralose, a modification of the biological molecule sucrose (table sugar), has chlorine atoms (green) in place of some of the hydroxyl groups of sucrose. This modification results in chemical properties that give it a sweet taste, but no nutritional value.

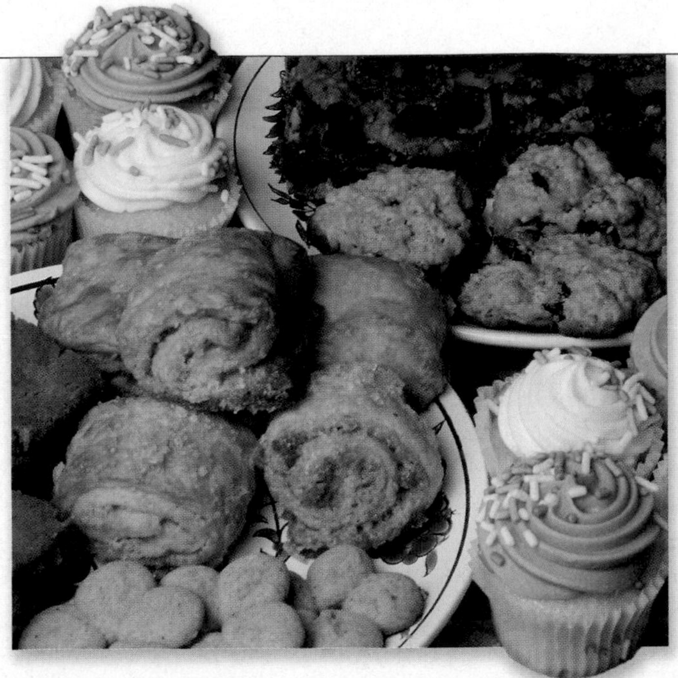

A Substitute for Sugar Sucralose can be used in place of sugar in baked goods, allowing dieters and people with diabetes to enjoy sweet treats.

CHAPTER OUTLINE

which also include proteins, lipids, and nucleic acids, differ in several significant ways from the small molecules and ions described in Chapter 2. First—not a surprise—they are larger, with molecular weights ranging from hundreds of Daltons (sucrose) to billions (some nucleic acids).

Second, these molecules all contain carbon atoms, and so belong to a group of what are known as *organic* chemicals. Third, they are held together largely by covalent bonds, which gives them important structural stability and forms the basis of some of their functions. And finally, carbohydrates, proteins, lipids, and nucleic acids are all unique to the living world. These molecular classes do not occur in inanimate nature. You won't find proteins in rocks—and if you do, you can be sure they came from some living organism.

IN THIS CHAPTER we will describe the chemical and biological properties of proteins, lipids, carbohydrates, and nucleic acids. We will see what these building blocks of life are made of, how they are assembled, and what roles they play in living organisms. We end the chapter with some ideas on how these molecules originated when life first began.

3.1 What Kinds of Molecules Characterize Living Things?

Four kinds of molecules are characteristic of living things: proteins, carbohydrates, lipids, and nucleic acids. Most of these *biological molecules* are large **polymers** (*poly*, "many"; *mer*, "unit") constructed by the covalent bonding of smaller molecules called **monomers** (**Table 3.1**). The monomers that make up each kind of biological molecule have similar chemical structures. Thus chains of chemically similar sugar monomers (*saccharides*) form the different carbohydrates; the thousands of different proteins are formed from combinations of a mere 20 *amino acids*, all of which share chemical similarities.

Polymers with molecular weights exceeding 1,000 are usually considered to be **macromolecules**. The proteins, polysaccharides (large carbohydrates), and nucleic acids of living systems certainly fall into this category. How these macromolecules function and interact with other molecules depends on properties of chemical groups in their monomers, the *functional groups*.

Functional groups give specific properties to molecules

Certain small groups of atoms, called **functional groups**, are consistently found together in a variety of different biological molecules. You will encounter several functional groups repeatedly in your study of biology (**Figure 3.1**). Each functional group has specific chemical properties, and when it is attached to a larger molecule, it confers those properties on the larger molecule. For example, the *hydroxyl group* is polar and attracts water molecules. Small molecules containing the hydroxyl group usually dissolve easily in water. The oxygen atom in the *keto group* is highly electronegative and can attract the hydrogen atom held by another electronegative atom (such as nitrogen), forming a hydrogen bond. The consistent chemical behavior of these functional groups helps us understand the properties of the molecules that contain them.

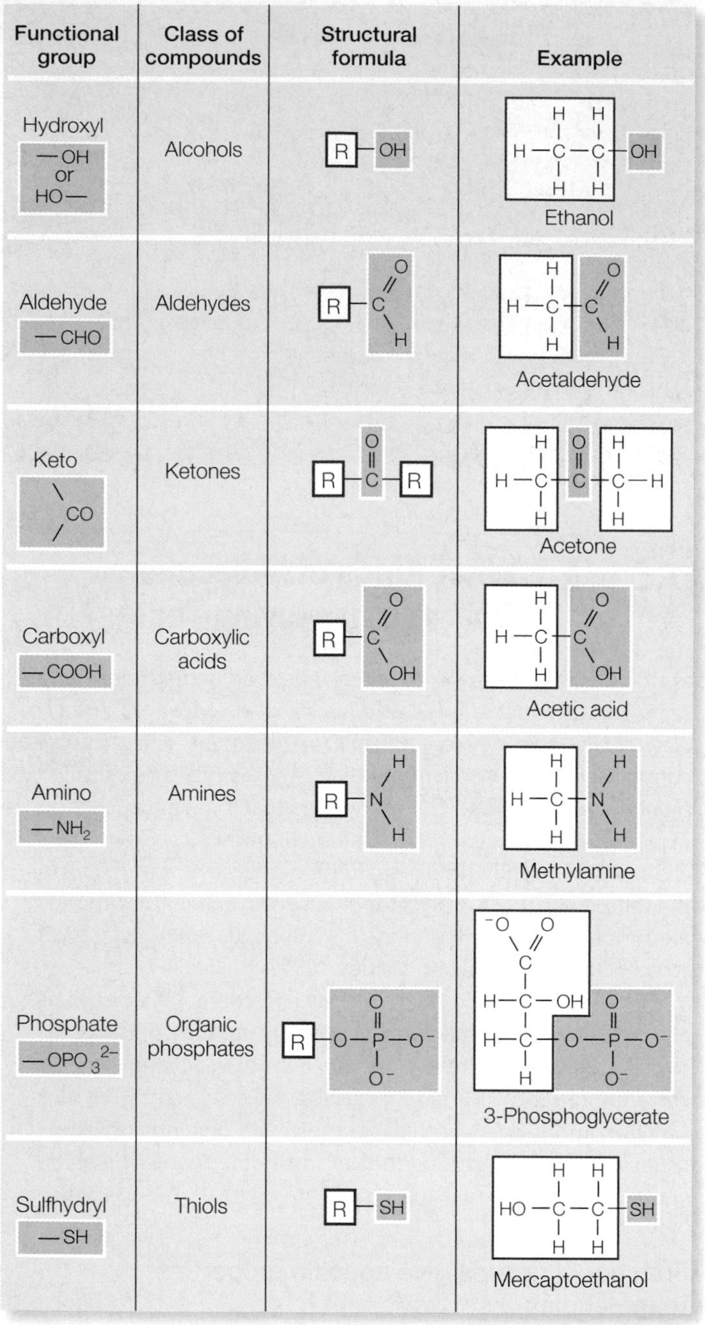

Functional group	Class of compounds	Structural formula	Example
Hydroxyl —OH or HO—	Alcohols	R—OH	H_3C—CH_2—OH Ethanol
Aldehyde —CHO	Aldehydes	R—C(=O)—H	H_3C—C(=O)—H Acetaldehyde
Keto \CO/	Ketones	R—C(=O)—R	H_3C—C(=O)—CH_3 Acetone
Carboxyl —COOH	Carboxylic acids	R—C(=O)—OH	H_3C—C(=O)—OH Acetic acid
Amino —NH_2	Amines	R—N(H)—H	H_3C—N(H)—H Methylamine
Phosphate —OPO_3^{2-}	Organic phosphates	R—O—P(=O)(O^-)—O^-	$^-$O—C(=O)—CH(OH)—CH_2—O—P(=O)(O^-)—O^- 3-Phosphoglycerate
Sulfhydryl —SH	Thiols	R—SH	HO—CH_2—CH_2—SH Mercaptoethanol

3.1 Some Functional Groups Important to Living Systems The seven functional groups shown here (highlighted in yellow) are the most common ones found in biologically important molecules. R represents the "remainder" of the molecule, which may be any of a large number of carbon skeletons or other chemical groups.

TABLE 3.1

The Building Blocks of Organisms

MONOMER	COMPLEX POLYMER (MACROMOLECULE)
Amino acid	Polypeptide (protein)
Monosaccharide (sugar)	Polysaccharide (carbohydrate)
Nucleotide	Nucleic acid

Isomers have different arrangements of the same atoms

Isomers are molecules that have the same chemical formula—the same kinds and numbers of atoms—but the atoms are arranged differently. (The prefix *iso-*, meaning "same," is encountered in many biological terms.) Of the different kinds of isomers, we will consider two: structural isomers and optical isomers.

Structural isomers differ in how their atoms are joined together. Consider two simple molecules, each composed of four carbon and ten hydrogen atoms bonded covalently, both with the formula C_4H_{10}. These atoms can be linked in two different ways, resulting in two different forms of the molecule:

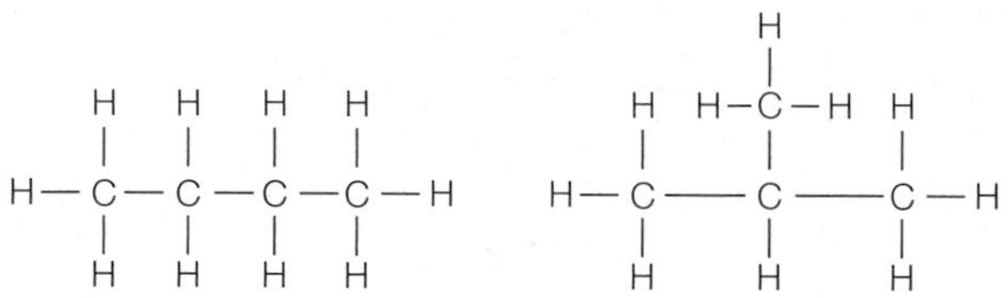

H_3C—CH_2—CH_2—CH_3
Butane

H_3C—CH(CH_3)—CH_3
Isobutane

The different bonding relationships in butane and isobutane are distinguished by their structural formulas, and the two molecules have different chemical properties.

Optical isomers occur when a carbon atom has four different atoms or groups of atoms attached to it. This pattern allows two different ways of making the attachments, each the mirror image of the other (**Figure 3.2**). Such a carbon atom is called an *asymmetrical carbon,* and the two resulting molecules are optical isomers of each other. You can envision your right and left hands as optical isomers. Just as a glove is specific for a particular hand, some biochemical molecules that can interact with one optical isomer of a carbon compound are unable to "fit" the other.

The structures of macromolecules reflect their functions

The four kinds of biological macromolecules are present in roughly the same proportions in all living organisms (**Figure 3.3**). A protein that has a certain function in an apple tree probably has a similar function in a human being because its chemistry is the same wherever it is found. One important advantage of this *biochemical unity* is that organisms can acquire needed biochemicals by eating other organisms. When you eat an apple, the molecules you take in include carbohydrates, lipids, and proteins that can be refashioned into the special varieties of those molecules needed by humans.

Each type of macromolecule performs some combination of functions, such as energy storage, structural support, protection, catalysis (speeding up a chemical reaction), transport, defense, regulation, movement, and information storage. These roles are not necessarily exclusive; for example, both carbohydrates and proteins can play structural roles, supporting and protecting tissues and organs. However, only the nucleic acids specialize in information storage. These macromolecules function as hereditary material, carrying the traits of both species and individuals from generation to generation.

The functions of macromolecules are directly related to their three-dimensional shapes and to the sequences and chemical prop-

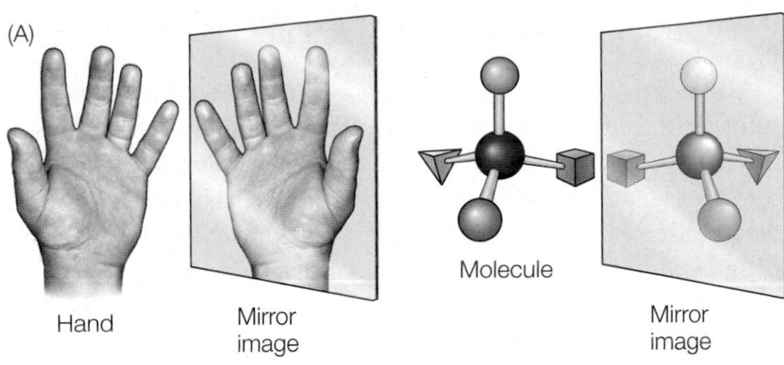

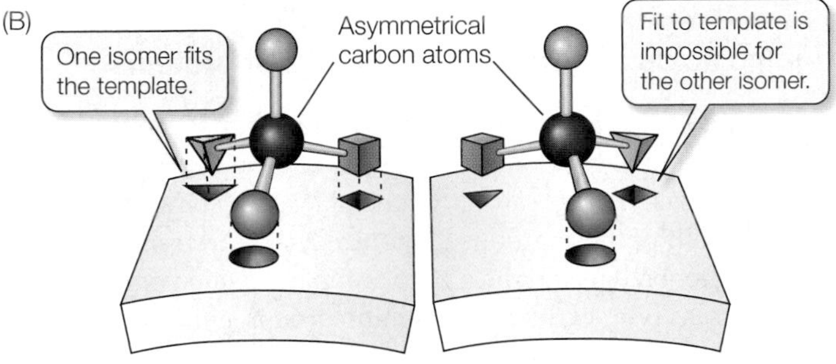

3.2 Optical Isomers (A) Optical isomers are mirror images of each other. (B) Molecular optical isomers result when four different atoms or groups are attached to a single carbon atom. If a template (representing a larger biological molecule) is laid out to match the groups on one carbon atom, the groups on that molecule's mirror-image isomer cannot be rotated to fit the same template.

erties of their monomers. Some macromolecules fold into compact spherical forms with surface features that make them water-soluble and capable of intimate interaction with other molecules. Some proteins and carbohydrates form long, fibrous systems that provide strength and rigidity to cells and organisms. Still other long, thin assemblies of proteins can contract and cause movement.

Because macromolecules are so large, they contain many different functional groups (see Figure 3.1). For example, a single large protein may contain hydrophobic, polar, and charged functional groups, each of which gives different specific properties to local sites on the macromolecule. As we will see, this diversity of properties determines the shapes of macromolecules and their interactions both with other macromolecules and with smaller molecules.

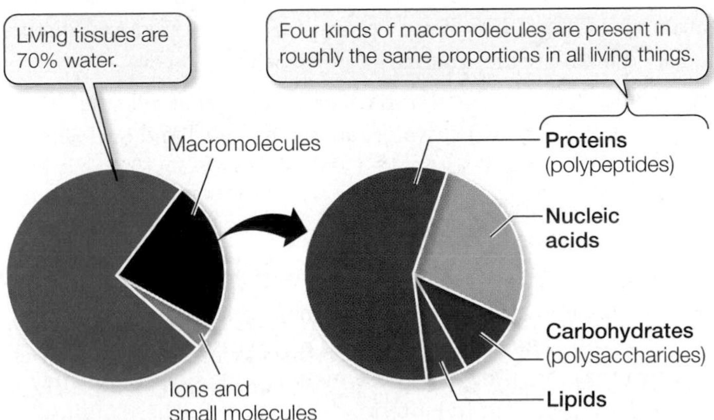

3.3 Substances Found in Living Tissues The substances shown here make up the nonmineral components of living tissue (bone would be an example of a mineral component).

Most macromolecules are formed by condensation and broken down by hydrolysis

Polymers are constructed from monomers by a series of reactions called **condensation reactions** (sometimes called *dehydration* reactions; both terms refer to the loss of water). Condensation reactions result in covalently bonded monomers. These reactions release a molecule of water for each covalent bond formed (**Figure 3.4A**). The condensation reactions that produce the different kinds of polymers differ in detail, but in all cases, polymers form only if energy is added to the system. In living systems, specific energy-rich molecules supply this energy.

The reverse of a condensation reaction is a **hydrolysis reaction** (*hydro*, "water"; *lysis*, "break"). Hydrolysis reactions digest polymers and produce monomers. Water reacts with the covalent bonds that link the polymer together, and the products are free

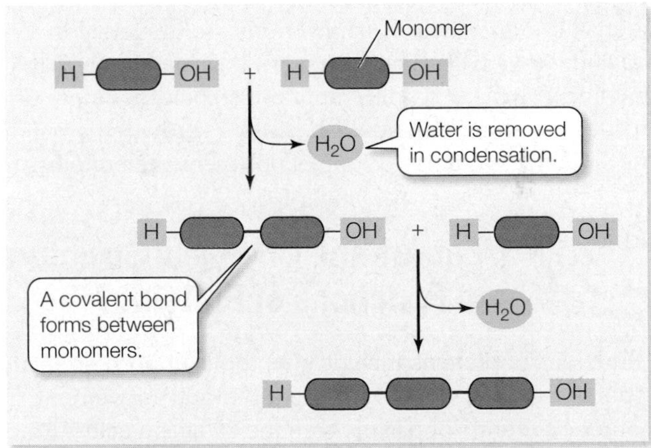

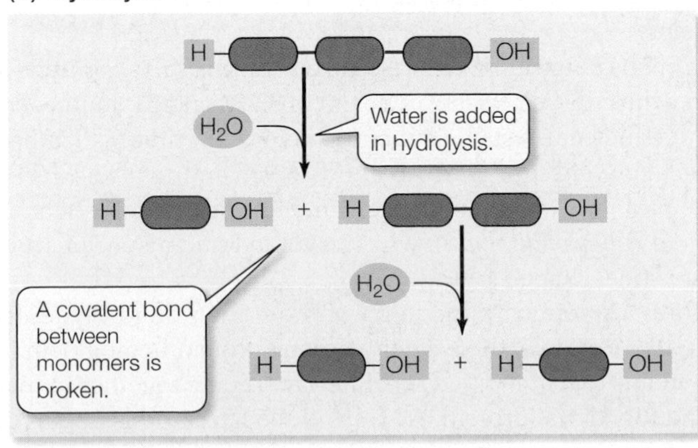

3.4 Condensation and Hydrolysis of Polymers (A) A condensation reaction links monomers into polymers. (B) A hydrolysis reaction breaks polymers into individual monomers.

monomers. The elements (H and O) of H_2O become part of the products (**Figure 3.4B**). The linkages between monomers thus can be formed and broken inside living tissues.

3.1 RECAP

The four kinds of large molecules that distinguish living tissues are proteins, lipids, carbohydrates, and nucleic acids. These biological molecules carry out a wide range of life-sustaining functions. Most of them are polymers, made up of linked monomeric subunits. Very large polymers are called macromolecules.

- Do you see how functional groups can affect the structure and function of macromolecules? (Keep this question in mind as you read the rest of this chapter.) See p. 39 and Figure 3.1
- Why is biochemical unity, as seen in the proportions of the four types of macromolecules present in all organisms, important for life? See p. 40 and Figure 3.3
- How do monomers link up to make polymers? How do they break down into monomers again? See pp. 41–42 and Figure 3.4

Condensation and hydrolysis reactions are universal in living things and, as we will see at the end of this chapter, their role in assembling polymers was an important step in the origin of life in an aqueous environment. We begin our study of the building blocks of life with a diverse group of polymers, the proteins.

3.2 What Are the Chemical Structures and Functions of Proteins?

The functions of **proteins** include structural support, protection, transport, catalysis, defense, regulation, and movement. The monomeric subunits of proteins are the 20 **amino acids**. Among the functions of macromolecules listed earlier, only two—energy storage and information storage—are not usually performed by proteins.

Proteins range in size from small ones such as the hormone insulin, which has a molecular weight of 5,733 and 51 amino acids, to huge molecules such as the muscle protein titin, which has a molecular weight of 2,993,451 and 26,926 amino acids. Each of these proteins consists of a single unbranched polymer of amino acids (a *polypeptide chain*), which is folded into a specific three-dimensional shape.

Many proteins are made up of more than one polypeptide chain. For example, the oxygen-carrying protein hemoglobin has four chains that are folded separately and come together to make up the functional protein. As we will see later in this book, proteins can associate with one another, forming "multi-protein machines" to carry out complex tasks such as DNA synthesis.

The *composition* of a protein refers to the relative amounts of the different amino acids it contains. Variation in the precise *sequence* of the amino acids in each polypeptide chain is the source of the diversity in protein structures and functions.

To understand the many functions of proteins that will be described throughout this book, we must first explore protein structure. First we will examine the properties of amino acids and see how they link together to form polypeptide chains. Then we will systematically examine protein structure and look at how a linear chain of amino acids is consistently folded into a specific, compact, three-dimensional shape. Finally, we will see how this three-dimensional structure provides a definitive physical and chemical environment that influences how other molecules can interact with the protein.

Amino acids are the building blocks of proteins

The amino acids have both a carboxyl functional group and an amino functional group (see Figure 3.1) attached to the same carbon atom, called the α carbon. At the pH values commonly found in cells, both of these groups are ionized: the carboxyl group has lost a hydrogen ion, and the amino group has gained one. Because they possess both carboxyl and amino groups, *amino acids are simultaneously acids and bases*.

Also attached to the α carbon atom are a hydrogen atom and a **side chain**, or **R group**, designated by the letter R:

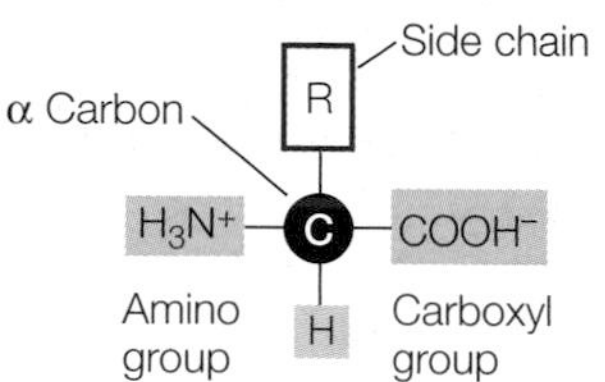

The α carbon in an amino acid is an asymmetrical carbon because it is bonded to four different atoms or groups or atoms. Therefore, amino acids exist in two isomeric forms, called D-amino acids and L-amino acids. D and L are abbreviations for the Latin terms for right (*dextro*) and left (*levo*). Only L-amino acids are commonly found in proteins in most organisms, and their presence is an important chemical "signature" of life.

The side chains of amino acids also contain functional groups, which are important in determining the three-dimensional structure and function of the protein macromolecule. The side chains are highlighted in white in **Table 3.2**. As the table shows, the 20 amino acids found in living organisms are grouped and distinguished by their side chains.

- The five amino acids that have electrically charged side chains (+1, –1) attract water (are hydrophilic) and oppositely charged ions of all sorts.
- The five amino acids that have polar side chains (δ^+, δ^-) tend to form hydrogen bonds with water and with other polar or charged substances. These amino acids are hydrophilic.
- Seven amino acids have side chains that are nonpolar hydrocarbons or very slightly modified hydrocarbons. In the watery environment of the cell, these hydrophobic side chains may cluster together in the interior of the protein. These amino acids are hydrophobic.

Three amino acids—cysteine, glycine, and proline—are special cases, although their side chains are generally hydrophobic.

TABLE 3.2

The Twenty Amino Acids

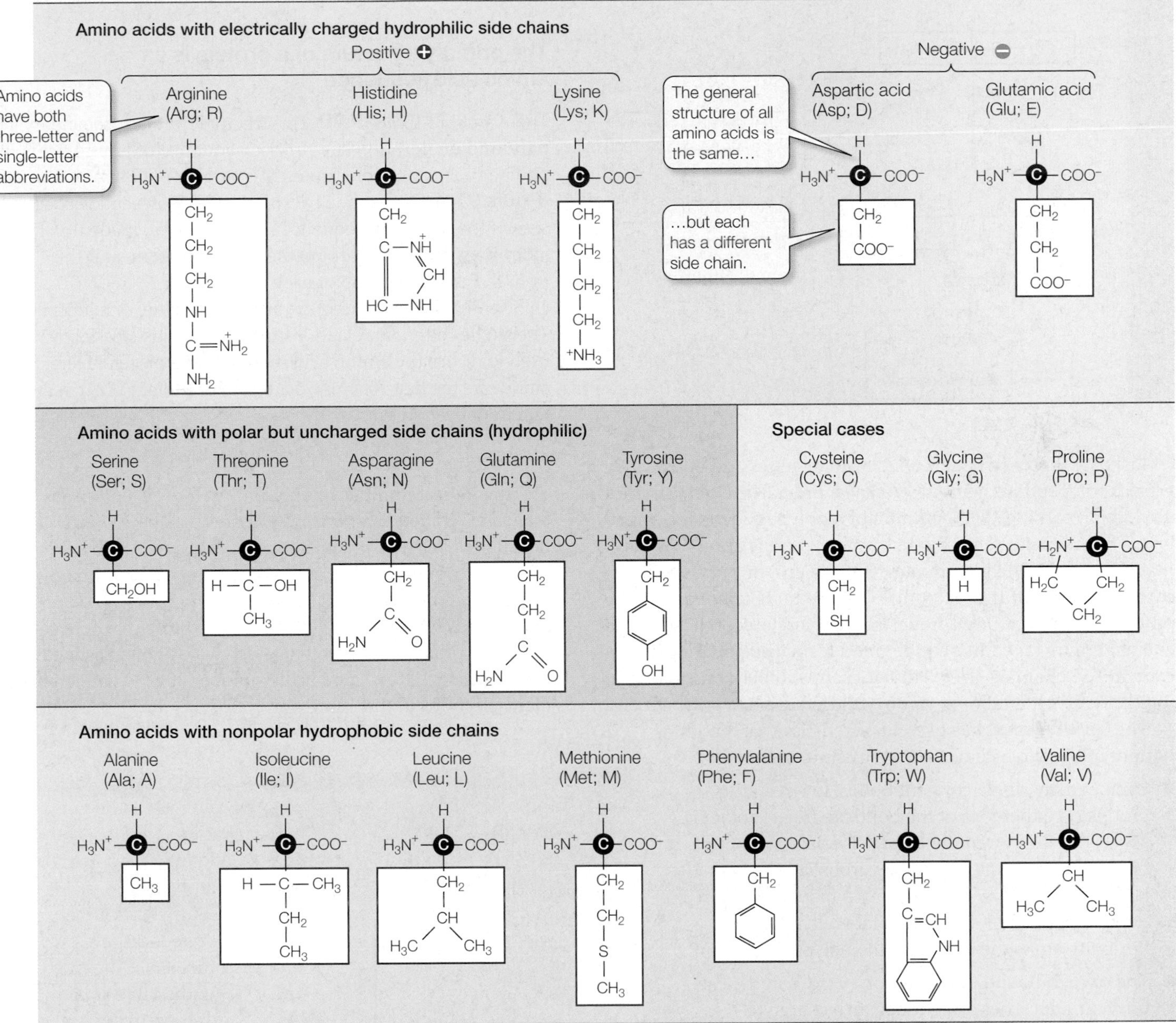

- The *cysteine* side chain, which has a terminal —SH group, can react with another cysteine side chain to form a covalent bond called a **disulfide bridge** (—S—S—) (**Figure 3.5**). Disulfide bridges help determine how a polypeptide chain folds. When cysteine is not part of a disulfide bridge, its side chain is hydrophobic.
- The *glycine* side chain consists of a single hydrogen atom and is small enough to fit into tight corners in the interior of a protein molecule, where a larger side chain could not fit.
- *Proline* differs from other amino acids because it possesses a modified amino group lacking a hydrogen on its nitrogen, which limits its hydrogen-bonding ability. In addition, the ring structure of proline limits rotation about its α carbon, so proline is often found at bends or loops in a protein.

Peptide bonds form the backbone of a protein

When amino acids polymerize, the carboxyl and amino groups attached to the α carbon are the reactive groups. The carboxyl group of one amino acid reacts with the amino group of another, undergoing a condensation reaction that forms a **peptide bond** (also called a *peptide linkage*). **Figure 3.6** gives a simplified description of this reaction. (In living systems, other molecules must activate the amino acids before this reaction proceeds, and there are intermediate steps in the process as well. We will examine these steps in Section 12.4.)

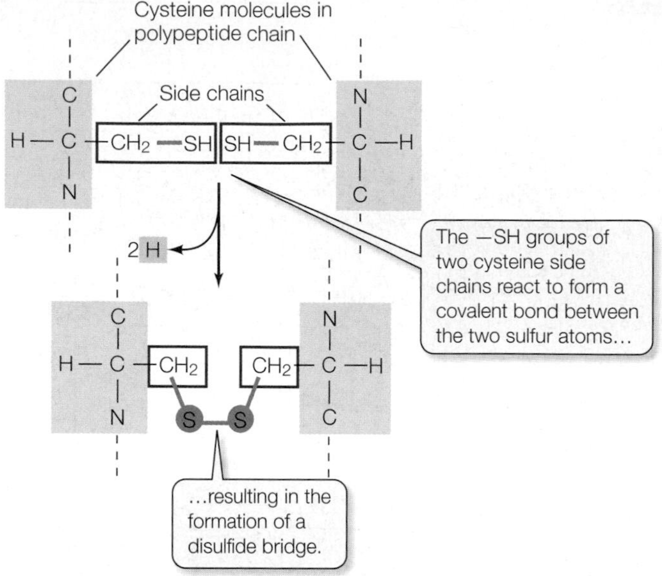

3.5 A Disulfide Bridge Two cysteine molecules in a polypeptide chain can form a disulfide bridge (—S—S—).

Just as a sentence begins with a capital letter and ends with a period, polypeptide chains have a linear order. The chemical "capital letter" marking the beginning of a polypeptide is the amino group of the first amino acid in the chain and is known as the *N terminus*. The "punctuation mark" for the end of the chain is the carboxyl group of the last amino acid—the *C terminus*. All the other amino and carboxyl groups in the chain (except those in side chains) are involved in peptide bond formation, so they do not exist in the chain as "free," intact groups. Biochemists refer to the "N → C," or "amino-to-carboxyl," orientation of polypeptides.

The peptide bond has two characteristics that are important in the three-dimensional structure of proteins:

- Unlike many single covalent bonds, in which the groups on either side of the bond are free to rotate in space, the C—N peptide linkage is relatively inflexible. The adjacent atoms (the α carbons of the two adjacent amino acids, shown in black in Figure 3.6) in a peptide bond are not free to rotate fully, and this inflexibility limits the folding of the polypeptide chain.
- The oxygen bound to the carbon (C═O) in the carboxyl group carries a slight negative charge (δ^-), whereas the hydrogen bound to the nitrogen (N—H) in the amino group is slightly positive (δ^+). This asymmetry of charge favors hydrogen bonding within the protein molecule itself and with other molecules, contributing to both the structure and the function of many proteins.

Before we explore the significance of these characteristics, however, we need to examine the importance of the order of amino acids.

The primary structure of a protein is its amino acid sequence

There are four levels of protein structure: primary, secondary, tertiary, and quaternary. The precise sequence of amino acids in a polypeptide chain constitutes the **primary structure** of a protein (**Figure 3.7A**). The peptide backbone of the polypeptide chain consists of the repeating sequence —N—C—C—, made up of the N atom from the amino group, the α carbon atom, and the C atom from the carboxyl group of each amino acid.

Scientists have deduced the primary structure of many proteins. The single-letter abbreviations for amino acids (see Table 3.2) are used to record the amino acid sequence of a protein. Here, for example, are the first 20 amino acids (out of a total of 124) in the protein ribonuclease from a cow:

KETAAAKFERQHMDSSTSAA

The theoretical number of different proteins is enormous. Since there are 20 different amino acids, there could be 20 × 20 = 400 distinct dipeptides (two linked amino acids), and 20 × 20 × 20 = 8,000 different tripeptides (three linked amino acids). Imagine this process of multiplying by 20 extended to a protein made up of 100 amino acids (which is considered a small protein). There could be 20^{100} (10^{130}) such small proteins, each with its own distinctive primary structure. How large is the number 20^{100}? There aren't that many electrons in the entire universe!

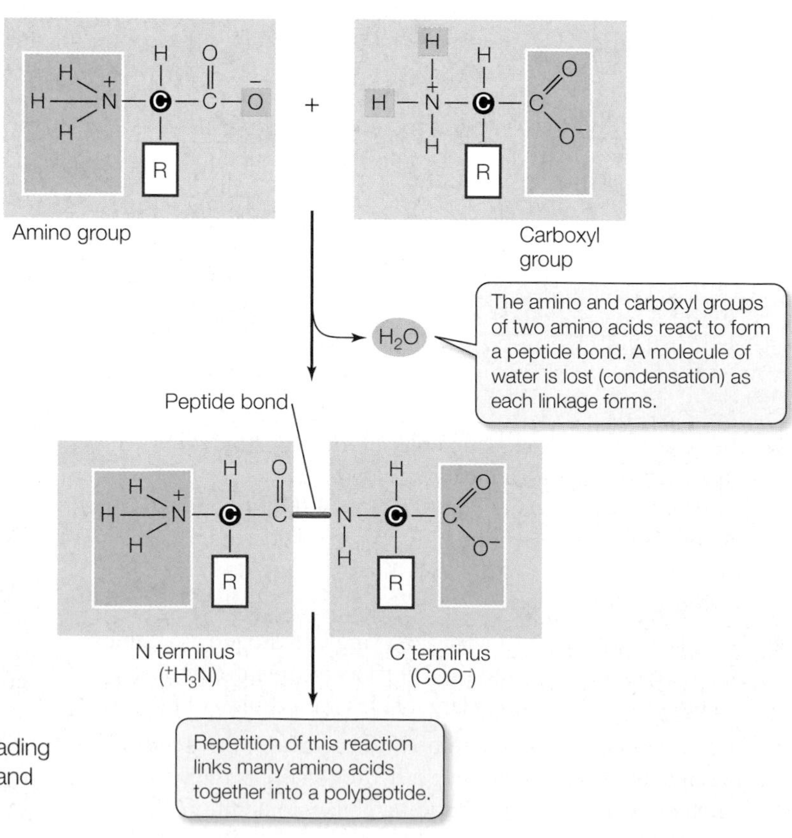

Amino group
Carboxyl group
The amino and carboxyl groups of two amino acids react to form a peptide bond. A molecule of water is lost (condensation) as each linkage forms.
H2O
Peptide bond
N terminus (+H3N)
C terminus (COO−)
Repetition of this reaction links many amino acids together into a polypeptide.

3.6 Formation of Peptide Bonds In living things, the reaction leading to a peptide bond has many intermediate steps, but the reactants and products are the same as those shown in this simplified diagram.

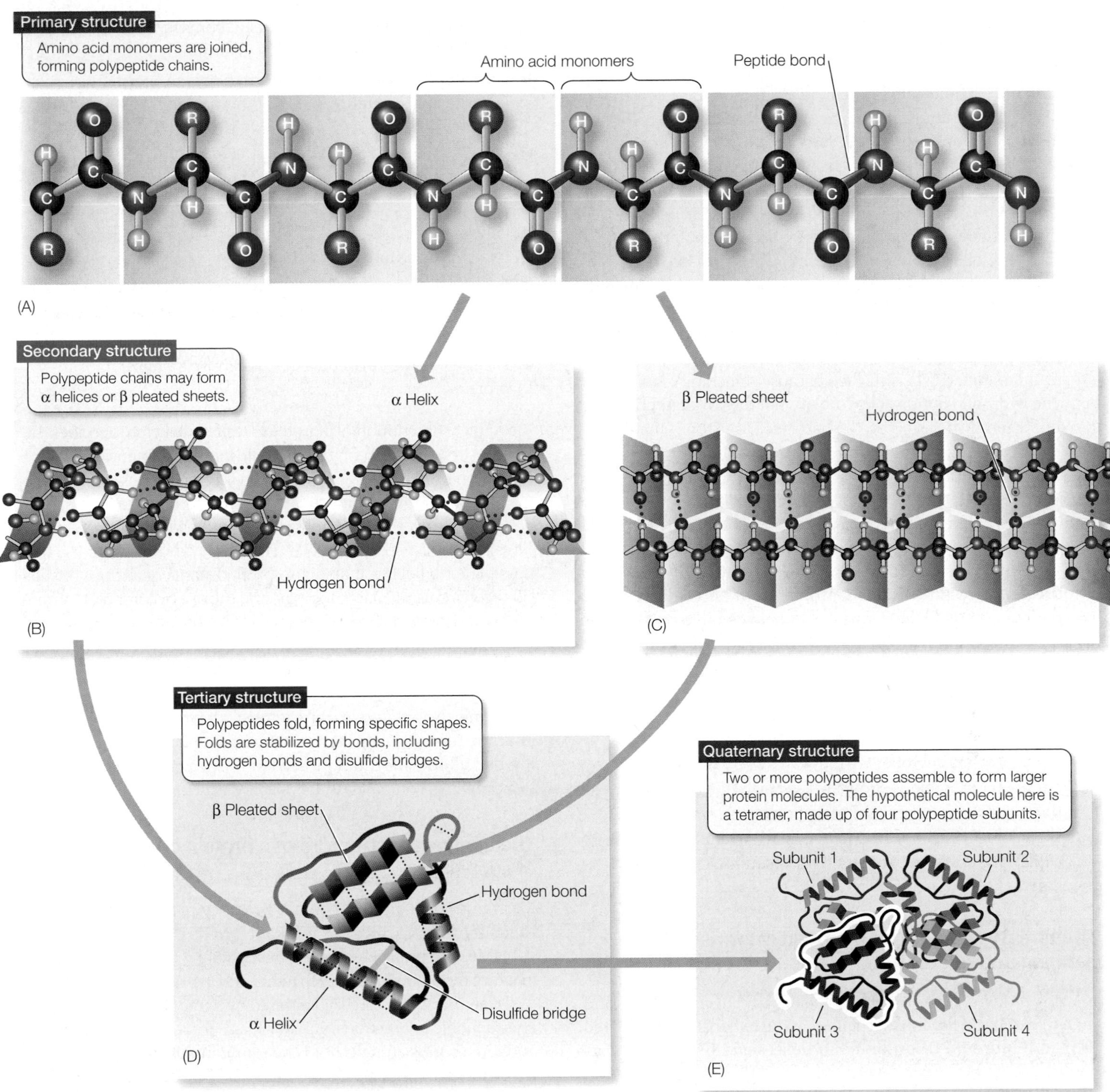
Primary structure
Amino acid monomers are joined, forming polypeptide chains.
Amino acid monomers
Peptide bond
(A)
Secondary structure
Polypeptide chains may form α helices or β pleated sheets.
α Helix
Hydrogen bond
(B)
β Pleated sheet
Hydrogen bond
(C)
Tertiary structure
Polypeptides fold, forming specific shapes. Folds are stabilized by bonds, including hydrogen bonds and disulfide bridges.
β Pleated sheet
Hydrogen bond
α Helix
Disulfide bridge
(D)
Quaternary structure
Two or more polypeptides assemble to form larger protein molecules. The hypothetical molecule here is a tetramer, made up of four polypeptide subunits.
Subunit 1
Subunit 2
Subunit 3
Subunit 4
(E)

3.7 The Four Levels of Protein Structure Secondary, tertiary, and quaternary structure all arise from the primary structure of the protein.

At the higher levels of protein structure, local coiling and folding give the molecule its final functional shape, but all of these levels derive from the primary structure—that is, the precise location of specific amino acids in the polypeptide chain. The properties associated with a precise sequence of amino acids determine how the protein can twist and fold, thus adopting a specific stable structure that distinguishes it from every other protein.

Primary structure is determined by covalent bonds. The next level of protein structure is determined by weaker hydrogen bonds.

The secondary structure of a protein requires hydrogen bonding

A protein's **secondary structure** consists of regular, repeated spatial patterns in different regions of a polypeptide chain. There are two basic types of secondary structure, both of them determined by hydrogen bonding between the amino acids that make up the primary structure.

THE α HELIX The α (alpha) **helix** is a right-handed coil that is "threaded" in the same direction as a standard wood screw (**Figure 3.7B**). The R groups extend outward from the peptide backbone of the helix. The coiling results from hydrogen bonds that form between the δ^+ hydrogen of the N—H of one amino acid and the δ^- oxygen of the C=O of another. When this pattern of hydrogen bonding is established repeatedly over a segment of the protein, it stabilizes the coil. The presence of amino acids with large R groups that distort the coil or otherwise prevent the formation of the necessary hydrogen bonds will keep an α helix from forming.

The α-helical secondary structure is common in the fibrous structural proteins called keratins, which make up hair, hooves, and feathers. Hair can be stretched because stretching requires that only the hydrogen bonds of the α helix, not the covalent bonds, be broken; when the tension on the hair is released, both the hydrogen bonds and the helix re-form.

THE β PLEATED SHEET A β (beta) **pleated sheet** is formed from two or more polypeptide chains that are almost completely extended and aligned. The sheet is stabilized by hydrogen bonds between the N—H groups on one chain and the C=O groups on the other (**Figure 3.7C**). A β pleated sheet may form between separate polypeptide chains, as in spider silk, or between different regions of a single polypeptide chain that is bent back on itself. Many proteins contain regions of both α helix and β pleated sheet in the same polypeptide chain.

When proteins get together, they can be very strong. Spider silk, for example, is stronger than Kevlar, the artificial polymer used in bulletproof vests. Potential uses for spider silk (if it could be purified in large enough quantities) include automobile airbags and surgical sutures.

The tertiary structure of a protein is formed by bending and folding

In many proteins, the polypeptide chain is bent at specific sites and then folded back and forth, resulting in the **tertiary structure** of the protein (**Figure 3.7D**). Although α helices and β pleated sheets contribute to the tertiary structure, usually only portions of the macromolecule have these secondary structures, and large regions consist of tertiary structure unique to a particular protein. The tertiary structure results in a macromolecule with a definite three-dimensional shape. The outer surfaces of the macromolecule present functional groups capable of chemically interacting with other molecules in the cell. These molecules might be other proteins (as in quaternary structure, as we will see below) or small molecules (as in enzymes; see Section 6.4).

While hydrogen bonding between the N—H and C=O groups within and between chains is responsible for secondary structure, the interactions between R groups—the amino acid side chains—determine tertiary structure. We described the various strong and weak interactions between atoms in Section 2.2. Many of these interactions are involved in determining tertiary structure.

- Covalent *disulfide bridges* can form between specific cysteine side chains (see Figure 3.5), holding a folded polypeptide in place.
- *Hydrophobic* side chains can aggregate together in the interior of the protein, away from water, folding the polypeptide in the process.
- *van der Waals forces* can stabilize the close interactions between hydrophobic side chains.
- *Ionic bonds* can form between positively and negatively charged side chains buried deep within a protein, away from water, forming a *salt bridge*.
- *Hydrogen bonds* between side chains also stabilize folds in proteins.

A complete description of a protein's tertiary structure specifies the location of every atom in the molecule in three-dimensional space in relation to all the other atoms. Such a description is available for the protein lysozyme (**Figure 3.8**). The first tertiary structures to be determined took years to figure out, but today dozens of new structures are published every week. The major advances making this possible have been our ability to produce large quantities of specific proteins using biotechnology, and the use of computers to analyze the atomic data.

Bear in mind that both tertiary structure and secondary structure derive from a protein's primary structure. If lysozyme is heated slowly, the heat energy will disrupt only the weak interactions and cause only the tertiary structure to break down. But the protein will return to its normal tertiary structure when it cools, demonstrating that all the information needed to specify the unique shape of a protein is contained in its primary structure.

The quaternary structure of a protein consists of subunits

Many functional proteins contain two or more polypeptide chains, called *subunits*, each of them folded into its own unique tertiary structure. The protein's **quaternary structure** results from the ways in which these subunits bind together and interact (see Figure 3.7E).

Quaternary structure is illustrated by hemoglobin (**Figure 3.9**). Hydrophobic interactions, van der Waals forces, hydrogen bonds, and ionic bonds all help hold the four subunits together to form the hemoglobin molecule. The function of hemoglobin is to carry oxygen in red blood cells. As hemoglobin binds one O_2 molecule, the four subunits shift their relative positions slightly, changing the quaternary structure. Ionic bonds are broken, exposing buried side chains that enhance the binding of additional O_2 molecules. The quaternary structure changes again when hemoglobin releases its O_2 molecules to the cells of the body.

Both shape and surface chemistry contribute to protein specificity

The specific shape and structure of a protein allows it to bind *non-covalently* to another molecule (often called a **ligand**), which in turn

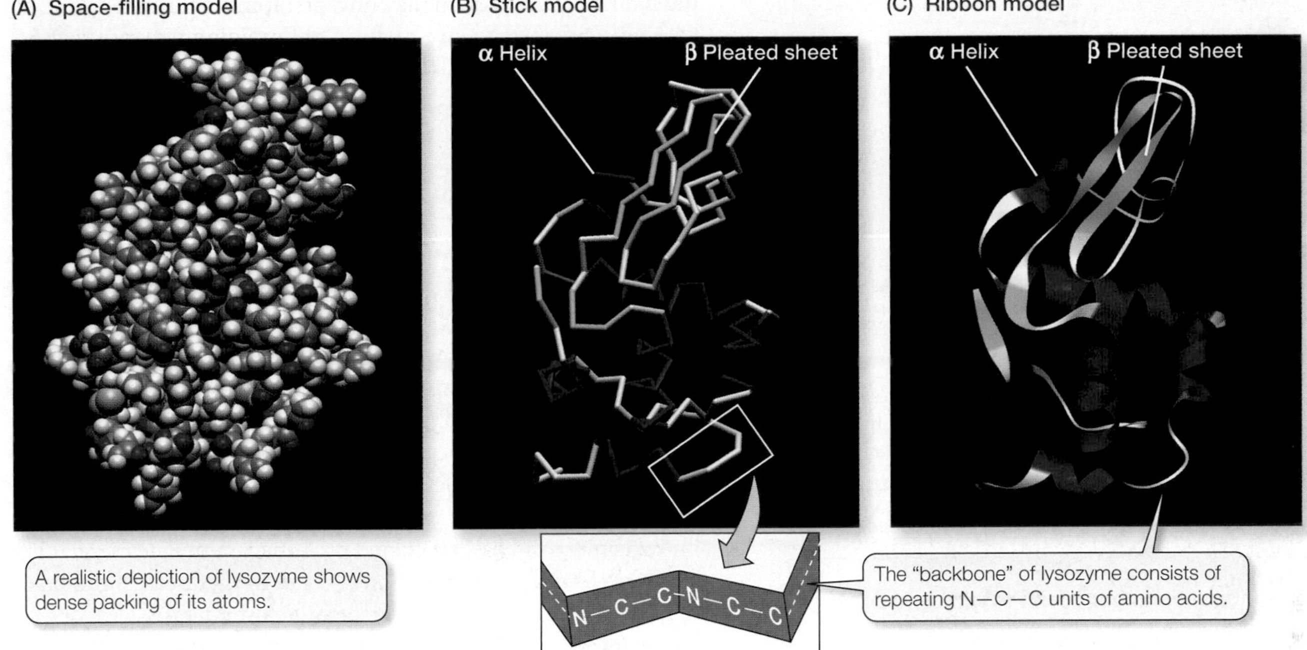

3.8 Three Representations of Lysozyme Different molecular representations of a protein emphasize different aspects of its tertiary structure. These three representations of lysozyme are similarly oriented.

allows other important biological events to occur. Here are just a few examples of how protein functions follow from protein structure:

- Two adjacent cells can stick together because proteins protruding from each of the cells interact with each other (cell junctions; see Section 5.2).
- A substance can enter a cell by binding to a carrier protein in the cell surface membrane (membrane transport; see Section 5.3).
- A chemical reaction can be speeded up when a type of protein called an *enzyme* binds to the reactants (enzyme–substrate interactions; see Section 6.3).
- Chemical signals such as hormones can bind to proteins on a cell's outer surface (receptor proteins; see Section 15.1 and Chapter 41).
- Defensive proteins called antibodies can recognize the shape of a virus coat and bind to it (immune response; see Chapter 18).

The biological specificity of protein function depends on two general properties of the protein: its shape, and the chemistry of its exposed surface groups.

- *Shape.* When a small molecule collides with and binds to a much larger protein, it is like a baseball being caught by a catcher's mitt: the mitt has a shape that binds to the ball and fits around it. Just as a hockey puck or a ping-pong ball is not designed to fit a baseball catcher's mitt, a given molecule will bind to a protein only if there is a general "fit" between their two three-dimensional shapes. The need for the right "fit" becomes even more specific after initial binding.
- *Chemistry.* The functional groups on the surface of a protein promote chemical interactions with other substances (**Figure 3.10**). These groups are the side chains of the exposed amino acids, and are therefore a property of the protein's primary structure.

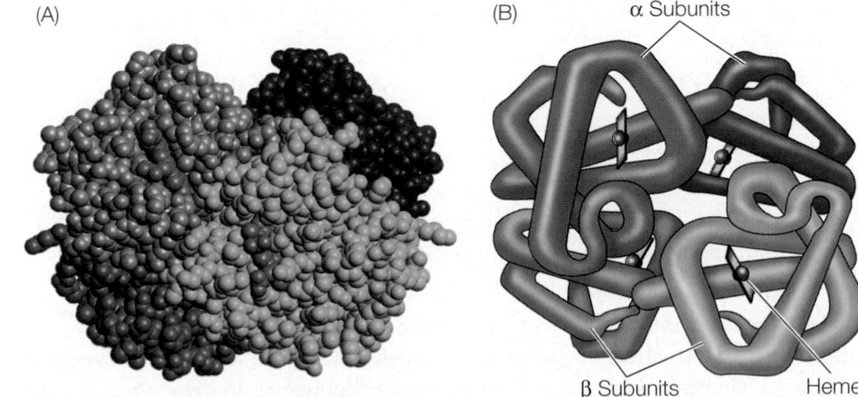

3.9 Quaternary Structure of a Protein Hemoglobin consists of four folded polypeptide subunits that assemble themselves into the quaternary structure shown here. In these two graphic representations, each type of subunit is a different color. The heme groups contain iron and are the oxygen-carrying sites.

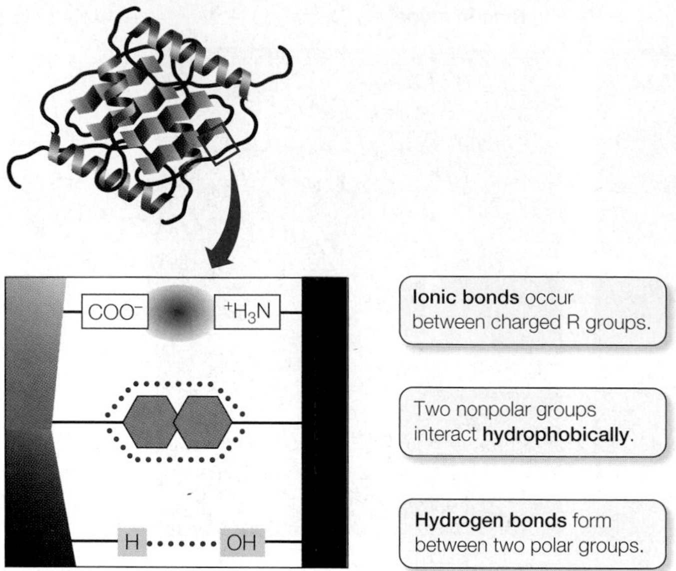

3.10 Noncovalent Interactions between Proteins and Other Molecules Noncovalent interactions allow a protein to bind tightly to another molecule with specific properties, or allow regions within a protein to interact with one another.

Examine the structures of the 20 amino acids in Table 3.2, noting the properties of the side chains. Exposed hydrophobic groups can bind to similarly nonpolar groups in the ligand. Charged R groups can bind to oppositely charged groups on the ligand. Polar R groups containing a hydroxyl (—OH) group can form a hydrogen bond with the ligand. These three types of interactions—hydrophobic, ionic, and hydrogen bonding—are each weak by themselves, but strong when all of them act together. So the exposure of appropriate amino acid R groups on the protein surface allows the binding of a specific ligand to occur.

Environmental conditions affect protein structure

Because it is determined by weak forces, the three-dimensional structure of proteins is sensitive to environmental conditions. Conditions that would not break covalent bonds, but can upset the weaker noncovalent interactions that determine secondary and tertiary structure, may affect a protein's shape and thus its function.

- *Increases in temperature* cause more rapid molecular movements and thus can break hydrogen bonds and hydrophobic interactions.
- *Alterations in pH* can change the pattern of ionization of carboxyl and amino groups in the R groups of amino acids, thus disrupting the pattern of ionic attractions and repulsions.
- *High concentrations of polar substances* such as urea can disrupt the hydrogen bonding that is crucial to protein structure. Nonpolar substances may also disrupt normal protein structure.

The loss of a protein's normal three-dimensional structure is called **denaturation**, and it is always accompanied by a loss of the normal biological function of the protein (**Figure 3.11**). Denaturation is usually irreversible because amino acids that were buried in the interior of the protein may now be exposed at the surface, and vice versa, causing a new structure to form or different molecules to bind to the protein. The boiling of an egg denatures its proteins and is, as you know, not reversible. However, as we saw earlier in the case of lysozyme, denaturation may be reversible in the laboratory. If the protein is allowed to cool or the denaturing chemicals are removed, the protein may return to its "native" shape and normal function.

Chaperonins help shape proteins

There are two occasions when a polypeptide chain is in danger of binding the wrong ligand. The first, as we have just seen, is following denaturation. In addition, when a protein has just been made and has not yet folded completely, it can present a surface that binds the wrong molecule. In the cell, a protein can sometimes fold incorrectly as it is made, with serious consequences. For example, in Alzheimer's disease, misfolded proteins accumulate in the brain and bind to one another, forming fibers in the areas of the brain that control memory, mood, and spatial awareness.

Living systems limit inappropriate protein interactions by making a class of proteins called, appropriately, **chaperonins** (recall the chaperones at school dances who tried to prevent "inappropriate interactions" among the students). Chaperonins were first identified in fruit flies as "heat shock" proteins, which prevented denaturing proteins from clumping together when the flies' temperatures were raised.

Some chaperonins work by trapping proteins that are in danger of inappropriate binding inside a molecular "cage" (**Figure 3.12**). This cage has several identical subunits, and is itself a good example of quaternary protein structure. Inside the cage, the targeted protein folds into the correct shape, and then is released at the appropriate time and place.

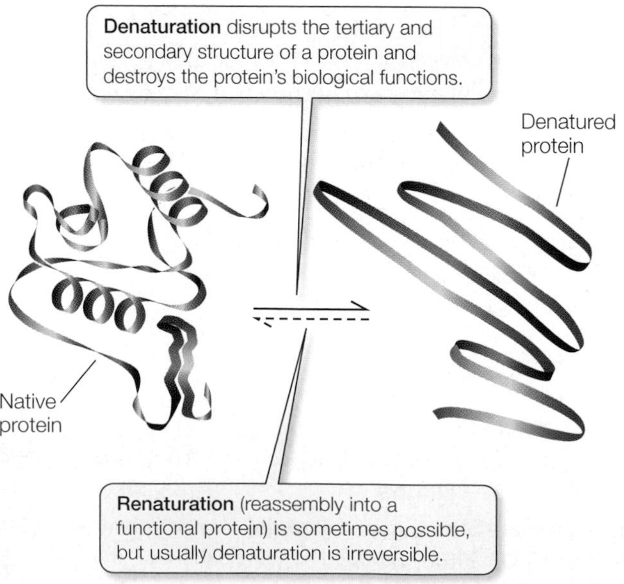

Denaturation disrupts the tertiary and secondary structure of a protein and destroys the protein's biological functions.
Denatured protein
Native protein
Renaturation (reassembly into a functional protein) is sometimes possible, but usually denaturation is irreversible.

3.11 Denaturation Is the Loss of Tertiary Protein Structure and Function Agents that can cause denaturation include high temperatures and certain chemicals.

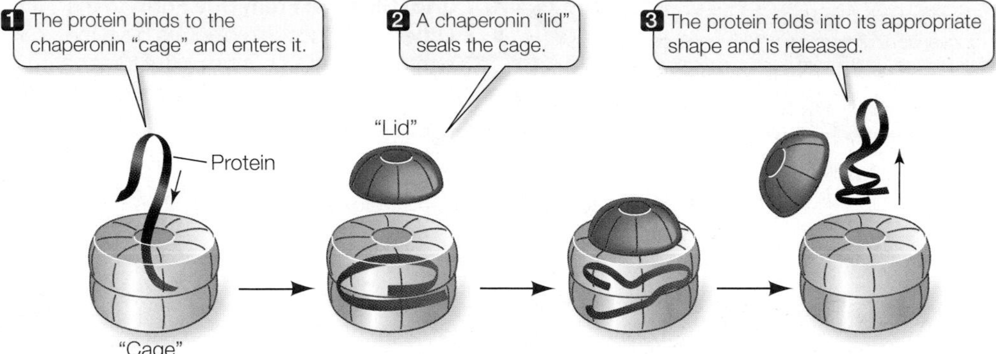

1 The protein binds to the chaperonin "cage" and enters it.
2 A chaperonin "lid" seals the cage.
3 The protein folds into its appropriate shape and is released.
Protein
"Lid"
"Cage"

3.12 Chaperonins Protect Proteins from Inappropriate Binding Chaperonins surround new or denatured proteins and prevent them from binding to the wrong ligand.

3.2 RECAP

Proteins are polymers of amino acids. The sequence of amino acids in a protein determines its primary structure; secondary, tertiary, and quaternary structures arise through interactions between the amino acids. A protein's three-dimensional shape ultimately determines its function.

- Can you describe the attributes of an amino acid's "R" group that would make it hydrophobic? Hydrophilic? See p. 42 and Table 3.2
- Can you sketch and explain how two amino acids link together to form a peptide bond? See p. 43 and Figure 3.6
- Can you describe the four levels of protein structure and how they are all ultimately determined by the protein's primary structure (i.e., its amino acid sequence)? See pp. 44–46 and Figure 3.7
- Do you understand the different ways environmental factors such as temperature and pH can affect the weak interactions that give a protein its specific shape and function? See p. 48

3.3 What Are the Chemical Structures and Functions of Carbohydrates?

Carbohydrates are molecules containing carbon atoms flanked by hydrogen atoms and hydroxyl groups (H—C—OH). They have two major biochemical roles:

- Carbohydrates are a source of energy that can be released in a form usable by body tissues.
- Carbohydrates serve as *carbon skeletons* that can be rearranged to form new molecules that are essential for biological structures and functions.

Some carbohydrates are relatively small, with molecular weights of less than 100. Others are true macromolecules, with molecular weights in the hundreds of thousands.

There are four categories of biologically important carbohydrates, which we will discuss in turn:

- **Monosaccharides** (*mono*, "one"; *saccharide*, "sugar"), such as glucose, ribose, and fructose, are simple sugars. They are the monomers from which the larger carbohydrates are constructed.
- **Disaccharides** (*di*, "two") consist of two monosaccharides linked together by covalent bonds.
- **Oligosaccharides** (*oligo*, "several") are made up of several (3–20) monosaccharides.
- **Polysaccharides** (*poly*, "many"), such as starch, glycogen, and cellulose, are large polymers composed of hundreds or thousands of monosaccharides.

The general formula for carbohydrates, CH_2O, gives the relative proportions of carbon, hydrogen, and oxygen in a monosaccharide (i.e., the proportions of these atoms are 1:2:1). In disaccharides, oligosaccharides, and polysaccharides, these proportions differ slightly from the general formula because two hydrogens and an oxygen are lost during each of the condensation reactions that form them.

Monosaccharides are simple sugars

Green plants produce monosaccharides through photosynthesis, and animals acquire them directly or indirectly from plants. All living cells contain the monosaccharide **glucose**. Cells use glucose as an energy source, breaking it down through a series of reactions that release stored energy and produce water and carbon dioxide. Glucose exists in two forms, the straight chain form and the ring form. The ring form predominates in more than 99 percent of biological circumstances because it is more stable under physiological conditions. There are two versions of the ring form, called α- and β-glucose, which differ only in the orientation of the —H and —OH attached to carbon 1 (**Figure 3.13**). The α and β forms interconvert and exist in equilibrium when dissolved in water.

Different monosaccharides contain different numbers of carbons. (The standard convention for numbering carbons in carbohydrates shown in Figure 3.13 is used throughout this book.) Most of the monosaccharides found in living systems belong to the D series of optical isomers. (Recall also that only L-amino acids occur in proteins—there is amazing specificity in biology!) Some monosaccharides are structural isomers, with the same kinds and numbers of atoms, but in different arrangements. For example, the **hexoses** (*hex*, "six"), a group of structural isomers, all have the for-

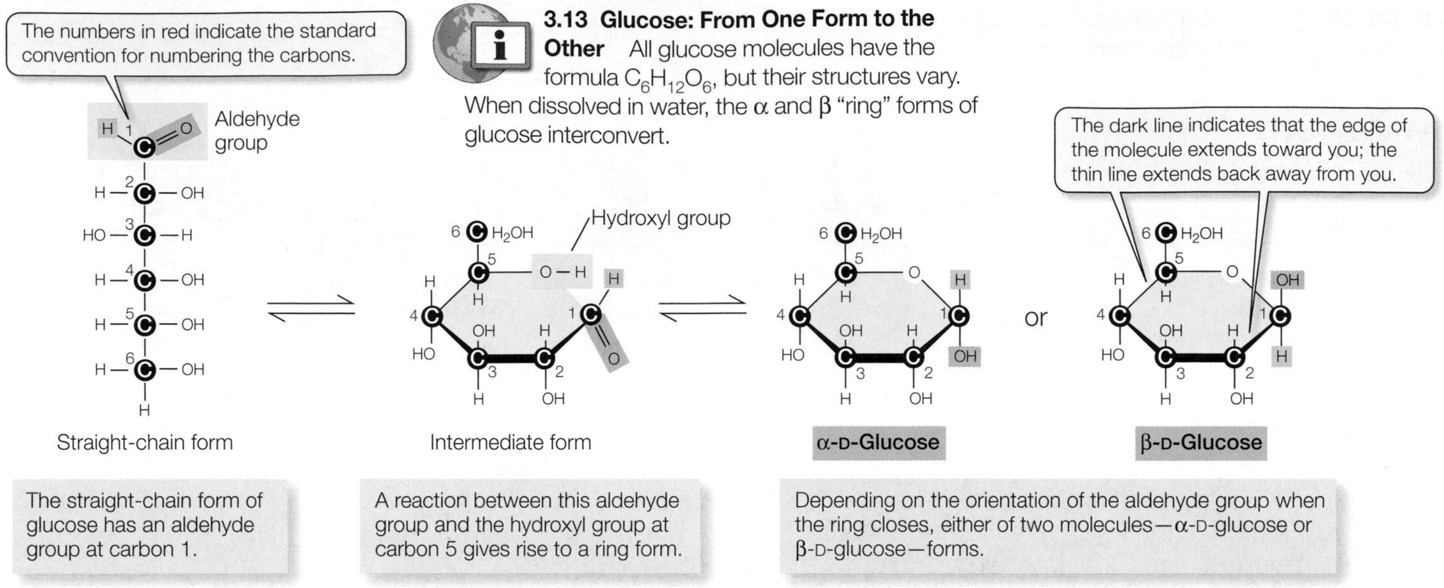

The numbers in red indicate the standard convention for numbering the carbons.
Aldehyde group
Hydroxyl group
The dark line indicates that the edge of the molecule extends toward you; the thin line extends back away from you.
or
Straight-chain form
Intermediate form
α-D-Glucose
β-D-Glucose
The straight-chain form of glucose has an aldehyde group at carbon 1.
A reaction between this aldehyde group and the hydroxyl group at carbon 5 gives rise to a ring form.
Depending on the orientation of the aldehyde group when the ring closes, either of two molecules—α-D-glucose or β-D-glucose—forms.

3.13 Glucose: From One Form to the Other All glucose molecules have the formula $C_6H_{12}O_6$, but their structures vary. When dissolved in water, the α and β "ring" forms of glucose interconvert.

mula $C_6H_{12}O_6$. Included among the hexoses are glucose, fructose (so named because it was first found in fruits), mannose, and galactose (**Figure 3.14**).

Pentoses (*pente*, "five") are five-carbon sugars. Two pentoses are of particular biological importance: ribose and deoxyribose form part of the backbones of the nucleic acids RNA and DNA, respectively (see Section 3.5). These two pentoses are not isomers; rather, one oxygen atom is missing from carbon 2 in deoxyribose (*de-*, "absent"). The absence of this oxygen atom is an important distinction between RNA and DNA.

Glycosidic linkages bond monosaccharides

The disaccharides, oligosaccharides, and polysaccharides are all constructed from monosaccharides that are covalently bonded together by condensation reactions that form **glycosidic linkages**. A single glycosidic linkage between two monosaccharides forms a disaccharide. For example, a molecule of sucrose is a disaccharide formed from a glucose molecule and a fructose molecule, while lactose (milk sugar) contains glucose and galactose. Sucrose, which is common table sugar in the human diet, is a major disaccharide in plants.

The disaccharide maltose contains two glucose molecules, but it is not the only disaccharide that can be made from two glucoses. When glucose molecules form a glycosidic linkage, the linkage will be one of two types, α or β, depending on whether the molecule that bonds its carbon 1 is α-D-glucose or β-D-glucose (see Figure 3.13). An α linkage with carbon 4 of a second glucose molecule produces maltose, whereas a β linkage gives cellobiose (**Figure 3.15**). Maltose and cellobiose are structural isomers, both having the formula $C_{12}H_{22}O_{11}$. However, they are different compounds with different properties. They undergo different chemical reactions and are recognized by different enzymes. For example, maltose can be hydrolyzed into its monosaccharides in the human

Three-carbon sugar

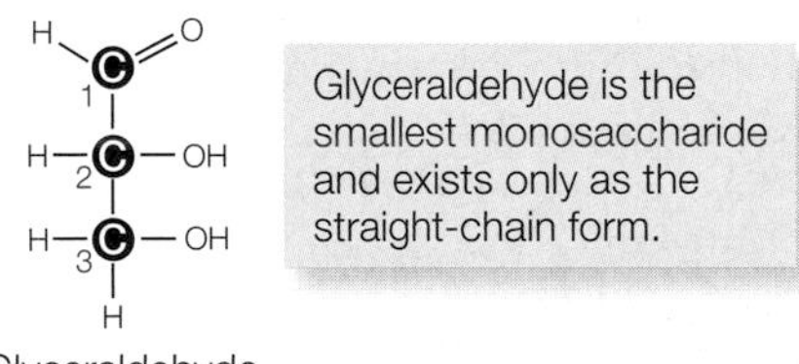

Glyceraldehyde is the smallest monosaccharide and exists only as the straight-chain form.
Glyceraldehyde

Five-carbon sugars (pentoses)

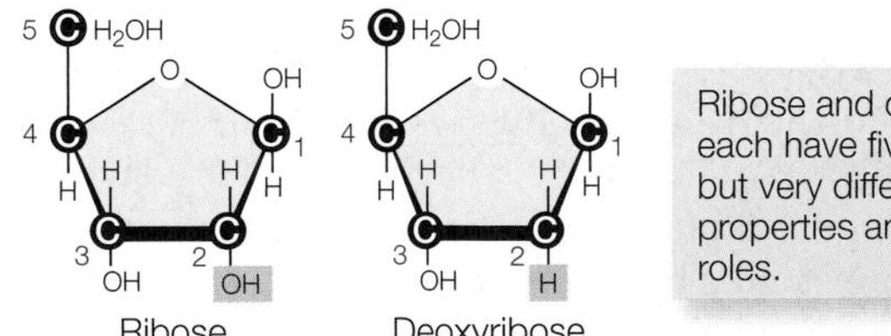

Ribose and deoxyribose each have five carbons, but very different chemical properties and biological roles.
Ribose
Deoxyribose

Six-carbon sugars (hexoses)

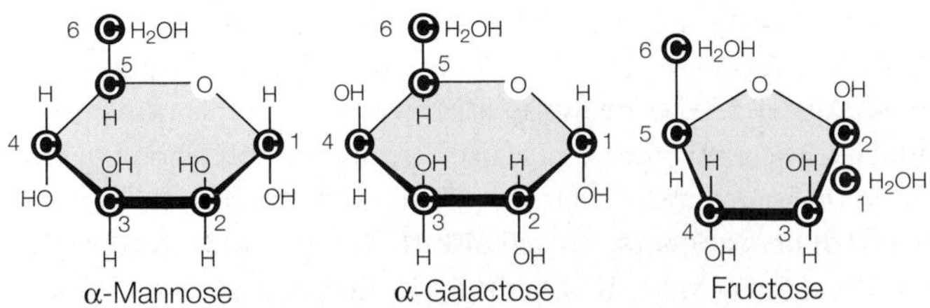

3.14 Monosaccharides Are Simple Sugars Monosaccharides are made up of varying numbers of carbons. Some hexoses are structural isomers that have the same kind and number of atoms, but the atoms are arranged differently. Fructose, for example, is a hexose, but forms a five-membered ring like the pentoses.

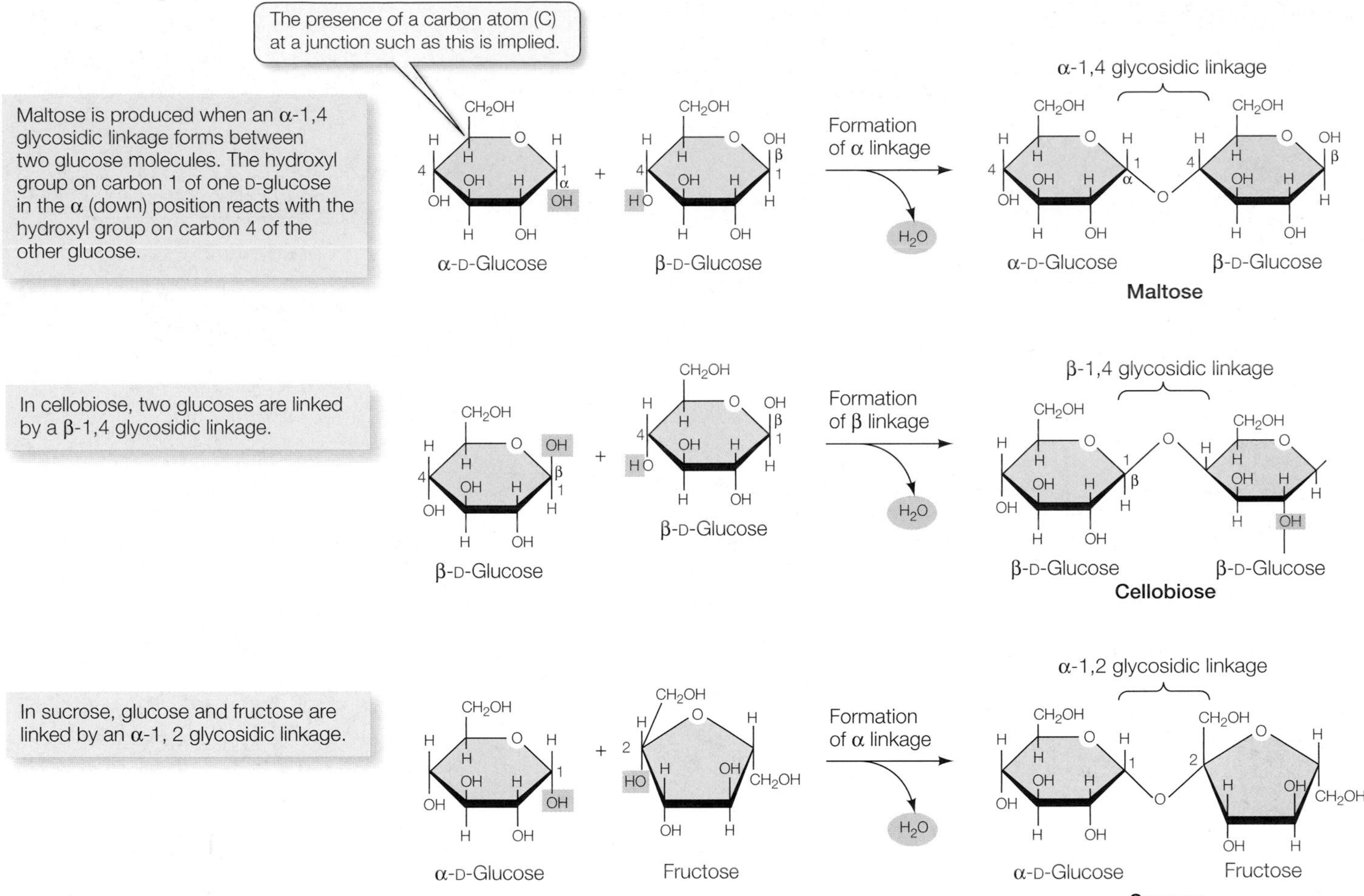

3.15 Disaccharides Are Formed by Glycosidic Linkages Glycosidic linkages between two monosaccharides can create many different disaccharides. Which disaccharide is formed depends on which monosaccharides are linked, and on the site (which carbon atom is linked) and form (α or β) of the linkage.

body, whereas cellobiose cannot. Certain microorganisms have the chemistry needed to break down cellobiose.

Oligosaccharides contain several monosaccharides bound by glycosidic linkages at various sites. Many oligosaccharides have additional functional groups, which give them special properties. Oligosaccharides are often covalently bonded to proteins and lipids on the outer cell surface, where they serve as recognition signals. The different human blood groups (such as the ABO blood types) get their specificity from oligosaccharide chains.

Polysaccharides store energy and provide structural materials

Polysaccharides are giant polymers of monosaccharides connected by glycosidic linkages (**Figure 3.16**):

- *Starch* is a polysaccharide of glucose with α-glycosidic linkages.
- *Glycogen* is a highly branched polysaccharide of glucose.
- *Cellulose* is also a polysaccharide of glucose, but its individual monosaccharides are connected by β-glycosidic linkages.

Starch comprises a family of giant molecules of broadly similar structure. While all starches are large polymers of glucose with α linkages (see Figure 3.16A), the different starches can be distinguished by the amount of branching that occurs at carbons 1 and 6 (see Figure 3.16B). Some plant starches are unbranched, such as plant amylose; others are moderately branched, such as plant amylopectin. Starch readily binds water, and when that water is removed, unbranched starch tends to form hydrogen bonds between the polysaccharide chains, which then aggregate.

Bread becomes hard and stale because when it dries out, the polysaccharide chains in starch aggregate. Adding water and gentle heat separates the chains and the bread becomes softer.

Glycogen stores glucose in animal livers and muscles. Starch and glycogen serve as energy storage compounds for plants and animals, respectively. Both of these polysaccharides are readily hydrolyzed into glucose monomers, which in turn can be further degraded to liberate their stored energy. But if it is glucose that is actually needed for fuel, why must it be stored as a polymer? The reason is that 1,000 glucose molecules would exert 1,000 times the osmotic pressure of a single glycogen molecule, causing water to

(A) Molecular structure

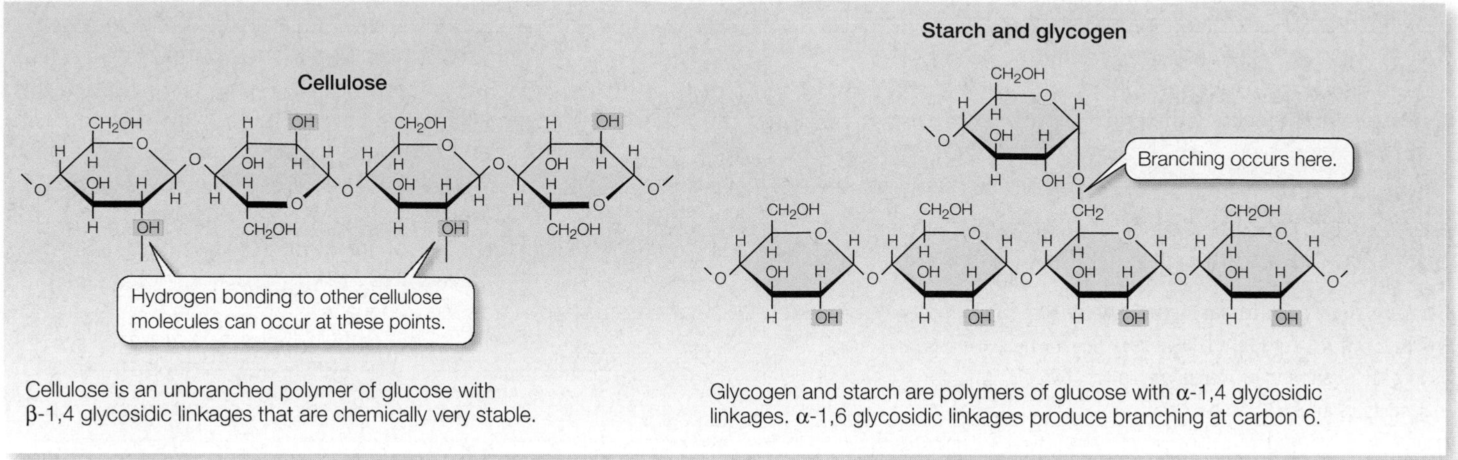

(B) Macromolecular structure

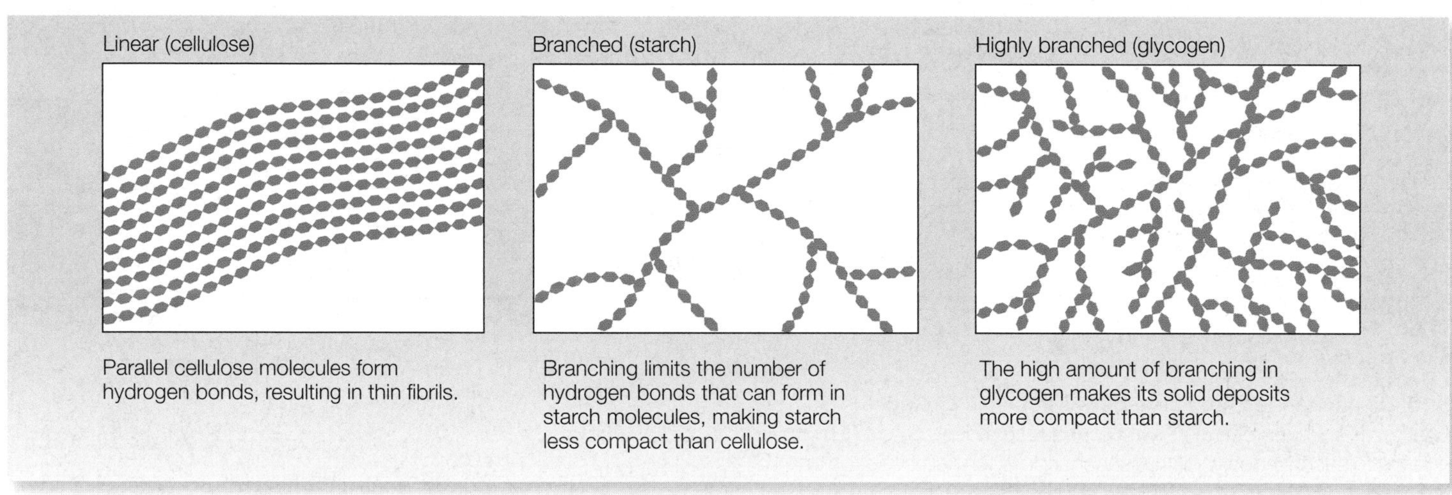

(C) Polysaccharides in cells

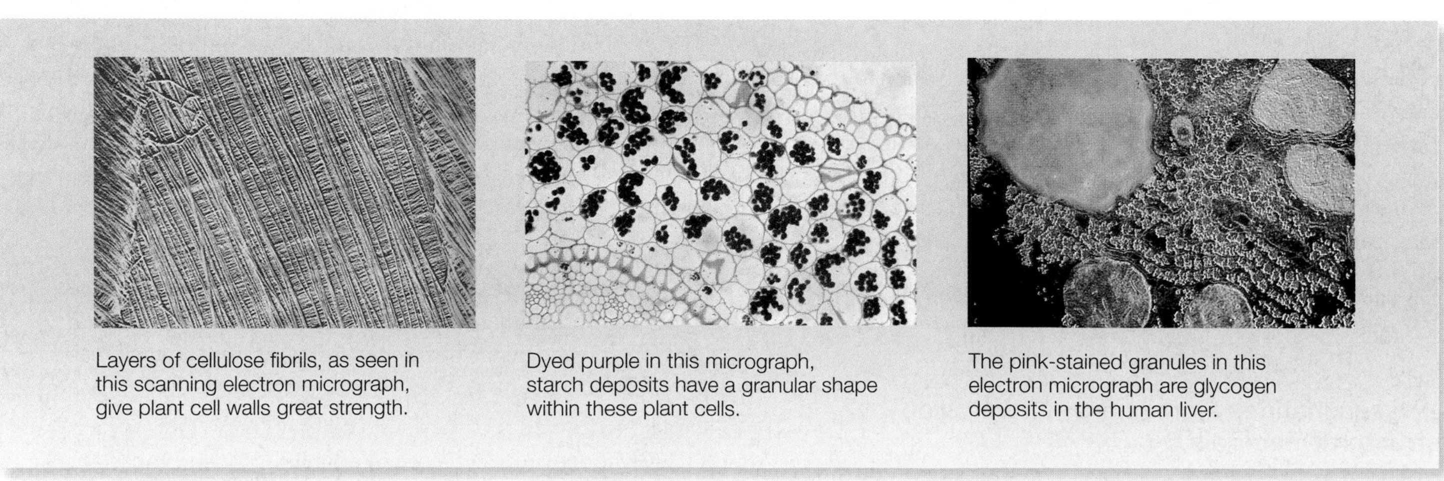

Layers of cellulose fibrils, as seen in this scanning electron micrograph, give plant cell walls great strength.
Dyed purple in this micrograph, starch deposits have a granular shape within these plant cells.
The pink-stained granules in this electron micrograph are glycogen deposits in the human liver.

3.16 Representative Polysaccharides Cellulose, starch, and glycogen demonstrate different levels of branching and compaction in polysaccharides.

enter the cells (see Section 5.3). If it were not for polysaccharides, many organisms would expend a lot of time and energy expelling excess water from their cells.

Cellulose is the predominant component of plant cell walls, and is by far the most abundant organic compound on Earth. Starch can be easily degraded by the actions of chemicals or enzymes. Cellulose, however, is chemically more stable because of its β-glycosidic linkages (see Figure 3.16A). Thus starch is a good storage medium that can be easily broken down to supply glucose for energy-producing reactions, while cellulose is an excellent structural material that can withstand harsh environmental conditions without changing.

Chemically modified carbohydrates contain additional functional groups

Some carbohydrates are chemically modified by the addition of functional groups, such as phosphate and amino groups (**Figure 3.17**). For example, carbon 6 in glucose may be oxidized from $—CH_2OH$ to a carboxyl group (—COOH), producing glucuronic acid. Or a phosphate group may be added to one or more of the —OH sites. Some of the resulting *sugar phosphates,* such as fructose 1,6-bisphosphate, are important intermediates in cellular energy reactions, which will be discussed in Chapter 7.

When an amino group is substituted for an —OH group, *amino sugars* such as glucosamine and galactosamine, are produced. These compounds are important in the extracellular matrix (see Section 4.4), where they form parts of glycoproteins involved in keeping tissues together. Galactosamine is a major component of cartilage, the material that forms caps on the ends of bones and stiffens the ears and nose. A derivative of glucosamine produces the polymer *chitin,* which is the principal structural polysaccharide in the skeletons of insects and many crustaceans (e.g., crabs and lobsters), as well as in the cell walls of fungi.

Taken together, fungi, insects, and crustaceans account for more than 80 percent of the species ever described. Thus the chitin that supports their bodies is one of the most abundant substances on Earth.

3.17 Chemically Modified Carbohydrates Added functional groups can modify the form and properties of a carbohydrate.

(A) Sugar phosphate

Fructose 1,6 bisphosphate is involved in the reactions that liberate energy from glucose. (The numbers in its name refer to the carbon sites of phosphate bonding; *bis-* indicates that two phosphates are present.)

Phosphate groups

Fructose

Fructose 1,6 bisphosphate

(B) Amino sugars

The monosaccharides glucosamine and galactosamine are amino sugars with an amino group in place of a hydroxyl group.

Amino group

Glucosamine **Galactosamine**

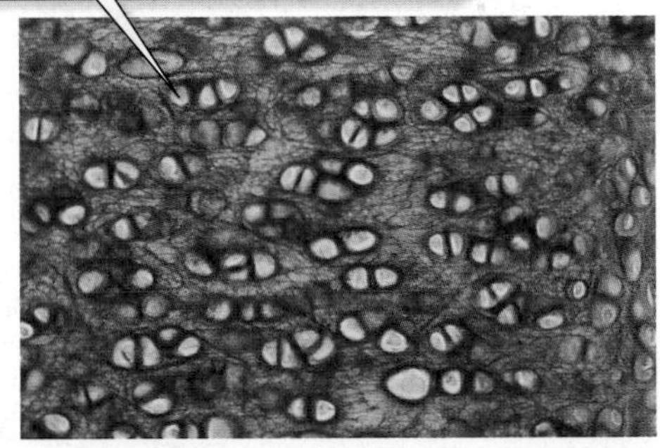

Galactosamine is an important component of cartilage, a connective tissue in vertebrates.

(C) Chitin

Chitin is a polymer of *N*-acetylglucosamine; *N*-acetyl groups provide additional sites for hydrogen bonding between the polymers.

Glucosamine

N-acetyl group

N-acetylglucosamine

Chitin

The external skeletons of insects are made up of chitin.

3.3 RECAP

Carbohydrates are composed of carbon, hydrogen, and oxygen in the general ratio of 1:2:1. They provide energy and structure to cells and are precursors of numerous important biological molecules. Monosaccharide monomers can be connected by glycosidic linkages to form disaccharides, oligosaccharides, and polysaccharides.

- Can you draw the chemical structure of a disaccharide formed by two monosaccharides? See Figure 3.15
- What qualities of the polysaccharides starch and glycogen make them useful for energy storage? See p. 51 and Figure 3.16
- From looking at the cellulose molecules in Figure 3.16A, can you see where a large number of hydrogen bonds are present in the linear structure of cellulose shown in Figure 3.16B? Do you understand why this structure is so strong?

We have now seen how amino acid monomers form protein polymers and how saccharide monomers form the polymers of carbohydrates. Now we will look at the lipids, which are unique among the four classes of large biological molecules in that they are not, strictly speaking, polymers.

3.4 What Are the Chemical Structures and Functions of Lipids?

Lipids are hydrocarbons that are insoluble in water because of their many nonpolar covalent bonds. As we saw in Section 2.2, nonpolar hydrocarbon molecules are hydrophobic and preferentially aggregate among themselves, away from water, which is polar. When these nonpolar hydrocarbons are sufficiently close together, weak but additive van der Waals forces hold them together. These huge macromolecular aggregations are not polymers in a strict chemical sense, because the individual lipid molecules are not covalently bonded. However, they can be considered to be polymers of individual lipid subunits.

There are several different types of lipids, and they play a number of roles in living organisms:

- Fats and oils store energy.
- Phospholipids play important structural roles in cell membranes.
- The carotenoids help plants capture light energy.
- Steroids and modified fatty acids play regulatory roles as hormones and vitamins.
- The fat in animal bodies serves as thermal insulation.
- A lipid coating around nerves provides electrical insulation.
- Oil or wax on the surfaces of skin, fur, and feathers repels water.

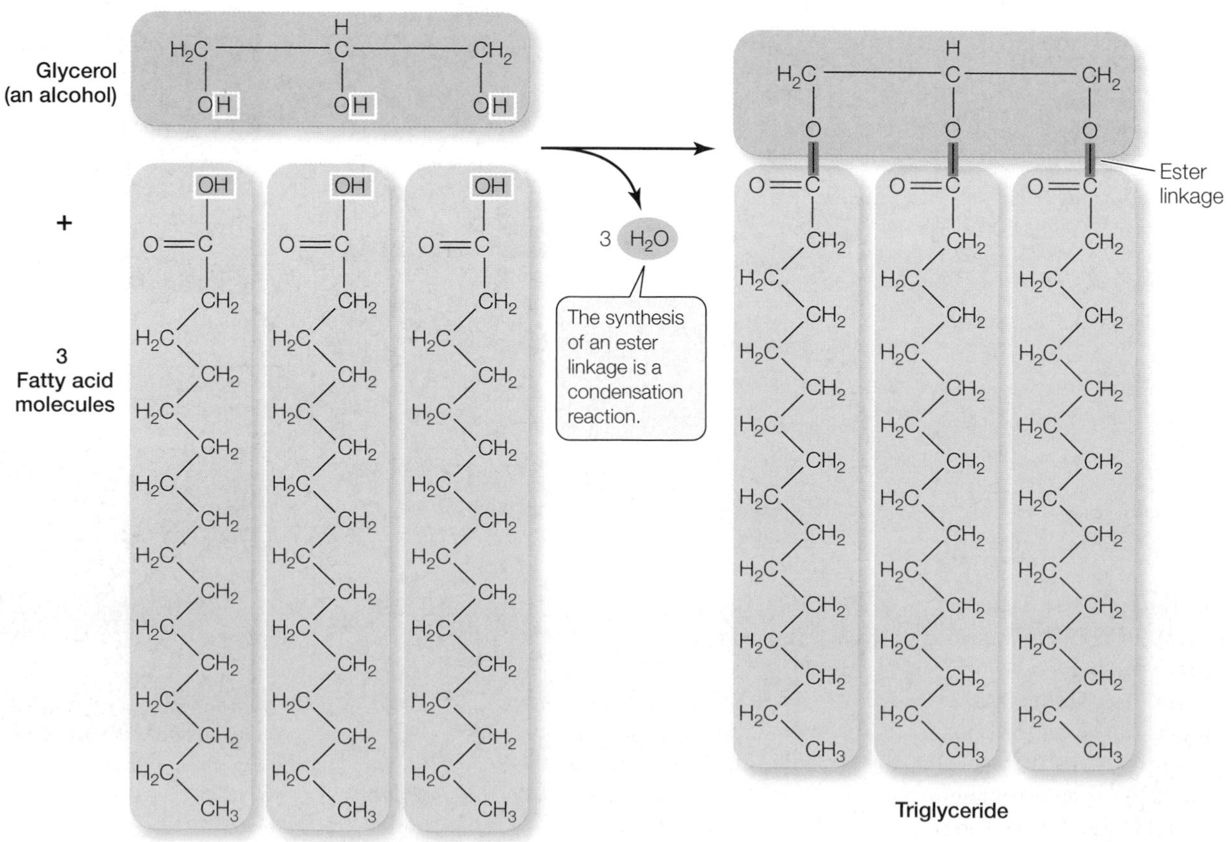

3.18 Synthesis of a Triglyceride In living things, the reaction that forms triglycerides is more complex, but the end result is as shown here.

Fats and oils store energy

Chemically, fats and oils are *triglycerides*, also known as *simple lipids*. Triglycerides that are solid at room temperature (20°C) are called **fats**; those that are liquid at room temperature are called **oils**. Triglycerides are composed of two types of building blocks: *fatty acids* and *glycerol*. **Glycerol** is a small molecule with three hydroxyl (—OH) groups (thus it is an alcohol). A **fatty acid** is made up of a long nonpolar hydrocarbon chain and a polar carboxyl group (—COOH). A **triglyceride** contains three fatty acid molecules and one molecule of glycerol. The carboxyl group of a fatty acid can bond with the hydroxyl group of glycerol, resulting in a covalent bond called an **ester linkage** and water (**Figure 3.18**).

The three fatty acids in a triglyceride molecule need not all have the same hydrocarbon chain length or structure:

- In **saturated** fatty acids, all the bonds between the carbon atoms in the hydrocarbon chain are single bonds—there are no double bonds. That is, all the bonds are saturated with hydrogen atoms (**Figure 3.19A**). These fatty acid molecules are relatively rigid and straight, and they pack together tightly, like pencils in a box.
- In **unsaturated** fatty acids, the hydrocarbon chain contains one or more double bonds. Oleic acid, for example, is a *monounsaturated* fatty acid that has one double bond near the middle of the hydrocarbon chain, which causes a kink in the molecule (**Figure 3.19B**). Some fatty acids have more than one double bond and have multiple kinks; they are *polyunsaturated* fatty acids. Kinks prevent the unsaturated fat molecules from packing together tightly.

The kinks in fatty acid molecules are important in determining the fluidity and melting point of a lipid. The triglycerides of animal fats tend to have many long-chain saturated fatty acids, packed tightly together; these fats are usually solids at room temperature and have a high melting point. The triglycerides of plants, such as corn oil, tend to have short or unsaturated fatty acids. Because of their kinks, these fatty acids pack together poorly and have a low melting point, and these triglycerides are usually liquids at room temperature.

Fats and oils are marvelous energy storehouses. In 1900, the German engineer Rudolf Diesel used peanut oil to power one of his early automobile engines, and there has been a recent resurgence of biodiesel technology. One successful application converts used cooking oil into "French fry fuel."

Phospholipids form biological membranes

Because lipids and water do not interact, when water and lipids are mixed together two distinct layers form. Many biologically important substances—such as ions, sugars, and free amino acids—that are soluble in water are insoluble in lipids.

Like triglycerides, **phospholipids** contain fatty acids bound to glycerol by ester linkages. In phospholipids, however, any one of several phosphate-containing compounds replaces one of the fatty acids (**Figure 3.20A**). The phosphate functional group has a negative electric charge, so this portion of the molecule is hydrophilic, attracting polar water molecules. But the two fatty acids are hydrophobic, so they tend to aggregate away from water.

In an aqueous environment, phospholipids line up in such a way that the nonpolar, hydrophobic "tails" pack tightly together and the phosphate-containing "heads" face outward, where they interact with water. The phospholipids thus form a **bilayer**: a sheet two molecules thick, with water excluded from the core (**Figure 3.20B**). Biological membranes have this kind of phospholipid bilayer structure, and we will devote all of Chapter 5 to their biological functions.

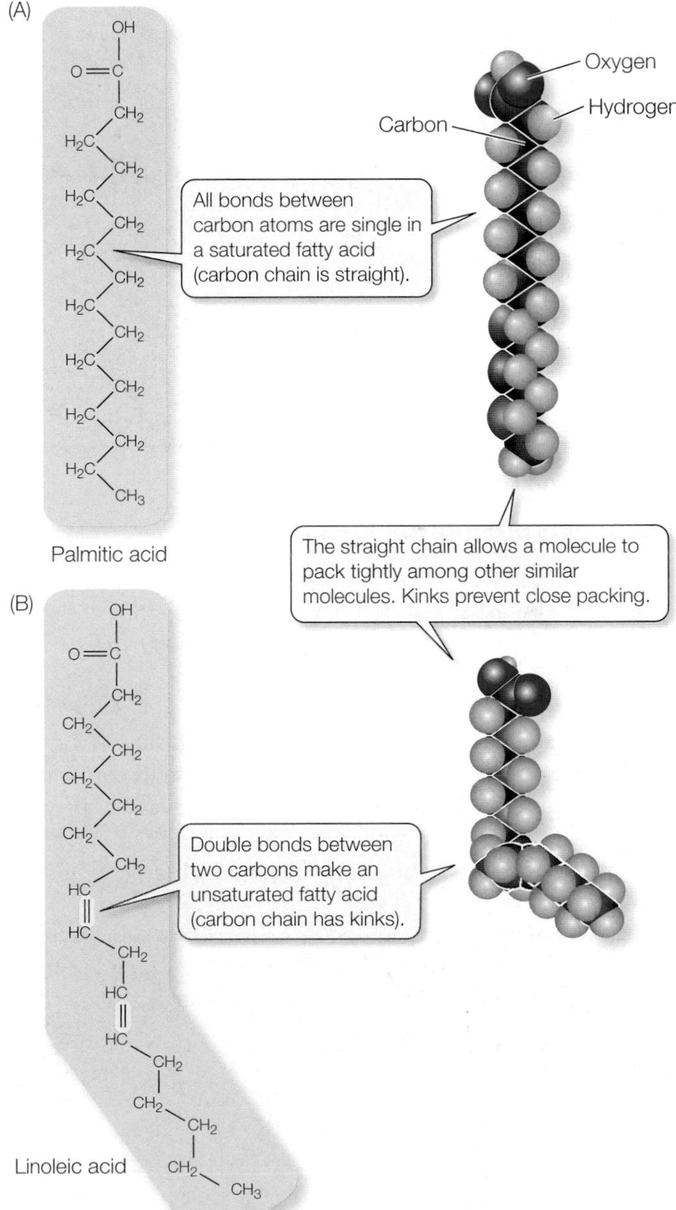

3.19 Saturated and Unsaturated Fatty Acids (A) The straight hydrocarbon chain of a saturated fatty acid allows the molecule to pack tightly among other similar molecules. (B) In unsaturated fatty acids, kinks in the chain prevent close packing. Note the convention in the space-filling molecular models shown here: gray, H; red, O; black, C.

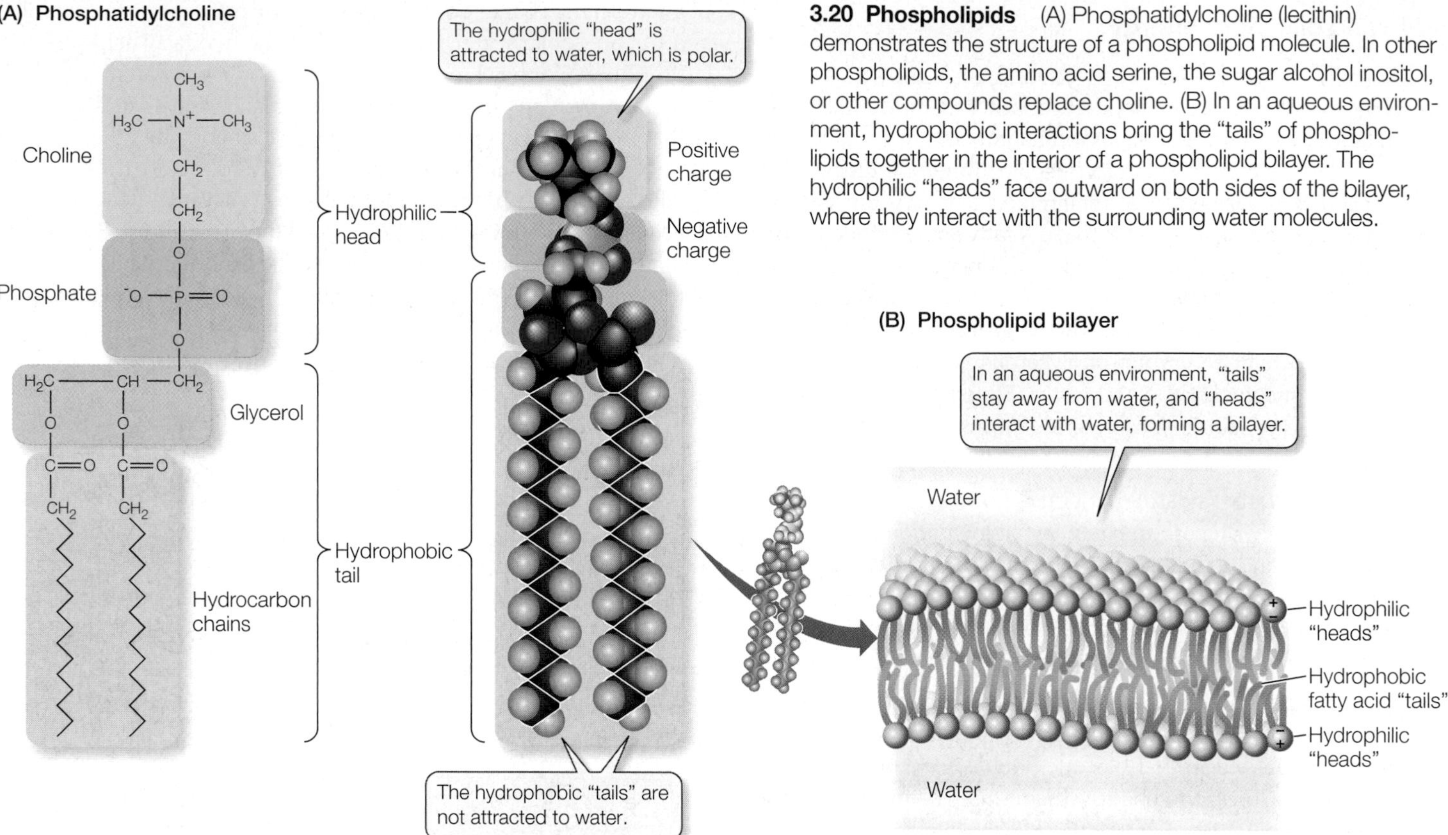

3.20 Phospholipids (A) Phosphatidylcholine (lecithin) demonstrates the structure of a phospholipid molecule. In other phospholipids, the amino acid serine, the sugar alcohol inositol, or other compounds replace choline. (B) In an aqueous environment, hydrophobic interactions bring the "tails" of phospholipids together in the interior of a phospholipid bilayer. The hydrophilic "heads" face outward on both sides of the bilayer, where they interact with the surrounding water molecules.

Not all lipids are triglycerides

A number of classes of lipids are not based on the glycerol-fatty acid structure we have just described. Yet because they are made up largely of carbon and hydrogen and are nonpolar, these molecules are still classified as lipids.

CAROTENOIDS The *carotenoids* are a family of light-absorbing pigments found in plants and animals. Beta-carotene (β-carotene) is one of the pigments that traps light energy in leaves during photosynthesis. In humans, a molecule of β-carotene can be broken down into two vitamin A molecules (**Figure 3.21**), from which we make the pigment rhodopsin, which is required for vision. Carotenoids are responsible for the colors of carrots, tomatoes, pumpkins, egg yolks, and butter.

STEROIDS The *steroids* are a family of organic compounds whose multiple rings share carbons (**Figure 3.22**). The steroid cholesterol is an important constituent of membranes. Other steroids function as hormones, chemical signals that carry messages from one part of the body to another (see Chapter 41). Cholesterol is synthesized in the liver and is the starting material for making testosterone and other steroid hormones, as well as the bile salts that help break down dietary fats so that they can be digested. Cholesterol is absorbed from foods such as milk, butter, and animal fats.

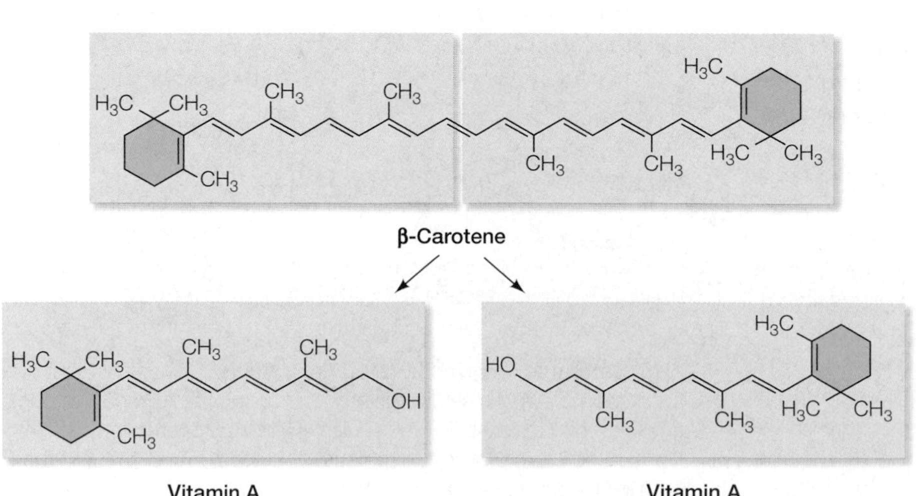

3.21 β-Carotene is the Source of Vitamin A The carotenoid β-carotene is symmetrical around its central double bond; when that bond is broken, the result is two vitamin A molecules. The simplified structural formula used here is standard chemical shorthand for large organic molecules with many carbon atoms. Structural formulas are simplified by omitting the C (indicating a carbon atom) at the intersections of the lines representing covalent bonds. Hydrogen atoms (H) to fill all the available bonding sites on each C are assumed.

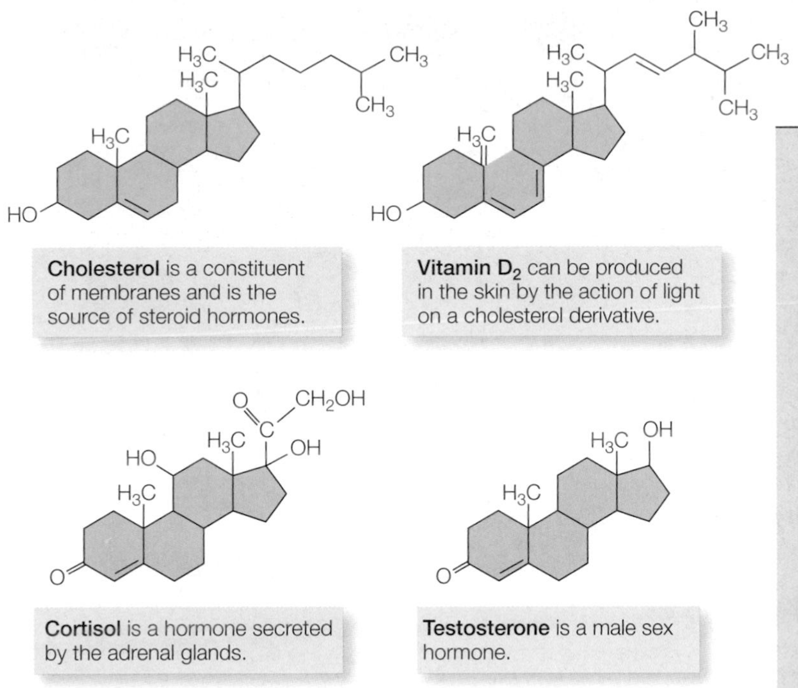

Cholesterol is a constituent of membranes and is the source of steroid hormones.

Vitamin D_2 can be produced in the skin by the action of light on a cholesterol derivative.

Cortisol is a hormone secreted by the adrenal glands.

Testosterone is a male sex hormone.

3.22 All Steroids Have the Same Ring Structure The steroids shown here, all important in vertebrates, are composed of carbon and hydrogen and are highly hydrophobic. However, small chemical variations, such as the presence or absence of a hydroxyl group, can produce enormous functional differences among these molecules.

VITAMINS **Vitamins** are small molecules that are not synthesized by the human body and so must be acquired from the diet (see Chapter 50). For example, *vitamin A* is formed from the β-carotene found in green and yellow vegetables (see Figure 3.21). In humans, a deficiency of vitamin A leads to dry skin, eyes, and internal body surfaces, retarded growth and development, and night blindness, which is a diagnostic symptom for the deficiency. Vitamins D, E, and K are also lipids.

WAXES The sheen on human hair is more than cosmetic. Glands in the skin secrete a waxy coating that repels water and keeps the hair pliable. Birds that live near water have a similar waxy coating on their feathers. The shiny leaves of holly plants, familiar during winter holidays, also have a waxy coating. Finally, bees make their honeycombs out of wax. All waxes have the same basic structure: they are formed by an ester linkage between a saturated, long-chain fatty acid and a saturated, long-chain alcohol. The result is a very long molecule, with 40–60 CH_2 groups. For example, here is the structure of beeswax:

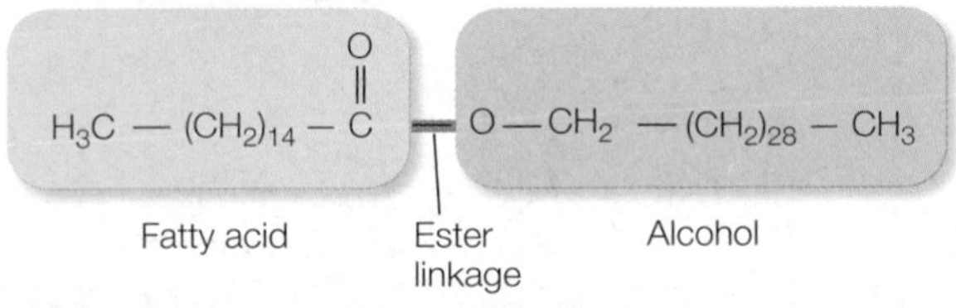

$$H_3C-(CH_2)_{14}-\overset{O}{\overset{\|}{C}}-O-CH_2-(CH_2)_{28}-CH_3$$

Fatty acid — Ester linkage — Alcohol

This highly nonpolar structure accounts for the impermeability of wax to water.

3.4 RECAP

Lipids are nonpolar molecules largely composed of carbon and hydrogen. They are important in energy storage and biological structures. Cell membranes contain phospholipids, which are composed of hydrophobic fatty acids linked to glycerol and a hydrophilic phosphate group.

- Can you draw the molecular structures of fatty acids and glycerol and show how they are linked to form a triglyceride? See p. 55 and Figure 3.18
- What is the difference between fats and oils? See p. 55
- Can you explain how the polar nature of phospholipids results in their forming a bilayer? See p. 55 and Figure 3.20
- Why are steroids and some vitamins classified as lipids? See pp. 56–57

Although all the large molecules discussed in this chapter are found only in living organisms, the final class of biological molecules we will discuss is especially linked to the living world. The function of the nucleic acids is nothing less than the transmission of life's "blueprint" to every new organism.

3.5 What Are the Chemical Structures and Functions of Nucleic Acids?

The **nucleic acids** are polymers specialized for the storage, transmission, and use of genetic information. There are two types of nucleic acids: **DNA** (deoxyribonucleic acid) and **RNA** (ribonucleic acid). DNA is a macromolecule that encodes hereditary information and passes it from generation to generation. Through an RNA intermediate, the information encoded in DNA is used to specify the amino acid sequence of proteins. Information flows from DNA to DNA in reproduction, but in the nonreproductive activities of the cell, information flows from DNA to RNA to proteins, which ultimately carry out life's functions.

Nucleotides are the building blocks of nucleic acids

Nucleic acids are composed of monomers called **nucleotides**, each of which consists of a pentose sugar, a phosphate group, and a nitrogen-containing **base**. The bases of the nucleic acids take one of two chemical forms: a single-ring structure called a **pyrimidine** or a fused-ring structure called a **purine** (**Figure 3.23**). (Molecules consisting of a pentose sugar and a nitrogenous base—but no phosphate group—are called *nucleosides*.) In DNA, the pentose sugar is **deoxyribose**, which differs from the **ribose** found in RNA by one oxygen atom (see Figure 3.14).

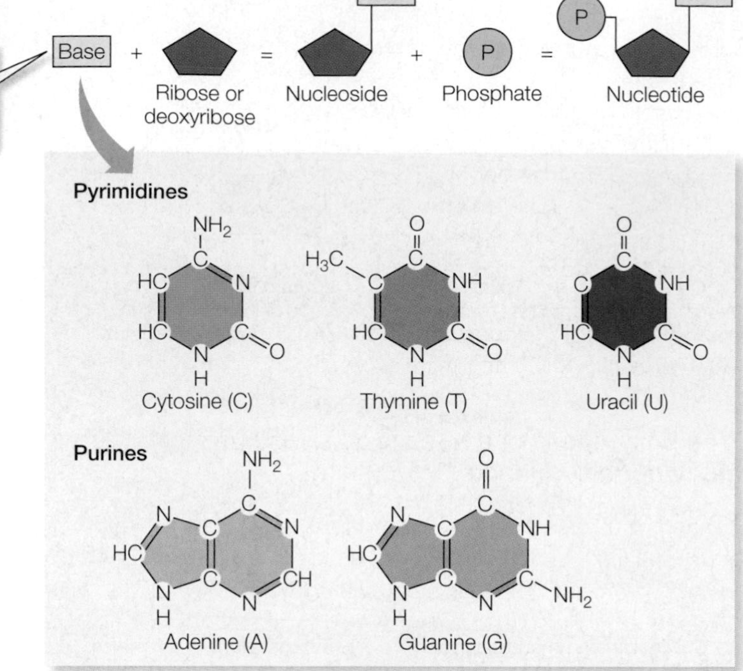
The base may be either a pyrimidine or a purine.
Base
Ribose or deoxyribose
Nucleoside
Phosphate
Nucleotide
Pyrimidines
Cytosine (C)
Thymine (T)
Uracil (U)
Purines
Adenine (A)
Guanine (G)

3.23 Nucleotides Have Three Components A nucleotide consists of a phosphate group, a pentose sugar (ribose or deoxyribose), and a nitrogen-containing base, all linked together by covalent bonds. The nitrogenous bases have two different chemical forms: purines have two fused rings, and the smaller pyrimidines have a single ring.

In both RNA and DNA, the backbone of the macromolecule consists of a chain of alternating pentose sugars and phosphate groups (sugar–phosphate–sugar–phosphate). The bases are attached to the sugars and project from the polynucleotide chain (**Figure 3.24**). The nucleotides are joined by **phosphodiester linkages** between the sugar of one nucleotide and the phosphate of the next (*diester* refers to the two covalent bonds formed by —OH groups reacting with acidic phosphate groups). The phosphate

3.24 Distinguishing Characteristics of DNA and RNA RNA is usually a single strand. DNA usually consists of two strands running in opposite directions (antiparallel).

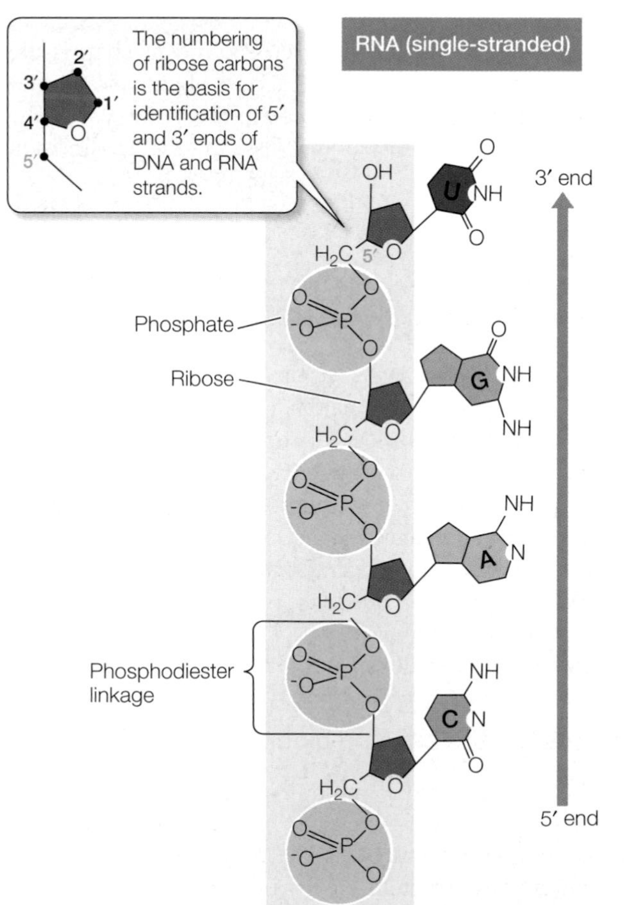
The numbering of ribose carbons is the basis for identification of 5′ and 3′ ends of DNA and RNA strands.
RNA (single-stranded)
3′ end
5′ end
Phosphate
Ribose
Phosphodiester linkage
In RNA, the bases are attached to ribose. The bases in RNA are the purines adenine (A) and guanine (G) and the pyrimidines cytosine (C) and uracil (U).

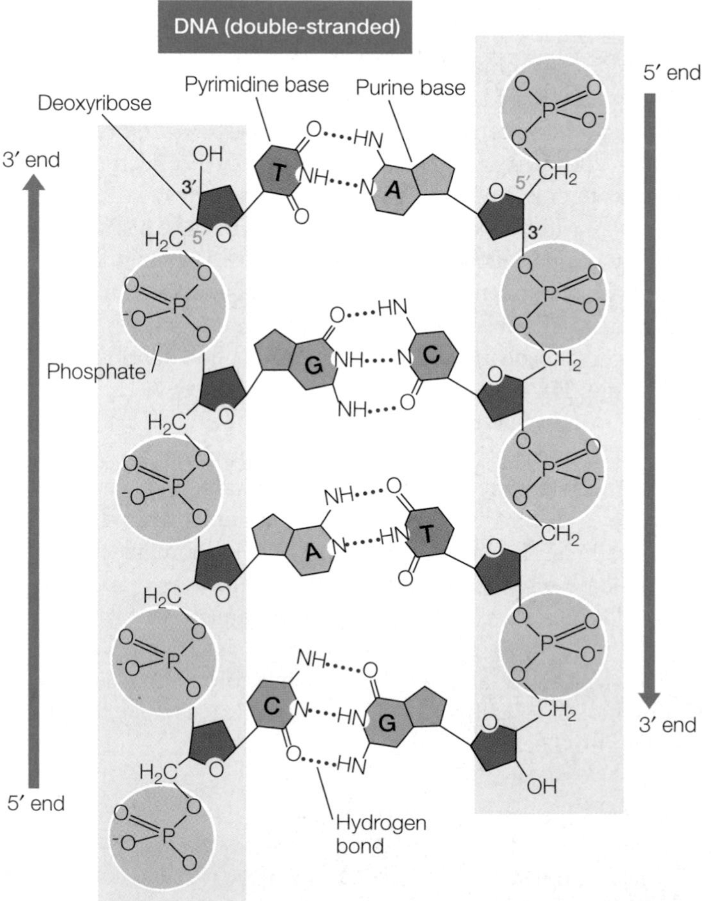
DNA (double-stranded)
Deoxyribose
Pyrimidine base
Purine base
3′ end
5′ end
Phosphate
Hydrogen bond
In DNA, the bases are attached to deoxyribose, and the base thymine (T) is found instead of uracil. Hydrogen bonds between purines and pyrimidines hold the two strands of DNA together.

groups link carbon 3′ in one pentose sugar to carbon 5′ in the adjacent sugar.

Most RNA molecules consist of only one polynucleotide chain. DNA, however, is usually double-stranded; its two polynucleotide chains are held together by hydrogen bonding between their nitrogenous bases. The two strands of DNA run in opposite directions. You can see what this means by drawing an arrow through a phosphate group from carbon 5 to carbon 3 in the next ribose. If you do this for both strands of the DNA in Figure 3.24, the arrows will point in opposite directions. This *antiparallel* orientation allows the strands to fit together in three-dimensional space.

The uniqueness of a nucleic acid resides in its nucleotide sequence

Only four nitrogenous bases—and thus only four nucleotides—are found in DNA. The DNA bases and their abbreviations are **adenine** (A), **cytosine** (C), **guanine** (G), and **thymine** (T). A key to understanding the structure and function of nucleic acids is the principle of **complementary base pairing**. In double-stranded DNA, adenine and thymine always pair (A-T), and cytosine and guanine always pair (C-G).

Base pairing is complementary because of three factors: the sites for hydrogen bonding on each base; the geometry of the sugar–phosphate backbone, which brings opposite bases near each other; and the molecular sizes of the paired bases. Adenine and guanine are both purines; thymine and cytosine are both pyrimidines. The pairing of a large purine with a smaller pyrimidine ensures stability and consistency in the double-stranded molecule of DNA.

RNA is also made up of four different monomers, but its nucleotides differ from those of DNA. In RNA the nucleotides are termed *ribonucleotides* (the ones in DNA are *deoxyribonucleotides*). They contain ribose rather than deoxyribose, and instead of the base thymine, RNA uses the base **uracil** (U). The other three bases are the same as those in DNA (**Table 3.3**).

Although RNA is generally single-stranded, complementary hydrogen bonding between ribonucleotides plays important roles in determining the three-dimensional shapes of some types of RNA molecules (**Figure 3.25**). Complementary base pairing can also take place between ribonucleotides and deoxyribonucleotides. In RNA, guanine and cytosine pair (G-C), as in DNA, but adenine pairs with uracil (A-U). Adenine in an RNA strand can pair either with uracil (in another RNA strand) or with thymine (in a DNA strand).

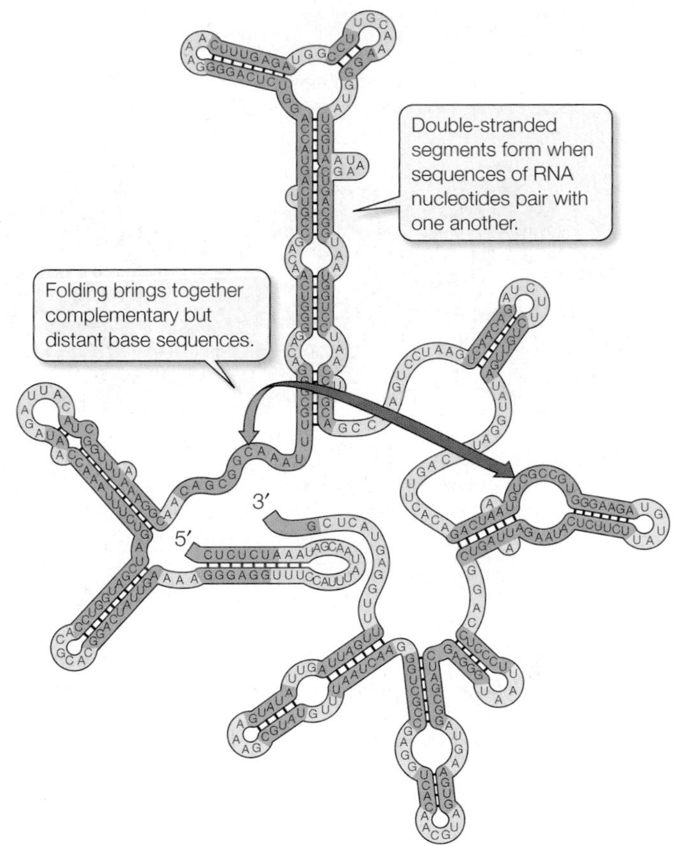
Double-stranded segments form when sequences of RNA nucleotides pair with one another.
Folding brings together complementary but distant base sequences.
3′
5′

3.25 Hydrogen Bonding in RNA When a single-stranded RNA folds in on itself, hydrogen bonds can stabilize it into a three-dimensional shape.

TABLE 3.3

Distinguishing RNA from DNA

NUCLEIC ACID	SUGAR	BASES
RNA	Ribose	Adenine
		Cytosine
		Guanine
		Uracil
DNA	Deoxyribose	Adenine
		Cytosine
		Guanine
		Thymine

DNA is a purely *informational* molecule. The information in DNA is encoded in the sequence of bases carried in its strands—the information encoded in the sequence TCAG is different from the information in the sequence CCAG. RNA uses the information carried in the nucleotide sequence of a DNA molecule to specify the amino acid sequence that in turn dictates the primary structure of a protein. This information can be read easily and reliably, in a specific order, as we will see in Chapter 12.

The three-dimensional physical appearance of DNA is strikingly uniform. The segment shown in **Figure 3.26** could be from any DNA molecule. The variations in DNA—the different sequences of bases—are strictly internal. Through hydrogen bonding, the two complementary polynucleotide strands pair and twist to form a **double helix**. When compared with the complex and varied tertiary structures of proteins, this uniformity is surprising. But this structural contrast makes sense in terms of the functions of these two classes of macromolecules. As we saw in Section 3.2, the different and unique shapes of proteins permit these macromolecules to recognize specific "target" molecules. The unique three-dimensional form of each protein matches at least a portion of the

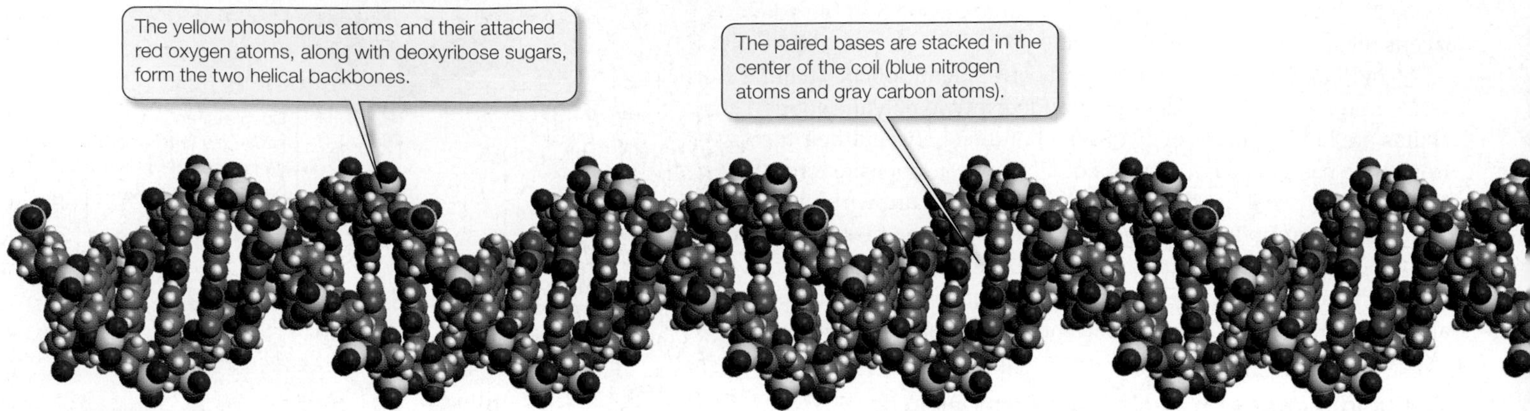

The yellow phosphorus atoms and their attached red oxygen atoms, along with deoxyribose sugars, form the two helical backbones.
The paired bases are stacked in the center of the coil (blue nitrogen atoms and gray carbon atoms).

3.26 The Double Helix of DNA The backbones of the two strands in a DNA molecule are coiled in a double helix. In this model, the small white atoms represent hydrogen.

surface of the target molecule. In other words, structural diversity in the molecules to which proteins bind requires corresponding diversity in the structure of the proteins themselves. Structural diversity is necessary in DNA as well. However, the diversity of DNA is found in the structure of its base sequence rather than in the physical shape of the molecule.

DNA reveals evolutionary relationships

Because DNA carries hereditary information between generations, a theoretical series of DNA molecules, with changes in base sequences, stretches back through the lineages of all organisms to the beginning of evolutionary time. Closely related living species should have more similar base sequences than species judged by other criteria to be more distantly related. The details of how scientists use this information are covered in Chapter 25.

The elucidation and examination of DNA base sequences has confirmed many of the evolutionary relationships that have been inferred from the more traditional comparisons of body structures, biochemistry, and physiology. For example, the closest living relative of humans (*Homo sapiens*) has usually been considered to be the chimpanzee (genus *Pan*), and in fact chimpanzee DNA shares more than 98 percent of its DNA base sequence with human DNA. Increasingly, scientists turn to DNA analyses to elucidate evolutionary relationships when studies of structure are not possible or are not conclusive. For example, DNA studies revealed a close relationship between starlings and mockingbirds that was not expected on the basis of their anatomy or behavior.

Nucleotides have other important roles

Nucleotides are more than just the building blocks of nucleic acids. As we will see in later chapters, there are several nucleotides with other functions:

- ATP (adenosine triphosphate) acts as an energy transducer in many biochemical reactions (see Section 6.2).
- GTP (guanosine triphosphate) serves as an energy source, especially in protein synthesis. It also has a role in the transfer of information from the environment to cells (see Section 15.2).
- cAMP (cyclic adenosine monophosphate), a special nucleotide in which an additional bond forms between the sugar and phosphate groups, is essential in many processes, including the actions of hormones and the transmission of information by the nervous system (see Section 15.3).

3.5 RECAP

The nucleic acids DNA and RNA are polymers of nucleotide monomers. The sequence of nucleotides in DNA carries the information used by RNA to specify primary protein structure. The genetic information in DNA is passed from generation to generation and can be studied to understand evolutionary relationships.

- Can you describe the key differences between DNA and RNA? Between purines and pyrimidines? See pp. 57–59, Figure 3.23, and Table 3.3
- Do you understand how purines and pyrimidines pair up in complementary bonding between nucleotides? See p. 59 and Figure 3.24
- Do you see how there can be immense diversity among DNA molecules even though they all appear to be structurally similar? See pp. 59–60

We have seen that the nucleic acids RNA and DNA carry the blueprint of life, and that the heritage of these macromolecules reaches back to the beginning of evolutionary time. But where did the nucleic acids come from? How did the building blocks of life originally arise?

3.6 How Did Life on Earth Begin?

As we saw in Chapter 2, living things are composed of the same atomic elements as the inanimate universe—the 92 naturally occurring elements of the periodic table (see Figure 2.2). But the arrangements of these atoms into molecules are unique in biological systems. You won't find biological molecules in inanimate matter (unless they came from a once-living organism).

How life began on Earth, or anywhere else, is impossible to know for certain. There are two prevailing scientific theories for the origin of life on Earth:

- The molecules of life arrived on Earth from extraterrestrial sources.
- Life is the result of chemical evolution on Earth.

Could life have come from outside Earth?

As mentioned in Chapter 2, comets are thought to have brought Earth most of its life-engendering water. Recently, it has become apparent that several meteorites from Mars have landed on Earth, and that some of these meteors carry some molecules that are possibly characteristic of life.

In 1984, a softball-sized rock was found on the ice in the Allan Hills region of Antarctica. ALH 84001, as it came to be called, was a meteorite that came from Mars (**Figure 3.27**). We know this because the composition of the gases trapped within the rock was identical to the Martian atmosphere, which is quite different from Earth's atmosphere. Radioactive dating and mineral analyses determined that ALH 84001 was 4.5 billion years old and was blasted off the Martian surface 16 million years ago, landing on Earth fairly recently, about 13,000 years ago.

Scientists found water trapped below the Martian meteorite's surface. This discovery was not surprising, considering that surface observations had shown that liquid water was once abundant on Mars (see Chapter 2). Because water is essential for life, scientists wondered whether the meteorite might contain other signs of life as well. Their analysis revealed two substances related to living systems. First, simple carbon-containing molecules called polycyclic aromatic hydrocarbons were present in small but unmistakable amounts; these substances can be formed by living organisms. Second, crystals of magnetite, an iron oxide mineral made by many living things on Earth, were found in the interior of the rock.

3.27 Was Life Once Here? The meteorite ALH 84001, which came from Mars and landed in Antarctica, contains chemical signatures of life.

ALH 84001 is not the only visitor from outer space that has been shown to contain chemical signatures of life. Fragments of a meteorite that fell around the town of Murchison, Australia, in 1969 were found to contain molecules that are unique to life, including purines, pyrimidines, and amino acids. Although the presence of such molecules in rocks may suggest that those rocks once harbored life, it does not prove that there were living things in the rocks when they landed on Earth.

Most scientists find it hard to believe that an organism in a meteorite could survive thousands of years of traveling through space, followed by intense heat as the meteorite passed through Earth's atmosphere. But there is some evidence that the heat inside some meteorites may not have been severe. When weakly magnetized rock is heated, it reorients its magnetic field to align with the magnetic field around it. In the case of ALH 84001, this would have been Earth's powerful magnetic field, which would have affected the meteorite as it approached our planet. Careful measurements indicate that, while reorientation did occur at the surface of the rock, it did not occur in the inside. The scientists who took these measurements, Benjamin Weiss and Joseph Kirschvink at the California Institute of Technology, concluded that the inside of ALH 84001 was never heated over 40°C on its trip to Antarctica. This evidence makes a long interplanetary trip by living organisms more plausible.

Did life originate on Earth?

Both Earth and Mars once had the water and other simple molecules that could, under the right conditions, form the large molecules unique to life. The second theory of the origin of life on Earth, **chemical evolution**, holds that conditions on primitive Earth led to the emergence of these molecules. Scientists have sought to reconstruct those primitive conditions.

Early in the twentieth century, researchers proposed that there was little oxygen gas (O_2) in Earth's first atmosphere. Oxygen gas is thought to have accumulated in quantity about 2.5 billion years ago as the by-product of photosynthesis by single-celled life forms (today, it constitutes 21 percent of our atmosphere). In the 1950s, Stanley Miller and Harold Urey set up an experimental "atmosphere" containing the gases they believed to have been present in Earth's early atmosphere: hydrogen gas, ammonia, methane gas, and water vapor. Through these gases, they passed a spark to simulate lightning, then cooled the system so the gases would condense and collect in a watery solution, or "ocean" (**Figure 3.28**). Within days, the system contained numerous complex molecules, including amino acids, purines, and pyrimidines—some of the building blocks of life.

In science, an experiment and its results must be constantly reinterpreted, repeated, and refined as more knowledge accumulates. The results of the Miller–Urey experiments have undergone several such refinements:

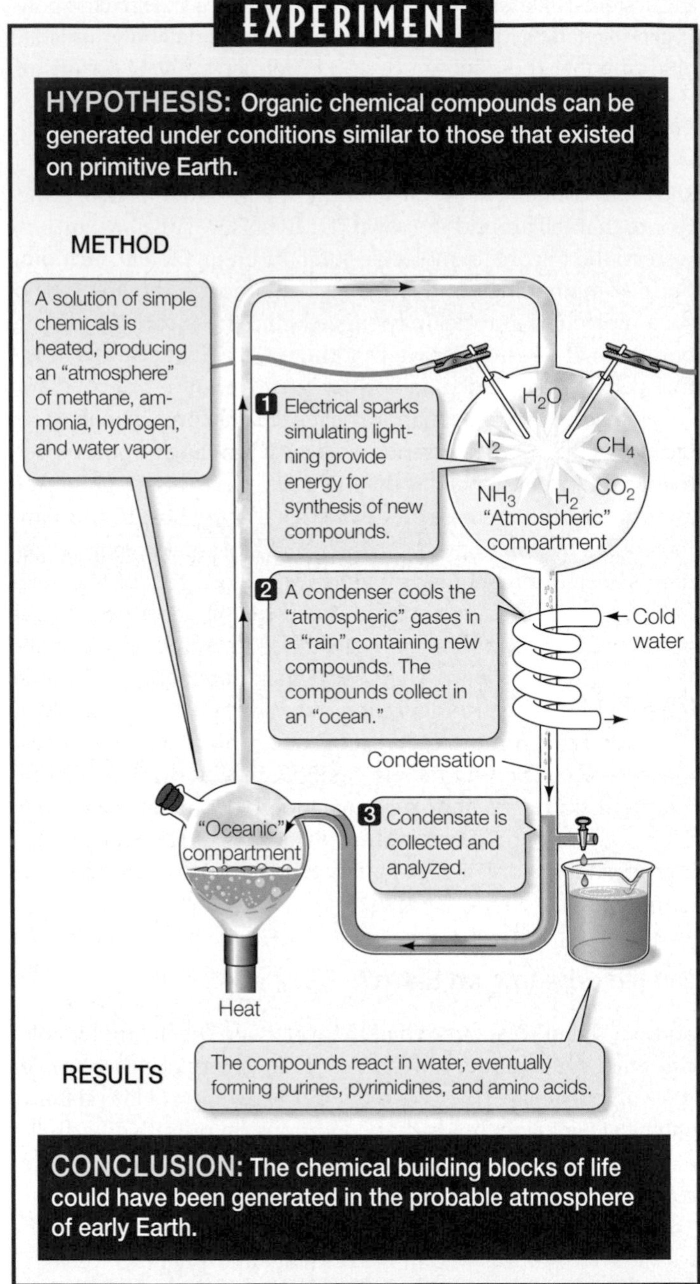

EXPERIMENT
HYPOTHESIS: Organic chemical compounds can be generated under conditions similar to those that existed on primitive Earth.
METHOD
A solution of simple chemicals is heated, producing an "atmosphere" of methane, ammonia, hydrogen, and water vapor.
1 Electrical sparks simulating lightning provide energy for synthesis of new compounds.
H_2O
N_2
CH_4
NH_3
H_2
CO_2
"Atmospheric" compartment
2 A condenser cools the "atmospheric" gases in a "rain" containing new compounds. The compounds collect in an "ocean."
Cold water
Condensation
"Oceanic" compartment
3 Condensate is collected and analyzed.
Heat
RESULTS
The compounds react in water, eventually forming purines, pyrimidines, and amino acids.
CONCLUSION: The chemical building blocks of life could have been generated in the probable atmosphere of early Earth.

3.28 Synthesis of Prebiotic Molecules in an Experimental Atmosphere The Miller–Urey experiment simulated possible atmospheric conditions on primitive Earth and obtained some of the molecular building blocks of biological systems. FURTHER RESEARCH: If O_2 were present in the "atmosphere" in this experiment, what results would you predict?

- The amino acids in living things are always L-isomers (see Figure 3.2 and p. 40). But a mixture of D- and L-isomers appeared in the amino acids formed in the Miller–Urey experiments. Recent experiments show that natural processes could have selected the L-amino acids from the mixture. Some minerals, especially calcite-based rocks, have unique crystal structures that selectively bind to D- or L-amino acids, separating the two. Such rocks were abundant on early Earth.
- Scientists' views of Earth's original atmosphere have changed since Miller and Urey did their experiment. There is abundant evidence of major volcanic eruptions 4 billion years ago, which would have released carbon dioxide (CO_2), nitrogen (N_2), hydrogen sulfide (H_2S), and sulfur dioxide (SO_2) into the atmosphere. Experiments using these gases in addition to the ones in the original experiment have produced more diverse molecules.

Chemical evolution may have led to polymerization

The Miller–Urey experiment and other experiments that followed it provided a plausible scenario for the formation of the building blocks of life. The next step would be the condensation of these monomers into polymers (see Figure 3.4). This poses a major problem, because in water, short polymers tend to hydrolyze back into monomers.

Scientists have used model systems to try to simulate conditions (most of them with a low water content) under which polymers could be made:

- *Solid mineral surfaces*, such as finely divided clays, have a large surface area, and the silicates within the minerals may have been catalytic (speeded up the reactions) for the early carbon-based molecules.
- *Hydrothermal vents* deep in the ocean, where hot water emerges from beneath Earth's crust, contain metals such as iron and nickel. These metals have been shown in laboratory experiments to catalyze polymerization of amino acids in the absence of oxygen.
- *Hot pools* at the edges of oceans may, through evaporation, have concentrated monomers to the point where polymerization was favored (the "primordial soup" hypothesis).

In whatever ways the earliest stages of chemical evolution occurred, they resulted in the emergence of monomers and polymers that have probably remained unchanged in their general structure and function for 3.8 billion years.

RNA may have been the first biological catalyst

The three-dimensional structure of a folded RNA molecule presents a unique surface to the external environment (see Figure 3.25). These surfaces are every bit as specific as those of proteins. Just as the shapes of proteins allow them to function as catalysts, speeding up reactions that would ordinarily take place too slowly to be biologically useful, the three-dimensional shapes and other chemical properties of certain RNA molecules allow them to function as catalysts.

The Miller–Urey experiment and other such experiments in prebiotic chemistry yielded both amino acids and nucleotides. Organisms can synthesize RNA and proteins from these monomers. But if protein synthesis requires DNA and RNA, and nucleic acid synthesis requires proteins (enzymes), we are confronted with a chicken-and-egg type of question: when life originated, which came first, the proteins or the nucleic acids? The discovery of cat-

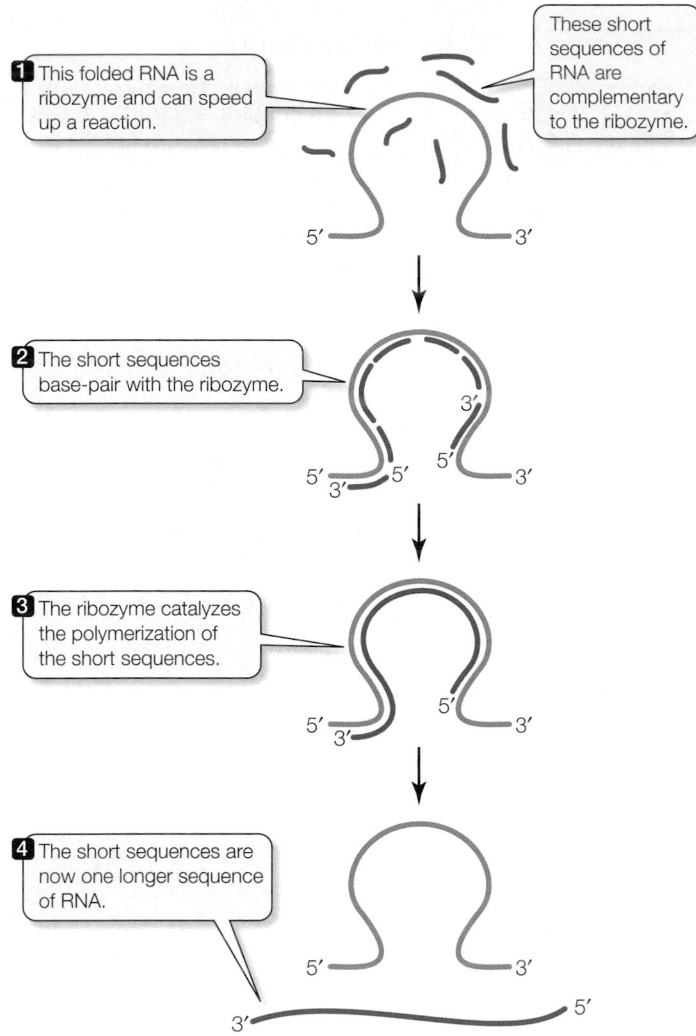

1 This folded RNA is a ribozyme and can speed up a reaction.
These short sequences of RNA are complementary to the ribozyme.
5′
3′
2 The short sequences base-pair with the ribozyme.
5′
3′
5′
3′
5′
3′
3 The ribozyme catalyzes the polymerization of the short sequences.
5′
3′
5′
3′
4 The short sequences are now one longer sequence of RNA.
5′
3′
3′
5′

3.29 An Early Catalyst for Life? This laboratory reconstruction shows that a ribozyme (a folded RNA molecule) can catalyze the polymerization of several short RNA strands into a longer molecule. Such a process could be a precursor for the copying of nucleic acids, which is essential for their duplication and expression.

alytic RNAs provided a solution to this dilemma. Catalytic RNAs, called **ribozymes**, can catalyze reactions on their own nucleotides as well as in other cellular substances.

Given that RNA can be both informational (in its nucleotide sequence) and catalytic (due to its ability to form unique three-dimensional shapes; see Figure 3.25), it has been hypothesized that early life existed in an "RNA world"—a world before DNA. It is thought that when RNA was first made, it could have acted as a catalyst for its own replication as well as for the synthesis of proteins. DNA could eventually have evolved from RNA. Some laboratory evidence supports this scenario:

- RNAs of different sequences have been put in a test tube and made to replicate on their own. Such self-replicating ribozymes speed up the synthesis of RNA 7 million times.
- In the test tube, a ribozyme can catalyze the assembly of short RNAs into a longer molecule—the beginning of priming of RNA for protein synthesis (**Figure 3.29**).
- In living organisms today, the formation of peptide bonds (see Figure 3.6) is catalyzed by a ribozyme.
- In certain viruses called retroviruses, there is an enzyme called reverse transcriptase that catalyzes the synthesis of DNA from RNA.

While this evidence suggests that RNA could have been the first polymer, scientists are a long way from a plausible explanation for the origins of the other large molecules characteristic of life, such as polysaccharides, proteins, and lipids.

Experiments disproved spontaneous generation of life

The idea that life originated from nonliving matter is not new. In fact, many cultures and religions have descriptions of such events. During the Renaissance (a period from about 1450 to 1700 AD, marked by the birth of modern science), most people thought that at least some forms of life arose repeatedly and directly from inanimate or decaying matter by *spontaneous generation*. For instance, it was suggested that mice arose from sweaty clothes placed in dim light, frogs came from moist soil, and flies were produced from meat. Scientists such as the Italian physician and poet Francesco Redi, however, doubted the assumptions of the time. Redi proposed that flies arose not by some mysterious transformation of decaying meat, but from other flies that laid their eggs on the meat. In 1668, Redi performed a scientific experiment—a relatively new concept at the time—to test his hypothesis. He set out several jars containing chunks of meat.

- One jar contained meat exposed both to air and to flies.
- A second jar contained meat in a container wrapped in a fine cloth so that the meat was exposed to air, but not to flies.
- The meat in the third jar was in a sealed container and thus was not exposed to either air or flies.

As he had hypothesized, Redi found maggots, which then hatched into flies, only in the first container. This finding demonstrated that maggots could occur only where flies were present. The idea that a complex organism like a fly could appear de novo from a nonliving substance in the meat, or from "something in the air," was laid to rest.

With the advent of the Leeuwenhook microscope in the 1660s, a vast new biological world was unveiled. Under microscopic observation, virtually every environment on Earth was found to be teeming with tiny organisms such as bacteria. Some scientists believed that these organisms arose spontaneously from their rich chemical environment. Experiments by the great French scientist Louis Pasteur disproved this idea, showing that microorganisms come only from other microorganisms, and that an environment without life remains lifeless unless contaminated by living creatures (**Figure 3.30**).

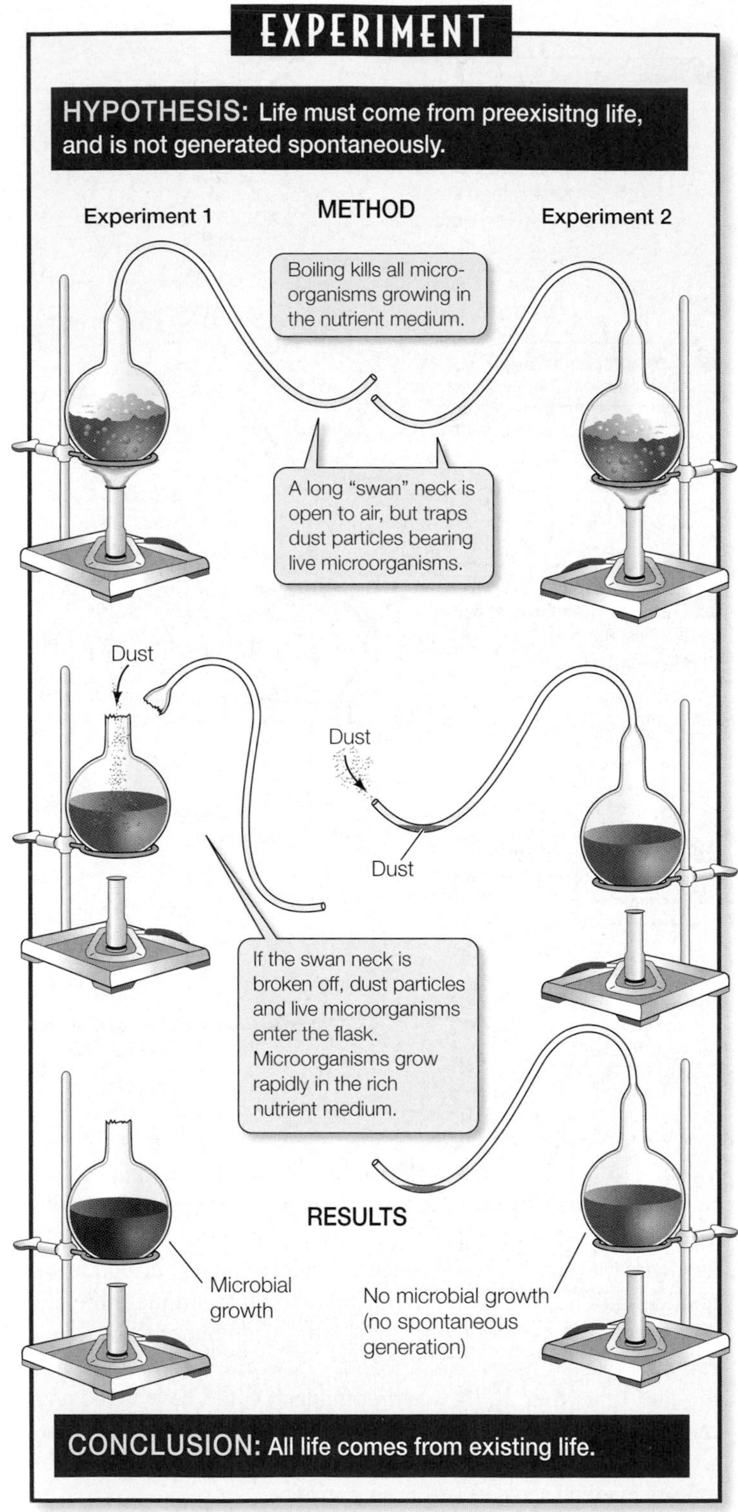

EXPERIMENT
HYPOTHESIS: Life must come from preexisitng life, and is not generated spontaneously.
Experiment 1
METHOD
Experiment 2
Boiling kills all micro-organisms growing in the nutrient medium.
A long "swan" neck is open to air, but traps dust particles bearing live microorganisms.
Dust
Dust
Dust
If the swan neck is broken off, dust particles and live microorganisms enter the flask. Microorganisms grow rapidly in the rich nutrient medium.
RESULTS
Microbial growth
No microbial growth (no spontaneous generation)
CONCLUSION: All life comes from existing life.

3.30 Disproving the Spontaneous Generation of Life Louis Pasteur's classic experiments showed that, under conditions existing on Earth today, an inanimate solution remains lifeless unless a living organism contaminates it.

Pasteur's experiments proved that life cannot arise from nonliving materials, but the Miller–Urey experiments indicate that it could have—at least at the molecular level. How can we reconcile the results of these two experiments? Bear in mind that the atmospheric and planetary constituents of Earth today are vastly different from those of prebiotic Earth. The oxygen in today's atmosphere would break down many molecules as they were formed, and the energy sources that might have propelled those early chemical reactions no longer dominate our planet.

3.6 RECAP

The chemicals of life could have originated elsewhere in the universe, or they could have evolved on Earth. Chemical signatures of life on meteorites from Mars make the first idea plausible. Laboratory experiments support the second hypothesis.

- What experimental evidence indicates that the chemical evolution of life could have taken place on Earth? See pp. 61–62 and Figure 3.28
- Do you understand what polymerization is? Reviewing the macromolecules described in this chapter, do you understand why this chemical process is so important? See p. 62 and Figure 3.29
- Francesco Redi's experiment, described on page 63, is one of the earliest known examples of the scientific method. Do you see all the elements of the method, as described in Section 1.3, in this experiment? Are these elements present in the experiments shown in Figures 3.28 and 3.30?

The diverse life forms of Earth are composed of chemical building blocks including atoms, small molecules, and large biological molecules. But from this molecular foundation emerges the cell, a structure that is the basis of life as we know it today. It is the cell that perpetuates life, and the cell is the basis for Pasteur's truism "all life from life." The next chapter describes the structure and function of living cells.

CHAPTER SUMMARY

3.1 What kinds of molecules characterize living things?

See Web/CD Tutorial 3.1

Macromolecules are **polymers** constructed by the formation of covalent bonds between smaller molecules called **monomers**. Macromolecules in living organisms include polysaccharides, proteins, and nucleic acids.

Functional groups are small groups of atoms that are consistently found together in a variety of different macromolecules. Functional groups have particular chemical properties that they confer on any larger molecule of which they are a part. Review Figure 3.1, Web/CD Activity 3.1

Structural and optical **isomers** have the same kinds and numbers of atoms, but differ in their structures and properties. Review Figure 3.2

The many functions of macromolecules are directly related to their three-dimensional shapes, which in turn is the result of the sequences and chemical properties of their monomers.

Monomers are joined by **condensation reactions**, which release a molecule of water for each bond formed. **Hydrolysis reactions** use water to break polymers into monomers. Review Figure 3.4

3.2 What are the chemical structures and functions of proteins?

See Web/CD Activity 3.2

The functions of proteins include support, protection, catalysis, transport, defense, regulation, and movement.

Amino acids are the monomers from which proteins are constructed. The properties of the amino acids depend on their **side chains**, or **R groups**, which may be charged, polar, or hydrophobic. Review Table 3.2

Peptide bonds covalently link amino acids into polypeptide chains. These bonds form by condensation reactions between the carboxyl and amino groups. Review Figure 3.6

The **primary structure** of a protein is the order of amino acids in the chain. This chain is folded into a **secondary structure**, which in different parts of the protein may be an **α helix** or a **β pleated sheet**. Review Figure 3.7A–C

Disulfide bonds and noncovalent interactions between amino acids cause the polypeptide chain to fold into a three-dimensional **tertiary structure** and allow multiple chains to interact in a **quaternary structure.** Review Figure 3.7D,E

The specific shape and structure of a protein allows it to bind noncovalently to other molecules, often called **ligands.**

Heat, alterations in pH, or certain chemicals can all result in protein **denaturation**, which involves the loss of tertiary and/or secondary structure as well as biological function. Review Figure 3.11

Chaperonins assist protein folding by preventing binding to inappropriate ligands. Review Figure 3.12

3.3 What are the chemical structures and functions of carbohydrates?

Carbohydrates contain carbon bonded to hydrogen and oxygen atoms in a ratio of 1:2:1, or $(CH_2O)_n$.

Monosaccharides are the monomers that make up carbohydrates. **Hexoses** such as **glucose** are six-carbon monosaccharides; **pentoses** have five carbons. Review Figure 3.14, Web/CD Activity 3.3

Glycosidic linkages, which have either an α or a β orientation in space, covalently link monosaccharides into larger units such as **disaccharides**, **oligosaccharides**, and **polysaccharides**. Review Figure 3.15

Starch stores energy in plants. Starch and **glycogen** are formed by α-glycosidic linkages between glucose monomers and are distinguished by the amount of branching they exhibit. They can be easily broken down to release stored energy.

Cellulose, a very stable glucose polymer, is the principal component of the cell walls of plants.

3.4 What are the chemical structures and functions of lipids?

Fats and **oils** are **triglycerides**, composed of three **fatty acids** covalently bonded to a molecule of **glycerol** by **ester linkages**. Review Figure 3.18

Saturated fatty acids have a hydrocarbon chain with no double bonds. The hydrocarbon chains of **unsaturated** fatty acids have one or more double bonds that bend the chain, making close packing less possible. Review Figure 3.19

Phospholipids have a hydrophobic hydrocarbon "tail" and a hydrophilic phosphate "head." In water, the interactions of the hydrophobic tails and hydrophilic heads of phospholipids generate a **phospholipid bilayer** that is two molecules thick. The heads are directed outward, where they interact with the surrounding water. The tails are packed together in the interior of the bilayer. Review Figure 3.20

3.5 What are the chemical structures and functions of nucleic acids?

The unique function of the **nucleic acids**—**DNA** and **RNA**—is information storage; they form the hereditary material that passes genetic information to the next generation.

Nucleic acids are polymers of nucleotides. A **nucleotide** consists of a phosphate group, a pentose sugar (**ribose** in RNA and **deoxyribose** in DNA), and a nitrogen-containing **base**. Review Figure 3.23, Web/CD Activity 3.4

In DNA the nucleotide bases are **adenine**, **guanine**, **cytosine**, and **thymine**. **Uracil** substitutes for thymine in RNA. The nucleotides are joined by **phosphodiester linkages** between the sugar of one nucleotide and the phosphate of the next.

RNA is single-stranded. DNA is a **double helix** in which there is **complementary base pairing** based on hydrogen bonds between adenine and thymine (A-T) and between guanine and cytosine (G-C). The two strands of the DNA double helix run in opposite directions. Review Figures 3.24 and 3.26, Web/CD Activity 3.5

The information content of DNA and RNA resides in their **base sequences**.

3.6 How did life on Earth begin?

Chemical evolution proposes that conditions on early Earth could have produced the macromolecules that distinguish living things. Review Figure 3.28, Web/CD Tutorial 3.2

Because it can form a three-dimensional shape, RNA could act as a **ribozyme**, an RNA surface on which chemical reactions proceed at a faster rate. Review Figure 3.29

Experiments have ruled out the continuous spontaneous generation of life. Review Figure 3.30, Web/CD Tutorial 3.3

SELF-QUIZ

1. The most abundant molecule in the cell is
   a. a carbohydrate.
   b. a lipid.
   c. a nucleic acid.
   d. a protein.
   e. water.
2. All lipids are
   a. triglycerides.
   b. polar.
   c. hydrophilic.
   d. polymers of fatty acids.
   e. more soluble in nonpolar solvents than in water.
3. All carbohydrates
   a. are polymers.
   b. are simple sugars.
   c. consist of one or more simple sugars.
   d. are found in biological membranes.
   e. are more soluble in nonpolar solvents than in water.
4. Which of the following is not a carbohydrate?
   a. Glucose
   b. Starch
   c. Cellulose
   d. Hemoglobin
   e. Deoxyribose
5. All proteins
   a. are enzymes.
   b. consist of one or more polypeptide chains.
   c. are amino acids.
   d. have quaternary structures.
   e. are more soluble in nonpolar solvents than in water.
6. Which of the following statements about the primary structure of a protein is not true?
   a. It may be branched.
   b. It is determined by the structure of the corresponding DNA.
   c. It is unique to that protein.
   d. It determines the tertiary structure of the protein.
   e. It is the sequence of amino acids in the protein.
7. The amino acid leucine
   a. is found in all proteins.
   b. cannot form peptide linkages.
   c. is hydrophobic.
   d. is hydrophilic.
   e. is identical to the amino acid lysine.
8. The quaternary structure of a protein
   a. consists of four subunits—hence the name quaternary.
   b. is unrelated to the function of the protein.
   c. may be either alpha or beta.
   d. depends on covalent bonding among the subunits.
   e. depends on the primary structures of the subunits.
9. All nucleic acids
   a. are polymers of nucleotides.
   b. are polymers of amino acids.
   c. are double-stranded.
   d. are double-helical.
   e. contain deoxyribose.
10. Which of the following statements about condensation reactions is *not* true?
    a. Protein synthesis results from them.
    b. Polysaccharide synthesis results from them.
    c. Nucleic acid synthesis results from them.
    d. They consume water as a reactant.
    e. Different condensation reactions produce different kinds of macromolecules.

FOR DISCUSSION

1. Suppose that, in a given protein, one lysine is replaced by aspartic acid (see Table 3.2). Does this change occur in the primary structure or in the secondary structure? How might it result in a change in tertiary structure? In quaternary structure?
2. If there are 20 different amino acids commonly found in proteins, how many different dipeptides are there? How many different tripeptides? How many different trinucleotides? How many different single-stranded RNAs composed of 200 nucleotides?
3. Why might RNA have preceded proteins in the evolution of biological macromolecules?

FOR INVESTIGATION

1. The Miller–Urey experiment (see Figure 3.28) showed that it was possible for amino acids to be formed from gases hypothesized to have been in Earth's early atmosphere. These amino acids were dissolved in water. Knowing what you do about the polymerization of amino acids into proteins (see Figure 3.6), how would you set up experiments to show that proteins can form under the conditions of early Earth? What properties would you expect of those proteins?
2. The interpretation of Pasteur's experiment (see Figure 3.30) depended on the inactivation of microorganisms by heat. We now know that some microorganisms can survive at very high temperatures. How would this change the interpretation of Pasteur's experiment? What experiments would you do to inactivate such microbes?

PART TWO
Cells and Energy

CHAPTER 4 Cells: The Working Units of Life

The oldest evidence of life?

Charles Darwin faced a dilemma. In his great book, *The Origin of Species*, he proposed the theory of natural selection to explain the gradual appearance and disappearance of different forms of organisms. But he realized that the fossil record, on which he based his theory, was incomplete, especially for the beginning of life. In Darwin's time—the middle of the nineteenth century—the oldest known fossils were complex organisms found in rocks dated at about 550 million years ago (the Cambrian period). Did simpler organisms exist before that time? If so, where were their fossils? These would surely provide a link to the origin of life.

Conditions on Earth were probably suitable for the emergence of life by 4 billion years ago, about 600 million years after Earth began to form. But at the turn of the twentieth century, the oldest known fossils were clumps of algae (simple aquatic photosynthetic organisms) close to 1 billion years old—still far short of the origin of life. Would it even be possible to find older fossils? Most ancient rocks are igneous—formed by high-temperature processes such as volcanic eruptions—and geologic action has altered them drastically over the millenia. It is unlikely that any cellular fossil could survive these extreme conditions.

It took until the 1990s to find older evidence of life. It was then that, in a few places on Earth's surface, scientists discovered a phenomenon—some relatively unchanged 3.5-billion-year-old rocks. In one of these rock samples from what is now Australia, geologist J. William Schopf saw chains and clumps of what looked tantalizingly like contemporary cyanobacteria ("blue-green" bacteria). Schopf needed to prove that these chains were once alive, not just the results of simple chemical reactions. He and his colleagues looked for chemical evidence of photosynthesis.

The use of carbon dioxide in photosynthesis is a hallmark of life and leaves a unique chemical signature—a specific ratio of isotopes of carbon (C^{13}: C^{12}) in the resulting carbohydrates. Schopf showed that the Australian material had this signature. Microscopic examination of the chains revealed substructures characteristic of living systems that were not likely to be the result of simple chemical reactions. Schopf's evidence suggests that the Australian sample is indeed the remains of a truly ancient living organism.

In addition to confirming the presence of the earliest life, Schopf's fossils suggest that life requires not just the right collection of macromolecules, but compartmentalization. The macromolecules of life perform unique functions because they are enclosed in

The Earliest Evidence of Life? This fossil from Western Australia is 3.5 billion years old. Its form is similar to that of modern filamentous cyanobacteria (inset).

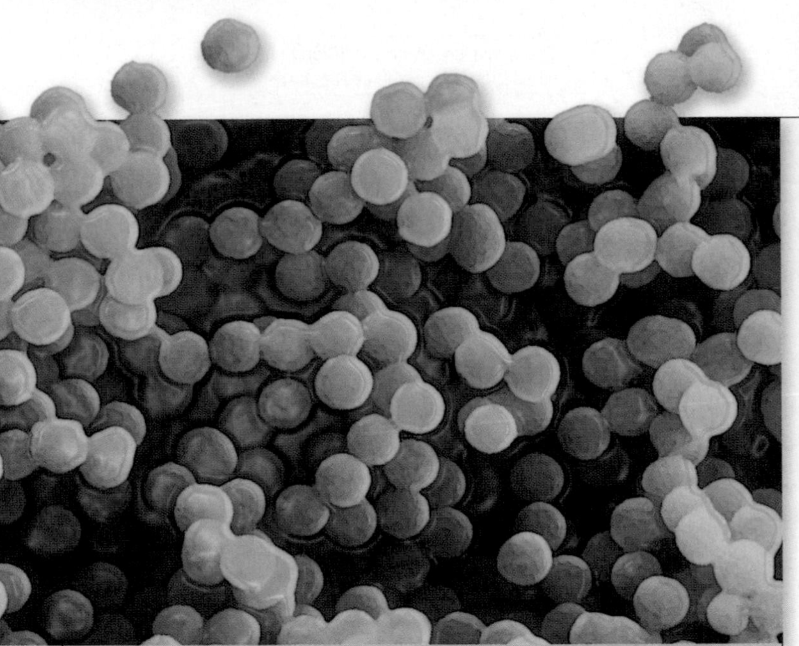

An Early "Cell" from the Laboratory Mixing amino acids and lipids formed in experiments simulating prebiotic conditions on Earth results in cell-like structures called proteinoids. They are surrounded by a membrane bilayer and can carry out some chemical reactions.

CHAPTER OUTLINE

structures that separate them from one another and from the external environment. This *compartmentalization* takes the form of cells and is seen in organelles that characterize eukaryotic cells. How did such compartmentalization arise?

Scientists have tried to model the origin of cells in the laboratory. In such modeling experiments, aggregates of molecules form round structures similar to cells. These chemical compartments can perform some biochemical reactions and can exchange materials with their environment. Taken together with the results of prebiotic chemistry experiments and the "RNA world" hypothesis described in Section 3.6, these experiments suggest that something similar to these aggregates might have been the first cells.

IN THIS CHAPTER we examine the structure and some of the functions of the "living compartment" known as the cell. We begin with the cell theory (the basis of cell biology), and then examine the simple cells of the single-celled organisms known as prokaryotes. We then tour the more complex eukaryotic cell and its various internal compartments, each of which performs specific functions for the cell.

4.1 What Features of Cells Make Them the Fundamental Unit of Life?

Just as atoms are the building blocks of chemistry, *cells are the building blocks of life*. The **cell theory** was described in Section 1.1 as the first unifying principle of biology. Recall the critical three tenets of the cell theory:

- Cells are the fundamental units of life.
- All organisms are composed of cells.
- All cells come from preexisting cells.

Cells contain water and the other small and large molecules we examined in the previous two chapters. Each cell contains at least 10,000 different types of molecules, most of them present in many copies. Cells use these molecules to transform matter and energy, to respond to their environment, and to reproduce themselves.

The cell theory has three important implications:

- Studying cell biology is in some sense the same as studying life. The principles that underlie the functions of the single cell of a bacterium are similar to those governing the approximately 60 trillion cells of your body.
- Life is continuous. All those cells in your body came from a single cell, a fertilized egg, which came from the fusion of two cells, a sperm and an egg from your parents, whose cells also came from fertilized eggs, and so on.
- The origin of life on Earth was marked by the origin of the first cells.

Cell size is limited by the surface area-to-volume ratio

Most cells are tiny. The volumes of cells range from 1 to 1,000 cubic micrometers (μm^3) (**Figure 4.1**). There are some exceptions: the eggs of birds are, relatively speaking, enormous, and individual cells of several types of algae and bacteria are large enough to be viewed with the unaided eye. And although neurons (nerve cells) have a volume that is within the "normal" range, they often have fine projections that may extend for meters, carrying signals from one part of a large animal to another.

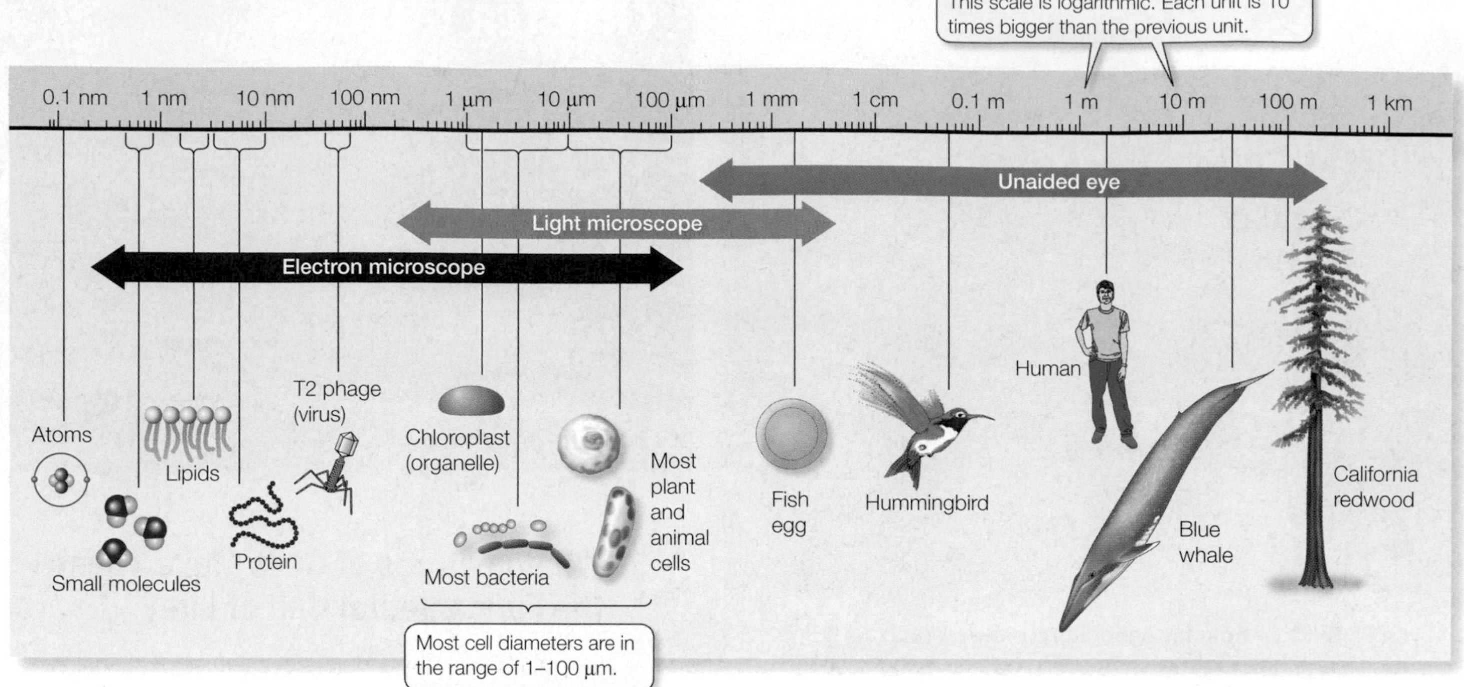

4.1 The Scale of Life This logarithmic scale shows the relative sizes of molecules, cells, and multicellular organisms.

Most cells are miniscule. About 2,000 human skin cells lined up in a row would fit across this page.

Small cell size is a practical necessity arising from the change in the **surface area-to-volume ratio** of any object as it increases in size. As an object increases in volume, its surface area also increases, but not to the same extent (**Figure 4.2**). This phenomenon has great biological significance for two reasons:

- The *volume* of a cell determines the amount of chemical activity it carries out per unit of time.
- The *surface area* of a cell determines the amount of substances the cell can take in from the outside environment and the amount of waste products it can release to the environment.

As a living cell grows larger, its chemical activity, and thus its rate of waste production and its need for resources, increase faster than its surface area. In addition, cells must often distribute substances from one place to another within the cell; the smaller the cell, the more easily this is accomplished. This explains why large organisms must consist of many small cells: cells must be small in volume in order to maintain a large enough surface area-to-volume ratio and an ideal internal volume. The large surface area represented by the myriad small cells of a multicellular organism enables it to carry out the many different functions required for survival.

Microscopes are needed to visualize cells

The smallest object a person can typically discern is about 0.2 mm (200 μm) in size. We refer to this measure as **resolution**, the distance apart two objects must be in order for the eye to distinguish them as separate; if they are closer together, they appear as a single blur. Most cells are much smaller than 200 μm, and thus are invisible to

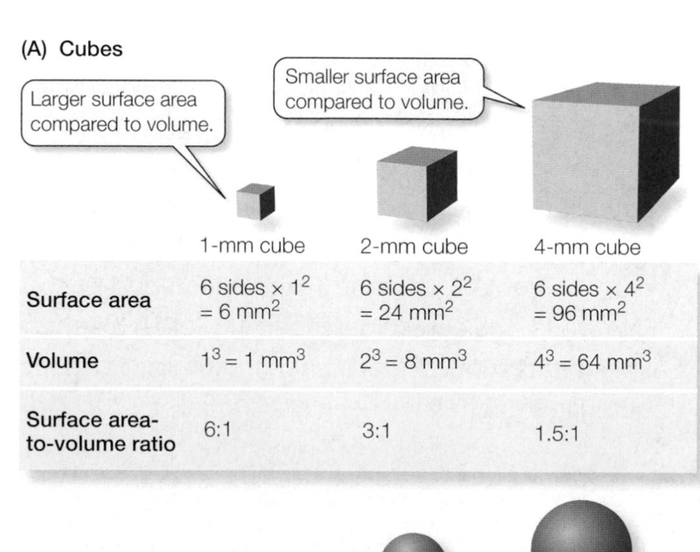

	1-mm cube	2-mm cube	4-mm cube
Surface area	6 sides × 1^2 = 6 mm^2	6 sides × 2^2 = 24 mm^2	6 sides × 4^2 = 96 mm^2
Volume	1^3 = 1 mm^3	2^3 = 8 mm^3	4^3 = 64 mm^3
Surface area-to-volume ratio	6:1	3:1	1.5:1

(B) Spheres

Diameter	1 μm	2 μm	3 μm
Surface area $4 \pi r^2$	3.14 μm^2	12.56 μm^2	28.26 μm^2
Volume $4/3 \pi r^3$	0.52 μm^3	4.19 μm^3	14.18 μm^3
Surface area-to-volume ratio	6:1	3:1	2:1

4.2 Why Cells Are Small Whether it is cuboid (A) or spheroid (B), as an object grows larger its volume increases more rapidly than its surface area. Cells must maintain a large surface area-to-volume ratio in order to function. This fact explains why large organisms must be composed of many small cells rather than a few huge ones.

RESEARCH METHOD

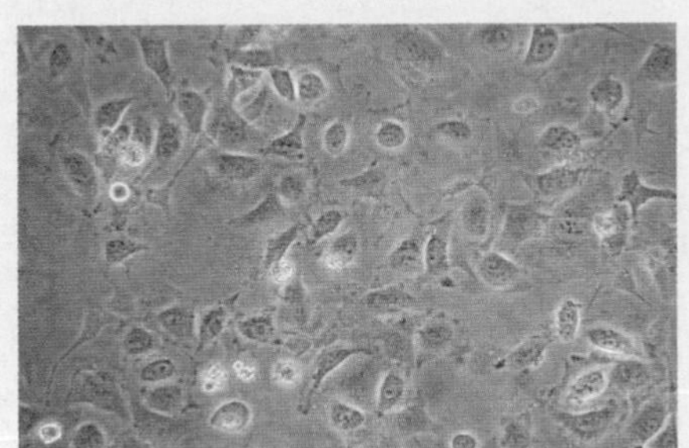

In **bright-field microscopy**, light passes directly through these human cells. Unless natural pigments are present, there is little contrast and details are not distinguished.

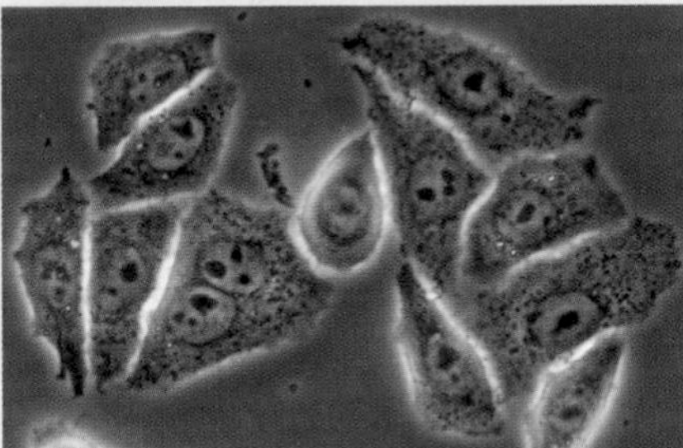

In **phase-contrast microscopy**, contrast in the image is increased by emphasizing differences in refractive index (the capacity to bend light), thereby enhancing light and dark regions in the cell.

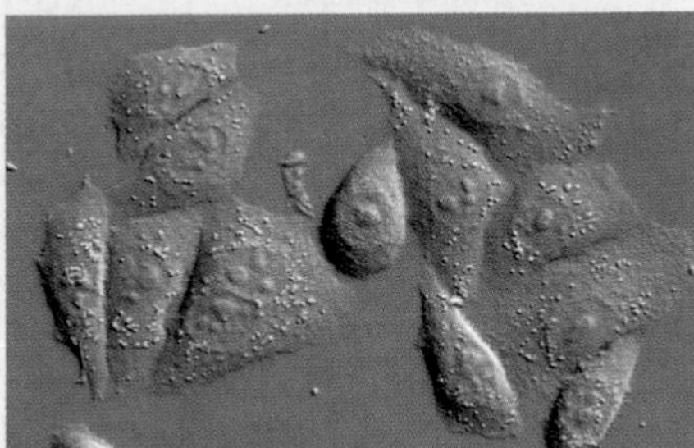

Differential interference-contrast microscopy uses two beams of polarized light. The combined images look as if the cell is casting a shadow on one side.

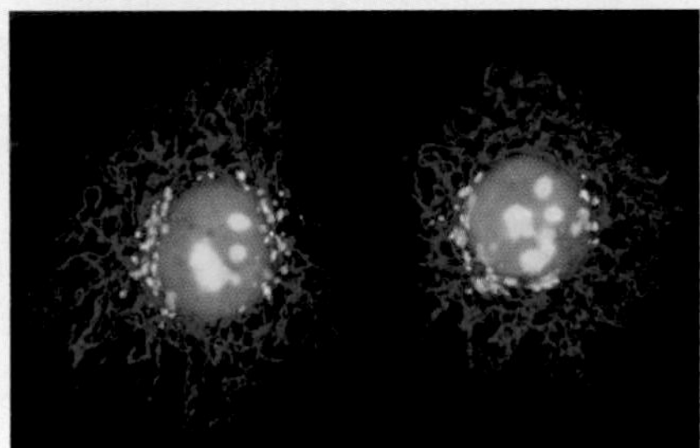

In **fluorescence microscopy**, a natural substance in the cell or a fluorescent dye that binds to a specific cell material is stimulated by a beam of light, and the longer-wavelength fluorescent light is observed coming directly from the dye.

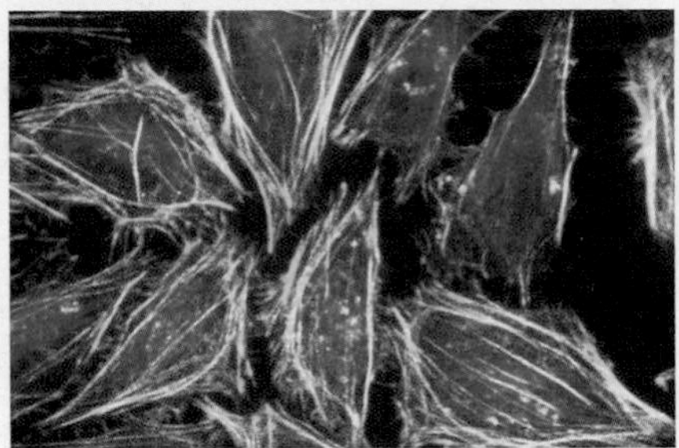

Confocal microscopy uses fluorescent materials but adds a system of focusing both the stimulating and emitted light so that a single plane through the cell is seen. The result is a sharper two-dimensional image than with standard fluorescence microscopy.

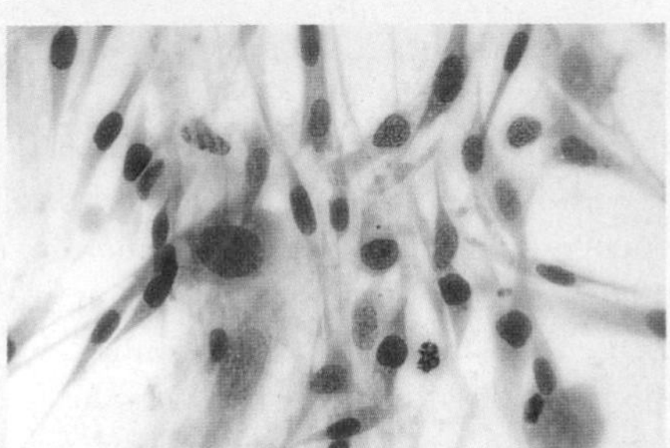

In **stained bright-field microscopy**, a stain added to preserve cells enhances contrast and reveals details not otherwise visible. Stains differ greatly in their chemistry and their capacity to bind to cell materials, so many choices are available.

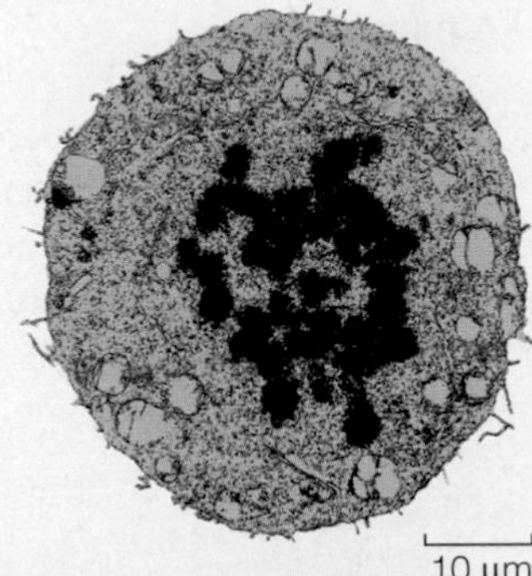

In **transmission electron microscopy** (TEM), a beam of electrons is focused on the object by magnets. Objects appear darker if they absorb the electrons. If the electrons pass through they are detected on a fluorescent screen.

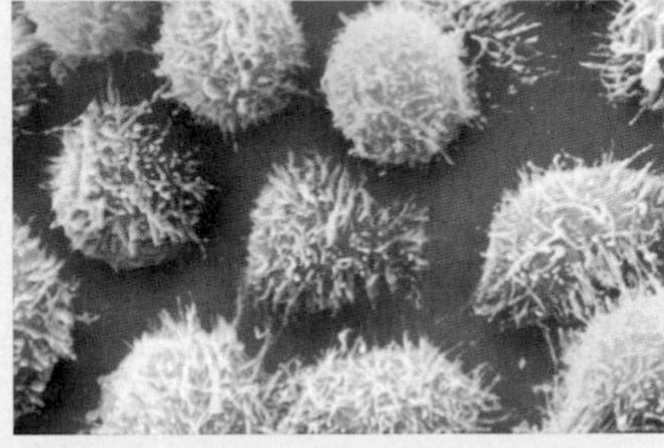

Scanning electron microscopy (SEM) directs electrons to the surface of the sample, where they cause other electrons to be emitted. These electrons are viewed on a screen. The three-dimensional surface of the object can be visualized.

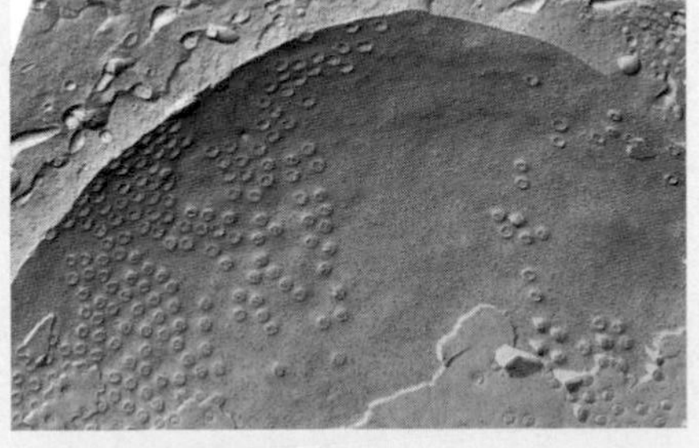

In **freeze-fracture microscopy**, cells are frozen and then a knife is used to crack them open. The crack often passes through the interior of plasma and internal membranes. The "bumps" that appear are usually large proteins embedded in the interior of the membrane.

4.3 Looking at Cells The top two rows show some techniques used in light microscopy. The lower three images were created using electron microscopes. All of these images are of a particular type of cultured cell known as HeLa cells. The story of these cells and their widespread use in research is told at the start of Chapter 9.

the human eye. *Microscopes* improve resolution so that cells and their internal structures can be seen (**Figure 4.3**).

There are two basic types of microscopes:

- A *light microscope* uses glass lenses and visible light to form a magnified image of an object. It has a resolution of about 0.2 μm, which is 1,000 times that of the human eye. It allows visualization of cell sizes and shapes and some internal cell structures. Internal structures are hard to see under visible light, so cells are often chemically treated and their components stained with various dyes to make certain structures stand out.

- An *electron microscope* uses electromagnets to focus an electron beam, much as a light microscope uses glass lenses to focus a beam of light. Since we cannot see electrons, the electron microscope directs them at a fluorescent screen or photographic film to create a visible image. The resolution of electron microscopes is about 0.2 nm, about 1,000,000 times greater than the human eye. This resolution permits the details of many subcellular structures to be distinguished.

Many techniques have been developed to enhance the views of cells we see under the light and electron microscopes.

Cells are surrounded by a plasma membrane

Each cell is surrounded by a membrane that separates it from its environment, creating a segregated (but not isolated) compartment. This **plasma membrane** is composed of a phospholipid bilayer, with the hydrophilic "heads" of the lipids facing the cell's aqueous interior on one side of the membrane and the extracellular environment on the other (see Figure 3.20). Proteins and other molecules are embedded in the lipids. We will devote most of Chapter 5 to detailing the structure and functions of the plasma membrane, but summarize its roles here:

- The plasma membrane allows the cell to maintain a more or less *constant internal environment*. A self-maintaining, constant internal environment (the state known as *homeostasis*) is a key characteristic of life that will be discussed in detail in Chapter 40.
- The plasma membrane acts as a *selectively permeable barrier*, preventing some substances from crossing it while permitting other substances to enter and leave the cell.
- As the cell's boundary with the outside environment, the plasma membrane is important in *communicating with adjacent cells* and *receiving signals from the environment*. We will describe this function in Chapter 15.
- The plasma membrane often has proteins protruding from it that are responsible for *binding and adhering to adjacent cells*.

Cells are prokaryotic or eukaryotic

As we learned in Section 1.2, biologists classify all living things into three domains: Archaea, Bacteria, and Eukarya. The organisms in Archaea and Bacteria are collectively called **prokaryotes** because they have in common a *prokaryotic* cell organization. Prokaryotic cells do not typically have membrane-enclosed internal compartments. The first cells were probably similar in organization to modern prokaryotes.

Eukaryotic cell organization, on the other hand, is particular to members of the domain Eukarya (**eukaryotes**), which includes the protists, plants, fungi, and animals. The genetic material (DNA) of eukaryotic cells is contained in a special membrane-enclosed compartment called the **nucleus**. Eukaryotic cells also contain other membrane-enclosed compartments in which specific chemical reactions take place.

4.1 RECAP

The cell theory is a unifying principle of biology. Surface area-to-volume ratios limit the size of cells. Both prokaryotic and eukaryotic cells are enclosed within a plasma membrane, but prokaryotic cells lack membrane-enclosed internal compartments.

- Why can we say that cell biology embodies all the principles of life? See p. 69
- Can you explain why cells are small? See p. 70 and Figure 4.2

Prokaryotes and eukaryotes alike have prospered for many hundreds of millions of years, and many evolutionary success stories have arisen from both types of cell organization. Let's look first at the organization of prokaryotic cells.

4.2 What Are the Characteristics of Prokaryotic Cells?

Prokaryotes can live on more different and diverse energy sources than any other living organisms, and they can inhabit greater environmental extremes, such as very hot springs and very salty water. As we examine prokaryotic cells in this section, bear in mind that there are vast numbers of prokaryotic species, and that the Bacteria and Archaea are distinguished in numerous ways. These differences, and the vast diversity of organisms in these two domains, will be the subject of Chapter 26.

Prokaryotic cells are generally smaller than eukaryotic cells, ranging from 0.25×1.2 µm to 1.5×4 µm. Each individual prokaryote is a single cell, but many types of prokaryotes are usually seen in chains, small clusters, or even clusters containing hundreds of individuals. In this section we will first consider the features that cells in the domains Bacteria and Archaea have in common. Then we will examine structural features that are found in some, but not all, prokaryotes.

Prokaryotic cells share certain features

All prokaryotic cells have the same basic structure (**Figure 4.4**):

- The plasma membrane encloses the cell, regulating the traffic of materials into and out of the cell and separating it from its environment.
- The **nucleoid** contains the hereditary material (DNA) of the cell.

The rest of the material enclosed in the plasma membrane is called the **cytoplasm**. The cytoplasm is composed of two components: the more fluid cytosol and insoluble suspended particles, including ribosomes.

- The **cytosol** consists mostly of water that contains dissolved ions, small molecules, and soluble macromolecules such as proteins.

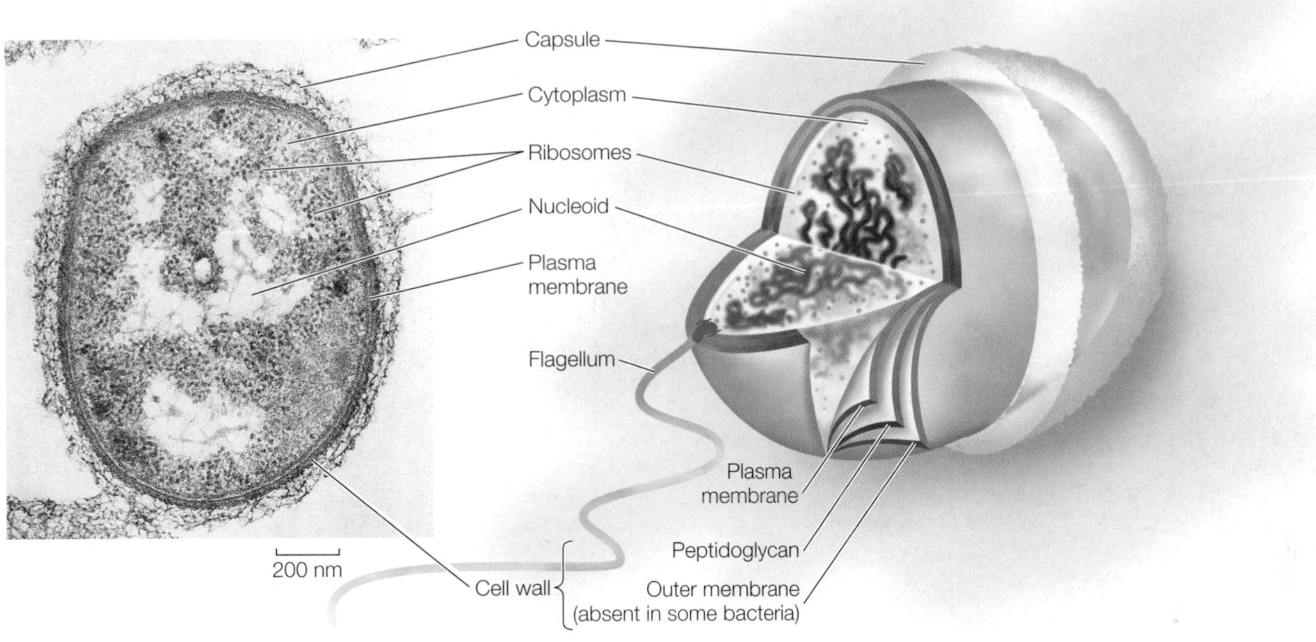

4.4 A Prokaryotic Cell The bacterium *Pseudomonas aeruginosa* illustrates the typical structures shared by all prokaryotic cells. This bacterium also has protective structures in an outer membrane that some, but not all, prokaryotes have. The flagellum and capsule are also structures found in some, but not all, prokaryotic cells.

- **Ribosomes** are complexes of RNA and proteins about 25 nm in diameter. They are the sites of protein synthesis.

The cytoplasm is not a static region. Rather, the substances in this environment are in constant motion. For example, a typical protein moves around the entire cell within a minute, and it encounters many molecules along the way.

Although structurally less complicated than eukaryotic cells, prokaryotic cells are functionally complex, carrying out thousands of biochemical reactions.

Some prokaryotic cells have specialized features

As they evolved, some prokaryotes developed specialized structures that gave a selective advantage to those cells that had them. These structures include a protective cell wall, an internal membrane for compartmentalization of some chemical reactions, and flagella for cell movement through the watery environment. These features are shown in Figures 4.4 and 4.5.

CELL WALLS Most prokaryotes have a cell wall located *outside* the plasma membrane. The rigidity of the cell wall supports the cell and determines its shape. The cell walls of most bacteria, but not archaea, contain *peptidoglycan*, a polymer of amino sugars, cross-linked by covalent bonds to form a single giant molecule around the entire cell. In some bacteria, another layer—the outer membrane (a polysaccharide-rich phospholipid membrane)—encloses the peptidoglycan layer. Unlike the plasma membrane, this outer membrane is not a major permeability barrier.

When pathogenic bacteria such *Salmonella*, *Shigella*, and *Neisseria* infect a human, lipopolysaccharide fragments from their membranes, called endotoxins, are released into the bloodstream. Among other unfortunate effects, these molecules cause fever and interfere with blood clotting, leading to excessive bleeding (hemorrhaging).

Enclosing the cell wall in some bacteria is a layer of slime, composed mostly of polysaccharides and referred to as a *capsule*. The capsules of some bacteria may protect them from attack by white blood cells in the animals they infect. The capsule helps keep the cell from drying out, and sometimes it helps the bacterium attach to other cells. Many prokaryotes produce no capsule, and those that do have capsules can survive even if they lose them, so the capsule is not essential to prokaryotic life.

As you will see later in this chapter, eukaryotic plant cells also have a cell wall, but it differs in composition and structure from the cell walls of prokaryotes.

INTERNAL MEMBRANES Some groups of bacteria—the cyanobacteria and some others—carry on photosynthesis. In these photosynthetic bacteria, the plasma membrane folds into the cytoplasm to form an internal membrane system that contains compounds needed for photosynthesis. The development of photosynthesis, which requires membranes, was an important event in the early evolution of life on Earth. Other prokaryotes have internal membrane folds that remain attached to the plasma membrane. These folds may function in cell division or in various energy-releasing reactions.

FLAGELLA AND PILI Some prokaryotes swim by using appendages called **flagella** (**Figure 4.5A**). A single flagellum, made of a protein called *flagellin*, looks at times like a tiny corkscrew. A complex mo-

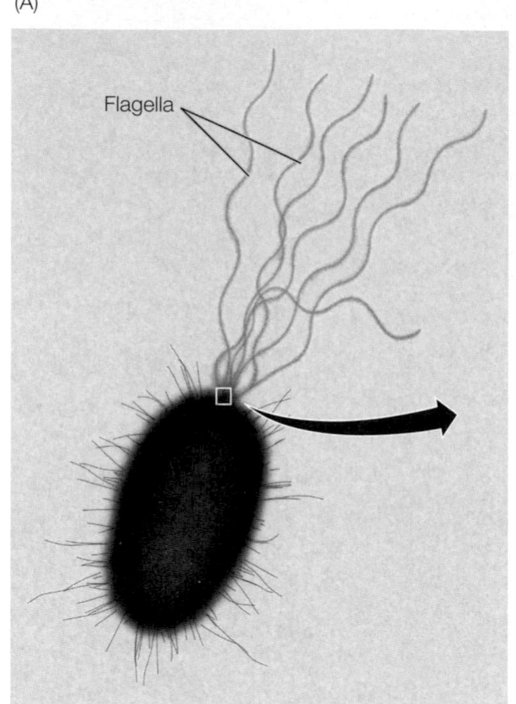

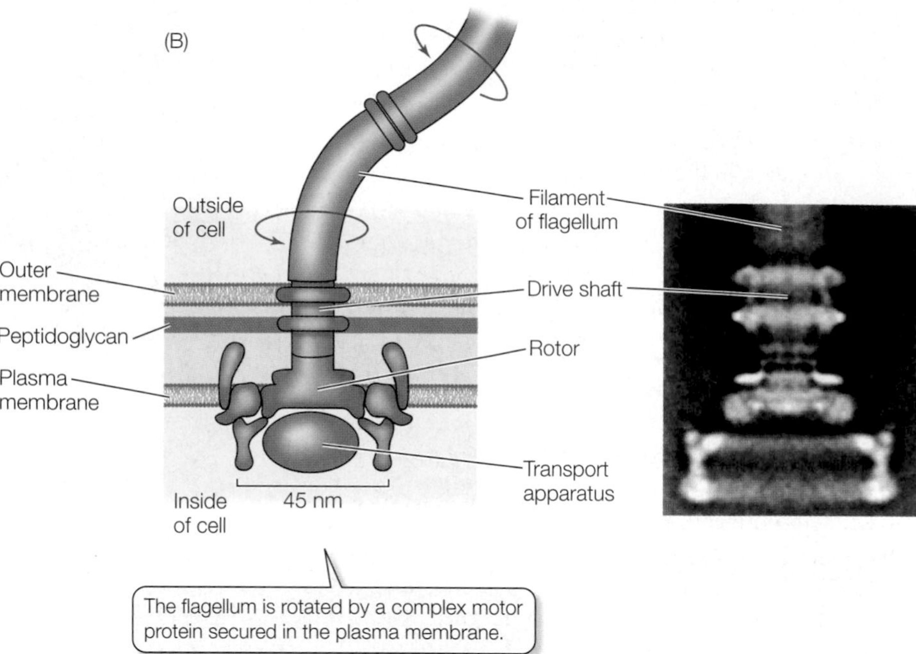

4.5 Prokaryotic Flagella (A) Flagella contribute to the movement and adhesion of prokaryotic cells. (B) Complex protein ring structures anchored in the plasma membrane form a motor unit that rotates the flagellum and propels the cell.

tor protein spins the flagellum on its axis like a propeller, driving the cell along. The motor protein is anchored to the plasma membrane and, in some bacteria, to the outer membrane of the cell wall (**Figure 4.5B**). We know that the flagella cause the motion of the cell because if they are removed, the cell does not move.

Pili project from the surfaces of some groups of bacteria. Shorter than flagella, these hairlike structures help bacteria adhere to one another when they exchange genetic material, as well as to animal cells for protection and food.

CYTOSKELETON Some prokaryotes, especially rod-shaped bacteria, have an internal filamentous helical structure just inside the plasma membrane. The proteins that make up this structure are similar in amino acid sequence to actin in eukaryotic cells, and since actin is part of the cytoskeleton in those cells (see Section 4.3), it has been suggested that the helical filaments in prokaryotes play a role in maintaining cell shape.

4.2 RECAP

Prokaryotic organisms can live on diverse energy sources and in extreme environments. Unlike eukaryotic cells, prokaryotic cells do not have internal compartments.

- What structures are present in all prokaryotic cells? See pp. 72–73 and Figure 4.4
- Describe the structure and function of a specialized prokaryotic cell feature, such as the cell wall, capsule, or flagellum. See pp. 73–74

The cells of animals, plants, fungi, and protists are usually larger than those of prokaryotic organisms, and they are structurally more complex. What specific structures characterize the eukaryotic cell?

4.3 What Are the Characteristics of Eukaryotic Cells?

Eukaryotic cells generally have dimensions up to 10 times greater than those of prokaryotes; for example, the spherical yeast cell has a diameter of 8 μm. Like prokaryotic cells, eukaryotic cells have a plasma membrane, cytoplasm, and ribosomes. But in addition to this shared organization, eukaryotic cells have compartments within the cytoplasm whose interiors are separated from the cytosol by a membrane.

Compartmentalization is the key to eukaryotic cell function

Some of the compartments in eukaryotic cells have been characterized as factories that make specific products. Others are like power plants that take in energy in one form and convert it into a more useful form. These membranous compartments, as well as other structures (such as ribosomes) that lack membranes but possess distinctive shapes and functions, are called **organelles**. Each of these organelles has specific roles in its particular cell. These roles are defined by the chemical reactions each organelle can carry out.

- The *nucleus* contains most of the cell's genetic material (DNA). The replication of the genetic material and the first steps in decoding genetic information take place in the nucleus.
- The *mitochondrion* is a power plant and industrial park, where energy stored in the bonds of carbohydrates and fatty acids is converted into a form more useful to the cell (ATP).
- The *endoplasmic reticulum* and *Golgi apparatus* are compartments in which some proteins synthesized by the ribosomes are packaged and sent to appropriate locations in the cell.
- *Lysosomes* and *vacuoles* are cellular digestive systems in which large molecules are hydrolyzed into usable monomers.
- *Chloroplasts* (found in only some cells) perform photosynthesis.

The membrane surrounding each organelle does two essential things. First, it keeps the organelle's molecules away from other molecules in the cell with which they might react inappropriately. Second, it acts as a traffic regulator, letting important raw materials into the organelle and releasing its products to the cytoplasm. The evolution of compartmentalization was an important development in the ability of eukaryotic cells to specialize, forming the organs and tissues of a complex multicellular body.

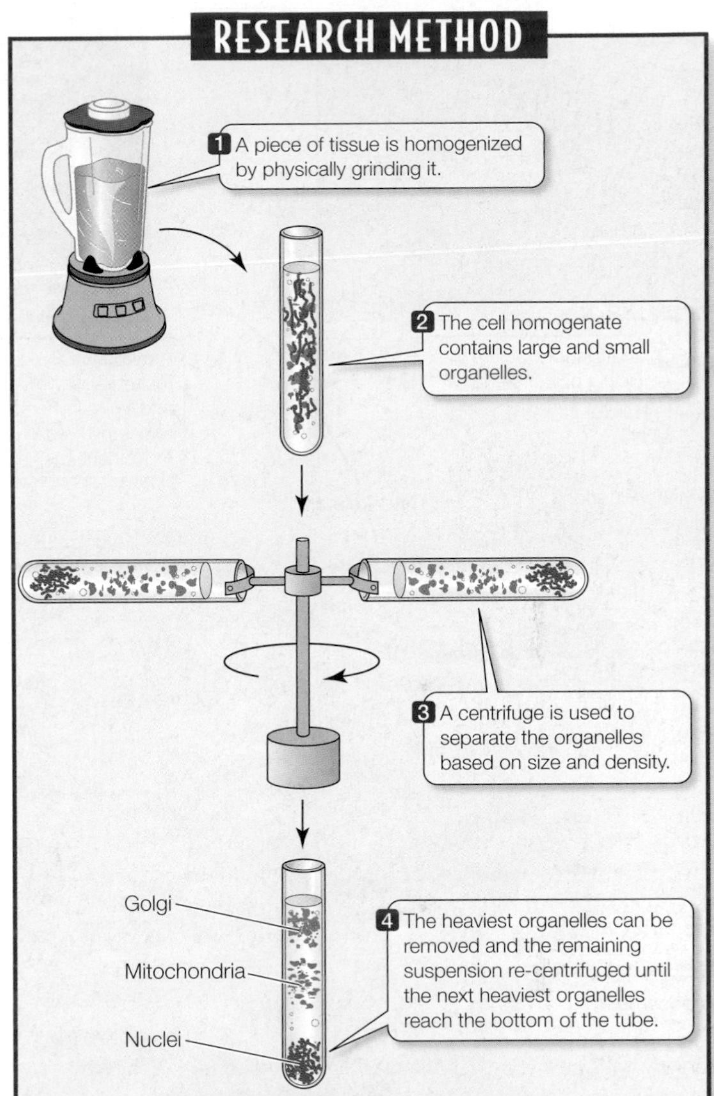

RESEARCH METHOD
1 A piece of tissue is homogenized by physically grinding it.
2 The cell homogenate contains large and small organelles.
3 A centrifuge is used to separate the organelles based on size and density.
Golgi
Mitochondria
Nuclei
4 The heaviest organelles can be removed and the remaining suspension re-centrifuged until the next heaviest organelles reach the bottom of the tube.

4.6 Cell Fractionation The organelles of cells can be separated from one another after the cells are broken open and centrifuged.

Organelles can be studied by microscopy or isolated for chemical analysis

Cell organelles were first detected by light and electron microscopy. The use of stains targeted to specific macromolecules has allowed cell biologists to determine the chemical compositions of organelles (see Figure 4.20, which shows a single cell stained for three different proteins).

Another way to look at cells is to take them apart. *Cell fractionation* begins with the destruction of the plasma membrane, which allows the cytoplasmic components to flow out into a test tube. The various organelles can then be separated from one another on the basis of size or density (**Figure 4.6**). Biochemical analyses can then be done on the isolated organelles. Microscopy and cell fractionation have complemented each other, giving us a complete picture of the structure and function of each organelle.

Microscopy of plant and animal cells reveals that many of the organelles are identical in both cell types. This can be seen in **Figure 4.7**, which also illustrates the prominent differences between eukaryotic cells and the prokaryotic cell diagrammed in Figure 4.4.

Some organelles process information

Living things depend on accurate, appropriate information—internal signals, environmental cues, and stored instructions—to respond appropriately to changing conditions and maintain a constant internal environment. In the cell, information is stored in the sequence of nucleotides in DNA molecules. Most of the DNA in eukaryotic cells resides in the nucleus. Information is translated from the language of DNA into the language of proteins at the ribosomes. This process is described in detail in Chapter 12.

NUCLEUS The single **nucleus** is usually the largest organelle in a cell (**Figure 4.8**; see also Figure 4.6). The nucleus of most animal cells is approximately 5 μm in diameter—substantially larger than most prokaryotic cells. The nucleus has several roles in the cell:

- It is the site of DNA replication.
- It is the site of genetic control of the cell's activities.
- A region within the nucleus, the **nucleolus**, begins the assembly of ribosomes from RNA and specific proteins.

The nucleus is surrounded by two membranes, which together form the *nuclear envelope*. The two membranes of the nuclear envelope are separated by 10–20 nm and are perforated by about 3,500 *nuclear pores* approximately 9 nm in diameter, which connect the interior of the nucleus with the cytoplasm. At these pores, the outer membrane of the nuclear envelope is continuous with the inner membrane. The pores are composed of over 100 different proteins, interacting hydrophobically. Each pore is surrounded by a pore complex made up of eight large protein aggregates arranged

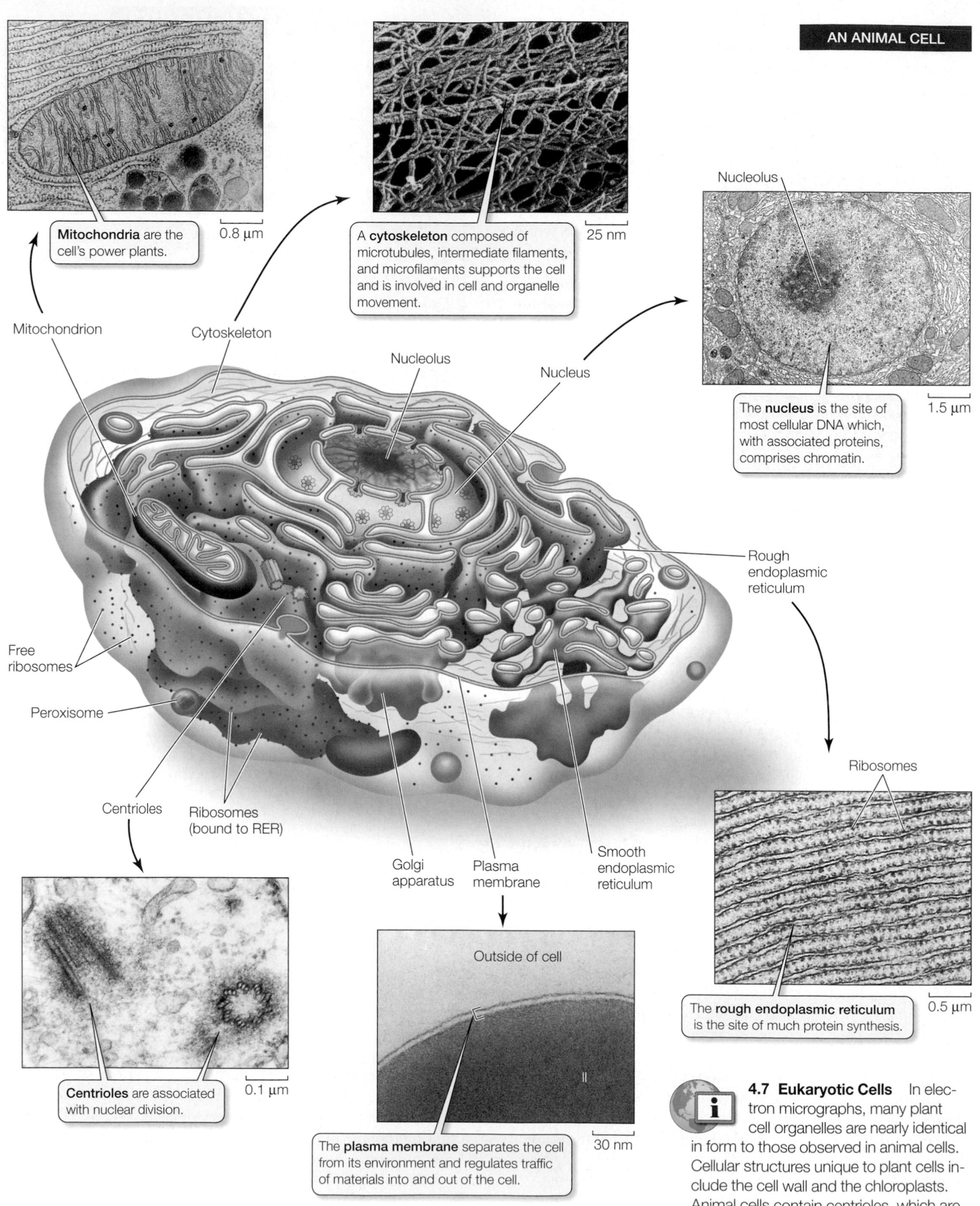
AN ANIMAL CELL
Mitochondria are the cell's power plants.
0.8 µm
A cytoskeleton composed of microtubules, intermediate filaments, and microfilaments supports the cell and is involved in cell and organelle movement.
25 nm
Nucleolus
The nucleus is the site of most cellular DNA which, with associated proteins, comprises chromatin.
1.5 µm
Mitochondrion
Cytoskeleton
Nucleolus
Nucleus
Rough endoplasmic reticulum
Free ribosomes
Peroxisome
Centrioles
Ribosomes (bound to RER)
Golgi apparatus
Plasma membrane
Smooth endoplasmic reticulum
Ribosomes
The rough endoplasmic reticulum is the site of much protein synthesis.
0.5 µm
Outside of cell
Centrioles are associated with nuclear division.
0.1 µm
The plasma membrane separates the cell from its environment and regulates traffic of materials into and out of the cell.
30 nm

4.7 Eukaryotic Cells In electron micrographs, many plant cell organelles are nearly identical in form to those observed in animal cells. Cellular structures unique to plant cells include the cell wall and the chloroplasts. Animal cells contain centrioles, which are not found in plant cells.

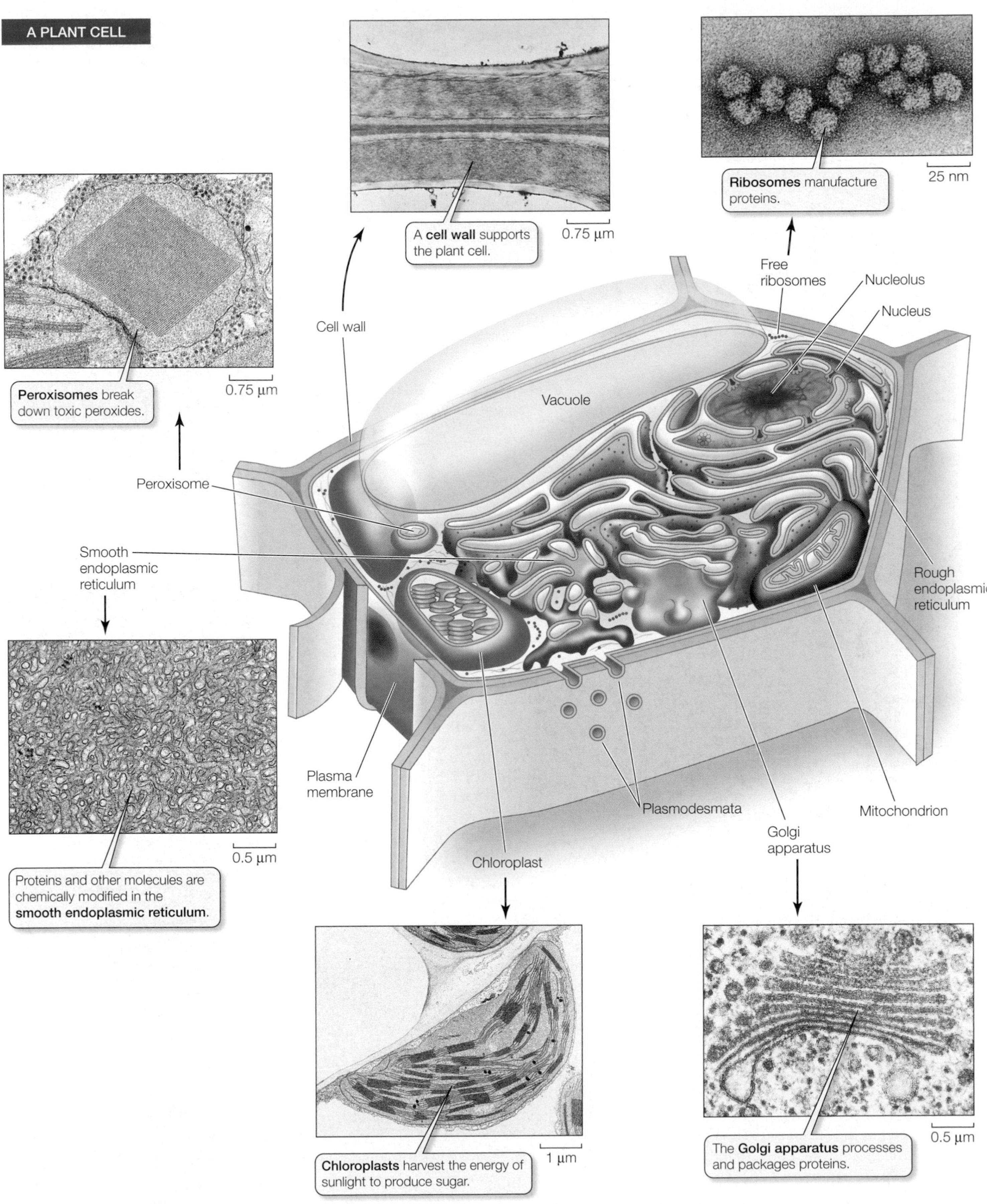

A PLANT CELL
A **cell wall** supports the plant cell.
0.75 µm
Ribosomes manufacture proteins.
25 nm
Peroxisomes break down toxic peroxides.
0.75 µm
Cell wall
Free ribosomes
Nucleolus
Nucleus
Vacuole
Peroxisome
Smooth endoplasmic reticulum
Rough endoplasmic reticulum
Plasma membrane
Plasmodesmata
Mitochondrion
Golgi apparatus
Chloroplast
0.5 µm
Proteins and other molecules are chemically modified in the **smooth endoplasmic reticulum**.
1 µm
Chloroplasts harvest the energy of sunlight to produce sugar.
0.5 µm
The **Golgi apparatus** processes and packages proteins.

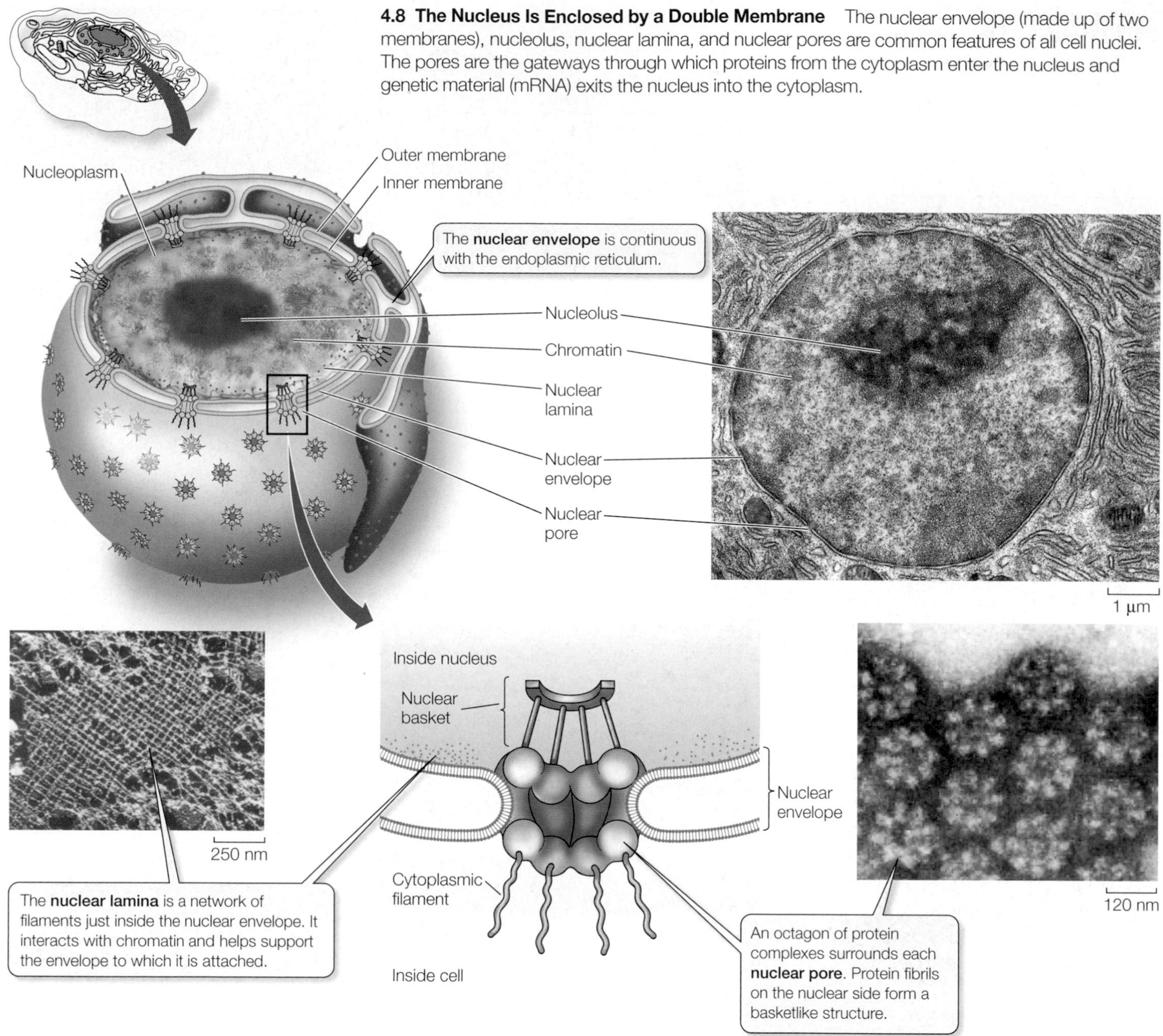
Outer membrane
Inner membrane
Nucleoplasm
The **nuclear envelope** is continuous with the endoplasmic reticulum.
Nucleolus
Chromatin
Nuclear lamina
Nuclear envelope
Nuclear pore
1 μm
250 nm
Inside nucleus
Nuclear basket
Nuclear envelope
Cytoplasmic filament
Inside cell
120 nm
The **nuclear lamina** is a network of filaments just inside the nuclear envelope. It interacts with chromatin and helps support the envelope to which it is attached.
An octagon of protein complexes surrounds each **nuclear pore**. Protein fibrils on the nuclear side form a basketlike structure.

4.8 The Nucleus Is Enclosed by a Double Membrane The nuclear envelope (made up of two membranes), nucleolus, nuclear lamina, and nuclear pores are common features of all cell nuclei. The pores are the gateways through which proteins from the cytoplasm enter the nucleus and genetic material (mRNA) exits the nucleus into the cytoplasm.

in an octagon where the inner and outer membranes merge (see Figure 4.8).

The nuclear pore works much like a turnstile gate at a sports event. Just as children can pass under the gate, small substances, such as ions and molecules with a molecular weight of less than 10,000 daltons, freely diffuse through the pore. Larger molecules (up to 50,000 Da) can also diffuse through, but they take longer. Still larger molecules, such as many proteins that are made in the cytoplasm and imported into the nucleus, are like adults at the gate: they cannot get through without a "ticket." In the case of imported proteins, the "ticket" is a short sequence of amino acids that is part of the protein. We know that this sequence is the *nuclear localization signal* from several lines of evidence:

- The signal sequence occurs in most nuclear proteins, but not in proteins that remain in the cytoplasm.
- If the signal sequence is removed from a protein, it stays in the cytoplasm.
- If the signal sequence is added to a protein that normally stays in the cytoplasm, that protein moves into the nucleus.
- Some viruses have a signal sequence that allows them to enter the nucleus; viruses without the signal sequence do not enter the nucleus as virus particles.

How does the signal sequence result in passage through the nuclear pore? Apparently, it has a three-dimensional structure that allows it to bind noncovalently to one of the proteins in the pore that acts as a *receptor*. Binding results in the receptor changing its three-dimensional shape so that it stretches the pore to let the large, imported protein pass through.

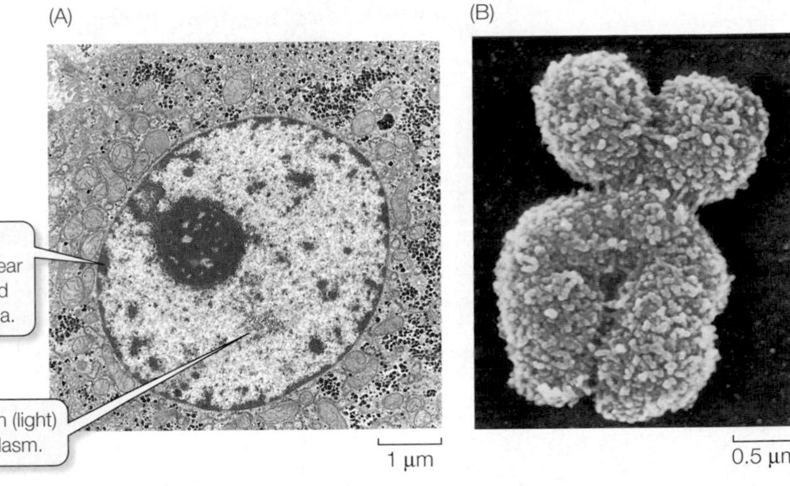
(A)
(B)
Dense chromatin (dark) near the nuclear envelope is attached to the nuclear lamina.
Diffuse chromatin (light) is in the nucleoplasm.
1 μm
0.5 μm

4.9 Chromatin and Chromosomes (A) When a cell is not dividing, the nuclear DNA is aggregated with proteins to form chromatin, which is dispersed throughout the nucleus. (B) The chromatin in a dividing cell is packed into dense bodies called chromosomes.

At certain sites, the outer membrane of the nuclear envelope folds outward into the cytoplasm and is continuous with the membrane of another organelle, the endoplasmic reticulum (which we will describe shortly). Inside the nucleus, DNA combines with proteins to form a fibrous complex called *chromatin*. Chromatin consists of exceedingly long, thin threads. Prior to cell division, the chromatin aggregates to form discrete, readily visible structures called *chromosomes* (**Figure 4.9**).

Surrounding the chromatin are water and dissolved substances collectively referred to as the *nucleoplasm*. Within the nucleoplasm, a network of apparently structural proteins called the *nuclear matrix* organizes the chromatin. At the periphery of the nucleus, the chromatin is attached to a protein meshwork, called the *nuclear lamina*, which is formed by the polymerization of proteins called *lamins* into filaments. The nuclear lamina maintains the shape of the nucleus by its attachment to both the chromatin and the nuclear envelope.

During most of a cell's life cycle, the nuclear envelope is a stable structure. When the cell reproduces, however, the nuclear envelope breaks down into pieces of membrane with attached pore complexes. The envelope re-forms when distribution of the replicated DNA to the daughter cells is completed (see Section 9.3). How this amazing self-assembly occurs is not known.

RIBOSOMES In prokaryotic cells, **ribosomes** float freely in the cytoplasm. In eukaryotic cells they are found in two places: in the cytoplasm, where they may be free or attached to the surface of the endoplasmic reticulum; and inside mitochondria and chloroplasts. In each of these locations, the ribosomes are the sites where proteins are synthesized under the direction of nucleic acids. Although they seem small in comparison to the cell in which they are contained, ribosomes are huge machines made up of several dozen kinds of molecules.

The ribosomes of prokaryotes and eukaryotes are similar in that both consist of two different-sized subunits. Eukaryotic ribosomes are somewhat larger, but the structure of prokaryotic ribosomes is better understood. Chemically, ribosomes consist of a special type of RNA called *ribosomal RNA (rRNA)*, to which more than 50 different protein molecules are noncovalently bound.

The endomembrane system is a group of interrelated organelles

Much of the volume of some eukaryotic cells is taken up by an extensive **endomembrane system**. This system includes two main components, the endoplasmic reticulum and the Golgi apparatus. Continuities between the nuclear envelope and the endomembrane system are visible under the electron microscope. Tiny, membrane-surrounded droplets called *vesicles* appear to shuttle between the various components of the endomembrane system. This system has various structures, but all of them are essentially compartments, closed off by their membranes from the cytoplasm. The membranes and materials in this system move between its various organelles.

ENDOPLASMIC RETICULUM Electron micrographs reveal a network of interconnected membranes branching throughout the cytoplasm of a eukaryotic cell, forming tubes and flattened sacs. These membranes are collectively called the **endoplasmic reticulum**, or **ER**. The interior compartment of the ER, referred to as the *lumen*, is separate and distinct from the surrounding cytoplasm (**Figure 4.10**). The ER can enclose up to 10 percent of the interior volume of the cell, and its foldings result in a surface area many times greater than that of the plasma membrane.

Rough endoplasmic reticulum (RER) is ER that is studded with ribosomes, which are temporarily attached to the outer faces of its flattened sacs.

- As a compartment, the RER segregates certain newly synthesized proteins away from the cytoplasm and transports them to other locations in the cell.
- While inside the RER, proteins can be chemically modified so as to alter their function and eventual destination.

The attached ribosomes are sites for the synthesis of proteins that function outside the cytosol—that is, proteins that are to be exported from the cell, incorporated into membranes, or moved into the organelles of the endomembrane system. These proteins enter the lumen of the RER as they are synthesized. As in nuclear localization, this is accomplished by a sequence of amino acids on the protein that acts as a RER localization signal. Once in the lumen of the RER, these proteins undergo several changes, including the formation of disulfide bridges and folding into their tertiary structures (see Figure 3.7D).

Some proteins gain carbohydrate groups in the RER, thus becoming *glycoproteins*. In the case of proteins directed to the lysosomes, the carbohydrate groups are part of an "addressing" system that ensures that the right proteins are directed to those organelles.

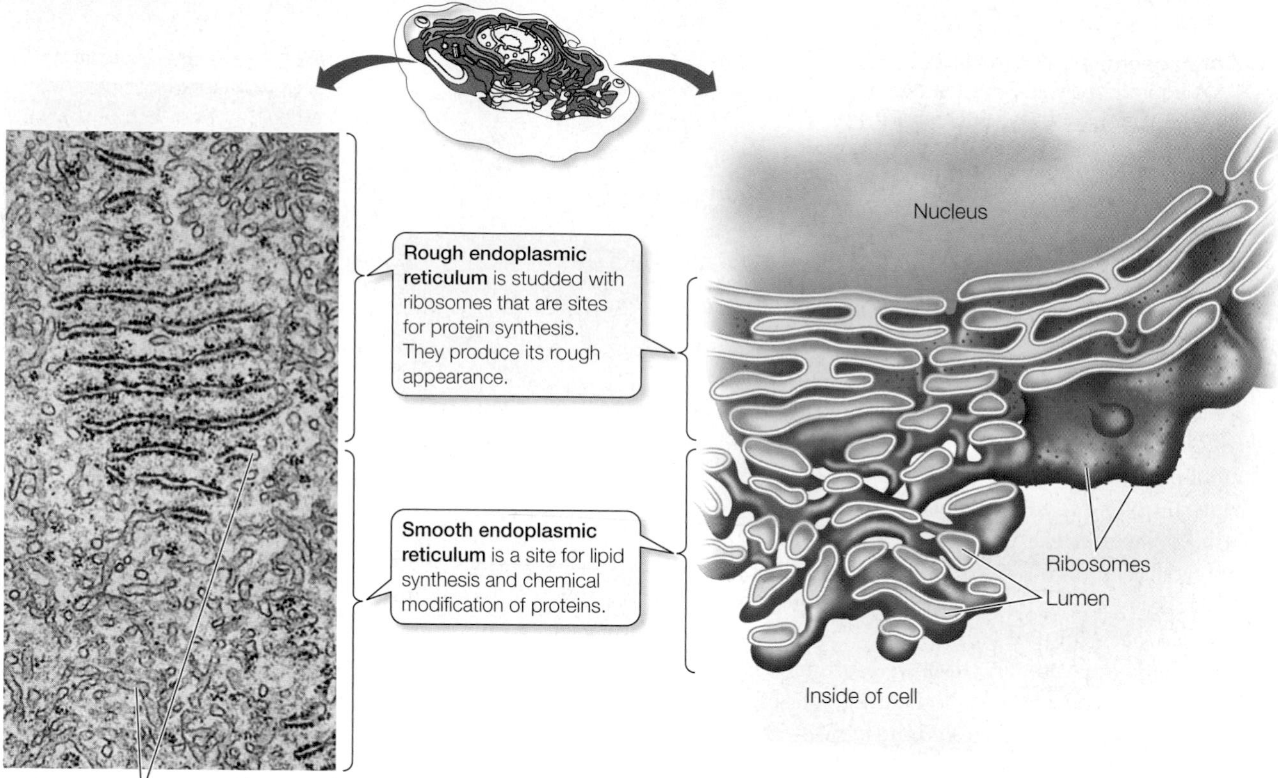

Rough endoplasmic reticulum is studded with ribosomes that are sites for protein synthesis. They produce its rough appearance.
Smooth endoplasmic reticulum is a site for lipid synthesis and chemical modification of proteins.
Nucleus
Ribosomes
Lumen
Inside of cell
Lumen
0.5 μm

4.10 Endoplasmic Reticulum The transmission electron micrograph on the left shows a two-dimensional slice through the three-dimensional structures depicted in the drawing. In normal living cells, the membranes of the ER never have open ends; they define closed compartments set off from the surrounding cytoplasm.

Smooth endoplasmic reticulum (**SER**) is more tubular (less like flattened sacs) than the RER and lacks ribosomes (see Figure 4.10). Within the lumen of the SER, some proteins that have been synthesized on the RER are chemically modified. In addition, the SER has three other important roles:

- It is responsible for chemically modifying small molecules taken in by the cell, especially drugs and pesticides.
- It is the site for the hydrolysis of glycogen in animal cells.
- It is the site for the synthesis of lipids and steroids.

Cells that synthesize a lot of protein for export are usually packed with endoplasmic reticulum. Examples include glandular cells that secrete digestive enzymes and white blood cells that secrete antibodies. In contrast, cells that carry out less protein synthesis (such as storage cells) contain less ER. Liver cells, which modify molecules that enter the body from the digestive system, have abundant smooth ER.

GOLGI APPARATUS The exact appearance of the **Golgi apparatus** (named for its discoverer, Camillo Golgi) varies from species to species, but it almost always consists of flattened membranous sacs called *cisternae* (singular *cisterna*), piled up like saucers, and small membrane-enclosed vesicles (**Figure 4.11**). The entire apparatus is about 1 μm long.

The Golgi apparatus has several roles:

- It receives proteins from the ER and may further modify them.
- It concentrates, packages, and sorts proteins before they are sent to their cellular or extracellular destinations.
- It is where some polysaccharides for the plant cell wall are synthesized.

In the cells of plants, protists, fungi, and many invertebrate animals, the stacks of cisternae are individual units scattered throughout the cytoplasm. In vertebrate cells, a few such stacks usually form a larger, single, more complex Golgi apparatus.

The Golgi apparatus appears to have three functionally distinct parts: a bottom, a middle, and a top. The bottom cisternae, constituting the *cis* region of the Golgi apparatus, lie nearest to the nucleus or a patch of RER (see Figure 4.11). The top cisternae, constituting the *trans* region, lie closest to the surface of the cell. The cisternae in the middle make up the *medial* region. (The terms *cis, trans,* and medial derive from Latin words meaning, respectively, "on the same side," "on the opposite side," and "in the middle.") These three parts of the Golgi apparatus contain different enzymes and perform different functions.

The Golgi apparatus receives proteins from the ER, packages them, and sends them on their way. Since there is often no direct membrane continuity between ER and Golgi apparatus, how does a protein get from one organelle to the other? The protein could simply leave the ER, travel across the cytoplasm, and enter the Golgi apparatus. But that would expose the protein to interactions with other molecules in the cytoplasm. On the other hand, segregation from the cytoplasm could be maintained if a piece of the ER could "bud off," forming a membranous vesicle that contains the protein—and that is exactly what happens (see Figure 4.11). As we will see in Section 5.1, many of the molecules in a mem-

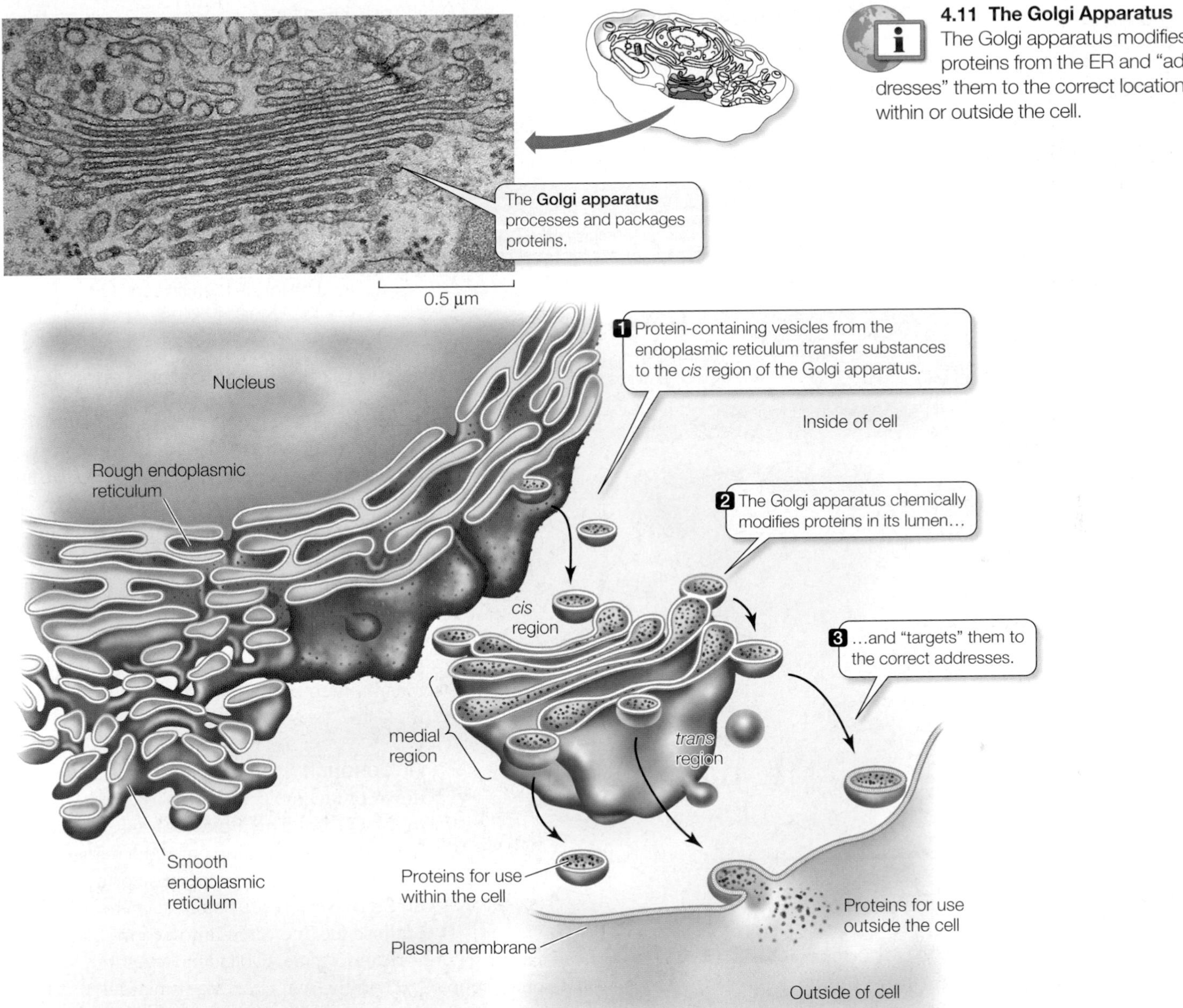
The Golgi apparatus processes and packages proteins.
0.5 μm
Nucleus
1 Protein-containing vesicles from the endoplasmic reticulum transfer substances to the cis region of the Golgi apparatus.
Inside of cell
Rough endoplasmic reticulum
2 The Golgi apparatus chemically modifies proteins in its lumen...
cis region
3 ...and "targets" them to the correct addresses.
medial region
trans region
Smooth endoplasmic reticulum
Proteins for use within the cell
Proteins for use outside the cell
Plasma membrane
Outside of cell

4.11 The Golgi Apparatus The Golgi apparatus modifies proteins from the ER and "addresses" them to the correct locations within or outside the cell.

brane are not bonded to one another (they "rub shoulders" rather than "hold hands"). Thus a biological membrane is not an ordered, solid structure, but a flexible, fluid boundary. Pieces of a membrane can easily bud off from or fuse with another membrane.

Proteins make the passage from ER to Golgi apparatus safely enclosed in a vesicle. Once it arrives, the vesicle fuses with the membrane of the Golgi apparatus, releasing its cargo. Other small vesicles may move between the cisternae, transporting proteins, and it appears that some proteins move from one cisterna to the next through tiny channels. Vesicles budding off from the *trans* region carry their contents away from the Golgi apparatus (see Figure 4.11).

LYSOSOMES Originating from the Golgi apparatus are organelles called **lysosomes**. They contain digestive enzymes, and they are the sites where macromolecules—proteins, polysaccharides, nucleic acids, and lipids—are hydrolyzed into their monomers (see Figure 3.4). Lysosomes are about 1 μm in diameter, are surrounded by a single membrane, and have a densely staining, featureless interior (**Figure 4.12**). There may be dozens of lysosomes in a cell, depending on its needs.

Lysosomes are sites for the breakdown of food and foreign objects taken up by the cell. These materials get into the cell by a process called *phagocytosis* (*phago*, "eat"; *cytosis*, "cellular"), in which a pocket forms in the plasma membrane and eventually deepens and encloses material from outside the cell. This pocket becomes a small vesicle that breaks free of the plasma membrane to move into the cytoplasm as a *phagosome* containing the food or other material (see Figure 4.12). The phagosome fuses with a *primary lysosome*, forming a *secondary lysosome*, in which digestion occurs.

The effect of this fusion is rather like releasing hungry foxes into a chicken coop: the enzymes in the secondary lysosome quickly hydrolyze the food particles. These reactions are enhanced by the mild acidity of the lysosome's interior, where the pH is lower than in the surrounding cytoplasm. The products of digestion diffuse through the membrane of the lysosome, providing raw materials for other cell processes. The "used" secondary lyso-

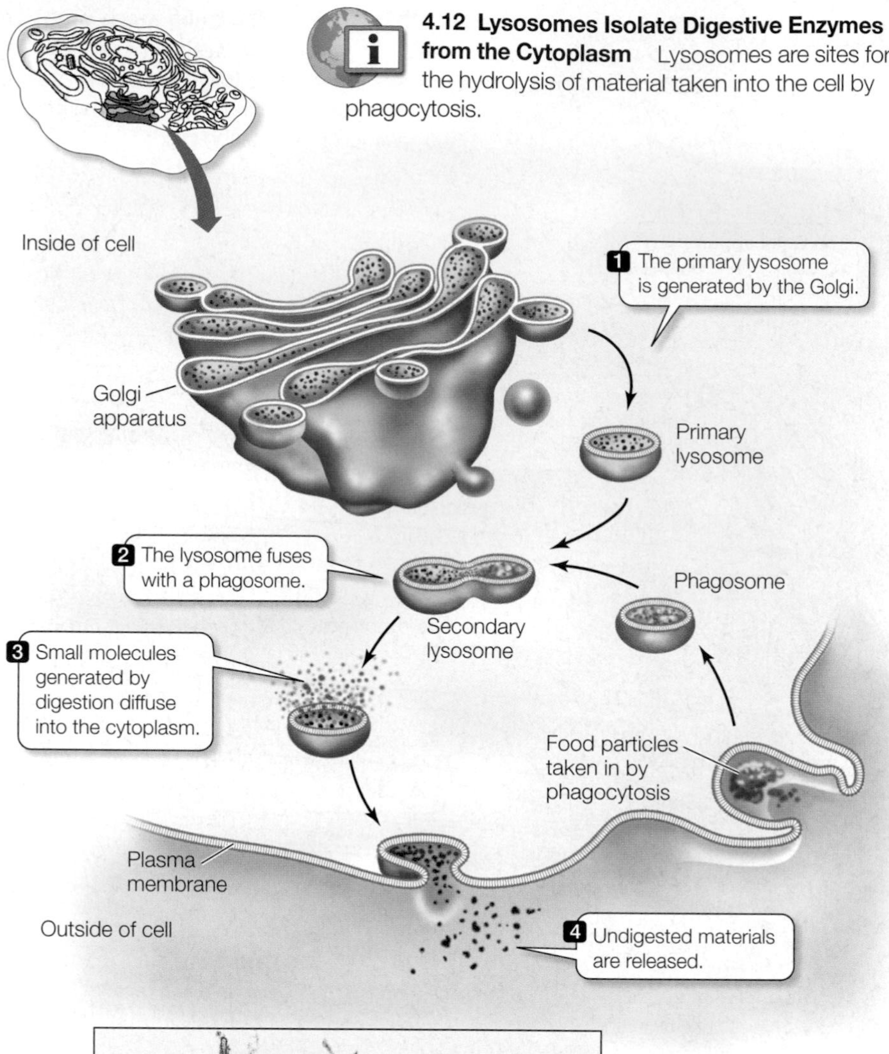

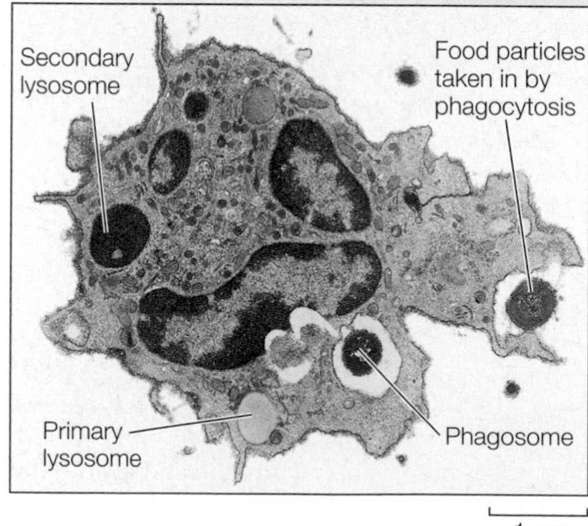

4.12 Lysosomes Isolate Digestive Enzymes from the Cytoplasm Lysosomes are sites for the hydrolysis of material taken into the cell by phagocytosis.

some, now containing undigested particles, then moves to the plasma membrane, fuses with it, and releases the undigested contents to the environment.

Lysosomes are also where the cell digests its own material in a process called *autophagy*. Autophagy is an ongoing process in which organelles such as mitochondria are engulfed by lysosomes and hydrolyzed into monomers, which pass out of the lysosome through its membrane into the cytoplasm for reuse.

Plant cells do not appear to contain lysosomes, but the central vacuole of a plant cell (which we will describe below) may function in an equivalent capacity because it, like lysosomes, contains many digestive enzymes.

Fatal lysosomal storage diseases can occur when lysosomes malfunction and molecules that should get broken down accumulate inside them instead.

Some organelles transform energy

A cell uses energy to synthesize the materials it needs for activities such as growth, reproduction, and movement. Energy is transformed from one form to another in mitochondria (found in eukaryotic cells) and in chloroplasts (found in eukaryotic cells that harvest energy from sunlight). In contrast, energy transformations in prokaryotic cells are associated with enzymes attached to the inner surface of the plasma membrane or to extensions of the plasma membrane that protrude into the cytoplasm.

MITOCHONDRIA In eukaryotic cells, the breakdown of fuel molecules such as glucose begins in the cytosol. The molecules that result from this partial degradation enter the **mitochondria** (singular *mitochondrion*), whose primary function is to convert the potential chemical energy of those fuel molecules into a form that the cell can use: the energy-rich molecule ATP (adenosine triphosphate). The production of ATP in the mitochondria using fuel molecules and molecular oxygen (O_2) is called *cellular respiration*.

Typical mitochondria are somewhat less than 1.5 μm in diameter and 2–8 μm in length—about the size of many bacteria. The number of mitochondria per cell ranges from one contorted giant in some unicellular protists to a few hundred thousand in large egg cells. An average human liver cell contains more than a thousand mitochondria. Cells that require the most chemical energy tend to have the most mitochondria per unit of volume.

Mitochondria have two membranes. The *outer membrane* is smooth and protective, and it offers little resistance to the movement of substances into and out of the mitochondrion. Immediately inside the outer membrane is an *inner membrane*, which folds inward in many places, and thus has a surface area much greater than that of the outer membrane (**Figure 4.13**). The folds tend to be quite regular, giving rise to shelflike structures called *cristae*.

Embedded in the inner mitochondrial membrane are many large protein complexes that participate in cellular respiration. The inner membrane exerts much more control over what enters and leaves the space it encloses than does the outer membrane. The

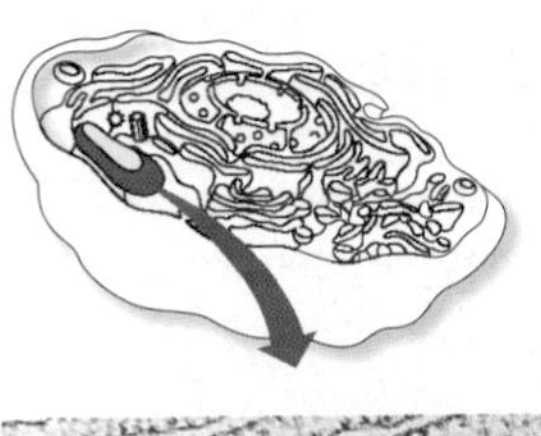

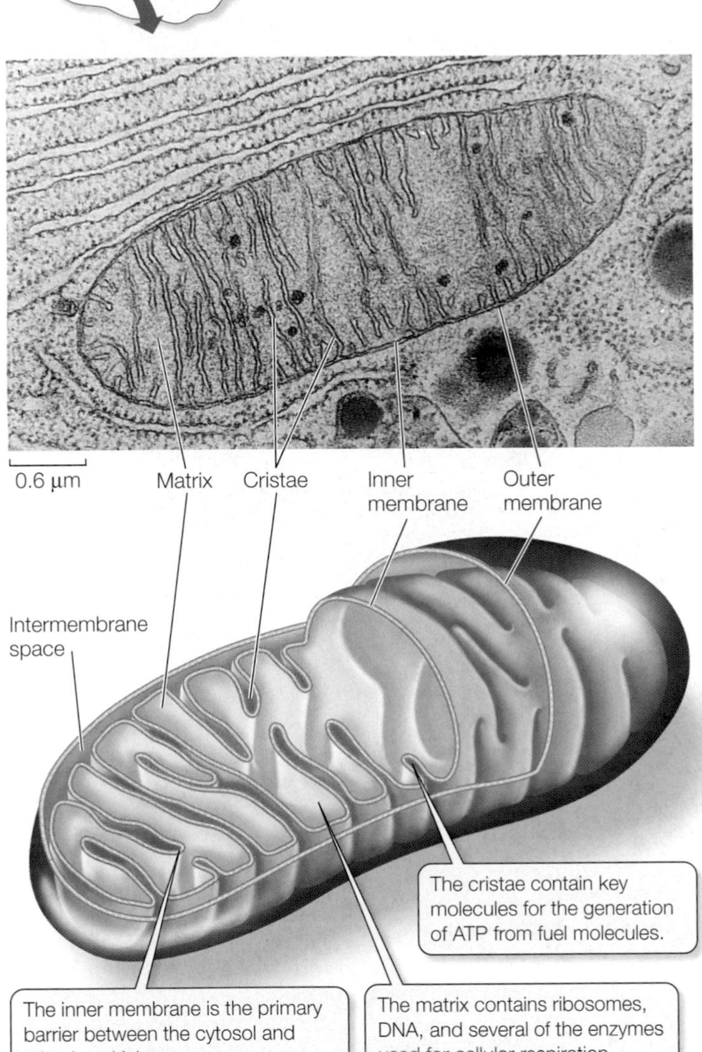
0.6 μm
Matrix
Cristae
Inner membrane
Outer membrane
Intermembrane space
The cristae contain key molecules for the generation of ATP from fuel molecules.
The inner membrane is the primary barrier between the cytosol and mitochondrial enzymes.
The matrix contains ribosomes, DNA, and several of the enzymes used for cellular respiration.

4.13 A Mitochondrion Converts Energy from Fuel Molecules into ATP The electron micrograph is a two-dimensional slice through a three-dimensional organelle. As the drawing emphasizes, the cristae are extensions of the inner mitochondrial membrane.

space enclosed by the inner membrane is referred to as the *mitochondrial matrix*. In addition to many enzymes, the matrix contains some ribosomes and DNA that are used to make some of the proteins needed for cellular respiration. We will see in Chapter 7 how the different parts of the mitochondrion work together in cellular respiration.

PLASTIDS One class of organelles—the *plastids*—is produced only in the cells of plants and certain protists. There are several types of plastids, with different functions.

Chloroplasts contain the green pigment *chlorophyll* and are the sites of photosynthesis (**Figure 4.14**). In photosynthesis, light energy is converted into the chemical energy of bonds between atoms. The molecules formed by photosynthesis provide food for the photosynthetic organisms, as well as for other organisms that eat them. Directly or indirectly, photosynthesis is the energy source for most of the living world.

Chloroplasts are variable in size and shape (**Figure 4.15**). Like a mitochondrion, a chloroplast is surrounded by two membranes. In addition, there is a series of internal membranes whose structure and arrangement vary from one group of photosynthetic organisms to another. Here we concentrate on the chloroplasts of the flowering plants. Even those chloroplasts show some variation, but the pattern shown in Figure 4.14 is typical.

The internal membranes of chloroplasts look like stacks of flat, hollow pita bread. These stacks, called *grana* (singular *granum*), consist of a series of flat, closely packed, circular compartments called **thylakoids**. In addition to phospholipids and proteins, the membranes of the thylakoids contain chlorophyll and other pigments that harvest light for photosynthesis (we will see how they do this in Section 8.2). The thylakoids of one granum may be connected to those of other grana, making the interior of the chloroplast a highly developed network of membranes, much like the ER.

The fluid in which the grana are suspended is called the *stroma*. Like the mitochondrial matrix, the chloroplast stroma contains ribosomes and DNA, which are used to synthesize some, but not all, of the proteins that make up the chloroplast.

Animal cells do not produce chloroplasts, but some do contain functional chloroplasts. These chloroplasts are either taken up from the partial digestion of green plants or contained within unicellular algae that live within the animal's tissues. The green color of some corals and sea anemones comes from chloroplasts in algae that live within those animals (see Figure 4.15C). The animals derive some of their nutrition from the photosynthesis that their chloroplast-containing "guests" carry out. Such an intimate relationship between two different organisms is called *symbiosis*.

Other types of plastids have functions different from those of chloroplasts:

- *Chromoplasts* contain red, orange, and/or yellow pigments and give color to plant organs such as flowers (**Figure 4.16A**). The chromoplasts have no known chemical function in the cell, but the colors they give to some petals and fruits probably encourage animals to visit flowers and thus aid in pollination, or to eat fruits and thus aid in seed dispersal. (On the other hand, carrot roots gain no apparent advantage from being orange.)
- *Leucoplasts* are storage depots for starches and fats (**Figure 4.16B**).

Several other organelles are surrounded by a membrane

Peroxisomes are organelles that collect the toxic peroxides (such as hydrogen peroxide, H_2O_2) that are the unavoidable by-products of cellular chemical reactions. These peroxides can be safely broken down inside the peroxisomes without mixing with other parts of the cell. Peroxisomes are small organelles, about 0.2 to 1.7 μm

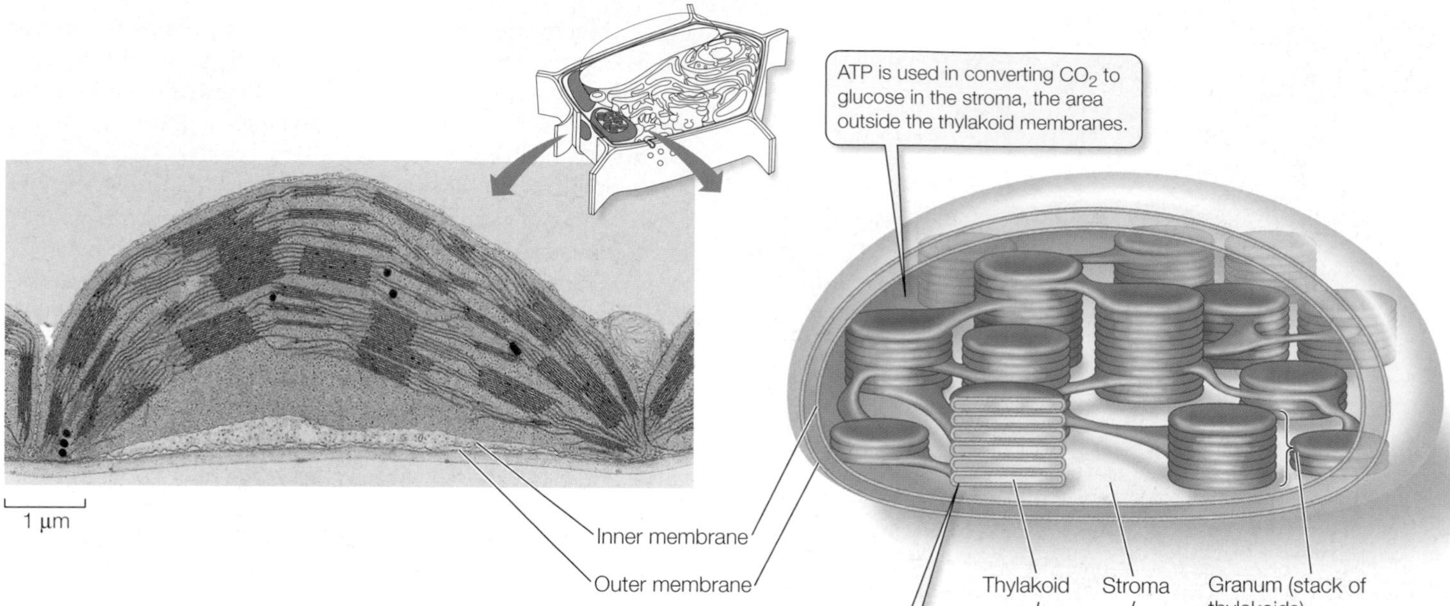

ATP is used in converting CO_2 to glucose in the stroma, the area outside the thylakoid membranes.

1 μm

Inner membrane

Outer membrane

Thylakoid

Stroma

Granum (stack of thylakoids)

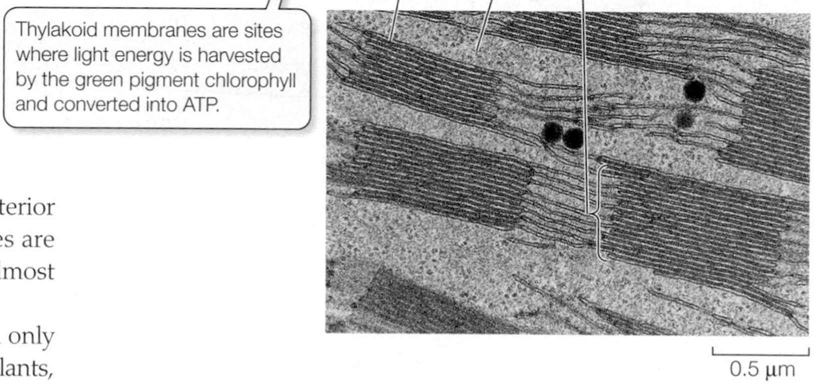

Thylakoid membranes are sites where light energy is harvested by the green pigment chlorophyll and converted into ATP.

0.5 μm

4.14 Chloroplasts Feed the World The electron micrograph shows a chloroplast from a leaf of corn. Chloroplasts are large compared with mitochondria and contain an extensive network of photosynthetic thylakoid membranes.

in diameter. They have a single membrane and a granular interior containing specialized enzymes (**Figure 4.17**). Peroxisomes are found at one time or another in at least some of the cells of almost every eukaryotic species.

A structurally similar organelle, the **glyoxysome**, is found only in plants. Glyoxysomes, which are most prominent in young plants, are the sites where stored lipids are converted into carbohydrates for transport to growing cells.

Many eukaryotic cells, but particularly those of plants and protists, contain membrane-enclosed **vacuoles** filled with aqueous

4.15 Being Green (A) In green plants, chloroplasts are concentrated in the leaf cells. (B) Green algae are photosynthetic and filled with chloroplasts. (C) No animal species produces its own chloroplasts, but in this symbiotic arrangement, a sea anemone is nourished by the chloroplasts of unicellular green algae living within its tissues.

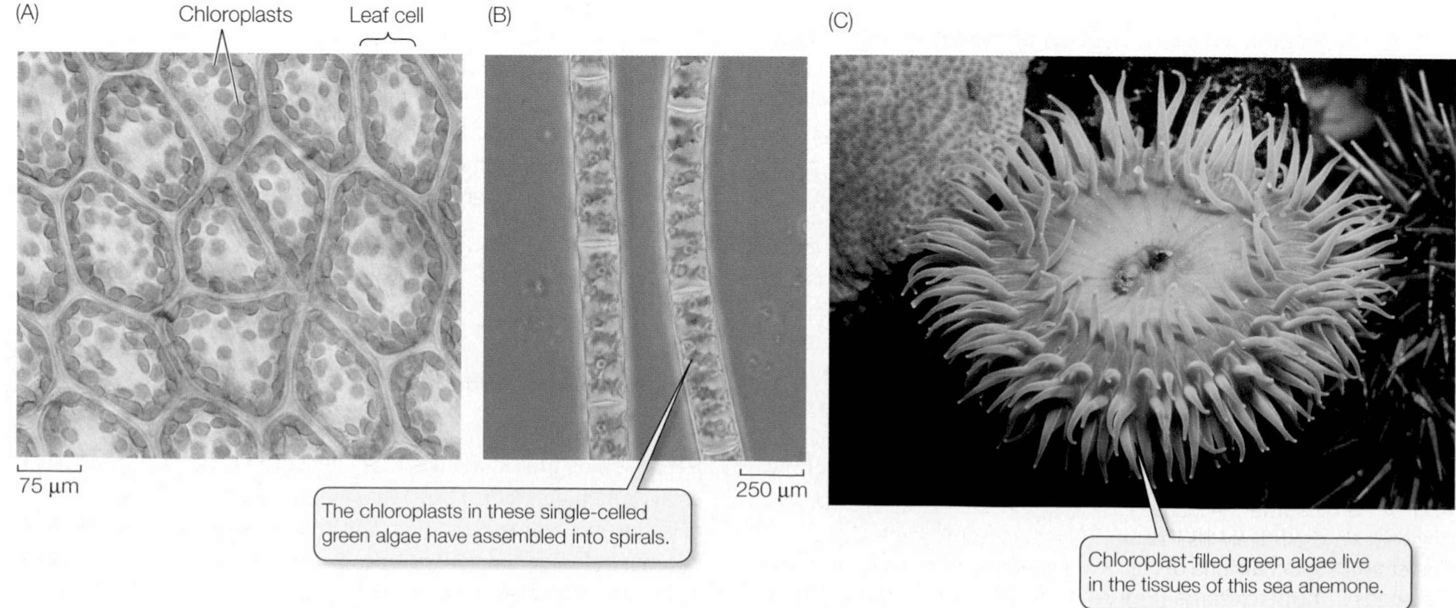
(A)
Chloroplasts
Leaf cell
75 μm
(B)
250 μm
The chloroplasts in these single-celled green algae have assembled into spirals.
(C)
Chloroplast-filled green algae live in the tissues of this sea anemone.

4.16 Chromoplasts and Leucoplasts (A) Colorful pigments stored in the chromoplasts of flowers like this poppy may help attract pollinating insects. (B) Leucoplasts in the cells of a potato are filled with white starch grains.

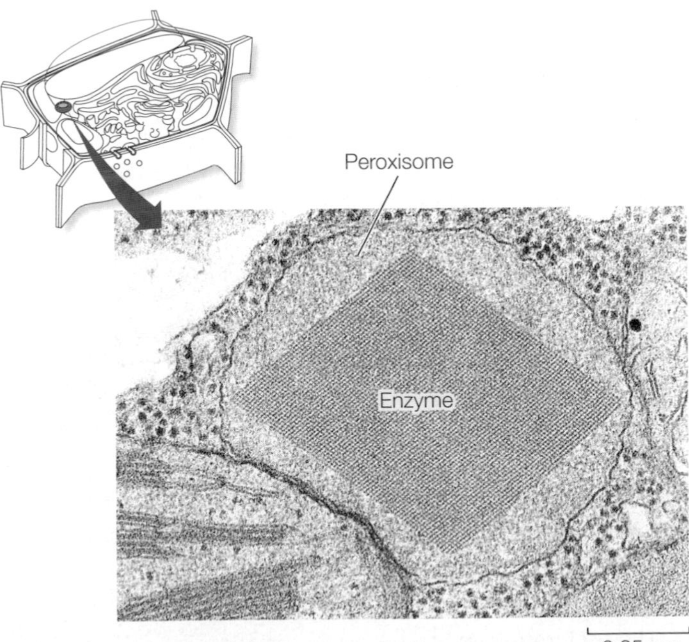

4.17 A Peroxisome A diamond-shaped crystal, composed of an enzyme, almost entirely fills this rounded peroxisome in a leaf cell. The enzyme catalyzes one of the reactions that breaks down toxic peroxides in the peroxisome.

solutions containing many dissolved substances (**Figure 4.18**). Plant vacuoles have several functions:

- *Storage*: Plant cells produce a number of toxic by-products and waste products, many of which are simply stored within vacuoles. Because they are poisonous or distasteful, these stored materials deter some animals from eating the plants, and may thus contribute to plant survival.

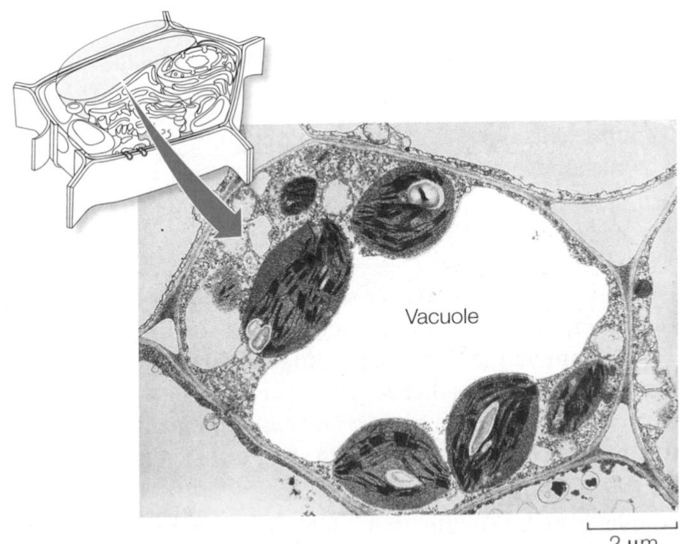

4.18 Vacuoles in Plant Cells Are Usually Large The large central vacuole in this cell is typical of mature plant cells. Smaller vacuoles are visible toward each end of the cell.

- *Structure*: In many plant cells, enormous vacuoles take up more than 90 percent of the cell volume and grow as the cell grows. The presence of dissolved substances in the vacuole causes water to enter, making the vacuole swell like a balloon. Plant cells have a rigid cell wall, which resists the swelling of the vacuole, providing *turgor*, or stiffness, that helps to support the plant.
- *Reproduction*: Some pigments (especially blue and pink ones) in the petals and fruits of flowering plants are contained in vacuoles. These pigments—the *anthocyanins*—are visual cues that help attract the animals that assist in pollination or seed dispersal.
- *Digestion*: In some plants, vacuoles in seeds contain enzymes that hydrolyze seed proteins into monomers that a developing plant embryo can use as food.

Food vacuoles are found in some simple and evolutionarily ancient groups of eukaryotes—single-celled protists and simple multicellular organisms such as sponges, for example—that have no digestive system. In these organisms, the cells engulf food particles by phagocytosis, generating a food vacuole. Fusion of this vacuole with a lysosome results in digestion, and small molecules leave the vacuole and enter the cytoplasm for use or distribution to other organelles.

Contractile vacuoles are found in many freshwater protists. Their function is to get rid of the excess water that rushes into the cell because of the imbalance in solute concentration between the interior of the cell and its freshwater environment. The contractile vacuole enlarges as water enters, then abruptly contracts, forcing the water out of the cell through a special pore structure.

The cytoskeleton is important in cell structure

In addition to its many membrane-enclosed organelles, the eukaryotic cytoplasm contains a set of long, thin fibers called the **cytoskeleton**. The cytoskeleton fills several important roles:

- It supports the cell and maintains its shape.
- It provides for various types of cellular movement.
- It positions organelles within the cell.
- Some of its fibers act as tracks or supports for motor proteins, which move organelles within the cell.
- It interacts with extracellular structures, helping to anchor the cell in place.

How do we know that the structural fibers of the cytoskeleton can achieve all these dynamic functions? Observing an individual structure under the microscope and observing a function in a living cell that contains that structure may be suggestive, but in science *mere correlation does not show cause and effect*. In cell biology, there are two ways to show that a structure or process "A" causes function "B":

- *Inhibition:* Use a drug that inhibits A and see if B still occurs. If it does not, then A is probably a causative factor for B. **Figure 4.19** shows an experiment with such a drug (an *inhibitor*) that demonstrates cause and effect in the case of the cytoskeleton and cell movement.

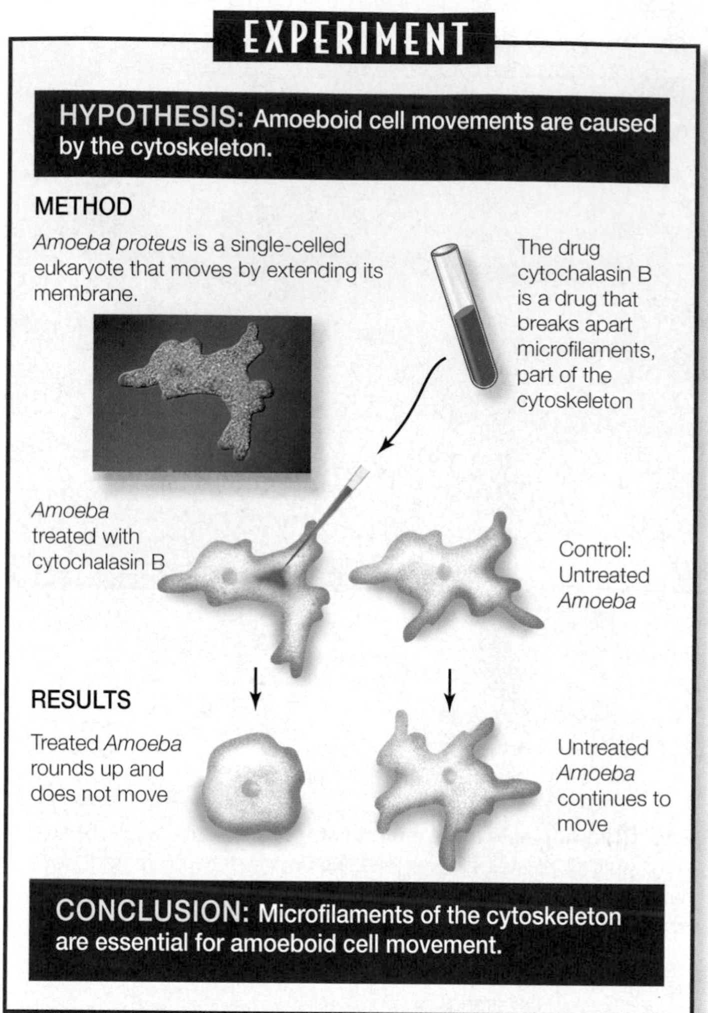

EXPERIMENT
HYPOTHESIS: Amoeboid cell movements are caused by the cytoskeleton.
METHOD
Amoeba proteus is a single-celled eukaryote that moves by extending its membrane.
The drug cytochalasin B is a drug that breaks apart microfilaments, part of the cytoskeleton
Amoeba treated with cytochalasin B
Control: Untreated Amoeba
RESULTS
Treated Amoeba rounds up and does not move
Untreated Amoeba continues to move
CONCLUSION: Microfilaments of the cytoskeleton are essential for amoeboid cell movement.

4.19 Showing Cause and Effect in Biology A substance known to inhibit a structure (in this case, cytochalasin B, a drug that inhibits microfilament formation) is applied in order to discover whether it also inhibits a function (in this case, cell movement in *Amoeba*). FURTHER RESEARCH: The drug colchicine breaks apart microtubules. How would you show that these components of the cytoskeleton are *not* involved in cell movement in *Amoeba*?

- *Mutation:* Look at a cell that lacks the gene (or genes) for A and see if B still occurs. If it does not, then A is probably a causative factor for B. Part Three of this book describes many experiments using this genetic approach.

MICROFILAMENTS **Microfilaments** can exist as single filaments, in bundles, or in networks. They are about 7 nm in diameter and up to several micrometers long. Microfilaments have two major roles:

- They help the entire cell or parts of the cell to move.
- They determine and stabilize cell shape.

Microfilaments are assembled from *actin*, a protein that exists in several forms and has many functions, especially in animals. The actin found in microfilaments (which are also known as *actin filaments*) is extensively folded and has distinct "plus" and "minus"

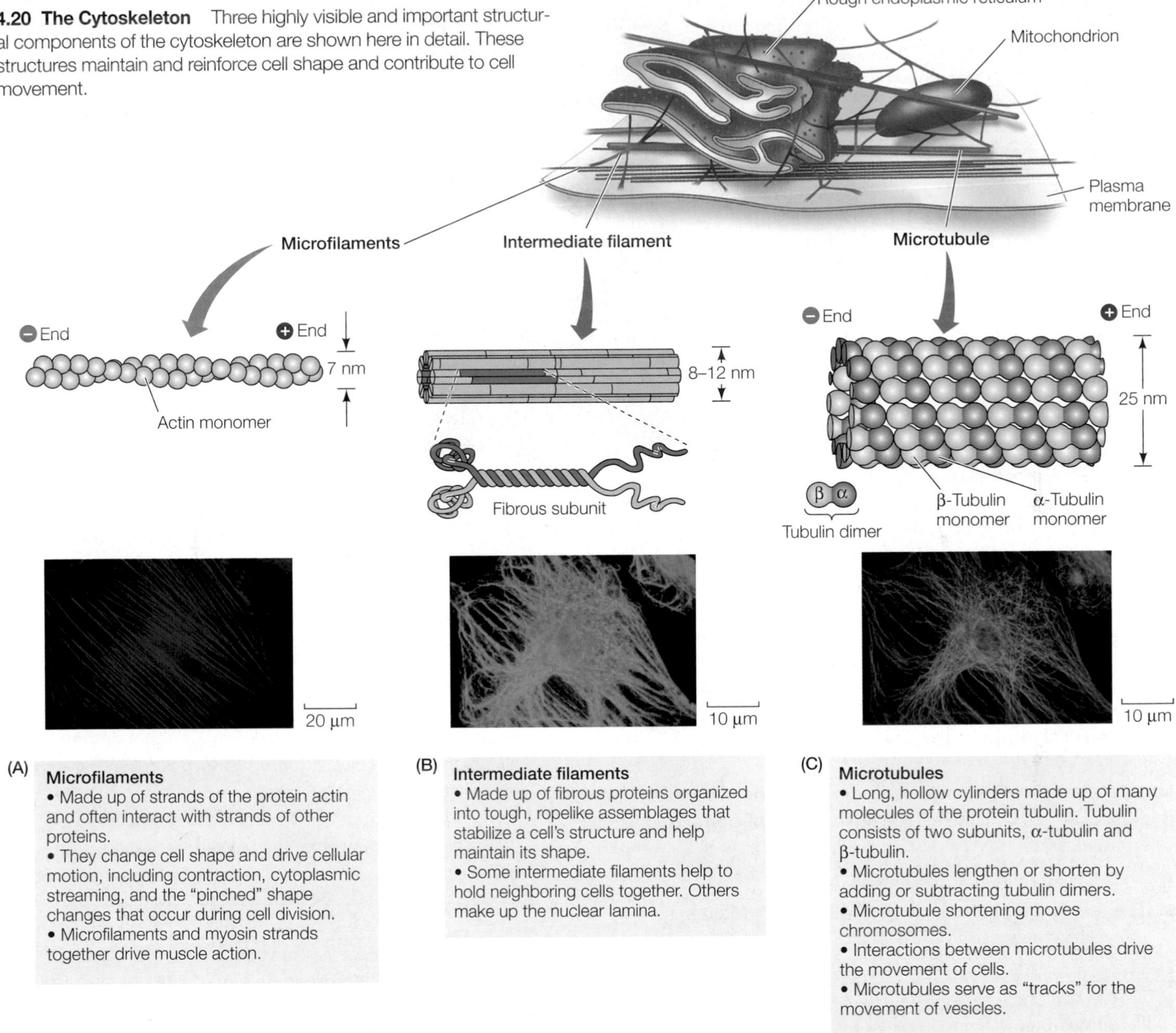

4.20 The Cytoskeleton Three highly visible and important structural components of the cytoskeleton are shown here in detail. These structures maintain and reinforce cell shape and contribute to cell movement.

(A) **Microfilaments**
- Made up of strands of the protein actin and often interact with strands of other proteins.
- They change cell shape and drive cellular motion, including contraction, cytoplasmic streaming, and the "pinched" shape changes that occur during cell division.
- Microfilaments and myosin strands together drive muscle action.

(B) **Intermediate filaments**
- Made up of fibrous proteins organized into tough, ropelike assemblages that stabilize a cell's structure and help maintain its shape.
- Some intermediate filaments help to hold neighboring cells together. Others make up the nuclear lamina.

(C) **Microtubules**
- Long, hollow cylinders made up of many molecules of the protein tubulin. Tubulin consists of two subunits, α-tubulin and β-tubulin.
- Microtubules lengthen or shorten by adding or subtracting tubulin dimers.
- Microtubule shortening moves chromosomes.
- Interactions between microtubules drive the movement of cells.
- Microtubules serve as "tracks" for the movement of vesicles.

ends. These ends interact with other actin monomers to form long, double helical chains (**Figure 4.20A**). The polymerization of actin into microfilaments is reversible, and they can disappear from cells, breaking down into monomers of free actin.

In the muscle cells of animals, actin filaments are associated with another protein, the "motor protein" *myosin*, and the interactions of these two proteins account for the contraction of muscles (described in Section 47.1). In non-muscle cells, actin filaments are associated with localized changes of shape in the cell. For example, microfilaments are involved in a flowing movement of the cytoplasm called *cytoplasmic streaming* and in the "pinching" contractions that divide an animal cell into two daughter cells. Microfilaments are also involved in the formation of cellular extensions called *pseudopodia* (*pseudo*, "false;" *podia*, "feet") that enable some cells to move.

In some cell types, microfilaments form a meshwork just inside the plasma membrane. Actin-binding proteins then cross-link the microfilaments to form a rigid structure that supports the cell. For example, microfilaments support the tiny microvilli that line the human intestine, giving it a larger surface area through which to absorb nutrients (**Figure 4.21**).

INTERMEDIATE FILAMENTS There are at least 50 different kinds of **intermediate filaments**, many of them specific to a few cell types. They generally fall into six molecular classes (based on amino acid sequence) that share the same general structure, being composed of fibrous proteins of the keratin family, similar to the protein that makes up hair and fingernails. These proteins are organized into tough, ropelike assemblages 8 to 12 nm in diameter (**Figure 4.20B**). Intermediate filaments have two major structural functions:

- They stabilize cell structure.
- They resist tension.

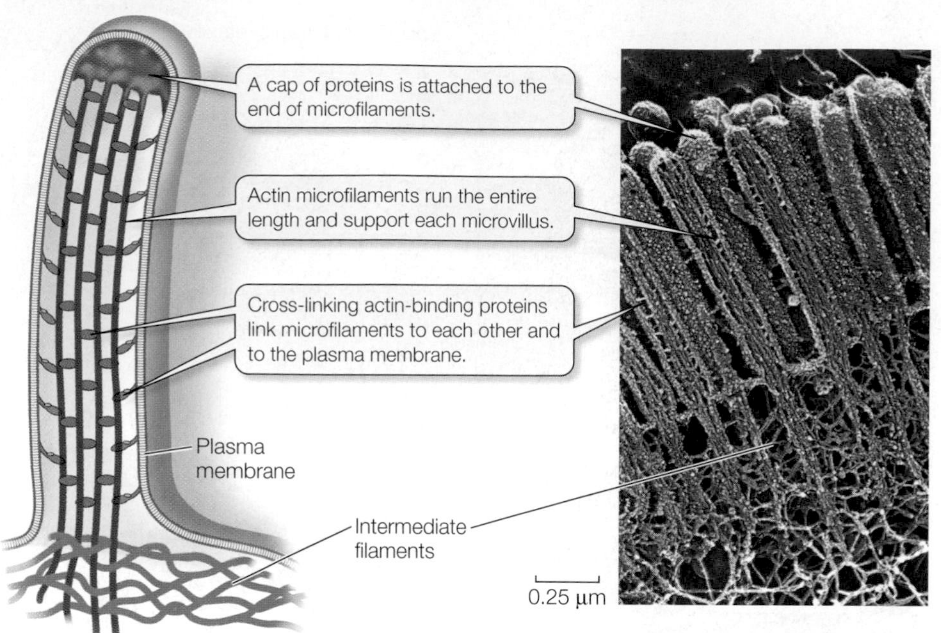
A cap of proteins is attached to the end of microfilaments.
Actin microfilaments run the entire length and support each microvillus.
Cross-linking actin-binding proteins link microfilaments to each other and to the plasma membrane.
Plasma membrane
Intermediate filaments
0.25 μm

4.21 Microfilaments for Support Cells that line the intestine are folded into tiny projections called microvilli, which are supported by microfilaments. The microvilli increase the surface area of the cells, facilitating their absorption of small molecules.

In some cells, intermediate filaments radiate from the nuclear envelope and may maintain the positions of the nucleus and other organelles in the cell. The lamins of the nuclear lamina are intermediate filaments. Other kinds of intermediate filaments help hold a complex apparatus of microfilaments in the microvilli in intestinal cells (see Figure 4.21). Still other kinds stabilize and help maintain rigidity in body surface tissues by connecting "spot welds" called *desmosomes* between adjacent cells (see Figure 5.7B).

MICROTUBULES **Microtubules** are long, hollow, unbranched cylinders about 25 nm in diameter and up to several micrometers long. Microtubules have two roles in the cell:

- They form a rigid internal skeleton for some cells.
- They act as a framework along which motor proteins can move structures within the cell.

Microtubules are assembled from molecules of the protein *tubulin*. Tubulin is a *dimer*—a molecule made up of two monomers. The polypeptide monomers that make up a tubulin dimer are known as α-tubulin and β-tubulin. Thirteen chains of tubulin dimers surround the central cavity of the microtubule (**Figure 4.20C**). The two ends of a microtubule are different: one end is designated the plus (+) end, the other the minus (–) end. Tubulin dimers can be rapidly added or subtracted, mainly at the plus end, lengthening or shortening the microtubule. This capacity to change length rapidly makes microtubules *dynamic structures*. This dynamic property of microtubules can be seen in animal cells, where they are often found in parts of the cell that are changing shape.

Many microtubules radiate from a region of the cell called the *microtubule organizing center*. Tubule polymerization results in a rigid structure, and tubule depolymerization leads to its collapse.

In plants, microtubules help control the arrangement of the cellulose fibers of the cell wall. Electron micrographs of plants frequently show microtubules lying just inside the plasma membrane of cells that are forming or extending their cell walls. Experimental alteration of the orientation of these microtubules leads to a similar change in the cell wall and a new shape for the cell.

In many cells, microtubules serve as tracks for **motor proteins**, specialized molecules that use energy to change their shape and move. Motor proteins bond to and move along the microtubules, carrying materials from one part of the cell to another. Microtubules are also essential in distributing chromosomes to daughter cells during cell division. They are also intimately associated with movable cell appendages: the cilia and flagella.

Because microtubules are crucial for proper chromosome alignment during cell division, drugs such as colchicine and taxol that disrupt microtubule dynamics also disrupt cell division. If these drugs can be targeted to cancer cells—whose division is rampant—they can be useful in treating the disease.

CILIA AND FLAGELLA Many eukaryotic cells possess flagella or cilia, or both. These whiplike organelles may push or pull the cell through its aqueous environment, or they may move surrounding fluid over the surface of the cell. Cilia and eukaryotic flagella (which are quite distinct from prokaryotic flagella) are both assembled from specialized microtubules and have identical internal structures, but differ in their length and pattern of beating:

- **Cilia** (singular *cilium*) are shorter than flagella and are usually present in great numbers (**Figure 4.22A**). They beat stiffly in one direction and recover flexibly in the other direction (like a swimmer's arm), so that the recovery stroke does not undo the work of the power stroke.
- Eukaryotic *flagella* are longer than cilia and are usually found singly or in pairs. Waves of bending propagate from one end of a flagellum to the other in snakelike undulation.

In cross section, a typical cilium or eukaryotic flagellum is surrounded by the plasma membrane and contains a "9 + 2" array of microtubules. As **Figure 4.22B** shows, nine fused pairs of microtubules—called *doublets*—form an outer cylinder, and one pair of unfused microtubules runs up the center. A spoke radiates from

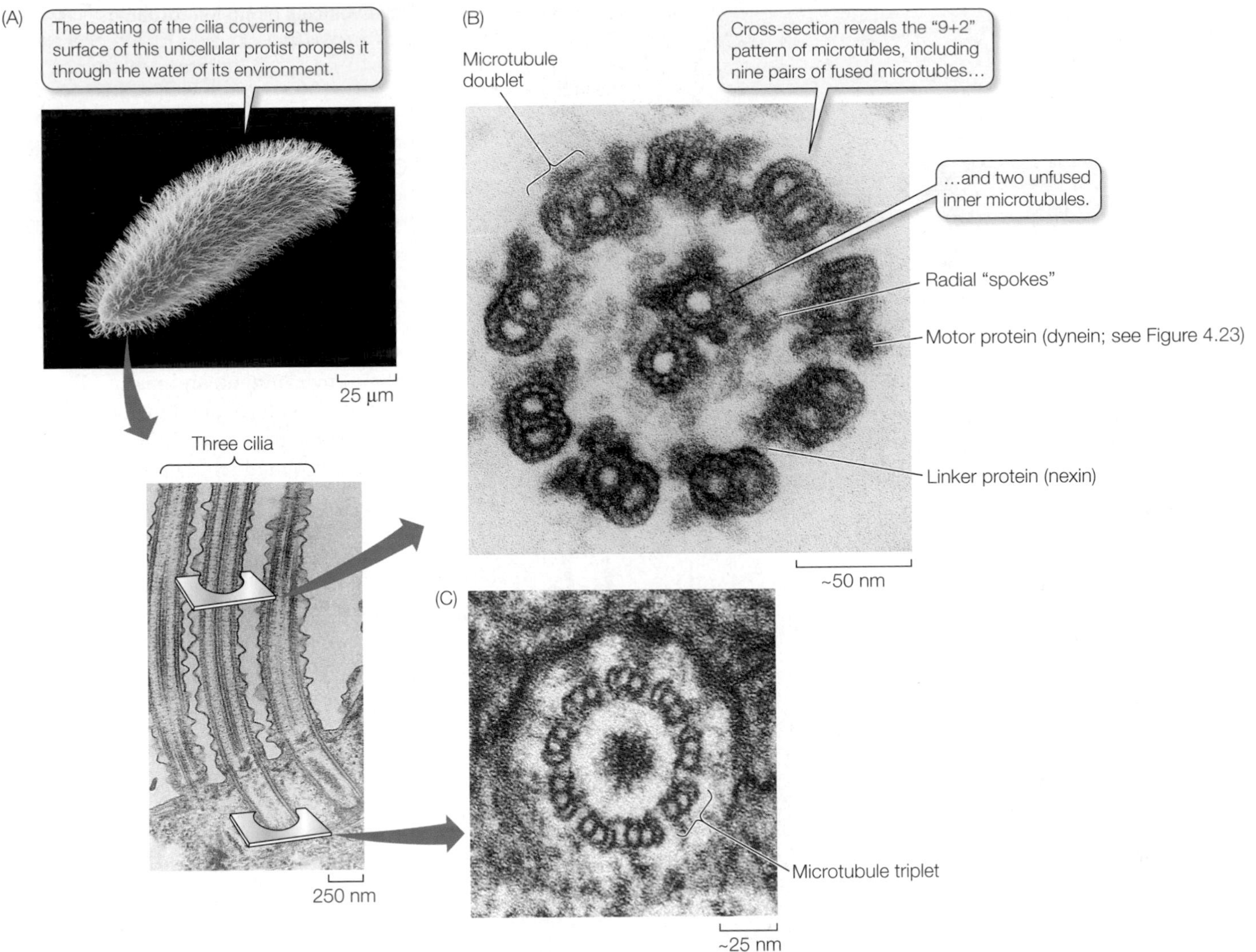
(A)
The beating of the cilia covering the surface of this unicellular protist propels it through the water of its environment.
25 μm
Three cilia
250 nm
(B)
Microtubule doublet
Cross-section reveals the "9+2" pattern of microtubles, including nine pairs of fused microtubles…
…and two unfused inner microtubules.
Radial "spokes"
Motor protein (dynein; see Figure 4.23)
Linker protein (nexin)
~50 nm
(C)
Microtubule triplet
~25 nm

4.22 Sliding Microtubules Cause Cilia to Bend (A) This unicellular eukaryotic organism (a ciliate protist) can coordinate the beating of its cilia, allowing rapid movement. (B) A cross section of a single cilium shows the arrangement of the microtubules and proteins. (C) Cross section of a basal body.

one microtubule of each doublet and connects the doublet to the center of the structure.

In the cytoplasm at the base of every eukaryotic flagellum or cilium is an organelle called a *basal body*. The nine microtubule doublets extend into the basal body. In the basal body, each doublet is accompanied by another microtubule, making nine sets of three microtubules. The central, unfused microtubules in the cilium do not extend into the basal body (**Figure 4.22C**).

Centrioles are almost identical to the basal bodies of cilia and flagella. Centrioles are found in the microtubule organizing centers of all eukaryotes except the seed plants and some protists. Under the light microscope, a centriole looks like a small, featureless particle, but the electron microscope reveals that it is made up of a precise bundle of microtubules arranged as nine sets of three fused microtubules each. Centrioles are involved in the formation of the mitotic spindle, to which the chromosomes attach (see Figure 9.9).

MOTOR PROTEINS The nine microtubule doublets of cilia and flagella are linked by proteins. The motion of cilia and flagella results from the sliding of the microtubule doublets past each other. This sliding is driven by a motor protein called *dynein*, which can change its three-dimensional shape. All motor proteins work by undergoing reversible shape changes powered by energy from ATP. Dynein molecules attached to one microtubule doublet bind to a neighboring doublet. As the dynein molecules change shape, they move the microtubule doublet past its neighbor (**Figure 4.23A**).

Another motor protein, *kinesin*, carries protein-laden vesicles from one part of the cell to another. Kinesin and similar motor proteins bind to a vesicle or other organelle, then "walk" it along a microtubule by changing their shape. This process is driven by a protein region of about 350 amino acids that is similar to myosin, the motor protein that binds to and moves actin microfilaments in muscle. Recall that microtubules have a plus end and a minus end. Dynein moves attached organelles toward the minus end, while kinesin moves them toward the plus end (**Figure 4.23B**).

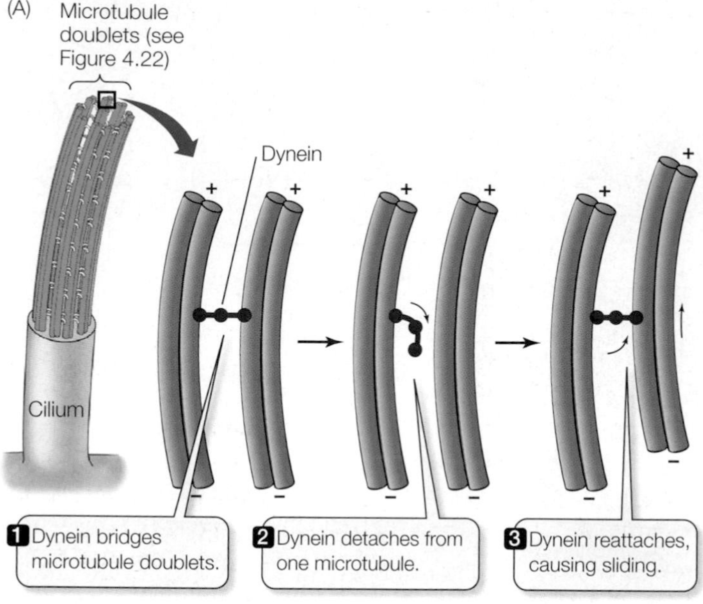
(A)
Microtubule doublets (see Figure 4.22)
Dynein
Cilium
1 Dynein bridges microtubule doublets.
2 Dynein detaches from one microtubule.
3 Dynein reattaches, causing sliding.

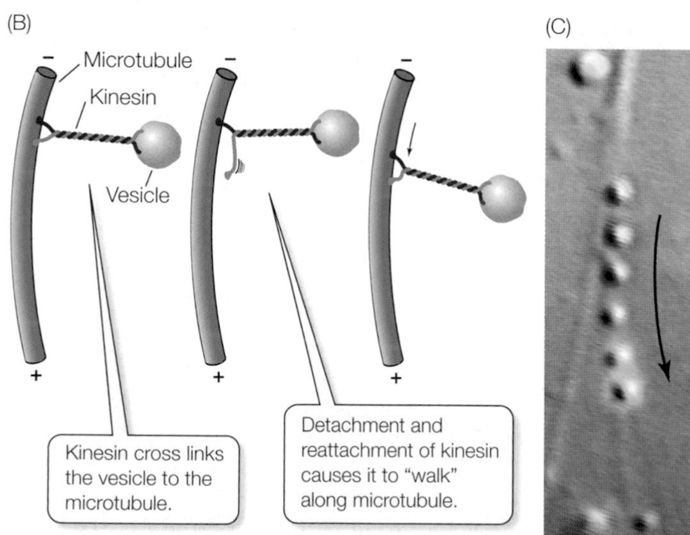
(B)
Microtubule
Kinesin
Vesicle
Kinesin cross links the vesicle to the microtubule.
Detachment and reattachment of kinesin causes it to "walk" along microtubule.
(C)

4.23 Motor Proteins Drive Vesicles along Microtubules (A) The motor protein dynein allows microtubule doublets to slide past one another in a cilium or flagellum. (B) Kinesin delivers vesicles or organelles to various parts of the cell by moving along microtubule "railroad tracks." Kinesin moves things from the minus toward the plus end of a microtubule; dynein works similarly, but moves from the plus toward the minus end. (C) Powered by kinesin, a vesicle (color) moves along a microtubule track in the protist *Dictyostelium.* The time sequence (time-lapse micrography at half-second intervals) is shown by the color gradient of purple to blue.

4.3 RECAP

The hallmark of eukaryotic cells is compartmentalization. Membrane-enclosed organelles process information, transform energy, form internal compartments for transporting proteins, and carry out intracellular digestion. An internal cytoskeleton plays several structural roles.

- What are some advantages of organelle compartmentalization? See pp. 74–75
- What are the differences between rough and smooth endoplasmic reticulum? See pp. 79–80 and Figure 4.10
- Describe how motor proteins and microtubules move materials within the cell. See pp. 88–90 and Figure 4.23

All cells interact with their environment, and many eukaryotic cells are part of a multicellular organism and must interact with other cells. The plasma membrane plays a crucial role in these interactions, but other structures outside that membrane are involved as well.

4.4 What Are the Roles of Extracellular Structures?

Although the plasma membrane is the functional barrier between the inside and the outside of a cell, many structures are produced by cells and secreted to the outside of the plasma membrane, where they play essential roles in protecting, supporting, or attaching cells. Because they are outside the plasma membrane, these structures are said to be *extracellular*. The peptidoglycan cell wall of bacteria is an example of such an extracellular structure. In eukaryotes, other extracellular structures—the cell walls of plants and the extracellular matrices found between the cells of animals—play similar roles. Both of these structures are made up of two components: a prominent fibrous macromolecule, and a gel-like medium.

The plant cell wall is an extracellular structure

The **cell wall** of plant cells is a semirigid structure outside the plasma membrane (**Figure 4.24**). It consists of cellulose fibers (see Figure 3.16) embedded in other complex polysaccharides and proteins. The cell wall has three major roles in plants:

- It provides support for the cell and limits its volume by remaining rigid.
- It acts as a barrier to infection by fungi and other organisms that can cause plant diseases.
- It contributes to plant form by growing as plant cells expand.

Because of their thick cell walls, plant cells viewed under a light microscope appear to be entirely isolated from one another. But electron microscopy reveals that this is not the case. The cytoplasm of adjacent plant cells is connected by numerous plasma membrane–lined channels, called **plasmodesmata**, that are about 20–40 nm in diameter and extend through the walls of adjoining cells (see Figure 15.20). Plasmodesmata permit the diffusion of water, ions, small molecules, and RNA and proteins between connected cells, ensuring uniform concentrations of these substances.

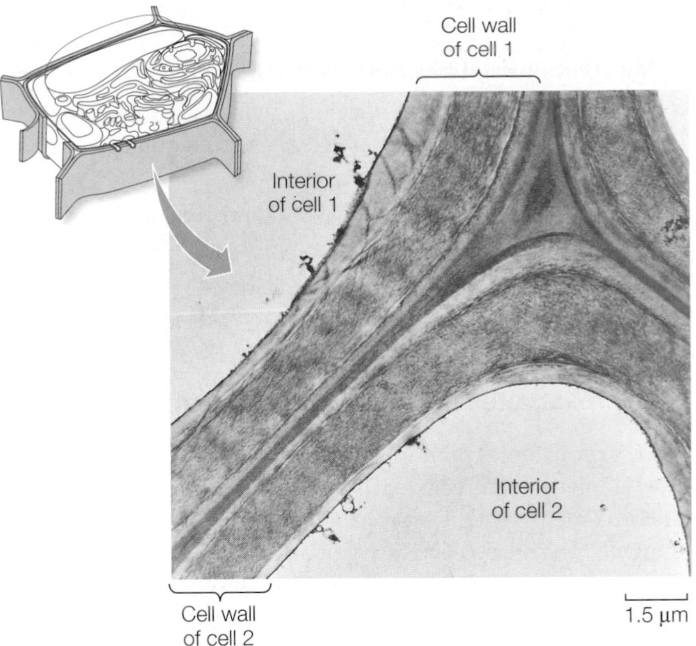

4.24 The Plant Cell Wall The semirigid cell wall provides support for plant cells.

The extracellular matrix supports tissue functions in animals

The cells of animals lack the semirigid cell wall that is characteristic of plant cells, but many animal cells are surrounded by, or are in contact with, an **extracellular matrix**. This matrix is composed of fibrous proteins such as **collagen** (the most abundant protein in mammals, constituting 25 percent of the protein in the human body), a matrix of glycoproteins termed **proteoglycans**, consisting primarily of sugars, and a third group of proteins that link the fibrous proteins and gel-like proteoglycan matrix together (**Figure 4.25**). These proteins, along with other substances that are specific to certain body tissues, are secreted by cells that are present in or near the matrix.

The functions of the extracellular matrix are many:

- It holds cells together in tissues.
- It contributes to the physical properties of cartilage, skin, and other tissues.
- It helps filter materials passing between different tissues.
- It helps orient cell movements during embryonic development and during tissue repair.
- It plays a role in chemical signaling from one cell to another.

In the human body, some tissues, such as those in the brain, have very little extracellular matrix; other tissues, such as bone and cartilage, have large amounts. Bone cells are embedded in an extracellular matrix that consists primarily of collagen and calcium phosphate. This matrix gives bone its familiar rigidity. Epithelial cells, which line body cavities, lie together in a sheet spread over a *basal lamina*, or *basement membrane*, a form of extracellular matrix (see Figure 4.25).

Some extracellular matrices are made up, in part, of an enormous proteoglycan. A single molecule of this proteoglycan consists of many hundreds of polysaccharide chains covalently attached to about a hundred proteins, all of which are attached in turn to one enormous polysaccharide. The molecular weight of this proteoglycan can exceed 100 million Da; the molecule takes up as much space as an entire prokaryotic cell.

4.4 RECAP

Extracellular structures are produced by cells and secreted outside the plasma membrane. Most are composed of a fibrous component in a gel-like medium.

- Do you understand the functions of the cell wall in plants and the extracellular matrix in animals?

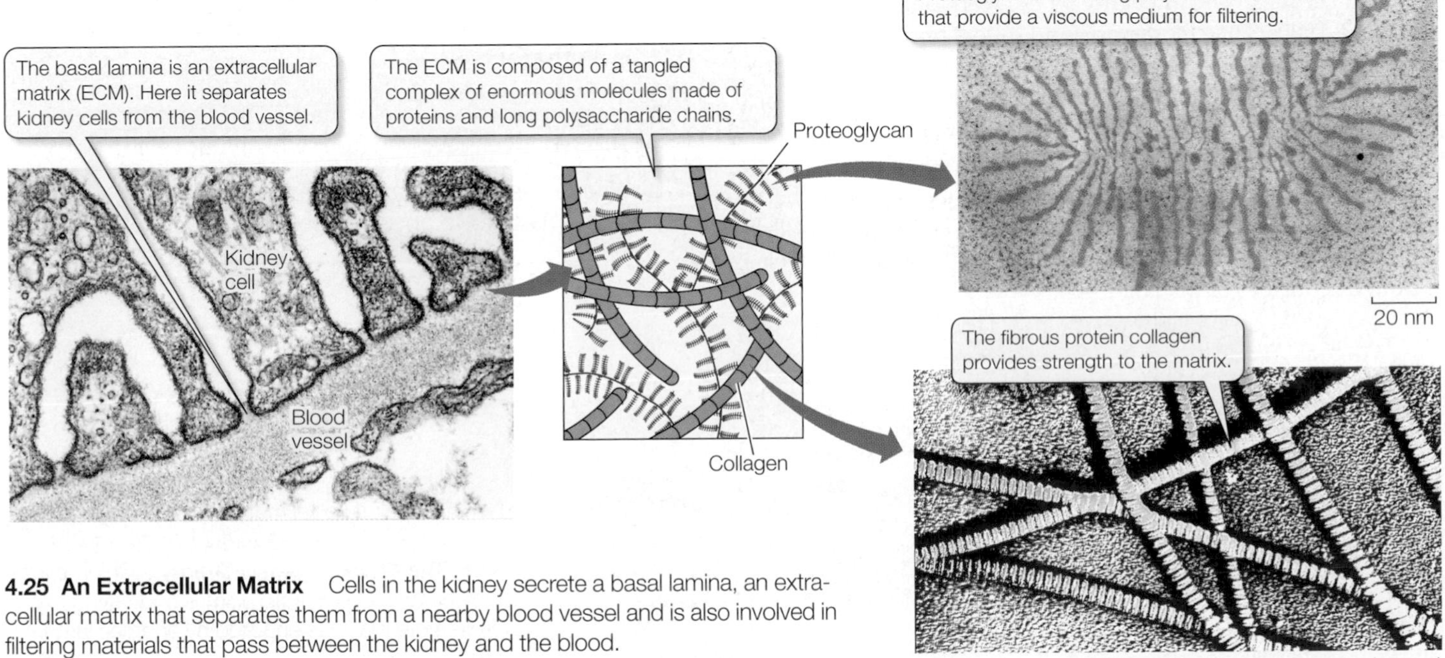

The basal lamina is an extracellular matrix (ECM). Here it separates kidney cells from the blood vessel.
The ECM is composed of a tangled complex of enormous molecules made of proteins and long polysaccharide chains.
Proteoglycans have long polysaccharide chains that provide a viscous medium for filtering.
Proteoglycan
Kidney cell
Blood vessel
Collagen
20 nm
The fibrous protein collagen provides strength to the matrix.
100 nm

4.25 An Extracellular Matrix Cells in the kidney secrete a basal lamina, an extracellular matrix that separates them from a nearby blood vessel and is also involved in filtering materials that pass between the kidney and the blood.

4.5 How Did Eukaryotic Cells Originate?

We began this chapter by describing some of the earliest evidence of prokaryotic cells on Earth, dating to about 3.5 billion years ago. The living world was entirely prokaryotic until about 1.5 billion years ago, when eukaryotic cells first appeared. This was a major event in the history of life, as the compartmentalization that is the hallmark of eukaryotic cells permitted new biochemical functions to evolve much more readily.

The endosymbiosis theory suggests how eukaryotes evolved

The first prokaryotes probably absorbed their food directly from the environment. Then some became photosynthetic (see Figure 1.10). Others, however, fed on smaller prokaryotes by engulfing them (**Figure 4.26**). Now suppose that a small, photosynthetic prokaryote was ingested by a larger one, but was not digested. Instead, the photosynthetic organism survived, trapped within the cytoplasm of the larger cell. Suppose further that the ingested prokaryote divided at about the same rate as the larger one, so successive generations of the larger cell contained the offspring of the smaller one. Such **endosymbiosis** (*endo*, "within"; *symbiosis*, "living together") would have benefited both partners: the smaller cell would have provided the larger one with monosaccharides made by photosynthesis, and the larger cell would have protected the smaller one. Over evolutionary time, the photosynthetic prokaryote might have evolved into the modern chloroplast.

Similar evidence and arguments support the proposition that mitochondria are the descendants of respiring prokaryotes engulfed by larger prokaryotes. The benefit of such an endosymbiotic relationship might have been the capacity of the engulfed prokaryote to detoxify molecular oxygen (O_2), which was increasing in Earth's atmosphere as a result of photosynthesis.

This **endosymbiosis theory** of eukaryote evolution was first suggested as early as the nineteenth century, but it was expounded and given substance and credence by the work of Lynn Margulis during the 1980s. The chloroplasts and mitochondria of eukaryotes are about the size of prokaryotic cells. These organelles also contain their own DNA and ribosomes and synthesize some of their own components. They are not independent of control by the nucleus, however. The vast majority of their proteins are encoded by nuclear DNA and synthesized in the cell's cytoplasm, then imported into the organelle. It is possible that over time the smaller prokaryotes that gave rise to organelles gradually lost much of their DNA to the nucleus of the larger cell.

Much circumstantial evidence favors the endosymbiosis theory:

- On an evolutionary time scale (i.e., millions of years), there is evidence of the movement of DNA between organelles in the eukaryotic cell.
- There are many biochemical similarities between chloroplasts and photosynthetic bacteria.
- DNA sequencing shows strong similarities between modern chloroplast DNA and that of a photosynthetic prokaryote.

Both prokaryotes and eukaryotes continue to evolve

Today's prokaryotes contain a number of the structures that are present in eukaryotic cells, such as a cytoskeleton, ribosomes, and a plasma membrane. It is probable that these structures evolved gradually. The shared chemistry of all living things, described in Chapter 3, suggests that eukaryotic cells evolved from prokaryotes. For example, both prokaryotes and eukaryotes:

- use nucleic acids as their genetic material;
- use the same 20 amino acids in their proteins; and
- use D sugars and L amino acids.

As described on page 73, many modern prokaryotes have folds in their internal membranes. It is believed that over evolutionary time such folds gradually closed off to form compartments, leading to the formation of the endoplasmic reticulum, Golgi apparatus, and lysosomes. Today's prokaryotic cells have their genetic material in a central region called the nucleoid (see Figure 4.4). During cell reproduction, this DNA becomes attached to the plasma membrane. If the membrane, aided by microfilaments, had folded over the DNA and sealed itself, the cell would have successfully compartmentalized its DNA in a nucleus.

4.5 RECAP

Eukaryotic cells arose long after prokaryotic cells. The endosymbiosis theory suggests how eukaryotic cells may have evolved from prokaryotic ancestors.

- Can you describe some of the evidence that suggests eukaryotic cells evolved from prokaryotic cells?
- What evidence suggests that mitochondria and chloroplasts were once independent prokaryotic cells?

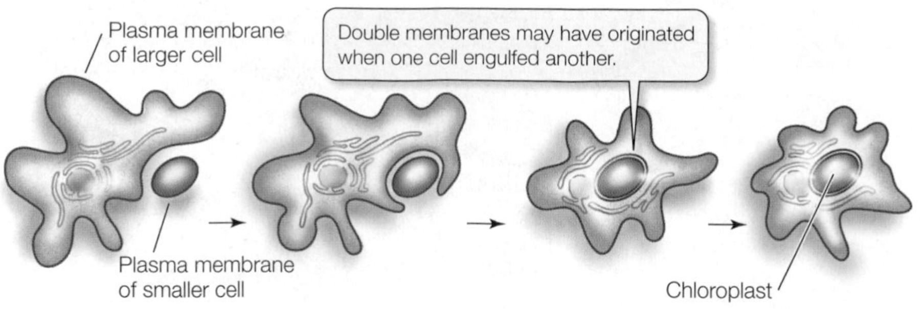

4.26 The Endosymbiosis Theory Chloroplasts may be descended from a small photosynthetic prokaryote that was engulfed by another, larger cell.

The fluidity of the plasma membrane undoubtedly played a major role in the compartmentalization of eukaryotic cells. Microfilaments of the cytoskeleton form a scaffold supporting the membrane, but recall that these microfilaments can contract. Contracting microfilaments attached to a membrane can deform and reshape it, like a rubber band. The double membrane that encloses mitochondria and chloroplasts could also have arisen through endosymbiosis. The outer membrane may have come from the engulfing cell's plasma membrane and the inner membrane from the engulfed cell's plasma membrane. This structure—the plasma membrane that surrounds and contains the cell—is so interesting and important that we devote all of Chapter 5 to explaining it.

CHAPTER SUMMARY

4.1 What features of cells make them the fundamental unit of life?

See Web/CD Activities 4.1 and 4.2

All cells come from preexisting cells.

Cells are small because a cell's **surface area** must be large compared with its **volume** to accommodate exchanges with its environment. Review Figure 4.2

All cells are enclosed by a selectively permeable plasma membrane that separates them from the external environment.

While certain biochemical processes, molecules, and structures are shared by all kinds of cells, two major categories of cells—**prokaryotes** and **eukaryotes**—are easily distinguished.

4.2 What are the characteristics of prokaryotic cells?

Prokaryotic cells have no internal compartments, but have a **nucleoid** region containing DNA and cytoplasm containing **cytosol**, **ribosomes**, proteins, and small molecules. Some prokaryotes have additional protective structures, including a cell wall, an outer membrane, and a capsule. Review Figure 4.4

Some prokaryotes have folded membranes that may be photosynthetic membranes, and some have flagella or pili. Review Figure 4.5

4.3 What are the characteristics of eukaryotic cells?

See Web/CD Tutorial 4.1

Eukaryotic cells are larger than prokaryotic cells and contain many membrane-enclosed **organelles**. The membranes that envelop organelles ensure compartmentalization of their functions. Review Figure 4.6

The **nucleus** contains most of the cell's DNA and controls protein synthesis. **Ribosomes** are sites of protein synthesis. Review Figure 4.8

The **endomembrane system**—consisting of the **endoplasmic reticulum** and **Golgi apparatus**—is a series of interrelated compartments enclosed by membranes. It segregates proteins and modifies them. **Lysosomes** contain many digestive enzymes. Review Figures 4.10, 4.11, and 4.12, Web/CD Activity 4.3, Web/CD Tutorial 4.2

Mitochondria and chloroplasts process energy. **Mitochondria** are present in most eukaryotic organisms and contain the enzymes needed for cellular respiration. The cells of photosynthetic eukaryotes contain **chloroplasts** that harvest light energy for photosynthesis. Review Figures 4.13 and 4.14

Vacuoles are prominent in many plant cells and consist of a membrane-enclosed compartment full of water and dissolved substances.

The **microfilaments**, **intermediate filaments**, and **microtubules** of the **cytoskeleton** provide the cell with shape, strength, and movement.

4.4 What are the roles of extracellular structures?

The plant **cell wall** consists principally of cellulose. Cell walls are pierced by **plasmodesmata** that join the cytoplasms of adjacent cells.

In animals, the **extracellular matrix** consists of different kinds of proteins, including **collagen** and **proteoglycans**. Review Figure 4.25

4.5 How did eukaryotic cells originate?

The **endosymbiosis theory** states that organelles such as mitochondria and chloroplasts originated when larger prokaryotes engulfed, but did not digest, smaller prokaryotes. Mutual benefits permitted this symbiotic relationship to be maintained, allowing the smaller cells to evolve into the eukaryotic organelles observed today. Review Figure 4.26

Infoldings of the plasma membrane could have led to the formation of some membrane-enclosed organelles, such as the endomembrane system and nucleus.

SELF-QUIZ

1. Which structure is present in both prokaryotic cells and eukaryotic plant cells?
   a. Chloroplasts
   b. Cell wall
   c. Nucleus
   d. Mitochondria
   e. Microtubules
2. The major factor limiting cell size is the
   a. concentration of water in the cytoplasm.
   b. need for energy.
   c. presence of membrane-enclosed organelles.
   d. ratio of surface area to volume.
   e. composition of the plasma membrane.
3. Which statement about mitochondria is *not* true?
   a. The inner mitochondrial membrane folds to form cristae.
   b. Mitochondria are usually 1 μm or smaller in diameter.
   c. Mitochondria are green because they contain chlorophyll.
   d. Fuel molecules from the cytosol are oxidized in mitochondria.
   e. ATP is synthesized in mitochondria.
4. Which statement about plastids is *true*?
   a. They are found in prokaryotes.
   b. They are surrounded by a single membrane.
   c. They are the sites of cellular respiration.
   d. They are found only in fungi.
   e. They contain several types of pigments or polysaccharides.

5. If all the lysosomes within a cell suddenly ruptured, what would be the most likely result?
   a. The macromolecules in the cytosol would begin to break down.
   b. More proteins would be made.
   c. The DNA within mitochondria would break down.
   d. The mitochondria and chloroplasts would divide.
   e. There would be no change in cell function.
6. The Golgi apparatus
   a. is found only in animals.
   b. is found in prokaryotes.
   c. is the appendage that moves a cell around in its environment.
   d. is a site of rapid ATP production.
   e. modifies and packages proteins.
7. Which organelle is *not* surrounded by one or more membranes?
   a. Ribosome
   b. Chloroplast
   c. Mitochondrion
   d. Peroxisome
   e. Vacuole
8. The cytoskeleton consists of
   a. cilia, flagella, and microfilaments.
   b. cilia, microtubules, and microfilaments.
   c. internal cell walls.
   d. microtubules, intermediate filaments, and microfilaments.
   e. calcified microtubules.
9. Microfilaments
   a. are composed of polysaccharides.
   b. are composed of actin.
   c. allow cilia and flagella to move.
   d. make up the spindle that aids the movement of chromosomes.
   e. maintain the position of the chloroplast in the cell.
10. Which statement about the plant cell wall is *not* true?
   a. Its principal chemical components are polysaccharides.
   b. It lies outside the plasma membrane.
   c. It provides support for the cell.
   d. It completely isolates adjacent cells from one another.
   e. It is semirigid.

FOR DISCUSSION

1. The drug vincristine is used to treat many cancers. It apparently works by causing microtubules to depolymerize. Vincristine use has many side effects, including loss of dividing cells and nerve problems. Explain why this might be so.
2. Through how many membranes would a molecule have to pass in moving from the interior of a chloroplast to the interior of a mitochondrion? From the interior of a lysosome to the outside of a cell? From one ribosome to another?
3. How does the possession of double membranes by chloroplasts and mitochondria relate to the endosymbiosis theory of the origins of these organelles? What other evidence supports the theory?
4. Compare the extracellular matrix of the animal cell with the plant cell wall with respect to composition of the fibrous and nonfibrous components, rigidity, and connectivity of cells.

FOR INVESTIGATION

The pathway of newly synthesized proteins through the cell can be followed by a "pulse-chase" experiment. Proteins are tagged with a radioactive isotope (the "pulse"), and the cell is allowed to process them for varying periods of time. The locations of the radioactive proteins are then determined by isolating cell organelles and quantifying their radioactivity. How would you use this method, and what results would you expect, for (A) a lysosomal enzyme? (B) a protein that is released from the cell?

CHAPTER 5 The Dynamic Cell Membrane

Disaster at the plasma membrane

On their first date, Paul and Ann shared an exotic lunch that included canned palm fruit. The next morning, they were suddenly stricken with the worst diarrhea, nausea, and vomiting either had ever experienced. They arrived at the hospital in shock, clinging to life with very low blood pressure and irregular heartbeat. Analysis of their stool samples quickly revealed the reason for their illness: *Vibrio cholerae*, the bacterium that causes cholera. Their infection was traced to the palm fruit, which was canned in El Salvador—a known location for the bacterium ever since a cholera epidemic devastated South and Central America in the early 1990s. Cholera spreads when an uninfected person ingests the bacterium, usually in water contaminated with the feces of infected individuals. In this case, contaminated water may have been used to irrigate the palm fruit.

Until late in the nineteenth century, people had no idea what caused this dreaded disease, nor did they understand exactly how it spread. Then, in 1854, the physician John Snow traced a cholera outbreak in central London to a single water pump. Once he removed the pump's handle, the outbreak subsided. Some 30 years later, in Berlin, Robert Koch examined cholera-contaminated water under a microscope and isolated *Vibrio cholerae*. These seemingly simple discoveries linking a disease to a specific cause helped usher in the medical science of *epidemiology*, the study of how diseases spread among populations and how they can be controlled.

Our current knowledge allows us to explain even more about cholera, including the physical events that occurred after Ann and Paul ate the contaminated palm fruit. Most of the bacteria they ingested died in the acidic environment of the stomach. But a few survived, bound themselves to the plasma membranes of cells in the small intestine, and released a toxic protein. This toxin entered healthy cells and did two things to their plasma membranes. It inactivated certain membrane proteins that "pump" sodium ions into the cell and, at the same time, it opened channels in the membrane that are normally closed, allowing chloride ions to escape from inside the cell into the intestine.

Healthy plasma membranes enclosing intestinal cells act as "gatekeepers," maintaining higher concentrations of Na^+ and Cl^- *inside* the cells than *outside*, in the intestine. This necessary imbalance was overturned, and Na^+ and Cl^- left the cells and accumulated in the intestine. Through a process called osmosis, this reversed

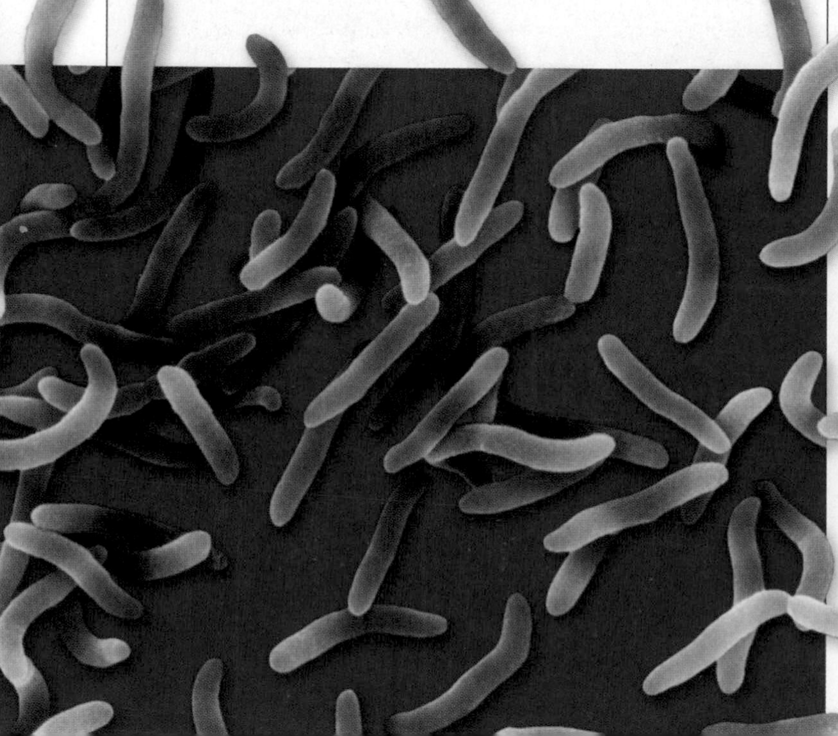

Vibrio cholerae This bacterium causes alterations in the membranes of cells in the human intestine, leading to cholera, a life-threatening but treatable disease.

Treating Cholera In the early 1990s, a cholera epidemic broke out in Peru and spread to other countries in Central and South America, leading to over a million cases in 3 years. Here a medic from the World Health Organization administers oral rehydration therapy to a Peruvian child.

ion imbalance caused water to be pulled out of other body cells and into the intestine, resulting in severe diarrhea and potentially fatal dehydration. Fortunately, treating cholera is straightforward: doctors administer oral rehydration therapy to replace the lost ions and water. Paul and Ann were given quantities of a specially balanced solution of NaCl (salt) and glucose (sugar), and both recovered fully.

Cholera is a serious threat in regions where sanitation is inadequate, and whenever a disaster (such as a hurricane or an earthquake) disrupts water supplies. But rehydration therapy is inexpensive, safe, rapid, and effective—and it is all the result of knowing how biological membranes work.

IN THIS CHAPTER we focus on the structure and functions of biological membranes. Membranes are dynamic structures that perform their vital physiological roles by allowing cells to interact with other cells and with molecules in the environment. We describe the structural aspects of those interactions here. Membranes also regulate which molecules and ions enter and leave the cell. The selective permeability of membranes is an important characteristic of life.

CHAPTER OUTLINE

5.1 What Is the Structure of a Biological Membrane?

The physical organization and functioning of all biological membranes depend on their constituents: lipids, proteins, and carbohydrates. The lipids establish the physical integrity of the membrane and create an effective barrier to the rapid passage of hydrophilic materials such as water and ions. In addition, the phospholipid bilayer serves as a lipid "lake" in which a variety of proteins "float" (**Figure 5.1**). This general design is known as the **fluid mosaic model**.

The proteins embedded in the phospholipid bilayer have a number of functions, including moving materials through the membrane and receiving chemical signals from the cell's external environment. Each membrane has a set of proteins suitable to the specialized function of the cell or organelle it surrounds.

The carbohydrates associated with membranes are attached either to the lipids or to protein molecules. In plasma membranes, they are located on the outside, where they protrude into the environment, away from the cell. Like some of the proteins, carbohydrates are crucial in recognizing specific molecules.

Lipids constitute the bulk of a membrane

The lipids in biological membranes are usually phospholipids. Recall from Section 2.2 that some compounds are hydrophilic ("water-loving") and others are hydrophobic ("water-hating"), and from Section 3.4 that a phospholipid molecule has regions of both kinds:

- *Hydrophilic regions*: The phosphorus-containing "head" of the phospholipid is electrically charged and hence associates with polar water molecules.
- *Hydrophobic regions*: The long, nonpolar fatty acid "tails" of the phospholipid associate with other nonpolar materials, but they do not dissolve in water or associate with hydrophilic substances.

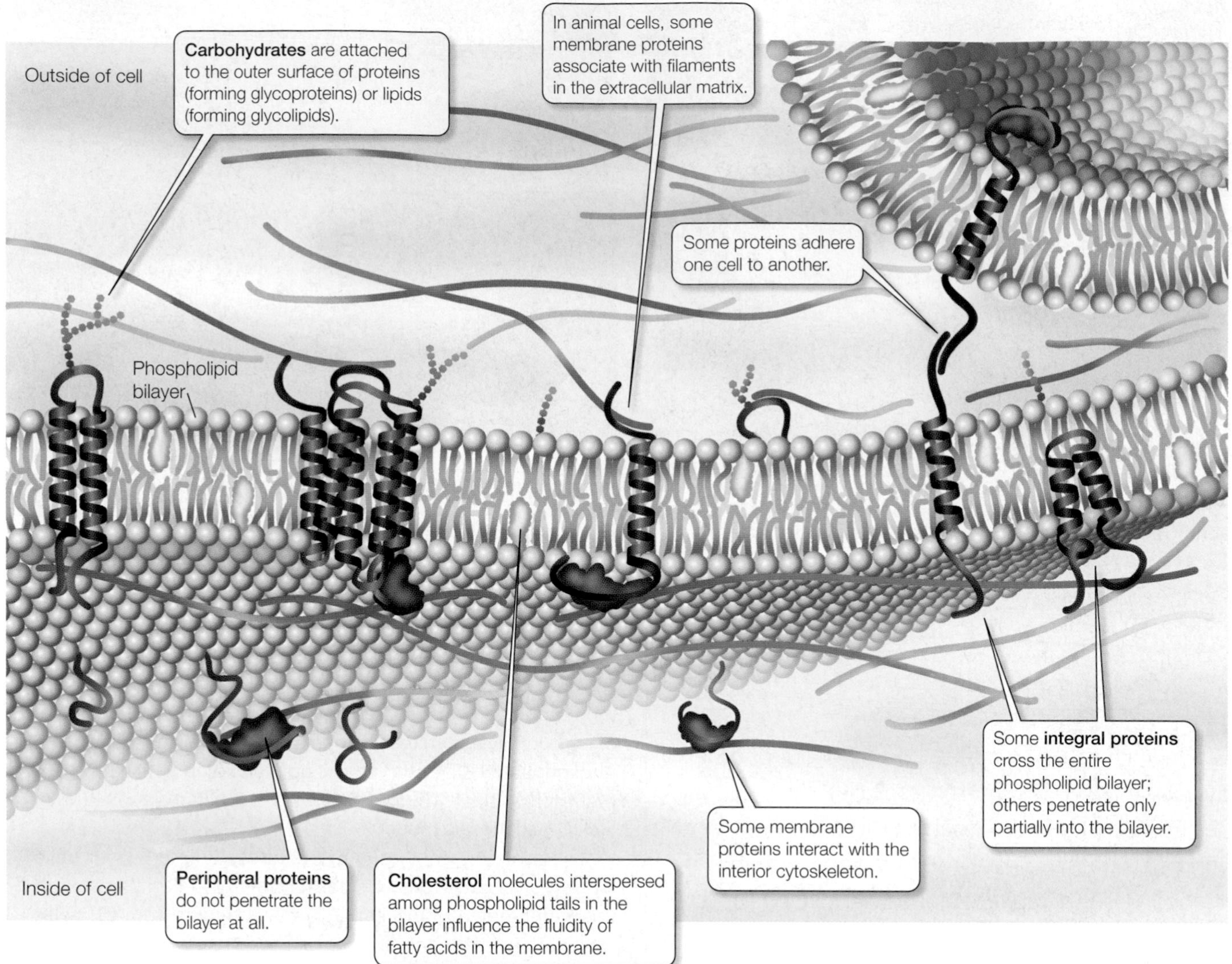
Outside of cell
Carbohydrates are attached to the outer surface of proteins (forming glycoproteins) or lipids (forming glycolipids).
In animal cells, some membrane proteins associate with filaments in the extracellular matrix.
Some proteins adhere one cell to another.
Phospholipid bilayer
Some **integral proteins** cross the entire phospholipid bilayer; others penetrate only partially into the bilayer.
Some membrane proteins interact with the interior cytoskeleton.
Inside of cell
Peripheral proteins do not penetrate the bilayer at all.
Cholesterol molecules interspersed among phospholipid tails in the bilayer influence the fluidity of fatty acids in the membrane.

5.1 The Fluid Mosaic Model The general molecular structure of biological membranes is a continuous phospholipid bilayer in which proteins are embedded. (The purple "ribbons" represent structural elements of the cell and extracellular matrix.)

Because of these properties, one way in which phospholipids can coexist with water is to form a *bilayer*, with the fatty acid "tails" of the two layers interacting with each other and the polar "heads" facing the outside aqueous environment (**Figure 5.2**).

In the laboratory, it is easy to make artificial bilayers with the same organization as natural membranes. In addition, small holes in a phospholipid bilayer seal themselves spontaneously. This capacity of lipids to associate with one another and maintain a bilayer organization helps biological membranes fuse during vesicle formation, phagocytosis, and related processes.

All biological membranes have a similar structure, but membranes from different cells or organelles may differ greatly in their lipid *composition*:

- Phospholipids can differ in terms of fatty acid chain length, degree of unsaturation (double bonds) in the fatty acids, and the polar (phosphate-containing) groups present.
- Up to 25 percent of the lipid content of a membrane may be cholesterol. When present, cholesterol is important to membrane integrity, and most cholesterol in membranes is not hazardous to your health. A molecule of cholesterol is commonly situated next to an unsaturated fatty acid (see Figure 5.1).

You can rinse mud off your hands with plain water, but you can't get rid of grease that way because grease isn't water-soluble. Soap molecules have one end that is water-soluble and one end that is fat-soluble, making it possible to wash grease off with soap and water.

The phospholipid bilayer stabilizes the entire membrane structure, but leaves it flexible, not rigid. At the same time, the fatty acids of the phospholipids make the hydrophobic interior of the membrane somewhat *fluid*—about as fluid as lightweight machine oil. This fluidity permits some molecules to move laterally (side to side) within the plane of the membrane. A given phospholipid molecule in the plasma membrane can travel from one end of the cell to the

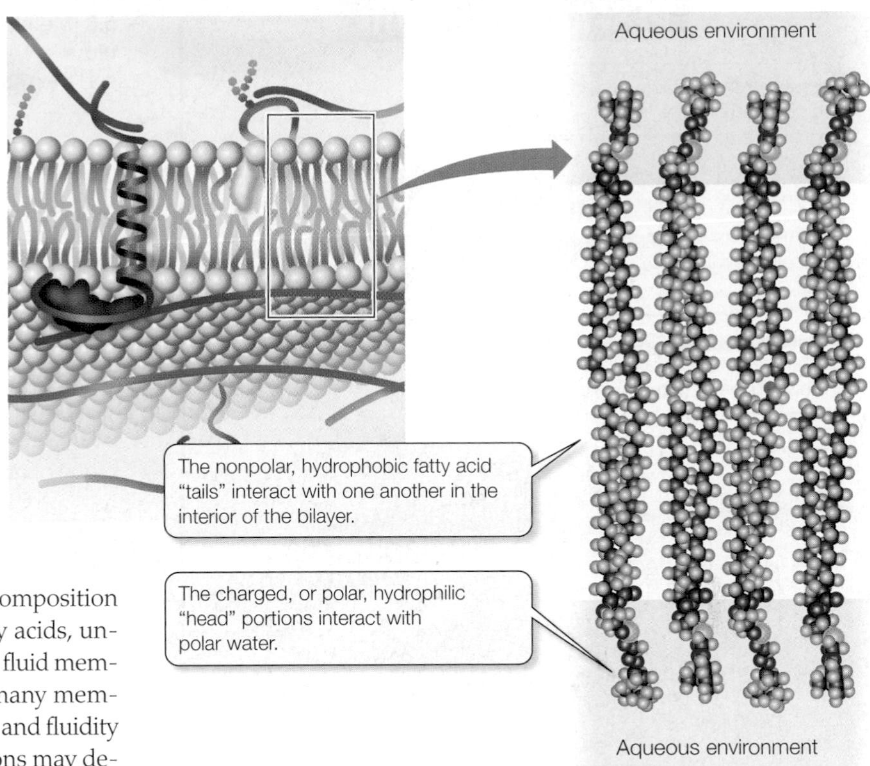

Aqueous environment
The nonpolar, hydrophobic fatty acid "tails" interact with one another in the interior of the bilayer.
The charged, or polar, hydrophilic "head" portions interact with polar water.
Aqueous environment

5.2 A Phospholipid Bilayer Separates Two Aqueous Regions The eight phospholipid molecules shown here represent a small cross section of a membrane bilayer.

other in a little more than a second! On the other hand, seldom does a phospholipid molecule in one half of the bilayer flip over to the other side and trade places with another phospholipid molecule. For such a swap to happen, the polar part of each molecule would have to move through the hydrophobic interior of the membrane. Since phospholipid flip-flops are rare, the inner and outer halves of the bilayer may be quite different in the kinds of phospholipids they contain.

The fluidity of a membrane is affected by its lipid composition and by its temperature. In general, shorter-chain fatty acids, unsaturated fatty acids, and less cholesterol result in more fluid membranes. Adequate membrane fluidity is essential for many membrane functions. Because molecules move more slowly and fluidity decreases at reduced temperatures, membrane functions may decline in organisms that cannot keep their bodies warm. To address this problem, some organisms simply change the lipid composition of their membranes under cold conditions, replacing saturated with unsaturated fatty acids and using fatty acids with shorter tails. Such changes play a part in the survival of plants and hibernating animals and bacteria during the winter.

Membrane proteins are asymmetrically distributed

All biological membranes contain proteins. Typically, plasma membranes have one protein molecule for every 25 phospholipid molecules. This ratio varies, however, depending on membrane function. In the inner membrane of the mitochondrion, which is specialized for energy processing, there is one protein for every 15 lipids. On the other hand, myelin, a membrane that encloses some neurons (nerve cells) and uses the properties of lipids to act as an electrical insulator, has only one protein per 70 lipids.

Many membrane proteins are embedded in, or extend across, the phospholipid bilayer (see Figure 5.1). Like phospholipids, these proteins have both hydrophilic and hydrophobic regions.

- *Hydrophilic regions*: Stretches of amino acids with hydrophilic side chains (see Table 3.2) give certain regions of the protein a polar character. Those regions, or *domains*, interact with water, sticking out into the aqueous extracellular environment or cytoplasm.
- *Hydrophobic regions*: Stretches of amino acids with hydrophobic side chains give other regions of the protein a nonpolar character. These domains interact with the fatty acid chains in the interior of the phospholipid bilayer, away from water.

A special preparation method for electron microscopy, called **freeze-fracturing**, reveals proteins embedded in the phospholipid bilayer of cellular membranes (**Figure 5.3**). The bumps that can be seen protruding from the interior of these membranes are not observed in pure lipid bilayers.

According to the fluid mosaic model, the proteins and lipids in a membrane are independent of each other and *interact only noncovalently*. The polar ends of proteins can interact with the polar ends of lipids, and the nonpolar regions of both molecules can interact hydrophobically.

There are two general types of membrane proteins:

- **Integral membrane proteins** have hydrophobic domains and penetrate the phospholipid bilayer. Many of these proteins have long, hydrophobic α-helical regions (see Section 3.2) that span the core of the bilayer. Their hydrophilic ends protrude into the aqueous environments on either side of the membrane (**Figure 5.4**).
- **Peripheral membrane proteins** lack hydrophobic domains and are not embedded in the bilayer. Instead, they have polar or charged regions that interact with similar regions on exposed parts of integral membrane proteins or with the polar heads of phospholipid molecules (see Figure 5.1).

Some membrane proteins are covalently attached to fatty acids or other lipid groups. These proteins can be classified as a special type of integral protein, as their hydrophobic lipid component allows them to insert themselves into the phospholipid bilayer.

Proteins are *asymmetrically distributed* on the inner and outer surfaces of a membrane. Integral membrane proteins that protrude on both sides of the membrane, known as **transmembrane proteins**, show different "faces" on the two membrane surfaces. Such proteins have certain specific domains on the outer side of the mem-

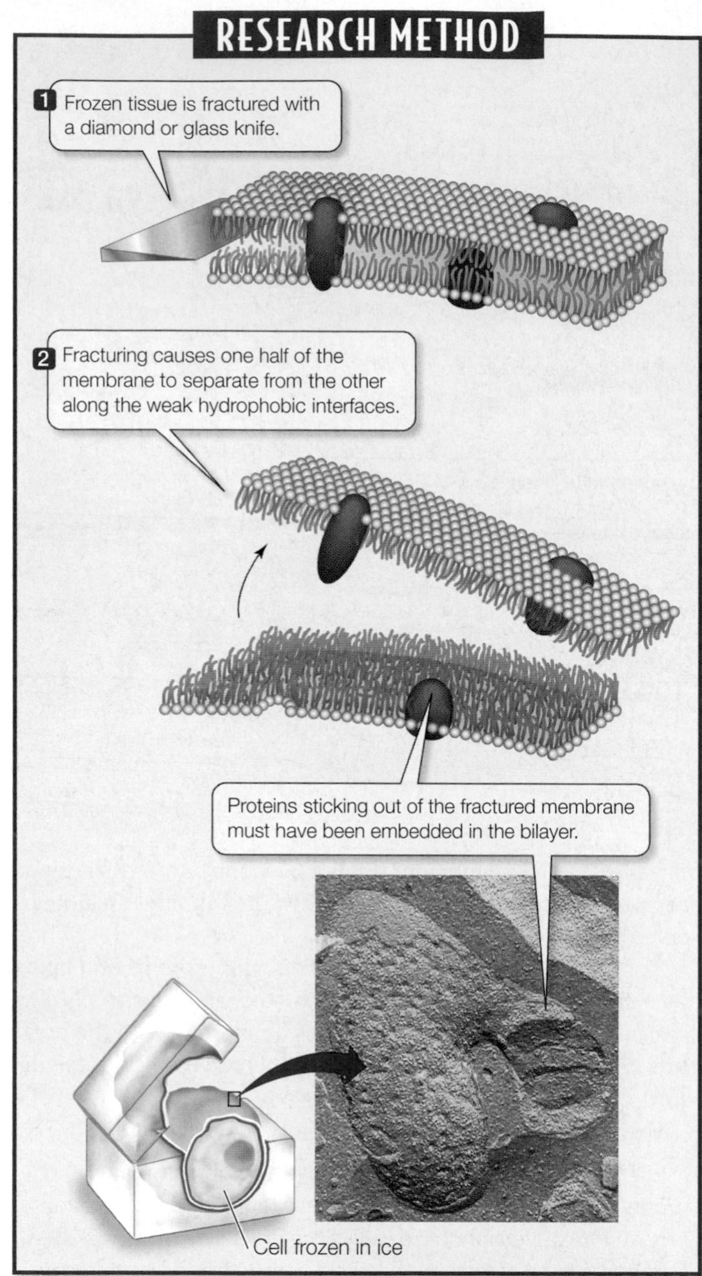

RESEARCH METHOD
1 Frozen tissue is fractured with a diamond or glass knife.
2 Fracturing causes one half of the membrane to separate from the other along the weak hydrophobic interfaces.
Proteins sticking out of the fractured membrane must have been embedded in the bilayer.
Cell frozen in ice

5.3 Membrane Proteins Revealed by the Freeze-Fracture Technique This membrane from a spinach chloroplast was first frozen and then separated so that the bilayer was split open.

brane, other domains within the membrane, and still other domains on the inner side of the membrane. Peripheral membrane proteins are localized on one side of the membrane or the other, but not both. This arrangement gives the two surfaces of the membrane different properties. As we will soon see, these differences have great functional significance.

Like lipids, many membrane proteins move around relatively freely within the phospholipid bilayer. Experiments using the technique of *cell fusion* illustrate this migration dramatically. When two cells are fused, a single continuous membrane forms and surrounds both cells, and some proteins from each cell distribute themselves uniformly around this membrane.

Although many proteins are free to migrate in the membrane, others are not, but rather appear to be "anchored" to a specific region of the membrane. These membrane regions are like a corral of horses on a farm: the horses are free to move around within the fenced area, but cannot get outside of it. For instance, the protein in the plasma membrane of a muscle cell that recognizes a chemical signal from neurons is normally found only at the site where a neuron meets the muscle cell. There are two ways in which the movement of proteins within a membrane can be restricted:

- The cytoskeleton may have components just below the inner face of the membrane that are attached to membrane proteins protruding into the cytoplasm.
- *Lipid rafts,* which are groups of lipids in a semisolid (not quite fluid) state, may trap proteins within a region. These lipids have a different composition than the surrounding phospholipids; for example, they may have very long fatty acid chains.

Membranes are dynamic

Membranes are constantly forming, transforming from one type to another, fusing with one another, and breaking down (**Figure 5.5**).

- In eukaryotes, *phospholipids* are synthesized on the surface of the smooth endoplasmic reticulum and are rapidly distributed to membranes throughout the cell.
- *Membrane proteins* are inserted into the rough endoplasmic reticulum as they form on ribosomes.
- *Functioning membranes* also move about within eukaryotic cells. Portions of the rough ER bud away as vesicles and join the *cis* face of the Golgi apparatus. Rapidly—often in less than an hour—these segments of membrane find themselves in

Hydrophilic R groups in exposed parts of the protein interact with aqueous environments.
Outside of cell (aqueous)
Hydrophobic interior of bilayer
Hydrophobic R groups interact with the hydrophobic core of the membrane, away from water.
Inside of cell (aqueous)

5.4 Interactions of Integral Membrane Proteins An integral membrane protein is held in the membrane by the distribution of the hydrophilic and hydrophobic side chains of its amino acids. The hydrophilic ends extend into the aqueous cell exterior and internal cytoplasm; the lipid core of the membrane is hydrophobic.

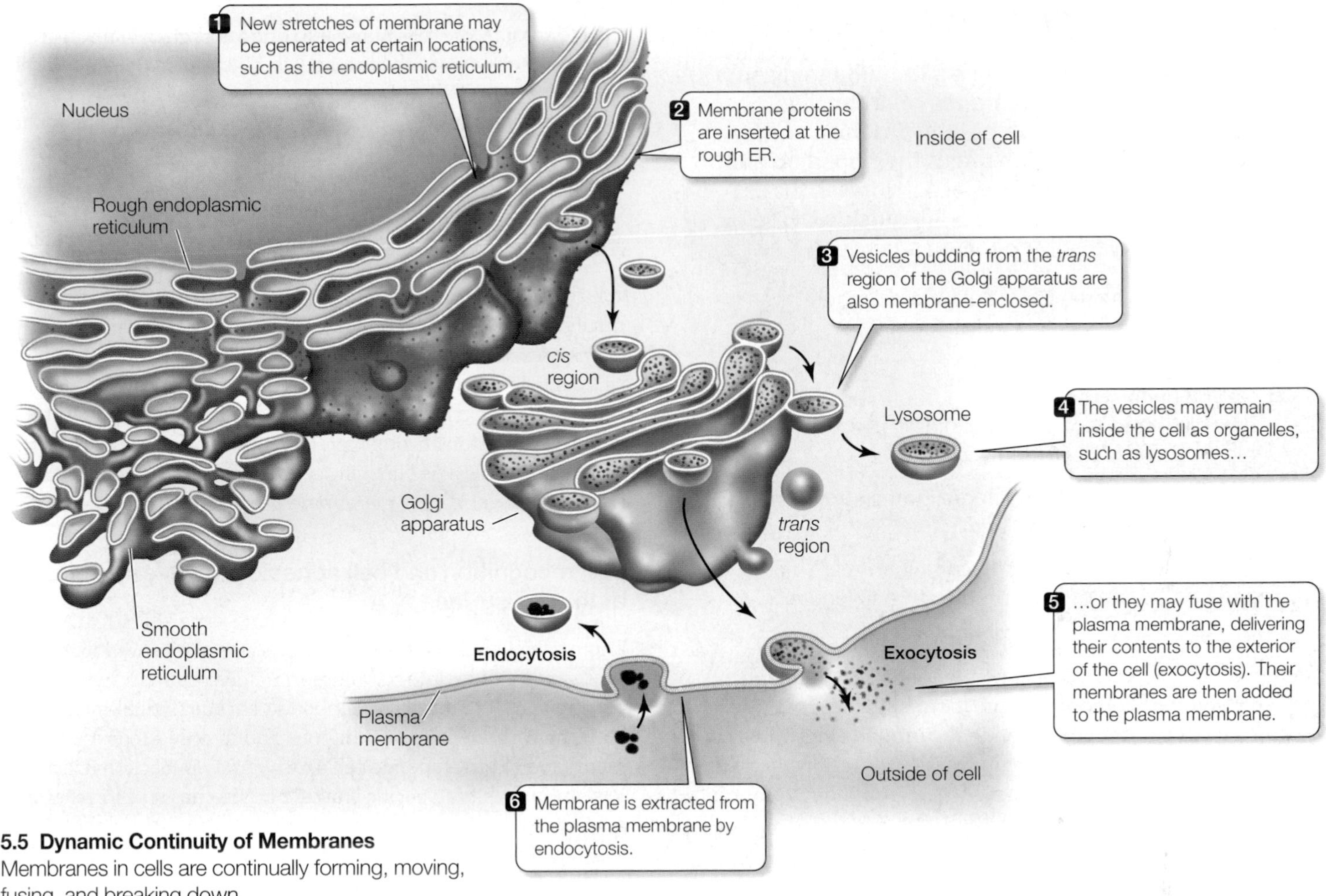

5.5 Dynamic Continuity of Membranes
Membranes in cells are continually forming, moving, fusing, and breaking down.

the *trans* regions of the Golgi apparatus, from which they bud away to join the plasma membrane.

- Additions to the plasma membrane from fusion with *vesicles* derived from the Golgi apparatus are largely balanced by the removal of membrane in processes such as phagocytosis, affording a recovery path by which internal membranes are replenished.

Because all membranes appear similar under the electron microscope, and because they interconvert readily, we might expect all subcellular membranes to be chemically identical. However, that is not the case, for there are major chemical differences among the membranes of even a single cell. Membranes are changed chemically when they form parts of certain organelles. In the Golgi apparatus, for example, the membranes of the *cis* face closely resemble those of the endoplasmic reticulum in chemical composition, but those of the *trans* face are more similar to the plasma membrane. As a vesicle is formed, the mix of proteins and lipids in its membrane is selected, just as its internal contents are selected, to correspond with the vesicle's target membrane.

Membrane carbohydrates are recognition sites

In addition to lipids and proteins, many membranes contain significant amounts of carbohydrates. The carbohydrates are located on the outer surface of the membrane and serve as recognition sites for other cells and molecules (see Figure 5.1).

Membrane-associated carbohydrates may be covalently bonded to lipids or to proteins:

- **Glycolipids** consist of a carbohydrate covalently bonded to a lipid. The carbohydrate units of glycolipids often extend to the outside of the plasma membrane, where they serve as recognition signals for interactions between cells. For example, the carbohydrate of some glycolipids changes when a cell becomes cancerous. This change may allow white blood cells to target cancer cells for destruction.
- **Glycoproteins** consist of a carbohydrate covalently bonded to a protein. The bound carbohydrates are oligosaccharide chains, usually not exceeding 15 monosaccharide units in length. Glycoproteins enable a cell to be recognized by other cells and proteins.

An "alphabet" of monosaccharides on membranes can be used to generate a diversity of messages. Recall from Section 3.3 that sugar molecules can be formed from three to seven carbons attached at different sites to one another, forming linear or branched oligosaccharides with many different three-dimensional shapes. An oligosaccharide of a specific shape on one cell can bind to a mirror-image shape on an adjacent cell. This binding is the basis of cell–cell adhesion.

5.1 RECAP

The fluid mosaic model applies to both the plasma membrane and the membranes of organelles. Membrane proteins have both hydrophilic and hydrophobic regions, which affect their position and function in the membrane. Carbohydrates that attach to lipids and proteins on the outside of the membrane serve as recognition sites.

- Do you see why the term "fluid mosaic model" is an apt description of biological membranes? See pp. 98–99
- Do you understand how the hydrophobic and hydrophilic regions of phospholipids cause a membrane bilayer to form? See Figures 5.1 and 5.2
- Can you differentiate an integral protein from a peripheral protein? See p. 98 and Figure 5.1
- Can you explain how plasma membranes are made and broken down? See pp. 100–101 and Figure 5.5

Now that we understand the structure of biological membranes, let's see how their components function. In the rest of this chapter we'll focus on the membrane that surrounds individual cells: the plasma membrane. We'll begin with a look at how the plasma membrane allows individual cells to be grouped together into tissues.

5.2 How Is the Plasma Membrane Involved in Cell Adhesion and Recognition?

Some organisms, such as bacteria, are *unicellular;* that is, the entire organism is a single cell. Others, such as plants and animals, are *multicellular*—composed of many cells. Often these cells exist in specialized blocks of cells with similar functions, called *tissues*. Your body has about 60 trillion cells, arranged in different kinds of tissues (such as muscle, nerve, or skin).

Two processes allow cells to arrange themselves in groups:

- **Cell recognition**, in which one cell specifically binds to another cell of a certain type
- **Cell adhesion**, in which the connection between the two cells is strengthened

Both processes involve the plasma membrane. They are most easily studied if the cells in a tissue are separated into individual cells, then allowed to adhere to one another again. Simple organisms provide a good model for the complex tissues of larger species. Studies of sponges, for example, have revealed how cells associate with one another.

A sponge is a multicellular marine animal with a simple body plan (see Section 31.5). The cells of the sponge are connected, but can be separated mechanically by passing the animal several times through a fine wire screen (**Figure 5.6A**). Through this process, what was a single animal becomes hundreds of individual cells suspended in seawater. Remarkably, if the cell suspension is shaken for a few hours, the cells bump into one another and stick together in the same shape as the original sponge! The *cells recognize and adhere to one another*.

There are many different species of sponges. If disaggregated sponge cells from two different species are placed in the same container, the cells float around and bump into one another, but the cells will stick only to other cells of the same species. Two different sponges form, just like the ones at the start of the experiment.

Such tissue-specific and species-specific cell recognition and cell adhesion are essential in the formation and maintenance of tissues and multicellular organisms. Think of your own body. What keeps muscle cells bound to muscle cells and skin to skin? Specific cell adhesion is so obvious a characteristic of complex organisms that it is easy to overlook. You will see many examples of specific cell adhesion throughout this book; here, we describe its general principles. As you will see, cell recognition and cell adhesion depend on plasma membrane proteins.

Cell recognition and cell adhesion involve proteins at the cell surface

The molecule responsible for cell recognition and adhesion in sponges is a huge integral membrane glycoprotein (which is 80 percent sugar) that is partly embedded in the plasma membrane, with the carbohydrate sticking out and exposed to the environment (and to other sponge cells). As we saw in Section 3.2, a protein not only has a specific shape, but also has specific chemical groups exposed on its surface where they can interact with other substances, including other proteins. Both of these features allow binding to other specific molecules. The cells of the ground-up sponge in Figure 5.6A find one another again through recognition of exposed chemical groups on their membrane glycoproteins. In the majority of plant cells, the plasma membrane is covered with a thick cell wall, but this structure, too, has adhesion proteins that allow cells to bind to one another.

In most cases, the binding of cells in a tissue is **homotypic**; that is, the same molecule sticks out of both cells, and the exposed surfaces bind to each other. But **heterotypic** binding (of cells with different proteins) can also occur. In this case, *different* chemical groups on *different* surface molecules have an affinity for one another. For example, when the mammalian sperm meets the egg, different proteins on the two types of cells have complementary binding surfaces. Similarly, some algae form similar-appearing male and female reproductive cells (analogous to sperm and eggs) that have flagella to propel them toward each other. Male and female cells can recognize each other by heterotypic proteins on their flagella (**Figure 5.6B**).

Three types of cell junctions connect adjacent cells

In a complex multicellular organism, cell recognition proteins allow specific types of cells to bind to one another. Often, both cells contribute material to additional membrane structures that connect them to one another. These specialized structures, called **cell junctions**, are most evident in electron micrographs of *epithelial tissues*, which are layers of cells that line body cavities or cover body surfaces. We will examine three types of cell junctions that enable

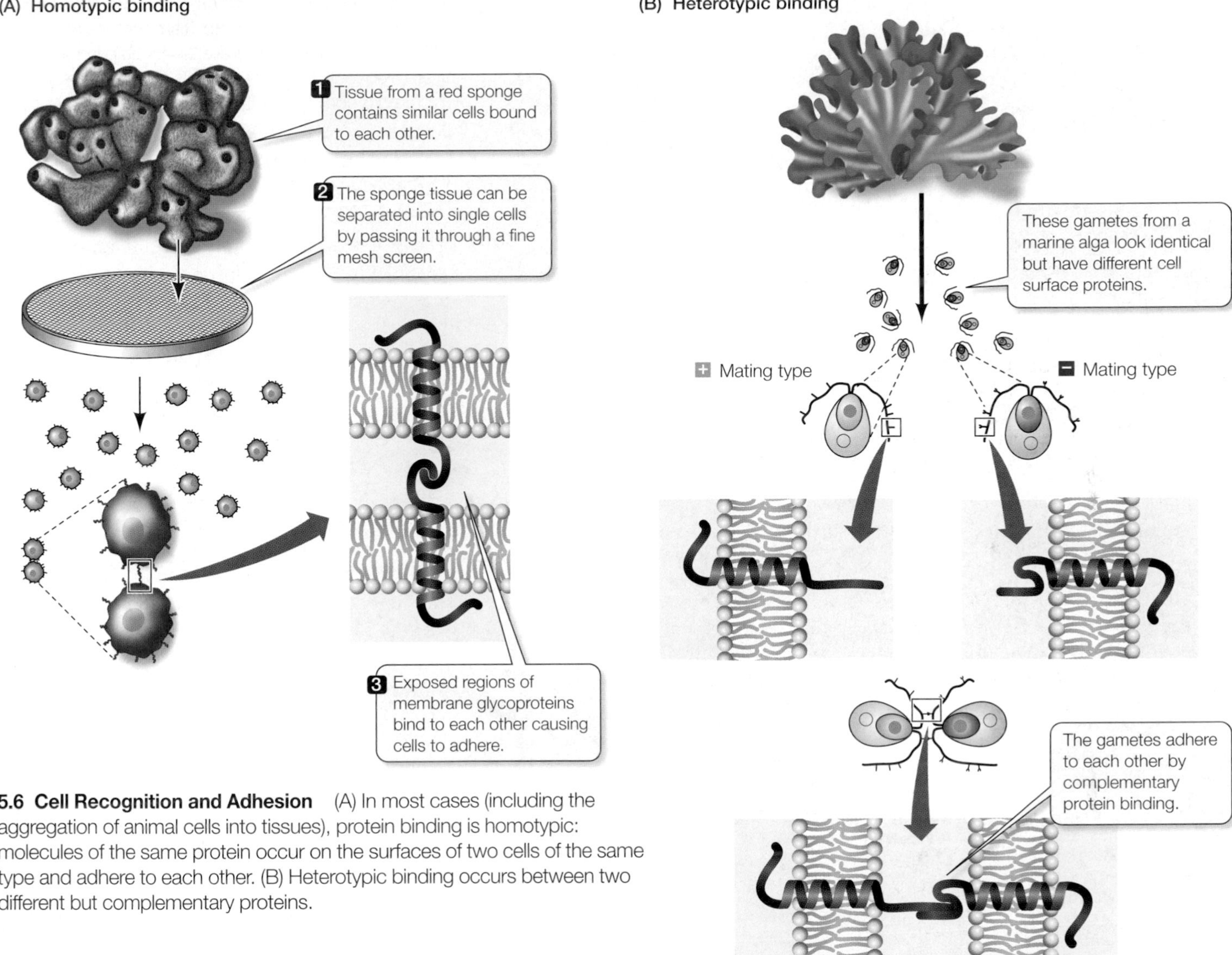

(A) Homotypic binding
1 Tissue from a red sponge contains similar cells bound to each other.
2 The sponge tissue can be separated into single cells by passing it through a fine mesh screen.
3 Exposed regions of membrane glycoproteins bind to each other causing cells to adhere.
(B) Heterotypic binding
These gametes from a marine alga look identical but have different cell surface proteins.
+ Mating type
− Mating type
The gametes adhere to each other by complementary protein binding.

5.6 Cell Recognition and Adhesion (A) In most cases (including the aggregation of animal cells into tissues), protein binding is homotypic: molecules of the same protein occur on the surfaces of two cells of the same type and adhere to each other. (B) Heterotypic binding occurs between two different but complementary proteins.

cells to make direct physical contact and connect to one another: tight junctions, desmosomes, and gap junctions.

TIGHT JUNCTIONS SEAL TISSUES **Tight junctions** are specialized structures that link adjacent epithelial cells. They result from the mutual binding of specific proteins in the plasma membranes of two epithelial cells, which form a series of joints encircling each cell (**Figure 5.7A**). Found in the lining of the lumen (cavity) of organs such as the intestine, tight junctions have two functions:

- They prevent substances from moving through the spaces between cells. Thus any substance entering the body from the lumen of the intestine must pass through the epithelial cells that form the tight junction.
- They define different functional regions of the membrane by restricting the migration of membrane proteins and phospholipids from one region of the cell to another. Thus the membrane proteins and phospholipids in the *apical* (tip) region of the cell (facing the lumen) can be different from those in the *basolateral* (*basal*, bottom; *lateral*, side) regions of the junction cells (facing the body cavity outside the lumen).

By forcing materials to enter certain cells, and by allowing different areas of the same cell to express different membrane proteins with different functions, tight junctions help ensure the directional movement of materials into the body.

DESMOSOMES HOLD CELLS TOGETHER **Desmosomes** connect adjacent plasma membranes. Desmosomes hold adjacent cells firmly together, acting like spot welds or rivets (**Figure 5.7B**). Each desmosome has a dense structure called a *plaque* on the cytoplasmic side of the plasma membrane. To this plaque are attached special cell adhesion molecules (CAMs) that stretch from the plaque through the plasma membrane of one cell, across the intercellular space, and through the plasma membrane of the adjacent cell, where they bind to the plaque proteins in that cell.

The plaque is also attached to fibers in the cytoplasm. These fibers, which are intermediate filaments of the cytoskeleton (see Figure 4.20), are made of a protein called *keratin*. They stretch from

(A)

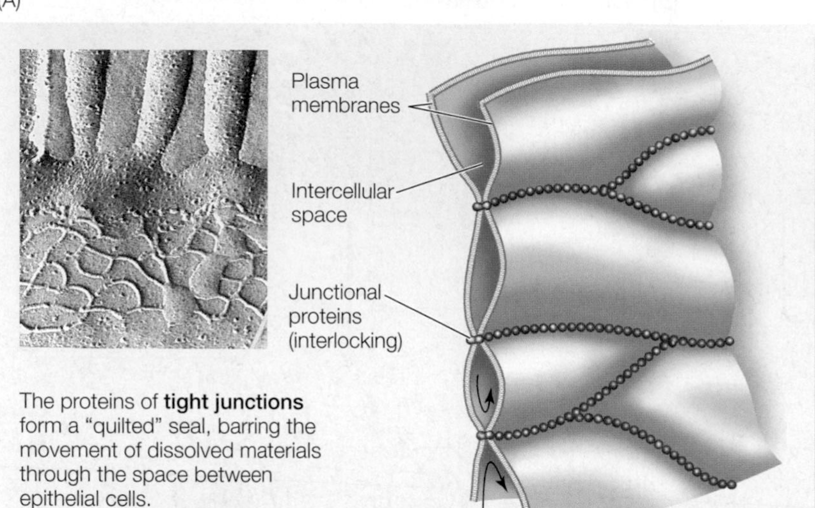

The proteins of **tight junctions** form a "quilted" seal, barring the movement of dissolved materials through the space between epithelial cells.

(B)

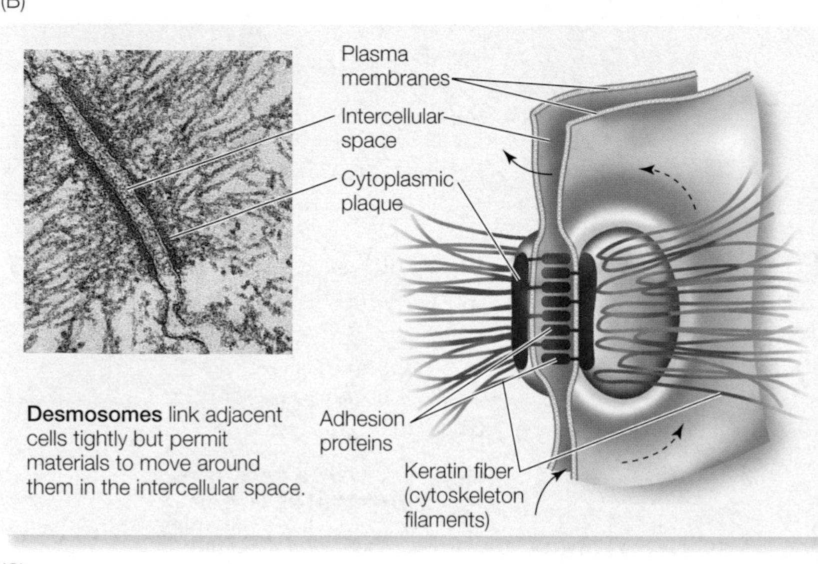

Desmosomes link adjacent cells tightly but permit materials to move around them in the intercellular space.

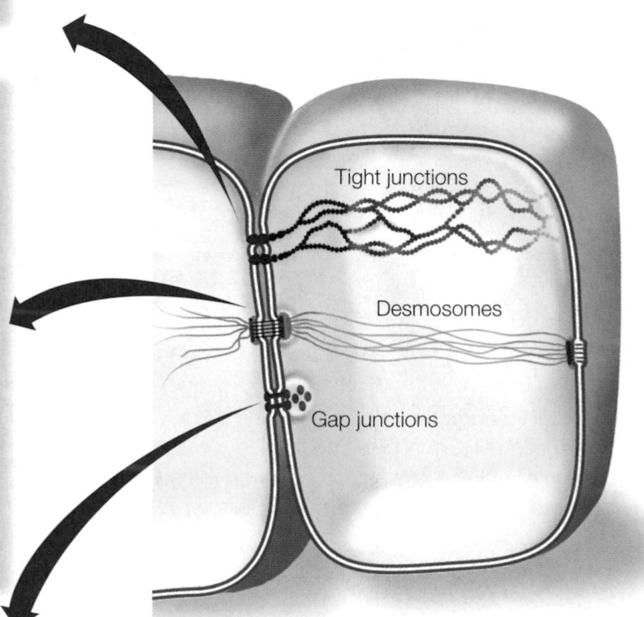

(C)

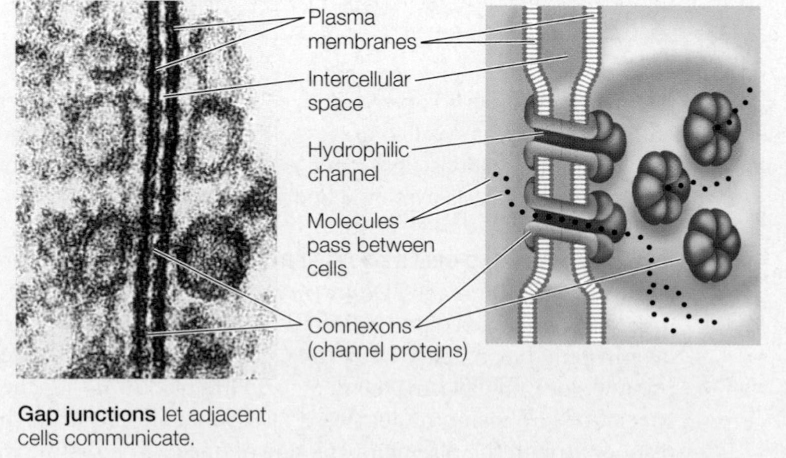

Gap junctions let adjacent cells communicate.

5.7 Junctions Link Animal Cells Together Tight junctions (A) and desmosomes (B) are abundant in epithelial tissues. Gap junctions (C) are also found in some muscle and nerve tissues, in which rapid communication between cells is important.

one cytoplasmic plaque across the cell to connect with another plaque on the other side of the cell. Anchored thus on both sides of the cell, these extremely strong fibers provide great mechanical stability to epithelial tissues, which often receive rough wear while protecting the integrity of the organism's body surface.

GAP JUNCTIONS ARE A MEANS OF COMMUNICATION

Whereas tight junctions and desmosomes have mechanical roles, **gap junctions** facilitate communication between cells. Each gap junction is made up of specialized channel proteins, called *connexons,* that span the plasma membranes of two adjacent cells and the intercellular space between them (**Figure 5.7C**). Dissolved small molecules and ions can pass from cell to cell through these junctions. We will describe their role in more detail, as well as that of *plasmodesmata,* which perform a similar role in plants, when we discuss cell communication in Chapter 15.

5.2 RECAP

In multicellular organisms, cells arrange themselves in groups by two processes, cell recognition and cell adhesion. Both processes are mediated by integral proteins in the plasma membrane.

- Can you explain the difference between cell recognition and cell adhesion? See p. 102
- The three types of cell junctions described here all have different effects for the passage of materials between cells and through intercellular space. Can you describe how each type deals with this molecular "traffic"? See pp. 103–104 and Figure 5.7

We have just examined how membrane structure accommodates one major membrane function: the binding of one cell to another. Now we turn to the second major function of membranes: regulating the substances that enter or leave a cell.

5.3 What Are the Passive Processes of Membrane Transport?

Biological membranes allow some substances, but not others, to pass through them. This characteristic of membranes is called **selective permeability**. Selective permeability allows the membrane to determine what substances enter or leave a cell or organelle.

There are two fundamentally different processes by which substances cross biological membranes:

- The processes of *passive transport* do not require any input of outside energy to drive them.
- The processes of *active transport* require the input of chemical energy from an outside source.

This section focuses on the passive processes by which substances enter and leave the cell. The energy for these processes is found in the substances themselves, and in the motive force generated by the difference in their concentrations on the two sides of the membrane. Passive transport processes include two types of *diffusion*: *simple diffusion* through the phospholipid bilayer, and *facilitated diffusion* through channel proteins or by means of carrier proteins.

Diffusion is the process of random movement toward a state of equilibrium

Nothing in this world is ever absolutely at rest. Everything is in motion, although the motions may be very small. An important consequence of all this random jiggling of molecules is that all the components of a solution tend eventually to become evenly distributed throughout the system. For example, if a drop of ink is allowed to fall into a container of water, the pigment molecules of the ink are initially very concentrated. Without human intervention, such as stirring, the pigment molecules of the ink move about at random, spreading slowly through the water until eventually the concentration of pigment—and thus the intensity of color—is exactly the same in every drop of liquid in the container. A solution in which the solute particles are uniformly distributed is said to be at *equilibrium* because there will be no future net change in their concentration. Equilibrium does not mean that the particles have stopped moving; it just means that they are moving in such a way that their overall distribution does not change.

Diffusion is the process of random movement toward a state of equilibrium. Although the motion of each individual particle is absolutely random, the *net* movement of particles is directional until equilibrium is reached. Diffusion is thus *net movement from regions of greater concentration to regions of lesser concentration* (**Figure 5.8**).

In a complex solution (one with many different solutes), the diffusion of each solute is independent of that of the others. How fast a substance diffuses depends on four factors:

- The *diameter* of the molecules or ions: smaller molecules diffuse faster
- The *temperature* of the solution: higher temperatures lead to faster diffusion because ions or molecules have more energy, and thus move more rapidly
- The *electric charge*, if any, of the diffusing material
- The *concentration gradient* in the system—that is, the change in solute concentration with distance in a given direction: the greater the concentration gradient, the more rapidly a substance diffuses

DIFFUSION WITHIN CELLS AND TISSUES Within cells, or wherever distances are very short, solutes distribute themselves rapidly by diffusion. Small molecules and ions may move from one end of an organelle to another in a millisecond (10^{-3} s, or one-thousandth of a second). However, the usefulness of diffusion as a transport mechanism declines drastically as distances become greater. In the absence of mechanical stirring, diffusion across more than a centimeter may take an hour or more, and diffusion across meters may take years! Diffusion would not be adequate to distribute materials over the length of the human body (much less that of larger organisms), but within our cells or across layers of one or two cells, diffusion is rapid enough to distribute small molecules and ions almost instantaneously.

DIFFUSION ACROSS MEMBRANES In a solution without barriers, all the solutes diffuse at rates determined by temperature, their physical properties, and the concentration gradient of each solute. If a biological membrane divides the solution into separate compartments, then the movement of the different solutes can be affected by the properties of the membrane. The membrane is said to be *permeable* to solutes that can cross it more or less easily, but *impermeable* to substances that cannot move across it.

Molecules to which the membrane is impermeable remain in separate compartments, and their concentrations are apt to be different on the two sides of the membrane. Molecules to which the membrane is permeable diffuse from one compartment to the other until their concentrations are equal on both sides of the membrane. When the concentrations of the diffusing substance on the two sides of the permeable membrane are identical, equilibrium is reached. Individual molecules continue to pass through the membrane after equilibrium is established, but equal numbers

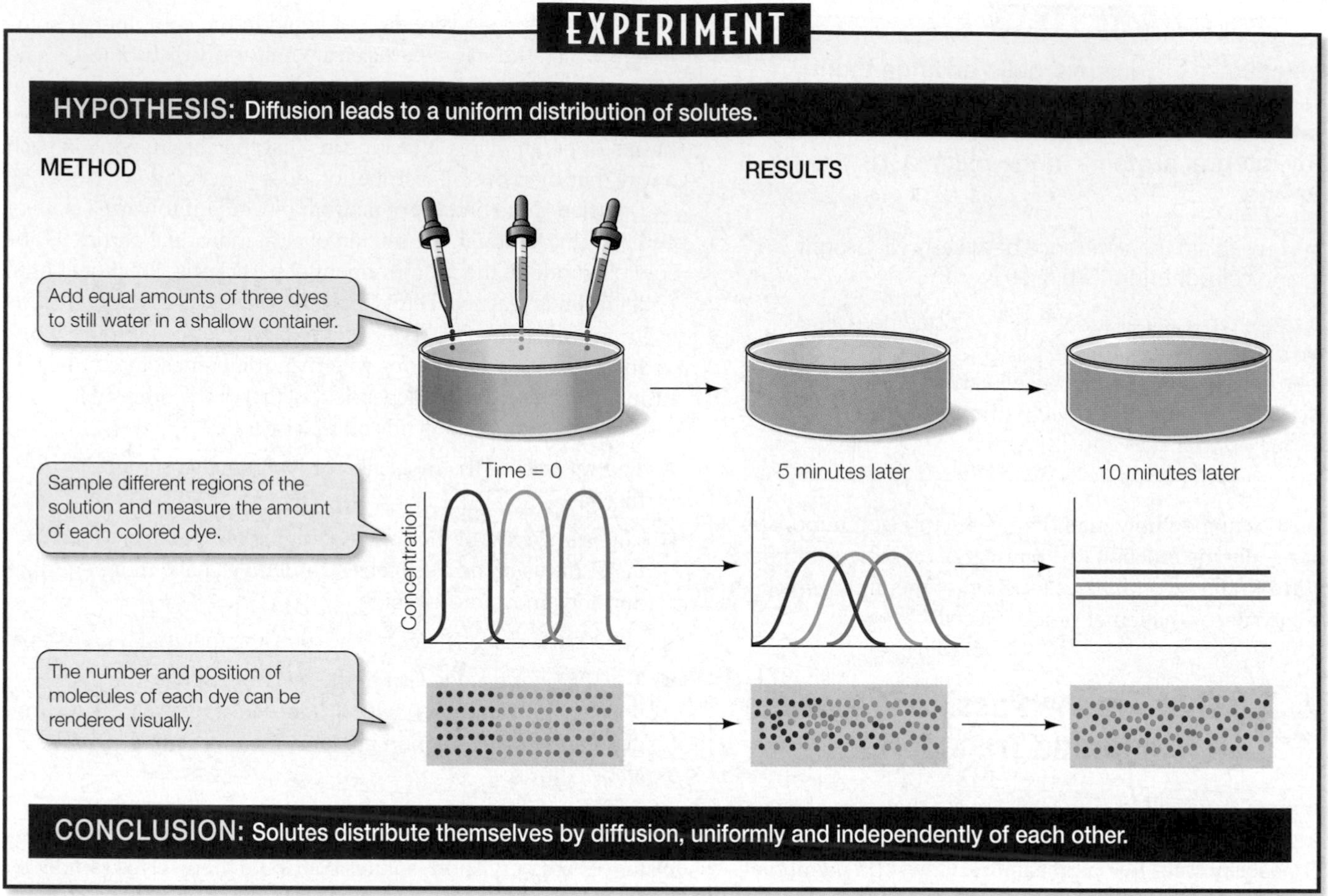

EXPERIMENT
HYPOTHESIS: Diffusion leads to a uniform distribution of solutes.
METHOD
RESULTS
Add equal amounts of three dyes to still water in a shallow container.
Sample different regions of the solution and measure the amount of each colored dye.
The number and position of molecules of each dye can be rendered visually.
Time = 0
5 minutes later
10 minutes later
Concentration
CONCLUSION: Solutes distribute themselves by diffusion, uniformly and independently of each other.

5.8 Diffusion Leads to Uniform Distribution of Solutes A simple experiment demonstrates that solutes move from regions of greater concentration to regions of lesser concentration until equilibrium is reached.

of molecules move in each direction, so there is *no net change* in concentration.

Simple diffusion takes place through the phospholipid bilayer

In **simple diffusion**, small molecules pass through the phospholipid bilayer of the membrane. A molecule that is itself hydrophobic, and hence is soluble in lipids, enters the membrane readily and is thus able to pass through it. The more lipid-soluble the molecule is, the more rapidly it diffuses through the membrane bilayer. This statement holds true over a wide range of molecular weights. Only water itself and the smallest of molecules seem to deviate from this rule, passing through bilayers much more rapidly than their lipid solubilities would predict.

On the other hand, electrically charged or polar molecules, such as amino acids, sugars, and ions, do not pass readily through a membrane, for two reasons:

- Cells are made up of, and exist in, water; polar substances form many hydrogen bonds with water and ions are surrounded by water molecules, thus preventing their "escape" to the membrane.
- The interior of the membrane is hydrophobic, and hydrophilic substances tend to be excluded from it.

Consider two types of molecules: a small protein made up of a few amino acids, and a cholesterol-based steroid of equivalent size. The protein, being polar, will diffuse slowly through the membrane, while the nonpolar steroid will diffuse through it readily.

Osmosis is the diffusion of water across membranes

Water molecules are abundant enough and small enough that they move through membranes by a diffusion process called **osmosis**. This completely passive process uses no metabolic energy and can be understood in terms of solute concentrations. Osmosis depends on the *number* of solute particles present, not on the kinds of particles. We will describe osmosis using red blood cells and plant cells as examples.

Red blood cells are normally suspended in a fluid called *plasma*, which contains salts, proteins, and other solutes. Examining a drop of blood under the light microscope reveals that these red cells have a characteristic doughnut shape. If pure water is added to the drop of blood, the red cells quickly swell and burst. Similarly, if slightly wilted lettuce is placed in pure water, it soon becomes crisp; by weighing it before and after, we can show that it has taken up water. If, on the other hand, red blood cells or crisp lettuce leaves are placed in a relatively concentrated solution of salt or sugar, the

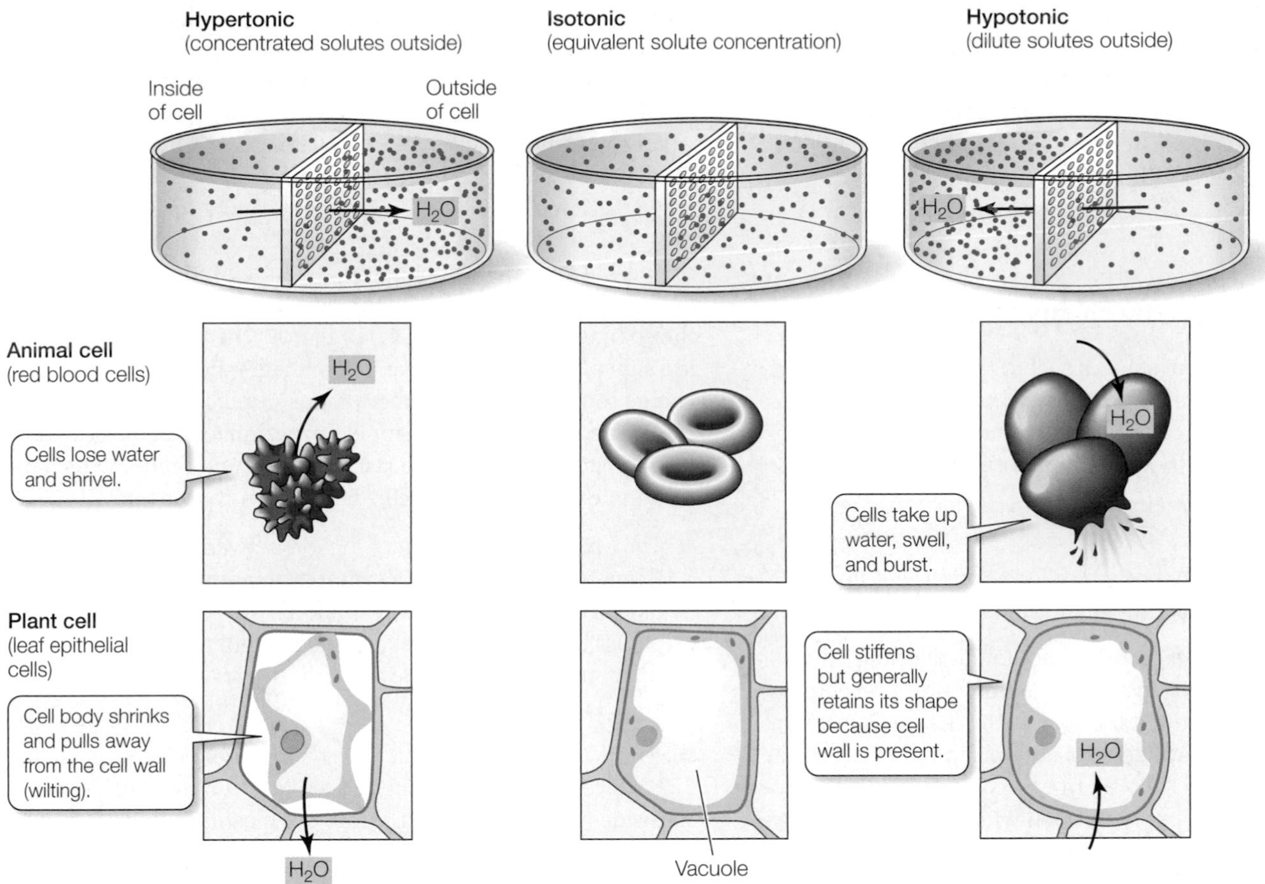

5.9 Osmosis Can Modify the Shapes of Cells In an isotonic solution (center column), plant and animal cells maintain consistent, characteristic shapes. In a solution that is hypotonic to the cells (right), water enters the cells; an environment that is hypertonic to the cells (left) draws water out of the cells.

leaves become limp (they wilt), and the red blood cells pucker and shrink (see Figure 5.9, left).

From analyses of such observations, we know that the difference in solute concentration between a cell and its surrounding environment determines whether water will move from the environment into the cell or out of the cell into the environment. Other things being equal, if two different solutions are separated by a membrane that allows water, but not solutes, to pass through, water molecules will move across the membrane toward the solution with a higher solute concentration. In other words, *water will diffuse from a region of its higher concentration* (with a lower concentration of solutes) *to a region of its lower concentration* (with a higher concentration of solutes).

Three terms are used to compare the solute concentrations of two solutions separated by a membrane (**Figure 5.9**):

- **Isotonic** solutions have equal solute concentrations.
- A **hypertonic** solution has a higher solute concentration than the other solution with which it is being compared.
- A **hypotonic** solution has a lower solute concentration than the other solution with which it is being compared.

Water moves from a hypotonic solution across a membrane to a hypertonic solution.

When we say that "water moves," bear in mind that we are referring to the *net* movement of water. Since it is so abundant, water is constantly moving across the plasma membrane into and out of cells. What concerns us here is whether the overall movement is greater in one direction or the other.

The concentration of solutes in the environment determines the direction of osmosis in all animal cells. A red blood cell takes up water from a solution that is hypotonic to the cell's contents. The cell bursts because its plasma membrane cannot withstand the swelling of the cell. The integrity of red blood cells (and other blood cells) is absolutely dependent on the maintenance of a constant solute concentration in the plasma in which they are suspended: the plasma must be isotonic to the blood cells if the cells are not to burst or shrink. Regulation of the solute concentration of body fluids is thus an important process for organisms without cell walls.

In contrast to animal cells, the cells of plants, archaea, bacteria, fungi, and some protists have cell walls that limit the volume of the cells and keep them from bursting. Cells with sturdy walls take up a limited amount of water, and in so doing they build up an internal pressure against the cell wall that prevents further water from entering. This pressure within the cell is called **turgor pressure.** Turgor pressure keeps plants upright and is the driving force for the enlargement of plant cells. It is a normal and essential component of plant growth. If enough water leaves the cells, turgor pressure drops and the plant wilts.

Turgor pressure in plant cells reaches about 100 pounds per square inch—several times greater than the pressure in automobile tires. This pressure is so great that the cells would slip by and detach from one another, were it not for adhesive molecules called pectins in the plant cell wall.

Diffusion may be aided by channel proteins

As we saw earlier, polar substances such as amino acids and sugars and charged substances such as ions do not readily diffuse across membranes. But they do cross the hydrophobic phospholipid bilayer passively (that is, without the input of energy) in two ways:

- Integral membrane proteins may form *channels* through which these substances can pass.
- Binding to a membrane protein called a *carrier protein* can speed up the diffusion of these substances.

Both of these processes are forms of **facilitated diffusion**.

Membrane **channel proteins** have a central pore lined with polar amino acids and water (to bind to polar or charged substances and allow them to pass through) and nonpolar amino acids on the outside of the protein (to keep them embedded in the phospholipid bilayer). The central pore can open when stimulated, allowing hydrophilic polar substances to pass through (**Figure 5.10**).

ION CHANNELS AND THE MEMBRANE POTENTIAL The best-studied channel proteins are the **ion channels**. As you will see in later chapters, the movement of ions into and out of cells is important in many biological processes, ranging from the electrical activity of the nervous system to the opening of the pores in leaves that allow gas exchange with the environment. Hundreds of ion channels have been identified, each of them specific for a particular ion. All of them show the same basic structure of a hydrophilic pore that allows a particular ion to move through it.

Just as a fence may have a gate that can be opened or closed, most ion channels are *gated*: they can be closed to ion passage or opened. A **gated channel** opens when something happens to change the three-dimensional shape of the protein. Depending on the channel, this stimulus can range from the binding of a chemical signal (a *ligand-gated* channel; see Figure 5.10) to an electrical charge caused by an imbalance of ions (a *voltage-gated* channel).

Once a voltage-gated channel opens, millions of ions can rush through it per second. How fast the ions move, and in which direction (into or out of the cell), depends on two factors:

- The *concentration gradient* of the ion between the inside and the outside of the cell. For example, in animal cells, because of active transport (discussed below), the concentration of potassium (K^+) ions is usually much higher inside the cell than outside, so K^+ will tend to *diffuse out of the cell* through an open potassium channel.
- An *electrochemical gradient* forms when there is an overall imbalance in the charged substances between the outside and inside of the cell—that is, across the plasma membrane. In animal cells, there is a higher concentration of Cl^- and other negatively charged ions (anions) inside the cell than outside. These negatively charged substances cannot get through the membrane, so there is a tendency for K^+ to *stay inside the cell* to balance out the negative charges and maintain neutrality.

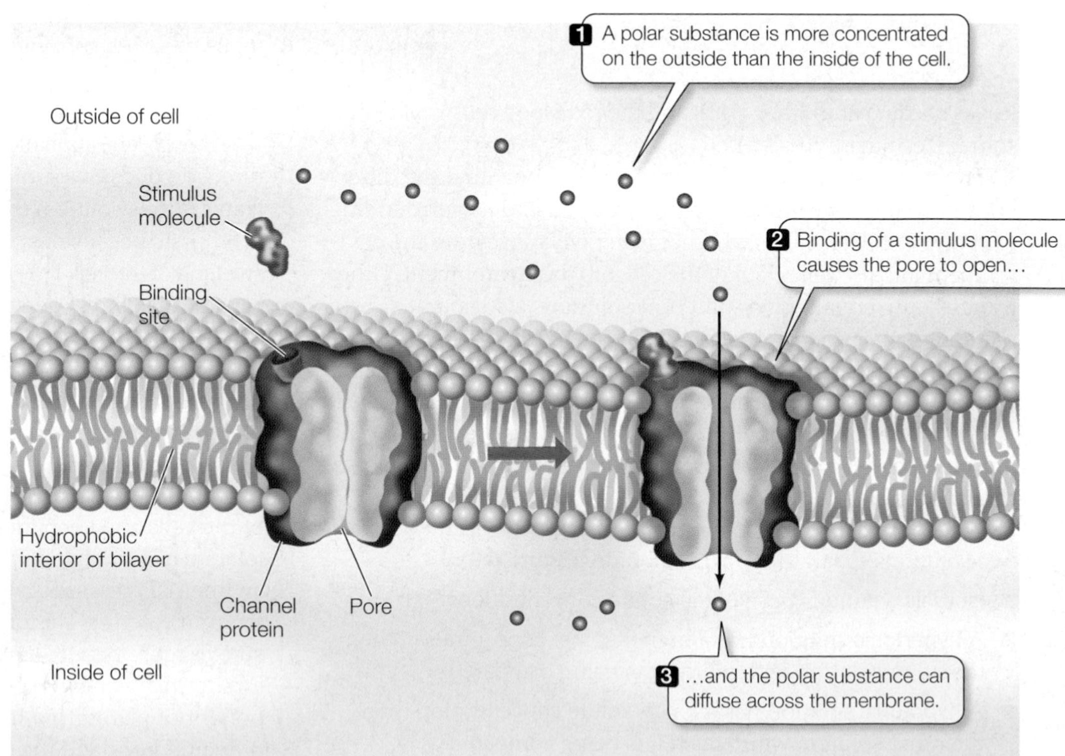

1 A polar substance is more concentrated on the outside than the inside of the cell.
Outside of cell
Stimulus molecule
Binding site
2 Binding of a stimulus molecule causes the pore to open...
Hydrophobic interior of bilayer
Channel protein
Pore
Inside of cell
3 ...and the polar substance can diffuse across the membrane.

5.10 A Gated Channel Protein Opens in Response to a Stimulus The channel protein has a pore of polar amino acids and water. It is anchored in the hydrophobic bilayer interior by its outer coating of nonpolar amino acids. The protein changes its three-dimensional shape when a stimulus molecule binds to it, opening the pore so that hydrophilic polar substances can pass through.

It is the totality of these two forces—simple diffusion based on concentration gradients, and electrochemical imbalances—that determines the direction in which an ion such as K^+ flows through a channel protein. Eventually, an equilibrium is reached at which the ion's rate of diffusion through the channel and out of the cell is balanced by the rate of entry through the channel due to electrical attraction. Obviously, the relative concentrations of K^+ on both sides of the membrane will not be equal, as we would expect if diffusion were the only force involved. Instead, the attraction of electrical charges keeps some extra K^+ inside the cell. This sets up a charge imbalance across the plasma membrane, as there is more K^+ on the inside than on the outside. This charge imbalance is called the **membrane potential**.

The membrane potential is related to the concentration imbalance of K^+ by the *Nernst equation*:

$$E_K = 2.3\frac{RT}{zF}\log\frac{[K]_o}{[K]_i}$$

where R is the gas constant, F is the Faraday constant (both familiar to chemistry students), T is the temperature, and z is the charge on the ion (+1). Solving for 2.3 RT/zF at 20°C ("room temperature"), the equation becomes much simpler:

$$E_K = 58\log\frac{[K]_o}{[K]_i}$$

where E_K is the membrane potential (in millivolts, mV) that results from the ratio of K^+ concentrations outside the cell $[K]_o$ and inside the cell $[K]_i$.

Actual measurements from animal cells give a total membrane potential of approximately –70 mV across the membrane, where the inside is negative with respect to the outside (see Figure 44.7). Cells have a tremendous amount of potential energy stored in their membrane potentials. In fact, the brain cells you are using to read this book have more potential energy—about 200,000 volts per centimeter—than the high-voltage electric lines powering your reading light, which carry about 2 volts per centimeter.

As you will see when you study plant and animal physiology, the energy stored in the membrane potential of cells is the basis of many important biological processes. We will return to the Nernst equation and its applications to biology in Chapter 44.

THE SPECIFICITY OF ION CHANNELS How does an ion channel allow one ion, but not another, to pass through? It is not simply a matter of charge or size. For example, a sodium ion (Na^+), with a radius of 0.095 nanometers, is smaller than K^+, at 0.130 nm; both carry the same positive charge. Yet the potassium channel lets only K^+ pass through the membrane, and not the smaller Na^+. The elegant explanation was recently discovered when Roderick MacKinnon determined the structure of a potassium channel from a bacterium (**Figure 5.11**).

Being charged, both Na^+ and K^+ are attracted to water molecules. They have water "shells" in solution, held by the attraction of their positive charges to the negatively charged oxygen atoms on the water molecules (see Figure 2.11). To get through a membrane channel, K^+ must let go of its water. The "naked" K^+ is now attracted to the oxygen atoms on the pore of the channel protein. In the potassium channel, oxygen atoms are located at the stem of a funnel-shaped protein region. The K^+ ion just fits the stem, and so can get into a position where it is more strongly attracted to the oxygen atoms there than to those of water. The smaller Na^+ ion, on the other hand, is kept a bit more distant from the oxygen atoms on the stem of the channel, and prefers to be surrounded by water. So Na^+ does not enter the potassium channel.

As we mentioned, water crosses the plasma membrane at a rate far in excess of expectations, given its polarity. One way that water can do this is by hydrating some ions as they pass through ion channels. Up to 12 water molecules may coat an ion as it traverses a channel. Another way that water enters cells rapidly is through water channels, called **aquaporins**. Channel proteins that allow water to pass through them have been characterized in many membranes, from the membrane of plant vacuoles, where they are important in maintaining turgor pressure, to the mammalian kidney, where they function to retain water that would otherwise be lost through urine.

Many plants, like some humans, follow the sun. When ion channels in cells joining the leaf to the stem open in response to sunlight, K^+ and Cl^- enter the cells by diffusion. Water follows by osmosis and the stem cells swell, causing the leaf to tilt toward the sun.

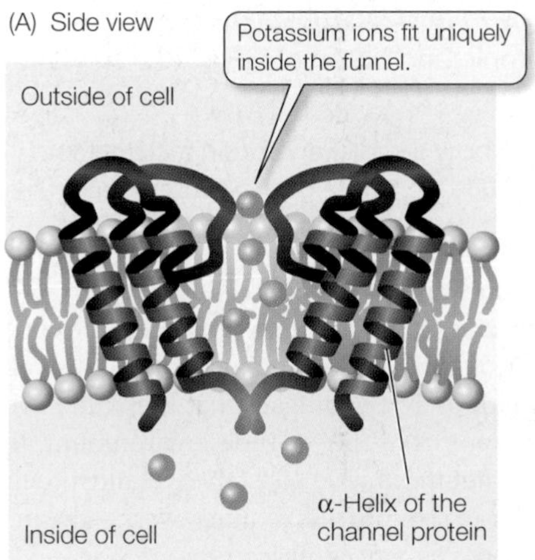

5.11 The Potassium Channel Roderick MacKinnon determined this structure for the selective K^+ channel in the bacterium *Streptomyces lividans*. (A) Potassium ions funnel through the channel in this side view, attracted by oxygen atoms on the α-helices of the channel protein. It is a "custom fit" for K^+; other ions do not pass through this channel. (B) View looking down from the top of the channel.

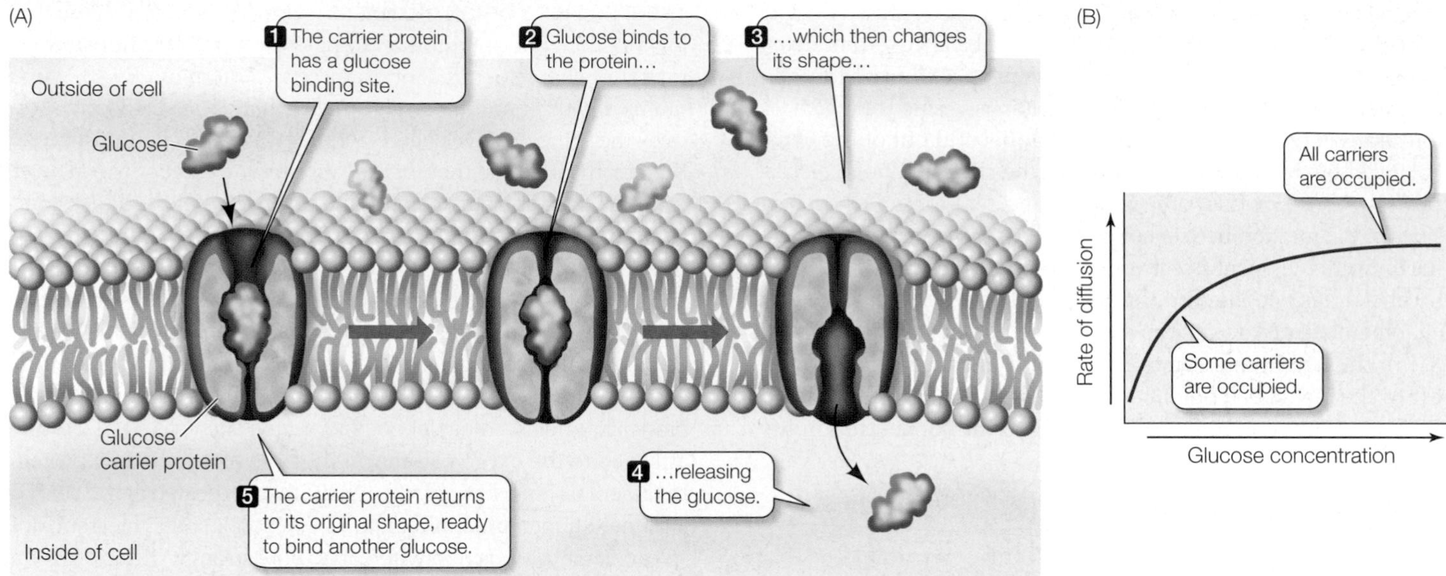

5.12 A Carrier Protein Facilitates Diffusion The glucose transporter is a carrier protein that allows glucose to enter the cell at a faster rate than would be possible by simple diffusion. (A) The transporter binds to glucose, brings it into the membrane interior, then changes shape, releasing glucose into the cell cytoplasm. (B) At lower glucose concentrations, not all the transporters are occupied and an increase in the number of glucose molecules increases the rate of diffusion. At high concentrations, all the transporter proteins are in use (the system is saturated), and the rate of diffusion plateaus.

Carrier proteins aid diffusion by binding substances

Another kind of facilitated diffusion involves not just the opening of a channel, but the actual binding of the transported substance to membrane proteins. These proteins are called **carrier proteins** and, like channel proteins, they allow diffusion both into and out of the cell. Carrier proteins transport polar molecules such as sugars and amino acids.

For example, glucose is the major energy source for most mammalian cells. The membranes of those cells contain a carrier protein—the glucose transporter—that facilitates glucose uptake into the cell. Binding of glucose to a specific three-dimensional site on the transporter protein causes the protein to change its shape and release glucose on the cytoplasmic side of the membrane (**Figure 5.12A**). Since glucose is broken down almost as soon as it enters a cell, there is almost always a strong concentration gradient favoring glucose entry, with a higher concentration outside the cell than inside. The transporter allows glucose molecules to cross the membrane and enter the cell much faster than they would by simple diffusion.

Transport by carrier proteins is different from simple diffusion. In both processes, the rate of movement depends on the concentration gradient across the membrane. However, in carrier-mediated transport, a point is reached at which increases in the concentration gradient are not accompanied by an increased rate of diffusion. At this point, the facilitated diffusion system is said to be *saturated* (**Figure 5.12B**). Because there are only a limited number of carrier protein molecules per unit of membrane area, the rate of diffusion reaches a maximum when all the carrier molecules are fully loaded with solute molecules. In other words, when the difference in solute concentration across the membrane is sufficiently high, not enough carrier molecules are free at a given moment to handle all the solute molecules.

5.3 RECAP

Diffusion is the movement of ions or molecules from a region of greater concentration to a region of lesser concentration. Water can diffuse through cell membranes by a process called osmosis. Channel proteins, which can be open or closed, and carrier proteins facilitate diffusion of charged and polar substances.

- What properties of a substance determine whether, and how fast, it will diffuse across a membrane? See p. 105
- Can you describe osmosis and explain the terms hypertonic, hypotonic, and isotonic? See p. 107 and Figure 5.9
- Do you understand how a channel protein facilitates diffusion? See pp. 108–109 and Figure 5.10

Passive transport allows substances to enter cells from the environment in such a way that, when equilibrium is reached, the concentrations of a substance inside the cell and just outside the cell are equal. But one hallmark of living things is that they can have a composition quite different from that of their environment. To achieve this they must sometimes move substances against their natural tendencies to diffuse. This process requires work—the input of energy—and is known as *active transport.*

5.4 How Do Substances Cross Membranes against a Concentration Gradient?

In many biological situations, an ion or small molecule must be moved across a membrane from a region of lower concentration to a region of higher concentration. In these cases, the substance cannot rush into or out of cells by diffusion. The movement of a substance across a biological membrane *against* a concentration gradient—called **active transport**—requires the expenditure of chemical energy. The differences between diffusion and active transport are summarized in **Table 5.1**.

TABLE 5.1

Membrane Transport Mechanisms

TRANSPORT MECHANISM	EXTERNAL ENERGY REQUIRED?	DRIVING FORCE	MEMBRANE PROTEIN REQUIRED?	SPECIFICITY
Simple diffusion	No	With concentration gradient	No	Not specific
Facilitated diffusion	No	With concentration gradient	Yes	Specific
Active transport	Yes	ATP hydrolysis (against concentration gradient)	Yes	Specific

Active transport is directional

Three types of membrane proteins are involved in active transport (**Figure 5.13**):

- **Uniports** move a single substance in one direction. For example, a calcium-binding protein found in the plasma membrane and endoplasmic reticulum of many cells actively transports Ca^{2+} to regions of its higher concentration either outside the cell or inside the ER.
- **Symports** move two substances in the same direction. For example, the uptake of amino acids from the intestine by the cells that line it requires the simultaneous binding of Na^+ and an amino acid to the same transport protein.
- **Antiports** move two substances in opposite directions, one into the cell and the other out of the cell. For example, many cells have a sodium–potassium pump that moves Na^+ out of the cell and K^+ into it.

Symports and antiports are known as *coupled transporters* because they move two substances at once.

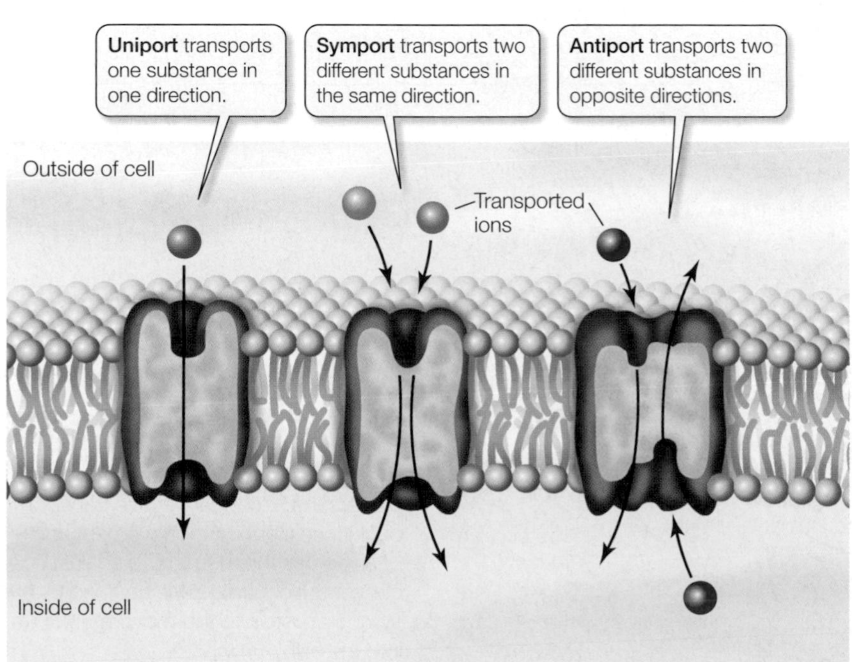

Uniport transports one substance in one direction.
Symport transports two different substances in the same direction.
Antiport transports two different substances in opposite directions.
Outside of cell
Transported ions
Inside of cell

5.13 Three Types of Proteins for Active Transport Note that in each of the three cases, transport is directional. Symport and antiport are examples of coupled transport.

Primary and secondary active transport rely on different energy sources

There are two basic types of active transport:

- **Primary active transport** requires the direct participation of the energy-rich molecule ATP.
- **Secondary active transport** does not use ATP directly; rather, its energy is supplied by an ion concentration gradient established by primary active transport.

In primary active transport, energy released by the hydrolysis of ATP drives the movement of specific ions against a concentration gradient. (We give the details of how ATP provides energy to cells in Section 6.2.) For example, we saw earlier that concentrations of potassium ions (K^+) inside a neuron are much higher than in the fluid bathing the nerve, whereas the concentration of sodium ions (Na^+) is much higher in the fluid outside. Nevertheless, a protein in the nerve cell membrane continues to pump Na^+ out of the neuron and K^+ in against these concentration gradients, ensuring that the gradients are maintained. This **sodium–potassium (Na^+–K^+) pump** is found in all animal cells. The pump is an integral membrane glycoprotein. It breaks down a molecule of ATP and uses the energy released to bring two K^+ ions into the cell and export three Na^+ ions (**Figure 5.14**). The Na^+–K^+ pump is thus an antiport.

In secondary active transport, the movement of a substance against its concentration gradient is accomplished using energy "regained" by letting ions move across the membrane *with* their concentration gradient. For example, once the sodium–potassium pump establishes a concentration gradient of sodium ions, the passive diffusion of some Na^+ back into the cell can provide energy for the secondary active transport of glucose into the cell (**Figure**

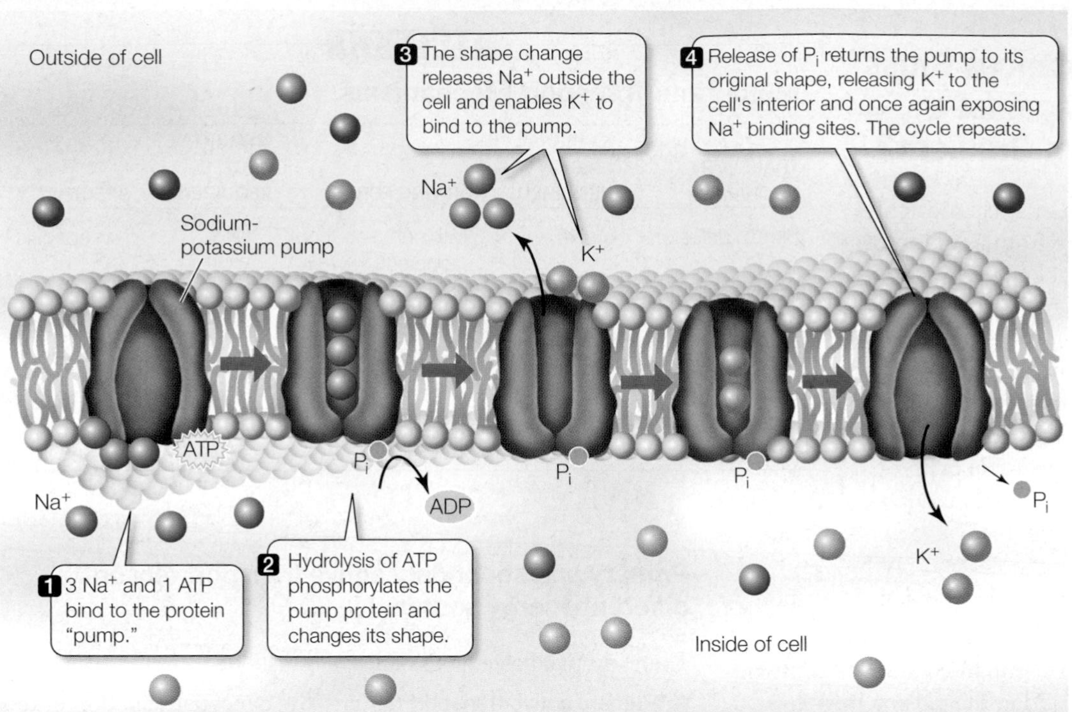

Outside of cell
3 The shape change releases Na^+ outside the cell and enables K^+ to bind to the pump.
4 Release of P_i returns the pump to its original shape, releasing K^+ to the cell's interior and once again exposing Na^+ binding sites. The cycle repeats.
Na^+
K^+
Sodium-potassium pump
ATP
P_i
ADP
Na^+
1 3 Na^+ and 1 ATP bind to the protein "pump."
2 Hydrolysis of ATP phosphorylates the pump protein and changes its shape.
K^+
Inside of cell

5.14 Primary Active Transport: The Sodium–Potassium Pump In active transport, energy is used to move a solute against its concentration gradient. Even though the Na^+ concentration is higher outside the cell and the K^+ concentration is higher inside the cell, for each molecule of ATP hydrolyzed, two K^+ are pumped into the cell and three Na^+ are pumped out of the cell. (The hydrolysis of ATP releases energy and splits the ATP molecule into a molecule of ADP and an inorganic phosphate ion, P_i; see Section 6.2.)

5.15). Secondary active transport aids in the uptake of amino acids and sugars, which are essential raw materials for cell maintenance and growth. Both types of coupled transport proteins—symports and antiports—are used for secondary active transport.

5.4 RECAP

Active transport across a membrane is directional and requires an input of energy to move substances against a concentration gradient. Active transport allows a cell to maintain small molecules and ions in concentrations very different from those of the surrounding environment.

- Why is energy required for active transport? See p. 111
- Do you understand the difference between primary active transport and secondary active transport? See p. 111
- Do you see why the sodium–potassium pump is an antiport? See pp. 111–112 and Figure 5.14

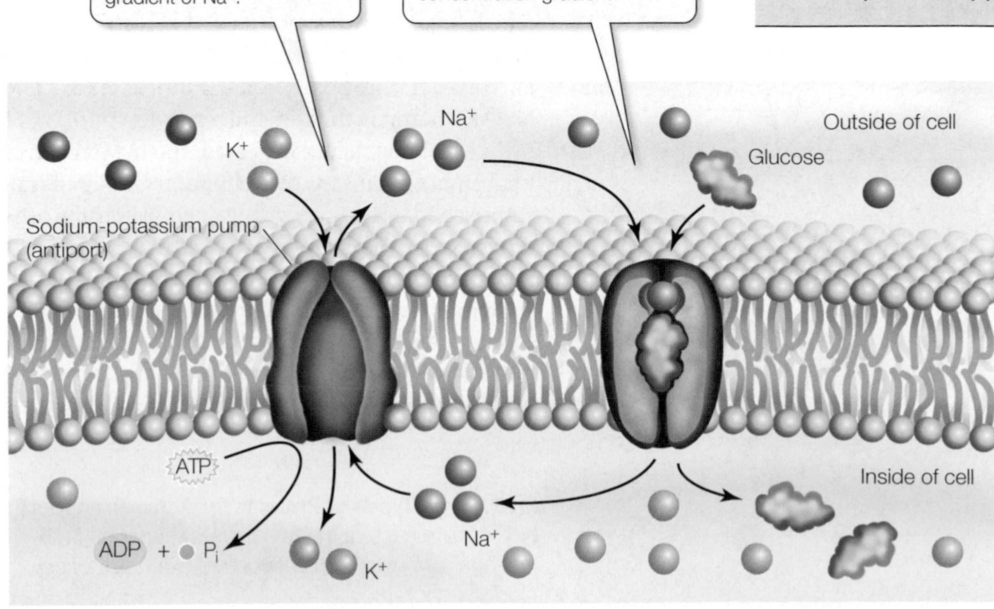

5.15 Secondary Active Transport The Na^+ concentration gradient established by primary active transport (left) powers the secondary active transport of glucose (right). The movement of glucose across the membrane against its concentration gradient is coupled by a symport protein to the movement of Na^+ into the cell.

We have examined a number of ways in which ions and small molecules can enter and leave cells. But what about large molecules? Their size means they diffuse very slowly, and their bulk makes it difficult for them to pass through the membrane directly. It takes a totally different mechanism to move large molecules across membranes.

5.5 How Do Large Molecules Enter and Leave a Cell?

Macromolecules such as proteins, polysaccharides, and nucleic acids are simply too large and too charged or polar to pass through biological membranes. This is actually a fortunate property—think of the consequences if such molecules diffused out of cells. A red blood cell would not retain its hemoglobin! On the other hand, cells must sometimes take up or secrete intact large molecules. As we saw in Section 4.3, this can be done by means of vesicles that either pinch off from the plasma membrane and enter the cell (*endocytosis*) or fuse with the plasma membrane and release their contents (*exocytosis*).

Macromolecules and particles enter the cell by endocytosis

Endocytosis is a general term for a group of processes that bring small molecules, macromolecules, large particles, and even small cells into the eukaryotic cell (**Figure 5.16A**). There are three types of endocytosis: *phagocytosis, pinocytosis,* and *receptor-mediated endocytosis*. In all three, the plasma membrane invaginates (folds inward) around materials from the environment, forming a small pocket. The pocket deepens, forming a vesicle. This vesicle separates from the plasma membrane and migrates with its contents to the cell's interior.

- In **phagocytosis** ("cellular eating"), part of the plasma membrane engulfs large particles or even entire cells. Phagocytosis is used by unicellular protists as a cellular feeding process and by some white blood cells that defend the body by engulfing foreign cells and substances. The food vacuole or phagosome that forms usually fuses with a lysosome, where its contents are digested (see Figure 4.12).
- In **pinocytosis** ("cellular drinking"), vesicles also form. However, these vesicles are smaller, and the process operates to bring small dissolved substances or fluids into the cell. Like phagocytosis, pinocytosis is relatively nonspecific as to what it brings into the cell. For example, pinocytosis goes on constantly in the *endothelium,* the single layer of cells that separates a tiny blood capillary from the surrounding tissue, allowing the cells to rapidly acquire fluids from the blood.
- In **receptor-mediated endocytosis**, specific reactions at the cell surface trigger the uptake of specific materials.

Let's take a closer look at this last process.

Receptor-mediated endocytosis is highly specific

Receptor-mediated endocytosis is used by animal cells to capture specific macromolecules from the cell's environment. This process depends on **receptor proteins**, integral membrane proteins that can bind to a specific molecule in the cell's environment. The uptake process is similar to that in nonspecific endocytosis. However, in receptor-mediated endocytosis, receptor proteins at particular sites on the extracellular surface of the plasma membrane bind to specific substances. These sites are called *coated pits* because they form a slight depression in the plasma membrane, and their cytoplasmic surfaces are coated by other proteins, such as *clathrin*.

When a receptor protein binds to its specific ligand, its coated pit invaginates and forms a *coated vesicle* around the bound macromolecule. Strengthened and stabilized by clathrin molecules, this vesicle carries the macromolecule into the cell (**Figure 5.17**). Once inside, the vesicle loses its clathrin coat and may fuse with a lysosome, where the engulfed material is processed and released into the cytoplasm. Because of its specificity for particular macromolecules, receptor-mediated endocytosis is a rapid and efficient method of taking up substances that may exist at low concentrations in the cell's environment.

Receptor-mediated endocytosis is the method by which cholesterol is taken up by most mammalian cells. Water-insoluble cholesterol and triglycerides are packaged by liver

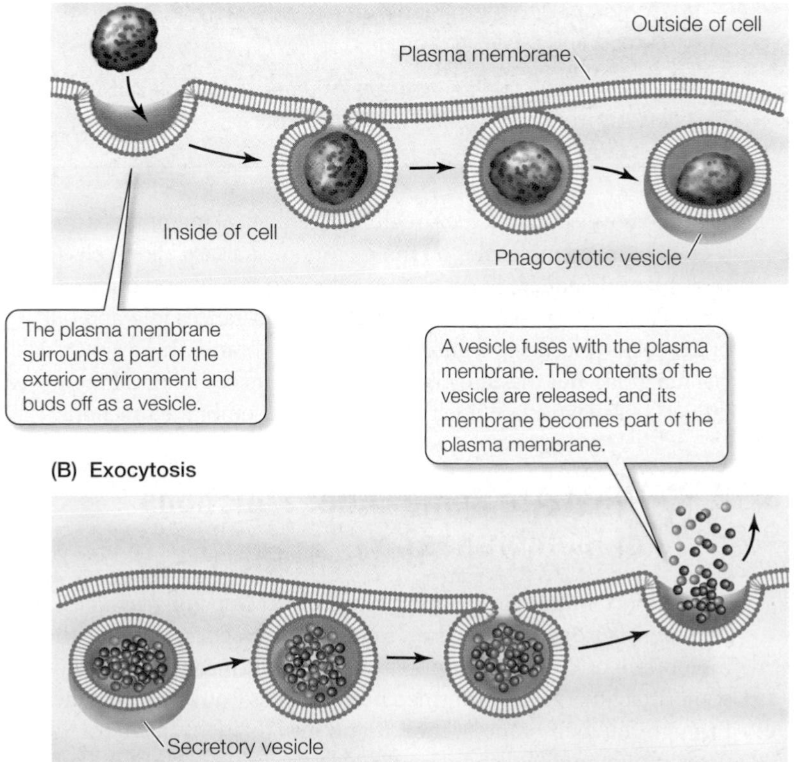
(A) Endocytosis
Outside of cell
Plasma membrane
Inside of cell
Phagocytotic vesicle
The plasma membrane surrounds a part of the exterior environment and buds off as a vesicle.
A vesicle fuses with the plasma membrane. The contents of the vesicle are released, and its membrane becomes part of the plasma membrane.
(B) Exocytosis
Secretory vesicle

5.16 Endocytosis and Exocytosis (A) Endocytosis and (B) exocytosis are used by eukaryotic cells to take up substances from and release substances to the outside environment.

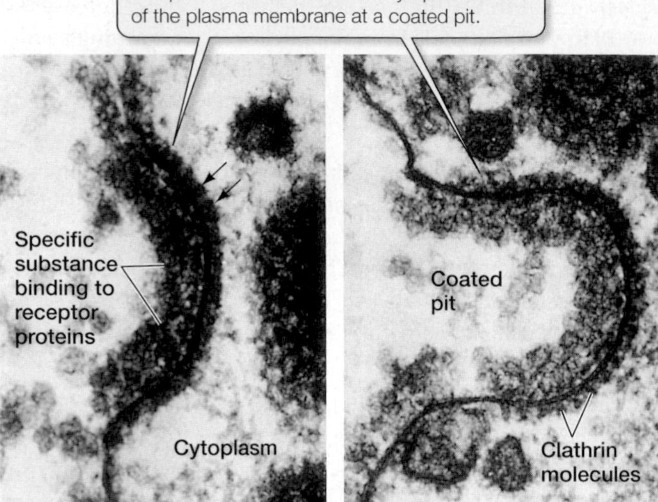
The protein clathrin coats the cytoplasmic side of the plasma membrane at a coated pit.
The endocytosed contents are surrounded by a clathrin-coated vesicle.
Specific substance binding to receptor proteins
Cytoplasm
Coated pit
Clathrin molecules
Coated vesicle

5.17 Formation of a Coated Vesicle In receptor-mediated endocytosis, the receptor proteins in a coated pit bind specific macromolecules, which are then carried into the cell by a coated vesicle.

cells into lipoprotein particles, which are then secreted into the bloodstream to provide body tissues with lipids. One type of lipoprotein particles, called *low-density lipoproteins* or LDLs, must be taken up by the liver for recycling. This uptake also occurs via receptor-mediated endocytosis, beginning with the binding of LDLs to specific receptor proteins on the liver cell surface. Once engulfed by endocytosis, the LDL particle is freed from the receptors. The receptors segregate to a region of the vesicle that buds off and forms a new vesicle, which is recycled to the plasma membrane. The freed LDL particle remains in the original vesicle, which fuses with a lysosome in which the LDL is digested and the cholesterol made available for cell use.

Persons with the inherited disease *familial hypercholesterolemia* have dangerously high levels of cholesterol in their blood because a deficient LDL receptor protein prevents receptor-mediated endocytosis of LDL in their liver.

Exocytosis moves materials out of the cell

Exocytosis is the process by which materials packaged in vesicles are secreted from a cell when the vesicle membrane fuses with the plasma membrane (**Figure 5.16B**). The initial event in this process is the binding of a membrane protein protruding from the cytoplasmic side of the vesicle with a membrane protein on the cytoplasmic side of the target site on the plasma membrane. The phospholipid bilayers of the two membranes merge, and an opening to the outside of the cell develops. The contents of the vesicle are released into the environment, and the vesicle membrane is smoothly incorporated into the plasma membrane.

In Chapter 4, we encountered exocytosis as the last step in the processing of material engulfed by phagocytosis: the secretion of indigestible materials to the environment. Exocytosis is also important in the secretion of many different substances, including digestive enzymes from the pancreas, neurotransmitters from neurons, and materials for the construction of the plant cell wall.

5.5 RECAP

Endocytosis and exocytosis transport molecules that are too large, charged, or polar to be transported through a membrane by passive or active transport. Endocytosis may be specific, mediated by a receptor protein in the membrane.

- Do you understand the difference between phagocytosis and pinocytosis? See p. 113
- Can you describe an example of receptor-mediated endocytosis? See p. 114 and Figure 5.17

We have now examined the structure of biological membranes and seen how macromolecules on the membrane surface allow cells to recognize and adhere to each other, allowing tissues and organs to form. We have also seen how the traffic of substances into and out of the cell is selectively regulated by the membrane. These are crucial functions, but they are not the only aspects of biological membranes. Let's take a brief look at a few other membrane functions.

5.6 What Are Some Other Functions of Membranes?

One important function of membranes is to *keep different materials separate* from one another. Recall from Section 4.3 that the membrane of the rough endoplasmic reticulum is a site for ribosome attachment. Newly formed proteins pass from the ribosomes through the membrane of the ER and into its interior, where the proteins are modified and then delivered to other parts of the cell.

This system requires a separate compartment to segregate these proteins from the rest of the cell.

The plasma membranes of certain types of cells, such as neurons, muscle cells, and some eggs, respond in different ways to the electric charges carried on ions. These membranes are thus *electrically excitable*, which gives them important special properties. For example, in neurons, the plasma membrane conducts nerve impulses from one end of the cell to the other.

Other biological activities and properties associated with membranes are discussed in the chapters that follow. These activities have been essential to the specialization of cells, tissues, and organisms throughout evolution. Three of these activities are especially important:

- *Some organelle membranes help transform energy.* The membranes of certain organelles are specialized for processing energy (**Figure 5.18A**). For example, the inner mitochondrial membrane helps convert the energy of fuel molecules to the energy of ATP, and the thylakoid membranes of chloroplasts participate in the conversion of light energy to the energy of chemical bonds. These important processes, vital to the life of most eukaryotic organisms, are discussed in detail in Chapters 7 and 8.
- *Some membrane proteins organize chemical reactions.* Many cellular processes depend on a series of enzyme-catalyzed reactions in which the products of one reaction serve as the reactants for the next. For such a series of reactions to occur, all the necessary molecules must come together. In a solution, for example, reactant and enzyme molecules are randomly distributed and collisions among them are random; thus a complete series of chemical reactions may occur only very slowly. However, if the different enzymes are bound to a membrane in sequential order, the product of one reaction can be released close to the enzyme for the next reaction. Such an "assembly line" allows reactions to proceed rapidly and efficiently (**Figure 5.18B**).
- *Some membrane proteins process information.* As we have seen, biological membranes may have protruding integral membrane proteins or attached carbohydrates that can bind to specific substances in the environment. The binding of a specific ligand can serve as a signal to initiate, modify, or turn off a cell function (**Figure 5.18C**). In this type of information processing, specificity in binding is essential.

We have seen the informational role of a specific receptor protein in the endocytosis of LDL and its cargo of cholesterol. Another example is the binding of a hormone, such as insulin, to specific receptors on a target cell, such as a liver cell, to elicit a response in that cell—in this case, the uptake of glucose. Chapter 15 will discuss many other examples of the role of membrane proteins in information processing, but one good example brings us back to the story that opened this chapter—that of the disease cholera.

(A) Energy transformation

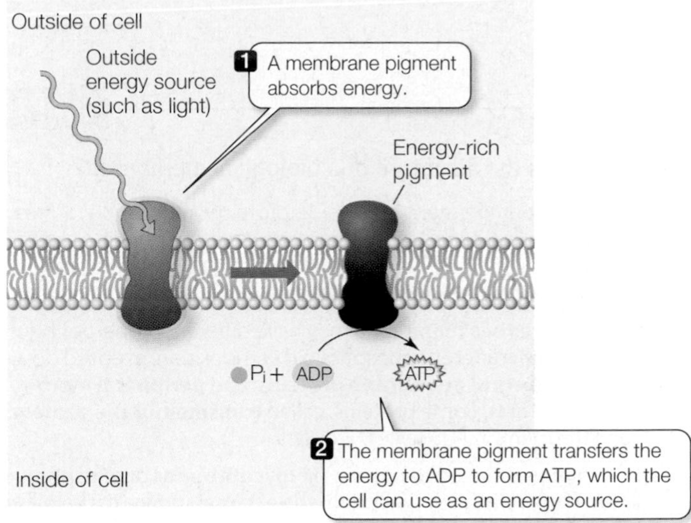

(B) Organizing chemical reactions

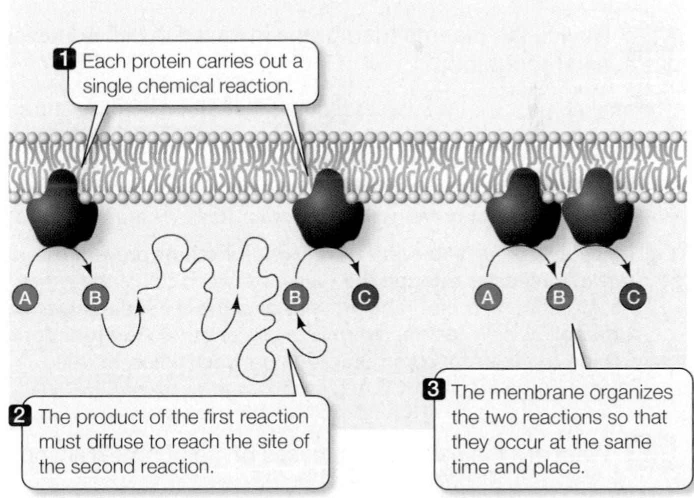

(C) Information processing

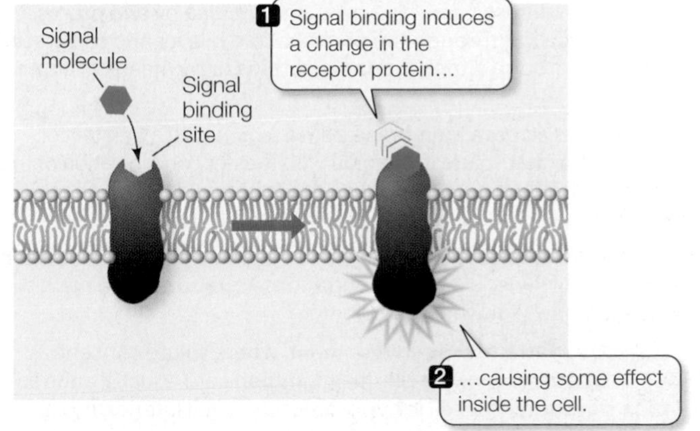

5.18 More Membrane Functions (A) The membranes of organelles such as mitochondria and chloroplasts are specialized for the transformation of energy. (B) When a series of biochemical reactions must take place in sequence, the membrane can sometimes arrange the needed enzymes in an "assembly line" to ensure that the reactions occur in proximity to one another. (C) Membrane proteins have an information-processing role. Receptors on membranes conduct signals from outside the cell that trigger changes inside the cell.

Cholera toxin disrupts cellular information processing. The toxin protein has two subunits, one of which binds to a cell surface glycolipid, causing a change in the three-dimensional shape of the toxin protein so that the second subunit can enter the cell. This subunit acts as an enzyme to modify a peripheral protein called adenylate cyclase on the inner surface of the cell's plasma membrane. The modified adenylate cyclase opens chloride channels in the membrane. Resulting accumulation of Cl^- and Na^+ in the intestine is followed by osmotic loss of water.

Whereas cholera was once almost invariably lethal, even in the developed world, it is now much less so. In 1997, a cholera epidemic broke out among 90,000 Rwandan refugees living in camps in the Congo. The majority of the subsequent 1,521 deaths occurred among people who were outside the health care network serving the camps. Everyone else was saved.

CHAPTER SUMMARY

5.1 What is the structure of a biological membrane?

Biological membranes consist of lipids, proteins, and carbohydrates. The **fluid mosaic model** of membrane structure describes a phospholipid bilayer in which proteins can move about laterally within the membrane. See Web/CD Activity 5.1

The two surfaces of a membrane may have different properties because of their different phospholipid composition, exposed domains of **integral membrane proteins**, and **peripheral membrane proteins**. Some proteins, called **transmembrane proteins**, span the membrane. Review Figure 5.1

Carbohydrates attached to proteins in **glycoproteins** or phospholipids in **glycolipids** project from the external surface of the plasma membrane and function as recognition signals.

Membranes are not static structures, but are constantly forming, exchanging and breaking down. Review Figure 5.5

5.2 How is the plasma membrane involved in cell adhesion and recognition?

The assembly of cells into tissues requires that they recognize and adhere to one another. **Cell recognition** and **cell adhesion** depend on integral membrane proteins that protrude from the cell surface. Binding can be between the same proteins from two cells (**homotypic**) or different proteins (**heterotypic**). Review Figure 5.6

Cell junctions connect adjacent cells. **Tight junctions** prevent the passage of molecules through the spaces between cells and restrict the migration of membrane proteins over the cell surface. **Desmosomes** allow cells to adhere firmly to one another. **Gap junctions** provide channels for communication between adjacent cells. Review Figure 5.7, Web/CD Activity 5.2

5.3 What are the passive processes of membrane transport?

See Web/CD Tutorial 5.1

Membranes exhibit **selective permeability**, regulating which substances pass through them.

Substances can diffuse passively across a membrane by two processes: **simple diffusion** through the phospholipid bilayer and **facilitated diffusion** through either **channel proteins** or by means of a carrier protein.

A solute diffuses across a membrane from a region with a greater concentration of that solute to a region with a lesser concentration of that solute. Equilibrium is reached when the concentrations of the solute are identical on both sides of the membrane. Review Figure 5.8

In **osmosis**, water diffuses from regions of higher water concentration to regions of lower water concentration. **Aquaporins** are membrane channels for water diffusion.

Most cells are in an **isotonic** environment, where solute concentrations on both sides of the plasma membrane are equal. If a the solution surrounding a cell is **hypotonic** to the cell interior, more water enters the cell than leaves it. In plant cells, this leads to **turgor pressure**. In a **hypertonic** solution, more water leaves the cell than enters it. Review Figure 5.9

Ion channels are membrane proteins that allow rapid facilitated diffusion of ions through membranes. These can be **gated channels**, where the channel can be opened or closed by certain conditions of chemicals. The opening or closing of channels can set up an **electrochemical gradient** with unequal charged species on different sides of a membrane. Review Figure 5.10

Carrier proteins bind to polar molecules such as sugars and amino acids and transport them across the membrane. The maximum rate of this type of facilitated diffusion is limited by the number of carrier (transporter) proteins in the membrane; once all carrier proteins are in use, the system is saturated and increases in solute concentration will not increase the rate of diffusion. Review Figure 5.12

5.4 How do substances cross membranes against a concentration gradient?

See Web/CD Tutorial 5.2

Active transport requires the use of chemical energy to move substances across a membrane against a concentration gradient. Active transport proteins may be uniports, symports, or antiports. Review Figure 5.13

In **primary active transport**, energy from the hydrolysis of ATP is used to move ions into or out of cells against their concentration gradients. The **sodium-potassium pump** is an important example. Review Figure 5.14

Secondary active transport couples the passive movement of one substance with its concentration gradient to the movement of another substance against its concentration gradient. Energy from ATP is used indirectly to establish the concentration gradient that results in the movement of the first substance. Review Figure 5.15

5.5 How do large molecules enter and leave a cell?

See Web/CD Tutorial 5.3

Endocytosis is the transport of macromolecules, large particles, and small cells into eukaryotic cells by means of engulfment by and vesicle formation from the plasma membrane. **Phagocytosis** and **pinocytosis** are types of endocytosis. Review Figure 5.16A

In **receptor-mediated endocytosis**, a specific **receptor protein** on the plasma membrane binds to a particular macromolecule.

In **exocytosis**, materials in vesicles are secreted from the cell when the vesicles fuse with the plasma membrane. Review Figure 5.16B

5.6 What are some other functions of membranes?

Membranes function as sites for energy transformations, for organizing chemical reactions and for recognition and initial processing of extracellular signals. Review Figure 5.18

SELF-QUIZ

1. Which statement about membrane phospholipids is *not* true?
   a. They associate to form bilayers.
   b. They have hydrophobic "tails."
   c. They have hydrophilic "heads."
   d. They give the membrane fluidity.
   e. They flip-flop readily from one side of the membrane to the other.
2. When a hormone molecule binds to a specific protein on the plasma membrane, the protein it binds to is called a
   a. ligand.
   b. clathrin.
   c. receptor protein.
   d. hydrophobic protein.
   e. cell adhesion molecule.
3. Which statement about membrane proteins is *not* true?
   a. They all extend from one side of the membrane to the other.
   b. Some serve as channels for ions to cross the membrane.
   c. Many are free to migrate laterally within the membrane.
   d. Their position in the membrane is determined by their tertiary structure.
   e. Some play roles in photosynthesis.
4. Which statement about membrane carbohydrates is *not* true?
   a. Most are bound to proteins.
   b. Some are bound to lipids.
   c. They are added to proteins in the Golgi apparatus.
   d. They show little diversity.
   e. They are important in recognition reactions at the cell surface.
5. Which statement about animal cell junctions is *not* true?
   a. Tight junctions are barriers to the passage of molecules between cells.
   b. Desmosomes allow cells to adhere firmly to one another.
   c. Gap junctions block communication between adjacent cells.
   d. Connexons are made of protein.
   e. The fibers associated with desmosomes are made of protein.
6. You are studying how the protein transferrin enters cells. When you examine cells that have taken up transferrin, you find it inside clathrin-coated vesicles. Therefore, the most likely mechanism for uptake of transferrin is
   a. facilitated diffusion.
   b. an antiport.
   c. receptor-mediated endocytosis.
   d. gap junctions.
   e. ion channels.
7. Which statement about ion channels is *not* true?
   a. They form pores in the membrane.
   b. They are proteins.
   c. All ions pass through the same type of channel.
   d. Movement through them is from high concentrations to low concentrations.
   e. Movement through them is by simple diffusion.
8. Facilitated diffusion and active transport both
   a. require ATP.
   b. require the use of proteins as carriers.
   c. carry solutes in only one direction.
   d. increase without limit as the concentration gradient increases.
   e. depend on the solubility of the solute in lipids.
9. Primary and secondary active transport both
   a. generate ATP.
   b. are based on passive movement of Na^+ ions.
   c. include the passive movement of glucose molecules.
   d. use ATP directly.
   e. can move solutes against their concentration gradients.
10. Which statement about osmosis is *not* true?
    a. It obeys the laws of diffusion.
    b. In animal tissues, water moves into cells if they are hypertonic to their environment.
    c. Red blood cells must be kept in a plasma that is hypotonic to the cells.
    d. Two cells with identical solute concentrations are isotonic to each other.
    e. Solute concentration is the principal factor in osmosis.

FOR DISCUSSION

1. Muscle function requires calcium ions (Ca^{2+}) to be pumped into a subcellular compartment against a concentration gradient. What types of molecules are required for this to happen?
2. Section 27.5 will describe the diatoms, protists that have complex glassy structures in their cell walls (see Figure 27.1). These structures form within the Golgi apparatus. How do these structures reach the cell wall without having to pass through a membrane?
3. Organisms that live in fresh water are almost always hypertonic to their environment. In what way is this a serious problem? How do some organisms cope with this problem?
4. Contrast nonspecific endocytosis and receptor-mediated endocytosis.
5. The emergence of the phospholipid membrane was important to the origin of cells. Describe the most important properties of membranes that might have allowed cells containing them to thrive in comparison with molecular aggregates without membranes.

FOR INVESTIGATION

Under certain laboratory conditions, cells of two different species can be induced to fuse, combining their two plasma membranes in much the same way that vesicles fuse with the plasma membrane in exocytosis. Antibodies labeled with fluorescent dyes can be used to stain specific proteins on the cell surface. How would you use cell fusion to investigate whether membrane proteins diffuse in the plane of the membrane?

CHAPTER 6 Energy, Enzymes, and Metabolism

Sensitivity to alcohol

The guests were at their tables as the bride and groom entered the room to enthusiastic applause. Champagne glasses were raised in the first of many toasts. But no sooner had Frank drunk his first small glass of bubbly than he started feeling ill. With his face flushed and his heart beating rapidly, he excused himself and left the reception. It was hours before he felt normal again. The same thing had happened to him before, also when he drank a small amount of an alcoholic beverage.

There are people who can drink a lot of alcohol without getting sick, and there are others, like Frank, who become ill after consuming only a small amount. These individual differences are a function of the biochemistry of alcohol once it enters the body. Typically, a series of two chemical reactions transforms ethyl alcohol (the alcohol in beverages) into acetate in the body: first ethyl alcohol is transformed into acetaldehyde, then acetaldehyde is transformed into acetate.

These reactions constitute a *metabolic pathway*, in which the product of the first reaction, acetaldehyde, serves as the raw material for the second. Neither reaction can occur to any appreciable extent without help. Each reaction needs to be "speeded up" by a different *catalyst*—a specific *enzyme*.

Enzymes are proteins, and as such, each has a specific amino acid sequence. The enzyme that speeds up the second reaction in the metabolism of alcohol is aldehyde dehydrogenase (ALDH), which has 517 amino acids. Most of the guests at Frank's table had the same amino acid sequence in their ALDH. In this "typical" sequence, the 487th amino acid is glutamic acid, whose side chain carries a negative charge (see Table 3.1). The other guests' ALDH folded into the correct three-dimensional structure, and the enzyme did its job—speeding up the conversion of acetaldehyde into acetate. In Frank's cells, however, the amino acid at position 487 is lysine instead of glutamic acid; lysine has a *positive* charge. This seemingly tiny difference is enough to change the three-dimensional structure of ALDH and interfere with its catalytic function.

Because of its structural defect, Frank's ALDH works, but only slowly. When he drank that small glass of champagne, the first metabolic reaction (the conversion of alcohol to acetaldehyde) happened quickly, but the second reaction (the conversion of acetaldehyde to acetate) did not. And, whereas alcohol has some pleasant effects on the body, the effects produced by acetaldehyde are not so pleasant. As

A Catalyst for Alcohol The enzyme aldehyde dehydrogenase catalyzes an important step in the breakdown of alcohol.

A Precursor to Trouble Some people have a weakly active form of the aldehyde dehydrogenase enzyme. When they drink alcohol, acetaldehyde accumulates in their bodies, with ill effects.

acetaldehyde accumulated in his body, Frank became ill. His facial flushing and accelerated heartbeat were typical short-term effects of acetaldehyde; in the longer term, it can cause brain abnormalities and even some cancers.

Frank's atypical ALDH sequence shows how individual differences in alcohol tolerance are not necessarily behavioral, but can be due to differences in basic biochemistry. (So think twice before kidding a person because he or she can't "hold his liquor.")

IN THIS CHAPTER we begin our study of biochemical transformations, focusing on the role of energy. After describing the physical principles that underlie energy transformations and how these principles apply to biology, we will see how the energy carrier ATP plays an important role in the cell. The rest of the chapter deals with the nature, activities, and regulation of enzymes, which speed up biochemical transformations and without which life would not be possible.

CHAPTER OUTLINE

6.1 What Physical Principles Underlie Biological Energy Transformations?

Metabolic reactions and catalysts are essential to the biochemical transformation of energy by living things. Whether it is a plant using light energy to produce carbohydrates or a cat transforming food energy so it can leap to a countertop (where it hopes to find food so it can obtain more energy), the transformation of energy is a hallmark of life.

Physicists define **energy** as the capacity to do work. *Work* is what occurs when a force operates on an object over a distance. In biochemistry, it is more useful to consider energy as *the capacity for change*. No cell creates energy; all living things must obtain energy from the environment. Indeed, one of the fundamental physical laws is that energy can neither be created nor destroyed. However, energy can be *transformed* from one type into another, and living cells carry out many such energy transformations. Energy transformations are linked to the chemical transformations that occur in cells—the breaking of chemical bonds, the movement of substances across membranes, and so forth.

There are two basic types of energy and of metabolism

Energy comes in many forms: chemical, electric, heat, light, and mechanical. But all forms of energy can be considered as one of two basic types:

- **Potential energy** is the energy of state or position—that is, stored energy. It can be stored in many forms: in chemical bonds, or as a concentration gradient, or even as an electric charge imbalance (as in the membrane potential we described in Section 5.3). Think of a crouching cat, muscles hunched taut, holding still as it prepares to pounce.
- **Kinetic energy** is the energy of movement—that is, the type of energy that does work. Think of the cat leaping as some of the potential energy stored in its taut muscles is converted into muscle contractions.

Just as potential and kinetic energy can interconvert, the *form* of that energy can also interconvert. The potential energy in the cat's muscles is in covalent bonds (chemical energy), while

Potential chemical energy is converted into kinetic mechanical energy when the cat leaps.
Potential chemical energy is stored in the muscles of the cat.

6.1 Energy Conversions and Work A leaping cat illustrates both the conversion between potential and kinetic energy and the conversion of energy from one form (chemical) to another (mechanical).

the kinetic energy of the pouncing cat is mechanical (**Figure 6.1**). You can think of many other such interconversions: reading this book, for example, involves light energy being converted into chemical energy in your eyes, and the chemical energy being converted into electric energy in the nerve cells that carry messages to your brain.

In any living organism, chemical reactions are occurring continuously. **Metabolism** is defined as the sum total of these reactions. Two types of metabolic reactions occur in all cells of all organisms:

- **Anabolic reactions** (*anabolism*) link simple molecules to form more complex molecules. The synthesis of a protein from amino acids is an anabolic reaction. Anabolic reactions require an input of energy and capture it in the chemical bonds that are formed.
- **Catabolic reactions** (*catabolism*) break down complex molecules into simpler ones and release the energy stored in chemical bonds.

Catabolic and anabolic reactions are often linked. The energy released in catabolic reactions is often used to drive anabolic reactions—that is, to do biological work, such as the contraction of a leaping cat's muscles or cellular activities such as active transport across a membrane.

Because living organisms are part of the physical universe, the **laws of thermodynamics** (*thermo*, "energy"; *dynamics*, "change") that operate in stars also apply to energy transformations in cells.

The first law of thermodynamics: Energy is neither created nor destroyed

The first law of thermodynamics states that in any conversion of energy, it is neither created nor destroyed. Another way of saying this is that *in any conversion of energy from one form to another, the total energy before and after the conversion is the same* (**Figure 6.2A**). As you will see in the next two chapters, potential energy in the chemical bonds of carbohydrates can be converted into potential energy in the form of ATP. This energy can then be converted into kinetic energy and used to do mechanical work, such as muscle contraction.

The second law of thermodynamics: Disorder tends to increase

The second law of thermodynamics states that, although energy cannot be created or destroyed, *when energy is converted from one form into another, some of that energy becomes unavailable to do work* (**Figure 6.2B**). In other words, no physical process or chemical reaction is 100 percent efficient, and not all the energy released can be converted into work. Some energy is lost to a form associated with disorder. Think of disorder as a kind of randomness. It takes energy to impose order on a system. Unless energy is applied to impose order on a system, that system will be randomly arranged or disordered. Your bedroom or dorm room is probably a good analogy. If you are like we were as students, your room is rather disordered, and it takes energy to organize it. The second law applies to all energy transformations, but we will focus here on chemical reactions in living systems.

NOT ALL ENERGY CAN BE USED In any system, the total energy includes the usable energy that can do work *and* the unusable energy that is lost to disorder:

$$\text{total energy} = \text{usable energy} + \text{unusable energy}$$

In biological systems, the total energy is called **enthalpy** (***H***). The usable energy that can do work is called **free energy** (***G***). Free energy is what cells require for all the chemical reactions of cell growth, cell division, and the maintenance of cell health. The unusable energy is represented by **entropy** (***S***), which is a measure of the disorder of the system, multiplied by the *absolute temperature* (T). Thus we can rewrite the word equation above more precisely as

$$H = G + TS$$

Because we are interested in usable energy, we rearrange this expression:

$$G = H - TS$$

Why do you get hot when you exercise? Your muscles convert the chemical energy you obtain from food to the mechanical energy of muscle contraction. Less than 20 percent of this energy conversion does the work of exercise; the rest is lost as heat.

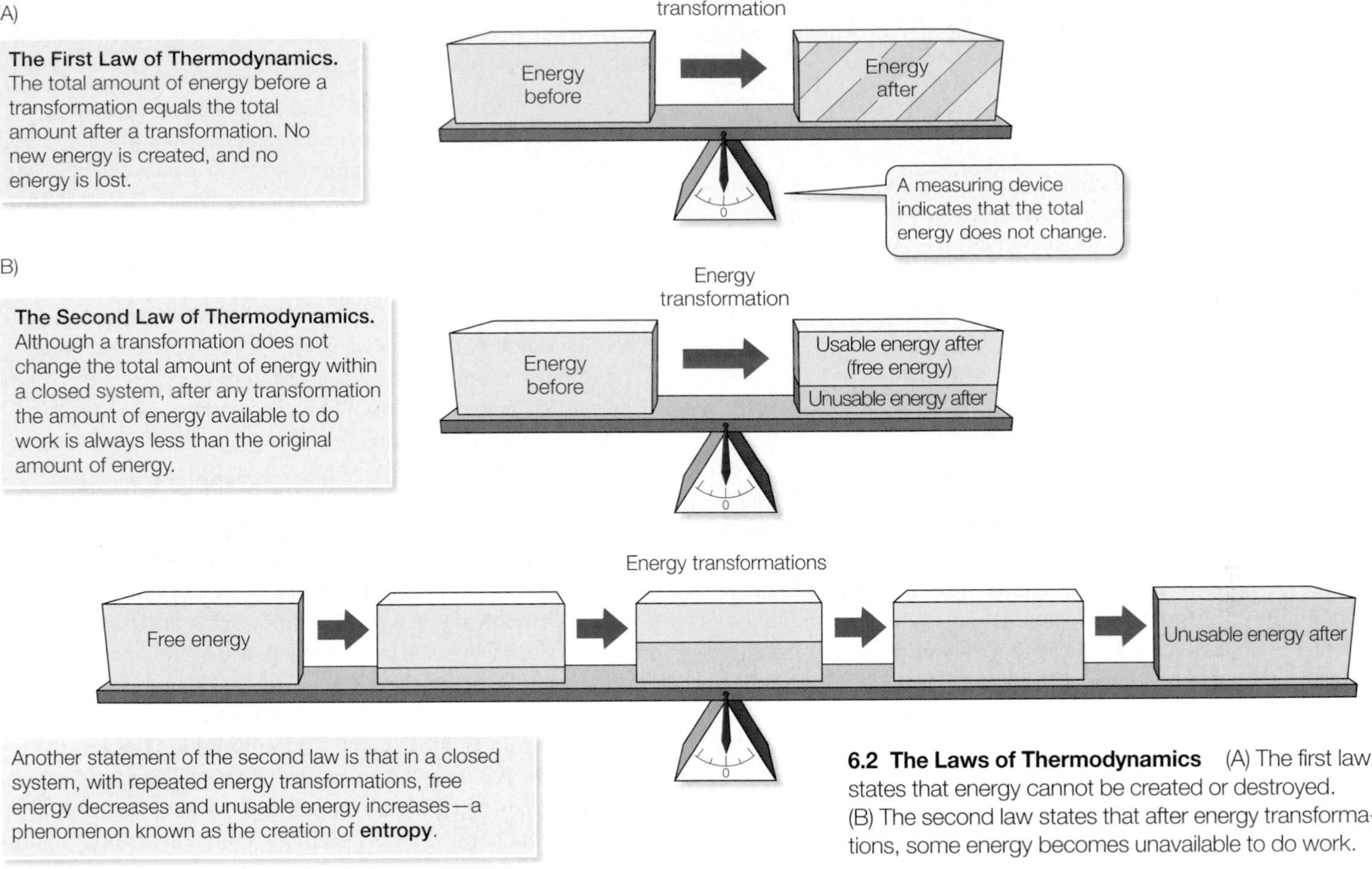

6.2 The Laws of Thermodynamics (A) The first law states that energy cannot be created or destroyed. (B) The second law states that after energy transformations, some energy becomes unavailable to do work.

Although we cannot measure *G*, *H*, or *S* absolutely, we can determine the *change* in each at a constant temperature. Such energy changes are measured in calories (cal) or joules (J).* A change in energy is represented by the Greek letter *delta* (Δ). The **change in free energy** (ΔG) of any chemical reaction is equal to the difference in free energy between the products and the reactants,

$$\Delta G_{\text{reaction}} = G_{\text{products}} - G_{\text{reactants}}$$

Such a change can be either positive or negative; that is, the free energy of the products can be more or less than the free energy of the reactants. If the products have more free energy, then there must have been some input of energy into the reaction (remember, energy cannot be created, so some must somehow have been added from an external source).

At a constant temperature, ΔG is defined in terms of the change in total energy (ΔH) and the change in entropy (ΔS):

$$\Delta G = \Delta H - T\Delta S$$

This equation tells us whether free energy is released or consumed by a chemical reaction:

- If ΔG is negative ($\Delta G < 0$), free energy is released.
- If ΔG is positive ($\Delta G > 0$), free energy is required (consumed).

If the necessary free energy is not available, the reaction does not occur. The sign and magnitude of ΔG depend on the two factors on the right of the equation:

- ΔH: In a chemical reaction, ΔH is the total amount of energy added to the system ($\Delta H > 0$) or released ($\Delta H < 0$).
- ΔS: Depending on the sign and magnitude of ΔS, the entire term, $T\Delta S$, may be negative or positive, large or small. In other words, in living systems at a constant temperature (no change in *T*), the magnitude and sign of ΔG can depend a lot on changes in entropy. Large changes in entropy make ΔG more negative in value, as shown by the minus sign in front of the $T\Delta S$ term.

If a chemical reaction increases entropy, its products are more disordered or random than its reactants. If there are more products than reactants, as in the hydrolysis of a protein to its amino acids, the products have considerable freedom to move around. The disorder in a solution of amino acids will be large compared with that in the protein, in which peptide bonds and other forces prevent free movement. So in hydrolysis, the change in entropy (ΔS) will be positive.

If there are fewer products, and they are more restrained in their movements than the reactants, ΔS will be negative. For example, a large protein linked by peptide bonds is less free in its move-

*A *calorie* is the amount of heat energy needed to raise the temperature of 1 gram of pure water from 14.5°C to 15.5°C. A *joule* is an energy measure in the commonly used SI system. 1 J = 0.239 cal; conversely, 1 cal = 4.184 J. Thus, for example, 486 cal = 2,033 J, or 2.033 kJ. Although defined in terms of heat, the calorie and the joule are measures of any form of energy—mechanical, electrical, or chemical. When you compare data on energy, always compare joules to joules and calories to calories.

ments than a solution of the hundreds or thousands of amino acids from which it was synthesized.

DISORDER TENDS TO INCREASE The second law of thermodynamics also predicts that, *as a result of energy transformations, disorder tends to increase*. Chemical changes, physical changes, and biological processes all tend to increase entropy and therefore tend toward disorder or randomness (see Figure 6.2B). This tendency for disorder to increase gives a directionality to physical processes and chemical reactions. It explains why some reactions proceed in one direction rather than another.

How does the second law apply to organisms? Consider the human body, with its highly complex structures constructed of simple molecules. This increase in complexity appears to be in conflict with the second law. But this is not the case! Constructing 1 kg of a human body requires that about 10 kg of biological materials be metabolized and in the process converted into CO_2, H_2O, and other simple molecules. This metabolism creates far more disorder than the order in 1 kg of flesh, and the conversions require a lot of energy. *Life requires a constant input of energy to maintain order*. There is no conflict with the second law of thermodynamics.

Having seen that the laws of thermodynamics apply to living things, we will now turn to a consideration of how these laws apply to biochemical reactions.

Chemical reactions release or consume energy

Recall that anabolic reactions tend to increase complexity (order) in the cell, while catabolic reactions break down complexity (create disorder). Accordingly, anabolic reactions require energy, while catabolic reactions release energy:

- Catabolic reactions may break down an ordered reactant, such as a protein molecule, into smaller, more randomly distributed products, such as amino acids. *Reactions that release free energy* ($-\Delta G$) *are called* **exergonic reactions** (**Figure 6.3A**). For example:

  complex molecules → free energy + small molecules

- Anabolic reactions may make a single product, such as a protein (a highly ordered substance), out of many smaller reactants, such as amino acids (less ordered). *Reactions that require or consume free energy* ($+\Delta G$) *are called* **endergonic reactions** (**Figure 6.3B**). For example:

  free energy + small molecules → complex molecules

In principle, chemical reactions can run both forward and backward. For example, if compound A can be converted into compound B (A → B), then B, in principle, can be converted into A (B → A), although at given concentrations of A and B, only one of these directions will be favored. Think of the overall reaction as resulting from competition between forward and reverse reactions ($A \rightleftharpoons B$). Increasing the concentration of A speeds up the forward reaction, and increasing the concentration of B favors the reverse reaction. At some concentration of A and B, the forward and reverse reactions take place at the same rate. At this concentration, no further net change in the system is observable, although individual molecules are still forming and breaking apart. This balance between forward and reverse reactions is known as **chemical equilibrium**. Chemical equilibrium is a static state, a state of no net change, and a state in which $\Delta G = 0$.

(A) Exergonic reaction

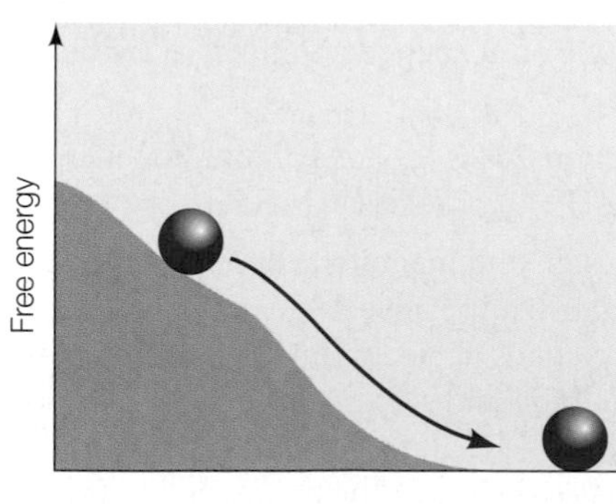
Free energy

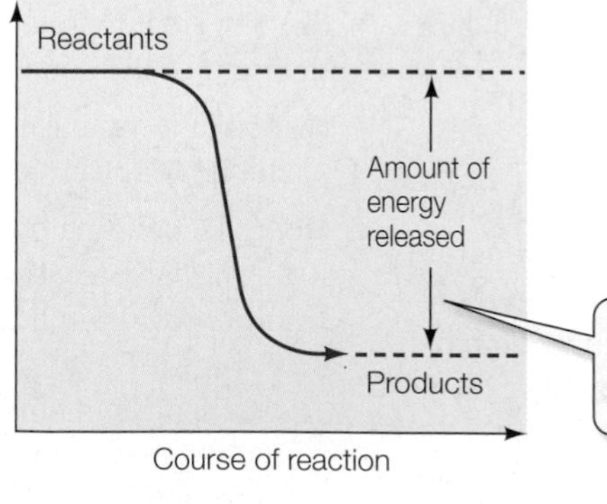
Reactants
Amount of energy released
Products
Course of reaction
In an exergonic reaction, energy is released as the reactants form lower-energy products. ΔG is negative.

(B) Endergonic reaction

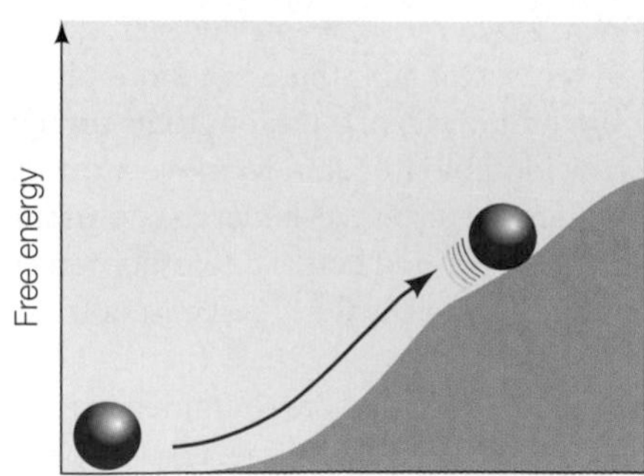
Free energy

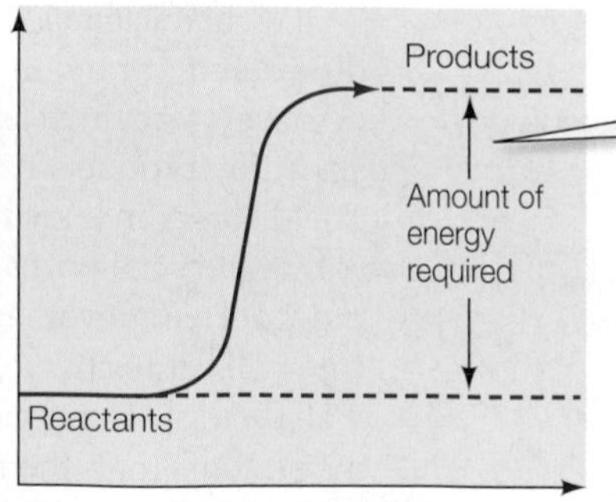
Products
Amount of energy required
Reactants
Course of reaction
Energy must be added for an endergonic reaction, in which reactants are converted to products with a higher energy level. ΔG is positive.

6.3 Exergonic and Endergonic Reactions (A) In an exergonic reaction, the reactants behave like a ball rolling down a hill, and energy is released. (B) A ball will not roll uphill by itself. Driving an endergonic reaction, like moving a ball uphill, requires adding free energy.

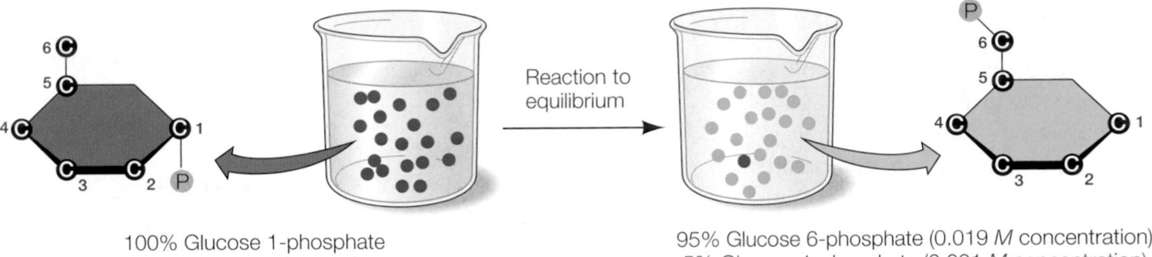

6.4 Chemical Reactions Run to Equilibrium No matter what quantities of glucose 1-phosphate and glucose 6-phosphate are dissolved in water, when equilibrium is attained, there will always be 95 percent glucose 6-phosphate and 5 percent glucose 1-phosphate.

Chemical equilibrium and free energy are related

Every chemical reaction proceeds to a certain extent, but not necessarily to completion. In other words, all the reactants present are not necessarily converted into products. Each reaction has a specific equilibrium point, and that equilibrium point is related to the free energy released by the reaction under specified conditions. To understand the principle of equilibrium, consider the following example.

Most cells contain glucose 1-phosphate, which is converted in the cell into glucose 6-phosphate. Imagine that we start out with an aqueous solution of glucose 1-phosphate that has a concentration of 0.02 *M*. (*M* stands for molar concentration; see Section 2.4). The solution is maintained under constant environmental conditions (25°C and pH 7). As the reaction proceeds slowly to equilibrium, the concentration of the product, glucose 6-phosphate, rises from 0 to 0.019 *M*, while the concentration of the reactant, glucose 1-phosphate, falls to 0.001 *M*. At this point, equilibrium is reached (**Figure 6.4**). From then on, the reverse reaction, from glucose 6-phosphate to glucose 1-phosphate, progresses at the same rate as the forward reaction.

At equilibrium, then, this reaction has a product-to-reactant ratio of 19:1 (0.019/0.001), so the forward reaction has gone 95 percent of the way to completion ("to the right," as written). Therefore, the forward reaction is an exergonic reaction. This result is obtained every time the experiment is run under the same conditions. The reaction is described by the equation

$$\text{glucose 1-phosphate} \rightleftharpoons \text{glucose 6-phosphate}$$

The change in free energy (ΔG) for any reaction is related directly to its point of equilibrium. The further toward completion the point of equilibrium lies, the more free energy is released. In an exergonic reaction, such as the conversion of glucose 1-phosphate to glucose 6-phosphate, ΔG is a *negative* number (in this example, $\Delta G = -1.7$ kcal/mol, or -7.1 kJ/mol).

A large, *positive* ΔG for a reaction means that it proceeds hardly at all to the right ($A \rightarrow B$). But if the product is present, such a reaction runs backward, or "to the left" ($A \leftarrow B$) (nearly all B is converted into A). A ΔG value near zero is characteristic of a readily reversible reaction: reactants and products have almost the same free energies.

6.1 RECAP

Biochemical energy transformations are governed by two laws of thermodynamics. These transformations can release or consume energy, and may not run to completion but instead end up at an equilibrium point.

- Do you understand the differences between potential energy and kinetic energy? Between anabolism and catabolism? See pp. 119–120
- Can you explain the laws of thermodynamics and how they relate to biology? See pp. 120–122 and Figure 6.2
- Do you understand the difference between endergonic and exergonic reactions and the importance of ΔG? See p. 120 and Figure 6.3

The principles of thermodynamics that we have been discussing apply to all energy transformations in the universe, so they are very powerful and useful. Next, we'll apply them to reactions in cells that involve the biological energy currency, ATP.

6.2 What Is the Role of ATP in Biochemical Energetics?

Cells rely on **adenosine triphosphate**, or **ATP**, for the capture and transfer of the free energy they need to do chemical work. ATP operates as a kind of "energy currency." That is, just as you may earn money from a job and then spend it on a meal, some of the free energy that is released by certain exergonic reactions is captured in ATP, which can then release free energy to drive endergonic reactions.

ATP is produced by cells in a number of ways (which we will describe in the next two chapters), and it is used in many ways. ATP is not an unusual molecule. In fact, it has another important use in the cell: it is a nucleotide that can be converted into a building block for nucleic acids. But two things about ATP make it especially useful to cells: it releases a relatively large amount of energy when hydrolyzed, and it can *phosphorylate* (donate a phosphate group to) many different molecules. We will examine these two properties in the discussion that follows.

ATP hydrolysis releases energy

An ATP molecule consists of the nitrogenous base adenine bonded to ribose (a sugar), which is attached to a sequence of three phosphate groups (**Figure 6.5A**). The hydrolysis of ATP yields free energy, as well as ADP (adenosine *di*phosphate) and an inorganic phosphate ion; the ion, HPO_4^{2-}, is commonly abbreviated P_i. Thus:

$$ATP + H_2O \rightarrow ADP + P_i + \text{free energy}$$

The important property of this reaction is that it is exergonic, releasing free energy. The change in free energy (ΔG) is about –7.3 kcal/mol (–30 kJ/mol) at a defined temperature, pH, and solute concentration.

Two characteristics of ATP account for the free energy released by the loss of one of its phosphate groups:

- The free energy of the P—O bond between phosphate groups (called a phosphoric acid anhydride bond) is much higher than the energy of the H—O bond that forms after hydrolysis. So some usable energy is released by hydrolysis.
- Because phosphates are negatively charged and so repel each other, it takes energy to get phosphates near enough to each other to make the covalent bond that links them together (e.g., to add a phosphate to ADP to make ATP).

Bioluminescence—the production of light by living organisms (**Figure 6.5B**)—is an example of an endergonic reaction driven by ATP hydrolysis as well as of the interconversion of energy forms (chemical to light). The chemical that becomes luminescent is called luciferin (after the devil, Lucifer):

$$\text{luciferin} + O_2 + ATP \xrightarrow{\text{luciferase}} \text{oxyluciferin} + AMP + PP_i + \text{light}$$

This reaction and the enzyme (luciferase) that catalyzes it occur in a wide variety of organisms besides the familiar firefly, ranging from marine organisms to mushrooms. The light is generally used to avoid predators or signal to mates.

Soft-drink companies use the firefly proteins luciferin and luciferase to detect bacterial contamination. Where there are living cells, there is ATP, and when the firefly proteins encounter ATP and oxygen, they give off light. Thus, soda that lights up is contaminated.

(A)

ATP (space-filling model)

ATP (structural formula)

Adenine

Phosphate groups

Ribose

Adenosine

AMP (Adenosine monophosphate)

ADP (Adenosine diphosphate)

ATP (Adenosine triphosphate)

(B)

6.5 ATP (A) ATP is richer in energy than its relatives ADP and AMP. The hydrolysis of ATP releases the energy stored in the P—O bond between the second and third phosphates. (B) Fireflies use ATP to initiate the oxidation of luciferin. This process converts chemical energy into light energy, emitting rhythmic flashes that signal the insect's readiness to mate. Very little of the energy in this conversion is lost as heat.

ATP couples exergonic and endergonic reactions

As we have just seen, the *hydrolysis of ATP is exergonic* and yields ADP, P_i, and free energy. The reverse reaction, the *formation of ATP from ADP and* P_i, *is endergonic* and consumes as much free energy as is released by the hydrolysis of ATP:

$$ADP + P_i + \text{free energy} \rightarrow ATP + H_2O$$

Many different exergonic reactions in the cell can provide the energy to convert ADP into ATP. In eukaryotes, the most important of these reactions is cellular respiration, in which some of the energy released from fuel molecules is captured in ATP. The formation and hydrolysis of ATP constitute what might be called an "energy-coupling cycle," in which ADP picks up energy from exergonic reactions to become ATP, which donates energy to endergonic reactions.

How does this ATP cycle trap and release energy? An exergonic reaction is coupled to the endergonic reaction that forms ATP from ADP and P_i (**Figure 6.6**). *Coupling of exergonic and endergonic reac-*

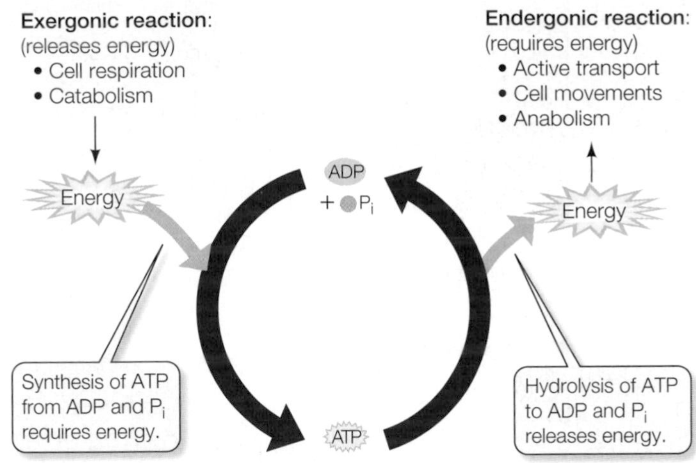

6.6 Coupling of Reactions Exergonic cellular reactions release the energy needed to make ATP from ADP. The energy released from the conversion of ATP back to ADP can be used to fuel endergonic reactions.

tions is very common in metabolism. When it forms, ATP captures free energy and retains it in the form of its P—O bond. ATP then diffuses to another site in the cell, where its hydrolysis releases free energy to drive an endergonic reaction.

A specific example of this energy-coupling cycle is shown in **Figure 6.7**. The formation of the amino acid glutamine has a positive ΔG (is endergonic) and will not proceed without the input of free energy from ATP hydrolysis, which has a negative ΔG (is exergonic). The overall ΔG for the coupled reactions (when the two ΔGs are added together) is negative. Hence the reactions proceed exergonically when they are coupled, and glutamine is synthesized.

An active cell requires millions of molecules of ATP per second to drive its biochemical machinery. An ATP molecule is consumed within a second of its formation, on average. At rest, an average person produces and hydrolyzes about 40 kg of ATP per day—as much as some people weigh. This means that each ATP molecule undergoes about 10,000 cycles of synthesis and hydrolysis every day.

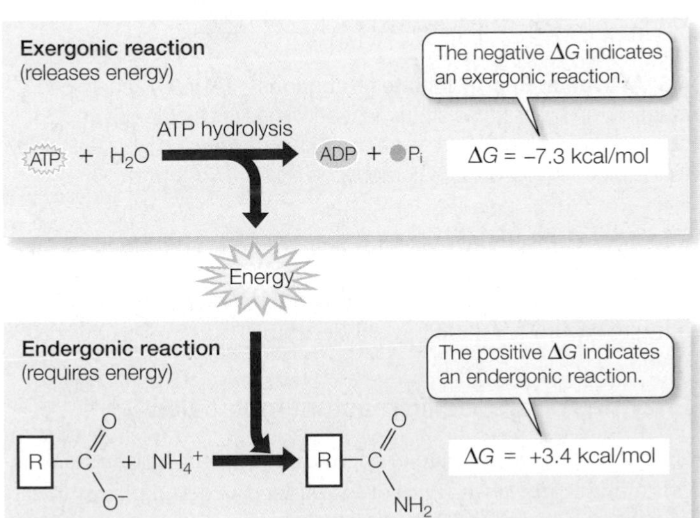

6.7 Coupling of ATP Hydrolysis to an Endergonic Reaction The synthesis of the amino acid glutamine from glutamate and an ammonium ion is an endergonic reaction that must be coupled with the exergonic hydrolysis of ATP.

6.2 RECAP

ATP is the "energy currency" of cells. ATP captures some of the free energy released by exergonic reactions, which can then be released by ATP hydrolysis and used to drive endergonic reactions.

- How does ATP store energy? See p. 124
- Can you explain the concept of coupled reactions? See pp. 124–125 and Figure 6.7

ATP is synthesized and used up very rapidly. But these biochemical reactions could not proceed so rapidly without the help of catalytic proteins called enzymes.

6.3 What Are Enzymes?

When we know the change in free energy (ΔG) of a reaction, we know where the equilibrium point of the reaction lies: the more negative ΔG is, the further the reaction proceeds toward completion. However, ΔG tells us nothing about the **rate** of a reaction—the speed at which it moves toward equilibrium. The reactions that occur in cells are so slow that they could not contribute to life unless the cells did something to speed them up. That is the role of **catalysts**: substances that speed up a reaction without being permanently altered by that reaction. A catalyst does not cause a reaction that would not take place eventually without it, but merely speeds up the rates of both forward and backward reactions, allowing equilibrium to be approached more rapidly.

Most biological catalysts are proteins called **enzymes**. Although we will focus here on proteins, some catalysts—perhaps the earliest ones in the origin of life—are RNA molecules called ribozymes (see Section 3.6). *A biological catalyst,* whether protein or RNA, *is a framework or scaffold in which chemical catalysis takes place.* Over time, proteins have evolved as catalysts, probably because of their great diversity in three-dimensional structure and variety of chemical functions.

In this section we will identify the energy barrier that controls the rate of chemical reactions. Then we'll focus on the role of enzymes: how they interact with specific reactants, how they lower the energy barrier, and how they permit reactions to proceed more quickly. In Section 6.4 we'll look at how enzymes contribute to the coupling of reactions.

For a reaction to proceed, an energy barrier must be overcome

An exergonic reaction may release a great deal of free energy, but the reaction may take place very slowly. Some reactions are slow because there is an *energy barrier* between reactants and products. Think about the propane stove we described in Section 2.3. The burning of propane ($C_3H_8 + 5\ O_2 \rightarrow 3\ CO_2 + 4\ H_2O$ + energy) is an exergonic reaction—energy is released in the form of heat and light. Once started, the reaction goes to completion: all of the propane reacts with oxygen to form carbon dioxide and water vapor.

Because burning propane liberates so much energy, you might expect this reaction to proceed rapidly whenever propane is exposed to oxygen. But this does not happen. Simply mixing propane with air produces no reaction. Propane will start burning only if a spark—an input of energy—is provided. (In the case of the stove, this energy is supplied by a match.) The need for this spark to start the reaction shows that there is an energy barrier between the reactants and the products.

In general, exergonic reactions proceed only after the reactants are pushed over the energy barrier by a small amount of added energy. The energy barrier thus represents the amount of energy needed to start the reaction, known as the **activation energy (E_a)** (**Figure 6.8A**). Recall the ball rolling down the hill in Figure 6.3. The ball has a lot of potential energy at the top of the hill. However, if the ball is stuck in a small depression, it will not roll down the hill, even though that action is exergonic. To start the ball rolling, a small amount of energy (activation energy) is needed to get the ball out of the depression (**Figure 6.8B**).

In a chemical reaction, the activation energy is the energy needed to change the reactants into unstable molecular forms called **transition-state species**. Transition-state species have higher free energies than either the reactants or the products. Their bonds may be stretched and hence unstable. Although the amount of activation energy needed for different reactions varies, it is often small compared with the change in free energy of the reaction. The activation energy that starts a reaction is recovered during the ensuing "downhill" phase of the reaction, so it is not a part of the net free energy change, ΔG (see Figure 6.8A).

Where does the activation energy come from? In any collection of reactants at room or body temperatures, some molecules are moving around, and they could use their kinetic energy of motion to overcome the energy barrier, enter the transition state, and react. However, at these temperatures, only a few molecules have enough energy to do this; most have insufficient kinetic energy for activation, so the reaction takes place slowly. If the system were heated, all the reactant molecules would move faster and have more kinetic energy. Since more molecules would have energy exceeding the required activation energy, the reaction would speed up.

However, adding enough heat to increase the average kinetic energy of the molecules would not work in living systems. Such a nonspecific approach would accelerate all reactions, including destructive ones, such as the denaturation of proteins (see Figure 3.11). A more effective way to speed up a reaction in a living system is to lower the energy barrier by bringing the reactants into close proximity to each other. In living cells, enzymes accomplish this task.

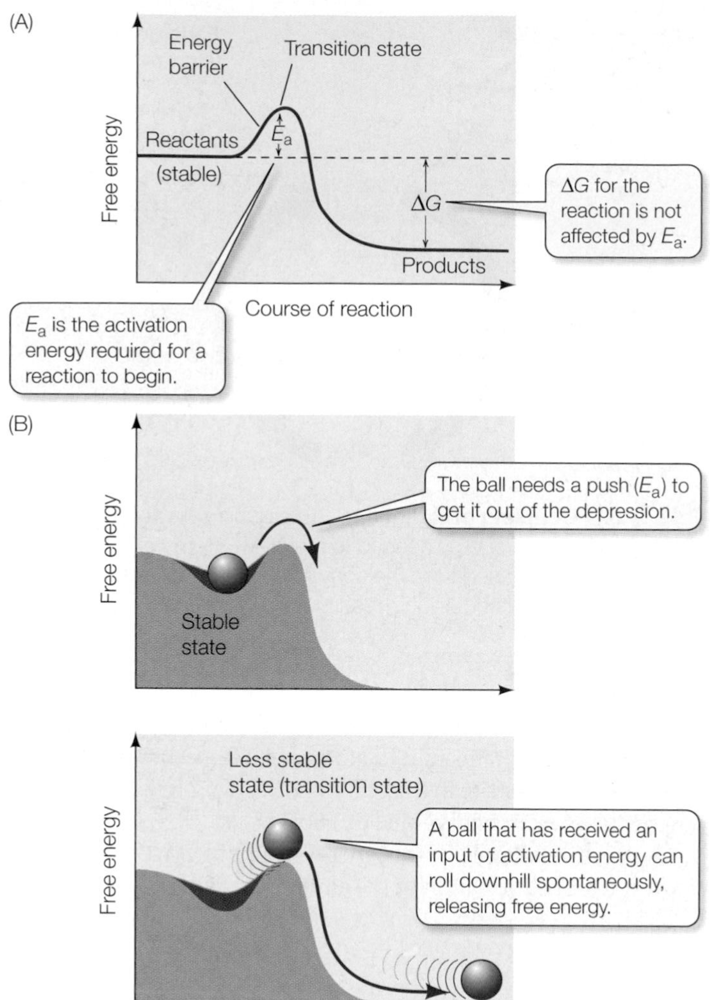

6.8 Activation Energy Initiates Reactions (A) In any chemical reaction, an initial stable state must become less stable before change is possible. (B) A ball on a hillside provides a physical analogy to the biochemical principle graphed in (A).

Enzymes bind specific reactant molecules

Catalysts increase the rate of chemical reactions. Most *nonbiological* catalysts are *nonspecific*. For example, powdered platinum catalyzes virtually any reaction in which molecular hydrogen (H_2) is a reactant. In contrast, most *biological* catalysts are *highly specific*. These complex molecules of protein (enzymes) or RNA (ribozymes) catalyze relatively simple chemical reactions. An enzyme or ribozyme usually recognizes and binds to only one or a few closely related reactants, and it catalyzes only a single chemical reaction. In the discussion that follows, we focus on enzymes, but remember that similar rules of chemical behavior apply to ribozymes as well.

In an enzyme-catalyzed reaction, the reactants are called **substrates**. Substrate molecules bind to a particular site on the en-

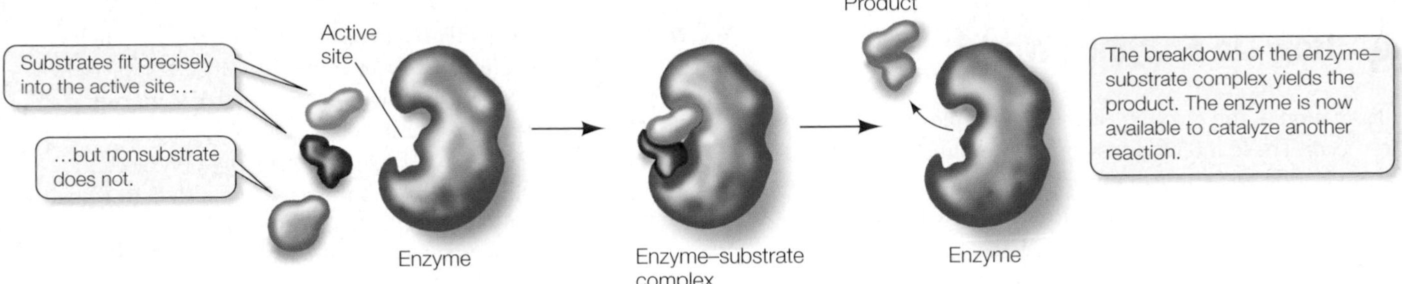

6.9 Enzyme and Substrate An enzyme is a protein catalyst with an active site capable of binding one or more substrate molecules.

zyme, called the **active site**, where catalysis takes place (**Figure 6.9**). The specificity of an enzyme results from the exact three-dimensional shape and structure of its active site, into which only a narrow range of substrates can fit. Other molecules—with different shapes, different functional groups, and different properties—cannot properly fit and bind to the active site.

The names of enzymes reflect the specificity of their functions and often end with the suffix "-ase." For example, the enzyme RNA polymerase catalyzes the formation of RNA, but not DNA, and the enzyme hexokinase accelerates the phosphorylation of hexose sugars, but not pentose sugars.

The binding of a substrate to the active site of an enzyme produces an **enzyme–substrate complex** (ES) held together by one or more means, such as hydrogen bonding, electrical attraction, or covalent bonding. The enzyme–substrate complex gives rise to product and free enzyme:

$$E + S \rightarrow ES \rightarrow E + P$$

where E is the enzyme, S is the substrate, P is the product, and ES is the enzyme–substrate complex. The free enzyme (E) is in the same chemical form at the end of the reaction as at the beginning. While bound to the substrate, it may change chemically, but by the end of the reaction it has been restored to its initial form.

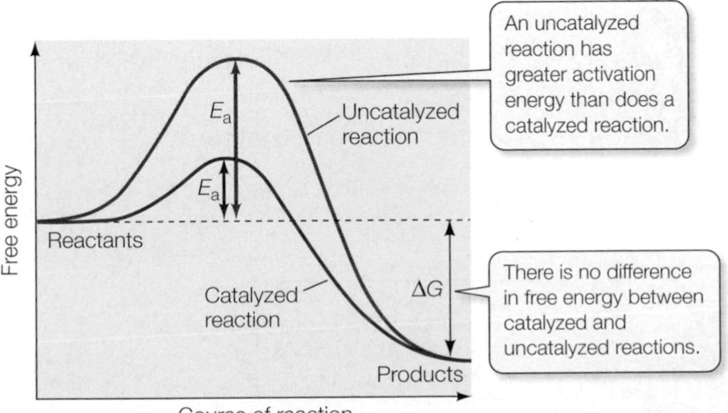

6.10 Enzymes Lower the Energy Barrier Although the activation energy is lower in an enzyme-catalyzed reaction than in an uncatalyzed reaction, the energy released is the same with or without catalysis. In other words, E_a is lower, but ΔG is unchanged.

Enzymes lower the energy barrier but do not affect equilibrium

When reactants are part of an enzyme–substrate complex, they require less activation energy than the transition-state species of the corresponding uncatalyzed reaction (**Figure 6.10**). Thus the enzyme lowers the energy barrier for the reaction—it offers the reaction an easier path. When an enzyme lowers the energy barrier, both the forward and the reverse reactions speed up, so the enzyme-catalyzed overall reaction proceeds toward equilibrium more rapidly than the uncatalyzed reaction. The final equilibrium is the same with or without the enzyme. Similarly, adding an enzyme to a reaction does not change the difference in free energy (ΔG) between the reactants and the products.

Each year over 500 tons of protease enzymes are added to laundry detergents to break down proteins such as the pizza stains on your shirt. These enzymes are produced by bacteria grown in huge stainless steel tanks.

Enzymes can change the rate of a reaction substantially. For example, if 600 molecules of a protein with arginine as its terminal amino acid just sit in solution, the protein molecules tend toward disorder, and the terminal peptide bonds break, releasing the arginines (ΔS increases). After 7 years, about half (300) of the proteins will have undergone this reaction. With the enzyme carboxypeptidase A catalyzing the reaction, however, the 300 arginines are released in half a second!

6.3 RECAP

A chemical reaction requires a "push" over the energy barrier to get started. Enzymes supply this activation energy by binding specific reactants (substrates).

- How does the structure of enzymes makes them specific? See pp. 126–127 and Figure 6.9
- Do you understand the relationship between enzymes and the equilibrium point of a reaction? See p. 127

Now that we understand the structure and specificity of enzymes, let's see how they work to speed up chemical reactions between the substrate molecules.

6.4 How Do Enzymes Work?

After formation of the enzyme–substrate complex, chemical interactions occur. These interactions contribute directly to the breaking of old bonds and the formation of new ones. In catalyzing a reaction, an enzyme may use one or more of the following mechanisms:

- *Enzymes orient substrates.* While free in solution, substrates are rotating and tumbling around and may not have the proper orientation to interact when they collide. Part of the activation energy needed to start a reaction is used to bring together the specific atoms between which bonds are to form (**Figure 6.11A**). For example, if acetyl CoA and oxaloacetate are to form citrate (a reaction that is part of the metabolism of glucose, as we will see in Section 7.2), the two substrates must be oriented so that the carbon atom of the methyl group of acetyl CoA can form a covalent bond with the carbon atom of the carbonyl group of oxaloacetate. The active site of the enzyme citrate synthase has just the right shape to bind these two molecules so that these atoms are adjacent.
- *Enzymes induce strain in the substrate.* Once a substrate has bound to its active site, an enzyme can cause bonds in the substrate to stretch, putting it in an unstable transition state (**Figure 6.11B**). For example, lysozyme is a protective enzyme that destroys invading bacteria by cleaving polysaccharide chains in their cell walls. Lysozyme's active site "stretches" the bonds of the bacterial polysaccharide—one of lysozyme's substrates. The stretching renders the polysaccharide bonds unstable and more reactive to lysozyme's other substrate, water.
- *Enzymes temporarily add chemical groups to substrates.* The side chains (R groups) of an enzyme's amino acids may be direct participants in making its substrates more chemically reactive (**Figure 6.11C**).
  - In *acid–base catalysis*, the acidic or basic side chains of the amino acids forming the active site may transfer H^+ to or from the substrate, destabilizing a covalent bond in the substrate and permitting it to break.
  - In *covalent catalysis*, a functional group in a side chain forms a temporary covalent bond with a portion of the substrate.

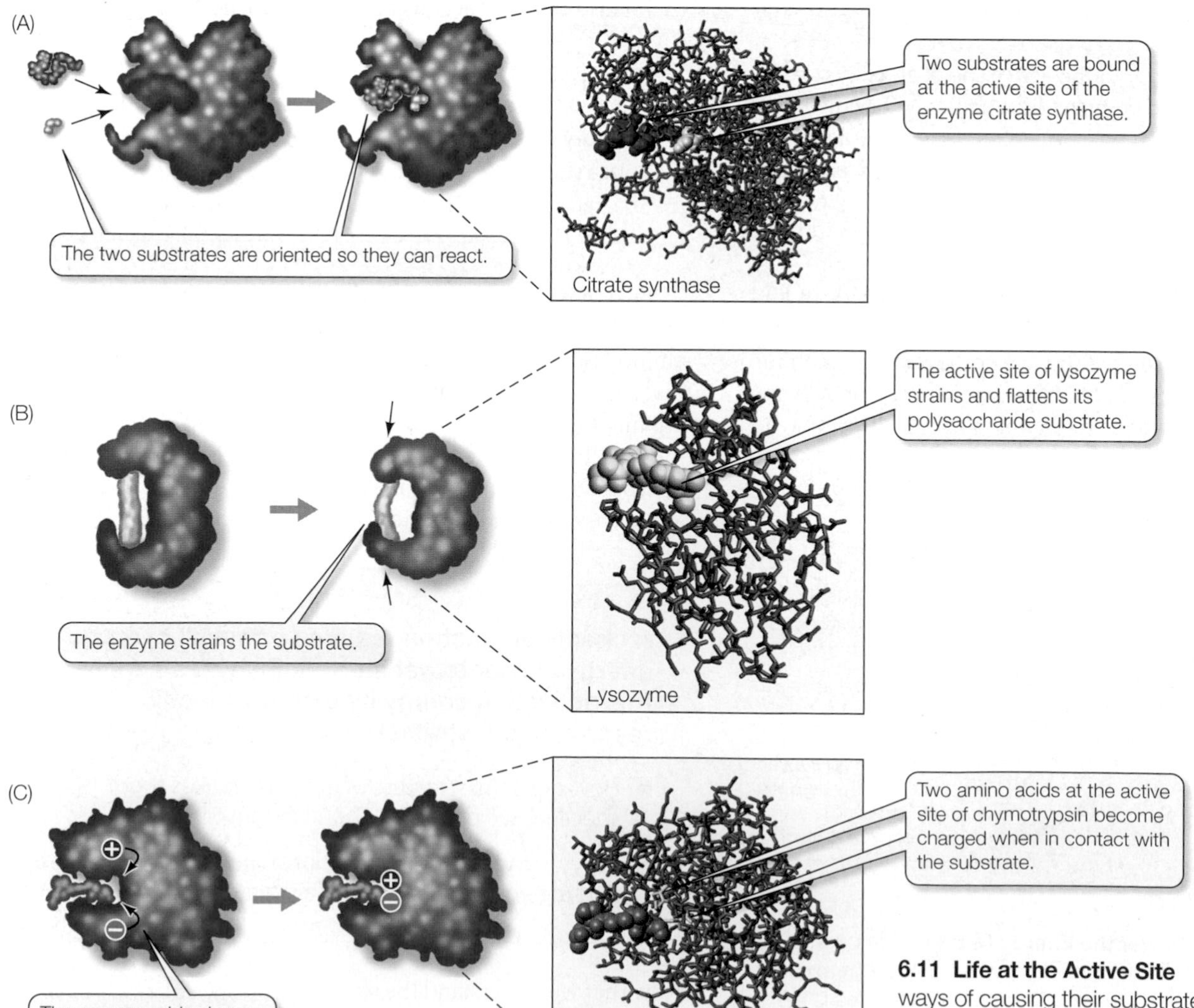
(A)
Two substrates are bound at the active site of the enzyme citrate synthase.
The two substrates are oriented so they can react.
Citrate synthase
(B)
The active site of lysozyme strains and flattens its polysaccharide substrate.
The enzyme strains the substrate.
Lysozyme
(C)
Two amino acids at the active site of chymotrypsin become charged when in contact with the substrate.
The enzyme adds charges to the substrate.
Chymotrypsin

6.11 Life at the Active Site Enzymes have several ways of causing their substrates to enter the transition state: (A) orientation, (B) physical strain, and (C) chemical charge.

- In *metal ion catalysis,* metal ions such as copper, iron, and manganese, which are firmly bound to side chains of the enzyme, can lose or gain electrons without detaching from the enzyme. This ability makes them important participants in oxidation–reduction reactions, which involve loss or gain of electrons.

Molecular structure determines enzyme function

Most enzymes are much larger than their substrates. An enzyme is typically a protein containing hundreds of amino acids and may consist of a single folded polypeptide chain or of several subunits. Its substrate is generally a small molecule. The active site of the enzyme is usually quite small, not more than 6–12 amino acids. Two questions arise from these observations:

- What features of the active site allow it to recognize and bind the substrate?
- What is the role of the rest of the huge protein?

THE ACTIVE SITE IS SPECIFIC TO THE SUBSTRATE The remarkable ability of an enzyme to select exactly the right substrate depends on a precise interlocking of molecular shapes and interactions of chemical groups at the active site. The binding of the substrate to the active site depends on the same kinds of forces that maintain the tertiary structure of the enzyme: hydrogen bonds, the attraction and repulsion of electrically charged groups, and hydrophobic interactions.

In 1894, the German chemist Emil Fischer compared the fit between an enzyme and its substrate to that of a lock and key. Fischer's model persisted for more than half a century with only indirect evidence to support it. The first direct evidence came in 1965, when David Phillips and his colleagues at the Royal Institution in London succeeded in crystallizing the enzyme lysozyme and determined its tertiary structure using the techniques of X-ray crystallography (described in Section 11.2). They observed a pocket in lysozyme that neatly fits its substrate (see Figure 6.11B).

AN ENZYME CHANGES SHAPE WHEN IT BINDS A SUBSTRATE As proteins, enzymes are not immutable structures. The three-dimensional shapes of many enzymes change when they bind to their substrates. These shape changes expose the active site (or sites) of the enzyme. A change in enzyme shape caused by substrate binding is called **induced fit**.

An example of induced fit can be seen in the enzyme hexokinase, which catalyzes the reaction

$$\text{glucose} + \text{ATP} \rightarrow \text{glucose 6-phosphate} + \text{ADP}$$

Induced fit brings reactive side chains from the hexokinase active site into alignment with the substrates (**Figure 6.12**), facilitating its catalytic mechanisms. Equally important, the folding of hexokinase to fit around the glucose substrate excludes water from the active site. This is essential, because the two molecules binding to the active site are glucose and ATP. If water were present, ATP could be hydrolyzed to ADP and phosphate. But since water is absent, the transfer of a phosphate from ATP to glucose is favored.

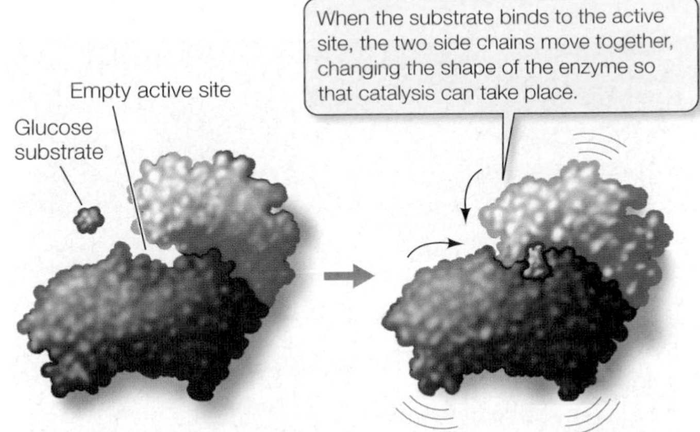

When the substrate binds to the active site, the two side chains move together, changing the shape of the enzyme so that catalysis can take place.
Empty active site
Glucose substrate

6.12 Some Enzymes Change Shape When Substrate Binds to Them Shape changes result in an induced fit between enzyme and substrate, improving the catalytic ability of the enzyme. Induced fit can be observed in the enzyme hexokinase seen with and without one of its substrates, glucose (its other substrate is ATP).

Induced fit at least partly explains why enzymes are so large. The rest of the macromolecule may have two roles:

- It provides a framework so that the amino acids of the active site are properly positioned in relation to the substrate.
- It participates in the small but significant changes in protein shape and structure that result in induced fit.

Some enzymes require other molecules in order to function

As large and complex as enzymes are, many of them require the presence of other, nonprotein molecular "partners" in order to function (**Table 6.1**):

TABLE 6.1

Some Examples of Nonprotein "Partners" of Enzymes

TYPE OF MOLECULE	ROLE IN CATALYZED REACTIONS
COFACTORS	
Iron (Fe^{2+} or Fe^{3+})	Oxidation/reduction
Copper (Cu^{+} or Cu^{2+})	Oxidation/reduction
Zinc (Zn^{2+})	Helps bind NAD
COENZYMES	
Biotin	Carries $-COO^{-}$
Coenzyme A	Carries $-CO-CH_3$
NAD	Carries electrons
FAD	Carries electrons
ATP	Provides/extracts energy
PROSTHETIC GROUPS	
Heme	Binds ions, O_2, and electrons; contains iron cofactor
Flavin	Binds electrons
Retinal	Converts light energy

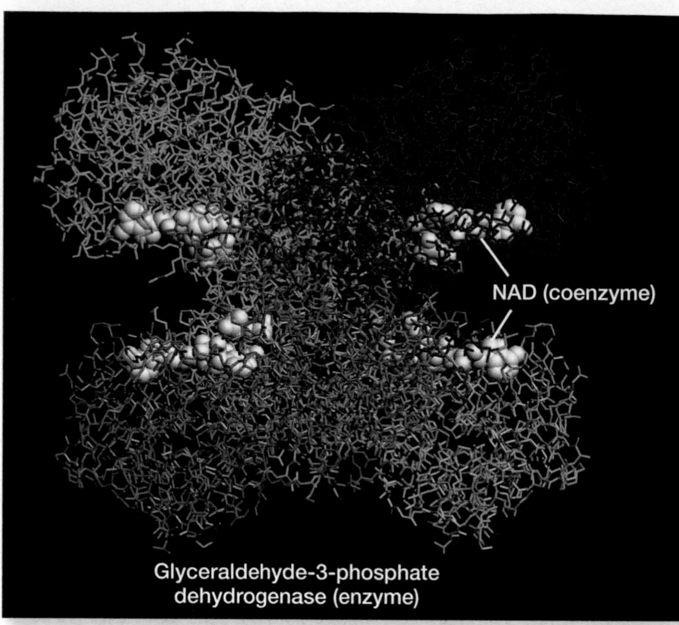
NAD (coenzyme)
Glyceraldehyde-3-phosphate dehydrogenase (enzyme)

6.13 An Enzyme with a Coenzyme Some enzymes require coenzymes in order to function. This illustration shows the relative sizes of the four subunits (red, yellow, green, and purple) of the enzyme triose phosphate dehydrogenase and its coenzyme, NAD (white).

- **Prosthetic groups** are distinctive, non-amino acid atoms or molecular groupings that are permanently bound to their enzymes.
- **Cofactors** are inorganic ions such as copper, zinc, or iron that bind to certain enzymes and are essential to their function.
- **Coenzymes** are carbon-containing molecules that are required for the action of one or more enzymes. Coenzymes are usually relatively small compared with the enzyme to which they temporarily bind (**Figure 6.13**).

Prosthetic groups include, for example, the flavin nucleotides that are bound to the mitochondrial enzyme succinate dehydrogenase, which plays an important role in cellular respiration (see Section 7.2). Although it is not an enzyme, hemoglobin provides another example of a protein bound to a prosthetic group—in this case, heme (see Figure 3.9).

Coenzymes move from enzyme molecule to enzyme molecule, adding or removing chemical groups from the substrate. Coenzymes are like substrates in that they are not permanently bound to the enzyme, but must collide with the enzyme and bind to its active site. In addition, a coenzyme changes chemically during the reaction and then separates from the enzyme to participate in other reactions.

ATP and ADP can be considered coenzymes because they are necessary for some reactions, are changed by those reactions, and bind to and detach from the enzymes that catalyze those reactions. In the next chapter we will encounter other coenzymes that function in energy-harvesting reactions by accepting or donating electrons or hydrogen atoms. In animals, some coenzymes are produced from *vitamins*—substances that must be obtained from food because they cannot be synthesized by the body. For example, the B vitamin niacin is used to make the coenzyme NAD.

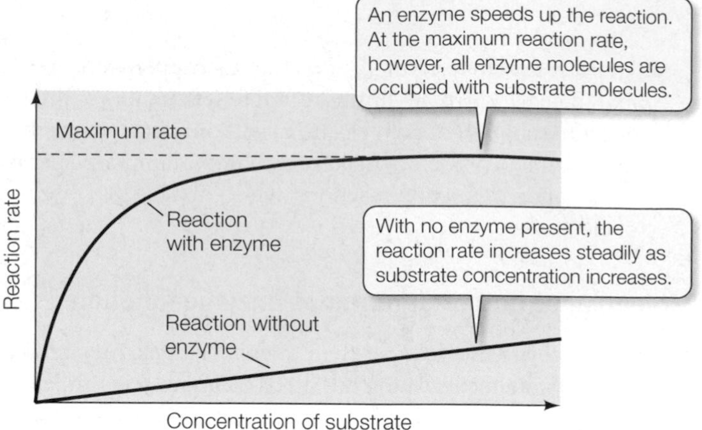
An enzyme speeds up the reaction. At the maximum reaction rate, however, all enzyme molecules are occupied with substrate molecules.
Maximum rate
Reaction with enzyme
With no enzyme present, the reaction rate increases steadily as substrate concentration increases.
Reaction without enzyme
Reaction rate
Concentration of substrate

6.14 Catalyzed Reactions Reach a Maximum Rate Because there is usually less enzyme than substrate present, the reaction rate levels off when the enzyme becomes saturated.

Substrate concentration affects reaction rate

For a reaction of the type A → B, the rate of the uncatalyzed reaction is directly proportional to the concentration of A. The higher the concentration of substrate, the more reactions per unit of time. Addition of the appropriate enzyme speeds up the reaction, of course, but it also changes the shape of a plot of rate versus substrate concentration (**Figure 6.14**). At first, the rate of the enzyme-catalyzed reaction increases as the substrate concentration increases, but then it levels off. When further increases in the substrate concentration do not significantly increase the reaction rate, the maximum rate is attained.

Since the concentration of an enzyme is usually much lower than that of its substrate, what we are seeing is a *saturation* phenomenon like the one that occurs in facilitated diffusion (see Figure 5.12B). When all the enzyme molecules are bound to substrate molecules, the enzyme is working as fast as it can—at its maximum rate. Nothing is gained by adding more substrate, because no free enzyme molecules are left to act as catalysts.

The maximum rate of a catalyzed reaction can be used to measure how efficient the enzyme can be—that is, how many molecules of substrate are converted into product per unit of time when there is an excess of substrate present. This turnover number ranges from 1 molecule every 2 seconds for lysozyme to an amazing 40 million molecules per second for the liver enzyme catalase.

6.4 RECAP

Enzymes change during the reaction they catalyze, often participating in the reaction itself by temporarily changing shape or donating electrons. Some enzymes require cofactors, coenzymes, or prosthetic groups in order to function.

- Can you describe three mechanisms of enzyme catalysis? See p. 128 and Figure 3.11
- Do you understand the chemical role of coenzymes in enzymatic reactions? See p. 130

Now that we understand more about how enzymes work, let's see what makes it possible for the myriad enzymes in a complex organism, each specific to just one or a few reactions, to work together.

6.5 How Are Enzyme Activities Regulated?

A major characteristic of life is *homeostasis,* the maintenance of stable internal conditions (see Chapter 40). How a cell maintains a relatively constant internal environment while thousands of chemical reactions are going on at once is an amazing story. These chemical reactions are organized into *metabolic pathways* in which the product of one reaction is a reactant for the next. The pathway for the metabolism of alcohol in humans described at the opening of this chapter is just one of many such pathways that regulate the internal environment and provide for such diverse functions as the catabolism of glucose and the anabolism of amino acids. These pathways do not exist in isolation, but interact extensively. And *each* reaction in each pathway is catalyzed by a specific enzyme. Within a cell or organism, enzymes control whether and how much a metabolic pathway functions. If even one enzyme in the pathway is inactive, that step, and all steps subsequent to it, shut down.

The "flow" of chemicals (for example, a carbon atom) through these interacting pathways can be studied, but gets complicated quickly, since each pathway influences the others. Mathematical algorithms for the computer are used to model these pathways and show how they mesh in an interdependent system (**Figure 6.15**); such models can help predict what will happen if the concentration of one molecule or another is altered. This new field of biology, which has numerous applications, is called **systems biology**.

Regulation of the rates at which thousands of different enzymes operate contributes to homeostasis within an organism. In this section, we will investigate the role of enzymes in organizing and regulating metabolic pathways. In living cells, enzymes can be activated or inhibited in various ways, so the presence of an enzyme alone does not ensure that it is functioning. There are also mechanisms that alter the rate at which some enzymes catalyze reactions, making enzymes the target points at which entire sequences of chemical reactions can be regulated. Finally, we will examine how the environment—particularly temperature and pH—affects enzyme activity.

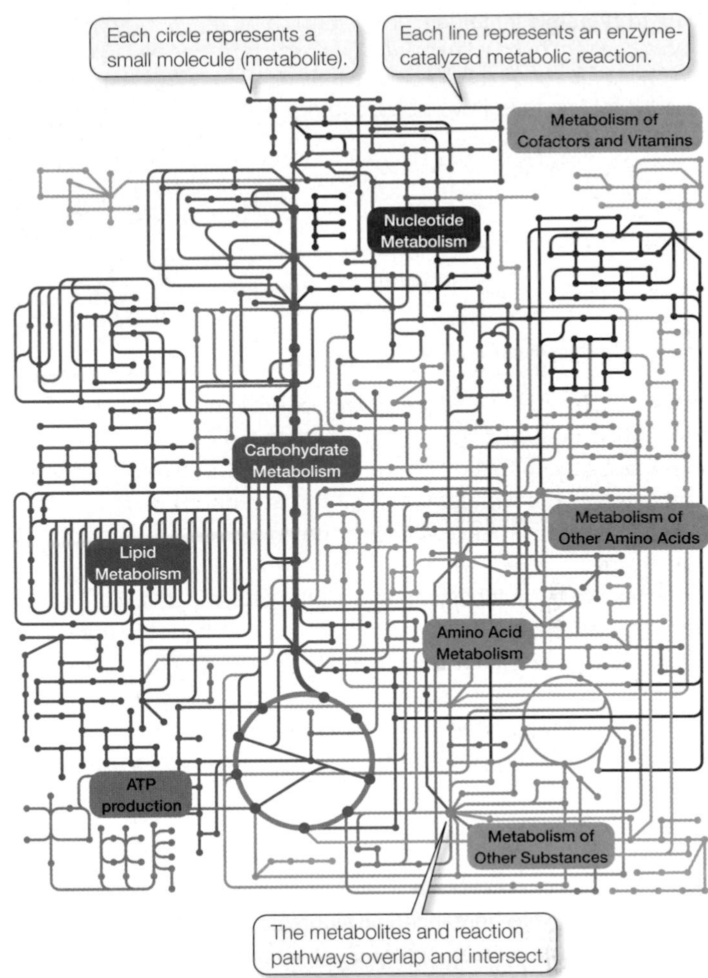

6.15 Metabolic Pathways The complex interactions of metabolic pathways can be modeled by the tools of systems biology. In cells, the main elements controlling these pathways are enzymes.

Enzymes can be regulated by inhibitors

Various **inhibitors** can bind to enzymes, slowing down the rates of enzyme-catalyzed reactions. Some inhibitors occur naturally in cells; others are artificial. Naturally occurring inhibitors regulate metabolism; artificial ones can be used to treat disease, to kill pests, or in the laboratory to study how enzymes work. Some inhibitors irreversibly inhibit the enzyme by permanently binding to it. Others have reversible effects; that is, they can become unbound from the enzyme. The removal of a natural reversible inhibitor increases an enzyme's rate of catalysis.

IRREVERSIBLE INHIBITION Some inhibitors covalently bond to certain side chains at the active site of an enzyme, thereby permanently inactivating the enzyme by destroying its capacity to interact with its normal substrate. An example of an irreversible inhibitor is DIPF (diisopropylphosphorofluoridate), which reacts with serine (**Figure 6.16**). DIPF is an irreversible inhibitor of acetylcholinesterase, whose operation is essential to the normal function of the nervous system. Because of their effect on acetylcholinesterase, DIPF and other similar compounds are classified as *nerve gases*. One of them, Sarin, was used in an attack on the Tokyo subway in 1995, resulting in a dozen deaths and hundreds hospitalized. The widely used insecticide malathion is a derivative of DIPF that inhibits only insect acetylcholinesterase, not the mammalian enzyme.

REVERSIBLE INHIBITION Not all inhibition is irreversible. Some inhibitors are similar enough to a particular enzyme's natural substrate to bind noncovalently to its active site, yet different enough that the enzyme catalyzes no chemical reaction. While such a molecule is bound to the enzyme, the natural substrate cannot enter the active site; thus the inhibitor effectively wastes the enzyme's time, preventing its catalytic action. Such molecules are called **com-**

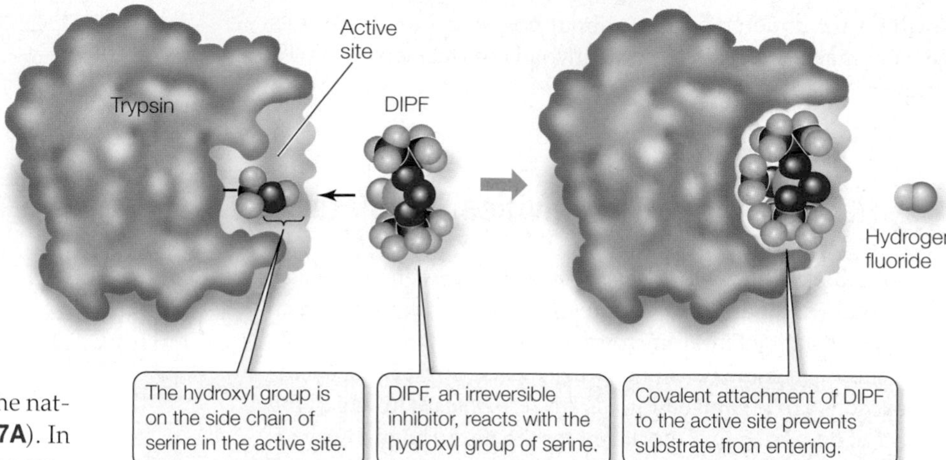
Active site
Trypsin
DIPF
Hydrogen fluoride
The hydroxyl group is on the side chain of serine in the active site.
DIPF, an irreversible inhibitor, reacts with the hydroxyl group of serine.
Covalent attachment of DIPF to the active site prevents substrate from entering.

6.16 Irreversible Inhibition DIPF forms a stable covalent bond with the side chain of the amino acid serine at the active site of the enzyme trypsin, thus irreversibly disabling the enzyme.

petitive inhibitors because they compete with the natural substrate for the active site (**Figure 6.17A**). In these cases, the inhibition is reversible. When the concentration of the competitive inhibitor is reduced, it detaches from the active site, and the enzyme is active again.

Noncompetitive inhibitors bind to the enzyme at a site distinct from the active site. Their binding can cause a change in the shape of the enzyme that alters the active site (**Figure 6.17B**). In this case, the active site may still bind substrate molecules, but the rate of product formation may be reduced. Noncompetitive inhibitors, like competitive inhibitors, can become unbound, so their effects are reversible.

Frank (see p. 118) got sick from a single glass of champagne because he has nonfunctional aldehyde dehydrogenase (ALDH). A drug called Antabuse is a competitive inhibitor of ALDH. Antabuse is used to treat alcoholism, since it causes alcoholics to get sick if they consume alcohol.

Allosteric enzymes control their activity by changing their shape

The change in enzyme shape due to noncompetitive inhibitor binding is an example of **allostery** (*allo*, "different"; *stery*, "shape"). In this case, the binding of the inhibitor *induces* the protein to change its shape. More common are enzymes that already exist in the cell in more than one possible shape (**Figure 6.18**):

- The *active* form of the enzyme has the proper shape for substrate binding.
- The *inactive* form of the enzyme has a shape that cannot bind the substrate but can bind an inhibitor. Binding of an inhibitor to a site separate from the active site (i.e., the site where substrate binds) stabilizes the inactive form, and it becomes less likely to convert to the active form.
- The active form can be stabilized by binding of an *activator* to a third site on the enzyme.

Like substrate binding, the binding of inhibitors and activators is highly specific.

Most (but not all) enzymes that are allosterically regulated are proteins with quaternary structure; that is, they are made up of multiple polypeptide subunits. The active site is present on one such subunit, called the **catalytic subunit**, while the regulatory site(s)

(A) Competitive inhibition

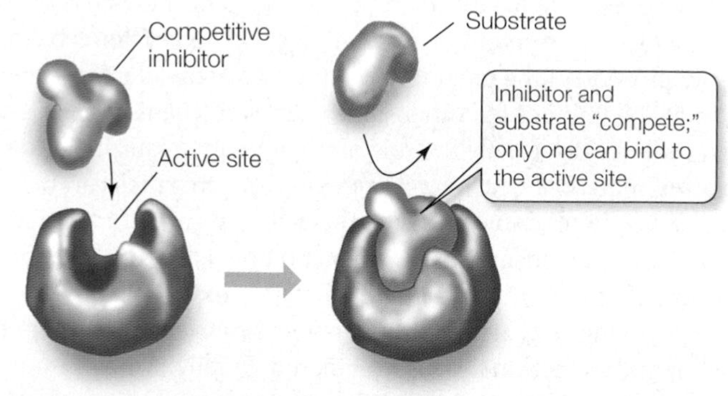
Competitive inhibitor
Substrate
Inhibitor and substrate "compete;" only one can bind to the active site.
Active site

(B) Noncompetitive inhibition

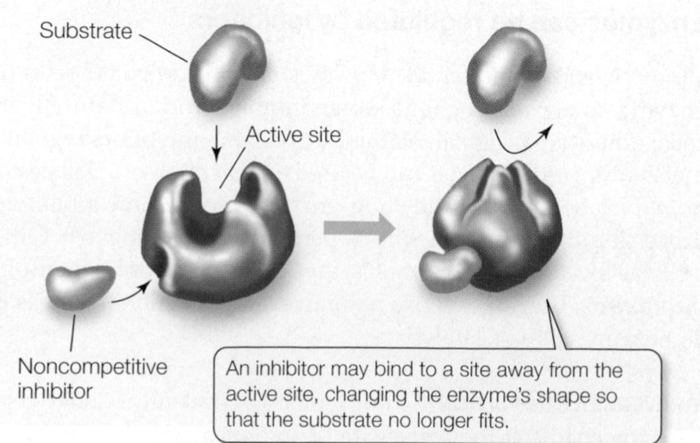
Substrate
Active site
Noncompetitive inhibitor
An inhibitor may bind to a site away from the active site, changing the enzyme's shape so that the substrate no longer fits.

6.17 Reversible Inhibition (A) A competitive inhibitor binds temporarily to the active site of an enzyme. (B) A noncompetitive inhibitor binds temporarily to the enzyme at a site away from the active site. In both instances, the enzyme's function is disabled for only as long as the inhibitor remains bound.

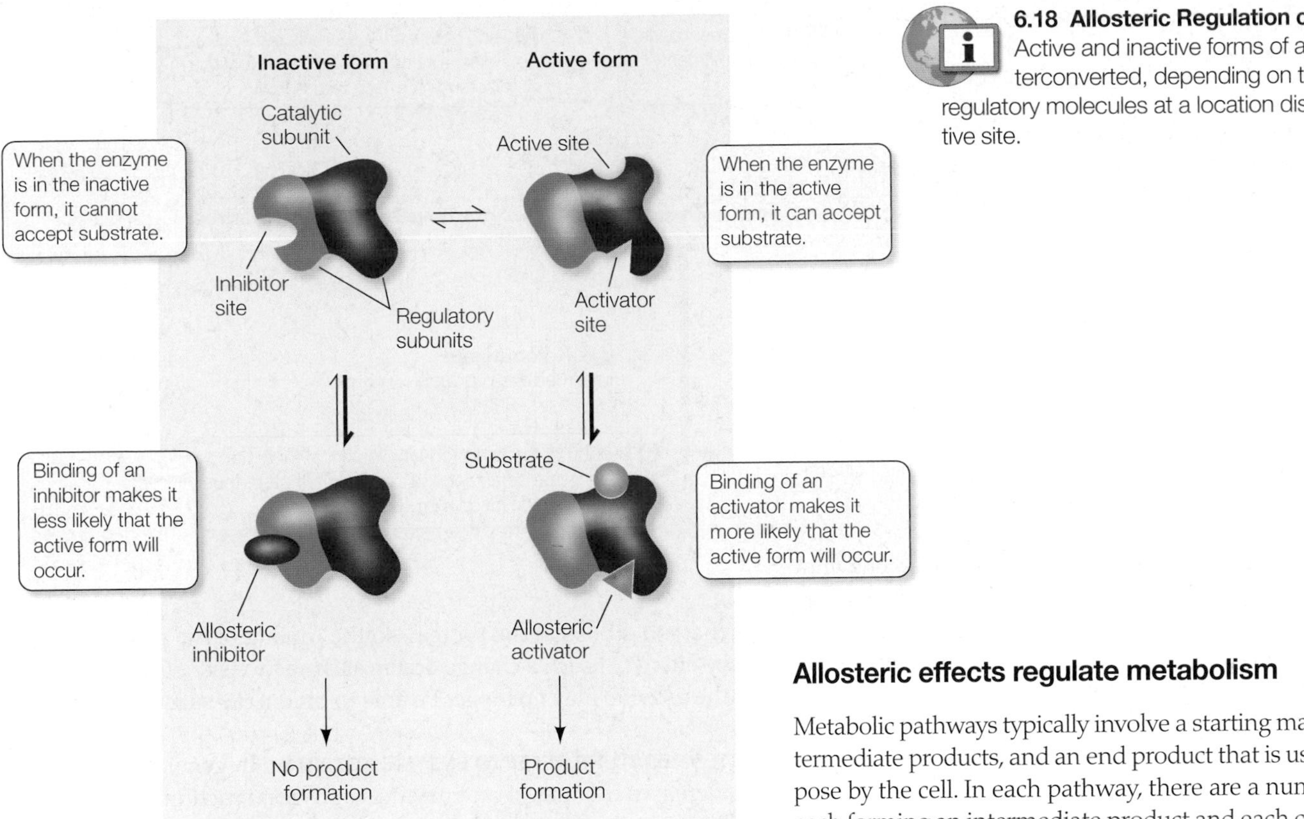
Inactive form
Active form
Catalytic subunit
Active site
When the enzyme is in the inactive form, it cannot accept substrate.
When the enzyme is in the active form, it can accept substrate.
Inhibitor site
Regulatory subunits
Activator site
Substrate
Binding of an inhibitor makes it less likely that the active form will occur.
Binding of an activator makes it more likely that the active form will occur.
Allosteric inhibitor
Allosteric activator
No product formation
Product formation

6.18 Allosteric Regulation of Enzymes Active and inactive forms of an enzyme are interconverted, depending on the binding of regulatory molecules at a location distant from the active site.

for activators and/or inhibitors are on different polypeptide sequences, called **regulatory subunits**.

Allosteric enzymes and nonallosteric enzymes differ greatly in their reaction rates when the substrate concentration is low. Graphs of reaction rate plotted against substrate concentration show this relationship. For a nonallosteric enzyme, the plot looks like that in **Figure 6.19A**. The reaction rate first increases very sharply with increasing substrate concentration, then tapers off to a constant maximum rate as the supply of enzyme becomes saturated with substrate. The plot for many allosteric enzymes is radically different, having a *sigmoid* (S-shaped) appearance (**Figure 6.19B**). The increase in reaction rate with increasing substrate concentration is slight at low substrate concentrations, but within a certain range, the reaction rate is extremely sensitive to relatively small changes in substrate concentration. Because of this sensitivity, allosteric enzymes are important in regulating entire metabolic pathways.

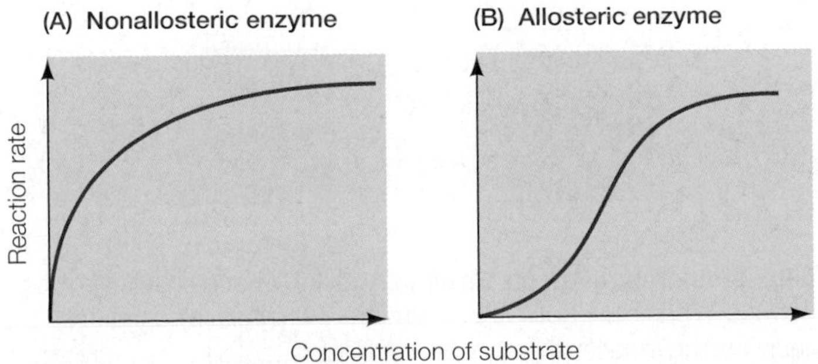
(A) Nonallosteric enzyme
(B) Allosteric enzyme
Reaction rate
Concentration of substrate

6.19 Allostery and Reaction Rate How the rate of an enzyme-catalyzed reaction changes with increasing substrate concentration depends on whether the enzyme is allosterically regulated.

Allosteric effects regulate metabolism

Metabolic pathways typically involve a starting material, various intermediate products, and an end product that is used for some purpose by the cell. In each pathway, there are a number of reactions, each forming an intermediate product and each catalyzed by a different enzyme. The first step in a pathway is called the **commitment step**, meaning that once this enzyme-catalyzed reaction occurs, the "ball is rolling," and the other reactions happen in sequence, leading to the end product. But what if the cell has no need for that product—for example, if that product is available from its environment in adequate amounts? It would be energetically wasteful for the cell to continue making something it does not need.

One way that cells solve this problem is to shut down the metabolic pathway by having the final product allosterically inhibit the enzyme that catalyzes the commitment step (**Figure 6.20**). This mechanism is known as **feedback inhibition** or **end-product inhibition**. When the end product is present at a high concentration, some of it binds to an allosteric site on the commitment step enzyme, thereby causing it to become inactive. We will describe many other examples of such allosteric interactions in later chapters.

Enzymes are affected by their environment

Enzymes enable cells to perform chemical reactions and carry out complex processes rapidly without using the extremes of temperature and pH employed by chemists in the laboratory. However, because of their three-dimensional structures and the chemistry of the side chains in their active sites, enzymes are highly sensitive to temperature and pH. We described the general effects of these environmental factors on proteins in Section 3.2. Here

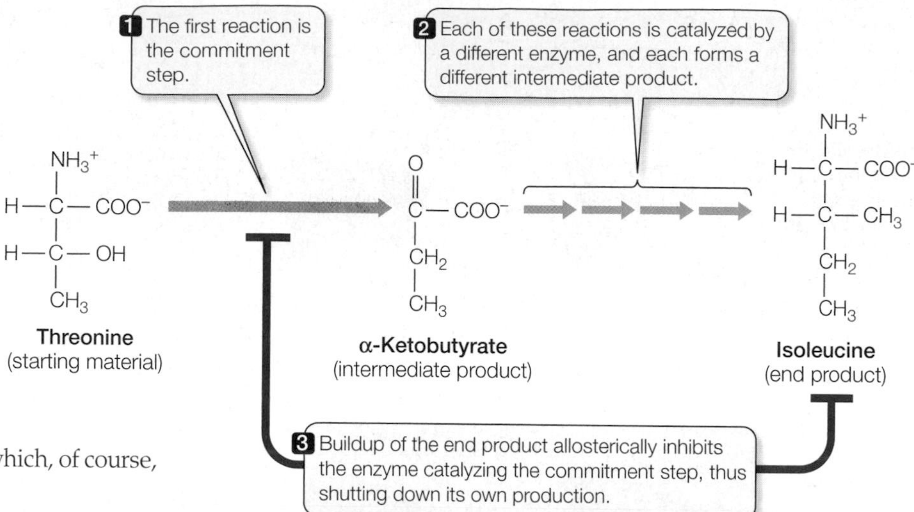

1 The first reaction is the commitment step.
2 Each of these reactions is catalyzed by a different enzyme, and each forms a different intermediate product.
Threonine
(starting material)
α-Ketobutyrate
(intermediate product)
Isoleucine
(end product)
3 Buildup of the end product allosterically inhibits the enzyme catalyzing the commitment step, thus shutting down its own production.

6.20 Feedback Inhibition of Metabolic Pathways The commitment step is catalyzed by an allosteric enzyme that can be inhibited by the end product of the pathway. The specific pathway shown here is the synthesis of isoleucine, an amino acid, from threonine in bacteria. This pathway is typical of many enzyme-catalyzed biological reactions.

we will examine their effects on enzyme function, which, of course, depends on enzyme structure and chemistry.

pH AFFECTS ENZYME ACTIVITY The rates of most enzyme-catalyzed reactions depend on the pH of the solution in which they occur. Each enzyme is most active at a particular pH; its activity decreases as the solution is made more acidic or more basic than its "ideal" (optimal) pH (**Figure 6.21**).

Several factors contribute to this effect. One is the ionization of carboxyl, amino, and other groups on either the substrate or the enzyme. In neutral or basic solutions, carboxyl groups (—COOH) release H^+ to become negatively charged carboxylate groups (—COO^-). Similarly, amino groups (—NH_2) accept H^+ in neutral or acidic solutions, becoming positively charged —NH_3^+ groups (see the discussion of acids and bases in Section 2.4). Thus, in a neutral solution, a molecule with an amino group is attracted electrically to another molecule that has a carboxyl group, because both groups are ionized and the two groups have opposite charges. If the pH changes, however, the ionization of these groups may change. For example, at a low pH (high H^+ concentration), the excess H^+ may react with —COO^- to form COOH. If this happens, the group is no longer charged and cannot interact with other charged groups in the protein, so the folding of the protein may be altered. If such a change occurs at the active site of an enzyme, the enzyme may no longer be able to bind to its substrate.

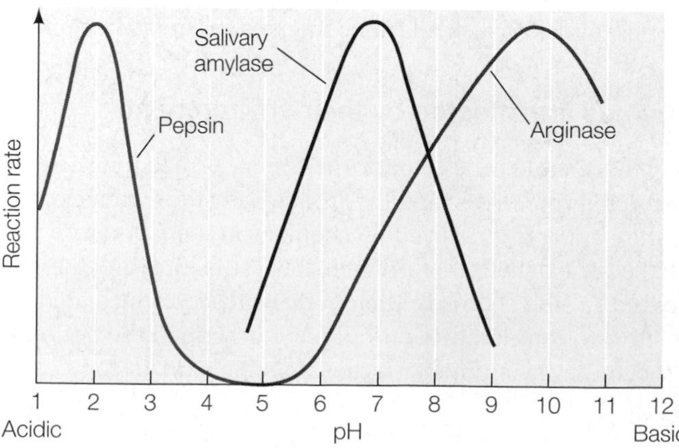

Salivary amylase
Pepsin
Arginase
Reaction rate
1 2 3 4 5 6 7 8 9 10 11 12
Acidic
pH
Basic

6.21 pH Affects Enzyme Activity Each enzyme catalyzes its reaction at a maximum rate at a particular pH. The activity curves peak at the pH at which each enzyme is most effective. For example, pepsin is a protease active in the acidic environment of the stomach.

TEMPERATURE AFFECTS ENZYME ACTIVITY In general, warming increases the rate of an enzyme-catalyzed reaction because at higher temperatures, a greater proportion of the reactant molecules have enough kinetic energy to provide the activation energy for the reaction (**Figure 6.22**). Temperatures that are too high, however, inactivate enzymes, because at high temperatures enzyme molecules vibrate and twist so rapidly that some of their noncovalent bonds break. When heat changes their tertiary structure, enzymes become denatured and lose their function. Some enzymes denature at temperatures only slightly above that of the human body, but a few are stable even at the boiling or freezing points of water. All enzymes, however, have an optimal temperature for activity.

Individual organisms adapt to changes in the environment in many ways, one of which is based on groups of enzymes, called **isozymes**, that catalyze the same reaction but have different chemical compositions and physical properties. Different isozymes

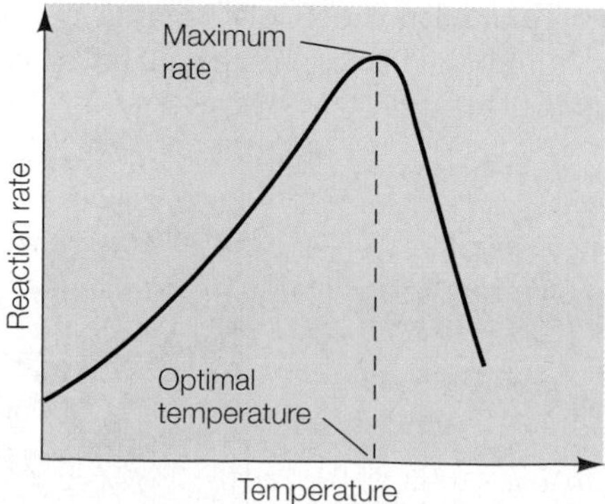

Maximum rate
Reaction rate
Optimal temperature
Temperature

6.22 Temperature Affects Enzyme Activity Each enzyme is most active at a particular optimal temperature. At higher temperatures, denaturation reduces the enzyme's activity.

within a given group may have different optimal temperatures. The rainbow trout, for example, has several isozymes of the enzyme acetylcholinesterase. If a rainbow trout is transferred from warm water to near-freezing water (2°C), the fish produces an isozyme of acetylcholinesterase that is different from the one it produces at the higher temperature. The new isozyme has a lower optimal temperature, allowing the fish's nervous system to perform normally in the colder water.

In general, enzymes adapted to warm temperatures fail to denature at those temperatures because their tertiary structures are held together largely by covalent bonds, such as disulfide bridges, instead of the more heat-sensitive weak chemical interactions. Most enzymes in humans are more stable at high temperatures than those of the bacteria that infect us, so that a moderate fever tends to denature bacterial enzymes, but not our own.

6.5 RECAP

The rates of most enzyme-catalyzed reactions are affected by chemicals (such as inhibitors) and by environmental factors (such as temperature and pH).

- Can you describe the difference between reversible and irreversible inhibitors of enzymes? See p. 131
- How are allosteric enzymes regulated? See p. 131–133 and Figure 6.18
- Can you explain the concept of feedback inhibition? Can you envision how the reactions shown in Figure 6.20 might fit into a systems diagram such as Figure 6.15?

CHAPTER SUMMARY

6.1 What physical principles underlie biological energy transformations?

Energy is the capacity to do work. In a biological system, the usable energy is called **free energy** (*G*). The unusable energy is **entropy**, a measure of the disorder of the system.

Potential energy is the energy of state or position; it includes the energy stored in chemical bonds. **Kinetic energy** is the energy of motion; it is the type of energy that can do work.

The **laws of thermodynamics** apply to living organisms. The first law states that energy cannot be created or destroyed. The second law states that energy transformations decrease the amount of energy available to do work (free energy) and increase disorder. Review Figure 6.2

The **change in free energy** (ΔG) of a reaction determines its point of chemical equilibrium, at which the forward and reverse reactions proceed at the same rate.

Exergonic reactions release free energy and have a negative ΔG. **Endergonic reactions** consume or require free energy and have a positive ΔG. Endergonic reactions proceed only if free energy is provided. Review Figure 6.3

Metabolism is the sum total of the biochemical (metabolic) reactions in an organism. **Catabolic reactions** are associated with breakdown of complex molecules and release energy (are exergonic). **Anabolic reactions** build complexity in the cell and are endergonic.

6.2 What is the role of ATP in biochemical energetics?

ATP (**adenosine triphosphate**) serves as an energy currency in cells. Hydrolysis of ATP releases a relatively large amount of free energy.

The ATP cycle couples exergonic and endergonic reactions, transferring free energy from the exergonic to the endergonic reaction. Review Figure 6.6, Web/CD Activity 6.1

6.3 What are enzymes?

The **rate** of a chemical reaction is independent of ΔG, but is determined by the **energy barrier**. **Enzymes** are protein catalysts that affect the rates of biological reactions by lowering the energy barrier, supplying the **activation energy** needed to initiate a reaction.

Substrates bind to the enzyme's **active site**—the site of **catalysis**—forming an **enzyme–substrate complex**. Enzymes are highly specific for their substrates.

At the active site, a substrate can be oriented correctly, chemically modified, or strained. As a result, the substrate readily forms its **transition state**, and the reaction proceeds. Review Figure 6.9

6.4 How do enzymes work?

Binding substrate causes many enzymes to change shape, exposing their active site(s) and allowing catalysis. The change in enzyme shape caused by substrate binding is known as **induced fit**. Review Figure 6.10, Web/CD Activity 6.2

Some enzymes require other substances, known as **cofactors**, to carry out catalysis. **Prosthetic groups** are permanently bound to the enzyme; **coenzymes** are not. Coenzymes can be considered substrates, as they are changed by the reaction and then released from the enzyme.

Substrate concentration affects the rate of an enzyme-catalyzed reaction.

6.5 How are enzyme activities regulated?

Metabolism is organized into pathways in which the product of one reaction is a reactant for the next reaction. Each reaction in the pathway is catalyzed by an enzyme.

Enzyme activity is subject to regulation. Some inhibitors react irreversibly with enzymes. Others react reversibly. See Web/CD Tutorial 6.1

Allosteric regulators bind to a site different from the active site and stabilize the active or inactive form of an enzyme. Review Figure 6.18, Web/CD Tutorial 6.2

The end product of a metabolic pathway may inhibit the allosteric enzyme that catalyzes the **commitment step** of that pathway. Review Figure 6.20

Enzymes are sensitive to their environment. Both pH and temperature affect enzyme activity. Review Figures 6.21 and 6.22

SELF-QUIZ

1. Coenzymes differ from enzymes in that coenzymes are
   a. only active outside the cell.
   b. polymers of amino acids.
   c. smaller molecules, such as vitamins.
   d. specific for one reaction.
   e. always carriers of high-energy phosphate.
2. Which statement about thermodynamics is true?
   a. Free energy is used up in an exergonic reaction.
   b. Free energy cannot be used to do work.
   c. The total amount of energy can change after a chemical transformation.
   d. Free energy can be kinetic but not potential energy.
   e. Entropy has a tendency to increase.
3. In a chemical reaction,
   a. the rate depends on the value of ΔG.
   b. the rate depends on the activation energy.
   c. the entropy change depends on the activation energy.
   d. the activation energy depends on the value of ΔG.
   e. the change in free energy depends on the activation energy.
4. Which statement about enzymes is *not* true?
   a. They usually consist of proteins.
   b. They change the rate of the catalyzed reaction.
   c. They change the ΔG of the reaction.
   d. They are sensitive to heat.
   e. They are sensitive to pH.
5. The active site of an enzyme
   a. never changes shape.
   b. forms no chemical bonds with substrates.
   c. determines, by its structure, the specificity of the enzyme.
   d. looks like a lump projecting from the surface of the enzyme.
   e. changes the ΔG of the reaction.
6. The molecule ATP is
   a. a component of most proteins.
   b. high in energy because of the presence of adenine.
   c. required for many energy-producing biochemical reactions.
   d. a catalyst.
   e. used in some endergonic reactions to provide energy.
7. In an enzyme-catalyzed reaction,
   a. a substrate does not change.
   b. the rate decreases as substrate concentration increases.
   c. the enzyme can be permanently changed.
   d. strain may be added to a substrate.
   e. the rate is not affected by substrate concentration.
8. Which statement about enzyme inhibitors is *not* true?
   a. A competitive inhibitor binds the active site of the enzyme.
   b. An allosteric inhibitor binds a site on the active form of the enzyme.
   c. A noncompetitive inhibitor binds a site other than the active site.
   d. Noncompetitive inhibition cannot be completely overcome by the addition of more substrate.
   e. Competitive inhibition can be completely overcome by the addition of more substrate.
9. Which statement about feedback inhibition of enzymes is *not* true?
   a. It is exerted through allosteric effects.
   b. It is directed at the enzyme that catalyzes the first committed step in a metabolic pathway.
   c. It affects the rate of reaction, not the concentration of enzyme.
   d. It acts very slowly.
   e. It is an example of irreversible inhibition.
10. Which statement about temperature effects is *not* true?
    a. Raising the temperature may reduce the activity of an enzyme.
    b. Raising the temperature may increase the activity of an enzyme.
    c. Raising the temperature may denature an enzyme.
    d. Some enzymes are stable at the boiling point of water.
    e. All enzymes have the same optimal temperature.

FOR DISCUSSION

1. What makes it possible for endergonic reactions to proceed in organisms?
2. Consider two proteins: one is an enzyme dissolved in the cytosol of a cell, the other is an ion channel in its plasma membrane. Contrast the structures of the two proteins, indicating at least two important differences.
3. Plot free energy versus the course of an endergonic reaction and that of an exergonic reaction. Include the activation energy in both plots. Label E_a and ΔG on both graphs.
4. Consider an enzyme that is subject to allosteric regulation. If a competitive inhibitor (not an allosteric inhibitor) is added to a solution containing such an enzyme, the ratio of enzyme molecules in the active form to those in the inactive form increases. Explain this observation.

FOR INVESTIGATION

In humans, hydrogen peroxide (H_2O_2) is a dangerous toxin produced as a by-product of several metabolic pathways. The accumulation of H_2O_2 is prevented by its conversion to harmless H_2O, a reaction catalyzed by the appropriately named enzyme, catalase. Air pollutants can inhibit this enzyme and leave individuals susceptible to tissue damage by H_2O_2. How would you investigate whether catalase has an allosteric or a nonallosteric mechanism, and whether the pollutants are acting as competitive or noncompetitive inhibitors?

CHAPTER 7 Pathways That Harvest Chemical Energy

Of mice and marathons

Like success in your biology course, winning a prestigious marathon comes only after a lot of hard work. The leg muscles of elite distance runners contain more mitochondria than most of us have. The chemical energy released by the hydrolysis of ATP in those mitochondria can be converted into mechanical energy to move the muscles.

The cells of muscle tissue associate into two types of muscle fibers. Most people have about equal proportions of each type. But in top marathon racers, 90 percent of the body's muscle is made up of so-called *slow-twitch* fibers. Cells of these fibers have lots of mitochondria and use oxygen to break down fats and carbohydrates, forming ATP. In contrast, the muscles of sprinters are about 80 percent *fast-twitch* fibers, which have fewer mitochondria. Fast-twitch fibers generate short bursts of ATP in the absence of O_2, but the ATP is soon used up. Extensive research with athletes has shown that training can improve the efficiency of the blood circulation to the muscle fibers, providing more oxygen, and even change the ratio of fast-twitch to slow-twitch fibers.

Now enter Marathon Mouse. No, this is not a cartoon character or a computer game, but a very real mouse that was genetically programmed by Ron Evans at the Salk Institute to express high levels of the protein PPARδ in its muscles. This protein normally controls the breakdown of fat in fat tissues, and it is also present in slow-twitch muscles, where it stimulates controlled fat breakdown to yield ATP. Evans's mouse was supposed to break down fats better, and thus be leaner—but there was an unexpected bonus. With high levels of PPARδ came an increase in slow-twitch fibers and a decrease in fast-twitch ones. It was as if the mouse had been in marathon training for a long time!

Marathon mice are leaner and meaner than ordinary mice. Leaner, because they are good at burning fat; and meaner in terms of their ability to run long distances. On an exercise wheel, a normal mouse can run for 90 minutes and about a half-mile (900 meters) before it gets tired. PPARδ-enhanced mice can run almost twice as long and twice as far—marks of true distance runners. Could we also manipulate genes to enhance performance (and fat burning) in humans?

The genetic engineering of people, if it is feasible, is probably far in the future. But implanting genetically altered muscle tissue is actually not such a far-fetched idea, and has already raised concerns over improper athletic en-

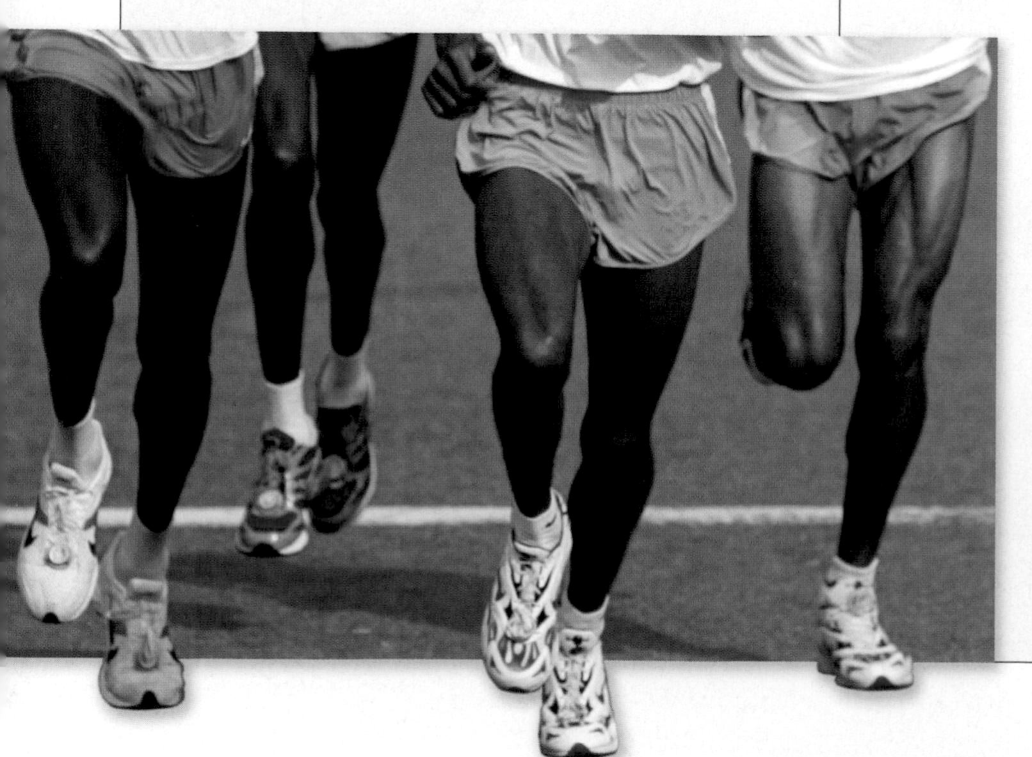

Marathon Men It takes a lot of training to run a marathon. One of the results of all that training is that the leg muscles become packed with slow-twitch muscle fibers with cells rich in energy-metabolizing mitochondria.

Marathon Mouse This mouse can run for much longer than a normal mouse because its energy metabolism has been genetically altered.

hancement. More likely in the near term is the use of an experimental drug called GW501516 (developed by the drug company Glaxo-Smith/Kline), which activates the PPARδ protein. When Evans and colleagues gave the drug to normal mice, they achieved the same results as with the genetically modified mice. Because this drug stimulates fat breakdown, it is being tested to treat obesity.

The energy stored in ATP is the energy you use all the time to fuel both conscious actions, like running a marathon or turning the pages of this book, and your body's automatic actions, such as breathing or contracting your heart muscles.

IN THIS CHAPTER we will discover how cells produce usable energy, usually in the form of ATP. We will describe the metabolic pathway by which glucose is oxidized both in the presence and the absence of O_2. We close the chapter with an overview of the relationships between the metabolic pathways that use and produce the four biologically important classes of molecules—carbohydrates, fats, proteins, and nucleic acids.

CHAPTER OUTLINE

7.1 How Does Glucose Oxidation Release Chemical Energy?

Fuels are molecules whose stored energy can be released for use. Wood burning in a campfire releases energy as heat and light. In cells, chemical fuels release chemical energy that is used to make ATP, which in turn can be used to drive endergonic reactions.

Photosynthetic organisms use energy from sunlight to synthesize their own fuels, as we will describe in Chapter 8. In nonphotosynthesizers, however, the most common chemical fuel is the sugar glucose ($C_6H_{12}O_6$). Other molecules, such as fats or proteins, can also supply energy, but to release it they must be converted either into glucose or into intermediate compounds in the various pathways of glucose metabolism.

In this section we explore how cells obtain energy from glucose by the chemical process of oxidation, which is carried out through a series of metabolic pathways. Several principles, some of which we noted in Section 6.5, govern metabolic pathways:

- Complex chemical transformations in the cell occur in a series of separate reactions that form a metabolic pathway.
- Each reaction in a metabolic pathway is catalyzed by a specific enzyme.
- Metabolic pathways are similar in all organisms, from bacteria to humans.
- Many metabolic pathways are compartmentalized in eukaryotes, with certain reactions occurring inside a specific organelle.
- Each metabolic pathway is regulated by key enzymes that can be inhibited or activated, thereby determining how fast the reactions will go.

Cells trap free energy while metabolizing glucose

As we saw in Section 2.3, the familiar process of combustion (burning) is very similar to the chemical processes that release energy in cells. If glucose is burned in a flame, it reacts with oxygen gas (O_2), forming carbon dioxide and water and releasing energy in the form of heat. The balanced equation for this combustion reaction is

$$C_6H_{12}O_6 + 6\ O_2 \rightarrow 6\ CO_2 + 6\ H_2O + \text{free energy}$$

The same equation applies to the metabolism of glucose in cells. The principles of metabolism cited above also apply to this process: the metabolism of glucose is a multistep pathway; each step is catalyzed by an enzyme; the process is compartmentalized; and the pathway is under enzymatic control.

The glucose metabolism pathway "traps" the stored energy of glucose in ATP molecules via the reaction

$$ADP + P_i + \text{free energy} \rightarrow ATP$$

The energy trapped in ATP can be used to do cellular work, such as movement of muscles or active transport across a membrane, just as heat energy captured from combustion can be used to do work.

The change in free energy (ΔG) resulting from the complete conversion of glucose and O_2 to CO_2 and water, whether by combustion or by metabolism, is –686 kcal/mol (–2,870 kJ/mol). Thus the overall reaction is highly exergonic and can drive the endergonic formation of a great deal of ATP from ADP and phosphate. The many steps of glucose metabolism allow this energy to be captured in ATP.

Three metabolic processes play important roles in the harvesting of energy from glucose: *glycolysis*, *cellular respiration*, and *fermentation* (**Figure 7.1**). All three involve metabolic pathways made up of many distinct chemical reactions.

- **Glycolysis** begins glucose metabolism in all cells and produces two molecules of the three-carbon product *pyruvate*. A small amount of the energy stored in glucose is captured in usable forms. Glycolysis does not use O_2.
- **Cellular respiration** uses O_2 (thus it is **aerobic**) from the environment and completely converts each pyruvate molecule into three molecules of CO_2 through a set of metabolic pathways. In the process, a great deal of the energy stored in the covalent bonds of pyruvate is released and transferred to ADP and phosphate to form ATP.
- **Fermentation** does not involve O_2 (it is **anaerobic**). Fermentation converts pyruvate into lactic acid or ethyl alcohol (ethanol), which are still relatively energy-rich molecules. Because the breakdown of glucose is incomplete, much less energy is released by fermentation than by cellular respiration.

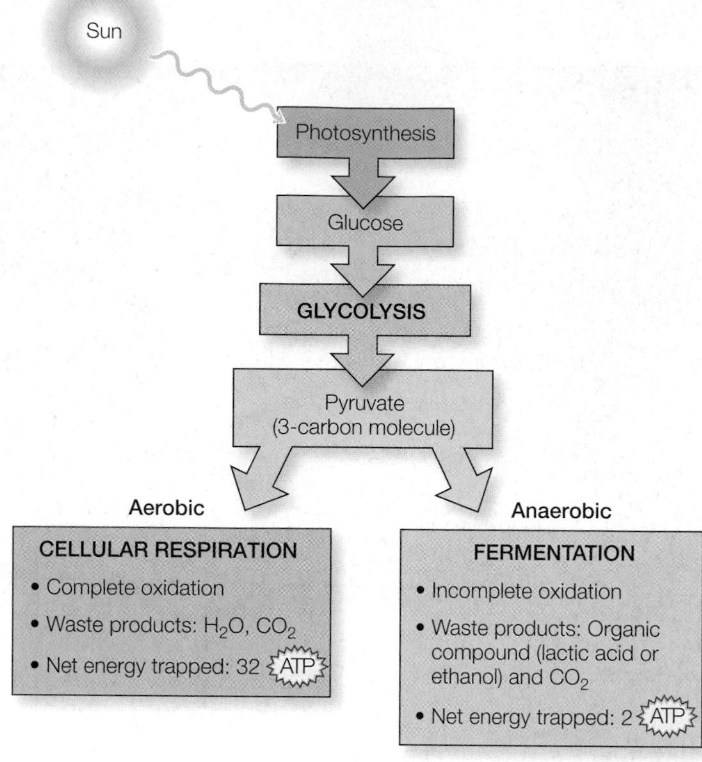

7.1 Energy for Life Living organisms obtain their energy from the food compounds produced by photosynthesis. They convert these compounds into glucose, which they metabolize by glycolysis to produce the 3-carbon compound pyruvate. Pyruvate molecules are then metabolized by either anaerobic fermentation or aerobic cellular respiration. The net result is energy "trapped" in ATP molecules that power the activities of living cells.

An overview: Harvesting energy from glucose

The energy-harvesting processes in cells use different combinations of metabolic pathways depending on the presence or absence of O_2:

- When O_2 is available as the final electron acceptor, four pathways operate (**Figure 7.2A**). **Glycolysis** takes place first, and is followed by the three pathways of cellular respiration: **pyruvate oxidation**, the **citric acid cycle** (also called the *Krebs cycle* or the *tricarboxylic acid cycle*), and the **electron transport chain** (also called the *respiratory chain*).
- When O_2 is unavailable, pyruvate oxidation, the citric acid cycle, and the electron transport chain do not function, and the pyruvate produced by glycolysis is further metabolized by **fermentation** (**Figure 7.2B**).

These five metabolic pathways, which we will consider one at a time, occur in different locations in the cell (**Table 7.1**).

Redox reactions transfer electrons and energy

As Section 6.2 described, the addition of a phosphate group to ADP to make ATP is an endergonic reaction that can extract and store energy from exergonic reactions. Another way of transferring energy is to transfer electrons. A reaction in which one substance transfers one or more electrons to another substance is called an *oxidation–reduction reaction*, or **redox reaction**.

- **Reduction** is the gain of one or more electrons by an atom, ion, or molecule.
- **Oxidation** is the loss of one or more electrons.

Although oxidation and reduction are always defined in terms of traffic in electrons, we may also think in these terms when hydrogen atoms (*not* hydrogen ions) are gained or lost, because transfers of hydrogen atoms involve transfers of electrons ($H = H^+ + e^-$). Thus, when a molecule loses hydrogen atoms, it becomes oxidized.

Oxidation and reduction *always occur together*: as one material is oxidized, the electrons it loses are transferred to another material, reducing that material. In a redox reaction, we call the reactant that becomes reduced an *oxidizing agent* and the one that be-

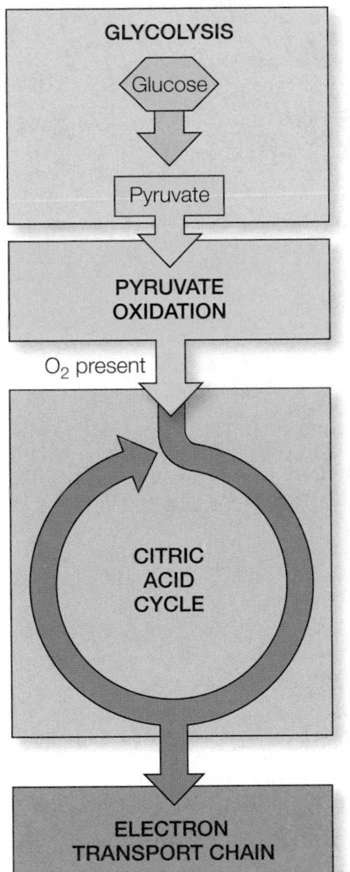

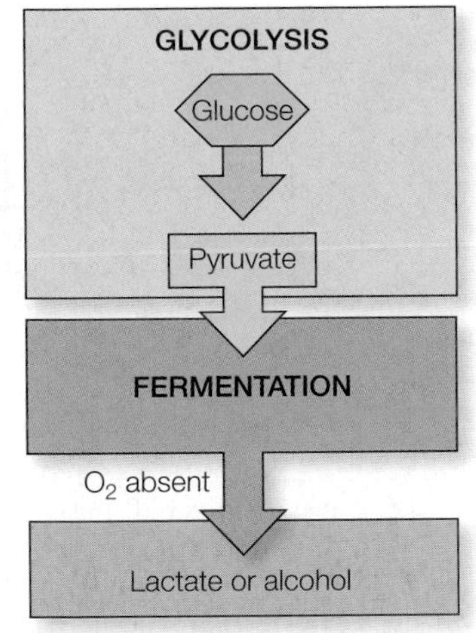

7.2 Energy-Producing Metabolic Pathways Energy-producing reactions can be grouped into five metabolic pathways: glycolysis, pyruvate oxidation, the citric acid cycle, the electron transport chain, and fermentation. (A) The three lower pathways occur only in the presence of O_2 and are collectively referred to as cellular respiration. (B) When O_2 is unavailable, glycolysis is followed by fermentation.

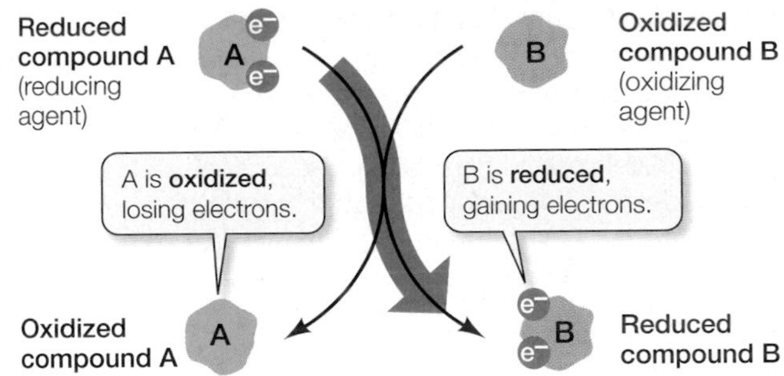

7.3 Oxidation and Reduction Are Coupled In a redox reaction, reactant A is oxidized and reactant B is reduced. In the process, A loses electrons and B gains electrons. Protons may be transferred along with electrons so that what is actually transferred (wide blue arrow) are hydrogen atoms: $AH_2 + B \rightarrow A + BH_2$.

comes oxidized a *reducing agent* (**Figure 7.3**). In both the combustion and the metabolism of glucose, glucose is the reducing agent (electron donor) and O_2 is the oxidizing agent (electron acceptor).

In a redox reaction, energy is transferred. Much of the energy originally present in the reducing agent becomes associated with the reduced product. (The rest remains in the reducing agent or is lost.) As we will see, some of the key reactions of glycolysis and cellular respiration are highly exergonic redox reactions.

The coenzyme NAD is a key electron carrier in redox reactions

Section 6.4 described the role of coenzymes, small molecules that assist in enzyme-catalyzed reactions. ADP acts as a coenzyme when it picks up energy released in an exergonic reaction and uses it to form ATP (an endergonic reaction). In a similar fashion, the coenzyme **NAD** (**nicotinamide adenine dinucleotide**) acts as an electron carrier, in this case in redox reactions (**Figure 7.4A**).

NAD exists in two chemically distinct forms, one oxidized (NAD^+) and the other reduced ($NADH + H^+$) (**Figure 7.4B**). Both forms participate in biological redox reactions. The reduction reaction

$$NAD^+ + 2\,H \rightarrow NADH + H^+$$

is formally equivalent to the transfer of two hydrogen atoms ($2\,H^+ + 2\,e^-$). However, what is actually transferred is a *hydride ion* (H^-, which is a proton and two electrons), leaving a free proton (H^+).

TABLE 7.1

Cellular Locations for Energy Pathways in Eukaryotes and Prokaryotes

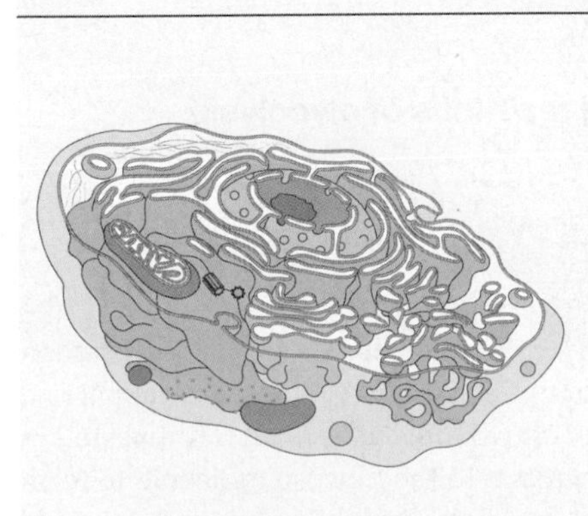

EUKARYOTES	PROKARYOTES
External to mitochondrion	**In cytoplasm**
Glycolysis	Glycolysis
Fermentation	Fermentation
	Citric acid cycle
Inside mitochondrion	**On plasma membrane**
Inner membrane	Pyruvate oxidation
Electron transport chain	Electron transport chain
Matrix	
Citric acid cycle	
Pyruvate oxidation	

7.4 NAD Is an Energy Carrier in Redox Reactions Thanks to its ability to carry free energy and electrons, NAD is a major energy carrier in redox reactions and a universal energy intermediary in cells. (A) Each curved black arrow represents either an oxidation or reduction reaction. The wide blue arrow shows the path of transferred electrons (compare with Figure 7.2). (B) NAD^+ is the oxidized form and NADH the reduced form of NAD^+. The unshaded portion of the molecule (left) remains unchanged by the redox reaction.

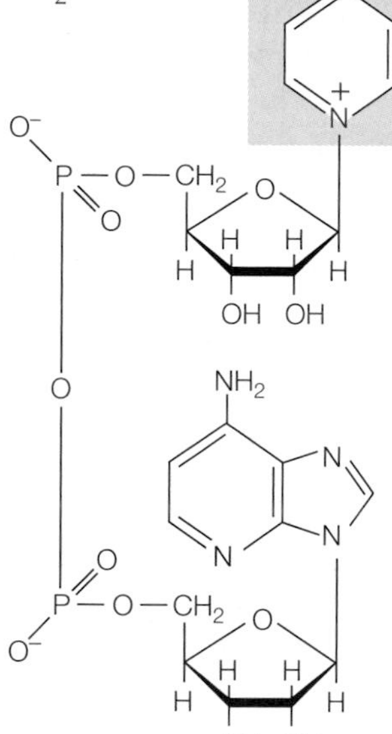

This notation emphasizes that reduction is accomplished by the addition of electrons.

Oxygen is highly electronegative and readily accepts electrons from NADH. The oxidation of NADH + H^+ by O_2,

$$NADH + H^+ + \tfrac{1}{2}O_2 \rightarrow NAD^+ + H_2O$$

is highly exergonic, with a ΔG of –52.4 kcal/mol (–219 kJ/mol). Note that the oxidizing agent appears here as "½ O_2" instead of "O." This notation emphasizes that it is molecular oxygen, O_2, that acts as the oxidizing agent.

Just as a molecule of ATP can be thought of as bundling or packaging about 12 kcal/mol (50 kJ/mol) of free energy, NAD can be thought of as packaging free energy in larger bundles (approximately 50 kcal/mol, or 200 kJ/mol). NAD is a common, but not the only, electron carrier in cells. As we will see, another carrier, **FAD (flavin adenine dinucleotide)**, also transfers electrons during glucose metabolism.

7.1 RECAP

When glucose is oxidized, the free energy released is trapped in ATP. The five metabolic pathways of the cell combine in different ways to produce ATP, which supplies the energy for myriad other reactions in living cells.

- What principles govern metabolic pathways in cells? See p. 139
- Do you understand how the coupling of oxidation and reduction can transfer energy from one molecule to another? See pp. 140–141 and Figure 7.2
- Do you understand the roles of NAD and O_2 with respect to electrons in a redox reaction? See pp. 141–142 and Figure 7.3

Now that we have an overview of the metabolic pathways that harvest energy from glucose, let's begin our more detailed examination of those pathways with a look at the three that begin the process when O_2 is available as an electron acceptor: glycolysis, pyruvate oxidation, and the citric acid cycle.

7.2 What Are the Aerobic Pathways of Glucose Metabolism?

Glycolysis takes place in the cytosol of cells. It converts glucose into pyruvate, produces a small amount of energy, and generates no CO_2. In glycolysis, some of the covalent bonds between carbon and hydrogen in the glucose molecule are oxidized, releasing some of the energy stored in this carbohydrate. After 10 enzyme-catalyzed reactions, the end products of glycolysis are 2 molecules of **pyruvate** (pyruvic acid), 4 molecules of ATP, and 2 molecules of NADH. Glycolysis can be divided into two stages: energy-investing reactions that use ATP, and energy-harvesting reactions that produce ATP (**Figure 7.5**).

The energy-investing reactions of glycolysis require ATP

Using Figure 7.5 as a guide, let's work our way through the glycolytic pathway.

The first five reactions of glycolysis are *endergonic*; that is, the cell is investing free energy in the glucose molecule, rather than releasing energy from it. In two separate reactions (*reactions 1 and 3* in Figure 7.5), the energy of two molecules of ATP is invested in attaching two phosphate groups to the glucose molecule to form fructose 1,6-bisphosphate, which has a free energy substantially

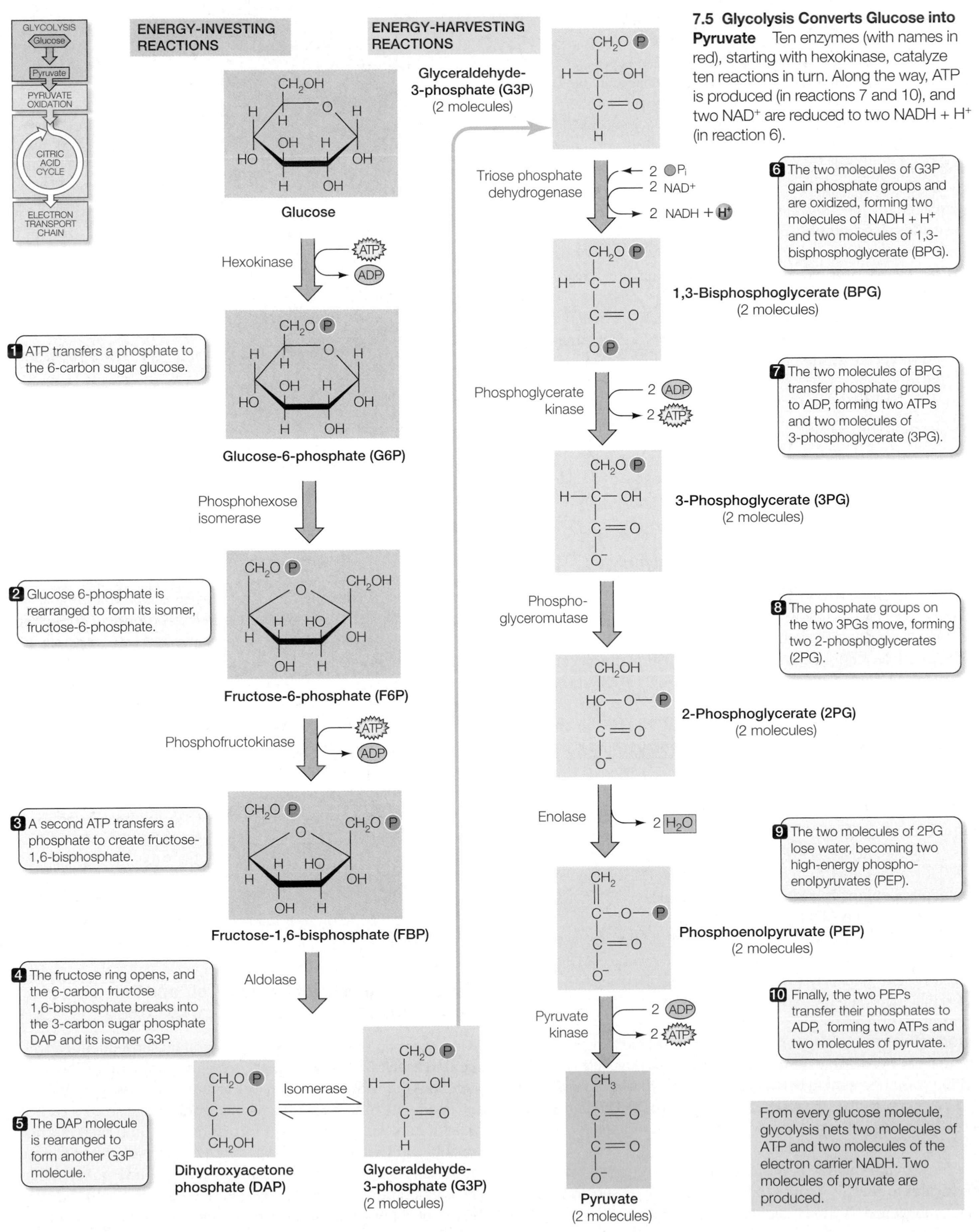

GLYCOLYSIS
Glucose
Pyruvate
PYRUVATE OXIDATION
CITRIC ACID CYCLE
ELECTRON TRANSPORT CHAIN
ENERGY-INVESTING REACTIONS
ENERGY-HARVESTING REACTIONS
Glucose
Hexokinase
ATP
ADP
1 ATP transfers a phosphate to the 6-carbon sugar glucose.
Glucose-6-phosphate (G6P)
Phosphohexose isomerase
2 Glucose 6-phosphate is rearranged to form its isomer, fructose-6-phosphate.
Fructose-6-phosphate (F6P)
Phosphofructokinase
ATP
ADP
3 A second ATP transfers a phosphate to create fructose-1,6-bisphosphate.
Fructose-1,6-bisphosphate (FBP)
Aldolase
4 The fructose ring opens, and the 6-carbon fructose 1,6-bisphosphate breaks into the 3-carbon sugar phosphate DAP and its isomer G3P.
Isomerase
5 The DAP molecule is rearranged to form another G3P molecule.
Dihydroxyacetone phosphate (DAP)
Glyceraldehyde-3-phosphate (G3P) (2 molecules)
Glyceraldehyde-3-phosphate (G3P) (2 molecules)
Triose phosphate dehydrogenase
2 Pi
2 NAD+
2 NADH + H+
6 The two molecules of G3P gain phosphate groups and are oxidized, forming two molecules of NADH + H+ and two molecules of 1,3-bisphosphoglycerate (BPG).
1,3-Bisphosphoglycerate (BPG) (2 molecules)
Phosphoglycerate kinase
2 ADP
2 ATP
7 The two molecules of BPG transfer phosphate groups to ADP, forming two ATPs and two molecules of 3-phosphoglycerate (3PG).
3-Phosphoglycerate (3PG) (2 molecules)
Phospho-glyceromutase
8 The phosphate groups on the two 3PGs move, forming two 2-phosphoglycerates (2PG).
2-Phosphoglycerate (2PG) (2 molecules)
Enolase
2 H2O
9 The two molecules of 2PG lose water, becoming two high-energy phospho-enolpyruvates (PEP).
Phosphoenolpyruvate (PEP) (2 molecules)
Pyruvate kinase
2 ADP
2 ATP
10 Finally, the two PEPs transfer their phosphates to ADP, forming two ATPs and two molecules of pyruvate.
Pyruvate (2 molecules)
From every glucose molecule, glycolysis nets two molecules of ATP and two molecules of the electron carrier NADH. Two molecules of pyruvate are produced.

7.5 Glycolysis Converts Glucose into Pyruvate Ten enzymes (with names in red), starting with hexokinase, catalyze ten reactions in turn. Along the way, ATP is produced (in reactions 7 and 10), and two NAD^+ are reduced to two NADH + H^+ (in reaction 6).

higher than that of glucose. Later, these phosphate groups will be transferred to ADP to make new molecules of ATP.

Although both of these steps of glycolysis use ATP as one of their substrates, each is catalyzed by a different, specific enzyme. The enzyme hexokinase catalyzes *reaction 1*, in which a phosphate group from ATP is attached to the six-carbon glucose molecule, forming glucose 6-phosphate. (A *kinase* is any enzyme that catalyzes the transfer of a phosphate group from ATP to another substrate.) In *reaction 2*, the six-membered glucose ring is rearranged into a five-membered fructose ring. In *reaction 3*, the enzyme phosphofructokinase adds a second phosphate (taken from another ATP) to the fructose ring, forming a six-carbon sugar, fructose 1,6-bisphosphate.

Reaction 4 opens up the six-carbon sugar ring and cleaves it into two different three-carbon sugar phosphates: dihydroxyacetone phosphate and glyceraldehyde 3-phosphate. In *reaction 5*, one of those products, dihydroxyacetone phosphate, is converted into a second molecule of the other one, glyceraldehyde 3-phosphate (G3P, a triose phosphate). In summary, by the halfway point of the glycolytic pathway, two things have happened:

- Two molecules of ATP have been invested.
- The six-carbon glucose molecule has been converted into two molecules of a three-carbon sugar phosphate, glyceraldehyde 3-phosphate (G3P).

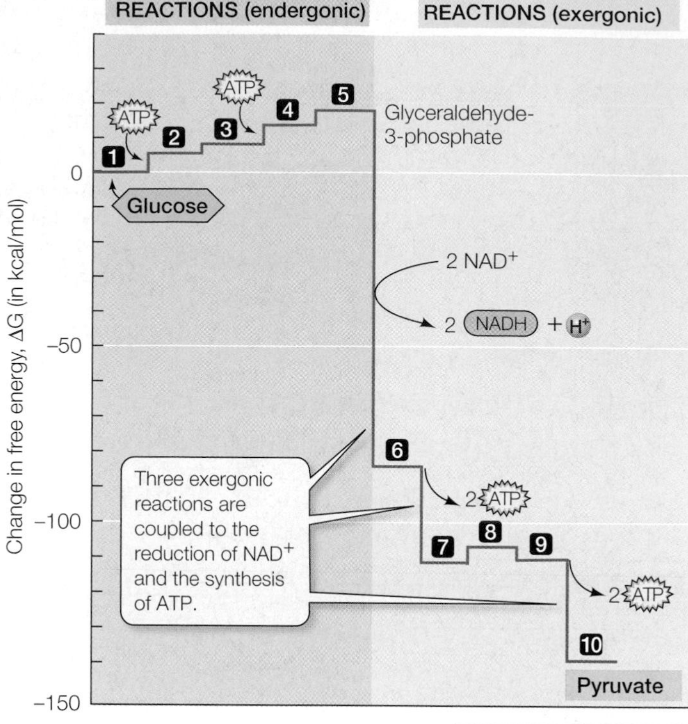

7.6 Changes in Free Energy During Glycolysis Each reaction of glycolysis results in a change in free energy. Refer to Figure 7.5 for the numbered reactions (e.g., reaction 1 is catalyzed by hexokinase).

The energy-harvesting reactions of glycolysis yield NADH + H^+ and ATP

In the discussion that follows, remember that *each reaction occurs twice for each glucose molecule* because each glucose molecule has been split into two molecules of G3P. It is the fate of G3P that now concerns us—its transformation will generate both NADH + H^+ and ATP.

PRODUCING NADH + H^+ *Reaction 6* is catalyzed by the enzyme triose phosphate dehydrogenase, and its end product is a phosphate ester, 1,3-bisphosphoglycerate (BPG). Reaction 6 is an *oxidation*, and it is accompanied by a large drop in free energy—more than 100 kcal of energy per mole of glucose is released in this exergonic reaction (**Figure 7.6**). If this big energy drop were simply a loss of heat, glycolysis would not provide useful energy to the cell. However, rather than being lost as heat, this energy is *stored* as chemical energy by reducing two molecules of NAD^+ to make two molecules of NADH + H^+.

Because NAD^+ is present in small amounts in the cell, it must be recycled to allow glycolysis to continue; if none of the NADH is oxidized back to NAD^+, glycolysis comes to a halt. The metabolic pathways that follow glycolysis carry out this oxidation, as we will see.

PRODUCING ATP In *reactions 7–10* of glycolysis, the two phosphate groups of BPG are transferred one at a time to molecules of ADP, with a rearrangement in between. More than 20 kcal (83.6 kJ/mol) of free energy is stored in ATP for every mole of BPG broken down. Finally, we are left with two moles of pyruvate for every mole of glucose that entered glycolysis.

The enzyme-catalyzed transfer of phosphate groups from donor molecules to ADP molecules to form ATP is called **substrate-level phosphorylation**. (*Phosphorylation* is the addition of a phosphate group to a molecule. *Substrate-level phosphorylation* is distinguished from the oxidative phosphorylation carried out by the electron transport chain, which we will discuss later in this chapter.) In glycolysis, energy released by oxidation is used to form ATP. An example of substrate-level phosphorylation occurs in *reaction 7*, in which phosphoglycerate kinase catalyzes the transfer of a phosphate group from BPG to ADP, forming ATP. Both reactions 6 and 7 are exergonic, even though a substantial amount of energy is consumed in the formation of ATP.

To summarize:

- The energy-investing steps of glycolysis use the energy of hydrolysis of two ATP molecules per glucose molecule.
- The energy-releasing steps of glycolysis produce four ATP molecules per glucose molecule.

If O_2 is present, glycolysis is followed by the three pathways of cellular respiration.

Pyruvate oxidation links glycolysis and the citric acid cycle

The oxidation of pyruvate to acetate and its subsequent conversion to **acetyl CoA** is the link between glycolysis and all the other reactions of cellular respiration (see Figure 7.7). Coenzyme A (CoA) is a complex molecule responsible for binding the two-carbon acetate molecule. Acetyl CoA formation is a multi-step reaction catalyzed by the *pyruvate dehydrogenase complex*, an enormous multi-enzyme complex that is attached to the inner mitochondrial membrane. Pyruvate diffuses into the mitochondrion, where a series of coupled reactions takes place:

1. Pyruvate is oxidized to a two-carbon acetyl group (acetate), and CO_2 is released.
2. Part of the energy from this oxidation is captured by the reduction of NAD^+ to NADH + H^+.
3. Some of the remaining energy is stored temporarily by the combining of the acetyl group with CoA, forming acetyl CoA:

$$\text{pyruvate} + NAD^+ + \text{CoA} \rightarrow \text{acetyl CoA} + NADH + H^+ + CO_2$$

Acetyl CoA has 7.5 kcal/mol (31.4 kJ/mol) more energy than simple acetate. Acetyl CoA can donate the acetyl group to acceptor molecules, much as ATP can donate phosphate groups to various acceptors. Acetyl CoA donates its acetyl group to the four-carbon compound oxaloacetate to form the six-carbon citrate, the compound that initiates the citric acid cycle, one of life's most important energy-harvesting pathways.

> Arsenic, the classic poison of rodent exterminators and murder mysteries, acts by inhibiting pyruvate dehydrogenase, thus decreasing acetyl CoA production. The lack of acetyl CoA stops the citric acid cycle and cells eventually "starve to death" for lack of ATP.

The citric acid cycle completes the oxidation of glucose to CO_2

Acetyl CoA is the starting point for the citric acid cycle. This pathway of eight reactions completely oxidizes the two-carbon acetyl group to two molecules of carbon dioxide. The free energy released from these reactions is captured by ADP and the electron carriers NAD and FAD. **Figure 7.7** shows the energetics of the pathway. In addition, recall that energy is released upon oxidation and stored in either ATP, $FADH_2$, or NADH + H^+.

The citric acid cycle is maintained in a *steady state*—that is, although the intermediate compounds in the cycle enter and leave it, the concentrations of those intermediates do not change much. Pay close attention to the numbered reactions in **Figure 7.8** as you read the next several paragraphs.

The energy temporarily stored in acetyl CoA drives the formation of citrate from oxaloacetate (*reaction 1*). During this reaction, the coenzyme A molecule is removed and can be reused. In *reaction 2*, the citrate molecule is rearranged to form isocitrate. In *reaction 3*, a CO_2 molecule and two hydrogen atoms are removed, converting isocitrate into α-ketoglutarate. This reaction produces a large drop in free energy, some of which is stored in NADH + H^+.

Reaction 4 of the citric acid cycle, in which α-ketoglutarate is oxidized to succinyl CoA, is similar to oxidation of pyruvate to acetyl CoA and, like that reaction, is catalyzed by a multi-enzyme complex. In *reaction 5*, some of the energy in succinyl CoA is harvested to make GTP (guanosine triphosphate) from GDP and P_i, which is another example of substrate-level phosphorylation. GTP is then used to make ATP from ADP.

Free energy is released in *reaction 6*, in which the succinate released from succinyl CoA in reaction 5 is oxidized to fumarate. In the process, two hydrogens are transferred to an enzyme that contains the carrier FAD. After a molecular rearrangement (*reaction 7*), one more NAD^+ reduction occurs, producing oxaloacetate from malate (*reaction 8*). These two reactions illustrate a common biochemical mechanism: water (H_2O) is added in reaction 7 to form an —OH group, and then the H from that —OH group is removed in reaction 8 to reduce NAD^+ to NADH + H^+. The final product, oxaloacetate, is ready to combine with another acetyl group from acetyl CoA and go around the cycle again. The citric acid cycle operates twice for each glucose molecule that enters glycolysis (once for each pyruvate that enters the mitochondrion).

To summarize:

- The *inputs* to the citric acid cycle are acetate (in the form of acetyl CoA), water, and oxidized electron carriers (NAD^+ and FAD).

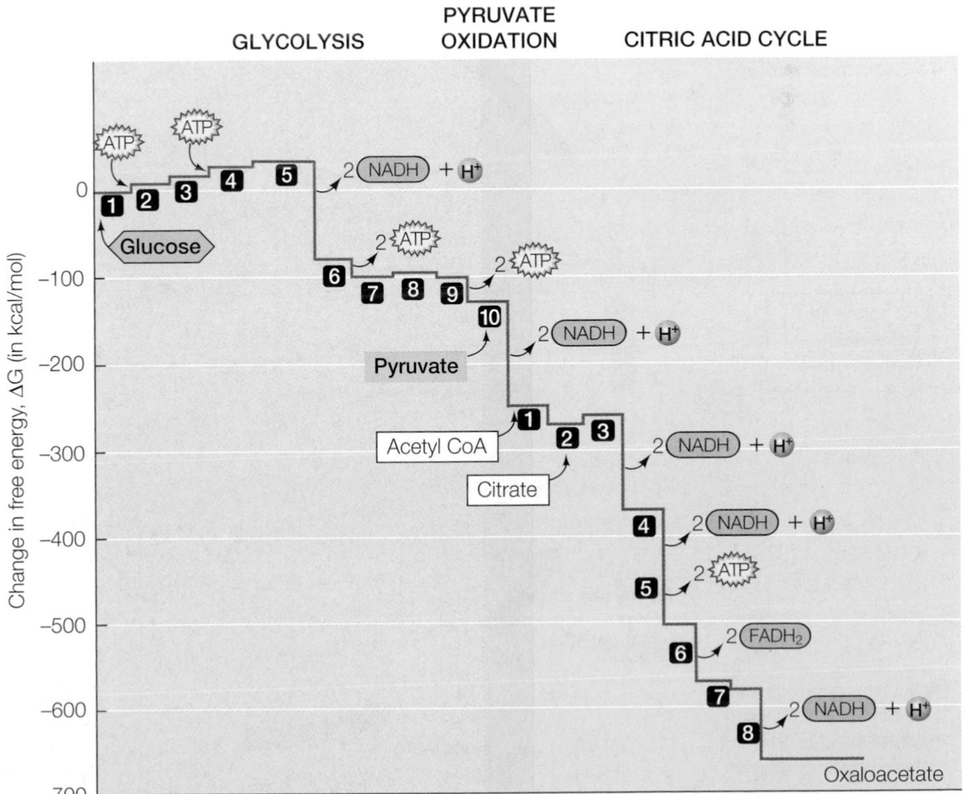

7.7 The Citric Acid Cycle Releases Much More Free Energy Than Glycolysis Does Electron carriers (NAD in glycolysis; NAD and FAD in the citric acid cycle) are reduced and ATP is generated in reactions coupled to other reactions, producing major drops in free energy as metabolism proceeds.

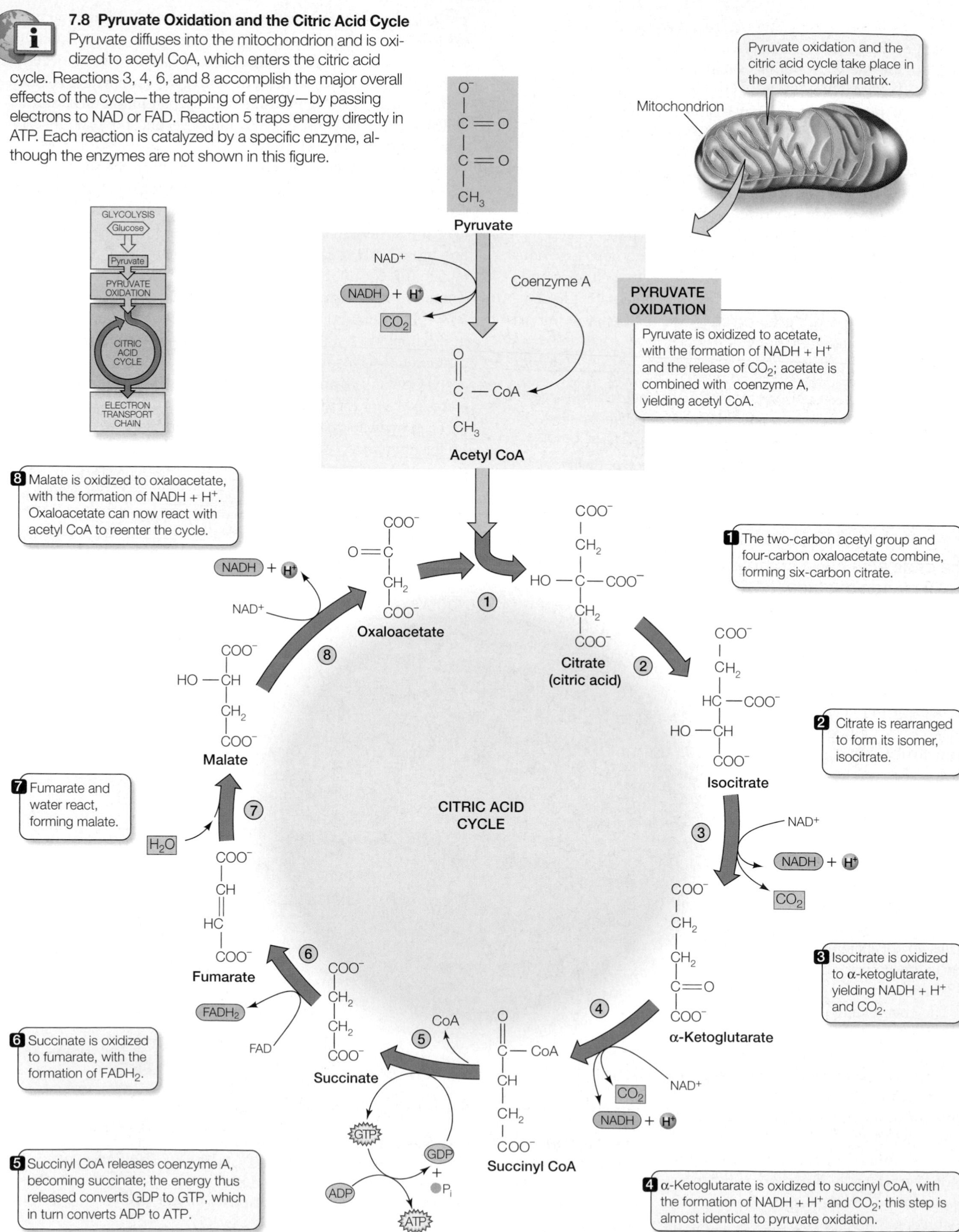
Pyruvate oxidation and the citric acid cycle take place in the mitochondrial matrix.
Mitochondrion
GLYCOLYSIS
Glucose
Pyruvate
PYRUVATE OXIDATION
CITRIC ACID CYCLE
ELECTRON TRANSPORT CHAIN
Pyruvate
NAD+
NADH + H+
CO2
Coenzyme A
Acetyl CoA
PYRUVATE OXIDATION
Pyruvate is oxidized to acetate, with the formation of NADH + H+ and the release of CO2; acetate is combined with coenzyme A, yielding acetyl CoA.
8 Malate is oxidized to oxaloacetate, with the formation of NADH + H+. Oxaloacetate can now react with acetyl CoA to reenter the cycle.
Oxaloacetate
Citrate (citric acid)
1 The two-carbon acetyl group and four-carbon oxaloacetate combine, forming six-carbon citrate.
2 Citrate is rearranged to form its isomer, isocitrate.
Isocitrate
Malate
7 Fumarate and water react, forming malate.
H2O
CITRIC ACID CYCLE
3 Isocitrate is oxidized to α-ketoglutarate, yielding NADH + H+ and CO2.
α-Ketoglutarate
Fumarate
FADH2
FAD
6 Succinate is oxidized to fumarate, with the formation of FADH2.
Succinate
CoA
Succinyl CoA
GTP
GDP
Pi
ADP
ATP
5 Succinyl CoA releases coenzyme A, becoming succinate; the energy thus released converts GDP to GTP, which in turn converts ADP to ATP.
4 α-Ketoglutarate is oxidized to succinyl CoA, with the formation of NADH + H+ and CO2; this step is almost identical to pyruvate oxidation.

7.8 Pyruvate Oxidation and the Citric Acid Cycle Pyruvate diffuses into the mitochondrion and is oxidized to acetyl CoA, which enters the citric acid cycle. Reactions 3, 4, 6, and 8 accomplish the major overall effects of the cycle—the trapping of energy—by passing electrons to NAD or FAD. Reaction 5 traps energy directly in ATP. Each reaction is catalyzed by a specific enzyme, although the enzymes are not shown in this figure.

- The *outputs* are carbon dioxide, reduced electron carriers (NADH + H^+ and $FADH_2$), and a small amount of ATP. Overall, for each acetyl group, the citric acid cycle removes two carbons as CO_2 and uses four pairs of hydrogen atoms to reduce electron carriers.

The citric acid cycle is regulated by concentrations of starting materials

We have seen how pyruvate, a three-carbon molecule, is completely oxidized to CO_2 by pyruvate dehydrogenase and the citric acid cycle. For the cycle to start anew, the starting molecules—acetyl CoA from oxidation of glucose, and oxidized electron carriers—must again all be in place. Because the electron carriers were reduced during the cycle (as well as in glycolysis—see reaction 6 in Figure 7.5), they are not able to accept electrons from the citric acid cycle reactions until they are *reoxidized*:

$$\text{NADH} \rightarrow \text{NAD}^+ + \text{H}^+ + e^-$$

$$\text{FADH}_2 \rightarrow \text{FAD} + 2\text{H}^+ + 2e^-$$

The oxidations of these electron carriers take place as coupled redox reactions, so that some other molecule (call it "X") gets reduced:

- NAD gets oxidized: $\text{NADH} \rightarrow \text{NAD}^+ + \text{H}^+ + e^-$
- An electron acceptor gets reduced: $\text{X} + \text{H}^+ + e^- \rightarrow \text{XH}$

Cells have two chemical ways of accomplishing this:

- *Fermentation*: If O_2 is absent, "X" is pyruvate, the end product of glycolysis. In the process of reoxidizing NADH formed by glycolysis, pyruvate is reduced to lactate or ethyl alcohol. These are the final products, so the citric acid cycle cannot occur.
- *Oxidative phosphorylation*: If O_2 is present, "X" is O_2. Pyruvate is fully oxidized to CO_2, and all the NADH and $FADH_2$ of the citric acid cycle is reoxidized. Energy is released by the oxidation of these electron carriers, and this energy is tapped to form ATP.

7.2 RECAP

The oxidation of glucose in the presence of O_2 involves glycolysis, pyruvate oxidation, and the citric acid cycle. In glycolysis, glucose is converted into pyruvate with some resulting energy capture. In the citric acid cycle, pyruvate is fully oxidized to CO_2, and more energy is captured in the form of reduced electron carriers.

- Can you describe the *net energy yield* of glycolysis in terms of energy invested and energy harvested? See p. 144 and Figure 7.6
- What role does pyruvate oxidation play in the citric acid cycle? See pp. 144–145 and Figure 7.8
- Why is reoxidation of NADH crucial for continuation of the citric acid cycle? See p. 147

The oxidation of pyruvate and the citric acid cycle require the presence of O_2 as an electron receptor. We will look now at fermentation, the metabolic pathway that processes pyruvate in the absence of oxygen. Section 7.4 will cover oxidative phosphorylation, which completes the electron transfers of aerobic cellular respiration

7.3 How Is Energy Harvested from Glucose in the Absence of Oxygen?

Fermentation, like glycolysis, occurs in the cytosol. With no O_2 present to serve as an electron acceptor, fermentation uses NADH + H^+ formed by glycolysis to reduce pyruvate (or one of its metabolic derivatives) and regenerate NAD^+. As we saw in Figure 7.5, NAD^+ is required for *reaction 6* of glycolysis, so once the cell has replenished its NAD^+ supply by fermentation, it can metabolize more glucose.

Prokaryotic organisms are known to use many different fermentation pathways. Best understood, however, are two short fermentation pathways found in many different cells: *lactic acid fermentation*, whose end product is lactic acid (lactate); and *alcoholic fermentation*, whose end product is ethyl alcohol.

Pyruvate serves as the electron acceptor in **lactic acid fermentation** (**Figure 7.9**), which takes place in many microorganisms. It can also take place in some muscle cells, especially during high activity. As you will see in Chapter 47, muscular contraction requires a lot of energy, which must be supplied by ATP. But the flow of

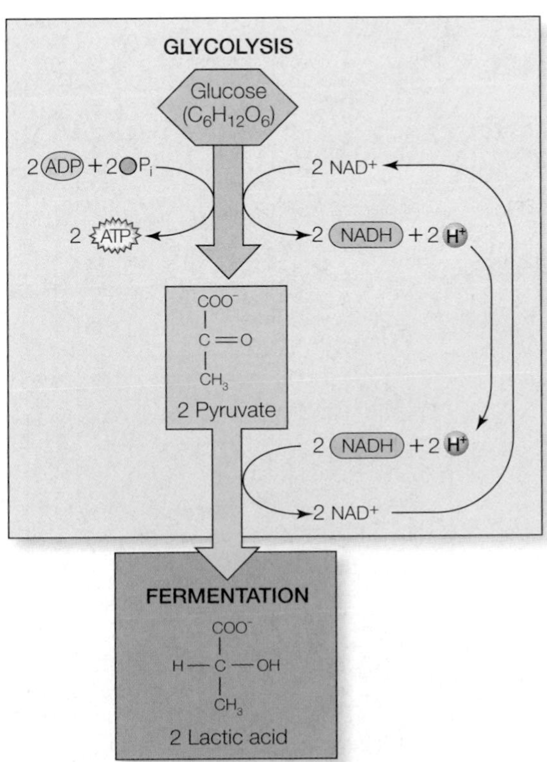

7.9 Lactic Acid Fermentation Glycolysis produces pyruvate, as well as ATP and NADH + H^+, from glucose. Lactic acid fermentation, using NADH + H^+ as a reducing agent, reduces pyruvate to lactic acid (lactate).

blood cannot supply enough O_2 during active contractions. To allow muscle tissue to continue to function in an anaerobic state, the muscle cells "switch" to lactic acid fermentation. Lactic acid buildup becomes a problem after prolonged periods because the acid ionizes, forming H^+ and lowering the pH of the cell, which reduces cellular activities (and causes muscle cramps).

Alcoholic fermentation takes place in certain yeasts and some plant cells under anaerobic conditions. This process requires two enzymes to metabolize pyruvate (**Figure 7.10**). First, carbon dioxide is removed from pyruvate, leaving the compound acetaldehyde. Second, the acetaldehyde is reduced by NADH + H^+, producing NAD^+ and ethyl alcohol (ethanol). Alcoholic beverages are made by the anaerobic fermentation of yeast cells using glucose from plant sources, such as grapes (wine) and barley (beer).

Billions and billions of prokaryotic microorganisms live by fermentation in the anaerobic environments of mammalian digestive organs such as the small intestine and the rumen (a specialized "stomach" found in cows and their relatives). These microorganisms may make crucial contributions to the "host" mammal's health and survival.

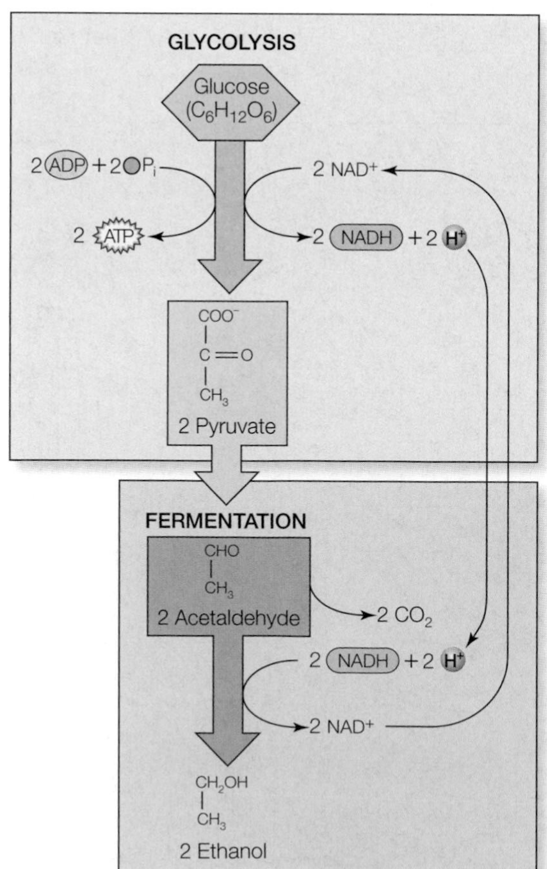

7.10 Alcoholic Fermentation In alcoholic fermentation, pyruvate from glycolysis is converted into acetaldehyde, and CO_2 is released. The NADH + H^+ from glycolysis acts as a reducing agent, reducing acetaldehyde to ethanol.

Fermentation enables glycolysis to produce a small amount of ATP through substrate-level phosphorylation. The net yield of two ATPs per glucose molecule is a minimal amount of usable energy, so it should not surprise you to learn that most organisms existing in anaerobic environments are small microbes that grow relatively slowly.

7.3 RECAP

In the absence of O_2, fermentation pathways use NADH + H^+ formed by glycolysis to reduce pyruvate or its derivatives and regenerate NAD^+. The energy yield of fermentation is low because glucose is only partially oxidized.

- Do you understand why replenishing NAD^+ is crucial to cellular metabolism? See p. 11

Fermentation pathways were the source of energy for much of life's early evolution, since Earth's early atmosphere lacked free oxygen. But when O_2 is present to accept electrons, the reoxidation of NADH and $FADH_2$ results in the synthesis of large amounts of ATP. Let's see how this process works.

7.4 How Does the Oxidation of Glucose Form ATP?

The overall process of ATP synthesis resulting from the reoxidation of electron carriers in the presence of O_2 is called **oxidative phosphorylation**. Two stages of the process can be distinguished:

1. *The electron transport chain.* The electrons from NADH and $FADH_2$ pass through a series of membrane-associated electron carriers. The flow of electrons along this electron transport chain accomplishes the active transport of protons across the inner mitochondrial membrane, out of the matrix, creating a proton concentration gradient.
2. *Chemiosmosis.* The protons diffuse back into the mitochondrial matrix through a proton channel, which couples this diffusion to the synthesis of ATP.

Before we proceed with the details of these two pathways, let's reflect on an important question: Why should the electron transport chain have so many components and complex processes? Why, for example, don't cells use the following single step?

$$NADH + H^+ + \tfrac{1}{2} O_2 \rightarrow NAD^+ + H_2O$$

The answer is that this reaction would be fundamentally untamable. It would be extremely exergonic—rather like setting off a stick of dynamite in the cell. There is no biochemical way to harvest that burst of energy efficiently and put it to physiological use (that is, no metabolic reaction is so endergonic as to consume a significant fraction of that energy in a single step). To control the release of energy during the oxidation of glucose in a cell, evolution has resulted in the lengthy electron transport chain: a series of reactions, each of which releases a small, manageable amount of energy.

The electron transport chain shuttles electrons and releases energy

The electron transport chain contains large integral proteins, smaller mobile proteins, and even a smaller lipid molecule (**Figure 7.11**):

- Four large protein complexes (I, II, III, and IV) containing electron carriers and associated enzymes are integral proteins of the inner mitochondrial membrane in eukaryotes (see Figure 4.13). Three of them are transmembrane proteins (extend through both sides of the membrane).
- **Cytochrome *c*** is a small peripheral protein that lies in the intermembrane space. It is loosely attached to the inner mitochondrial membrane.
- A nonprotein component called **ubiquinone** (abbreviated **Q**) is a small, nonpolar molecule that moves freely within the hydrophobic interior of the phospholipid bilayer of the inner mitochondrial membrane.

As illustrated in **Figure 7.12**, NADH + H^+ passes electrons to the first large protein complex (I), called NADH-Q reductase, which in turn passes the electrons to Q. The second complex (II), succinate dehydrogenase, passes electrons to Q from $FADH_2$ during the formation of fumarate from succinate in reaction 6 of the citric acid cycle (see Figure 7.8). These electrons enter the chain later than those from NADH.

The third complex (III), cytochrome *c* reductase, receives electrons from Q and passes them to cytochrome *c*. The fourth complex (IV), cytochrome *c* oxidase, receives electrons from cytochrome *c* and passes them to oxygen, which, along with these extra electrons ($\frac{1}{2} O_2^-$) picks up two hydrogen ions (H^+) to form H_2O.

The electron carriers of the electron transport chain (including those contained in the three transmembrane protein complexes) differ as to how they change when they become reduced. NAD^+, for example, accepts H^- (a hydride ion—one proton and two electrons), leaving the proton from the other hydrogen atom to float free: that is, the result is NADH + H^+ (see Figure 7.3B). Other carriers, including Q, bind both protons and both electrons, becoming, for example, QH_2. Beyond Q, the chain transports electrons only. Electrons, not protons, are passed from Q to cytochrome *c*. An electron from QH_2 reduces the cytochrome's Fe^{3+} to Fe^{2+}.

What happens to the protons? Electron transport within each of the three transmembrane protein complexes results, as we'll see, in the pumping of protons across the inner mitochondrial membrane, and the return of those protons across the membrane is coupled to the formation of ATP. Thus the energy originally contained

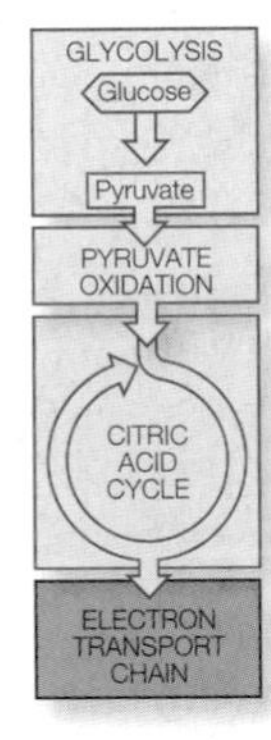

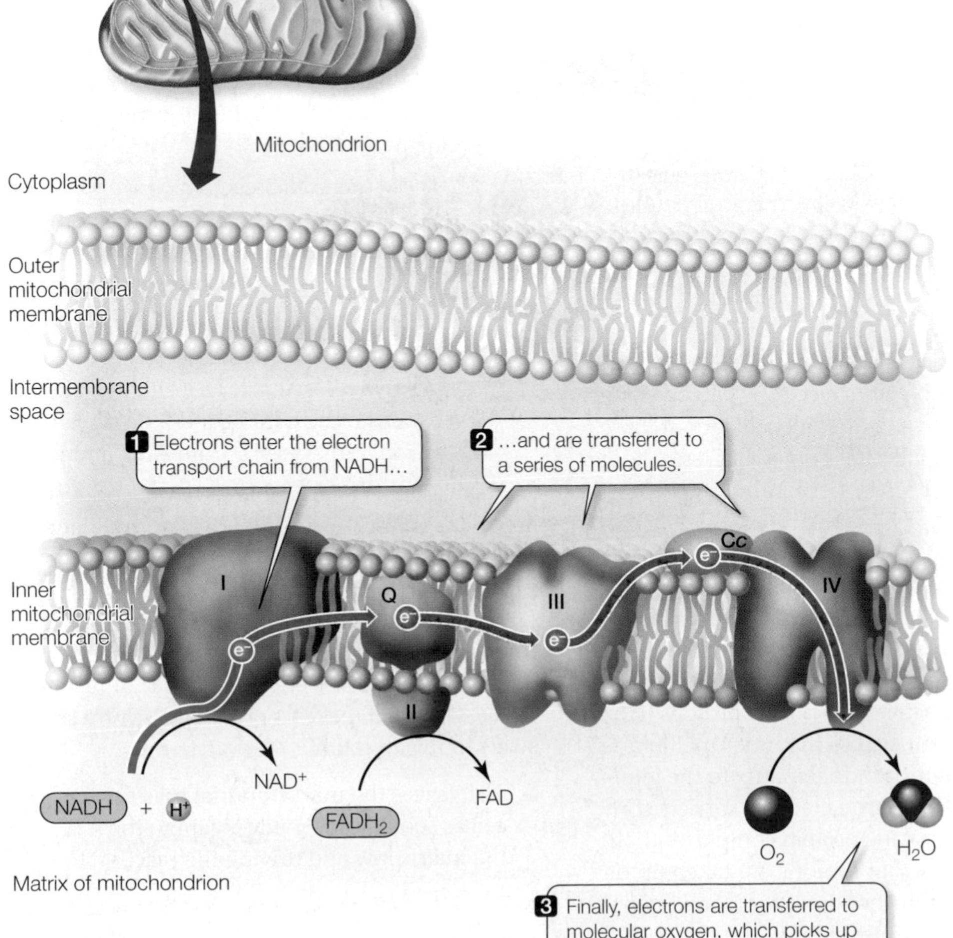

7.11 The Oxidation of NADH + H^+ Electrons from NADH + H^+ are passed through the electron transport chain, a series of protein complexes in the inner mitochondrial membrane containing electron carriers and enzymes. The carriers gain free energy when they become reduced and release free energy when they are oxidized.

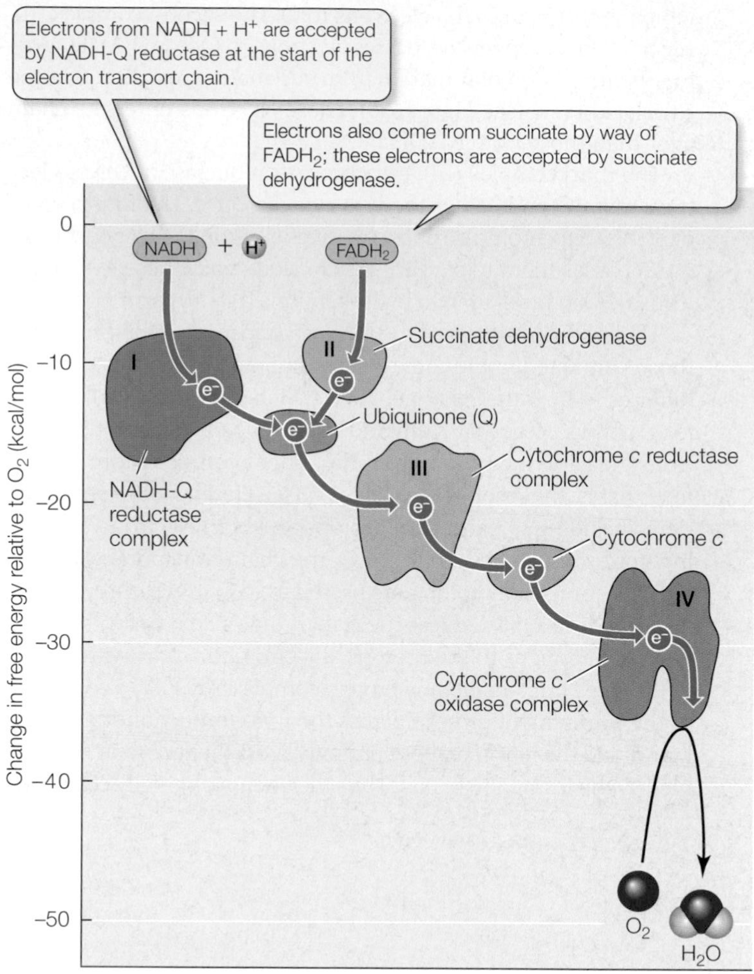

7.12 The Complete Electron Transport Chain Electrons enter the chain from two sources, but they follow the same pathway from Q onward.

in glucose and other fuel molecules is finally captured in the cellular energy currency, ATP. For each pair of electrons passed along the chain from NADH + H^+ to oxygen, theoretically about three molecules of ATP are formed. Actually, it is somewhat less, about 2.5 ATP per NADH oxidized and 1.5 ATP per FADH oxidized.

Proton diffusion is coupled to ATP synthesis

As shown in Figure 7.11, all the electron carriers and enzymes of the electron transport chain except cytochrome *c* are embedded in the inner mitochondrial membrane. The operation of the electron transport chain results in the active transport of protons (H^+), against their concentration gradient, across the inner membrane of the mitochondrion from the mitochondrial matrix to the intermembrane space. This occurs because the electron carriers contained in the three large transmembrane protein complexes (I, III, and IV) are arranged in such a way that protons are taken up on one side of the membrane (in the mitochondrial matrix) and transported, along with electrons, to the other side (to the intermembrane space) (**Figure 7.13**). Thus the transmembrane protein complexes of the electron transport chain act as *proton pumps*. Because of the positive charge on the protons (H^+), this pumping creates not only a difference in proton concentration, but also a difference in electric charge, across the inner mitochondrial membrane, making the mitochondrial matrix more negative than the intermembrane space. (In other words, the intermembrane space becomes more acidic than the mitochondrial matrix.)

Together, the proton concentration gradient and the charge difference constitute a source of potential energy called the **proton-motive force**. This force tends to drive the protons back across the membrane, just as the charge on a battery drives the flow of electrons, discharging the battery.

The conversion of the proton-motive force into kinetic energy is prevented by the fact that protons cannot cross the hydrophobic phospholipid bilayer of the inner mitochondrial membrane by simple diffusion. However, they can diffuse across the membrane by passing through a specific proton channel, called **ATP synthase**, which couples proton movement to the synthesis of ATP. This coupling of proton-motive force and ATP synthesis is called the *chemiosmotic mechanism*, or **chemiosmosis**.

THE CHEMIOSMOTIC MECHANISM FOR ATP SYNTHESIS The chemiosmotic mechanism uses ATP synthase to couple proton diffusion to ATP synthesis. This mechanism has three parts:

1. The flow of electrons from one electron carrier to another in the electron transport chain is a series of exergonic reactions that occurs in the inner mitochondrial membrane.
2. These exergonic reactions drive the endergonic pumping of H^+ out of the mitochondrial matrix and across the inner membrane into the intermembrane space. This pumping establishes and maintains a H^+ gradient.
3. The potential energy of the H^+ gradient, or proton-motive force, is harnessed by ATP synthase. This protein has two roles: it acts as a channel allowing H^+ to diffuse back into the matrix, and it uses the energy of that diffusion to make ATP from ADP and P_i.

ATP synthesis is a reversible reaction, and ATP synthase can also act as an ATPase, hydrolyzing ATP to ADP and P_i:

$$ATP \rightleftharpoons ADP + P_i + \text{free energy}$$

If the reaction goes to the right, free energy is released and is used to pump H^+ out of the mitochondrial matrix. If the reaction goes to the left, it uses free energy from H^+ diffusion into the matrix to make ATP. What makes it prefer ATP synthesis? There are two answers to this question:

- ATP leaves the mitochondrial matrix for use elsewhere in the cell as soon as it is made, keeping the ATP concentration in the matrix low and driving the reaction toward the left.
- The H^+ gradient is maintained by electron transport and proton pumping. (The electrons, you will recall, come from the oxidation of NADH and $FADH_2$, which are themselves re-

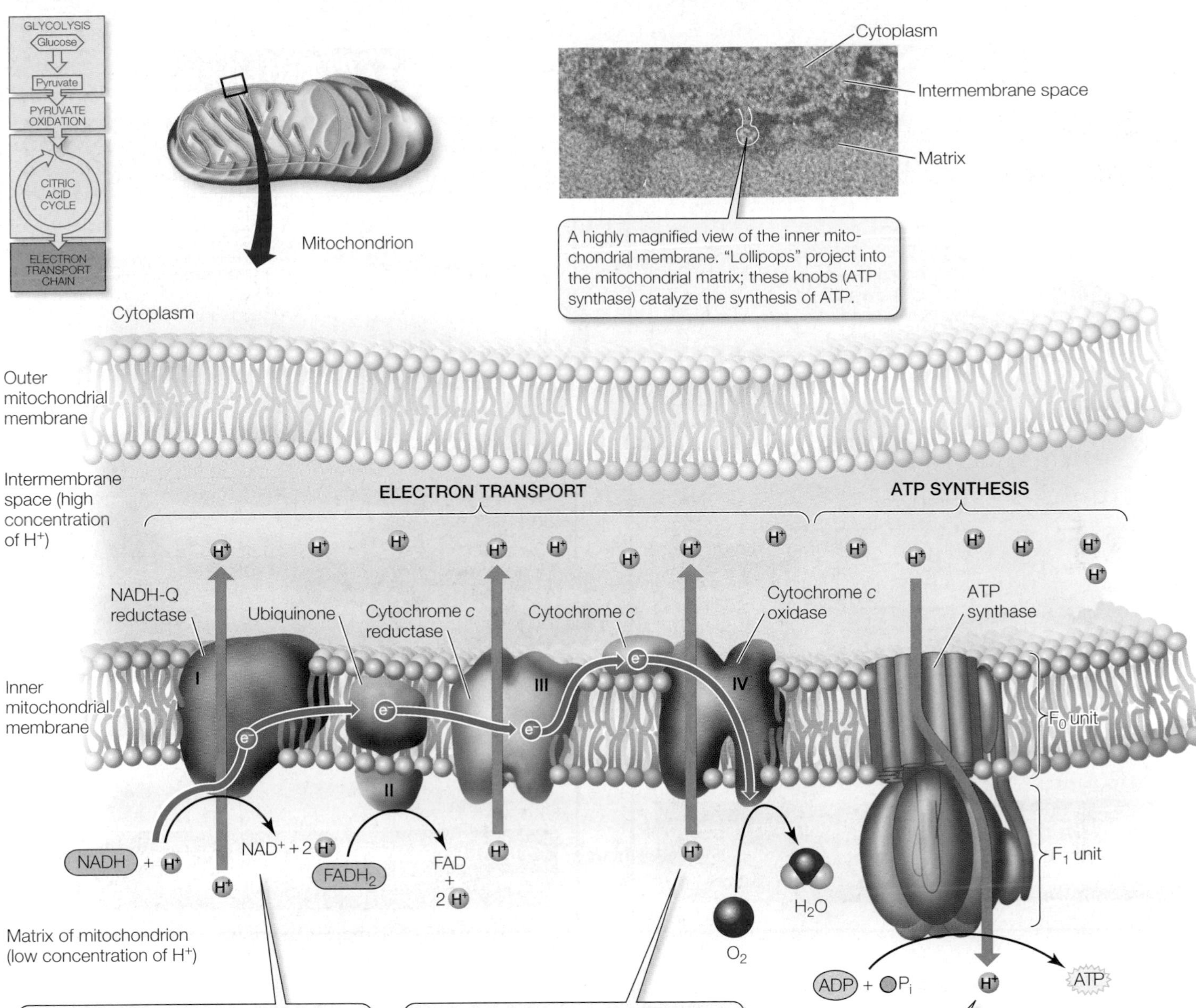

7.13 A Chemiosmotic Mechanism Produces ATP As electrons pass through the transmembrane protein complexes in the electron transport chain, protons are pumped from the mitochondrial matrix into the intermembrane space. As the protons return to the matrix through ATP synthase, ATP is formed.

duced by the oxidations of glycolysis and the citric acid cycle. So, one reason you eat food is to replenish the H^+ gradient!)

Every day a person hydrolyzes about 10^{25} ATP molecules to ADP. The vast majority of this ADP is "recycled"—converted back to ATP—using free energy from the oxidation of glucose.

EXPERIMENTS DEMONSTRATE CHEMIOSMOSIS Two key experiments demonstrated (1) that a proton (H^+) gradient across a membrane can drive ATP synthesis; and (2) that the enzyme ATP synthase is the catalyst for this reaction (**Figure 7.14**):

- *Experiment 1* tested the hypothesis that ATP synthesis is driven by the H^+ gradient across the inner mitochondrial membrane. In this experiment, mitochondria without a food source were "fooled" into making ATP by raising the H^+ concentration in their environment. Isolated mitochondria were exposed to a low H^+ concentration, then suddenly placed in a medium with a high H^+ concentration. The outer mitochondrial membrane, unlike the inner one, is freely permeable to H^+, so H^+ rapidly diffused into the intermembrane space. This

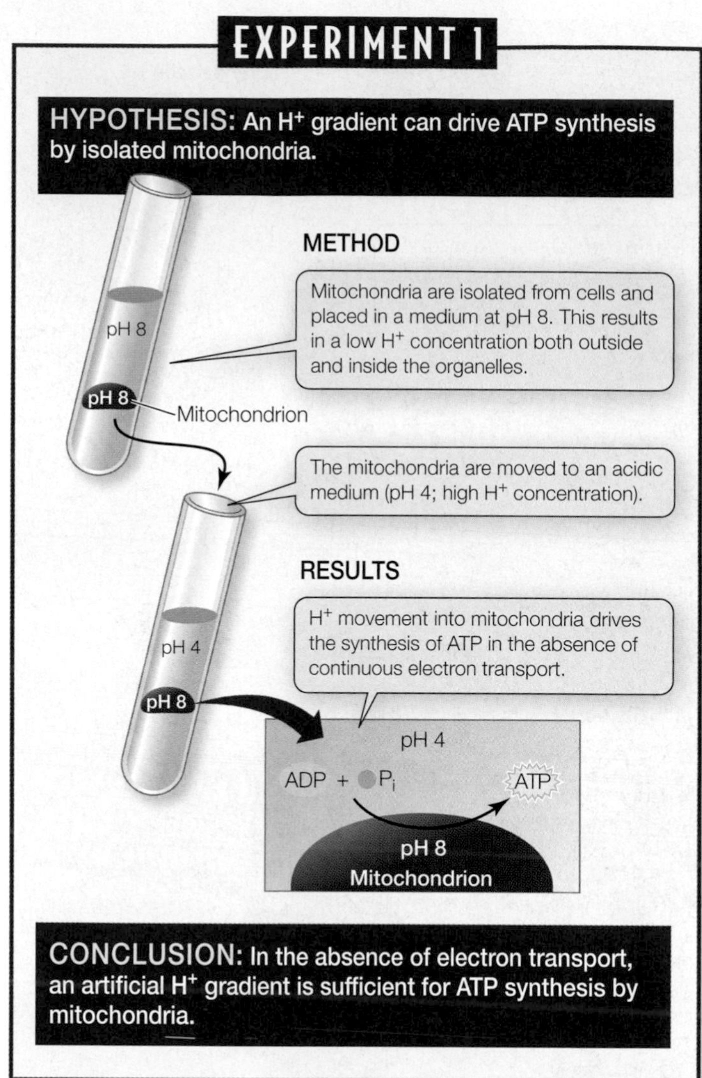

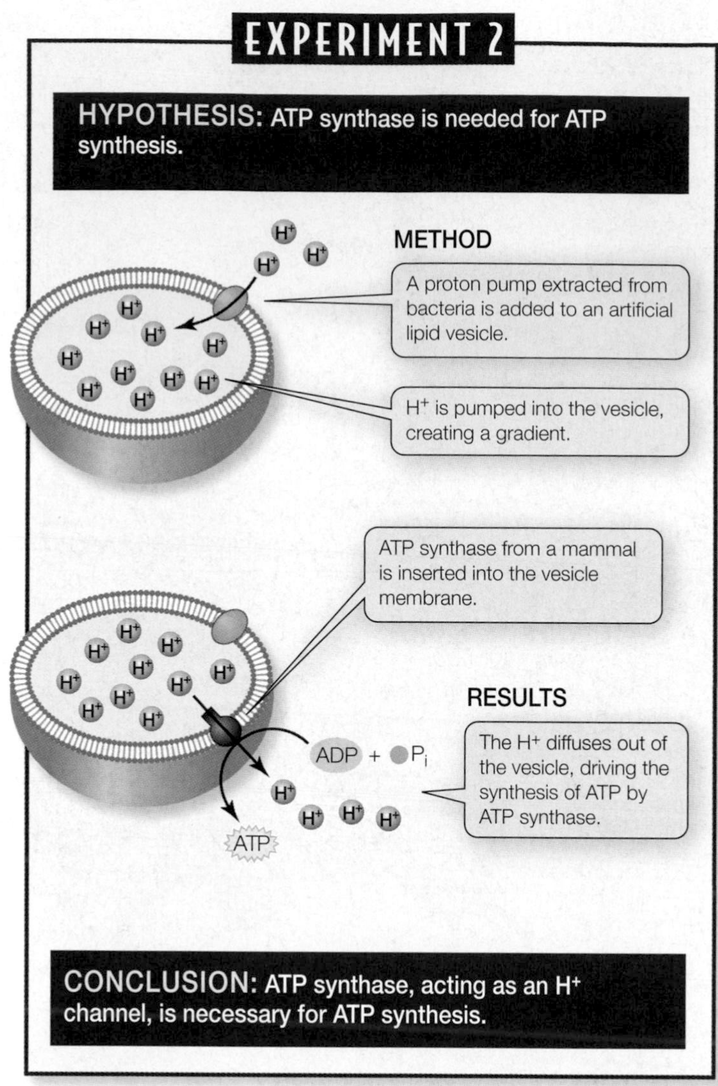

7.14 Two Experiments Demonstrate the Chemiosmotic Mechanism An H^+ gradient across a membrane is all that is needed to drive the synthesis of ATP by the enzyme ATP synthase. It doesn't matter whether the H^+ gradient is produced artificially, as in these experiments, or by the electron transport chain found in nature. FURTHER RESEARCH: What would happen in Experiment 2 if a second ATP synthase, oriented in the opposite way to the one originally inserted in the membrane, was added?

created an artificial gradient across the inner membrane, which the mitochondria used to make ATP from ADP and P_i. This result supported the hypothesis, providing strong evidence for the chemiosmotic mechanism.

- *Experiment 2* tested the hypothesis that the enzyme ATPase couples a proton gradient to ATP synthesis. A proton pump isolated from a bacterium was added to artificial membrane vesicles. When an appropriate energy source was provided, H^+ was pumped into the vesicles, creating a proton gradient. If mammalian ATP synthase was then inserted into the membranes of these vesicles and the energy source removed, the vesicles made ATP, even in the absence of the usual electron carriers. Again, this result supported the hypothesis, showing that ATP synthase is the coupling factor.

UNCOUPLING PROTON DIFFUSION FROM ATP PRODUCTION Another way to demonstrate the chemiosmotic mechanism is to show that the diffusion of H^+ and the formation of ATP must be *tightly coupled;* that is, that protons must pass only through the ATP synthase channel in order to move into the mitochondrial matrix. If a second type of H^+ diffusion channel (not ATP synthase) is inserted into the mitochondrial membrane, the energy of the H^+ gradient is released as heat, rather than being coupled to the synthesis of ATP. Such uncoupling molecules are deliberately used by some organisms to generate heat instead of ATP. For example, the natural uncoupling protein thermogenin plays an important role in regulating the temperature of some mammals under some circumstances, especially newborn human infants, who lack the hair to keep warm, and hibernating animals. We will describe this process in more detail in Section 40.4.

HOW ATP SYNTHASE WORKS: A MOLECULAR MOTOR Now that we have established that the H^+ gradient is needed for ATP synthesis, a question remains: How does the enzyme actually make ATP from ADP and P_i? This is certainly a fundamental question in biology, as it underlies energy harvesting in most cells. Look at the structure of ATP synthase in Figure 7.13. It is composed of two parts: the F_0 unit, a transmembrane region that is the H^+ channel, and the F_1 unit, the "lollipop" of interacting subunits that constitute the active site for ATP synthesis. It is believed that ATP synthase turns potential energy from the H^+ gradient into the kinetic energy of movement. The subunits of F_1 rotate, exposing the active site for ATP synthesis.

7.4 RECAP

Energy from the reoxidation of electron carriers is released in small steps by the electron transport chain. Through chemiosmosis, this energy is used to synthesize ATP.

- Can you describe roles of oxidation and reduction in the electron transport chain? See Figures 7.11–7.13
- Do you understand the proton motive force and how it drives chemiosmosis? See p. 150 and Figure 7.13
- Do you see how the two experiments described in Figure 7.14 demonstrate the chemiosmotic mechanism? See pp. 151–152

Oxidative phosphorylation captures a great deal of energy in ATP. What makes it so much more efficient than fermentation?

7.5 Why Does Cellular Respiration Yield So Much More Energy Than Fermentation?

The total net energy yield from glycolysis via fermentation is 2 molecules of ATP per molecule of glucose oxidized. The maximum yield of ATP that can be harvested from a molecule of glucose through glycolysis followed by cellular respiration is much greater—about 32 molecules of ATP (**Figure 7.15**). (See Figures 7.5, 7.8, and 7.13 to review where the ATP molecules come from.)

Why is so much more ATP produced by the metabolic pathways that operate in the presence of O_2? Recall that glycolysis is only a partial oxidation of glucose, as is fermentation. Much more energy remains in the end products of fermentation, lactic acid and ethanol than in the end product of cellular respiration, CO_2. In cellular respiration, carriers (mostly NAD^+) are reduced in pyruvate oxidation and the citric acid cycle, then oxidized by the electron transport chain, with the accompanying production of ATP by chemiosmosis (2.5 ATP for each NADH + H^+ and 1.5 ATP for each $FADH_2$). In an aerobic environment, a cell or organism capable of aerobic metabolism will be at an advantage (in terms of energy availability per glucose molecule) over one limited to fermentation.

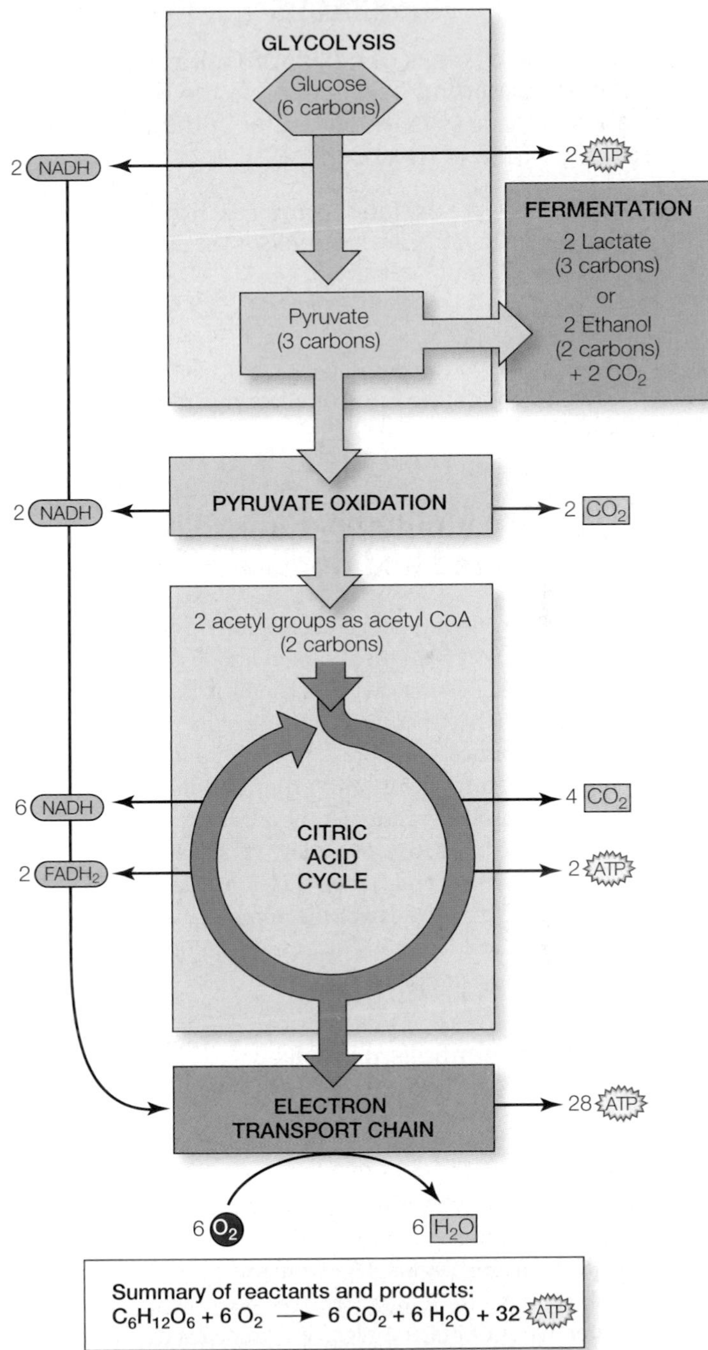

7.15 Cellular Respiration Yields More Energy Than Glycolysis Does Electron carriers are reduced in pyruvate oxidation and the citric acid cycle, then oxidized by the electron transport chain. These reactions produce ATP via chemiosmosis.

The total gross yield of ATP from one molecule of glucose processed through glycolysis and cellular respiration is 32. However, we may subtract two from that gross—for a net yield of 30 ATP—because in some animal cells the inner mitochondrial membrane is impermeable to NADH, and a "toll" of one ATP must be paid for each NADH produced in glycolysis that is shuttled into the mitochondrial matrix.

7.5 RECAP

The electron carriers of cellular respiration allow for the full oxidation of glucose, so the energy yield from glucose is much higher in the presence of O_2 than in its absence.

- Can you calculate the total energy yield from glucose in human cells in the presence versus the absence of O_2? See p. 153 and Figure 7.15

Now that we've seen how cells harvest energy, let's see how that energy moves through the interconnections among metabolic pathways in the cell.

7.6 How Are Metabolic Pathways Interrelated and Controlled?

Glycolysis and the pathways of cellular respiration do not operate in isolation from the rest of metabolism. Rather, there is an interchange, with biochemical traffic flowing both into these pathways and out of them, to and from the synthesis and breakdown of amino acids, nucleotides, fatty acids, and other building blocks of life. Carbon skeletons enter these pathways from other molecules that are broken down to release their energy (catabolism), and carbon skeletons leave these pathways to form the major macromolecular constituents of the cell (anabolism). These relationships are summarized in **Figure 7.16**.

Catabolism and anabolism involve interconversions of biological monomers

A hamburger or veggie burger contains three major sources of carbon skeletons for the person who eats it: carbohydrates, mostly as starch (a polysaccharide); lipids, mostly as triglycerides (three fatty acids attached to glycerol); and proteins (polymers of amino acids). Looking at Figure 7.16, you can see how each of these three types of macromolecules can be used in catabolism or anabolism.

CATABOLIC INTERCONVERSIONS Polysaccharides, lipids, and proteins can all be broken down to provide energy:

- *Polysaccharides* are hydrolyzed to glucose. Glucose then passes through glycolysis and cellular respiration, where its energy is captured in NADH and ATP.
- *Lipids* are broken down into their constituents, glycerol and fatty acids. Glycerol is converted into dihydroxyacetone phosphate (DAP), an intermediate in glycolysis, and fatty acids are converted into acetyl CoA in the mitochondria. In both cases, further oxidation to CO_2 and release of energy occur.
- *Proteins* are hydrolyzed to their amino acid building blocks. The 20 different amino acids feed into glycolysis or the citric acid cycle at different points. A specific example is shown in **Figure 7.17**, in which the amino acid glutamate is converted into α-ketoglutarate, an intermediate in the citric acid cycle.

ANABOLIC INTERCONVERSIONS Many catabolic pathways can operate in reverse. Glycolytic and citric acid cycle intermediates, instead of being oxidized to form CO_2, can be reduced and used to form glucose in a process called **gluconeogenesis** (which means "new formation of glucose"). Likewise, acetyl CoA can be used to form fatty acids. The most common fatty acids have an even number of carbons: 14, 16, or 18. These molecules are formed by adding two-carbon acetyl CoA "units" one at a time until the ap-

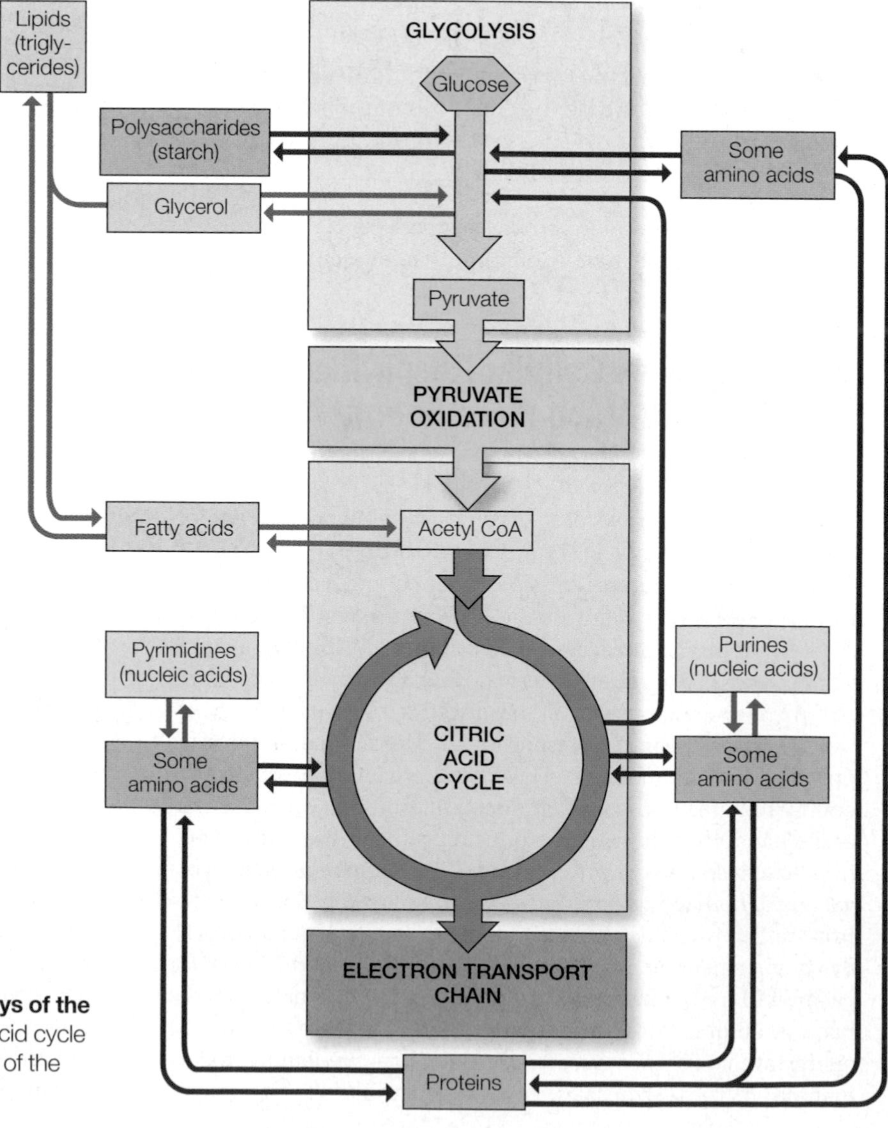

7.16 Relationships among the Major Metabolic Pathways of the Cell Note the central position of glycolysis and the citric acid cycle in this network of metabolic pathways. Also note that many of the pathways can operate in reverse.

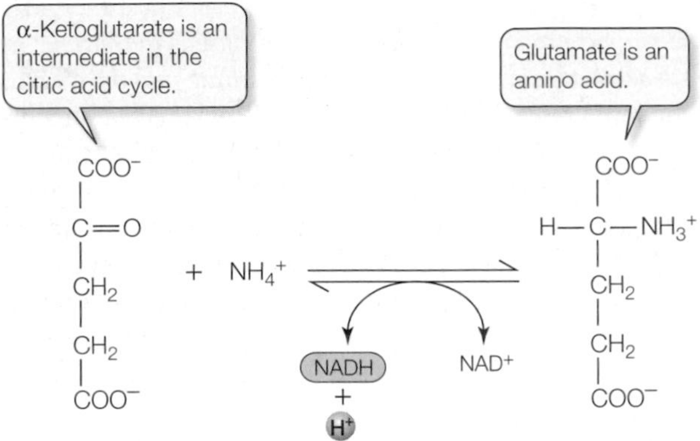

7.17 Coupling Metabolic Pathways This reaction, in which glutamate and α-ketoglutarate are interconverted, is catalyzed by the enzyme glutamate dehydrogenase.

propriate carbon chain length is reached. Amino acids can be formed by reversible reactions such as the one shown in Figure 7.17, and can then be polymerized into proteins.

Some intermediates in the citric acid cycle are used in the synthesis of various cellular constituents. For example, α-ketoglutarate is a starting point for purines and oxaloacetate for pyrimidines, both constituents of the nucleic acids DNA and RNA. α-Ketoglutarate is also a starting point for chlorophyll synthesis. Acetyl CoA is a building block for various pigments, plant growth substances, rubber, and the steroid hormones of animals, among other molecules.

Catabolism and anabolism are integrated

A carbon atom from a protein in your burger can end up in DNA or fat or CO_2, among other fates. How does the cell "decide" which metabolic pathway to follow? With all of these possible interconversions, you might expect that cellular concentrations of various biochemical molecules would vary widely. For example, you might think that the level of oxaloacetate in your cells would depend on what you eat (some food molecules form oxaloacetate) and on whether oxaloacetate is used up (in the citric acid cycle or in forming the amino acid aspartate). Remarkably, the levels of these substances in what is called the "metabolic pool"—the sum total of all the biochemical molecules in a cell—are quite constant. The cell regulates the enzymes of catabolism and anabolism so as to maintain a balance. This *metabolic homeostasis* gets upset only in unusual circumstances. Let's look one such unusual circumstance: undernutrition.

Glucose is an excellent source of energy. From Figure 7.16, you can see that lipids and proteins can also serve as energy sources. Any one, or all three, could be used to provide the energy your body needs. In reality, things are not so simple. Proteins, for example, have essential roles in your body as enzymes and structural elements; they are not stored for energy, and using them for energy might deprive you of a catalyst for a vital reaction.

Polysaccharides and fats (trigylcerides) have no such catalytic roles. But polysaccharides, because they are somewhat polar, can bind a lot of water. Because they are nonpolar, fats do not bind as much water as polysaccharides do. So, in water, fats weigh less than polysaccharides. In addition, fats are more reduced than carbohydrates (have more C—H bonds as opposed to C—OH bonds) and have more energy stored in their bonds. For these two reasons, fats are a better means of storing energy than polysaccharides. It is not surprising, then, that a typical person has about one day's worth of food energy stored as glycogen (a polysaccharide), a week's food energy as usable proteins in blood, and over a month's food energy stored as fats.

What happens if a person does not eat enough to produce sufficient ATP and NADH for anabolism and biological activities? This situation can be the result of a deliberate decision to lose weight, but for too many people, it is forced upon them because not enough food is available. In either case, the first energy stores in the body to be used are the glycogen stores in muscle and liver cells. These stores do not last long, and next come the fats.

The level of acetyl CoA rises as fatty acids are broken down. However, a problem remains: because fatty acids cannot cross from the blood to the brain, the brain can use only glucose as its energy source. With glucose already depleted, the body must convert something else to make glucose for the brain. This gluconeogenesis uses mostly amino acids, largely from the breakdown of proteins. Without sufficient food intake, both proteins (for glucose) and fats (for energy) are used up. After several weeks of starvation, fat stores become depleted, and the only energy source left is proteins. At this point, essential proteins, such as muscle proteins and the antibodies used to fight off infections, get broken down. The loss of such proteins can lead to severe illness and eventual death.

Seasonal migrations between Canada and South America made by the tiny blackpoll warbler are fueled by 10–12 grams of fat, which is about equal to their lean body mass. The glycogen required to supply the same amount of energy would weigh 10 times more. The birds would not be able to fly (or even walk).

Metabolic pathways are regulated systems

We have described the relationships between metabolic pathways and noted that these pathways work together to provide homeostasis in the cell and organism. But how does the cell regulate interconversions between these pathways to maintain constant metabolic pools? In its most simplified sense, this is a problem of systems biology (see Figure 6.15).

Consider what happens to the starch in your burger bun. In the digestive system, starch is hydrolyzed to glucose, which enters the blood for distribution to the rest of the body. Before this happens, however, a regulatory check must be made: is there already enough glucose in the blood to supply the body's needs? If there is, the excess glucose is converted into glycogen and stored in the liver. If not enough glucose is supplied by food, that glycogen is broken down, or other molecules are used to make glucose by gluconeogenesis.

The end result is that the level of glucose in the blood is remarkably constant. We will describe the details of how this happens in Part Eight of this book. For now, it is important to realize

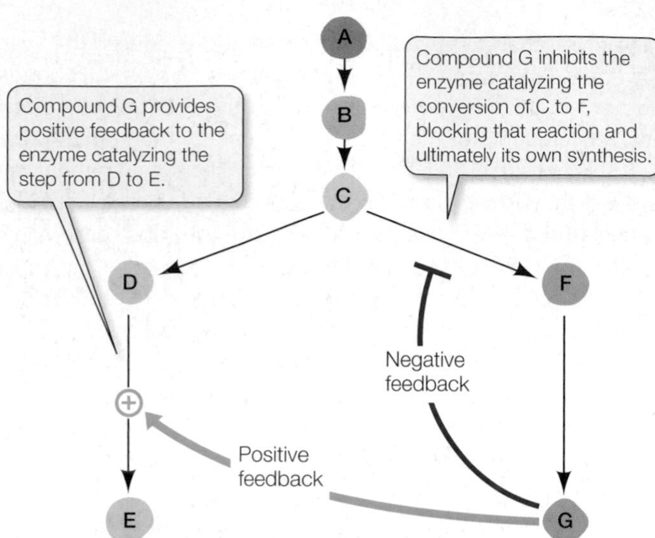

7.18 Regulation by Negative and Positive Feedback Allosteric regulation plays an important role in metabolic pathways. Excess accumulation of some products can shut down their synthesis or stimulate the synthesis of other products that require the same raw materials.

that the interconversions of glucose involve many steps, each catalyzed by an enzyme, and it is here that controls often reside.

Glycolysis, the citric acid cycle, and the electron transport chain are regulated by *allosteric* control of the enzymes involved. As described in Section 6.5, in metabolic pathways, a high concentration of the products of a later reaction can suppress the action of enzymes that catalyze an earlier reaction. On the other hand, an excess of the product of one pathway can speed up reactions in another pathway, diverting raw materials away from synthesis of the first product (**Figure 7.18**). These *negative and positive feedback control mechanisms* are used at many points in the energy-harvesting pathways, which are summarized in **Figure 7.19**.

- *The main control point in glycolysis is the enzyme phosphofructokinase* (reaction 3 in Figure 7.5). This enzyme is allosterically inhibited by ATP and activated by ADP or AMP. As long as fermentation proceeds, yielding a relatively small amount of ATP, phosphofructokinase operates at full efficiency. But when cellular respiration begins producing 16 times more ATP than fermentation does, the abundant ATP allosterically inhibits the enzyme, and the conversion of fructose 6-phosphate into fructose 1,6-bisphosphate declines, as does the rate of glucose utilization.

- *The main control point in the citric acid cycle is the enzyme isocitrate dehydrogenase, which converts isocitrate into α-ketoglutarate* (reaction 3 in Figure 7.8). NADH + H^+ and ATP are feedback inhibitors of this reaction; ADP and NAD^+ are activators. If too much ATP is accumulating, or if NADH + H^+ is being produced faster than it can be used by the electron transport chain, the conversion of isocitrate is slowed, and the citric acid cycle is essentially shut down. A shutdown of the citric acid cycle would cause large amounts of isocitrate and citrate to accumulate if the conversion of acetyl CoA to citrate were not also slowed by abundant ATP and NADH + H^+. An excess of citrate acts as an additional feedback inhibitor to slow the fructose 6-phosphate reaction early in glycolysis. Consequently, if the citric acid cycle has been slowed down because of abundant ATP (and not because of a lack of oxygen), glycolysis is shut down as well. Both processes resume when the

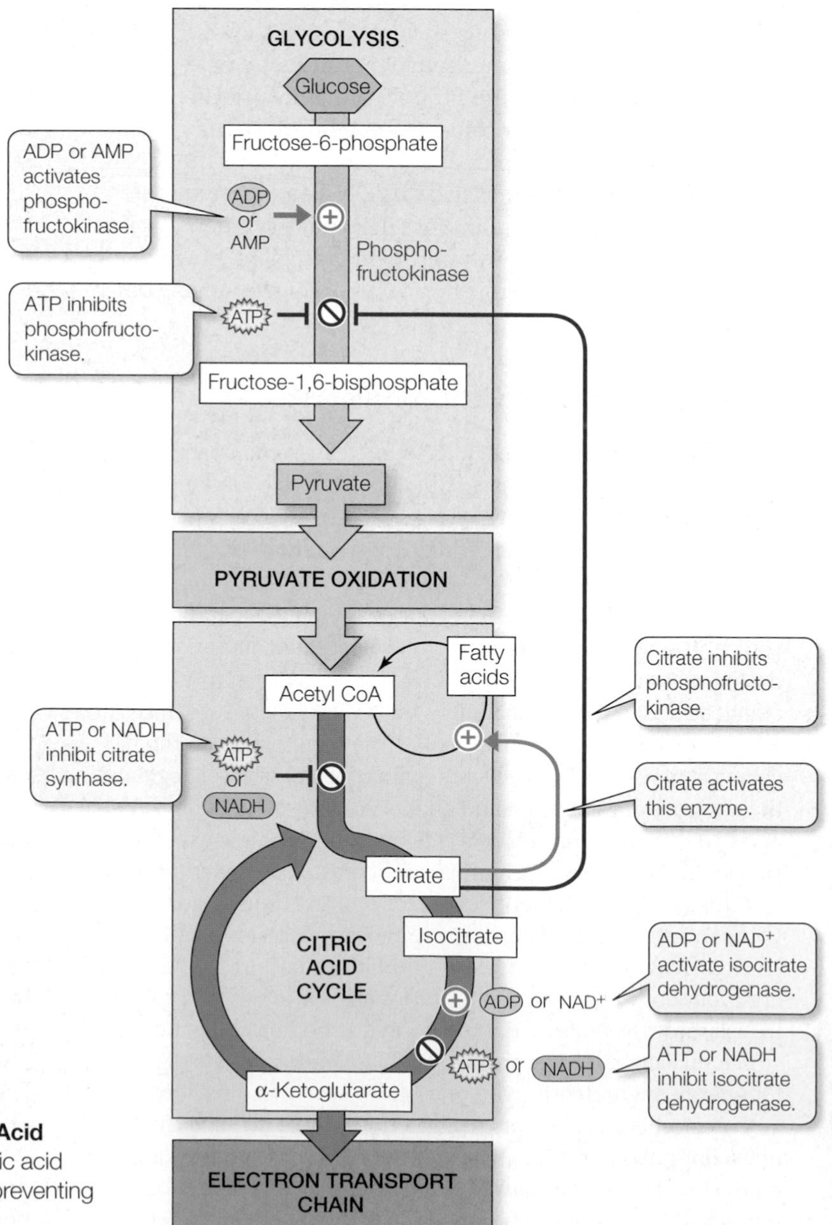

7.19 Allosteric Regulation of Glycolysis and the Citric Acid Cycle Allosteric feedback controls glycolysis and the citric acid cycle at crucial early steps, increasing their efficiency and preventing the excessive buildup of intermediates.

ATP level falls and they are needed again. Allosteric control keeps these processes in balance.

- *Another control point involves acetyl CoA.* If too much ATP is being made and the citric acid cycle shuts down, the accumulation of citrate diverts acetyl CoA to the synthesis of fatty acids for storage. That is one reason why people who eat too much accumulate fat. These fatty acids may be metabolized later to produce more acetyl CoA.
- *A final control point comes with cell differentiation.* For example, as illustrated in the opening of this chapter, a single protein, PPARδ, appears to control the proliferation of slow-twitch muscle fibers. These cells, which are full of mitochondria, catabolize fats and carbohydrates aerobically. The result is a steady release of ATP, whose hydrolysis can be used to power muscles in sustained activity. Long-distance running is a good way to combat obesity.

7.6 RECAP

Glucose can be made from intermediates in glycolysis and the citric acid cycle by a process called gluconeogenesis. The metabolic pathways of lipids and amino acids are tied to those of glucose metabolism. Key enzymes link and regulate the various pathways.

- Can you give an example of a catabolic interconversion of a lipid? Of an anabolic interconversion of a protein? See p. 154 and Figure 7.16
- Can you explain how phosphofructokinase serves as a control point for glycolysis? See p. 156 and Figure 7.19
- What would happen if there were no allosteric mechanism for sensing the level of acetyl CoA?

CHAPTER SUMMARY

7.1 How does glucose oxidation release chemical energy?

As a material is **oxidized**, the electrons it loses are transferred to another material, which is thereby **reduced**. Such **redox reactions** transfer large amounts of energy. Review Figure 7.2, Web/CD Activities 7.1, 7.2

The coenzyme **NAD** is a key electron carrier in biological redox reactions. It exists in two forms, one oxidized (NAD^+) and the other reduced ($NADH + H^+$).

Glycolysis operates in the presence or absence of O_2. Under **aerobic** conditions, **cellular respiration** continues the process of breaking down glucose. Under **anaerobic** conditions, **fermentation** occurs. Review Figure 7.4

The three pathways of cellular respiration are **pyruvate oxidation**, the **citric acid cycle**, and the **electron transport chain**.

7.2 What are the aerobic pathways of glucose metabolism?

Glycolysis consists of 10 enzyme-catalyzed reactions that occur in the cell cytosol. Two **pyruvate** molecules are produced for each partially oxidized molecule of glucose, providing the starting material for both cellular respiration and fermentation. Review Figure 7.5

The first five reactions of glycolysis require an investment of energy; the last five produce energy. The net gain is two molecules of ATP.

The enzyme-catalyzed transfer of phosphate groups to ADP is called **substrate-level phosphorylation**.

Pyruvate oxidation follows glycolysis and links glycolysis to the citric acid cycle. In this pathway, pyruvate is converted into **acetyl CoA**.

Acetyl CoA is the starting point of the citric acid cycle. It reacts with oxaloacetate to produce citrate. A series of eight enzyme-catalyzed reactions oxidize citrate and regenerate oxaloacetate, continuing the cycle. Review Figure 7.8, Web/CD Activity 7.3

7.3 How is energy harvested from glucose in the absence of oxygen?

In the absence of O_2, glycolysis is followed by fermentation. Together, these pathways partially oxidize glucose and generate end products such as **lactic acid** or **ethanol** along with a small amount of ATP. Review Figures 7.9 and 7. 10

7.4 How does the oxidation of glucose form ATP?

Oxidation of electron carriers in the presence of O_2 releases energy that can be used to form ATP in a process called **oxidative phosphorylation**.

NADH and $FADH_2$ from glycolysis, pyruvate oxidation, and the citric acid cycle are oxidized by the electron transport chain, regenerating NAD^+ and FAD. Oxygen (O_2) is the final acceptor of electrons and protons, forming water (H_2O). Review Figure 7.11, Web/CD Activity 7.4, Web/CD Tutorial 7.1

The electron transport chain not only moves electrons across its carriers, but also pumps protons across the inner mitochondrial membrane, creating the **proton-motive force**.

The protons constituting the proton-motive force can return to the mitochondrial matrix via **ATP synthase**, which couples this movement of protons to the synthesis of ATP. This process is called **chemiosmosis**. Review Figure 7.13, Web/CD Tutorial 7.2

7.5 Why does cellular respiration yield so much more energy than fermentation?

For each molecule of glucose used, fermentation yields 2 molecules of ATP. In contrast, glycolysis operating with pyruvate oxidation, the citric acid cycle, and the electron transport chain yields up to 32 molecules of ATP per molecule of glucose. Review Figure 7.15, Web/CD Activity 7.5

7.6 How are metabolic pathways interrelated and controlled?

The **catabolic pathways** of carbohydrates, fats, and proteins feed into the energy-harvesting metabolic pathways. Review Figure 7.16

Anabolic pathways use intermediate components of the energy-harvesting pathways to synthesize fats, amino acids, and other essential building blocks.

The formation of glucose from glycolytic and citric acid cycle intermediates is called **gluconeogenesis**.

The rates of glycolysis and the citric acid cycle are regulated by allosteric feedback, by storage of excess acetyl CoA, and by cell differentiation. See Figure 7.19, Web/CD Activity 7.6

SELF-QUIZ

1. The role of oxygen gas in our cells is to
   a. catalyze reactions in glycolysis.
   b. produce CO_2.
   c. form ATP.
   d. accept electrons from the electron transport chain.
   e. react with glucose to split water.
2. Oxidation and reduction
   a. entail the gain or loss of proteins.
   b. are defined as the loss of electrons.
   c. are both endergonic reactions.
   d. always occur together.
   e. proceed only under aerobic conditions.
3. NAD^+ is
   a. a type of organelle.
   b. a protein.
   c. present only in mitochondria.
   d. a part of ATP.
   e. formed in the reaction that produces ethanol.
4. Glycolysis
   a. takes place in the mitochondrion.
   b. produces no ATP.
   c. has no connection with the electron transport chain.
   d. is the same thing as fermentation.
   e. reduces two molecules of NAD^+ for every glucose molecule processed.
5. Fermentation
   a. takes place in the mitochondrion.
   b. takes place in all animal cells.
   c. does not require O_2.
   d. requires lactic acid.
   e. prevents glycolysis.
6. Which statement about pyruvate is *not* true?
   a. It is the end product of glycolysis.
   b. It becomes reduced during fermentation.
   c. It is a precursor of acetyl CoA.
   d. It is a protein.
   e. It contains three carbon atoms.
7. The citric acid cycle
   a. takes place in the mitochondrion.
   b. produces no ATP.
   c. has no connection with the electron transport chain.
   d. is the same thing as fermentation.
   e. reduces two NAD^+ for every glucose processed.
8. The electron transport chain
   a. is located in the mitochondrial matrix.
   b. includes integral membrane proteins.
   c. always produces ATP.
   d. reoxidizes reduced coenzymes.
   e. operates simultaneously with fermentation.
9. Compared with fermentation, the aerobic pathways of glucose metabolism produce
   a. more ATP.
   b. pyruvate.
   c. fewer protons for pumping in mitochondria.
   d. less CO_2.
   e. more oxidized coenzymes.
10. Which statement about oxidative phosphorylation is *not* true?
   a. It is the formation of ATP by the electron transport chain.
   b. It is brought about by chemiosmosis.
   c. It requires aerobic conditions.
   d. It takes place in mitochondria.
   e. Its functions can be served equally well by fermentation.

FOR DISCUSSION

1. Trace the sequence of chemical changes that occurs in mammalian tissue when the oxygen supply is cut off. The first change is that the cytochrome *c* oxidase system becomes totally reduced, because electrons can still flow from cytochrome *c*, but there is no oxygen to accept electrons from cytochrome *c* oxidase. What are the remaining steps?
2. Some cells that use the aerobic pathways of glucose metabolism can also thrive by using fermentation under anaerobic conditions. Given the lower yield of ATP (per molecule of glucose) in fermentation, how can these cells function so efficiently under anaerobic conditions?
3. The drug antimycin A blocks electron transport in mitochondria. Explain what would happen if experiment 1 in Figure 7.14 were repeated in the presence of this drug.
4. You eat a burger that contains polysaccharides, proteins, and lipids. Using your knowledge of the integration of biochemical pathways, explain how the amino acids in the proteins and the glucose in the polysaccharides can end up as fats.

FOR INVESTIGATION

A protein present in the fat of newborn babies uncouples the synthesis of ATP from electron transport and instead generates heat. How would you investigate the hypothesis that this uncoupling protein adds a second proton channel to the mitochondrial membrane?

CHAPTER 8 Photosynthesis: Energy from Sunlight

The ultimate energy source

If all the carbohydrates produced by photosynthesis in a year were in the form of sugar cubes, there would be 300 quadrillion of them. Lined up, these cubes would extend from Earth to the planet Pluto. That's a lot of photosynthesis!

Imagine what would happen if photosynthesis on Earth were reduced. That catastrophe actually happened about 65 million years ago, when a large meteorite slammed into Earth in what is now southern Mexico. Geological evidence suggests that the impact raised a huge cloud of dust, blocking the sun and curtailing photosynthesis—and hence plant growth, and thus the survival of species that depended on plants. The impact may have led to the extinction of the dinosaurs (among other species). Their extinction, however, benefited the early mammals, who thrived in the absence of competition from the great reptiles. (You might not even be here if photosynthesis had not been so drastically scaled back those many million years ago.)

Powered by sunlight, green plants use the reactions of photosynthesis to convert simple chemicals in the environment—carbon dioxide and water—into carbohydrates. The emergence of photosynthesis was a key event in the evolution of life, integrating an external energy source—the energy of sunlight—into the living world. Photosynthesizing organisms using solar energy to make their own food provide an entry point to the biosphere for chemical energy. Most other organisms depend on photosynthesizers for the raw materials of metabolism, such as glucose, as well as for atmospheric oxygen.

Globally, more than 10 billion tons of carbon is *fixed* by plants every year. By this we mean carbon molecules are converted from being part of a simple gas (CO_2) into more complex, reduced molecules (carbohydrates), making carbon available for use as food. As you might guess, Earth's "food chain" requires a lot of photosynthesis. On the African plain, it takes 6 acres of grassland to convert enough CO_2 into plant matter that supports the growth of a single grazing gazelle.

Humans consume a huge amount of Earth's photosynthetic output. To determine how much, Mark Imhoff and colleagues at the NASA Center in Maryland estimated the net productivity of Earth's photosynthetic organisms—that is, the amount of carbon dioxide they fix minus the amount they use for their own growth and reproduction. Having estimated available fixed carbon, the scientists then estimated

Primary Producers Covering less than 2 percent of Earth's surface, rainforests are photosynthetic dynamos, producing about one-fifth of the oxygen gas in the atmosphere.

A Photosynthetic Disaster An artist's conception of the presumed meteorite impact of 65 million years ago, which raised a dust cloud that choked the atmosphere and severely reduced photosynthetic output across the planet.

human carbohydrate consumption. Direct consumption included all carbohydrates consumed as food, fuel, fiber, or timber. Indirect consumption included disposal of crops we do not use, fires set to clear the land, and land cleared for cities.

The astounding conclusion: humans appropriate one-third of all the carbon fixed each year, leaving two-thirds for the entire remainder of the biosphere. This is by far the greatest consumption ratio for any single species in known evolutionary history. Can we keep consuming this much photosynthetic output indefinitely? The 2002 United Nations Conference on Sustainability concluded that we need to take steps to protect our photosynthetic future. An important first step in examining ecological sustainability is to understand photosynthesis.

IN THIS CHAPTER we first explore how photosynthesis converts light energy into chemical energy in the form of reduced electron carriers and ATP. Then we'll see how these two sources of chemical energy are used to drive the synthesis of carbohydrates from carbon dioxide. We'll show how the relationships between these two processes are imperative for plant growth.

CHAPTER OUTLINE

8.1 What Is Photosynthesis?

Photosynthesis (literally, "synthesis from light") is a metabolic process by which the energy of sunlight is captured and used to convert carbon dioxide (CO_2) and water (H_2O) into carbohydrate sugars ($C_6H_{12}O_6$) and oxygen gas (O_2) (**Figure 8.1**). By early in the nineteenth century, scientists had grasped these broad outlines of photosynthesis and had established several things about the way the process worked:

- The water for photosynthesis in land plants comes primarily from the soil and must travel from the roots to the leaves.
- Plants take in carbon dioxide, and release water and O_2, through tiny openings in leaves called *stomata* (singular *stoma*) (see Figure 8.1).
- Light is absolutely necessary for the production of oxygen and sugars.

By 1804, scientists had summarized photosynthesis as follows:

carbon dioxide + water + light energy → sugar + oxygen

In molecular terms, this equation seems to be the *reverse* of the overall equation for cellular respiration (see Section 7.1). More precisely, it can be written as:

$$6\ CO_2 + 6\ H_2O \rightarrow C_6H_{12}O_6 + 6\ O_2$$

While they have proved to be essentially correct, these equations left much more to know about the process of photosynthesis, which, in fact, is *not* the reverse of cellular respiration. What are the reactions of photosynthesis? What role does light play in these reactions? How do carbons become linked to form sugars? And where does the oxygen gas come from: from CO_2 or H_2O?

It took almost another century to determine the source of the O_2 released during photosynthesis. One of the first biological experiments to utilize radioisotopes traced the flow of oxygen in plants. Samuel Ruben and Martin Kamen allowed two groups of plants to carry on photosynthesis (**Figure 8.2**). Plants in the first group were supplied with water containing the oxygen isotope ^{18}O and with CO_2 containing only the common

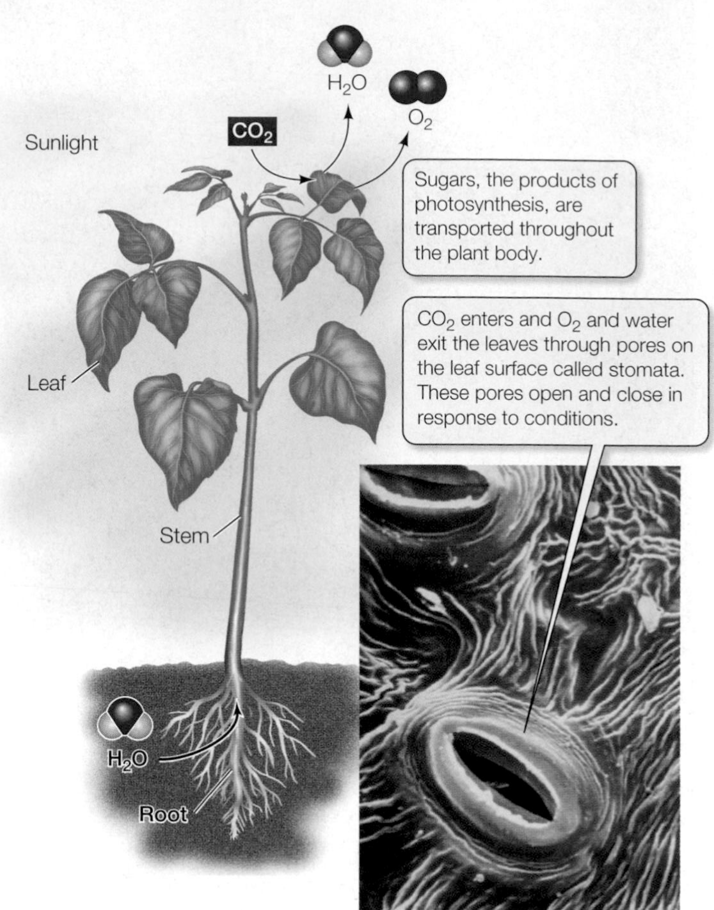

8.1 The Ingredients for Photosynthesis A typical terrestrial plant uses light from the sun, water from the soil, and carbon dioxide from the atmosphere to form organic compounds by photosynthesis.

oxygen isotope ^{16}O; plants in the second group were supplied with CO_2 labeled with ^{18}O and water containing only ^{16}O. Oxygen gas from each group of plants was collected and analyzed. Oxygen gas containing ^{18}O was produced in abundance by the plants that had been given ^{18}O-labeled water, but *not* by those given ^{18}O-labeled CO_2.

These results showed that all the oxygen gas produced during photosynthesis comes from water, as reflected in the revised balanced equation:

$$6\ CO_2 + 12\ H_2O \rightarrow C_6H_{12}O_6 + 6\ O_2 + 6\ H_2O$$

Water appears on both sides of the equation because water is both used as a reactant (the twelve molecules on the left) and released as a product (the six new ones on the right). This revised equation accounts for all water molecules needed for all the oxygen gas produced.

Photosynthesis involves two pathways

Our equation summarizes the overall process of photosynthesis, but not the stages by which it is completed. Like glycolysis and the other metabolic pathways that harvest energy in cells, photosynthesis consists not of a single reaction, but of many reactions. The reactions of photosynthesis are commonly divided into two main pathways:

- The **light reactions** are driven by light energy. This pathway converts light energy into chemical energy in the form of ATP and a reduced electron carrier (NADPH + H^+).
- The **light-independent reactions** do not use light directly, but instead use ATP, NADPH + H^+ (made by the light reactions), and CO_2 to produce sugars. There are three different forms of the light-independent pathway that reduces CO_2: the *Calvin cycle*, *C_4 photosynthesis*, and *crassulacean acid metabolism*.

The light-independent reactions are sometimes called the *dark reactions* because they do not directly require light energy. However, both the light-dependent and light-independent reactions stop in the dark because ATP synthesis and $NADP^+$ reduction require light. The reactions of both pathways proceed within the chloroplast, but they reside in different parts of that organelle (**Figure 8.3**). The two pathways are linked by the exchange of ATP and ADP and of $NADP^+$ and NADPH, and the rate of each set of reactions depends on the rate of the other.

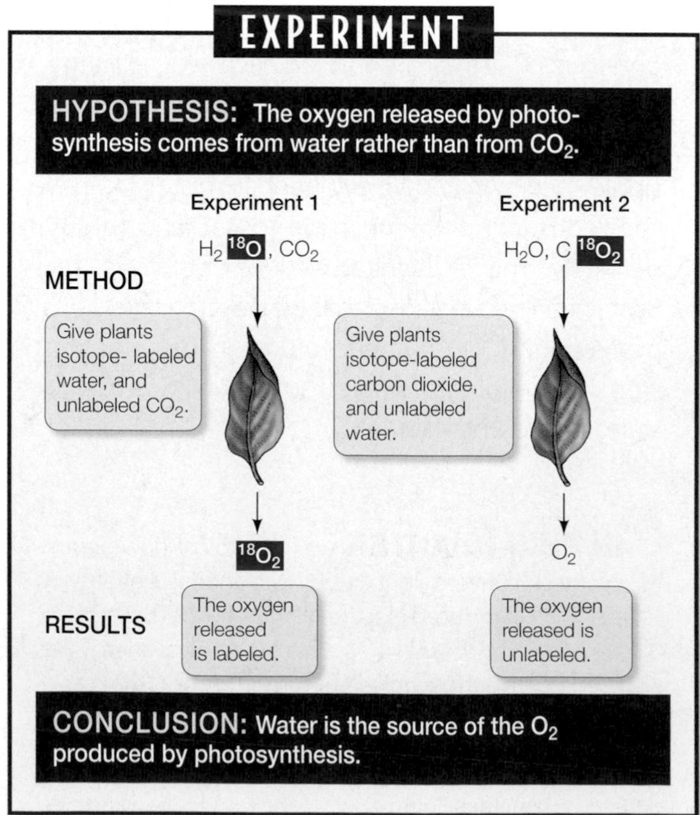

EXPERIMENT
HYPOTHESIS: The oxygen released by photosynthesis comes from water rather than from CO_2.
Experiment 1
Experiment 2
METHOD
$H_2{}^{18}O$, CO_2
H_2O, $C^{18}O_2$
Give plants isotope- labeled water, and unlabeled CO_2.
Give plants isotope-labeled carbon dioxide, and unlabeled water.
$^{18}O_2$
O_2
RESULTS
The oxygen released is labeled.
The oxygen released is unlabeled.
CONCLUSION: Water is the source of the O_2 produced by photosynthesis.

8.2 Water Is the Source of the Oxygen Produced by Photosynthesis Because only plants given isotope-labeled water released isotope-labeled O_2, this experiment showed that water is the source of the oxygen released during photosynthesis.

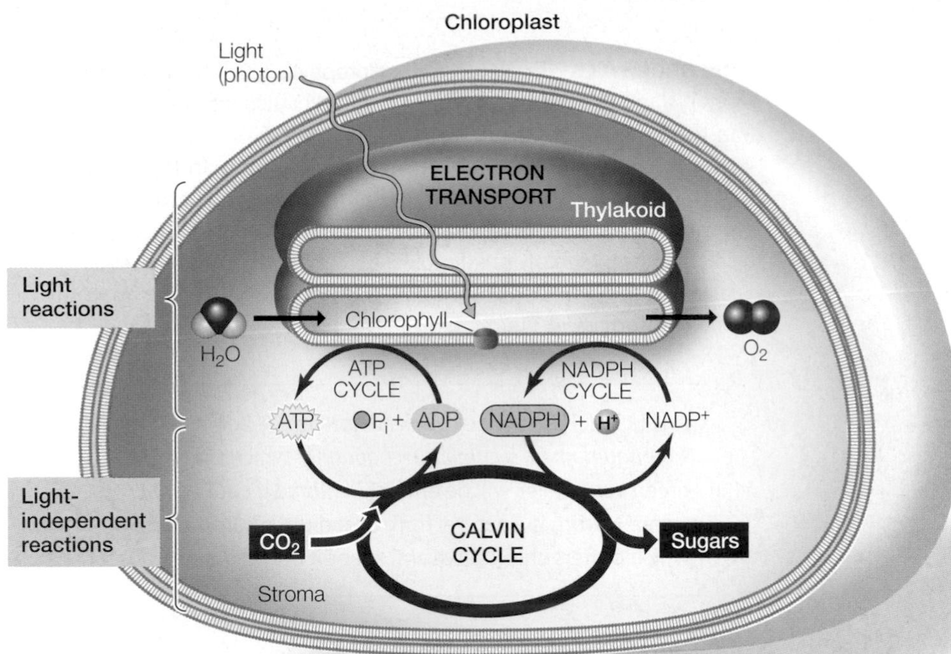

8.3 An Overview of Photosynthesis Photosynthesis comprises two pathways: the light reactions and the light-independent reactions. These reactions take place in the thylakoids and the stroma of chloroplasts, respectively.

8.1 RECAP

The light reactions of photosynthesis convert light energy into chemical energy. The light-independent reactions use that chemical energy to reduce CO_2 to carbohydrates.

- Can you explain how we know that water is the source of the O_2 required by photosynthesis? See p. 162 and Figure 8.2
- What is the relationship between the light reactions and the light-independent reactions of photosynthesis? See p. 162 and Figure 8.3

We will describe the light reactions and the light-independent reactions separately and in detail. But since these two photosynthetic pathways are powered by the energy of sunlight, let's begin by discussing the physical nature of light and the specific photosynthetic molecules that capture its energy.

8.2 How Does Photosynthesis Convert Light Energy into Chemical Energy?

Light is a source of both energy and information. Later chapters will explore the many roles of light in the transmission of information. Here our focus is on light as a source of energy.

Light behaves as both a particle and a wave

Light is a form of **electromagnetic radiation**. It comes in discrete packets called **photons**. Light also behaves as if it were propagated in waves. The amount of energy contained in a single photon is inversely proportional to its **wavelength**: the shorter the wavelength, the greater the energy of the photons. To be active in a biological process, a photon must be absorbed by a receptive molecule, and it must have sufficient energy to perform the chemical work required.

Absorbing a photon excites a pigment molecule

When a photon meets a molecule, one of three things happens:

- The photon may bounce off the molecule—it may be *scattered* or *reflected*.
- The photon may pass through the molecule—it may be *transmitted*.
- The photon may be *absorbed* by the molecule.

Neither of the first two outcomes causes any change in the molecule. In the case of **absorption**, however, the photon disappears. Its energy, however, cannot disappear, because, according to the first law of thermodynamics, energy is neither created nor destroyed. Instead, when a molecule absorbs a photon, that molecule acquires the energy of the photon. It is thereby raised from a **ground state** (with lower energy) to an **excited state** (with higher energy) (**Figure 8.4A**).

The difference in free energy between the molecule's excited state and its ground state is approximately equal to the free energy of the absorbed photon (a small amount is lost to entropy). The increase in energy boosts one of the electrons within the molecule into a shell farther from its nucleus; this electron is now held less firmly (**Figure 8.4B**), making the molecule more chemically reactive.

Absorbed wavelengths correlate with biological activity

The electromagnetic spectrum (**Figure 8.5**) encompasses the wide range of wavelengths (and hence energy levels) that photons can have. The specific wavelengths absorbed by a particular molecule are characteristic of that type of molecule. Molecules that absorb wavelengths in the *visible spectrum*—that region of the spectrum that is visible to humans as light—are called **pigments**.

When a beam of white light (light containing visible light of *all* wavelengths) falls on a pigment, certain wavelengths of the light are absorbed. The remaining wavelengths, which are scattered or transmitted, make the pigment appear to us to be colored. For example, if a pigment absorbs both blue and red light—as chlorophyll does—what we see is the remaining light, which is primarily green.

If we plot the wavelengths of the light absorbed by a purified pigment, the result is an **absorption spectrum** for that pigment. If we plot the biological activity of a photosynthetic organism as a function of the wavelengths of light to which the organism is ex-

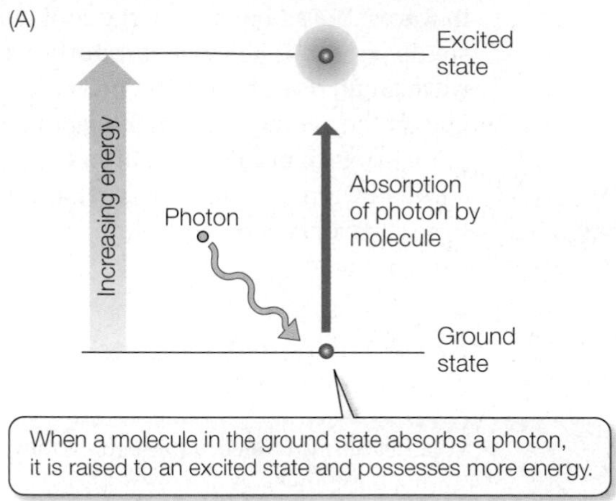

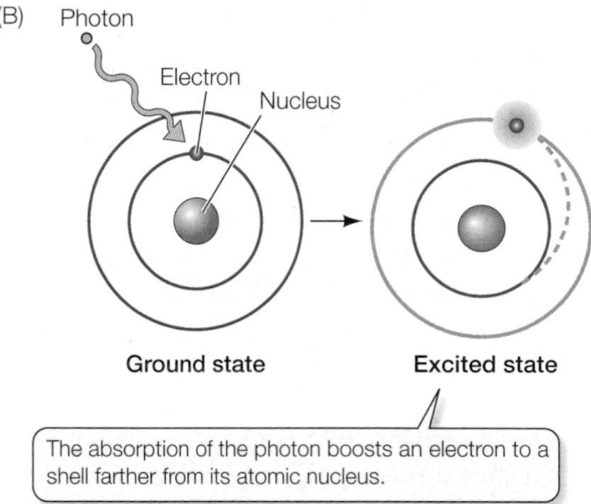

8.4 Exciting a Molecule (A) When a molecule absorbs the energy of a photon, it is raised from a ground state to an excited state. (B) In the excited state, an electron is boosted to a more distant shell, where it is held less firmly.

posed, the result is an **action spectrum**. **Figure 8.6** shows the absorption spectrum for a pigment, chlorophyll *a,* isolated from the leaves of a plant and the action spectrum for photosynthetic activity for the same plant. A comparison of the two spectra shows that the wavelengths at which photosynthesis is maximal are the same wavelengths at which chlorophyll *a* absorbs light.

Photosynthesis uses energy absorbed by several pigments

The light energy used for photosynthesis is not absorbed by just one type of pigment. Instead, several different pigments with different absorption spectra absorb the energy that is eventually used for photosynthesis. In photosynthetic organisms of all kinds (plants, protists, and bacteria), these pigments include *chlorophylls, carotenoids,* and *phycobilins.*

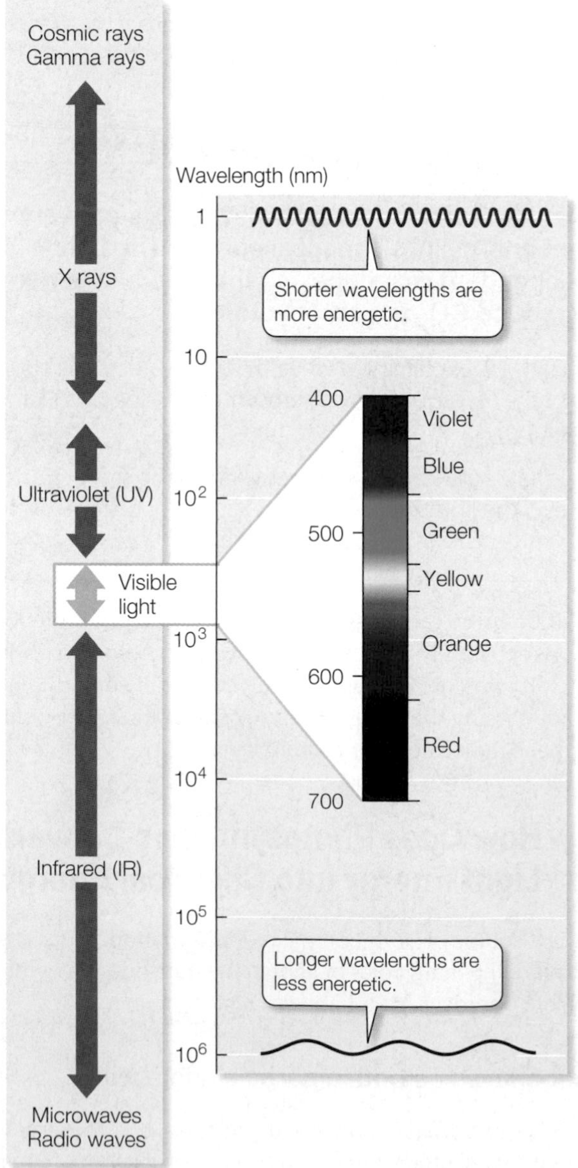

8.5 The Electromagnetic Spectrum The portion of the electromagnetic spectrum that is visible to humans as light is shown in detail at the right.

CHLOROPHYLLS In plants, two **chlorophylls** predominate: chlorophyll *a* and chlorophyll *b*. These two molecules differ only slightly in their molecular structure. Both have a complex ring structure similar to that of the heme group of hemoglobin. In the center of each chlorophyll ring is a magnesium atom, and attached at a peripheral location on the ring is a long hydrocarbon "tail," which can anchor the chlorophyll molecule to integral proteins in the thylakoid membrane of a chloroplast (**Figure 8.7**; review the anatomy of a chloroplast in Figure 4.14).

An ancient Roman cookbook advised *omne holus smaragdinum fit, si cum nitro coquatur*—which means, "All green vegetables will be emerald-colored if cooked with nitrum." Nitrum is a natural form of sodium bicarbonate, which buffers chlorophyll and keeps it bright green.

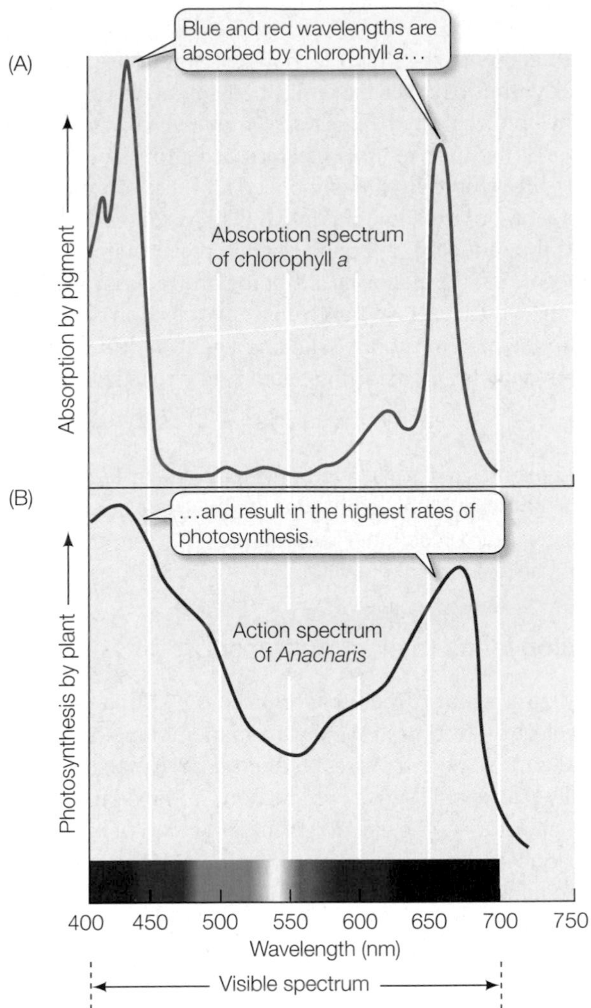

8.6 Absorption and Action Spectra The absorption spectrum (A) of the purified pigment chlorophyll *a* from the aquatic plant *Anacharis* is similar to the action spectrum (B) obtained when different wavelengths of light are shone on the intact plant and the rate of photosynthesis is measured.

ACCESSORY PIGMENTS We see in Figure 8.6 that chlorophyll absorbs blue and red wavelengths, which are near the two ends of the visible spectrum. Thus, if only chlorophyll were active in photosynthesis, much of the visible spectrum would go unused. However, all photosynthetic organisms possess **accessory pigments**, which absorb photons intermediate in energy between the red and the blue wavelengths and then transfer a portion of that energy to the chlorophylls. Among these accessory pigments are **carotenoids**, such as β-carotene, which absorb photons in the blue and blue-green wavelengths and appear deep yellow. The **phycobilins**, which are found in red algae and in cyanobacteria, absorb various yellow-green, yellow, and orange wavelengths.

Light absorption results in photochemical change

Any pigment molecule can become excited when its absorption spectrum matches the energies of incoming photons. After a pigment molecule absorbs a photon and enters an excited state (see Figure 8.4), that molecule returns to the ground state. When this happens, some of the absorbed energy may be given off as heat, and the rest may be given off as light energy, or *fluorescence*. Because some of the absorbed light energy is lost as heat, the fluorescence has less energy and longer wavelengths than the absorbed light. When there is fluorescence, there are no permanent chemical changes or biological functions—no chemical work is done. If fluorescence does not occur, the pigment molecule may pass the absorbed energy along to another molecule, provided that the target molecule is very near, has the right orientation, and has the appropriate structure to receive the energy.

The pigments in photosynthetic organisms are arranged into energy-absorbing **antenna systems**. In these systems, the pigments are packed together and attached to thylakoid membrane proteins in such a way that the excitation energy from an absorbed photon can be passed along from one pigment molecule in the sys-

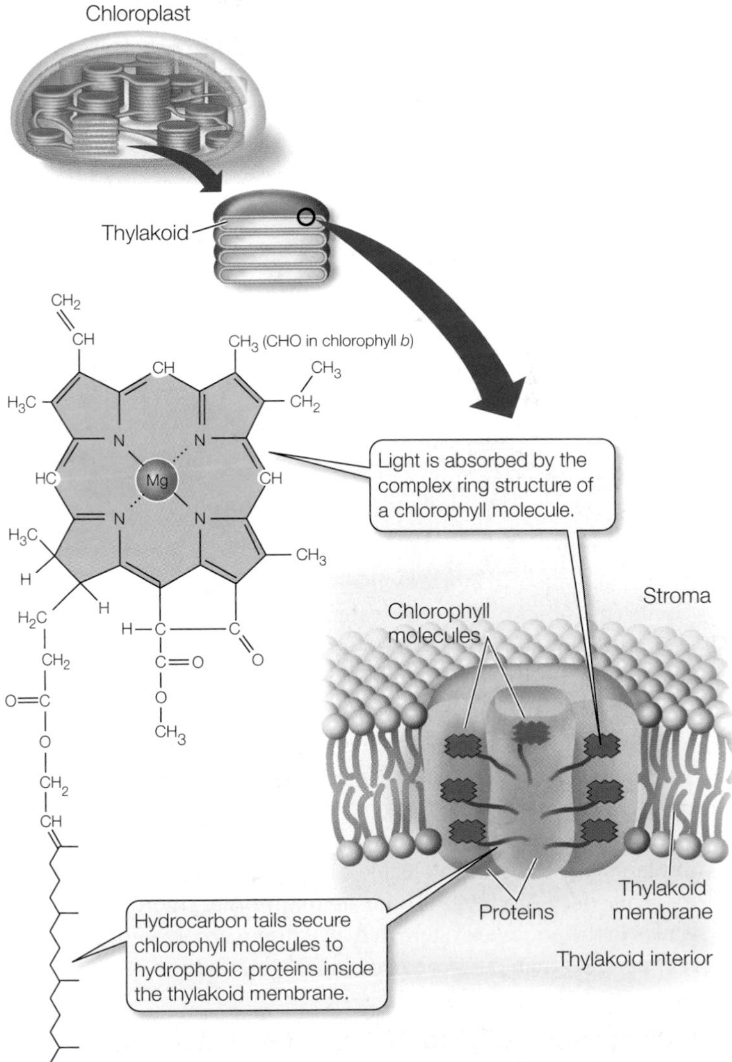

8.7 The Molecular Structure of Chlorophyll Chlorophyll consists of a complex ring structure (green area) with a magnesium atom at the center, plus a hydrocarbon "tail." The "tail" anchors the chlorophyll molecule to an integral protein in the thylakoid membrane. Chlorophyll *a* and chlorophyll *b* are identical except for the replacement of a methyl group (—CH_3) with an aldehyde group (—CHO) at the upper right.

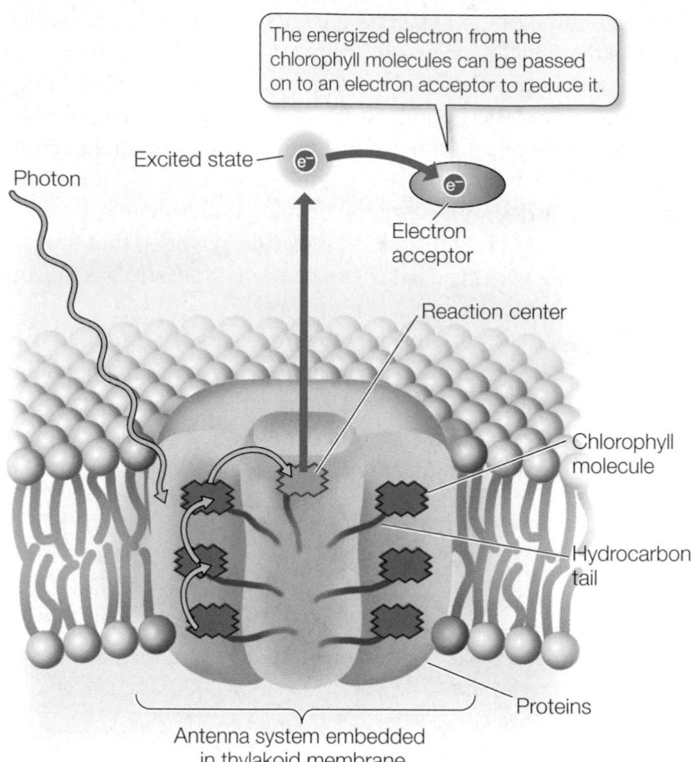

8.8 Energy Transfer and Electron Transport Rather than being lost as fluorescence, energy from a photon may be transferred from one pigment molecule to another. In an antenna system, an excited pigment molecule can transfer energy through a series of other pigment molecules to a pigment molecule in the reaction center. That molecule may become sufficiently excited that it gives up its excited electron, which can then be passed on to an electron acceptor.

tem to another (**Figure 8.8**). Excitation energy moves from pigments that absorb shorter wavelengths (higher energy) to pigments that absorb longer wavelengths (lower energy). Thus the excitation ends up in the one pigment molecule in the antenna system that absorbs the longest wavelengths; this molecule is in the **reaction center** of the antenna system.

It is the reaction center that converts the absorbed light energy into chemical energy. It is in the reaction center that a pigment molecule absorbs sufficient energy that it actually gives up its excited electron (is chemically oxidized) and becomes positively charged. In plants, the pigment molecule in the reaction center is always a molecule of chlorophyll *a*. There are many other chlorophyll *a* molecules in the antenna system, but all of them absorb light at shorter wavelengths than does the molecule in the reaction center.

Excited chlorophyll in the reaction center acts as a reducing agent

Chlorophyll has two vital roles in photosynthesis:

- It absorbs light energy and transforms it into chemical energy in the form of electrons.
- It transfers those electrons to other molecules.

We have dealt with the first role; now we turn to the second.

Photosynthesis harvests chemical energy by using the excited chlorophyll molecule in the reaction center as a *reducing agent* (*electron donor*) to reduce a stable electron acceptor (see Figure 8.8). Ground-state chlorophyll (symbolized Chl) is not much of a reducing agent, but excited chlorophyll (Chl^*) is a good one. To understand the reducing capability of Chl^*, recall that in an excited molecule, one of the electrons is zipping around in a shell farther away from its nucleus. Less tightly held, this electron can be passed on in a redox reaction to an oxidizing agent. Thus Chl^* (but not Chl) can react with an oxidizing agent A in a reaction like this:

$$Chl^* + A \rightarrow Chl^+ + A^-$$

This, then, is the first consequence of light absorption by chlorophyll: the chlorophyll becomes a reducing agent (Chl^*) and participates in a redox reaction. (The resulting Chl^+ is a strong oxidizing agent, as described below.)

Reduction leads to electron transport

The oxidizing agent A that gets reduced by Chl^* is the first in a chain of electron carriers in the thylakoid membrane of the chloroplast that participate in a process termed *electron transport*. This energetically "downhill" series of reductions and oxidations is similar to what occurs in the electron transport chain of mitochondria (see Section 7.4). The final electron acceptor is **$NADP^+$ (nicotinamide adenine dinucleotide phosphate)**, which gets reduced:

$$NADP^+ + e^- \rightarrow NADPH + H^+$$

The energy-rich NADPH + H^+ is a stable, reduced coenzyme. Its oxidized form is $NADP^+$. Just as NAD^+ couples the metabolic pathways of cellular respiration, $NADP^+$ couples the two photosynthetic pathways. $NADP^+$ is identical to NAD^+ except that the former has an additional phosphate group attached to each ribose (see Figure 7.4). Whereas NAD^+ participates in catabolism, $NADP^+$ is used in anabolic (synthetic) reactions, such as carbohydrate synthesis from CO_2, that require energy from reducing power.

There are two different systems of electron transport in photosynthesis:

- *Noncyclic electron transport* produces NADPH + H^+ and ATP.
- *Cyclic electron transport* produces only ATP.

We'll consider these two systems before considering the role of chemiosmosis in photophosphorylation—a process that is very similar to oxidative phosphorylation in mitochondria.

Noncyclic electron transport produces ATP and NADPH

In **noncyclic electron transport**, light energy is used to oxidize water, forming O_2, H^+, and electrons.

When chlorophyll loses electrons upon excitation by light, it has an "electron hole" and thus a strong tendency to "grab" electrons from another molecule to replenish those it lost. In chemical terms, then, Chl^+ (see above) is a strong oxidizing agent. The replenishing electrons come from water, splitting the H–O–H bonds. As the electrons are passed from water to chlorophyll, and ultimately to

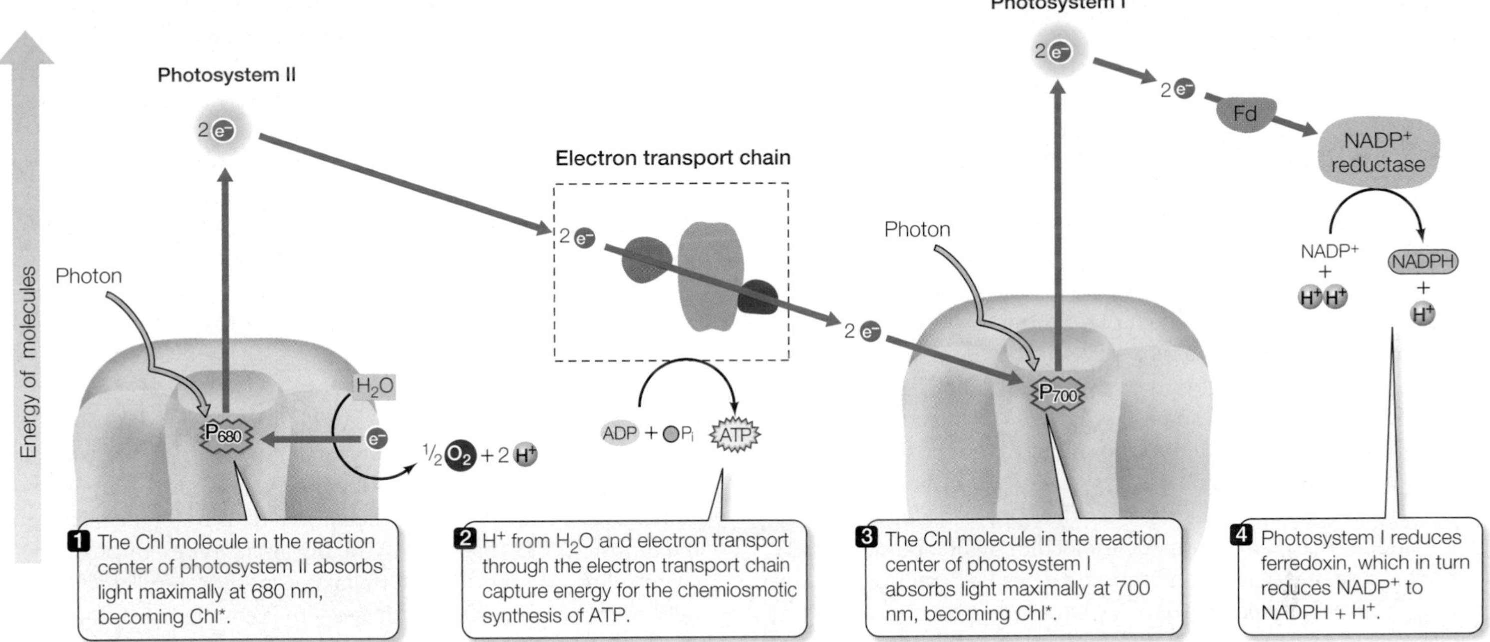

Photosystem II
Photosystem I
Electron transport chain
Energy of molecules
Photon
2 e⁻
Fd
NADP⁺ reductase
NADP⁺ + H⁺ H⁺
NADPH + H⁺
H₂O
P680
P700
½ O₂ + 2 H⁺
ADP + Pi
ATP
1 The Chl molecule in the reaction center of photosystem II absorbs light maximally at 680 nm, becoming Chl*.
2 H⁺ from H₂O and electron transport through the electron transport chain capture energy for the chemiosmotic synthesis of ATP.
3 The Chl molecule in the reaction center of photosystem I absorbs light maximally at 700 nm, becoming Chl*.
4 Photosystem I reduces ferredoxin, which in turn reduces NADP⁺ to NADPH + H⁺.

8.9 Noncyclic Electron Transport Uses Two Photosystems The energy of excited chlorophyll molecules in the reaction centers of photosystems I and II allows energized electrons to reduce carriers, setting up electron transport. The term "Z scheme" describes the path (blue arrows) of electrons as they travel through the two photosystems, using energy level as the *y*-axis of the "graph."

$NADP^+$, they pass through a chain of electron carriers in the thylakoid membrane. These redox reactions are exergonic, and some of the free energy released is ultimately used to form ATP by chemiosmosis.

TWO PHOTOSYSTEMS ARE REQUIRED Noncyclic electron transport requires the participation of two different **photosystems** —light-driven molecular units in the thylakoid membrane, each of which consists of many chlorophyll molecules and accessory pigments bound to proteins in *separate* energy-absorbing antenna systems.

- **Photosystem I** uses light energy to reduce $NADP^+$ to NADPH + H^+.
- **Photosystem II** uses light energy to oxidize water molecules, producing electrons, protons (H^+), and O_2.

The reaction center for photosystem I contains a chlorophyll *a* molecule called P_{700} because it can best absorb light with a wavelength of 700 nm. The reaction center for photosystem II contains a chlorophyll *a* molecule called P_{680} because it absorbs light maximally at 680 nm. Thus photosystem II requires photons that are somewhat more energetic (i.e., have shorter wavelengths) than those required by photosystem I. To keep noncyclic electron transport going, both photosystems must be constantly absorbing light, thereby boosting electrons to higher shells from which they may be captured by specific oxidizing agents. Photosystems I and II complement each other, interacting in a way that has been described in a model called the **Z scheme** (because the path of the electrons, when placed along an axis of rising energy level, resembles a sideways letter Z; **Figure 8.9**).

ELECTRON TRANSPORT: THE Z SCHEME In the Z scheme model describing the reactions of noncyclic electron transport from water to $NADP^+$, photosystem II comes before photosystem I. When photosystem II absorbs photons, electrons pass from P_{680} to the primary electron acceptor—the first carrier in the electron transport chain—and P_{680} is oxidized to $P_{680}{}^+$. Electrons from the oxidation of water are passed to $P_{680}{}^+$, reducing it once again to P_{680}, which can then absorb more photons. The electrons from photosystem II pass through a series of exergonic reactions in the electron transport chain that are indirectly coupled to proton pumping across the thylakoid membrane (described in Figure 8.11). This *chemiosmotic* pumping creates a proton gradient that produces energy for ATP synthesis.

In photosystem I, the reaction center containing P_{700} becomes excited to $P_{700}{}^*$, which leads to the reduction of an oxidizing agent called **ferredoxin** (Fd) and the production of $P_{700}{}^+$. $P_{700}{}^+$ returns to the ground state by accepting electrons passed through the electron transport chain from photosystem II.

With this accounting for the source of the electrons entering photosystem II, we can now consider the fate of the electrons from photosystem I. These electrons are used in the last step of noncyclic electron transport, in which two electrons and two protons are used to reduce a molecule of $NADP^+$ to NADPH + H^+.

In summary:

- Noncyclic electron transport extracts electrons from water and passes them ultimately to NADPH + H^+, utilizing photons absorbed by photosystems I and II and resulting in ATP synthesis.
- Noncyclic electron transport yields NADPH + H^+, ATP, and O_2.

Cyclic electron transport produces ATP but no NADPH

Noncyclic electron transport produces ATP and NADPH + H^+. However, as we will see, the light-independent reactions of photosynthesis use more ATP than NADPH + H^+. **Cyclic electron transport** occurs in some organisms when the ratio of NADPH + H^+ to $NADP^+$ in the chloroplast is high. This process, which produces only ATP, is called *cyclic* because an electron passed from an excited chlorophyll molecule at the outset cycles *back to the same chlorophyll molecule* at the end of the chain of reactions (**Figure 8.10**).

Before cyclic electron transport begins, P_{700}, the reaction center chlorophyll molecule of photosystem I, is in the ground state. It absorbs a photon and becomes P_{700}^*. The P_{700}^* then reacts with oxidized ferredoxin (Fd_{ox}) to produce reduced ferredoxin (Fd_{red}). The reaction is exergonic, releasing free energy. Reduced ferredoxin (Fd_{red}) passes its added electron to a different oxidizing agent, **plastoquinone** (**PQ**, a small organic molecule), which pumps two H^+ back across the thylakoid membrane. Thus Fd_{red} reduces PQ, and PQ_{red} passes the electron to the electron transport chain by way of **plastocyanin** (**PC**) until it completes its cycle by returning to P_{700}^+, resulting in a restoration of its uncharged form, P_{700}. By the time the electron from P_{700}^* travels through the electron transport chain and comes back to reduce P_{700}^+, all the energy from the original photon has been released. This cycle is a series of redox reactions, each exergonic, and the released energy is stored in the form of a proton gradient that can be used to produce ATP.

Chemiosmosis is the source of the ATP produced in photophosphorylation

Section 7.4 considers the chemiosmotic mechanism for ATP formation in the mitochondrion. A similar chemiosmotic mechanism operates in **photophosphorylation**, the light-driven production of ATP from ADP and P_i in the chloroplast. In chloroplasts, electron transport through the electron transport chain is coupled to the transport of protons (H^+) across the thylakoid membrane, which results in a proton gradient across the membrane (**Figure 8.11**).

About 60 percent of the amino acid sequence in chloroplast ATP synthase is the same as that in human mitochondrial ATP synthase—a remarkable similarity, given that plants and animals had their most recent common ancestor more than a billion years ago.

The electron carriers in the thylakoid membrane are oriented so that protons move from the stroma—the interior matrix of the chloroplast—into the lumen of the thylakoid. Thus the lumen becomes acidic with respect to the stroma. This difference leads to the diffusion of H^+ back out of the thylakoid lumen through specific protein channels in the thylakoid membrane. These channels are enzymes—ATP synthases—that couple the diffusion of protons to the formation of ATP, just as in mitochondria (see Figure 7.14). The mechanisms of the two enzymes are also similar. What is different is their orientation: in plants, protons flow through the ATP synthase out of the thylakoid lumen, but in animals they flow *into* the mitochondrial matrix.

8.2 RECAP

Conversion of light energy into chemical energy occurs when pigments absorb photons. Light energy is transferred to electrons, which act to reduce a series of molecules in the chloroplast.

- How does chlorophyll absorb and transfer light energy? See pp. 165–166 and Figure 8.8
- Do you understand how electrons are produced in photosystem II and then flow to photosystem I? See pp. 166–167 and Figure 8.9
- How does cyclic electron transport in photosystem I result in the production of ATP? See p. 168 and Figure 8.10

We have seen how light energy drives the synthesis of ATP and NADPH + H^+. We now turn to the light-independent reactions of photosynthesis, which use these two energy-rich coenzymes to reduce CO_2 and form carbohydrates.

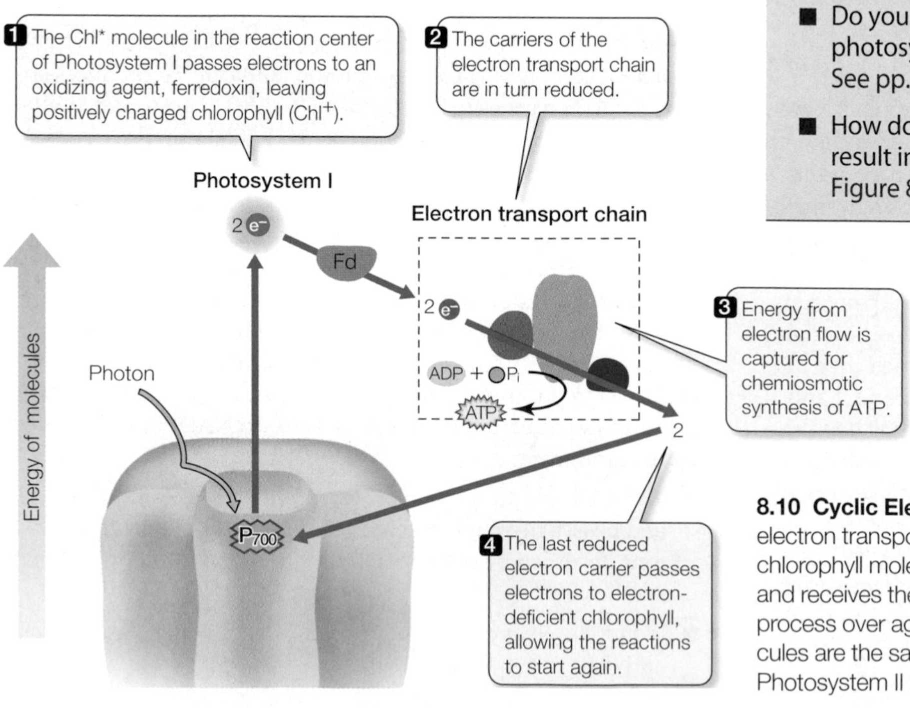

1 The Chl* molecule in the reaction center of Photosystem I passes electrons to an oxidizing agent, ferredoxin, leaving positively charged chlorophyll (Chl^+).

2 The carriers of the electron transport chain are in turn reduced.

Photosystem I

Electron transport chain

2e⁻

Fd

Energy of molecules

Photon

ADP + Pi

ATP

P700

3 Energy from electron flow is captured for chemiosmotic synthesis of ATP.

4 The last reduced electron carrier passes electrons to electron-deficient chlorophyll, allowing the reactions to start again.

8.10 Cyclic Electron Transport Traps Light Energy as ATP Cyclic electron transport produces ATP, but no NADPH + H^+. The same chlorophyll molecule passes on the electrons that start the reactions and receives the electrons at the end of the reactions to start the process over again. Photosystem I and the electron transport molecules are the same as in noncyclic electron transport (see Figure 8.9). Photosystem II is not involved in cyclic electron transport.

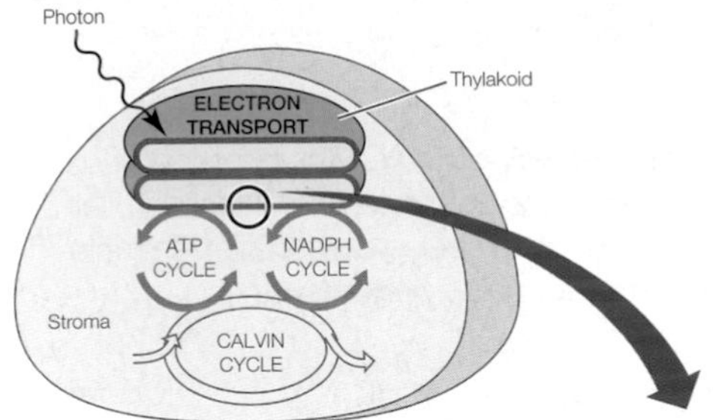

8.11 Chloroplasts Form ATP Chemiosmotically Protons (H^+) pumped across the thylakoid membrane from the stroma during electron transport make the lumen of the thylakoid more acidic than the stroma. Driven by this pH difference, the protons diffuse back to the stroma through ATP synthase channels, which couple the energy of proton diffusion to the formation of ATP from ADP + P_i.

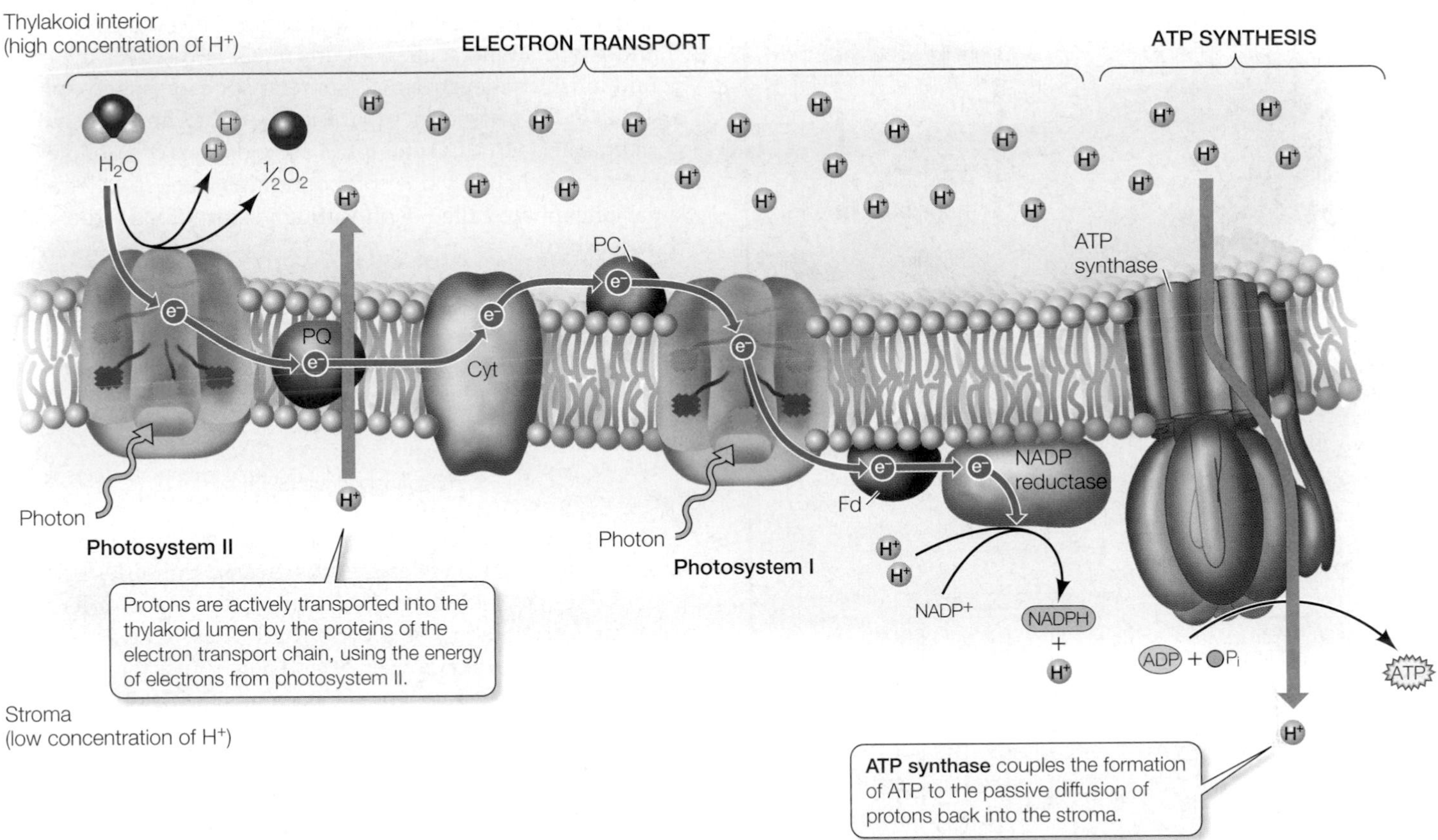

8.3 How Is Chemical Energy Used to Synthesize Carbohydrates?

Most of the enzymes that catalyze the reactions of CO_2 fixation are dissolved in the stroma of the chloroplast, and that is where those reactions take place. However, these enzymes use the energy in ATP and NADPH, produced in the thylakoids by the light reactions, to reduce CO_2 to carbohydrates. Because there is no stockpiling of these energy-rich coenzymes, those "light-independent" reactions take place *only in the light*, when these coenzymes are being generated.

Radioisotope labeling experiments revealed the steps of the Calvin cycle

To identify the sequence of reactions by which the carbon from CO_2 ends up in carbohydrates, scientists found a way to label CO_2 so that it could be followed after being taken up by a photosynthetic cell. In the 1950s, an experiment performed by Melvin Calvin, Andrew Benson, and their colleagues used radioactively labeled CO_2 in which some of the carbon atoms were not the normal ^{12}C, but its radioisotope ^{14}C. Although ^{14}C is distinguished by its emission of radiation, chemically it behaves virtually identically to nonradioactive ^{12}C. In general, enzymes do not distinguish between isotopes of an element in their substrates, so photosynthesizing cells utilize $^{14}CO_2$ no differently than $^{12}CO_2$.

Calvin and his colleagues exposed cultures of the unicellular green alga *Chlorella* to $^{14}CO_2$ for 30 seconds. They then rapidly killed the cells and extracted their organic compounds. They separated the different compounds from one another by *paper chromatography* (**Figure 8.12**), a technique that had been invented just a few years previously by two British scientists. The algal extracts were dissolved in alcohol, and then applied to a sheet of filter paper, where they formed hydrogen bonds with the cellulose of the paper. The paper was put into a solvent called phenol-water, which crept up the paper by capillary action, much like water being absorbed by a paper towel. The various molecules in the algal extract were dissolved in the phenol-water and carried along. But as the solvent moved, some became less and less attracted to the solvent molecules compared with their attraction with the paper. Finally, they came out of solution and stayed where they were. Different molecules had different properties in this regard, and the use of a second solvent moving in a different direction provided even more separation. An X-ray film was exposed to the filter paper to reveal the positions of radioactive compounds.

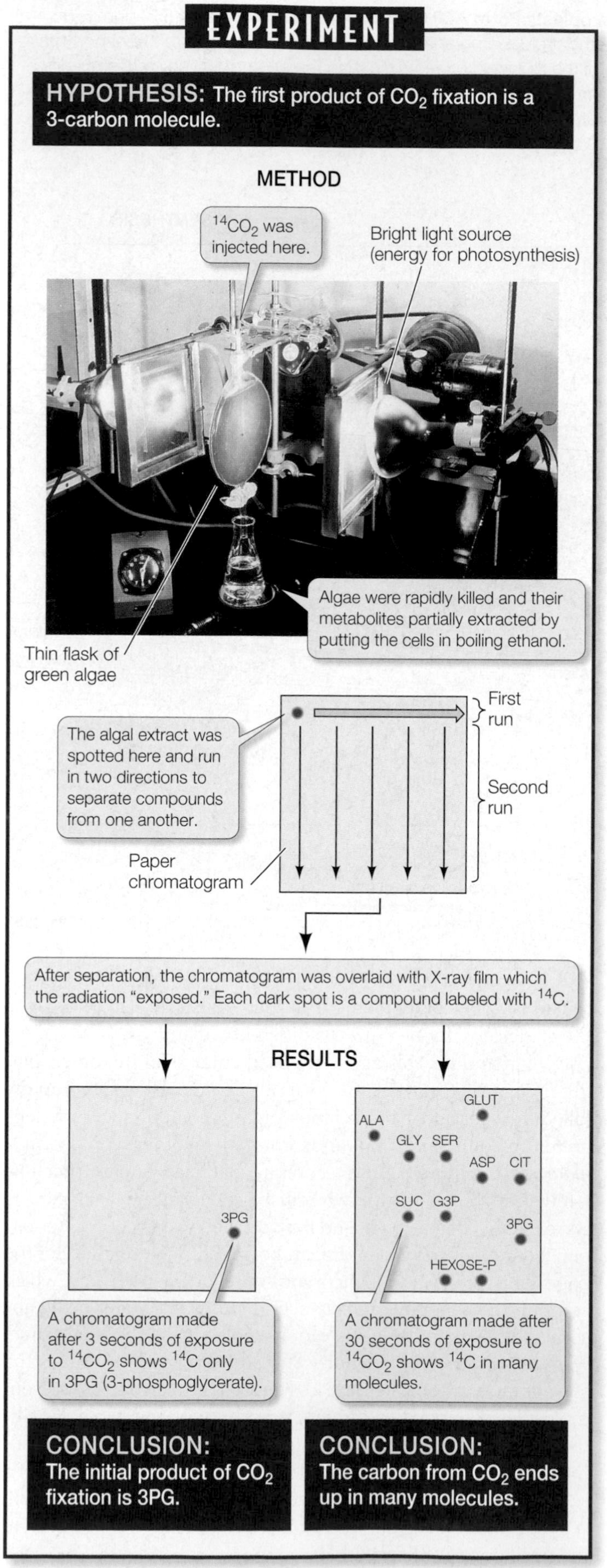

EXPERIMENT

HYPOTHESIS: The first product of CO_2 fixation is a 3-carbon molecule.

METHOD

$^{14}CO_2$ was injected here.

Bright light source (energy for photosynthesis)

Algae were rapidly killed and their metabolites partially extracted by putting the cells in boiling ethanol.

Thin flask of green algae

The algal extract was spotted here and run in two directions to separate compounds from one another.

First run

Second run

Paper chromatogram

After separation, the chromatogram was overlaid with X-ray film which the radiation "exposed." Each dark spot is a compound labeled with ^{14}C.

RESULTS

3PG

ALA GLY SER GLUT ASP CIT SUC G3P 3PG HEXOSE-P

A chromatogram made after 3 seconds of exposure to $^{14}CO_2$ shows ^{14}C only in 3PG (3-phosphoglycerate).

A chromatogram made after 30 seconds of exposure to $^{14}CO_2$ shows ^{14}C in many molecules.

CONCLUSION: The initial product of CO_2 fixation is 3PG.

CONCLUSION: The carbon from CO_2 ends up in many molecules.

8.12 Tracing the Pathway of CO_2 The historical photograph at the top shows the apparatus Calvin and his colleagues used to follow radiolabeled carbon dioxide molecules ($^{14}CO_2$) as they were transformed by photosynthesis.

But many compounds in the algal extract, including monosaccharides and amino acids, contained ^{14}C. To discover the compound in which the labeled carbon first appears (suggesting the first step in the pathway of CO_2 fixation), Calvin and his team exposed the algae to $^{14}CO_2$ for just 3 seconds. This 3-second exposure revealed that only one compound was labeled—a 3-carbon sugar phosphate called 3-phosphoglycerate (3PG) (the ^{14}C is shown in red):

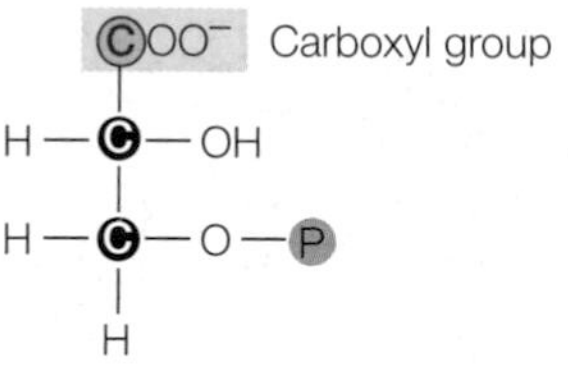

3-Phosphoglycerate (3PG)

By tracing the steps with successive, increasingly long exposures, Calvin and his colleagues discovered the series of compounds through which the carbon taken up in CO_2 flows. Its pathway was discovered to be a cycle that "fixes" CO_2 in a larger molecule, produces a carbohydrate, and regenerates the initial CO_2 acceptor. This cycle was appropriately named the **Calvin cycle** (**Figure 8.13**).

The initial reaction in the Calvin cycle adds the 1-carbon CO_2 to an acceptor molecule, the 5-carbon compound **ribulose 1,5-bisphosphate** (**RuBP**). The product is an intermediate 6-carbon compound, which quickly breaks down and forms two 3-carbon molecules of 3PG (as Calvin and colleagues observed; **Figure 8.14**). The enzyme that catalyzes this fixation reaction, **ribulose bisphosphate carboxylase/oxygenase** (**rubisco**), is the most abundant protein in the world, constituting up to 50 percent of all the protein in every plant leaf.

The Calvin cycle is made up of three processes

The Calvin cycle uses the high-energy coenzymes made in the thylakoids during the light reactions (ATP and NADPH) to reduce CO_2 to a carbohydrate in the stroma. Three processes make up the cycle:

- *Fixation of CO_2.* As we have seen, this reaction is catalyzed by rubisco, and its product is 3PG.
- *Reduction of 3PG to form glyceraldehyde 3-phosphate (G3P).* This series of reactions involves a phosphorylation (using the ATP made in the light reactions) and a reduction (using the NADPH made in the light reactions).
- *Regeneration of the CO_2 acceptor, RuBP.* Most of the G3P ends up as RuMP (ribulose monophosphate), and ATP is used to convert this compound into RuBP. So for every "turn" of the cycle, with one CO_2 fixed, the CO_2 acceptor is regenerated.

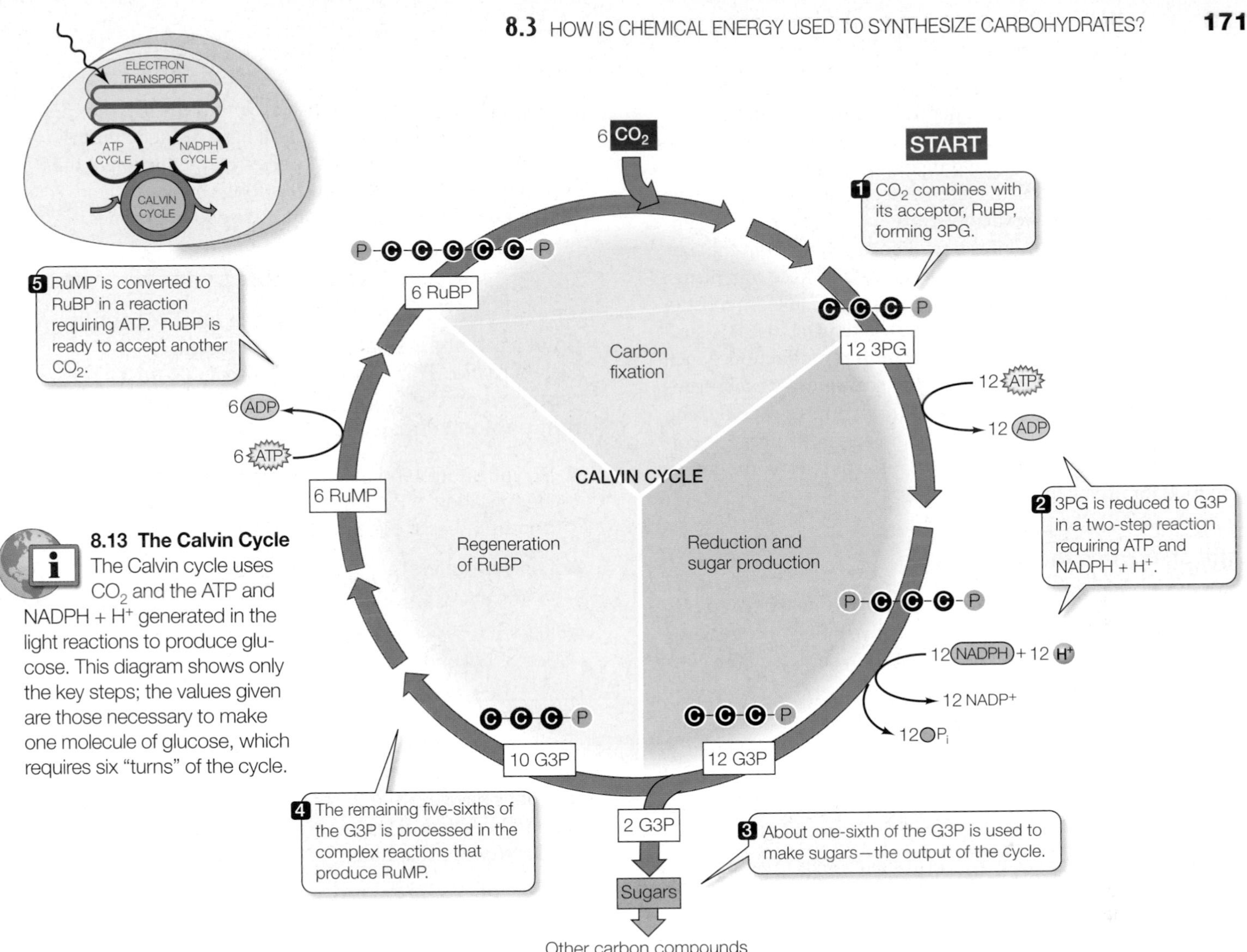
ELECTRON TRANSPORT
ATP CYCLE
NADPH CYCLE
CALVIN CYCLE
6 CO_2
START
1 CO_2 combines with its acceptor, RuBP, forming 3PG.
12 3PG
12 ATP
12 ADP
2 3PG is reduced to G3P in a two-step reaction requiring ATP and NADPH + H+.
12 NADPH + 12 H+
12 NADP+
12 Pi
12 G3P
10 G3P
2 G3P
Sugars
Other carbon compounds
3 About one-sixth of the G3P is used to make sugars—the output of the cycle.
4 The remaining five-sixths of the G3P is processed in the complex reactions that produce RuMP.
6 RuMP
6 ATP
6 ADP
6 RuBP
5 RuMP is converted to RuBP in a reaction requiring ATP. RuBP is ready to accept another CO_2.
Carbon fixation
Reduction and sugar production
Regeneration of RuBP
CALVIN CYCLE

8.13 The Calvin Cycle The Calvin cycle uses CO_2 and the ATP and NADPH + H+ generated in the light reactions to produce glucose. This diagram shows only the key steps; the values given are those necessary to make one molecule of glucose, which requires six "turns" of the cycle.

The product of this cycle is glyceraldehyde 3-phosphate (G3P), which is a 3-carbon sugar phosphate, also called triose phosphate:

Glyceraldehyde 3-phosphate (G3P)

In a typical leaf, five-sixths of the G3P is recycled into RuBP. There are two fates for the remaining G3P:

- One-third of it ends up in the polysaccharide *starch*, which is stored in the chloroplast.
- Two-thirds of it is converted in the cytosol into the disaccharide *sucrose*, which is transported out of the leaf to other organs in the plant, where it is hydrolyzed to its constituent monosaccharides: *glucose* and *fructose*.

The fate of the carbon atom in CO_2 is followed in red.

The enzyme rubisco catalyzes the reaction of CO_2 with RuBP.

The reaction intermediate splits into two molecules of 3-phosphoglycerate (3PG).

Carbon dioxide

Ribulose 1,5-bisphosphate (RuBP)

Rubisco

Six-carbon skeleton of reaction intermediate

8.14 RuBP Is the Carbon Dioxide Acceptor CO_2 is added to a 5-carbon compound, RuBP. The resulting 6-carbon compound immediately splits into two molecules of the sugar phosphate 3PG.

These carbohydrates are subsequently used by the plant to make other compounds. Their carbon molecules are incorporated into amino acids, lipids, and the building blocks of nucleic acids.

The products of the Calvin cycle are of crucial importance to the entire biosphere, for the covalent bonds of the carbohydrates generated by the cycle represent the total energy yield from the harvesting of light by photosynthetic organisms. These organisms, which are also called *autotrophs* ("self-feeders"), release most of this energy by glycolysis and cellular respiration and use it to support their own growth, development, and reproduction. Much plant matter ends up being consumed by *heterotrophs* ("other-feeders"), such as animals, which cannot photosynthesize and depend on autotrophs for both raw materials and energy sources. Glycolysis and cellular respiration in heterotroph cells release free energy from food for use by the heterotrophs.

Light stimulates the Calvin cycle

The Calvin cycle uses NADPH and ATP, which, as we have seen, are made through photophosphorylation. Two other processes connect the light reactions with this CO_2 fixation pathway. Both connections are indirect, but significant:

- *Light-induced pH changes in the stroma activate some enzymes in the Calvin cycle.* Proton pumping from the stroma into the thylakoids increases the pH of the stroma from 7 to 8 (a tenfold decrease in H^+ concentration). This favors the activation of rubisco.
- *Light-induced electron flow reduces disulfide bonds to activate four Calvin cycle enzymes* (**Figure 8.15**). When ferredoxin is reduced in photosystem I (see Figure 8.9), it passes some electrons to a small, soluble protein called *thioredoxin*. This protein in turn passes electrons to four enzymes in the CO_2 fixation pathway. These enzymes all have disulfide bridges (see Figure 3.5) near their active sites, and reduction of the sulfurs breaks the bridges. The resulting change in their three-dimensional shape activates these four enzymes.

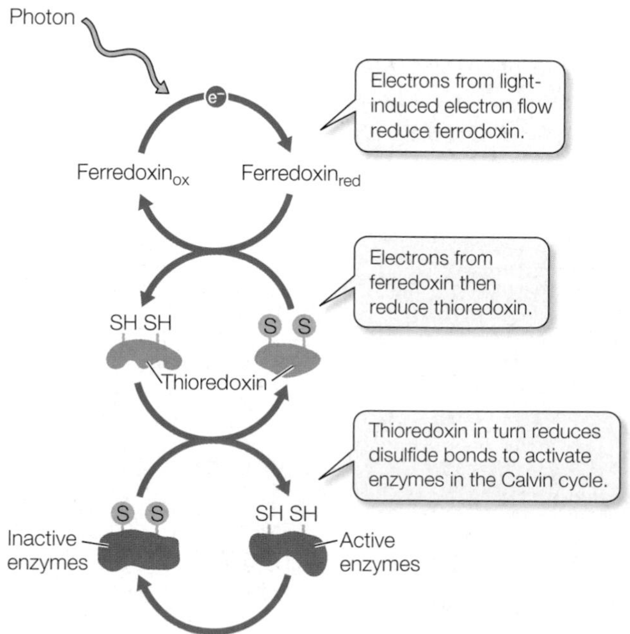
Photon
e⁻
Electrons from light-induced electron flow reduce ferrodoxin.
Ferredoxin$_{ox}$
Ferredoxin$_{red}$
Electrons from ferredoxin then reduce thioredoxin.
SH SH
S S
Thioredoxin
Thioredoxin in turn reduces disulfide bonds to activate enzymes in the Calvin cycle.
S S
SH SH
Inactive enzymes
Active enzymes

8.15 The Photochemical Reactions Stimulate the Calvin Cycle By reducing (breaking) disulfide bridges, electrons from the light reactions activate enzymes in CO_2 fixation.

8.3 RECAP

ATP and NADPH produced in the light reactions power the synthesis of carbohydrates by the Calvin cycle. This cycle produces G3P and also regenerates the initial CO_2 acceptor, RuBP, so that photosynthesis can continue.

- Do you understand the experiments that led to the identification of RuBP as the initial CO_2 acceptor in photosynthesis? See pp. 169–170 and Figure 8.12
- Can you outline the three processes of the Calvin cycle? See p. 170 and Figure 8.13
- In what ways does light stimulate the Calvin cycle? See p. 172 and Figure 8.15

Although all green plants carry out the Calvin cycle, in some plant groups variations on (or additional steps in) the light-independent reactions have evolved in response to certain environmental conditions. Let's look at these environmental limitations and the metabolic bypasses that have evolved to circumvent them.

8.4 How Do Plants Adapt to the Inefficiencies of Photosynthesis?

One major limitation of rubisco is its tendency to react with O_2 instead of CO_2. This reaction leads to a process called photorespiration, which lowers the overall rate of CO_2 fixation. After examining this problem, we'll look at some biochemical pathways and features of plant anatomy that compensate for the limitations of rubisco.

Rubisco catalyzes RuBP reaction with O_2 as well as with CO_2

As its full name indicates, rubisco is an **oxygenase** as well as a **carboxylase**; that is, it can add O_2 to the acceptor molecule RuBP instead of CO_2. These two reactions compete with each other, so if RuBP reacts with O_2, it cannot react with CO_2. This reaction reduces the overall amount of CO_2 that is converted into carbohydrates, and therefore limits plant growth.

When O_2 is added to RuBP, one of the products is a 2-carbon compound, phosphoglycolate:

$$\text{RuBP} + O_2 \rightarrow \text{phosphoglycolate} + \text{3PG}$$

Plants have evolved a metabolic pathway that can partially recover the carbon that has been channeled away from the Calvin cycle into phosphoglycolate. The phosphoglycolate forms glycolate, which diffuses into membrane-enclosed organelles called *peroxi-*

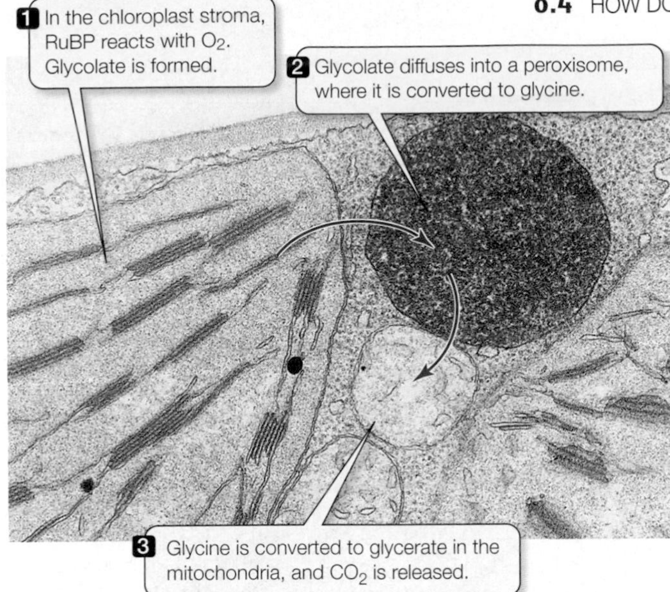

1 In the chloroplast stroma, RuBP reacts with O_2. Glycolate is formed.
2 Glycolate diffuses into a peroxisome, where it is converted to glycine.
3 Glycine is converted to glycerate in the mitochondria, and CO_2 is released.

8.16 Organelles of Photorespiration The reactions of photorespiration take place in the chloroplasts, peroxisomes, and finally, in the mitochondria.

somes (**Figure 8.16**). There, a series of reactions converts it into the amino acid glycine:

$$\text{glycolate} \rightarrow \text{glycine}$$

The glycine then diffuses into a mitochondrion, where two glycine molecules are converted into glycerate (a 3-carbon molecule) and CO_2:

$$2 \text{ glycine} \rightarrow \text{glycerate} + CO_2$$

This pathway is called **photorespiration** because it consumes O_2 and releases CO_2. It uses ATP and NADPH produced in the light reactions, just like the Calvin cycle. The net effect is to take two 2-carbon molecules and make one 3-carbon molecule. So one carbon of the four is released as CO_2, and three of the carbons (75 percent) are recovered as fixed carbon. In other words, photorespiration reduces the net carbon fixed by the Calvin cycle by 25 percent.

How does rubisco "decide" whether to act as an oxygenase or a carboxylase? There are three factors involved:

- Rubisco has ten times more affinity for CO_2 than for O_2, and so favors CO_2 fixation.
- In the leaf, the relative concentrations of CO_2 and O_2 vary. If O_2 is relatively abundant, rubisco acts as an oxygenase, and photorespiration ensues. If CO_2 predominates, rubisco fixes it, and the Calvin cycle occurs.
- Photorespiration is more likely at high temperatures. On a hot, dry day, the stomata that allow water to evaporate from the leaf close to prevent water loss (see Figure 8.1). But this also prevents gases from entering and leaving the leaf. The CO_2 concentration in the leaf falls because CO_2 is being used up by photosynthetic reactions, and the O_2 concentration rises because of these same reactions. As the ratio of CO_2 to O_2 in the leaf falls, the oxygenase activity of rubisco is favored, and photorespiration proceeds.

C_4 plants can bypass photorespiration

In plants such as roses, wheat, and rice, the palisade **mesophyll** cells, which lie just below the surface of the leaf, are full of chloroplasts that contain abundant rubisco (**Figure 8.17A**). On a hot day, these leaves close their stomata to conserve water. The level of CO_2 in the air spaces of the leaves falls, and that of O_2 rises, as photosynthesis goes on. Under these conditions, rubisco acts as an oxygenase, and photorespiration occurs. Because the first product of CO_2 fixation in these plants is the 3-carbon molecule 3PG, they are called **C_3 plants**.

Corn, sugarcane, and other tropical grasses (**Figure 8.17B**) also close their stomata on a hot day, but their rate of photosynthesis does not fall, nor does photorespiration occur. They keep the ratio of CO_2 to O_2 around rubisco high so that rubisco continues to act as a carboxylase. They do this in part by making a 4-carbon compound, *oxaloacetate*, as the first product of CO_2 fixation, and so are called **C_4 plants**.

C_4 plants perform the normal Calvin cycle, but they have an additional early reaction that fixes CO_2 without losing carbon to photorespiration. Because this initial CO_2 fixation step can function

(A) Arrangement of cells in a C_3 leaf

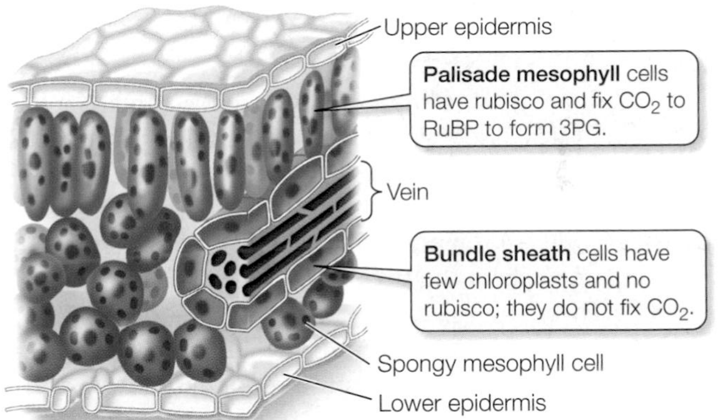

(B) Arrangement of cells in a C_4 leaf

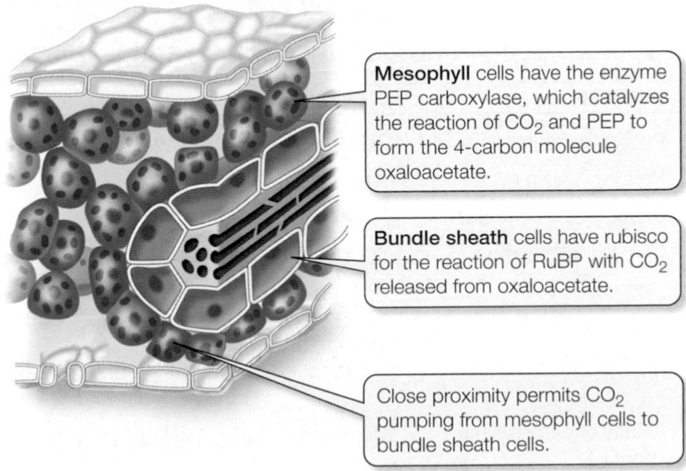

Mesophyll cells have the enzyme PEP carboxylase, which catalyzes the reaction of CO_2 and PEP to form the 4-carbon molecule oxaloacetate.
Bundle sheath cells have rubisco for the reaction of RuBP with CO_2 released from oxaloacetate.
Close proximity permits CO_2 pumping from mesophyll cells to bundle sheath cells.

8.17 Leaf Anatomy of C_3 and C_4 Plants Carbon dioxide fixation occurs in different organelles and cells of the leaves in (A) C_3 plants and (B) C_4 plants.

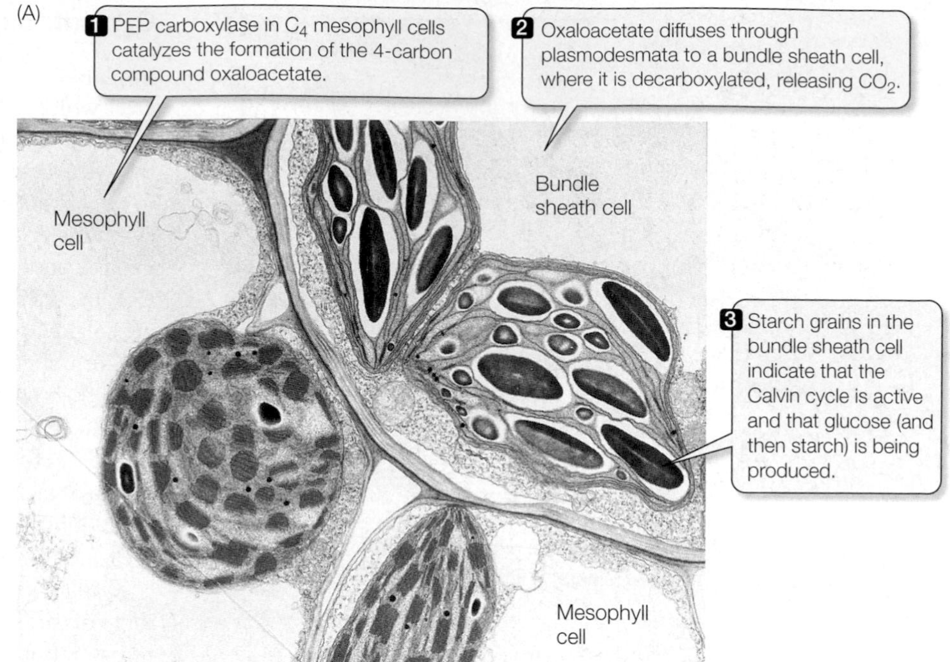

(A)
1 PEP carboxylase in C_4 mesophyll cells catalyzes the formation of the 4-carbon compound oxaloacetate.
2 Oxaloacetate diffuses through plasmodesmata to a bundle sheath cell, where it is decarboxylated, releasing CO_2.
Bundle sheath cell
Mesophyll cell
3 Starch grains in the bundle sheath cell indicate that the Calvin cycle is active and that glucose (and then starch) is being produced.
Mesophyll cell

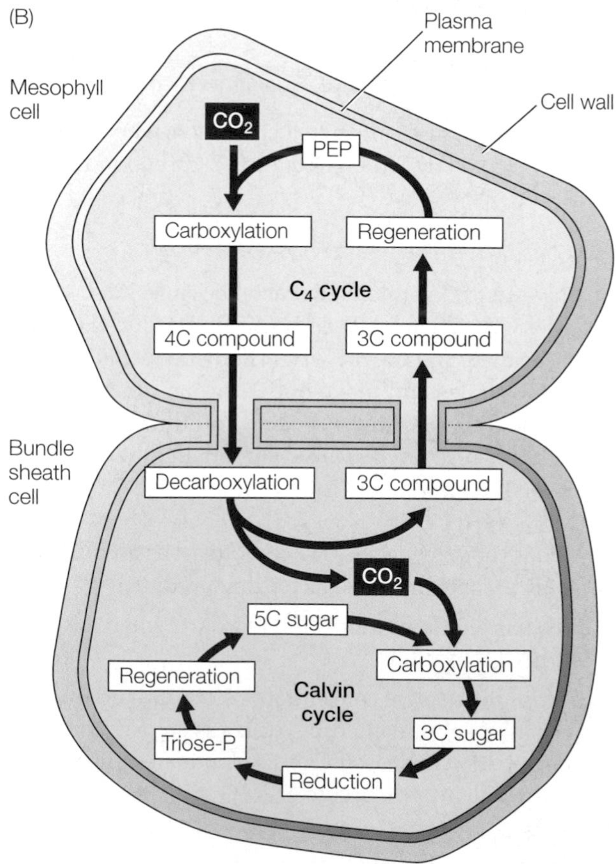

8.18 The Anatomy and Biochemistry of C_4 Carbon Fixation (A) Carbon dioxide is fixed initially in the mesophyll cells, but enters the Calvin cycle in the bundle sheath cells. (B) The two cell types share an interconnected biochemical pathway for CO_2 assimilation.

even at low levels of CO_2 and high temperatures, C_4 plants very effectively optimize photosynthesis under conditions that inhibit it in C_3 plants. C_4 plants have two separate enzymes for CO_2 fixation, located in two different parts of the leaf (**Figure 8.18**; see also Figure 8.17B). The first enzyme, which is present in the cytosol of mesophyll cells near the surface of the leaf, fixes CO_2 to a 3-carbon acceptor compound, **phosphoenolpyruvate** (**PEP**), to produce the 4-carbon fixation product, oxaloacetate. This enzyme, **PEP carboxylase**, has two advantages over rubisco:

- It does not have oxygenase activity.
- It fixes CO_2 even at very low CO_2 levels.

So even on a hot day when the stomata are closed, the CO_2 concentration in the leaf is low, and the O_2 concentration is high, PEP carboxylase just keeps on fixing CO_2.

Oxaloacetate diffuses out of the mesophyll cells and into the **bundle sheath cells** (see Figure 8.17B), located in the interior of the leaf. The chloroplasts in the bundle sheath cells contain abundant rubisco. There, the 4-carbon oxaloacetate loses one carbon (is *decarboxylated*), forming CO_2 and regenerating the 3-carbon acceptor compound, PEP, in the mesophyll cells. Thus the role of PEP is to bind CO_2 from the air in the leaf and carry it to the bundle sheath cells, where it is "dropped off" at rubisco. This process essentially pumps up the CO_2 concentration around rubisco, so that it acts as a carboxylase and begins the Calvin cycle.

Kentucky bluegrass, a C_3 plant, thrives on lawns in April and May. But in the heat of summer, it does not do as well, and

TABLE 8.1

Comparison of Photosynthesis in C_3 and C_4 Plants

VARIABLE	C_3 PLANTS	C_4 PLANTS
Photorespiration	Extensive	Minimal
Perform Calvin cycle?	Yes	Yes
Primary CO_2 acceptor	RuBP	PEP
CO_2-fixing enzyme	Rubisco (RuBP carboxylase/oxygenase)	PEP carboxylase and rubisco
First product of CO_2 fixation	3PG (3-carbon compound)	Oxaloacetate (4-carbon compound)
Affinity of carboxylase for CO_2	Moderate	High
Photosynthetic cells of leaf	Mesophyll	Mesophyll + bundle sheath
Classes of chloroplasts	One	Two

Bermuda grass (crabgrass), a C_4 plant, takes over the lawn. The same is true on a global scale for crops: C_3 plants, such as soybeans, rice, wheat, and barley, have been adapted for human food production in temperate climates, while C_4 plants, such as corn and sugarcane, originated and are grown in the tropics. **Table 8.1** compares C_3 and C_4 photosynthesis.

C_3 plants are certainly more ancient than C_4 plants. While C_3 photosynthesis appears to have begun about 3.5 billion years ago, C_4 plants appeared about 12 million years ago. A possible factor in the emergence of the C_4 pathway is the decline in atmospheric CO_2. When dinosaurs dominated Earth 100 million years ago, the concentration of CO_2 in the atmosphere was four times what it is now. As CO_2 levels declined thereafter, the more efficient C_4 plants would have had an advantage over their C_3 counterparts. In the last two hundred years, however, CO_2 levels began increasing. Measurement of atmospheric CO_2 at sites far away from industrial sources of the gas, such as the top of a volcano in Hawaii and bubbles in Antarctic ice, show that the level of this gas has risen significantly, from 250 parts per million in 1800 to 370 ppm now. This rise may be affecting photosynthesis and plant growth. Currently, the level of CO_2 is not enough for maximal activity of CO_2 fixation by rubisco, so photorespiration occurs and reduces the growth of C_3 plants, favoring C_4 plants. If CO_2 in the atmosphere increases even more, the reverse will occur, and C_3 plants will have a comparative advantage.

Climate scientists predict that atmospheric CO_2 could rise to 600 ppm by 2100. If this happens, the overall growth of crops such as rice and wheat should increase. This may or may not translate into more food, given that other effects of the human-spurred CO_2 increase (such as global warming) will also alter Earth's ecosystem.

CAM plants also use PEP carboxylase

Other plants besides the C_4 plants use PEP carboxylase to fix and accumulate CO_2. Such plants include some water-storing plants (called *succulents*) of the family Crassulaceae, many cacti, pineapples, and several other kinds of flowering plants. The CO_2 metabolism of these plants is called **crassulacean acid metabolism**, or **CAM**, after the family of succulents in which it was discovered. CAM is much like the metabolism of C_4 plants in that CO_2 is initially fixed into a 4-carbon compound. In CAM plants, however, the processes of initial CO_2 fixation and the Calvin cycle are separated in time, rather than in space.

- *At night*, when it is cooler and water loss is minimized, the stomata open. CO_2 is fixed in mesophyll cells to form the 4-carbon compound oxaloacetate, which is converted into malic acid.
- *During the day*, when the stomata close to reduce water loss, the accumulated malic acid is shipped to the chloroplasts, where its decarboxylation supplies the CO_2 for operation of the Calvin cycle, and the light reactions supply the necessary ATP and NADPH + H^+.

8.4 RECAP

Rubisco catalyzes the fixation of CO_2 to RuBP but can also fix O_2 to that molecule. This diversion of rubisco decreases net CO_2 fixation. C_4 photosynthesis and crassulacean acid metabolism allow plants to get around this problem.

- Can you describe how photorespiration recovers some of the carbon that is channeled away from the Calvin cycle? See p. 173
- What do C_4 plants do to keep the concentration of CO_2 around rubisco high, and why? See pp. 173–174
- Can you describe CO_2 fixation in CAM plants? See p. 175

Now that we understand how photosynthesis produces carbohydrates, let's see how those carbohydrates fuel the metabolism of plants, and how the pathways of photosynthesis are connected to other metabolic pathways.

8.5 How Is Photosynthesis Connected to Other Metabolic Pathways in Plants?

Green plants are autotrophs and can synthesize all the molecules they need from simple starting materials: CO_2, H_2O, and phosphate, sulfate, and ammonium ions (NH_4^+). The NH_4^+ is needed for amino acids and comes either from the conversion of nitrogen-containing molecules in soil water taken up by the plant's roots, or from the conversion of N_2 gas from the atmosphere by bacteria, as we'll see in Chapter 26.

Plants use the carbohydrates generated by photosynthesis to provide energy for processes such as active transport and anabolism. Both cellular respiration and fermentation can occur in plants, although the former is far more common. Plant cellular respiration, unlike photosynthesis, takes place both in the light and in the dark. Because glycolysis occurs in the cytosol, respiration in the mitochondria, and photosynthesis in the chloroplasts, all these processes can proceed simultaneously.

Photosynthesis and respiration are closely linked through the Calvin cycle (**Figure 8.19**). The partitioning of G3P is particularly important:

- Some G3P from the Calvin cycle is part of the glycolysis pathway and can be converted into pyruvate. This pyruvate can be used in cellular respiration for energy, or its carbon skeletons can be used anabolically to make lipids, proteins, and other carbohydrates (see Figure 7.17).
- Some G3P can enter a pathway that is the reverse of glycolysis (*gluconeogenesis*; see Section 7.6). In this case, hexose-phosphates and then sucrose are formed and transported to the nonphotosynthetic tissues of the plant (such as the root).

Energy flows from sunlight to reduced carbon in photosynthesis to ATP in respiration. Energy can also be stored in the bonds of macromolecules such as polysaccharides, lipids, and proteins. For a plant to grow, energy storage (as body structures) must exceed

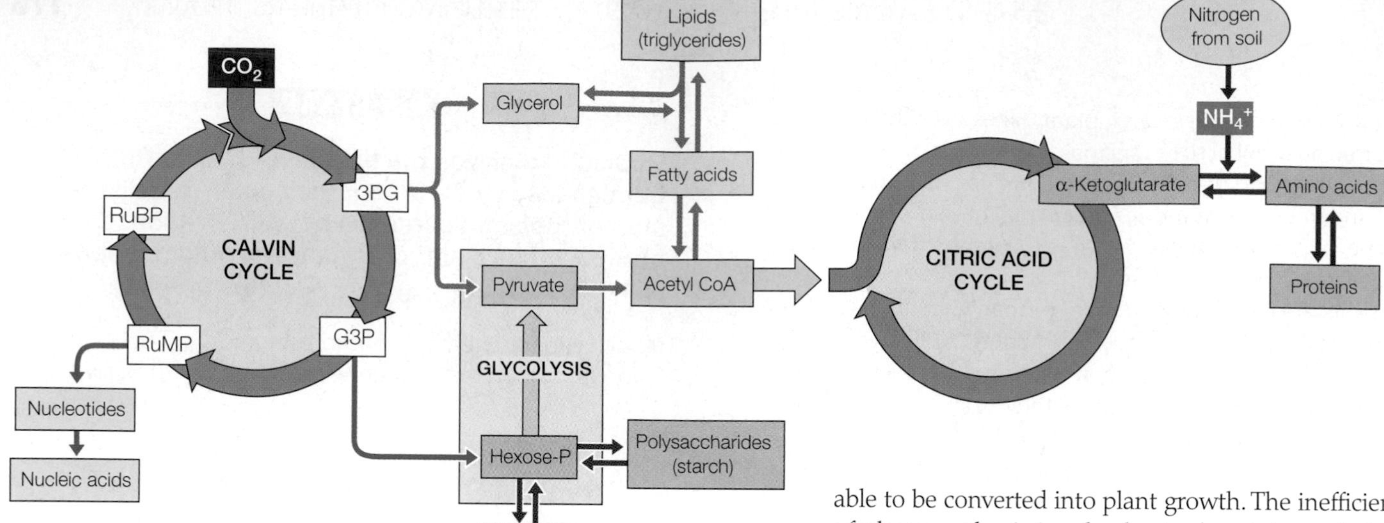

8.19 Metabolic Interactions in a Plant Cell The products of the Calvin cycle are used in the reactions of cellular respiration (glycolysis and the citric acid cycle).

energy release; that is, overall carbon fixation by photosynthesis must exceed respiration. This principle is the basis of the ecological food chain, as we will see in later chapters.

The opening of this chapter makes clear the extent to which people are dependent on photosynthesis. Given the uncertainties of the photosynthetic future (such as climate change), it would be wise to seek ways to reduce our dependence on photosynthesis or to improve photosynthetic efficiency. **Figure 8.20** shows the various ways in which solar energy is utilized and lost. In essence, only 5 percent of the sunlight that reaches Earth is available to be converted into plant growth. The inefficiencies of photosynthesis involve basic chemistry and physics (some light energy is not absorbed by photosynthetic pigments) as well as biology (plant anatomy and leaf exposure, photorespiration, inefficiencies in the metabolic pathways). While it is hard to change chemistry and physics, biologists can use their knowledge of plants to improve on the basic biology of photosynthesis. This in turn should result in more efficient use of resources and better food production.

8.5 RECAP

The products of photosynthesis are utilized in glycolysis and the citric acid cycle as well as in synthesizing lipids, proteins, and other carbohydrates.

- Do you understand how the pathways of glycolysis and the citric acid cycle are linked to photosynthesis in plant cells? See p. 175 and Figure 8.19

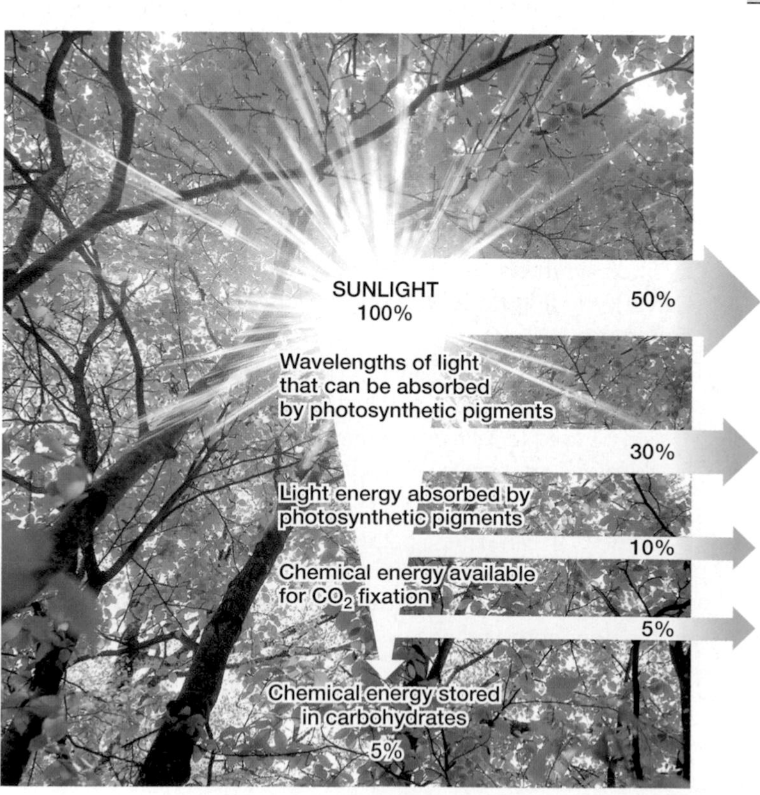

8.20 Energy Losses During Photosynthesis As we face an increasingly uncertain photosynthetic future, understanding the inefficiencies of photosynthesis becomes increasingly important. The pathways of photosynthesis preserve only about 5 percent of the sun's energy output as chemical energy in carbohydrates.

CHAPTER SUMMARY

8.1 What is photosynthesis?

In the process of **photosynthesis**, plants and other organisms take in CO_2, water, and light energy, producing O_2 and carbohydrates. See Web/CD Tutorial 8.1

The **light reactions** of photosynthesis convert light energy into chemical energy. They produce ATP and reduce $NADP^+$ to NADPH + H^+. Review Figure 8.3

The **light-independent reactions** do not use light directly but instead use ATP and NADPH + H^+ to reduce CO_2, forming carbohydrates.

8.2 How does photosynthesis convert light energy into chemical energy?

Light is a form of **electromagnetic radiation**. It is emitted in particle-like packets called **photons** but has wavelike properties.

Molecules that absorb light in the visible spectrum are called **pigments**. Photosynthetic organisms have several pigments, most notably **chlorophylls**, but also **accessory pigments** such as **carotenoids** and **phycobilins**.

Absorption of a photon puts a pigment molecule in an **excited state** that has more energy than its **ground state**. Review Figure 8.4

Each compound has a characteristic **absorption spectrum**. An **action spectrum** reflects the biological activity of a photosynthetic organism for a given wavelength of light. Review Figure 8.6

The pigments in photosynthetic organisms are arranged into **antenna systems** that absorb energy from light and funnel this energy to a single chlorophyll *a* molecule in the **reaction center**. Chlorophyll acts as a reducing agent, transferring excited electrons to other molecules. Review Figure 8.8

Noncyclic electron transport uses **photosystems I** and **II** to produces ATP, NADPH + H^+, and O_2. **Cyclic electron transport** uses only photosystem I and produces only ATP. Review Figures 8.9 and 8.10

Chemiosmosis is the mechanism of ATP production in **photophosphorylation**. Review Figure 8.11, Web/CD Tutorial 8.2

8.3 How is chemical energy used to synthesize carbohydrates?

The **Calvin cycle** makes carbohydrates from CO_2. The cycle consists of three processes: fixation of CO_2, reduction and carbohydrate production, and regeneration of RuBP. See Web/CD Tutorial 8.3

RuBP is the initial CO_2 acceptor, and **3PG** is the first stable product of CO_2 fixation. The enzyme **rubisco** catalyzes the reaction of CO_2 and RuBP to form 3PG. Review Figure 8.13, Web/CD Activity 8.1

ATP and NADPH formed by the light reactions are used in the reduction of 3PG to form G3P. Light stimulates enzymes in the Calvin cycle, further integrating the two pathways.

8.4 How do plants adapt to the inefficiencies of photosynthesis?

Rubisco can catalyze a reaction between O_2 and RuBP in addition to the reaction between CO_2 and RuBP. At high temperatures and low CO_2 concentrations, the **oxygenase** function of rubisco is favored over its **carboxylase** function.

When rubisco functions as an oxygenase, the result is **photorespiration**, which significantly reduces the efficiency of photosynthesis.

In **C_4 plants**, CO_2 is fixed to PEP in mesophyll cells. The 4-carbon product releases its CO_2 to rubisco in the interior of the leaf, at the bundle sheath cells. Review Figure 8.17, Web/CD Activity 8.2

CAM plants operate much like C_4 plants, but their initial CO_2 fixation by **PEP carboxylase** is temporally separated from the Calvin cycle, rather than spatially separated as in C_4 plants.

8.5 How is photosynthesis connected to other metabolic pathways in plants?

Photosynthesis and cellular respiration are linked through the Calvin cycle, the citric acid cycle, and glycolysis. Review Figure 8.19

To survive, a plant must photosynthesize more than it respires.

Photosynthesis utilizes only a small portion of the energy of sunlight. Review Figure 8.20

SELF-QUIZ

1. In noncyclic photosynthetic electron transport, water is used to
   a. excite chlorophyll.
   b. hydrolyze ATP.
   c. reduce chlorophyll.
   d. oxidize NADPH.
   e. synthesize chlorophyll.
2. Which statement about light is true?
   a. An absorption spectrum is a plot of biological effectiveness versus wavelength.
   b. An absorption spectrum may be a good means of identifying a pigment.
   c. Light need not be absorbed to produce a biological effect.
   d. A given kind of molecule can occupy any energy level.
   e. A pigment loses energy as it absorbs a photon.
3. Which statement about chlorophylls is *not* true?
   a. Chlorophylls absorb light near both ends of the visible spectrum.
   b. Chlorophylls can accept energy from other pigments, such as carotenoids.
   c. Excited chlorophyll can either reduce another substance or fluoresce.
   d. Excited chlorophyll may be an oxidizing agent.
   e. Chlorophylls contain magnesium.
4. In cyclic electron transport,
   a. oxygen gas is released.
   b. ATP is formed.
   c. water donates electrons and protons.
   d. NADPH + H^+ forms.
   e. CO_2 reacts with RuBP.
5. Which of the following does *not* happen in noncyclic electron transport?
   a. Oxygen gas is released.
   b. ATP forms.
   c. Water donates electrons and protons.
   d. NADPH + H^+ forms.
   e. CO_2 reacts with RuBP.

6. In the chloroplasts,
   a. light leads to the pumping of protons out of the thylakoids.
   b. ATP forms when protons are pumped into the thylakoids.
   c. light causes the stroma to become more acidic than the thylakoids.
   d. protons return passively to the stroma through protein channels.
   e. proton pumping requires ATP.
7. Which statement about the Calvin cycle is *not* true?
   a. CO_2 reacts with RuBP to form 3PG.
   b. RuBP forms by the metabolism of 3PG.
   c. ATP and NADPH + H^+ form when 3PG is reduced.
   d. The concentration of 3PG rises if the light is switched off.
   e. Rubisco catalyzes the reaction of CO_2 and RuBP.
8. In C_4 photosynthesis,
   a. 3PG is the first product of CO_2 fixation.
   b. rubisco catalyzes the first step in the pathway.
   c. 4-carbon acids are formed by PEP carboxylase in bundle sheath cells.
   d. photosynthesis continues at lower CO_2 levels than in C_3 plants.
   e. CO_2 released from RuBP is transferred to PEP.
9. Photosynthesis in green plants occurs only during the day. Respiration in plants occurs
   a. only at night.
   b. only when there is enough ATP.
   c. only during the day.
   d. all the time.
   e. in the chloroplast after photosynthesis.
10. Photorespiration
   a. takes place only in C_4 plants.
   b. includes reactions carried out in peroxisomes.
   c. increases the yield of photosynthesis.
   d. is catalyzed by PEP carboxylase.
   e. is independent of light intensity.

FOR DISCUSSION

1. Both photosynthetic electron transport and the Calvin cycle stop in the dark. Which specific reaction stops first? Which stops next? Continue answering the question "Which stops next?" until you have explained why both pathways have stopped.
2. In what principal ways are the reactions of electron transport in photosynthesis similar to the reactions of oxidative phosphorylation discussed in Section 7.4?
3. Differentiate between cyclic and noncyclic electron transport in terms of (1) the products and (2) the source of electrons for the reduction of oxidized chlorophyll.
4. What two experimental techniques made it possible to elucidate the Calvin cycle? How were these techniques used in the investigation?
5. If water labeled with ^{18}O is added to a suspension of photosynthesizing chloroplasts, which of the following compounds will first become labeled with ^{18}O: ATP, NADPH, O_2, or 3PG? If water labeled with ^{3}H is added to a suspension of photosynthesizing chloroplasts, which of the same compounds will first become radioactive? If CO_2 labeled with ^{14}C is added to a suspension of photosynthesizing chloroplasts, which of those compounds will first become radioactive?
6. The Viking lander was sent to Mars in 1976 to detect signs of life. Explain the rationale behind the following experiments this unmanned probe performed:
   a. A scoop of dirt was inserted into a container and $^{14}CO_2$ was added. After a while during the Martian day, the $^{14}CO_2$ was removed and the dirt was heated to a high temperature. Scientists monitoring the experiment back on Earth looked for the release of $^{14}CO_2$ as a sign of life.
   b. The same experiment was performed, except that the dirt was heated to a high temperature for 30 minutes and then allowed to cool to Martian temperature right after scooping and *before* the $^{14}CO_2$ was added. If experiment *a* released $^{14}CO_2$, then this experiment should not release it, if living things were present.

FOR INVESTIGATION

Calvin's experiment (see Figure 8.12) laid the foundations for a full description of the pathway for fixation of CO_2. Given the metabolic interrelationships between pathways in plants, how would you do an experiment to follow the fate of fixed carbon through photosynthesis to proteins?

PART THREE
Heredity and the Genome

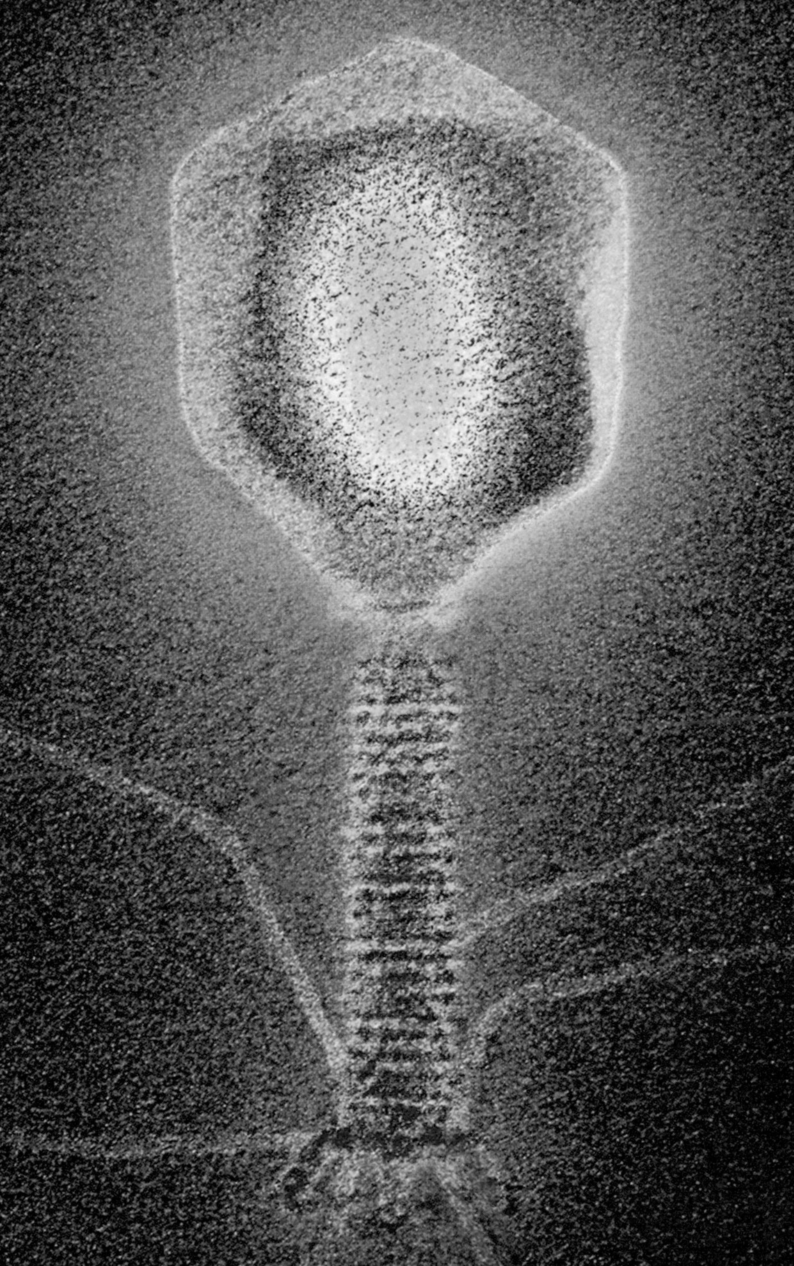

CHAPTER 9 Chromosomes, the Cell Cycle, and Cell Division

The immortal cells of Henrietta Lacks

On January 28, 1951, 31-year-old Henrietta Lacks found blood spotting her underwear. Sensing that something was wrong, the mother of five children convinced her husband to take her to nearby Johns Hopkins Hospital in Baltimore, Maryland. An examination of her cervix revealed the reason for the blood spots: a tumor the size of a quarter. Her doctor sent a piece of the tumor to a pathologist in the clinical laboratory, who confirmed that the tumor was malignant.

A week later, Henrietta was back in the hospital, where physicians treated her tumor with radium to try to kill it. Before the treatment began, however, they removed a small sample of cells from the tumor and sent them to the research laboratory of George and Margaret Gey, two scientists at Johns Hopkins who had been trying for 20 years to coax human cells to live and multiply outside the body, or in vitro. They were attempting this in the belief that if they could get human cells to thrive in vitro, they could use those cells to find a cure for cancer. They hit paydirt with Henrietta's tumor cells, which grew more vigorously than any cells the Geys had previously cultured.

Unfortunately, the tumor cells also grew rapidly in Henrietta Lacks's body. Within a few months, cancerous cells had spread to almost all of her organs. She died on October 4, 1951. On that same day, George Gey appeared on national television displaying a test tube containing her cells—which he named HeLa cells—and saying that a cure for cancer was near.

Because of their robust ability to reproduce themselves, HeLa cells became a staple of much important basic and applied biomedical research. In controlled settings, the cells could be infected with viruses, and they were instrumental in developing the supply of poliovirus that led to the first vaccine against that disease. Although Henrietta herself had never been outside of Virginia and Maryland, her cells have traveled all over the world. HeLa cells even went into space aboard the space shuttle. Over the past half-century, tens of thousands of research articles have been published using information obtained from Henrietta's cells. But the hope that HeLa cells would lead to a quick cure for cancer has proved to be unfounded.

Cancer remains the second leading cause of death (after heart disease) in most of the world's developed nations. However, Henrietta Lacks would

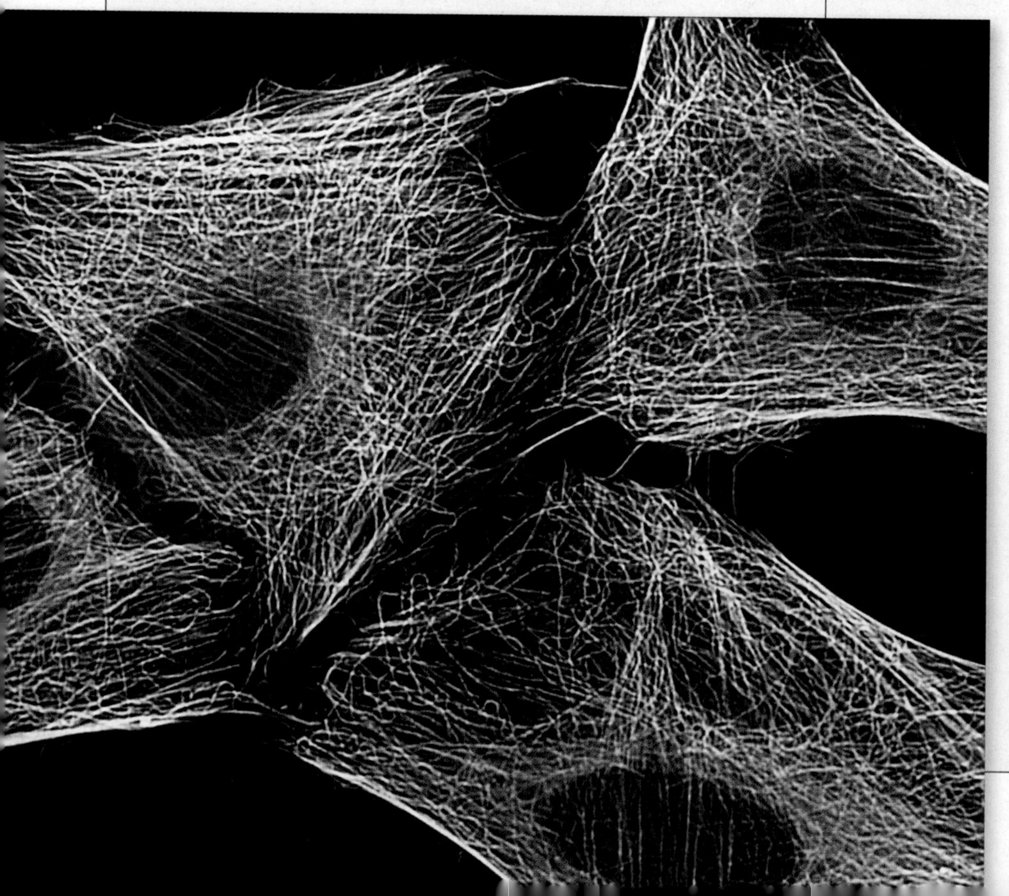

HeLa Cells These rapidly reproducing cancer cells have been cultured in many laboratories and have contributed greatly to biomedical research.

CHAPTER OUTLINE

Henrietta Lacks Mrs. Lacks, shown here in front of her home in Baltimore, Maryland, died of cancer in 1951. She left a legacy in the form of cultured cells from the tumor that killed her.

probably have lived much longer if she had had access to a simple medical test that was first used in 1941. This test, called the Pap test, can detect precancerous cells in a woman's cervix, usually allowing them to be removed before they become cancerous. In the United States, the Pap test has prevented an estimated 90 percent of deaths from cervical cancer. Had such testing been performed in time, HeLa cells would never have developed.

In normal tissues, cell division (cell "birth") is offset by cell loss (cell "death"). Unlike most normal cells, most cancer cells, including HeLa cells, keep growing because they have a genetic imbalance that heavily favors cell division over cell death. Cancer treatments using radiation or drugs aim to alter this balance in favor of cell death.

IN THIS CHAPTER we will see how cells give rise to more cells. We will describe how prokaryotic cells produce two new organisms from an original single-celled organism. Then we will describe the two types of eukaryotic cell and nuclear division—mitosis and meiosis—and relate them to asexual and sexual reproduction in eukaryotic organisms. Finally, to balance our discussion of cell proliferation through division, we will describe the important process of programmed cell death, also known as apoptosis.

9.1 How Do Prokaryotic and Eukaryotic Cells Divide?

Unicellular organisms use cell division primarily to reproduce themselves, whereas in multicellular organisms it also plays important roles in growth and in the repair of tissues (**Figure 9.1**).

In order for any cell to divide, four events must occur:

- There must be a *reproductive signal*. This signal, which may come from either inside or outside the cell, initiates cell division.
- **Replication** of DNA (the genetic material) and other vital cell components must occur so that each of the two new cells will have identical genes and complete cell functions.
- The cell must distribute the replicated DNA to each of the two new cells. This process is called **segregation**.
- New material must be added to the cell membrane (and the cell wall, in organisms that have one) in order to separate the two new cells by a process called **cytokinesis**.

These four events proceed somewhat differently in prokaryotes and eukaryotes.

Prokaryotes divide by binary fission

In prokaryotes, cell division results in the reproduction of the entire single-celled organism. The cell grows in size, replicates its DNA, and then essentially divides into two new cells, a process called **binary fission**.

REPRODUCTIVE SIGNALS The reproductive rates of many prokaryotes respond to conditions in the environment. The bacterium *Escherichia coli*, a species commonly used in genetic studies, is a "cell division machine"; essentially, it divides continuously. Typically, cell division in *E. coli* takes 40 minutes at 37°C. But if abundant sources of carbohydrates and mineral nutrients are available, the division cycle speeds up, and the cells may divide in as little as 20 minutes. Another bacterium, *Bacillus subtilis*, stops dividing when food supplies are low, then resumes dividing when conditions improve. These observations

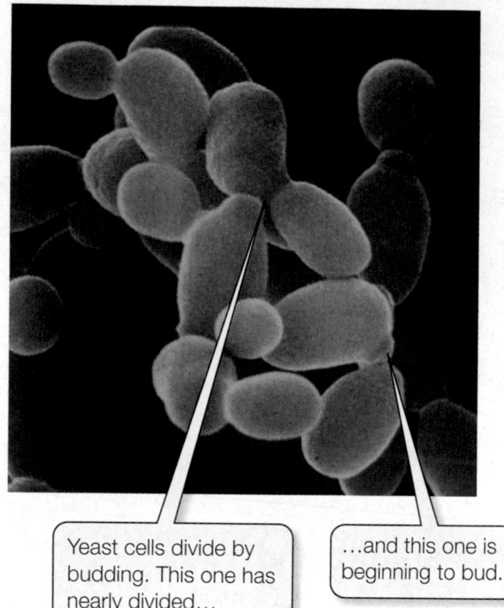

(A) Reproduction
Yeast cells divide by budding. This one has nearly divided...
...and this one is beginning to bud.

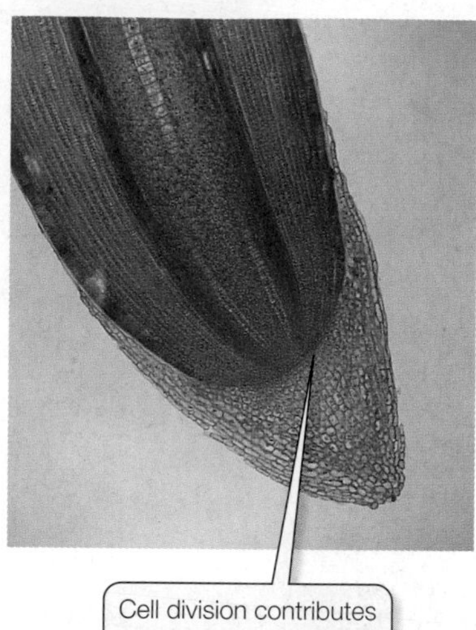

(B) Growth
Cell division contributes to the growth of this root tissue.

(C) Regeneration
Cell division contributes to the regeneration of a lizard's tail.

9.1 Important Consequences of Cell Division Cell division is the basis for (A) reproduction, (B) growth, and (C) repair and regeneration of tissues.

suggest that external factors, such as environmental conditions and nutrient concentrations, are signals for the initiation of cell division in prokaryotes.

REPLICATION OF DNA As we saw in Section 4.3, a **chromosome** is a DNA molecule containing genetic information. When a cell divides, all of its chromosomes must be replicated, and each of the two resulting copies must find its way into one of the two new cells.

Most prokaryotes have only one chromosome, a single long DNA molecule with proteins bound to it. In the bacterium *E. coli*, the DNA is a continuous molecule, often called a *circular chromosome*. Although often drawn as circles, circular chromosomes are not perfectly round. If the bacterial DNA were arranged in an actual circle, it would be about 1.6 million nm (1.6 mm) in circumference. The bacterium itself is only about 1μm (1,000 nm) in diameter and about 4 μm long. Thus if the bacterial DNA were fully extended, it would form a circle over 100 times larger than the cell! To fit into the cell, the DNA must be compacted. The DNA molecule accomplishes some of this packaging by folding in on itself. Positively charged (basic) proteins bound to negatively charged (acidic) DNA contribute to this folding. Circular chromosomes are characteristic of almost all prokaryotes, as well as some viruses, and are also found in the chloroplasts and mitochondria of eukaryotic cells.

Two regions of the prokaryotic chromosome play a functional role in cell reproduction:

- *ori*: the site where replication of the circle starts (the *ori*gin of replication)
- *ter*: the site where replication ends (the *ter*minus of replication)

Chromosome replication takes place as the DNA is threaded through a "replication complex" of proteins near the center of the cell. (These proteins include the enzyme DNA polymerase, whose important role in replication will be discussed further in Section 11.3.) During the process of prokaryotic DNA replication, the cell grows, and the two daughter DNA's are segregated from one another at opposite ends of the cell.

SEGREGATION OF DNA The two DNA's are replicated at the center of the cell and as they do so, the *ori* regions move towartd opposite ends of the cell. DNA adjacent to the *ori* region binds proteins that are essential for this segregation. This is an active process, since these binding proteins hydrolyze ATP (**Figure 9.2**). The prokaryotic cytoskeleton (see Section 4.2) may be involved in DNA segregation, either actively moving the DNA along, or passively acting as a "railroad track" along which DNA moves.

CYTOKINESIS Cell separation, or cytokinesis, begins after chromosome replication is finished. The first event of cytokinesis is a pinching in of the plasma membrane to form a ring similar to a purse string. Fibers composed of a protein similar to eukaryotic tubulin (which makes up microtubules) are major components of this ring. As the membrane pinches in, new cell wall materials are synthesized, which finally separate the two cells.

Under optimal environmental conditions, a population of *E. coli* cells doubles in size every 20 minutes. Theoretically, in about one week a single *E. coli* cell could produce a ball of bacteria the size of the Earth! Thankfully for other organisms, the *E. coli* would run out of nutrients long before that happened.

Eukaryotic cells divide by mitosis or meiosis

Many complex eukaryotes, such as humans and flowering plants, originate from a single cell, the fertilized egg. This cell derives from

9.2 Prokaryotic Cell Division (A) The process of cell division in a bacterium. (B) These two cells of the bacterium *Pseudomonas aeruginosa* have almost completed cytokinesis.

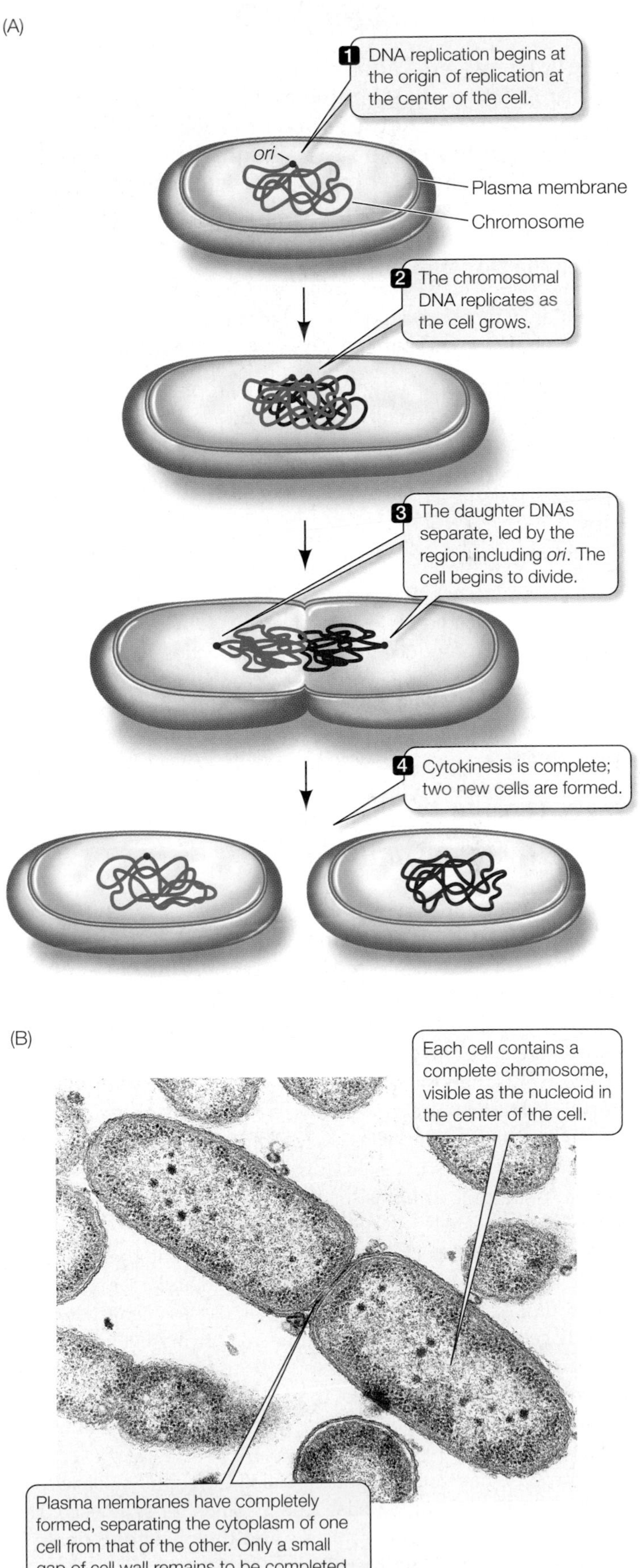

(A)
1 DNA replication begins at the origin of replication at the center of the cell.
ori
Plasma membrane
Chromosome
2 The chromosomal DNA replicates as the cell grows.
3 The daughter DNAs separate, led by the region including ori. The cell begins to divide.
4 Cytokinesis is complete; two new cells are formed.
(B)
Each cell contains a complete chromosome, visible as the nucleoid in the center of the cell.
Plasma membranes have completely formed, separating the cytoplasm of one cell from that of the other. Only a small gap of cell wall remains to be completed.

the union of two sex cells, called **gametes**, from the organism's parents—that is, a sperm and an egg—and thus contains genetic material from both parents. Specifically, the fertilized egg contains one set of chromosomes from the male parent and one set from the female parent.

The formation of a multicellular organism from a fertilized egg is called *development*. Development involves both cell reproduction and cell specialization. For example, an adult human has several trillion cells, all ultimately derived from a fertilized egg, yet many of those cells have specialized roles. How these cells become specialized for different functions is the subject of Chapter 43; here our focus is on cell reproduction.

As in prokaryotes, cell reproduction in eukaryotes entails reproductive signals, DNA replication, segregation, and cytokinesis. The details, however, are quite different:

- Unlike prokaryotes, eukaryotic cells do not constantly divide whenever environmental conditions are adequate. In fact, eukaryotic cells that are part of a multicellular organism and have become specialized seldom divide. In a eukaryotic organism, the signals for cell division are related not to the environment of a single cell, but to the needs of the entire organism.
- While most prokaryotes have a single main chromosome, eukaryotes usually have many (humans have 46), so the processes of replication and segregation, while basically the same as in prokaryotes, are more intricate. In eukaryotes, the newly replicated chromosomes are closely associated with each other (they are thus known as *sister chromatids*), and a different mechanism, called **mitosis**, is used to segregate them into two new nuclei.
- Eukaryotic cells have a distinct nucleus, which has to be divided into two new nuclei, each containing an identical set of chromosomes. Thus, in eukaryotes, cytokinesis is distinct from segregation of the genetic material and can happen only after duplication of the entire nucleus.
- Cytokinesis proceeds differently in plant cells (which have a cell wall) than in animal cells (which do not).

A second mechanism of nuclear division, **meiosis**, occurs only in cells that produce the gametes involved in sexual reproduction. That is, meiosis occurs only in cells that produce the sperm and eggs that will be contributed to a new organism. While the two products of mitosis are genetically identical to the cell that produced them—they both have the same DNA—the products of meiosis are not. As we will see in Section 9.5, meiosis generates diversity by shuffling the genetic material, resulting in new gene combinations. Meiosis plays a key role in sexual life cycles.

9.1 RECAP

Four events are required for cell division: a reproductive signal, replication of the genetic material (DNA), segregation of replicated DNA, and separation of the two daughter cells (cytokinesis). In prokaryotes, cell division is rapid; in eukaryotes, the process is more involved, and the nucleus must be duplicated before cell division can occur.

- Can you describe the type of reproductive signal that leads the bacterium *Bacillus subtilis* to divide? See p. 181
- Can you explain why DNA must be replicated and segregated *before* a cell can divide? See p. 182
- Do you understand the difference between cell division in prokaryotes (binary fission) and mitosis in eukaryotes? See pp. 182–183

What determines whether a cell will divide? How does mitosis lead to identical cells, and meiosis to diversity? Why do most eukaryotic organisms reproduce sexually? In the sections that follow, we will describe the details of the two eukaryotic cell division processes, mitosis, and meiosis, and their roles in heredity, development, and evolution.

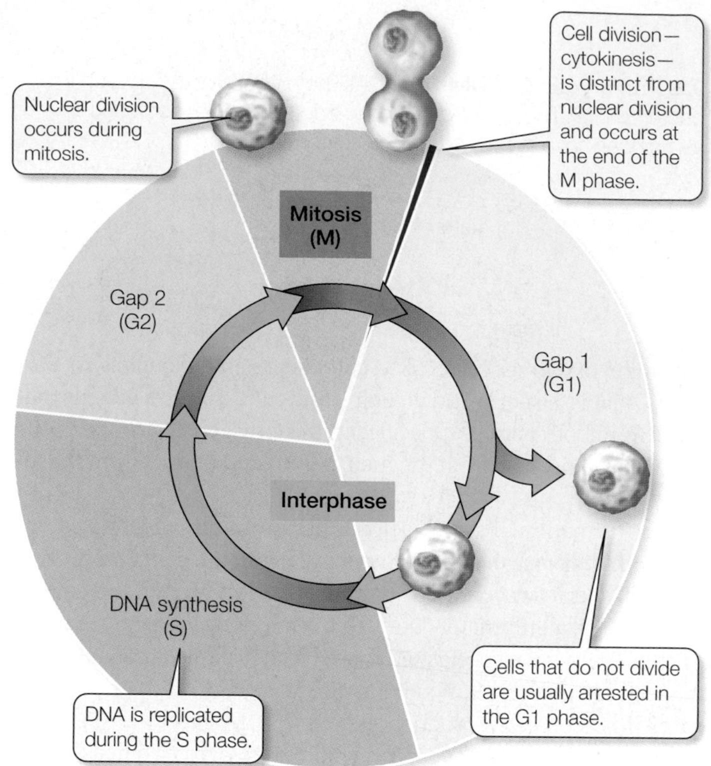

9.3 The Eukaryotic Cell Cycle The cell cycle consists of a mitotic (M) phase, during which mitosis and cytokinesis take place, and a long period of growth known as interphase. Interphase has three subphases (G1, S, and G2) in cells that divide.

9.2 How Is Eukaryotic Cell Division Controlled?

A cell lives and functions until it divides or dies. Or, if it is a gamete (egg or sperm), it lives until it fuses with another gamete. Some types of cells, such as red blood cells, lose the capacity to divide as they mature. Other cell types, such as cortical cells in plant stems, divide only rarely. Some cells, like the cells in a developing embryo, are specialized for rapid division.

The events that occur to produce two eukaryotic cells from one are referred to as the **cell cycle**. Between divisions—that is, for most of its life—a eukaryotic cell is in a condition called **interphase**. For most types of eukaryotic cells, the cell cycle has two phases: mitosis and interphase. In this section we will describe the events of interphase, especially those that trigger mitosis.

A given cell lives for one turn of the cell cycle and then becomes two cells. The cell cycle, when repeated again and again, is a constant source of new cells. However, even in tissues engaged in rapid growth, cells spend most of their time in interphase. Examination of any collection of dividing cells, such as the tip of a root or a slice of liver, will reveal that most of the cells are in interphase most of the time; only a small percentage of the cells will be in mitosis at any given moment.

Interphase has three subphases, called G1, S, and G2. The cell's DNA replicates during **S phase** (the S stands for synthesis). The period between the end of mitosis and the onset of S phase is called **G1**, or Gap 1. Another gap phase—**G2**—separates the end of S phase and the beginning of mitosis. Mitosis and cytokinesis are referred to as the **M phase** of the cell cycle (**Figure 9.3**).

Let's look at the events of interphase in more detail:

- *G1 phase.* During G1, a cell is preparing for S phase, so at this stage each chromosome is a single, unreplicated structure. G1 is quite variable in length in different cell types. Some rapidly dividing embryonic cells dispense with it entirely, while other cells may remain in G1 for weeks or even years. In many cases, these cells enter a resting phase, called G0. Special internal and external signals are needed to prompt a cell to leave G0 and reenter the cell cycle at G1.
- *The G1-to-S transition.* It is at the G1-to-S transition that the commitment to cell division (and thus another cell cycle) is made.
- *S phase.* During S phase, the process of DNA replication, which we will describe in detail in Section 11.3, is completed. Where there was formerly one chromosome, there are now two sister chromatids joined together and awaiting segregation into two new cells by mitosis or meiosis.
- *G2 phase.* During G2, the cell makes preparations for mitosis—for example, by synthesizing components of the microtubules that will move the chromatids to opposite ends of the dividing cell.

Cyclins and other proteins trigger events in the cell cycle

How are appropriate decisions to enter the S or M phases made? A first indication that there were substances that control these tran-

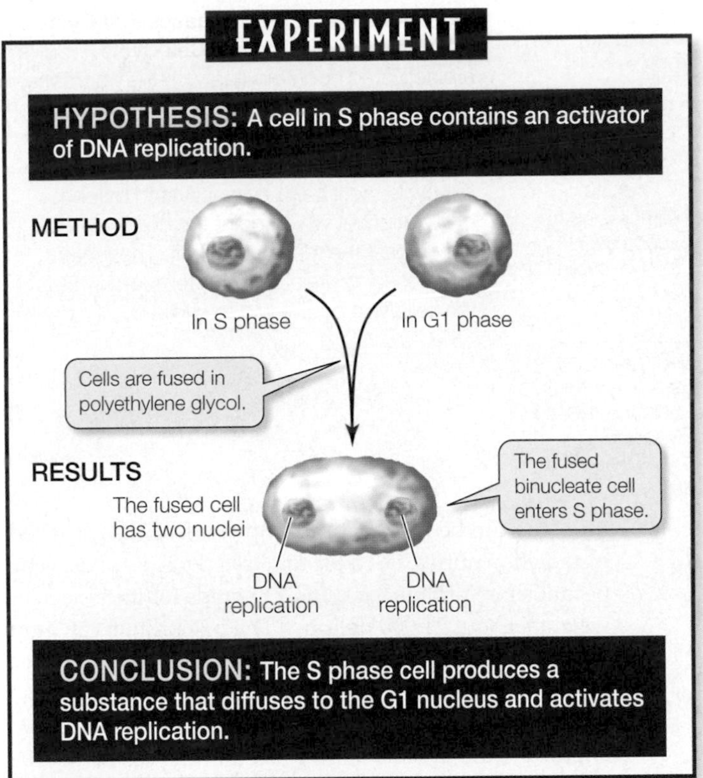

EXPERIMENT
HYPOTHESIS: A cell in S phase contains an activator of DNA replication.
METHOD
In S phase
In G1 phase
Cells are fused in polyethylene glycol.
RESULTS
The fused cell has two nuclei
The fused binucleate cell enters S phase.
DNA replication
DNA replication
CONCLUSION: The S phase cell produces a substance that diffuses to the G1 nucleus and activates DNA replication.

9.4 Regulation of the Cell Cycle Cells can be induced to fuse by certain substances (such as the sugar alcohol polyethylene glycol) that disaggregate plasma membranes. Such fusion initially produces a binucleate cell. If an S phase cell is fused with an early G1 cell, the latter is stimulated by the former to enter S phase. FURTHER RESEARCH: How would you use this method to show that a cell in M phase produces an activator of mitosis?

sitions came from experiments using *cell fusion.* Fusing mammalian cells at different phases of the cell cycle showed that a cell in S phase produces a substance that activates DNA replication (**Figure 9.4**). Similar experiments point to a molecular activator for entry into M phase.

These transitions—from G1 to S and from G2 to M—depend on the activation of a type of protein called **cyclin-dependent kinase**, or **Cdk**. Recall from Section 7.2 that a *kinase* is an enzyme that catalyzes the transfer of a phosphate group from ATP to another molecule; this phosphate transfer is called *phosphorylation.*

$$\text{protein} + \text{ATP} \xrightarrow{\text{kinase}} \text{protein—P} + \text{ADP}$$

What does phosphorylation do to a protein? As discussed in Section 3.2, proteins have both hydrophilic regions (which tend to interact with water on the outside of the protein macromolecule) and hydrophobic regions (which tend to interact with one another on the inside of the macromolecule). These regions are important in giving a protein its three-dimensional shape. Phosphate groups are charged, so an amino acid with such a group tends to be on the outside of the protein. Thus phosphorylation changes the shape and function of a protein by changing its charges.

By catalyzing the phosphorylation of certain target proteins, Cdk's play important roles in initiating the steps of the cell cycle. The discovery that Cdk's induce cell division is a beautiful example of how research on different organisms and different cell types can converge on a single mechanism. One group of scientists, led by James Maller at the University of Colorado, was studying immature sea urchin eggs, trying to find out how they are stimulated to divide and form the precursor cells of the mature egg. A protein called *maturation promoting factor* was purified from maturing eggs, which by itself prodded immature eggs into division. Meanwhile, at the University of Washington, Leland Hartwell, who was studying the cell cycle in yeast (a single-celled eukaryote), found a strain that was stalled at the G1–S boundary because it lacked a Cdk. This yeast Cdk was discovered to be very similar to the sea urchin's maturation promoting factor. Similar Cdk's were soon found to control the G1-to-S transition in many other organisms, including humans.

Cdk's are not active by themselves. Binding to a second type of protein, called **cyclin**, activates a Cdk. This binding—an example of allosteric regulation (see Section 6.5)—activates the Cdk by altering its shape and exposing its active site (**Figure 9.5**). It is the cyclin–Cdk complex that acts as a protein kinase and triggers

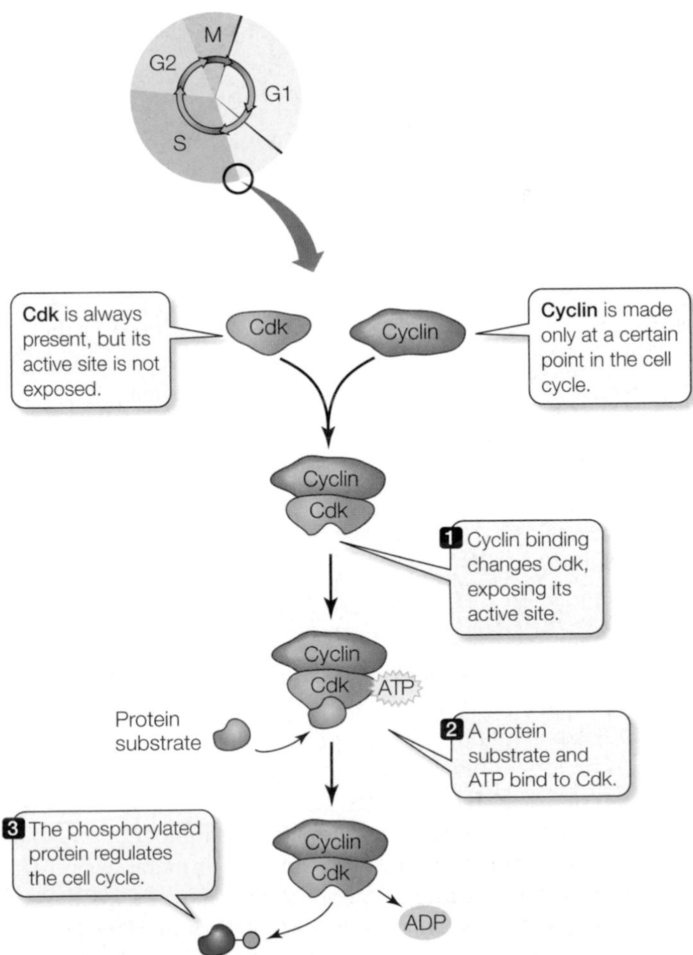

Cdk is always present, but its active site is not exposed.

Cyclin is made only at a certain point in the cell cycle.

Protein substrate
2 A protein substrate and ATP bind to Cdk.

9.5 Cyclin Binding Activates Cdk Binding of a cyclin changes the three-dimensional structure of an inactive Cdk, making it a protein kinase.

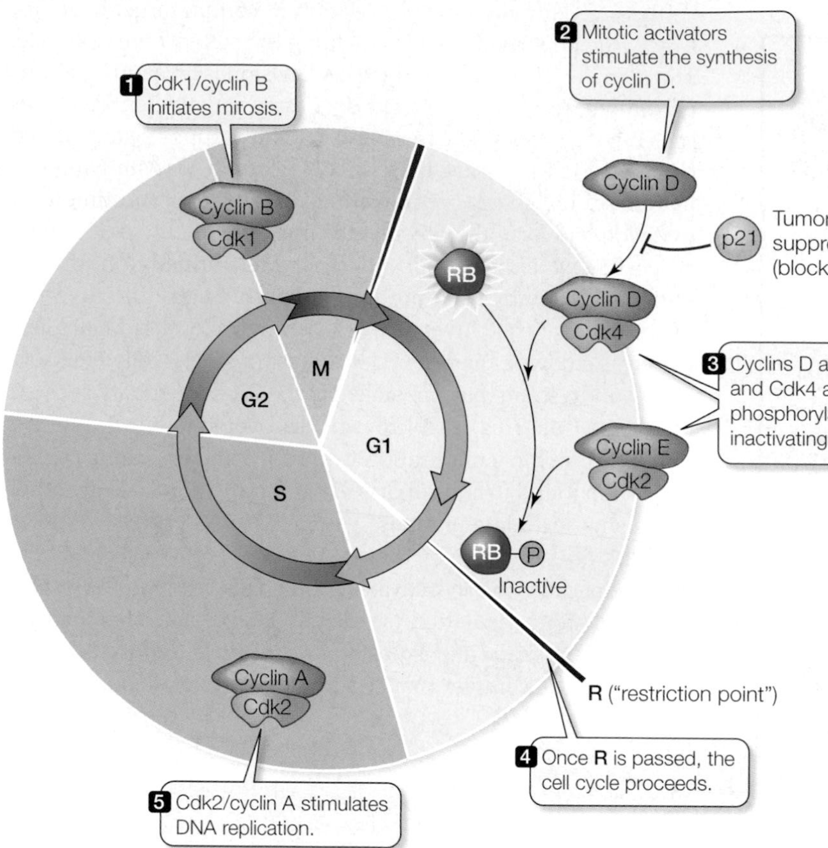
1 Cdk1/cyclin B initiates mitosis.
2 Mitotic activators stimulate the synthesis of cyclin D.
Cyclin D
Cyclin B
Cdk1
p21
Tumor suppressor (blocks cyclin D)
RB
Cyclin D
Cdk4
3 Cyclins D and E and Cdk4 and Cdk2 phosphorylate RB, inactivating it.
M
G2
G1
S
Cyclin E
Cdk2
RB
P
Inactive
Cyclin A
Cdk2
R ("restriction point")
4 Once R is passed, the cell cycle proceeds.
5 Cdk2/cyclin A stimulates DNA replication.

9.6 Cyclin-Dependent Kinases and Cyclins Trigger Transitions in the Cell Cycle Cdk's are activated upon binding with the appropriate cyclin. There are four cyclin/Cdk controls during the typical cell cycle in humans. RB is a tumor suppressor that has the potential to inhibit the cell cycle but it can be deactivated by phosphorylation, at which point the cell cycle proceeds past the restriction point. The tumor suppressor protein p21 can temporarily stop the cell cycle by binding to cyclin D.

the transition from G1 to S phase. Then cyclin breaks down and the Cdk becomes inactive.

Several different cyclin–Cdk combinations act at various stages of the mammalian cell cycle (**Figure 9.6**):

- Cyclin D–Cdk4 acts during the middle of G1. It moves the cell past the **restriction point** (**R**), a key decision point beyond which the rest of the cell cycle is normally inevitable.
- Cyclin E–Cdk2 also acts in the middle of G1; it works in concert with Cyclin D–Cdk4 to move the cell cycle past the restriction point.
- Cyclin A–Cdk2 acts during the S phase to stimulate DNA replication.
- Cyclin B–Cdk1 acts at the G2–M boundary, initiating the transition to mitosis.

The key to progress past the restriction point is a protein called **RB** (**retinoblastoma protein**). RB normally inhibits the cell cycle. But when RB is phosphorylated by a protein kinase, it becomes inactive and no longer blocks the restriction point, and the cell progresses past G1 into S phase. (Note the double negative here—a cell function happens because an inhibitor is inhibited. This phenomenon is common in the control of cellular metabolism.) The enzymes that catalyze RB phosphorylation are Cdk4 and Cdk2. So what is needed for a cell to pass the restriction point is the synthesis of cyclins D and E, which activate Cdk4 and Cdk2, which phosphorylate RB, which becomes inactivated.

The cyclin–Cdk complexes act as *checkpoints*, points at which a cell cycle's progress can be monitored to determine whether the next step can be taken. For example, if DNA is damaged by radiation during G1, a protein called p21 is made. (The p stands for "protein" and the 21 stands for its molecular weight—about 21,000 daltons.) The p21 protein can bind to the two G1 Cdk's, preventing their activation by cyclins. So the cell cycle stops while repairs are made to DNA. The p21 protein breaks down after the DNA is repaired, allowing the cyclins to bind to Cdk's so that the cell cycle proceeds. There are checkpoints at several other points in the cell cycle. For example, at the end of S phase, there is a checkpoint for complete DNA replication; if it is not complete, the cell cycle stops before mitosis.

Because cancer results from inappropriate cell division, it is not surprising that these cyclin–Cdk controls are disrupted in cancer cells. For example, some fast-growing breast cancers have too much cyclin D, which overstimulates Cdk4 and thus cell division. For example, a protein called p53 prevents normal cells from dividing by stimulating the synthesis of p21 and thereby inhibiting Cdk's. More than half of all human cancers contain defective p53, resulting in the absence of cell cycle controls. Proteins such as p53, p21, and RB that normally block the cell cycle are known as *tumor suppressors*.

The RB protein was named for retinoblastoma, a tumor that arises from uncontrolled cell division in the embryonic cells that produce the retina of the eye. This cancer affects about one infant in 20,000. Fortunately, the tumor can be removed, although the individual can lose sight in the affected eye.

Growth factors can stimulate cells to divide

Cyclin–Cdk complexes provide cells with an internal control on their progress through the cell cycle. Not all cells in an organism go through the cell cycle on a regular basis. Some cells either no longer go through the cell cycle, or go through it slowly and divide infrequently. If such cells are to divide, they must be stimulated by external chemical signals called **growth factors**. For example, when you cut yourself and bleed, specialized cell fragments called *platelets* gather at the wound to initiate blood clotting. The platelets produce and release a protein called *platelet-derived growth factor* that diffuses to the adjacent cells in the skin and stimulates them to divide and heal the wound.

Other growth factors include *interleukins*, which are made by one type of white blood cell and promote cell division in other cells that are essential for the body's immune system defenses. *Erythropoietin*, made by the kidney, stimulates the division of bone marrow cells and the production of red blood cells. In addition, many hormones promote division in specific cell types.

We will describe the physiological roles of growth factors in later chapters, but all of them act in a similar way. They bind to their target cells via specialized receptor proteins on the target cell surface. This specific binding triggers events within the target cell that initiate the cell cycle. Cancer cells often divide inappropriately because they make their own growth factors, or because they no longer require growth factors to start cycling.

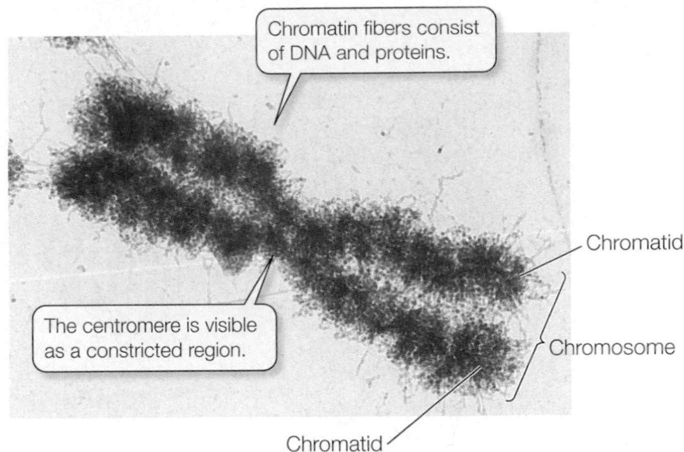

9.7 Chromosomes, Chromatids, and Chromatin A human chromosome in metaphase, shown as the cell prepares to divide.

9.2 RECAP

The eukaryotic cell cycle is under both external and internal controls. A series of kinases, themselves under control by cyclins, controls the eukaryotic cell cycle. External signals such as growth factors can initiate the cell cycle.

- Can you draw a cell cycle diagram showing the various phases of interphase? See p. 184 and Figure 9.3
- Do you understand how cyclins and RB protein control the progress of the cell cycle? See pp. 185–186 and Figure 9.6
- Do you understand the differences between external and internal controls of the cell cycle? See pp. 186–187

Having outlined the events of the eukaryotic cell cycle, now let's take a closer look at the events of the M phase—mitosis.

9.3 What Happens during Mitosis?

The third essential step in the process of cell division— segregation of the replicated DNA—occurs during mitosis. Segregation is accomplished by packaging the huge DNA molecules and their associated proteins into very compact chromosomes. After segregation by mitosis, cytokinesis separates the two cells. Let's now look at these steps more closely.

Eukaryotic DNA is packed into very compact chromosomes

The eukaryotic chromosome shown in **Figure 9.7** consists of two gigantic, linear, double-stranded molecules of DNA complexed with many proteins to form a dense material called **chromatin**. Before S phase, each chromosome contains only one such double-stranded DNA molecule. After the DNA molecule replicates during S phase, however, there are two double-stranded DNA molecules, known as **sister chromatids**. The sister chromatids are held together along most of their length by a protein complex called **cohesin**. They stay this way until mitosis, when most of the cohesin is removed, except in a region called the **centromere** at which the chromatids remain held together (see Figures 9.7 and 9.11). After replication, a second group of proteins called **condensins** coats the DNA molecules and makes them more compact.

If all the DNA in a typical human cell were put end to end, it would be nearly 2 meters long. Yet the nucleus is only 5 μm (0.000005 meters) in diameter. So, although the DNA in an interphase nucleus is "unwound," it is still impressively packed, as illustrated in **Figure 9.8**. This packing is achieved largely by proteins associated closely with the chromosomal DNA.

Chromosomes contain large quantities of proteins called **histones** (*histos*, "web" or "loom"). There are five classes of histones. All of them have a positive charge at cellular pH levels because of their high content of the basic amino acids lysine and arginine. These positive charges attract the negative phosphate groups on DNA. These DNA-histone interactions, as well as histone-histone interactions, result in the formation of beadlike units called **nucleosomes**. Each nucleosome contains the following components:

- Eight histone molecules, two each of four of the histone classes, united to form a core or spool
- 146 base pairs of DNA, 1.65 turns of it wound around the histone core
- Histone H1 (the remaining histone class) on the outside of the DNA, which may clamp it to the histone core

During interphase, a chromosome consists of a single DNA molecule running around vast numbers of nucleosomes like beads on a string. Between the nucleosomes stretches a variable amount of non-nucleosomal "linker" DNA. During this time, the DNA is exposed to the nuclear environment, so it is accessible to proteins involved in its replication and in the regulation of its expression, as we will see in Chapter 14.

During both mitosis and meiosis, the chromatin becomes ever more tightly coiled and condensed as the nucleosomes pack together. Further coiling of the chromatin continues up to the time at which the chromatids begin to move apart.

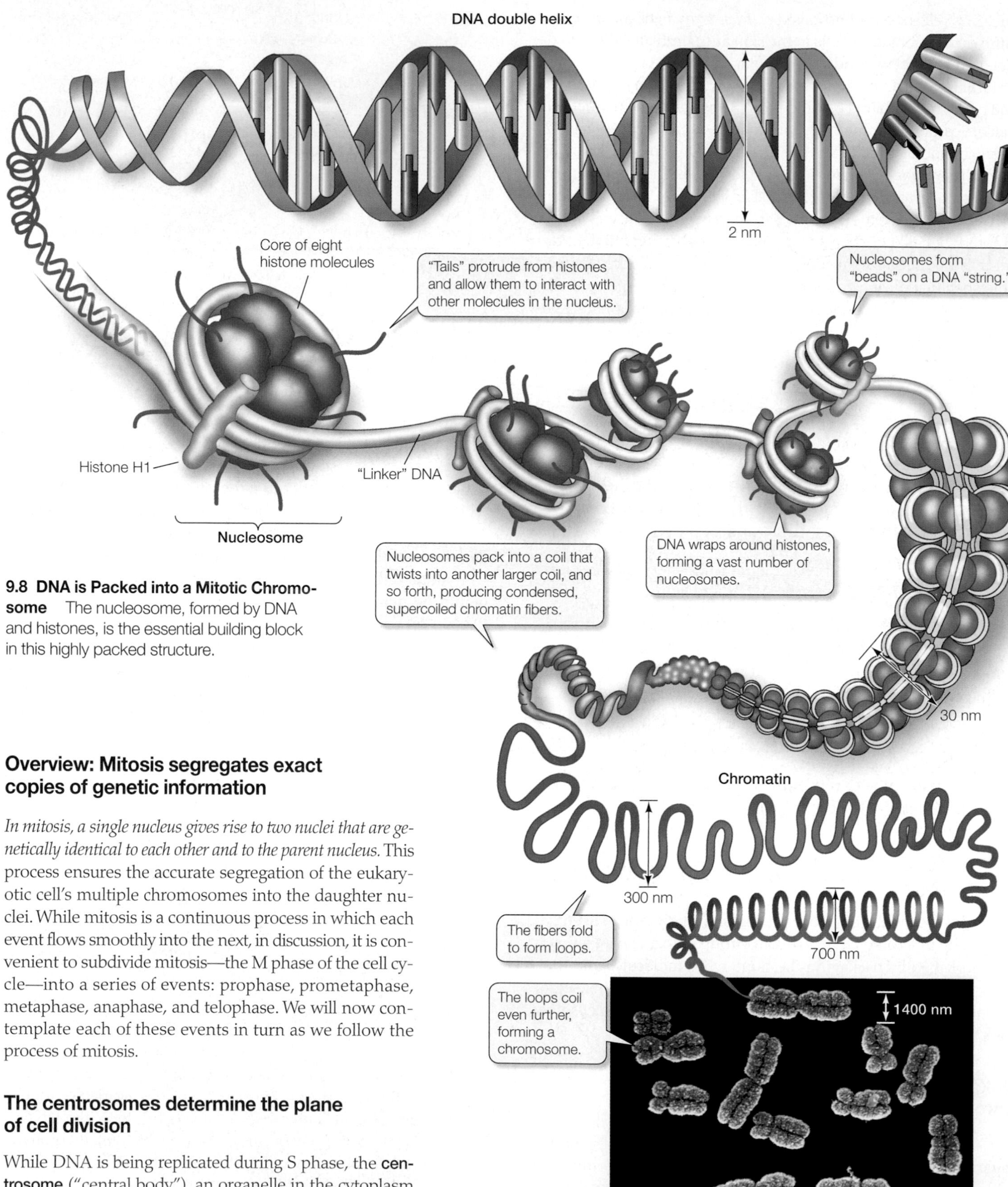

DNA double helix
2 nm
Core of eight histone molecules
"Tails" protrude from histones and allow them to interact with other molecules in the nucleus.
Nucleosomes form "beads" on a DNA "string."
Histone H1
"Linker" DNA
Nucleosome
Nucleosomes pack into a coil that twists into another larger coil, and so forth, producing condensed, supercoiled chromatin fibers.
DNA wraps around histones, forming a vast number of nucleosomes.
30 nm
Chromatin
300 nm
The fibers fold to form loops.
700 nm
The loops coil even further, forming a chromosome.
1400 nm
Metaphase chromosomes

9.8 DNA is Packed into a Mitotic Chromosome The nucleosome, formed by DNA and histones, is the essential building block in this highly packed structure.

Overview: Mitosis segregates exact copies of genetic information

In mitosis, a single nucleus gives rise to two nuclei that are genetically identical to each other and to the parent nucleus. This process ensures the accurate segregation of the eukaryotic cell's multiple chromosomes into the daughter nuclei. While mitosis is a continuous process in which each event flows smoothly into the next, in discussion, it is convenient to subdivide mitosis—the M phase of the cell cycle—into a series of events: prophase, prometaphase, metaphase, anaphase, and telophase. We will now contemplate each of these events in turn as we follow the process of mitosis.

The centrosomes determine the plane of cell division

While DNA is being replicated during S phase, the **centrosome** ("central body"), an organelle in the cytoplasm near the nucleus, doubles, forming a pair of centrosomes. In many organisms, each centrosome consists of a pair of **centrioles**, each one a hollow tube lined with nine microtubules. The two tubes are at right angles to each other.

At the G2-to-M transition, the two centrosomes separate from each other, moving to opposite ends of the nuclear envelope. The orientation of the centrosomes determines the plane at which the cell will divide, and therefore the spatial relationship of the two new cells to the parent cell. This relationship may be of little consequence to single free-living cells such as yeasts, but it is important for cells that make up part of a body tissue.

Surrounding the centrioles is a high concentration of tubulin dimers, and these proteins initiate the formation of microtubules, which will orchestrate chromosomal movement. (In plant cells, which lack centrosomes, a distinct microtubule organizing center at either end of the cell plays the same role.) The formation of microtubules will lead to the formation of the spindle structure that is required for the orderly segregation of the chromosomes.

Chromatids become visible and the spindle forms during prophase

During interphase, only the nuclear envelope, the nucleoli, and a barely discernible tangle of chromatin are visible under the light microscope. The appearance of the nucleus changes as the cell enters **prophase**—the beginning of mitosis. Most of the cohesin that has held the two products of DNA replication together since S phase is removed, so the individual chromatids become visible. They are still held together by a small amount of cohesin at the centromere. Late in prophase, specialized three-layered structures called **kinetochores** develop in the centromere region, one on each chromatid. These structures will be important in chromosome movements.

Each of the two centrosomes serves as a *mitotic center*, or *pole*, toward which the chromosomes will move (**Figure 9.9A**). Microtubules form between each pole and the chromosomes to make up a **spindle**, which serves both as a structure to which the chromosomes will attach and as a framework keeping the two poles apart. The spindle is actually two half-spindles: each microtubule runs from one pole to the middle of the spindle, where it overlaps with microtubules extending from the other half-spindle. The microtubules are initially unstable, constantly forming and falling apart, until they contact microtubules from the other half-spindle and become more stable.

There are two types of microtubules in the spindle:

- *Polar microtubules* are the microtubules we have just described; they form the framework of the spindle. They have abundant tubulin around the centrioles. Tubulin dimers aggregate to form long fibers that extend into the middle region of the cell.
- *Kinetochore microtubules*, which form later, attach to the kinetochores on the chromosomes. The two sister chromatids in each chromosome pair will become attached by their kinetochore to kinetochore microtubules in opposite halves of the spindle (**Figure 9.9B**). This ensures that one chromatid of the pair will eventually move to one pole while the other chromatid moves to the opposite pole. Movement of the chromatids is the central feature and accomplishment of mitosis.

Chromosome movements are highly organized

It is during the next three phases of mitosis—prometaphase, metaphase, and anaphase—that the chromosomes move (**Figure 9.10**). During these phases, the centromeres holding the two chromatids together separate, and the former sister chromatids move away from each other in opposite directions.

PROMETAPHASE **Prometaphase** is marked by the disappearance of the nuclear envelope and the nucleoli. The material that composes them remains in the cytoplasm, however, to be reassembled when the daughter nuclei form. In prometaphase, the chromosomes begin to move toward the poles, but this movement is counteracted by two factors:

- A repulsive force from the poles pushes the chromosomes toward the middle region, or **equatorial plate** (metaphase plate), of the cell.

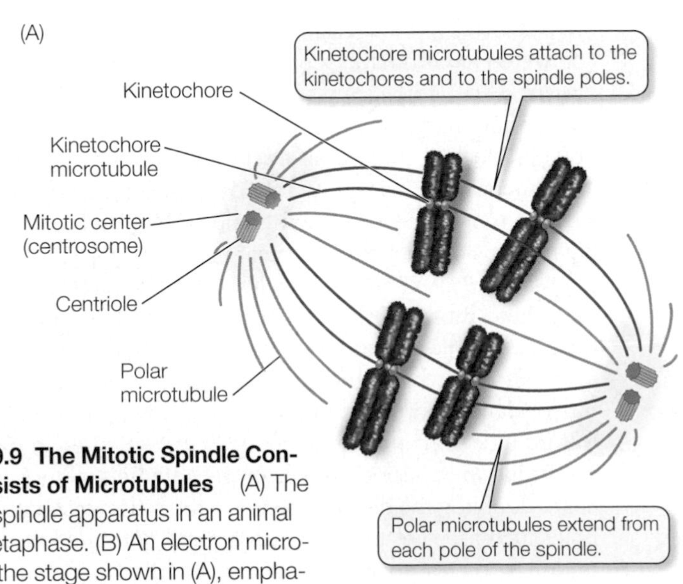
(A)
Kinetochore microtubules attach to the kinetochores and to the spindle poles.
Kinetochore
Kinetochore microtubule
Mitotic center (centrosome)
Centriole
Polar microtubule
Polar microtubules extend from each pole of the spindle.

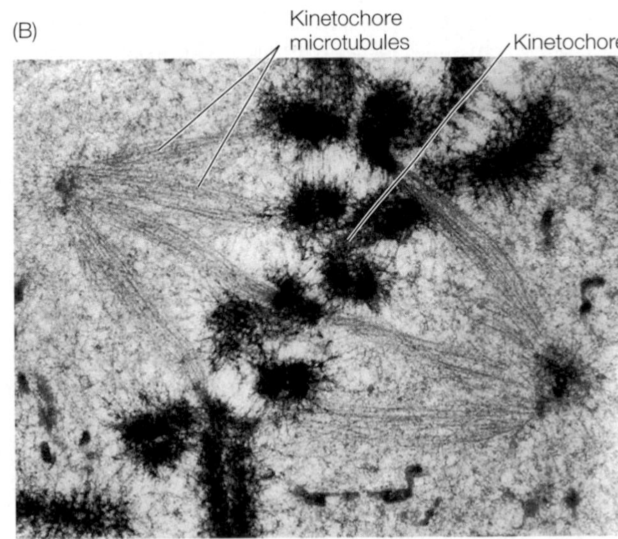
(B)
Kinetochore microtubules
Kinetochore

9.9 The Mitotic Spindle Consists of Microtubules (A) The spindle apparatus in an animal cell at metaphase. (B) An electron micrograph of the stage shown in (A), emphasizing the kinetochore microtubules.

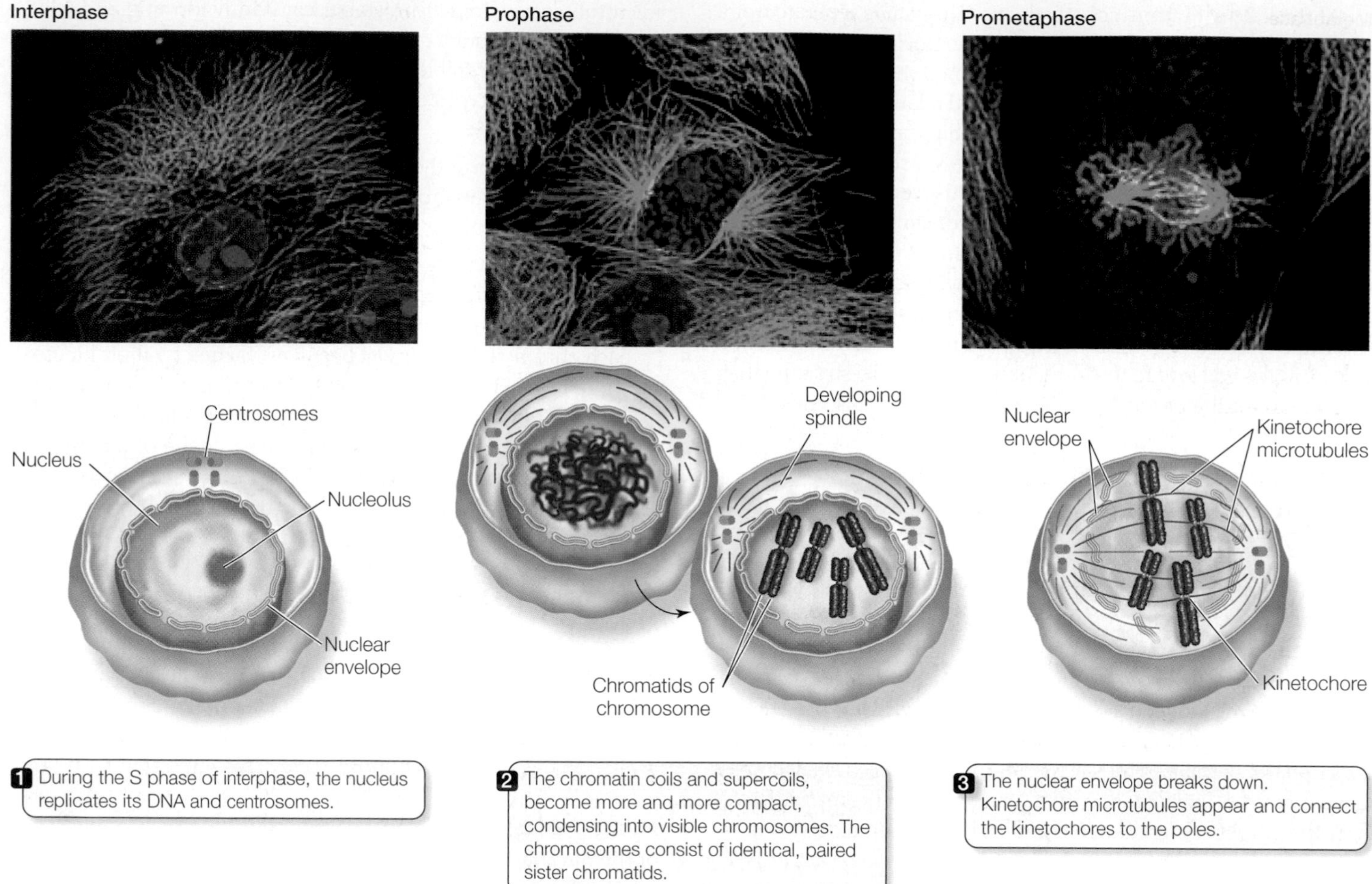

Interphase
Prophase
Prometaphase
Centrosomes
Nucleus
Nucleolus
Nuclear envelope
Developing spindle
Chromatids of chromosome
Nuclear envelope
Kinetochore microtubules
Kinetochore
1 During the S phase of interphase, the nucleus replicates its DNA and centrosomes.
2 The chromatin coils and supercoils, become more and more compact, condensing into visible chromosomes. The chromosomes consist of identical, paired sister chromatids.
3 The nuclear envelope breaks down. Kinetochore microtubules appear and connect the kinetochores to the poles.

9.10 Mitosis Mitosis results in two new nuclei that are genetically identical to each other and to the nucleus from which they were formed. In the micrographs, the green dye stains microtubules (and thus the spindle); the red dye stains the chromosomes. The chromosomes in the diagrams are stylized to emphasize the fates of the individual chromatids.

- The two chromatids are still held together at the centromere by cohesin.

Thus, during prometaphase, the chromosomes appear to move aimlessly back and forth between the poles and the middle of the spindle. Gradually, the centromeres approach the equatorial plate.

METAPHASE The cell is said to be in **metaphase** when all the centromeres arrive at the equatorial plate. Metaphase is the best time to see the sizes and shapes of chromosomes because they are maximally condensed. The chromatids are now clearly connected to one pole or the other by microtubules. At the end of metaphase, all of the chromatid pairs separate simultaneously.

ANAPHASE Separation of the chromatids marks the beginning of **anaphase**, during which the two sister chromatids move to opposite ends of the spindle. Each chromatid contains one double-stranded DNA molecule and is now referred to as a **daughter chromosome**.

This separation occurs because the cohesin holding the sister chromatids together is hydrolyzed by a specific protease, appropriately called *separase*. Until this point, separase has been present but inactive, because it has been bound to an inhibitory subunit called *securin*. Once all the chromatids are connected to the spindle, securin is hydrolyzed, allowing separase to catalyze cohesin breakdown (**Figure 9.11**). In this way, chromosome alignment is connected to chromatid separation. This process, called the *spindle checkpoint*, apparently senses whether there are any kinetochores still unattached to the spindle. If there are, securin breakdown is blocked, and the sister chromatids stay together.

What propels this highly organized mass migration? Two things seem to move the chromosomes along. First, at the kinetochores are proteins that act as "molecular motors." These proteins, called *cytoplasmic dynein*, have the ability to hydrolyze ATP to ADP and phosphate, thus releasing energy to move the chromosomes along the microtubules toward the poles. These motor proteins account for about 75 percent of the force of motion. Second, the kinetochore microtubules shorten from the poles, drawing the chromosomes toward them, accounting for about 25 percent of the motions.

During anaphase the poles of the spindle are pushed farther apart, doubling the distance between them. The distance between poles increases because the overlapping polar microtubules extending from opposite ends of the spindle contain motor proteins that cause them to slide past each other, in much the same

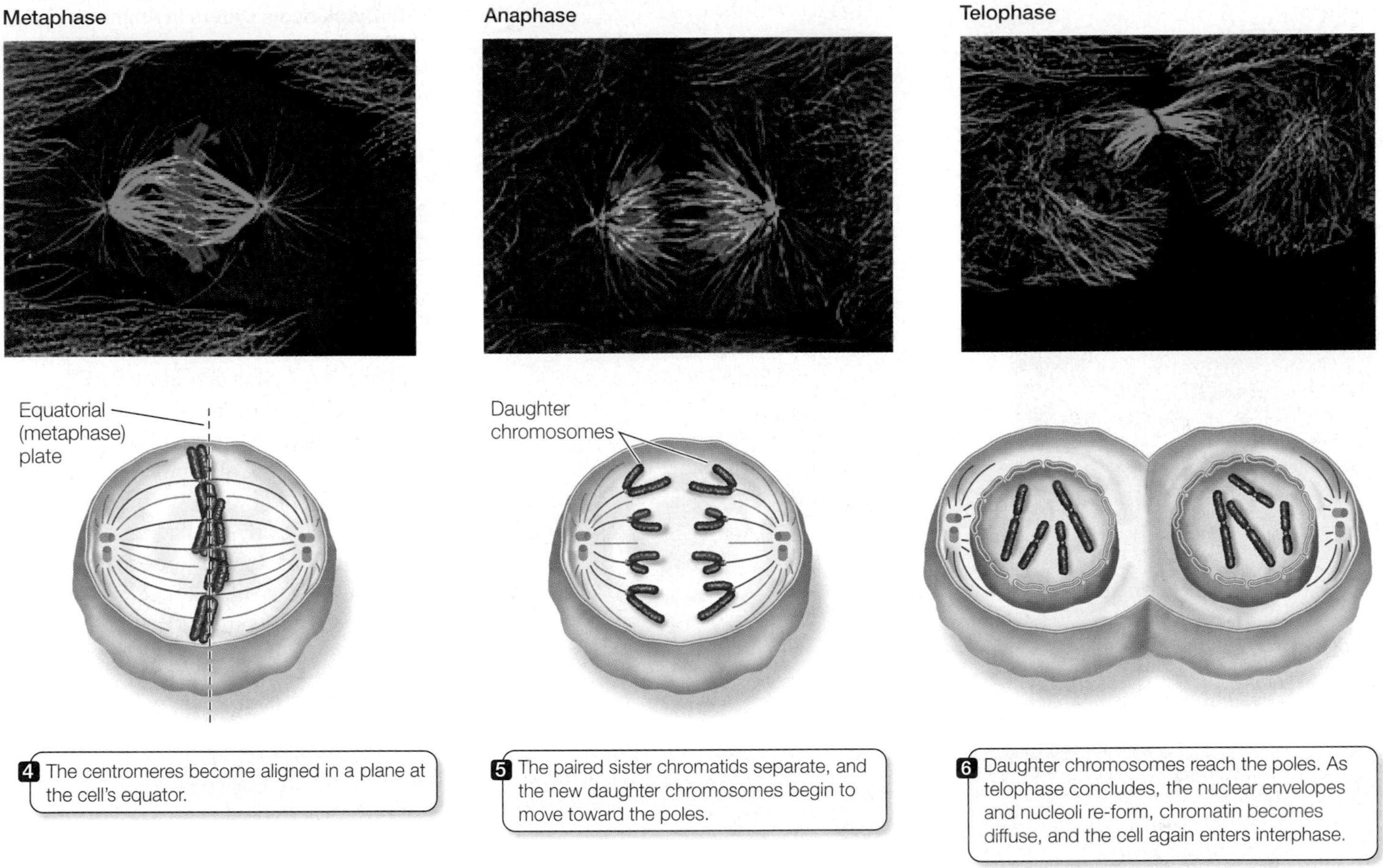

Metaphase

Equatorial (metaphase) plate

4 The centromeres become aligned in a plane at the cell's equator.

Anaphase

Daughter chromosomes

5 The paired sister chromatids separate, and the new daughter chromosomes begin to move toward the poles.

Telophase

6 Daughter chromosomes reach the poles. As telophase concludes, the nuclear envelopes and nucleoli re-form, chromatin becomes diffuse, and the cell again enters interphase.

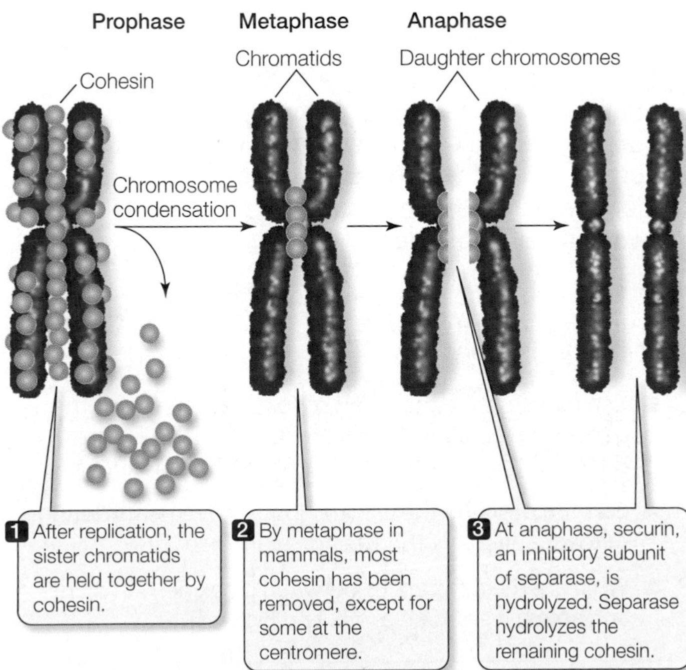

Prophase
Metaphase
Anaphase
Cohesin
Chromatids
Daughter chromosomes
Chromosome condensation
1 After replication, the sister chromatids are held together by cohesin.
2 By metaphase in mammals, most cohesin has been removed, except for some at the centromere.
3 At anaphase, securin, an inhibitory subunit of separase, is hydrolyzed. Separase hydrolyzes the remaining cohesin.

9.11 Chromatid Attachment and Separation Cohesin protein holds sister chromatids together. The enzyme separase hydrolyzes cohesin at the onset of anaphase, allowing the chromatids to separate into daughter chromosomes.

way that microtubules slide in cilia and flagella (see Figure 4.23A). This polar separation further separates one set of daughter chromosomes from the other.

Chromosomes move slowly, even in cellular terms. At about 1 μm per minute, it takes them 10–60 minutes to complete their journey to the poles—roughly equivalent to a person taking 9 million years to travel across the United States. This slow speed may ensure that the chromosomes segregate accurately.

Nuclei re-form during telophase

When the chromosomes stop moving at the end of anaphase, the cell enters **telophase**. Two sets of chromosomes (formerly referred to as daughter chromosomes), containing identical DNA and thus carrying identical sets of hereditary instructions, are now at the opposite ends of the spindle, which begins to break down. The chromosomes uncoil until they are once again the diffuse tangle of chromatin that is characteristic of interphase. The nuclear envelopes and nucleoli, which were disaggregated during prophase, coalesce and re-form their respective structures. When these and other changes are complete, telophase—and mitosis—is at an end, and each of the daughter nuclei enters another interphase.

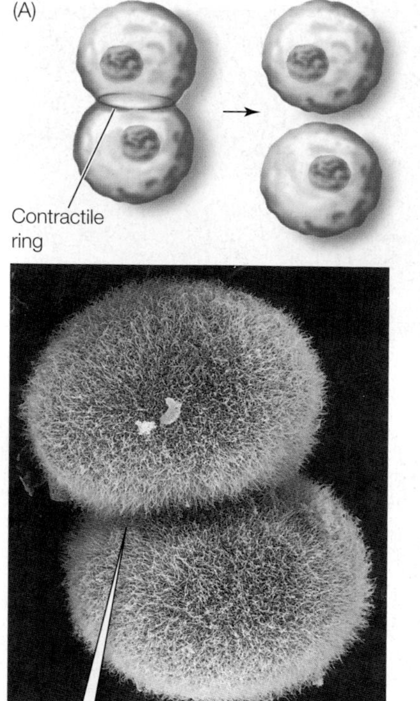

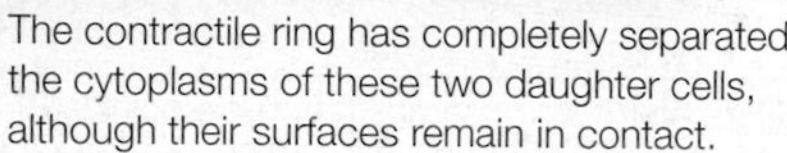

The contractile ring has completely separated the cytoplasms of these two daughter cells, although their surfaces remain in contact.

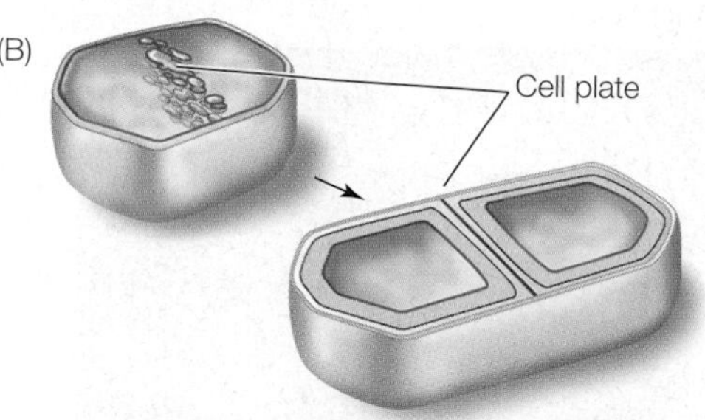

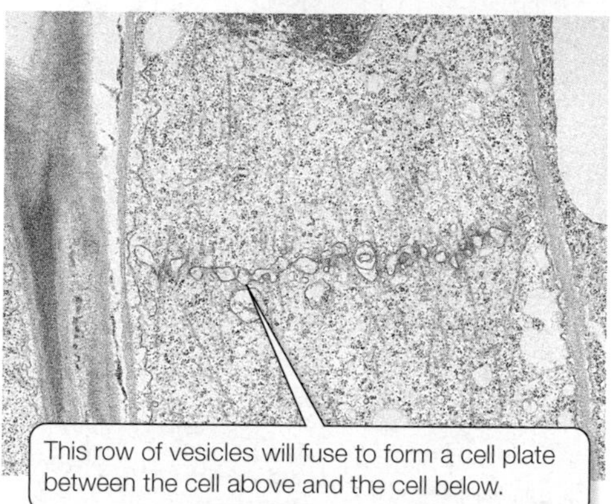

This row of vesicles will fuse to form a cell plate between the cell above and the cell below.

9.12 Cytokinesis Differs in Animal and Plant Cells Plant cells must divide differently from animal cells because they have cell walls. (A) A sea urchin zygote (fertilized egg) that has just completed cytokinesis at the end of the first cell division of its development into an embryo. (B) A dividing plant cell in late telophase. The resulting two cells will be separated not by space, but by a solid cell wall.

Mitosis is beautifully precise. Its result is two nuclei that are identical to each other and to the parent nucleus in chromosomal makeup, and hence in genetic constitution. Next, the two nuclei must be isolated in separate cells, which requires the division of the cytoplasm.

Cytokinesis is the division of the cytoplasm

Mitosis refers only to the division of the nucleus. The division of the cell's cytoplasm, which follows mitosis, is accomplished by cytokinesis. In different organisms, cytokinesis may be accomplished in different ways. The differences between the process in plants and in animals are substantial.

Animal cells usually divide by a furrowing of the plasma membrane, as if an invisible thread were cinching the cytoplasm between the two poles (**Figure 9.12A**). The "invisible thread" is actually microfilaments of actin and myosin (see Figure 4.20A) in what is called a *contractile ring* just beneath the plasma membrane. These two proteins interact to produce a contraction, just as they do in muscles, thus pinching the cell in two. These microfilaments assemble rapidly from actin monomers that are present in the interphase cytoskeleton. Their assembly appears to be under the control of calcium ions released from storage sites in the center of the cell.

Plant cell cytoplasm divides differently because plants have cell walls. In plant cells, as the spindle breaks down after mitosis, membranous vesicles derived from the Golgi apparatus appear at the equatorial plate, roughly midway between the two daughter nuclei. Propelled along microtubules by the motor protein kinesin, these vesicles fuse to form new plasma membrane and contribute their contents to a *cell plate*, which is the beginning of a new cell wall (**Figure 9.12B**).

Following cytokinesis, both daughter cells contain all the components of a complete cell. A precise distribution of chromosomes is ensured by mitosis. In contrast, organelles such as ribosomes, mitochondria, and chloroplasts need not be distributed equally between daughter cells as long as some of each are present in both cells; accordingly, there is no mechanism with a precision comparable to that of mitosis to provide for their equal allocation to daughter cells. As we will see in Chapter 43, the unequal distribution of cytoplasmic components during development can have functional significance for the two new cells.

9.3 RECAP

Mitosis is the division of the nucleus of a eukaryotic cell into two nuclei identical to each other and to the parent nucleus. The process of mitosis, while continuous, can be viewed as a series of events (prophase, prometaphase, metaphase, anaphase, and telophase) during which sister chromatids are separated into daughter chromosomes, around which nuclei re-form. Once two identical nuclei have formed, the cell divides into two cells by cytokinesis.

- What is the difference between a chromosome, a chromatid, and a daughter chromosome? See p. 187 and Figures 9.7 and 9.11
- Can you describe the levels of "packing" by which the genetic information contained in linear DNA is condensed into chromosomes? See p. 187 and Figure 9.8
- Can you describe the chromosome movements during mitosis? See pp. 189–190 and Figure 9.10
- What are the differences between cytokinesis in plant and animal cells? See p. 192 and Figure 9.12

The intricate process of mitosis results in two cells that are genetically identical. But, as mentioned earlier, there is another eukaryotic cell division process, called meiosis, that results in genetic diversity. What is the role of that process?

9.4 What Is the Role of Cell Division in Sexual Life Cycles?

The mitotic cell cycle repeats itself over and over. By this process, a single cell can give rise to a vast number of other cells. Meiosis, on the other hand, produces just four daughter cells, which may not undergo further duplications. Mitosis and meiosis are both involved in reproduction, but they have different reproductive roles.

Reproduction by mitosis results in genetic constancy

Asexual reproduction, sometimes called *vegetative reproduction*, is based on mitotic division of the nucleus. A cell undergoing mitosis may be an entire single-celled organism reproducing itself with each cell cycle, or it may be a cell in a multicellular organism that breaks off a piece to produce a new multicellular organism. Some multicellular organisms can reproduce themselves by releasing cells derived from mitosis and cytokinesis or by having a multicellular piece break away and grow on its own (**Figure 9.13**). In asexual reproduction, offspring are **clones** of the parent organism; that is, the offspring are *genetically identical* to the parent. If there is any variation among the offspring, it is likely to be due to *mutations*, or changes, in the genetic material. Asexual reproduction is a rapid and effective means of making new individuals, and it is common in nature.

Reproduction by meiosis results in genetic diversity

Unlike asexual reproduction, **sexual reproduction** results in an organism that is not identical to the parent organism. Sexual reproduction requires gametes created by meiosis; two parents each contribute one gamete to each of their offspring. Meiosis produces gametes—and thus offspring—that differ genetically not only from each parent, but from one another as well. Because of this genetic variation, some offspring may be better adapted than others to survive and reproduce in a particular environment. Meiosis thus generates the genetic diversity that is the raw material for natural selection and evolution.

In most multicellular organisms, **somatic cells**—those body cells that are *not* specialized for reproduction—each contain two sets of chromosomes, which are found in pairs. One chromosome of each pair comes from each of the organism's two parents. The members of such a **homologous pair** are similar in size and appearance (except for the sex chromosomes found in some species, as we will see in Section 10.4). The two chromosomes (the **homologs**) of a homologous pair bear corresponding, though generally not identical, genetic information.

Gametes, on the other hand, contain only a single set of chromosomes—that is, one homolog from each pair. The number of chromosomes in a gamete is denoted by n, and the cell is said to be **haploid**. Two haploid gametes fuse to form a new organism, the **zygote**, in a process called **fertilization**. The zygote thus has two sets of chromosomes, just as somatic cells do. Its chromosome number is denoted by $2n$, and the zygote is said to be **diploid**.

All sexual life cycles have certain hallmarks:

- There are two parents, each of which provides chromosomes to the offspring in the form of a gamete produced by meiosis.
- Each gamete is haploid; that is, it contains a single set of chromosomes.
- The two gametes—often identifiable as a female egg and a male sperm—fuse to produce a single cell, the zygote, or fertilized egg. The zygote thus contains two sets of chromosomes (is diploid).

(A)

9.13 Asexual Reproduction (A) Some cactus, such as this cholla, have brittle stems that break off easily. Fragments on the ground set down roots and mitotically grow an entire new plant that is genetically identical to the plant it came from. (B) These spool-shaped cells are asexual spores formed by a fungus. Each spore contains a nucleus produced by a mitotic division. A spore is genetically identical to the parent that fragmented to produce it.

(B)

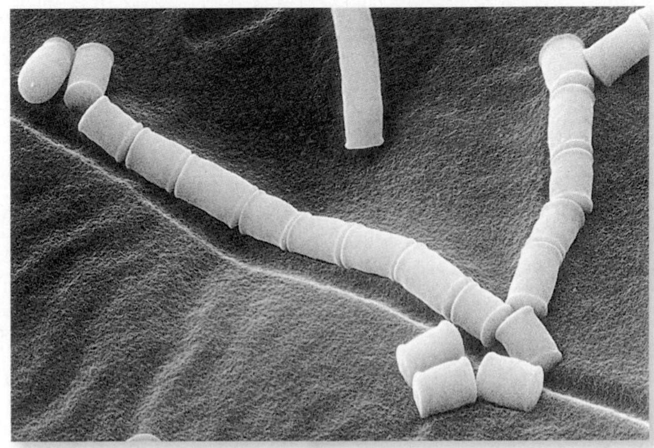

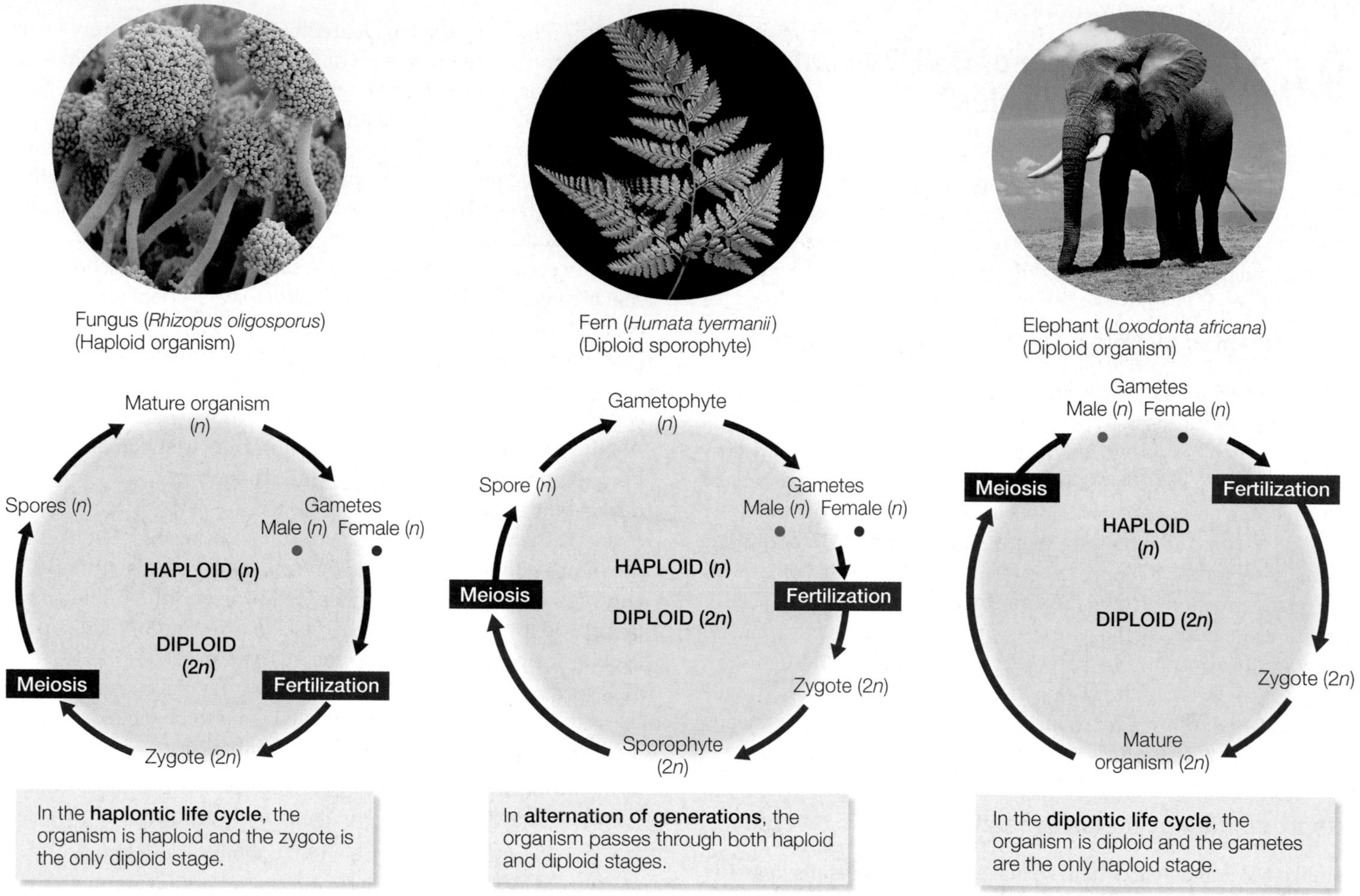

Fungus (*Rhizopus oligosporus*)
(Haploid organism)

Mature organism (n)
Spores (n)
Gametes Male (n) Female (n)
HAPLOID (n)
DIPLOID ($2n$)
Meiosis
Fertilization
Zygote ($2n$)

In the **haplontic life cycle**, the organism is haploid and the zygote is the only diploid stage.

Fern (*Humata tyermanii*)
(Diploid sporophyte)

Gametophyte (n)
Spore (n)
Gametes Male (n) Female (n)
HAPLOID (n)
DIPLOID ($2n$)
Meiosis
Fertilization
Zygote ($2n$)
Sporophyte ($2n$)

In **alternation of generations**, the organism passes through both haploid and diploid stages.

Elephant (*Loxodonta africana*)
(Diploid organism)

Gametes Male (n) Female (n)
Meiosis
Fertilization
HAPLOID (n)
DIPLOID ($2n$)
Zygote ($2n$)
Mature organism ($2n$)

In the **diplontic life cycle**, the organism is diploid and the gametes are the only haploid stage.

9.14 Fertilization and Meiosis Alternate in Sexual Reproduction In sexual reproduction, haploid (n) cells or organisms alternate with diploid ($2n$) cells or organisms.

As shown in **Figure 9.14**, after zygote formation, different kinds of sexual life cycles are observed:

- In **haplontic** organisms, such as most protists and many fungi, the tiny zygote is the only diploid cell in the life cycle; the mature organism is haploid. The zygote undergoes meiosis to produce haploid cells, or **spores**. These spores form the new organism, which may be single-celled or multicellular, by mitosis. The mature haploid organism produces gametes by mitosis, which fuse to form the diploid zygote.
- Most plants and some protists display **alternation of generations**, in which meiosis does not give rise to gametes, but to haploid spores. The spores divide by mitosis to form an alternate, haploid life stage (the *gametophyte*). It is this haploid stage that forms gametes by mitosis. The gametes fuse to form a diploid zygote, which divides by mitosis to become the diploid *sporophyte*.
- In **diplontic** organisms, which include animals and some plants, the gametes are the only haploid cells in the life cycle, and the mature organism is diploid. Gametes are formed by meiosis, which fuse to form a diploid zygote. The zygote divides by mitosis to form the mature organism.

These life cycles are described in greater detail in Part Six. In this chapter we will focus on the role of sexual reproduction in generating diversity among individual organisms.

TABLE 9.1

Numbers of Pairs of Chromosomes in Some Plant and Animal Species

COMMON NAME	SPECIES	NUMBER OF CHROMOSOME PAIRS
Mosquito	*Culex pipiens*	3
Housefly	*Musca domestica*	6
Toad	*Bufo americanus*	11
Rice	*Oryza sativa*	12
Frog	*Rana pipiens*	13
Alligator	*Alligator mississippiensis*	16
Rhesus monkey	*Macaca mulatta*	21
Wheat	*Triticum aestivum*	21
Human	*Homo sapiens*	23
Potato	*Solanum tuberosum*	24
Donkey	*Equus asinus*	31
Horse	*Equus caballus*	32
Dog	*Canis familiaris*	39
Carp	*Cyprinus carpio*	52

(A)

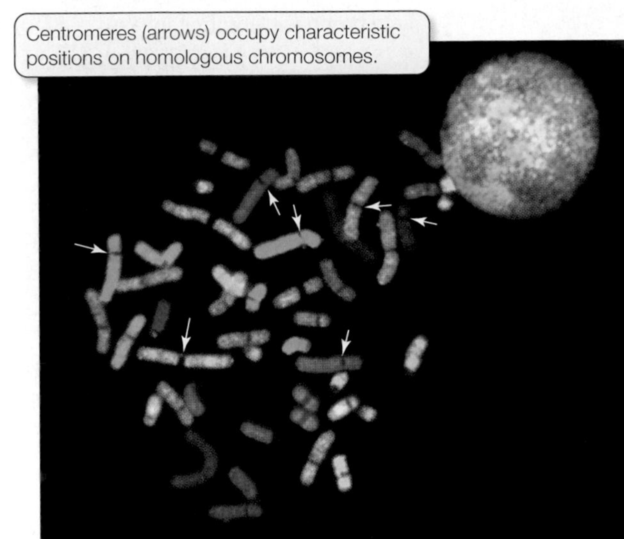

Centromeres (arrows) occupy characteristic positions on homologous chromosomes.

(B)

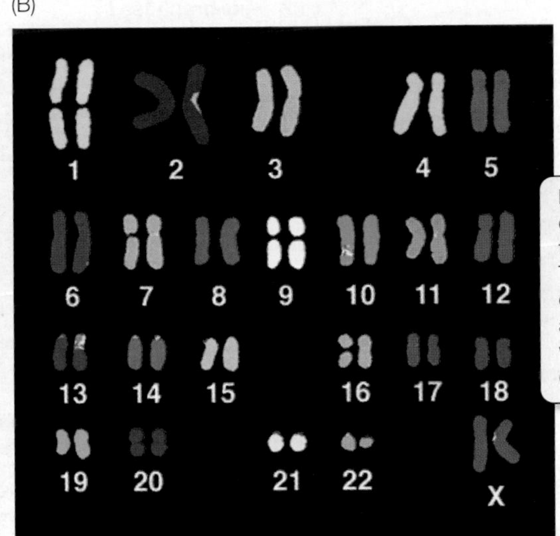

9.15 The Human Karyotype (A) Chromosomes from a human cell in metaphase of mitosis. The DNA of each chromosome has a specific nucleotide sequence that is stained by a specific colored dye, so that each homologous pair shares a distinctive color. Each chromosome at this stage is composed of two chromatids, but they cannot be distinguished by this "chromosome painting" technique. The multicolored globe is an interphase nucleus. (B) This image, produced by computerized analysis of the image on the left, with homologous pairs lined up together and numbered, clearly reveals the human karyotype.

The essence of sexual reproduction is the random selection of half of a parent's diploid chromosome set to make a haploid gamete, followed by the fusion of two such haploid gametes to produce a diploid cell containing genetic information from both gametes. Both of these steps contribute to a shuffling of genetic information in the population, so that no two individuals have exactly the same genetic constitution. The diversity provided by sexual reproduction opens up enormous opportunities for evolution.

The number, shapes, and sizes of the metaphase chromosomes constitute the karyotype

When cells are in metaphase of mitosis, it is often possible to count and characterize their individual chromosomes. This is a relatively simple process in some organisms, thanks to techniques that can capture cells in metaphase and spread out their chromosomes. A photomicrograph of the entire set of chromosomes can then be made, and the images of the individual chromosomes can be placed in an orderly arrangement. Such a rearranged photomicrograph reveals the number, shapes, and sizes of the chromosomes in a cell, which together constitute its **karyotype** (**Figure 9.15**).

Individual chromosomes can be recognized by their lengths, the positions of their centromeres, and characteristic banding that is visible when they are stained and observed at high magnification. When the cell is diploid, the karyotype consists of homologous pairs of chromosomes—23 pairs for a total of 46 chromosomes in humans, and greater or smaller numbers of pairs in other diploid species. There is no simple relationship between the size of an organism and its chromosome number (**Table 9.1**).

9.4 RECAP

Meiosis is necessary for sexual reproduction, in which haploid gametes fuse to produce a diploid zygote. Sexual reproduction results in genetic diversity, the foundation of evolution.

- What is the difference, in terms of genetics, between asexual and sexual reproduction? See p. 193
- Can you describe how fertilization produces a diploid organism? See p. 193
- What general features do all sexual life cycles have in common? See p. 194 and Figure 9.14

Meiosis, unlike mitosis, results in daughter cells that are genetically different from, and have only half as many chromosomes as, the parent cell. How are these changes accomplished?

9.5 What Happens When a Cell Undergoes Meiosis?

Meiosis consists of *two* nuclear divisions that reduce the number of chromosomes to the haploid number in preparation for sexual reproduction. Although the *nucleus divides twice* during meiosis, the *DNA is replicated only once*. Unlike the products of mitosis, the products of meiosis are different both from one another and from the parent cell. To understand the process of meiosis and its specific details, it is useful to keep in mind the overall functions of meiosis:

- To reduce the chromosome number from diploid to haploid
- To ensure that each of the haploid products has a complete set of chromosomes
- To promote genetic diversity among the products

Throughout this discussion, refer to **Figure 9.16** to help you visualize each step.

MEIOSIS I

Early prophase I

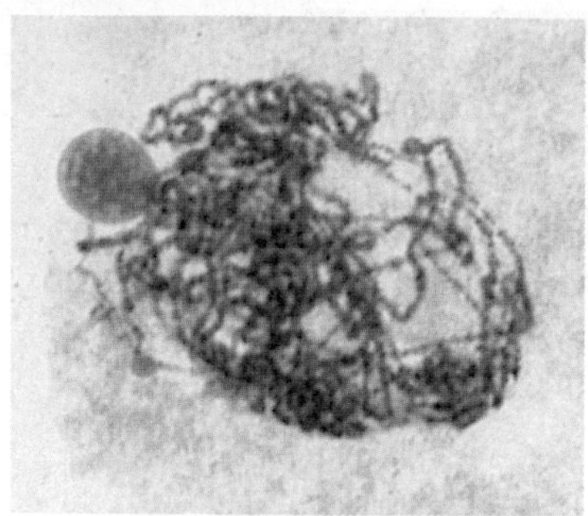

1 The chromatin begins to condense following interphase.

Mid-prophase I

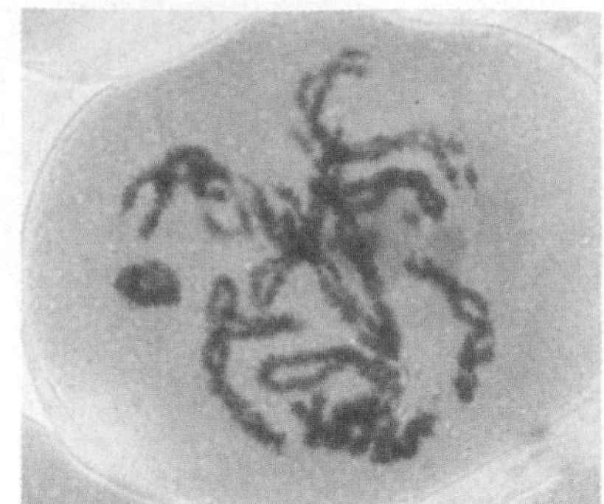

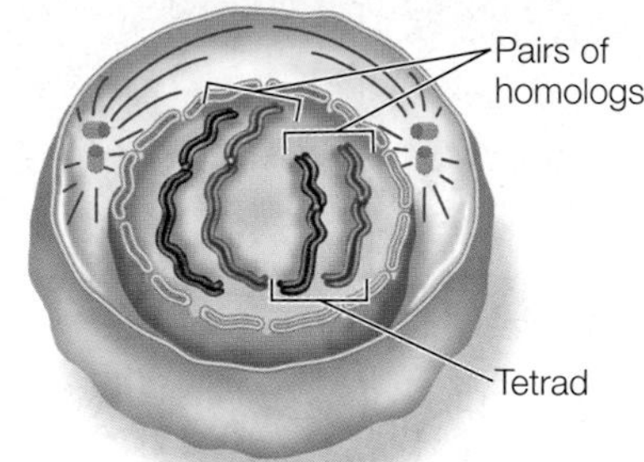

2 Synapsis aligns homologs, and chromosomes condense further.

Late prophase I–prometaphase

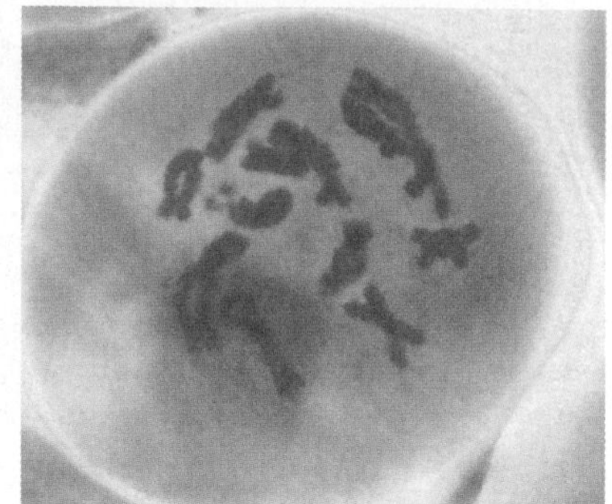

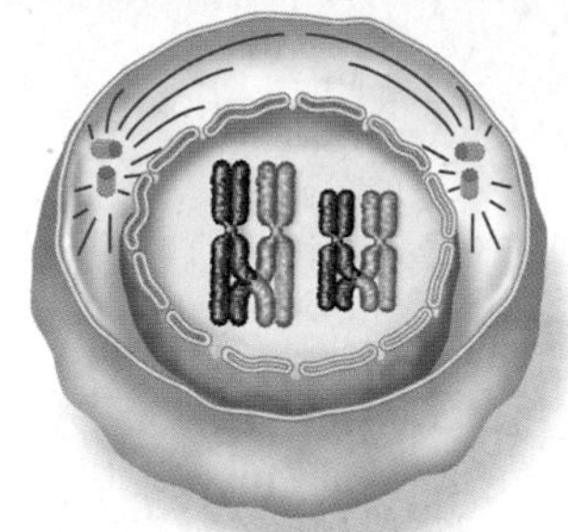

3 The chromosomes continue to coil and shorten. Crossing over results in an exchange of genetic material. In prometaphase the nuclear envelope breaks down.

MEIOSIS II

Prophase II

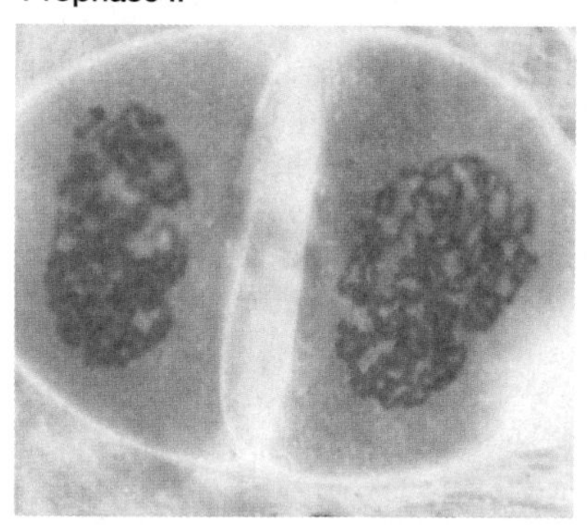

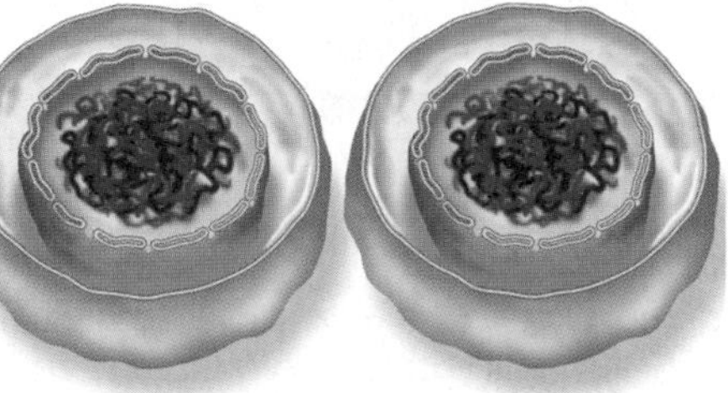

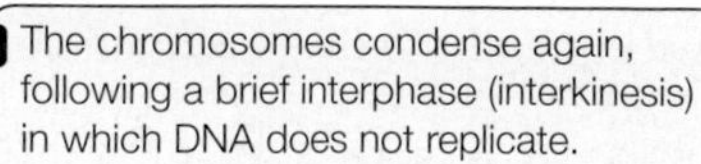

7 The chromosomes condense again, following a brief interphase (interkinesis) in which DNA does not replicate.

Metaphase II

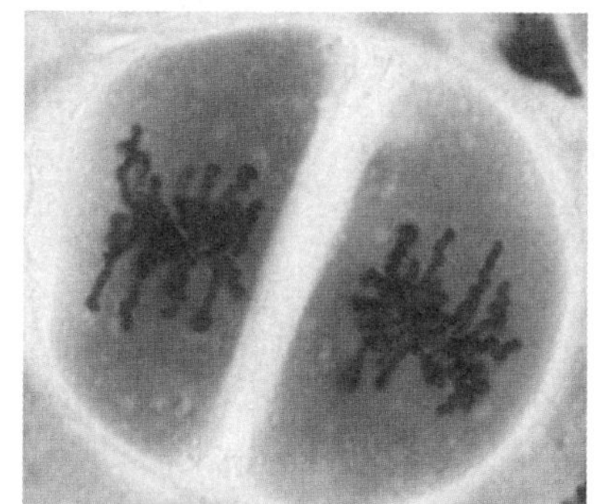

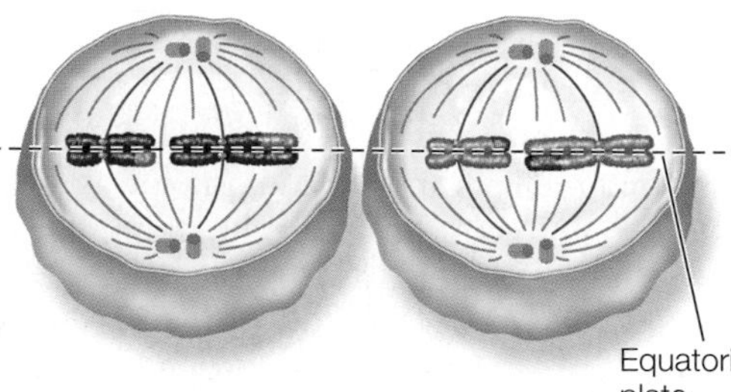

8 The centrosomes of the paired chromatids line up at the equatorial plates of each cell.

Anaphase II

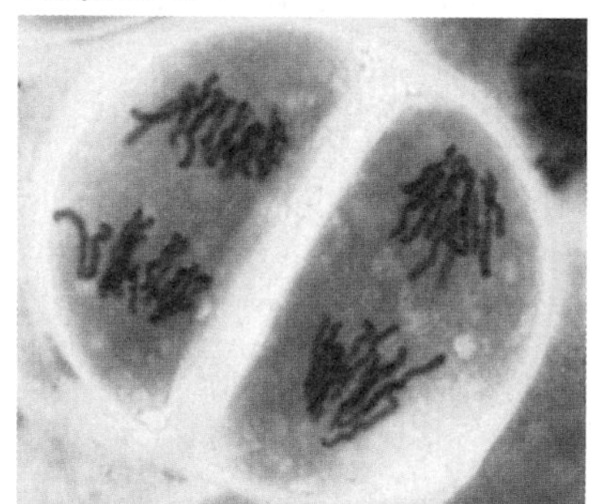

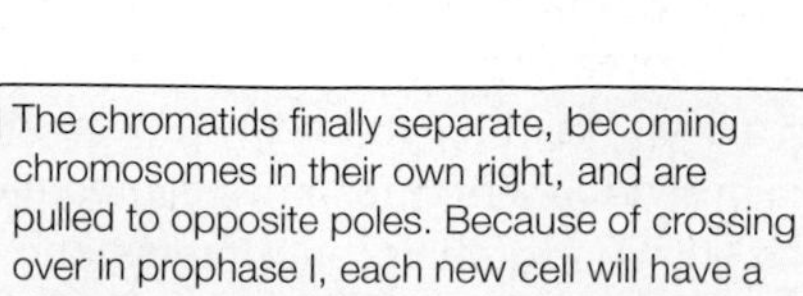

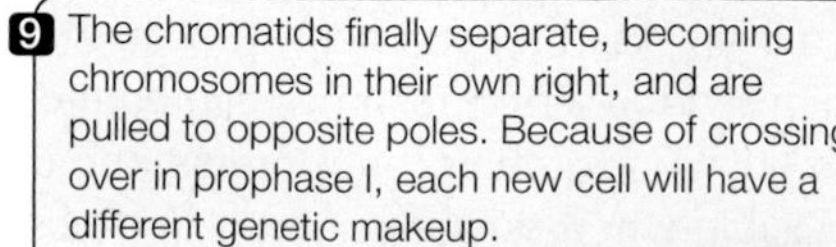

9 The chromatids finally separate, becoming chromosomes in their own right, and are pulled to opposite poles. Because of crossing over in prophase I, each new cell will have a different genetic makeup.

9.16 Meiosis In meiosis, two sets of chromosomes are divided among four nuclei, each of which then has half as many chromosomes as the original cell. Four haploid cells are the result of two successive nuclear divisions. The micrographs show meiosis in the male reproductive organ of a lily; the diagrams show corresponding phases in an animal. (For instructional purposes, the chromosomes from one parent are colored blue and those from the other parent are red.)

Metaphase I

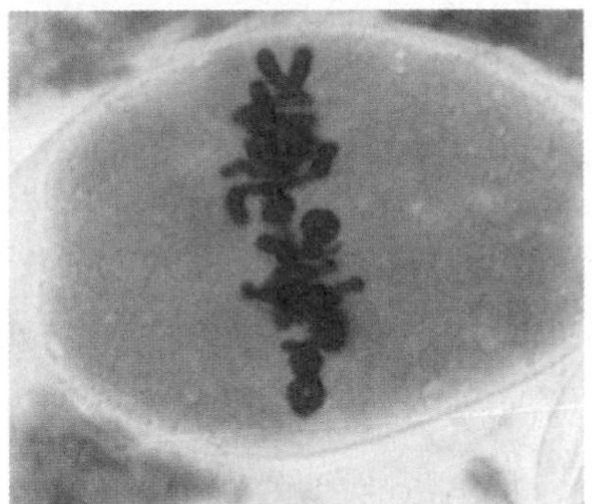

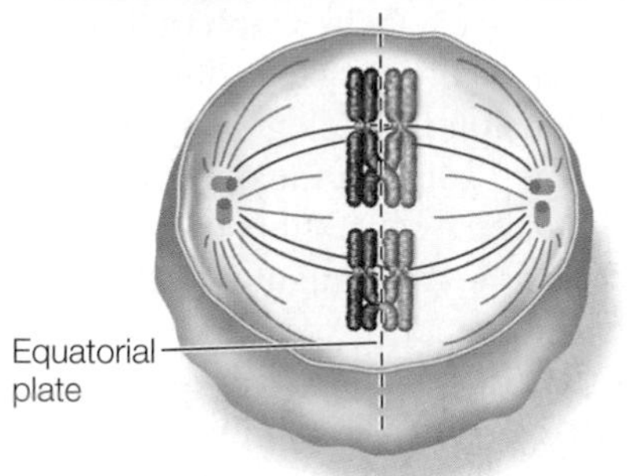

4 The homologous pairs line up on the equatorial (metaphase) plate.

Anaphase I

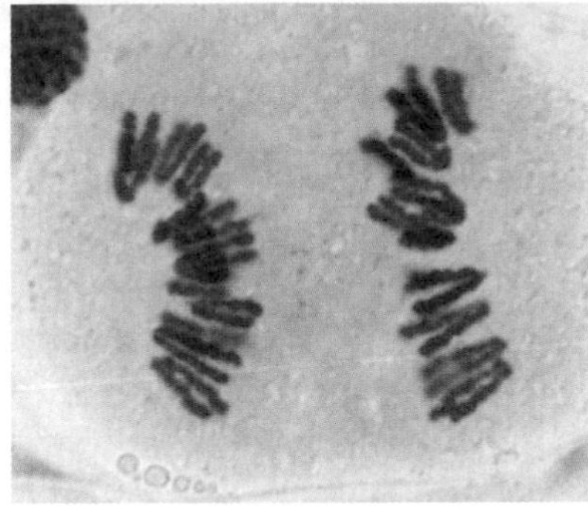

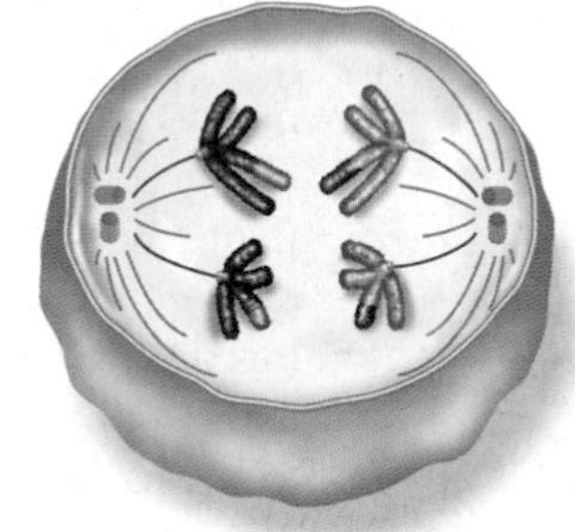

5 The homologous chromosomes (each with two chromatids) move to opposite poles of the cell.

Telophase I

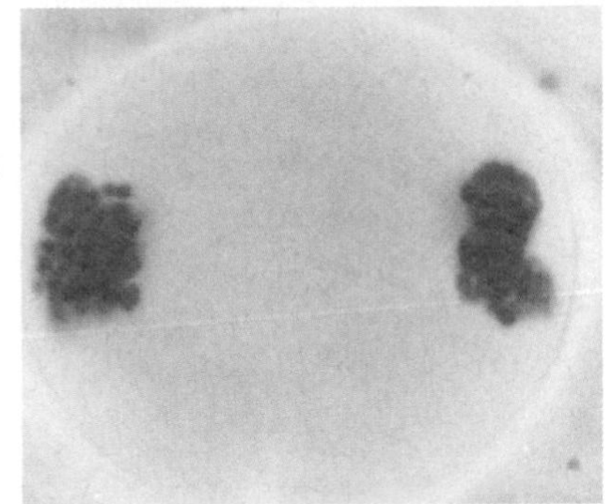

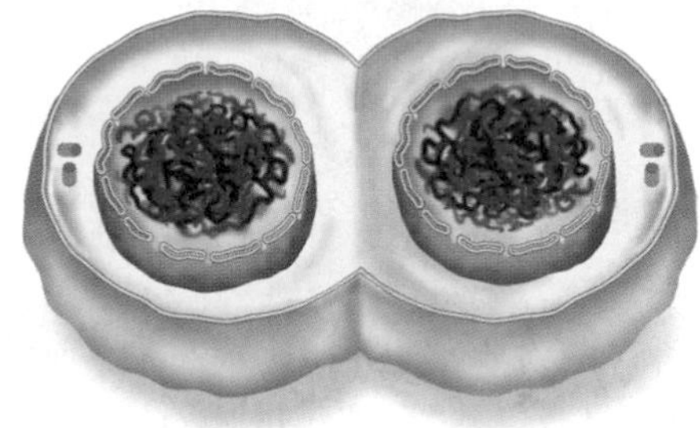

6 The chromosomes gather into nuclei, and the original cell divides.

Telophase II

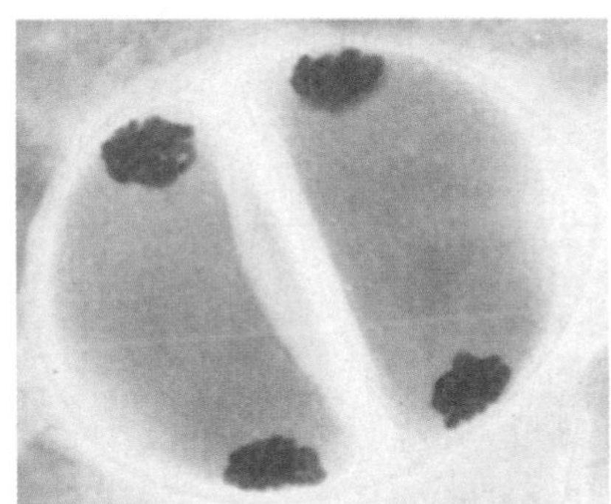

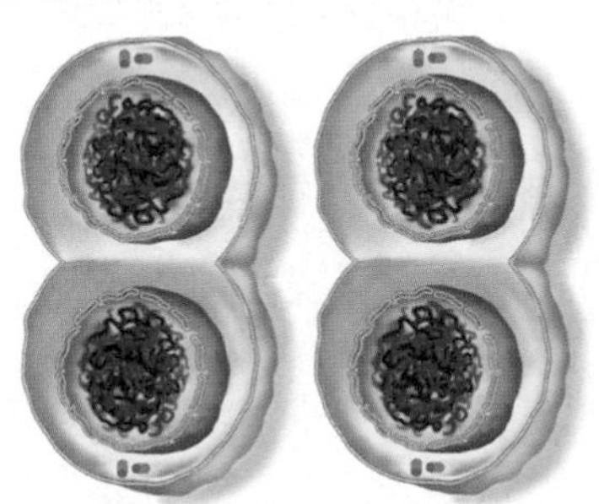

10 The chromosomes gather into nuclei, and the cells divide.

Products

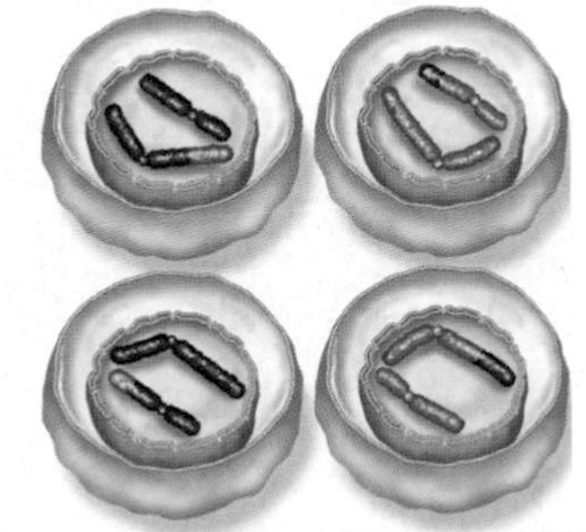

11 Each of the four cells has a nucleus with a haploid number of chromosomes.

The first meiotic division reduces the chromosome number

Two unique features characterize the first of the two meiotic divisions, **meiosis I**. The first is that homologous chromosomes come together to pair along their entire lengths. No such pairing occurs in mitosis. The second is that after metaphase I, the homologous chromosomes separate. The individual chromosomes, each consisting of two sister chromatids, remain intact until the end of metaphase II in the second meiotic division.

Like mitosis, meiosis I is preceded by an interphase with an S phase, during which each chromosome is replicated. As a result, each chromosome consists of two sister chromatids, held together by cohesin proteins.

Meiosis I begins with a long prophase I (the first three frames of Figure 9.16), during which the chromosomes change markedly. The homologous chromosomes pair by adhering along their lengths, a process called **synapsis**. This pairing process lasts from prophase I to the end of metaphase I.

By the time chromosomes can be clearly seen under the light microscope, the two homologs are already tightly joined. This joining begins at the telomeres and is mediated by the recognition of homologous DNA sequences on homologous chromosomes. In addition, a special group of proteins may form a scaffold called the *synaptonemal complex*, which runs lengthwise along the homologous chromosomes and appears to join them together.

The four chromatids of each pair of homologous chromosomes form what is called a **tetrad**, or *bivalent*. In other words, a tetrad consists of four chromatids, two from each of two homologous chromosomes. For example, there are 46 chromosomes in a hu-

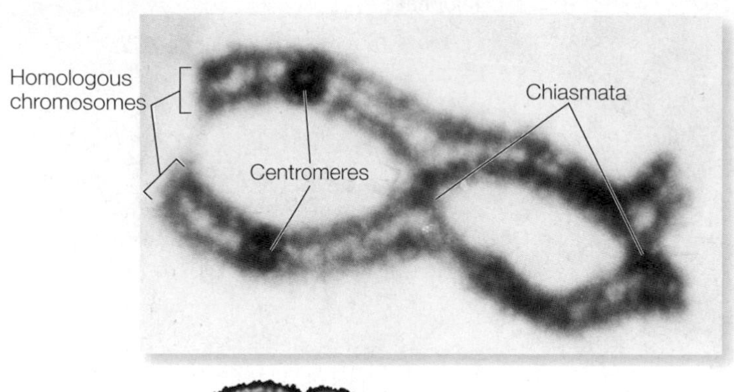

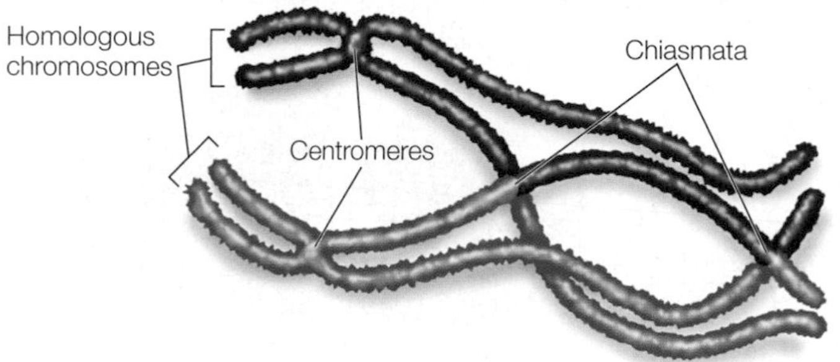

9.17 Chiasmata: Evidence of Exchange between Chromatids This micrograph shows a pair of homologous chromosomes, each with two chromatids, during prophase I of meiosis in a salamander. Two chiasmata are visible.

man diploid cell at the beginning of meiosis, so there are 23 homologous pairs of chromosomes, each with two chromatids (that is, 23 tetrads), for a total of 92 chromatids during prophase I.

Throughout prophase I and metaphase I, the chromatin continues to coil and compact, so that the chromosomes appear ever thicker. At a certain point, the homologous chromosomes seem to repel each other, especially near the centromeres, but they are held together by physical attachments mediated by cohesins. These cohesins are different from the ones holding the two sister chromatids together. Regions having these attachments take on an X-shaped appearance (**Figure 9.17**) and are called **chiasmata** (singular *chiasma*, "cross").

A chiasma reflects an exchange of genetic material between nonsister chromatids on homologous chromosomes—what geneticists call **crossing over** (**Figure 9.18**). The chromosomes usually begin exchanging material shortly after synapsis begins, but chiasmata do not become visible until later, when the homologs are repelling each other. Crossing over increases genetic variation among the products of meiosis by reshuffling genetic information among the homologous pairs. We will have a great deal to say about crossing over and its genetic consequences in the coming chapters.

There seems to be plenty of time for the complicated events of prophase I to occur. Whereas mitotic prophase is usually measured in minutes, and all of mitosis seldom takes more than an hour or two, meiosis can take much longer. In human males, the cells in the testis that undergo meiosis take about a week for prophase I and about a month for the entire meiotic cycle. In the cells that will become eggs, prophase I begins long before a woman's birth, during her early fetal development, and ends as much as decades later, during the monthly ovarian cycle (see Figure 42.13).

Prophase I is followed by prometaphase I, during which the nuclear envelope and the nucleoli disaggregate. A spindle forms, and microtubules become attached to the kinetochores of the chromosomes. In meiosis I, the kinetochores of both chromatids in each chromosome become attached to the same half-spindle. Thus the entire chromosome, consisting of two chromatids, will migrate to one pole. Which member of a homologous chromosome pair becomes attached to each half-spindle, and thus which member will go to which pole, is random. By metaphase I, all the chromosomes have moved to the equatorial plate. Up to this point, homologous pairs are held together by chiasmata.

The homologous chromosomes separate in anaphase I, when the individual chromosomes, each still consisting of two chromatids, are pulled to the poles, with one homolog of a pair going to one pole and the other homolog going to the opposite pole. (Note that this process differs from the separation of *chromatids* during mitotic anaphase.) Each of the two daughter nuclei from this division thus contains only one set of chromosomes, not the two sets that were present in the original diploid nucleus. However, because they consist of two chromatids rather than just one, each of these chromosomes has twice the mass as that of a chromosome at the end of a mitotic division.

In some organisms, there is a telophase I, with the reaggregation of nuclear envelopes. When there is a telophase I, it is followed

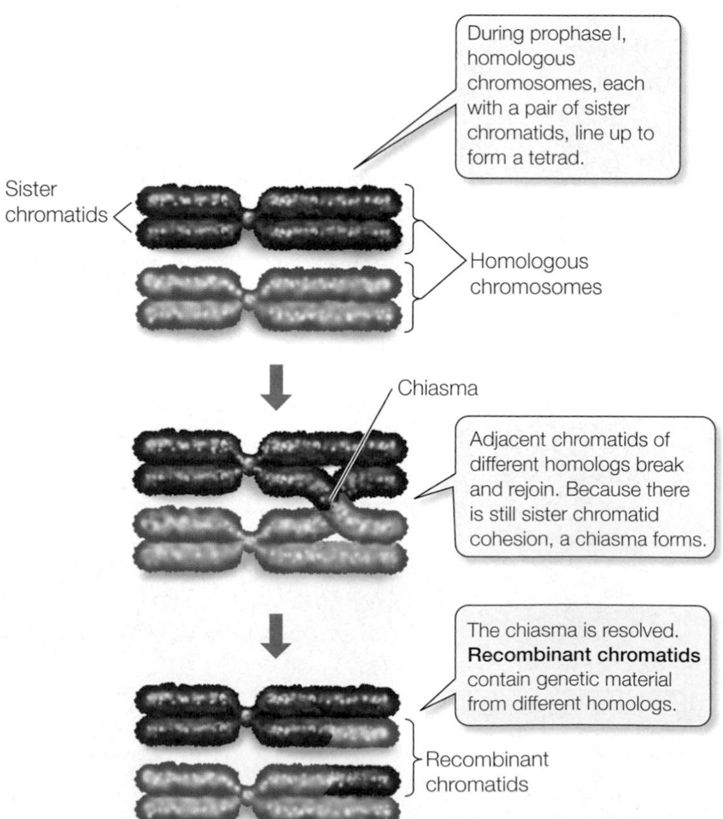

During prophase I, homologous chromosomes, each with a pair of sister chromatids, line up to form a tetrad.
Sister chromatids
Homologous chromosomes
Chiasma
Adjacent chromatids of different homologs break and rejoin. Because there is still sister chromatid cohesion, a chiasma forms.
The chiasma is resolved. **Recombinant chromatids** contain genetic material from different homologs.
Recombinant chromatids

9.18 Crossing Over Forms Genetically Diverse Chromosomes The exchange of genetic material by crossing over may result in new combinations of genetic information on the recombinant chromosomes. The two different colors distinguish the chromosomes contributed by the male and female parent.

by an interphase, called **interkinesis**, similar to the mitotic interphase. During interkinesis the chromatin is partially uncoiled; however, there is no replication of the genetic material, because each chromosome already consists of two chromatids. Furthermore, the sister chromatids at interkinesis are generally not genetically identical, because crossing over in prophase I has reshuffled genetic material between the maternal and paternal chromosomes. In other organisms, the chromosomes move directly into the second meiotic division.

The second meiotic division separates the chromatids

Meiosis II is similar to mitosis in many ways. In each of the two nuclei produced by meiosis I, the chromosomes line up at the equatorial plate at metaphase II. The centromeres of the sister chromatids separate because of cohesin breakdown, and the daughter chromosomes move to the poles in anaphase II.

There are three major differences between meiosis II and mitosis:

- DNA replicates before mitosis, but not before meiosis II.
- In mitosis, the sister chromatids that make up a given chromosome are identical. In meiosis II, they may differ over part of their length if they participated in crossing over during prophase I.
- The number of chromosomes on the equatorial plate in meiosis II is half the number in the mitotic nucleus.

The result of meiosis is four nuclei; each nucleus is haploid and has a single set of unreplicated chromosomes that differs from those of the other nuclei in its exact genetic composition. The differences among the four haploid nuclei result from crossing over during prophase I and from the random segregation of homologous chromosomes during anaphase I.

The activities and movements of chromosomes during meiosis result in genetic diversity

What are the consequences of the synapsis and segregation of homologous chromosomes during meiosis? In mitosis, each chromosome behaves independently of its homolog; its two chromatids are sent to opposite poles at anaphase. If a mitotic division begins with *x* chromosomes, we end up with *x* chromosomes in each daughter nucleus, and each chromosome consists of one chromatid. Each of the two sets of chromosomes (one of paternal and one of maternal origin) is divided equally and distributed equally to each daughter cell. In meiosis, things are very different. **Figure 9.19** compares the two processes.

In meiosis, chromosomes of maternal origin pair with their paternal homologs during synapsis. Segregation of the homologs during meiotic anaphase I ensures that each pole receives one member of each homologous pair. For example, at the end of meiosis I in humans, each daughter nucleus contains 23 of the original 46 chromosomes. In this way, the chromosome number is decreased from diploid to haploid. Furthermore, meiosis I guarantees that each daughter nucleus gets one full set of chromosomes.

The products of meiosis I are genetically diverse for two reasons:

- Synapsis during prophase I allows the maternal chromosome in each homologous pair to exchange segments with the paternal homolog by crossing over. The resulting **recombinant** chromatids contain some genetic material from each parent.
- It is a matter of chance which member of a homologous pair goes to which daughter cell at anaphase I. For example, if there are two homologous pairs of chromosomes in the diploid parent nucleus, a particular daughter nucleus could get paternal chromosome 1 and maternal chromosome 2, or paternal 2 and maternal 1, or both maternal, or both paternal. It all depends on the way in which the homologous pairs line up at metaphase I. This phenomenon is termed **independent assortment**.

Note that of the four possible chromosome combinations just described, only two produce daughter nuclei that are the same as one of the parental types (except for any material exchanged by crossing over). The greater the number of chromosomes, the less probable that the original parental combinations will be reestablished, and the greater the potential for genetic diversity. Most species of diploid organisms do indeed have more than two pairs of chromosomes. In humans, with 23 chromosome pairs, 2^{23} (8,388,608) different combinations can be produced just by the mechanism of independent assortment. Taking the extra genetic shuffling afforded by crossing over into account, the number of possible combinations is virtually infinite.

Meiotic errors lead to abnormal chromosome structures and numbers

In the complex process of cell division, things occasionally go wrong. A pair of homologous chromosomes may fail to separate during meiosis I, or sister chromatids may fail to separate during meiosis II or during mitosis. This phenomenon is called **nondisjunction**. Conversely, homologous chromosomes may fail to remain together. Either problem can result in the production of *aneuploid* cells. **Aneuploidy** is a condition in which one or more chromosomes are either lacking or present in excess (**Figure 9.20**).

One cause of aneuploidy may be a lack of cohesins. Recall that in meiosis, these proteins, formed after DNA replication, hold the two homologous chromosomes together into metaphase I (see Figure 9.11). They ensure that when the chromosomes line up at the equatorial plate, one homolog will face one pole and the other homolog will face the other pole. Without this "glue," the two homologs may line up randomly at metaphase I, just like chromosomes during mitosis, and there is a 50 percent chance that both will go to the same pole. If, for example, during the formation of a human egg, both members of the chromosome 21 pair go to the same pole during anaphase I, the resulting eggs will contain either two of chromosome 21 or none at all. If an egg with two of these chromosomes is fertilized by a normal sperm, the resulting zygote will have three copies of the chromosome: it will be **trisomic** for chromosome 21. A child with an extra chromosome 21 has the symptoms of *Down syndrome*: impaired intelligence; characteristic abnormalities of the hands, tongue, and eyelids; and an increased susceptibility to cardiac abnormalities and diseases such as leukemia. If an egg that did not receive chromosome 21 is fertilized by a normal sperm, the zygote will have only one copy: it will be **monosomic** for chromosome 21.

Other abnormal chromosomal events can also occur. In a process called **translocation**, a piece of a chromosome may break away and become attached to another chromosome. For example, a particular large part of one chromosome 21 may be translocated to another chromosome. Individuals who inherit this translocated piece along with two normal chromosomes 21 will have Down syndrome.

Trisomies (and the corresponding monosomies) are surprisingly common in human zygotes, with 10–30 percent of all conceptions showing aneuploidy. But most of the embryos that develop from such zygotes do not survive to birth, and those that do often die before the age of 1 year. Trisomies and monosomies for most chromosomes other than chromosome 21 are lethal to the embryo. At least one-fifth of all recognized pregnancies are spontaneously terminated (miscarried) during the first 2 months, largely because of such trisomies and monosomies. (The actual proportion of spontaneously terminated pregnancies is certainly higher, because the earliest ones often go unrecognized.)

Polyploids can have difficulty in cell division

As mentioned in Section 9.4, both diploid and haploid nuclei can divide by mitosis. Multicellular diploid and multicellular haploid individuals both develop from single-celled beginnings by mitotic divisions. Likewise, mitosis may proceed in diploid organisms even when a chromosome is missing from one of the haploid sets or when there is an extra copy of one of the chromosomes (as in people with Down syndrome).

Organisms with complete extra sets of chromosomes may sometimes be produced by artificial breeding or by natural acci-

9.19 Mitosis and Meiosis: A Comparison Meiosis differs from mitosis chiefly by the synapsis of homologs and by the failure of the centromeres to separate at the end of metaphase I.

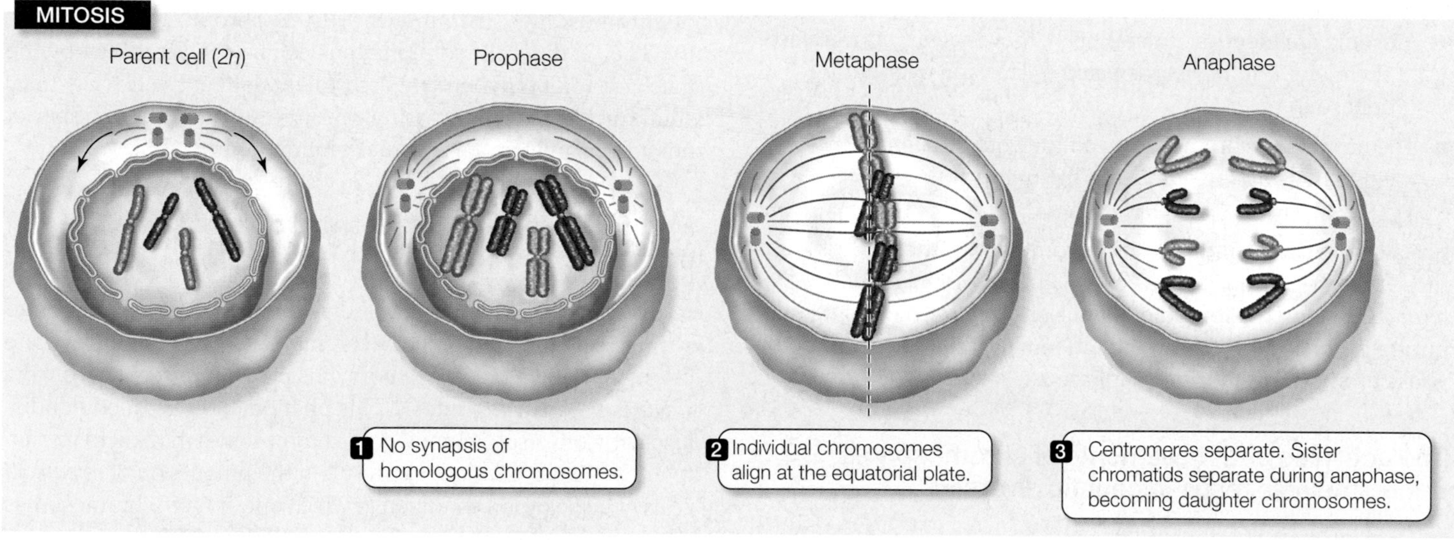
MITOSIS
Parent cell (2*n*)
Prophase
Metaphase
Anaphase
1 No synapsis of homologous chromosomes.
2 Individual chromosomes align at the equatorial plate.
3 Centromeres separate. Sister chromatids separate during anaphase, becoming daughter chromosomes.

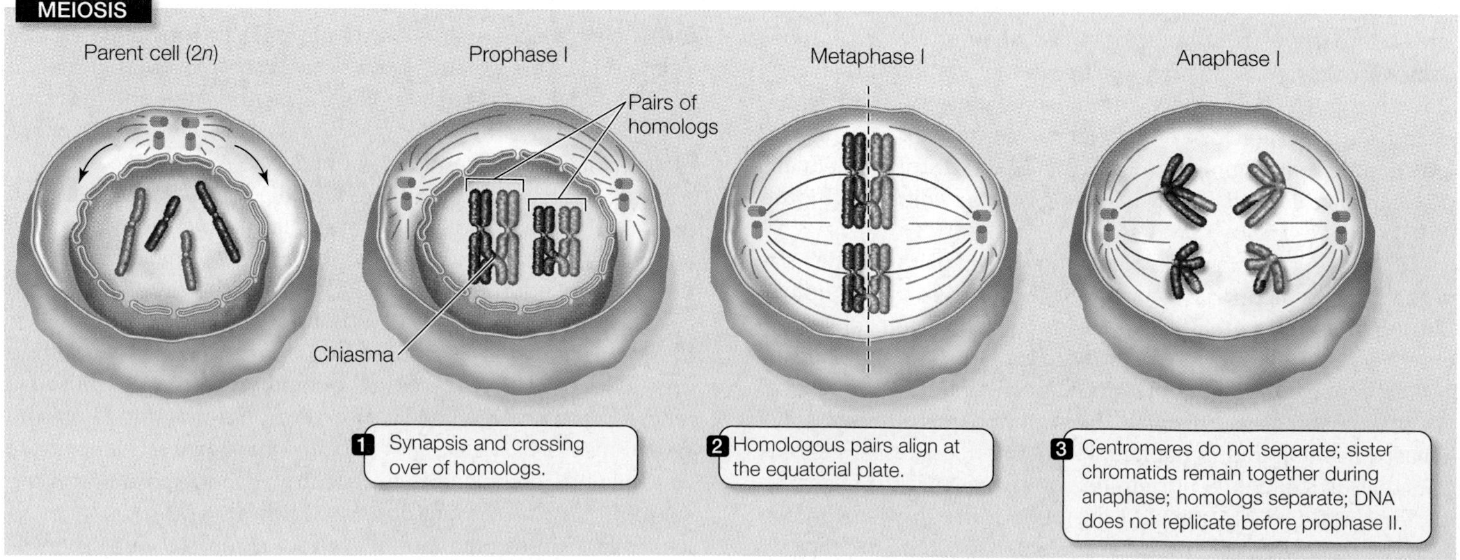
MEIOSIS
Parent cell (2*n*)
Prophase I
Pairs of homologs
Chiasma
Metaphase I
Anaphase I
1 Synapsis and crossing over of homologs.
2 Homologous pairs align at the equatorial plate.
3 Centromeres do not separate; sister chromatids remain together during anaphase; homologs separate; DNA does not replicate before prophase II.

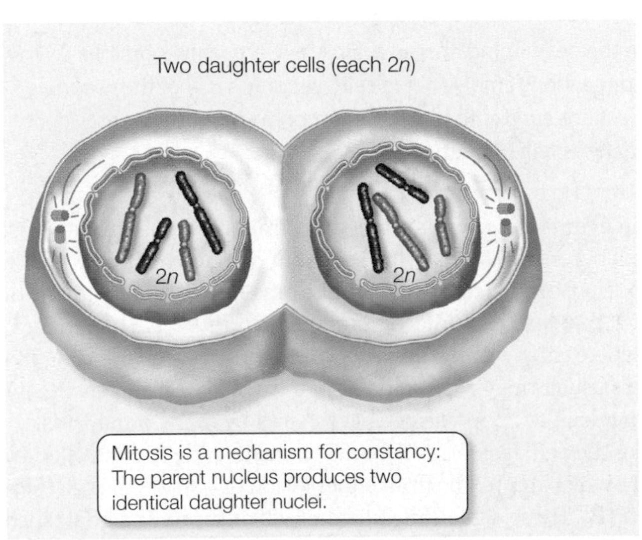

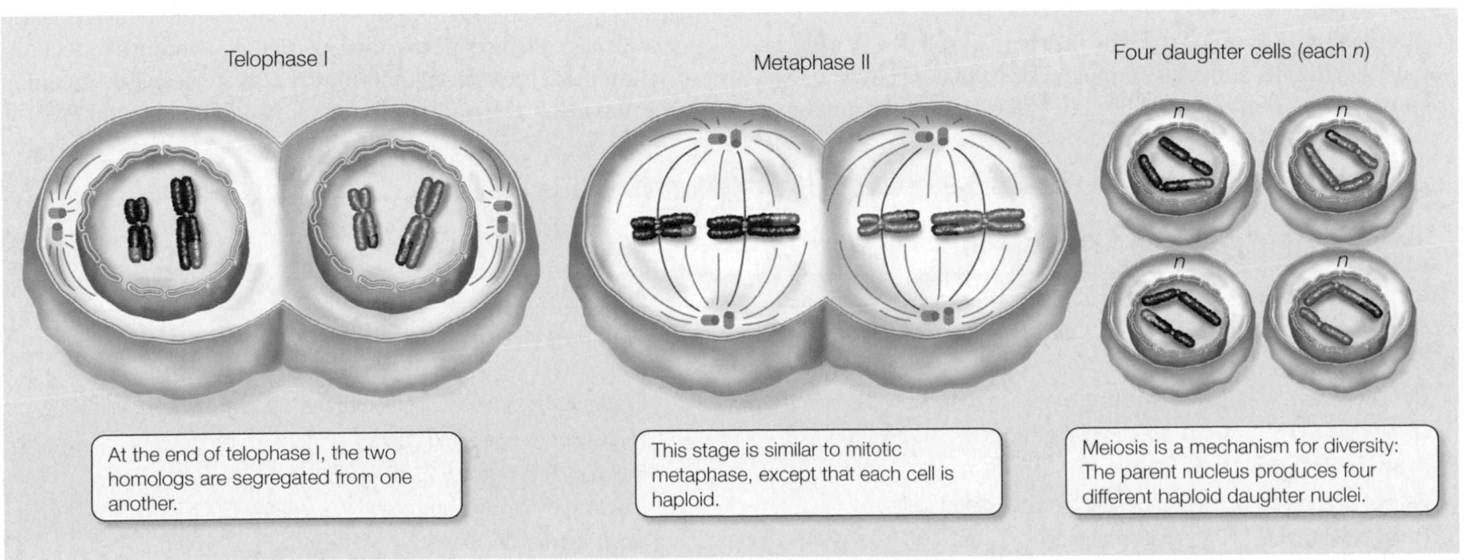

dents. Under some circumstances, triploid ($3n$), tetraploid ($4n$), and higher-order **polyploid** nuclei may form. Each of these *ploidy levels* represents an increase in the number of complete sets of chromosomes present.

Telophase I
Metaphase II
Four daughter cells (each *n*)
n
n
n
n
At the end of telophase I, the two homologs are segregated from one another.
This stage is similar to mitotic metaphase, except that each cell is haploid.
Meiosis is a mechanism for diversity: The parent nucleus produces four different haploid daughter nuclei.

If a nucleus has one or more extra full sets of chromosomes, its abnormally high ploidy in itself does not prevent mitosis, because in mitosis each chromosome behaves independently of the others. In meiosis, however, homologous chromosomes must synapse to begin division. If even one chromosome has no homolog, anaphase I cannot send representatives of that chromosome to both poles. A diploid nu-

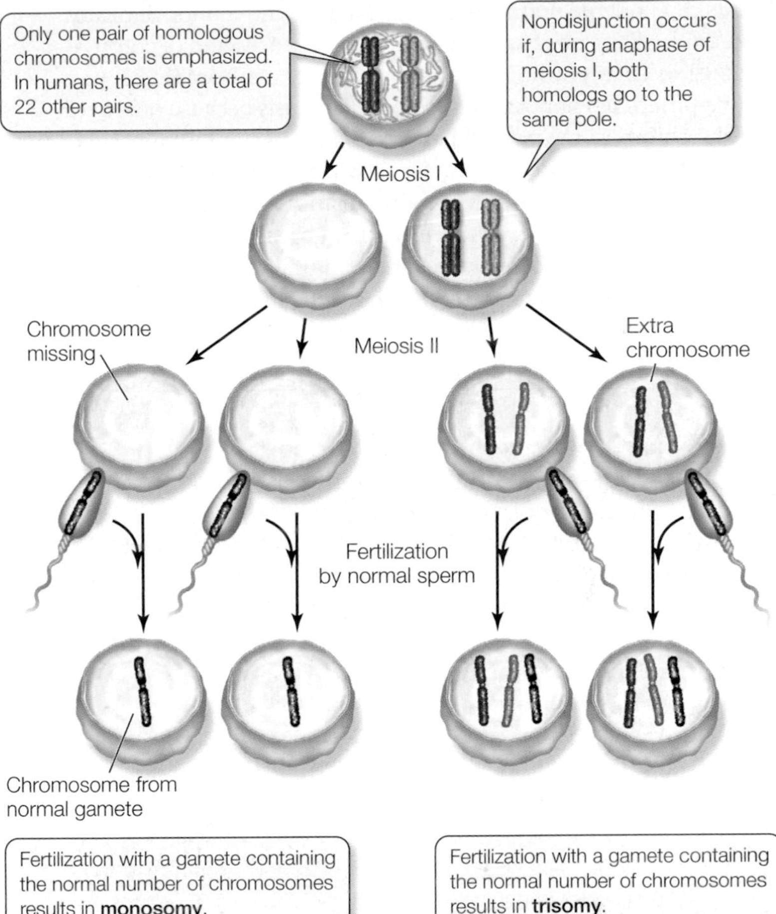

9.20 Nondisjunction Leads to Aneuploidy Nondisjunction occurs if homologous chromosomes fail to separate during meiosis I. The result is aneuploidy: one or more chromosomes are either lacking or present in excess.

cleus can undergo normal meiosis; a haploid one cannot. Similarly, a tetraploid nucleus has an even number of each kind of chromosome, so each chromosome can pair with its homolog. But a triploid nucleus cannot undergo normal meiosis because one-third of the chromosomes would lack partners. This limitation has important consequences for the fertility of triploid, tetraploid, and other chromosomally unusual organisms. Such organisms are common in modern agriculture. Modern bread wheat plants, for example, are hexaploids, the result of naturally occurring crosses between three different grasses, each having its own diploid set of 14 chromosomes.

9.5 RECAP

Meiosis produces four daughter cells in which the chromosome number is reduced from diploid to haploid. Because of the independent assortment of chromosomes and the crossing over of homologous chromatids, the four products of meiosis are not genetically identical. Meiotic errors, such as failure of a homologous chromosome pair to separate, can lead to abnormal numbers of chromosomes.

- Can you describe how crossing over and independent assortment result in unique daughter nuclei? See p. 198 and Figure 9.18
- Do you understand the differences between meiosis and mitosis? See p. 199 and Figure 9.19
- What is aneuploidy, and how can it arise from nondisjunction during meiosis? See p. 199 and Figure 9.20

As mentioned at the start of this chapter, an essential role of cell division in complex eukaryotes is to replace cells that die. What happens to those cells?

9.6 How Do Cells Die?

Cells die in one of two ways. The first type of cell death, **necrosis**, occurs when cells either are damaged by toxins or starved of oxygen or essential nutrients. These cells usually swell up and burst, releasing their contents into the extracellular environment. This process often results in inflammation (see Section 18.2). The scab that forms around a wound is a familiar example of necrotic tissue.

More typically, cell death is due to **apoptosis** (Greek, "falling apart"). Apoptosis is a genetically programmed series of events that result in cell death. These two modes of cell death are compared in **Table 9.2**.

Why would a cell initiate apoptosis, which is essentially "cell suicide"? There are two possible reasons:

- *The cell is no longer needed by the organism.* For example, before birth, a human fetus has weblike hands, with connective tissue between the fingers. As development proceeds, this unneeded tissue disappears as its cells undergo apoptosis (see Figure 19.14).
- *The longer cells live, the more prone they are to genetic damage that could lead to cancer.* This is especially true of cells in the blood and in the epithelia lining organs such as the intestine, which are exposed to high levels of toxic substances. Such cells normally die after only days or weeks.

In the developing human brain, there are many possible connections among nerve cells, yet only a few of these connections survive to birth. The rest of the brain cells—as many as half of them—die.

The events of apoptosis are very similar in most organisms. The cell becomes isolated from its neighbors, cuts up its chromatin into nucleosome-sized pieces, and forms membranous lobes, or "blebs," that break up into cell fragments (**Figure 9.21A**). In a remarkable example of the economy of nature, the surrounding living cells usually ingest the remains of the dead cell. The genetic signals that lead to apoptosis are also shared by many organisms.

Like the cell division cycle, the cell death cycle is controlled by signals, which may come from either inside or outside the cell (**Figure 9.21B**). These signals include the lack of a mitotic signal (such as a growth factor) and the recognition of damaged DNA. External signals (or a lack of them) can cause a receptor protein in the plasma membrane to change its shape and in turn activate a class of enzymes called **caspases**. Internal signals can cause mitochondria to release molecules that in turn activate caspases. In either case, the result is that the cell turns into a killing machine as the

TABLE 9.2

Two Different Ways for Cells to Die

	NECROSIS	APOPTOSIS
Stimuli	Low O_2, toxins, ATP depletion, damage	Specific, genetically programmed physiological signals
ATP required	No	Yes
Cellular pattern	Swelling, organelle disruption, tissue death	Chromatin condensation, membrane blebbing, single-cell death
DNA breakdown	Random fragments	Nucleosome-sized fragments
Plasma membrane	Bursts	Blebbed (see Figure 9.21A)
Fate of dead cells	Ingested by white blood cells	Ingested by neighboring cells
Reaction in tissue	Inflammation	No inflammation

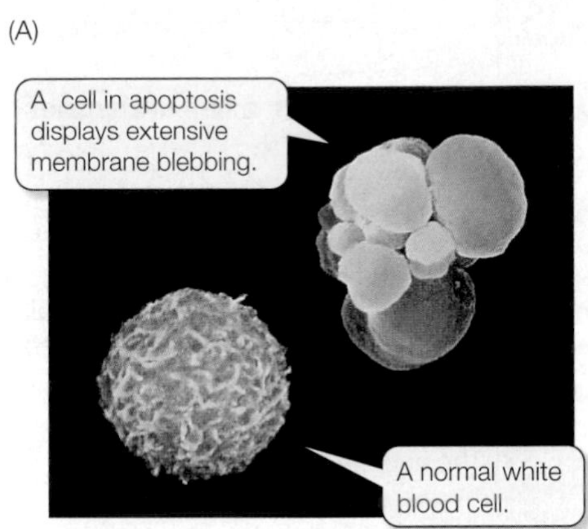

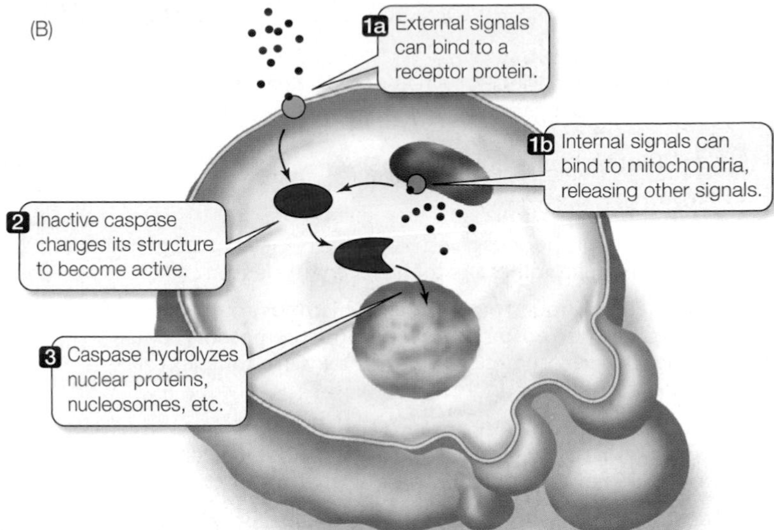

(B)
1a External signals can bind to a receptor protein.
1b Internal signals can bind to mitochondria, releasing other signals.
2 Inactive caspase changes its structure to become active.
3 Caspase hydrolyzes nuclear proteins, nucleosomes, etc.

9.21 Apoptosis: Programmed Cell Death (A) Many cells are genetically programmed to "self-destruct" when they are no longer needed, or when they have lived long enough to accumulate a burden of DNA damage that might harm the organism. (B) Both external and internal signals stimulate caspases, the enzymes that break down specific cell constituents, resulting in apoptosis.

caspases hydrolyze proteins of the nuclear envelope, nucleosomes, and plasma membrane. As we will see in Section 17.5, many of the drugs used to treat diseases of excess cell proliferation, such as cancer, work through these signals.

9.6 RECAP

The growth of a population of cells depends on the relative rates of cell reproduction and cell death. Cell death can occur either by necrosis or by apoptosis. Apoptosis is governed by precise molecular controls.

- What are some differences between apoptosis and necrosis? See Table 9.2
- Can you think of some situations in which apoptosis is necessary? See p. 202
- How is apoptosis regulated? See Figure 9.21

We have now looked at the cell cycle and how cells divide by fission, by mitosis, and by meiosis. We have seen how meiosis produces the genetic complement that will be passed on to the next generation. We will spend the next five chapters in the "realm of the genome," looking at heredity as it was explicated by Gregor Mendel in the nineteenth century and at the exponential amount of knowledge that has been acquired since the elucidation of the genetic code. Virtually no area of modern life is untouched by the field of genetics.

CHAPTER SUMMARY

9.1 How do prokaryotic and eukaryotic cells divide?

Cell division is necessary for the reproduction, growth, and repair of organisms.

Cell division must be initiated by a reproductive signal. Before a cell can divide, the genetic material (DNA) must be **replicated** and **segregated** to separate portions of the cell. **Cytokinesis** then divides the cytoplasm into two cells.

In prokaryotes, most cellular DNA is a single molecule, usually in the form of a circular **chromosome**. Prokaryotes reproduce by **binary fission.** Review Figure 9.2

In eukaryotes, cells divide by either **mitosis** or **meiosis**. Eukaryotic cell division follows the same general pattern as binary fission, but with significant differences in detail. For example, eukaryotic cells have a distinct nucleus (not just a chromosome) that must be replicated. Replicated chromosomes, called **sister chromatids**, must be separated by mitosis.

Cells that produce gametes undergo a special kind of nuclear division called meiosis; the two nuclei produced by meiosis are not genetically identical.

9.2 How is eukaryotic cell division controlled?

The eukaryotic **cell cycle** has two main phases: **interphase** (during which cells are not dividing) and mitosis or **M phase** (when cells are dividing).

During most of the cell cycle, the cell is in interphase, which is divided into three subphases: **S, G1**, and **G2**. DNA is replicated during the S phase. Mitosis and cytokinesis take place during the M phase. Review Figure 9.3

Cyclin/Cdk complexes regulate the passage of cells through checkpoints in the cell cycle. The suppressor protein **RB** inhibits the cell cycle. Cdk4 and Cdk2 act in concert to inactivate RB and allow the cell cycle to progress beyond the **restriction point**. Review Figure 9.6

CHAPTER SUMMARY

Controls external to the cell, such as **growth factors** and hormones, can also stimulate the cell to begin a division cycle.

9.3 What happens during mitosis?

See Web/CD Tutorial 9.1

In mitosis, a single nucleus gives rise to two nuclei that are genetically identical to each other and to the parent nucleus.

DNA is wrapped around proteins called **histones**, forming beadlike units called **nucleosomes**. A eukaryotic chromosome contains strings of nucleosomes bound to proteins in a complex called **chromatin**. Review Figure 9.8

At mitosis, the replicated chromatids are held together at the **centromere**. Each chromatid consists of one double-stranded DNA molecule. Review Figure 9.9, Web/CD Activity 9.1

Mitosis can be divided into several phases, called **prophase**, **prometaphase**, **metaphase**, **anaphase**, and **telophase**.

During mitosis, sister chromatids, attached by **cohesin** protein, line up at the **equatorial plate**, and attach to the **spindle**. The chromatids separate (becoming **daughter chromosomes**) and migrate to opposite ends of the cell. Review Figure 9.10, Web/CD Activity 9.2

Nuclear division is usually followed by cytokinesis. Animal cell cytoplasm usually divides by a contractile ring made up of actin microfilaments. In plant cells, cytokinesis is accomplished by vesicles that fuse to form a cell plate. Review Figure 9.12

9.4 What is the role of cell division in sexual life cycles?

Asexual reproduction produces a **clone**, a new organism that is genetically identical to the parent. Any genetic variation is the result of mutations.

In **sexual reproduction**, two **haploid** gametes—one from each parent—unite in **fertilization** to form a genetically unique, **diploid zygote**. There are several patterns of sexual life cycles: **haplontic** life cycles, **alternation of generations**, and **diplontic** life cycles. Review Figure 9.14, Web/CD Activity 9.3

In sexually reproducing organisms, certain cells in the adult undergo meiosis, a process by which a diploid cell produces haploid gametes. Each gamete contains a random selection of one of each **pair of homologous chromosomes** from the parent.

The numbers, shapes, and sizes of the chromosomes constitute the **karyotype** of an organism.

9.5 What happens when a cell undergoes meiosis?

See Web/CD Tutorial 9.2

Meiosis consists of two nuclear divisions, **meiosis I** and **meiosis II**, that collectively reduce the chromosome number from diploid to haploid. It ensures that each haploid cell contains one member of each chromosome pair, and results in four genetically diverse haploid cells, usually gametes. Review Figure 9.16, Web/CD Activity 9.4

In anaphase I, entire chromosomes, each with two chromatids, migrate to the poles. By the end of meiosis I, there are two nuclei, each with the haploid number of chromosomes.

In meiosis II, the sister chromatids separate. No DNA replication precedes this division, which in other aspects is similar to mitosis.

During prophase I of the first meiotic division, homologous chromosomes undergo **synapsis** to form pairs in a **tetrad**. Chromatids can form junctions called **chiasmata** and genetic material may be exchanged between the two homologs by **crossing over**. Review Figure 9.18

Both crossing over during prophase I and the **independent assortment** of the homologs that migrate to each pole during anaphase I ensure that the genetic composition of each haploid gamete is different from that of the parent cell and from that of the other gametes.

In **nondisjunction**, one member of a homologous pair of chromosomes fails to separate from the other, and both go to the same pole. Pairs of homologous chromosomes may also fail to stick together when they should. Both problems can lead to one gamete having an extra chromosome and another lacking that chromosome. Review Figure 9.20

The union of a gamete with an abnormal chromosome number with a normal haploid gamete at fertilization results in **aneuploidy**. Such genetic abnormalities are invariably harmful or lethal to the organism.

9.6 How do cells die?

Cells may die by **necrosis**, or they may self-destruct by **apoptosis**, a genetically programmed series of events that includes the detachment of the cell from its neighbors and the fragmentation of its nuclear DNA. Review Table 9.2

Apoptosis is regulated by external and internal signals. These signals result in activation of a class of enzymes called **caspases** that hydrolyze proteins of the cell, leading to its destruction. Review Figure 9.21

SELF-QUIZ

1. Which statement about eukaryotic chromosomes is *not* true?
   a. They sometimes consist of two chromatids.
   b. They sometimes consist only of a single chromatid.
   c. They normally possess a single centromere.
   d. They consist only of proteins.
   e. They are clearly visible as defined bodies under the light microscope.
2. Nucleosomes
   a. are made of chromosomes.
   b. consist entirely of DNA.
   c. consist of DNA wound around a histone core.
   d. are present only during mitosis.
   e. are present only during prophase.
3. Which statement about the cell cycle is *not* true?
   a. It consists of mitosis and interphase.
   b. The cell's DNA replicates during G1.
   c. A cell can remain in G1 for weeks or much longer.
   d. DNA is not replicated during G2.
   e. Cells enter the cell cycle as a result of internal or external signals.

4. Which statement about mitosis is *not* true?
   a. A single nucleus gives rise to two identical daughter nuclei.
   b. The daughter nuclei are genetically identical to the parent nucleus.
   c. The centromeres separate at the onset of anaphase.
   d. Homologous chromosomes synapse in prophase.
   e. The centrosomes organize the microtubules of the spindle fibers.
5. Which statement about cytokinesis is true?
   a. In animals, a cell plate forms.
   b. In plants, it is initiated by furrowing of the membrane.
   c. It follows mitosis.
   d. In plant cells, actin and myosin play an important part.
   e. It is the division of the nucleus.
6. Apoptosis
   a. occurs in all cells.
   b. involves the formation of the plasma membrane.
   c. does not occur in an embryo.
   d. is a series of programmed events resulting in cell death.
   e. is the same as necrosis.
7. In meiosis,
   a. meiosis II reduces the chromosome number from diploid to haploid.
   b. DNA replicates between meiosis I and meiosis II.
   c. the chromatids that make up a chromosome in meiosis II are identical.
   d. each chromosome in prophase I consists of four chromatids.
   e. homologous chromosomes separate from one another in anaphase I.
8. In meiosis,
   a. a single nucleus gives rise to two daughter nuclei.
   b. the daughter nuclei are genetically identical to the parent nucleus.
   c. the centromeres separate at the onset of anaphase I.
   d. homologous chromosomes synapse in prophase I.
   e. no spindle forms.
9. A plant has a diploid chromosome number of 12. An egg cell of that plant has 5 chromosomes. The most probable explanation is
   a. normal mitosis.
   b. normal meiosis.
   c. nondisjunction in meiosis I.
   d. nondisjunction in meiosis I and II.
   e. nondisjunction in mitosis.
10. The number of daughter chromosomes in a human cell in anaphase II of meiosis is
   a. 2.
   b. 23.
   c. 46.
   d. 69.
   e. 92.

FOR DISCUSSION

1. The story of HeLa cells raises some important issues in the ethics of biomedical research. Henrietta Lacks was not asked for permission to use her cells, and neither she nor her surviving family benefited in any direct way from the many commercial uses of her cells. What types of regulations are needed to ensure ethical handling of opportunities such as that presented by Mrs. Lacks's cells?
2. Compare the roles of cohesins in mitosis, meiosis I, and meiosis II.
3. Compare and contrast mitosis (and subsequent cytokinesis) in animals and plants.
4. Contrast mitotic prophase and prophase I of meiosis. Contrast mitotic anaphase and anaphase I of meiosis.
5. Compare the sequence of events in the mitotic cell cycle with the sequence of events in apoptosis.

FOR INVESTIGATION

1. Suggest two ways in which, with the help of a microscope, one might determine the relative duration of the various phases of mitosis.
2. Describing the events and controls of the cell cycle is much easier if the cells under investigation are synchronous; that is, if a population of cells are all in the same stage of the cell cycle. This can be accomplished with various chemicals. But some populations of cells are naturally synchronous. The anther (male sex organ) of a lily plant contains cells that become pollen grains (male gametes). As anthers develop in the flower, their length correlates precisely with the stage of the meiotic cycle in those cells. These stages each take many days, so that an anther that is 1.5 millimeters long, for example, contains cells in early prophase I. How would you use lily anthers to investigate the roles of cyclins and Cdk's in the meiotic cell cycle?

CHAPTER 10 Genetics: Mendel and Beyond

The wisdom of the rabbis

In the Middle Eastern desert of 1,800 years ago, the rabbi faced a dilemma. A Jewish woman had given birth to a son. As required by the laws set down by God's commandment to Abraham almost 2,000 years previously and later reiterated by Moses, the mother brought her 8-day-old son to the rabbi for ritual penile circumcision. The rabbi knew that the woman's two previous sons had bled to death when their foreskins were cut. Yet the biblical commandment remained: unless he was circumcised, the boy could not be counted among those with whom God had made His solemn covenant. After consultation with other rabbis, it was decided to exempt this, the third son.

Almost a thousand years later, in the twelfth century, the physician and biblical commentator Moses Maimonides reviewed this and numerous other cases in the rabbinical literature and stated that in such instances the third son should not be circumcised. Furthermore, the exemption should apply whether the mother's son was "from her first husband or from her second husband." The bleeding disorder, he reasoned, was clearly carried by the mother and passed on to her sons.

Without any knowledge of our modern concepts of genes and genetics, the rabbis had linked a human disease (which we now know as hemophilia A) to a pattern of inheritance (which we know as sex linkage). Only in the past few decades have the precise biochemical nature of hemophilia A and its genetic determination been worked out.

Humans normally have two copies each of 22 of the 23 chromosomes in the human karyotype, as described in Chapter 9. Thus, even if a given gene on one of the chromosomes is mutant, the normal gene on the second copy of that chromosome can usually produce a functional protein. But one pair of chromosomes is different. In the case of X and Y chromosomes, males receive only one copy of each; females receive two copies of the X chromosome (but no Y chromosome). The genetic mutation that causes the blood clotting malfunction of hemophilia is located on the X chromosome, and males carrying the mutation have no "back-up" normal gene. Color blindness, a physical condition with only minor ramifications for most individuals who suffer from it, has a similar pattern of transmission.

How do we account for and predict such patterns of inheritance? Much about inheritance was intuited even before scientists and

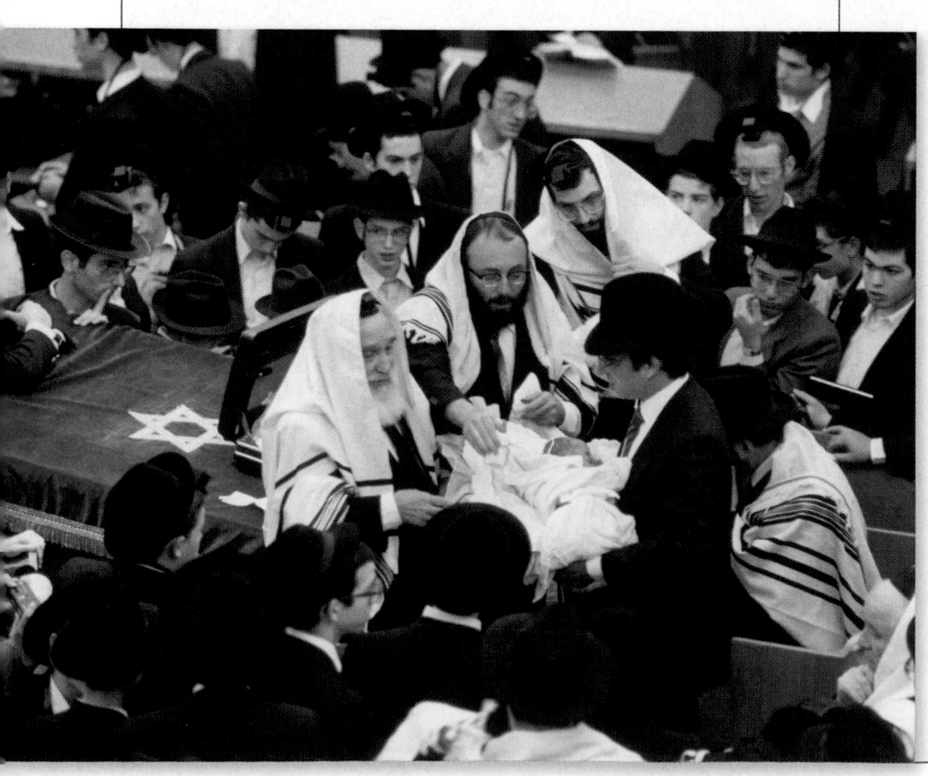

An Ancient Ritual A male infant undergoes ritual circumcision in accordance with Jewish laws. Sons of Jewish mothers who carry the gene for hemophilia may be exempted from this ritual.

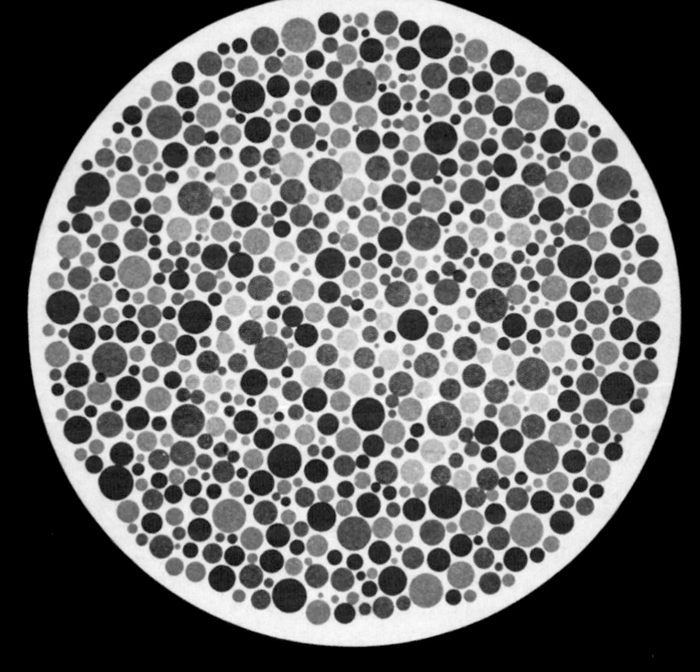

Test for a Sex-Linked Trait Like hemophilia, the mutant allele for red-green color blindness is carried on the X chromosome. Unlike hemophilia, however, this condition is not usually deleterious. In the simple test shown here, a person with normal color vision sees the number 74; people with the most typical type of color blindness see 21; and severely color blind people cannot distinguish any numeral.

CHAPTER OUTLINE

scholars knew that genes and chromosomes existed—as proven by the ruling of that wise rabbi almost two thousand years ago. Indeed, the foundations of the science of inheritance and genetic transmission were laid in the 1860s by some of the most amazing experiments and feats of data analysis in the history of biological science. It was almost 50 years before the significance of these experiments and their analyses by Gregor Mendel was recognized by the scientific community. Once that recognition was finally achieved, however, natural science and medicine began to move forward at an unprecedented pace.

IN THIS CHAPTER we will discuss how the units of inheritance—genes—are transmitted from generation to generation. We will show that many of the rules that govern inheritance can be explained by the behavior of chromosomes during meiosis. We will describe the interactions of genes with one another and with the environment, and we will see how the specific positions of genes on chromosomes affect diversity.

10.1 What Are the Mendelian Laws of Inheritance?

Much of the early study of biological inheritance was done with plants and animals of economic importance. Records show that people were deliberately cross-breeding date palm trees and horses as early as 5,000 years ago. By the early nineteenth century plant breeding was widespread, especially for ornamental flowers such as tulips. Plant breeders of that time were operating under two key assumptions about how inheritance worked. Only one of those assumptions turned out to be correct.

- *Each parent contributes equally to offspring (correct).* In the 1770s, the German botanist Josef Gottlieb Kölreuter studied the offspring of **reciprocal crosses**, in which plants are *crossed* (mated with each other) in opposite directions. For example, in one cross, males that have white flowers are mated with females that have red flowers, while in a complementary cross, red-flowered males and white-flowered females are mated. In Kölreuter's studies, such reciprocal crosses always gave identical results, showing that both parents contributed equally to the offspring.
- *Hereditary determinants blend in offspring (incorrect).* Kölreuter and others proposed that there were hereditary determinants in the egg and sperm cells. When these determinants came together in a single cell after mating, they were believed to blend together. If a plant that had one form of a characteristic (say, red flowers) was crossed with one that had a different form of that characteristic (blue flowers), the offspring would have a blended combination of the two parents' characteristics (purple flowers). According to the blending theory, it was thought that once heritable elements were combined, they could not be separated again (like inks of different colors mixed together). The red and blue hereditary determinants were thought to be forever blended into the new purple one.

In his experiments in the 1860s, Gregor Mendel confirmed the first of these two assumptions, but refuted the second.

Mendel brought new methods to experiments on inheritance

Gregor Mendel was an Austrian monk, not an academic scientist (**Figure 10.1**). He was well qualified, however, to undertake scientific investigations. After his 1850 failure in an examination for a teaching certificate in natural science, he undertook intensive studies in physics, chemistry, mathematics, and various aspects of biology at the University of Vienna. His studies in physics and mathematics strongly influenced his use of experimental and quantitative methods in his studies of heredity, and it was those quantitative experiments that were key to his successful deductions.

Over the seven years he spent working out the principles of inheritance in plants, Mendel made crosses between and noted the resulting characteristics of 24,034 plants. Analysis of his meticulously gathered data suggested to him a new theory of how inheritance might work. His work culminated in a public lecture in 1865 and a detailed written publication in 1866. Mendel's paper appeared in a journal that was received by 120 libraries, and he sent reprinted copies (of which he had obtained 40) to several distinguished scholars. However, his theory was not readily accepted. In fact, it was mostly ignored.

One reason Mendel's paper received so little attention was that most prominent biologists of his time were not in the habit of thinking in mathematical terms, even the simple terms Mendel used. Even Charles Darwin, whose theory of evolution by natural selection was predicated on heritable variation among individuals, failed to understand the significance of Mendel's findings. In

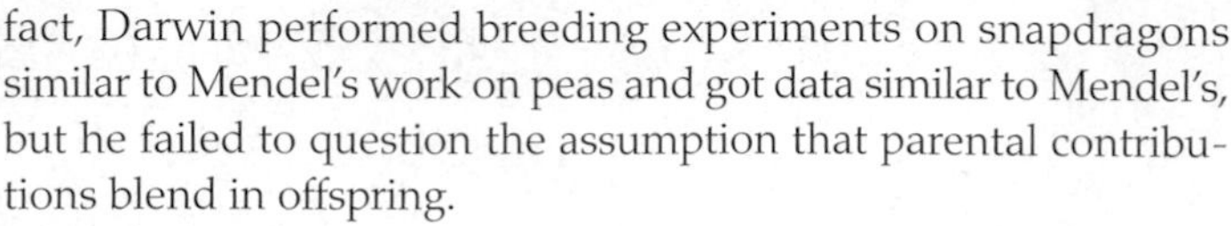
fact, Darwin performed breeding experiments on snapdragons similar to Mendel's work on peas and got data similar to Mendel's, but he failed to question the assumption that parental contributions blend in offspring.

Mendel's work may have gone unnoticed in part because he had little credibility as a biologist. In fact his lowest test scores were in biology. Whatever the reasons, Mendel's pioneering paper had no discernible influence on the scientific community for more than 30 years.

By 1900, the events of meiosis had been observed and described, and Mendel's discoveries burst into sudden prominence as a result of independent experiments by three plant geneticists: Hugo DeVries, Carl Correns, and Erich von Tschermak. Each carried out crossing experiments, each published his principal findings in 1900, and each cited Mendel's 1866 paper. These three men realized that chromosomes and meiosis provided a physical explanation for the theory that Mendel had proposed to explain the data from his crosses.

That Mendel was able to achieve his remarkable insights before the discovery of genes and meiosis was largely due to his experimental methods. His work is a definitive example of extensive preparation, fortunate choice of experimental subject, meticulous execution, and imaginative yet logical interpretation. Let's take a closer look at these experiments and the conclusions and hypotheses that emerged.

10.1 Gregor Mendel and His Garden The Austrian monk Gregor Mendel (left) did his pathbreaking experiments in genetics in a garden of the monastery at Brno, in what is now the Czech Republic.

Mendel devised a careful research plan

Mendel chose to study the common garden pea because of its ease of cultivation, the feasibility of controlled pollination, and the availability of varieties with contrasting traits. He controlled pollination, and thus fertilization, of his parent plants by manually moving pollen from one plant to another (**Figure 10.2**). Thus he knew the parentage of the offspring in his experiments. The pea plants Mendel studied produce male and female sex organs and gametes in the same flower. If untouched, they naturally *self-pollinate*—that is, the female organ of each flower receives pollen from the male organs of the same flower. Mendel made use of this natural phenomenon in some of his experiments.

Mendel began by examining different varieties of peas in a search for heritable characters and traits suitable for study:

- A **character** is an observable physical feature, such as flower color.
- A **trait** is a particular form of a character, such as purple flowers or white flowers.
- A **heritable trait** is one that is passed from parent to offspring.

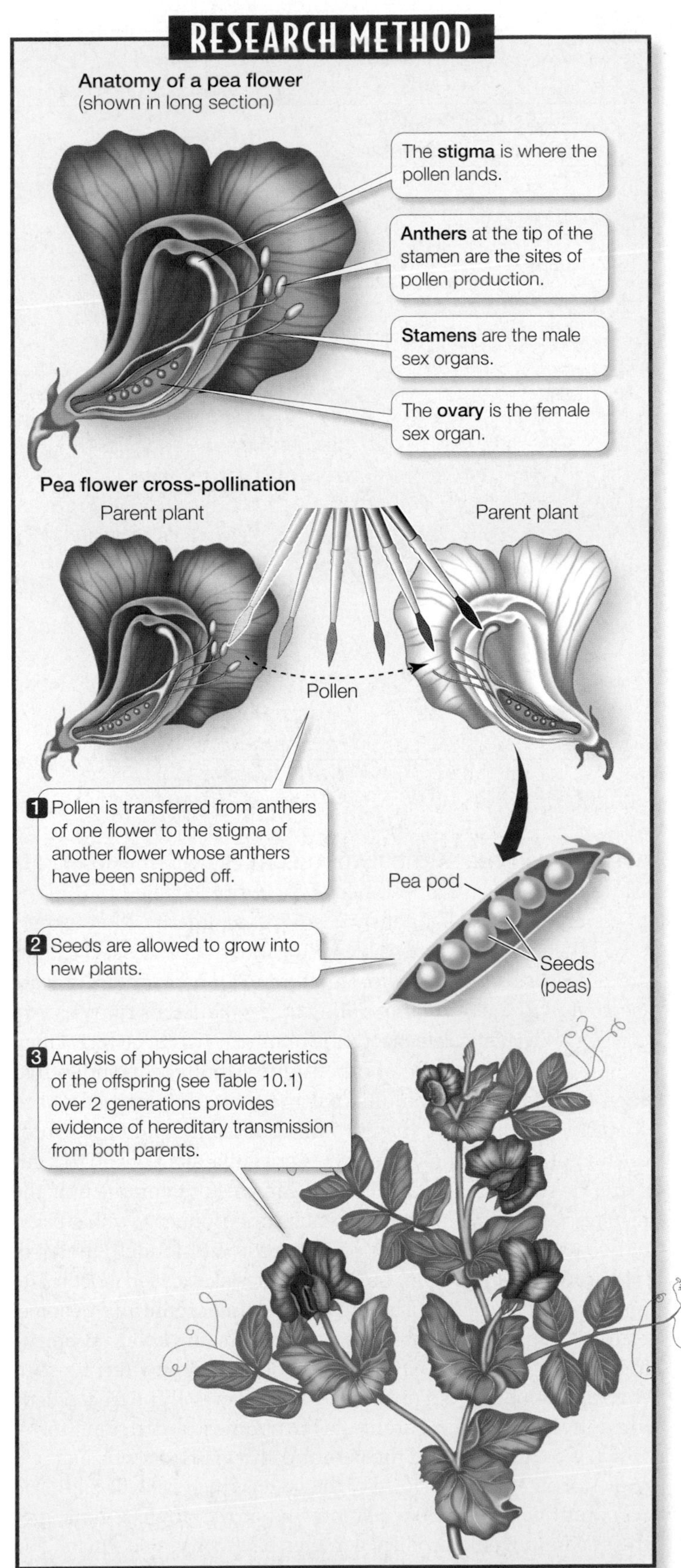
RESEARCH METHOD
Anatomy of a pea flower
(shown in long section)
The **stigma** is where the pollen lands.
Anthers at the tip of the stamen are the sites of pollen production.
Stamens are the male sex organs.
The **ovary** is the female sex organ.
Pea flower cross-pollination
Parent plant
Parent plant
Pollen
1 Pollen is transferred from anthers of one flower to the stigma of another flower whose anthers have been snipped off.
Pea pod
2 Seeds are allowed to grow into new plants.
Seeds (peas)
3 Analysis of physical characteristics of the offspring (see Table 10.1) over 2 generations provides evidence of hereditary transmission from both parents.

10.2 A Controlled Cross between Two Plants Plants were widely used in early genetic studies because it is easy to control which individuals mate with which. Mendel used the garden pea (*Pisum sativum*) in many of his experiments.

Mendel looked for characters with well-defined, contrasting alternative traits, such as purple flowers versus white flowers. Furthermore, these traits had to be **true-breeding**, meaning that the observed trait was the only form present for many generations. In other words, if they were true-breeding, peas with white flowers, when crossed with one another, would give rise only to progeny with white flowers for many generations; tall plants bred to tall plants would produce only tall progeny.

Mendel isolated each of his true-breeding strains by repeated inbreeding (done by crossing of sibling plants that were seemingly identical or by allowing individuals to self-pollinate) and selection. In most of his work, Mendel concentrated on the seven pairs of contrasting traits shown in **Table 10.1**. Before performing any experimental cross, he made sure that each potential parent was from a true-breeding strain—an essential point in his analysis of his experimental results.

Mendel then performed his crosses in the following manner:

- He collected pollen from one parental strain and placed it on the stigma (female organ) of flowers of the other strain whose anthers (male organs) had been removed (so that the recipient plant could not fertilize itself). The plants providing and receiving the pollen were the **parental generation**, designated **P**.
- In due course, seeds formed and were planted. The seeds and the resulting new plants constituted the **first filial generation**, or $\mathbf{F_1}$. Mendel and his assistants examined each F_1 plant to see which traits it bore and then recorded the number of F_1 plants expressing each trait.
- In some experiments the F_1 plants were allowed to self-pollinate and produce a **second filial generation**, $\mathbf{F_2}$. Again, each F_2 plant was characterized and counted.

Mendel's first experiments involved monohybrid crosses

The term *hybrid* refers to offspring of crosses between organisms differing in one or more traits. In Mendel's first experiment, he crossed two true-breeding parental (P) lineages differing in just *one* trait, producing *monohybrids* (the F_1 generation). He subsequently planted the F_1 seeds and allowed the resulting plants to self-pollinate to produce the F_2 generation. This technique is referred to as a **monohybrid cross**, even though in this case, the monohybrid plants were not literally crossed, but self-pollinated.

Mendel performed the same experiment for all seven pea-plant traits. His method is illustrated in **Figure 10.3**, using the seed shape trait as an example. He took pollen from pea plants of a true-breeding strain with wrinkled seeds and placed it on the stigmas of flowers of a true-breeding strain with spherical seeds. He also performed the complementary cross, in which the parental source of each trait is reversed: he placed pollen from the spherical-seeded strain on the stigmas of flowers of the wrinkled-seeded strain. In all cases, all F_1 seeds produced by crosses of the P plants were spherical—it was as if the wrinkled seed trait had disappeared completely.

The following spring, Mendel grew 253 F_1 plants from these spherical seeds. Each of these plants was allowed to self-polli-

TABLE 10.1

Mendel's Results from Monohybrid Crosses

PARENTAL GENERATION PHENOTYPES		F_2 GENERATION PHENOTYPES			
DOMINANT	RECESSIVE	DOMINANT	RECESSIVE	TOTAL	RATIO
Spherical seeds ×	Wrinkled seeds	5,474	1,850	7,324	2.96:1
Yellow seeds ×	Green seeds	6,022	2,001	8,023	3.01:1
Purple flowers ×	White flowers	705	224	929	3.15:1
Inflated pods ×	Constricted pods	882	299	1,181	2.95:1
Green pods ×	Yellow pods	428	152	580	2.82:1
Axial flowers ×	Terminal flowers	651	207	858	3.14:1
Tall stems (1 m) ×	Dwarf stems (0.3 m)	787	277	1,064	2.84:1

nate to produce F_2 seeds. In all, 7,324 F_2 seeds were produced, of which 5,474 were spherical and 1,850 wrinkled (Figure 10.3 and Table 10.1).

Mendel observed that the wrinkled seed trait was never expressed in the F_1 generation, even though it reappeared in the F_2 generation. This lead him to conclude that the spherical seed trait was **dominant** to the wrinkled seed trait, which he called **recessive**. In each of the other six pairs of traits Mendel studied, one trait proved to be dominant over the other trait. The trait that disappears in the F_1 generation of true-breeding crosses is always the recessive trait.

Mendel also observed that the ratio of the two traits in the F_2 generation was always the same—approximately 3:1—for each of the seven pea-plant traits he studied. That is, *three-fourths of the F_2 generation showed the dominant trait and one-fourth showed the recessive trait* (see Table 10.1). For example, Mendel's monohybrid cross for seed shape produced a ratio of 5,474:1,850 = 2.96:1. The two reciprocal crosses in the parental generation yielded similar outcomes in the F_2; it did not matter which parent contributed the pollen, just as Kölreuter had shown.

REJECTION OF THE BLENDING THEORY Mendel's monohybrid cross experiments showed that inheritance cannot be the result of a blending phenomenon. According to the blending theory, Mendel's F_1 seeds should have had an appearance intermediate between those of the two parents—in other words, they should have been slightly wrinkled. Furthermore, the blending theory offered no explanation for the reappearance of the wrinkled trait in the F_2 seeds after its apparent absence in the F_1 seeds.

SUPPORT FOR THE PARTICULATE THEORY Given the absence of blending and the reappearance of the wrinkled seed trait in the F_2 generation of his monohybrid cross experiments, Mendel proposed that the units responsible for the inheritance of specific traits are present as *discrete particles* that occur in pairs and segregate (separate) from one another during the formation of gametes. According to his **particulate theory**, the units of inheritance retain their integrity in the presence of other units. Mendel concluded that each pea plant has two units (particles) of inheritance for each character, one from each parent. He proposed that during the production of gametes, only one of these paired units is given to a gamete. He concluded that while each gamete contains one unit, the resulting zygote contains two, because it is produced by the fusion of two gametes. This conclusion is the core of Mendel's model of inheritance. Mendel's unit of inheritance is now called a **gene**. The totality of all the genes of an organism is that organism's **genome**.

Mendel reasoned that in his experiments, the two true-breeding parent plants had different forms of the gene affecting seed shape (although he did not use the term "gene"). The true-breeding spherical-seeded parent had two genes of the same form, which we will call *S*, and the parent with wrinkled seeds had two copies of an alternative form of the gene, which we will call *s*. The *SS* parent would produce gametes having a single *S* gene, and the *ss* parent would produce gametes having a single *s* gene. The cross producing the F_1 generation would donate an *S* from one parent and an *s* from the other to each seed; the F_1 offspring would thus be *Ss*. We say that *S* is dominant over *s* because the trait specified by *s* is not evident—is not *expressed*—when both forms of the gene are present.

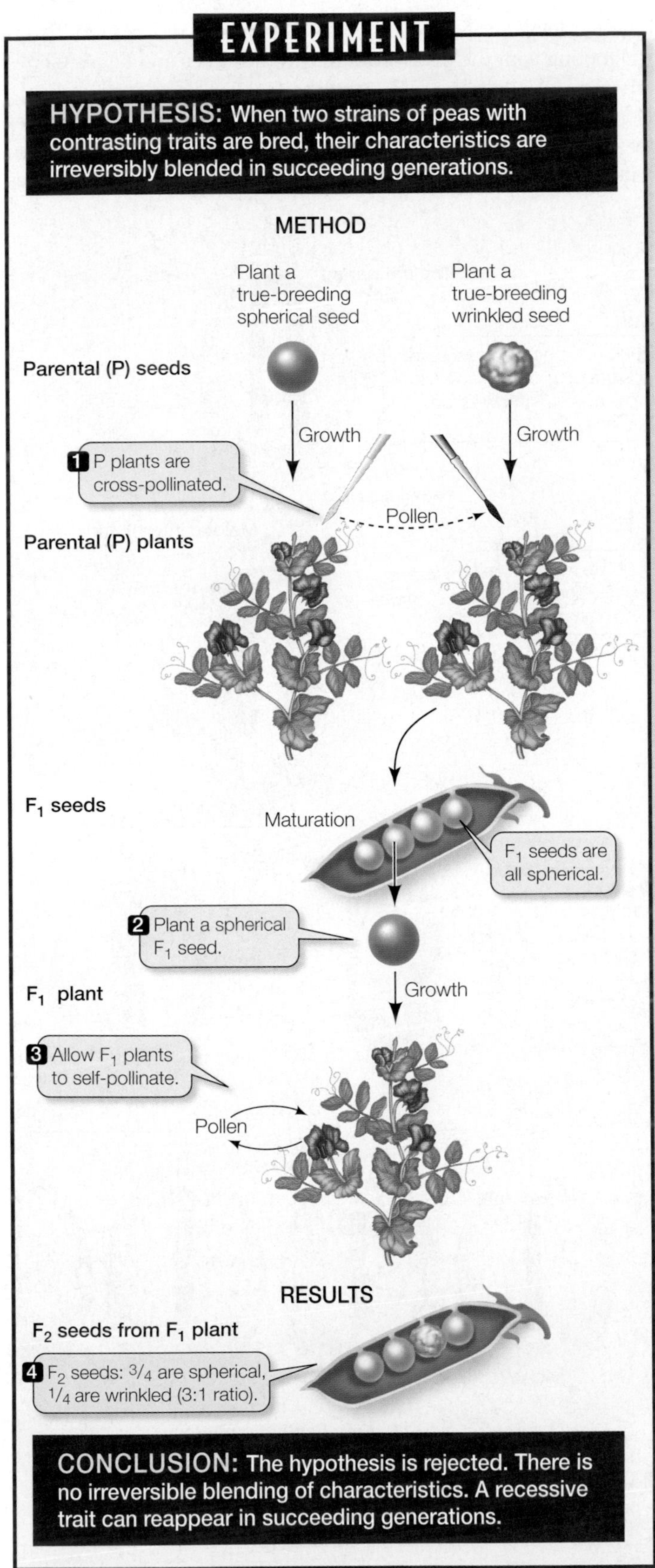

EXPERIMENT
HYPOTHESIS: When two strains of peas with contrasting traits are bred, their characteristics are irreversibly blended in succeeding generations.
METHOD
Plant a true-breeding spherical seed
Plant a true-breeding wrinkled seed
Parental (P) seeds
Growth
Growth
1 P plants are cross-pollinated.
Pollen
Parental (P) plants
F_1 seeds
Maturation
F_1 seeds are all spherical.
2 Plant a spherical F_1 seed.
F_1 plant
Growth
3 Allow F_1 plants to self-pollinate.
Pollen
RESULTS
F_2 seeds from F_1 plant
4 F_2 seeds: 3/4 are spherical, 1/4 are wrinkled (3:1 ratio).
CONCLUSION: The hypothesis is rejected. There is no irreversible blending of characteristics. A recessive trait can reappear in succeeding generations.

10.3 Mendel's Monohybrid Experiments The pattern Mendel observed in the F_2 generation—3/4 of the seeds spherical, 1/4 wrinkled—was the same no matter which strain contributed the pollen in the parental generation.

Alleles are different forms of a gene

The different forms of a gene (*S* and *s* in this case) are called **alleles**. Individuals that are true-breeding for a trait contain two copies of the same allele. For example, all the individuals in a population of a strain of true-breeding peas with wrinkled seeds must have the allele pair *ss*; if the dominant *S* allele were present, the plants would produce spherical seeds.

We say that the individuals that produce wrinkled seeds are **homozygous** for the allele *s*, meaning that they have two copies of the same allele (*ss*). Some peas with spherical seeds—the ones with the genotype *SS*—are also homozygous. However, not all plants with spherical seeds have the *SS* genotype. Some spherical-seeded plants, like Mendel's F_1, are **heterozygous**: they have two different alleles of the gene in question (in this case, *Ss*). An individual that is homozygous for a character is sometimes called a *homozygote*; a *heterozygote* is heterozygous for the character in question.

As a somewhat more complex example of inheritance, let's consider three gene pairs. An individual with the alleles *AABbcc* is homozygous for the *A* and *C* genes, because it has two *A* alleles and two *c* alleles, but heterozygous for the *B* gene, because it contains the *B* and *b* alleles.

The physical appearance of an organism is its **phenotype**. Mendel correctly supposed the phenotype to be the result of the **genotype**, or genetic constitution, of the organism showing the phenotype. Spherical seeds and wrinkled seeds are two phenotypes, which are the result of *three* genotypes: the wrinkled seed phenotype is produced by the genotype *ss*, whereas the spherical seed phenotype is produced by two genotypes, *SS* and *Ss*.

What's in a name—of a gene? Names such as *tall*, *short*, *spherical*, and *wrinkled* describe a trait. So what do *Drosophila* geneticists name a fruit fly that is impaired in learning experiments? Why, *dunce*, of course! A mutation that prevents heart formation? Call it *tinman*. A gene that produces extra bristles on the face is *groucho*; flies with no external genitalia are *ken* and *barbie*.

Mendel's first law says that the two copies of a gene segregate

How does Mendel's model of inheritance explain the ratios of traits seen in the F_1 and F_2 generations? Consider first the F_1, in which all progeny have the spherical seed phenotype. According to Mendel's model, when any individual produces gametes, the two copies of a gene separate, so that each gamete receives only one copy. This is *Mendel's first law*, the **law of segregation**. Thus from each parent of the P generation, every individual in the F_1 inherits one gene copy, and has the genotype *Ss* (**Figure 10.4**).

Now let's consider the composition of the F_2 generation. Half of the gametes produced by the F_1 generation have the *S* allele and the other half the *s* allele. Since both *SS* and *Ss* plants produce spherical seeds while *ss* produces wrinkled seeds, in the F_2 generation there are three ways to get a spherical-seeded plant, but

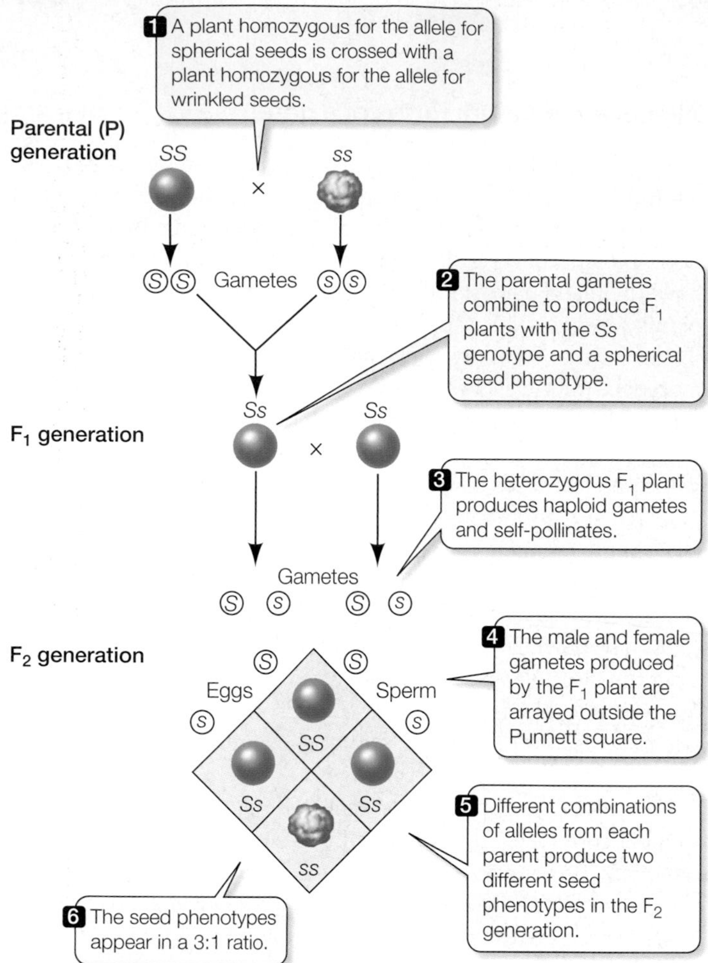

1 A plant homozygous for the allele for spherical seeds is crossed with a plant homozygous for the allele for wrinkled seeds.
Parental (P) generation
SS
ss
×
Gametes
2 The parental gametes combine to produce F_1 plants with the Ss genotype and a spherical seed phenotype.
F_1 generation
Ss
Ss
3 The heterozygous F_1 plant produces haploid gametes and self-pollinates.
Gametes
F_2 generation
Eggs
Sperm
4 The male and female gametes produced by the F_1 plant are arrayed outside the Punnett square.
SS
Ss
Ss
ss
5 Different combinations of alleles from each parent produce two different seed phenotypes in the F_2 generation.
6 The seed phenotypes appear in a 3:1 ratio.

10.4 Mendel's Explanation of Inheritance Mendel concluded that inheritance depends on discrete factors from each parent that do not blend in the offspring.

only one way to get a wrinkled-seeded plant (*s* from both parents)—predicting a 3:1 ratio remarkably close to the values Mendel found experimentally for all seven of the traits he compared (see Table 10.1).

The allele combinations that will result from a cross can be predicted using a **Punnett square**, a method devised in 1905 by the British geneticist Reginald Crundall Punnett. This device ensures that we consider all possible combinations of gametes when calculating expected genotype frequencies. A Punnett square looks like this:

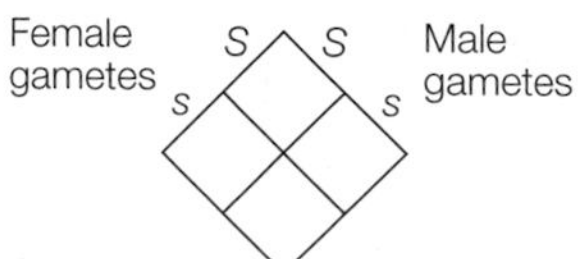

It is a simple grid with all possible male gamete (haploid sperm) genotypes shown along one side and all possible female gamete (haploid egg) genotypes along another side. The grid is completed by filling in each square with the diploid genotype that can be generated from each combination of gametes (see Figure 10.4). In this example, to fill in, say, the rightmost square, we put in the *S* from the egg (female gamete) and the *s* from the pollen (male gamete), yielding *Ss*.

Mendel did not live to see his theory placed on a sound physical footing with the discoveries of chromosomes and DNA. Genes are now known to be regions of the DNA molecules in chromosomes. More specifically, a gene is a sequence of DNA that resides at a particular site on a chromosome, called a **locus** (plural **loci**), and encodes a particular character. Genes are expressed in the pheno-

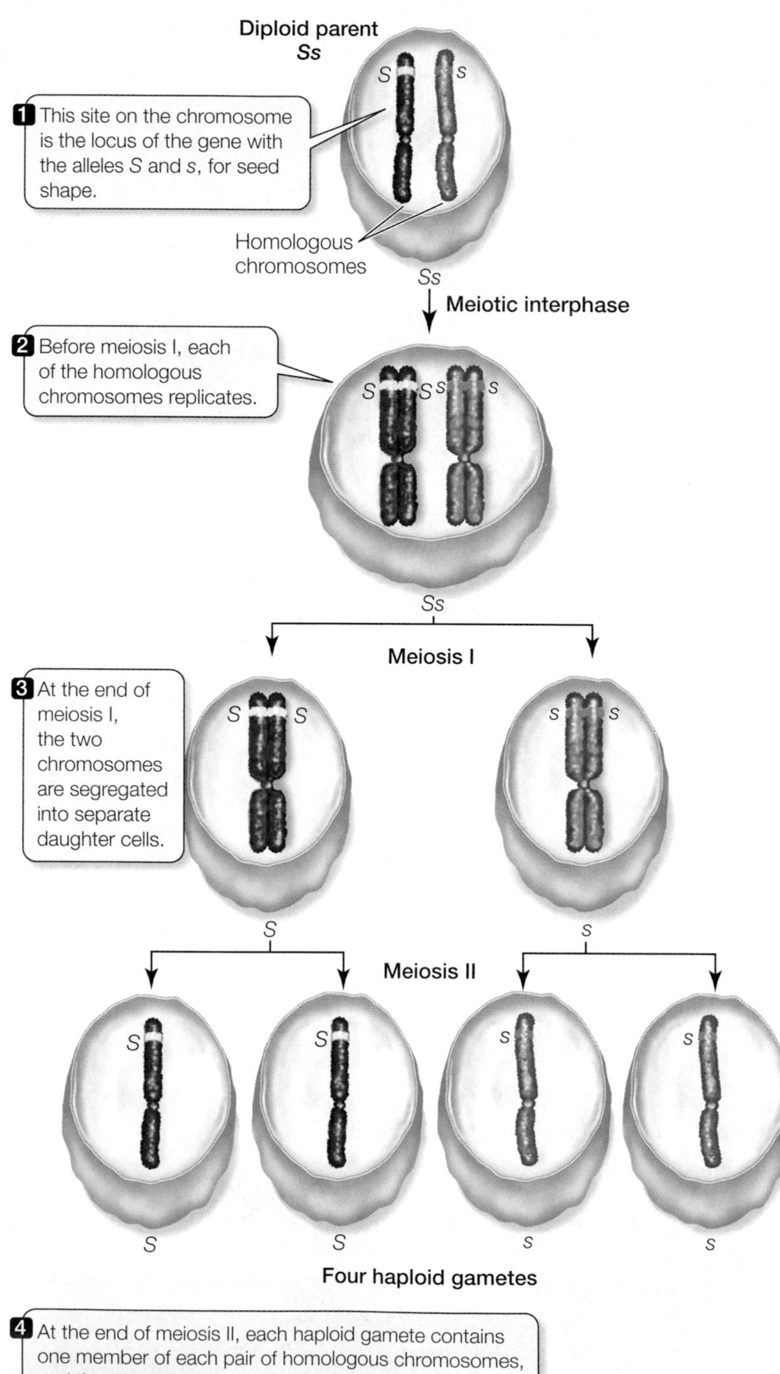

10.5 Meiosis Accounts for the Segregation of Alleles Although Mendel had no knowledge of chromosomes or meiosis, we now know that a pair of alleles resides on homologous chromosomes, and that meiosis segregates those alleles.

type mostly as proteins with particular functions, such as enzymes. So a dominant gene can be thought of as a region of DNA that is expressed as a functional enzyme, while a recessive gene typically expresses a nonfunctional enzyme. Mendel arrived at his law of segregation with no knowledge of chromosomes or meiosis, but today we can picture the different alleles of a gene segregating as chromosomes separate in meiosis I (**Figure 10.5**).

Mendel verified his hypothesis by performing a test cross

Mendel set out to test his hypothesis that there were two possible allele combinations (*SS* and *Ss*) in the spherical-seeded F_1 generation. He did so by performing a **test cross**, which is a way of finding out whether an individual showing a dominant trait is homozygous or heterozygous. In a test cross, the individual in question is crossed with an individual known to be homozygous for the recessive trait—an easy individual to identify, because in order to have the recessive phenotype, it must be homozygous for the recessive trait.

For the seed shape gene that we have been considering, the recessive homozygote used for the test cross is *ss*. The individual being tested may be described initially as *S_* because we do not yet know the identity of the second allele. We can predict two possible results:

- If the individual being tested is homozygous dominant (*SS*), all offspring of the test cross will be *Ss* and show the dominant trait (spherical seeds) (**Figure 10.6, left**).
- If the individual being tested is heterozygous (*Ss*), then approximately half of the offspring of the test cross will be heterozygous and show the dominant trait (*Ss*), but the other half will be homozygous for, and will show, the recessive trait (*ss*) (**Figure 10.6, right**).

The second prediction matches the results that Mendel obtained; thus Mendel's hypothesis accurately predicted the results of his test cross.

With his first hypothesis confirmed, Mendel went on to ask another question: How do different pairs of genes behave in crosses when considered together?

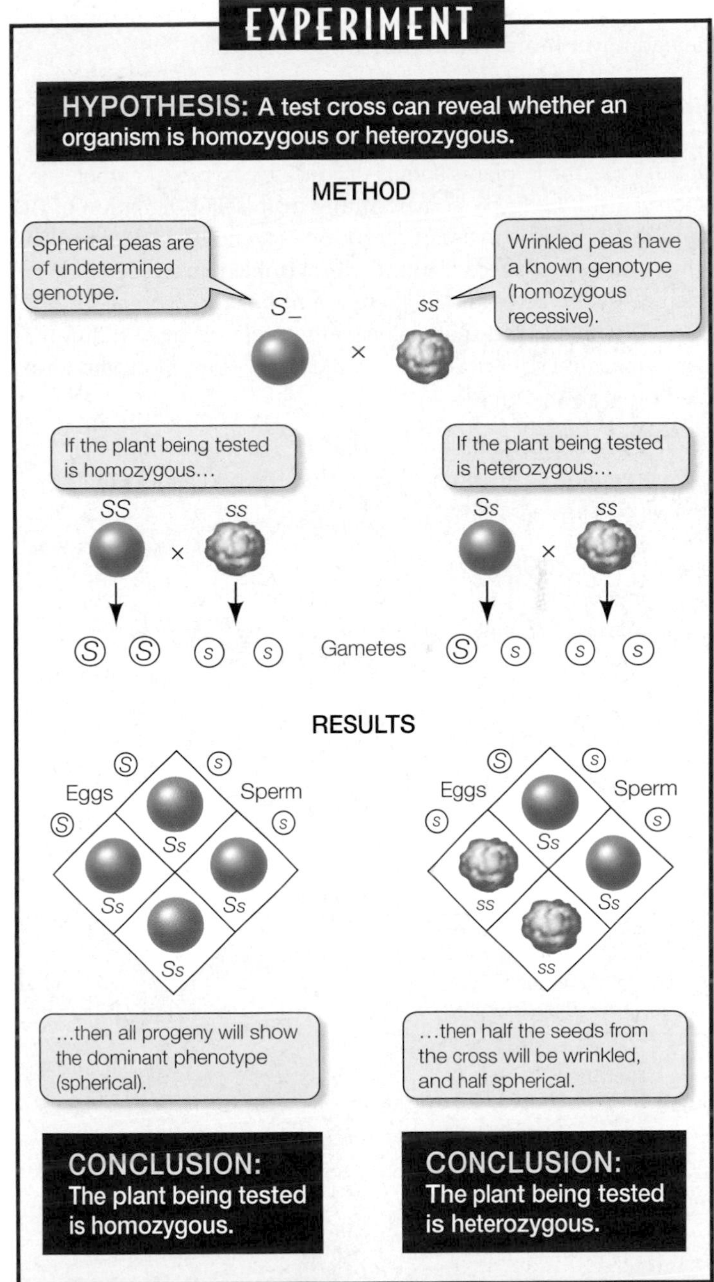

EXPERIMENT
HYPOTHESIS: A test cross can reveal whether an organism is homozygous or heterozygous.
METHOD
Spherical peas are of undetermined genotype.
Wrinkled peas have a known genotype (homozygous recessive).
S_
ss
×
If the plant being tested is homozygous...
If the plant being tested is heterozygous...
SS
ss
Ss
ss
Gametes
S S s s
S s s s
RESULTS
Eggs
Sperm
Ss
Ss
Ss
Ss
ss
ss
...then all progeny will show the dominant phenotype (spherical).
...then half the seeds from the cross will be wrinkled, and half spherical.
CONCLUSION: The plant being tested is homozygous.
CONCLUSION: The plant being tested is heterozygous.

10.6 Homozygous or Heterozygous? An individual with a dominant phenotype may be homozygous or heterozygous. Its genotype can be determined by crossing it with a homozygous recessive individual and observing the phenotypes of the progeny produced. This procedure is known as a test cross. FURTHER RESEARCH: What would be the result if the "tester" plant was homozygous for spherical instead of wrinkled seeds?

Mendel's second law says that copies of different genes assort independently

Consider an organism that is heterozygous for two genes (*SsYy*), in which the *S* and *Y* alleles came from its mother and *s* and *y* came from its father. When this organism produces gametes, do the alleles of maternal origin (*S* and *Y*) go together to one gamete and those of paternal origin (*s* and *y*) to another gamete? Or can a single gamete receive one maternal and one paternal allele, *S* and *y* (or *s* and *Y*)?

To answer these questions, Mendel performed another series of experiments. He began with peas that differed in two seed characters: seed shape and seed color. One true-breeding parental strain produced only spherical, yellow seeds (*SSYY*), and the other produced only wrinkled, green ones (*ssyy*). A cross between these two strains produced an F_1 generation in which all the plants were *SsYy*. Because the *S* and *Y* alleles are dominant, the F_1 seeds were all spherical and yellow.

Mendel continued this experiment to the F_2 generation by performing a **dihybrid cross** (a cross between individuals that are identical double heterozygotes) with F_1 plants (although again, in this case, this was done by allowing the F_1 plants to self-pollinate). There are two possible ways in which such doubly heterozygous plants might produce gametes, as Mendel saw it (remember that he had never heard of chromosomes or meiosis):

1. The alleles could maintain the associations they had in the parental generation (that is, they could be *linked*).

In this case, the F_1 plants should produce two types of gametes (*SY* and *sy*), and the F_2 progeny resulting from self-pollination of the F_1 plants should consist of three times as many plants bearing spherical, yellow seeds as ones with wrinkled, green seeds. Were such results to be obtained, there might be no reason to suppose that seed shape and seed color were regulated by two different genes, because spherical seeds would always be yellow and wrinkled ones always green.

2. The segregation of *S* from *s* could be independent of the segregation of *Y* from *y* (that is, that the two genes could be *unlinked*).

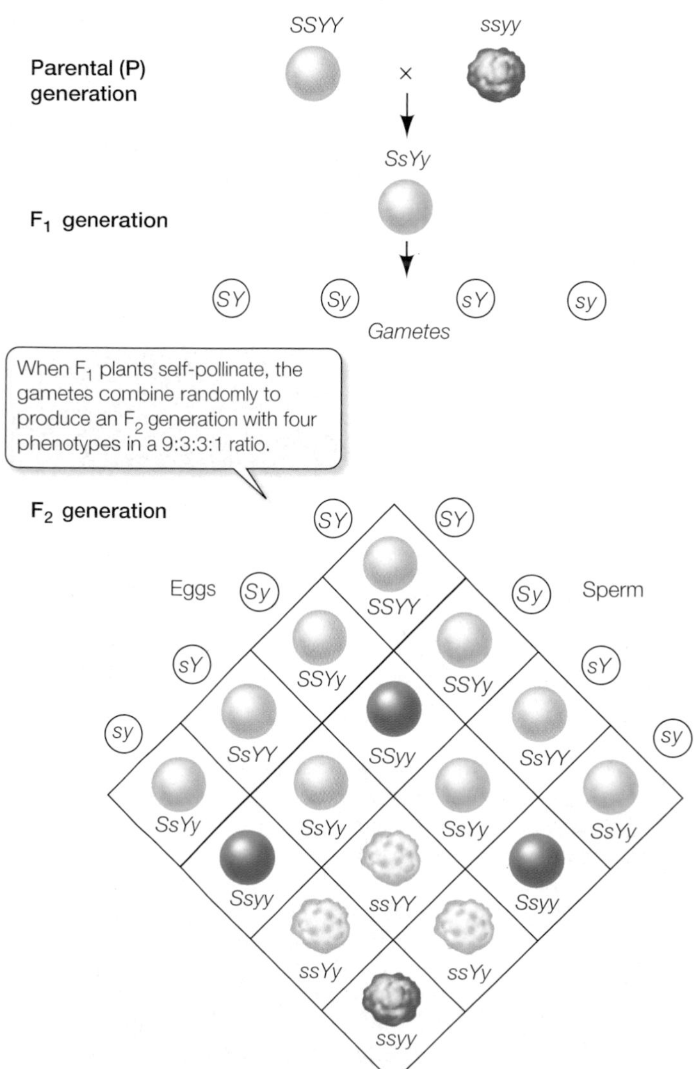

10.7 Independent Assortment The 16 possible combinations of gametes in this dihybrid cross result in 9 different genotypes. Because *S* and *Y* are dominant over *s* and *y*, respectively, the 9 genotypes result in four phenotypes in a ratio of 9:3:3:1. These results show that the two genes segregate independently.

In this case, four kinds of gametes should be produced by the F_1 in equal numbers: *SY*, *Sy*, *sY*, and *sy*. When these gametes combine at random, they should produce an F_2 having nine different genotypes. The F_2 progeny could have any of three possible genotypes for shape (*SS*, *Ss*, or *ss*) and any of three possible genotypes for color (*YY*, *Yy*, or *yy*). The combined nine genotypes should produce four phenotypes (spherical yellow, spherical green, wrinkled yellow, wrinkled green). Putting these data into a Punnett square, we can predict that these four phenotypes will occur in a ratio of 9:3:3:1 (**Figure 10.7**).

Mendel's dihybrid crosses supported the second prediction: four different phenotypes appeared in the F_2 in a ratio of about 9:3:3:1. The parental traits appeared in new combinations (spherical green and wrinkled yellow) in some progeny. Such new combinations are called **recombinant** phenotypes.

These results led Mendel to the formulation of what is now known as *Mendel's second law*: Alleles of different genes assort independently of one another during gamete formation. That is, the segregation of the alleles of gene A is independent of the segregation of the alleles of gene B. We now know that this **law of independent assortment** is not as universal as the law of segregation, because it applies to genes located on separate chromosomes, but not always to those located on the same chromosome, as we will see in Section 10.4. However, it is correct to say that *chromosomes* segregate independently during the formation of gametes, and so do any two genes on separate homologous chromosome pairs (**Figure 10.8**).

One of Mendel's major contributions to the science of genetics was his use of the rules of statistics and probability to analyze his masses of data from hundreds of crosses producing thousands of plants. His mathematical analyses revealed clear patterns in the data that allowed him to formulate his hypotheses. Ever since Mendel, geneticists have used simple mathematics in the same ways that Mendel did.

Punnett squares or probability calculations: A choice of methods

Punnett squares provide one way of solving problems in genetics, and probability calculations provide another. Many people find it easiest to use the principles of probability, some of which are intuitive and familiar. For example, when we flip a coin, the law of probability states that it has an equal probability of landing "heads" or "tails." For any given toss of a fair coin, the probability of heads is independent of what happened in all the previous tosses. A run of ten straight heads implies nothing about the next toss. No "law of averages" increases the likelihood that the next toss will come up tails, and no "momentum" makes an eleventh occurrence of heads any more likely. On the eleventh toss, the odds of getting heads are still 50-50.

The basic conventions of probability are simple:

- If an event is absolutely certain to happen, its probability is 1.
- If it cannot possibly happen, its probability is 0.
- All other events have a probability between 0 and 1.

A coin toss results in heads approximately half the time, so the probability of heads is ½—as is the probability of tails.

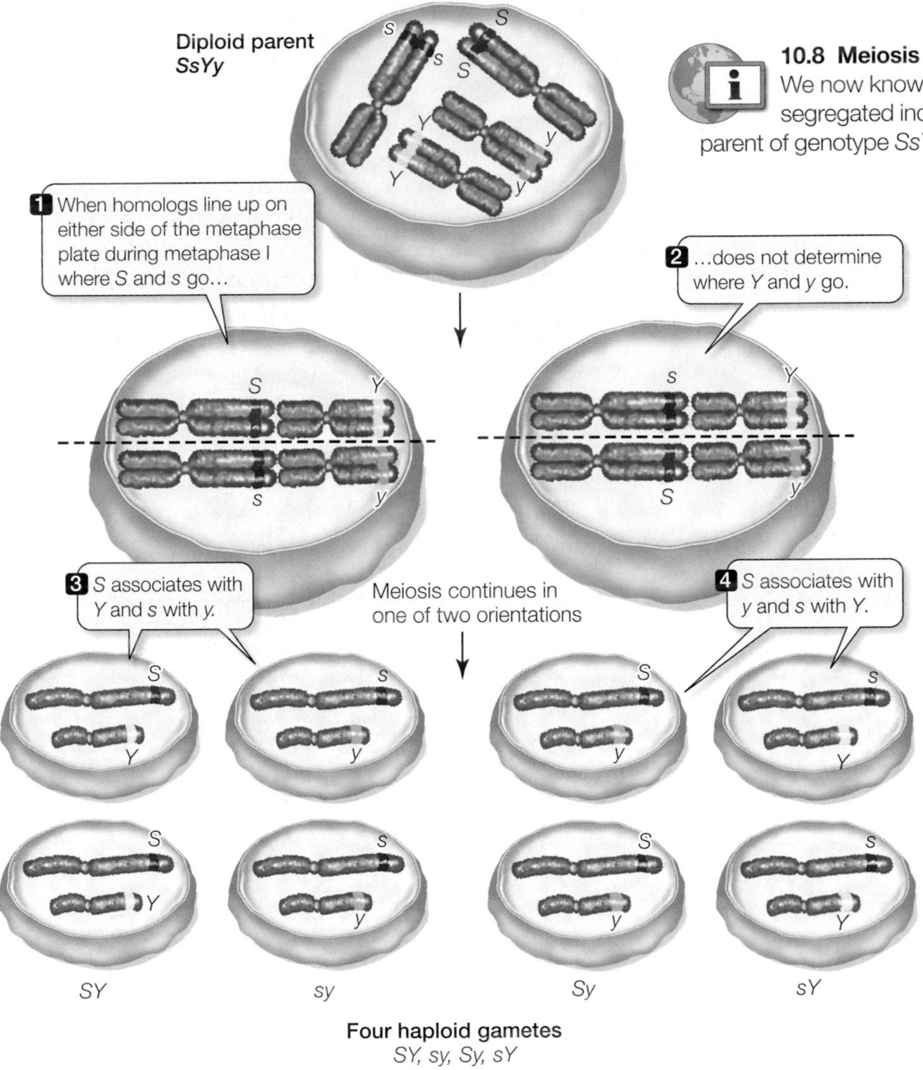

Diploid parent
SsYy
1 When homologs line up on either side of the metaphase plate during metaphase I where S and s go…
2 …does not determine where Y and y go.
3 S associates with Y and s with y.
Meiosis continues in one of two orientations
4 S associates with y and s with Y.
SY
sy
Sy
sY
Four haploid gametes
SY, sy, Sy, sY

10.8 Meiosis Accounts for Independent Assortment of Alleles We now know that copies of genes on different chromosomes are segregated independently during metaphase I of meiosis. Thus a parent of genotype *SsYy* can form gametes with four different genotypes.

MULTIPLYING PROBABILITIES How can we determine the probability of two independent events happening together? If two coins (a penny and a dime, say) are tossed, each acts independently of the other. What, then, is the probability of both coins coming up heads? Half the time, the penny comes up heads; of that fraction, half the time the dime also comes up heads. Therefore, the *joint probability* of both coins coming up heads is half of one-half, or $\frac{1}{2} \times \frac{1}{2} = \frac{1}{4}$. To find the joint probability of independent events, then, we multiply the probabilities of the individual events (**Figure 10.9**). How does this method apply to genetics?

To see how joint probability is calculated in genetics problems, let's consider the monohybrid cross. The probabilities of two events are involved: gamete formation and random fertilization.

Calculating the probabilities involved in gamete formation is straightforward. A homozygote can produce only one type of gamete, so, for example, the probability of an *SS* individual producing gametes with the genotype *S* is 1. The heterozygote *Ss* produces *S* gametes with a probability of $\frac{1}{2}$, and *s* gametes with a probability of $\frac{1}{2}$.

Now let's see how the rules of probability might predict the ratio of the F_2 progeny of the cross in Figure 10.4. These plants are obtained by self-pollination of F_1 plants of genotype *Ss*. The probability that an F_2 plant will have the genotype *SS* must be $\frac{1}{2} \times \frac{1}{2} = \frac{1}{4}$, because there is a 50-50 chance that the sperm will have the genotype *S*, and that chance is independent of the 50-50 chance that the egg will have the genotype *S*. Similarly, the probability of *ss* offspring is $\frac{1}{2} \times \frac{1}{2} = \frac{1}{4}$.

ADDING PROBABILITIES How are probabilities calculated when an event can happen in different ways? The probability of an F_2 plant getting an *S* allele from the sperm and an *s* allele from the egg is $\frac{1}{4}$, but remember that the same genotype can also result from an *s* from the sperm and an *S* from the egg, also with a probability of $\frac{1}{4}$. *The probability of an event that*

1 Two coin tosses are **independent events**, each with an outcome probability (*P*) of $\frac{1}{2}$.

2 This outcome is the result of two independent events. The joint probability is $\frac{1}{2} + \frac{1}{2} = \frac{1}{4}$ (**multiplication rule**).

$P = \frac{1}{2}$ $P = \frac{1}{2}$ $P = \frac{1}{2}$ $P = \frac{1}{2}$

S S S S S s S s s S s s

$\frac{1}{2} \times \frac{1}{2} = \frac{1}{4}$ $\frac{1}{2} \times \frac{1}{2} = \frac{1}{4}$ $\frac{1}{2} \times \frac{1}{2} = \frac{1}{4}$ $\frac{1}{2} \times \frac{1}{2} = \frac{1}{4}$

Because there are two ways to arrive at a heterozygote, we add the probabilities of the two individual outcomes: $\frac{1}{4} + \frac{1}{4} = \frac{1}{2}$ (**addition rule**).

10.9 Using Probability Calculations in Genetics Like the results of a coin toss, the probability of any given combination of alleles from a sperm and an egg appearing in the offspring of a cross can be obtained by multiplying the probabilities of each event. Since a heterozygote can be formed in two ways, these two probabilities are added together.

can occur in two or more different ways is the sum of the individual probabilities of those ways. Thus the probability that an F_2 plant will be a heterozygote is equal to the sum of the probabilities of the two ways of forming a heterozygote: $1/4 + 1/4 = 1/2$ (see Figure 10.9). The three genotypes are therefore expected in the ratio $1/4$ *SS*:$1/2$ *Ss*:$1/4$ *ss*—hence the 1:2:1 ratio of genotypes and the 3:1 ratio of phenotypes seen in Figure 10.4.

PROBABILITY AND THE DIHYBRID CROSS If F_1 plants heterozygous for two independent characters self-pollinate, the resulting F_2 plants express four different phenotypes. The proportions of these phenotypes are easily determined by probability calculations. Let's see how this works for the experiment shown in Figure 10.7.

Using the principles described above, we can calculate that the probability that an F_2 seed will be spherical is $3/4$: the probability of an *Ss* heterozygote ($1/2$) plus the probability of an *SS* homozygote ($1/4$) = $3/4$. By the same reasoning, the probability that a seed will be yellow is also $3/4$. The two characters are determined by separate genes and are independent of each other, so the joint probability that a seed will be both spherical and yellow is $3/4 \times 3/4 = 9/16$. What is the probability of F_2 seeds being both wrinkled and yellow? The probability of being yellow is again $3/4$; the probability of being wrinkled is $1/2 \times 1/2 = 1/4$. The joint probability that a seed will be both wrinkled and yellow, then, is $1/4 \times 3/4 = 3/16$. The same probability applies, for similar reasons, to spherical, green F_2 seeds. Finally, the probability that F_2 seeds will be both wrinkled and green is $1/4 \times 1/4 = 1/16$. Looking at all four phenotypes, we see they are expected in the ratio of 9:3:3:1.

Probability calculations and Punnett squares give the same results. Learn to do genetics problems both ways, and then decide which method you prefer.

Mendel's laws can be observed in human pedigrees

How are Mendel's laws of inheritance applied to humans? Mendel worked out his laws by performing many planned crosses and counting many offspring. Neither of these approaches is possible with humans, so human geneticists rely on **pedigrees**: family trees

(A) Dominant inheritance
Generation I (parents)
Generation II
Generation III
Every affected individual has an affected parent.
About 1/2 of the offspring (of both sexes) of an affected parent are affected.
Oldest Youngest
Siblings
(B) Recessive inheritance
Generation I (parents)
Generation II
Generation III
Generation IV
1 One parent is heterozygous...
2 ...and the recessive allele is passed on to 1/2 of the phenotypically normal offspring.
3 These cousins are both heterozygous.
4 Mating of heterozygous recessive parents may produce homozygous recessive (affected) offspring.
Unaffected
Affected
Heterozygote (unaffected phenotype)
Female
Male
Mating
Mating between relatives

10.10 Pedigree Analysis and Inheritance (A) This pedigree represents a family affected by Huntington's disease, which results from a rare dominant allele. Everyone who inherits this allele is affected. (B) The family in this pedigree carries the allele for albinism, a recessive trait. Because the trait is recessive, heterozygotes do not have the albino phenotype, but they can pass the allele on to their offspring. Affected persons must inherit the allele from two heterozygous parents or (rarely) from one homozygous and one heterozygous parent. In this family, the heterozygous parents are cousins, but the same result could occur if the parents were unrelated but heterozygous.

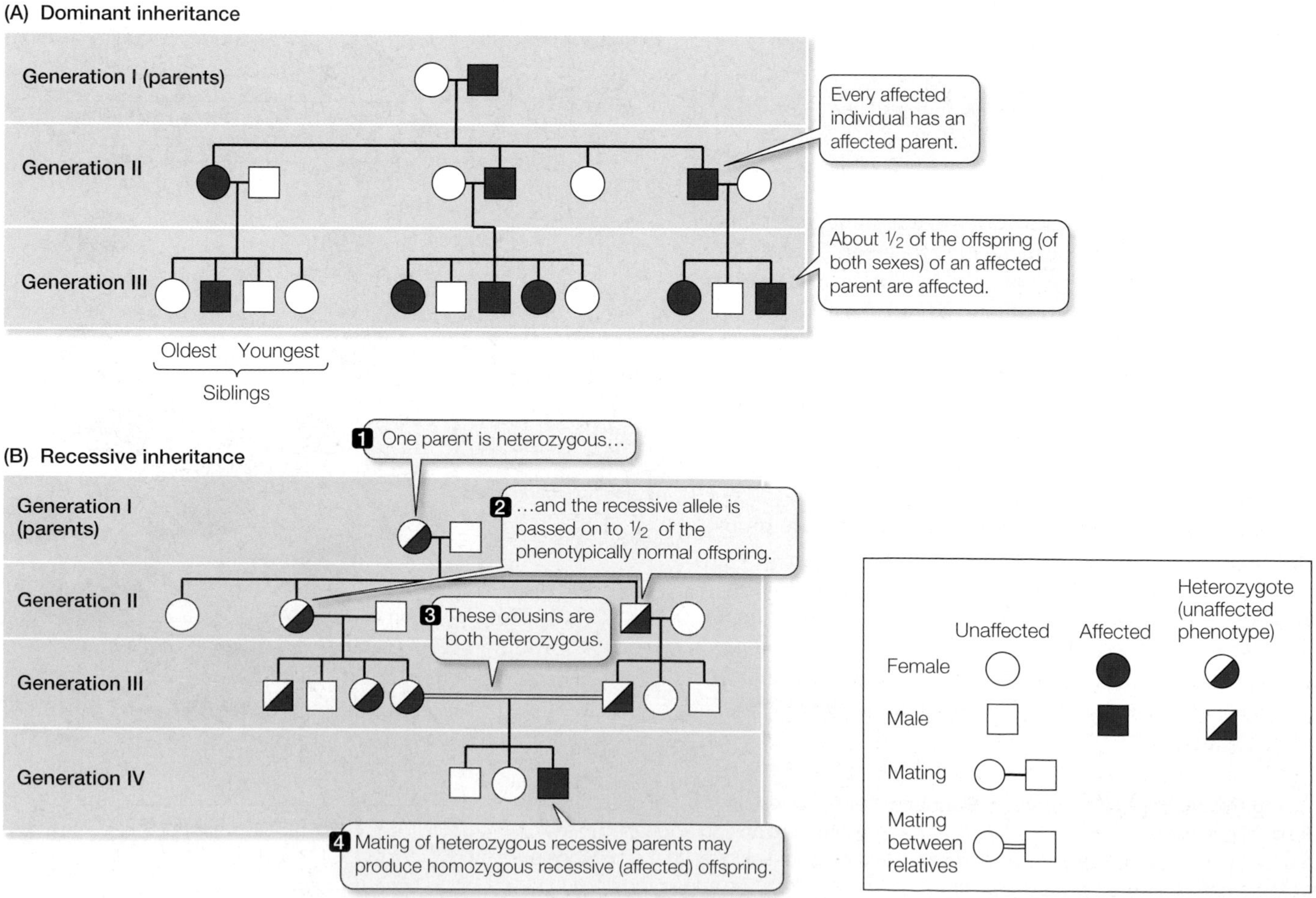

that show the occurrence of phenotypes (and alleles) in several generations of related individuals.

Because humans have such small numbers of offspring, human pedigrees do not show the clear proportions of offspring phenotypes that Mendel saw in his pea plants. For example, when a man and a woman who are both heterozygous for a recessive allele (say, *Aa*) have children together, each child has a 25 percent probability of being a recessive homozygote (*aa*). Thus if this couple were to have dozens of children, one-fourth of them would be recessive homozygotes (*aa*). But the offspring of a single couple are likely to be too few to show the exact one-fourth proportion. In a family with only two children, for example, both could easily be *aa* (or *Aa*, or *AA*).

What if we want to know whether a recessive allele is carried by both the mother and the father? Human geneticists assume that any allele that causes an abnormal phenotype (such as a genetic disease) is rare in the human population. This means that if some members of a given family have a rare allele, it is highly unlikely that an outsider marrying into that family will have that same rare allele.

Human geneticists may wish to know whether a particular rare allele that causes an abnormal phenotype is dominant or recessive. **Figure 10.10A** is a pedigree showing the pattern of inheritance of a rare *dominant allele*. The following are the key features to look for in such a pedigree:

- Every affected person has an affected parent.
- About half of the offspring of an affected parent are also affected.
- The phenotype occurs equally in both sexes.

Compare this pattern with **Figure 10.10B**, which shows the pattern of inheritance of a rare *recessive* allele:

- Affected people usually have two parents who are not affected.
- In affected families, about one-fourth of the children of unaffected parents are affected.
- The phenotype occurs equally in both sexes.

In pedigrees showing inheritance of a recessive phenotype, it is not uncommon to find a marriage of two relatives. This observation is a result of the rarity of recessive alleles that give rise to abnormal phenotypes. For two phenotypically normal parents to have an affected child (*aa*), the parents must both be heterozygous (*Aa*). If a particular recessive allele is rare in the general population, the chance of two people marrying who are both carrying that allele is quite low. On the other hand, if that allele is present in a family, two cousins might share it (see Figure 10.10B). This is why studies on populations isolated either culturally (by religion, as with the Amish in the United States) or geographically (as on islands) have been so valuable to human geneticists. People in these groups tend to have large families, or to marry among themselves, or both.

Because the major use of pedigree analysis is in the clinical evaluation and counseling of patients with inherited abnormalities, a single pair of alleles is usually followed. However, just as pedigree analysis shows the segregation of alleles, it also can show independent assortment if two different allele pairs are considered.

10.1 RECAP

Mendel showed that genetic determinants are particulate and do not "blend" or disappear when the genes from two gametes combine. Mendel's first law states that the two copies of a gene segregate during gamete formation. His second law states that genes assort independently during gamete formation. The frequencies with which different allele combinations will be expressed in offspring can be calculated with a Punnett square or using probability theory.

- What results seen in the F_1 and F_2 generations of Mendel's monohybrid cross experiments refuted the blending theory of inheritance? See p. 210 and Figures 10.3 and 10.4
- Can you explain Mendel's experiment on segregation of alleles in terms of meiosis? See pp. 211–212 and Figure 10.5
- Can you explain, in terms of meiosis, how Mendel's dihybrid cross experiments suggested independent assortment of alleles? See pp. 213–214 and Figures 10.7 and 10.8
- Draw human pedigrees for dominant and recessive inheritance? See pp. 216–217 and Figure 10.10

The laws of inheritance as articulated by Mendel remain valid today; his discoveries laid the groundwork for all future studies of genetics. Inevitably, however, we have learned that things are more complicated. Let's take a look at some of these complications, beginning with the interactions between alleles at different loci.

10.2 How Do Alleles Interact?

In many cases, alleles do not show the simple relationships between dominance and recessiveness that we have described. Existing alleles are subject to mutation, and thus may give rise to new alleles, so there can be many alleles for a single character. Furthermore, a single allele may have multiple phenotypic effects.

New alleles arise by mutation

Different alleles of a gene exist because genes are subject to **mutations**, which are rare, stable, and inherited changes in the genetic material. In other words, an allele can mutate to become a different allele. Mutation, which will be discussed in detail in Section 12.6, is a random process; different copies of the same allele may be changed in different ways.

Geneticists usually define one particular allele of a gene as the **wild type**; this allele is the one that is present in most individuals in nature ("the wild") and gives rise to an expected trait or phenotype. Other alleles of that gene, often called *mutant alleles*, may pro-

Possible genotypes	CC, Cc^{ch}, Cc^{h}, Cc	$c^{ch}c^{ch}$	$c^{ch}c^{h}$, $c^{ch}c$	$c^{h}c^{h}$, $c^{h}c$	cc
Phenotype	Dark gray	Chinchilla	Light gray	Point restricted	Albino

10.11 Inheritance of Coat Color in Rabbits There are four alleles of the gene for coat color in these Netherlands dwarf rabbits. Different combinations of two alleles give different coat colors. The dominance hierarchy is $C > c^{ch} > c^{h} > c$.

duce a different phenotype. The wild-type and mutant alleles reside at the same locus and are inherited according to the rules set forth by Mendel. A genetic locus with a wild-type allele that is present less than 99 percent of the time (the rest of the alleles being mutant) is said to be **polymorphic** (Greek *poly*, "many," and *morph*, "form").

Many genes have multiple alleles

Because of random mutations, more than two alleles of a given gene may exist in a group of individuals. (Any one individual has only two alleles—one from its mother and one from its father.) In fact, there are many examples of such multiple alleles.

Coat color in rabbits, for example, is determined by one gene with four alleles. Any rabbit with the *C* allele (paired with any of the four) is dark gray, and a rabbit with *cc* is albino. The intermediate colors result from the different allele combinations shown in **Figure 10.11**.

Multiple alleles increase the number of possible phenotypes. In Mendel's monohybrid cross, there was just one pair of alleles (*Ss*) and two possible phenotypes (resulting from *SS* or *Ss* and *ss*). The four alleles of the rabbit coat color gene produce five different phenotypes.

Dominance is not always complete

In the single pairs of alleles studied by Mendel, dominance is *complete* when an individual is heterozygous. That is, an *Ss* individual always expresses the *S* phenotype. However, many genes have alleles that are not dominant or recessive to one another. Instead, the heterozygotes show an intermediate phenotype—at first glance, like that predicted by the old blending theory of inheritance. For example, if a true-breeding red snapdragon is crossed with a true-breeding white one, all the F_1 flowers are pink. That this phenomenon can still be explained in terms of Mendelian genetics, rather than blending, is readily demonstrated by a further cross.

The blending theory predicts that if one of the pink F_1 snapdragons is crossed with a true-breeding white one, all the offspring should be a still lighter pink. In fact, approximately half of the offspring are white, and half are the same shade of pink as the F_1 parent. When the F_1 pink snapdragons are allowed to self-pollinate, the resulting F_2 plants show a ratio of 1 red:2 pink:1 white (**Figure 10.12**). Clearly the hereditary particles—

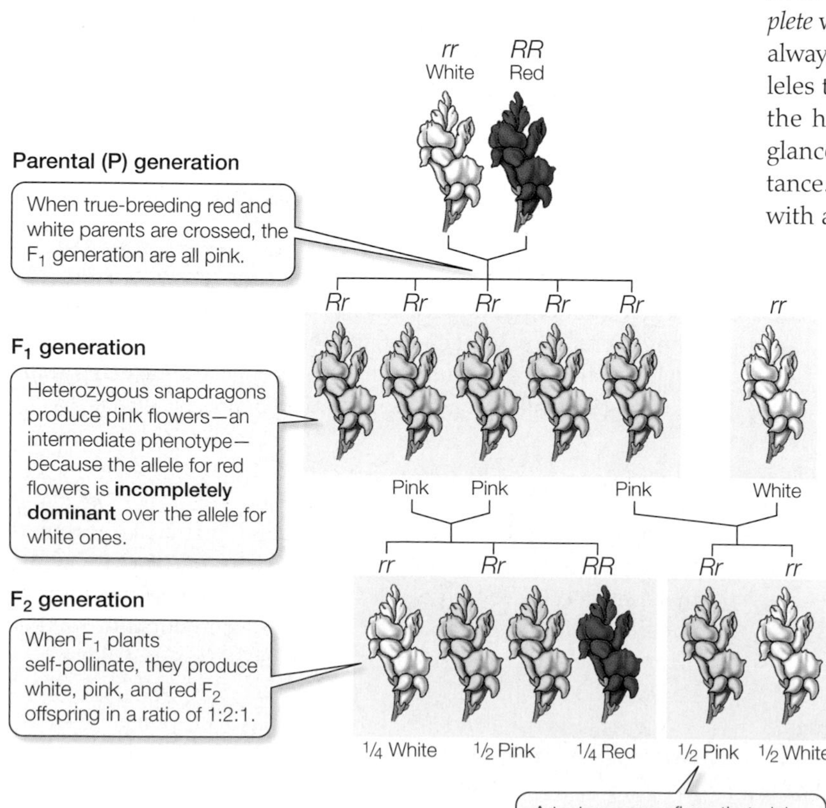
rr
White
RR
Red
Parental (P) generation
When true-breeding red and white parents are crossed, the F1 generation are all pink.
Rr Rr Rr Rr Rr
rr
F1 generation
Heterozygous snapdragons produce pink flowers—an intermediate phenotype—because the allele for red flowers is incompletely dominant over the allele for white ones.
Pink Pink Pink White
rr Rr RR Rr rr
F2 generation
When F1 plants self-pollinate, they produce white, pink, and red F2 offspring in a ratio of 1:2:1.
1/4 White 1/2 Pink 1/4 Red 1/2 Pink 1/2 White
A test cross confirms that pink snapdragons are heterozygous.

10.12 Incomplete Dominance Follows Mendel's Laws An intermediate phenotype can occur in heterozygotes when neither allele is dominant. The heterozygous phenotype (here, pink flowers) may give the appearance of a blended trait, but the traits of the parental generation reappear in their original forms in succeeding generations, as predicted by Mendel's laws of inheritance.

10.13 ABO Blood Reactions Are Important in Transfusions This graph shows the results of mixing red blood cells of types A, B, AB, and O with serum containing anti-A or anti-B antibodies. As you look down the columns, note that each of the types, when mixed separately with anti-A and with anti-B, gives a unique pair of results; this is the basic method by which blood is typed. People with type O blood are good blood donors because O cells do not react with either anti-A or anti-B antibodies. People with type AB blood are good recipients, since they make neither type of antibody.

Blood type of cells	Genotype	Antibodies made by body	Reaction to added antibodies	
			Anti-A	Anti-B
A	I^AI^A or I^Ai^O	Anti-B		
B	I^BI^B or I^Bi^O	Anti-A		
AB	I^AI^B	Neither anti-A nor anti-B		
O	i^Oi^O	Both anti-A and anti-B		

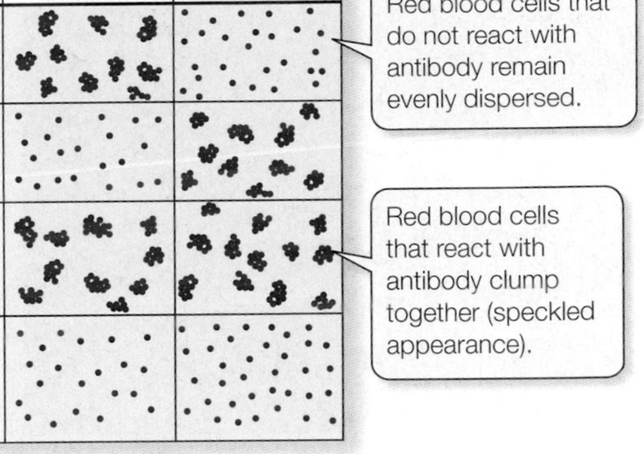

Red blood cells that do not react with antibody remain evenly dispersed.

Red blood cells that react with antibody clump together (speckled appearance).

the genes—have not blended; they are readily sorted out in the F_2 generation.

We can understand these results in terms of the Mendelian laws of inheritance. When heterozygotes show a phenotype intermediate between those of the two homozygotes, the gene is said to be governed by **incomplete dominance**. In other words, neither of the two alleles is dominant. Incomplete dominance is common in nature. In fact, Mendel's study of seven pea-plant traits is unusual in that all seven traits happened to be characterized by complete dominance.

In codominance, both alleles at a locus are expressed

Sometimes the two alleles at a locus produce two different phenotypes that *both* appear in heterozygotes, a phenomenon called **codominance**. A good example of codominance is seen in the ABO blood group system in humans.

Early attempts at blood transfusion frequently killed the patient. Around 1900, the Austrian scientist Karl Landsteiner mixed blood cells and *serum* (blood from which cells have been removed) from different individuals. He found that only certain combinations of blood are compatible. In other combinations, the red blood cells from one individual form clumps in the presence of serum from the other individual. This discovery led to our ability to administer compatible blood transfusions that do not kill the recipient.

Clumps form in incompatible transfusions because specific proteins in the serum, called *antibodies*, react with foreign, or "nonself," cells. The antibodies react with proteins on the surface of nonself cells, called *antigens*. (We will learn much more about the function of antibodies and antigens in Chapter 18.) Blood compatibility is determined by a set of three alleles (I^A, I^B, and i^O) at one locus, which determine the antigens on the surface of red blood cells. Different combinations of these alleles in different people produce four different blood types, or phenotypes: A, B, AB, and O (**Figure 10.13**). The AB phenotype found in individuals of genotype I^AI^B is an example of codominance—these individuals produce cell surface antigens of both the A and B types.

Some alleles have multiple phenotypic effects

Mendel's principles were further extended when it was discovered that a single allele can influence more than one phenotype. When a single allele has more than one distinguishable phenotypic effect, we say that the allele is **pleiotropic**. A familiar example of pleiotropy involves the allele responsible for the coloration pattern (light body, darker extremities) of Siamese cats. The same allele is also responsible for the characteristic crossed eyes of Siamese cats. Although these effects appear to be unrelated, both result from the same protein produced under the influence of the allele.

10.2 RECAP

Genes are subject to random mutations that give rise to new alleles; thus many genes have more than two alleles. Dominance is not necessarily an all-or-nothing phenomenon.

- Can you explain how the experiment in Figure 10.12 demonstrates incomplete dominance? See pp. 218–219
- How does the AB blood type result from codominance? See p. 219 and Figure 10.13

Thus far we have treated the phenotype of an organism, with respect to a given character, as a simple result of the alleles of a single gene. In many cases, however, several genes interact to determine a phenotype. To complicate things further, the physical environment may interact with the genetic constitution of an individual in determining the phenotype.

10.3 How Do Genes Interact?

Epistasis occurs when the phenotypic expression of one gene is affected by another gene. For example, two genes determine coat color in Labrador retrievers:

- Allele *B* (black pigment) is dominant to *b* (brown)
- Allele *E* (pigment deposition in hair) is dominant to *e* (no deposition, so hair is yellow)

So a dog with *BB* or *Bb* is black; one with *bb* is brown; and one with *ee* is yellow regardless of the *B*/*b* alleles present. Clearly, gene *E* de-

A dog with alleles B and E is black.

(A) Black labrador (*B_E_*)

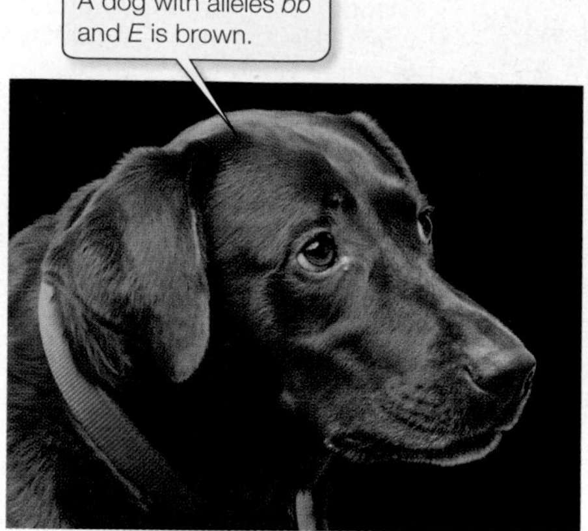

A dog with alleles bb and E is brown.

(B) Chocolate labrador (*bbE_*)

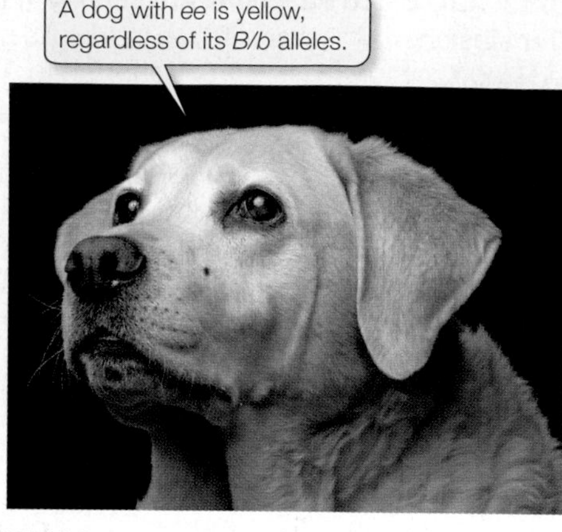

(C) Yellow labrador (*_ _ee*)

10.14 Genes May Interact Epistatically Epistasis occurs when one gene alters the phenotypic effect of another gene. In Labrador retrievers, the *E*/*e* gene determines the expression of the *B*/*b* gene.

termines the expression of *B*/*b* (**Figure 10.14**). If two dogs that are *BbEe* are mated, the phenotypic ratio among the puppies will be $^9/_{16}$ black: $^3/_{16}$ brown: $^4/_{16}$ yellow. Can you show why?

The song *Camptown Races* by Stephen Foster laments that "Somebody bet on the bay." A bay horse is dark brown. Palominos are blonde, chestnuts are reddish brown, and so on. Horses come in a wide array of colors and patterns, the result of epistasis involving multiple alleles of at least seven genes. Skin color in humans is likewise determined by multiple alleles and genes.

Hybrid vigor results from new gene combinations and interactions

Early in the twentieth century a paper called "The composition of a field of maize" by G. H. Shull had a lasting impact on the field of applied genetics. Farmers have known for centuries that matings among close relatives (known as **inbreeding**) can result in offspring of lower quality than matings between unrelated individuals. The problems with inbreeding arise because close relatives tend to have the same recessive alleles, some of which may be harmful, as we saw in our discussion of human pedigrees in Section 10.1. In fact, it has long been known that if one crosses two different true-breeding, homozygous genetic strains of a plant or animal, the offspring are phenotypically much stronger, larger, and in general more "vigorous" than either of the parents (**Figure 10.15**).

Shull began his experiment with two of the thousands of existing varieties of corn (maize). Both varieties produced about 20 bushels of corn per acre. But when he crossed them, the yield of their offspring was an astonishing 80 bushels per acre. This phenomenon is known as **heterosis** (short for *heterozygosis*), or *hybrid vigor*. The cultivation of hybrid corn spread rapidly in the United States and all over the world, quadrupling grain production. The practice of hybridization has spread to many other crops and animals used in agriculture. For example, beef cattle that are crossbred are larger and live longer than cattle bred within their own genetic strain.

The mechanism by which heterosis works is not known. A widely accepted hypothesis is *overdominance*, in which the heterozygous condition in certain important genes is superior to either homozygote. Another hypothesis is that the homozygotes have alleles that inhibit growth, and these are less active or absent in the heterozygote.

The environment affects gene action

The phenotype of an individual does not result from its genotype alone. Genotype and environment interact to determine the phenotype of an organism. Environmental variables such as light, temperature, and nutrition can affect the expression of a genotype as a phenotype.

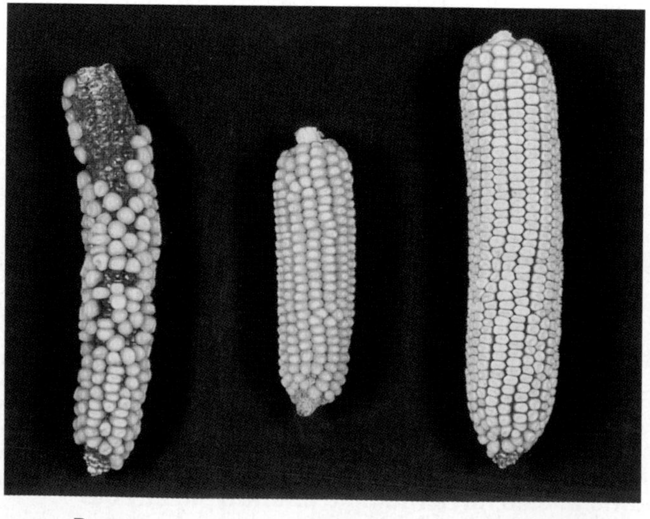

10.15 Hybrid Vigor in Corn The heterozygous F_1 offspring is larger and more vigorous than either of its homozygous parents.

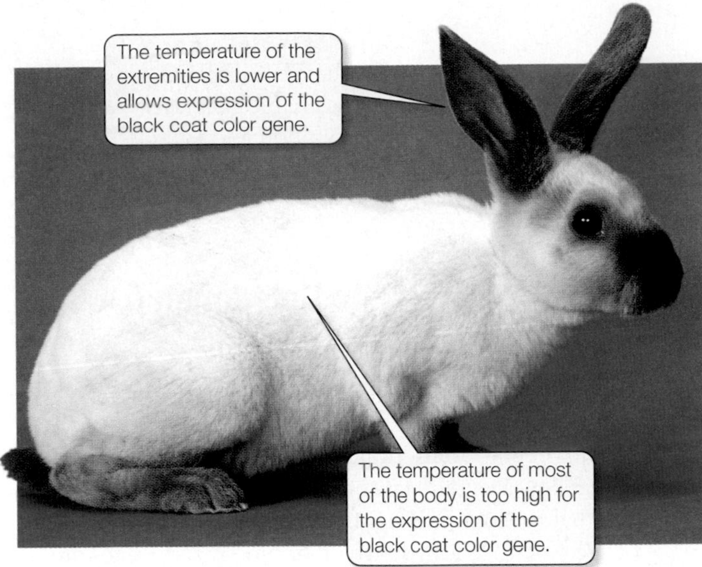

The temperature of the extremities is lower and allows expression of the black coat color gene.
The temperature of most of the body is too high for the expression of the black coat color gene.

10.16 The Environment Influences Gene Expression This rabbit expresses a coat pattern known as "chocolate point." Its genotype specifies dark fur, but the enzyme for dark fur is inactive at normal body temperature, so only the rabbit's extremities—the coolest regions of the body—express this phenotype.

A familiar example of this phenomenon involves "point restriction" coat patterns found in Siamese cats and certain rabbit breeds (**Figure 10.16**). These animals have a genotype that should result in dark fur all over the body. However, an enzyme that produces the dark fur has a mutation that renders it inactive at temperatures above a certain point (usually around 35°C). The animals maintain a body temperature above this point, and so their fur is mostly light. However, the extremities—feet, ears, nose, and tail—are cooler, about 25°C, so the fur on these regions is dark.

A simple experiment shows that the dark fur is temperature-dependent. If a patch of white fur on a point-restricted rabbit's back is removed and an ice pack is placed on the skin where the patch was, the fur that grows back will be dark. This indicates that the gene for dark fur was there all along; it's the environment that inhibited its expression.

Two parameters describe the effects of genes and environment on the phenotype:

- **Penetrance** is the proportion of individuals in a group with a given genotype that actually show the expected phenotype.
- **Expressivity** is the degree to which a genotype is expressed in an individual.

For an example of environmental effects on expressivity, consider how Siamese cats kept indoors or outdoors in different climates might look.

Most complex phenotypes are determined by multiple genes and the environment

The differences between individual organisms in simple characters, such as those that Mendel studied in pea plants, are discrete and **qualitative**. For example, the individuals in a population of pea plants are either short or tall. For most complex characters, however, such as height in humans, the phenotype varies more or less continuously over a range. Some people are short, others are tall, and many are in between the two extremes. Such variation within a population is called **quantitative**, or *continuous*, variation (**Figure 10.17**).

Sometimes this variation is largely genetic. For instance, much of human eye color is the result of a number of genes controlling the synthesis and distribution of dark melanin pigment. Dark eyes have a lot of it, brown eyes less, and green, gray, and blue eyes even less. In the latter cases, the distribution of other pigments in the eye is what determines light reflection and color.

In most cases, however, quantitative variation is due to *both genes and environment*. Height in humans certainly falls into this category. If you look at families, you often see that parents and their offspring all tend to be tall or short. However, nutrition also plays a role in height: American 18-year-olds today are about 20 percent

10.17 Quantitative Variation Quantitative variation is produced by the interaction of genes and environment. These students (women in white on the left; men in blue on the right) show continuous variation in height that is the result of interactions between many alleles and the environment.

taller than their great-grandparents were at the same age, a difference that is certainly not genetic.

Geneticists call the genes that together determine such complex characters **quantitative trait loci**. Identifying these loci is a major challenge, and an important one. For example, the amount of grain that a variety of rice produces in a growing season is determined by many interacting genetic factors. Crop plant breeders have worked hard to decipher these factors in order to breed higher-yielding rice strains. In a similar way, human characteristics such as disease susceptibility and behavior are caused in part by quantitative trait loci.

10.3 RECAP

In epistasis, one gene affects the expression of another. Perhaps the most challenging problem for genetics is the explanation of complex phenotypes that are caused by many interacting genes and the environment.

- Can you explain the difference between penetrance and expressivity? See p. 221
- How is quantitative variation different from qualitative variation? See p. 221

In the next section we'll see how the discovery that genes occupy specific positions on chromosomes enabled Mendel's successors not only to provide a physical explanation for his model of inheritance, but also to provide an explanation for those cases where Mendel's second law does not apply.

10.4 What Is the Relationship between Genes and Chromosomes?

The observation that genes located on the same chromosome do not always follow Mendel's law of independent assortment raised questions that led to new insights: What is the pattern of inheritance of such genes? How do we determine where genes are located on a chromosome, and the distances between them?

The answers to these and many other genetic questions were worked out in studies of the fruit fly *Drosophila melanogaster*. Its small size, the ease with which it could be bred, and its short generation time made this animal an attractive experimental subject. Beginning in 1909, Thomas Hunt Morgan and his students pioneered the study of *Drosophila* in Columbia University's famous "fly room," where they discovered the phenomena described in this section. *Drosophila* remains extremely important in studies of

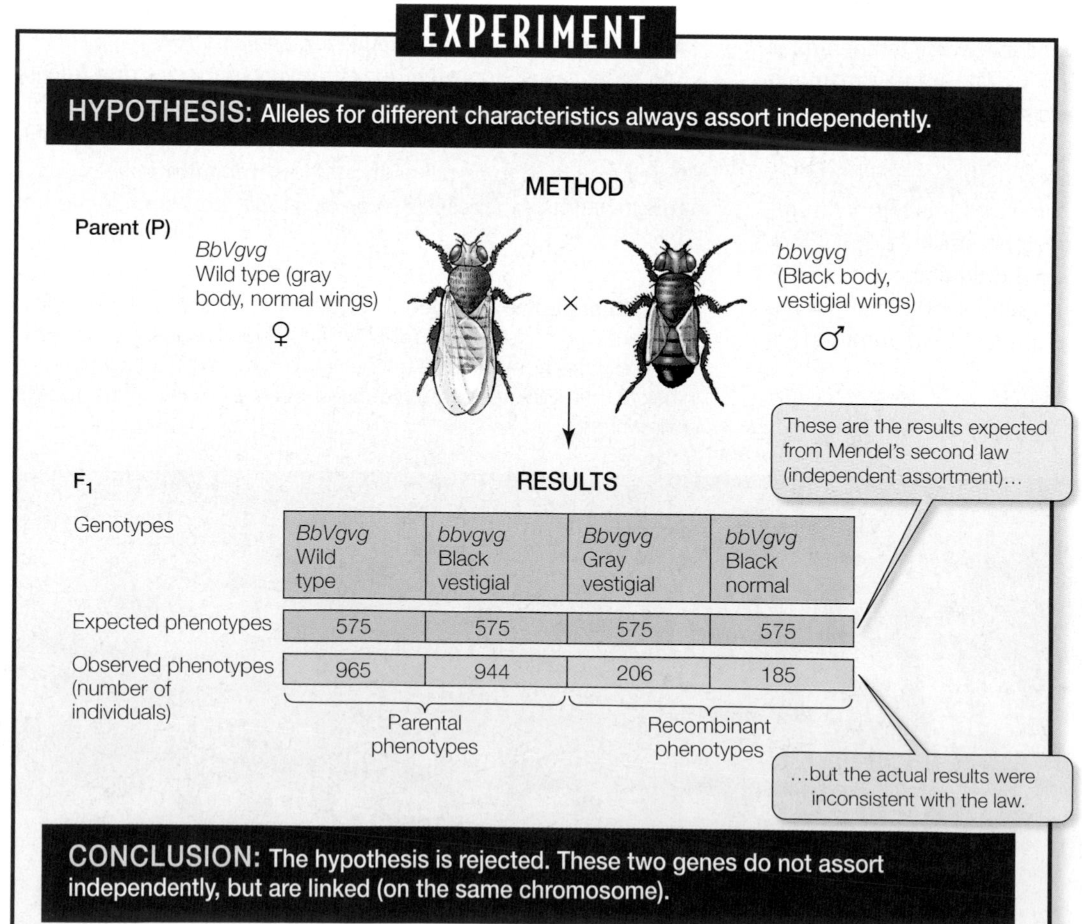

EXPERIMENT
HYPOTHESIS: Alleles for different characteristics always assort independently.
METHOD
Parent (P)
BbVgvg
Wild type (gray body, normal wings)
♀
×
bbvgvg
(Black body, vestigial wings)
♂
These are the results expected from Mendel's second law (independent assortment)...
F1
RESULTS
Genotypes
BbVgvg Wild type
bbvgvg Black vestigial
Bbvgvg Gray vestigial
bbVgvg Black normal
Expected phenotypes
575
575
575
575
Observed phenotypes (number of individuals)
965
944
206
185
Parental phenotypes
Recombinant phenotypes
...but the actual results were inconsistent with the law.
CONCLUSION: The hypothesis is rejected. These two genes do not assort independently, but are linked (on the same chromosome).

10.18 Some Alleles Do Not Assort Independently Morgan's studies showed that the genes for body color and wing size in *Drosophila* are linked, so that their alleles do not assort independently. Linkage accounts for the departure of the phenotype ratios Morgan observed from those predicted by Mendel's law of independent assortment. FURTHER RESEARCH: Look again at Mendel's dihybrid cross (Figure 10.7). If the genes for seed shape and seed color were linked, what would these results be?

chromosome structure, population genetics, the genetics of development, and the genetics of behavior.

Genes on the same chromosome are linked

Some of the crosses Morgan performed with fruit flies yielded phenotypic ratios that were not in accord with those predicted by Mendel's law of independent assortment. Morgan crossed *Drosophila* with two known genotypes, *BbVgvg* × *bbvgvg*,* for two different characters, body color and wing shape:

- *B* (wild-type gray body), is dominant over *b* (black body)
- *Vg* (wild-type wing) is dominant over *vg* (vestigial, a very small wing)

Morgan expected to see four phenotypes in a ratio of 1:1:1:1, but that is not what he observed. The body color gene and the wing size gene were not assorting independently; rather, they were, for the most part, inherited together (**Figure 10.18**).

These results became understandable to Morgan when he considered the possibility that the two loci are on the same chromosome—that is, that they might be linked. After all, since the number of genes in a cell far exceeds the number of chromosomes, each chromosome must contain many genes. We now say that the full set of loci on a given chromosome constitutes a **linkage group**. The number of linkage groups in a species equals its number of homologous chromosome pairs.

Suppose, now, that the *Bb* and *Vgvg* loci are indeed located on the same chromosome. Why, then, didn't *all* of Morgan's F_1 flies have the parental phenotypes—that is, why did his cross result in anything other than gray flies with normal wings (wild-type) and black flies with vestigial wings? If linkage were *absolute*—that is, if chromosomes always remained intact and unchanged—we would expect to see just those two types of progeny. However, this is not always what happens.

Genes can be exchanged between chromatids

Absolute linkage is extremely rare. If linkage were absolute, Mendel's law of independent assortment would apply only to loci on different chromosomes. What actually happens is more complex, and therefore more interesting. Because chromosomes can break, recombination of genes can occur. That is, genes at different loci on the same chromosome do sometimes separate from one another during meiosis.

Genes may recombine when two homologous chromosomes physically exchange corresponding segments during prophase I of meiosis—that is, by crossing over (**Figure 10.19**; see also Figure 9.18). As described in Section 9.5, DNA is replicated during the S phase, so that by prophase I, when homologous chromosome pairs come together to form tetrads, each chromosome consists of two chromatids. The exchange event involves only two of the four chromatids in a tetrad, one from each member of the homologous pair, and can occur at any point along the length of the chromosome. The chromosome segments involved are exchanged reciprocally, so both chromatids involved in crossing over become recombinant (that is, each chromatid ends up with genes from both of the organism's parents). Usually several exchange events occur along the length of each homologous pair.

When crossing over takes place between two linked genes, not all the progeny of a cross have the parental phenotypes. Instead, recombinant offspring appear as well, as they did in Morgan's cross. They appear in proportions called **recombinant frequencies**, which are calculated by dividing the number of recombinant prog-

*Do you recognize this type of cross? It is a *test cross* for the two gene pairs; see Figure 10.6.

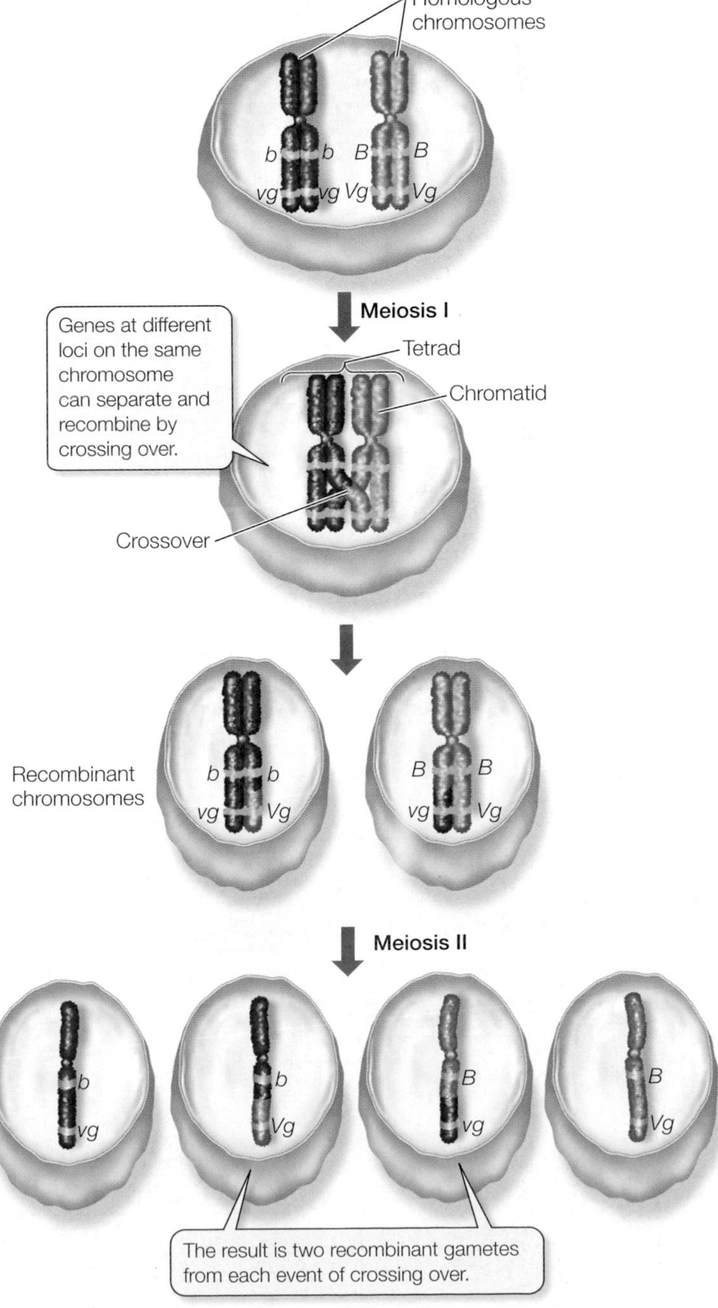

10.19 Crossing Over Results in Genetic Recombination Genes at different loci on the same chromosome can be separated from one another and recombined by crossing over. Such recombination occurs during prophase I of meiosis.

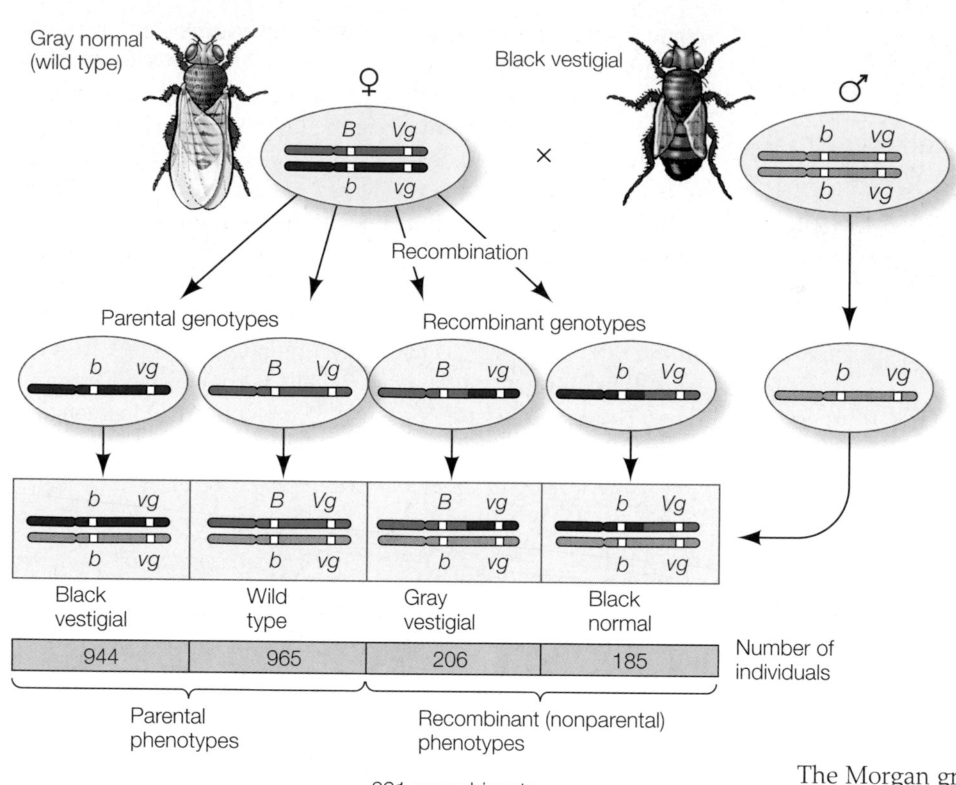

10.20 Recombinant Frequencies The frequency of recombinant offspring (those with a phenotype different from either parent) can be calculated.

eny by the total number of progeny (**Figure 10.20**). Recombinant frequencies will be greater for loci that are farther apart on the chromosome than for loci that are closer together because an exchange event is more likely to occur between genes that are far apart than between genes that are close together.

Geneticists can make maps of chromosomes

If two loci are very close together on a chromosome, the odds of crossing over between them are small. In contrast, if two loci are far apart, crossing over could occur between them at many points. This pattern is a consequence of the mechanism of crossing over: the farther apart two genes are, the more places there are in the chromosome for breakage and reunion of chromatids to occur. In a population of cells undergoing meiosis, a greater proportion of the cells will undergo recombination between two loci that are far apart than between two loci that are close together. In 1911, Alfred Sturtevant, then an undergraduate student in T. H. Morgan's fly room, realized how this simple insight could be used to show where different genes lie on a chromosome in relation to one another.

The Morgan group had determined recombinant frequencies for many pairs of linked *Drosophila* genes. Sturtevant used those recombinant frequencies to create **genetic maps** that showed the arrangement of genes along the chromosome (**Figure 10.21**). Ever since Sturtevant demonstrated this method, geneticists have mapped the chromosomes of eukaryotes, prokaryotes, and viruses, assigning distances between genes in **map units**. A map unit corresponds to a recombinant frequency of 0.01; it is also referred to as a **centimorgan** (**cM**), in honor of the founder of the fly room. You, too, can work out a genetic map (**Figure 10.22**).

10.21 Steps toward a Genetic Map Because the chance of a recombinant genotype occurring increases with the distance between two loci on a chromosome, Sturtevant was able to derive this partial map of a *Drosophila* chromosome from the Morgan group's data on the recombinant frequencies of five recessive traits. He used an arbitrary unit of distance—the map unit, or centimorgan (cM)—equivalent to a recombinant frequency of 0.01.

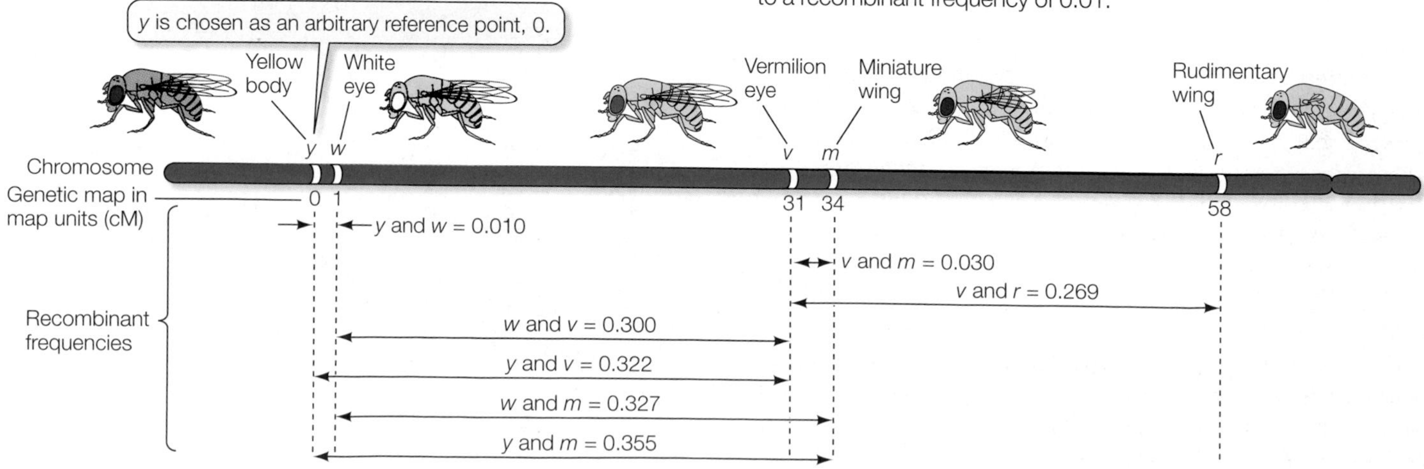

10.22 Map These Genes The object of this exercise is to determine the order of three loci (*a*, *b*, and *c*) on a chromosome, as well as the map distances (in cM) between them.

1 At the outset, we have no idea of the individual distances between the genes, and there are several possible sequences (*a-b-c*, *a-c-b*, *b-a-c*).

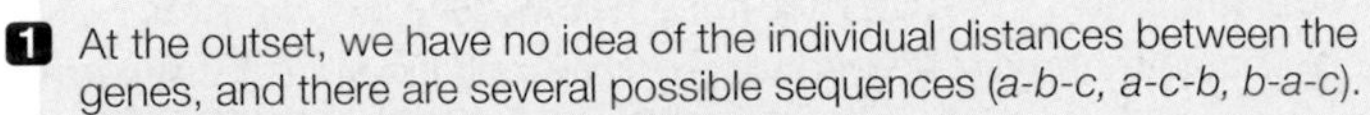

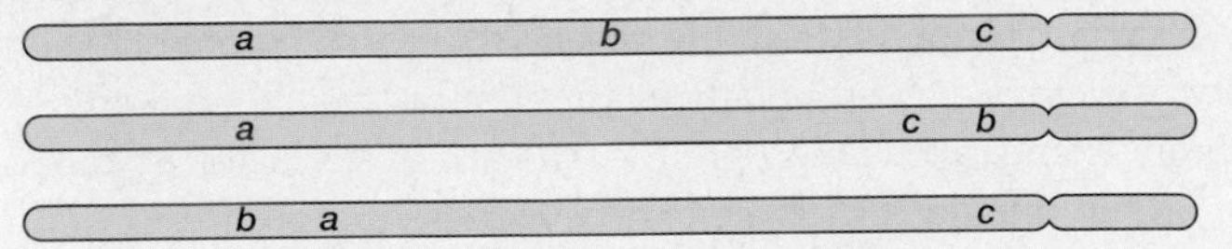

We make a cross *AABB* × *aabb*, and obtain an F_1 generation with a genotype *AaBb*. We test cross these *AaBb* individuals with *aabb*. Here are the genotypes of the first 1,000 progeny:

450 *AaBb*, 450 *aabb*, 50 *Aabb*, and 50 *aaBb*.
(parental types) (recombinant types)

2 How far apart are the *a* and *b* genes?

What is the recombinant frequency? Which are the recombinant types, and which are the parental types?

Recombinant frequency (*a* to *b*) = (50 + 50)/1,000 = 0.1
So the map distance is

Map distance = 100 × recombinant frequency =
100 × 0.1 = 10 cM

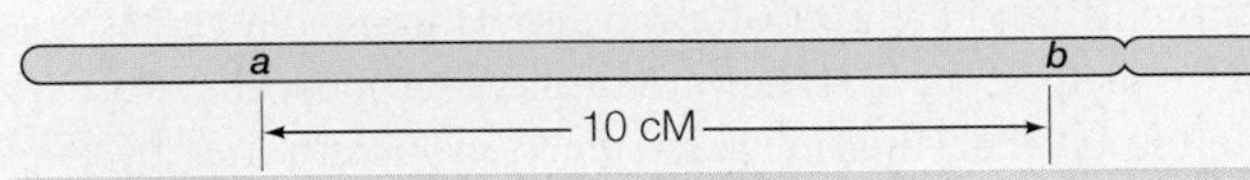

3 How far apart are the *a* and *c* genes?

Now we make a cross *AACC* × *aacc*, obtain an F_1 generation, and test cross it, obtaining

460 *AaCc*, 460 *aacc*, 40 *Aacc*, and 40 *aaCc*

Recombinant frequency (*a* to *c*) = (40 + 40)/1,000 = 0.08

Map distance = 100 × recombinant frequency =
100 × 0.08 = 8 cM

4 How far apart are the *b* and *c* genes?

We make a cross *BBCC* × *bbcc*, obtain an F_1 generation, and test cross it, obtaining

490 *BbCc*, 490 *bbcc*, 10 *Bbcc*, and 10 *bbCc*

Recombinant frequency (*b* to *c*) = (10 + 10)/1,000 = 0.02

Map distance = 100 × recombinant frequency =
100 × 0.02 = 2 cM

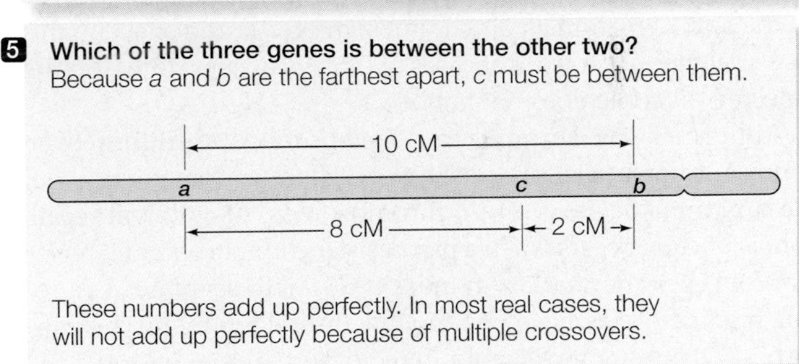

5 Which of the three genes is between the other two?
Because *a* and *b* are the farthest apart, *c* must be between them.

These numbers add up perfectly. In most real cases, they will not add up perfectly because of multiple crossovers.

Linkage is revealed by studies of the sex chromosomes

In Mendel's work, reciprocal crosses always gave identical results; it did not matter, in general, whether a dominant allele was contributed by the mother or by the father. But in some cases, the parental origin of a chromosome does matter. For example, human males inherit a bleeding disorder called hemophilia from their mother, not from their father. To understand the types of inheritance in which the parental origin of an allele is important, we must consider the ways in which sex is determined in different species.

SEX DETERMINATION BY CHROMOSOMES In corn, every diploid adult has both male and female reproductive structures. The tissues in these two types of structures are genetically identical, just as roots and leaves are genetically identical. Plants such as corn, in which the same individual produces both male and female gametes, are said to be *monoecious* (Greek, "one house"). Other plants, such as date palms and oak trees, and most animals are *dioecious* ("two houses"), meaning that some individuals can produce only male gametes and the others can produce only female gametes. In other words, dioecious organisms have two sexes.

In most dioecious organisms, sex is determined by differences *in the chromosomes*, but such determination operates in different ways in different groups of organisms. For example, in many animals, including humans, sex is determined by a single **sex chromosome**, or by a pair of them. Both males and females have two copies of each of the rest of the chromosomes, which are called **autosomes**.

The sex chromosomes of female mammals consist of a pair of X chromosomes. Male mammals, on the other hand, have one X chromosome and a sex chromosome that is not found in females: the Y chromosome. Females may be represented as XX and males as XY:

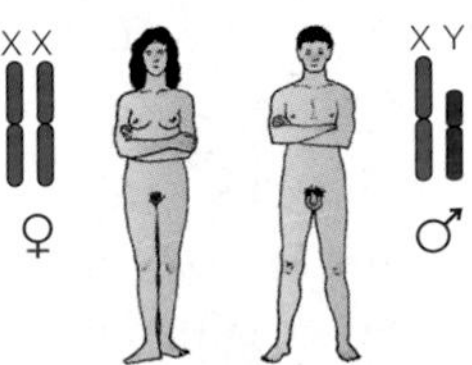

Male mammals produce *two* kinds of gametes. Each gamete has a complete set of autosomes, but half the gametes carry an X chromosome and the other half carry a Y. When an X-bearing sperm fertilizes an egg, the resulting XX zygote is female; when a Y-bearing sperm fertilizes an egg, the resulting XY zygote is male.

The situation is different in birds, in which males are XX and females are XY (to avoid confusion, these chromosomes are called ZZ and ZW):

In these organisms, the female produces two types of gametes, carrying Z or W. Whether the egg is Z or W determines the sex of the offspring, in contrast to humans and fruit flies, in which the sperm, carrying either X or Y, determines the sex.

SEX CHROMOSOME ABNORMALITIES REVEALED THE GENE THAT DETERMINES SEX There must be genes on the Y (or W) chromosome that determine sex (male or female, respectively). But how can we be sure? As previously pointed out, one way to determine cause (in the case of mammals, a Y chromosome gene) and effect (in this case, maleness) is to look at cases of biological error, in which the expected outcome does not happen.

Abnormal sex chromosome constitutions resulting from nondisjunction in meiosis (see Section 9.5) tell us something about the functions of the X and Y chromosomes. As you will recall, nondisjunction occurs when a pair of sister chromosomes (in meiosis I) or sister chromatids (in meiosis II) fail to separate. As a result, a gamete may have one too few or one too many chromosomes. Assuming fertilization by another gamete with the full haploid chromosome set, the resulting offspring are aneuploid, with fewer or more chromosomes than normal.

In humans, XO individuals sometimes appear. (The O implies that a chromosome is missing—that is, individuals that are XO have only one sex chromosome.) Human XO individuals are females who are physically moderately abnormal but mentally normal; usually they are also sterile. The XO condition in humans is called *Turner syndrome.* It is the only known case in which a person can survive with only one member of a chromosome pair (here, the XY pair), although most XO conceptions are spontaneously terminated early in development. XXY individuals also occur; this condition, which affects males, is called *Klinefelter syndrome,* and results in overlong limbs and sterility.

These observations suggested that the gene that determines maleness is located on the Y chromosome. Observations of people with other types of chromosomal abnormalities helped researchers to pinpoint the location of that gene:

- Some XY individuals are phenotypically women but lack a small portion of the Y chromosome.
- Some men are genetically XX but have a small piece of the Y chromosome attached to another chromosome.

It was clear that the Y fragments that are respectively missing and present in these two cases contained the maleness-determining gene, which was named *SRY* (*s*ex-determining *r*egion on the *Y* chromosome).

The *SRY* gene encodes a protein involved in **primary sex determination**—that is, the determination of the kinds of gametes that an individual will produce and the organs that will make them. In the presence of functional SRY protein, an embryo develops sperm-producing testes. (Notice that *italic type* is used for the name of a gene, but roman type is used for the name of a protein.) If the embryo has no Y chromosome, the *SRY* gene is absent, and thus the SRY protein is not made. In the absence of the SRY protein, the embryo develops egg-producing ovaries. In this case, a gene on the X chromosome called *DAX1* produces an anti-testis factor. So the role of *SRY* in a male is to inhibit the maleness inhibitor encoded by *DAX1*. The SRY protein does this in male cells, but since it is not present in females, *DAX1* can act to inhibit maleness.

Primary sex determination is not the same as **secondary sex determination**, which results in the outward manifestations of maleness and femaleness (such as body type, breast development, body hair, and voice). These outward characteristics are not determined directly by the presence or absence of the Y chromosome. Rather, they are determined by genes scattered on the autosomes and X chromosome that control the actions of hormones, such as testosterone and estrogen.

Superficially, *Drosophila melanogaster* follows the same pattern of sex determination seen in mammals—females are XX and males are XY. However, XO *Drosophila* are males (rather than females, as in mammals) and are almost always indistinguishable from normal XY males except that they are sterile. XXY individuals are normal, fertile *females*. Thus, in *Drosophila*, sex is determined by the *ratio of X chromosomes to autosome sets*. If there is one X chromosome for each set of autosomes, the individual is a female; if there is only one X chromosome for the two sets of autosomes, the individual is a male. The Y chromosome plays no sex-determining role in *Drosophila*, but it is needed for male fertility.

Genes on sex chromosomes are inherited in special ways

Genes on sex chromosomes do not show the Mendelian patterns of inheritance. In *Drosophila* and in humans, the Y chromosome carries few known genes, but a substantial number of genes, affecting a great variety of characters, are carried on the X chromosome. Any such gene is present in two copies in females but in only one copy in males. Therefore, males will always be **hemizygous** for genes on the X chromosome—they will have only one copy of each, and it will be expressed. Thus reciprocal crosses do not give identical results for characters whose genes are carried on the sex chromosomes, and these characters do not show the usual Mendelian ratios for the inheritance of genes located on autosomes.

The first, and still one of the best, examples of inheritance of characters governed by loci on the sex chromosomes (**sex-linked** inheritance) is that of eye color in *Drosophila*. The wild-type eye color of these flies is red. In 1910, Morgan discovered a mutation that causes white eyes. He experimented by crossing flies of the wild-type and mutant phenotypes. His results demonstrated that the eye color locus is on the X chromosome.

- When a homozygous red-eyed female was crossed with a (hemizygous) white-eyed male, all the sons and daughters had red eyes, because red is dominant over white and all the progeny had inherited a wild-type X chromosome from their mothers (**Figure 10.23A**).

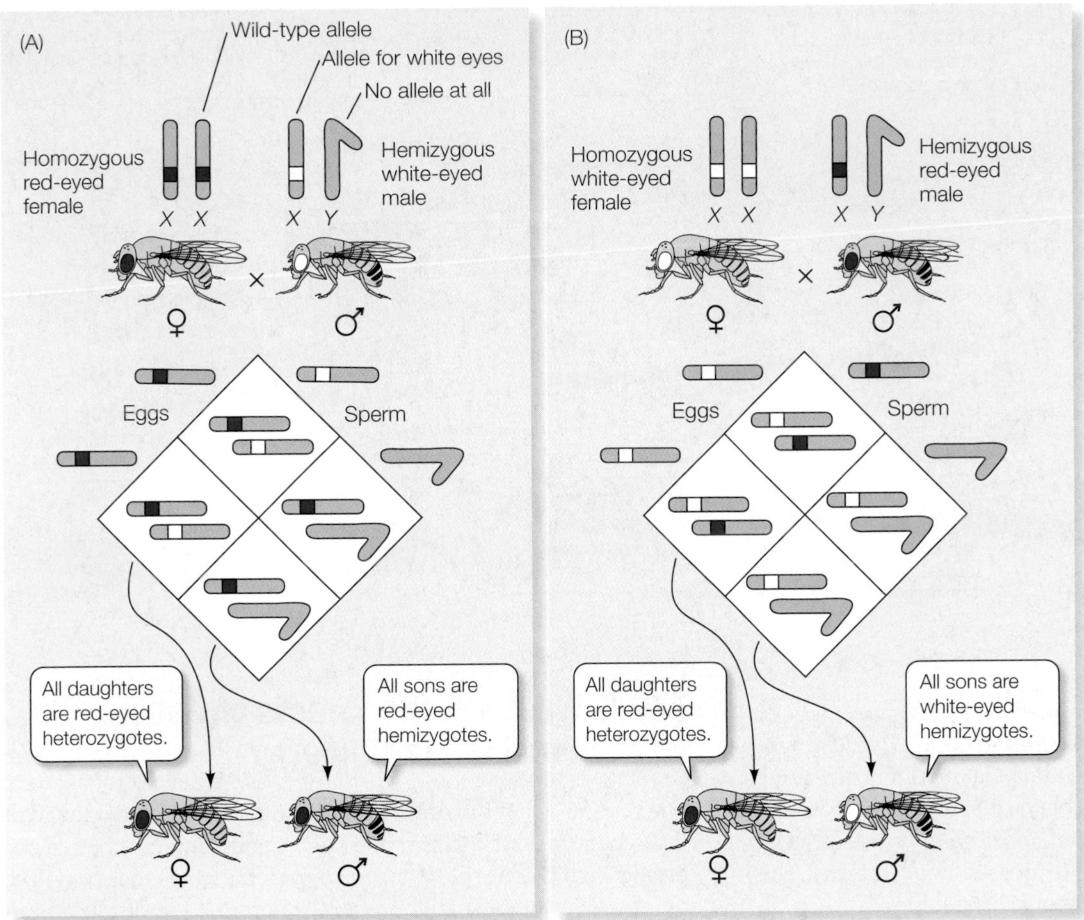

10.23 Eye Color Is a Sex-Linked Trait in *Drosophila* Morgan demonstrated that a mutant allele that causes white eyes in *Drosophila* is carried on the X chromosome. Note that in this case, the reciprocal crosses do not have the same results.

- In the reciprocal cross, in which a white-eyed female was mated with a red-eyed male, all the sons were white-eyed and all the daughters were red-eyed (**Figure 10.23B**).

The sons from the reciprocal cross inherited their only X chromosome from their white-eyed mother; the Y chromosome they inherited from their father does not carry the eye color locus. The daughters, on the other hand, got an X chromosome bearing the white allele from their mother and an X chromosome bearing the red allele from their father; they were therefore red-eyed heterozygotes.

- When heterozygous females were mated with red-eyed males, half their sons had white eyes, but all their daughters had red eyes.

Together, these three results showed that eye color was carried on the X chromosome and not on the Y.

Humans display many sex-linked characters

The human X chromosome carries about 2,000 known genes. The alleles at these loci follow the same pattern of inheritance as those for white eyes in *Drosophila*. One gene on the human X chromosome, for example, has a mutant recessive allele that leads to red-green color blindness, a hereditary disorder, as described at the start of this chapter. Red-green color blindness appears in individuals who are homozygous or hemizygous for the recessive mutant allele.

Pedigree analysis of X-linked recessive phenotypes (**Figure 10.24**) reveals the following patterns:

- The phenotype appears much more often in males than in females, because only one copy of the rare allele is needed for its expression in males, while two copies must be present in females.
- A male with the mutation can pass it on only to his daughters; all his sons get his Y chromosome.
- Daughters who receive one mutant X chromosome are heterozygous **carriers**. They are phenotypically normal, but they can pass the mutant X to both sons and daughters (but do so only half of the time, on average, since half of their X chromosomes carry the normal allele).
- The mutant phenotype can skip a generation if the mutation passes from a male to his daughter (who will be phenotypically normal) and thus to her son.

Red-green color blindness (see p. 207) is an X-linked recessive phenotype, as are several important human diseases. Human mutations inherited as X-linked dominant phenotypes are rarer than X-linked recessives because dominant phenotypes appear in every generation, and because people carrying the harmful mutation, even as heterozygotes, often fail to survive and reproduce. (Look at the four points above and try to determine what would happen if the mutation were dominant.)

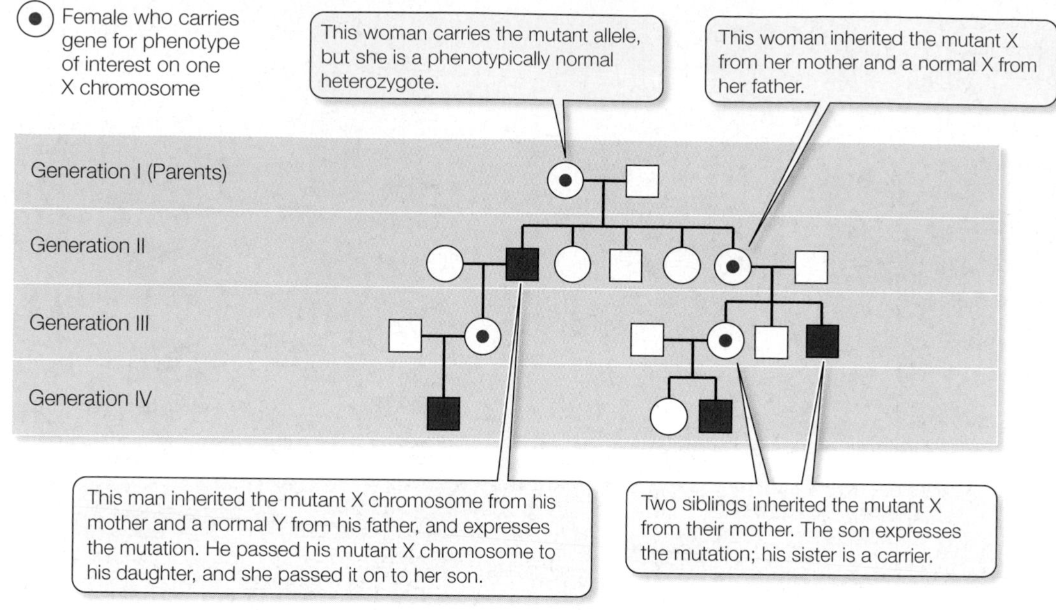

Female who carries gene for phenotype of interest on one X chromosome
This woman carries the mutant allele, but she is a phenotypically normal heterozygote.
This woman inherited the mutant X from her mother and a normal X from her father.
Generation I (Parents)
Generation II
Generation III
Generation IV
This man inherited the mutant X chromosome from his mother and a normal Y from his father, and expresses the mutation. He passed his mutant X chromosome to his daughter, and she passed it on to her son.
Two siblings inherited the mutant X from their mother. The son expresses the mutation; his sister is a carrier.

10.24 Red-Green Color Blindness Is a Sex-Linked Trait in Humans The mutant allele for red-green color blindness is inherited as an X-linked recessive.

The small human Y chromosome carries several dozen genes. Among them is the maleness determinant, *SRY*. Interestingly, for some genes on the Y, there are similar, but not identical, genes on the X. For example, one of the proteins that make up ribosomes is encoded by a gene on the Y that is expressed only in male cells, while the X-linked counterpart is expressed in both sexes. This means that there are "male" and "female" ribosomes; the significance of this phenomenon is unknown. Y-linked alleles are passed only from father to son. (You can verify this with a Punnett square.)

10.4 RECAP

Simple Mendelian ratios are not observed when genes are linked on the same chromosome. Linkage is indicated by atypical frequencies of gametes evidenced by offspring in a test cross. Sex linkage in humans refers to genes on the X chromosome that have no counterpart on the Y.

- Can you explain the concept of linkage and its implications for the results of genetic crosses? See p. 223 and Figures 10.19 and 10.20
- How does a sex-linked gene behave differently in genetic crosses than a gene on an autosome? See pp. 226–227 and Figure 10.23

The genes we've discussed so far in this chapter are all in the cell nucleus. But other organelles, including mitochondria and plastids, also carry genes. What are they, and how are they inherited?

10.5 What Are the Effects of Genes Outside the Nucleus?

The nucleus is not the only organelle in a eukaryotic cell that carries genetic material. As described in Section 4.3, mitochondria and plastids, which may have arisen from prokaryotes that colonized other cells, contain small numbers of genes. For example, in humans, there are about 24,000 genes in the nuclear genome and 37 in the mitochondrial genome. Plastid genomes are about five times larger than those of mitochondria. In any case, several of the genes of cytoplasmic organelles are important for organelle assembly and function, so it is not surprising that mutations of these genes can have profound effects on the organism.

The inheritance of organelle genes differs from that of nuclear genes for several reasons:

- In most organisms, mitochondria and plastids are inherited only from the mother. As you will learn in Chapter 43, eggs contain abundant cytoplasm and organelles, but the only part of the sperm that survives to take part in the union of haploid gametes is the nucleus. So you have inherited your mother's mitochondria (with their genes), but not your father's.
- There may be hundreds of mitochondria or plastids in a cell. So a cell is not diploid for organelle genes; rather, it is highly polyploid.
- Organelle genes tend to mutate at much faster rates than nuclear genes, so there are multiple alleles of organelle genes.

The phenotypes of mutations in the DNA of organelles reflect the organelles' roles. For example, in plants and some photosynthetic protists, certain plastid gene mutations affect the proteins that assemble chlorophyll molecules into photosystems and result in a phenotype that is essentially white instead of green. Mitochondrial gene mutations that affect one of the complexes in the electron transport chain result in less ATP production. These mutations have especially noticeable effects in tissues with a high energy requirement, such as the nervous system, muscles, and kidneys. In 1995, Greg LeMond, a professional cyclist who had won the famous Tour de France three times, was forced to retire because of muscle weakness caused by a mitochondrial mutation.

10.5 RECAP

Genes in the genomes of organelles, specifically plastids and mitochondria, do not behave in a Mendelian fashion.

- Can you explain why genes carried in the organelle genomes are usually inherited only from the mother?

CHAPTER SUMMARY

10.1 What are the Mendelian laws of inheritance?

Physical features of organisms, or **characters**, can exist in different forms, or **traits**. A **heritable trait** is one that can be passed from parent to offspring. A **phenotype** is the physical appearance of an organism; a **genotype** is the genetic constitution of the organism.

The different forms of a **gene** are called **alleles**. Organisms that have two identical alleles for a trait are called **homozygous**; organisms that have two different alleles for a trait are called **heterozygous**. A gene resides at a particular site on a chromosomes called a **locus**.

Mendel's experiments included **reciprocal crosses** and **monohybrid** crosses of **true-breeding** pea plants. Analysis of his meticulously tabulated data led Mendel to propose a **particulate theory** of inheritance stating that discrete units (now called genes) are responsible for the inheritance of specific traits, to which both parents contribute equally.

Mendel's first law, the **law of segregation**, states that when any individual produces gametes, the two copies of a gene separate, so that each gamete receives only one member of the pair. Thus every individual in the F_1 inherits one copy from each parent. Review Figures 10.4 and 10.5

Mendel used a **test cross** to find out whether an individual showing a dominant phenotype is homozygous or heterozygous. Review Figure 10.6, Web/CD Activity 10.1

Mendel's use of **dihybrid crosses** to study the inheritance of two characters led to his second law: the **law of independent assortment**. The independent assortment of genes in meiosis leads to **recombinant** phenotypes. Review Figures 10.7 and 10.8, Web/CD Tutorial 10.1

Probability calculations and **pedigrees** help geneticists trace Mendelian inheritance patterns. Review Figures 10.9 and 10.10

10.2 How do alleles interact?

New alleles arise by random **mutation**. Many genes have multiple alleles. A **wild-type** allele gives rise to the predominant form of a trait. When the wild-type allele is present at a locus less than 99 percent of the time, the locus is said to be **polymorphic**. Review Figure 10.11

In **incomplete dominance**, neither of two alleles is dominant. The heterozygous phenotype is intermediate between the homozygous phenotypes. Review Figure 10.12

Codominance exists when two alleles at a locus produce two different phenotypes that both appear in heterozygotes.

An allele that affects more than one trait is said to be **pleiotropic**.

10.3 How do genes interact?

In **epistasis**, one gene affects the expression of another. Review Figure 10.14

Environmental conditions can affect the expression of a genotype.

Penetrance is the proportion of individuals in a group with a given genotype that show the expected phenotype. **Expressivity** is the degree to which a genotype is expressed in an individual.

Variation in phenotype can be **qualitative** (discrete) or **quantitative** (graduated, continuous). Most quantitative traits are the result of the effects of several genes and the environment. Genes that together determine quantitative characters are called **quantitative trait loci**.

10.4 What is the relationship between genes and chromosomes?

See Web/CD Tutorial 10.2

Each chromosome carries many genes. Genes on the same chromosome are referred to as a **linkage group**.

Genes on the same chromosome can recombine by crossing over. The resulting recombinant chromosomes have new combinations of alleles. Review Figures 10.19 and 10.20

Sex chromosomes carry genes that determine whether the organism will produce male or female gametes. All other chromosomes are called **autosomes**. The specific functions of X and Y chromosomes differ among groups of organisms.

Primary sex determination in mammals is usually a function of the presence or absence of the *SRY* gene. **Secondary sex determination** results in the outward manifestations of maleness or femaleness.

In fruit flies and mammals, the X chromosome carries many genes, but the Y chromosome has only a few. Males have only one allele (are **hemizygous**) for X-linked genes, so rare recessive alleles show up phenotypically more often in males than in females. Females may be unaffected **carriers** of such alleles.

10.5 What are the effects of genes outside the nucleus?

Cytoplasmic organelles such as plastids and mitochondria contain small numbers of genes. In many organsisms, cytoplasmic genes are inherited only from the mother because male gametes contribute only their nucleus (i.e., no cytoplasm) to the zygote at fertilization.

See Web/CD Activities 10.2 and 10.3 for a concept review of this chapter.

SELF-QUIZ

1. In a simple Mendelian monohybrid cross, tall plants are crossed with short plants, and the F_1 plants are allowed to self-pollinate. What fraction of the F_2 generation are both tall *and* heterozygous?
   a. 1/8
   b. 1/4
   c. 1/3
   d. 2/3
   e. 1/2
2. The phenotype of an individual
   a. depends at least in part on the genotype.
   b. is either homozygous or heterozygous.
   c. determines the genotype.
   d. is the genetic constitution of the organism.
   e. is either monohybrid or dihybrid.
3. The ABO blood groups in humans are determined by a multiple-allele system in which I^A and I^B are codominant and dominant to i^O. A newborn infant is type A. The mother is type O. Possible genotypes of the father are
   a. A, B, or AB
   b. A, B, or O
   c. O only
   d. A or AB
   e. A or O
4. Which statement about an individual that is homozygous for
   an allele is *not* true?
   a. Each of its cells possesses two copies of that allele.
   b. Each of its gametes contains one copy of that allele.
   c. It is true-breeding with respect to that allele.
   d. Its parents were necessarily homozygous for that allele.
   e. It can pass that allele to its offspring.
5. Which statement about a test cross is *not* true?
   a. It tests whether an unknown individual is homozygous or heterozygous.
   b. The test individual is crossed with a homozygous recessive individual.
   c. If the test individual is *heterozygous*, the progeny will have a 1:1 ratio.
   d. If the test individual is *homozygous*, the progeny will have a 3:1 ratio.
   e. Test cross results are consistent with Mendel's model of inheritance.
6. Linked genes
   a. must be immediately adjacent to one another on a chromosome.
   b. have alleles that assort independently of one another.
   c. never show crossing over.
   d. are on the same chromosome.
   e. always have multiple alleles.
7. In the F_2 generation of a dihybrid cross
   a. four phenotypes appear in the ratio 9:3:3:1 if the loci are linked.
   b. four phenotypes appear in the ratio 9:3:3:1 if the loci are unlinked.
   c. two phenotypes appear in the ratio 3:1 if the loci are unlinked.
   d. three phenotypes appear in the ratio 1:2:1 if the loci are unlinked.
   e. two phenotypes appear in the ratio 1:1 whether or not the loci are linked.
8. The genetic sex of a human is determined by
   a. ploidy, with the male being haploid.
   b. the Y chromosome.
   c. X and Y chromosomes, the male being XX.
   d. the number of X chromosomes, the male being XO.
   e. Z and W chromosomes, the male being ZZ.
9. In epistasis
   a. nothing changes from generation to generation.
   b. one gene alters the effect of another.
   c. a portion of a chromosome is deleted.
   d. a portion of a chromosome is inverted.
   e. the behavior of two genes is entirely independent.
10. In humans, spotted teeth are caused by a dominant sex-linked gene. A man with spotted teeth whose mother had normal teeth marries a woman with normal teeth. Therefore,
    a. all of their daughters will have normal teeth.
    b. all of their daughters will have spotted teeth.
    c. all of their children will have spotted teeth.
    d. half of their sons will have spotted teeth.
    e. all of their sons will have spotted teeth.

GENETIC PROBLEMS

1. Using the Punnett squares below, show that for typical dominant and recessive autosomal traits, it does not matter which parent contributes the dominant allele and which the recessive allele. Cross true-breeding tall plants (*TT*) with true-breeding dwarf plants (*tt*).

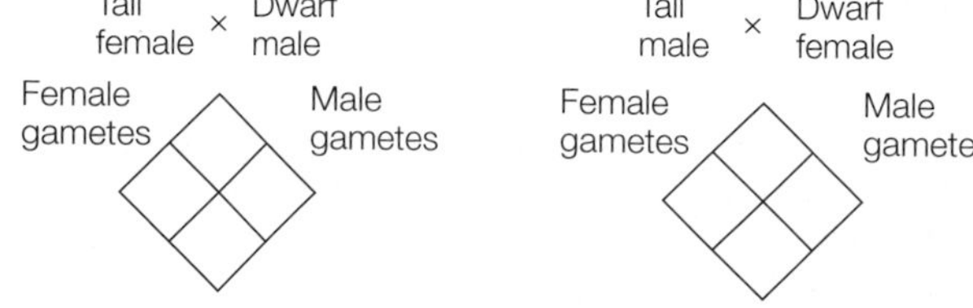

2. Show diagrammatically what occurs when the F_1 offspring of the cross in Question 1 self-pollinate.

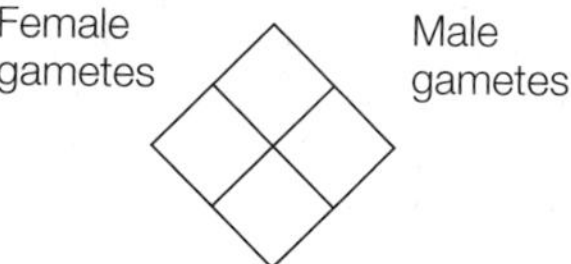

3. A new student of genetics suspects that a particular recessive trait in fruit flies (dumpy wings, which are somewhat smaller and more bell-shaped than the wild type) is sex-linked. A single mating between a fly with dumpy wings (*dp*; female) and a fly with wild-type wings (*Dp*; male) produces three dumpy-winged females and two wild-type males in the F_1 generation. On the basis of these data, is the trait sex-linked or autosomal? What were the genotypes of the parents? Explain how these conclusions can be reached on the basis of so few data.

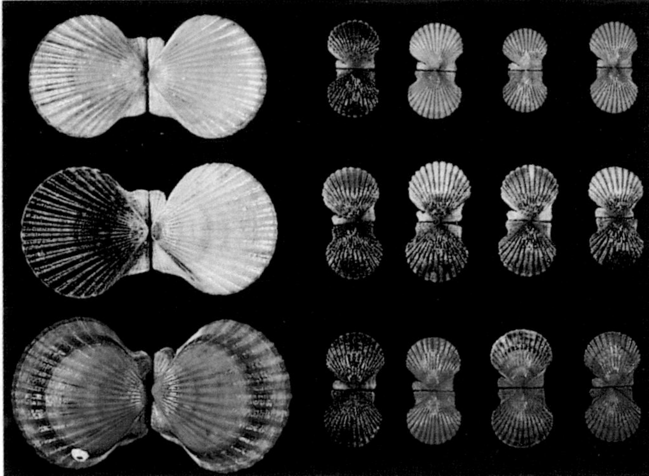

4. The photograph shows the shells of 15 bay scallops, *Argopecten irradians*. These scallops are hermaphroditic; that is, a single individual can reproduce sexually, as did the pea plants of the F_1 generation in Mendel's experiments. Three color schemes are evident: yellow, orange, and black and white. The color-determining gene has three alleles. The top row shows a yellow scallop and a representative sample of its offspring, the middle row shows a black-and-white scallop and its offspring, and the bottom row shows an orange scallop and its offspring. Assign a suitable symbol to each of the three alleles participating in color control; then determine the genotype of each of the three parent individuals and tell what you can about the genotypes of the different offspring. Explain your results carefully.
5. The sex of some fishes is determined by the same XY system as in humans. An allele of one locus on the Y chromosome of the fish *Lebistes* causes a pigmented spot to appear on the dorsal fin. A male fish that has a spotted dorsal fin is mated with a female fish that has an unspotted fin. Describe the phenotypes of the F_1 and the F_2 generations from this cross.
6. In *Drosophila melanogaster*, the recessive allele *p*, when homozygous, determines pink eyes. *Pp* or *PP* results in wild-type eye color. Another gene, on another chromosome, has a recessive allele, *sw*, that produces short wings when homozygous. Consider a cross between females of genotype *PPSwSw* and males of genotype *ppswsw*. Describe the phenotypes and genotypes of the F_1 generation and of the F_2 generation produced by allowing the F_1 progeny to mate with one another.
7. On the same chromosome of *Drosophila melanogaster* that carries the *p* (pink eyes) locus, there is another locus that affects the wings. Homozygous recessives, *byby*, have blistery wings, while the dominant allele *By* produces wild-type wings. The *P* and *By* loci are very close together on the chromosome; that is, the two loci are tightly linked. In answering these questions, assume that no crossing over occurs.
   a. For the cross *PPByBy* × *ppbyby*, give the phenotypes and genotypes of the F_1 and of the F_2 generations produced by interbreeding of the F_1 progeny.
   b. For the cross *PPbyby* × *ppByBy*, give the phenotypes and genotypes of the F_1 and of the F_2 generations.
   c. For the cross of Question 7*b*, what further phenotype(s) would appear in the F_2 generation if crossing over occurred?
   d. Draw a nucleus undergoing meiosis at the stage in which the crossing over (Question 7*c*) occurred. In which generation (P, F_1, or F_2) did this crossing over take place?
8. Consider the following cross of *Drosophila melanogaster* (alleles are as described in Question 6): Males with genotype *Ppswsw* are crossed with females of genotype *ppSwsw*. Describe the phenotypes and genotypes of the F_1 generation.
9. In the Andalusian fowl, a single pair of alleles for one gene controls the color of the feathers. Three colors are observed: blue, black, and splashed white. Crosses among these three types yield the following results:

PARENTS	PROGENY
Black × blue	Blue and black (1:1)
Black × splashed white	Blue
Blue × splashed white	Blue and splashed white (1:1)
Black × black	Black
Splashed white × splashed white	Splashed white

   a. What progeny would result from the cross blue × blue?
   b. If you wanted to sell eggs, all of which would yield blue fowl, how should you proceed?
10. In *Drosophila melanogaster*, white (*w*), eosin (w^e), and wild-type red (w^+) are multiple alleles at a single locus for eye color. This locus is on the X chromosome. A female that has eosin (pale orange) eyes is crossed with a male that has wild-type eyes. All the female progeny are red-eyed; half the male offspring have eosin eyes, and half have white eyes.
   a. What is the order of dominance of these alleles?
   b. What are the genotypes of the parents and progeny?
11. Red-green color blindness is a recessive trait. Two people with normal vision have two sons, one color-blind and one with normal vision. If the couple also has daughters, what proportion of them will have normal vision? Explain.
12. A mouse with an agouti coat is mated with an albino mouse of genotype *aabb*. Half of the offspring are albino, one-fourth are black, and one-fourth are agouti. What are the genotypes of the agouti parents and of the various kinds of offspring? (*Hint*: See the section on epistasis.)
13. The disease Leber's optic neuropathy is caused by a mutation in a gene carried on mitochondrial DNA. What would be the phenotype of their first child if a man with this disease married a woman who did not have the disease? What would be the result if the wife had the disease and the husband did not?

FOR INVESTIGATION

Sometimes scientists get lucky. Consider Mendel's dihybrid cross (see Figure 10.7). Peas have a haploid number of 7 chromosomes, so many of their genes are linked. What would Mendel's results have been if the genes for seed color and seed shape were *linked* with a map distance of 10 units? Now, consider Morgan's fruit flies (see Figure 10.21). Suppose that the genes for body color and wing shape were *not* linked? What results would Morgan have obtained?

CHAPTER 11 DNA and Its Role in Heredity

A structure for our times

In Michael Crichton's novel *Jurassic Park* and its film counterpart, fictional scientists were depicted using biotechnology to produce living dinosaurs for display in a theme park. In the story, the scientists isolated the DNA of dinosaurs from fossilized insects that had sucked the reptiles' blood. The insects, which had been preserved intact in amber (fossilized tree resin), yielded DNA that could be used to produce living individuals of long-extinct organisms such as *Tyrannosaurus rex*.

The premise of Crichton's novel was based on an actual scientific paper that claimed to show reptilian DNA sequences in a fossil insect. Unfortunately, the scientific report was not upheld; the "preserved" DNA turned out to be a contaminant from modern organisms.

Despite the fact that the preservation of intact DNA over millions of years is highly improbable, the popular success of *Jurassic Park* did bring the idea of DNA as the genetic material to the attention of millions of readers and viewers. Indeed, even before the novel and movie, the image of the DNA double helix was a familiar secular icon.

The double helix was first proposed by James Watson and Francis Crick in a short paper in the scientific journal *Nature*. A drawing of the structure made by Crick's wife, Odile, accompanied the article, and its simplicity and elegance made it an instant hit, not just with scientists, but with the general public. As Watson put it later, "A structure this pretty just had to exist."

Deoxyribonucleic acid—DNA—and its double-helical structure has become one of the great symbols of science of our era. It is not just trumpeted on the covers of newsmagazines as the "secret of life," but has moved from academic obscurity to common speech. One sees advertisements about a company whose customers get "into the DNA of business." A perfume is named "DNA" and is advertised as "The Essence of Life." A digital media software system is called the "DNA Server."

This is not the first time such a powerful symbol has emerged from science. Think of the mushroom cloud of a nuclear explosion and the Bohr model of the atom, with its electrons whizzing around the nucleus. Salvador Dali was the first well-known artist to use the DNA double helix in his whimsical creations. A portrait of Sir John Sulston, a Nobel prize-

Resurrecting the Rex Scientists and artists have been creating inanimate reconstructions of dinosaurs for more than 100 years. Michael Crichton's novel *Jurassic Park* was based on the fictional premise that DNA retrieved from fossils could produce living dinosaurs such as *Tyrannosaurus rex.*

CHAPTER OUTLINE

Bejeweled with DNA The double helix of DNA has become an iconic symbol of modern science and culture. Artists and designers make use of the widely recognized shape in many ways.

winning geneticist, is made of tiny bacterial colonies, each containing a piece of Sulston's DNA. The Brazilian artist Eduardo Kac translated a sentence from the Bible into a DNA nucleotide base sequence and incorporated this DNA into bacteria. The viewer can turn on an ultraviolet lamp to change the DNA sequence and the biblical verse it represents. DNA sculptures abound, and jewelry made with the double helix motif is called the "strands of life" collection.

But it is not only DNA's structure that stirs our society. It is what that structure symbolizes, which is nothing less than the promise and perils of our rapidly expanding knowledge of genetics.

IN THIS CHAPTER we will first describe the key experiments that led to the determination that the genetic material is DNA. We will then describe the structure of the DNA molecule and how this structure determines its function. We will describe the processes by which DNA is replicated, repaired, and maintained. Finally, we present two practical applications arising from our knowledge of DNA replication: the polymerase chain reaction and DNA sequencing.

11.1 What Is the Evidence that the Gene is DNA?

By the early twentieth century, geneticists had associated the presence of genes with chromosomes. Research began to focus on exactly which chemical component of chromosomes comprised this genetic material.

By the 1920s, scientists knew that chromosomes were made up of DNA and proteins. At this time a new dye was developed that could bind specifically to DNA and turned red in direct proportion to the amount of DNA present in a cell. This technique provided circumstantial evidence that DNA was the genetic material:

- *It was in the right place.* DNA was confirmed to be an important component of the nucleus and the chromosomes, which were known to carry genes.
- *It varied among species.* When cells from different species were stained with the dye and their color intensity measured, each species appeared to have its own specific amount of nuclear DNA.
- *It was present in the right amounts.* The amount of DNA in somatic cells (body cells not specialized for reproduction) was twice that in reproductive cells (eggs or sperm)—as might be expected for diploid and haploid cells, respectively.

But circumstantial evidence is *not* a scientific demonstration of cause and effect. After all, proteins are also present in cell nuclei. Science relies on experiments to test hypotheses. The convincing demonstration that DNA is the genetic material came from two sets of experiments, one on bacteria and the other on viruses.

DNA from one type of bacterium genetically transforms another type

The history of biology is filled with incidents in which research on one specific topic has—with or without answering the question originally asked—contributed richly to another, apparently unrelated area. Such a case of serendipity is seen in the work of Frederick Griffith, an English physician.

In the 1920s, Griffith was studying the bacterium *Streptococcus pneumoniae*, or pneumococcus, one of the agents that cause pneumonia in humans. He was trying to develop a vaccine against this devastating illness (antibiotics had not yet been discovered). Griffith was working with two strains of pneumococcus:

- Cells of the S strain produced colonies that looked smooth (S). Covered by a polysaccharide capsule, these cells were protected from attack by a host's immune system. When S cells were injected into mice, they reproduced and caused pneumonia (the strain was *virulent*).
- Cells of the R strain produced colonies that looked rough (R), lacked the protective capsule, and were not virulent.

Griffith inoculated some mice with heat-killed S pneumococci. These heat-killed bacteria did not produce infection. However, when Griffith inoculated other mice with a mixture of living R bacteria and heat-killed S bacteria, to his astonishment, the mice died of pneumonia (**Figure 11.1**). When he examined blood from the hearts of these mice, he found it full of living bacteria—many of them with characteristics of the virulent S strain! Griffith concluded that, in the presence of the dead S pneumococci, some of the living R pneumococci had been transformed into virulent S strain organisms.

Did this transformation of the bacteria depend on something that happened in the mouse's body? No. It was shown that simply incubating living R and heat-killed S bacteria together in a test tube yielded the same transformation. Years later, another group of scientists discovered that a cell-free extract of heat-killed S cells also transformed R cells. (A *cell-free extract* contains all the contents of ruptured cells, but no intact cells.) This result demonstrated that some substance—called at the time a chemical **transforming principle**—from the dead S pneumococci could cause a heritable change in the affected R cells. This was an extraordinary discovery: treatment with a chemical substance permanently changed an inherited characteristic. Now it remained to identify the chemical structure of this substance.

The transforming principle is DNA

Identifying the transforming principle was a crucial step in the history of biology. It was accomplished over a period of years by Oswald Avery and his colleagues at what is now Rockefeller University. They treated samples known to contain the pneumococcal transforming principle in a variety of ways to destroy different types of molecules—proteins, nucleic acids, carbohydrates, and lipids—and tested the treated samples to see if they had retained transforming activity.

11.1 Genetic Transformation of Nonvirulent Pneumococci Frederick Griffith's experiments demonstrated that something in the virulent S strain could transform nonvirulent R strain bacteria into a lethal form, even when the S strain bacteria had been killed by high temperatures. FURTHER RESEARCH: How would you show that heat-killed R strain bacteria can transform living S strain bacteria?

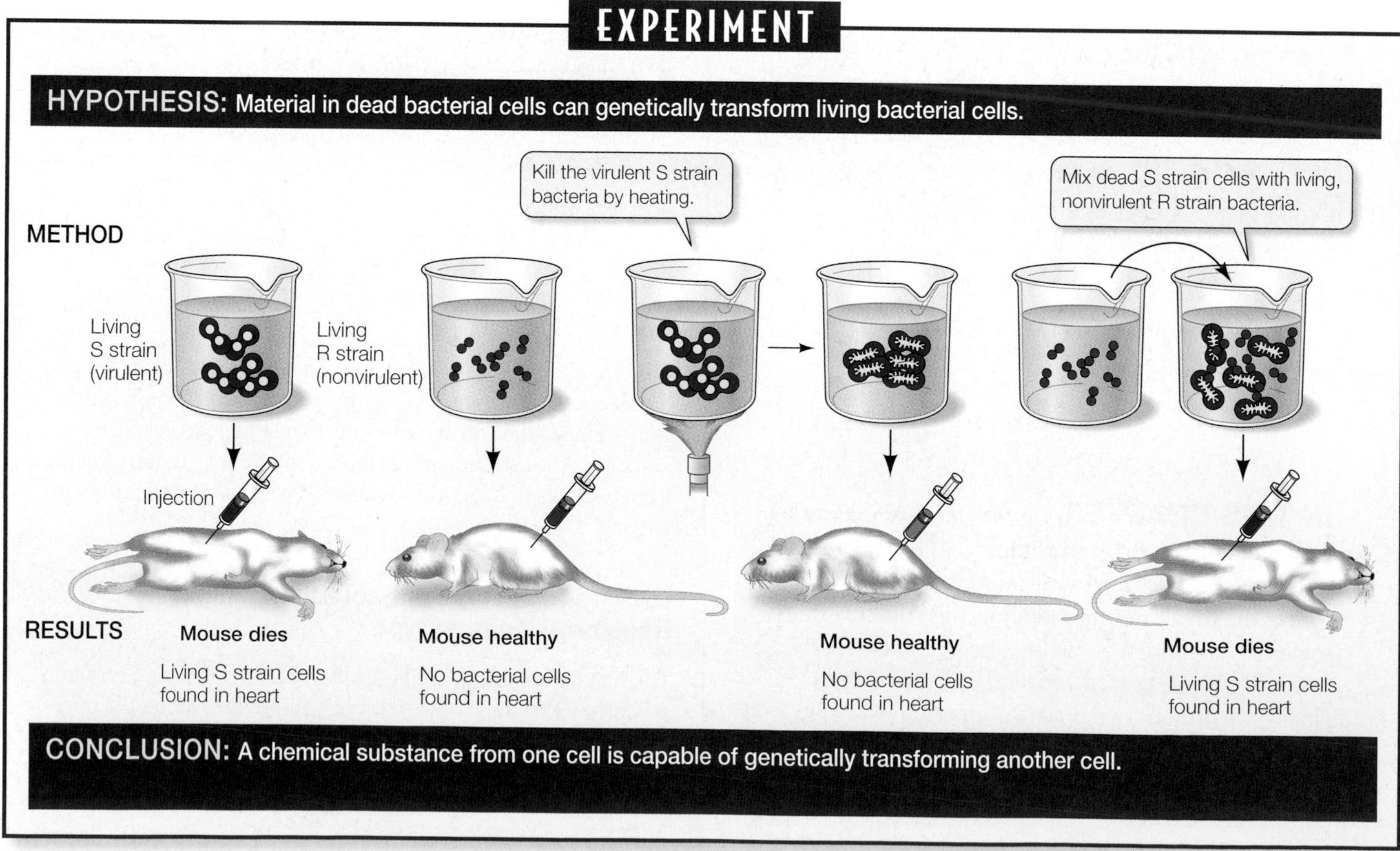
EXPERIMENT
HYPOTHESIS: Material in dead bacterial cells can genetically transform living bacterial cells.
METHOD
Kill the virulent S strain bacteria by heating.
Mix dead S strain cells with living, nonvirulent R strain bacteria.
Living S strain (virulent)
Living R strain (nonvirulent)
Injection
RESULTS
Mouse dies
Living S strain cells found in heart
Mouse healthy
No bacterial cells found in heart
Mouse healthy
No bacterial cells found in heart
Mouse dies
Living S strain cells found in heart
CONCLUSION: A chemical substance from one cell is capable of genetically transforming another cell.

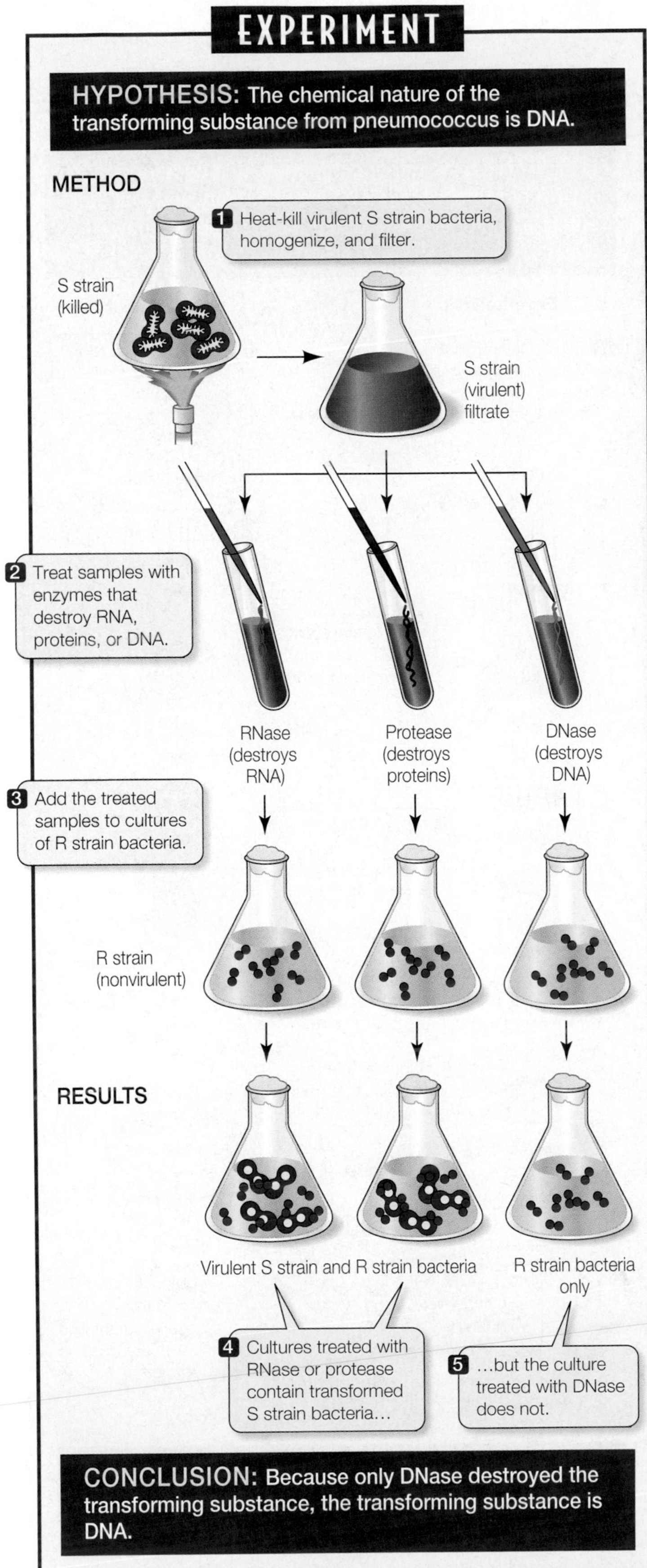

EXPERIMENT
HYPOTHESIS: The chemical nature of the transforming substance from pneumococcus is DNA.
METHOD
1 Heat-kill virulent S strain bacteria, homogenize, and filter.
S strain (killed)
S strain (virulent) filtrate
2 Treat samples with enzymes that destroy RNA, proteins, or DNA.
RNase (destroys RNA)
Protease (destroys proteins)
DNase (destroys DNA)
3 Add the treated samples to cultures of R strain bacteria.
R strain (nonvirulent)
RESULTS
Virulent S strain and R strain bacteria
R strain bacteria only
4 Cultures treated with RNase or protease contain transformed S strain bacteria...
5 ...but the culture treated with DNase does not.
CONCLUSION: Because only DNase destroyed the transforming substance, the transforming substance is DNA.

11.2 Genetic Transformation by DNA Experiments by Avery, MacLeod, and McCarty showed that DNA from virulent S strain pneumococci was responsible for transformation in Griffith's experiments (see Figure 11.1).

The answer was always the same: if the DNA in the sample was destroyed, transforming activity was lost, but there was no loss of activity when proteins, carbohydrates, or lipids were destroyed (**Figure 11.2**). As a final step, Avery, with Colin MacLeod and Maclyn McCarty, isolated virtually pure DNA from a sample containing pneumococcal transforming principle and showed that it caused bacterial transformation. (We now know that the gene encoding the enzyme that catalyzes the synthesis of the pneumococcal polysaccharide capsule was transferred during transformation.)

The work of Avery, MacLeod, and McCarty was a milestone in establishing that DNA is the genetic material in bacterial cells. However, when it was first published (in 1944), it had little impact, for two reasons. First, most scientists did not believe that DNA was chemically complex enough to be the genetic material, especially given the much greater chemical complexity of proteins. Second, and perhaps more important, bacterial genetics was a new field of study—it was not yet clear that bacteria even *had* genes.

Viral replication experiments confirmed that DNA is the genetic material

The questions about bacteria were soon resolved as researchers identified genes and mutations. Bacteria and viruses seemed to undergo genetic processes similar to those in fruit flies and pea plants. Experiments with these relatively simple systems were designed to discover the nature of the genetic material.

In 1952, Alfred Hershey and Martha Chase of the Carnegie Laboratory of Genetics published a paper that had a much greater immediate impact than Avery's 1944 paper. The Hershey–Chase experiment, which sought to determine whether DNA or protein was the genetic material, was carried out with a virus that infects bacteria. This virus, called bacteriophage T2, consists of little more than a DNA core packed inside a protein coat (**Figure 11.3**). The virus is thus made of the two materials that were, at the time, the leading candidates for the genetic material.

When bacteriophage T2 attacks a bacterium, part (but not all) of the virus enters the bacterial cell. About 20 minutes later, the cell bursts, releasing dozens of viruses. Clearly the virus is somehow able to replicate itself inside the bacterium. Hershey and Chase deduced that the entry of some viral component affects the genetic program of the host bacterial cell, transforming it into a bacteriophage factory. They set out to determine which part of the virus—protein or DNA—enters the bacterial cell. To trace the two components of the virus over its life cycle, Hershey and Chase labeled each component with a specific radioisotope:

- Proteins contain some *sulfur* (in the amino acids cysteine and methionine), an element not present in DNA. Sulfur has a radioactive isotope, ^{35}S. Hershey and Chase grew bacteriophage T2 in a bacterial culture in the presence of ^{35}S, so the proteins of the resulting viruses were labeled with the radioisotope.
- The deoxyribose–phosphate "backbone" of DNA is rich in *phosphorus* (see Figure 3.24), an element that is not present in most proteins. Phosphorus also has a radioisotope, ^{32}P. The researchers grew another batch of T2 in a bacterial culture in the presence of ^{32}P, thus labeling the viral DNA with ^{32}P.

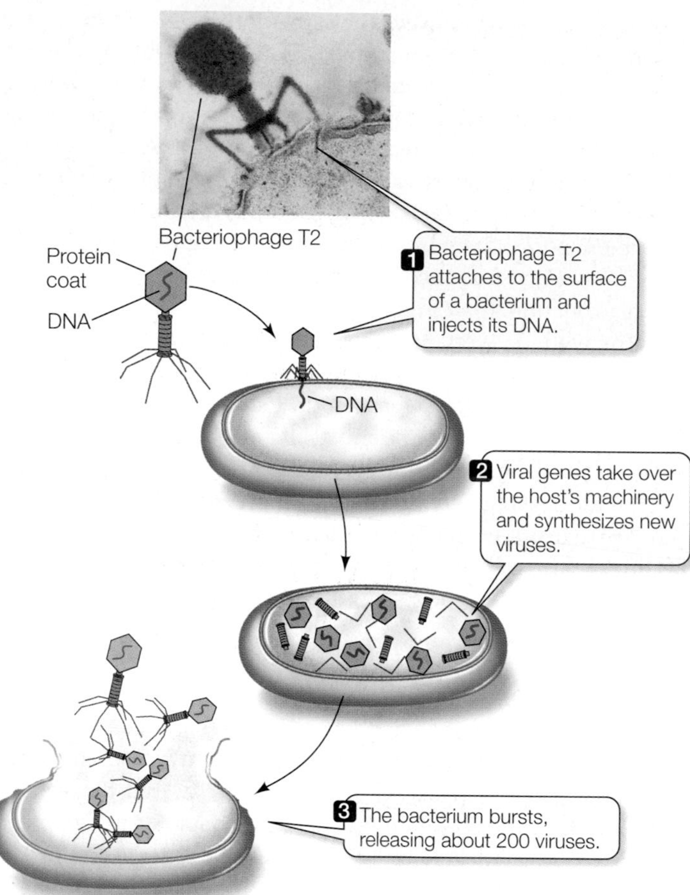

11.3 Bacteriophage T2: Reproduction Cycle Bacteriophage T2 is parasitic on *E. coli*, depending on the bacterium to produce new viruses. The external structures of bacteriophage T2 consist entirely of protein; its DNA is injected into the host bacterium.

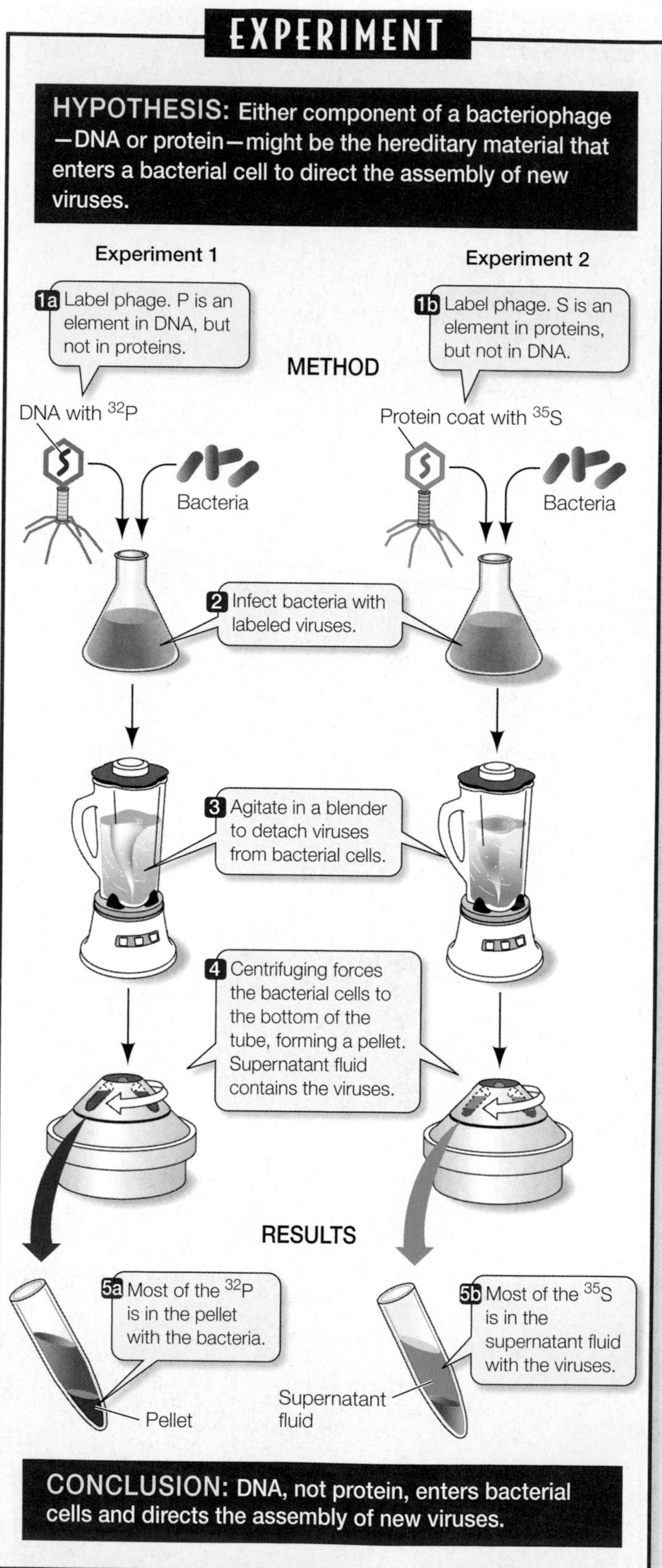

11.4 The Hershey–Chase Experiment This classic experiment demonstrated that DNA, not protein, is the genetic material. When bacterial cells were infected by radiolabeled T2 bacteriophage, only labeled DNA was found in the bacteria; labeled protein remained in the viral matter.

Using these radioactively labeled viruses, Hershey and Chase performed their revealing experiments (**Figure 11.4**). In one experiment, they allowed ^{32}P-labeled bacteriophage to infect bacteria; in the other, the bacteria were infected by ^{35}S-labeled bacteriophage. After a few minutes, they agitated each mixture of infected bacteria vigorously in a kitchen blender, which (without bursting the bacteria) stripped away the parts of the virus that had not penetrated the bacteria. Then they separated the bacteria from the rest of the material in a *centrifuge*. Spinning solutions or suspensions at high speed in a centrifuge causes solutes or particles to separate and form a gradient according to their density. The lighter remains of the viruses (those parts that had not penetrated the bacteria) were captured in the supernatant fluid, while the heavier bacterial cells segregated in a "pellet" in the bottom of the container. The scientists found that the supernatant fluid contained most of the ^{35}S (and thus the viral protein), while most of the ^{32}P (and thus the viral DNA) had stayed with the bacteria. These results suggested that it was DNA that had been transferred to the bacteria, and that DNA therefore was the compound responsible for redirecting the genetic program of the bacterial cell.

Hershey and Chase performed other similar but longer-range experiments, allowing a progeny (offspring) generation of viruses to be collected. The resulting viruses contained almost none of the original ^{35}S and none of the parental viral protein. They did, however, contain about one-third of the original ^{32}P—and thus, presumably, one-third of the original DNA. Because DNA was carried over in the virus from generation to generation but protein was not, the logical conclusion was that the hereditary information of the virus is contained in the DNA.

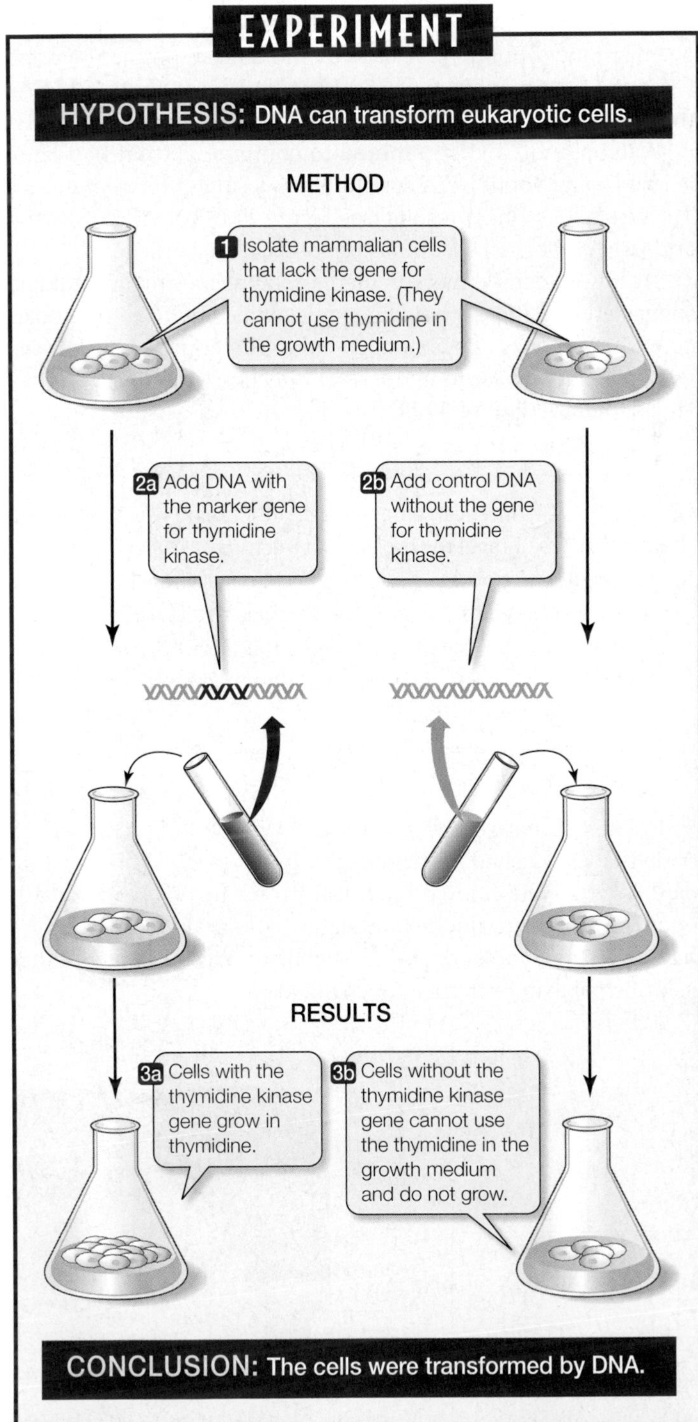

EXPERIMENT
HYPOTHESIS: DNA can transform eukaryotic cells.
METHOD
1 Isolate mammalian cells that lack the gene for thymidine kinase. (They cannot use thymidine in the growth medium.)
2a Add DNA with the marker gene for thymidine kinase.
2b Add control DNA without the gene for thymidine kinase.
RESULTS
3a Cells with the thymidine kinase gene grow in thymidine.
3b Cells without the thymidine kinase gene cannot use the thymidine in the growth medium and do not grow.
CONCLUSION: The cells were transformed by DNA.

11.5 Transfection in Eukaryotic Cells The use of a marker gene demonstrates that mammalian cells can be genetically transformed by DNA.

Eukaryotic cells can also be genetically transformed by DNA

With the publication of the evidence for DNA as the genetic material in bacteria and viruses, the question arose as to whether DNA could be similarly shown to be the genetic material in complex eukaryotes. Some simple experiments, such as injecting a white duck with DNA from a brown duck and reporting the recipient turning brown, or feeding flatworms DNA from worms that had learned a simple task and then seeing the recipients immediately get smarter, were reported, but the results were dubious because no one could duplicate them. It would be impossible for a large molecule such as DNA to get into all the cells of the body, let alone avoid hydrolysis into nucleotides in the digestive system.

However, genetic transformation of eukaryotic cells by DNA (called **transfection**) *can* be demonstrated. The key is to use a genetic **marker**, a gene whose presence in the recipient cell confers an observable phenotype. In the experiments with pneumococcus, these phenotypes were the smooth polysaccharide capsule and virulence. In eukaryotes, researchers usually use a nutritional or antibiotic resistance marker gene that permits the growth of transformed recipient cells but not of nontransformed cells. For example, in the absence of the gene that codes for thymidine kinase, an enzyme needed to make use of thymidine, mammalian cells do not grow. When DNA containing the thymidine kinase marker gene is added to a culture of mammalian cells lacking this gene, however, some cells will grow, demonstrating that they have been transfected with the gene (**Figure 11.5**). Any cell can be transfected in this way, even an egg cell. In this case, a whole new genetically transformed organism can result; such an organism is referred to as *transgenic*. Transformation in eukaryotes is the final line of evidence for DNA as the genetic material.

11.1 RECAP

Experiments on bacteria and on viruses demonstrated that DNA is the genetic material.

- At the time of Griffith's experiments in the 1920s, what circumstantial evidence suggested to scientists that DNA might be the genetic material? See p. 233
- Why were the experiments of Avery, MacLeod, and McCarty definitive evidence that DNA was the genetic material? See pp. 234–235 and Figure 11.2
- What attributes of bacteriophage T2 were key to the Hershey–Chase experiments demonstrating that DNA is the genetic material? See pp. 235–236 and Figure 11.4

As soon as scientists became convinced that the genetic material was DNA, they began efforts to learn its precise three-dimensional chemical structure. In determining the structure of DNA, scientists hoped to find the answers to two questions: How is DNA replicated between nuclear divisions, and how does it direct the synthesis of specific proteins? They were eventually able to answer both questions.

11.2 What Is the Structure of DNA?

The structure of DNA was deciphered only after many types of experimental evidence and theoretical considerations were considered together. The crucial evidence was obtained by *X-ray crystallography*. Some chemical substances, when they are isolated and purified, can be made to form crystals. The positions of atoms in a crystallized substance can be inferred from the pattern of diffraction of X-rays passed through it (**Figure 11.6A**). The attempt to characterize DNA would have been impossible without the crystallographs prepared in the early 1950s by the English chemist Rosalind Franklin (**Figure 11.6B**). Franklin's work, in turn, depended on the success of the English biophysicist Maurice Wilkins, who prepared a sample containing very uniformly oriented DNA fibers. These DNA preparations provided samples for diffraction that were far better than previous ones, and the crystallographs Franklin prepared from them suggested a spiral or helical molecule.

The chemical composition of DNA was known

The chemical composition of DNA also provided important clues to its structure. Biochemists knew that DNA was a polymer of *nucleotides*. Each nucleotide of DNA consists of a molecule of the sugar deoxyribose, a phosphate group, and a nitrogen-containing base (see Figures 3.23 and 3.24). The only differences among the four nucleotides of DNA are their nitrogenous bases: the purines **adenine** (**A**) and **guanine** (**G**), and the pyrimidines **cytosine** (**C**) and **thymine** (**T**).

In 1950, Erwin Chargaff at Columbia University reported some observations of major importance. He and his colleagues found that DNA from many different species—and from different sources within a single organism—exhibits certain regularities. In almost all DNA, the following rule holds: The amount of adenine equals the amount of thymine (A = T), and the amount of guanine equals the amount of cytosine (G = C) (**Figure 11.7**). As a result, the total abundance of purines (A + G) equals the total abundance of pyrimidines (T + C). The structure of DNA could not have been worked out without this observation, now known as *Chargaff's rule*, yet its significance was overlooked for at least three years.

Watson and Crick described the double helix

The solution to the puzzle of the structure of DNA was accelerated by *model building*: the assembly of three-dimensional representations of possible molecular structures using known relative molecular dimensions and known bond angles. This technique, originally exploited in structural studies by the American biochemist Linus Pauling, was used by the English physicist Francis Crick and the American geneticist James D. Watson (**Figure 11.8A**), then both at the Cavendish Laboratory of Cambridge University.

Watson and Crick attempted to combine all that had been learned so far about DNA structure into a single coherent model. The crystallographers' results (see Figure 11.6) convinced Watson and Crick that the DNA molecule is **helical** (cylindrically spiral). The results of density measurements and previous model building suggested that there are two polynucleotide chains in the molecule. Modeling studies had also led to the conclusion that the two chains in DNA run in opposite directions—that is, that they are **antiparallel** (**Figure 11.8B**).

> In 1952, the American biochemist Linus Pauling was also working to unravel the molecular structure of DNA. He attempted to travel to a meeting in London, where he hoped to convince the British researchers to allow him to view their crystallography. However, the U.S. Department of State revoked Pauling's passport based on his stance against the Korean War.

In late February of 1953, Crick and Watson built a model out of tin that established the general structure of DNA. This structure explained all the known chemical properties of DNA, and it opened the door to understanding its biological functions. There have been minor amendments to that first published structure, but its principal features remain unchanged.

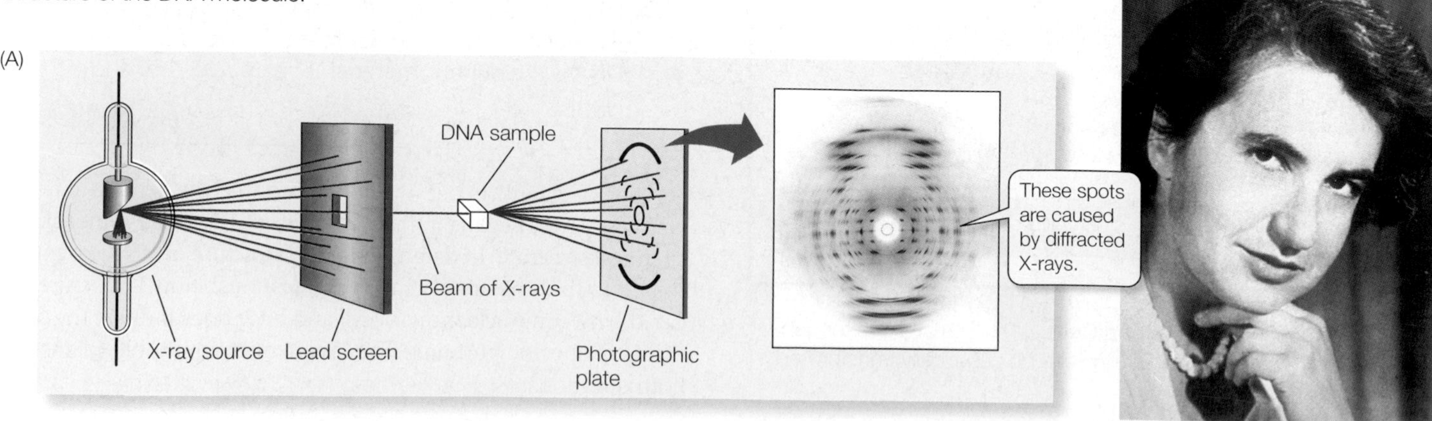

11.6 X-Ray Crystallography Helped Reveal the Structure of DNA (A) The positions of atoms in a crystallized chemical substance can be inferred by the pattern of diffraction of X-rays passed through it. The pattern of DNA is both highly regular and repetitive. (B) Rosalind Franklin's crystallography helped scientists to visualize the helical structure of the DNA molecule.

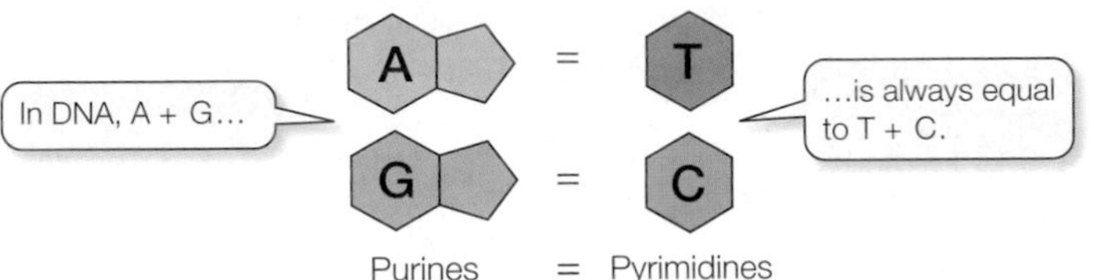

11.7 Chargaff's Rule In DNA, the total abundance of purines is equal to the total abundance of pyrimidines.

Four key features define DNA structure

Four features summarize the molecular architecture of the DNA molecule:

- It is a *double-stranded helix* of uniform diameter.
- It is *right-handed*. [Hold your right hand with the thumb pointing up (the axis of the helix) and the fingers curled around (the sugar–phosphate backbone) and you have the idea.]
- It is *antiparallel* (the two strands run in opposite directions).
- The outer edges of the nitrogenous bases are *exposed* in the major and minor grooves.

THE HELIX The sugar–phosphate "backbones" of the polynucleotide chains coil around the outside of the helix, and the nitrogenous bases point toward the center. The two chains are held together by hydrogen bonding between specifically paired bases (**Figure 11.9**). Consistent with Chargaff's rule,

- Adenine (A) pairs with thymine (T) by forming two hydrogen bonds.
- Guanine (G) pairs with cytosine (C) by forming three hydrogen bonds.

Every *base pair* consists of one purine (A or G) and one pyrimidine (T or C). This pattern is known as **complementary base pairing**.

Because the AT and GC pairs are of equal length, they fit into a fixed distance between the two chains (like rungs on a ladder), and the diameter of the helix is thus uniform. The base pairs are flat, and their stacking in the center of the molecule is stabilized by hydrophobic interactions (see Section 2.2), contributing to the overall stability of the double helix.

ANTIPARALLEL STRANDS What does it mean to say that the two DNA strands are *antiparallel*? The direction of each strand is determined by examining the bonds between the alternating phosphate groups and sugars that make up the backbones of each strand. Look closely at the five-carbon sugar (deoxyribose) mole-

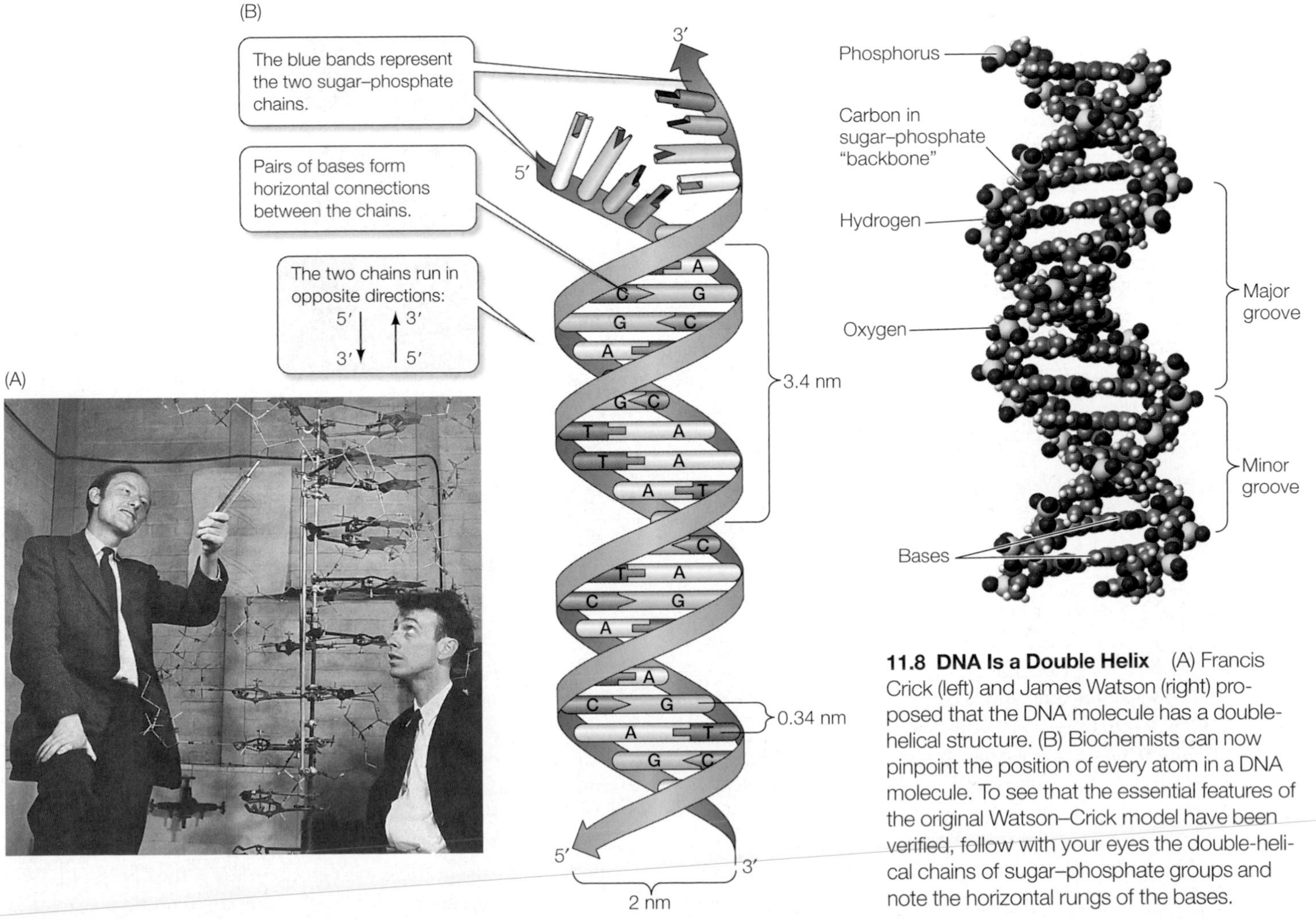

11.8 DNA Is a Double Helix (A) Francis Crick (left) and James Watson (right) proposed that the DNA molecule has a double-helical structure. (B) Biochemists can now pinpoint the position of every atom in a DNA molecule. To see that the essential features of the original Watson–Crick model have been verified, follow with your eyes the double-helical chains of sugar–phosphate groups and note the horizontal rungs of the bases.

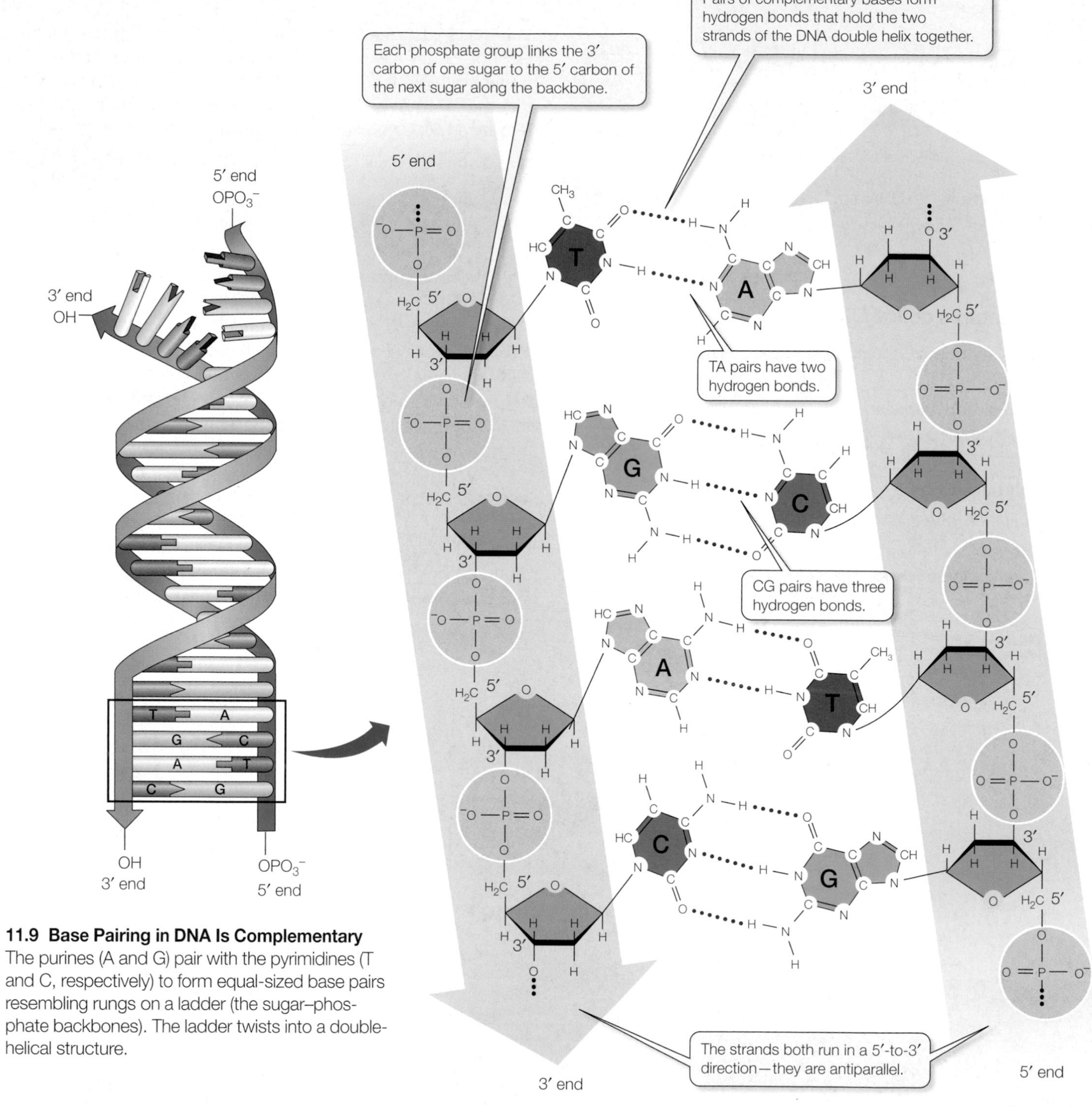

11.9 Base Pairing in DNA Is Complementary The purines (A and G) pair with the pyrimidines (T and C, respectively) to form equal-sized base pairs resembling rungs on a ladder (the sugar–phosphate backbones). The ladder twists into a double-helical structure.

cules in Figure 11.9. The number followed by a prime (′) designates the position of a carbon atom in the sugar. In the sugar–phosphate backbone of DNA, the phosphate groups connect to the 3′ carbon of one deoxyribose molecule and the 5′ carbon of the next, linking successive sugars together.

Thus the two ends of a polynucleotide chain differ. At one end of a chain is a free (not connected to another nucleotide) 5′ phosphate group (—OPO_3^-); this end is called the 5′ end. At the other end is a free 3′ hydroxyl group (—OH); this end is called the 3′ end. In a DNA double helix, the 5′ end of one strand is paired with the 3′ end of the other strand, and vice versa. In other words, were you to draw an arrow for each strand running from 5′ to 3′, the arrows would point in opposite directions.

BASE EXPOSURE AT THE GROOVES Looking back at Figure 11.8B, note the major and minor grooves in the helix. From these grooves, the exposed outer edges of the flat, hydrogen-bonded base pairs are accessible for potential hydrogen bonding. As seen in Figure 11.8, two hydrogen bonds join each AT base pair, while three hydrogen bonds join each GC base pair. Hydrogen-bonding opportunities also exist at the C=O group in T and the "N" group in A; the GC base pair offers additional hydrogen bonding possibilities.

Thus the surfaces of the AT and GC base pairs offer slightly different, chemically distinct surfaces that another molecule, such as a protein, could recognize and bind to. Access to the exposed base-pair sequences in the major and minor grooves is the key to protein–DNA interactions in the replication and expression of the genetic information in DNA.

The double-helical structure of DNA is essential to its function

The genetic material performs four important functions, and the DNA structure proposed by Watson and Crick was elegantly suited to three of them.

- *The genetic material stores an organism's genetic information.* With its millions of nucleotides, the base sequence of a DNA molecule could encode and store an enormous amount of information and could account for species and individual differences. DNA fits this role nicely.
- *The genetic material is susceptible to mutation*, or permanent changes in the information it encodes. For DNA, mutations might be simple changes in the linear sequence of base pairs.
- *The genetic material is precisely replicated* in the cell division cycle. Replication could be accomplished by complementary base pairing, A with T and G with C. In the original publication of their findings in the journal *Nature* in 1953, Watson and Crick coyly pointed out, "It has not escaped our notice that the specific pairing we have postulated immediately suggests a possible copying mechanism for the genetic material."
- *The genetic material is expressed as the phenotype.* This function is not obvious in the structure of DNA. However, as we will see in the next chapter, the nucleotide sequence of DNA is copied into RNA, which is in turn converted into a linear sequence of amino acids—a protein. The folded forms of proteins provide much of the phenotype of an organism.

11.2 RECAP

DNA is a double helix made up of two antiparallel polynucleotide chains. The two chains are joined by hydrogen bonds between the nucleotide bases, which pair specifically, A with T and G with C. Chemical groups of the bases that are exposed in the grooves of the helix can be recognized by other molecules.

- Can you describe some of the evidence that Watson and Crick used to come up with the double helix model for DNA? See pp. 238–239
- Do you understand how the double-helical structure of DNA relates to its function?

Once the structure of DNA was understood, it was possible to discover how DNA replicates itself. Let's examine the experiments that taught us how this elegant process works.

11.3 How Is DNA Replicated?

The mechanism of DNA replication that had suggested itself to Watson and Crick was soon confirmed. First, experiments showed that single strands of DNA could be replicated in a test tube containing simple substrates and an enzyme. Then a truly classic experiment showed that each of the two strands of the double helix can serve as a template for a new strand of DNA.

Three modes of DNA replication appeared possible

The prediction that the DNA molecule contains the information needed for its own replication was confirmed by the work of Arthur Kornberg, then at Washington University in St. Louis. He showed that DNA with the same base composition as parental DNA can be synthesized in a test tube containing three substances:

- The substrates, deoxyribonucleoside triphosphates dATP, dCTP, dGTP, and dTTP
- A **DNA polymerase** enzyme
- DNA, which serves as a **template** to guide the incoming nucleotides

Recall that a nucleoside is a nitrogen base attached to a sugar. The four deoxyribonucleoside triphosphates each consist of a nitrogen base attached to deoxyribose, which in turn is attached to three phosphate groups.

The next question was which of three possible replication patterns was occurring:

- *Semiconservative replication*, in which each parent strand serves as a template for a new strand, and the two new DNA molecules each have one old and one new strand (**Figure 11.10A**)
- *Conservative replication*, in which the original double helix serves as a template for, but does not contribute to, a new double helix (**Figure 11.10B**)
- *Dispersive replication*, in which fragments of the original DNA molecule serve as templates for assembling two new molecules, each containing old and new parts, perhaps at random (**Figure 11.10C**)

Watson and Crick's original paper suggested that DNA replication was semiconservative, but Kornberg's experiment did not provide a basis for choosing among these three models.

Meselson and Stahl demonstrated that DNA replication is semiconservative

The work of Matthew Meselson and Franklin Stahl convinced the scientific community that the pattern seen in DNA is **semiconservative replication**. Working at the California Institute of Technology, Meselson and Stahl devised a simple way to distinguish old parent strands of DNA from newly copied ones: *density labeling*.

The key to their experiment was the use of a "heavy" isotope of nitrogen. Heavy nitrogen (^{15}N) is a rare, nonradioactive isotope that makes molecules containing it more dense than chemically identical molecules containing the common isotope, ^{14}N. Meselson, Stahl, and Jerome Vinograd grew two cultures of the bacterium *Escherichia coli* for many generations:

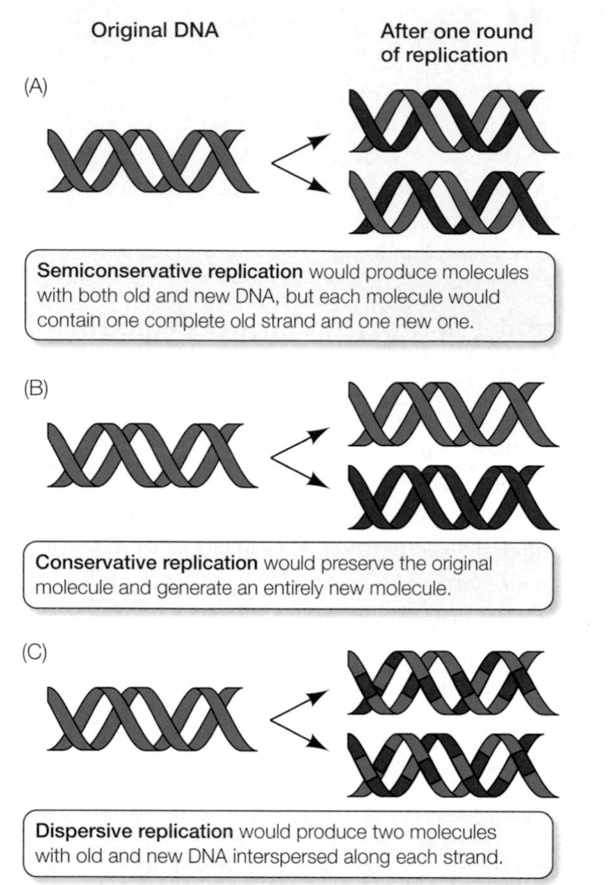

Original DNA
After one round of replication
(A)
Semiconservative replication would produce molecules with both old and new DNA, but each molecule would contain one complete old strand and one new one.
(B)
Conservative replication would preserve the original molecule and generate an entirely new molecule.
(C)
Dispersive replication would produce two molecules with old and new DNA interspersed along each strand.

11.10 Three Models for DNA Replication In each model, original DNA is shown in blue and newly synthesized DNA in red.

- One culture was grown in a medium whose nitrogen source (ammonium chloride, NH_4Cl) was made with ^{15}N instead of ^{14}N. As a result, all the DNA in the bacteria was "heavy."
- Another culture was grown in a medium containing ^{14}N, and all the DNA in these bacteria was "light."

When extracts from the two cultures were combined and centrifuged, two separate DNA bands formed, showing that this method could distinguish DNA samples of slightly different densities.

Next, the researchers grew another *E. coli* culture on ^{15}N medium, then transferred it to normal ^{14}N medium and allowed the bacteria to continue growing (**Figure 11.11**). Under the conditions they used, *E. coli* cells divide, replicating their DNA every 20 minutes. Meselson and Stahl collected some of the bacteria after each division and extracted DNA from the samples. They found that the density gradient was different in each bacterial generation:

- At the time of the transfer to the ^{14}N medium, the DNA was uniformly labeled with ^{15}N, and hence was relatively dense.
- After one generation in the ^{14}N medium, when the DNA had been duplicated once, all the DNA was of an intermediate density.
- After two generations, there were two equally large DNA bands: one of low density and one of intermediate density.
- In samples from subsequent generations, the proportion of low-density DNA increased steadily.

The results of this experiment can be explained only by the semiconservative model of DNA replication. In the first round of DNA replication in the ^{14}N medium, the strands of the double helix—both heavy with ^{15}N—separated. Each strand then acted as the template for a second strand, which contained only ^{14}N and hence was less dense. Each double helix then consisted of one ^{15}N strand and one ^{14}N strand, and was of intermediate density. In the second replication, the ^{14}N-containing strands directed the synthesis of partners with ^{14}N, creating low-density DNA, and the ^{15}N strands formed new ^{14}N partners (see Figure 11.10).

The crucial observation demonstrating the semiconservative model was that intermediate-density DNA (^{15}N–^{14}N) appeared in the first generation and continued to appear in subsequent generations. With the other models, the results would have been quite different (see Figure 11.10):

- In conservative replication, the first generation would have had both high-density DNA (^{15}N–^{15}N) and low-density DNA (^{14}N–^{14}N), but no intermediate-density DNA.
- In dispersive replication, the density of the new DNA would have been half that of the parent DNA, but DNA of this density would not continue to appear in subsequent generations.

The Meselson–Stahl experiment, called by some scientists among the most elegant ever done by biologists, was an excellent example of the scientific method. It began with three hypotheses—the three models of DNA replication—and was designed so that the results could differentiate between them.

We all began life as a fertilized egg, with only one set of double-stranded DNA molecules from our parents. Given semiconservative replication, do we still have those original parental strands? There is some evidence that we may. During mitosis, stem cells in the adult body preferentially retain DNA with the "old" strands. A similar mechanism may operate during early development.

There are two steps in DNA replication

Semiconservative DNA replication in the cell involves a number of different enzymes and other proteins. It takes place in two steps:

- The DNA double helix is unwound to separate the two template strands and make them available for new base pairing.
- New nucleotides are joined by phosphodiester linkages to each growing new strand in a sequence determined by complementary base pairing with the bases on the template strand.

A key observation is that *nucleotides are added to the growing new strand at the 3′ end*—the end at which the DNA strand has a free

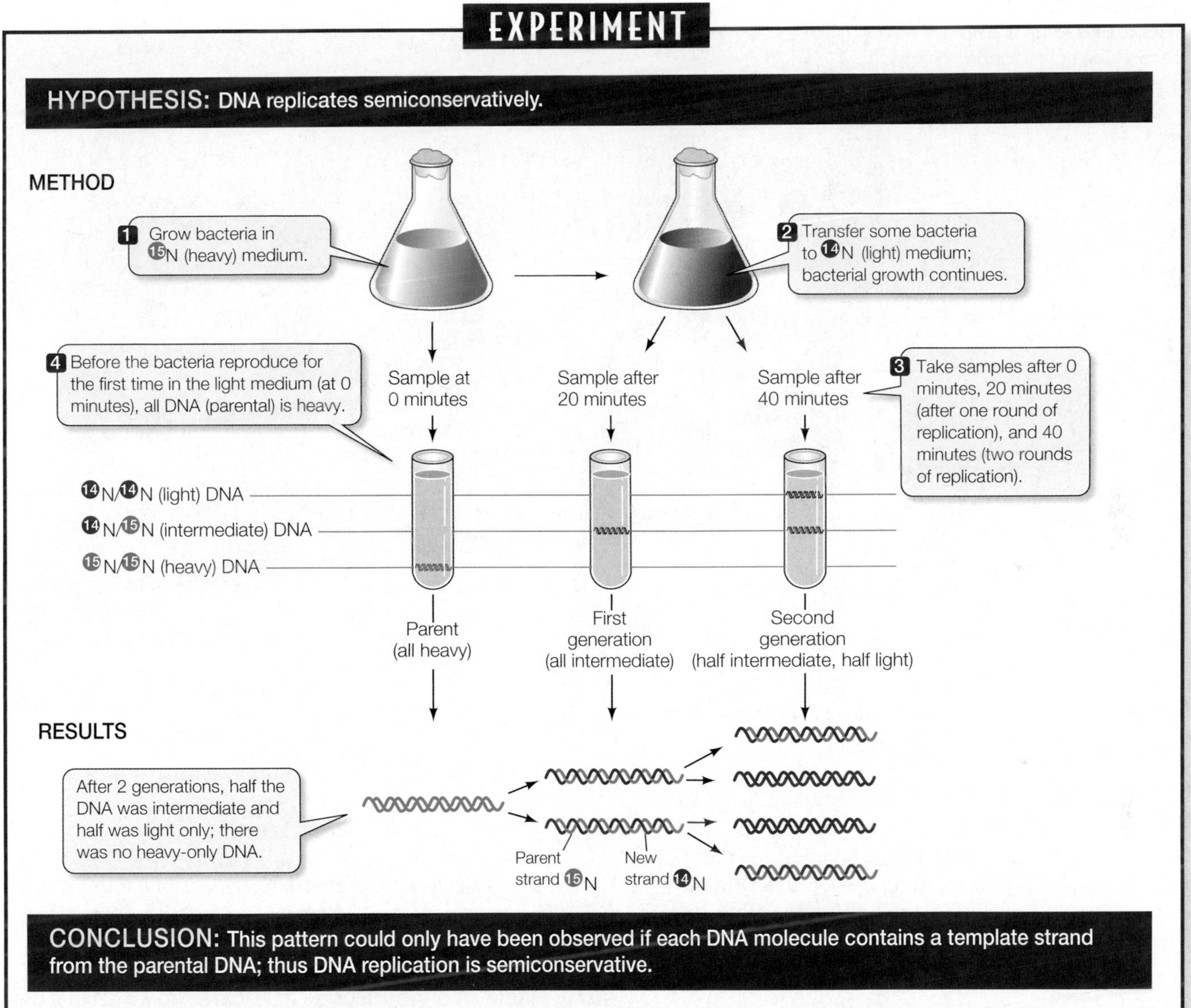

EXPERIMENT
HYPOTHESIS: DNA replicates semiconservatively.
METHOD
1 Grow bacteria in 15N (heavy) medium.
2 Transfer some bacteria to 14N (light) medium; bacterial growth continues.
4 Before the bacteria reproduce for the first time in the light medium (at 0 minutes), all DNA (parental) is heavy.
Sample at 0 minutes
Sample after 20 minutes
Sample after 40 minutes
3 Take samples after 0 minutes, 20 minutes (after one round of replication), and 40 minutes (two rounds of replication).
14N/14N (light) DNA
14N/15N (intermediate) DNA
15N/15N (heavy) DNA
Parent (all heavy)
First generation (all intermediate)
Second generation (half intermediate, half light)
RESULTS
After 2 generations, half the DNA was intermediate and half was light only; there was no heavy-only DNA.
Parent strand 15N
New strand 14N
CONCLUSION: This pattern could only have been observed if each DNA molecule contains a template strand from the parental DNA; thus DNA replication is semiconservative.

11.11 The Meselson–Stahl Experiment A centrifuge was used to separate DNA molecules labeled with isotopes of different densities. This experiment revealed a pattern that supports the semiconservative model of DNA replication. FURTHER RESEARCH: If you continued this experiment for two more generations (as Meselson and Stahl actually did), what would be the composition (in terms of low- and intermediate-density) of the fourth generation DNA?

hydroxyl (—OH) group on the 3′ carbon of its terminal deoxyribose (**Figure 11.12**). One of the three phosphate groups in a deoxyribonucleoside triphosphate is attached to the 5′ position of the sugar. The bonds linking the other two phosphate groups to the nucleotide are broken, releasing energy for the reaction.

DNA is threaded through a replication complex

DNA is replicated through the interaction of the template strand with a huge protein complex called the **replication complex**, which catalyzes the reactions involved. All chromosomes have at least one base sequence, called the **origin of replication** (***ori***), to which this replication complex initially binds. As noted above, this binding is based on the recognition of the different nucleotide bases by proteins. DNA replicates *in both directions* from the origin of replication, forming two **replication forks**. Both of the separated strands of the parent molecule act as templates simultaneously, and the formation of the new strands is guided by complementary base pairing.

Until recently, DNA replication was depicted as a locomotive (the replication complex) moving along a railroad track (the DNA) (**Figure 11.13A**). The current view is that this model may not be correct. Instead, the replication complex seems to be stationary, attached to nuclear structures, and it is the DNA that moves, essentially threading through the complex as single strands and emerging as double strands (**Figure 11.13B**). All replication complexes contain several proteins with different roles in DNA replication; we will describe these proteins as we examine the steps of the process.

11.12 Each New DNA Strand Grows from Its 5′ End to Its 3′ End The DNA strand at the right (blue) is the template for the synthesis of the complementary strand that is growing at the left (pink).

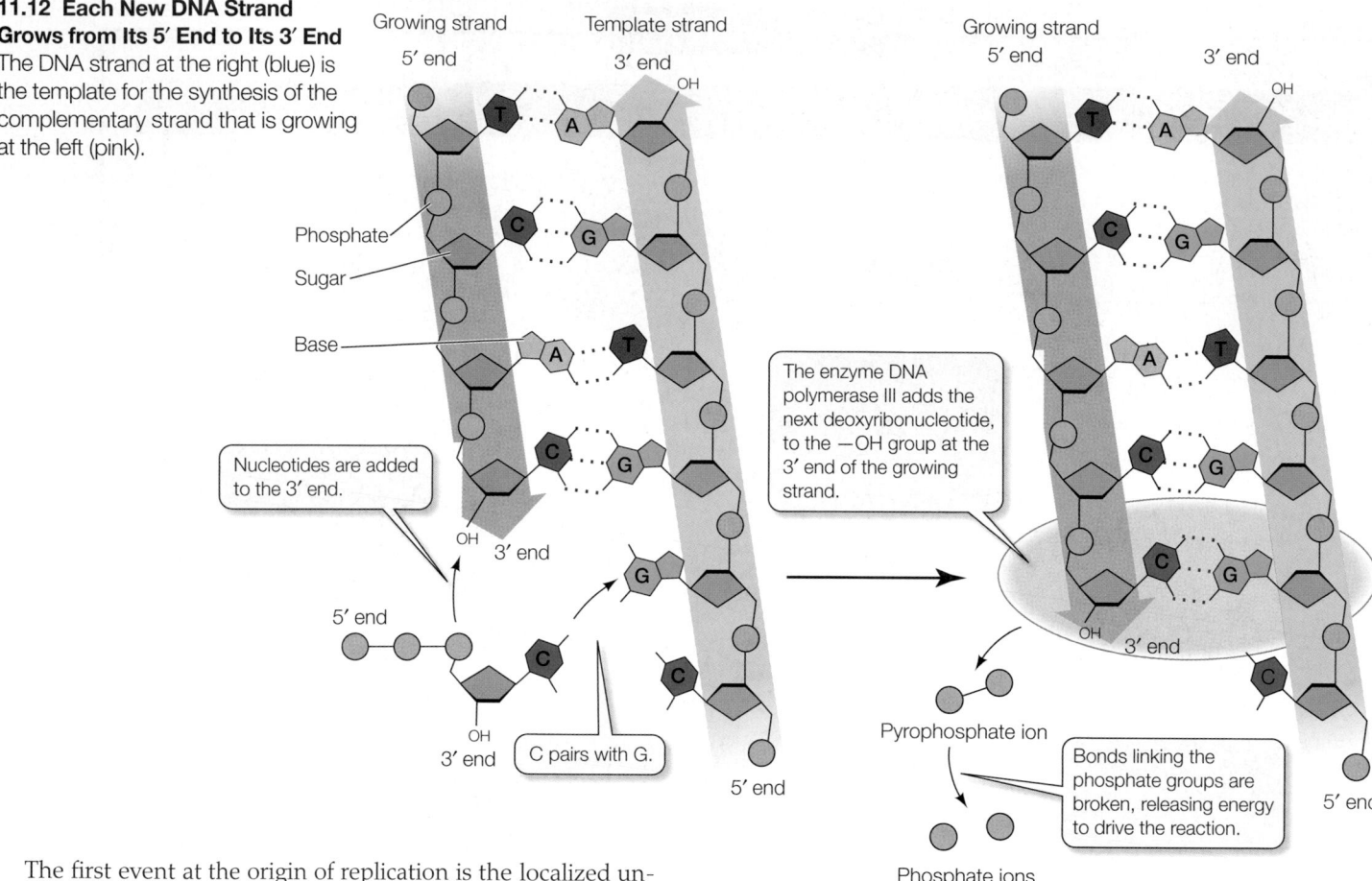

The first event at the origin of replication is the localized unwinding (denaturation) of DNA. There are several forces that hold the two strands together, including hydrogen bonding and the hydrophobic interactions of the bases. An enzyme called **DNA helicase** uses energy from ATP hydrolysis to unwind the DNA, and special proteins called **single-strand binding proteins** bind to the unwound strands to keep them from reassociating into a double helix. This process makes each of the two template strands available for complementary base pairing.

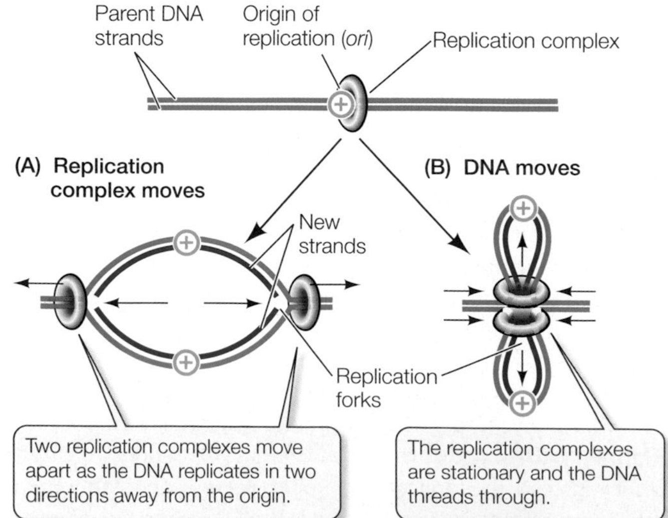

11.13 Two Views of DNA Replication (A) It was once thought that the replication complex moved along DNA like a locomotive moving along a railroad track. (B) More recent evidence suggests that the DNA is threaded through the stationary replication complex.

SMALL CIRCULAR CHROMOSOMES REPLICATE FROM A SINGLE ORIGIN Small circular chromosomes, such as the 1–4-million-base-pair DNA of bacteria, have a single origin of replication. As the DNA moves through the replication complex, the replication forks grow around the circle (**Figure 11.14A**). Two interlocking circular DNAs are formed, and they are separated by an enzyme called **DNA topoisomerase**.

DNA polymerases are very fast. In the bacterium *E. coli*, replication can be as fast as 1,000 bases per second, and it takes 20–40 minutes to replicate the bacterium's 4.7 million base pairs. Human polymerases are slower (50 bases per second), and human chromosomes are much larger (about 80 million base pairs). In this case, to get the job done in an hour, it takes many polymerases working at many replication forks.

LARGE LINEAR CHROMOSOMES HAVE MANY ORIGINS In large linear chromosomes, such as a human chromosome, there are hundreds of origins of replication. Origins of replication that are adjacent to one another along the linear chromosome can be bound by replication complexes at the same time and replicated simultaneously. So there are many replication forks in eukaryotic DNA (**Figure 11.14B**).

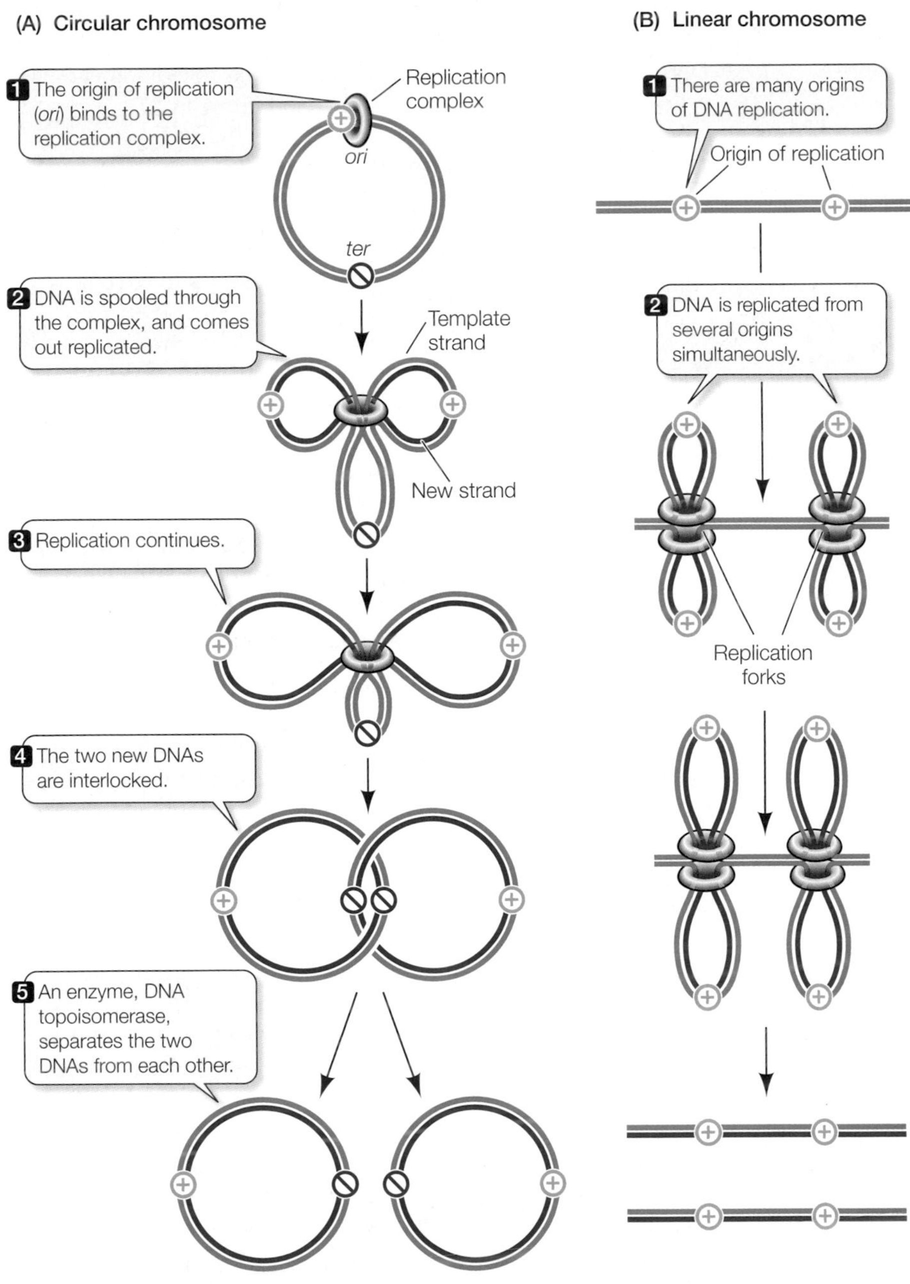

(A) Circular chromosome
1 The origin of replication (ori) binds to the replication complex.
Replication complex
ori
ter
2 DNA is spooled through the complex, and comes out replicated.
Template strand
New strand
3 Replication continues.
4 The two new DNAs are interlocked.
5 An enzyme, DNA topoisomerase, separates the two DNAs from each other.
(B) Linear chromosome
1 There are many origins of DNA replication.
Origin of replication
2 DNA is replicated from several origins simultaneously.
Replication forks

11.14 Replication in Small Circular and Large Linear Chromosomes (A) Small circular chromosomes have a single origin (*ori*) and terminus (*ter*) of replication. (B) Larger linear chromosomes have many origins of replication.

DNA polymerases add nucleotides to the growing chain

DNA polymerases are much larger than their substrates, the deoxyribonucleoside triphosphates, and the template DNA, which is very thin (**Figure 11.15A**). Molecular models of the enzyme–substrate–template complex from bacteria show that the enzyme is shaped like an open right hand with a palm, a thumb, and fingers (**Figure 11.15B**). The palm holds the active site of the enzyme and brings together the substrate and the template. The finger regions rotate inward and have precise shapes that can recognize the different shapes of the four nucleotide bases.

NO DNA BEGINS WITHOUT A PRIMER DNA polymerases can elongate a polynucleotide strand by covalently linking new nucleotides to a previously existing strand, but they cannot start a strand from scratch. Therefore, a "starter" strand, called a **primer**, is required. In DNA replication, the primer is a short single strand of RNA (**Figure 11.16**). This RNA strand, complementary to the DNA template strand, is synthesized one nucleotide at a time by an enzyme called **primase**. DNA polymerase then adds nucleotides to the 3′ end of the primer and continues until the replication of that section of DNA has been completed. Then the RNA

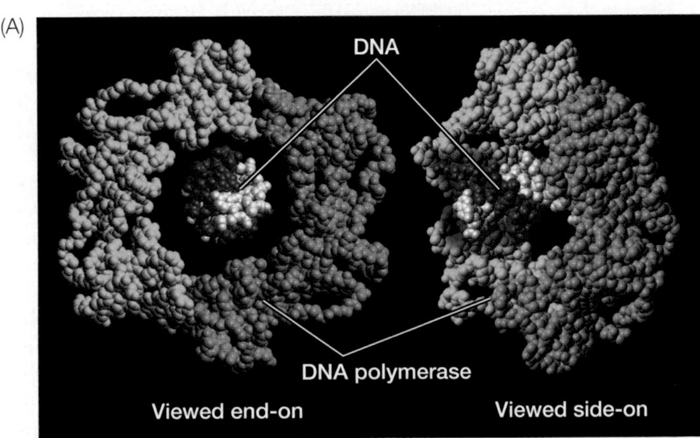

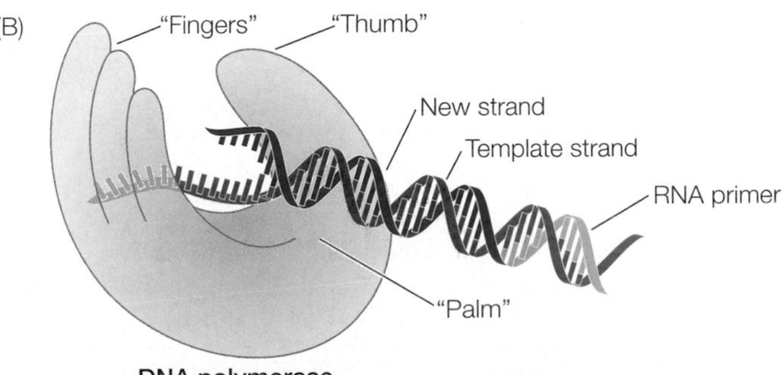

11.15 DNA Polymerase Binds to the Template Strand (A) The DNA polymerase enzyme (blue and green) is much larger than the DNA molecule (red and white). (B) DNA polymerase is shaped like a hand, and in this side-on view, its "fingers" can be seen curling around the DNA. These "fingers" can recognize the distinctive shapes of the four bases.

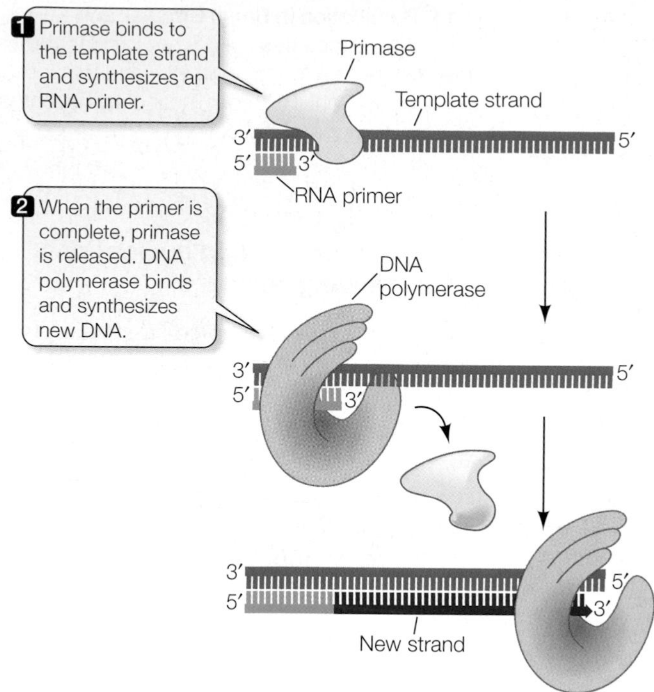

1 Primase binds to the template strand and synthesizes an RNA primer.
Primase
Template strand
3′
5′
5′
3′
RNA primer
2 When the primer is complete, primase is released. DNA polymerase binds and synthesizes new DNA.
DNA polymerase
3′
5′
5′
3′
3′
5′
5′
3′
New strand

11.16 No DNA Forms without a Primer DNA polymerases require a primer—a "starter" strand of DNA or RNA to which they can add new nucleotides.

primer is degraded, DNA is added in its place, and the resulting DNA fragments are connected by the action of other enzymes. When DNA replication is complete, each new strand consists only of DNA.

CELLS CONTAIN SEVERAL DIFFERENT DNA POLYMERASES Most cells contain more than one DNA polymerase, but only one of them is responsible for chromosomal DNA replication. The others are involved in primer removal and DNA repair. Fourteen DNA polymerases have been identified in humans; the one catalyzing most replication is DNA polymerase δ. In the bacterium *E. coli* there are five DNA polymerases; the one responsible for replication is DNA polymerase III. Various other proteins play roles in other replication tasks; some of these are shown in **Figure 11.17**.

THE TWO DNA STRANDS GROW DIFFERENTLY As Figure 11.17 shows, the DNA at the replication fork opens up like a zipper in one direction. Study **Figure 11.18** and try to imagine what is happening over a short period of time. Remember that the two DNA strands are antiparallel; that is, the 3′ end of one strand is paired with the 5′ end of the other.

- One newly replicating strand (the **leading strand**) is pointing in the "right" direction to grow continuously at its 3′ end as the fork opens up.
- The other new strand (the **lagging strand**) is pointing in the "wrong" direction: as the fork opens up further, its exposed 3′ end gets farther and farther away from the fork, and an unreplicated gap is formed, which would get bigger and bigger if there were not a special mechanism to overcome this problem.

Synthesis of the lagging strand requires working in relatively small, discontinuous stretches (100 to 200 nucleotides at a time in eukaryotes; 1,000 to 2,000 at a time in prokaryotes). These discontinuous stretches are synthesized just as the leading strand is, by the addition of new nucleotides one at a time to the 3′ end of the new strand, but the synthesis of this new strand moves in the direction opposite to that in which the replication fork is moving. These stretches of new DNA for the lagging strand are called **Okazaki fragments**, after their discoverer, the Japanese biochemist Reiji Okazaki. While the leading strand grows continuously "forward," the lagging strand grows in shorter, "backward" stretches with gaps between them.

A single primer suffices for synthesis of the leading strand, but each Okazaki fragment requires its own primer. In bacteria, DNA polymerase III synthesizes Okazaki fragments by adding nucleotides to a primer until it reaches the primer of the previous fragment. At this point, DNA polymerase I (discovered by Arthur Kornberg) removes the old primer and replaces it with DNA. Left behind is a tiny nick—the final phosphodiester linkage between the adjacent Okazaki fragments is missing. The enzyme **DNA ligase** catalyzes the formation of that bond, linking the fragments and making the lagging strand whole (**Figure 11.19**).

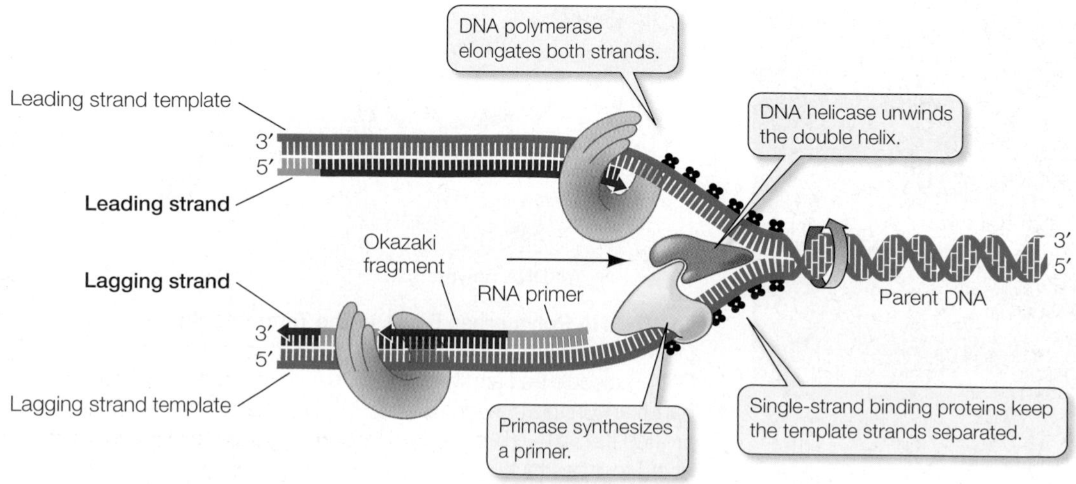

DNA polymerase elongates both strands.
Leading strand template
3′
5′
Leading strand
DNA helicase unwinds the double helix.
Okazaki fragment
Lagging strand
RNA primer
3′
5′
Parent DNA
3′
5′
Lagging strand template
Primase synthesizes a primer.
Single-strand binding proteins keep the template strands separated.

11.17 Many Proteins Collaborate in the Replication Complex Several proteins in addition to DNA polymerase are involved in DNA replication. The two molecules of DNA polymerase shown here are actually part of the same complex.

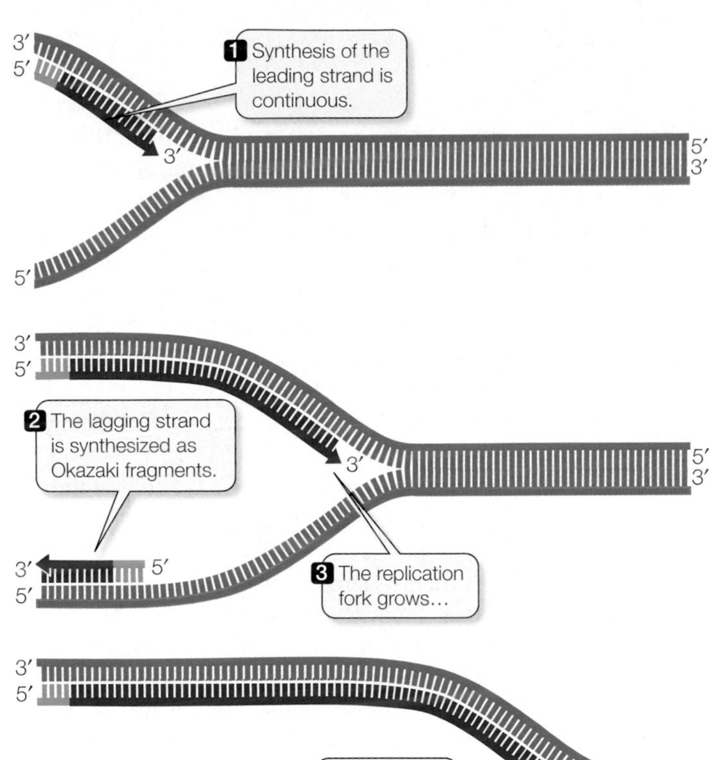

11.18 The Two New Strands Form in Different Ways As the parent DNA unwinds, both new strands are synthesized in the 5′-to-3′ direction, although their template strands are antiparallel. The leading strand grows continuously forward, but the lagging strand grows in short discontinuous stretches called Okazaki fragments. Eukaryotic Okazaki fragments are hundreds of nucleotides long, with gaps between them.

Working together, DNA helicase, the two DNA polymerases, primase, DNA ligase, and the other proteins of the replication complex do the job of DNA synthesis with a speed and accuracy that are almost unimaginable. In *E. coli*, the replication complex makes new DNA at a rate in excess of 1,000 base pairs per second, committing errors in fewer than one base in a million.

A SLIDING DNA CLAMP MAKES DNA POLYMERASE PROCESSIVE How do DNA polymerases work so fast? We saw in Section 6.4 that an enzyme catalyzes a chemical reaction:

substrate binds to enzyme → one product is formed →
enzyme is released → cycle repeats

It is hard to envision a reaction so fast that it would be possible to go through such a cycle for each nucleotide added to DNA. Instead, DNA polymerases are **processive**—that is, they catalyze many polymerizations each time they bind to a DNA molecule:

substrates bind to one enzyme → many products are formed →
enzyme is released → cycle repeats

The newly replicated strand is stabilized by a **sliding DNA clamp** (**Figure 11.20**). This protein has multiple identical subunits assembled into a doughnut shape. The doughnut's "hole" is just large enough to encircle the DNA double helix, along with a single layer of water molecules for lubrication. The clamp binds to DNA just behind DNA polymerase, keeping it associated tightly with the newly replicated DNA. If the clamp is absent, DNA polymerase dissociates from DNA after 20–100 polymerizations. With the clamp, it can polymerize up to 50,000 nucleotides before it detaches.

Telomeres are not fully replicated

As we have just seen, replication of the lagging strand occurs by the addition of Okazaki fragments to RNA primers. When the ter-

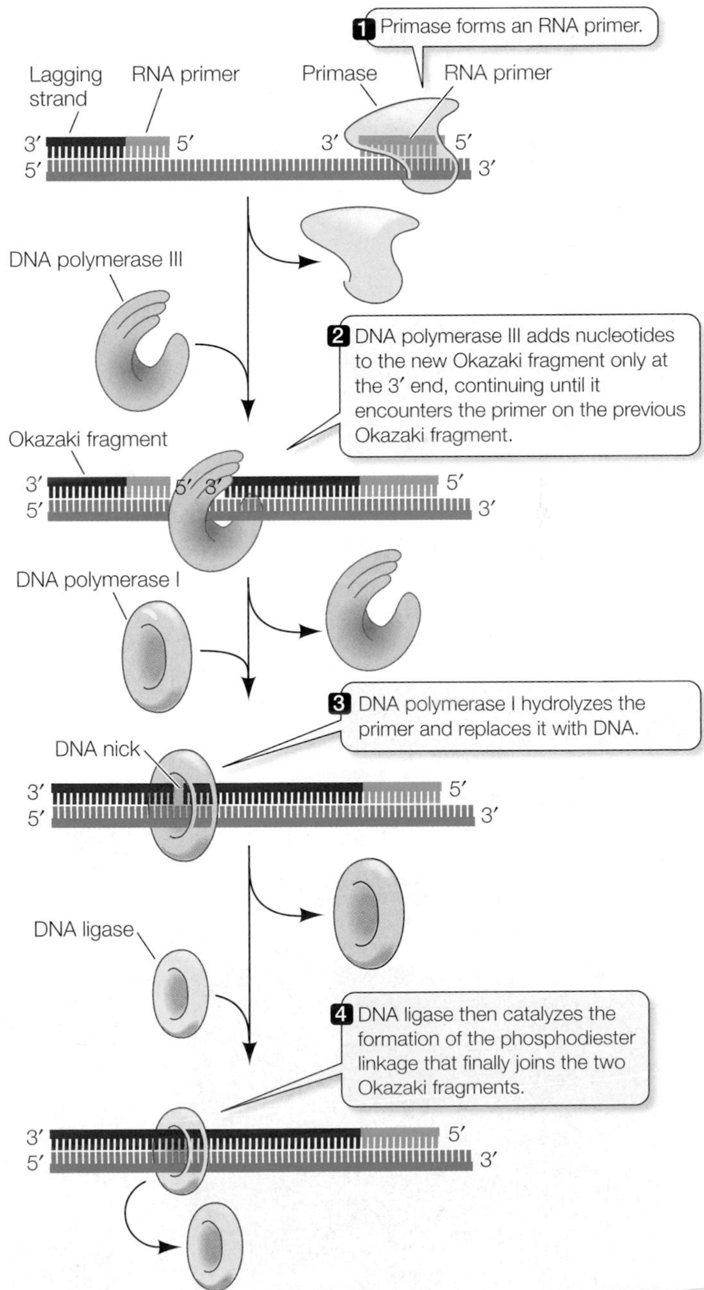

11.19 The Lagging Strand Story In bacteria, DNA polymerase I and DNA ligase cooperate with DNA polymerase III to complete the complex task of synthesizing the lagging strand.

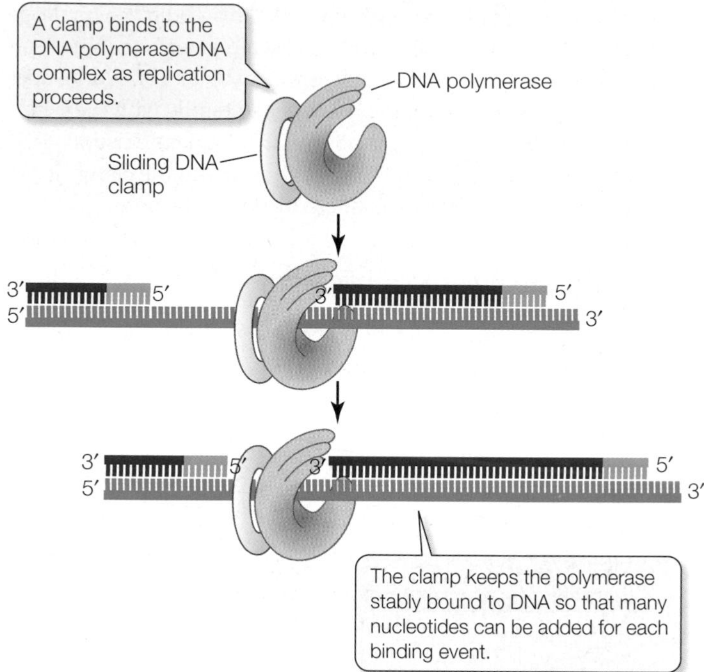

A clamp binds to the DNA polymerase-DNA complex as replication proceeds.
DNA polymerase
Sliding DNA clamp
3′
5′
5′
3′
5′
3′
The clamp keeps the polymerase stably bound to DNA so that many nucleotides can be added for each binding event.

11.20 A Sliding DNA Clamp Increases the Efficiency of DNA Polymerization The clamp keeps DNA polymerase bound to DNA, so that thousands of nucleotides can be polymerized every time the enzyme binds to a template strand.

minal RNA primer is removed, no DNA can be synthesized to replace it because there is no DNA 3′ end to extend (that is, there is no complementary DNA strand). So the new chromosome formed by DNA replication has a bit of single-stranded DNA at each end (**Figure 11.21A**). This situation activates mechanisms that cut off the single-stranded region, along with some of the intact double-stranded end. Thus the chromosome becomes slightly shorter with each cell division.

In many eukaryotes, there are repetitive sequences at the ends of chromosomes called **telomeres**. In humans, the telomere sequence is TTAGGG, and it is repeated about 2,500 times. These repeats bind special proteins that maintain the stability of chromosome ends. Each human chromosome can lose 50–200 base pairs of telomeric DNA after each round of DNA replication and cell division. After 20–30 divisions, the chromosomes are unable to take part in cell division, and the cell dies.

This phenomenon explains in part why cells do not last the entire lifetime of the organism: their telomeres are lost. Yet constantly dividing cells, such as bone marrow stem cells and gamete-producing cells, maintain their telomeric DNA. An enzyme, appropriately called **telomerase**, catalyzes the addition of any lost telomeric sequences (**Figure 11.21B**). Telomerase contains an RNA sequence that acts as a template for the telomeric repeat sequence.

Telomerase is expressed in more than 90 percent of human cancers and may be an important factor in the ability of cancer cells to divide continuously. Since most normal cells do not have this ability, telomerase is an attractive target for drugs designed to attack tumors specifically.

There is also interest in telomerase and aging. When a gene expressing high levels of telomerase is added to human cells in culture, their telomeres do not shorten. Instead of dying after 20–30

11.21 Telomeres and Telomerase (A) Removal of the RNA primer at the 3′ end of the lagging strand leaves a region of DNA—the telomere—unreplicated. (B) The enzyme telomerase binds to the 3′ end and extends the lagging strand of DNA. An RNA sequence embedded in telomerase provides a template so that, overall, the DNA does not get shorter. (C) Bright fluorescent staining marks the telomeric regions on these blue-stained human chromosomes.

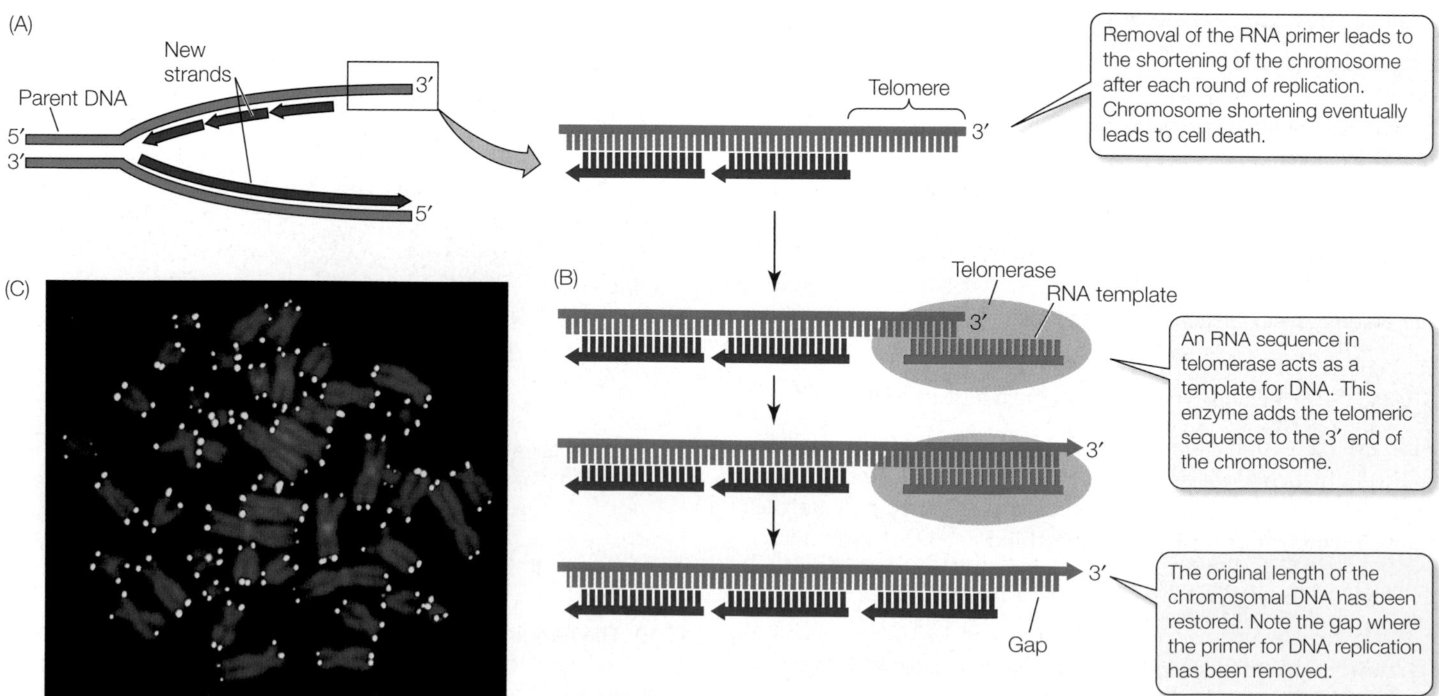

cell generations, the cells become immortal. It remains to be seen how this finding relates to the aging of a large organism.

11.3 RECAP

Meselson and Stahl showed that DNA replication is semiconservative: each parent strand serves as a template for a new strand. A complex of proteins, most notably DNA polymerases, is involved in replication. New DNA is polymerized in one direction only, but since the two strands are antiparallel, one strand is made continuously and the other in Okazaki fragments that are eventually joined.

- Do you see how the Meselson–Stahl experiment differentiated between the three models for DNA replication? See pp. 242–243 and Figures 11.10 and 11.11
- Can you recognize five enzymes needed for DNA replication and explain the role of each? See pp. 244–245 and Figures 11.15, 11.16, 11.17, and 11.21
- Do you undersand how the leading strand of DNA be replicated continuously while the lagging strand must be replicated in fragments? See p. 246 and Figure 11.18

The complex process of DNA replication is amazingly accurate, but it is not perfect. What happens when things go wrong?

11.4 How Are Errors in DNA Repaired?

DNA is accurately replicated and faithfully maintained. The price of failure can be great: the transmission of genetic information is at stake, as is the functioning and even the life of a cell or multicellular organism. Yet the replication of DNA is not perfectly accurate, and the DNA of nondividing cells is subject to damage by natural chemical alterations of the bases as well as by environmental agents. In the face of these threats, how has life gone on so long?

The preservers of life are DNA repair mechanisms. DNA polymerases initially make a significant number of mistakes in assembling polynucleotide strands. The observed error rate of one for every 10^5 bases replicated would result in about 60,000 mutations every time a human cell divided. Fortunately, our cells have at least three DNA repair mechanisms at their disposal:

- A **proofreading** mechanism corrects errors in replication as DNA polymerase makes them.
- A **mismatch repair** mechanism scans DNA immediately after it has been replicated and corrects any base-pairing mismatches.
- An **excision repair** mechanism removes abnormal bases that have formed because of chemical damage and replaces them with functional bases.

Every time it introduces a new nucleotide into a growing polynucleotide strand, DNA polymerase performs a **proofreading** function (**Figure 11.22A**). When a DNA polymerase recognizes a mispairing of bases, it removes the improperly introduced nucleotide and tries again. (Other proteins of the replication complex also play roles in proofreading.) The error rate for this process is only about 1 in 10,000 base pairs, and it lowers the overall error rate for replication to about one base in every 10^{10} bases replicated.

After DNA has been replicated, a second set of proteins surveys the newly replicated molecule and looks for mismatched base pairs that were missed in proofreading (**Figure 11.22B**). For example, this **mismatch repair** mechanism might detect an AC base pair instead of an AT pair. But how does the repair mechanism "know" whether the AC pair should be repaired by removing the C and replacing it with T or by removing the A and replacing it with G?

Individuals who suffer from a condition known as xeroderma pigmentosum lack an excision repair mechanism that normally corrects the damage caused by ultraviolet radiation. They can develop skin cancers after even a brief exposure to sunlight.

The mismatch repair mechanism can detect the "wrong" base because a DNA strand is chemically modified some time after replication. In prokaryotes, methyl groups ($—CH_3$) are added to some adenines. Immediately after replication, methylation has not yet occurred on the newly replicated strand, so the new strand is "marked" (distinguished by being unmethylated) as the one in which errors should be corrected.

When mismatch repair fails, DNA sequences are altered. One form of colon cancer arises in part from a failure of mismatch repair.

DNA molecules can also be damaged during the life of a cell (for example, when it is in G1). High-energy radiation, chemicals from the environment, and random spontaneous chemical reactions can all damage DNA. **Excision repair** mechanisms deal with these kinds of damage.

Certain enzymes constantly "inspect" the cell's DNA (**Figure 11.22C**). When they find mispaired bases, chemically modified bases, or points at which one strand has more bases than the other (with the result that one or more bases of one strand form an unpaired loop), these enzymes cut the defective strand. Another enzyme cuts away the bases adjacent to and including the offending base, and DNA polymerase and DNA ligase synthesize and seal up a new (usually correct) base sequence to replace the excised one.

11.4 RECAP

DNA replication is not perfect; in addition, DNA may be naturally altered or damaged. Repair mechanisms exist that detect and repair mismatched or damaged DNA.

- Do you understand the roles of DNA proofreading, mismatch repair, and excision repair? See Figure 11.22

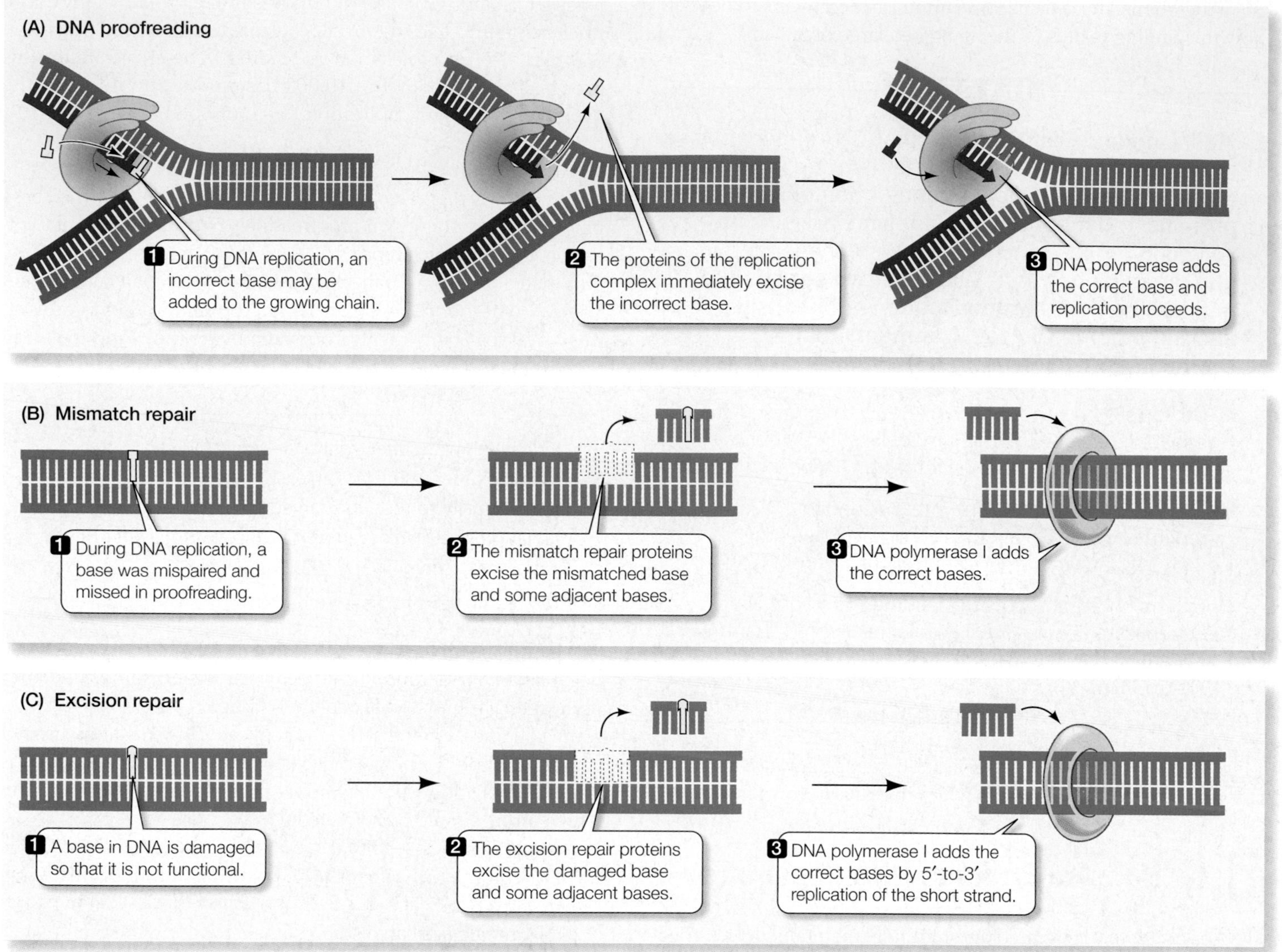

11.22 DNA Repair Mechanisms The proteins of the replication complex also play roles in the life-preserving DNA repair mechanisms, helping to ensure the exact replication of template DNA and repair any damage that occurs.

Understanding how DNA is replicated and repaired has allowed scientists to develop techniques for studying genes. We'll look at just two of those techniques next.

11.5 What Are Some Applications of Our Knowledge of DNA Structure and Replication?

The principles underlying DNA replication in cells have been used to develop two laboratory techniques that have been vital in analyzing genes and genomes. The first technique allows researchers to make multiple copies of short DNA sequences, and the second allows them to determine the base sequence of a DNA molecule.

The polymerase chain reaction makes multiple copies of DNA

Since DNA can be replicated in the laboratory, it is possible to make multiple copies of a DNA sequence. The **polymerase chain reaction** (**PCR**) technique essentially automates this process by copying a short region of DNA many times in a test tube.

PCR is a cyclic process in which a sequence of steps is repeated over and over again (**Figure 11.23**):

- Double-stranded fragments of DNA are separated into single strands by heating (*denatured*).
- A short, artificially synthesized primer is added to the mixture, along with the four deoxyribonucleoside triphosphates (dATP, dGTP, dCTP, and dTTP) and DNA polymerase.
- DNA polymerase catalyzes the production of complementary new strands.

A single cycle takes a few minutes to double the amount of DNA, leaving the new DNA in the double-stranded state. Repeating the cycle many times leads to an exponential increase in the number

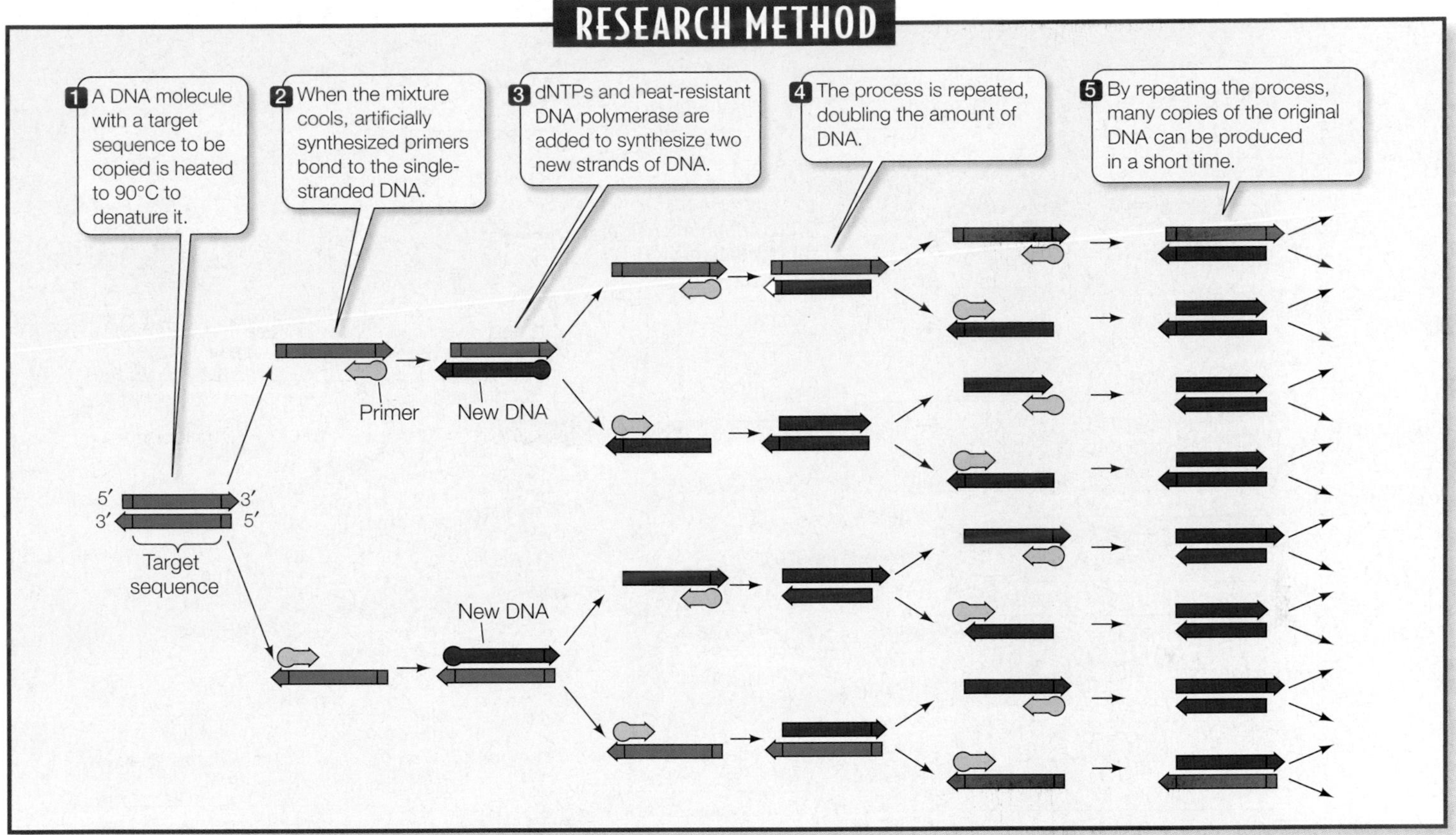

RESEARCH METHOD
1 A DNA molecule with a target sequence to be copied is heated to 90°C to denature it.
2 When the mixture cools, artificially synthesized primers bond to the single-stranded DNA.
3 dNTPs and heat-resistant DNA polymerase are added to synthesize two new strands of DNA.
4 The process is repeated, doubling the amount of DNA.
5 By repeating the process, many copies of the original DNA can be produced in a short time.
Primer
New DNA
5′
3′
3′
5′
Target sequence
New DNA

11.23 The Polymerase Chain Reaction The steps in this cyclic process are repeated many times to produce multiple copies of a DNA fragment.

of copies of the DNA sequence; this process is referred to as *amplifying* the sequence.

The PCR technique requires that the base sequences at the 3′ end of each strand of the target DNA sequence be known so that complementary primers, usually 15–20 bases long, can be made in the laboratory. Because of the uniqueness of DNA sequences, usually only two primers of this length will bind to only one region of DNA in an organism's genome. This specificity in the face of the incredible diversity of target DNA is a key to the power of PCR.

One initial problem with PCR was its temperature requirements. To denature the DNA, it must be heated to more than 90°C—a temperature that destroys most DNA polymerases. The PCR technique would not be practical if new polymerase had to be added after denaturation in each cycle.

This problem was solved by nature: in the hot springs at Yellowstone National Park, as well as in other high-temperature locations, lives a bacterium called, appropriately, *Thermus aquaticus*. The means by which this organism survives temperatures up to 95°C was investigated by Thomas Brock and his colleagues. They discovered that *T. aquaticus* has an entire metabolic machinery that is heat-resistant, including DNA polymerase that does not denature at these high temperatures.

Scientists pondering the problem of copying DNA by PCR read Brock's basic research articles and got a clever idea: Why not use *T. aquaticus* DNA polymerase in the PCR technique? It could withstand the 90°C denaturation temperature and would not have to be added during each cycle. The idea worked, and it earned biochemist Kerry Mullis a Nobel prize. PCR has had an enormous impact on genetic research. Some of its most striking applications will be described in Chapters 13 through 17.

The nucleotide sequence of DNA can be determined

Another important technique allows researchers to determine the base sequence of a DNA molecule. This **DNA sequencing** technique relies on the use of artificially altered nucleosides. As we saw earlier in this chapter, the deoxyribonucleoside triphosphates (dNTPs) that are the normal substrates for DNA replication contain the sugar deoxyribose. If that sugar is replaced with 2,3-dideoxyribose, the resulting *di*deoxyribonucleoside triphosphate (ddNTP) will still be added by DNA polymerase to a growing polynucleotide chain. However, because ddNTPs lack a hydroxyl group (—OH) at the 3′ position, the next nucleotide cannot be added (**Figure 11.24A**). Thus synthesis stops at the position where ddNTP has been incorporated into the growing end of a DNA strand.

To determine the sequence of DNA, a fragment of DNA (usually no more than 700 base pairs long) is denatured. The resulting single strands of DNA are placed in a test tube and mixed with

- DNA polymerase, to synthesize the complementary strand
- Short, artificially synthesized primers appropriate for the DNA sequence
- The four dNTPs (dATP, dGTP, dCTP, and dTTP)

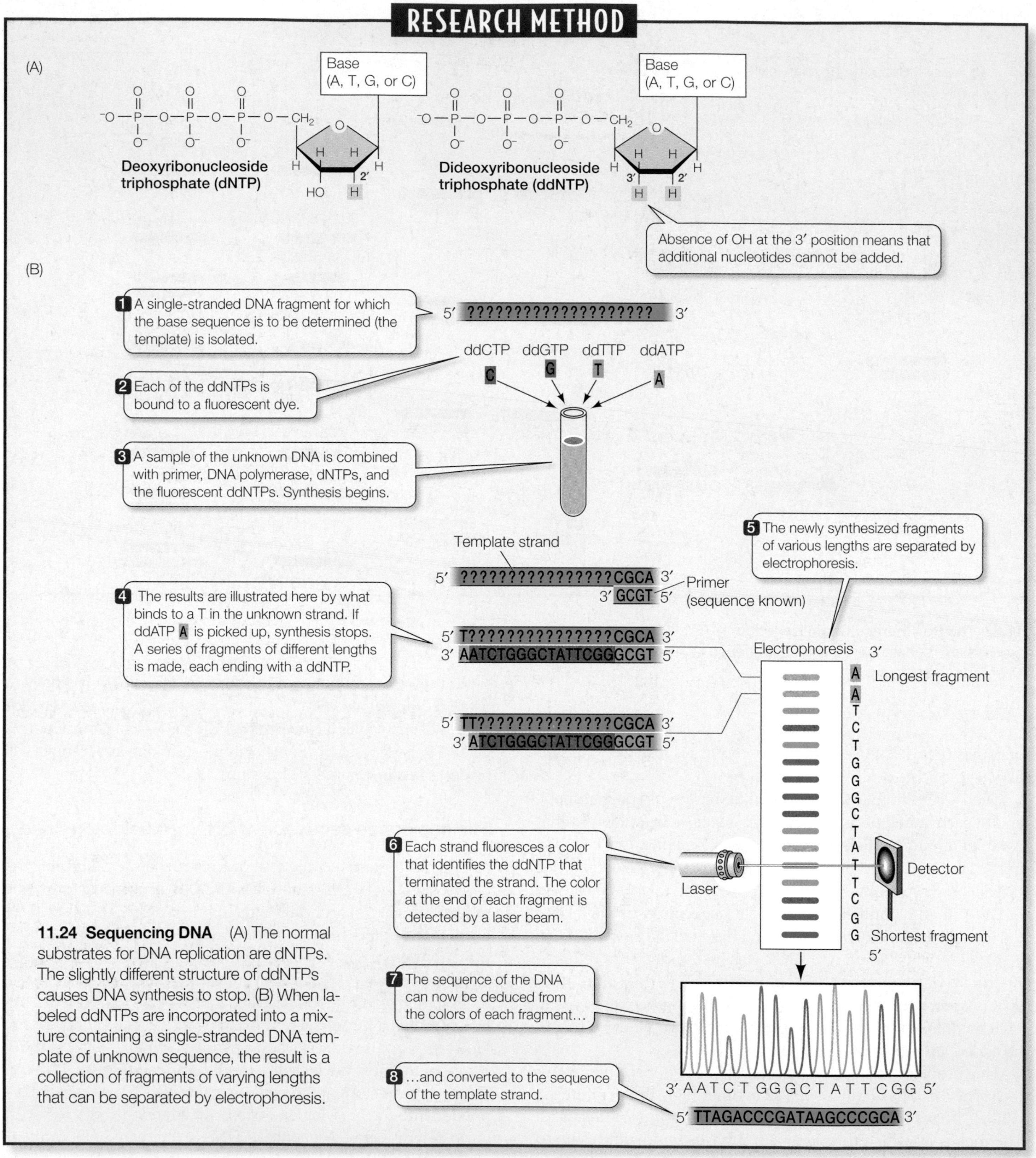
RESEARCH METHOD
(A)
Base
(A, T, G, or C)
Deoxyribonucleoside triphosphate (dNTP)
HO
H
2′
Base
(A, T, G, or C)
Dideoxyribonucleoside triphosphate (ddNTP)
3′
2′
Absence of OH at the 3′ position means that additional nucleotides cannot be added.
(B)
1 A single-stranded DNA fragment for which the base sequence is to be determined (the template) is isolated.
5′ ???????????????????? 3′
ddCTP ddGTP ddTTP ddATP
C G T A
2 Each of the ddNTPs is bound to a fluorescent dye.
3 A sample of the unknown DNA is combined with primer, DNA polymerase, dNTPs, and the fluorescent ddNTPs. Synthesis begins.
Template strand
5′ ????????????????CGCA 3′
3′ GCGT 5′
Primer (sequence known)
4 The results are illustrated here by what binds to a T in the unknown strand. If ddATP A is picked up, synthesis stops. A series of fragments of different lengths is made, each ending with a ddNTP.
5′ T???????????????CGCA 3′
3′ AATCTGGGCTATTCGGGCGT 5′
5′ TT??????????????CGCA 3′
3′ ATCTGGGCTATTCGGGCGT 5′
5 The newly synthesized fragments of various lengths are separated by electrophoresis.
Electrophoresis
3′
A
A
T
C
T
G
G
G
C
T
A
T
T
C
G
G
Longest fragment
Shortest fragment
5′
6 Each strand fluoresces a color that identifies the ddNTP that terminated the strand. The color at the end of each fragment is detected by a laser beam.
Laser
Detector
7 The sequence of the DNA can now be deduced from the colors of each fragment…
8 …and converted to the sequence of the template strand.
3′ AATCTGGGCTATTCGG 5′
5′ TTAGACCCGATAAGCCCGCA 3′

11.24 Sequencing DNA (A) The normal substrates for DNA replication are dNTPs. The slightly different structure of ddNTPs causes DNA synthesis to stop. (B) When labeled ddNTPs are incorporated into a mixture containing a single-stranded DNA template of unknown sequence, the result is a collection of fragments of varying lengths that can be separated by electrophoresis.

- Small amounts of the four ddNTPs, each bonded to a fluorescent "tag" that emits a different color of light

DNA replication proceeds, and the test tube soon contains a mixture of the template DNA strands and shorter, new complementary strands. The new strands, each ending with a fluorescent ddNTP, are of varying lengths. For example, each time a T is reached on the template strand, DNA polymerase adds either a dATP or a ddATP to the growing complementary strand. If dATP is added, the strand continues to grow. If ddATP is added, growth stops.

After DNA replication has been allowed to proceed for a while, the new DNA fragments are denatured from their templates. The fragments are then subjected to *electrophoresis* (see Figure 16.2). This technique sorts the DNA fragments by length and can detect differences in fragment length as short as one base. During the electrophoresis run, the fragments pass in order of increasing length through a laser beam that excites the fluorescent tags. The light emitted is then detected, and the resulting information—that is, which color of fluorescence, and therefore which ddNTP, is at the end of a strand of which length—is fed into a computer. The computer processes this information and prints out the DNA sequence of the fragment (**Figure 11.24B**). DNA sequencing has formed the basis of the new science of genomics.

11.5 RECAP

Knowledge of the mechanisms of DNA replication led to the development of techniques for making multiple copies of DNA sequences and for determining the nucleotide sequence of DNA molecules.

- Do you understand the role of primers in PCR? See pp. 250–251 and Figure 11.23
- Can you explain why dideoxyribonucleosides are used in DNA sequencing? See pp. 251–252 and Figure 11.24

CHAPTER SUMMARY

11.1 What is the evidence that the gene is DNA?

Griffith's experiments in the 1920s demonstrated that some substance in cells—then called a **transforming principle**—can cause heritable change in other cells. Review Figure 11.1

The location and quantity of DNA in the cell suggested that DNA might be the genetic material. Experiments by Avery, MacLeod, and McCarty isolated the transforming principle from bacteria and identified it as DNA. Review Figure 11.2

The Hershey–Chase experiment established conclusively that DNA (and not protein) is the genetic material by tracing the DNA of radiolabeled viruses with which they infected bacterial cells. Review Figure 11.4

Genetic transformation of eukaryotic cells is called **transfection.** Transformation and transfection can be studied with the aid of a **marker** gene that confers a known and observable phenotype. Review Figure 11.5

11.2 What is the structure of DNA?

Chargaff's rule states that the amount of **adenine** in DNA is equal to the amount of **thymine**, and that the amount of **guanine** is equal to the amount of **cytosine**; thus the total abundance of purines (A + G) equals the total abundance of pyrimidines (T + C).

X-ray crystallography showed that the DNA molecule is **helical**. Watson and Crick proposed that DNA is a double-stranded helix in which the strands are **antiparallel**. Review Figure 11.8

Complementary base pairing between A and T and between G and C accounts for Chargaff's rule. The bases are held together by hydrogen bonding. Review Figure 11.9

11.3 How is DNA replicated?

See Web/CD Tutorial 11.1

Meselson and Stahl showed the replication of DNA to be **semiconservative**. Each parent strand acts as a **template** for the synthesis of a new strand; thus the two replicated DNA molecules each contain one parent strand and one newly synthesized strand. Review Figure 11.11, Web/CD Tutorial 11.2

In DNA replication, the enzyme **DNA polymerase** catalyzes the addition of nucleotides to the 3′ end of each strand. Which nucleotides are added is determined by complementary base pairing with the template strand. Review Figure 11.12

The **replication complex** is a huge protein complex that attaches to the chromosome at the **origin of replication** (***ori***).

Replication proceeds from the origin of replication on both strands in the 5′-to-3′ direction, forming two **replication forks**.

Many proteins assist in DNA replication. **DNA helicase** separates the strands, and **single-strand binding proteins** keep the strands from reassociating.

In prokaryotes, two interlocking circular DNAs are formed; they are separated by an enzyme called **DNA topoisomerase**. Review Figure 11.14

Primase catalyzes the synthesis of a short RNA **primer** to which nucleotides are added by DNA polymerase. Review Figure 11.16

The **leading strand** is synthesized continuously and the **lagging stand** in pieces called **Okazaki fragments**. The fragments are joined together by **DNA ligase**. Review Figures 11.18 and 11.19, Web/CD Tutorial 11.3

The speed with which DNA polymerization proceeds is attributed to the **processive** nature of DNA polymerases, which can catalyze many polymerizations at a time. A **sliding DNA clamp** helps ensure the stability of this process. Review Figure 11.20

DNA replication leaves a short, unreplicated sequence, the **telomere**, at the 3′ end of the chromosome. Unless the enzyme **telomerase** is present, the sequence is removed. After mutiple cell cycles, the telomeres shorten, leading to chromosome instability and cell death. Review Figure 11.21

11.4 How are errors in DNA repaired?

DNA polymerases make about one error in 10^5 bases replicated. DNA is also subject to natural alteration and chemical damage. DNA can be repaired by three different mechanisms: **proofreading**, **mismatch repair**, and **excision repair**. Review Figure 11.22

11.5 What are some applications of our knowledge of DNA structure and replication?

The **polymerase chain reaction** technique uses DNA polymerase to make multiple copies of DNA in the laboratory. Review Figure 11.23

DNA sequencing techniques use the principles of DNA replication to determine the nucleotide sequence of DNA. Review Figure 11.24

SELF-QUIZ

1. Griffith's studies of *Streptococcus pneumoniae*
   a. showed that DNA is the genetic material of bacteria.
   b. showed that DNA is the genetic material of bacteriophages.
   c. demonstrated the phenomenon of bacterial transformation.
   d. proved that prokaryotes reproduce sexually.
   e. proved that protein is not the genetic material.
2. In the Hershey–Chase experiment,
   a. DNA from parent bacteriophages appeared in progeny bacteriophages.
   b. most of the phage DNA never entered the bacteria.
   c. more than three-fourths of the phage protein appeared in progeny phages.
   d. DNA was labeled with radioactive sulfur.
   e. DNA formed the coat of the bacteriophages.
3. Which statement about complementary base pairing is *not* true?
   a. It plays a role in DNA replication.
   b. In DNA, T pairs with A.
   c. Purines pair with purines, and pyrimidines pair with pyrimidines.
   d. In DNA, C pairs with G.
   e. The base pairs are of equal length.
4. In semiconservative replication of DNA,
   a. the original double helix remains intact and a new double helix forms.
   b. the strands of the double helix separate and act as templates for new strands.
   c. polymerization is catalyzed by RNA polymerase.
   d. polymerization is catalyzed by a double-helical enzyme.
   e. DNA is synthesized from amino acids.
5. Which of the following does *not* occur during DNA replication?
   a. Unwinding of the parent double helix
   b. Formation of short pieces that are connected by DNA ligase
   c. Complementary base pairing
   d. Use of a primer
   e. Polymerization in the 3′-to-5′ direction
6. The primer used for DNA replication
   a. is a short strand of RNA added to the 3′ end.
   b. is present only once on the leading strand.
   c. remains on the DNA after replication.
   d. ensures that there will be a free 5′ end to which nucleotides can be added.
   e. is added to only one of the two template strands.
7. One strand of DNA has the sequence 5′-ATTCCG-3′. The complementary strand for this is
   a. 5′-TAAGGC-3′
   b. 5′-ATTCCG-3′
   c. 5′-ACCTTA-3′
   d. 5′-CGGAAT-3′
   e. 5′-GCCTTA-3′
8. The role of DNA ligase in DNA replication is to
   a. add more nucleotides to the growing strand one at a time.
   b. open up the two DNA strands to expose template strands.
   c. ligate base to sugar to phosphate in a nucleotide.
   d. bond Okazaki fragments to one another.
   e. remove incorrectly paired bases.
9. The polymerase chain reaction
   a. is a method for sequencing DNA.
   b. is used to transcribe specific genes.
   c. amplifies specific DNA sequences.
   d. does not require DNA replication primers.
   e. uses a DNA polymerase that denatures at 55°C.
10. What is the correct order for the following events in excision repair of DNA?
    (1) Base-paired DNA is made complementary to the template
    (2) Damaged bases are recognized
    (3) DNA ligase seals the new strand to existing DNA
    (4) Part of a single strand is excised
    a. 1234
    b. 2134
    c. 2413
    d. 3421
    e. 4231

FOR DISCUSSION

1. Suppose that Meselson and Stahl had continued their experiment on DNA replication for another ten bacterial generations. Would there still have been any ^{14}N–^{15}N hybrid DNA present? Would it still have appeared in the centrifuge tube? Explain.
2. If DNA replication were conservative rather than semiconservative, what results would Meselson and Stahl have observed? Diagram the results using the conventions of Figure 11.10.
3. Using the following information, calculate the number of origins of DNA replication on a human chromosome: DNA polymerase adds nucleotides at 3,000 base pairs per minute in one direction; replication is bidirectional; S phase lasts 300 minutes; there are 120 million base pairs per chromosome. With a typical chromosome 3 μm long, how many origins are there per μm?
4. The drug dideoxycytidine, used to treat certain viral infections, is a nucleotide made with 2′,3′-dideoxyribose. This sugar lacks —OH groups at both the 2′ and the 3′ positions. Explain why this drug stops the growth of a DNA chain when added to DNA.

FOR INVESTIGATION

Outline a series of experiments using radioactive isotopes to show that bacterial DNA and not protein enters the host cell and is responsible for bacterial transformation.

CHAPTER 12 From DNA to Protein: Genotype to Phenotype

Toxic avenger at the ribosome

In 1978, Georgi Markov, a Bulgarian journalist who had written articles critical of the then-Communist government of Bulgaria, was living in exile in London. As he stood one evening at a bus stop near Waterloo Station, a man—possibly a Bulgarian secret agent—brushed up against him and, seemingly by accident, poked him with an umbrella. Markov felt a sharp pain. Within a few hours, he started to feel weak. A high temperature, vomiting, and more severe symptoms soon followed. Two days later he was dead.

Police investigators found a tiny perforated pellet embedded in Markov's leg, and in that pellet was a small amount of ricin, a highly toxic molecule isolated from the seeds of the tropical castor bean plant, *Ricinus communis*. The seeds of *Ricinus* have been used for centuries as a source of castor oil, a natural product once frequently administered to children to "clean out" the digestive tract. Castor oil is used today in the plastics industry. The toxin ricin is a protein that is not present in the seed oil, and people found out the hard way that it is one of the most poisonous substances made by any organism. About 1 milligram (an amount the size of the head of a pin) can kill a human.

Markov's murder is not the only case of deliberate use of ricin. Small amounts of ricin were found in caves in Afghanistan occupied by the terrorist group Al-Qaeda, and the poison may have been used in the war between Iran and Iraq that raged throughout much of the 1980s. In the 1990s, four members of an anti-tax group were arrested for plotting to use home-grown ricin to kill a U.S. government official. And on February 3, 2004, the U.S. Senate offices were closed when ricin was found in a mailroom.

Much has been written about the possible ways ricin might be used in a terrorist attack. That is unlikely, however, because relatively large amounts of it would be needed to harm a significant number of people. Unlike bacteria such as those that cause anthrax, ricin molecules are proteins; they do not reproduce.

Ricin enters cells by binding to membrane glycoproteins and glycolipids that contain the sugar galactose. Since many cell surface molecules contain this sugar, ricin can bind to most cells. After being endocytosed and released into the cell cytoplasm, ricin kills the cell by blocking protein synthesis. More specifically, it catalyzes the modifi-

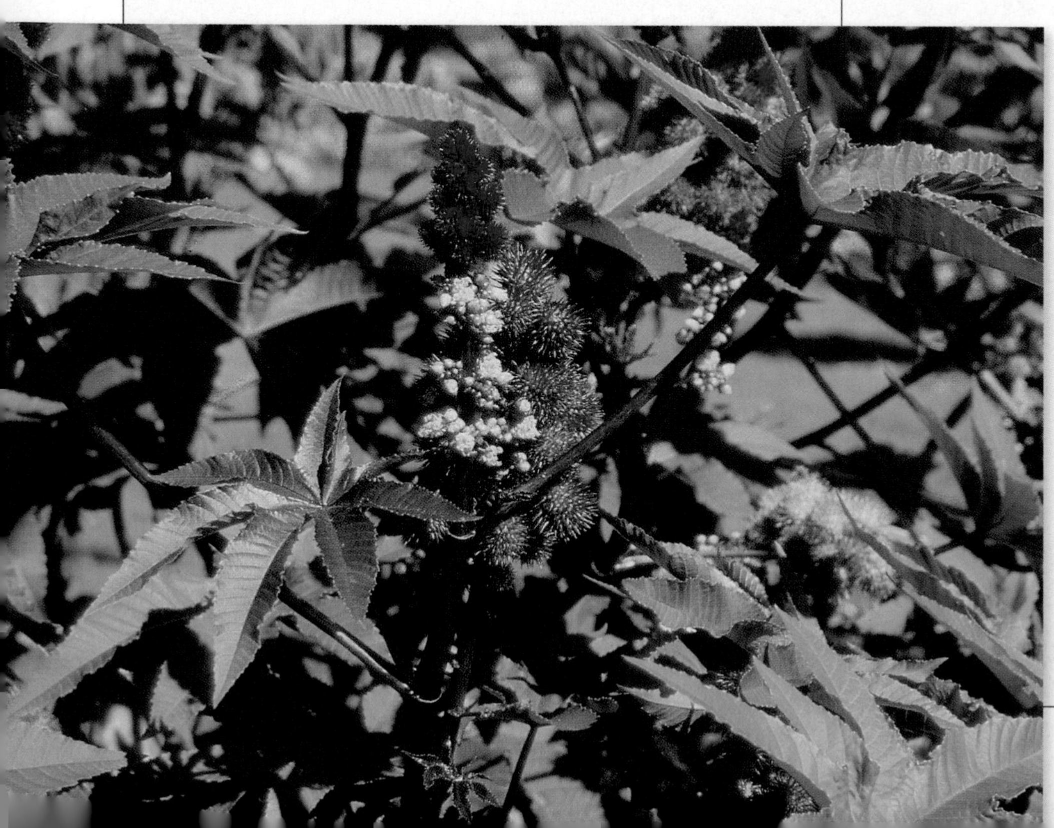

***Ricinus communis*, the Castor Bean Plant** This brightly colored tropical plant produces ricin, a lethal toxin that inhibits protein synthesis at the ribosome.

Ricin's Target Ricin inactivates the ribosome, which is the site of protein synthesis. Ribosomes are large aggregates of macromolecules, containing dozens of proteins and ribosomal RNA in two subunits (violet and gray) and three molecules of transfer RNA (gold).

cation and cleavage of one of the large RNA molecules that make up the eukaryotic ribosome—the "workbench" of protein synthesis. A single ricin molecule in the cytoplasm can modify 1500 ribosomes, killing the cell in minutes.

Proteins are the major phenotypic expression of the genotype—the genetic information encoded in a cell's DNA. Ricin inhibits the cell's ability to express the genotype as phenotype through protein synthesis, and therefore ricin-poisoned cells cannot survive.

IN THIS CHAPTER we will see how genes are expressed as proteins. We begin with evidence for the relationship between genes and proteins, and then fill in some of the details of transcription (the copying of the gene sequence of DNA into a sequence of RNA) and translation (the use of a sequence of RNA to make a polypeptide with a defined sequence of amino acids). Finally, we will define mutations and their phenotypes in specific molecular terms.

CHAPTER OUTLINE

12.1 What Is the Evidence that Genes Code for Proteins?

In Chapter 11, we defined genes as sequences of DNA and learned that genes are expressed as physical characteristics known as the *phenotype*. Here, we define the phenotype as proteins. What is the evidence for this definition?

The molecular basis of phenotypes was actually discovered before it was known that DNA was the genetic material. Scientists had studied the chemical differences between individuals carrying wild-type and mutant alleles in organisms as diverse as humans and bread molds. They found that the major phenotypic differences were the result of differences in specific proteins.

Experiments on bread mold established that genes determine enzymes

Because of life's basis in the cell theory, scientists investigating a biological phenomenon can reasonably assume that what is found in one organism applies to others. Thus they often search for a *model organism,* one that is easy to grow in the laboratory or observe in the field and which shows the phenomenon to be studied. In previous chapters we have seen several examples of model organisms, including:

- Pea plants (*Pisum sativum*) used by Mendel in his genetics experiments
- Fruit flies (*Drosophila*) used by Morgan in his genetics experiments
- *E. coli* used by Meselson and Stahl to study DNA replication

To this list we now add the common bread mold, *Neurospora crassa*. *Neurospora* is a type of fungus known as an ascomycete (see Chapter 30). This mold is haploid for most of its life, making its genetics straightforward (since there are no dominant–recessive relationships). It is simple to culture and grows well in the laboratory. In the 1940s, George W. Beadle and Edward L. Tatum at Stanford University undertook studies to chemically define the phenotype in *Neurospora*.

The roles of enzymes in biochemistry were being described at the time Beadle and Tatum began their work. They hypothesized that the expression of a gene as a phenotype could occur through an enzyme. They grew *Neurospora* on a minimal nutritional medium containing sucrose, minerals, and a vitamin. Using this medium, the enzymes of wild-type *Neurospora* could catalyze the metabolic reactions needed to make all the chemical constituents of their cells, including proteins. These wild-type strains are called *prototrophs* ("original eaters").

Mutations provide a powerful way to determine cause and effect in biology. Nowhere has this been so evident as in the elucidation of biochemical pathways. Beadle and Tatum treated wild-type *Neurospora* with X rays, which act as a *mutagen* (something known to cause mutations). When they examined the treated molds, they found that some mutant strains could no longer grow on the minimal medium, but grew only if they were supplied with additional nutrients. The scientists hypothesized that these *auxotrophs* ("increased eaters") must have suffered mutations in genes that coded for the enzymes used to synthesize the nutrients they now needed to obtain from their environment. For each auxotrophic strain, Beadle and Tatum were able to find a single compound that, when added to the minimal medium, supported the growth of that strain. This result suggested that mutations have simple effects, and that each mutation causes a defect in only one enzyme in a metabolic pathway. These conclusions became known as the **one-gene, one-enzyme hypothesis** (**Figure 12.1**).

One group of auxotrophs, for example, could grow only if the minimal medium was supplemented with the amino acid arginine. (Wild-type *Neurospora* makes its own arginine.) These mutant strains were designated *arg* mutants. Beadle and Tatum found several different *arg* mutant strains. They proposed two alternative hypotheses to explain why these different genetic strains had the same phenotype:

- The different *arg* mutants could have mutations in *the same gene*, as in the case of the different eye color alleles of fruit flies. In this case, the gene might code for an enzyme involved in arginine synthesis.
- The different *arg* mutants could have mutations in *different genes*, each coding for a separate function that leads to arginine production. These independent functions might be different enzymes along the same biochemical pathway.

Some of the *arg* mutant strains fell into each of the two categories. Genetic crosses showed that some of the mutations were at the same chromosomal locus, and were different alleles of the same gene. Other mutations were at different loci, or on different chromosomes, and so were not alleles of the same gene. Beadle and Tatum concluded that these different genes participated in governing a single biosynthetic pathway—in this case, the pathway leading to arginine synthesis (see the Interpretation in Figure 12.1).

By growing different *arg* mutants in the presence of various compounds suspected to be intermediates in the biosynthetic pathway for arginine, Beadle and Tatum were able to classify each mutation as affecting one enzyme or another and to order the compounds along the pathway. Then they broke open the wild-type and mutant cells and examined them for enzyme activities.

12.1 One Gene, One Enzyme Beadle and Tatum studied several *arg* mutants of *Neurospora*. The different *arg* mutant strains required the addition of different compounds to their growth medium in order to synthesize the arginine required for their growth. Step through the figure to follow the reasoning that upheld the "one-gene, one-enzyme" hypothesis. FURTHER RESEARCH: If a diploid *Neurospora* spore were made from two haploid cells, one with mutant 3 and the other with mutant 2, what would be its phenotype?

The results confirmed their hypothesis: each mutant strain was indeed missing a single active enzyme in the pathway.

The gene–enzyme connection had been proposed 40 years earlier by the Scottish physician Archibald Garrod, who studied the inherited human disease alkaptonuria. In 1908, Garrod linked the biochemical phenotype of alkaptonuria to a missing enzyme, and thus to an abnormal gene. Today we know of hundreds of examples of such hereditary diseases.

One gene determines one polypeptide

The gene–enzyme relationship has undergone several modifications in light of our current knowledge of molecular biology. Many proteins, including many enzymes, are composed of more than one polypeptide chain, or subunit (that is, they have a quaternary structure; see Section 3.2). Look at the illustration of hemoglobin in Figure 3.9: this protein has four polypeptides, two of each of type of chain. In the case of hemoglobin, each polypeptide chain is specified by its own separate gene. Thus it is more correct to speak of a **one-gene, one-polypeptide relationship**.

In other words, *the function of a gene is to control the production of a single, specific polypeptide*. This statement remains true despite the fact that we know now of genes that code for forms of RNA that are not translated into polypeptides, and have discovered still other gene sequences that do not themselves produce physical polypeptides but instead are involved in controlling which *other* DNA sequences are expressed (i.e., produce polypeptides).

12.1 RECAP

Beadle and Tatum's studies of mutations in bread molds led to our understanding of the one-gene, one-polypeptide relationship: the function of a gene is to code for a specific polypeptide.

- What is a model organism, and why was *Neurospora* a good model for studying biochemical genetics? See p. 257
- How were Beadle and Tatum's experiments on *Neurospora* set up to determine, on the basis of phenotypes of mutant strains, the order of a biochemical pathway? See p. 258 and Figure 12.1
- Do you understand the distinction between the phrases "one gene, one protein" and "one gene, one polypeptide"?

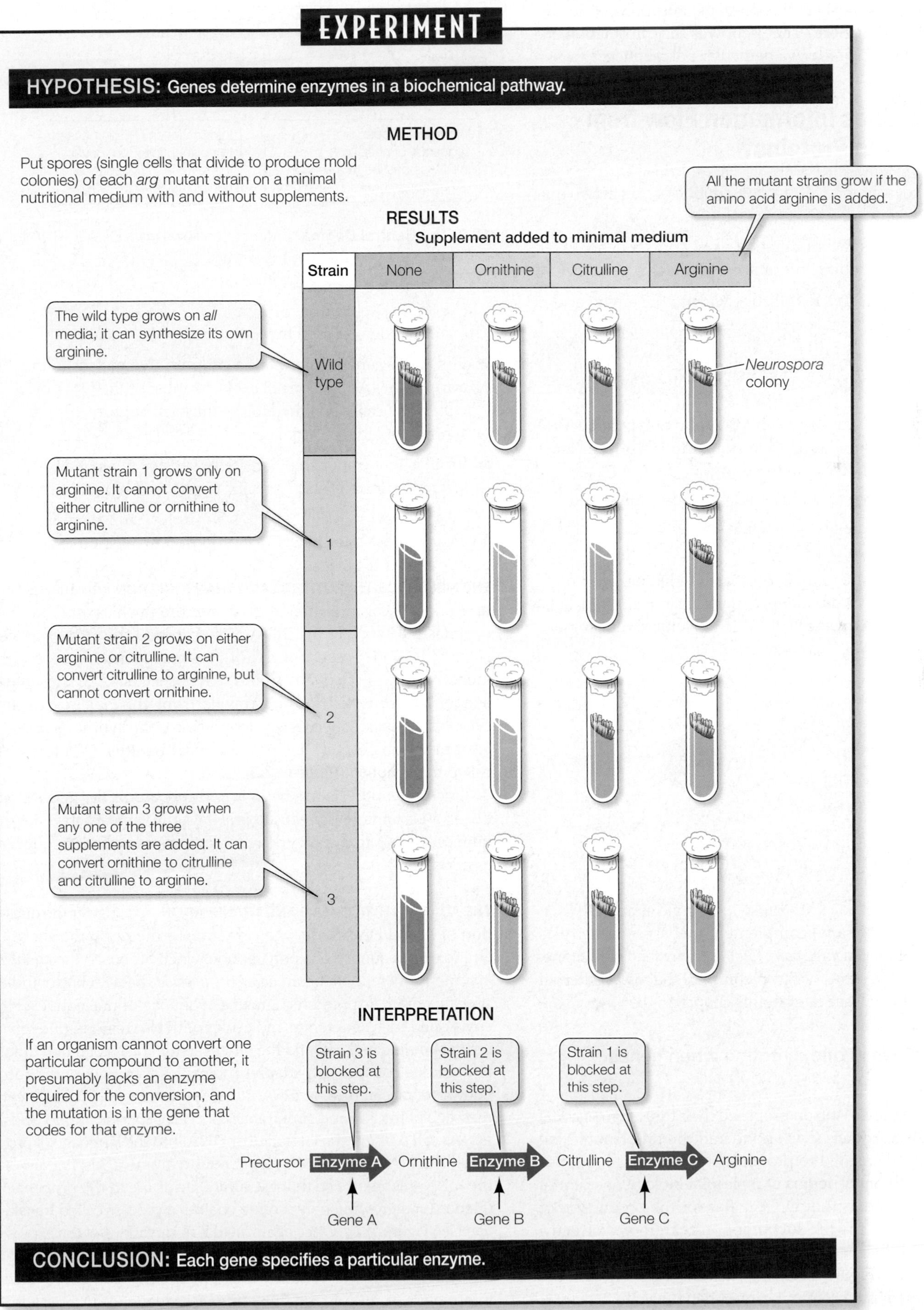
EXPERIMENT
HYPOTHESIS: Genes determine enzymes in a biochemical pathway.
METHOD
Put spores (single cells that divide to produce mold colonies) of each *arg* mutant strain on a minimal nutritional medium with and without supplements.
RESULTS
Supplement added to minimal medium
Strain
None
Ornithine
Citrulline
Arginine
All the mutant strains grow if the amino acid arginine is added.
The wild type grows on *all* media; it can synthesize its own arginine.
Wild type
Neurospora colony
Mutant strain 1 grows only on arginine. It cannot convert either citrulline or ornithine to arginine.
1
Mutant strain 2 grows on either arginine or citrulline. It can convert citrulline to arginine, but cannot convert ornithine.
2
Mutant strain 3 grows when any one of the three supplements are added. It can convert ornithine to citrulline and citrulline to arginine.
3
INTERPRETATION
If an organism cannot convert one particular compound to another, it presumably lacks an enzyme required for the conversion, and the mutation is in the gene that codes for that enzyme.
Strain 3 is blocked at this step.
Strain 2 is blocked at this step.
Strain 1 is blocked at this step.
Precursor
Enzyme A
Ornithine
Enzyme B
Citrulline
Enzyme C
Arginine
Gene A
Gene B
Gene C
CONCLUSION: Each gene specifies a particular enzyme.

Now that we have established the one-gene, one-polypeptide relationship, how does it work? That is, how is the information encoded in DNA used to produce a particular polypeptide?

12.2 How Does Information Flow from Genes to Proteins?

The expression of a gene to form a polypeptide occurs in two major steps:

- *Transcription* copies the information of a DNA sequence (a gene) into corresponding information in an RNA sequence.
- *Translation* converts this RNA sequence into the amino acid sequence of a polypeptide.

RNA differs from DNA

RNA is a key intermediary between DNA and polypeptide. **RNA (ribonucleic acid)** is a polynucleotide similar to DNA (see Figure 3.24), but it differs from DNA in three ways:

- RNA generally consists of only one polynucleotide strand.
- The sugar molecule found in RNA is ribose, rather than the deoxyribose found in DNA.
- Although three of the nitrogenous bases (adenine, guanine, and cytosine) in RNA are identical to those in DNA, the fourth base in RNA is **uracil** (**U**), which is similar to thymine but lacks the methyl (—CH_3) group.

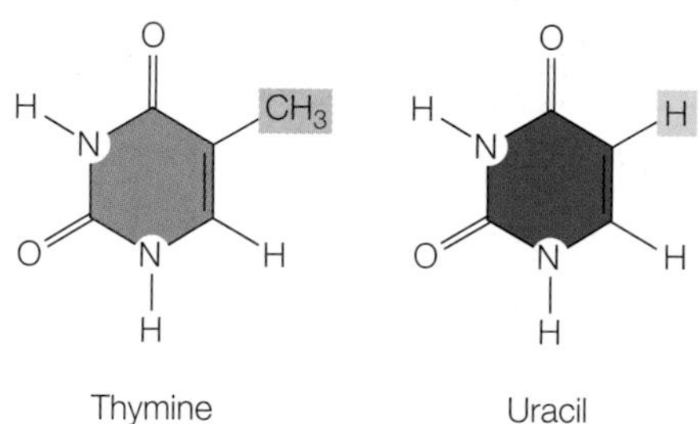

The bases in RNA can pair with those in a single strand of DNA. This pairing obeys the same complementary base-pairing rules as in DNA, except that *adenine pairs with uracil* instead of thymine. Single-stranded RNA can fold into complex shapes by internal base pairing, as we will see later in this chapter.

Information flows in one direction when genes are expressed

Soon after he and James Watson proposed their three-dimensional structure for DNA, Francis Crick pondered the problem of how DNA is functionally related to proteins. This led him to propose what he called the **central dogma** of molecular biology. The central dogma, simply stated, is that DNA codes for the production of RNA, RNA codes for the production of protein (more correctly, polypeptide), and protein does not code for the production of protein, RNA, or DNA (**Figure 12.2**). In Crick's words, "once 'information' has passed into protein it cannot get out again."

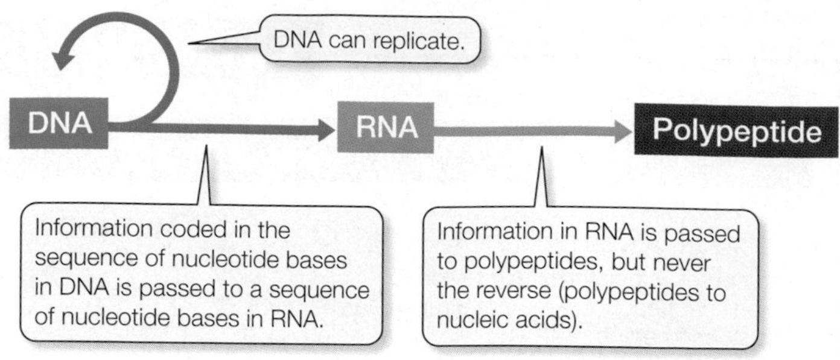

DNA can replicate.
DNA
RNA
Polypeptide
Information coded in the sequence of nucleotide bases in DNA is passed to a sequence of nucleotide bases in RNA.
Information in RNA is passed to polypeptides, but never the reverse (polypeptides to nucleic acids).

12.2 The Central Dogma Information flows from DNA to RNA to polypeptide, as indicated by the arrows.

The central dogma raised two questions:

- How does genetic information get from the nucleus to the cytoplasm? (As Section 4.3 explains, most of the DNA of a eukaryotic cell is confined to the nucleus, but proteins are synthesized in the cytoplasm.)
- What is the relationship between a specific nucleotide sequence in DNA and a specific amino acid sequence in a protein?

To answer these questions, Crick proposed two hypotheses.

THE MESSENGER HYPOTHESIS AND TRANSCRIPTION To answer the question of how information gets from the nucleus and into the cytoplasm, Crick and his colleagues proposed the *messenger hypothesis*. They proposed that an RNA molecule forms as a complementary copy of one DNA strand of a particular gene. This **messenger RNA**, or **mRNA**, then travels from the nucleus to the cytoplasm, where it serves as a template for the synthesis of proteins at the ribosomes. The process by which this RNA forms is called **transcription** (**Figure 12.3**).

Crick's hypothesis has been tested repeatedly for genes that code for proteins, and the result is always the same: each gene sequence in DNA that codes for a protein is expressed as a sequence in mRNA.

THE ADAPTER HYPOTHESIS AND TRANSLATION To answer the question of how a DNA sequence gets transformed into the specific amino acid sequence of a polypeptide, Crick proposed the *adapter hypothesis*: there must be an *adapter molecule* that can both bind a specific amino acid and recognize a sequence of nucleotides. He envisioned such adapters as molecules with two regions, one serving the binding function and the other serving the recognition function. In due course, such adapter molecules were found: they are known as **transfer RNA**, or **tRNA**. Because they recognize the genetic message of mRNA and simultaneously carry specific amino acids, tRNAs can *translate* the language of DNA into the language of proteins. The tRNA adapters, carrying bound amino acids, line up on the mRNA sequence so that the amino acids are in the proper sequence for a growing polypeptide chain—a process called **translation** (see Figure 12.3). Once again, actual observations of the expression of thousands of genes have confirmed the hypothesis that tRNA acts as the intermediary between the nucleotide sequence information in mRNA and the amino acid sequence in a protein.

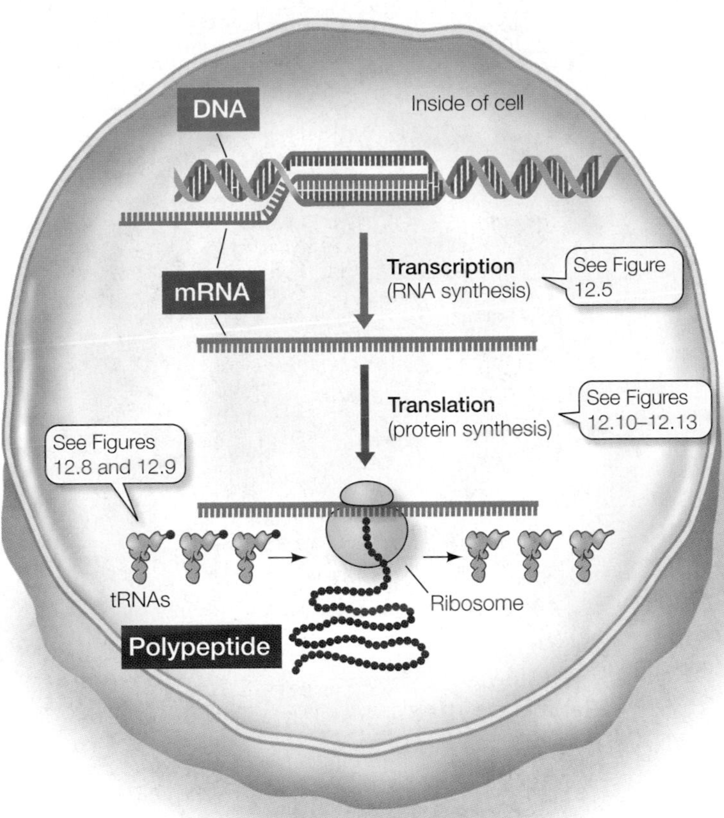

12.3 From Gene to Protein This diagram summarizes the processes of gene expression in prokaryotes. In eukaryotes, the processes are somewhat more complex.

Summarizing the main features of the central dogma, the messenger hypothesis, and the adapter hypothesis, we may say that a given gene is transcribed to produce a messenger RNA (mRNA) molecule complementary to one of the DNA strands, and that transfer RNA (tRNA) molecules translate the sequence of bases in the mRNA into the appropriate sequence of linked amino acids during protein synthesis.

RNA viruses are exceptions to the central dogma

Certain viruses present exceptions to the central dogma. As we saw in Section 11.1, *viruses* are acellular infectious particles that reproduce inside cells. Many viruses, such as the tobacco mosaic virus, influenza viruses, and poliovirus, have RNA rather than DNA as their genetic material. With its nucleotide sequence, RNA could potentially act as an information carrier and be expressed as protein. But if RNA is usually single-stranded, how does it replicate? The viruses generally solve this problem by transcribing from RNA to RNA, making an RNA strand that is complementary to their genome. This "opposite" strand is then used to make multiple copies of the viral genome by transcription:

Human immunodeficiency viruses (HIV) and certain rare tumor viruses also have RNA as their genome, but do not replicate it as RNA-to-RNA. Instead, after infecting a host cell, they make a DNA copy of their genome and use it to make more RNA. This RNA is then used both as a template for making more copies of the viral genome and as mRNA to produce viral proteins.

Synthesis of DNA from RNA is called **reverse transcription**, and not surprisingly, such viruses are called **retroviruses**.

12.2 RECAP

The central dogma of molecular biology states that DNA codes for the production of RNA and RNA codes for the production of protein (polypeptides). Proteins do *not* code for the production of protein, RNA, or DNA. Transcription is the process that copies a DNA sequence into mRNA. Translation is the process by which this information is converted into protein. Transfer RNA recognizes the genetic information in messenger RNA and brings the appropriate amino acid into position in a growing polypeptide chain.

- Do you understand the central dogma of molecular biology? See p. 260 and Figure 12.2
- Can you describe the roles of mRNA and tRNA in gene expression? See p. 260 and Figure 12.3

Let's look at the physical processes underlying the central dogma in more detail. We'll begin by describing how the information in DNA is transcribed to produce RNA.

12.3 How Is the Information Content in DNA Transcribed to Produce RNA?

In normal prokaryotic and eukaryotic cells, RNA synthesis is directed by DNA. Transcription—the formation of a specific RNA from a specific DNA—requires several components:

- A DNA template for complementary base pairing
- The appropriate ribonucleoside triphosphates (ATP, GTP, CTP, and UTP) to act as substrates
- An *RNA polymerase* enzyme

Within each gene, only one of the two strands of DNA—the **template strand**—is transcribed. The other, complementary DNA strand, referred to as the *non-template strand*, remains untranscribed. For different genes in the same DNA molecule, different strands may be transcribed. That is, the strand that is the non-template strand in one gene may be the template strand in another.

Not only mRNA is produced by transcription. The same process is responsible for the synthesis of tRNA and ribosomal RNA (rRNA), whose important roles in protein synthesis will be described below. Like polypeptides, these RNAs are encoded by specific genes.

RNA polymerases share common features

RNA polymerases from both prokaryotes and eukaryotes catalyze the synthesis of RNA from template DNA. There is only one RNA polymerase in bacteria, while there are three in eukaryotes. Yet all share a common structure that resembles a crab claw (**Figure 12.4**). Catalysis occurs in several steps:

- The enzyme recognizes certain bases within the DNA double helix and binds to them.
- Once the template DNA has bound to the enzyme, the "pincers" close, keeping DNA in a double-stranded form called a *closed complex*.
- A conformational change in the RNA polymerase occurs, denaturing a short (10 base pairs) stretch of DNA and forming an *open complex*.
- The open complex makes the unpaired bases within DNA available to pair with ribonucleotides, and RNA synthesis begins.

Like DNA polymerases, RNA polymerases are *processive*; that is, a single enzyme–template binding event results in the polymerization of hundreds of RNA bases. Unlike DNA polymerases, RNA polymerases do not require a primer.

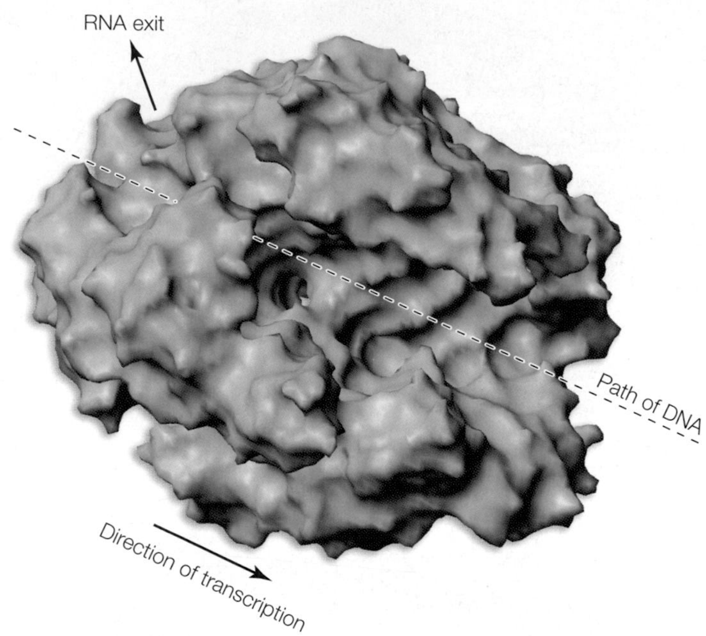

12.4 RNA Polymerase This enzyme from yeast is similar to most other RNA polymerases.

Transcription occurs in three steps

Transcription can be divided into three distinct processes: *initiation*, *elongation*, and *termination*.

INITIATION **Initiation**, which begins transcription, requires a **promoter**, a special sequence of DNA to which RNA polymerase binds very tightly (**Figure 12.5A**). There is at least one promoter for each gene (or, in prokaryotes, each set of genes). Promoters are important control sequences that "tell" the RNA polymerase three things:

- Where to start transcription
- Which strand of DNA to transcribe
- The direction to take from the start

A promoter, which is a specific sequence in the DNA that reads in a particular direction, orients the RNA polymerase and thus "aims" it at the correct strand to use as a template. Promoters function somewhat like the punctuation marks that determine how a sequence of words is to be read as a sentence. Part of each promoter is the **initiation site**, where transcription begins. Groups of nucleotides lying "upstream" from the initiation site (5′ on the non-template strand and 3′ on the template strand) help the RNA polymerase bind.

Although every gene has a promoter, not all promoters are identical. Some promoters are more effective at transcription initiation than others. Furthermore, there are differences between transcription initiation in prokaryotes and in eukaryotes (which will be explored in Chapters 13 and 14).

ELONGATION Once RNA polymerase has bound to the promoter, it begins the process of **elongation** (**Figure 12.5B**). RNA polymerase unwinds the DNA about 10 base pairs at a time and reads the template strand in the 3′-to-5′ direction. Like DNA polymerase, RNA polymerase adds new nucleotides to the 3′ end of the growing strand, but does not require a primer to get this process started. The new RNA elongates from the first base, which forms its 5′ end, to its 3′ end. The RNA transcript is thus antiparallel to the DNA template strand.

Unlike DNA polymerases, RNA polymerases do not proofread and correct their work (see Section 11.4). Transcription errors occur at a rate of one for every 10^4 to 10^5 bases. Because many copies of RNA are made, however, and because they often have only a relatively short life span, these errors are not as potentially harmful as mutations in DNA.

TERMINATION What tells RNA polymerase to stop adding nucleotides to a growing RNA transcript? Just as initiation sites in the DNA template strand specify the starting point for transcription, particular base sequences specify its **termination** (**Figure 12.5C**). The mechanisms of termination are complex and of more than one kind. For some genes, the newly formed transcript falls away from the DNA template and the RNA polymerase. For others, a helper protein pulls the transcript away.

In prokaryotes, which have no nuclear envelope and can have ribosomes located near the chromosome, translation often begins near the 5′ end of the mRNA before transcription of the mRNA molecule is complete. In eukaryotes, the situation is more complicated. First, there is a spatial separation of transcription (in the nucleus) and translation (in the cytoplasm). Second, the first product of transcription is a pre-mRNA that is longer than the final mRNA and must undergo considerable processing before it can be translated. The advantages of this processing and its mechanisms will be discussed in Section 14.3.

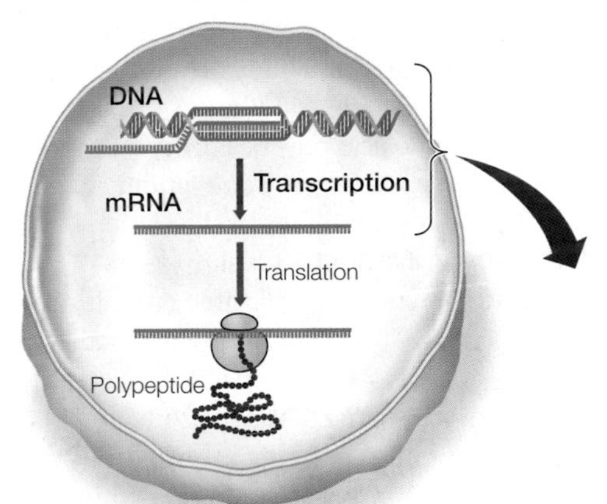

The information for protein synthesis lies in the genetic code

The **genetic code** relates genes (DNA) to mRNA and mRNA to the amino acids that make up proteins. The genetic code specifies which amino acids will be used to build a protein. You can think of the genetic information in an mRNA molecule as a series of sequential, nonoverlapping three-letter "words." Each sequence of

12.5 DNA Is Transcribed to Form RNA DNA is partially unwound by RNA polymerase to serve as a template for RNA synthesis. The RNA transcript is formed and then peels away, allowing the DNA that has already been transcribed to rewind into a double helix. Three distinct processes—initiation, elongation, and termination—constitute DNA transcription. RNA polymerase is much larger in reality than indicated here, covering about 50 base pairs.

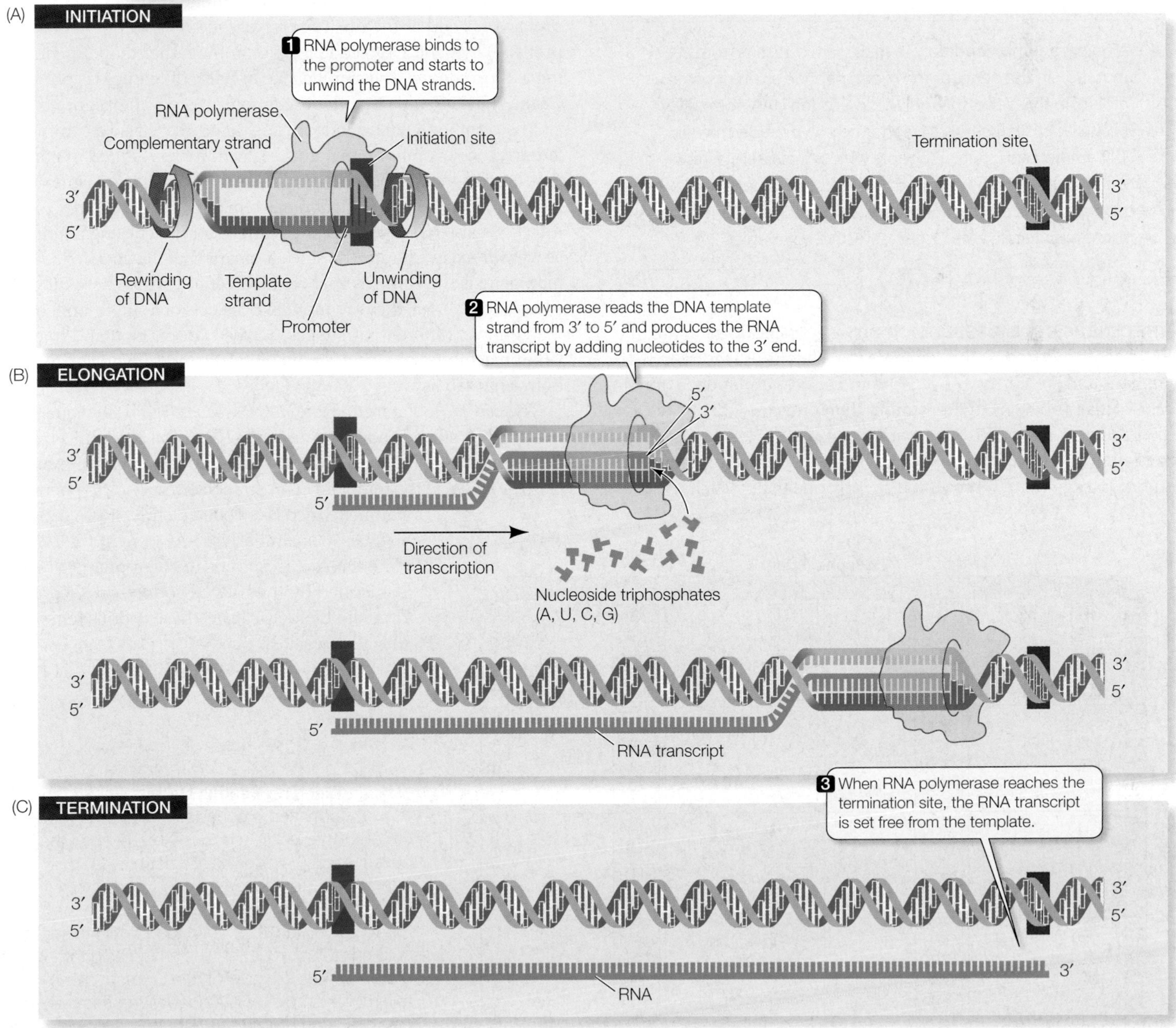

(A) INITIATION

1 RNA polymerase binds to the promoter and starts to unwind the DNA strands.

RNA polymerase
Complementary strand
Initiation site
Termination site
3′
5′
Rewinding of DNA
Template strand
Promoter
Unwinding of DNA

2 RNA polymerase reads the DNA template strand from 3′ to 5′ and produces the RNA transcript by adding nucleotides to the 3′ end.

(B) ELONGATION

Direction of transcription
Nucleoside triphosphates (A, U, C, G)
RNA transcript

3 When RNA polymerase reaches the termination site, the RNA transcript is set free from the template.

(C) TERMINATION

RNA

three nucleotide bases (the three "letters") along the mRNA polynucleotide chain specifies a particular amino acid. Each three-letter "word" is called a **codon**. Each codon is complementary to the corresponding triplet of bases in the DNA molecule from which it was transcribed. Thus the genetic code is the means of relating codons to their specific amino acids.

The complete genetic code is shown in **Figure 12.6**. Notice that there are many more codons than there are different amino acids in proteins. Combinations of the four available "letters" (the bases) give 64 (4^3) different three-letter codons, yet these codons determine only 20 amino acids. AUG, which codes for methionine, is also the **start codon**, the initiation signal for translation. Three of the codons (UAA, UAG, UGA) are **stop codons**, or termination signals for translation; when the translation machinery reaches one of these codons, translation stops, and the polypeptide is released from the translation complex.

A severe anemic condition, α-thalassemia, results from a mutation in the gene for the α-polypeptide chain of hemoglobin. In this gene, the mRNA stop codon UAA normally occurs at base position 142, and the polypeptide chain is 141 amino acids long. In people with α-thalassemia, however, position 142 is mutated to GAA (glutamine). The next stop codon doesn't occur until position 173, resulting in a protein molecule with larger, defective α subunits.

THE GENETIC CODE IS REDUNDANT BUT NOT AMBIGUOUS After the start and stop codons, the remaining 60 codons are far more than enough to code for the other 19 amino acids—and indeed, for almost all amino acids, there is more than one codon. Thus we say that the genetic code is *redundant*. For example, leucine is represented by six different codons (see Figure 12.6). Only methionine and tryptophan are represented by only one codon each.

The term *redundant* should not be confused with *ambiguous*. If the code were ambiguous, a single codon could specify either of two (or more) different amino acids, and there would be doubt about which amino acid should be incorporated into a growing polypeptide chain. Redundancy in the code simply means that there is more than one clear way to say, "Put leucine here." The genetic code is *not* ambiguous: a given amino acid may be encoded by more than one codon, but a codon can code for only one amino acid.

THE GENETIC CODE IS (NEARLY) UNIVERSAL Over 40 years of experiments on thousands of organisms from all three domains reveal that the genetic code is nearly *universal*, applying to all the species on our planet. That is, for almost every species, the codons that specify the amino acids are the same. Thus the code must be an ancient one that has been maintained intact throughout the evolution of living organisms. Exceptions are known: within mitochondria and chloroplasts, the code differs slightly from that in prokaryotes and in the nuclei of eukaryotic cells; in one group of protists, UAA and UAG code for glutamine rather than functioning as stop codons. The significance of these differences is not yet clear. What is clear is that the exceptions are few and slight.

The common genetic code means that there is also a common language for evolution. As natural selection has resulted in gradual changes in the genomes of different types of organisms, the raw material of genetic variation has remained the same. The common code also has profound implications for genetic engineering, as we will see in Chapter 16, since it means that the code for a human gene is the same as that for a bacterial gene. Where differences in code exist, they are more like dialects of a single language than entirely different languages. So the transcription and translation machinery of a bacterium could theoretically utilize genes from a human as well as its own genes.

The codons in Figure 12.6 are *mRNA codons*. The base sequence of the DNA strand that is transcribed to produce the mRNA is complementary and antiparallel to these codons. Thus, for example, 3′-AAA-5′ in the template DNA strand corresponds to phenylalanine (which is encoded by the mRNA codon 5′-UUU-3′); similarly, 3′-ACC-5′ in the template DNA corresponds to tryptophan (which is encoded by the mRNA codon 5′-UGG-3′). How did biologists learn these codons for specific amino acids?

First letter	Second letter: U	Second letter: C	Second letter: A	Second letter: G	Third letter
U	UUU Phenyl-alanine	UCU Serine	UAU Tyrosine	UGU Cysteine	U
	UUC	UCC	UAC	UGC	C
	UUA Leucine	UCA	UAA Stop codon	UGA Stop codon	A
	UUG	UCG	UAG Stop codon	UGG Tryptophan	G
C	CUU Leucine	CCU Proline	CAU Histidine	CGU Arginine	U
	CUC	CCC	CAC	CGC	C
	CUA	CCA	CAA Glutamine	CGA	A
	CUG	CCG	CAG	CGG	G
A	AUU Isoleucine	ACU Threonine	AAU Asparagine	AGU Serine	U
	AUC	ACC	AAC	AGC	C
	AUA	ACA	AAA Lysine	AGA Arginine	A
	AUG Methionine; start codon	ACG	AAG	AGG	G
G	GUU Valine	GCU Alanine	GAU Aspartic acid	GGU Glycine	U
	GUC	GCC	GAC	GGC	C
	GUA	GCA	GAA Glutamic acid	GGA	A
	GUG	GCG	GAG	GGG	G

12.6 The Genetic Code Genetic information is encoded in mRNA in three-letter units—codons—made up of the bases uracil (U), cytosine (C), adenine (A), and guanine (G). To decode a codon, find its first letter in the left column, then read across the top to its second letter, then read down the right column to its third letter. The amino acid the codon specifies is given in the corresponding row. For example, AUG codes for methionine, and GUA codes for valine.

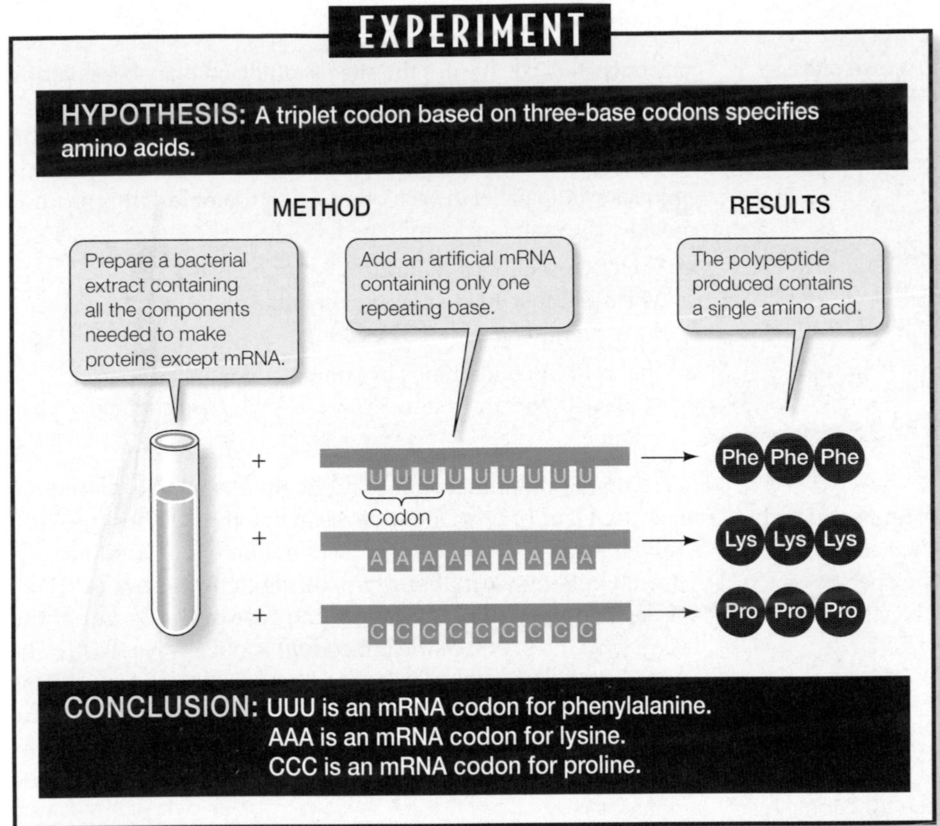

EXPERIMENT
HYPOTHESIS: A triplet codon based on three-base codons specifies amino acids.
METHOD
RESULTS
Prepare a bacterial extract containing all the components needed to make proteins except mRNA.
Add an artificial mRNA containing only one repeating base.
The polypeptide produced contains a single amino acid.
U U U U U U U U U
Codon
A A A A A A A A A
C C C C C C C C C
Phe Phe Phe
Lys Lys Lys
Pro Pro Pro
CONCLUSION: UUU is an mRNA codon for phenylalanine.
AAA is an mRNA codon for lysine.
CCC is an mRNA codon for proline.

12.7 Deciphering the Genetic Code Nirenberg and Matthaei used a test-tube protein synthesis system to determine the amino acids specified by synthetic mRNAs of known codon composition. FURTHER RESEARCH: What would be the result if the artificial mRNA were poly G?

Biologists used artificial messengers to decipher the genetic code

Molecular biologists "broke" the genetic code in the early 1960s. The problem they addressed was perplexing: how could more than 20 "code words" be written with an "alphabet" consisting of only four "letters"? In other words, how could four bases (A, U, G, and C) code for 20 different amino acids?

A triplet code, based on three-letter codons, was considered likely. Since there are only four letters (A, G, C, U), a one-letter code clearly could not unambiguously encode 20 amino acids; it could encode only four of them. A two-letter code could unambiguously code for only $4 \times 4 = 16$ codons—still not enough. But a triplet code could account for up to $4 \times 4 \times 4 = 64$ codons, more than enough to encode the 20 amino acids.

Marshall W. Nirenberg and J. H. Matthaei, at the U.S. National Institutes of Health, made the first decoding breakthrough in 1961 when they realized that they could use a simple artificial polynucleotide instead of a complex natural mRNA as a messenger. They could then identify the polypeptide that the artificial messenger encoded. They prepared an artificial mRNA in which all the bases were uracil (this artificial mRNA was aptly named *poly U*). When poly U was added to a test tube containing all the ingredients necessary for bacterial protein synthesis (ribosomes, all the amino acids, activating enzymes, tRNAs, and other factors), a polypeptide formed. This polypeptide chain was composed of only one kind of amino acid: phenylalanine (Phe). Poly U coded for poly Phe! Accordingly, UUU appeared to be the mRNA code word—the codon—for phenylalanine. Following up on this success, Nirenberg and Matthaei soon showed that CCC codes for proline and AAA for lysine (**Figure 12.7**). (Poly G presented some chemical problems and was not tested initially.) UUU, CCC, and AAA were three of the easiest codons; different approaches were required to work out the rest.

Other scientists later found that simple artificial mRNAs only three nucleotides long—each amounting to a codon—could bind to a ribosome, and that the resulting complex could then cause the binding of the corresponding tRNA with its specific amino acid. Thus, for example, simple UUU caused the tRNA carrying phenylalanine to bind to the ribosome. After this discovery, complete deciphering of the genetic code was relatively simple. To discover which amino acid a codon represents, Nirenberg repeated his experiment using a sample of artificial mRNA for that codon and observed which amino acid became bound to it.

12.3 RECAP

Transcription, which is catalyzed by an RNA polymerase, proceeds in three steps: initiation, elongation, and termination. The genetic code relates the information in mRNA (as a linear sequence of codons) to protein (a linear sequence of amino acids).

- Describe the steps of gene transcription that produce mRNA. See p. 262 and Figure 12.5
- How do RNA polymerases work? See p. 262
- Can you explain how the genetic code was elucidated? See pp. 264–265 and Figure 12.7

We now turn to the second step in gene expression, the translation of the nucleotide sequence in mRNA into an amino acid sequence in a polypeptide chain.

12.4 How Is RNA Translated into Proteins?

As Crick's adapter hypothesis proposed, the translation of mRNA into proteins requires a molecule that links the information contained in mRNA codons with specific amino acids in proteins. That function is performed by tRNA. Two key events must take place to ensure that the protein made is the one specified by mRNA:

- tRNA must read mRNA codons correctly.
- tRNA must deliver the amino acids that correspond to the mRNA codons it has read.

Once the tRNAs "decode" the mRNA and deliver the appropriate amino acids, components of the ribosome catalyze the formation of peptide bonds between amino acids. We now turn to these two steps.

Transfer RNAs carry specific amino acids and bind to specific codons

The codon in mRNA and the amino acid in a protein are related by way of an *adapter*—a specific tRNA with an attached amino acid. For each of the 20 amino acids, there is at least one specific type ("species") of tRNA molecule. The tRNA molecule has three functions:

- It carries (is "charged with") an amino acid.
- It associates with mRNA molecules.
- It interacts with ribosomes.

Its molecular structure relates clearly to all of these functions. A tRNA molecule has about 75 to 80 nucleotides. It has a *conformation* (a three-dimensional shape) that is maintained by complementary base pairing (hydrogen bonding) within its own sequence (**Figure 12.8**).

The conformation of a tRNA molecule is exquisitely suited for its interaction with specific binding sites on ribosomes. At the 3′ end of every tRNA molecule is its *amino acid attachment site*: a site to which its specific amino acid binds covalently. At about the midpoint of the tRNA sequence is a group of three bases, called the **anticodon**, that constitutes the site of complementary base pairing (via hydrogen bonding) with mRNA. Each tRNA species has a unique anticodon, which is complementary to the mRNA codon for that tRNA's amino acid. At contact, the codon and the anticodon are antiparallel to each other. As an example of this process, consider the amino acid arginine:

- The DNA sequence that codes for arginine is 3′-GCC-5′, which is transcribed, by complementary base pairing, to produce the mRNA codon 5′-CGG-3′.
- That mRNA codon binds by complementary base pairing to a tRNA with the anticodon 3′-GCC-5′, which is charged with arginine.

Recall that 61 different codons encode the 20 amino acids in proteins (see Figure 12.6). Does this mean that the cell must produce 61 different tRNA species, each with a different anticodon? No. The cell gets by with about two-thirds that number of tRNA species because the specificity for the base at the 3′ end of the codon (and the 5′ end of the anticodon) is not always strictly observed. This phenomenon, called *wobble*, allows the alanine codons GCA, GCC, and GCU, for example, all to be recognized by the same tRNA. Wobble is allowed in some matches, but not in others; of most importance, it does not allow the genetic code to be ambiguous.

It took 3 years for Cornell University chemist Robert Holley and his team to sequence the 80 nucleotides of alanine-tRNA from yeast, an effort for which Holley won the 1968 Nobel prize in medicine. With today's technology, sequencing a nucleic acid with 80 nucleotides takes a few seconds.

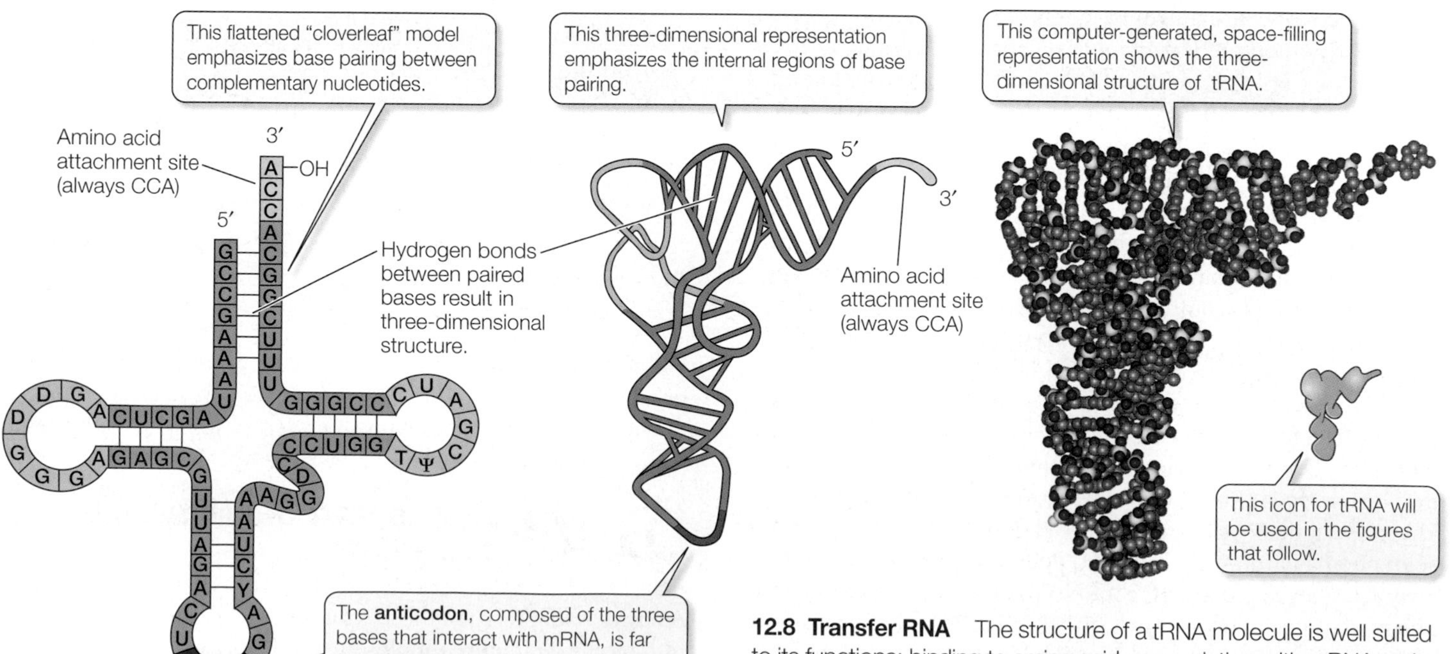
This flattened "cloverleaf" model emphasizes base pairing between complementary nucleotides.
This three-dimensional representation emphasizes the internal regions of base pairing.
This computer-generated, space-filling representation shows the three-dimensional structure of tRNA.
Amino acid attachment site (always CCA)
3′
5′
–OH
Hydrogen bonds between paired bases result in three-dimensional structure.
Amino acid attachment site (always CCA)
This icon for tRNA will be used in the figures that follow.
The **anticodon**, composed of the three bases that interact with mRNA, is far from the amino acid attachment site.

12.8 Transfer RNA The structure of a tRNA molecule is well suited to its functions: binding to amino acids, associating with mRNA molecules, and interacting with ribosomes.

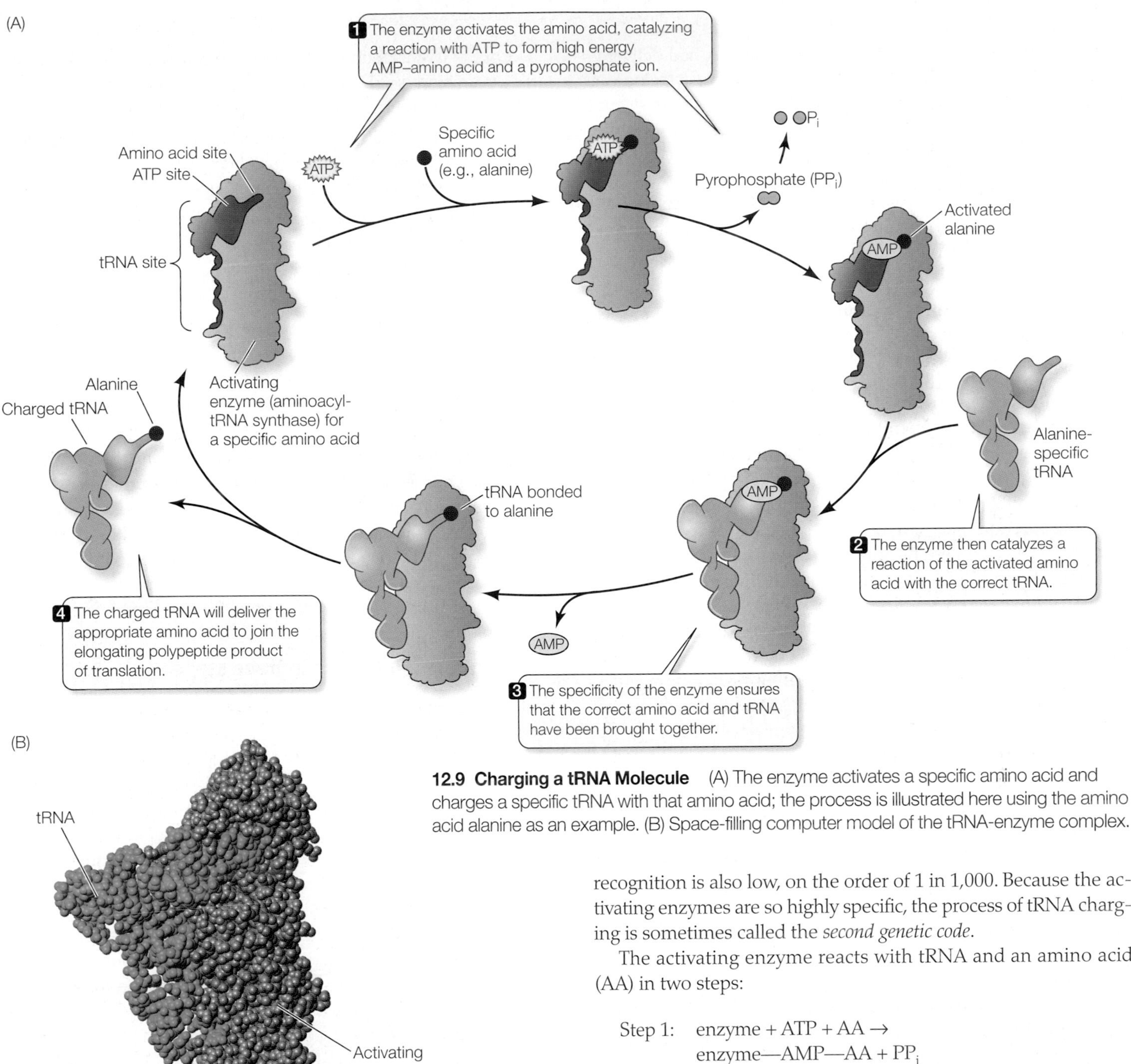
(A)
1 The enzyme activates the amino acid, catalyzing a reaction with ATP to form high energy AMP–amino acid and a pyrophosphate ion.
Amino acid site
ATP site
tRNA site
ATP
Specific amino acid (e.g., alanine)
ATP
Pi
Pyrophosphate (PPi)
Activated alanine
AMP
Alanine
Charged tRNA
Activating enzyme (aminoacyl-tRNA synthase) for a specific amino acid
Alanine-specific tRNA
tRNA bonded to alanine
AMP
2 The enzyme then catalyzes a reaction of the activated amino acid with the correct tRNA.
4 The charged tRNA will deliver the appropriate amino acid to join the elongating polypeptide product of translation.
AMP
3 The specificity of the enzyme ensures that the correct amino acid and tRNA have been brought together.
(B)
tRNA
Activating enzyme

12.9 Charging a tRNA Molecule (A) The enzyme activates a specific amino acid and charges a specific tRNA with that amino acid; the process is illustrated here using the amino acid alanine as an example. (B) Space-filling computer model of the tRNA-enzyme complex.

Activating enzymes link the right tRNAs and amino acids

The charging of each tRNA with its correct amino acid is achieved by a family of activating enzymes, known more formally as *aminoacyl-tRNA synthetases* (**Figure 12.9**). Each activating enzyme is specific for one amino acid and for its corresponding tRNA. The enzyme has a three-part active site that recognizes three smaller molecules: a specific amino acid, ATP, and a specific tRNA. Since tRNA is large and has a complex three-dimensional structure, the activating enzyme's recognition of tRNA is quite specific and has a very low error rate. Remarkably, the error rate for amino acid recognition is also low, on the order of 1 in 1,000. Because the activating enzymes are so highly specific, the process of tRNA charging is sometimes called the *second genetic code*.

The activating enzyme reacts with tRNA and an amino acid (AA) in two steps:

Step 1: enzyme + ATP + AA $\rightarrow$
enzyme—AMP—AA + PP_i

Step 2: enzyme—AMP—AA + tRNA $\rightarrow$
enzyme + AMP + tRNA—AA

The amino acid is attached to the 3′ end of the tRNA (to a free OH group on the ribose) with an energy-rich bond, forming charged tRNA. This bond will provide the energy for the synthesis of the peptide bond that will join adjacent amino acids.

A clever experiment by Seymour Benzer and his colleagues at Purdue University demonstrated the importance of the specificity of the attachment of tRNA to its amino acid. In their laboratory, the amino acid cysteine, already properly attached to its tRNA, was chemically modified to become a different amino acid, alanine. Which component—the amino acid or the tRNA—would be recognized when this hybrid charged tRNA was put into a protein-synthesizing system? The answer was: the tRNA. Everywhere in the synthesized protein where cysteine was supposed to be, alanine appeared instead. The cysteine-specific tRNA had delivered its cargo (alanine) to every mRNA "address" where cysteine was called for. This experiment showed that the protein synthesis machinery recognizes the anticodon of the charged tRNA, not the amino acid attached to it. If activating enzymes in nature did what Benzer did in the laboratory and charged tRNAs with the wrong amino acids, those amino acids would be inserted into proteins at inappropriate places, leading to alterations in protein shape and function.

The ribosome is the workbench for translation

The **ribosome** is the molecular workbench where the task of translation is accomplished. Its structure enables it to hold mRNA and charged tRNAs in the right positions, thus allowing a polypeptide chain to be assembled efficiently. A given ribosome does not specifically produce just one kind of protein. A ribosome can use any mRNA and all species of charged tRNAs, and thus can be used to make many different polypeptide products. The mRNA, as a linear sequence of codons, specifies the polypeptide sequence to be made.

Although ribosomes are small in contrast to other cellular organelles, their mass of several million daltons makes them large in comparison with charged tRNAs. Each ribosome consists of two subunits, a large one and a small one (**Figure 12.10**). In eukaryotes, the large subunit consists of three different molecules of **ribosomal RNA** (**rRNA**) and about 45 different protein molecules, arranged in a precise pattern. The small subunit consists of one rRNA molecule and 33 different protein molecules. When not active in the translation of mRNA, the ribosomes exist as separated subunits.

The ribosomes of prokaryotes are somewhat smaller than those of eukaryotes, and their ribosomal proteins and RNAs are different. Mitochondria and chloroplasts also contain ribosomes, some of which are similar to those of prokaryotes.

The different proteins and rRNAs in a ribosomal subunit are held together by ionic and hydrophobic forces, not covalent bonds. If these forces are disrupted by a detergent, for example, the proteins and rRNAs separate from one another. When the detergent is removed, the entire complex structure self-assembles. This is like separating the pieces of a jigsaw puzzle and having them fit together again without human hands to guide them!

On the large subunit of the ribosome are three sites to which tRNA can bind (see Figure 12.10). A charged tRNA traverses these three sites in order:

- The *A* (amino acid) *site* is where the charged tRNA anticodon binds to the mRNA codon, thus lining up the correct amino acid to be added to the growing polypeptide chain.
- The *P* (polypeptide) *site* is where the tRNA adds its amino acid to the growing polypeptide chain.
- The *E* (exit) *site* is where the tRNA, having given up its amino acid, resides before being released from the ribosome and going back to the cytosol to pick up another amino acid and begin the process again.

An important role of the ribosome is to make sure that the mRNA–tRNA interactions are accurate; that is, that a charged tRNA with the correct anticodon (e.g., 3′-UAC-5′) binds to the appropriate codon in mRNA (e.g., 5′-AUG-3′). When this occurs, hydrogen bonds form between the base pairs. But these hydrogen bonds are not enough to hold the tRNA in place. The rRNA of the small ribosomal subunit plays a role in validating the three-base-pair match. If hydrogen bonds have not formed between all three base pairs, the tRNA must be the wrong one for that mRNA codon, and that tRNA is ejected from the ribosome.

Translation takes place in three steps

Like transcription, translation occurs in three steps: initiation, elongation, and termination.

INITIATION The translation of mRNA begins with the formation of an **initiation complex**, which consists of a charged tRNA bearing what will be the first amino acid of the polypeptide chain and a small ribosomal subunit, both bound to the mRNA (**Figure 12.11**). The rRNA of the small ribosomal subunit first binds to a complementary ribosome binding site (known as the *Shine–Dalgarno sequence*) on the mRNA. This sequence is "upstream" (toward the 5′ end) of the actual start codon that begins translation.

Recall that the mRNA start codon in the genetic code is AUG (see Figure 12.6). The anticodon of a methionine-charged tRNA binds to this start codon by complementary base pairing to complete the initiation complex. Thus the first amino acid in a polypeptide chain is always methionine. Not all mature proteins have methionine as their N-terminal amino acid, however. In many cases, the initiator methionine is removed by an enzyme after translation.

After the methionine-charged tRNA has bound to the mRNA, the large subunit of the ribosome joins the complex. The methionine-charged tRNA now lies in the P

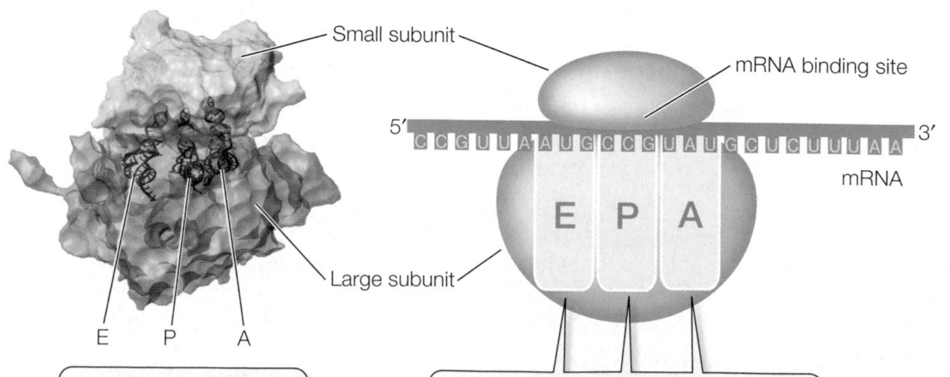
Small subunit
mRNA binding site
5′
CCGUUAAUGCCGUAUGCUCUUUAA
3′
mRNA
E P A
Large subunit
E P A
Ribosomes are irregularly shaped and composed of two subunits.
There are 3 sites for tRNA binding. Codon–anticodon interactions between tRNA and mRNA occur only at the P and A sites.

12.10 Ribosome Structure Each ribosome consists of a large and a small subunit. The subunits remain separate when they are not in use for protein synthesis.

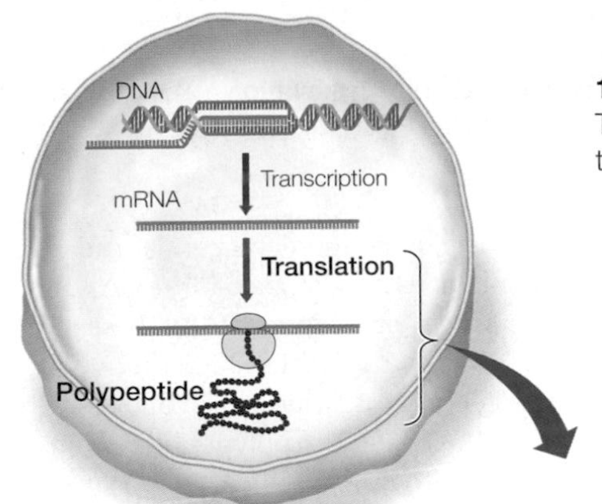
DNA
Transcription
mRNA
Translation
Polypeptide

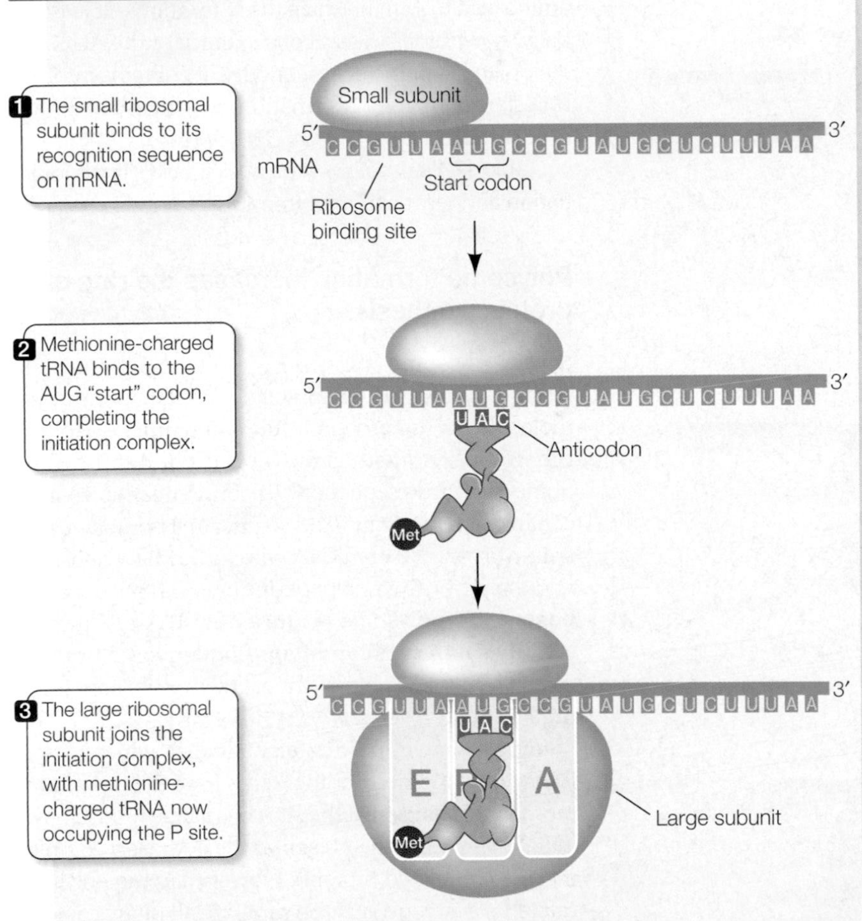
INITIATION
1 The small ribosomal subunit binds to its recognition sequence on mRNA.
Small subunit
5′
3′
C C G U U A A U G C C G U A U G C U C U U U A A
mRNA
Start codon
Ribosome binding site
2 Methionine-charged tRNA binds to the AUG "start" codon, completing the initiation complex.
U A C
Anticodon
Met
3 The large ribosomal subunit joins the initiation complex, with methionine-charged tRNA now occupying the P site.
E P A
Large subunit

12.11 The Initiation of Translation Translation begins with the formation of an initiation complex.

site of the ribosome, and the A site is aligned with the second mRNA codon. These ingredients—mRNA, two ribosomal subunits, and methionine-charged tRNA—are put together properly by a group of proteins called *initiation factors.*

The prokaryotic ribosome is smaller and has a different collection of proteins than the eukaryotic ribosome. Some antibacterial antibiotics work by binding and inhibiting specific ribosomal proteins that are essential to the bacterium but do not exist in the eukaryotic ribosome.

ELONGATION A charged tRNA whose anticodon is complementary to the second codon of the mRNA now enters the open A site of the large ribosomal subunit (**Figure 12.12**). The large subunit then catalyzes two reactions:

- It breaks the bond between the tRNA in the P site and its amino acid.
- It catalyzes the formation of a peptide bond between that amino acid and the one attached to the tRNA in the A site.

Because the large subunit performs these two actions, it is said to have *peptidyl transferase activity.* In this way, methionine (the amino acid in the P site) becomes the N terminus of the new protein. The second amino acid is now bound to methionine, but remains attached to its tRNA at the A site.

How does the large ribosomal subunit catalyze this binding? Harry Noller and his colleagues at the University of California at Santa Cruz found that if they removed almost all the proteins from the large subunit, it still catalyzed peptide bond formation. But if the RNA was destroyed, so was peptidyl transferase activity. Part of the rRNA in the large subunit interacts with the end of the charged tRNA where the amino acid is attached. Thus RNA appears to be the catalyst, probably tRNA. This situation is unusual, because proteins are the usual catalysts in biological systems. The recent purification and crystallization of ribosomes has allowed scientists to examine their structure in detail, and the catalytic role of rRNA in peptidyl transferase activity has been confirmed, supporting the hypothesis that RNA, and catalytic RNA in particular, evolved before DNA (see Section 3.6).

After the first tRNA releases its methionine, it moves to the E site and is then dissociated from the ribosome, returning to the cytosol to become charged with another methionine. The second tRNA, now bearing a *dipeptide* (a two-amino-acid chain), is shifted to the P site as the ribosome moves one codon along the mRNA in the 5′-to-3′ direction.

The elongation process continues, and the polypeptide chain grows, as the steps are repeated:

- The next charged tRNA enters the open A site, where its anticodon binds with the mRNA codon.
- Its amino acid forms a peptide bond with the amino acid chain in the P site, so that it picks up the growing polypeptide chain from the tRNA in the P site.
- The tRNA in the P site is transferred to the E site and then released. The ribosome shifts one codon, so that the entire tRNA–polypeptide complex moves to the newly vacated P site.

All these steps are assisted by proteins called *elongation factors.*

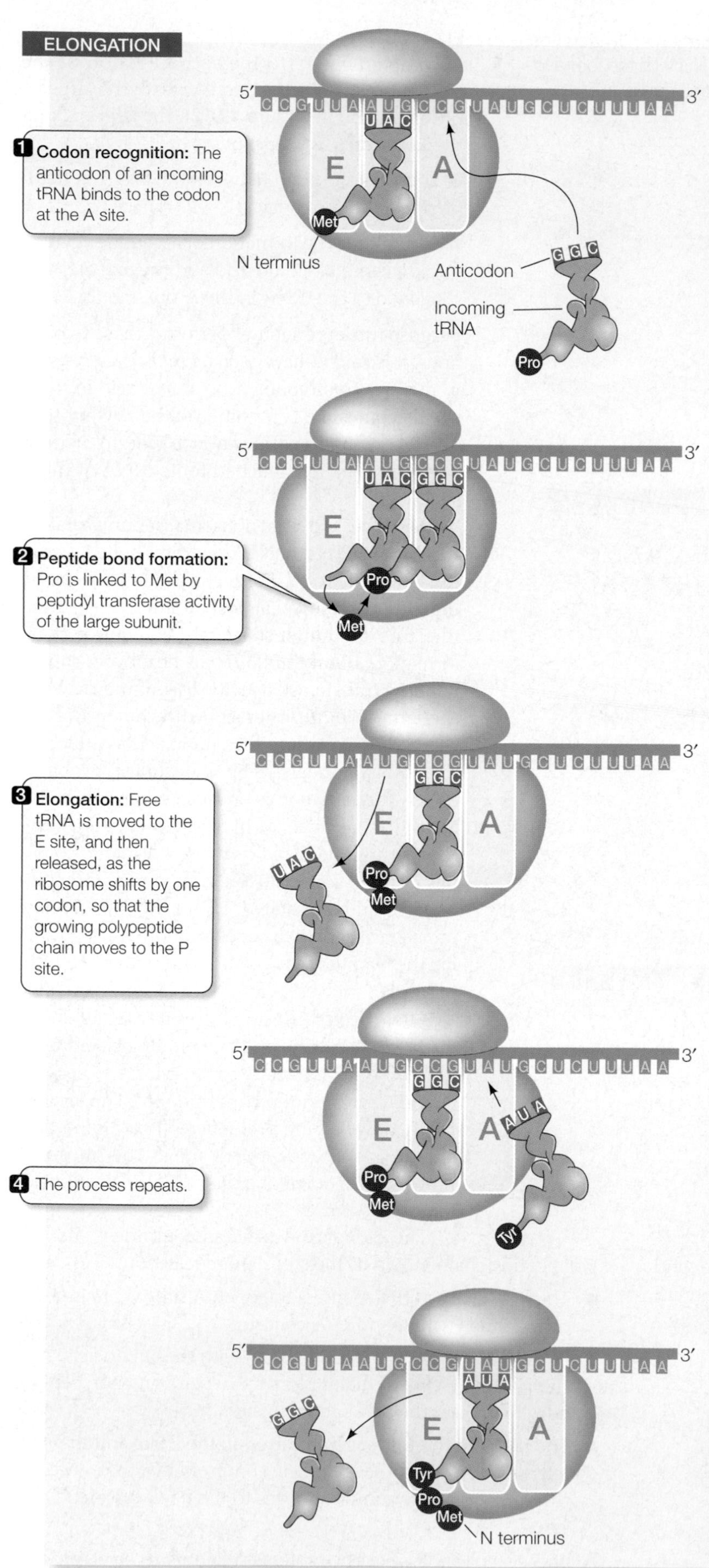

ELONGATION
5′
3′
E
A
1 Codon recognition: The anticodon of an incoming tRNA binds to the codon at the A site.
Met
N terminus
Anticodon
Incoming tRNA
Pro
2 Peptide bond formation: Pro is linked to Met by peptidyl transferase activity of the large subunit.
3 Elongation: Free tRNA is moved to the E site, and then released, as the ribosome shifts by one codon, so that the growing polypeptide chain moves to the P site.
4 The process repeats.
Tyr
N terminus

12.12 The Elongation of Translation The polypeptide chain elongates as the mRNA is translated.

TERMINATION The elongation cycle ends, and translation is terminated, when a stop codon—UAA, UAG, or UGA—enters the A site (**Figure 12.13**). These codons encode no amino acids, nor do they bind tRNA. Rather, they bind a protein *release factor*, which allows hydrolysis of the bond between the polypeptide chain and the tRNA in the P site.

The newly completed polypeptide thereupon separates from the ribosome. Its C terminus is the last amino acid to join the chain. Its N terminus, at least initially, is methionine, as a consequence of the AUG start codon. In its amino acid sequence, it contains information specifying its conformation, as well as its ultimate cellular destination.

Table 12.1 summarizes the nucleic acid signals for initiation and termination of transcription and translation.

Polysome formation increases the rate of protein synthesis

Several ribosomes can work simultaneously at translating a single mRNA molecule, producing multiple molecules of the protein at the same time. As soon as the first ribosome has moved far enough from the Shine–Dalgarno sequence, a second initiation complex can form, then a third, and so on. An assemblage consisting of a strand of mRNA with its beadlike ribosomes and their growing polypeptide chains is called a **polyribosome**, or **polysome** (**Figure 12.14**). Cells that are actively synthesizing proteins contain large numbers of polysomes and few free ribosomes or ribosomal subunits.

A polysome is like a cafeteria line, in which patrons follow one another, adding items to their trays. At any moment, the person at the start has a little food (a newly initiated protein); the person at the end has a complete meal (a completed protein). However, in the polysome cafeteria, everyone gets the same meal: many copies of the same protein are made from a single mRNA.

TABLE 12.1

Signals that Start and Stop Transcription and Translation

	TRANSCRIPTION	TRANSLATION
Initiation	Promoter sequence in DNA	AUG start codon in mRNA
Termination	Terminator sequence in DNA	UAA, UAG, or UGA stop codon in mRNA

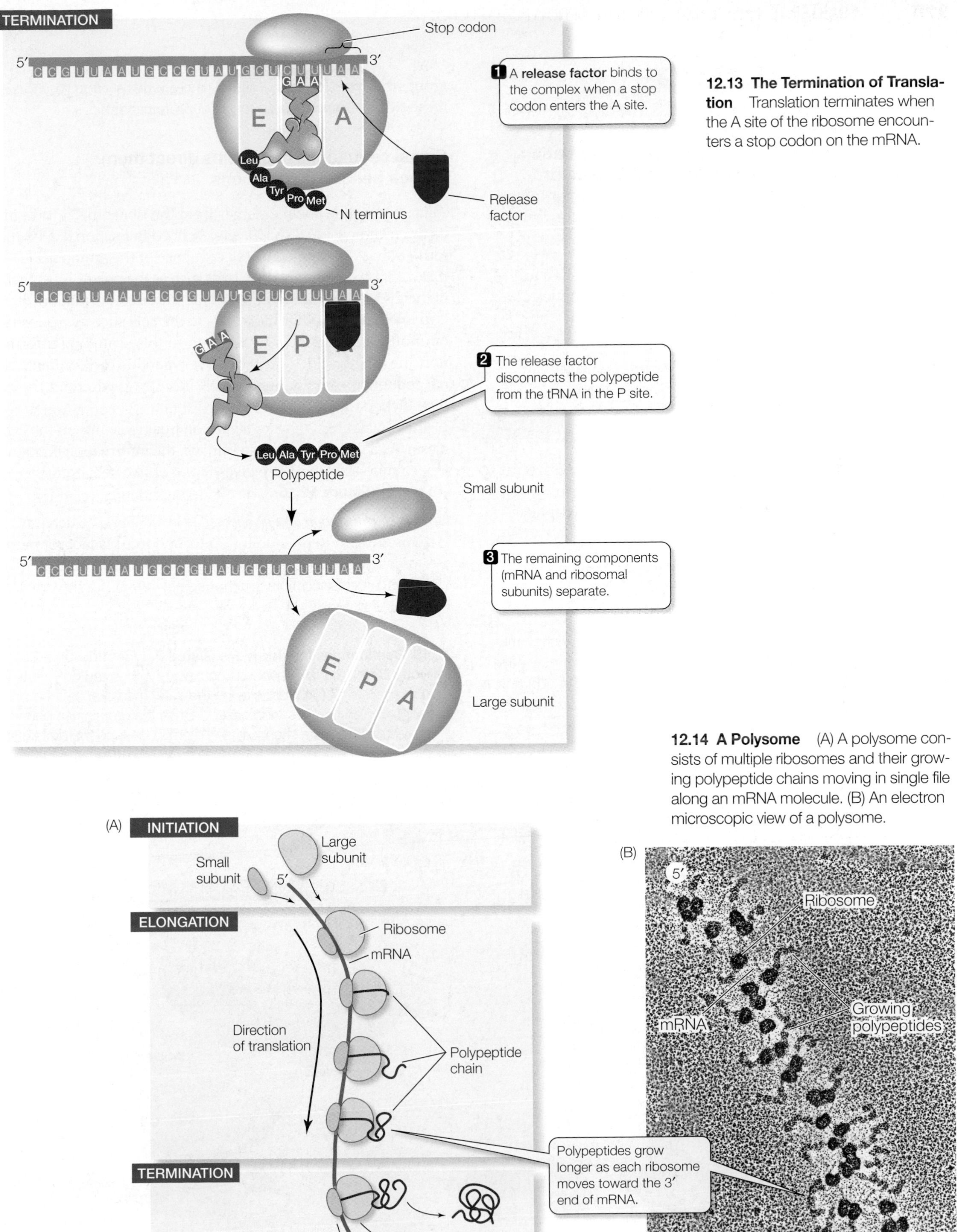

12.13 The Termination of Translation Translation terminates when the A site of the ribosome encounters a stop codon on the mRNA.

12.14 A Polysome (A) A polysome consists of multiple ribosomes and their growing polypeptide chains moving in single file along an mRNA molecule. (B) An electron microscopic view of a polysome.

12.4 RECAP

A key step in protein synthesis is the attachment of an amino acid to its proper tRNA, which is carried out by an activating enzyme. Translation of the genetic information from mRNA into protein occurs at the ribosome. Multiple ribosomes may act on a single mRNA to make multiple copies of the protein for which it codes.

- Do you understand how an amino acid is attached to a specific tRNA, and why the term "second genetic code" is associated with this process? See p. 267 and Figure 12.9
- Describe the events of initiation, elongation, and termination of translation. See pp. 268-270 and Figures 12.11, 12.12, and 12.13

The polypeptide chain that is released from the ribosome is not necessarily a functional protein. Let's look at some of the posttranslational changes that affect the fate and function of polypeptides.

12.5 What Happens to Polypeptides after Translation?

Especially in eukaryotic cells, the site of a polypeptide's function may be far away from its point of synthesis in the cytoplasm; it may need to be moved into an organelle, or even out of the cell. In addition, polypeptides are often modified by the addition of new chemical groups that have functional significance. In this section we examine these two *posttranslational* aspects of protein synthesis.

Signal sequences in proteins direct them to their cellular destinations

As a polypeptide chain emerges from the ribosome, it folds into its three-dimensional shape. As described in Section 3.2, its conformation is determined by the sequence of the amino acids that make up the protein, and by factors such as the polarity and charge of their R groups. Ultimately, a polypeptide's conformation allows it to interact with other molecules in the cell, such as a substrate or another polypeptide. In addition to this structural information, the amino acid sequence of a polypeptide can contain a **signal sequence**—an "address label" indicating where in the cell the polypeptide belongs.

Protein synthesis always begins on free ribosomes in the cytoplasm. As a polypeptide chain is made, the information contained in its amino acid sequence gives it one of two sets of further instructions (**Figure 12.15**):

- *"Finish translation and be released to an organelle."* Such proteins are sent to the nucleus, mitochondria, plastids, or peroxisomes, depending on the address in their instructions; or, lacking such specific instructions, they remain in the cytosol.

12.15 Destinations for Newly Translated Polypeptides in a Eukaryotic Cell Signal sequences on newly synthesized polypeptides bind to specific receptor proteins on the outer membrane of the organelle to which they are "addressed." Once the protein has bound to it, the receptor forms a channel in the membrane, and the protein enters the organelle.

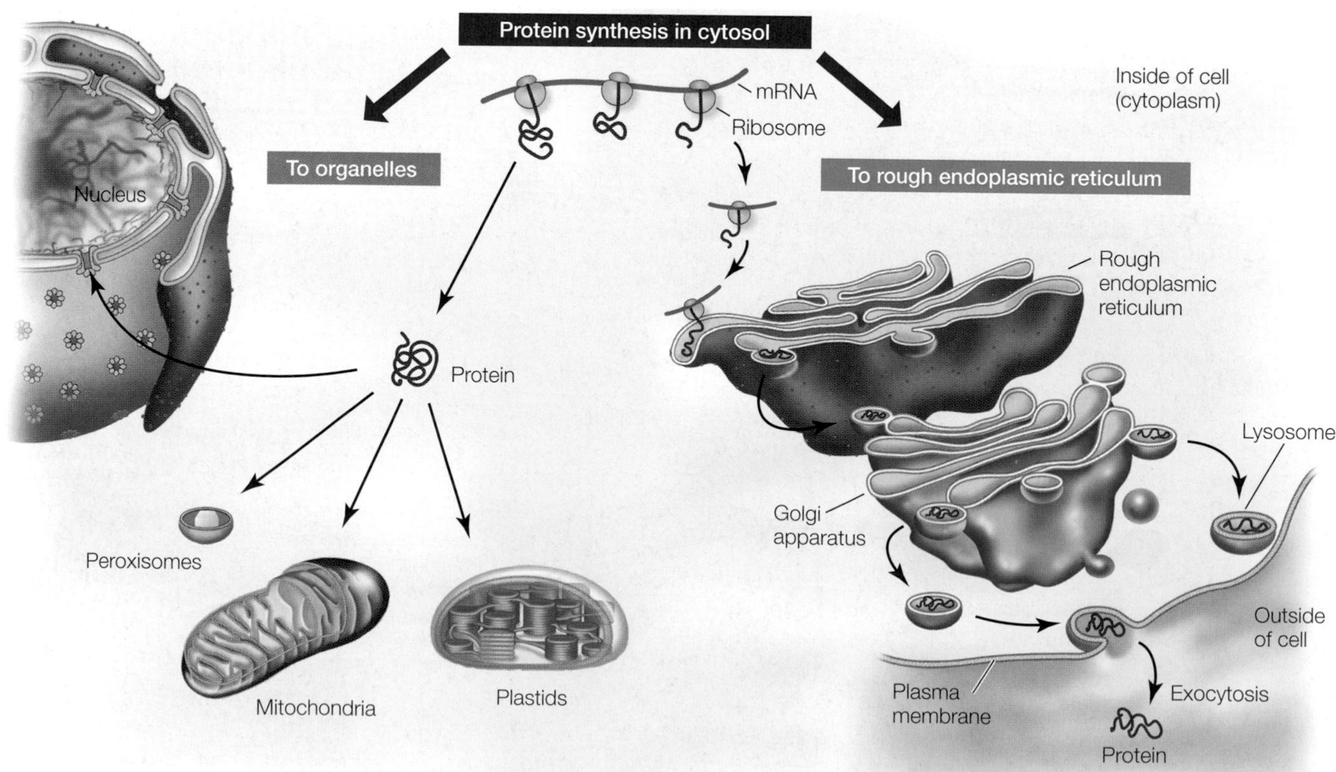

- *"Stop translation, go to the endoplasmic reticulum, and finish synthesis there."* After protein synthesis is completed, such proteins may be retained in the ER, and sent to the Golgi apparatus. From there, they may be sent to the lysosomes, to the plasma membrane, or, lacking such specific instructions, they may be secreted from the cell via vesicles that fuse with the plasma membrane.

Proteins destined to remain in the cytoplasm by default have no signal.

DESTINATION: NUCLEUS, MITOCHONDRION, OR CHLOROPLAST After translation, some folded polypeptides have a short exposed sequence of amino acids that acts like a postal "zip code," directing them to an organelle. These signal (or *localization*) sequences are either at the N terminus or in the interior of the amino acid chain. For example, the following sequence directs a protein to the nucleus:

—Pro—Pro—Lys—Lys—Lys—Arg—Lys—Val—

This amino acid sequence would occur in the histone proteins associated with nuclear DNA, but not in citric acid cycle enzymes, which are addressed to the mitochondria.

Signal sequences have a conformation that allows them to bind to a specific receptor protein, appropriately called a **docking protein**, on the outer membrane of the appropriate organelle. Once the protein has bound to it, the receptor forms a channel in the membrane, allowing the protein to pass through to its organelle destination. In this process, the protein is usually unfolded by a chaperonin (see Figure 3.12) so that it can pass through the channel, then refolds into its normal conformation.

DESTINATION: ENDOPLASMIC RETICULUM If a specific hydrophobic sequence of 15–30 amino acids occurs at the N terminus of a polypeptide chain, the polypeptide is sent initially to the ER and then to the Golgi, where it can be modified for eventual transport to the lysosomes, the plasma membrane, or out of the cell. In the cytoplasm, before translation is finished and while the polypeptide is still attached to a ribosome, the signal sequence binds to a **signal recognition particle** composed of protein and RNA (**Figure 12.16**). This binding blocks further protein synthesis until the ribosome becomes attached to a specific receptor protein in the membrane of the rough ER. Once again, the receptor protein is converted into a channel, through which the growing polypeptide passes. The elongating polypeptide may be retained in the ER membrane itself, or it may enter the interior space—the lumen—of the ER. In either case, an enzyme in the lumen of the ER removes the signal sequence from the polypeptide chain.

At this point, protein synthesis resumes, and the chain grows longer until its sequence is completed. If the finished protein enters the ER lumen, it can be transported to its appropriate location—to other cellular compartments or to the outside of the cell—via the ER and the Golgi apparatus without mixing with other molecules in the cytoplasm.

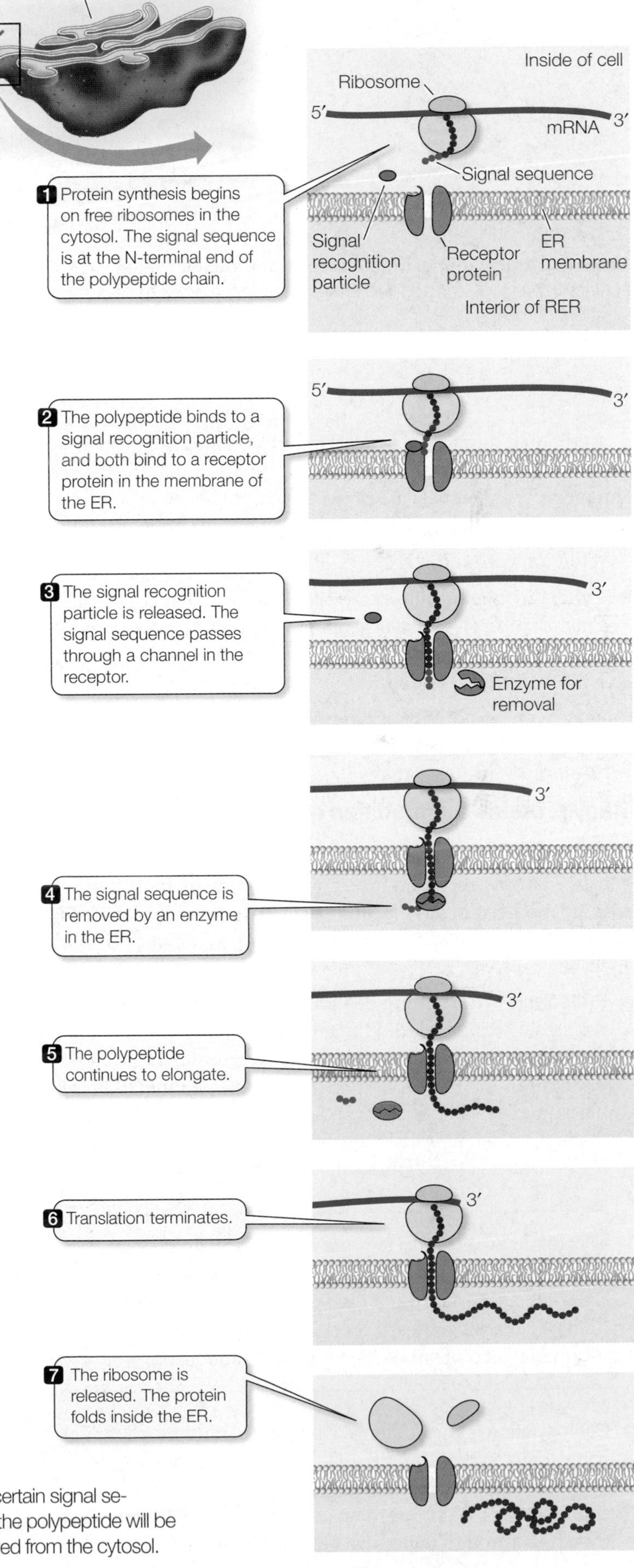

12.16 A Signal Sequence Moves a Polypeptide into the ER When a certain signal sequence of amino acids is present at the beginning of a polypeptide chain, the polypeptide will be taken into the endoplasmic reticulum. The finished protein is thus segregated from the cytosol.

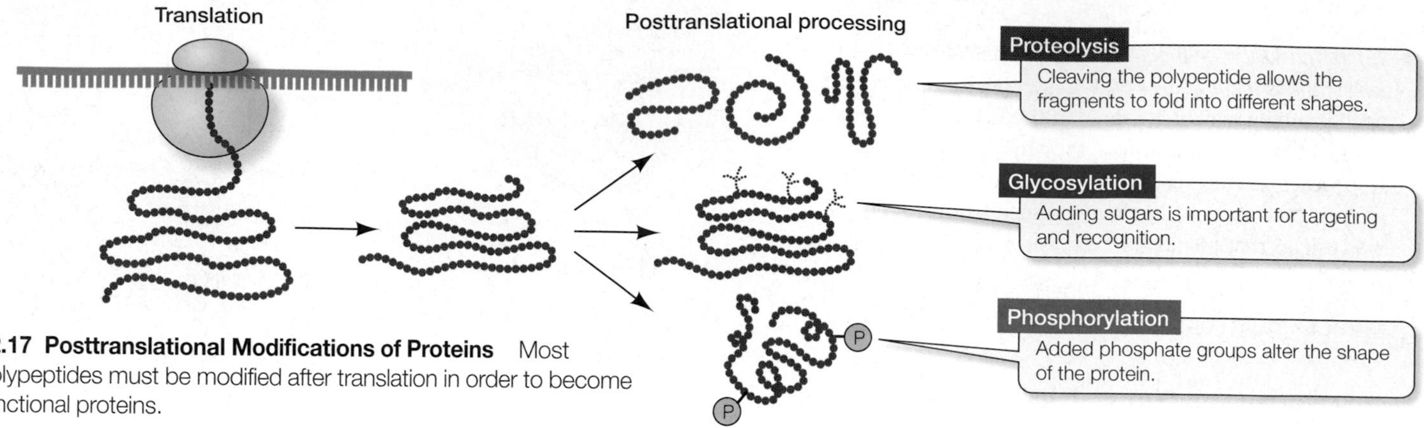

12.17 Posttranslational Modifications of Proteins Most polypeptides must be modified after translation in order to become functional proteins.

Additional signals are needed to further direct the protein (remember that the signal sequence that sent it to the ER has been removed). These signals are of two kinds:

- Some are sequences of amino acids that allow the protein's retention within the ER.
- Others are sugars, which are added in the Golgi apparatus. The resulting *glycoproteins* end up either at the plasma membrane or in a lysosome (or plant vacuole), depending on which sugars are added.

Proteins with no additional signals pass from the ER through the Golgi apparatus and are secreted from the cell.

Many proteins are modified after translation

Most finished proteins are not identical to the polypeptide chains translated from mRNA on the ribosomes. Instead, most polypeptides are modified in any of a number of ways after translation (**Figure 12.17**). These modifications are essential to the final functioning of the protein.

- **Proteolysis** is the cutting of a polypeptide chain. Cleavage of the signal sequence from the growing polypeptide chain in the ER is an example of proteolysis; the protein might move back out of the ER through the membrane channel if the signal sequence were not cut off. Some proteins are actually made from *polyproteins* (long polypeptides) that are cut into final products by enzymes called *proteases*.

Proteases are essential to some viruses, including HIV, because the large viral polyprotein cannot fold properly unless it is cut. Certain drugs used to treat AIDS work by inhibiting the HIV protease, thereby preventing the formation of proteins needed for viral reproduction.

- **Glycosylation** is the addition of sugars to proteins to form glycoproteins. In both the ER and the Golgi apparatus, resident enzymes catalyze the addition of various sugars or short sugar chains to certain amino acid R groups on proteins as they pass through. One such type of "sugar coating" is essential for addressing proteins to lysosomes, as mentioned above. Other types are important in the conformation and the recognition functions of proteins at the cell surface. Still other attached sugars help to stabilize proteins stored in vacuoles in plant seeds.
- **Phosphorylation**, the addition of phosphate groups to proteins, is catalyzed by protein kinases. The charged phosphate groups change the conformation of a protein, often exposing the active site of an enzyme or a binding site for another protein.

12.5 RECAP

Signal sequences in polypeptides "address" them to their appropriate destinations inside or outside the cell. Many polypeptides are modified after translation.

- Do you understand how signal sequences determine where a protein will go after it is made? See pp. 272–273 and Figure 12.16
- Can you explain some ways in which posttranslational modifications alter protein structure and function? See p. 274 and Figure 12.17

All of the processes we have just described result in a functional protein only if the amino acid sequence of that protein is correct. If the sequence is not correct, cellular dysfunction may result. Changes in the DNA—mutations—are a major source of errors in amino acid sequences.

12.6 What Are Mutations?

In Chapter 10, we described mutations as inherited changes in genes, and we saw that the new alleles that result may produce altered phenotypes (short pea plants instead of tall, for example). Now that we understand the chemical nature of genes and how they are expressed as phenotypes, we will return to the concept of mutations for a more specific definition.

Errors in DNA replication can occur in any cell undergoing the cell cycle, and these errors are passed on to the daughter cells.

Mutations in multicellular organisms can be divided into two types:

- **Somatic mutations** are those that occur in *somatic* (body) cells. These mutations are passed on to the daughter cells after mitosis, and to the offspring of those cells in turn, but are not passed on to sexually produced offspring. A mutation in a single human skin cell, for example, could result in a patch of skin cells, all with the same mutation, but would not be passed on to a person's children.
- **Germ line mutations** are those that occur in the cells of the *germ line*—the specialized cells that give rise to gametes. A gamete with the mutation passes it on to a new organism at fertilization.

Some mutations cause their phenotypes only under certain *restrictive* conditions. They are not detectable under other, *permissive* conditions. These phenotypes are known as **conditional mutants**. Many conditional mutants are temperature-sensitive; that is, they show the altered phenotype only at a certain temperature (recall the rabbit in Figure 10.16). The mutant allele in such an organism may code for an enzyme with an unstable tertiary structure that is altered at the restrictive temperature.

All mutations are alterations in the nucleotide sequence of DNA. At the molecular level, we can divide mutations into two categories:

- **Point mutations** are mutations of single base pairs and so are limited to single genes: one allele (usually dominant) becomes another allele (usually recessive) because of an alteration (gain, loss, or substitution) of a single nucleotide (which, after DNA replication, becomes a mutant base pair).
- **Chromosomal mutations** are more extensive alterations than point mutations. They may change the position or orientation of a DNA segment without actually removing any genetic information, or they may cause a segment of DNA to be irretrievably lost or duplicated.

Point mutations change single nucleotides

Point mutations result from the addition or subtraction of a nucleotide base, or the substitution of one base for another, in the DNA. Point mutations can be the result of errors in DNA replication that are not corrected in proofreading, or they can be caused by environmental mutagens such as chemicals and radiation.

Point mutations in DNA usually result in changes in mRNA, but changes in mRNA may or may not result in changes in the protein. *Silent mutations* have no effect on the protein. *Missense* and *nonsense* mutations result in changes in the protein, some of them drastic.

SILENT MUTATIONS Because of the redundancy of the genetic code, some base substitutions result in no change in amino acids when the altered mRNA is translated; for this reason, they are called **silent mutations**. For example, there are four mRNA codons that code for proline: CCA, CCC, CCU, and CCG (see Figure 12.6). If the template strand of DNA has the sequence 5′-CGG-3′, it will be transcribed as 5′-CCG-3′ in mRNA, and proline-charged tRNA will bind to it at the ribosome. If there is a mutation such that the codon in the template DNA now reads AGG, the mRNA codon will be CCU—and the tRNA that binds it will still carry proline:

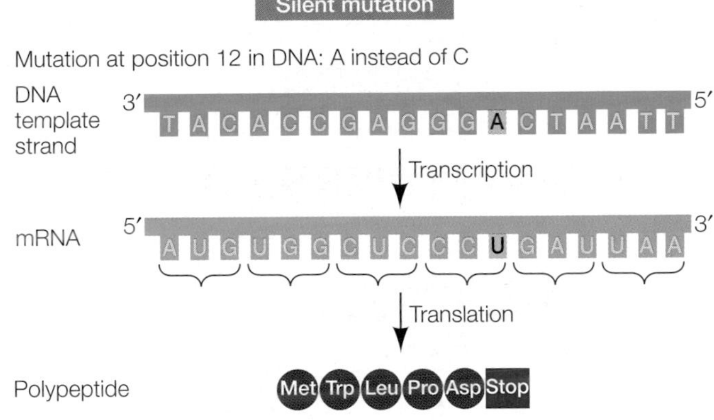

Silent mutations are quite common, and they result in genetic diversity that is not expressed as phenotypic differences.

MISSENSE MUTATIONS In contrast to silent mutations, some base substitutions change the genetic message such that one amino acid substitutes for another in the protein. These changes are called **missense mutations**:

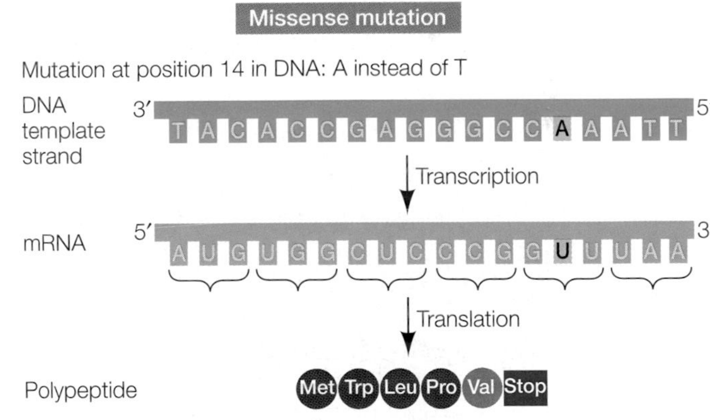

A specific example of a missense mutation is the sickle allele for human β-globin. Sickle-cell disease results from a defect in hemoglobin, a protein in human red blood cells that carries oxygen. The sickle allele of the gene that codes for β-globin subunits of hemoglobin differs from the normal allele by one base, and thus codes for a polypeptide that differs by one amino acid from the normal protein. Individuals who are homozygous for this recessive allele have defective, sickle-shaped red blood cells (**Figure 12.18**) that cause abnormalities in blood circulation and lead to serious illness.

A missense mutation may cause a protein not to function, but often its effect is only to reduce the functional efficiency of the protein. Therefore, individuals carrying missense mutations may survive, even though the affected protein is essential to life. Through evolution, some missense mutations can even improve functional efficiency.

NONSENSE MUTATIONS **Nonsense mutations**, another type of mutation in which one base is substituted for another, are more often disruptive than missense mutations. In a nonsense mutation, the base substitution causes a stop codon, such as UAG, to form in the mRNA product:

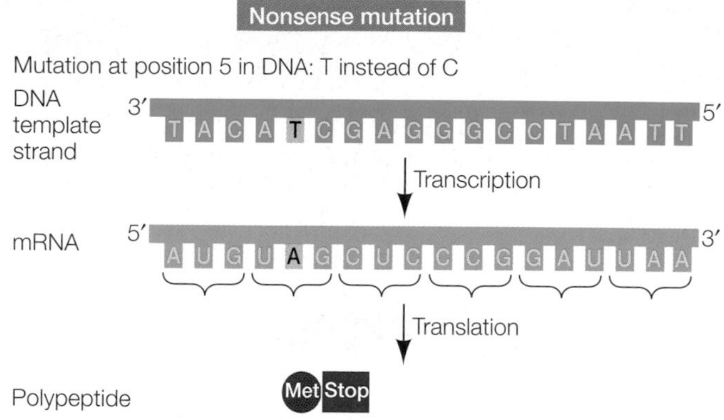

Result: Only one amino acid translated; no protein made

A nonsense mutation results in a shortened protein, since translation does not proceed beyond the point where the mutation occurred. Such short proteins are usually not functional.

FRAME-SHIFT MUTATIONS Not all point mutations are base substitutions. Single base pairs may be inserted into or deleted from DNA. Such mutations are known as **frame-shift mutations** because they interfere with the decoding of the genetic message by throwing it out of register:

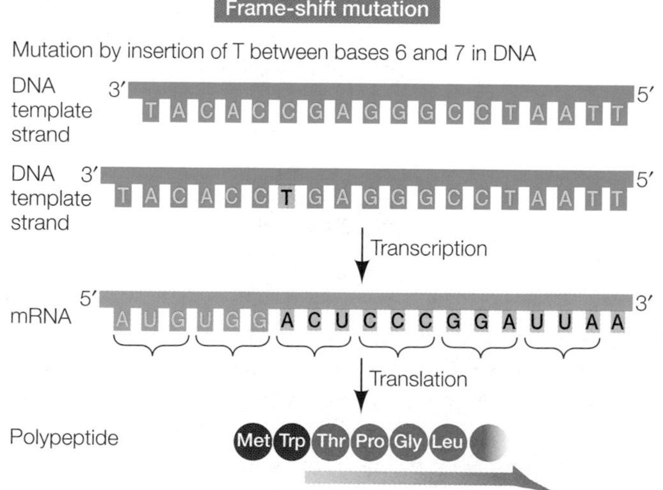

Result: All amino acids changed beyond the insertion

Think again of codons as three-letter words, each corresponding to a particular amino acid. Translation proceeds codon by codon; if a base is added to the mRNA or subtracted from it, translation proceeds perfectly until it comes to the one-base insertion or deletion. From that point on, the three-letter words in the genetic message are one letter out of register. In other words, such mutations shift the "reading frame" of the message. Frame-shift mutations almost always lead to the production of nonfunctional proteins.

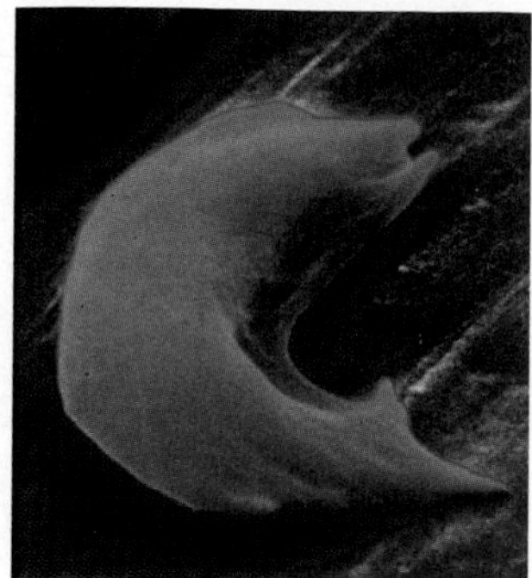

Sickle-cell phenotype

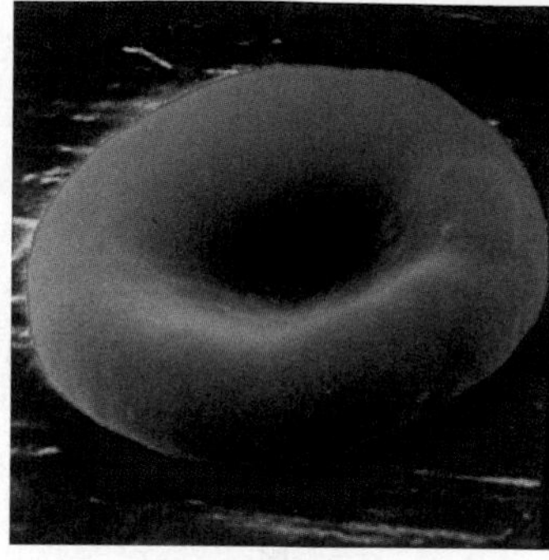

Normal phenotype

12.18 Sickled and Normal Red Blood Cells The misshapen red blood cell on the left is caused by a missense mutation and an incorrect amino acid in one of the two polypeptides of hemoglobin.

Chromosomal mutations are extensive changes in the genetic material

Changes in single nucleotides are not the most dramatic changes that can occur in the genetic material. Whole DNA molecules can break and rejoin, grossly disrupting the sequence of genetic information. There are four types of such chromosomal mutations: *deletions, duplications, inversions,* and *translocations*. These mutations can be caused by severe damage to chromosomes resulting from mutagens or by drastic errors in chromosome replication.

- **Deletions** remove part of the genetic material (**Figure 12.19A**). Like frame-shift point mutations, their consequences can be severe unless they affect unnecessary genes or are masked by the presence, in the same cell, of normal alleles of the deleted genes. It is easy to imagine one mechanism that could produce deletions: a DNA molecule might break at two points, and the two end pieces might rejoin, leaving out the DNA between the breaks.
- **Duplications** can be produced at the same time as deletions (**Figure 12.19B**). Duplication would arise if homologous chromosomes broke at different positions and then reconnected to the wrong partners. One of the two chromosomes produced by this mechanism would lack a segment of DNA (it would have a deletion), and the other would have two copies (a duplication) of the segment that was deleted from the first chromosome.
- **Inversions** can also result from breaking and rejoining of chromosomes. A segment of DNA may be removed and reinserted into the same location in the chromosome, but "flipped" end over end so that it runs in the opposite direction (**Figure 12.19C**). If the break site includes part of a DNA segment that codes for a protein, the resulting protein will be drastically altered and almost certainly nonfunctional.
- **Translocations** result when a segment of DNA breaks off, moves from its chromosome, and is inserted into a different chromosome. Translocations may be reciprocal, as in **Figure 12.19D**, or nonreciprocal, as the mutation involving duplication and deletion in Figure 12.19B illustrates. Translocations often lead to duplications and deletions and may result in sterility if normal chromosome pairing in meiosis cannot occur.

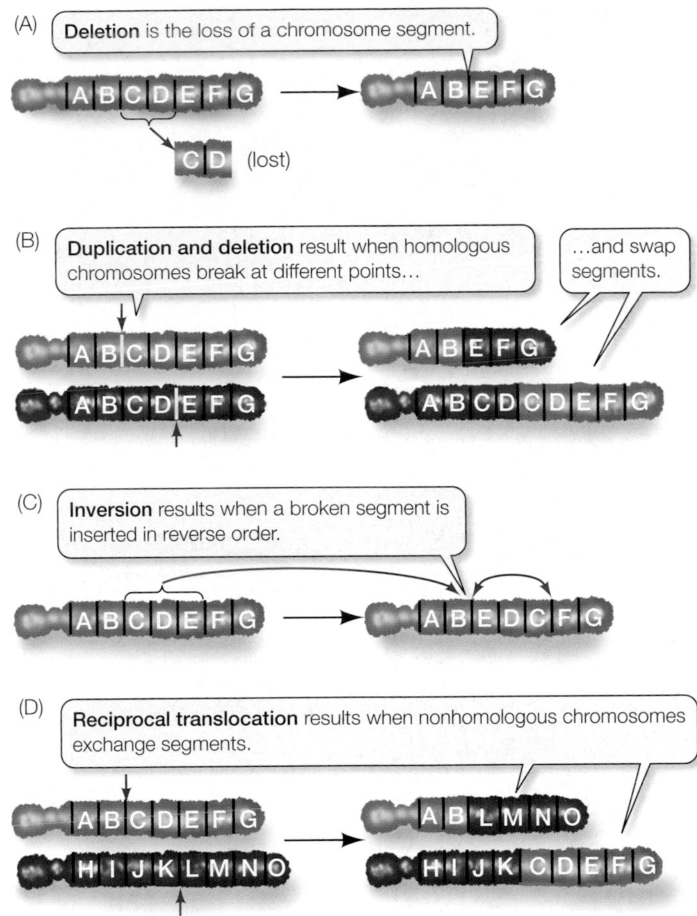

12.19 Chromosomal Mutations Chromosomes may break during replication, and parts of chromosomes may then rejoin incorrectly

Mutations can be spontaneous or induced

It is useful to distinguish two types of mutations in terms of their causes:

- **Spontaneous mutations** are permanent changes in the genetic material that occur without any outside influence. In other words, they occur simply because the machinery of the cell is imperfect.
- **Induced mutations** occur when some agent outside the cell—a **mutagen**—causes a permanent change in DNA.

Spontaneous mutations may occur by several mechanisms:

- *The four nucleotide bases of DNA are somewhat unstable.* They can exist in two different forms (called *tautomers*), one of which is common and one rare. When a base temporarily forms its rare tautomer, it can pair with the wrong base. For example, C normally pairs with G, but if C is in its rare tautomer at the time of DNA replication, it pairs with (and DNA polymerase will insert) A. The result is a point mutation: G → A (**Figure 12.20A,C**).
- *Bases may change because of a chemical reaction.* For example, loss of an amino group in cytosine (a reaction called *deamination*) forms uracil. When DNA replicates, instead of a G opposite what was C, DNA polymerase adds an A (base-pairs with U).
- *DNA polymerase can make errors in replication* (see Section 11.4)—for example, inserting a T opposite a G. Most of these errors are repaired by the proofreading function of the replication complex, but some errors escape detection and become permanent.
- *Meiosis is not perfect.* Nondisjunction—failure of homologous chromosomes to separate during meiosis—can occur, leading to one too many or one too few chromosomes (aneuploidy; see Figure 9.20). Random chromosome breakage and rejoining can produce deletions, duplications, and inversions, or, when involving nonhomologous chromosomes, translocations.

Mutagens can alter DNA by several mechanisms, thus inducing mutations.

- *Some chemicals can alter the nucleotide bases.* For example, nitrous acid (HNO_2) and its relatives can turn cytosine in DNA into uracil by deamination: they convert an amino group on cytosine (—NH_2) into a keto group (—C=O). This alteration has the same result as a spontaneous deamination: instead of a G, DNA polymerase inserts an A (**Figure 12.20B,C**).
- *Some chemicals add groups to the bases.* For instance, benzpyrene, a component of cigarette smoke, adds a large chemical group to guanine, making it unavailable for base pairing. When DNA polymerase reaches such a modified guanine, it inserts any of the four bases; of course, three-fourths of the time the inserted base will not be cytosine, and a mutation results.
- *Radiation damages the genetic material.* Radiation can damage DNA in two ways. First, ionizing radiation (X-rays) produces highly reactive chemical species called *free radicals*, which can change bases in DNA to unrecognizable (by DNA polymerase) forms. It can also break the sugar–phosphate backbone of DNA, causing chromosomal abnormalities. Second, ultraviolet radiation from the sun (or a tanning lamp) is absorbed by thymine in DNA, causing it to form interbase covalent bonds with adjacent nucleotides. This, too, plays havoc with DNA replication.

Mutations have both costs and benefits. The costs are obvious, since mutations often produce an organism that is less fit for its current environment. Somatic mutations can lead to cancer—an effect we will return to in Chapter 17. But germ line mutations are also essential to life, because they provide the genetic diversity on which the forces of evolution act.

Mutations are the raw material of evolution

Without mutation, there would be no evolution. As we will see in Part Five of this book, mutation does not drive evolution, but it provides the genetic diversity on which natural selection and other agents of evolution act.

All mutations are rare events, but mutation frequencies vary from organism to organism and from gene to gene within a given organism. The frequency of mutation is usually much lower than one mutation per 10^4 base pairs per DNA replication, and some-

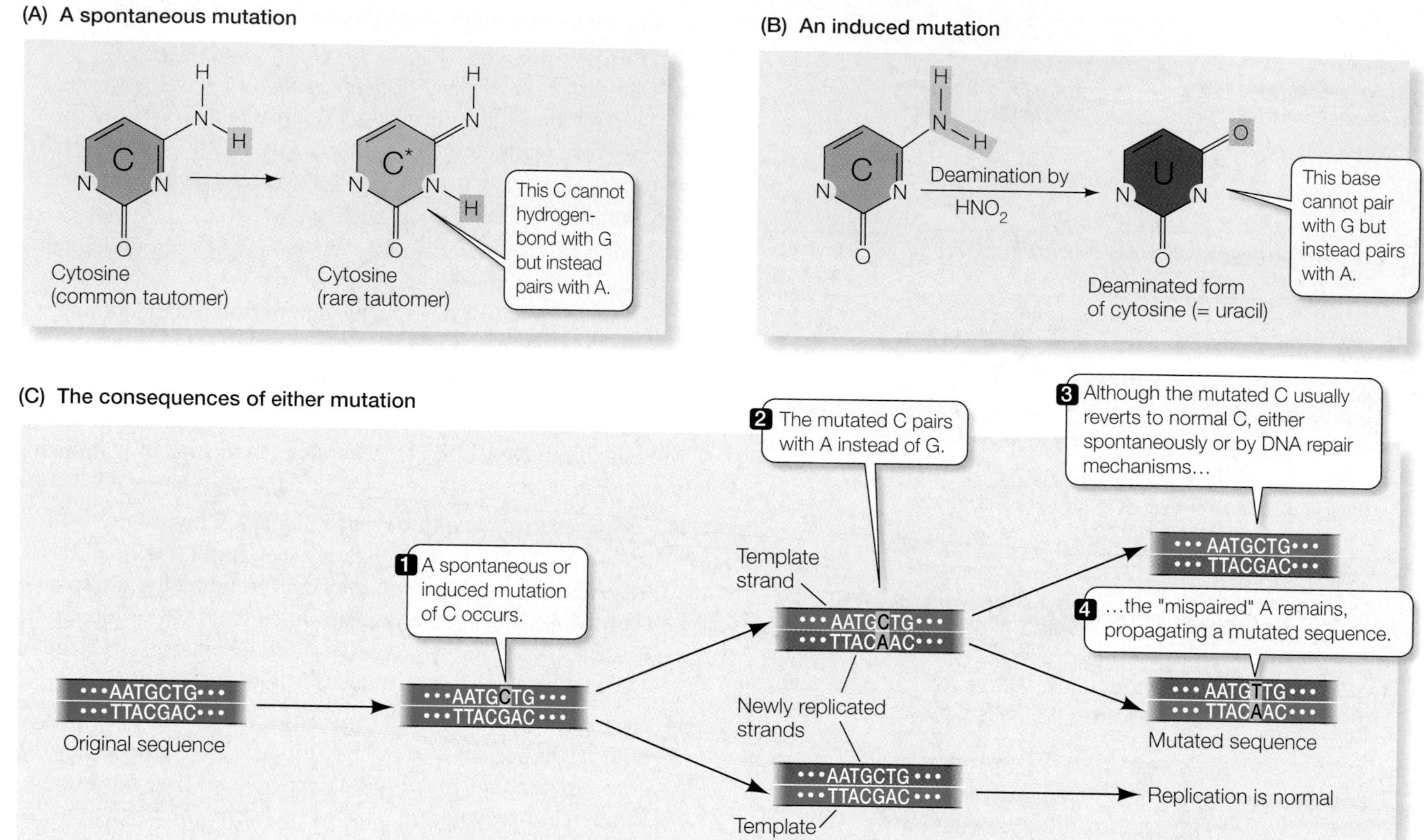

12.20 Spontaneous and Induced Mutations (A) All four nitrogenous bases in DNA exist in both a prevalent (common) form and a rare form. When a base spontaneously forms its rare tautomer, it can pair with a different base. (B) Mutagenic chemicals such as nitrous acid can induce changes in the bases. (C) In both spontaneous and induced mutations, the result is a permanent change in the DNA sequence following replication.

times as low as one mutation per 10^9 base pairs per replication. Most mutations are point mutations in which one nucleotide is substituted for another during the synthesis of a new DNA strand.

Mutations can harm the organism that carries them, or they can be neutral (have no effect on the organism's ability to survive or produce offspring). Once in a while, a mutation improves an organism's adaptation to its environment, or it becomes favorable when environmental conditions change.

Most of the complex creatures living on Earth have more genes than the simpler creatures do. Humans, for example, have 20 times more genes than prokaryotes have. How did these new genes arise? Whole genes might be duplicated, and the bearer of the duplication would have a surplus of genetic information that might be turned to good use. Subsequent mutations in one of the two copies of the gene might not have an adverse effect on survival because the other copy of the gene would continue to produce functional protein. The extra gene might mutate over and over again without ill effect because its original function would be fulfilled by the original copy.

If the random accumulation of mutations in the extra gene led to the production of a useful protein (for example, an enzyme with an altered specificity for the substrates it binds, allowing it to catalyze different—but related—reactions), natural selection would tend to perpetuate the existence of this new gene. New copies of genes may also arise through the activity of *transposable elements*, which are discussed in Chapters 13 and 14.

12.6 RECAP

Mutations are alterations in the nucleotide sequence of DNA. They may take the form of changes in single nucleotides or extensive rearrangements of the genetic material in chromosomes. If they occur in somatic cells, they will be passed on to daughter cells; if they occur in germ line cells, they will be passed on to offspring.

- Can you distinguish the various kinds of chromosomal mutations: deletions, duplications, inversions, and translocations? See p. 276 and Figure 12.19
- Do you understand the difference between spontaneous and induced mutations? Can you give an example of each? See p. 277 and Figure 12.20

CHAPTER SUMMARY

12.1 What is the evidence that genes code for proteins?

Beadle and Tatum's experiments on metabolic enzymes in the bread mold *Neurospora* led to the **one-gene, one-enzyme hypothesis**. We now know that there is a **one-gene, one-polypeptide relationship**. Review Figure 12.1

12.2 How does information flow from genes to proteins?

The **central dogma** of molecular biology states that DNA codes for RNA and RNA codes for protein. Protein does not code for protein, RNA, or DNA. Review Figure 12.2

The process by which the information in DNA is copied to RNA is called **transcription**. The process by which a protein is built from the information in RNA is called **translation**. Review Figure 12.3

Certain RNA viruses are exceptions to the central dogma. These **retroviruses** synthesize DNA from RNA in **reverse transcription**.

The product of transcription, **messenger RNA** (**mRNA**), travels from the nucleus to the cytoplasm. **Transfer RNA** (**tRNA**) molecules translate the genetic information in mRNA into a corresponding sequence of amino acids to produce a polypeptide.

12.3 How is the information content in DNA transcribed to produce RNA?

In a given gene, only one of the two strands of DNA (the **template strand**) acts as a template for transcription. **RNA polymerase** is the catalyst for transcription.

RNA transcription from DNA proceeds in three steps: **initiation**, **elongation**, and **termination**. Review Figure 12.5, Web/CD Tutorial 12.1

Initiation requires a **promoter**, to which DNA polymerase binds. Part of each promoter is the **initiation site**, where transcription begins.

Elongation of the RNA molecule proceeds from the 5′ to 3′ end.

Particular base sequences specify termination, at which point transcription ends and the RNA transcript separates from the DNA template.

The **genetic code** is a "language" of triplets of mRNA nucleotide bases (**codons**) corresponding to 20 specific amino acids; there are **start** and **stop codons** as well. The code is redundant (an amino acid may be represented by more than one codon), but not ambiguous (no single codon represents more than one amino acid). Review Figure 12.6, Web/CD Activity 12.1, Web/CD Tutorial 12.2

12.4 How is RNA translated into proteins?

See Web/CD Tutorial 12.3

In translation, amino acids are linked in an order specified by the codons in mRNA. This task is achieved by tRNAs, which bind to (are charged with) specific amino acids.

Each tRNA species has an amino acid attachment site as well as an **anticodon** complementary to a specific mRNA codon. A specific activating enzyme charges each tRNA with its specific amino acid. Review Figures 12.8 and 12.9

The **ribosome** is the molecular workbench where translation takes place. It has one large and one small subunit, both made of **ribosomal RNA** and proteins.

Three sites on the large subunit of the ribosome interact with tRNA anticodons. The A site is where the charged tRNA anticodon binds to the mRNA codon; the P site is where the tRNA adds its amino acid to the growing polypeptide chain; and the E site is where the tRNA is released. Review Figure 12.10

Translation occurs in three steps: **initiation**, **elongation**, and **termination**.

The **initiation complex** consists of tRNA bearing the first amino acid, the small ribosomal subunit, and mRNA. The mRNA binds to a specific complementary sequence on rRNA. Review Figure 12.11

The growing polypeptide chain is elongated by the formation of peptide bonds between amino acids, catalyzed by RNA. Review Figure 12.12

When a stop codon reaches the A site, it terminates translation by binding a release factor. Review Figure 12.13

In a **polysome**, more than one ribosome moves along a strand of mRNA at one time. Review Figure 12.14

12.5 What happens to polypeptides after translation?

Signal sequences of amino acids direct polypeptides to their cellular destinations. Review Figure 12.15

Destinations in the cytoplasm include organelles, which proteins enter upon recognizing surface receptors called **docking proteins**.

Proteins "addressed" to the ER bind to a **signal recognition particle**. Review Figure 12.16

Posttranslational modifications of polypeptides include **proteolysis**, in which a polypeptide is cut into smaller fragments; **glycosylation**, in which sugars are added; and **phosphorylation**, in which phosphate groups are added. Review Figure 12.17

12.6 What are mutations?

Mutations are heritable changes in DNA. **Somatic mutations** are passed on to daughter cells, but only **germ line mutations** are passed on to sexually produced offspring.

Point mutations result from alterations in single base pairs of DNA. **Silent mutations** result in no change in amino acids when the altered mRNA is translated into a polypeptide. **Missense**, **nonsense**, and **frame-shift** mutations do cause changes in the amino acids produced. Review pp. 275–276

Chromosomal mutations (**deletions**, **duplications**, **inversions**, or **translocations**) involve large regions of a chromosome. Review Figure 12.19

Spontaneous mutations occur because of instabilities in DNA or chromosomes. **Induced** mutations occur when a **mutagen** damages DNA. Review Figure 12.20

Mutations, although often detrimental to an individual organism, are the raw material of evolution.

SELF-QUIZ

1. Which of the following is *not* a difference between RNA and DNA?
   a. RNA has uracil; DNA has thymine.
   b. RNA has ribose; DNA has deoxyribose.
   c. RNA has five bases; DNA has four.
   d. RNA is a single polynucleotide strand; DNA is a double strand.
   e. RNA is relatively smaller than human chromosomal DNA.
2. Normally, *Neurospora* can synthesize all 20 amino acids. A certain strain of this mold cannot grow in minimal nutritional medium, but grows only when the amino acid leucine is added to the medium. This strain
   a. is dependent on leucine for energy.
   b. has a mutation affecting the biochemical pathway leading to the synthesis of proteins.
   c. has a mutation affecting the biochemical pathway leading to the synthesis of all 20 amino acids.
   d. has a mutation affecting the biochemical pathway leading to the synthesis of leucine.
   e. has a mutation affecting the biochemical pathways leading to the syntheses of 19 of the 20 amino acids.
3. An mRNA has the sequence 5′-AUGAAAUCCUAG-3′. What is the template DNA strand for this sequence?
   a. 5′-TACTTTAGGATC-3′
   b. 5′-ATGAAATCCTAG-3′
   c. 5′-GATCCTAAAGTA-3′
   d. 5′-TACAAATCCTAG-3′
   e. 5′-CTAGGATTTCAT-3′
4. The adapters that allow translation of the four-letter nucleic acid language into the 20-letter protein language are called
   a. aminoacyl-tRNA synthetases.
   b. transfer RNAs.
   c. ribosomal RNAs.
   d. messenger RNAs.
   e. ribosomes.
5. At a certain location in a gene, the non-template strand of DNA has the sequence GAA. A mutation alters the triplet to GAG. This type of mutation is called
   a. silent.
   b. missense.
   c. nonsense.
   d. frame-shift.
   e. translocation.
6. Transcription
   a. produces only mRNA.
   b. requires ribosomes.
   c. requires tRNAs.
   d. produces RNA growing from the 5′ end to the 3′ end.
   e. takes place only in eukaryotes.
7. Which statement about translation is *not* true?
   a. It is RNA-directed polypeptide synthesis.
   b. An mRNA molecule can be translated by only one ribosome at a time.
   c. The same genetic code operates in almost all organisms and organelles.
   d. Any ribosome can be used in the translation of any mRNA.
   e. There are both start and stop codons.
8. Which statement about RNA is *not* true?
   a. Transfer RNA functions in translation.
   b. Ribosomal RNA functions in translation.
   c. RNAs are produced by transcription.
   d. Messenger RNAs are produced on ribosomes.
   e. DNA codes for mRNA, tRNA, and rRNA.
9. The genetic code
   a. is different for prokaryotes and eukaryotes.
   b. has changed during the course of recent evolution.
   c. has 64 codons that code for amino acids.
   d. has more than one codon for many amino acids
   e. is ambiguous.
10. A mutation that results in the codon UAG where there had been UGG is
    a. a nonsense mutation.
    b. a missense mutation.
    c. a frame-shift mutation.
    d. a large-scale mutation.
    e. unlikely to have a significant effect.

FOR DISCUSSION

1. How is it possible that a point mutation, consisting of the replacement of a single nucleotide base in DNA by a different base, might not result in an error in protein production?
2. Har Gobind Khorana at the University of Wisconsin synthesized artificial mRNAs such as poly CA (CACA …) and poly CAA (CAACAACAA …). He found that poly CA codes for a polypeptide consisting of threonine (Thr) and histidine (His), in alternation (His–Thr–His–Thr …). There are two possible codons in poly CA, CAC and ACA. One of these must code for histidine and the other for threonine—but which is which? The answer comes from results with poly CAA, which produces three different polypeptides: poly Thr, poly Gln (glutamine), and poly Asn (asparagine). (An artificial mRNA can be read, inefficiently, beginning at any point in the chain; there is no specific initiation signal. Thus poly CAA can be read as a polymer of CAA, of ACA, or of AAC.) Compare the results of the poly CA and poly CAA experiments, and determine which codon codes for threonine and which for histidine.
3. Look back at Question 2. Using the genetic code in Figure 12.6 as a guide, deduce what results Khorana would have obtained had he used poly UG and poly UGG as artificial messengers. In fact, very few such artificial messengers would have given useful results. For an example of what could happen, consider poly CG and poly CGG. If poly CG were the messenger, a mixed polypeptide of arginine and alanine (Arg–Ala–Ala–Arg …) would be obtained; poly CGG would give three polypeptides: poly Arg, poly Ala, and poly Gly (glycine). Can any codons be determined from only these data? Explain.
4. Errors in transcription occur about 100,000 times as often as do errors in DNA replication. Why can this high rate be tolerated in RNA synthesis but not in DNA synthesis?

FOR INVESTIGATION

Beadle and Tatum's experiments showed that a biochemical pathway could be deduced from mutant strains. In bacteria, the biosynthesis of the amino acid tryptophan (T) from the precursor chorismate (C) involves four intermediate chemical compounds, which we will call D, E, F, and G. Here are the phenotypes of various mutant strains, where each strain has a mutation in a gene for a different enzyme, + means growth with the stated addition to the medium, and 0 means no growth. Based on these data, order the compounds (C, D, E, F, G, and T) and enzymes (1, 2, 3, 4, and 5) in a biochemical pathway.

	ADDITION TO MEDIUM					
MUTANT STRAIN	C	D	E	F	G	T
1	0	0	0	0	+	+
2	0	+	+	0	+	+
3	0	+	0	0	+	+
4	0	+	+	+	+	+
5	0	0	0	0	0	+

CHAPTER 13 The Genetics of Viruses and Prokaryotes

Mutation of a bird virus results in human infection

On May 9, 1997, a 3-year-old boy in Hong Kong developed a cough and fever. His physician treated the boy with antibiotics and aspirin, but the fever got worse. The boy was hospitalized on May 15. Unfortunately, his fever progressed to lung failure, and he died 6 days later.

In an attempt to find the cause of the boy's death, researchers took fluid drawn from his lungs prior to his death and added it to mammalian kidney cells in a laboratory culture. Two days later, the cells were dying, with each cell releasing hundreds of influenza virus particles. Public health professionals at Hong Kong Hospital searched for signs that the boy's flu was one of the strains that had infected people earlier that winter. They tested for glycoproteins on the virus surface that would allow it to attach to human cells. None of the tests worked, indicating that the virus was not typical human flu. What was this deadly new virus?

By August they were able to determine that the little boy had been infected with H5N1, a flu virus previously known to infect only chickens. The boy's day care provider had kept chicks for the children to play with, and several of the chicks died. Genetic tests showed that the nucleotide sequences of the viruses in the birds and in the boy matched. Comparison of this sequence with that of flu virus confined to birds revealed that a mutation in the gene for the viral surface glycoprotein had made the avian virus capable of binding to and infecting human cells.

By December, more human cases of "bird flu" appeared in Hong Kong; of 18 people infected, 6 died. Although there was no clear connection between the infected individuals and diseased birds, all of the victims had visited live poultry markets in the week before developing symptoms. Tests of live chickens from these markets revealed a high rate of H5N1 infection. Hong Kong health officials immediately closed the border to mainland China and directed the slaughter of every chicken on the island. Within days, over 1.5 million birds were killed and, in all likelihood, a major epidemic was avoided.

But the bird flu virus was not wiped out by this action. H5N1 has been found in birds other than chickens, and on continents other than Asia. There have been more cases of human infection, but so far prompt action to kill infected birds has prevented widespread occurrence of the disease.

Chicken Farming With the spread of bird flu throughout the Far East, raising chickens there has become a hazardous occupation.

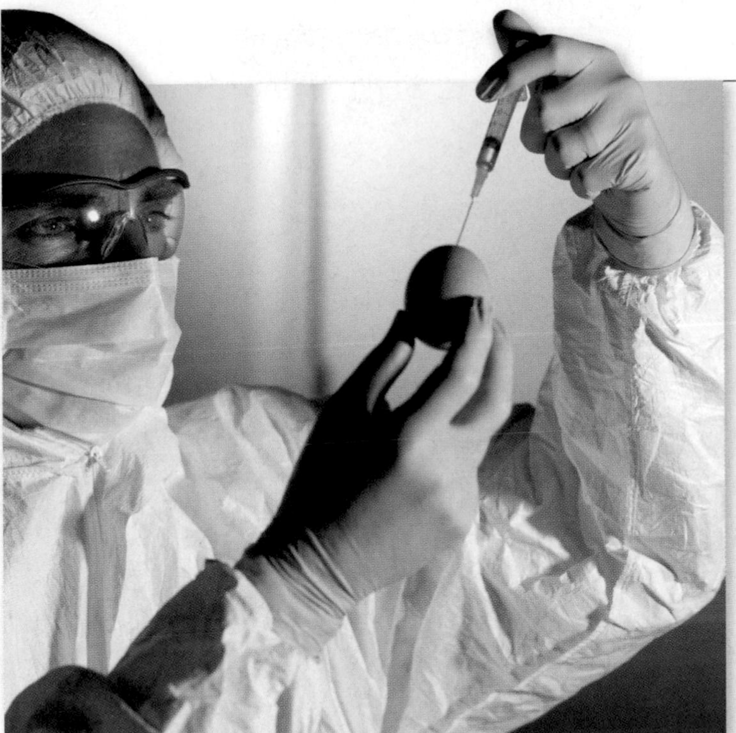

Searching for a Vaccine Avian flu virus is injected into an egg in part of the procedure to develop a vaccine. The development of flu vaccines is always a race against time. Because viruses evolve so rapidly, new vaccines are constantly needed.

We haven't always been so lucky. The "Spanish flu" epidemic of 1918, which may have started with a single soldier, spread to Europe with U.S. troops fighting in World War I. The resulting pandemic led to 40 million deaths worldwide. Flu pandemics in 1957 and 1968 killed a million people each. In all three pandemics, a single gene mutation made it possible for an animal influenza virus to infect people.

When will the next flu pandemic occur? The answers lie in the molecular genetics of viruses and their evolution.

IN THIS CHAPTER we will describe the growth and reproduction of viruses and bacteria, and we will see how these organisms transmit their genes. First we'll examine the nature of viruses and see how they infect, reproduce in, and express their genes in host cells. Next we'll see how prokaryotes can exchange genes. We will also describe how the expression of prokaryotic genes is regulated and what DNA sequencing has revealed about the prokaryotic genome.

CHAPTER OUTLINE

13.1 How Do Viruses Reproduce and Transmit Genes?

Many prokaryotes and viruses have served as model organisms for studying the structure, function, and transmission of genes (**Figure 13.1**). They are excellent model organisms compared with more complex eukaryotes for several reasons:

- *Their genomes are small*. A typical bacterium contains about a thousandth as much DNA as a single human cell, and a typical bacteriophage contains about a hundredth as much DNA as a bacterium.
- *They reproduce quickly*. A single milliliter of growth medium can contain more than 10^9 cells of the bacterium *Escherichia coli*, and its numbers can double every 25 minutes.
- *They are usually haploid*, which makes genetic analyses easier.

Viruses are not cells

Unlike the organisms that make up the three domains of the living world, **viruses** are *acellular*; that is, they are not cells and do not consist of cells. Most viruses are composed of only nucleic acid and a few proteins. Viruses do not carry out two of the basic functions of cellular life: they do not regulate the transport of substances into and out of themselves by membranes, and they perform no metabolic functions—that is, they do not take in nutrients or expel wastes. But they can reproduce in systems that do perform these functions: living cells.

Most viruses are much smaller than even the smallest bacteria (**Table 13.1**). Viruses have become well understood only within the last half century, but the first step on this path of discovery was taken by the Russian botanist Dmitri Ivanovsky in 1892. He was trying to find the cause of tobacco mosaic disease, which results in the destruction of photosynthetic tissues in plants and can devastate a tobacco crop. Ivanovsky passed an extract of diseased tobacco leaves through a fine porcelain filter, a technique that had been used previously by physicians and veterinarians to isolate disease-causing bacteria.

To Ivanovsky's surprise, the disease agent in this case was not retained on the filter. It passed through, and the liquid fil-

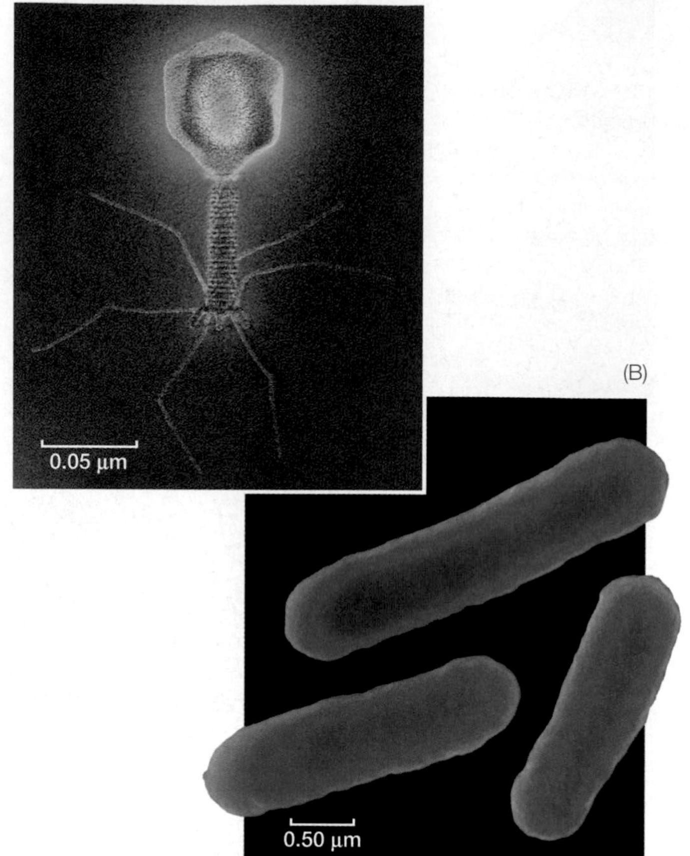

13.1 Model Organisms Viruses and bacteria are valuable model organisms for studying genetics and molecular biology. (A) Bacteriophage T4, a commonly studied virus, is about 10 times smaller than (B) *Escherichia coli*, a commonly studied bacterium.

trate still caused tobacco mosaic disease. But instead of concluding that the agent was smaller than a bacterium, he assumed that his filter was faulty. Louis Pasteur's recent demonstration that bacteria could cause disease was the dominant idea at the time, and Ivanovsky chose not to challenge it. But, as often happens in science, someone soon came along who did. In 1898, the Dutch microbiologist Martinus Beijerinck repeated Ivanovsky's experiment and also showed that the tobacco mosaic disease agent could diffuse through an agar gel. He called the tiny agent *contagium vivum fluidum*, which later became shortened to *virus*.

Almost 40 years later, the disease agent was crystallized by Wendell Stanley (who won the Nobel prize for his efforts). The crystalline viral preparation became infectious again when it was dissolved. It was soon shown that crystallized viral preparations consist of proteins and nucleic acids. Still, it was not until the electron microscope allowed direct observation of viruses in the 1950s that it became clear how different they are from bacteria and other organisms.

Viruses reproduce only with the help of living cells

Whole viruses never arise directly from preexisting viruses. Viruses are *obligate intracellular parasites*; that is, they develop and reproduce only within the cells of specific hosts. The cells of animals, plants, fungi, protists, and prokaryotes (both bacteria and archaea) can serve as hosts to viruses. Viruses use the host's DNA replication and protein synthesis machinery to reproduce themselves, usually destroying the host cell in the process. The host cell releases progeny viruses, which then infect new hosts.

Viruses outside of host cells exist as individual particles called **virions**. The virion, the basic unit of a virus, consists of a central core of either DNA or RNA (but not both) surrounded by a **capsid**, or coat, composed of one or more proteins. Because they lack the distinctive cell wall and ribosomal biochemistry of bacteria, *viruses are not affected by antibiotics* that target these structures.

TABLE 13.1

Relative Sizes of Microorganisms

MICROORGANISM	TYPE	TYPICAL SIZE RANGE (μm^3)
Protists	Eukaryote	5,000–50,000
Photosynthetic bacteria	Prokaryote	5–50
Spirochetes	Prokaryote	0.1–2.0
Mycoplasmas	Prokaryote	0.01–0.1
Poxviruses	Virus	0.01
Influenza virus	Virus	0.0005
Poliovirus	Virus	0.00001

Viruses are classified according to several characteristics:

- Whether the genome is DNA or RNA
- Whether the nucleic acid is single-stranded or double-stranded
- Whether the shape of the virion is simple or complex
- Whether the virion is surrounded by a membrane

Some of these variations are shown in **Figure 13.2**.

Another important characteristic of a virus is the type of organism it infects and the manner in which the infection occurs. Most viruses simply infect and immediately replicate in their host cells. Others can infect a host cell but postpone reproduction, remaining inactive in the host cell until conditions for replication are favorable.

Bacteriophage reproduce by a lytic cycle or a lysogenic cycle

Viruses that infect bacteria are known as **bacteriophage** or **phage** (Greek *phagos*, "one that eats"; note that "phage" is both singular and plural). They recognize their prospective hosts by means of proteins in the capsid, which bind to specific receptor proteins or carbohydrates in the host's cell wall. The virions, whose nucleic acids must penetrate the cell wall for successful infection, are often equipped with tail assemblies that inject the phage's nucleic acid through the cell wall into the host bacterium. After the nucleic acid has entered the host, one of two things happens, depending on the type of phage:

- The virus reproduces immediately and kills the host cell.
- The virus postpones reproduction by integrating its nucleic acid into the host cell's genome.

13.2 Virions Come in Various Shapes (A) The tobacco mosaic virus (a plant virus) consists of an inner helix of RNA covered with a helical array of protein molecules. (B) Many animal viruses, such as this adenovirus, have a capsid as an outer shell. Inside the capsid is a spherical mass of proteins and DNA. (C) In some viruses, such as this herpes virus, a membrane envelope surrounds the capsid.

The Hershey–Chase experiment (see Figure 11.4) involved the first type of viral reproductive cycle, called the **lytic cycle**, so named because the infected bacterium *lyses* (bursts), releasing progeny phage. The alternative fate is the **lysogenic cycle**, in which the infected bacterium does not lyse, but instead harbors the viral nucleic acid in its genome and passes it along over many generations until conditions trigger a lytic cycle. Most viruses reproduce only by the lytic cycle; others undergo both types of reproductive cycles (**Figure 13.3**).

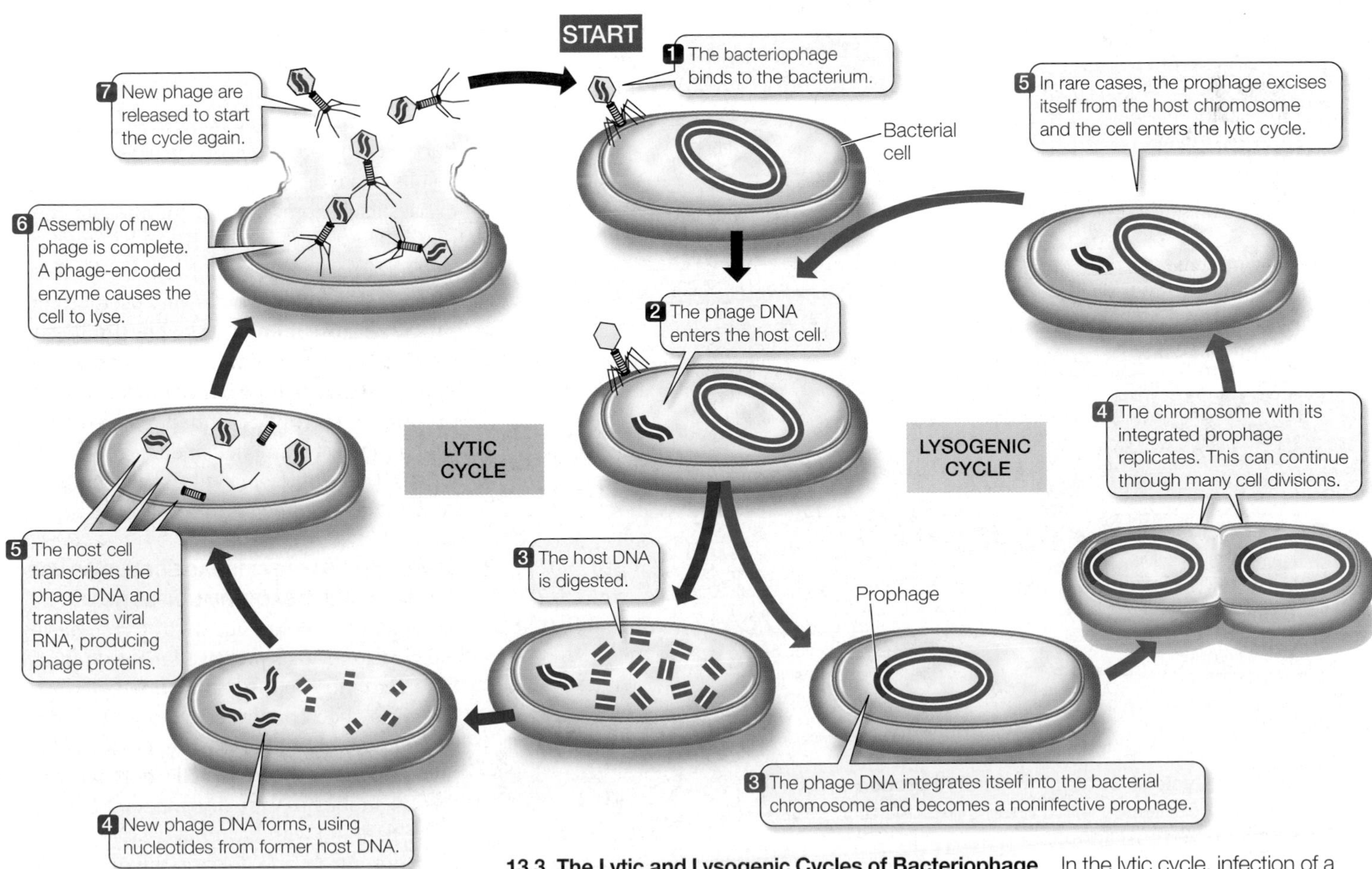

START
1 The bacteriophage binds to the bacterium.
Bacterial cell
2 The phage DNA enters the host cell.
LYTIC CYCLE
3 The host DNA is digested.
4 New phage DNA forms, using nucleotides from former host DNA.
5 The host cell transcribes the phage DNA and translates viral RNA, producing phage proteins.
6 Assembly of new phage is complete. A phage-encoded enzyme causes the cell to lyse.
7 New phage are released to start the cycle again.
LYSOGENIC CYCLE
Prophage
3 The phage DNA integrates itself into the bacterial chromosome and becomes a noninfective prophage.
4 The chromosome with its integrated prophage replicates. This can continue through many cell divisions.
5 In rare cases, the prophage excises itself from the host chromosome and the cell enters the lytic cycle.

13.3 The Lytic and Lysogenic Cycles of Bacteriophage In the lytic cycle, infection of a bacterium by viral DNA leads directly to the multiplication of the virus and lysis of the host cell. In the lysogenic cycle, an inactive prophage is replicated as part of the host's chromosome.

If bacteria are everywhere, then bacteriophage are everywhere, and in much larger numbers. There are up to 100 million viruses in a milliliter of seawater; similar numbers are found in soil water. Scientists estimate that in nature, there are some 10 phage per every bacterium.

THE LYTIC CYCLE A virus that reproduces only by the lytic cycle is called a **virulent** virus. Once a virulent phage has bound to and injected its nucleic acid into a bacterium, that nucleic acid takes over the host's synthetic machinery. It does so in two stages (**Figure 13.4**):

- The viral genome contains promoter sequences that attract host RNA polymerase. In the *early stage*, viral genes that lie adjacent to this promoter are transcribed. These *early genes* often code for proteins that shut down host transcription, stimulate viral genome replication, and stimulate viral gene transcription. Viral nuclease enzymes digest the host's chromosome, providing nucleotides for the synthesis of viral genomes.
- In the *late stage*, viral *late genes*, which code for the proteins of the viral capsid and those that lyse the host cell to release the new virions, are transcribed.

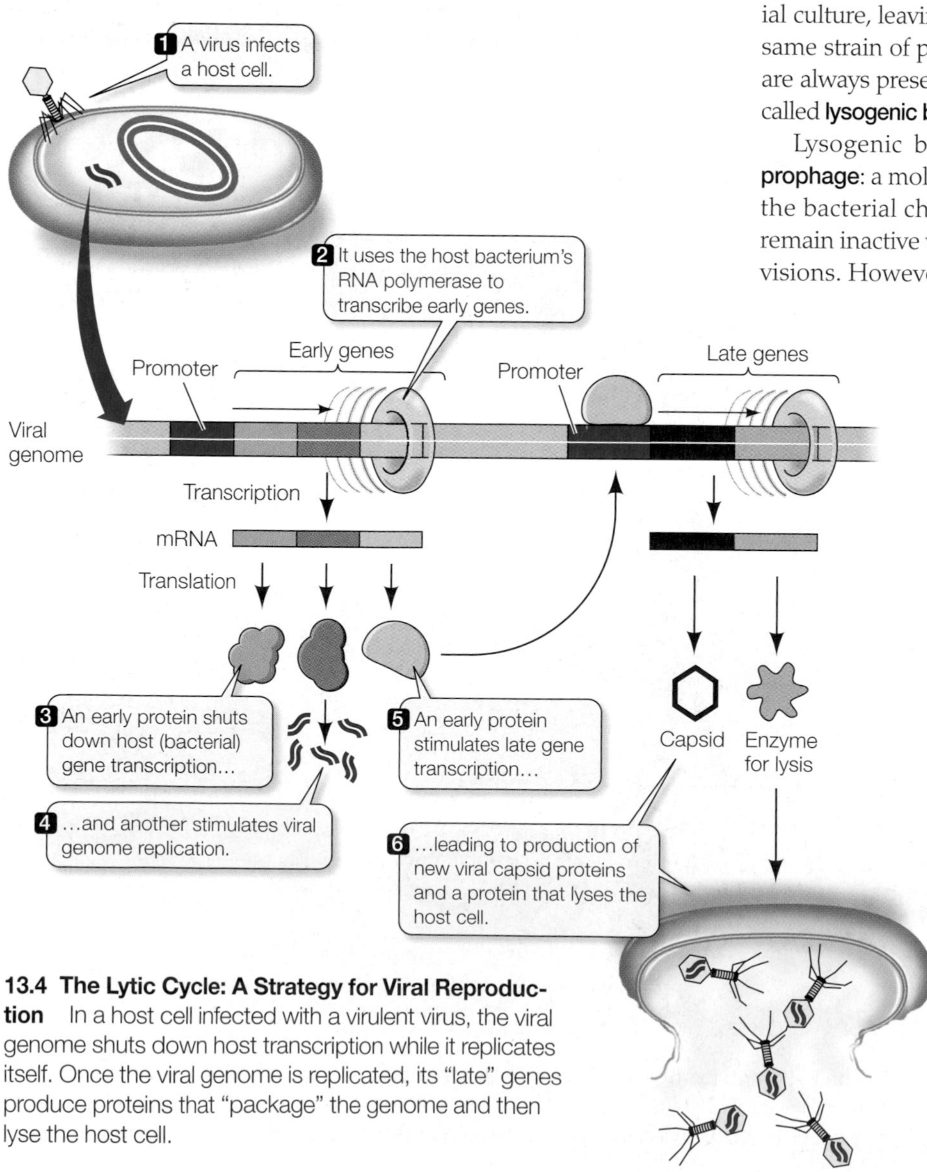

The whole process—from binding and infection to release of phage by lysis of the host cell—takes about half an hour. However, this sequence of transcriptional events is carefully controlled: premature lysis of the host cell before virions are assembled and ready for release would stop the infection.

Two viruses can infect a cell at the same time. This is an unusual event because once a lytic cycle is under way, there is usually not enough time for an additional infection. In addition, an early viral protein may prevent further infections. When two different viral genomes are in the same host cell, however, there exists the possibility of genetic recombination by crossing over (as in prophase I of meiosis in eukaryotes; see Figure 9.18). This phenomenon enables genetically different but related viruses to swap genes and create new strains.

THE LYSOGENIC CYCLE Viral infection does not always result in lysis of the host cell. Some phage seem to disappear from a bacterial culture, leaving the bacteria "immune" to further attack by the same strain of phage. In such cultures, however, a few free phage are always present. Bacteria harboring viruses that are not lytic are called **lysogenic bacteria**, and the viruses are called **temperate** viruses.

Lysogenic bacteria contain a noninfective entity called a **prophage**: a molecule of phage DNA that has been integrated into the bacterial chromosome (see Figure 13.3). The prophage can remain inactive within the bacterial genome through many cell divisions. However, an occasional lysogenic bacterium can be induced to activate its prophage. This activation results in a lytic cycle, in which the prophage excises itself from the host chromosome and reproduces.

The capacity to switch between the lysogenic and the lytic cycle is very useful to the phage because it enhances opportunities to produce the maximum number of progeny viruses. When its host cell is growing and reproducing rapidly, the phage is lysogenic. When the host is stressed or damaged by mutagens, the prophage is released from its inactive state, and the lytic cycle proceeds.

USING LYTIC BACTERIOPHAGE COULD BE USEFUL IN TREATING BACTERIAL INFECTIONS Because lytic bacteriophage destroy their bacterial hosts, scientists perceived that they might be useful in treating infectious diseases caused by bacteria. Indeed, one of the early discoverers of phage, the French-Canadian microbiologist Felix D'Herelle, noted in 1917 (before antibiotics were discovered) that when some patients with bacterial dysentery were recovering from the disease, the quantity of phage near the bacteria was much higher than when the disease was at its peak.

13.4 The Lytic Cycle: A Strategy for Viral Reproduction In a host cell infected with a virulent virus, the viral genome shuts down host transcription while it replicates itself. Once the viral genome is replicated, its "late" genes produce proteins that "package" the genome and then lyse the host cell.

D'Herelle tried using phage to control infections of chickens by the bacterium *Salmonella gallinarum*. To do this, he divided chickens into two groups, one that was given phage and another that was not. Then he exposed both groups to the infectious bacteria. The phage-protected group did not get the bacterial disease. Later, he used phage successfully to treat people in Egypt infected with plague-causing bacteria and people in India with infectious cholera.

The emergence of antibiotics and of phage-resistant bacteria reduced interest in phage therapy. However, interest has revived now that bacterial resistance to antibiotics is becoming common. Bacteriophage are even being investigated as a means of treating edible fruits and vegetables to prevent bacterial contamination. In addition to advancing our understanding of fundamental biological processes, the study of bacteriophage has opened the door to investigations of viruses that infect eukaryotes.

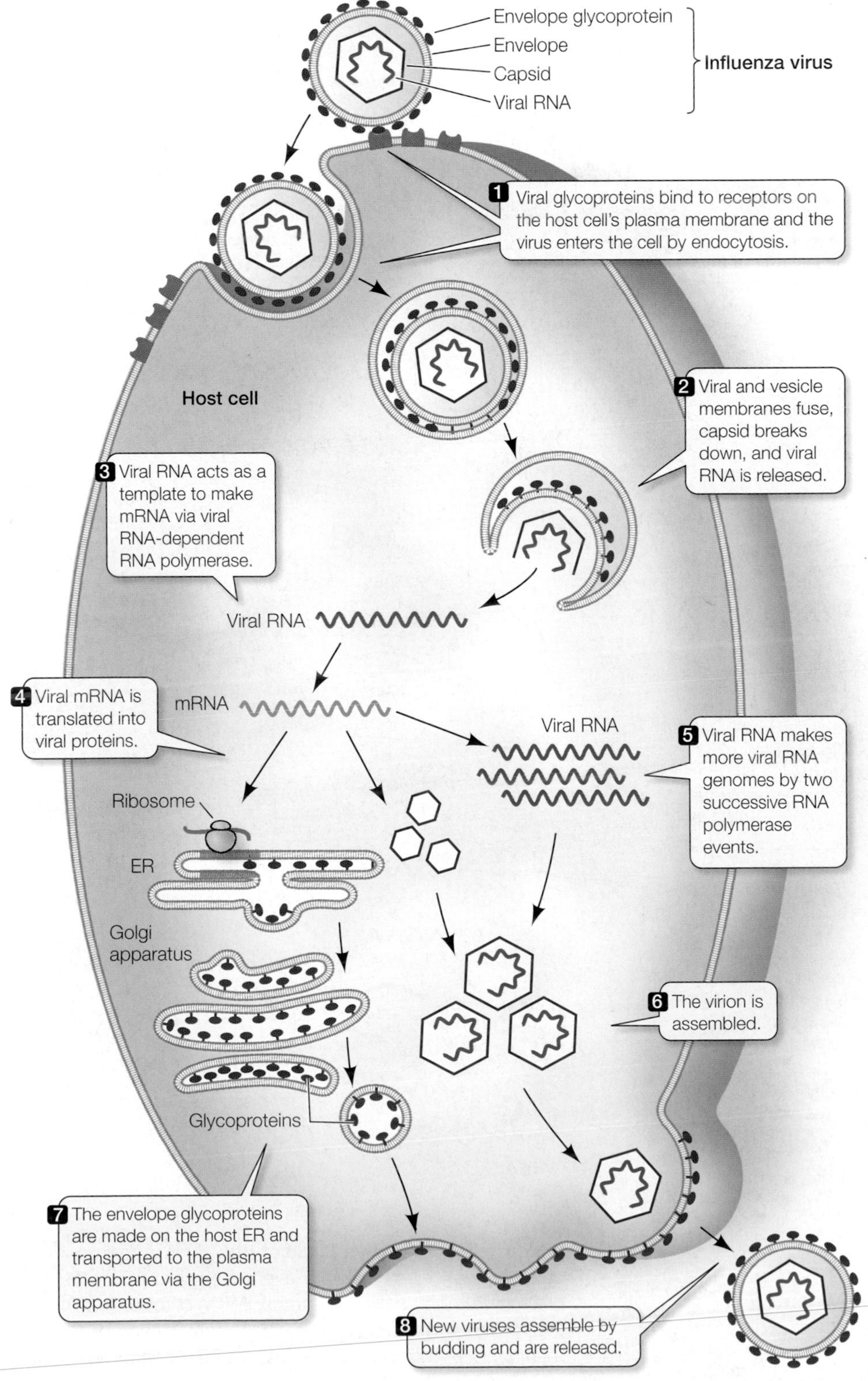

Envelope glycoprotein
Envelope
Capsid
Viral RNA
Influenza virus
1 Viral glycoproteins bind to receptors on the host cell's plasma membrane and the virus enters the cell by endocytosis.
Host cell
2 Viral and vesicle membranes fuse, capsid breaks down, and viral RNA is released.
3 Viral RNA acts as a template to make mRNA via viral RNA-dependent RNA polymerase.
Viral RNA
4 Viral mRNA is translated into viral proteins.
mRNA
Viral RNA
5 Viral RNA makes more viral RNA genomes by two successive RNA polymerase events.
Ribosome
ER
Golgi apparatus
6 The virion is assembled.
Glycoproteins
7 The envelope glycoproteins are made on the host ER and transported to the plasma membrane via the Golgi apparatus.
8 New viruses assemble by budding and are released.

13.5 The Reproductive Cycle of the Influenza Virus The enveloped influenza virus is taken into the host cell by endocytosis. Once inside, fusion of the vesicle and viral membranes releases the viral genome, which replicates and assembles new virions.

Animal viruses have diverse reproductive cycles

Almost all vertebrates are susceptible to viral infections, but among invertebrates, such infections are common only in arthropods (the group that includes insects and crustaceans). One group of viruses, called *arboviruses* (short for "arthropod-borne viruses"), is transmitted to vertebrates through insect bites. Although they are carried within the arthropod host's cells, arboviruses apparently do not harm that host; they affect only the bitten and infected vertebrate. The arthropod acts as a **vector**—an intermediate carrier—by transmitting the virus from one vertebrate host to another.

Animal viruses are very diverse. Some are just particles consisting of proteins surrounding a nucleic acid. Others have a membrane derived from the host cell's plasma membrane and are called *enveloped* viruses. Some animal viruses have DNA as their genetic material; others have RNA. In most cases, the viral genome is small, coding for only a few proteins.

Like that of bacteriophage, the lytic cycle of animal viruses can be divided into early and late stages (see Figure 13.4). Animal viruses enter cells in one of three ways:

- A naked virion (without a membrane envelope) is taken up by endocytosis, which traps it within a membranous vesicle inside the host cell. The membrane of the vesicle breaks down, releasing the virion into the cytoplasm, and the host cell digests the protein capsid, liberating the viral nucleic acid, which takes charge of the host cell.
- Enveloped viruses may also be taken up by endocytosis (see Figure 13.5) and released from a vesicle. In these viruses,

the viral membrane is studded with glycoproteins that bind to receptors on the host cell's plasma membrane.

- More commonly, the membranes of the host and the enveloped virus fuse, releasing the rest of the virion into the cell (see Figure 13.6).

Following viral reproduction, enveloped viruses usually escape from the host cell by a budding process in which they acquire a membrane envelope from the host cell's plasma membrane.

Both influenza virus and human immunodeficiency virus (HIV) are single-stranded RNA viruses, yet their life cycles illustrate two very different strategies of infection and genome replication. Influenza virus is taken up into a membrane vesicle by endocytosis (**Figure 13.5**). Fusion of the viral and vesicle membranes releases the virion into the cell. The virus carries its own enzyme to replicate its RNA genome. This enzyme is an RNA-dependent RNA polymerase using RNA as a template (as opposed to the DNA-dependent RNA polymerases that use a DNA template; see Section 12.3). This newly synthesized viral RNA strand is then used as mRNA to make, by complementary base pairing, more copies of the viral genome.

Retroviruses such as HIV have a more complex reproductive cycle (**Figure 13.6**). The virus enters a host cell by direct fusion of the viral envelope and the host plasma membrane. A distinctive feature of the retroviral life cycle is RNA-directed DNA synthesis. This process, driven by the viral enzyme **reverse transcriptase**, produces a DNA **provirus** consisting of cDNA (complementary DNA transcribed from the RNA genome), which is the form of the viral genome that gets integrated into the host's DNA. The provirus resides in the host chromosome permanently and is occasionally activated to produce new virions. When this happens, the provirus is transcribed as mRNA, which is then translated into viral proteins using the host cell's protein-synthesizing machinery. Viral glycoproteins are inserted into the host

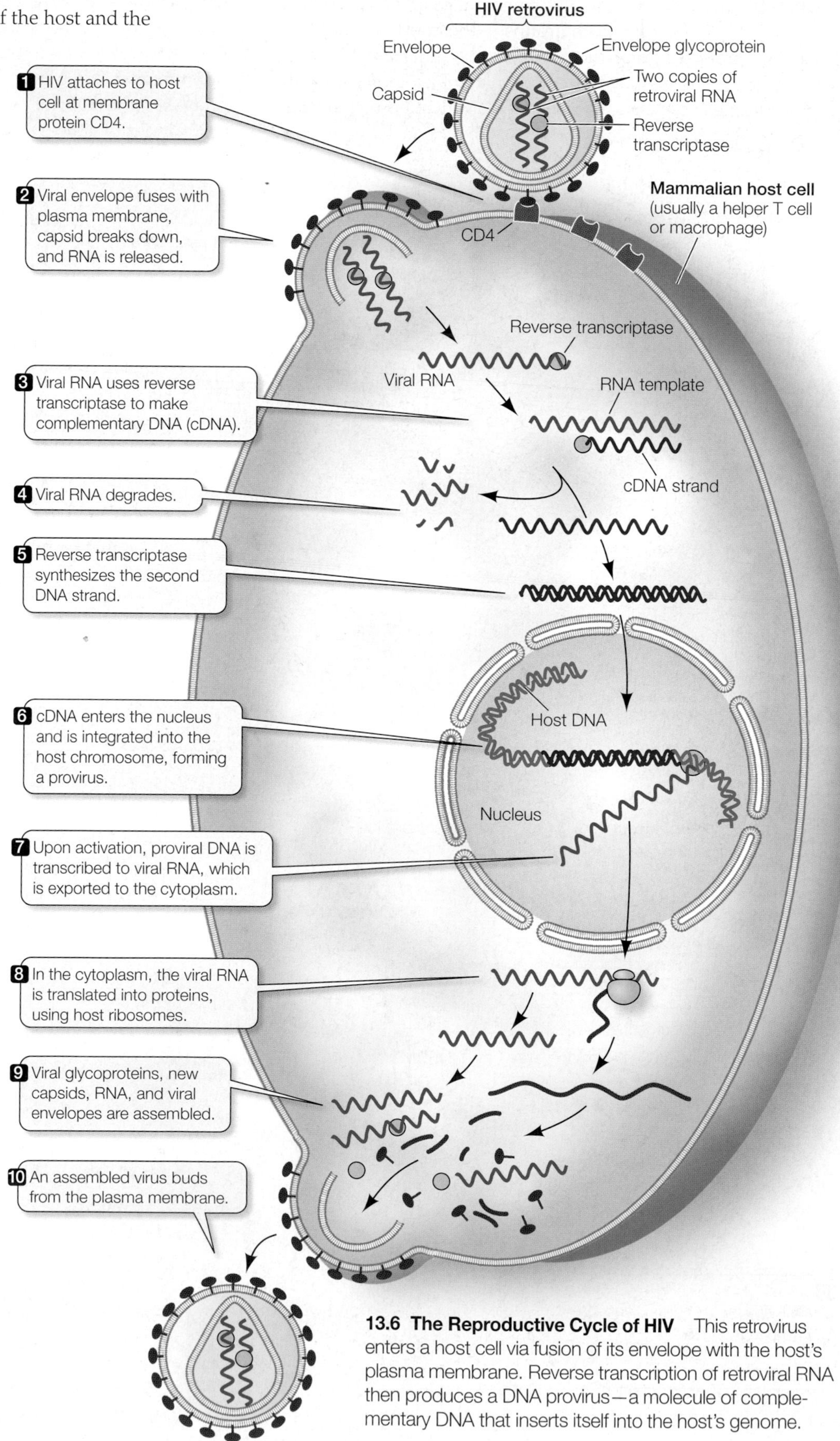
HIV retrovirus
Envelope
Envelope glycoprotein
Capsid
Two copies of retroviral RNA
Reverse transcriptase
Mammalian host cell (usually a helper T cell or macrophage)
CD4
1 HIV attaches to host cell at membrane protein CD4.
2 Viral envelope fuses with plasma membrane, capsid breaks down, and RNA is released.
Reverse transcriptase
Viral RNA
RNA template
3 Viral RNA uses reverse transcriptase to make complementary DNA (cDNA).
cDNA strand
4 Viral RNA degrades.
5 Reverse transcriptase synthesizes the second DNA strand.
Host DNA
6 cDNA enters the nucleus and is integrated into the host chromosome, forming a provirus.
Nucleus
7 Upon activation, proviral DNA is transcribed to viral RNA, which is exported to the cytoplasm.
8 In the cytoplasm, the viral RNA is translated into proteins, using host ribosomes.
9 Viral glycoproteins, new capsids, RNA, and viral envelopes are assembled.
10 An assembled virus buds from the plasma membrane.

13.6 The Reproductive Cycle of HIV This retrovirus enters a host cell via fusion of its envelope with the host's plasma membrane. Reverse transcription of retroviral RNA then produces a DNA provirus—a molecule of complementary DNA that inserts itself into the host's genome.

(A)

(B)

13.7 Wheat Streak Mosaic Virus (A) A tiny mite is the vector for wheat streak mosaic virus. (B) Photosynthetic tissues damaged by the virus are visible on these wheat plants as yellow streaks.

cell's plasma membrane, which will become the viral envelope. Other viral proteins form capsids, which enclose viral RNA molecules. Release of virions from the cell is by a process of budding very similar to exocytosis. Almost every step in this complex cycle can, in principle, be attacked by therapeutic drugs; this fact is used by researchers in their quest to conquer AIDS, the deadly condition caused by HIV infection in humans, as is discussed further in Section 18.7.

Animal viruses, including human viruses, take a severe toll on human and animal health. But our well-being is also challenged by plant viruses and the diseases they cause.

Many plant viruses spread with the help of vectors

Viral diseases of flowering plants are common. Plant viruses can be transmitted *horizontally*, from one plant to another, or *vertically*, from parent to offspring. To infect a plant cell, viruses must pass through a cell wall as well as a plasma membrane. Most plant viruses accomplish this through their association with vectors, which are often insects. When an insect vector penetrates a cell wall with its proboscis (snout), virions can move from the insect into the plant. Once inside a plant cell, the virus reproduces and spreads to other cells in the plant. Within a structure such as a leaf, the virus spreads through the plasmodesmata, the cytoplasmic connections between cells (see Figure 15.20).

An example of a virus that causes an economically important plant disease is the wheat streak mosaic virus (**Figure 13.7**). It enters the leaf of a wheat plant via a tiny (1 mm long) insect, the mite *Aceria tosichella*. Yellow streaks are seen in the leaves as the infection spreads and destroys photosynthetic tissues. Without the chemical energy from photosynthesis, the plant's production of wheat grains can be severely reduced. This disease was first detected in Canada in 1960 and in the Unites States in 1964. The only way to control it is to eliminate the host insects (or plants) to break the viral life cycle. There is serious concern about the use of this virus as a bioterrorism agent, as its spread could wreak havoc on food supplies.

13.1 RECAP

Viruses are not cells. They consist of nucleic acid and a few proteins, and require a host cell to reproduce. In the lytic cycle, the viral genome directs the host cell to generate new virions along with proteins that cause the host cell to lyse and release them. In the lysogenic cycle, viruses may remain inactive in host cells for long periods.

- How can bacteria and viruses make good model organisms? See p. 283
- Do you understand the lytic and lysogenic cycles of bacteriophage? See pp. 284–285 and Figure 13.3
- Describe the HIV life cycle. See p. 288 and Figure 13.6

How does a phage manage to switch between the lysogenic and the lytic cycle? How does it know when it is to its advantage to make this switch? Let's examine these questions and other aspects of the regulation of viral gene expression.

13.2 How Is Gene Expression Regulated in Viruses?

The mechanisms used by viruses within a host cell for the regulation of gene expression are similar to those used by their hosts. Even a "simple" biological agent such as a virus is faced with complicated molecular decisions when its genome enters a cell. For example, the viral genome must direct the shutdown of host transcription and translation, then redirect the host's protein synthesis machinery to virus production and host cell lysis. All the genes involved in this process must be activated in the right order. In temperate viruses, which can insert their genome (or a DNA copy) into the host chromosome, an additional issue arises: When should the provirus leave the host chromosome and undergo a lytic cycle?

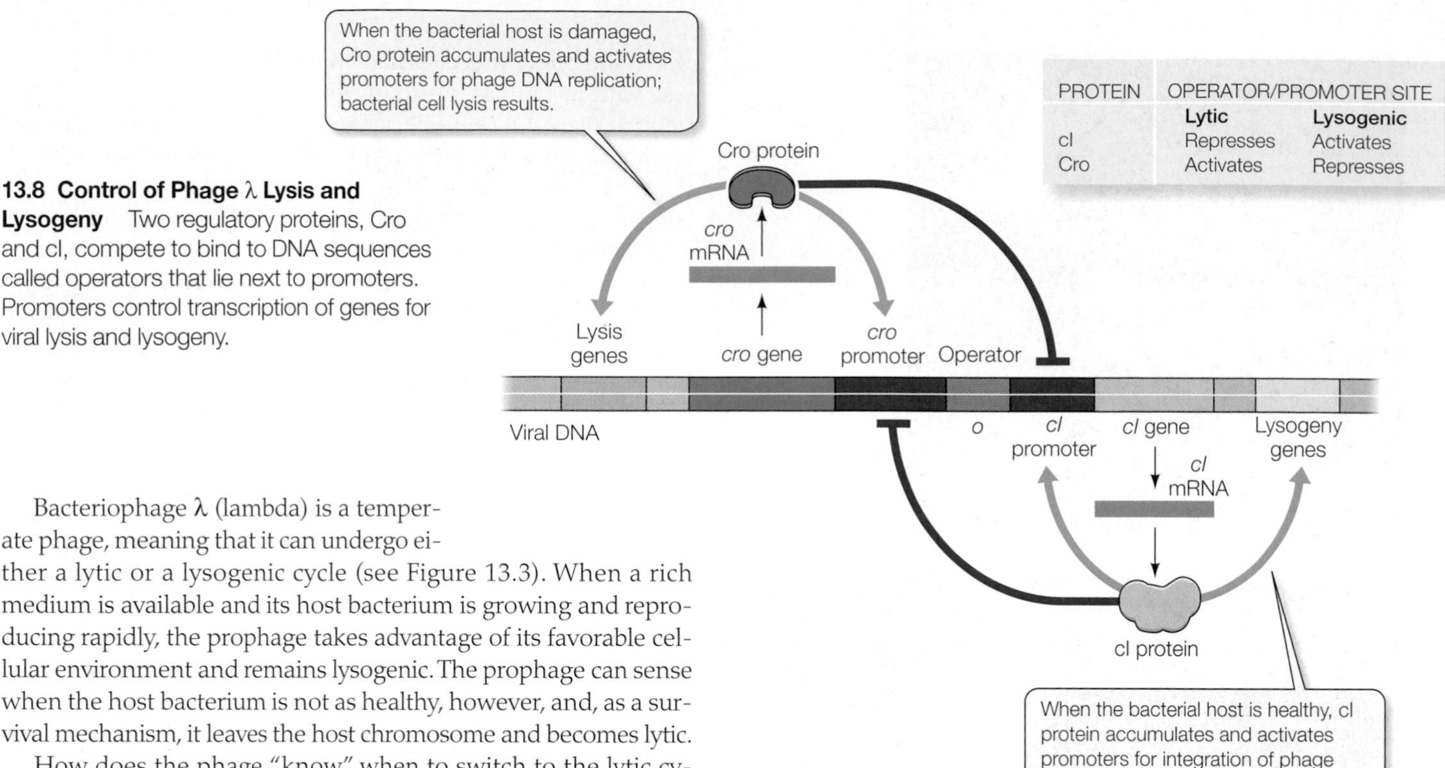
When the bacterial host is damaged, Cro protein accumulates and activates promoters for phage DNA replication; bacterial cell lysis results.
Cro protein
cro mRNA
Lysis genes
cro gene
cro promoter
Operator
Viral DNA
o
cI promoter
cI gene
cI mRNA
Lysogeny genes
cI protein
When the bacterial host is healthy, cI protein accumulates and activates promoters for integration of phage DNA into the host chromosome. The phage enters the lysogenic cycle.

PROTEIN	OPERATOR/PROMOTER SITE	
	Lytic	**Lysogenic**
cI	Represses	Activates
Cro	Activates	Represses

13.8 Control of Phage λ Lysis and Lysogeny Two regulatory proteins, Cro and cI, compete to bind to DNA sequences called operators that lie next to promoters. Promoters control transcription of genes for viral lysis and lysogeny.

Bacteriophage λ (lambda) is a temperate phage, meaning that it can undergo either a lytic or a lysogenic cycle (see Figure 13.3). When a rich medium is available and its host bacterium is growing and reproducing rapidly, the prophage takes advantage of its favorable cellular environment and remains lysogenic. The prophage can sense when the host bacterium is not as healthy, however, and, as a survival mechanism, it leaves the host chromosome and becomes lytic.

How does the phage "know" when to switch to the lytic cycle? A kind of "genetic switch" is involved in sensing conditions within the host. Two viral regulatory proteins, cI and Cro, compete for two promoters on phage DNA. The two promoters control the transcription of the viral genes involved in the lytic and the lysogenic cycles, respectively, and the two regulatory proteins have opposite effects on the two promoters (**Figure 13.8**). Additional DNA sequences called operators can bind proteins that affect RNA polymerase binding to adjacent promoters.

Phage infection is essentially a "race" between these two regulatory proteins. In a healthy *E. coli* host cell, Cro synthesis is low, so cI "wins," and the phage enters a lysogenic cycle. If the host cell is damaged by mutagens or other stress, Cro synthesis is high, promoters for phage DNA and viral capsid proteins are activated, and bacterial lysis ensues. The two regulatory proteins are made very early in phage infection, and each has a binding site for a specific DNA sequence.

The life cycle of phage λ, which has been greatly simplified here, is a paradigm for viral infections throughout the biological world. The lessons learned from transcriptional controls in this system have been applied again and again to other viruses, including HIV.

13.2 RECAP

Special viral proteins that interact with host and viral DNA sequences are the keys to the regulation of viral gene expression.

- Describe the genetic switch that regulates the bacteriophage λ life cycle between lytic and lysogenic. See Figure 13.8

Now that we've seen how viruses reproduce, transmit, and regulate the expression of their genes, let's see how these processes work in their hosts—the prokaryotes.

13.3 How Do Prokaryotes Exchange Genes?

In contrast to viruses, prokaryotes (organisms in the domains Bacteria and Archaea) are living cells that carry out all the basic functions of life. Prokaryotes usually reproduce asexually, but nonetheless have several ways of recombining their genes. Whereas in eukaryotes, genetic recombination occurs between the genomes of two parents, recombination in prokaryotes results from the interaction of the genome of one cell with a much smaller sample of genes—a DNA fragment—from another cell.

The reproduction of prokaryotes gives rise to clones

Most prokaryotes reproduce by the division of a single cell into two identical offspring (see Figure 9.2). In this way, a single cell gives rise to a **clone**—a population of genetically identical individuals. Prokaryotes can reproduce very rapidly. A population of *E. coli*, as we saw above, can double every 20 minutes as long as conditions remain favorable. That is one of the reasons that this bacterium is used so widely in research as a model organism.

Simple, reliable methods exist for isolating single bacterial cells and rapidly growing them into clones for identification and study. Pure cultures of *E. coli* or other bacteria can be grown on the surface of a solid *minimal nutrient medium* containing a sugar, minerals, a nitrogen source such as ammonium chloride (NH_4Cl), and a solidifying agent such as agar (**Figure 13.9**). If the number of cells spread on the medium is small, each cell will give rise to a small, rapidly growing *bacterial colony*. If a large number of cells is spread on the medium, their growth will produce one continuous layer—

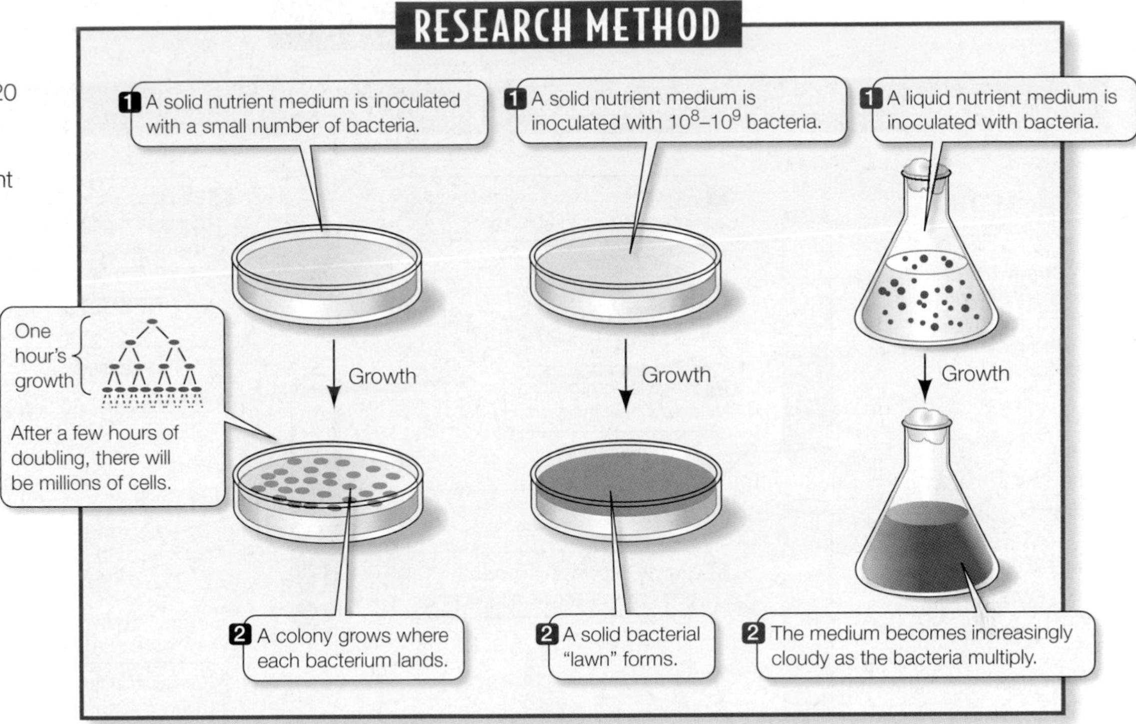
RESEARCH METHOD
1 A solid nutrient medium is inoculated with a small number of bacteria.
1 A solid nutrient medium is inoculated with 10^8–10^9 bacteria.
1 A liquid nutrient medium is inoculated with bacteria.
One hour's growth
Growth
Growth
Growth
After a few hours of doubling, there will be millions of cells.
2 A colony grows where each bacterium lands.
2 A solid bacterial "lawn" forms.
2 The medium becomes increasingly cloudy as the bacteria multiply.

13.9 Growing Bacteria in the Laboratory A population of *Escherichia coli* doubles every 20 minutes in laboratory culture. The three techniques of culture shown here are used for different applications.

a *bacterial lawn*. Bacteria can also be grown in a liquid medium. We'll see examples of all these techniques in this chapter.

Bacteria have several ways of recombining their genes

The existence and heritability of mutations in bacteria has attracted the attention of geneticists. If there were no form of exchange of genetic information between individuals, bacteria would not be useful for genetic analysis. But how can these asexually reproducing organisms exchange genetic information? In eukaryotes, genetic recombination occurs between homologous chromosomes from two parents during meiosis. While prokaryotes usually reproduce asexually, they still have several ways of recombining their genes.

CONJUGATION The most important way in which bacteria recombine their genes is through the interaction of the genome of one cell with a sample of genes—a DNA fragment—from another cell. In 1946, Joshua Lederberg and Edward Tatum demonstrated that such exchanges do occur, although they are rare events.

Initially, Lederberg and Tatum grew two auxotrophic (nutrient-requiring) mutant strains of *E. coli*. Like the *Neurospora* studied by Beadle and Tatum (see Figure 12.1), these strains could not grow on a minimal nutritional medium, but required supplementation with nutrients that they could not synthesize for themselves because of an enzyme defect.

- Strain 1 required the amino acid methionine and the vitamin biotin for growth; it could make its own threonine and leucine. So its phenotype (and genotype) is given as $met^-bio^-thr^+leu^+$.
- Strain 2 required neither methionine nor biotin, but could not grow without the amino acids threonine and leucine. Its phenotype is given as $met^+bio^+thr^-leu^-$.

Lederberg and Tatum mixed these two mutant strains and cultured them together for several hours in a liquid medium supplemented with methionine, biotin, threonine, and leucine, so that both strains could grow. The bacteria were then removed from the medium by centrifugation, washed, and transferred to minimal medium, which lacked all four supplements. Neither strain 1 nor strain 2 had been able to grow on this medium before because of their nutritional requirements. Nevertheless, after the two strains had been cultured together, a few bacterial colonies did appear on the minimal medium (**Figure 13.10**). Because they were growing in the minimal medium, these colonies must have consisted of bacteria that were $met^+bio^+thr^+leu^+$; that is, they must have been prototrophic. These colonies appeared at a rate of approximately one for every 10 million cells originally placed on the plates (10^{-7}).

Where did these prototrophic colonies come from? Lederberg and Tatum were able to rule out mutation, and other investigators ruled out transformation (a process we discussed in Section 11.1 and which we'll look at in more detail shortly). A third possibility is that the two strains of *E. coli* had exchanged genetic material, producing some cells containing met^+ and bio^+ alleles from strain 2 and thr^+ and leu^+ alleles from strain 1. Later experiments showed that such an exchange, during **conjugation**, had indeed occurred. One bacterial cell—the recipient—had received DNA from another cell—the donor—that included the two wild-type (⁺) alleles that were missing in the recipient. Recombination had then created a genotype with four wild-type alleles.

The physical contact between two bacteria required for conjugation can be observed under the electron microscope (**Figure 13.11**). It is initiated by a thin projection called a **sex pilus** (plural *pili*). Once the sex pili bring the two cells into proximity, the actual transfer of DNA occurs through a thin cytoplasmic bridge called a **conjugation tube** that forms between the cells. Since the bacter-

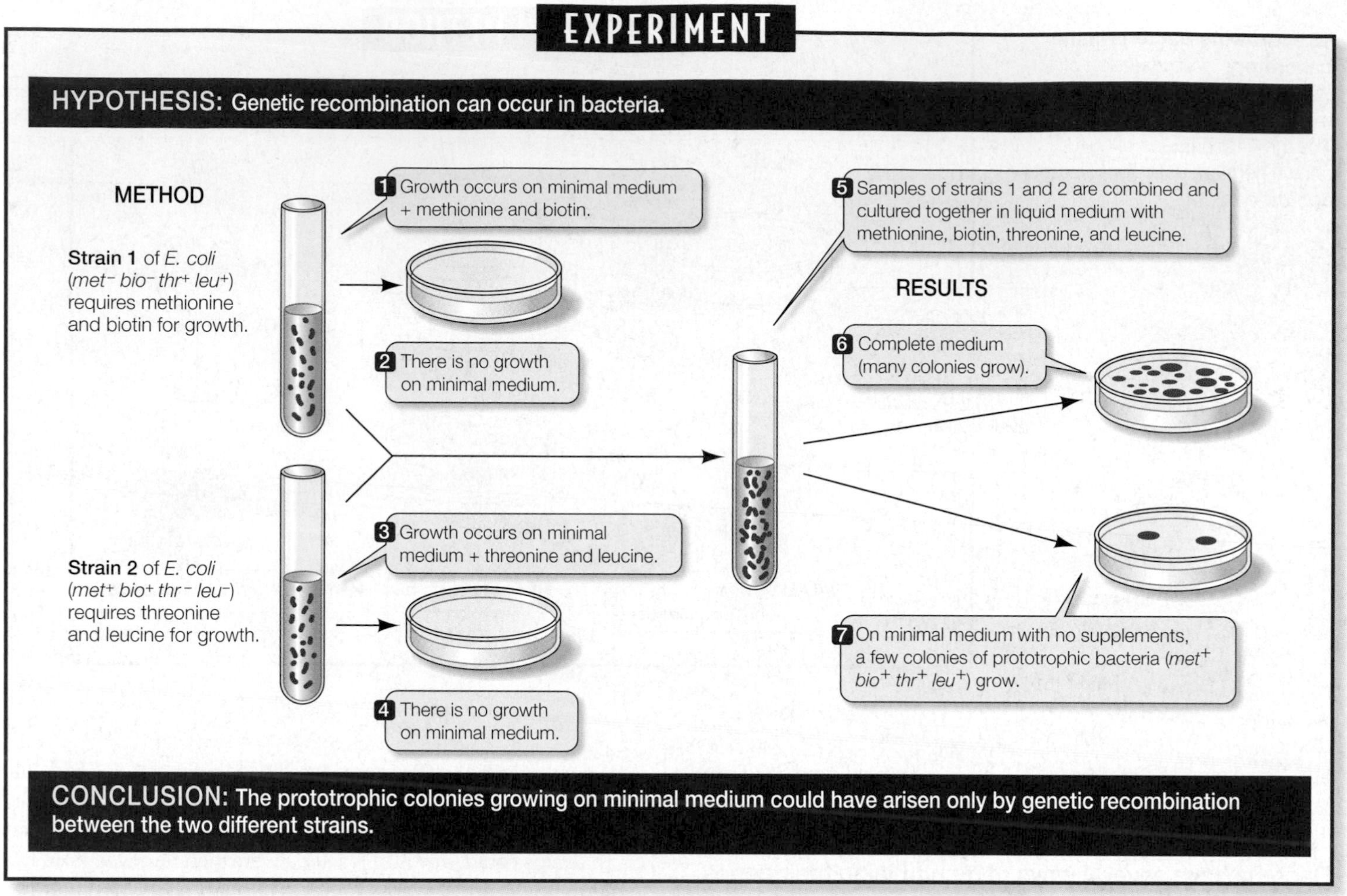

EXPERIMENT

HYPOTHESIS: Genetic recombination can occur in bacteria.

METHOD

Strain 1 of *E. coli* (*met$^-$ bio$^-$ thr$^+$ leu$^+$*) requires methionine and biotin for growth.

1 Growth occurs on minimal medium + methionine and biotin.

2 There is no growth on minimal medium.

Strain 2 of *E. coli* (*met$^+$ bio$^+$ thr$^-$ leu$^-$*) requires threonine and leucine for growth.

3 Growth occurs on minimal medium + threonine and leucine.

4 There is no growth on minimal medium.

5 Samples of strains 1 and 2 are combined and cultured together in liquid medium with methionine, biotin, threonine, and leucine.

RESULTS

6 Complete medium (many colonies grow).

7 On minimal medium with no supplements, a few colonies of prototrophic bacteria (*met$^+$ bio$^+$ thr$^+$ leu$^+$*) grow.

CONCLUSION: The prototrophic colonies growing on minimal medium could have arisen only by genetic recombination between the two different strains.

13.10 Lederberg and Tatum's Experiment After being grown together, a mixture of two complementary auxotrophic strains of *E. coli* contained a few cells that gave rise to new prototrophic colonies. This experiment proved that genetic recombination takes place in prokaryotes. FURTHER RESEARCH: How would you set up experiments to distinguish reversion mutations (back to wild type; for example, *met$^-$ bio$^-$ thr$^+$ leu$^+$* to *met$^+$ bio$^+$ thr$^+$ leu$^+$*) from recombination in these experiments?

ial chromosome is circular, it must be made linear (cut) before it can pass through the tube. Contact between the cells is brief—only rarely long enough for the entire donor genome to enter the recipient cell. Therefore, the recipient cell usually receives only a portion of the donor DNA.

Once the donor DNA fragment is inside the recipient cell, it can recombine with the recipient cell's genome. In much the same way that chromosomes pair up, gene for gene, in prophase I of meiosis, the donor DNA can line up beside its homologous genes in the recipient, and crossing over can occur. Enzymes that can cut and rejoin DNA molecules are active in bacteria, so gene(s) from the donor can become integrated into the genome of the recipient, thus changing the recipient's genetic constitution (**Figure 13.12**), even though only about half the transferred genes become integrated in this way.

TRANSFORMATION Frederick Griffith obtained the first evidence for the transfer of prokaryotic genes more than 75 years ago when he discovered the transforming principle (see Figure 11.1). We can now explain Griffith's results: DNA had leaked from dead cells of

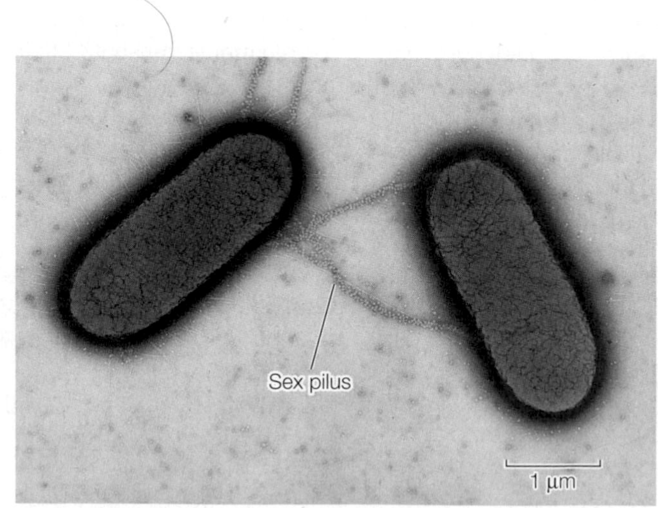

13.11 Bacterial Conjugation Sex pili draw two bacteria into close contact, so that a cytoplasmic conjugation tube can form. DNA is transferred from one cell to the other via the conjugation tube.

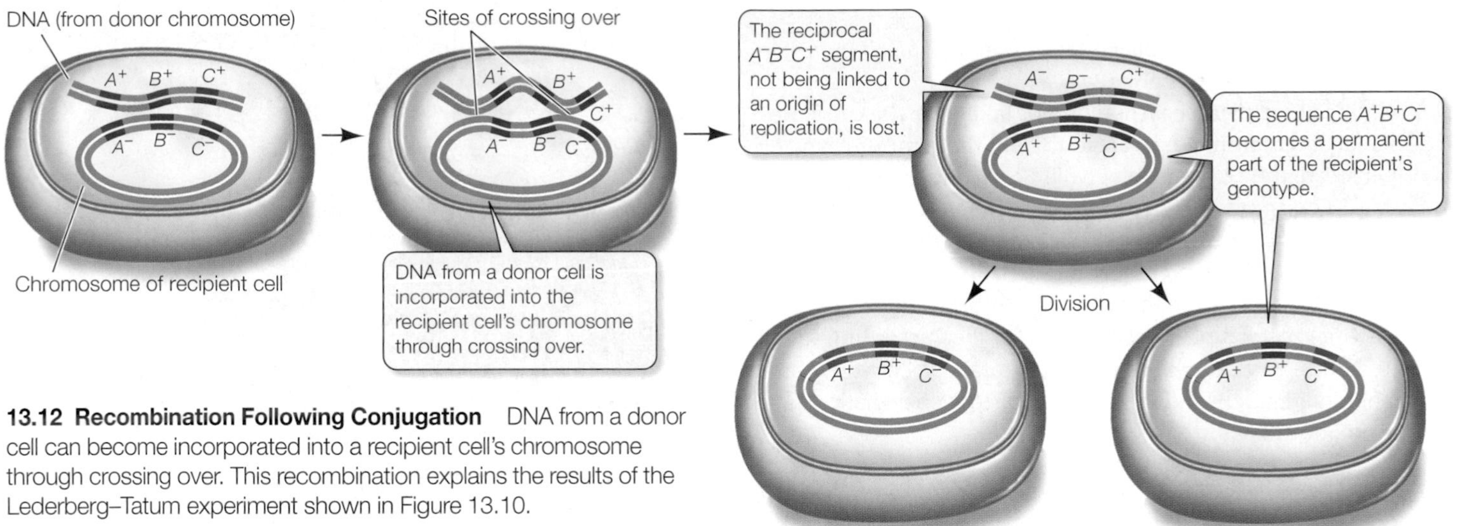

13.12 Recombination Following Conjugation DNA from a donor cell can become incorporated into a recipient cell's chromosome through crossing over. This recombination explains the results of the Lederberg–Tatum experiment shown in Figure 13.10.

pathogenic pneumococci and was taken up as free DNA by living nonvirulent pneumococci, which became virulent as a result. This phenomenon, called **transformation**, occurs in nature in some species of bacteria when cells die and their DNA leaks out (**Figure 13.13A**). Once transforming DNA is inside a host cell, an event very similar to recombination occurs, and new genes can be incorporated into the host chromosome.

TRANSDUCTION When bacteriophage undergo a lytic cycle, they package their DNA in capsids, as we saw in Section 13.1. These capsids generally form before the viral DNA is inserted into them. Sometimes bacterial DNA fragments are inserted into an empty phage capsid instead of, or along with, the phage DNA (**Figure 13.13B**). So, when the new virion infects another bacterium, the bacterial DNA is injected into the new host cell. This mechanism of DNA transfer is called **transduction**. Needless to say, it does not result in a productive viral infection. Instead, the incoming DNA fragment can recombine with the host chromosome, resulting in the replacement of host cell genes with bacterial genes from the virus's former host.

Plasmids are extra chromosomes in bacteria

In addition to their main chromosome, many bacteria harbor additional smaller, circular chromosomes called **plasmids**. They contain at most a few dozen genes and, most important, an origin of replication (*ori*, the sequence where DNA replication starts), which defines them as chromo-

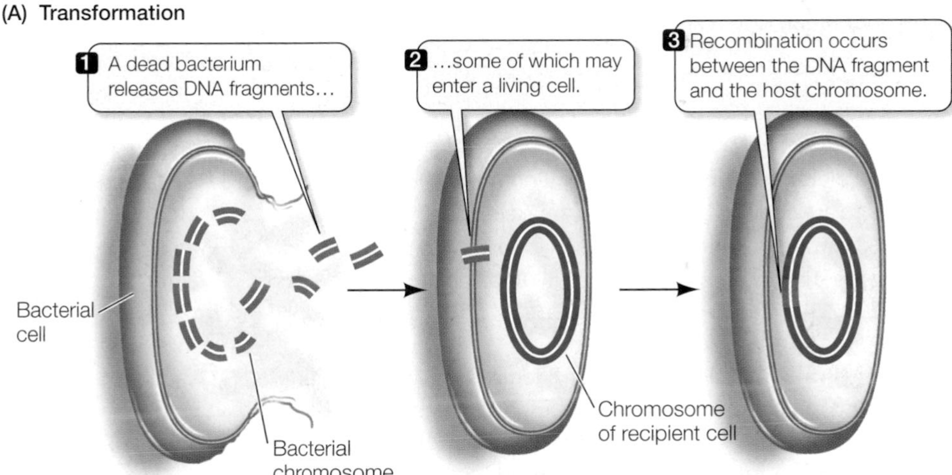

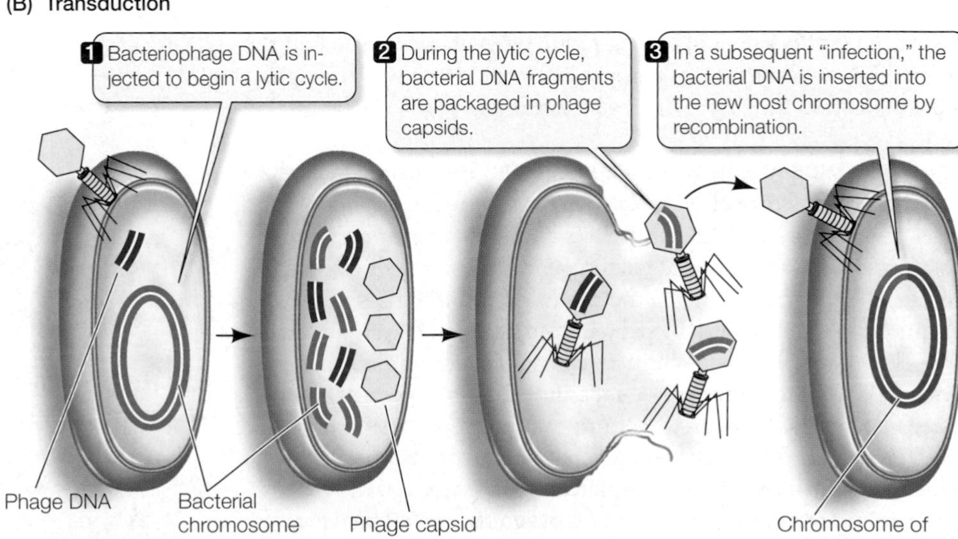

13.13 Transformation and Transduction When a new DNA fragment enters a bacterial cell, recombination can occur. (A) Transforming DNA can leak from dead bacterial cells and be taken up by a living bacterium, which may incorporate the new genes into its chromosome. (B) In transduction, some viruses carry bacterial DNA fragments from one cell to another.

somes. Plasmids usually replicate at the same time as the main bacterial chromosome, but not always.

Plasmids are *not* viruses. They do not take over the cell's molecular machinery or make a capsid to help them move from cell to cell. Instead, they can move between cells during conjugation, thereby adding some new genes to the recipient bacterium (**Figure 13.14**). Because plasmids exist independently of the main chromosome, they do not need to recombine with the main chromosome to add their genes to the recipient cell's genome.

There are several types of plasmids, classified according to the kinds of genes they carry. Some code for catabolic enzymes, others enable conjugation, and still others code for genes that circumvent antibiotic attack.

SOME PLASMIDS CARRY SPECIAL GENES Some plasmids, called **metabolic factors**, carry genes that allow their recipients to carry out unusual metabolic functions. For example, some bacteria can actually thrive on unusual hydrocarbons found in oil spills, and can use them as a carbon source. Plasmids carry the genes for the enzymes that break down such hydrocarbons.

Bacteria such as *Phenylobacterium immobile* can break down oil from oil spills to less harmful CO_2. In a process called bioremediation, nutrients that promote bacterial growth are added to the polluted area, along with the bacteria. Bioremediation has been used to help clean up major oil spills, as well as smaller releases in urban areas.

Plasmids called **fertility factors**, or **F factors**, encode the genes needed for conjugation. F factors have approximately 25 genes, including the ones that make both the sex pilus and the conjugation tube. A cell harboring an F factor is referred to as F^+. It can transfer a copy of the F factor to an F^- cell, making the recipient F^+. Sometimes the F factor integrates itself into the main chromosome (at which point it is no longer a plasmid); when it does, it can bring along other genes from that chromosome when it moves through the conjugation tube from one cell to another.

SOME PLASMIDS ARE RESISTANCE FACTORS **Resistance factors**, also called *R factors* or *R plasmids*, may carry genes coding for proteins that destroy or modify antibiotics. Other R factors provide resistance to heavy metals that bacteria encounter in their environment. R factors first came to the attention of biologists in 1957 during an epidemic of dysentery in Japan, when it was discovered that some strains of the *Shigella* bacterium, which causes dysentery, were resistant to several antibiotics. Researchers found that resistance to the entire spectrum of antibiotics in use could be transferred by conjugation even when no genes on the main chromosome were transferred. Eventually it was shown that the genes for antibiotic resistance are carried on plasmids. Each R factor carries one or more genes conferring resistance to particular antibiotics, as well as genes that code for proteins involved in the conjugation with a recipient bacterium. As far as biologists can determine, R factors providing resistance to naturally occurring antibiotics existed long before antibiotics were discovered and used by humans. However, R factors seem to have become more abundant in modern times, possibly because the heavy use of antibiotics in hospitals selects for bacterial strains bearing them.

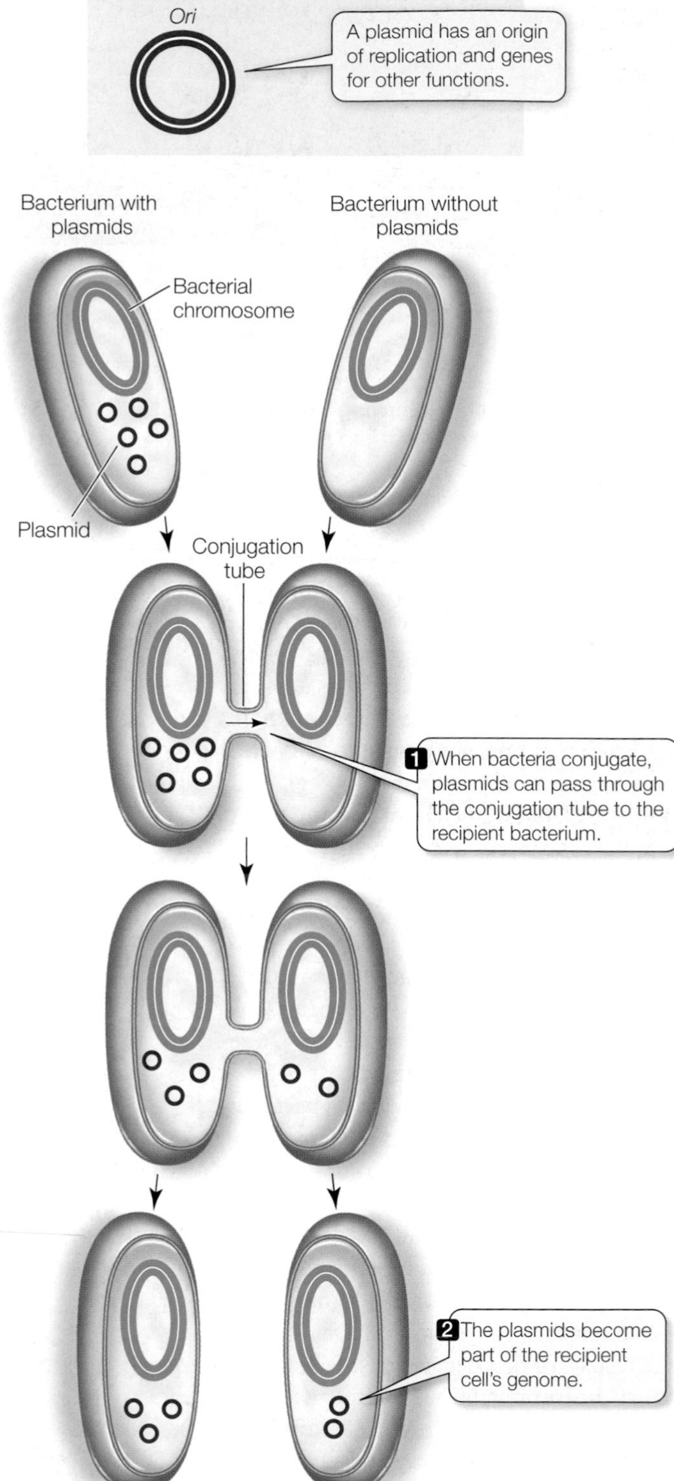
Ori
A plasmid has an origin of replication and genes for other functions.
Bacterium with plasmids
Bacterium without plasmids
Bacterial chromosome
Plasmid
Conjugation tube
1 When bacteria conjugate, plasmids can pass through the conjugation tube to the recipient bacterium.
2 The plasmids become part of the recipient cell's genome.

13.14 Gene Transfer by Plasmids When plasmids enter a cell via conjugation, their genes can be expressed in the recipient cell.

Antibiotic resistance poses a serious threat to human health, and the inappropriate use of antibiotics contributes to this prob-

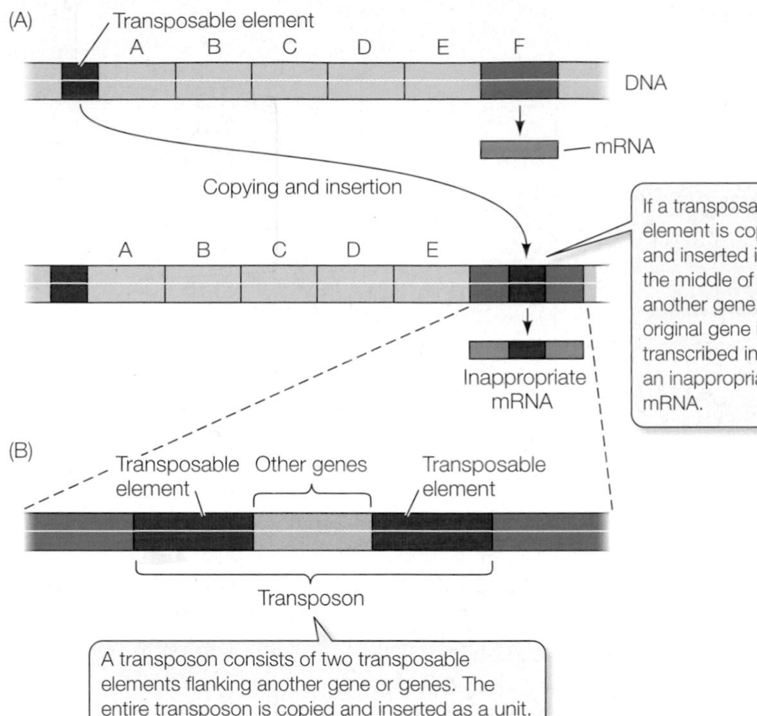
(A)
Transposable element
A B C D E F
DNA
mRNA
Copying and insertion
A B C D E
If a transposable element is copied and inserted into the middle of another gene, the original gene is transcribed into an inappropriate mRNA.
Inappropriate mRNA
(B)
Transposable element
Other genes
Transposable element
Transposon
A transposon consists of two transposable elements flanking another gene or genes. The entire transposon is copied and inserted as a unit.

13.15 Transposable Elements and Transposons (A) Transposable elements are segments of DNA that can be inserted at new locations, either on the same chromosome or on a different chromosome. (B) Transposons consist of transposable elements combined with other genes.

lem. You have probably gone to a physician because of a sore throat, which can have either a viral or a bacterial cause. The best way to determine the causative agent is for the doctor to take a small sample from your inflamed throat, culture it, and identify any bacteria that are present. But perhaps you feel you cannot wait another day for the results. Impatient, you ask the doctor to give you something to make you feel better. The doctor prescribes an antibiotic, which you take. The sore throat gradually gets better, and you think that the antibiotic did the job.

But suppose the infection is viral. In that case, the antibiotic has done nothing to combat the infection, which just runs its normal course. However, it may do something harmful: by killing many normal bacteria in your body, the antibiotic may select for bacteria harboring R factors. These bacteria may survive and reproduce in the presence of the antibiotic, and may soon become quite numerous. You may continue to harbor these bacteria or pass them on to other people. The next time you get a bacterial infection, there may be a ready supply of resistant bacteria in your body, and antibiotics may be ineffective.

Acquisition of antibiotic resistance in pathogenic bacteria provides an example of evolution in action. In the years after they were first discovered in the twentieth century, antibiotics were very successful in combating diseases that had plagued humans for millennia, such as cholera, tuberculosis, and leprosy. But over time, resistant bacteria have appeared. This is classic natural selection: genetic variation exists among bacteria, and those that have survived the onslaught of antibiotics must have a genetic constitution that allows them to do so.

Transposable elements move genes among plasmids and chromosomes

As we have seen, plasmids, viruses, and even phage capsids (in the case of transduction) can transport genes from one bacterial cell to another. There is another type of "gene transport" that occurs within the individual cell. It relies on segments of DNA that can be inserted either at a new location on the same chromosome or into another chromosome. These DNA sequences are called **transposable elements**. Their insertion often produces phenotypic effects by disrupting the genes into which they are inserted (**Figure 13.15A**).

Some transposable elements are relatively short sequences, 1,000–2,000 base pairs long. Such sequences are found at many sites on the *E. coli* main chromosome. In one mechanism of transposition, the transposable element replicates independently of the rest of the chromosome. The copy then inserts itself at other, seemingly random sites on the chromosome. The genes encoding the enzymes necessary for this insertion are found within the transposable element itself. Other transposable elements are cut from their original sites and inserted elsewhere without replication. Longer transposable elements (about 5,000 base pairs) carry one or more additional genes and are called **transposons** (**Figure 13.15B**).

Transposable elements have contributed to the evolution of plasmids. R factors probably originally gained their genes for antibiotic resistance through the activity of transposable elements. One piece of evidence for this conclusion is that each resistance gene in an R factor was originally part of a transposon.

As we have seen, rapid asexual reproduction can produce huge clones of prokaryotes. These genetically identical cells are all equally vulnerable to any change in the environment. Recombination by means of conjugation, transformation, and transduction, or the acquisition of new genes by means of plasmids and transposable elements, introduces genetic diversity into bacterial populations, and this diversity may allow at least some cells to survive under changing conditions.

13.3 RECAP

Bacteria can exchange genetic information through conjugation, transformation, and transduction. Plasmids are small chromosomes that can be transferred between bacteria. Transposable elements and transposons are small pieces of DNA that can move from one place to another in the genome.

- Describe the experiment that demonstrated that genetic recombination occurs in bacteria. See p. 291 and Figure 13.10
- How is DNA exchanged between bacteria during transformation? Transduction? See pp. 292–293 and Figure 13.13
- Do you understand the importance of plasmids in spreading antibiotic resistance? See pp. 294–295

Antibiotics are not the only threats to the survival of prokaryotes. Such threats are numerous, and include changes in environmental temperature and scarcity of resources. Prokaryotes may respond to changes in their environment not only by exchanging genes, but also by regulating the expression of their genes.

13.4 How Is Gene Expression Regulated in Prokaryotes?

Prokaryotes conserve energy and resources by making proteins only when they are needed. The protein content of a bacterium can change rapidly when conditions warrant. There are several ways in which a prokaryotic cell could shut off the supply of an unneeded protein. The cell can:

- downregulate the transcription of mRNA for that protein
- hydrolyze the mRNA after it is made and prior to translation
- prevent translation of the mRNA at the ribosome
- hydrolyze the protein after it is made
- inhibit the function of the protein

Whichever mechanism is used, it must do two things:

- It must respond to environmental signals. (If you find yourself feeling hungry, you might go to the kitchen.)
- It must be efficient. (Why wander through every room in the house if all you want is a midnight snack in the kitchen?)

The earlier the cell intervenes in the process of protein synthesis, the less energy it has to expend. Selective blocking of transcription is far more efficient than transcribing the gene, translating the message, and then degrading or inhibiting the protein. While examples of all five mechanisms for regulating protein levels are found in nature, prokaryotes generally use the most efficient one: transcriptional regulation.

Regulating gene transcription conserves energy

As a normal inhabitant of the human intestine, *E. coli* must be able to adjust to sudden changes in its chemical environment. Its host may present it with one foodstuff one hour and another the next. This variation presents the bacterium with a metabolic challenge. Glucose is its preferred energy source, and is the easiest sugar to metabolize, but not all of its host's foods contain an abundant supply of glucose. For example, the bacterium may suddenly be deluged with milk, whose predominant sugar is lactose. Lactose is a β-galactoside—a disaccharide containing galactose β-linked to glucose (see Section 3.3). To be taken up and metabolized by *E. coli*, lactose must be acted on by three proteins:

- *β-galactoside permease* is a carrier protein in the bacterial plasma membrane that moves the sugar into the cell.
- *β-galactosidase* is an enzyme that catalyzes the hydrolysis of lactose to glucose and galactose.
- *β-galactoside transacetylase* is an enzyme that transfers acetyl groups from acetyl CoA to certain β-galactosides. Its role in the metabolism of lactose is not clear.

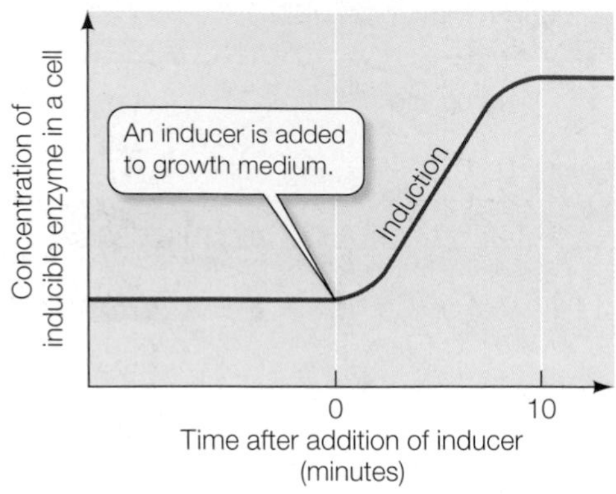

13.16 An Inducer Stimulates the Synthesis of an Enzyme It is most efficient for a cell to produce an enzyme only when it is needed. Some enzymes are induced by the presence of the substance they act upon (for example, β-galactosidase is induced by the presence of lactose).

When *E. coli* is grown on a medium that contains glucose, but no lactose or other β-galactosides, the levels of these three proteins are extremely low—the cell does not waste energy and materials making the unneeded enzymes. If, however, the environment changes such that lactose is the predominant sugar available and very little glucose is present, the bacterium promptly begins making all three enzymes, and they increase rapidly in abundance. For example, there are only two molecules of β-galactosidase present in an *E. coli* cell when glucose is present in the medium. But when glucose is absent, the presence of lactose can induce the synthesis of 3,000 molecules of β-galactosidase per cell!

If lactose is removed from *E. coli*'s environment, synthesis of the three enzymes that process it stops almost immediately. The enzyme molecules that have already formed do not disappear; they are merely diluted during subsequent cell divisions until their concentration falls to the original low level within each bacterium.

Compounds that stimulate the synthesis of a protein (such as lactose in our example) are called **inducers** (**Figure 13.16**). The proteins that are produced are called **inducible** proteins, whereas proteins that are made all the time at a constant rate are called **constitutive** proteins.

We have now seen two basic ways of regulating the rate of a metabolic pathway. Section 6.5 described allosteric regulation of enzyme activity (the rate of enzyme-catalyzed reactions); this mechanism allows rapid fine-tuning of metabolism. Regulation of protein synthesis—that is, regulation of the concentration of enzymes—is slower, but produces a greater savings of energy. **Figure 13.17** compares these two modes of regulation.

A single promoter can control the transcription of adjacent genes

The genes that encode the synthesis of the three enzymes that process lactose in *E. coli* are called **structural genes**, indicating that they specify the primary structure (the amino acid sequence) of a protein molecule. In other words, structural genes are genes that can be transcribed into mRNA.

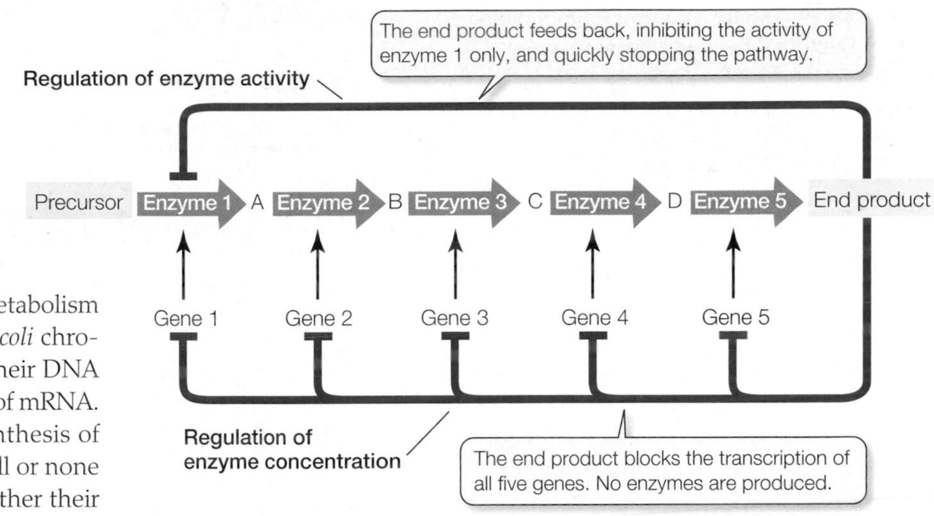

13.17 Two Ways to Regulate a Metabolic Pathway Feedback from the end product of a metabolic pathway can block enzyme activity (allosteric regulation), or it can stop the transcription of genes that code for the enzymes in the pathway (transcriptional regulation).

The three structural genes involved in the metabolism of lactose lie adjacent to one another on the *E. coli* chromosome. This arrangement is no coincidence: their DNA is transcribed into a single, continuous molecule of mRNA. Because this particular mRNA governs the synthesis of all three lactose-metabolizing enzymes, either all or none of these enzymes are made, depending on whether their common message—their mRNA—is present in the cell.

The three genes share a single promoter. As we saw in Section 12.3, a *promoter* is a DNA sequence to which RNA polymerase binds to initiate transcription. We also noted that some promoters are more effective at transcription initiation than others. The promoter for these three *E. coli* structural genes can be very effective, so the maximum rate of mRNA synthesis can be high. There is, however, a mechanism to shut down mRNA synthesis when the enzymes are not needed. That mechanism is called an *operon*, and was elegantly worked out by François Jacob and Jacques Monod.

Operons are units of transcription in prokaryotes

Prokaryotes shut down transcription by placing an obstacle between the promoter and the structural genes it regulates. A short stretch of DNA called the **operator** lies in this position. It can bind very tightly with a special type of protein molecule, called a **repressor**, to create such an obstacle:

- When the repressor protein is bound to the operator, it blocks the transcription of mRNA.
- When the repressor is not attached to the operator, mRNA synthesis proceeds constitutively.

The whole unit, consisting of the closely linked structural genes and the DNA sequences that control their transcription, is called an **operon**. An operon always consists of a promoter, an operator, and two or more structural genes (**Figure 13.18**). The promoter and operator are binding sites on DNA and are not transcribed.

E. coli has numerous mechanisms to control the transcription of operons; we will focus on three of them here. Two of these control mechanisms depend on interactions of a repressor protein with the operator, and the third depends on interactions of other proteins with the promoter.

Operator–repressor control induces transcription in the *lac* operon

The operon containing the genes for the three lactose-metabolizing proteins of *E. coli* is called the *lac operon* (see Figure 13.18). As we have just seen, RNA polymerase can bind to the promoter, and a repressor protein can bind to the operator.

The repressor protein has two binding sites: one for the operator and the other for inducers. The inducers of the *lac* operon, as we know, are molecules of lactose. Binding with an inducer changes the shape of the repressor protein (by allosteric modification; see Section 6.3). This change in shape prevents the repressor from binding to the operator (**Figure 13.19**). As a result, RNA polymerase can bind to the promoter and start transcribing the structural genes of the *lac* operon. The mRNA transcribed from these genes is translated on ribosomes to synthesize the three proteins required for metabolizing lactose.

What happens if the concentration of lactose drops? As the lactose concentration decreases, the inducer (lactose) molecules separate from the repressor. Free of lactose molecules, the repressor

13.18 The *lac* Operon of *E. coli* The *lac* operon of *E. coli* is a segment of DNA that includes a promoter, an operator, and the three structural genes that code for lactose-metabolizing enzymes.

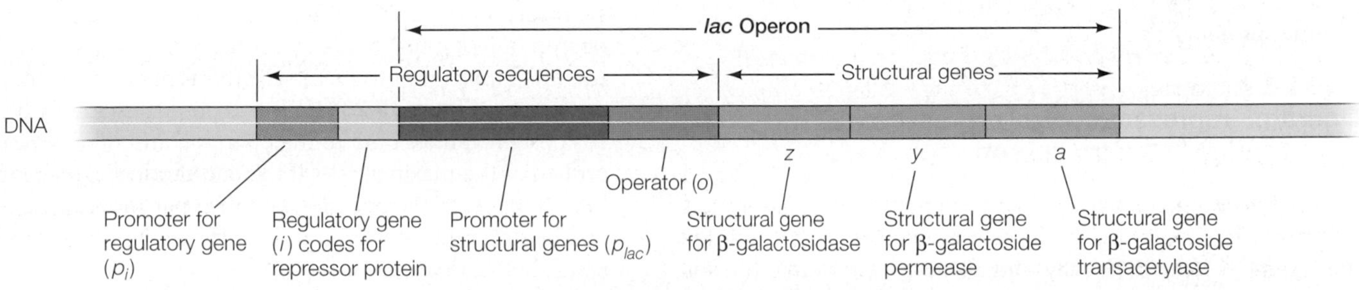

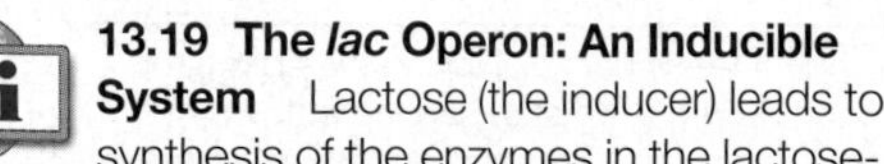

13.19 The *lac* Operon: An Inducible System Lactose (the inducer) leads to synthesis of the enzymes in the lactose-metabolizing pathway by preventing the repressor protein (which would have stopped transcription) from binding to the operator.

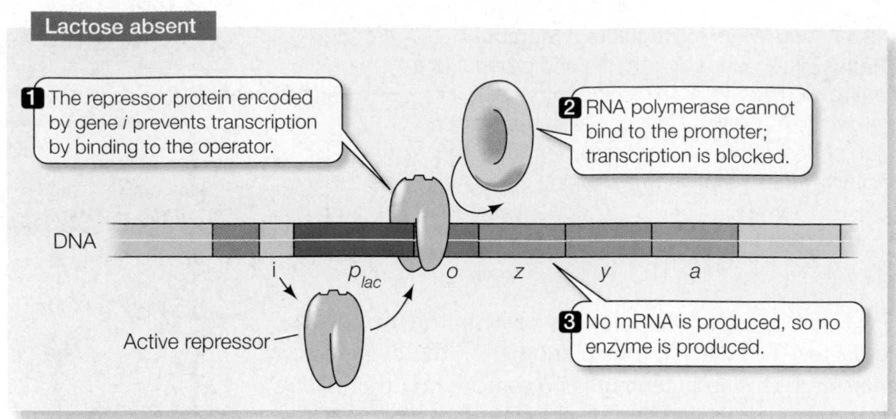

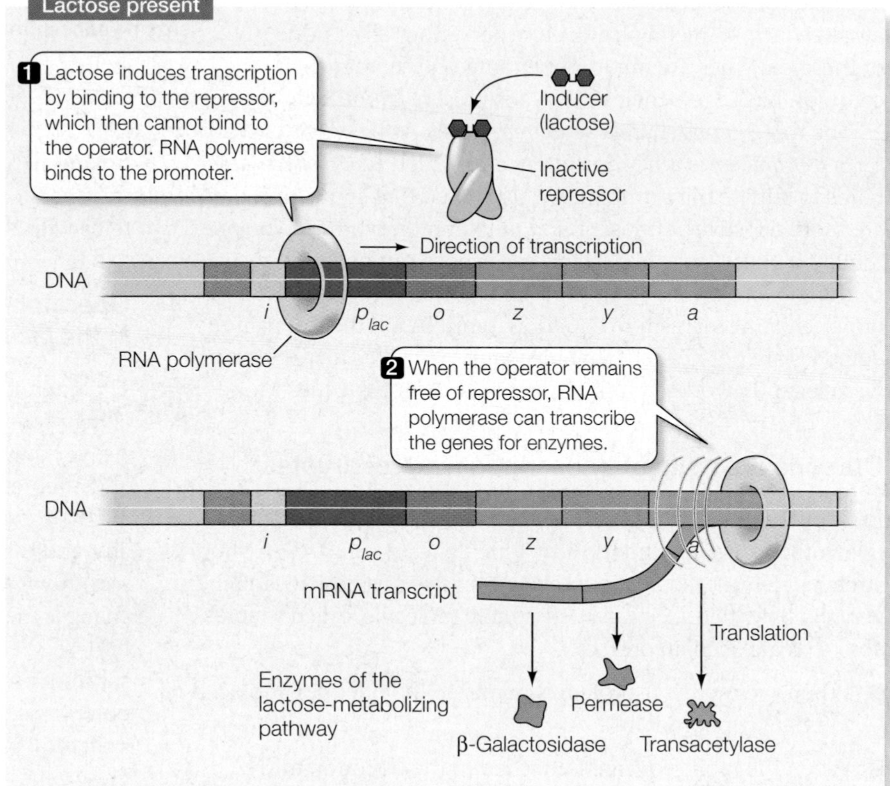

returns to its original shape and binds to the operator, and transcription of the *lac* operon stops. Translation stops soon thereafter because the mRNA that is already present breaks down quickly. Thus it is the presence or absence of lactose—the inducer—that regulates the binding of the repressor to the operator, and therefore the synthesis of the proteins needed to metabolize it. In other words, the *lac* operon is an *inducible system*.

Repressor proteins are encoded by **regulatory genes.** The regulatory gene that codes for the repressor of the *lac* operon is called the *i* (inducibility) gene. The *i* gene happens to lie close to the operon that it regulates (see Figure 13.18), but some regulatory genes are distant from their operons. Like all other genes, the *i* gene itself has a promoter, which can be designated p_i. Because this promoter does not bind RNA polymerase very effectively, only enough mRNA to synthesize about ten molecules of repressor protein per cell per generation is produced. This quantity of the repressor is enough to regulate the operon effectively—producing more would be a waste of energy. There is no operator between p_i and the *i* gene. Therefore, the repressor of the *lac* operon is a constitutive protein; that is, it is made at a constant rate that is not subject to environmental control.

Let's review the important features of inducible systems such as the *lac* operon:

- In the absence of inducer, the operon is turned off.
- Control is exerted by a regulatory protein—the repressor—that turns the operon off.
- Adding inducer changes the repressor and turns the operon on.
- Regulatory genes produce proteins whose sole function is to regulate the expression of other genes.
- Certain other DNA sequences (operators and promoters) do not code for proteins, but are binding sites for regulatory or other proteins.

Operator–repressor control represses transcription in the *trp* operon

We have seen how *E. coli* benefits from having an inducible system for lactose metabolism: only when lactose is present does the system switch on. Equally valuable to a bacterium is the ability to switch off the synthesis of certain enzymes in response to the excessive accumulation of their end products. For example, when the amino acid tryptophan, an essential constituent of proteins, is present in ample concentrations, it is advantageous to stop making the enzymes for tryptophan synthesis. When the synthesis of a protein can be turned off in response to such a biochemical cue, that protein is said to be **repressible.**

In *repressible systems*, such as the *trp* operon that controls the synthesis of tryptophan, the repressor protein cannot shut off its operon unless it first binds to a **corepressor**, which may be either the metabolic end product itself (tryptophan in this case) or an analog of it (**Figure 13.20**). If the end product is absent, the repressor protein cannot bind to the operator, and the operon is transcribed at the maximal rate. If the end product is present, the repressor binds to the operator, and the operon is turned off.

The difference between inducible and repressible systems is small, but significant:

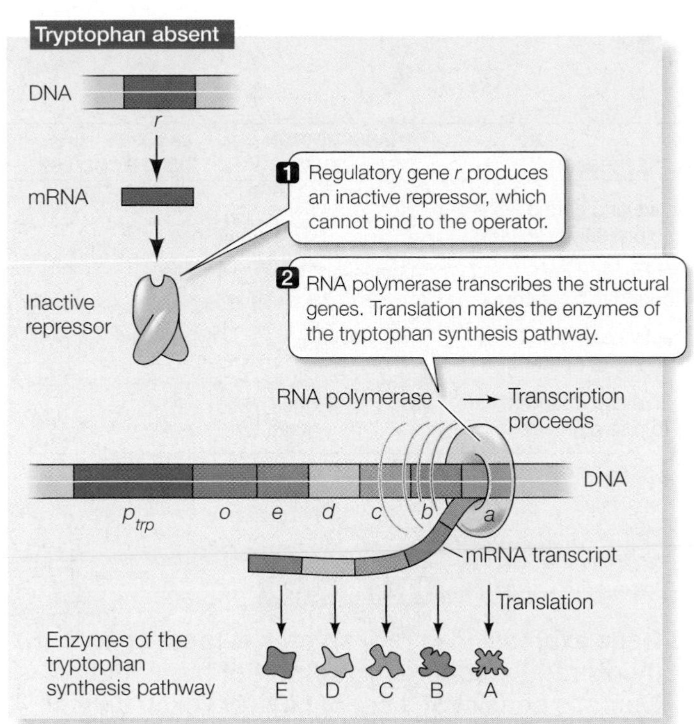

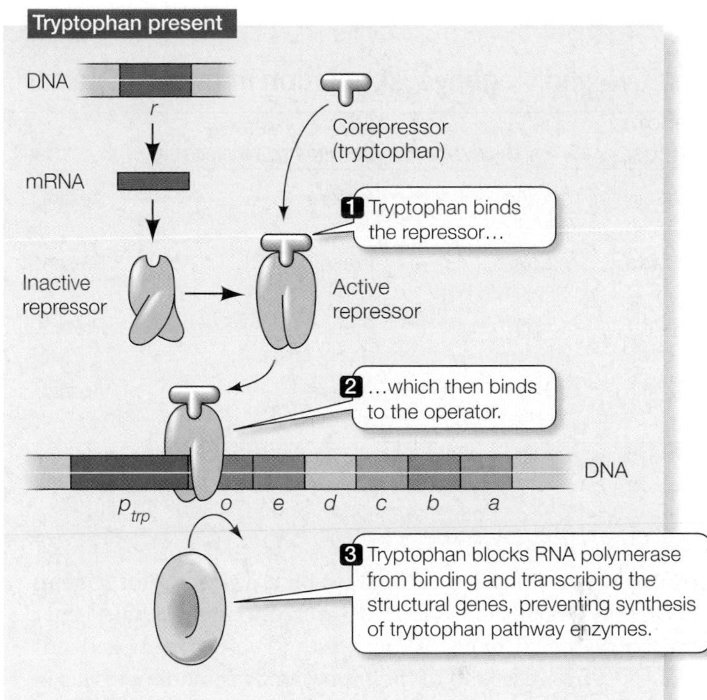

13.20 The *trp* Operon: A Repressible System Because tryptophan activates an otherwise inactive repressor, it is called a corepressor.

- In inducible systems, the *substrate* of a metabolic pathway (the inducer) interacts with a regulatory protein (the repressor), rendering the repressor incapable of binding to the operator and thus *allowing* transcription.
- In repressible systems, the *product* of a metabolic pathway (the corepressor) interacts with a regulatory protein to make it capable of binding to the operator, thus *blocking* transcription.

In general, inducible systems control catabolic pathways (which are turned on only when the substrate is available), whereas repressible systems control anabolic pathways (which are turned off until the product becomes unavailable). In both systems, the regulatory molecule functions by binding to the operator. Next we will consider an example of control by binding to the promoter.

Protein synthesis can be controlled by increasing promoter efficiency

Suppose an *E. coli* cell lacks a sufficient supply of glucose, its preferred energy source, but has access to lactose, another sugar. The *lac* operon encodes enzymes that catabolize lactose, and can increase transcription of those enzymes by increasing the efficiency of the promoter (**Figure 13.21**).

In the *lac* operon and other similar operons, the promoter binds RNA polymerase in a series of steps. First,

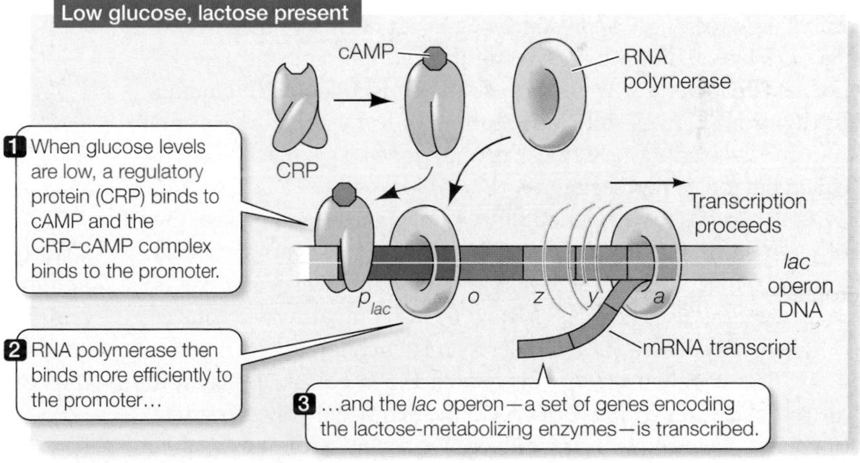

Low glucose, lactose present
cAMP
RNA polymerase
CRP
1 When glucose levels are low, a regulatory protein (CRP) binds to cAMP and the CRP–cAMP complex binds to the promoter.
Transcription proceeds
lac operon DNA
P lac
o z y a
mRNA transcript
2 RNA polymerase then binds more efficiently to the promoter…
3 …and the lac operon—a set of genes encoding the lactose-metabolizing enzymes—is transcribed.

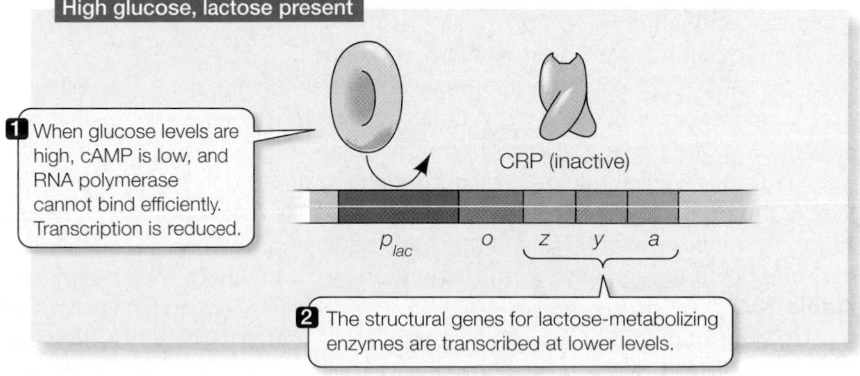

High glucose, lactose present
1 When glucose levels are high, cAMP is low, and RNA polymerase cannot bind efficiently. Transcription is reduced.
CRP (inactive)
P lac
o z y a
2 The structural genes for lactose-metabolizing enzymes are transcribed at lower levels.

13.21 Catabolite Repression Regulates the *lac* Operon The promoter for the *lac* operon does not function efficiently in the absence of cAMP, as occurs when glucose levels are high. High glucose levels thus repress the catabolites (metabolic enzymes) for lactose.

TABLE 13.2

Positive and Negative Regulation in the *lac* Operon[a]

GLUCOSE	cAMP LEVELS	RNA POLYMERASE BINDING TO PROMOTER	LACTOSE	*lac* REPRESSOR	TRANSCRIPTION OF *lac* GENES?	LACTOSE USED BY CELLS?
Present	Low	Absent	Absent	Active and bound to operator	No	No
Present	Low	Present, not efficient	Present	Inactive and not bound to operator	Low level	No
Absent	High	Present, very efficient	Present	Inactive and not bound to operator	High level	Yes
Absent	High	Absent	Absent	Active and bound to operator	No	No

[a]Negative regulators are in red type.

a regulatory protein called CRP (short for *c*AMP *r*eceptor *p*rotein) binds the nucleotide adenosine 3′,5′-cyclic monophosphate, better known as cyclic AMP, or cAMP. Next, the CRP–cAMP complex binds to DNA just upstream (5′) of the promoter. This binding results in more efficient binding of RNA polymerase to the promoter, and thus an elevated level of transcription of the structural genes.

When glucose becomes abundant in the medium, the bacterium does not need to break down alternative food molecules, so synthesis of the enzymes that catabolize these molecules diminishes or ceases. The presence of glucose decreases the synthesis of these enzymes by lowering the cellular concentration of cAMP. The lower cAMP concentration leads to less CRP binding to the promoter, less efficient binding of RNA polymerase, and reduced transcription of the structural genes. This mechanism is called **catabolite repression**.

CRP controls the efficiency of RNA polymerase binding in many other operons, such as those that metabolize the sugars arabinose and galactose. In general, when an adequate level of glucose is present, *E. coli* uses the CRP system to turn off pathways that use other energy sources. As you will see in later chapters of this book, cAMP is a widely used signaling molecule in eukaryotes as well as in prokaryotes. The use of cAMP in such widely diverse situations as a bacterium sensing glucose levels and a human sensing hunger demonstrates the prevalence of common themes in biochemistry and natural selection.

The inducible *lac* and repressible *trp* systems of *E. coli*—the two operator–repressor systems—are examples of **negative control** of transcription because the regulatory protein (the repressor) in each case prevents transcription. The catabolite repression mechanism is an example of **positive control** of transcription because the regulatory molecule (the CRP–cAMP complex) is a transcriptional activator. The relationships between these positive and negative regulators in the *lac* operon are summarized in **Table 13.2**.

The control of gene expression by regulatory proteins is not unique to prokaryotes. It occurs in viruses, as we saw earlier, and as we will see in the next chapter, it is important in eukaryotes as well. Throughout nature, genomes contain not only sequences that are transcribed, but nontranscribed sequences that, by binding proteins, can determine whether a particular gene will be transcribed.

13.4 RECAP

Gene expression in prokaryotes is most commonly regulated through control of transcription. Operons consist of a set of closely linked genes and DNA sequences that control their transcription. Operons can be regulated by both negative and positive controls.

- What are the key differences between an inducible system and a repressible system? See pp. 298–299 and Figures 13.19 and 13.20
- Describe the difference between positive and negative control of transcription. See pp. 299–300 and Table 13.2

Clearly, an amazing amount of knowledge about prokaryotic genomes has been recently acquired, and advanced sequencing and genomic techniques promise to deliver more information about specific prokaryotic genes and their functions. To what uses might we hope to put this information?

13.5 What Have We Learned from the Sequencing of Prokaryotic Genomes?

When DNA sequencing first became possible in the late 1970s, the first organisms to be sequenced were the simplest viruses. Soon, over 150 viral genomes, including those of important animal and plant pathogens, had been sequenced. Information on how these viruses infect their hosts and reproduce came quickly as a result. But the manual sequencing techniques used on viruses were not up to the task of elucidating the genomes of prokaryotes and eukaryotes, the smallest of which are a hundred times larger than those of a bacteriophage. In the past decade, however, the automated sequencing techniques described in Section 11.5 have rapidly added many prokaryotic sequences to biologists' store of knowledge.

In 1995, a team led by Craig Venter and Hamilton Smith determined the first sequence of a free-living cellular organism, the bacterium *Haemophilus influenzae*. Many more prokaryotic sequences have followed. These sequences have revealed not only how prokaryotes apportion their genes to perform different cellular functions, but also how their specialized functions are carried out. A beginning has even been made on the provocative question of what the minimal requirements for a living cell might be.

Three types of information can be obtained from a genomic sequence, including:

- *Open reading frames*, which are the coding regions of genes. For protein-coding genes, these regions can be recognized by the start and stop codons for translation.
- *Amino acid sequences of proteins*. These sequences can be deduced from the DNA sequences of open reading frames by applying the genetic code (see Figure 12.6).
- *Regulatory sequences*, such as promoters and terminators for transcription.

Perhaps the record for speedy genome sequencing was that of the virus that causes SARS (Severe Adult Respiratory Syndrome). It took less than two weeks for a team led by Steven Jones at the British Columbia Cancer Center in Vancouver to sequence the viral genome.

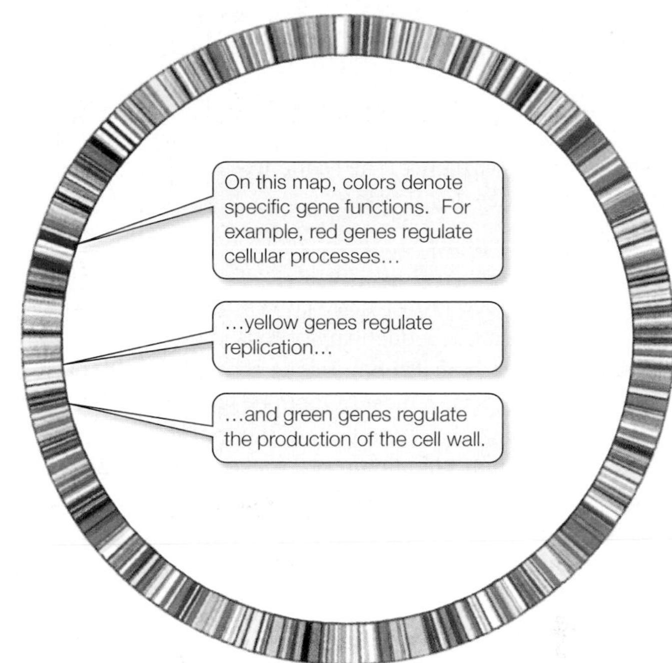

On this map, colors denote specific gene functions. For example, red genes regulate cellular processes...
...yellow genes regulate replication...
...and green genes regulate the production of the cell wall.

13.22 Functional Organization of the Genome of *H. influenzae* The entire DNA sequence has 1,830,137 base pairs.

FUNCTIONAL GENOMICS **Functional genomics** is the assignment of functions to the products of genes described by genomic sequencing. This field, less than a decade old, is now a major occupation of biologists. Let's see how its methods were applied to the bacterium *H. influenzae* once its sequence was known.

The only host for *H. influenzae* is humans. It lives in the upper respiratory tract and can cause ear infections or, more seriously, meningitis in children. Its single circular chromosome has 1,830,137 base pairs (**Figure 13.22**). In addition to its origin of replication and the genes coding for rRNAs and tRNAs, this bacterial chromosome has 1,743 regions containing amino acid codons as well as the transcriptional (promoter) and translational (start and stop codons) information needed for protein synthesis—that is, regions that are likely to be genes that code for proteins.

When this sequence was first announced, only 1,007 (58 percent) of the bacterium's genes had amino acid sequences that corresponded to proteins with known functions—in other words, only 58 percent were genes that the researchers, based on their knowledge of the functions of bacteria, expected to find. The remaining 42 percent of its genes coded for proteins that were unknown to researchers. The roles of most of the unknown proteins have been identified since that time by a process known as **annotation**.

Most annotation of genes and proteins with known functions confirmed a century of biochemical description of bacterial enzymatic pathways. For example, genes for enzymes making up entire pathways of glycolysis, fermentation, and electron transport were found. Some of the remaining gene sequences for unknown proteins may code for membrane proteins, including those involved in active transport. Another important finding was that highly infective strains of *H. influenzae* have genes coding for surface proteins that attach the bacterium to the human respiratory tract, while noninfective strains lack those genes.

COMPARATIVE GENOMICS Soon after the sequence of *H. influenzae* was announced, smaller (*Mycoplasma genitalium*, 580,070 base pairs) and larger (*E. coli*, 4,639,221 base pairs) prokaryotic sequences were completed. Thus began a new era in biology, the era of **comparative genomics**, in which the genome sequences of different organisms are compared to see what genes one organism has that are missing in another, in order to relate the results to physiology.

M. genitalium, for example, lacks the enzymes needed to synthesize amino acids, which the other two prokaryotes possess. This finding reveals that *M. genitalium* is a parasite, which must obtain all its amino acids from its environment, the human urogenital tract. *E. coli* has 55 regulatory genes coding for transcriptional activators and 58 for repressors; *M. genitalium* has only 3 genes for activators. Comparisons such as these have led to the formulation of specific questions about how an organism lives the way it does. We'll see many more applications of comparative genomics in the next chapter.

The sequencing of prokaryotic genomes has many potential benefits

Prokaryotic genome sequencing promises to provide insights into microorganisms that cause human diseases as well as those involved in global ecological cycles (see Chapter 56).

- *Chlamydia trachomatis* causes the most common sexually transmitted disease in the United States. Because it is an intracellular parasite, it has been very hard to study. Among its

900 genes are several for ATP synthesis—something scientists used to think this bacterium could not do.

- *Rickettsia prowazekii* causes typhus; it infects people bitten by louse vectors. Of its 634 genes, 6 code for proteins that are essential for its virulence. The sequences of these genes are being used to develop vaccines.
- Not a lot is known about the Sargasso Sea, a floating mass of warm water in the North Atlantic Ocean where a large mat of *Sargassum* seaweed grows, but relatively few other life forms have been observed. Sequencing of cells in this water has shown that it harbors its own unique ecosystem, with over 1,000 bacterial species, most of them never described before, that are specially adapted to live among the *Sargassum* mats. Earth's oceans may contain up to 2 million prokaryotic species, and the soils another 4 million. The roles of 99 percent of them in Earth's ecology remain obscure. Sequencing their genomes may provide some answers.
- *Mycobacterium tuberculosis* causes tuberculosis. It has a large genome for a prokaryote, coding for 4,000 proteins. Over 250 of these proteins are used to metabolize lipids, so this may be the main way that the bacterium gets its energy. Some of its genes code for previously unidentified cell surface proteins; these genes are targets for potential vaccines.
- *Streptomyces coelicolor* and its close relatives produce two-thirds of all the antibiotics currently in clinical use, including streptomycin, tetracycline, and erythromycin. The genome sequence of this bacterium reveals 22 clusters of genes responsible for antibiotic production, of which only 4 were previously known. This finding may lead to more and better antibiotics to combat resistant pathogens.
- In addition to the well-known carbon dioxide, another important gas contributing to the atmospheric "greenhouse effect" and global warming is methane (CH_4; see Figure 2.8). Some bacteria, such as *Methanococcus*, produce methane in the stomachs of cows. Others, such as *Methylococcus*, remove methane from the air and use it as a carbon–energy source. The genomes of both of these bacteria have been sequenced. Understanding the genes involved in methane production and oxidation may help us meet the challenge of global warming.
- *E. coli* strain O157:H7 in hamburger can cause severe illness when ingested, as happens to at least 70,000 people a year in the United States. Its genome has 5,416 genes, of which 1,387 are different from those in the familiar (and harmless) laboratory strains of this bacterium. Remarkably, many of these unique genes are also present in other pathogenic bacteria, such as *Salmonella* and *Shigella*. This finding suggests that there is extensive genetic exchange between these species, and that "superbugs" may be on the horizon.

Will defining the genes required for cellular life lead to artificial life?

When the genomes of prokaryotes and eukaryotes are compared, a striking conclusion arises: certain genes are present in all organisms (*universal genes*). There are also some (nearly) universal gene segments—one that codes for an ATP binding site, for example—

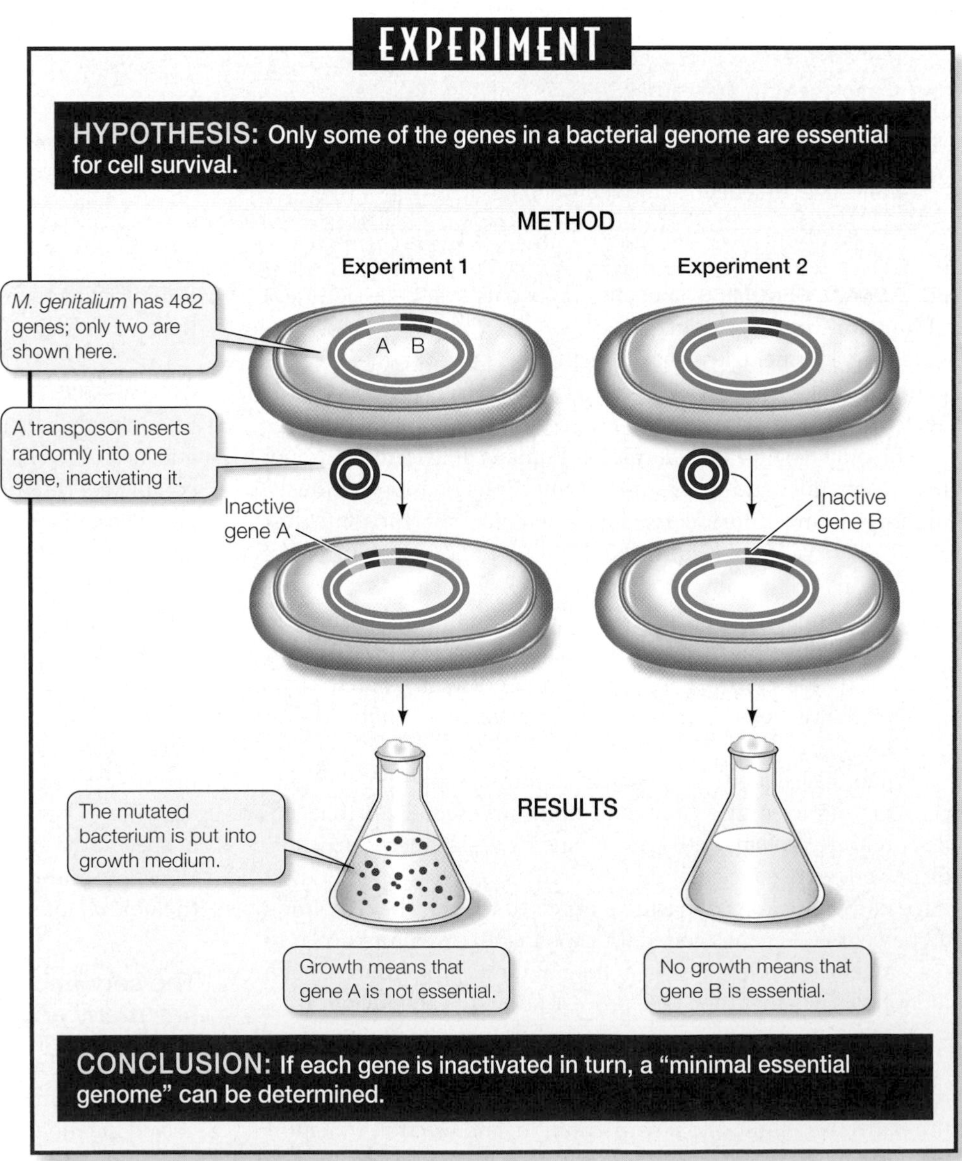

EXPERIMENT
HYPOTHESIS: Only some of the genes in a bacterial genome are essential for cell survival.
METHOD
Experiment 1
Experiment 2
M. genitalium has 482 genes; only two are shown here.
A B
A transposon inserts randomly into one gene, inactivating it.
Inactive gene A
Inactive gene B
The mutated bacterium is put into growth medium.
RESULTS
Growth means that gene A is not essential.
No growth means that gene B is essential.
CONCLUSION: If each gene is inactivated in turn, a "minimal essential genome" can be determined.

13.23 Using Transposon Mutagenesis to Determine the Minimal Genome By inactivating the genes in a bacterial genome one by one, scientists can determine which genes are essential for the cell's survival.

that are present in many genes in many organisms. These findings suggest that there is some ancient, minimal set of DNA sequences common to all cells. One way to identify these sequences is to look for them (or, more realistically, to have a computer look for them).

Another way to define the minimal genome is to take the organism with the simplest genome, deliberately mutate one gene at a time, and see what happens. *Mycoplasma genitalium* has the smallest known genome—only 482 genes. Even so, some of its genes are dispensable under some circumstances. It has genes for metabolizing both glucose and fructose. In the laboratory, the organism can survive on a medium supplying only one of those sugars, making the genes for metabolizing the other sugar unnecessary. But what about other genes? Experiments using transposons as mutagens have addressed this question. When the bacterium is exposed to transposons, they insert themselves into a gene at random, mutating and inactivating it (**Figure 13.23**). The mutated bacteria are sequenced to determine which gene was mutated, and then tested for growth and survival.

The astonishing result of these studies is that *M. genitalium* can survive in the laboratory without the services of 100 of its genes, leaving a minimal genome of 382 genes! But is this really what it takes to make a viable organism? The only way to find out would be to build the genome, gene by gene. A team led by Craig Venter at a private company is trying to do this. Since it is possible to synthesize artificial DNA molecules, and since the DNA sequences of all of the genes of *M. genitalium* are known, making and then assembling the genome in the laboratory should be straightforward. The genome will then be inserted into an empty bacterial cell. If it starts transcribing mRNA and making proteins, it may result in a viable cell. If so, human-created life will be a reality.

In addition to the obvious technical feat of creating artificial life, this technique could have many important applications. New microbes could be made with entirely new abilities, such as degrading oil spills, making synthetic fibers, reducing tooth decay, or converting cellulose to ethanol for use as fuel. On the other hand, fears of the misuse or mishandling of this knowledge are not unfounded. It would be equally possible to develop synthetic bacteria with detrimental properties, such as species toxic to people, plants, or animals, and use them as agents of biological warfare or bioterrorism. As described at the beginning of this chapter, the mutation of avian flu into a lethal human pathogen suggests the danger that new species, however small, can sometimes pose. The "genomics genie" is, for better or worse, already out of the bottle.

13.5 RECAP

DNA sequencing is used to study the genomes of prokaryotes of importance to humans and ecosystems. Functional genomics uses these sequences to determine the functions of gene products; comparative genomics uses sequence similarities with genes with known roles in other organisms to help identify functions.

- Can you describe how open reading frames are recognized in a genomic sequence? What kind of information can be derived from open reading frames? See p. 301
- Can you give some examples of prokaryotic genomes that have been sequenced and what the sequence has shown? See p. 302
- How are selective inactivation studies being used to determine the minimal genome? See pp. 302–303 and Figure 13.23

CHAPTER SUMMARY

13.1 How do viruses reproduce and transmit genes?

Prokaryotes and viruses are useful model organisms for the study of genetics and molecular biology because they contain much less DNA than eukaryotes, grow and reproduce rapidly, and are usually haploid.

Viruses are not cells, and rely on host cells to reproduce.

The basic unit of a virus is a **virion**, which consists of a nucleic acid genome (DNA or RNA) and a protein coat, called a **capsid**.

Bacteriophage are viruses that infect bacteria.

Virulent viruses undergo a **lytic cycle**, which causes the host cell to burst, releasing new virions.

Temperate viruses can also undergo a **lysogenic cycle**, in which a molecule of their DNA, called a **prophage**, is inserted into the host chromosome, where it replicates for generations. Such hosts are referred to as **lysogenic bacteria**. Review Figure 13.3

Some viruses have promoters that bind host RNA polymerase, which they use to transcribe their own genes and proteins. Review Figure 13.4

Many animal and plant viruses are spread by **vectors**, such as insects.

The viruses that infect animals include both RNA and DNA viruses. Enveloped viruses have a membrane derived from the host's plasma membrane. Review Figure 13.5

A **retrovirus** uses reverse transcriptase to generate a cDNA **provirus** from its RNA genome. The provirus is incorporated into the animal host's DNA and can be activated to produce new virions. Review Figure 13.6

13.2 How is gene expression regulated in viruses?

Whether a temperate phage undergoes a lytic or a lysogenic cycle depends on the state of cellular environment offered by the host, which is assessed by regulatory proteins that compete for promoters on phage DNA. Review Figure 13.8

13.3 How do prokaryotes exchange genes?

Most prokaryotes reproduce by the division of single cells, giving rise to **clones** of genetically identical cells. Allowed to multiply on a

CHAPTER SUMMARY

solidified nutrient medium, they will produce bacterial colonies or a bacterial lawn. Review Figure 13.9

An experiment by Lederberg and Tatum demonstrated that bacteria can exchange genetic information. Review Figure 13.10

In **conjugation**, two bacterial cells are brought into proximity by **sex pili**. A portion of the donor cell's chromosome moves into the recipient cell through a **conjugation tube** and may become incorporated into the recipient cell's chromosome.

In **transformation**, bacteria take up free DNA released by dead bacterial cells.

In **transduction**, phage capsids carry bacterial DNA from one bacterium to another. Review Figure 13.13

Plasmids are small bacterial chromosomes that are independent of the main chromosome and may be transferred between cells. Plasmids may act as **metabolic factors**, **fertility factors**, or **resistance factors**. Review Figure 13.14

Transposable elements are short DNA sequences that can move from one place to another in the bacterial genome, often interrupting gene sequences. Transposable elements that carry other genes are called **transposons**. Review Figure 13.15

13.4 How is gene expression regulated in prokaryotes?

In prokaryotes, the synthesis of some proteins is regulated so that they are made only when they are needed. Proteins that are made only in the presence of a particular compound—an **inducer**—are **inducible** proteins. Proteins that are made at a constant rate regardless of conditions are **constitutive** proteins.

An **operon** consists of a promoter, an **operator**, and two or more **structural genes**. Promoters and operators do not code for proteins, but serve as binding sites for regulatory proteins. Review Figure 13.18

Regulatory genes code for regulatory proteins, such as **repressors**. When a repressor binds to an operator, transcription of the structural gene is inhibited. Review Figure 13.19, Web/CD Tutorial 13.1

The *lac* operon is an example of an inducible system, in which the presence of an inducer keeps the repressor from binding the operator and allows transcription.

A **repressible** protein is one whose synthesis is turned off by the presence of a particular compound—a **corepressor**. The *trp* operon is an example of a repressible system, in which the presence of a corepressor causes the repressor to bind to the operator and stops transcription. Review Figure 13.20, Web/CD Tutorial 13.2

In catabolite repression transcription is enhanced by the binding of a regulatory protein to the promoter. Review Figure 13.21

13.5 What have we learned from the sequencing of prokaryotic genomes?

Functional genomics relates gene sequences to protein functions.

In **comparative genomics**, the genome sequences of different organisms are compared.

Sequencing studies have revealed universal genes that are present in all organisms and unique genes associated with organism functions.

By mutating individual genes in a small genome, scientists can determine the minimal genome required for cellular life. These studies could lead to the creation of artificial cells. Review Figure 13.23

See Web/CD Activity 13.1 for a concept review of this chapter.

SELF-QUIZ

1. Which of the following statements about the *lac* operon is *not* true?
   a. When lactose binds to the repressor, the repressor can no longer bind to the operator.
   b. When lactose binds to the operator, transcription is stimulated.
   c. When the repressor binds to the operator, transcription is inhibited.
   d. When lactose binds to the repressor, the shape of the repressor is changed.
   e. When the repressor is mutated, one possibility is that it does not bind to the operator.
2. Which of the following is *not* a type of viral reproduction?
   a. DNA virus in a lytic cycle
   b. DNA virus in a lysogenic cycle
   c. RNA virus by a double-stranded RNA intermediate
   d. RNA virus by reverse transcription to make cDNA
   e. RNA virus by acting as tRNA
3. In the lysogenic cycle of bacteriophage λ,
   a. a repressor, cI, blocks the lytic cycle.
   b. a bacteriophage carries DNA between bacterial cells.
   c. both early and late phage genes are transcribed.
   d. the viral genome is made into RNA, which stays in the host cell.
   e. many new viruses are made immediately, regardless of host health.
4. An operon is
   a. a molecule that can turn genes on and off.
   b. an inducer bound to a repressor.
   c. a series of regulatory sequences controlling transcription of protein-coding genes.
   d. any long sequence of DNA.
   e. a promoter, an operator, and a group of linked structural genes.
5. Which statement is true of both transformation and transduction?
   a. DNA is transferred between viruses and bacteria.
   b. Neither occurs in nature.
   c. Small fragments of DNA move from one cell to another.
   d. Recombination between the incoming DNA and host cell DNA does not occur.
   e. A conjugation tube is used to transfer DNA between cells.
6. Plasmids
   a. are circular protein molecules.
   b. are required by bacteria.
   c. are tiny bacteria.
   d. may confer resistance to antibiotics.
   e. are a form of transposable element.

7. The minimal genome can be estimated for a prokaryote
   a. by counting the total number of genes.
   b. by comparative genomics.
   c. as about 5,000 genes.
   d. by transposon mutagenesis, one gene at a time.
   e. by leaving out genes coding for tRNA.
8. When tryptophan accumulates in a bacterial cell,
   a. it binds to the operator, preventing transcription of adjacent genes.
   b. it binds to the promoter, allowing transcription of adjacent genes.
   c. it binds to the repressor, causing it to bind to the operator.
   d. it binds to the genes that code for enzymes.
   e. it binds to RNA and initiates a negative feedback loop to reduce transcription.
9. The promoter in the *lac* operon is
   a. the region that binds the repressor.
   b. the region that binds RNA polymerase.
   c. the gene that codes for the repressor.
   d. a structural gene.
   e. an operon.
10. The CRP–cAMP system
   a. produces many catabolites.
   b. requires ribosomes.
   c. operates by an operator–repressor mechanism.
   d. is an example of positive control of transcription.
   e. relies on operators.

FOR DISCUSSION

1. Viruses sometimes carry DNA from one cell to another by transduction. Sometimes a segment of bacterial DNA is incorporated into a phage protein coat without any phage DNA. These particles can infect a new host. Would the new host become lysogenic if the phage originally came from a lysogenic host? Why or why not?
2. Compare the life cycles of the viruses that cause influenza and AIDS (Figures 13.5 and 13.6) with respect to:
   a. how the virus enters the cell.
   b. how the virion is released in the cell.
   c. how the viral genome is replicated.
   d. how new viruses are produced.
3. Compare promoters adjacent to early and late genes in the bacteriophage lytic cycle.
4. The repressor protein that acts on the *lac* operon of *E. coli* is encoded by a regulatory gene. The repressor protein is made in small quantities and at a constant rate per cell. Would you surmise that the promoter for this repressor protein is efficient or inefficient? Is synthesis of the repressor constitutive, or is it under environmental control?
5. A key characteristic of a repressible enzyme system is that the repressor protein must react with a corepressor (typically, the end product of a pathway) before it can bind to the operator of an operon to shut the operon off. How is this system different from an inducible enzyme system?

FOR INVESTIGATION

A patient was admitted to the hospital with a urinary tract infection, and antibiotic-sensitive *Klebsiella* bacteria were identified as the cause. He was treated with ampicillin and kanamycin. However, during his hospital stay, the bacteria became resistant to both antibiotics. The patient in the bed next to him had *E. coli* that were resistant to these two antibiotics. Three possible explanations for the development of resistance in *Klebsiella* are (A) spontaneous mutation, (B) bacterial conjugation and transfer of resistance genes on the main chromosome, and (C) transfer of a plasmid containing the resistance genes. How would you investigate these possibilities?

CHAPTER 14 The Eukaryotic Genome and Its Expression

Endangered genomes

The derivations of the common and scientific names of the cheetah, *Acinonyx jubatus*, tell the story:

- The word "cheetah" is from the Hindi *chiita*, meaning "spotted." The small black spots on yellow fur are unmistakable in this animal.
- *Acinonyx* in Greek means "no-move claw." Alone among the cats, cheetahs cannot fully retract their claws—an advantage in running fast and in hunting.
- *Jubatus* in Latin means "maned"; manes are characteristic of cheetah cubs.

This sleek, muscular cat is a solitary hunter, preying on smaller mammals such as gazelles and hares. It stalks its victim to within about 10–30 meters, then chases the prey at speeds up to 110 kilometers (70 miles) per hour. The chase is usually over in a minute or less.

There are only about 12,000 cheetahs in the world today, almost all of them in Africa. Some of the decline in their numbers is due to human intervention. For instance, farmers have hunted them down because they were (mistakenly) blamed for killing cattle. But something else is involved. When the sequences of the two parental strands of DNA from an individual cheetah are compared, there is a high degree of homozygosity: both alleles for protein-coding genes are usually the same. In addition, the *same sequences are present in almost all cheetahs*. It's as if they all came from a single set of parents!

Fossil remains of cheetahs and of several related extinct cat species provide part of the explanation for this remarkable genetic homogeneity. The modern cheetah probably evolved in Africa about 15 million years ago, then migrated to Asia and North America. Until the end of the last glaciation (about 10,000 years ago), cheetahs were widespread. Then, something (we're not sure what) happened. Many related species—such as the saber-toothed tiger—died out. But a few cheetahs apparently survived, and we presume that the genomes of all the cheetahs alive today are descended from those few. As we will learn in Chapter 22, such an event is called a *bottleneck*.

With their homogeneous genomes, cheetahs lack the genetic variability that can help a species survive adverse conditions. For example, cheetahs are very susceptible to diseases because they lack the genetic diversity that might allow them to mount an attack on new pathogens.

The cheetah is not the only species with high genomic homogeneity. The few Florida panthers that remain are likewise homogeneous in their DNA sequences. Like the cheetah, they are susceptible to diseases and genetic defects. In the Florida panther's case, however, their decline in numbers is recent and

Fast Cat Cheetahs are the swiftest of all the world's land animals. Most of the 12,000 cheetahs living in the wild carry almost identical gene sequences, primarily the result of an evolutionary event some 10,000 years ago that wiped out all but a few individuals.

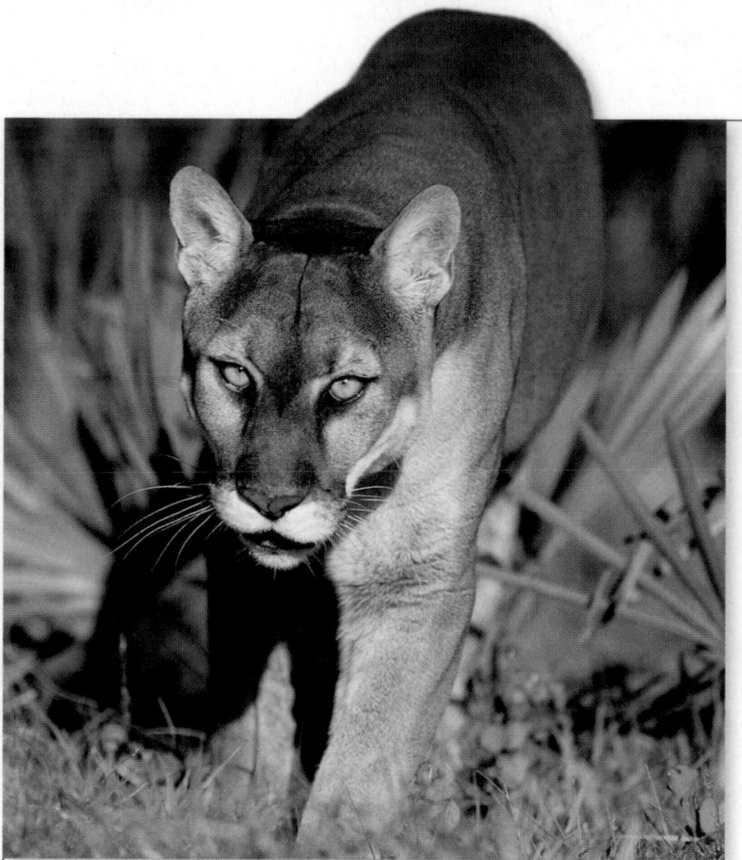

Endangered Cat The Florida panther, *Felis concolor coryi*, is a highly endangered North American cougar. The U.S. Fish and Wildlife Service estimates the number of living adult Florida panthers to be between 30 and 50.

is due solely to human intervention. Overhunted during the nineteenth century, their habitat in the southeastern United States was almost totally eliminated during the twentieth century, having been cleared to make way for human habitation.

When a variety of alleles is present in a population, it means there is a lot of room for new variations of the protein (or proteins) the gene transcribes. The complex process of turning a gene into a protein is the subject of this chapter.

IN THIS CHAPTER we will look at the organization and expression of eukaryotic genomes, which are significantly different from prokaryotic genomes. We'll begin by highlighting some of these differences and describing some of the model organisms that have made our understanding of eukaryotic genomes possible. In the rest of this chapter we look at the many points at which the complex process of eukaryotic gene expression can be regulated before, during, and after transcription and translation.

CHAPTER OUTLINE

14.1 What Are the Characteristics of the Eukaryotic Genome?

As genomes have been sequenced and their expression studied, and as the roles of the proteins they encode are annotated, a number of major differences have emerged between eukaryotic genomes and their prokaryotic counterparts (**Table 14.1**). Key differences between eukaryotic and prokaryotic genomes include:

- In terms of haploid DNA content, *eukaryotic genomes are larger than those of prokaryotes*. This difference is not surprising, given that in multicellular organisms there are many cell types and many jobs to do; many proteins—all encoded by DNA—are needed to do those jobs. A typical virus contains enough DNA to code for only a few proteins—about 10,000 base pairs (bp). The most thoroughly studied prokaryote, *E. coli*, has sufficient DNA (over 4.6 million bp) to make several thousand different proteins and regulate their synthesis. Humans have considerably more genes and regulatory sequences: some 6 billion bp (2 meters of DNA) are crammed into each diploid human cell. However, the amount of DNA in a organism does not always correlate with its complexity. For example, the lily (which produces beautiful flowers each spring, but produces fewer proteins than a human does) has 18 times more DNA than a human.
- *Eukaryotic genomes have more regulatory sequences*—and many more regulatory proteins that bind to them—than prokaryotic genomes do. The great complexity of eukaryotes requires a great deal of regulation, which is evident in the many points of control associated with the expression of the eukaryotic genome.
- *Much of eukaryotic DNA is noncoding.* Interspersed throughout many eukaryotic genomes are various kinds of DNA sequences that are not transcribed into mRNA. And coding regions of genes sometimes contain sequences that do not appear in the mRNA that is translated at the ribosome.
- *Eukaryotes have multiple chromosomes.* The genomic "encyclopedia" of a eukaryote is separated into multiple

TABLE 14.1

A Comparison of Prokaryotic and Eukaryotic Genes and Genomes

CHARACTERISTIC	PROKARYOTES	EUKARYOTES
Genome size (base pairs)	10^4–10^7	10^8–10^{11}
Repeated sequences	Few	Many
Noncoding DNA within coding sequences	Rare	Common
Transcription and translation separated in cell	No	Yes
DNA segregated within a nucleus	No	Yes
DNA bound to proteins	Some	Extensive
Promoters	Yes	Yes
Enhancers/silencers	Rare	Common
Capping and tailing of mRNA	No	Yes
RNA splicing required (spliceosomes)	Rare	Common
Number of chromosomes in genome	One	Many

"volumes." This separation requires that each chromosome have, at a minimum, three defining DNA sequences that we have described in previous chapters: an origin of replication (*ori*) recognized by the DNA replication machinery; a centromere region that holds the replicated chromosomes together before mitosis; and a telomeric sequence at each end of the chromosome.

- *In eukaryotes, transcription and translation are physically separated.* The nuclear envelope separates DNA and its transcription (inside the nucleus) from the sites where mRNA is translated into polypeptides (the ribosomes in the cytoplasm). This separation allows for many points of regulation before translation begins: in the synthesis of a primary (pre-mRNA) transcript, in its processing into mature mRNA, and in its transport to the cytoplasm for translation (**Figure 14.1**).

Model organisms reveal the characteristics of eukaryotic genomes

Most of the lessons learned from eukaryotic genomics have come from several simple model organisms that have been studied extensively: the yeast *Saccharomyces cerevisiae*, the nematode (roundworm) *Caenorhabditis elegans*, the fruit fly *Drosophila melanogaster*, and—representing plants—the thale cress, *Arabidopsis thaliana*.

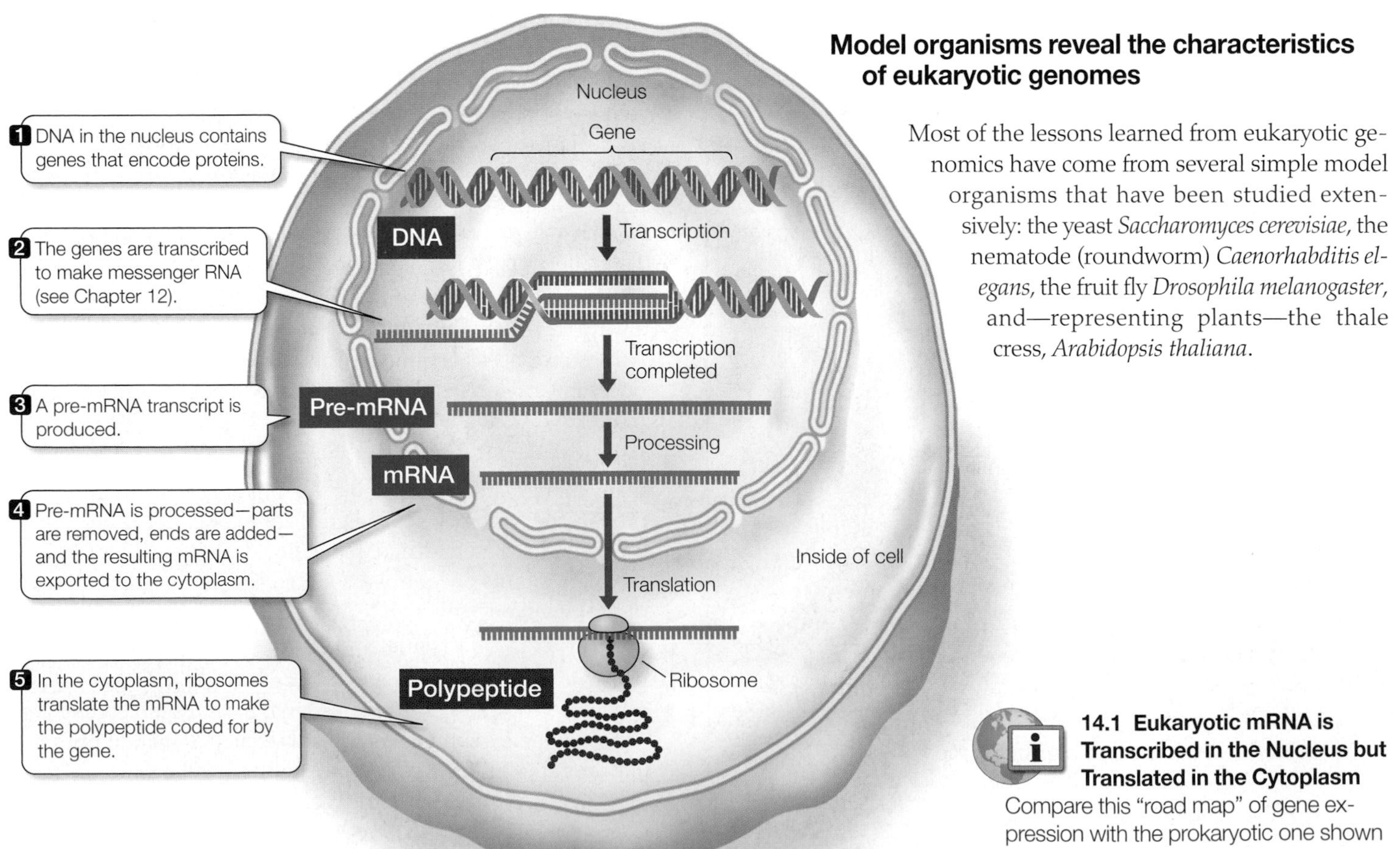

1 DNA in the nucleus contains genes that encode proteins.
2 The genes are transcribed to make messenger RNA (see Chapter 12).
3 A pre-mRNA transcript is produced.
4 Pre-mRNA is processed—parts are removed, ends are added—and the resulting mRNA is exported to the cytoplasm.
5 In the cytoplasm, ribosomes translate the mRNA to make the polypeptide coded for by the gene.
Nucleus
Gene
DNA
Transcription
Transcription completed
Pre-mRNA
Processing
mRNA
Translation
Inside of cell
Ribosome
Polypeptide

14.1 Eukaryotic mRNA is Transcribed in the Nucleus but Translated in the Cytoplasm Compare this "road map" of gene expression with the prokaryotic one shown in Figure 12.3.

The genome of the puffer fish *Fugu rubripes* is the most compact known among vertebrates. It has about the same number of genes as the human genome in about one-eighth as much DNA. The two genomes are so similar that, as Sydney Brenner put it, "the *Fugu* genome is the *Reader's Digest* version of the Book of Man."

YEAST: THE BASIC EUKARYOTIC MODEL Yeasts are single-celled eukaryotes. Like most eukaryotes, they have membrane-enclosed organelles, such as a nucleus and endoplasmic reticulum, as well as a life cycle with alternation of haploid and diploid generations (see Figure 9.14).

In comparison with the prokaryote *E. coli*, whose genome has about 4.6 million bp on a single circular chromosome, the genome of budding yeast (*Saccharomyces cerevisiae*) has 16 linear chromosomes and a haploid content of more than 12 million bp. More than 600 scientists around the world collaborated in mapping and sequencing the yeast genome. When they began, they knew of about 1,000 yeast genes coding for RNAs or proteins. The final sequence revealed 5,800 genes, and annotation has assigned probable roles to most of them. Some of these genes are homologous to genes found in prokaryotes, but many are not.

The most striking difference between the yeast genome and that of *E. coli* is in the number of genes for protein targeting (**Table 14.2**). Both of these single-celled organisms appear to use about the same numbers of genes to perform the basic functions of cell survival. It is the compartmentalization of the eukaryotic yeast cell into organelles that requires it to have so many more genes. This finding is direct, quantitative confirmation of something we have known for a century: the eukaryotic cell is structurally more complex than the prokaryotic cell.

Genes encoding several other types of proteins are present in the yeast and other eukaryotic genomes, but have no homologs in prokaryotes:

- Genes encoding histones that package DNA into nucleosomes
- Genes encoding cyclin-dependent kinases that control cell division
- Genes encoding proteins involved in the processing of mRNA

THE NEMATODE: UNDERSTANDING EUKARYOTIC DEVELOPMENT In 1965, Sydney Brenner, fresh from being part of the team that first isolated mRNA, looked for a simple organism in which to study multicellularity. He settled on *Caenorhabditis elegans*, a millimeter-long nematode (roundworm) that normally lives in the soil. But it also lives in the laboratory, where it has become a favorite model organism of developmental biologists (see Section 19.4). The nematode has a transparent body, which scientists can watch over 3 days as a fertilized egg divides and forms an adult worm made up of nearly 1,000 cells. In spite of its small number of cells, the nematode has a nervous system, digests food, reproduces sexually, and ages. So it is not surprising that an intense effort was made to sequence the genome of this model organism.

The *C. elegans* genome is eight times larger than that of yeast (97 million bp) and has four times as many protein-coding genes (19,099). Once again, sequencing revealed far more genes than expected: when the sequencing effort began, researchers estimated that the worm would have about 6,000 genes and about that many proteins. Clearly, it has far more. About 3,000 genes in the worm have direct homologs in yeast; these genes code for basic eukaryotic cell functions. What do the rest of the genes—the bulk of the worm genome—do?

In addition to surviving, growing, and dividing, as single-celled organisms do, multicellular organisms must have genes for holding cells together to form tissues, for cell differentiation to divide up tasks among those tissues, and for intercellular communication pathways to coordinate their activities (**Table 14.3**). Many of the genes so far identified in *C. elegans* that are not present in yeast perform these roles, which will be described in the remainder of this chapter and the next one.

TABLE 14.2

Comparison of the Genomes of *E. coli* and Yeast

	E. COLI	YEAST
Genome length (base pairs)	4,640,000	12,068,000
Number of protein-coding genes	4,300	5,800
Proteins with roles in:		
Metabolism	650	650
Energy production/storage	240	175
Membrane transport	280	250
DNA replication/repair/recombination	120	175
Transcription	230	400
Translation	180	350
Protein targeting/secretion	35	430
Cell structure	180	250

***DROSOPHILA MELANOGASTER*: RELATING GENETICS TO GENOMICS** The fruit fly *Drosophila melanogaster* is a famous model organism, as its study elucidated many principles of genetics (see Section 10.4). Over 2,500 mutations of *Drosophila melanogaster* had been described by the 1990s, when genome sequencing began, and this fact alone was a good reason for sequencing its DNA. The fruit fly is a much larger organism than *C. elegans*, both in size (it has 10 times more cells) and complexity, and it undergoes a complicated developmental transformation from egg to larva to pupa to adult.

Not surprisingly, the fly's genome is larger than that of *C. elegans* (about 180 million bp). But as we mentioned earlier, genome size does not necessarily correlate with the number of genes encoded. In this case, the larger fruit fly genome

TABLE 14.3

C. elegans Genes Essential to Multicellularity

FUNCTION	PROTEIN/DOMAIN	NUMBER OF GENES
Transcription control	Zinc finger; homeobox	540
RNA processing	RNA binding domains	100
Nerve impulse transmission	Gated ion channels	80
Tissue formation	Collagens	170
Cell interactions	Extracellular domains; glycotransferases	330
Cell–cell signaling	G protein-linked receptors; protein kinases; protein phosphatases	1,290

contains fewer genes (13,449) than does the smaller nematode genome. Annotation has given us a snapshot of the functions of the *Drosophila* genes so far characterized (**Figure 14.2**). The distribution of their functions is typical of complex eukaryotes.

Several additional findings from the fruit fly genome are notable:

- 18,941 different mRNAs are transcribed from the 13,449 genes. This means that the fruit fly genome codes for more proteins than it has genes—a discovery of significance in understanding other genomes, including our own.
- An additional 514 genes code for RNAs that are not translated into proteins. These include genes for tRNAs and rRNAs (see Chapter 12), but there are also 123 genes coding for small RNAs that remain in the nucleus. We will look at their possible roles later in this chapter.

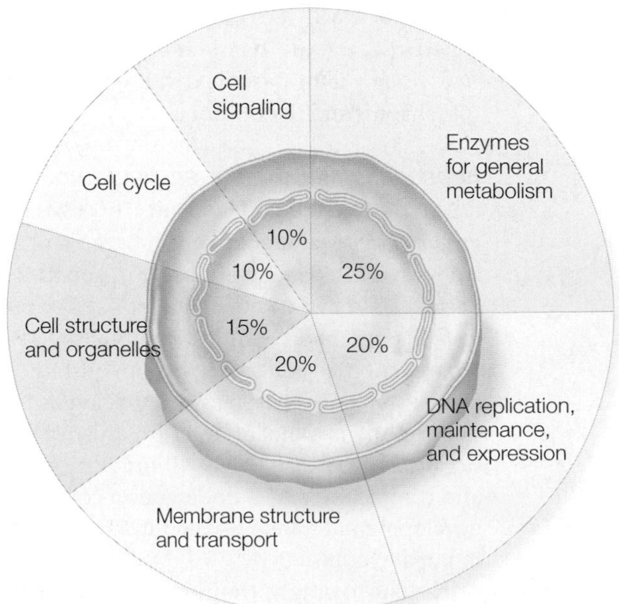

14.2 Functions of the Eukaryotic Genome The functions of the protein-coding genes in the genome of *Drosophila melanogaster* show a pattern that is typical of many complex organisms.

- Comparative genomics has revealed many protein-coding sequences that are similar in fruit flies and other organisms. Not surprisingly, the similarities are strongest with other *Drosophila* species. But a survey of genes involved in human diseases shows that about half have sequences similar to sequences found in fruit flies. Finding out the function of a gene in fruit flies may lead to finding out its function in humans.

ARABIDOPSIS: STUDYING THE GENOMES OF PLANTS

About 250,000 species of flowering plants dominate land and fresh water. But in the history of life, the flowering plants are fairly young, having evolved only about 200 million years ago. Given the pace of mutation and other genetic changes, the differences among these plants are likely to be relatively small. So, although it is the genomes of the plants used by people as food and fiber that hold the greatest interest for us, it is not surprising that instead of sequencing the huge genomes of wheat (16 billion bp) or corn (3 billion bp), scientists first chose to sequence a simpler flowering plant.

Arabidopsis thaliana, the thale cress, is a member of the mustard family and has long been a favorite model organism for study by plant biologists. It is small (hundreds could grow and reproduce in the space occupied by this page), is easy to manipulate, and has a small (119 million bp) genome.

The *Arabidopsis* genome has about 26,000 protein-coding genes, but remarkably, many of these genes are duplicates and probably originated by chromosomal rearrangements. When these duplicate genes are subtracted from the total, about 15,000 unique genes are left—a number similar to the numbers of genes found in the fruit fly and nematode genomes. Indeed, many of the genes found in these animals have homologs in the plant, suggesting that plants and animals have a common ancestor.

But *Arabidopsis* has some genes that distinguish it as a plant (**Table 14.4**). These genes include those involved in photosynthesis, in the transport of water into the root and throughout the plant, in the assembly of the cell wall, in the uptake and metabolism of inorganic substances from the environment, and in the synthesis of specific molecules used for defense against herbivores (organisms that eat plants).

TABLE 14.4

***Arabidopsis* Genes Unique to Plants**

FUNCTION	NUMBER OF GENES
Cell wall and growth	420
Water channels	300
Photosynthesis	139
Defense and metabolism	94

TABLE 14.5

Comparison of the Rice and *Arabidopsis* Genomes

FUNCTION	PERCENTAGE OF GENOME	
	RICE	*ARABIDOPSIS*
Cell structure	9	10
Enzymes	21	20
Ligand binding	10	10
DNA binding	10	10
Signal transduction	3	3
Membrane transport	5	5
Cell growth and maintenance	24	22
Other functions	18	20

Justifying its position as a model plant, these "plant" genes in *Arabidopsis* were also found in the genome of rice, the first major crop plant whose sequence has been determined. Rice (*Oryza sativa*) is the world's most important crop; it is the staple diet for 3 billion people. In fact, two *O. sativa* sequences of have been deciphered: that of *O. sativa indica*, the rice subspecies grown in China and most of tropical Asia, and that of the subspecies *japonica*, which is grown in Japan and other temperate climates (such as the United States). Both genomes are about the same size (430 million bp), yet in this much larger genome is a set of genes remarkably similar to that of *Arabidopsis* (**Table 14.5**). And many of the genes in rice are also present in the much larger genomes of corn and wheat.

Of course, rice as a whole, and each subspecies, has its own particular set of genes that make it unique. The *indica* subspecies is estimated to have 46,000–55,000 genes, while *japonica* has some 32,000–50,000—both higher numbers of genes than in *Arabidopsis*. These "extra" genes include genes for characters that are specific to rice, such as the physiology that allows rice to grow for part of the season submerged in water; the nutrient-packed seeds that sustain human lives; and resistance to certain plant diseases, such as viruses and fungi. Analyses of these and other rice genes will no doubt lead to significant improvements in this crop, and to the improvement of the other grain crops as well.

Eukaryotic genomes contain many repetitive sequences

Studies of the genomes of eukaryotic model organisms have revealed that they contain numerous repetitive DNA sequences that do not code for polypeptides.

HIGHLY REPETITIVE SEQUENCES Two types of *highly repetitive sequences* are found in eukaryotic genomes:

- *Minisatellites* are 10–40 base pairs long and are repeated up to several thousand times. Because DNA polymerase tends to make errors in copying these sequences, the number of copies present varies among individuals. For example, one person might have 300 minisatellites at a particular locus, whereas another person will have 500 minisatellites at that same locus. This variation provides a set of molecular genetic markers that can be used to identify an individual, as we will see in Section 16.1.
- *Microsatellites* are very short (1–3 bp) sequences, present in small clusters of 15–100 copies. They are scattered all over the genome and are also known as simple sequence repeats.

Why call these noncoding repetitive sequences "satellites"? The term conveys that the character of these sequences is different from that of the rest of the genome. For example, the ratio of GC base pairs to AT base pairs may set a satellite apart from the rest of the genome. Recall that there are three hydrogen bonds between G and C, and only two between A and T. This makes the GC base pair somewhat more tightly held than the AT pair. Therefore GC-rich DNA is denser than AT-rich DNA. So, if a satellite has a different proportion of GC base pairs from the rest of DNA, it will have a different density. DNAs of different densities can be separated in a centrifuge, so it is easy to separate some satellites from the rest of the genome.

These highly repetitive sequences are not represented in mature mRNA transcripts. While laboratory scientists have made use of these sequences in genetic studies, their roles in eukaryotes remain unclear.

MODERATELY REPETITIVE SEQUENCES While highly repetitive sequences are not transcribed into mRNA, some *moderately repetitive DNA* sequences are transcribed. These sequences code for tRNAs and rRNAs, which are used in protein synthesis.

The cell makes tRNAs and rRNAs constantly, but even at the maximum rate of transcription, single copies of the DNA sequences coding for them would be inadequate to supply the large amounts of these molecules needed by most cells; hence the genome has multiple copies of these sequences. Since these moderately repetitive sequences are transcribed into RNA, they are properly termed "genes"; that is, there are genes that code for rRNA and tRNA.

In mammals, four different rRNA molecules make up the ribosome: the 18S, 5.8S, 28S, and 5S rRNAs. (The S stands for *Svedberg unit*, a measure of how a substance behaves in a centrifuge that is related to the size of a molecule.) The 18S, 5.8S, and 28S rRNAs are transcribed from a repeated DNA sequence as a single precursor RNA molecule, which is twice the size of the three ultimate products (**Figure 14.3**). Several posttranscriptional steps cut this precursor into the final three rRNA products and discard the noncoding "spacer" RNA. The sequence encoding these RNAs is moderately repetitive in humans: a total of 280 copies of the sequence are located in clusters on five different chromosomes.

TRANSPOSONS Outside of the moderately repetitive DNA sequences that code for rRNA, most moderately repetitive sequences are not stably integrated into the genome. Instead, these sequences can move from place to place in the genome, and thus are called *transposable elements* or **transposons**. Transposons make up over 40 percent of the human genome, far more than the 3–10 percent found in the other sequenced eukaryotes.

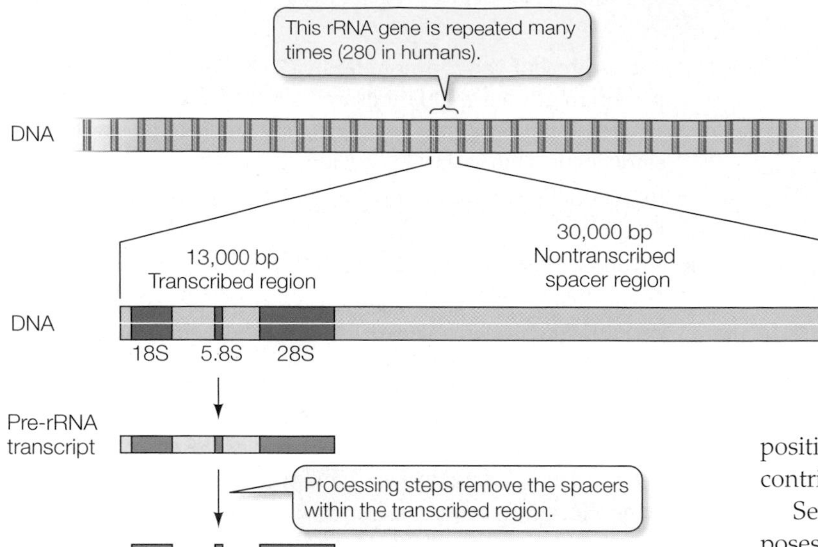

14.3 A Moderately Repetitive Sequence Codes for rRNA This rRNA gene, along with its nontranscribed spacer region, is repeated 280 times in the human genome, with clusters on five chromosomes. Once this gene has been transcribed, posttranscriptional processing removes the spacers within the transcribed region and separates the primary transcript into the three final rRNA products.

There are four main types of transposons in eukaryotes:

1. *SINEs* (short *in*terspersed *e*lements) are up to 500 bp long and are transcribed, but not translated. There are about 1.5 million of them in the human genome, making up 15 percent of human DNA. A single type, the 300 bp *Alu* element, accounts for 11 percent of the human genome; it is present in a million copies scattered over all the chromosomes.
2. *LINEs* (*l*ong *in*terspersed *e*lements) are up to 7,000 bp long, and some are transcribed and translated into proteins. They constitute about 17 percent of the human genome.

Both of these elements are present in more than 100,000 copies. They move about the genome in a distinctive way: they make an RNA copy of themselves, which acts as a template for new DNA, which then inserts itself at a new location in the genome. In this "copy and paste" mechanism, the original sequence stays where it is, and the copy inserts itself at a new location.

3. *Retrotransposons* also make an RNA copy of themselves when they move about the genome. They constitute about 8 percent of the human genome. Some of them code for some of the proteins necessary for their own transposition, and others do not.
4. *DNA transposons* do not use an RNA intermediate, but actually move to a new spot in the genome without replicating (**Figure 14.4**).

What role do these moving sequences play in the cell? There are few answers to this question. The best answer so far seems to be that transposons are cellular parasites that simply replicate themselves. These replications can lead to the insertion of a transposon at a new location, which can have important consequences. For example, the insertion of a transposon into the coding region of a gene results in a mutation (see Figure 14.4). This phenomenon accounts for a few rare forms of several human genetic disease, including hemophilia and muscular dystrophy. If the insertion of a transposon takes place in the germ line, a gamete with a new mutation results. If the insertion takes place in a somatic cell, cancer may result.

If a transposon replicates not just itself but also an adjacent gene, the result may be a gene duplication. A transposon can carry a gene, or a part of it, to a new location in the genome, shuffling the genetic material and creating new genes. Clearly, transposition stirs the genetic pot in the eukaryotic genome and thus contributes to genetic variation.

Section 4.5 described the theory of endosymbiosis, which proposes that chloroplasts and mitochondria are the descendants of once free-living prokaryotes. Transposons may have played a role in endosymbiosis. In living eukaryotes, although chloroplasts and mitochondria contain some DNA, the nucleus contains most of the genes that encode the organelles' proteins. If the organelles were once independent, they must originally have contained all of those genes. How did the genes move to the nucleus? They may have done so by DNA transpositions between organelles and nucleus, which still occur today. The DNA that remains in the organelles may be the remnants of more complete prokaryotic genomes.

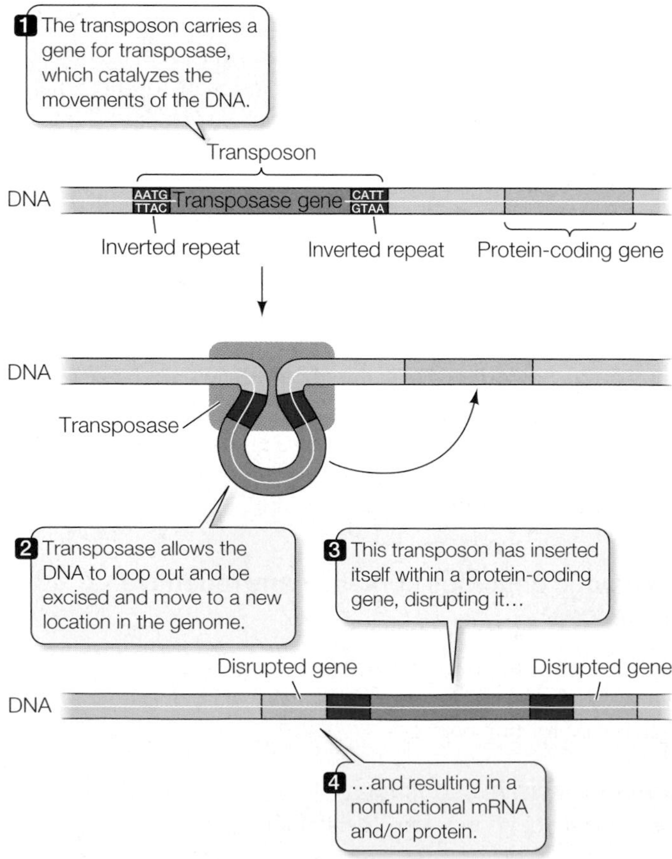

14.4 DNA Transposons and Transposition At the end of each DNA transposon is an inverted repeat sequence that helps in the transposition process.

14.1 RECAP

The sequencing of the genomes of model organisms has demonstrated common features of the eukaryotic genome, which include the presence of repetitive sequences and transposons.

- Can you describe some of the major differences between prokaryotic and eukaryotic genomes? See pp. 307–308 and Table 14.1
- What is one function of genes found in *C. elegans* that has no counterpart in the genome of yeast? See p. 309
- What purpose is served by having multiple copies of sequences coding for rRNA in the mammalian genome? See p. 311
- What effects can transposons have on a genome? See p. 312 and Figure 14.4

Studies of eukaryotic genomes have revealed that they contain coding as well as noncoding sequences. We turn next to the subject at the heart of molecular genetics: genes that code for proteins, and the surprising finding that noncoding regions exist even within these genes.

14.2 What Are the Characteristics of Eukaryotic Genes?

Like their prokaryotic counterparts, many protein-coding genes in eukaryotes exist in only one copy per haploid genome. But eukaryotic genes have two distinctive characteristics that are uncommon among prokaryotes:

- They contain noncoding internal sequences.
- They form gene families—groups of structurally and functionally related "cousins" within the genome.

Protein-coding genes contain noncoding sequences

The structure and transcription of a typical eukaryotic gene is diagrammed in **Figure 14.5**. Preceding the coding region of a eukaryotic gene is a *promoter*, to which an RNA polymerase binds to begin the transcription process. Unlike the prokaryotic enzyme, however, a eukaryotic RNA polymerase does not recognize the promoter sequence by itself, but requires help from other molecules, as we'll see below. At the other end of the gene, after the coding region, is a DNA sequence appropriately called the **terminator**, which signals the end of transcription. It is important that you distinguish the terminator from the stop codon:

- The *terminator* sequence is usually after the stop codon and signals the end of transcription by RNA polymerase.
- The *stop codon* is within the coding region and, when transcribed into mRNA, signals the end of translation at the ribosome.

Eukaryotic protein-coding genes may also contain noncoding base sequences, called **introns**. One or more introns may be interspersed with the coding sequences, which are called **exons**. Transcripts of the introns appear in the primary mRNA transcript, called **pre-mRNA**, but they are removed by the time the mature mRNA—the final mRNA that will be translated—leaves the nucleus. Pre-mRNA processing involves cutting introns out of the pre-mRNA transcript and splicing together the remaining exon transcripts.

How can we locate introns within a eukaryotic gene? The easiest way is by **nucleic acid hybridization**, the method that originally

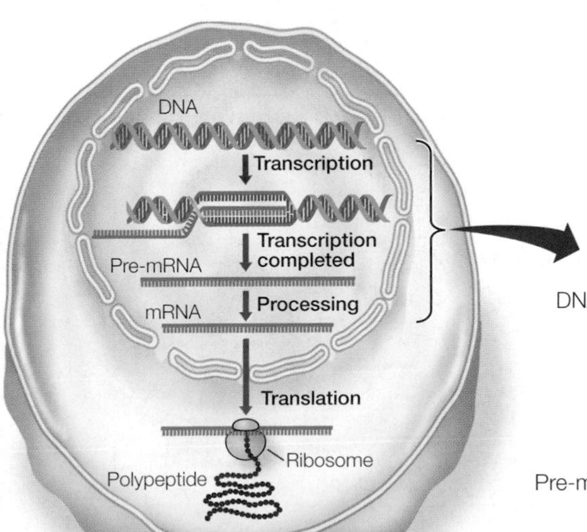

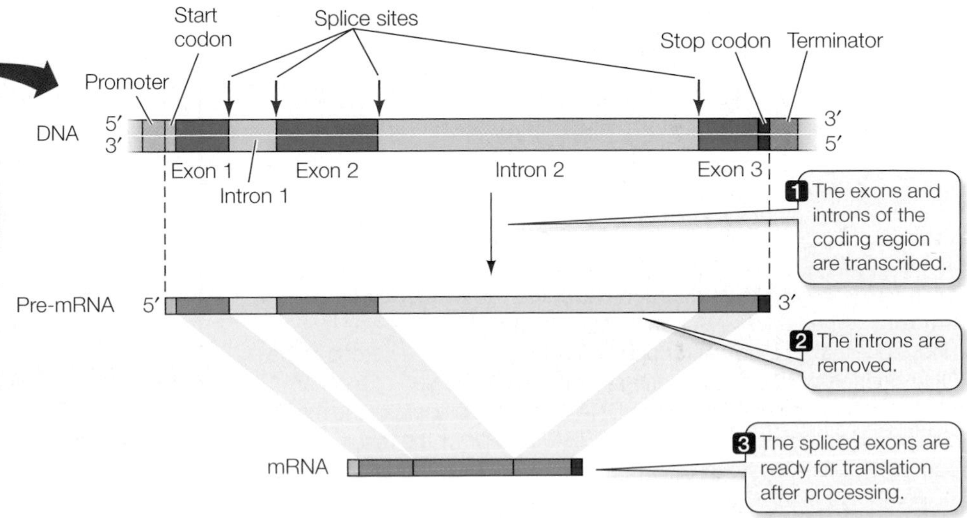

14.5 Transcription of a Eukaryotic Gene The β-globin gene diagrammed here is about 1,600 bp long. The three exons—the protein-coding sequences—contain 441 base pairs (codons for 146 amino acids plus a stop codon). The two introns—noncoding sequences of DNA containing almost 1,000 bp between them—are initially transcribed, but are spliced out of the pre-mRNA transcript.

DNA
Transcription
Pre-mRNA
Transcription completed
mRNA
Processing
Translation
Ribosome
Polypeptide

Start codon
Splice sites
Stop codon
Terminator
Promoter
DNA 5′ 3′
3′ 5′
Exon 1
Intron 1
Exon 2
Intron 2
Exon 3
1 The exons and introns of the coding region are transcribed.
Pre-mRNA 5′ 3′
2 The introns are removed.
mRNA
3 The spliced exons are ready for translation after processing.

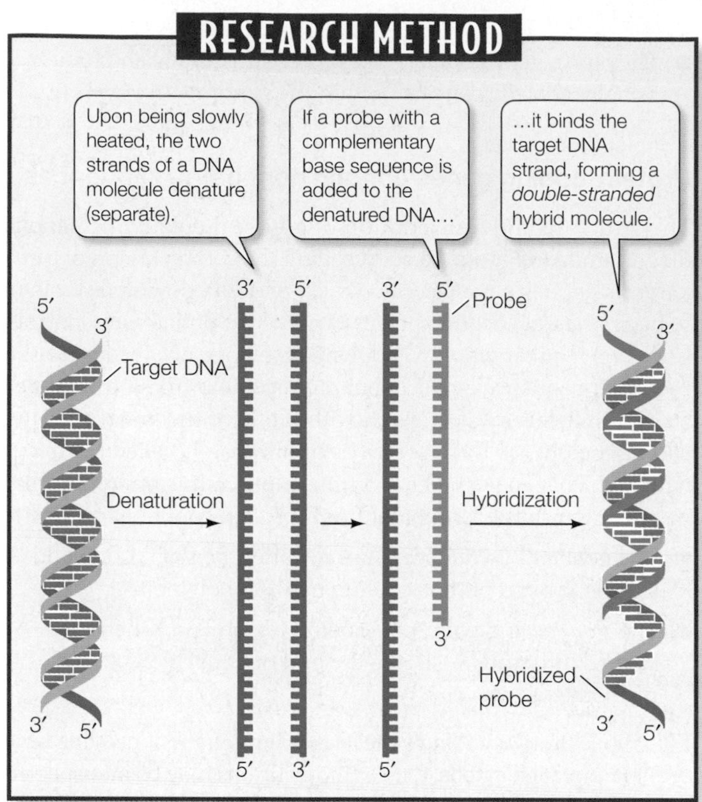
RESEARCH METHOD
Upon being slowly heated, the two strands of a DNA molecule denature (separate).
If a probe with a complementary base sequence is added to the denatured DNA...
...it binds the target DNA strand, forming a *double-stranded* hybrid molecule.
5′
3′
Target DNA
3′
5′
3′
5′
Probe
5′
3′
Denaturation
Hybridization
3′
Hybridized probe
3′
5′
3′
5′
5′
3′
5′

14.6 Nucleic Acid Hybridization Base pairing permits the detection of a sequence complementary to the probe.

revealed the existence of introns. This method, outlined in **Figure 14.6**, has been crucial for studying the relationship between eukaryotic genes and their transcripts. It involves two steps:

- The target DNA is denatured to break the hydrogen bonds between the base pairs and separate the two strands.
- A single-stranded nucleic acid from another source (called a **probe**) is incubated with the denatured DNA. If the probe has a base sequence complementary to the target DNA, a probe–target double helix forms by hydrogen bonding between the bases. Because the two strands are from different sources, the resulting double-stranded molecule is called a *hybrid*.

Biologists used nucleic acid hybridization to examine the β-globin gene, which encodes one of the globin proteins that make up he-

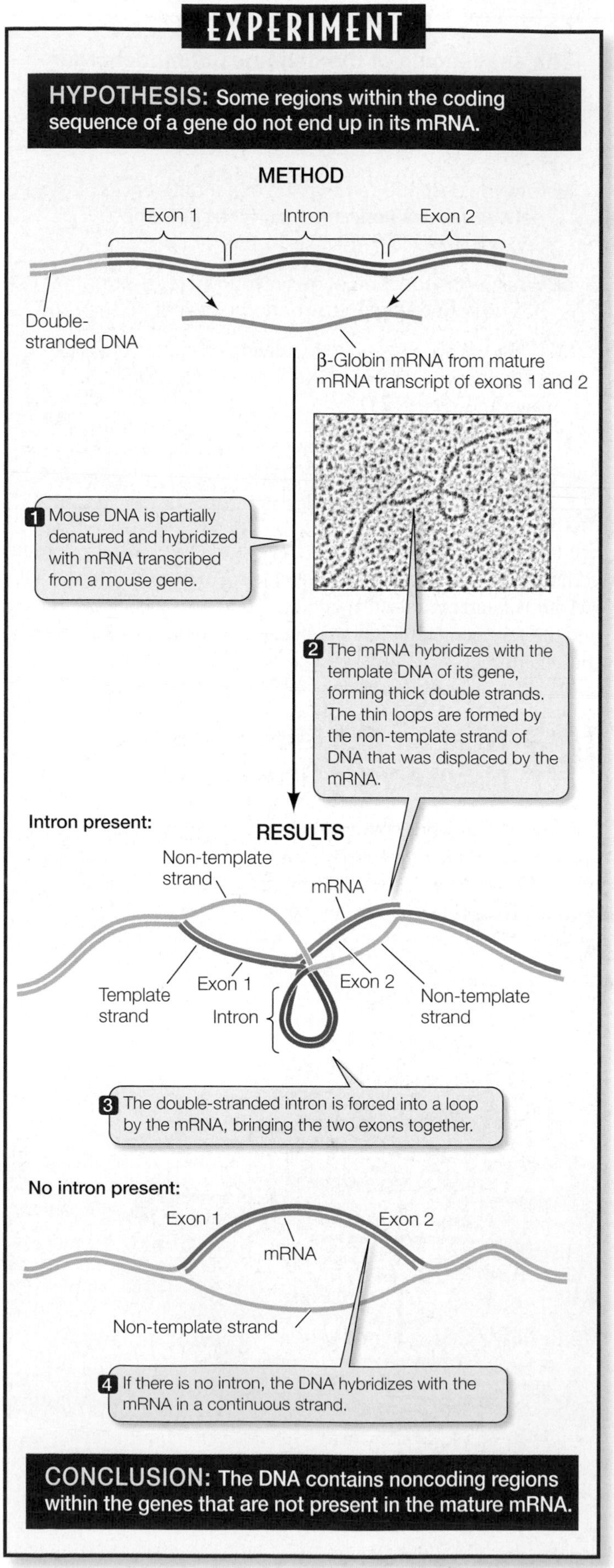

EXPERIMENT

HYPOTHESIS: Some regions within the coding sequence of a gene do not end up in its mRNA.

METHOD

Exon 1
Intron
Exon 2
Double-stranded DNA
β-Globin mRNA from mature mRNA transcript of exons 1 and 2

1 Mouse DNA is partially denatured and hybridized with mRNA transcribed from a mouse gene.

2 The mRNA hybridizes with the template DNA of its gene, forming thick double strands. The thin loops are formed by the non-template strand of DNA that was displaced by the mRNA.

RESULTS

Intron present:
Non-template strand
mRNA
Template strand
Exon 1
Intron
Exon 2
Non-template strand

3 The double-stranded intron is forced into a loop by the mRNA, bringing the two exons together.

No intron present:
Exon 1
Exon 2
mRNA
Non-template strand

4 If there is no intron, the DNA hybridizes with the mRNA in a continuous strand.

CONCLUSION: The DNA contains noncoding regions within the genes that are not present in the mature mRNA.

14.7 Nucleic Acid Hybridization Revealed the Existence of Introns When an mRNA transcript of the β-globin gene was hybridized with the double-stranded DNA of that gene, the introns in the DNA "looped out," demonstrating that the coding region of a eukaryotic gene can contain noncoding DNA that is not present in the mature mRNA transcript. FURTHER RESEARCH: Draw the result assuming that there were three exons and two introns.

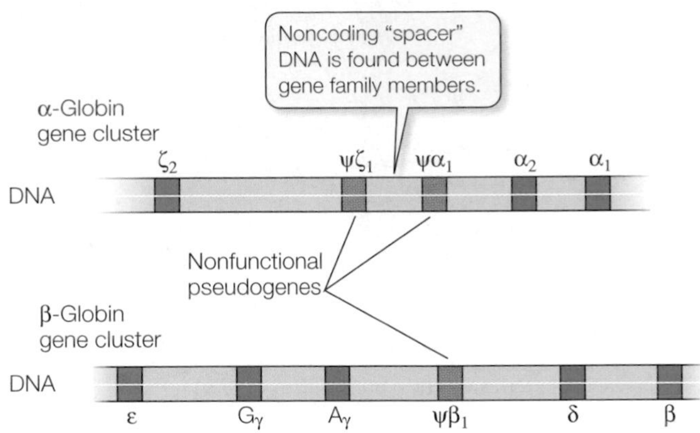

14.8 The Globin Gene Family The α-globin and β-globin clusters of the human globin gene family are located on different chromosomes. The genes of each cluster are separated by noncoding "spacer" DNA. The nonfunctional pseudogenes are indicated by the Greek letter psi (ψ). The γ gene has two variants, A_γ and G_γ.

moglobin. Follow the experiment in **Figure 14.7** carefully as we describe what they did and what happened.

The researchers first denatured DNA containing the β-globin gene by heating it slowly, then added previously isolated, mature β-globin mRNA. As expected, the mRNA bound to the DNA by complementary base pairing. The researchers expected to obtain a linear (1:1) matchup of the mRNA to the coding DNA. That expectation was only partly met: there were indeed stretches of RNA–DNA hybridization, but some looped structures were also visible. These loops were the introns, stretches of DNA that did not have complementary bases on the mature mRNA.

Later studies would show complete hybridization of *pre*-mRNA to DNA, revealing that the introns were indeed part of the pre-mRNA transcript. Somewhere on the path from primary transcript (pre-mRNA) to mature mRNA, the introns had been removed, and the exons had been spliced together. We will examine this splicing process in the next section.

Most (but not all) vertebrate genes contain introns, as do many other eukaryotic genes (and even a few prokaryotic ones). The largest human gene, for the muscle protein called titin, has 363 exons, which together code for 38,138 amino acids. Introns interrupt, but do not scramble, the DNA sequence that codes for a polypeptide chain. The base sequence of the exons, taken in order, is complementary to that of the mature mRNA product. In some cases, the separated exons code for different functional regions, or *domains*, of the protein. For example, the globin polypeptides that make up hemoglobin each have two domains: one for binding to a nonprotein pigment called heme, and another for binding to the other globin subunits. These two domains are encoded by different exons in the globin genes.

Gene families are important in evolution and cell specialization

About half of all eukaryotic protein-coding genes exist in only one copy in the haploid genome (two copies in somatic cells). The rest are present in multiple copies. Over evolutionary time, the copies of a gene may undergo separate mutations, giving rise to a group of closely related genes called a **gene family**. Some gene families, such as the genes encoding the globin proteins that make up hemoglobin, contain only a few members; other families, such as the genes encoding the immunoglobulins that make up antibodies, have hundreds of members.

Like the members of any family, the DNA sequences in a gene family are usually different from one another. As long as one member retains the original DNA sequence and thus codes for the proper protein, the other members can mutate slightly, extensively, or not at all. The availability of such "extra" genes is important for "experiments" in evolution: If the mutated gene is useful, it may be selected for in succeeding generations. If the mutated gene is a total loss, the functional copy is still there to save the day.

The gene family encoding the globins is a good example of the gene families found in vertebrates. These proteins are found in hemoglobin as well as in myoglobin (an oxygen-binding protein present in muscle). The globin genes all arose from a single common ancestor gene long ago. In humans, there are three functional members of the alpha-globin (α-globin) cluster and five in the beta-globin (β-globin) cluster (**Figure 14.8**). In adults, each hemoglobin molecule is a tetramer containing two identical α-globin subunits, two identical β-globin subunits, and four heme pigments (each held inside a globin subunit) (see Figure 3.9).

During human development, different members of the globin gene cluster are expressed at different times and in different tissues (**Figure 14.9**). This *differential gene expression* has great physiological significance. For example, γ-globin, a subunit found in the hemoglobin of the human fetus, binds O_2 more tightly than adult hemoglobin does. This specialized form of hemoglobin ensures that in the placenta, where the maternal and fetal circulation come close to one another, O_2 will be transferred from the mother's blood to the developing fetus's blood. Just before birth, the synthesis of fetal hemoglobin proteins in the liver stops, and the bone marrow cells take over, making the adult forms. Thus hemoglobins with different binding affinities for O_2 are provided at different stages of human development.

In addition to genes that encode proteins, many gene families include nonfunctional **pseudogenes**, designated with the Greek letter psi (ψ) (see Figure 14.8). These pseudogenes are the "black sheep" of a gene family: they result from mutations that cause a loss of function, rather than an enhanced or new function. The DNA sequence of a pseudogene may not differ greatly from that of other family members. It may simply lack a promoter, for example, and thus fail to be transcribed. Or it may lack the recognition sites needed for the removal of introns (a process we will describe in the next section) and thus be transcribed into pre-mRNA, but not correctly processed into a useful mature mRNA. In some gene families, pseudogenes outnumber functional genes. Because some members of the family are functional, there appears to be little selection pressure for evolution to eliminate pseudogenes.

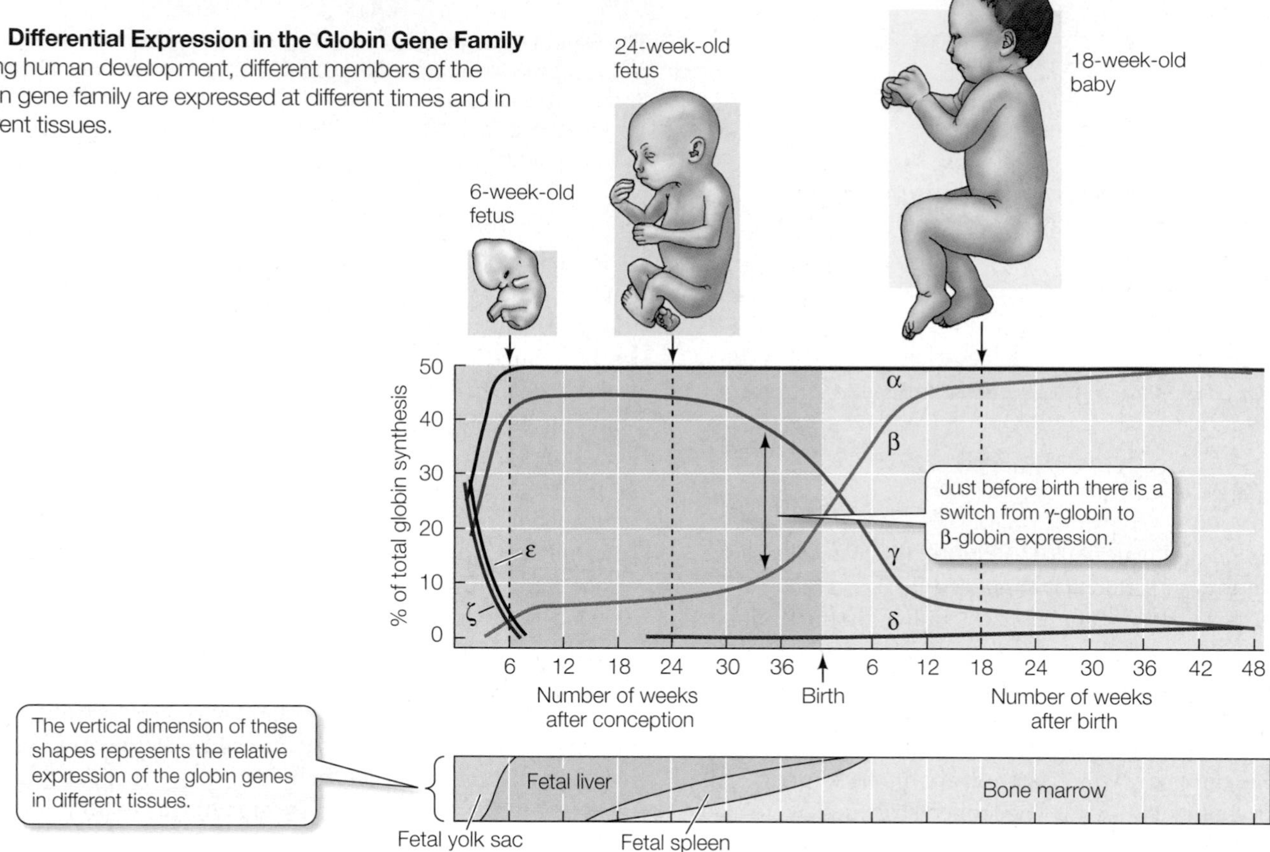

14.9 Differential Expression in the Globin Gene Family During human development, different members of the globin gene family are expressed at different times and in different tissues.

14.2 RECAP

Most eukaryotic genes contain noncoding sequences called introns, which are removed from the pre-mRNA transcript. Many eukaryotic genes belong to groups of closely related genes called gene families.

- Can you describe the method of nucleic acid hybridization? See p. 314 and Figure 14.6
- Can you describe the experiment that showed that the β-globin gene contains introns? See p. 315 and Figure 14.7
- What are gene families, and how might they be important in evolution? See p. 315

We now know that eukaryotic protein-coding genes contain some sequences that do not appear in the mature mRNA that is translated into proteins. How are these noncoding sequences eliminated from the pre-mRNA transcript?

14.3 How Are Eukaryotic Gene Transcripts Processed?

The primary transcript of a eukaryotic gene is modified in two ways before it leaves the nucleus: both ends of the pre-mRNA are modified, and the introns are removed.

The primary transcript of a protein-coding gene is modified at both ends

Two steps in the processing of pre-mRNA take place in the nucleus, one at each end of the molecule (**Figure 14.10**):

- A **G cap** is added to the 5′ end of the pre-mRNA as it is transcribed. The G cap is a chemically modified molecule of guanosine triphosphate (GTP). It apparently facilitates the binding of mRNA to the ribosome for translation and protects the mRNA from being digested by *ribonucleases* that break down RNAs.
- A **poly A tail** is added to the 3′ end of pre-mRNA at the end of transcription. Near the 3′ end of pre-mRNA, and after the last codon, is the sequence AAUAAA. This sequence acts as a signal for an enzyme to cut the pre-mRNA. Immediately after this cleavage, another enzyme adds 100 to 300 adenine bases ("poly A") to the 3′ end of the pre-mRNA. This "tail" may assist in the export of the mRNA from the nucleus and is important for mRNA stability.

Splicing removes introns from the primary transcript

The next step in the processing of eukaryotic pre-mRNA within the nucleus is deletion of the introns. If these RNA sequences were not removed, an mRNA translating to a very different amino acid sequence, and possibly a nonfunctional protein, would result. A process called **RNA splicing** removes the introns and splices the exons together.

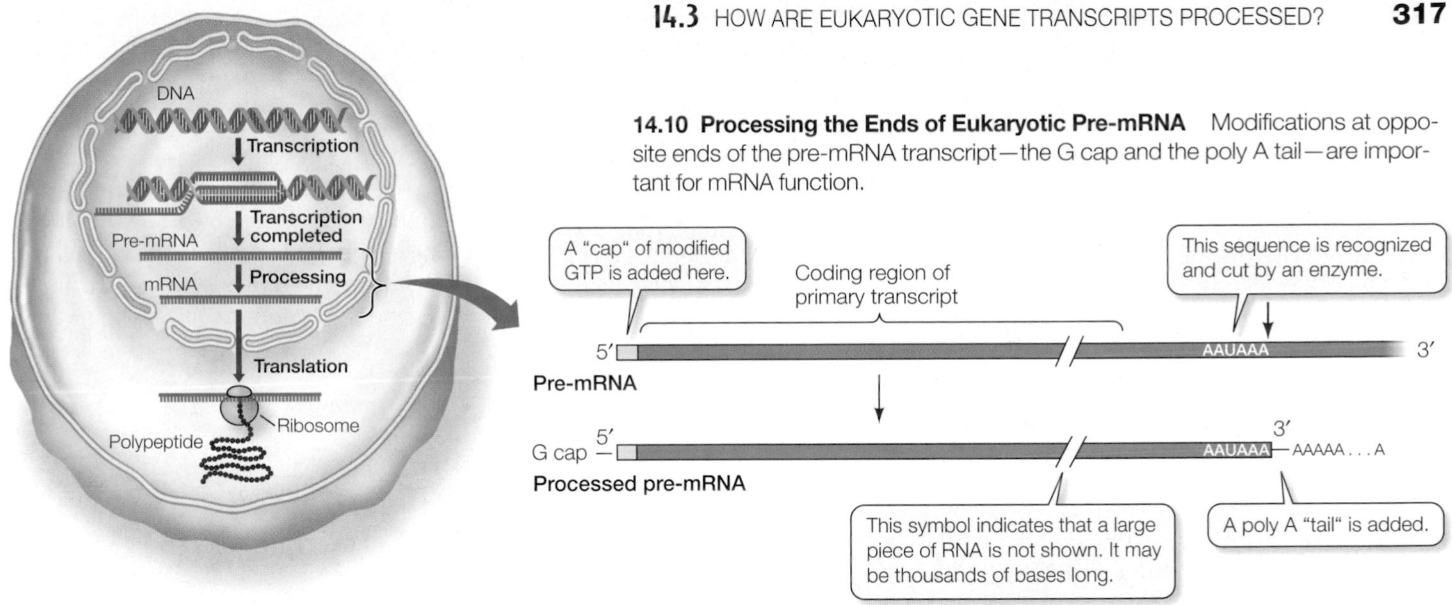

14.10 Processing the Ends of Eukaryotic Pre-mRNA Modifications at opposite ends of the pre-mRNA transcript—the G cap and the poly A tail—are important for mRNA function.

As soon as the pre-mRNA is transcribed, it is bound by several **small nuclear ribonucleoprotein particles** (**snRNPs**, commonly pronounced "snurps"). There are several types of these RNA–protein particles in the nucleus.

At the boundaries between introns and exons are **consensus sequences**—short stretches of DNA that appear, with little variation ("consensus"), in many different genes. The RNA in one of the snRNPs has a stretch of bases complementary to the consensus sequence at the 5′ exon–intron boundary, and it binds to the pre-mRNA by complementary base pairing. Another snRNP binds to the pre-mRNA near the 3′ intron–exon boundary (**Figure 14.11**).

Next, using energy from ATP, proteins are added to form a large RNA–protein complex called a **spliceosome**. This complex cuts the pre-mRNA, releases the introns, and joins the ends of the exons together to produce mature mRNA.

Molecular studies of human genetic diseases have provided insights into consensus sequences and splicing machinery. People with a genetic disease called beta thalassemia, for example, make inadequate amounts of the β-globin subunit of hemoglobin. These people suffer from severe anemia because they have an inadequate supply of red blood cells. In some cases, the genetic mutation that causes the disease occurs at a consensus sequence in the β-globin gene. Consequently, β-globin pre-mRNA cannot be spliced correctly, and nonfunctional β-globin mRNA is made.

This finding offers an excellent example of how mutations can elucidate cause-and-effect relationships in biology. In the logic of science, merely linking two phenomena (for example, consensus sequences and splicing) does not prove that one is necessary for the other. In an experiment, the scientist alters one phenomenon (for example, the bases of the consensus sequence) to see whether the other (for example, splicing) occurs. In beta thalassemia, nature has done this experiment for us.

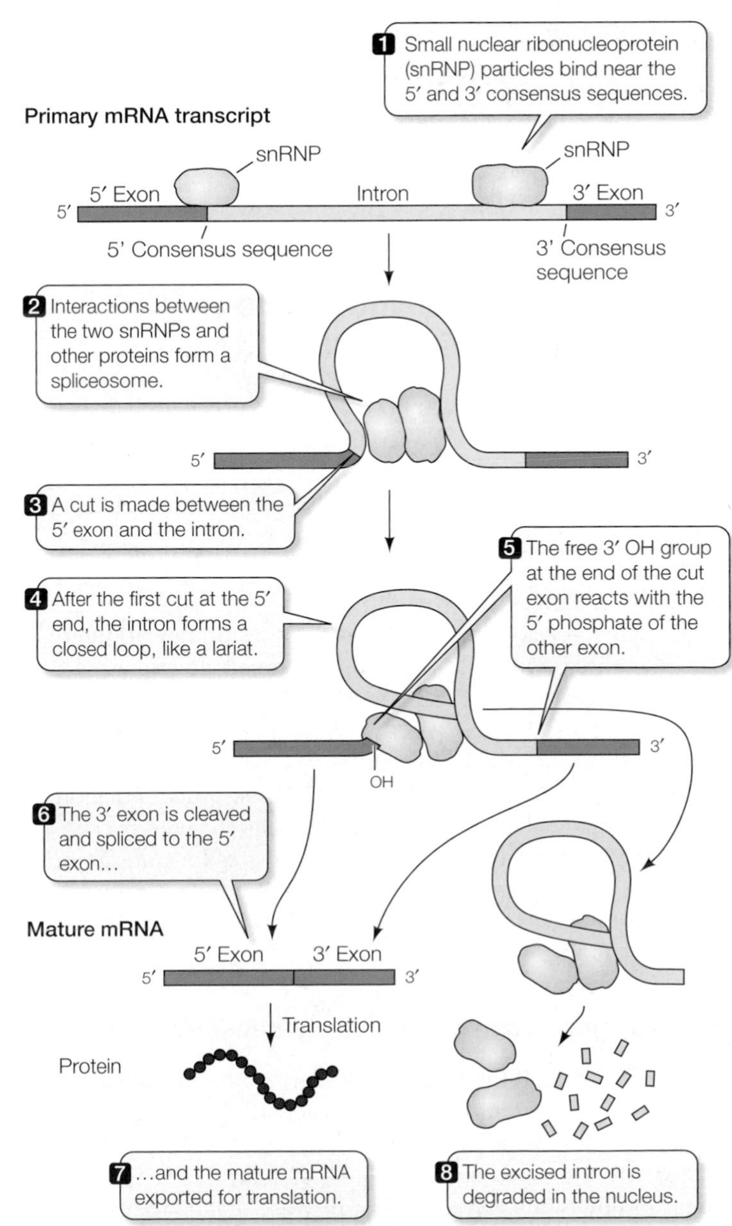

1 Small nuclear ribonucleoprotein (snRNP) particles bind near the 5′ and 3′ consensus sequences.
Primary mRNA transcript
snRNP
snRNP
5′ Exon
Intron
3′ Exon
5′ Consensus sequence
3′ Consensus sequence
2 Interactions between the two snRNPs and other proteins form a spliceosome.
3 A cut is made between the 5′ exon and the intron.
4 After the first cut at the 5′ end, the intron forms a closed loop, like a lariat.
5 The free 3′ OH group at the end of the cut exon reacts with the 5′ phosphate of the other exon.
OH
6 The 3′ exon is cleaved and spliced to the 5′ exon…
Mature mRNA
5′ Exon
3′ Exon
Translation
Protein
7 …and the mature mRNA exported for translation.
8 The excised intron is degraded in the nucleus.

14.11 The Spliceosome: An RNA Splicing Machine The binding of snRNPs to consensus sequences bordering the introns on the pre-mRNA lines up the splicing machinery. After the snRNPs bind to the pre-mRNA, other proteins join the complex to form a spliceosome. This mechanism determines the exact position of each cut in the pre-mRNA with great precision.

After processing is completed in the nucleus, the mature mRNA leaves that organelle through the nuclear pores. A protein called TAP binds to the 5′ end of processed mRNA. This protein in turn binds to others, which are recognized by a receptor at the nuclear pore. Unprocessed or incompletely processed pre-mRNAs remain in the nucleus.

14.3 RECAP

The primary transcript of a eukaryotic gene is modified while still in the nucleus. First, its 5′ and 3′ ends are modified, and then its introns are spliced out.

- Can you describe how the pre-mRNA transcript is modified at the 5′ and 3′ ends? See p. 316 and Figure 14.10
- How does RNA splicing happen? What are the consequences if it does not happen correctly? See p. 317 and Figure 14.11

While each of a multicellular organism's somatic cells contains a complete set of genes, no cell expresses all of those genes. Each cell type usually expresses only those genes that it needs for its own development and function. How a cell controls gene expression is the subject of the next section.

14.4 How Is Eukaryotic Gene Transcription Regulated?

For development to proceed normally, and for each cell in a multicellular organism to acquire and maintain its proper specialized function, certain proteins must be synthesized at just the right times and in just the right cells. Thus the expression of eukaryotic genes must be precisely regulated. Unlike DNA replication, which is regulated in every cell on an all-or-none basis, gene expression is highly selective.

Gene expression can be regulated at a number of different points in the process of transcribing and translating the gene into a protein (**Figure 14.12**). In this section we will describe the mechanisms that result in the selective transcription of specific genes. Some of these mechanisms involve nuclear proteins that alter chromosome function or structure. In other cases, the regulation of transcription involves changes in the DNA itself: genes may be selectively replicated to provide more templates for transcription, or even rearranged on the chromosome. The following two sections will examine the regulation of eukaryotic gene expression after transcription.

Specific genes can be selectively transcribed

The brain cells and the liver cells of a mouse have some proteins in common and others that are characteristic of each cell type. Yet both cells have the same DNA sequences and, therefore, the same genes. Are the differences in protein content due to differential transcription of the genes? Or is it the case that all the genes are transcribed in both cell types, and some mechanism that acts after transcription is responsible for the differences in proteins?

These two alternatives— *transcriptional regulation* and *posttranscriptional regulation*—can be distinguished by examining the actual mRNA sequences made within the nucleus of each cell type. Such analyses indicate that for some proteins, the mechanism of regulation is differential gene transcription. Both brain and liver

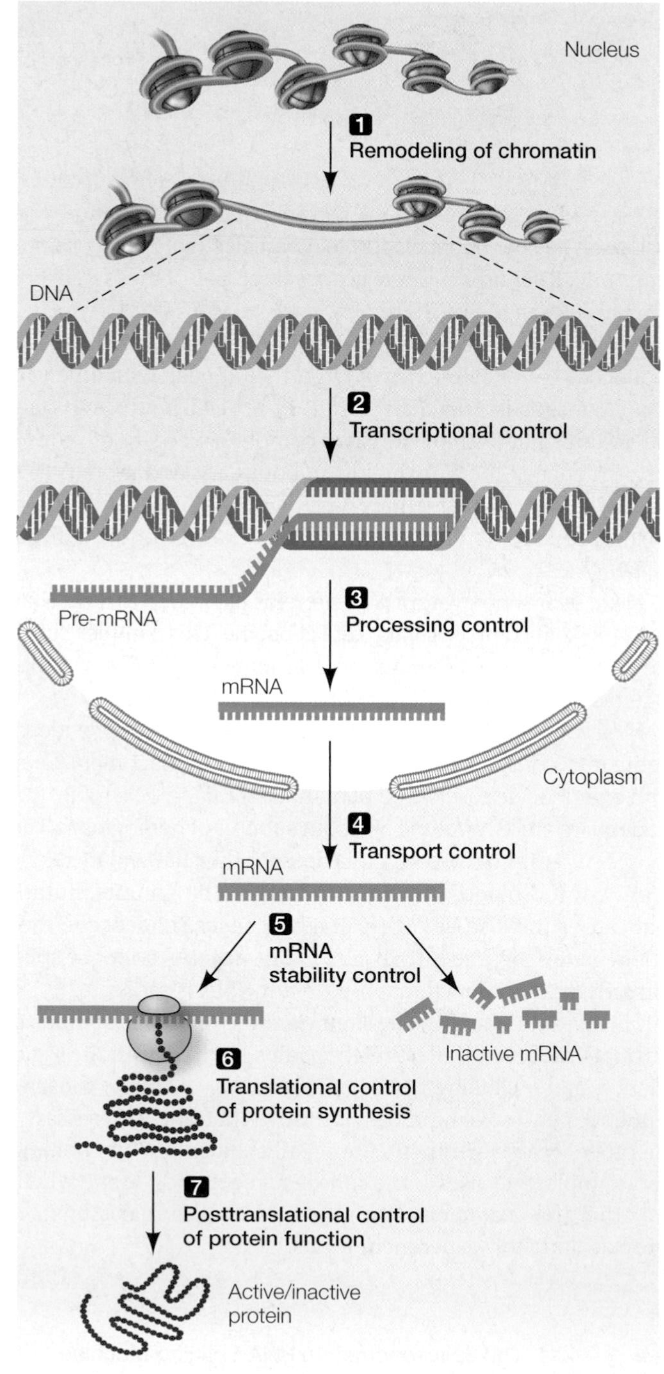

14.12 Potential Points for the Regulation of Gene Expression Gene expression can be regulated before transcription (1), during transcription (2, 3), after transcription but before translation (4, 5), at translation (6), or after translation (7).

cells, for example, transcribe "housekeeping" genes—those that encode proteins involved in the basic metabolic processes that occur in every living cell, such as glycolysis enzymes. But liver cells transcribe some genes for liver-specific proteins, and brain cells transcribe some genes for brain-specific proteins. And neither cell type transcribes the genes for proteins that are characteristic of muscle, blood, bone, or the other specialized cell types in the body.

CONTRASTING EUKARYOTES AND PROKARYOTES Unlike prokaryotes, in which functionally related genes are often grouped into operons that are transcribed as a unit, eukaryotes tend to have solitary genes. Thus the regulation of several genes at once requires common control elements in each of the genes that allow all of the genes to respond to the same signal.

In contrast to the single RNA polymerase in bacteria, eukaryotes have three different RNA polymerases. Each eukaryotic polymerase catalyzes the transcription of a specific type of gene. Only one (RNA polymerase II) transcribes protein-coding genes. The other two transcribe the DNA that codes for rRNA (polymerase I) and for tRNA and small nuclear RNAs (polymerase III).

The diversity of eukaryotic polymerases is reflected in the diversity of eukaryotic promoters, which tend to be much more varied in their sequences than prokaryotic promoters. Furthermore, most eukaryotic genes have additional sequences that can regulate the rate of their transcription. Whether a eukaryotic gene is transcribed depends on the sum total of the effects of all of these DNA and protein elements; thus there are many points of possible regulation.

Finally, the initiation of transcription in eukaryotes is very different from that in prokaryotes, in which RNA polymerase directly recognizes the promoter. In eukaryotes, many proteins are involved in initiating transcription. We will confine the following discussion to RNA polymerase II, which catalyzes the transcription of most protein-coding genes, but the mechanisms for the other two RNA polymerases are similar.

TRANSCRIPTION FACTORS As Section 13.4 described, the prokaryotic promoter is a sequence of DNA near the 5′ end of the coding region of a gene or operon where RNA polymerase begins transcription. A prokaryotic promoter has two essential sequences. One is the **recognition sequence**—the sequence recognized by RNA polymerase. The second, closer to the initiation site, is the **TATA box** (so called because it is rich in AT base pairs), where DNA begins to denature so that the template strand can be exposed.

Some plants can attract bacteria and fungi to their roots to provide them with important nutrients. A chemical signal generated by the plant causes the microbes to make oligosaccharides called "nod factors." Nod factors, in turn, cause the plant to activate two transcription factors that increase the expression of genes that result in a fruitful symbiosis.

Things are different in eukaryotes. Eukaryotic RNA polymerase II cannot simply bind to the promoter and initiate transcription. Rather, it does so only after various regulatory proteins, called **transcription factors**, have assembled on the chromosome (**Figure 14.13**). First, the protein TFIID ("TF" stands for transcription factor) binds to the TATA box. Its binding changes both its own shape and that of the DNA, presenting a new surface that attracts the binding of other transcription factors to form a *transcription complex*. RNA polymerase II does not bind until several other proteins have bound to this complex.

Some DNA sequences, such as the TATA box, are common to the promoters of many eukaryotic genes and are recognized by transcription factors that are found in all the cells of an organism.

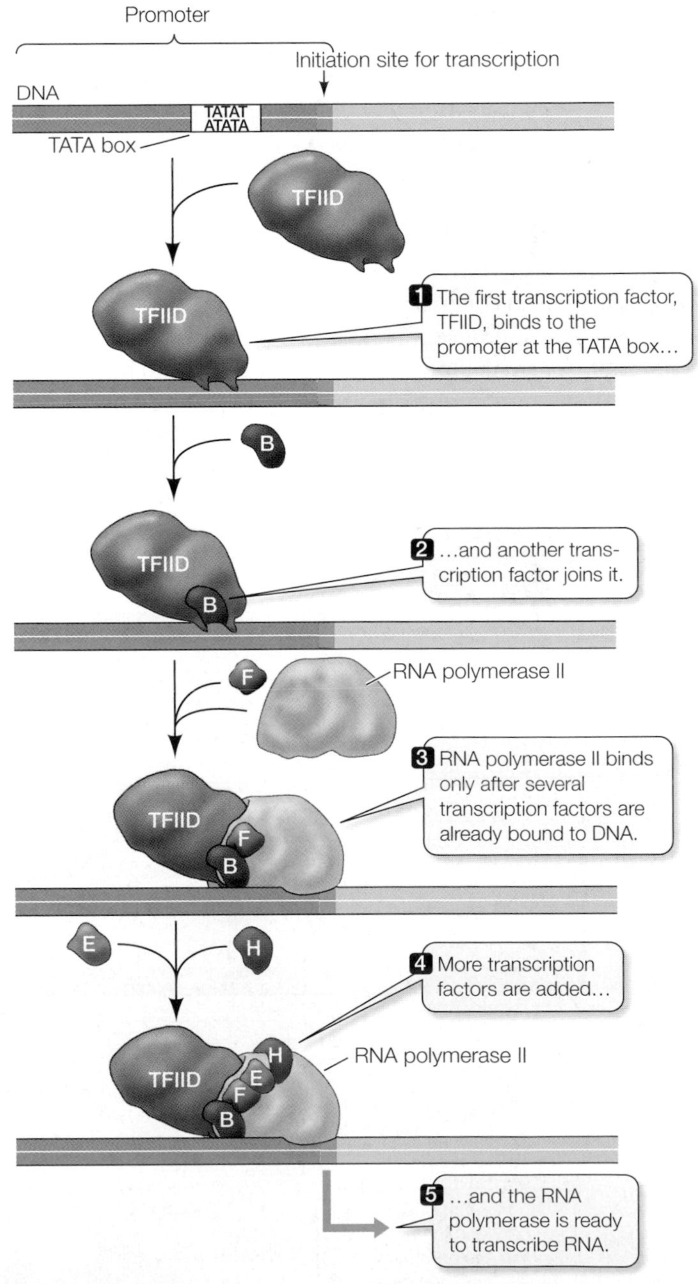

Promoter
Initiation site for transcription
DNA
TATAT
ATATA
TATA box
TFIID
1 The first transcription factor, TFIID, binds to the promoter at the TATA box...
B
2 ...and another transcription factor joins it.
F
RNA polymerase II
3 RNA polymerase II binds only after several transcription factors are already bound to DNA.
E
H
4 More transcription factors are added...
RNA polymerase II
5 ...and the RNA polymerase is ready to transcribe RNA.

14.13 The Initiation of Transcription in Eukaryotes Except for TFIID, which binds to the TATA box, each transcription factor in this transcription complex has binding sites only for the other proteins in the complex, and does not bind directly to DNA. B, E, F, and H are transcription factors.

Other sequences found in promoters are specific to only a few genes and are recognized by transcription factors found only in certain tissues. These specific transcription factors play an important role in *differentiation,* the specialization of cells during development.

REGULATORS, ENHANCERS, AND SILENCERS IN DNA In addition to the promoter, two other types of regulatory DNA sequences bind proteins that activate RNA polymerase. The recently discovered **regulator sequences** are clustered just upstream of the promoter. Various *regulator proteins* (seven for the β-globin gene) may bind to these regulator sequences (**Figure 14.14**). The resulting complexes bind to the adjacent transcription complex and activate it.

Much farther away—up to 20,000 bp away—from the promoter are **enhancer sequences**. Enhancer sequences bind *activator* proteins, and this binding strongly stimulates the transcription complex. How enhancers exert their influence is not clear. In one proposed model, the DNA bends (it is known to do so) so that the activator protein is in contact with the transcription complex.

In addition, there are *negative* regulatory sequences on DNA, called **silencer** sequences, that have the opposite effect from enhancers. Silencers turn off transcription by binding proteins appropriately called *repressor* proteins.

How do these proteins and DNA sequences—transcription factors, regulators, enhancers, activators, silencers, and repressors—regulate transcription? Apparently, in most tissues, a small amount of RNA is transcribed from all genes, but the combination of these factors determines the rate of transcription. In the immature red blood cells in bone marrow, for example, which make a large amount of β-globin, transcription of the β-globin gene is stimulated by the binding of seven regulator proteins and six activator proteins. But in white blood cells in the same bone marrow, these thirteen proteins are not made and do not bind to the regulator and enhancer sequences adjacent to the β-globin gene; consequently, the β-globin gene is hardly transcribed at all.

PROTEIN STRUCTURES INVOLVED IN PROTEIN–DNA INTERACTIONS The regulation and coordination of gene expression requires the binding of many specialized proteins to DNA. Among DNA-binding proteins, there are four common structural themes in the protein domains that bind to DNA. These themes, called *motifs,* consist of different combinations of structural elements and special components: *helix-turn-helix, zinc finger, leucine zipper,* and *helix-loop-helix* (**Figure 14.15**). DNA-binding proteins with specific motifs in their binding domains are involved in the activation and inactivation of certain types of genes, both during development and in the adult organism.

Let's look at how just one of these motifs works. As pointed out in Section 11.2, the complementary bases in DNA not only form hydrogen bonds with each other, but can form additional hydrogen bonds with proteins, particularly at points exposed in the major and minor grooves (see Figure 11.8B). In this way, an intact DNA double helix can be recognized by a protein whose structure:

- fits into the major or minor groove
- has amino acids that can project into the interior of the double helix
- has amino acids that can form hydrogen bonds with the interior bases

Proteins with a helix-turn-helix motif, in which two α-helices are connected via a non-helical turn, fit these three criteria. The interior-facing "recognition" helix is the one whose amino acids interact with the bases inside the DNA. The exterior-facing helix sits on the sugar–phosphate backbone, ensuring that the interior helix is presented to the bases in the correct configuration. Many repressor proteins have this helix-turn-helix configuration in their structure. Binding of these repressors to DNA prevents other proteins from interacting with that DNA, and this is what inhibits transcription.

COORDINATING THE EXPRESSION OF SEVERAL GENES How do eukaryotic cells coordinate the regulation of several genes whose transcription must be turned on at

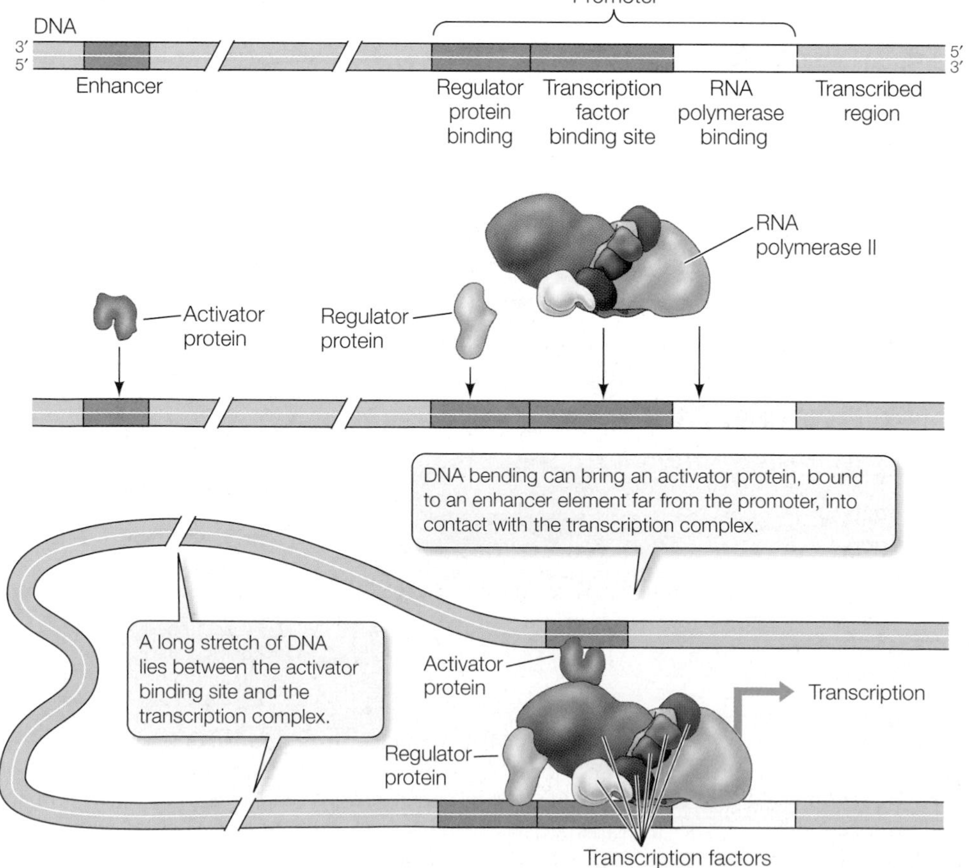
Promoter
DNA
3′
5′
5′
3′
Enhancer
Regulator protein binding
Transcription factor binding site
RNA polymerase binding
Transcribed region
RNA polymerase II
Activator protein
Regulator protein
DNA bending can bring an activator protein, bound to an enhancer element far from the promoter, into contact with the transcription complex.
A long stretch of DNA lies between the activator binding site and the transcription complex.
Activator protein
Transcription
Regulator protein
Transcription factors

14.14 Transcription Factors, Regulators, and Activators The actions of many proteins determine whether and where RNA polymerase II will transcribe DNA.

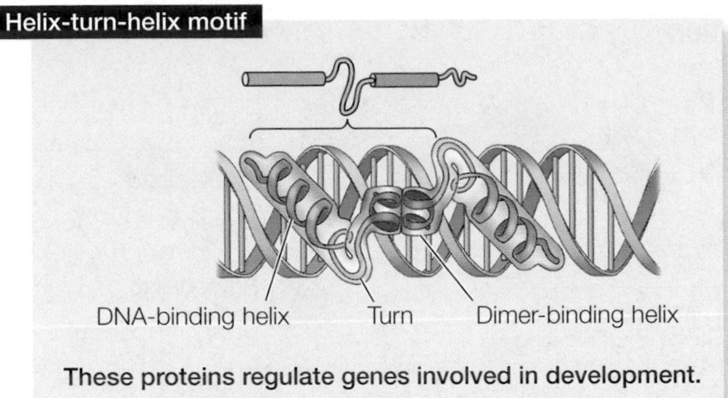
Helix-turn-helix motif
DNA-binding helix
Turn
Dimer-binding helix
These proteins regulate genes involved in development.

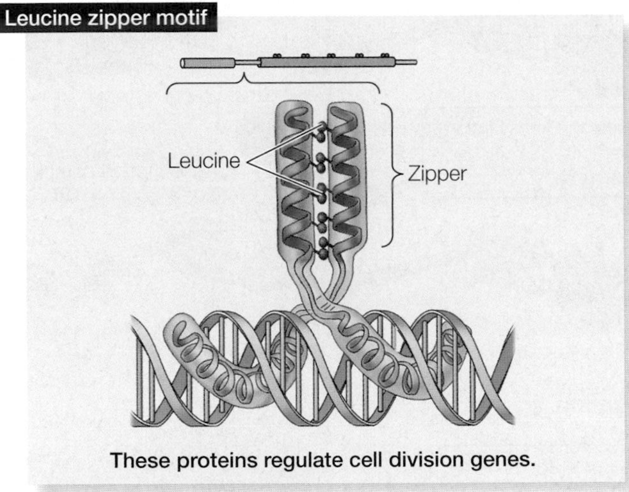
Leucine zipper motif
Leucine
Zipper
These proteins regulate cell division genes.

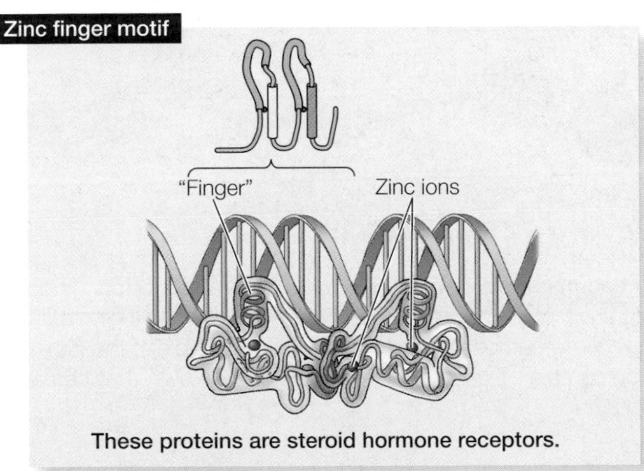
Zinc finger motif
"Finger"
Zinc ions
These proteins are steroid hormone receptors.

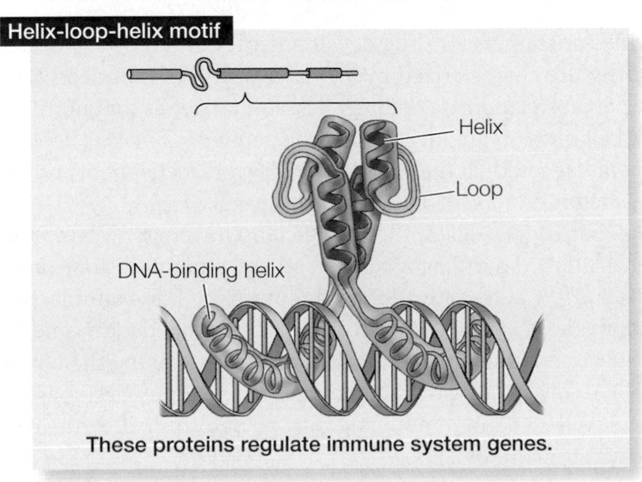
Helix-loop-helix motif
Helix
Loop
DNA-binding helix
These proteins regulate immune system genes.

14.15 Protein–DNA Interactions The DNA-binding domains of most regulatory proteins contain one of four structural motifs.

the same time? In prokaryotes, in which related genes are linked together in an operon, a single regulatory system can regulate several adjacent genes. But in eukaryotes, the several genes whose regulation requires coordination may be far apart on a chromosome, or even on different chromosomes.

In such a case, regulation can be achieved if the various genes all have the same regulator sequences, which bind the same regulator proteins. One of the many examples of this phenomenon is provided by the response of an organism to a stressor—for example, that of plants to drought. Under conditions of drought stress, a plant must synthesize a number of proteins, but the genes for those proteins are scattered throughout the genome. However, each of these genes has a specific regulator sequence near its promoter, called the *stress response element* (SRE). The binding of a regulator protein to this element stimulates RNA synthesis (**Figure 14.16**). The proteins made from these genes are involved not only in water conservation, but also in protecting the plant against ex-

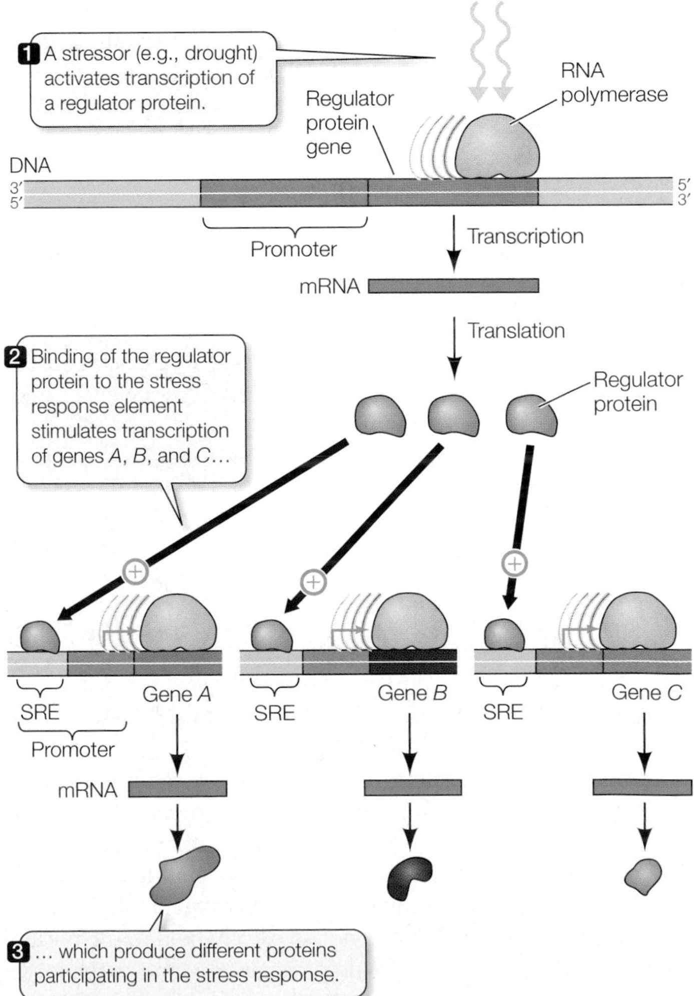
1 A stressor (e.g., drought) activates transcription of a regulator protein.
Regulator protein gene
RNA polymerase
DNA
3′
5′
5′
3′
Promoter
Transcription
mRNA
Translation
2 Binding of the regulator protein to the stress response element stimulates transcription of genes A, B, and C...
Regulator protein
SRE
Gene A
Promoter
SRE
Gene B
SRE
Gene C
mRNA
3 ... which produce different proteins participating in the stress response.

14.16 Coordinating Gene Expression A single environmental signal, such as drought stress, causes the synthesis of a transcriptional regulator protein that acts on many genes.

cess salt in the soil and against freezing. This finding has considerable importance for agriculture, in which crops are often grown under less than optimal conditions.

Gene expression can be regulated by changes in chromatin structure

Other mechanisms that regulate transcription act on the structure of chromatin and chromosomes. As Section 9.3 describes, chromatin contains a number of proteins as well as DNA. The packaging of DNA into nucleosomes by these nuclear proteins can make DNA physically inaccessible to RNA polymerase and the rest of the transcription apparatus, much as the binding of a repressor to the operator in the prokaryotic *lac* operon prevents transcription (see Section 13.4). Chromatin structure at both the local and whole-chromosome levels affects transcription.

CHROMATIN REMODELING Recall that DNA is wound around proteins called histones to form a structure called a nucleosome. Nucleosomes block both the initiation and elongation steps of transcription. In a process called **chromatin remodeling**, two types of remodeling proteins inactivate these two blocks (**Figure 14.17**). To allow initiation, the first remodeling protein binds upstream of the initiation site, disaggregating the nucleosomes so that the transcription complex can bind and RNA polymerase can begin transcription. To allow elongation, the second remodeling protein binds once transcription is under way, allowing the transcription complex to move through the nucleosomes.

THE HISTONE CODE How are nucleosomes disassembled to allow transcription (and then reassembled)? Each histone protein has an approximately 20–amino acid "tail" at its N terminus that sticks out of the compact structure. This tail has certain amino acids (notably lysine) that are positively charged. These amino acids are targets for enzymes that add acetyl groups to the positively charged amino acids, thus changing their charges:

$$\cdots\mathrm{N(H)}-\mathrm{C(H)}[(\mathrm{CH_2})_3-{}^{+}\mathrm{NH_3}]-\mathrm{C(=O)}\cdots + \mathrm{CoA}-\mathrm{S}-\mathrm{C(=O)}-\mathrm{CH_3} \rightarrow \cdots\mathrm{N(H)}-\mathrm{C(H)}[(\mathrm{CH_2})_3-\mathrm{HN}-\mathrm{C(=O)}-\mathrm{CH_3}]-\mathrm{C(=O)}\cdots + \mathrm{CoA}$$

Lysine in histone — Acetyl-CoA — Acetyl-lysine

Reducing the positive charge of the histone tails reduces the affinity of the histones for DNA, opening up the compact nucleosome. Ordinarily, because histone proteins are positively charged and DNA is negatively charged (owing to its phosphate groups), the attachment of these two molecules is electrostatic. Without this electrostatic attraction, DNA is not so closely held to the nucleosome, and chromatin remodeling proteins can bind to the looser nucleosome–DNA complex.

While gene activation calls for enzymes (*histone acetyltransferases*) to add acetyl groups, gene repression requires other enzymes (*histone deacetylases*) to *remove* the acetyl groups. The latter enzymes are targets for drug development. Certain diseases such as cancer are characterized by a greater proportion of deacetylation than acetylation at certain genes, so that genes that normally block cell division are inactive. A drug acting as a *histone deacetylase inhibitor* could tilt the activity ratio toward acetylation, and the genes might be activated, turning off the cell cycle.

In what David Allis of the Rockefeller University in New York City has dubbed the "histone code," several types of histone modification affect gene activation and repression. For example, histones are subject to methylation, which is associated with gene inactivation, and phosphorylation, as well as acetylation. All of these effects are reversible. So whether a eukaryotic gene becomes activated by chromatin remodeling may be determined by the pattern of histone modification.

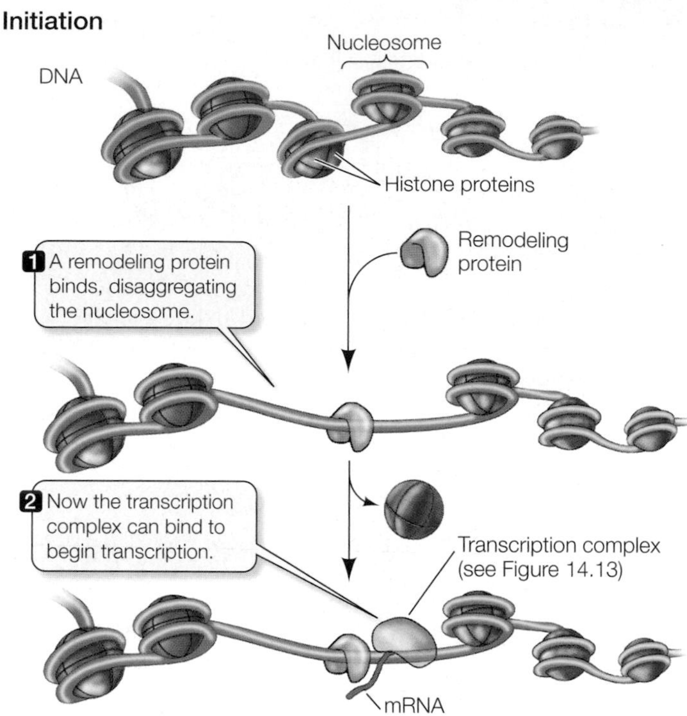

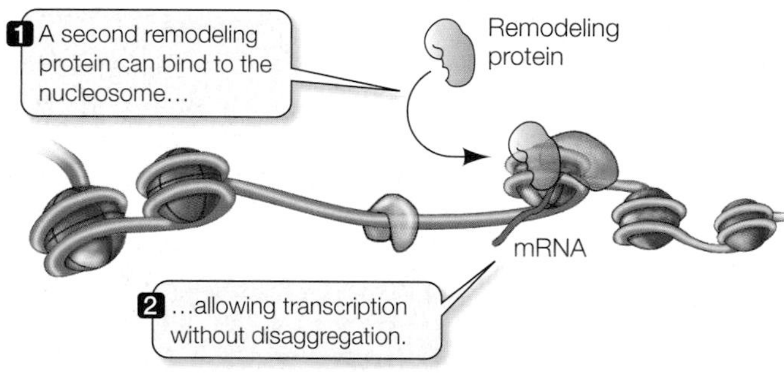

14.17 Local Remodeling of Chromatin for Transcription Initiation of transcription requires that nucleosomes change their structure, becoming less compact. This makes DNA accessible to the transcription complex. During elongation of RNA, however, they can remain intact.

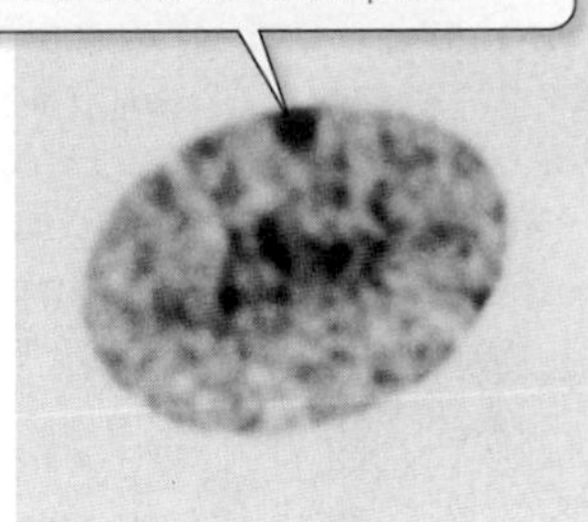

The Barr body is the condensed, inactive member of a pair of X chromosomes in the cell. The other X is not condensed and is active in transcription.

14.18 A Barr Body in the Nucleus of a Female Cell The number of Barr bodies per nucleus is equal to the number of X chromosomes minus one. Thus normal males (XY) have no Barr body, whereas normal females (XX) have one.

WHOLE-CHROMOSOME EFFECTS Some transcriptional regulation mechanisms act on entire chromosomes. Under a microscope, two kinds of chromatin can be distinguished in the stained interphase nucleus: *euchromatin* and *heterochromatin*. Euchromatin is diffuse and stains lightly; it contains the DNA that is transcribed into mRNA. Heterochromatin is condensed and stains darkly; any genes it contains are generally not transcribed.

Perhaps the most dramatic example of heterochromatin is seen in the inactive X chromosome of mammals. A normal female mammal has two X chromosomes; a normal male has an X and a Y. The X and Y chromosomes probably arose from a pair of autosomes about 300 million years ago. Over time, mutations in the Y chromosome resulted in maleness-determining genes (see Section 10.4), and the Y chromosome gradually lost most of the genes it once shared with its X homolog. As a result, there is a great difference between females and males in the "dosage" of X-linked genes. Each female cell has two copies of the genes on the X chromosome, and therefore has the potential to produce twice as much of the protein products of these genes as a male cell has. Nevertheless, for 75 percent of genes on the X, transcription is generally the same in males and in females. How does this happen?

Mary Lyon, Liane Russell, and Ernest Beutler independently suggested in 1961 that one of the X chromosomes in each cell of a female is, to a significant extent, transcriptionally inactivated early in embryonic development. That is, they proposed that a copy of the X remains inactive in each embryonic cell, and in all the cells arising from it. In a given embryonic cell, the "choice" of which X in the pair of Xs to inactivate is random. Recall that one X in a female comes from her father and one from her mother. Thus, in one embryonic cell, the paternal X might be the one remaining transcriptionally active, but in a neighboring cell, the maternal X might be active.

During interphase, a single, stainable nuclear body, called a *Barr body* (after its discoverer, Murray Barr), can be seen in cells of human females under the light microscope (**Figure 14.18**). This clump of heterochromatin, which is not present in males, is the inactivated X chromosome. The number of Barr bodies in a nucleus is equal to the number of X chromosomes minus one (the one represents the X chromosome that remains transcriptionally active). So a female with the normal two X chromosomes will have one Barr body, a rare female with three Xs will have two, an XXXX female will have three, and an XXY male will have one. These observations suggest that the interphase cells of each person, male or female, have a single active X chromosome, making the dosage of the expressed X chromosome genes constant across both sexes.

Condensation of the inactive X chromosome makes its DNA sequences physically unavailable to the transcriptional machinery. One mechanism of condensation is the addition of a methyl group ($—CH_3$) to the 5′ position of cytosine on DNA. Methylation of cytosines is associated with transcriptionally inactive genes. For example, many cytosines of the DNA of the inactive X chromosome are methylated, while few cytosines on the active X are methylated. Methylated DNA appears to bind certain chromosomal proteins that may be responsible for heterochromatin formation.

The otherwise inactive X chromosome has one gene that is only lightly methylated and is transcriptionally active, called *Xist* (for *X i*nactivation-*s*pecific *t*ranscript). *Xist* is heavily methylated on, and not transcribed from, the other, "active" X chromosome. The RNA transcribed from *Xist* does not leave the nucleus and is not an mRNA. Instead, it appears to bind to the X chromosome from which it is transcribed, and this binding somehow leads to a spreading of inactivation along the chromosome. This RNA transcript is known as an **interference RNA** (**Figure 14.19**).

How does the transcriptionally active X overcome the effects of *Xist* RNA? Apparently, there is an anti-*Xist* gene, appropriately called *Tsix*. This gene codes for an RNA that binds by complementary base pairing to *Xist* RNA at the active X chromosome.

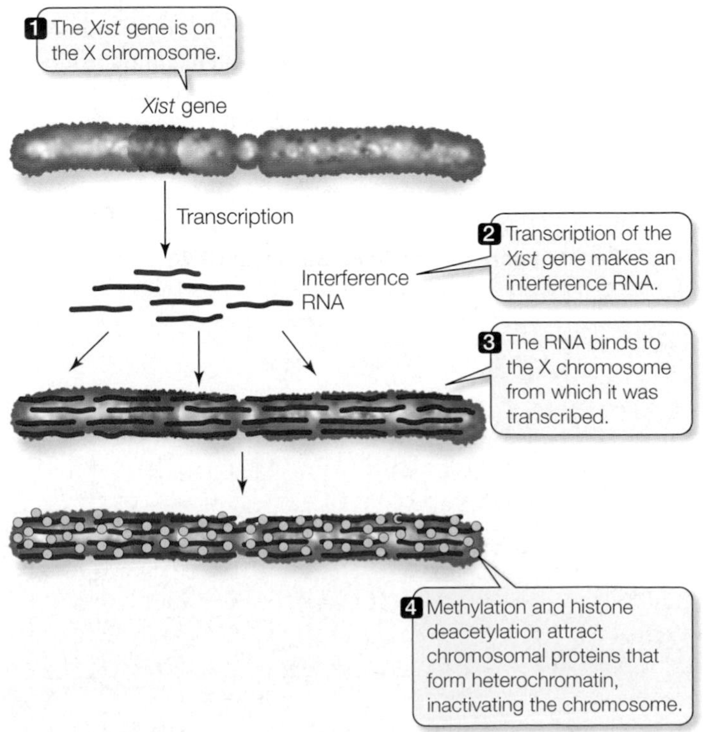

14.19 A Model for X Chromosome Inactivation Interference RNA and chromosomal proteins combine to inactivate the X chromosome.

Selective gene amplification results in more templates for transcription

Another way for one cell to make more of a certain gene product than another cell does is to make more copies of the appropriate gene and transcribe them all. The process of creating more copies of a gene in order to increase its transcription is called **gene amplification**.

As described earlier, the genes that code for three of the four human ribosomal RNAs are linked together in a unit, and this unit is repeated several hundred times in the genome to provide multiple templates for rRNA synthesis (rRNA is the most abundant kind of RNA in the cell). In some circumstances, however, even this moderate repetition is not enough to satisfy the demands of the cell.

The mature eggs of frogs and fishes, for example, have up to a trillion ribosomes. These ribosomes are used for the massive protein synthesis that follows fertilization. The precursor cell that will differentiate into the egg contains fewer than 1,000 copies of the rRNA gene cluster, and would take 50 years to make a trillion ribosomes if it transcribed those rRNA genes at peak efficiency. How does the egg end up with so many ribosomes (and so much rRNA)?

The cell solves this problem by selectively amplifying its rRNA gene clusters until there are more than a million copies (**Figure 14.20**). In fact, this gene complex goes from being 0.2 percent of the total genome to 68 percent. These million copies, transcribed at the maximum rate, are just enough to make the necessary trillion ribosomes in a few days.

The mechanism for selective amplification of a single gene is not clearly understood, but it has important medical implications. In some cancers, a cancer-causing gene called an *oncogene* becomes amplified (see Section 17.4). In addition, when some tumors are treated with a drug that targets a single protein, amplification of the gene for the target protein leads to an excess of that protein, and the cell becomes resistant to the prescribed dose of the drug.

14.4 RECAP

A number of transcription factors must bind to a eukaryotic promoter before RNA polymerase will bind to it and begin transcription, providing a number of ways to increase or decrease transcription. Proteins that bind to chromatin, chemical modification of histones and nucleosomes, and selective gene amplification can influence transcription as well.

- Can you describe some of the different ways in which transcription factors regulate gene transcription? See pp. 319–320 and Figure 14.13
- What is the histone code? See p. 322
- Do you understand why and how X chromosome inactivation occurs? See p. 323

There are many ways in which gene expression can be regulated even after the gene has been transcribed. The processing of pre-mRNA described in Section 14.3 is merely one opportunity for doing so.

14.5 How Is Eukaryotic Gene Expression Regulated After Transcription?

As we have seen, pre-mRNA is processed by cutting out the introns and splicing the exons together. If exons are selectively deleted from the pre-mRNA, different proteins can be synthesized. The longevity of mRNA in the cytoplasm can also be regulated: the longer an mRNA exists in the cytoplasm, the more of its protein can be made.

Different mRNAs can be made from the same gene by alternative splicing

Most primary mRNA transcripts contain several introns (see Figure 14.5). We have seen how the splicing mechanism recognizes the boundaries between exons and introns. What would happen if the β-globin pre-mRNA, which has two introns, were spliced from the start of the first intron to the end of the second? Not only

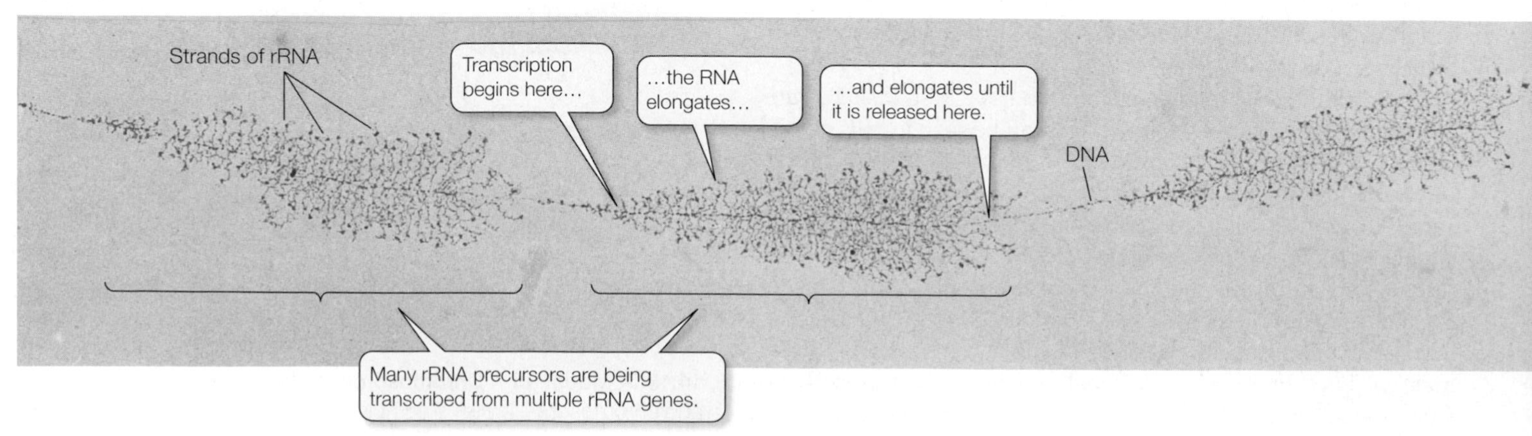
Strands of rRNA
Transcription begins here...
...the RNA elongates...
...and elongates until it is released here.
DNA
Many rRNA precursors are being transcribed from multiple rRNA genes.

14.20 Transcription from Multiple Genes for rRNA Elongating strands of rRNA transcripts form arrowhead-shaped regions, each centered on a DNA sequence that codes for three of the four ribosomal subunits.

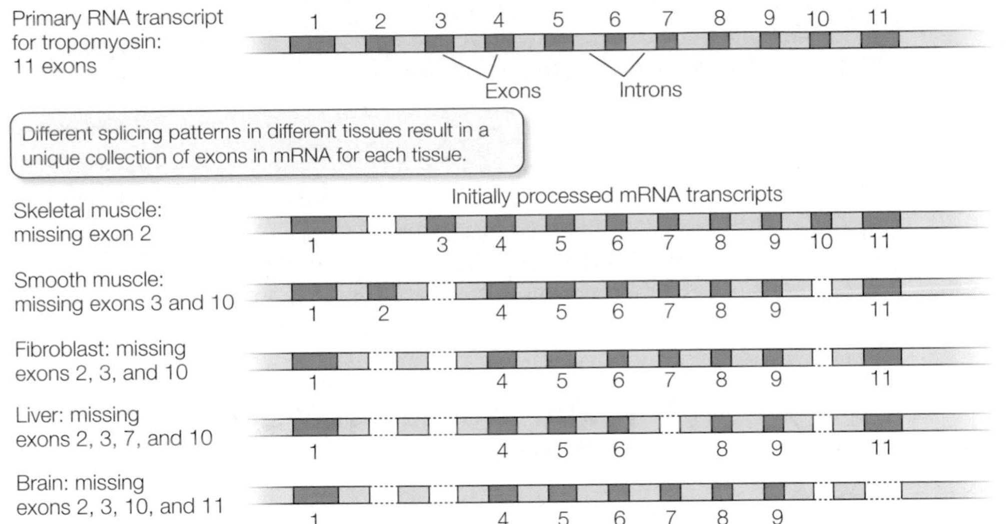

14.21 Alternative Splicing Results in Different Mature mRNAs and Proteins In mammals, the protein tropomyosin is encoded by a gene that has 11 exons. Tropomyosin pre-mRNA is spliced differently in different tissues, resulting in five different forms of the protein.

the two introns, but also the middle exon, would be spliced out. An entirely new protein (certainly not a β-globin) would be made, and the functions of normal β-globin would be lost.

Such **alternative splicing** can be a deliberate mechanism for generating a family of different proteins from a single gene. In mammals, for example, a single pre-mRNA for the structural protein tropomyosin is spliced differently in five different tissues to give five different mature mRNAs. These mRNAs are translated into the five different forms of tropomyosin found in skeletal muscle, smooth muscle, fibroblast, liver, and brain (**Figure 14.21**).

Before the sequencing of the human genome began, most scientists estimated that they would find between 100,000 and 150,000 genes. You can imagine their surprise when the actual sequence revealed only about 24,000 genes! In fact, there are many more human mRNAs than there are human genes, and most of this variation comes from alternative splicing. Indeed, recent surveys show that half of all human genes are alternatively spliced. Alternative splicing may be a key to the differences in levels of complexity among organisms.

The stability of mRNA can be regulated

DNA, as the genetic material, must remain stable, and as we saw in Section 11.3, there are elaborate mechanisms for repairing it if it becomes damaged. RNA, however, has no such repair mechanisms. After it arrives in the cytoplasm, mRNA is subject to breakdown catalyzed by ribonucleases, which exist both in the cytoplasm and in lysosomes. The less time an mRNA spends in the cytoplasm, the less of its protein can be translated. But not all eukaryotic mRNAs have the same life span. Differences in the stabilities of mRNAs provide another mechanism for posttranscriptional regulation of protein synthesis.

Specific AU-rich nucleotide sequences within some mRNAs mark them for rapid breakdown by a ribonuclease complex called the **exosome**. Signaling molecules such as growth factors, for example, are made only when needed, and then break down rapidly. Their mRNAs are highly unstable because they contain an AU-rich sequence.

Small RNAs can break down mRNAs

Very small RNAs—about 20 bases long—that are complementary to a region in mRNA can bind by base pairing to that mRNA before it gets to the ribosome. This binding causes the target mRNA to break down, but even if it survives, its translation is inhibited because tRNA cannot bind to the already base-paired region. Although only recently discovered, these **micro RNAs** are a common mechanism of posttranscriptional regulation. There are about 250 genes in the human genome that code for micro RNAs.

Small RNAs begin as 70-nucleotide, double-stranded molecules. A protein complex appropriately named *dicer* cuts the RNAs down to "micro" size and directs them to their target mRNAs (**Figure 14.22**). They have been implicated in controlling genes with a wide range of functions, ranging from the development of the nervous system in the roundworm, to apoptosis in the fruit fly, to flower development in plants, to development of the blood system in humans. As we will see in Chapter 16, small RNAs are under development as drugs to block the expression of certain human genes in diseases.

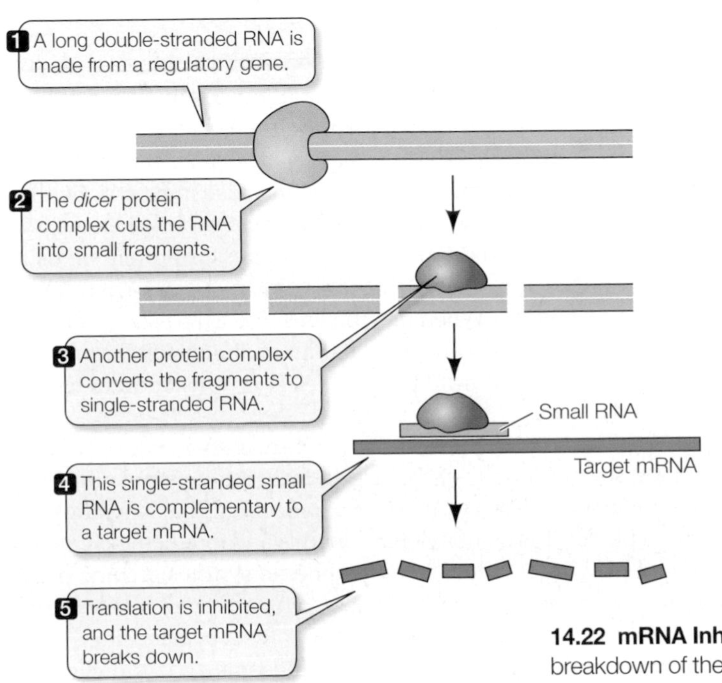

1 A long double-stranded RNA is made from a regulatory gene.
2 The *dicer* protein complex cuts the RNA into small fragments.
3 Another protein complex converts the fragments to single-stranded RNA.
Small RNA
Target mRNA
4 This single-stranded small RNA is complementary to a target mRNA.
5 Translation is inhibited, and the target mRNA breaks down.

14.22 mRNA Inhibition by Small RNAs Small RNAs result in inhibition of translation and breakdown of the target mRNA.

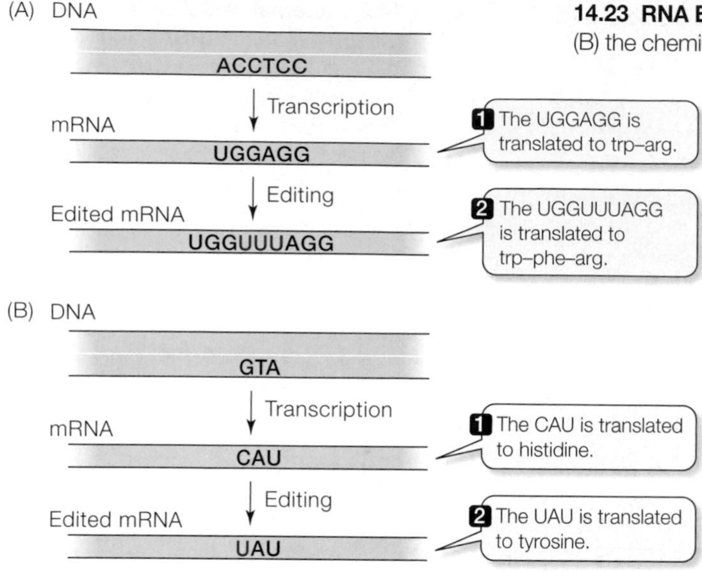

14.23 RNA Editing RNA can be edited by (A) the insertion of new nucleotides or by (B) the chemical alteration of existing nucleotides.

RNA can be edited to change the encoded protein

The sequence of mRNA can be changed after transcription and splicing by **RNA editing**. This editing can occur in two ways (**Figure 14.23**):

- *Insertion of nucleotides.* In the parasitic protist *Trypanosoma brucei,* certain mRNAs have been found that have a longer base sequence than predicted by the gene coding for them. Stretches of uracil are added after transcription, changing the protein that is made.
- *Alteration of nucleotides.* An enzyme can catalyze the deamination of cytosine, forming uracil. This process can affect a membrane channel protein in the mammalian nervous system that normally allows calcium and sodium to pass through. Editing of a certain cytosine in the mRNA for this protein to uracil changes the amino acid at that position in the polypeptide chain from histidine to tyrosine, and the channel protein no longer allows the passage of calcium.

14.5 RECAP

One of the most important means of posttranscriptional regulation is alternative RNA splicing, which allows more than one protein to be made from a gene. The stability of mRNA in the cytoplasm can also be regulated. Micro RNAs and RNA editing are two recently discovered mechanisms of regulation.

- Do you understand how a single pre-mRNA sequence can encode several different proteins? See pp. 324–325 and Figure 14.21
- How do micro RNAs regulate gene expression? See p. 325 and Figure 14.22
- What is RNA editing? See p. 326 and Figure 14.23

Even after mature mRNA is produced and transported to the cytoplasm, the presence of mRNA does not necessarily mean that it will be translated into a functional protein. Eukaryotes have several mechanisms for regulating gene expression during and after translation.

14.6 How Is Gene Expression Controlled During and After Translation?

Is the amount of a protein in a cell determined by the amount of its mRNA? Recently, the relationships between mRNAs and proteins in yeast cells were examined. For about a third of the dozens of genes surveyed, a clear relationship between mRNA and protein held: more of one led to more of the other. But for two-thirds of the proteins, no apparent relationship was observed. The concentrations of these proteins in the cell must therefore be determined by factors acting after the mRNA is made.

The initiation and extent of translation can be regulated

One way to regulate translation is through the G cap on mRNA. As we saw in Section 14.3, mRNA is capped at its 5′ end by a modified guanosine triphosphate molecule (see Figure 14.10). An mRNA that is capped with an unmodified GTP molecule is not translated. For example, stored mRNA in the egg cells of the tobacco hornworm moth has a G cap at its 5′ end, but the GTP molecule is not modified and the stored mRNA is not translated. After the egg is fertilized, however, the cap *is* modified, allowing the mRNA to be translated to produce the proteins needed for early embryonic development.

Conditions within a cell can influence translational processes. Within mammalian cells, for example, free iron ions (Fe^{2+}) are bound by a storage protein called *ferritin*. When iron is present in excess, ferritin synthesis rises dramatically. Yet the amount of ferritin mRNA remains constant. The increase in ferritin synthesis is due to an increased rate of mRNA translation. When the iron level in the cell is low, a translational repressor protein binds to ferritin mRNA and prevents its translation by blocking its attachment to a ribosome. When the iron level rises, the excess iron ions bind to the repressor and alter its three-dimensional structure, causing it to detach from the mRNA, and translation of ferritin proceeds.

Translational control can be used to keep a proper balance where several subunits associate to form a functional unit, as in hemoglobin molecules. As we saw in Section 14.2, a hemoglobin molecule consists of four globin subunits and four heme pigments. If globin synthesis does not equal heme synthesis, some heme stays free in the cell, waiting for a globin partner. Excess heme increases the rate of translation of globin mRNA by removing a block to the initiation of translation at the ribosome.

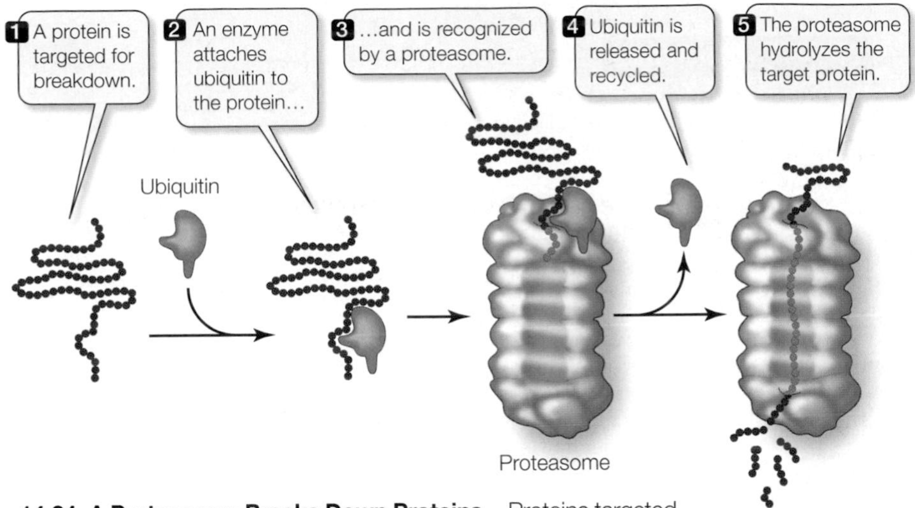

1 A protein is targeted for breakdown.
2 An enzyme attaches ubiquitin to the protein…
3 …and is recognized by a proteasome.
4 Ubiquitin is released and recycled.
5 The proteasome hydrolyzes the target protein.
Ubiquitin
Proteasome

14.24 A Proteasome Breaks Down Proteins Proteins targeted for breakdown are bound to ubiquitin, which "leads" the target protein to a proteasome (a complex composed of many polypeptides).

Posttranslational controls regulate the longevity of proteins

Most gene products—proteins—are modified after translation. Some of these are covalent changes, such as the addition of sugars (glycosylation), the addition of phosphate groups, or the removal of a signal sequence after a protein has crossed a membrane (see Figure 12.5).

One way to regulate the action of a protein in a cell is to regulate its lifetime in the cell. In many cases, an enzyme attaches a 76-amino acid protein called **ubiquitin** (so named because it is *ubiquitous*, or widespread) to a lysine in a protein targeted for breakdown. Other ubiquitin chains then attach to the primary one, forming a *polyubiquitin complex*. The protein–polyubiquitin complex then binds to a huge protein complex called a **proteasome** (**Figure 14.24**). The entryway to this "molecular chamber of doom" is a hollow cylinder. This part of the complex uses energy from ATP to cut off the ubiquitin for recycling and unfold its targeted protein "victim." The protein then passes by three different proteases (thus the name of the complex), which digest it into small peptides and amino acids.

The cellular concentrations of many proteins are determined not by differential expression of their genes, but by their degradation in proteasomes. Cyclins, for example, are degraded at just the right time during the cell cycle (see Section 9.2), and transcriptional regulators are broken down after they are used, lest the affected genes be always "on." Viruses can hijack this system. Human papillomavirus, which causes cervical cancer, targets the cell division inhibitory protein p53 for proteasomal degradation, resulting in unregulated cell division—i.e., cancer.

14.6 RECAP

Proteins in the cell can be targeted for breakdown by ubiquitin and then hydrolyzed in proteasomes.

- Do you understand the role of translational repressors? See p. 326
- Can you describe how a proteasome works? See p. 327 and Figure 14.24

CHAPTER SUMMARY

14.1 What are the characteristics of the eukaryotic genome?

There are many differences between prokaryotic and eukaryotic genomes and their mechanisms of expression. Review Table 14.1

Unlike prokaryotic DNA, eukaryotic DNA is contained within a nucleus, so that transcription and translation are physically separated. Review Figure 14.1, Web/CD Activity 14.1

Various types of highly repetitive sequences, known as satellites, characterize eukaryotic DNA. Most are not transcribed.

Some moderately repetitive DNA sequences are transcribed; many of these code for tRNA and rRNA, of which many copies are needed. Review Figure 14.3

Transposons are moderately repetitive sequences that are able to move about the genome. Review Figure 14.4

14.2 What are the characteristics of eukaryotic genes?

A typical eukaryotic protein-coding gene is flanked by promoter and **terminator** sequences and contains noncoding internal sequences, called **introns**. The coding sequences of such a gene are called **exons**. Review Figure 14.5

Introns are transcribed to the primary mRNA transcript (**pre-mRNA**), but are later removed and do not appear in the mature mRNA transcript.

In **nucleic acid hybridization**, DNA is denatured and incubated with a single-stranded **probe** that binds to the DNA by complementary base pairing. Review Figure 14.6

Some eukaryotic genes exist as members of **gene families**. Proteins may be made from these closely related genes at different times and in different tissues. Some members of gene families may be nonfunctional **pseudogenes**. Review Figure 14.8

14.3 How are eukaryotic gene transcripts processed?

The transcribed pre-mRNA is altered by the addition of a **G cap** at the 5′ end and a **poly A tail** at the 3′ end. Review Figure 14.10

The introns are removed from pre-mRNA by **RNA splicing**. A complex of **snRNPs** and enzymes, called a **spliceosome**, forms at the **consensus sequences** that lie between introns and exons. The spliceosome cuts out the introns and splices the exons together. Review Figure 14.11, Web/CD Tutorial 14.1

CHAPTER SUMMARY

14.4 How is eukaryotic gene transcription regulated?

Eukaryotic gene expression can be regulated before or during transcription, during pre-mRNA processing, and during or after translation. Review Figure 14.12, Web/CD Activity 14.2

Eukaryotes have three different RNA polymerases. RNA polymerase II transcribes protein-coding genes.

For transcription to occur, the protein TFIID must bind to the **TATA box** on the promoter, and other **transcription factors** must assemble on that protein, before RNA polymerase can bind at the promoter. Review Figure 14.13, Web/CD Tutorial 14.2

Other regulatory DNA sequences include **regulator** sequences, which bind regulator proteins and activate transcription, **enhancer** sequences, which bind activator proteins and stimulate transcription, and **silencer** sequences, which bind repressor proteins and turn off transcription. Review Figure 14.14

The DNA-binding domains of most DNA-binding proteins have one of four structural motifs.

Chromatin remodeling allows the transcription complex to bind to DNA and to move through the nucleosomes. Review Figure 14.17

Interference RNA, transcribed from the *Xist* gene, is important in inhibiting transcription of the inactive X chromosome. Review Figure 14.19

Some genes are selectively **amplified** in some cells. The extra copies of these genes result in increased transcription of their protein product.

14.5 How is eukaryotic gene expression regulated after transcription?

Alternative splicing of pre-mRNA can produce different proteins. Review Figure 14.21

Not all RNAs have the same life span. Regulating the stability of mRNA in the cytoplasm is a posttranscriptional mechanism for regulating protein synthesis. Specific AU-rich sequences can, for example, be rapidly broken down by an **exosome**. **Small RNAs** can base-pair with target mRNA sequences, preventing their translation and breaking them down. Review Figure 14.22

mRNA can be **edited** by the addition of new nucleotides or by the chemical alteration of existing nucleotides. Review Figure 14.23

14.6 How is gene expression controlled during and after translation?

Translational repressors can inhibit the translation of mRNA.

Proteasomes can degrade proteins that have been targeted for breakdown by attachment of **ubiquitin**. Review Figure 14.24

SELF-QUIZ

1. Eukaryotic protein-coding genes differ from their prokaryotic counterparts in that eukaryotic genes
   a. are double-stranded.
   b. are present in only a single copy.
   c. contain introns.
   d. have a promoter.
   e. transcribe mRNA.
2. Comparison of the genomes of yeast and bacteria shows that only yeast has many genes for
   a. energy metabolism.
   b. cell wall synthesis.
   c. intracellular protein targeting.
   d. DNA-binding proteins.
   e. RNA polymerase.
3. The genomes of the fruit fly and the nematode are similar to that of yeast, except that the former organisms have many genes for
   a. intercellular signaling.
   b. synthesis of polysaccharides.
   c. cell cycle regulation.
   d. intracellular protein targeting.
   e. transposable elements.
4. Which of the following does *not* occur after mRNA is transcribed?
   a. Binding of RNA polymerase II to the promoter
   b. Capping of the 5′ end
   c. Addition of a poly A tail to the 3′ end
   d. Splicing out of the introns
   e. Transport to the cytosol
5. Which statement about RNA splicing is *not* true?
   a. It removes introns.
   b. It is performed by small nuclear ribonucleoprotein particles (snRNPs).
   c. It always removes the same introns.
   d. It is usually directed by consensus sequences.
   e. It shortens the RNA molecule.
6. Eukaryotic transposons
   a. always use RNA for replication.
   b. are approximately 50 bp long.
   c. are made up of either DNA or RNA.
   d. do not contain genes coding for transposition.
   e. make up about 40 percent of the human genome.
7. Which statement about selective gene transcription in eukaryotes is *not* true?
   a. Different classes of RNA polymerase transcribe different parts of the genome.
   b. Transcription requires transcription factors.
   c. Genes are transcribed in groups called operons.
   d. Both positive and negative regulation occur.
   e. Many proteins bind at the promoter.
8. Heterochromatin
   a. contains more DNA than does euchromatin.
   b. is transcriptionally inactive.
   c. is responsible for all negative transcriptional control.
   d. clumps the X chromosome in human males.
   e. occurs only during mitosis.
9. Translational control
   a. is not observed in eukaryotes.
   b. is a slower form of regulation than transcriptional control.
   c. can be achieved by only one mechanism.
   d. requires that mRNA be uncapped.
   e. ensures that heme synthesis equals globin synthesis.
10. Control of gene expression in eukaryotes includes all of the following *except*
    a. alternative splicing of RNA transcripts.
    b. binding of proteins to DNA.
    c. transcription factors.
    d. feedback inhibition of enzyme activity by allosteric control.
    e. DNA methylation.

FOR DISCUSSION

1. In rats, a gene 1,440 bp long codes for an enzyme made up of 192 amino acids. Discuss this apparent discrepancy. How long would the initial and final mRNA transcripts be?
2. The genomes of rice, wheat, and corn are similar to one other and to that of *Arabidopsis* in many ways. Discuss how these plants might nevertheless have very different proteins.
3. The activity of the enzyme dihydrofolate reductase (DHFR) is high in some tumor cells. This activity makes the cells resistant to the anticancer drug methotrexate, which targets DHFR. Assuming that you had the complementary DNA for the gene that encodes DHFR, how would you show whether this increased activity was due to increased transcription of the single-copy DHFR gene or to amplification of the gene?
4. Describe the steps in the production of a mature, translatable mRNA from a eukaryotic gene that contains introns. Compare this to the situation in prokaryotes (see Section 13.3).
5. A protein-coding gene has three introns. How many different proteins can be made from alternative splicing of the pre-mRNA transcribed from this gene?

FOR INVESTIGATION

Nucleic acid hybridization has had many uses in biological research and clinical medicine. It is possible to make a DNA probe that fluoresces under ultraviolet light. Suppose you want to determine whether the fetus a woman is carrying has an extra copy of chromosome 21 (and so has Down syndrome). You take a sample of cells from the fetus. These cells are mostly in interphase. Assuming you have isolated the DNA for several genes that map on chromosome 21, outline the experiments you would do to determine whether the fetus has Down syndrome.

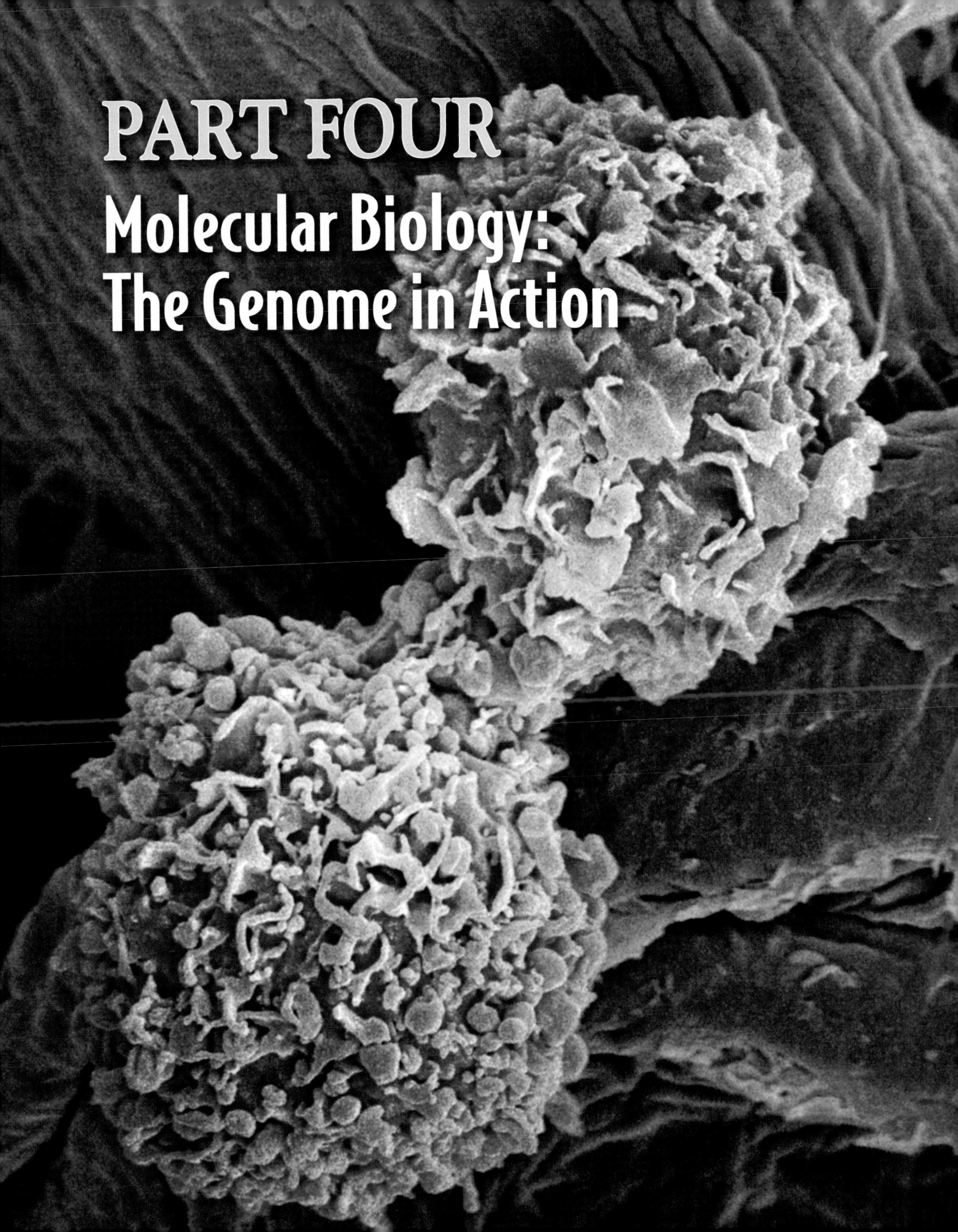

PART FOUR
Molecular Biology: The Genome in Action

CHAPTER 15 Cell Signaling and Communication

Have a cup of signals

It's probably happened to you: it's late, you have a paper due tomorrow, and you've put it off until the last minute. You're exhausted, but you need to stay awake and alert so that you can get your work done. What do you do? Well, for starters, you might have a cup (or several cups) of coffee. Many people turn to coffee when they need to wake themselves up or give themselves an energy boost.

Legend has it that the energy-inducing effects of coffee were first noticed a thousand years ago in what is now Ethiopia. A goat herder named Kaldi is said to have noticed that his goats became very frisky after they ate the berries of a certain plant. His curiosity aroused, Kaldi ate some of the berries himself, and thoroughly enjoyed the result. Word of his discovery spread throughout the region. Soon monks at a nearby monastery found that eating the berries kept them awake during their late-night prayers. The monks came up with the idea of drying the berries for storage and transport, and subsequently learned that pulverizing the dried berries and heating the powder in water resulted in an enjoyable beverage. Coffee shops were not far behind.

On a typical day, at least 90 percent of North Americans and Europeans consume caffeine in some form—in tea (which may contain up to 90 mg of caffeine per 8-ounce cup), cola (50 mg), chocolate (20 mg in a dark bar), as well as coffee (up to 180 mg per cup). Caffeine is our most popular drug, and like many drugs it is a *signal molecule.* To understand the effects of caffeine, we must first understand the pathways by which the body's cells respond to signals in their environment.

A cell's response to any signal molecule takes place in three sequential steps. First, the signal binds to a receptor protein in the cell, often on the outside surface of the plasma membrane. Second, the binding of the signal causes a message to be conveyed to the inside of the cell and amplified. Third, the cell changes its activity in response to the signal.

Caffeine acts in different ways in different tissues. A tired person's brain produces adenosine molecules that bind to specific receptor proteins, resulting in decreased brain activity and increased drowsiness. Caffeine's molecular structure is similar to that of adenosine, so it occupies the adenosine receptors without inhibiting brain cell function, much like a competitive enzyme inhibitor, and alertness is restored. Adenosine also opens up, or *dilates*, the blood vessels supplying the brain, causing headaches; thus caffeine is a common ingredient in headache remedies.

A Signal to the Body The caffeine in coffee signals cells in the body of this coffee drinker. The effects of these signals help him stay alert.

Traditional Ethiopian Coffee Ceremony Coffee plays an important ceremonial role in the traditional culture of Ethiopia, where (according to legend) people originally discovered its energizing effects.

In heart and liver cells, caffeine indirectly stimulates the same signaling pathway normally stimulated by epinephrine, the "fight-or-flight" hormone. The result is an increased heartbeat rate. Muscles tighten up and the liver is stimulated to convert glycogen into glucose and release it into the bloodstream.

How can a single molecule from outside the body have so many biological effects?

IN THIS CHAPTER we describe types of signals that affect cells, which include both chemicals produced by other cells and substances from outside the body, as well as physical and environmental factors such as light. We'll learn that whatever the signal, it affects only those cells that have the appropriate receptor protein to respond to the specific signal. We'll trace the steps of signal transduction by which the receptor communicates that a signal has been received, thus effecting a change in cell function.

CHAPTER OUTLINE

15.1 What Are Signals, and How Do Cells Respond to Them?

Both prokaryotic and eukaryotic cells process information from their environment. This information can be in the form of a physical stimulus, such as the light reaching your eyes as you read this book, or chemicals that bathe a cell, such as lactose in the medium surrounding *E. coli*. It may come from outside the organism, such as the scent of a female moth seeking a mate in the dark, or from a neighboring cell within the organism, as in the heart, where thousands of muscle cells contract in unison by transmitting signals to one another.

Of course, the mere presence of a signal does not mean that a cell will respond to it, just as you do not pay close attention to every sound in your environment as you study. To respond to a signal, the cell must have a specific *receptor* that can detect it. This section provides examples of some types of cellular signals and one model *signal transduction pathway*—the series of steps that lead to a cell's response to a signal. After discussing signals in this section, we will consider their receptors in Section 15.2.

Cells receive signals from the physical environment and from other cells

The physical environment is full of signals. Our sense organs allow us to respond to light, odors and tastes (chemical signals), temperature, touch, and sound. Bacteria and protists respond to even minute chemical changes in their environment. Plants respond to light as a signal. For example, at sunset, at night, or in the shade, not only the amount but also the wavelengths of the light reaching Earth's surface differ from that of full sunlight in the daytime. These variations act as signals that affect plant growth and reproduction. Some plants also respond to temperature: when the weather gets cold, they respond either by becoming tolerant to cold or by accelerating flowering. Even magnetism can be a signal: some bacteria and birds orient themselves to Earth's magnetic poles, like a needle on a compass.

A cell deep inside a large multicellular organism is far away from the exterior environment. Instead, its environment consists of other cells and extracellular fluids. Cells receive their

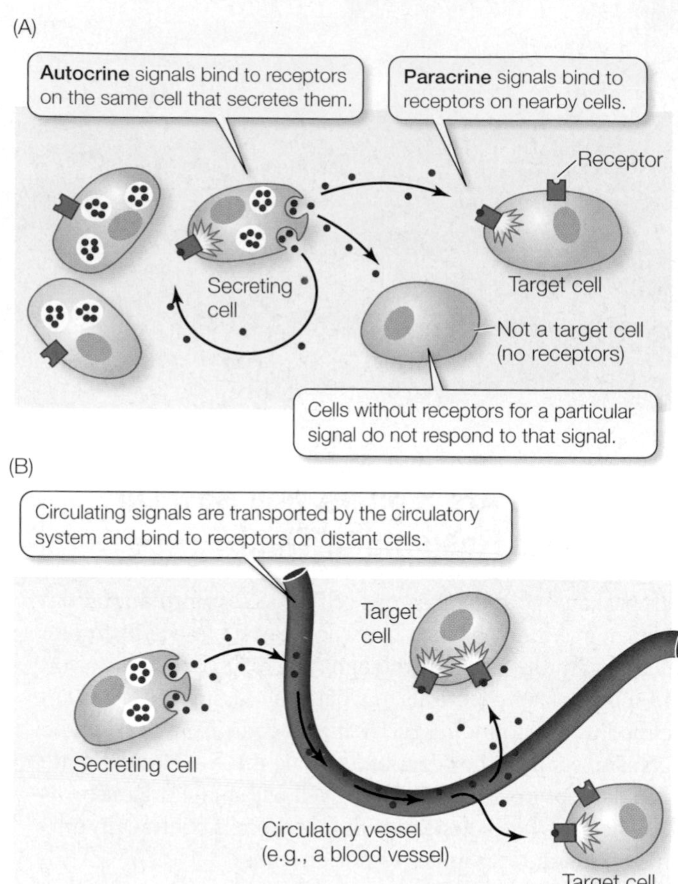
(A)
Autocrine signals bind to receptors on the same cell that secretes them.
Paracrine signals bind to receptors on nearby cells.
Receptor
Secreting cell
Target cell
Not a target cell (no receptors)
Cells without receptors for a particular signal do not respond to that signal.
(B)
Circulating signals are transported by the circulatory system and bind to receptors on distant cells.
Target cell
Secreting cell
Circulatory vessel (e.g., a blood vessel)
Target cell

15.1 Chemical Signaling Systems (A) A signal molecule can diffuse to act on the cell that produces it, or on a nearby cell. (B) Many signals act on distant cells, to which they must be transported by the organism's circulatory system.

nutrients from, and pass their wastes into, extracellular fluids. Cells also receive signals—mostly chemical signals—from their extracellular fluid environment. Most of these chemical signals come from other cells. Cells also respond to chemical signals coming from the environment via the digestive and respiratory systems. And cells can respond to concentrations of certain chemicals, such as CO_2 and H^+, whose presence in the extracellular fluids results from the metabolic activities of other cells.

Inside a large multicellular organism, chemical signals made by the body itself reach a target cell by local diffusion or by circulation within the blood. **Autocrine** signals diffuse to and affect the cells that make them, while **paracrine** signals diffuse to and affect nearby cells (**Figure 15.1A**). Signals to distant cells are called *hormones* and usually travel through the circulatory system (**Figure 15.1B**).

In all cases, in order for a signal to be transmitted, the target cell must be able to receive or sense the signal and respond to it, and the response must have some effect on the function of the cell. Depending on the target cell and the signal, these effects range from the cell's entering the cell cycle to heal a wound, to moving to a new location in the embryo to form a tissue, to releasing enzymes that digest food, to sending messages to the brain about the book you are reading.

A signal transduction pathway involves a signal, a receptor, transduction, and effects

The entire signaling process—from the signal's affecting a receptor, to conveying the message to the cytoplasm, to the cell's final response—is called a **signal transduction pathway**. Let's look at an example of such a pathway in *E. coli*. In Section 13.4, we saw that this bacterium responds to changes in the nutrient content of its environment by altering its transcription of certain genes, such as those in the *lac* operon. The bacterium must also be able to sense and respond to other kinds of changes in its environment, such as changes in solute concentration.

In the human intestine, where *E. coli* lives, the solute concentration around the bacterium often rises far above that inside the cell. The principle of diffusion tells us that when this happens, water will diffuse out of the cell, and solutes will move into the cell. But the bacterium must maintain homeostasis, so it must perceive and respond to this environmental signal (**Figure 15.2, step 1**). The pathway by which *E. coli* does so has much in common with signal transduction pathways in more complex multicellular eukaryotes. The pathway involves two major components: a receptor and a responder.

RECEPTOR A **receptor** is the first component of a signal transduction pathway. The *E. coli* receptor protein for changes in solute concentration is called EnvZ. EnvZ is a transmembrane protein that extends through the bacterium's plasma membrane into the space between the plasma membrane and the highly porous outer membrane, which forms a complex with the cell wall. When the solute concentration of the extracellular environment rises, so does the solute concentration in the space between the two membranes. This change in its aqueous medium causes the part of the receptor protein sticking into the intermembrane space to undergo a change in *conformation* (its three-dimensional shape).

As we saw in Section 6.5, changing the tertiary structure of one part of a protein often leads to changes in distant parts of the protein. In the case of the bacterial EnvZ receptor, the conformational change in the intermembrane domain of the protein is transmitted to the domain that lies in the cytoplasm, initiating the events of signal transduction. This conformational change exposes an active site, so that EnvZ becomes a *protein kinase*, an enzyme that catalyzes the addition of a phosphate group from ATP to one of EnvZ's own histidine molecules. In other words, EnvZ phosphorylates itself (**Figure 15.2, step 2**).

RESPONDER A **responder** is the second component of a signal transduction pathway. The charged phosphate group added to the histidine causes the cytoplasmic domain of the EnvZ protein to change its shape again. It now binds to a second protein, OmpR, which takes the phosphate group from EnvZ. This phosphorylation changes the shape of OmpR in turn (**Figure 15.2, step 3**). This change in a responder is a key event in signaling for three reasons:

- The signal on the outside of the cell has now been *transduced* to a protein that lies totally within the cell's cytoplasm.
- The altered responder can *do something*. In the case of the phosphorylated OmpR, that "something" is to bind to a promoter on *E. coli* DNA adjacent to the sequence that codes for

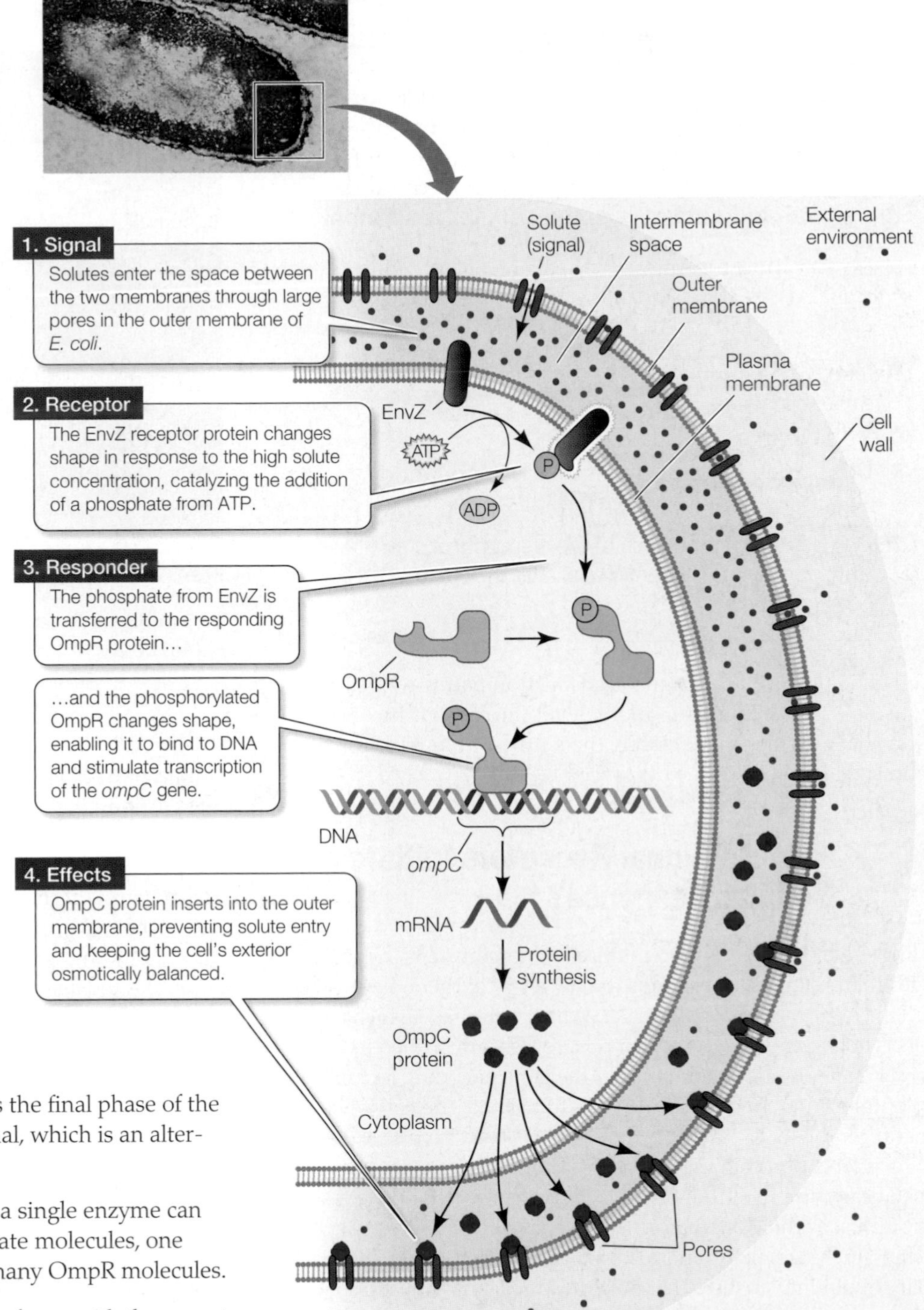

1. Signal
Solutes enter the space between the two membranes through large pores in the outer membrane of E. coli.
2. Receptor
The EnvZ receptor protein changes shape in response to the high solute concentration, catalyzing the addition of a phosphate from ATP.
3. Responder
The phosphate from EnvZ is transferred to the responding OmpR protein...
...and the phosphorylated OmpR changes shape, enabling it to bind to DNA and stimulate transcription of the ompC gene.
4. Effects
OmpC protein inserts into the outer membrane, preventing solute entry and keeping the cell's exterior osmotically balanced.
Solute (signal)
Intermembrane space
External environment
Outer membrane
Plasma membrane
Cell wall
EnvZ
ATP
ADP
P
OmpR
DNA
ompC
mRNA
Protein synthesis
OmpC protein
Cytoplasm
Pores

15.2 A Model Signal Transduction Pathway *E. coli* responds to the signal of an increase in solute concentration in its environment. The basic steps of such signal transduction pathways occur in all living organisms.

the protein OmpC. This binding begins the final phase of the signaling pathway: the *effect* of the signal, which is an alteration in cell function.

- The signal has been *amplified*. Because a single enzyme can catalyze the conversion of many substrate molecules, one EnvZ molecule alters the structure of many OmpR molecules.

Phosphorylated OmpR is a transcription factor with the correct three-dimensional structure to bind to the promoter of the *ompC* gene, resulting in an increase in the transcription of that gene. Translation of *ompC* mRNA results in the production of OmpC protein, which leads to the response that regulates *E. coli*'s solute concentration (**Figure 15.2, step 4**). The OmpC protein is inserted into the outer membrane of the bacterial cell, where it blocks pores and prevents solutes from entering the intermembrane space. As a result, the solute concentration in the intermembrane space is lowered, and homeostasis is restored. Thus the *E. coli* cell can go on behaving just as if the external environment had a normal solute concentration.

Many of the same elements we have highlighted in this prokaryotic signal transduction system will reappear in many other signal transduction pathways in eukaryotic organisms:

- A receptor protein *changes its conformation* upon interacting with a signal.
- A conformational change in the receptor protein gives it *protein kinase* activity, resulting in the transfer of a phosphate group from ATP to a target protein.
- This *phosphorylation* alters the function of a responder protein.
- The signal is *amplified*.
- A *transcription factor* is activated.
- The *synthesis of a specific protein* is turned on.
- The action of the protein *alters cell activity*.

15.1 RECAP

Cells are constantly exposed to molecular signals, both from the external environment and from within the body itself. To respond to such a signal, a cell must have a receptor that detects the signal and activates some cellular response.

- Do you know the difference between an autocrine signal, a paracrine signal, and a hormone? See p. 334 and Figure 15.1
- Describe the roles of the two major components in a signal transduction pathway. See p. 334 and Figure 15.2
- Do you understand why it is usually important for a signal to be amplified? See p. 335
- Do you understand each of the elements of signal transduction described in the list at the close of this section?

The general features of signal transduction pathways described in this section will recur in more detail throughout the chapter. First let's consider more closely the nature of the receptors that bind signal molecules.

15.2 How Do Signal Receptors Initiate a Cellular Response?

Although any given cell in a multicellular organism is bombarded with many signals, it responds to only a few of them, because no cell makes receptors for all signals. Which cells make which receptors is determined by the regulatory processes described in Chapter 14; in short, if a cell transcribes the gene encoding a particular receptor and the resulting mRNA is translated, the cell will have that receptor.

A receptor protein that binds to a chemical signal does so very specifically, in much the same way as an enzyme binds to a substrate. Just as there are many types of enzymes with diverse specificities, there are many kinds of signal receptor proteins. This specificity of binding ensures that only those cells that make a specific receptor will respond to a given signal.

Receptors have specific binding sites for their signals

A specific chemical signal molecule fits into a three-dimensional site on its receptor (**Figure 15.3**). A molecule that binds to a receptor site in another molecule in this way is called a **ligand**. As you saw with the example in *E. coli*, binding of the signaling ligand causes the receptor protein to change its three-dimensional shape, and that conformational change initiates a cellular response. The ligand does not contribute further to this response. In fact, the ligand signal usually is not metabolized into useful products. Its role is purely to "knock on the door." (This is in sharp contrast to the enzyme–substrate interactions described in Chapter 6, whose whole purpose is to change the substrate into a useful product.)

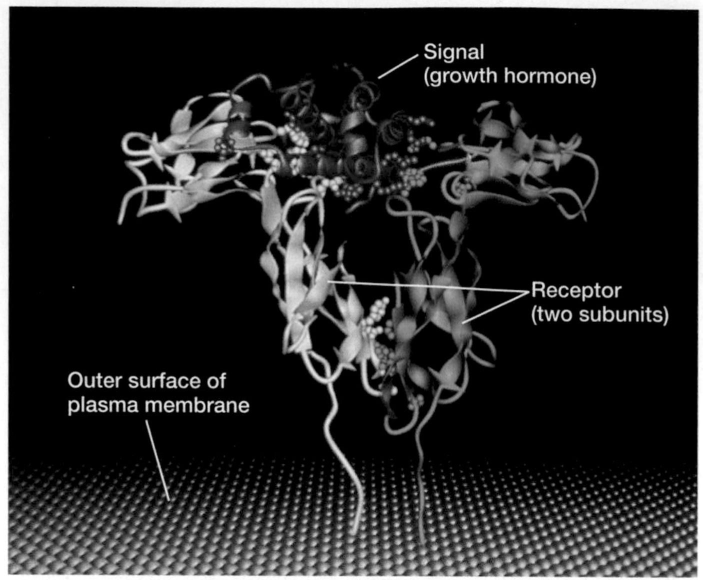

15.3 A Signal Bound to Its Receptor Human growth hormone is shown bound to its receptor, a transmembrane protein. Only the extracellular regions of the receptor are shown.

Receptors bind to their ligands according to chemistry's law of mass action:

$$R + L \rightleftharpoons RL$$

This means that the binding is reversible, although for most ligand–receptor complexes, the equilibrium point is far to the right—that is, binding is favored. Reversibility is important, however, because if the ligand were never released, the receptor would be continuously stimulated.

As with enzymes, inhibitors can bind to the ligand binding site on a receptor protein. Both natural and artificial inhibitors of receptor binding are important in medicine. For example, over two-thirds of the drugs that alter human behavior bind to specific receptors in the brain.

Receptors can be classified by location

Receptors can be classified by their location in the cell, which largely depends on the nature of their ligands. The chemistry of signal molecules is quite variable, but they can be divided into two classes (**Figure 15.4**):

- *Ligands with cytoplasmic receptors:* Small or nonpolar ligands can diffuse across the phospholipid bilayer of the plasma membrane and enter the cell. Estrogen, for example, is a lipid-soluble steroid hormone that can easily diffuse across the plasma membrane and enter the cell; it binds to a receptor in the cytoplasm.
- *Ligands with plasma membrane receptors:* Large or polar ligands cannot cross the plasma membrane. Insulin, for example, is a protein hormone that cannot diffuse through the plasma membrane; instead, it binds to a receptor that is a transmembrane protein with an extracellular binding domain.

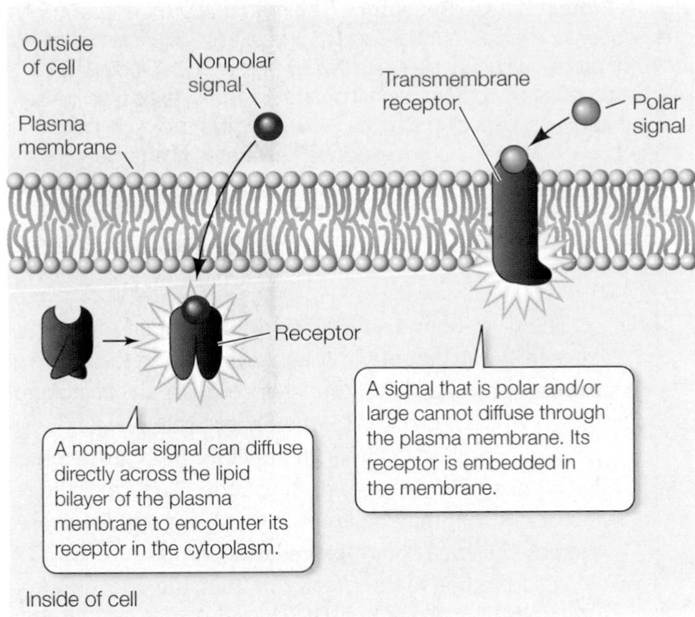

15.4 Two Locations for Receptors Receptors can be located in the plasma membrane or in the cytoplasm of the cell.

In complex eukaryotes such as mammals, there are three well-studied types of plasma membrane receptors that bear examination: *ion channels, protein kinases,* and *G protein-linked receptors*.

ION CHANNEL RECEPTORS As described in Section 5.3, the plasma membranes of many types of cells contain channel proteins that can be open or closed. These **ion channels** act as "gates," allowing ions such as Na^+, K^+, Ca^{2+}, or Cl^- to enter or leave the cell (see Figure 5.10). The gate-opening mechanism is an alteration in the three-dimensional shape of the channel protein upon ligand binding. Some ion channels are plasma membrane receptors for signal molecules; others act later in signal transduction pathways. Each type of ion channel receptor has its own signal. These signals include sensory stimuli such as light, sound, and electric charge differences across the plasma membrane, as well as chemical ligands such as hormones and neurotransmitters.

The acetylcholine receptor, which is located in the plasma membrane of vertebrate skeletal muscle cells, is an example of a gated ion channel. This receptor protein is a sodium channel that binds the ligand *acetylcholine*, which is a *neurotransmitter*, a chemical signal released from neurons (nerve cells) (**Figure 15.5**). When *two* molecules of acetylcholine bind to the receptor, it opens for about a thousandth of a second. That is enough time for Na^+, which is more concentrated outside the cell than inside, to rush into the cell. The change in Na^+ concentration in the cell initiates a series of events that result in muscle contraction.

Acetylcholine receptors are present in brain cells as well as skeletal muscle. Nicotine binds strongly to these receptors and activates many pathways in the brain. The effects of nicotine may be transiently pleasurable, but ultimately result in addiction.

PROTEIN KINASE RECEPTORS Like the EnvZ receptor protein of *E. coli*, some eukaryotic receptor proteins become **protein kinases** when they are activated: that is, they catalyze the transfer of a phosphate group from ATP to a specific protein, referred to as the *target protein*. This phosphorylation can alter the conformation and activity of the target protein.

The receptor for insulin is an example of a protein kinase receptor. Insulin is a protein hormone made by the mammalian pancreas. Its receptor has two copies each of two different polypeptide subunits (**Figure 15.6**). As with acetylcholine, two molecules of insulin must bind to the receptor. When insulin binds to its extracellular subunits, the receptor protein changes its shape to expose a protein kinase active site in the cytoplasm. Like the *E. coli* receptor described above, the insulin receptor *autophosphorylates* (phosphorylates itself). Then, as a protein kinase, it catalyzes the phosphorylation of certain cytoplasmic proteins, appropriately called insulin response substrates. These proteins then initiate many cellular responses, including the insertion of glucose transporters (see Figure 5.12) into the plasma membrane.

G PROTEIN-LINKED RECEPTORS A third category of eukaryotic plasma membrane receptors is the *seven-transmembrane-spanning G protein-linked receptors*. This long name identifies a fascinating group of receptors, all of which are composed of a single protein with seven transmembrane regions. These seven regions pass through the phospholipid bilayer, separated by short loops that extend either outside or inside the

15.5 A Gated Ion Channel The acetylcholine receptor (AChR) is a gated ion channel for sodium ions. It is made up of five polypeptide subunits. When acetylcholine molecules (ACh) bind to two of the subunits, the gate opens and Na^+ flows into the cell.

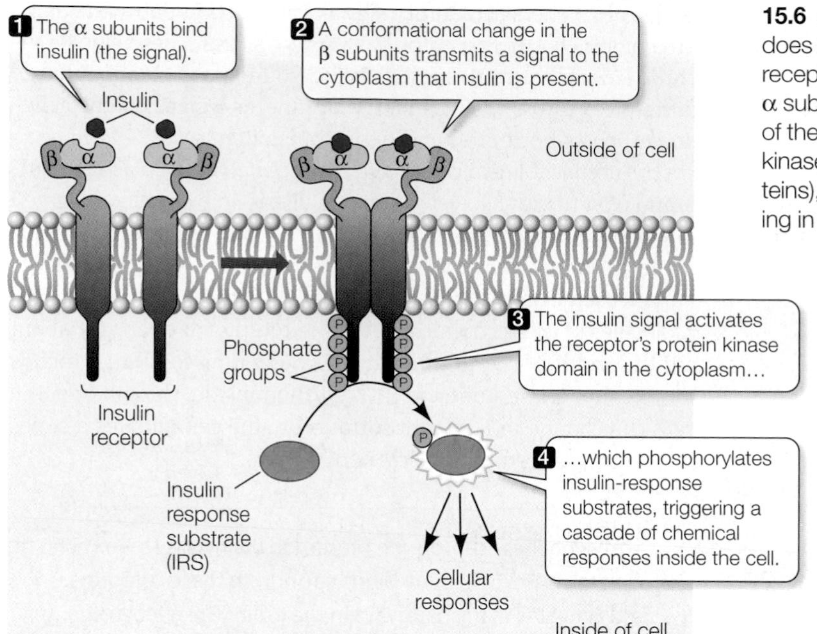

15.6 A Protein Kinase Receptor The mammalian hormone insulin does not enter the cell, but is bound by the extracellular domain of a receptor protein with four subunits (two α and two β). Binding to the α subunit causes a conformational change in the cytoplasmic domain of the β subunits, exposing a protein kinase active site. This protein kinase phosphorylates insulin response substrates (the target proteins), triggering further responses within the cell and eventually resulting in the transport of glucose across the membrane into the cell.

cell. Ligand binding on the extracellular side of the receptor protein changes the shape of its cytoplasmic region, exposing a binding site for a mobile membrane **G protein**.

The term *G protein* describes a class of signaling proteins with three polypeptide subunits that are characterized by their ability to bind GDP and GTP (guanosine diphosphate and triphosphate, respectively; these are nucleoside phosphates like ADP and ATP). A G protein has two important binding sites: one for the G protein-linked receptor and the other for GDP/GTP. The inactive G protein binds GDP to one of its subunits (**Figure 15.7A**). When the G protein binds to an activated receptor protein, GDP is exchanged for GTP (**Figure 15.7B**). At the same time, the ligand is released from the extracellular side of the receptor.

The GTP-bound subunit of the G protein then separates from the parent G protein, diffusing in the plane of the phospholipid bilayer until it encounters an **effector protein** to which it can bind. An effector protein is just what its name implies: it causes an effect in the cell. The binding of the GTP-bearing G protein subunit activates the effector—which may be an enzyme or an ion channel—thereby causing changes in cell function (**Figure 15.7C**).

After binding to the effector protein, the GTP on the G protein is hydrolyzed to GDP. The now inactive G protein subunit separates from the effector protein. The G protein subunit must form a complex with other subunits before binding to yet another activated receptor. When an activated receptor is bound, the G protein exchanges its GDP for GTP, and the cycle begins again.

By means of their diffusing subunits, G proteins can either activate or inhibit an effector protein. An example of an *activating* response involves the receptor for epinephrine (adrenaline), a hor-

15.7 A G Protein-Linked Receptor (A) The inactive G protein has binding sites for its linked receptor and for GDP/GTP. (B) Binding of an extracellular signal—in this case, a hormone—activates the G protein-linked receptor. The activated receptor binds the G protein, which is then activated by the exchange of GDP for GTP. (C) The GTP-bound subunit separates from the parent G protein and activates an effector protein—in this case, an enzyme that catalyzes a reaction in the cytoplasm, amplifying the signal. This figure is a generalized diagram that could apply to any member of the large family of G proteins and the signals they react to.

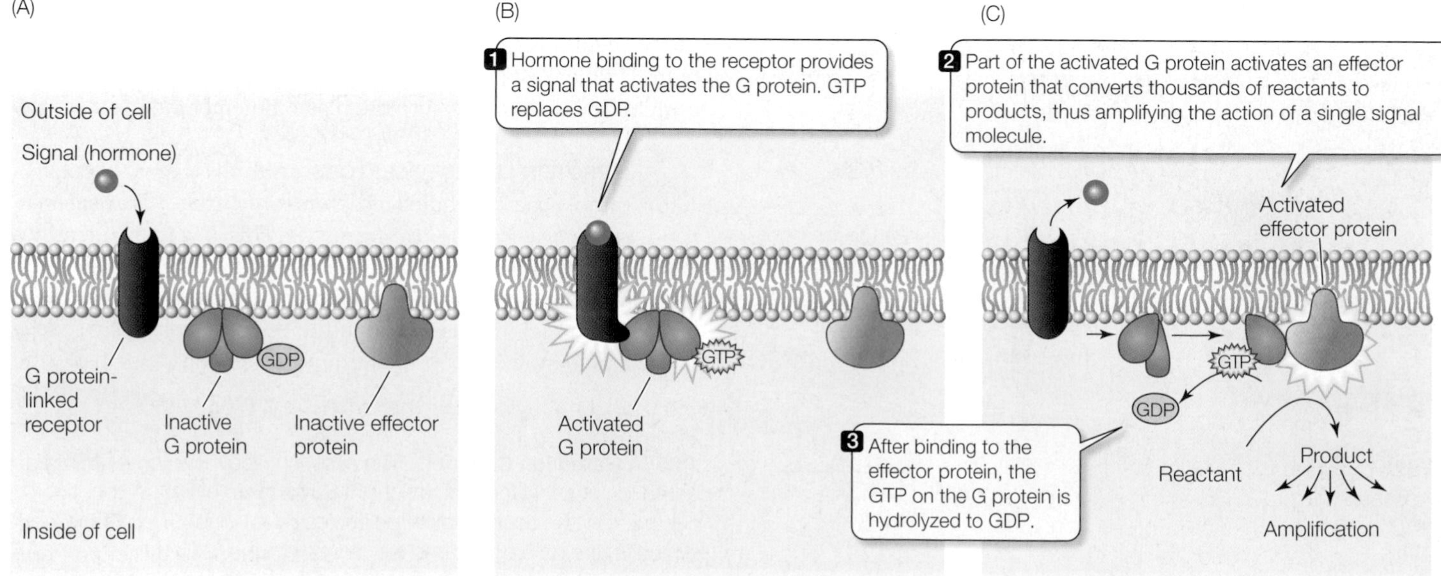

mone made by the adrenal gland in response to stress or heavy exercise. In heart muscle, this hormone binds to its G protein-linked receptor, activating a G protein. The GTP-bound subunit then activates a membrane-bound enzyme to produce a small molecule, cyclic AMP that has many effects on the cell (as we will see below), including glucose mobilization for energy and muscle contraction.

G protein-mediated *inhibition* occurs when the same hormone, epinephrine, binds to its receptor in the smooth muscle cells surrounding blood vessels lining the digestive tract. Again, the epinephrine-bound receptor changes its shape and activates a G protein, and the GTP-bound subunit binds to a target enzyme. But in this case, the enzyme is inhibited instead of being activated. As a result, the muscles relax and the blood vessel diameter increases, allowing more nutrients to be carried away from the digestive system to the rest of the body. Thus the same signal and initial signaling mechanism can have different consequences in different cells, depending on the nature of the responding cell.

CYTOPLASMIC RECEPTORS **Cytoplasmic receptors** are located inside the cell and bind to signals that can diffuse across the plasma membrane. Binding to the signaling ligand causes the receptor to change its shape so that it can enter the cell nucleus, where it acts as a transcription factor affecting gene expression. But this general view is somewhat simplified. The receptor for the hormone cortisol, for example, is normally bound to a chaperone protein, which blocks it from entering the nucleus. Binding of the hormone causes the receptor to change its shape so that the chaperone is released (**Figure 15.8**). This allows the receptor to fold into an appropriate conformation for entering the nucleus and initiating transcription of mRNA.

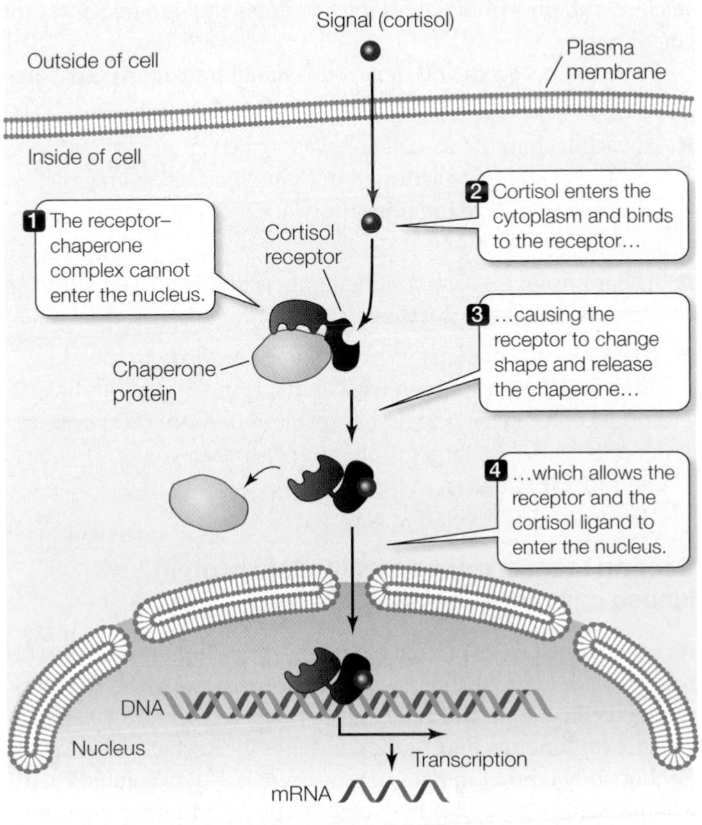

15.2 RECAP

Receptors for signals are proteins that bind, or are changed by, specific signals and then initiate a response in the cell. These receptors may be at the plasma membrane or inside the cell.

- Do you understand the specificity of receptors for ligands? See p. 336
- What are the three different kinds of plasma membrane receptors seen in complex eukaryotes? See pp. 337–338

Now that we have discussed signals and receptors, let's examine the characteristics of the transducers that mediate between the receptor and the cellular response.

15.8 A Cytoplasmic Receptor The receptor for cortisol is bound to a chaperone protein. Binding of the signal releases the chaperone and allows the receptor protein to enter the cell's nucleus, where it functions as a transcription factor.

15.3 How Is a Response to a Signal Transduced through the Cell?

As we have just seen, the same signal may produce different responses in different tissues. When epinephrine, for example, binds to receptors on heart muscle cells, it stimulates muscle contraction, but when it binds to receptors on smooth muscle cells in the blood vessels of the digestive system, it slows muscle contraction. These different responses to the same signal–receptor complex are mediated by the events of signal transduction. These events, which are critical to the cell's response, may be either direct or indirect:

- **Direct transduction** is a function of the receptor itself and occurs at the plasma membrane (**Figure 15.9A**).
- In **indirect transduction**, which is more common, another molecule, termed a **second messenger**, mediates a further interaction between the receptor and the cell's response (**Figure 15.9B**).

In both cases, the signal initiates a **cascade** of events, in which proteins interact with other proteins until the final responses are achieved. Through such a cascade, a weak initial signal can be both *amplified* and *distributed* to cause several different responses in the target cell.

Protein kinase cascades amplify a response to ligand binding

We have seen that when a signal binds to a protein kinase receptor, the receptor's conformation changes, exposing a protein kinase active site, which catalyzes the phosphorylation of target proteins.

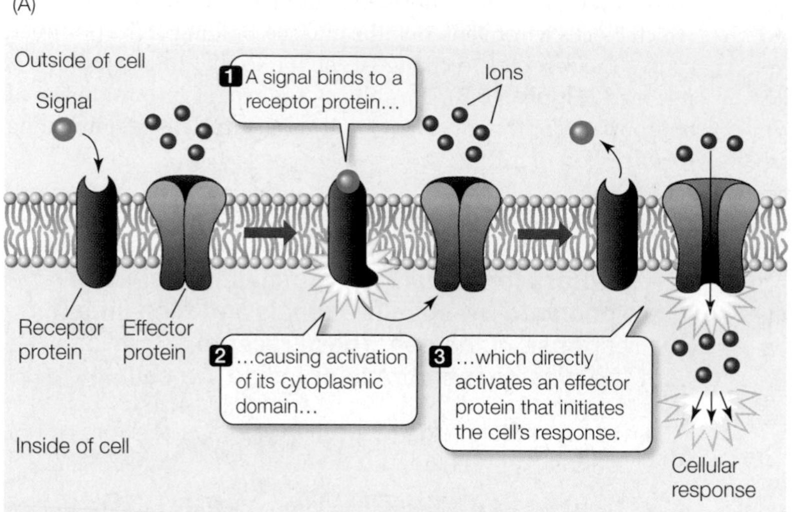

15.9 Direct and Indirect Signal Transduction (A) All the events of direct transduction occur at the plasma membrane at or near the receptor. (B) In indirect transduction, a second messenger mediates the events.

This process is an example of direct signal transduction. Protein kinase receptors are important in binding signals that stimulate cell division in both plants and animals. For example, the growth factors described in Section 9.2 that serve as external inducers of the cell cycle work by binding to protein kinase receptors.

The complete signal transduction pathway that occurs after a protein kinase receptor binds a growth factor was worked out through studies on a cell that went wrong. Many human bladder cancers contain an abnormal form of a protein called Ras (so named because it was first isolated from a *ra*t *s*arcoma tumor). Investigations of these bladder cancers showed that this Ras protein was a G protein, but was always active because it was permanently bound to GTP, and thus caused continuous cell division. If the cancer cells' Ras protein was inhibited, they stopped dividing. This discovery has led to a major effort to develop specific Ras inhibitors for cancer treatment.

What does Ras do in normal, noncancerous cells? Researchers knew that cells must be stimulated by growth factors (signals) in order to enter the cell cycle and divide (see Section 9.2). One hypothesis was that Ras was an intermediary between the binding of a growth factor to its receptor and the ultimate response of cell division. To investigate this hypothesis, the researchers treated cells in a culture dish with both a Ras inhibitor and a growth factor. Cell division did not occur, confirming their hypothesis.

After this discovery, the next step was to work out what the activated growth factor receptor did to Ras, and what Ras did to stimulate further events in signal transduction. This signaling pathway has been worked out, and it is an example of a more general phenomenon, called a **protein kinase cascade** (**Figure 15.10**). Such cascades are key to the external regulation of many cellular activities. Indeed, the eukaryotic genome codes for hundreds, even thousands, of such kinases.

The unbound receptors for growth factors exist in the plasma membrane as separate polypeptide chains (subunits). When the growth factor signal binds to a subunit, it associates with another subunit to form a dimer, which changes its shape to expose a protein kinase active site. Its kinase activity sets off a series of events, activating several other protein kinases in turn. The final phosphorylated, activated protein kinase—MAP kinase—moves into the nucleus and phosphorylates target proteins that are necessary for cell division.

Protein kinase cascades are useful signal transducers for three reasons:

- At each step in the cascade of events, the signal is *amplified*, because each newly activated protein kinase is an enzyme, which can catalyze the phosphorylation of many target proteins.
- The information from a signal that originally arrived at the plasma membrane is *communicated* to the nucleus.
- The multitude of steps provides some *specificity* to the process. As we have seen with epinephrine, signal binding and receptor activation do not result in the same response in all cells. Different target proteins at each step in the cascade can provide variation in the response.

Second messengers can stimulate protein kinase cascades

As we have just seen, protein kinase receptors initiate a protein kinase cascade right at the plasma membrane. However, the stimulation of events in the cell is more often indirect. In a series of clever experiments, Earl Sutherland, Edwin Krebs, and Edmond Fischer showed that in many cases, a small, water-soluble chemical messenger mediates between the plasma membrane receptor and cytoplasmic events. These researchers were investigating the activation of the liver enzyme *glycogen phosphorylase* by the hormone epinephrine. Glycogen phosphorylase catalyzes the hydrolysis of glycogen stored in the liver so that the resulting glucose

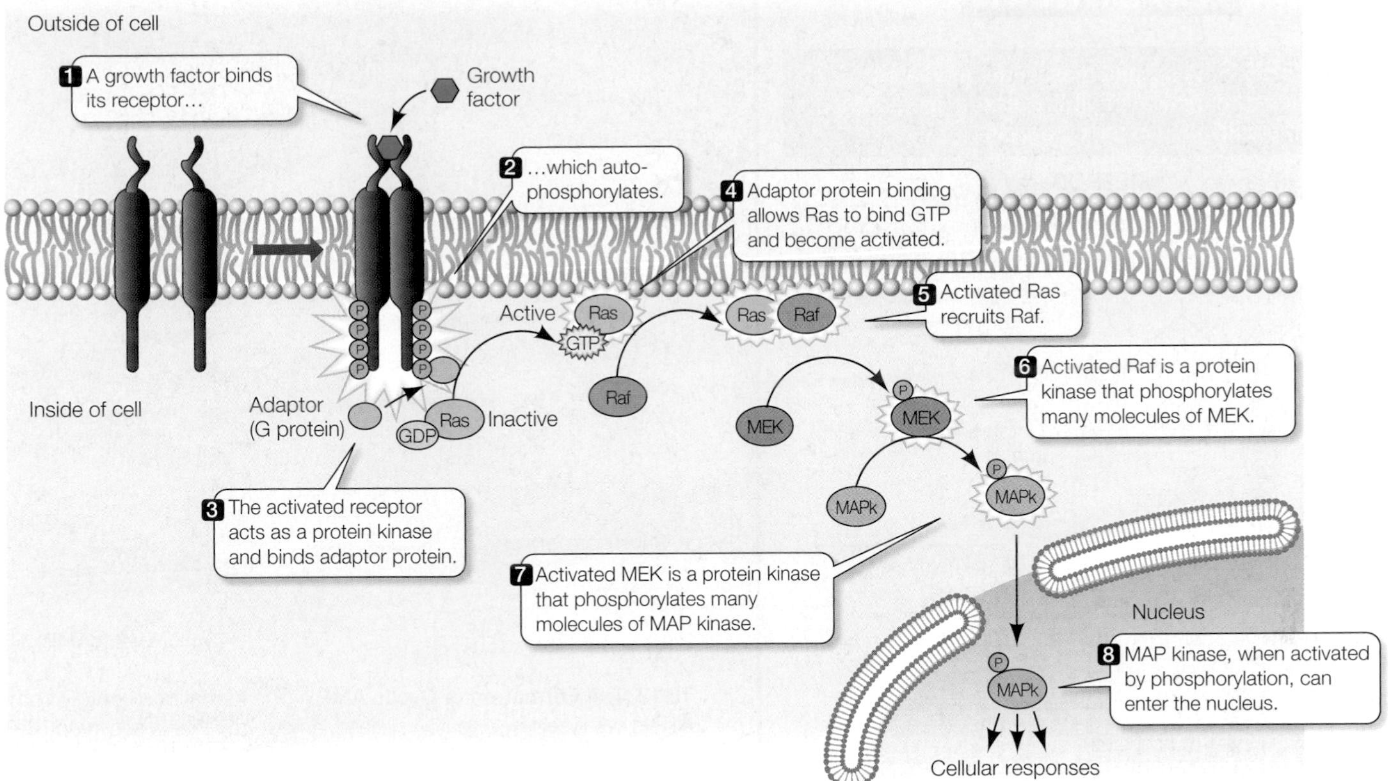

Outside of cell
1 A growth factor binds its receptor…
Growth factor
2 …which autophosphorylates.
4 Adaptor protein binding allows Ras to bind GTP and become activated.
5 Activated Ras recruits Raf.
Active
Ras
GTP
Raf
P
Inside of cell
Adaptor (G protein)
GDP
Inactive
6 Activated Raf is a protein kinase that phosphorylates many molecules of MEK.
MEK
MAPk
3 The activated receptor acts as a protein kinase and binds adaptor protein.
7 Activated MEK is a protein kinase that phosphorylates many molecules of MAP kinase.
Nucleus
8 MAP kinase, when activated by phosphorylation, can enter the nucleus.
Cellular responses

15.10 A Protein Kinase Cascade In a protein kinase cascade, a series of proteins are sequentially activated. In this example, the growth factor receptor protein stimulates the G protein Ras, which mediates a cascading series of reactions. The final product of the cascade, MAP kinase (MAPk), enters the nucleus and causes changes in transcription.

molecules can be released to the blood to fuel the fight-or-flight response (see Figure 41.4). It is normally present in liver cell cytoplasm, but is inactive except in the presence of epinephrine.

The researchers found that glycogen phosphorylase could be activated in liver cells that had been broken open, but only if the entire cell contents, including the plasma membrane fragments, were present. Under these circumstances epinephrine bound to the plasma membranes, but the active phosphorylase was present in the cytoplasm. The researchers hypothesized that there must be some chemical messenger that transmits the message of epinephrine binding (at the membrane) to the phosphorylase (in the cytoplasm). To investigate the production of this message, they separated the plasma membrane fragments from the cytoplasm of broken liver cells followed the sequence of steps described in **Figure 15.11**. This experiment confirmed their hypothesis that hormone binding to the membrane receptor had caused the production of a small, water-soluble molecule that then diffused into the cytoplasm, where it activated the enzyme. This small molecule was later determined to be *cyclic AMP* (*cAMP*), the same molecule involved in the *lac* operon regulatory system in *E. coli* (see Section 13.4). Here, cAMP was working as a **second messenger**.

Second messengers are substances that are released into the cytoplasm after the first messenger—the signal—binds to its receptor. In contrast to the specificity of receptor binding, second messengers affect many processes in the cell, and they allow a cell to respond to a single event at the plasma membrane with many events inside the cell. Like protein kinase cascades, second messengers amplify the signal: a single epinephrine molecule, for example, leads to the production of many molecules of cAMP, which then activate many enzyme targets. Second messengers themselves do not have enzymatic activity; rather, they act as *cofactors* or *allosteric regulators* of target enzymes (see Chapter 6).

Cyclic AMP is a common and widely used second messenger. *Adenylyl cyclase*, the effector enzyme that catalyzes the formation of cAMP from ATP, is located on the cytoplasmic surface of the plasma membrane of target cells (**Figure 15.12**). Usually it is activated by the binding of G proteins, which are themselves activated by receptors (see above).

Cyclic AMP has two major target types. In many kinds of sensory cells, cAMP binds to *ion channels* and thus opens them. Cyclic AMP may also bind to a *protein kinase* in the cytoplasm, whose active site is exposed as a result. A protein kinase cascade (see Figure 15.10) ensues, leading to the final effects in the cell.

Second messengers can be derived from lipids

Phospholipids, in addition to their roles as structural components of the plasma membrane, are involved in signal transduction. When certain phospholipids are hydrolyzed into their component parts by enzymes called *phospholipases*, second messengers are formed.

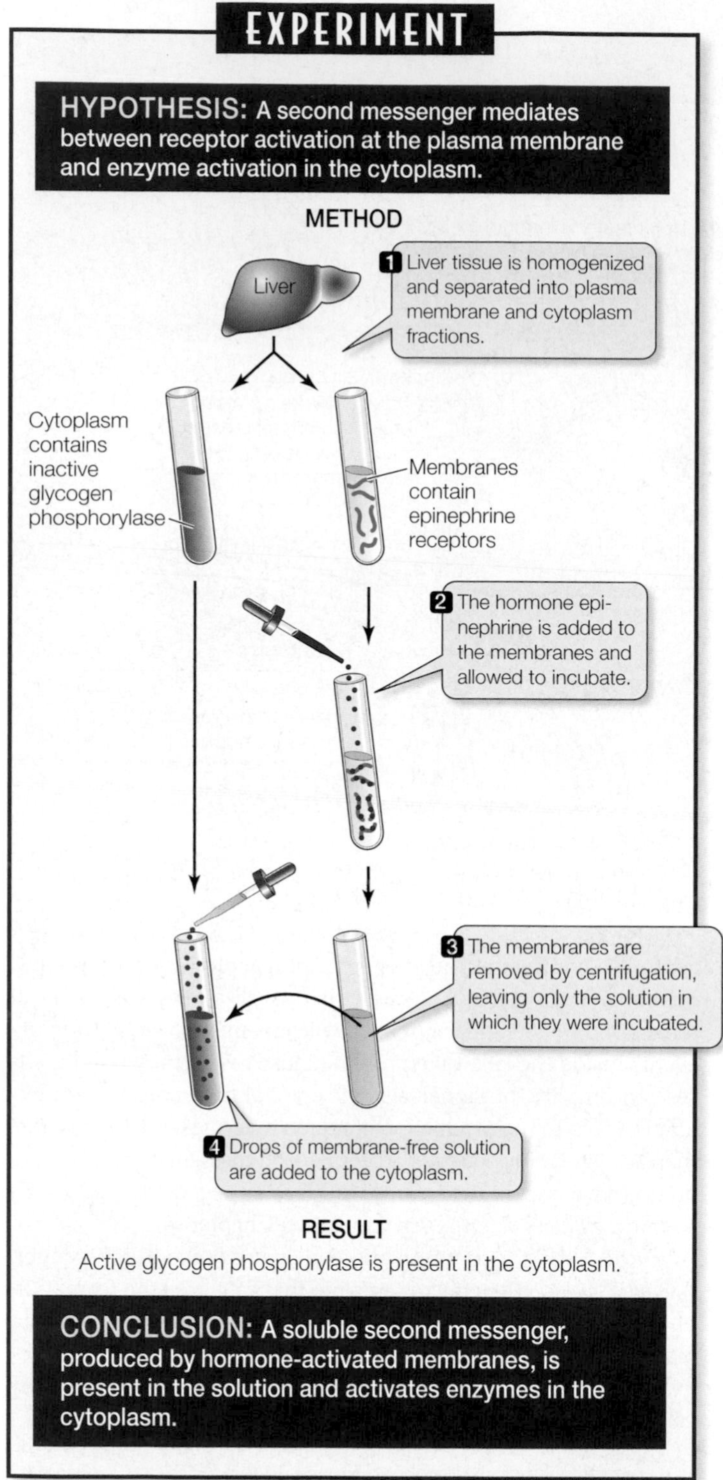

EXPERIMENT
HYPOTHESIS: A second messenger mediates between receptor activation at the plasma membrane and enzyme activation in the cytoplasm.
METHOD
Liver
1 Liver tissue is homogenized and separated into plasma membrane and cytoplasm fractions.
Cytoplasm contains inactive glycogen phosphorylase
Membranes contain epinephrine receptors
2 The hormone epinephrine is added to the membranes and allowed to incubate.
3 The membranes are removed by centrifugation, leaving only the solution in which they were incubated.
4 Drops of membrane-free solution are added to the cytoplasm.
RESULT
Active glycogen phosphorylase is present in the cytoplasm.
CONCLUSION: A soluble second messenger, produced by hormone-activated membranes, is present in the solution and activates enzymes in the cytoplasm.

15.11 The Discovery of a Second Messenger Sutherland and colleagues separated the components of a cell signaling pathway. In this manner, they were able to show that a soluble second messenger was present. FURTHER RESEARCH: The soluble molecule produced in this experiment was later identified as cAMP. How would you show that cAMP, and not ATP, is the second messenger in this system?

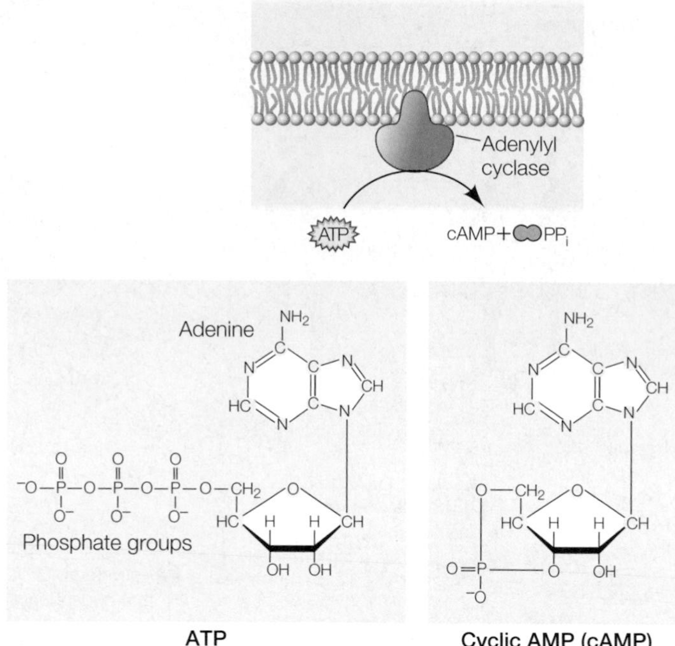

15.12 The Formation of Cyclic AMP The formation of cAMP from ATP is catalyzed by adenylyl cyclase, an enzyme that is activated by G proteins.

The best-studied of the lipid-derived second messengers come from hydrolysis of the **phosphatidyl inositol-bisphosphate** (**PIP2**), which, like all phospholipids, has a hydrophobic portion (two fatty acid tails attached to a molecule of glycerol, which together form **diacylglycerol**, or **DAG**) embedded in the plasma membrane; and a hydrophilic portion (**inositol triphosphate**, or **IP$_3$**) projecting into the cytoplasm.

As with cAMP, the receptors involved in this second-messenger system are often G protein-linked receptors. The G protein subunit activated by the receptor diffuses within the plasma membrane and activates an enzyme, phospholipase C. This enzyme cleaves off the IP$_3$ from PIP2, leaving the glycerol and the two attached fatty acids (DAG) in the phospholipid bilayer:

$$\underset{\text{in membrane}}{\text{PIP2}} \xrightarrow{\text{Phospholipase C}} \underset{\text{released to cytoplasm}}{\text{IP}_3} + \underset{\text{in membrane}}{\text{DAG}}$$

IP$_3$ and DAG are both second messengers and have different modes of action that build on each other. DAG activates a membrane-bound enzyme, protein kinase C (PKC). PKC is dependent on Ca^{2+} (hence the "C"), and that is where IP$_3$ plays an essential role. IP$_3$ diffuses through the cytoplasm to the smooth endoplasmic reticulum, where it opens an ion channel, releasing Ca^{2+} into the cytoplasm. There, in combination with DAG, the Ca^{2+} causes PKC to become active. PKC can then phosphorylate a wide variety of proteins, leading to the ultimate response of the cell (**Figure 15.13**).

This pathway is apparently the target for lithium, used clinically for many years as a psychoactive drug to treat bipolar (manic-depressive) disorder. This serious illness occurs in about 1 person in 100, in whom an overactive IP$_3$/DAG signal transduction pathway in the brain leads to excessive brain activity in certain regions.

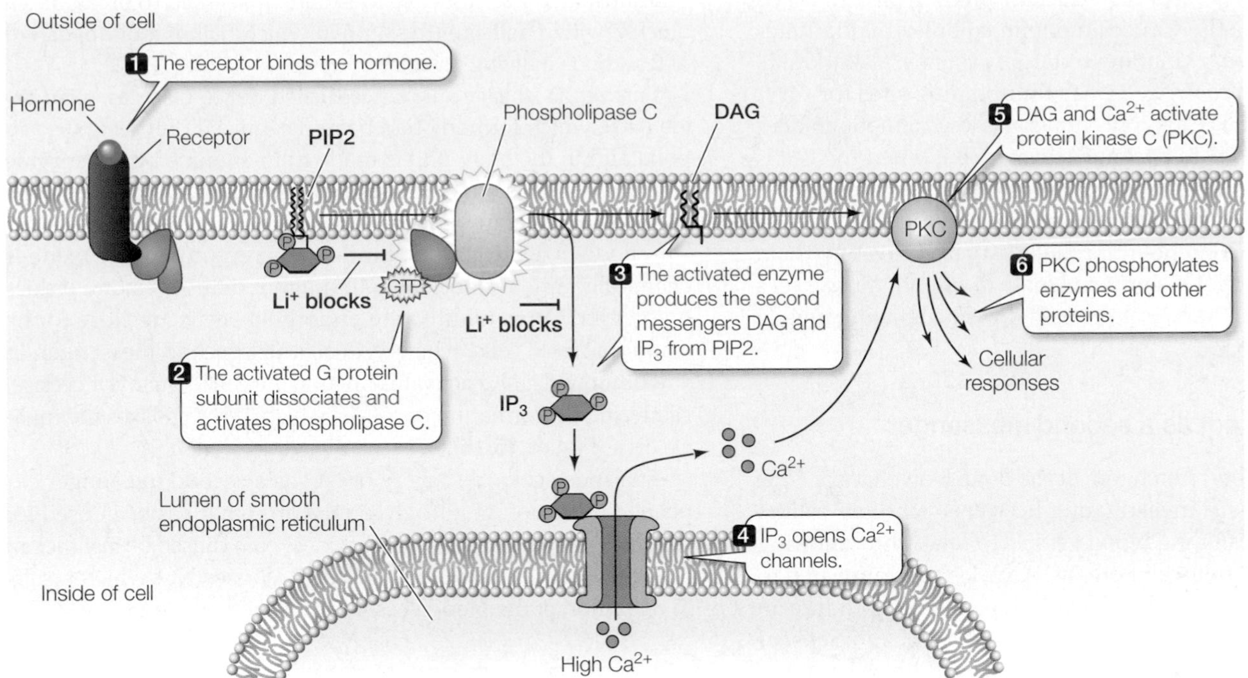

15.13 The IP_3/DAG Second-Messenger System Phospholipase C hydrolyzes the phospholipid PIP2 into its components, IP_3 and DAG, both of which are second messengers. Lithium ions (Li^+), used to treat bipolar disorder, block this pathway (red type).

Lithium ions (Li^+) "tone down" this pathway in two ways, as indicated by the red notations in Figure 15.13. Li^+ inhibits G protein activation of phospholipase C, and also inhibits the synthesis of IP_3. The overall result is that brain activity returns to normal.

Calcium ions are involved in many signal transduction pathways

Calcium ions are scarce inside most cells, which have a cytoplasmic Ca^{2+} concentration of only about 0.1 m*M*; Ca^{2+} concentrations outside the cell and within the endoplasmic reticulum are usually much higher. This concentration difference is maintained by active transport proteins at the plasma and ER membranes that pump Ca^{2+} out of the cytoplasm. In contrast to cAMP and the lipid-derived second messengers, intracellular Ca^{2+} concentration cannot be increased by making more Ca^{2+}. Instead, the opening and closing of ion channels and the action of membrane pumps regulate ion levels in a cellular compartment.

There are many signals that can cause calcium channels to open, including IP_3 (as we saw in the previous section). The entry of a sperm into an egg is a very important signal that causes a massive opening of Ca^{2+} channels; (**Figure 15.14**). Whatever the signal, the open calcium channels result in a dramatic increase in cytoplasmic Ca^{2+} concentration, up to a hundredfold within a fraction of a second. As we saw earlier, this increase activates protein kinase C. In addition, Ca^{2+} controls other ion channels and stimulates secretion by exocytosis.

A distinctive aspect of Ca^{2+} signaling is that the ion can stimulate its own release from intracellular stores. For example, in some plant leaf cells, the hormone abscisic acid binds to gated calcium channels in the plasma membrane and opens them, causing the ion to rush into the cells. This influx is not enough to trigger the cell's response, however. The ion binds to calcium channels in the endoplasmic reticulum and in the membranes of vacuoles, causing those organelles to release their Ca^{2+} stores as well.

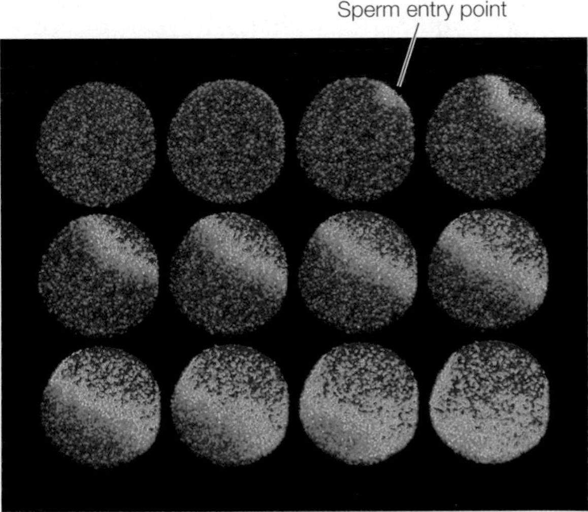

15.14 Calcium Ions as a Second Messenger The concentration of Ca^{2+} can be measured by a dye that fluoresces and turns red when it binds the ion. Here, fertilization causes a wave of Ca^{2+}, photographed at 5-second intervals, to pass through the egg of a starfish. Calcium signaling is used by virtually all animal groups to deliver the message that fertilization is complete and the mitotic divisions, or *cleavages*, that will result in a new organism can begin.

In some cases, Ca^{2+} acts via a calcium-binding protein called *calmodulin,* and it is the Ca^{2+}/calmodulin complex that initiates the cellular response by binding to target proteins. Calmodulin, which is present in many cells, has four binding sites for Ca^{2+}. When the cytoplasmic Ca^{2+} concentration is low, calmodulin does not bind enough Ca^{2+} to become activated. But when the cell is stimulated by a signal that causes a rise in the Ca^{2+} concentration, all four binding sites are filled. The calmodulin then changes its shape and binds to a number of cellular targets, activating them in turn. One such target is a protein kinase in smooth muscle cells that phosphorylates the muscle protein myosin, initiating muscle contraction.

Nitric oxide can act as a second messenger

Pharmacologist Robert Furchgott, at the State University of New York in Brooklyn, was investigating how acetylcholine causes the smooth muscles lining blood vessels to relax, thus allowing more blood to flow to certain organs. Acetylcholine appeared to stimulate the IP_3/DAG signal transduction pathway to produce an influx of Ca^{2+}, leading to an increase in the level of another second messenger, *cyclic GMP* (cGMP). Cyclic GMP then bound to a protein kinase, which in turn stimulated a protein kinase cascade leading to muscle relaxation. So far, the pathway seemed straightforward.

While the signal transduction pathway Furchgott was studying seemed to work in intact animals, it did not work on isolated strips of artery tissue. When Furchgott switched to tubular sections of artery, however, signal transduction did occur. There turned out to be a crucial difference between these two tissue preparations: in the strips, the delicate inner layer of cells that lines blood vessels had been lost. Furchgott hypothesized that this layer, the *endothelium,* was producing some chemical that diffused into the smooth muscle cells and was needed for their response to acetylcholine. The substance was not easy to isolate. It seemed to break down quickly, with a half-life (the time in which half of it disappeared) of 5 seconds in living tissue.

Furchgott's elusive substance turned out to be a gas, **nitric oxide** (**NO**), which formerly had been recognized only as a toxic air pollutant! In the body, NO is made from arginine by the enzyme *NO synthase*. This enzyme is activated by Ca^{2+}, which enters the endothelial cells through a channel opened by PIP2, which is released when acetylcholine binds to its receptor. Nitric oxide is chemically very unstable, and although it diffuses readily, it does not get far. Conveniently, the endothelial cells are close to the smooth muscle cells, where NO acts as a second messenger. In smooth muscle, NO activates an enzyme called guanylyl cyclase, catalyzing the formation of cGMP, which in turn relaxes the muscle cells (**Figure 15.15**).

The spectacular discovery of NO as a second messenger explained the action of nitroglycerin, a drug that has been used for over a century to treat angina, the chest pain caused by insufficient blood flow to the heart. Nitroglycerin releases NO, which results in relaxation of the blood vessels and increased blood flow.

Developed to treat angina by the NO pathway, the drug Viagra was only modestly useful for that purpose. However, men taking it reported more pronounced penile erections. During sexual stimulation, NO acts as a second messenger, causing an increase in cGMP and ensuing relaxation of the smooth muscles in the corpus cavernosum of the penis, which fills with blood, resulting in an erection.

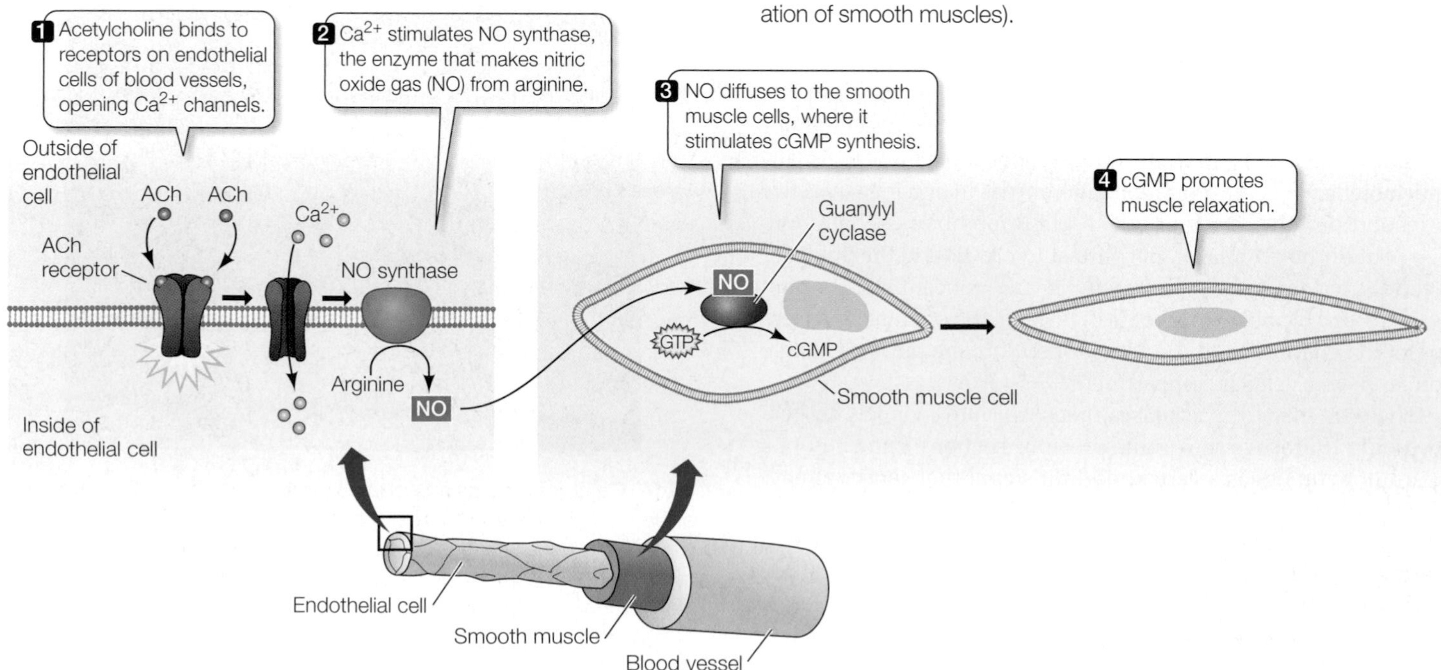
1 Acetylcholine binds to receptors on endothelial cells of blood vessels, opening Ca^{2+} channels.
2 Ca^{2+} stimulates NO synthase, the enzyme that makes nitric oxide gas (NO) from arginine.
3 NO diffuses to the smooth muscle cells, where it stimulates cGMP synthesis.
4 cGMP promotes muscle relaxation.
Outside of endothelial cell
ACh
ACh
ACh receptor
Ca^{2+}
NO synthase
Arginine
NO
Inside of endothelial cell
Guanylyl cyclase
NO
GTP
cGMP
Smooth muscle cell
Endothelial cell
Smooth muscle
Blood vessel

15.15 Nitric Oxide as a Second Messenger Nitric oxide (NO) is an unstable gas, which nevertheless serves as a second messenger mediating between a signal (acetylcholine) and its effect (the relaxation of smooth muscles).

Signal transduction is highly regulated

There are several ways in which cells can regulate the activity of a transducer. The concentration of NO, which breaks down quickly, can be regulated only by how much of it is made. The concentration of Ca^{2+}, on the other hand, is determined by both membrane pumps and ion channels, as we have seen. To regulate protein kinase cascades, G proteins, and cAMP, there are enzymes that convert the activated transducer back to its inactive precursor (**Figure 15.16**).

The balance between these two enzyme activities (making and breaking down active molecules) is what determines the ultimate cellular response to a signal. Cells can alter this balance in several ways:

- *Synthesis or breakdown of the enzymes involved*. For example, synthesis of adenylyl cyclase and breakdown of phosphodiesterase (which breaks down cAMP) would tilt the balance in favor of more cAMP in the cell.
- *Activation or inhibition of the enzymes by other molecules*. Simple examples include signal binding to a receptor activating G proteins, and caffeine inhibiting phosphodiesterase in muscle cells. Viagra inhibits the phosphodiesterase that normally breaks down cGMP.

Because cell signaling is so important in diseases such as cancer, a search is under way for new drugs that can modulate the activities of the enzymes regulating the concentrations of molecules in signal transduction pathways.

15.16 Regulation of Signal Transduction Some signals lead to the production of active transducers such as (A) enzymes, (B) G proteins, and (C) cAMP. Other enzymes (red type) inactivate these transducers.

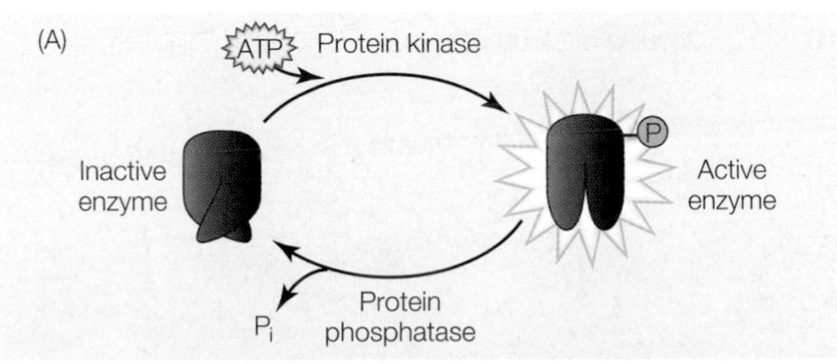

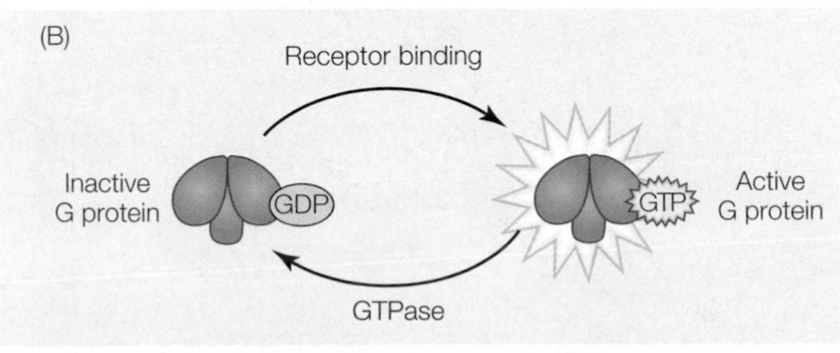

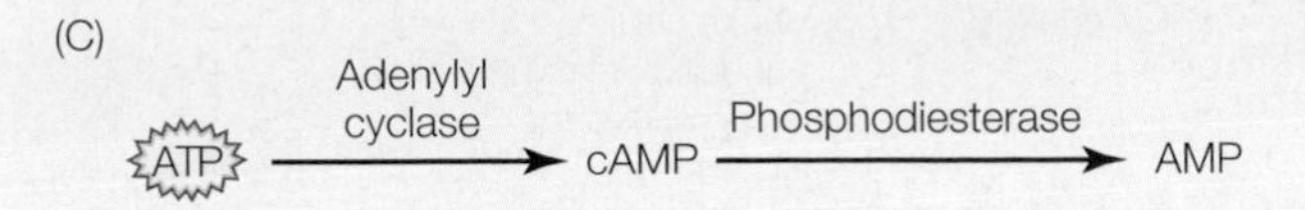

15.3 RECAP

Signal transduction is the step between the binding of a signal to a receptor and the ultimate cellular response. In direct signal transduction, the receptor is the effector; more commonly, a second messenger acts to amplify the signaling to the effector. Protein kinase cascades amplify and regulate signaling.

- Can you explain how a protein kinase cascade amplifies a signal's message inside the cell? See p. 340 and Figure 15.10
- Describe the role of cAMP as a second messenger. See p. 341
- How are signal transduction cascades regulated? See p. 345 and Figure 15.16

We have seen how the binding of a signal to its receptor initiates the response of a cell to the signal, and how the direct or indirect transduction of the signal to the inside of the cell amplifies the signal. In the next section we will consider the third and final step in the signal transduction process, the actual effects of the signal on cell function.

15.4 How Do Cells Change in Response to Signals?

The effects of a signal on cell function take three primary forms: the opening of ion channels, changes in the activities of enzymes, or differential gene transcription.

Ion channels open in response to signals

The opening of ion channels is a key step in the response of the nervous system to signals. Sensory neurons in the sense organs, for example, become stimulated through the opening of ion channels. We will focus here on one such signal transduction pathway, that for the sense of smell, which responds to gaseous molecules in the environment.

The sense of smell is well developed in mammals, some of which have more than 1,000 genes for odor signal receptors—making this the largest gene family known. Each of the thousands of neurons in the nose expresses one of these receptors. The identification of which chemical signal, or *odorant*, activates which receptor is just getting under way.

The average human is born with about 950 genes for odor signal receptor proteins, but very few people express more than 400 of them; some express far fewer, which may explain why you are able to smell certain things your roommate cannot, or vice versa.

When an odorant molecule binds to its receptor, a G protein becomes activated, which in turn activates adenylyl cyclase, which catalyzes the formation of the second messenger cAMP. This molecule then binds to ion channels, causing them to open (**Figure 15.17**). The resulting influx of Na^+ and Ca^{2+} causes the neuron to become stimulated so that it sends a signal to the brain that a particular odor is present.

Enzyme activities change in response to signals

Proteins will change their shape, and thus their function, if they are modified either covalently or noncovalently. We have seen examples of both types of modification in our description of signal transduction. Protein kinases add phosphate groups to a target protein, and this covalent change alters the protein's conformation. Cyclic AMP binds to target proteins allosterically, and this noncovalent interaction changes the protein's conformation. In both cases, previously inaccessible active sites are exposed, and the target protein goes on to perform a new cellular role.

The G protein-mediated protein kinase cascade stimulated by epinephrine in liver cells results in the phosphorylation of two key enzymes in glycogen metabolism, with opposite effects:

- *Inhibition.* Glycogen synthase, which catalyzes the joining of glucose molecules to synthesize the energy-storing molecule glycogen, is inactivated by phosphorylation. Thus the epinephrine signal prevents glucose from being stored in glycogen (**Figure 15.18, step 1**).
- *Activation.* Phosphorylase kinase is activated when a phosphate group is added to it. It goes on to stimulate a protein kinase cascade that ultimately leads to the activation by phosphorylation of glycogen phosphorylase, the other key enzyme in glucose metabolism. This enzyme liberates glucose molecules from glycogen (**Figure 15.18, steps 2 and 3**).

Thus the same signaling pathway inhibits the storage of glucose as glycogen (by inhibiting glycogen synthase) and promotes the release of glucose through glycogen breakdown (by activating glycogen phosphorylase). As we mentioned earlier, the released glucose fuels the fight-or-flight response to epinephrine.

The *amplification* of the signal in this pathway is impressive; as detailed in Figure 15.18, each single molecule of epinephrine that arrives at the plasma membrane ultimately results in *10,000 molecules* of blood glucose:

1	molecule of epinephrine bound to the membrane activates
20	molecules of cAMP, which activate
20	molecules of protein kinase A, which activate
100	molecules of phosphorylase kinase, which activate
1,000	molecules of glycogen phosphorylase, which produce
10,000	molecules of glucose 1-phosphate, which produce
10,000	molecules of blood glucose

15.17 A Signal Transduction Pathway Leads to the Opening of Ion Channels In the signal transduction pathway for the sense of smell, the final effect is the opening of ion channels. The resulting influx of Na^+ and Ca^{2+} stimulates the transmission of a scent message to a specific region of the brain.

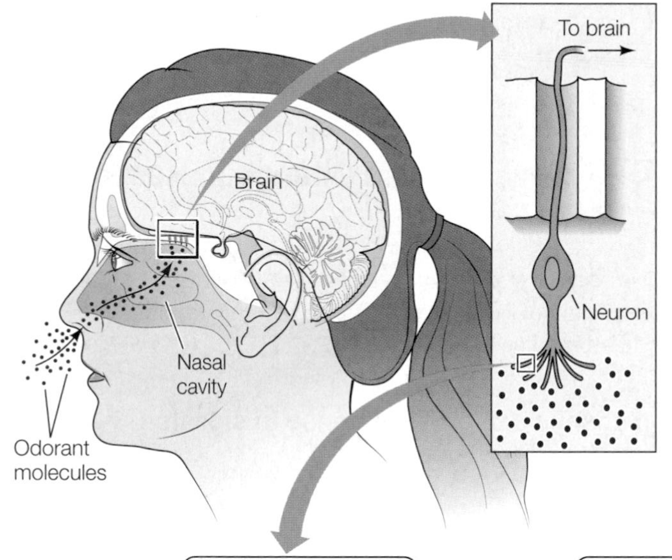

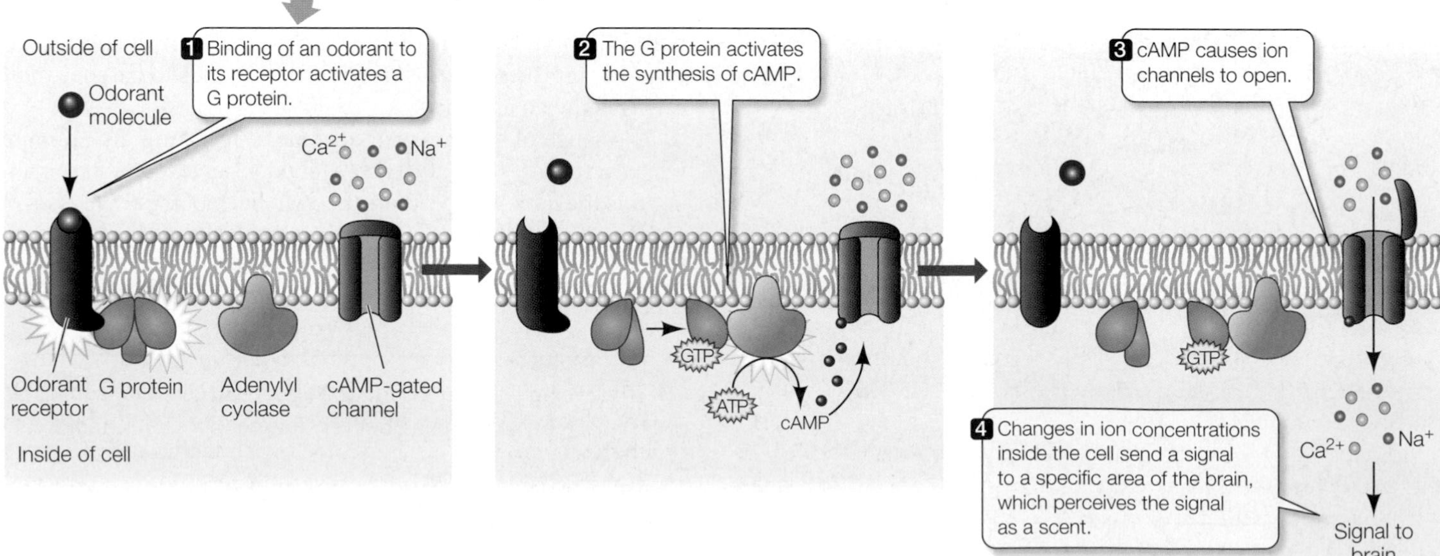

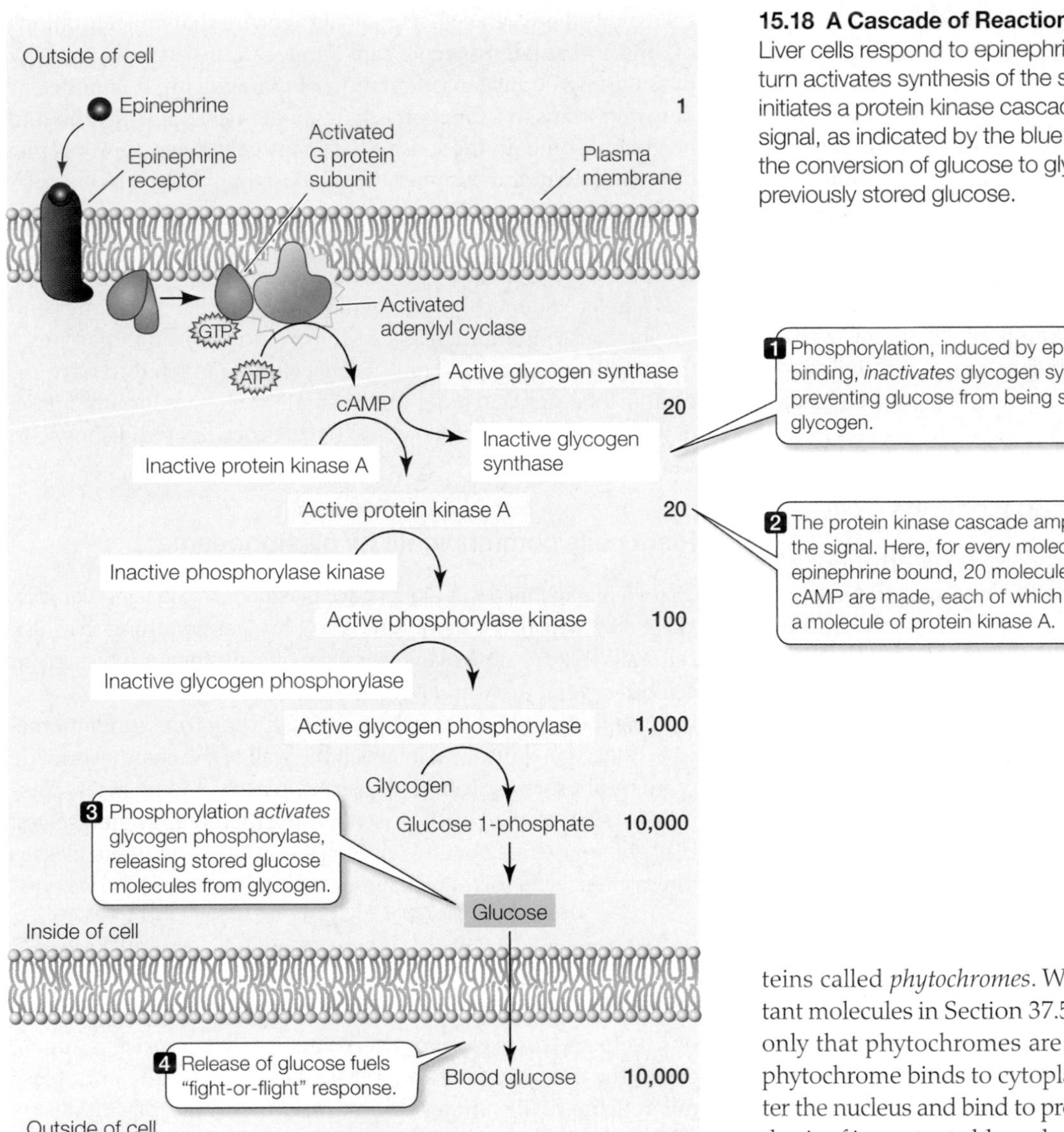

15.18 A Cascade of Reactions Leads to Altered Enzyme Activity Liver cells respond to epinephrine by activating G proteins, which in turn activates synthesis of the second messenger cAMP. Cyclic AMP initiates a protein kinase cascade, greatly amplifying the epinephrine signal, as indicated by the blue numbers. The cascade both inhibits the conversion of glucose to glycogen and stimulates the release of previously stored glucose.

Signals can initiate gene transcription

Plasma membrane receptors are involved in activating a broad range of gene expression responses. The Ras signaling pathway, for example, ends in the nucleus (see Figure 15.10). The final protein kinase in the Ras signaling cascade, MAPk, enters the nucleus and phosphorylates a transcription factor, which stimulates the transcription of a number of genes involved in cell proliferation.

As described in Section 15.2, lipid-soluble hormones can diffuse through the plasma membrane and meet their receptors in the cytoplasm. In this case, binding of the ligand allows the ligand–receptor complex to enter the nucleus, where it binds to hormone-responsive elements at the promoters of a number of genes (see Figure 15.8). In some cases, transcription is stimulated, and in others it is inhibited.

In plants, light acts as a signal to initiate the formation of chloroplasts. Between the light signal and its effect on the chloroplast is a transcription-mediated signal transduction pathway. In bright sunlight, red wavelengths of light are absorbed by receptor proteins called *phytochromes*. We will say more about these important molecules in Section 37.5, but for now it is important to know only that phytochromes are activated by red light. An activated phytochrome binds to cytoplasmic regulatory proteins, which enter the nucleus and bind to promoters of genes involved in the synthesis of important chloroplast proteins. Synthesis of these chloroplast proteins is the key to plant "greening."

15.4 RECAP

Cells respond to signal transduction by activating enzymes, opening membrane channels, or initiating gene transcription.

- Describe the role that cAMP plays in the sense of smell. See pp. 345–346 and Figure 15.17
- Can you explain how amplification of a signal occurs and why it is important in a cell's response to changes in its environment? See p. 346 and Figure 15.18

We have described how signals from a cell's environment can influence that cell. But the environment of a cell in a multicellular organism is more than the extracellular medium—it includes neighboring cells as well. In the next section we'll look at specialized junctions between cells that allow them to signal one another directly.

15.5 How Do Cells Communicate Directly?

Most cells are in contact with their neighbors. Section 5.2 described various ways in which cells adhere to one another, such as recognition proteins protruding from the cell surface, tight junctions, and desmosomes. However, as we know from our own experience with our neighbors (and roommates), just being in proximity does not necessarily mean that there is functional communication. Neither tight junctions nor desmosomes are specialized for intercellular communication. However, many multicellular organisms have specialized cell junctions that allow their cells to communicate directly. In animals, these structures are *gap junctions*; in plants, they are *plasmodesmata*.

Animal cells communicate by gap junctions

Gap junctions are channels between adjacent cells that occur in many animals, occupying up to 25 percent of the area of the plasma membrane (**Figure 15.19**). Gap junctions traverse the narrow space between the plasma membranes of two cells (the "gap") by means of channel proteins called **connexons**. The walls of these channels are composed of six subunits of an integral membrane protein. In two cells close to each other, two connexons come together, forming a channel that links the two cytoplasms. There may be hundreds of these channels between a cell and its neighbors. The channel pores are about 1.5 nm in diameter—far too narrow for the passage of large molecules such as proteins. But they are wide enough to allow small signal molecules and ions to pass between the cells. Experiments in which a labeled signal molecule or ion is injected into one cell show that it can readily pass into the adjacent cells if the cells are connected by gap junctions.

Gap junctions permit metabolic cooperation among the linked cells. Such cooperation ensures the sharing of important small molecules such as ATP, metabolic intermediates, amino acids, and coenzymes between cells. It may also ensure that concentrations of ions and small molecules are similar in linked cells, thereby maintaining equivalent regulation of metabolism. It is not clear how important this function is in many tissues, but it is known to be vital in some. In the lens of the mammalian eye, for example, only the cells at the periphery are close enough to the blood supply to allow diffusion of nutrients and wastes. But because lens cells are connected by large numbers of gap junctions, material can diffuse between them rapidly and efficiently.

There is evidence that signal molecules such as hormones and second messengers such as cAMP can move through gap junctions. If this is true, then only a few cells would need to have receptors for a signal in order for the signal to spread throughout the tissue. In this way, a tissue could have a coordinated response to the signal.

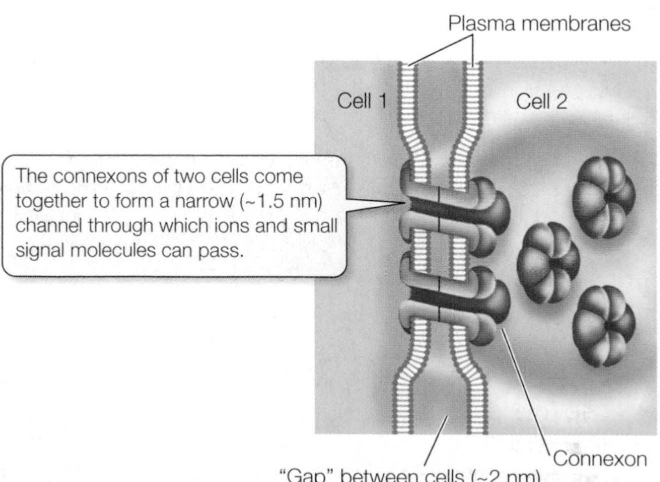
Plasma membranes
Cell 1
Cell 2
The connexons of two cells come together to form a narrow (~1.5 nm) channel through which ions and small signal molecules can pass.
Connexon
"Gap" between cells (~2 nm)

15.19 Gap Junctions Connect Animal Cells An animal cell may contain hundreds of gap junctions connecting it to neighboring cells. Gap junctions are too small for proteins, but small molecules such as ATP, metabolic intermediates, amino acids, and coenzymes can pass through them.

Plant cells communicate by plasmodesmata

Instead of gap junctions, plants have **plasmodesmata** (singular *plasmodesma*), which are membrane-lined bridges spanning the thick cell walls that separate plant cells from one another. A typical plant cell has several thousand plasmodesmata.

Plasmodesmata differ from gap junctions in one fundamental way: unlike gap junctions, in which the wall of the channel is made of integral proteins from the adjacent plasma membranes, plasmodesmata are lined by the fused plasma membranes themselves. Plant biologists are so familiar with the notion of a tissue as cells interconnected in this way that they refer to these continuous cytoplasms as a *symplast* (see Section 35.1).

The diameter of a plasmodesma is about 6 nm, far larger than a gap junction channel. But the actual space available for diffusion is about the same—1.5 nm. A look at the interior of the plasmodesma gives the reason for this reduction in pore size: a tubule called the **desmotubule**, apparently derived from the endoplasmic reticulum, fills up most of the opening of the plasmodesma (**Figure 15.20**). So, typically, only small metabolites and ions move between plant cells. This fact is important physiologically to plants, which lack the tiny circulatory vessels (capillaries) many animals use to bring gases and nutrients to every cell.

Diffusion from cell to cell through plasma membranes is probably inadequate for hormonal responses in plants. Instead, they rely on more rapid diffusion through plasmodesmata to ensure that all cells of a tissue respond to a signal at the same time. In C_4 plants (see Section 8.4), abundant plasmodesmata help to move the carbon fixed in the mesophyll rapidly to the bundle sheath cells. A similar transport system, found at the junctions of nonvascular tissues and phloem, conducts organic solutes throughout the plant.

Plasmodesmata are not merely passive channels, but can be regulated. Plant viruses may infect cells at one location, then spread rapidly through a plant organ by plasmodesmata until they reach the plant's vascular tissue (circulatory system). These viruses, and even their RNA, would appear to be many times too large to pass through the desmotubules, but they get through, apparently by making "movement proteins" that increase the pore size temporarily while attached to the viral genome. Similar movement proteins made by the plants themselves are involved in transporting mRNAs and even proteins such as transcription factors between

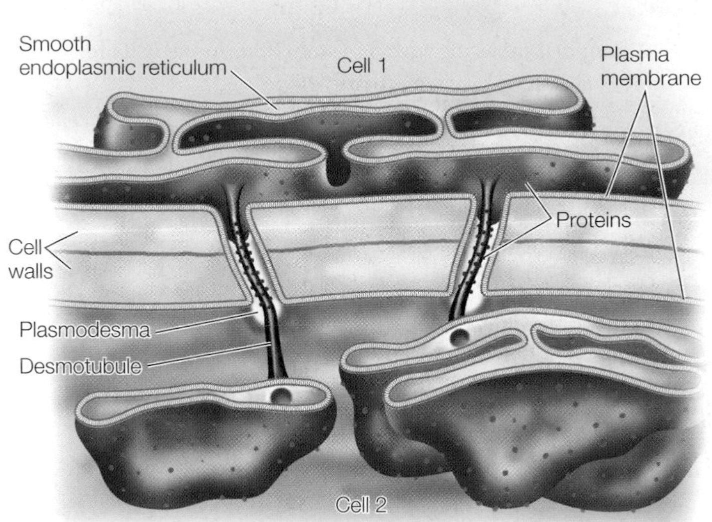

15.20 Plasmodesmata Connect Plant Cells The desmotubule, derived from the smooth endoplasmic reticulum, fills up most of the space inside a plasmodesma, leaving a tiny gap through which small metabolites and ions can pass.

plant cells. This finding opens up the possibility of long-distance regulation of transcription and translation.

15.5 RECAP

Cells can communicate with their neighbors through specialized cell junctions. In animals, these structures are gap junctions; in plants, they are plasmodesmata.

- What roles do gap junctions and plasmodesmata play in cell signaling?

CHAPTER SUMMARY

15.1 What are signals, and how do cells respond to them?

Cells receive many signals from the physical environment and from other cells. **Autocrine** signals affect the cells that make them; **paracrine** signals diffuse to and affect nearby cells. Review Figure 15.1

A **signal transduction pathway** involves the interaction of a signal molecule with a **receptor**; the transduction of the signal via a **responder** within the cell; and an effect on the function of the cell. Review Figure 15.2, Web/CD Activity 15.1

15.2 How do signal receptors initiate a cellular response?

Cells respond to signals only if they have specific receptor proteins that can bind those signals. Depending on the nature of its signaling **ligand**, a receptor may be located in the plasma membrane or in the cytoplasm of the target cell. Review Figure 15.4

Receptors located in the plasma membrane include **ion channels**, **protein kinases**, and **G protein-linked receptors**.

Ion channel receptors are "gated": the gate "opens" when the three-dimensional structure of the channel protein is altered by ligand binding. Review Figure 15.5

G proteins have two important binding sites: one for a G protein-linked receptor, and another for GDP/GTP. G proteins can either activate or inhibit an **effector protein**. Review Figure 15.7, Web/CD Tutorial 15.1

When bound by a ligand, **cytoplasmic receptors** change their shape and enter the cell nucleus. Review Figure 15.8

15.3 How is a response to a signal transduced through the cell?

Direct transduction is a function of the receptor itself and occurs at the plasma membrane. **Indirect transduction** involves a **second messenger**. Review Figure 15.9

Protein kinase cascades amplify a response to receptor binding. Review Figure 15.10

Second messengers include **cyclic AMP** (**cAMP**); **inositol triphosphate** (**IP_3**) and **diacylglycerol** (**DAG**), which are derived from the phospholipid **phosphatidyl inositol-bisphosphate** (**PIP2**); calcium ions; and the gas **nitric oxide** (**NO**).

Signal transduction can be regulated in several ways. The balance between making and breaking down the enzymes involved determines the ultimate cellular response to a signal.

15.4 How do cells change in response to signals?

The ultimate cellular response to a signal may be the opening of ion channels, the alteration of enzyme activities, or changes in gene transcription. Review Figure 15.17

Protein kinases covalently add phosphate groups to target proteins; cAMP allosterically binds target proteins. Whether by covalent or noncovalent action, changes in the target protein's conformation expose previously inaccessible active sites.

Activated enzymes may either inhibit or activate other enzymes in a signal transduction pathway, leading to impressive amplification of a signal. Review Figure 15.18

Lipid-soluble signals, such as steroid hormones, can diffuse through the plasma membrane and meet their receptors in the cytoplasm; the ligand–receptor complex may then enter the nucleus to affect gene transcription.

15.5 How do cells communicate directly?

Most animal cells can communicate with one another directly through small pores in their plasma membranes called **gap junctions.** Channel proteins called **connexons** form thin channels between two adjacent cells through which small signal molecules and ions can pass. Review Figure 15.19

Plant cells are connected by somewhat larger pores called **plasmodesmata**, which traverse both plasma membranes and cell walls. The **desmotubule** narrows the opening of the plasmodesma. Review Figure 15.20

See Web/CD Activity 15.2 for a concept review of this chapter.

SELF-QUIZ

1. What is the correct order for the following events in the interaction of a cell with a signal? (1) alteration of cell function; (2) signal binds to receptor; (3) signal released from source; (4) signal transduction.
   a. 1234
   b. 2314
   c. 3214
   d. 3241
2. Why do some signals ("first messengers") trigger a "second messenger" to activate a target cell?
   a. The first messenger requires activation by ATP.
   b. The first messenger is not water soluble.
   c. The first messenger binds to many types of cells.
   d. The first messenger cannot cross the plasma membrane.
   e. There are no receptors for the first messenger.
3. Steroid hormones such as estrogen act on target cells by
   a. initiating second messenger activity.
   b. binding to membrane proteins.
   c. initiating DNA transcription.
   d. activating enzymes.
   e. binding to membrane lipids.
4. The major difference between a cell that responds to a signal and one that does not is the presence of a
   a. DNA sequence that binds to the signal.
   b. nearby blood vessel.
   c. receptor.
   d. second messenger.
   e. transduction pathway.
5. Which of the following is *not* a consequence of signal binding to a receptor?
   a. Activation of receptor enzyme activity
   b. Diffusion of receptor in the plasma membrane
   c. Change in conformation of the receptor protein
   d. Breakdown of the receptor to amino acids
   e. Release of the signal from the receptor
6. A nonpolar molecule such as a steroid hormone usually binds to a
   a. cytoplasmic receptor.
   b. protein kinase.
   c. ion channel.
   d. phospholipid.
   e. second messenger.
7. Which of the following is *not* a common type of receptor?
   a. Ion channel
   b. Protein kinase
   c. G protein-linked
   d. Transcription factor
   e. Adenylyl cyclase
8. Which of the following is *not* true of the protein kinase cascade?
   a. The signal is amplified.
   b. A second messenger is formed.
   c. Target proteins are phosphorylated.
   d. The cascade ends up in the nucleus.
   e. The cascade begins at the plasma membrane.
9. Which of the following is *not* a second messenger for signal transduction?
   a. Calcium ions
   b. Nitric oxide gas
   c. ATP
   d. Cyclic AMP
   e. Diacylglycerol
10. Plasmodesmata and gap junctions
    a. allow small molecules and ions to pass rapidly between cells.
    b. are both membrane-lined channels.
    c. are channels about 1 mm in diameter.
    d. are present only once per cell.
    e. are involved in cell recognition in signaling.

FOR DISCUSSION

1. Like the Ras protein itself, the various components of the Ras signaling pathway were discovered when cancer cells showed mutations in one or another of their components. What might be the biochemical consequences of mutations in the genes coding for (A) Raf and (B) MAP kinase that resulted in rapid cell division?
2. Cyclic AMP is a second messenger in many different responses. How can the same messenger act in different ways in different cells?
3. Compare direct communication via plasmodesmata or gap junctions with receptor-mediated communication between cells. What are the advantages of one method over the other?
4. The tiny invertebrate *Hydra* has an apical region, which has tentacles, and a long, slender body. *Hydra* can reproduce asexually when cells on the body wall differentiate and form a bud, which then breaks off as a new organism. Buds form only at certain distances from the apex, leading to the idea that the apex releases a signal molecule that diffuses down the body and, at high concentrations (i.e., near the apex), inhibits bud formation. *Hydra* lacks a circulatory system, so the inhibitor must diffuse from cell to cell. If you had an antibody that binds to connexons to plug up the gap junctions, how would you show that *Hydra*'s inhibitory factor passes *through* these junctions?

FOR INVESTIGATION

Endosymbiotic bacteria in the marine invertebrate *Begula neritina* synthesize bryostatins, a name derived from *Begula*'s animal group Ectoprocta, once known as *bryozoans* ("moss animals"); and *stat* (stop). Bryostatins curtail cell division in many cell types, including several cancers. It is proposed that bryostatins inhibit protein kinase C (see Figure 15.13). How would you investigate this hypothesis, and how would you relate this inhibition to cell division?

CHAPTER

16 Recombinant DNA and Biotechnology

Baby 81

The tsunami of December 26, 2004, struck the coastal town of Kalmunai, Sri Lanka, with such force that 4-month-old Abilass Jeyarajah was torn from his mother's arms and swept away. Hours later, while his parents desperately searched the devastated town, their tiny son washed up on the beach a kilometer away, alive. A local schoolteacher found him and brought him to the hospital—the eighty-first patient admitted that day. The hospital was overwhelmed with 1,000 bodies, many of them children. Since Abilass was alive and healthy, he was dubbed "Baby 81, the miracle baby" and became an instant celebrity among the staff as they went about their grim duties.

Meanwhile, the parents kept looking. Two days later, they met the schoolteacher, who told them about the baby he had found. Rushing to the hospital, the Jeyarajahs were elated to find their son but were in for a rude shock. Eight other couples who had also lost infants were claiming Baby 81 as theirs. The baby remained in the hospital while the case went to court.

Judge M. P. Mohaideen faced a situation not unlike King Solomon, who 3,000 years ago was asked to decide which of two women was the mother of an infant. Solomon's method of determining parentage is told in a famous biblical passage (I Kings 3:16–28). The Sri Lankan judge had a different method: he called in molecular biologists.

With 6 billion base pairs of DNA packaged in 46 chromosomes, each one of us is unique. Although our protein-coding sequences are similar (after all, our phenotypes are similar), only a few percent of the total base pairs in the human genome actually code for genes. As explained in Chapter 14, the eukaryotic genome contains many repeated sequences, and between individuals the repeat frequency may differ, offering one way to differentiate individuals. Differences in a single base pair due to DNA replication errors or random mutations also distinguish individuals from one another, and these differences are inherited.

After the Tsunami In December of 2004, a tsunami originating in the Indian Ocean struck over a broad region that encompassed many nations in Southeast Asia. The result was an unprecedented humanitarian disaster that left almost a quarter of a million people dead and many more homeless.

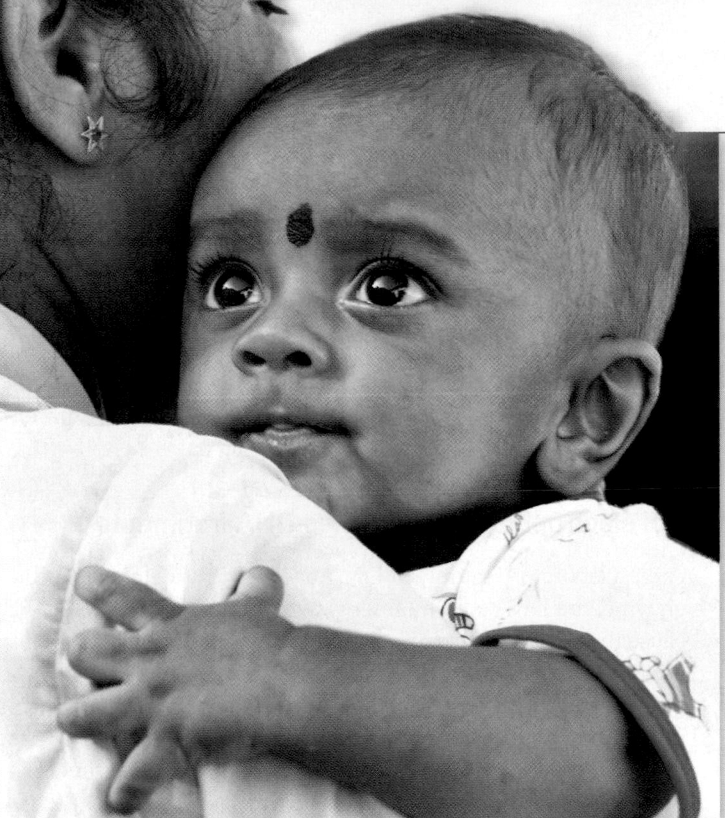

"Baby 81" Abilass Jeyarajah survived the tsunami and was reunited with his parents by court order after DNA testing proved that he is indeed their son.

CHAPTER OUTLINE

It is now possible to analyze these differences in DNA (amplified by PCR) to identify people. When DNA from the nine sets of contesting parents was analyzed and compared with that of Baby 81, the only parents whose sequences were consistent with being passed on to the baby were the Jeyarajas. On February 14, 2005, the judge ruled that the Jeyarajas are the true parents, and Baby 81 got his real name and parents back.

IN THIS CHAPTER we will describe some of the techniques that are used to manipulate DNA. We begin with a description of how DNA molecules are cut into smaller fragments and how these fragments are spliced together to create recombinant DNA. We will then describe how recombinant (or any other) DNA is introduced into a suitable host cell. Then we'll look at some other techniques for copying, storing, and altering DNA. Finally, we'll see how some of these DNA manipulation techniques have been applied by scientists.

16.1 How Are Large DNA Molecules Analyzed?

Scientists have long realized that the chemical reactions that living cells employ for one purpose may be applied in the laboratory for other, novel purposes. Similarly, the manipulation of DNA molecules is based on an understanding of the chemical structure of DNA, the properties of certain enzymes, and the rules of complementary base pairing.

In this section we will see how some of the numerous naturally occurring enzymes that cleave and repair DNA are used in the laboratory to manipulate and recombine DNA. We'll learn how enzymes are used to cut DNA into fragments and how these fragments are analyzed.

Restriction enzymes cleave DNA at specific sequences

All organisms, including bacteria, must have ways of dealing with their enemies. As Section 13.1 describes, bacteria are attacked by viruses called bacteriophage that inject their genetic material into the host cell. Some bacteria defend themselves against such invasions by producing **restriction enzymes** (also known as *restriction endonucleases*), which cut double-stranded DNA molecules—such as those injected by phage—into smaller, noninfectious fragments (**Figure 16.1**). These enzymes break the bonds of the DNA backbone between the 3′ hydroxyl group of one nucleotide and the 5′ phosphate group of the next nucleotide. This cutting process is called **restriction digestion**.

There are many such restriction enzymes, each of which cleaves DNA at a specific sequence of bases, called a **recognition sequence** or a **restriction site**. Most recognition sequences are 4–6 base pairs long. The sequence is recognized through the principles of protein–DNA interactions (see Section 11.2): the base pairs inside the double helix of DNA have small differences, so that they fit differently into the three-dimensional structure of an enzyme.

Why doesn't a restriction enzyme cut the DNA of the bacterial cell that makes it? One way that the cell protects itself is

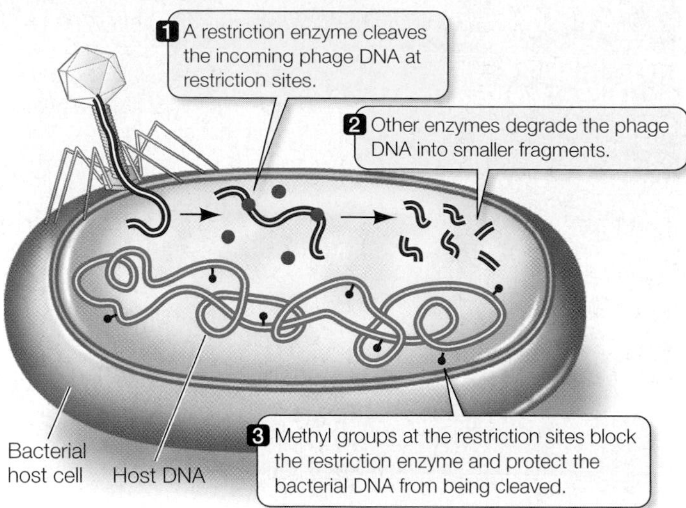

1 A restriction enzyme cleaves the incoming phage DNA at restriction sites.
2 Other enzymes degrade the phage DNA into smaller fragments.
3 Methyl groups at the restriction sites block the restriction enzyme and protect the bacterial DNA from being cleaved.
Bacterial host cell
Host DNA

16.1 Bacteria Fight Invading Viruses with Restriction Enzymes Bacteria produce restriction enzymes that degrade phage DNA by cleaving it into fragments. Methylation protects the bacteria's own DNA from being cleaved.

by modifying the restriction site on its own DNA after the sequence has been made during DNA replication. Specific modifying enzymes called *methylases* add methyl (—CH_3) groups to certain bases at the restriction sites of the host's DNA after it has been replicated. The methylation of the host's bases makes the recognition sequence unrecognizable to the restriction enzyme. But unmethylated phage DNA is efficiently recognized and cleaved.

Bacterial restriction enzymes can be isolated from the cells that make them and used as biochemical reagents in the laboratory. If DNA from any organism is incubated in a test tube with a restriction enzyme, that DNA will be cut wherever the restriction site occurs. A specific sequence of bases defines each restriction site. For example, the enzyme *Eco*RI (named after its source, a strain of the bacterium *E. coli*) cuts DNA only where it encounters the following paired sequence in the DNA double helix:

$$
\begin{array}{c}
\downarrow \\
5'\ldots G\,AATTC\ldots 3' \\
3'\ldots CTTAA\,G\ldots 5' \\
\uparrow
\end{array}
$$

Notice that this sequence reads the same in the 5′-to-3′ direction on both strands. It is *palindromic*, like the word "mom," in the sense that it is the same in both directions from the 5′ end. The *Eco*RI enzyme has two identical active sites on its two subunits, which cleave the two strands simultaneously between the G and the A of each strand.

The *Eco*RI recognition sequence occurs, on average, about once in every 4,000 base pairs in a typical prokaryotic genome, or about once per four prokaryotic genes. So *Eco*RI can chop a large piece of DNA into smaller pieces containing, on average, just a few genes. Using *Eco*RI in the laboratory to cut small genomes, such as those of viruses that have tens of thousands of base pairs, may result in a few fragments. For a huge eukaryotic chromosome with tens of millions of base pairs, a very large number of fragments will be created.

Of course, "on average" does not mean that the enzyme cuts all stretches of DNA at regular intervals. The *Eco*RI recognition sequence does not occur even once in the 40,000 base pairs of the genome of a phage called T7—a fact that is crucial to the survival of this virus, since its host is *E. coli*. Fortunately for *E. coli*, the *Eco*RI recognition sequence does appear in the DNA of other phage.

Hundreds of restriction enzymes have been purified from various microorganisms. In the test tube, different restriction enzymes that recognize different restriction sites can be used to cut the same sample of DNA. Thus restriction enzymes can be used as "knives" for genetic "surgery" to cut a sample of DNA in many different, specific places.

Gel electrophoresis separates DNA fragments

After a laboratory sample of DNA has been cut with a restriction enzyme, the DNA is in fragments, which must be separated. Because the recognition sequence does not occur at regular intervals, the fragments are not all the same size, and this property provides a way to separate them from one another. Separating the fragments is necessary to determine the number and molecular sizes (in base pairs) of the fragments produced or to identify and purify an individual fragment of particular interest.

A convenient way to separate or purify DNA fragments is by **gel electrophoresis** (**Figure 16.2**). A mixture of DNA fragments is placed in a well in a porous gel, and an electric field (with positive and negative ends) is applied across the gel. Because of its phosphate groups, DNA is negatively charged at neutral pH; therefore, because opposite charges attract, the DNA fragments move toward the positive end of the field. The porous gel acts like a sieve, and the smaller molecules move faster than the larger ones. After a fixed time, and while all the fragments are still in the gel, the electric power is shut off. The separated fragments can be visualized by staining them with a fluorescent dye. Under ultraviolet light, they can then be seen as bars or spots in the gel and can be examined or removed individually. Electrophoresis gives us two types of information:

- *The sizes of the fragments*. DNA fragments of known size are often placed in a well in the gel next to the sample to provide a standard of comparison.
- *The presence of specific DNA sequences*. A specific DNA sequence can be revealed with a single-stranded DNA probe (**Figure 16.3**). The sample DNA is denatured (unwound and separated into single strands) while still in the gel, then the gel is treated so that the DNA is transferred to a nylon filter to make a "blot" (called a *Southern blot*, after the scientist who developed the method). The filter is then exposed to a single-stranded DNA probe with a sequence complementary to the one that is being sought. If the sequence of interest is present in the sample DNA, the probe will hybridize with it. The probe can be labeled with a radioisotope. After hybridization, spots of radioactivity indicate that the probe has hybridized with its target sequence at that location. Unbound probes stay in solution.

The gel region containing the desired fragment (in size or sequence) can be cut out as a lump of gel, and the pure DNA fragment can then be removed from the gel by diffusion into a small volume of water.

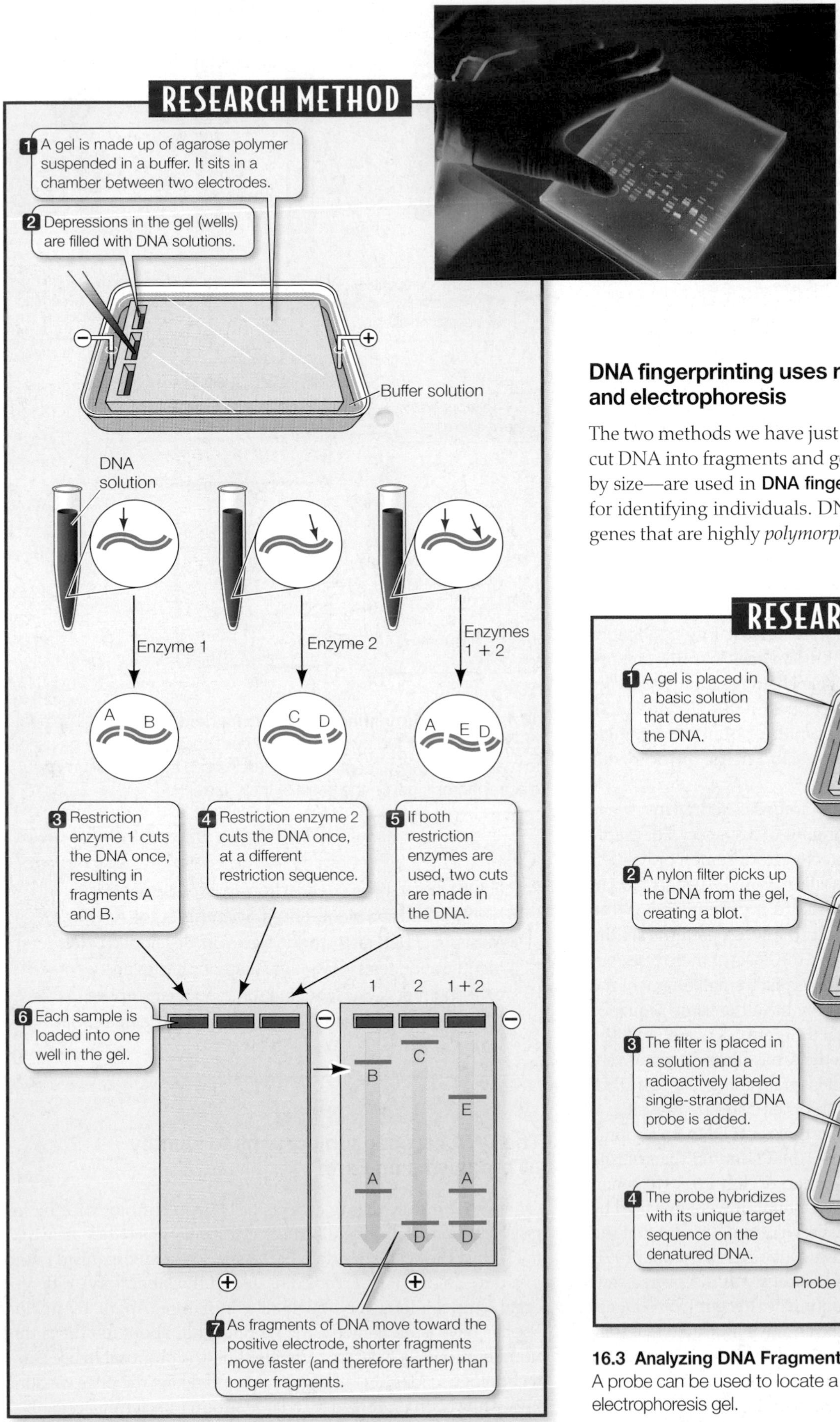

RESEARCH METHOD
1 A gel is made up of agarose polymer suspended in a buffer. It sits in a chamber between two electrodes.
2 Depressions in the gel (wells) are filled with DNA solutions.
Buffer solution
DNA solution
Enzyme 1
Enzyme 2
Enzymes 1 + 2
A B
C D
A E D
3 Restriction enzyme 1 cuts the DNA once, resulting in fragments A and B.
4 Restriction enzyme 2 cuts the DNA once, at a different restriction sequence.
5 If both restriction enzymes are used, two cuts are made in the DNA.
6 Each sample is loaded into one well in the gel.
1 2 1+2
B C E A A D D
7 As fragments of DNA move toward the positive electrode, shorter fragments move faster (and therefore farther) than longer fragments.

16.2 Separating Fragments of DNA by Gel Electrophoresis
A mixture of DNA fragments is placed in a gel and an electric field is applied across the gel. The negatively charged DNA moves toward the positive end of the field, with smaller molecules moving faster than larger ones. When the electric power is shut off, the separated fragments can be analyzed.

DNA fingerprinting uses restriction analysis and electrophoresis

The two methods we have just described—restriction digestion to cut DNA into fragments and gel electrophoresis to separate them by size—are used in **DNA fingerprinting**, a DNA-based technique for identifying individuals. DNA fingerprinting works best with genes that are highly *polymorphic*—that is, genes that have multi-

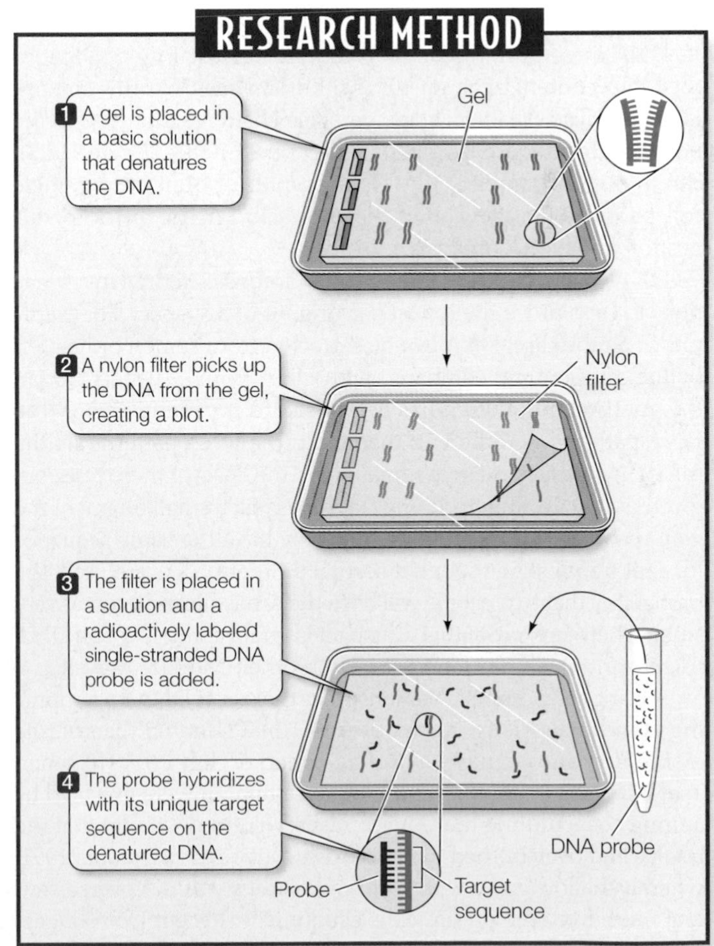

RESEARCH METHOD
1 A gel is placed in a basic solution that denatures the DNA.
Gel
2 A nylon filter picks up the DNA from the gel, creating a blot.
Nylon filter
3 The filter is placed in a solution and a radioactively labeled single-stranded DNA probe is added.
4 The probe hybridizes with its unique target sequence on the denatured DNA.
DNA probe
Probe
Target sequence

16.3 Analyzing DNA Fragments by Southern Blotting
A probe can be used to locate a specific DNA fragment on an electrophoresis gel.

ple alleles and are therefore likely to be different in different individuals. Two types of polymorphisms are especially informative:

- **Single nucleotide polymorphisms (SNPs)** are inherited variations involving a single nucleotide base. These polymorphisms have been mapped for many organisms. If one parent is homozygous for A at a certain point on the genome, for example, and the other parent has a G at that point, the offspring will be heterozygous: one chromosome will have A at that point and the other, G.
- **Short tandem repeats (STRs)** are short, moderately repetitive DNA sequences that occur side by side on the chromosomes. These repeat patterns are also inherited. For example, an individual might inherit a chromosome 15 with a short sequence repeated six times from her mother and a chromosome 15 with the same sequence repeated two times from her father.

STRs are easily detectable if they lie between two recognition sequences for a restriction enzyme. If the DNA from the heterozygous individual described above is cut with a restriction enzyme, it will form two different-sized fragments: one larger (the one from the mother) and the other smaller (the one from the father). These patterns can be easily seen by the use of gel electrophoresis (**Figure 16.4**). With several different STRs (as many as eight are used, each with numerous alleles), an individual's unique pattern becomes apparent.

DNA fingerprinting methods require at least 1 μg of DNA, or the DNA content of about 100,000 human cells, but this amount is not always available. The polymerase chain reaction (PCR), the powerful technique described in Section 11.5 (see Figure 11.23), can multiply ("amplify") the DNA of interest from even a single cell, producing in a few hours the necessary 1 μg for restriction digestion and electrophoresis.

DNA fingerprinting can be used in forensics (crime investigation) to help prove the innocence or guilt of a suspect. For example, in a rape case, DNA can be extracted from semen or hair left by the attacker and compared with DNA from a suspect. So far, this method more often has been used to prove innocence (the DNA patterns are different) than guilt (the DNA patterns are the same). It is easy to exclude someone on the basis of these tests, but given that DNA fingerprinting examines just a small fragment the genome, two people could theoretically have the same sequence for that fragment, although if several different STRs are used, the probability that two people will have the same alleles becomes very small. Therefore, proof that a suspect is guilty cannot rest on DNA fingerprinting alone, but must rely on other evidence as well.

A fascinating example demonstrates the use of DNA fingerprinting in the analysis of historical events. Three hundred years of rule by the Romanov dynasty in Russia ended on July 16, 1918, when Tsar Nicholas II, his wife, and their five children were executed by a firing squad during the Communist revolution. A report that the bodies had been burned to ashes was never questioned until 1991, when a shallow grave with several skeletons was discovered several miles from the presumed execution site. Recent DNA fingerprinting of bone fragments found in this grave indicated that they came from an older man and woman and three female children, who were clearly related to one another (**Figure 16.5**) and were also related to several living descendants of the Tsar.

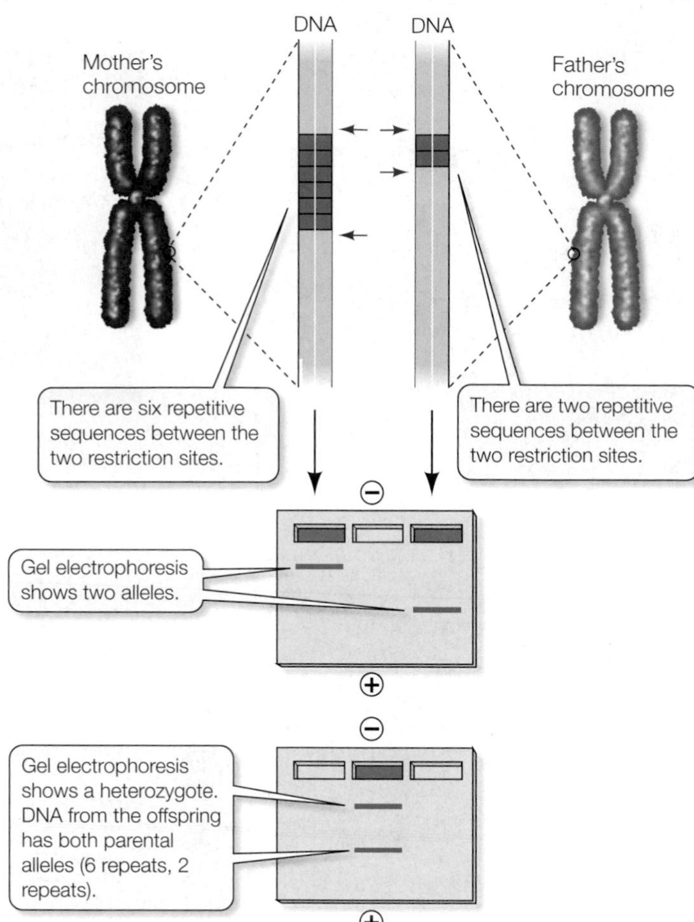

16.4 DNA Fingerprinting with Short Tandem Repeats The number of STRs inherited by an individual from each parent can be used to make a DNA fingerprint. The two alleles can be identified in an electrophoresis gel on the basis of their sizes.

Beluga caviar, the roe (eggs) from Beluga sturgeon, is highly prized and highly priced. Scientists at the American Museum of Natural History in New York developed DNA identification tests for various sturgeon species and went on a gourmet caviar shopping spree. To their surprise, they found that a quarter of the tins labeled Beluga caviar, weren't.

The DNA barcode project aims to identify all organisms on Earth

One of the most exciting aspects of DNA technology for biologists is its potential to identify the species with which they are working. While they can be certain of the species of an organism raised in a pure culture (say, *E. coli* cultivated in the laboratory), different organisms can look very much alike in nature. About 1.7 million species have been named and described, but about ten times that number probably have yet to be identified. A proposal to use DNA technology to identify known species and detect the ones we don't know has been endorsed by a large group of scientific organiza-

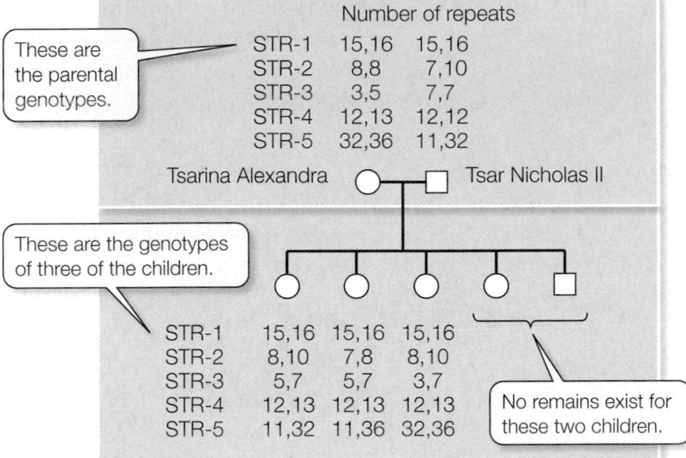

16.5 DNA Fingerprinting the Russian Royal Family The skeletal remains of Tsar Nicholas II, his wife Alexandra, and three of their children were found in 1991 and subjected to DNA fingerprinting. Five STRs were tested. The results can be interpreted as follows: Using STR-2 as an example, the parents had genotypes 8,8 (homozygous) and 7,10 (heterozygous). The three children all inherited type 8 from the Tsarina and either type 7 or type 10 from the Tsar.

tions known as the Consortium for the Barcode of Life (CBOL).

Paul Hebert at the University of Guelph in Ontario, Canada, has proposed to identify each species with a "DNA barcode" using a short sequence from a single gene. The gene he chose is the cytochrome oxidase gene, which is present in most cells. Because the gene mutates readily, there should be many allelic differences between species. A fragment of 650–750 base pairs in this gene is being sequenced for all organisms, and so far, sufficient variation has been detected to make it diagnostic of each species (**Figure 16.6**). Once the DNA of the targeted gene fragment has been sequenced for all species, a simple device for conducting field analyses can be developed. The barcode project has the potential to advance biological research on evolution, to track species diversity in ecologically significant areas, to help identify new species, and even to detect undesirable microbes or bioterrorism agents.

16.1 RECAP

Large DNA molecules can be cut into smaller pieces by restriction digestion and then sorted by gel electrophoresis. DNA fingerprinting uses these two techniques to analyze DNA polymorphisms for the purpose of identifying individuals. Scientists hope to identify species using DNA analyses.

- Do you understand how a restriction enzyme recognizes a restriction site on DNA? See p. 355
- On what principle does gel electrophoresis separate DNA fragments? See p. 355 and Figure 16.2
- What are STRs and how are they used to identify individuals? See p. 357

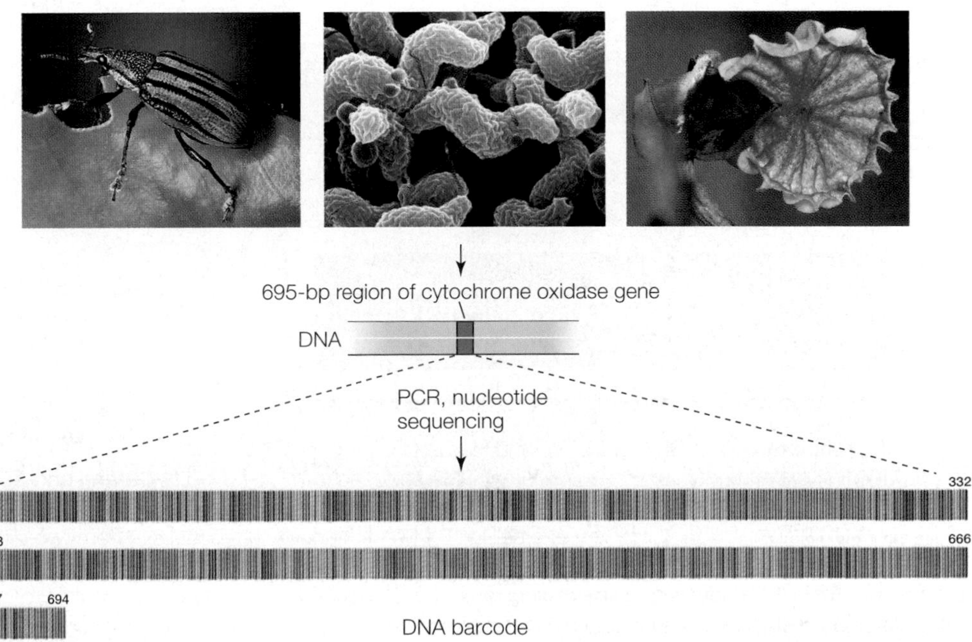

16.6 A DNA Barcode DNA from a 650–750-base pair region of the cytochrome oxidase gene from any organism can be amplified using PCR and then sequenced to identify a species. Each color represents one of the four DNA bases.

We've just seen how scientists can cut DNA into fragments, and how they can use those fragments to identify polymorphisms. One of the most exciting ways of using DNA fragments, however, is to put them together in different combinations—to create recombinant DNA.

16.2 What Is Recombinant DNA?

Restriction enzymes cut the DNA backbone into fragments. These fragments can be joined by a second type of enzyme, DNA ligase (the same enzyme that joins Okazaki fragments during DNA replication; see Section 11.3). Once they had isolated restriction enzymes and DNA ligase, scientists realized that they could use these enzymes to generate and splice together any two DNA sequences. Stanley Cohen and Herbert Boyer did just that in 1973. They used restriction enzymes to cut and then DNA ligase to join DNA sequences from two *E. coli* plasmids (see Figure 13.14) containing different antibiotic resistance genes. The resulting plasmid, when inserted into new *E. coli* cells, gave those cells resistance to both antibiotics (**Figure 16.7**). The era of **recombinant DNA** was born.

Some restriction enzymes cleave DNA cleanly, cutting both strands exactly opposite one another. Others, such as *Eco*RI, make staggered cuts in a palindromic recognition sequence, cutting one strand of the double helix several bases away from where they cut the other (**Figure 16.8**). After *Eco*RI makes its two cuts in the complementary strands, the ends of the strands are held together only by the hydrogen bonds between four base pairs. These hydrogen bonds are too weak to persist at warm temperatures (above room temperature), so the fragments of DNA separate when they are warmed. As a result, there are single-stranded "tails" at the location of each cut. These tails are called **sticky ends** because they have a specific base sequence that can bind by base pairing with complementary sticky ends. If n restriction sites for a given restriction enzyme are present in a linear DNA molecule, then $n + 1$ fragments will be made, all with the same complementary sequences at their sticky ends.

After a DNA molecule has been cut with a restriction enzyme, complementary sticky ends can form hydrogen bonds with one

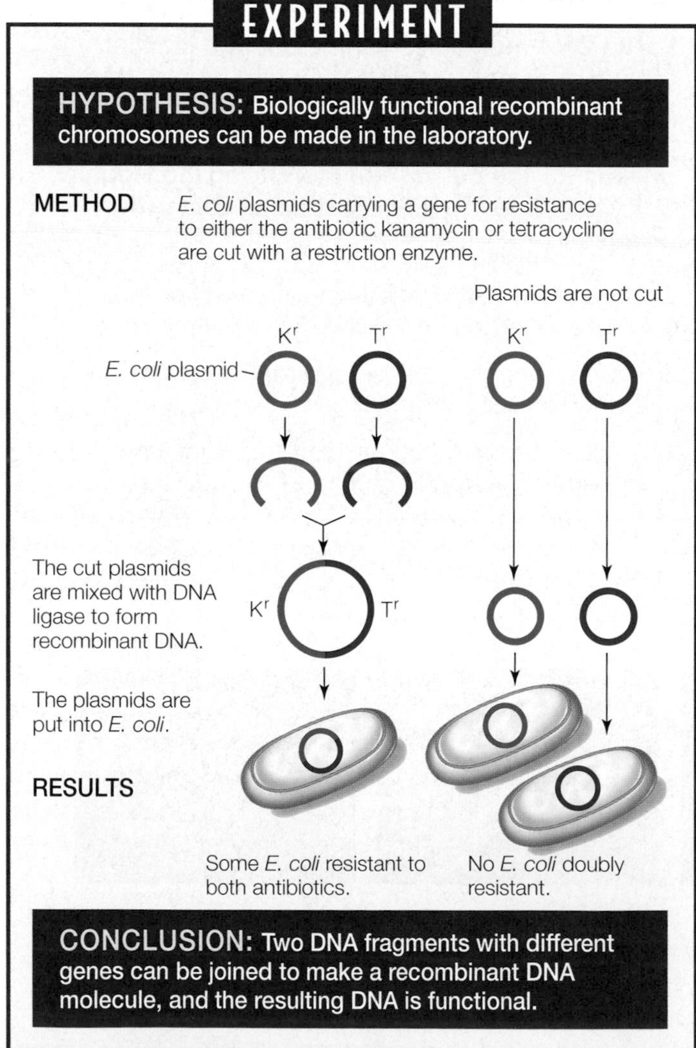

16.7 Recombinant DNA Stanley Cohen and Herbert Boyer performed the first experiment in which two different DNA sequences were combined in the laboratory to make a new, functional DNA molecule. FURTHER RESEARCH: Only one cell in 10,000 took up the plasmid in the experiment. The spontaneous mutation rate to T^r or K^r is one cell in 10^6. How would you distinguish between genetic transformation and spontaneous mutation in this experiment?

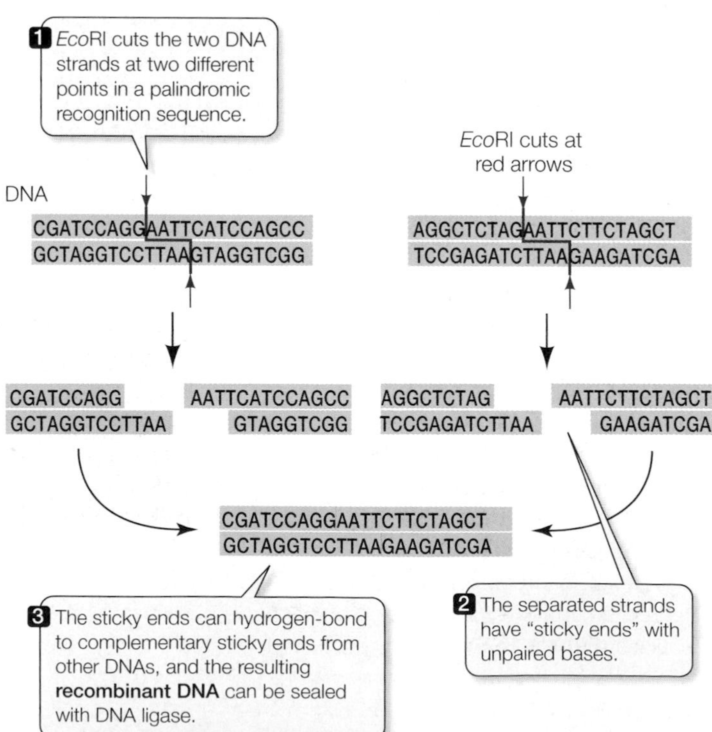

1 *Eco*RI cuts the two DNA strands at two different points in a palindromic recognition sequence.
*Eco*RI cuts at red arrows
DNA
CGATCCAGGAATTCATCCAGCC
GCTAGGTCCTTAAGTAGGTCGG
AGGCTCTAGAATTCTTCTAGCT
TCCGAGATCTTAAGAAGATCGA
CGATCCAGG
GCTAGGTCCTTAA
AATTCATCCAGCC
GTAGGTCGG
AGGCTCTAG
TCCGAGATCTTAA
AATTCTTCTAGCT
GAAGATCGA
CGATCCAGGAATTCTTCTAGCT
GCTAGGTCCTTAAGAAGATCGA
3 The sticky ends can hydrogen-bond to complementary sticky ends from other DNAs, and the resulting **recombinant DNA** can be sealed with DNA ligase.
2 The separated strands have "sticky ends" with unpaired bases.

16.8 Cutting and Splicing DNA Some restriction enzymes (*Eco*RI is shown here) make staggered cuts in DNA. *Eco*RI can be used to cut DNA from two different sources (blue and yellow). The cuts leave sticky ends, exposed bases that can hybridize with complementary fragments. Sticky ends from different DNAs can bind to each other, forming recombinant DNA.

another. The original ends may rejoin, or an end may pair with a complementary end from another fragment. Furthermore, because the ends of all fragments cut by the same restriction enzyme are the same, fragments from one source, such as a human, can be joined to fragments from another source, such as a bacterium.

When the temperature is lowered, the fragments *anneal* (come together by hydrogen bonding) at their sticky ends, but these associations are unstable because they are held together by only a few hydrogen bonds. The associated sticky ends can be permanently spliced together by DNA ligase.

Many restriction enzymes do not produce sticky ends. Instead, they cut both DNA strands at the same base pair within the recognition sequence, making "blunt" ends. DNA ligase can also connect blunt-ended fragments.

With these two enzyme tools—restriction enzymes and DNA ligase—scientists can cut and rejoin different DNA molecules from *any and all sources* to form recombinant DNA.

16.2 RECAP

DNA fragments from different biological sources can be linked together to make recombinant DNA.

- Describe how Cohen and Boyer made the first recombinant DNA. See Figure 16.7
- Do you understand how a staggered cut in DNA creates a "sticky end"? See pp. 358–359 and Figure 16.8

Recombinant DNA has no biological significance until it is installed inside a living cell, which can replicate and transcribe the transplanted genetic information. How can that be accomplished?

16.3 How Are New Genes Inserted into Cells?

One goal of recombinant DNA technology is to **clone** (produce many copies of) a particular gene, either for analysis or to produce its protein product in quantity. If recombinant DNA is to be used to make a protein, it must be inserted, or **transfected**, into host cells. Such altered hosts are referred to as **transgenic** cells or organisms. The choice of a host cell—prokaryotic or eukaryotic—is important to this endeavor.

Once the host species has been selected, the recombinant DNA is brought together with a population of host cells and, under specific conditions, enters some of them. Because all the host cells proliferate—not just the few that receive the recombinant DNA—the scientist must be able to determine which cells actually contain the sequence of interest. One common method of identifying cells with recombinant DNA is to tag the inserted sequence with **reporter genes**, whose phenotypes are easily observed. These phenotypes serve as *genetic markers* for the sequence of interest. Antibiotic resistance genes were the markers used in Cohen and Boyer's experiment (see Figure 16.7)

Genes can be inserted into prokaryotic or eukaryotic cells

The initial successes of recombinant DNA technology were achieved using bacteria as hosts. As we have seen in preceding chapters, bacterial cells are easily grown and manipulated in the laboratory. Much of their molecular biology is known, especially for certain well-studied bacteria such as *E. coli*, and they have numerous genetic markers that can be used to select for cells harboring the recombinant DNA. Furthermore, bacteria contain plasmids, small circular chromosomes that are easily manipulated to carry recombinant DNA into the cell.

In some important ways, however, bacteria are not ideal organisms for studying and expressing eukaryotic genes. Consider how differently the processes of transcription and translation proceed in prokaryotes and eukaryotes, and recall that DNA often contains the signals for these specific functions. Bacteria, for example, lack the splicing machinery to excise introns from the initial RNA transcript of a eukaryotic gene. Recall also that many eukaryotic proteins are extensively modified after translation by processes such as glycosylation and phosphorylation, and that these modifications are often essential for the protein's activity.

When the expression of a new gene in a eukaryote is the point of the experiment—that is, its aim is to produce a transgenic organism—a eukaryotic host is selected. This host may be a mouse, a wheat plant, a yeast, or even a human. Yeasts such as *Saccharomyces* are common eukaryotic hosts for recombinant DNA studies.

The advantages of using yeasts include rapid cell division (a life cycle completed in 2–8 hours), ease of growth in the laboratory, and a relatively small genome size (about 12 million base pairs and 6,000 genes). The yeast genome is several times larger than that of *E. coli* (see Table 14.2), but has only one-fourth as many genes as the human genome. Nevertheless, yeasts have most of the characteristics of other eukaryotes, except for those characteristics involved in multicellularity (see Table 14.3).

Plant cells can also be used as hosts, especially if the desired result is a transgenic plant. The property that makes plant cells good hosts is their *totipotency*—that is, the ability of a differentiated plant cell to act like a fertilized egg and produce an entire new organism. Isolated plant cells grown in culture can take up recombinant DNA, and by manipulation of the growth medium, these transgenic cells can be induced to form an entire new plant, which can then be reproduced naturally in the field.

Whatever host is chosen, a vehicle for carrying the DNA into the host cell is needed.

Vectors carry new DNA into host cells

The challenge of inserting new DNA into a cell lies not just in getting it into the host cell, but in getting it to replicate as the host cell divides. DNA polymerase, the enzyme that catalyzes DNA replication, does not bind to just any sequence of DNA. If the new DNA is to be replicated, it must become part of a segment of DNA that contains an origin of replication, called a **replicon**, or *replication unit*.

There are two general ways in which the newly introduced DNA can become part of a replicon:

- It can be inserted near an origin of replication in a host chromosome after entering the host cell. Although such insertion is often a random event, it is nevertheless a common method of integrating a new gene into a host cell.
- It can enter the host cell as part of a carrier DNA sequence—a **vector**— that already has the appropriate origin of replication.

A vector should have four characteristics:

- The ability to *replicate independently* in the host cell
- A *recognition sequence* for a restriction enzyme that will allow the vector to be cut and combined with the new DNA
- A *reporter gene* that will announce its presence in the host cell
- A *small size* in comparison to the host chromosomes

Several types of vectors fit this profile, including plasmids, viruses, and artificial chromosomes.

PLASMIDS AS VECTORS Plasmids have all four of the characteristics of a useful vector (**Figure 16.9A**). First, they are small (an *E. coli* plasmid has 2,000–6,000 base pairs, as compared with the main *E. coli* chromosome, which has more than 4.6 million base pairs). Furthermore, because they are so small, many plasmids have only a single recognition sequence for a given restriction enzyme. When such a plasmid is cut with a restriction enzyme, it is transformed into a linear molecule with sticky ends. The sticky ends of another DNA fragment cut with the same restriction enzyme can pair with the sticky ends of the plasmid, resulting in a circular plasmid containing the new DNA at a known location.

Two other characteristics make plasmids good vectors. As we have seen, many plasmids contain genes that confer resistance to antibiotics, which can serve as reporter genes (genetic markers). Finally, plasmids have an origin of replication (*ori*) and can replicate independently of the host chromosome. It is not uncommon for a bacterial cell with a single main chromosome to contain hundreds of copies of a recombinant plasmid.

The plasmids commonly used as vectors in the laboratory have been extensively altered by recombinant DNA technology, and most are combinations of genes and other sequences from several sources. Many of these plasmids have a single marker for antibiotic resistance.

VIRUSES AS VECTORS Constraints on plasmid replication limit the size of the new DNA that can be inserted into a plasmid to about 10,000 base pairs. Although many prokaryotic genes may be smaller than this, 10,000 base pairs is much smaller than most eukaryotic genes, with their introns and extensive flanking sequences. A vector that accommodates larger DNA inserts is needed.

Both prokaryotic and eukaryotic viruses are often used as vectors for eukaryotic DNA. Bacteriophage λ, which infects *E. coli*, has a DNA genome of about 45,000 base pairs. If the genes that cause the host cell to die and lyse—about 20,000 base pairs—are eliminated, the virus can still attach to a host cell and inject its DNA. The deleted 20,000 base pairs can be replaced with DNA from another organism.

Because viruses infect cells naturally, they offer a great advantage over plasmids, which often require artificial means to coax them to enter host cells. As we will see in Section 17.5, viruses are important vectors in human gene therapy.

ARTIFICIAL CHROMOSOMES AS VECTORS Bacterial plasmids are not good vectors for eukaryotic hosts such as yeasts because prokaryotic and eukaryotic DNA sequences use different origins of replication. To remedy this problem, scientists have created a "minimalist chromosome" called the **yeast artificial chromosome**, or **YAC** (**Figure 16.9B**). This artificial DNA molecule contains not only the yeast origin of replication, but the yeast centromere and telomere sequences as well, making it a true eukaryotic chromosome. YACs also contain artificially synthesized restriction sites and useful reporter genes (for yeast nutritional requirements). YACs are only about 10,000 base pairs in size, but can accommodate 50,000 to 1.5 million base pairs of inserted DNA. These artificial chromosomes carry out eukaryotic DNA replication and gene expression normally in yeast cells.

PLASMID VECTORS FOR PLANTS An important vector for carrying new DNA into many types of plants is a plasmid found in *Agrobacterium tumefaciens*. This bacterium lives in the soil and causes a plant disease called crown gall, which is characterized by the pres-

16.9 Vectors for Carrying DNA into Cells (A) A plasmid with reporter genes for antibiotic resistance can be incorporated into an *E. coli* cell. (B) A DNA molecule synthesized in the laboratory constitutes an artificial chromosome that can carry its inserted DNA into yeasts. (C) The Ti plasmid, isolated from the bacterium *Agrobacterium tumefaciens*, is used to insert DNA into many types of plants.

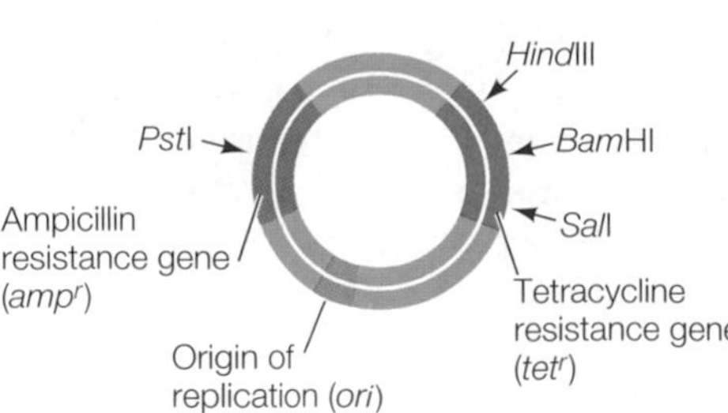

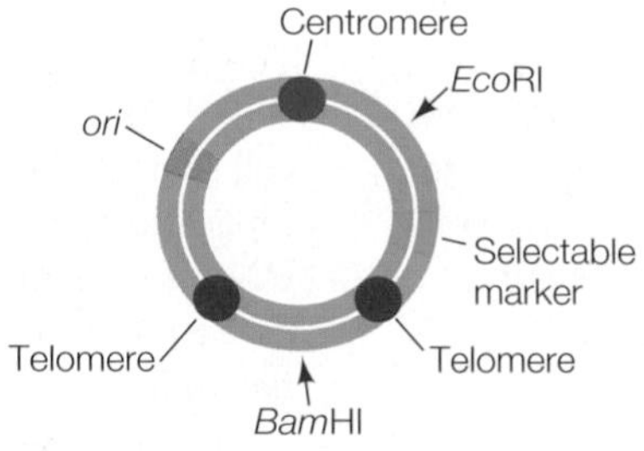

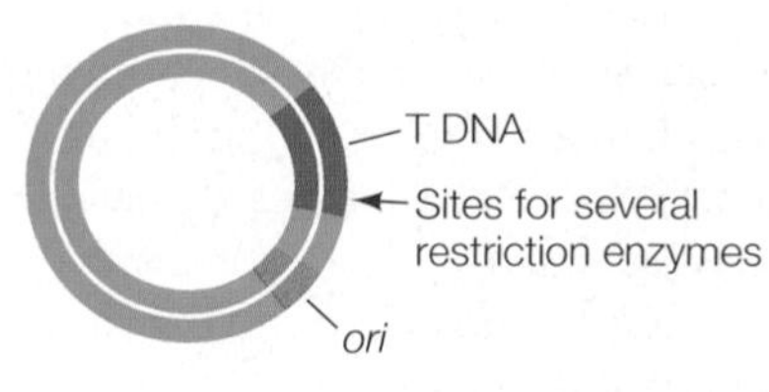

ence of growths, or tumors, in the plant. *A. tumefaciens* contains a plasmid called Ti (for tumor-inducing) (**Figure 16.9C**). The Ti plasmid contains a region called T DNA, which inserts copies of itself into the chromosomes of infected plant cells. The T DNA contains recognition sequences for several restriction enzymes, so that new DNA can be inserted into it. When the T DNA is thus altered, the plasmid no longer produces tumors, but the transposon, with its new DNA, can still be inserted into the host plant cell's chromosomes. A plant cell containing this DNA can then be grown in culture or induced to form a new, transgenic plant.

Whatever vector is effective, the problem of identifying those host cells which have actually taken up the recombinant DNA remains.

Reporter genes identify host cells containing recombinant DNA

Even when a population of host cells interacts with an appropriate vector, only a small proportion of the cells actually take up the vector. In addition, since the process of making recombinant DNA is far from perfect, only a few of the vectors that have moved into the host cells will actually contain the DNA sequence of interest. How can we select only the host cells that contain that sequence?

One common procedure uses *E. coli* bacteria as hosts and the pBR322 plasmid (see Figure 16.9A), which carries the genes for resistance to the antibiotics ampicillin and tetracycline, as a vector. When the plasmid is incubated with the restriction enzyme *Bam*HI, the enzyme encounters its recognition sequence, GGATCC, only once, at a site within the gene for tetracycline resistance. If foreign DNA is inserted at this restriction site, the presence of those "extra" base pairs within the tetracycline resistance gene inactivates it. So plasmids containing the inserted DNA will carry an intact gene for ampicillin resistance, but not for tetracycline resistance (**Figure 16.10**). This difference is the key to the selection of the host bacteria that contain the recombinant plasmid.

The cutting and insertion process results in three types of DNA, all of which can be taken up by host bacteria:

- The recombinant plasmid—the one we want—turns out to be the rarest type of DNA. Its uptake confers resistance to ampicillin, but not to tetracycline, on host *E. coli*.
- More common are bacteria that take up plasmids that have sealed their own ends back together. These plasmids retain intact genes for resistance to both ampicillin and tetracycline.
- Even more common are bacteria that take up the foreign DNA sequence alone, without the plasmid; since it is not part of a replicon, it does not survive as the bacteria divide. These host cells remain susceptible to both antibiotics.

The vast majority (more than 99.9 percent) of host cells take up no DNA at all and remain susceptible to both antibiotics. So the unique drug-resistant phenotype of the cells with recombinant DNA (tetracycline-sensitive and ampicillin-resistant) marks them in a way that can be detected by simply adding ampicillin and/or tetracycline to the medium surrounding the cells.

In addition to genes for antibiotic resistance, several other reporter genes are used to detect recombinant DNA in host cells:

- Artificial vectors include restriction sites within the *lac* operon (see Figure 13.18). When the *lac* operon is inactivated by the insertion of foreign DNA, the vector no longer carries its function into the host cell.
- Green fluorescent protein, which normally occurs in the jellyfish *Aequopora victoriana*, does not require a substrate, but

16.10 Marking Recombinant DNA by Inactivating a Gene Scientists can inactivate reporter genes within plasmid vectors to mark the host cells that have incorporated recombinant DNA. The host bacteria in this experiment could display any of the three phenotypes indicated in the table.

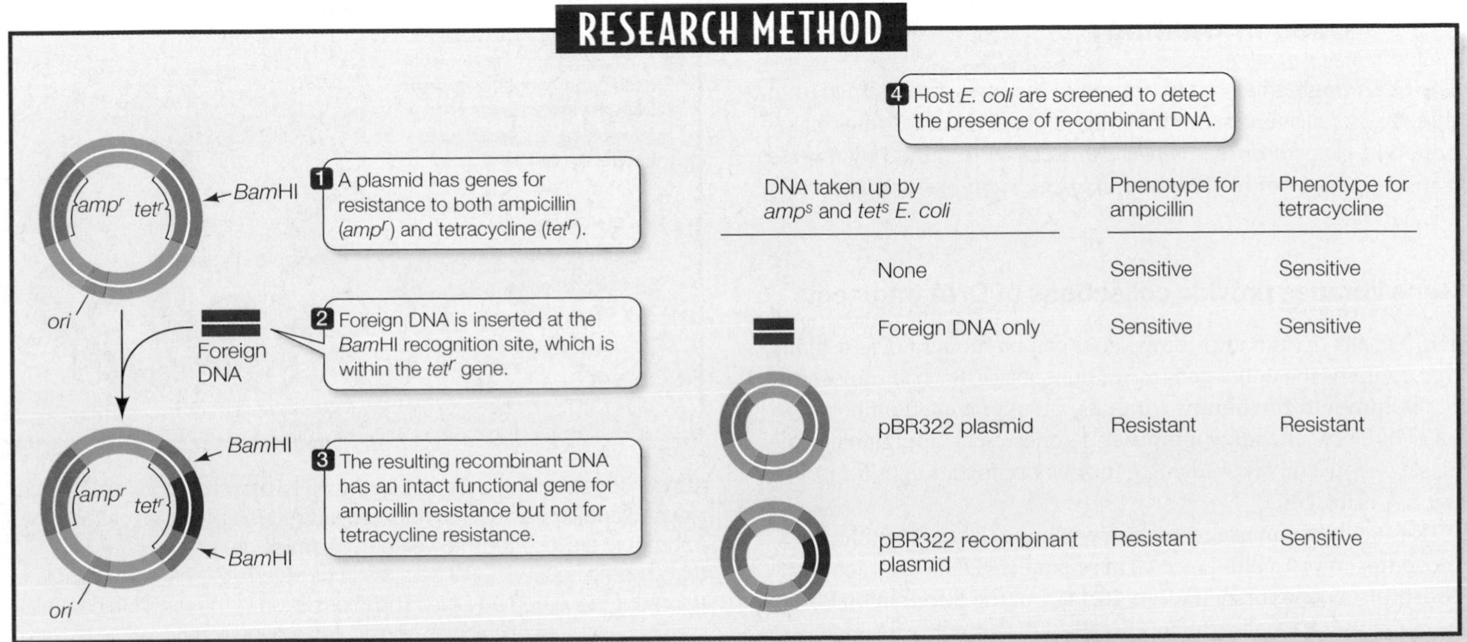
RESEARCH METHOD
ampʳ
tetʳ
BamHI
ori
1 A plasmid has genes for resistance to both ampicillin (ampʳ) and tetracycline (tetʳ).
Foreign DNA
2 Foreign DNA is inserted at the BamHI recognition site, which is within the tetʳ gene.
BamHI
ampʳ
tetʳ
BamHI
ori
3 The resulting recombinant DNA has an intact functional gene for ampicillin resistance but not for tetracycline resistance.
4 Host E. coli are screened to detect the presence of recombinant DNA.

DNA taken up by amp^s and tet^s *E. coli*	Phenotype for ampicillin	Phenotype for tetracycline
None	Sensitive	Sensitive
Foreign DNA only	Sensitive	Sensitive
pBR322 plasmid	Resistant	Resistant
pBR322 recombinant plasmid	Resistant	Sensitive

emits visible light when exposed to ultraviolet light. It is now widely used as a reporter gene.

After exposure to the vector, the host cells are grown on a solid medium. If the concentration of cells dispersed on the solid medium is low, each cell will divide and grow into a distinct bacterial colony. The colonies that contain recombinant DNA can be identified by reporter gene expression and removed from the medium, then grown in large amounts in liquid culture. A quick examination of a plasmid can confirm whether the cells of the colony actually have the recombinant DNA. The power of bacterial transformation to amplify a gene is indicated by the fact that a 1-liter culture of bacteria harboring the human β-globin gene in the pBR322 plasmid has as many copies of that gene as the sum total of all the cells in a typical adult human being (10^{14}).

16.3 RECAP

Recombinant DNA can be cloned by using a vector to insert it into a suitable host cell. The vector often has genetic markers that give the host cell a phenotype by which recombinant cells can be identified.

- What are the characteristics of a good vector for introducing new DNA into a host cell? See p. 360
- Do you understand how cells harboring a vector that carries recombinant DNA can be selected? See p. 361 and Figure 16.10

Now that we have described how DNA can be cut, inserted into a vector, and transfected into host cells, and how host cells carrying recombinant DNA can be identified, let's pause briefly to consider where the genes or DNA fragments used in these procedures come from.

16.4 What Are the Sources of DNA Used in Cloning?

The DNA fragments used in cloning procedures are obtained from three principal sources: random fragments of chromosomes maintained as *gene libraries*, *complementary DNA* obtained by reverse transcription from mRNA, and *artificial synthesis* or *mutation* of DNA.

Gene libraries provide collections of DNA fragments

The 23 pairs of human chromosomes can be thought of as a library that contains the entire genome of our species. Each chromosome, or "volume" in the library, contains, on average, 80 million base pairs of DNA, encoding a thousand genes. Such a huge molecule is not very useful for studying genomic organization or for isolating a specific gene.

Researchers can use restriction enzymes to break human chromosomes into smaller pieces. These smaller DNA fragments still constitute a **gene library** (**Figure 16.11**), but the information is now in many more, smaller "volumes." Each fragment is inserted into a vector, which is then taken up by a host cell. When bacteria are used as hosts, proliferation of one cell produces a colony of recombinant cells, each of which harbors many copies of the same fragment of human DNA.

When plasmids are used as vectors, about 200,000 separate fragments are required to make a library of the human genome. By using phage λ, which can carry four times as much DNA as a plasmid, the number of volumes can be reduced to about 50,000.

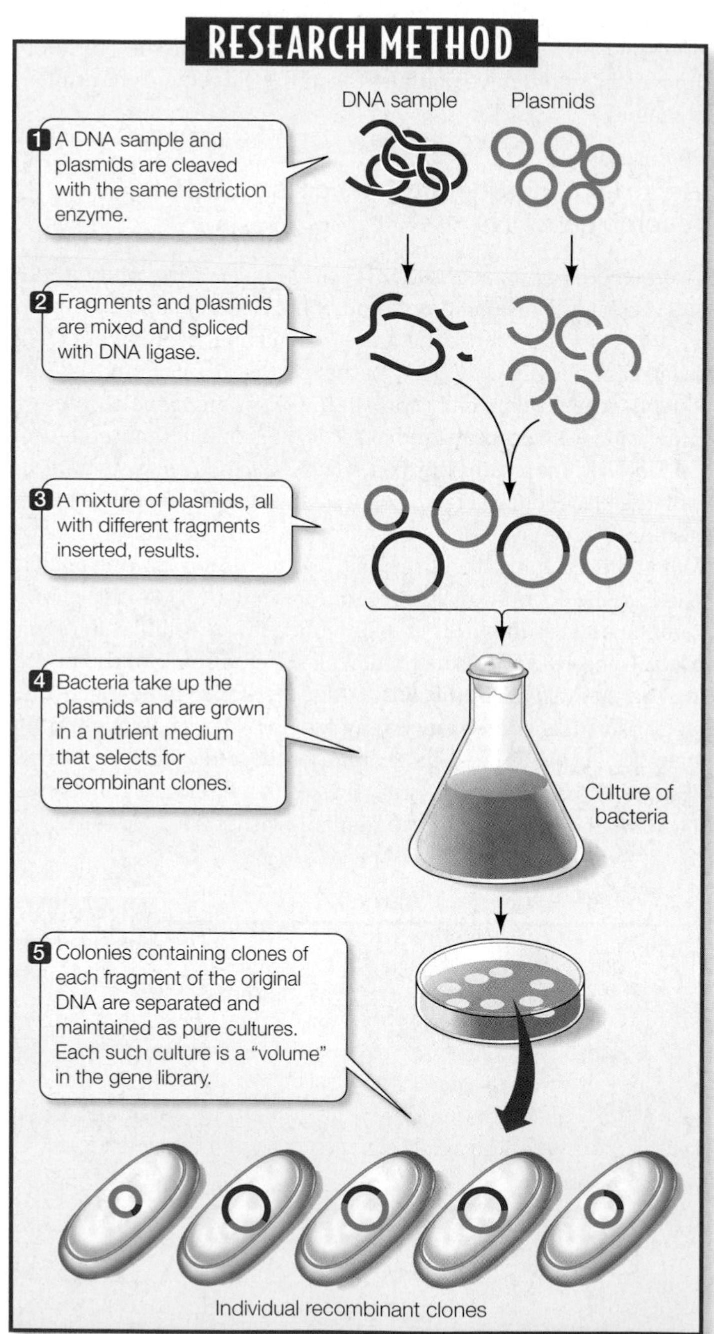

RESEARCH METHOD
DNA sample
Plasmids
1 A DNA sample and plasmids are cleaved with the same restriction enzyme.
2 Fragments and plasmids are mixed and spliced with DNA ligase.
3 A mixture of plasmids, all with different fragments inserted, results.
4 Bacteria take up the plasmids and are grown in a nutrient medium that selects for recombinant clones.
Culture of bacteria
5 Colonies containing clones of each fragment of the original DNA are separated and maintained as pure cultures. Each such culture is a "volume" in the gene library.
Individual recombinant clones

16.11 Constructing a Gene Library Human chromosomal DNA is isolated and broken up into fragments using restriction enzymes. The fragments are inserted into vectors (plasmids are shown here) and taken up by host bacterial cells, each of which then harbors a single fragment of the human DNA. The information in the resulting bacterial cultures and sets of colonies constitutes a gene library.

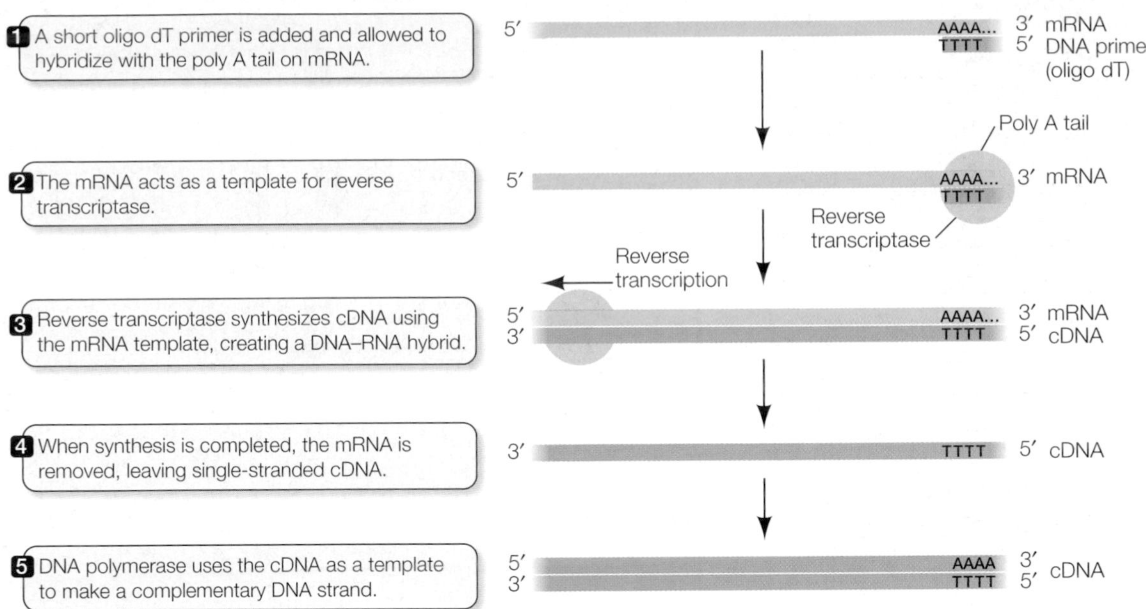
1 A short oligo dT primer is added and allowed to hybridize with the poly A tail on mRNA.
5′
AAAA... 3′ mRNA
TTTT 5′ DNA primer (oligo dT)
Poly A tail
2 The mRNA acts as a template for reverse transcriptase.
5′
AAAA... 3′ mRNA
TTTT
Reverse transcriptase
Reverse transcription
3 Reverse transcriptase synthesizes cDNA using the mRNA template, creating a DNA–RNA hybrid.
5′
3′
AAAA... 3′ mRNA
TTTT 5′ cDNA
4 When synthesis is completed, the mRNA is removed, leaving single-stranded cDNA.
3′
TTTT 5′ cDNA
5 DNA polymerase uses the cDNA as a template to make a complementary DNA strand.
5′
3′
AAAA 3′
TTTT 5′ cDNA

16.12 Synthesizing Complementary DNA Gene libraries that include only genes transcribed in a particular tissue at a particular time can be made from complementary DNA. cDNA synthesis is especially useful for identifying mRNAs that are present only in a few copies, and is often a starting point for gene cloning.

Although this seems like a large number, a single petri plate can hold up to 80,000 phage colonies, or *plaques*, and is easily screened for the presence of a particular DNA sequence by denaturing the phage DNA and applying a particular probe.

cDNA libraries are constructed from mRNA transcripts

A much smaller DNA library—one that includes only the genes transcribed in a particular tissue—can be made from **complementary DNA**, or **cDNA** (**Figure 16.12**). Recall that most eukaryotic mRNAs have a poly A tail—a string of adenine bases at their 3′ end (see Figure 14.10). The first step in cDNA production is to extract mRNA from a tissue and allow its poly A tail to hybridize with a molecule called *oligo dT*, which consists of a string of thymine bases. The oligo dT serves as a primer, and the mRNA as a template, for the enzyme reverse transcriptase, which synthesizes DNA from RNA. In this way, a cDNA strand complementary to the mRNA is formed.

A collection of cDNAs from a particular tissue at a particular time in the life cycle of an organism is called a *cDNA library*. Messenger RNAs do not last long in the cytoplasm and are often present in small amounts, so a cDNA library is a "snapshot" that preserves the transcription pattern of the cell. Complementary DNA libraries have been invaluable in comparisons of gene expression in different tissues at different stages of development. Their use has shown, for example, that up to one-third of all the genes of an animal are expressed only during prenatal development. Complementary DNA is also a good starting point for the cloning of eukaryotic genes. It is especially useful for cloning genes expressed at low levels in only a few cell types.

DNA can be synthesized chemically in the laboratory

If we know the amino acid sequence of a protein, we can use the genetic code to figure out what DNA sequence codes for each amino acid and assemble the corresponding fragments of DNA. Such artificial DNA synthesis is now fully automated, and a special service laboratory can make short to medium-length sequences overnight for any number of investigators.

Determining the base sequence of the gene that codes for a given protein is just one step in the design of a synthetic gene. Other sequences must be added, such as flanking sequences for transcription initiation, termination, and regulation and start and stop codons for translation initiation and termination. Of course, these noncoding DNA sequences must be the ones actually recognized by the host cell if the synthetic gene is to be transcribed. It does no good to have a prokaryotic promoter sequence near a gene if that gene is to be inserted into a yeast cell for expression. Appropriate selection of the codon for a given amino acid is another important consideration: many amino acids are encoded by more than one codon (see Figure 12.6), and host organisms vary in their use of synonymous codons.

DNA mutations can be created in the laboratory

Mutations that occur in nature have been important in demonstrating cause-and-effect relationships in biology. Recombinant DNA technology allows us to ask "What if?" questions without having to find mutations in nature. Because synthetic DNA can be made in any desired sequence, it can be manipulated to create specific mutations, the consequences of which can be observed when the mutant DNA is expressed by a host cell. These *mutagenesis techniques* have revealed many cause-and-effect relationships. For example, it was hypothesized that the signal sequence at the beginning of a secreted protein is essential to its passage through the membrane of the endoplasmic reticulum (see Section 12.5). A gene coding for such a protein, but with the codons for the signal sequence deleted, was synthesized. Sure enough, when this gene was expressed in yeast cells, the protein did not cross the ER membrane. When the signal sequence codons were added to an unrelated gene encoding a soluble cytoplasmic protein, that protein did cross the ER membrane.

16.4 RECAP

DNA for cloning can be obtained from gene libraries, cDNA made from mRNA, and artificially synthesized DNA fragments. Gene function can be investigated by intentionally introducing mutations into natural or synthetic genes.

- Do you understand how a gene library can be made? See p. 362 and Figure 16.11
- Do you understand how reverse transcriptase is used to make cDNA? See p. 363 and Figure 16.12

As we have just seen, artificial mutations provide an excellent means of investigating questions about the role of a gene in cell function. Let's look at some ways in which the tools described in this section can be used to study the effects of a gene, beginning with two techniques for blocking a gene's function altogether.

16.5 What Other Tools Are Used to Manipulate DNA?

Section 11.5 describes DNA sequencing and the polymerase chain reaction, two important techniques arising from our understanding of DNA replication. In this section we will examine three additional techniques for manipulating DNA:

- *Knockout experiments*: the use of genetic recombination to create an inactive, or "knocked-out," gene
- *Gene silencing*: the creation of artificial antisense RNA and interference RNA that can block the translation of specific mRNAs
- *DNA chips*: microarrays that detect the presence of many different sequences simultaneously

Genes can be inactivated by homologous recombination

A technique called *homologous recombination* can be used to replace a gene inside a cell with an inactivated form of that gene to see what happens to a living organism lacking that gene. Such a manipulation is called a **knockout** experiment.

Mice are frequently used in knockout experiments (**Figure 16.13**). The normal allele of the mouse gene to be tested is inserted into a plasmid. Restriction enzymes are then used to insert a fragment containing a reporter gene into the middle of the normal gene. This addition of extra DNA plays havoc with the targeted gene's transcription and translation; a functional mRNA is seldom made from a gene whose sequence has been thus interrupted. Next, the plasmid is transfected into a stem cell from an early mouse embryo. (A **stem cell** is an undifferentiated cell that divides and differentiates into specialized cells.)

Because much of the targeted gene is still present in the plasmid (although in two separated regions), homologous sequence recognition takes place between the inactive allele on the plasmid and the active (normal) allele in the mouse genome. The sequence in the plasmid lines up with the homologous sequence in the

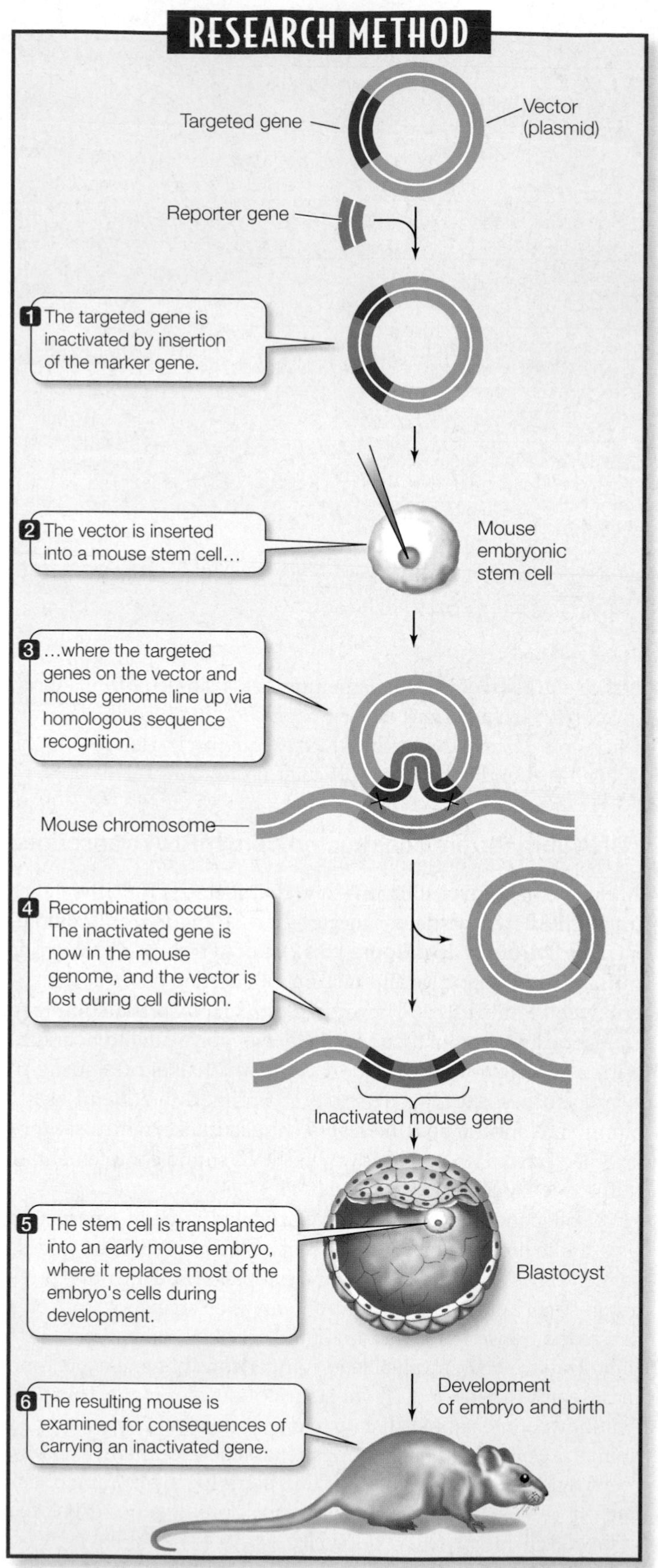

RESEARCH METHOD
Targeted gene
Vector (plasmid)
Reporter gene
1 The targeted gene is inactivated by insertion of the marker gene.
2 The vector is inserted into a mouse stem cell...
Mouse embryonic stem cell
3 ...where the targeted genes on the vector and mouse genome line up via homologous sequence recognition.
Mouse chromosome
4 Recombination occurs. The inactivated gene is now in the mouse genome, and the vector is lost during cell division.
Inactivated mouse gene
5 The stem cell is transplanted into an early mouse embryo, where it replaces most of the embryo's cells during development.
Blastocyst
Development of embryo and birth
6 The resulting mouse is examined for consequences of carrying an inactivated gene.

16.13 Making a Knockout Mouse Homologous recombination is used to replace a normal mouse gene with an inactivated copy of that gene, thus "knocking out" the gene. Discovering what happens to a mouse with an inactive gene tells us much about the normal role of that gene.

mouse chromosome, and sometimes recombination occurs, in which case the plasmid's inactive allele can be "swapped" with the functional allele in the host cell. Neither allele can be expressed, however; the allele inserted into the mouse chromosome is an interrupted fragment of the normal gene, and the gene inserted into the plasmid usually lacks its promoter. The reporter gene in the insert is, however, functional, and it is used to identify those stem cells carrying the inactivated gene.

A transfected stem cell is now transplanted into an early mouse embryo, and (through some clever tricks beyond the scope of this discussion) a knockout mouse carrying the inactivated gene in homozygous form is produced. The mutant mouse can then be observed for phenotypic changes providing clues to the function of the non-inactivated gene in the normal, wild-type animal. The knockout technique has been important in assessing the roles of certain genes during development.

Antisense RNA and interference RNA can prevent the expression of specific genes

The expression of a gene can also be blocked by stopping the translation of mRNA. As is often the case, this technique is an example of scientists imitating nature. As described in Section 14.5 (see Figure 14.22), gene expression is occasionally controlled by the production of a small RNA molecule (micro RNA) that is complementary to mRNA. Such a complementary molecule is called **antisense** RNA because it binds by base pairing to the "sense" bases on the mRNA that code for a protein. The resulting double-stranded RNA hybrid inhibits translation of the mRNA, and the hybrid tends to be broken down rapidly in the cytoplasm. Although the gene continues to be transcribed, translation does not take place. After determining the sequence of a gene and its mRNA in the laboratory, scientists can make a specific antisense RNA and add it to a cell to prevent translation of that gene's mRNA (**Figure 16.14**, left).

A related technique takes advantage of **interference RNA (RNAi)**, a rare natural mechanism for inhibiting mRNA translation. In this case, a short (about 20 nucleotides) double-stranded RNA is unwound to single strands by a protein complex that guides this RNA to a complementary region on mRNA. The protein complex catalyzes the breakdown of the targeted mRNA.

Armed with this knowledge, scientists can synthesize a *small interfering RNA* (siRNA) to inhibit the translation of *any* known gene (Figure 16.14, right). Because these double-stranded siRNAs are more stable than antisense RNAs, RNAi is preferred over antisense RNA as a means of blocking RNA translation.

Antisense RNA and RNAi have been widely used to test cause-and-effect relationships. For example, when antisense RNA was used to block the synthesis of a protein essential for the growth of cancer cells, the cells reverted to a normal phenotype. Such gene silencing techniques offer great potential for the development of drugs to treat diseases that are the result of the inappropriate expression of specific genes.

In macular degeneration, near-blindness results when blood vessels proliferate in the eye. The signaling molecule for vessel proliferation is a growth factor. An RNAi has been developed that targets this growth factor's mRNA. Without the signal to turn it on, vessel growth stops, and sometimes reverses.

DNA chips can reveal DNA mutations and RNA expression

The emerging science of genomics must deal with two major quantitative realities. First, there are a very large number of genes in eukaryotic genomes. Second, the pattern of gene expression in different tissues at different times is quite distinctive. For example, a cell from a skin cancer at its early stage may have a unique mRNA "fingerprint" that differs from that of both normal skin cells and the cells of a more advanced skin cancer.

To find such patterns, scientists could isolate the mRNA from a cell and test it by hybridization with each gene in the genome, one gene at a time. But it would be far simpler to do these hybridizations all in one step. This is possible with **DNA chip** technology, which provides large arrays of sequences for hybridization experiments.

The development of DNA chips was inspired by methods used for decades by the semiconductor industry. You may be familiar with the silicon microchip, in which an array of microscopic electric circuits is etched onto a tiny chip. In the same way, a series of DNA sequences can be attached to a glass slide in a precise order

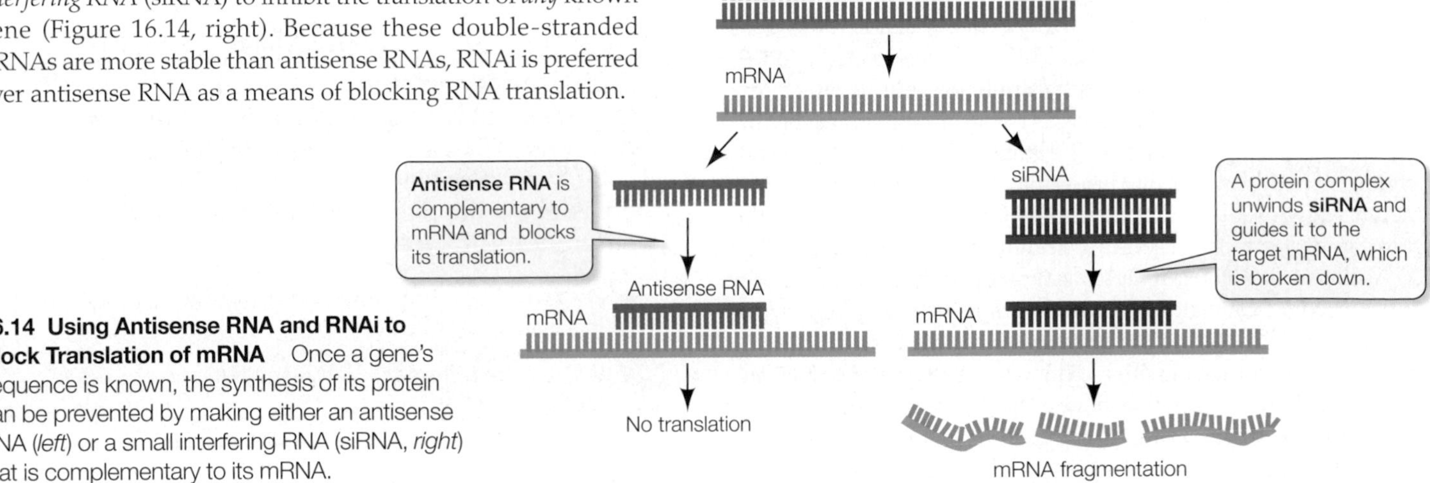

16.14 Using Antisense RNA and RNAi to Block Translation of mRNA Once a gene's sequence is known, the synthesis of its protein can be prevented by making either an antisense RNA (*left*) or a small interfering RNA (siRNA, *right*) that is complementary to its mRNA.

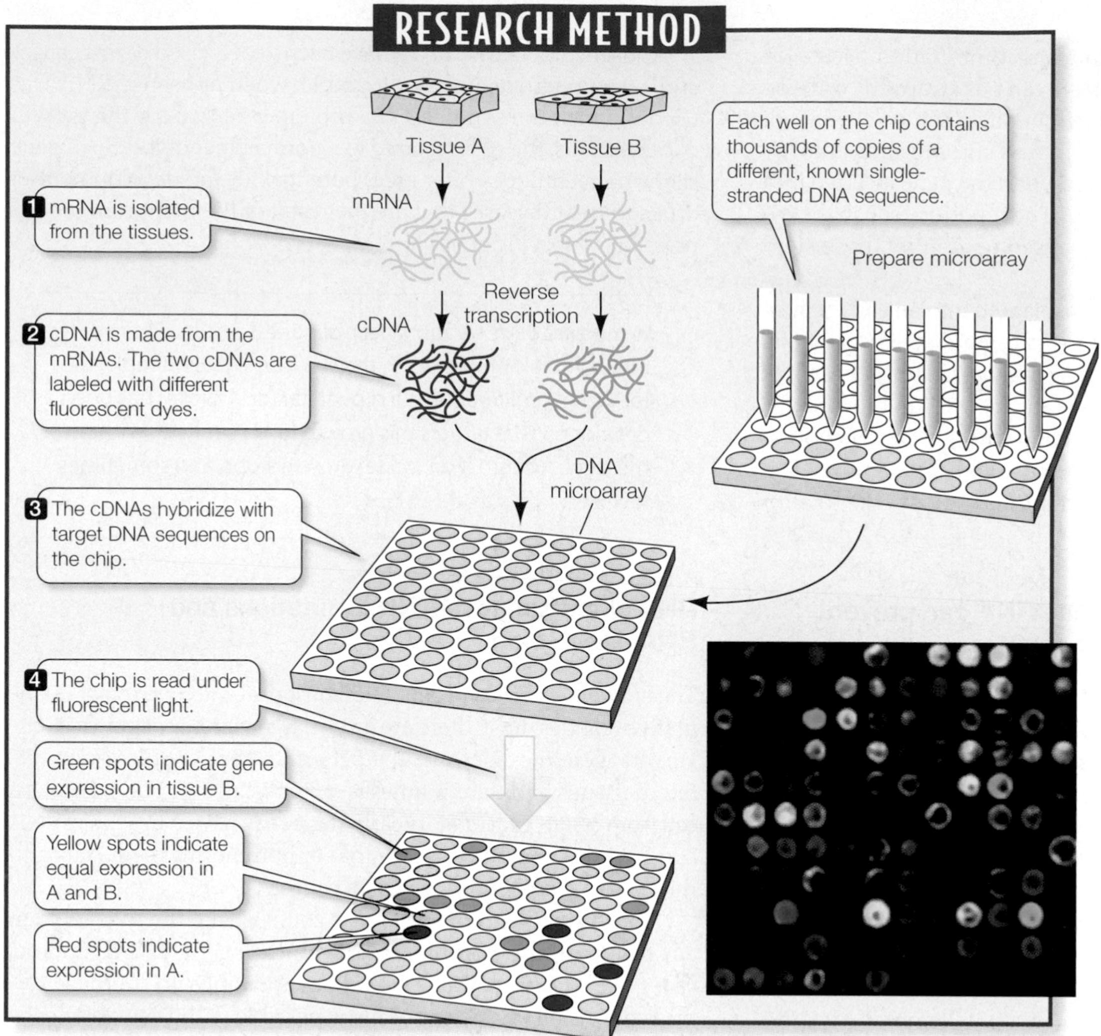
RESEARCH METHOD
Tissue A
Tissue B
mRNA
1 mRNA is isolated from the tissues.
Each well on the chip contains thousands of copies of a different, known single-stranded DNA sequence.
Prepare microarray
Reverse transcription
cDNA
2 cDNA is made from the mRNAs. The two cDNAs are labeled with different fluorescent dyes.
DNA microarray
3 The cDNAs hybridize with target DNA sequences on the chip.
4 The chip is read under fluorescent light.
Green spots indicate gene expression in tissue B.
Yellow spots indicate equal expression in A and B.
Red spots indicate expression in A.

16.15 DNA on a Chip Thousands of known DNA sequences can be attached to a glass slide in an organized grid pattern and hybridized with cDNA derived from mRNA from tissue samples to find out what genes are being expressed in the tissues.

(**Figure 16.15**). The slide is divided into a grid of microscopic spots, or "wells," each of which contains thousands of copies of a particular sequence up to 20 nucleotides long. A computer controls the addition of the sequences in a predetermined pattern. Each 20-base-long sequence can hybridize with only one genomic DNA (or cDNA or RNA) sequence, and thus is a unique identifier of a gene. Up to 60,000 different sequences can be placed on a single chip.

If cellular mRNA is to be analyzed, it is usually incubated with reverse transcriptase (RT) to make cDNA (see Figure 16.12), and the cDNA is amplified by the polymerase chain reaction (PCR) prior to hybridization (see Figure 11.23). This technique, called **RT-PCR**, ensures that mRNA sequences naturally present in only a few copies (or in a small sample, such as a cancer biopsy) will be numerous enough to form a signal. The amplified cDNA is tagged with a fluorescent dye and used to probe the DNA on the chip. Complementary DNA sequences that form hybrids with the DNA on the chip can be located by a sensitive scanner under fluorescent light.

A clinical use of DNA chips was developed by Laura van'T Veer and her colleagues at the Netherlands Cancer Institute. Most women with breast cancer are treated with surgery to remove the tumor, and then treated with radiation soon afterward to kill cancer cells that the surgeon may have missed. In some patients, however, overlooked breast cancer cells eventually form tumors, either in the breast or elsewhere in the body. The challenge for physicians is to develop criteria to identify those patients and treat them aggressively with tumor-killing chemotherapy. Over the years, doctors have used such characteristics as original tumor size, tumor cell morphology, and tumor spread to lymph nodes (see Section 17.4) to predict whether a patient's breast cancer will recur. These methods have been only moderately successful.

Enter DNA chips. van'T Veer's group looked at the expression of about 1,000 genes in tumors from patients whose prognosis they already knew (the tumors were stored specimens). They found 70 genes whose expression differed dramatically between tumors from patients with poor prognoses (that is, their cancers recurred) and with good prognoses, developing what is called a "gene expression signature." This "signature on a chip" is useful in clinical decision making: patients with a good signature can avoid unnecessary chemotherapy, while those with a poor signature can receive aggressive treatment.

16.5 RECAP

Researchers can study the function of a gene by knocking out that gene in a living organism. Antisense RNA and RNAi silence genes by selectively blocking mRNA translation. DNA chips allow the simultaneous analysis of many different mRNA transcripts.

- Do you understand how a gene can be "knocked out" in a living organism? See p. 364 and Figure 16.13
- What is the difference between RNAi and antisense RNA? See p. 365

Now that we've seen how DNA can be fragmented, recombined, manipulated, and put back into living organisms, let's look at some examples of how these techniques can be put together to make useful products.

16.6 What Is Biotechnology?

Biotechnology is the use of living cells to produce materials useful to people, such as foods, medicines, and chemicals. People have been doing this for a very long time. For example, the use of yeasts to brew beer and wine dates back at least 8,000 years, and the use of bacterial cultures to make cheese and yogurt is a technique many centuries old. For a long time, however, people were not aware of the cellular bases of these biochemical transformations.

About 100 years ago, thanks largely to Louis Pasteur's work, it became clear that specific bacteria, yeasts, and other microbes could be used as biological converters to make certain products. Alexander Fleming's discovery that the mold *Penicillium* makes the antibiotic penicillin led to the large-scale commercial culture of microbes to produce antibiotics as well as other useful chemicals. Today, microbes are grown in vast quantities to make much of the industrial-grade alcohol, glycerol, butyric acid, and citric acid that are used by themselves or as starting materials in the manufacture of other products.

But the harvesting of proteins, including hormones and enzymes, was limited to the minuscule amounts that could be extracted from organisms that produce them naturally. Yields were low, and purification was difficult and costly. All this has changed with the advent of gene cloning. The ability to insert almost any gene into bacteria or yeasts, along with methods to induce the gene to make its product in large amounts and export it from the cells, has turned these microbes into versatile factories for important products. Key to this boom in biotechnology has been the development of specialized vectors that not only carry genes into cells, but make those cells express them at high levels.

Expression vectors can turn cells into protein factories

If a eukaryotic gene is inserted into a typical plasmid (see Figure 16.9) and transformed into *E. coli*, little, if any, of the product of the gene will be made by the host cell unless other key sequences are also included. The bacterial promoter for RNA polymerase binding, the terminator for transcription, and a special sequence on mRNA that is necessary for ribosome binding are all needed if the gene is to be expressed and its product synthesized in the bacterial cell.

To solve this kind of problem, scientists make **expression vectors** that have all the characteristics of typical vectors as well as the extra sequences needed for the foreign gene (also called a *transgene*) to be expressed in the host cell. For bacterial hosts, these additional sequences include the elements named above (**Figure 16.16**); for eukaryotes, they include the poly A addition sequence, transcription factor binding sites, and enhancers. Once these sequences are placed at the appropriate location in the vector, a gene

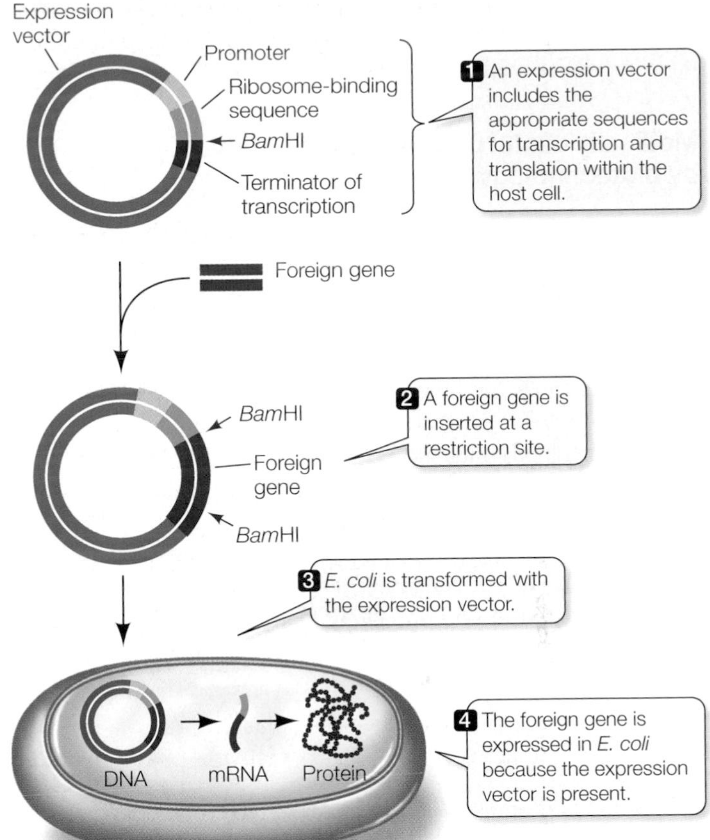
Expression vector
Promoter
Ribosome-binding sequence
BamHI
Terminator of transcription
1 An expression vector includes the appropriate sequences for transcription and translation within the host cell.
Foreign gene
BamHI
Foreign gene
BamHI
2 A foreign gene is inserted at a restriction site.
3 E. coli is transformed with the expression vector.
DNA
mRNA
Protein
4 The foreign gene is expressed in E. coli because the expression vector is present.

16.16 An Expression Vector Allows a Transgene to Be Expressed in a Host Cell A transformed eukaryotic gene may not be expressed in *E. coli* if it lacks the necessary bacterial sequences for promotion, termination, and ribosome binding. Expression vectors contain these additional sequences, enabling the eukaryotic protein to be synthesized in the prokaryotic cell.

transfected by that vector can be expressed in almost any kind of host cell.

An expression vector can be modified in various ways.

- An *inducible promoter*, which responds to a specific signal, can be made part of an expression vector. For example, a specific promoter that responds to hormonal stimulation can be used so that the transgene will transcribe its mRNA when the hormone is added. An enhancer that responds to hormonal stimulation can also be added so that transcription and protein synthesis will occur at high rates—a goal of obvious importance in the manufacture of an industrial product.
- A *tissue-specific promoter*, which is expressed only in a certain tissue at a certain time, can be used if localized expression is desired. For example, many seed proteins are expressed only in the plant embryo. Coupling a gene to a seed-specific promoter will allow the gene to be expressed only as a seed protein.
- *Signal sequences* can be added to the expression vector so that the product of the gene is directed to an appropriate destination. For example, when yeast or bacterial cells making a pro-

tein are to be maintained in a large vessel, it is economical to include a signal directing the protein to be secreted into the extracellular medium for easier recovery.

Medically useful proteins can be made by biotechnology

Many medically useful products are being made by biotechnology (**Table 16.1**), and hundreds more are in various stages of development. The manufacture of tissue plasminogen activator provides a good illustration of a medical application of biotechnology.

When a wound begins bleeding, a blood clot soon forms to stop the flow. Later, as the wound heals, the clot dissolves. How does the blood perform these conflicting functions at the right times? Mammalian blood contains an enzyme called plasmin that catalyzes the dissolution of the clotting proteins. But plasmin is not always active; if it were, a blood clot would dissolve as soon as it formed. Instead, plasmin is "stored" in the blood in an inactive form called plasminogen. The conversion of plasminogen to plasmin is activated by an enzyme, appropriately called *tissue plasminogen activator* (TPA), that is produced by cells lining the blood vessels:

$$\underset{\text{(inactive)}}{\text{plasminogen}} \xrightarrow{\text{TPA}} \underset{\text{(active)}}{\text{plasmin}}$$

Heart attacks and strokes can be caused by blood clots that form in major blood vessels leading to the heart or the brain, respectively. During the 1970s, a bacterial enzyme called streptokinase was found to stimulate the dissolution of clots in some patients. Treatment with this enzyme saved lives, but its use had side effects. Streptokinase was a protein foreign to the body, so patients' immune systems reacted against it. More important, the drug sometimes prevented clotting throughout the entire circulatory system, leading to an almost hemophilia-like condition in some patients.

The discovery of TPA and its isolation from human tissues led to the hope that this enzyme might be used to treat heart attack and stroke victims—that it would bind specifically to clots, and that it would not provoke an immune reaction. But the amounts

TABLE 16.1

Some Medically Useful Products of Biotechnology

PRODUCT	USE
Colony-stimulating factor	Stimulates production of white blood cells in patients with cancer and AIDS
Erythropoietin	Prevents anemia in patients undergoing kidney dialysis and cancer therapy
Factor VIII	Replaces clotting factor missing in patients with hemophilia A
Growth hormone	Replaces missing hormone in people of short stature
Insulin	Stimulates glucose uptake from blood in people with insulin-dependent (Type I) diabetes
Platelet-derived growth factor	Stimulates wound healing
Tissue plasminogen activator	Dissolves blood clots after heart attacks and strokes
Vaccine proteins: Hepatitis B, herpes, influenza, Lyme disease, meningitis, pertussis, etc.	Prevent and treat infectious diseases

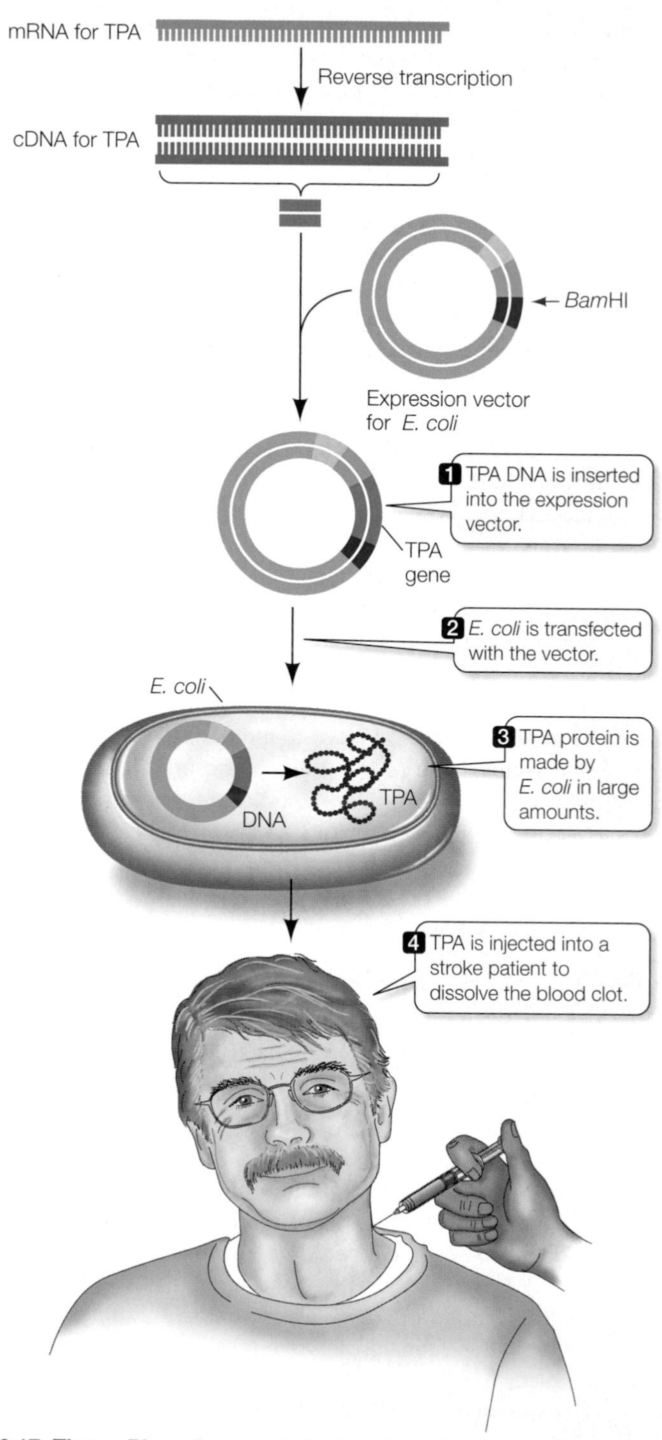

16.17 Tissue Plasminogen Activator: From Protein to Gene to Drug TPA is a naturally occurring human protein involved in dissolving blood clots. Its manufacture and use for treating patients suffering from blood clotting in the heart or brain—in other words, heart attacks or strokes—was made possible by recombinant DNA technology.

of TPA that could be harvested from human tissues were tiny, certainly not enough to inject at the site of a clot in the emergency room.

Recombinant DNA technology solved this problem. TPA mRNA was isolated and used to make a cDNA copy, which was then inserted into an expression vector and transfected into *E. coli* (**Figure 16.17**). The transgenic bacteria made the protein in quantity, and it soon became available commercially. This drug has had considerable success in dissolving blood clots in people undergoing heart attacks and, especially, strokes.

Another way of making medically useful products in large amounts is what has come to be called **pharming**: the production of proteins in milk. A transgene coding for a useful protein product can be transfected into the eggs of a female domestic animal, such as a sheep, goat, or cow, next to the promoter for lactoglobulin, a protein made in large amounts in milk. The resulting transgenic animal secretes large amounts of the protein in its milk. These natural "bioreactors" can produce a large supply of the protein, which can be easily separated from the other components of the milk (**Figure 16.18**). Products being produced by pharming include blood clotting factors for treating hemophilia and antibodies for treating colon cancer.

DNA manipulation is changing agriculture

The cultivation of plants and husbanding of animals that constitute *agriculture* give us the world's oldest examples of biotechnology, dating back more than 8,000 years. Over the centuries, people have adapted crops and farm animals to their needs. Through cultivation and selective breeding (artificial selection) of these organisms, desirable characteristics, such as ease of cooking seeds or fat content of meat, have been imparted and improved. In addition, people have developed crops with desirable growth characteristics, such as high yield, a reliable ripening season, and resistance to diseases.

Until recently, the most common way to improve crop plants and farm animals was to select and breed varieties with desirable phenotypes that existed in nature through mutational variation. The advent of genetics a century ago was followed by its application to plant and animal breeding. A crop plant or animal with desirable genes could be identified, and through deliberate crosses, those genes could be introduced into a widely used variety of that organism.

Despite some spectacular successes, such as the breeding of "supercrops" of wheat, rice, and corn, such deliberate crossing remains a hit-or-miss affair. Many desirable traits are complex in their genetics, and it is hard to predict the results of a cross or to maintain a prized combination as a pure-breeding variety year after year. In sexual reproduction, combinations of unlinked genes are quickly separated in meiosis during gamete formation. Moreover, traditional crop plant breeding takes a long time: many plants can reproduce only once or twice a year—a far cry from the rapid reproduction of bacteria.

Modern recombinant DNA technology has several advantages over traditional methods of breeding:

- *The ability to target specific genes.* Allowing a breeder to select for specific genes makes the breeding process more precise and less likely to fail as a result of the incorporation of unforeseen genes.
- *The ability to introduce any gene from any organism into a plant or animal species.* This ability, combined with mutagenesis techniques, vastly expands the range of possible new traits.
- *The ability to generate new organisms quickly.* Manipulating cells in the laboratory and regenerating a whole plant by cloning is much faster than traditional breeding.

Consequently, recombinant DNA technology has found many applications in agriculture (**Table 16.2**). We will describe a few examples here to demonstrate the approaches that have been used.

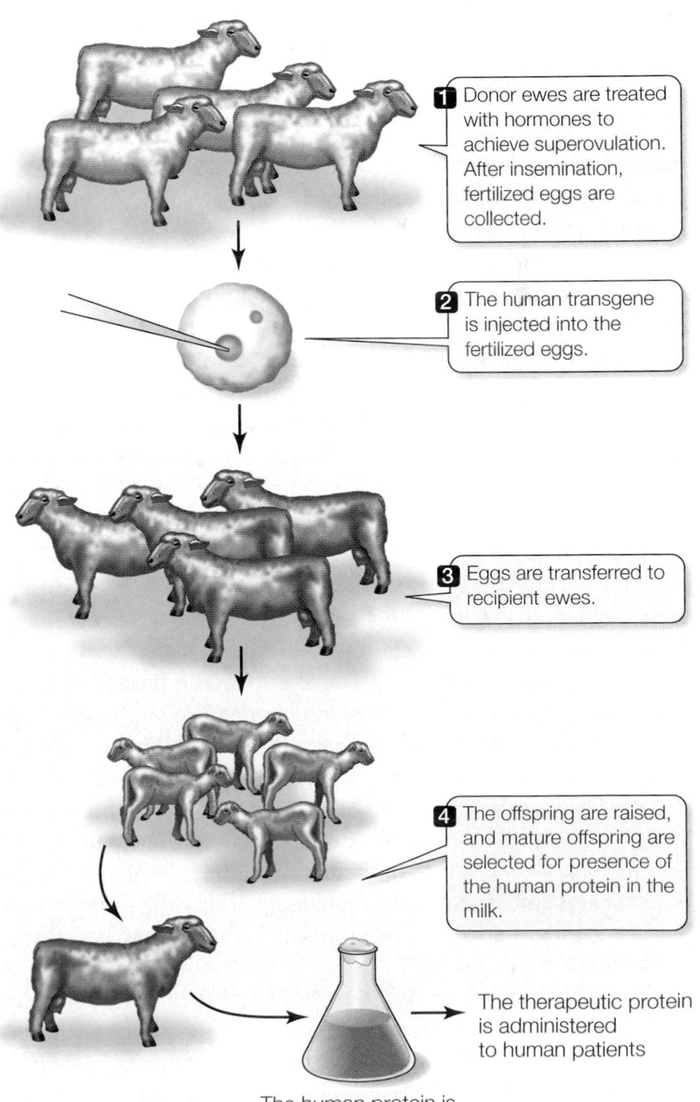

1 Donor ewes are treated with hormones to achieve superovulation. After insemination, fertilized eggs are collected.
2 The human transgene is injected into the fertilized eggs.
3 Eggs are transferred to recipient ewes.
4 The offspring are raised, and mature offspring are selected for presence of the human protein in the milk.
The therapeutic protein is administered to human patients
The human protein is extracted from the milk

16.18 Pharming An expression vector carrying a desired gene can be put into an animal egg and, using reproductive technology, the egg can be made to form an early embryo, which is implanted into a surrogate mother. The transgenic offspring produce the new protein in their milk.

TABLE 16.2
Agricultural Applications of Biotechnology under Development

PROBLEM	TECHNOLOGY/GENES
Improving the environmental adaptations of plants	Genes for drought tolerance, salt tolerance
Improving nutritional traits	High-lysine seeds
Improving crops after harvest	Delay of fruit ripening; sweeter vegetables
Using plants as bioreactors	Plastics, oils, and drugs produced in plants
Controlling crop pests	Herbicide tolerance; resistance to viruses, bacteria, fungi, insects

On the site of a nineteenth-century hat factory in Connecticut, where toxic mercury used in the curing process still contaminates the soil, 160 transgenic cottonwood trees have been planted to clean up the soil. The trees contain a bacterial gene called *MerA*, whose enzyme product converts toxic ionic mercury to volatile and less harmful metallic mercury.

PLANTS THAT MAKE THEIR OWN INSECTICIDES Humans are not the only species that consumes crop plants. Plants are subject to infections by viruses, bacteria, and fungi, but probably the most important crop pests are herbivorous insects. From the locusts of biblical (and modern) times to the cotton boll weevil, insects have continually eaten the crops people grow.

The development of insecticides has improved the situation somewhat, but insecticides have their own problems. Most, such as the organophosphates, are relatively nonspecific, killing not only pests in the field but beneficial insects in the broader ecosystem as well. Some even have toxic effects on other groups of organisms, including people. What's more, many insecticides persist in the environment for a long time.

Some bacteria have solved their own pest problem by producing proteins that can kill insects. For example, there are dozens of strains of *Bacillus thuringiensis*, each of which produces a protein toxic to the insect larvae that prey on it. The toxicity of this protein is 80,000 times that of the usual commercial insecticides. When a hapless larva eats the bacteria, the toxin becomes activated, binding specifically to the insect's gut to produce holes, and the insect starves to death.

Dried preparations of *B. thuringiensis* have been sold for decades as a safe insecticide that breaks down rapidly in the environment. But this *biodegradation* is their limitation, because it means that the dried bacteria must be applied repeatedly during the growing season. A more permanent approach would be to have the crop plants make the toxin themselves.

The toxin genes from different strains of *B. thuringiensis* have been isolated and cloned, and they have been extensively modified by the addition of plant promoters and terminators, plant poly A addition sequences, and plant regulatory elements. These modified genes have been introduced into plant cells in the laboratory using the Ti plasmid vector (see Figure 16.9C), and transgenic plants have been grown and tested for insect resistance in the field. Corn, cotton, soybeans, tomatoes, and other crops are being grown successfully with this added gene. Pesticide usage by farmers growing these transgenic crops is greatly reduced.

CROPS THAT ARE RESISTANT TO HERBICIDES Herbivorous insects are not the only threat to agriculture. Weeds may grow in fields and compete with crop plants for water and soil nutrients. Glyphosate (Roundup®) is a widely used and effective *herbicide*, or weed killer. It works only on plants, by inhibiting an enzyme system in the chloroplast that is involved in the synthesis of amino acids. Glyphosate is truly a "miracle herbicide" killing 76 of the world's 78 most prevalent weeds. Unfortunately, it also kills crop plants, so it is best used to rid a field of weeds before the crop plant starts to grow. But, as any gardener knows, when the crop begins to grow, the weeds reappear. If the crop were not affected by the herbicide, the herbicide could be applied to the field at any time and would kill only the weeds.

Scientists have used expression vectors to make plants that synthesize so much of the target enzyme for glyphosate that they are unaffected by it. The gene has been inserted into corn, cotton, and soybean plants, making them resistant to glyphosate. This technology expanded so rapidly in the late 1990s that half of the U.S. crops of these three plants now contain this high-expressing gene.

GRAINS WITH IMPROVED NUTRITIONAL CHARACTERISTICS To remain healthy, humans must eat foods (or supplements) containing an adequate amount of β-carotene, which the body converts into vitamin A (see Figure 3.21). About 400 million people worldwide suffer from vitamin A deficiency, which makes them susceptible to infections and blindness. One reason is that rice grains, which do not contain β-carotene, but only a precursor molecule for it, make up a large part of their diet. Other organisms, such as the bacterium *Erwinia* and daffodil plants, have enzymes that can convert the precursor into β-carotene. The genes for this biochemical pathway are present in the bacterial and daffodil genomes, but not in the rice genome.

Scientists isolated one of the genes for the β-carotene pathway from the bacterium and the other two from daffodil plants. They added promoter signals for expression in the developing rice grain, and then added each gene to rice plants by using the Ti plasmid vector from *Agrobacterium tumefaciens* (see Figure 16.9C). The resulting rice plants produce grains that look yellow because of their high β-carotene content (**Figure 16.19**). About 300 grams of this cooked rice a day can supply all the β-carotene a person needs. This new transgenic strain has been crossed with more locally adapted strains in the hope of improving the diets of millions of people.

CROPS THAT ADAPT TO THE ENVIRONMENT Throughout human history, agriculture has depended on ecological management—tailoring the environment to the needs of crop plants. A farm field is an unnatural, human-designed system, and when conditions in

16.19 Transgenic Rice Is Rich in β-Carotene (A) The grains from a new transgenic strain of rice are yellow because they make the pigment β-carotene, which is converted by humans into vitamin A. (B) Normal rice does not contain β-carotene.

that field become intolerable, the crops die. The Fertile Crescent, the region between the Tigris and Euphrates rivers in the Middle East where agriculture probably originated 10,000 years ago, is no longer fertile. It is now a desert, largely because the soil has a high salt concentration. Few plants can grow on salty soils, primarily because the environment is hypertonic to plant roots, and water leaves them, resulting in wilting.

Recently, a gene was discovered in *Arabidopsis thaliana* that allows this tiny weed to thrive in salty soils. The gene codes for a protein that transports sodium ions into the central vacuole. When this gene was added to tomato plants, they grew in soils four times as salty as the normal lethal level (**Figure 16.20**). This finding raises the prospect of growing useful crops on what were previously unproductive soils.

More important, this example illustrates what could become a fundamental shift in the relationship between crop plants and the environment. Instead of manipulating the environment to suit the plant, biotechnology may allow us to adapt the plant to the environment. As a result, some of the negative effects of agriculture, such as water pollution, could be lessened.

There is public concern about biotechnology

Concerns have been raised by the general public about the safety and wisdom of genetically modifying crops. These concerns are centered on three claims:

- Genetic manipulation is an unnatural interference with nature.
- Genetically altered foods are unsafe to eat.
- Genetically altered crop plants are dangerous to the environment.

Advocates of biotechnology tend to agree with the first claim. However, they point out that *all* crops are unnatural in the sense that they come from artificially bred plants growing in a manipulated environment (a farmer's field). Recombinant DNA technology just adds another level of sophistication to these techniques.

Biotechnology advocates counter the concern about whether genetically engineered crops are safe for human consumption by pointing out that only single genes are added and that these genes are specific for plant function. For example, the *B. thuringiensis* toxin produced by transgenic plants has no effect on people. However, as plant biotechnology moves from adding genes to improve plant growth to adding genes that affect human nutrition, such concerns will become more pressing.

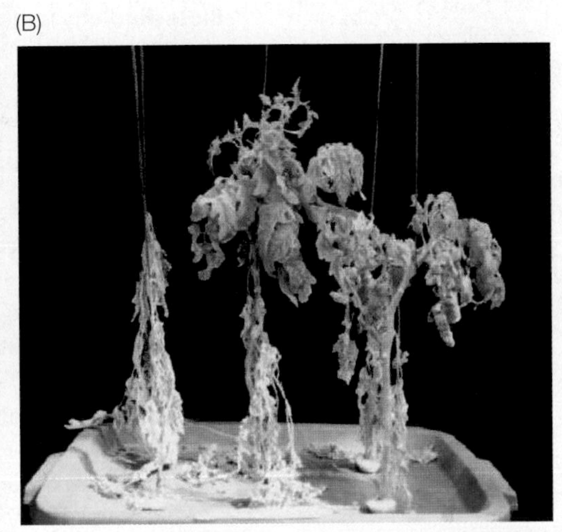

16.20 Salt-Tolerant Tomato Plants Transgenic plants containing a gene for salt tolerance thrive in salty soils (A), while plants without the transgene die (B).

The third concern, about environmental effects, centers on the possible "escape" of transgenes from crops to other species. If the gene for herbicide resistance, for example, were inadvertently transferred from a crop plant to a nearby weed, that weed could thrive in herbicide-treated areas. Or beneficial insects could eat plant materials containing *B. thuringiensis* toxin and die. Transgenic plants undergo extensive field testing before they are approved for use, but the complexity of the biological world makes it impossible to predict all potential environmental effects of transgenic organisms. Because of the potential benefits of agricultural biotechnology (see Table 16.2), scientists believe that it is wise to proceed with caution.

16.6 RECAP

Expression vectors maximize the expression of transgenes inserted into host cells. Biotechnology has been used to produce medicines and to develop transgenic plants with improved agricultural and nutritional characteristics.

- Do you understand how expression vectors work? See p. 367 and Figure 16.16
- What are some of the concerns that people might have about agricultural biotechnology? See pp. 371–372

CHAPTER SUMMARY

16.1 How are large DNA molecules analyzed?

Restriction enzymes, which are made by bacteria as a defense against viruses, bind to and cut DNA at specific **recognition sequences** (also called **restriction sites**). These enzymes can be used to produce small fragments of DNA for study, a technique known as **restriction digestion**. Review Figure 16.1

DNA fragments can be separated by size using **gel electrophoresis**. Review Figure 16.2, Web/CD Tutorial 16.1

Specific DNA sequences can be identified in a gel by probes with a complementary sequence in a procedure known as Southern blotting. Review Figure 16.3

DNA fingerprinting can distinguish between specific individuals, or reveal which individuals are most closely related, by detecting polymorphisms in their genes, particularly **single nucleotide polymorphisms** (**SNPs**) and **short tandem repeats** (**STRs**). Review Figure 16.4

The DNA barcoding project uses sequencing of a single region of DNA to identify species.

16.2 What is recombinant DNA?

Recombinant DNA is formed by the combination of two DNA sequences from different sources. Review Figure 16.7

Many restriction enzymes make staggered cuts in the two strands of DNA, creating fragments that have **sticky ends** with unpaired bases.

DNA fragments with sticky ends can be used to create recombinant DNA if DNA molecules from different sources are cut with the same restriction enzyme and spliced together with DNA ligase. Review Figure 16.8

16.3 How are new genes inserted into cells?

One goal of recombinant DNA technology is to **clone** a particular gene, either for analysis or to produce its protein product in quantity.

Bacteria, yeasts, and cultured plant cells are commonly used as hosts for recombinant DNA. Host cells into which recombinant DNA is inserted, or **transformed**, are called **transgenic cells**.

To identify host cells that have taken up a foreign gene, the inserted sequence can be tagged with **reporter genes**, genetic markers with easily identifiable phenotypes.

Expression of the foreign gene in the host cell requires that it become part of a segment of DNA that contains a **replicon** (origin and terminus of replication).

Vectors are DNA sequences that can carry new DNA into host cells. Plasmids, viruses, and **yeast artificial chromosomes** are all used as vectors. Review Figure 16.9

16.4 What are the sources of DNA used in cloning?

DNA fragments from a genome can be inserted in host cells to create a **gene library**. Review Figure 16.11

The mRNAs produced in a certain tissue at a certain time can be extracted and used to create **complementary DNA** (**cDNA**) by reverse transcription. Review Figure 16.12

Synthetic DNA containing any desired sequence can be made and mutated in the laboratory.

16.5 What other tools are used to manipulate DNA?

Homologous recombination can be used to "**knock out**" a gene in a living organism. Review Figure 16.13

Gene silencing techniques can be used to inactivate the mRNA transcript of a gene, which may provide clues to the gene's function. Artificially created **antisense** RNA or **interference RNA** (**RNAi**) can be added to a cell to prevent translation of a specific mRNA. Review Figure 16.14

DNA chip technology permits the screening of mRNA for thousands of sequences at the same time. Review Figure 16.15, Web/CD Tutorial 16.2

In a technique called **RT-PCR**, mRNA from a specific tissue is made into cDNA by reverse transcription and then amplified. This cDNA can be used as a probe on a DNA chip to explore the pattern of gene expression in the tissue.

16.6 What is biotechnology?

Biotechnology is the use of living cells to produce materials useful to people. Recombinant DNA technology has resulted in a boom in biotechnology.

Expression vectors allow a transgene to be expressed in a host cell. Review Figure 16.16, Web/CD Activity 16.1

Recombinant DNA techniques have been used to make medically useful proteins. Review Figure 16.17

Pharming uses transgenic animals that produce useful products in their milk. Review Figure 16.18

Because recombinant DNA technology has several advantages over traditional agricultural biotechnology, it is being extensively applied to agriculture.

Transgenic crop plants can be adapted to their environment, instead of vice versa.

There is public concern about the application of recombinant DNA technology to food production.

SELF-QUIZ

1. Restriction enzymes
   a. play no role in bacteria.
   b. cleave DNA at highly specific recognition sequences.
   c. are inserted into bacteria by bacteriophage.
   d. are made only by eukaryotic cells.
   e. add methyl groups to specific DNA sequences.
2. When fragments of DNA of different sizes are placed in an electric field,
   a. the smaller pieces move most rapidly toward the positive pole.
   b. the larger pieces move most rapidly toward the positive pole.
   c. the smaller pieces move most rapidly toward the negative pole.
   d. the larger pieces move most rapidly toward the negative pole.
   e. the smaller and larger pieces move at the same rate.
3. From the list below, select the sequence of steps for inserting a piece of foreign DNA into a plasmid vector, introducing the plasmid into bacteria, and verifying that the plasmid and the foreign gene are present:
   (1) Transfect host cells.
   (2) Select for the lack of plasmid reporter gene 1 function.
   (3) Select for the plasmid reporter gene 2 function.
   (4) Digest vector and foreign DNA with a restriction enzyme, which inactivates plasmid reporter gene 1.
   (5) Ligate the digested plasmid together with the foreign DNA.
   a. 45132
   b. 45123
   c. 13425
   d. 32145
   e. 13254
4. Possession of which feature is not desirable in a vector for gene cloning?
   a. An origin of DNA replication
   b. Genetic markers for the presence of the vector
   c. Multiple recognition sequences for the restriction enzyme to be used
   d. One recognition sequence each for one to several different restriction enzymes
   e. Genes other than the target for transfection
5. RNA interference (RNAi) inhibits
   a. DNA replication.
   b. transcription of specific genes.
   c. recognition of the promoter by RNA polymerase.
   d. transcription of all genes.
   e. translation of specific mRNAs.
6. Complementary DNA (cDNA)
   a. is produced from ribonucleoside triphosphates.
   b. is produced by reverse transcription.
   c. is the "other strand" of single-stranded DNAs in a virus.
   d. requires no template for its synthesis.
   e. cannot be placed into a vector because it has the opposite base sequence of the vector DNA.
7. In a gene library of frog DNA in *E. coli* bacteria,
   a. all bacterial cells have the same sequences of frog DNA.
   b. all bacterial cells have different sequences of frog DNA.
   c. each bacterial cell has a random fragment of frog DNA.
   d. each bacterial cell has many fragments of frog DNA.
   e. the frog DNA is transcribed into mRNA in the bacterial cells.
8. An expression vector requires all of the following except
   a. genes for ribosomal RNA.
   b. a reporter gene.
   c. a promoter of transcription.
   d. an origin of DNA replication.
   e. restriction enzyme recognition sequences.
9. "Pharming" is a term that describes
   a. the use of animals in transgenic research.
   b. plants making genetically altered foods.
   c. synthesis of recombinant drugs by bacteria.
   d. large-scale production of cloned animals.
   e. synthesis of a drug by a transgenic animal in its milk.
10. In DNA fingerprinting,
   a. a positive identification can be made.
   b. a gel blot is all that is required.
   c. multiple restriction enzymes generate unique fragments.
   d. the polymerase chain reaction amplifies finger DNA.
   e. the variation in repeated sequences between two restriction sites is evaluated.

FOR DISCUSSION

1. In the recombinant DNA experiment in Figure 16.7, would you expect any *E. coli* to be doubly resistant (to both antibiotics) if the plasmids carrying antibiotic resistance genes were introduced into the cell at the same time but not recombined in the test tube before the experiment?
2. Compare PCR (see Section 11.5) and cloning as methods to amplify a gene. What are the requirements, benefits, and drawbacks of each method?
3. As specifically as you can, outline the steps you would take to (A) insert and express the gene for a new, nutritious seed protein in wheat, and (B) insert and express a gene for a human enzyme in sheep's milk.
4. Compare traditional genetic methods with molecular methods for producing genetically altered plants. For each case, describe (*a*) sources of new genes; (*b*) numbers of genes transferred; and (*c*) how long the process takes.

FOR INVESTIGATION

Green fluorescent protein (GFP) from a jellyfish can be incorporated into a vector as a reporter gene to signal the presence of the vector in a host cell (the cell glows under UV light). How would you alter the technique in Figure 16.10 to substitute GFP for one (or both) of the antibiotic resistance markers?

CHAPTER 17 Genome Sequencing, Molecular Biology, and Medicine

Genomes of the founders of Quebec

In 1535, the French explorer Jacques Cartier sailed up Canada's St. Lawrence River, arriving at a town called Stadacona by the native Iroquois already living there. When Samuel de Champlain arrived with 28 farmers 70 years later, the Algonquin then residing there called it "Quebecq," for "the place where the river narrows."

Over the next 150 years, 15,000 people came to join the first settlers. Half returned home because of the harsh conditions, and thousands more left for greener pastures in the south and west. About 2,600 stayed, had children, and formed the basis of the vibrant French-Canadian society of what is now Quebec Province of Canada. Cultural and religious traditions were very strong, and French Canadians tended to marry within their own society. Even today, about two-thirds of the current population of 6 million Quebecois is descended from that founding group of 2,600. For instance, the marriage of Pierre and Anne Tremblay in 1657 produced 12 children, and there are now 280,000 of their descendants living in Quebec!

French Canadians are a genomic gold mine. Because so many people in this population came from just a few ancestors, alleles that were present in the founding fathers and mothers are more common than in a population derived from a larger set of founders. In addition, because Quebec has a socialized health care system and a centralized database, getting medical records is straightforward.

Scientists are mapping the genomes of thousands of French Canadians, focusing on people whose four grandparents are all French Canadian. This "genome prospecting" involves looking for single nucleotide polymorphisms (SNPs). Scientists are trying to correlate the presence of these SNPs with diseases that may have been inherited from the founders. Such diseases would be indicated by a high frequency in the population along with the presence of a particular SNP. Linkage of genotype and phenotype may suggest genes involved in either the disease itself or susceptibility to it. So far, the researchers have evidence for several genes with some involvement in Crohn's disease, a serious inflammation of the intestines, as well as psoriasis, a skin disorder. Having identified a gene associated with a disease, the next steps include determining its function and seeking possible cures.

Until now, treatment of diseases has been—and largely continues to be—a "top-down" affair, with alle-

Samuel de Champlain In 1608, Champlain led a small group of French settlers to what is now Quebec.

Les Fêtes de la Nouvelle France This family has dressed in period costume for the annual celebration of the founding of Quebec. Scientists are mapping the genomes of many French Canadians.

CHAPTER OUTLINE

viating symptoms the primary goal. For some infectious diseases, knowledge of the microbe causing the disease and the availability of specific antibiotics has led to a cure. But for many complex diseases that still plague humans, such as cancer and heart disease, cures are elusive. For many diseases there can be more than one cause. Molecular medicine aims to identify causes of disease through better understanding of the interactions of genes, proteins, and the environment. The "bottom-up" approach taken in Quebec—identifying genes first, then causes—represents a new era in the understanding, prevention, and cure of disease.

IN THIS CHAPTER we first discuss the kinds of abnormal proteins that can result from an abnormal allele of a gene, whether that allele is inherited or has its origin in a mutation. We then consider how abnormal proteins can cause human genetic diseases, including cancer, and how the alleles that produce them can be detected. Then we'll see how this knowledge has been applied to the development of new treatments. The end of the chapter discusses the promise of molecular biology and medicine embodied in the sequencing of the human genome.

17.1 How Do Defective Proteins Lead to Diseases?

Biochemical genetics, the science that relates genotype (DNA) and phenotype (proteins), has been clearly described in viruses, prokaryotes, and eukaryotic model organisms, but it applies to humans, as well.

Genetic mutations may make proteins dysfunctional

Genetic mutations are often expressed phenotypically as proteins that differ from normal (wild-type) proteins. In principle, a mutation in any gene encoding a protein could result in a genetic disease. Abnormalities in enzymes, receptor proteins, transport proteins, structural proteins, and most of the other functional classes of proteins have all been implicated in genetic diseases.

DYSFUNCTIONAL ENZYMES In 1934, the urine of two mentally retarded young siblings was found to contain phenylpyruvic acid, an unusual by-product of the metabolism of the amino acid phenylalanine. It was not until two decades later, however, that the complex clinical phenotype of the disease that afflicted these children, called *phenylketonuria* (PKU), was traced back to its molecular phenotype. The disease resulted from an abnormality in a single enzyme, phenylalanine hydroxylase (**Figure 17.1**). This enzyme normally catalyzes the conversion of dietary phenylalanine to tyrosine, but it was not active in PKU patients' livers. Lack of this conversion led to excess phenylalanine in the blood and explained the accumulation of phenylpyruvic acid. Later, the amino acid sequences of phenylalanine hydroxylase in normal people were compared with those in individuals with PKU. Many people with PKU had tryptophan instead of arginine in position 408 of this long polypeptide chain of 451 amino acids.

The exact cause of the mental retardation in PKU remains elusive, although, as we will see later in this chapter, it can be prevented. We can, however, understand why most people with

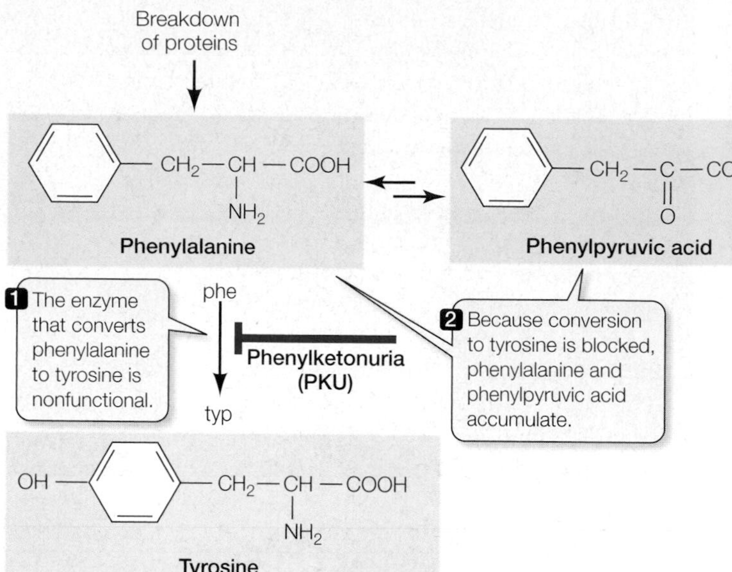

17.1 One Gene, One Enzyme Phenylketonuria is caused by an abnormality in a specific enzyme in the metabolic pathway that metabolizes the amino acid phenylalanine. Knowing the molecular causes of such single-gene, single-enzyme metabolic diseases can aid researchers in developing screening tests as well as treatments.

PKU have light skin and hair color. The pigment melanin, which is responsible for dark skin and hair, is made from tyrosine, which people with PKU cannot synthesize adequately.

Hundreds of human genetic diseases that result from enzyme abnormalities have been discovered, many of which lead to mental retardation and premature death. Most of these diseases are rare; PKU, for example, shows up in one newborn out of every 12,000. But these diseases are just the tip of the iceberg. Some mutations result in amino acid changes that have no obvious clinical effects. In fact, at least 30 percent of all proteins whose sequences are known show detectable amino acid differences among individuals. Thus polymorphism does not necessarily mean disease. There can be numerous normal alleles of a gene, each producing normally functioning forms of its protein.

ABNORMAL HEMOGLOBIN The first human genetic disease known to be caused by an amino acid abnormality was *sickle-cell disease.* This blood disorder most often afflicts people whose ancestors came from the tropics or from the Mediterranean. About 1 in 655 African-Americans are homozygous for the sickle allele and have the disease. The abnormal allele produces abnormal hemoglobin that results in sickle-shaped red blood cells (see Figure 12.18). These cells tend to block narrow blood capillaries, especially when the oxygen concentration of the blood is low. The result is tissue damage and eventually death by organ failure.

Recall that human hemoglobin is a protein with quaternary structure, containing four globin chains—two α chains and two β chains—as well as the pigment heme (see Figure 3.9). In sickle-cell disease, one of the 146 amino acids in the β-globin chain is abnormal: at position 6, the normal glutamic acid has been replaced by valine. This replacement changes the charge of the protein (glutamic acid is negatively charged and valine is neutral), causing it to form long, needle-like aggregates in the red blood cells. The result is *anemia,* a deficiency of normal red blood cells and an impaired ability of the blood to carry oxygen.

Because hemoglobin is easy to isolate and study, its variations in the human population have been extensively documented (**Figure 17.2**). Hundreds of single amino acid alterations in β-globin have been reported. For example, at the same position that is mutated in sickle-cell disease, the normal glutamic acid may be replaced by lysine, causing hemoglobin C disease. In this case, the resulting anemia is usually not severe. Many alterations of hemoglobin have no effect on the protein's function, and thus no clinical phenotype. That is fortunate, because about 5 percent of all humans are carriers for one of these variants.

ALTERED MEMBRANE PROTEINS Some of the most common human genetic diseases show their phenotypes as altered membrane receptors or transport proteins. About one person in 500 is born with *familial hypercholesterolemia* (FH), in which levels of cholesterol in the blood are several times higher than normal. The excess cholesterol can accumulate on the inner walls of blood vessels (a condition called *atherosclerosis*), leading to complete blockage if a blood clot forms. If a clot forms in a major vessel serving the heart, the heart becomes starved of oxygen, and a heart attack results. If a clot forms in the brain, the result is a stroke. People with FH often die of heart attacks before the age of 45.

Unlike PKU, which is characterized by the inability to convert phenylalanine to tyrosine, the problem in FH is not an inability to convert cholesterol to other products. People with FH have all the machinery needed to metabolize cholesterol. The problem is that they are unable to transport cholesterol into the liver and other cells that use it.

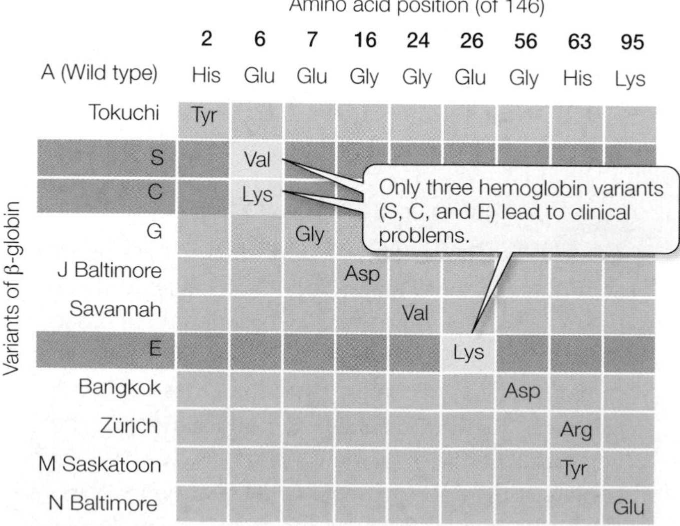

17.2 Hemoglobin Polymorphism Each of these mutant alleles changes a single amino acid in the 146–amino acid chain of β-globin. Only three of the hundreds of known variants of β-globin are known to lead to clinical abnormalities.

Cholesterol travels through the bloodstream in protein-containing particles called *lipoproteins*. One type of lipoprotein, low-density lipoprotein (LDL), carries cholesterol to the liver cells (**Figure 17.3A**). After binding to a specific receptor on the plasma membrane of a liver cell, LDL is taken up by endocytosis and delivers its cholesterol to the interior of the cell. People with FH lack a functional version of the receptor protein. Of the 840 amino acids that make up the receptor, only one may be abnormal, but that is enough to change its structure so that it cannot bind to LDL.

Among Caucasians, about one baby in 2,500 is born with *cystic fibrosis*. The clinical phenotype of this genetic disease is an unusually thick and dry mucus that lines surface tissues such as the airways of the respiratory system and the ducts of glands. In the respiratory passageways, this thick mucus obstructs the passage of air and also prevents the cilia on the surfaces of the epithelial cells from working efficiently to clear out the bacteria and fungal spores that we all take in with every breath. The results are recurrent infections as well as liver, pancreatic, and digestive failures, causing malnutrition and poor growth. People with cystic fibrosis often die in their thirties.

The cause of the thick mucus is a lack of a functional membrane protein, the chloride transporter (**Figure 17.3B**). In normal cells, this ion channel opens to release Cl^- to the outside of an epithelial cell. The resulting imbalance, in which there are more Cl^- ions outside the cell than inside, causes water to leave the cell by osmosis, resulting in a moist, thin mucus outside the cell. A single amino acid change in the channel protein prevents it from reaching the plasma membrane and functioning properly. As a consequence, water does not lubricate the mucus, and thick mucus and associated clinical problems are the result.

ALTERED STRUCTURAL PROTEINS In some genetic diseases, the defective protein is not an enzyme or receptor, but one involved in biological structure. *Duchenne muscular dystrophy* and *hemophilia* are cases in point.

About one boy in 3,000 is born with Duchenne muscular dystrophy. People with this disease show progressive muscle weakness and are wheelchair-bound by their teenage years. They usually die in their twenties, when the muscles that serve their respiratory system fail. Normal people have a protein in their skeletal muscles called *dystrophin*, which connects the actin filaments of the muscle cells to the extracellular matrix. People with Duchenne muscular dystrophy do not have a working copy of dystrophin, so their muscle cells become structurally disorganized, and the muscles stop working.

Hemophilia is the consequence of the absence of one of the blood clotting proteins. In normal people, inactive blood clotting proteins are always present in the blood and become active only at a wound. Some people with hemophilia risk death from even minor cuts, since they cannot stop bleeding.

We have thus far described disease-causing alterations in wild-type genes. Some abnormal proteins that result in disease, however, arise from other causes.

(A) Hypercholesterolemia

Normal liver cell: Cholesterol, as part of low-density lipoprotein (LDL), enters the cell after LDL binds to a receptor.

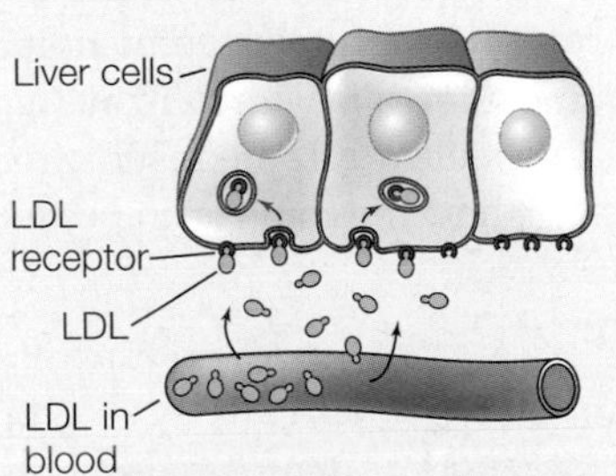

Familial hypercholesterolemia: Absence of a functional LDL receptor prevents cholesterol from entering the cells, and it accumulates in the blood.

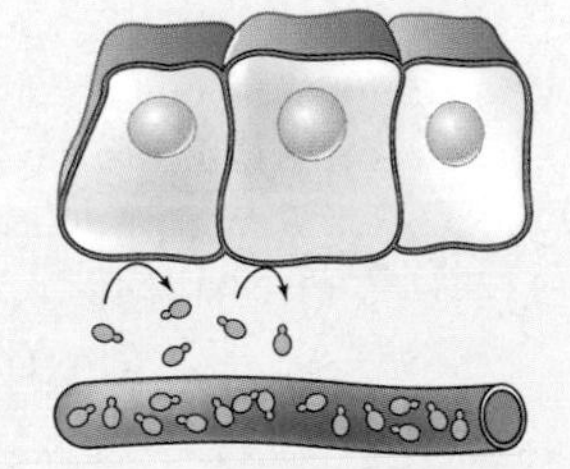

(B) Cystic fibrosis

Normal cell lining the airway: Cl^- leaves the cell through an ion channel. Water follows by osmosis. The resulting moist thin mucus allows cilia to beat and sweep away foreign particles, including bacteria.

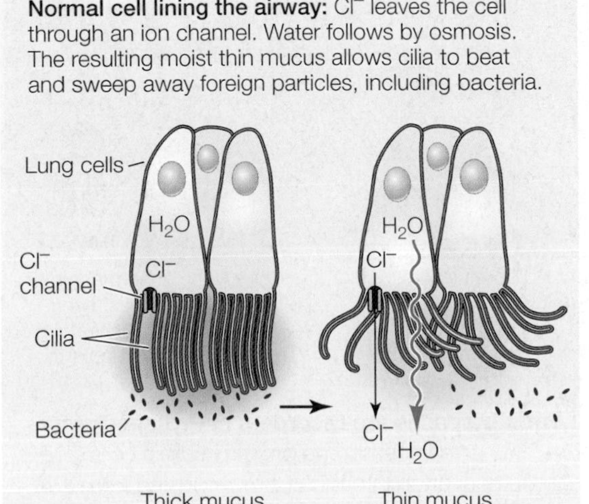

Cystic fibrosis: Lack of a Cl^- channel causes a thick, viscous mucus to form. Cilia cannot beat properly and remove bacteria; infections can easily take hold.

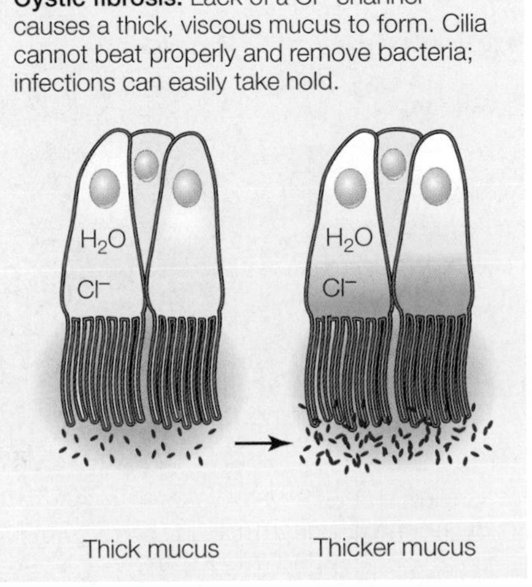

17.3 Genetic Diseases of Membrane Proteins The two left panels illustrate normal cell function, while the two right panels show the abnormalities caused by (A) familial hypercholesterolemia and (B) cystic fibrosis.

King George III of England suffered from porphyria, a metabolic disorder resulting in nonfunctioning enzymes in the porphyrin pathway that synthesizes heme, the pigment in hemoglobin. Its symptoms—intermittent bouts of high fever, pain, and delirium—may have led to George III's poor administrative decisions, which resulted in the American Revolution and England's loss of its North American colonies.

Prion diseases are disorders of protein conformation

Transmissible spongiform encephalopathies (TSEs) are degenerative brain diseases that occur in many mammals, including humans, in which the brain gradually develops holes, leaving it looking like a sponge. Scrapie, a TSE that causes affected sheep and goats to rub the wool off their bodies, has been known for 250 years. Chronic wasting disease is a TSE that causes elk and deer to become emaciated. In the 1980s, a TSE that appeared in cows in Britain was traced to the cows having eaten products from sheep that had scrapie. These cows would shake and rub their bodies against fences, and their staggering led farmers to dub them "mad cows."

It turns out that TSEs are caused by defective proteins, but the defects arise *not from mutations* in the genes that express them, but from errors in *conformation*—the folding of proteins into the proper three-dimensional shape. This was discovered in the 1990s after some people who had eaten beef from cows with bovine spongiform encephalopathy (BSE) got a human version of the disease (dubbed "mad cow disease" by the media). There was a time delay of several years between when the cows got the disease and its emergence in humans; thus there is clearly a period of several years between the consumption of meat and the development of symptoms (**Figure 17.4**).

The disease *kuru* offers another example of how humans can acquire a TSE by consuming an infective agent. Kuru is a TSE that was observed in the 1950s among members of the Fore tribe of New Guinea who had consumed the brains of people who had died of it. When this ritual cannibalism stopped, so did the epidemic of kuru.

Researchers found that TSEs can be transmitted from one animal species to another via brain extracts from a diseased animal. At first, a virus was suspected. But when Tikva Alper at Hammersmith Hospital, London, treated infectious extracts with high doses of ultraviolet light to inactivate nucleic acids, they still caused TSEs. She proposed that the causative agent for TSEs was a protein, not a virus. Later, Stanley Prusiner at the University of California purified the protein responsible and showed it to be free of DNA or RNA. He called it a *proteinaceous infective particle*, or **prion**.

Normal brain cells contain a membrane protein called PrP^c. A protein with the same amino acid sequence is present in TSE-affected brain tissues, but that protein, called PrP^{sc}, has a different three-dimensional shape (**Figure 17.5**). Thus TSEs are not caused by a mutated gene (the primary structures of the two proteins are the same), but are somehow caused by an alteration in protein conformation. The altered three-dimensional structure of the protein has profound effects on its function in the cell. PrP^{sc} is insoluble, and it piles up as fibers in brain tissue, causing cell death.

How can the exposure of a normal cell to material containing PrP^{sc} result in a TSE? The abnormal PrP^{sc} protein seems to induce a conformational change in the normal PrP^c protein so that it, too, becomes abnormal, just as one rotten apple results in a whole barrel full of rotten apples. Just how the conversion occurs, and how it causes a TSE, are unclear.

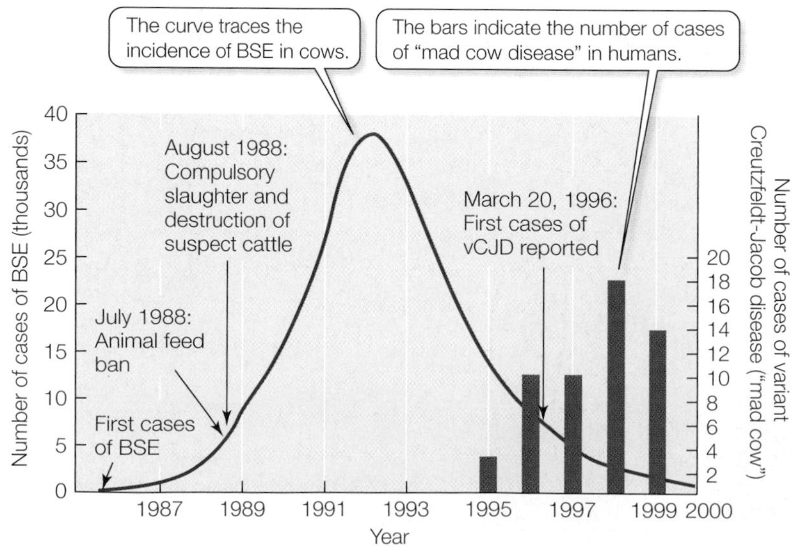

17.4 Mad Cow Disease in Britain There was a lag time of several years between the development of bovine spongiform encephalopathy (BSE) in cows and its equivalent in humans (vCJD).

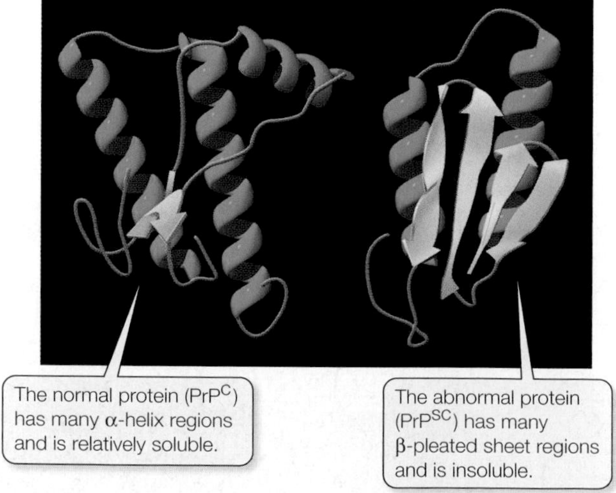

17.5 Prion Diseases are Disorders of Protein Conformation A normal membrane protein in brain cells (PrP^C, left) can be converted into the disease-causing form (PrP^{SC}, right), which has a different three-dimensional structure.

Prions appear to represent a highly unusual phenomenon in human disease. The vast majority of inherited diseases are understood in terms of proteins that are products of functional or dysfunctional genes. But the expression of these genes, like that of all genes, is influenced by the environment.

Most diseases are caused by both genes and environment

The human diseases for which clinical phenotypes can be traced to a single altered protein and its altered gene may number in the thousands, and in most cases they are dramatic evidence of a one-gene, one-polypeptide relationship. Taken together, these diseases have a frequency of about 1 percent in the total human population.

Far more common, however, are diseases that are **multifactorial**; that is, diseases that are caused by the interactions of many genes and proteins and the environment. Although we tend to call individuals either normal (wild-type) or abnormal (mutant), the totality of our genetic makeup is what determines, for example, who among us can eat a high-fat diet and not experience a heart attack and which of us will succumb to disease when exposed to infectious bacteria. Estimates suggest that up to 60 percent of all people are affected by diseases that are genetically influenced.

Human genetic diseases have several patterns of inheritance

As in any human genetic system, the alleles that cause genetic diseases may be inherited in a dominant or recessive pattern, and may be carried on autosomes or on sex chromosomes (see Section 10.4). In addition, some human diseases are caused by more extensive chromosomal abnormalities (see Section 9.5). Different inheritance patterns can be seen when genetic diseases are followed over several human generations.

AUTOSOMAL RECESSIVE PATTERN PKU, sickle-cell disease, and cystic fibrosis are all caused by autosomal recessive mutant alleles. Typically, both parents of an affected person are carriers (with a normal phenotype and a heterozygous genotype). Each time they conceive a child, they have a 25 percent (one in four) chance of having an affected (homozygous) son or daughter.

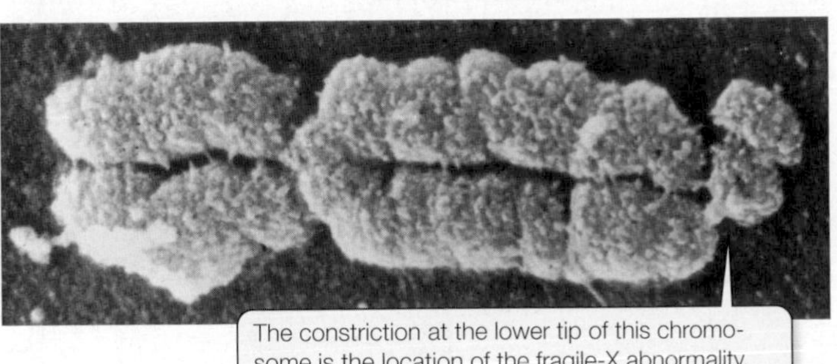

The constriction at the lower tip of this chromosome is the location of the fragile-X abnormality.

In the cells of a person who is homozygous for a harmful autosomal recessive mutant allele, a nonfunctional, mutant version of the protein it encodes is made. Thus a biochemical pathway or important cell function is disrupted, and disease results. Heterozygotes, with one normal and one mutant allele, often have 50 percent of the normal level of functional protein. For example, people who are heterozygous for the PKU allele have half as many active molecules of phenylalanine hydroxylase in their liver cells as individuals who carry two normal alleles for this enzyme, but this 50 percent suffices for normal cellular function.

AUTOSOMAL DOMINANT PATTERN Familial hypercholesterolemia is caused by an abnormal autosomal dominant allele. In this case, the presence of only one mutant allele is enough to produce the clinical phenotype. In people who are heterozygous for familial hypercholesterolemia, having half the normal number of functional receptors for low-density lipoprotein on the surfaces of liver cells is simply not enough to clear cholesterol from the blood. In autosomal dominance, direct transmission from an affected parent to offspring is the rule.

X-LINKED RECESSIVE PATTERN Hemophilia is an X-linked recessive condition (see pp. 206–207); that is, the gene responsible is on the X chromosome. Thus a son who inherits a mutant allele on the X chromosome from his mother will have the disease, because his Y chromosome does not contain a normal allele. However, a daughter who inherits one mutant allele will be an unaffected heterozygous carrier, since she has two X chromosomes, and hence two alleles. Because, until recently, few males with these diseases lived to reproduce, the most common pattern of inheritance has been from carrier mother to son, and all rare X-linked diseases are much more common in males than in females.

CHROMOSOMAL ABNORMALITIES Chromosomal abnormalities also cause human diseases. Such abnormalities include a gain or loss of one or more chromosomes (aneuploidy; see Figure 9.20), loss of a piece of a chromosome (deletions), and the transfer of a piece of one chromosome to another chromosome (translocations). About one newborn in 200 is born with a chromosomal abnormality. While some of these abnormalities are inherited, many are the result of meiotic events such as nondisjunction.

One common cause of mental retardation is *fragile-X syndrome* (**Figure 17.6**). About one male in 1,500 and one female in 2,000 is affected. These people have a constriction near the tip of the X chromosome that tends to break during preparation for microscopy, giving the syndrome its name. Although the basic pattern of inheritance is that of an X-linked recessive trait, there are departures from this pattern. Not all people with the fragile-X chromosomal abnormality are mentally retarded, as we will see in Section 17.2.

17.6 A Fragile-X Chromosome at Metaphase The chromosomal abnormality associated with fragile-X syndrome shows up under the microscope as a constriction in the chromosome.

17.1 RECAP

Many genetic mutations are expressed as nonfunctional enzymes, structural proteins, or membrane proteins. Human genetic diseases may be inherited in dominant, recessive, or X-linked patterns.

- Can you describe an example of an abnormal protein in humans that results from a genetic mutation ? See pp. 376–377
- How is the brain cell membrane protein PrP^C related to diseases caused by prions? See p. 378
- How are autosomal recessive and autosomal dominant human genetic diseases distinguished? See p. 379

What kinds of changes in DNA result in nonfunctional proteins? An important task of molecular medicine is to find and identify such genetic changes.

17.2 What Kinds of DNA Changes Lead to Diseases?

The isolation and description of human mutations has proceeded rapidly since the modern techniques described in Chapter 16 were developed. When the protein phenotype is known, as in the case of abnormal hemoglobins, cloning the gene responsible has been straightforward, although time-consuming. In other cases, such as Duchenne muscular dystrophy, a chromosome deletion associated with the disease in a patient has pointed the way to the missing gene. In still other cases, such as cystic fibrosis, only a subtle molecular marker was available to lead investigators to the defective gene. In both of the latter examples, the primary phenotype—the defective protein—was unknown; only when the gene was isolated was the protein found.

One way to identify a gene is to start with its protein

The primary phenotype for sickle-cell disease was described in the 1950s as a single amino acid change in β-globin. On the basis of the clinical picture of sickled red blood cells, β-globin was certainly the right protein to examine. By the 1970s, researchers were able to isolate β-globin mRNA from immature red blood cells, which transcribe the globins as their major gene product. A cDNA copy of this mRNA was made and used to probe a human gene library to find the β-globin gene (**Figure 17.7A**). DNA sequencing was

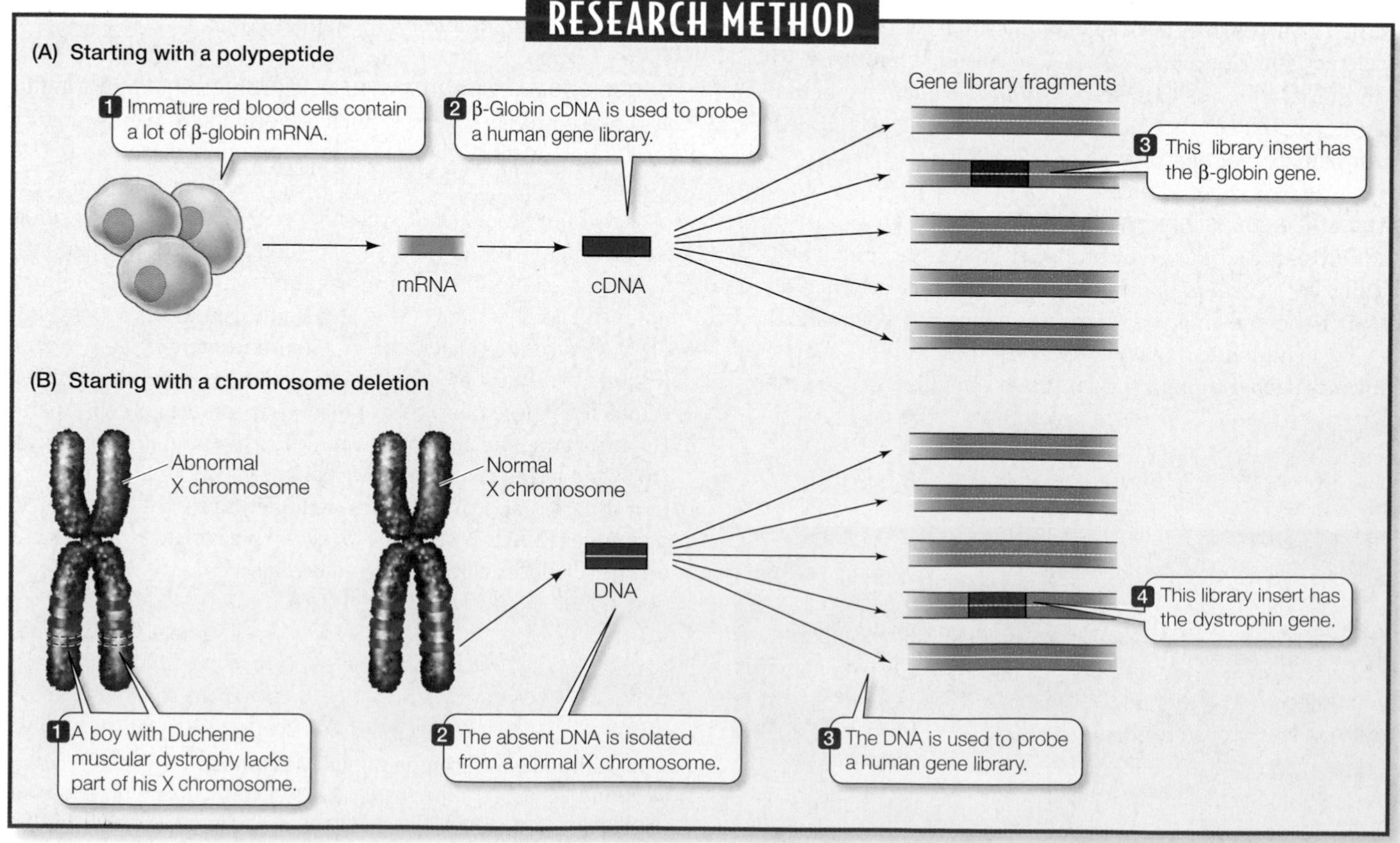
RESEARCH METHOD
(A) Starting with a polypeptide
1 Immature red blood cells contain a lot of β-globin mRNA.
2 β-Globin cDNA is used to probe a human gene library.
Gene library fragments
3 This library insert has the β-globin gene.
mRNA
cDNA
(B) Starting with a chromosome deletion
Abnormal X chromosome
Normal X chromosome
DNA
4 This library insert has the dystrophin gene.
1 A boy with Duchenne muscular dystrophy lacks part of his X chromosome.
2 The absent DNA is isolated from a normal X chromosome.
3 The DNA is used to probe a human gene library.

17.7 Two Strategies for Isolating Human Genes (A) Starting with the normal β-globin protein, researchers were able to isolate the mRNA from which it was translated. Once cDNA was made by reverse transcription from the isolated mRNA, it could be used to probe a gene library to isolate the gene. (B) Researchers found a chromosome deletion in some patients affected with Duchenne muscular dystrophy. They then compared the affected persons' chromosomes with normal chromosomes to locate the DNA sequence that was missing.

then used to compare the normal gene with the gene from people with sickle-cell disease. As previously described, it was found that a point mutation had changed only one base pair in the entire β-globin gene.

Chromosome deletions can lead to gene and then protein isolation

The inheritance pattern of Duchenne muscular dystrophy is consistent with an X-linked recessive trait. But until the late 1980s, neither the abnormal protein involved nor the gene encoding it had been described. This failure was not from lack of effort: almost every known muscle protein had been tested without success. Then several boys with the disease were found to have a small deletion in their X chromosome. Comparison of the affected X chromosomes with normal ones made possible the isolation of the gene that was missing in the boys (**Figure 17.7B**).

Genetic markers can point the way to important genes

In cases in which no candidate protein nor chromosome deletion has been identified, a technique called **positional cloning** has been invaluable. To understand this method, imagine an astronaut looking down from space, trying to find her son on a park bench on Chicago's North Shore. The astronaut first picks out reference points—landmarks that will lead her to the park. She recognizes the shape of North America, then moves to Lake Michigan, the Sears Tower, and so on. Once she has zeroed in on the North Shore park, she can use advanced optical instruments to find her son.

The reference points for positional cloning are genetic markers. These markers can be located anywhere in the DNA; the only requirement is that they be polymorphic (have more than one allele).

RFLPs As Section 16.1 describes, restriction enzymes cut DNA molecules at specific recognition sequences. On a particular human chromosome, a given restriction enzyme may make hundreds of cuts, producing many DNA fragments. The enzyme *Eco*RI, for example, cuts DNA at

5′... GAATTC ... 3′

Suppose this recognition sequence exists in a certain stretch of human chromosome 7. The restriction enzyme will cut this stretch once and make two fragments of DNA. Now suppose that, in some people, this sequence is mutated as follows:

5′... GAGTTC ... 3′

This sequence will not be recognized by the restriction enzyme; thus it will remain intact and yield one larger fragment of DNA.

Differences in the DNA sequence of designated segments of homologous chromosomes are called **restriction fragment length polymorphisms**, or **RFLPs** (**Figure 17.8**). They can be easily seen as bands on an electrophoresis gel. A RFLP band pattern is inherited in a Mendelian fashion and can be followed through a pedigree. Thousands of such markers have been described for the human genome.

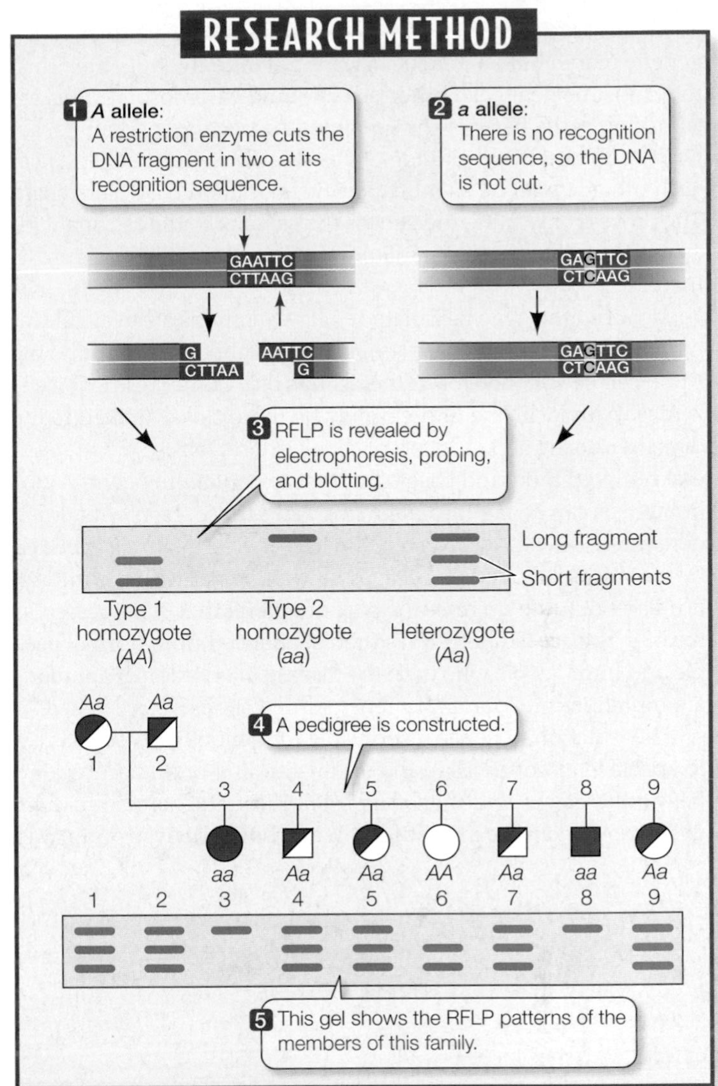

17.8 RFLP Mapping Restriction fragment length polymorphisms are differences in DNA sequences that serve as genetic markers. Thousands of such markers have been described for the human genome.

SNPs As noted in Section 16.1, single nucleotide polymorphisms (SNPs) are widespread in the eukaryotic genome. SNP maps indicate that there is roughly one SNP for every 1,330 base pairs in the human genome. SNPs can be detected by direct sequence comparisons or by special chemical methods such as mass spectrometry (see Section 17.6). They differ from RFLPs in that they do not necessarily alter the recognition sequence for a restriction enzyme.

Genetic markers such as RFLPs and SNPs can be used as landmarks to find genes of interest if the genes, too, are polymorphic. The key to this method is the well-known observation that if two genes are located near each other on the same chromosome, they are usually passed on together from parent to offspring. The same holds true for any pair of genetic markers.

To narrow down the location of a gene, a scientist must find a marker and a gene that are *always inherited together*. To do this, family medical histories are taken and pedigrees are constructed. If a genetic marker and a genetic disease are inherited together, then they must be near each other on the same chromosome. This assumption is the basis of the genome prospecting in Quebec that we described at the beginning of this chapter. Unfortunately, "near each other" might be as much as several million base pairs apart. The process of locating the gene is thus similar to the astronaut focusing on Chicago: the first landmarks lead to only an approximate location.

How can the gene be isolated? Many methods are available for narrowing the search. For example, the neighborhood around the RFLP can be screened for further RFLPs using other restriction enzymes. With luck, one of them may be more closely linked to the disease-causing gene. Once a relatively short DNA sequence (several hundred thousand bases) thought to contain the gene is pinpointed, it can be cut into fragments, and those fragments, when denatured, can be used to probe mRNA expressed in cells affected by the disease. If one of the fragments hybridizes with the mRNA, it means that the fragment is part of a gene that is expressed as mRNA. The candidate gene is then sequenced from normal people and from people who have the disease in question. If appropriate mutations are found, the gene of interest has been isolated.

The isolation of genes responsible for genetic diseases has led to spectacular advances in the understanding of human biology. Before the genes, and then the proteins, for Duchenne muscular dystrophy and for cystic fibrosis were isolated, dystrophin and the chloride transporter had never been described. Thus the identification of mutant genes has opened up new vistas in our understanding of how the human body works.

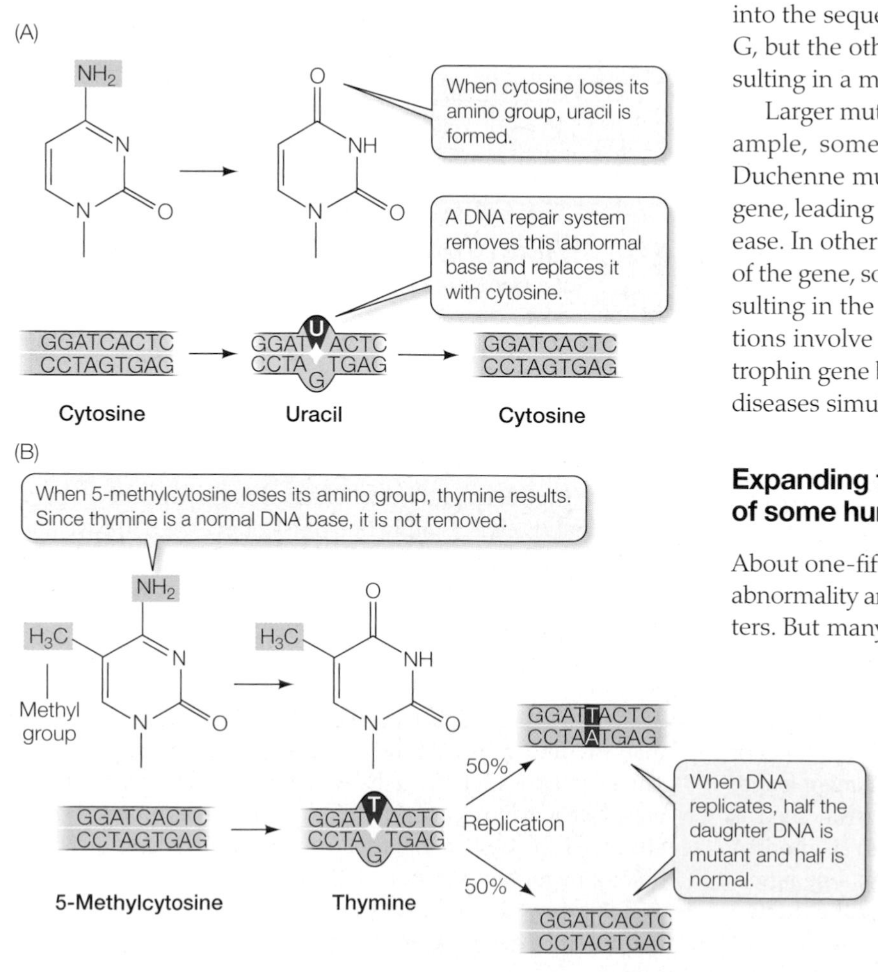

Disease-causing mutations may involve any number of base pairs

Disease-causing mutations may involve a single base pair (as we saw in the case of hemophilia), a long stretch of DNA, multiple segments of DNA, or even entire chromosomes. Some mutations have no effect at all, as we saw in Chapter 12, while others have many deleterious effects.

DNA sequencing has revealed that mutations occur most often at certain base pairs. These "hot spots" are often located where cytosine has been methylated to 5-methylcytosine (see Section 14.4). Either spontaneously or with chemical prodding, *un*methylated cytosine can lose its amino group and form uracil (**Figure 17.9A**). This type of error, however, is usually detected by the cell and repaired. A DNA repair mechanism recognizes this uracil as inappropriate for DNA (after all, uracil occurs only in RNA) and replaces the uracil with cytosine.

When 5-methylcytosine loses its amino group, however, the product is thymine, a natural base for DNA. The repair mechanism that recognized the inappropriateness of uracil ignores this thymine (**Figure 17.9B**). The mismatch repair mechanism, however, recognizes that GT is a mismatched pair (which should be GC). It cannot tell, though, which base was incorrectly inserted into the sequence. So, half of the time, it matches a new C to the G, but the other half of the time, it matches a new A to the T, resulting in a mutation.

Larger mutations may involve many base pairs of DNA. For example, some deletions in the X chromosome that result in Duchenne muscular dystrophy cover only part of the dystrophin gene, leading to an incomplete protein and a mild form of the disease. In other cases, however, deletions span the entire sequence of the gene, so that the protein is missing entirely from muscle, resulting in the severe form of the disease. In yet other cases, deletions involve millions of base pairs and cover not only the dystrophin gene but adjacent genes as well; the result may be several diseases simultaneously.

Expanding triplet repeats demonstrate the fragility of some human genes

About one-fifth of all males that have the fragile-X chromosomal abnormality are phenotypically normal, as are most of their daughters. But many of those daughters' sons are mentally retarded. In

(A)

When cytosine loses its amino group, uracil is formed.

A DNA repair system removes this abnormal base and replaces it with cytosine.

Cytosine

Uracil

Cytosine

(B)

When 5-methylcytosine loses its amino group, thymine results. Since thymine is a normal DNA base, it is not removed.

Methyl group

Replication

When DNA replicates, half the daughter DNA is mutant and half is normal.

5-Methylcytosine

Thymine

17.9 5-Methylcytosine in DNA Is a "Hot Spot" for Mutations (A) Cytosine can lose an amino group either spontaneously or because of exposure to certain chemical mutagens. Such mutations are usually repaired. (B) If the cytosine has been methylated to 5-methylcytosine, however, the mutation is unlikely to be repaired and a C-G base pair is replaced with a T-A pair.

a family in which the fragile-X syndrome appears, later generations tend to show earlier onset and more severe symptoms of the disease. It is almost as if the abnormal allele itself is changing—and getting worse. And that's exactly what is happening.

The gene responsible for fragile-X syndrome (*FMR1*) contains a repeated triplet, CGG, at a certain point in the promoter region. In normal people, this triplet is repeated 6 to 54 times (the average is 29). In mentally retarded people with fragile-X syndrome, the CGG sequence is repeated 200 to 2,000 times.

Males carrying a moderate number of repeats (55–200) show no symptoms and are called *premutated*. These repeats become more numerous as the daughters of these men pass the chromosome on to their children (**Figure 17.10**). With more than 200 repeats, increased methylation of the cytosines in the CGG triplets is likely, accompanied by transcriptional inactivation of the *FMR1* gene. The normal role of the protein product of this gene is to bind to mRNAs involved in neuron function and regulate their translation at the ribosome. When the FMR1 protein is not made in adequate amounts, these mRNAs are not properly translated, and nerve cells die. Their loss often results in mental retardation.

This phenomenon of **expanding triplet repeats** has been found in over a dozen other diseases, such as myotonic dystrophy (involving repeated CTG triplets) and Huntington's disease (in which CAG is repeated). Such repeats, which may be found within a protein-coding region or outside it, appear to be present in many other genes without causing harm. How the repeats expand is not known; one theory is that DNA polymerase may slip after copying a repeat and then fall back to copy it again.

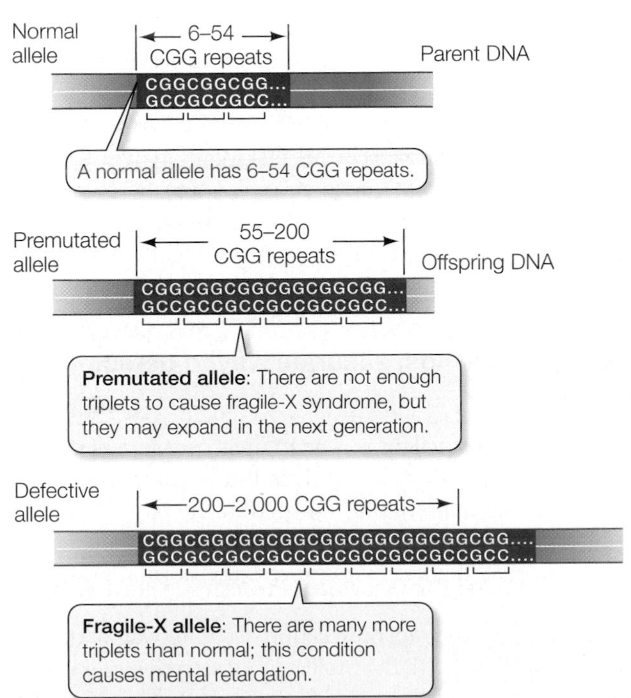

17.10 The CGG Repeats in the *FMR1* Gene Expand with Each Generation The genetic defect in fragile-X syndrome is caused by 200 or more repeats of the CGG triplet.

DNA changes in males and females can have different consequences

Just after fertilization in a mammalian egg, before the nuclei from the egg and from the sperm have fused, there are two haploid *pronuclei*—one from the sperm and the other from the egg. The two pronuclei can be distinguished from each other, and they can be carefully removed with a micropipette and placed in other eggs. It is possible to make mouse zygotes in the laboratory with two male or two female pronuclei, but these diploid cells do not go on to develop into mice. Invariably, if the two sets of chromosomes come from only one sex, development begins, but is quickly terminated. The same thing happens in those rare instances when this occurs in humans—for instance, if two sperm enter an enucleated egg. Again, a fetus never develops.

In addition to showing the obvious need for two sexes, these observations raise the possibility that the male and female genomes are *not functionally equivalent*. In fact, there are groups of genes that differ in their phenotypic effects depending on which parent they came from. This phenomenon is called **genomic imprinting**.

A dramatic example of genomic imprinting is the inheritance and phenotype of a certain small deletion on human chromosome 15:

- A deletion on the mother's chromosome 15 results in a thin child with a wide mouth and prominent jaw (*Angelman syndrome*).
- The same deletion on the father's chromosome 15 results in a short and obese child with small hands and feet (*Prader-Willi syndrome*).

The other alleles for this region on chromosome 15 are not deleted in these people; that is, they are heterozygotes, with one normal and one deleted allele. But the "normal" alleles must differ in males and females in these cases. They must be imprinted in very different ways in the two sexes to result in such different phenotypes. How this happens is not clear.

17.2 RECAP

Genes involved in disease have been identified by first detecting the abnormal DNA sequence and then the protein that the wild-type allele encodes. Linkage between genetic markers (such as RFLPs and SNPs) and a gene of interest is also useful in isolating human genes. Unusual features such as expanding triplet repeats have been detected in the human genome.

- How have gene libraries been instrumental in identifying abnormal gene sequences associated with diseases? See p. 380 and Figure 17.7
- How are genetic markers such as RFLPs and SNPs used to narrow the search for genes of interest, and why must these genes be polymorphic? See p. 381 and Figure 17.8
- Why do many genetic mutations involve GC base pairing? See p. 382

The determination of the precise molecular phenotypes and genotypes of various human genetic diseases has made it possible to diagnose these diseases even before symptoms first appear. Let's take a detailed look at some of these genetic screening techniques.

17.3 How Does Genetic Screening Detect Diseases?

Genetic screening is the use of a test to identify people who have, are predisposed to, or are carriers of a genetic disease. It can be done at many times of life and used for many purposes.

- *Prenatal screening* can identify an embryo or fetus with a disease so that medical intervention can be applied or decisions about continuing the pregnancy can be made.
- *Newborn babies* can be screened so that proper medical intervention can be initiated quickly for those babies who need it.
- *Asymptomatic people who have a relative with a genetic disease* can be screened to determine whether they are carriers of the disease or are likely to develop the disease themselves.

Genetic screening can be done at the level of either the phenotype or the genotype.

Screening for disease phenotypes can make use of protein expression

At the level of the phenotype, genetic screening involves examining a protein for abnormal structure or function. Since many proteins are enzymes, low enzyme activity is strongly suggestive of a mutation, as we saw in Section 17.1. Perhaps the best example of this kind of protein screening is a test for phenylketonuria that has made it possible to identify the disease in newborns so that treatments that prevent the development of mental retardation can be started.

Babies born with phenylketonuria have a normal phenotype because excess phenylalanine in their blood before birth diffuses across the placenta to the mother's circulatory system. Since the mother is almost always heterozygous, and therefore has adequate phenylalanine hydroxylase activity, her body metabolizes the excess phenylalanine from the fetus. After birth, however, the baby begins to consume protein-rich food (milk) and to break down some of its own proteins. Phenylalanine enters the baby's blood and accumulates. After a few days, the phenylalanine level in its blood may be ten times higher than normal. Within days, the developing brain is damaged, and untreated children with PKU become severely mentally retarded. If detected early, PKU can be treated with a special diet low in phenylalanine to avoid the brain damage that would otherwise result. Thus early detection is imperative.

A simple and inexpensive screening test for PKU was devised in 1963 by Robert Guthrie (**Figure 17.11**). This elegant method uses auxotrophic bacteria that require phenylalanine to grow well. If blood samples from newborns are added to a plate containing these bacteria, bacterial growth will be observed around those samples that are high in phenylalanine. The test can be automated so that a screening laboratory can process many samples in a day. This test is gradually being replaced by more accurate chemical tests that measure phenylalanine directly.

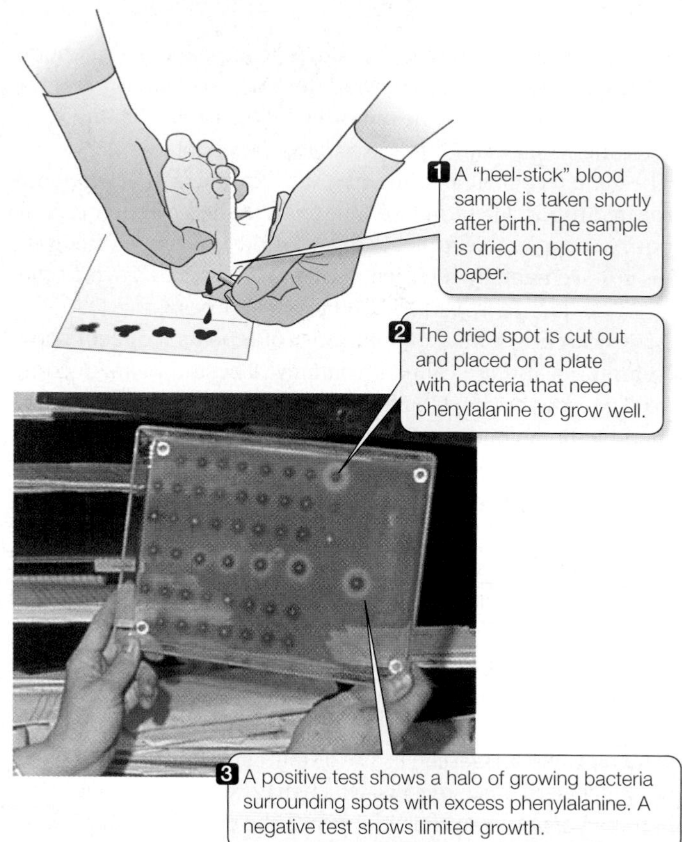

1 A "heel-stick" blood sample is taken shortly after birth. The sample is dried on blotting paper.
2 The dried spot is cut out and placed on a plate with bacteria that need phenylalanine to grow well.
3 A positive test shows a halo of growing bacteria surrounding spots with excess phenylalanine. A negative test shows limited growth.

17.11 Genetic Screening of Newborns for Phenylketonuria A simple test devised by Robert Guthrie in 1963 is used to screen newborns for phenylketonuria. Early detection means that the symptoms of the condition can be prevented by putting the baby on a therapeutic diet.

Genetic screening using newborn babies' blood is now done for up to 25 diseases, some rare (occurring in 1 infant in over 100,000) and others more common (occurring in 1 in 3,500). With early intervention, many of these infants can be successfully treated. So it is not surprising that newborn screening is legally mandatory in many countries, including the United States and Canada.

DNA testing is the most accurate way to detect abnormal genes

The blood level of phenylalanine is an indirect measure of phenylalanine hydroxylase activity in the liver. But how can we screen for genetic diseases that are not reflected in the blood? What if blood is difficult to obtain, as it is in a fetus? How are genetic abnormalities in heterozygotes, who express the normal protein at some level, identified?

DNA testing offers the most direct and accurate way of detecting an abnormal allele. With the molecular description of so many of the genetic mutations responsible for human diseases, it has become possible to examine directly any cell in the body at any time during the life span for mutations. These methods work best for diseases caused by only one or a few different mutations. However, with the amplification power of PCR, only one or a few cells are needed for testing.

Consider, for example, two parents who are both heterozygous for the cystic fibrosis allele, have had a child with the disease, and want a normal child. If treated with the appropriate hormones, the mother can be induced to "superovulate," releasing several eggs. One of the eggs can be injected with a single sperm from her husband and the resulting zygote allowed to divide to the 8-cell stage. If one of these embryonic cells is removed, it can be tested for the presence of the cystic fibrosis allele(s). If the test is negative, the remaining 7-cell embryo can be implanted in the mother's womb and go on to develop normally.

Such *preimplantation screening* is performed only rarely. More typical are analyses of fetal cells after implantation in the womb. Fetal cells can be analyzed at about the tenth week of pregnancy by *chorionic villus sampling* or during the thirteenth to seventeenth weeks by *amniocentesis*, two sampling methods described in Section 43.5. In either case, only a few fetal cells are required.

Newborns can also be screened for genetic mutations. The blood samples used for screening for PKU and other disorders contain enough of the baby's blood cells to permit extraction of the DNA, its amplification by PCR, and testing. Pilot studies of screening methods for sickle-cell disease and cystic fibrosis are under way, and other diseases will surely follow.

DNA testing is also widely used to test adults for heterozygosity. For example, a sister or female cousin of a boy with Duchenne muscular dystrophy can determine whether she is a carrier of the X chromosome deletion that results in the disease.

Of the numerous methods of DNA testing, two are the most widespread. We will describe their use to detect the mutation in the β-globin gene that results in sickle-cell disease.

SCREENING FOR ALLELE-SPECIFIC CLEAVAGE DIFFERENCES There is a difference between the normal and sickle alleles of the β-globin gene with respect to a restriction enzyme recognition sequence. Around codon position 6 in the normal gene is the sequence

5′... CCTGAGGAG... 3′

This sequence is recognized by the restriction enzyme *Mst*II, which will cleave DNA at

5′... CCTNAGGAG... 3′

where *N* is any base.

In the sickle allele, the DNA sequence is

5′... CCTGTGGAG... 3′

The point mutation at codon position 6 makes this sequence unrecognizable by *Mst*II. When *Mst*II fails to make the cut in the mutant allele, gel electrophoresis detects a larger DNA fragment (**Figure 17.12**).

This *allele-specific cleavage* method of DNA testing is the same as the use of RFLPs in positional cloning (see Figure 17.8). It works only if a restriction enzyme exists that can recognize the sequence of either the normal or the mutant allele.

SCREENING BY ALLELE-SPECIFIC OLIGONUCLEOTIDE HYBRIDIZATION The *allele-specific oligonucleotide hybridization* method uses short artificial DNA strands called *oligonucleotides* that will hybridize either (in this example) with the denatured normal β-globin DNA sequence or with the sickle mutant sequence. Usually, an oligonucleotide probe of at least a dozen bases is needed to form a stable double helix with the target DNA. If the probe is radioactively or fluorescently labeled, hybridization can be readily detected (**Figure 17.13**). This method is easier and faster than allele-specific cleavage, and will work no matter what the sequence of the normal or mutant allele is.

17.12 DNA Testing by Allele-Specific Cleavage Allele-specific cleavage, a technique identical to RFLP analysis, can be used to detect mutations such as the one that causes sickle-cell disease.

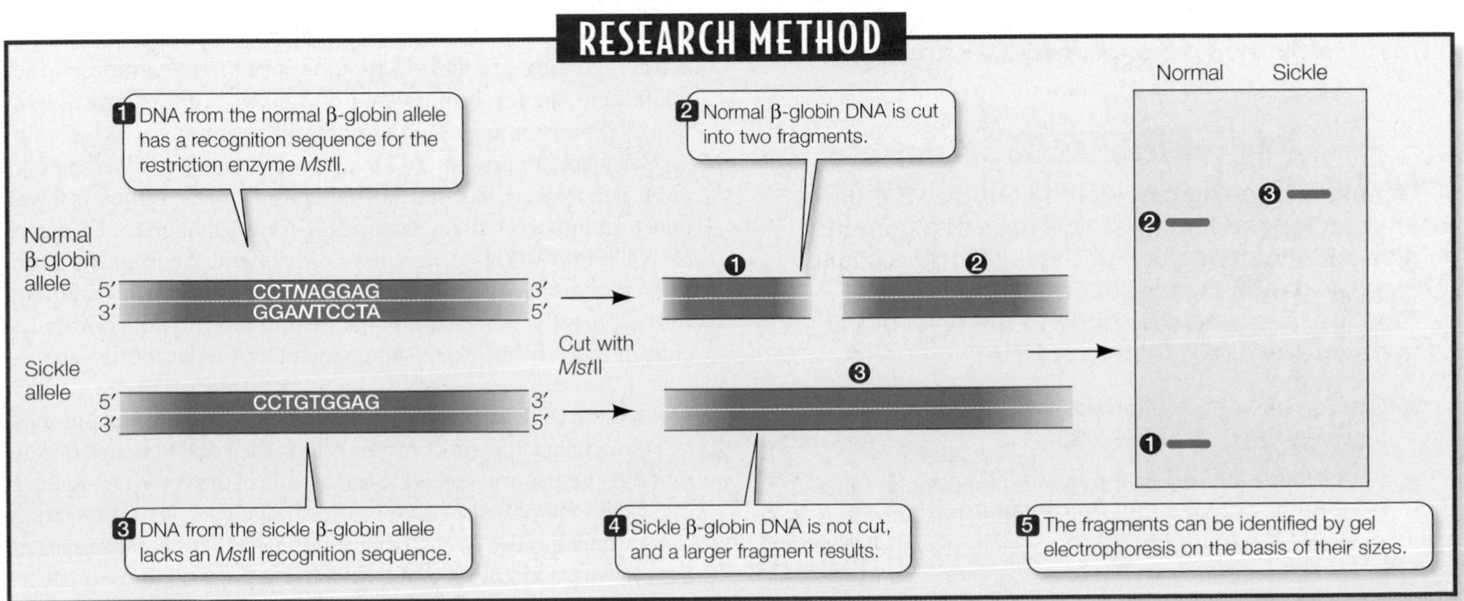

RESEARCH METHOD
1 DNA from the normal β-globin allele has a recognition sequence for the restriction enzyme MstII.
2 Normal β-globin DNA is cut into two fragments.
Normal β-globin allele
5′ CCTNAGGAG 3′
3′ GGANTCCTA 5′
Cut with MstII
Sickle allele
5′ CCTGTGGAG 3′
3′ 5′
Normal
Sickle
3 DNA from the sickle β-globin allele lacks an MstII recognition sequence.
4 Sickle β-globin DNA is not cut, and a larger fragment results.
5 The fragments can be identified by gel electrophoresis on the basis of their sizes.

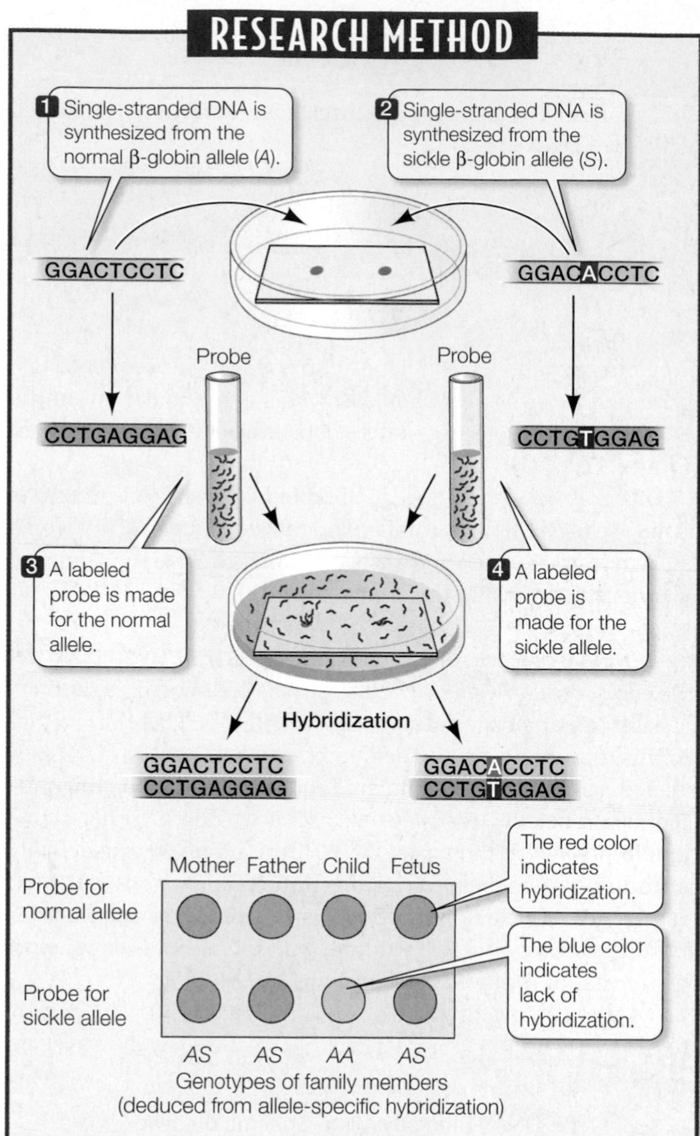
RESEARCH METHOD
1 Single-stranded DNA is synthesized from the normal β-globin allele (A).
2 Single-stranded DNA is synthesized from the sickle β-globin allele (S).
GGACTCCTC
GGACACCTC
Probe
Probe
CCTGAGGAG
CCTGTGGAG
3 A labeled probe is made for the normal allele.
4 A labeled probe is made for the sickle allele.
Hybridization
GGACTCCTC
CCTGAGGAG
GGACACCTC
CCTGTGGAG
Mother Father Child Fetus
Probe for normal allele
Probe for sickle allele
The red color indicates hybridization.
The blue color indicates lack of hybridization.
AS AS AA AS
Genotypes of family members
(deduced from allele-specific hybridization)

17.13 DNA Testing by Allele-Specific Oligonucleotide Hybridization Testing of this family reveals that three of them are heterozygous carriers of the sickle allele. The first child, however, has inherited two normal alleles and is neither affected by the disease nor a carrier.

17.3 RECAP

Genetic screening can identify people who have, are predisposed to, or are carriers of a genetic disease. It can be done at the level of the phenotype by identifying an abnormal protein or its function. It can also be done at the level of the genotype by direct testing of DNA.

- Can you describe how newborn babies are screened for phenylketonuria? See p. 384
- What is the advantage of screening for genetic mutations by allele-specific oligonucleotide hybridization relative to screening for allele-specific cleavage differences? See p. 385

We usually think of genetic diseases as inherited, but we'll now turn our attention to a genetic disease that most often affects somatic cells: cancer.

17.4 What Is Cancer?

Perhaps no malady affecting people in the industrialized world instills more fear than cancer. One in three Americans will have some form of cancer in their lifetime, and at present, one in four will die of it. With a million new cases and half a million deaths in the United States annually, cancer ranks second only to heart disease as a killer. Cancer was less common a century ago; then, as now in many regions of the world, people died of infectious diseases and did not live long enough to get cancer. Cancer tends to be a disease of the later years of life; children are much less frequently afflicted.

Since the U.S. government declared "war on cancer" in 1970, a tremendous amount of information on cancer cells—on their growth and spread and on their molecular changes—has been obtained. Perhaps the most remarkable discovery is that cancer is caused primarily by genetic changes. These changes are mostly mutations in the DNA of somatic cells that are propagated by mitosis.

Cancer cells differ from their normal counterparts

Cancer cells differ from the normal cells from which they originate in two major ways.

CANCER CELLS LOSE CONTROL OVER CELL DIVISION Most cells in the body divide only if they are exposed to extracellular influences, such as growth factors or hormones. Cancer cells do not respond to these controls, and instead divide more or less continuously, ultimately forming **tumors** (large masses of cells). By the time a physician can feel a tumor or see one on an X ray or CAT scan, it already contains millions of cells.

Benign tumors resemble the tissue they came from, grow slowly, and remain localized where they develop. A lipoma, for example, is a benign tumor of fat cells that may arise in the armpit and remain there. Benign tumors are not cancers, but they must be removed if they impinge on an important organ, such as the brain.

Malignant tumors, on the other hand, do not look like their parent tissue at all. A flat, specialized epithelial cell in the lung wall may turn into a relatively featureless, round, malignant lung cancer cell (**Figure 17.14**). Malignant cells often have irregular structures, such as variable nucleus sizes and shapes. Many malignant cells express the gene for telomerase and thus do not shorten the ends of their chromosomes after each DNA replication.

CANCER CELLS SPREAD TO OTHER TISSUES The second, and most fearsome, characteristic of cancer cells is their ability to invade surrounding tissues and spread to other parts of the body. This spreading, called **metastasis**, occurs in several stages. First, the cancer cells extend into the tissue that surrounds them by secreting digestive enzymes that disintegrate the surrounding cells and extracellular materials, working their way toward a blood vessel. Then, some

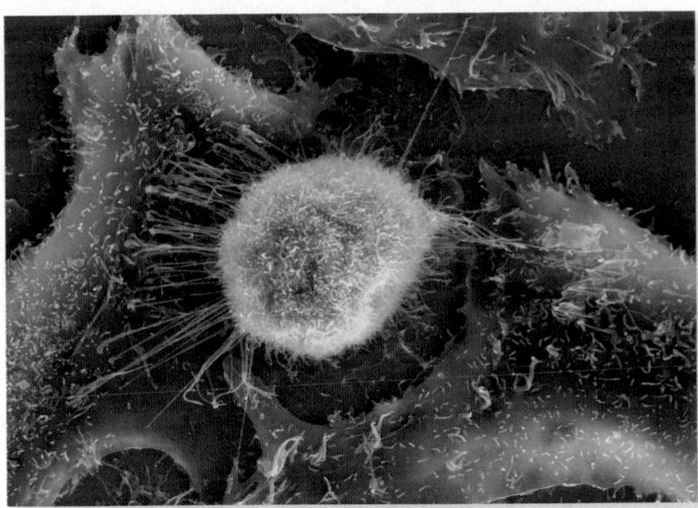

17.14 A Cancer Cell with Its Normal Neighbors This small-cell lung cancer cell (yellow-green) is quite different from the surrounding lung epithelial cells from which it came. This particular form of cancer is very lethal, with a 5-year survival rate of 10 percent. Most cases are caused by tobacco smoking.

of the cancer cells enter the bloodstream or the lymphatic system. The journey through these vessels is perilous, and few of the cancer cells survive—perhaps one in 10,000. When by chance a cancer cell arrives at an organ suitable for its further growth, it expresses cell surface proteins that allow it to bind to and invade the new host tissue. Finally, the tumor at the new site secretes chemical signals that cause blood vessels to grow to the tumor and supply it with oxygen and nutrients. This process is called *angiogenesis*.

Different forms of cancer affect different parts of the body. About 85 percent of all human tumors are *carcinomas*—cancers that arise in surface tissues such as the skin and the epithelial cells that line the organs. Lung cancer, breast cancer, colon cancer, and liver cancer are all carcinomas. *Sarcomas* are cancers of tissues such as bone, blood vessels, and muscle. *Leukemias* and *lymphomas* affect the cells that give rise to blood cells.

Some cancers are caused by viruses

In 1909, Peyton Rous, a young physician beginning his research career at the Rockefeller University in New York City, received a phone call from panicked chicken farmers on Long Island. Their chickens were coming down with a strange illness that caused their muscles to deteriorate before they died. The disease was spreading rapidly through the flocks: when one chicken in a coop got it, soon the others did, too. After some medical—or, more correctly, veterinary—detective work, Rous diagnosed the disease as a sarcoma—a muscle tumor—and the cause as a virus. This was a key discovery, because it showed that cancer could be a transmissible disease.

Unfortunately, Rous's discovery was not immediately followed up, and a search for human tumor viruses was initially fruitless. It was only when scientists returned to tumor virology in animals in the 1960s that the importance of Rous's work was fully appreciated. He was awarded the Nobel prize in 1966, 57 years after his discovery! For a time in the 1960s, it was thought that much of human cancer came from viruses. But careful investigations have shown this not to be the case. About 15 percent of human cancers are virally induced. At least five types of human cancer are probably caused by viruses (**Table 17.1**).

TABLE 17.1

Human Cancers Known To Be Caused by Viruses

CANCER	ASSOCIATED VIRUS
Liver cancer	Hepatitis B virus
Lymphoma, nasopharyngeal cancer	Epstein–Barr virus
T cell leukemia	Human T cell leukemia virus (HTLV-I)
Anogenital cancers	Papillomavirus
Kaposi's sarcoma	Kaposi's sarcoma herpesvirus

Hepatitis B, a liver disease that affects people all over the world, is caused by the hepatitis B virus, which contaminates blood or is carried from mother to child during birth. The viral infection can be long-lasting and may flare up numerous times. The hepatitis B virus is associated with liver cancer, especially in Asia and Africa, where millions of people are infected. But it does not cause cancer by itself. Some gene mutations that are necessary for tumor formation occur in the infected cells of Asians and Africans, although apparently not in those of Europeans and North Americans.

An important group of virally induced cancers among North Americans and Europeans is the various anogenital cancers caused by papillomaviruses. The genital and anal warts that these viruses cause often develop into tumors. These viruses seem to be able to act on their own, not needing mutations in host cells for tumors to arise. Normally, the viral genome is circular, and it undergoes a lytic cycle, producing more viruses. Occasionally, the circle is broken and the virus integrates itself into the chromosome of a host cell in the uterine cervix, disrupting a gene that normally stimulates viral replication and blocks cell division and thus stimulating the cell cycle. Sexual transmission of these papillomaviruses is unfortunately widespread.

Cancers caused by viruses can be prevented and treated with antiviral vaccines. Large vaccination programs in Asia are already reducing the incidence of liver cancer caused by hepatitis B. Recently, an effective vaccine was developed for the papillomaviruses that cause cervical cancer.

Most cancers are caused by genetic mutations

What causes the 85 percent of cancers that are not caused by viruses? Because most cancers develop in older people, it is reasonable to assume that one must live long enough for a series of events to occur. This assumption turns out to be correct, and the events are genetic mutations.

DNA can be damaged in many ways. As Section 12.6 describes, spontaneous mutations arise because of chemical changes in the nucleotides. In addition, certain mutagens, called **carcinogens**, can cause mutations that lead to cancer. Familiar carcinogens include the chemicals that are present in tobacco smoke and meat preservatives, ultraviolet light from the sun, and ionizing radiation from sources of radioactivity. Less familiar, but just as harmful, are thousands of chemicals that are naturally present in the foods people eat. According to one estimate, these "natural" carcinogens account for well over 80 percent of human exposure.

Both natural and synthetic carcinogens damage DNA, usually by causing changes from one base to another. In somatic cells that divide often, such as epithelial and bone marrow stem cells, there is less time for DNA repair mechanisms to work before replication occurs again. Therefore, such cells are especially susceptible to cancer.

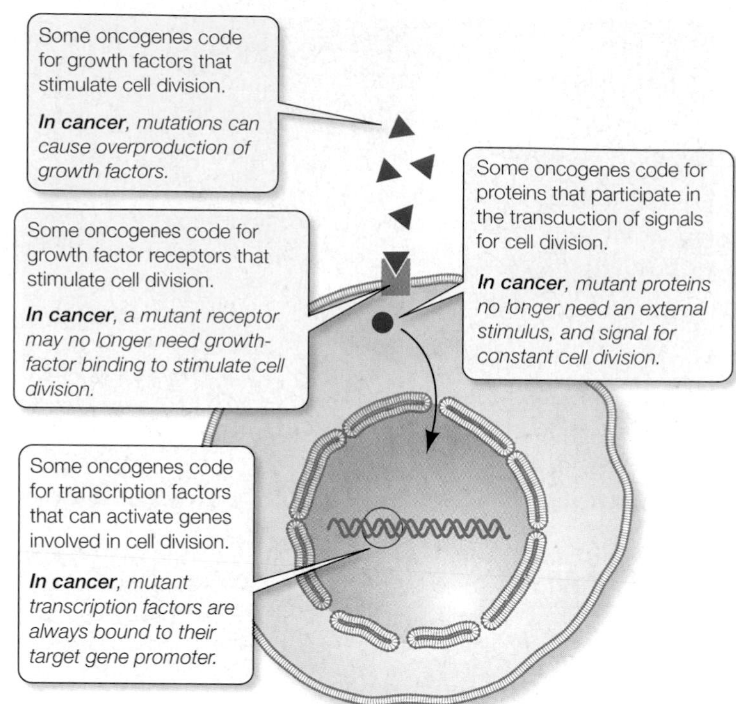

Some oncogenes code for growth factors that stimulate cell division.
In cancer, mutations can cause overproduction of growth factors.
Some oncogenes code for growth factor receptors that stimulate cell division.
In cancer, a mutant receptor may no longer need growth-factor binding to stimulate cell division.
Some oncogenes code for proteins that participate in the transduction of signals for cell division.
In cancer, mutant proteins no longer need an external stimulus, and signal for constant cell division.
Some oncogenes code for transcription factors that can activate genes involved in cell division.
In cancer, mutant transcription factors are always bound to their target gene promoter.

17.15 Oncogene Products Stimulate Cell Division Mutations can affect any of the several ways in which oncogenes can stimulate cell division, thus causing cancer.

Two kinds of genes are changed in many cancers

The changes in the control of cell division that lie at the heart of cancer can be likened to the controls of an automobile. To make a car move, two things must happen: the gas pedal must be pressed, and the brake must be released. In the human genome, some genes act as *oncogenes*, which "press the gas pedal" to stimulate cell division, and some act as *tumor suppressor genes*, which "put the brake on" to inhibit it.

ONCOGENES The first hint that **oncogenes** (from the Greek *onco*, "mass") were necessary for cells to become cancerous came with the identification of animal cancers induced by viruses. In many cases, these viruses bring a new gene into their host cell that stimulates cell division when it is expressed in the viral genome. It soon became apparent that the viral oncogenes had counterparts that are not usually transcribed in the genomes of host cells. So the search for genes that are damaged by carcinogens quickly zeroed in on these cellular oncogenes. Several dozen such genes were soon found.

Oncogenes are genes that have the capacity to stimulate cell division but are normally "turned off" in differentiated, nondividing cells. Many of them are involved in the pathways by which growth factors stimulate cell division (**Figure 17.15**). Some remarkable oncogenes control *apoptosis* (programmed cell death; see Section 9.6). Activation of these genes by mutation causes them to prevent apoptosis, allowing cells that normally die to continue dividing.

Some oncogenes can be activated by point mutations, others by chromosome changes such as translocations, and still others by gene amplification. Whatever the mechanism, the result is the same: the oncogene becomes activated, and the "gas pedal" for cell division is pressed.

TUMOR SUPPRESSOR GENES About 10 percent of all cancers are inherited. Often the inherited form of a cancer is clinically similar to a noninherited form that occurs later in life, called the *sporadic* form. However, the inherited form strikes much earlier in life, and it usually shows up as multiple tumors.

In 1971, Alfred Knudson used these observations to predict that for a cancer to occur, a **tumor suppressor gene**, which normally acts as a "brake" on cell division, must be inactivated. But in contrast to oncogenes, in which one mutant allele is all that is needed for activation, the full inactivation of a tumor suppressor gene requires that both alleles be turned off, which requires two mutational events. It takes a long time for both alleles in a single cell to mutate and cause sporadic cancer. But people with inherited cancer are born with one mutant allele for a tumor suppressor gene, and need just one more mutational event for its full inactivation (**Figure 17.16**).

The isolation of various tumor suppressor genes has confirmed Knudson's "two-hit" hypothesis. Some of these genes are involved in inherited forms of rare childhood cancers such as retinoblastoma (a tumor of the eye) and Wilms' tumor of the kidney as well as in inherited breast and prostate cancers.

An inherited form of breast cancer demonstrates the effect of tumor suppressor genes. The 9 percent of women who inherit one mutant allele of the gene *BRCA1* have a 60 percent chance of having breast cancer by age 50 and an 82 percent chance of developing it by age 70. The comparable figures for women who inherit two normal alleles of the gene are 2 percent and 7 percent, respectively.

How do tumor suppressor genes act in the cell? Like the oncogenes, they are normally involved in cell division (**Figure 17.17**). Some regulate progress through the cell cycle (which we described in Section 9.2). The *Rb* gene, which was first described for its contribution to retinoblastoma, is active during the G1 phase. In its active form, it encodes a protein that binds to and inactivates tran-

(A)

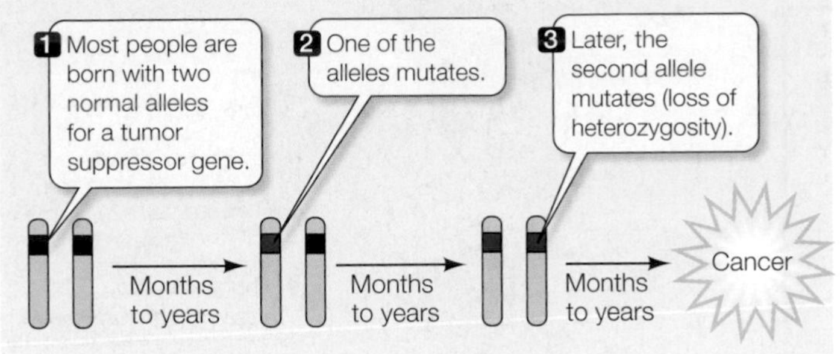

(B)

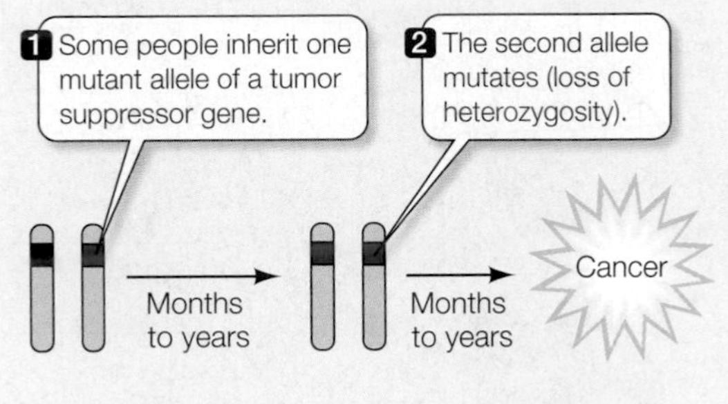

17.16 The "Two-Hit" Hypothesis for Cancer (A) Although a single mutation can activate an oncogene, two mutations are needed to inactivate a tumor suppressor gene. (B) An inherited predisposition to cancer occurs in people born with one allele already mutated.

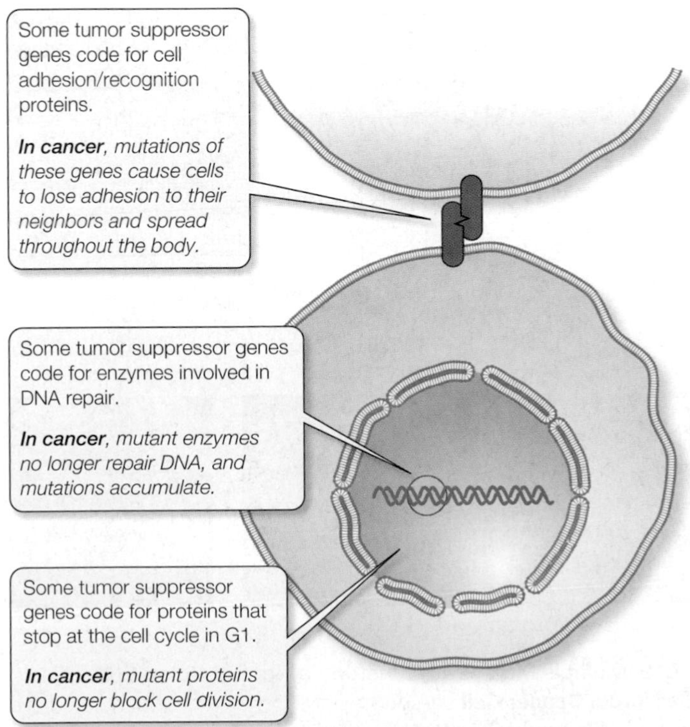

17.17 Tumor Suppressor Gene Products Inhibit Cell Division Mutations can affect any of the several ways in which tumor suppressor genes inhibit cell division, allowing cells to divide and form a tumor.

scription factors that are necessary for progress to the S phase and the rest of the cell cycle. In nondividing cells, *Rb* remains active, preventing cell division until the proper growth factor signals are present. When *Rb* is inactivated by mutation, the cell cycle moves forward independently of growth factors.

The protein product of another widespread tumor suppressor gene, *p53*, also stops the cell cycle at G1. It does this by acting as a transcription factor, stimulating the production of (among other things) a protein that blocks the interaction of a cyclin and a protein kinase needed for moving the cell cycle beyond G1. This gene is mutated in many types of cancers, including lung cancer and colon cancer.

Several events must occur to turn a normal cell into a malignant cell

The "gas pedal" and "brake" analogies we have been using for oncogenes and tumor suppressor genes, respectively, are elegant but simplified. There are many oncogenes and tumor suppressor genes, some of which act only in certain cells at certain times. Therefore, a complex sequence of events must occur before a normal cell becomes malignant, as demonstrated by an experiment in which mutations of oncogenes and tumor suppressor genes were introduced in different combinations (**Figure 17.18**). When normal rat cells growing in culture were transfected with an expression vector containing a mutant form of an oncogene (such as *ras*, which encodes a protein involved in signaling cell division), this alone was not enough to turn them into cancer cells. Likewise, if an expression vector containing a mutant form of a tumor suppressor gene (such as *p53*) was introduced in such a way that it overshadowed the normal form already in the cells, that alone was not enough to turn normal cells into cancer cells. *Both* an active gas pedal and a mutant brake are needed for malignant transformation.

More than two gene mutations are usually needed for full-blown cancer. Because colon cancer progresses to full malignancy slowly, it is possible to describe the oncogene and tumor suppressor gene mutations at each stage in great molecular detail. **Figure 17.19** outlines the progress of this form of cancer. At least four tumor suppressor genes and one oncogene must be mutated in sequence for an epithelial cell in the colon to become metastatic. Although the occurrence of all these events in a single cell might appear unlikely, remember that the colon has millions of cells, that the cells giving rise to colon epithelial cells are constantly dividing, and that these changes take place over many years of exposure to natural and synthetic carcinogens as well as spontaneous mutations.

The characterization of the molecular changes in tumor cells has opened up the possibility of genetic diagnosis and screening for cancer. Many cancers are now commonly diagnosed in part by specific oligonucleotide probes for oncogene or tumor suppressor gene alterations. It is also possible to detect early in life whether an individual has inherited a mutant tumor suppressor gene. A person who inherits mutated copies of the tumor suppressor genes involved in colon cancer, for example, normally would have a high probability of developing this cancer by age 40. Surgical removal of the colon would prevent a metastatic tumor from arising.

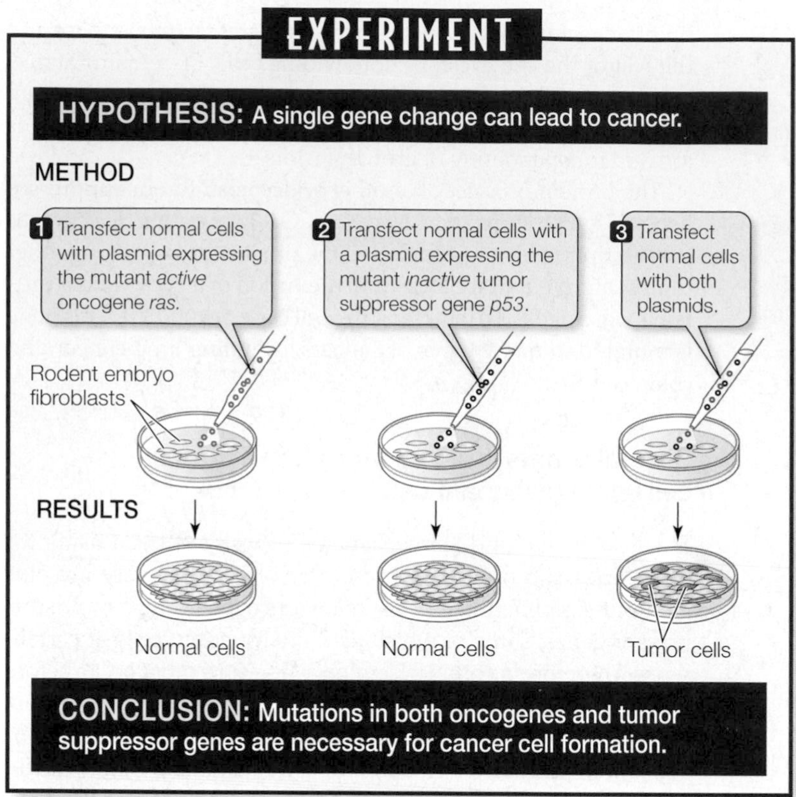

17.18 Cancer Is the Result of Multiple Genetic Alterations A complex series of events leads to malignancy in a normal cell. FURTHER RESEARCH: What would happen if you performed this experiment using the *normal* alleles for *ras* and *p53*?

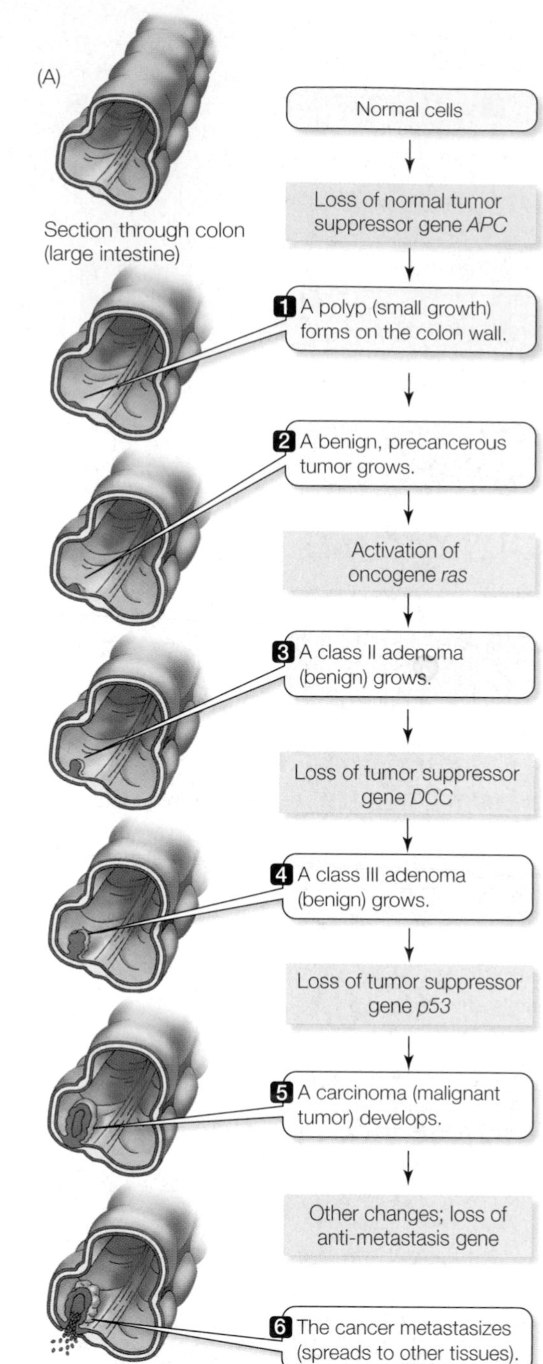

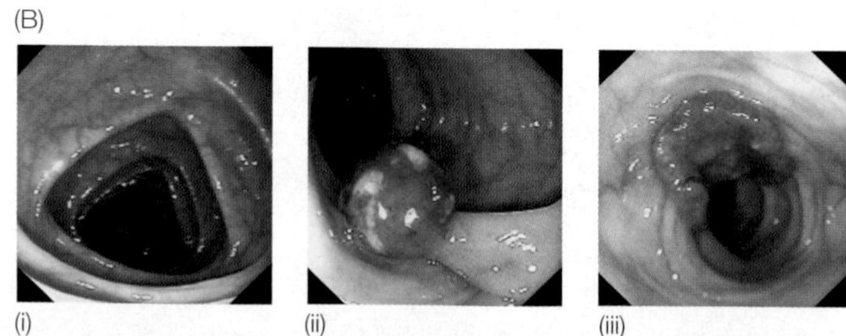

17.19 Multiple Mutations Transform a Normal Colon Epithelial Cell into a Cancer Cell (A) At least five genes must be mutated in a single cell to produce metastatic colon cancer. (B) Colonoscopy is the best screening test for colon cancer. These views reveal (i) normal colon tissue, (ii) a benign adenoma (stalked polyp), and (iii) adenocarcinoma (a malignant tumor).

17.4 RECAP

Uncontrolled cell division and metastasis (spread) are the hallmarks of cancer. Although some cancers are caused by viruses or inherited, most are due to the accumulation of genetic mutations in two classes of genes: oncogenes and tumor suppressor genes.

- What are some viruses that can cause cancer? See p. 387 and Table 17.1
- Can you describe the functions of oncogenes and tumor suppressor genes? See p. 388 and Figures 17.15 and 17.17
- What is the two-hit hypothesis? See p. 388 and Figure 17.16

Ongoing research has resulted in the development of increasingly accurate diagnostic tests and a better understanding of various genetic diseases at the molecular level. This knowledge is now being applied to develop new treatments for genetic diseases. In the next section we will survey various approaches to treatment, ranging from modifications of the mutant phenotype to gene therapy, in which the normal version of a mutant gene is supplied.

17.5 How Are Genetic Diseases Treated?

Most treatments for genetic diseases simply try to alleviate the symptoms that affect the patient. But to effectively treat these diseases—whether they affect all cells, as in inherited disorders such as PKU, or only somatic cells, as in cancer—physicians must be able to diagnose the disease accurately, understand how the disease works at the molecular level, and intervene early, before the disease ravages or kills the individual. There are two main approaches to treating genetic diseases: modifying the disease phenotype or replacing the defective gene.

Genetic diseases can be treated by modifying the phenotype

Altering the phenotype of a genetic disease so that it no longer harms an individual is commonly done in one of three ways: by restricting the substrate of a deficient enzyme, by inhibiting a harmful metabolic reaction, or by supplying a missing protein product.

RESTRICTING THE SUBSTRATE Restricting the substrate of a deficient enzyme is the approach taken when a newborn is diagnosed with PKU. In this case, the deficient enzyme is phenylalanine hydroxylase, and the substrate is phenylalanine. The infant's inability to break down the phenylalanine in food leads to a buildup of the substrate, which causes the clinical symptoms. So the infant is immediately put on a special diet that contains only enough phenylalanine for immediate use. Lofenelac, a milk-based product that is low in phenylalanine, is fed to these infants just like formula. Later, certain fruits, vegetables, cereals, and noodles low in phenylalanine can be added to the diet. Meat, fish, eggs, dairy products, and bread, which contain high amounts of phenylalanine, must be avoided, especially during childhood, when brain development is most rapid. The artificial sweetener aspartame must also be avoided because it is made of two amino acids, one of which is phenylalanine.

People with PKU are generally advised to stay on a low-phenylalanine diet for life. Although maintaining these dietary restrictions may be difficult, it is effective. Numerous follow-up studies since newborn screening was initiated have shown that people with PKU who stay on the diet are no different from the rest of the population in terms of mental ability. This is an impressive achievement in public health, given the severity of mental retardation in untreated patients.

METABOLIC INHIBITORS As described in Section 17.1, people with familial hypercholesterolemia accumulate dangerous levels of cholesterol in their blood. These people are not only unable to metabolize dietary cholesterol, but also synthesize a lot of it. One effective treatment for people with this disease is a statin drug, which blocks the patient's own cholesterol synthesis. Patients who receive this drug need only worry about cholesterol in the diet, and not about the cholesterol their cells are making.

Metabolic inhibitors also form the basis of chemotherapy for cancer. The strategy is to kill rapidly dividing cells, since rapid cell division is the hallmark of malignancy. But such a strategy is not

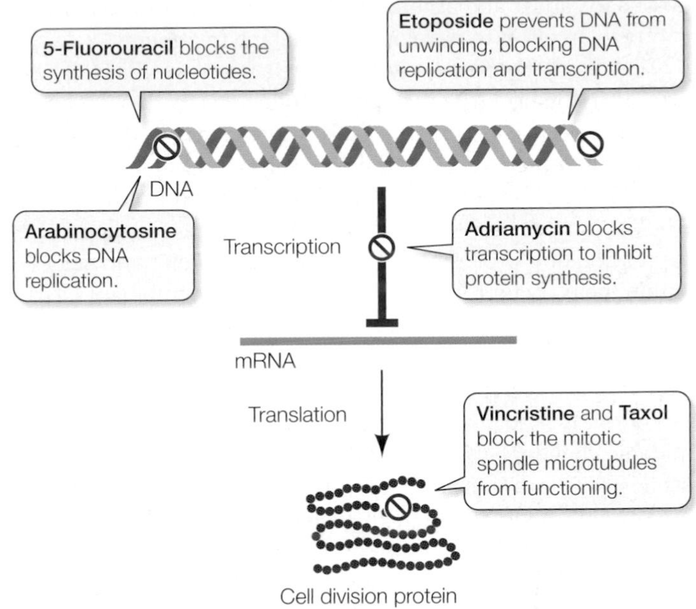

17.20 Strategies for Killing Cancer Cells The medications used in chemotherapy attack rapidly dividing cancer cells in several ways. Unfortunately, most of them also affect rapidly dividing noncancerous cells.

selective for tumor cells. Many drugs can kill cancer cells (**Figure 17.20**), but most of those drugs also damage other, noncancerous, dividing cells in the body. Therefore, it is not surprising that people undergoing chemotherapy suffer side effects such as loss of hair (due to damage to the skin epithelium), digestive upsets (gut epithelial cells), and anemia (bone marrow stem cells). The effective dose of these highly toxic drugs for treating the cancer is often just below the dose that would kill the patient, so they must be used with utmost care. Often they can control the spread of cancer, but not cure it.

SUPPLYING THE MISSING PROTEIN An obvious way to treat a disease phenotype in which a functional protein is missing is to supply that protein. This approach is the basis of treatment of hemophilia, in which the missing blood clotting protein is supplied in pure form. The production of human clotting proteins by recombinant DNA technology has made it possible to provide a pure protein instead of blood products, which can be contaminated with viruses or other pathogens.

Unfortunately, the phenotypes of many diseases caused by genetic mutations are very complex. Simple interventions like those we have described do not work for most such diseases. Indeed, a recent survey showed that current therapies for 351 diseases caused by single-gene mutations improved patients' life spans by only 15 percent.

Gene therapy offers the hope of specific treatments

Clearly, if a cell lacks a functional allele, it would be optimal to provide that allele, which is the aim of gene therapy. Diseases ranging from the rare inherited disorders caused by single-gene mutations to cancer and atherosclerosis are under intensive investigation in an effort to develop gene therapy treatments.

The object of **gene therapy** is to insert a new gene that will be expressed in the host. The new DNA is often attached to a promoter that will be active in human cells. The physicians who are developing such treatments are confronted by all the challenges of recombinant DNA technology: they must find effective vectors and ensure efficient uptake, precise insertion into the host DNA, appropriate expression and processing of mRNA and protein, and selection within the body for the cells that contain the recombinant DNA.

Which human cells should be the targets of gene therapy? The best approach would be to replace the nonfunctional allele with a functional one in every cell of the body. But vectors that could do this are simply not available, and delivery to every cell poses a formidable challenge. Until recently, attempts at gene therapy have used *ex vivo* techniques. That is, physicians have taken cells from the patient's body, added the new gene to those cells in the laboratory, and then returned the cells to the patient in the hope that the correct gene product would be made (**Figure 17.21**). Two examples demonstrate this technique:

- *Adenosine deaminase* is needed for maturation of white blood cells, and people without this enzyme have severe immune system deficiencies. A functional gene for adenosine deaminase was introduced via a viral vector into the white blood cells of a girl with a genetic deficiency of this enzyme. Unfortunately, mature white blood cells were used, and although they survived for a time in the girl and provided some therapeutic benefit, they eventually died, as is the normal fate of such cells. Further clinical trials have used bone marrow stem cells, which constantly divide to produce white blood cells.
- People with *hemophilia* had some skin cells removed and transfected with a plasmid containing a normal allele of the gene coding for the blood clotting protein they were missing. The cells were then reintroduced into the patients' body fat, where they produced adequate protein for normal clotting.

The other approach to gene therapy is to insert the gene directly into cells in the body of the patient. This *in vivo* approach is being attempted for various types of cancer. Lung cancer cells, for example, are accessible to such treatment if the DNA or vector is given as an aerosol through the respiratory system. Vectors carrying functional alleles of the tumor suppressor genes that are mutated in the patient's tumors, as well as vectors expressing antisense RNAs targeting oncogene mRNAs, have been successfully introduced in this way into patients with lung cancer, with some clinical improvement.

Several thousand patients, over half of them with cancer, have undergone gene therapy. Most of these clinical trials have been at a preliminary level, in which people are given the therapy to see whether it has any toxicity and whether the new gene is actually incorporated into the patient's genome. More ambitious trials are under way, in which a larger number of patients will receive the therapy with the hope that their disease will disappear, or at least improve.

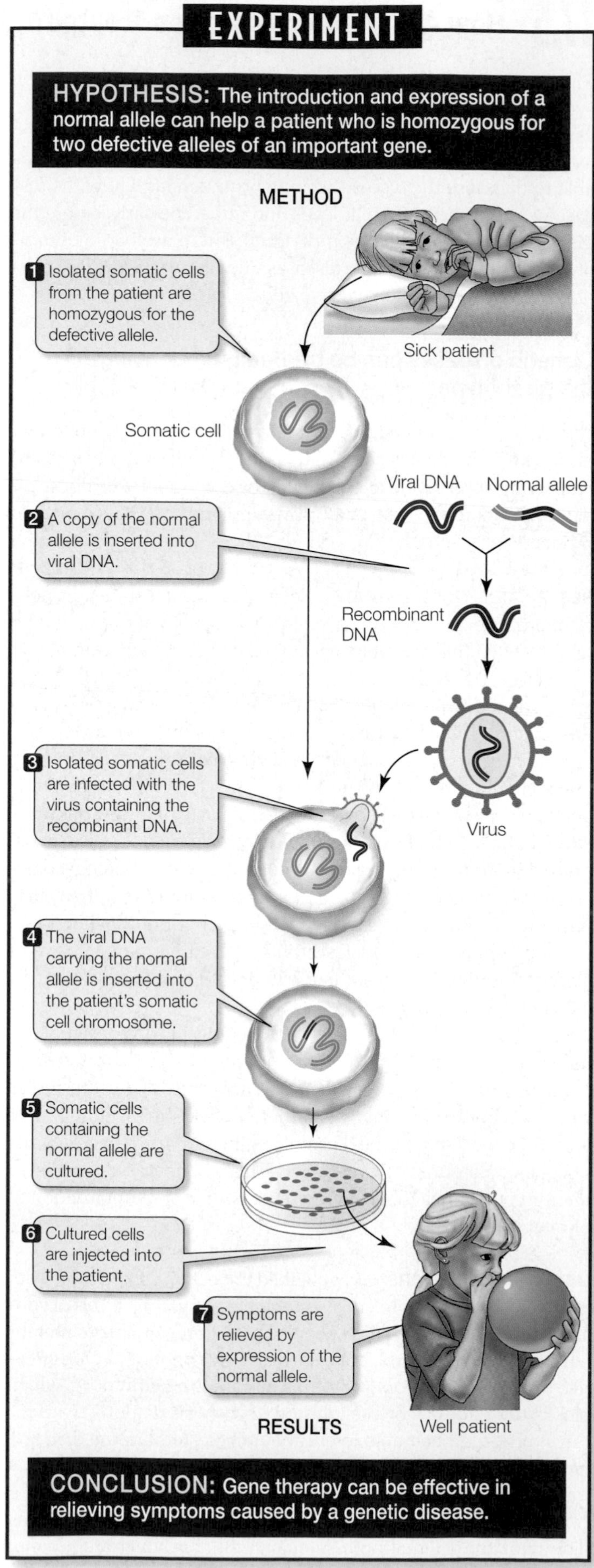

EXPERIMENT
HYPOTHESIS: The introduction and expression of a normal allele can help a patient who is homozygous for two defective alleles of an important gene.
METHOD
1 Isolated somatic cells from the patient are homozygous for the defective allele.
Sick patient
Somatic cell
Viral DNA
Normal allele
2 A copy of the normal allele is inserted into viral DNA.
Recombinant DNA
3 Isolated somatic cells are infected with the virus containing the recombinant DNA.
Virus
4 The viral DNA carrying the normal allele is inserted into the patient's somatic cell chromosome.
5 Somatic cells containing the normal allele are cultured.
6 Cultured cells are injected into the patient.
7 Symptoms are relieved by expression of the normal allele.
RESULTS
Well patient
CONCLUSION: Gene therapy can be effective in relieving symptoms caused by a genetic disease.

17.21 Gene Therapy: The *Ex Vivo* Approach New genes are added to somatic cells taken from a patient's body. These transgenic cells are then returned to the body to make the missing gene product.

17.5 RECAP

Treatment of human genetic diseases may involve interventions that attempt to modify the abnormal phenotype, such as restricting the substrate of a deficient enzyme, inhibiting a harmful metabolic reaction, or supplying a missing protein. Gene therapy aims to address a genetic defect by inserting normal alleles into a patient's cells.

- Can you describe how metabolic inhibitors used in chemotherapy function in treating cancer? See p. 391 and Figure 17.20
- Can you explain how *ex vivo* gene therapy works and give an example? See p. 392 and Figure 17.21

Therapies for simple genetic diseases remain a challenge. But most diseases are far more complex, involving many genes interacting with the environment. Knowledge of these genes is coming from the Human Genome Project.

17.6 What Have We Learned from the Human Genome Project?

In 1984, the United States government sponsored a conference on the detection of DNA damage in people exposed to significant levels of radiation, such as those who had survived the atomic bombs in Japan 39 years earlier. Scientists attending this conference quickly realized that the ability to detect such damage would also be useful in evaluating environmental mutagens. But in order to detect changes in the human genome, scientists first needed to know its normal sequence.

In 1986, Renato Dulbecco, who had won the Nobel prize for his pioneering work on cancer-causing viruses, suggested that determining the normal sequence of human DNA could also be a boon to cancer research. He proposed that the scientific community be mobilized for the task. The result was the publicly funded **Human Genome Project**, an international effort. In the 1990s, private industry launched its own sequencing effort.

There are two approaches to genome sequencing

Because of their differing sizes, the 46 human chromosomes can be separated from one another and identified (see Figure 9.15). So it is possible to isolate the DNA of each chromosome for sequencing. The straightforward approach would be to start at one end of a chromosome and simply sequence the entire 120 million base pairs. Unfortunately, this approach is not practical, since only about 700 base pairs can be sequenced at a time (see Figure 11.24 for a description of the DNA sequencing technique).

To sequence an entire genome, chromosomal DNA is first cut into fragments about 500 base pairs long, then each fragment is sequenced. For the haploid human genome, which has about 3.2 billion base pairs, there are more than 6 million such fragments. The problem then becomes putting these millions of fragments back together. This task is accomplished by assembling smaller portions of DNA sequence. These portions of DNA sequence almost always overlap and must be aligned so that the positions of all of the bases are properly accounted for. Two methods were used to align the DNA fragments: *hierarchical sequencing* and *shotgun sequencing*.

HIERARCHICAL SEQUENCING The publicly funded sequencing team used a method known as **hierarchical sequencing**. First, they systematically identified short marker sequences along the chromosomes, ensuring that every fragment of DNA to be sequenced would contain a marker (**Figure 17.22A**). This method can be compared to making a road map, showing towns with the mileage separating them. The "towns" are the marker sequences, and the "mileage" is in base pairs. The simplest markers are the recognition sequences for restriction enzymes.

Some restriction enzymes recognize 8–12 base pairs in DNA, not just the usual 4–6 base pairs. A DNA molecule with several million base pairs will have relatively few of these larger sites, and thus the enzyme will generate a small number of relatively large fragments. These large fragments can be added to a vector called a **bacterial artificial chromosome** (**BAC**), which can carry about 250,000 base pairs of inserted DNA, and inserted into bacteria to create a gene library.

The fragments in this library are arranged in the proper order along the chromosome map by using the marker sequences. To arrange the DNA fragments on the map, libraries made with different restriction enzymes are compared. If two large fragments of DNA cut with different enzymes have the same marker, they must overlap. This method works, but is slow.

SHOTGUN SEQUENCING Instead of finding markers, fragmenting the DNA, and then sequencing it, the **shotgun sequencing** method cuts the DNA at random into small, sequencing-ready fragments and lets powerful computers search for markers that overlap (**Figure 17.22B**). The fragments can then be aligned.

The shotgun approach, which has been used by private industry, is much faster than the hierarchical approach because there is no need to make a map. At first this method was greeted with considerable skepticism. One concern was that without rigorous prior mapping of marker sites on the chromosomes, the computer might pick out repetitive sequences common to many DNA fragments and line the fragments up incorrectly. But the rapid development of sophisticated computers and software for DNA sequencing and analysis (the advent of **bioinformatics**) allowed the shotgun method to be refined so that inaccurate alignment is not a major problem. The entire 180 million-base-pair fruit fly genome was sequenced by the shotgun method in little over a year. This success proved that the shotgun method might work for the much larger human genome, and in fact, it did.

The sequence of the human genome contained many surprises

The two teams of scientists announced a draft human genome sequence in June 2000 to great fanfare, and published their data simultaneously in February 2001. By the start of 2005, the final

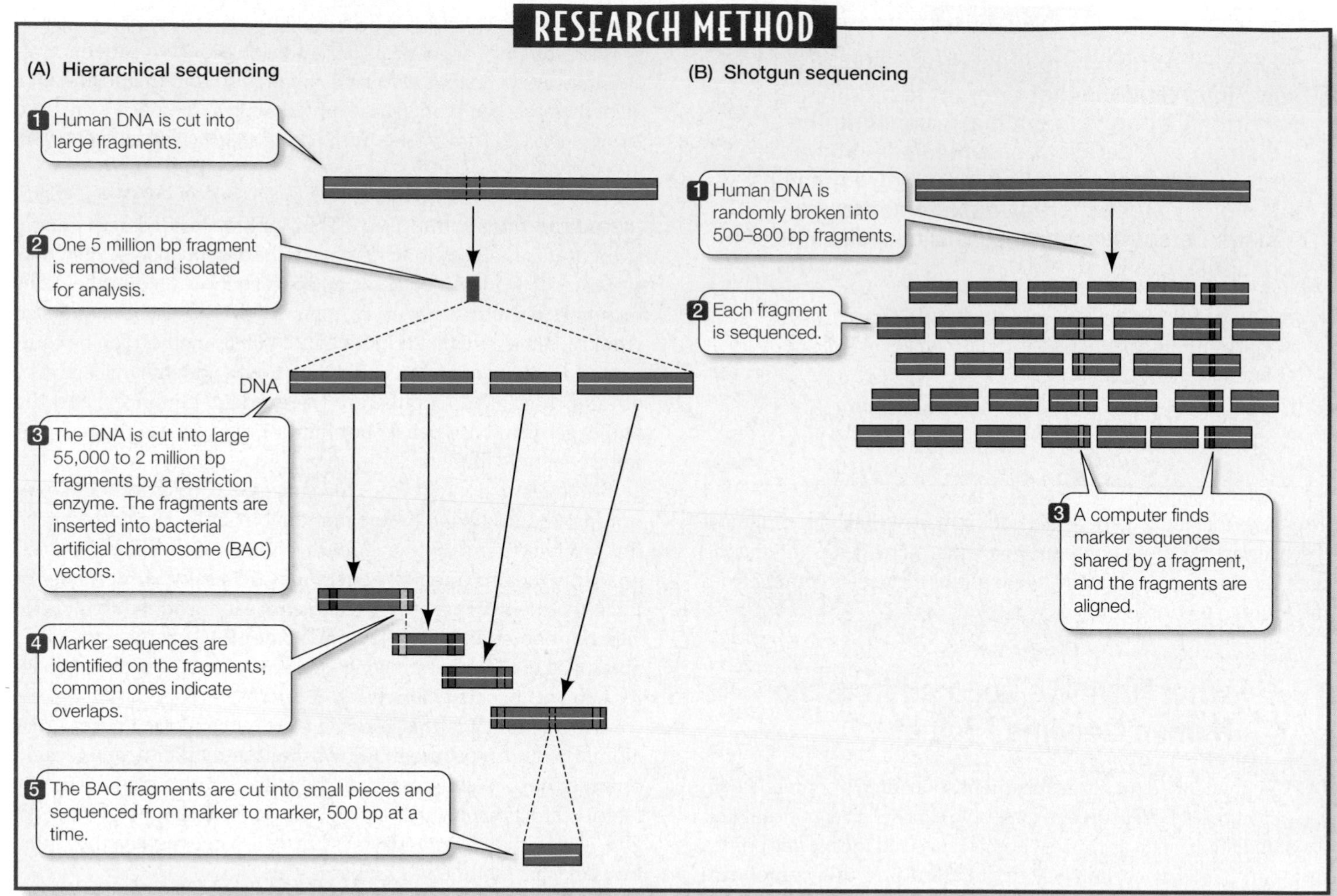
RESEARCH METHOD
(A) Hierarchical sequencing
1 Human DNA is cut into large fragments.
2 One 5 million bp fragment is removed and isolated for analysis.
DNA
3 The DNA is cut into large 55,000 to 2 million bp fragments by a restriction enzyme. The fragments are inserted into bacterial artificial chromosome (BAC) vectors.
4 Marker sequences are identified on the fragments; common ones indicate overlaps.
5 The BAC fragments are cut into small pieces and sequenced from marker to marker, 500 bp at a time.
(B) Shotgun sequencing
1 Human DNA is randomly broken into 500–800 bp fragments.
2 Each fragment is sequenced.
3 A computer finds marker sequences shared by a fragment, and the fragments are aligned.

17.22 Two Approaches to Sequencing DNA (A) In the hierarchical approach to genome sequencing, genetic markers are mapped, and DNA fragments are then aligned by matching overlapping sites with the same markers. (B) In the shotgun approach, the DNA is fragmented and a computer is used to find overlapping markers.

sequence was completed, two years ahead of the schedule set more than a decade previously and well under budget.

Many interesting and surprising facts about the human genome were revealed upon its sequencing.

- *Of its 3.2 billion base pairs, fewer than 2 percent make up coding regions, containing a total of about 24,000 genes.* Before sequencing began, estimates of the number of human genes ranged from 80,000 to 100,000. This lower number of genes—not many more than the fruit fly's—means that the observed diversity of proteins, which led to the 100,000 estimate, must be produced posttranscriptionally. That is, the average human gene must code for several different proteins.
- *The average gene has 27,000 base pairs.* Gene sizes vary greatly, from 1,000 to 2.4 million base pairs. Variation in gene size is to be expected, given that human proteins (and RNAs) vary in size. Human proteins range from 100 to about 5,000 amino acids per polypeptide chain.
- *Virtually all human genes have many introns.* As we noted previously, only 2 percent of the human genome is coding DNA.
- *Over 50 percent of the genome is made up of highly repetitive sequences.* Repetitive sequences near genes are GC-rich, while those farther away from genes are AT-rich.
- *Almost all (99.9 percent) of the genome is the same in all people.* Despite this apparent homogeneity, there are, of course, many individual differences. Scientists have mapped over 2 million single nucleotide polymorphisms (SNPs)—bases that differ in at least 1 percent of people.
- *Genes are not evenly distributed over the genome.* The small chromosome 19 is packed densely with genes, while chromosome 8 has long stretches of "gene desert," with no coding regions. The Y chromosome has the fewest genes (231), while chromosome 1 has the most (2,968).
- *Many genes exist whose functions are unknown.* There are 740 genes coding for RNAs that are not translated into proteins. Of these RNAs, several dozen are tRNAs, and a few are rRNAs and splicing RNAs. The roles of the rest are not clear. Nor are the roles of the hundreds of genes encoding protein kinases, although it is a good bet that they are involved in cell signaling.

The ENCODE project (*Encyclopedia of DNA Elements*) has been set up to identify all of the functional sequences in the human genome—not just the protein-coding sequences, but others as well, such as those coding for small RNAs involved in gene regulation. This project will make use of comparative data from closely related species, such as the chimpanzee.

The human genome sequence has many applications

Reading the human "book of life" is an achievement that ranks high among other recent great events in scientific exploration. The human genome sequence, and the tools developed to read it, are changing biology in many ways.

- Mapping technology and SNPs have made the isolation of human genes by positional cloning much easier because of the huge number of genetic markers now available. Many disease-related genes have been identified in this way.
- Genetic variation in drug metabolism has been a medical problem for a long time. The emerging field of **pharmacogenomics** is identifying the genes responsible for this variation and developing tests to predict who will react best to which medications.
- DNA chips (see Figure 16.15) are being used to analyze the expression of thousands of genes in different cells in different biochemical states. For example, a Cancer Genome Anatomy Project is seeking to make an mRNA "fingerprint" of a tumor at each stage of its development. Finding out which genes are expressed at which stage will be important not only in diagnosis, but also in identifying targets for therapy.
- "Genome prospecting" refers to the search for important polymorphisms in specific human populations. For example, the Pima Indians in Arizona have a high frequency of extreme obesity and diabetes. A search of their genomes might reveal genes predisposing them to these conditions.

The end result of all this knowledge may be a new approach to medical care, in which each person's genome will be used to prescribe lifestyle changes and treatments that can maximize that person's genetic potential.

The use of genetic information poses ethical questions

When the genetic defect that causes cystic fibrosis was discovered, many people predicted a "tidal wave" of genetic testing for heterozygous carriers. Everyone, it was thought, would want the test—especially the relatives of people with the disease. But this tidal wave has not developed. To find out why, a team of psychologists, ethicists, and geneticists interviewed 20,000 people in the United States. What the researchers found surprised them. Most people are simply not very interested in their genetic makeup, unless they have a close relative with a genetic disease and are involved in a decision about pregnancy.

There are other people, however, who might be very interested in the results of genetic testing. For example, people who test positive for genetic abnormalities, from hypercholesterolemia to cancer, might be denied employment or health insurance. Consequently, there are laws that prohibit discrimination on the basis of genetic information.

The search for valuable genes in diverse human populations has raised many concerns about exploitation and commercialization of a person's DNA sequence. Is a gene that confers resistance to cancer, for example, the property of an individual, an ethnic group in which it may be frequent, the pharmaceutical company that finds it, or humanity at large?

We described at the beginning of this chapter how the French Canadian population of Quebec is being used to discover genes relevant to human diseases. People there have been happy to cooperate with the laboratories doing the investigations. Several other populations that came from a small group of founders are being used by human geneticists for genome analyses:

- 2.5 million Costa Ricans who came from about 4,000 founders 12 generations ago
- 10 million Ashkenazic Jews (from Eastern Europe) who came from about 1,500 founders 30 generations ago
- 1.6 million Sardinians who came from about 500 founders 400 generations ago

Perhaps the best-known population under intensive study is in Iceland, where 280,000 people are descended from about 10,000 settlers who came from Europe 1,100 years ago. A company has been set up, with government support, to sell the knowledge that comes from analyzing the genomes of Iceland's people. The company's approach to mining this genetic lode is illustrated by its search for genes that predispose people to asthma, a respiratory disease:

- The names of Icelanders with asthma were searched for in a genealogy database.
- One group of 104 patients was descended from a single ancestor, born in 1710 (11 generations ago). It was considered likely that the same genes for predisposition to asthma were present in all 104 patients.
- Marker sequences were sought that would identify alleles that all 104 patients had in common, and a small number of such genes were subsequently identified.

The characterization of these genes will lead to a greater understanding of how a group of genes interacts to produce a complex phenotype.

The proteome is more complex than the genome

As mentioned above, many genes encode more than a single protein (**Figure 17.23A**). Alternative splicing leads to different combinations of exons in the mature mRNAs transcribed from a single gene (see Figure 14.21). Posttranslational modifications also increase the number of proteins that derive from one gene (see Figure 12.17). Therefore, the sum total of the proteins produced by an organism—its **proteome**—is more complex than its genome.

Two technologies are commonly used to analyze the proteome:

- Because of their unique amino acid compositions (primary structures), most proteins have unique electric charges and sizes. On the basis of these two properties, they can be separated by *two-dimensional gel electrophoresis*. Thus isolated, individual proteins can be analyzed, sequenced, and studied (**Figure 17.23B**).

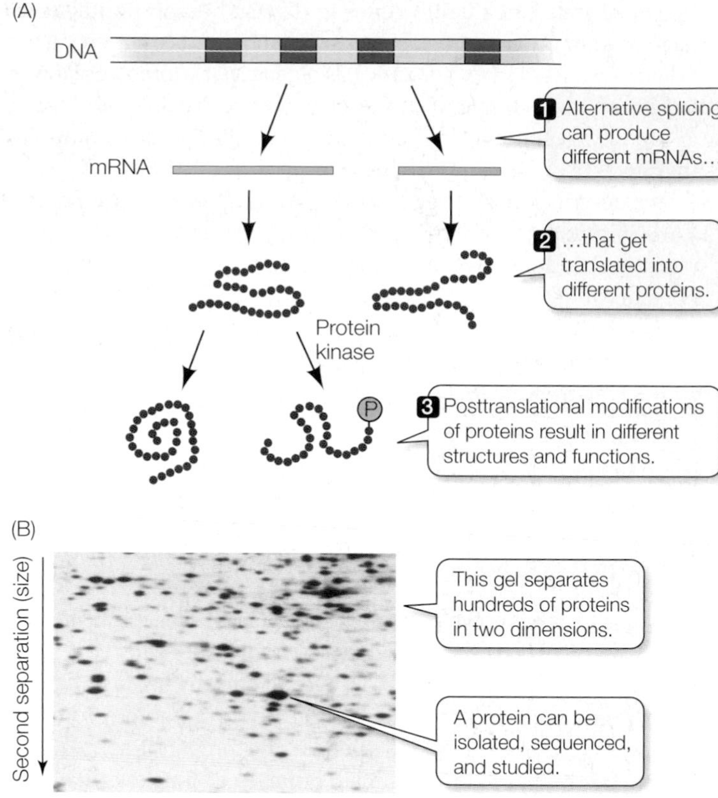
(A)
DNA
mRNA
1 Alternative splicing can produce different mRNAs...
2 ...that get translated into different proteins.
Protein kinase
P
3 Posttranslational modifications of proteins result in different structures and functions.
(B)
Second separation (size)
This gel separates hundreds of proteins in two dimensions.
A protein can be isolated, sequenced, and studied.
First separation (charge)

17.23 Proteomics (A) A small number of genes can code for a large number of proteins. (B) A cell's proteins can be separated on the basis of charge and size by gel electrophoresis. The two separations can distinguish most proteins from one another.

- *Mass spectrometry* uses electromagnets to identify proteins by the masses of their atoms and displays them as peaks on a graph.

The ultimate aim of proteomics is just as ambitious as that of genomics. While genomics seeks to describe the genome and its expression, proteomics seeks to describe the phenotypes of the expressed proteins.

An amazing example of proteomics, combined with DNA chip technology, is the recent comparison of brain proteins in chimpanzees and humans. Comparative DNA sequencing has shown that humans and chimpanzees differ by no more than 3 percent at the DNA level. Svante Pääbo and his colleagues examined gene expression in the "thinking" part (the cortex) of the brains of three chimpanzees and three humans who had died of natural causes. Of 12,000 DNA sequences tested for expression as mRNA, only 175 (1.4 percent) differed between the two species—a truly humbling result. But proteomics showed that the *specific proteins* expressed by those sequences differed by 7.4 percent, probably due to alternative splicing. The *amounts* of those proteins differed even more (31.4 percent). So what makes our brains different from a chimpanzee's is more quantitative than qualitative. This finding suggests that the control of gene expression may be the key to human evolution.

Systems biology integrates data from genomics and proteomics

With the ease of generating DNA sequences, mRNA expression profiles, and proteomic profiles, a huge amount of data about biological systems is accumulating. How does it all fit together? These kinds of studies are essentially *reductionist*, dissecting biology into ever smaller parts, as has been most of biological science up to now. In contrast, Charles Darwin looked at all of nature, and fit it all together by taking a "top-down" approach in his theory of natural selection. Can we do the same with our molecular data?

Systems biology aims to integrate molecular data into a coherent picture of life. A *system* is a group of parts that interact with one another, forming a whole that is often greater than the sum of the parts. An electric light bulb provides an analogy. You can take three parts—a filament made of tungsten, a metal cup, and a thin glass bowl—and study them individually, but only when the three are put together, with the filament inside the glass bowl and the metal cup at the end, does a "system," with properties that none of the parts possess by themselves, emerge.

Systems biologists strive to discover such *emergent properties* by studying the interactions in a biological system. If the interactions within a system are understood, predictions of how events will proceed in that system under new physiological conditions may be made. For example, when a drug is designed to inhibit one enzyme in a biological pathway, systems biology may predict what the effects will be on the many other pathways that interact with that one. This approach could make drug side effects more predictable and better controlled, and even, if care is taken, eliminate them.

Figure 17.24 is a diagram of a metabolic pathway. It shows interactions between and amounts of some mRNA transcripts, proteins, and metabolites involved in fat metabolism in two different genetic strains of mice. One strain has a mutation that results in a tendency to get atherosclerosis and the other (the wild type) does not. The redder a dot, the more of the molecule it represents in the mutant strain compared with the wild type; the greener a dot, the less of a molecule in the mutant strain. Looking at the left of the diagram, you can see that protein A is clearly up-regulated in the mouse strain that gets atherosclerosis, while protein B is down-regulated in that mutant strain.

Systems biologists can try to use such information to understand how increasing or decreasing the amounts of one molecule might affect the levels of others. For example, what if a drug were designed to increase the level of protein B in the mutant strain of mice, so that it was now at the level seen in normal mice? Looking at Figure 17.24, you can see that protein B has several lines of interaction with other molecules, which in turn interact with still others. A whole new field of computational biology is being developed to help scientists make these kinds of predictions.

The objective of molecular medicine, as in all medicine, is to improve the health of the patient. Scientists have made great progress in understanding and treating diseases caused by single molecular events, such as phenylketonuria. The new approach of systems biology holds promise for understanding and treating the more complex diseases, such as heart disease and cancer, that continue to plague humankind.

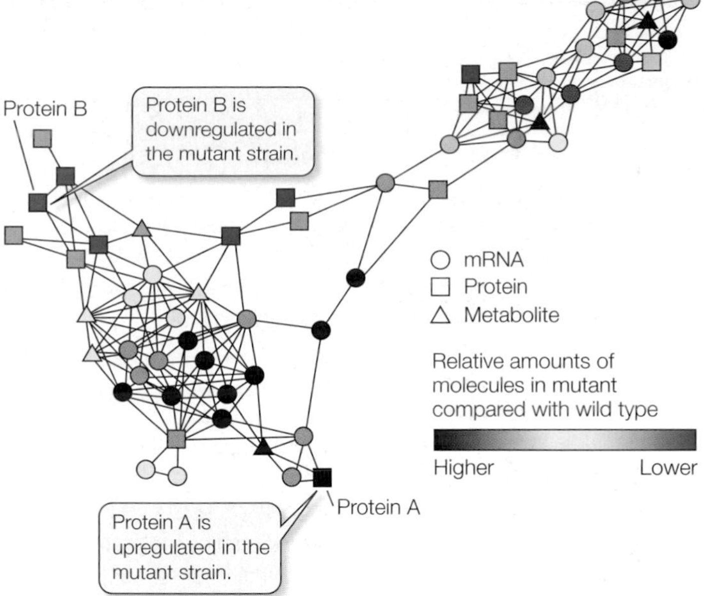

17.24 Applying Systems Biology This diagram compares the levels and interactions of mRNA transcripts, proteins, and metabolites in the pathways involving fat metabolism in two strains of mice. One strain (mutant) tends to get atherosclerosis; the other (wild type) does not. The intensities of the colors indicate amounts of molecules in the mutant strain compared with the wild-type strain (redder means increased and greener means decreased). Lines indicate interactions between molecules.

17.6 RECAP

Advances in DNA sequencing technology made it possible to map the entire human genome. Having a map of the human genome makes it possible to isolate specific genes, to compare genomic sequences between individuals and between species, and to identify mutations that cause disease. Next steps include analysis of the proteome and integrating molecular knowledge into the framework of systems biology.

- Can you explain the difference between hierarchical and shotgun DNA sequencing? See p. 393 and Figure 17.22
- What is the proteome and how does it relate to the genome? See pp. 395–396 and Figure 17.23
- What is systems biology and what is its importance in molecular medicine? See p. 396

CHAPTER SUMMARY

17.1 How do defective proteins lead to diseases?

Abnormalities in nearly all classes of proteins, including enzymes, transport proteins, receptor proteins, and structural proteins, have been implicated in genetic diseases.

While a single amino acid difference can be the cause of disease, amino acid variations have been detected in many functional proteins. Review Figure 17.2

Transmissible spongiform encephalopathies (TSEs) are degenerative brain diseases that can be transmitted from one animal to another by consumption of infected tissues. The infective agent is a **prion**, a protein with an abnormal conformation.

Multifactorial diseases are caused by the interactions of many genes and proteins with the environment. They are much more common than diseases caused by mutations in a single gene.

Predictable patterns of inheritance are associated with some human genetic diseases. Autosomal recessive, autosomal dominant, and X-linked patterns are common.

17.2 What kinds of DNA changes lead to human diseases?

It is possible to isolate both the mutant genes and the abnormal proteins responsible for human diseases.

Positional cloning is a method of locating disease-causing genes by finding polymorphic genetic markers that are inherited with the disease.

Restriction fragment length polymorphisms (**RFLPs**) and single nucleotide polymorphisms (SNPs) are commonly used as genetic markers in positional cloning. Review Figure 17.8, Web/CD Activity 17.1

Mutations occur often where cytosine has been methylated to 5-methylcytosine. Review Figure 17.9

The effects of fragile-X syndrome worsen with each generation. This pattern is the result of an **expanding triplet repeat**. Review Figure 17.10

Genomic imprinting results in a gene being expressed differently depending on the sex of the parent from which it comes.

17.3 How does genetic screening detect human diseases?

Genetic screening can detect human genetic diseases, alleles predisposing people to those diseases, or carriers of those diseases.

Genetic screening can be done by looking for abnormal protein expression.

DNA testing is the direct identification of mutant alleles by sequence comparison. Any cell can be tested at any time in the life cycle.

The two predominant methods of DNA testing are the allele-specific cleavage method and allele-specific oligonucleotide hybridization method. Review Figures 17.12 and 17.13, Web/CD Tutorial 17.1

17.4 What is cancer?

Cancer cells fail to respond to the normal controls on cell division, and divide continuously.

Tumors may be **benign** or **malignant**. Malignant tumors can spread to other parts of the body through a process called **metastasis**.

Some types of human cancers are caused by viruses, but 85 percent of human cancers are caused by genetic mutations of somatic cells. These mutations occur most commonly in dividing cells. **Carcino-**

CHAPTER SUMMARY

gens can cause mutations that lead to cancer, but some mutations arise spontaneously.

Normal cells contain **oncogenes** that stimulate cell division. When mutated, these genes may become active in a cell in which they are normally turned off. Review Figure 17.15

Normal cells also contain **tumor suppressor genes** that inhibit cell division. When mutated, they may become inactive. Review Figure 17.17

The two-hit hypothesis describes the difference between inherited cancer, in which an individual inherits one mutant allele of a tumor suppressor gene and a somatic mutation occurs later in the second allele, and sporadic cancer, in which two mutational events must occur in the same somatic cell to produce cancer. Review Figure 17.16

In most cases, mutations must activate several oncogenes and inactivate several tumor suppressor genes to produce a malignant tumor. Review Figure 17.19

17.5 How are genetic diseases treated?

There are three ways to modify the phenotype of a genetic disease: restrict the substrate of a deficient enzyme, inhibit a harmful metabolic reaction, or supply a missing protein.

In **gene therapy**, a mutant gene is replaced with a normal gene. Both *ex vivo* and *in vivo* therapies are being developed. Review Figure 17.21

17.6 What have we learned from the Human Genome Project?

The entire human genome has been sequenced using both **hierarchical** and **shotgun** methods. Review Figure 17.22, Web/CD Tutorial 17.2

The human genome has only about 24,000 genes.

Humans make many more proteins than predicted by their number of genes. Thus the **proteome** is more complex than the genome. Review Figure 17.23

Systems biology attempts to unify data from genomics and proteomics into a coherent, predictive whole. Review Figure 17.24

See Web/CD Activity 17.2 for a concept review of this chapter.

SELF-QUIZ

1. Phenylketonuria is an example of a genetic disease in which
   a. a single enzyme is not functional.
   b. inheritance is sex-linked.
   c. two parents without the disease cannot have a child with the disease.
   d. mental retardation always occurs, regardless of treatment.
   e. a transport protein does not work properly.
2. Mutations of the gene for β-globin
   a. are usually lethal.
   b. occur only at amino acid position 6.
   c. number in the hundreds.
   d. always result in sickling of red blood cells.
   e. can always be detected by gel electrophoresis.
3. Multifactorial (complex) diseases
   a. are less common than single-gene diseases.
   b. involve the interaction of many genes with the environment.
   c. affect less than 1 percent of humans.
   d. involve the interactions of several mRNAs.
   e. are exemplified by sickle-cell disease.
4. In fragile-X syndrome,
   a. females are affected more severely than males.
   b. a short sequence of DNA is repeated many times to create the fragile site.
   c. both the X and Y chromosomes tend to break when prepared for microscopy.
   d. all people who carry the gene that causes the syndrome are mentally retarded.
   e. the basic pattern of inheritance is autosomal dominant.
5. Most genetic diseases are rare because
   a. each person is unlikely to be a carrier for harmful alleles.
   b. genetic diseases are usually sex-linked and so uncommon in females.
   c. genetic diseases are always dominant.
   d. two parents probably do not carry the same recessive alleles.
   e. mutation rates in humans are low.
6. Mutational "hot spots" in human DNA
   a. always occur in genes that are transcribed.
   b. are common at cytosines that have been modified to 5-methylcytosine.
   c. involve long stretches of nucleotides.
   d. occur where there are long repeats.
   e. are very rare in genes that code for proteins.
7. Newborn genetic screening for PKU
   a. is very expensive.
   b. detects phenylketones in urine.
   c. has not led to the prevention of mental retardation resulting from this disorder.
   d. must be done during the first day of an infant's life.
   e. uses bacterial growth to detect excess phenylalanine in blood.
8. Genetic diagnosis by DNA testing
   a. detects only mutant and not normal alleles.
   b. can be done only on eggs or sperm.
   c. involves hybridization to rRNA.
   d. utilizes restriction enzymes and a polymorphic site.
   e. cannot be done with PCR.
9. Most human cancers
   a. are caused by viruses.
   b. are in blood cells or their precursors.
   c. involve mutations of somatic cells.
   d. spread through solid tissues rather than by the blood or lymphatic system.
   e. are inherited.
10. Current treatments for genetic diseases include all of the following *except*
    a. restricting a dietary substrate.
    b. replacing the mutant gene in all cells.
    c. alleviating the patient's symptoms.
    d. inhibiting a harmful metabolic reaction.
    e. supplying a protein that is missing.

FOR DISCUSSION

1. How do oncogenes and tumor suppressor genes and their functions change in tumor cells? Propose targets for cancer therapy involving their gene products.
2. In the past, it was common for people with phenylketonuria (PKU) who were placed on a low-phenylalanine diet after birth to be allowed to return to a normal diet during their teenage years. Although the levels of phenylalanine in their blood were high, their brains were thought to be beyond the stage of being harmed. If a woman with PKU becomes pregnant, however, a problem arises. Typically, the fetus is heterozygous, but is unable at early stages of development to metabolize the high levels of phenylalanine that arrive from the mother's blood. Why is the fetus heterozygous? What do you think would happen to the fetus during this "maternal PKU" situation? What would be your advice to a woman with PKU who wants to have a child?
3. Cystic fibrosis is an autosomal recessive disease in which thick mucus is produced in the lungs and airways. The gene responsible for this disease codes for a protein composed of 1,480 amino acids. In most patients with cystic fibrosis, the protein has 1,479 amino acids: a phenylalanine is missing at position 508. A baby is born with cystic fibrosis. He has an older brother. How would you test the DNA of the older brother to determine whether he is a carrier for cystic fibrosis? How would you design a gene therapy protocol to "cure" the cells in the younger brother's lungs and airways?
4. A number of efforts are under way to identify human genetic polymorphisms that correlate with multifactorial diseases such as diabetes, heart disease, and cancer. What would be the uses of such information? What concerns do you think are being raised by the people whose genomes are being analyzed?

FOR INVESTIGATION

We have seen that human cells in culture can be transformed from normal to tumor cells (see Figure 17.18), and that human genes can be isolated by positional cloning (see Figure 17.8). A group of related prostate cancer patients do not have mutations in any of the known tumor suppressor genes. You suspect that a new, as yet undiscovered tumor suppressor gene has been mutated. How would you use these methods to isolate the new gene from these patients?

CHAPTER 18 Immunology: Gene Expression and Natural Defense Systems

The most dangerous foe

On January 6, 1777, George Washington, commander of the Revolutionary army of the fledgling United States, wrote to his chief physician: "Finding smallpox to be spreading much, and fearing that no precaution can prevent it from running through the whole of our army, I have determined that the troops shall be inoculated. Should the disease rage with its usual virulence, we should have more to dread from it than the sword of the enemy."

Washington was speaking from experience. He himself had survived smallpox as a teenager, in 1751. And during the year 1776 his army lost 1,000 men in battle—and 10,000 men to smallpox. After Washington's men were inoculated, the death rate due to smallpox in the Revolutionary army plummeted.

Inoculation in Washington's army meant introducing a small amount of fluid obtained from a smallpox pustule on a recent victim to a healthy person. For reasons not understood at the time, most people inoculated in this way became immune to the full ravages of smallpox—as Washington himself, having survived the disease, was immune.

Two decades later, the English country doctor Edward Jenner observed that milkmaids frequently contracted a mild infection called cowpox, and that these women seemed *not* to be affected by the smallpox epidemics that ravaged England. Reasoning that there was some kind of cross-resistance between the two infections, Jenner performed an experiment that modern medical science would forbid as unethical.

Jenner rubbed powdered remnants from cowpox lesions into scratches on the arm of a young boy named James Phipps. Not unexpectedly, James came down with a mild case of cowpox. Six weeks later, Jenner infected James with the residue of smallpox lesions, discovering that, as Jenner expected (and, we presume, hoped), the boy did not contract smallpox.

Because of the connection with cows, for which the Latin word is *vacca*, the term coined for Jenner's procedure was *vaccination*. A much safer approach than inoculation, the practice of vaccination spread quickly. More potent vaccines were developed, and massive programs were instituted to vaccinate much of the world's population. By 1980, smallpox had literally disappeared.

Why was George Washington immune to smallpox? Why did inoculation save his soldiers? How could a vaccination program rid the world of smallpox? The answers lie in the cells and molecules of

Vaccination An artist's depiction of the first smallpox vaccination. Edward Jenner used a solution containing a related virus—cowpox—to confer immunity against smallpox.

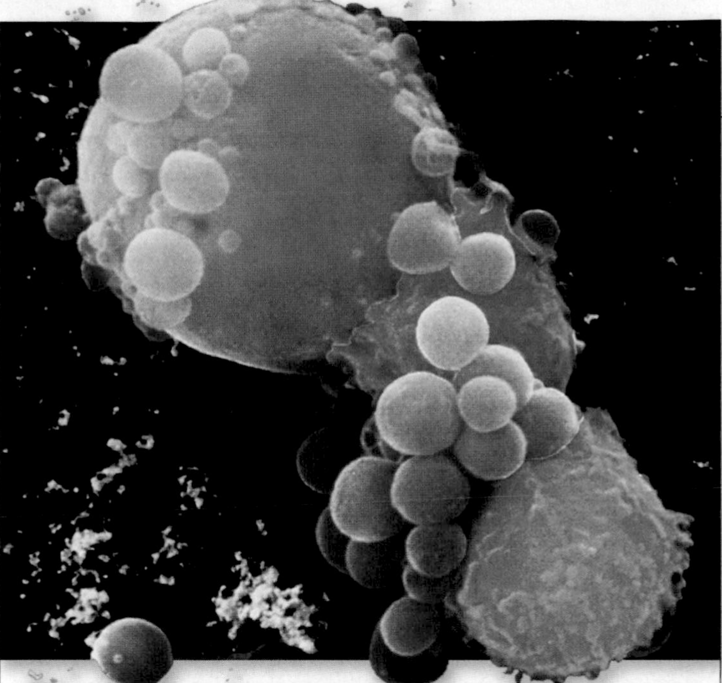

T Cells in Action Two cytotoxic T cells (orange) have come into contact with virus-infected cells (pink). The infected cell at the top left has begun to die, as indicated by membrane blisters. The process is complete in the cell in the center, which has disintegrated into small blebs.

CHAPTER OUTLINE

our immune systems. When Washington caught smallpox in 1751, specialized white blood cells called macrophages engulfed and destroyed some of the smallpox viruses and displayed fragments of the viruses on their surfaces. Other specialized white blood cells, T cells, recognized the fragments and become "activated." Descendants of activated T cells attacked Washington's virus-infected cells, preventing the lethal spread of the disease. Other descendants of the T cells persisted in his body as "memory cells," which defended him when he was exposed to the disease later. Inoculation and vaccination against a pathogen stimulate the formation of these memory cells, which then protect the body against infection.

IN THIS CHAPTER we'll see how the nonspecific immune system attempts to prevent pathogens from entering the body. We then describe how the specific immune system targets invaders such as smallpox virus for destruction. We'll see how reshuffling genetic material helps animals fight off a mind-boggling diversity of potential invaders. We conclude by considering what happens when this complex and crucial system malfunctions.

18.1 What Are the Major Defense Systems of Animals?

Animals have a number of ways of defending themselves against **pathogens**—harmful organisms and viruses that can cause disease. These defense systems are based on the distinction between *self*—the animal's own molecules—and *nonself*, or foreign, molecules. In this section we will consider the mechanisms by which animals recognize and mount defenses against nonself molecules. Many of these mechanisms are based on the principles of genetics and molecular biology that have been discussed in earlier chapters.

There are two general types of defense mechanisms:

- **Nonspecific defenses**, or *innate defenses*, are inherited mechanisms that protect the body from many kinds of pathogens. Nonspecific defenses, which typically act very rapidly, include barriers such as the skin, molecules that are toxic to invaders, and phagocytic cells that ingest invaders. Most animals—including both invertebrates and vertebrates—as well as plants have nonspecific defenses.
- **Specific defenses** are *adaptive* mechanisms aimed at a specific pathogen. For example, these defense systems can make an antibody protein that will recognize, bind to, and destroy a certain virus if that virus ever enters the bloodstream. Specific defense mechanisms are found in vertebrate animals. They are typically slow to develop and long-lasting.

In animals that have both kinds of defense mechanisms, nonspecific and specific defenses operate together as a coordinated defense system. Because the specific defenses often require days or even weeks to become effective, the nonspecific defenses are the body's first line of defense. Considering that a bacterium can produce 20 million progeny a day, the nonspecific defenses are tremendously important in keeping disease at bay.

Blood and lymph tissues play important roles in defense systems

The components of the mammalian defense system are dispersed throughout the body and interact with almost all of its other tissues and organs. The *lymphoid tissues*, which include the thymus, bone marrow, spleen, and lymph nodes, are essential parts of the defense system (**Figure 18.1**). Central to their functioning are the blood and lymph.

Blood and lymph are both fluid tissues that consist of water, dissolved solutes, and cells.

- **Blood plasma** is a yellowish solution containing ions, small molecular solutes, and soluble proteins. Suspended in the plasma are red blood cells, white blood cells, and platelets (cell fragments essential to blood clotting). While red blood cells are normally confined to the *closed circulatory system* (the heart, arteries, capillaries, and veins), white blood cells and platelets are also found in the lymph.
- **Lymph** is a fluid derived from the blood and other tissues that accumulates in intercellular spaces throughout the body. From these spaces, the lymph moves slowly into the vessels of the *lymphatic system*. Tiny lymph capillaries conduct this fluid to larger ducts that eventually join together, forming one large vessel, the *thoracic duct*, which joins a major vein (the left subclavian vein) near the heart. By this system of vessels, the lymph is eventually returned to the blood and the circulatory system.

At many sites along the lymph vessels are small, roundish structures called **lymph nodes**, which contain a variety of white blood cells. As lymph passes through a lymph node, it is filtered and "inspected" for nonself materials by these defensive cells.

White blood cells play many defensive roles

One milliliter of human blood typically contains about 5 billion red blood cells and 7 million of the larger white blood cells. All of these cells originate from *pluripotent stem cells* in the bone marrow, which constantly divide and can differentiate into a wide variety of blood cell types (**Figure 18.2**). **White blood cells** (also called *leukocytes*) have nuclei and are colorless, unlike mammalian red blood cells, which lose their nuclei during development. White blood cells can leave the closed circulatory system and enter intercellular spaces where nonself cells or substances are present. The number of white blood cells in the blood and lymph may rise sharply in response to invading pathogens, providing medical professionals with a useful clue for detecting an infection.

Several types of white blood cells are important in the body's defenses. White blood cells fall into two broad groups:

- **Granular cells** include histamine-producing signaling cells as well as **phagocytes**, which engulf and digest foreign cells and cellular debris. Among the most important phagocytes are dendritic cells and **macrophages**. In addition to engulfing nonself materials, macrophages have the important function of presenting partly digested nonself materials to the T cells.
- **Lymphocytes** participate in specific defenses against nonself or altered cells, such as virus-infected cells and tumor cells. There are two types of lymphocytes, B cells and T cells.

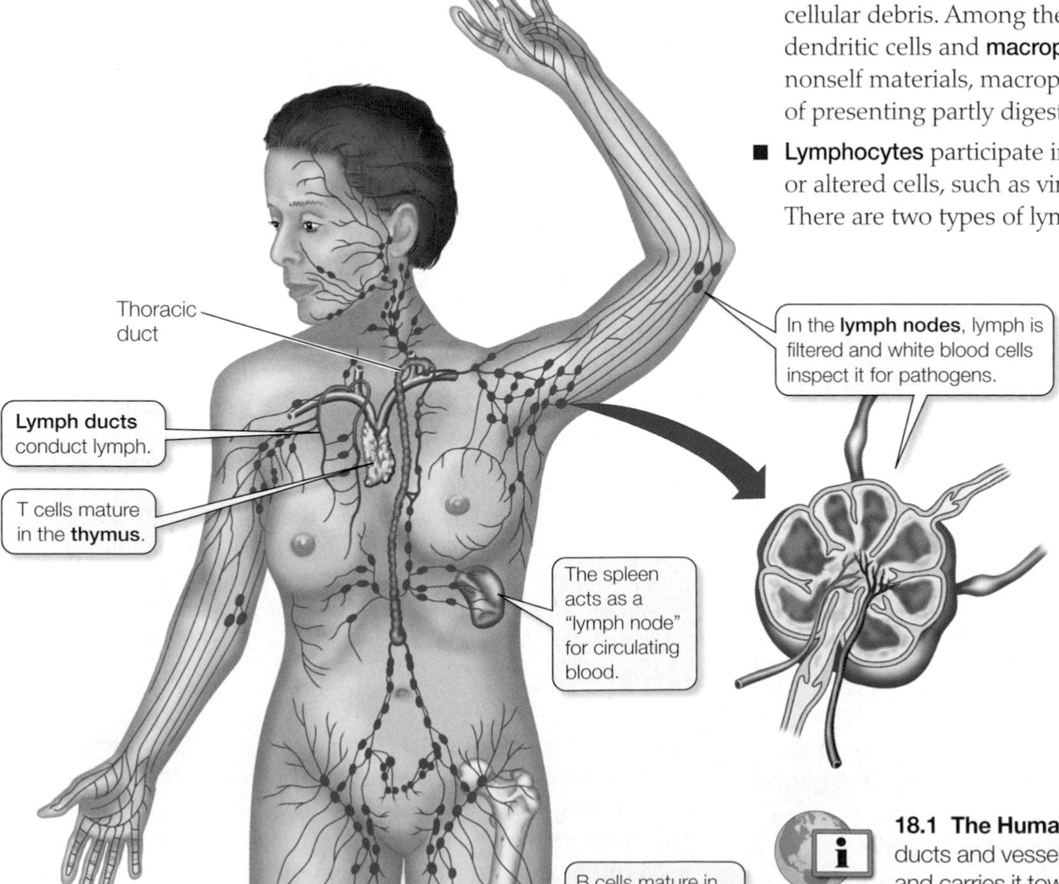

Thoracic duct
Lymph ducts conduct lymph.
T cells mature in the thymus.
In the lymph nodes, lymph is filtered and white blood cells inspect it for pathogens.
The spleen acts as a "lymph node" for circulating blood.
B cells mature in the bone marrow.

18.1 The Human Lymphatic System A network of ducts and vessels collects lymph from body tissues and carries it toward the heart, where it mixes with blood to be pumped back to the tissues. Other lymphoid tissues, including the thymus, spleen, and bone marrow, are also essential to the body's defense system.

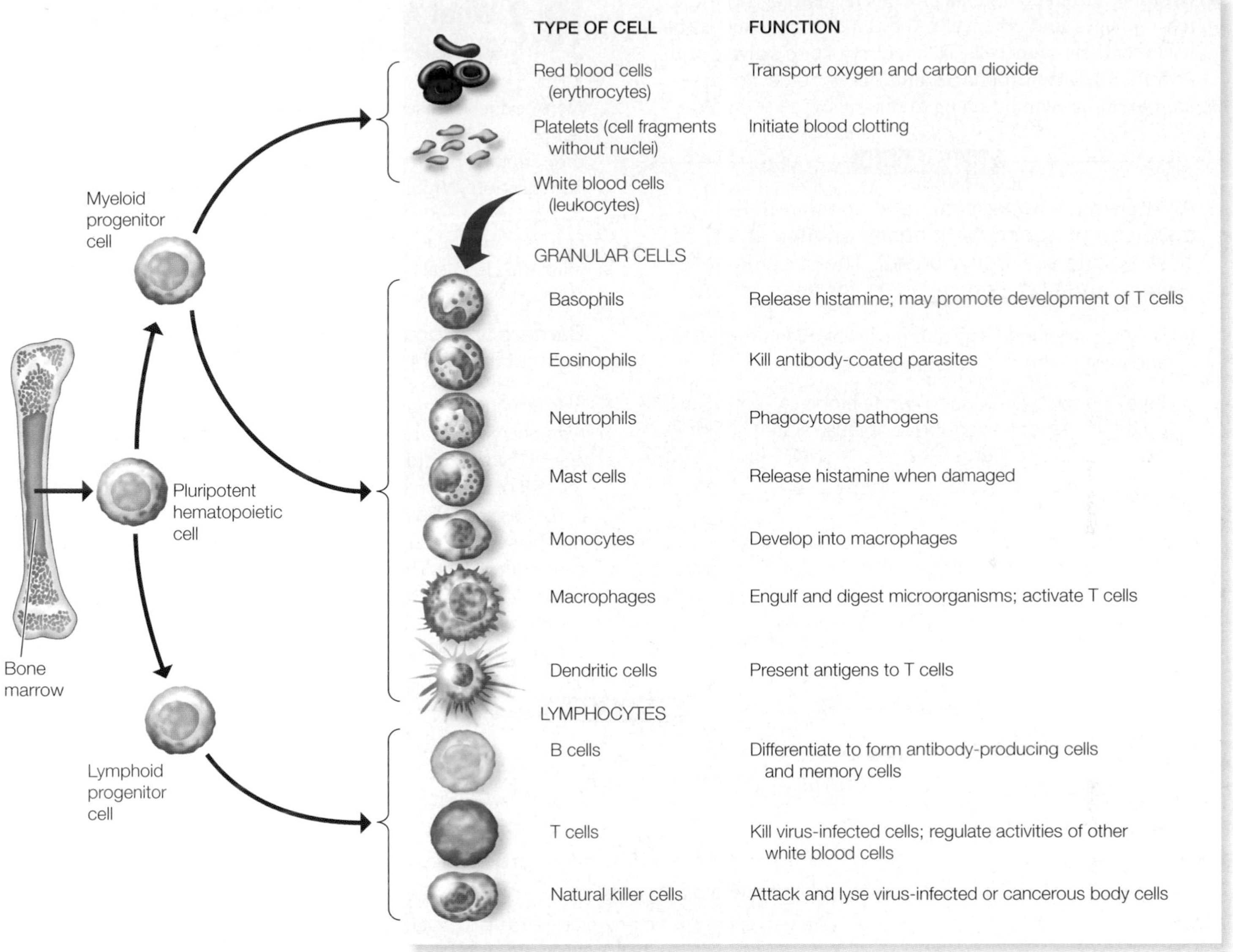

Myeloid progenitor cell

Pluripotent hematopoietic cell

Bone marrow

Lymphoid progenitor cell

TYPE OF CELL	FUNCTION
Red blood cells (erythrocytes)	Transport oxygen and carbon dioxide
Platelets (cell fragments without nuclei)	Initiate blood clotting
White blood cells (leukocytes)	
GRANULAR CELLS	
Basophils	Release histamine; may promote development of T cells
Eosinophils	Kill antibody-coated parasites
Neutrophils	Phagocytose pathogens
Mast cells	Release histamine when damaged
Monocytes	Develop into macrophages
Macrophages	Engulf and digest microorganisms; activate T cells
Dendritic cells	Present antigens to T cells
LYMPHOCYTES	
B cells	Differentiate to form antibody-producing cells and memory cells
T cells	Kill virus-infected cells; regulate activities of other white blood cells
Natural killer cells	Attack and lyse virus-infected or cancerous body cells

18.2 Blood Cells Pluripotent stem cells in the bone marrow can differentiate into red blood cells, platelets, and the various types of white blood cells.

- Immature **T cells** migrate from the bone marrow via the blood to the thymus, where they mature.
- Mature **B cells** leave the bone marrow and circulate through the blood and lymph vessels. B cells make specialized proteins called *antibodies* that enter the blood and bind to nonself substances.

Fundamental to the interactions, control, and defensive functioning of these white blood cells are defensive proteins and other signals.

Immune system proteins bind pathogens or signal other cells

The cells that defend mammalian bodies work together like cast members in a drama, interacting with one another and with the cells of invading pathogens. These cell–cell interactions are accomplished by a variety of key proteins, including receptors, other cell surface proteins, and signaling molecules. While these proteins will be discussed in more detail later in the chapter, four of the major players warrant mention here:

- **Antibodies** are proteins that bind specifically to certain substances identified by the immune system as nonself or altered self, thereby denaturing the invading nonself substance. They are secreted by B cells as defensive weapons.
- **T cell receptors** are integral membrane proteins on the surfaces of T cells. They recognize and bind to nonself substances on the surfaces of other cells.
- **Major histocompatibility complex** (**MHC**) proteins protrude from the surfaces of most cells in the mammalian body. They are important self-identifying labels and play major parts in coordinating interactions between lymphocytes and macrophages.

- **Cytokines** are soluble signal proteins released by T cells, macrophages, and other cells. They bind to and alter the behavior of their target cells. Different cytokines activate or inactivate B cells, macrophages, and T cells. Some cytokines limit tumor growth by killing tumor cells.

18.1 RECAP

Animals have nonspecific and specific defenses against pathogens, both based on their ability to differentiate self from nonself. The specific defenses target specific invaders for destruction.

- Do you understand the differences between specific and nonspecific defenses? See p. 401
- What are the two classes of white blood cells, and how do they function in vertebrate defense systems? See pp. 402–403 and Figure 18.2

The role that the specific defenses may be called upon to play in fighting disease often depends on the success of the nonspecific responses to invading pathogens. We turn now to these nonspecific defenses that protect vertebrates from disease.

18.2 What Are the Characteristics of the Nonspecific Defenses?

Nonspecific defenses are general protection mechanisms that attempt to stop pathogens from invading the body. As noted above, they are the first lines of the body's defense system, in both time and location. While the specific defense systems of vertebrates evolved about 500 million years ago, the nonspecific defenses are far older. In humans, they include physical barriers as well as cellular and chemical defenses (**Table 18.1**).

Barriers and local agents defend the body against invaders

Skin is a primary nonspecific defense against invasion. Fungi, bacteria, and viruses rarely penetrate healthy, unbroken skin. But damage to the skin or to the internal surface tissues greatly increases the risk of infection by pathogens.

The bacteria and fungi that normally live and reproduce in great numbers on our body surfaces without causing disease are referred to as **normal flora**. These natural occupants of our bodies compete with pathogens for space and nutrients and are thus a form of nonspecific defense.

TABLE 18.1

Human Nonspecific Defenses

DEFENSIVE MECHANISM	FUNCTION
Surface barriers	
Skin	Prevents entry of pathogens and foreign substances
Acid secretions	Inhibit bacterial growth on skin
Mucus	Prevents entry of pathogens; produces defensins that kill pathogens
Mucous secretions	Trap bacteria and other pathogens in digestive and respiratory tracts
Nasal hairs	Filter bacteria in nasal passages
Cilia	Move mucus and trapped materials away from respiratory passages
Gastric juice	Concentrated HCl and proteases destroy pathogens in stomach
Acid in vagina	Limits growth of fungi and bacteria in female reproductive tract
Tears, saliva	Lubricate and cleanse; contain lysozyme, which destroys bacteria
Nonspecific cellular, chemical, and coordinated defenses	
Normal flora	Compete with pathogens; may produce substances toxic to pathogens
Fever	Body-wide response inhibits microbial multiplication and speeds body repair processes
Coughing, sneezing	Expels pathogens from upper respiratory passages
Inflammatory response (involves leakage of blood plasma and phagocytes from capillaries)	Limits spread of pathogens to neighboring tissues; concentrates defenses; digests pathogens and dead tissue cells; released chemical mediators attract phagocytes and lymphocytes to site
Phagocytes (macrophages and neutrophils)	Engulf and destroy pathogens that enter body
Natural killer cells	Attack and lyse virus-infected or cancerous body cells
Antimicrobial proteins	
Interferons	Released by virus-infected cells to protect healthy tissue from viral infection; mobilize specific defenses
Complement proteins	Lyse microorganisms, enhance phagocytosis, and assist in inflammatory and antibody responses

Skin is a major organ of the vertebrate body. In an "average" adult human, skin accounts for about 15 percent of the body's weight. A typical square inch of living human skin holds over 500 sweat glands, 20 blood vessels, and more than a thousand nerve endings.

The mucous membranes found at the surfaces of the visual, respiratory, digestive, excretory, and reproductive systems have other defenses against pathogens. Tears, nasal mucus, and saliva contain an enzyme called **lysozyme** that attacks the cell walls of many bacteria. Mucus in the nose traps airborne microorganisms, and most of those that get past this filter end up trapped in mucus deeper in the respiratory tract. Mucus and trapped pathogens are removed by the beating of cilia in the respiratory passageway, which continuously move a sheet of mucus and the debris it contains up toward the nose and mouth. Sneezing is another way to remove microorganisms from the respiratory tract.

Finally, the mucous membranes produce **defensins**, peptides that consist of 29–42 amino acids and contain hydrophobic domains. They are toxic to a wide range of pathogens, including bacteria, microbial eukaryotes, and enveloped viruses. Defensins insert themselves into the plasma membranes of these organisms and, by some unknown mechanism, kill the invaders. They are also produced in phagocytes, where they kill pathogens trapped by phagocytosis.

Pathogens that reach the digestive tract (stomach, small intestine, and large intestine) are met by other defenses. The gastric juice in the stomach is a deadly environment for many bacteria because of the hydrochloric acid and proteases (protein-digesting enzymes) that are secreted into it. The intact lining of the small intestine is not normally penetrated by bacteria, and some pathogens are killed by bile salts secreted into this part of the digestive tract. The large intestine harbors many bacteria, which multiply freely; however, they are usually removed quickly with the feces. Most of the bacteria in the large intestine are normal flora that provide benefits to their host. We probably add to this beneficial flora when we eat foods such as active-culture yogurt and various cheeses.

All of these barriers and local agents are *nonspecific* defenses because they act on all invading pathogens in the same way. More complex nonspecific defenses await any pathogens that manage to elude this first line of defense.

Other nonspecific defenses include specialized proteins and cellular processes

Pathogens that penetrate the body's outer and inner surfaces encounter more complex nonspecific defenses that involve the secretion of various defensive proteins as well as defensive cells.

COMPLEMENT PROTEINS Vertebrate blood contains about 30 different antimicrobial proteins that make up the **complement system**. These proteins, in different combinations, provide three types of defenses. In each type of defense, the complement proteins act in a characteristic sequence, or *cascade*, with each protein activating the next:

- First they attach to microbes, which helps phagocytes recognize and destroy them.
- Then they activate the inflammation response and attract phagocytes to the site of infection.
- Finally, they lyse (burst) invading cells (such as bacteria).

INTERFERONS When cells are infected by a virus, they produce small amounts of antimicrobial proteins called **interferons** that increase the resistance of neighboring cells to infection by the same *or other* viruses. Interferons have been found in many vertebrates and are one of the body's first lines of nonspecific defense against the internal spread of viral infection.

Interferons differ from species to species, and each vertebrate species produces at least three different interferons. All interferons are glycoproteins (proteins with attached carbohydrate groups) consisting of about 160 amino acids. By binding to receptors in the plasma membranes of uninfected cells, interferons stimulate a signaling pathway that inhibits viral reproduction inside the infected cells. They also stimulate lysosome activity that digests viral proteins into peptides, which, when brought to the cell surface, stimulate the specific immune system (see Section 18.5).

PHAGOCYTES Some phagocytes travel freely in the circulatory and lymphatic systems; others can move out of blood vessels and adhere to certain tissues. Pathogenic cells, viruses, or fragments of these invaders can become attached to the plasma membrane of a phagocyte (**Figure 18.3**), which ingests them by phagocytosis. Defensins inside these phagocytes then kill the pathogens.

NATURAL KILLER CELLS One class of lymphocytes, known as **natural killer cells**, can distinguish virus-infected cells and some tumor cells from their normal counterparts and initiate the lysis of these target cells. In addition to this nonspecific action, natural killer cells interact with the specific defenses.

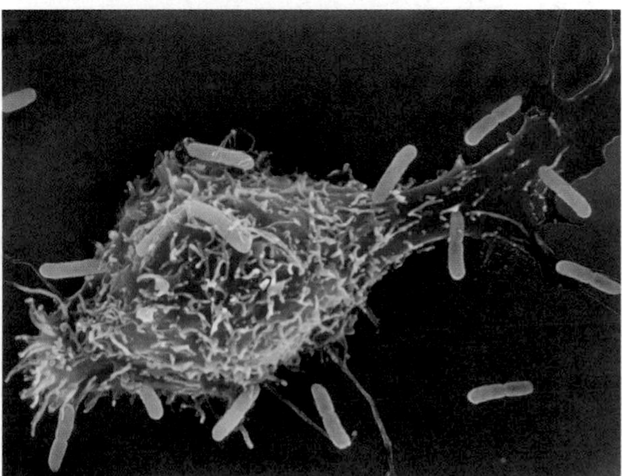

18.3 A Phagocyte and Its Bacterial Prey Several bacteria (yellow in this artificially colored micrograph) have become attached to the surface of a phagocyte in the human bloodstream. The bacteria will be engulfed by the phagocyte and destroyed. A single phagocyte can digest many bacteria.

Inflammation is a coordinated response to infection or injury

The body employs the **inflammation** response in dealing with infection or with any other process that causes tissue injury, either on the surface of the body or internally. The damaged body cells initiate the inflammation response by releasing small molecules and enzymes. Cells adhering to the skin and the linings of organs, called **mast cells**, release a chemical signal, called **histamine**, when they are damaged, as do white blood cells called **basophils**.

You have no doubt experienced the symptoms of inflammation: redness and swelling, accompanied by heat and pain. The redness and heat of inflammation result from histamine-induced dilation of blood vessels in the infected or injured area (**Figure 18.4**). Histamine also causes the capillaries (the smallest blood vessels) to become leaky, allowing blood plasma, along with complement proteins and phagocytes, to escape into the tissue, causing the characteristic swelling. The pain of inflammation results from increased pressure (from the swelling) and from the action of leaked enzymes on sensory nerve endings.

In damaged or infected tissue, complement proteins and other chemical signals attract phagocytes—neutrophils first, and then monocytes, which develop into macrophages. The phagocytes, which engulf the invaders and dead tissue cells, are responsible for most of the healing associated with inflammation. They produce several cytokines, which (among other functions) signal the brain to produce a fever. This rise in body temperature inhibits the growth of the invading pathogens.

Cytokines may also stimulate the cells lining the blood vessels to produce cell adhesion molecules that allow phagocytes in the capillaries to attach to the lining, leave the vessel, and enter the site of the injury. The arrival of phagocytes initiates a specific response to the pathogen.

Calor, dolor, rubor, tumor—the diagnosis of inflammation remains unchanged since 40 A.D. when the Roman medical scribe Aulus Cornelius Celsus described the syndrome in his text *De medicina*.

In some severe bacterial infections, the inflammation response does not remain local. Instead, it is disseminated throughout the bloodstream in a condition called *sepsis*. As in a local infection or injury, blood vessels dilate, but they do so throughout the body. This situation is a medical emergency and can be lethal.

Following inflammation, *pus* may accumulate. Pus is a collection of dead cells (bacteria, neutrophils and the damaged body cells) and leaked fluid. A normal result of inflammation, pus is gradually consumed and digested by macrophages.

18.4 Interactions of Cells and Chemical Signals Result in Inflammation The histamine-induced swelling of the inflammation response is accompanied by redness, heat, and pain. The chemical signals associated with inflammation attract the phagocytes that digest the pathogens and damaged cells.

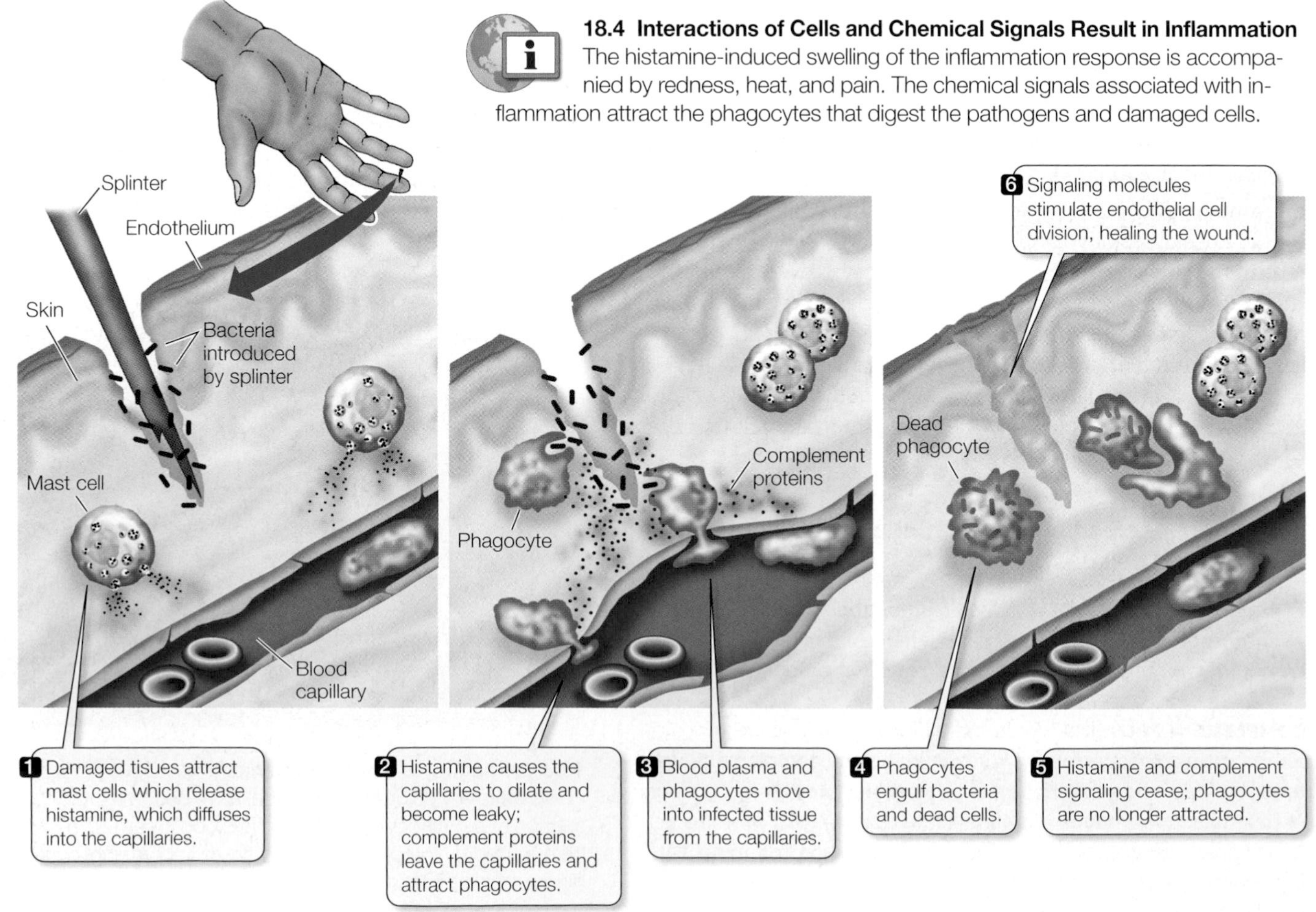

Splinter
Endothelium
Skin
Bacteria introduced by splinter
Mast cell
Blood capillary
Complement proteins
Phagocyte
Dead phagocyte
6 Signaling molecules stimulate endothelial cell division, healing the wound.
1 Damaged tisues attract mast cells which release histamine, which diffuses into the capillaries.
2 Histamine causes the capillaries to dilate and become leaky; complement proteins leave the capillaries and attract phagocytes.
3 Blood plasma and phagocytes move into infected tissue from the capillaries.
4 Phagocytes engulf bacteria and dead cells.
5 Histamine and complement signaling cease; phagocytes are no longer attracted.

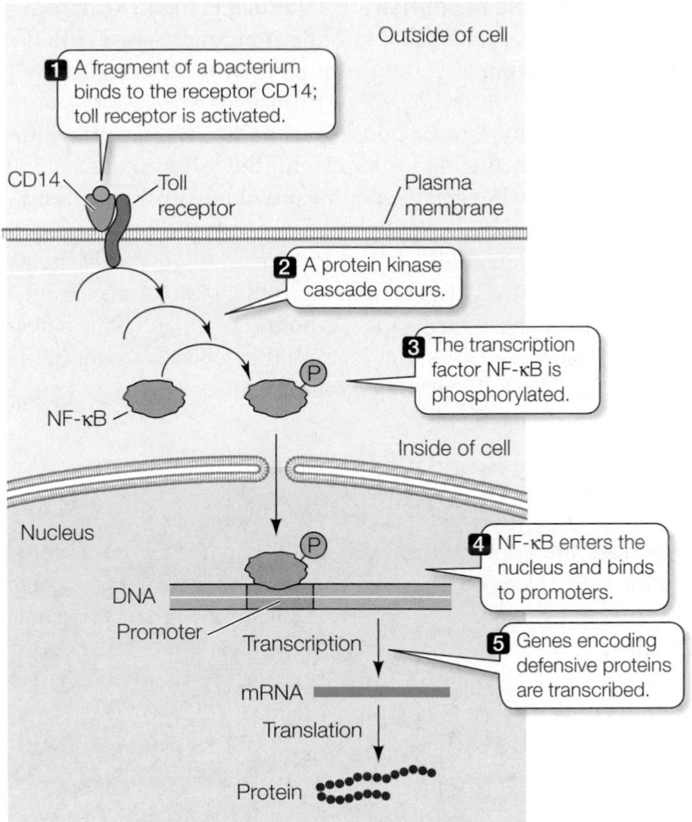
Outside of cell
1 A fragment of a bacterium binds to the receptor CD14; toll receptor is activated.
CD14
Toll receptor
Plasma membrane
2 A protein kinase cascade occurs.
3 The transcription factor NF-κB is phosphorylated.
P
NF-κB
Inside of cell
Nucleus
4 NF-κB enters the nucleus and binds to promoters.
DNA
Promoter
Transcription
5 Genes encoding defensive proteins are transcribed.
mRNA
Translation
Protein

18.5 Cell Signaling and Defense Binding of a pathogenic molecule to the toll receptor initiates a signal transduction pathway which results in the transcription of genes whose products are involved in specific and nonspecific defenses. CD14 is expressed on the surface of white blood cells, including macrophages and monocytes.

A cell signaling pathway stimulates the body's defenses

An invading pathogen such as a bacterium can be regarded as a signal. In response to that signal, the body produces molecules such as complement proteins, interferons, and cytokines that regulate phagocytosis and other defense processes. Not surprisingly, the link between signal and response is a signal transduction pathway, similar to the ones we considered in Section 15.3. The receptor in this pathway is a membrane protein called **toll**. This receptor was originally discovered in fruit flies, in which it plays an essential role in sensing infection by fungi. Comparative genomics has revealed at least ten similar receptors in humans.

Toll is part of a protein kinase cascade that ultimately results in the transcription of at least 40 genes involved in both nonspecific and specific defenses (**Figure 18.5**). The molecules that stimulate this pathway are made only by microbes and include some bacterial and fungal cell wall fragments. Binding of these molecules to toll sets in motion a cascade that results in the phosphorylation of the transcription factor NF-κB. As a result, the transcription factor's conformation changes, allowing it to enter the nucleus, bind to the promoters of genes encoding defensive proteins, and activate their transcription.

18.2 RECAP

Nonspecific defenses are the first line of defense against pathogens. Barriers such as the skin, defensive proteins, and coordinated responses such as inflammation are important nonspecific defenses.

- How do complement proteins and interferons defend the body against microbes? See p. 405
- Can you describe the inflammation response? See p. 406 and Figure 18.4

How does the body deal with pathogens that get by these nonspecific defenses? The next section describes the development of immunity to specific pathogens.

18.3 How Does Specific Immunity Develop?

The body's nonspecific defenses are numerous and effective, but some invaders elude them. Vertebrates animals deal with these pathogens by means of defenses targeted against specific threats. In this section we outline the main features of the specific immune system and consider the two major types of specific responses: the *humoral immune response*, which produces antibodies, and the *cellular immune response*, which destroys infected cells.

The specific immune system has four key traits

The four characteristic features of the specific immune system are specificity; the ability to respond to an enormous diversity of foreign molecules and organisms; the ability to distinguish self from nonself; and immunological memory.

SPECIFICITY Lymphocytes (B cells and T cells) are crucial to specific immunity. T cell receptors and the antibodies produced by B cells recognize and bind to specific nonself or altered-self substances (**antigens**), and this interaction initiates a specific immune response. The specific sites on antigens that the immune system recognizes are called **antigenic determinants** or *epitopes*:

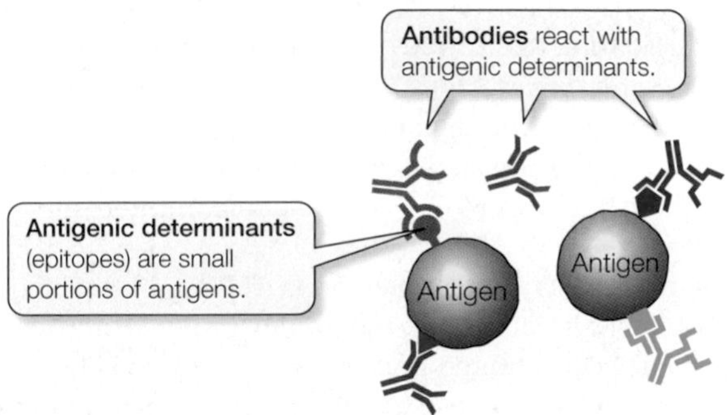
Antibodies react with antigenic determinants.
Antigenic determinants (epitopes) are small portions of antigens.
Antigen
Antigen

Chemically, an antigenic determinant is a specific portion of a large molecule, such as a certain sequence of amino acids that may be present in several proteins. A large antigen, such as a whole cell, may have many different antigenic determinants on its surface, each capable of being bound by a specific antibody or T cell. Even a single protein has multiple, different antigenic determinants. Some epitopes provoke a more powerful immune response than others and are referred to as *immunodominant*. The host animal responds to the presence of an antigen by producing highly specific defenses: T cells or antibodies that are complementary to, or fit, the antigenic determinants of that antigen. Each T cell and each antibody is *specific for a single antigenic determinant.*

DISTINGUISHING SELF FROM NONSELF The human body contains tens of thousands of different proteins, each with a specific three-dimensional structure capable of generating immune responses. Thus every cell in the body bears a tremendous number of antigenic determinants. A crucial requirement of an individual's immune system is that it recognize the body's own antigenic determinants and not attack them.

DIVERSITY Challenges to the immune system are numerous. Pathogens take many forms: individual foreign molecules, viruses, bacteria, protists, fungi, and multicellular parasites. Furthermore, each pathogenic species usually exists in many subtly differing genetic strains, and each strain possesses multiple surface features. Estimates vary, but a reasonable guess is that humans can respond *specifically* to 10 million different antigenic determinants. Upon recognizing an antigenic determinant, the immune system responds by activating lymphocytes of the appropriate specificity.

IMMUNOLOGICAL MEMORY After responding to a particular type of pathogen once, the immune system "remembers" that pathogen and can usually respond more rapidly and powerfully to the same threat in the future. This **immunological memory** usually saves us from repeats of childhood diseases such as chicken pox. Vaccination against specific diseases works because the immune system "remembers" the antigenic determinants that are introduced into the body.

Two types of specific immune responses interact

The specific immune system mounts two types of responses against invaders: the *humoral immune response* and the *cellular immune response*. These two responses operate in concert—simultaneously and cooperatively, sharing many mechanisms.

HUMORAL IMMUNE RESPONSE In the **humoral immune response** (from the Latin *humor*, "fluid"), antibodies react with antigenic determinants on pathogens in blood, lymph, and tissue fluids. An animal can produce a staggering diversity of antibodies capable of binding to almost any conceivable antigen the animal encounters.

Some antibodies are soluble and travel free in the blood and lymph; others exist as integral membrane proteins on B cells. The first time a specific antigen invades the body, it may be detected and bound by a B cell whose membrane antibody recognizes one of its antigenic determinants. This binding activates the B cell, which makes and secretes multiple copies of an antibody with the same specificity as its membrane antibody.

CELLULAR IMMUNE RESPONSE The **cellular immune response** is directed against antigens that have become established within a cell of the host animal. It detects and destroys virus-infected or mutated cells.

The cellular immune response is carried out by T cells within the lymph nodes, the bloodstream, and the intercellular spaces. These T cells have integral membrane proteins—T cell receptors—that recognize and bind to antigenic determinants. T cell receptors are rather similar to antibodies in structure and function, each including specific molecular configurations that bind to specific antigenic determinants. Once a T cell is bound to an antigenic determinant, it initiates an immune response that typically results in the total destruction of the nonself or altered self cell.

Genetic changes and clonal selection generate the specific immune response

How does the tremendous diversity of the specific immune response arise? How do lymphocytes specific for certain antigenic determinants proliferate? The answers lie in the process of **clonal selection**.

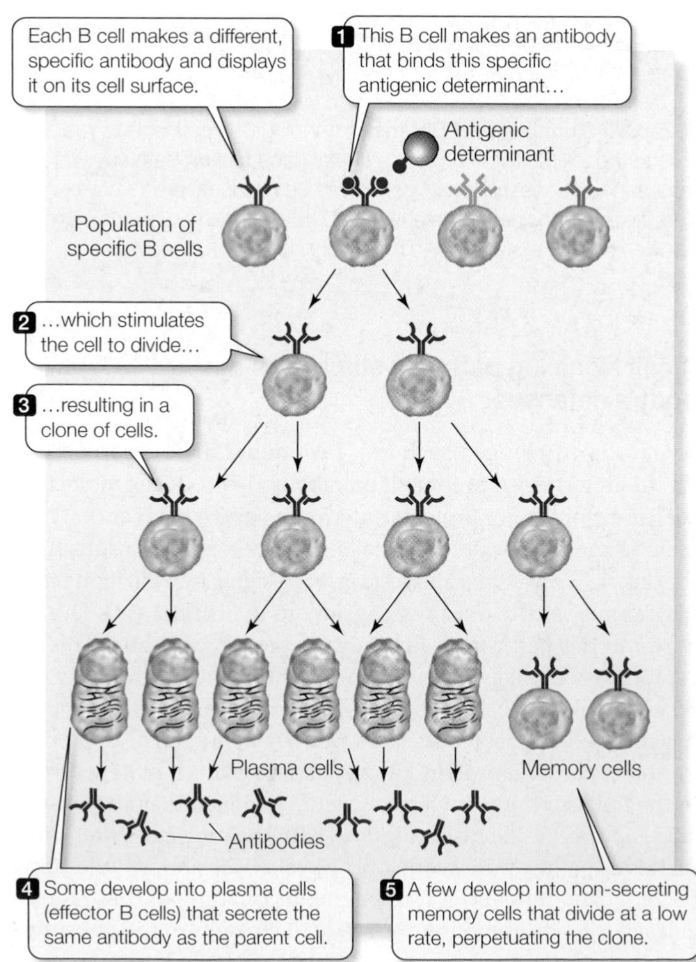

Each B cell makes a different, specific antibody and displays it on its cell surface.
1 This B cell makes an antibody that binds this specific antigenic determinant...
Antigenic determinant
Population of specific B cells
2 ...which stimulates the cell to divide...
3 ...resulting in a clone of cells.
Plasma cells
Memory cells
Antibodies
4 Some develop into plasma cells (effector B cells) that secrete the same antibody as the parent cell.
5 A few develop into non-secreting memory cells that divide at a low rate, perpetuating the clone.

18.6 Clonal Selection in B Cells The binding of an antigen to a specific antibody on the surface of a B cell stimulates that cell to divide, producing a clone of genetically identical cells to fight that invader.

- *Diversity is generated primarily by DNA changes*—chromosomal rearrangements and mutations—that occur just after the B and T cells are formed in the bone marrow. Each B cell is able to produce *only one kind of antibody*. Thus there are millions of different B cells, each one producing a particular antibody. Similarly, there are millions of different T cells, each of which has a unique type of T cell receptor that binds to a particular antigenic determinant on a target cell.
- *Antigen binding "selects" a particular B or T cell for proliferation.* When an antigen that fits the surface antibody on a B cell binds to it, that B cell is activated. It divides to form a *clone* of cells (a genetically identical group derived from a single cell), all of them producing that particular antibody (**Figure 18.6**). In the same way, a foreign or abnormal cell *selects* for the proliferation of a T cell expressing a particular T cell receptor on its surface (hence, *clonal selection*).

Immunity and immunological memory result from clonal selection

An activated lymphocyte produces two types of daughter cells, *effector cells* and *memory cells*.

- **Effector cells** carry out the attack on the antigen. Effector B cells, called **plasma cells**, secrete antibodies. Effector T cells release cytokines, which initiate reactions that destroy nonself or altered cells. Effector cells live only a few days.
- **Memory cells** are long-lived cells that retain the ability to start dividing on short notice to produce more effector and more memory cells. Memory B and possibly T cells may survive in the body for decades, dividing at a low rate.

Between them, these two types of lymphocytes can respond to an antigen in two different ways:

- When the body first encounters a particular antigen, a **primary immune response** is activated, in which the "naive" lymphocytes that recognize that antigen proliferate to produce clones of effector and memory cells. The effector cells destroy the invaders and then die, but one or more clones of memory cells have now been added to the immune system and provide immunological memory.
- After a primary immune response to a particular antigen, subsequent encounters with the same antigen will trigger a much more rapid and powerful **secondary immune response**. The memory cells that bind with that antigen proliferate, launching a huge army of plasma cells and effector T cells.

The first time a vertebrate animal is exposed to a particular antigen, there is a time lag (usually several days) before the number of antibody molecules and T cells specific to that antigen slowly increases. But for years afterward—sometimes for life—the immune system "remembers" that particular antigen. The secondary immune response is characterized by a shorter lag time, a greater rate of antibody production, and a larger total production of antibodies or T cells than the primary immune response.

TABLE 18.2

Some Human Pathogens for Which Vaccines are Available

INFECTIOUS AGENT	DISEASE	VACCINATED POPULATION
Bacteria		
Bacillus anthracis	Anthrax	Those at risk in biological warfare
Bordetella pertussis	Whooping cough	Children and adults
Clostridium tetani	Tetanus	Children and adults
Corynebacterium diphtheriae	Diphtheria	Children
Haemophilus influenzae	Meningitis	Children
Mycobacterium tuberculosis	Tuberculosis	All people
Salmonella typhi	Typhoid fever	Areas exposed to agent
Streptococcus pneumoniae	Pneumonia	Elderly
Vibrio cholerae	Cholera	People in areas exposed to agent
Viruses		
Adenovirus	Respiratory disease	Military personnel
Hepatitis A	Liver disease	Areas exposed to agent
Hepatitis B	Liver disease, cancer	All people
Influenza virus	Flu	All people
Measles virus	Measles	Children and adolescents
Mumps virus	Mumps	Children and adolescents
Poliovirus	Polio	Children
Rabies virus	Rabies	Persons exposed to agent
Rubella virus	German measles	Children
Vaccinia virus	Smallpox	Laboratory workers, military personnel
Varicella-zoster virus	Chicken pox	Children

Vaccines are an application of immunological memory

Thanks to immunological memory, recovery from many diseases, such as chicken pox, provides a *natural immunity* to those diseases. However, it is possible to provide *artificial immunity* against many life-threatening diseases by *inoculation*: the introduction of antigenic determinants into the body. **Immunization** is inoculation with antigenic proteins, pathogen fragments, or other molecular antigens. **Vaccination** is inoculation with whole pathogens that have been modified so that they cannot cause disease.

Immunization or vaccination initiates a primary immune response, generating memory cells without making the person ill. Later, if the same or very similar pathogens attack, specific memory cells already exist. They recognize the antigen and quickly overwhelm the invaders with a massive production of lymphocytes and antibodies.

Because the antigens used for immunization or vaccination are either parts of or toxic proteins produced by a pathogenic organism, they must be altered so that they cannot cause disease but are still able to provoke an immune response. There are four principal ways to do this:

- *Inactivation* involves treating the antigen—in this case, usually a whole organism such as a bacterium—with heat or chemicals so that it is killed.
- *Attenuation* involves reducing the virulence of a virus by repeatedly infecting cells with it in the laboratory.
- *Recombinant DNA technology* can be used to produce peptide fragments that bind to and activate lymphocytes but do not have the harmful part of a protein toxin.
- *DNA vaccines* are being developed that will introduce a gene encoding an antigen into the body.

Although smallpox vaccination programs have seemingly rid the world of this virus (with everyone vaccinated, the virus simply had nowhere to go), it still exists in several national laboratories under strict security. Because other supplies of the virus may exist and might be used in biological warfare, the known supplies of the virus are being used to generate new vaccines.

For most of the 70 bacteria, viruses, fungi, and parasites known to cause serious human diseases, vaccines are already available or will be in the next few years (**Table 18.2**). Vaccination has completely or almost completely wiped out some deadly diseases, such as smallpox, diphtheria, and polio, in industrialized countries.

Animals distinguish self from nonself and tolerate their own antigens

Normally, the body is tolerant of its own molecules—the same molecules that would generate an immune response in another individual. **Immunological tolerance** is based on two mechanisms: *clonal deletion* and *clonal anergy*.

CLONAL DELETION **Clonal deletion** removes certain immature B and T cells from the immune system early in their differentiation. Immature B cells in the bone marrow and T cells in the thymus may encounter self antigens. Any immature B or T cell that shows the potential to mount an immune response against self antigens undergoes programmed cell death (apoptosis) within a short time.

CLONAL ANERGY **Clonal anergy** is the suppression of the immune response to self antigens and occurs after lymphocytes mature. A mature T cell, for example, may encounter and recognize a self antigen on the surface of a body cell. But before it sends out the cytokines

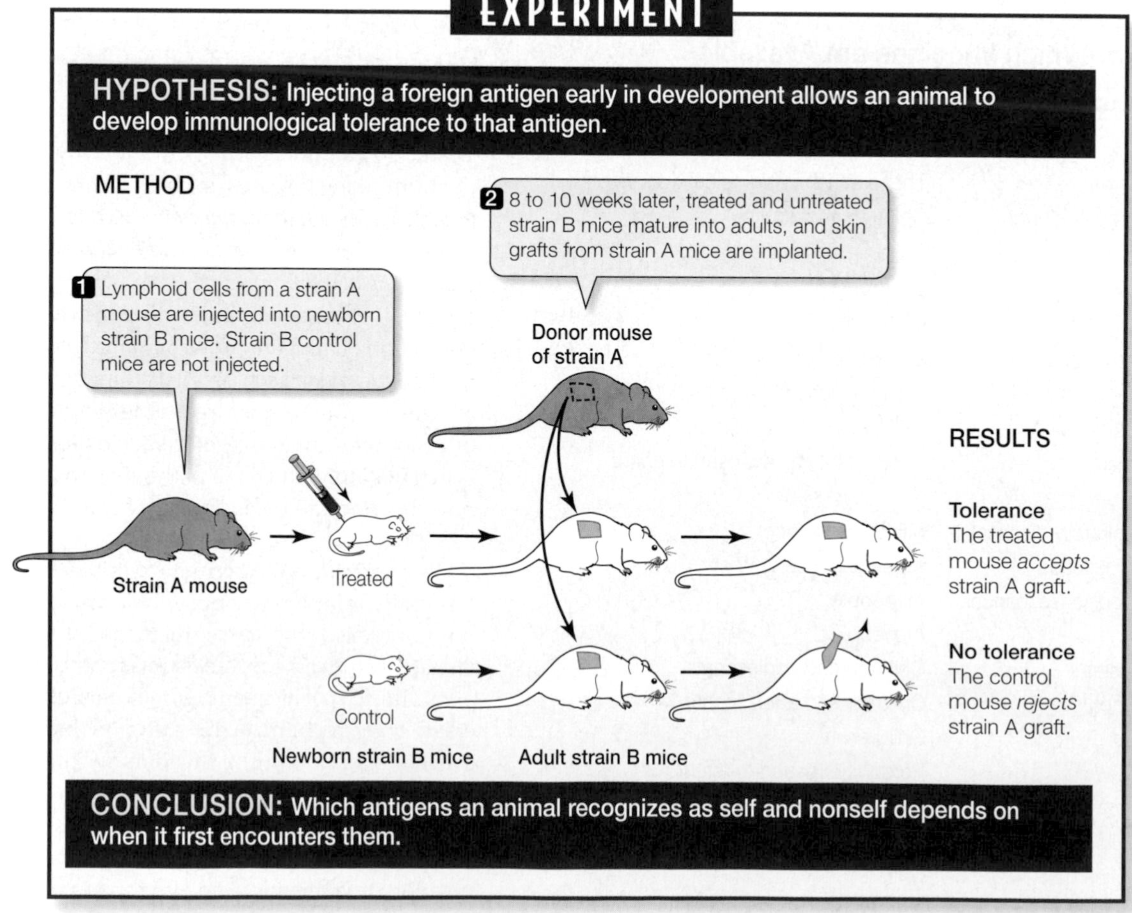

EXPERIMENT
HYPOTHESIS: Injecting a foreign antigen early in development allows an animal to develop immunological tolerance to that antigen.
METHOD
1 Lymphoid cells from a strain A mouse are injected into newborn strain B mice. Strain B control mice are not injected.
2 8 to 10 weeks later, treated and untreated strain B mice mature into adults, and skin grafts from strain A mice are implanted.
Donor mouse of strain A
Strain A mouse
Treated
Control
Newborn strain B mice
Adult strain B mice
RESULTS
Tolerance
The treated mouse *accepts* strain A graft.
No tolerance
The control mouse *rejects* strain A graft.
CONCLUSION: Which antigens an animal recognizes as self and nonself depends on when it first encounters them.

18.7 Immunological Tolerance A mouse can tolerate a skin graft from a genetically different mouse if lymphoid cells from the second mouse are injected into the first one as a newborn. FURTHER RESEARCH: What would happen if the lymphoid cells were heated in a boiling water bath prior to injection into strain B mice?

that signal the initiation of an immune response, the T cell must encounter not only an antigen, but also a second molecule, CD28, on the cell surface. Most body cells lack CD28 and thus will not be attacked by the cellular immune response. CD28 is an example of a *co-stimulatory signal* that is expressed only on certain antigen-presenting cells. Such cells "present" nonself antigens on their surfaces, thus stimulating the cellular immune response, as we will see in Section 18.5. They include the macrophages that wander through the body's fluids and the *dendritic cells* that reside among the linings of the respiratory and digestive tracts, as well as B cells.

Immunological tolerance was discovered through the observation that some *nonidentical* twin cattle with different blood types contained some of each other's red blood cells. Why didn't these "foreign" blood cells cause immune responses resulting in their elimination? One hypothesis was that the blood cells had passed between the fetal animals in the womb before the lymphocytes had matured, so that each calf regarded the other's red blood cells as self. This hypothesis was confirmed when it was shown that injecting a foreign antigen into an animal early in its fetal development caused that animal henceforth to recognize that antigen as self (**Figure 18.7**).

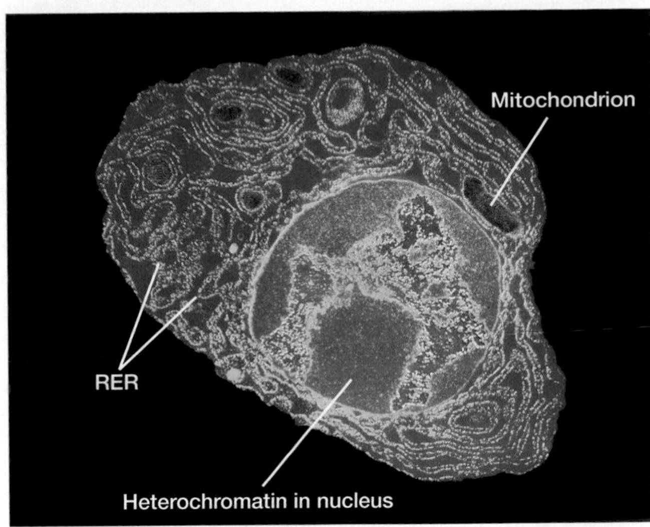

18.8 A Plasma Cell The prominent nucleus with large amounts of heterochromatin (orange) and the cytoplasm (bright blue) crowded with rough endoplasmic reticulum (RER) are features of a cell that is actively synthesizing and exporting proteins—in this case, a specific antibody. Whole blocks of genes not needed for this specialized function are kept turned off in the heterochromatin.

18.3 RECAP

The specific immune system reacts against nonself or altered-self substances called antigenic determinants. The system generates amazing diversity, distinguishes self from nonself, and has immunological memory. Immune system diversity arises from clonal selection.

- Can you explain how an antigenic determinant initiates a specific immune response? See pp. 407–408
- What is clonal selection and how does it account for the specific immune system's diversity and immunological memory? See p. 408 and Figure 18.6
- How do vaccines make use of immunological memory? See pp. 409–410

Now that we understand the general features of the specific immune system, let's focus in more detail on the B lymphocytes and the humoral response.

18.4 What Is the Humoral Immune Response?

Every day, billions of B cells survive the test of clonal deletion and are released from the bone marrow into the circulation. B cells are the basis for the humoral immune response.

Some B cells develop into plasma cells

B cells begin by making an antibody that is expressed as a receptor protein on the cell surface. As we have seen, if a B cell is activated by antigen binding to this receptor, it becomes a plasma cell, making an antibody protein that is secreted into the bloodstream, and gives rise to a clone of plasma cells as well as memory cells (see Figure 18.6).

Usually, for a B cell to develop into an antibody-secreting plasma cell, a *helper T cell* (T_H) with the same specificity must also bind to the antigen. Thus the B cell also functions as an antigen-presenting cell, as we will see in Section 18.5. The division and differentiation of the B cell is stimulated by the receipt of chemical signals from the T_H cell.

As plasma cells develop, the number of ribosomes and the amount of endoplasmic reticulum in their cytoplasm increase greatly (**Figure 18.8**). These increases allow the cells to synthesize and secrete large amounts of antibody proteins, up to 2,000 per second! All the plasma cells arising from a given B cell produce antibodies that are specific for the antigenic determinant that originally bound to the parent B cell. Thus antibody specificity is maintained as B cells proliferate.

Different antibodies share a common structure

Antibodies belong to a class of proteins called **immunoglobulins**. There are several types of immunoglobulins, but all contain a tetramer consisting of four polypeptide chains. In each immunoglobulin molecule, two of these polypeptides are identical *light chains*, and two are identical *heavy chains*. Disulfide bonds hold the chains together. Each polypeptide chain has a *constant region* and a *variable region* (**Figure 18.9**):

- The amino acid sequences of the **constant regions** are similar among the immunoglobulins. They determine the destination and function—the *class*—of the antibody.

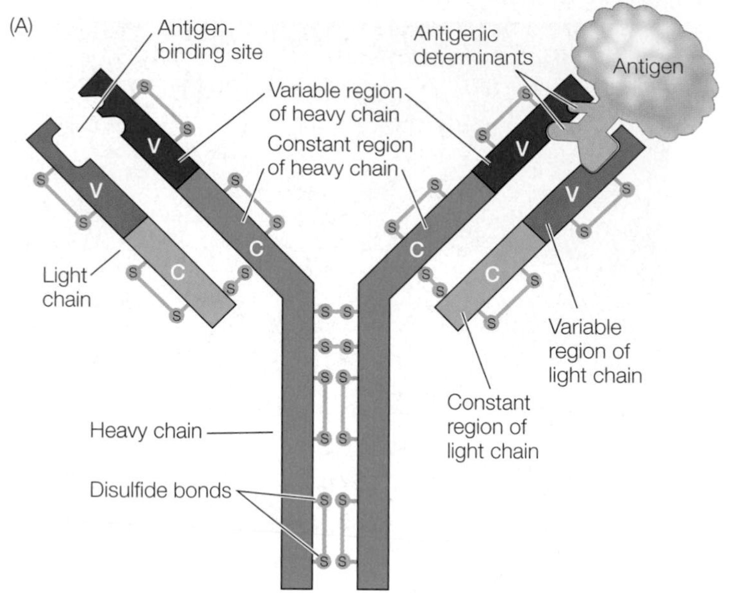

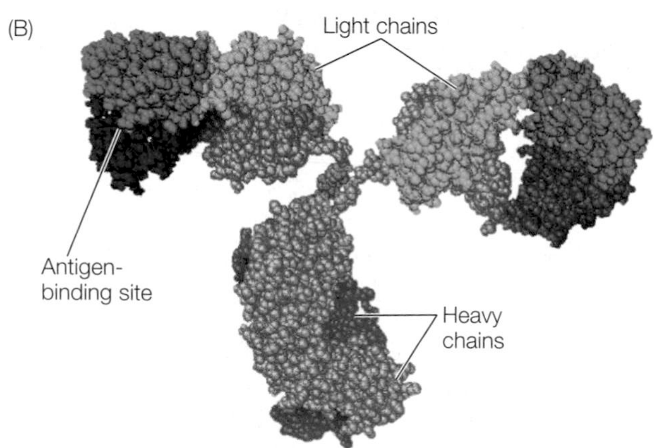

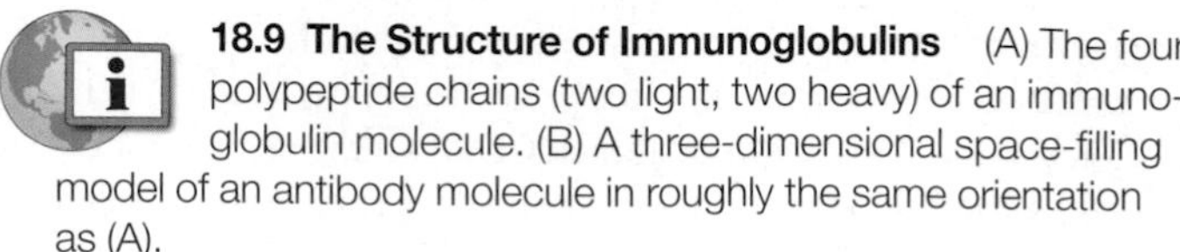

18.9 The Structure of Immunoglobulins (A) The four polypeptide chains (two light, two heavy) of an immunoglobulin molecule. (B) A three-dimensional space-filling model of an antibody molecule in roughly the same orientation as (A).

- The amino acid sequences of the **variable regions** are different for each specific immunoglobulin. Differences in their secondary structure result in differences in the three-dimensional antigen-binding site of each immunoglobulin and are responsible for antibody specificity.

The two antigen-binding sites on each immunoglobulin molecule are identical, making the antibody *bivalent* (*bi*, "two"; *valent*, "binding"). This ability to bind two antigen molecules at once permits the antibody to form a large complex with antigen and other antibody molecules. Such a complex is an easy target for ingestion and breakdown by phagocytes and complement.

There are five classes of immunoglobulins

While the variable regions are responsible for the *specificity* of an immunoglobulin, the constant regions of the heavy chain determine the *class* of the antibody—for example, whether it will be an integral membrane receptor or a soluble antibody that is secreted into the bloodstream. The five immunoglobulin classes are described in **Table 18.3**. The most abundant immunoglobulin class is IgG; these soluble antibody proteins make up about 80 percent of the total immunoglobulin content of the bloodstream. They are made in greatest quantity during a secondary immune response. IgG defends the body in several ways. For example, after some IgG

TABLE 18.3

Antibody Classes

CLASS	GENERAL STRUCTURE	LOCATION	FUNCTION
IgG	Monomer	Free in blood plasma; about 80 percent of circulating antibodies	Most abundant antibody in primary and secondary immune responses; crosses placenta and provides passive immunization to fetus
IgM	Pentamer	Surface of B cell; free in blood plasma	Antigen receptor on B cell membrane; first class of antibodies released by B cells during primary response
IgD	Monomer	Surface of B cell	Cell surface receptor of mature B cell; important in B cell activation
IgA	Dimer	Saliva, tears, milk, and other body secretions	Protects mucosal surfaces; prevents attachment of pathogens to epithelial cells
IgE	Monomer	Secreted by plasma cells in skin and tissues lining gastrointestinal and respiratory tracts	When bound to antigens, binds to mast cells and basophils to trigger release of histamine that contributes to inflammation and some allergic responses

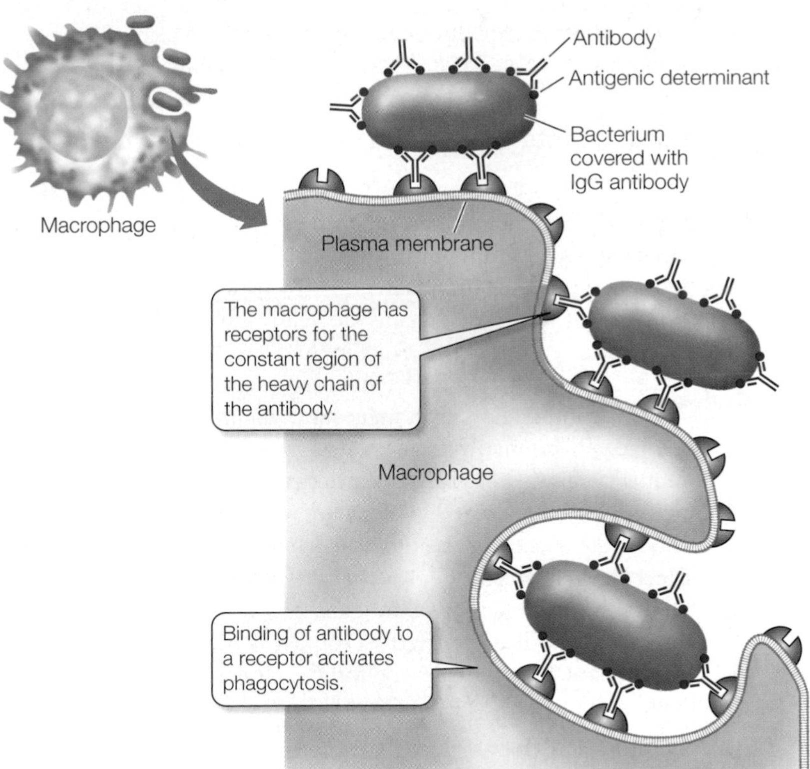
Antibody
Antigenic determinant
Bacterium covered with IgG antibody
Macrophage
Plasma membrane
The macrophage has receptors for the constant region of the heavy chain of the antibody.
Macrophage
Binding of antibody to a receptor activates phagocytosis.

18.10 IgG Antibodies Promote Phagocytosis When IgG antibodies bind to a bacterium, receptors on a macrophage can recognize and bind to those antibodies.

molecules bind to antigens, they become attached by their heavy chains to macrophages. This attachment permits the macrophages to destroy the antigens by phagocytosis (**Figure 18.10**).

Monoclonal antibodies have many uses

The specificity of antibodies suggested to scientists that they might be useful for detecting a specific substance in a fluid. An initial challenge to scientists seeking to accomplish this was that the immune response to a complex antigen is *polyclonal*—that is, because most antigens carry many different antigenic determinants, a single antigen will produce a complex mixture of antibodies, each made by a different clone of B cells.

Suppose that a physician wishes to measure the level of the hormone estrogen in a woman's blood. This could be done by adding an antibody specific for estrogen to a sample of her blood and observing how much antigen–antibody complex formed. But, as emphasized in our study of biochemistry, many biological molecules share regions of similar structure—all human steroid hormones, for example, have a similar multi-ring structure (see Figure 3.22). A polyclonal group of antibodies targeted to estrogen would be uninformative because the antibodies would bind not just to estrogen, but to any steroid hormone present in the blood sample. What is needed is a clone of B cells to produce large amounts of an antibody that binds to *only one* antigenic determinant—a **monoclonal antibody**. How could such a clone be produced?

A clone of cells that produce a single antibody can be made artificially by fusing a B cell (which has a finite lifetime and makes a lot of antibody) with a tumor cell (which has an infinite lifetime and can be grown in culture). The resulting hybrid cell, called a **hybridoma**, makes a specific monoclonal antibody and proliferates in culture (**Figure 18.11**).

A Canadian company called "Toxin Alert" has developed a plastic food wrap impregnated with antibodies against *Listeria*, *E. coli*, and *Salmonella* bacteria, which together account for 80 percent of food-related deaths in the developed world. The antibodies are distributed in an "X" pattern on the wrap, so if bacteria are present, the antibodies bind and a colored "X" appears on the package.

RESEARCH METHOD

Antigen
Myeloma cell culture
1 Inoculate a mouse with antigen.
2 Isolate B cells from the spleen.
3 Myeloma cells grow well in culture.
4 B cells produce antibodies but do not proliferate in culture.
B cells
Myeloma cells
5 A myeloma cell is fused with a B cell to form a hybridoma.
Hybridoma
6 A single hybridoma cell is isolated and grown in culture, and assayed for antibody.
7 Antibody-producing hybridomas proliferate indefinitely in culture.

18.11 Creating Hybridomas for the Production of Monoclonal Antibodies Cancerous myeloma cells and normal B cells can be fused so that the proliferative properties of the myeloma cells are merged with the specificity of the antibody-producing B cells.

Monoclonal antibodies have many practical applications:

- *Immunoassays* use the specificity of monoclonal antibodies to detect tiny amounts of molecules in tissues and fluids. This technique is used, for example, in pregnancy tests to detect the hormone made by the developing embryo.
- *Immunotherapy* uses monoclonal antibodies targeted against antigenic determinants on the surfaces of cancer cells. The coupling of a radioactive ligand or toxin to the antibody makes it into a medical "smart bomb." In some cases, binding of the antibody itself is enough to trigger a cellular immune response that destroys the cancer. (This is the case with Herceptin,® a monoclonal antibody that binds to a growth factor receptor on breast cancer cells.
- *Passive immunization* is inoculation with an immediately acting, but not long-lasting, monoclonal antibody. This approach is necessary when therapy must be effective quickly (within hours). Examples of such life-threatening situations include the early symptoms of rabies infection, rattlesnake bites, and babies born with hepatitis B virus infection—all cases in which the toxic nature of the infection is so serious that there is not enough time to allow the person's immune system to mount its own defense.

18.4 RECAP

The humoral immune response is based on the synthesis by B cells of specific antibodies directed against specific antigens. The specificity of an antibody derives from the amino acid sequence of its variable regions. Monoclonal antibodies are specific to one antigenic determinant and can be produced artificially for use in diagnostics and therapy.

- Can you describe the B cell's response to an antigen? See p. 411
- How are the structure and function of an antibody molecule related? See p. 412 and Figure 18.9
- Can you explain what a monoclonal antibody is and how are they used? See pp. 413–414 and Figure 18.11

Both T cells and B cells are involved in the humoral immune response. T cells are also the effectors of the cellular immune response, which we explore in the next section.

18.5 What Is the Cellular Immune Response?

The effector molecules of the humoral immune response are the antibodies secreted by plasma cells that develop from activated B cells. The *cellular immune response*, which is directed against any factor (e.g., a virus or mutation) that changes a normal cell into an abnormal cell, is directed by T cells.

Two types of effector T cells (*helper T cells* and *cytotoxic T cells*) are involved in the cellular immune response, along with the *MHC* (*major histocompatibility complex*) proteins, which underlie the immune system's tolerance for the body's own cells.

T cell receptors are found on two types of T cells

Like B cells, T cells possess specific membrane receptors. T cell receptors are not immunoglobulins, however, but glycoproteins with molecular weights about half that of an IgG. They are made up of two polypeptide chains, each encoded by a separate gene (**Figure 18.12**). Thus the two chains are nearly always different in their amino acid sequence, especially in their variable regions.

The genes that code for T cell receptors are similar to those that code for immunoglobulins, suggesting that both are derived from a single, evolutionarily more ancient group of genes. Like the immunoglobulins, T cell receptors include both variable and constant regions, and the variable regions are responsible for their specificity. There is one major difference: whereas antibodies bind to an intact antigen, T cell receptors bind to a piece of an antigen displayed on the surface of an antigen-presenting cell.

When a T cell is activated by contact with a specific antigenic determinant, it proliferates and forms a clone. Its descendants differentiate into two types of effector T cells:

- **Cytotoxic T cells**, or T_C cells, recognize virus-infected or mutated cells and kill them by inducing lysis (see page 401).
- **Helper T cells**, or T_H cells, assist both the cellular and humoral immune responses.

As mentioned in Section 18.4, a specific T_H cell must bind to an antigenic determinant presented on a B cell before that B cell can become activated. The helper T cell becomes the "conductor" of the "immunological orchestra" as it sends out chemical signals that not only result in its own proliferation and that of the B cell, but also set in motion the actions of cytotoxic T cells, as we will see shortly.

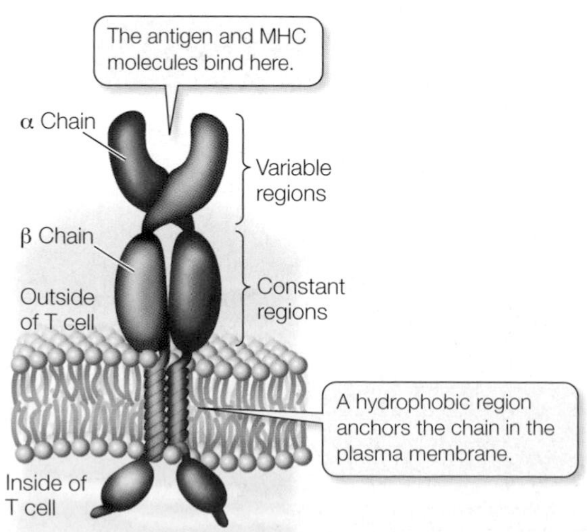

18.12 A T Cell Receptor In both T cell receptors and immunoglobulins, the specificity of each antigen-binding site is determined by two polypeptide chains. T cell receptors are bound more firmly to the plasma membrane of the T cell that produces them than antibodies are to B cells.

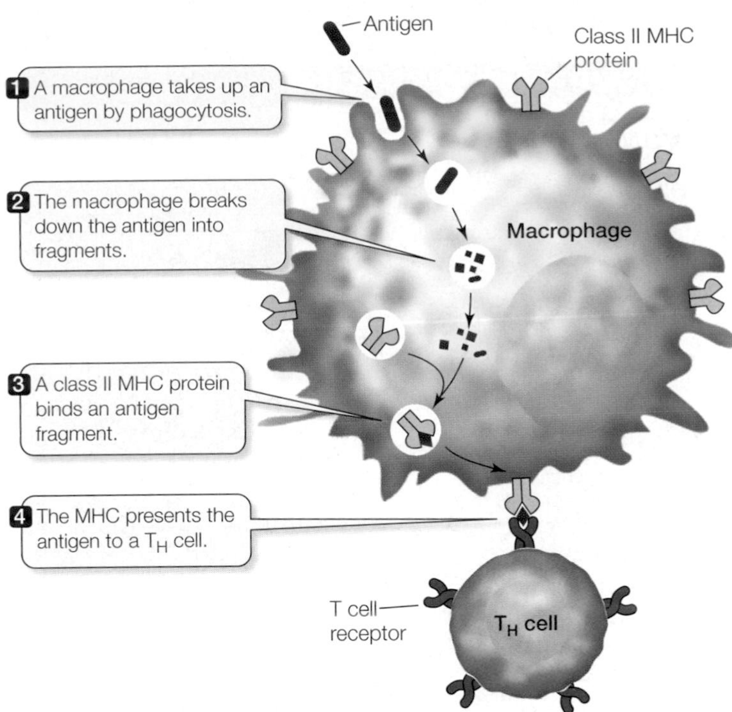

Antigen
Class II MHC protein
1 A macrophage takes up an antigen by phagocytosis.
2 The macrophage breaks down the antigen into fragments.
Macrophage
3 A class II MHC protein binds an antigen fragment.
4 The MHC presents the antigen to a T_H cell.
T cell receptor
T_H cell

18.13 Macrophages Are Antigen-Presenting Cells A fragment of an antigen is displayed by MHC II on the surface of a macrophage. T cell receptors on a specific helper T cell can then bind to and interact further with the antigen–MHC II complex.

Now that we are familiar with the two types of T cells, we can address the question of how T cells recognize their antigenic determinants and the role the MHC proteins play in the process.

The MHC encodes proteins that present antigens to the immune system

We have seen that an animal's immune system recognizes the body's own cells by their surface proteins. Several types of mammalian cell surface proteins are involved in this process, but we will focus here on the products of a cluster of genes called the **major histocompatibility complex**, or **MHC**. These proteins have important roles in the cellular and humoral immune responses as well as in self-tolerance.

The MHC gene products are plasma membrane glycoproteins. In humans, the MHC proteins are called *human leukocyte antigens* (HLA), while in mice they are called *H-2 proteins*. Their major role is to present antigens to a T cell receptor in such a way that it can distinguish between self and nonself antigens. There are two classes of MHC proteins:

- *Class I MHC proteins* are present on the surface of every nucleated cell in the animal body. When cellular proteins are degraded into small peptide fragments by a proteasome (see Section 14.6), an MHC I protein may bind to a fragment and travel to the plasma membrane. There, the MHC I protein "presents" the cellular peptide to T_C cells. The T_C cells have a surface protein called CD8 that recognizes and binds to MHC I.
- *Class II MHC proteins* are found mostly on the surfaces of B cells, macrophages, and other antigen-presenting cells. When an antigen-presenting cell ingests a nonself antigen, such as a virus, the antigen is broken down in a phagosome. An MHC II molecule may bind to one of the fragments and carry it to the cell surface, where it is presented to a T_H cell (**Figure 18.13**). T_H cells have a surface protein called CD4 that recognizes and binds to MHC II.

To accomplish their roles in antigen presentation, both MHC I and MHC II proteins have an antigen-binding site, which can hold a peptide of about 10–20 amino acids (**Figure 18.14**). The T cell receptor recognizes not just the antigenic fragment, but the fragment *bound to an MHC I or MHC II molecule*. The table in Figure 18.14 summarizes the relationships of T cells and antigen-presenting cells.

In humans, there are three genetic loci for MHC I and three for MHC II; each of these six loci has as many as 100 different alleles.

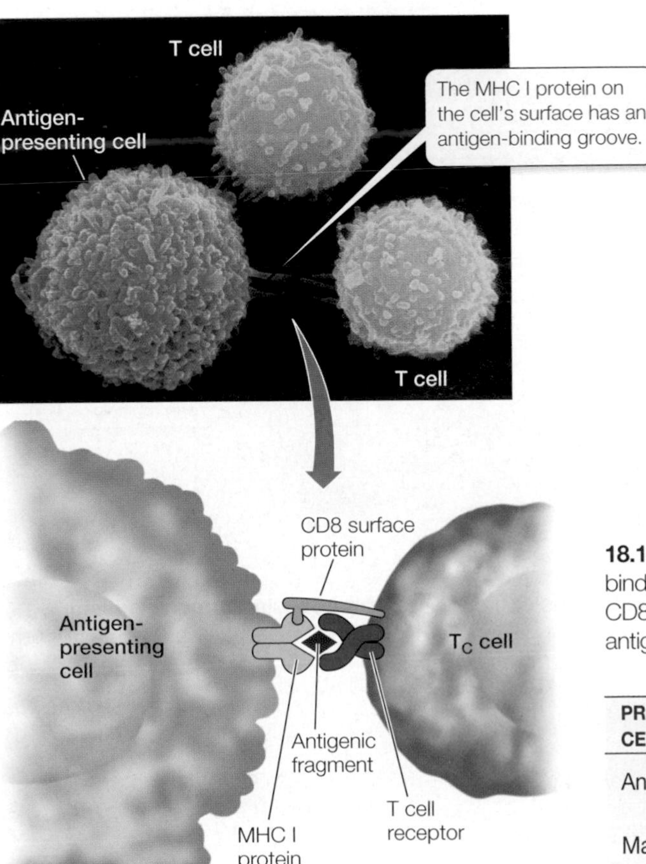

T cell
Antigen-presenting cell
The MHC I protein on the cell's surface has an antigen-binding groove.
T cell
CD8 surface protein
Antigen-presenting cell
T_C cell
Antigenic fragment
T cell receptor
MHC I protein

18.14 The Interaction between T Cells and Antigen-Presenting Cells An antigen-binding site in the MHC I protein holds an antigen, which it presents to cytotoxic T cells. CD8 surface proteins on T_C cells bind to MHC I, and specific T cell receptors bind to the antigen. The binding of MHC II protein by T_H cells works in a similar manner.

PRESENTING CELL TYPE	ANTIGEN PRESENTED	MHC CLASS	T CELL TYPE	T CELL SURFACE PROTEIN
Any cell	Intracellular protein fragment	Class I	Cytotoxic T cell (T_C)	CD8
Macrophages and B cells	Fragments from extracellular proteins	Class II	Helper T cell (T_H)	CD4

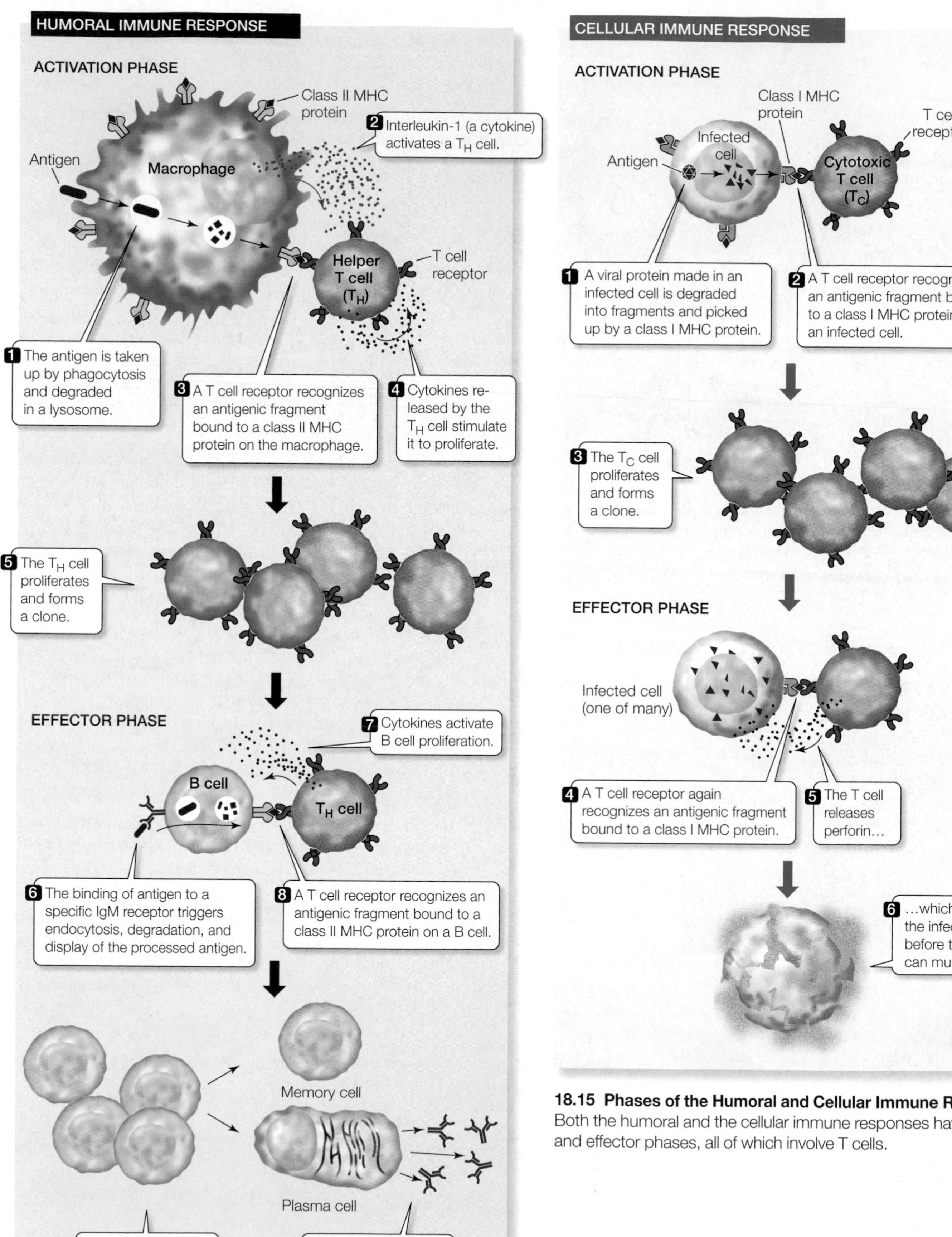

18.15 Phases of the Humoral and Cellular Immune Responses Both the humoral and the cellular immune responses have activation and effector phases, all of which involve T cells.

With so many possible allele combinations, it is not surprising that different people are very likely to have different MHC genotypes. Similarities in base sequences between the MHC genes and the genes coding for antibodies and T cell receptors suggest that all three may have descended from the same ancestral genes and are part of a gene "superfamily." Thus major aspects of the immune system in vertebrates seem to be woven together by a common evolutionary thread.

Helper T cells and MHC II proteins contribute to the humoral immune response

When a T_H cell binds to an antigen-presenting macrophage, the T_H cell releases cytokines, which activate the T_H cell to produce a clone of T_H cells with the same specificity. The steps to this point constitute the *activation phase* of the humoral immune response, and they occur in the lymphoid tissues. Next comes the *effector phase*, in which the T_H cells activate B cells with the same specificity to produce antibodies (**Figure 18.15, left**).

B cells are also antigen-presenting cells. B cells take up antigens bound to their surface immunoglobulin receptors by endocytosis, break them down, and display antigenic fragments on class II MHC proteins. When a T_H cell binds to the displayed antigen–MHC II complex, it releases cytokines, which cause the B cell to produce a clone of plasma cells. Finally, the plasma cells secrete antibodies, completing the effector phase of the humoral immune response.

Cytotoxic T cells and MHC I proteins contribute to the cellular immune response

Class I MHC proteins play a role in the cellular immune response that is similar to the role played by class II MHC proteins in the humoral immune response. In a virus-infected or mutated cell, foreign or abnormal proteins or peptide fragments combine with MHC I molecules. The resulting complex is displayed on the cell surface and presented to T_C cells. When a T_C cell recognizes and binds to this –antigen–MHC I complex, it is activated to proliferate (**Figure 18.15, right**).

In the effector phase of the cellular immune response, T_C cells recognize and bind to cells bearing the same antigen–MHC I complex. These T_C cells produce a substance called perforin, which lyses the target cell. In addition, the T_C cells can bind to a specific receptor (called Fas) on the target cell that initiates apoptosis in that cell. These two mechanisms, cell lysis and programmed cell death, work in concert to eliminate the altered host cell.

Because T_C cells recognize self MHC proteins complexed with *nonself* antigens, they help rid the body of its own virus-infected cells. Because they also recognize MHC proteins complexed with *altered self* antigens (altered as a result of mutations), they help eliminate tumor cells, since most tumor cells have been altered by mutations.

In addition to the binding of an antigen–MHC complex to their receptors, T cells must receive a second signal for activation. This co-stimulatory signal occurs after the initial specific binding and involves the interaction of additional proteins on the T cell with the CD28 protein on certain antigen-presenting cells, as we saw above. This second binding event leads to T cell activation, including cytokine production and proliferation. It also sets in motion the production of an *inhibitor* of these events, so that the response is appropriately terminated. This inhibitor, a cell surface protein called CTLA4 competes with CD28, blocking the activation process, especially for self antigens.

MHC proteins underlie the tolerance of self

MHC proteins play a key role in establishing self-tolerance, without which an animal would be destroyed by its own immune system. Throughout the animal's life, developing T cells are tested in the thymus. This "test" consists of two "questions":

1. Can this cell recognize the body's MHC proteins? A T cell unable to recognize self MHC proteins would be useless to the animal because it could not participate in any immune reactions. Such a T cell fails the test and dies within about 3 days.
2. Does this cell bind to self MHC proteins *and* to one of the body's own antigens? A T cell that satisfied both of these criteria would be harmful or lethal to the animal; it also fails the test and undergoes apoptosis.

T cells that survive this test mature into either T_C cells or T_H cells.

In humans, one consequence of the major histocompatibility complex became important with the development of organ transplant surgery. Because the proteins produced by the MHC are specific to each individual, they act as nonself antigens if transplanted into another individual. An organ or a piece of tissue transplanted from one person to another is recognized as nonself and soon provokes an immune response; the tissue is then killed, or "rejected," by the host's cellular immune system. But if the transplant is performed immediately after birth, or if it comes from a genetically identical person (an identical twin), the material is recognized as self and is not rejected.

The rejection problem can be overcome by treating a patient with drugs, such as *cyclosporin*, that suppress the immune system. Cyclosporin works by blocking the activation of a transcription factor essential for T cell development. This approach, however, compromises the ability of transplant recipients to defend themselves against pathogens. These risks must be managed by the use of antibiotics and other drugs.

18.5 RECAP

The cellular immune response acts against antigens expressed on the surfaces of virus-infected or mutated body cells. Specific receptors on T cells bind to antigens displayed on the cell surface by MHC proteins. MHC proteins are also involved in self-tolerance.

- What are the roles of a T cell receptor in cellular immunity? See p. 414
- Can you describe the events of the cellular immune response to a virus-infected cell? See p. 415 and Figure 18.15
- What is the role of MHC proteins in the cellular immune response? See p. 417

We have alluded to genes that encode various components of the immune system and to the tremendous diversity in the immune response. We will now consider the genetic mechanisms that make this diversity possible.

18.6 How Do Animals Make So Many Different Antibodies?

Each cell of a newborn mammal possesses a full set of genetic information for immunoglobulin synthesis. At each of the loci coding for the heavy and light antibody chains, it has one allele from its mother and one from its father. Throughout the animal's life, each of its cells begins with the same full set of immunoglobulin genes. However, as B cells develop, their genomes become modified in such a way that each mature B cell can produce one—and only one—specific type of immunoglobulin. In other words, different B cells develop *slightly different genomes* encoding different antibody specificities. How can a single organism produce millions of different genomes?

One hypothesis was that mammals simply have millions of antibody genes. However, a simple calculation (the number of base pairs needed per antibody gene multiplied by millions of antibodies) shows that if this were true, the entire human genome would be taken up by antibody genes! More than 30 years ago, an alternative hypothesis was proposed: a relatively small number of genes recombine to produce many unique combinations, and it is this shuffling of the genetic deck, plus the random pairing of light and heavy chains, that produces antibody diversity. This second hypothesis is now the accepted molecular genetic theory.

In this section we will describe the unusual genetic events that generate the enormous antibody diversity that normally characterizes each individual mammal. Then we will see how similar events produce five classes of antibodies with slightly different functions.

Antibody diversity results from DNA rearrangement and other mutations

Each gene encoding an immunoglobulin chain is in reality a "supergene" assembled from several clusters of smaller genes scattered along part of a chromosome (**Figure 18.16**). Every cell in the body has hundreds of genes, located in separate clusters, that are potentially capable of participating in the synthesis of the variable and constant regions of immunoglobulin chains. In most body cells and tissues, these genes remain intact and separated from one another. During B cell development, however, these genes are cut out, rearranged, and joined together. One gene from each cluster is chosen randomly for joining, and the others are deleted (**Figure 18.17**).

In this manner, a unique antibody supergene is assembled from randomly selected "parts." Each B cell precursor in the animal assembles its own two specific antibody supergenes, one for a specific heavy chain and the other, assembled independently, for a specific light chain. This remarkable example of essentially irreversible cell differentiation generates an enormous diversity of antibody specificities from the same starting genome, one for each individual B cell.

In both humans and mice, the gene clusters coding for immunoglobulin heavy chains are on one pair of chromosomes and those for light chains are on others. The variable region of the light chain is encoded by two families of genes; the variable region of the heavy chain is encoded by three families.

Figure 18.16 illustrates the gene families coding for the heavy-chain constant and variable regions in mice. There are multiple genes coding for each of the four kinds of regions in the polypeptide chain: 100 *V*, 30 *D*, 6 *J*, and 8 *C*. Each B cell that becomes committed to making an antibody randomly selects *one* gene for each of these clusters to make the final heavy-chain coding sequence, *VDJC*. So the number of *different* heavy chains that can be made through this random recombination process is quite large (144,000 possible combinations in mice).

Now consider that the light chains are similarly constructed, with a similar amount of diversity made possible by random recombination. If we assume that light-chain diversity is the same as heavy-chain diversity, the number of possible combinations of light and heavy chains is 144,000 different light chains × 144,000 different heavy chains = 21 *billion* possibilities! But there are other mechanisms that generate even more diversity:

- When the DNA sequences coding for the *V*, *J*, and *C* regions are rearranged so that they are next to one another, the re-

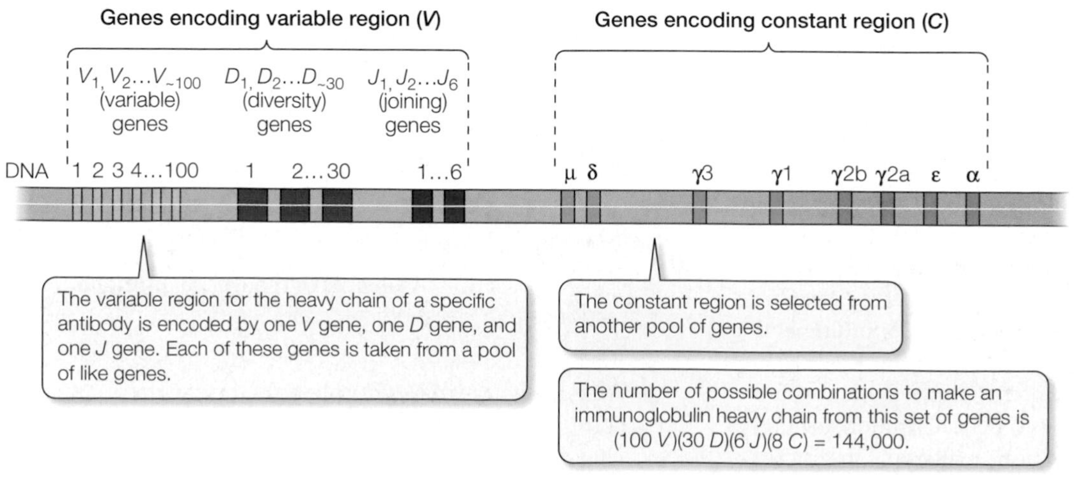

18.16 Heavy-Chain Genes Mouse immunoglobulin heavy chains have four regions, each of which is coded for by one of a number of possible genes selected from a cluster of similar genes.

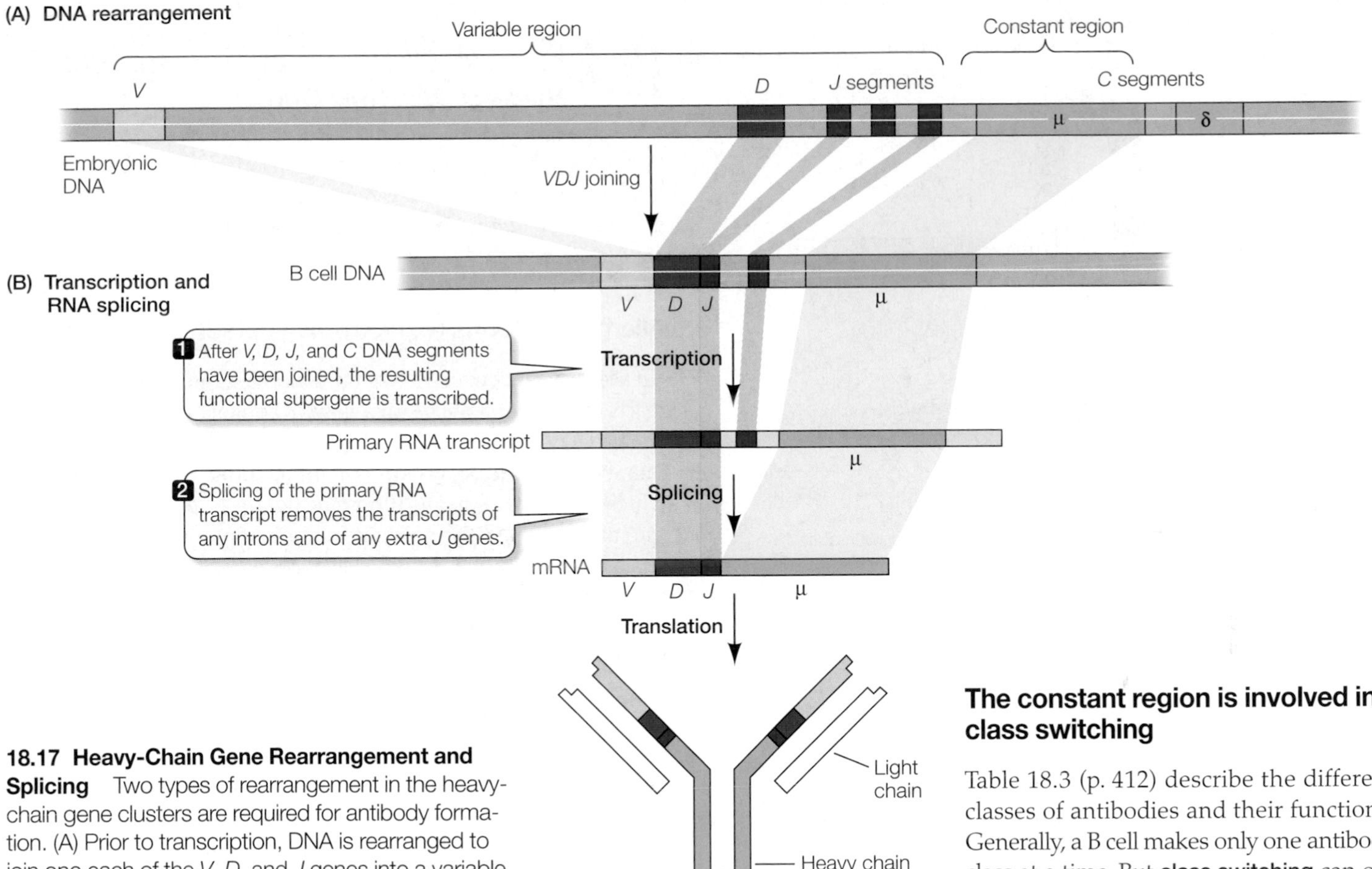

18.17 Heavy-Chain Gene Rearrangement and Splicing Two types of rearrangement in the heavy-chain gene clusters are required for antibody formation. (A) Prior to transcription, DNA is rearranged to join one each of the *V*, *D*, and *J* genes into a variable region supergene. (B) After transcription, RNA splicing joins the *VDJ* region to the constant region.

combination event is not precise, and errors occur at the junctions. This *imprecise recombination* can create new codons at the junctions, with resulting amino acid changes.

- After the DNA sequences are cut out and before they are joined, an enzyme, *terminal transferase*, often adds some nucleotides to the free ends of the DNAs. These additional bases create *insertion mutations*.
- There is a relatively high spontaneous *mutation rate* in immunoglobulin genes. Once again, this process creates many new alleles and adds to antibody diversity.

When we add these possibilities to the billions of combinations that can be made by random DNA rearrangements, it is not surprising that the immune system can mount a response to almost any natural or artificial substance.

Once this pretranscriptional processing in completed, pre-mRNA can be transcribed from each supergene. Posttranscriptional processing removes the remaining introns, so that the mature mRNA contains a continuous coding sequence for an immunoglobulin light chain or heavy chain. Translation then produces the polypeptide chains, which combine to form an active antibody protein.

This genetic system is capable of still other kinds of changes, as seen when a B cell or plasma cell switches the immunoglobulin class it produces, but retains its antibody specificity.

The constant region is involved in class switching

Table 18.3 (p. 412) describe the different classes of antibodies and their functions. Generally, a B cell makes only one antibody class at a time. But **class switching** can occur, in which a B cell changes the antibody class it synthesizes. For example, a B cell making IgM can switch to making IgG.

Early in its life, a B cell produces IgM molecules, which are the receptors responsible for its recognition of a specific antigenic determinant. At this time, the constant region of the antibody's heavy chain is encoded by the first constant region gene, the μ gene (see Figure 18.17). If the B cell later becomes a plasma cell during a humoral immune response, another deletion commonly occurs in the cell's DNA, positioning the heavy-chain variable region genes (consisting of the same *V*, *D*, and *J* genes) next to a constant region gene farther down the original DNA (**Figure 18.18**). Such a DNA deletion results in the production of an antibody with a different constant region of the heavy chain, and therefore a different function. However, the antibody produced has *the same variable regions*, and therefore the same antigen specificity, as before. The new antibody falls into one of the four other immunoglobulin classes (IgA, IgD, IgE, or IgG), depending on which of the constant region genes is placed adjacent to the variable region genes.

After switching classes, the plasma cell cannot go back to making the previous immunoglobulin class, because that part of the DNA has been lost. On the other hand, if additional constant region genes are still present, the cell may switch classes again.

What triggers class switching, and what determines the class to which a given B cell will switch? T_H cells direct the course of an immune response and determine the nature of the attack on the antigen. These T cells induce class switching by sending cytokine signals. The cytokines bind to receptors on the target B cells,

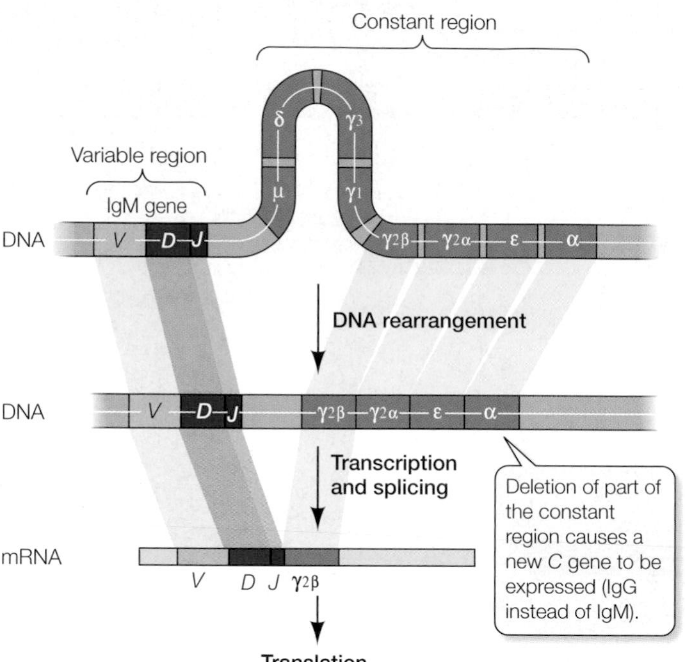

18.18 Class Switching The supergene produced by joining *V*, *D*, *J*, and *C* genes (see Figure 18.19) may later be modified, causing a different *C* region to be transcribed. This modification, known as class switching, is accomplished by deletion of part of the constant region gene cluster. Shown here is class switching from IgM to IgG.

generating a signal transduction cascade that results in altered transcription of the immunoglobulin genes.

18.6 RECAP

The immune system can make millions of antibodies with different specificities by rearranging the *V*, *D*, and *J* genes that code for variable regions in the immature B cell. Additional posttranscriptional processing and mRNA splicing result in a unique immunoglobulin chain. The class of the immunoglobulin molecule can be changed by deletion of a gene coding for the constant region of the heavy chain.

- How can millions of antibodies with different specificities be generated from a relatively small number of genes? See pp. 418–419 and Figures 18.16 and 18.17
- What is the role of the constant region of the immunoglobulin in class switching? See p. 419 and Figure 18.18

Given the numerous and complex cellular interactions that activate the immune system and generate antibody diversity, you may perceive many points at which the immune system can fail. We now turn to several situations in which one or more components of this complex system malfunction.

18.7 What Happens When the Immune System Malfunctions?

Sometimes the immune system fails us in one way or another. It may overreact, as in an *allergic reaction*; it may attack self antigens, as in an *autoimmune disease*; or it may function weakly or not at all, as in an *immune deficiency* disease.

Allergic reactions result from hypersensitivity

An **allergic reaction** arises when the human immune system overreacts to (is *hypersensitive* to) a dose of antigen. Although the antigen itself may present no danger to the host, the inappropriate immune response may produce inflammation and other symptoms, which can cause serious illness or death. Allergic reactions are the most familiar examples of this phenomenon. There are two types of allergic reactions: *immediate hypersensitivity* and *delayed hypersensitivity*.

IMMEDIATE HYPERSENSITIVITY **Immediate hypersensitivity** arises when an individual exposed to an antigen in a food, pollen, or the venom of an insect, referred to as an *allergen*, makes large amounts of IgE. When this happens, mast cells in tissues and basophils in blood bind the constant end of the IgE. If that individual is exposed to that allergen again, binding of the allergen to the IgE causes the mast cells and basophils to rapidly release a large amount of histamine (**Figure 18.19**). This results in symptoms such as dilation of blood vessels, inflammation, and difficulty breathing. If not treated with antihistamines, a severe allergic reaction can lead to death. Why an initial exposure to an allergen stimulates IgE production in some people is not known. There is some evidence for genetic factors predisposing people to an allergic response.

Allergy to pollen can be treated by a process called *desensitization*. The process involves injecting small amounts of the allergen (typically just an extract of the offending plant tissue) into the skin—enough to stimulate IgG production but not enough to stimulate IgE production. So the next time the person is exposed to the allergen, IgG binds to it, tying it up before IgE can bind it and exert its harmful effects. Desensitization does not work for food allergens because the IgE response to those substances is so strong that even a small amount of antigen provokes it.

One of the most common food allergens is the peanut. Some people are so sensitive that kissing someone who has recently eaten a peanut can cause a life-threatening allergic reaction.

DELAYED HYPERSENSITIVITY **Delayed hypersensitivity** is an allergic reaction that does not begin until hours after exposure to an antigen. In this case, the antigen is taken up by antigen-presenting cells and a T cell response is initiated. An example is the rash that develops after exposure to poison ivy.

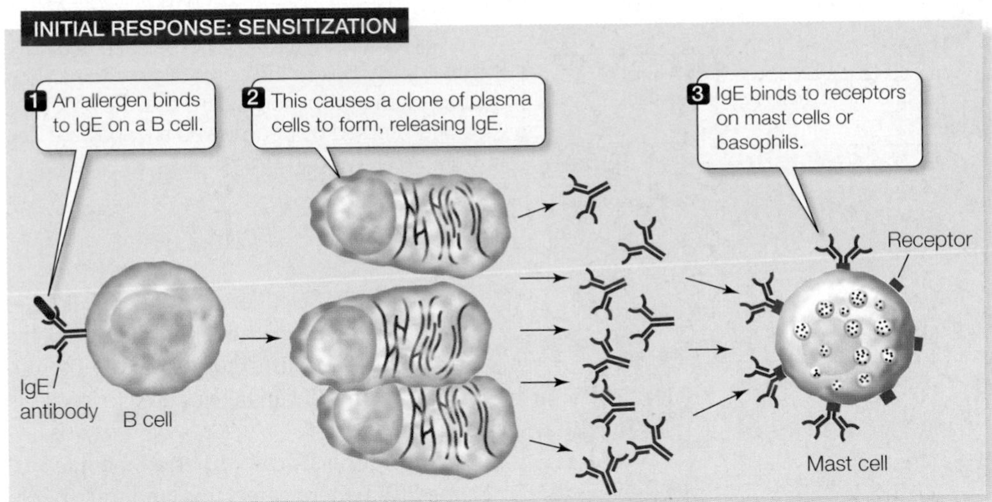

INITIAL RESPONSE: SENSITIZATION
1 An allergen binds to IgE on a B cell.
2 This causes a clone of plasma cells to form, releasing IgE.
3 IgE binds to receptors on mast cells or basophils.
Receptor
IgE antibody
B cell
Mast cell

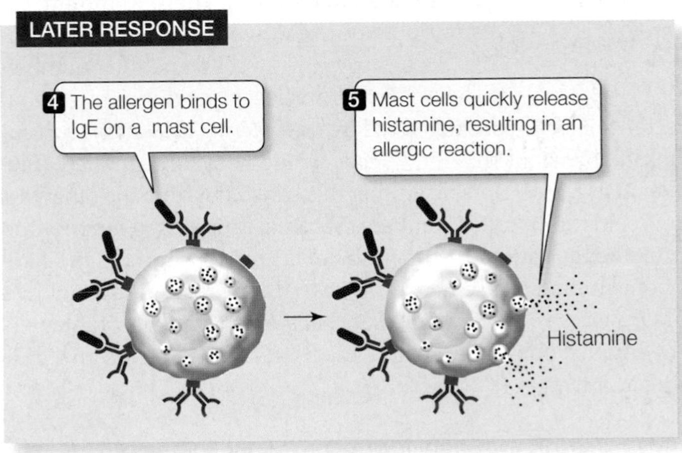

18.19 An Allergic Reaction An allergen is an antigen that stimulates B cells to make large amounts of IgE antibodies that bind to mast cells and basophils. When the body encounters the allergen again, these cells produce large amounts of histamine that have harmful physiological effects.

Autoimmune diseases are caused by reactions against self antigens

Sometimes, people produce one or more "forbidden clones" of B and T cells directed against self antigens. Although the precise origin of **autoimmunity** is not known, there are several hypotheses:

- *Failure of clonal deletion:* A clone of lymphocytes making antibodies against self antigens that should have been destroyed by apoptosis is not.
- *Viral infection:* A virus that has an antigenic determinant that resembles a self antigen infects a person, and because of the polyclonal response to complex antigens, the body generates some antibodies against the self antigen.
- *Molecular mimicry:* T cells that recognize a nonself antigen also recognize something on the self that has a similar structure.

Analyses of human pedigrees show that autoimmune diseases tend to "run in families," indicating a genetic component. Genome scans for SNPs in patients with these diseases have found a transcription factor involved in B cell development, called RUNX1, that may be involved. Some alleles of *MHC II* are strongly linked to certain autoimmune diseases.

Autoimmunity does not always result in disease, but a number of autoimmune diseases are common:

- People with *systemic lupus erythematosis* (SLE) have antibodies to many cellular components, including DNA and nuclear proteins. These antinuclear antibodies can cause serious damage when they bind to normal tissue antigens to form large circulating antigen–antibody complexes, which become stuck in tissues and provoke inflammation.
- People with *rheumatoid arthritis* have difficulty in shutting down a T cell response. We mentioned earlier that the inhibitor CTLA4 blocks T cells from reacting to self antigens. People with rheumatoid arthritis may have low CTLA4 activity, which results in inflammation of the joints due to the infiltration of excess white blood cells.
- *Hashimoto's thyroiditis* is the most common autoimmune disease in women over 50. Immune cells attack thyroid secretions resulting in fatigue, depression, weight gain, and other symptoms.
- *Insulin-dependent diabetes mellitus*, or type I diabetes, occurs most often in children. It is caused by an immune reaction against several proteins in the cells of the pancreas that manufacture the protein hormone insulin. This reaction kills the insulin-producing cells, so people with type I diabetes must take insulin daily in order to survive.

AIDS is an immune deficiency disorder

There are a number of inherited and acquired *immune deficiency disorders*. In some individuals, T or B cells never form; in others, B cells lose the ability to give rise to plasma cells. In either case, the affected individual is unable to mount an immune response and thus lacks a major line of defense against pathogens.

Because of its essential roles in both the humoral and cellular immune responses, the T_H cell is perhaps the most central of all the components of the immune system—a significant cell to lose to an immune deficiency disorder. This cell is the target of **HIV** (*h*uman *i*mmunodeficiency *v*irus), the retrovirus that eventually results in **AIDS** (*a*cquired *i*mmune *d*eficiency *s*yndrome).

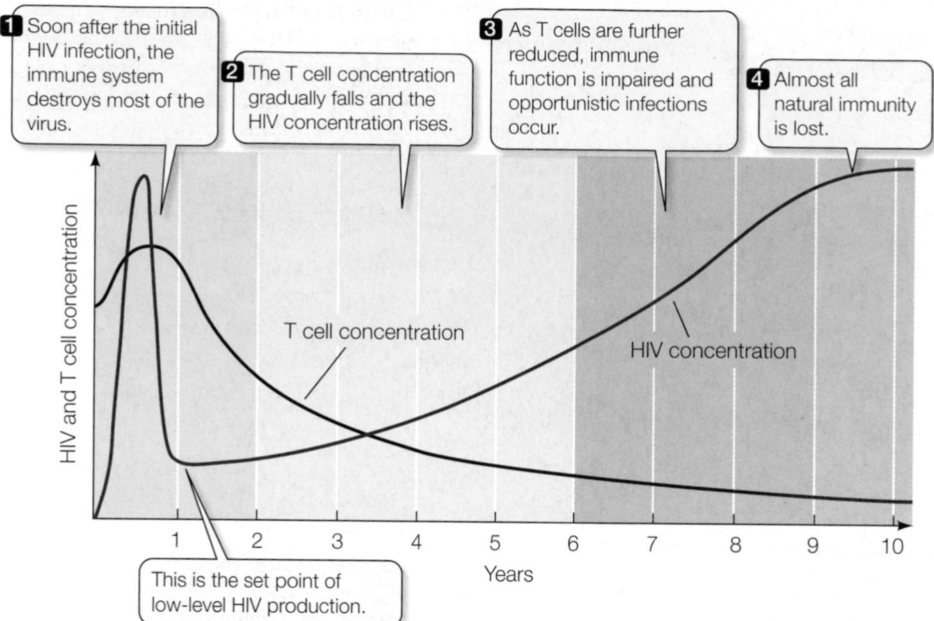

18.20 The Course of an HIV Infection An HIV infection may be carried, unsuspected, for many years before the onset of symptoms. This long "dormant" period means that the infection is often spread by people who are unaware that they are carrying the virus.

HIV can be transmitted from person to person in several ways:

- Through blood, such as by a needle contaminated with the virus after being used to inject an infected individual
- Through the exposure of broken skin, an open wound, or mucous membranes to body fluids, such as blood or semen, from an infected individual
- Through the blood of an infected mother to her baby during birth

HIV initially infects macrophages, T_H cells, and dendritic cells in blood and tissues. These infected cells carry the virus to the lymph nodes and spleen, where T and B cells are present. Normally, the dendritic cells present their captured antigens to T_H cells in the lymph nodes, and this causes the T_H cells to proliferate and form a clone (see Figure 18.15). But HIV preferentially infects activated—not resting—T_H cells. So the HIV arriving in the lymph nodes proceeds to infect the many activated T_H cells that are already responding to other antigens there. These two processes—the transport of the virus to the nodes and the presence in the nodes of cells already receptive to viral infection—combine to ensure that HIV reproduces vigorously. Up to 10 billion viruses are made every day during this initial phase of infection. The numbers of T_H cells quickly drop, and infected people show symptoms similar to mononucleosis, such as enlarged lymph nodes and fever.

These symptoms abate within 3 weeks, however, as T cells recognize infected cells, an immune response is mounted, and antibodies specific to HIV appear in the blood (**Figure 18.20**). By this time, the patient has a high level of circulating HIV complexed with antibodies, which is gradually removed by the action of dendritic cells over the next several months. But before they are removed, these antibody-complexed viruses can still infect T_H cells that come into contact with them. This secondary infection process reaches a low, steady-state level called the "set point." This point varies among individuals and is a strong predictor of the rate of progression of the disease. For most people, it takes 8–10 years, even without treatment, for the more severe manifestations of AIDS to develop. In some, it can take as little as a year; in others, 20 years.

During this dormant period, people carrying HIV generally feel fine, and their T_H cell levels are adequate for them to mount immune responses. Eventually, however, the virus destroys the T_H cells, and their numbers fall to dangerous levels. At this point, the infected person is considered to have *full-blown AIDS* and is susceptible to infections that the T_H cells would normally eliminate (**Figure 18.21**). Most notable among these infections are the otherwise rare skin tumor called Kaposi's sarcoma, caused by a herpesvirus; pneumonia, caused by the fungus *Pneumocystis carinii*; and lymphoma tumors, caused by the Epstein–Barr virus. These conditions are called *opportunistic infections* because they take advantage of the crippled immune system of the host. They lead to death within a year or two.

HIV INFECTION AND REPLICATION As the AIDS epidemic has grown, so has our knowledge of the molecular biology of HIV infection. We described the life cycle of HIV in Section 13.1 (see Figure 13.6). Briefly, as an enveloped retrovirus, the viral core of HIV is surrounded by a membrane, and its genome is RNA.

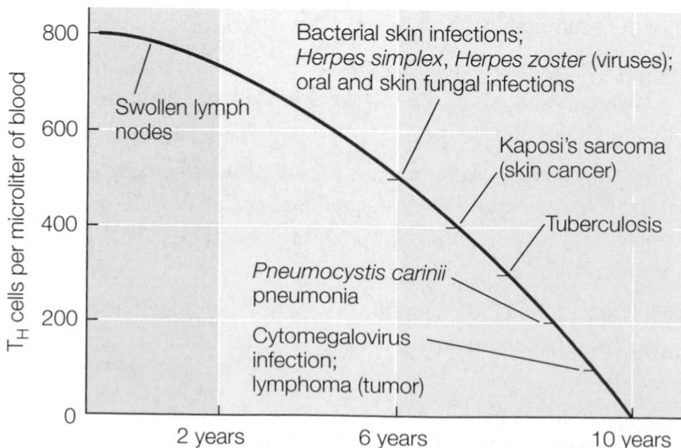

18.21 Relationship between T_H Cell Count and Opportunistic Infections As HIV kills more and more T_H cells, the immune system is less and less able to defend the body against various pathogens, including many that are not usually infectious to healthy people.

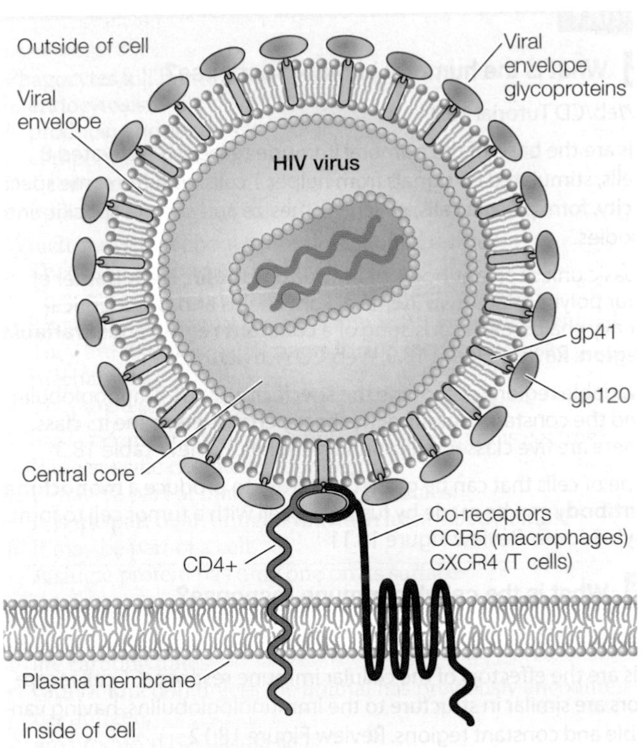

18.22 Two Receptors for HIV Two proteins on the cell surface are needed for HIV to enter a cell. Some people lack the co-receptor CCR5 and so are resistant to HIV infection.

Several virally encoded proteins are especially vital for HIV infection and replication:

- The virus uses the membrane proteins *gp120* and *gp41* to attach to human cells (*gp* stands for *glycoprotein*). These proteins target two proteins on the host cell surfaces, CD4 and a co-receptor (CCR5 in macrophages and CXCR4 in T cells) (**Figure 18.22**). Binding of the virus is followed by membrane fusion and the viral genome with attached proteins enters the cell.
- *Reverse transcriptase* catalyzes the synthesis of cDNA from viral RNA. Unfortunately, it lacks the proofreading function of many DNA polymerases, and makes about 10 errors in every round of cDNA synthesis. This creates a pool of mutant viruses.
- *Integrase* catalyzes the insertion of cDNA into the host chromosome.
- *Protease* is necessary to complete the formation of individual viral proteins from larger initial products of translation.

Additional proteins encoded by HIV are needed to complete the transcription of its cDNA (Tat protein) and the splicing of RNA (Rev protein).

TREATING HIV INFECTION Armed with a staggering amount of detailed knowledge of HIV, scientists have been attempting to develop treatments for AIDS. The general therapeutic strategy is to try to block stages in the viral life cycle without damaging the host cell. Potential therapeutic agents that interfere with the major steps of the life cycle are being tested. It is crucial to block only steps that are unique to the virus so that drug therapies do not harm the patient by blocking a step in the patient's own metabolism.

*H*ighly *a*ctive *a*nti*r*etroviral *t*herapy (**HAART**), developed in the late 1990s, has had considerable success in delaying the onset of AIDS symptoms in people infected with HIV, and in prolonging the lives of people with AIDS. The logic of HAART comes from cancer treatment: employ a combination of drugs that act at different parts of the viral life cycle. Generally, the HAART regimen uses a protease inhibitor and two reverse transcriptase inhibitors. The two reverse transcriptase inhibitors can be given as a single pill. These drug regimens may eliminate HIV entirely in some people, especially in those treated within the first few days after infection, before the virus has arrived in the lymph nodes. Most patients, however, face a lifetime of anti-HIV therapy.

Unfortunately, many patients who take HAART develop mutant strains of HIV that are resistant to this regimen; this is a consequence of the error rate of reverse transcriptase. Thus there is a never-ending race to modify HAART by adding new and/or different drug combinations, and today there are around 150 different HAART treatments. In short, we seem trapped in an evolutionary struggle with HIV. How can we gain a lasting advantage? The greatest hope is for the development of a vaccine against HIV. The first clinical trials of vaccines directed against the HIV membrane protein gp120 were not successful, but other vaccines are under development.

The worldwide AIDS epidemic began in the early 1980s and continues to expand, especially in Africa and Southeast Asia. About 40 million people are currently infected with HIV. Some 4 million new infections occur each year, along with 3 million deaths. At least half of the people who have been infected since the epidemic began are already dead. What can be done until biomedical science can bring the epidemic to an end?

Above all, people must recognize that they are in danger whenever they have sex with a partner whose *total* sexual history they do not know. The danger rises as the number of sexual partners rises, and the danger is much greater if partners participating in sexual intercourse are not protected by a latex condom. The danger that heterosexual intercourse will transmit HIV rises significantly if either partner has another sexually transmitted disease.

18.7 RECAP

Failures of the immune system include allergic reactions (caused by hypersensitivity to antigens), autoimmune diseases (caused by reactions against self antigens), and immune deficiency disorders.

- How does immediate hypersensitivity develop? See p. 420 and Figure 18.19
- What is an autoimmune disease? Give an example. See p. 421
- Can you describe the course of events in the human immune system during HIV infection? See pp. 422–423 and Figure 18.20

CHAPTER 19 Differential Gene Expression in Development

Stem cells from fat

Dr. Marc Hedrick was concerned about fat—other people's fat. As a plastic surgeon in Los Angeles, part of his medical practice involved performing liposuction, a technique to remove unwanted fat. He wondered if this excess fat could be put to some use. Looking at globs of fat under the microscope, Hedrick was surprised to see cells that resembled bone cells. How did bone cells come to be found in excised fat tissue?

Hedrick hypothesized that fat stores harbor a population of *stem cells*—dividing, unspecialized cells that have the potential to produce many different cell types on demand. In this case, the potential of the stem cells would be limited to types of cells whose origins in the embryo are related to fat; this would include bone, cartilage, blood vessel, and muscle cells.

The surgeon set out to isolate these stem cells. There was no shortage of discarded fat from Los Angeles cosmetic surgery practices, and soon he had a cell population that, when provided with growth factors and hormones, could differentiate into several types of cells in a laboratory dish.

But can these fat-derived adult stem cells differentiate inside an animal? Experiments on rats and mice have now shown that these stem cells can be implanted into organs and that they differentiate into cells appropriate for the tissue they are in. Thus they could be used to treat damaged hearts, injured bones, and damaged blood vessels. An advantage of using fat stem cells in human medicine is that the patient's own cells can be used. The immune system, as we saw in Chapter 18, recognizes and rejects nonself tissues. With fat stem cells, there would be no rejection implanted tissue.

Hedrick has developed the machinery to separate stem cells from fat quickly enough to be done in the operating room while patients await treatment for a damaged heart or other tissue. Two hundred million stem cells are enough for therapy, and they can be obtained from 450 grams (about a pound) of fat. Recently, three women in Japan had stem cells from their own fat implanted to help reconstruct and heal wounded breast tissue following mastectomy. The era of personalized medicine has begun.

Behind all of this impressive medical technology lies a great deal of basic developmental genetics. The processes that underlie the differentiation of stem cells in an adult are the same ones that occur in the embryo. Much of our knowledge of developmental

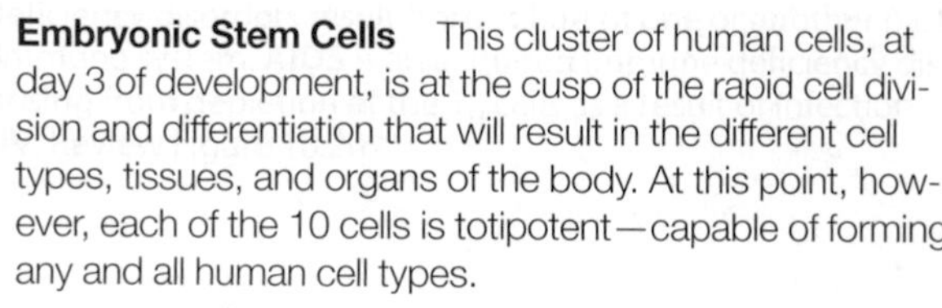

Embryonic Stem Cells This cluster of human cells, at day 3 of development, is at the cusp of the rapid cell division and differentiation that will result in the different cell types, tissues, and organs of the body. At this point, however, each of the 10 cells is totipotent—capable of forming any and all human cell types.

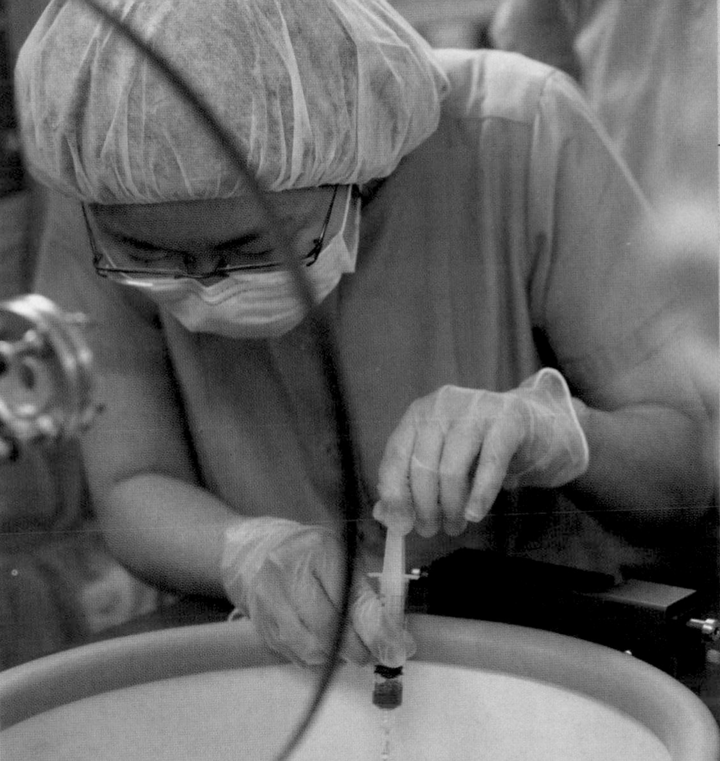

Fat As a Source of Stem Cells This centrifuge separates dense fatty tissues from the lighter stem cells. Stem cells from fat have been found to be capable of differentiating into several specialized cell types.

genetics has come from studies on certain model organisms, such as the fruit fly *Drosophila melanogaster*, the nematode *Caenorhabditis elegans*, frogs, sea urchins, and a flowering plant, the thale cress, *Arabidopsis thaliana*. As we saw in Section 14.1, the genomes of all eukaryotes are surprisingly similar, and the cellular and molecular principles underlying their development also turn out to be similar. Thus discoveries from one organism aid us in understanding other organisms, including ourselves.

IN THIS CHAPTER we will see that every body cell in a multicellular organism contains all of the genes present in the zygote that gave rise to that organism—even though only some of those genes are expressed in any particular cell type. We will see how cellular changes during development result from the differential expression of genes. We will describe the various mechanisms of transcriptional control and chemical signaling that work together to produce a complex organism.

CHAPTER OUTLINE

19.1 What Are the Processes of Development?

Development is the process by which a multicellular organism undergoes a series of progressive changes, taking on the successive forms that characterize its life cycle (**Figure 19.1**). In its earliest stages of development, a plant or animal is called an **embryo**. Sometimes the embryo is contained within a protective structure, such as a seed coat, an eggshell, or a uterus. An embryo does not photosynthesize or feed actively; instead, it obtains its food from its mother directly or indirectly (by way of nutrients stored in a seed or egg). A series of embryonic stages may precede the birth of the new, independent organism. Most organisms continue to develop throughout their life cycle; development ceases only with death.

Development proceeds via determination, differentiation, morphogenesis, and growth

Four processes are responsible for the developmental changes an organism undergoes as it progresses from an embryo to mature adulthood:

- **Determination** sets the developmental *fate* of a cell—what type of cell it will become—even before any characteristics of that cell type are observable.
- **Differentiation** is the process by which different types of cells arise; that is, differentiation leads to cells with the specific structures and functions of that cell's determined fate.
- **Morphogenesis** (Greek for "origin of form") is the shaping of differentiated cells into the multicellular body and its organs.
- **Growth** is the increase in size of the body and its organs by cell division and cell expansion.

Mitosis produces daughter cells that are chromosomally and genetically identical to the cell that divides to produce them (see Section 9.1). Why, then, are the cells of a multicellular organism not all identical in structure or function? Cells differ from one another because different genes are expressed in different cells, a phenomenon called **differential gene expression**.

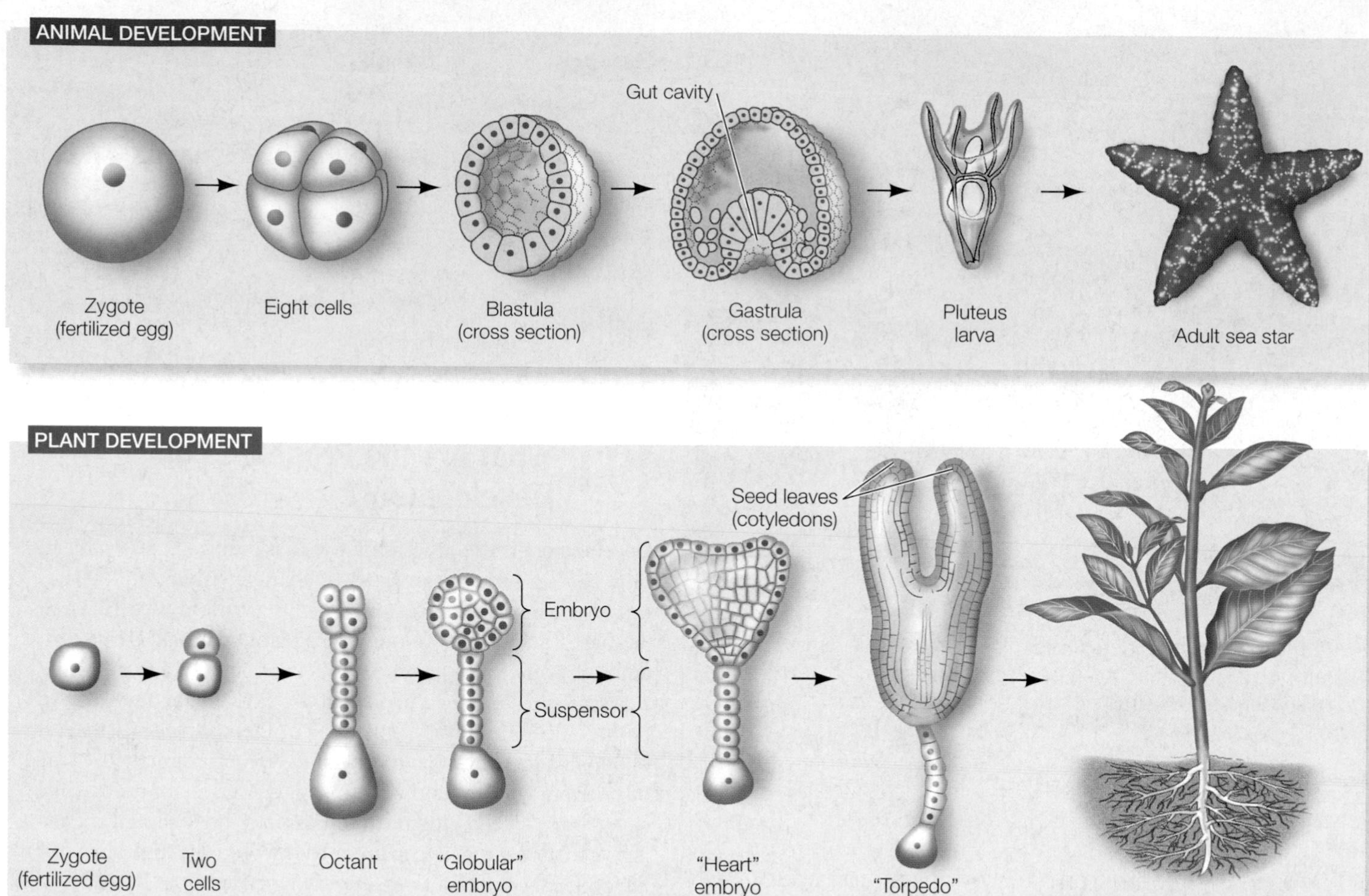

19.1 From Fertilized Egg to Adult Stages of development from zygote to maturity are shown for an animal and for a plant.

In an early embryo that consists of only a few cells, each cell has the potential to develop in many different ways. As development proceeds, regulation of the expression of genes results in the production of different proteins and thus in cells with progressively different features and functions.

Morphogenesis, the process by which recognizable tissues and organs develop in an organism, can occur in several ways. In plant development, cells are constrained by cell walls and do not move around the body, so organized division and expansion of cells are the major processes that build the plant body. In animals, cell movements are very important in morphogenesis, as we will see in Section 43.2. And in both plants and animals, apoptosis (programmed cell death) is essential to orderly development. Like differentiation, morphogenesis results ultimately from the closely regulated activities of genes and their products as well as from the interplay of signals secreted by one group of cells and their effects on other (target) cells.

In all multicellular organisms, repeated mitotic divisions generate the multicellular body. Increases in the sizes of cells also contribute to growth. In plants, cell expansion begins shortly after the first divisions of the fertilized egg, or *zygote*. In animals, on the other hand, cell expansion is often slow to begin: the animal embryo may consist of thousands of cells before it becomes larger than the original zygote. Growth continues throughout the individual's life in some species, but reaches a more or less stable end point in others.

Cell fates become more and more restricted

A cell's **fate**—that is, the type of cell into which it will ultimately differentiate—is a function of both differential gene expression and morphogenesis. The role of morphogenesis in determining cell fate can be revealed in experiments in which specific cells of an early embryo are grafted into new positions on another embryo. The cells are marked with stains so that their development into adult structures can be traced.

For instance, we know that the blue-shaded area of the early frog embryo shown in **Figure 19.2** normally becomes part of the skin of the tadpole (the larva of a frog). However, if we cut out a piece from this region and transplant it to another location on another early frog embryo, it does not become skin. The type of tissue it does become is determined by its new location in the embryo. The developmental potential of these early embryonic cells—that is, their range of possible fates—is thus greater than their actual fate.

The developmental potential of cells becomes restricted fairly early in normal development. Tissue from a later-stage frog embryo, for example, if taken from a region that normally develops

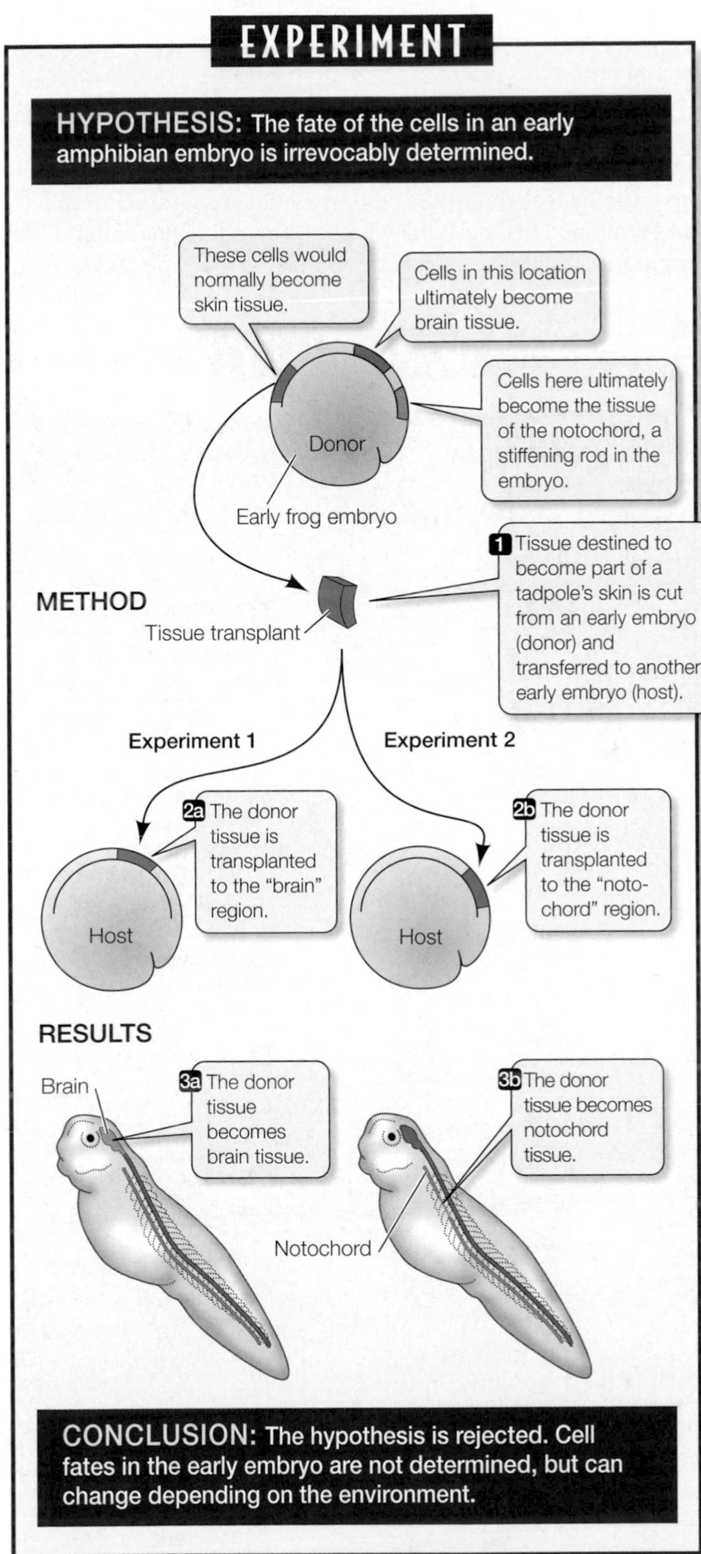

EXPERIMENT
HYPOTHESIS: The fate of the cells in an early amphibian embryo is irrevocably determined.
These cells would normally become skin tissue.
Cells in this location ultimately become brain tissue.
Cells here ultimately become the tissue of the notochord, a stiffening rod in the embryo.
Donor
Early frog embryo
METHOD
Tissue transplant
1 Tissue destined to become part of a tadpole's skin is cut from an early embryo (donor) and transferred to another early embryo (host).
Experiment 1
Experiment 2
2a The donor tissue is transplanted to the "brain" region.
2b The donor tissue is transplanted to the "noto-chord" region.
Host
Host
RESULTS
Brain
3a The donor tissue becomes brain tissue.
3b The donor tissue becomes notochord tissue.
Notochord
CONCLUSION: The hypothesis is rejected. Cell fates in the early embryo are not determined, but can change depending on the environment.

19.2 Developmental Potential in Early Frog Embryos Cells that would normally form one type of tissue can form completely different tissue types when they are experimentally moved to another location. In this experiment, epithelial (skin) tissue from an early-stage frog embryo is transplanted from a donor to a host embryo. The tissue that develops in the host tadpole is not skin, but is consistent with the location to which the tissue is transplanted. FURTHER RESEARCH: What would happen if tissue from an adult were transplanted into an early embryo?

into the brain, becomes brain tissue even if transplanted to a part of an early-stage embryo destined to become another structure.

Determination is influenced by the action of both the extracellular environment and the contents of the cell on the cell's genome. Determination is not something that is visible under the microscope—cells do not change their appearance when they become determined. Determination is followed by differentiation—the actual changes in biochemistry, structure, and function that result in cells of different types. Differentiation often involves a change in appearance as well as function. *Determination is a commitment; the final realization of that commitment is differentiation.*

19.1 RECAP

Development takes place via the processes of determination, differentiation, morphogenesis, and growth. Cells in the early embryo have not yet had their differentiated fates determined; as development proceeds, their potential fates become more and more restricted.

- Can you name and describe the four processes of development? See p. 427
- Do you understand how the experiment in Figure 19.2 illuminates how cell fates become determined? See pp. 428–429

Is a photosynthetic cell in the mesophyll of a leaf or a liver cell in a human being irrevocably committed to that specialization? Under the right experimental circumstances, differentiation is reversible in many cells. The next section describes how the genome of a cell can be induced to express a different pattern of differentiation.

19.2 Is Cell Differentiation Irreversible?

A zygote has the ability to give rise to every type of cell in the adult body; in other words, it is **totipotent**. Its genome contains instructions for all of the structures and functions that will arise throughout the life cycle of the organism. Later in development, the cellular descendants of the zygote lose their totipotency and become determined. These determined cells then differentiate into specific types of specialized cells. A liver cell in a human or a mesophyll cell in the leaf of a green plant generally retains its differentiated form and function throughout its life. But this does not necessarily mean that these cells have irrevocably lost their totipotency.

Most of the differentiated cells of an animal or plant have nuclei containing the entire genome and certainly have the genetic capacity for totipotency. We explore here several examples of how this capacity has been demonstrated experimentally.

Plant cells are usually totipotent

A food storage cell in a carrot root normally faces a dark future. It cannot photosynthesize or give rise to new carrot plants. However, if we isolate that cell from the root, maintain it in a suitable nutrient medium, and provide it with appropriate chemical cues, we can "fool" the cell into acting as if it were a zygote. It can divide and give rise to a mass of undifferentiated cells, called a *callus*, and eventually to a complete plant (**Figure 19.3**). Since the new plant is genetically identical to the cell from which it came, we call the plant a **clone**.

The ability of scientists to clone an entire carrot plant from a differentiated root cell indicates that the cell contains the entire carrot genome and that it can express the appropriate genes in the right sequence. Many types of cells from other plant species show similar behavior in the laboratory. This ability to generate a whole plant from a single cell has been invaluable in agricultural biotechnology. For example, forest products companies harvest trees for use in making paper, lumber, and other products. To replace the trees reliably, they remove fragments of leaves and, in a procedure just like the one used on carrot root cells, form embryos and then small plants, which can be planted in the forest. The yield of trees from these clones is higher and more predictable than that from seeds produced by fertilization.

Among animals, the cells of early embryos are totipotent

Animal somatic cells cannot be fooled into acting like zygotes as easily as plant cells can, but nuclear transplantation experiments have demonstrated that animal somatic cells retain their totipotency. Such experiments were first done on frogs by Robert Briggs and Thomas King, who asked whether the nuclei of early frog embryos had retained the ability to do what the totipotent zygote nucleus could do. They first removed the nucleus from an unfertilized egg, forming an *enucleated* egg. Then, with a very fine glass tube, they punctured a cell from an early embryo and drew up part of its contents, including the nucleus, which they injected into the enucleated egg. They stimulated the eggs to divide, and many went on to form embryos, tadpoles, and eventually, frogs. These experiments led to two important conclusions:

- No information is lost from the nuclei of cells as they pass through the early stages of embryonic development. This fundamental principle of developmental biology is known as **genomic equivalence**.
- The cytoplasmic environment around a nucleus can modify its fate.

Similar experiments have been performed on rhesus monkeys, in which a single cell can be removed from an 8-cell embryo and fused with an enucleated egg. This *cell fusion* technique causes the nucleus of the embryonic cell to enter the egg cytoplasm. The resulting cell acts like a zygote, forming an embryo, which can be implanted into a foster mother, who ultimately gives birth to a normal monkey. Each of the remaining 7 cells from the original embryo can give rise to offspring by the same cell fusion technique.

In humans, the totipotency of early embryonic cells permits both genetic screening (see Section 17.3) and certain assisted reproductive technologies (see Section 42.4). An 8-cell human embryo can be isolated in the laboratory and a single cell removed and examined to determine whether a harmful genetic condition is present. The cells can then be stimulated to divide and form an

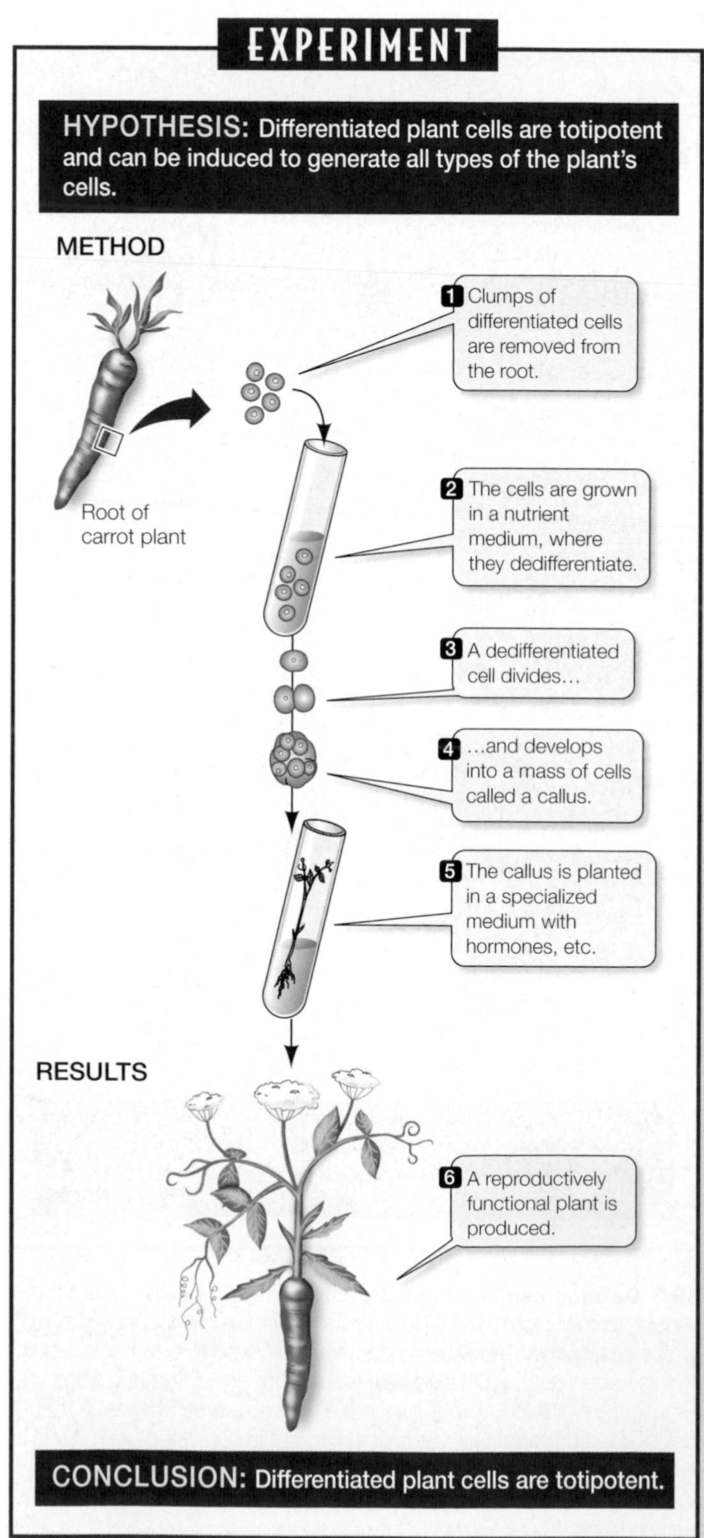

EXPERIMENT
HYPOTHESIS: Differentiated plant cells are totipotent and can be induced to generate all types of the plant's cells.
METHOD
Root of carrot plant
1 Clumps of differentiated cells are removed from the root.
2 The cells are grown in a nutrient medium, where they dedifferentiate.
3 A dedifferentiated cell divides...
4 ...and develops into a mass of cells called a callus.
5 The callus is planted in a specialized medium with hormones, etc.
RESULTS
6 A reproductively functional plant is produced.
CONCLUSION: Differentiated plant cells are totipotent.

19.3 Cloning a Plant Differentiated, specialized food storage cells from the root of a carrot can be induced to dedifferentiate by placing them in a new chemical environment. These cells can then act like early embryonic cells and form a new plant.

embryo, which can be implanted into the mother's uterus, where it develops into an infant.

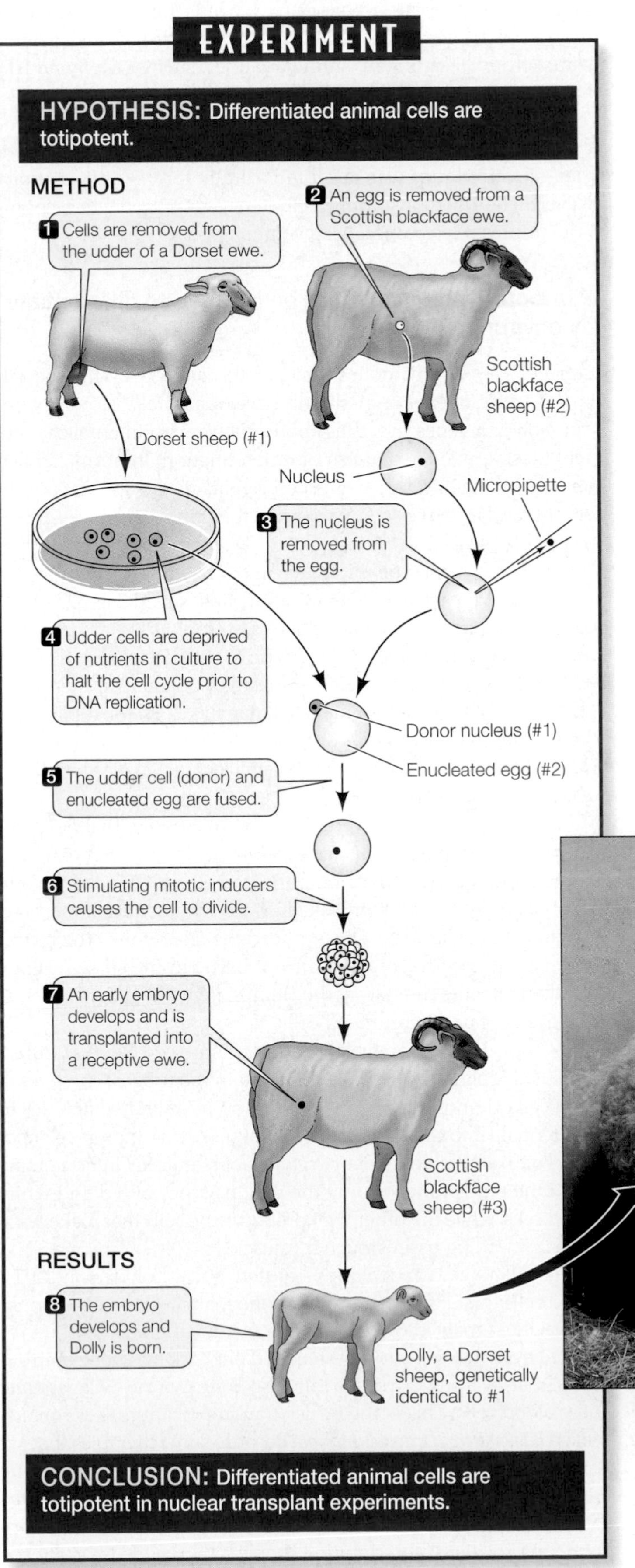

EXPERIMENT
HYPOTHESIS: Differentiated animal cells are totipotent.
METHOD
1 Cells are removed from the udder of a Dorset ewe.
2 An egg is removed from a Scottish blackface ewe.
Scottish blackface sheep (#2)
Dorset sheep (#1)
Nucleus
Micropipette
3 The nucleus is removed from the egg.
4 Udder cells are deprived of nutrients in culture to halt the cell cycle prior to DNA replication.
Donor nucleus (#1)
Enucleated egg (#2)
5 The udder cell (donor) and enucleated egg are fused.
6 Stimulating mitotic inducers causes the cell to divide.
7 An early embryo develops and is transplanted into a receptive ewe.
Scottish blackface sheep (#3)
RESULTS
8 The embryo develops and Dolly is born.
Dolly, a Dorset sheep, genetically identical to #1
CONCLUSION: Differentiated animal cells are totipotent in nuclear transplant experiments.

The somatic cells of adult animals retain the complete genome

The experimental reproductive cloning of animals from adult somatic cells was revolutionized in the late 1990s when Ian Wilmut and his colleagues at a biotechnology company in Scotland used the cell fusion technique to clone sheep. Previous attempts to produce mammals by this method had worked, as in the rhesus monkey case, only if the donor nucleus was from an early embryo. Problems had arisen when mammalian donor cells that were in the G2 phase of the cell cycle (see Figure 9.3) were fused with the cytoplasm of eggs that were also in G2; extra DNA replication took place that created havoc with the cell cycle in the egg when it attempted to divide. Wilmut found a way to ensure that the cells used in animal cloning experiments were in the G1 phase of the cell cycle. His procedure is outlined in **Figure 19.4**.

Wilmut took differentiated cells from a ewe's udder and starved them of nutrients for a week, thus halting the cells in G1 phase of the cell cycle. One of these cells was fused with an enucleated egg from a different breed of ewe. When mitotic inducers in the egg cytoplasm were stimulated, the donor nucleus entered S phase, and the rest of the cell cycle proceeded normally. After several cell divisions, the resulting early embryo was transplanted into the womb of a surrogate mother.

19.4 Cloning a Mammal In 1996, the experimental procedure described here produced the first cloned mammal, a Dorset sheep named Dolly (shown on the left in the photo). As an adult, Dolly mated and subsequently gave birth to a "normal" offspring (the lamb on the right), thus proving the genetic viability of cloned mammals.

Out of 277 successful attempts to fuse adult cells with enucleated eggs, one lamb survived to be born; she was named Dolly, and she became world-famous overnight. DNA analyses confirmed that Dolly's nuclear genes were identical to those of the ewe from whose udder the donor nucleus had been obtained. Dolly grew to adulthood, mated, and produced offspring in the classic manner (see Figure 19.4), thus proving her status as a fully functioning adult animal. Dolly died in 2003 at the age of six—middle age for a sheep. Whether her cloned genome contributed to her relatively early death is still being debated.

One goal of Wilmut's experiments was to develop a method of cloning transgenic sheep—sheep into which genes with therapeutic properties have been introduced. For example, transgenic sheep have been developed that are capable of expressing pharmaceutical products in their milk (see Section 16.6). The hope was that the cloning procedure could make multiple, identical transgenic sheep that are all reliable producers of drugs.

The trick of starving donor cells for cloning has been applied to other mammals. Mice have been cloned using the somatic cells surrounding the egg as a source of donor nuclei (**Figure 19.5**). Cattle have been cloned to preserve a rare breed in New Zealand. Genetically engineered goats that produce several useful proteins in their milk have now been cloned to expand the numbers of these valuable animals. The recent successful cloning of horses may be a prelude to the cloning of valuable racing and show animals.

Cloning might be useful in preserving endangered species. More than 25 years ago, geneticists at the San Diego Zoo began freezing cells from endangered species, creating a modern-day Noah's Ark in anticipation of the emergence of new knowledge about animal development and its application to reproductive cloning. With that knowledge now at hand, they have begun thawing out cells and using them as nuclear donors to expand the populations of endangered species by cloning. The banteng, an endangered relative of the cow, was the first animal cloned in this way, using a cow enucleated egg and a cow surrogate mother. Meanwhile, in China, scientists are using rabbits as surrogate mothers for cloned pandas (which are only about 6 centimeters long when they are born). Even pets such as cats and dogs have been cloned. The evidence that clones are genetically identical to their nuclear donor "parent" comes from DNA fingerprinting analyses, such as SNP and STR typing (see Section 16.1).

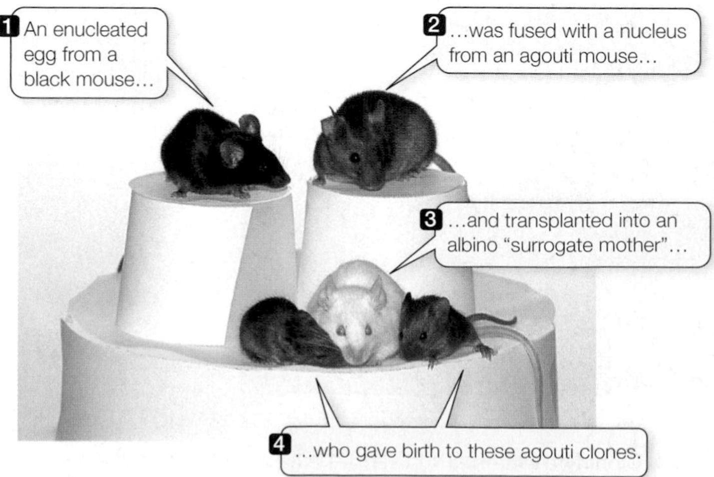

1 An enucleated egg from a black mouse...
2 ...was fused with a nucleus from an agouti mouse...
3 ...and transplanted into an albino "surrogate mother"...
4 ...who gave birth to these agouti clones.

19.5 Cloned Mice Because so much is known about mouse genetics and molecular biology, cloned mice may be useful in studies of basic biology.

The advent of cloning has been surrounded by controversy and ethical concerns, but cloning is not a new scientific concept. The idea of totipotency was accepted long before Dolly was born. Nevertheless, demonstrating totipotency via reproductive cloning is an impressive technical achievement.

Pluripotent stem cells can be induced to differentiate by environmental signals

Genomic equivalence implies that differentiated cells stay specialized because of environment and developmental history, not because of their genes, and thus appropriate environmental changes could result in a new pattern of differentiation. In normal development, a complex series of timed signals results in the patterns of differentiation we see in a newborn organism. If these signals could be described in enough detail, we should be able to understand how any cell type might become any other.

In plants, the growing regions at the tips of the roots and stems contain *meristems,* which are clusters of undifferentiated, rapidly dividing cells. These cells can give rise to the specialized cell types that make up roots and stems, respectively. Plants have far fewer (15–20) cell types than animals (as many as 200). Most plant cell types differ in the structure of their cell walls, whereas most animal cell types have specific cytoplasmic characteristics and many cell-specific proteins.

In mammals, **stem cells** are found in adult tissues that need frequent cell replacement, such as the skin, the inner lining of the intestine, and the blood system. Canadians Ernest McCulloch and James Till discovered stem cells in 1960 while doing bone marrow transplants in mice. They noticed that sometimes the recipient mice developed small clumps of tissue in the spleen. When they looked more carefully at the clumps, they found that each was composed of undifferentiated cells.

As they divide, stem cells produce daughter cells that differentiate to replace dead cells and maintain the tissues. These adult stem cells are not totipotent, but they do have limited abilities to differentiate into certain cell types; in other words, they are **pluripotent**. For example, there are two types of stem cells in bone marrow. One type produces only the various types of red and white blood cells, while the other type produces the cells that make bone and surrounding tissues, such as muscle.

The differentiation of pluripotent stem cells is "on demand." The blood cells that differentiate in the bone marrow do so in response to specific signals known as growth factors. If blood cells are removed from the circulatory system and put back into bone marrow, the signals will still be present and the bone marrow will generate new blood cells. This is the basis of an important cancer therapy called *bone marrow transplantation*. Because some therapies that kill cancer cells also kill other dividing cells (see Section 17.4), bone marrow stem cells in patients will die if exposed to the therapy agents. To prevent this, stem cells are removed from the patient's bone marrow and stored during therapy, then added back to the bone marrow when therapy is over. The stem cells retain their ability to differentiate in the bone marrow environment.

Adjacent cells can also influence stem cell differentiation. For example, the bone marrow stem cells that can form muscle will do so if implanted into the heart (**Figure 19.6**). This has been demonstrated in animal experiments, in which the stem cells were used to repair a damaged heart, and clinical uses in human patients are beginning.

Embryonic stem cells are potentially powerful therapeutic agents

As stated earlier, totipotent stem cells are found only in early embryos. In laboratory mice, these embryonic stem cells can be removed from an early embryo (the *blastocyst*; see Figure 43.4) and grown almost indefinitely. When injected back into a mouse blastocyst, the stem cells mix with the resident cells and differentiate to form all the cell types of the mouse. This kind of experiment shows that blastocyst cells do not lose any of their developmental potential while growing in the laboratory.

Embryonic stem cells growing in the laboratory also can be induced to differentiate in a particular way if the right signal is provided (**Figure 19.7**). For example, treatment of mouse embryonic stem cells with a derivative of vitamin A causes them to form neu-

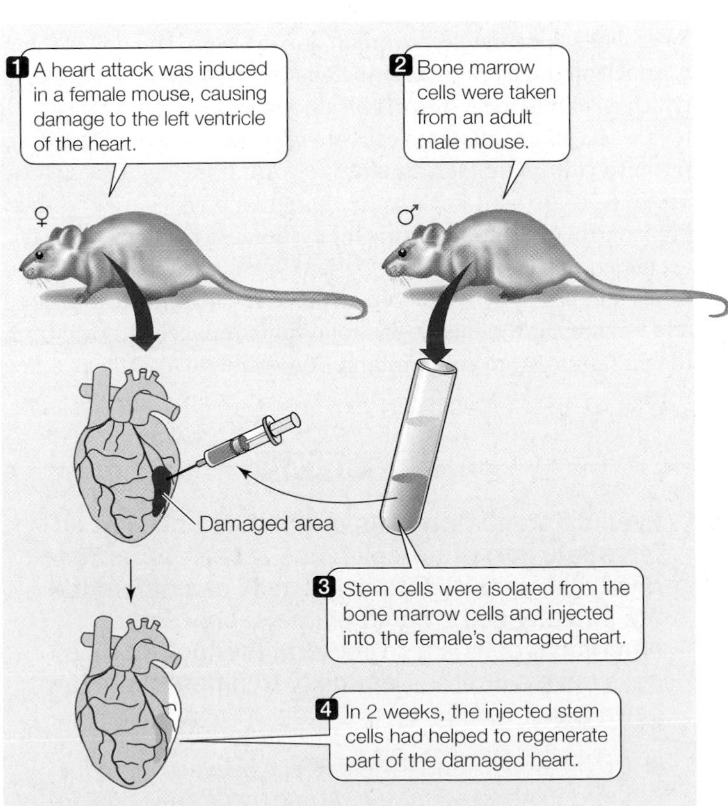

19.6 Repairing a Damaged Heart Pluripotent stem cells from the bone marrow of a male mouse were used successfully to repair the damaged heart of a female mouse. Mice of different sexes were used so that the Y chromosome could act as a genetic marker for the donor cells.

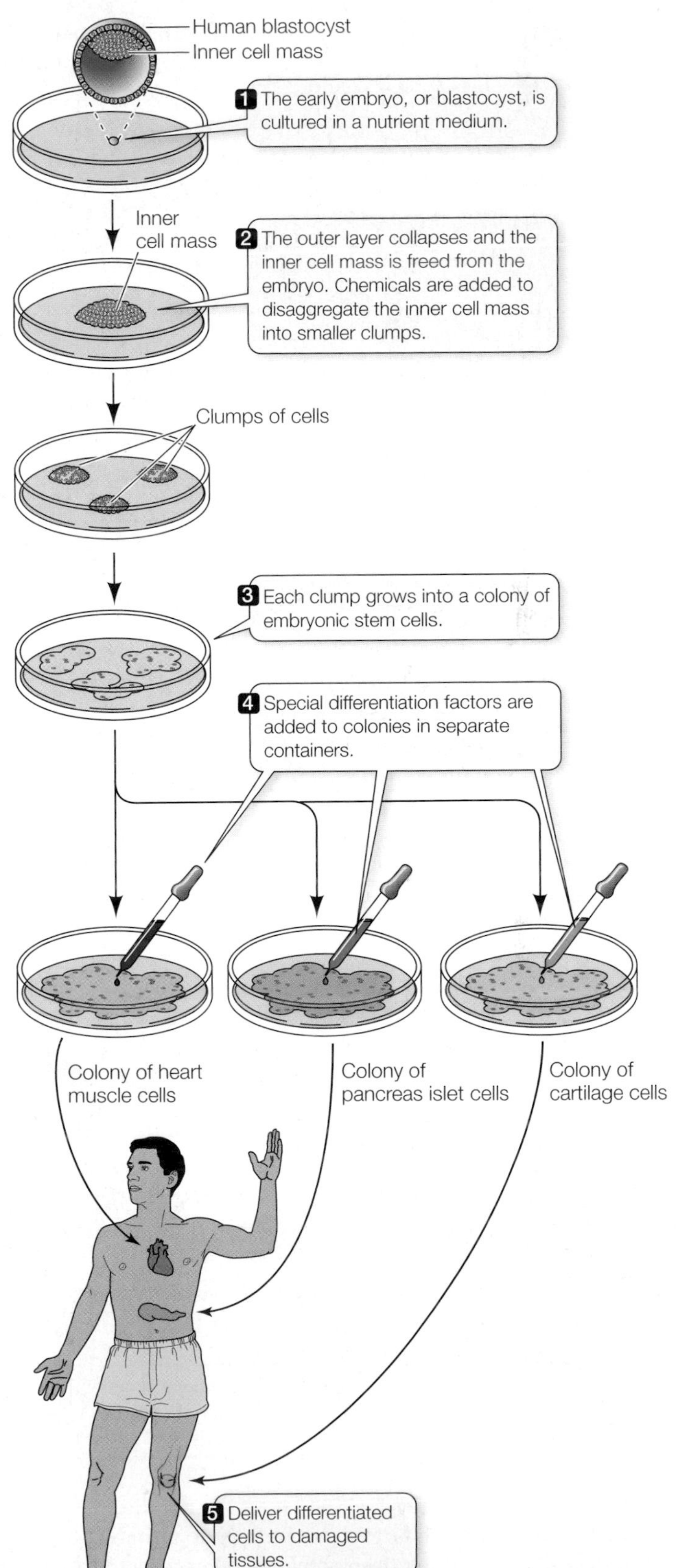

19.7 The Potential Use of Embryonic Stem Cells in Medicine Totipotent human embryonic stem cells can be cultured in the laboratory and induced to differentiate into a particular cell type. The use of these cells to replace tissues damaged by injury or disease is under intensive investigation.

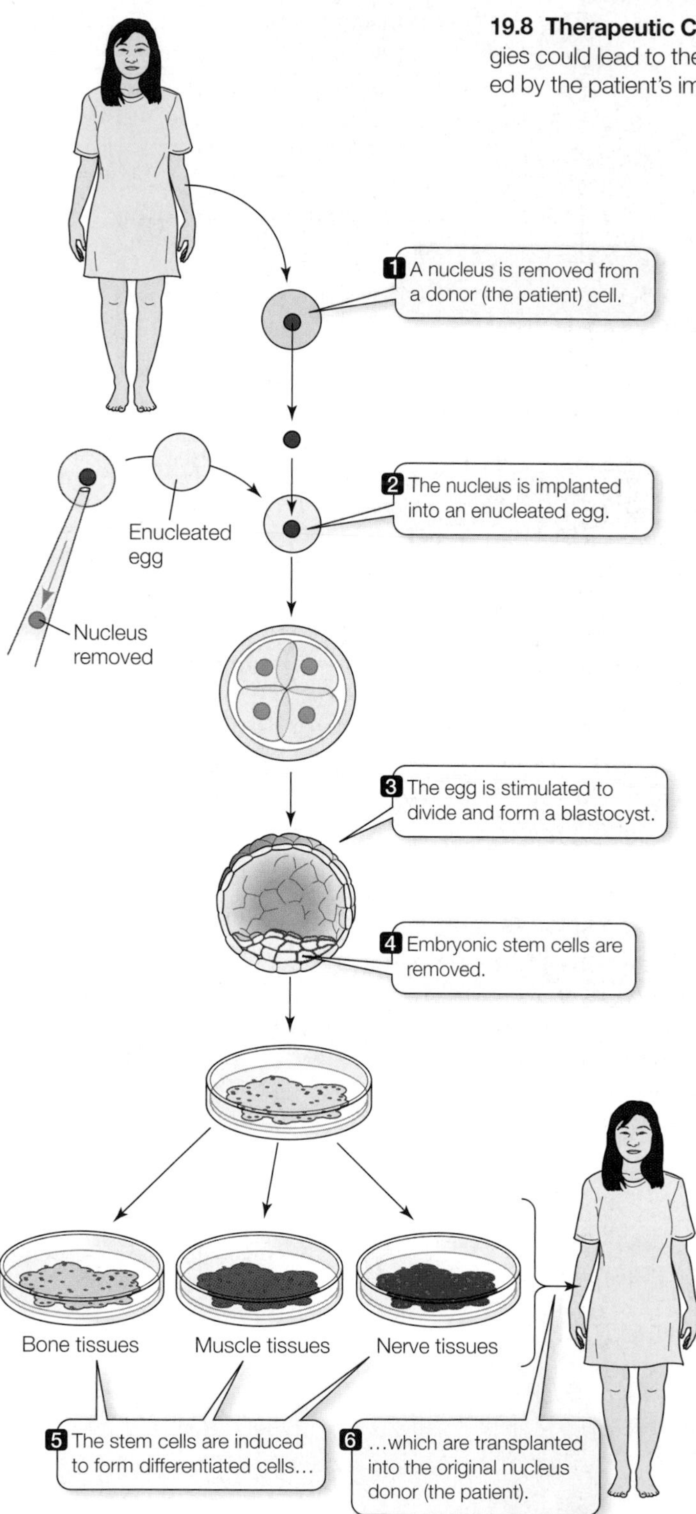
1 A nucleus is removed from a donor (the patient) cell.
Enucleated egg
Nucleus removed
2 The nucleus is implanted into an enucleated egg.
3 The egg is stimulated to divide and form a blastocyst.
4 Embryonic stem cells are removed.
Bone tissues
Muscle tissues
Nerve tissues
5 The stem cells are induced to form differentiated cells…
6 …which are transplanted into the original nucleus donor (the patient).

19.8 Therapeutic Cloning The combination of nuclear transplantation and stem cell technologies could lead to the production of cells and tissues for transplantation that would not be rejected by the patient's immune system.

rons, while other growth factors induce them to form blood cells, again demonstrating their developmental potential and the roles of environmental signals. This finding raises the possibility of using embryonic stem cell cultures as sources of differentiated cells for clinical medicine. A key advance toward this use has been the ability to grow human embryonic stem cells in the laboratory.

Embryonic stem cells could be harvested from human embryos made for in vitro fertilization with the consent of the donors. *In vitro fertilization* is a medical procedure used by couples who want a child but cannot conceive naturally. In the procedure, up to 10 eggs are taken from the mother's ovaries and exposed to the father's sperm, with the hope that some early embryos will form. Only a few of these embryos are implanted into the mother's uterus for development; the remaining embryos could be used for stem cell research.

A problem would arise, however, if such embryonic stem cells were induced to differentiate to form a tissue for transplantation—say, pancreatic tissue for a patient with diabetes. The cells and the recipient would be genetically different, so the recipient's immune system would form T cell receptors that would initiate an immune response, leading to rejection of the implanted cells (see Section 18.5). The problem of tissue rejection has led to the idea of **therapeutic cloning**, in which nuclear transplantation and stem cell implantation technologies would be combined.

If stem cells were to be derived from an embryo formed by nuclear transplantation using the patient's own nuclei, the cells would be genetically identical to the patient's own cells (**Figure 19.8**). A plausible scenario based on this technology is as follows: A patient suffers bone damage in an accident. Some of his skin cells are removed and brought to a laboratory, where the nuclei are isolated. Meanwhile, a woman donates eggs, and one of them is enucleated. The patient's nucleus is implanted into the egg, which is stimulated to form a blastocyst in a laboratory dish. After a week, embryonic stem cells are removed from the blastocyst, put into culture, and stimulated to form bone cell precursors. These cells are sent back to the patient's physician, who then implants them at the site of the injury. Because the implanted cells are genetically identical to the patient's cells, they are not rejected by his immune system. The cells proceed to differentiate into bone and help repair the injury. The total time from cell removal from the patient to stem cell implantation could be as little as a few weeks.

19.2 RECAP

Even differentiated cells retain their ability to differentiate into other cell types, given appropriate chemical signals. Totipotent cells can differentiate into any cell type; in animals, only early embryonic cells are totipotent. Pluripotent stem cells have a limited capability to differentiate into certain cell types.

- Do you understand the difference between reproductive cloning (see Figure 19.4) and therapeutic cloning? Why is in vitro fertilization *not* the same thing as reproductive cloning? See pp. 431–434
- How are stem cells found in adult body tissues different from embryonic stem cells? See pp. 432–433

Cloning experiments and observations of stem cells have shown that a differentiated cell still has all of the genes of every other cell type. But not all genes are expressed in every cell. What turns gene expression on and off as cells differentiate? In the next section we explore several of the controls of gene expression that lead to cell differentiation.

19.3 What Is the Role of Gene Expression in Cell Differentiation?

Although every cell contains all the genes needed to produce every protein encoded by its genome, each cell synthesizes only selected proteins. For example, certain cells in our hair follicles continuously produce keratin, the protein that makes up hair, while other cell types in the body do not produce keratin. What determines whether a cell will produce keratin or not?

Chapter 14 describes a number of ways in which cells regulate gene expression—and hence the production of proteins. These pathways included transcriptional, translational, and posttranslational controls. The major controls of the gene expression that results in cell differentiation are transcriptional.

Differential gene transcription is a hallmark of cell differentiation

The gene for β-globin, one of the protein components of hemoglobin, is expressed in red blood cells as they form in the bone marrow of mammals. That this same gene is also present—but unexpressed—in neurons in the brain (which do not make hemoglobin) can be demonstrated by nucleic acid hybridization. Recall that in nucleic acid hybridization, a probe made of single-stranded DNA or RNA of known sequence is added to denatured DNA to reveal complementary coding regions on the DNA template strand (see Figure 14.6). A probe for the β-globin gene can be applied to DNA from both brain cells and immature red blood cells (recall that mature mammalian red blood cells lose their nuclei during development). In both cases, the probe finds its complement, showing that the β-globin gene is present in both types of cells. On the other hand, if the probe is applied to mRNA, rather than DNA, from the two cell types, it finds β-globin mRNA only in the red blood cells, not in the brain cells. This result shows that the gene is expressed in only one of the two cell types.

What leads to this differential gene expression? One well-studied example of cell differentiation is the conversion of undifferentiated muscle precursor cells, called *myoblasts*, into the large, multinucleated *muscle fibers* that make up mammalian skeletal muscles. The key event that starts this conversion is the expression of a gene called *MyoD* (*myo*blast-*d*etermination gene). The protein product of this gene is a *transcription factor* (MyoD) with a helix-loop-helix domain (see Figure 14.15), which not only binds to the promoters of muscle-determining genes to stimulate their transcription, but also acts on its own promoter to keep its levels high in the myoblasts and in their descendants.

Strong evidence for the controlling role of *MyoD* in muscle fiber differentiation comes from experiments in which an artificial DNA sequence containing an active promoter adjacent to *MyoD* was transfected into the precursors of other cell types. For example, when this sequence was added to fat cell precursors, the fat cells were reprogrammed to become muscle cells. Genes such as *MyoD* that direct the most fundamental decisions in development (often by regulating other genes on other chromosomes) usually encode transcription factors. Such genes, which act as a kind of molecular on-off switch, are called *developmental genes*.

Tools of molecular biology are used to investigate development

As the *MyoD* system demonstrates, transcription factors are important regulators of cell differentiation in the embryo. Determination and differentiation are not carried out by the actions of single genes, however, but rather by a complex series of interactions of many genes and their products. Eric Davidson at the California Institute of Technology leads a team of scientists who have explored these interactions in the early stages of development in the sea urchin, which has long been a favorite model organism of developmental biologists.

Davidson and his colleagues have described dozens of genes that are turned on and off during sea urchin development. Indeed, they estimate that at least one-third of the eukaryotic genome is used *only* during development. Besides studying transcription using DNA chips, they inactivated the expression of single genes using RNAi (see Section 16.5). The researchers looked at the effects of "silenced" genes not only on overall phenotype, but also on other proteins. For example, for a single gene called *endo16* that codes for a cell membrane protein, they were able to describe where and when in the embryo it is transcribed; how much of the protein was made; how long the transcript and protein lasted in the cell; and what molecules interacted with the *endo16* promoter to regulate its transcription during development. The result is a complex network of hundreds of RNAs and proteins interacting. The computational tools of systems biology are important in setting up and analyzing this network.

19.3 RECAP

Differentiation involves selective gene expression. In some cases, a single transcription factor can cause a cell to differentiate in a certain way. In others, complex interactions between genes and proteins determine a sequence of transcriptional events that leads to differential gene expression.

- Do you understand how differential gene expression underlies cell differentiation? See pp. 435
- What is a transcription factor? Do you understand the role of transcription factors in differentiation? See pp. 435

We have seen how cell differentiation involves the extensive transcriptional regulation of genes. But what causes a cell to express one set of genes, and not some other set? In other words, how is a cell's fate determined?

19.4 How Is Cell Fate Determined?

The intricate networks of transcriptional controls that lead to cell differentiation are stimulated by chemical signals. In general, there are two mechanisms for producing such signals:

- **Cytoplasmic segregation.** A factor within an egg, zygote, or precursor cell may be unequally distributed in the cytoplasm. After cell division, the factor ends up in some daughter cells or regions of cells, but not others.
- **Induction.** A factor is actively produced and secreted by certain cells to induce other cells to differentiate.

Cytoplasmic segregation can determine polarity and cell fate

Some differences in patterns of gene expression are the result of *cytoplasmic* differences between cells. The emergence of **polarity**—the difference between the "top" and "bottom" ends of an organism or structure—is one such phenomenon. Polarity is obvious throughout development. Our heads are distinct from our rear ends, and the distal ends of our arms and legs (wrists, ankles, fingers, toes) differ from the proximal ends (shoulders and hips). Polarity may develop early; even within the fertilized egg, yolk and other factors are often distributed asymmetrically. During early development, polarity is specified by an *animal pole* at the top ("north pole") of the zygote and a *vegetal pole* at the bottom (the "south pole").

A famous series of experiments by Hans Driesch demonstrated the effects of cytoplasmic segregation on development (**Figure 19.9**). Very early development in sea urchins occurs by equal mitotic divisions of the fertilized egg; there is no increase in size at this stage. If an 8-cell embryo is cut vertically, both halves develop into normal (albeit small) embryos. But if an 8-cell embryo is cut horizontally, the top half does not develop at all, while the bottom half develops into a small, abnormal embryo.

Clearly, then, there must be at least one factor essential for development that is segregated in the vegetal half of the sea urchin egg, such that the bottom cells of the 8-cell embryo have it and the top cells do not. This and many other experiments have established that certain materials, called **cytoplasmic determinants**, are distributed unequally in the egg cytoplasm. These materials play a role in directing the embryonic development of many organisms (**Figure 19.10**).

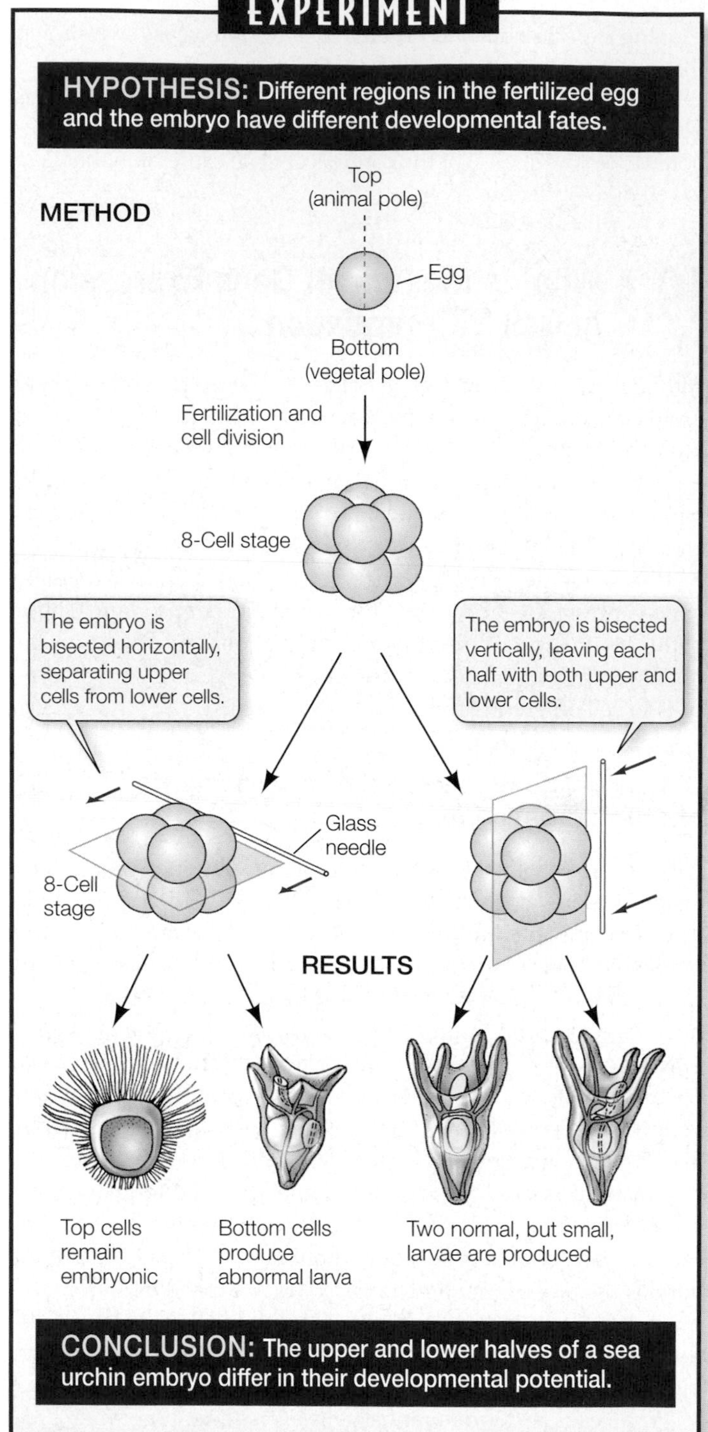
EXPERIMENT
HYPOTHESIS: Different regions in the fertilized egg and the embryo have different developmental fates.
METHOD
Top (animal pole)
Egg
Bottom (vegetal pole)
Fertilization and cell division
8-Cell stage
The embryo is bisected horizontally, separating upper cells from lower cells.
The embryo is bisected vertically, leaving each half with both upper and lower cells.
Glass needle
8-Cell stage
RESULTS
Top cells remain embryonic
Bottom cells produce abnormal larva
Two normal, but small, larvae are produced
CONCLUSION: The upper and lower halves of a sea urchin embryo differ in their developmental potential.

19.9 Asymmetry in the Early Sea Urchin Embryo The top and bottom halves of an 8-cell sea urchin differ in the cytoplasmic determinants they contain. Cells from both halves are necessary to produce a normal larva.

In the fertilized eggs of many animals, the nucleus is positioned near the animal pole, and yolk molecules that will nourish the embryo accumulate in the vegetal half. The presence of yolk can slow down cell division, so the top halves of such eggs undergo many more cell divisions than the bottom halves do.

The cytoskeleton contributes to the asymmetrical distribution of cytoplasmic determinants in the egg. Recall from Section 4.3 that an important function of the microtubules and microfilaments in the cytoskeleton is to help move materials in the cell. Two properties allow these structures to accomplish this:

- Microtubules and microfilaments have polarity— they grow by adding subunits to the plus (+) end.
- Cytoskeletal elements can bind specific proteins.

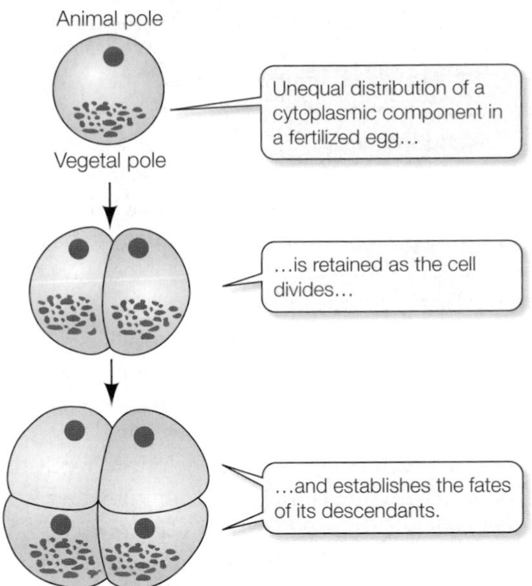

19.10 The Principle of Cytoplasmic Segregation The unequal distribution of some component in the cytoplasm of a cell may determine the fates of its descendants.

For example, in the sea urchin egg, a protein binds to both the growing (+) end of a microfilament and to an mRNA encoding a cytoplasmic determinant. As the microfilament grows toward one end of the cell, it carries the mRNA along with it. This asymmetrical distribution of the mRNA leads to a similar distribution of the protein it encodes.

Inducers passing from one cell to another can determine cell fates

Experimental work on developing embryos has established that in many cases, the fates of particular cells and tissues are determined by interactions with other specific tissues in the embryo. Many such instances of induction, in which one tissue causes an adjacent tissue to develop in a particular manner, have been observed in developing animal embryos. These effects are mediated by intercellular communication—that is, by chemical signals and signal transduction mechanisms. We will describe two examples of differentiation by embryonic induction: one in the developing vertebrate eye, and the other in a developing reproductive structure in the nematode *C. elegans*.

LENS DIFFERENTIATION IN THE VERTEBRATE EYE The development of the lens of the vertebrate eye is a classic example of induction. In a frog embryo, the developing forebrain bulges out at both sides to form the *optic vesicles*, which expand until they come into contact with the cells at the surface of the head (**Figure 19.11**). The surface tissue in the region of contact with the optic vesicles thickens, forming a *lens placode*. The lens placode bends inward, folds over on itself, and ultimately detaches from the surface tissue to produce a structure that will develop into the lens. If the growing optic vesicle is cut away before it contacts the surface cells, no lens forms. Placing an impermeable barrier between the optic vesicle and the surface cells also prevents the lens from forming. These observations suggest that the surface tissue begins to develop into a lens when it receives a signal—an **inducer**—from the optic vesicle.

A chain of inductive interactions results in the development of the eye. There is a "dialogue" between the developing optic vesicle and the surface tissue. The optic vesicle induces lens development, and the developing lens determines the size of the *optic cup* that forms from the optic vesicle. If head surface tissue from a frog species with small eyes is grafted over the optic vesicle of one with large eyes, both lens and optic cup will have an intermediate size. The developing lens also induces the surface tissue over it to develop into a *cornea*, a specialized layer that allows light to pass through and enter the eye.

As this example shows, tissues do not induce themselves; rather, different tissues interact and induce one another. Embryonic inducers trigger a sequence of gene expression in the responding cells. How cells switch on different sets of genes that govern development and direct the formation of body plans is a subject of great interest to both developmental and evolutionary biologists. We will look at embryonic induction in more detail in Chapter 43.

VULVAL DIFFERENTIATION IN THE NEMATODE The tiny nematode *Caenorhabditis elegans* is a favorite model organism for studying

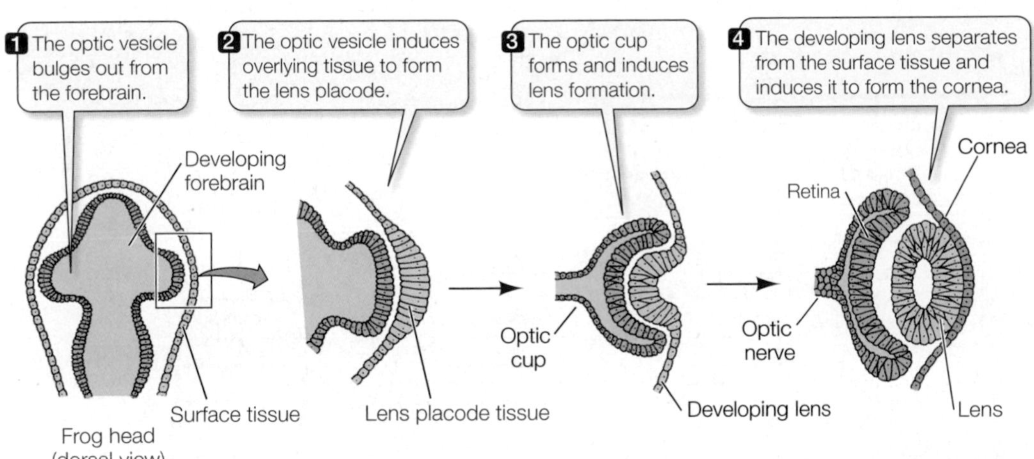

19.11 Embryonic Inducers in the Vertebrate Eye The eye of a frog develops as different tissues take their turns inducing one another.

development. It normally lives in the soil, where it feeds on bacteria, but can also grow in the laboratory if supplied with its food source. The process of development from fertilized egg to larva takes only about 8 hours, and the worm reaches the adult stage in just 3.5 days. The process is easily observed using a low-magnification dissecting microscope because the body covering is transparent (**Figure 19.12A**). The development of *C. elegans* does not vary from one individual to the next, so it has been possible to identify the source of each of the 959 somatic cells of the adult worm.

The adult nematode is *hermaphroditic*, containing both male and female reproductive organs. It lays eggs through a pore called the *vulva* on the ventral (belly) surface. During development, a single cell, called the *anchor cell*, induces the vulva to form. If the anchor cell is destroyed by laser surgery, no vulva forms; the eggs are fertilized inside the parent, and ultimately the baby worms consume the parent. This striking phenotype has allowed geneticists to identify the genes that have a role in the development of the vulva.

The anchor cell controls the fates of six cells on the developing worm's ventral surface through two molecular signals, the *primary inducer* and the *secondary inducer*. Each of these cells has three possible fates: it may become a primary vulval precursor cell, a secondary vulval precursor cell, or simply become part of the worm's surface—an epidermal cell (**Figure 19.12B**).

The anchor cell produces the primary inducer, which diffuses out of the cell and interacts with adjacent cells. Cells that receive enough primary inducer become vulval precursor cells; cells slightly farther from the anchor cell become epidermal cells. Thus the anchor cell, by releasing the primary inducer, determines whether a cell takes the "track" toward becoming part of the vulva or the track toward becoming part of the epidermis.

The cell closest to the anchor cell, having received the most primary inducer, differentiates into the primary vulval precursor cell. It produces its own inducer (the secondary inducer), which acts on the two neighboring cells and directs them to become secondary vulval precursor cells. Thus the primary vulval precursor cell produces a second signal, determining whether a vulval precursor cell will take the primary track or the secondary track. The two inducers control the activation or inactivation of specific genes through a signal transduction cascade in the responding cells (**Figure 19.13**).

Nematode development illustrates the important observation that *much of development is controlled by molecular switches that allow a cell to proceed down one of two alternative tracks.* One challenge for developmental biologists is to find these switches and determine how they work. The primary inducer released by the *C. elegans* anchor cell appears to be a growth factor homologous to a mammalian

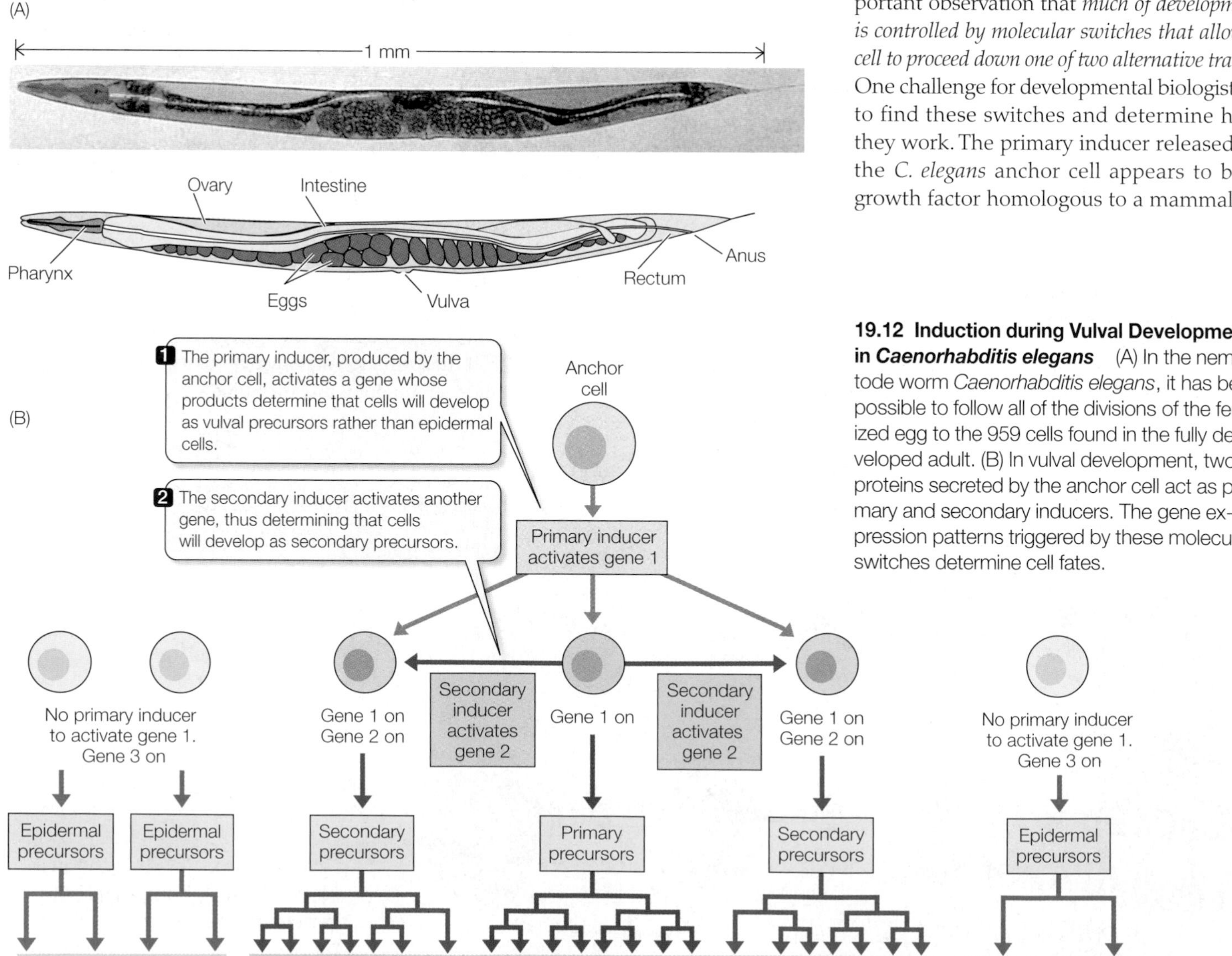

19.12 Induction during Vulval Development in *Caenorhabditis elegans* (A) In the nematode worm *Caenorhabditis elegans*, it has been possible to follow all of the divisions of the fertilized egg to the 959 cells found in the fully developed adult. (B) In vulval development, two proteins secreted by the anchor cell act as primary and secondary inducers. The gene expression patterns triggered by these molecular switches determine cell fates.

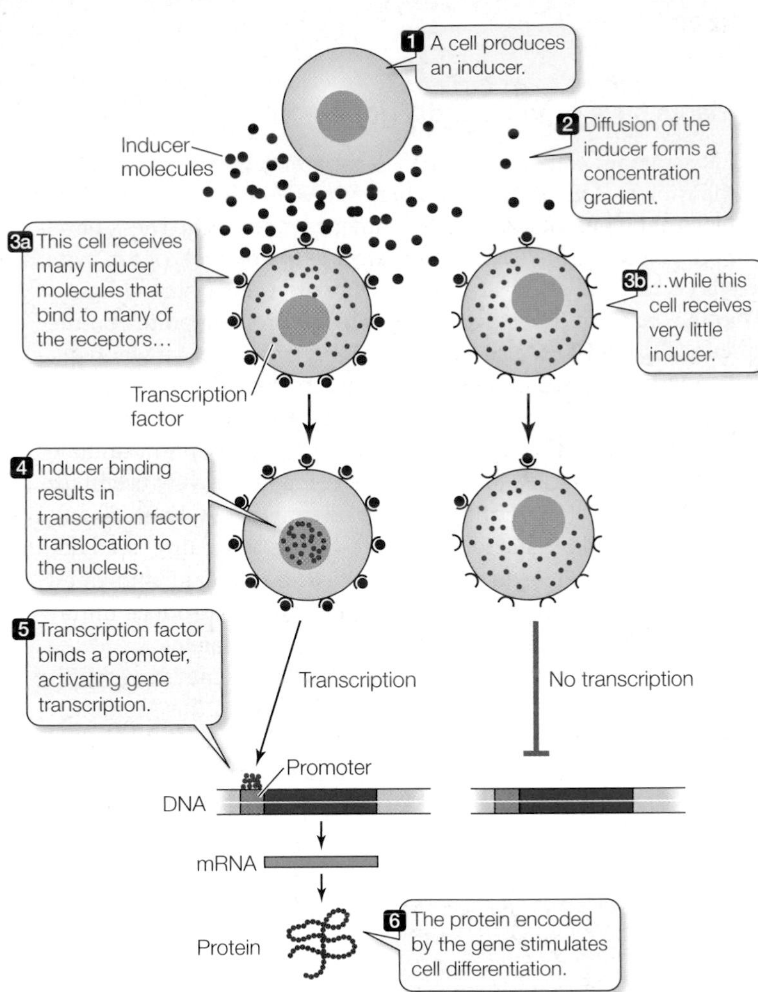
1 A cell produces an inducer.
2 Diffusion of the inducer forms a concentration gradient.
Inducer molecules
3a This cell receives many inducer molecules that bind to many of the receptors...
3b ...while this cell receives very little inducer.
Transcription factor
4 Inducer binding results in transcription factor translocation to the nucleus.
5 Transcription factor binds a promoter, activating gene transcription.
Transcription
No transcription
Promoter
DNA
mRNA
Protein
6 The protein encoded by the gene stimulates cell differentiation.

19.13 Embryonic Induction The concentration of an inducer directly affects the degree to which a transcription factor is activated. The inducer acts by binding to a receptor on the target cell. This binding is followed by signal transduction involving transcription factor translocation from the cytoplasm to the nucleus.

growth factor called EGF (*e*pidermal *g*rowth *f*actor). The nematode growth factor, called LIN-3, binds to a receptor on the surface of a potential vulval precursor cell, setting in motion a signal transduction cascade involving the Ras protein and MAP kinases (see Figure 15.10). The end result is increased transcription of the genes involved in the differentiation of vulval cells.

19.4 RECAP

Cytoplasmic segregation is the unequal distribution of molecular signaling factors in the egg, zygote, or early embryo. Embryonic induction occurs when one cell or tissue sends a chemical signal to another. Both types of signaling trigger transcription factor activity, and thus cellular differentiation.

- Can you describe how cytoplasmic segregation results in polarity in a fertilized egg, and how polarity affects cell differentiation? See p. 436 and Figure 19.10
- Do you understand how embryonic induction forms the tissues of a vertebrate's eyes? See p. 437 and Figure 19.11
- How do inducer molecules interact with transcription factors to produce differentiated cells? See p. 438 and Figure 19.13

The differentiation of cells is beginning to be understood in terms of molecular and cellular events, but how do these events result in the organization of multitudes of cells into specific body parts, such as a leaf, a flower, a shoulder blade, or a tear duct? We now turn to the final phase of development, the formation of organs.

19.5 How Does Gene Expression Determine Pattern Formation?

Pattern formation is the process that results in the spatial organization of a tissue or organism. It is inextricably linked to morphogenesis, the creation of body form. You might expect morphogenesis to involve a lot of cell division, followed by differentiation—and it does. But what you might not expect is the degree of programmed cell death—apoptosis—that occurs during morphogenesis.

Some genes determine programmed cell death during development

We noted in Section 9.6 that apoptosis is used extensively during development to "sculpt" organs. For example, in the early human embryo, the hands and feet look like tiny paddles: the fingers and toes are linked by connective tissue. Between days 41 and 56 of development, the cells between the digits die, freeing the individual fingers and toes (**Figure 19.14**). As we will see when we discuss mammalian development in Chapter 43, many structures (such as the primitive structure in the back of the human embryo called the notochord) form and then disappear; this is also an example of apoptosis.

Model organisms have been very useful in studying the genes involved in apoptosis. For example, the nematode *C. elegans* produces precisely 1,090 somatic cells as it develops from a fertilized egg into an adult, but 131 of those cells die (leaving 959 cells to develop into an adult worm, as described in the previous section). The sequential expression of two genes, called *ced-4* and *ced-3* (for *ce*ll *d*eath), appears to control this cell death.

In the nematode nervous system, 302 neurons come from 405 precursors; thus 103 neural precursor cells undergo apoptosis. If the protein encoded by either *ced-3* or *ced-4* is nonfunctional, all 405 cells form neurons, resulting in abnormal brain development. Experiments have identified a third gene, *ced-9*, that codes for an inhibitor of apoptosis—that is, it codes for a protein that blocks the function of the *ced-3* and *ced-4* genes. Where apoptosis is required, *ced-3* and *ced-4* are active and *ced-9* is inactive; if apoptosis is not appropriate, *ced-9* is active and blocks *ced-3* and *ced-4*.

A similar system of cell death genes acts in humans. The proteins—a class of enzymes called *caspases*—that stimulate apopto-

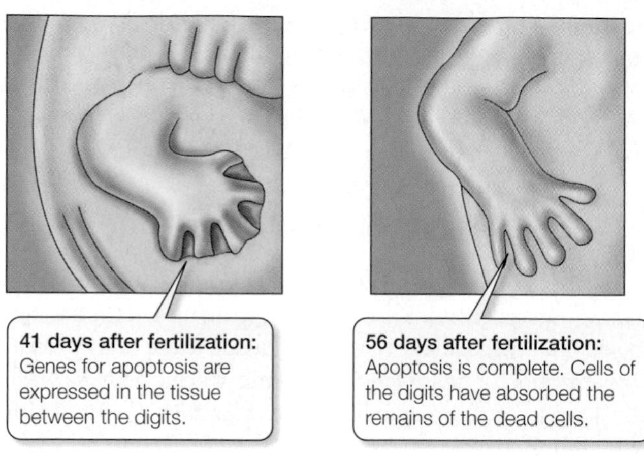

41 days after fertilization:
Genes for apoptosis are expressed in the tissue between the digits.
56 days after fertilization:
Apoptosis is complete. Cells of the digits have absorbed the remains of the dead cells.

19.14 Apoptosis Removes the Tissue between Human Fingers Early in the second month of human development, the tissue connecting the fingers is removed by apoptosis, freeing the individual fingers.

sis are similar in amino acid sequence to the protein encoded by *ced-3*, and a human protein (BCL-2) that inhibits apoptosis is similar to the product of *ced-9*. So humans and nematodes, two creatures separated by more than 600 million years of evolutionary history, have similar genes controlling programmed cell death.

Plants have organ identity genes

Like animals, plants have organs—for example, leaves and roots. Many plants form flowers, and many flowers are composed of four types of organs: sepals, petals, stamens, and carpels. These floral organs occur in *whorls*, which are groups of each organ type stacked around a central axis. The whorls develop from meristems in the shape of domes, which develop at growing points on the stem (**Figure 19.15A**). How is the identity of a particular whorl determined? A group of genes, called **organ identity genes**, code for proteins that act in combination to produce specific whorl features.

Organ identity genes have been well described in the thale cress, *Arabidopsis thaliana*. This model organism is very useful for studies of development because of its small size (about 25 cm), abundant seed production (over 1,000 seeds per plant), rapid development (from seed to plant to seed in 6 weeks), and small genome (see Section 14.1). Finally, it is easy to produce mutations in this plant by treating the seeds with mutagens.

The development of the flower begins with the meristem, which contains about 700 undifferentiated cells. Within this seemingly homogeneous cell population, individual cells "sense" their position and differentiate into the whorls. This happens through the expression of three classes of organ identity genes, which code for proteins that act in combination with one another:

- Genes in class A are expressed in whorls 1 and 2 (which form sepals and petals, respectively).

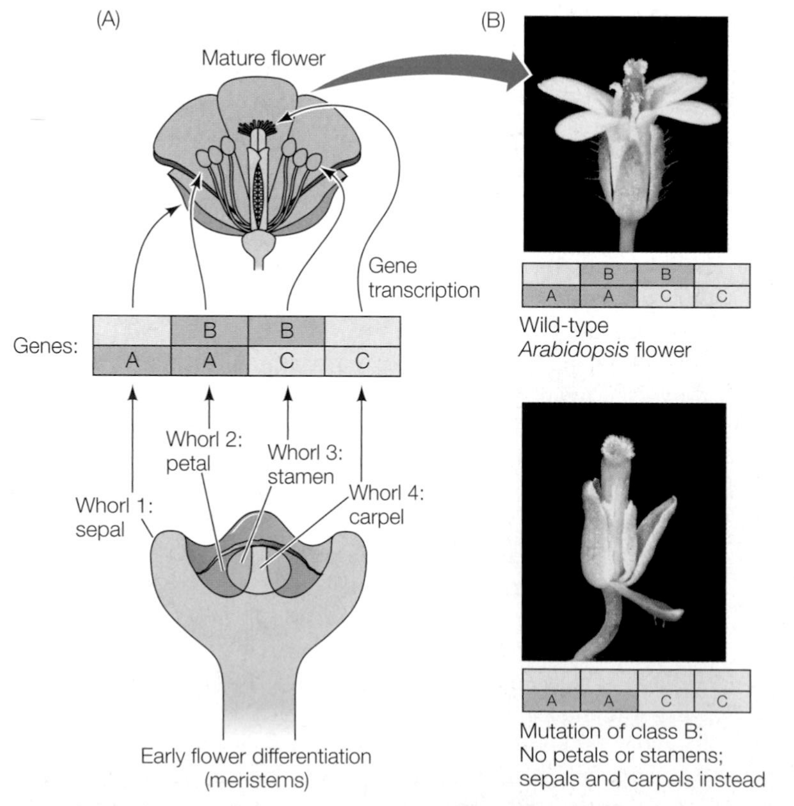

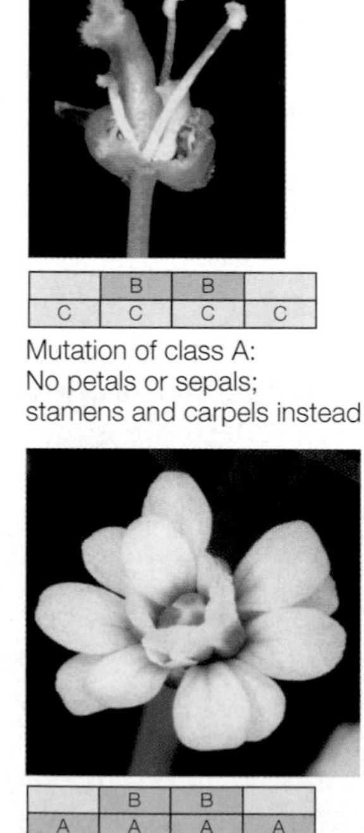

B B
C C C C
Mutation of class A:
No petals or sepals;
stamens and carpels instead

B B
A A A A
Mutation of class C:
No stamens or carpels;
petals and sepals instead

19.15 Organ Identity Genes in *Arabidopsis* Flowers (A) The four organs of a flower—carpels (yellow), stamens (green), petals (purple), and sepals (pink)—grow in whorls that develop from meristems. (B) When a mutation in one of the three organ identity genes occurs, one type of organ replaces another. Such mutations helped scientists decipher the pattern of gene expression that gives rise to normal flowers.

- Genes in class B are expressed in whorls 2 and 3 (which form petals and stamens, respectively).
- Genes in class C are expressed in whorls 3 and 4 (which form stamens and carpels, respectively).

There are two lines of experimental evidence for this model of organ identity gene function (**Figure 19.15B**):

- *Loss-of-function mutations:* for example, a mutation in gene class A results in no sepals or petals.
- *Gain-of-function mutations:* for example, the promoter for gene class C can be artificially coupled to gene class A. In this case, A is expressed in all four whorls, resulting in only sepals and petals.

Gene classes A, B, and C code for subunits of transcription factors, which are active as dimers. Gene regulation in these cases is *combinatorial*—that is, the composition of the dimer determines which genes will be activated by the transcription factor. For example, a dimer made up of two monomers of transcription factor A would activate transcription of the genes that make sepals; a dimer made up of A and B monomers would result in petals, and so forth. A common feature of the A, B, and C proteins, as well as many other plant transcription factors, is a DNA-binding domain called the **MADS box**. These 200-amino-acid proteins also have domains that can bind to other proteins in a *transcription initiation complex.*

The MADS box gets its name from homologous amino acid sequences found in four proteins in widely divergent species: *MCMI* (from yeast), *AGAMOUS* (*Arabidopsis*), *DEFICIENS* (snapdragons), and *SRF* (humans).

In addition to being fascinating to biologists, plant organ identity genes have caught the eye of horticultural and agricultural scientists. Flowers filled with petals instead of stamens and carpels often have mutations of the C genes. Many of the foods that make up the human diet, such as the grains of wheat, rice, and corn, come from fruits and seeds. These fruits and seeds form from the carpels (the female reproductive organs) of the flower. Genetically modifying the number of carpels on a particular plant could increase the amount of grain a crop could produce.

A gene called *leafy* codes for a protein that controls the transcription of the organ identity genes. Plants with a mutation that causes the underexpression of *leafy* are just that—they make leaves, but no flowers. The protein product of this gene acts as a transcription factor, stimulating gene classes A, B, and C so that they produce flowers (**Figure 19.16**). This finding, too, has practical applications. It usually takes 6–20 years for a citrus tree to produce flowers and fruits. Scientists have made orange trees transgenic for *leafy* coupled to a strongly expressed promoter; the transgenic trees flower and fruit years earlier than normal trees.

Morphogen gradients provide positional information

During development, the key cellular question, "What am I (or what will I be)?" is often answered in part by "Where am I?" Think of a cell in the apical meristem of a developing *Arabidopsis*: it needs to "know" in what whorl it is located. This spatial "sense" is called **positional information**. Positional information usually comes in the form of a signal, called a **morphogen**, that diffuses from one group of cells down a body axis, setting up a concentration gradient. There are two requirements for a signal to be considered a morphogen:

- It must directly affect target cells, rather than triggering a secondary signal that affects target cells.
- Different concentrations of the signal must cause different effects.

The development of the vertebrate limb provides us with an example of a morphogen in action. The limb develops from a paddle-shaped *limb bud*. The cells that become the bones and muscles of the limb must receive positional information. If they do not, the limbs will be totally disorganized (imagine a hand with only thumbs or only little fingers). A group of cells at the posterior base of the limb bud, just where it joins the body wall, is called the *zone of polarizing activity* and makes a morphogen called BMP2, which forms a gradient. This gradient determines the posterior–anterior ("little finger to thumb") axis of the developing limb. The cells getting the highest dose of BMP2 make the little finger, and those getting the lowest dose make the thumb.

Wild type

Leafy mutant

19.16 A Nonflowering Mutant Mutations in the *leafy* gene of *Arabidopsis* prevent the transcription of the organ identity genes, and the resulting plant does not produce any flowers.

Different concentrations of morphogens act through differential regulation of gene expression in their target cells. The model organism most often used for studying this process has been the fruit fly.

In the fruit fly, a cascade of transcription factors establishes body segmentation

Insects such as the fruit fly *Drosophila melanogaster* develop a highly modular body composed of different types of segments. Complex interactions of different sets of genes underlie pattern formation in segmented bodies.

Unlike the body segments of segmented worms such as earthworms, which are all essentially alike, the segments of the *Drosophila* body are clearly different from one another. The adult fly has an anterior *head* (composed of several fused segments), three different *thoracic* segments, and eight *abdominal* segments at the posterior end. In the *Drosophila* larva, the thoracic and abdominal segments all appear to be similar, but at this point it has *already been determined* that they will form these specialized adult segments. Several types of genes are expressed sequentially in the embryo to define these segments:

- First, a set of genes from the mother's cells adjacent to the egg sets up anterior–posterior and dorsal–ventral axes in the egg.
- Next, a series of genes in the embryo successively define the position of each cell in a segment relative to these axes. The end result is that a cell "knows" precisely where it is in the embryo; for example, that it is part of the head at the very front.
- Finally, a set of genes called *Hox genes* control the ultimate identity of each segment; for example, determining that the cells at the very front of the head will make antennae.

The genes involved in each of these steps code for transcription factors, which in turn control the synthesis of transcription factors acting on the next set of genes. This cascade of events may remind you of a signal transduction cascade (see Section 15.3), only in this case it is a cascade of events over time and location, rather than in a single cell. The final expressed genes are the ones familiar to you: they code for protein kinases, receptors, and other proteins that carry out the functions of the cell.

The description of these events in fruit fly development is one of the great achievements in modern biology. We will only skim the surface of the process here, but keep in mind the basic principle of a transcriptional cascade. As we will see in Chapter 20, the fruit fly has been a true model organism in this case, as the basic pattern of events is similar in many other organisms, including humans.

MATERNAL EFFECT GENES Like those of the sea urchin, *Drosophila* eggs and larvae are characterized by unevenly distributed cytoplasmic determinants (see Figure 19.10). These molecules, which include both mRNAs and proteins, are the products of specific **maternal effect genes**. These genes are transcribed in the cells of the mother's ovary that surround what will be the anterior portion of the egg. Two maternal effect genes, called *bicoid* and *nanos*, determine the anterior–posterior axis of the egg. Other maternal effect genes that will not be described here determine the dorsal–ventral axis.

The mRNAs for *bicoid* and *nanos* diffuse from the mother's cells into what will be the anterior end of the egg through cytoplasmic bridges. The *bicoid* mRNA stays where it enters the egg and is translated to produce Bicoid protein, which diffuses away from the anterior end, establishing a gradient in the egg (**Figure 19.17**). Where it is present in sufficient concentration, Bicoid acts as a transcription factor to stimulate the transcription of the *hunchback* gene in the egg. A gradient of the latter protein establishes the head, or anterior, region.

Meanwhile, *nanos* mRNA is transported by the cytoskeleton to the posterior end of the egg, where it enters the egg and is translated and its protein forms a gradient. The Nanos protein is not a morphogen; instead, it is an inhibitor of the translation of *hunchback* mRNA. It is the low level of *hunchback* expression in the posterior because of this "double whammy"—inhibition of its translation by Nanos and a lack of stimulation of its transcription by Bicoid—that results in the establishment of the posterior region.

How do we know all of this? Throughout this book, we have described two approaches to demonstrating cause and effect: we study mutations that *prevent* an event from occurring, and we do experiments to *cause* that event to occur. Scientists demonstrated that maternal effect genes specify the axes of the *Drosophila* embryo by causing mutations in those genes, and by experiments in which cytoplasm was transferred from one egg to another:

- Females that are homozygous for a particular mutation of *bicoid* produce larvae with no head and no thorax; thus the Bicoid protein must be needed for the anterior structures to develop.
- If the eggs of these mutant females are inoculated at the anterior end with cytoplasm from the anterior region of a wild-type egg, the "rescued" eggs develop into normal larvae; this experiment also shows that Bicoid protein is involved in the development of anterior structures.
- If Bicoid protein from a wild-type egg is injected into the posterior region of an egg, anterior structures develop there. The degree of induction depends on how much cytoplasm is injected.
- Eggs from homozygous *nanos* mutant females develop into larvae with missing abdominal segments.
- Injecting cytoplasm from the posterior region of a wild-type egg into a *nanos* mutant egg allows normal development.

The events involving *bicoid*, *nanos*, and *hunchback* begin before fertilization and continue after it. Before the *Drosophila* egg is fertilized, *bicoid* and *nanos* mRNAs are made by the maternal cells and enter the egg. After the egg is fertilized and laid, nuclear divisions begin. In *Drosophila*, cytokinesis does not begin right away; until the thirteenth nuclear division, the embryo is a single, multinucleated cell called a *syncytium*. At this early stage, *bicoid* and *nanos* mRNAs are translated and establish gradients.

After the axes of the embryo are determined, the next step in pattern formation is the determination of the locations of the segments.

SEGMENTATION GENES The number, boundaries, and polarity of the larval segments are determined by proteins encoded by the **segmentation genes**. These genes are expressed when there are about 6,000 nuclei in the embryo. These nuclei all look the same, but in terms of gene expression, they are not equivalent. Their fates are sealed early in development.

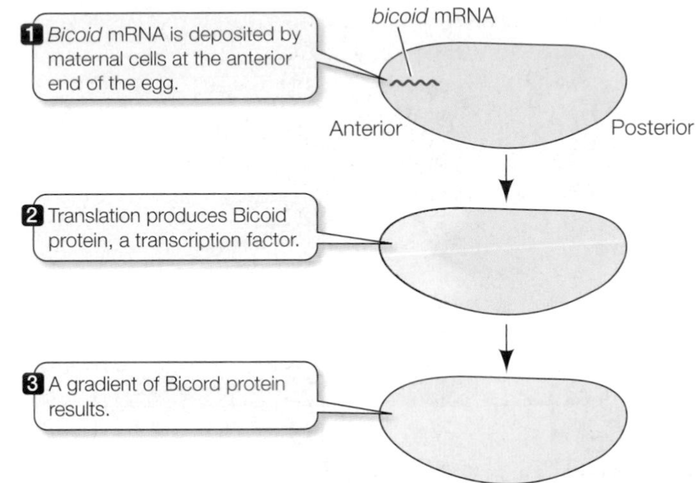

bicoid mRNA
1 Bicoid mRNA is deposited by maternal cells at the anterior end of the egg.
Anterior
Posterior
2 Translation produces Bicoid protein, a transcription factor.
3 A gradient of Bicord protein results.

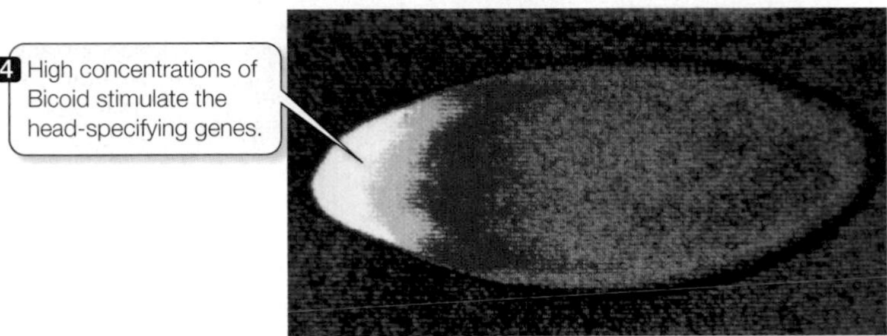

4 High concentrations of Bicoid stimulate the head-specifying genes.

19.17 Bicoid Protein Provides Positional Information The anterior–posterior axis of *Drosophila* arises from the gradient of a morphogen encoded by *bicoid*, a maternal effect gene. Bicoid protein acts as a transcription factor to activate a gene that specifies that this region will produce the structures of the head. Other maternal effect genes in the posterior portion of the embryo inhibit Bicoid, thus limiting its region.

Three classes of segmentation genes act one after the other to regulate finer and finer details of the segmentation pattern (**Figure 19.18**):

- **Gap genes** organize broad areas along the anterior–posterior axis. Mutations in gap genes result in gaps in the body plan—the omission of several larval segments.
- **Pair rule genes** divide the embryo into units of two segments each. Mutations in pair rule genes result in embryos missing every other segment.
- **Segment polarity genes** determine the boundaries and anterior–posterior organization of the individual segments. Mutations in segment polarity genes can result in segments in which posterior structures are replaced by reversed (mirror-image) anterior structures.

The expression of these genes is sequential. The maternal effect protein Bicoid, which begins the cascade, acts as a morphogen and transcription factor to stimulate the expression in the egg of genes such as *hunchback* that set up the anterior–posterior axis. As a result, a nucleus in the egg "knows" where it is. The Hunchback protein stimulates gap gene transcription; the products of the gap genes stimulate pair rule genes; and the products of the pair rule genes stimulate segment polarity genes. By the end of this cascade,

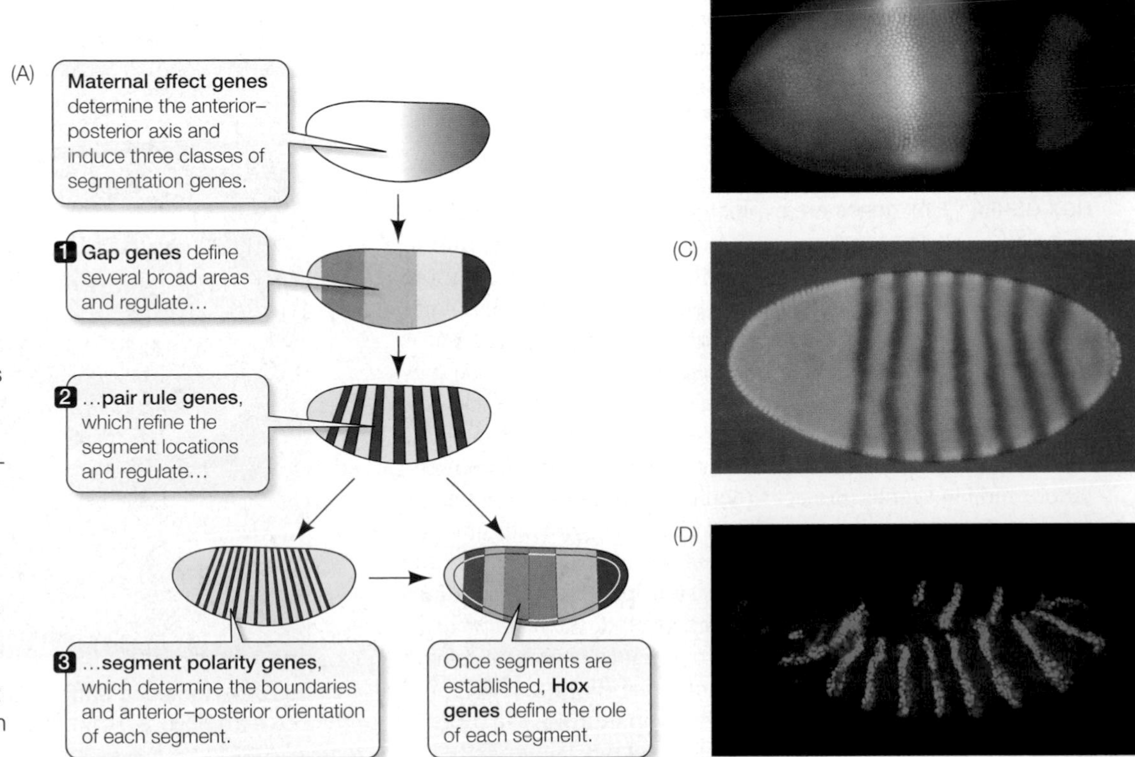

1 Gap genes define several broad areas and regulate...

19.18 A Gene Cascade Controls Pattern Formation in the *Drosophila* Embryo (A) Maternal effect genes (see Figure 19.17) induce gap, pair rule, and segment polarity genes—collectively referred to as segmentation genes. (B) Two gap genes, *hunchback* (orange) and *Krüppel* (green) overlap; both genes are transcribed in the yellow area. (C) The pair rule gene *fushi tarazu* is transcribed in the dark blue areas. (D) The segment polarity gene *engrailed* (bright green) is seen here at a slightly more advanced stage than is depicted in (A).

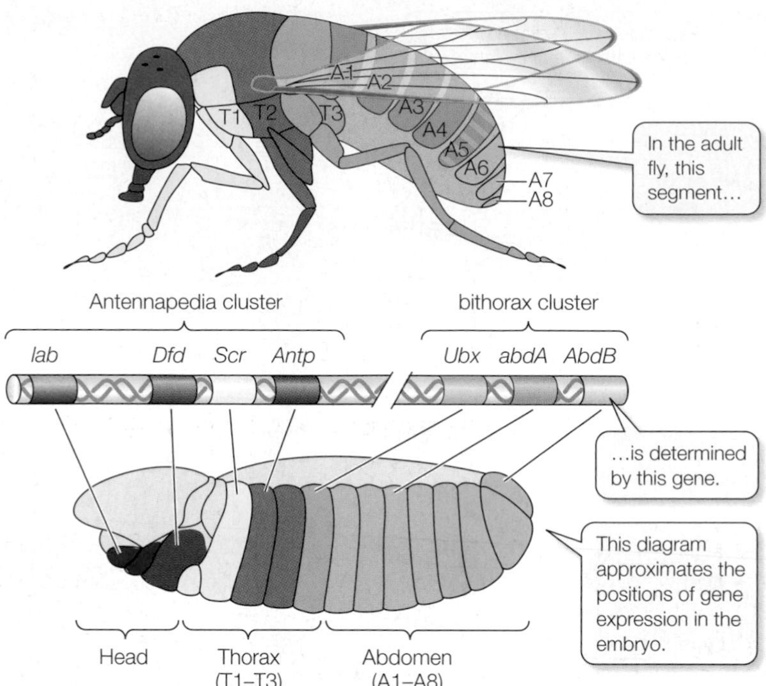

19.19 Hox Genes in *Drosophila* Two clusters of genes on chromosome 3 (center) determine segment function in the adult fly (top). These genes are expressed in the embryo (bottom) long before the structures of the segments actually appear.

a group of nuclei at the very anterior of the egg, for example, "knows" that they are going to become part of the first anterior segment in the adult fly.

The next set of genes in the cascade determines the form and function of each segment.

Drosophila genes are often named for their mutant phenotype. Thus the *hunchback* gene received its name because flies who lack this gene have deformed or missing heads. Likewise, *fushi tarazu* is Japanese for "not enough parts"—and, indeed, mutants for this pair rule gene have too few segments.

HOX GENES **Hox genes** are expressed in different combinations along the length of the embryo and tell each segment what to become. Hox gene expression tells the cells of a segment in the head to make eyes, those of a segment in the thorax to make wings, and so on. Remarkably, the *Drosophila* Hox genes map on chromosome 3, in the same order as the segments whose function they determine (**Figure 19.19**). By the time the fruit fly larva hatches, its segments are completely determined. Hox genes are analogous to the organ identity genes of plants. The maternal effect, segmentation, and Hox genes interact to "build" a *Drosophila* larva step by step, beginning with the unfertilized egg.

Once again, how do we know that the Hox genes determine segment identity? An important clue came from bizarre mutations observed in *Drosophila* that were called *homeotic* (Greek *homeos*, "a being resembling something else"). In the *Antennapedia* mutation, legs grow on the head in place of antennae (**Figure 19.20**), and in the *bithorax* mutation, an extra pair of wings grows in a thoracic segment where wings do not normally grow (see Figure 20.2). Edward Lewis at Caltech found that *Antennapedia* and *bithorax* mutations resulted from changes in Hox genes.

The first cluster of Hox genes, the bithorax cluster, specifies anterior segments, starting with genes for the different head segments and ending with thoracic segments. The second cluster, the Antennapedia cluster, begins with a gene specifying the last thoracic segment, followed by a gene for the anterior abdominal segments, and ends with a gene for the posterior abdominal segments. Lewis hypothesized that all of the *Drosophila*

(A)

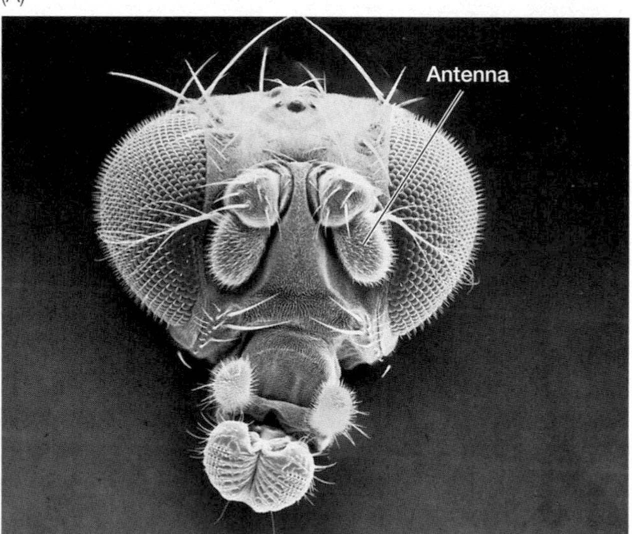

(B)

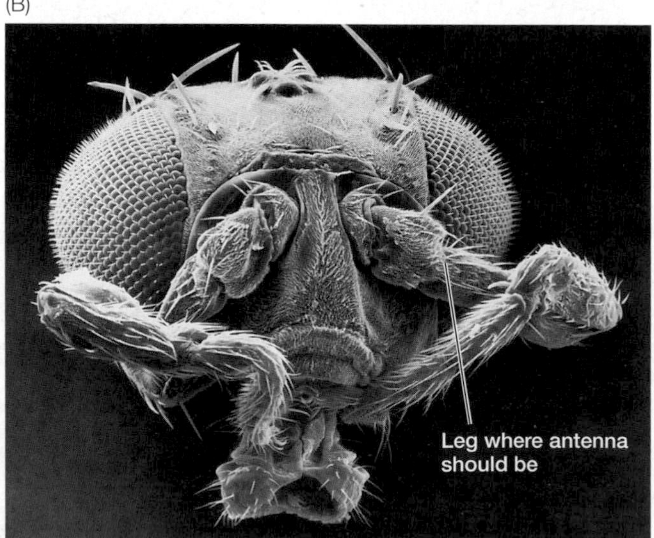

19.20 A Homeotic Mutation in *Drosophila* Mutations of the Hox genes cause body parts to form on inappropriate segments. (A) A wild-type fruit fly. (B) An *Antennapedia* mutant fruit fly.

Hox genes might have come from the duplication of a single gene in an ancestral, unsegmented organism.

Homeobox-containing genes encode transcription factors

Molecular biologists confirmed Lewis's hypothesis using nucleic acid hybridization. Several scientists found that a probe for a sequence found in one of the genes in the bithorax cluster bound not only to its own gene, but also to the adjacent genes in its cluster and to the genes in the Antennapedia cluster. In other words, this DNA sequence is common to all the Hox genes in both clusters. It is also found in several of the segmentation genes, as well as other genes that encode transcription factors.

This 180-base-pair DNA sequence is called the **homeobox**. It encodes a 60-amino acid sequence, called the *homeodomain*, that binds to DNA. The homeodomain recognizes a specific DNA sequence in the promoter of its target genes, but that recognition is usually not sufficient to allow the transcription factor to bind fully to a promoter and turn the target gene on or off. Other transcription factors are also involved.

Genes containing the homeobox are found in many animals, including humans. They play a role in development similar to the role the MADS box genes play in plants. The evolutionary significance of these common pathways for development will be discussed in the next chapter.

19.5 RECAP

A cascade of transcription factors governs pattern formation and the subsequent development of animal and plant organs. In plants, organ identity genes code for transcription factors that determine which organ a group of cells will form. In animals, morphogen gradients provide positional information.

- Do you see how apoptosis is crucial in shaping the developing embryo? See p. 440
- Can you describe how organ identity genes act in *Arabidopsis*? See p. 441 and Figure 19.15
- What are the key attributes of a morphogen? How does Bicoid protein fit this definition? See pp. 441–442 and Figure 19.17

CHAPTER SUMMARY

19.1 What are the processes of development?

A multicellular organism begins its **development** as an **embryo**. A series of embryonic stages may precede the birth of an independent organism. Review Figure 19.1

The processes of development are **determination**, **differentiation**, **morphogenesis**, and **growth**.

Differential gene expression is responsible for the differences between cell types. A cell's **fate** is determined by environmental factors, such as its position in the embryo, as well as by intracellular influences.

Determination is followed by differentiation, the actual changes in biochemistry, structure, and function that result in cells of different types. Determination is a commitment; differentiation is the realization of that commitment.

19.2 Is cell differentiation irreversible?

See Web/CD Tutorial 19.1

The zygote is **totipotent**; it is capable of forming every type of cell in the adult body.

The ability to create **clones** from differentiated cells demonstrates the principle of **genomic equivalence**. Review Figures 19.3 and 19.4

Stem cells produce daughter cells that differentiate when provided with appropriate intercellular signals. Some **pluripotent** stem cells in the adult body can differentiate into a limited number of cell types to replace dead cells and maintain tissues.

Embryonic stem cells are totipotent and can be cultured in the laboratory. With suitable environmental stimulation, these cells can be induced to differentiate. A technique called **therapeutic cloning** could combine nuclear transplantation and stem cell technologies to replace cells or tissues damaged by injury or disease. Review Figures 19.7 and 19.8

19.3 What is the role of gene expression in cell differentiation?

Differential gene expression results in cell differentiation. Transcription factors are especially important in regulating gene expression during differentiation.

Complex interactions of many genes and their products are responsible for differentiation during development.

19.4 How is cell fate determined?

Cytoplasmic segregation—the unequal distribution of **cytoplasmic determinants** in the egg, zygote, or early embryo—can establish **polarity** and lead to determination of a cell's descendants. Review Figures 19.9 and 19.10, Web/CD Tutorial 19.2

Induction is a process by which embryonic animal tissues direct the development of neighboring cells and tissues by secreting chemical signals, called **inducers**. Review Figure 19.11

The induction of the vulva in the nematode *Caenorhabditis elegans* offers an example of how inducers act as molecular switches to direct a cell down one of two differentiation paths. Review Figures 19.12 and 19.13

19.5 How does gene expression determine pattern formation?

Pattern formation is the process that results in the spatial organization of a tissue or organism.

During development, selective elimination of cells by apoptosis results from the expression of specific genes.

Combinatorial interactions of transcription factors coded for by **organ identity genes** in plants causes the formation of sepals, petals, stamens, and carpels. Review Figure 19.15

CHAPTER SUMMARY

The transcription factors encoded by floral organ identity genes contain an amino acid sequence called the **MADS box** that can bind to DNA.

Both plants and animals use **positional information** as a basis for pattern formation. Positional information usually comes in the form of a signal called a **morphogen**. Different concentrations of the morphogen cause different effects.

In the fruit fly *Drosophila melanogaster*, a cascade of transcription factors sets up the axes of the embryo and then causes the cells of each body segment to differentiate into particular organs. The cascade involves the sequential expression of maternal effect genes, gap genes, pair rule genes, segment polarity genes, and Hox genes. Review Figures 19.17 and 19.18, Web/CD Tutorial 19.3

The **homeobox** is a DNA sequence found in Hox genes and other genes that code for transcription factors. The sequence of amino acids encoded by the homeobox is called the homeodomain. Review Figure 19.19

SELF-QUIZ

1. Which statement about determination is true?
   a. Differentiation precedes determination.
   b. All cells are determined after two cell divisions in most organisms.
   c. A determined cell will keep its determination no matter where it is placed in an embryo.
   d. A cell changes its appearance when it becomes determined.
   e. A differentiated cell has the same pattern of transcription as a determined cell.
2. Cloning experiments on sheep, frogs, and mice have shown that
   a. nuclei of adult cells are totipotent.
   b. nuclei of embryonic cells can be totipotent.
   c. nuclei of differentiated cells have different genes than zygote nuclei have.
   d. differentiation is fully reversible in all cells of a frog.
   e. differentiation involves permanent changes in the genome.
3. The term "induction" describes a process in which a cell or group of cells
   a. influences the development of another group of cells.
   b. triggers the cell movements in an embryo.
   c. stimulates the transcription of their own genes.
   d. organizes the egg cytoplasm before fertilization.
   e. inhibits the movement of the embryo.
4. The term "therapeutic cloning" describes a method for
   a. modification of a clone by a transgene.
   b. combining nuclear transplantation and stem cell technologies.
   c. making clones that produce useful drugs.
   d. producing embryonic stem cells for transplantation.
   e. making many identical copies of an organism.
5. Which statement about cytoplasmic determinants in *Drosophila* is *not* true?
   a. They specify the dorsal–ventral and anterior–posterior axes of the embryo.
   b. Their positions in the embryo are determined by cytoskeletal action.
   c. Some are products of specific genes in the mother fruit fly.
   d. They do not produce gradients.
   e. They have been studied by the transfer of cytoplasm from egg to egg.
6. In fruit flies, the following genes are used to determine segment polarity: (k) gap genes; (l) Hox genes; (m) maternal effect genes; (n) pair rule genes. In what order are these genes expressed during development?
   a. klmn
   b. lknm
   c. mknl
   d. nkml
   e. nmkl
7. Which statement about embryonic induction is *not* true?
   a. One group of cells induces adjacent cells to develop in a certain way.
   b. It triggers a sequence of gene expression in target cells.
   c. Single cells cannot form an inducer.
   d. A tissue may be induced as well as an inducer.
   e. The chemical identification of specific inducers has not been achieved.
8. In the process of pattern formation in the *Drosophila* embryo,
   a. the first steps are specified by Hox genes.
   b. mutations in pair rule genes result in embryos missing every other segment.
   c. mutations in gap genes result in the insertion of extra segments.
   d. segment polarity genes determine the dorsal–ventral axes of segments.
   e. segmentation is the same as in earthworms.
9. Homeotic mutations
   a. are often severe and result in structures at inappropriate places.
   b. cause subtle changes in the forms of larvae or adults.
   c. occur only in prokaryotes.
   d. do not affect the animal's DNA.
   e. are confined to the zone of polarizing activity.
10. Which statement about the homeobox is *not* true?
   a. It is transcribed and translated.
   b. It is found only in animals.
   c. Proteins containing the homeodomain bind to DNA.
   d. It is a sequence of DNA shared by many genes.
   e. It occurs only in Hox genes.

FOR DISCUSSION

1. Molecular biologists can attach genes to active promoters and insert them into cells (see Section 16.3). What would happen if the following were inserted and overexpressed? Explain your answers.
   a. ced-9 in embryonic neuron precursors in *C. elegans*
   b. MyoD1 in undifferentiated myoblasts
   c. the gene for BMP2 in a chick limb bud
   d. nanos at the anterior end of the *Drosophila* embryo
2. A powerful method to test for the function of a gene in development is to generate a "knockout" organism, in which the gene in question is inactivated (see Section 16.5). What do you think would happen in each of the following cases?
   a. a knocked-out *ced-9* in *C. elegans*
   b. a knocked-out *nanos* in *Drosophila*
3. Look at the chart of organ identity mutations in Figure 19.15. What pattern do you perceive in the results of these mutations, and what might this pattern mean?
4. During development, the potential of a cell becomes ever more limited, until, in the normal course of events, its potential is the same as its original prospective fate. On the basis of what you have learned in this chapter, discuss possible mechanisms for the progressive limitation of the cell's potential.
5. How were biologists able to obtain such a complete accounting of all the cells in *Caenorhabditis elegans*? What major conclusions came from these studies?

FOR INVESTIGATION

Cloning involves considerable reprogramming of gene expression in a differentiated cell so that it acts like an egg cell. How would you investigate this reprogramming?

CHAPTER 20

Development and Evolutionary Change

The eyes have it

Eyes are not essential for survival; many animals and all plants get by just fine without them. However, over 90 percent of all animals *do* have eyes or some type of light-sensing organs, and having eyes can confer a selective advantage.

About a dozen different kinds of eyes are found among the different animals, including the camera-like eyes of humans and the compound eyes of insects, with their thousands of individual units. In trying to understand the origin of this variety, scientists—starting with Charles Darwin—proposed that eyes evolved independently many times in different animal groups, and that each improvement in the ability of eyes to gather light and form images conferred a selective advantage on their possessor.

Our understanding of the evolution of eyes remained at this level until 1915, when a mutant fruit fly without eyes was found and the gene involved, appropriately called *eyeless*, was mapped on one of its chromosomes. This mutant fly remained a laboratory curiosity until the 1990s, when the Swiss developmental biologists Rebecca Quiring and Walter Gehring began looking for transcription factors that might be involved in fly development. They found one, and when they mapped its gene, it was at the *eyeless* locus. The product of the *eyeless* gene is a transcription factor that controls the formation of the eye. Quiring and Gehring demonstrated this by transplanting early embryonic tissue that was overexpressing the *eyeless* gene to other places in the embryo. It did not matter where they placed the tissue; even on legs, as long as the gene was active, an eye developed.

When they did comparative genomic studies, Quiring and Gehring were in for a big surprise: the *eyeless* gene sequence was quite similar to that of *Pax6*, a gene in mice that, when mutated, leads to the development of very small eyes. Could the very different eyes of flies and mice just be variations on a common developmental theme?

To test the similarity in function between the insect and mammalian genes, Quiring and Gehring inserted the mouse gene *Pax6* into the fly genome and repeated their transplantation experiments. Once again, eyes developed. A gene whose expression normally leads to the development of a mammalian "camera" eye now led to the development of

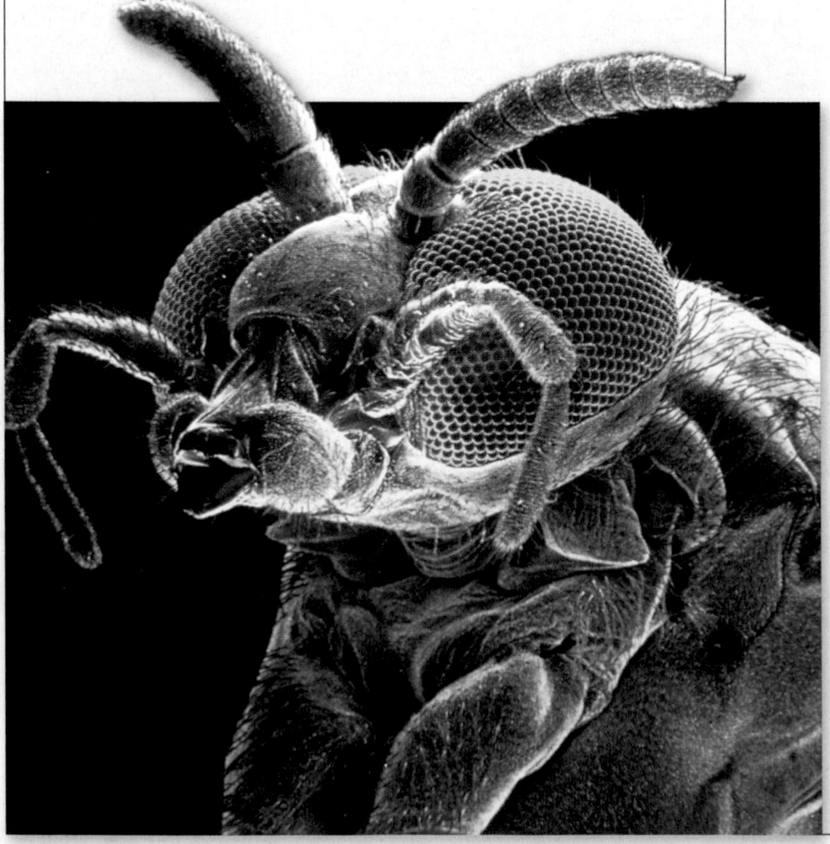

Eye of the Fly Unlike the single-lensed eyes of vertebrates, the compound eyes of flies and some other insects are composed of thousands of individual lenses, or ommatidia.

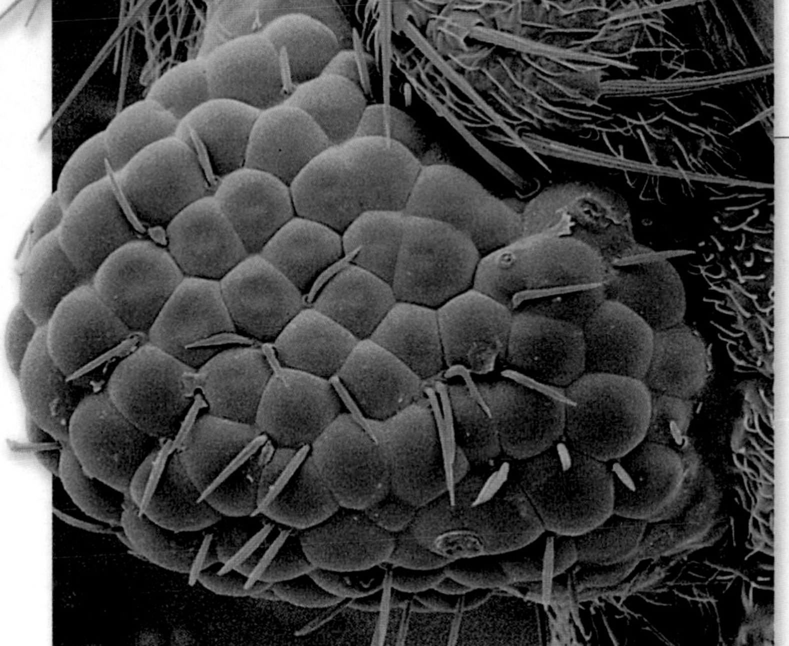

A Mouse Gene Can Produce a Fly's Eye When the mouse *Pax6* eye-specifying gene was implanted in the limb disk of a fruit fly, ommatidia emerged in place of a leg.

CHAPTER OUTLINE

an insect's "compound" eye—a very different eye type. Thus a single transcription factor appears to function as a molecular switch that turns on eye development. Although eyes evolved many times during animal evolution, all of them may depend on the same gene. The special features of the many different eyes in diverse animals all evolved from a common developmental process.

The discovery that the same genes govern development in a wide variety of animals led to the rapid proliferation of a new discipline: evolutionary developmental biology, often known as "evo-devo." Evolutionary developmental biologists compare the genes that regulate development in many different multicellular organisms to understand how a single gene can do so many different things.

IN THIS CHAPTER we will see that the genes that control pattern formation are shared by a diverse array of organisms. We will discover how changes can occur in some parts of an organism without causing undesirable changes in other parts, and how a common set of genes can produce such a great variety of body forms. We will also see how some organisms can modulate their development by responding to signals from their environment. Finally, we will see how developmental processes constrain evolution.

20.1 How Does a Molecular Tool Kit Govern Development?

Biologists have long known that a fruit fly looks like a fruit fly and a mouse looks like a mouse largely because of their genes; clearly, certain genes are unique to each type of organism. But we have also discussed how the genomes of complex organisms share numerous homologous coding sequences and proteins (see Section 14.1). When developmental biologists began to describe the events that specify differentiation, morphogenesis, and pattern formation at the molecular level, they found common themes that encompassed both phenomena.

Developmental genes in diverse organisms are similar, but have different results

As we saw in Chapter 19, many of the genes that control development encode transcription factors that act on proteins in a cascade of events. These *developmental genes* were not discovered until the mid 1980s because geneticists had previously concentrated their attention on the transmission of inherited characteristics from adult organisms to their offspring. Until that time, genetics was almost exclusively the study of that part of DNA encoding structural proteins and enzymes. The processes of development remained a "molecular black box."

The discovery of developmental mutants in *Drosophila* eventually led to the identification of genes and gene products that are responsible for some of normal insect development (see Section 19.5). Using the tools of comparative genomics, scientists discovered that a similar set of genes exists in vertebrates. This discovery of a common set of developmental genes in organisms as evolutionarily distant as fruit flies and mice led to a major conclusion of evolutionary developmental biology: *the amazing diversity of organisms is produced by a modest number of regulatory genes.* Differences in body form result from differences in where and when these genes are turned on and off.

The transcription factors and extracellular signals that govern pattern formation in multicellular organisms, and the genes that encode them, can be thought of as a **molecular tool kit**, in the same sense that a few tools in a carpenter's tool kit can be used to build many different structures.

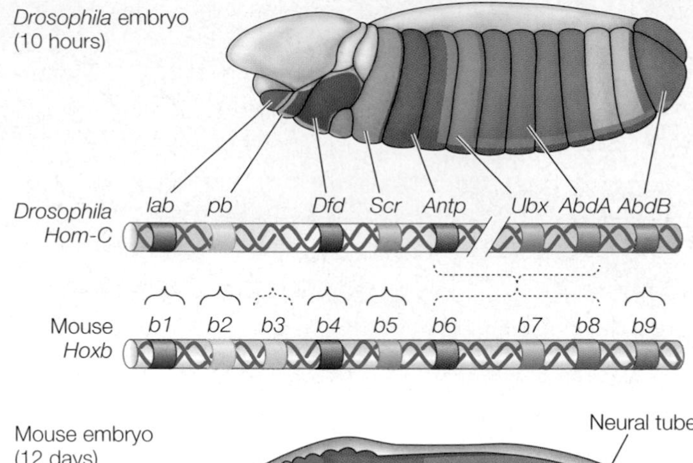

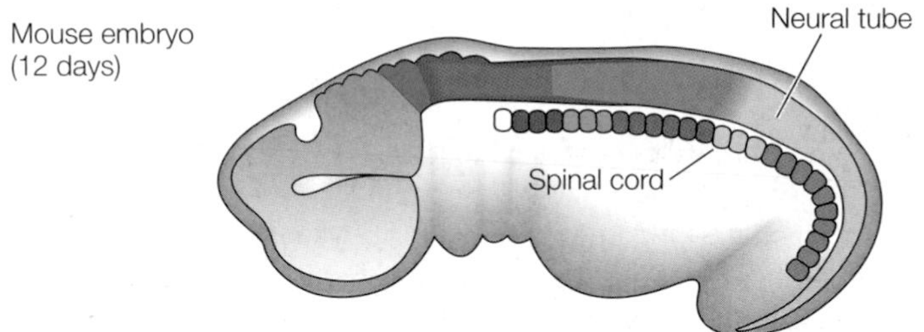

20.1 Regulatory Genes Show Similar Expression Patterns Homologous genes encoding similar transcription factors are expressed in similar patterns along the anterior–posterior axes of both insects and vertebrates.

Plants and animals share many regulatory genes. We saw in Section 19.5 that members of two families of genes that encode transcription factors, the MADS box genes of plants and the Hox genes of fruit flies, regulate important developmental processes. In *Drosophila*, these genes govern the development of structures such as legs, antennae, and wings, whereas in plants they govern the development of flowers, fruits, and seeds. So even though the structures produced are unique and dissimilar, the "rules" governing cell differentiation during development are quite similar in animals and plants. This similarity is remarkable, given that the evolutionary paths of these two lineages diverged in the far distant past. The regulatory genes have been *conserved*—that is, they have changed little—over many millions of years.

Most animals that move through their environment under their own power have bilaterally symmetrical bodies with a head (anterior) and a tail (posterior) end; the bodies of many of them are divided into segments, although these segments may not be visible externally (see Chapter 31). The same kinds of homeobox-containing genes encoding transcription factors provide positional information to cells along the anterior–posterior axis of the body in both insect and mammalian embryos. For example, certain *Drosophila* genes and the homologous genes of vertebrates both are expressed in the anterior regions of the brain (**Figure 20.1**). When certain Hox genes are mutated in *Drosophila*, the segments differentiate in the wrong way (see Figure 19.20). The *bithorax* mutation, which is caused by a deletion of the *Ubx* gene, causes the developing insect to form two sets of forewings rather than the normal one pair (**Figure 20.2A**). In vertebrates, altering the expression patterns of some Hox genes can change lumbar (abdominal) vertebrae into thoracic (ribbed) vertebrae (**Figure 20.2B**). Altering the expression of other genes can replace neck bones with duplications of the ear bones and jaw.

20.1 RECAP

A molecular tool kit consisting of highly conserved regulatory genes encoding transcription factors and chemical signals governs pattern formation in multicellular organisms.

- How does the story of the mouse *Pax6* gene demonstrate the existence of common developmental genes in different organisms? See pp. 448–449
- Do you understand the effects of mutations in developmental genes, and why the existence of mutant phenotypes led to the discovery of these genes? See pp. 449–450

(A)

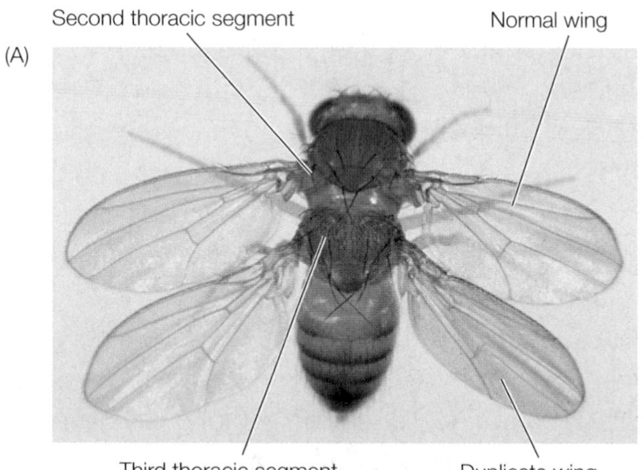

(B)

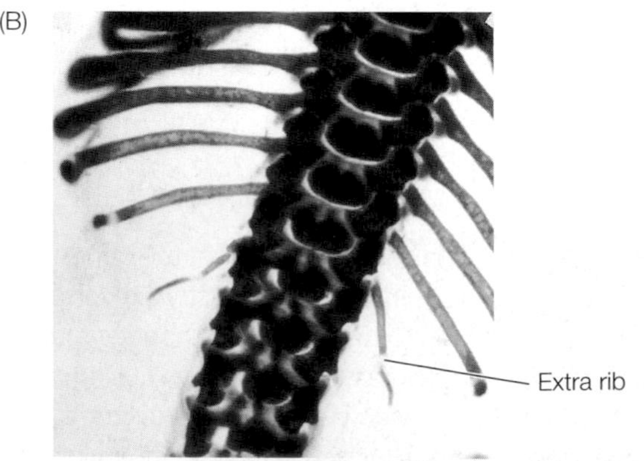

20.2 Altering Homeotic Genes Changes Morphology (A) Deletion of the *Ubx* gene in *Drosophila* converts the third thoracic segment, which does not normally bear wings, into a duplication of the second thoracic (forewing-bearing) segment. (B) Deletion of the *Hoxc-8* gene in mice transforms a lumbar (abdominal) vertebra into a copy of a thoracic (ribbed) vertebra.

The striking abnormalities that first alerted geneticists to the existence of regulatory genes in *Drosophila* transformed one body part into another part normally found elsewhere. Surprisingly, these mutations affected only single structures. A fly with a leg coming out of its head was, in all other respects, normal. How was it possible that a mutation with such a massive effect on one structure did not cause deformities in any other part of the fly's body?

20.2 How Can Mutations with Large Effects Change Only One Part of the Body?

It was by studying homeotic mutations that geneticists discovered that embryos, like adults, are made up of **modules**—functional entities encompassing genes and the various signaling pathways the genes stimulate or inhibit, as well as the physical structures that result from these signaling cascades.

The form of each module in an organism may be changed independently of the other modules because many developmental genes exert their effects on only a single module. The form of a developing animal's heart, for example, can change independently of changes in its limbs because the genes that govern heart formation do not affect limb formation, and vice versa. If this were not true, a single mutation in a developmental gene would likely result in an adult with multiple, widely different deformities. Such an adult would be unlikely to function well in any environment.

Genetic switches govern how the molecular tool kit is used

Different structures can evolve within a single organism using a common set of genetic instructions because there are components of DNA, called *genetic switches*, that control how the molecular tool kit is used. These DNA sequences and the signal cascades that converge and act on these sequences determine when and where genes will be turned on and off. Multiple switches control each gene by influencing its expression at different times and in different places. In turn, most elements of the molecular tool kit influence more than one developmental process. They are able to do so because multiple switches allow them to be used in many contexts.

Genetic switches integrate positional information in the developing embryo and play a key role in making different modules develop differently. We have seen that different Hox genes are expressed in different segments and appendages of developing fruit flies. The pattern and functioning of each segment depends on the unique Hox gene or combination of Hox genes that are expressed in it. Genetic switches control this activity by activating each Hox gene in different zones of the body.

As an example, consider the formation of *Drosophila* wings. A fruit fly has three thoracic segments, the first of which bears no wings. The second segment bears the large forewings, and the third segment bears small hind wings that function as balancing organs. Hox proteins are not expressed in forewing cells, but all hind wing cells express the Hox gene *Ultrabithorax* (*Ubx*) because a set of genetic switches activates the *Ubx* gene in the third thoracic segment. *Ubx* turns off genes that promote the formation of the veins and other structures of the forewing, and it turns on genes that promote the formation of hind wing features (**Figure 20.3**). In butterflies, on the other hand, *Ubx* does not bind in the third-segment cells, so full wings develop.

Modularity allows differences in the timing and spatial pattern of gene expression

Modularity also allows the relative *timing* of different developmental processes to shift independently of one another, a process called **heterochrony**. That is, the genes regulating the development of one module (say, the eyes of vertebrates) may be expressed at different times in different species, relative to the genes regulating the development of other modules.

Two salamander species of the genus *Bolitoglossa* illustrate how heterochrony can result in new morphology. The webbing between the toes of the larvae of most salamander species, including *Bolitoglossa rostratus* (**Figure 20.4A**), disappears as the animals mature, resulting in feet with separated toes, well suited to moving on the ground. But what if a mutation slowed or stopped expression of the genes that trigger apoptosis in the webbing? The resulting "juvenile" webbed feet would act like suction cups, allowing the animal to adhere to tree branches (**Figure 20.4B**). This is exactly what happened to *Bolitoglossa occidentalis*, opening up a new, arboreal way of life for this salamander species.

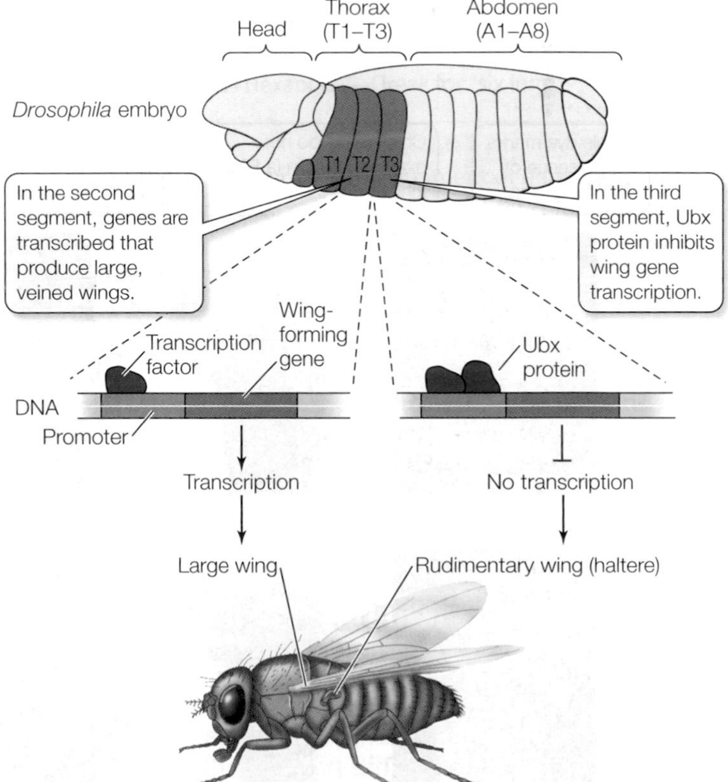

20.3 Segments Differentiate under Control of Genetic Switches The binding of a single protein, Ultrabithorax, determines whether a thoracic segment produces full wings or the reduced wings known as halteres (balancers).

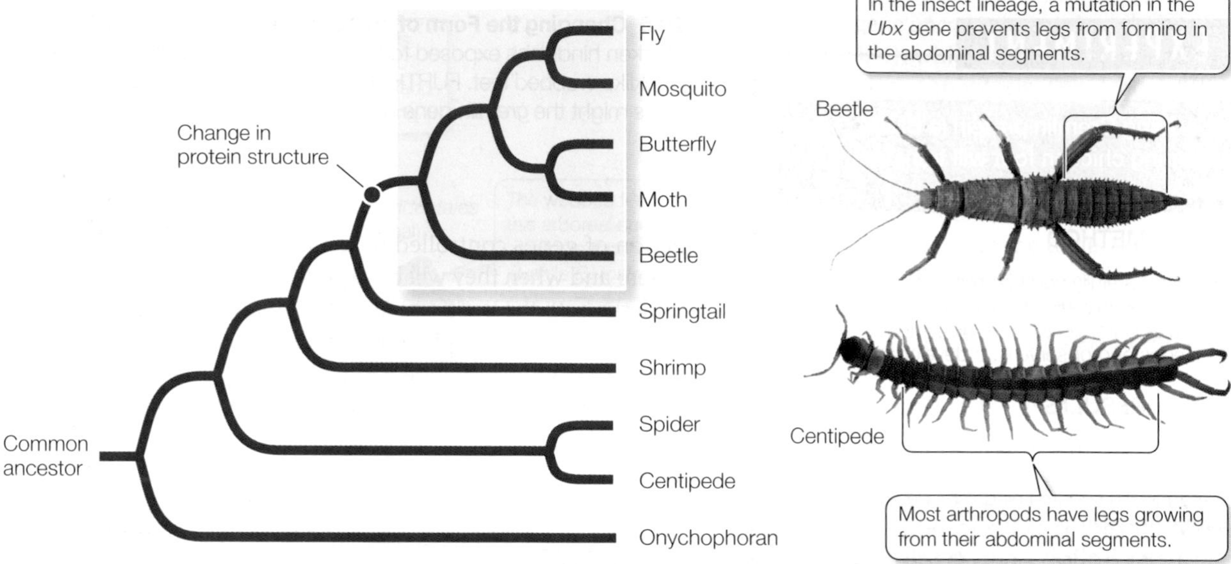
Fly
Mosquito
Butterfly
Moth
Beetle
Springtail
Shrimp
Spider
Centipede
Onychophoran
Change in protein structure
Common ancestor
In the insect lineage, a mutation in the Ubx gene prevents legs from forming in the abdominal segments.
Beetle
Centipede
Most arthropods have legs growing from their abdominal segments.

20.7 Mutation in a Hox Gene Changed the Number of Legs in Insects In the insect lineage (blue) of the arthropods, a mutation in the *Ubx* gene resulted in a protein that inhibits the *dll* gene, which is required for legs to form. Because insects express *Ubx* in their abdominal segments, no legs grow from these segments. Other arthropods, such as centipedes, do grow legs from their abdominal segments.

Our discussion so far might suggest that the form of an adult organism is determined entirely by its genes. Yet we know that the form of an adult organism is also influenced by environmental conditions during its development. We also know that no single body form is best for all environments. How are developmental processes modified so that the adult organism is adapted for the environment in which it will live?

20.4 How Does the Environment Modulate Development?

The environment in which an individual will live may differ from the one in which its parents lived. It would be advantageous for a developing individual to "sense" the kind of environment in which it will function as an adult and modify its development accordingly. As it turns out, the development of many organisms is modified by environmental conditions. The ability of an organism to modify its development in response to environmental conditions is called **developmental plasticity**. In other words, a single genotype may produce a range of phenotypes, and signals from the environment may determine which phenotype is expressed. But how can organisms respond to signals from the environment?

No single way of responding to signals from the environment would always result in adaptation because environmental signals tell organisms many different things. We can divide signals from the environment into two major types, based on their significance and how organisms should respond to them:

- *Some environmental signals are accurate predictors of future conditions.* Some of these signals always occur, but organisms may develop without ever encountering others. In either case, the developmental processes of organisms should respond adaptively to these signals when they occur.
- *Some environmental signals are poorly correlated with future conditions.* Organisms should fail to respond to such signals.

Do developing organisms respond to these different types of signals as we expect them to?

Organisms respond to signals that accurately predict the future

Organisms *do* respond to signals that are accurate predictors of future environmental conditions. For example, seasonal changes in day length occur reliably at the same time every year. Increasing day length accurately predicts the approach of summer; decreasing day length signals winter is coming. In most tropical regions, wet and dry seasons alternate in a regular pattern during the year. Developing organisms respond to these signals in such a way that the adults they become are adapted to the future conditions.

The West African butterfly *Bicyclus anynana* has two color forms. The dry-season form matches the dead brown leaves on the dry-season forest floor, where the butterfly typically rests. The more active wet-season form has a white line along the wing and conspicuous ventral eyespots on its hind wings. These eyespots deceive predatory lizards and birds into attacking the wing, rather than the butterfly's actual eye, increasing the butterfly's chances of escape.

Temperature during pupation determines the color form of the adult butterfly. Pupae that develop in the soil under cooler temperatures (less than 20°C) produce the dry-season color form; temperatures above 24°C produce the wet-season color form. In the late larval stages, transcription of the *distal-less* gene is restricted to several small areas of the wing that have the potential to become the centers of eyespots. During pupal development, the area over which distal-less protein is expressed increases with temper-

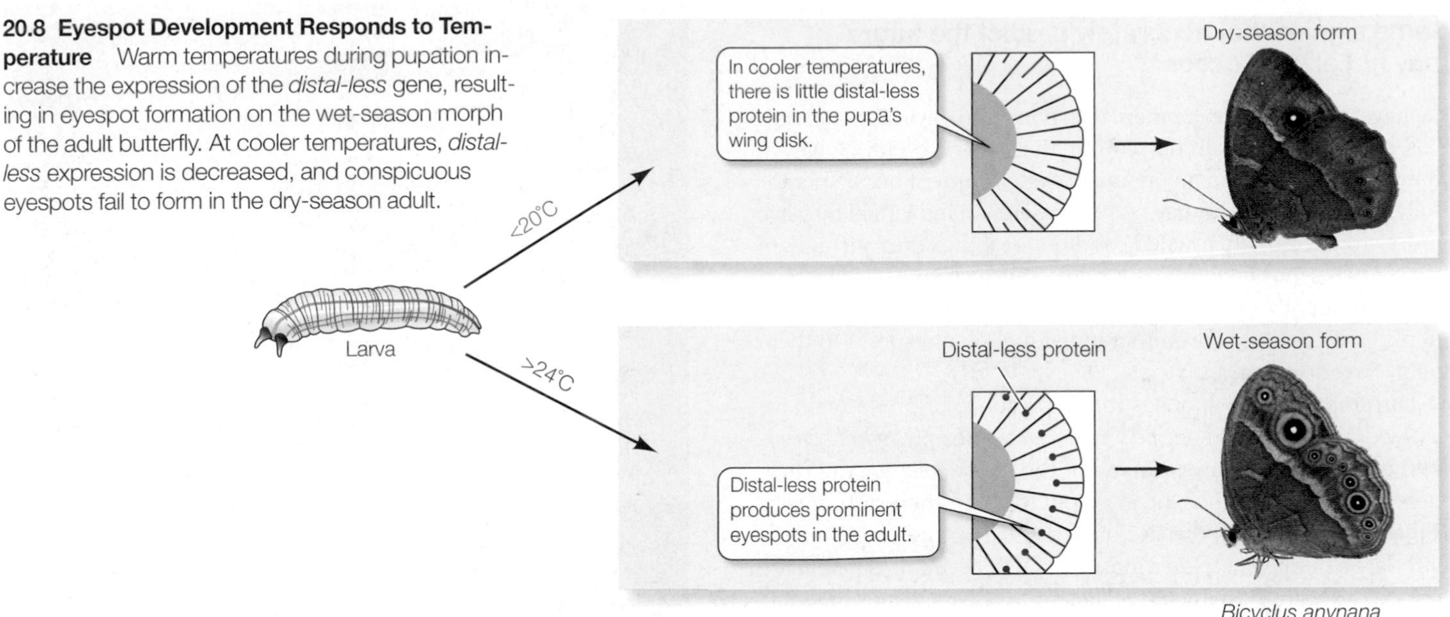

20.8 Eyespot Development Responds to Temperature Warm temperatures during pupation increase the expression of the *distal-less* gene, resulting in eyespot formation on the wet-season morph of the adult butterfly. At cooler temperatures, *distal-less* expression is decreased, and conspicuous eyespots fail to form in the dry-season adult.

ature, resulting in conspicuous eyespots on adults that pupate while exposed to warm temperatures (**Figure 20.8**). Thus, by responding to an environmental signal—temperature—the pupae develop into adults with a form that is adapted to the conditions under which they will live.

Another example of developmental plasticity in response to seasonal changes occurs in the larvae of the moth *Nemoria arizonaria*. This moth produces two generations each year. Larvae (caterpillars) that hatch from eggs in spring feed on oak flowers (catkins). These larvae complete their development, form pupae, and transform themselves into adult moths in summer. These moths lay their eggs on oak trees. The larvae that hatch from these eggs in summer eat oak leaves, complete their development, and lay their eggs on oak branches. These eggs overwinter and hatch the following spring. The spring caterpillars resemble the catkins on which they feed (**Figure 20.9A**); the summer caterpillars that feed on oak leaves resemble small, year-old oak branches (**Figure 20.9B**). Both types of caterpillars are well camouflaged in the environment in which they feed. An experimenter was able to convert spring caterpillars into summer caterpillars by feeding them oak leaves. A chemical in the oak leaves induces them to develop the twiglike summer form.

Another example of developmental plasticity resulting in phenotypic change in response to the environment is found among the venomous tiger snakes (genus *Notechis*) of Australia. On the Australian mainland, tiger snakes eat small frogs and mice. On nearby Carnac Island, they eat much larger prey, the chicks of silver gulls. Island snakes grow larger, and have larger heads relative to their body length, than mainland snakes. In the laboratory, juvenile island snakes fed large mice developed larger heads and larger jaws than their siblings that ate only small prey. The heads and jaws of juvenile mainland snakes changed much less with diet. Thus head development in island snakes shows marked developmental plasticity, but that of mainland snakes does not.

20.9 Spring and Summer Forms of a Caterpillar Differ (A) Spring caterpillars of the moth *Nemoria arizonaria* resemble the oak catkins on which they feed. (B) Summer caterpillars of the same species resemble oak twigs.

Some signals that accurately predict the future may not always occur

We have discussed developmental responses to environmental signals, such as changes in day length, that always occur. However, many other changes in an organism's environment occur sporadically: predators may or may not be present; an individual may live under crowded or uncrowded conditions; or the sexes and ages of its associates may change. Nevertheless, if such changes have occurred frequently during the evolution of a species, developmental plasticity may evolve so that individuals can respond to them when they do occur.

Developmental responses to the presence of predators have evolved in numerous species. When water fleas (*Daphnia cucullata*) encounter predatory larvae of the fly *Chaoborus*, the "helmets" on the top of their heads grow to twice their normal size (**Figure 20.10**). The fly larvae have difficulty ingesting *Daphnia* with large helmets. Helmet induction also occurs if *Daphnia* are exposed to water in which fly larvae have been swimming. Offspring that develop in the abdomens of mothers with induced large helmets are born with large helmets. Why don't all individuals develop large helmets just in case? Because there is a cost in terms of allocation of limited energy: *Daphnia* with large helmets produce fewer eggs than do *Daphnia* with small helmets.

20.11 Light Seekers The bean plants on the left were grown experimentally under low light levels. The plant's cells have elongated in response to low light, and the plants have become spindly. The control plants on the right were grown under normal light conditions.

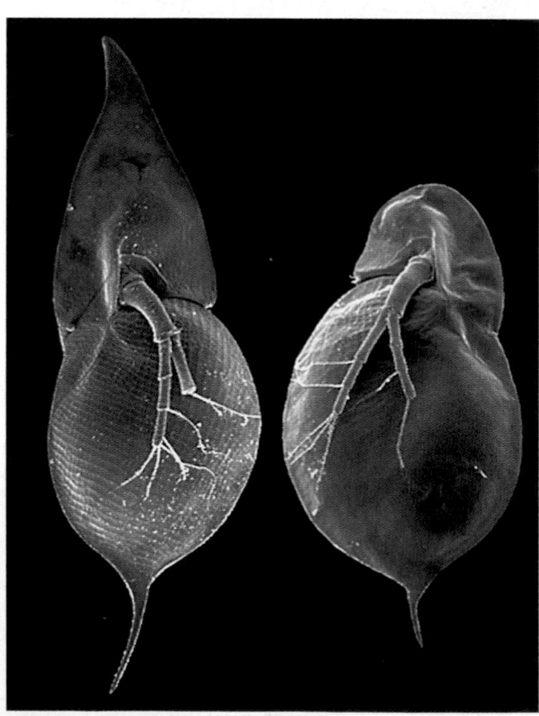

Daphnia cucullata

20.10 Predator-Induced Developmental Plasticity in *Daphnia* This scanning electron micrograph shows the predator-induced form of *Daphnia* (left), with an enlarged helmet, and the normal form of this tiny crustacean (right). These two *Daphnia* are genetically identical individuals from a single asexually produced clone.

Light exerts a powerful influence on plant development. Low light levels stimulate the elongation of cells, so that plants growing in the shade become spindly (**Figure 20.11**). This developmental plasticity is adaptive because a spindly plant is more likely to reach a patch of brighter light than a plant that remains compact. And because they have undifferentiated tissues called meristems, plants can continue to respond to light as long as they grow.

Among some animals, learning is a developmental response to environmental change. As you know from your struggles to absorb the contents of this book, learning is costly. Learning takes much effort and time, during which an individual cannot do other useful things. But learning can continue throughout adult life, allowing an individual to adjust its behavior to the physical, biological, and social environment in which it matures. As we will see in Chapter 53, learning is particularly important in species with complex social systems. Individuals of these species must learn the identities and individual characteristics of many associates and adjust their behavior accordingly. Meanwhile, bear in mind that, as difficult as learning may be, ignorance is even more costly.

Organisms do not respond to signals that are poorly correlated with future conditions

Just as we expect organisms to respond to environmental signals that accurately predict the future, we expect them to ignore signals that are poorly correlated with future conditions, because responding to such signals would not help them adapt to the environment in which they will live. Consider, for example, seed production by plants. Plants respond to changing environmental conditions by varying their size, shape, and number of their flowers, as well as the *number* of seeds they produce; but the *size* of the seeds they produce remains nearly constant.

The amount of energy a growing plant has available to allocate to seed production depends, among other things, on temperature, rainfall, and the sizes and numbers of its neighbors. But these environmental conditions in any one year are poor predictors of what those conditions will be in the next year. A seedling that germinates from a large seed will survive better under conditions of intense competition than a seedling that germinates from a small seed because it can grow larger using the greater energy reserves stored within the seed. But, for a given amount of energy, a mature plant can produce far fewer large seeds than small seeds. So if the next generation of plants grows under more favorable conditions, plants that produced a larger number of smaller seeds in the previous year will have more surviving offspring. Thus plants do not change the sizes of the seeds they produce in response to the conditions under which they grow.

Instead, many plants have genes for seed dormancy, so that the seeds made in one year germinate at staggered intervals over future years that will have different and unknowable rainfall patterns and densities of neighbors. Seed size has evolved in response to the average conditions encountered by plants over many generations, and thus remains relatively consistent.

Organisms may lack appropriate responses to new environmental signals

Although organisms can respond adaptively to environmental signals that have occurred frequently during their recent evolutionary histories, organisms typically lack adaptive responses to environmental signals that they have not encountered before. Chapter 1 of this book gave an example in describing the effects of a particular parasite on frog limb development. The lack of useful developmental responses to newly encountered environmental events is an important problem today because human societies have changed the environment in so many ways.

For example, humans today release thousands of chemical compounds into the environment, some of which disrupt normal development. In 1962, Rachel Carson's classic book *Silent Spring* focused attention on the devastating effects the widely used chemical pesticide DDT had on bird populations, in part by interfering with the development of the eggshell. Research over the years has confirmed the malignant effects of DDT on the reproductive systems and development of many birds and mammals.

Another event that made headlines in 1962 underscored the unforeseen effects of environmental agents, this time affecting humans. More than 7,000 infants with missing or underdeveloped limbs were born to women who had taken a drug called thalidomide. Nobody in the medical profession had expected thalidomide, which was believed to be a safe mild sedative, to affect the expression of limb development genes in the fetus.

20.4 RECAP

Developmental plasticity enables developing organisms to adjust their forms to fit the environment in which they live. Organisms respond to environmental signals that are accurate predictors of future conditions, but fail to respond to signals that are poorly correlated with future conditions.

- Describe several examples of how an organism's phenotype can be a response to environmental signals. See pp. 454–455 and Figure 20.8
- Can you explain how learning allows an individual to adjust its behavior to the physical and social environment in which it matures? See p. 456
- Why do organisms typically lack adaptive responses to environmental signals that their ancestors never encountered? See p. 457

Appropriate responses to new environmental conditions are likely to evolve over time, but what are the limits of such evolution? Do developmental genes dictate what structures and forms are possible?

20.5 How Do Developmental Genes Constrain Evolution?

Three decades ago, the French geneticist François Jacob made the analogy that evolution works like a tinker, assembling new structures by combining and modifying the available materials, and not like an engineer, who is free to develop dramatically different designs (say, a jet engine to replace a propeller-driven engine). The evolution of form has not been governed by the appearance of radically new genes, but by modifications of what the existing genes do. Developmental genes and their expression constrain evolution in two major ways:

- Nearly all evolutionary innovations are modifications of previously existing structures.
- The genes that control development are highly conserved; that is, the regulatory genes themselves do not greatly change over the course of evolution.

Evolution proceeds by changing what's already there

The features of organisms almost always evolve from pre-existing features in their ancestors. To cite a classic example, insects, birds, and bats did not "invent" their wing genes. Wings arose as modifications of an existing structure. Wings evolved independently in insects and vertebrates—only once in the insects, but in three equally independent instances among the vertebrates (**Figure 20.12**). But in all four cases, the wings are modified limbs.

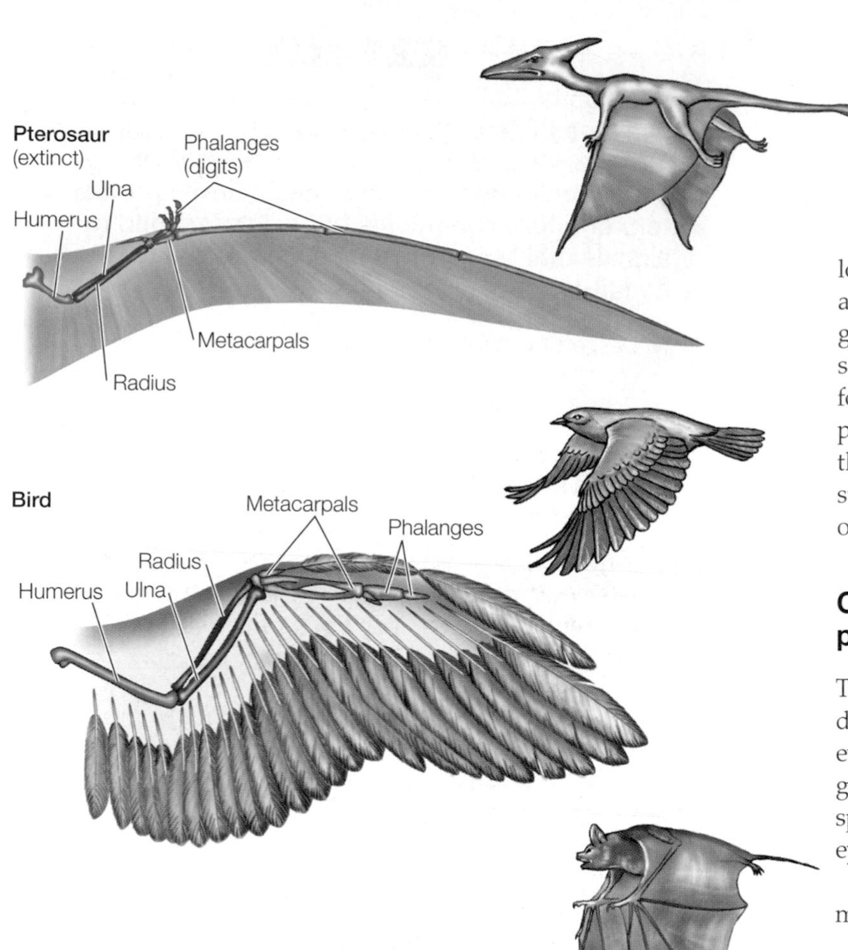

20.12 Wings Evolved Three Times in Vertebrates The wings of pterosaurs, birds, and bats are all modified forelimbs constructed of the same skeletal components. However, the components have different forms in the different groups.

Dr. Seuss was not alone in conceiving animals that evolution never produced. Storytellers and writers have envisioned many imaginary beings with wings sprouting from their backs—think of fairies, flying horses, and angels. Such beings are likely to remain confined to the world of human imagination, because wings probably cannot evolve from any structure other than an already existing limb.

Developmental controls also influence how organisms lose structures. The ancestors of snakes lost their limbs as a result of changes in the body segments in which the Hox genes that suppress limb formation express themselves. The same process, involving the same genes, was responsible for the dramatic evolutionary changes in the number of appendages arthropods have and on which body segments they are borne (see Figure 20.7). But in the case of snakes, suppression is not necessarily complete, and some snakes occasionally develop rudimentary limbs (**Figure 20.13**).

Conserved developmental genes can lead to parallel evolution

The nucleotide sequences of many of the genes that govern development have been highly conserved throughout the evolution of multicellular organisms—in other words, these genes exist in similar form across a broad spectrum of species. We made mention of this fact in relation to the *Pax6* eye-specifying gene described at the start of this chapter.

The existence of highly conserved developmental genes makes it likely that similar traits will evolve repeatedly, es-

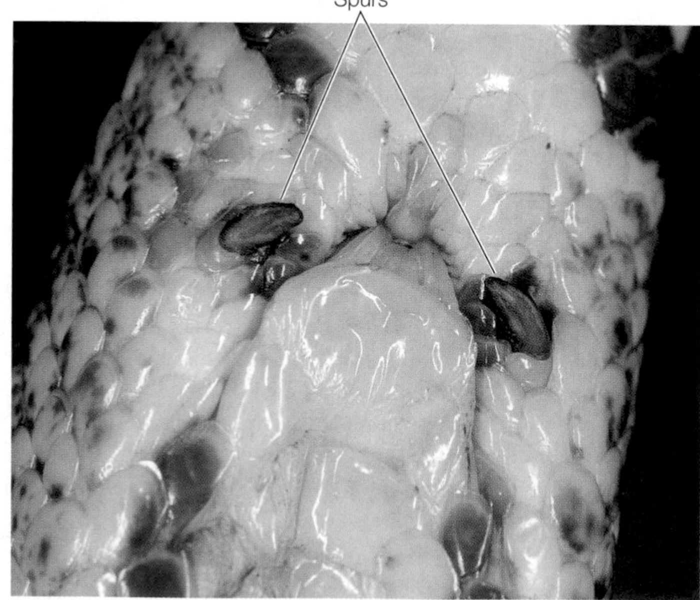

20.13 Developmental Genes and Lost Structures In snakes, Hox gene expression in the developing vertebral column suppresses limb formation. However, in a number of snake species, mutations occasionally result in the development of tiny legs. These spurs, or claws, are the outward projections of a royal python's vestigial hind legs.

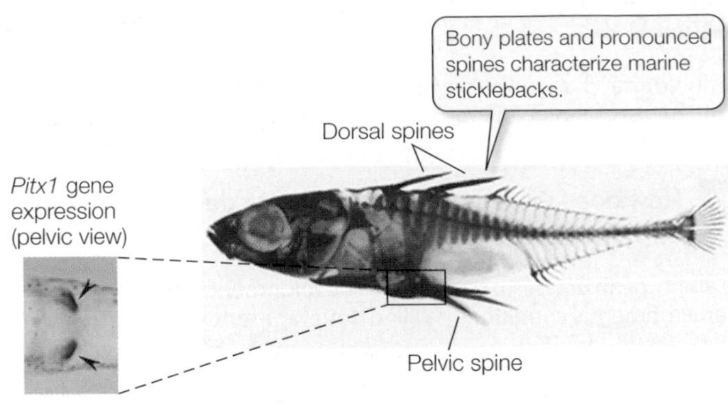
Bony plates and pronounced spines characterize marine sticklebacks.
Dorsal spines
Pitx1 gene expression (pelvic view)
Pelvic spine

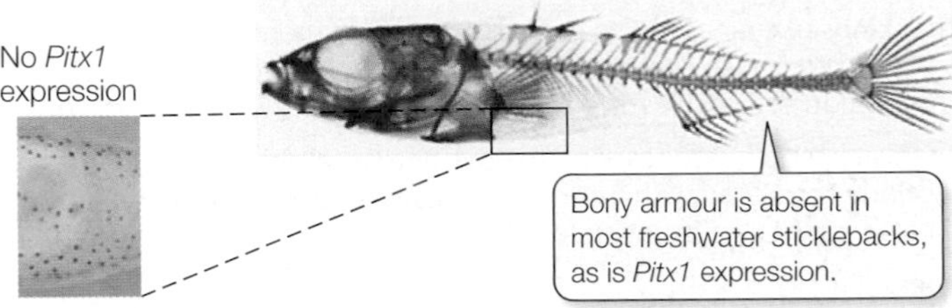
No Pitx1 expression
Bony armour is absent in most freshwater sticklebacks, as is Pitx1 expression.

20.14 Parallel Phenotypic Evolution in Sticklebacks Three-spined sticklebacks are small fish (often no more than 3 cm) that are widespread in the oceans. Marine sticklebacks (top) have sharp spines and a bony skeleton. In the many independent cases in which freshwater sticklebacks have arisen from marine ancestors, the spines and bony armor are greatly reduced or disappear; the phenotypic change from marine to freshwater forms seems to be consistently parallel, the result of the absence of *Ptx1* gene expression.

pecially among closely related species—a process called **parallel phenotypic evolution**. A good example of parallel phenotypic evolution is provided by a small fish, the three-spined stickleback (*Gasterosteus aculeatus*)

Sticklebacks are widely distributed across the Atlantic and Pacific Oceans and are found in many freshwater lakes. Marine populations of this species spend most of their lives at sea but return to fresh water to breed. Members of freshwater populations live in lakes and never journey to salt water. Genetic evidence shows that freshwater populations have arisen independently many times from adjacent marine ones. Marine sticklebacks have several structures that protect them from predators: well-developed pelvic bones, long dorsal and pelvic spines, and bony plates. In the freshwater populations descended from them, this body armor is greatly reduced, and dorsal and pelvic spines are much shorter or even lacking (**Figure 20.14**).

When researchers began to investigate the genetics of stickleback evolution, they expected that different genes would govern the changes in different populations. Instead, wherever they looked—Japan, British Columbia, California, Iceland—investigators have found that the same single gene, or small set of nearby genes, is responsible for the loss of armor in freshwater populations, as well as for changes in a suite of other bony parts, including jaws and bones that protect the gills. That gene, *Pitx1*, is inactive in the bone-forming regions of freshwater stickleback embryos, but active in those regions in marine sticklebacks. (Note that *modularity* is important to the results of gene expression in this example; the *Pitx1* gene is expressed in other developing organs in both marine and freshwater sticklebacks; it is only in the developing bone that structural differences arise.)The consistency of these changes in gene expression that have arisen multiple times in freshwater populations strongly suggests that the changes are adaptive. Most of the lakes in which freshwater sticklebacks live formed less than 10,000 years ago, after the glaciers that covered them retreated. These lakes lack predatory fish, but they contain predatory dragonfly larvae that can more easily capture sticklebacks with long spines. In addition, calcium is typically in short supply in freshwater lakes, making armor more costly for the fish to produce, and freshwater lakes have abundant vegetation in which sticklebacks can hide, so they have less need of an armored defense.

20.5 RECAP

Developmental controls constrain evolution because nearly all evolutionary innovations are modifications of previously existing structures, and because the conservation of many genes makes it likely that similar traits will evolve repeatedly.

- Do you understand how diverse body forms have evolved by means of modifications in the functioning of existing genes? See p. 457 and Figure 20.12
- Do you understand how the differences between marine and freshwater sticklebacks exemplify phenotypic evolution by gene regulation? See p. 459 and Figure 20.14

Many novel traits have arisen that failed to persist beyond a single generation. Part Five of this book examines the processes of evolution—the powerful forces that influence the different survival and reproductive success of various life forms—and how different adaptations become prevalent in different environments, resulting over evolutionary time in the luminous diversity of life.

CHAPTER SUMMARY

20.1 How does a molecular tool kit govern development?

Evolutionary diversity is produced using a modest number of regulatory genes. Review Figure 20.1

The transcription factors and chemical signals that govern pattern formation in the bodies of multicellular organisms, and the genes that encode them, can be thought of as a "molecular tool kit."

Regulatory genes have been highly conserved during evolution. Review Figure 20.2

20.2 How can mutations with large effects change only one part of the body?

See Web/CD Tutorial 20.1

The bodies of both adults and embryos are organized into self-contained units called **modules** that can be modified independently. Modularity allows the timing of different developmental processes to shift independently, a process called **heterochrony**. Review Figure 20.4

Alterations in the spatial expression patterns of regulatory genes can also result in evolutionary changes. Review Figures 20.5 and 20.6

20.3 How can differences among species evolve?

The action of genes controlled by genetic switches that determine where and when they will be expressed or suppressed underlies both the transformation of an individual from an egg to an adult and the evolution of differences among species.

Morphological changes in species can evolve through mutations in the genes that regulate the differentiation of segments. Review Figure 20.7

20.4 How does the environment modulate development?

See Web/CD Activity 20.1

The ability of an organism to modify its development in response to environmental conditions is called **developmental plasticity**.

During development, organisms respond to environmental signals that are accurate predictors of future conditions. Some of these signals, such as changes in day length, always occur. Review Figure 20.8

Other kinds of changes in an organism's environment may or may not occur, but if such changes have occurred frequently during the evolution of a species, the ability to respond to them may evolve.

Organisms tend not to respond to environmental signals that are poorly correlated with future conditions, or to environmental signals that they have not encountered before.

20.5 How do developmental genes constrain evolution?

Virtually all evolutionary innovations are modifications of preexisting structures. Review Figure 20.12

Because many genes that govern development have been highly conserved, similar traits are likely to evolve repeatedly, especially among closely related species, a process called **parallel phenotypic evolution**.

SELF-QUIZ

1. Plants and animals share many regulatory genes that
   a. govern only the early development of the embryo.
   b. are inherited from a shared multicellular ancestor.
   c. govern the development of similar structures in both groups.
   d. govern the development of very different structures in the two groups.
   e. operate by very different rules.
2. Geneticists discovered regulatory genes only recently because
   a. geneticists did not believe that development was genetically controlled.
   b. geneticists concentrated their attention on genes that produced small effects.
   c. geneticists competed with developmental biologists for research funds and therefore de-emphasized the importance of development for understanding evolution.
   d. regulatory genes were particularly difficult to discover.
   e. geneticists concentrated their attention on the transmission of inherited characteristics from adults to their offspring.
3. Which of the following is *not* true of genetic switches?
   a. They control how a molecular tool kit is used.
   b. They integrate positional information in an embryo.
   c. A single switch controls each gene.
   d. They allow different structures to evolve within an individual organism.
   e. They determine when and where a gene is turned on or off.
4. Ducks have webbed feet and chickens don't because
   a. ducks need webbed feet to swim, whereas terrestrial chickens do not.
   b. both duck and chicken embryos express BMP4 in the webbing between the toes, but the *gremlin* gene is expressed in the webbing cells only in ducks.
   c. both duck and chicken embryos express BMP4 in the webbing between the toes, but the *gremlin* gene is expressed in the webbing cells only in chickens.
   d. only duck embryos express BMP4 in the webbing between the toes.
   e. only chicken embryos express BMP4 in the webbing between the toes.
5. Modularity is important for development because it
   a. guarantees that all units of a developing embryo will change in a coordinated way.
   b. coordinates the establishment of the anterior–posterior axis of the developing embryo.
   c. allows changes in the genes to change one part of the body without affecting other parts.
   d. guarantees that the timing of gene expression is the same in all parts of a developing embryo.
   e. allows organisms to be built up one module at a time.
6. Organisms often respond developmentally to regularly occurring environmental signals that accurately predict future conditions by
   a. stopping development until the signal changes.
   b. altering their development such that the resulting adult is adapted to the future environment.
   c. altering their development such that the resulting adult can produce offspring adapted to the future environment.
   d. producing new mutants.
   e. developing normally because the predicted conditions may not last long.

7. The process whereby changes in the relative timing of different developmental events can change the form of an organism is called
   a. heterochrony.
   b. developmental plasticity.
   c. adaptation.
   d. modularity.
   e. mutation.
8. *Daphnia* with large helmets are more difficult for some predators to capture and eat, but not all *Daphnia* produce large helmets because
   a. individuals with large helmets cannot feed efficiently.
   b. individuals with large helmets have trouble mating.
   c. individuals with large helmets produce fewer eggs than individuals with small helmets.
   d. individuals with large helmets become ensnared in vegetation.
   e. some individuals lack the genes that govern helmet formation.
9. Which of the following plant structures does *not* change in response to the conditions under which a plant grows?
   a. Roots
   b. Seeds
   c. Leaves
   d. Stems
   e. Branches
10. Parallel phenotypic evolution is common because
    a. closely related organisms typically face similar problems.
    b. the conservation of regulatory genes during evolution means that similar traits are likely to evolve repeatedly.
    c. many different phenotypes can be produced by a given genotype.
    d. phenotypic plasticity, which generates parallel phenotypic evolution, is widespread.
    e. evolutionary biologists have looked especially hard to find evidence of it.

FOR DISCUSSION

1. What components of environmental influences on development would probably be missed if investigations were confined to simple organisms such as bacteria and single-celled eukaryotes?
2. A spadefoot toad tadpole that develops in a rapidly drying pond is likely to eat many of its brothers and sisters. How can eating its siblings, which share half of an individual's genes, be favored by natural selection?
3. If evolutionary innovations can result from rather simple changes in the timing of expression of a few genes, why have such innovations arisen relatively infrequently during evolution?
4. François Jacob stated that evolution was more like tinkering than engineering. Does the observation that developmental genes have changed little over evolutionary time support his assertion? Why?
5. Despite their major differences, plants and animals share many of the genes that regulate development. What are the implications of this observation for the ways in which humans can respond to the adverse effects of the many substances we release into the environment that cause developmental abnormalities in plants and animals? What kinds of substances are most likely to have such effects? Why?

FOR INVESTIGATION

Figure 20.6 describes an experiment in which the protein Gremlin, which inhibits expression of the *BMP4* gene, was introduced into the foot of a developing chicken. What result would you expect from introducing Gremlin protein into other parts of the body of a developing chicken? Why? Into what other body parts would it be most informative to introduce Gremlin? If you were particularly interested in parallel phenotypic evolution, into what other organisms would you want to introduce Gremlin?

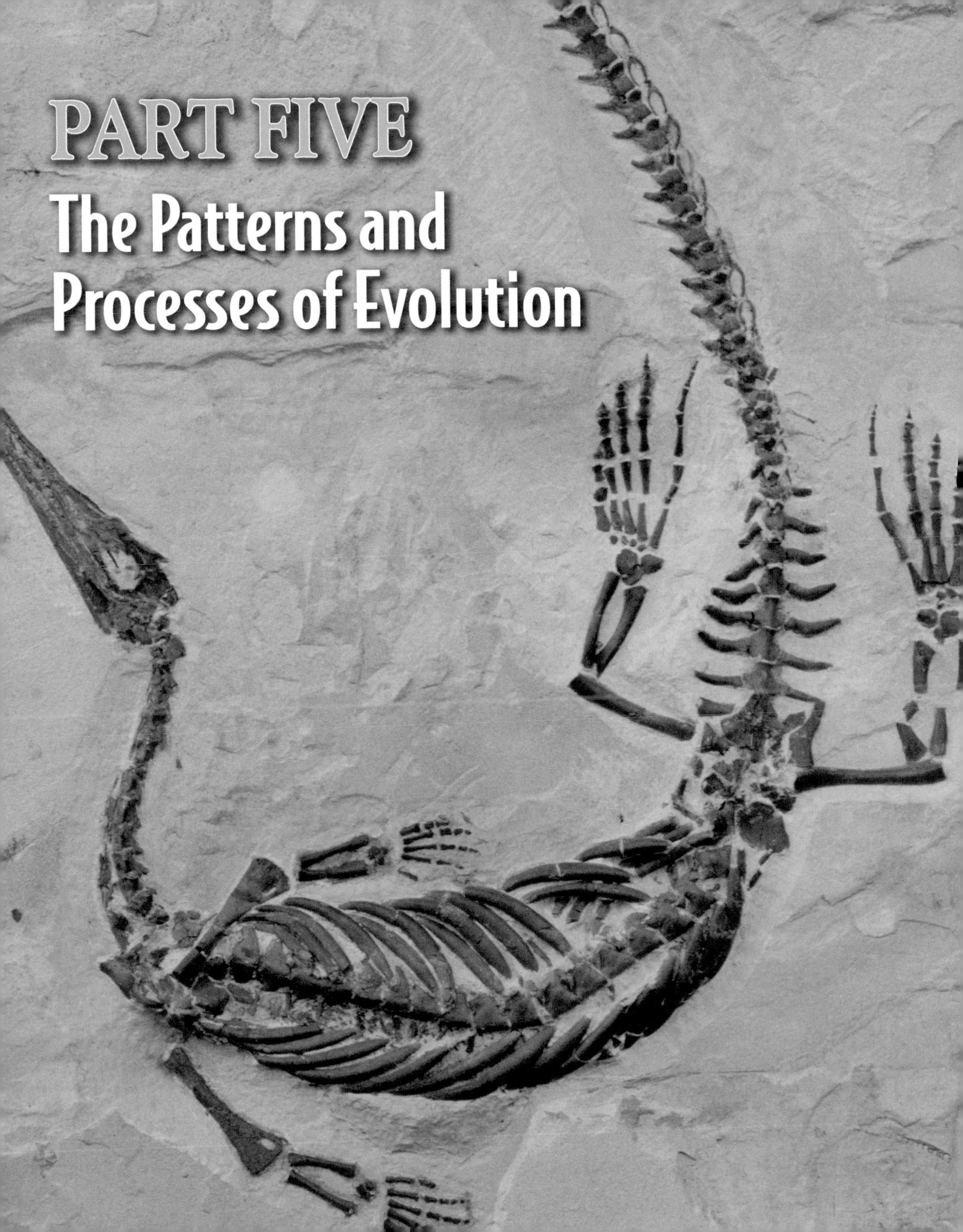

PART FIVE

The Patterns and Processes of Evolution

CHAPTER 21 The History of Life on Earth

Giant rodents

There are more species of rodents than any other group of mammals, and most of them are very small. House mice, for example, weigh about 30 grams, Norway rats weigh about 300 grams, and most squirrels weigh between 300 and 600 grams. South America, however, is home to a diverse group of rodents, including the familiar guinea pigs and chinchillas, that are, on average, significantly larger than rodents elsewhere. The largest living rodent, the capybara, weighs in at about 50 *kilograms* and lives in the marshes of South America.

Large rodents have been a distinctive feature of South America's evolutionary history. During the Miocene epoch (about 10 million years ago), the continent was home to a rodent the size of a buffalo. *Phoberomys pattersoni*, as it has been named, weighed about 700 kilograms—at least 10 times more than a capybara. The shapes of its fossilized teeth told paleontologists that *Phoberomys* was indeed a rodent, and also indicated that it fed on the grass of marshes and swamps, as the capybara does today. Paleontologists determined the weight of this giant rodent by applying a known, mathematically constant relationship between body mass and limb-bone diameter in living rodents to the fossilized bones.

How do biologists know *Phoberomys* lived 10 million years ago? The *Phoberomys* fossils were found in association with fossils of other mammals that no longer live today, in a specific layer of rock that was laid down a long time ago. But how long ago? The stratigraphic layering of different rocks allows us to tell their ages relative to each other, but does not indicate a given layer's absolute age.

One of the remarkable achievements of twentieth-century science was the development of sophisticated techniques that use the decay rates of various radioisotopes, changes in Earth's magnetic field, and the presence or absence of certain molecules to infer conditions and events in the remote past, and to date them accurately. It is those methods that provided the ages of the rocks in which the fossils of *Phoberomys* were found.

We are so accustomed to having time-measuring devices all around us that we forget how recently those devices were invented. When Galileo studied the motion of a ball rolling down an inclined plane about 400 years ago, he used his pulse to mark off equal intervals of time. The development of the science of biology is inti-

The World's Largest Rodents The South American capybara (*Hydrochaeris hydrochaeris*) is the largest extant rodent species. Adult capybaras, such as the one shown here with two young offspring, can weigh as much as an adult human.

Younger Rocks Lie on Top of Older Rocks In the Grand Canyon, the Colorado River has cut through and exposed many strata of ancient rocks. The oldest rocks visible here formed about 540 million years ago. The youngest rocks, at the top, are about 500 million years old. Fossilized remains of organisms that existed during the same evolutionary timeframe are found together in the same stratum.

mately linked to changing concepts of time, especially of the age of Earth. Biology as we know it could not and did not develop until about 150 years ago, when geologists first provided solid evidence that Earth was ancient. Before 1850, most people believed that Earth was no more than a few thousand years old. Charles Darwin could not have developed his theory of evolution by natural selection if he had not known that Earth was very old and that millions of years were available for life's evolution.

IN THIS CHAPTER we first describe how scientists assign dates to events in the distant evolutionary past. We then review the major changes in physical conditions on Earth during the past 4 billion years and look at how those changes affected life. We will describe the major patterns in the evolution of life and explain why rates of evolution change over time within and among groups of organisms. Finally we discuss the importance of evolutionary processes that are operating today.

CHAPTER OUTLINE

21.1 How Do Scientists Date Ancient Events?

Many evolutionary changes happen rapidly enough to be studied directly and manipulated experimentally. Plant and animal breeding by agriculturalists and the evolution of resistance to pesticides are good examples of rapid, short-term evolution. Other changes, such as the appearance of new species and evolutionary lineages, usually take place over much longer time frames.

To understand the long-term patterns of evolutionary change that we will trace throughout Parts Five and Six of this book, we must think in time frames spanning many millions of years, and we must imagine events and conditions very different from those we observe today. Earth of the distant past is, to us, a foreign planet inhabited by strange organisms. The continents were not where they are today, and climates were sometimes dramatically different from those of today.

As the opening of this chapter shows, **fossils**—the preserved remains of ancient organisms—can tell us a great deal about the body form, or *morphology*, of organisms that lived long ago, as well as how and where they lived. But to understand patterns of evolutionary change, we must also understand how life changed over time.

Earth's history is largely recorded in its rocks. We cannot tell the ages of rocks just by looking at them, but we can determine the ages of rocks relative to one another. The first person to formally recognize that this could be done was the seventeenth-century Danish physician Nicolaus Steno. Steno realized that in undisturbed *sedimentary* rock (rocks formed by the accumulation of grains on the bottoms of bodies of water), the oldest layers, or **strata** (singular *stratum*), lie at the bottom, and successively higher strata are progressively younger.

Geologists, particularly the eighteenth-century English scientist William Smith, subsequently combined Steno's insight with their observations of fossils contained within sedimentary rocks. They concluded that:

- Fossils of similar organisms were found in widely separated places on Earth.
- Certain organisms were always found in younger rocks than certain other organisms.

- Organisms found in higher, more recent strata were more similar to modern organisms than were those found in lower, more ancient strata.

These patterns revealed much about the relative ages of sedimentary rocks as well as patterns in the evolution of life. But the geologists still could not tell how old the rocks actually were. A method of dating rocks did not become available until after radioactivity was discovered at the beginning of the twentieth century.

Radioisotopes provide a way to date rocks

Radioactive isotopes of atoms (see Section 2.1) decay in a predictable pattern over long time periods. During each successive time interval, or **half-life**, half of the remaining radioactive material of the radioisotope decays, either changing into another element or becoming the stable isotope of the same element (**Figure 21.1**).

Each radioisotope has a characteristic half-life (**Table 21.1**). To use a radioisotope to date a past event, we must know or estimate the concentration of the isotope at the time of that event. In the case of carbon, the production of new ^{14}C in the upper atmosphere (by the reaction of neutrons with ^{14}N) just balances the natural radioactive decay of ^{14}C. Therefore, the ratio of ^{14}C to its stable isotope, ^{12}C, is relatively constant in living organisms and their environment. However, as soon as an organism dies, it ceases to exchange carbon compounds with its environment. Its decaying ^{14}C is no longer replenished, and the ratio of ^{14}C to ^{12}C in its remains decreases through time. *Paleontologists* (scientists who study fossils) can use the ratio of ^{14}C to ^{12}C in fossil organisms to date fossils that are less than 50,000 years old (and thus the sedimentary rocks that contain those fossils) with a fair degree of certainty. After that time so little ^{14}C remains that the limits of detection are reached.

TABLE 21.1

Half-Lives of Some Radioisotopes

RADIOISOTOPE	HALF-LIFE
Phosphorus-32 (^{32}P)	14.3 days
Tritium (^{3}H)	12.3 years
Carbon-14 (^{14}C)	5,700 years
Potassium-40 (^{40}K)	1.3 billion years
Uranium-238 (^{238}U)	4.5 billion years

TABLE 21.2

Earth's Geological History

RELATIVE TIME SPAN	ERA	PERIOD	ONSET	MAJOR PHYSICAL CHANGES ON EARTH
	Cenozoic	Quaternary	1.8 mya	Cold/dry climate; repeated glaciations
		Tertiary	65 mya	Continents near current positions; climate cools
	Mesozoic	Cretaceous	145 mya	Northern continents attached; Gondwana begins to drift apart; meteorite strikes Yucatán Peninsula
		Jurassic	200 mya	Two large continents form: Laurasia (north) and Gondwana (south); climate warm
		Triassic	251 mya	Pangaea begins to slowly drift apart; hot/humid climate
	Paleozoic	Permian	297 mya	Continents aggregate into Pangaea; large glaciers form; dry climates form in interior of Pangaea
		Carboniferous	359 mya	Climate cools; marked latitudinal climate gradients
		Devonian	416 mya	Continents collide at end of period; meteorite probably strikes Earth
		Silurian	444 mya	Sea levels rise; two large continents form; hot/humid climate
		Ordovician	488 mya	Massive glaciation, sea level drops 50 meters
		Cambrian	542 mya	O_2 levels approach current levels
Precambrian	Precambrian		600 mya	O_2 level at >5% of current level
			1.5 bya	O_2 level at >1% of current level
			3.8 bya	O_2 first appears in atmosphere
			4.5 bya	

Note: mya, million years ago; bya, billion years ago.

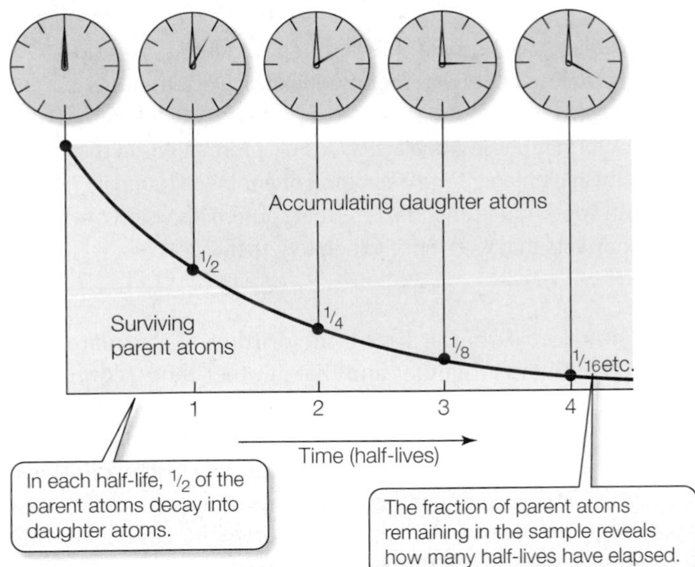

21.1 Radioactive Isotopes Allow Us to Date Ancient Rocks The decay of radioactive "parent" atoms into stable "daughter" isotopes happens at a steady rate known as the half-life. Different radioisotopes have different but characteristic half-lives that allow us to measure how much time has elapsed since the rocks containing such isotopes were laid down.

MAJOR EVENTS IN THE HISTORY OF LIFE
Humans evolve; many large mammals become extinct
Diversification of birds, mammals, flowering plants, and insects
Dinosaurs continue to diversify; flowering plants and mammals diversify; mass extinction at end of period (≈76% of species disappear)
Diverse dinosaurs; radiation of ray-finned fishes
Early dinosaurs; first mammals; marine invertebrates diversify; first flowering plants; mass extinction at end of period (≈65% of species disappear)
Reptiles diversify; amphibians decline; mass extinction at end of period (≈96% of species disappear)
Extensive "fern" forests; first reptiles; insects diversify
Fishes diversify; first insects and amphibians; mass extinction at end of period (≈75% of species disappear)
Jawless fishes diversify; first ray-finned fishes; plants and animals colonize land
Mass extinction at end of period (≈75% of species disappear)
Most animal phyla present; diverse photosynthetic protists
Ediacaran fauna
Eukaryotes evolve; several animal phyla appear
Origin of life; prokaryotes flourish

Radioisotope dating methods have been expanded and refined

Sedimentary rocks are formed from materials that existed for varying lengths of time before being transported, sometimes over long distances, to the site of their deposition. Therefore, the isotopes in a sedimentary rock do not contain reliable information about the date of its formation. Dating rocks more ancient than 50,000 years requires estimating isotope concentrations in *igneous* rocks (rocks formed when molten material cools). To date sedimentary rocks, geologists search for places where volcanic ash or lava flows have intruded into beds of those rocks.

A preliminary estimate of the age of the igneous rock determines which isotope is used to date it. The decay of potassium-40 to argon-40 has been used to date most of the ancient events in the evolution of life. Fossils in the adjacent sedimentary rock that are similar to those in other rocks of known ages provide additional clues.

Radioisotope dating of rocks, combined with fossil analysis, is the most powerful method of determining geological age. But in places where sedimentary rocks do not contain suitable igneous intrusions and few fossils are present, paleontologists turn to other dating methods. One method, known as *paleomagnetic dating*, relates the ages of rocks to patterns in Earth's magnetism, which change over time. Earth's magnetic poles move and occasionally reverse themselves. Because both sedimentary and igneous rocks preserve a record of Earth's magnetic field at the time they were formed, paleomagnetism helps determine the ages of those rocks. Other dating methods, which we will describe in later chapters, use continental drift, sea level changes, and molecular clocks.

Using these methods, geologists divided the history of life into geological *eras*, which in turn are subdivided into *periods* (**Table 21.2**). The boundaries between these divisions are based on the striking differences scientists observed in the assemblages of fossil organisms contained in successive layers of rocks; hence the suffix *zoic* ("of life") for the eras. Geologists established and named these divisions before they knew the ages of the eras and periods, and we are constantly refining the dates used for these boundaries as new discoveries are made.

21.1 RECAP

Fossils in sedimentary rocks enabled geologists to determine the relative ages of organisms, but absolute dating was not possible until the discovery of radioactivity. Geologists divide the history of life into eras and periods, based on assemblages of fossil organisms found in successive layers of rocks.

- What observations about fossils suggested to geologists that they could be used to determine the relative ages of rocks? See p. 465
- How is the rate of decay of radioisotopes used to estimate the absolute ages of rocks? See pp. 466–467 and Figure 21.1

The scale at the left of Table 21.2 gives a relative sense of geological time, especially the vast expanse of the Precambrian era, during which early life evolved amid stupendous physical changes. Earth continued to undergo physical changes that influenced the evolution of life, and these physical events and important milestones are listed in the table. Now we will describe the most important of these changes in more detail.

21.2 How Have Earth's Continents and Climates Changed over Time?

The maps and globes that adorn our walls, shelves, and books give an impression of a static Earth. It would be easy for us to assume that the continents have always been where they are. But we would be wrong. The idea that Earth's land masses have changed position over the millennia, and that they continue to do so, was first put forth in 1912 by the German meteorologist and geophysicist Alfred Wegener. His book, *The Origin of Continents and Oceans*, was initially met with skepticism and resistance. By the 1960s, however, physical evidence and increased understanding of the geophysics of *plate tectonics* had convinced virtually all geologists of the reality of Wegener's vision.

Earth's crust consists of a number of solid *plates* approximately 40 kilometers thick, which collectively make up the *lithosphere*. The lithospheric plates float on a fluid layer of molten rock, or *magma* (**Figure 21.2**). The magma circulates because heat produced by radioactive decay deep in Earth's core sets up convection currents in the fluid. The plates move because magma rises and exerts tremendous pressure. Where plates are pushed together, either they move sideways past each other, or one plate slides under the other, pushing up mountain ranges and carving deep *rift valleys*. (When they occur under water, such valleys are known as *trenches*.) Where plates are pushed apart, ocean basins may form between them. The movement of the lithospheric plates and the continents they contain is known as **continental drift**.

The idea of continental drift captured Wegener as he studied the close fit between the coastlines of western Africa and eastern South America. Geological evidence linking rock formations in Appalachia to similar formations in the Scottish highlands, and geological phenomena found in both South Africa and Brazil, further spurred his vision of continents that were once joined together.

We now know that at times, the drifting of the plates has brought continents together and that at other times, continents have drifted apart. The positions and sizes of the continents influence oceanic circulation patterns, sea levels, and global climates. Mass extinctions of species, particularly marine organisms, have usually accompanied major drops in sea level, which exposed vast areas of the continental shelves, killing the marine organisms that lived in the shallow seas that had covered them (**Figure 21.3**).

Oxygen has steadily increased in Earth's atmosphere

The continents have moved irregularly over Earth's surface, but some physical changes, such as the increase in atmospheric oxygen, have been largely unidirectional. The atmosphere of early Earth probably contained little or no free oxygen gas (O_2). The increase in atmospheric oxygen came in two big steps more than a billion years apart. The first step occurred about 2.4 billion years ago, when certain bacteria evolved the ability to use water as the source of hydrogen ions for photosynthesis. By chemically splitting H_2O, these bacteria generated atmospheric O_2 as a waste product. They also made electrons available for reducing CO_2 to form organic compounds (see Section 8.3).

One group of oxygen-generating bacteria, the *cyanobacteria*, formed rocklike structures called *stromatolites*, which are abun-

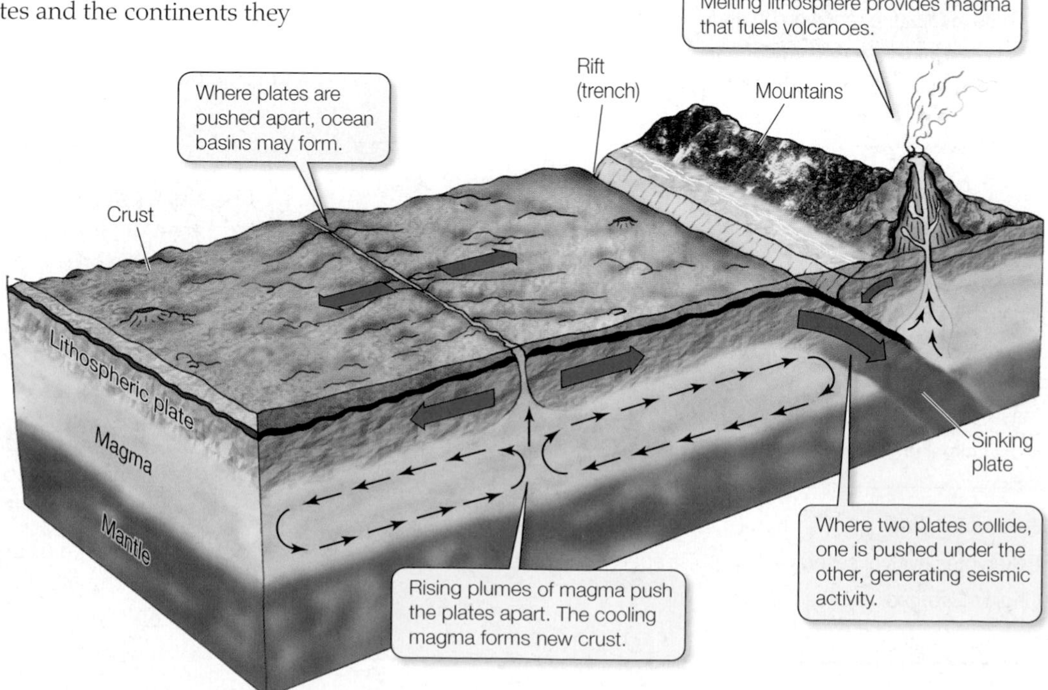

Melting lithosphere provides magma that fuels volcanoes.
Rift (trench)
Mountains
Where plates are pushed apart, ocean basins may form.
Crust
Lithospheric plate
Magma
Mantle
Sinking plate
Where two plates collide, one is pushed under the other, generating seismic activity.
Rising plumes of magma push the plates apart. The cooling magma forms new crust.

21.2 Plate Tectonics and Continental Drift The heat of Earth's core generates convection currents (arrows) that push the lithospheric plates, along with the land masses lying on them, together or apart. When lithospheric plates collide, one slides under the other. The resulting seismic activity can create mountains and deep rift valleys (oceanic trenches).

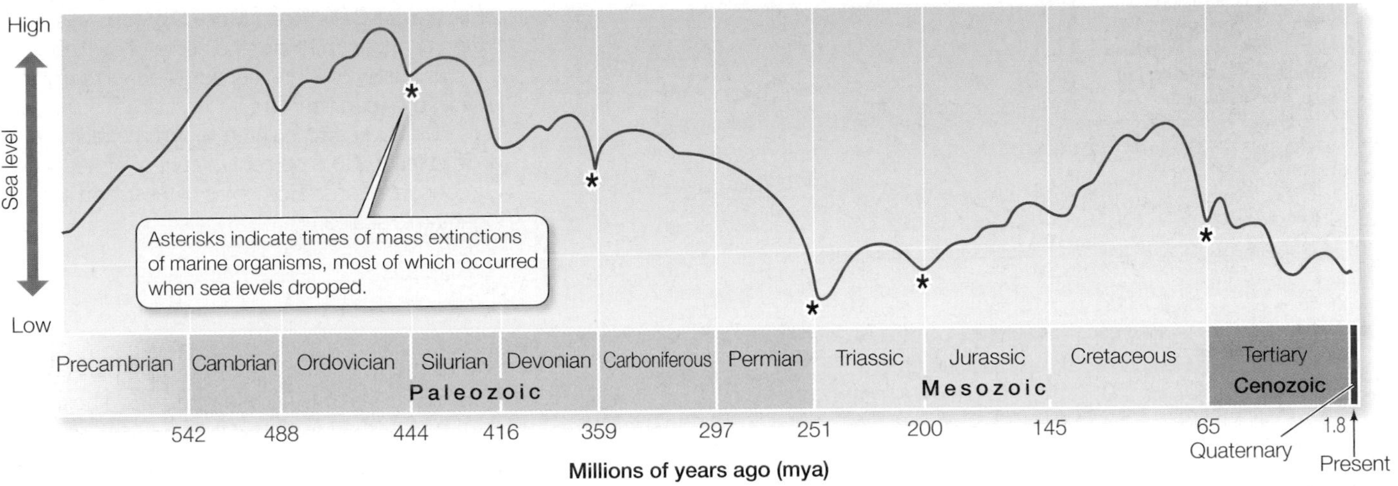

21.3 Sea Levels Have Changed Repeatedly Most mass extinctions (indicated by asterisks) have coincided with periods of low sea levels.

dantly preserved in the fossil record. Cyanobacteria are still forming stromatolites today in a few very salty places on Earth (**Figure 21.4**). Cyanobacteria liberated enough O_2 to open the way for the evolution of oxidation reactions as the energy source for the synthesis of ATP (see Section 7.1). Their ability to split water doubtless contributed to their extraordinary success.

Thus the evolution of life irrevocably changed the physical nature of Earth. Those physical changes, in turn, influenced the evolution of life. When it first appeared in the atmosphere, oxygen was poisonous to the anaerobic prokaryotes that inhabited Earth at the time. Those prokaryotes that evolved the ability to metabolize O_2 not only survived, but also gained a number of advantages. Aerobic metabolism proceeds at more rapid rates and harvests energy more efficiently than anaerobic metabolism (see Section 7.5). Consequently, organisms with aerobic metabolism replaced anaerobes in most of Earth's environments.

An atmosphere rich in O_2 also made possible larger cells and more complex organisms. Small unicellular aquatic organisms can obtain enough O_2 by simple diffusion even when O_2 concentrations are very low. Larger unicellular organisms have lower surface area-to-volume ratios (see Figure 4.2). In order to obtain enough O_2 by simple diffusion, they must live in an environment with a relatively high concentration of O_2. Bacteria can thrive on 1 percent of the current atmospheric O_2 levels; eukaryotic cells require oxygen levels that are at least 2 to 3 percent of current atmospheric concentrations. (For concentrations of dissolved O_2 in the oceans to reach these levels, much higher atmospheric concentrations were needed.)

Probably because it took many millions of years for Earth to develop an oxygenated atmosphere, only unicellular prokaryotes lived on Earth for more than 2 billion years.

(A)

(B)

21.4 Stromatolites (A) A vertical section through a fossil stromatolite. (B) These rocklike structures are living stromatolites that thrive in the very salty waters of Shark Bay, Western Australia. Layers of cyanobacteria are found in the uppermost parts of the structures.

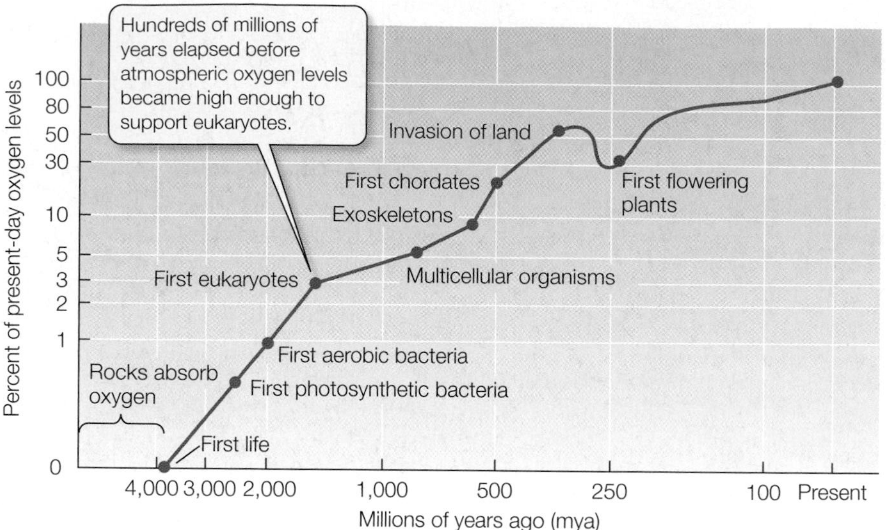

21.5 Larger Cells Need More Oxygen As oxygen concentrations in the atmosphere rose, the complexity of life increased. Although aerobic prokaryotes can flourish with less, larger eukaryotic cells with lower surface area-to-volume ratios require at least 2 to 3 percent of current atmospheric O_2 concentrations. (Both axes of the graph are logarithmic scales.)

About a billion years ago (1 bya), atmospheric O_2 concentrations became high enough for large eukaryotic cells to flourish (**Figure 21.5**). Further increases in atmospheric O_2 levels 700 to 570 million years ago (mya) enabled multicellular organisms to evolve.

In contrast to this largely unidirectional change in atmospheric O_2 concentration, most physical conditions on Earth have oscillated in response to the planet's internal processes, such as volcanic activity and continental drift. Extraterrestrial events, such as collisions with meteorites, have also left their mark. In some cases, as we will see later in this chapter, these events caused **mass extinctions**, during which a large proportion of the species living at the time disappeared. After each mass extinction, the diversity of life rebounded, but recovery took millions of years.

Earth's climate has shifted between hot/humid and cold/dry conditions

Through much of its history, Earth's climate was considerably warmer than it is today, and temperatures decreased more gradually toward the poles. At other times, however, Earth was colder than it is today. Large areas were covered with glaciers during the end of the Precambrian and during parts of the Carboniferous and Permian periods. These cold periods were separated by long periods of milder climates (**Figure 21.6**). For Earth to be in a cold, dry state, atmospheric CO_2 levels had to have been unusually low, but scientists do not know what caused such low concentrations. Because we are living in one of the colder periods in the history of Earth, it is difficult for us to imagine the mild climates that were found at high latitudes during much of the history of life. During the Quaternary period there have been a series of glacial advances, interspersed with warmer interglacial intervals during which the glaciers retreated.

Weather often changes rapidly; climates usually change slowly. Major climatic shifts have taken place over periods as short as 5,000 to 10,000 years, however, primarily as a result of changes in Earth's orbit around the sun. A few climatic shifts have been even more rapid. For example, during one Quaternary interglacial period, the Antarctic Ocean changed from being ice-covered to being nearly ice-free in less than a hundred years. Such rapid changes are usually caused by sudden shifts in ocean currents. Some climate changes have been so rapid that the extinctions caused by them appear "instantaneous" in the fossil record.

21.6 Hot/Humid and Cold/Dry Conditions Have Alternated over Earth's History Throughout Earth's history, periods of cold climates and glaciations (white depressions) have been separated by long periods of milder climates.

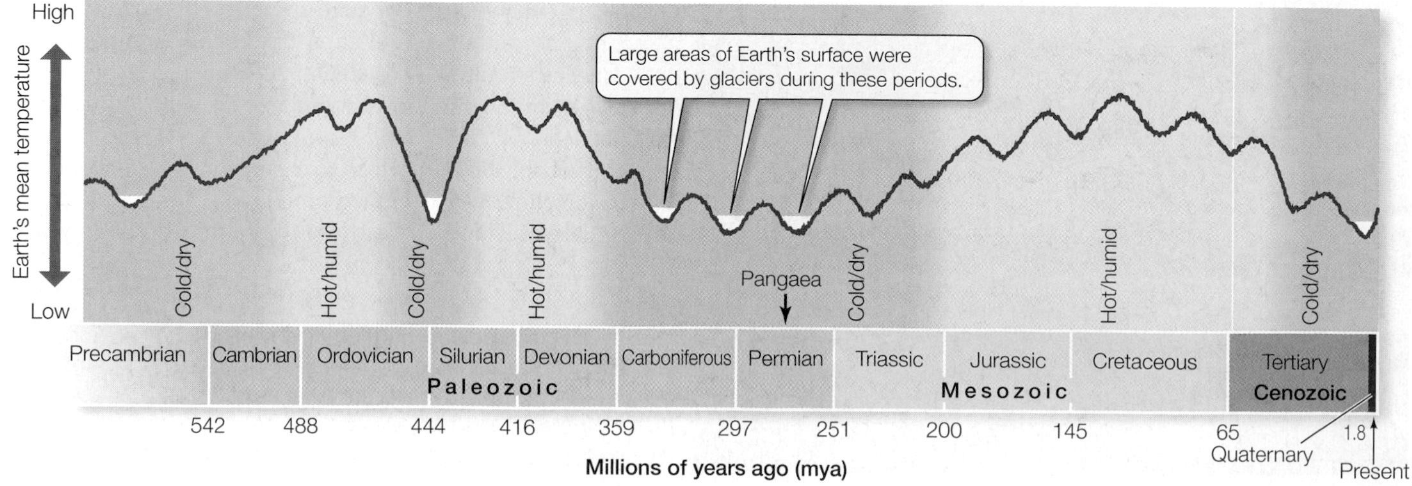

We are living today in a time of rapid climate change caused by a buildup of atmospheric CO_2, primarily from the burning of fossil fuels. The current concentration of atmospheric CO_2 is greater than at any time in many thousands of years, except for one warm interval 5,000 years ago, when the level may have been slightly higher than it is today. A doubling of the atmospheric CO_2 concentration, which may happen during the current century unless major efforts are made to reduce human consumption of fossil fuels, would probably increase the average temperature of Earth, cause droughts in the central regions of continents, increase precipitation in coastal areas, melt glaciers and ice caps, and result in sea level rises that would flood coastal cities and agricultural land. The possible consequences of such climate changes will be discussed in Chapters 56 and 57.

Volcanoes have occasionally changed the history of life

Most volcanic eruptions produce only local or short-lived effects, but a few very large volcanic eruptions have had major consequences for life. The collision of continents during the Permian period (about 275 mya) to form a single, gigantic land mass, called **Pangaea**, caused massive volcanic eruptions. The ash ejected by volcanoes into the atmosphere reduced the penetration of sunlight to Earth's surface, lowering temperatures, reducing photosynthesis, and triggering massive glaciation. Massive volcanic eruptions also occurred as the continents drifted apart during the late Triassic period and at the end of the Cretaceous.

Extraterrestrial events have triggered changes on Earth

At least 30 meteorites between the sizes of baseballs and soccer balls hit Earth each year. Collisions with large meteorites are rare, but large meteorites have probably been responsible for several mass extinctions. Several types of evidence tell us about these collisions. Their craters, and the dramatically disfigured rocks that resulted from their impact, are found in many places. Geologists have also discovered giant molecules that contain trapped helium and argon with isotope ratios characteristic of meteorites, which are very different from the ratios found on Earth.

A meteorite caused or contributed to a mass extinction at the end of the Cretaceous period (about 65 mya). The first clue that a meteorite was responsible came from the abnormally high concentrations of the element iridium in a thin layer separating rocks deposited during the Cretaceous from those deposited during the Tertiary (**Figure 21.7**). Iridium is abundant in some meteorites, but is exceedingly rare on Earth's surface. Subsequently, scientists discovered a circular crater 180 kilometers in diameter buried beneath the northern coast of the Yucatán Peninsula of Mexico (see p. 161). When it collided with Earth, the meteorite released energy equivalent to that of 100 million megatons of high explosives, creating great tsunamis. A massive plume of debris swelled to a diameter of up to 200 kilometers, spread around Earth, and descended. The descending debris heated the atmosphere to several hundred degrees, ignited massive fires, and blocked the sun, preventing plants from photosynthesizing. The settling debris formed the iridium-rich layer. About a billion tons of soot, which has a composition that matches smoke from forest fires, was also deposited. Many fossil species, particularly dinosaurs, found in Cretaceous rocks are not found in the Tertiary rocks of the next layer.

A thin band rich in iridium marks the boundary between rocks deposited in the Cretaceous and Tertiary periods.

21.7 Evidence of a Meteorite Impact Iridium is a metal common in some meteorites, but rare on Earth. Its high concentration in sediments deposited about 65 million years ago suggests the impact of a large meteorite.

21.2 RECAP

Conditions on Earth have changed dramatically over time. Some changes, such as increases in atmospheric concentrations of oxygen, have been primarily unidirectional, but other factors, such as climate, have repeatedly oscillated.

- Can you describe how increases in atmospheric concentrations of oxygen affected the evolution of multicellular organisms? See pp. 469–470 and Figure 21.5
- How have volcanic eruptions and collisions with meteorites influenced the course of life's evolution? See p. 471

The many dramatic physical events of Earth's history have influenced the nature and timing of evolutionary changes among Earth's living organisms. We now will look more closely at some of the major events that characterize the history of life on Earth.

21.3 What Are the Major Events in Life's History?

Life first evolved on Earth about 3.8 billion years ago. By about 1.5 billion years ago, eukaryotic organisms had evolved (see Figure 21.5). The fossil record of organisms that lived prior to 550 million years ago is fragmentary, but it is good enough to show that the total number of species and individuals increased dramatically in late Precambrian times. As we saw above, pre-Darwinian geologists divided geological history into eras and periods based on their distinct fossil assemblages. Biologists refer to the assemblage of all organisms of all kinds living at a particular time or place as the **biota** of that time or place. All of the plants living at a particular time or place are its **flora**; all of the animals are its **fauna**. Table 21.2 describes some of the physical and biological changes, such as mass extinctions and dramatic increases in the diversity of major groups of organisms, associated with each unit of time.

About 300,000 species of fossil organisms have been described, and the number is growing steadily. However, this number is only a tiny fraction of the species that have ever lived. We do not know how many species lived in the past, but we have ways of making reasonable estimates. Of the present-day biota, approximately 1.7 million species have been named. The actual number of living species is probably at least 10 million, because most species of insects and mites (the animal groups with the largest numbers of species; see Chapter 32) have not yet been described. So the number of described fossil species is less than 2 percent of the probable minimum number of living species. Life has existed on Earth for about 3.8 billion years. Species last, on average, less than 10 million years; therefore, Earth's biota must have turned over many times during geological history. So the total number of species that have lived over evolutionary time must vastly exceed the number living today. Why have so few species been described from fossils?

Several processes contribute to the paucity of fossils

Only a tiny fraction of organisms ever become fossils, and only a tiny fraction of those fossils are ever studied by paleontologists. Most organisms live and die in oxygen-rich environments in which they quickly decompose. They are not likely to become fossils unless they are transported by wind or water to sites that lack oxygen, where decomposition proceeds slowly or not at all. Furthermore, geological processes often transform rocks, destroying the fossils they contain, and many fossil-bearing rocks are deeply buried and inaccessible. Paleontologists have studied only a tiny fraction of the sites that contain fossils, but they find and describe many new fossils every year.

The number of known fossils is especially large for marine animals that had hard skeletons (which resist decomposition). Among the nine major animal groups with hard-shelled members, approximately 200,000 species have been described from fossils—roughly twice the number of living marine species in these same groups. Paleontologists lean heavily on these groups in their interpretations of the evolution of life. Insects and spiders are also relatively well represented in the fossil record (**Figure 21.8**). The fossil record, though incomplete, is good enough to demonstrate clearly that organisms of particular types are found in rocks of specific ages and that newer organisms appear sequentially in younger rocks.

By combining information about geological changes during Earth's history with evidence from the fossil record, scientists have composed portraits of what Earth and its inhabitants may have looked like at different times. We know in general where the continents were and how life changed over time, but many of the details are poorly known, especially for events in the more remote past. In this section we will provide an overview of how life has changed during its history on Earth. In Part Six of this book we will discuss the evolutionary history of particular groups of organisms in more detail.

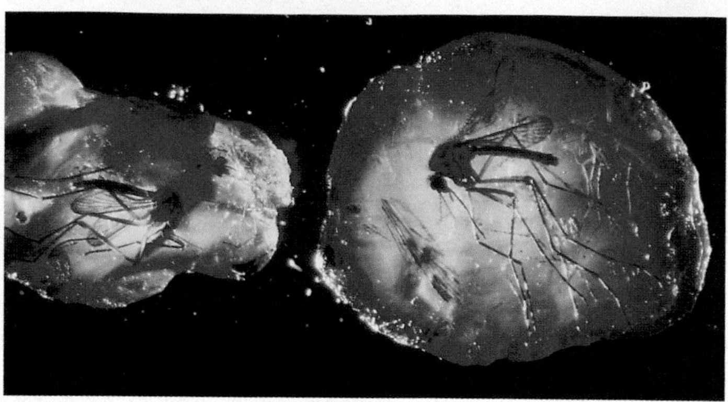

21.8 Insect Fossils These chunks of amber—fossilized tree resin—contain insects that were preserved when they were trapped in the sticky resin some 50 million years ago.

Precambrian life was small and aquatic

For most of its history, life was confined to the oceans, and all organisms were small. Over the long ages of the Precambrian—more than three billion years—the shallow seas slowly began to teem with life. For most of the **Precambrian**, life consisted of microscopic prokaryotes; eukaryotes probably evolved about two-thirds of the way through the era. Unicellular eukaryotes and small multicellular animals fed on floating photosynthetic microorganisms. Small floating organisms, known collectively as *plankton*, were eaten by slightly larger animals that filtered them from the water. Other animals ingested sediments on the seafloor and digested the remains of organisms within them. By the late Precambrian (about 650 mya), many kinds of multicellular soft-bodied animals had evolved. Some of them were very different from any animals living today, and may be members of groups that have no living descendants (**Figure 21.9**).

Life expanded rapidly during the Cambrian period

The **Cambrian** period (542–488 mya) marks the beginning of the **Paleozoic** era. The O_2 concentration in the Cambrian atmosphere was approaching its current level, and the continents had come together to form several large land masses. The largest of these land masses was called *Gondwana* (**Figure 21.10A**). A rapid diversification of life took place that we now refer to as the **Cambrian explosion** (although in fact it began before the Cambrian). Most of the major groups of animals that have species living today appeared during the Cambrian.

For the most part, fossils tell us only about the hard parts of organisms, but in three Cambrian fossil beds—the Burgess Shale in British Columbia, Sirius Passet in northern Greenland, and the Chengjiang site in southern China—the soft parts of many ani-

Spriggina floundersi

Mawsonites

21.9 Ediacaran Animals These fossils of soft-bodied invertebrates, excavated at Ediacara in southern Australia, were formed 600 million years ago. They illustrate the diversity of life that evolved in the Precambrian era.

mals were preserved (**Figure 21.10B**). Arthropods (crabs, shrimps, and their relatives) are the most diverse group in the Chinese fauna; some of them were large carnivores. Trilobites, members of an arthropod group that was abundant and diverse during the Cambrian (see Figure 32.21), suffered a major reduction at the end of the Cambrian, but they recovered and continued to be abundant until the end of the Permian, when they became extinct.

Many groups of organisms diversified

Geologists divide the remainder of the Paleozoic era into the Ordovician, Silurian, Devonian, Carboniferous, and Permian periods (see Table 21.2). Each period is characterized by the diversification of specific groups of organisms. Mass extinctions marked the ends of the Ordovician, Devonian, and Permian.

THE ORDOVICIAN (488–444 MYA) During the **Ordovician** period, the continents, which were located primarily in the Southern Hemisphere, still lacked multicellular plants. Evolutionary radiation of marine organisms was spectacular during the early Ordovician, especially among animals, such as brachiopods and mollusks, that lived on the seafloor and filtered small prey from the water. At the end of the Ordovician, as massive glaciers formed over Gondwana, sea levels were lowered about 50 meters, and ocean temperatures dropped. About 75 percent of the animal species became extinct, probably because of these major environmental changes.

THE SILURIAN (444–416 MYA) During the **Silurian** period, the northernmost continents coalesced, but the general positions of the continents did not change much. Marine life rebounded from the mass extinction at the end of the Ordovician. Animals able to swim and feed above the ocean bottom appeared for the first time, but no new major groups of marine organisms evolved. The tropical sea was uninterrupted by land barriers, and most marine organisms were widely distributed. On land, the first vascular plants appeared late in the Sil-

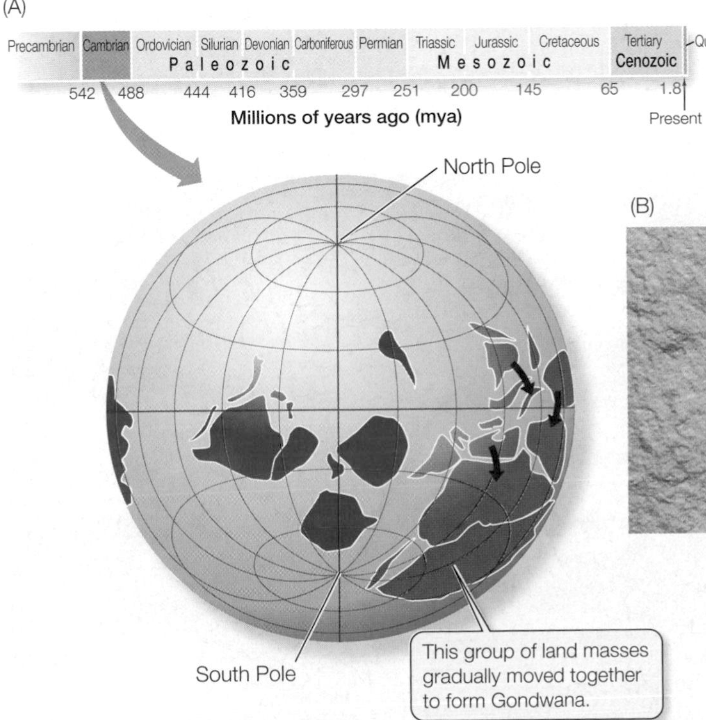

(B)

21.10 Cambrian Continents and Fauna (A) Positions of the continents during the middle of the Cambrian period (542–488 mya). This view of Earth has been distorted so that you can see both poles. (B) Fossil beds in China have yielded well-preserved remains of Cambrian animals such as this one, called *Jianfangia*.

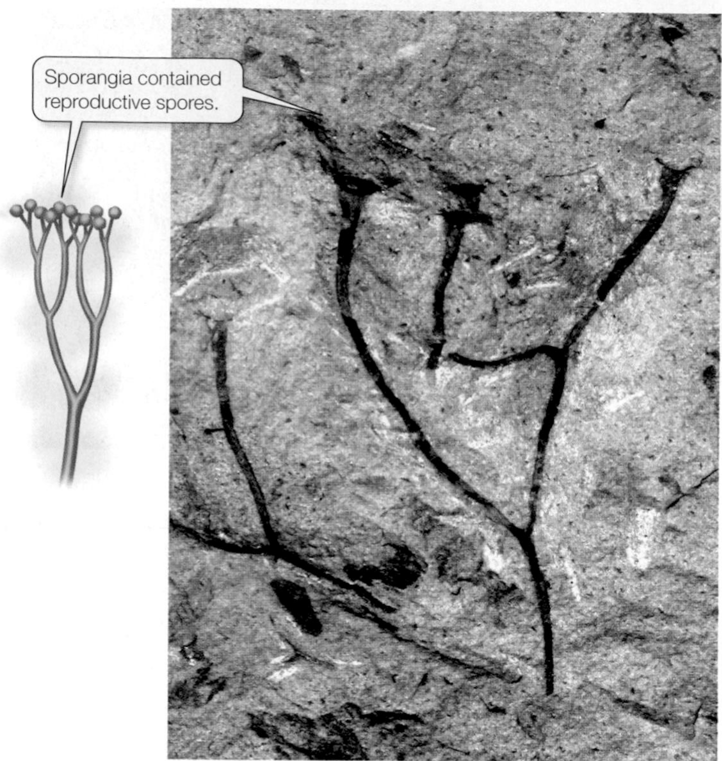

21.11 *Cooksonia*, the Earliest Known Vascular Plant These early plants were small and very simple in structure. However, they were true vascular plants (tracheophytes) with internal water-conducting cells (tracheids), well equipped to make the move from the aquatic to the terrestrial environment. This fossil of *Cooksonia pertoni* is from the Silurian period (about 420 mya).

urian period (about 420 mya). These plants were less than 50 cm tall and lacked roots and leaves (**Figure 21.11**). The first terrestrial arthropods—scorpions and millipedes—appeared at about the same time.

THE DEVONIAN (416–359 MYA) Rates of evolutionary change accelerated in many groups of organisms during the **Devonian** period. The northern land mass (called *Laurasia*) and the southern land mass (Gondwana) moved slowly toward each other (**Figure 21.12A**). There were great evolutionary radiations of corals and shelled squidlike cephalopods (**Figure 21.12B**). Fishes diversified as jawed forms replaced jawless ones, and heavy armor gave way to the less rigid outer coverings of modern fishes. All current major groups of fishes were present by the end of the period.

Terrestrial communities also changed dramatically during the Devonian. Club mosses, horsetails, and tree ferns became common toward the end of the Devonian; some attained the size of trees. Their deep roots accelerated the weathering of rocks, resulting in the development of the first forest soils. Distinct floras evolved on Laurasia and Gondwana toward the end of the period. A wind-pollinated precursor of seed-bearing plants, *Runcaria*, was found in sediments in Belgium laid down 385 million years ago. The ancestors of gymnosperms, the first plants to produce seeds, appeared later in the Devonian. The first known fossils of centipedes, spiders, mites, and insects date to this period. Fishlike amphibians began to occupy the land.

An extinction of about 75 percent of all marine species marked the end of the Devonian. Paleontologists are uncertain about the cause of this mass extinction, but two large meteorites that collided with Earth at that time, one in present-day Nevada and the other in Western Australia, may have been responsible, or at least a contributing factor.

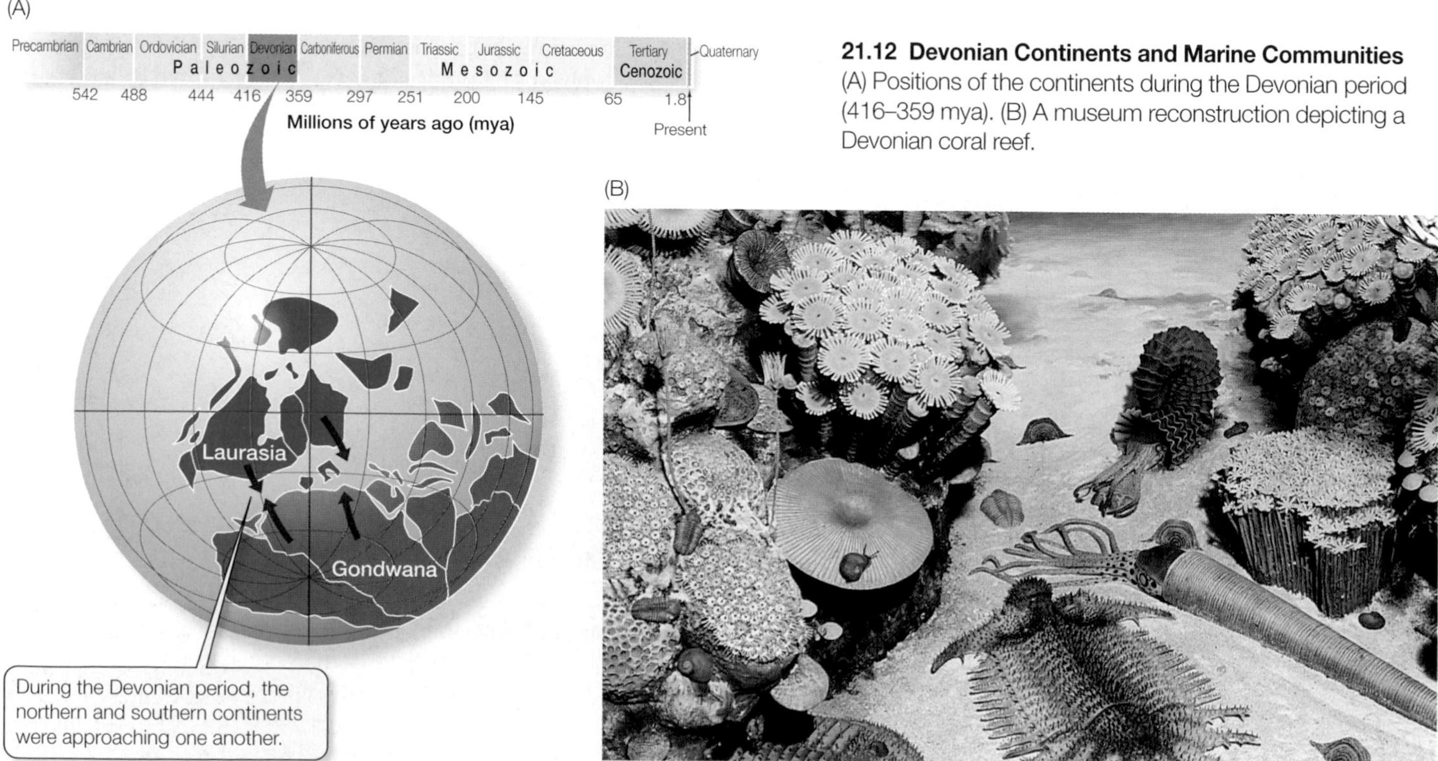

21.12 Devonian Continents and Marine Communities (A) Positions of the continents during the Devonian period (416–359 mya). (B) A museum reconstruction depicting a Devonian coral reef.

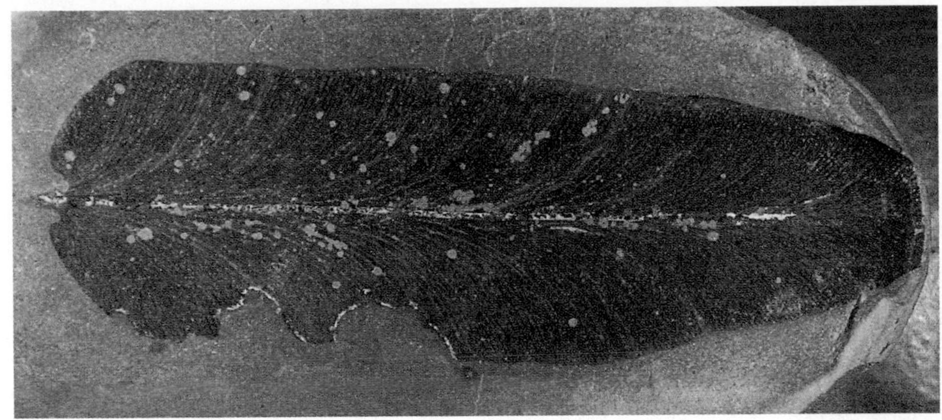

21.13 Evidence of Insect Diversification The margins of this fossil fern leaf from the Carboniferous period have been chewed by insects.

THE CARBONIFEROUS (359–297 MYA) Large glaciers formed over high-latitude Gondwana during the **Carboniferous** period, but extensive swamp forests grew on the tropical continents. These forests were not made up of the kinds of trees we know today, but were dominated by giant tree ferns and horsetails with small leaves (see Figure 28.8). Fossilized remains of those trees formed the coal we now mine for energy.

The diversity of terrestrial animals increased greatly during the Carboniferous. Snails, scorpions, centipedes, and insects were abundant and diverse. Insects evolved wings, becoming the first animals to fly. Flight gave them access to tall plants; plant fossils from this period show evidence of chewing by insects (**Figure 21.13**). Amphibians became larger and better adapted to terrestrial existence after splitting from the lineage leading to the *amniotes*, vertebrates with well-protected eggs that can be laid in dry places. In the seas, crinoids (sea lilies and feather stars) reached their greatest diversity, forming "meadows" on the seafloor (**Figure 21.14**).

THE PERMIAN (297–251 MYA) During the **Permian** period, the continents coalesced into the supercontinent *Pangaea* (**Figure 21.15**). Permian rocks contain representatives of most modern groups of insects. By the end of the period, one amniote group, the reptiles, greatly outnumbered the amphibians. Late in the period, the lineage leading to mammals diverged from one reptilian group. In fresh waters, the Permian period was a time of extensive diversification of ray-finned fishes.

Conditions for life deteriorated toward the end of the Permian. Massive volcanic eruptions resulted in outpourings of lava that covered large areas of Earth. The ash the volcanoes produced blocked sunlight and cooled the climate, resulting in the largest glaciers in Earth's history. Atmospheric oxygen concentrations gradually dropped from about 30 percent to about 12 percent. At such low oxygen concentrations, most animals would have been unable to survive at elevations above 500 meters; thus about half of the Permian land area would have been uninhabitable. The com-

21.14 A Carboniferous "Crinoid Meadow" Crinoids—the flowerlike organisms—were dominant marine animals during the Carboniferous period and may have formed communities similar to this one.

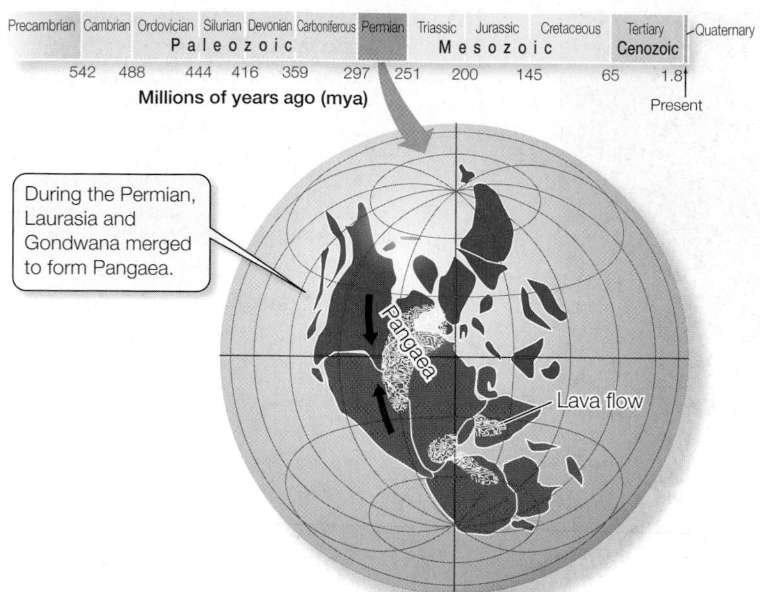

21.15 Pangaea Formed in the Permian Period At the end of the Permian period, massive lava flows spread over Earth, and the largest glaciers in Earth's history formed.

bination of these changes resulted in the most drastic mass extinction event in Earth's history. Many species became extinct simultaneously at the end of the period; others disappeared gradually over a period of several million years.

The Permian extinction may have come perilously close to wiping out life on Earth. Scientists estimate that about 96 percent of all species became extinct at that time.

Geographic differentiation increased during the Mesozoic era

The few organisms that survived the Permian mass extinction found themselves in a relatively empty world at the start of the **Mesozoic** era (251 mya). As Pangaea slowly separated into individual continents, the oceans rose and reflooded the continental shelves, forming huge, shallow inland seas. Atmospheric oxygen concentrations gradually rose to their former levels. Life again proliferated and diversified, but different groups of organisms came to dominate Earth. The three groups of *phytoplankton* (floating photosynthetic organisms) that dominate today's oceans—dinoflagellates, coccolithophores, and diatoms—became ecologically important at this time. New seed-bearing plants replaced the trees that had dominated the Permian forests.

During the Mesozoic, Earth's biota, which until that time had been relatively homogeneous, became increasingly **provincialized**; that is, distinct terrestrial biotas evolved on each continent. The biotas of the shallow waters bordering the continents also diverged from one another. The provincialization that began during the Mesozoic continues to influence the geography of life today. By the end of the era, the continents were close to their present positions, and many organisms looked similar to those living today.

The Mesozoic era is divided into three periods: the Triassic, Jurassic, and Cretaceous. The Triassic and Cretaceous were terminated by mass extinctions, probably caused by meteorite impacts.

THE TRIASSIC (251–200 MYA) Pangaea began to break apart during the **Triassic** period. Many invertebrate groups became more species-rich, and many burrowing animals evolved from groups living on the surfaces of seafloor sediments. On land, conifers and pteridosperms became the dominant trees. The first frogs and turtles appeared. A great radiation of reptiles began, which eventually gave rise to crocodilians, dinosaurs, and birds. The end of the Triassic was marked by a mass extinction that eliminated about 65 percent of the species on Earth. A large meteorite that crashed into what is now Quebec may have been responsible.

THE JURASSIC (200–145 MYA) During the **Jurassic** period, the land was once again divided in two large continents—Laurasia in the north, and Gondwana in the south. Ray-finned fishes began the great radiation that led to their dominance of the oceans. The first salamanders and lizards appeared, and flying reptiles (pterosaurs)

21.16 Jurassic Parkland The dinosaurs of the Mesozoic have captured human imaginations ever since their fossils were first discovered. This illustrations depicts dinosaurs that lived some 160 million years ago (the Jurassic period) on what are now the western plains of North America. In the foreground, a *Ceratosaurus* and two small *Coelurus* feed on the carcass of an *Apatosaurus.* In the background are (left to right) two *Camptosaurus*, *Stegosaurus*, *Brachiosaurus*, and another *Apatosaurus*.

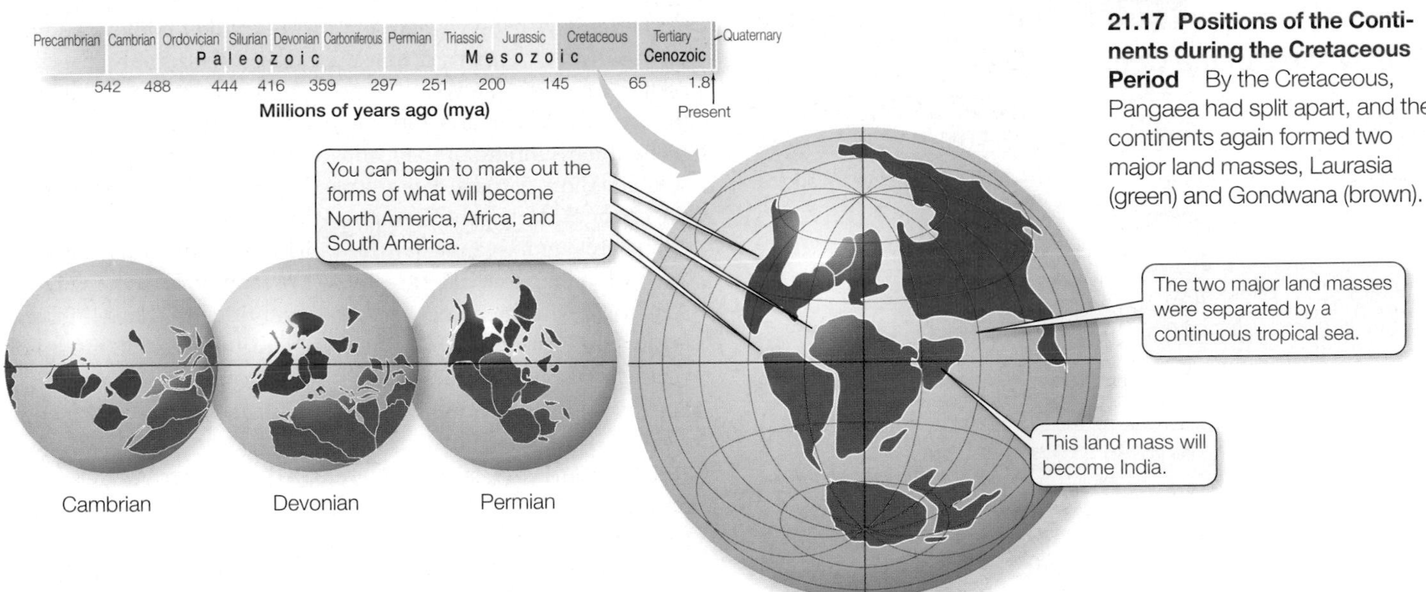

542 488 444 416 359 297 251 200 145 65 1.8

You can begin to make out the forms of what will become North America, Africa, and South America.

This land mass will become India.

21.17 Positions of the Continents during the Cretaceous Period By the Cretaceous, Pangaea had split apart, and the continents again formed two major land masses, Laurasia (green) and Gondwana (brown).

evolved. Dinosaur lineages evolved into predators that walked on two legs and large herbivores that walked on four legs (**Figure 21.16**). Several groups of mammals first appeared during this time. Plant evolution continued with the emergence of the flowering plants that dominate Earth's vegetation today.

THE CRETACEOUS (145–65 MYA) By the early **Cretaceous** period, Laurasia was completely separate from Gondwana, which was beginning to break apart. A continuous sea encircled the tropics (**Figure 21.17**). Sea levels were high, and Earth was warm and humid. Life proliferated both on land and in the oceans. Marine invertebrates increased in diversity and in number of species. On land, dinosaurs continued to diversify. The first snakes appeared during the Cretaceous, but the modern groups with the most species resulted from a later radiation. Early in the Cretaceous, flowering plants began the radiation that led to their current dominance on land. Fossils of the earliest known flowering plants, dated at 124 million years ago, were recently discovered in Liaoning Province in northeastern China (**Figure 21.18**). By the end of the period, many groups of mammals had evolved. Most of them were small, but one species recently discovered in China, *Repenomamus giganticus*, was large enough to capture and eat young dinosaurs.

As described earlier in this chapter, another meteorite-caused mass extinction took place at the end of the Cretaceous period. In the seas, many planktonic organisms and bottom-dwelling invertebrates became extinct. On land, all animals larger than about 25 kilograms in body weight apparently became extinct. Many species of insects died out, perhaps because the growth of their food plants was greatly reduced following the impact. Some species survived in the northern parts of North America and Eurasia, areas that were not subjected to the devastating fires that engulfed most low-latitude regions.

The modern biota evolved during the Cenozoic era

By the early **Cenozoic** era (65 mya), the positions of the continents resembled those of today, but Australia was still attached to Antarctica, and the Atlantic Ocean was much narrower. The Cenozoic era was characterized by an extensive radiation of mammals, but other groups were also undergoing important changes.

Flowering plants diversified extensively and came to dominate world forests, except in cool regions. Mutations of two genes in one group of plants allowed them to use atmospheric N_2 directly by forming symbioses with a few species of nitrogen-fixing bacteria (see Section 36.4). The evolution of this symbiosis between certain early Cenozoic plants and these specialized bacteria was

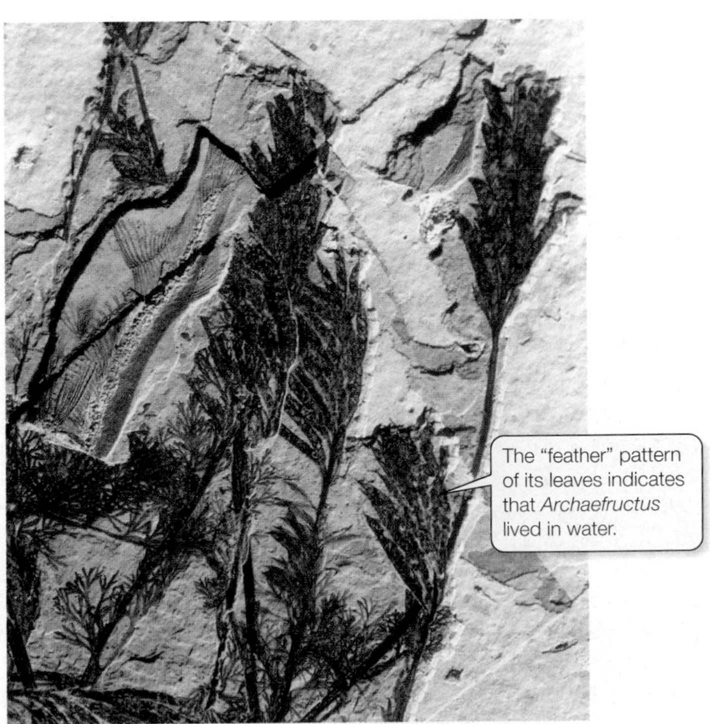

The "feather" pattern of its leaves indicates that *Archaefructus* lived in water.

21.18 Flowering Plants of the Cretaceous These fossils of *Archaefructus* are the earliest known examples of flowering plants, the type of plants most prevalent on Earth today.

TABLE 21.3

Subdivisions of the Cenozoic Era

PERIOD	EPOCH ONSET (MYA)	EPOCH ONSET (MYA)
Quaternary	Holocene[a]	0.01 (~10,000 years ago)
	Pleistocene	1.8
Tertiary	Pliocene	5.3
	Miocene	23
	Oligocene	34
	Eocene	55.8
	Paleocene	65

[a] The Holocene is also known as the Recent.

the first "green revolution" and dramatically increased the amount of nitrogen available for terrestrial plant growth.

The Cenozoic era is divided into two periods, the Tertiary and the Quaternary. Because both the fossil record and our subsequent knowledge of evolutionary history become more extensive closer to our own time, paleontologists have subdivided these periods into *epochs* (**Table 21.3**).

THE TERTIARY (65–1.8 MYA) During the **Tertiary** period, Australia began its northward drift. By 20 million years ago it had nearly reached its current position. The early Tertiary was a hot and humid time, during which the ranges of many plants shifted latitudinally. The tropics were probably too hot for rainforests, and were clothed in low-lying vegetation instead. In the middle of the Tertiary, however, Earth's climate became considerably drier and cooler. Many lineages of flowering plants evolved herbaceous (nonwoody) forms; grasslands spread over much of Earth.

By the beginning of the Cenozoic era, invertebrate faunas resembled those of today. It is among the vertebrates that evolutionary changes during the Tertiary period were most rapid. Snakes and lizards underwent extensive radiations during this period, as did birds and mammals. Three waves of mammals dispersed from Asia to North America across the land bridge that has intermittently connected the two continents during the past 55 million years ago. Rodents, marsupials, primates, and hoofed mammals appeared in North America for the first time.

THE QUATERNARY (1.8 MYA TO PRESENT) The current geological period, the **Quaternary**, is subdivided into two epochs, the *Pleistocene* and the *Holocene* (also known as the *Recent*). The Pleistocene was a time of drastic cooling and climate fluctuations. During four major and about 20 minor "ice ages," massive glaciers spread across the continents, and the ranges of animal and plant populations shifted toward the equator. The last of these glaciers retreated from temperate latitudes less than 15,000 years ago. Organisms are still adjusting to these changes. Many high-latitude ecological communities have occupied their current locations for no more than a few thousand years. Interestingly, relatively few species became extinct during these climate fluctuations.

21.19 Evolutionary Faunas Representatives of the three great evolutionary faunas are shown, together with a graph illustrating the number of major groups in each fauna over time.

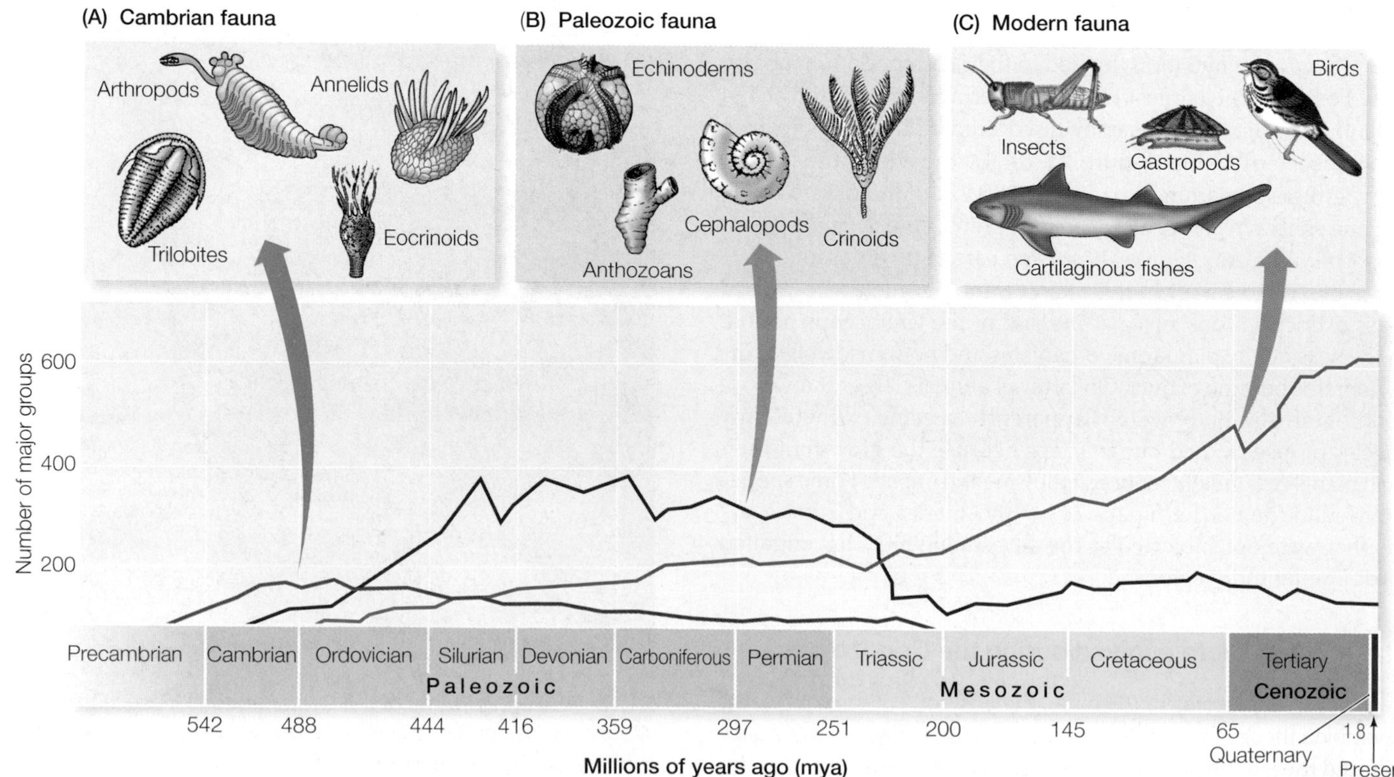

The Pleistocene was the time of hominoid evolution and radiation, resulting in the species *Homo sapiens*—modern humans (see Section 33.5). Many large bird and mammal species became extinct in Australia and in the Americas when *H. sapiens* arrived on those continents about 40,000 and 15,000 years ago, respectively. These extinctions were probably the result of hunting by humans, although the existing evidence does not convince all paleontologists.

Woolly mammoths survived on Siberia's Wrangel Island as recently as 4,000 years ago, long after hunters exterminated them from the mainland. The mammoths may have maintained their dry steppe environment on the island by trampling tundra mosses and recycling nitrogen, whereas the mainland environment became a moist tundra.

Three major faunas have dominated life on Earth

The fossil record reveals three great evolutionary radiations, each of which resulted in the evolution of major new faunas (**Figure 21.19**). The first one, the Cambrian explosion, actually began before the beginning of the Cambrian period. The second, about 60 million years later, resulted in the Paleozoic fauna. The great Permian extinctions 300 million years later were followed by the third great radiation, called the Triassic explosion, which led to our modern fauna.

During the Cambrian explosion, organisms ancestral to most of the present-day animal groups appeared, along with a number of groups that subsequently became extinct. The Paleozoic and Triassic explosions resulted in considerable diversification of the existing major groups of animals, but all of them were modifications of body plans that were already present when these great biological diversifications began (see Chapters 31–33).

21.3 RECAP

Life evolved in the Precambrian oceans. Life diversified as atmospheric oxygen approached its current level and the continents came together to form several large land masses. Numerous climate changes and rearrangements of the continents, as well as meteorite impacts, contributed to five mass extinctions.

- Why have so few of the multitudes of organisms that have existed over millenia become fossilized? See p. 472
- What do we mean when we refer to the "Cambrian explosion"? See pp. 472–473
- Can you identify the five mass extinctions and their possible causes? See pp. 473–477 and Table 21.2

The fossil record reveals broad patterns in life's evolution. It shows that many species changed very little over many millions of years, others changed only gradually, and still others underwent rapid changes that were followed by long periods of slow change. In short, the rate of evolutionary change has differed greatly at different times and among different lineages. Let's look at some examples of evolutionary patterns to determine why rates of evolutionary change are so variable.

21.4 Why Do Evolutionary Rates Differ among Groups of Organisms?

Fossils can reveal information about rates of change within particular lineages of organisms. Evolutionary change in a lineage may incorrectly appear to be rapid if its fossil record is very incomplete, but some rapid changes are well documented by excellent series of fossils.

Changes in the physical and biological environment are likely to stimulate evolutionary change. Organisms that live in environments that are changing are likely to evolve more rapidly than organisms living in relatively constant environments. When climates change, the ranges of some organisms may shift, and other organisms may find themselves with different predators or competitors. Similarly, predators may change in response to changes in their prey. In contrast, the morphology of organisms that live in relatively unchanging environments often changes slowly, if at all.

"Living fossils" exist today

Species whose morphology has changed little over millions of years are known as " living fossils." For example, the horseshoe crabs living today are almost identical in appearance to those that lived 300 million years ago (see Figure 32.30B). The sandy coastlines where horseshoe crabs spawn feature extremes in temperatures and salt concentrations that are lethal to many organisms. These harsh environments have changed relatively little over millennia, and the horseshoe crabs likewise remain relatively unchanged as they maintain the very specific adaptations that allow them to survive.

Similarly, the chambered nautiluses of the late Cretaceous are nearly indistinguishable from living species (see Figure 32.15F). Chambered nautiluses spend their days in deep, dark ocean waters, ascending to feed in food-rich surface waters only under the protective cover of darkness. Their intricate shells provide little protection against today's visually hunting fish; they survive, however, because they are adapted to a stringent, relatively unchanging environment in which recently evolved potential predators cannot survive.

The leaf shapes of many plants have changed little over time; fossilized leaves of *Ginkgo* trees from the Triassic, for example, are very similar to those of living trees (**Figure 21.20**). This may be because the physical nature of sunlight is unchanging, and the intricate photosynthetic mechanisms that harvest sunlight (see Chapter 8), once evolved, have remained relatively constant over millions of years.

Evolutionary changes have been gradual in most groups

The most striking feature of life's evolution is that rates of change are, on average, very slow. The fossil record contains many series of fossils that demonstrate gradual change in lineages over time. A good example is the series of fossils showing changes in the

21.20 "Living Fossils" Fossilized *Ginkgo* leaves from the Triassic appear very similar to the leaves of living ginkgo trees.

number of ribs on the exoskeleton in eight lineages of trilobites during the Ordovician (**Figure 21.21**). Rates of change differed among the lineages, and they did not all change at the same time, but all of the changes were gradual.

Why does slow, gradual change appear to dominate the fossil record? A likely reason is that climates have usually changed slowly. In addition, as climates changed, the ranges of most organisms shifted accordingly, so that the environments in which individuals lived actually changed very little. For example, as the climate warmed and glaciers retreated about 10,000 years ago, many species expanded their ranges northward (**Figure 21.22**).

Rates of evolutionary change are sometimes rapid

If the physical or biological environment changes rapidly, some lineages may also change rapidly. Good examples of rapid evolutionary change are provided by species that have been introduced into new regions that differ strikingly from the environment from which they came. For example, as recently as 1939, the house finch was confined to the arid and semiarid parts of western North America. That year, some captive finches were released in New York City. Many of them survived to form a small breeding population in the immediate vicinity of the city. During the early 1960s, that population began to grow and increase its range. By the 1990s, the house finch had spread across all of the eastern United States and southern Canada (**Figure 21.23**), colonizing regions with climates that differ dramatically from those of its original western range. Remarkably, by 2000, birds in different eastern finch populations that had been separated for only a few decades were as different in size as birds in western finch populations that had been separated for thousands of years.

Rates of extinction have also varied greatly

More than 99 percent of the species that have ever lived are extinct. Species have become extinct throughout the history of life, but extinction rates have fluctuated dramatically. Section 21.3 described at least five major extinction events that severely reduced the planet's biota. These events were often followed by high rates of evolution, as surviving organisms responded to a new environment with a different composition of predators, prey, and competitors.

Some groups have had high extinction rates while others were proliferating. For example, among the mollusks of the Atlantic

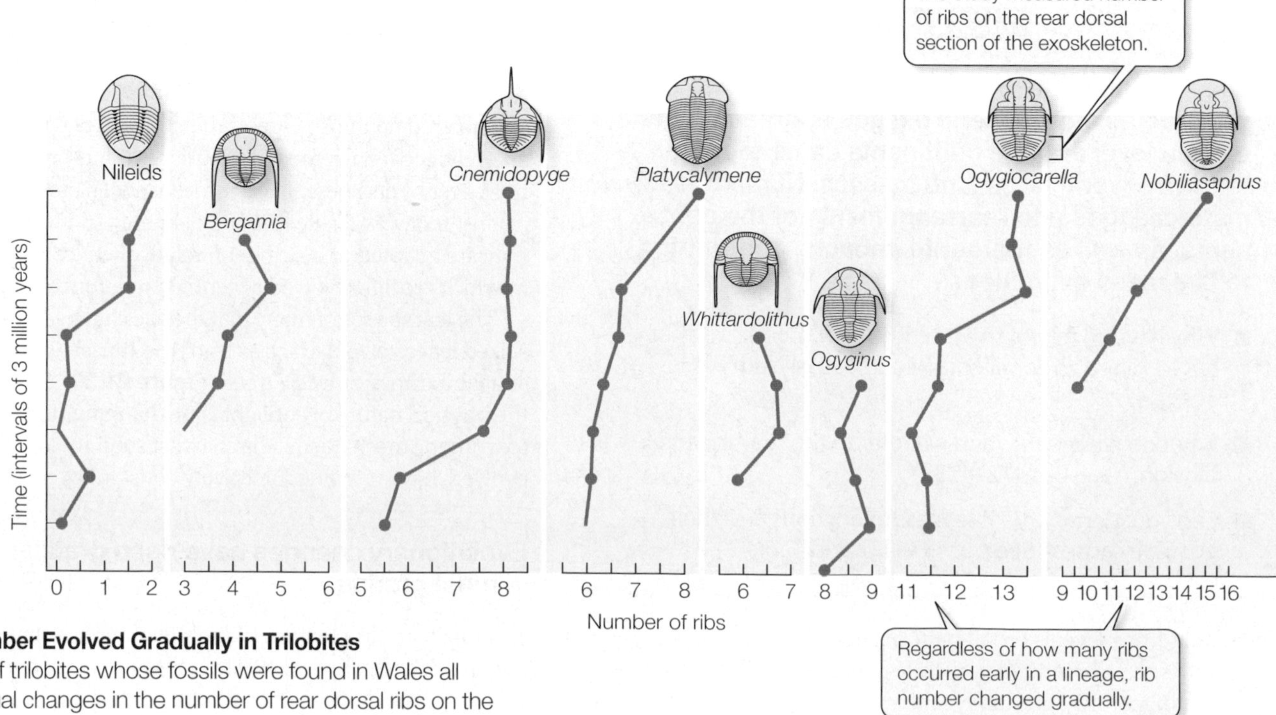

21.21 Rib Number Evolved Gradually in Trilobites
Eight lineages of trilobites whose fossils were found in Wales all displayed gradual changes in the number of rear dorsal ribs on the exoskeleton.

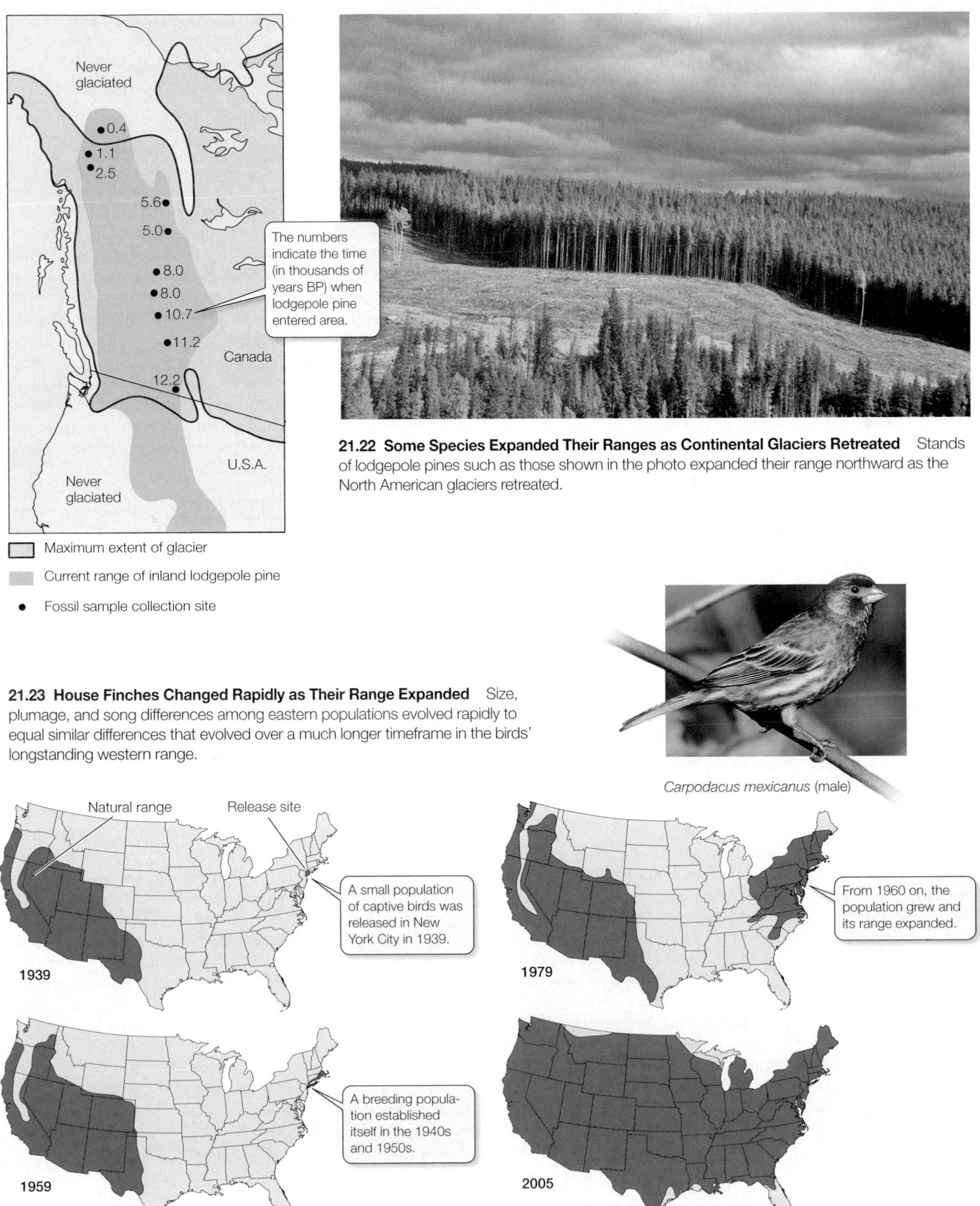

21.22 Some Species Expanded Their Ranges as Continental Glaciers Retreated Stands of lodgepole pines such as those shown in the photo expanded their range northward as the North American glaciers retreated.

21.23 House Finches Changed Rapidly as Their Range Expanded Size, plumage, and song differences among eastern populations evolved rapidly to equal similar differences that evolved over a much longer timeframe in the birds' longstanding western range.

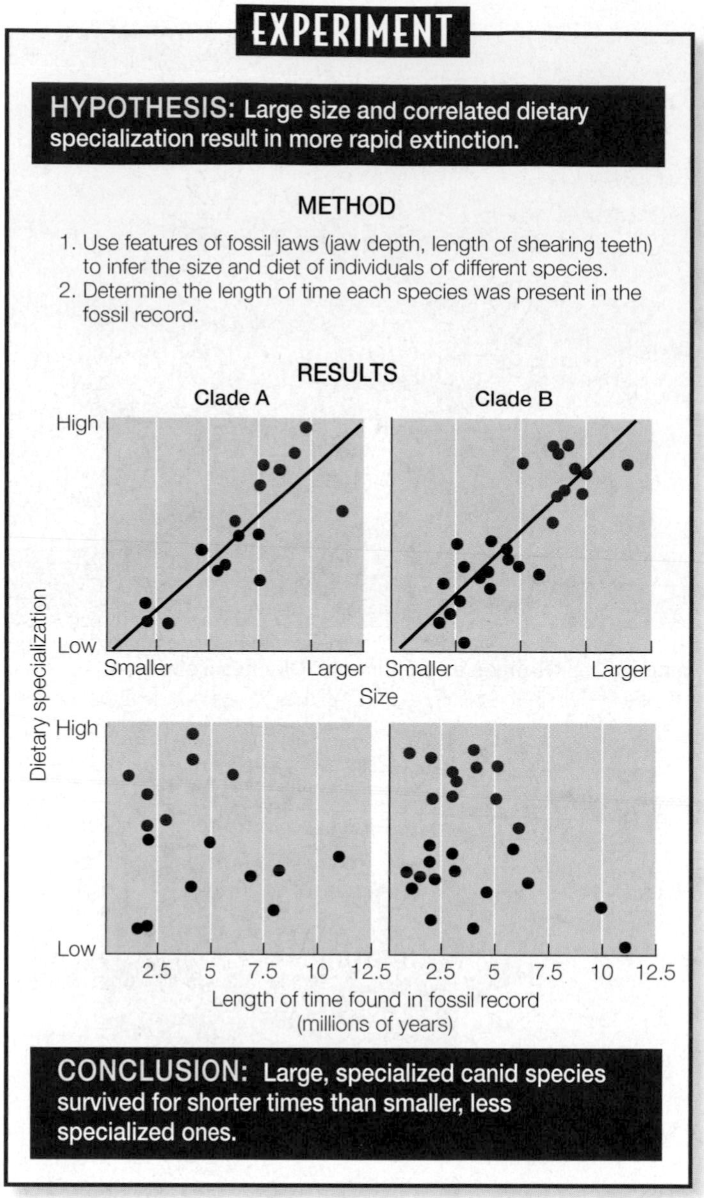

EXPERIMENT

HYPOTHESIS: Large size and correlated dietary specialization result in more rapid extinction.

METHOD

1. Use features of fossil jaws (jaw depth, length of shearing teeth) to infer the size and diet of individuals of different species.
2. Determine the length of time each species was present in the fossil record.

RESULTS

Clade A
Clade B
High
Low
Smaller
Larger
Smaller
Larger
Size
Dietary specialization
High
Low
2.5 5 7.5 10 12.5 2.5 5 7.5 10 12.5
Length of time found in fossil record (millions of years)

CONCLUSION: Large, specialized canid species survived for shorter times than smaller, less specialized ones.

21.24 Large, Specialized Canids Survived Shorter Times Each dot represents a canid species from the North American fossil records of two clades. In these two successive groups of canids, larger species are linked to more specialized diets (top graphs) and were more likely to become extinct (bottom graphs).

coastal plain of North America, species with broad geographic ranges were less likely to become extinct during normal times (when no mass extinctions were taking place) than were species with small geographic ranges. During the mass extinction at the end of the Cretaceous, however, groups of closely related mollusk species with large geographic ranges survived better than groups with small ranges, even if the individual species within the group had small ranges.

The diets of organisms may also influence their extinction rates. The fossil record shows that natural selection often favors larger size in carnivores (flesh-eating animals). However, larger carnivores typically have more specialized diets than smaller carnivores, and animals with specialized diets may be more vulnerable to loss of their food supply than are those with more generalized diets. A possible consequence is that large, specialized carnivores would be more likely to become extinct than smaller species with more generalized diets. An impressive fossil record of successive lineages of canids (dogs, wolves, and their relatives) enabled Blaire Van Valkenburgh and her colleagues to use the comparative method (see Section 1.3) to test and confirm this hypothesis (**Figure 21.24**).

At the end of the Cretaceous period, extinction rates on land were much higher among large vertebrates than among small ones. The same was true during the Pleistocene epoch, when extinction rates were high only among large mammals and large birds. During some mass extinctions, marine organisms were heavily hit, but terrestrial organisms survived well. Other mass extinctions affected organisms living in both environments. These differences are not surprising, given that major changes on land and in the oceans did not always coincide.

21.4 RECAP

Rates of evolutionary change vary greatly among lineages of organisms. Changes in the physical and biological environments typically induce evolutionary changes. Extinction rates have also fluctuated dramatically over evolutionary history.

- Why do slow, gradual changes dominate the fossil record? See pp. 479–480
- How can the diets of organisms influence their extinction rates? See p. 482 and Figure 21.24

Although the agents of evolutionary change are operating today as they have been since life first appeared on Earth, major changes are being driven by the dramatic increase of Earth's human population. Predation (hunting) by humans has caused the extinction of many large mammals in the recent past, and continues doing so today. Humans are also changing the physical and biological environment by dramatically altering Earth's vegetation, converting forests and grasslands to crops and pastures. Deliberately or inadvertently, we are moving thousands of species around the globe, reducing the provincialization of Earth's biota that began during the Mesozoic era. Humans have also taken charge of the evolution of certain species by means of artificial selection and biotechnology. As we saw in Chapter 16, modern molecular methods enable us to modify organisms by moving genes among even distantly related species. In short, humans have become a dominant agent of evolutionary change. How we wield our massive influence will powerfully affect the future of life on Earth.

CHAPTER SUMMARY

21.1 How do scientists date ancient events?

The relative ages of organisms can be determined by the dating of **fossils** and the **strata** of sedimentary rocks in which they are found.

Paleontologists use a variety of radioisotopes with different **half-lives** to date events at different times in the remote past. Review Figure 21.1

Geologists divide the history of life into eras and periods, based on major differences in the assemblages of fossils found in successive layers of rocks. Review Table 21.2

21.2 How have Earth's continents and climates changed over time?

See Web/CD Tutorial 21.1

Earth's crust consists of solid lithospheric plates that float on fluid magma. **Continental drift** caused by convection currents in the magma moves these plates and the continents they contain. Review Figure 21.2

Conditions on Earth have changed dramatically over time. Some changes, such as increases in atmospheric concentrations of oxygen, have been primarily unidirectional. Other changes, such as changes in climate, have undergone repeated oscillations. Review Figures 21.5 and 21.6

Oxygen-generating cyanobacteria liberated enough O_2 to open the door to oxidation reactions in metabolic pathways. The aerobic prokaryotes were able to harvest more energy than anaerobic organisms and began to proliferate. Increases in atmospheric O_2 levels supported the evolution of large eukaryotic cells.

Major physical events on Earth, such as the collision of continents that formed the supercontinent **Pangaea**, have affected Earth's surface, climate, and atmosphere. In addition, extraterrestrial events such as meteorite strikes created sudden and dramatic environmental shifts. All these changes have affected the history of life.

21.3 What are the major events in life's history?

Paleontologists use fossils and evidence of geological changes to determine what Earth and its **biota** may have looked like at different times.

During most of its history, life was confined to the oceans. Multicellular life diversified during the **Cambrian explosion**. Review Figure 21.10

The periods of the **Paleozoic** era were each characterized by the diversification of specific groups of organisms. Amniotes—vertebrates whose eggs can be laid in dry places—first appeared during the Carboniferous period.

During the **Mesozoic** era, Earth's biota became increasingly **provincialized**, as distinct terrestrial **biotas** evolved on each continent.

Five episodes of **mass extinction** punctuated the history of life in the Paleozoic and Mesozoic eras.

Three major **faunas** have dominated life on Earth, stemming from evolutionary radiations occurring in the Cambrian, Paleozoic, and Mesozoic eras. Review Figure 21.19

Earth's **flora** has been dominated by flowering plants since the **Cenozoic** era.

21.4 Why do evolutionary rates differ among groups of organisms?

Most lineages evolve slowly and gradually over time; the morphology of many species has changed little over many millions of years. Review Figure 21.21

Changes in the physical and biological environments typically induce evolutionary changes.

Extinction rates have fluctuated dramatically during the history of life, but some groups have had high extinction rates while others were proliferating.

See Web/CD Activity 21.1 for a concept review of this chapter.

SELF-QUIZ

1. The number of species of fossil organisms that have been described is about
   a. 50,000.
   b. 100,000.
   c. 200,000.
   d. 300,000.
   e. 500,000.
2. In undisturbed strata of sedimentary rock,
   a. the oldest rocks lie at the top.
   b. the oldest rocks lie at the bottom.
   c. the oldest rocks are in the middle.
   d. the oldest rocks are distributed among the strata of younger rocks.
   e. None of the above
3. Carbon-14 can be used to determine the ages of fossil organisms because
   a. all organisms contain many carbon compounds.
   b. carbon-14 has a regular rate of decay to carbon-12.
   c. the ratio of ^{14}C to ^{12}C in living organisms is always the same as that in the atmosphere.
   d. the production of new ^{14}C in the atmosphere just balances the natural radioactive decay of ^{14}C.
   e. All of the above
4. An important, generally unidirectional change in Earth during its history is a
   a. steady increase in volcanic activity.
   b. gradual coming together of the continents.
   c. steady increase in the oxygen content of the atmosphere.
   d. gradual warming of the climate.
   e. steady increase in Earth's precipitation.
5. The total of all species of organisms in a given region is known as the region's
   a. biota.
   b. flora.
   c. fauna.
   d. flora and fauna.
   e. diversity.
6. The coal beds we now mine for energy are the remains of
   a. trees that grew in swamps during the Carboniferous period.
   b. trees that grew in swamps during the Devonian period.
   c. trees that grew in swamps during the Permian period.
   d. small plants that grew in swamps during the Carboniferous period.
   e. None of the above

7. The cause of the mass extinction at the end of the Ordovician period was probably
   a. the collision of Earth with a large meteorite.
   b. massive volcanic eruptions.
   c. massive glaciation in Gondwana.
   d. the uniting of all continents to form Pangaea.
   e. changes in Earth's orbit.
8. The cause of the mass extinction at the end of the Mesozoic era probably was
   a. continental drift.
   b. the collision of Earth with a large meteorite.
   c. changes in Earth's orbit.
   d. massive glaciation.
   e. changes in the salt concentration of the oceans.
9. The times during the history of life when many new evolutionary lineages appeared were the
   a. Precambrian, Cambrian, and Triassic.
   b. Precambrian, Cambrian, and Tertiary.
   c. Cambrian, Paleozoic, and Triassic.
   d. Cambrian, Triassic, and Devonian.
   e. Paleozoic, Triassic, and Tertiary.
10. At which of the following times was there *no* mass extinction?
    a. The end of the Cretaceous period
    b. The end of the Devonian period
    c. The end of the Permian period
    d. The end of the Triassic period
    e. The end of the Silurian period

FOR DISCUSSION

1. Some groups of organisms have evolved to contain large numbers of species; other groups have produced only a few species. Is it meaningful to consider the former groups more successful than the latter? What does the word "success" mean in evolution? How does your answer influence your thinking about *Homo sapiens*, the only surviving representative of a clade that never had many species in it?
2. Scientists date ancient events using a variety of methods, but nobody was present to witness or record those events. Accepting those dates requires us to believe in the accuracy and appropriateness of indirect measurement techniques. What other basic scientific concepts are also based on the results of indirect measurement techniques?
3. Why is it useful to be able to date past events absolutely as well as relatively?
4. If we are living during one of the cooler periods in Earth's history, why should we be concerned about human activities that are contributing to global climate warming?
5. Large meteorites colliding with Earth have caused massive climate and evolutionary changes in the past. Should we attempt to take steps to prevent future impacts? What actions might we undertake? What adverse effects might such actions trigger?

FOR INVESTIGATION

The experiment in Figure 21.24 showed that, over evolutionary history, large canid species with specialized diets survived for shorter times than did smaller but less specialized species. Large herbivores also survived for shorter times that smaller ones. What hypotheses would you propose to explain these facts? How would you test your hypotheses?

CHAPTER 22 The Mechanisms of Evolution

Snake eats poisonous newt—and lives!

Newts and other salamanders move slowly, so they are easy prey for garter snakes and other predators. But some of these amphibians have evolved chemical defenses against predation—in a word, they are poisonous. The rough-skinned newt, *Taricha granulosa*, is a salamander that lives on the Pacific Coast of North America. The newt sequesters in its skin a potent neurotoxin called tetrodotoxin (TTX). TTX paralyzes nerves and muscles by blocking sodium channels (see Section 5.3). Most vertebrates, including predatory garter snakes, will die if they eat a rough-skinned newt.

But some snakes can eat rough-skinned newts and survive. In certain populations of the garter snake *Thamnophis sirtalis*, most individuals have TTX-resistant sodium channels in their nerves and muscles—but they pay a price for this attribute. TTX-resistant snakes can move only slowly for several hours after eating a newt, and they never move as fast as nonresistant snakes. Thus TTX-resistant snakes are more vulnerable to their own predators than are TTX-sensitive snakes that simply don't encounter poisonous newts.

Pufferfish, octopuses, tunicates, and some species of frogs also use TTX as a defensive chemical. Other animals and many plants use a variety of chemicals to defend themselves against predators, and many predators have evolved resistance to those chemicals. Both the production of and resistance to defensive chemicals are evolutionary *adaptations*. But adaptations that benefit an individual in one way may reduce the ability of the organism to carry out other activities, as happens with the decreased ability of TTX-resistant garter snakes to move quickly. And adaptations such as the ability to produce neurotoxins in the skin may be energetically costly for an organism to develop and maintain. In other words, to improve its performance in one area, an organism typically experiences reduced performance in some other area—there is a *trade-off* between the benefits of an adaptation and what it costs the organism that expresses it.

One tool of evolutionary biologists is to try to identify and measure the trade-offs that different adaptations impose, because the nature and strength of those trade-offs influence the extent to which different adaptations succeed, and thus how populations of organisms evolve. If TTX resistance had no cost, we might expect to find resistant garter snakes everywhere, even in environments where there are no toxic salamanders. This is not the case. If toxic defensive chemicals had no cost, individuals

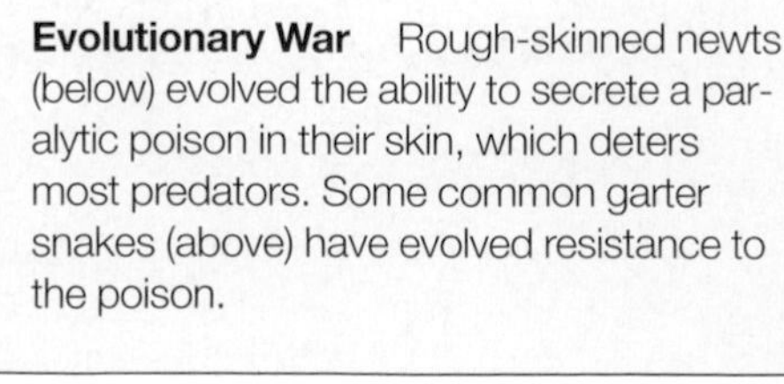

Evolutionary War Rough-skinned newts (below) evolved the ability to secrete a paralytic poison in their skin, which deters most predators. Some common garter snakes (above) have evolved resistance to the poison.

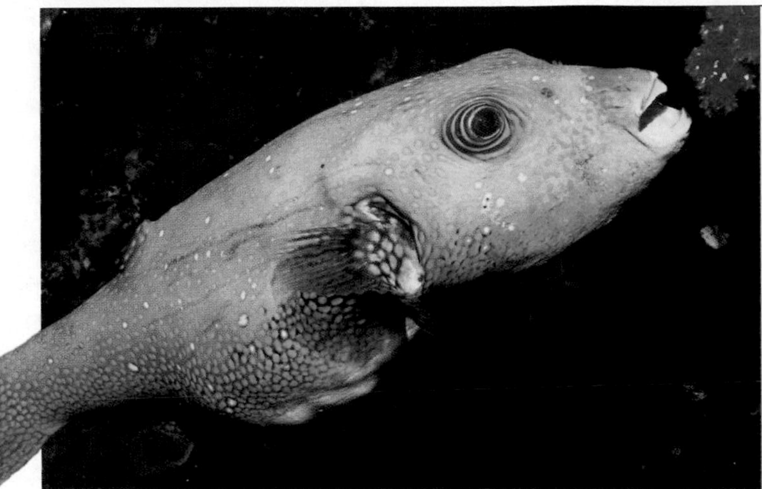

A Workable Plan Shows Up Elsewhere The slow-swimming map pufferfish (*Arothron mappa*), shown here feeding in the seas off Thailand, produces TTX in several of its organs, notably the liver. The level of toxicity appears to vary seasonally, and does not stop people from regarding the flesh of these fish as a dangerous delicacy.

of many prey species would probably be poisonous. They are not.

Preeminent among Charles Darwin's many contributions to biology was putting forth a plausible and testable hypothesis for a mechanism that could result in the adaptation of organisms to their environments. In effect, Darwin offered a mechanistic explanation for the evolution of all forms of life, including humans. Some people find it difficult to accept that the same mechanistic processes that determined the evolutionary pathways of plants, insects, and bacteria also guided human evolution. But as Darwin noted, "there is grandeur in this view of life."

IN THIS CHAPTER we will see how Charles Darwin developed his ideas, and then turn to the advances in our understanding of evolutionary mechanisms since Darwin's time. We will discuss the genetic basis of evolution and show how genetic variation within populations is measured. We will describe the mechanisms of evolution and show how biologists design studies to investigate them. Finally, we will discuss constraints on the pathways evolution can take.

CHAPTER OUTLINE

22.1 What Facts Form the Base of Our Understanding of Evolution?

Today, a rich array of geological, morphological, and molecular data support and enhance the factual basis of evolution; but when Charles Darwin was a youth, it was not evident to him (or to almost anyone else) that life had evolved. Darwin was passionately interested in both geology and natural history—the scientific study of how different organisms function and carry out their lives in nature. Despite having these interests, he had planned to become a doctor, but he was nauseated by surgery conducted without anesthesia. He gave up medicine to study at Cambridge University for a career as a clergyman of the Church of England. Darwin was more interested in science than in theology, however, and he became a companion of scientists on the faculty, especially the botanist John Henslow. In 1831, Henslow recommended Darwin for a position on the H.M.S. *Beagle*, which was preparing for a survey voyage around the world (**Figure 22.1**).

Whenever possible during the 5-year-long voyage, Darwin (who was often seasick) went ashore to study rocks and to observe and collect specimens of plants and animals. He noticed how strikingly the species he saw in South America differed from those of Europe. He observed that the species of the temperate regions of South America (Argentina and Chile) were more similar to those of tropical South America (Brazil) than they were to temperate European species. When he explored the Galápagos Islands, west of Ecuador, he noted that most of its animal species were found nowhere else, but were similar to those of mainland South America. Darwin also recognized that the animals of the archipelago differed from island to island. He postulated that some animals had come to the archipelago from mainland South America and then undergone different changes on each of the islands. But what mechanism could account for these changes?

When he returned to England in 1836, Darwin continued to ponder his observations. Within a decade he had developed the

(A)

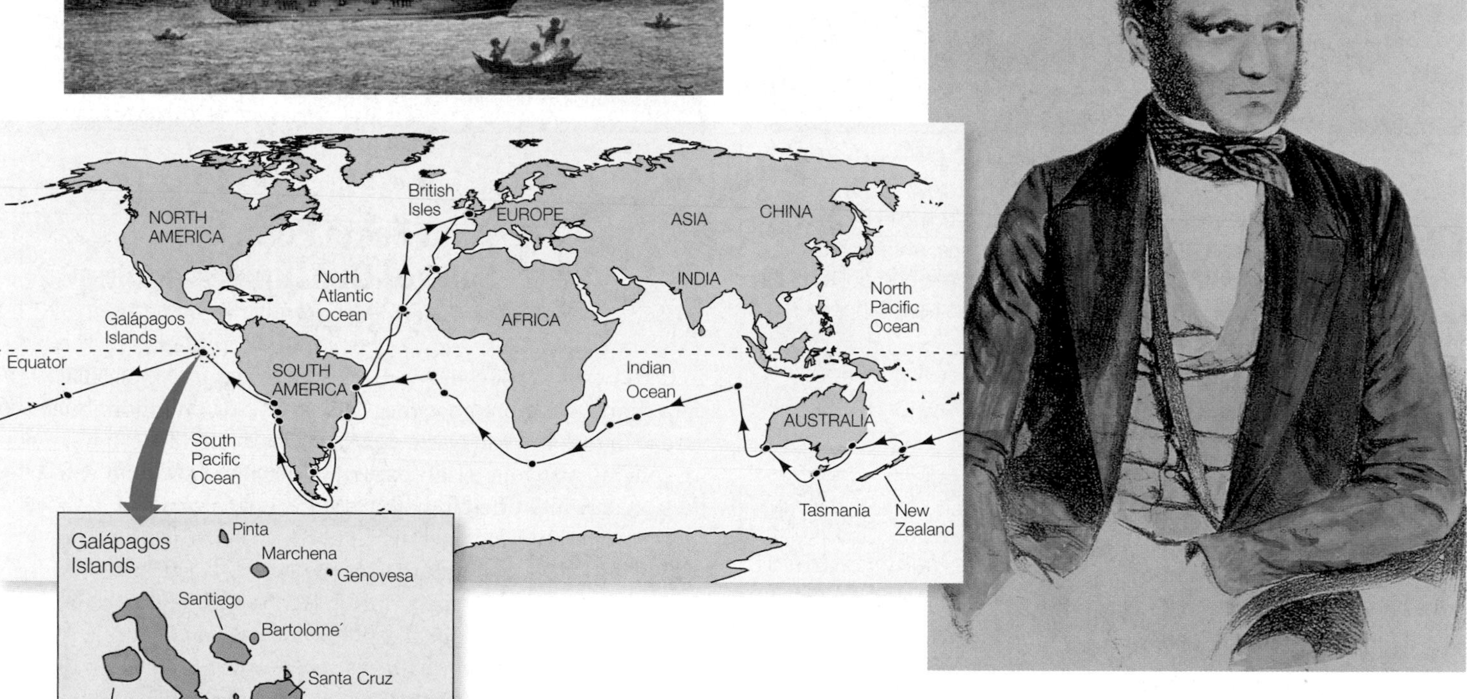

22.1 Darwin and the Voyage of the *Beagle* (A) The mission of H.M.S. *Beagle* was to chart the oceans and collect oceanographic and biological information from around the world. The map indicates the ship's path, and the inset shows the Galápagos Islands, whose organisms were an important source of Darwin's ideas on natural selection. (B) Charles Darwin at age 24, shortly after the *Beagle* returned to England.

major features of an explanatory theory for *evolutionary change* based on two major propositions:

- Species are not immutable; they change over time.
- The process that produces these changes is *natural selection*.

Darwin asserted that evolution is a historical fact that can be demonstrated to have taken place (his first proposition). In 1844, he wrote a long essay on natural selection, the process he described as the cause of evolution (his second proposition), but, despite urging from his wife and colleagues, he was reluctant to publish it, preferring to assemble more evidence first.

Darwin's hand was forced in 1858 when he received a letter and manuscript from another traveling naturalist, Alfred Russel Wallace, who was studying the biota of the Malay Archipelago. Wallace asked Darwin to evaluate the manuscript, in which Wallace proposed a theory of natural selection almost identical to Darwin's. At first Darwin was dismayed, believing that Wallace had preempted his idea. But parts of Darwin's 1844 essay, together with Wallace's manuscript, were presented to the Linnaean Society of London on July 1, 1858, thereby giving credit for the idea to both men. Darwin then worked quickly to finish his own book, *The Origin of Species*, which was published the next year.

Although both Darwin and Wallace independently articulated the concept of natural selection, Darwin developed his ideas first. Furthermore, *The Origin of Species* provided exhaustive evidence from many fields to support both natural selection and evolution itself; thus both concepts are more closely associated with Darwin than with Wallace.

The facts that Darwin used to conceive and develop his theory of evolution by natural selection were familiar to most contemporary biologists. His insight was to perceive the significance of relationships among them. Both Darwin and Wallace were influenced by the ideas of the economist Thomas Malthus, who in 1838 published *An Essay on the Principle of Population*. Malthus argued that because the rate of human population growth is greater than the rate of increase in food production, unchecked growth inevitably leads to famine. Darwin saw parallels throughout nature.

He recognized that populations of all species have the potential for rapid increases in numbers. To illustrate this point, he used the following example:

> Suppose ... there are eight pairs of birds, and that only four pairs of them annually ... rear only four young, and that these go on rearing their young at the same rate, then at the end of seven years ... there will be 2048 birds instead of the original sixteen.

Yet such rates of increase are rarely seen in nature. Therefore, Darwin reasoned that death rates in nature must also be high. Without high death rates, even the most slowly reproducing species would quickly reach enormous population sizes.

Darwin also observed that, although offspring tend to resemble their parents, the offspring of most organisms are not identical to one another or to their parents. He suggested that slight variations among individuals affect the chance that a given individual will survive and reproduce. Darwin called this differential survival and reproduction of individuals **natural selection**. Natural selection is formally defined as *the differential contribution of offspring to the next generation by various genetic types belonging to the same population.*

Darwin may have used the words "natural selection" because he was familiar with the **artificial selection** of individuals with certain desirable traits by animal and plant breeders. Many of Darwin's observations on the nature of variation came from domesticated plants and animals. Darwin was a pigeon breeder, and he knew firsthand the astonishing diversity in color, size, form, and behavior that pigeon breeders could achieve (**Figure 22.2**). He recognized close parallels between selection by breeders and selection in nature. As he argued in *The Origin of Species,*

> How can it be doubted, from the struggle each individual has to obtain subsistence, that any minute variation in structure, habits or instincts, adapting that individual better to the new conditions, would tell upon its vigour and health? In the struggle it would have a better chance of surviving; and those of its offspring which inherited the variation, be it ever so slight, would have a better chance.

That statement, written almost 150 years ago, still stands as a good expression of the process of evolution by natural selection.

It is important to remember, as Darwin clearly understood, that *individuals do not evolve; populations do.* A **population** is a group of individuals of a single species that live and interbreed in a particular geographic area at the same time. A major consequence of the evolution of populations is that their members become adapted to the environments in which they live. But what do biologists mean when they say that an organism is adapted to its environment?

Adaptation has two meanings

In evolutionary biology, the term **adaptation** refers both to the *processes* by which characteristics that appear to be useful to their bearers evolve—that is, the evolutionary mechanisms that produce them—and to the *characteristics* themselves. With respect to characteristics, an adaptation is a phenotypic characteristic that has helped an organism adjust to conditions in its environment. We will discuss the processes that result in adaptation in great detail in this chapter.

Biologists regard an organism as being adapted to a particular environment when they can demonstrate that a slightly different organism reproduces and survives less well in that environment. To understand adaptation, biologists compare the performance of individuals that differ in their traits. For example, biologists could assess the adaptive role of the chemical defenses of the rough-skinned newts described at the opening of this chapter by comparing the survival and reproductive rates of newts that had different concentrations of TTX in their skins.

When Darwin proposed his theory of evolution by natural selection, he could point to many examples of evolutionary mechanisms operating in nature, but none were supported by experiments. Since then, many observational and experimental studies of evolutionary mechanisms have been conducted. Biologists have also documented changes over time in the genetic composition of many populations, and our understanding of the mechanisms of inheritance has improved enormously.

Population genetics provides an underpinning for Darwin's theory

For a population to evolve, its members must possess heritable genetic variation, which is the raw material on which mechanisms of evolution act. In everyday life, we cannot directly observe the genetic compositions of organisms. What we do see in nature are *phenotypes*, the physical expressions of organisms' genes. The features

22.2 Many Types of Pigeons Have Been Produced by Artificial Selection Charles Darwin raised pigeons as a hobby, and he noted similar forces at work in artificial and natural selection. The pigeons shown here are just some of the more than 300 varieties that have been artificially selected by breeders to display different forms of characters such as color, size, and feather distribution.

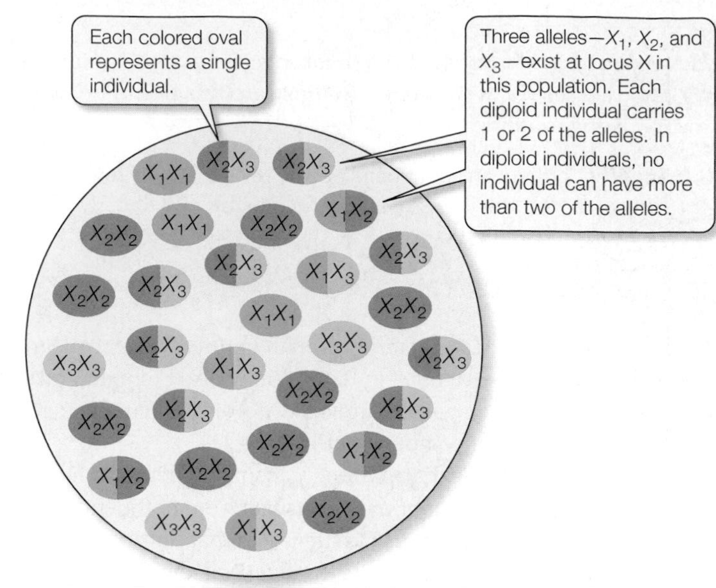

Each colored oval represents a single individual.
Three alleles—X_1, X_2, and X_3—exist at locus X in this population. Each diploid individual carries 1 or 2 of the alleles. In diploid individuals, no individual can have more than two of the alleles.
X_1X_1
X_2X_3
X_2X_2
X_1X_2
X_1X_3
X_3X_3

22.3 A Gene Pool A gene pool is the sum of all the alleles found in a population. The gene pool for only one locus, *X*, is shown in this figure. The allele frequencies in this gene pool are 0.20 for X_1, 0.50 for X_2, and 0.30 for X_3.

of a phenotype are its *characters*—eye color, for example. The specific form of a character, such as brown eyes, is a *trait*. A **heritable** trait is a characteristic of an organism that is at least partly determined by its genes. The genetic constitution that governs a character is called its *genotype*. *A population evolves when individuals with different genotypes survive or reproduce at different rates.*

The rediscovery of Gregor Mendel's publications in the early 1900s (see Section 10.1) paved the way for the development of the field of **population genetics**. Population genetics has three main goals:

- To explain the origin and maintenance of genetic variation
- To explain the patterns and organization of genetic variation
- To understand the mechanisms that cause changes in allele frequencies in populations

The perspective of population genetics complements the insights into evolutionary processes provided by developmental biology, which we discussed in Chapter 20.

As described in Section 10.1, different forms of a gene, called *alleles*, may exist at a particular locus. At any particular locus, a single individual has only some of the alleles found in the population to which it belongs (**Figure 22.3**). The sum of all copies of all alleles at all loci found in a population constitutes its **gene pool**. (We can also refer to the "gene pool" for a particular locus or loci.) The gene pool contains the genetic variation that produces the phenotypic traits on which natural selection acts. To understand evolution and the role of natural selection, we need to know how much genetic variation populations have, the sources of that genetic variation, and how genetic variation changes in populations over space and time.

Most populations are genetically variable

Nearly all populations have genetic variation for many characters. Artificial selection on different characters in a single European species of wild mustard produced many important crop plants (**Figure 22.4**). Agriculturalists could achieve these results because the original mustard population had genetic variation for the characters of interest.

Laboratory experiments also demonstrate the existence of considerable genetic variation in populations. In one such experiment, investigators attempted to breed populations of fruit flies (*Drosophila melanogaster*) with high or low numbers of bristles on their abdomens from an initial population with intermediate numbers of bristles. After 35 generations, all

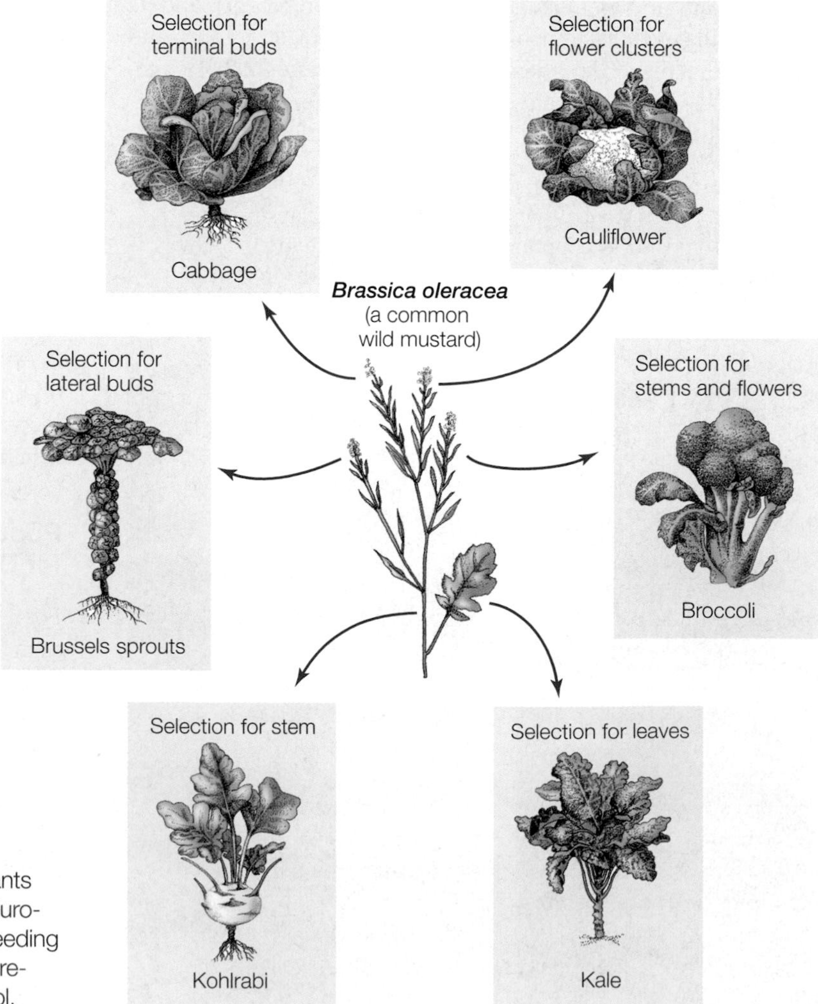

22.4 Many Vegetables from One Species All of the crop plants shown have been derived from a single wild mustard species. European agriculturalists produced these crops by choosing and breeding plants with unusually large buds, stems, leaves, or flowers. The results illustrate the vast amount of variation present in a gene pool.

22.5 Artificial Selection Reveals Genetic Variation In artificial selection experiments with *Drosophila melanogaster*, bristle numbers evolved rapidly. The graphs show the number of flies with different numbers of bristles after 35 generations of artificial selection.

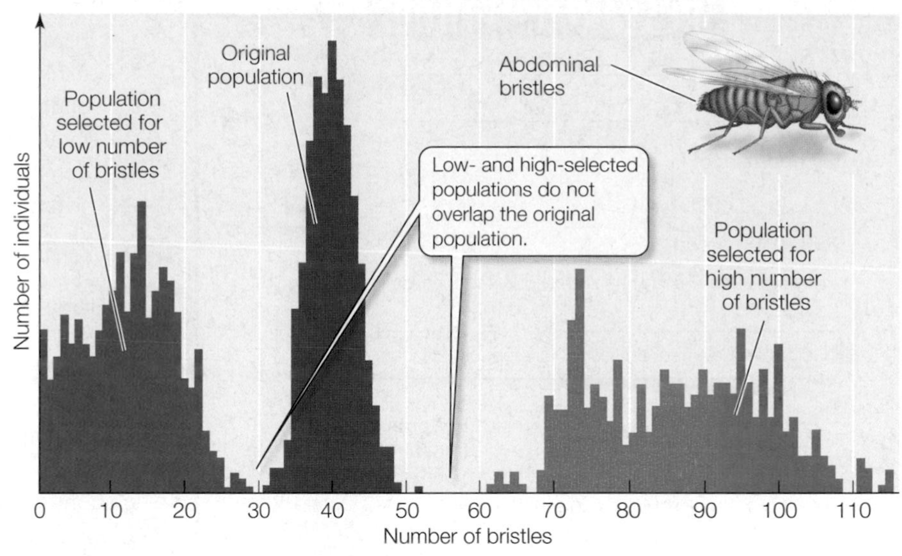

flies in both the high-bristle and low-bristle lineages had bristle numbers that fell well outside the range found in the original population (**Figure 22.5**). Thus there must have been considerable genetic variation in the original fruit fly population for selection to act on.

The study of the genetic basis of natural selection is difficult because genotypes alone do not uniquely determine all phenotypes. With dominance, for example, a particular phenotype can be produced by more than one genotype (e.g., *AA* and *Aa* individuals may be phenotypically identical). Similarly, as we described in detail in Section 20.4, different phenotypes can be produced by a given genotype, depending on the environment encountered during development. For example, the cells of all the leaves on a tree or shrub are usually genetically identical, yet leaves of the same plant often differ in shape and size.

Evolutionary change can be measured by allele and genotype frequencies

Allele frequencies are usually estimated in locally interbreeding groups, called **Mendelian populations**, within a geographic population of a species. To measure allele frequencies in a Mendelian population precisely, we would need to count every allele at every locus in every individual in it. By doing so, we could determine the frequencies of all alleles in the population. The word **frequency** in population genetics means *proportion*, so the frequency of a given allele or genotype is simply the proportion of the gene pool at that locus.

Fortunately, we do not need to make complete measurements, because we can reliably estimate allele frequencies for a given locus by counting alleles in a sample of individuals from the population. The sum of all allele frequencies at a locus is equal to 1, so measures of allele frequency range from 0 to 1.

An allele's frequency is calculated using the following formula:

$$p = \frac{\text{number of copies of the allele in the population}}{\text{sum of alleles in the population}}$$

If only two alleles (we'll call them *A* and *a*) for a given locus are found among the members of a diploid population, they may combine to form three different genotypes: *AA*, *Aa*, and *aa*. Such a population is said to be *polymorphic* at that locus, since there is more than one allele. Using the formula above, we can calculate the relative frequencies of alleles *A* and *a* in a population of *N* individuals as follows:

- Let N_{AA} be the number of individuals that are homozygous for the *A* allele (*AA*).
- Let N_{Aa} be the number that are heterozygous (*Aa*).
- Let N_{aa} be the number that are homozygous for the *a* allele (*aa*).

Note that $N_{AA} + N_{Aa} + N_{aa} = N$, the total number of individuals in the population, and that the total number of copies of both alleles present in the population is $2N$ because each individual is diploid. Each *AA* individual has two copies of the *A* allele, and each *Aa* individual has one copy of the *A* allele. Therefore, the total number of *A* alleles in the population is $2N_{AA} + N_{Aa}$. Similarly, the total number of *a* alleles in the population is $2N_{aa} + N_{Aa}$.

If *p* represents the frequency of *A*, and *q* represents the frequency of *a*, then

$$p = \frac{2N_{AA} + N_{Aa}}{2N}$$

and

$$q = \frac{2N_{aa} + N_{Aa}}{2N}$$

To show how this formula works, **Figure 22.6** calculates allele frequencies in two hypothetical populations, each containing 200 diploid individuals. Population 1 has mostly homozygotes (90 *AA*, 40 *Aa*, and 70 *aa*), whereas population 2 has mostly heterozygotes (45 *AA*, 130 *Aa*, and 25 *aa*).

The calculations in Figure 22.6 demonstrate two important points. First, notice that for each population, $p + q = 1$. If $p + q = 1$, then $q = 1 - p$. So when there are only two alleles at a given locus in a population, we can calculate the frequency of one allele and then easily obtain the second allele's frequency by subtraction. If there is only one allele at a given locus in a population, its frequency is 1: the population is then *monomorphic* at that locus, and the allele is said to be *fixed*.

The second thing to notice is that population 1 (consisting mostly of homozygotes) and population 2 (consisting mostly of heterozygotes) have the same allele frequencies for *A* and *a*. Thus they have the same gene pool for this locus. However, because the alleles in the gene pool are distributed differently among individuals, the *genotype frequencies* of the two populations differ. Geno-

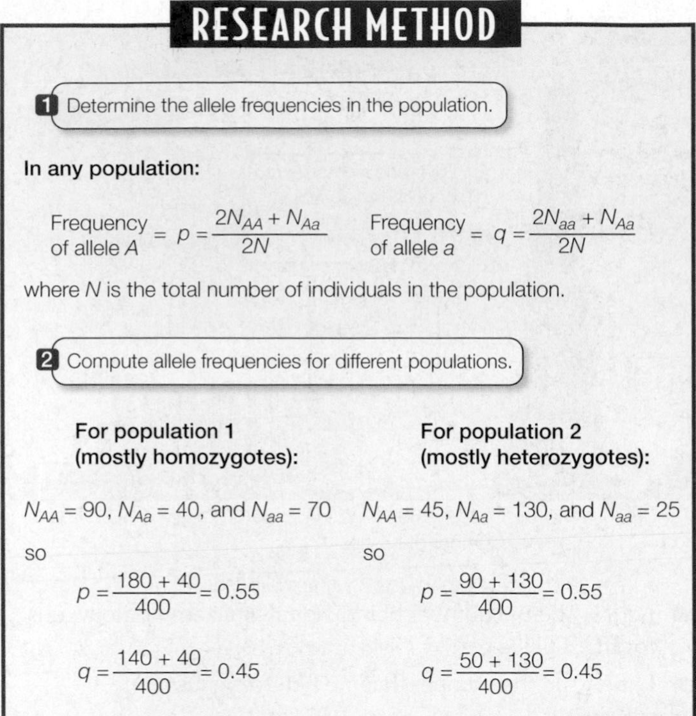

RESEARCH METHOD
1 Determine the allele frequencies in the population.
In any population:
Frequency of allele A $= p = \frac{2N_{AA} + N_{Aa}}{2N}$
Frequency of allele a $= q = \frac{2N_{aa} + N_{Aa}}{2N}$
where N is the total number of individuals in the population.
2 Compute allele frequencies for different populations.
For population 1 (mostly homozygotes):
$N_{AA} = 90$, $N_{Aa} = 40$, and $N_{aa} = 70$
so
$p = \frac{180 + 40}{400} = 0.55$
$q = \frac{140 + 40}{400} = 0.45$
For population 2 (mostly heterozygotes):
$N_{AA} = 45$, $N_{Aa} = 130$, and $N_{aa} = 25$
so
$p = \frac{90 + 130}{400} = 0.55$
$q = \frac{50 + 130}{400} = 0.45$

22.6 Calculating Allele Frequencies The gene pool and allele frequencies are the same in the two different populations, but the alleles are distributed differently between heterozygous and homozygous genotypes. In all cases, $p + q$ must equal 1.

type frequencies are calculated as the number of individuals that have a given genotype divided by the total number of individuals in the population. In population 1 in Figure 22.6, the genotype frequencies are 0.45 *AA*, 0.20 *Aa*, and 0.35 *aa*.

The frequencies of different alleles at each locus and the frequencies of different genotypes in a Mendelian population describe that population's **genetic structure**. Allele frequencies measure the amount of genetic variation in a population; genotype frequencies show how a population's genetic variation is distributed among its members. With these measurements, it becomes possible to consider how the genetic structure of a population changes or remains the same over generations—that is, to measure evolutionary change.

The genetic structure of a population does not change over time if certain conditions exist

In 1908, the British mathematician Godfrey Hardy and the German physician Wilhelm Weinberg independently deduced the conditions that must prevail if the genetic structure of a population is to remain the same over time. Hardy's equations explain why dominant alleles do not necessarily replace recessive alleles in populations, as well as other features of the genetic structure of populations: *If an allele is not advantageous, its frequency remains constant from generation to generation*; its frequency will not increase even if the allele is dominant.

Hardy wrote his equations in response to a question posed to him by the geneticist Reginald C. Punnett at the Cambridge University faculty club. Punnett wondered why, even though the allele for brachydactyly (short, stubby fingers) is dominant and the allele for normal-length fingers is recessive, most people in Britain have normal-length fingers.

Hardy–Weinberg equilibrium is the cornerstone of population genetics. The equation describes a model situation in which allele frequencies do not change across generations and genotype frequencies can be predicted from allele frequencies (**Figure 22.7**). The principles of Hardy–Weinberg equilibrium apply to sexually reproducing organisms. Several conditions must be met for a population to be at Hardy–Weinberg equilibrium:

- *Mating is random.* Individuals do not preferentially choose mates with certain genotypes.
- *Population size is infinite.* The larger a population, the smaller will be the effect of *genetic drift*—random (chance) fluctuations in allele frequencies.
- *There is no gene flow.* There is no migration either into or out of the population.
- *There is no mutation.* There is no change to alleles *A* and *a*, and no new alleles are added to change the gene pool.
- *Natural selection does not affect the survival of particular genotypes.* There is no differential survival of individuals with different genotypes.

If these ideal conditions hold, two major consequences follow. First, the frequencies of alleles at a locus remain constant from generation to generation. Second, following one generation of random mating, the genotype frequencies occur in the following proportions:

Genotype	*AA*	*Aa*	*aa*
Frequency	p^2	$2pq$	q^2

Consider generation 1 in Figure 22.7, in which the frequency of the *A* allele (p) is 0.55. Because we assume that individuals select mates at random, without regard to their genotype, gametes carrying *A* or *a* combine at random—that is, as predicted by the frequencies p and q. The probability that a particular sperm or egg in this example will bear an *A* allele rather than an *a* allele is 0.55. In other words, 55 out of 100 randomly sampled sperm or eggs will bear an *A* allele. Because $q = 1 - p$, the probability that a sperm or egg will bear an *a* allele is $1 - 0.55 = 0.45$. (You may wish to review the discussion of probability in Section 10.1.)

To obtain the probability of two *A*-bearing gametes coming together at fertilization, we multiply the two independent probabilities of their occurring separately:

$$p \times p = p^2 = (0.55)^2 = 0.3025$$

Therefore, 0.3025, or 30.25 percent, of the offspring in the next generation will have the *AA* genotype. Similarly, the probability of bringing together two *a*-bearing gametes is

$$q \times q = q^2 = (0.45)^2 = 0.2025$$

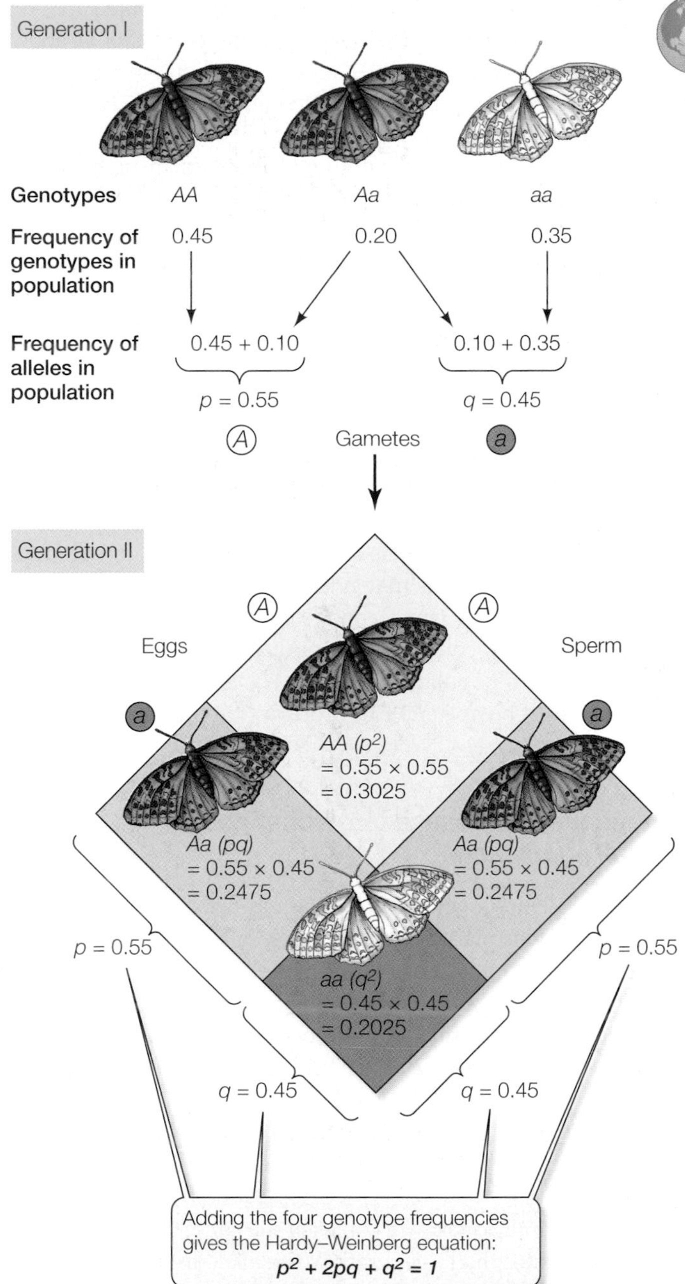

22.7 Calculating Hardy–Weinberg Genotype Frequencies The areas within the rectangle are proportional to the expected frequencies if mating is random with respect to genotype. Because there are two ways of producing a heterozygote, the probability of this event occurring is the sum of the two *Aa* squares. This example assumes that (1) the organism in question is diploid, (2) its generations do not overlap, (3) the gene under consideration has two alleles, and (4) allele frequencies are identical in males and females. Hardy–Weinberg equilibrium also applies if the gene has more than two alleles and generations overlap, but in those cases the mathematics is more complicated.

Thus 20.25 percent of the next generation will have the *aa* genotype.

Figure 22.7 also shows that there are two ways of producing a heterozygote: an *A* sperm may combine with an *a* egg, the probability of which is $p \times q$; or an *a* sperm may combine with an *A* egg, the probability of which is $q \times p$. Consequently, the overall probability of obtaining a heterozygote is $2pq$.

It is now easy to show that the allele frequencies p and q remain constant for each generation. If the frequency of *A* alleles in a randomly mating population is $p^2 + pq$, this frequency becomes $p^2 + p(1 - p) = p^2 + p - p^2 = p$. The original allele frequencies are unchanged, and the population is at *Hardy–Weinberg equilibrium*, as described by the **Hardy–Weinberg equation**:

$$p^2 + 2pq + q^2 = 1$$

If the genotype frequencies in the parental generation were to be altered (say, by emigration of a large number of *AA* individuals from the population), then the allele frequencies in the next generation would also be altered. However, based on the new allele frequencies, a single generation of random mating is sufficient to restore the genotype frequencies to Hardy–Weinberg (H-W) equilibrium, based on the new allele frequencies.

Deviations from Hardy–Weinberg equilibrium show that evolution is occurring

You may already have realized that populations in nature never meet the stringent conditions necessary to maintain them at H-W equilibrium. Why, then, is this model considered so important for the study of evolution? There are two reasons. First, the equation is useful for predicting the approximate genotype frequencies of a population from its allele frequencies.

Second, and crucially, the model describes the conditions that would result if there were *no* evolution in a population: the Hardy–Weinberg equation shows that *allele frequencies will remain the same from generation to generation unless some mechanism acts to change them*. Since the model's conditions are never met completely, allele frequencies in all populations do in fact deviate from Hardy–Weinberg equilibrium. In other words, *there are mechanisms acting to change allele frequencies, and these mechanisms drive evolution*. The patterns of deviation from Hardy–Weinberg equilibrium can help us identify specific mechanisms of evolutionary change.

22.1 RECAP

Laboratory experiments and the results of artificial selection demonstrate the existence of genetic variation in populations and provide evidence that natural selection can and does produce evolutionary change.

- Can you articulate the principle of natural selection? The two meanings of adaptation? See p. 489
- Do you see how calculating allele frequencies allows us to measure evolutionary change? See pp. 491–492 and Figure 22.6
- Why is the concept of Hardy–Weinberg equilibrium important even though the assumptions on which it is based are nearly always violated in nature? See pp. 492–493

We have briefly outlined Charles Darwin's vision of natural selection and adaptation and explained the mathematical basis of Hardy–Weinberg equilibrium and its importance for studying evolution. Now let's take a look at some of the forces that cause populations to deviate from equilibrium—the mechanisms of evolutionary change.

22.2 What Are the Mechanisms of Evolutionary Change?

Evolutionary mechanisms are forces that change the genetic structure of a population. Hardy–Weinberg equilibrium is a null hypothesis that assumes that those forces are absent. The known evolutionary mechanisms include mutation, gene flow, genetic drift, nonrandom mating, and natural selection. To understand evolutionary processes we need to discuss each of these mechanisms before considering natural selection in detail.

Mutations generate genetic variation

The origin of genetic variation is mutation. A mutation, as we saw in Section 12.6, is any change in an organism's DNA. Mutations appear to be random with respect to the adaptive needs of organisms. Most mutations are harmful to their bearers or are neutral, but if environmental conditions change, previously harmful or neutral alleles may become advantageous. In addition, mutations can restore to populations alleles that other evolutionary processes have removed. Thus mutations both create and help maintain genetic variation within populations.

Mutation rates are very low for most loci that have been studied. Rates as high as one mutation per locus in a thousand zygotes per generation are rare; one in a million is more typical. Nonetheless, these rates are sufficient to create considerable genetic variation because each of a large number of genes may mutate, chromosomal rearrangements may change many genes simultaneously, and populations often contain large numbers of individuals. For example, if the probability of a point mutation (an addition, subtraction, or substitution of a single base) were 10^{-9} per base pair per generation, then in each human gamete, the DNA of which contains 3×10^9 base pairs, there would be an average of three new point mutations ($3 \times 10^9 \times 10^{-9} = 3$). Therefore, each zygote would carry, on average, six new mutations. The current human population of about 6.5 billion people would be expected to carry about 40 billion new mutations that were not present one generation earlier.

One condition for Hardy–Weinberg equilibrium is that there be no mutation. Although this condition is never strictly met, the rate at which mutations arise at a single locus is usually so low that mutations by themselves result in only very small deviations from Hardy–Weinberg equilibrium. If large deviations are found, it is appropriate to dismiss mutation as the cause and to look for evidence of other evolutionary mechanisms acting on the population.

Gene flow may change allele frequencies

Few populations are completely isolated from other populations of the same species. Migration of individuals and movements of gametes between populations, referred to as **gene flow**, are common. If the arriving individuals or gametes survive and reproduce in their new location, they may add new alleles to the gene pool of the population, or they may change the frequencies of alleles already present if they come from a population with different allele frequencies. For a population to be at Hardy–Weinberg equilibrium, there must be no gene flow from populations with different allele frequencies.

Genetic drift may cause large changes in small populations

In small populations, **genetic drift**—random changes in allele frequencies—may produce large changes in allele frequencies from one generation to the next. Harmful alleles may increase in frequency and rare advantageous alleles may be lost. Even in large populations, genetic drift can influence the frequencies of alleles that do not influence the survival and reproductive rates of their bearers.

As an example, suppose we perform a cross of *Aa* × *Aa* fruit flies to produce an F_1 population in which $p = q = 0.5$ and in which the genotype frequencies are 0.25 *AA*, 0.50 *Aa*, and 0.25 *aa*. If we randomly select 4 individuals (= 8 copies of the gene) from the F_1 population to produce the F_2 generation, the allele frequencies in this small sample population may differ markedly from $p = q = 0.5$. If, for example, we happen by chance to draw 2 *AA* homozygotes and 2 heterozygotes (*Aa*), the allele frequencies in the sample will be $p = 0.75$ (6 out of 8) and $q = 0.25$ (2 out of 8). If we replicate this sampling experiment 1,000 times, one of the two alleles will be missing entirely from about 8 of the 1,000 sample populations.

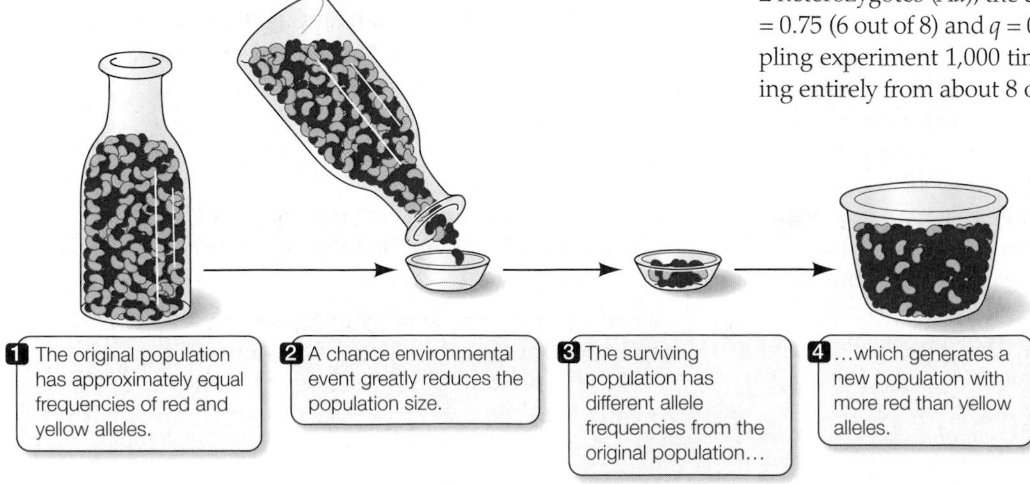

1 The original population has approximately equal frequencies of red and yellow alleles.
2 A chance environmental event greatly reduces the population size.
3 The surviving population has different allele frequencies from the original population...
4 ...which generates a new population with more red than yellow alleles.

22.8 A Population Bottleneck Population bottlenecks occur when only a few individuals survive a random event, resulting in a shift in allele frequencies within the population.

The same principles operate when a population suffers large losses. Populations that are normally large may pass through occasional periods when only a small number of individuals survive. During these **population bottlenecks**, genetic variation can be reduced by genetic drift. How this works is illustrated in **Figure 22.8**, in which red and yellow beans represent two different alleles of a gene. Most of the "surviving" beans in the small sample taken from the original bean population are, just by chance, red, so the new population has a much higher frequency of red beans than the previous generation had. In a real population, the allele frequencies would be described as having "drifted."

A population forced through a bottleneck is likely to lose much of its genetic variation. Millions of greater prairie chickens lived in the prairies of North America when Europeans first arrived. As a result of both hunting and habitat destruction, the Illinois population of prairie chickens plummeted from about 100 million birds in 1900 to fewer than 50 individuals in the 1990s (**Figure 22.9A**). A comparison of DNA from birds collected in Illinois during the middle of the twentieth century with DNA from the surviving population in the 1990s showed that Illinois prairie chickens had lost most of their genetic diversity. The remaining population is experiencing low reproductive success. Similarly, the California fan palm (*Washingtonia filifera*) was once widespread in California and Mexico; today it is restricted to a few oases in extreme southern California and adjacent Mexico (**Figure 22.9B**). The species has little genetic variation: an average individual is heterozygous at only 0.009 of its loci.

South African cheetah populations have very low genetic variation, suggesting an extreme bottleneck occurred in the past. These animals are so close genetically that skin grafts from one cheetah to another are accepted as "self" and do not provoke an immune response.

(A)

Tympanuchus cupido (male)

(B)

Washingtonia filifera

22.9 Species with Low Genetic Variation (A) Prairie chickens in Illinois lost most of their genetic variation when the population crashed from millions to fewer than 100 individuals. (B) The California fan palm, whose range has been reduced to a small area of southern California and neighboring Mexico, has little genetic variation.

Genetic drift can have similar effects when a few pioneering individuals colonize a new region. Because of its small size, the colonizing population is unlikely to have all the alleles found among members of its source population. The resulting change in genetic variation, called a **founder effect**, is equivalent to that in a large population reduced by a bottleneck. For example, the current population of the pitcher plant *Sarracenia purpurea* on a small island arose from a single individual that was planted there in 1912. Today the population has only one polymorphic locus in its entire genome.

Scientists were given an opportunity to study the genetic composition of founding populations when *Drosophila subobscura*, a well-studied species of fruit fly native to Europe, was discovered near Puerto Montt, Chile (in 1978), and at Port Townsend, Washington (in 1982). The *D. subobscura* founders probably reached Chile from Europe on a ship. A few flies carried north from Chile on another ship founded the North American population. In both South and North America, populations of the flies grew rapidly and expanded their ranges. Today in North America, *D. subobscura* ranges from British Columbia, Canada, to central California. In Chile it has spread across 15° of latitude (**Figure 22.10**).

European populations of *D. subobscura* have 80 chromosomal inversions (see Section 12.6), but the North and South American populations have only 20 such inversions—and they are the same 20 on both continents. North and South American populations also have lower allele diversity at certain enzyme-producing genes than European populations do. Only those alleles that have a frequency higher than 0.10 in European populations are present in the Americas. Thus, as expected for a small founding population, only a small part of the total genetic variation found in Europe reached the Americas. Geneticists estimate that somewhere between 4 and 100 flies founded the North and South American populations.

Nonrandom mating changes genotype frequencies

Mating patterns may alter genotype frequencies if individuals in a population choose other individuals of particular genotypes as mates (a phenomenon known as **nonrandom mating**). For exam-

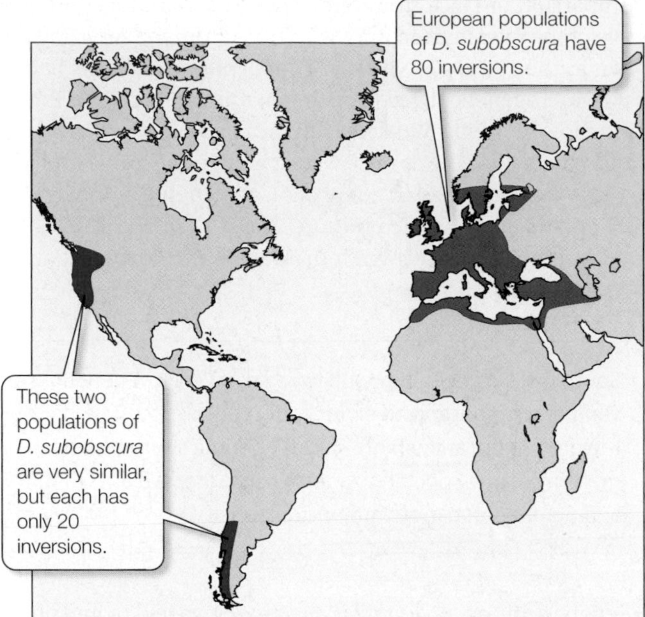

22.10 A Founder Effect Populations of the fruit fly *Drosophila subobscura* in North and South America contain less genetic variation than the European populations from which they came, as measured by the number of chromosome inversions in each population. Within two decades of arriving in the New World, *D. subobscura* populations had increased dramatically and spread widely in spite of their reduced genetic variation.

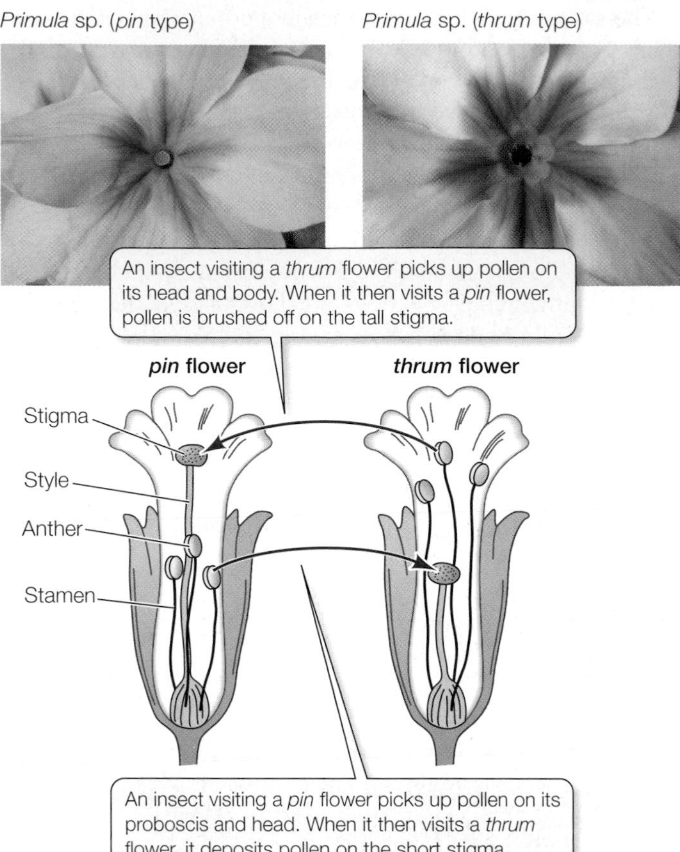

Primula sp. (pin type)
Primula sp. (thrum type)
An insect visiting a thrum flower picks up pollen on its head and body. When it then visits a pin flower, pollen is brushed off on the tall stigma.
pin flower
thrum flower
Stigma
Style
Anther
Stamen
An insect visiting a pin flower picks up pollen on its proboscis and head. When it then visits a thrum flower, it deposits pollen on the short stigma.

22.11 Flower Structure Fosters Nonrandom Mating Differing floral structure within the same plant species, as illustrated by this primrose, ensures that pollination usually occurs between individuals of different genotypes.

ple, if they mate preferentially with individuals of the same genotype, then homozygous genotypes will be overrepresented and heterozygous genotypes underrepresented, relative to Hardy–Weinberg expectations. Alternatively, individuals may mate primarily or exclusively with individuals of different genotypes.

Nonrandom mating is seen in some plant species, such as primroses (genus *Primula*), in which individual plants bear flowers of only one of two different types. One type, known as *pin*, has a long style (female reproductive organ) and short stamens (male reproductive organs). The other type, known as *thrum*, has a short style and long stamens (**Figure 22.11**). In many species with this reciprocal arrangement, pollen from one flower type can fertilize only flowers of the other type. Pollen grains from *pin* and *thrum* flowers are deposited on different parts of the bodies of insects that visit the flowers. When the insects visit other flowers, pollen grains from *pin* flowers are most likely to come into contact with stigmas of *thrum* flowers, and vice versa.

Self-fertilization (*selfing*), another form of nonrandom mating, is common in many groups of organisms, especially plants. Selfing reduces the frequencies of heterozygous individuals from Hardy–Weinberg equilibrium and increases the frequencies of homozygotes, but it does not change allele frequencies, and thus does not result in adaptation.

Sexual selection is a particularly important form of nonrandom mating that *does* change allele frequencies and often results in adaptations. We will discuss this important evolutionary mechanism in detail in the next section.

22.2 RECAP

Evolutionary mechanisms are forces that change the genetic structure of a population. The known evolutionary mechanisms include mutation, gene flow, genetic drift, nonrandom mating, and natural selection.

- Why do mutations, by themselves, result in only very small deviations from Hardy–Weinberg equilibrium? See p. 494
- Can you explain how genetic drift can cause large changes in small populations? See pp. 494–495 and Figure 22.8
- Why is it that some types of nonrandom mating alter genotype frequencies but do not change allele frequencies? See p. 496

The evolutionary mechanisms discussed so far influence the frequencies of alleles and genotypes in populations. Although all of these processes influence the course of biological evolution, none result in adaptations. For adaptation to occur, individuals that differ in heritable traits must survive and reproduce with different degrees of success. What evolutionary mechanisms have these effects?

22.3 What Evolutionary Mechanisms Result in Adaptation?

Adaptation occurs when some individuals in a population contribute more offspring to the next generation than others, such that *allele frequencies in the population change in a way that adapts individuals to the environment that influenced such reproductive success.* This is the mechanism that Darwin called "natural selection."

Although natural selection is formally defined as "the differential contribution of offspring to the next generation by various genotypes belonging to the same population," natural selection in fact acts on the *phenotype*—the physical features expressed by an organism with a given genotype—rather than acting directly on the genotype. The reproductive contribution of a phenotype to subsequent generations relative to the contributions of other phenotypes is called its **fitness**.

The absolute number of offspring produced by an individual does not influence the genetic structure of a population. Changes in absolute numbers of offspring are responsible for increases and decreases in the *size* of a population, but only changes in the *relative* success of different phenotypes within a population lead to changes in allele frequencies from one generation to another—that is, to adaptation. The fitness of individuals of a particular phenotype is a function of the probability of those individuals surviving multiplied by the average number of offspring they produce over their lifetimes. In other words, the *fitness of a phenotype is determined by the average rates of survival and reproduction of individuals with that phenotype*.

Natural selection produces variable results

To simplify our discussion until now, we have considered only characters influenced by alleles at a single locus. However, as we described in Section 10.3, most characters are influenced by alleles at more than one locus. Such characters are likely to show quantitative rather than qualitative variation. For example, the distribution of the body sizes of individuals in a population, a character that is influenced by genes at many loci as well as by the environment, is likely to resemble the bell-shaped curves shown in right-hand column of Figure 22.12.

Natural selection can act on characters with quantitative variation in any one of several different ways, producing quite different results:

- *Stabilizing selection* preserves the average characteristics of a population by favoring average individuals.
- *Directional selection* changes the characteristics of a population by favoring individuals that vary in one direction from the mean of the population.
- *Disruptive selection* changes the characteristics of a population by favoring individuals that vary in opposite directions from the mean of the population.

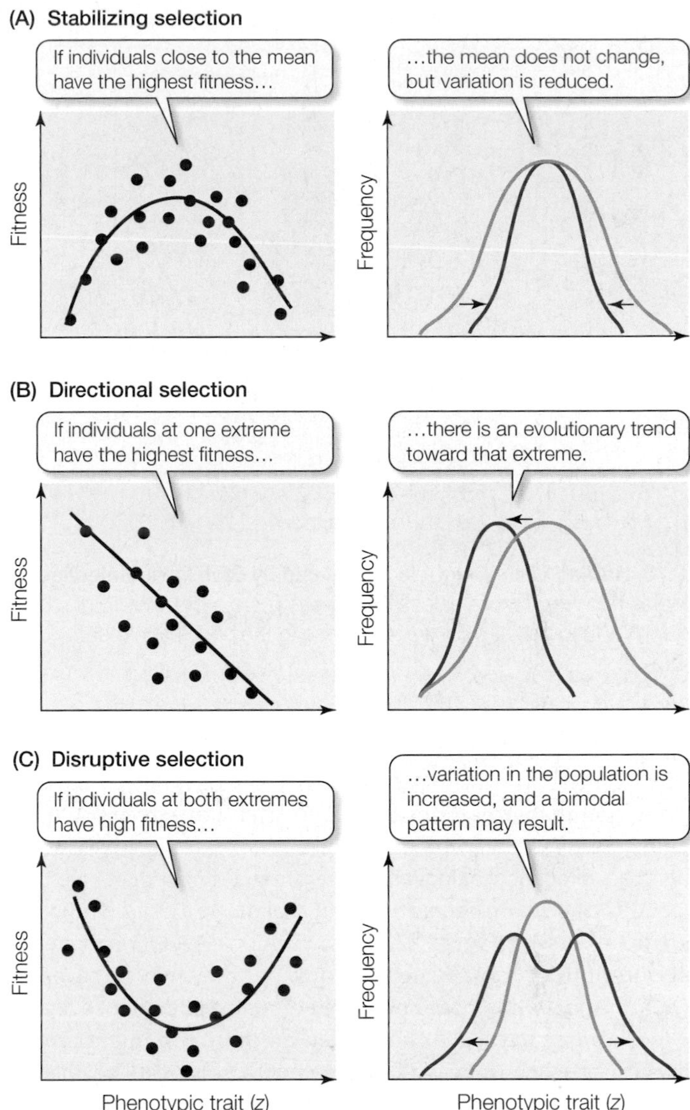

22.12 Natural Selection Can Operate on Quantitative Variation in Several Ways The graphs in the left-hand column show the fitnesses of individuals with different phenotypes of the same trait. Graphs at the right show the distribution of the phenotypes in the population before (light green) and after (dark green) the influence of selection.

STABILIZING SELECTION If the individuals in a population with both the smallest and the largest body sizes contribute relatively fewer offspring to the next generation than those closer to the average size do, then **stabilizing selection** is operating (**Figure 22.12A**). Stabilizing selection reduces variation in populations, but does not change the mean. Natural selection frequently acts in this way, countering increases in variation brought about by sexual recombination, mutation, or migration. Rates of evolution are typically very slow because natural selection is usually stabilizing. Stabilizing selection operates, for example, on human birth weight. Babies born lighter or heavier than the population mean die at higher rates than babies whose weights are close to the mean (**Figure 22.13**).

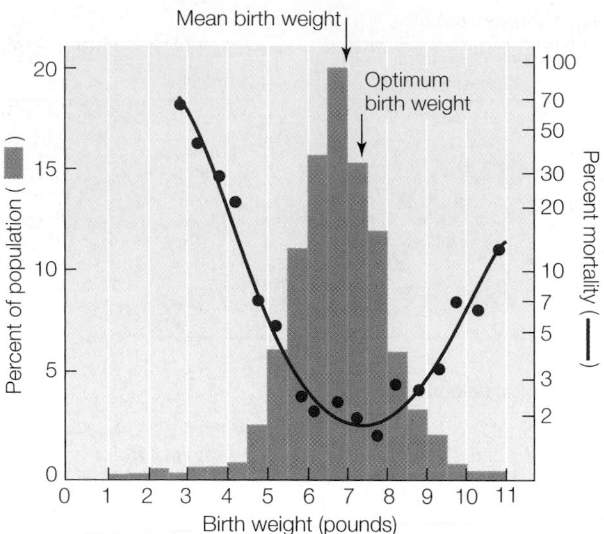

22.13 Human Birth Weight Is Influenced by Stabilizing Selection Babies that weigh more or less than average are more likely to die soon after birth than babies with weights close to the population mean.

DIRECTIONAL SELECTION If individuals at one extreme of a character distribution contribute relatively more offspring to the next generation than other individuals do, then the average value of that character in the population will shift toward that extreme. In this case, **directional selection** is operating. If directional selection operates over many generations, an *evolutionary trend* within the population is seen (**Figure 22.12B**). Directional evolutionary trends often continue for many generations, but they may be reversed when the environment changes and different phenotypes are favored, or they may be halted when an optimum phenotype is reached, or when trade-offs oppose further change. The character then falls under stabilizing selection.

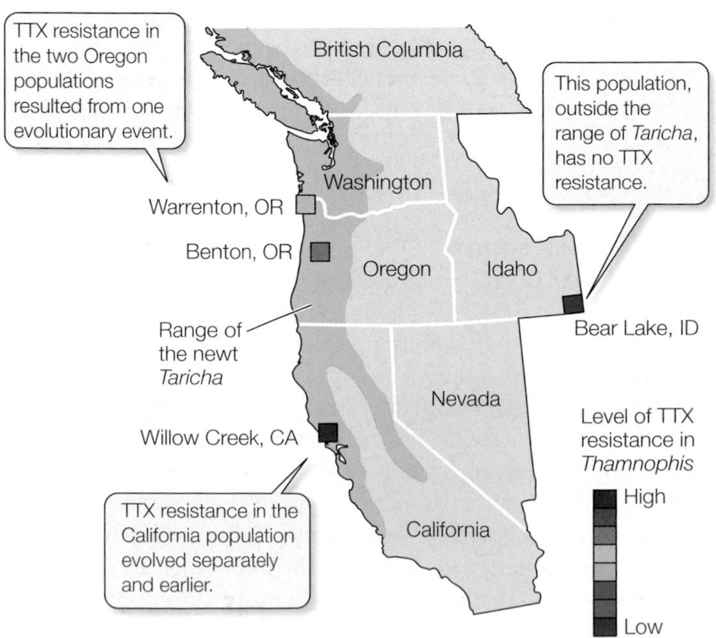

Directional selection produced the resistance to tetrodotoxin (TTX) in garter snakes discussed at the beginning of this chapter. TTX resistance evolved because, in some individual snakes, mutations in the allele (i.e., form of the gene) for the sodium channel proteins in their nerves and muscles resulted in a phenotype (i.e., physical protein) that could continue to function even when exposed to TTX. In areas where toxic newts abounded, such "mutant" snakes survived better than other individuals and left more offspring (who inherited their TTX-resistant allele for the sodium channel protein). TTX resistance has evolved independently several times within *Thamnophis sirtalis* populations in western North America via several different mutations in the same gene (**Figure 22.14**).

DISRUPTIVE SELECTION When **disruptive selection** operates, individuals at opposite extremes of a character distribution contribute more offspring to the next generation than do those close to the mean, increasing variation in the population (**Figure 22.12C**).

The strikingly bimodal (two-peaked) distribution of bill sizes in the black-bellied seedcracker (*Pyrenestes ostrinus*), a West African finch (**Figure 22.15**), illustrates how disruptive selection can influence populations in nature. The seeds of two types of sedges (marsh plants) are the most abundant food source for these finches during part of the year. Birds with large bills can readily crack the hard seeds of the sedge *Scleria verrucosa*. Birds with small bills can crack *S. verrucosa* seeds only with difficulty; however, they feed more efficiently on the soft seeds of *S. goossensii* than do birds with larger bills.

Young finches whose bills deviate markedly from the two predominant bill sizes do not survive as well as finches whose bills are close to one of the two sizes represented by the distribution peaks. Because there are few abundant food sources in the environment, and because the seeds of the two sedges do not overlap in hardness, birds with intermediate-sized bills are less efficient in using either one of the principal food sources. Disruptive selection therefore maintains a bimodal bill size distribution.

Sexual selection influences reproductive success

Sexual selection is a special type of natural selection that acts on characteristics that determine reproductive success. In *The Origin of Species*, Darwin devoted only a few pages to sexual selection, but he subsequently wrote a book about it—*The Descent of Man, and Selection in Relation to Sex* (1871). Sexual selection was Darwin's explanation for the evolution of apparently useless, and in some cases deleterious, but conspicuous characters, such as bright colors, long tails, horns, antlers, and elaborate courtship displays, in males of many species. He hypothesized that these features either improved the ability of their bearers to compete for access to mates (*intrasexual selection*) or made their bearers more attractive to members of the opposite sex (*intersexual selection*). The concept of sexual

22.14 Resistance to TTX Is Associated with the Presence of Newts Garter snakes (*Thamnophis sirtalis*) of the Pacific Coast have evolved resistance to the neurotoxin TTX, which is produced by a prey species found there, the newt *Taricha granulosa*. TTX resistance has evolved at least twice. The range of the newt is shown in blue.

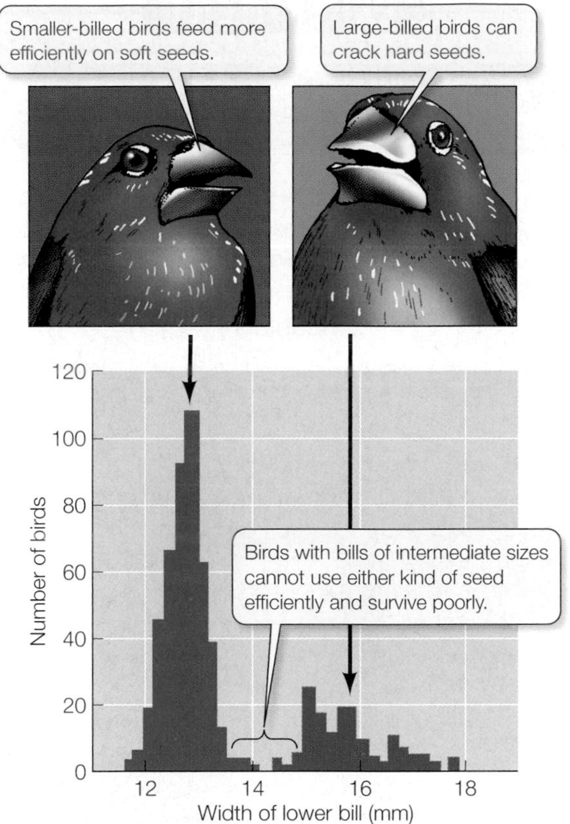

22.15 Disruptive Selection Results in a Bimodal Distribution The bimodal distribution of bill sizes in the black-bellied seedcracker of West Africa is a result of disruptive selection, which favors individuals with larger and smaller bill sizes over individuals with intermediate-sized bills.

selection was either ignored or questioned by Darwin's contemporaries, and for many decades afterward, but recent investigations have demonstrated its importance.

Darwin devoted an entire book to sexual selection because he recognized that, whereas natural selection typically favors traits that enhance the survival of their bearers or their descendants, sexual selection is primarily about success in reproduction. Of course, an animal must survive to reproduce, but if it survives and fails to reproduce, it makes no contribution to the next generation. Thus sexual selection may favor traits that enhance the bearer's chances of reproduction, but reduce its chances of survival. Such costly traits reliably demonstrate the quality of their possessors as mates because they enable the choosing sex (typically females) to distinguish between genuinely fit individuals and exaggerators. If females chose mates on the base of traits that could be easily faked, they would gain no fitness benefit.

One example of a trait that Darwin attributed to sexual selection is the remarkable tail of the male African long-tailed widowbird, which is longer than the bird's head and body combined. To demonstrate that sexual selection drove the evolution of widowbird tails, Malte Andersson, a behavioral ecologist at Gothenburg University, Sweden, captured some male widowbirds. He shortened the tails of some males by cutting them, and lengthened the tails of others by gluing on additional feathers. He cut and reglued the tail feathers of still other males, which served as controls. Male widowbirds normally select, and defend from other males, a territory where they perform courtship displays to attract females. Both short-tailed and long-tailed males successfully defended their display territories, indicating that a long tail does not confer an advantage in male–male competition. However, males with artificially elongated tails attracted about four times more females than did males with shortened tails (**Figure 22.16**).

Why do female widowbirds prefer males with long tails? The ability to grow and maintain a costly feature such as a long tail may indicate that the male bearing it is vigorous and healthy, even though the tail impairs the bird's ability to fly. Even though the manipulated males in Malte Andersson's investigation did not have to pay the price of growing and supporting (except briefly) artificially long tails, the hypothesis that having well-developed orna-

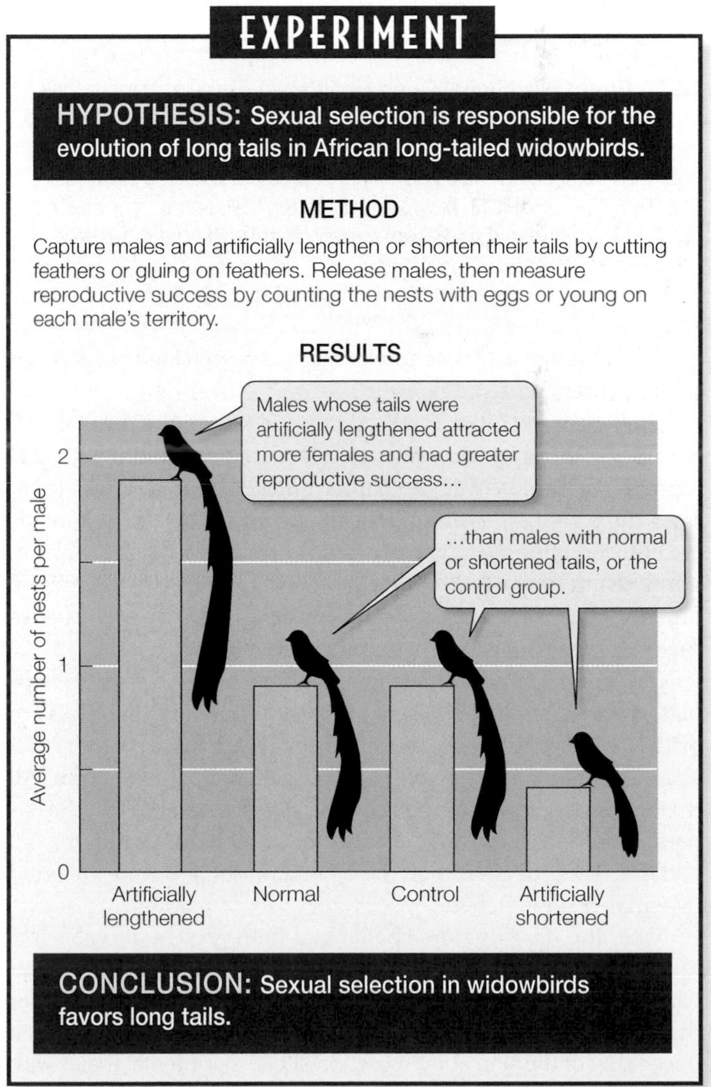

22.16 The Longer the Tail, the Better the Male Male widowbirds with shortened tails defended their display sites successfully, but attracted fewer females (and thus fathered fewer nests of eggs) than did males with normal and lengthened tails.

(A)

(B)

EXPERIMENT

HYPOTHESIS: Having a bright red bill signals good health in a male zebra finch.

METHOD

Provide carotenoids in drinking water for experimental, but not for control males. Challenge all males immunologically and measure response.

RESULTS

Experimental males responded more strongly to the immunological challenge. They also developed brighter bills than control males.

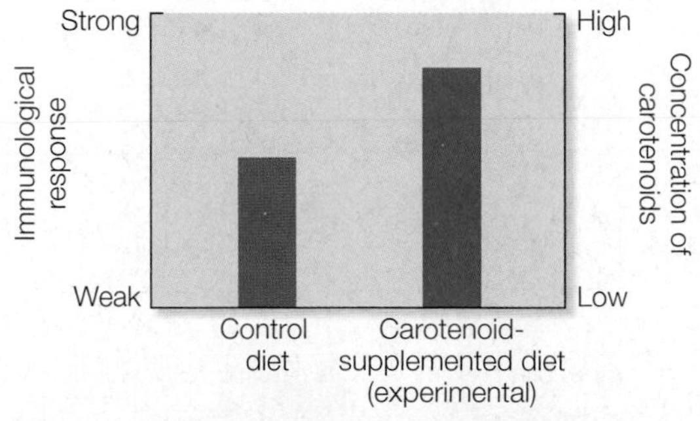

CONCLUSION: Males with high carotenoid levels have bright bills and are immunologically strong. Therefore, bill color is an indication of the health of a male zebra finch.

22.17 Bright Bills Signal Good Health (A) Female zebra finches preferentially choose mates with the brightest bill color, thus choosing the healthiest males. (B) This laboratory experiment demonstrated that bright bill color in the male zebra finch indicates a healthy individual. FURTHER RESEARCH: How would you test this same hypothesis in the field? What would constitute experimental and control birds?

mental traits signals vigor and health has been tested experimentally with captive zebra finches.

The bright red bills of male zebra finches are the result of red and yellow carotenoid pigments. Zebra finches (and most other animals) cannot synthesize carotenoids and must obtain them from their food. In addition to influencing bill color, carotenoids are antioxidants and components of the immune system. Males in good health may need to allocate fewer carotenoids to immune function than males in poorer health. If so, then females can use the brightness of his bill to assess a male's health.

Tim Birkhead and his colleagues at Sheffield University manipulated blood levels of carotenoids in genetically similar male zebra finches by giving experimental males drinking water with added carotenoids; they gave control males only distilled water. All the males had access to the same food. After one month, the experimental males had higher levels of carotenoids in their blood, had much brighter bills than the control males, and were preferred by female zebra finches.

Next, the investigators challenged both groups of males immunologically by injecting phytohemagglutinin (PHA) into their wings. PHA induces a response by T lymphocytes (see Chapter 18) that results in an accumulation of white blood cells and thus a thickening of the skin at the injection site. Experimental males with enhanced carotenoid levels developed thicker skins because they responded more strongly to PHA than control males did, indicating that they had a stronger immune system (**Figure 22.17**).

This experiment showed that when a female chooses a male with a bright red bill, she probably gets a mate with a healthy immune system. Such males are less likely to become infected with parasites and diseases, so they are less likely to pass on infections to their mates. Healthier males are also better able to assist with parental care than are males with duller bills.

22.3 RECAP

Variation in genotype leads to variation in fitness. Fitness is the reproductive contribution of a phenotype to subsequent generations relative to the contributions of other phenotypes in a population. Natural selection acts on phenotypes with quantitative variation in several different ways that can result in adaptation.

- Do you understand the connection between genotype and phenotype, and why natural selection that acts on the phenotype results in changes in genotype frequencies? See p. 497
- Can you describe the differences between stabilizing, directional, and disruptive selection? See pp. 497–498 and Figure 22.12
- Can you explain why Darwin devoted an entire book to sexual selection? See pp. 498–499

Genetic drift, stabilizing selection, and directional selection all tend to reduce genetic variation within populations. Nevertheless, as we have seen, most populations have considerable genetic variation. What processes produce and maintain genetic variation within populations?

22.4 How Is Genetic Variation Maintained within Populations?

We have seen that genetic variation is the raw material on which mechanisms of evolution act. Several forces—neutral mutations, sexual recombination, and frequency-dependent selection—operate to maintain genetic variation in populations despite the action of other forces that reduce it (such as genetic drift and many types of selection). In addition, as we will show, genetic variation may be maintained over geographic space.

Neutral mutations may accumulate within populations

As we saw in Section 12.6, some mutations do not affect the function of the proteins encoded by the mutated genes. An allele that does not affect the fitness of an organism—that is, an allele that is no better or worse than alternative alleles at the same locus—is called a **neutral allele**. Neutral alleles are unaffected by natural selection. Even in large populations, neutral alleles may be lost or their frequencies may increase, purely by genetic drift. Neutral alleles tend to accumulate in a population over time, providing it with considerable genetic variation.

Much of the *phenotypic* variation that we are able to observe is not neutral. However, modern techniques enable us to measure neutral variation at the *molecular* level and provide the means by which to distinguish it from adaptive variation. In Section 24.2 we will see how these techniques enable us to discriminate among alleles and how variation in neutral molecular traits can be used to estimate rates of evolution.

Sexual recombination amplifies the number of possible genotypes

In asexually reproducing organisms, each new individual is genetically identical to its parent unless there has been a mutation. When organisms reproduce sexually, however, offspring differ from their parents, due to crossing over and independent assortment of chromosomes during meiosis as well as the combination of genetic material from two different gametes, as described in Chapter 9. Sexual recombination generates an endless variety of genotypic combinations that increases the evolutionary potential of populations—a long-term advantage of sex.

The evolution of the mechanisms of meiosis and sexual recombination were crucial events in the history of life. Exactly how these attributes arose is puzzling, however, because sex has at least three striking disadvantages in the short term:

- Recombination breaks up adaptive combinations of genes.
- Sex reduces the rate at which females pass genes on to their offspring.
- Dividing offspring into separate genders greatly reduces the overall reproductive rate.

To see why this last disadvantage exists, consider an asexual female that produces the same number of offspring as a sexual female. Let's assume that both females produce two offspring, but that the sexual female must produce 50 percent males. In the next generation, both asexual F_1 females will produce two more offspring, but there is only one sexual F_1 female to produce offspring. The evolutionary problem is to identify the advantages of sex that can overcome such short-term disadvantages.

A number of hypotheses have been proposed for the existence of sex, none of them mutually exclusive. One is that sexual recombination facilitates repair of damaged DNA, because breaks and other errors in DNA on one chromosome can be repaired by copying the intact sequence from the homologous chromosome.

Another advantage of sexual reproduction is that it permits the elimination of deleterious mutations. As we saw in Section 11.4, DNA replication is not perfect. Errors are introduced in every generation, and many or most of these errors result in lower fitness. Asexual organisms have no mechanism to eliminate deleterious mutations. Hermann J. Muller noted that the accumulation of mutations in a non-recombining genome is like a genetic ratchet. The mutations accumulate—"ratchet up"—at each replication: that is, a mutation occurs and is passed on when the genome replicates, then two new mutations occur in the next replication, so three mutations are passed on, and so on. Deleterious mutations cannot be eliminated except by the death of the lineage. This accumulation of deleterious mutations in asexual lineages is known as *Muller's ratchet*.

In sexual species, on the other hand, genetic recombination produces some individuals with more of these deleterious mutations, and some with fewer. The individuals with fewer deleterious mutations are more likely to survive. Therefore, sexual reproduction allows natural selection to eliminate deleterious mutations from the population over time.

Another explanation for the existence of sex is that the great variety of genetic combinations created each generation by sexual recombination may be especially valuable as a defense against pathogens and parasites. Most pathogens and parasites have much shorter life cycles than their hosts, and thus can potentially evolve counteradaptations to host defenses rapidly. Sexual recombination might give the host's defenses a chance to keep up.

Sexual recombination does not influence the frequencies of alleles; rather, *sexual recombination generates new combinations of alleles on which natural selection can act*. It expands variation in a character influenced by alleles at many loci by creating new genotypes. That is why artificial selection for bristle number in *Drosophila* (see Figure 22.5) resulted in flies with more bristles than any flies in the initial population had.

Frequency-dependent selection maintains genetic variation within populations

Natural selection often preserves variation as a polymorphism. A polymorphism may be maintained when the fitness of a genotype (or phenotype) depends on its frequency in a population. This phenomenon is known as **frequency-dependent selection**.

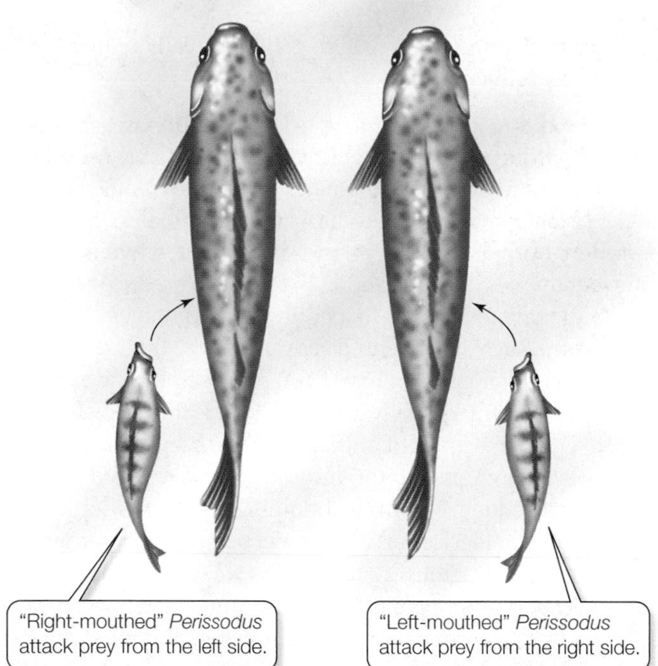

"Right-mouthed" *Perissodus* attack prey from the left side.
"Left-mouthed" *Perissodus* attack prey from the right side.

22.18 A Stable Polymorphism Frequency-dependent selection maintains equal proportions of left-mouthed and right-mouthed individuals of the scale-eating fish *Perissodus microlepis*.

A small fish that lives in Lake Tanganyika, in East Africa, provides an example of frequency-dependent selection. The mouth of this scale-eating fish, *Perissodus microlepis*, opens either to the right or to the left as a result of an asymmetrical jaw joint; the direction of opening is genetically determined (**Figure 22.18**). The scale-eater approaches its prey (another fish) from behind and dashes in to bite off several scales from its flank. "Right-mouthed" individuals always attack from the victim's left; "left-mouthed" individuals always attack from the victim's right. The distorted mouth enlarges the area of teeth in contact with the prey's flank, but only if the scale-eater attacks from the appropriate side.

Prey fish are alert to approaching scale-eaters, so attacks are more likely to be successful if the prey must watch both flanks. Vigilance by prey favors equal numbers of right-mouthed and left-mouthed scale-eaters, because if attacks from one side were more common than the other, prey fish would pay more attention to potential attacks from that side. Over an 11-year study of the scale-eaters in Lake Tanganyika, the polymorphism was found to be stable: the two forms of *P. microlepis* remained at about equal frequencies.

Environmental variation favors genetic variation

Environments vary greatly over time. A night is dramatically different from the preceding day. A cold, cloudy day differs from a clear, hot one. Day lengths and temperatures change seasonally. No single genotype is likely to perform well under all these conditions.

Colias butterflies of the Rocky Mountains live in environments where dawn temperatures often are too cold, and afternoon temperatures too hot, for the butterflies to fly. Populations of this butterfly are polymorphic for the enzyme phosphoglucose isomerase (PGI), which influences how well the butterfly flies at different temperatures. Some PGI genotypes can fly better during the cold hours of early morning; others perform better during midday heat. The optimal body temperature for flight is 35°C to 39°C, but some butterflies can fly with body temperatures as low as 29°C or as high as 40°C. During spells of unusually hot weather, heat-tolerant genotypes are favored; during spells of unusually cool weather, cold-tolerant genotypes are favored.

Heterozygous *Colias* butterflies can fly over a greater temperature range than homozygous individuals, which should give them an advantage in foraging and finding mates. Thus heterozygous males should be more successful in inseminating females than homozygous males. Ward Watt tested this prediction; His results are shown in **Figure 22.19**.

EXPERIMENT

HYPOTHESIS: Heterozygous male *Colias* butterflies should have better mating sucess than homozygous males because they can fly farther under a broader range of temperatures.

METHOD

Capture butterflies in the field, transfer them to the laboratory, and determine their genotypes. Allow females to lay eggs, and determine the genotypes of the offspring (and thus paternity and mating success) of males.

RESULTS

For both species, the proportion of heterozygous males that mated successfully was higher than the proportion of all males seeking females ("flying").

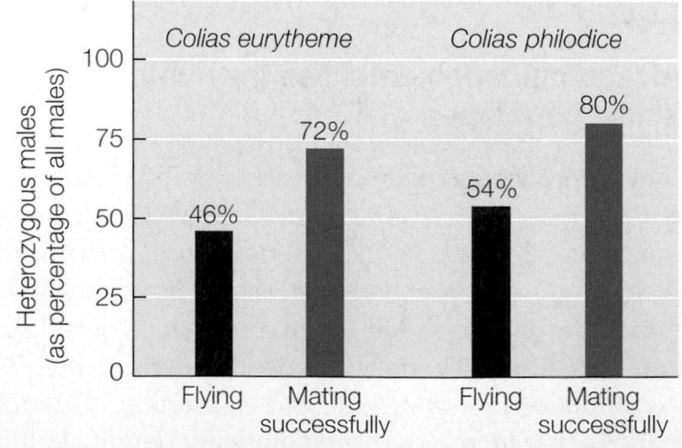

CONCLUSION: Heterozygous *Colias* males have a mating advantage over homozygous males.

22.19 A Heterozygote Mating Advantage Heterozygous male *Colias* butterflies can fly farther than homozygous males under a broader range of temperatures; thus heterozygotes are more successful in inseminating females.

Much genetic variation is maintained in geographically distinct subpopulations

Much of the genetic variation in large populations is preserved as differences among members living in different places (subpopulations). Subpopulations often vary genetically because they are subjected to different selective pressures in different environments. Environments may vary significantly over short distances. For example, in the Northern Hemisphere, temperature and soil moisture differ dramatically between north-facing and south-facing mountain slopes. In the Rocky Mountains of Colorado, the proportion of ponderosa pines (*Pinus ponderosa*) that are heterozygous for a particular peroxidase enzyme is particularly high on south-facing slopes, where temperatures fluctuate dramatically, often on a daily basis. This heterozygous genotype performs well over a broad range of temperatures. On north-facing slopes and at higher elevations, where temperatures are cooler and fluctuate less strikingly, a peroxidase homozygote, which has a lower optimum temperature, is much more frequent.

Plant species may also vary geographically in the chemicals they synthesize to defend themselves against herbivores. Some individuals of the white clover (*Trifolium repens*) produce the poisonous chemical cyanide. Poisonous individuals are less appealing to herbivores—particularly mice and slugs—than are nonpoisonous individuals. However, clover plants that produce cyanide are more likely to be killed by frost, because freezing damages cell membranes and releases cyanide into the plant's own tissues.

In European subpopulations of *Trifolium repens*, the frequency of cyanide-producing individuals increases gradually from north to south and from east to west (**Figure 22.20**). Poisonous plants make up a large proportion of clover subpopulations only in areas where winters are mild. Cyanide-producing individuals are rare where winters are cold, even though herbivores graze clovers heavily in those areas.

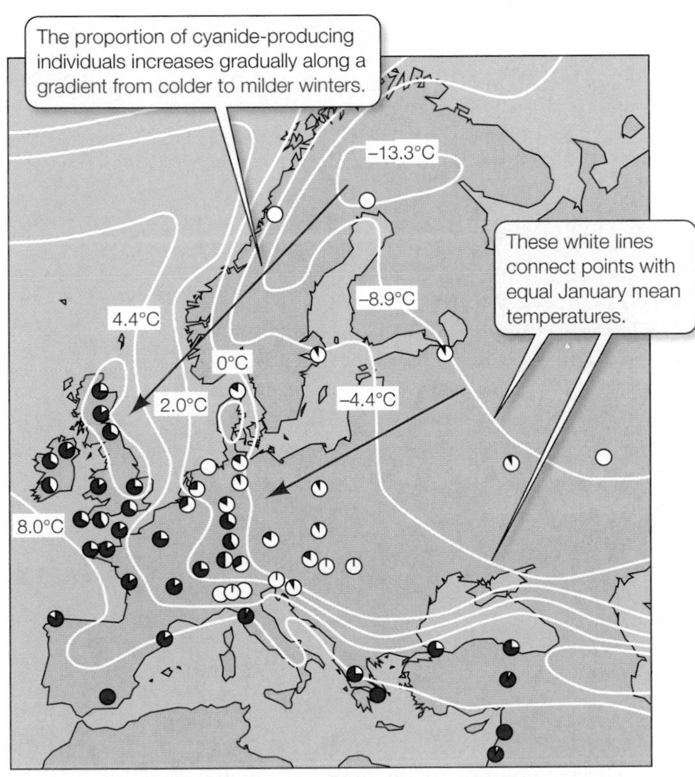
The proportion of cyanide-producing individuals increases gradually along a gradient from colder to milder winters.
−13.3°C
These white lines connect points with equal January mean temperatures.
−8.9°C
4.4°C
0°C
2.0°C
−4.4°C
8.0°C
Red indicates proportion of plants producing cyanide
White indicates proportion of plants not producing cyanide

22.20 Geographic Variation in a Defensive Chemical The frequency of cyanide-producing individuals in each subpopulation of white clover (*Trifolium repens*) in Europe depends on the winter temperature of its environment.

22.4 RECAP

Neutral mutations, sexual recombination, and frequency-dependent selection all act to maintain considerable genetic variation within most populations. Variation is also maintained among geographically distinct, genetically variable subpopulations.

- Do you understand why sexual recombination is so prevalent in nature, despite having at least three short-term evolutionary disadvantages? See p. 501
- How does frequency-dependent selection act to maintain genetic variation within a population? See pp. 501–502

The mechanisms of evolution have produced a remarkable variety of organisms, some of which are adapted to most environments that exist on Earth. This natural variation, and the success of breeders attempting to produce desired traits in domesticated plants and animals, suggests that natural selection can favor most traits that might be adaptive. But is this impression correct?

22.5 What Are the Constraints on Evolution?

We would be mistaken to assume that natural selection always results in adaptive traits. There are obviously constraints on evolution. Lack of appropriate genetic variation, for example, could prevent the development of many potentially favorable traits. (That is, if the allele for a given trait does not exist in a population, that trait cannot evolve even if it would be highly favored by natural selection.) Evolutionary biologists have long debated whether lack of genetic variation has seriously constrained the rate and direction of evolutionary change, but nobody knows for sure how important this constraint has been. Evidence is accumulating, however, for other types of constraints on evolution—indeed, imagine how many different organisms might have evolved if there were none.

Earlier in the book we learned of constraints imposed on organisms by the dictates of physics and chemistry. The size of cells, for example, is constrained by the stringencies of surface-area-to-volume ratios (see Section 2.1). The ways in which proteins can fold are limited by the bonding capacities of their constituent molecules (see Section 3.2). And the energy transfers that fuel life must operate within the laws of thermodynamics (see Section 6.1). Keep in mind that evolution works within the boundaries of these universal constraints, as well those mentioned below.

Developmental processes constrain evolution

As we saw in Section 20.5, developmental constraints on evolution are paramount because all evolutionary innovations are modifications of previously existing structures. Human engineers seeking to power an airplane may be able to design a completely new type of engine (jet) that can replace a previous type (propeller), but evolutionary changes cannot happen that way.

A striking example of such developmental constraints is provided by the evolution of fish that spend most of their time on the sea bottom. One lineage, the bottom-dwelling skates and rays, share a common ancestor with sharks, whose bodies were already somewhat ventrally flattened, and whose skeletal frame is made of flexible cartilage. Skates and rays evolved a body plan that further flattened their bellies, allowing them to swim along the ocean floor (**Figure 22.21A**).

Plaice, sole, and flounder, on the other hand, are bottom-dwelling descendants of deep-bellied, laterally flattened ancestors with bony skeletons. The only way such a fish can lie flat is to flop over onto its side. Their ability to swim is thus curtailed, but their bodies can lie still and are well camouflaged. During development, the eyes of these flatfishes are grotesquely twisted so that both eyes are one side of the body (**Figure 22.21B**). Small shifts in the position of one eye probably helped ancestral flatfishes see better, eventually resulting in the body form found today.

(A) *Taeniura lymma*

(B) *Bothus lunatus*

22.21 Two Solutions to a Single Problem (A) Stingrays, whose ancestors were dorsoventrally flattened, lie on their bellies. Their bodies are symmetrical around the dorsal backbone. (B) Flounders, whose ancestors were laterally flattened, lie on their sides. (The backbone of this fish is at the right.) Their eyes migrate during development so that both eyes are on the same side of the body.

Trade-offs constrain evolution

As we saw in the case of the garter snakes described at the opening of this chapter, all adaptations impose both fitness costs and fitness benefits. For an adaptation to evolve, the fitness benefits it confers must exceed the fitness costs it imposes—in other words, the trade-off must be worthwhile. Garter snakes pay for their ability to eat rough-skinned newts by being unable to move rapidly after having eaten a newt. Sufficient benefits apparently exist to override that cost *only* in areas where newts are common and are potentially major items in the snakes' diet.

The costs of developing and maintaining certain conspicuous features used by males to compete with other males for access to females have been estimated using the comparative method. In some mammals, such as deer, lions, and baboons, one male controls reproductive access to many females. Species in which males have multiple mates are said to be *polygynous* (the term *monogamous* applies to species that have just one mate). Males of many polygynous species are significantly larger than the females and often bear large weapons (such as horns, antlers, and large canine teeth). Such dramatic differences between the sexes are known as *sexual dimorphism*.

In polygynous species, large size and weaponry are needed to defend a male's multiple mates against other males of the species. Developing and maintaining these structures is energetically costly, and males are often injured during their contests. Their life span is often considerably shorter than that of females of the same species. However, since during his "time at the top" a polygynous male enjoys access to multiple females, his reproductive success will usually be greater than that of males he chases away, or that females reject as "inferior."

Short-term and long-term evolutionary outcomes sometimes differ

The short-term changes in allele frequencies within populations that we have emphasized in this chapter are an important focus of study for evolutionary biologists. These changes can be observed directly, they can be manipulated experimentally, and they demonstrate the actual processes by which evolution occurs. By themselves, however, they do not enable us to predict—or, more properly, "postdict" (because they have already happened)—the kinds of long-term evolutionary changes described in Section 21.3.

Patterns of evolutionary change can be strongly influenced by events that occur so infrequently (a meteorite impact, for example) or so slowly (continental drift) that they are unlikely to be observed during short-term studies. In addition, the ways in which evolutionary processes act may change over time; even among the descendants of a single ancestral species, different lineages may evolve in different directions. Therefore, additional types of evidence, demonstrating the effects of rare and unusual events on trends in the fossil record, must be gathered if we wish to understand the course of evolution over billions of years. In subsequent chapters, we will discuss the methods that biologists use to study long-term evolutionary changes and infer the processes that led to them.

22.5 RECAP

Developmental processes constrain evolution because all evolutionary innovations are modifications of previously existing structures. An adaptation can evolve only if the fitness benefits it confers exceed the fitness costs it imposes.

- How do trade-offs constrain the evolution of adaptations? See p. 504
- Do you see why the presence of a great deal of genetic variation within a population could increase the chances that some members of the population would survive an unprecedented environmental change? Do you also see why there is no guarantee that this would be the case?

Human beings have not altered the mechanisms of evolution. Humans have, however, influenced evolutionary events.

22.6 How Have Humans Influenced Evolution?

Evolutionary processes are operating today just as they have been since life first appeared on Earth. Like all living things, humans are still evolving, but as humans have changed their environment, the selective forces affecting us have also changed, and now differ from those that dominated most of human prehistory. Today few humans are killed by predators, and deaths due to bad weather, although not infrequent, are rare compared with deaths caused by diseases, accidents, wars, and murders. Today, some of the differences in survival and reproductive success among humans are related to genes that confer defenses against diseases like malaria, as well as health conditions related to the stresses of modern industrial life, such as hypertension, diabetes, and AIDS.

People in most parts of the world cannot digest lactose, the sugar in dairy products. Enzymes that digest lactose began to be selected for between 5,000 and 10,000 years ago when human populations in certain regions began to consume milk from their domesticated animals.

Many evolutionary changes are taking place all around us, and humans are influencing some of those changes. For example, our attempts to control populations of species we consider pests and to increase populations of those we consider desirable make us powerful agents of evolutionary change. In addition to producing the results we desire, these efforts often cause undesirable consequences, such as the evolution of resistance to antibiotics by pathogens and to pesticides by pests. Medicine and agriculture can respond creatively to the evolutionary changes they are causing only if their practitioners understand how and why those changes happen.

Ovis canadensis (trophy male)

Ovis canadensis (male)

22.22 Trophy Hunting Selects for Smaller Males Hunters preferentially kill large "trophy" rams rather than their less impressive counterparts.

Many current human-caused evolutionary changes are inadvertent. For example, sport hunters often seek large trophy animals to display on the walls of their homes. By preferentially shooting male mammals with particularly large ornaments, trophy hunters have inadvertently favored males with smaller ornaments. Trophy hunting is big business: one hunter paid more than a million Canadian dollars in 1998 and 1999 for permits to hunt trophy bighorn rams in Alberta. Since 1975, most rams (45 of 57) shot at Ram Mountain, Alberta, have been trophy rams: large individuals with large horns (**Figure 22.22**). The result has been a steady decline in the size of rams shot by hunters and in the size of their horns). Nearly all of these rams were shot before they had achieved most of the reproductive success that would be expected over their lifetimes. Ironically, selectively harvesting trophy rams is making trophy animals scarce. Unrestricted trophy hunting of other species has similarly resulted in diminished populations of animals with the valued traits.

Humans have influenced evolution in a number of far-reaching ways. We have moved thousands of species around the globe and modified organisms using biotechnology. Human activities have changed the climate and greatly increased the rate of extinction of other species. We will continue to see examples of the effects of *Homo sapiens* on the evolution of other species throughout this book.

CHAPTER SUMMARY

22.1 What facts form the base of our understanding of evolution?

See Web/CD Tutorial 22.1

Charles Darwin attributed changes in species over time to the possession of advantageous traits by some individuals. He understood that *individuals* do not evolve, but *populations* evolve when individuals with different heritable genotypes survive and reproduce at different rates.

Adaptation refers both to characteristics of organisms and the way those characteristics are acquired via **natural selection**.

The sum of all copies of all alleles at all loci found in the population constitutes its **gene pool** and represents the **genetic variation** that results in different phenotypic traits upon which natural selection can act. Review Figure 22.3

Artificial selection and laboratory experiments demonstrate the existence of considerable genetic variation in most populations. Review Figure 22.5

Allele frequencies measure the amount of genetic variation in a population; **genotype frequencies** show how a population's genetic variation is distributed among its members. Review Figure 22.6

Hardy–Weinberg equilibrium predicts the allele frequencies in populations in the absence of evolution. Deviation from these frequencies indicates the work of evolutionary mechanisms. Review Figure 22.7, Web/CD Tutorial 22.2

22.2 What are the mechanisms of evolutionary change?

Migration of individuals between populations results in **gene flow**.

In small populations, **genetic drift**—the random loss of individuals and the alleles they possess—may produce large changes in allele frequencies from one generation to the next and greatly reduce genetic variation. Review Figure 22.8

Population bottlenecks occur when only a few individuals survive a random event, resulting in a drastic shift in allele frequencies within the population and the loss of variation. Similarly, a population established by a small number of individuals colonizing a new region may lose variation via a **founder effect**.

Nonrandom mating may result in genotype frequencies that deviate from Hardy–Weinberg equilibrium.

22.3 What evolutionary mechanisms result in adaptation?

See Web/CD Tutorial 22.3

Fitness is the reproductive contribution of a phenotype to subsequent generations relative to the contributions of other phenotypes.

Changes in numbers of offspring are responsible for changes in the absolute size of a population, but only changes in the relative success of different phenotypes within a population lead to changes in allele frequencies.

Natural selection can act on traits with quantitative variation in several different ways, resulting in **stabilizing**, **directional**, or **disruptive selection**. Review Figure 22.12

Sexual selection is primarily about success in reproduction, not about success in survival. Review Figures 22.16 and 22.17

22.4 How is genetic variation maintained within populations?

Although genetic drift, stabilizing selection, and directional selection all tend to reduce genetic variation within populations, most populations have considerable genetic variation.

Neutral mutations, sexual recombination, and frequency-dependent selection can maintain genetic variation within populations.

Neutral alleles do not affect the fitness of an organism, are not affected by natural selection, and may accumulate or be lost by genetic drift.

Sexual reproduction generates countless genotypic combinations that increase the evolutionary potential of populations despite short-term disadvantages.

A polymorphism may be maintained by **frequency-dependent selection** when the fitness of a genotype depends on its frequency in a population.

Genetic variation may be maintained by the existence of genetically distinct subpopulations over geographic space. Review Figure 22.20

22.5 What are the constraints on evolution?

Developmental processes constrain evolution because all evolutionary innovations are modifications of previously existing structures.

Most adaptations impose costs. An adaptation can evolve only if the benefits it confers exceed the costs it imposes, a situation that leads to **trade-offs**.

22.6 How have humans influenced evolution?

Humans have become major agents of evolution as they attempt to control pests and diseases, move species around the globe, and modify organisms via biotechnology. Human activities are changing the climate and have greatly increased the rates of extinction of other species.

SELF-QUIZ

1. Garter snakes that are resistant to TTX move more slowly than those that are susceptible to the toxin. Their reduced speed is an example of
   a. an adaptation.
   b. genetic drift.
   c. natural selection.
   d. a trade-off.
   e. none of the above
2. Which of the following is *not* true?
   a. Darwin and Wallace were both influenced by Malthus.
   b. Wallace proposed a theory of evolution by natural selection that was similar to Darwin's.
   c. Malthus claimed that because human population growth would outstrip any increases in food production, famine was a likely result.
   d. Darwin realized that all populations had the capacity for rapid increases in numbers.
   e. All of the above are true.
3. The phenotype of an organism is
   a. the type specimen of its species in a museum.
   b. its genetic constitution, which governs its traits.
   c. the chronological expression of its genes.
   d. the physical expression of its genotype.
   e. the form it achieves as an adult.

4. The appropriate unit for defining and measuring genetic variation is the
   a. cell.
   b. individual.
   c. population.
   d. community.
   e. ecosystem.
5. Which statement about allele frequencies is *not* true?
   a. The sum of all allele frequencies at a locus is always 1.
   b. If there are two alleles at a locus and we know the frequency of one of them, we can obtain the frequency of the other by subtraction.
   c. If an allele is missing from a population, its frequency in that population is 0.
   d. If two populations have the same gene pool for a locus, they will have the same proportion of homozygotes at that locus.
   e. If there is only one allele at a locus, its frequency is 1.
6. Which of the following is *not* required for a population at Hardy–Weinberg equilibrium ?
   a. There is no migration between populations.
   b. Natural selection is not acting on the alleles in the population.
   c. Mating is random.
   d. The frequency of one allele must be greater than 0.7.
   e. All of the above conditions must be met.
7. The fitness of a genotype is a function of the
   a. average rates of survival and reproduction of individuals with that genotype.
   b. individuals that have the highest rates of both survival and reproduction.
   c. individuals that have the highest rates of survival.
   d. individuals that have the highest rates of reproduction.
   e. average reproductive rate of individuals with that genotype.
8. Laboratory selection experiments with fruit flies have demonstrated that
   a. bristle number is not genetically controlled.
   b. bristle number is not genetically controlled, but changes in bristle number are caused by the environment in which the fly is raised.
   c. bristle number is genetically controlled, but there is little variation on which natural selection can act.
   d. bristle number is genetically controlled, but selection cannot result in flies having more bristles than any individual in the original population had.
   e. bristle number is genetically controlled, and selection can result in flies having more bristles than any individual in the original population had.
9. Disruptive selection maintains a bimodal distribution of bill size in the West African seedcracker because
   a. bills of intermediate shapes are difficult to form.
   b. the birds' two major food sources differ markedly in size and hardness.
   c. males use their large bills in displays.
   d. migrants introduce different bill sizes into the population each year.
   e. older birds need larger bills than younger birds.
10. Which of the following is *not* a reason why trade-offs constrain evolution?
    a. All adaptations impose both fitness costs and benefits.
    b. Sexual selection for large size and weaponry shortens the life span of their possessors.
    c. Changes in allele frequencies may be influenced by events that occur very infrequently.
    d. Ability to consume toxic prey may reduce mobility.
    e. Adaptations can evolve only if the fitness benefits they confer exceed the costs they impose.

FOR DISCUSSION

1. In what ways does artificial selection by humans differ from natural selection? Was Darwin wise to base so much of his argument for natural selection on the results of artificial selection?
2. In nature, mating among individuals in a population is never truly random, immigration and emigration are common, and natural selection is seldom totally absent. Why, then, is Hardy–Weinberg equilibrium, which is based on assumptions known generally to be false, so useful in our study of evolution? Can you think of other models in science that are based on false assumptions? How are such models used?
3. As far as we know, natural selection cannot adapt organisms to future events. Yet many organisms appear to respond to natural events before they happen. For example, many mammals go into hibernation while it is still quite warm. Similarly, many birds leave the temperate zone for their southern wintering grounds long before winter has arrived. How can such "anticipatory" behaviors evolve?
4. Populations of most of the thousands of species that have been introduced to areas where they were previously not found, including those that have become pests, began with a few individuals. They should therefore have begun with much less genetic variation than the parent populations have. If genetic variation is advantageous, why have so many of these species been successful in their new environments?
5. Why is it important that the ways in which males advertise their health and vigor to females reliably indicate their status?
6. How does selection on humans today differ from selection on humans in the preindustrial past (i.e., prior to 1700)?
7. As more humans live longer, many people face degenerative conditions such as Alzheimer's disease that (in most cases) are linked to advancing age. Assuming that some individuals may be genetically predisposed to successfully combat these conditions, is it likely that natural selection alone would act to favor such a predisposition in human populations? Why or why not?

FOR INVESTIGATION

During the past 50 years, more than 200 species of insects that attack crop plants have become highly resistant to DDT and other pesticides. Using your recently acquired knowledge of evolutionary processes, explain the rapid and widespread evolution of resistance. What proposals concerning pesticide use would you make in order slow down the rate of evolution of resistance? Explain why you think your proposals could work and how you might test them.

CHAPTER 23 Species and Their Formation

Sex stimulates speciation (among other things)

Charles Darwin proposed that sexual selection was responsible for the evolution of such bright and conspicuous traits as the large antlers of male deer and the greatly enlarged tail feathers of male peacocks. Those traits are especially exaggerated in species in which individuals of one sex—usually males—compete for opportunities to mate with individuals of the opposite sex—usually females. Exaggerated traits evolve if they confer an advantage, either in competition among males for access to females, or in stimulating and attracting discriminating females.

In addition to leading to conspicuous male ornaments such as antlers and plumage, recent evidence suggests that sexual selection may also increase the rate at which new species form. Evidence for this effect of sexual selection comes from comparing the number of species found in sister clades, that is, clades that share a common ancestor. Because they share a common ancestor, sister clades have been evolving independently from one another for the same length of time. The rate at which species have formed in the two clades can be estimated by comparing the number of species in them today.

Birds with promiscuous mating systems offer some of the best examples of both of these effects of sexual selection. In many of these species males assemble in display grounds and females come there to choose with whom to copulate. After mating, the females of many species build their nests, lay their eggs, and raise their offspring with no help from the males. And in most of these species, the males have evolved bright plumage and often ornaments such as long tail feathers. In contrast, in monogamous species that form pair bonds and share the responsibilities of raising young, individuals of both sexes tend to have dull plumage and to look alike.

As an example of this phenomenon, there are about 320 species of hummingbirds in the Americas, all of which are promiscuous. However, there are only 103 species of swifts (sister clade to the hummingbirds), even though swifts are found worldwide. Male and female swifts form monogamous pair bonds and look alike.

Why does sexual selection stimulate the divergence of a lineage into many species? A likely reason is that random mutations result in different plumage elaborations in different parts of the range of a species. Thus, local populations of a species may develop unique male plumage patterns, each of which is favored by sexual selection in that area. Females may not respond positively to an immigrant male whose

A Male Hummingbird More than 300 species of hummingbirds are found in the Americas. Hummingbird males are promiscuous. They compete with other males for mates, and males with the most successful display will usually inseminate the most females and thus sire the most offspring.

CHAPTER OUTLINE

A Male Swift Swifts are the sister clade of hummingbirds. Swifts form pair bonds and the male and female, who look much alike, raise the young together. Thus the most successful males are those that help their mates raise the most offspring. There are only 103 species of swifts, even though they are found worldwide.

plumage differs from the pattern to which they are normally attracted.

The origin of new species—the splitting and divergence of a single lineage into distinct species—is one of the most important phenomena in biological science. Charles Darwin recognized its preeminence when he chose the title of his most famous book. But without the underlying knowledge supplied by modern genetics, Darwin was primarily viewing the consequences of speciation, not its underlying causes, and he recognized that as well. We are still looking for the answers to many questions about speciation, a process Darwin referred to as "the mystery of mysteries."

IN THIS CHAPTER we will describe what species are and discuss how Earth's millions of species came into being. We will examine the mechanisms by which a population splits into new species and how such separations are maintained. We will look at different factors that can make speciation a rapid or a very slow process. Finally, we will look at the conditions that give rise to the great species diversifications known as evolutionary radiations.

23.1 What Are Species?

The word *species* means, literally, "kinds." But how do biologists interpret "kinds"? Answering this question is complicated because species are the result of processes that unfold over time. In studying those processes, biologists pursue several different questions about species. How can we recognize and identify species? How do species form over time? How do species remain separate? Various definitions of species will emerge as we address these questions.

We can recognize and identify many species by their appearance

Someone who is knowledgeable about a group of organisms, such as lizards or birds, usually can distinguish the different species found in a particular area simply by looking at them. Standard field guides to birds, mammals, insects, and flowering plants are possible only because many species change little in appearance over large geographic distances. We can easily recognize male red-winged blackbirds from New York and from California, for example, as members of the same species (**Figure 23.1A**).

More than 200 years ago, the Swedish biologist Carolus Linnaeus originated the *binomial* system of Latinate nomenclature by which species are known today (see Section 1.2). Linnaeus described hundreds of species, and because he knew nothing about genetics or the mating behavior of the organisms he was naming, he classified them on the basis of their appearance; that is, he used a **morphological species concept**. Members of many of the groups that he classified as species by their appearance look alike because they share many of the alleles that code for their body structures. In many groups of organisms for which genetic data are unavailable, species are still recognized by their morphological traits.

But not all members of a species must look alike. For example, males, females, and young individuals may not resemble one another closely (**Figure 23.1B**). Scientists must use information other than appearance to decide whether such easily distinguished individuals are members of the same or different species. To understand the types of information they use, we need to understand how species form.

(A)

Agelaius phoeniceus (male, NY)

Agelaius phoeniceus (male, CA)

(B)

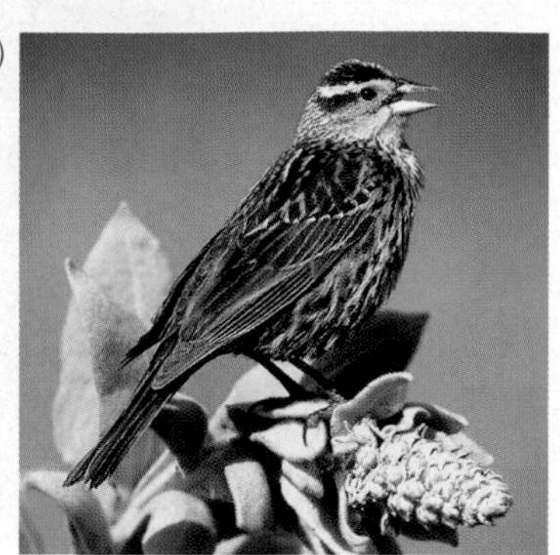

Agelaius phoeniceus (female)

23.1 Members of the Same Species Look Alike—or Not (A) Both of these male red-winged blackbirds are obviously members of the same species, even though one is from the eastern United States and the other is from California. (B) Because red-winged blackbirds are sexually dimorphic, the female of the species looks quite different from the male.

Species form over time

Evolutionary biologists think of species as branches on the tree of life. Each species has a history that starts at a speciation event and ends at either extinction or another speciation event, at which it produces two daughter species. This process is often gradual (**Figure 23.2**). **Speciation** is thus the process by which one species splits into two or more daughter species, which thereafter evolve as distinct lineages. The gradual nature of most speciation guarantees that in many cases, two populations at various stages in the process of becoming new species will exist. In these cases it is impractical to try to decide whether the individuals belong to species A or to species B. However, it *is* important to understand the processes that lead to the splitting of a single species into two species.

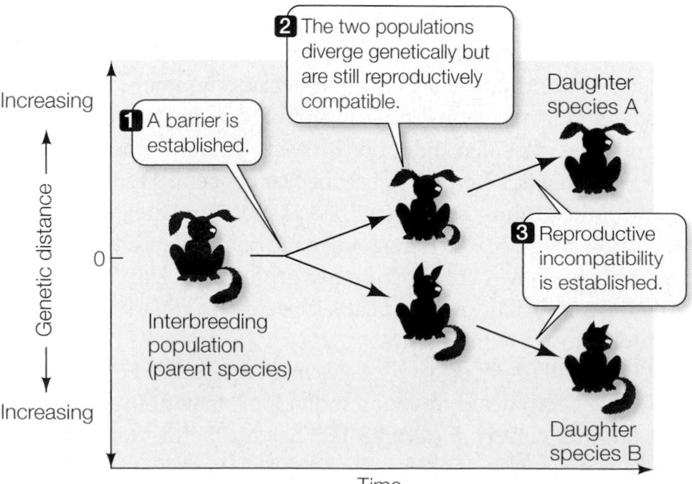

23.2 Speciation May Be a Gradual Process In this hypothetical example, genetic divergence between two separated populations begins before reproductive incompatibility evolves.

An important component of the process of speciation is the development of **reproductive isolation**. If individuals of a population mate with one another, but not with individuals of other populations, they constitute a distinct group within which genes recombine; that is, they are independent evolutionary units—separate branches on the tree of life. In 1940 Ernst Mayr proposed the following definition of species, known as the **biological species concept**: "Species are groups of actually or potentially interbreeding natural populations which are reproductively isolated from other such groups." The terms "actually" and "potentially" are important elements of the definition. "Actually" says that the individuals live in the same area and interbreed with one another. "Potentially" says that although the individuals do not live in the same area, and therefore cannot interbreed, other information suggests that they would do so if they did get together. This widely used species definition does not, however, apply to organisms that reproduce asexually.

Mayr's definition of species asserts that reproductive isolation is the most important criterion for identifying species, but how does reproductive isolation develop? In other words, how do species arise? We will devote much attention in this chapter to this important question.

23.1 RECAP

The biological species concept defines species in terms of reproductive isolation of groups. Speciation is usually a gradual process, and populations can exist at various stages in the process of becoming reproductively isolated from one another.

- Do you understand the importance of the biological species concept relative to the morphological species concept? See pp. 509–510
- Do you see why the morphological species concept is still a useful tool? See p. 510
- Do you see why the biological species concept is difficult to apply to asexually reproducing organisms? (Read Mayr's definition carefully.)

Although Charles Darwin titled his groundbreaking book *The Origin of Species*, it did not extensively discuss the processes of speciation. He devoted most of his attention to demonstrating that species are altered by natural selection over time. We will next discuss the many things that biologists have learned about "the mystery of mysteries" since Darwin's time.

23.2 How Do New Species Arise?

Not all evolutionary changes result in new species. A single lineage may change over time without giving rise to a new species. If evolutionary changes result in one species splitting into two or more daughter species, however, then speciation has occurred. Subsequently, within each isolated gene pool, allele and genotype frequencies may change as a result of the action of evolutionary mechanisms. If two populations are isolated from each other, and sufficient differences in their genetic structure accumulate during the period of isolation, then the two populations may not exchange genes when they come together again. As we will see, the amount of genetic difference that is needed to prevent gene exchange is highly variable.

Thus speciation requires that gene flow within a population whose members formerly exchanged genes be interrupted. Gene flow may be interrupted in two major ways, each of which characterizes a mode of speciation.

Time
A single species is distributed over a broad range.
A barrier separates two populations. Populations adapt to differing environments on opposite sides of the barrier.
The barrier is removed. The populations recolonize the intervening area and mingle, but do not interbreed.
Range of overlap

23.3 Allopatric Speciation Allopatric speciation may result when a population is divided into two separate populations by a physical barrier, such as rising sea levels.

Allopatric speciation requires almost complete genetic isolation

Speciation that results when a population is divided by a physical barrier is known as **allopatric speciation** (*allo*, "different"; *patris*, "country") or *geographic speciation* (**Figure 23.3**). Allopatric speciation is thought to be the dominant mode of speciation among most groups of organisms. The physical barrier that divides the range of a species may be a water body or a mountain range for terrestrial organisms, or dry land for aquatic organisms. Barriers can form when continents drift, sea levels rise, glaciers advance and retreat, and climates change. These processes continue to generate physical barriers today. The populations separated by such barriers are often, but not always, initially large. They evolve differences for a variety of reasons, including gene drift, but especially because the environments in which they live are, or become, different.

Allopatric speciation may also result when some members of a population cross an existing barrier and found a new, isolated population. The 14 species of finches of the Galápagos archipelago, 1,000 kilometers off the coast of Ecuador, were generated by allopatric speciation. Darwin's finches (as they are usually called, because Darwin was the first scientist to study them) arose in the Galápagos from a single South American species that colonized the islands. Today the 14 species of Darwin's finches differ strikingly from their closest mainland relative (**Figure 23.4**). The islands of the Galápagos archipelago are sufficiently distant from one another that finches infrequently disperse among them. In addition, environmental conditions differ among the islands. Some are relatively flat and arid; others have forested mountain slopes. Populations of finches on different islands have differentiated enough over millions of years that when occasional immigrants arrive from other islands, they either do not breed with the residents or, if they do, the resulting offspring often do not survive as well as those produced by pairs composed of island residents. The genetic distinctness and cohesiveness of the different finch species are thus maintained.

Many of the more than 800 species of the fruit fly genus *Drosophila* in the Hawaiian Islands are restricted to a single island. We know that these species are the descendants of new populations founded by individuals dispersing among the islands because the closest relative of a species on one island is often a species on a neighboring island rather than a species on the same island. Biologists who have studied the chromosomes of picture-winged species of *Drosophila* believe that speciation in this group of flies has resulted from at least 45 such founder events (**Figure 23.5**).

A physical barrier's effectiveness at preventing gene flow depends on the size and mobility of the species in question. The eight-lane highway that is an almost impenetrable barrier to a snail

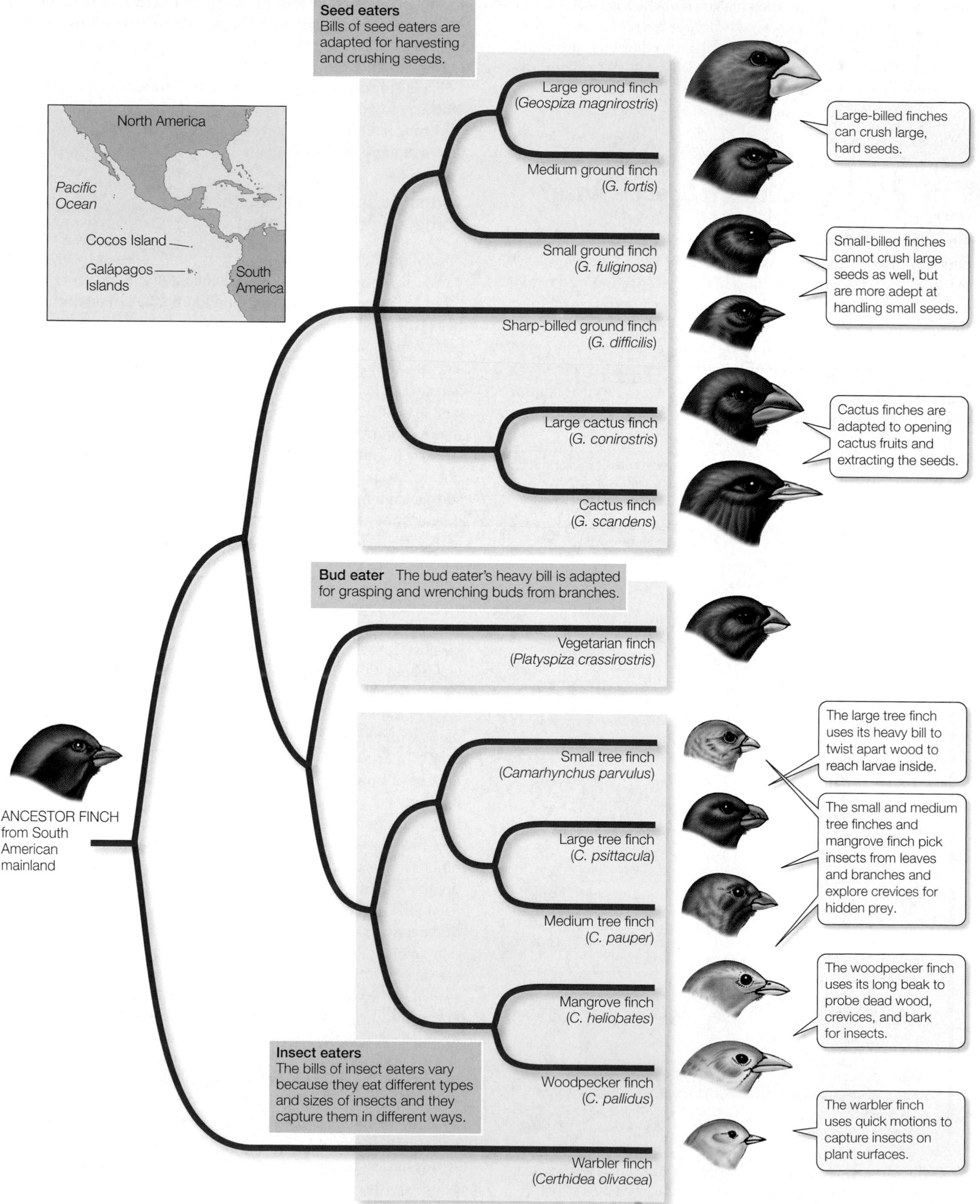

Seed eaters
Bills of seed eaters are adapted for harvesting and crushing seeds.
North America
Pacific Ocean
Cocos Island
Galápagos Islands
South America
Large ground finch (Geospiza magnirostris)
Medium ground finch (G. fortis)
Small ground finch (G. fuliginosa)
Sharp-billed ground finch (G. difficilis)
Large cactus finch (G. conirostris)
Cactus finch (G. scandens)
Large-billed finches can crush large, hard seeds.
Small-billed finches cannot crush large seeds as well, but are more adept at handling small seeds.
Cactus finches are adapted to opening cactus fruits and extracting the seeds.
Bud eater The bud eater's heavy bill is adapted for grasping and wrenching buds from branches.
Vegetarian finch (Platyspiza crassirostris)
Small tree finch (Camarhynchus parvulus)
Large tree finch (C. psittacula)
Medium tree finch (C. pauper)
Mangrove finch (C. heliobates)
Woodpecker finch (C. pallidus)
Warbler finch (Certhidea olivacea)
The large tree finch uses its heavy bill to twist apart wood to reach larvae inside.
The small and medium tree finches and mangrove finch pick insects from leaves and branches and explore crevices for hidden prey.
The woodpecker finch uses its long beak to probe dead wood, crevices, and bark for insects.
The warbler finch uses quick motions to capture insects on plant surfaces.
Insect eaters
The bills of insect eaters vary because they eat different types and sizes of insects and they capture them in different ways.
ANCESTOR FINCH from South American mainland

23.4 Allopatric Speciation among Darwin's Finches The descendants of the ancestral finch that colonized the Galápagos archipelago several million years ago evolved into 14 different species whose members are variously adapted to feed on seeds, buds, and insects. (The fourteenth species, not pictured here, lives in Cocos Island, farther north in the Pacific Ocean.)

is no barrier at all to a butterfly or a bird. Populations of wind-pollinated plants are isolated at the maximum distance pollen can be blown by the wind, but individual plants are effectively isolated at much shorter distances. Among animal-pollinated plants, the width of the barrier is the distance that animals can travel while carrying pollen or seeds. Even animals with great powers of dispersal are often reluctant to cross narrow strips of unsuitable habitat. For animals that cannot swim or fly, narrow water-filled gaps may be effective barriers. Gene flow can sometimes be interrupted, however, even in the absence of physical barriers.

Sympatric speciation occurs without physical barriers

Although physical isolation is usually required for speciation, under some circumstances speciation can occur without it. A partition of a gene pool without physical isolation is called **sympatric speciation** (*sym*, "together with"). But if speciation is usually a gradual process, how can reproductive isolation develop when individuals have frequent opportunities to mate with one another? What is required is some form of disruptive selection in which certain genotypes have high fitness on one or the other of two different resources, as we described for black-bellied seedcrackers (see Figure 22.15).

Disruptive selection in seedcrackers has not yet resulted in the formation of two species, but sympatric speciation via disruptive selection may be happening in a fruit fly (*Rhagoletis pomonella*) in New York State. Until the mid-1800s, *Rhagoletis* fruit flies courted, mated, and deposited their eggs only on hawthorn fruits. The larvae learned the odor of hawthorn as they fed on the fruits. When they emerged from their pupae as adults, they used this cue to locate other hawthorn plants on which to mate and lay their eggs. About 150 years ago, large commercial apple orchards were planted in the Hudson River valley. Apple trees are closely related to hawthorns. A few female *Rhagoletis* laid their eggs on apples, perhaps by mistake. Their larvae did not grow as well as the larvae on hawthorn fruits, but many did survive. These larvae recognized the odor of apples, so when they emerged as adults, they sought out apple trees, where they mated with other flies reared on apples.

Today there are two groups of *Rhagoletis pomonella* in the Hudson River valley that may be on the way to becoming distinct species (**Figure 23.6**). One feeds primarily on hawthorn fruits, the other on apples. The two incipient species are partly reproductively isolated because they mate primarily with individuals raised on the same fruit and because they emerge from their pupae at different times of the year. In addition, the apple-feeding flies have evolved so that they now grow more rapidly on apples than they originally did.

Sympatric speciation via ecological isolation (on different resources), as is happening in *Rhagoletis pomonella*, may be widespread among insects, many of which feed on only a single plant species, but the most common means of sympatric speciation is **polyploidy**, the production within an individual of duplicate sets of chromosomes. Polyploidy can arise either from chromosome duplication in a single species (**autopolyploidy**) or from the combining of the chromosomes of two different species (**allopolyploidy**).

An autopolyploid individual originates when (for example) cells that are normally diploid (with two sets of chromosomes) acciden-

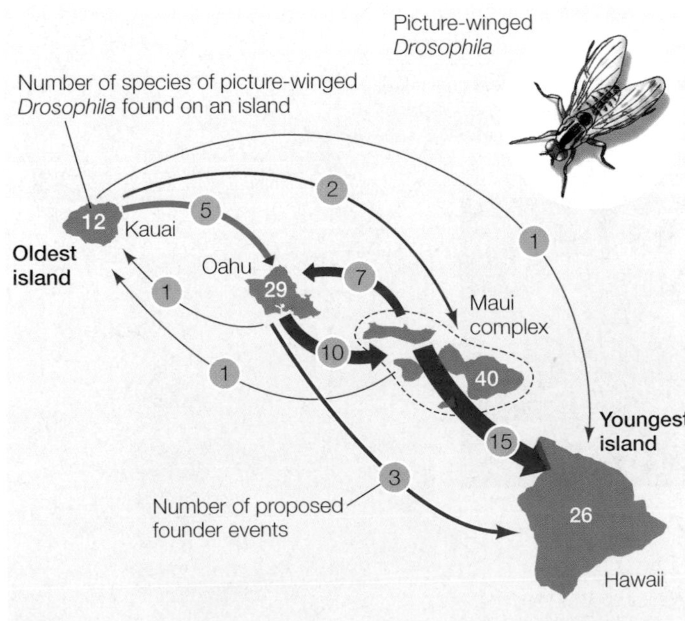

23.5 Founder Events Lead to Allopatric Speciation The large number of species of picture-winged *Drosophila* in the Hawaiian Islands is the result of founder events: the founding of new populations by individuals dispersing among the islands. The islands, which were formed in sequence as Earth's crust moved over a volcanic "hot spot," vary in age.

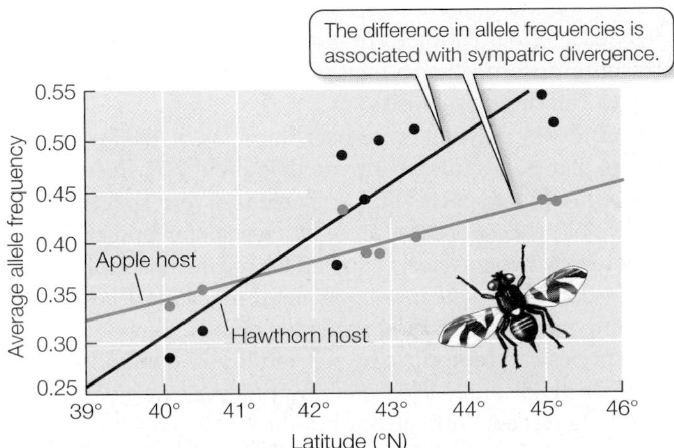

23.6 Sympatric Speciation May Be Underway in *Rhagoletis pomonella* Genetic differentiation (measured by the average frequencies of alleles at several loci) is increasing among *R. pomonella* flies who have developed preferences for different host plants.

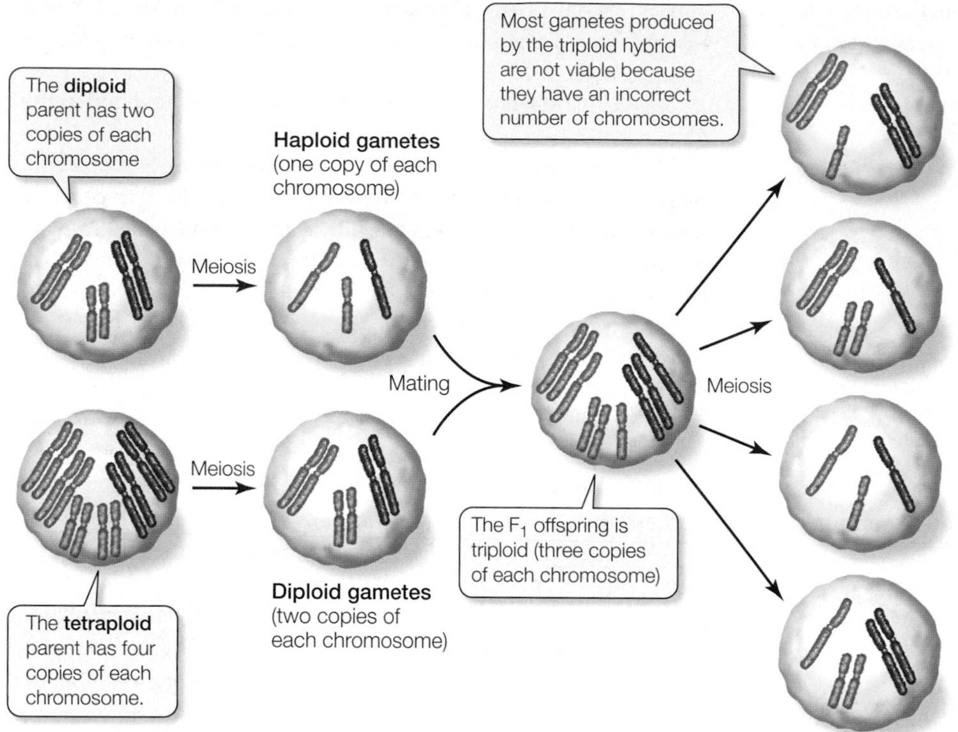

Diploid gametes (two copies of each chromosome)

The F_1 offspring is triploid (three copies of each chromosome)

Most gametes produced by the triploid hybrid are not viable because they have an incorrect number of chromosomes.

23.7 Tetraploids Are Soon Reproductively Isolated from Diploids Even if the triploid offspring of a diploid and a tetraploid parent survives and reaches sexual maturity, most of the gametes it produces have aneuploid numbers of chromosomes. Such triploid individuals are effectively sterile. (For simplicity, the diagram shows only three chromosomes; most species have many more than that.)

tally duplicate their chromosomes, resulting in a tetraploid individual (with four sets of chromosomes). Tetraploid and diploid plants of the same species are reproductively isolated because their hybrid offspring are triploid and are usually sterile. They cannot produce viable gametes because their chromosomes do not synapse correctly during meiosis (**Figure 23.7**). So a tetraploid plant cannot produce viable offspring by mating with a diploid individual—but it *can* do so if it self-fertilizes or mates with another tetraploid. Thus polyploidy can result in complete reproductive isolation in two generations—an important exception to the general rule that speciation is a gradual process.

Allopolyploids may also be produced when individuals of two different (but closely related) species interbreed, or **hybridize**. Allopolyploids are often fertile because each of the chromosomes has a nearly identical partner with which to pair during meiosis.

Speciation by polyploidy has been important in the evolution of plants. Botanists estimate that about 70 percent of flowering plant species and 95 percent of fern species are polyploids. Most of these arose as a result of hybridization between two species, followed by self-fertilization. New species arise by means of polyploidy much more easily among plants than among animals because plants of many species can reproduce by self-fertilization. In addition, if polyploidy arises in several offspring of a single parent, the siblings can fertilize one another.

How easily allopolyploidy can produce new species is illustrated by the salsifies (*Tragopogon*), relatives of sunflowers. Salsifies are weedy plants that thrive in disturbed areas around towns. People have inadvertently spread them around the world from their ancestral ranges in Eurasia. Three diploid species of salsify—*Tragopogon porrifolius, T. pratensis,* and *T. dubius*—were introduced into North America early in the twentieth century. Two tetraploid hybrids—*T. mirus* and *T. miscellus*—between the original three diploid species were first discovered in 1950. The hybrids have spread since their discovery and today are more widespread than their diploid parents (**Figure 23.8**).

Studies of their genetic material show that both salsify hybrids have formed more than once. Some populations of *T. miscellus*—a hybrid of *T. pratensis* and *T. dubius*—have the chloroplast genome of *T. pratensis;* other populations have the chloroplast genome of *T. dubius*. Genetic differences among local populations of *T. miscellus* show that this allopolyploid has formed independently at least 21 times. *T. mirus,* a hybrid of *T. porrifolius* and *T. dubius,* has formed 12 times. Scientists seldom know the dates and locations of species formation so well. *T. porri-*

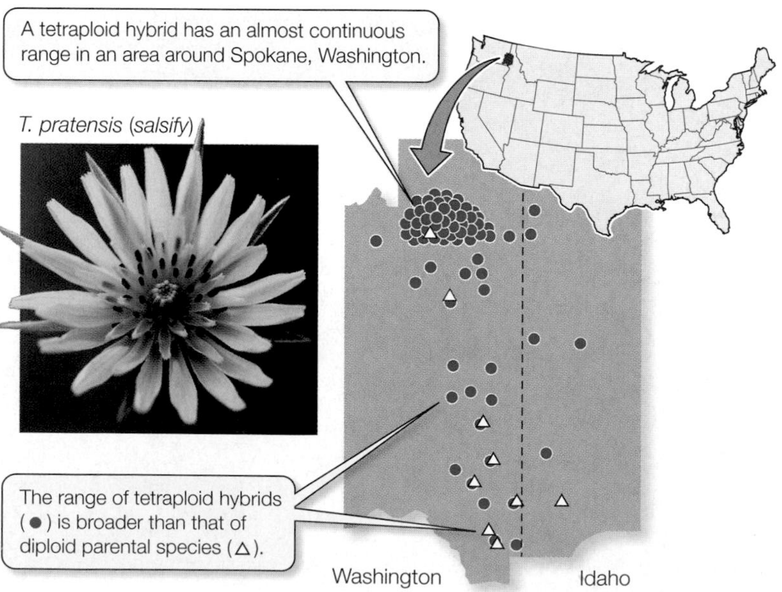

A tetraploid hybrid has an almost continuous range in an area around Spokane, Washington.
T. pratensis (salsify)
The range of tetraploid hybrids (●) is broader than that of diploid parental species (△).
Washington
Idaho

23.8 Polyploids May Outperform Their Parent Species *Tragopogon* species (salsifies) are members of the sunflower family. The map shows the distribution of the three diploid parent species and of the two tetraploid hybrid species of *Tragopogon* in eastern Washington and adjacent Idaho.

folius prefers wet, shady places; *T. dubius* prefers dry, sunny places. *T. mirus*, however, can grow in partly shaded environments where neither parent species does well. The success of these newly formed hybrid species of salsifies illustrates how so many species of flowering plants could have originated as polyploids.

23.2 RECAP

Allopatric speciation results from the separation of populations by geographic barriers; it is the dominant mode of speciation among most groups. Among plants and some animals, sympatric speciation via polyploidy has been frequent.

- Can you explain why an effective barrier to gene flow for one species may not effectively isolate another species? See pp. 511, 513
- How can speciation via polyploidy happen in two generations? See p. 514 and Figure 23.7

Polyploidy, as we have just seen, can result in a new species that is completely reproductively isolated from its parent species in two generations, but most populations separated by a physical barrier become reproductively isolated only very slowly. Let's see how reproductive isolation may become established once two populations have separated from each other.

23.3 What Happens when Newly Formed Species Come Together?

Once a barrier to gene flow is established, the separated populations may diverge genetically through the action of the evolutionary mechanisms we described in Section 22.2. Over many generations, differences may accumulate that reduce the probability that members of the two populations could mate and produce viable offspring. In this way, reproductive isolation can evolve as an incidental by-product of genetic changes in allopatric populations.

Partial reproductive isolation has evolved in this way in strains of *Phlox drummondii*. In 1835, Thomas Drummond, after whom this plant is named, collected seeds in Texas and distributed them to nurseries in Europe. Over the next 80 years, these nurseries established more than 200 true-breeding strains of this phlox, which differed in flower size and color and plant growth form. These plant breeders did not select directly for reproductive incompatibility between strains, but in subsequent experiments, in which the fertilization rates of flowers of various strains by pollen of other strains were measured and compared, it was discovered that reproductive compatibility between strains had been reduced from 14 to 50 percent, depending on the strains.

Geographic isolation does not necessarily lead to reproductive isolation, however, because genetic divergence does not cause reproductive isolation to appear as a byproduct. Therefore, populations that have been isolated from one another for millions of years may still be reproductively compatible. For example, American sycamores and European sycamores (also known as plane trees) have been geographically isolated from one another for at least 20 million years. Nevertheless, they are morphologically very similar (**Figure 23.9**), and they can form fertile hybrids, even though they never have an opportunity to do so in nature.

Reproductive incompatibility may arise by many different mechanisms, but it is useful to group them into two major types: prezygotic and postzygotic reproductive barriers.

Prezygotic barriers operate before fertilization

Several mechanisms that operate before fertilization—**prezygotic reproductive barriers**—may prevent individuals of different species or populations from interbreeding:

(A) *Platanus occidentalis* (American sycamore)

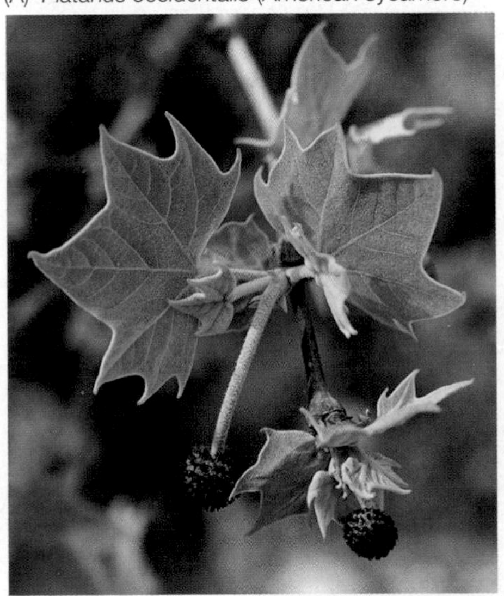

(B) *Platanus hispanica* (European sycamore)

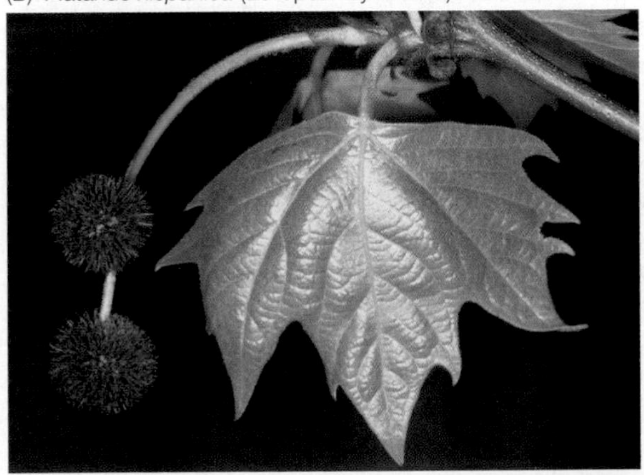

23.9 Geographically Separated, Morphologically Similar Although they have been separated by the Atlantic Ocean for at least 20 million years, American and European sycamores have diverged very little in appearance.

- *Habitat isolation*. Individuals of different species may select different habitats in the same general area in which to live or to mate. As a result, they may never come into contact during their respective mating periods. This is what is happening with *Rhagoletis* flies in the Hudson River valley.
- *Temporal isolation*. Many organisms have mating periods that are as short as a few hours or days. If the mating periods of two species do not overlap, they will be reproductively isolated by time, as the *Rhagoletis* flies are.
- *Mechanical isolation*. Differences in the sizes and shapes of reproductive organs may prevent the union of gametes from different species. Males of many insects, for example, have elaborate copulatory organs (penises) that may prevent them from inseminating females of other species.
- *Gametic isolation*. Sperm of one species may not attach to the eggs of another species because the eggs do not release the appropriate attractive chemicals, or the sperm may be unable to penetrate the egg because the two gametes are chemically incompatible.
- *Behavioral isolation*. Individuals of a species may reject, or fail to recognize, individuals of other species as mating partners. Behavioral isolation may have stimulated speciation in birds of paradise.

Sometimes the mate choice of one species is mediated by the behavior of individuals of other species. For example, whether two plant species hybridize may depend on the food preferences of their pollinators. The floral traits of plants can enhance reproductive isolation either by influencing which pollinators are attracted to the flowers or by altering where pollen is deposited on the bodies of pollinators.

The evolution of floral traits that generate reproductive isolation has been studied in columbines of the genus *Aquilegia*. Columbines have undergone recent and very rapid speciation. At the same time, they have evolved long floral nectar spurs—tubular outgrowths of petals that produce nectar at their tips. Animals pollinate these flowers while probing the spurs to collect nectar. The length of the spurs and the orientation of the flowers influence how efficiently pollinators can extract nectar. Two species, *Aquilegia formosa* and *A. pubescens*, that grow in the mountains of California can produce fertile hybrids. *A. formosa* has pendant (hanging) flowers and short spurs (**Figure 23.10A**); it is pollinated by hummingbirds. *A. pubescens* has upright flowers and long spurs (**Figure 23.10B**); it is pollinated by hawkmoths.

M. Fulton and S. Hodges tested discrimination among these flowers by hawkmoths by turning *A. formosa* flowers so that they were upright. Hawkmoths still visited mostly *A. pubescens* flowers (**Figure 23.10C**), probably because the flowers of the two species differ strongly in the color of light they reflect. Thus, although these two species are interfertile, hybrids rarely form in nature because the two species attract different pollinators.

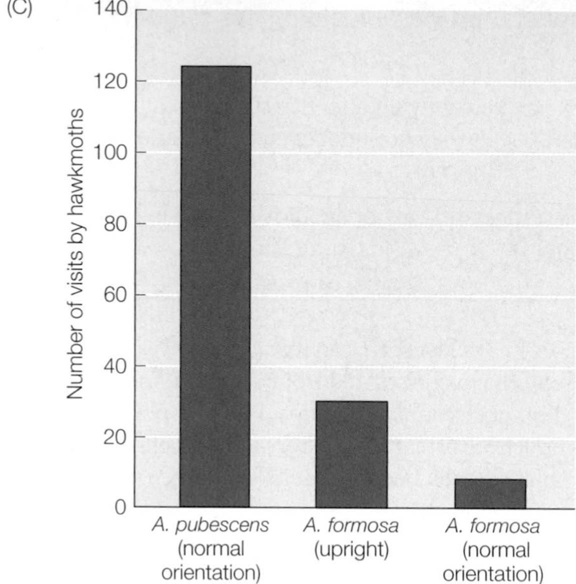

23.10 Hawkmoths Favor Flowers of One Columbine Species
(A) Flowers of *Aquilegia formosa* are normally pendant. (B) Flowers of *A. pubescens* are normally upright. (C) The hawkmoths that pollinate *A. pubescens* distinguish between flowers of the two species, even when *A. formosa* flowers are experimentally modified to be upright.

Postzygotic barriers operate after fertilization

If individuals of two different populations lack complete prezygotic reproductive barriers, **postzygotic reproductive barriers** may still prevent gene exchange. Genetic differences that accumulate while the populations are isolated from each other may reduce the survival and reproduction of hybrid offspring in any of several ways:

- *Low hybrid zygote viability*. Hybrid zygotes may fail to mature normally, either dying during development or developing such severe abnormalities that they cannot mate as adults.
- *Low hybrid adult viability*. Hybrid offspring may simply survive less well than offspring resulting from matings within populations.
- *Hybrid infertility*. Hybrids may mature normally, but be infertile when they attempt to reproduce. For example, the offspring of matings between horses and donkeys—mules—are healthy, but they are sterile; they produce no descendants.

Although natural selection does not directly favor the evolution of postzygotic reproductive barriers, if hybrid offspring survive poorly, natural selection may favor the evolution of prezygotic barriers. This happens because individuals that mate with individuals of the

EXPERIMENT

HYPOTHESIS: ***Phlox drummondii*** **has red flowers only where it is sympatric with pink-flowered** ***P. cuspidata*** **because having red flowers decreases interspecific hybridization.**

METHOD

1. Introduce equal numbers of red- and pink-flowered *P. drummondii* individuals into an area with many pink-flowered *P. cuspidata*.
2. After the flowering season ends, assess the genetic composition of the seeds produced by *P. drummondii* plants of both colors.

RESULTS

Of the seeds produced by pink-flowered *P. drummondii*, 38% were hybrids with *P. cuspidata*. Only 13% of the seeds produced by red-flowered individuals were genetic hybrids.

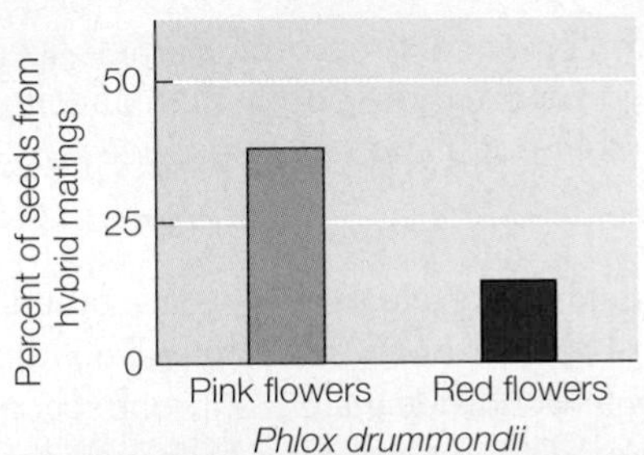

CONCLUSION: **For** ***Phlox drummondii*****, having flowers that differ in color from those of** ***P. cuspidata*** **reduces the amount of interspecific hybridization.**

23.11 Prezygotic Reproductive Barriers Most *Phlox drummondii* plants are pink, but in regions where they are sympatric with *P. cuspidata*, which is also pink, most individuals are red. An experiment by Donald Levin showed that the red color functions as a prezygotic reproductive barrier, because pollinators tend to visit flowers of only one color. FURTHER RESEARCH: This experiment did not address the probable reproductive advantages for individual *Phlox* plants of donating and receiving primarily intraspecific pollen. What experiments could be designed to measure such advantages?

other population will leave fewer surviving descendants than individuals that mate only within their own population. Such strengthening of prezygotic barriers is known as **reinforcement**.

Donald Levin of the University of Texas noticed that individuals of *Phlox drummondii* over most of the range of the species in Texas have pink flowers. However, where *P. drummondii* is sympatric with the pink-flowered *P. cuspidata*, they have red flowers. No other species of *Phlox* has red flowers. Levin performed an experiment whose results showed that reinforcement might explain the evolution of red flowers where the two species are sympatric (**Figure 23.11**).

Reinforcement can also be detected by using the comparative method. If reinforcement is occurring, then sympatric pairs of closely related species should evolve prezygotic reproductive barriers more rapidly than allopatric pairs of species. An investigation of related sympatric and allopatric species of *Agrodiaetus* butterflies is one of many studies that have demonstrated reinforcement. The colors of the wings of males, which females use to choose mates, have diverged much faster in sympatric than in allopatric populations.

Many closely related species in nature form hybrids in areas where their ranges overlap, and they may continue to do so for many years. Let's examine what happens when reproductive barriers do not completely prevent individuals from different populations from mating and producing offspring.

Hybrid zones may form if reproductive isolation is incomplete

If contact is reestablished between formerly isolated populations before complete reproductive isolation has developed, members of the two populations may interbreed. Three outcomes of such interbreeding are possible:

- If hybrid offspring are as fit as those resulting from matings within each population, hybrids may spread through both populations and reproduce with other individuals. The gene pools are then combined, and no new species result from the period of isolation.
- If hybrid offspring are less fit, complete reproductive isolation may evolve as reinforcement strengthens prezygotic reproductive barriers.
- Even if hybrid offspring are at some disadvantage, a narrow **hybrid zone** may exist if reinforcement does not happen, or the zone may persist for a long time while reinforcement may be developing.

Hybrid zones are excellent natural laboratories for the study of speciation. When a hybrid zone first forms, most hybrids are offspring of crosses between purebred individuals of the two species. However, subsequent generations include a variety of individuals with different proportions of their genes derived from the original two populations. Thus hybrid zones contain recombinant individuals resulting from many generations of hybridization. Detailed genetic studies can tell us much about why hybrid zones may be narrow and stable for long periods of time.

The hybrid zone between two species of European toads of the genus *Bombina* has been studied intensively. The fire-bellied toad (*B. bombina*) lives in eastern Europe. The closely related yellow-bellied toad (*B. variegata*) lives in western and southern Europe. The ranges of the two species meet in a narrow zone stretching 4,800 kilometers from eastern Germany to the Black Sea (**Figure 23.12**). Hybrids between the two species suffer from a range of defects, many of which are lethal. Those that survive often have skeletal abnormalities, such as misshapen mouths, ribs that are fused to vertebrae, and a reduced number of vertebrae.

By following the fates of thousands of toads from the hybrid zone, investigators found that a hybrid toad is on average only half as fit as a purebred individual. The hybrid zone is narrow because there is strong selection against hybrids, and because adult toads do not move over long distances. It has persisted for hundreds of years, however, because many purebred individuals move only a

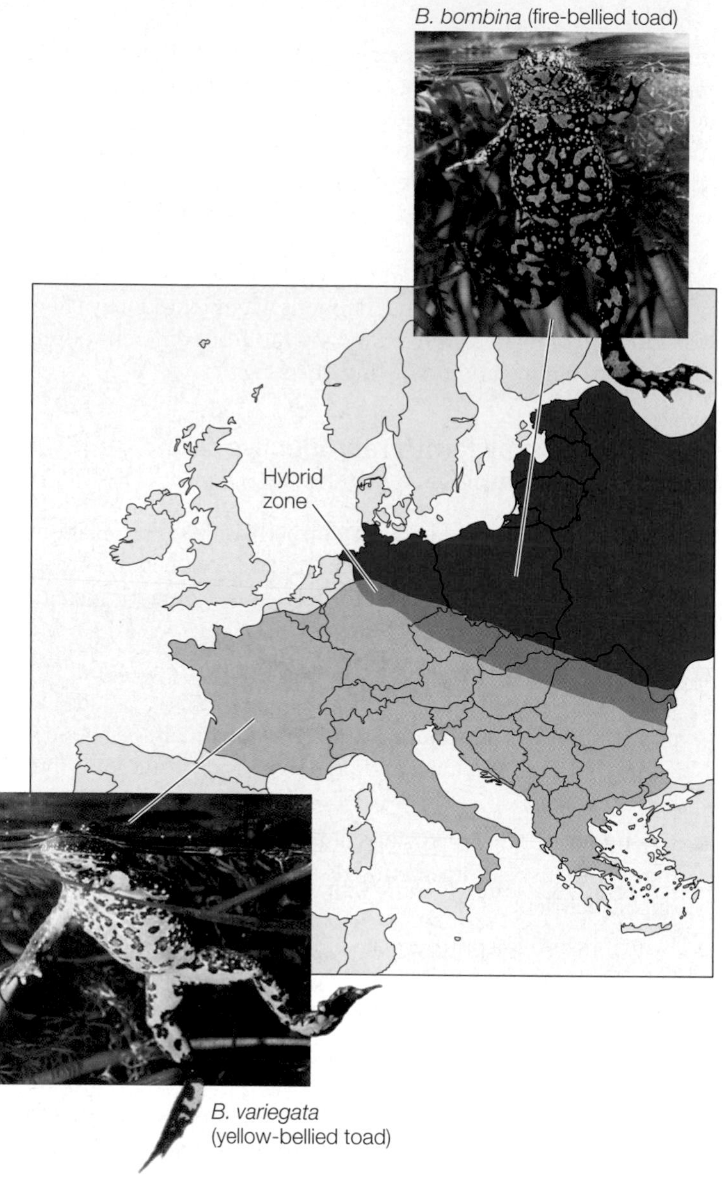

23.12 Hybrid Zones May Be Long and Narrow The narrow zone in Europe where fire-bellied toads meet and hybridize with yellow-bellied toads stretches across Europe. This hybrid zone has been stable for hundreds of years, but has never expanded, and no reinforcement has evolved.

short distance into the zone. They have not previously encountered individuals of the other species, so there has been no opportunity for reinforcement to evolve.

Human activity may result in new hybrid zones. One example is found among the many protead plants that are adapted to the dry, nutrient-poor soils of southwestern Australia. In undisturbed sand plain vegetation in this region, two related species of protead shrubs, *Banksia hookeriana* and *Banksia prionotes,* do not hybridize because their flowering seasons do not overlap. However, where human disturbances have disrupted their habitat, their flowering seasons expand and overlap, and fully fertile hybrids between the two species are now common.

The protead plant group is both distinctive and ancient, dating from the Mesozoic. Proteads were isolated on Gondwana when Pangaea broke apart, and no members of the group are native to the Northern Hemisphere. Modern proteads are most species-rich in southwest Australia (≈550 species) and the Cape region of South Africa (≈320 species)

23.3 RECAP

Reproduction isolation may result from prezygotic or postzygotic reproductive barriers. If reproductive isolation is incomplete, hybrid zones may form when previously separated populations come into contact.

- Can you describe the various kinds of prezygotic and postzygotic reproductive barriers? See pp. 515–516
- Why is reinforcement of prezygotic barriers likely if hybrid offspring survive more poorly than offspring produced by within-population matings? See p. 517

Some groups of organisms have many species; others have only a few. Hundreds of species of *Drosophila* evolved rapidly in the Hawaiian Islands, but worldwide there is only one species of horseshoe crab, even though its ancestry dates back more than 300 million years. Why do different groups of organisms have such different rates of speciation?

23.4 Why Do Rates of Speciation Vary?

Rates of speciation vary greatly because many factors influence the likelihood that a lineage will split to form two or more species. The larger the number of species in a group, the larger the number of opportunities for new species to form. For speciation by polyploidy, the more species in a group, the more species are available to hybridize with one another. For allopatric speciation, the larger the number of different species living in an area, the larger the number of species whose ranges will be bisected by a given physical barrier. For all these reasons, "the rich get richer": groups that are already species-rich are likely to speciate faster than species-poor groups.

Speciation rates are likely to be higher in species with poor dispersal abilities than in species with good dispersal abilities, because even a narrow barriers can be effective in dividing a species whose members are highly sedentary. The Hawaiian Islands have about a thousand species of land snails, many of which are restricted to a single valley. Because snails move only short distances, the high ridges that separate the valleys are effective barriers to their dispersal.

Populations of species that have specialized diets are more likely to diverge than are populations that have generalized diets. To investigate the effects of diet on rates of speciation, C. Mitter and colleagues compared species richness in some closely related groups of true bugs (hemipterans). The common ancestor of these

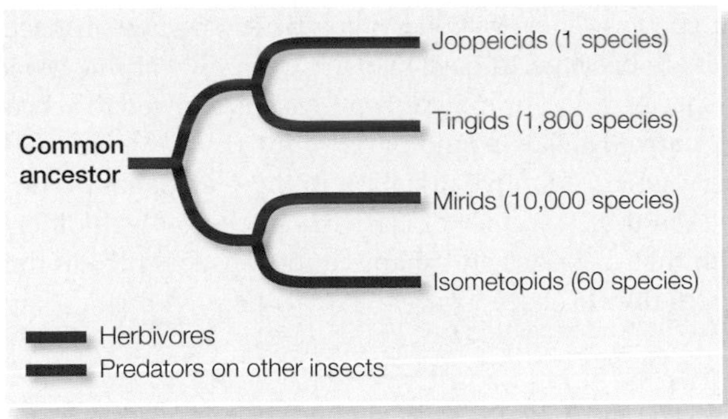

23.13 Dietary Shifts Can Promote Speciation Herbivorous groups of hemipteran insects have speciated several times faster than closely related predatory groups.

groups was a predator that fed on other insects, but a dietary shift to herbivory (eating plants) evolved at least twice in the groups under study. The herbivorous groups have many more species than those that are predatory (**Figure 23.13**). Herbivorous bugs typically specialize on one or a few closely related species of plants, whereas predatory bugs tend to feed on many different species of insects.

Speciation rates in plants are faster in animal-pollinated than wind-pollinated plants. Animal-pollinated groups have, on average, 2.4 times as many species as related groups pollinated by wind. Among animal-pollinated plants, speciation rates are correlated with pollinator specialization. In columbines (*Aquilegia*), the rate of evolution of new species has been about three times faster in lineages that have long nectar spurs as in closely related lineages that lack spurs. Why do spurs increase the speciation rate? Apparently it is because having longer spurs restricts the number of pollinator species that visit the flowers, thus increasing opportunities for reproductive isolation (see Figure 23.10).

The mechanisms of sexual selection (see Section 22.3) also appear to result in increased rates of speciation. Some of the most striking examples of sexual selection are found in birds with promiscuous mating systems.

Bird-watchers travel thousands of miles to Papua New Guinea to witness the mating displays of male birds of paradise, which have long, brightly colored tail feathers and are very distinct in appearance from the females (*sexual dimorphism*). In many of these species, males assemble at display grounds, and females come there to choose a male with whom to copulate. After mating, the females leave the display grounds, build their nests, lay their eggs, and feed their offspring with no help from the males. The males remain to court more females. There are 33 species of birds of paradise (**Figure 23.14A**).

The closest relatives of the birds of paradise are the manucodes. Male and female manucodes differ only slightly in size and plumage. They form monogamous pair bonds and both sexes contribute to raising the young. There are only 5 species of manucodes (**Figure 23.14B**).

Animals with complex sexually selected behaviors are likely to form new species at a high rate because they make sophisticated discriminations among potential mating partners. They distinguish members of their own species from members of other species, and they make subtle discriminations among members of their own species on the basis of size, shape, appearance, and behavior (see Figures 22.16 and 22.17). Such discriminations can greatly influence which individuals are most successful in mating and producing offspring, and may lead to rapid evolution of differences among populations.

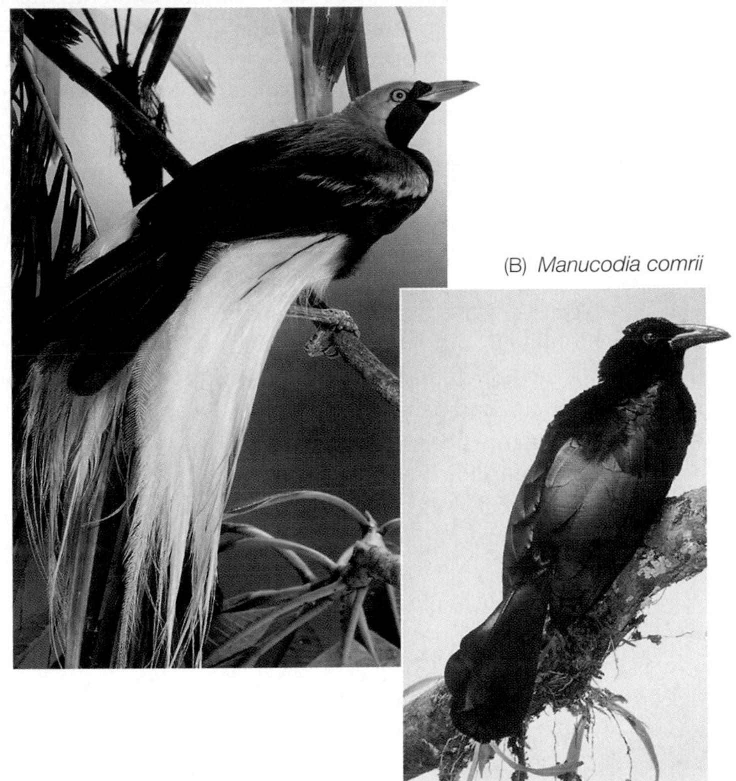

23.14 Sexual Selection in Birds Can Lead to Higher Speciation Rates (A) Birds of paradise and (B) manucodes are closely related bird groups of the South Pacific. However, speciation rates are much higher among the sexually dimorphic, polygynous birds of paradise (33 species) than among manucodes (5 species).

23.4 RECAP

The existing species richness of a group, dispersal ability, dietary specialization, type of pollination, and sexual selection are among the many factors that influence speciation rates.

- Do you understand why speciation rates tend to be higher in groups that already have a lot of species? See p. 518
- Do you understand why pollinator specialization in plants and sexual selection in animals typically stimulate speciation? See p. 519

The fossil record reveals that, at certain times in certain groups, speciation rates have been much higher than extinction rates, resulting in the proliferation of a large number of daughter species. Let's examine why such evolutionary radiations happen.

23.5 Why Do Adaptive Radiations Occur?

The proliferation of a large number of daughter species from a single ancestor is called an **evolutionary radiation**. If the rapid proliferation of species results in an array of species that live in a variety of environments and differ in the characteristics they use to exploit those environments, the radiation is said to be *adaptive*. If a radiation is not accompanied by any observed ecological differentiation among the species, it is said to be nonadaptive.

An **adaptive radiation** begins when genetic differentiation between populations evolves in response to differences in the environments they inhabit and the resources they use. Such differentiation is likely to occur in environments with abundant resources. A population is likely to encounter underutilized resources when it colonizes a new environment that contains relatively few species. That is why, as we saw in Chapter 21, adaptive radiations have frequently followed mass extinctions. Similarly, islands lack many groups of organisms found on the mainland, so the ecological opportunities that exist on islands may stimulate rapid evolutionary changes when a new species colonizes them. Water barriers also restrict gene flow among the islands in an archipelago, so populations on the different islands may evolve adaptations to their local environments.

Remarkable adaptive radiations have occurred in the Hawaiian Islands. The native biota of the Hawaiian Islands includes 1,000 species of flowering plants, 10,000 species of insects, 1,000 land snails, and more than 100 bird species. However, there were no amphibians, no terrestrial reptiles, and only one native terrestrial mammal—a bat—on the islands until humans introduced additional species. The 10,000 known native species of insects on Hawaii are believed to have evolved from only about 400 immigrant species; only 7 immigrant species are believed to account for all the native Hawaiian land birds. Similarly, as we saw earlier in this chapter, an adaptive radiation in the Galápagos archipelago resulted in the 14 species of Darwin's finches, which differ strikingly in their bill sizes and shapes and, consequently, in the food resources they use (see Figure 23.4).

The Hawaiian archipelago is Earth's most isolated group of islands. The Hawaiian Islands lie 4,000 kilometers from the nearest continental land mass and 1,600 kilometers from the nearest group of islands.

23.15 Rapid Evolution among Hawaiian Silverswords The Hawaiian silverswords, three closely related genera of the sunflower family, are believed to have descended from a single common ancestor (a plant similar to the tarweed) that colonized Hawaii from the Pacific coast of North America. The four plants shown here are more closely related than they appear to be based on their morphology.

Madia sativa (tarweed)

Argyroxiphium sandwicense

Wilkesia hobdyi

Dubautia menziesii

Adaptive radiations have been frequent among plants on the Hawaiian Islands. More than 90 percent of the 1,000 plant species on the Hawaiian Islands are *endemic*—that is, they are found nowhere else. Several groups of flowering plants have more diverse forms and life histories on the islands, and live in a wider variety of habitats, than do their close relatives on the mainland. An outstanding example is the 28 species of Hawaiian sunflowers called silverswords (genera *Argyroxiphium, Dubautia*, and *Wilkesia*). Chloroplast DNA sequences show that these species share a relatively recent common ancestor with a species of tarweed from the Pacific coast of North America (**Figure 23.15**). Whereas all mainland tarweeds are small, upright herbs (nonwoody plants), the silverswords include prostrate and upright herbs, shrubs, trees, and vines. Silversword species occupy nearly all the habitats of the Hawaiian Islands, from sea level to above timberline in the mountains. Despite their extraordinary morphological diversification, the silverswords are genetically very similar.

The island silverswords are more diverse in size and shape than the mainland tarweeds because the original colonizers arrived on islands that had very few plant species. In particular, there were few trees and shrubs, because such large-seeded plants rarely disperse to oceanic islands. Trees and shrubs have evolved from nonwoody ancestors on many oceanic islands. On the mainland, however, tarweeds live in ecological communities that contain many tree and shrub species in lineages with long evolutionary histories. In those environments, opportunities to exploit the "tree" way of life had already been preempted.

Adaptive radiations are common on islands but are not confined to them. Genetic analyses of the 1,563 species of ice plants, nearly all of which are endemic to southern Africa, show that the group radiated in that region within the last 9 million years. Ice plants are *succulents*, plants whose cells utilize the processes of crassulacean acid metabolism (CAM; see Section 8.4) to withstand extremely dry conditions. Their prominence in their desert home has given the region its name—the Succulent Karoo.

23.5 RECAP

Evolutionary radiation is the rapid proliferation of species from a single ancestor.

- Can you describe the difference between adaptive and nonadaptive radiation? See p. 520
- Do you understand why adaptive radiations are particularly common on islands? See pp. 520–521

The processes we have discussed in this chapter, operating over billions of years, have produced a world in which life is organized into millions of species, each adapted to live in a particular environment and to use environmental resources in a particular way. How these millions of species are organized into ecological communities is explored in Part Nine of this book.

CHAPTER SUMMARY

23.1 What are species?

Speciation is the process by which one species splits into two or more daughter species, which thereafter evolve as distinct lineages. Review Figure 23.2

The **morphological species concept** distinguishes species on the basis of physical similarities.

The **biological species concept** distinguishes species on the basis of **reproductive isolation**; that is, a species are considered to be made up of populations that can interbreed with each other. Asexual species cannot be defined with the biological species concept.

23.2 How do new species arise?

See Web/CD Tutorial 23.1

Speciation requires that gene flow within a population that once exchanged genes be interrupted.

Allopatric speciation, which results when populations are separated by a physical barrier, is the dominant mode of speciation. This type of speciation may follow from founder events, in which some members of a population cross a barrier and found a new, isolated population. Review Figure 23.3, Web/CD Tutorial 23.2

Sympatric speciation results when the genomes of two groups diverge in the absence of physical isolation. It can result when groups are ecologically isolated (i.e., depend on different resources). Review Figure 23.6

In plants, sympatric speciation can occur within two generations via **polyploidy**, an increase in the number of chromosomes. Polyploidy may arise from chromosome duplications within a species (autopolyploidy) or from **hybridization** that results in combining the chromosomes of two species (**allopolyploidy**). Review Figure 23.7

23.3 What happens when newly formed species come together?

Prezygotic barriers to reproduction operate before fertilization; **postzygotic reproductive barriers** operate after fertilization. Prezygotic barriers may be favored by natural selection if postzygotic barriers are incomplete.

Hybrid zones may form when previously separated populations come into contact if reproductive isolation is incomplete.

23.4 Why do rates of speciation vary?

Species richness, dispersal ability, dietary specialization, type of pollination, and sexual selection all influence speciation rates. Review Figure 23.13

23.5 Why do adaptive radiations occur?

An **evolutionary radiation**, a rapid proliferation of species from a common ancestor, may result in an array of species that live in a variety of environments. Radiation may also be stimulated by sexual selection.

Adaptive radiation, during which daughter species become ecologically differentiated, is likely to occur in environments where there is an array of underutilized resources.

See Web/CD Activity 23.1 for a concept review of this chapter.

SELF-QUIZ

1. The biological species concept defines a species as a group of
   a. actually interbreeding natural populations that are reproductively isolated from other such groups.
   b. potentially interbreeding natural populations that are reproductively isolated from other such groups.
   c. actually or potentially interbreeding natural populations that are reproductively isolated from other such groups.
   d. actually or potentially interbreeding natural populations that are reproductively connected to other such groups.
   e. actually interbreeding natural populations that are reproductively connected to other such groups.
2. Which of the following is *not* a condition that favors allopatric speciation?
   a. Continents drift apart and separate previously connected lineages.
   b. A mountain range separates formerly connected populations.
   c. Different environments on two sides of a barrier cause populations to diverge.
   d. The range of a species is separated by loss of intermediate habitat.
   e. Tetraploid individuals arise in one part of the range of a species.
3. Finches speciated in the Galápagos Islands because
   a. the Galápagos Islands are not far from the mainland.
   b. the Galápagos Islands are arid.
   c. the Galápagos Islands are small.
   d. the islands of the Galápagos archipelago are sufficiently isolated from one another that there is little migration among them.
   e. the islands of the Galápagos archipelago are close enough to one another that there is considerable migration among them.
4. Which of the following is *not* a potential prezygotic reproductive barrier?
   a. Temporal segregation of breeding seasons
   b. Differences in chemicals that attract mates
   c. Hybrid infertility
   d. Spatial segregation of mating sites
   e. Sperm that cannot penetrate an egg
5. A common means of sympatric speciation is
   a. polyploidy.
   b. hybrid infertility.
   c. temporal segregation of breeding seasons.
   d. spatial segregation of mating sites.
   e. imposition of a geographic barrier.
6. Narrow hybrid zones may persist for long times because
   a. hybrids are always at a disadvantage.
   b. hybrids have an advantage only in narrow zones.
   c. hybrid individuals never move far from their birthplaces.
   d. individuals that move into the zone have not previously encountered individuals of the other species, so reinforcement of reproductive barriers has not occurred.
   e. Narrow hybrid zones are artifacts because biologists generally restrict their studies to contact zones between species.
7. Which statement about speciation is *not* true?
   a. It always takes thousands of years.
   b. Reproductive isolation may develop slowly between diverging lineages.
   c. Among animals, it usually requires a physical barrier.
   d. Among plants, it often happens as a result of polyploidy.
   e. It has produced the millions of species living today.
8. Speciation is often rapid within groups whose species have complex behavior because
   a. individuals of such species make fine discriminations among potential mating partners.
   b. such species have short generation times.
   c. such species have high reproductive rates.
   d. such species have complex relationships with their environments.
   e. such species are particularly abundant.
9. Evolutionary radiations
   a. often happen on continents, but rarely on island archipelagoes.
   b. characterize birds and plants, but not other groups of organisms.
   c. have happened on continents as well as on islands.
   d. require major reorganizations of the genome.
   e. never happen in species-poor environments.
10. Speciation is an important component of evolution because it
    a. generates the variation on which natural selection acts.
    b. generates the variation on which genetic drift and mutations act.
    c. enabled Charles Darwin to perceive the mechanisms of evolution.
    d. generates the high extinction rates that drive evolutionary change.
    e. has resulted in a world with millions of species, each adapted for a particular way of life.

FOR DISCUSSION

1. The North American snow goose has two distinct color forms, blue and white. Matings between the two color forms are common. However, blue individuals pair with blue individuals and white individuals pair with white individuals much more frequently than would be expected by chance. Suppose that 75 percent of all mated pairs consisted of two individuals of the same color. What would you conclude about speciation processes in these geese? If 95 percent of pairs were the same color? If 100 percent of pairs were the same color?
2. Suppose pairs of snow geese of mixed colors were found only in a narrow zone within the broad Arctic breeding range of the geese. Would your answer to Question 1 remain the same? Would your answer change if mixed-color pairs were widely distributed across the breeding range of the geese?
3. Although many butterfly species are divided into local populations among which there is little gene flow, these species often show relatively little morphological variation among populations. Describe the studies you would conduct to determine what maintains this morphological similarity.
4. Evolutionary radiations are common and easily studied on oceanic islands, but in what types of *mainland* situations would you expect to find major evolutionary radiations? Why?
5. Fruit flies of the genus *Drosophila* are distributed worldwide, but 30–40 percent of the species in the genus are found on the Hawaiian Islands. What might account for this distribution pattern?
6. Evolutionary radiations take place when speciation rates exceed extinction rates. What factors can cause extinction rates to exceed

speciation rates in a clade? Name some clades in which human activities are increasing extinction rates without increasing speciation rates.

7. If it is true that natural selection does not directly favor lower viability of hybrids, why is it that hybrid individuals so often have lowered viability?

FOR INVESTIGATION

The experiment in Figure 23.10 changed only the orientation of the flowers. Although the flowers in the experiment were similarly oriented, they still differed in color, and probably in odor as well. What experiments could you design to determine the traits that the bees use to distinguish among the flowers of different *Aquilegia* species?

CHAPTER 24 The Evolution of Genes and Genomes

Molecular evolution and the conquest of polio

Prior to the late 1950s, thousands of children every year died or were left paralyzed by poliomyelitis (polio). This disease is caused by several strains of poliovirus, which infect humans through the mouth and multiply in the intestines. There are no symptoms in most infected individuals, but sometimes the virus invades the nervous system, resulting in rapid paralysis.

In 1955, Jonas Salk developed a vaccine against this nightmare disease. The vaccine, IPV (*i*nactivated *p*olio *v*accine), was based on killed virus. An injection of IPV produces antibodies to poliovirus in the blood (*serum immunity*) and prevents the spread of poliovirus into the nervous system. IPV does not prevent infection of the intestine, however, so infected individuals can still spread the virus to others. Thus, even though IPV prevented many cases of polio, the disease persisted, especially in developing countries where it was difficult to vaccinate everyone.

In 1958, Albert Sabin introduced an alternative: a live-virus vaccine that could be administered orally and provided a local immune reaction in the intestine as well as serum immunity. Sabin's discovery required selection of mutant strains of poliovirus, called *attenuated viruses*, that did not cause the disease.

Sabin could not have known the molecular details back in 1958, but today we know a small number of nucleotide substitutions present in the attenuated strains prevent them from infecting the nervous system. Antibodies produced in the lining of the intestine in response to attenuated viruses also prevent the multiplication of wild-type poliovirus, and the wild-type cannot infect the intestine of an immunized individual; thus Sabin's vaccine—called OPV, for *o*ral *p*olio *v*accine—prevents person-to-person transmission. Moreover, the attenuated strains of OPVs can spread to other individuals (especially in regions with poor sanitation), so an entire local population can become protected, even if not everyone receives the vaccine.

But OPVs have one major disadvantage: the attenuated strains continue to evolve. Very rarely, an attenuated virus undergoes a simple back substitution that results in its reversion to a virulent strain. Therefore, in regions where poliovirus has been virtually eliminated, the IPV vaccine is once again preferred: since it is based on a dead virus, it

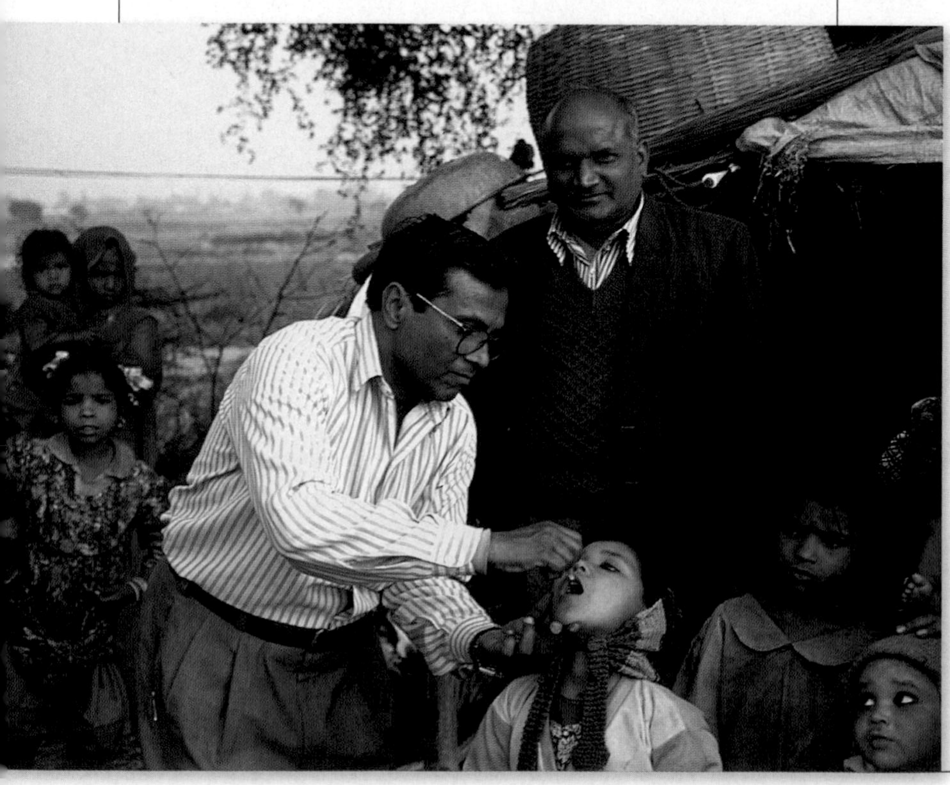

Combating Paralysis Mutant strains of poliovirus were the key to developing an effective vaccine that is helping to eliminate polio worldwide. Here a government health official administers the oral vaccine to children in rural India.

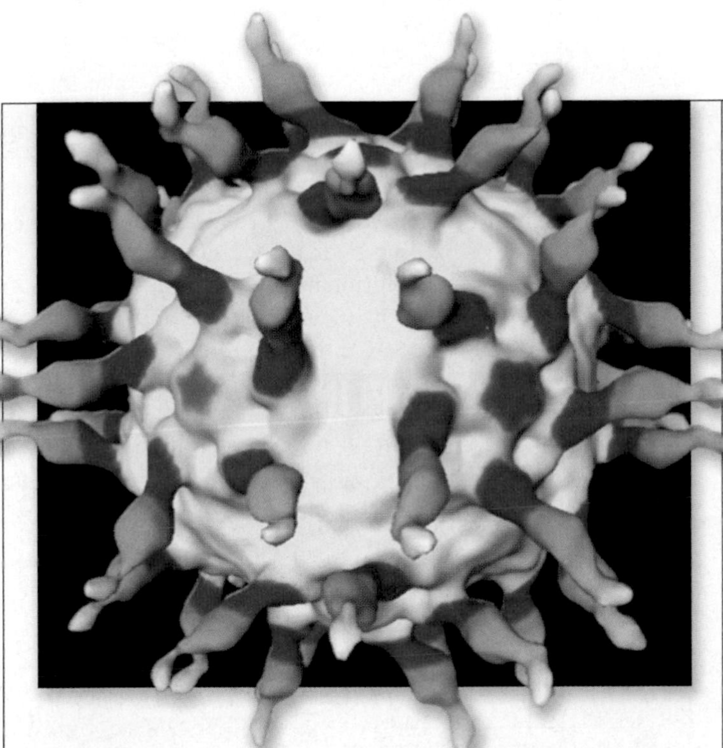

Killing a Killer Poliovirus infects its host by binding to receptors on the host cell surfaces. The binding sites on this inactivated virus are shown in red; antibodies have inactivated the sites and "killed" the virus. Injection of a vaccine made from inactivated poliovirus (IPV) stops the virus from affecting the nervous system.

cannot evolve. And now that entire poliovirus genomes have been sequenced, the molecular basis of attenuation is known—which is useful in selecting attenuated strains that are less subject to reversion to use in OPVs.

Today, when a polio outbreak occurs anywhere in the world, DNA sequencing can be used to determine the evolutionary relationships of the viruses involved and the preferred vaccination strategy. This medical success story is a result of our growing understanding of molecular evolution.

IN THIS CHAPTER we will see how molecular biologists use nucleic acids and proteins to infer both the patterns and the causes of molecular evolution. We will explore how the functions of molecules change, where new genes come from, and how genomes change in size. Finally, we will see how knowledge of the patterns of molecular evolution can help answer other biological questions.

CHAPTER OUTLINE

24.1 What Can Genomes Reveal about Evolution?

An organism's **genome** is the full set of genes it contains, as well as any noncoding regions of the DNA (or, in the case of some viruses, RNA). Most of the genes of eukaryotic organisms are found on chromosomes in the nucleus, but genes are also present in chloroplasts and mitochondria. In organisms that reproduce sexually, both males and females transmit nuclear genes, but mitochondrial and chloroplast genes usually are transmitted only via the cytoplasm of eggs, as we saw in Section 10.5.

Genomes must be replicated to be transmitted from parents to offspring. DNA replication does not occur without error, however. Mistakes in DNA replication—mutations—provide the raw material for evolutionary change. Mutations are essential for the long-term survival of life, for without the genetic variation they provide, organisms could not evolve in response to changes in their environment.

A particular copy of a gene will not be passed on to successive generations unless an individual with that copy survives and reproduces. Therefore, the capacity to cooperate with different combinations of other genes is likely to increase a particular allele's probability of fixation in a population. Moreover, the degree and timing of a gene's expression are affected by its location in the genome. For these reasons, the genes of an individual organism can be viewed as interacting members of a group, among which there are divisions of labor, but also strong interdependencies.

A genome, then, is not simply a random collection of genes in random order along chromosomes. Rather, it is a complex set of integrated genes and their regulatory sequences, as well as vast stretches of noncoding DNA that apparently have little direct function. The positions of genes, as well as their sequences, are subject to evolutionary change, as are the extent and location of noncoding DNA. All of these changes can affect the phenotype of an organism. Biologists have now sequenced the complete genomes of a large number of organisms (including humans), and this information is helping us to understand how and why organisms differ, how they function, and how they have evolved.

Evolution of genomes results in biological diversity

The field of **molecular evolution** investigates the mechanisms and consequences of the evolution of macromolecules. Molecular evolutionists study relationships between the structures of genes and proteins and the functions of organisms. They also use molecular variation to reconstruct evolutionary history and study the mechanisms and consequences of evolution. The molecules of special interest to molecular evolutionists are nucleic acids (DNA and RNA) and proteins. Students of this field ask questions such as, How do proteins acquire new functions? Why are the genomes of different organisms so variable in size? How has enlargement of genomes been accomplished? They also investigate the evolution of particular nucleic acids and proteins and use their findings to reconstruct the evolutionary histories of genes and the organisms that carry them. In this way, molecular evolutionists seek to understand the molecular basis for the biological diversity that we now observe in the world around us.

The evolution of nucleic acids and proteins depends on genetic variation introduced by mutations (for a review of the various kinds of mutations, refer to Section 12.6). One of several ways in which genes evolve is by means of **nucleotide substitutions**. In genes that encode proteins, some of these nucleotide substitutions can result in **amino acid replacements** in the encoded proteins. Changes in the amino acid sequence of a protein can change the charge, as well as the secondary (two-dimensional) and tertiary (three-dimensional) structure, of the protein. All of these phenotypic changes affect the way the protein functions in the organism.

Evolutionary changes in genes and proteins can be identified by comparing the nucleotide or amino acid sequences among different organisms. The longer two sequences have been evolving separately, the more differences they accumulate (although different genes in the same species evolve at different rates). Evolutionary analysis by such comparisons is essential for determining the timing of evolutionary changes in molecular characters, and knowing the timing of such changes is usually the first step in inferring their causes. Conversely, knowledge of the pattern and rate of evolutionary change in a given macromolecule is useful in reconstructing the evolutionary history of groups of organisms.

Before biologists can compare genes or proteins across different organisms, they must have a method for identifying homologous parts of these molecules. As we will see in Section 25.1, any features shared by two or more species that have been inherited from a common ancestor are said to be *homologous*. For example, the forelimbs of all mammals are homologous, although they differ greatly in their form and function (consider the fins of whales and the arms of humans). The concept of homology extends down to the level of particular nucleotide positions in genes. Therefore, one of the first steps in studying the evolution of genes or proteins is to align homologous positions in the nucleotide or amino acid sequences of interest.

Genes and proteins are compared through sequence alignment

Once the DNA or amino acid sequences of molecules from different organisms have been determined, they can be compared. Homologous positions can be identified only if we first pinpoint the locations of deletions and insertions that have occurred in the molecules of interest in the time since the organisms diverged from a common ancestor. A simple hypothetical example illustrates this **sequence alignment** technique. In **Figure 24.1** we compare two amino acid sequences (1 and 2) from homologous proteins in different organisms. The two sequences at first appear to differ in the number and identity of their amino acids, but if we insert a gap after the first amino acid in sequence 2 (after leucine), similarities in the two sequences become evident. This gap indicates the occurrence of one of two evolutionary events: an insertion of an amino acid in the longer protein, or a deletion of an amino acid

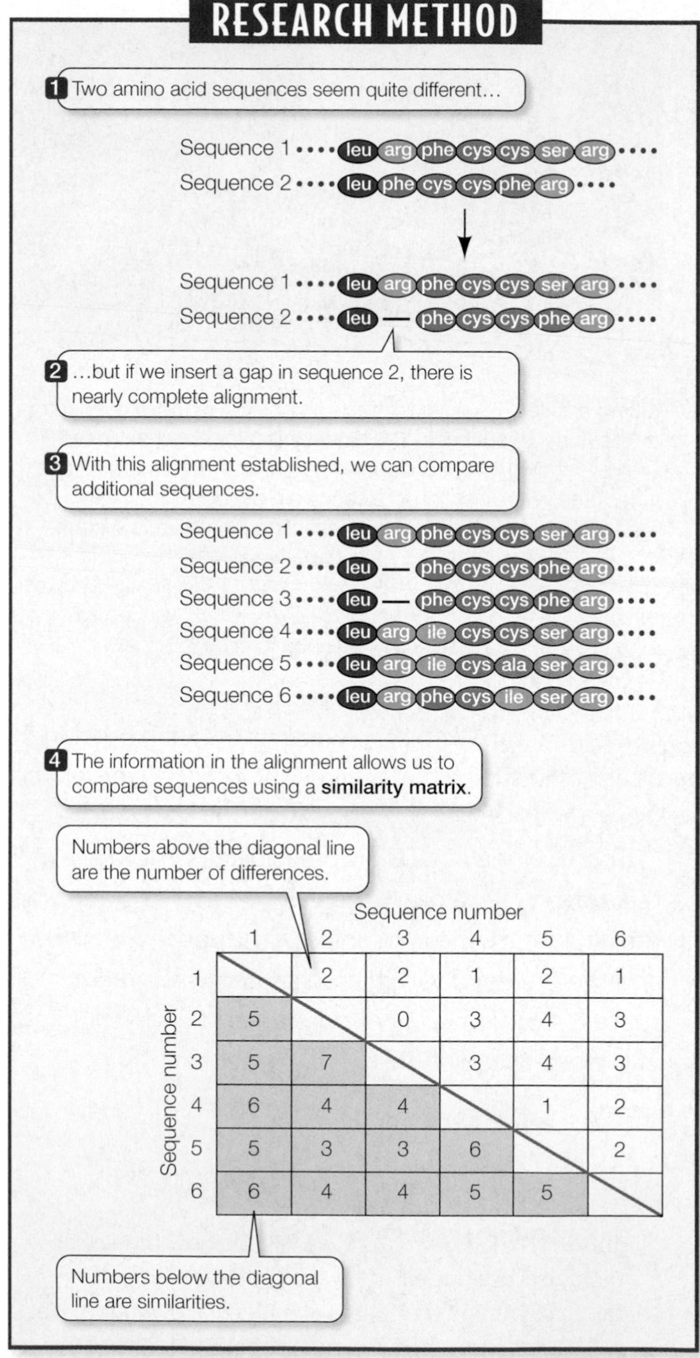

	1	2	3	4	5	6
1		2	2	1	2	1
2	5		0	3	4	3
3	5	7		3	4	3
4	6	4	4		1	2
5	5	3	3	6		2
6	6	4	4	5	5	

24.1 Amino Acid Sequence Alignment Insertion of a gap (—) allows us to align two homologous amino acid squences so that we can compare them. Once the alignment is established, sequences from more organisms can be added and compared. A similarity matrix sums similarities and differences between each pair of organisms.

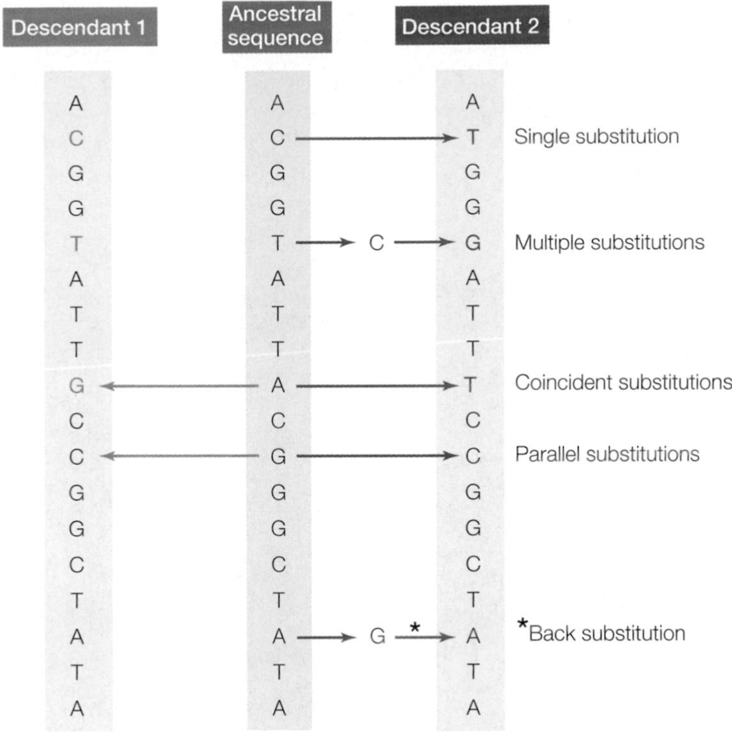

24.2 Multiple Substitutions Are Not Reflected in Pairwise Sequence Comparisons An ancestral sequence (center) gives rise to two descendant sequences through a series of substitutions. Although the two descendant sequences exhibit only three nucleotide differences (colored letters), these three differences resulted from a total of nine substitutions (arrows).

in the shorter protein. Having adjusted for this gap, we can see that the two sequences differ by only one amino acid at position 6 (serine or phenylalanine). Adding a single gap—that is, identifying a deletion or an insertion—*aligns* these sequences. Longer sequences and those that have diverged more extensively require more elaborate adjustments based on explicit models (computer algorithms) for the relative costs of deletions, insertions, and particular amino acid replacements.

Having aligned the sequences, we can compare them by counting the number of nucleotides or amino acids that differ between them. If we add more sequences to our original example and sum the number of similar and different amino acids in each pair of sequences, we can construct a **similarity matrix**, which gives us a measure of the minimum number of changes that have occurred during the divergence between each pair of organisms (see Figure 24.1).

Models of sequence evolution are used to calculate evolutionary divergence

The sequence comparison procedure illustrated in Figure 24.1 gives a simple count of the minimum number of changes between two species. In the context of two aligned DNA sequences, we can count the number of differences at homologous nucleotide positions, and this count indicates the minimum number of nucleotide substitutions that must have occurred between the two sequences.

However, this simple count of differences almost certainly underestimates the number of substitutions that have actually occurred since the sequences diverged from a common ancestor. When more than one substitution occurs between the ancestor and the descendants, as illustrated in **Figure 24.2**, the number of changes is not captured by a simple count of differences in a similarity matrix. Several phenomena can account for this, including:

- *Multiple substitutions:* More than one change occurs at a given position between the ancestor and a descendant.
- *Coincident substitutions:* At a given position, different substitutions occur between the ancestor and each descendant.
- *Parallel substitutions:* The same substitution occurs independently between the ancestor and each descendant.
- *Back substitutions:* In a variation on multiple substitutions, after a change at a given position, a subsequent substitution changes the position back to the ancestral state.

To correct for undercounting of substitutions, molecular evolutionists have developed mathematical models that describe how DNA (and protein) sequences evolve. These models take into account the rates of change from each nucleotide to another; for example, *transitions* (changes between two purines or between two pyrimidines) are more frequent than *transversions* (changes between a purine and pyrimidine). They also include parameters such as the different rates of substitution across different parts of a gene and the proportions of each nucleotide present in a given sequence. These parameters are estimated for a particular set of sequences, and then the models are used to correct for multiple substitutions, coincident substitutions, parallel substitutions, and back substitutions. The end result is a revised estimate of the total number of substitutions likely to have occurred between two sequences, which is almost always greater than the observed number of differences.

As sequence information becomes available for more and more proteins in an ever-expanding database, sequence alignments can be extended across multiple homologous sequences, and the minimum number of insertions, deletions, and substitutions can be summed across homologous genes or proteins of an entire group of organisms. **Figure 24.3** shows this kind of aligned data for cytochrome *c* sequences in a wide variety of animals, plants, and fungi. This type of information is used extensively in determining the evolutionary relationships among species.

Experimental studies examine molecular evolution directly

Although molecular evolutionists are often interested in naturally evolved sequences and proteins, molecular and phenotypic evolution can also be observed directly in the laboratory. Increasingly, evolutionary biologists are studying evolution experimentally. Because substitution rates are related to generation rate rather than to absolute time, most of these experiments use unicellular organisms or viruses with short generations. Viruses, bacteria, and unicellular eukaryotes (such as the yeasts) can be cultured in large populations in the laboratory, and many of these organisms can evolve

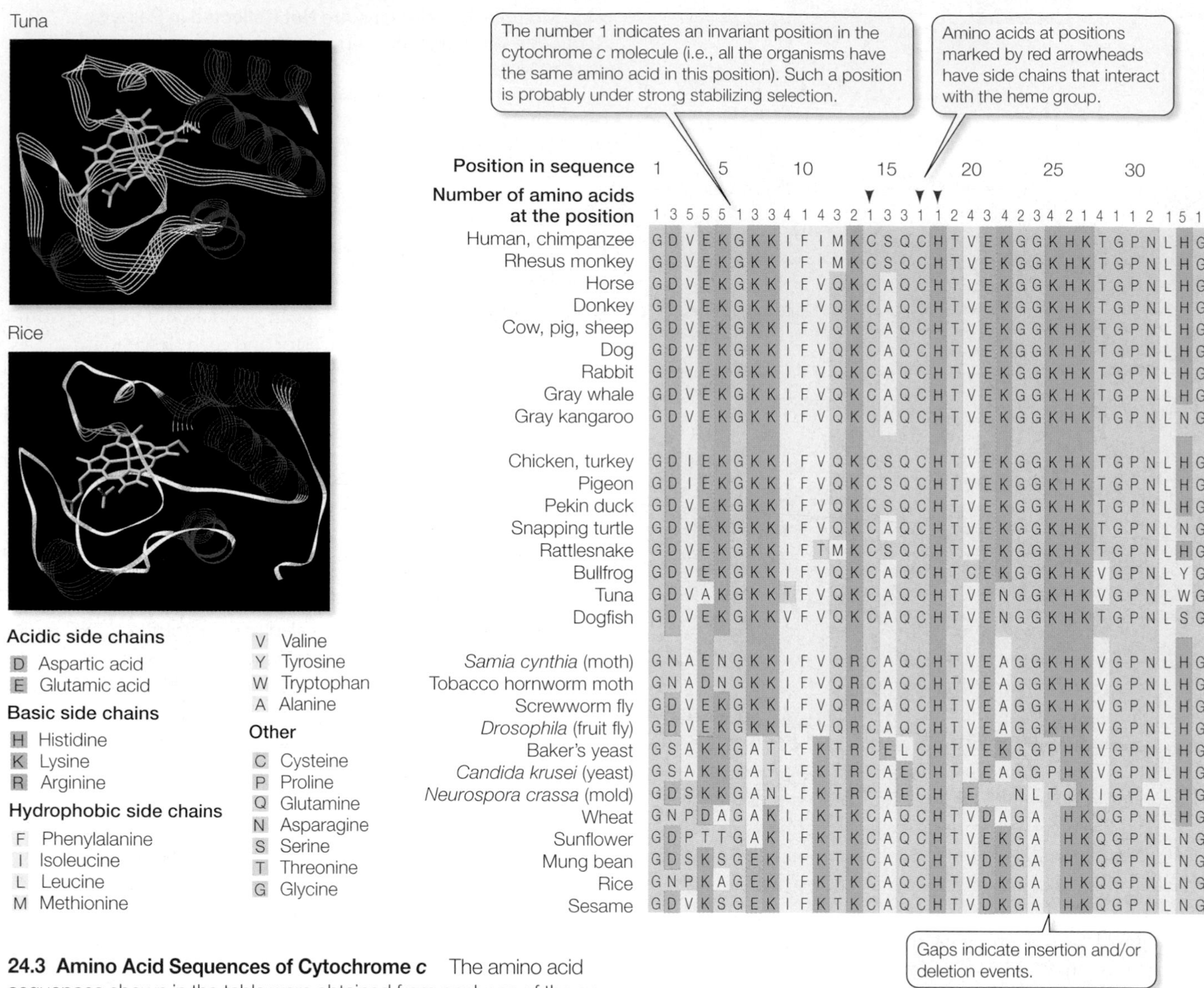

24.3 Amino Acid Sequences of Cytochrome *c* The amino acid sequences shown in the table were obtained from analyses of the enzyme cytochrome *c* from 33 species of plants, fungi, and animals. Note the lack of variation across the sequences at positions 70–80, suggesting that this region is under strong stabilizing selection, and changing its amino acid sequence would impair the protein's function. The computer graphics at the upper left are created from these sequences and show the three-dimensional structures of tuna and rice cytochrome *c*. Alpha helixes are in red, and the molecule's heme group is shown in yellow.

rapidly. In the case of some RNA viruses, the natural substitution rate may be as high as 10^{-3} substitutions per position per generation. Therefore, in a virus of a few thousand nucleotides, one or more substitutions are expected (on average) every generation, and these changes can easily be determined by sequencing the entire genome (because of its small size). Generation time may be only tens of minutes (rather than years or decades, as in humans), so biologists can directly observe substantial molecular evolution in a controlled population over the course of days, weeks, or months.

An example of an experimental evolutionary study is shown in **Figure 24.4**. Paul Rainey and Michael Travisano wanted to examine a potential cause of adaptive radiations, which are a major source of biological diversity. For instance, near the beginning of the Cenozoic era, mammals rapidly diversified into species as diverse as elephants, moles, whales, and bats. While Rainey and Travisano clearly couldn't experimentally manipulate mammals over many millions of years, they could test the idea that heterogenous environments lead to adaptive radiation by experimentally manipulating a lineage of bacteria.

Rainey and Travisano inoculated a number of flasks containing culture medium with the same strain of the bacterium *Pseudomonas fluorescens*. They then shook some of the cultures to maintain a constantly uniform environment, and left others alone (static cultures), allowing them to develop a spatially distinct structure. For instance, in the static cultures the environment on the surface film of the medium differed from that on the walls of the flasks and from parts of the culture not touching any surfaces.

When the cultures were started, the ancestral phenotype of the bacterium produced a smooth colony, which the investigators called a "smooth morph." Within just a few days, however, the static cultures consistently and independently developed two other

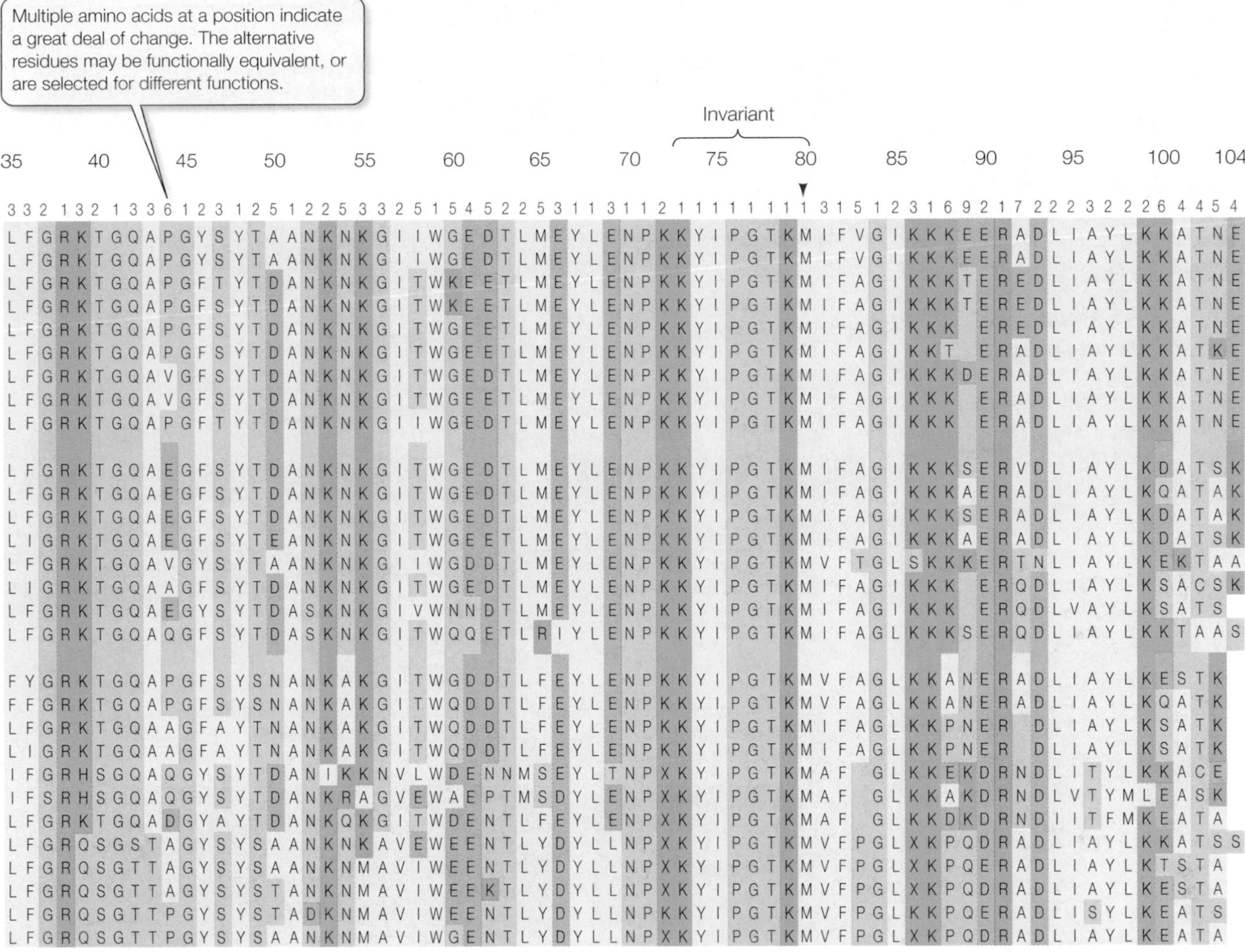
Multiple amino acids at a position indicate a great deal of change. The alternative residues may be functionally equivalent, or are selected for different functions.
Invariant
35 40 45 50 55 60 65 70 75 80 85 90 95 100 104

morphs: a "wrinkly spreader" and a "fuzzy spreader." The researchers determined that the two new morphs had a genetic basis and were adaptively superior in some of the environments of the static cultures. For example, the "wrinkly spreader" cells adhered firmly to one another as well as to surfaces, and were thus able to form a mat across the surface of the medium, where they could compete successfully for oxygen.

Molecular investigations of the genomes of these morphs showed that they had evolved repeatedly, and that many different substitutions could produce the same phenotypes. In contrast, the homogeneous shaken cultures showed no evolution. The same mutations occurred in the shaken cultures, but did not persist in those populations because the novel phenotypes that they produced were selectively disadvantageous under the "shaken" environmental conditions.

Experimental molecular evolutionary studies are used for a wide variety of purposes and have greatly expanded the ability of evolutionary biologists to test evolutionary concepts and principles. Biologists now routinely study evolution in the laboratory and, as we will see later in this chapter, use in vitro evolutionary techniques to produce novel molecules to perform new functions for industrial and pharmaceutical uses.

24.1 RECAP

A genome is the sum of all the genetic material of an organism—including both sequences coding for functional genes, and noncoding sequences. The genomes of all organisms evolve over time.

- Can you explain the relationship between a nucleotide substitution and an amino acid replacement? See p. 526
- Can you describe how biologists align nucleotide and amino acid sequences they wish to compare and how they estimate the number of changes that have occurred between pairs of aligned sequences? See p. 526–527 and Figure 24.1

Molecular evolutionists can directly observe the evolution of genomes over time, and they can compare the genomes of different organisms and reconstruct the changes that have occurred during their evolution. Let's turn now to the question of how genomes change and examine some of the consequences of those changes.

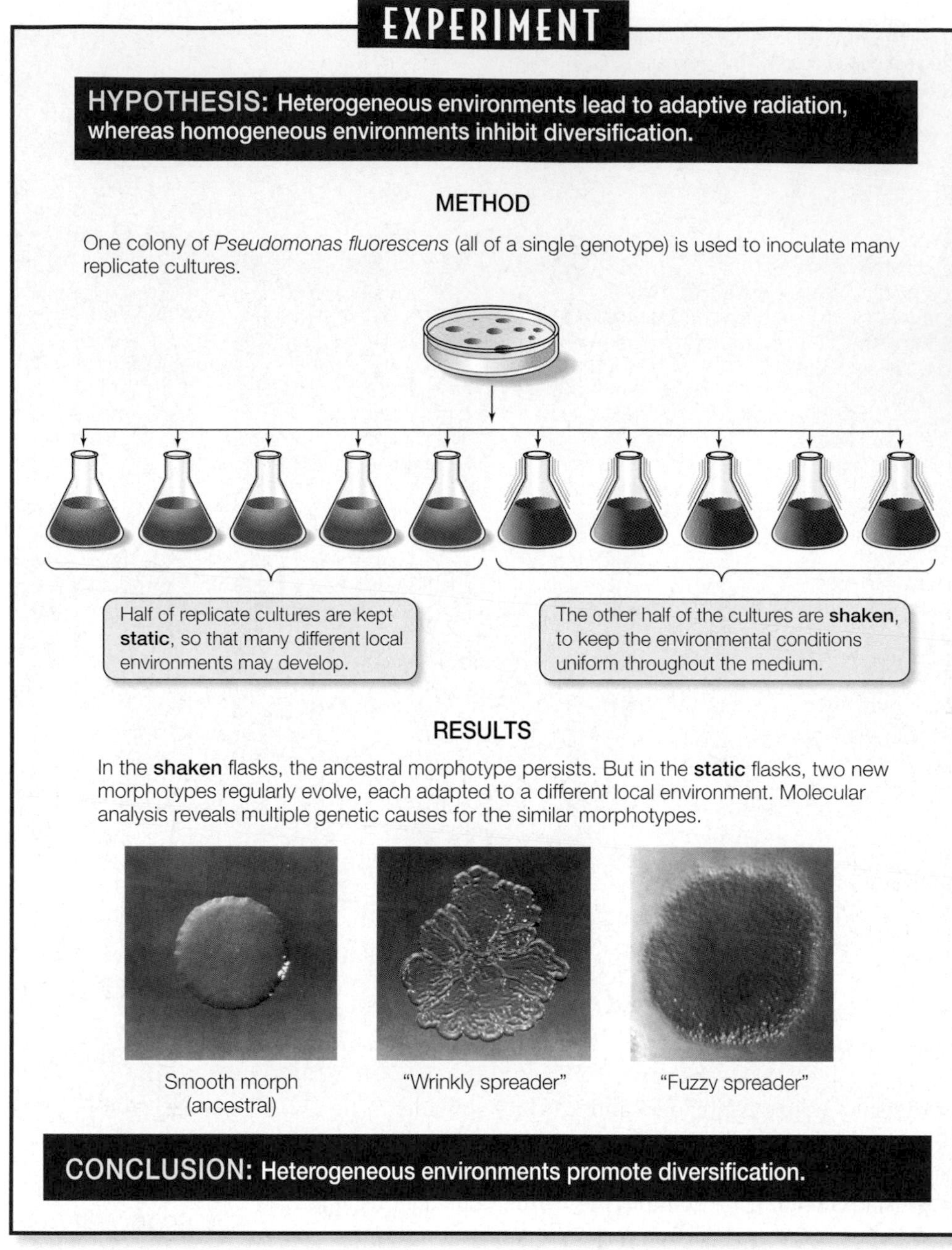

EXPERIMENT
HYPOTHESIS: Heterogeneous environments lead to adaptive radiation, whereas homogeneous environments inhibit diversification.
METHOD
One colony of *Pseudomonas fluorescens* (all of a single genotype) is used to inoculate many replicate cultures.
Half of replicate cultures are kept **static**, so that many different local environments may develop.
The other half of the cultures are **shaken**, to keep the environmental conditions uniform throughout the medium.
RESULTS
In the **shaken** flasks, the ancestral morphotype persists. But in the **static** flasks, two new morphotypes regularly evolve, each adapted to a different local environment. Molecular analysis reveals multiple genetic causes for the similar morphotypes.
Smooth morph (ancestral)
"Wrinkly spreader"
"Fuzzy spreader"
CONCLUSION: Heterogeneous environments promote diversification.

24.4 A Heterogeneous Environment Spurs Adaptive Radiation Rainey and Travisano's studies used a rapidly reproducing prokaryote species to model adaptive radiation. Their experiments indicated that phenotypic change is enhanced in a heterogeneous environment. The uniform environment of the shaken cultures showed no evolution. Molecular analysis revealed that same genetic mutations occurred in the shaken cultures, but did not persist because the novel phenotypes they produced were selectively disadvantageous under homogeneous environmental conditions. FURTHER RESEARCH: Once they have arisen in a heterogeneous environment, would the three evolved phenotypes compete successfully in a homogeneous environment?

24.2 What Are the Mechanisms of Molecular Evolution?

A *mutation*, as we saw in Chapter 12, is any change in the genetic material. A nucleotide substitution is one type of mutation. Many nucleotide substitutions have no effect on phenotype, even if the change occurs in a gene that encodes a protein, because most amino acids are specified by more than one codon (see Figure 12.6). A substitution that does not change the amino acid that is specified is known as a **synonymous** or **silent substitution** (**Figure 24.5A**). Synonymous substitutions do not affect the functioning of a protein (and hence the organism) and are therefore unlikely to be influenced by natural selection.

A nucleotide substitution that *does* change the amino acid sequence encoded by a gene is known as a **nonsynonymous substitution** (**Figure 24.5B**). In general, nonsynonymous substitutions are likely to be deleterious to the organism. But not every amino acid replacement alters a protein's shape and charge (and hence its functional properties). Therefore, some nonsynonymous substitutions may also be selectively neutral, or nearly so. Conversely, an amino acid replacement that confers an advantage to the organism would result in positive selection for the corresponding nonsynonymous substitution.

Enough analyses of mammalian genes have been performed to show that the rate of nonsynonymous nucleotide substitutions varies from nearly zero to about 3×10^{-9} substitutions per position per year. Synonymous substitutions in the protein-coding regions of genes have occurred about five times more rapidly than nonsynonymous substitutions. In other words, substitution rates are highest at nucleotide positions that *do not change the amino acid be-*

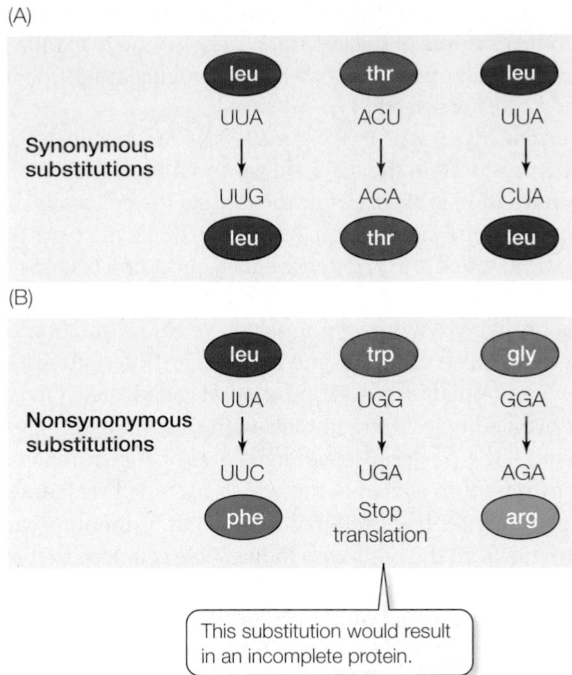

24.5 When One Nucleotide Does or Doesn't Make a Difference (A) Synonymous substitutions do not change the amino acid specified and do not affect protein function; such substitutions are less likely to be subject to natural selection. (B) Nonsynonymous substitutions do change the amino acid sequence and are likely to have an effect (often deleterious) on protein function; such substitutions are targets for natural selection.

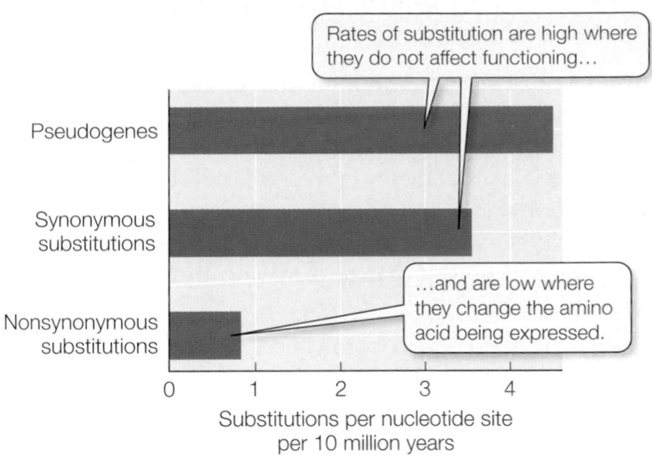

24.6 Rates of Substitution Differ Rates of nonsynonymous substitution typically are much lower than rates of synonymous substitution and the substitution rate in pseudogenes. This pattern reflects differing levels of functional constraints.

ing expressed (**Figure 24.6**). The rate of substitution is even higher in **pseudogenes**, which are duplicate copies of genes that have undergone one or more mutations that eliminate their ability to be expressed.

As we saw in Chapter 22, most natural populations of organisms harbor much more genetic variation than we would expect if genetic variation were influenced by natural selection alone. This discovery, combined with the knowledge that many mutations do not change molecular function, stimulated the development of the neutral theory of molecular evolution.

Much of evolution is neutral

In 1968, Motoo Kimura proposed the *neutral theory of molecular evolution*. Kimura suggested that, at the molecular level, the majority of mutations we observe in populations are selectively neutral; that is, they confer neither an advantage nor a disadvantage on their bearers. Therefore, these neutral mutations accumulate through genetic drift, rather than through positive directional selection (as described in Section 22.3).

The rate of fixation of neutral mutations by genetic drift is independent of population size. To see why this is so, consider a population of size N and a neutral mutation rate μ (mu) per gamete per generation at a locus. The number of new mutations would, on average, be $\mu \times 2N$, because $2N$ gene copies are available to mutate in a population of diploid organisms. According to genetic drift theory, the probability that a mutation will be fixed by drift alone is its frequency, p, which equals $1/(2N)$ for a newly arisen mutation. Therefore, the number of neutral mutations that arise per generation that are likely to become fixed in a given population is $2N\mu \times 1/(2N) = \mu$. Therefore, the rate of fixation of neutral mutations depends only on the neutral mutation rate (μ), and is independent of population size (N). A given mutation is more likely to appear in a large population than in a small one, but given that it does appear, it is more likely to become fixed in a small population. These two influences of population size exactly cancel each other out.

Therefore, the rate of fixation of neutral mutations is equal to the mutation rate. So if most mutations of macromolecules are neutral, and if the underlying mutation rate is constant, macromolecules evolving in different populations should diverge from one another at a constant rate. Empirically, the rate of evolution of particular genes and proteins is often relatively constant over time, and in effect can be used as a "molecular clock." (More will be said about this in Section 25.2, where we will see how molecular clocks can be used to calculate evolutionary divergence times between species.)

Although much of the genetic variation we observe in populations is the result of neutral evolution, that does not mean that most mutations have no effect on the organism. Many mutations are never observed in populations because they are lethal or strongly detrimental to the organism and are thus removed from the population through natural selection. Similarly, mutations that confer a selective advantage tend to be quickly fixed in populations, so they do not result in variation at the population level either. Nonetheless, if we compare homologous proteins from different populations or species, some amino acid positions will stay constant under stabilizing selection, others will vary through neutral genetic drift, and still others will vary as a result of positive selection for change. How can these evolutionary processes be distinguished?

Positive and stabilizing selection can be detected in the genome

As we have just seen, substitutions in a protein-coding gene can be either synonymous or nonsynonymous, depending on whether they change the resulting amino acid sequence of the protein. According to the neutral theory of molecular evolution, the relative rates of synonymous and nonsynonymous substitutions are expected to differ in regions of genes that are evolving neutrally, under positive selection for change, or staying unchanged under stabilizing selection.

- If a given amino acid in a protein can be one of many alternatives (without changing the protein's function), then an amino acid replacement is *neutral* with respect to the fitness of an organism. In this case, the rates of synonymous and nonsynonymous substitutions in the corresponding DNA sequences are expected to be very similar, so the ratio of the two rates would be close to 1.
- If a given amino acid position is under *strong stabilizing selection,* then the rate of synonymous substitutions in the corresponding DNA sequences is expected to be much higher than the rate of nonsynonymous substitutions.
- If a given amino acid position is under *strong selection for change,* the rate of nonsynonymous substitutions is expected to exceed the rate of synonymous substitutions in the corresponding DNA sequences.

By comparing the gene sequences that encode proteins from many species, the history and timing of synonymous and nonsynonymous substitutions can be determined (look back at Figure 24.2 for an example). This information can be mapped on a *phylogenetic tree,* a diagram of evolutionary relationships (the construction of phylogenetic trees will be discussed in more detail in Chapter 25).

Genes, or regions of genes, that are evolving under neutral, stabilizing, or positive selection can be identified by comparing the nature and rates of substitutions across the phylogenetic tree. Let's consider the example of the evolution of lysozyme to explore how and why particular positions of a gene sequence might be under different modes of selection.

The enzyme lysozyme (see Figure 3.8) is found in almost all animals. It is produced in the tears, saliva, and milk of mammals and in the whites of bird eggs. Lysozyme digests the cell walls of bacteria, rupturing and killing them. As a result, lysozyme plays an important role as a first line of defense against invading bacteria. Most animals defend themselves against bacteria by digesting them, which is probably why most animals have lysozyme. Some animals, however, also use lysozyme in the digestion of food.

Among mammals, a mode of digestion called *foregut fermentation* has evolved twice. In mammals with this mode of digestion, the foregut—the posterior esophagus and/or the stomach—has been converted into a chamber in which bacteria break down ingested plant matter by fermentation. Foregut fermenters can obtain nutrients from the otherwise indigestible cellulose that makes up a large proportion of the plant body. Foregut fermentation evolved independently in ruminants (a group of hoofed mammals that includes cattle) and in certain leaf-eating monkeys, such as langurs (**Figure 24.7A**). We know that these evolutionary events were independent because both langurs and ruminants have close relatives that are not foregut fermenters.

In both foregut-fermenting lineages, the enzyme lysozyme has been modified to play a new, nondefensive role. This lysozyme ruptures some of the bacteria that live in the foregut, releasing nutrients metabolized by the bacteria, which the mammal then absorbs. How many changes in the lysozyme molecule were needed to allow it to perform this function amid the digestive enzymes and acidic conditions of the mammalian foregut? To answer this question, molecular evolutionists compared the lysozyme-coding sequences in foregut fermenters with those of several of their nonfermenting relatives. They determined which amino acids differed and which were shared among the species (**Table 24.1**), and they also determined the rates of synonymous and nonsynonymous substitutions in the lysozyme genes across the evolutionary history of the sampled species.

For many of the amino acid positions of lysozyme, the rate of synonymous substitutions (in the corresponding gene) is much higher than the rate of nonsynonymous substitutions. This ob-

(A) *Presbytis entellus*

(B) *Opisthocomus hoazin*

24.7 Convergent Molecular Evolution Foregut-fermenting mammals such as the Hanuman langur (A) have been evolving independently from the hoatzin (B) for hundreds of millions of years, but they have independently evolved similar modifications to the enzyme lysozyme.

TABLE 24.1

Similarity Matrix for Lysozyme in Mammals

SPECIES	LANGUR	BABOON	HUMAN	RAT	CATTLE	HORSE
Langur*		14	18	38	32	65
Baboon	0		14	33	39	65
Human	0	1		37	41	64
Rat	0	1	0		55	64
Cattle*	5	0	0	0		71
Horse	0	0	0	0	1	

Shown above the diagonal line is the number of amino acid sequence *differences* between the two species being compared; below the line are the number of changes uniquely *shared* by the two species.

Asterisks (*) indicate foregut-fermenting species.

servation indicates that many of the amino acids that make up lysozyme are evolving under stabilizing selection. In other words, there is selection against change in the protein at these positions, and the observed amino acids must therefore be critical for lysozyme function. At other positions, several different amino acids function equally well, and the corresponding regions of the genes have similar rates of synonymous and nonsynonymous substitutions. The most striking finding is that amino acid replacements in lysozyme happened at a much higher rate in the lineage leading to langurs than in any other primates. The high rate of nonsynonymous substitutions in the langur lysozyme gene shows that lysozyme went through a period of rapid change in adapting to the stomachs of langurs. Moreover, the lysozymes of langurs and cattle share five amino acid replacements, all of which lie on the surface of the lysozyme molecule, well away from the enzyme's active site. Several of these shared replacements involve changes from arginine to lysine, which makes the proteins more resistant to attack by the pancreatic enzyme trypsin. By understanding the functional significance of amino acid replacements, molecular evolutionists can explain the observed changes in amino acid sequences in terms of changes in the functioning of the protein.

A large body of fossil, morphological, and molecular evidence shows that langurs and cattle do not share a recent common ancestor. However, langur and ruminant lysozymes share several amino acids that neither mammal shares with the lysozymes of its own closer relatives. The lysozymes of these two mammals have converged at some amino acid positions despite their very different ancestry. (We will see other examples of *convergent evolution* under similar selection pressures in Chapter 25.) The amino acids they share give these lysozymes the ability to lyse the bacteria that ferment plant material in the foregut.

An even more remarkable story emerges in the case of lysozyme in the crop of the hoatzin, a unique leaf-eating South American bird and the only known avian foregut fermenter (**Figure 24.7B**). Many birds have an enlarged esophageal chamber called a *crop*. Hoatzins have a crop that contains bacteria and acts as a fermenting chamber. Many of the amino acid replacements that occurred in the adaptation of hoatzin crop lysozyme are identical to the changes that evolved in ruminants and langurs. Thus, even though the hoatzin and the foregut-fermenting mammals have not shared a common ancestor in hundreds of millions of years, they have all evolved similar adaptations in their lysozymes that enable them to recover nutrients from their fermenting bacteria in a highly acidic environment.

Genome size and organization also evolve

We know that genome size varies tremendously among organisms. Across broad taxonomic categories, there is some correlation between genome size and organismal complexity. The genome of the tiny bacterium *Mycoplasma genitalium* has only 470 genes. *Rickettsia prowazekii*, the bacterium that causes typhus, has 634 genes. *Homo sapiens*, on the other hand, has about 23,000 protein-coding genes. **Figure 24.8** shows the relative number of genes for several prokaryotic and eukaryotic organisms.

Additional genomic comparisons reveal, however, that a larger genome does not always indicate greater complexity. It is not surprising that more complex genetic instructions are needed for building and maintaining a large, multicellular organism than a small, single-celled bacterium. What is surprising is that some organisms, such as lungfishes, some salamanders, and lilies, have about 40 times as much DNA as humans do. Structurally, a lungfish or a lily is not 40 times more complex as a human. So why does genome size vary so much?

Differences in genome size are not so great if we take into account only the portion of DNA that actually encodes RNAs or proteins. The organisms with the largest total amounts of nuclear DNA (some ferns and flowering plants) have 80,000 times as much DNA as the bacteria with the smallest genomes, but no species has more than about 100 times as many protein-coding genes as a bacterium. Therefore, much of the variation in genome size lies not in the number of functional genes, but in the amount of noncoding DNA (**Figure 24.9**).

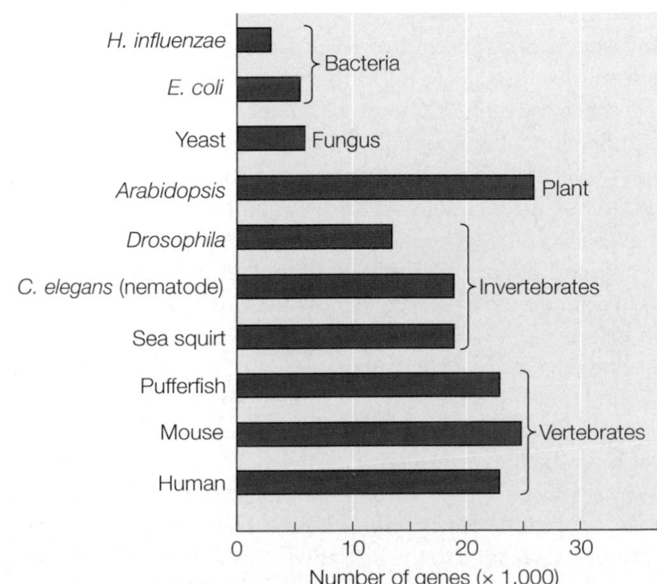

24.8 Genome Size Varies Widely The number of genes in the genome has been measured or estimated in a variety of organisms, ranging from single-celled prokaryotes to vertebrates.

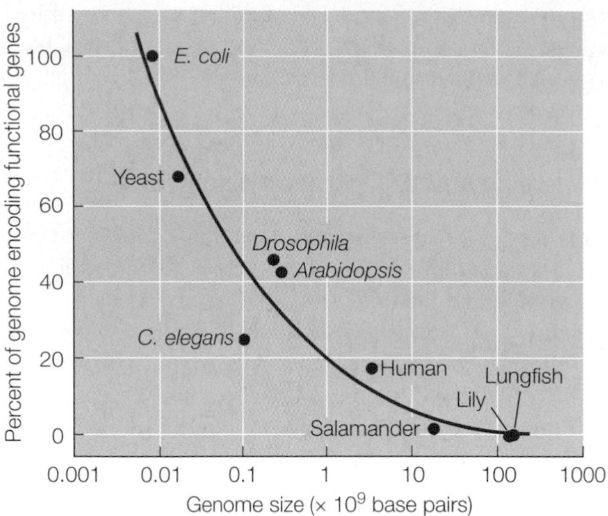

24.9 A Large Proportion of DNA Is Noncoding Most of the DNA of bacteria and yeasts encodes RNAs or proteins, but a large percentage of the DNA of multicellular species is noncoding.

Why do the cells of most organisms have so much noncoding DNA? Does this noncoding DNA have a function, or is it "junk"? Although much of this DNA does not appear to have a direct function, it can alter the expression of the surrounding genes. The degree or timing of gene expression can be changed dramatically depending on the gene's position relative to noncoding sequences. Other regions of noncoding DNA consist of pseudogenes that are simply carried in the genome because the cost of doing so is very small. These pseudogenes may become the raw material for the evolution of new genes with novel functions. Still other noncoding sequences consist of parasitic transposable elements that spread through populations because they reproduce faster than the host genome.

Investigators can use retrotransposons to estimate the rates at which species lose DNA. Retrotransposons are transposable elements that copy themselves through an RNA intermediate, as we saw in Section 14.1. The most common type of retrotransposon carries duplicated sequences at each end, called long terminal repeats (LTRs). Occasionally, LTRs recombine in the host genome, so that the DNA between them is excised. When this happens, one recombined LTR is left behind. The number of such "orphaned" LTRs in a genome is a measure of how many retrotransposons have been lost. By comparing the number of LTRs in the genomes of Hawaiian crickets (*Laupala*) and fruit flies (*Drosophila*), investigators found that *Laupala* loses DNA more than 40 times more slowly than *Drosophila*. As a result, the genome of *Laupala* is 11 times larger than that of *Drosophila*.

Why do species differ so greatly in the rate at which they gain or lose apparently functionless DNA? One hypothesis is that genome size is related to the rate at which the organism develops, which may be under selection pressure. Large genomes can slow down the rate of development and thus alter the relative timing of expression of particular genes. As we illustrated and discussed in Section 20.2, such changes in the timing of gene expression (*heterochrony*) can produce major changes in phenotype. Thus, although some noncoding DNA sequences may have no direct function, they may still affect the development of the organism.

Another hypothesis is that the proportion of noncoding DNA is related primarily to population size. Noncoding sequences that are only slightly deleterious are likely to be purged by selection primarily in species with large population sizes. In species with small populations, the effects of genetic drift can overwhelm selection against noncoding sequences that have small deleterious consequences. Therefore, selection against the accumulation of noncoding sequences is most effective in species with large populations, so such species (such as bacteria or yeasts) have relatively little noncoding DNA compared with species with small populations (see Figure 24.9).

New functions can arise by gene duplication

Gene duplication is another way in which proteins can acquire new functions. When a gene is duplicated, one copy of that gene is potentially freed from having to perform its original function. The identical copies of a duplicated gene can have any one of four different fates:

- Both copies of the gene may retain their original function (which can result in a change in the amount of gene product that is produced by the organism).
- Both copies of the gene may retain the ability to produce the original gene product, but the expression of the genes may diverge in different tissues or at different times of development.
- One copy of the gene may be incapacitated by the accumulation of deleterious substitutions and become a functionless pseudogene.
- One copy of the gene may retain its original function, while the second copy accumulates enough substitutions that it can perform a different function.

How often do gene duplications arise, and which of these four outcomes is most likely? Investigators have found that rates of gene duplication are fast enough for a yeast or *Drosophila* population to acquire several hundred duplicate genes over the course of a million years. They have also found that most of the duplicated genes in these organisms are very young. Many extra genes are lost from a genome within 10 million years (which is rapid on an evolutionary time scale).

Many gene duplications affect only one or a few genes at a time, but entire genomes are often duplicated in *polyploid* organisms (including many plants). When all the genes are duplicated, there are massive opportunities for new functions to evolve. That is exactly what appears to have happened in the evolution of vertebrates. The genomes of most jawed vertebrates appear to have four ancient copies of many major genes, which leads biologists to believe that two genome-wide duplication events occurred in the ancestor of these species. These duplications have allowed considerable specialization of individual vertebrate genes, many of which are now highly tissue-specific in their expression.

A novel function that evolved as a result of gene duplication is the ability of some fish—electric eels, among others—to produce electric signals. This function has evolved independently several times in different species, always through similar changes in duplicated sodium channel genes.

Although many extra genes disappear rapidly, some duplication events lead to the evolution of genes with new functions. Several successive rounds of duplication and mutation may result in a *gene family*, a group of homologous genes with related functions, often arrayed in tandem along a chromosome. An example of this process is provided by the *globin gene family* (see Figure 14.8) The globins were among the first proteins to be sequenced and compared. Comparisons of their amino acid sequences strongly suggest that the different globins arose via gene duplications. These comparisons can also tell us how long the globins have been evolving separately because differences among these proteins have accumulated with time.

Hemoglobin, a tetramer (four-subunit molecule) consisting of two α-globin and two β-globin polypeptide chains, carries oxygen in blood. Myoglobin, a monomer, is the primary oxygen storage protein in muscle. Myoglobin's affinity for O_2 is much higher than that of hemoglobin, but hemoglobin has evolved to be more diversified in its role. Hemoglobin binds O_2 in the lungs or gills, where the O_2 concentration is relatively high, transports it to deep body tissues, where the O_2 concentration is low, and releases it in those tissues. With its more complex tetrameric structure, hemoglobin is able to carry four molecules of O_2, as well as hydrogen ions and carbon dioxide, in the blood.

To estimate the time of the globin gene duplication that gave rise to the α- and β-globin gene clusters, we can create a *gene tree*, a phylogenetic tree based on the gene sequences that encode the various globins (**Figure 24.10**). The rate of molecular evolution of globin genes has been estimated from other studies, using the divergence times of groups of vertebrates that are well documented in the fossil record. These studies indicate an average rate of divergence for globin genes of about 1 substitution every 2 million years. Applying this rate to the gene tree, the two globin gene clusters are estimated to have split about 450 million years ago.

Some gene families evolve through concerted evolution

Although the members of the globin gene family have diversified in form and function, the members of many other gene families do not evolve independently of one another. For instance, almost all organisms have many (up to thousands of) copies of the ribosomal RNA genes. Ribosomal RNA is the principal structural element of the ribosome, and as such, has a primary role in protein synthesis. Every living species needs to synthesize proteins, often in large amounts (especially during early development). Having many copies of the ribosomal RNA genes ensures that organisms can rapidly produce many ribosomes and thereby maintain a high rate of protein synthesis.

Like all portions of the genome, ribosomal RNA genes evolve, and differences accumulate in the ribosomal RNA genes of different species. But within any one species, the multiple copies of ribosomal RNA genes are very similar, both structurally and functionally. This similarity makes sense, because ideally every ribosome within a species should synthesize proteins in the same way. In other words, the multiple copies of these genes within a species

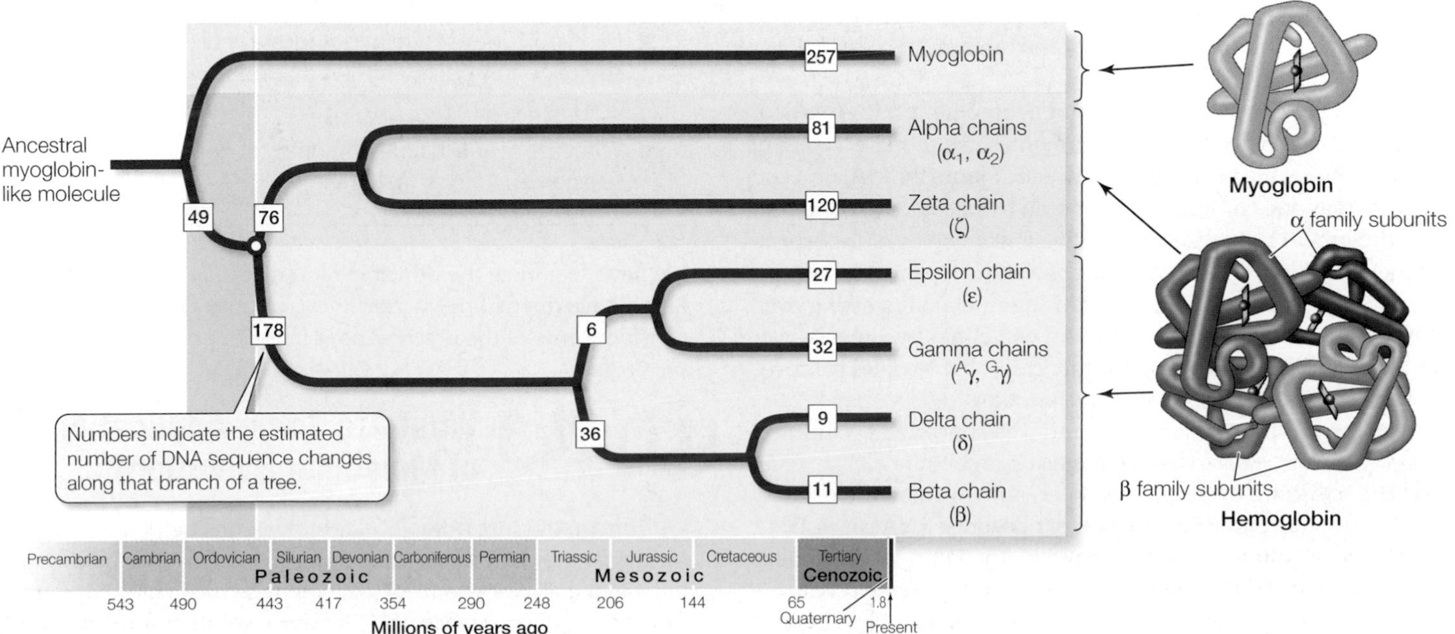

24.10 A Globin Family Gene Tree This gene tree suggests that the α-globin and β-globin gene clusters diverged about 450 million years ago (open circle), soon after the origin of the vertebrates.

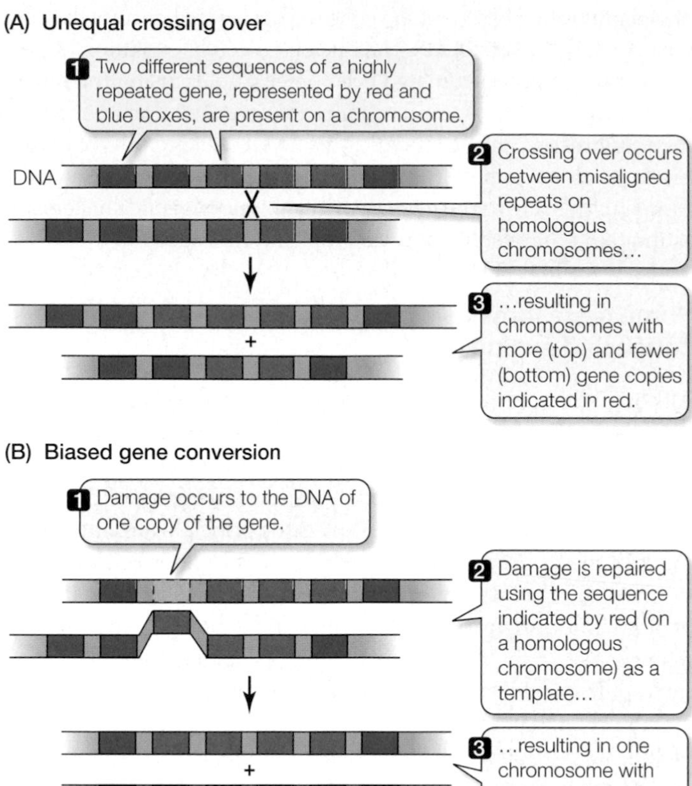
(A) Unequal crossing over
1 Two different sequences of a highly repeated gene, represented by red and blue boxes, are present on a chromosome.
DNA
2 Crossing over occurs between misaligned repeats on homologous chromosomes...
3 ...resulting in chromosomes with more (top) and fewer (bottom) gene copies indicated in red.
(B) Biased gene conversion
1 Damage occurs to the DNA of one copy of the gene.
2 Damage is repaired using the sequence indicated by red (on a homologous chromosome) as a template...
3 ...resulting in one chromosome with more copies of the red sequence.

24.11 Concerted Evolution Two mechanisms can produce concerted evolution of highly repeated genes. (A) Unequal crossing over results in deletions and duplications of a repeated gene. (B) Biased gene conversion can rapidly spread a mutation across multiple copies of a repeated gene.

are evolving in concert with one another, a phenomenon called **concerted evolution**.

How does concerted evolution occur? There must be one or more mechanisms to cause a substitution in one copy to spread to other copies within a species so that all of the copies remain similar. In fact, two different mechanisms appear to be responsible for concerted evolution. The first of these is *unequal crossing over*. When DNA is replicated during meiosis in a diploid species, the homologous chromosome pairs align and recombine by crossing over (see Section 9.5).

However, in the case of highly repeated genes, it is easy for genes to become displaced in alignment, since so many copies of the same genes are present in the repeats (**Figure 24.11A**). The end result is that one chromosome will gain extra copies of the repeat and the other chromosome will have fewer copies of the repeat. If a new substitution arises in one copy of the repeat, it can spread to new copies (or be eliminated) through unequal crossing over. Thus, over time, a novel substitution will either become fixed or lost entirely from the repeat. In either case, all the copies of the repeat will remain very similar to one another.

The second mechanism that produces concerted evolution is *biased gene conversion*. This mechanism can be much faster than unequal crossing over, and has been shown to be the primary mechanism for concerted evolution of ribosomal RNA genes. DNA strands break often, and are repaired by the DNA repair systems of cells (see Section 11.4). At many times during the cell cycle, the ribosomal RNA genes are clustered together in close proximity. If damage occurs to one of the genes, another copy of the RNA gene on another chromosome may be used to repair the damaged copy, and the sequence that is used as a template can thereby replace the original sequence (**Figure 24.11B**). In many cases, this repair system appears to be biased in favor of using particular sequences as templates for repair, and thus the favored sequence rapidly spreads across all the copies of the gene. In this way, changes may appear in a single copy and then rapidly spread to all the other copies.

Regardless of the mechanism responsible, the net result of concerted evolution is that the copies of a highly repeated gene do not evolve independently of one another. Mutations still occur, but once they arise in one copy, they either spread rapidly across all the copies or are lost from the genome completely. This process allows the products of each copy to remain similar through time in both sequence and function.

24.2 RECAP

By examining the relative rates of synonymous and nonsynonymous substitutions in genes across evolutionary history, biologists can distinguish the evolutionary mechanisms acting on individual genes. This knowledge, in turn, can lead to an understanding of protein function.

- Can you describe how the ratio of synonymous to nonsynonymous substitutions can be used to determine whether a gene is evolving neutrally, under positive selection, or under stabilizing selection? See pp. 530–532 and Figure 24.6
- Can you contrast two hypotheses for the wide diversity of genome sizes among different organisms? See pp. 533–534
- What are four possible outcomes of gene duplication? See p. 534

We have seen how the principles and methods of molecular evolution have opened new vistas in evolutionary science. Now let's consider some of the practical applications of this field.

24.3 What Are Some Applications of Molecular Evolution?

Our understanding of molecular biology has helped reveal how biological molecules function as well as how they diversify. Such knowledge allows scientists to create new molecules with novel functions in the laboratory, and to understand and treat disease.

Molecular sequence data are used to determine the evolutionary history of genes

A **gene tree** shows the evolutionary relationships of a single gene in different species or of the members of a gene family (as in Figure 24.10). The methods for constructing a gene tree are the same as those that will be presented in Section 25.2 for building phylogenetic trees. The process involves identifying differences between genes and using those differences to reconstruct the evolutionary history of the genes. Gene trees are often used to infer species phylogenetic trees, but the two types of trees are not necessarily equivalent. Processes such as gene duplication can give rise to differences between the phylogenetic trees of genes and species. From a gene tree, biologists can reconstruct the history and timing of gene duplication events and learn how gene diversification has resulted in the evolution of new protein functions.

All of the genes of a particular gene family have similar sequences because they have a common ancestry. As we will see in Chapter 25, features that are similar as a result of common ancestry are referred to as *homologs* of one another. However, when discussing gene trees, we usually need to distinguish between two forms of homology. Genes found in different species, and whose divergence we can trace to the speciation events that gave rise to various species, are called **orthologs**. Genes in the same or different species that are related through gene duplication events are called **paralogs**. When we examine a gene tree, the questions we wish to address determine whether we should compare orthologous or paralogous genes. If we wish to reconstruct the evolutionary history of the species that contain the genes, then our comparison should be restricted to orthologs (because they will reflect the history of speciation events). On the other hand, if we are interested in the changes in function that have resulted from gene duplication events, then the appropriate comparison is among paralogs (because they will reflect the history of gene duplication events). If our focus is on the diversification of a gene family through both processes, then we will want to include both paralogs and orthologs in our analysis.

Figure 24.12 depicts a gene tree for the members of a gene family called *engrailed* (its members encode transcription factors that regulate development). At least three gene duplications have occurred in this family, resulting in up to four different *engrailed* genes in some vertebrates (such as the zebrafish). All of the *engrailed* genes (*En*) are homologs because they have a common ancestor. Gene duplication events have generated paralogous *engrailed* genes in some lineages of vertebrates. We could compare the orthologous sequences of the *En1* group of genes to reconstruct the history of the bony vertebrates (i.e., all the vertebrates in Figure 24.12 except the lamprey), or we could use the orthologous sequences of the *En2* group of genes and expect the same answer (because

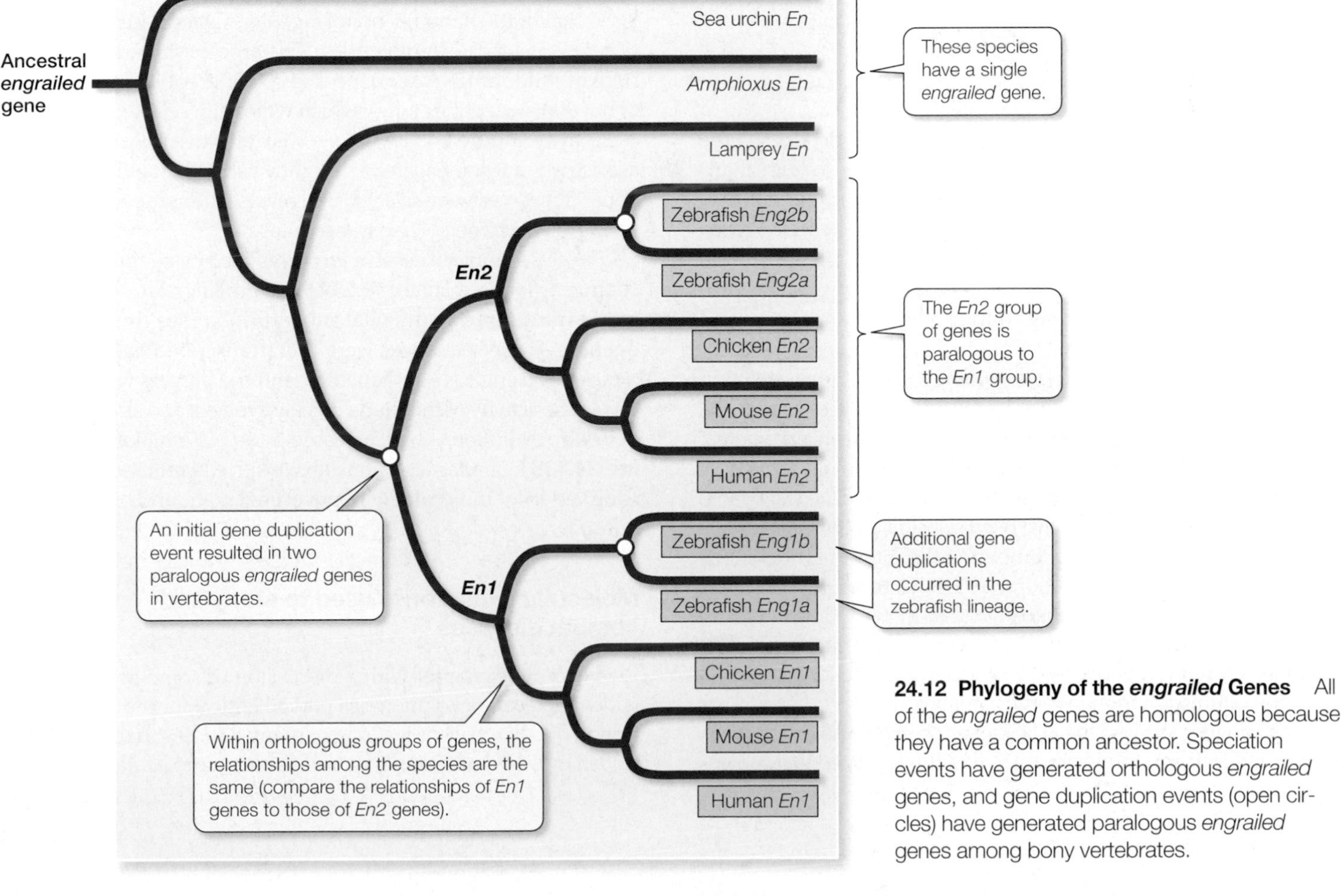

24.12 Phylogeny of the *engrailed* Genes All of the *engrailed* genes are homologous because they have a common ancestor. Speciation events have generated orthologous *engrailed* genes, and gene duplication events (open circles) have generated paralogous *engrailed* genes among bony vertebrates.

there is only one history of the underlying speciation events). All bony vertebrates have both of these groups of *engrailed* genes because the two groups arose from a gene duplication event in the common ancestor of bony vertebrates. If we wanted to focus on the diversification that occurred as a result of this gene duplication, then the appropriate comparison would be between the paralogous genes of the *En1* versus *En2* groups.

Gene evolution is used to study protein function

Earlier in this chapter we discussed the ways in which biologists can detect regions of genes that are under positive selection for change. What are the practical uses of this information? Consider the evolution of the family of gated sodium channel genes. Sodium channels have many functions, including the control of nerve impulses in the nervous system. Sodium channels can become blocked by various toxins, such as the tetrodotoxin that is present in newts (see the opening of Chapter 22), pufferfish, and many other animals. If a human eats those tissues of a pufferfish that contain tetrodotoxin, they can become paralyzed and die, because the tetrodotoxin blocks sodium channels and prevents nerves and muscles from functioning.

Despite its potential toxicity, pufferfish sushi (fugu) is considered a delicacy in some cultures. Only those tissues with minimal amounts of tetrodotoxin can be consumed safely, and errors in preparation lead quickly to the death of the consumer. The danger in its consumption is part of this sushi's appeal.

Pufferfish themselves have sodium channels; why doesn't the tetrodotoxin cause paralysis in the pufferfish? The sodium channels of pufferfish (and other animals that sequester tetrodotoxin) have evolved to become resistant to the toxin. Nucleotide substitutions in the pufferfish genome have resulted in changes to the proteins that make up sodium channels, and those changes prevent tetrodotoxin from binding to the sodium channel pore and blocking it.

Many other changes that have nothing to do with the evolution of tetrodotoxin resistance have occurred in these genes as well. Biologists who study the function of sodium channels can learn a great deal about how the channels work (and about neurological diseases that are caused by mutations in the sodium channel genes) by understanding which changes have been selected for tetrodotoxin resistance. They can do this by comparing the rates of synonymous and nonsynonymous substitutions across the genes in various lineages that have evolved tetrodotoxin resistance (including the garter snakes mentioned in Chapter 22). In a similar manner, molecular evolutionary principles are used to understand function and diversification of function in many other proteins.

As biologists studied the relationship between selection, evolution, and function in macromolecules, they realized that molecular evolution could be used in a controlled laboratory environment to produce new molecules with novel and useful functions. Thus were born the applications of in vitro evolution.

In vitro evolution produces new molecules

Living organisms produce thousands of compounds that humans have found useful. The search for such naturally occurring compounds, which can be used for pharmaceutical, agricultural, or industrial purposes, has been termed *bioprospecting*. These compounds are the result of millions of years of molecular evolution across millions of species of living organisms. Yet biologists can imagine molecules that could have evolved but have not, lacking the right combination of selection pressures and opportunities.

For instance, we might like to have a molecule that binds a particular environmental contaminant so that it can be easily isolated and extracted from the environment. But if the environmental contaminant is synthetic (not produced naturally), then it is unlikely that any living organism would have evolved a molecule with the function we desire. This problem was the inspiration for the field of **in vitro evolution**, in which new molecules are produced in the laboratory to perform novel and useful functions.

The principles of in vitro evolution are based on the principles of molecular evolution that we have learned from the natural world. Consider the evolution of a new RNA molecule that was produced in the laboratory using some of the techniques described in Chapter 16. This molecule's intended function was to join two other RNA molecules (acting as a ribozyme with a function similar to that of the naturally occurring DNA ligase described in Section 11.3, but for RNA molecules). The process started with a large pool of random RNA sequences (10^{15} different sequences, each about 300 nucleotides long), which were then selected for any ligase activity (**Figure 24.13A**). None were effective ribozymes for ligase activity, but some were very slightly better than others. The best of the ribozymes were selected and reverse-transcribed into cDNA (using the enzyme reverse transcriptase).

The cDNA molecules were then amplified using the polymerase chain reaction (see Figure 11.23). PCR amplification is not perfect, and it introduced many new substitutions into the pool of sequences. These sequences were then transcribed back into RNA molecules using RNA polymerase, and the process was repeated. The ligase activity of the RNAs quickly evolved, and after 10 rounds of in vitro evolution, it had increased by about 7 million times (**Figure 24.13B**). Similar techniques have since been used to create a wide variety of molecules with novel enzymatic and binding functions.

Molecular evolution is used to study and combat diseases

We began this chapter with a discussion of some of the ways in which molecular evolution has proved critical for the global eradication of polio. Polio is just one of many diseases that can be controlled only by understanding and applying molecular evolution-

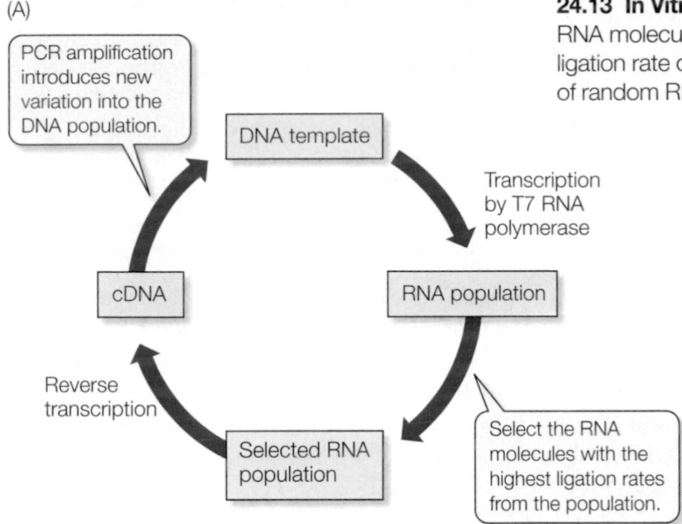

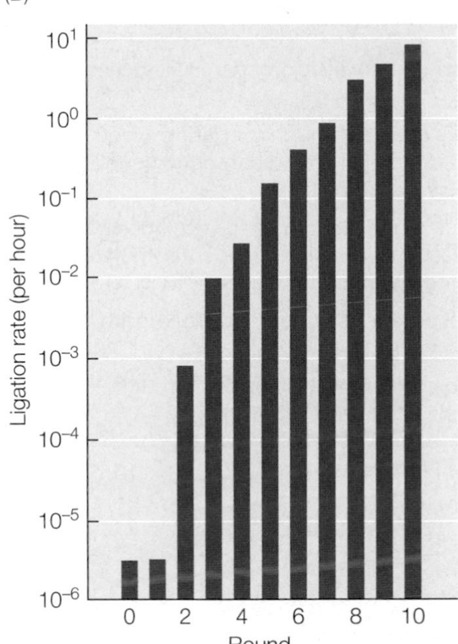

24.13 In Vitro Evolution of a Ribozyme (A) Methodology for in vitro evolution of new RNA molecules with the novel ability to ligate other RNA molecules. (B) Improvement of ligation rate over 10 rounds of in vitro evolution; the experiment started with a large pool of random RNA sequences.

ary principles. Many of humankind's most problematic diseases are caused by living, evolving organisms, and those organisms present a moving target. Their control depends on techniques that can track their evolution through time.

During the last century, transportation advances have allowed humans to move around the world with unprecedented speed and with increasing frequency. Unfortunately, this mobility has allowed pathogens to be transmitted among human populations at much higher rates, which has led to the global emergence of many "new" diseases. Most of these emerging diseases are caused by viruses, and virtually all new viral diseases have been identified by evolutionary comparison of their genomes with those of known viruses. In recent years, for example, rodent-borne hantaviruses have been identified as the source of widespread respiratory illnesses, and the virus (and its host) that causes Sudden Acute Respiratory Syndrome (SARS) has been identified using evolutionary comparisons of genes. Studies of the origins, the timing of emergence, and the global diversity of the human immunodeficiency virus (HIV) all depend on the principles of molecular evolution, as do the efforts to develop a vaccine for HIV. Global patterns of flu change as the influenza virus evolves each year, and new vaccines must be developed to track the evolution of these viruses.

In the future, molecular evolution will become even more critical to the identification of human (and other) diseases. Once biologists have collected data on the genomes of enough organisms, it will be possible to identify an infection by sequencing a portion of the infecting organism's genome and then comparing this sequence with other sequences on an evolutionary tree. At present, it is difficult to identify many common viral infections (those that cause "colds," for instance). As genomic databases and evolutionary trees increase, however, automated methods of sequencing and rapid phylogenetic comparison of the sequences will allow us to identify and treat a much wider array of human illnesses.

24.3 RECAP

Molecular evolutionary studies have provided biologists with new tools to understand the functions of macromolecules and how those functions can change through time. Molecular evolution is used to develop synthetic molecules for industrial and pharmaceutical uses and to identify and combat human diseases.

- Can you explain why a biologist might limit a particular investigation to orthologous (as opposed to paralogous) genes? See pp. 537–538
- Do you understand how gene evolution can be used to study protein function? See p. 538
- Can you describe the process of in vitro evolution? See p. 538 and Figure 24.13

Now that we have discussed how organisms and biological molecules evolve, we are ready to consider how evolutionary history is studied and why evolution is the principal organizing framework for biology. Chapter 25 describes some of the principles and uses of phylogenetic analysis: the study of evolutionary relationships.

CHAPTER SUMMARY

24.1 What can genomes reveal about evolution?

See Web/CD Activity 24.1

The field of **molecular evolution** concerns relationships between the structures of genes and proteins and the functions of organisms.

A **genome** is an organism's full set of genes and noncoding DNA. In eukaryotes, the genome includes genetic material in the nucleus of the cell as well as in mitochondria and chloroplasts (where present).

Nucleotide substitutions may or may not result in **amino acid replacements** in the encoded proteins.

The estimated number of substitutions between sequences can be calculated from a **similarity matrix** using models of sequence evolution that account for changes that cannot be observed directly. Review Figure 24.1, Web/CD Activity 24.2

The concept of homology (similarity that results from common ancestry) extends down to the level of particular positions in nucleotide or amino acid sequences. **Sequence alignments** from different organisms allow us to compare the sequences and identify homologous positions. Review Figure 24.3

24.2 What are the mechanisms of molecular evolution?

Nonsynonymous substitutions of nucleotides result in amino acid replacements in proteins, but **synonymous substitutions** do not. Review Figure 24.5

Rates of synonymous substitution are typically higher than rates of nonsynonymous substitution in protein-coding genes (a result of stabilizing selection). Review Figure 24.6

Much of the molecular change in nucleotide sequences is a result of neutral evolution. The rate of fixation of neutral mutations is independent of population size and is equal to the mutation rate.

Positive selection for change in a protein-coding gene may be detected by a higher rate of nonsynonymous than synonymous substitutions.

Genome size evolves by the addition or deletion of genes and noncoding DNA. The total size of genomes varies much more widely across living organisms than does the number of functional genes. Review Figures 24.8 and 24.9

Even though many noncoding regions of the genome may not have direct functions, these regions can affect the phenotype of an organism by influencing gene expression. Functionless **pseudogenes** can serve as the raw material for the evolution of new genes.

Gene duplications can result in increased production of the gene's product, in pseudogenes, or in new gene functions.

Some highly repeated genes evolve by **concerted evolution**: multiple copies within an organism maintain high similarity, while the genes continue to diverge between species.

24.3 What are some applications of molecular evolution?

See Web/CD Activity 24.3

Gene trees describe the evolutionary history of particular genes or gene families.

Orthologs are genes that are related through speciation events, whereas **paralogs** are genes that are related through gene duplication events. Review Figure 24.12

Protein function can be studied by examining gene evolution. Detection of positive selection can be used to identify molecular changes that have resulted in functional changes.

In vitro evolution is used to produce synthetic molecules with particular desired functions.

Many diseases are identified, studied, and combated through molecular evolutionary investigations.

SELF-QUIZ

1. A higher rate of synonymous than nonsynonymous substitutions in a protein-coding gene is expected under
   a. stabilizing selection.
   b. positive selection.
   c. neutral evolution.
   d. concerted evolution.
   e. none of the above
2. Before nucleotide and amino acid sequences can be compared in an evolutionary framework, they must be aligned in order to account for
   a. deletions and insertions.
   b. selection and neutrality.
   c. parallelisms and convergences.
   d. gene families.
   e. all of the above
3. Models of nucleotide sequence evolution, developed by biologists to estimate sequence divergence, include parameters that account for
   a. substitution rates between different nucleotides.
   b. differences in substitution rates across different positions in a gene.
   c. differences in nucleotide frequencies.
   d. all of the above
   e. none of the above
4. The rate of fixation of neutral mutations is
   a. independent of population size.
   b. higher in small populations than in large populations.
   c. higher in large populations than in small populations.
   d. slower than the rate of fixation of deleterious mutations.
   e. none of the above
5. Genome size differs widely among different species. What is the greatest contributing cause for these differences?
   a. The number of protein-coding genes
   b. The amount of noncoding DNA
   c. The number of duplicated genes
   d. The degree of concerted evolution
   e. The amount of positive selection for change in protein-coding genes
6. Which of the following is *not* true of concerted evolution?
   a. Concerted evolution refers to the nonindependent evolution of some repeated genes within a species.
   b. Unequal crossing over may produce concerted evolution.

c. Biased gene conversion may produce concerted evolution.
d. Ribosomal RNA genes are an example of a gene family that has undergone concerted evolution.
e. Concerted evolution results in divergence of members of a gene family within an organism.

7. When a gene is duplicated, which of the following outcomes may occur?
   a. Production of the gene's product may increase.
   b. The two copies may become expressed in different tissues.
   c. One copy of the gene may accumulate deleterious substitutions and become functionless.
   d. The two copies may diverge and acquire different functions.
   e. All of the above
8. Paralogous genes are genes that trace back to a common
   a. speciation event.
   b. substitution event.
   c. insertion event.
   d. deletion event.
   e. duplication event.
9. Which of the following is true of in vitro evolution?
   a. In vitro evolution refers to bioprospecting for naturally occurring macromolecules.
   b. In vitro evolution can produce new molecular sequences not known from nature.
   c. In vitro evolution can only produce new proteins.
   d. In vitro evolution only selects for changes that were present in the starting pool of molecules, and does not introduce any new mutations.
   e. All of the above
10. Which of the following is true of the use of molecular evolutionary studies of human disease?
   a. Molecular evolutionary studies are useful for identifying many diseases.
   b. Molecular evolutionary studies are often used to determine the origin of emerging diseases.
   c. Molecular evolutionary studies are important for developing vaccines against diseases.
   d. Molecular evolutionary studies are used to determine whether outbreaks of polio are the result of naturally occurring viruses or viruses that have evolved from attenuated viruses.
   e. All of the above

FOR DISCUSSION

1. Rates of evolutionary change differ among different molecules, and different species differ widely in generation times and population sizes. How does this variation limit how and in what ways we can use the concept of a molecular clock to help us answer questions about the evolution of both molecules and organisms?
2. One hypothesis proposed to explain the existence of large amounts of noncoding DNA is that the cost of maintaining that DNA is so small that natural selection is too weak to reduce it. How could you test this hypothesis against the hypothesis that genome size is functionally related to developmental rate?
3. If fossil evidence and molecular evidence disagree on the date of a major lineage split, which of the two kinds of evidence would you favor? Why?
4. Soon scientists will be able to produce and release into the wild genetically modified mosquitoes that are unable to harbor and transmit malarial parasites. What ethical issues need to be discussed before such releases are permitted?

FOR INVESTIGATION

Many groups of organisms that inhabit caves have evolved to become eyeless. For instance, although surface-dwelling crayfishes have functional eyes, several species that are restricted to underground habitats lack eyes. Opsins are a group of light-sensitive proteins known to have an important function in vision; opsin genes are expressed in eye tissues. Although cave-dwelling crayfishes lack eyes, opsin genes are still present in their genomes. Two alternative hypotheses are (1) the opsin genes are no longer experiencing stabilizing selection (because there is no longer selection for function in vision); or (2) the opsin genes are experiencing selection for a function other than vision. How would you investigate these alternatives using the sequences of the opsin genes in various species of crayfishes?

CHAPTER 25 Reconstructing and Using Phylogenies

Phylogenetic trees in the courtroom

Transmitting HIV, while irresponsible, is not usually prosecuted as a crime. But in one true-crime case, a woman we'll call "April" went to the police immediately upon learning she was HIV-positive. April believed she was the victim of an attempted murder by "Victor," a physician and her former boyfriend, who had repeatedly threatened violence when she tried to break up with him. April's contention was that Victor, under the pretense of administering vitamin therapy, had injected her with blood from one of his HIV-infected patients.

Police investigators discovered that Victor had drawn blood from one of his HIV-positive patients just before giving April the injection. This blood draw had no clinical purpose, and Victor had tried to hide the records of it. The police were convinced that he might indeed have committed the alleged crime.

The district attorney, however, had to show that April's HIV infection had come from Victor's patient, and from no other source. To reconstruct the history of the infection, the district attorney turned to *phylogenetic analysis*—the study of the evolutionary relationships among a group of organisms.

The district attorney's task was complicated by the nature of HIV. HIV is a retrovirus, in which poor repair of replication errors leads to a very high rate of evolution. Once a person is infected with HIV, the virus not only replicates quickly, but evolves quickly, so that the infected individual is soon host to a genetically diverse population of viruses. Thus, when one person transmits HIV to another, typically very few viral particles (often only one) initiate the infective event. But the person who is the source of the infection may be host to a large, genetically diverse population of viruses—not just the variant they transmit to the recipient.

Enter molecular phylogeny. Samples of HIV from an infected individual can be sequenced to trace their evolutionary lineages back to the originally transmitted virus. The virus that is passed to the recipient will be very closely related to some of the viruses in the source individual and more distantly related to others. A reconstruction of the evolutionary history of the viruses in both individuals is needed to reveal not only whether the two individuals' viruses are closely related, but also who infected whom.

To prove attempted murder, the district attorney needed to demonstrate that April's HIV was more closely related to that of Victor's patient than

Human Immunodeficiency Virus Human immunodeficiency virus (HIV) is the cause of acquired immunodeficiency syndrome, or AIDS. The biology of this retrovirus was discussed extensively in Chapter 13; however, to combat AIDS it is also essential to understand the phylogeny of HIV.

A Source of the Virus AIDS is a zoonotic disease, the virus having been transferred to humans from another animal. Phylogenetic analyses of immunodeficiency viruses show that humans acquired HIV-1 from chimpanzees (see Figure 25.8). Other forms of the virus have been passed to humans by different simians.

to other HIV variants in her community. Samples of HIV were isolated from the blood of the patient, from April, and from other HIV-positive individuals in the community. Phylogenetic analysis revealed that April's HIV was indeed closely related to a subset of the patient's HIV, and more distantly related to the other HIV sources in the community. Given this fact along with the other evidence in the case, Victor was convicted of attempted murder.

IN THIS CHAPTER we will examine the field of systematics, the scientific study of the diversity of life. We will see how phylogenetic methods are used to reconstruct evolutionary history and to study diversity across genes, populations, species, and larger groups of organisms. We will see how systematists reconstruct the past and predict the course of evolution. We end the chapter with a look at taxonomy, the theory and practice of classifying organisms.

CHAPTER OUTLINE

25.1 What Is Phylogeny?

A **phylogeny** is a description of the evolutionary history of relationships among organisms (or their parts). A **phylogenetic tree** is a diagram that portrays a reconstruction of that history. Phylogenetic trees are commonly used to depict the evolutionary history of species, populations, and genes. Each split (or *node*) in a phylogenetic tree represents a point at which lineages diverged in the past. In the case of species, these nodes represent past speciation events, when one lineage split into two. Thus a phylogenetic tree can be used to trace the evolutionary relationships from the ancient common ancestor of a group of species, through the various speciation events when lineages split, up to the present populations of the organisms.

A phylogenetic tree may portray the evolutionary history of all life forms; of a major evolutionary group (such as the insects); of a small group of closely related species; or in some cases, even the history of individuals, populations, or genes within a species. The common ancestor of all the organisms in the tree forms the *root* of the tree. The phylogenetic trees in this book depict time flowing from left (earliest) to right (most recent) (**Figure 25.1A**); it is equally common practice to draw trees with the earliest times at the bottom.

The timing of separations between lineages of organisms is shown by the positions of nodes on a time or divergence axis; this axis may have an explicit scale or simply show the relative timing of divergence events. In this book, the positions of nodes along the horizontal (time) axis have meaning, but the vertical distance between the branches does not. Vertical distances are adjusted for legibility and clarity of presentation; they do not correlate with the degree of similarity or difference between groups. Note too that lineages can be rotated around nodes in the tree, so the vertical order of taxa is also largely arbitrary (**Figure 25.1B**).

Any group of species that we designate or name is called a **taxon** (plural *taxa*). Some examples of familiar taxa include humans, primates, mammals, and vertebrates (note that in this series, each taxon in the list is also a member of the next, more inclusive taxon). Any taxon that consists of all the evolutionary descendants of a common ancestor is called a **clade**. Clades can be identified by picking any point on a phylogenetic tree and then tracing all the descendant lineages to the tips of the terminal branches. Two species that are each other's closest rela-

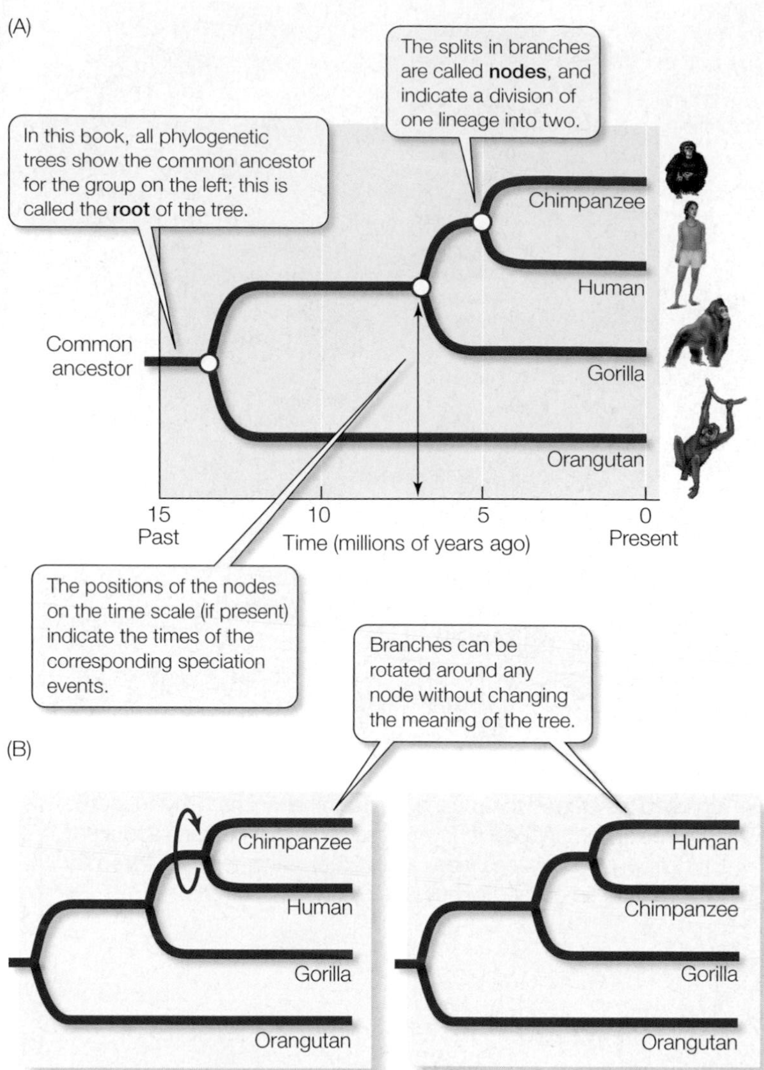

25.1 How to Read a Phylogenetic Tree (A) A phylogenetic tree displays the evolutionary relationships among organisms. Such trees can be produced with time scales, as shown here, or with no indication of time. If no time scale is shown, then the branch lengths show relative rather than absolute times of divergence. (B) Lineages can be rotated around a given node, so the vertical order of taxa is also largely arbitrary.

tives are called **sister species**; similarly, two clades that are each other's closest relatives are called **sister clades**.

Before the 1980s, phylogenetic trees were mostly seen in the literature on evolutionary biology, especially in **systematics**, the study of biodiversity. But today, in almost any scientific journal in the life sciences published in the last few years, you are likely to find articles that contain phylogenetic trees. Trees are widely used in studies of molecular biology, biomedicine, physiology, behavior, ecology, and virtually all other fields of biology. Why have phylogenetic studies become so important to biology?

All of life is connected through evolutionary history

In biology, we study life at all levels of organization—from genes, cells, organisms, populations, and species to the major divisions of life. In most cases, however, no individual gene or organism (or other unit of study) is exactly like any other gene or organism that we investigate.

Consider the individuals in your biology class. We recognize each person as an individual human, but we know that no two are exactly alike. If we knew everyone's family tree in detail, we could predict the genetic similarity of any pair of students. If we had this information, we would find that more closely related students would have many more traits in common (such as susceptibility or resistance to diseases). In a similar manner, biologists use phylogenies to make comparisons and predictions about shared traits across genes, populations, and species.

One of the greatest unifying concepts in biology is that all of life is connected through its evolutionary history. The complete, nearly 4-billion-year evolutionary history of life is known as the "Tree of Life." Biologists estimate that there are tens of millions of species on Earth. Of these, only 1.7 million have been formally described and named. New species are being discovered and named, and phylogenetic analyses reviewed and revised, all the time, but our knowledge of the Tree of Life is far from complete, even for known species. Nonetheless, knowledge of evolutionary relationships is essential for making comparisons in biology, so biologists build phylogenies for particular groups of interest as the need arises.

Any claim of an evolutionary association between a trait and a group of organisms is actually a statement about when during the history of the group the trait first arose, and about the maintenance of the trait since its first appearance. For example, the statement that the cytoskeleton is a trait possessed by all eukaryotes is an assertion that the cytoskeleton is an ancestral trait of eukaryotes that has been maintained during the subsequent evolution of all surviving eukaryote lineages.

Comparisons among species require an evolutionary perspective

When biologists make comparisons among species, they observe traits that differ within the group of interest, and try to ascertain when these traits evolved. In many cases, investigators are interested in how the evolution of a trait is dependent on environmental conditions or selective pressures. For instance, phylogenetic analyses have been used to discover changes in the genome of HIV that confer resistance to particular drug treatments. The association of a particular genetic change in HIV with a particular treatment provides a hypothesis about resistance that can be tested experimentally.

Any features shared by two or more species that have been inherited from a common ancestor are said to be **homologous**. As we noted in Section 24.1, homologous features may be any heritable traits, including DNA sequences, protein structures, anatomical structures, and even some behavior patterns. Traits that are shared by most or all of the organisms in a group of interest are likely to have been inherited from a common ancestor. For example, all living vertebrates have a vertebral column, all known fossil vertebrates had a vertebral column, and all vertebrates are descended from the same common ancestor. Therefore, the vertebral column is judged to be homologous in all vertebrates.

A trait that differs from its ancestral form is called a **derived trait**. Conversely, a trait that was present in the ancestor of a group is known as an **ancestral trait** for that group. Derived traits that are shared among a group of organisms, and are viewed as evidence of the common ancestry of the group, are called **synapomorphies** (*syn* means "shared," *apo* means "derived," and *morphy* refers to the "form" of a trait). Thus the vertebral column is considered a synapomorphy of the vertebrates.

Not all similar traits are evidence of relatedness, however. Similar traits in unrelated groups of organisms can develop for either of the following reasons:

- Independently evolved traits subjected to similar selection pressures may become superficially similar; this phenomenon is called **convergent evolution**. For example, although the bones of the wings of bats and birds are homologous, having been inherited from a common ancestor, the wings of bats and the wings of birds are not homologous because they evolved independently from the forelimbs of different nonflying ancestors (**Figure 25.2**).
- A character may revert from a derived state back to an ancestral state. Such a change is called an **evolutionary reversal**. For example, most frogs lack teeth in the lower jaw, but the ancestor of frogs did have such teeth. Teeth have been regained in the lower jaw of one frog genus, *Amphignathodon*, and thus represent an evolutionary reversal.

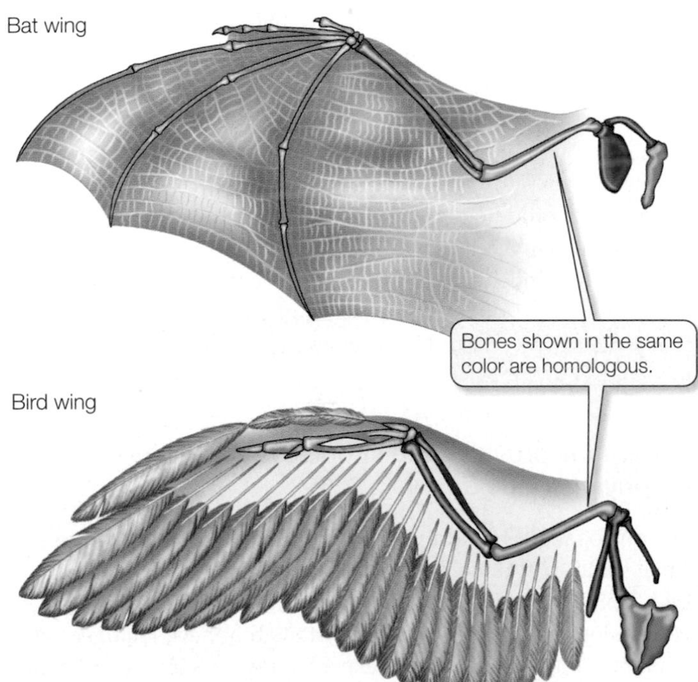

25.2 The Bones Are Homologous; the Wings Are Not The supporting bone structures of both bat wings and bird wings are derived from a common four-limbed ancestor and are thus homologous. However, the wings themselves—an adaptation for flight—evolved independently in the two groups.

Convergent evolution and evolutionary reversals generate traits that are similar for reasons other than inheritance from a common ancestor. Such traits are called *homoplastic traits* or **homoplasies**.

A particular trait may be ancestral or derived, depending on our point of reference in a phylogeny. For example, all birds have feathers, which are highly modified scales. We infer from this that feathers were present in the common ancestor of modern birds. Therefore, we consider the presence of feathers to be an *ancestral* trait for birds. However, feathers are not present in any other living vertebrates (or in any other animal species, for that matter). If we were reconstructing a phylogeny of all living vertebrates, the presence of feathers would be a *derived* trait that is found only among birds (and thus a synapomorphy of the birds).

25.1 RECAP

A phylogeny is a description of evolutionary relationships: how a group of genes, populations, or species have evolved from a common ancestor. All living organisms share a common ancestor and are related through the phylogenetic Tree of Life.

- Do you understand the different elements of a phylogenetic tree? See p. 543 and Figure 25.1
- Can you explain the difference between an ancestral and a derived trait? See p. 545
- Do you see how similar traits might arise in species that are not descended from a common ancestor? See p. 545 and Figure 25.2

Phylogenetic analyses have become increasingly important to many types of biological research in recent years, and they are the basis for the comparative nature of biology. But for the most part, evolutionary history cannot be observed directly. How, then, do biologists reconstruct the past?

25.2 How Are Phylogenetic Trees Constructed?

To illustrate how a phylogenetic tree is constructed, let's consider eight vertebrate animals: a lamprey, perch, pigeon, chimpanzee, salamander, lizard, mouse, and crocodile. We will assume initially that a given derived trait evolved only once during the evolution of these animals (that is, there has been no convergent evolution), and that no derived traits were lost from any of the descendant groups (there has been no evolutionary reversal). For simplicity, we have selected traits that are either present (+) or absent (–) (**Table 25.1**).

TABLE 25.1

Eight Vertebrates Ordered According to Unique Shared Derived Traits

TAXON	DERIVED TRAIT[a] JAWS	LUNGS	CLAWS OR NAILS	GIZZARD	FEATHERS	FUR	MAMMARY GLANDS	KERATINOUS SCALES
Lamprey (outgroup)	–	–	–	–	–	–	–	–
Perch	+	–	–	–	–	–	–	–
Salamander	+	+	–	–	–	–	–	–
Lizard	+	+	+	–	–	–	–	+
Crocodile	+	+	+	+	–	–	–	+
Pigeon	+	+	+	+	+	–	–	+
Mouse	+	+	+	–	–	+	+	–
Chimpanzee	+	+	+	–	–	+	+	–

[a]A plus sign indicates the trait is present, a minus sign that it is absent.

As we will see in Chapter 33, a group of jawless fishes called the lampreys is thought to have separated from the lineage leading to the other vertebrates before the jaw arose. Therefore, we will choose the lamprey as the *outgroup* for our analysis. An outgroup can be any species or group of species outside the group of interest; it follows, then, that we call the group of primary interest the *ingroup*. The outgroup is used to determine which traits of the ingroup are derived (evolved in the ingroup) and which are ancestral (evolved before the origin of the ingroup). In this case, derived traits are those that have been acquired by other members of the vertebrate lineage since they separated from the lamprey, whereas any trait that is present in both the lamprey and the other vertebrates is judged to be ancestral. The root of the tree is determined by the relationship of the ingroup to the outgroup.

We begin by noting that the chimpanzee and mouse share two derived traits: mammary glands and fur. Those traits are absent in both the outgroup and the other species of the ingroup. Therefore, we infer that mammary glands and fur are derived traits that evolved in a common ancestor of chimpanzees and mice after that lineage separated from the ones leading to the other vertebrates. In other words, we provisionally assume that mammary glands and fur evolved only once among the animals in our ingroup. These characters are thus synapomorphies that unite chimpanzees and mice (as well as all other mammals, but we have not included other mammalian species in this example).

By the same reasoning, we can infer that the other shared derived traits are synapomorphies for the various groups in which they are expressed. For instance, keratinous scales are a synapomorphy of the crocodile, pigeon, and lizard, meaning that we infer that these species inherited this trait from a common ancestor.

The pigeon has one unique trait in our list: feathers. As before, we provisionally assume that feathers evolved only once, in the ancestor of birds. Feathers are a synapomorphy of birds, but since we only have one bird in this example, the presence of feathers does not provide any clues about the relationships among the eight species of vertebrates we have sampled. On the other hand, gizzards are found in birds and crocodilians, and so this trait is evidence of the close relationship between birds and crocodilians.

By combining information about the various synapomorphies, we can construct a phylogenetic tree. We infer, for example, that mice and chimpanzees, the only two animals that share fur and mammary glands in our example, share a more recent common ancestor with each other than they do with pigeons and crocodiles. Otherwise, we would need to assume that the ancestors of pigeons and crocodiles also had fur and mammary glands, but subsequently lost them—unnecessary additional assumptions.

Figure 25.3 shows a phylogenetic tree for these eight vertebrates, based on the traits we used and the assumption that each derived trait evolved only once. This particular phylogenetic tree was easy to construct because the animals and traits we used fulfilled the assumptions that derived traits appeared only once in the tree and that they were never lost after they appeared. Had we included a snake in the group, our second assumption would have been violated, because we know that the lizard ancestors of snakes had limbs that were subsequently lost (along with their claws). We would need to examine additional traits to determine that the lineage leading to snakes separated from the one leading to lizards long after the lineage leading to lizards separated from the others. In fact, the analysis of a number of traits shows that snakes evolved from burrowing lizards that became adapted to a subterranean existence.

Parsimony provides the simplest explanation for phylogenetic data

The phylogenetic tree shown in Figure 25.3 is based on only a very small sample of traits. Typically, biologists construct phylogenetic trees using hundreds or thousands of traits. With larger data sets, we would expect to observe some traits that have changed more than once, and thus we would expect to see some convergence and evolutionary reversal. How do we determine which traits are synapomorphies and which are homoplasies?

One approach is to use the **parsimony** principle. In its most general form, the parsimony principle states that the preferred explanation of the observed data is the simplest explanation. Its application to the reconstruction of phylogenies means minimizing the

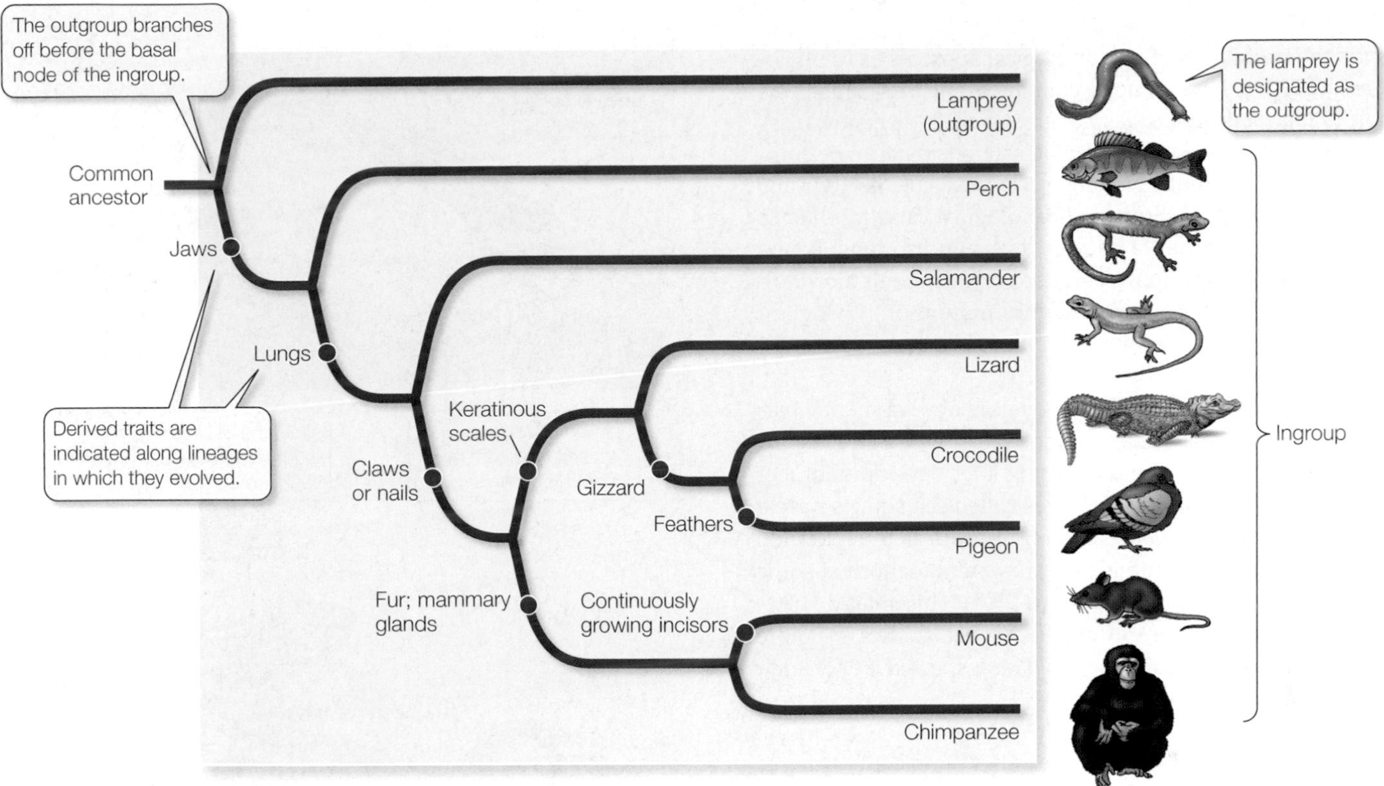

25.3 Inferring a Phylogenetic Tree This phylogenetic tree was constructed from the information given in Table 25.1 using the parsimony principle. Each clade in the tree is supported by at least one shared derived trait, or synapomorphy.

number of evolutionary changes that need to be assumed over all characters in all groups in the tree. In other words, the best hypothesis under the parsimony principle is one that requires the fewest homoplasies. This application of parsimony is a specific case of a general principle of logic called Occam's razor: the idea that the best explanation is the one that fits the data best and makes the fewest assumptions.

Using the parsimony principle is appropriate not because all evolutionary changes occurred parsimoniously, but because it is logical to adopt the simplest explanation that can account for the observed data. More complicated explanations are accepted only when the evidence requires them. Phylogenetic trees represent our best estimates about evolutionary relationships. They are continually modified as additional evidence becomes available.

Phylogenies are reconstructed from many sources of data

Naturalists constructed various forms of phylogenetic trees as far back as classical times. Tree construction has been revolutionized, however, by the advent of computer software for trait analysis and tree construction, allowing us to consider amounts of data far more vast than could ever before be processed.

Any trait that is genetically determined, and therefore heritable, can be used in a phylogenetic analysis. Evolutionary relationships can be revealed through studies of morphology, development, the fossil record, behavioral traits, and molecular traits such as DNA and protein sequences. All of these are used by modern systematists, and computer-driven tree construction and gene sequencing taken together have led to an explosion of new approaches to phylogeny.

Although the construction of phylogenetic trees is conceptually straightforward, the computations involved with large sets of traits and taxa can be challenging. For perspective, the number of possible phylogenetic trees for just 50 species greatly exceeds the number of atoms in the universe!

MORPHOLOGY An important source of phylogenetic information is *morphology*—that is, the presence, size, shape, and other attributes of body parts. Since living organisms have been studied for centuries, we have a wealth of recorded morphological data as well as extensive museum and herbarium collections of organisms whose traits can be measured. New technological tools, such as the electron microscope and computed tomography (CT) scans, enable systematists to examine and analyze the structures of organisms at much finer scales than was formerly possible. Most species of living organisms have been described and are known primarily from morphological data, so morphology provides the most comprehensive data set available for many taxa. The features of morphology that are important for phylogenetic analysis are often specific to a particular group of organisms. For example, the presence, development, shape, and size of various features of the skeletal system are important for the study of vertebrate phylogeny, whereas floral structures are important for studying the relationships among flowering plants.

Despite the usefulness of morphological approaches to phylogenetic analysis, they have some limitations. Some taxa exhibit very little morphological diversity, despite great species diversity. At the other extreme, there are few morphological traits that can be compared between very distantly related species (consider bacteria and vertebrates, for instance). Some morphological variation has an environmental (rather than a genetic) basis and so must be excluded from phylogenetic analyses. Therefore, although morphology provides an important source of information on phylogeny, additional sources of information are often necessary.

DEVELOPMENT Observations of similarities in developmental patterns may also reveal evolutionary relationships. Similarities in early developmental stages may be lost during later development. For example, the larvae of marine creatures called sea squirts have a rod in the back—the *notochord*—that disappears as they develop into adults. All vertebrate animals also have a notochord at some time during their development (**Figure 25.4**). This shared structure is one of the reasons for inferring that sea squirts are more closely related to vertebrates than would be suspected if only adult sea squirts were examined.

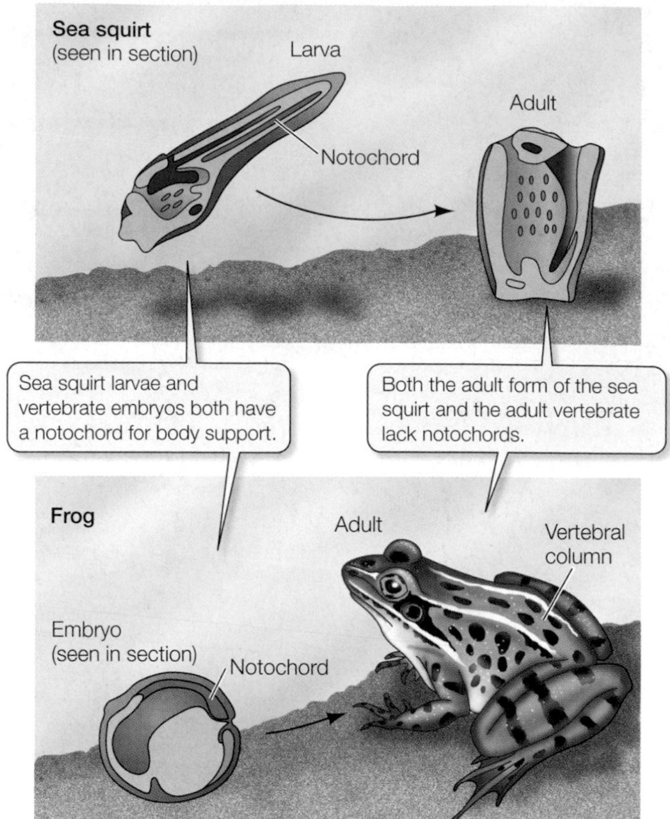

25.4 A Larva Reveals Evolutionary Relationships Sea squirt larvae, but not adults, have a well-developed notochord (orange) that reveals their evolutionary relationship to vertebrates, all of which have a notochord at some time during their life cycle. In adult vertebrates, the vertebral column replaces the notochord as the support structure.

PALEONTOLOGY The fossil record is another important source of information on evolutionary history. Fossils show us where and when organisms lived in the past and give us an idea of what they looked like. Fossils provide important evidence that helps us distinguish ancestral from derived traits. The fossil record can also reveal when lineages diverged and began their independent evolutionary histories. Furthermore, in groups with few species that have survived to the present, information on extinct species is often critical to an understanding of the large divergences between the surviving species. The fossil record does have limitations, however. Few or no fossils have been found for some groups whose phylogenies we may wish to determine, and the fossil record for many groups is fragmentary.

BEHAVIOR Some behavioral traits are culturally transmitted and others are inherited. If a particular behavior is culturally transmitted, then it may not accurately reflect evolutionary relationships (but may nonetheless reflect cultural connections). Bird songs, for instance, are often learned and may be inappropriate traits for phylogenetic analysis. Frog calls, on the other hand, are genetically determined and appear to be acceptable sources of information for reconstructing phylogenies.

MOLECULAR DATA All heritable variation is encoded in DNA, and so the complete genome of an organism contains an enormous set of traits (the individual nucleotide bases of DNA) that can be used in phylogenetic analyses. In recent years, DNA sequences have become one of the most widely used sources for constructing phylogenetic trees. Comparisons of nucleotide sequences are not limited to the DNA in the cell nucleus. Eukaryotes have genes in their mitochondria as well as in their nuclei; plant cells also have genes in their chloroplasts. The chloroplast genome (cpDNA), which is used extensively in phylogenetic studies of plants, has changed slowly over evolutionary time, so it is often used to study relatively ancient phylogenetic relationships. Most animal mitochondrial DNA (mtDNA) has changed more rapidly, so mitochondrial genes have been used extensively for studies of evolutionary relationships among closely related animal species. Many nuclear gene sequences are also commonly analyzed, and now that a number of whole genomes have been sequenced, they are also used to construct phylogenetic trees. Information on gene products (such as the amino acid sequences of proteins) is also widely used for phylogenetic analyses, as we discussed in Chapter 24.

Mathematical models expand the power of phylogenetic reconstruction

As biologists began to use DNA sequences to infer phylogenies in the 1970s and 1980s, they developed explicit mathematical models describing how DNA sequences change over time. These models account for multiple changes at a given position in a DNA sequence, and they also take into account different rates of change at different positions in a gene, at different positions in a codon, and among different nucleotides (see Section 24.1). For example, transitions (changes between two purines or between two pyrimidines) are usually more likely than are transversions (changes between a purine and pyrimidine).

Such mathematical models can be used to compute **maximum likelihood** solutions for phylogenetic estimation. A likelihood score of a tree is based on the probability of the observed data evolving on the specified tree, given an explicit mathematical model of evolution for the characters. The maximum likelihood solution, then, is the most likely tree given the observed data. Maximum likelihood methods can be used for any kind of characters, but they are most often used with molecular data, for which explicit mathematical models of evolutionary change are easier to develop. The principal advantages to maximum likelihood analyses are that they incorporate more information about evolutionary change than do parsimony methods, and they are easier to treat in a statistical framework. The principal disadvantages are that they are computationally intensive and require explicit models of evolutionary change (which may not be available for some kinds of character change).

The accuracy of phylogenetic methods can be tested

If phylogenetic trees represent reconstructions of past events, and if many of these events occurred before any humans were around to witness them, how can we test the accuracy of phylogenetic methods? Biologists have conducted experiments both in living organisms and with computer simulations that have demonstrated the effectiveness and accuracy of phylogenetic methods.

In one experiment that was designed to test the accuracy of phylogenetic analysis, a single viral culture of bacteriophage T7 was used as a starting point, and lineages were allowed to evolve from this ancestral virus in the laboratory (**Figure 25.5**). The initial culture was split into two separate lineages, one of which became the ingroup for analysis, and the other became the outgroup for rooting the tree. In the ingroup, the lineages were split in two after every 400 generations, and samples of the virus were saved for analysis at each branching point. The lineages were allowed to

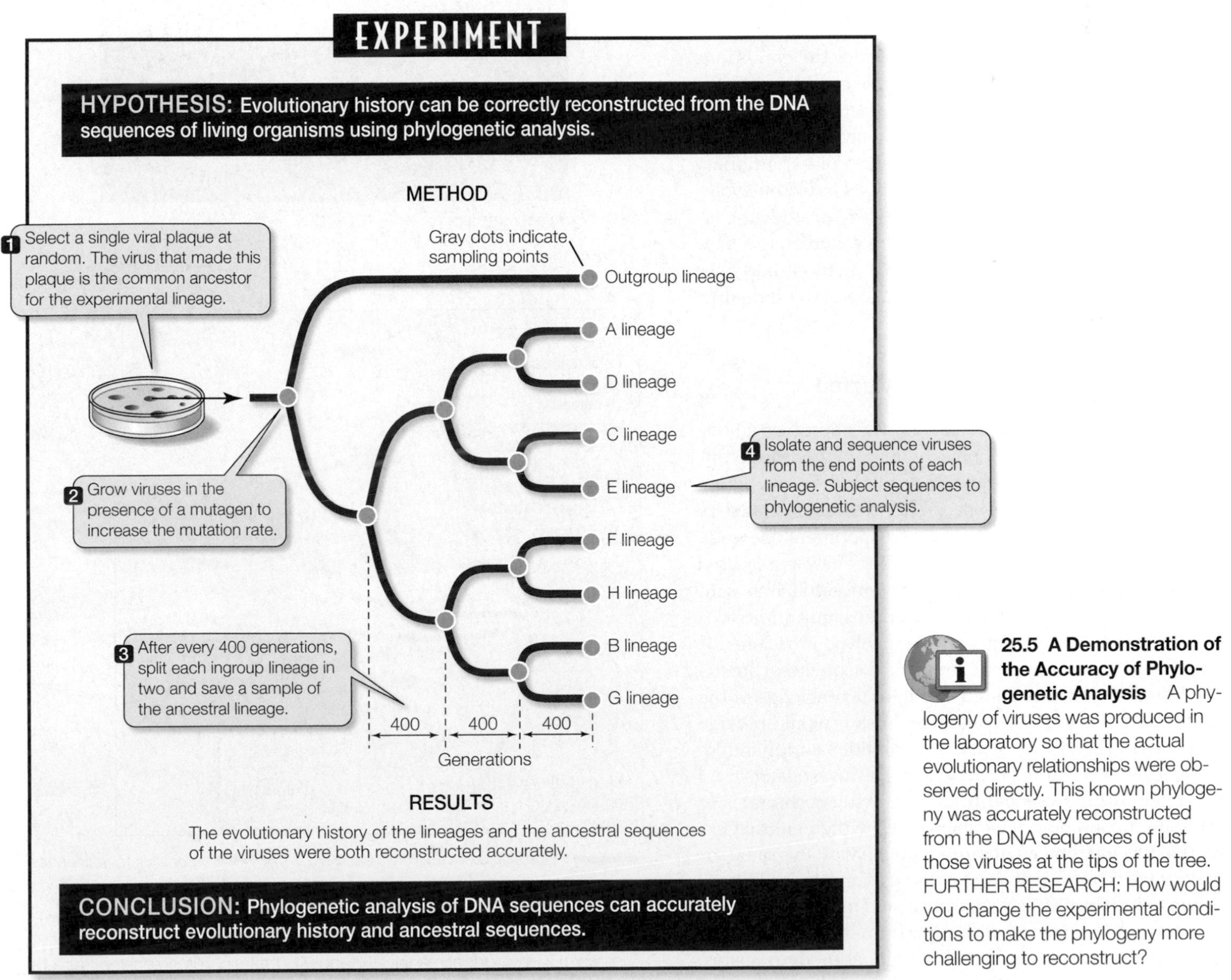

EXPERIMENT
HYPOTHESIS: Evolutionary history can be correctly reconstructed from the DNA sequences of living organisms using phylogenetic analysis.
METHOD
1 Select a single viral plaque at random. The virus that made this plaque is the common ancestor for the experimental lineage.
Gray dots indicate sampling points
Outgroup lineage
A lineage
D lineage
C lineage
E lineage
F lineage
H lineage
B lineage
G lineage
2 Grow viruses in the presence of a mutagen to increase the mutation rate.
4 Isolate and sequence viruses from the end points of each lineage. Subject sequences to phylogenetic analysis.
3 After every 400 generations, split each ingroup lineage in two and save a sample of the ancestral lineage.
400
400
400
Generations
RESULTS
The evolutionary history of the lineages and the ancestral sequences of the viruses were both reconstructed accurately.
CONCLUSION: Phylogenetic analysis of DNA sequences can accurately reconstruct evolutionary history and ancestral sequences.

25.5 A Demonstration of the Accuracy of Phylogenetic Analysis A phylogeny of viruses was produced in the laboratory so that the actual evolutionary relationships were observed directly. This known phylogeny was accurately reconstructed from the DNA sequences of just those viruses at the tips of the tree. FURTHER RESEARCH: How would you change the experimental conditions to make the phylogeny more challenging to reconstruct?

evolve until there were eight lineages in the ingroup and one in the outgroup. Mutagens were added to the viral cultures to increase the mutation rate so that the amount of change and the degree of homoplasy would be typical of the organisms analyzed in average phylogenetic analyses. The investigators then sequenced samples from the end points of the eight lineages, as well as from the ancestors at the branching points. They then gave the sequences from the end points of the lineages to other investigators to analyze, without revealing the known history of the lineages or the sequences of the ancestral viruses.

After the phylogenetic analysis had been completed, the investigators asked two questions: Did phylogenetic methods reconstruct the known history correctly, and were the sequences of the ancestral viruses reconstructed accurately? The answer in both cases was yes: the branching order of the lineages was reconstructed exactly as it had occurred, over 98 percent of the nucleotide positions of the ancestral viruses were reconstructed correctly, and 100 percent of the amino acid changes in the viral proteins were reconstructed correctly.

The experiment shown in Figure 25.5 demonstrated that phylogenetic analysis was accurate under the conditions tested, but it did not examine all possible conditions. Other experimental studies have taken other factors into account, such as the sensitivity of phylogenetic analysis to convergent environments and highly variable rates of evolutionary change. In addition, computer simulations based on evolutionary models have been used extensively to study the effectiveness of phylogenetic analysis. These studies, too, have confirmed the accuracy of phylogenetic methods and have also been used to refine those methods and extend them to new applications.

Ancestral states can be reconstructed

In addition to inferring the evolutionary relationships among lineages, biologists can use phylogenetic methods to reconstruct the morphology, behavior, or nucleotide and amino acid sequences of ancestral species (as was demonstrated for the ancestral sequences of T7 in the experiment shown in Figure 25.5). For instance, a phylogenetic analysis was used to reconstruct an opsin protein in the ancestral archosaur (the last common ancestor of birds, dinosaurs, and crocodiles). Opsins are pigment proteins involved in vision; different opsins (with different amino acid sequences) are excited by different wavelengths of light. Knowledge of the opsin sequence in the ancestral archosaur would provide clues about the animal's visual capabilities and therefore about some of its probable behaviors. The investigators used phylogenetic analysis of opsin from living vertebrates to estimate the amino acid sequence of the pigment that existed in the ancestral archosaur. A protein with this same sequence was then constructed in the laboratory. The investigators tested the reconstructed opsin and found a significant shift toward the red end of the spectrum in the light sensitivity of this protein compared with most modern opsins. Modern species that exhibit similar sensitivity are adapted for nocturnal vision, so the investigators inferred that the ancestral archosaur might have been active at night. Thus, reminiscent of the movie *Jurassic Park*, phylogenetic analyses are being used to reconstruct extinct species, one protein at a time!

(A)

Harpagochromis sp.

Ptyochromis sp.

(B)

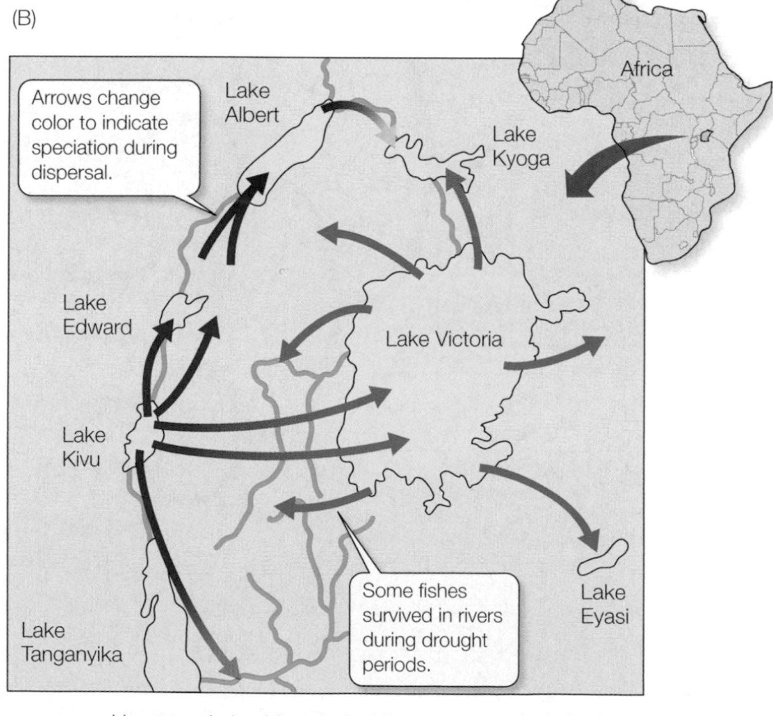

25.6 Origins of the Cichlid Fishes of Lake Victoria (A) These photographs show only two of the hundreds of cichlid species found in Lake Victoria. (B) Phylogenetic analysis suggests that cichlids colonized Lake Victoria by the routes shown on the map.

Molecular clocks add a dimension of time

For many applications, biologists want to know not only the order in which evolutionary lineages split, but also the timing of those splits. In 1965, Emile Zuckerkandl and Linus Pauling hypothesized that rates of molecular change were constant enough that they could be used to predict evolutionary divergence times—an idea that has become known as the *molecular clock hypothesis*.

Of course, different genes evolve at different rates, and there are also differences in evolutionary rates among species related to differing generation times, environments, efficiencies of DNA repair systems, and other biological factors. Nonetheless, among closely related species, a given gene usually evolves at a reasonably constant rate and can be used as a metric to gauge the time of divergence for a particular split in the phylogeny. These **molecular clocks** must be calibrated using independent data, such as the fossil record, known times of divergence, or biogeographic dates (such as the dates for separations of continents). Using such calibrations, times of divergence have been estimated for many groups of species.

Studies of cichlid fishes provide an example of the use of molecular clocks. The spectacular evolutionary radiation that produced more than 500 species in one group of cichlid fishes in Lake Victoria, in eastern Africa, was initially assumed to have occurred over a period of about 750,000 years, the presumed age of the lake basin. Recently discovered geological data suggest, however, that Lake Victoria dried up between 15,600 and 14,700 years ago. Biologists judged that the hundreds of morphologically diverse cichlids in Lake Victoria could not have evolved in such a short time, so they considered other hypotheses. One hypothesis assumed that the lake did not dry up completely. Another postulated that some of the fish species survived in rivers, from which they subsequently recolonized the lake. But how could they test these possibilities? Phylogenetic analyses based on molecular data helped resolve the problem.

Using mitochondrial DNA sequences from each of 300 species, investigators constructed a phylogenetic tree of the cichlid fishes of Lake Victoria and other lakes in the region. Their phylogenetic tree suggests that the ancestors of the Lake Victoria cichlids came from the geologically much older Lake Kivu. Today, Lake Kivu is home to only 15 species of cichlids, but the phylogenetic tree suggests that fishes from Lake Kivu colonized Lake Victoria on two different occasions (**Figure 25.6**). A molecular clock analysis also indicated that some of the cichlid lineages that are found only in Lake Victoria, and which therefore probably evolved there, split at least 100,000 years ago. Thus the inferred phylogeny of these fishes strongly suggests that Lake Victoria did not completely dry up about 15,000 years ago, and that many fish species survived in rivers and in pockets of water that remained in the deepest part of the lake throughout the most recent dry period.

Biologists once thought that HIV-1 moved from chimpanzees to humans shortly before AIDS became recognized as a disease in the early 1980s. However, molecular clock analysis of the main HIV-1 group of viruses demonstrated that HIV-1 in humans dates back at least to the 1920s or 1930s, and that other primate immunodeficiency viruses have moved into human populations several times over the past century.

25.2 RECAP

Phylogenetic trees can be constructed by using the parsimony principle to find the simplest explanation for the evolution of traits. Maximum likelihood methods incorporate more explicit models of evolutionary change to reconstruct evolutionary history.

- Do you understand how a phylogenetic tree is constructed? See pp. 547–548 and Figure 25.3
- Is there a way to test whether phylogenetic trees provide accurate reconstructions of evolutionary history? See pp. 549–550 and Figure 25.5
- How do molecular clocks add a time dimension to phylogenetic trees? See p. 551

Biologists in many fields now routinely reconstruct phylogenetic relationships. Let's examine some of the many uses of these phylogenetic trees.

25.3 How Do Biologists Use Phylogenetic Trees?

Information about the evolutionary relationships among organisms is useful to scientists investigating a wide variety of biological questions. In this section we will illustrate how phylogenetic trees can be used to ask questions about the past, to compare organisms in the present, and to make predictions about the future.

Phylogenies help us reconstruct the past

Most flowering plants reproduce by mating with another individual—a process called *outcrossing*. Many outcrossing species have mechanisms to prevent self-fertilization, and so are referred to as *self-incompatible*. Individuals of some species, however, regularly fertilize themselves with their own pollen; they are termed *selfing* species, which of course requires that they be *self-compatible*. How can we tell how often self-compatibility has evolved in a group of plants? We can do so by conducting a phylogenetic analysis of outcrossing and selfing species and testing the species for self-compatibility.

The evolution of fertilization mechanisms was examined in *Linanthus* (a genus in the phlox family), a group of plants with a diversity of breeding systems and pollination mechanisms. The outcrossing species of *Linanthus* have long petals and are pollinated by long-tongued flies; these species are self-incompatible. The self-compatible species, in contrast, all have short petals. The investigators reconstructed a phylogeny for 12 species in the genus using nuclear ribosomal DNA sequences (**Figure 25.7**). They determined whether each species was self-compatible by artificially pollinating flowers with the plant's own pollen or with pollen from other individuals and observing whether viable seeds formed.

Several lines of evidence suggest that self-incompatibility is the ancestral state in *Linanthus*. Multiple origins of self-incompatibil-

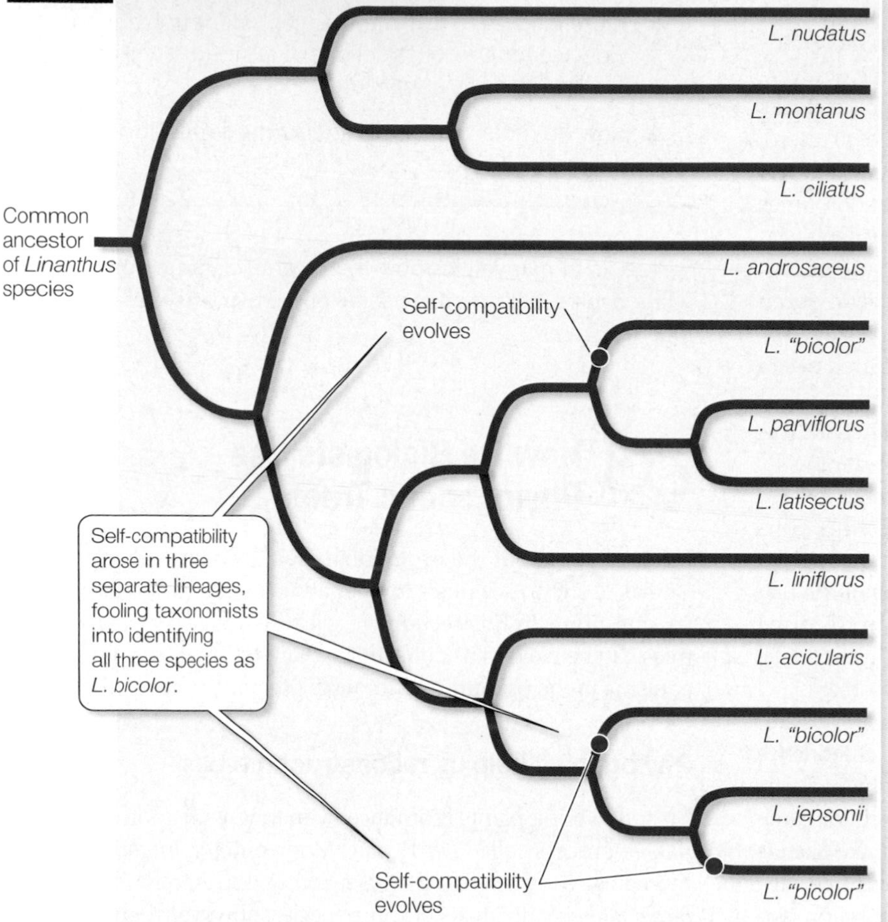

25.7 Phylogeny of a Section of the Plant Genus *Linanthus* Self-compatibility apparently evolved three times in this group. Because the form of the flowers converged in the three selfing lineages, taxonomists mistakenly thought that they were all varieties of a single species.

ity have not been found in any other flowering plant family. Self-incompatibility depends on physiological mechanisms in both the pollen and the stigma (the female organ on which pollen lands) and requires the presence of at least three different alleles. Therefore, a change from self-incompatibility to self-compatibility would be easier than the reverse change. In addition, in all self-incompatible species of *Linanthus*, the site of pollen rejection is the stigma, even though sites of pollen rejection vary greatly among other plant families.

Assuming that self-incompatibility is the ancestral state, the reconstructed phylogeny suggests that self-compatibility has evolved three times within this group of *Linanthus* (see Figure 25.7). The change to self-compatibility has been accompanied by the evolution of reduced petal size. Interestingly, the striking similarity of the flowers in the self-compatible groups once led to their being classified as members of a single species. The phylogenetic analysis using ribosomal DNA showed them to be members of three distinct lineages, however.

Reconstructing the past is important for understanding many biological processes. In the case of *zoonotic diseases* (diseases caused by infectious organisms that have been transferred to humans from another animal host), it is important to understand when, where, and how the disease first entered human populations. Human immunodeficiency virus (HIV) is the cause of such a zoonotic disease: acquired immunodeficiency syndrome, or AIDS. A phylogenetic analysis of immunodeficiency viruses shows that humans acquired these viruses from two different hosts: HIV-1 from chimpanzees, and HIV-2 from sooty mangabeys (**Figure 25.8**).

HIV-1 is the common form of the virus in human populations in central Africa, where chimpanzees are hunted for food, and HIV-2 is the common form of the virus in human populations in western Africa, where sooty mangabeys are hunted for food. Thus it seems likely that these viruses entered human populations through hunters who cut themselves while skinning chimpanzees and sooty mangabeys. The relatively recent global pandemic of AIDS occurred when these infections in local African populations rapidly spread through human populations around the world.

Phylogenies allow us to compare and contrast living organisms

Male swordtails (a group of fishes within the genus *Xiphophorus*) have a long, colorful extension of the tail (**Figure 25.9A**). The reproductive success of male swordtails is closely associated with these appendages. Males with long swords are more likely to mate successfully than are males with short swords (an example of *sexual selection*; see Chapters 22 and 23). Several explanations have been advanced for the evolution of these structures, including the hypothesis that the sword simply exploits a preexisting bias in the sensory system of the females. This *sensory exploitation hypothesis* suggests that female swordtails had a preference for males with long tails even before the tails evolved (perhaps because females assess the size of males by their total length, including the tail). Long tails became a sexually selected trait because of the preexisting preference of the females.

A phylogeny was used to identify the closest relatives of swordtails that had split from their lineage before the evolution of swords. These turned out to be the platyfishes, another group of *Xiphophorus* (**Figure 25.9B**). To test the sensory exploitation hypothesis, researchers attached artificial swordlike structures to the tails of some male platyfishes. Even though male platyfishes do not normally have such structures, female platyfishes preferred the males with the artificial swords, thus providing support for the hypothesis that

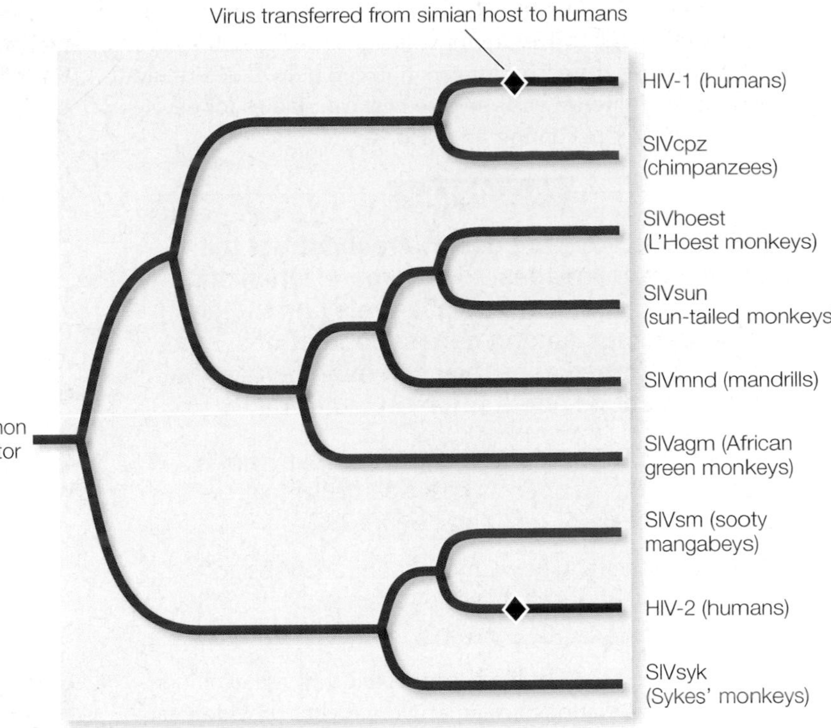

25.8 Phylogenetic Tree of Immunodeficiency Viruses Immunodeficiency viruses have been transmitted to humans from two different simian hosts (HIV-1 from chimpanzees, and HIV-2 from sooty mangabeys).

female *Xiphophorus* had a preexisting sensory bias that favored swords even before the trait evolved.

Biologists use phylogenies to predict the future

Influenza kills many people every year, and each year, many susceptible individuals receive a flu vaccine to reduce their chance of contracting the disease. Why do we need a new flu vaccine every year, when a single dose of vaccine for many other diseases provides many years of protection? The answer is that the rate of evolution of the influenza virus is sufficiently high that the flu viruses in circulation each year are often substantially different from the flu viruses that circulated in previous years.

A phylogenetic analysis of strains of influenza virus indicates that there is strong selection by the human immune system against most strains. Only those strains with the greatest number of amino acid replacements in particular positions of the protein hemagglutinin (a protein on the surface of influenza viruses that is recognized by the human immune system; **Figure 25.10**) are likely to leave descendant lineages in subsequent years. Thus, by conducting a phylogenetic analysis of hemagglutinin

25.9 The Origin of a Sexually Selected Trait in the Fish Genus *Xiphophorus* (A) A male swordtail, showing the large tail that evolved through sexual selection. (B) A male platyfish, a related species. Phylogenetic analysis reveals that the platyfish split from the swordtails before the evolution of the sword. But experimental analysis demonstrates that female platyfish prefer males with artificial swords. This finding supports the idea that the sword of swordtails evolved as a result of a preexisting preference in the females.

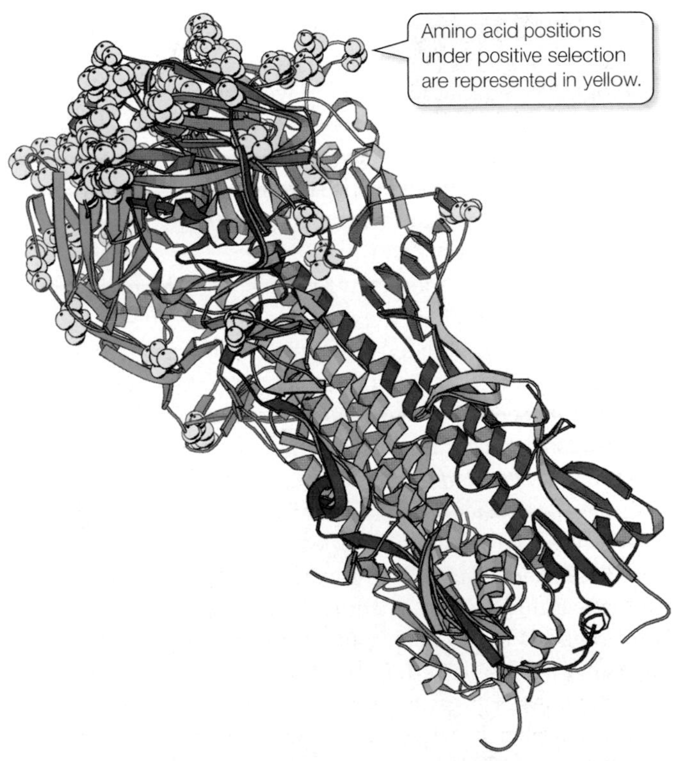

Amino acid positions under positive selection are represented in yellow.

25.10 Model of Hemagglutinin, a Surface Protein of Influenza The yellow balloon-shaped structures represent amino acids in the protein that are under strong selection for change. Changes in these amino acids produce proteins that are more likely to escape detection by the human immune system.

sequences, biologists can predict which of the currently circulating strains of influenza virus are most likely to survive and leave descendants in the future. This information can help to determine which strains of influenza virus are the best candidates for developing flu vaccine for upcoming flu seasons.

25.3 RECAP

Phylogenetic trees are used to reconstruct the past history of lineages, to determine when and where traits arose, and to make relevant biological comparisons among genes, populations, and species. In some cases they can even be used to make predictions about future evolution.

- Can you explain how phylogenetic trees can help us determine the number of times a particular trait evolved? See pp. 551–552 and Figure 25.7
- Do you understand how phylogenetic analysis can help produce more effective flu vaccines? See p. 553

As we have now seen, all of life is connected through evolutionary history, and the relationships among organisms provide a natural basis for making biological comparisons. For these reasons, biologists use phylogenetic relationships as the basis for organizing life into a coherent classification system.

25.4 How Does Phylogeny Relate to Classification?

The biological classification system in widespread use today is derived from a system developed by the Swedish biologist Carolus Linnaeus and has been used since the mid-1700s. Linnaeus's system, referred to as **binomial nomenclature**, allows scientists throughout the world to refer unambiguously to the same organisms by the same names (**Figure 25.11**).

Linnaeus gave each species two names, one identifying the species itself and the other the genus to which it belongs. A **genus** (plural, *genera;* adjectival form *generic*) is a group of closely related species. Optionally, the name of the taxonomist who first proposed the species name may be added at the end. Thus *Homo sapiens* Linnaeus is the name of the modern human species. *Homo* is the genus to which the species belongs, and *sapiens* identifies the particular species in the genus *Homo;* Linnaeus proposed the species name *Homo sapiens.* You can think of the generic name *Homo* as equivalent to your surname and the specific name *sapiens* as equivalent to your first name. The generic name is always capitalized; the name identifying the species always lowercased. Both names are italicized, whereas common names of organisms are not. Rather than repeating a generic name when it is used several times in the same discussion, biologists often spell it out only once and abbreviate it to the initial letter thereafter (for example, *D. melanogaster* is the abbreviated form of *Drosophila melanogaster*).

As we noted earlier, any group of organisms that is treated as a unit in a biological classification system, such as the genus *Drosophila,* or all insects, is called a *taxon.* In the Linnaean system, species and genera are further grouped into a hierarchical system of higher taxonomic categories. The taxon above the genus in the Linnaean system is the **family.** The names of animal families end in the suffix "-idae." Thus Formicidae is the family that contains all ant species, and the family Hominidae contains humans and our recent fossil relatives, as well as our closest living relatives, the chimpanzees and gorillas. Family names are based on the name of a member genus; Formicidae is based on the genus *Formica,* and Hominidae is based on *Homo.* Plant classification follows the same procedures, except that the suffix "-aceae" is used with family names instead of "-idae." Thus Rosaceae is the family that includes the genus of roses (*Rosa*) and its close relatives. Families, in turn, are grouped into **orders,** orders into **classes,** and classes into **phyla** (singular *phylum*), and phyla into **kingdoms.** However, the application of these various ranked levels of classification is subjective,

(A) *Campanula rotundifolia*

(B) *Endymion non-scriptus*

(C) *Mertensia virginica*

25.11 Many Different Plants Are Called Bluebells All three of these unrelated plant species are commonly called "bluebells." Binomial nomenclature allows us to communicate exactly what is being described. (A) *Campanula rotundifolia*, found on the North American Great Plains, belongs to a larger group of bellflowers. (B) *Endymion non-scriptus*, the English bluebell, is related to hyacinths. (C) *Mertensia virginica*, the Virginia bluebell, belongs in a very different group of plants known as borages.

and they are used largely for convenience. Although families are always grouped within orders, orders within classes, and so forth, there is nothing that makes a family in one group equivalent (in age, for instance) to a family in another group.

Linnaeus developed his system before evolutionary thought had become widespread in biology, but he still recognized the overwhelming hierarchy of life. Although Linnaeus did not know the basis of this hierarchy, biologists today universally recognize the common Tree of Life as the organizational basis for biological classification. Today, some biologists de-emphasize the use of ranked classifications, since it is a largely subjective decision whether a particular taxon is considered, say, an order or a class. But regardless of whether modern biologists explicitly use these ranked categories, they use evolutionary relationships as the basis for recognizing all biological taxa.

Phylogeny is the basis for modern biological classification

Biological classification systems are used to express relationships among organisms. The kind of relationship we wish to express influences which features we use to classify organisms. If, for instance, we were interested in a system that would help us decide what plants and animals were desirable as food, we might devise a classification based on tastiness, ease of capture, and the type of edible parts each organism possessed. Early Hindu classifications of organisms were designed according to these criteria. Biologists do not use such systems in formal classification today, but those systems served the needs of the people who developed them.

Taxonomists today use biological classifications to express the evolutionary relationships of organisms. Therefore, taxa in biological classifications are expected to be **monophyletic**, meaning that the taxon contains an ancestor and all descendants of that ancestor, and no other organisms (**Figure 25.12**). In other words, it is a historical group of related species, or a complete branch on the Tree of Life (also known as a *clade*). Although biologists seek to name monophyletic taxa, the detailed phylogenetic information needed to do so is not always available. A group that does not include its common ancestor is called a **polyphyletic** group. A group that does not include all the descendants of a common ancestor is called a **paraphyletic** group.

A true monophyletic group (or clade) can be removed from a phylogenetic tree by a single "cut" in the tree, as shown in Figure 25.12. Note that there are many monophyletic groups on any phylogenetic tree, and that these groups are successively smaller subsets of larger monophyletic groups. This hierarchy of biological taxa, with all of life as the most inclusive taxon and many smaller taxa within larger taxa, down to the individual species, is the modern basis for biological classification.

Virtually all taxonomists now agree that polyphyletic and paraphyletic groups are inappropriate as taxonomic units. The classifications used today still contain such groups because some organisms have not been evaluated phylogenetically. As mistakes in prior classifications are detected, taxonomic names are revised and polyphyletic and paraphyletic groups are eliminated from the classifications.

Several codes of biological nomenclature govern the use of scientific names

Several sets of explicit rules govern the use of scientific names. Biologists around the world follow these rules voluntarily to facilitate communication and dialogue. Although there may be dozens of common names for an organism in many different languages, the rules of biological nomenclature are designed so that there is only one correct scientific name for any single recognized taxon and (ideally) a given scientific name applies only to a single taxon (that is, each scientific name is unique). Sometimes the

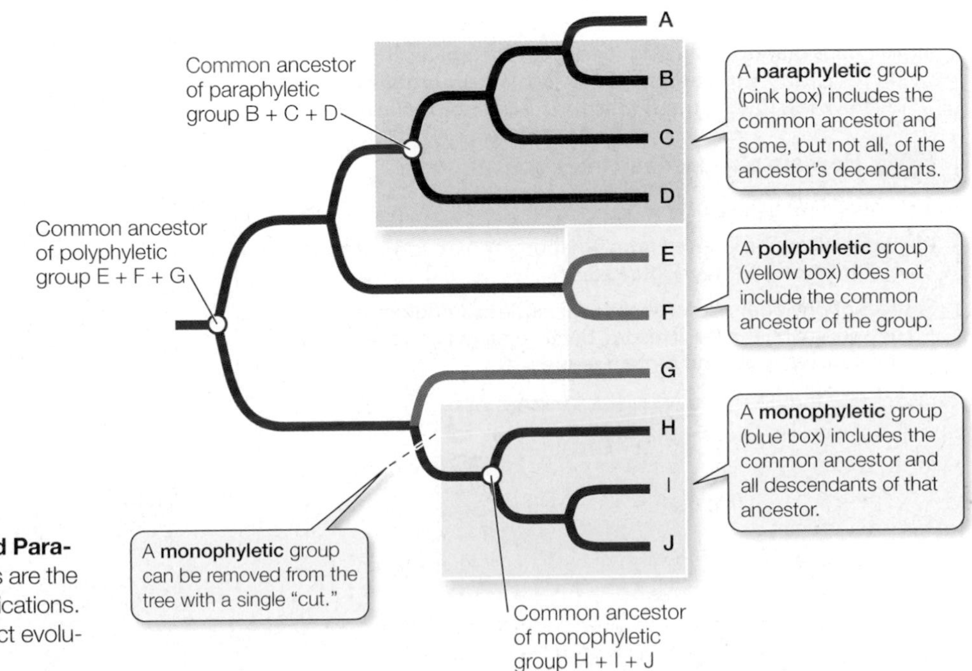

A
B
C
D
E
F
G
H
I
J
Common ancestor of paraphyletic group B + C + D
Common ancestor of polyphyletic group E + F + G
A paraphyletic group (pink box) includes the common ancestor and some, but not all, of the ancestor's decendants.
A polyphyletic group (yellow box) does not include the common ancestor of the group.
A monophyletic group (blue box) includes the common ancestor and all descendants of that ancestor.
A monophyletic group can be removed from the tree with a single "cut."
Common ancestor of monophyletic group H + I + J

25.12 Monophyletic, Polyphyletic, and Paraphyletic Groups Monophyletic groups are the basis of biological taxa in modern classifications. Polyphyletic and paraphyletic groups do not reflect evolutionary history.

same species is named more than once (when more than one taxonomist has taken up the task); the rules specify that the valid name is the first name proposed. If the same name is inadvertently given to two different species, then a replacement name must be given to the species that was named second.

Because of the historical separation of the fields of zoology, botany (including, originally, the study of fungi), and microbiology, different sets of taxonomic rules were developed for each of these groups. Yet another set of rules for classifying viruses emerged later. This has resulted in many duplicated names between groups that are governed by different sets of rules: *Drosophila*, for instance, is both a genus of fruit flies and a genus of fungi, and there are species within both groups that have identical names. Until recently these duplicated names caused little confusion, since traditionally biologists who studied fruit flies were unlikely to read the literature on fungi (and vice versa). Today, however, given the use of large, universal biological databases (such as GenBank, which includes DNA sequences from across all life), it is increasingly important that each taxon have a unique name. Taxonomists are now working to develop common sets of rules that can be applied across all living organisms.

25.4 RECAP

Biologists organize and classify life by identifying and naming monophyletic groups. Several sets of rules govern the use of scientific names so that each species and higher taxon can be identified and named unambiguously.

- Can you explain the difference between monophyletic, paraphyletic, and polyphyletic groups? See p. 555 and Figure 25.12
- Do you understand why biologists prefer monophyletic groups for use in formal classifications? See p. 555

Now that we have seen how evolution occurs and how it can be studied, we are ready to consider the products of evolutionary history. The vastness and diversity of the living world are what first excite many future biologists and convince them to study biology. In Part Six we will examine the great diversity of life and describe some of the reasons that biologists turn their attention to particular groups of organisms.

CHAPTER SUMMARY

25.1 What is phylogeny?

A **phylogeny** is a description of the history of descent of a group of organisms from their common ancestor. Groups of evolutionarily related species are represented as related branches in a **phylogenetic tree**. Review Figure 25.1

A group of species that consists of all the evolutionary descendants of a common ancestor is called a **clade**. Named clades and species are called **taxa**.

Homologies are similar traits that have been inherited from a common ancestor. Review Figure 25.2

A **derived** trait is one that differs from its form in the common ancestor of a lineage. Derived traits that support the common relationship among a group of species are called **synapomorphies**.

Similar traits may occur among species that do not result from common ancestry. **Convergent evolution** and **evolutionary reversals** can give rise to such traits, which are called **homoplasies**.

25.2 How are phylogenetic trees constructed?

See Web/CD Activity 25.1

Phylogenies can be inferred from synapomorphies using the principle of **parsimony**. Review Figure 25.3

Sources of phylogenetic information include morphology, patterns of development, the fossil record, behavioral traits, and molecular traits such as DNA and protein sequences.

Phylogenies can also be inferred by finding **maximum likelihood** solutions, using explicit evolutionary models of character change.

Biologists can use phylogenetic trees to reconstruct ancestral states. See Web/CD Tutorial 25.1

Phylogenetic trees may include estimates of times of divergence of lineages, as determined by a **molecular clock** analysis.

25.3 How do biologists use phylogenetic trees?

Phylogenetic trees are used to reconstruct the past and understand the origin of traits. Review Figure 25.7

Phylogenetic trees are used to make appropriate evolutionary comparisons among living organisms.

Phylogenetic trees can sometimes be used to predict future evolution.

25.4 How does phylogeny relate to classification?

Taxonomists organize biological diversity on the basis of evolutionary history.

Taxa in modern classifications are expected to be **monophyletic** groups. **Paraphyletic** and **polyphyletic** groups are not considered appropriate taxonomic units. Review Figure 25.12, Web/CD Activity 25.2

Several sets of rules govern the use of scientific names, with the goal of providing unique and universal names for biological taxa.

SELF-QUIZ

1. A clade is
   a. a type of phylogenetic tree.
   b. a group of evolutionarily related species that share a common ancestor.
   c. a tool for constructing phylogenetic trees.
   d. an extinct species.
   e. an ancestral species.
2. Phylogenetic trees may be reconstructed for
   a. genes.
   b. species.
   c. major evolutionary groups.
   d. viruses.
   e. All of the above
3. A shared derived trait, used as the basis for inferring a monophyletic group, is called
   a. a synapomorphy.
   b. a homoplasy.
   c. a parallel trait.
   d. a convergent trait.
   e. a phylogeny.
4. The parsimony principle can be used to infer phylogenies because
   a. evolution is nearly always parsimonious.
   b. it is logical to adopt the simplest hypothesis capable of explaining the known facts.
   c. once a trait changes, it never reverses condition.
   d. all species have an equal probability of evolving.
   e. closely related species are always very similar to one another.
5. Convergent evolution and evolutionary reversal are two sources of
   a. homology.
   b. parsimony.
   c. synapomorphy.
   d. monophyly.
   e. homoplasy.
6. Which of the following are commonly used to infer phylogenetic relationships among plants but not among animals?
   a. Nuclear genes
   b. Chloroplast genes
   c. Mitochondrial genes
   d. Ribosomal RNA genes
   e. Protein-coding genes
7. Which of the following is *not* true of maximum likelihood or parsimony methods for inferring phylogeny?
   a. The maximum likelihood method requires an explicit model of evolutionary character change.
   b. The parsimony method is computationally easier than the maximum likelihood method.
   c. The maximum likelihood method is easier to treat in a statistical framework.
   d. The maximum likelihood method is most often used with molecular data.
   e. Parsimony is usually used to infer time on a phylogenetic tree.
8. Taxonomists strive to include taxa in biological classifications that are
   a. monophyletic.
   b. paraphyletic.
   c polyphyletic.
   d. homoplastic.
   e monomorphic.
9. Which of the following groups have separate sets of rules for nomenclature?
   a. Animals
   b. Plants and fungi
   c. Bacteria
   d. Viruses
   e. All of the above
10. If two scientific names are proposed for the same species, how do taxonomists decide which name should be used?
    a. The name that provides the most accurate description of the organism is used.
    b. The name that was proposed most recently is used.
    c. The name that was used in the most recent taxonomic revision is used.
    d. The first name to be proposed is used, unless that name was previously used for another species.
    e. Taxonomists use whichever name they prefer.

FOR DISCUSSION

1. Why are taxonomists concerned with identifying species that share a single common ancestor?
2. How are fossils used to identify ancestral and derived traits of organisms?
3. The parsimony principle is often used to reconstruct phylogenetic trees. Given that evolutionary processes are not always parsimonious, why is it used as a guiding principle?
4. A student of the evolution of frogs has proposed a strikingly new classification of frogs based on an analysis of a few mitochondrial genes from about 10 percent of frog species. Should frog taxonomists immediately accept the new classification? Why or why not?
5. Linnaeus developed his system of classification before Darwin developed his theory of evolution by natural selection, and most classifications of organisms initially were proposed by nonevolutionists. Yet many parts of these classifications are still used today, with minor modifications, by most evolutionary taxonomists. Why?
6. Classification systems summarize much information about organisms and enable us to remember the traits of many organisms. From our general knowledge, how many traits can you associate with the following names: conifer, fern, bird, mammal?

FOR INVESTIGATION

West Nile virus causes mortality in many species of birds, and can cause fatal encephalitis (inflammation of the brain) in humans and horses. It was first isolated in Africa (where it is thought to be endemic) in the 1930s, and by the 1990s it had been found throughout much of Eurasia. West Nile virus was not found in North America until 1999, but since then, it has quickly spread across most of the United States. The genome of West Nile virus evolves quickly. How could you use phylogenetic analysis to investigate the origin of the West Nile virus that was introduced into North America in 1999?

PART SIX

The Evolution of Diversity

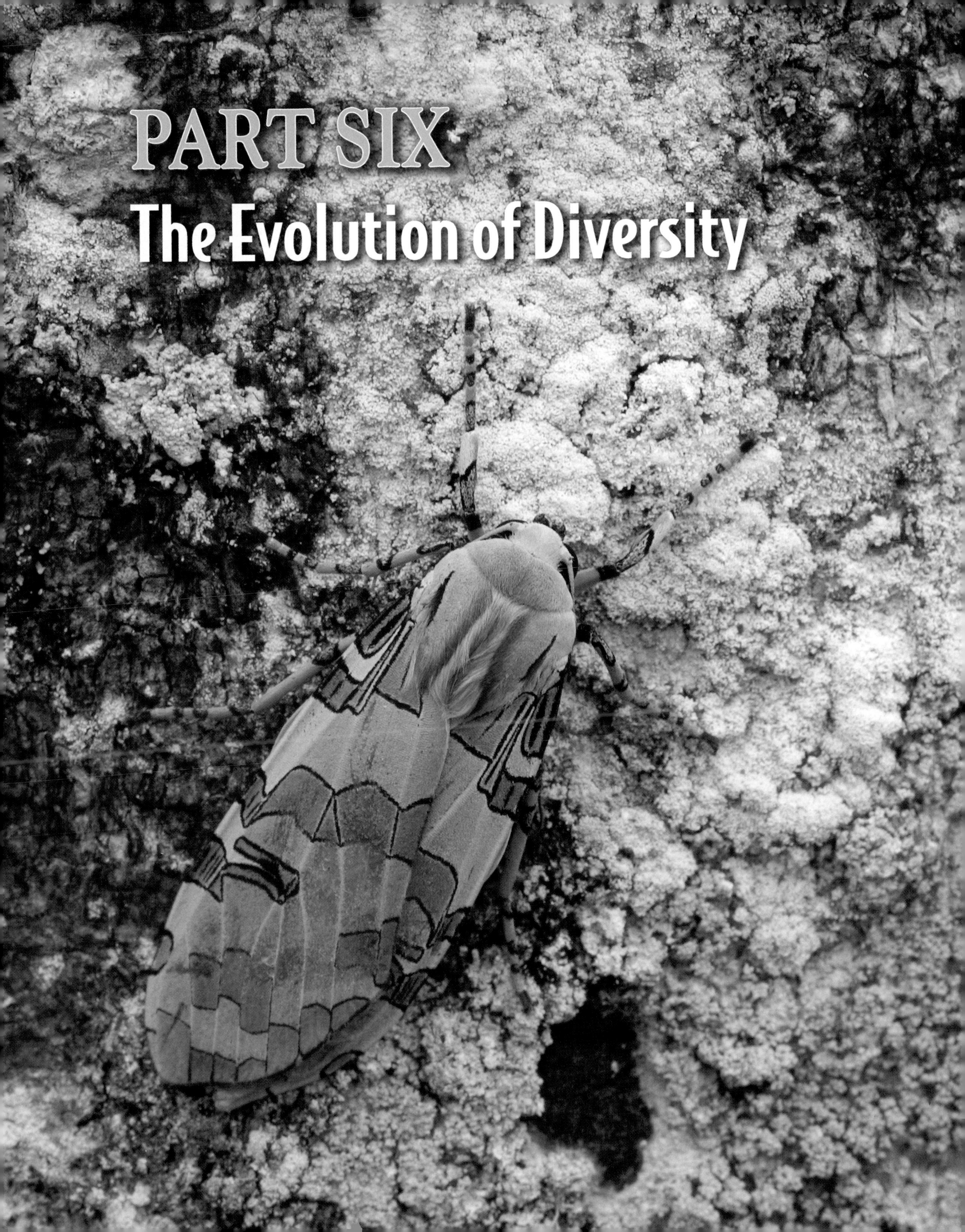

CHAPTER 26 Bacteria and Archaea: The Prokaryotic Domains

Life on the Red Planet?

The ancient Phoenicians called it the "river of fire." Today, Spanish astrobiologist Ricardo Amils Pibernat calls Spain's Río Tinto ("painted river") a possible model for the scene of the origin of life that may have existed on Mars. The Río Tinto wends its way through a huge deposit of iron pyrite—"fool's gold." The river's deep color comes about because prokaryotes in the river and in the acidic soil from which it arises convert iron pyrite into sulfuric acid and dissolved iron.

The Río Tinto has a pH of 2 and exceptionally high concentrations of heavy metals, especially iron. Oxygen concentrations in the river and in its source soil are extremely low. Amils believes this soil resembles the kind of environment in which life could have arisen on Mars. Whatever the truth of that speculation, the Río Tinto represents one of the most unusual habitats for life on Earth.

Is there life—presumably microscopic—on Mars today? Most scientists are skeptical about the possibility. However, at a meeting in 2005, Italian space scientist Vittorio Formisano reported some suggestive findings. In particular, there appears to be formaldehyde, a breakdown product of methane, in the Martian atmosphere.

In order to account for the observed concentration of formaldehyde in the Martian atmosphere, some 2.5 million tons of methane would have to be produced every year. Formisano argues that there are only three explanations for such a high rate of methane production: solar radiation-induced chemical reactions at the planetary surface; chemical reactions within the planet; or biochemical reactions in living organisms. There is no geological or chemical evidence for the first two explanations. Could living microorganisms similar to the many known methane-producing prokaryotes on Earth be responsible?

Just how likely is it that populations of simple organisms could be living in the Martian crust? We know there are enormous numbers of prokaryotes living deep in Earth's subsurface in rocks, mines, oil reservoirs, ice sheets, and sediments as much as 90 meters below the ocean floor. Some of these populations have been living and metabolizing underground for millions of years.

Prokaryotes are relatively simple in structure, but don't make the mistake of underestimating their capabilities. The prokaryotes are by far the most numer-

Earth or Ancient Mars? Spain's Río Tinto owes its rusty red color—and its extreme acidity—to the action of prokaryotes on iron pyrite-rich soil.

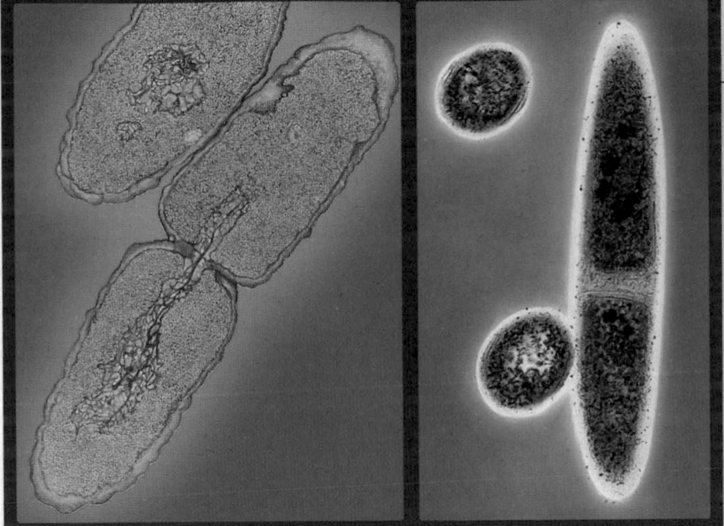

Very Different Prokaryotes *Salmonella typhimurium* (left) is a member of the domain Bacteria; *Methanospirillum hungatii* (right) is classified in the Archaea, and is probably more closely related to the eukaryotes than to the bacteria. The cells in both images are dividing, but are not shown to the same scale; the archaeal cells are actually about one-tenth the size of the *Salmonella* cells.

CHAPTER OUTLINE

ous organisms on Earth, where they are found in more different habitats than are the eukaryotes. There are more prokaryotes living on and inside your body than you have human cells. The prokaryotes are masters of metabolic ingenuity, having developed more ways to obtain energy from the environment than the eukaryotes have. And they have been around longer than other organisms.

Late in the twentieth century, it became apparent to many microbiologists that certain prokaryotes differ so fundamentally from others in their metabolic processes that they should be considered members of a distinct evolutionary lineage. Furthermore, the two prokaryotic lineages diverged very early in life's evolution. We refer to these lineages as the domains Bacteria and Archaea.

IN THIS CHAPTER we will discuss the distribution of prokaryotes and examine their remarkable metabolic diversity. We will describe the impediments to the resolution of evolutionary relationships among the prokaryotes and survey the surprising diversity of organisms within each domain. Finally, we will discuss the effects of prokaryotes on their environments.

26.1 How Did the Living World Begin to Diversify?

What does it mean to be *different*? You and the person nearest you look very different—certainly you appear more different than the two cells shown above. But the two of you are members of the same species, while these two tiny organisms that look so much alike actually are classified in entirely separate domains. Still, you—in the domain Eukarya—and those two prokaryotes in the domains Bacteria and Archaea have a lot in common. All three of you:

- conduct glycolysis
- replicate DNA semiconservatively
- have DNA that encodes polypeptides
- produce those polypeptides by transcription and translation using the same genetic code
- have plasma membranes and ribosomes in abundance

Despite these commonalities among life's domains, there are also major differences. Let's first distinguish between the Eukarya and the two prokaryotic domains. Note that "domain" is a subjective term used for the largest groups of life. There is no objective definition of a domain, any more than of a kingdom or a family.

The three domains differ in significant ways

Prokaryotic cells differ from eukaryotic cells in three important ways:

- Prokaryotic cells lack a cytoskeleton, and in the absence of organized cytoskeletal proteins, they lack mitosis. Prokaryotic cells divide by their own method, *binary fission*, after replicating their DNA (see Figure 9.2).
- The organization and replication of the genetic material differs. The DNA of the prokaryotic cell is not organized within a membrane-enclosed nucleus. DNA molecules in prokaryotes (both bacteria and archaea) are usually circu-

TABLE 26.1

The Three Domains of Life on Earth

CHARACTERISTIC	BACTERIA	DOMAIN ARCHAEA	EUKARYA
Membrane-enclosed nucleus	Absent	Absent	*Present*
Membrane-enclosed organelles	Absent	Absent	*Present*
Peptidoglycan in cell wall	*Present*	Absent	Absent
Membrane lipids	Ester-linked Unbranched	*Ether-linked* *Branched*	Ester-linked Unbranched
Ribosomes[a]	70S	70S	*80S*
Initiator tRNA	*Formylmethionine*	Methionine	Methionine
Operons	Yes	Yes	*No*
Plasmids	Yes	Yes	*Rare*
RNA polymerases	One	One[b]	*Three*
Ribosomes sensitive to chloramphenicol and streptomycin	*Yes*	No	No
Ribosomes sensitive to diphtheria toxin	*No*	Yes	Yes
Some are methanogens	No	*Yes*	No
Some fix nitrogen	Yes	Yes	*No*
Some conduct chlorophyll-based photosynthesis	Yes	*No*	Yes

[a]70S ribosomes are smaller than 80S ribosomes.
[b]Archaeal RNA polymerase is similar to eukaryotic polymerases.

lar; in the best-studied prokaryotes, there is a single chromosome, but there are often plasmids as well (see Section 13.3).

- Prokaryotes have none of the membrane-enclosed cytoplasmic organelles that modern eukaryotes have—mitochondria, Golgi apparatus, and others. However, the cytoplasm of a prokaryotic cell may contain a variety of infoldings of the plasma membrane and photosynthetic membrane systems not found in eukaryotes.

A glance at **Table 26.1** will show you that there are also major differences (most of which cannot be seen even under the microscope) between the two prokaryotic domains. In some ways the archaea are more like eukayrotes; in other ways they are more like bacteria. The architectures of prokaryotic and eukaryotic cells were compared in Chapter 4. The basic unit of the archaea and bacteria is the prokaryotic cell, which contains a full complement of genetic and protein-synthesizing systems, including DNA, RNA, and all the enzymes needed to transcribe and translate the genetic information into proteins. The prokaryotic cell also contains at least one system for generating the ATP it needs.

Genetic studies clearly indicate that all three domains had a single common ancestor, and that the present-day Archaea share a more recent common ancestor with the Eukarya than they do with the Bacteria (**Figure 26.1**). To treat all the prokaryotes as a single domain within a two-domain classification of organisms would result in a domain that is paraphyletic. That is, a single domain "Prokaryotes" would not include all the descendants of their common ancestor. (See Section 25.4 for a discussion of paraphyletic groups.)

The common ancestor of all three domains had DNA as its genetic material; its machinery for transcription and translation produced RNAs and proteins, respectively. It probably had a circular chromosome.

The Archaea, Bacteria, and Eukarya are all products of billions of years of mutation, natural selection, and genetic drift, and they are all well adapted to present-day environments. None is "primitive." The common ancestor of the Archaea and the Eukarya probably lived more than 2 billion years ago, and the common ancestor of the Archaea, the Eukarya, and the Bacteria probably lived more than 3 billion years ago. The earliest prokaryotic fossils date back at least 3.5 billion years, and those ancient fossils indicate that there was considerable diversity among the prokaryotes even during the earliest days of life.

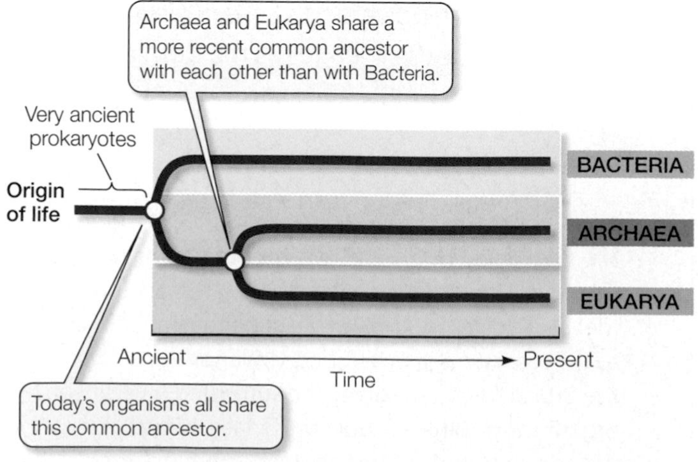

26.1 The Three Domains of the Living World Most biologists believe that all three domains share a common prokaryotic ancestor.

26.1 RECAP

Prokaryotic bacteria and archaea are as different from each other as they are from the eukaryotes. Eukaryotes share a more recent common ancestor with archaea than with bacteria.

- Do you understand the principal differences between the prokaryotes and the eukaryotes? See pp. 561–562 and Table 26.1
- Can you explain why we do not group the Bacteria and the Archaea together in a single domain? See p. 562 and Figure 26.1

The prokaryotes were alone on Earth for a very long time, adapting to new environments and to changes in existing environments. They have survived to this day—and in massive numbers. Prokaryotes are around us everywhere.

26.2 Where Are Prokaryotes Found?

Although we cannot see them with the naked eye, the prokaryotes are the most successful of all creatures on Earth, if success is measured by numbers of individuals. Bacteria and archaea in the oceans number more than 3×10^{28}. This stunning number is perhaps 100 million times as great as the number of stars in the visible universe.

> The bacteria living in a single human intestinal tract outnumber all the humans who have ever lived.

Prokaryotes are found in every type of environment on the planet, from the coldest to the hottest, from the most acidic to the most alkaline and the saltiest. Some live where oxygen is abundant, others where there is no oxygen at all. They have established themselves at the bottom of the seas, in rocks more than 2 kilometers deep in solid crust, in the clouds, and on and inside other organisms, large and small. Their effects on our environment are diverse and profound.

Three shapes are particularly common among the bacteria: spheres, rods, and curved or helical forms (**Figure 26.2**). Among their other shapes are long filaments and branched filaments. A spherical bacterium is called a *coccus* (plural *cocci*). Cocci may live singly or may associate in two- or three-dimensional arrays as chains, plates, blocks, or clusters of cells. A rod-shaped bacterium is called a *bacillus* (plural *bacilli*). Helical forms (like a corkscrew) are the third main bacterial shape. Bacilli and helical forms may be single, they may form chains, or they may gather in regular clusters. Less is known about the shapes of archaea because many of these organisms have never been seen; they are known only from samples of DNA from the environment, as we will describe in Section 26.4.

Prokaryotes are almost all unicellular, although some multicellular ones are known. Associations such as chains or clusters do not signify multicellularity because each individual cell is fully viable and independent. These associations arise as cells adhere to one another after reproducing by binary fission. Associations in the form of chains are called **filaments**. Some filaments become enclosed within delicate tubular sheaths.

Prokaryotes generally form complex communities

Prokaryotic cells and their associations do not usually live in isolation. Rather, they live in communities of many different species of organisms, often including microscopic eukaryotes. (Microscopic organisms are sometimes collectively referred to as *microbes*.) Some microbial communities form layers in sediments, and others form clumps a meter or more in diameter. While some microbial communities are harmful to humans, others provide us with important services. They help us digest our food, they break down municipal waste, and they recycle organic matter in the environment.

Many microbial communities tend to form dense **biofilms**. Upon contacting a solid surface, the cells lay down a gel-like polysaccharide matrix that then traps other cells, forming a biofilm (**Figure**

26.2 Bacterial Cell Shapes (A) These acid-producing cocci grow in the mammalian gut. (B) The bacillus *E. coli*, a resident of the human gut, is one of the most thoroughly studied organisms on Earth. (C) This helical bacterium belongs to a genus of human pathogens that cause leptospirosis, an infection spread by contaminated water. This particular strain was isolated in 1915 from the blood of a soldier serving in World War I.

(A) *Enterococcus* sp.

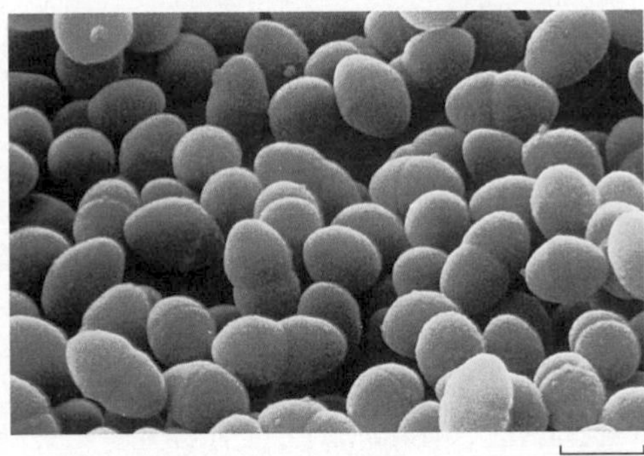

(B) *Escherichia coli*

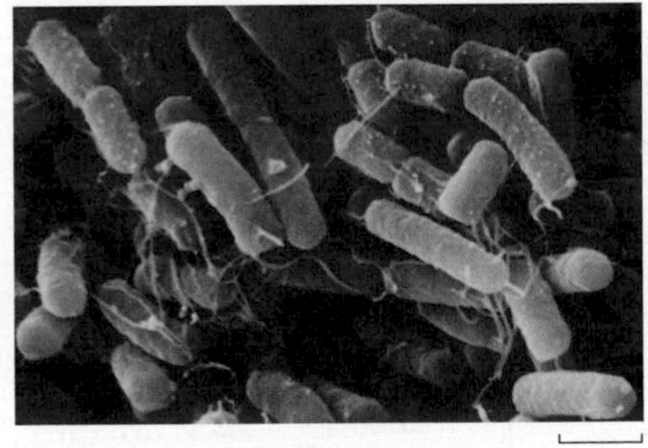

(C) *Leptospira interrogans*

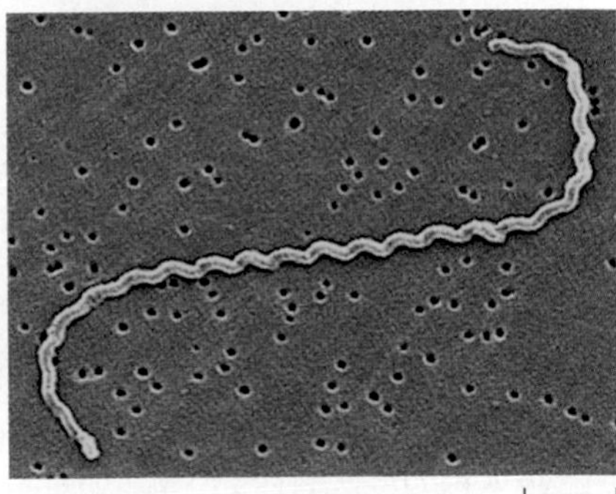

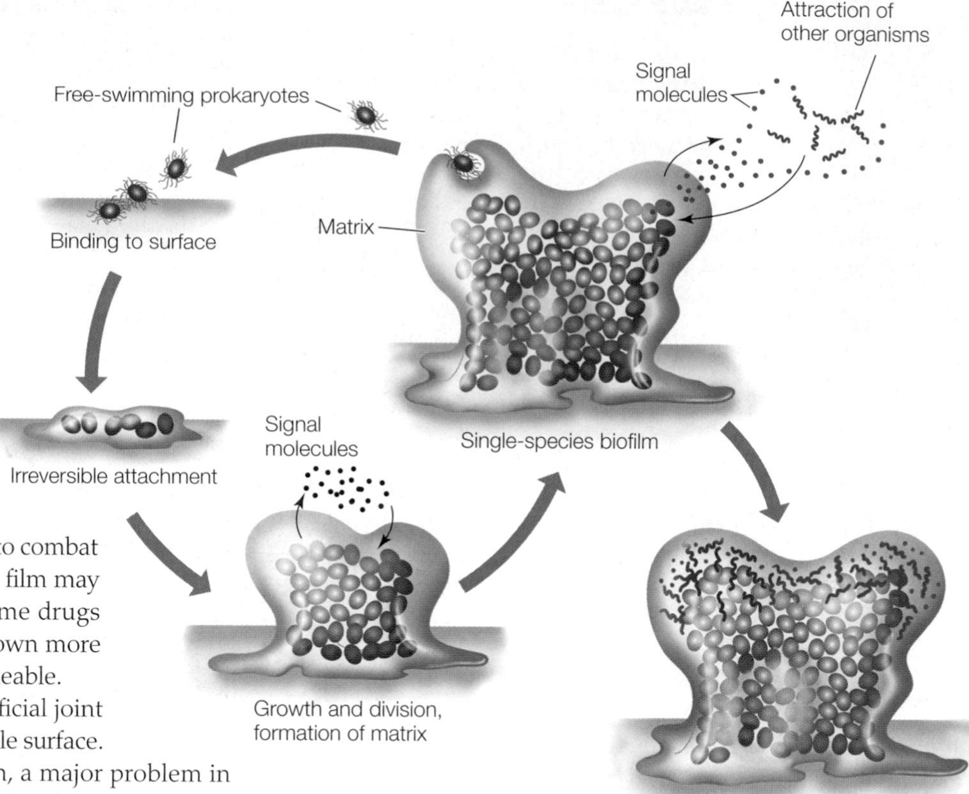

26.3 Forming a Biofilm Free-swimming microorganisms such as bacteria and archaea readily attach themselves to surfaces and form films stabilized and protected by a surrounding matrix. Once the population size is great enough, the developing biofilm can send chemical signals that attract other microorganisms.

26.3). Once a biofilm forms, it is difficult to kill the cells. Pathogenic (disease-causing) bacteria are hard for the immune system—and modern medicine—to combat once they form a biofilm. For example, the film may be impermeable to antibiotics. Worse, some drugs stimulate the bacteria in a biofilm to lay down more matrix, making the film even more impermeable.

Biofilms form on contact lenses, on artificial joint replacements, and on just about any available surface. They foul metal pipes and cause corrosion, a major problem in steam-driven electricity generation plants. Fossil stromatolites—large, rocky structures with alternating layers of fossilized microbial biofilm and calcium carbonate—are the oldest remnants of early life on Earth (see Figure 21.4). Stromatolites still form today in some parts of the world.

Biofilms are the object of much current research. For example, some biologists are studying the chemical signals used by bacteria in biofilms to communicate with one another. By blocking the signals that lead to the production of the matrix polysaccharides, they may be able to prevent biofilms from forming.

(A)

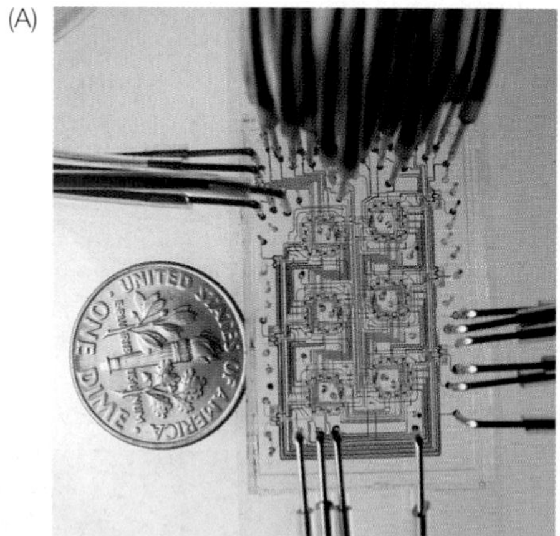

(B)

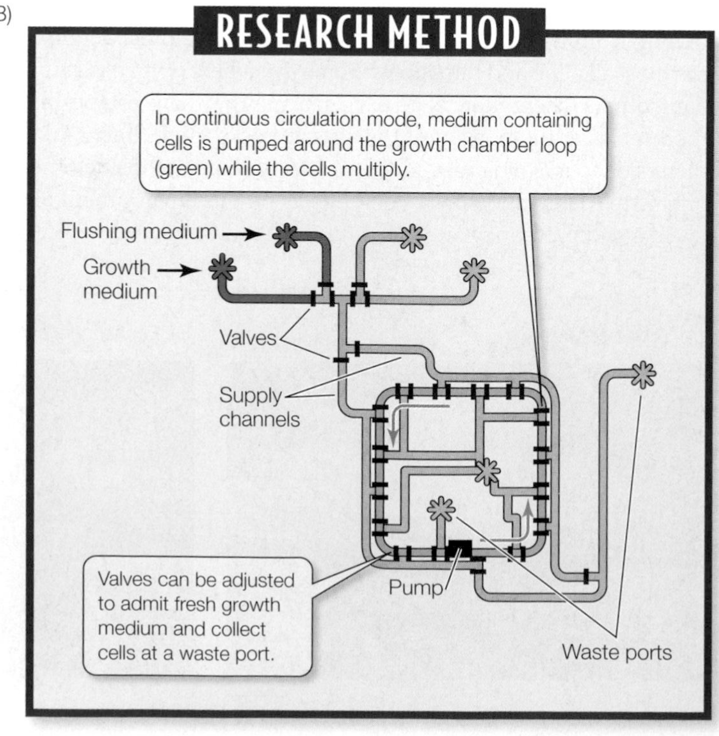

RESEARCH METHOD
In continuous circulation mode, medium containing cells is pumped around the growth chamber loop (green) while the cells multiply.
Flushing medium
Growth medium
Valves
Supply channels
Valves can be adjusted to admit fresh growth medium and collect cells at a waste port.
Pump
Waste ports

26.4 Microchemostats Allow Us to Study Microbial Dynamics (A) Six microchemostats on a single chip. Biologists and engineers monitor microbial population dynamics in six different chambers simultaneously by the application of microfluidic techniques. (B) A microchemostat is equipped with input ports for growth medium and flushing medium, and output ports for cells and waste. Tiny valves, controlled by a computer, direct flow. Two modes are shown here. A growth medium containing a bacterial population circulates in a loop with a total volume of 16 nanoliters.

That stain on your teeth is a biofilm called dental plaque, a coating of bacteria and hard matrix on and between the teeth. A softer bacterial biofilm forms on your tongue if you don't brush it regularly. Both biofilms can cause bad breath.

A team of bioengineers and chemical engineers recently devised a sophisticated technique that enables them to monitor biofilm development in extremely small populations of bacteria, cell by cell. They developed a tiny chip housing six separate growth chambers, or "microchemostats" (**Figure 26.4A**). The techniques of *microfluidics* use microscopic tubes and computer-controlled valves to direct fluid flow through complex "plumbing circuits" in the growth chambers (**Figure 26.4B**).

26.2 RECAP

Prokaryotes have established themselves everywhere on Earth. They often live in communities called biofilms.

- Can you describe the three most common forms of bacterial cells? See p. 563 and Figure 26.2
- Do you understand how biofilms form and why they are of special interest to researchers? See p. 563 and Figure 26.3

The prokaryotes are not only the most numerous living organisms, but also the most widely dispersed. To what do they owe their spectacular success?

26.3 What Are Some Keys to the Success of Prokaryotes?

Features of the prokaryotes that have contributed to their success include unique cell surfaces and modes of locomotion, communication, nutrition, and reproduction. These characteristics vary from group to group and even within groups of prokaryotes.

Prokaryotes have distinctive cell walls

Many prokaryotes have a thick and relatively stiff cell wall. This cell wall is quite different from those of land plants and algae, which contain cellulose and other polysaccharides, and from those of fungi, which contain chitin. Almost all bacteria have cell walls containing **peptidoglycan** (a polymer of amino sugars). Archaeal cell walls are of differing types, but most contain significant amounts of protein. One group of archaea has *pseudopeptidoglycan* in its cell wall; as you have probably already guessed from the prefix *pseudo*, pseudopeptidoglycan is similar to, but distinct from, the peptidoglycan of bacteria. The monomers making up pseudopeptidoglycan differ from and are differently linked than those of peptidoglycan. Peptidoglycan is a substance unique to bacteria; its absence from the walls of archaea is a key difference between the two prokaryotic domains.

To appreciate the complexity of some bacterial cell walls, consider the reactions of bacteria to a simple staining process. The **Gram stain** separates most types of bacteria into two distinct groups, Gram-positive and Gram-negative. A smear of cells on a microscope slide is soaked in a violet dye and treated with iodine; it is then washed with alcohol and counterstained with safranine (a red dye). **Gram-positive** bacteria retain the violet dye and appear blue to purple (**Figure 26.5A**). The alcohol washes the vi-

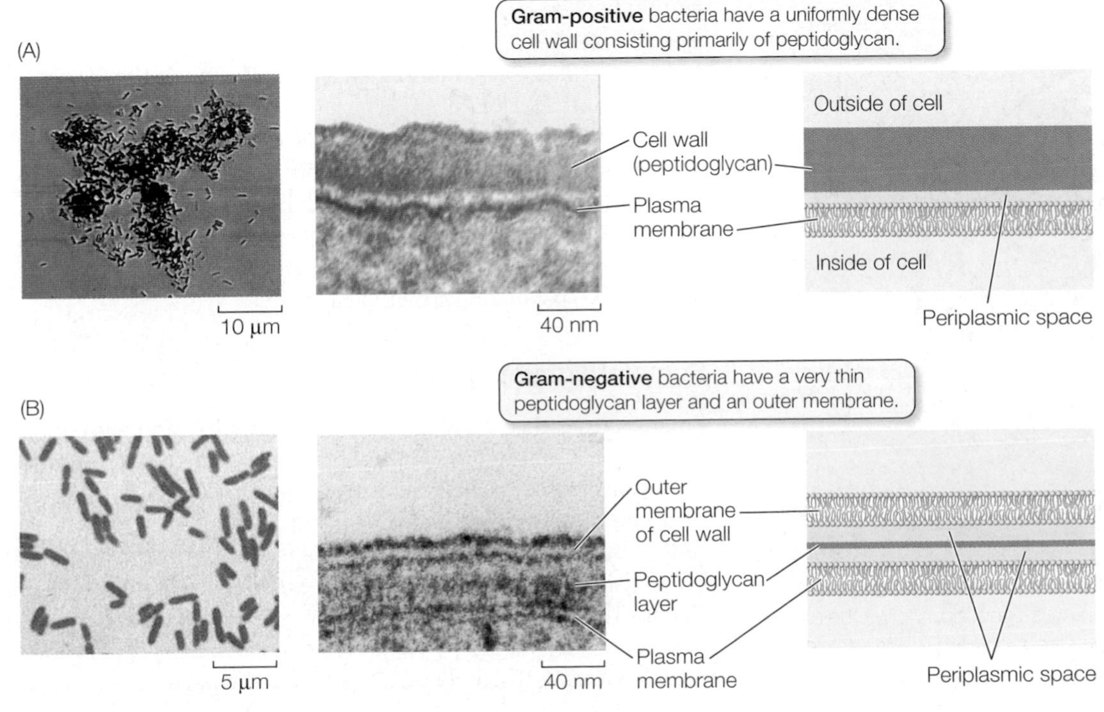

26.5 The Gram Stain and the Bacterial Cell Wall When treated with Gram stain, the cell wall components of different bacteria react in one of two ways. (A) Gram-positive bacteria have a thick peptidoglycan cell wall that retains the violet dye and appears deep blue or purple. (B) Gram-negative bacteria have a thin peptidoglycan layer that does not retain the violet dye, but picks up the counterstain and appears pink to red.

olet stain out of **Gram-negative** cells; these cells then pick up the safranine counterstain and appear pink to red (**Figure 26.5B**).

For many bacteria, the Gram-staining results correlate roughly with the structure of the cell wall. A Gram-negative cell wall usually has a thin peptidoglycan layer, and outside the peptidoglycan layer the cell is surrounded by a second, outer membrane quite distinct in chemical makeup from the plasma membrane (see Figure 26.5B). Between the inner (plasma) and outer membranes of Gram-negative bacteria is a *periplasmic space*. This space contains proteins that are important in digesting some materials, transporting others, and detecting chemical gradients in the environment.

A Gram-positive cell wall usually has about five times as much peptidoglycan as a Gram-negative wall. This thick peptidoglycan layer is a meshwork that may serve some of the same purposes as the periplasmic space of the Gram-negative cell wall.

The consequences of the different features of prokaryotic cell walls are numerous and relate to the disease-causing characteristics of some bacteria. Indeed, the cell wall is a favorite target in medical combat against pathogenic bacteria because it has no counterpart in eukaryotic cells. Antibiotics such as penicillin and ampicillin, as well as other agents that specifically interfere with the synthesis of peptidoglycan-containing cell walls, tend to have little, if any, effect on the cells of humans and other eukaryotes.

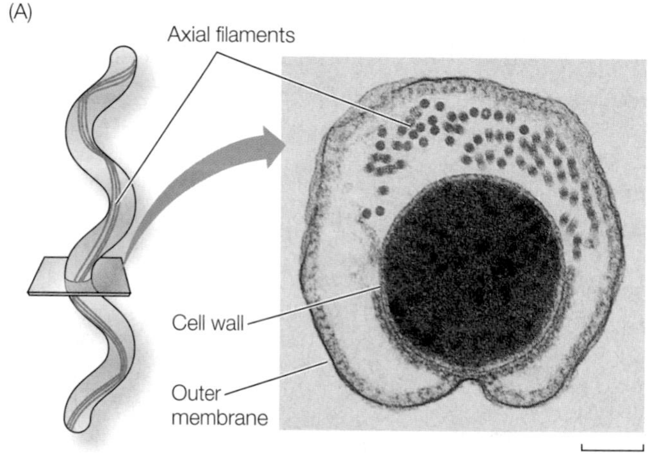

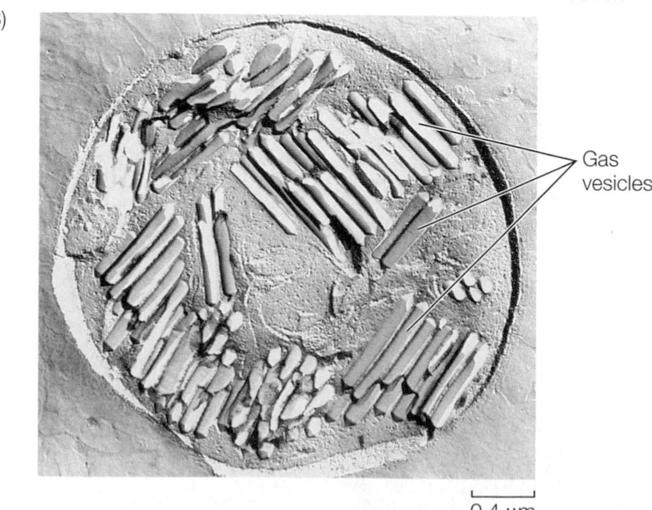

26.6 Structures Associated with Prokaryote Motility (A) A spirochete from the gut of a termite, seen in cross section, shows the axial filaments used to produce a corkscrew-like motion. (B) Gas vesicles in a cyanobacterium, visualized by the freeze-fracture technique.

Prokaryotes have distinctive modes of locomotion

Although many prokaryotes cannot move, others are *motile*. These organisms move by one of several means. Some helical bacteria, called *spirochetes*, use a corkscrew-like motion made possible by modified flagella, called *axial filaments*, running along the axis of the cell beneath the outer membrane (**Figure 26.6A**). Many cyanobacteria and a few other groups of bacteria use various poorly understood gliding mechanisms, including rolling. Various aquatic prokaryotes, including some cyanobacteria, can move slowly up and down in the water by adjusting the amount of gas in *gas vesicles* (**Figure 26.6B**). By far the most common type of locomotion in prokaryotes, however, is that driven by flagella.

Prokaryotic **flagella** are slender filaments that extend singly or in tufts from one or both ends of the cell or are randomly distributed all around it (**Figure 26.7**). A prokaryotic flagellum consists of a single fibril made of the protein *flagellin*, projecting from the cell surface, plus a hook and basal body responsible for motion (see Figure 4.5). In contrast, the flagellum of eukaryotes is enclosed by the plasma membrane and usually contains a circle of nine pairs of microtubules surrounding two central microtubules, all containing the protein tubulin, along with many other associated proteins. The prokaryotic flagellum rotates about its base, rather than beating as a eukaryotic flagellum or cilium does.

Prokaryotes reproduce asexually, but genetic recombination can occur

Prokaryotes reproduce by **binary fission**, an asexual process. Recall, however, that there are also processes—transformation, conjugation, and transduction—that allow the exchange of genetic information between some prokaryotes quite apart from reproduction (see Chapter 13).

Some prokaryotes multiply very rapidly. One of the fastest is the bacterium *Escherichia coli*, which under optimal conditions has a generation time of about 20 minutes. The shortest known prokaryote generation times are about 10 minutes. Generation times of 1 to 3 hours are common for others; some extend to days. Bacteria living deep in Earth's crust may suspend their growth for more than a century without dividing and then multiply for a few days before suspending growth again.

Some prokaryotes communicate

Prokaryotes can communicate with one another and with other organisms. One communication channel that they employ is chemical. Another is physical, with light as the medium.

Some bacteria release chemical signals—substances that are sensed by other bacteria of the same species. They can announce their availability for conjugation, for example, by means of such signals. They can also monitor the size of their population. As the number of bacteria in a particular region increases, the concentration of a chemical signal builds up. When the bacteria sense that

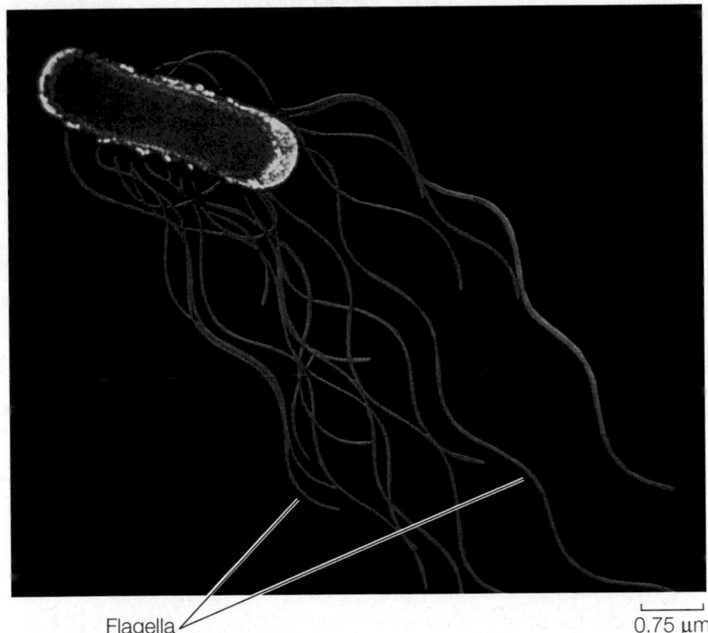

26.7 Some Prokaryotes Use Flagella for Locomotion Multiple flagella propel this *Salmonella* bacillus.

their population has become sufficiently large, they can commence activities that smaller numbers could not manage, such as forming a biofilm (see Figure 26.3). This "counting" technique is called **quorum sensing**.

Like many other organisms, such as fireflies, some bacteria can emit light by a process called **bioluminescence**. A complex, enzyme-catalyzed reaction requiring ATP causes the emission of light, but not heat. Often such bacteria luminesce only when a quorum has been sensed.

How could bioluminescence be useful to a prokaryote? One fairly well understood case is that of some bacteria of the genus *Vibrio*. These bacteria can live freely, but thrive when inside the guts of fish. Inside the fish, they may attach to food particles and be expelled as waste along with the particulate matter. Reproducing on the particles, their population increases until the glowing particle attracts another fish, which ingests it along with the bacteria—giving them a new home and food source for a while. In this case, *Vibrio* is communicating with another species and enhancing its own nutritional status. *Vibrio* in the Indian Ocean, off the eastern coast of Africa, provide the only instance in which bioluminescence is visible from space. These bacteria become concentrated over an area of several thousand square kilometers (**Figure 26.8**).

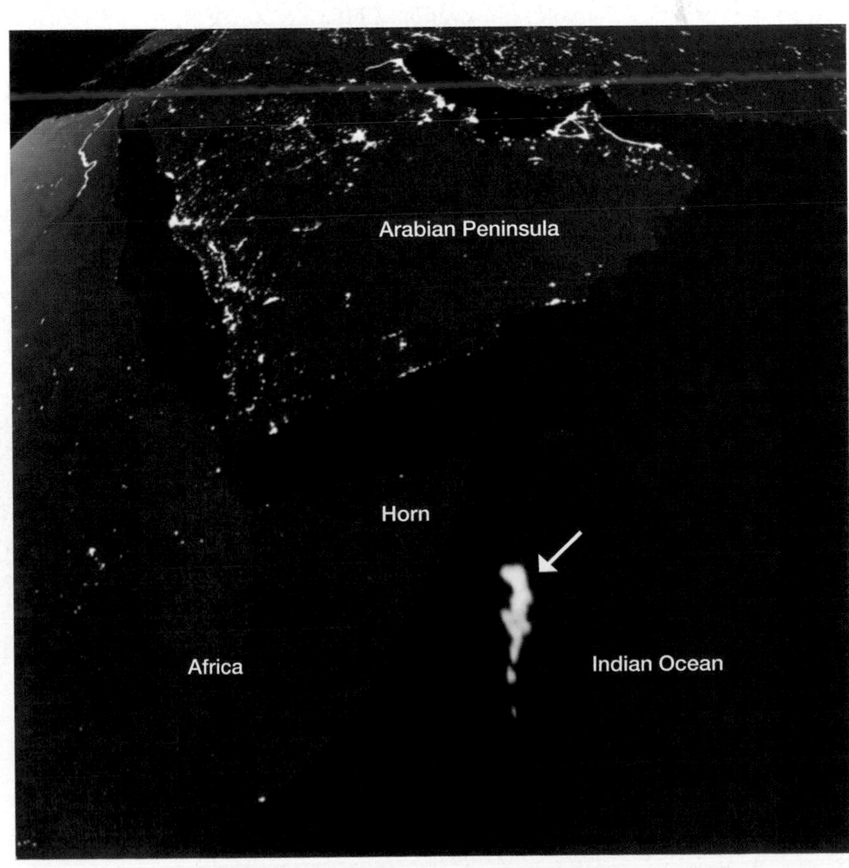

26.8 Bioluminescent Bacteria Seen from Space In this satellite photo, legions of *Vibrio harveyi* form a glowing patch (arrow) some thousands of square kilometers in area in the Indian Ocean, off the Horn of Africa. Compare their blue glow with the white light of cities in eastern Africa and the Middle East.

Some fish have "headlights"—colonies of bioluminescent bacteria that glow together. Other fish are attracted to the light and then eaten by the host fish. This arrangement, profitable to the host, also probably provides the bacteria with nutrients released as the fish eats.

Some soil-dwelling bacteria bioluminesce, producing eerily glowing patches of ground at night. What is the adaptive value of this particular case of bacterial bioluminescence? We're not certain, but one hypothesis links the phenomenon to the time in evolutionary history when most organisms went extinct because their metabolisms could not deal with increasing levels of oxygen in the atmosphere. The distant ancestor of bioluminescent soil bacteria may have avoided oxidative reactions (and subsequent death) by dissipating excess energy, emitting it as light.

Prokaryotes have amazingly diverse metabolic pathways

The members of the two prokaryotic domains outdo all other groups in metabolic diversity. Eukaryotes, while much more diverse in size and shape, draw on fewer metabolic mechanisms for their energy needs. In fact, much of the energy metabolism of eukaryotes is carried out in organelles—mitochondria and chloroplasts—that are descended from bacteria, as we saw in Section 4.5.

The long evolutionary history of the bacteria and archaea, during which they have had time to explore a wide variety of habitats, has led to the extraordinary diversity of their metabolic "lifestyles"—their use or nonuse of oxygen, their energy sources, their sources of carbon atoms, and the materials they release as waste products.

ANAEROBIC VERSUS AEROBIC METABOLISM Some prokaryotes can live only by anaerobic metabolism because molecular oxygen is poisonous to them. These oxygen-sensitive organisms are called **obligate anaerobes**.

Other prokaryotes can shift their metabolism between anaerobic and aerobic modes (see Chapter 7) and thus are called **facultative anaerobes**. Many facultative anaerobes alternate between anaerobic metabolism (such as fermentation) and cellular respiration as conditions dictate. **Aerotolerant anaerobes** cannot conduct cellular respiration, but are not damaged by oxygen when it is present. By definition, an anaerobe does not use oxygen as an electron acceptor for its respiration.

At the other extreme from the obligate anaerobes, some prokaryotes are **obligate aerobes**, unable to survive for extended periods in the *absence* of oxygen. They require oxygen for cellular respiration.

NUTRITIONAL CATEGORIES Biologists recognize four broad nutritional categories of organisms: photoautotrophs, photoheterotrophs, chemolithotrophs, and chemoheterotrophs. Prokaryotes are represented in all four groups (**Table 26.2**).

TABLE 26.2

How Organisms Obtain Their Energy and Carbon

NUTRITIONAL CATEGORY	ENERGY SOURCE	CARBON SOURCE
Photoautotrophs (found in all three domains)	Light	Carbon dioxide
Photoheterotrophs (some bacteria)	Light	Organic compounds
Chemolithotrophs (some bacteria, many archaea)	Inorganic substances	Carbon dioxide
Chemoheterotrophs (found in all three domains)	Organic compounds	Organic compounds

Photoautotrophs perform photosynthesis. They use light as their source of energy and carbon dioxide as their source of carbon. Like the photosynthetic eukaryotes, the cyanobacteria, a group of photoautotrophic bacteria, use chlorophyll *a* as their key photosynthetic pigment and produce oxygen as a by-product of noncyclic electron transport (see Section 8.1).

In contrast, the other photosynthetic bacteria use *bacteriochlorophyll* as their key photosynthetic pigment, and they do not release oxygen gas. Some of these photosynthesizers produce particles of pure sulfur instead because hydrogen sulfide (H_2S), rather than H_2O, is their electron donor for photophosphorylation. Bacteriochlorophyll absorbs light of longer wavelengths than the chlorophyll used by all other photosynthesizing organisms. As a result, bacteria using this pigment can grow in water beneath fairly dense layers of algae, using light of wavelengths that are not absorbed by the algae (**Figure 26.9**).

Photoheterotrophs use light as their source of energy, but must obtain their carbon atoms from organic compounds made by other organisms. They use compounds such as carbohydrates, fatty acids, and alcohols as their organic "food." For example, compounds released from plant roots (as in rice paddies) or from decomposing cyanobacteria in hot springs are taken up by photoheterotrophs and metabolized to form building blocks for other compounds; sunlight provides the necessary ATP through photophosphorylation (see Section 8.2). The purple nonsulfur bacteria, among others, are photoheterotrophs.

Chemolithotrophs (also called chemoautotrophs) obtain their energy by oxidizing inorganic substances, and they use some of that energy to fix carbon dioxide. Some chemolithotrophs use reactions identical to those of the typical photosynthetic cycle (see Chapter 8), but others use other pathways to fix carbon dioxide. Some bacteria oxidize ammonia or nitrite ions to form nitrate ions. Others oxidize hydrogen gas, hydrogen sulfide, sulfur, and other materials. Many archaea are chemolithotrophs.

Deep-sea hydrothermal vent ecosystems are based on chemolithotrophic prokaryotes that are incorporated into large communities of crabs, mollusks, and giant worms, all living at a depth of 2,500 meters, below any hint of light from the sun. These bacteria obtain energy by oxidizing hydrogen sulfide and other substances released in the near-boiling water that flows from volcanic vents in the ocean floor.

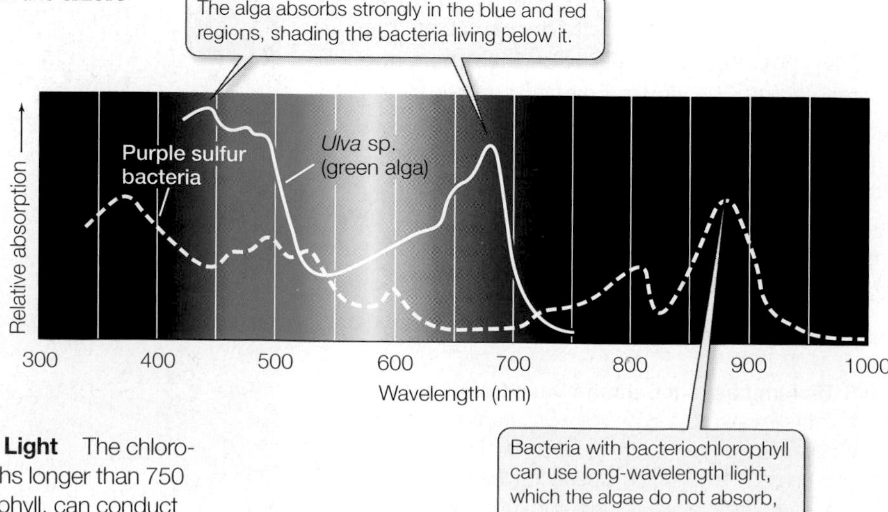
The alga absorbs strongly in the blue and red regions, shading the bacteria living below it.
Purple sulfur bacteria
Ulva sp. (green alga)
Relative absorption
300
400
500
600
700
800
900
1000
Wavelength (nm)
Bacteria with bacteriochlorophyll can use long-wavelength light, which the algae do not absorb, for their photosynthesis.

26.9 Bacteriochlorophyll Absorbs Long-Wavelength Light The chlorophyll in *Ulva*, a green alga, absorbs no light of wavelengths longer than 750 nm. Purple sulfur bacteria, which contain bacteriochlorophyll, can conduct photosynthesis using longer wavelengths.

Finally, **chemoheterotrophs** obtain both energy and carbon atoms from one or more complex organic compounds. Most known bacteria and archaea are chemoheterotrophs—as are all animals and fungi and many protists.

NITROGEN AND SULFUR METABOLISM Key metabolic reactions in many prokaryotes involve nitrogen or sulfur. For example, some bacteria carry out respiratory electron transport without using oxygen as an electron acceptor. These organisms use oxidized inorganic ions such as nitrate, nitrite, or sulfate as electron acceptors. Examples include the **denitrifiers**, bacteria that release nitrogen to the atmosphere as nitrogen gas (N_2). These normally aerobic bacteria, mostly species of the genera *Bacillus* and *Pseudomonas*, use nitrate (NO_3^-) as an electron acceptor in place of oxygen if they are kept under anaerobic conditions:

$$2\,NO_3^- + 10\,e^- + 12\,H^+ \rightarrow N_2 + 6\,H_2O$$

Nitrogen fixers convert atmospheric nitrogen gas into a chemical form usable by the nitrogen fixers themselves as well as by other organisms. They convert nitrogen gas into ammonia:

$$N_2 + 6\,H \rightarrow 2\,NH_3$$

All organisms require nitrogen for their proteins, nucleic acids, and other important compounds. Nitrogen fixation is thus vital to life as we know it; this all-important biochemical process is carried out by a wide variety of archaea and bacteria (including cyanobacteria), but by no other organisms. We'll discuss this vital process in detail in Chapter 36.

Ammonia is oxidized to nitrate in soil and in seawater by chemolithotrophic bacteria called **nitrifiers**. Bacteria of two genera, *Nitrosomonas* and *Nitrosococcus*, convert ammonia to nitrite ions (NO_2^-), and *Nitrobacter* oxidizes nitrite to nitrate (NO_3^-).

What do the nitrifiers get out of these reactions? Their metabolism is powered by the energy released by the oxidation of ammonia or nitrite. For example, by passing the electrons from nitrite through an electron transport chain, *Nitrobacter* can make ATP, and using some of this ATP, it can also make NADH. With this ATP and NADH, the bacterium can convert CO_2 and H_2O to glucose.

26.3 RECAP

Prokaryotes have distinctive cell walls and modes of locomotion, communication, reproduction, and nutrition.

- Can you describe bacterial cell wall architecture? See p. 565 and Figure 26.5
- How are the four nutritional categories of prokaryotes distinguished? See p. 568 and Table 26.2
- Can you see why nitrogen metabolism in the prokaryotes is vital to other organisms? See p. 569

We noted earlier that only very recently have scientists appreciated the huge distinctions between Bacteria and Archaea. How do researchers approach the classification of organisms they can't even see?

26.4 How Can We Determine Prokaryote Phylogeny?

As detailed in Chapter 25, there are three primary motivations for classification schemes: to identify unknown organisms, to reveal evolutionary relationships, and to provide universal names. Classifying bacteria and archaea is of particular importance to humans, because scientists and medical technologists must be able to identify bacteria quickly and accurately—when the bacteria are pathogenic, lives may depend on it. In addition, many emerging biotechnologies (see Chapter 16) depend on a thorough knowledge of prokaryote biochemistry, and understanding an organism's phylogeny can contribute to such knowledge.

Size complicates the study of prokaryote phylogeny

Until recently, taxonomists based their classification schemes for the prokaryotes on readily observable phenotypic characters such as color, motility, nutritional requirements, antibiotic sensitivity, and reaction to the Gram stain. Although such schemes have facilitated the identification of prokaryotes, they have not provided insights into how these organisms evolved—a question of great interest to microbiologists and to all students of evolution. The prokaryotes and the microbial eukaryotes (protists; Chapter 27) have presented major challenges to those who attempted phylogenetic classifications.

Nobody had *seen* an individual prokaryote until about 300 years ago. Prokaryotes are so small that they were invisible before the invention of the first simple microscope. Even under the best light microscopes, we still don't learn much about them.

When biologists learned how to grow bacteria in pure culture on nutrient media, some useful information began to flow in. A great deal was learned about the genetics, nutrition, and metabolism of prokaryotes. With little but cell shape, size, and nutritional requirements for criteria, however, microbiologists were unable to deduce classification schemes that made sense in evolutionary terms. Only recently have systematists had the right tools for tackling this task.

The nucleotide sequences of prokaryotes reveal their evolutionary relationships

Analyses of the nucleotide sequences of ribosomal RNA have provided us with the first apparently reliable measures of evolutionary distance among taxonomic groups. Ribosomal RNA (rRNA) is particularly useful for evolutionary studies of living organisms for several reasons:

- rRNA is evolutionarily ancient.
- No living organism lacks rRNA.
- rRNA plays the same role in translation in all organisms.
- rRNA has evolved slowly enough that sequence similarities between groups of organisms are easily found.

Comparisons of short stretches of rRNA from a great many organisms have revealed recognizable base sequences that are characteristic of particular taxonomic groups. More recent investigations have focused on longer stretches of DNA, even entire genes (especially rRNA genes).

These data are very helpful, but things aren't as simple as we might wish. When biologists examined more genes, contradictions began to appear and new questions arose. In some groups of prokaryotes, analyses of different nucleotide sequences suggested different phylogenetic patterns. How could such a situation have arisen?

Lateral gene transfer may complicate phylogenetic studies

It is now clear that, from early in evolution to the present day, genes have been moving among many prokaryotic species by **lateral gene transfer**. As we have seen, a gene from one species can become incorporated into the genome of another. For example, the 1,869 genes of *Thermotoga maritima,* a bacterium that can survive extremely high temperatures, have all been sequenced. In comparing the sequences of *T. maritima* genes against sequences of genes for the same proteins in other species, it was found that nearly 20 percent of this bacterium's genes have their closest match not with other bacterial species, but with archaeal species.

Mechanisms of lateral gene transfer include transfer by plasmids and viruses and uptake of DNA by transformation. Such transfers are well documented, not just between species in the prokaryotic domains, but between prokaryotes and eukaryotes.

A gene that has been transferred will be inherited by the recipient's progeny and in time will be recognized as part of the normal genome of its descendants. When phylogenies are inferred using gene trees, the presence of a laterally transferred gene can result in mistaken assumptions about relationships (**Figure 26.10A,B**). Biologists are still assessing the extent of lateral gene transfer among prokaryotes and its implications for phylogeny, especially at the early stages of evolution.

It is unclear whether lateral gene transfer has seriously complicated our attempts to resolve the tree of prokaryotic life. Recent work suggests that it has not—while it complicates studies within some individual species, it need not present problems at higher levels. It is now possible to make nucleotide sequence comparisons involving entire genomes, and many scientists feel this work will reveal a *stable core* of crucial genes that are uncomplicated by lateral gene transfer. Gene trees using this stable core should reveal more accurate phylogenetic relationships (**Figure 26.10C**). The problem remains that only a very small proportion of the prokaryotic world has been described and studied.

The great majority of prokaryote species have never been studied

Some prokaryotes have defied all attempts to grow them in pure culture, causing biologists to wonder how many species, and possibly even important clades, we might be missing. A window onto this problem was opened with the introduction of a new way to look at nucleic acid sequences. Unable to work with the whole genome of a single species, biologists instead examine sequences in individual genes collected from a random sample of the environment.

Norman Pace, of the University of Colorado, isolated individual rRNA gene sequences from extracts of environmental samples such as soil or seawater. Comparison of such sequences with previously known ones revealed an extraordinary number of new sequences, implying that they came from previously unrecognized species. Biologists have described only about 5,000 species of bacteria and archaea. The results of Pace's studies strongly suggest that there may be as many as half a million prokaryote species out there. This finding presents a great challenge, but also a great opportu-

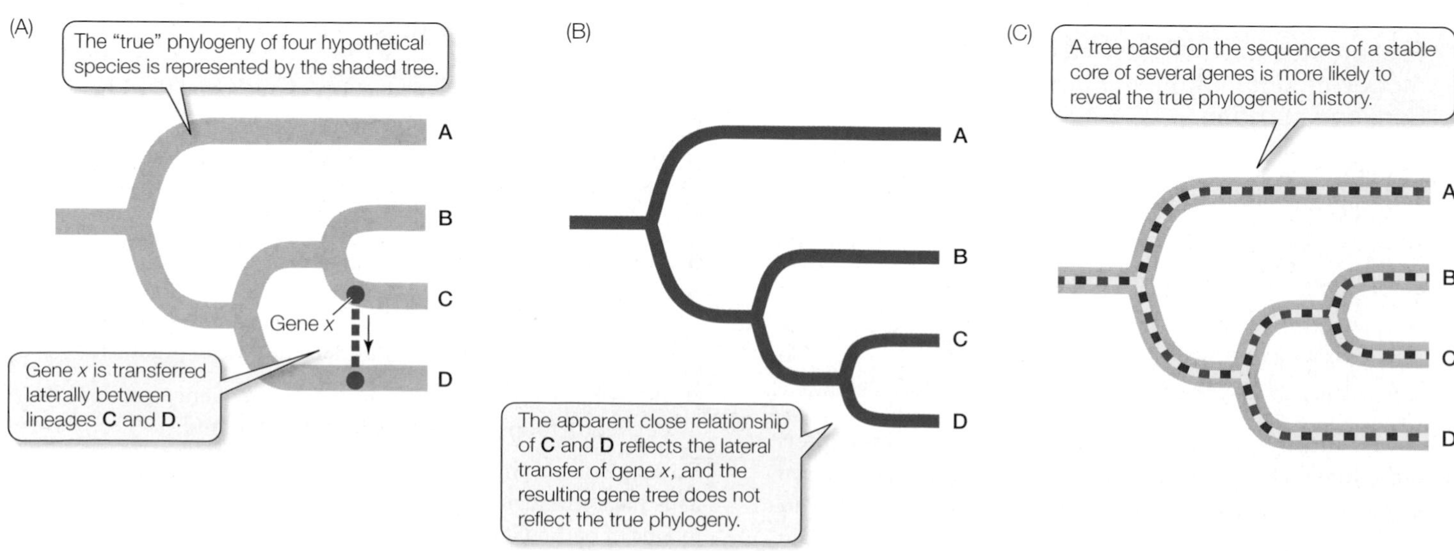
(A)
The "true" phylogeny of four hypothetical species is represented by the shaded tree.
A
B
C
D
Gene x
Gene x is transferred laterally between lineages C and D.
(B)
A
B
C
D
The apparent close relationship of C and D reflects the lateral transfer of gene x, and the resulting gene tree does not reflect the true phylogeny.
(C)
A tree based on the sequences of a stable core of several genes is more likely to reveal the true phylogenetic history.
A
B
C
D

26.10 Lateral Gene Transfer Complicates Phylogenetic Relationships (A) The phylogeny of four hypothetical species is shown as a shaded tree. (B) In a tree based on gene *x*, the true relationship is obscured. (C) Many systematists studying prokaryote phylogeny believe that eventually we will establish a "stable core" of prokaryote genes that is resistant to lateral transfer. Such a tree, however, remains in the future, awaiting the sequencing of many more prokaryotic genomes.

nity to explore prokaryotic diversity and better understand the phylogenetic tree of life.

Mutations are a major source of prokaryotic variation

Assuming that the prokaryote groups we are about to describe do indeed represent clades, these groups are amazingly complex. A single clade of bacteria or archaea may contain extraordinarily diverse species; on the other hand, a species in one group may be phenotypically almost indistinguishable from one or many species in another group. What are the sources of these phylogenetic patterns?

Although prokaryotes can acquire new alleles by transformation, transduction, or conjugation, the most important sources of genetic variation in populations of prokaryotes are probably mutation and genetic drift (described in Section 22.2). Mutations, especially recessive mutations, are slow to make their presence felt in populations of humans and other diploid organisms. In contrast, a mutation in a prokaryote, which is haploid, has immediate consequences for that organism. If it is not lethal, it will be transmitted to and expressed in the organism's daughter cells—and in their daughter cells, and so on. Thus a beneficial mutant allele spreads rapidly.

The rapid multiplication of many prokaryotes, coupled with mutation, natural selection, genetic drift, and lateral gene transfer, allows rapid phenotypic changes within their populations. Important changes, such as loss of sensitivity to an antibiotic, can occur over broad geographic areas in just a few years. Think how many significant metabolic changes could have occurred over even modest time spans, let alone over the entire history of life on Earth. When we introduce the proteobacteria, the largest group of bacteria, you will see that its different subgroups have easily and rapidly adopted and abandoned metabolic pathways under selective pressure from their environments.

26.4 RECAP

The study of prokaryote phylogeny is complicated by the organisms' small size, our inability to grow some of them in pure culture, and lateral gene transfer. Nucleotide sequences are especially helpful in distinguishing prokaryote clades.

- How did biologists classify bacteria before it became possible to determine nucleotide sequences? See p. 569
- Can you explain why rRNA is useful for evolutionary studies? See p. 569
- Do you see how lateral gene transfer could complicate evolutionary studies? See p. 570 and Figure 26.10

In spite of the difficulties described here, biologists have identified many clades of prokaryotes. We identify the characteristics of some of them in the next section.

26.5 What Are the Major Known Groups of Prokaryotes?

The Bacteria are the better-studied of the two prokaryote domains, and here we use a widely accepted classification scheme that enjoys considerable support from nucleotide sequence data. More than a dozen clades have been proposed under this scheme; we will describe just of a few of them here. We pay the closest attention to six groups: the spirochetes, chlamydias, high-GC Gram-positives, cyanobacteria, low-GC Gram-positives, and proteobacteria (**Figure 26.11**). First, however, we'll mention one property that is shared by members of three other groups.

Three of the bacterial groups once thought to have branched out earliest during bacterial evolution are all **thermophiles** (heat-lovers), as are the most ancient of the archaea. This observation supported the hypothesis that the first living organisms were thermophiles that appeared in much hotter environments than are predominant today. Recent nucleic acid-based evidence suggests, however, that those clades may have arisen more recently than did the spirochetes and chlamydias.

Spirochetes move by means of axial filaments

Spirochetes are Gram-negative, motile, chemoheterotrophic bacteria characterized by unique structures called axial filaments, which are modified flagella running through the periplasmic space (see Figure 26.6A). The cell body is a long cylinder coiled into a he-

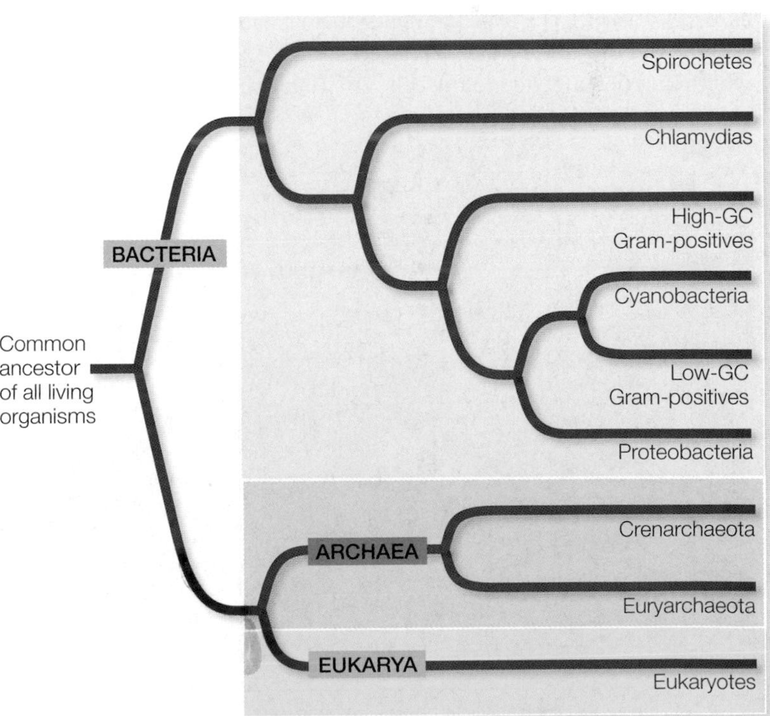

26.11 Two Domains: A Brief Overview This abridged summary classification of the domains Bacteria and Archaea shows their relationships to each other and to the Eukarya. The relationships among the many clades of bacteria, not all of which are listed here, are incompletely resolved at this time.

26.12 A Spirochete This corkscrew-shaped bacterium causes syphilis in humans.

lix (**Figure 26.12**). The axial filaments begin at either end of the cell and overlap in the middle, and there are typical protein motors where they are attached to the cell wall. These structures rotate, as they do in other prokaryotic flagella (see Figure 4.5) Many spirochetes live in humans as parasites; a few are pathogens, including those that cause syphilis and Lyme disease. Others live free in mud or water.

Chlamydias are extremely small parasites

Chlamydias are among the smallest bacteria (0.2–1.5 μm in diameter). They can live only as parasites within the cells of other organisms. It was once believed that this obligate parasitism resulted from an inability of chlamydias to produce ATP—that chlamydias were "energy parasites." Genomic sequencing results from the end of the twentieth century indicate, however, that chlamydias have the genetic capability to produce at least some ATP. They can augment this capacity by using an enzyme called a translocase, which allows them to take up ATP from the cytoplasm of their host in exchange for ADP from their own cells.

These tiny Gram-negative cocci are unique prokaryotes because of their complex life cycle, which involves two different forms of cells, *elementary bodies* and *reticulate bodies* (**Figure 26.13**). In humans, various strains of chlamydias cause eye infections (especially trachoma), sexually transmitted diseases, and some forms of pneumonia.

Some high-GC Gram-positives are valuable sources of antibiotics

High-GC Gram-positives, also known as *actinobacteria*, derive their name from the relatively high G+C/A+T ratio of their DNA. They develop an elaborately branched system of filaments (**Figure 26.14**). The shapes of these bacteria closely resemble the filamentous growth habit of fungi at a reduced scale. Some high-GC

1 **Elementary bodies** are taken into a eukaryotic cell by phagocytosis...

2 ...where they develop into thin-walled **reticulate bodies**, which grow and divide.

3 Reticulate bodies reorganize into elementary bodies, which are liberated by the rupture of the host cell.

Chlamydia psittaci

0.2 μm

26.13 Chlamydias Change Form during Their Life Cycle Elementary bodies and reticulate bodies are the two major phases of the chlamydia life cycle.

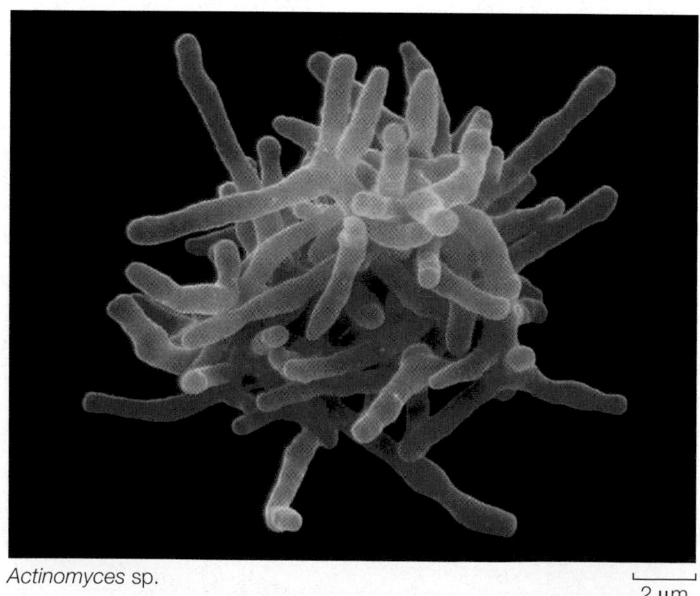

26.14 Filaments of a High-GC Gram-Positive The branching filaments seen in this scanning electron micrograph are typical of this medically important bacterial group.

Gram-positives reproduce by forming chains of spores at the tips of the filaments. In species that do not form spores, the branched, filamentous growth ceases, and the structure breaks up into typical cocci or bacilli, which then reproduce by binary fission.

The high-GC Gram-positives include several medically important bacteria. *Mycobacterium tuberculosis* causes tuberculosis, which kills 3 million people each year. Genetic data suggest that this bacterium arose 3 million years ago in East Africa, making it the most ancient known human bacterial affliction. *Streptomyces* produce streptomycin as well as hundreds of other antibiotics. We derive most of our antibiotics from members of the high-GC Gram-positives.

Cyanobacteria are important photoautotrophs

Cyanobacteria, sometimes called *blue-green bacteria* because of their pigmentation, are photoautotrophs that require only water, nitrogen gas, oxygen, a few mineral elements, light, and carbon dioxide to survive. They use chlorophyll *a* for photosynthesis and release oxygen gas; many species also fix nitrogen. Their photosynthesis was the basis of the "oxygen revolution" that transformed Earth's atmosphere (see Section 21.2).

Cyanobacteria carry out the same type of photosynthesis that is characteristic of eukaryotic photosynthesizers. They contain elaborate and highly organized internal membrane systems called *photosynthetic lamellae* or *thylakoids*. The chloroplasts of photosynthetic eukaryotes are derived from an endosymbiotic cyanobacterium.

Cyanobacteria may live free as single cells or associate in colonies. Depending on the species and on growth conditions, colonies of cyanobacteria may range from flat sheets one cell thick to filaments to spherical balls of cells.

Some filamentous colonies of cyanobacteria differentiate into three cell types: vegetative cells, spores, and heterocysts (**Figure 26.15**). **Vegetative cells** photosynthesize, **spores** are resting stages that can survive harsh environmental conditions and eventually develop into new filaments, and **heterocysts** are cells specialized for nitrogen fixation. All of the known cyanobacteria with heterocysts fix nitrogen. Heterocysts also have a role in reproduction: when filaments break apart to reproduce, the heterocyst may serve as a breaking point.

Not all low-GC Gram-positives are Gram-positive

The **low-GC Gram-positives**, as their name suggests, have a lower G+C/A+T ratio than do the high-GC Gram-positives. They are also sometimes called *firmicutes*. Some of the low-GC Gram-positives are in fact Gram-negative, and some have no cell wall at all. Nonetheless, this group is a clade.

Some low-GC Gram-positives produce **endospores** (**Figure 26.16**)—heat-resistant resting structures—when a key nutrient such as nitrogen or carbon becomes scarce. The bacterium replicates its DNA and encapsulates one copy, along with some of its cytoplasm, in a tough cell wall heavily thickened with peptidogly-

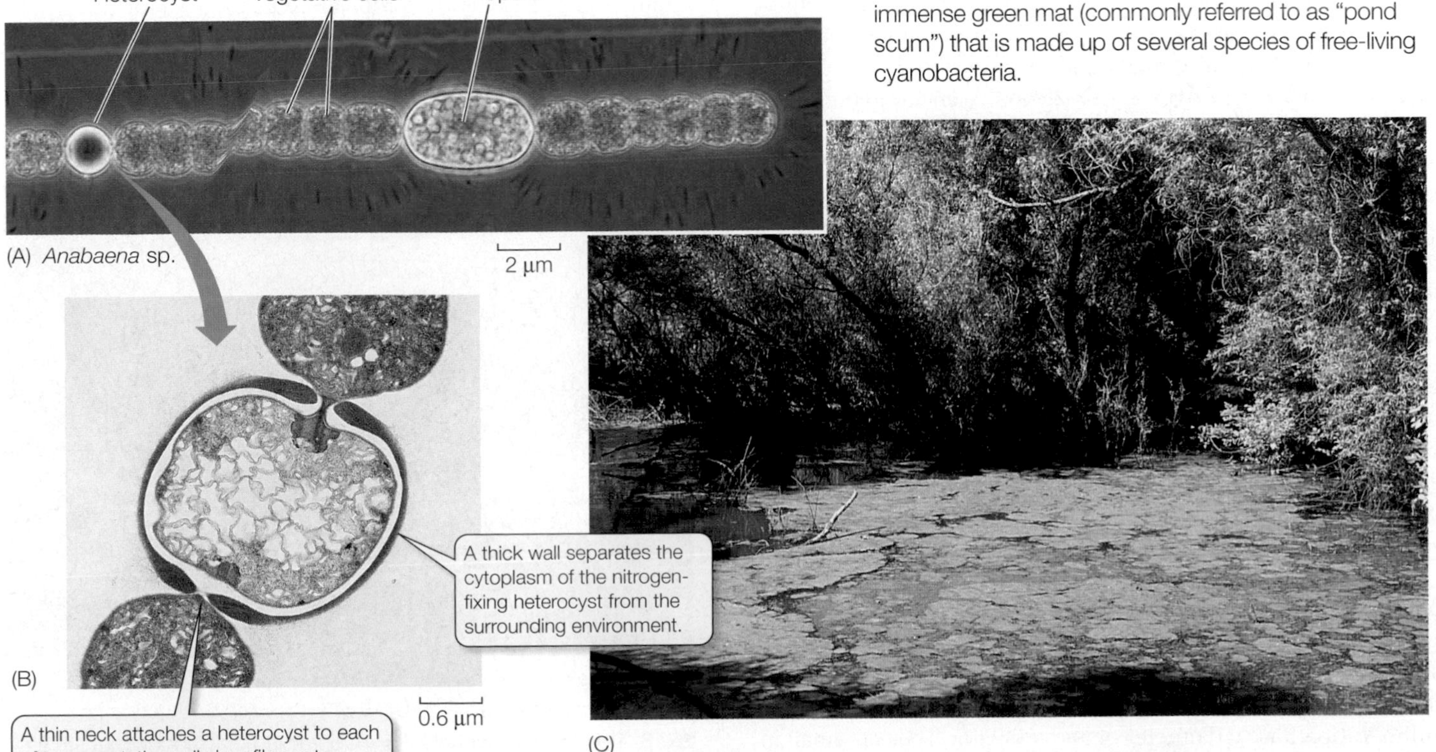
Heterocyst
Vegetative cells
Spore
(A) *Anabaena* sp.
2 μm
A thick wall separates the cytoplasm of the nitrogen-fixing heterocyst from the surrounding environment.
(B)
0.6 μm
A thin neck attaches a heterocyst to each of two vegetative cells in a filament.
(C)

26.15 Cyanobacteria (A) *Anabaena* is a genus of cyanobacteria that form filamentous colonies containing three cell types. (B) Heterocysts are specialized for nitrogen fixation and serve as a breaking point when filaments reproduce. (C) Cyanobacteria appear in enormous numbers in some environments. This California pond has experienced eutrophication: phosphorus and other nutrients generated by human activity have accumulated in the pond, feeding an immense green mat (commonly referred to as "pond scum") that is made up of several species of free-living cyanobacteria.

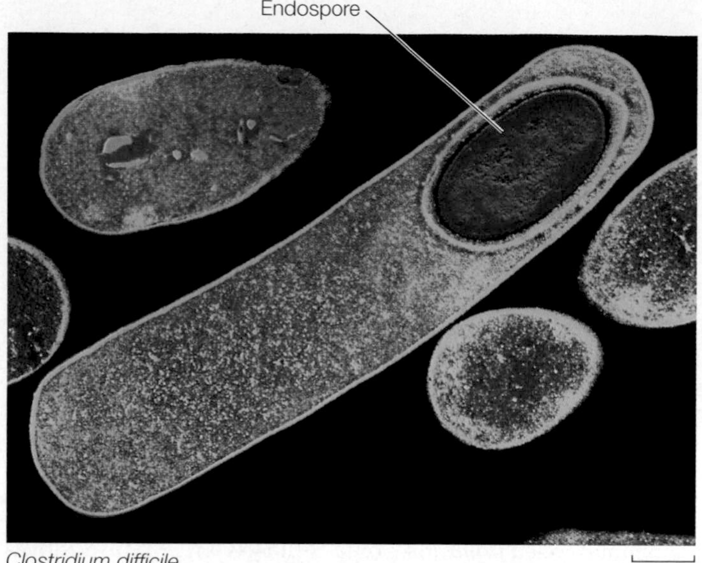

26.16 A Structure for Waiting Out Bad Times This low-GC Gram-positive bacterium, which can cause severe colitis in humans, produces endospores as resistant resting structures.

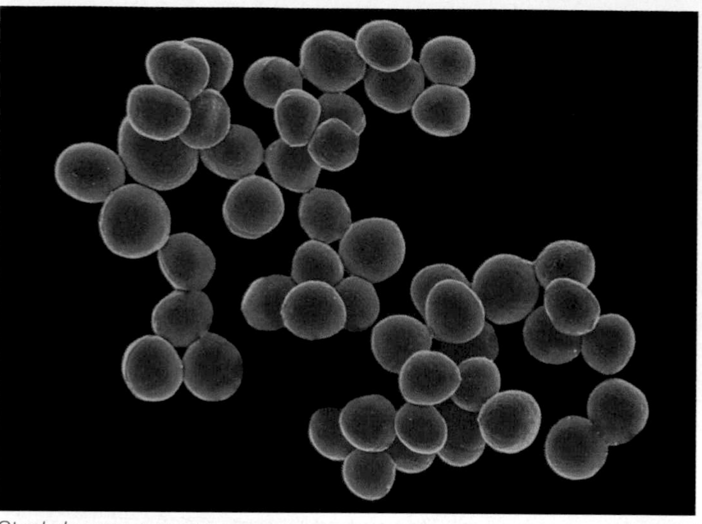

26.17 Low-GC Gram-Positives "Grape clusters" are the usual arrangement of staphylococci.

can and surrounded by a spore coat. The parent cell then breaks down, releasing the endospore. Endospore production is not a reproductive process; the endospore merely replaces the parent cell. The endospore, however, can survive harsh environmental conditions that would kill the parent cell, such as high or low temperatures or drought, because it is *dormant*—its normal activity is suspended. Later, if it encounters favorable conditions, the endospore becomes metabolically active and divides, forming new cells like the parent.

Dormant endospores of *Bacillus anthracis*, the causal agent of anthrax, germinate upon sensing specific molecules in the cytoplasm of blood cells called macrophages. Endospores of nonpathogenic *Bacillus* species do not germinate in this environment. Some endospores can be reactivated after more than a thousand years of dormancy. There are credible claims of reactivation of *Bacillus* endospores after millions of years. Members of this endospore-forming group of low-GC Gram-positives include the many species of *Clostridium* and *Bacillus*. The toxins produced by *C. botulinum* are among the most poisonous ever discovered; the lethal dose for humans is about one-millionth of a gram (1 μg).

The genus *Staphylococcus*—the staphylococci—includes low-GC Gram-positives that are abundant on the human body surface; they are responsible for boils and many other skin problems (**Figure 26.17**). *Staphylococcus aureus* is the best-known human pathogen in this genus; it is found in 20 to 40 percent of normal adults (and in 50 to 70 percent of hospitalized adults). It can cause respiratory, intestinal, and wound infections in addition to skin diseases.

Another interesting group of low-GC Gram-positives, the **mycoplasmas**, lack cell walls, although some have a stiffening material outside the plasma membrane. Some of them are the smallest cellular creatures ever discovered—they are even smaller than chlamydias (**Figure 26.18**). The smallest mycoplasmas capable of multiplication have a diameter of about 0.2 μm. They are small in another crucial sense as well: they have less than half as much DNA as most other prokaryotes—but they still can grow autonomously. It has been speculated that the amount of DNA in a mycoplasma may be the minimum amount required to encode the essential properties of a living cell.

The Proteobacteria are a large and diverse group

By far the largest group of bacteria, in terms of numbers of described species, is the **proteobacteria**, sometimes referred to as the *purple bacteria*. Among the proteobacteria are many species of Gram-negative, bacteriochlorophyll-containing, sulfur-using photoautotrophs. However, the proteobacteria also include dramatically diverse bacteria that bear no resemblance to those species

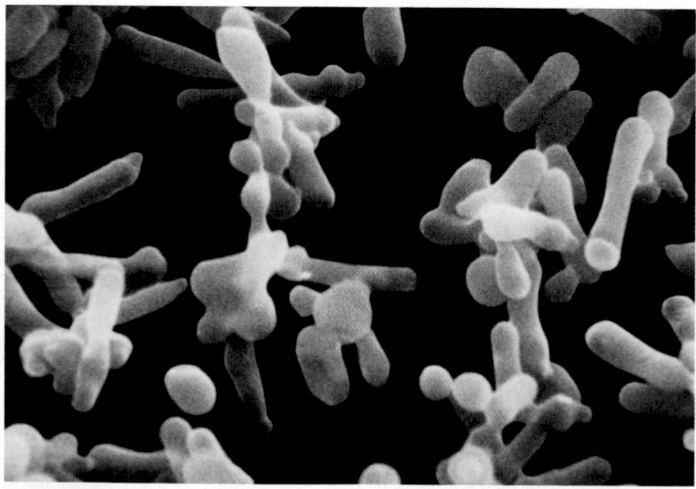

26.18 The Tiniest Living Cells Containing only about one-fifth as much DNA as *E. coli*, mycoplasmas are the smallest known bacteria.

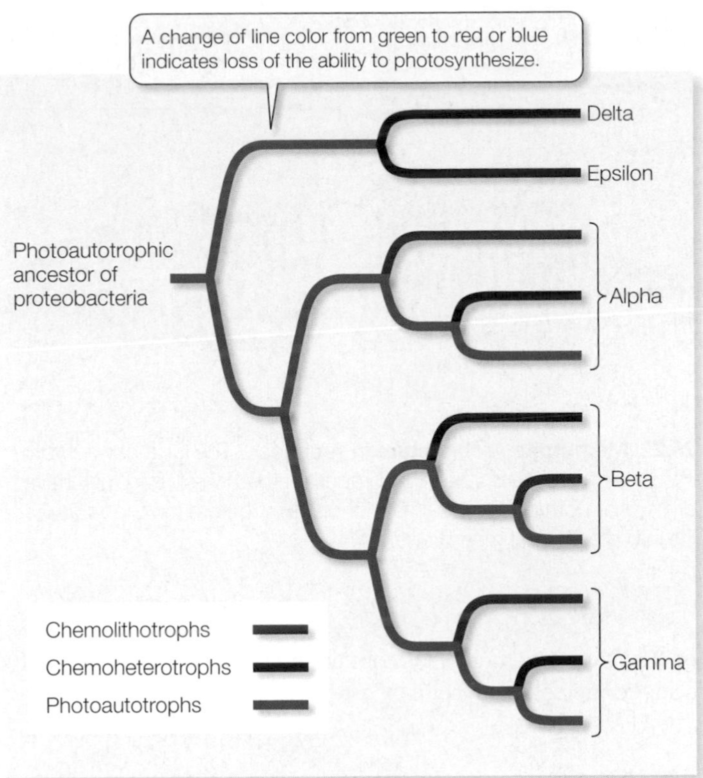

26.19 Modes of Nutrition in the Proteobacteria The common ancestor of all proteobacteria was probably a photoautotroph. As they encountered new environments, the delta and epsilon proteobacteria lost the ability to photosynthesize. In the other three groups, some evolutionary lineages became chemolithotrophs or chemoheterotrophs.

in phenotype. The mitochondria of eukaryotes were derived from a proteobacterium by endosymbiosis, as we'll see in Section 27.2.

No characteristic demonstrates the diversity of the proteobacteria more clearly than their metabolic pathways (**Figure 26.19**). There are five groups of proteobacteria: alpha, beta, gamma, delta, and epsilon. The common ancestor of all the proteobacteria was a photoautotroph. Early in evolution, two of the groups of proteobacteria lost their ability to photosynthesize and have been chemoheterotrophs ever since. The other three groups still have photoautotrophic members, but in *each* group, some evolutionary lines have abandoned photoautotrophy and taken up other modes of nutrition. There are chemolithotrophs and chemoheterotrophs in all three groups. Why? One possibility is that each of the trends shown in Figure 26.19 was an evolutionary response to selective pressures encountered as these bacteria colonized new habitats that presented new challenges and opportunities. Lateral gene transfer may have played a role in these responses.

Among the proteobacteria are some nitrogen-fixing genera, such as *Rhizobium* (see Figure 36.7), and other bacteria that contribute to the global nitrogen and sulfur cycles. *Escherichia coli*, one of the most studied organisms on Earth, is a proteobacterium. So, too, are many of the most famous human pathogens, such as *Yersinia pestis* (plague), *Vibrio cholerae* (cholera), and *Salmonella typhimurium* (gastrointestinal disease).

26.20 A Crown Gall This colorful tumor (the large mass at the left) growing on the stem of a geranium plant is caused by the proteobacterium *Agrobacterium tumefaciens*.

Fungi cause most plant diseases, and viruses cause others, but about 200 known plant diseases are of bacterial origin. *Crown gall*, with its characteristic tumors (**Figure 26.20**), is one of the most striking. The causal agent of crown gall is *Agrobacterium tumefaciens*, a proteobacterium that harbors a plasmid used in recombinant DNA studies as a vehicle for inserting genes into new plant hosts.

We have discussed six clades of bacteria in some detail, but other bacterial clades are well known, and there probably are dozens more waiting to be discovered. This conservative estimate is based on the fact that many bacteria have never been cultured in the laboratory.

Archaea differ in several important ways from bacteria

Archaea are well known for living in extreme habitats such as those with high salinity (salt content), low oxygen concentrations, high temperatures, or high or low pH (**Figure 26.21**). However, many archaea live in habitats that are not extreme. Perhaps the largest number of archaea live in the ocean depths.

One current classification scheme divides Archaea into two principal groups, **Euryarchaeota** and **Crenarchaeota**. Less is known about two more recently discovered groups, **Korarchaeota** and **Nanoarchaeota**. In fact, we know relatively little about the phylogeny of archaea, in part because the study of archaea is still in its early stages. Two characteristics shared by all archaea are the absence of peptidoglycan in their cell walls and the presence of lipids of distinctive composition in their cell membranes (see Table 26.1). Their separation from the Bacteria and Eukarya was supported when biologists sequenced the first archaeal genome. It consisted of 1,738 genes, more than half of which were unlike any genes ever found in the other two domains.

The unusual lipids in the membranes of archaea are found in all archaea, and in no bacteria or eukaryotes. Most bacterial and

EXPERIMENT

HYPOTHESIS: Some prokaryotes can grow and multiply at temperatures above 120°C.

METHOD

1. Seal samples of unidentified microorganisms taken from the vicinity of a thermal vent in tubes with medium containing Fe^{3+} as an electron acceptor. Control tubes contain the electron acceptor but no cells.
2. Hold experimental and control tubes for 10 hours in a sterilizer at 121°C. Reduction of the Fe^{3+} produces Fe^{2+} as magnetite, indicating the presence of living cells (left-hand photo).
3. In a second experiment, isolate and test for growth at various temperatures.

RESULTS

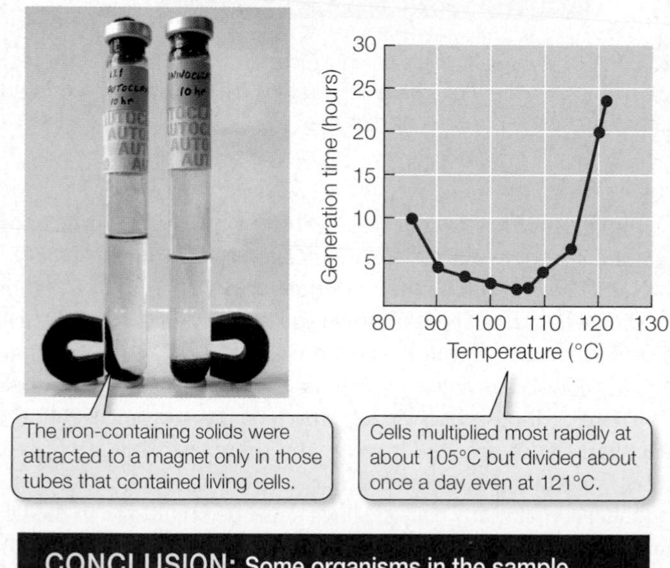

Generation time (hours)
30
25
20
15
10
5
80
90
100
110
120
130
Temperature (°C)
The iron-containing solids were attracted to a magnet only in those tubes that contained living cells.
Cells multiplied most rapidly at about 105°C but divided about once a day even at 121°C.

CONCLUSION: Some organisms in the sample multiplied at 121°C, the highest temperature yet known to allow growth of an organism.

26.21 What Is the Highest Temperature an Organism Can Tolerate? Kazem Kashefi and Derek Lovley isolated an unidentified prokaryotic organism from water samples taken near a hydrothermal vent in the northeastern Pacific Ocean. Tests to establish growth and metabolism were run at temperatures from 85°C to 130°C in sealed tubes with Fe^{3+} as an electron acceptor (as Fe_2O_3, or common rust).* The organism continued growing at temperatures as high as 121°C—a sterilizing temperature, known to destroy all previously described microorganisms. Genes from the surviving organism (dubbed "Strain 121") were sequenced, and comparison of those sequences with genes from known archaean species indicated that Strain 121 is probably also an archaean. FURTHER RESEARCH: Strain 121 did not grow during a 2-hour exposure to 130°C, but it did not die, either. How would you demonstrate that it was still alive?

*Some prokaryotes metabolize Fe^{3+} to Fe^{2+}, converting rust to magnetite. Living, growing cells thus produce a material that is attracted to magnets, whereas dead cells do not.

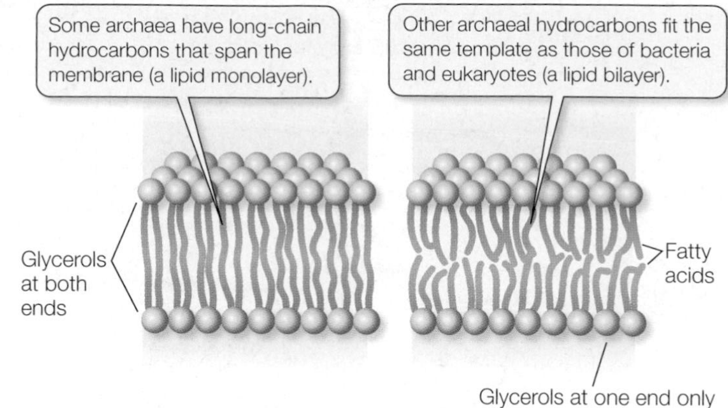

26.22 Membrane Architecture in Archaea The long-chain hydrocarbons of many archaeal membranes are branched, and may have glycerol at both ends. This lipid monolayer structure (on the left) still fits into a biological membrane.

eukaryotic membrane lipids contain unbranched long-chain fatty acids connected to glycerol by ester linkages:

$$-\overset{\overset{\displaystyle O}{\|}}{C}-O-\underset{\underset{\displaystyle H}{|}}{\overset{\overset{\displaystyle H}{|}}{C}}-$$

In contrast, some archaeal membrane lipids contain long-chain hydrocarbons connected to glycerol by *ether linkages*:

$$-\underset{\underset{\displaystyle H}{|}}{\overset{\overset{\displaystyle H}{|}}{C}}-O-\underset{\underset{\displaystyle H}{|}}{\overset{\overset{\displaystyle H}{|}}{C}}-$$

In addition, the long-chain hydrocarbons of the archaea are branched. One class of these lipids, with hydrocarbon chains 40 carbon atoms in length, contains glycerol at *both* ends of the hydrocarbons (**Figure 26.22**). This *lipid monolayer* structure, unique to Archaea, still fits in a biological membrane because the lipids are twice as long as the typical lipids in the bilayers of other membranes. Lipid monolayers and bilayers are both found among the archaea. The effects, if any, of these structural features on membrane performance are unknown. In spite of this striking difference in their membrane lipids, the membranes seen in all three domains have similar overall structures, dimensions, and functions.

Many Crenarchaeota live in hot, acidic places

Most known Crenarchaeota are both thermophilic (heat-loving) and **acidophilic** (acid-loving). Members of the genus *Sulfolobus* live in hot sulfur springs at temperatures of 70°C–75°C. They die of "cold" at 55°C (131°F). Hot sulfur springs are also extremely acidic. *Sulfolobus* grows best in the range from pH 2 to pH 3, but it readily tolerates pH values as low as 0.9. One species of the genus *Ferroplasma* lives at a pH near 0. Some acidophilic hyperthermophiles maintain an internal pH near 7 (neutral) in spite of their

26.23 Some Would Call It Hell; These Archaea Call It Home Masses of heat- and acid-loving archaea form an orange mat inside a volcanic vent on the island of Kyushu, Japan. Sulfurous residue is visible at the edges of the archaeal mat.

26.24 Extreme Halophiles Commercial seawater evaporating ponds, such as these in San Francisco Bay, are attractive homes for salt-loving archaea, which are easily visible here because of their carotenoid pigments.

acidic environment. These and other hyperthermophiles thrive where very few other organisms can even survive (**Figure 26.23**).

The Euryarchaeota live in many surprising places

Some species of Euryarchaeota share the property of producing methane (CH_4) by reducing carbon dioxide. All of these **methanogens** are obligate anaerobes, and methane production is the key step in their energy metabolism. Comparison of rRNA nucleotide sequences has revealed a close evolutionary relationship among all these methanogens, which were previously assigned to several unrelated bacterial groups.

Methanogens release approximately 2 billion tons of methane gas into Earth's atmosphere each year, accounting for 80 to 90 percent of the methane in the atmosphere, including that associated with mammalian belching. Approximately a third of this methane comes from methanogens living in the guts of grazing herbivores such as cattle, sheep, and deer. Methane is increasing in Earth's atmosphere by about 1 percent per year and is a contributor to the greenhouse effect. Part of the increase is due to increases in cattle and rice farming and the methanogens associated with both.

One methanogen, *Methanopyrus*, lives on the ocean bottom near boiling hot hydrothermal vents. *Methanopyrus* can survive and grow at 110°C. It grows best at 98°C and not at all at temperatures below 84°C.

Another group of Euryarchaeota, the **extreme halophiles** (salt lovers), lives exclusively in very salty environments. Because they contain pink carotenoid pigments, they can be easily seen under some circumstances (**Figure 26.24**). Halophiles grow in the Dead Sea and in brines of all types: pickled fish may sometimes show reddish pink spots that are colonies of halophilic archaea. Few other organisms can live in the saltiest of the homes that the extreme halophiles occupy; most would "dry" to death, losing too much water to the hypertonic environment. Extreme halophiles have been found in lakes with pH values as high as 11.5—the most alkaline environment inhabited by living organisms, and almost as alkaline as household ammonia.

Some of the extreme halophiles have a unique system for trapping light energy and using it to form ATP—without using any form of chlorophyll—when oxygen is in short supply. They use the pigment *retinal* (also found in the vertebrate eye) combined with a protein to form a light-absorbing molecule called *bacteriorhodopsin*, and they form ATP by a chemiosmotic mechanism of the kind described in Figure 7.13.

Another member of the Euryarchaeota, *Thermoplasma*, has no cell wall. It is thermophilic and acidophilic, its metabolism is aerobic, and it lives in coal deposits. It has the smallest genome among the archaea, and perhaps the smallest (along with the mycoplasmas) of any free-living organism—1,100,000 base pairs.

Korarchaeota and Nanoarchaeota are less well known

The Korarchaeota are known only by evidence derived from DNA isolated directly from hot springs. No korarchaeote has been successfully grown in pure culture.

Another archaean has been discovered at a deep-sea hydrothermal vent off the coast of Iceland. It is the first representative of a group christened Nanoarchaeota because of their minute size. This organism lives attached to cells of *Ignicoccus*, a crenarchaeote. Because of their association, the two species can be grown together in culture (**Figure 26.25**).

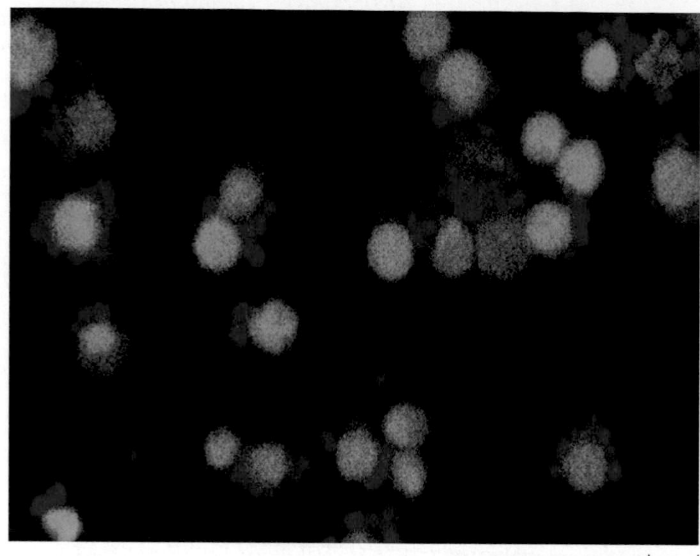

26.25 A Nanoarchaeote Growing in Mixed Culture with a Crenarchaeote *Nanoarchaeum equitans* (red), discovered living near deep-ocean hydrothermal vents, is the only representative of the nanoarchaeote group so far discovered. This tiny organism lives attached to cells of the crenarchaeote *Ignicoccus* (green). For this confocal laser micrograph, the two species were visually differentiated by fluorescent dye "tags" that are specific to their separate gene sequences.

26.5 RECAP

The relationships among the groups of Bacteria and Archaea are only partially understood. Each group is diverse in form and metabolism.

- Can you explain how metabolic diversity could have become so great in the proteobacteria? See p. 575 and Figure 26.19
- What makes the membranes of archaea unique? See p. 576 and Figure 26.22

Because prokaryotes have so many different metabolic and nutritional capabilities, and because they can live in so many environments, it is reasonable to expect that they affect their environments in many ways. As we are about to see, prokaryotes directly affect humans—in ways both beneficial and harmful.

26.6 How Do Prokaryotes Affect Their Environments?

Prokaryotes live in and exploit all kinds of environments and are part of all ecosystems. In this section we'll examine the roles of prokaryotes that live in soils, in water, and even in other organisms, where they may exist in a neutral, beneficial, or parasitic relationship with their host's tissues. The roles of some prokaryotes living in extreme environments have yet to be determined.

Remember that in spite of our frequent mention of prokaryotes as human pathogens, only a small minority of the known prokaryotic species are pathogenic. Many more prokaryotes play positive roles in our lives and in the biosphere. We make direct use of many bacteria and a few archaea in such diverse applications as cheese production, sewage treatment, and the industrial production of an amazing variety of antibiotics, vitamins, organic solvents, and other chemicals.

Prokaryotes are important players in element cycling

Many prokaryotes are **decomposers**, organisms that metabolize organic compounds in dead organisms and other organic material and return the products to the environment as inorganic substances. Prokaryotes, along with fungi, return tremendous quantities of organic carbon to the atmosphere as carbon dioxide, thus carrying out a key step in the carbon cycle. Prokaryotic decomposers also return inorganic nitrogen and sulfur to the environment.

Animals depend on plants and other photosynthetic organisms for their food, directly or indirectly. But plants depend on other organisms—prokaryotes—for their own nutrition. The extent and diversity of life on Earth would not be possible without nitrogen fixation by prokaryotes. Nitrifiers are crucial to the biosphere because they convert the products of nitrogen fixation into nitrate ions, the form of nitrogen most easily used by many plants (see Figure 36.8). Plants, in turn, are the source of nitrogen compounds for animals and fungi. Denitrifiers also play a key role in keeping the nitrogen cycle going. Without denitrifiers, which convert nitrate ions back into nitrogen gas, all forms of nitrogen would leach from the soil and end up in lakes and oceans, making life on land impossible. Other prokaryotes—both bacteria and archaea—contribute to a similar cycle of sulfur.

In the ancient past, the cyanobacteria had an equally dramatic effect on life: their photosynthesis generated oxygen, converting Earth's atmosphere from an anaerobic to an aerobic environment. A major result was the wholesale loss of obligate anaerobic species that could not tolerate the O_2 generated by the cyanobacteria. Only those anaerobes that were able to adapt to aerobic conditions or colonize environments that remained anaerobic (such as very wet ones) survived. However, this transformation to aerobic environments made possible the evolution of cellular respiration and the subsequent explosion of eukaryotic life.

Ten trillion tons of methane gas lie deep under the ocean floor. Is it possible we could experience a disastrous "big burp" as this gas escapes to the atmosphere? Fortunately, legions of archaea also live below the ocean bottom, where they metabolize the methane as it rises. Virtually none of the gas gets even as far as the ocean deeps.

Prokaryotes live on and in other organisms

Prokaryotes work together with eukaryotes in many ways. As we have seen, mitochondria and chloroplasts are descended from what were once free-living bacteria. Much later in evolutionary history, some plants became associated with bacteria to form cooperative nitrogen-fixing nodules on their roots (see Figure 36.9).

Many animals harbor a variety of bacteria and archaea in their digestive tracts. Cattle depend on prokaryotes to perform important steps in digestion. Like most animals, cattle cannot produce cellulase, the enzyme needed to start the digestion of the cellulose that makes up the bulk of their plant food. However, bacteria living in a special section of the gut, called the rumen, produce enough cellulase to process the daily diet for the cattle.

Humans use some of the metabolic products—especially vitamins B_{12} and K—of bacteria living in our large intestine. These and other bacteria and archaea line our intestines with a dense biofilm that is in intimate contact with the lining of the gut. This biofilm facilitates nutrient transfer from the intestine into the body and induces immunity to the gut contents. The biofilm in the gut is a major part of an "organ" consisting of prokaryotes that is essential to our health. Its makeup varies from time to time and from region to region of the intestinal tract, and it has a complex ecology that scientists have just begun to explore in detail.

We are heavily populated, inside and out, by bacteria. Although very few of them are agents of disease, popular notions of bacteria as "germs" arouse our curiosity about those few.

A small minority of bacteria are pathogens

The late nineteenth century was a productive era in the history of medicine—a time during which bacteriologists, chemists, and physicians proved that many diseases are caused by microbial agents. During this time the German physician Robert Koch laid down a set of four rules for establishing that a particular microorganism causes a particular disease:

- The microorganism is always found in individuals with the disease.
- The microorganism can be taken from the host and grown in pure culture.
- A sample of the culture produces the disease when injected into a new, healthy host.
- The newly infected host yields a new, pure culture of microorganisms identical to those obtained in the second step.

These rules, called **Koch's postulates**, were very important in a time when it was not widely understood that microorganisms cause disease. Although medical science today has more powerful diagnostic tools, the postulates remain useful on occasion. For example, physicians were taken aback in the 1990s when stomach ulcers—long accepted and treated as the result of excess stomach acid—were proven by Koch's postulates to be caused by the bacterium *Helicobacter pylori* (see Figure 50.14).

Only a tiny percentage of all prokaryotes are **pathogens** (disease-producing organisms), and of those that are known, all are in the domain Bacteria. For an organism to be a successful pathogen, it must overcome several hurdles:

- It must arrive at the body surface of a potential host.
- It must enter the host's body.
- It must evade the host's defenses.
- It must multiply inside the host.
- It must infect a new host.

Failure to overcome any of these hurdles ends the reproductive career of a pathogenic organism. However, in spite of the many defenses available to potential hosts (see Chapter 18), some bacteria are very successful pathogens.

For the host, the consequences of a bacterial infection depend on several factors. One is the **invasiveness** of the pathogen—its ability to multiply within the body of the host. Another is its **toxigenicity**—its ability to produce chemical substances (*toxins*) that are harmful to the tissues of the host. *Corynebacterium diphtheriae*, the agent that causes diphtheria, has low invasiveness and multiplies only in the throat, but its toxigenicity is so great that the entire body is affected. In contrast, *Bacillus anthracis*, which causes anthrax (a disease primarily of cattle and sheep, but which is also sometimes fatal in humans), has low toxigenicity, but an invasiveness so great that the entire bloodstream ultimately teems with the bacteria. Both are low-GC Gram-positives.

There are two general types of bacterial toxins: exotoxins and endotoxins. **Endotoxins** are released when certain Gram-negative bacteria grow or lyse (burst). These toxins are lipopolysaccharides (complexes consisting of a polysaccharide and a lipid component) that form part of the outer bacterial membrane (see Figure 26.5). Endotoxins are rarely fatal; they normally cause fever, vomiting, and diarrhea. Among the endotoxin producers are some strains of the gamma proteobacteria *Salmonella* and *Escherichia*.

Exotoxins are usually soluble proteins released by living, multiplying bacteria, and they may travel throughout the host's body. They are highly toxic—often fatal—to the host, but do not produce fevers. Exotoxin-induced human diseases include tetanus (from *Clostridium tetani*), botulism (from *Clostridium botulinum*), cholera (from *Vibrio cholerae*), and plague (from *Yersinia pestis*). Anthrax results from three exotoxins produced by *Bacillus anthracis*.

Pathogenic bacteria are often surprisingly difficult to combat, even with today's arsenal of antibiotics. One source of this difficulty is the ability of prokaryotes to form biofilms.

26.6 RECAP

Prokaryotes play key roles in the cycling of Earth's elements. While many prokaryotes are beneficial and even necessary to other forms of life, some are pathogens.

- Can you describe the roles of bacteria in the nitrogen cycle? See p. 569 and p. 578
- Do you understand the challenges facing a pathogen? See p. 579

CHAPTER SUMMARY

26.1 How did the living world begin to diversify?

See Web/CD Tutorial 26.1

Two of life's three domains, **Bacteria** and **Archaea**, are prokaryotic. They are distinguished from Eukarya in several ways, including the cell's lack of membrane-enclosed organelles. Review Table 26.1

Archaea and Eukarya share a common ancestor not shared by Bacteria. The common ancestor of all three domains probably lived more than 3 billion years ago, and the common ancestor of the Archaea and Eukarya at least 2 billion years ago. Review Figure 26.1

26.2 Where are prokaryotes found?

Prokaryotes are the most numerous organisms on Earth. They occupy an enormous variety of habitats, including inside other organisms and deep in Earth's crust.

The three most common bacterial body forms are **cocci** (spheres), **bacilli** (rods), and **helices** (spirals). The cells of some bacteria aggregate, forming **filaments** and other structures.

Prokaryotes form complex communities, some of which become dense films called **biofilms**. Review Figure 26.3

26.3 What are some keys to the success of prokaryotes?

Most prokaryotes have cell walls. Almost all bacterial cell walls contain **peptidoglycan**. Review Figure 26.5, Web/CD Activity 26.1

Bacteria can be classified into two groups by the **Gram stain**.

Prokaryotes move by a variety of means, including axial filaments, gas vesicles, and flagella.

Prokaryotes undergo genetic recombination but reproduce asexually.

Prokaryote metabolism is very diverse. Some prokaryotes are anaerobic, others are aerobic, and yet others can shift between these modes. Prokaryotes are classified as **photoautotrophs**, **photoheterotrophs**, **chemolithotrophs**, or **chemoheterotrophs**. Review Table 26.2

The metabolic pathways of some prokaryotes involve sulfur or nitrogen. **Nitrogen fixers** convert nitrogen gas into a form organisms can metabolize.

26.4 How can we determine prokaryote phylogeny?

Early attempts to classify prokaryotes were hampered by their small size and difficulties growing them in pure culture. Phylogenetic classification of prokaryotes is now based on rRNA and other nucleotide sequences.

Lateral gene transfer has occurred throughout evolutionary history, but it may not complicate the elucidation of prokaryote phylogeny. Review Figure 26.10

Only a tiny percentage of all prokaryote species have been described.

26.5 What are the major known groups of prokaryotes?

Several clades of prokaryotes have been recognized. The members of a prokaryote clade often differ profoundly from one another. Review Figures 26.11 and 26.19

Of the clades of Bacteria, the **proteobacteria** embrace the largest number of species. Other important groups include the **cyanobacteria**, the **spirochetes**, the **chlamydias**, and the **low-GC Gram-positives**. Some **high-GC Gram positives** produce important antibiotics.

The cell walls of archaeans lack peptidoglycan, and archaeal membrane lipids differ from those of bacteria and eukaryotes.

The best-studied groups of Archaea are the **Euryarchaeota** and the **Crenarchaeota**.

26.6 How do prokaryotes affect their environments?

Prokaryotes play key roles in the cycling of elements such as nitrogen, oxygen, sulfur, and carbon. One such role is as **decomposers** of dead organisms.

Nitrogen-fixing bacteria fix the nitrogen needed by all other organisms. Nitrifiers convert that nitrogen into forms that can be used by plants, and denitrifiers ensure that nitrogen is returned to the atmosphere.

Oxygen production by early photosynthetic cyanobacteria reconfigured Earth's atmosphere, which made aerobic forms of life possible.

Prokaryotes inhabiting the guts of many animals help them digest their food.

Koch's postulates establish the criteria by which an organism may be classified as a **pathogen**. Relatively few bacteria—and no archaea—are known to be pathogens.

SELF-QUIZ

1. Most prokaryotes
   a. are agents of disease.
   b. lack ribosomes.
   c. evolved from the most ancient eukaryotes.
   d. lack a cell wall.
   e. are chemoheterotrophs.
2. The division of the living world into three domains
   a. is strictly arbitrary.
   b. is based on the morphological differences between archaea and bacteria.
   c. emphasizes the greater importance of eukaryotes.
   d. was proposed by the early microscopists.
   e. is strongly supported by data on rRNA sequences.
3. Which statement about the archaeal genome is true?
   a. It is much more similar to the bacterial genome than to eukaryotic genomes.
   b. Many of its genes are genes that are never observed in bacteria or eukaryotes.
   c. It is much smaller than the bacterial genome.
   d. It is housed in the nucleus.
   e. No archaeal genome has yet been sequenced.
4. Which statement about nitrogen metabolism is *not* true?
   a. Certain prokaryotes reduce atmospheric N_2 to ammonia.
   b. Some nitrifiers are soil bacteria.
   c. Denitrifiers are obligate anaerobes.
   d. Nitrifiers obtain energy by oxidizing ammonia and nitrite.
   e. Without the nitrifiers, terrestrial organisms would lack a nitrogen supply.

5. All photosynthetic bacteria
   a. use chlorophyll *a* as their photosynthetic pigment.
   b. use bacteriochlorophyll as their photosynthetic pigment.
   c. release oxygen gas.
   d. produce particles of sulfur.
   e. are photoautotrophs.
6. Gram-negative bacteria
   a. appear blue to purple following Gram staining.
   b. are the most abundant of the bacterial groups.
   c. are all either bacilli or cocci.
   d. contain no peptidoglycan in their cell walls.
   e. are all photosynthetic.
7. Endospores
   a. are produced by viruses.
   b. are reproductive structures.
   c. are very delicate and easily killed.
   d. are resting structures.
   e. lack cell walls.
8. Chlamydias
   a. are among the smallest archaea.
   b. live on the surface of human skin.
   c. are never pathogenic to humans.
   d. are Gram-negative.
   e. have a very simple life cycle.
9. Which statement about mycoplasmas is *not* true?
   a. They lack cell walls.
   b. They are the smallest known cellular organisms.
   c. They contain the same amount of DNA as do other prokaryotes.
   d. They cannot be killed with penicillin.
   e. Some are pathogens.
10. Archaea
   a. have cytoskeletons.
   b. have distinctive lipids in their plasma membranes.
   c. survive only at moderate temperatures and near neutrality.
   d. all produce methane.
   e. have substantial amounts of peptidoglycan in their cell walls.

FOR DISCUSSION

1. Why do systematic biologists find rRNA sequence data more useful than data on metabolism or cell structure for classifying prokaryotes?
2. Why does lateral gene transfer make it so difficult to arrive at agreement on prokaryote phylogeny?
3. Differentiate among the members of the following sets of related terms:
   a. prokaryotic/eukaryotic
   b. obligate anaerobe/facultative anaerobe/obligate aerobe
   c. photoautotroph/photoheterotroph/chemolithotroph/chemoheterotroph
   d. Gram-positive/Gram-negative
4. Why are the endospores of low-GC Gram-positives not considered to be reproductive structures?
5. Originally, the cyanobacteria were called "blue-green algae" and were not grouped with the bacteria. Suggest several reasons for this (abandoned) tendency to separate the cyanobacteria from the bacteria. Why are the cyanobacteria now grouped with the other bacteria?
6. The high-GC Gram-positives are of great commercial interest. Why?
7. Thermophiles are of great interest to molecular biologists and biochemists. Why? What practical concerns might motivate that interest?
8. How can biologists discuss the Korarchaeota when they have never seen one?

FOR INVESTIGATION

Kashefi and Lovley were able to grow an unnamed archaean at temperatures over 120°C only because they used Fe^{3+} as an electron acceptor—no other electron acceptor that they tried allowed growth (see Figure 26.21). How might you explore the same or other high-temperature environments for other hyperthermophilic organisms not detected by Kashefi and Lovley using Fe^{3+}?

CHAPTER 27 # The Origin and Diversification of the Eukaryotes

A tale of three trypanosomes

Among the most deadly organisms on Earth are the trypanosomes, single-celled microscopic organisms that cause several grim diseases, mainly in developing countries. In central Africa, tsetse flies carry *Trypanosoma brucei* and its relatives, which cause African sleeping sickness. There is no vaccine to prevent the infection, and only one drug is currently available to treat it. That drug—melarsoprol—kills about 5 percent of the patients who take it and is without effect on another 30 percent, but it is the only treatment available. There are 300,000 to 500,000 cases of sleeping sickness and more than 50,000 deaths from the disease each year.

Assassin bugs carry another trypanosome, *Trypanosoma cruzi*, which causes Chagas' disease. This disease affects 16 to 18 million people, primarily in Central and South America. Again, there is no vaccine and no effective drug, and 20,000 to 50,000 people die from Chagas' disease each year. Still another trypanosome, *Leishmania major*, causes a family of often fatal human diseases collectively called leishmaniasis. This organism is transmitted by sand flies. There are about 2 million cases and an estimated 60,000 deaths from leishmaniasis each year, and—you guessed it—there is no vaccine and no good treatment.

Research and development of a new medicine typically requires about a billion dollars and a dozen years. The income from sales must compensate the developers and yield enough profit to make production a viable commercial enterprise. Trypanosome diseases mostly strike the poorest of the poor in developing countries, and so offer little financial incentive for drug and vaccine developers.

International health organizations rank the trypanosome diseases among the "most neglected diseases." Their enormous toll is barely mentioned in the Western news media, in stark contrast to the copious coverage given to any proposed new cancer treatment or the latest brouhaha over a flu vaccine.

Some developing countries, such as India, Cuba, Brazil, and South Africa, have the technical expertise and manufacturing capacity for pharmaceutical research and production. But even with such expertise, preventing and treating trypanosome diseases presents obstacles. Trypanosomes can evade recognition and destruction by the human immune system, vaccines, and drugs by constantly changing the cell surface recognition molecules of their own or infected host cells.

In addition, unlike the prokaryotic bacteria, these unicellular microbes are eukaryotes—

Thugs in a Huddle Sometimes trypanosomes like these *Leishmania major* form clusters held together by a tangle of mucilage secreted around their flagella; nobody is sure yet why they behave this way.

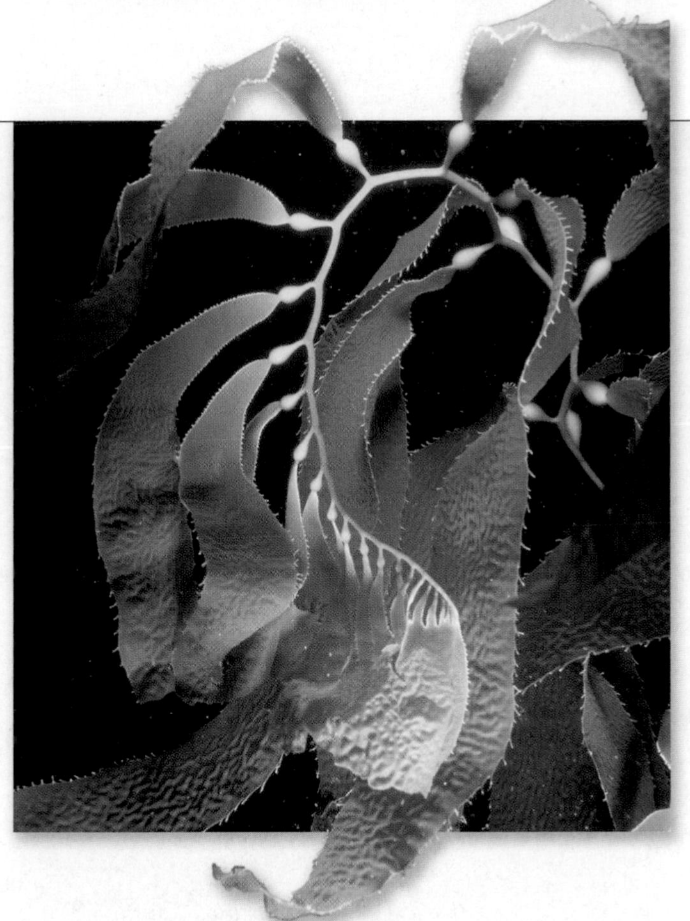

A Giant among Microbial Eukaryotes "Microbial" certainly does not describe *Macrocystis pyrifera*, the giant kelp, which can reach 60 meters in length. Despite this, they are often placed with the microbial eukaryotes in a large, paraphyletic grouping.

their cells are similar to ours, and remedies that attack them often damage our own cells as well. Trypanosomes are one group among the legions of *microbial eukaryotes*. Several different microbial eukaryote lineages were the ancestors of all multicellular life. Indeed, some of these organisms are multicellular themselves, and hardly microbial. They are a diverse lot, and difficult to categorize.

IN THIS CHAPTER we look at some of the many effects that microbial eukaryotes have on their environment. We then describe the origin and early diversification of the eukaryotes and the complexity achieved by some single cells. We then explore some of the diversity of microbial eukaryote body forms and adaptations and present the developing current view of the evolutionary relationships among some of the microbial eukaryotes.

CHAPTER OUTLINE

27.1 How Do Microbial Eukaryotes Affect the World Around Them?

Many modern members of the Eukarya (such as trees, mushrooms, and dogs, not to mention human beings) are familiar to us. We have no problem recognizing these organisms as land plants, fungi, and animals, respectively. However, trypanosomes and a dazzling assortment of other eukaryotes—mostly microscopic organisms—do not fit into any of these three groups. Eukaryotes that are neither land plants, animals, nor fungi have traditionally been grouped into a category called **protists**, and for the sake of clarity and convenience, other chapters in this book have used that term. In this chapter, however, we will refer to them as **microbial eukaryotes** (even though not all of them are actually microbial) to emphasize that these organisms do not constitute a clade, but are *paraphyletic* (see Figure 25.12).

The microbial eukaryotes are extremely diverse, and their effects on other organisms and on the physical environment are almost as diverse. Some microbial eukaryotes are food for marine animals, while others poison the sea; some are packaged as nutritional supplements, and some are pathogens; the remains of some form the sands of many modern beaches, and others are a major source of today's ever more expensive crude oil.

The phylogeny and morphology of the microbial eukaryotes both illustrate their diversity

The true phylogeny of these organisms is still the subject of research and debate, but we do know that *the microbial eukaryotes are not a clade*. Some groups of microbial eukaryotes are more closely related to the animals than they are to other microbial eukaryotes, while others are closely related to the land plants (**Table 27.1**; see also Figure 27.17). Some microbial eukaryotes are motile, while others do not move; some are photosynthetic, others heterotrophic; most are unicellular, but some are multicellular. Most are microscopic, but a few are huge: giant kelps, for example, sometimes achieve lengths greater than that of a football field (see above left). Many unicellular eukaryotes are constituents of the **plankton** (free-floating microscopic aquatic organisms). Photosynthetic members of the plankton are called **phytoplankton**.

TABLE 27.1

Major Eukaryote Clades

CLADE	ATTRIBUTES	EXAMPLE (GENUS)
Chromalveolates		
Haptophytes	Unicellular, often with calcium carbonate scales	*Emiliania*
Alveolates	Sac-like structures beneath plasma membrane	
Apicomplexans	Apical complex for penetration of host	*Plasmodium*
Dinoflagellates	Pigments give golden-brown color	*Gonyaulax*
Ciliates	Cilia; two types of nuclei	*Paramecium*
Stramenopiles	Hairy and smooth flagella	
Brown algae	Multicellular; marine; photosynthetic	*Macrocystis*
Diatoms	Unicellular; photosynthetic; two-part cell walls	*Thalassiosira*
Oomycetes	Mostly coenocytic; heterotrophic	*Saprolegnia*
Plantae		
Glaucophytes	Peptidoglycan in chloroplasts	*Cyanophora*
Red algae	No flagella; chlorophyll *a* and *c*; phycoerythrin	*Chondrus*
Chlorophytes	Chlorophyll *a* and *b*	*Ulva*
*Land plants (Chs. 28–29)	Chlorophyll *a* and *b*; protected embryo	*Ginkgo*
Charophytes	Chlorophyll *a* and *b*; mitotic spindle oriented as in land plants	*Chara*
Excavates		
Diplomonads	No mitochondria; two nuclei; flagella	*Giardia*
Parabasalids	No mitochondria; flagella and undulating membrane	*Trichomonas*
Heteroloboseans	Can transform between amoeboid and flagellate stages	*Naegleria*
Euglenids	Flagella; spiral strips of protein support cell surface	*Euglena*
Kinetoplastids	Kinetoplast within mitochondrion	*Trypanosoma*
Rhizaria		
Cercozoans	Threadlike pseudopods	*Cercomonas*
Foraminiferans	Long, branched pseudopods; calcium carbonate shells	*Globigerina*
Radiolarians	Glassy endoskeleton; thin, stiff pseudopods	*Astrolithium*
Unikonts		
Opisthokonts	Single, posterior flagellum	
*Fungi (Ch. 30)	Heterotrophs that feed by absorption	*Penicillium*
Choanoflagellates	Resemble sponge cells; heterotrophic; with flagella	*Choanoeca*
*Animals (Chs. 31–33)	Heterotrophs that feed by ingestion	*Drosophila*
Amoebozoans	Amoebas with lobe-shaped pseudopods	
Loboseans	Feed individually	*Amoeba*
Plasmodial slime molds	Form coenocytic feeding bodies	*Physarum*
Cellular slime molds	Cells retain their identity in pseudoplasmodium	*Dictyostelium*

*Clades marked with an asterisk are made up of multicellular organisms and are discussed in the chapters indicated. All other groups listed are treated here as *microbial eukaryotes* (often known as *protists*).

Phytoplankton are the primary producers of the marine food web

A single microbial eukaryote clade, the *diatoms*, is responsible for about a fifth of all of the photosynthetic carbon fixation on Earth—about the same amount of photosynthesis performed by all of Earth's rainforests. These spectacular unicellular organisms (**Figure 27.1**) are the predominant members of the phytoplankton, but other microbial eukaryote clades also include important phytoplanktonic species that contribute heavily to global photosynthesis. Like green plants on land, the phytoplankton serve as a gateway for energy from the sun into the living world; in other words, they are *primary producers*. In turn, they are eaten by heterotrophs, including animals and other microbial eukaryotes. Those consumers are, in turn, eaten by other consumers. Most aquatic heterotrophs

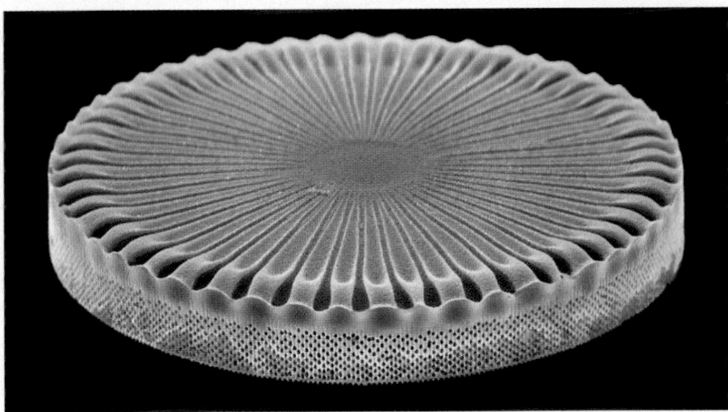

27.1 Architecture in Miniature: A Photosynthetic Diatom This artificially colored scanning electron micrograph shows the intricate patterning of diatom cell walls. These spectacular unicellular eukaryotes are photosynthetic and dominate the aquatic phytoplankton community.

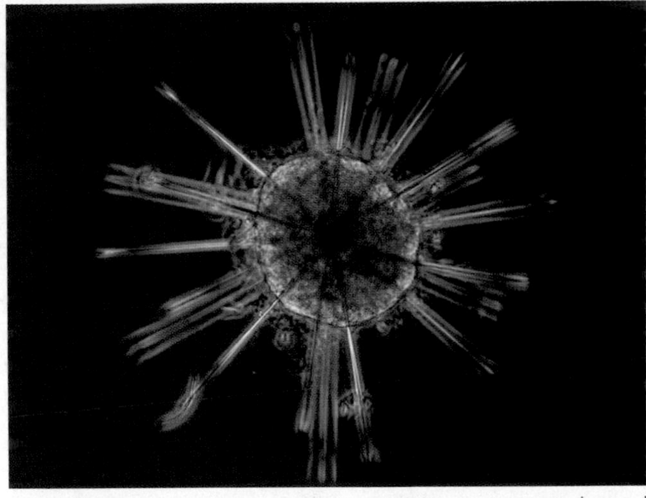

27.2 Two Microbial Eukaryotes in an Endosymbiotic Relationship Photosynthetic dinoflagellates (see Figure 27.18) are living as endosymbionts within this radiolarian, providing organic nutrients for the radiolarian and imparting the golden brown pigmentation seen at the center of its glassy skeleton.

(with the exception of some species existing in the deep ocean) depend on the photosynthesis performed by phytoplankton.

Some microbial eukaryotes are endosymbionts

Endosymbiosis is the condition in which two organisms live together, one inside the other (see Figure 4.15C). Endosymbiosis is very common among the microbial eukaryotes, many of which live within the cells of animals. Members of the *dinoflagellates* are common microbial eukaryote symbionts in both animals and in other microbial eukaryotes; most but not all dinoflagellate endosymbiont species are photosynthetic. Many *radiolarians,* for example, harbor photosynthetic endosymbionts (**Figure 27.2**). As a result, the radiolarians, which are not photosynthetic themselves, appear greenish or golden, depending on the type of endosymbiont they contain. This arrangement is often mutually beneficial: the radiolarian can make use of the organic nutrients produced by its photosynthetic guest, and the guest may in turn make use of metabolites made by the host or receive physical protection. In some cases, however, the guest is exploited for its photosynthetic products while receiving no benefit itself.

Some dinoflagellates live endosymbiotically in the cells of corals, contributing products of their photosynthesis to the partnership. The importance to the coral is demonstrated when the dinoflagellates are attacked by certain bacteria; the coral is ultimately damaged or destroyed when its nutrient supply is reduced.

Some microbial eukaryotes are deadly

The best-known pathogenic microbial eukaryotes are members of the genus *Plasmodium,* a highly specialized group of *apicomplexans* that spend part of their life cycle as parasites within human red blood cells, where they are the cause of malaria (**Figure 27.3**). In terms of the number of people affected, malaria is one of the world's three most serious infectious diseases, and it kills more than a million people each year. Every 30 seconds malaria kills someone somewhere—usually in sub-Saharan Africa, although malaria occurs in more than 100 countries. About 600 million people suffer from this disease.

Female mosquitoes of the genus *Anopheles* transmit *Plasmodium* to humans. In other words, *Anopheles* is the *vector* for malaria. The parasite enters the human circulatory system when an infected *Anopheles* mosquito penetrates the human skin in search of blood. The parasites find their way to cells in the liver and the lymphatic system, change their form, multiply, and reenter the bloodstream, attacking red blood cells.

The parasites multiply inside the red blood cells, which then burst, releasing new swarms of parasites. If another *Anopheles* bites the victim, the mosquito takes in *Plasmodium* cells along with blood. Some of the ingested cells are gametes that formed in human cells. The gametes unite in the mosquito, forming zygotes that lodge in the mosquito's gut, divide several times, and move into its salivary glands, from which they can be passed on to another human host. Thus *Plasmodium* is an extracellular parasite in the mosquito vector and an intracellular parasite in the human host.

Plasmodium has proved to be a singularly difficult pathogen to attack. The complex *Plasmodium* life cycle is best broken by the removal of stagnant water, in which mosquitoes breed. The use of insecticides to reduce the *Anopheles* population can be effective, but their benefits must be weighed against the ecological, economic, and health risks posed by the insecticides themselves.

The genomes of one malarial parasite, *Plasmodium falciparum,* and one of its vectors, *Anopheles gambiae,* have been sequenced and published. These advances should lead to a better understand-

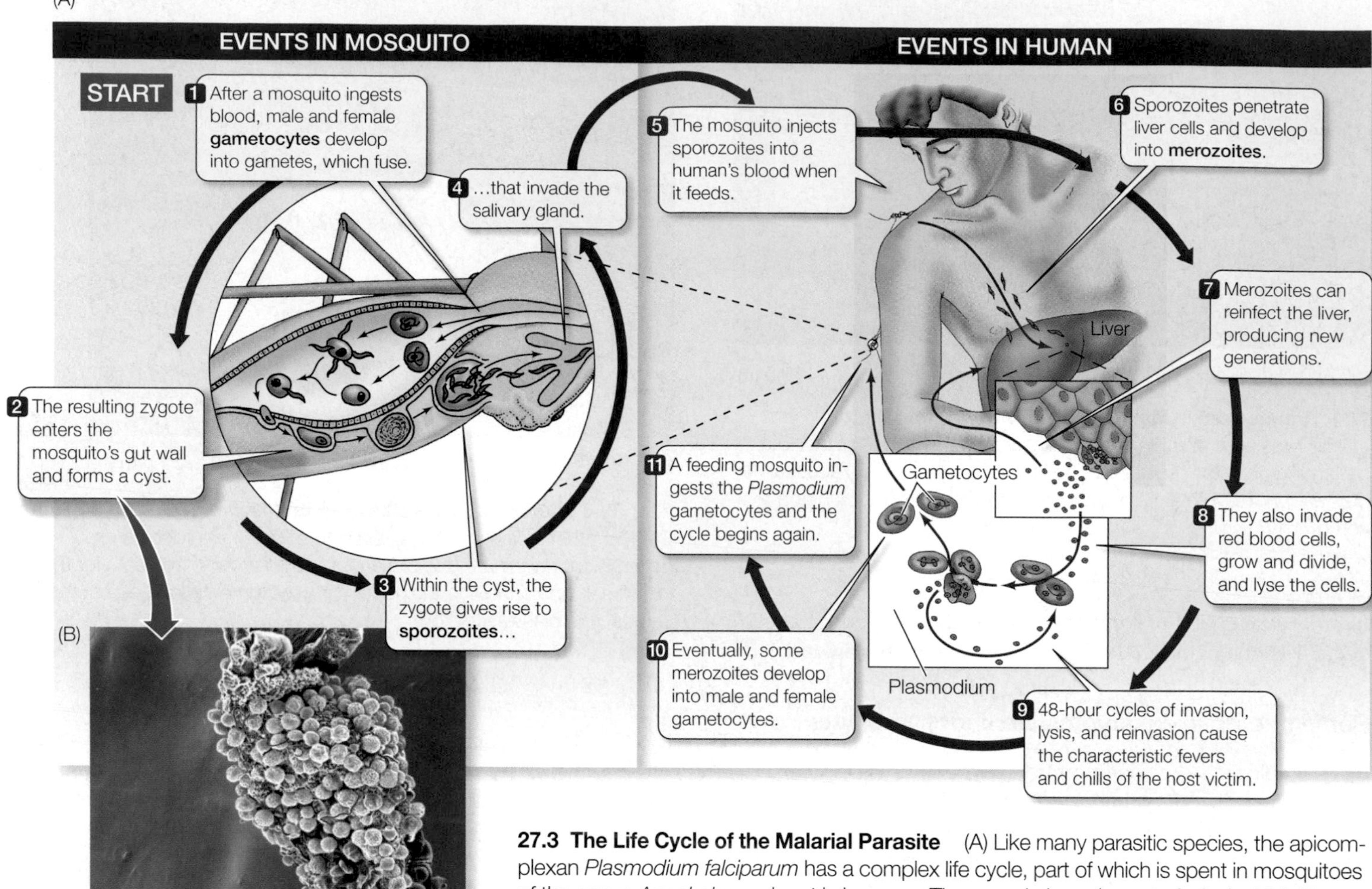
(A)
EVENTS IN MOSQUITO
EVENTS IN HUMAN
START
1 After a mosquito ingests blood, male and female **gametocytes** develop into gametes, which fuse.
2 The resulting zygote enters the mosquito's gut wall and forms a cyst.
3 Within the cyst, the zygote gives rise to **sporozoites**...
4 ...that invade the salivary gland.
5 The mosquito injects sporozoites into a human's blood when it feeds.
6 Sporozoites penetrate liver cells and develop into **merozoites**.
7 Merozoites can reinfect the liver, producing new generations.
Liver
8 They also invade red blood cells, grow and divide, and lyse the cells.
Gametocytes
Plasmodium
9 48-hour cycles of invasion, lysis, and reinvasion cause the characteristic fevers and chills of the host victim.
10 Eventually, some merozoites develop into male and female gametocytes.
11 A feeding mosquito ingests the *Plasmodium* gametocytes and the cycle begins again.
(B)
170 µm

27.3 The Life Cycle of the Malarial Parasite (A) Like many parasitic species, the apicomplexan *Plasmodium falciparum* has a complex life cycle, part of which is spent in mosquitoes of the genus *Anopheles* and part in humans. The sexual phase (gamete fusion) of this life cycle takes place in the insect, and the zygote is the only diploid stage. (B) Encysted *Plasmodium* zygotes (artificially colored blue) cover the stomach wall of a mosquito. Invasive sporozoites will hatch from the cysts and be transmitted to a human, in whom the parasite causes malaria.

ing of the biology of malaria and to the possible development of drugs, vaccines, or other means of dealing with this pathogen or its insect vectors. In the opening of Chapter 30 we describe a novel treatment of mosquito netting with fungi that attack mosquitoes, first reported in 2005.

Some *kinetoplastids* are human pathogens, such as the trypanosomes discussed at the opening of this chapter. Recall from the opening of this chapter that trypanosomes cause sleeping sickness, leishmaniasis, and Chagas' disease. The genomes of all three of these trypanosomes were sequenced in 2005.

Some *chromalveolates*, including diatoms, dinoflagellates, and haptophytes, reproduce in enormous numbers in warm and somewhat stagnant waters. The result can be a "red tide," so called because of the reddish color of the sea that results from the pigments of dinoflagellates (**Figure 27.4A**). During a dinoflagellate red tide, the concentration of cells may reach 60 million per liter of ocean water. Some red tide species produce a potent nerve toxin that can kill tons of fish. The genus *Gonyaulax* produces a toxin that can accumulate in shellfish in amounts that, although not fatal to the shellfish, may kill a person who eats the shellfish.

The haptophyte *Emiliania huxleyi* is one of the smallest unicellular eukaryotes, but it can form tremendous blooms in ocean waters. This *coccolithophore* ("sphere of stone") has an armored coating that makes the surface water more reflective (**Figure 27.4B**). This reflectivity cools the deeper layers of water below the bloom by reducing the amount of sunlight that penetrates. At the same time, it is possible that *E. huxleyi* contributes to global warming, because its metabolism increases the amount of dissolved CO_2 in ocean waters.

We continue to rely on the products of ancient marine microbial eukaryotes

Diatoms are lovely to look at, as we saw in Figure 27.1, but their importance to us goes far beyond aesthetics. They store oil as an

27.4 Chromalveolates Can Bloom in the Oceans (A) By reproducing in astronomical numbers, the dinoflagellate *Gonyaulax tamarensis* can cause toxic red tides, such as this one along the coast of Baja California. (B) Massive blooms of this coccolithophore, a tiny haptophyte, can reduce the amount of sunlight that is able to penetrate to the waters below.

energy reserve and to help them float at the correct depth in the ocean. Over millions of years, diatoms have died and sunk to the ocean floor, ultimately undergoing chemical changes and becoming a major source of petroleum and natural gas, two of our most important energy supplies and political concerns.

Other marine microbial eukaryotes have also contributed to today's world. Some *foraminiferans,* for example, secrete cells of calcium carbonate. After they reproduce (by mitosis and cytokinesis), the daughter cells abandon the parent shell and make new shells of their own. The discarded shells of ancient foraminiferans make up extensive limestone deposits in various parts of the world, forming a layer hundreds to thousands of meters deep over millions of square kilometers of ocean bottom. Foraminiferan shells also make up the sand of some beaches. A single gram of such sand may contain as many as 50,000 foraminiferan shells and shell fragments.

The shells of individual foraminiferans are easily preserved as fossils in marine sediments. The shells of foraminiferan species have distinctive shapes (**Figure 27.5**), and each geological period has a distinctive assemblage of foraminiferan species. For this reason, and because they are so abundant, the remains of foraminiferans are especially valuable in classifying and dating sedimentary rocks, as well as in oil prospecting. Studies of foraminiferan fossil shells are also used in determining the global temperatures prevalent at the time of their existence.

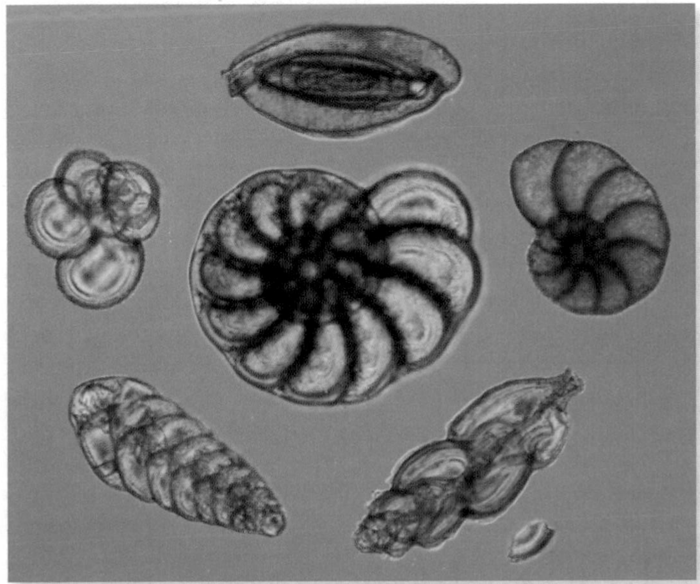

27.5 Foraminiferan Shells Are Building Blocks Foraminiferan shells are made of protein hardened with calcium carbonate. Over millions of years their remains have formed limestone deposits and sandy beaches. Several species are shown in this micrograph.

27.1 RECAP

Many different groups of species are treated here as "microbial eukaryotes," also known as "protists." Most but not all of these organisms are unicellular. These organisms have many effects, both positive and negative, on other organisms and on global ecosystems. Some species are primary producers, many are endosymbionts, and some are pathogens.

- Do you appreciate the full extent to which these organisms are *not* monophyletic? See p. 583 and Table 27.1; glance ahead to Figure 27.17
- Describe the role of female mosquitoes of the genus *Anopheles* as the vector for malaria. See p. 585 and Figure 27.3

This section has presented a very brief overview of the many diverse types of microbial eukaryotes. Indeed, perhaps the only thing all these organisms clearly have in common is that they are all eukaryotes. As we work to appreciate their origins and diversity, we are also working to understand the origin of the eukaryotic cell itself.

27.2 How Did the Eukaryotic Cell Arise?

The eukaryotic cell differs in many ways from the prokaryotic cell. Given the nature of evolutionary processes, these many differences cannot all have arisen simultaneously. We can make some reasonable inferences about the most important events that led to the evolution of a new cell type, bearing in mind that the global environment underwent an enormous change—from anaerobic to aerobic—during the course of these events (see Section 21.2). Keep in mind as you read that these inferences, although reasonable and grounded, are still conjectural; the hypothesis we pursue here is one of a few under current consideration. We present it as a framework for thinking about this challenging problem.

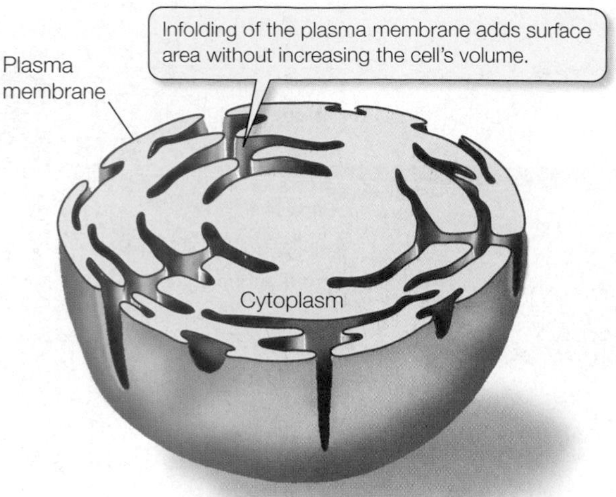

Infolding of the plasma membrane adds surface area without increasing the cell's volume.
Plasma membrane
Cytoplasm

27.6 Membrane Infolding The loss of the rigid prokaryotic cell wall may have allowed the plasma membrane to fold inward and create more surface area.

The modern eukaryotic cell arose in several steps

Several events preceded the origin of the modern eukaryotic cell:

- The origin of a flexible cell surface
- The origin of a cytoskeleton
- The origin of a nuclear envelope
- The appearance of digestive vesicles, or *vacuoles*
- The endosymbiotic acquisition of certain organelles

RAMIFICATIONS OF A FLEXIBLE CELL SURFACE Many ancient fossil prokaryotes look like rods, and we presume that they, like most present-day prokaryotic cells, had firm cell walls. The first step toward the eukaryotic condition was the loss of the cell wall by an ancestral prokaryotic cell. This wall-less condition is present in some present-day prokaryotes, although many others have developed new types of cell walls. Let's consider the possibilities open to a flexible cell without a wall.

First, think of cell size. As a cell grows larger, its surface area-to-volume ratio decreases (see Figure 4.2). Unless the surface area can be increased, the cell volume will reach an upper limit. If the cell's surface is flexible, it can fold inward and elaborate itself, creating more surface area for gas and nutrient exchange (**Figure 27.6**).

With a surface flexible enough to allow infolding, the cell can exchange materials with its environment rapidly enough to sustain a larger volume and more rapid metabolism. Furthermore, a flexible surface can pinch off bits of the environment, bringing them into the cell by endocytosis (**Figure 27.7, steps 1–3**).

CHANGES IN CELL STRUCTURE AND FUNCTION Other early steps in the evolution of the eukaryotic cell are likely to have included three advances: the appearance of a cytoskeleton; the formation of ribosome-studded internal membranes, some of which surrounded the DNA; and the evolution of digestive vesicles (**Figure 27.7, steps 3–7**).

A cytoskeleton made up of microfilaments and microtubules would support the cell and allow it to manage changes in shape, to distribute daughter chromosomes, and to move materials from one part of the now much larger cell to other parts. The presence of microtubules in the cytoskeleton could have evolved in some cells to give rise to the characteristic eukaryotic flagellum. The origin of the cytoskeleton is becoming clearer, as homologs of the genes that encode many cytoskeletal proteins have been found in modern prokaryotes.

The DNA of a prokaryotic cell is attached to a site on its plasma membrane. If that region of the plasma membrane were to fold into the cell, the first step would be taken toward the evolution of a *nucleus*, a primary feature of the eukaryotic cell.

From an intermediate kind of cell, the next step was probably *phagocytosis*—the ability to consume other cells by engulfing and digesting them. The first true eukaryote possessed a cytoskeleton and a nuclear envelope. It may have had an associated endoplasmic reticulum and Golgi apparatus, and perhaps one or more flagella of the eukaryotic type.

ENDOSYMBIOSIS AND ORGANELLES While the processes already outlined were taking place, the cyanobacteria were busy generating oxygen gas as a product of photosynthesis. The increasing O_2 levels in the atmosphere had disastrous consequences for most other living things because most organisms of the time (archaea and bacteria) were unable to tolerate the newly aerobic, oxidizing environment. But some prokaryotes managed to cope with these changes, and—fortunately for us—so did some of the new phagocytic eukaryotes.

At about this time endosymbiosis might have come into play (**Figure 27.7, steps 8 and 9**). Recall that the theory of endosymbiosis proposes that certain organelles are the descendants of prokaryotes engulfed, but not digested, by ancient eukaryotic cells (see Section 4.5). A crucial endosymbiotic event in the history of the Eukarya was the incorporation of a proteobacterium that evolved into the mitochondrion. Initially, the new organelle's primary function was probably to detoxify O_2 by reducing it to water.

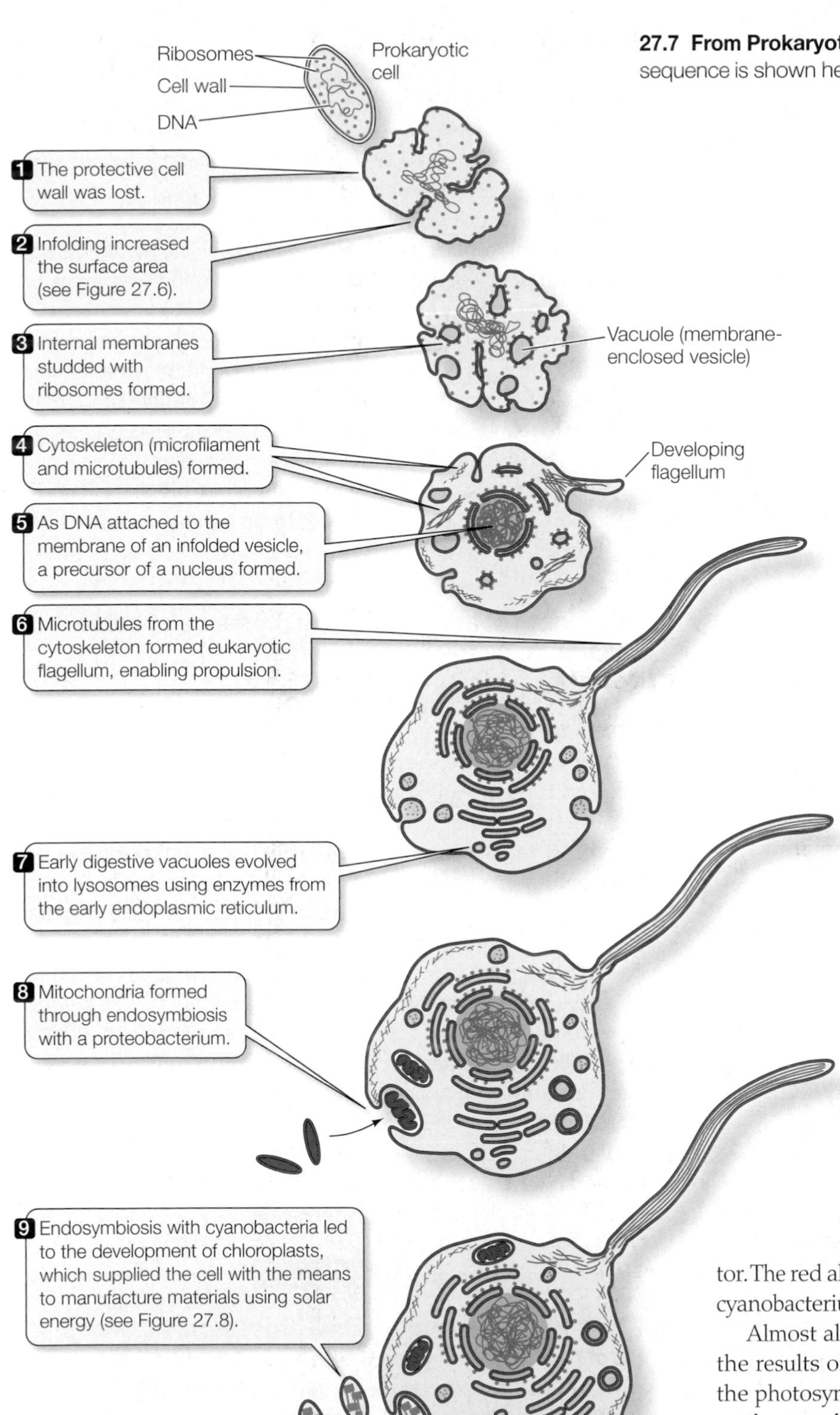
Ribosomes
Cell wall
DNA
Prokaryotic cell
1 The protective cell wall was lost.
2 Infolding increased the surface area (see Figure 27.6).
3 Internal membranes studded with ribosomes formed.
Vacuole (membrane-enclosed vesicle)
4 Cytoskeleton (microfilament and microtubules) formed.
Developing flagellum
5 As DNA attached to the membrane of an infolded vesicle, a precursor of a nucleus formed.
6 Microtubules from the cytoskeleton formed eukaryotic flagellum, enabling propulsion.
7 Early digestive vacuoles evolved into lysosomes using enzymes from the early endoplasmic reticulum.
8 Mitochondria formed through endosymbiosis with a proteobacterium.
9 Endosymbiosis with cyanobacteria led to the development of chloroplasts, which supplied the cell with the means to manufacture materials using solar energy (see Figure 27.8).

27.7 From Prokaryotic Cell to Eukaryotic Cell One possible evolutionary sequence is shown here.

Later, this reduction became coupled with the formation of ATP—respiration. Upon completion of this step, the basic modern eukaryotic cell was complete.

Some important eukaryotes are the result of yet another endosymbiotic step, the incorporation of a prokaryote related to today's cyanobacteria, which became the chloroplast.

Chloroplasts are a study in endosymbiosis

Eukaryotes in several different groups possess chloroplasts, and groups with chloroplasts appear in several distantly related clades. Some of these groups differ in the photosynthetic pigments their chloroplasts contain. And we'll see that not all chloroplasts have a pair of surrounding membranes—in some microbial eukaryotes, they are surrounded by *three or more* membranes. We now understand these observations in terms of a remarkable series of endosymbioses, supported by extensive evidence from electron microscopy and nucleic acid sequencing.

All chloroplasts trace their ancestry back to the engulfment of one cyanobacterium by a larger eukaryotic cell (**Figure 27.8A**). This event, the step that gave rise to the photosynthetic eukaryotes, is known as **primary endosymbiosis**. The cyanobacterium, a Gram-negative bacterium, had both an inner and an outer membrane. Thus the original chloroplasts had two surrounding membranes—the inner and outer membranes of the cyanobacterium. What about the peptidoglycan-containing wall of the bacterium? It is still represented today by a bit of peptidoglycan between the chloroplast membranes of *glaucophytes*, the first microbial eukaryote group to branch off following primary endosymbiosis.

Primary endosymbiosis gave rise to the chloroplasts of the "green algae" (including *chlorophytes* and *charophytes*) and the *red algae*. It is almost certain that both trace back to a single primary endosymbiosis, and that the divergence of these two distinct lineages occurred later. The photosynthetic land plants would arise from a green algal ancestor. The red algal chloroplast retains certain pigments of the original cyanobacterium that are absent in green algal chloroplasts.

Almost all remaining photosynthetic microbial eukaryotes are the results of secondary or tertiary endosymbiosis. For example, the photosynthetic *euglenids* derived their chloroplasts from **secondary endosymbiosis** (**Figure 27.8B**). Their ancestor took up a unicellular chlorophyte, retaining the endosymbiont's chloroplast and eventually losing the rest of its constituents. This history explains why the photosynthetic euglenids have the same photosynthetic pigments as the chlorophytes and land plants. It also accounts for the third membrane of the euglenoid chloroplast, which is derived from the euglenid's plasma membrane (as a result of endocytosis). Other evidence for secondary endosymbiosis comes from the observation that certain chloroplast products of secondary endosymbiosis contain traces of the nuclei of the cells that were engulfed.

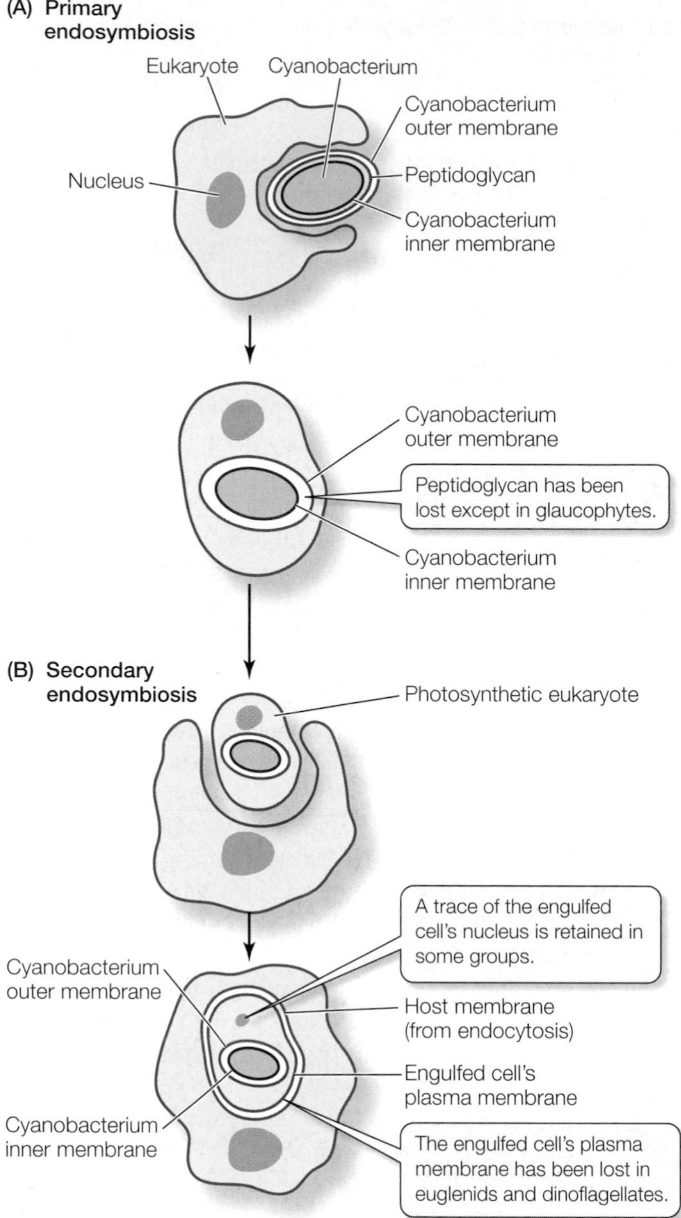

27.8 Endosymbiotic Events in the Family Tree of Chloroplasts (A) A single instance of primary endosymbiosis ultimately gave rise to all of today's chloroplasts. A eukaryotic cell engulfed a cyanobacterium, but did not digest it. (B) Secondary endosymbiosis—the uptake and retention of a chloroplast-containing cell by another eukaryotic cell—took place a few times. Tertiary endosymbiosis and sequential secondary endosymbiosis have also occurred.

Members of another photosynthetic microbial eukaryote group have chloroplasts derived by secondary endosymbiosis with a unicellular red alga. The chloroplasts of cryptophytes (one of the clades of chromalveolates) contain reduced red algal nuclei and appear to be sister to all other chromalveolate chloroplasts. The red clade of the chloroplasts may have been involved in only one secondary endosymbiosis, but the green clade of chloroplasts appears to have been involved in more than one secondary endosymbiosis.

Some dinoflagellates took up other partners by *tertiary endosymbiosis*. For example, one dinoflagellate apparently lost its chloroplast and took up a haptophyte (itself the result of secondary endosymbiosis). The result is the dinoflagellate *Karenia brevis*. There has been at least one case of *sequential secondary endosymbiosis*—another dinoflagellate lost its red algal chloroplast and engulfed a chlorophyte, thus becoming a new dinoflagellate species with a green algal chloroplast with two membranes. The actual number of endosymbiotic events in the course of evolution was small, but the results were diverse and profound.

Although euglenoid chloroplasts are descendants of a chlorophyte and haptophyte chloroplasts are descendants of a red alga, this does not mean that euglenids themselves are descendants of a chlorophyte, nor are haptophytes themselves descendants of a red alga. The ancestors that took up green or red algae in secondary endosymbioses had their own evolutionary histories. Thus the nuclear and chloroplast genomes of these organisms have different histories.

We cannot yet account for the presence of some prokaryotic genes in eukaryotes

Several uncertainties remain about the origins of eukaryotic cells. Lateral gene transfer complicates the study of eukaryote origins, just as it complicates the study of relationships among prokaryotes. And it seems unlikely that lateral gene transfer could have been extensive enough to account for the increasing numbers of genes of bacterial origin that are being found in eukaryotes by ongoing genetic analyses.

An endosymbiotic origin of mitochondria and chloroplasts accounts for the presence of bacterial genes encoding enzymes for energy metabolism (respiration and photosynthesis) in eukaryotes, but it does not explain the presence of some other bacterial genes. The eukaryotic genome clearly is a mixture of genes with different origins. A recent suggestion is that the Eukarya might have arisen from the mutualistic fusion of a Gram-negative bacterium and an archaean.

Many interesting ideas about eukaryotic origins await additional data and analysis. We can expect that these questions and others will eventually yield to additional research.

27.2 RECAP

The modern eukaryotic cell arose from an ancestral prokaryote in several steps, one of which involved endosymbiosis. The exact origins of eukaryotic cells are uncertain.

- Can you explain the importance of a flexible cell surface in the origin of eukaryotes? See p. 588 and Figure 27.6
- Can you identify some of the probable events involved in the evolution of the eukaryotic cell from a prokaryotic cell? See p. 588 and Figure 27.7
- Do you understand the difference between the origin of chloroplasts by primary and secondary endosymbiosis? See p. 589 and Figure 27.8

Having considered some of the known and suspected steps that led from the prokaryotic to the eukaryotic condition, let's now see what use the microbial eukaryotes made of their new features.

27.3 How Did the Microbial Eukaryotes Diversify?

The eukaryotic cell possesses some very useful features. The cytoskeleton allows for various means of locomotion, and also manages the controlled movement of cellular constituents (notably the mitotic and meiotic chromosomes). The specialized organelles of eukaryotes support a variety of activities. Given these tools, the microbial eukaryotes have been able to explore many environments and have exploited a variety of nutrient sources.

Microbial eukaryotes have different lifestyles

Most microbial eukaryotes are aquatic. Some live in marine environments, others in fresh water, and still others in the body fluids of other organisms. Many unicellular aquatic eukaryotes are plankton floating freely in the water. The *slime molds* inhabit damp soil and the moist, decaying bark of rotting trees. Other microbial eukaryotes also live in soil water, and some of them contribute to the global nitrogen cycle by preying on soil bacteria and recycling their nitrogen compounds into nitrates.

Some microbial eukaryotes are photosynthetic autotrophs, some are heterotrophs, and some switch with ease between the autotrophic and heterotrophic modes of nutrition. Some of the heterotrophs ingest their food; others, including many parasites, absorb nutrients from their environment.

Certain microbial eukaryotes, formerly classified as animals, are sometimes referred to as *protozoans,* although biologists increasingly believe that term lumps together too many phylogenetically distant groups. Most protozoans are ingestive heterotrophs. Similarly, there are several kinds of photosynthetic microbial eukaryotes that some biologists still refer to as *algae* (singular *alga*). Although these two terms are useful in some contexts, they do not correspond with our current understanding of phylogeny, and we generally avoid them in this chapter, except as parts of descriptive names such as "red algae."

Microbial eukaryotes have diverse means of locomotion

Although a few microbial eukaryote groups consist entirely of nonmotile organisms, most groups include cells that move, either by amoeboid motion, by ciliary action, or by means of flagella. Each of these types of motion is based on activities of the cytoskeleton.

In *amoeboid motion,* the cell forms **pseudopods** ("false feet") that are extensions of its constantly changing cell shape. Cells such as the one shown in **Figure 27.9** simply extend a pseudopod and then flow into it. Regions of the cytoplasm alternate between a more liquid state and a stiffer state, and a network of cytoskeletal microfilaments squeezes the more liquid cytoplasm forward.

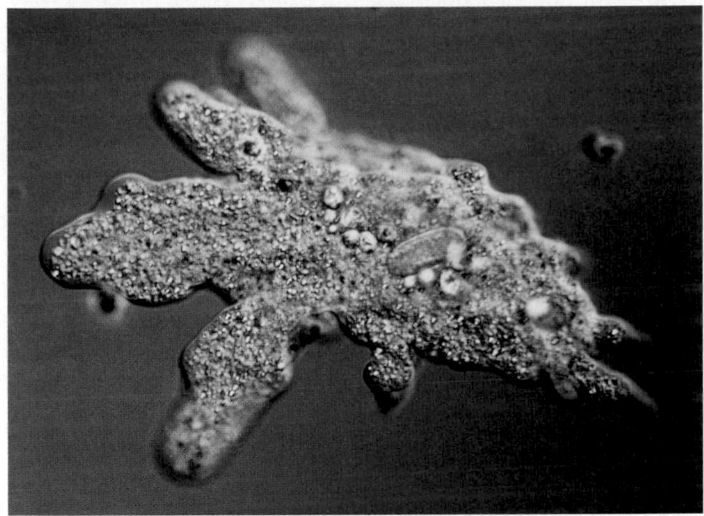

27.9 An Amoeba The flowing pseudopods are constantly changing shape as the amoeba moves and feeds.

As shown in Figure 27.7, the proteins of the eukaryotic cytoskeleton form microtubules that allowed the evolution of different means of locomotion. *Cilia* are tiny, hairlike organelles that beat in a coordinated fashion to move the cell forward or backward. Some ciliated organisms can change direction rapidly in response to their environment. A eukaryotic *flagellum* moves like a whip; some flagella *push* the cell forward, others *pull* the cell forward. Cilia and eukaryotic flagella are identical in cross section, with a "9 + 2" arrangement of microtubules (see Figure 4.22); they differ only in length.

Microbial eukaryotes employ vacuoles in several ways

Most unicellular organisms are microscopic. As we noted above, an important reason that cells are small is that they need enough membrane surface area in relation to their volume to support the exchange of materials required for their existence. Many relatively large unicellular eukaryotes minimize this problem by having membrane-enclosed *vacuoles* of various types that increase their effective surface area.

Organisms living in fresh water are hypertonic to their environment (see Section 5.3). Many freshwater microbial eukaryotes such as *Paramecium* address this problem by means of specialized vacuoles that excrete the excess water they constantly take in by osmosis. Members of several groups have such **contractile vacuoles**. The excess water collects in the contractile vacuole, which then expels the water from the cell (**Figure 27.10**).

A second important type of vacuole found in *Paramecium* and many other microbial eukaryotes is the **food vacuole**. These organisms engulf solid food by endocytosis, forming a vacuole within which the food is digested (**Figure 27.11**). Smaller vesicles containing digested food pinch away from the food vacuole and enter the cytoplasm. These tiny vesicles provide a large surface area across which the products of digestion may be absorbed by the rest of the cell.

27.10 Contractile Vacuoles Bail Out Excess Water Contractile vacuoles remove the water that constantly enters freshwater microbial eukaryotes by osmosis.

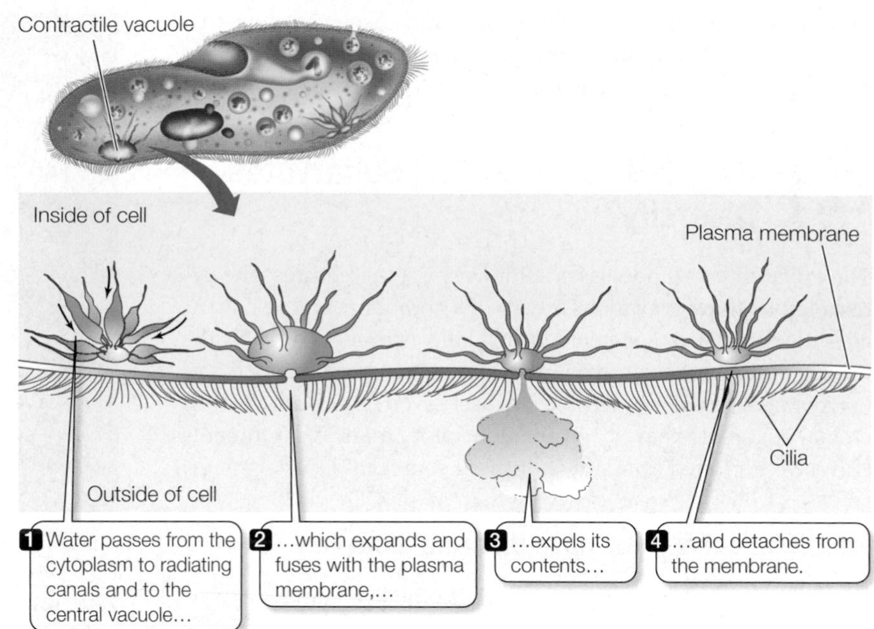
Contractile vacuole
Inside of cell
Plasma membrane
Cilia
Outside of cell
1 Water passes from the cytoplasm to radiating canals and to the central vacuole...
2 ...which expands and fuses with the plasma membrane,...
3 ...expels its contents...
4 ...and detaches from the membrane.

The cell surfaces of microbial eukaryotes are diverse

A few microbial eukaryotes, such as the amoeba shown in Figure 27.9, are surrounded by only a plasma membrane, but most have stiffer surfaces that maintain the structural integrity of the cell. Many have cell walls, which are often complex in structure, outside the plasma membrane. Other microbial eukaryotes that lack cell walls have a variety of ways of strengthening their surfaces.

Paramecium has proteins in its cell surface—known as a *pellicle* in this genus—that make it flexible but resilient. Other groups have external "shells," which the organism either produces itself, as foraminiferans do (see Figure 27.5), or makes from bits of sand and thickenings immediately beneath the plasma membrane, as some amoebas do (**Figure 27.12A**). The complex cell walls of diatoms are glassy, based on silicon (**Figure 27.12B**; see also Figure 27.1). Biologists recently measured, at a microscopic scale, the forces needed to break single, living diatoms. They discovered that the glassy cell walls are exceptionally strong. Evolution of these walls by natural selection may have given diatoms an enhanced defense against predators, and thus an edge over competitors.

EXPERIMENT

HYPOTHESIS: Paramecium digests its food by making food vacuoles acidic.

METHOD Paramecia are fed yeast cells that are stained with Congo red, a pH indicator.

RESULTS

1 A food vacuole forms around yeast cells.

Stained yeast cells

Oral groove

2 The change in color shows that the vacuole has become acidic, which helps digest the yeast cells.

3 As products of digestion move into the cytosol, the pH increases in the vacuole. The dye becomes red again.

4 Waste material is expelled.

CONCLUSION: Acidification of the food vacuoles assists digestion.

27.11 Food Vacuoles Handle Digestion and Excretion An experiment with *Paramecium* demonstrates the function of food vacuoles. *Paramecium* ingests food by way of the oral groove shown at the left. The dye Congo red turns green at an acidic pH and red at neutral or basic pH. Acid has a digestive function in other organisms (as in the human stomach), so its presence in food vacuoles suggests that digestion is taking place there. FURTHER RESEARCH: How might you determine whether food vacuoles in *Paramecium* contain digestive enzymes?

27.3 RECAP

Microbial eukaryotes are diverse in their habitat, nutrition, locomotion, and body form.

- Can you explain the roles of the cytoskeleton in the locomotion of microbial eukaryotes? See p. 591
- Do you understand the operation of contractile and food vacuoles in *Paramecium*? See p. 591 and Figures 27.10 and 27.11

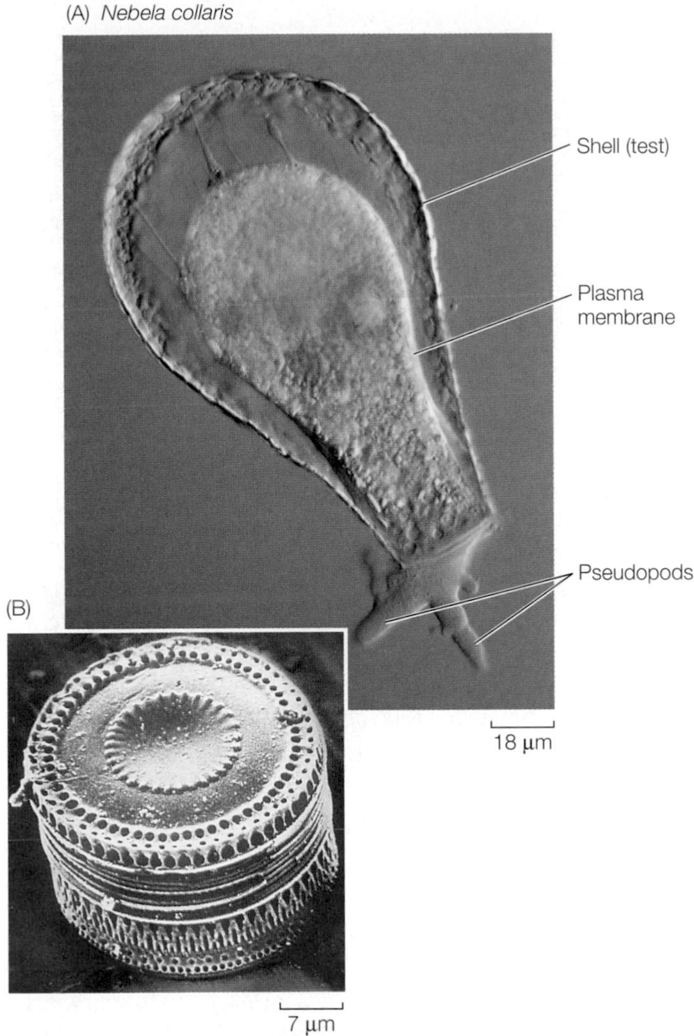

27.12 Cell Surfaces in the Microbial Eukaryotes (A) This testate amoeba has built a lightbulb-shaped shell, or test, by gluing sand grains together. Its pseudopods extend through the single aperture in the test (compare with Figure 27.9). (B) The artificial color added to this scanning electron micrograph shows the intricate patterning of the silicate cell surface of a diatom.

The diversity of body form, habitat, nutrition and locomotion found among the microbial eukaryotes reflects the diversity of avenues pursued during the early evolution of eukaryotes. The diverse reproductive modes and life cycles that have evolved among the microbial eukaryotes are a case in point.

27.4 How Do Microbial Eukaryotes Reproduce?

Although most microbial eukaryotes practice both asexual and sexual reproduction, some groups lack sexual reproduction. As we will see, some microbial eukaryotes separate the acts of sex and reproduction, so that the two are not directly linked.

The asexual reproductive processes found among the microbial eukaryotes include

- *Binary fission*: equal splitting of the cell, with mitosis followed by cytokinesis
- *Multiple fission*: splitting into more than two cells
- *Budding*: the outgrowth of a new cell from the surface of an old one
- *Spores*: the formation of specialized cells that are capable of developing into new organisms

Sexual reproduction in microbial eukaryotes takes various forms. In some microbial eukaryotes, as in animals, the gametes are the only haploid cells. In others, the zygote is the only diploid cell. In still others, by contrast, both diploid and haploid cells undergo mitosis, giving rise to *alternation of generations*.

Some microbial eukaryotes have reproduction without sex, and sex without reproduction

Like all groups in the *ciliate* clade, members of the genus *Paramecium* possess two types of nuclei, commonly a single *macronucleus* and, within the same cell, from one to several *micronuclei*. The micronuclei, which are typical eukaryotic nuclei, are essential for genetic recombination. The macronucleus is derived from micronuclei. Each macronucleus contains many copies of the genetic information, packaged in units containing very few genes each. The macronuclear DNA is transcribed and translated to regulate the life of the cell. In asexual reproduction, all of the nuclei are copied before the cell divides.

Paramecia also have an elaborate sexual behavior called **conjugation**, in which two paramecia line up tightly against each other and fuse in the oral groove region of the body. Nuclear material is extensively reorganized and exchanged over the next several hours (**Figure 27.13**). As a result of this process, each cell ends up with two haploid micronuclei, one of its own and one from the other cell, which fuse to form a new diploid micronucleus. A new macronucleus develops from the micronucleus through a series of dramatic chromosomal rearrangements. The exchange of nuclei is fully reciprocal—each of the two paramecia gives and receives an equal amount of DNA. The two organisms then separate and go their own ways, each equipped with new combinations of alleles.

Conjugation in *Paramecium* is a *sexual* process of genetic recombination, but it is not a *reproductive* process. The same two cells that begin the process are there at the end, and no new cells are created. As a rule, each asexual clone of paramecia must periodically conjugate. Experiments have shown that if some species are not permitted to conjugate, the clones can live through no more than approximately 350 cell divisions before they die out.

Many microbial eukaryote life cycles feature alternation of generations

What do we mean by **alternation of generations**, the type of life cycle found in many multicellular microbial eukaryotes and all land plants? In alternation of generations, a multicellular, diploid, spore-producing organism gives rise to a multicellular, haploid, gamete-producing organism. When two haploid gametes fuse (a process

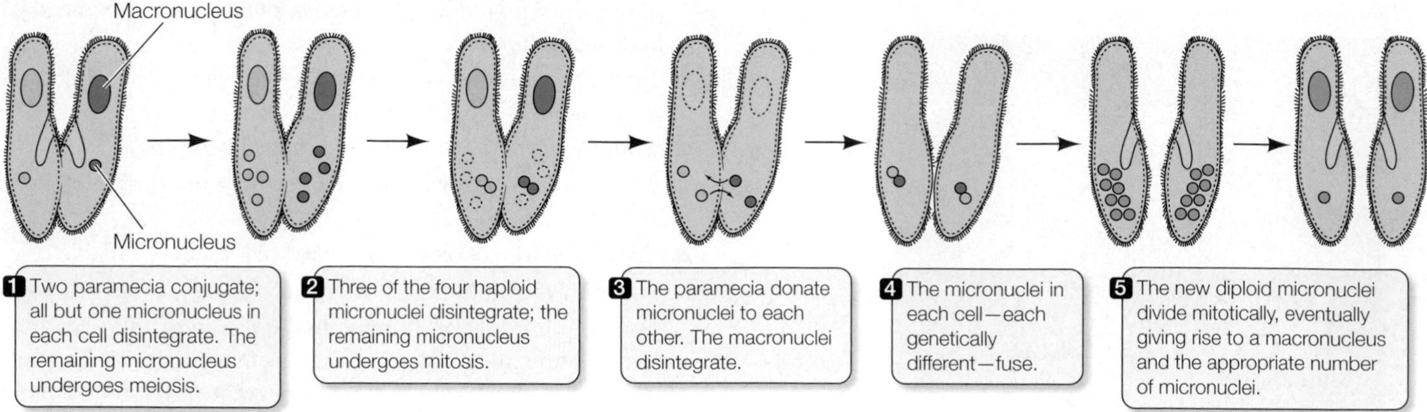

Macronucleus
Micronucleus
1 Two paramecia conjugate; all but one micronucleus in each cell disintegrate. The remaining micronucleus undergoes meiosis.
2 Three of the four haploid micronuclei disintegrate; the remaining micronucleus undergoes mitosis.
3 The paramecia donate micronuclei to each other. The macronuclei disintegrate.
4 The micronuclei in each cell—each genetically different—fuse.
5 The new diploid micronuclei divide mitotically, eventually giving rise to a macronucleus and the appropriate number of micronuclei.

27.13 Paramecia Achieve Genetic Recombination by Conjugating The exchange of micronuclei by conjugating *Paramecium* individuals results in genetic recombination. After conjugation, the cells separate and continue their lives as two individuals.

called *fertilization*, or *syngamy*), a diploid organism is formed (**Figure 27.14**). The haploid organism, the diploid organism, or both may also reproduce asexually.

The two alternating generations (spore-producing and gamete-producing) differ genetically (one has diploid cells and the other has haploid cells), but they may or may not differ morphologically. In **heteromorphic** alternation of generations, the two generations differ morphologically; in **isomorphic** alternation of generations, they do not, despite their genetic difference.

Examples of both heteromorphic and isomorphic alternation of generations are found in both brown algae and green algae. In discussing the life cycles of land plants and multicellular photosynthetic microbial eukaryotes, we will use the terms **sporophyte** ("spore plant") and **gametophyte** ("gamete plant") to refer to the multicellular diploid and haploid generations, respectively.

Gametes are not produced by meiosis because the gametophyte generation is already haploid. Instead, specialized cells of the diploid sporophyte, called **sporocytes**, divide meiotically to produce four haploid spores. The spores may eventually germinate and divide mitotically to produce multicellular haploid gametophytes, which produce gametes by mitosis and cytokinesis.

Gametes, unlike spores, can produce new organisms only by fusing with other gametes. The fusion of two gametes produces a diploid zygote, which then undergoes mitotic divisions to produce a diploid organism: the sporophyte generation. The sporocytes of the sporophyte generation then undergo meiosis and produce haploid spores, starting the cycle anew.

Chlorophytes provide examples of several life cycles

We can use different chlorophyte species to summarize the major features of microbial eukaryote life cycles. Let's begin with the sea lettuce *Ulva lactuca*. Like many chlorophytes, sea lettuce exhibits alternation of generations. The diploid sporophyte of this common multicellular seashore organism is a broad sheet only two cells thick. Some of its cells (sporocytes) differentiate and undergo meiosis and cytokinesis, producing motile haploid spores. These *zoospores* swim away, each propelled by four flagella, and some eventually find a suitable place to settle. The zoospores then lose their flagella and begin to divide mitotically, producing a thin filament that develops into a broad sheet only two cells thick. The gametophyte thus produced looks just like the sporophyte—in other words, *Ulva lactuca* has an isomorphic life cycle (**Figure 27.15**).

In most species of *Ulva*, the female and male gametes are also structurally indistinguishable, making those species **isogamous**—having gametes of identical appearance. Other chlorophytes, including some other species of *Ulva*, are **anisogamous**—having female gametes that are distinctly larger than the male gametes.

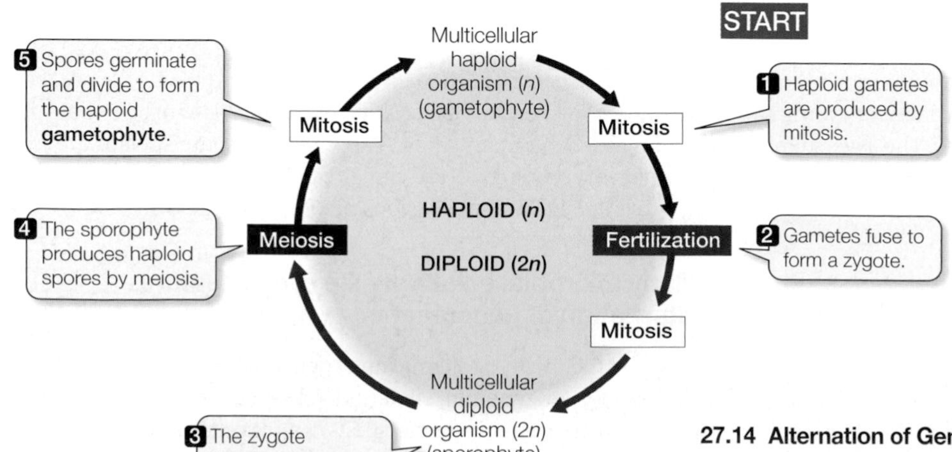

27.14 Alternation of Generations In many multicellular photosynthetic eukaryotes and all land plants, a diploid generation that produces spores alternates with a haploid generation that produces gametes.

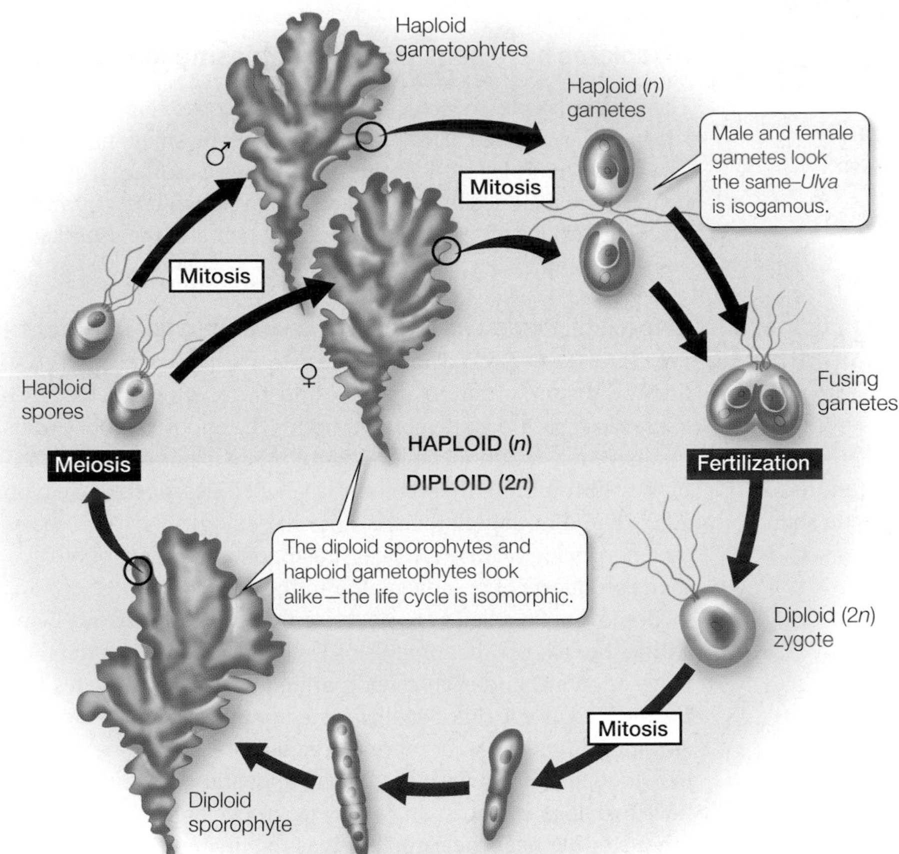

27.15 An Isomorphic Life Cycle The sexual life cycle of *Ulva lactuca* (sea lettuce) is an example of isomorphic alternation of generations.

pletes only part of its life cycle in the human host, and the other part in an insect (see Figure 27.3). Many other microbial eukaryote life cycles require the participation of two different host species.

What could be the advantage of a life cycle with two hosts? This remains an intriguing question. It may be relevant that in the human pathogens described above, the sexual phase of the organism's life cycle—the fusion of gametes into a zygote—takes place in the insect vector. Could this imply that the human host is nothing but a copying machine for the products of sexual reproduction in the vector?

The life cycles of many other chlorophytes do not feature alternation of generations. Some chlorophytes have a **haplontic** life cycle, in which a multicellular haploid individual produces gametes that fuse to form a zygote. The zygote functions directly as a sporocyte, undergoing meiosis to produce spores, which in turn produce a new haploid individual. In the entire haplontic life cycle, only one cell—the zygote—is diploid. The filamentous organisms of the genus *Ulothrix* are examples of haplontic chlorophytes (**Figure 27.16**).

Some other chlorophytes have a **diplontic** life cycle like that of many animals. In a diplontic life cycle, meiosis of diploid sporocytes produces haploid gametes directly; the gametes fuse, and the resulting diploid zygote divides mitotically to form a new multicellular diploid sporophyte. In such organisms, all cells except the gametes are diploid. Between these two extremes are chlorophytes in which the gametophyte and sporophyte generations are both multicellular, but one generation (usually the sporophyte) is much larger and more prominent than the other.

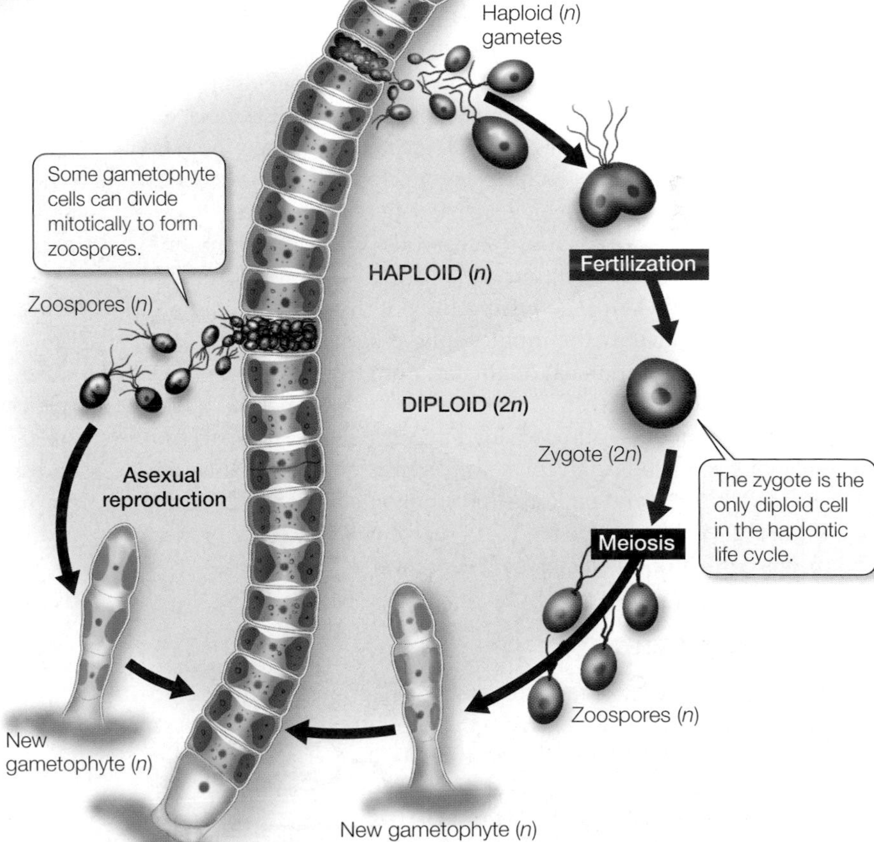

27.16 A Haplontic Life Cycle In the life cycle of *Ulothrix*, a filamentous, multicellular haploid gametophyte generation alternates with a diploid sporophyte generation consisting of a single cell (the zygote). *Ulothrix* gametophytes can also reproduce asexually.

The life cycles of some microbial eukaryotes require more than one host species

The three trypanosome diseases discussed at the opening of this chapter share a striking feature with malaria: in each case, the eukaryote pathogen com-

27.4 RECAP

Microbial eukaryotes reproduce both asexually and sexually, and their life cycles are diverse.

- Why is conjugation between paramecia considered a sexual process, but not a reproductive process? See p. 593 and Figure 27.13
- Can you explain the difference between the diplontic human life cycle and a life cycle with alternation of generations? See p. 595

The success of the microbial eukaryotes' diverse adaptations for nutrition, locomotion, and reproduction is evident from the abundance and diversity of eukaryotes living today. In the next section we will survey that diversity.

27.5 What Are the Major Groups of Eukaryotes?

Biologists used to classify the microbial eukaryotes on the basis of features such as those we have described in Sections 27.3 and 27.4. However, scientists using the advanced technologies of electron microscopy and gene sequencing have revealed many new patterns of evolutionary relatedness. New molecular biological techniques, such as rRNA gene sequencing (see Section 26.4), are making it possible to explore evolutionary relationships among the microbial eukaryotes in ever greater detail and with greater confidence. Today we recognize great diversity within many microbial eukaryote clades, whose members have explored a great variety of lifestyles.

The phylogeny of microbial eukaryotes is an area of exciting, challenging research. Their marvelous diversity of body forms and nutritional lifestyles alone justify study of these organisms, and questions about how the multicellular eukaryotic groups (land plants, fungi, and animals) originated from the microbial groups stimulate further interest.

Most eukaryotes can be divided into five major groups: chromalveolates (including alveolates and stramenopiles), Plantae, excavates, Rhizaria, and unikonts (opisthokonts and amoebozoans) (**Figure 27.17**; also see Table 27.1). As we will see, some of these groups consist of organisms with very diverse body plans.

CHROMALVEOLATES

We'll begin our tour of microbial eukaryote groups with the **chromalveolates**, a group that includes the haptophytes, the cryptophytes, and two other large clades: the alveolates and the stramenopiles. The **haptophytes** are unicellular organisms with flagella; many are "armored" with elaborate scales (see Figure 27.4B). The role of the **cryptophytes** in the story of chloroplast evolution was described in Section 27.2.

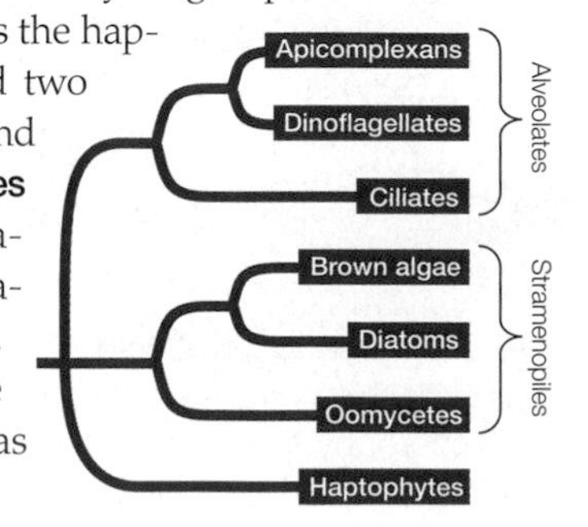

Alveolates have sacs under their plasma membrane

The synapomorphy that characterizes the **alveolate** clade is the possession of sacs called *alveoli* just below their plasma membranes. The alveoli may play a role in supporting the cell surface. These organisms are all unicellular, but they are diverse in body form. The alveolate groups we'll consider in detail here are the dinoflagellates, apicomplexans, and ciliates.

DINOFLAGELLATES ARE UNICELLULAR MARINE ORGANISMS WITH TWO FLAGELLA Most **dinoflagellates** are marine organisms. A distinctive mixture of photosynthetic and accessory pigments gives their chloroplasts a golden brown color. (The endosymbiotic events that gave rise to dinoflagellates with different numbers of membranes surrounding their chloroplasts were described in Section 27.2.) The dinoflagellates are of great ecological, evolutionary, and morphological interest. They are important primary photosynthetic producers of organic matter in the oceans.

Some dinoflagellates are photosynthetic endosymbionts living within the cells of other organisms, including various invertebrates (such as corals) and even other marine microbial eukaryotes (see Figure 27.2). Some dinoflagellates are nonphotosynthetic and live as parasites within other marine organisms.

Dinoflagellates have a distinctive appearance. They generally have two flagella, one in an equatorial groove around the cell, the other starting near the same point as the first and passing down a longitudinal groove before extending into the surrounding medium (**Figure 27.18**). Some dinoflagellates, notably *Pfiesteria piscicida*, can take on different forms, including amoeboid ones, depending on environmental conditions. It has been claimed that *P. piscicida* can occur in at least two dozen distinct forms, although this claim is highly controversial. In any case, this remarkable dinoflagellate is harmful to fish and can, when present in great numbers, both stun and feed on them.

ALL APICOMPLEXANS ARE PARASITES Exclusively parasitic organisms, the **apicomplexans** derive their name from the *apical complex*, a mass of organelles contained within the *apical* end (the tip) of a cell. These organelles help the apicomplexan invade its host's tissues. For example, the apical complex enables the merozoites and sporozoites of *Plasmodium*, the causal agent of malaria, to enter their target cells in the human body (see Figure 27.3).

Like many obligate parasites, apicomplexans have elaborate life cycles featuring asexual and sexual reproduction by a series of very dissimilar life stages. Often these stages are associated with two different types of host organisms, as is the case with *Plasmodium*.

The apicomplexan *Toxoplasma* alternates between cats and rats to complete its life cycle. A rat infected with *Toxoplasma* loses its fear of cats, making it more likely to be eaten by, and thus transfer the parasite to, a cat.

Apicomplexans lack contractile vacuoles. They contain a much-reduced, nonfunctional chloroplast (derived, like all chromalveolate chloroplasts, from secondary endosymbiosis of a red alga). This chloroplast might be a target for a future antimalarial drug.

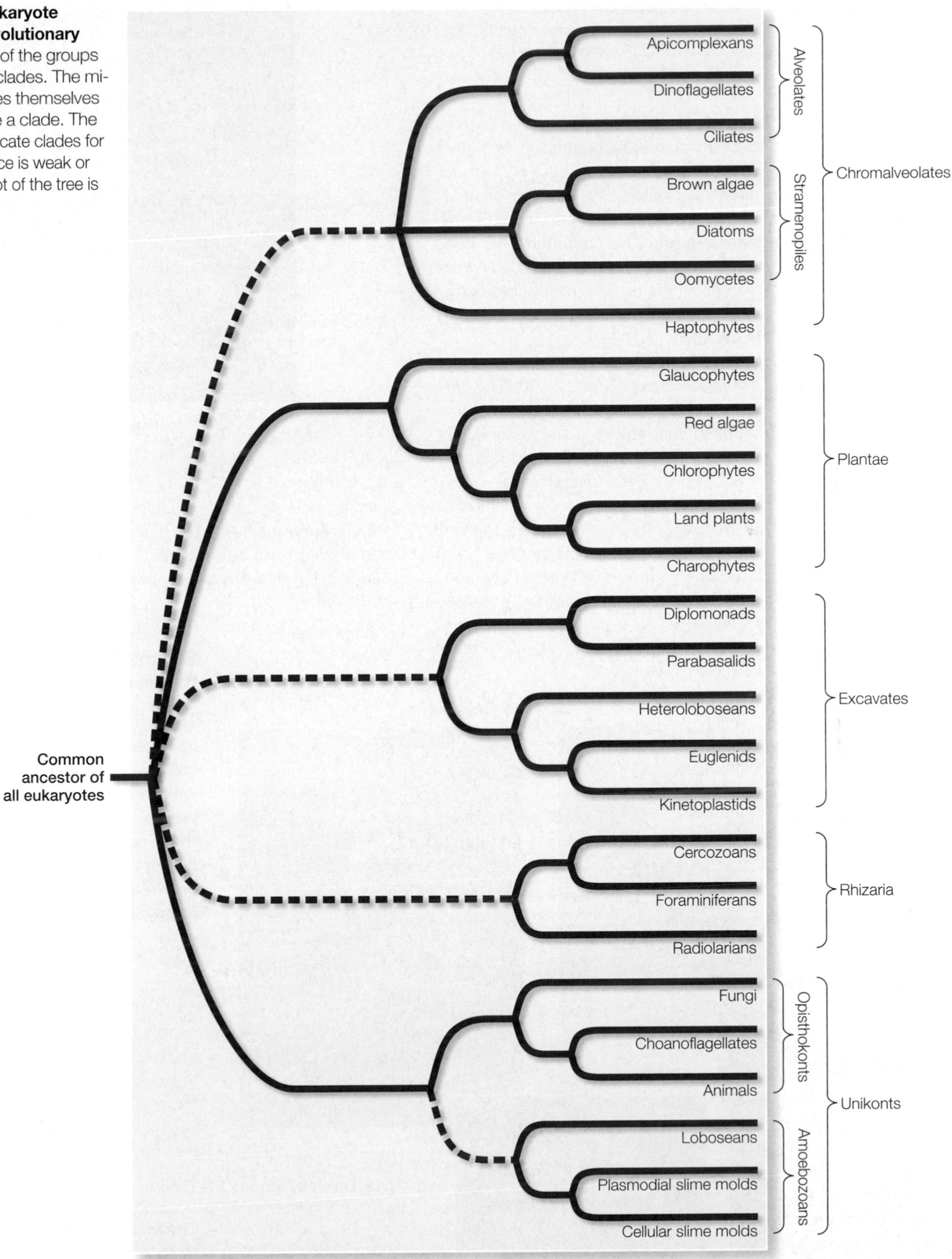

27.17 Major Eukaryote Groups in an Evolutionary Context Most of the groups shown here are clades. The microbial eukaryotes themselves do not constitute a clade. The dashed lines indicate clades for which the evidence is weak or disputed. The root of the tree is uncertain.

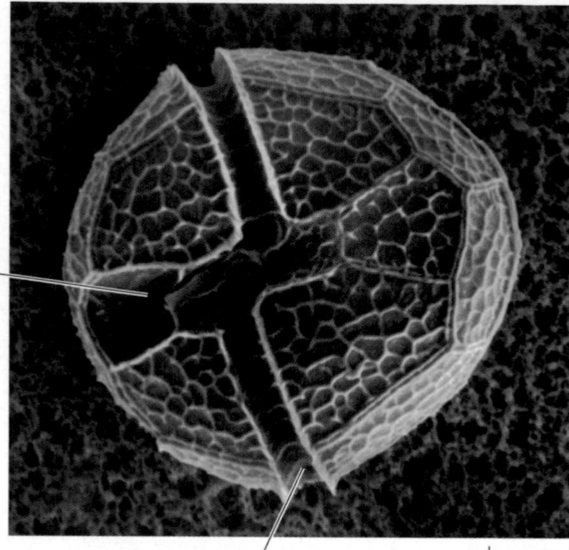

27.18 A Dinoflagellate The dinoflagellates are an important group of alveolates. Most of them are photosynthetic and are a crucial component of the world's phytoplankton. They are often endosymbiotic (see Figure 27.2) and can be the agents of deadly ocean "blooms" (see Figure 27.4A).

CILIATES HAVE TWO TYPES OF NUCLEI The **ciliates** are so named because they characteristically have numerous hairlike cilia shorter than, but otherwise identical to, eukaryotic flagella. This group is noteworthy for its diversity and ecological importance (**Figure 27.19**). Almost all ciliates are heterotrophic (although a few contain photosynthetic endosymbionts), and they are much more complex in body form than are most other unicellular eukaryotes. The definitive characteristic of ciliates is the possession of two types of nuclei (see Figure 27.13).

Paramecium, a frequently studied ciliate genus, exemplifies the complex structure and behavior of ciliates (**Figure 27.20**). The slipper-shaped cell is covered by an elaborate pellicle, a structure composed principally of an outer membrane and an inner layer of closely packed, membrane-enclosed sacs (the alveoli) that surround the bases of the cilia. Defensive organelles called *trichocysts* are also present in the pellicle. In response to a threat, a microscopic explosion expels the trichocysts in a few milliseconds, and they emerge as sharp darts, driven forward at the tip of a long, expanding filament.

The cilia provide a form of locomotion that is generally more precise than locomotion by flagella or pseudopods. A paramecium can coordinate the beating of its cilia to propel itself either forward or backward in a spiraling manner. It can also back off swiftly when it encounters a barrier or a negative stimulus. The coordination of ciliary beating is probably the result of a differential distribution of ion channels in the plasma membrane near the two ends of the cell.

Stramenopiles have two unequal flagella, one with hairs

The synapomorphy that defines the **stramenopiles** is the possession of rows of tubular hairs on the longer of their two flagella. Some stramenopiles lack flagella, but they are descended from ancestors that possessed flagella. The stramenopiles include the diatoms and the brown algae, which are photosynthetic, and the oomycetes and slime nets, which are not. Most golden algae are photosynthetic, but nearly all of them become heterotrophic when light intensity is limiting or when there is a plentiful food supply; some even feed on diatoms or bacteria. The slime nets, formerly thought to be close relatives of the slime molds,

(A) *Paramecium bursaria* — Cilia — 10 μm

(B) *Vorticella* sp. — Cilia — 10 μm

(C) *Paracineta* sp. — Tentacles — 20 μm

(D) *Euplotes* sp. — Cirri — Oral groove — 25 μm

27.19 Diversity among the Ciliates (A) A free-swimming organism, this paramecium belongs to a ciliate group whose members have many cilia of uniform length. (B) Members of this group have cilia on their mouthparts. (C) In this group, tentacles replace cilia as development proceeds. (D) This ciliate "walks" on bundled cilia, called cirri, which project from its body. Other cilia are grouped into flat sheets that sweep food particles into its oral groove; this individual has ingested some green algae.

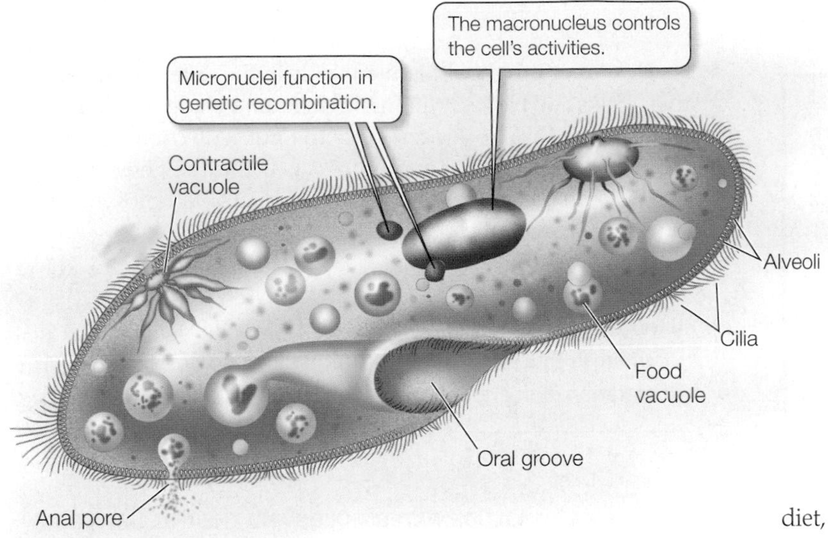
The macronucleus controls the cell's activities.
Micronuclei function in genetic recombination.
Contractile vacuole
Alveoli
Cilia
Food vacuole
Oral groove
Anal pore

27.20 Anatomy of *Paramecium* This diagram shows the complex structure of a typical paramecium.

are unicellular organisms that produce networks of filaments along which the cells move.

DIATOMS ABOUND IN THE OCEANS **Diatoms** are unicellular organisms, although some species associate in filaments. Many have sufficient carotenoids in their chloroplasts to give them a yellow or brownish color. All make *chrysolaminarin* (a carbohydrate) and oils as photosynthetic storage products. Diatoms lack flagella except in male gametes.

Architectural magnificence on a microscopic scale is the hallmark of the diatoms. As mentioned earlier, almost all diatoms deposit silicon in their cell walls. The cell wall is constructed in two pieces, with the top overlapping the bottom like the top and bottom of a petri plate (see Figure 27.1). The silicon-impregnated walls have intricate patterns unique to each species (**Figure 27.21**). Despite their remarkable morphological diversity, however, all diatoms are symmetrical—either bilaterally (with "right" and "left" halves) or radially (with the type of symmetry possessed by a circle).

Diatoms reproduce both sexually and asexually. Asexual reproduction is by binary fission and is somewhat constrained by the stiff, silica-containing cell wall. Both the top and the bottom of the "petri plate" become tops of new "plates" without changing appreciably in size; as a result, the new cell made from the former bottom is smaller than the parent cell. If this process continued indefinitely, one cell line would simply vanish, but sexual reproduction largely solves this potential problem. Gametes are formed, shed their cell walls, and fuse. The resulting zygote then increases substantially in size before a new cell wall is laid down.

Diatoms are found in all the oceans and are frequently present in great numbers, making them major photosynthetic producers in coastal waters. Diatoms are also common in fresh water and even occur on the wet surfaces of terrestrial mosses.

Patches of ocean are occasionally populated by dense plankton blooms dominated by diatoms. Surprisingly, these diatom blooms are not heavily grazed by the copepods (tiny planktonic crustaceans) that are their usual predators. Thus much of the diatom population dies and sinks to the seafloor ungrazed. Later, as the bloom dissipates, the copepod population increases.

Why doesn't the copepod population peak and take advantage of the apparent riches of the diatom bloom? It had been suspected that this "failure" might result from a slow response by the copepod life cycle. Recent experimental data suggest, however, that the explanation lies in the effects of the diatoms on the reproductive success of the copepods when diatoms make up too great a proportion of their diet (**Figure 27.22**). While diatoms constitute a good food source for copepods when they make up a moderate proportion of the diet, they are toxic in large quantities because they release a toxic compound when they are crushed.

Because the silicon-containing walls of dead diatom cells resist decomposition, certain sedimentary rocks are composed almost entirely of diatom skeletons that sank to the seafloor over time. Diatomaceous earth, which is obtained from such rocks, has many industrial uses, such as insulation, filtration, and metal polishing. It has also been used as an "Earth-friendly" insecticide that clogs the tracheae (breathing structures) of insects.

THE BROWN ALGAE ARE MULTICELLULAR All the **brown algae** are multicellular, and some are large: giant kelps, such as those of the genus *Macrocystis*, may be up to 60 meters long (see p. 583). The brown algae obtain their namesake color from the carotenoid *fucoxanthin*, which is abundant in their chloroplasts. The combination of this yellow-orange pigment with the green of chlorophylls *a* and *c* yields a brownish tinge.

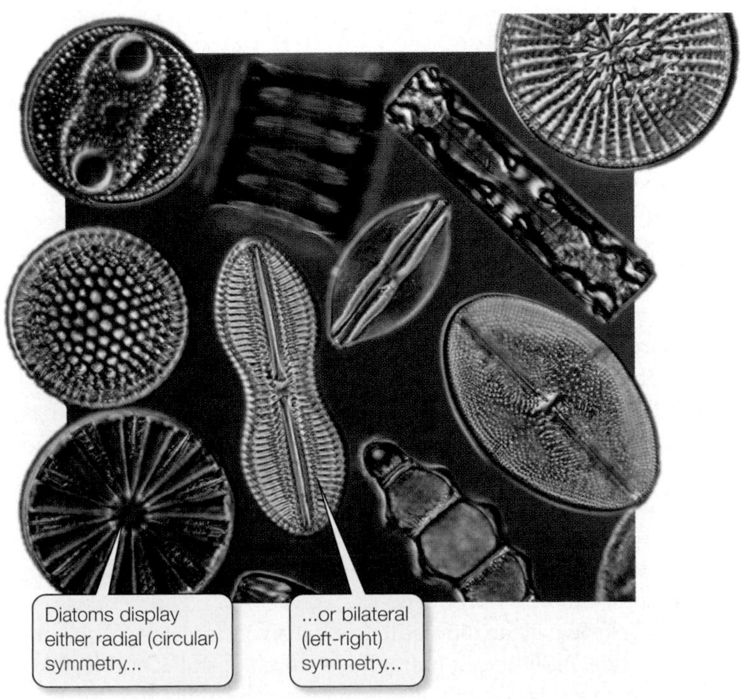
Diatoms display either radial (circular) symmetry...
...or bilateral (left-right) symmetry...

27.21 Diatom Diversity Diatoms exhibit a splendid variety of species-specific forms.

EXPERIMENT

HYPOTHESIS: Copepods fail to flourish during a diatom bloom because of the toxicity of the diatoms.

METHOD

1. Female copepods are fed for 10 days on either the diatom *Calanus helgolandicus* or a nontoxic diet of dinoflagellates (control).
2. Eggs and newly hatched larvae are counted each day.

RESULTS

Although almost all eggs of the control group hatched each day, fewer and fewer of the eggs from the diatom-fed females hatched as the experiment progressed.

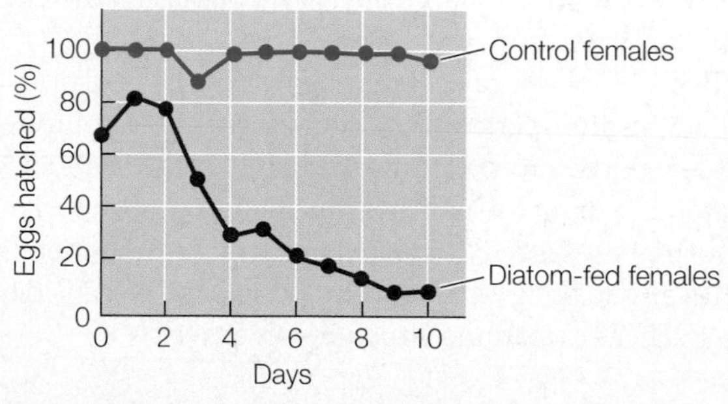

CONCLUSION: Toxicity is a plausible explanation for the failure of copepods to flourish during a diatom bloom.

27.22 Why Don't Copepods Flourish During Diatom Blooms? Adrianna Ianora and her colleagues tested the hypothesis that diatoms impair the reproduction of copepods when they constitute too great a proportion of the diet. FURTHER RESEARCH: How might you further test this hypothesis?

The brown algae are almost exclusively marine. They are composed either of branched filaments (**Figure 27.23A,B**) or of leaflike growths (**Figure 27.23C**). Some float in the open ocean; the most famous example is the genus *Sargassum,* which forms dense mats in the Sargasso Sea in the mid-Atlantic. Most brown algae, however, are attached to rocks near the shore. A few thrive only where they are regularly exposed to heavy surf; a notable example is the sea palm *Postelsia palmaeformis* of the Pacific coast. All of the attached forms develop a specialized structure, called a *holdfast,* that literally glues them to the rocks (**Figure 27.23D**). The "glue" of the holdfast is *alginic acid,* a gummy polymer of sugar acids found in the walls of many brown algal cells. In addition to its function in holdfasts, alginic acid cements algal cells and filaments together, and is harvested and used by humans as an emulsifier in ice cream, cosmetics, and other products.

Some brown algae differentiate extensively into specialized organs. Some, like the sea palm, have stemlike stalks and leaflike blades. Some develop gas-filled cavities or bladders that serve as floats. In addition to organ differentiation, the larger brown algae also exhibit considerable tissue differentiation. Most of the giant kelps have photosynthetic filaments only in the outermost regions of their stalks and blades. Within the stalks and blades lie filaments of tubular cells that closely resemble the nutrient-conducting tissue of land plants. Called *trumpet cells* because they have flaring ends, these tubes rapidly conduct the products of photosynthesis through the body of the organism.

The specialized flotation bladders of brown algae often contain as much as 5 percent carbon monoxide—a concentration high enough to kill a human.

THE OOMYCETES INCLUDE WATER MOLDS AND THEIR RELATIVES A nonphotosynthetic stramenopile group called the **oomycetes** consists in large part of the water molds and their terrestrial relatives, such as the downy mildews. Water molds are filamentous and stationary, and they are *absorptive heterotrophs*—that is, they secrete enzymes that digest large food molecules into smaller molecules that the water mold can absorb. If you have seen a whitish, cottony mold growing on dead fish or dead insects in water, it was probably a water mold of the common genus *Saprolegnia* (**Figure 27.24**).

Don't be confused by the *mycete* in the name of this group. That term means "fungus," and it is there because these organisms were once classified as fungi. However, we now know that the oomycetes are unrelated to the fungi.

Some oomycetes are **coenocytes**: that is, they have many nuclei enclosed in a single plasma membrane. Their filaments have no cross-walls to separate the many nuclei into discrete cells. Their cytoplasm is continuous throughout the body of the organism, and there is no single structural unit with a single nucleus, except in certain reproductive stages. A distinguishing feature of the oomycetes is their flagellated reproductive cells. Oomycetes are diploid throughout most of their life cycle and have cellulose in their cell walls.

The water molds, such as *Saprolegnia,* are all aquatic and **saprobic** (they feed on dead organic matter). Some other oomycetes are terrestrial. Although most of the terrestrial oomycetes are harmless or helpful decomposers of dead matter, a few are serious plant parasites that attack crops such as avocados, grapes, and potatoes.

Although their presumed chromalveolate ancestors had chloroplasts and were photosynthetic, the oomycetes lack chloroplasts. The next major group we'll consider, the Plantae, is predominantly photosynthetic.

PLANTAE

Glaucophytes
Red algae
Chlorophytes
Land plants
Charophytes

The **Plantae** consists of several major clades, including glaucophytes, red algae, chlorophytes, charophytes, and the land plants, all of which probably trace their chloroplasts back to a single incidence of endosymbiosis (see Section 27.2). It is for this reason that the small clade known as the **glaucophytes**, unicellular organisms that live in fresh water, is of great interest to students of evolution.

(A) *Pelvetia canaliculata*

(B) *Ectocarpus* sp.

(C) *Postelsia palmaeformis*

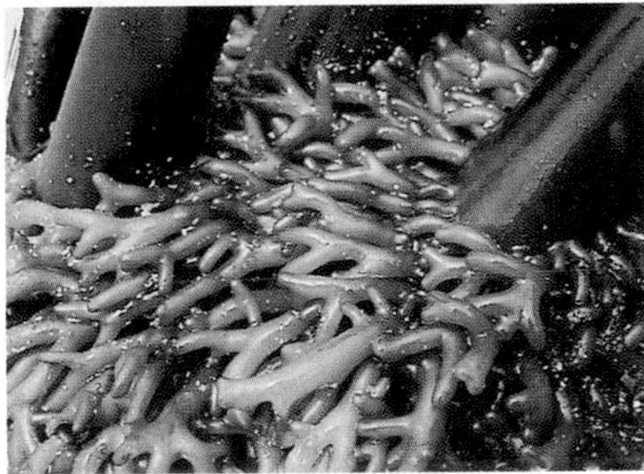

(D) *Postelsia palmaeformis*

27.23 Brown Algae (A) Channeled wrack seaweed illustrates the filamentous growth form of brown algae. (B) Filaments of the microscopic brown alga *Ectocarpus* seen through a light microscope. (C) Sea palms exemplify the leaflike growth form. Sea palms grow in the intertidal zone, where they take a tremendous pounding by the surf. (D) Sea palms and many other brown algal species are "glued" to the substratum by tough, branched structures called holdfasts.

The glaucophytes were likely the first group to diverge after the primary endosymbiosis event. Their chloroplast is unique in containing a small amount of peptidoglycan between its inner and outer membranes—the same arrangement as that found in cyanobacteria (see Figure 27.8A). The presence of peptidoglycan, the characteristic cell wall component of bacteria, suggests that the glaucophytes resemble the common ancestor of all Plantae.

Saprolegnia sp.

27.24 An Oomycete The filaments of a water mold radiate from the carcass of an insect.

Red algae have a distinctive accessory photosynthetic pigment

Almost all **red algae** are multicellular (**Figure 27.25**). Their characteristic color is a result of the accessory photosynthetic pigment *phycoerythrin,* which is found in relatively large amounts in the chloroplasts of many species. In addition to phycoerythrin, red algae contain phycocyanin, carotenoids, and chlorophyll *a*.

The red algae include species that grow in the shallowest tide pools as well as the photosynthesizers found deepest in the ocean (as deep as 260 meters if nutrient conditions are right and the water is clear enough to permit light to penetrate). Very few red algae inhabit fresh water. Most grow attached to a substratum by a holdfast.

In a sense, the red algae are misnamed. They have the capacity to change the relative amounts of their various photosynthetic pigments depending on the light conditions where they are growing. Thus the leaflike *Chondrus crispus,* a common North Atlantic red alga, may appear bright green when it is growing at or near the surface of the water and deep red when growing at greater depths. The ratio of pigments present depends to a remarkable degree on the intensity of the light that reaches the alga. In deep water,

27.25 Red Algae (A) Under the light microscope, reproductive structures can be seen in this red alga. (B) A coralline red alga grows along the coast of central Oregon.

where the light is dimmest, the alga accumulates large amounts of phycoerythrin. The algae in deep water have as much chlorophyll as the green ones near the surface, but the accumulated phycoerythrin makes them look red.

In addition to being the only photosynthetic eukaryotes with phycoerythrin among their pigments, the red algae have two other distinctive characteristics:

- They store the products of photosynthesis in the form of *floridean starch*, which is composed of very small, branched chains of approximately 15 glucose monomers.
- They produce no motile, flagellated cells at any stage of their life cycle. The male gametes lack cell walls and are slightly amoeboid; the female gametes are completely immobile.

Some red algal species enhance the formation of coral reefs (see Figure 27.25B). Like coral animals, they possess the biochemical machinery for secreting calcium carbonate, which they deposit both in and around their cell walls. After the deaths of corals and algae, the calcium carbonate persists, sometimes forming substantial rocky masses.

Some red algae produce large amounts of mucilaginous polysaccharide substances, which contain the sugar galactose with a sulfate group attached. This material readily forms solid gels and is the source of agar, a substance widely used in the laboratory for making a solid aqueous medium on which tissue cultures and many microorganisms can be grown.

A red alga became the ancestor of the distinctive chloroplasts of the photosynthetic chromalveolates by secondary endosymbiosis, as we saw in Section 27.2.

Chlorophytes, charophytes, and land plants contain chlorophylls *a* and *b*

One major clade of "green algae" consists of the **chlorophytes**. A sister group to the chlorophytes contains another green algal clade (the **charophytes**, or **charales**) along with the *land plants* (see Section 28.1). The green algae share several characters that distinguish them from other microbial eukaryotes: like the land plants, they contain chlorophylls *a* and *b*, and their reserve of photosynthetic products is stored as starch in chloroplasts. Through secondary endosymbiosis, a chlorophyte became the chloroplast of the euglenids.

There are more than 17,000 species of chlorophytes. Most are aquatic—some are marine, but more are freshwater forms—but others are terrestrial, living in moist environments. The chlorophytes range in size from microscopic unicellular forms to multicellular forms many centimeters in length.

CHLOROPHYTES VARY IN SHAPE AND CELLULAR ORGANIZATION We find among the chlorophytes an incredible variety in shape and body form. *Chlamydomonas* is an example of the simplest type: unicellular and flagellated.

Surprisingly large and well-formed colonies of cells are found in such freshwater groups as the genus *Volvox* (**Figure 27.26A**). The cells in these colonies are not differentiated into specialized tissues and organs, as in land plants and animals, but the colonies show vividly how the preliminary step of this great evolutionary innovation might have been taken. In *Volvox*, the origins of cell specialization can be seen in certain cells within the colony that are specialized for reproduction.

While *Volvox* is colonial and spherical, *Oedogonium* is multicellular and filamentous, and each of its cells has only one nucleus. *Cladophora* is multicellular, but each cell is multinucleate. *Bryopsis* is tubular and coenocytic, forming cross-walls only when reproductive structures form. *Acetabularia* is a single, giant uninucleate cell a few centimeters long that becomes multinucleate only at the end of its reproductive stage. *Ulva lactuca* is a thin, membranous sheet a few centimeters across; its distinctive appearance justifies its common name: sea lettuce (**Figure 27.26B**).

As we mentioned above, the chlorophytes are the largest clade of green algae, but there are other green algal clades as well. Those clades are branches of a clade that also includes the land plants, which will be described in the next chapter.

EXCAVATES

Diplomonads
Parabasalids
Heteroloboseans
Euglenids
Kinetoplastids

The **excavates** includes several diverse clades, several of which lack mitochondria. This absence of mitochondria once led to the view that these groups might be the basal clades of the eukaryotes. However, the trait seems to be a derived condition, judging in part from the presence of

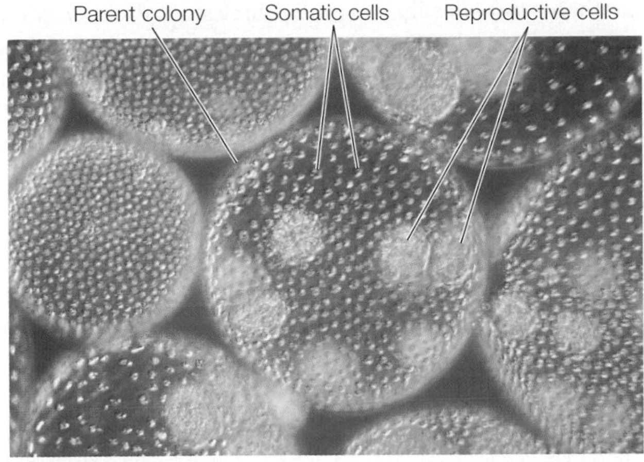

(A) *Volvox* sp.

(B) *Ulva lactuca*

27.26 Chlorophytes (A) *Volvox* colonies are precisely spaced arrangements of cells. Specialized reproductive cells produce daughter colonies, which will eventually release new individuals. (B) A stand of sea lettuce exposed by low tide.

nuclear genes normally associated with mitochondria. Ancestors of these organisms probably possessed mitochondria that were lost or reduced in the course of evolution. The existence of such organisms today shows that eukaryotic life is feasible without mitochondria, and for that reason, these groups are the focus of much attention.

Diplomonads and parabasalids are excavates that lack mitochondria

The **diplomonads** and **parabasalids**, all unicellular, lack mitochondria. *Giardia lamblia*, a diplomonad, is a familiar parasite that contaminates water supplies and causes the intestinal disease giardiasis (**Figure 27.27A**). This tiny organism contains two nuclei bounded by nuclear envelopes, and it has a cytoskeleton and multiple flagella.

Trichomonas vaginalis is a parabasalid responsible for a sexually transmitted disease in humans (**Figure 27.27B**). Infection of the male urethra, where it may occur without symptoms, is less common than infection of the vagina. In addition to flagella and a cytoskeleton, the parabasalids have undulating membranes that also contribute to the cell's locomotion.

Heteroloboseans alternate between amoeboid forms and forms with flagella

The amoeboid body form appears in several microbial eukaryote groups—including the loboseans and heteroloboseans—that are only distantly related to one another. These groups belong, respectively, to the unikonts and excavates. Amoebas of the free-living heterolobosean genus *Naegleria*, some of which can enter humans and cause a fatal disease of the nervous system, usually have a two-stage life cycle, in which one stage has amoeboid cells and the other flagellated cells.

Euglenids and kinetoplastids have distinctive mitochondria and flagella

The euglenids and kinetoplastids, both of which are excavate clades, together constitute a clade of unicellular organisms with flagella. Their mitochondria contain distinctive, disc-shaped cristae,

(A) *Giardia* sp.

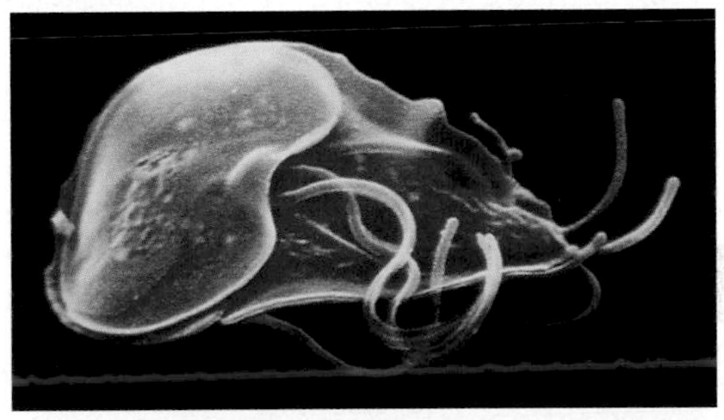

(B) *Trichomonas vaginalis*

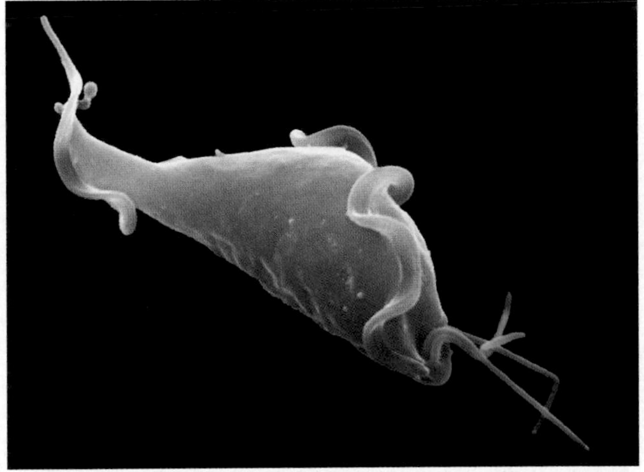

27.27 Some Excavate Groups Lack Mitochondria (A) *Giardia*, a diplomonad, has flagella and two nuclei. (B) *Trichomonas*, a parabasalid, has flagella and undulating membranes. Neither of these organisms possesses mitochondria.

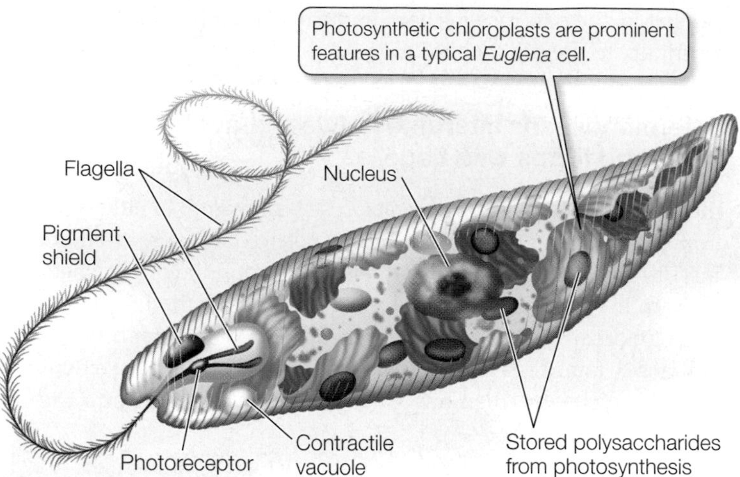

27.28 A Photosynthetic Euglenid Several *Euglena* species possess flagella. In this species, the second flagellum is rudimentary.

and their flagella contain a crystalline rod not found in other organisms. They reproduce asexually by binary fission.

EUGLENIDS HAVE SUPPORTING STRIPS OF PROTEIN The **euglenids** possess flagella arising from a pocket at the anterior end of the cell. Spiraling strips of proteins under their plasma membranes control the cell's shape. Some members of the group are photosynthetic. Euglenids used to be claimed by the zoologists as animals and by the botanists as plants.

Figure 27.28 depicts a cell of the genus *Euglena*. Like most other euglenids, this common freshwater organism has a complex cell structure. It propels itself through the water with the longer of its two flagella, which may also serve as an anchor to hold the organism in place. The second flagellum is often rudimentary.

Euglenids have very diverse nutritional requirements. Many species are always heterotrophic. Other species are fully autotrophic in sunlight, using chloroplasts to synthesize organic compounds through photosynthesis. The chloroplasts of euglenids are surrounded by three membranes as a result of secondary endosymbiosis (see Figure 27.8B). When kept in the dark, these euglenids lose their photosynthetic pigment and begin to feed exclusively on dissolved organic material in the water around them. Such a "bleached" *Euglena* resynthesizes its photosynthetic pigment when it is returned to the light and becomes autotrophic again. But *Euglena* cells treated with certain antibiotics or mutagens lose their photosynthetic pigment completely; neither they nor their descendants are ever autotrophs again. However, those descendants function well as heterotrophs.

KINETOPLASTIDS HAVE MITOCHONDRIA THAT EDIT THEIR OWN RNA The **kinetoplastids** are unicellular parasites with two flagella and a single, large mitochondrion. That mitochondrion contains a *kinetoplast*—a unique structure housing multiple, circular DNA molecules and associated proteins. Some of these DNA molecules encode "guides" that edit messenger RNA within the mitochondrion.

The trypanosomes discussed at the opening of this chapter are kinetoplastids. Recall that they are able to change their cell surface recognition molecules frequently, which allows them to evade our best attempts to kill them and eradicate the diseases they cause (**Table 27.2**)

Rhizaria

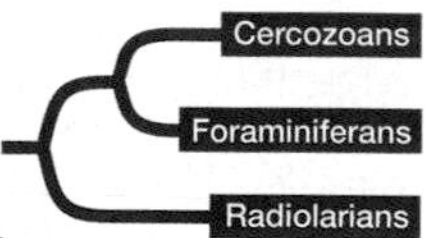

Three closely related groups of **Rhizaria** are unicellular aquatic eukaryotes. Foraminiferans, radiolarians, and cercozoans typically have long, thin pseudopodia that contrast with the broader, lobelike pseudopodia of the familiar amoebas. These groups have contributed to ocean sediments, some of which have become terrestrial features in the course of geological history.

The **cercozoans** are a diverse group, with many forms and habitats. Some are amoeboid, while others have flagella. Some are aquatic; others live in soil. One group of cercozoans possesses chloroplasts derived from a green alga by secondary endosymbiosis—and that chloroplast contains a trace of the green alga's nucleus.

Foraminiferans have created vast limestone deposits

Foraminiferans secrete external shells of calcium carbonate (see Figure 27.5). Some foraminiferans live as plankton, and many others live at the bottom of the sea. Living foraminiferans have been

TABLE 27.2

A Comparison of Three Kinetoplastid Trypanosomes

	TRYPANOSOMA BRUCEI	*TRYPANOSOMA CRUZI*	*LEISHMANIA MAJOR*
Human disease	Sleeping sickness	Chagas' disease	Leishmaniasis
Insect vector	Tsetse fly	Assassin bug	Sand fly
Vaccine or effective cure	None	None	None
Strategy for survival	Changes surface recognition molecules frequently	Causes changes in surface recognition molecules on host cell	Reduces effectiveness of macrophage hosts
Site in human body	Bloodstream; attacks nerve tissue in final stages	Enters cells, especially muscle cells	Enters cells, primarily macrophages
Deaths per year	>50,000	43,000	60,000 (?)

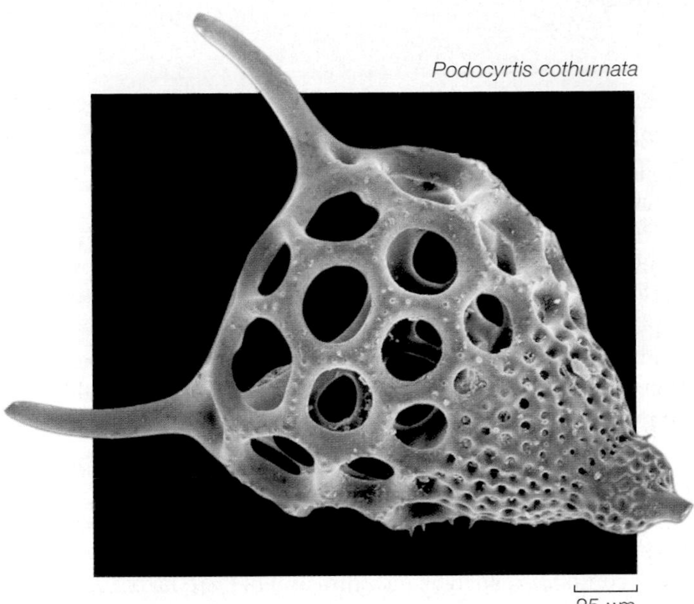

27.29 A Radiolarian's Glass House Radiolarians secrete intricate glassy skeletons such as the one shown here. A living radiolarian is shown in Figure 27.2.

found at the deepest point in the world's oceans—10,896 meters down in the Challenger Deep, in the western Pacific. At that depth, they cannot secrete a normal shell because the surrounding water is too poor in calcium carbonate.

Long, threadlike, branched pseudopods reach out through numerous microscopic pores in the shell and interconnect to create a sticky net, which planktonic foraminifera use to catch smaller plankton. The pseudopods provide locomotion in some species.

Radiolarians have thin, stiff pseudopods

The **radiolarians** are recognizable by their thin, stiff pseudopods, which are reinforced by microtubules. These pseudopods play important roles:

- They greatly increase the surface area of the cell for exchange of materials with the environment.
- They help the cell float in its marine environment.

Found exclusively in marine environments, radiolarians may be the most beautiful of all microorganisms (see Figure 27.2). Almost all radiolarian species secrete glassy *endoskeletons* (internal skeletons). A central capsule lies within the cytoplasm. The skeletons of the different species are as varied as snowflakes, and many have elaborate geometric designs (**Figure 27.29**). A few radiolarians are among the largest of the unicellular eukaryotes, measuring several millimeters across.

Unikonts

We now consider a large clade that may be close to the root of the eukaryote tree: the **unikonts**, eukaryotes whose flagella, if present, are single. (The name *unikont* derives from "single cone.") The animals and fungi are both believed to have arisen from a common ancestor within the **opisthokont** clade. Recent extensive work with multiple gene phylogenies (see Section 26.4) have led many scientists to the conclusion that the *amoebozoans* are sister to the opisthokonts.

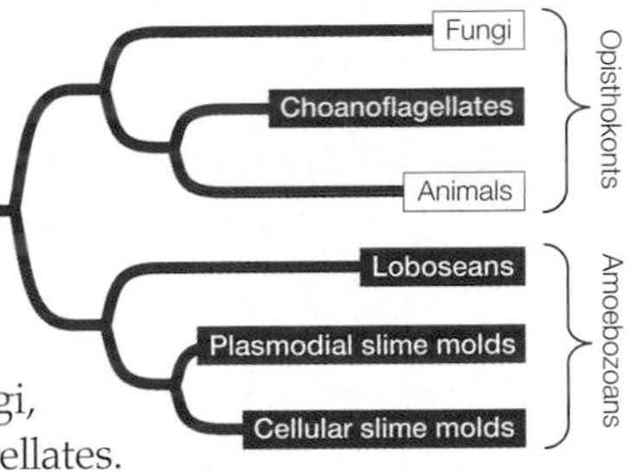

The synapomorphy of the opisthokonts is that their flagellum, if present, is posterior, as in animal sperm. The flagella of all other eukaryotes are anterior. Major groups of opisthokonts include the fungi, the animals, and the choanoflagellates. Fungi and animals will be discussed in Chapters 30–33. The choanoflagellates, or collar flagellates, are sister to the animals, and the animal–choanoflagellate clade is sister to the fungi.

Some choanoflagellates are colonial (**Figure 27.30A**). They bear a striking resemblance to the most characteristic type of cell found in the sponges (compare **Figure 27.30B** with Figure 31.7).

Amoebozoans use lobe-shaped pseudopods for locomotion

The lobe-shaped pseudopods used by **amoebozoans** are a hallmark of the amoeboid body plan. The amoebozoan pseudopod differs in form and function from the slender pseudopods of Rhizaria. We'll consider three amoebozoan groups here: the loboseans and two clades of slime molds.

LOBOSEAN CELLS LIVE INDEPENDENTLY OF ONE ANOTHER A **lobosean**, such as the *Amoeba proteus* shown in Figure 27.9, consists of a single cell. Unlike the cells of the slime molds, loboseans

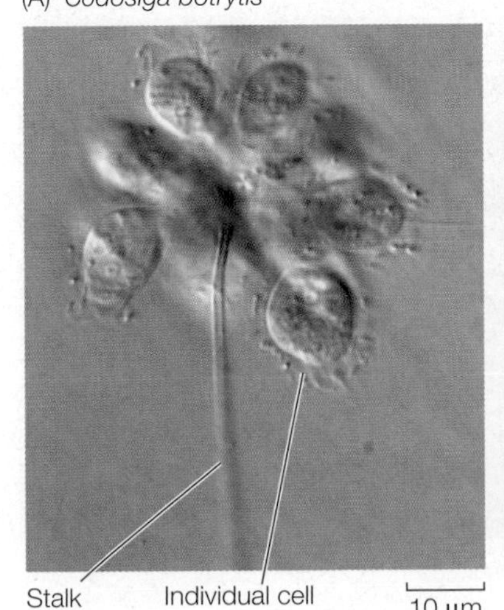

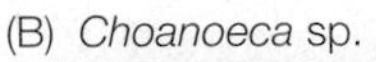

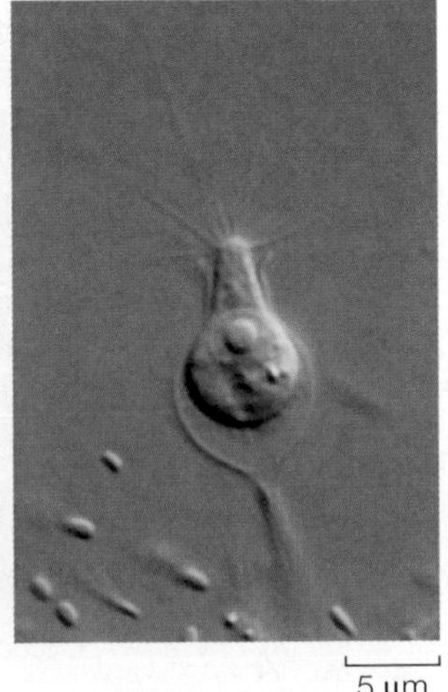

27.30 A Link to the Animals Choanoflagellates are sister to the animals. (A) The formation of colonies by unicellular organisms, as in this choanoflagellate species, is one route to the evolution of multicellularity. (B) A solitary choanoflagellate illustrates the similarity of this microbial eukaryote group to a cell type present in the multicellular sponges (see Figure 31.7).

do not aggregate. A lobosean feeds on small organisms and particles of organic matter by phagocytosis, engulfing them with its pseudopods. Many are adapted for life on the bottoms of lakes, ponds, and other bodies of water. Their creeping locomotion and their manner of engulfing food particles fit them for life close to a relatively rich supply of sedentary organisms or organic particles. Most loboseans exist as predators, parasites, or scavengers.

Some loboseans are shelled, living in casings of sand grains glued together (see Figure 27.12A). Others have shells secreted by the organism itself.

SLIME MOLDS RELEASE SPORES FROM ERECT FRUITING BODIES The two major groups of *slime molds* share only general characteristics. All are motile, all ingest particulate food by endocytosis, and all form spores on erect structures called *fruiting bodies*. They undergo striking changes in organization during their life cycles, and one stage consists of isolated cells that take up food particles by endocytosis. Some slime molds may cover areas of 1 meter or more in diameter while in their less aggregated stage. Such a large slime mold may weigh more than 50 grams. Slime molds of both types favor cool, moist habitats, primarily in forests. They range from colorless to brilliantly yellow and orange.

PLASMODIAL SLIME MOLDS FORM MULTINUCLEATE MASSES If the nucleus of an amoeba began rapid mitotic division, accompanied by a tremendous increase in cytoplasm and organelles but no cytokinesis, the resulting organism might resemble the **plasmodial slime molds**. During its vegetative (feeding) phase, a plasmodial slime mold is a wall-less mass of cytoplasm with numerous diploid nuclei. This mass streams very slowly over its substratum in a remarkable network of strands called a *plasmodium*.* The plasmodium of such a slime mold is another example of a coenocyte, with many nuclei enclosed in a single plasma membrane. The outer cytoplasm of the plasmodium (closest to the environment) is normally less fluid than the interior cytoplasm and thus provides some structural rigidity (**Figure 27.31A**).

Plasmodial slime molds provide a dramatic example of movement by **cytoplasmic streaming**. The outer cytoplasmic region of the plasmodium becomes more fluid in places, and cytoplasm rushes into those areas, stretching the plasmodium. This streaming somehow reverses its direction every few minutes as cytoplasm rushes into a new area and drains away from an older one, moving the plasmodium over its substratum. Sometimes an entire wave of plasmodium moves across the substratum, leaving strands behind. Microfilaments and a contractile protein called *myxomyosin* interact to produce the streaming movement. As it moves, the plasmodium engulfs food particles by endocytosis—predominantly bacteria, yeasts, spores of fungi, and other small organisms, as well as decaying animal and plant remains.

A plasmodial slime mold can grow almost indefinitely in its plasmodial stage, as long as the food supply is adequate and other conditions, such as moisture and pH, are favorable. However, one of two things can happen if conditions become unfavorable. First, the plasmodium can form an irregular mass of hardened cell-like components called a *sclerotium*. This resting structure rapidly becomes a plasmodium again when favorable conditions are restored.

Alternatively, the plasmodium can transform itself into spore-bearing fruiting structures (**Figure 27.31B**). These stalked or branched structures rise from heaped masses of plasmodium. They derive their rigidity from walls that form and thicken between their nuclei. The diploid nuclei of the plasmodium divide by meiosis as the fruiting structure develops. One or more knobs, called *sporangia*, develop on the end of the stalk. Within a sporangium, haploid nuclei become surrounded by walls and form spores. Eventually, as the fruiting body dries, it sheds its spores.

*Do not confuse the plasmodium of a plasmodial slime mold with the genus *Plasmodium*, the apicomplexan that is the cause of malaria.

27.31 Plasmodial Slime Molds (A) Plasmodia of the yellow slime mold *Physarum* cover a rock in Nova Scotia. (B) The fruiting structures of *Physarum*.

(A) *Physarum polycephalum*

(B) *Physarum polycephalum*

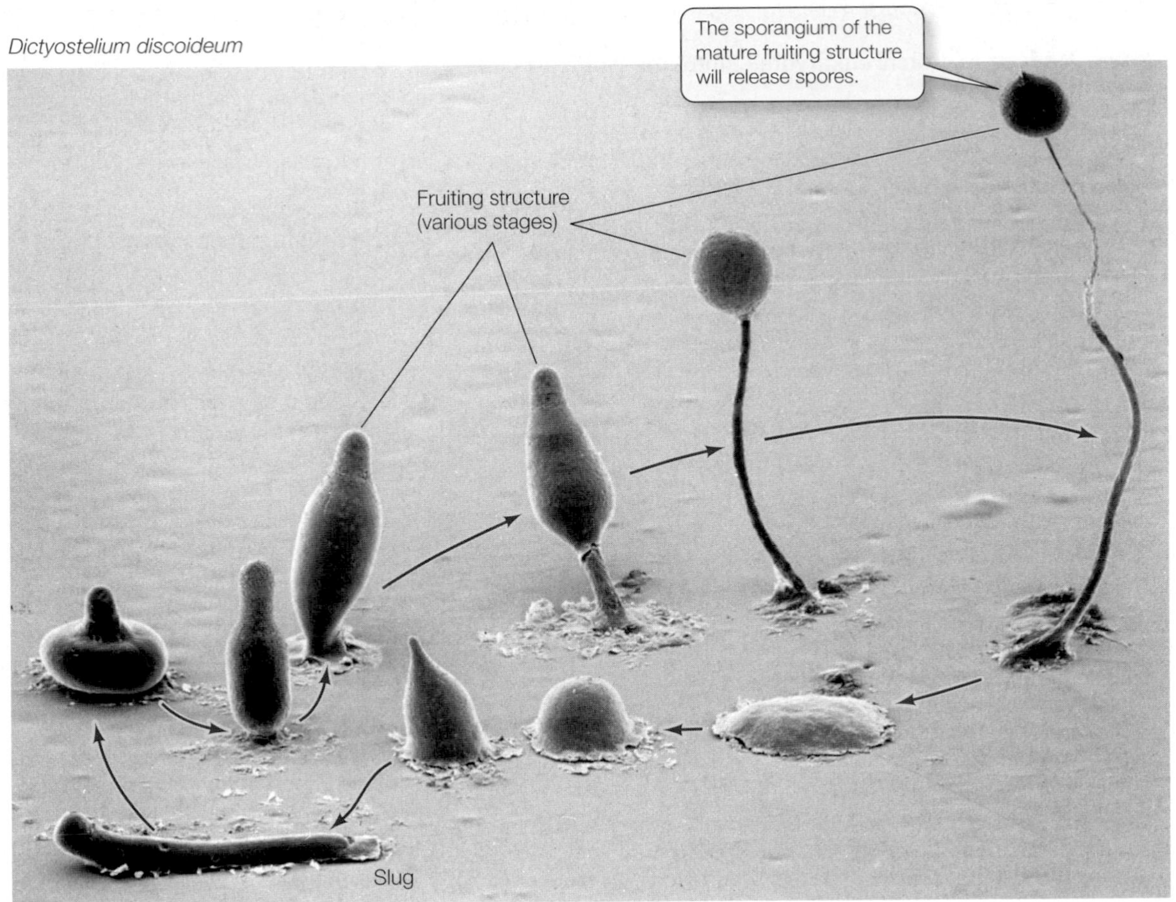

27.32 A Cellular Slime Mold The life cycle of the slime mold *Dictyostelium* is shown here in a composite micrograph.

The spores germinate into wall-less, haploid cells called *swarm cells*, which can either divide mitotically to produce more haploid swarm cells or function as gametes. Swarm cells can live as separate individual cells that move by means of flagella or pseudopods, or they can become walled and resistant resting cysts when conditions are unfavorable; when conditions improve again, the cysts release swarm cells. Two swarm cells can also fuse to form a diploid zygote, which divides by mitosis (but without a wall forming between the nuclei) and thus forms a new, coenocytic plasmodium.

CELLS RETAIN THEIR IDENTITY IN THE CELLULAR SLIME MOLDS

Whereas the plasmodium is the basic vegetative (feeding, nonreproductive) unit of the plasmodial slime molds, an amoeboid cell is the vegetative unit of the **cellular slime molds**. Large numbers of cells called *myxamoebas*, which have single haploid nuclei, engulf bacteria and other food particles by endocytosis and reproduce by mitosis and fission. This simple life cycle stage, consisting of swarms of independent, isolated cells, can persist indefinitely as long as food and moisture are available.

When conditions become unfavorable, however, the cellular slime molds aggregate and form fruiting structures, as do their plasmodial counterparts. The individual myxamoebas aggregate into a mass called a *slug* or *pseudoplasmodium* (**Figure 27.32**). Unlike the true plasmodium of the plasmodial slime molds, this structure is not simply a giant sheet of cytoplasm with many nuclei; the individual myxamoebas retain their plasma membranes and, therefore, their identity.

A slug may migrate over its substratum for several hours before becoming motionless and reorganizing to construct a delicate, stalked fruiting structure. Cells at the top of the fruiting structure develop into thick-walled spores, which are eventually released. Later, under favorable conditions, the spores germinate, releasing myxamoebas.

The cycle from myxamoebas through slug and spores to new myxamoebas is asexual. Cellular slime molds also have a sexual cycle, in which two myxamoebas fuse. The product of this fusion develops into a spherical structure that ultimately germinates, releasing new haploid myxamoebas.

In subsequent chapters we will explore the three classic groups of multicellular eukaryotes that arose within the opisthokonts. Chapters 28 and 29 describe the rise and diversification of the land plants, Chapter 30 presents the fungi, and Chapters 31–33 describe the animals. All three of these groups arose from microbial eukaryote ancestors.

CHAPTER SUMMARY

27.1 How do microbial eukaryotes affect the world around them?

The **microbial eukaryotes** (also called **protists**) embrace several diverse groups of mostly unicellular eukaryotic organisms. They constitute a paraphyletic group, not a clade.

The diatoms, part of the **plankton**, are responsible for up to a fifth of all carbon fixation on Earth. **Phytoplankton** are the primary producers in the marine environment.

Endosymbiosis is common among microbial eukaryotes and often helpful to both partners. Pathogenic microbial eukaryotes include species of *Plasmodium* and trypanosomes. Review Figure 27.3

27.2 How did the eukaryotic cell arise?

Several events led to the evolution of the modern eukaryotic cell from an ancestral prokaryote. Probable early events include the loss of the cell wall and infolding of the plasma membrane. Review Figure 27.6

The development of a cytoskeleton gave the evolving cell increasing control over its shape and distribution of daughter chromosomes. Review Figure 27.7

Some organelles were acquired by endosymbiosis. Mitochondria evolved from a proteobacterium. **Primary endosymbiosis** of a eukaryote and a cyanobacterium gave rise to the chloroplasts, beginning with those of glaucophytes, red algae, and green algae. Review Figure 27.8A, Web/CD Tutorial 27.1

Secondary endosymbiosis of eukaryotes with other eukaryotes already equipped with chloroplasts gave rise to the chloroplasts of euglenids, stramenopiles, and other groups. Review Figure 27.8B

27.3 How did the microbial eukaryotes diversify?

Some microbial eukaryotes are photosynthetic autotrophs, some are heterotrophs, and some are both.

The cytoskeleton allows for various means of locomotion. Most microbial eukaryotes are motile, moving by amoeboid motion with **pseudopods** or by means of cilia or flagella.

Some microbial eukaryote cells contain **contractile vacuoles** that pump out excess water or **food vacuoles** where food is digested. Review Figures 27.10 and 27.11, Web/CD Tutorial 27.2

Many microbial eukaryotes have protective cell surfaces such as cell walls, external "shells," or shells constructed from sand.

27.4 How do microbial eukaryotes reproduce?

Most microbial eukaryotes reproduce both asexually and sexually.

Conjugation in paramecia is a sexual process but not a reproductive one. Review Figure 27.13

Alternation of generations is a feature of many microbial eukaryote life cycles, which include a multicellular diploid phase and a multicellular haploid phase. Review Figure 27.14

Alternation of generations may be **heteromorphic** or **isomorphic**. The alternating generations are the (diploid) **sporophyte** and (haploid) **gametophyte**. Specialized cells of the sporophyte, called **sporocytes**, divide meiotically to produce haploid spores.

Depending on whether their gametes appear identical or dissimilar, species are termed **isogamous** or **anisogamous**.

In a **haplontic** life cycle, the zygote is the only diploid cell. In a **diplontic** life cycle, the gametes are the only haploid cells. Review Figures 27.15 and 27.16, Web/CD Activities 27.1 and 27.2

Some microbial eukaryote life cycles involve more than one host species.

27.5 What are the major groups of eukaryotes?

Most eukaryotes can be divided into five major groups: chromalveolates, Plantae, excavates, Rhizaria, and unikonts. Review Figure 27.17

The **chromalveolates** include the haptophytes, cryptophytes, alveolates, and stramenopiles. The **cryptophytes** represent a key link in the story of the evolution of chloroplasts.

Alveolates are unicellular organisms with sacs (alveoli) beneath their plasma membranes. Alveolate clades include the marine **dinoflagellates**; the parasitic **apicomplexans**; and diverse, highly motile **ciliates.** Web/CD Activity 27.3

Stramenopiles typically have two flagella of unequal length, the longer bearing rows of tubular hairs. Included among the stramenopiles are the unicellular **diatoms**, the multicellular **brown algae**, and the nonphotosynthetic **oomycetes** including the water molds and downy mildews.

Plantae is a clade containing several other clades, including **glaucophytes**, **red algae**, **chlorophytes**, **land plants**, and **charophytes**. All are photosynthetic and contain chloroplasts. Glaucophyte chloroplasts contain peptidoglycan between their inner and outer membranes.

The **excavates** include the diplomonads, parabasalids, heterolobosean, euglenids, and kinetoplastids. The **diplomonads** and **parabasalids** lack mitochondria, having apparently lost them during their evolution. **Heteroloboseans** are amoebas with a two-stage life cycle. **Euglenids** are often photosynthetic and have anterior flagella and strips of protein that support their cell surface. **Kinetoplastids** have a single, large mitochondrion, in which mitochondrial messenger RNA is edited.

Rhizaria are unicellular and aquatic; most are amoeboid. This group includes the **foraminiferans** whose shells have contributed to great limestone deposits; the **radiolarians** with thin, stiff pseudopods and glassy endoskeletons; and the **cercozoans**, which take many forms and live in diverse habitats.

The **unikonts** encompass organisms with single flagella on their flagellated cells (if any). They can be divided into two subgroups: the opisthokonts and the amoebozoans. In the **opisthokonts**, the flagellum (if any) is posterior. The opisthokont subgroups are the fungi, the choanoflagellates, and the animals. **Choanoflagellates** resemble the cells of sponges and are sister to the animal clade.

The **amoebozoans** move by means of lobe-shaped pseudopodia. They comprise the loboseans, plasmodial slime molds, and cellular slime molds. A **lobosean** consists of a single cell; these cells do not aggregate. **Plasmodial slime molds** are amoebozoans whose feeding phase is coenocytic. In the feeding phase, movement is by **cytoplasmic streaming**. In **cellular slime molds**, the individual cells maintain their identity at all times, but aggregate to form fruiting bodies.

SELF-QUIZ

1. Microbial eukaryotes with flagella
   a. appear in several clades.
   b. are all algae.
   c. all have pseudopods.
   d. are all colonial.
   e. are never pathogenic.
2. Which statement about eukaryotic phytoplankton is *not* true?
   a. Some are important primary producers.
   b. Some contributed to the formation of petroleum.
   c. Some form toxic "red tides."
   d. Some are food for marine animals.
   e. They constitute a clade.
3. Apicomplexans
   a. possess flagella.
   b. possess a glassy shell.
   c. are all parasitic.
   d. are algae.
   e. include the trypanosomes that cause sleeping sickness.
4. The ciliates
   a. move by means of flagella.
   b. use amoeboid movement.
   c. include *Plasmodium*, the agent of malaria.
   d. possess both a macronucleus and micronuclei.
   e. are autotrophic.
5. The chloroplasts of photosynthetic microbial eukaryotes
   a. are structurally identical.
   b. gave rise to mitochondria.
   c. are all descended from a once free-living cyanobacterium.
   d. all have exactly two surrounding membranes.
   e. are all descended from a once free-living red alga.
6. Which statement about the brown algae is *not* true?
   a. They are all multicellular.
   b. They use the same photosynthetic pigments as do land plants.
   c. They are almost exclusively marine.
   d. A few are many meters in length.
   e. They are stramenopiles.
7. Which statement about the chlorophytes is *not* true?
   a. They use the same photosynthetic pigments as do land plants.
   b. Some are unicellular.
   c. Some are multicellular.
   d. All are microscopic in size.
   e. They display a great diversity of life cycles.
8. The red algae
   a. are mostly unicellular.
   b. are mostly marine.
   c. owe their red color to a special form of chlorophyll.
   d. have flagella on their gametes.
   e. are all heterotrophic.
9. The plasmodial slime molds
   a. form a plasmodium that is a coenocyte.
   b. lack fruiting bodies.
   c. consist of large numbers of myxamoebas.
   d. consist at times of a mass called a pseudoplasmodium.
   e. possess flagella.
10. The cellular slime molds
   a. possess apical complexes.
   b. lack fruiting bodies.
   c. form a plasmodium that is a coenocyte.
   d. have haploid myxamoebas.
   e. possess flagella.

FOR DISCUSSION

1. For each type of organism below, give a single characteristic that may be used to differentiate it from the other, related organism(s) named in parentheses.
   a. Foraminiferans (radiolarians)
   b. *Euglena* (*Volvox*)
   c. *Trypanosoma* (*Giardia*)
   d. Plasmodial slime molds (cellular slime molds)
2. In what sense are sex and reproduction independent of each other in the ciliates? What does that suggest about the role of sex in biology?
3. Why are dinoflagellates and apicomplexans placed in one group of microbial eukaryotes and brown algae and oomycetes in another?
4. Unlike many microbial eukaryotes, apicomplexans lack contractile vacuoles. Why don't apicomplexans need a contractile vacuole?
5. Giant seaweeds (mostly brown algae) have "floats" that aid in keeping their fronds suspended at or near the surface of the water. Why is it important that the fronds be suspended in this way?
6. Why are algal pigments so much more diverse than those of land plants?
7. Consider the chloroplasts of chlorophytes, euglenids, and red algae. For each of these groups, indicate how many membranes surround their chloroplasts, and offer a reasonable explanation in each case. Why do some dinoflagellates have more membranes around their chloroplasts than other dinoflagellates?

FOR INVESTIGATION

During a bloom of diatoms, their predators, the copepods, do not show a corresponding increase in population. The copepod population rises only later, after the diatom bloom has declined. When the diatoms are at a normal (non-bloom) population level, what proportion of the copepods' diet do they constitute?

CHAPTER 28 Plants without Seeds: From Sea to Land

What surprises lurk in a rock?

John William Dawson, later Sir William Dawson, was an outspoken critic of Charles Darwin and evolutionary theory. Ironically, he made one of the first contributions to our understanding of the early evolution of plants. While surveying the geology of Nova Scotia, Dawson, a Canadian scientist, found a puzzling fossil fragment on the Gaspé Peninsula. It was 1859—the very year in which Darwin published *The Origin of Species*.

What Dawson found was the remains of a remarkable plant, which he named *Psilophyton*, meaning "naked plant." The fossilized plant appeared to have no roots and no leaves, and its stem grew both below and above the ground. The aboveground part of the stem, which branched and ended in spore cases, was about 50 centimeters tall. Could this fossil plant represent one of the stages in the momentous transition of plants from an aquatic existence to the forests that cover the land today? Dawson presented *Psilophyton* and other finds in a major lecture delivered in 1870, but his audience was unreceptive, with his fellow scientists chuckling that *Psilophyton* grew nowhere but in Dawson's imagination. His published drawing of the fossil plant was regarded as an amusing curiosity.

Decades later, Dawson's interpretation of *Psilophyton* was vindicated by the work of two British botanists who studied plant fossils. The year 1915 found Robert Kidston and William Lang in the hills near Rhynie, Scotland, where they discovered evidence that a marsh had existed there 400 million years ago. In the intervening hundreds of millions of years, that Devonian marsh had become a flinty, fossil-laden rock called chert. Although the rock now lay in a hilly site some 50 kilometers from the sea, its original location must have been near the shore.

The most startling feature of the Rhynie chert was the great abundance of fossils of a small plant with spore cases but neither roots nor leaves—a plant, in fact, that looked strikingly like Dawson's "imaginary" *Psilophyton*! Kidston and Lang described their find in detail and gave it the genus name *Rhynia* in honor of the site of its discovery.

Rhynia is just one of several ancient groups of plants that lack seeds and other features of more modern plants. The glory

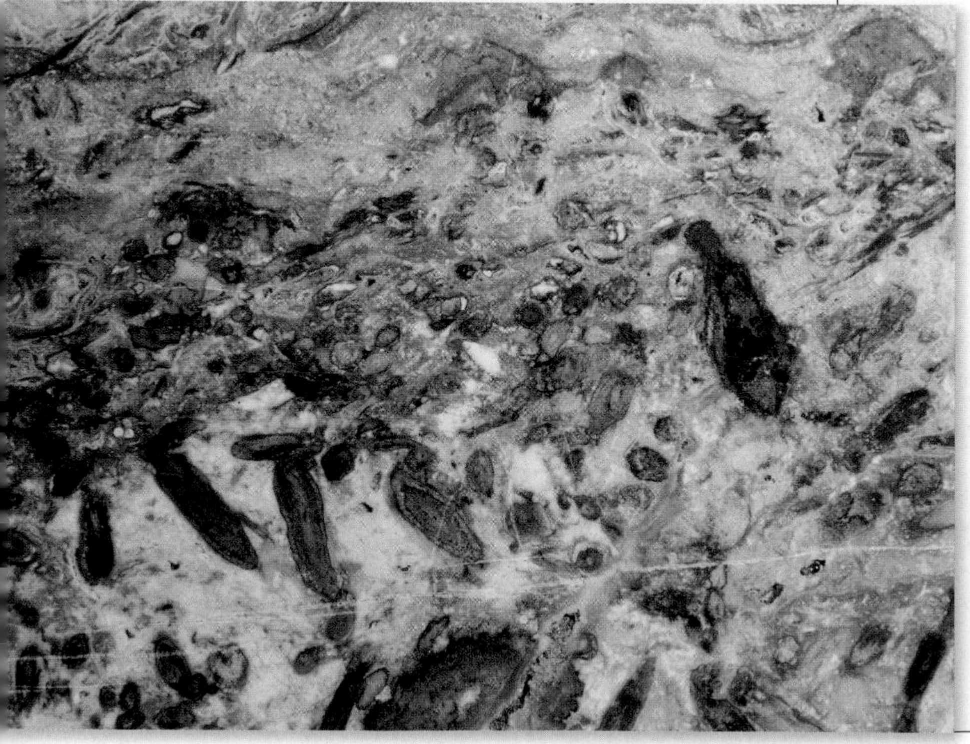

Making It on Land A sample of Rhynie chert shows fragments of fossilized *Rhynia gwynne-vaughani*, a vascular plant abundant during the Devonian. These early land plants were 30–60 centimeters tall and had no leaves and no roots. The dark, spindle-shaped objects seen against the lightest rock are glancing cuts through stems; the circles and ovals are stem cross-sections.

Green Mansions Seedless plants—mosses and ferns—dominate the understory of this pristine rainforest on the island of Maui, Hawaii. The only seed plant we see is the branching tree in the background.

days of the seedless plants are long past, but many of them—most notably the mosses and ferns—are still abundant. We rely on these surviving plants, as well as on fossils, to help us understand the evolution of plants from aquatic algae growing at the edge of a sea or marsh. Much of the fossilized history, and the great bulk of the mass, of seedless plants was preserved as coal, which today is an economically important substance that we burn as fuel and to produce electricity. What a library of botanical history the fossils in the world's coal beds provide!

IN THIS CHAPTER we will see how plants invaded the land, how land plants evolved, and how plant clades diversified, resulting in plants ever better equipped to face the challenges of terrestrial environments. Our descriptions here will concentrate on those land plants that lack seeds. Chapter 29 completes our survey of plants by considering the seed plants, which dominate the terrestrial scene today.

CHAPTER OUTLINE

28.1 How Did the Land Plants Arise?

The question "Where did plants come from?" embraces two questions: "What were the ancestors of plants?" and "Where did those ancestors live?" We will consider both questions in this section.

The **land plants** are monophyletic: all land plants descend from a single common ancestor and form a branch of the evolutionary tree of life. One of the key shared derived traits, or synapomorphies, of the land plants is development from an embryo protected by tissues of the parent plant. For this reason, land plants are sometimes referred to as **embryophytes** (*phyton*, "plant"). Land plants retain the derived features they share with the "green algae" described in Chapter 27: the use of chlorophylls *a* and *b* in photosynthesis, and the use of starch as a photosynthetic storage product. Both land plants and green algae have cellulose in their cell walls.

There are several ways to define "plant" and still refer to a clade (**Figure 28.1**). For example, if the land plants are included with a certain paraphyletic group of green algae, we define a monophyletic group (the **streptophytes**) with several synapomorphies, including the retention of the egg in the parent body. The addition of the remainder of the green algae to the streptophytes gives another clade, commonly called the **green plants**, with synapomorphies including the possession of chlorophyll *b*. Other groups, such as stramenopiles and red algae, are also called "plants" by some people. Green plants, streptophytes, and land plants have each been called "the plant kingdom" by different authorities. There are no objective criteria for defining a kingdom (or any other taxonomic rank); definitions of plant groups are thus somewhat subjective. In this book we use the unmodified common name "plants" to refer to the embryophytes, or monophyletic land plants.

There are ten major groups of land plants

The land plants of today fall naturally into ten major clades (**Table 28.1**). Members of seven of those clades possess well-developed vascular systems that transport materials throughout the plant body. We call these seven groups, collectively, the **vascular plants**, or *tracheophytes*, because they all possess conducting cells called **tracheids**. Together, the seven groups of vascular plants constitute a clade.

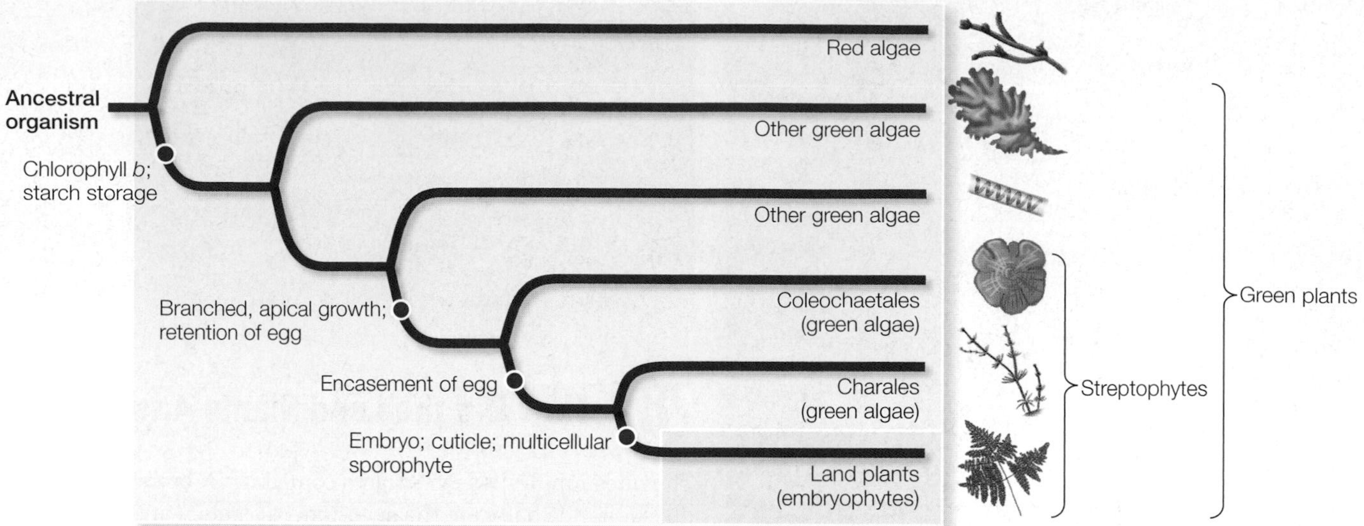

28.1 What Is a Plant? There are various definitions of the term "plant." In this book, we use the most restrictive definition: plants as embryophytes. To include the green algae, we use the broader term "green plants."

The remaining three clades (liverworts, hornworts, and mosses) lack tracheids. These three groups are sometimes collectively called *bryophytes*, but in this text we reserve that term for their most familiar members, the mosses. Instead, we refer to them as **nonvascular plants**, but note that, collectively, *these three groups are not a clade*. Some nonvascular plants have conducting cells, but none have tracheids.

Where did the plants come from? To determine which living organisms are most closely related to plants, biologists considered several of the synapomorphies of land plants and looked for their origins in various other groups.

The land plants arose from a green algal clade

Several synapomorphies, backed by clear-cut evidence from molecular studies, indicate that the sister clade of the plants is a group of aquatic green algae called the **Charales** (see Figure 28.1):

- Plasmodesmata joining the cytoplasm of adjacent cells (see Figure 15.20)
- Branching, apical (tip) growth
- Retention of the egg in the parent organism

TABLE 28.1

Classification of Land Plants

GROUP	COMMON NAME	CHARACTERISTICS
NONVASCULAR PLANTS		
Hepatophyta	Liverworts	No filamentous stage; gametophyte flat
Anthocerophyta	Hornworts	Embedded archegonia; sporophyte grows basally (from the ground)
Bryophyta	Mosses	Filamentous stage; sporophyte grows apically (from the tip)
VASCULAR PLANTS		
Lycophyta	Club mosses and allies	Microphylls in spirals; sporangia in leaf axils
Pteridophyta	Horsetails, whisk ferns, ferns	Differentiation between main stem and side branches (overtopping growth)
SEED PLANTS		
Gymnosperms		
Cycadophyta	Cycads	Compound leaves; swimming sperm; seeds on modified leaves
Ginkgophyta	Ginkgo	Deciduous; fan-shaped leaves; swimming sperm
Gnetophyta	Gnetophytes	Vessels in vascular tissue; opposite, simple leaves
Coniferophyta	Conifers	Seeds in cones; needle-like or scale-like leaves
Angiosperms	Flowering plants	Endosperm; carpels; gametophytes much reduced; seeds within fruit

Note: No extinct groups are included in this classification.

28.2 The Closest Relatives of Land Plants The land plants probably evolved from a common ancestor shared with the Charales, a green algal group. (A) Molecular evidence seems to favor stoneworts of the genus *Chara* as sister group to the plants. (B) Evidence from morphology indicates that the clade including this coleochaete alga is closely related to the land plants.

Stoneworts of the genus *Chara* are members of the Charales that resemble plants in terms of their rRNA and DNA sequences, peroxisome contents, mechanics of mitosis and cytokinesis, and chloroplast structure (**Figure 28.2A**). Strong evidence from morphology-based cladistic analysis suggests that the Coleochaetales, a group of green algae that includes the genus *Coleochaete* (**Figure 28.2B**), are also fairly closely related to the land plants. *Coleochaete*-like algae have several features found in plants, such as a flattened form in contrast to the branched-filament form of *Chara*.

28.1 RECAP

Land plants are photosynthetic organisms that develop from embryos protected by parent plant tissue.

- Can you explain the different possible uses of the term "plant"? See p. 611 and Figure 28.1
- Do you recognize the key difference between the vascular plants and the other three clades of land plants? See pp. 611–612 and Table 28.1
- What evidence supports the phylogenetic relationship between land plants and Charales? See pp. 612–613

The green algal ancestors of the plants lived at the margins of ponds or marshes, ringing them with a green mat. It was from such a marginal habitat, which was sometimes wet and sometimes dry, that early plants made the transition onto land.

28.2 How Did Plants Colonize and Thrive on Land?

Land plants, or their immediate ancestors in those ancient green mats, first appeared in the terrestrial environment between 400 and 500 million years ago. How did they survive in an environment that differed so dramatically from the aquatic environment of their ancestors? While the water essential for life is everywhere in the aquatic environment, it is hard to obtain and retain water in the terrestrial environment.

Adaptations to life on land distinguish land plants from green algae

No longer bathed in fluid, organisms on land faced potentially lethal desiccation (drying). Large terrestrial organisms had to develop ways to transport water to body parts distant from the source. And whereas water provides aquatic organisms with support against gravity, a plant living on land must either have some other support system or sprawl unsupported on the ground. A land plant must also use different mechanisms for dispersing its gametes and progeny than its aquatic relatives, which can simply release them into the water. Survival on land required numerous adaptations. The first colonists—the *nonvascular plants*—met these challenges.

Most of the characteristics that distinguish land plants from green algae are evolutionary adaptations to life on land:

- The *cuticle*, a waxy covering that retards water loss (desiccation)
- *Gametangia*, cases that enclose plant gametes and prevent them from drying out
- *Embryos*, which are young plants contained within a protective structure
- Certain *pigments* that afford protection against the mutagenic ultraviolet radiation that bathes the terrestrial environment
- Thick *spore walls* containing *sporopollenin*, a polymer that protects the spores from desiccation and resists decay
- A *mutually beneficial association with a fungus* that promotes nutrient uptake from the soil

The **cuticle** may be the most important—and earliest—of these features. Composed of several unique waxy lipids (see Section 3.4) that coat the leaves and stems of land plants, the cuticle has several functions, the most obvious and important of which is to keep water (which the plant's ancestors obtained from their environment) from evaporating away from the plant body.

The lotus is revered as a symbol of purity. Its leaves are never dirty, even when they emerge from muddy water. The specific pattern of wax deposition in the lotus cuticle allows water to roll off the leaf surface, pushing dirt particles away. This "lotus effect" keeps the plant clean (and thus better able to trap light for photosynthesis).

Ancient plants also helped themselves and their descendants by contributing to the formation of soil. Acid secreted by plants helps break down rock, and the organic compounds produced by the breakdown of dead plants contribute to soil structure. Such effects are repeated today as plants grow in new areas.

The nonvascular plants usually live where water is available

The nonvascular plants—today's liverworts, hornworts, and mosses—are thought to be similar to the earliest land plants. Most of these plants grow in dense mats, usually in moist habitats (**Figure 28.3**). Even the largest nonvascular plants are only about half a meter tall, and most are only a few centimeters tall or long. Why have they not evolved to be taller? The probable answer is that they lack an efficient vascular system for conducting water and minerals from the soil to distant parts of the plant body.

The nonvascular plants lack the leaves, stems, and roots that characterize the vascular plants, although they have structures analogous to each. Their growth pattern allows water to move through the mats of plants by capillary action. They have leaflike structures that readily catch and hold any water that splashes onto them. They are small enough that minerals can be distributed throughout their bodies by diffusion. As in all land plants, layers of maternal tissue protect their embryos from desiccation. They also have a cuticle, although it is often very thin (or even absent in some species) and thus is not highly effective in retarding water loss.

Most nonvascular plants live on the soil or on other plants, but some grow on bare rock, dead and fallen tree trunks, and even on buildings. The ability to grow on such marginal surfaces results from a mutualistic association with the glomeromycetes, a clade of fungi. The earliest land plants were already colonized by these fungi; the association dates back at least 460 million years. The fungal association probably promoted the absorption of water and minerals, especially phosphorus, from the first "soils."

Nonvascular plants are widely distributed over six continents and even exist (albeit very locally) on the coast of the seventh, Antarctica. They are successful and well adapted to their environments. Most are terrestrial. Some live in wetlands. Although a few nonvascular plant species live in fresh water, these aquatic forms are descended from terrestrial ones. There are no marine nonvascular plants.

Land plants and green algae differ not just in structure, but with respect to their life cycles. Differences in the life cycles of land plants and those of their ancestors are also a function of the relative dependence of plants on a water supply.

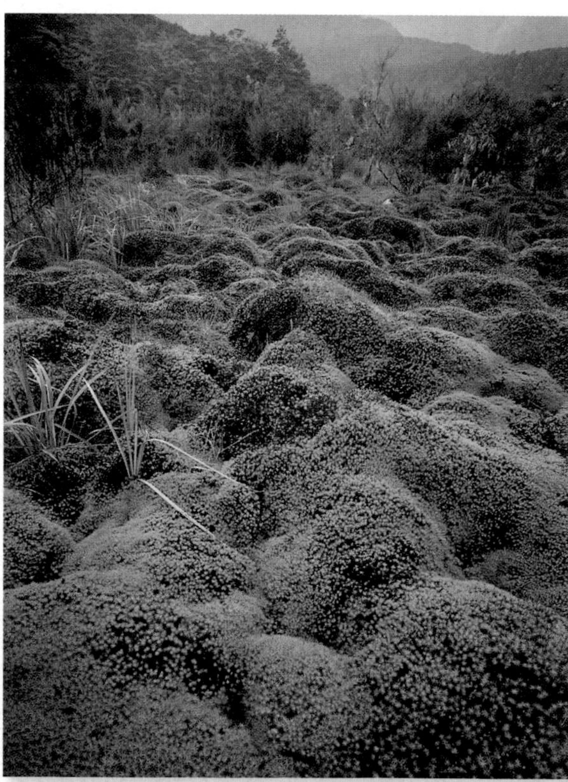

28.3 Mosses Form Dense Mats Dense moss forms hummocks in a valley on New Zealand's South Island.

Life cycles of land plants feature alternation of generations

A universal feature of the life cycles of land plants is alternation of generations (**Figure 28.4**). Recall from Section 27.4 the two hallmarks of alternation of generations:

- The life cycle includes both multicellular diploid individuals and multicellular haploid individuals.
- Gametes are produced by mitosis, not by meiosis. Meiosis produces spores that develop into multicellular haploid individuals.

If we begin looking at the plant life cycle at the single-cell stage—the diploid zygote—then the first phase of the cycle is the formation, by mitosis and cytokinesis, of a multicellular embryo, which eventually grows into a mature diploid plant. This multicellular diploid plant is the **sporophyte** ("spore plant").

Cells contained in **sporangia** (singular *sporangium*, "spore vessel") of the sporophyte undergo meiosis to produce haploid, uni-

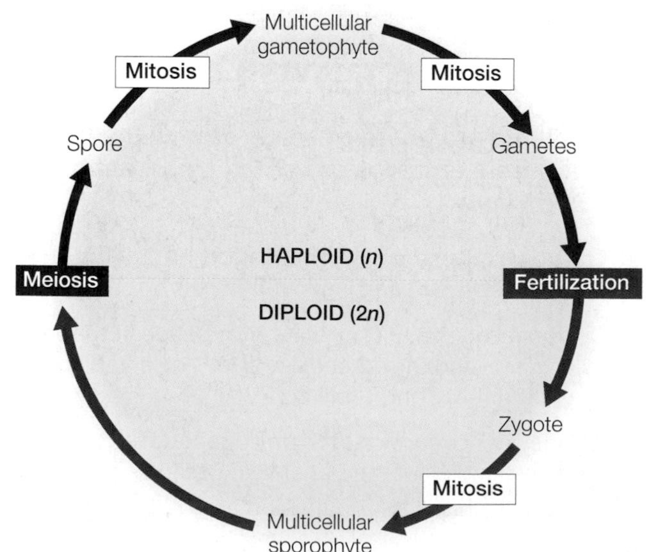

28.4 Alternation of Generations in Plants A diploid sporophyte generation that produces spores by meiosis alternates with a haploid gametophyte generation that produces gametes by mitosis.

cellular spores. By mitosis and cytokinesis, a spore forms a haploid plant. This multicellular haploid plant, called the **gametophyte** ("gamete plant"), produces haploid gametes by mitosis. The fusion of two gametes (*syngamy*, or *fertilization*) forms a single diploid cell—the zygote—and the cycle is repeated (**Figure 28.5**).

The *sporophyte generation* extends from the zygote through the adult multicellular diploid plant; the *gametophyte generation* extends from the spore through the adult multicellular haploid plant to the gametes. The transitions between the generations are accomplished by fertilization and meiosis. In all plants, the sporophyte and gametophyte differ genetically: the sporophyte has diploid cells, and the gametophyte has haploid cells.

Reduction of the gametophyte generation is a major theme in plant evolution. In the nonvascular plants, the gametophyte is larger, longer-lived, and more self-sufficient than the sporophyte. However, in those groups that appeared later in plant evolution, the sporophyte generation is the larger, longer-lived, and more self-sufficient one. In the seed plants, this evolutionary trend has led to a condition in which water is not required for the sperm to reach the egg.

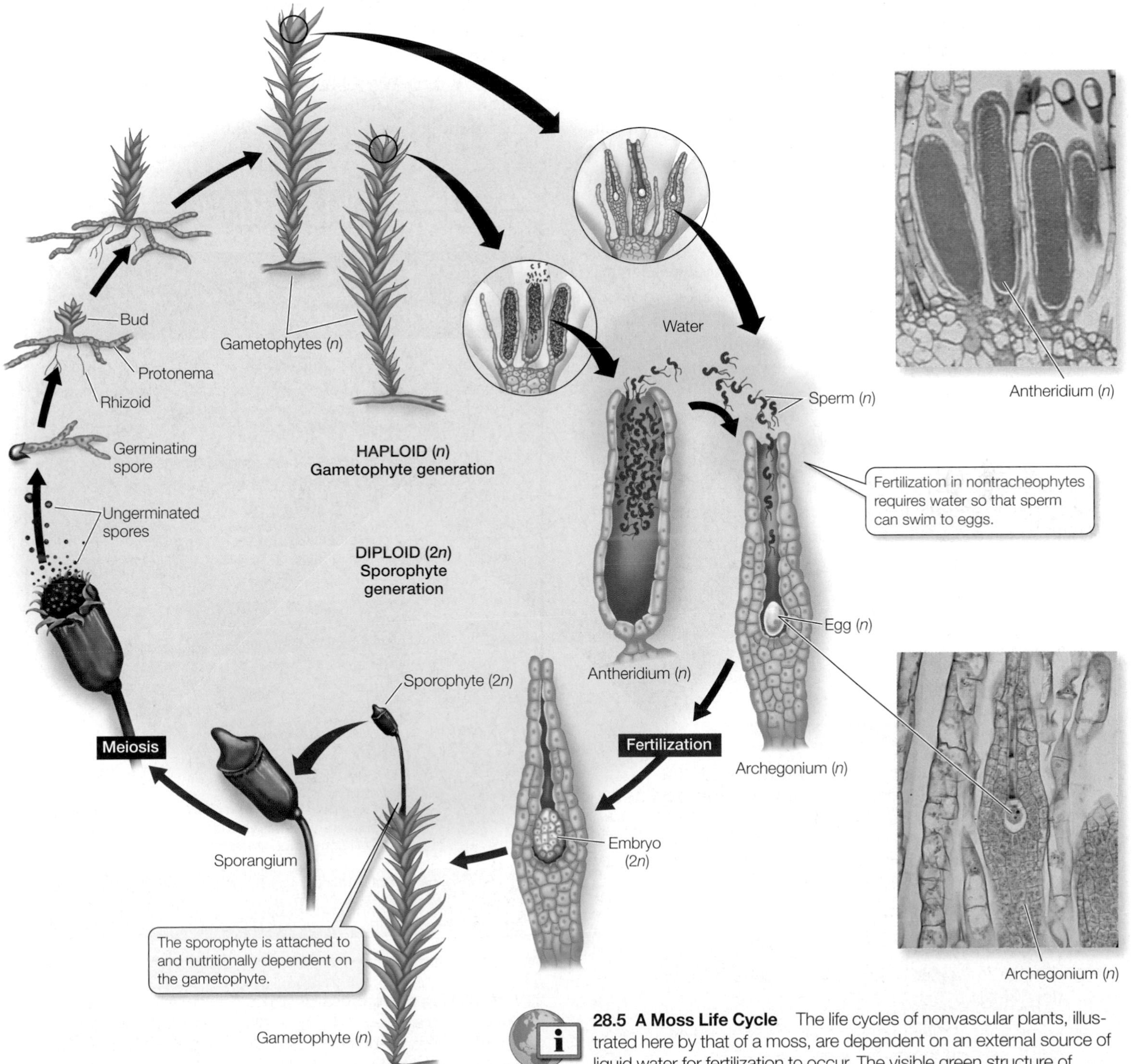

28.5 A Moss Life Cycle The life cycles of nonvascular plants, illustrated here by that of a moss, are dependent on an external source of liquid water for fertilization to occur. The visible green structure of such plants is the gametophyte.

The sporophytes of nonvascular plants are dependent on gametophytes

In nonvascular plants, the conspicuous green structure visible to the naked eye is the gametophyte (see Figure 28.5), in contrast to the familiar forms of vascular plants, such as ferns and seed plants, which are sporophytes. The gametophyte of liverworts, hornworts, and mosses is photosynthetic and is therefore nutritionally independent, whereas the sporophyte may or may not be photosynthetic, but is always nutritionally dependent on the gametophyte and remains permanently attached to it.

A nonvascular plant sporophyte produces unicellular haploid spores as products of meiosis within a sporangium. A spore germinates, giving rise to a multicellular haploid gametophyte whose cells contain chloroplasts and are thus photosynthetic. Eventually gametes form within specialized sex organs, the **gametangia**. The **archegonium** is a multicellular, flask-shaped female sex organ with a long neck and a swollen base, which produces a single egg. The **antheridium** is a male sex organ in which sperm, each bearing two flagella, are produced in large numbers (see the insets in Figure 28.5).

Once released from the antheridium, the sperm must swim or be splashed by raindrops to a nearby archegonium on the same or a neighboring plant. The sperm are aided in this task by chemical attractants released by the egg or the archegonium. Before sperm can enter the archegonium, certain cells in the neck of the archegonium must break down, leaving a water-filled canal through which the sperm swim to complete their journey. Note that *all of these events require liquid water*.

On arrival at the egg, the nucleus of a sperm fuses with the egg nucleus to form a diploid zygote. Mitotic divisions of the zygote produce a multicellular, diploid sporophyte embryo. The base of the archegonium grows to protect the embryo during its early development. Eventually, the developing sporophyte elongates sufficiently to break out of the archegonium, but it remains connected to the gametophyte by a "foot" that is embedded in the parent tissue and absorbs water and nutrients from it. In liverworts, hornworts, and mosses, the sporophyte remains attached to the gametophyte throughout its life. The sporophyte produces a sporangium, within which meiotic divisions produce spores and thus the next gametophyte generation.

28.2 RECAP

Terrestrial organisms must obtain and retain water for their life processes, for support against gravity, and for reproduction.

- Describe several adaptations of plants to the terrestrial environment. See p. 613
- Can you explain the cycle of alternation of generations? See pp. 614–615 and Figure 28.5
- Why do some plants require liquid water for fertilization? See p. 616
- In the nonvascular plants, how is the sporophyte dependent on the gametophyte? See p. 616

Further adaptations to the terrestrial environment appeared as plants continued to evolve. One of the most important of these later adaptations was the appearance of vascular tissues.

28.3 What Features Distinguish the Vascular Plants?

We now know that nonvascular plants evolved tens of millions of years before the earliest vascular plants, even though vascular plants—which fossilize more readily because of the chemical makeup of their vascular tissues—appear earlier in the fossil record. Persuasive DNA and structure-based evidence for the earlier appearance of nonvascular plants was further supported by recent experimental evidence suggesting that ancient microfossils predating the first appearance of nonvascular plants are actually fragments of ancient liverworts (**Figure 28.6**).

EXPERIMENT

HYPOTHESIS: Ancient microfossils, predating the earliest known nonvascular plant fossils, could be fragments of ancient liverworts.

METHOD

1. Investigators allowed liverworts to rot in soil or subjected them to high-temperature acid treatment, then examined the degraded material by light and scanning electron microscopy.
2. They compared images of the degraded material with those of microfossils from ancient rocks.

RESULTS

Sheets of decay-resistant cells from the lower surface of the degraded liverworts resembled some cell-sheet microfossils, showing rosette-like groupings of cells around "pores."

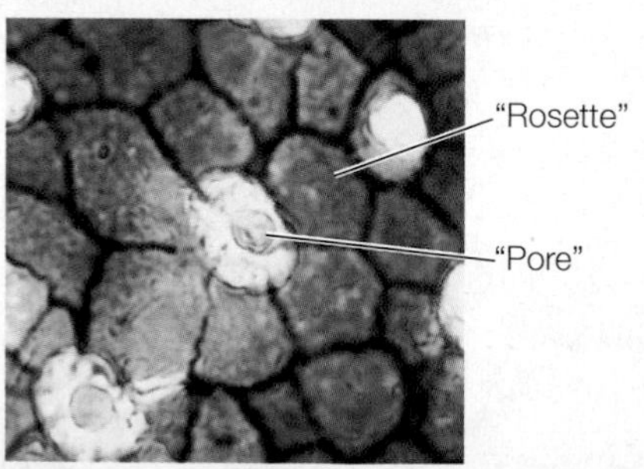

Resistant fragments of liverwort rhizoids resembled some tubular microfossils.

CONCLUSION: The ancient microfossils may be fragments of liverworts.

28.6 Mimicking a Microfossil? Linda Graham and her collaborators studied two species of decay-resistant liverworts. These plants yielded degradation fragments closely similar in appearance to microfossils characteristic of rocks from the Cambrian through the Devonian, suggesting that those microfossils were also from liverworts.

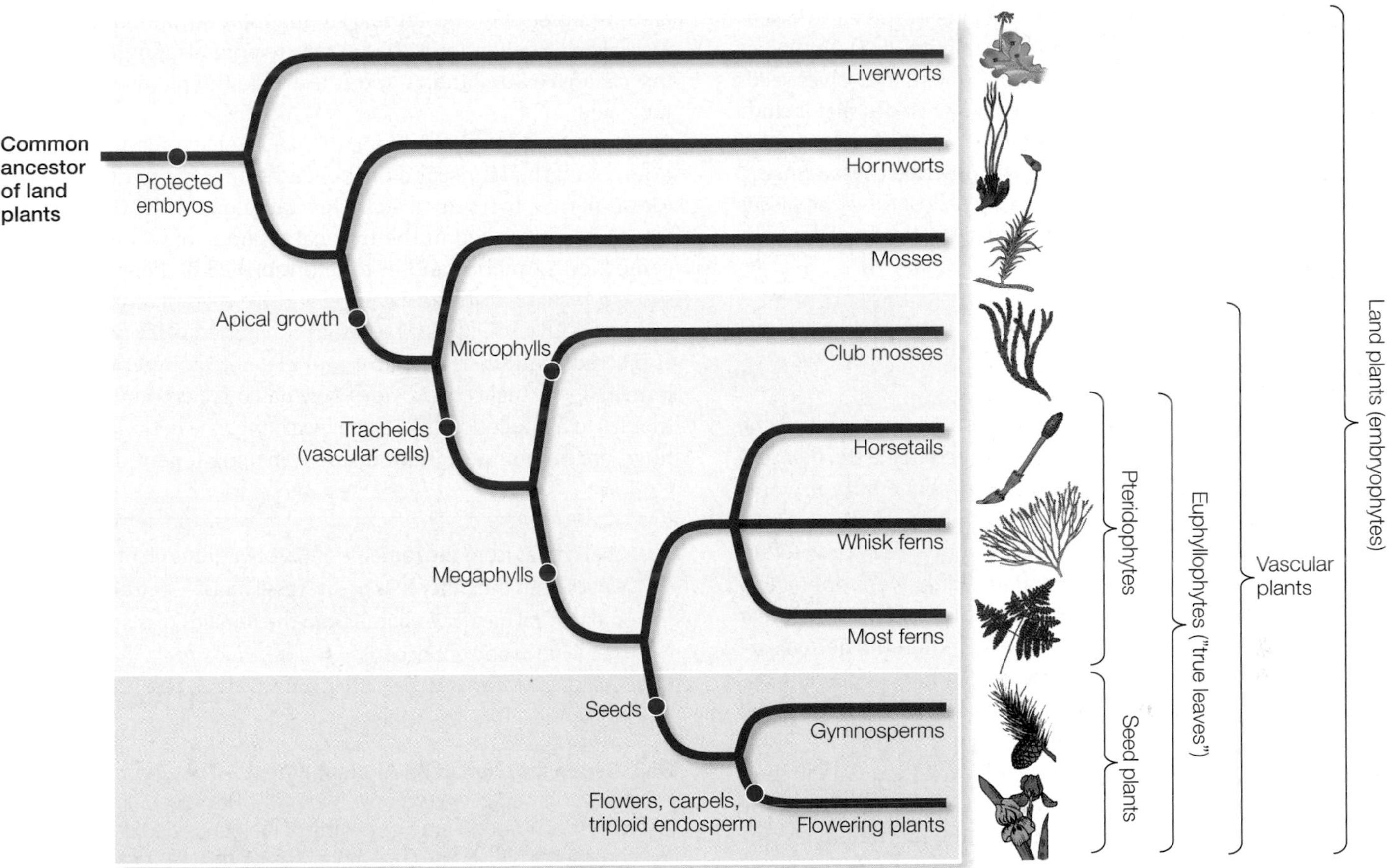

28.7 The Evolution of Today's Plants Three key characteristics that emerged during plant evolution—protected embryos, vascular tissues, and seeds—are adaptations to life in a terrestrial environment.

The first land plants were nonvascular, lacking both water-conducting and food-conducting tissue. The first true vascular plants, possessing specialized tracheids, arose later (**Figure 28.7**).

Vascular tissues transport water and dissolved materials

Vascular plants differ from the other land plants in crucial ways, one of which is the possession of a well-developed **vascular system** consisting of tissues specialized for the transport of materials from one part of the plant to another. One type of vascular tissue, the **xylem**, conducts water and minerals from the soil to aerial parts of the plant; because some of its cell walls are stiffened by a substance called *lignin*, xylem also provides support in the terrestrial environment. The other type of vascular tissue, the **phloem**, conducts the products of photosynthesis from sites where they are produced or released to sites where they are used or stored.

Familiar vascular plants include the club mosses, ferns, conifers, and angiosperms (flowering plants). Although they are an extraordinarily large and diverse group, the vascular plants can be said to have been launched by a single evolutionary event. Sometime during the Paleozoic era, probably well before the Silurian period (440 mya), the sporophyte generation of a now long-extinct plant produced a new cell type, the tracheid. The tracheid is the principal water-conducting element of the xylem in all vascular plants except the angiosperms, and even in the angiosperms, tracheids persist alongside a more specialized and efficient system of vessels and fibers derived from them.

The evolution of tracheids had two important consequences. First, it provided a pathway for transport of water and mineral nutrients from a source of supply to regions of need in the plant body. Second, the stiff cell walls of tracheids provided something almost completely lacking—and unnecessary—in the largely aquatic green algae: rigid structural support. Support is important in a terrestrial environment because it allows plants to grow upward as they compete for sunlight to power photosynthesis. A taller plant can receive direct sunlight and photosynthesize more readily than a shorter plant, whose leaves may be shaded by the taller one. Increased height also improves the dispersal of spores. Thus the tracheid set the stage for the complete and permanent invasion of land by plants.

The vascular plants featured another evolutionary novelty: a branching, independent sporophyte. A branching sporophyte can produce more spores than an unbranched body, and it can develop in complex ways. The sporophyte of a vascular plant is nutritionally independent of the gametophyte at maturity. Among the vascular plants, the sporophyte is the large and obvious plant that one normally notices in nature, in contrast to the sporophyte of nonvascular plants, which is attached to, dependent on, and usually much smaller than the gametophyte.

The present-day evolutionary descendants of the early vascular plants belong to six major groups (see Figure 28.7). Two types of life cycles are seen in the vascular plants: one that involves seeds and another that does not. The life cycles of the clades that include the club mosses and the ferns and their relatives, the horsetails and whisk ferns, do not involve seeds. We will describe these nonseed vascular plant groups in detail after taking a closer look at vascular plant evolution. The major groups of seed plants will be described in Chapter 29.

Vascular plants have been evolving for almost half a billion years

The evolution of an effective cuticle and of protective layers for the gametangia (archegonia and antheridia) helped make the first vascular plants successful, as did the initial absence of herbivores (plant-eating animals) on land. By the late Silurian period, vascular plants were being preserved as fossils that we can study today. During the Silurian, the largest vascular plants were only a few centimeters tall, yet fossils uncovered in Wales in 2004 gave clear evidence for the earliest known wildfire, which burned vigorously even in the Silurian atmosphere, which had 14 percent less oxygen than today's. The small plants must have been abundant to sustain fire in such an atmosphere.

Two groups of vascular plants that still exist today made their first appearances during the Devonian period (409–354 mya): the lycophytes (club mosses and relatives) and the pteridophytes (including horsetails and ferns). Their proliferation made the terrestrial environment more hospitable to animals. Amphibians and insects arrived on land at about the time the plants became established.

Trees of various kinds appeared in the Devonian period and dominated the landscape of the Carboniferous period. Mighty forests of lycophytes up to 40 meters tall, along with horsetails and tree ferns, flourished in the tropical swamps of what would become North America and Europe (**Figure 28.8**). Plant parts from those forests sank in the swamps and were gradually covered by sediment. Over millions of years, as the buried plant material was subjected to intense pressure and elevated temperatures, coal formed. Today that coal provides over half our electricity—and contributes to air pollution and global warming. The deposits, although huge, are not infinite, and they are not being renewed.

Coal comes from the remains of Carboniferous plants, but what about the other two great "fossil fuels"—petroleum and natural gas? They come from the remains of plankton that lived in ancient oceans.

28.8 Reconstruction of an Ancient Forest This Carboniferous forest once thrived in what is now Michigan. The "trees" to the left and in the background are lycophytes of the genus *Lepidodendron*; abundant ferns are visible to the right. The plant in the foreground is a relative of the horsetails.

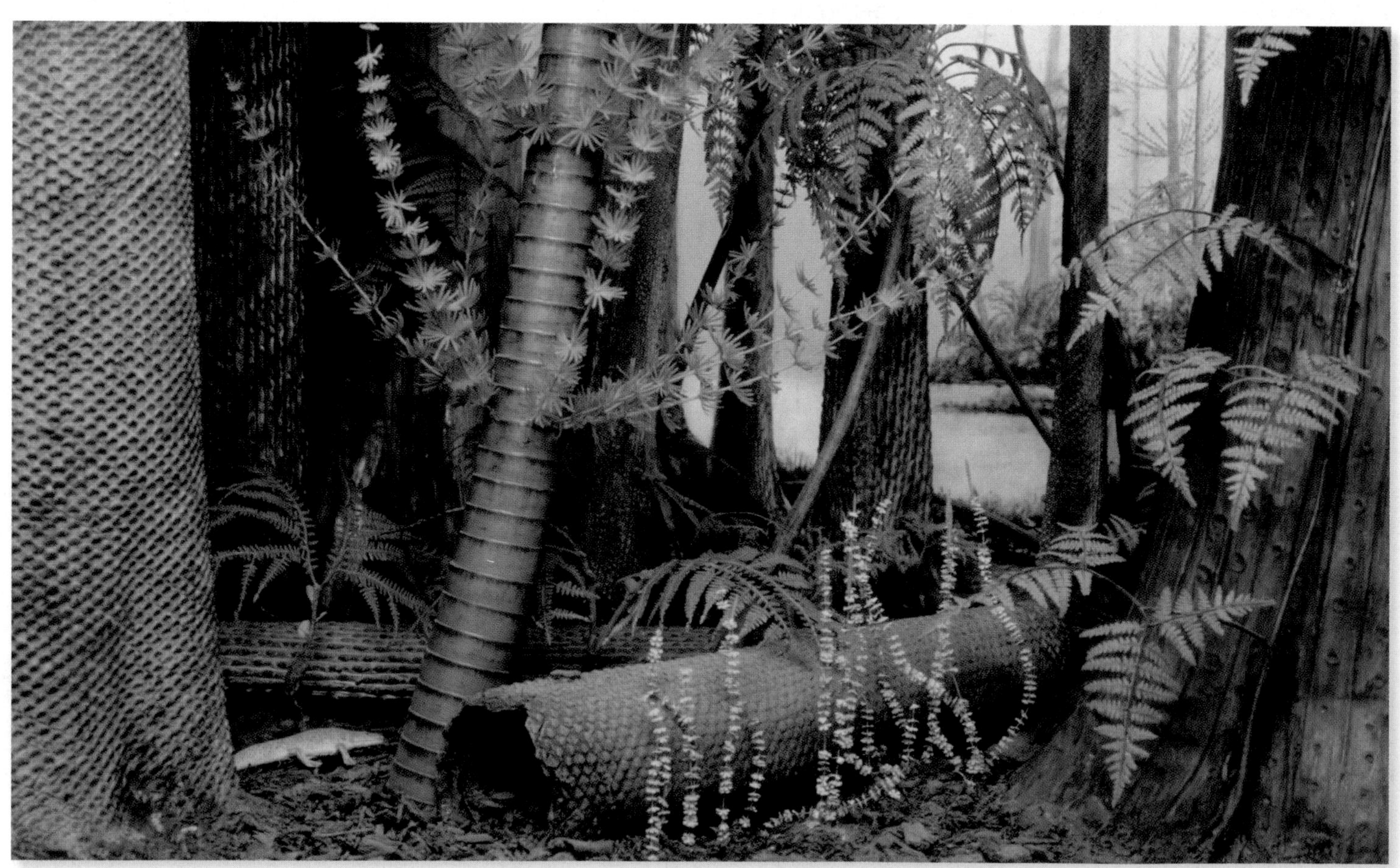

In the subsequent Permian period, the continents came together to form a single gigantic land mass called Pangaea. The continental interior became warmer and drier, but late in the period glaciation was extensive. The 200-million-year reign of the lycophyte–fern forests came to an end as they were replaced by forests of seed plants (gymnosperms), which were prevalent until a different group of seed plants (angiosperms) overtook the landscape less than 80 million years ago.

The earliest vascular plants lacked roots and leaves

The earliest known vascular plants belonged to the now-extinct group called **rhyniophytes**. The rhyniophytes were one of a very few types of vascular plants in the Silurian period. The landscape at that time probably consisted of bare ground, with stands of rhyniophytes in low-lying moist areas. Early versions of the structural features of all the other vascular plant groups appeared in the rhyniophytes of that time. These shared features strengthen the case for the origin of all vascular plants from a common nonvascular plant ancestor.

At the beginning of this chapter we described the discovery of some important fossils in Devonian rocks near Rhynie, Scotland. The preservation of these plants was remarkable, considering that the rocks were more than 395 million years old. These fossil plants had a simple vascular system of phloem and xylem, but not all had the tracheids characteristic of today's vascular plants.

These plants also lacked roots. Like most modern ferns and lycophytes, they were apparently anchored in the soil by horizontal portions of stem, called **rhizomes**, which bore water-absorbing unicellular filaments called **rhizoids**. These rhizomes also bore aerial branches, and sporangia—homologous to the sporangia of mosses—were found at the tips of those branches. Their branching pattern was *dichotomous;* that is, the apex (tip) of the shoot divided to produce two equivalent new branches, each pair diverging at approximately the same angle from the original stem (**Figure 28.9**). Scattered fragments of such plants had been found earlier, but never in such profusion or so well preserved as those discovered by Kidston and Lang.

Although they were apparently ancestral to the other vascular plant groups, the rhyniophytes themselves are long gone. None of their fossils appear anywhere after the Devonian period.

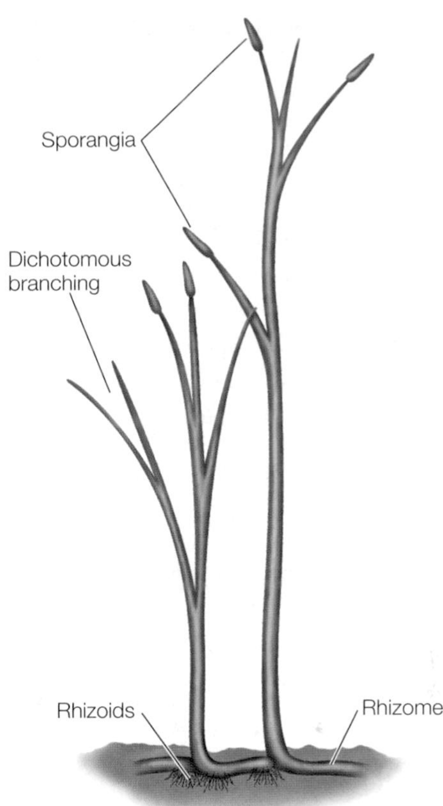

28.9 An Ancient Vascular Plant Relative This extinct plant, *Aglaophyton major* (a rhyniophyte), lacked roots and leaves. It had a central column of xylem running through its stems, but no true tracheids. The rhizome is a horizontal underground stem, not a root. The dichotomously branching aerial stems were less than 50 centimeters tall, and some were topped by sporangia. Other very similar rhyniophytes, such as *Rhynia*, did have tracheids.

The vascular plants branched out

A new group of vascular plants—the **lycophytes** (club mosses and their relatives)—also appeared in the Silurian period. Another—the **pteridophytes** (ferns and fern allies)—appeared during the Devonian period. These two groups, both still with us today, arose from rhyniophyte-like ancestors. These new groups featured specializations not found in the rhyniophytes, including true roots, true leaves, and a differentiation between two types of spores. The pteridophytes and seed plants constitute a clade called the **euphyllophytes**.

An important synapomorphy of the euphyllophytes is **overtopping** growth, a pattern in which one branch differentiates from and grows beyond the others. Overtopping growth would have given these plants an advantage in the competition for light for photosynthesis, enabling them to shade their dichotomously growing competitors. And, as we'll see, the overtopping growth of the euphyllophytes enabled a new type of leaf to evolve.

Roots may have evolved from branches

The rhyniophytes had only rhizoids arising from a rhizome with which to gather water and minerals. How, then, did subsequent groups of vascular plants come to have the complex roots we see today? Roots probably arose separately in the lycophytes and euphyllophytes.

It is probable that roots had their evolutionary origins as a branch, either of a rhizome or of the aboveground portion of a stem. That branch presumably penetrated the soil and branched further. The underground portion could anchor the plant firmly, and even in this primitive condition, it could absorb water and minerals. The discovery of several fossil plants from the Devonian period, all having horizontal stems (rhizomes) with both underground and aerial branches, supported this hypothesis.

Underground and aboveground branches, growing in sharply different environments, were subjected to very different selection pressures during the succeeding millions of years. Thus the two parts of the plant body—the aboveground shoot system and the underground root system—diverged in structure and evolved distinct internal and external anatomies. In spite of these differences, scientists believe that the root and shoot systems of vascular plants are homologous—that they were once part of the same organ.

Pteridophytes and seed plants have true leaves

Thus far we have used the term "leaf" rather loosely. In the strictest sense, a *leaf* is a flattened photosynthetic structure emerging laterally from a stem or branch and possessing true vascular tissue. Using this precise definition as we take a closer look at true leaves in the vascular plants, we see that there are two different types of leaves, very likely of different evolutionary origins.

The first leaf type, the **microphyll**, is usually small and only rarely has more than a single vascular strand, at least in plants alive today. Club mosses (lycophytes), of which only a few genera survive, have such simple leaves. Some biologists believe that microphylls had their evolutionary origins as sterile sporangia (**Figure 28.10A**). The principal characteristic of this type of leaf is a vascular strand that departs from the vascular system of the stem in such a way that the structure of the stem's vascular system is scarcely disturbed. This was true even in the lycophyte trees of the Carboniferous period, many of which had leaves many centimeters long.

The other leaf type is found in pteridophytes and seed plants. This larger, more complex leaf is called a **megaphyll**. The megaphyll is thought to have arisen from the flattening of a dichotomously branching stem system with overtopping growth. This change was followed by the development of photosynthetic tissue between the members of overtopped groups of branches (**Figure 28.10B**), which had the advantage of increasing the photosynthetic surface area of those branches. Megaphylls evolved more than once, in different clades of euphyllophytes with overtopping growth.

The first megaphylls, which were very small, appeared in the Devonian period. We might expect that evolution should have led swiftly to the appearance of more and larger megaphylls because of their greater photosynthetic capacity. However, it took some 50 million years, until the Carboniferous period, for large megaphylls to become common. Why should this have been so, especially given that other advances in plant structure were taking place during that time?

According to one theory, the high concentration of CO_2 in the atmosphere during the Devonian period restricted the development of the tiny pores, called *stomata*, that allow a leaf to take up CO_2 for use in photosynthesis. With more CO_2 available, fewer stomata were needed. Today, when stomata are open, they allow water vapor to escape the leaf and CO_2 to enter. In the Devonian, larger leaves would have absorbed heat from sunlight, but would

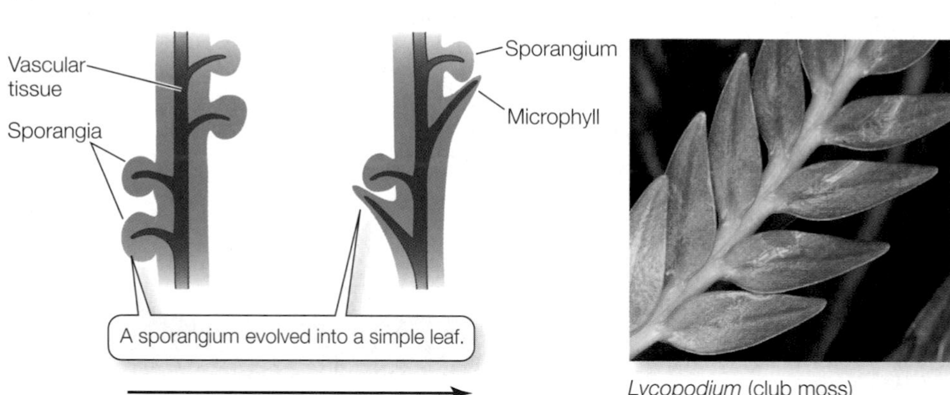
(A)
Vascular tissue
Sporangia
Sporangium
Microphyll
A sporangium evolved into a simple leaf.
Time
Lycopodium (club moss)

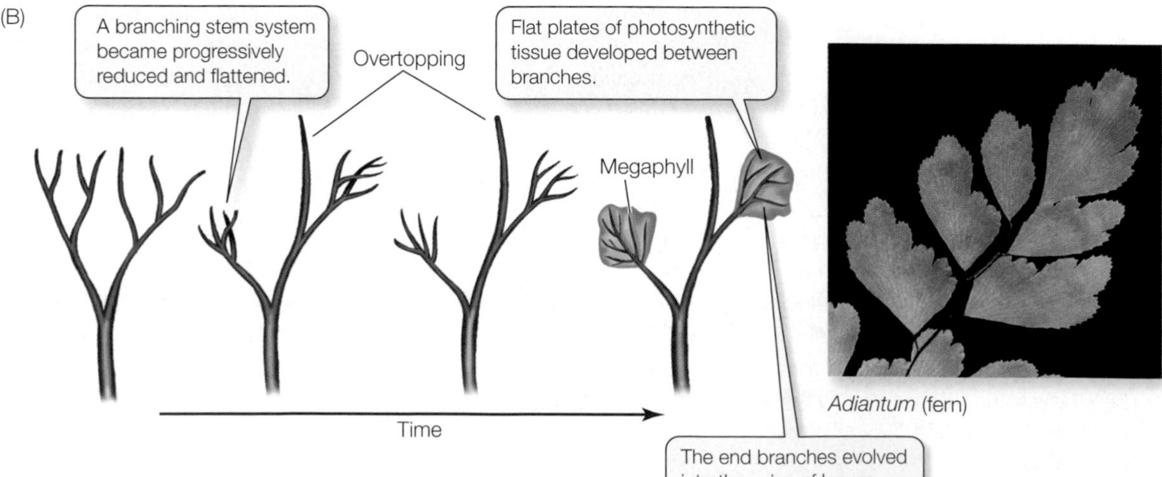
(B)
A branching stem system became progressively reduced and flattened.
Overtopping
Flat plates of photosynthetic tissue developed between branches.
Megaphyll
Time
The end branches evolved into the veins of leaves.
Adiantum (fern)

28.10 The Evolution of Leaves (A) Microphylls are thought to have evolved from sterile sporangia. (B) The megaphylls of pteridophytes and seed plants may have arisen as photosynthetic tissue developed between branch pairs that were "left behind" as dominant branches overtopped them.

EXPERIMENT

HYPOTHESIS: High concentrations of CO_2 in the Devonian atmosphere delayed the increase in leaf size.

METHOD

1. Scientists analyzed 300 plant fossils from the Devonian and Carboniferous periods and calculated the sizes of the leaves.
2. They compared the pattern of change in leaf size over time with that of change in atmosphere CO_2 concentrations.

RESULTS

Among these fossils, a rise in leaf size corresponded in time with a decline in the estimated concentration of CO_2 in the atmosphere.

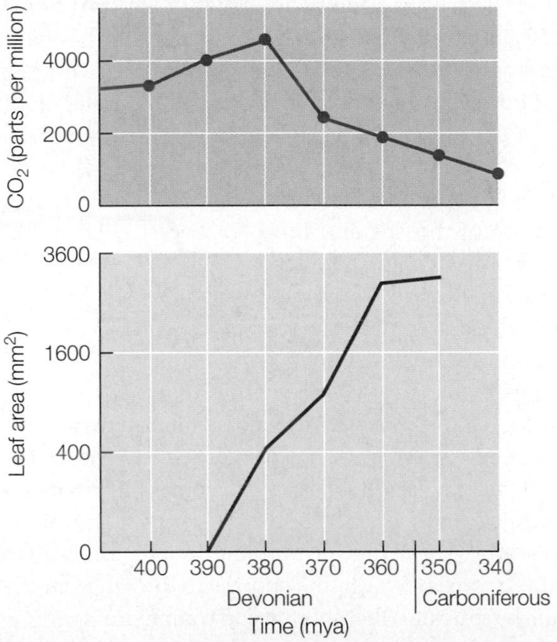

CONCLUSION: Leaf sizes increased as CO_2 concentrations decreased.

28.11 Decreasing CO_2 Levels and the Evolution of Megaphylls How can biologists test hypotheses relating to events that happened hundreds of millions of years ago? C. P. Osborne and colleagues gathered and measured the leaves of plant fossils from the Devonian and Carboniferous periods. They then compared the sizes of the leaves with estimates of atmospheric CO_2 concentrations under which the plants had lived. FURTHER RESEARCH: What sort of experiment would you do to determine the effects of stomata on overheating of fern leaves today?

have been unable to lose heat fast enough by evaporation of water through their limited number of stomata. The resulting overheating would have been lethal. Recent research has supported this theory, indicating that larger megaphylls evolved only as CO_2 concentrations dropped over millions of years (**Figure 28.11**).

Heterospory appeared among the vascular plants

In the most ancient of the present-day vascular plants, the gametophyte and the sporophyte are independent, and both are usually photosynthetic. Spores produced by the sporophyte are of a single type and develop into a single type of gametophyte that bears both female and male reproductive organs. The female organ is a multicellular archegonium containing a single egg. The male organ is an antheridium, producing many sperm. Such plants, which bear a single type of spore, are said to be **homosporous** (**Figure 28.12A**).

A system with two distinct types of spores evolved somewhat later. Plants of this type are said to be **heterosporous** (**Figure 28.12B**). In heterospory, one type of spore—the **megaspore**—develops into a specifically female gametophyte (a **megagametophyte**) that produces only eggs. The other type, the **microspore**, is smaller and develops into a male gametophyte (a **microgametophyte**) that produces only sperm. The sporophyte produces megaspores in small numbers in **megasporangia**, and microspores in large numbers in **microsporangia**. Heterospory affects not only the spores and the gametophyte, but also the sporophyte itself, which must develop two types of sporangia.

The most ancient vascular plants were all homosporous, but heterospory evidently evolved several times in the early descendants of the rhyniophytes. The fact that heterospory evolved repeatedly suggests that it affords selective advantages. Subsequent evolution in the land plants featured ever greater specialization of the heterosporous condition. All seed plants are heterosporous.

The nonseed vascular plants, like the nonvascular plants, require liquid water at a key stage in their life cycles: sperm reach eggs only by means of liquid water. As we will see in the next chapter, the seed plants have taken the next evolutionary step and can unite sperm and egg without relying on water.

28.3 RECAP

A new type of cell, the tracheid, marked the origin of the vascular plants. Later evolutionary events included the appearance of roots and leaves.

- How do the vascular tissues xylem and phloem serve the vascular plants? See p. 617
- Do you understand the difference between the two leaf types, microphylls and megaphylls? See p. 620 and Figure 28.10
- Can you explain the concept of heterospory? See p. 621 and Figure 28.12

The liverworts, hornworts, mosses, lycophytes, and pteridophytes have come a long way from the aquatic environment to meet the challenges of life on dry land. Let's look at the diversity within these groups.

28.12 Homospory and Heterospory (A) Homosporous plants bear a single type of spore. Each gametophyte has two types of sex organs, antheridia (male) and archegonia (female). (B) Heterosporous plants bear two types of spores that develop into distinctly male and female gametophytes.

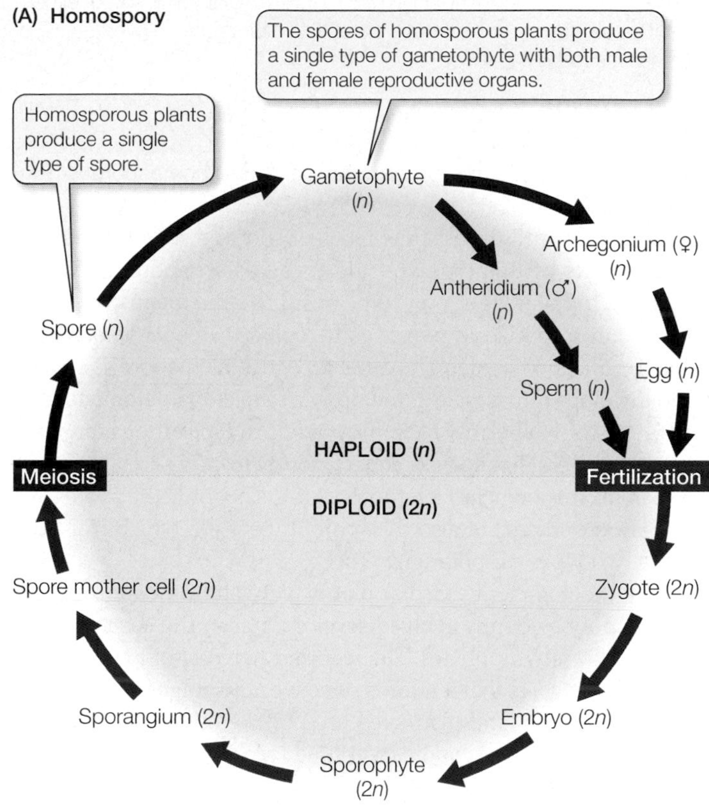

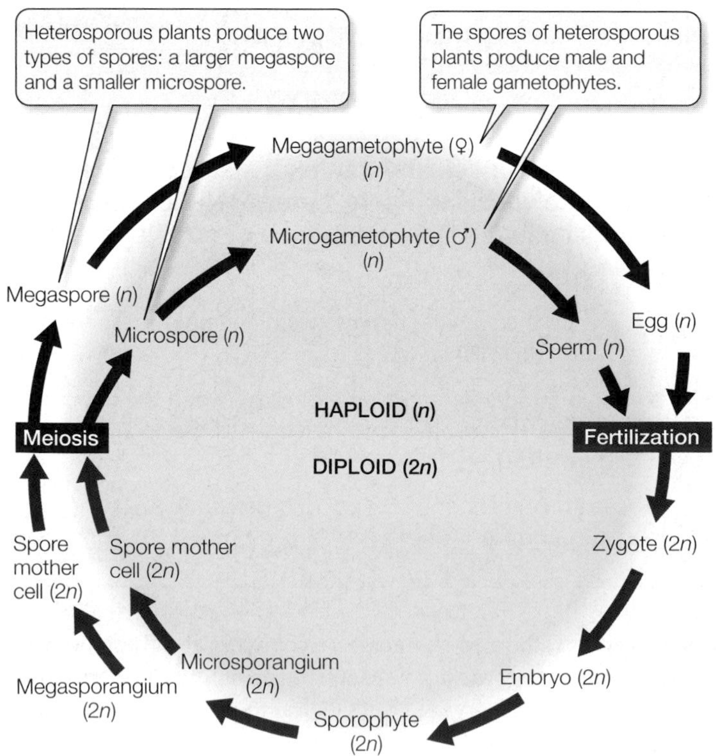

28.4 What Are the Major Clades of Seedless Plants?

Three clades of land plants lack tracheids; these nonvascular plants are the liverworts, hornworts, and mosses. Seedless vascular plants include three clades—club mosses, horsetails and whisk ferns—and the ferns, which are not a clade. The structure and growth pattern of the sporophyte differ among the three nonvascular land plant groups.

Liverworts may be the most ancient surviving plant clade

There are about 9,000 species of **liverworts** (**Hepatophyta**). Most liverworts have leafy gametophytes (**Figure 28.13A**). Some have *thalloid* gametophytes—green, leaflike layers that lie flat on the ground (**Figure 28.13B**). The simplest liverwort gametophytes, however, are flat plates of cells, a centimeter or so long, that produce antheridia or archegonia on their upper surfaces and rhizoids on their lower surfaces.

Liverwort sporophytes are shorter than those of mosses and hornworts, rarely exceeding a few millimeters. The liverwort sporophyte has a stalk that connects sporangium and foot. In most species, the stalk elongates by expansion of cells throughout its length. This elongation raises the sporangium above ground level, allowing the spores to be dispersed more widely. The sporangia of liverworts are simple: a globular sporangium wall surrounds a mass of spores. In some species of liverworts, spores are not released by the sporophyte until the surrounding sporangium wall rots. In other liverworts, however, the spores are thrown from the sporangium by structures that shorten and compress a "spring" as they dry out. When the stress becomes sufficient, the compressed spring snaps back to its resting position, throwing spores in all directions.

Liverworts
Hornworts
Mosses
Vascular plants

Among the most familiar thalloid liverworts are species of the genus *Marchantia*. *Marchantia* is easily recognized by the characteristic structures on which its male and female gametophytes bear their antheridia and archegonia (**Figure 28.13C**). Like most liverworts, *Marchantia* also reproduces asexually by simple fragmentation of the gametophyte. *Marchantia* and some other liverworts and mosses also reproduce asexually by means of *gemmae* (singular *gemma*), which are lens-shaped clumps of cells. In a few liverworts, the gemmae are loosely held in structures called *gemmae cups*, which promote dispersal of the gemmae by raindrops (see Figure 28.13C).

Hornworts have stomata, distinctive chloroplasts, and sporophytes without stalks

The group **Anthocerophyta** comprises the approximately 100 species of **hornworts**, so named because their sporophytes look like little horns (**Figure 28.14**). Hornworts appear at first glance to be liverworts with very simple gametophytes. Their gametophytes are flat plates of cells a few cells thick.

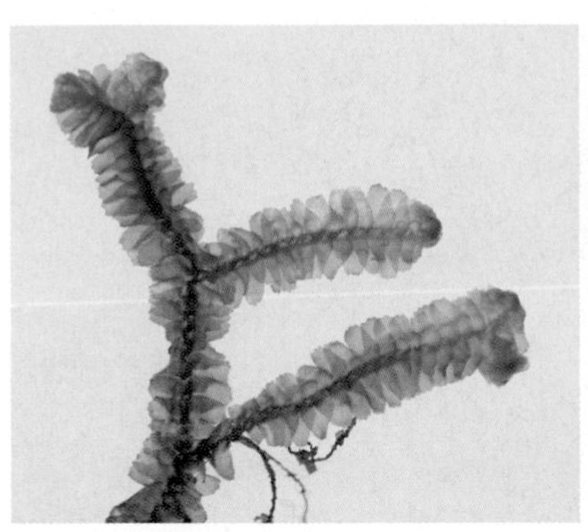

(A) *Bazzania trilobata*

(B) *Marchantia* sp.

These cups contain gemmae—small, lens-shaped outgrowths of the plant body, each capable of developing into a new plant.
The banana-like structures bear archegonia.

(C) *Marchantia* sp.

28.13 Liverwort Structures Liverworts display various characteristic structures. (A) The gametophyte of a leafy liverwort. (B) Gametophytes of a thalloid liverwort. (C) This thalloid liverwort bears archegonia in the structures that look like bunches of bananas. It also bears gemmae cups containing gemmae.

The hornworts, along with the mosses and vascular plants, share an advance over the liverwort clade in their adaptation to life on land: they have stomata. Stomata may be a shared derived trait of hornworts and all other land plants except liverworts, although hornwort stomata do not close, as do those in mosses and vascular plants, and may have evolved independently.

Hornworts have two characteristics that distinguish them from both liverworts and mosses. First, the cells of hornworts each contain a single large, platelike chloroplast, whereas the cells of the other two groups contain numerous small, lens-shaped chloroplasts. Second, of the sporophytes in all three of these groups, those of the hornworts come closest to being capable of growth without a set limit. Liverwort and moss sporophytes have a stalk that stops growing as the sporangium matures, so elongation of the sporophyte is strictly limited. The hornwort sporophyte, however, has no stalk. Instead, a basal region of the sporangium remains capable of indefinite cell division, continuously producing new spore-bearing tissue above. The sporophytes of some hornworts growing in mild and continuously moist conditions can become as tall as 20 centimeters. Eventually the sporophyte's growth is limited by the lack of a transport system.

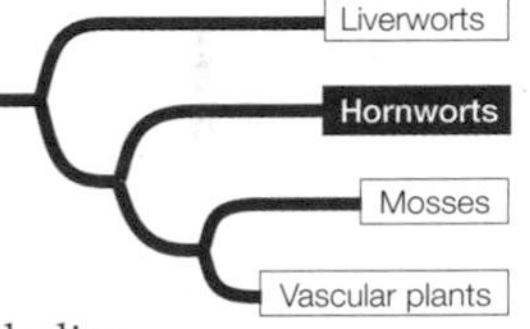

To support their metabolism, the hornworts need access to nitrogen. Hornworts have internal cavities filled with mucilage; these cavities are often populated by cyanobacteria that convert atmospheric nitrogen gas into a form usable by their host plant.

We have presented the hornworts as sister to the clade consisting of mosses and vascular plants, but this is only one possible interpretation of the current data. The exact evolutionary status of the hornworts is still unclear, and in some analyses they are placed as the sister group to the remaining land plants.

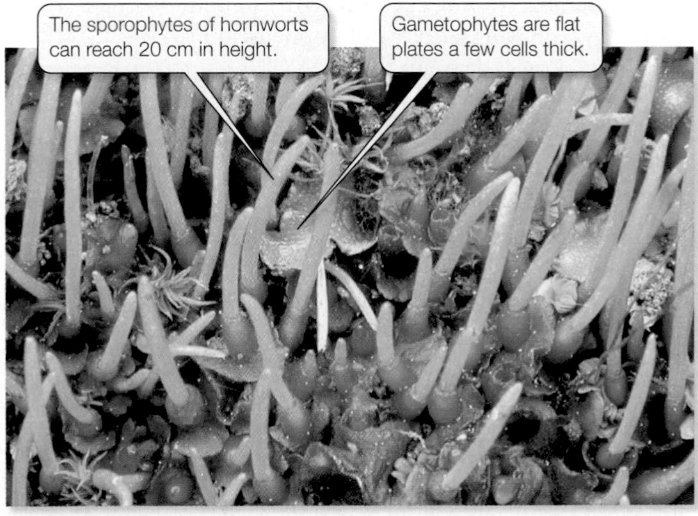

Anthoceros sp.

28.14 A Hornwort The sporophytes of many hornworts resemble little horns.

Water and sugar transport mechanisms emerged in the mosses

The most familiar of the land plants without tracheids are the **mosses** (**Bryophyta**). There are about 15,000 species of mosses—more than of liverworts and hornworts combined—and these hardy little plants are found in almost every terrestrial environment. They are often found on damp, cool ground, where they form thick mats (see Figure 28.3). The mosses are probably sister to the vascular plants (see Figure 28.7).

In mosses, the gametophyte begins its development following spore germination as a branched, filamentous structure called a *protonema* (see Figure 28.5). Although the protonema looks a bit like a filamentous green alga, it is unique to the mosses. Some of the filaments contain chloroplasts and are photosynthetic; others, called rhizoids, are nonphotosynthetic and anchor the protonema to the substratum. After a period of linear growth, cells close to the tips of the photosynthetic filaments divide rapidly in three dimensions to form *buds*. The buds eventually develop a distinct tip, or apex, and produce the familiar leafy moss shoot with leaflike structures arranged spirally. These leafy shoots produce antheridia or archegonia (see Figure 28.5).

Sporophyte development in most mosses follows a precise pattern, resulting ultimately in the formation of an absorptive foot anchored to the gametophyte, a stalk, and, at the tip, a swollen sporangium. In contrast to liverworts and hornworts, the sporophytes of mosses and vascular plants grow by **apical cell division**, in which a region at the growing tip provides an organized pattern of cell division, elongation, and differentiation. This growth pattern allows extensive and sturdy vertical growth of sporophytes. Apical cell division is a synapomorphy of mosses and vascular plants.

Some moss gametophytes are so large that they could not transport enough water solely by diffusion. Gametophytes and sporophytes of many mosses contain a type of cell called a *hydroid*, which dies and leaves a tiny channel through which water can travel. The hydroid may be the progenitor of the tracheid, the characteristic water-conducting cell of the vascular plants, but it lacks lignin and the cell wall structure found in tracheids. The possession of hydroids and of a limited system for transport of sugar by some mosses (via cells called *leptoids*) shows that the old term "nonvascular plant" is somewhat misleading when applied to mosses. Despite their simple system of internal transport, however, the mosses are not vascular plants because they lack true xylem and phloem.

Mosses of the genus *Sphagnum* often grow in swampy places, where the plants begin to decompose in the water after they die. Rapidly growing upper layers of moss compress the deeper-lying, decomposing layers. Partially decomposed plant matter is called *peat*. In some parts of the world, people derive the majority of their fuel from peat bogs (**Figure 28.15**). *Sphagnum*-dominated peatlands cover an area approximately half as large as the United States—more than 1 percent of Earth's surface. Long ago, continued compression of peat composed primarily of other nonseed plants gave rise to coal.

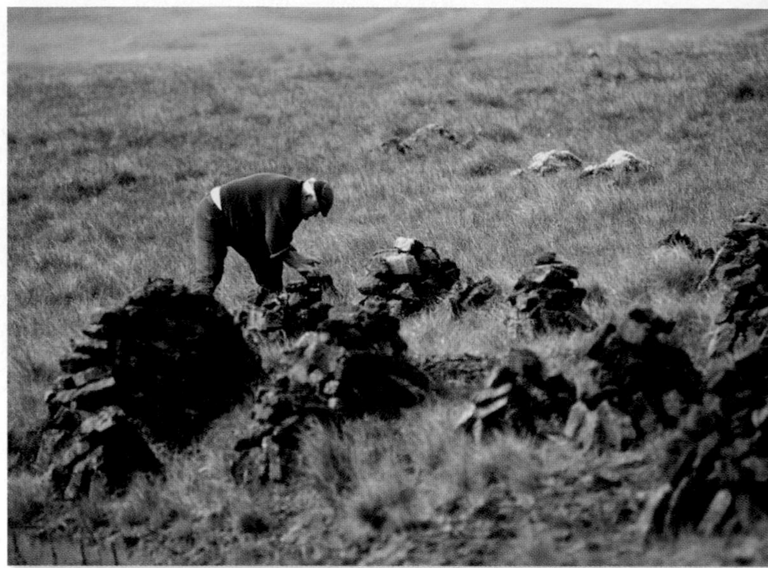

28.15 Harvesting Peat from a Bog A farmer extracts peat, formed from decomposing *Sphagnum* mosses. Peat bogs are an important source of fuel in some areas of the world, such as the southern Ireland location shown here.

Some vascular plants have vascular tissue but not seeds

The earliest vascular plant clades that survive to this day do not have seeds. These plants have a large, independent sporophyte and a small gametophyte that is independent of the sporophyte. The gametophytes of the surviving nonseed vascular plants are rarely more than 1 or 2 centimeters long and are short-lived, whereas their sporophytes are often highly visible and long-lived; the sporophyte of a tree fern, for example, may be 15 or 20 meters tall and may live for many years.

The most prominent resting stage in the life cycle of the seedless vascular plants is the single-celled spore. A spore may "rest" for some time before developing further. This feature makes their life cycle similar to those of the fungi, the green algae, and the nonvascular plants, but not, as we will see in the next chapter, to that of the seed plants. Nonseed vascular plants must have an aqueous environment for at least one stage of their life cycle because fertilization is accomplished by flagellated, swimming sperm.

The ferns are the most abundant and diverse group of seedless vascular plants today, but the club mosses and horsetails were once dominant elements of Earth's vegetation. A fourth group, the whisk ferns, contains only two genera. Let's look at the characteristics of these four groups and at some of the evolutionary advances that appeared in them.

The club mosses are sister to the other vascular plants

The **club mosses** and their relatives, the spike mosses and quillworts (together called lycophytes), diverged earlier than all other living vascular plants; that is, the remaining vascular plants share an ancestor that was not ancestral to the lycophytes. There are relatively few surviving species of lycophytes— just over 1,200.

The lycophytes have roots that branch dichotomously. The arrangement of vascular tissue in their stems is simpler than in the other vascular plants. They bear only microphylls, and these simple leaves are arranged spirally on the stem. Growth in club mosses comes entirely from apical cell division, and branching in the stems is also dichotomous, by a division of the apical cluster of dividing cells.

Liverworts
Hornworts
Mosses
Club mosses
Horsetails
Whisk ferns
Most ferns
Seed plants

The sporangia of many club mosses are aggregated in conelike structures called *strobili* (singular *strobilus*; **Figure 28.16**). A strobilus is a cluster of spore-bearing leaves inserted on an axis (linear supporting structure). Other club mosses lack strobili and bear their sporangia on (or adjacent to) the upper surfaces of leaves called *sporophylls*. This placement contrasts with the terminal spo-

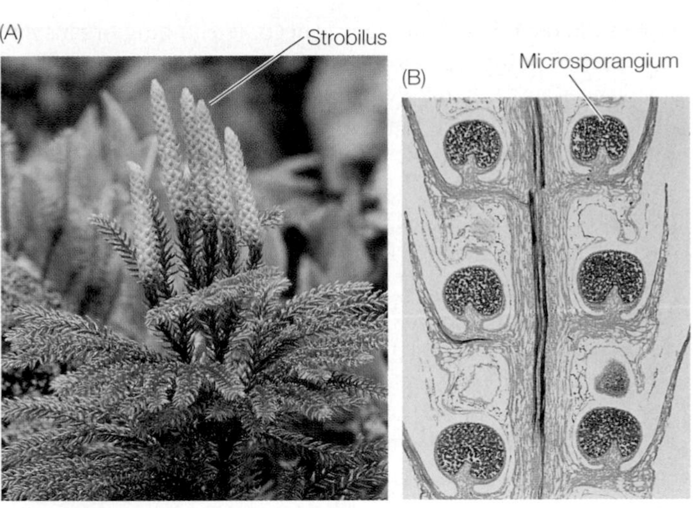

28.16 Club Mosses (A) Strobili are visible at the tips of this club moss. Club mosses have microphylls arranged spirally on their stems. (B) A thin section through a strobilus of a club moss, showing microsporangia.

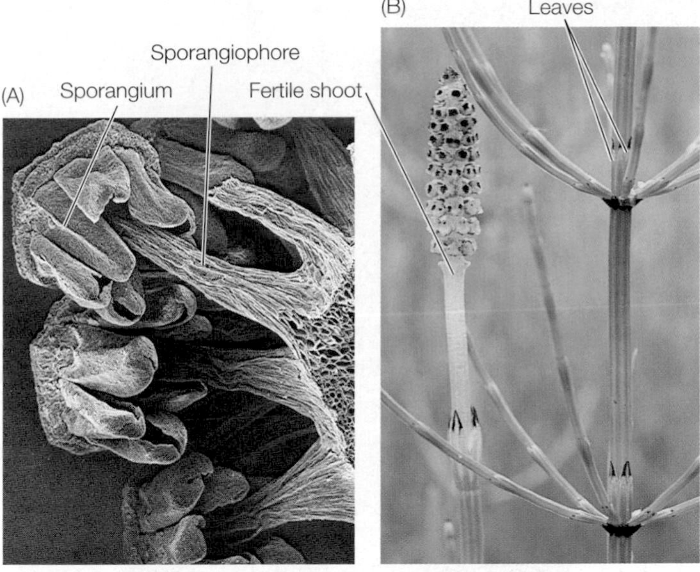

28.17 Horsetails (A) Sporangia and sporangiophores of a horsetail. (B) Vegetative and fertile shoots of the marsh horsetail. Reduced megaphylls can be seen in whorls on the stem of the vegetative shoot on the right. The sporangia on the fertile shoot (left) are ready to disperse their spores.

rangia of the rhyniophytes. There are both homosporous species and heterosporous species of club mosses.

Although they are only a minor element of present-day vegetation, the lycophytes are one of two groups that appear to have been the dominant vegetation during the Carboniferous period. One type of coal (cannel coal) is formed almost entirely from fossilized spores of the tree lycophyte *Lepidodendron*—which gives us an idea of the abundance of this genus in the forests of that time (see Figure 28.8). Other major elements of Carboniferous vegetation included horsetails and ferns.

Horsetails, whisk ferns, and ferns constitute a clade

Once thought to be only distantly related, the horsetails, whisk ferns, and ferns form a clade, the pteridophytes, or "ferns and fern allies." Within that clade, the whisk ferns and the horsetails are both monophyletic; the ferns are not. However, most ferns do belong to a single clade, the *leptosporangiate ferns*. In the pteridophytes—and in all seed plants—there is differentiation between the main stem and side branches (overtopping). This pattern contrasts with the dichotomous branching characteristic of the lycophytes and rhyniophytes (see Figure 28.9).

HORSETAILS The **horsetails** are represented by only about 15 present-day species. All are in a single genus, *Equisetum*. These plants are sometimes called "scouring rushes" because silica deposits found in their cell walls made them useful for cleaning. They have true roots that branch irregularly. Their sporangia curve back toward the stem on the ends of short stalks called *sporangiophores* (**Figure 28.17A**). Horsetails have a large sporophyte and a small gametophyte, both independent.

The small leaves of horsetails are reduced megaphylls and form in distinct whorls (circles) around the stem (**Figure 28.17B**). Growth in horsetails originates to a large extent from discs of dividing cells just above each whorl of leaves, so each segment of the stem grows from its base. Such basal growth is uncommon in plants, although it is found in the grasses, a major group of flowering plants, as well as in the hornworts.

WHISK FERNS There once was some disagreement about whether the rhyniophytes are entirely extinct. The confusion arose because of the existence today of about 15 species in two genera of rootless, spore-bearing plants, *Psilotum* and *Tmesipteris*, collectively called the **whisk ferns**. *Psilotum flaccidum* (**Figure 28.18**) has only minute scales instead of true leaves, but plants of the genus *Tmesipteris* have flattened photosynthetic organs—reduced megaphylls—with well-developed vascular tissue. Are these two genera the living relics of the rhyniophytes, or do they have more recent origins?

Psilotum and *Tmesipteris* once were thought to be evolutionarily ancient descendants of anatomically simple ancestors. That hypothesis was weakened by an enormous hole in the fossil record between the rhyniophytes that apparently became extinct more than 300 million years ago, and *Psilotum* and *Tmesipteris*, which are modern plants. DNA sequence data finally settled the question in favor of a more modern origin of the whisk ferns from fernlike ancestors. The whisk ferns are a clade of highly specialized plants that evolved fairly recently from anatomically more complex ancestors by loss or reduction of megaphylls and true roots. Whisk fern gametophytes live below the surface of the ground and lack chlorophyll. They depend on fungal partners for their nutrition.

28.18 A Whisk Fern *Psilotum* was once considered by some to be a surviving rhyniophyte and by others to be a fern. It is now included in the pteridophytes, and it is widespread in the tropics and subtropics.

FERNS The **ferns** constitute a group that first appeared during the Devonian period and today consists of more than 12,000 species. The ferns are not monophyletic, although the clade of leptosporangiate ferns includes about 97 percent of fern species. The leptosporangiate ferns differ from other ferns in having sporangia with walls only one cell thick, borne on a stalk.

The sporophytes of the ferns, like those of the seed plants, have true roots, stems, and leaves. Ferns are characterized by large leaves with branching vascular strands (**Figure 28.19A**). During its development, the fern leaf unfurls from a tightly coiled "fiddlehead" (**Figure 28.19B**). Some fern leaves become climbing organs and may grow to be as much as 30 meters long. A few species have small leaves as a result of evolutionary reduction, but even these small leaves have more than one vascular strand, and are thus megaphylls (**Figure 28.19C**).

THE FERN LIFE CYCLE In all ferns, *spore mother cells* inside the sporangia undergo meiosis to form haploid spores. Once shed, the spores may be blown great distances by the wind and eventually germinate to form independent gametophytes far from the parent sporophyte. A case in point: Old World climbing fern, *Lygodium microphyllum*, is currently spreading disastrously through the Florida Everglades, choking off the growth of other plants. This rapid spread is testimony to the effectiveness of windborne spores. Another example is the remarkable diversity of ferns that have spread through the isolated Hawaiian Islands.

Fern gametophytes have the potential to produce both antheridia and archegonia, although not necessarily at the same time or on the same gametophyte. Sperm swim through water to archegonia—often to those on other gametophytes—where they unite with an egg. The resulting zygote develops into a new sporophyte embryo. The young sporophyte sprouts a root and can thus grow independently of the gametophyte. In the alternating generations of a fern, the gametophyte is small, delicate, and short-lived, but the sporophyte can be very large and can sometimes survive for hundreds of years (**Figure 28.20**).

Because they require water for the transport of the male gametes to the female gametes, most ferns inhabit shaded, moist woodlands and swamps. Tree ferns can reach heights of 20 meters.

28.19 Fern Leaves Take Many Forms (A) The leaves of northern maidenhair fern form a pattern in this photograph. (B) The "fiddlehead" (developing leaf) of a common forest fern will unfurl and expand to give rise to a complex adult leaf such as those in (A). (C) The leaves of two species of water ferns.

Tree ferns are not as rigid as woody plants and they have poorly developed root systems. Thus they do not grow in sites exposed directly to strong winds, but rather in ravines or beneath trees in forests. The sporangia of ferns are found on the undersurfaces of the leaves, sometimes covering the whole undersurface and sometimes only at the edges. In most species the sporangia are found in clusters called *sori* (singular *sorus*) (see the inset in Figure 28.20).

Most ferns are homosporous. However, two groups of aquatic ferns, the Marsileaceae and the Salviniaceae (see Figure 28.19C), are derived from a common ancestor that evolved heterospory. The megaspores and microspores of these plants (which germinate to produce female and male gametophytes, respectively) are produced in different sporangia (megasporangia and microsporangia), and the microspores are always much smaller and greater in number than the megaspores.

A few genera of ferns produce a tuberous, fleshy gametophyte instead of the characteristic flattened, photosynthetic structure produced by most ferns. These tuberous gametophytes depend on a mutualistic fungus for nutrition; in some genera, even the sporophyte embryo must become associated with the fungus before its development can proceed. In Section 30.2 we will see that there are many other important plant–fungus mutualisms.

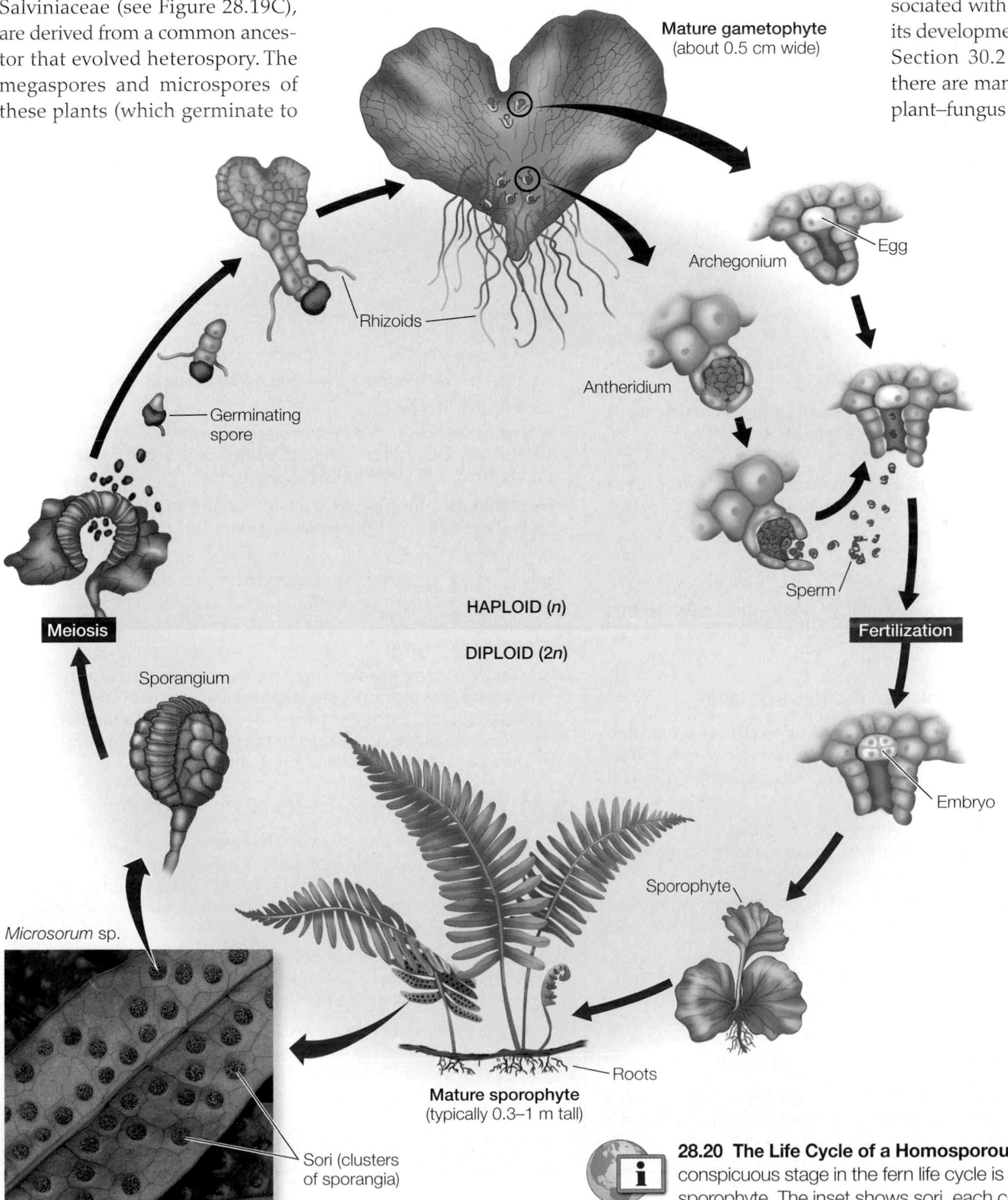

28.20 The Life Cycle of a Homosporous Fern The most conspicuous stage in the fern life cycle is the mature diploid sporophyte. The inset shows sori, each containing many spore-producing sporangia, on the underside of a leaf of a fern.

28.4 RECAP

Three clades of land plants lack true vascular systems (liverworts, hornworts, and mosses). The seedless vascular plants are the club mosses, horsetails, whisk ferns, and ferns. The ferns are not a clade.

- What is the difference between the branching patterns of lycophytes and pteridophytes? See p. 625
- Why was it once thought that whisk ferns were close relatives of the rhyniophytes? See p. 625
- Why do most ferns live in shady, moist areas? See p. 626

The seedless vascular plants, and especially the ferns, were long considered an evolutionary *cul-de-sac*—that is, a group with great diversity in the fossil record but less diversity in the present. However, recent DNA-based research has suggested that the diversification of today's ferns took place much more recently than previously thought. The expansion of flowering plants and their dominance of forests actually predates the diversification of extant ferns, which presumably took advantage of the new environments created by those forests.

All the vascular plants we have discussed thus far disperse themselves by spores. In the next chapter we will discuss the plants that dominate most of Earth's vegetation today—the seed plants, whose seeds afford new sporophytes protection unavailable to those of the other vascular plants. Even without such protection, ancient vascular plants were so successful and abundant that they formed vast forests, some of the remains of which are with us to this day as coal. Their modern representatives have come a long way from the plants discovered by Dawson and by Kidston and Lang.

CHAPTER SUMMARY

28.1 How did the land plants arise?

Land plants, sometimes referred to as **embryophytes**, are photosynthetic eukaryotes that develop from embryos protected by parental tissue. Review Figure 28.1

Streptophytes include the land plants and certain green algae. **Green plants** include the streptophytes and the remaining green algae.

Land plants arose from an aquatic green algal ancestor related to today's **Charales**.

There are ten major groups of living land plants. Seven groups (the **vascular plants**) have well developed water-conducting tissues with cells including **tracheids**; three groups (the **nonvascular plants**) do not. Review Table 28.1

28.2 How did plants colonize and thrive on land?

The acquisition of a **cuticle**, gametangia, a protected embryo, protective pigments, thick spore walls with a protective polymer, and a mutualistic association with a fungus are all adaptations to terrestrial life.

All land plant life cycles feature alternation of generations, in which a multicellular **sporophyte** alternates with a multicellular **gametophyte**. Review Figure 28.4

Spores form in **sporangia**; gametes form in **gametangia**. In nonvascular plants the female and male gametangia are, respectively, an **archegonium** and an **antheridium**.

The sporophyte of liverworts, hornworts, and mosses is smaller than the gametophyte and depends on it for water and nutrition. Review Figure 28.5, Web/CD Tutorial 28.1

28.3 What features distinguish the vascular plants?

A **vascular system**, consisting of **xylem** and **phloem**, conducts water, minerals, and products of photosynthesis through the bodies of vascular plants. Review Figure 28.7

In vascular plants, the sporophyte is larger than and independent of the gametophyte.

The **rhyniophytes**, the earliest vascular plants, are known to us only in fossil form. They lacked roots and leaves but possessed **rhizomes** and **rhizoids**. Review Figure 28.9

Lycophytes (club mosses and relatives) and **pteridophytes** (ferns and allies) appeared later. **Euphyllophytes** include the pteridophytes and seed plants.

Roots may have evolved from rhizomes or from branches. **Microphylls** probably evolved from sterile sporangia, and **megaphylls** may have resulted from the flattening and reduction of an **overtopping**, branching stem system. Review Figure 28.10

Many seedless vascular plants are **homosporous**, but **heterospory**—the production of distinct **megaspores** and **microspores**—evolved several times. Megaspores develop into **megagametophytes**; microspores develop into **microgametophytes**. Review Figure 28.12, Web/CD Activities 28.1 and 28.2

28.4 What are the major clades of seedless plants?

The nonvascular plant clades are the **liverworts** (Hepatophyta), the **hornworts** (Anthocerophyta), and the **mosses** (Bryophyta). The nonseed vascular plant groups are the **club mosses** and their relatives and the pteridophytes (**horsetails**, **whisk ferns**, and **ferns**). Review Figure 28.7

Hornworts, mosses, and vascular plants all have surface pores (stomata) in their leaves. In mosses and vascular plants, the sporophytes grow by **apical cell division**.

The ferns are not a clade, although 97 percent of fern species do constitute a clade called the leptosporangiate ferns. Ferns have megaphylls with branching vascular strands. See Web/CD Activity 28.3

SELF-QUIZ

1. Land plants differ from photosynthetic protists in that only the plants
   a. are photosynthetic.
   b. are multicellular.
   c. possess chloroplasts.
   d. have multicellular embryos protected by the parent.
   e. are eukaryotic.
2. Which statement about alternation of generations in land plants is *not* true?
   a. The gametophyte and sporophyte differ in appearance.
   b. Meiosis occurs in sporangia.
   c. Gametes are always produced by meiosis.
   d. The zygote is the first cell of the sporophyte generation.
   e. The gametophyte and sporophyte differ in chromosome number.
3. Which statement is not evidence for the origin of plants from the green algae?
   a. Some green algae have multicellular sporophytes and multicellular gametophytes.
   b. Both plants and green algae have cellulose in their cell walls.
   c. The two groups have the same photosynthetic pigments.
   d. Both plants and green algae produce starch as their principal storage carbohydrate.
   e. All green algae produce large, stationary eggs.
4. Liverworts, hornworts, and mosses
   a. lack a sporophyte generation.
   b. grow in dense masses, allowing capillary movement of water.
   c. possess xylem and phloem.
   d. possess true leaves.
   e. possess true roots.
5. Which statement is *not* true of the mosses?
   a. The sporophyte is dependent on the gametophyte.
   b. Sperm are produced in archegonia.
   c. There are more species of mosses than of liverworts and hornworts combined.
   d. The sporophyte grows by apical cell division.
   e. Mosses are probably sister to the vascular plants.
6. Megaphylls
   a. probably evolved only once.
   b. are found in all the vascular plant groups.
   c. probably arose from sterile sporangia.
   d. are the characteristic leaves of club mosses.
   e. are the characteristic leaves of horsetails and ferns.
7. The rhyniophytes
   a. lacked tracheids.
   b. possessed true roots.
   c. possessed sporangia at the tips of stems.
   d. possessed leaves.
   e. lacked branching stems.
8. Club mosses and horsetails
   a. have larger gametophytes than sporophytes.
   b. possess small leaves.
   c. are represented today primarily by trees.
   d. have never been a dominant part of the vegetation.
   e. produce fruits.
9. Which statement about ferns is *not* true?
   a. The sporophyte is larger than the gametophyte.
   b. Most are heterosporous.
   c. The young sporophyte can grow independently of the gametophyte.
   d. The leaf is a megaphyll.
   e. The gametophytes produce archegonia and antheridia.
10. The leptosporangiate ferns
   a. are not a monophyletic group.
   b. have sporangia with walls more than one cell thick.
   c. constitute a minority of all ferns.
   d. are pteridophytes.
   e. produce seeds.

FOR DISCUSSION

1. Mosses and ferns share a common trait that makes water droplets a necessity for sexual reproduction. What is that trait?
2. Are the mosses well adapted to terrestrial life? Justify your answer.
3. Ferns display a dominant sporophyte generation (with large leaves). Describe the major advance in anatomy that enables most ferns to grow much larger than mosses.
4. What features distinguish club mosses from horsetails? What features distinguish these groups from rhyniophytes? From ferns?
5. Why did some botanists once believe that the whisk ferns should be classified together with the rhyniophytes?
6. Contrast microphylls with megaphylls in terms of structure, evolutionary origin, and occurrence among plants.

FOR INVESTIGATION

Osborne's findings on the evolution of megaphylls (see Figure 28.11) support the concept that large megaphylls became common only after the atmospheric CO_2 level had dropped, so that more stomata were produced, allowing water to evaporate and cool larger leaves. How might you extend that work to confirm the involvement of temperature as a factor limiting leaf size?

CHAPTER 29 The Evolution of Seed Plants

A seed from Biblical times germinates in 2005

The Judean date was once much prized. The Prophet Muhammad admired its nutritional and medicinal properties; in the Koran it is associated with heaven and described as a symbol of goodness. The Judean date was the source of the "honey" in the Biblical "land of milk and honey." Today that ancient strain of Judean date is gone. Or is it?

Around 2,000 years ago, at the beginning of the Common Era, a seed developed in a fruit on a Judean date palm. The fruit that contained that seed found its way to a storeroom in the fortress Masada in Judea. In 73 C.E. a band of 960 Jewish Zealots involved in a religious revolt against Rome fled to this refuge with their families. Roman legions followed, and the ensuing siege lasted more than two years. In the end, rather than be killed or enslaved by the Roman soldiers, the Zealots are said to have killed themselves and their families in a dramatic mass suicide.

Twenty centuries later, archeologists working in Masada discovered the seed of that overlooked Judean date and confirmed its age. The previous record for seed survival and germination was 1,200 years, held by lotus seeds that recently germinated under the care of scientists in China. But botanist Elaine Solowey succeeded in making the 2,000-year-old date seed germinate! The resulting seedling has continued to thrive and grow. Perhaps the ancient Egyptians were right to place date seeds in the tombs of the Pharaohs as symbols of immortality.

Seeds are important structures for the evolutionary survival of plants. They protect the plant embryo within from environmental extremes through what may be a very long and stressful resting period—in the case of the Judean date, many centuries in a harsh desert. Seeds of the coconut palm remain dormant for years as they float across vast expanses of ocean, finally washing up on a distant shore, where they germinate and grow. Such hardiness is one of the properties that have contributed to making seed plants the predominant plants on Earth. All of today's forests are dominated by seed plants.

A Refuge As a precaution, King Herod of Judea fortified Masada and stocked it with water and food (including Judean dates).

The Hardy Seed This coconut seed has arrived on a beach, where it germinated successfully. The evolution of seeds was a major factor in the eventual dominance of the seed plants.

CHAPTER OUTLINE

So will the seedling growing under Dr. Solowey's care serve as the parent of a new population of Judean dates, thus resurrecting that genotype from extinction? Unfortunately, it cannot, because date trees are of two different sexes. It will be many years before we learn the sex of this plant (if it survives). And regardless of its sex, it cannot reproduce alone.

In fact many seed plant species do not need a partner in order to reproduce sexually, but the sexuality of date palms and other plants was recognized from the earliest days of agriculture. The life-giving attributes of the seed plants are indispensable, and humans have always sought to understand and augment plant reproductive cycles.

IN THIS CHAPTER we will first describe the defining characteristics of the seed plants as a group. Then we will describe the flowers and fruits that are characteristic of their most dominant group, the flowering plants or angiosperms. We will examine some of the unsolved problems in seed plant evolution, concluding with a survey of the diversity of living seed plants.

29.1 How Did Seed Plants Become Today's Dominant Vegetation?

By the late Devonian period, more than 360 million years ago, Earth was home to a great variety of land plants, many of which we discussed in the previous chapter. These plants shared the hot, humid terrestrial environment with insects, spiders, centipedes, and fishlike amphibians. The plants and animals affected one another, acting as agents of natural selection.

Late in this period an innovation appeared: some plants developed extensively thickened woody stems, which resulted from the proliferation of xylem. This type of growth in the diameter of stems and roots is called **secondary growth**. The first plants with this adaptation were seedless vascular plants called *progymnosperms*, all species of which are now extinct.

The seed plants are the most recent group of vascular plants to appear. The earliest fossil evidence of seed plants is found in Devonian rocks. Like the progymnosperms, these *seed ferns* were woody. They possessed fernlike foliage, but had seeds attached to their leaves. By the Carboniferous period, new lineages of seed plants had evolved (**Figure 29.1**).

The several clades of seed ferns are known only as fossils. Two of those clades are basal to the surviving seed plants, which fall into two groups, the **gymnosperms** (such as pines and cycads) and the **angiosperms** (flowering plants). There are four living groups of gymnosperms and one of angiosperms (**Figure 29.2**). The phylogenetic relationships among these five groups have not yet been resolved; we will discuss some of the issues involved later in this chapter. All living gymnosperms and many angiosperms show secondary growth. The life cycles of all seed plants also share major features, as we are about to see.

Features of the seed plant life cycle protect gametes and embryos

In Section 28.2, we noted a trend in plant evolution: the sporophyte became less dependent on the gametophyte, which became smaller in relation to the sporophyte. This trend continued with the appearance of the seed plants, whose gametophyte generation is reduced even further than it is in the ferns

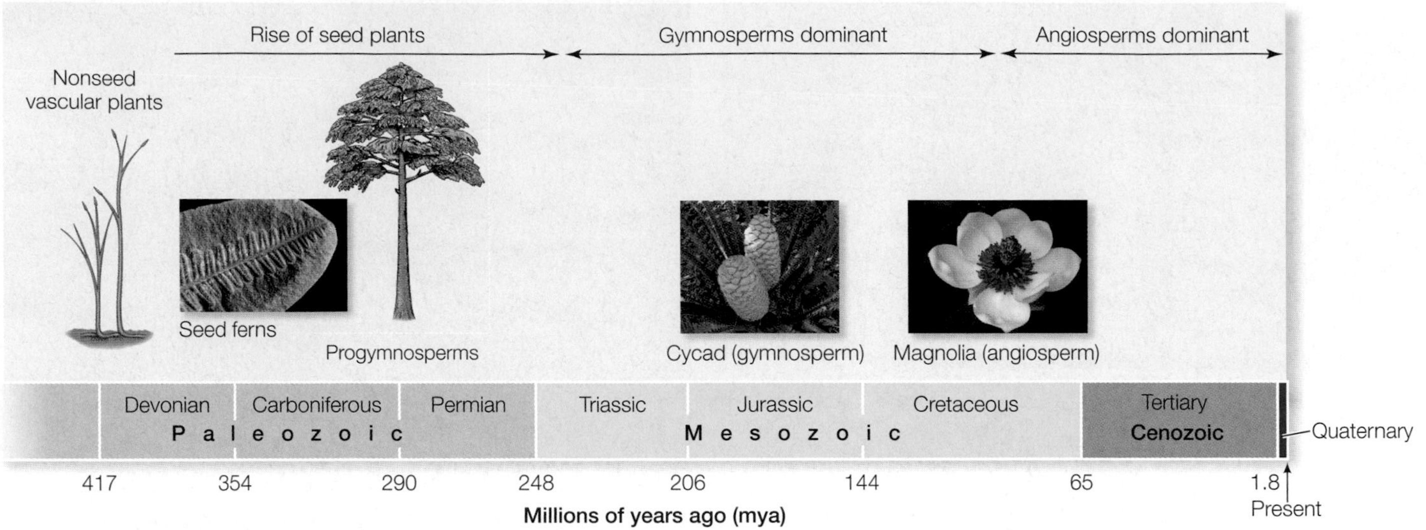

29.1 Highlights in the History of Seed Plants Woody growth evolved in the seedless progymnosperms. The now-extinct seed ferns had woody growth, fernlike foliage, and seeds attached to their leaves. New lineages of seed plants arose during the Carboniferous.

(**Figure 29.3**). The haploid gametophyte develops partly or entirely while attached to and nutritionally dependent on the diploid sporophyte.

Among the seed plants, only the earliest groups of gymnosperms (and their few survivors) had swimming sperm. Later groups of gymnosperms and the angiosperms evolved other means of bringing eggs and sperm together. The culmination of this striking evolutionary trend in seed plants was independence from the liquid water that earlier plants needed for sexual reproduction. This gave the seed plants a major advantage as they spread over the terrestrial environment.

Seed plants are heterosporous (see Figure 28.12B); that is, they produce two types of spores, one that becomes the male gametophyte and one that becomes the female gametophyte. They form separate megasporangia and microsporangia on structures that are grouped on short axes, such as the cones and strobili of conifers and the flowers of angiosperms.

As in other land plants, the spores of seed plants are produced by meiosis within the sporangia, but in seed plants, the megaspores are not shed. Instead, they develop into female gametophytes within the megasporangia. These megagametophytes are dependent on the sporophyte for food and water.

In most seed plant species, only one of the meiotic products in a megasporangium survives. The surviving haploid nucleus divides mitotically, and the resulting cells divide again to produce a multicellular female gametophyte. This megagametophyte is retained within the megasporangium, where it matures and produces an egg by mitosis. The megagametophyte, in turn, houses the early development of the next sporophyte generation when the egg is fertilized. The megasporangium is surrounded by sterile sporophytic structures, which form an **integument** that protects the megasporangium and its contents. Together, the megasporangium and integument constitute the **ovule**, which develops into a seed.

Within the microsporangium, the meiotic products are microspores, which divide mitotically within the spore wall one or a few times to form a male gametophyte called a **pollen grain**. Pollen

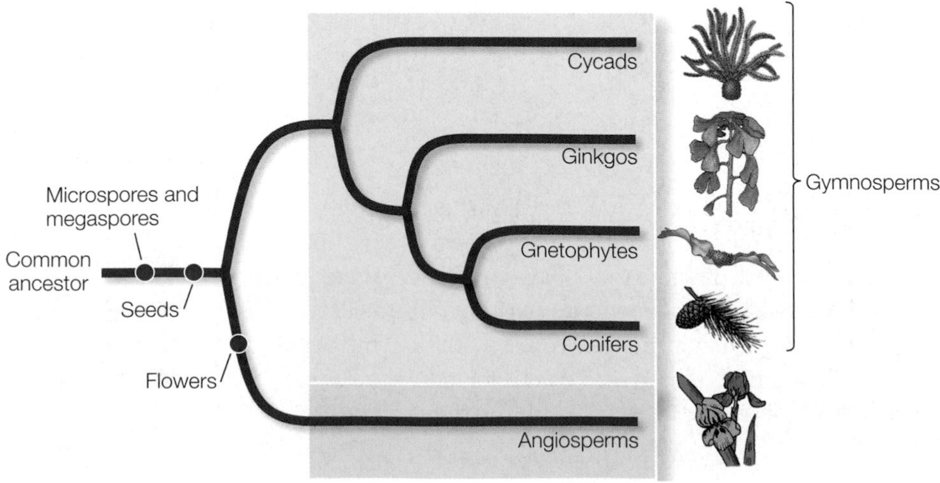

29.2 The Major Groups of Living Seed Plants There are four groups of gymnosperms and one of angiosperms. Their exact evolutionary relationship is still uncertain, but this cladogram represents one current interpretation.

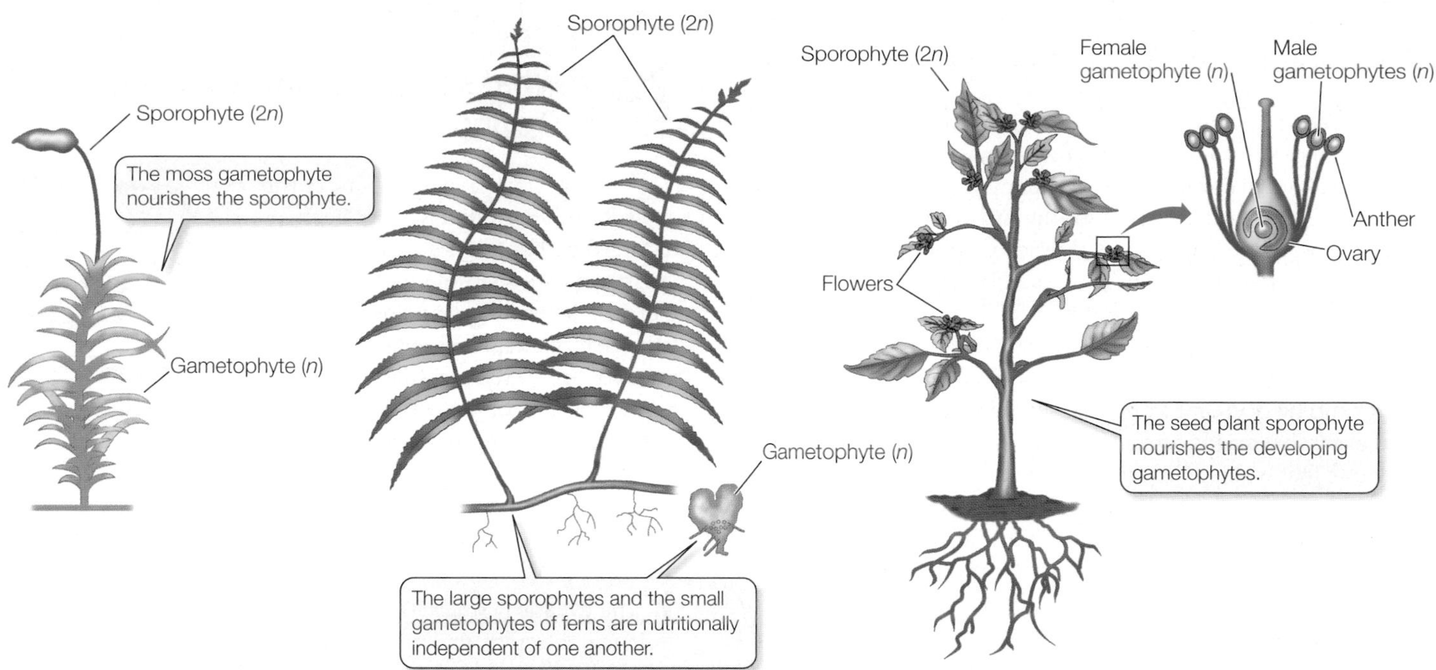

29.3 The Relationship between Sporophyte and Gametophyte Has Evolved In the course of plant evolution, the gametophyte has been reduced and the sporophyte has become more prominent.

grains are released from the microsporangium to be distributed by wind, an insect, a bird, or a plant breeder (**Figure 29.4**). The wall of the pollen grain contains *sporopollenin*, the most chemically resistant biological compound known, which protects the pollen grain against dehydration and chemical damage—another advantage in terms of survival in the terrestrial environment. Recall that sporopollenin in spore walls also contributed to the successful colonization of the terrestrial environment by the earliest land plants.

The arrival of a pollen grain at an appropriate landing point, close to a female gametophyte on a sporophyte of the same species, is called **pollination**. A pollen grain that reaches this point develops further. It produces a slender **pollen tube** that elongates and digests its way through the sporophytic tissue toward the megagametophyte. When the tip of the pollen tube reaches the megagametophyte, sperm are released from the tube, and fertilization occurs.

The resulting diploid zygote divides repeatedly, forming an embryonic sporophyte. After a period of embryonic development, growth is temporarily suspended (the embryo enters a *dormant* stage). The end product at this stage is a multicellular **seed**.

The seed is a complex, well-protected package

A seed may contain tissues from three generations. A *seed coat* develops from the tissues of the diploid sporophyte parent that surround the megasporangium (the integument). Within the megasporangium is the haploid female gametophytic tissue from the next generation, which contains a supply of nutrients for the developing embryo. (This tissue is fairly extensive in most gymnosperm seeds. In angiosperm seeds its place is taken by a tissue called endosperm, which we will describe below.) In the center

Betula pendula

29.4 Pollen Grains Pollen grains are the male gametophytes of seed plants. The pollen of this silver birch is dispersed by the wind, and grains may land near the female gametophytes of the same or other silver birch trees.

of the seed is the third generation, the embryo of the new diploid sporophyte.

The seed of a gymnosperm or an angiosperm is a well-protected resting stage. The seeds of some species may remain *viable* (capable of growth and development) for many years, germinating only when conditions are favorable for the growth of the sporophyte, as did the date seed from Biblical times mentioned at the beginning of this chapter. In contrast, the embryos of nonseed plants develop directly into sporophytes, which either survive or die, depending on environmental conditions; there is no dormant stage in the life cycle.

During the dormant stage, the seed coat protects the embryo from excessive drying and may also protect it against potential predators that would otherwise eat the embryo and its nutrient reserves. Many seeds have structural adaptations that promote their dispersal by wind or, more often, by animals. When the young sporophyte resumes growth, it draws on the food reserves in the seed. The possession of seeds is a major reason for the enormous evolutionary success of the seed plants, which are the dominant life forms of Earth's modern terrestrial flora in most areas. But there is another reason for their dominance: secondary growth.

A change in anatomy enabled seed plants to grow to great heights

The most ancient seed plants produced **wood**—extensively proliferated xylem—which gave them the support to grow taller than other plants around them, thus capturing more light for photosynthesis. The younger portion of wood is well adapted for water transport, while older wood becomes clogged with resins or other materials. Although no longer functional in transport, the older wood continues to provide support for the plant.

Not all seed plants are woody. In the course of seed plant evolution, many seed plants lost the woody growth habit; however, other advantageous attributes helped them become established in an astonishing variety of places.

29.1 RECAP

Pollen, seeds, and wood are major evolutionary innovations of the seed plants. Protection of the gametes and embryos is a hallmark of the seed plants.

- Can you distinguish between the roles of the megagametophyte and the pollen grain? See p. 632
- Do you understand the importance of pollen in freeing seed plants from dependence on liquid water? See p. 633
- What are some of the advantages afforded by seeds? By wood? See pp. 633–634

The seed ferns and progymnosperms have long been extinct, but the surviving seed plants have been remarkable successes. Let's look at the most ancient seed plants that survive today: the gymnosperms.

29.2 What Are the Major Groups of Gymnosperms?

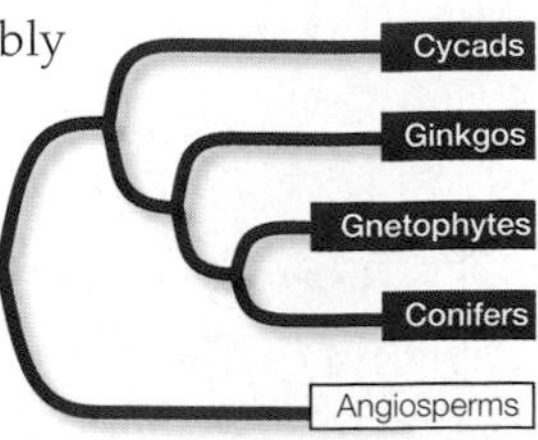

The extant gymnosperms are probably a clade, although that has not been established beyond a doubt. The gymnosperms are seed plants that do not form flowers. Gymnosperms derive their name (which means "naked-seeded") from the fact that their ovules and seeds are not protected by ovary or fruit tissue. Although there are probably fewer than 850 species of living gymnosperms, these plants are second only to the angiosperms in their dominance of the terrestrial environment.

The four major groups of living gymnosperms bear little superficial resemblance to one another:

- The **cycads** (*Cycadophyta*) are palmlike plants of the tropics and subtropics, growing as tall as 20 meters (**Figure 29.5A**). Of the present-day gymnosperms, the cycads are probably the earliest-diverging clade. There are 140 species of cycads. Their tissues are often highly toxic to humans.
- **Ginkgos** (*Ginkgophyta*), which were common during the Mesozoic era, are represented today by a single genus and species, *Ginkgo biloba*, the maidenhair tree (**Figure 29.5B**). There are both male (microsporangiate) and female (megasporangiate) maidenhair trees. The difference is determined by X and Y sex chromosomes, as in humans; few other plants have sex chromosomes.
- **Gnetophytes** (*Gnetophyta*) number about 90 species in three very different genera, which share certain characteristics analogous to ones found in the angiosperms, as we will see. One of the gnetophytes is *Welwitschia* (**Figure 29.5C**), a long-lived desert plant with just two straplike leaves that sprawl on the sand and can grow as long as 3 meters.
- **Conifers** (*Coniferophyta*) are by far the most abundant of the gymnosperms. There are about 600 species of these cone-bearing plants, including the pines and redwoods (**Figure 29.5D**).

All living gymnosperms except for the gnetophytes have only tracheids as water-conducting and support cells within the xylem. In angiosperms, cells called vessel elements and fibers, which are specialized for water conduction and support, respectively, are found alongside tracheids. While the gymnosperm water transport and support system may thus seem somewhat less efficient than that of the angiosperms, it serves some of the largest trees known. The coastal redwoods of California are the tallest gymnosperms; the largest are well over 100 meters tall.

During the Permian, as environments became warmer and dryer, the conifers and cycads flourished. Gymnosperm forests changed over time as the gymnosperm groups evolved. Gymnosperms dominated the Mesozoic era, during which the continents drifted apart and dinosaurs strode the Earth. They were the principal trees in all forests until less than 100 million years ago, and even down to the present day, conifers are the dominant trees in many forests.

(A) *Encephalartos villosus*

(B) *Ginkgo biloba*

(C) *Welwitschia mirabilis*

(D) *Sequoiadendron giganteum*

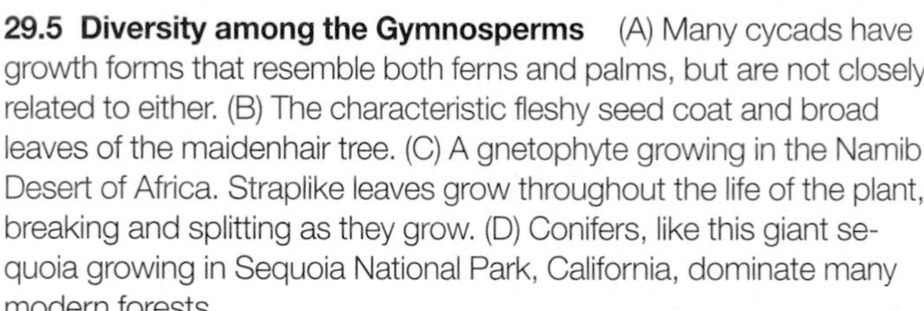

29.5 Diversity among the Gymnosperms (A) Many cycads have growth forms that resemble both ferns and palms, but are not closely related to either. (B) The characteristic fleshy seed coat and broad leaves of the maidenhair tree. (C) A gnetophyte growing in the Namib Desert of Africa. Straplike leaves grow throughout the life of the plant, breaking and splitting as they grow. (D) Conifers, like this giant sequoia growing in Sequoia National Park, California, dominate many modern forests.

The oldest living thing on Earth is a gymnosperm in California—a bristlecone pine called "Methuselah"—that began its life about 4,800 years ago, at about the time the ancient Egyptians were just starting to develop writing.

The relationship between gnetophytes and conifers is a subject of continuing research

Although the cycads, ginkgo, and gnetophytes are all clades, the status of the conifers is uncertain. There is evidence favoring a hypothesis—the "gnetifer" hypothesis—that the gnetophytes and conifers are sister clades (**Figure 29.6A**). However, there is also evidence favoring an alternative hypothesis—the "gnepine" hypothesis—that the gnetophytes are nested within a paraphyletic conifer group (**Figure 29.6B**). Why should this confusion exist, and how will it be resolved?

DNA data make it clear that the gnetophytes and conifers are closely related—but how? Most studies based on chloroplast and mitochondrial genes have supported the gnepine hypothesis. However, other studies based on rDNA (DNA that codes for rRNA) have favored the gnetifer hypothesis, as have some studies of chloroplast genes. An important source of confusion may be the use of different (usually small) sets of species for comparison; another is the use of different DNA sources.

It is widely agreed that what is needed are studies including more species and more genes. It may also be helpful to use new groups as outgroups (see Section 25.2) in future analyses. It is reasonable to expect that such studies will clarify the situation.

Whether or not they are a clade, the conifers are the most abundant gymnosperms. Let's look at them in more detail.

(A) Gnetifer hypothesis

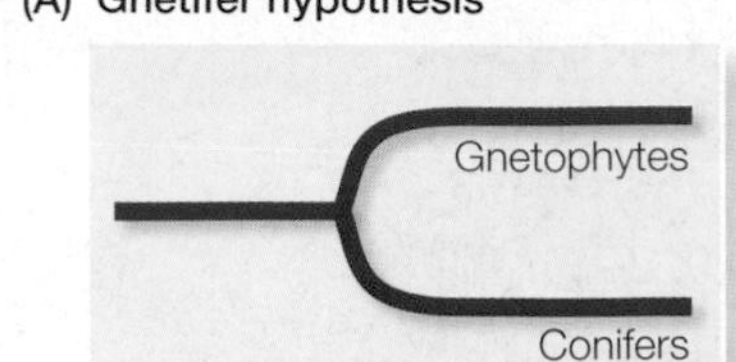

(B) Gnepine hypothesis

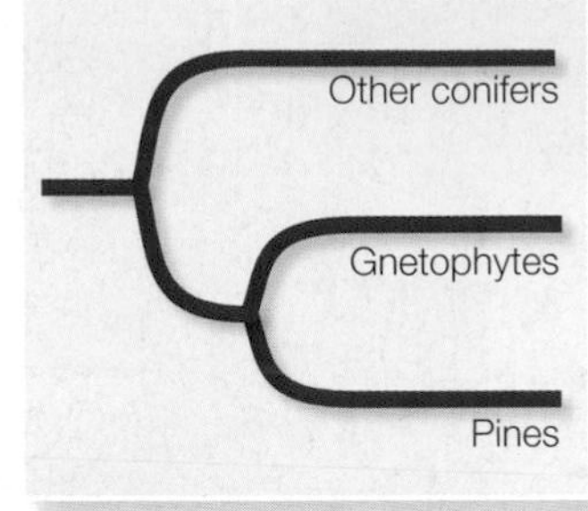

29.6 Two Interpretations of Conifer Phylogeny (A) According to the gnetifer hypothesis, the conifers are a clade. (B) According to the gnepine hypothesis, the conifers are paraphyletic.

Conifers have cones but no motile gametes

The great Douglas fir and cedar forests of the northwestern United States and the massive boreal forests of pine, fir, and spruce found in northern regions of Eurasia and North America, as well as on the upper slopes of mountain ranges everywhere, rank among the great vegetation formations of the world. All these trees belong to one group of gymnosperms, Coniferophyta—the conifers, or cone-bearers. A **cone** is a short axis (a modified stem) bearing a tight cluster of scales, which are reduced *branches* specialized for reproduction (**Figure 29.7A**). A **strobilus** is a conelike cluster of scales that are modified *leaves* inserted on an axis (**Figure 29.7B**). Megaspores, and thus seeds, are produced in cones, and microspores are produced in strobili. Cones are usually much larger than strobili.

We will use the life cycle of a pine to illustrate reproduction in gymnosperms (**Figure 29.8**). The production of male gametophytes in the form of pollen grains frees the plant completely from its dependence on liquid water for fertilization. Wind, rather than water, assists conifer pollen grains in their first stage of travel from the strobilus to the female gametophyte inside a cone (see Figure 29.4). The pollen tube provides the sperm with the means for the last stage of travel by elongating and digesting its way through maternal sporophytic tissue. When it reaches the female gametophyte, it releases two sperm, one of which degenerates after the other unites with an egg. Union of sperm and egg results in a zygote; mitotic divisions and further development of the zygote result in an embryo.

The megasporangium, in which the female gametophyte will form, is enclosed in a layer of sporophytic tissue—the integument—that will eventually develop into the seed coat that protects the embryo. The integument, the megasporangium inside it, and the tissue attaching it to the maternal sporophyte constitute the ovule. The pollen grain enters through a small opening in the integument at the tip of the ovule, the **micropyle**.

Most conifer ovules are borne exposed on the upper surfaces of the modified branches that form the scales of the cone. The only protection of the ovules comes from the scales, which are tightly pressed against one another within the cone. Some pines, such as the lodgepole pine, have such tightly closed cones that only fire suffices to split them open and release the seeds.

About half of all conifer species have soft, fleshy fruitlike tissues associated with their seeds; examples are the fleshy cones or "berries" of juniper and yew. Animals may eat these tissues and then disperse the seeds in their feces, often carrying them considerable distances from the parent plant.

29.7 Cones and Strobili (A) The scales of cones are modified branches. (B) The spore-bearing structures in strobili are modified leaves.

29.8 The Life Cycle of a Pine Tree In conifers and other gymnosperms, the gametophytes are microscopically small and nutritionally dependent on the sporophyte generation.

The sporophyte is enormous.

The same plant has both pollen-producing strobili and egg-producing cones.

Immature cone

Scale of cone

Section through scale

Ovule

Megasporocyte

Megasporangium

Meiosis

Functional megaspore

Pollen chamber

Strobili

Sporophyte (about 10–100 m)

Scale of strobilus

Section through scale

Meiosis

Microspores

Micropyle

Pollen grain

Pollen grain

Seed coat

Female gametophyte

Embryo

DIPLOID (2*n*) Sporophyte generation

HAPLOID (*n*) Gametophyte generation

Winged seed

Female gametophyte

Archegonium

Egg

Sperm

Male gametophyte (germinating pollen grain)

Fertilization

The gametophytes are tiny.

Wing

Seed

Scale of cone

Developing embryos

Mature cone

Zygote

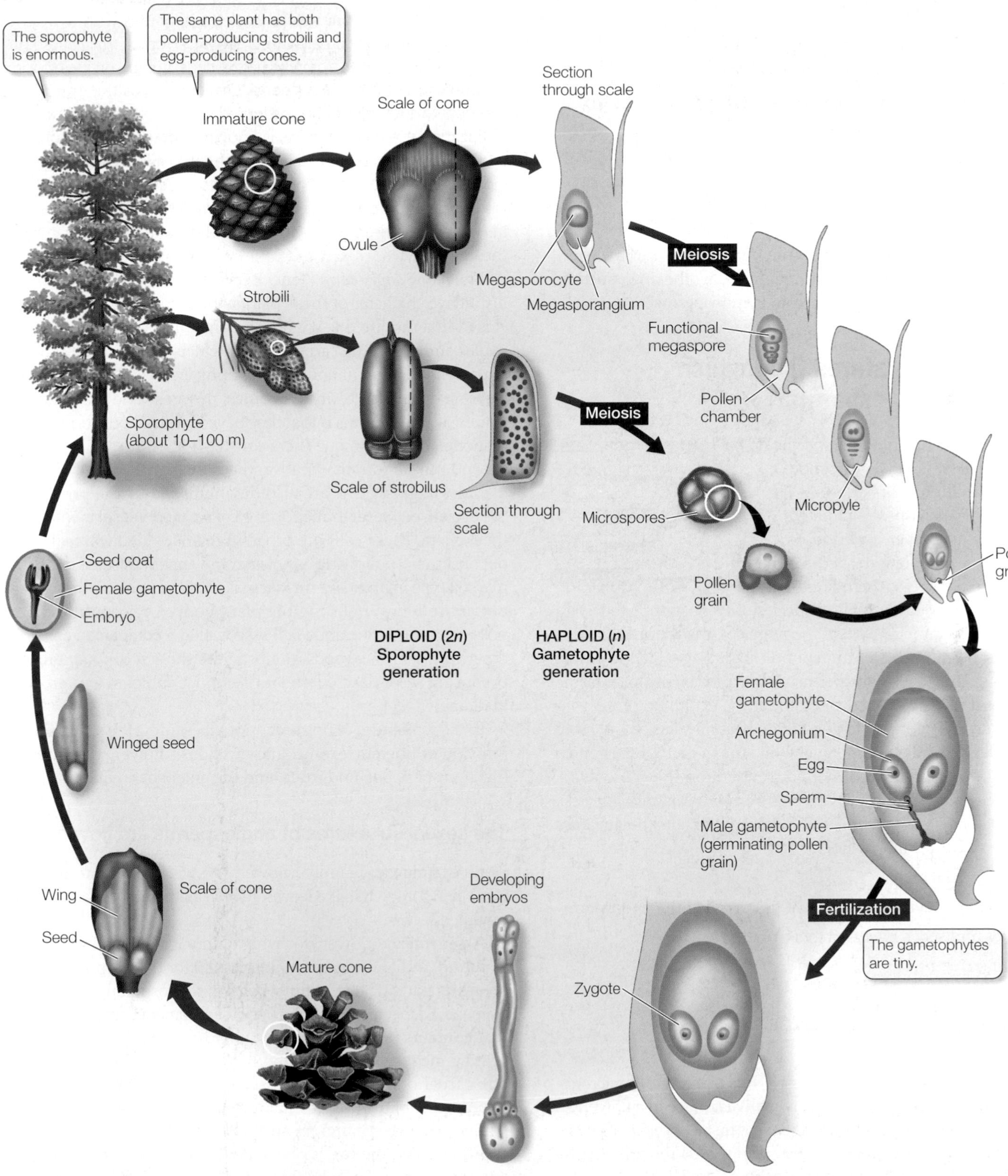

29.2 RECAP

The four groups of gymnosperms are woody and have naked seeds. Their phylogenetic relationships are still uncertain.

- What distinguishes the gnetifer from the gnepine hypothesis? See p. 635 and Figure 29.6
- Can you explain the difference between a cone and a strobilus? See p. 636 and Figure 29.8
- Do you understand the role of the integument? See pp. 632 and 636

Juniper and yew "berries" are not true fruits, which are characteristic of the plant group that is dominant today: the angiosperms. How else do angiosperms differ from gymnosperms?

29.3 What Features Distinguish the Angiosperms?

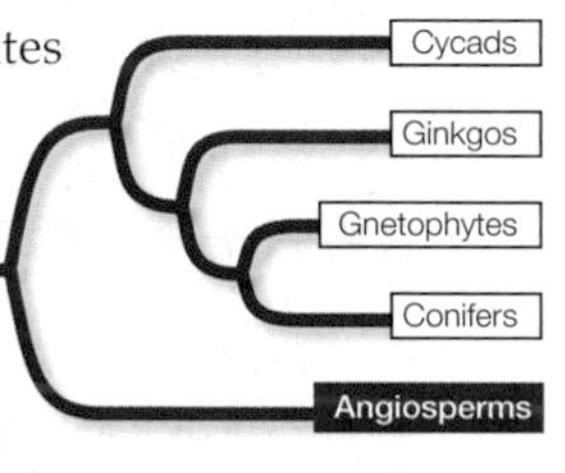

The oldest evidence of angiosperms dates back to the early Cretaceous period, about 140 million years ago (see Figure 29.1). The angiosperms radiated explosively and, over a period of only about 60 million years, became the dominant plant life of the planet. There are more than a quarter million species of angiosperms today.

The female gametophyte of the angiosperms is even more reduced than that of the gymnosperms, usually consisting of just seven cells. Thus the angiosperms represent the current extreme of the trend we have traced throughout the evolution of the vascular plants: the sporophyte generation becomes larger and more independent of the gametophyte, while the gametophyte generation becomes smaller and more dependent on the sporophyte. What else sets the angiosperms apart from other plants?

The major synapomorphies (shared derived traits) that characterize the angiosperms include:

- Double fertilization
- Production of a triploid nutritive tissue called the endosperm
- Ovules and seeds enclosed in a carpel
- Flowers
- Fruits
- Xylem with vessel elements and fibers
- Phloem with companion cells

In the angiosperms, pollination consists of the arrival of a microgametophyte—a pollen grain—on a receptive surface in a flower. As in the gymnosperms, pollination is just the first in a series of events that result in the formation of a seed. The next is the growth of a pollen tube extending to the megagametophyte. The third event is a fertilization process which, in detail, is unique to the angiosperms.

Double fertilization was long considered the single most reliable distinguishing characteristic of the angiosperms. *Two* male gametes, contained within a single microgametophyte, participate in fertilization events within the megagametophyte of an angiosperm. The nucleus of one sperm combines with that of the egg to produce a diploid zygote, the first cell of the sporophyte generation. In most angiosperms, the other sperm nucleus combines with two other haploid nuclei of the female gametophyte to form a *triploid* ($3n$) nucleus (see Figure 29.14 below). That nucleus, in turn, divides to form a triploid tissue, the **endosperm**, that nourishes the embryonic sporophyte during its early development. This process, in which two fertilization events take place, is known as **double fertilization**.

Double fertilization occurs in nearly all present-day angiosperms. We are not sure when and how it evolved because there is no known fossil evidence on this point. It may have first resulted in two diploid embryos, as it does in the three existing genera of the gnetophytes..

The name *angiosperm* ("enclosed seed") is drawn from another distinctive character of these plants: the ovules and seeds are enclosed in a modified leaf called a **carpel**. Besides protecting the ovules and seeds, the carpel often interacts with incoming pollen to prevent self-pollination, thus favoring cross-pollination and increasing genetic diversity. Of course, the most evident diagnostic feature of angiosperms is that they have **flowers**. Production of **fruits** is another of their unique characteristics. As we will see, both flowers and fruits afford major advantages to angiosperms.

Most angiosperms are also distinguished by the possession of specialized water-transporting cells called **vessel elements** in their xylem. These cells are broad in diameter and connect without obstruction, allowing easy water movement. A second distinctive cell type in angiosperm xylem is the **fiber**, which plays an important role in supporting the plant body. Angiosperm phloem possesses another unique cell type, called a **companion cell**. Like the gymnosperms, woody angiosperms show secondary growth, producing secondary xylem and secondary phloem and growing in diameter.

In the remainder of this section we'll examine the structure and function of flowers, evolutionary trends in flower structure, the functions of pollen and fruits, and the angiosperm life cycle.

The sexual structures of angiosperms are flowers

If you examine any familiar flower, you will notice that the outer parts look somewhat like leaves. In fact, all the parts of a flower *are* modified leaves.

A generalized flower (for which there is no exact counterpart in nature) is diagrammed in **Figure 29.9** for the purpose of identifying its parts. The structures bearing microsporangia are called **stamens**. Each stamen is composed of a **filament** bearing an **anther** that contains pollen-producing microsporangia. The structures bearing megasporangia are the carpels. A structure composed of one carpel or two or more fused carpels is called a **pistil**. The swollen base of the pistil, containing one or more ovules (each containing a megasporangium surrounded by its protective integument), is called the **ovary**. The apical stalk of the pistil is the **style**, and the terminal surface that receives pollen grains is the **stigma**.

In addition, a flower often has several specialized sterile (non-spore-bearing) leaves. The inner ones are called **petals** (collectively, the **corolla**) and the outer ones **sepals** (collectively, the

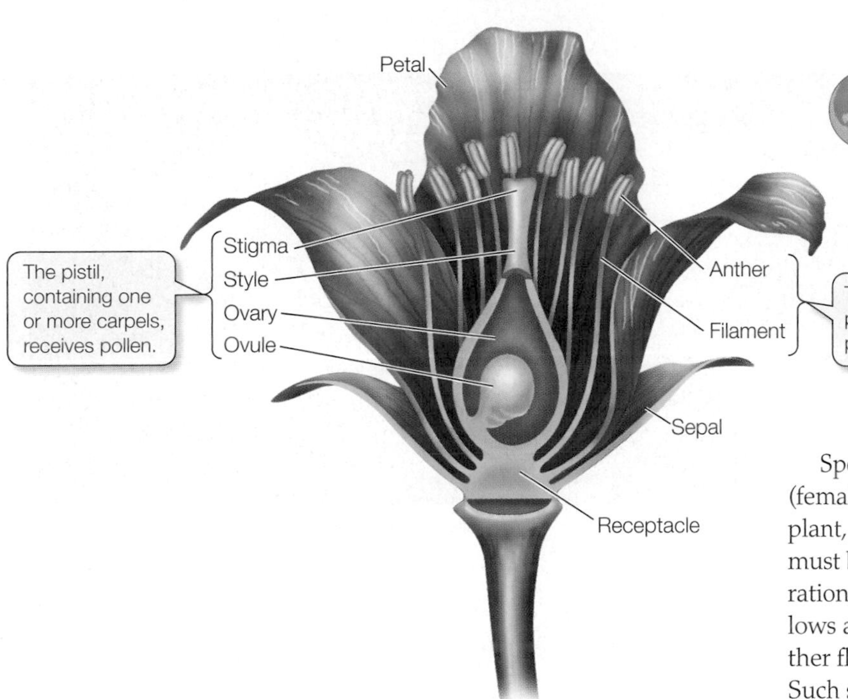

29.9 A Generalized Flower Not all flowers possess all the structures shown here, but they must possess a stamen (bearing microsporangia), a pistil (containing megasporangia), or both in order to play their role in reproduction. Flowers that have both, as this one does, are referred to as perfect.

calyx). The corolla and calyx, which can be quite showy, often play roles in attracting animal pollinators to the flower. The calyx more commonly protects the immature flower in bud. From base to apex, the sepals, petals, stamens, and carpels (which are referred to as the *floral organs*; see Figure 19.15) are usually positioned in circular arrangements or whorls and attached to a central stalk called the **receptacle**.

The generalized flower shown in Figure 29.9 has both megasporangia and microsporangia; such flowers are referred to as **perfect**. Many angiosperms produce two types of flowers, one with only megasporangia and the other with only microsporangia. Consequently, either the stamens or the carpels are nonfunctional or absent in a given flower, and the flower is referred to as **imperfect**.

Species such as corn or birch, in which both megasporangiate (female) and microsporangiate (male) flowers occur on the same plant, are said to be **monoecious** (meaning "one-housed"—but, it must be added, one house with separate rooms). Complete separation is the rule in some other angiosperm species, such as willows and date palms; in these species, a given plant produces either flowers with stamens or flowers with pistils, but never both. Such species are said to be **dioecious** ("two-housed").

Flowers come in an astonishing variety of forms, as you will realize if you think of some of the flowers you recognize. The generalized flower shown in Figure 29.9 has distinct petals and sepals arranged in distinct whorls. In nature, however, petals and sepals sometimes are indistinguishable. Such appendages are called **tepals** (see Figure 29.11A). In other flowers, petals, sepals, or tepals are completely absent.

Flowers may be single, or they may be grouped together to form an **inflorescence**. Different families of flowering plants have their own, characteristic types of inflorescences, such as the compound umbels of the carrot family, the heads of the aster family, and the spikes of many grasses (**Figure 29.10**).

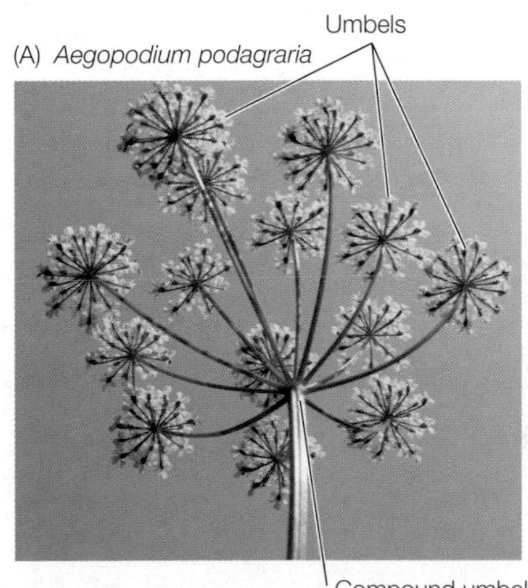

29.10 Inflorescences (A) The inflorescence of bishop's goutweed (a member of the carrot family) is a compound umbel. Each umbel bears flowers on stalks that arise from a common center. (B) Sunflowers are members of the aster family; their inflorescence is a head. In a head, each of the long, petal-like structures is a ray flower; the central portion of the head consists of dozens to hundreds of disk flowers. (C) Grasses such as this fountain grass have inflorescences called spikes.

(A) *Magnolia watsonii* (B) *Paphiopedilum maudiae*

29.11 Flower Form and Evolution (A) A magnolia flower shows the major features of early flowers: it is radially symmetrical, and the individual tepals, carpels, and stamens are separate, numerous, and attached at their bases. (B) Orchids, such as this ladyslipper, have a bilaterally symmetrical structure that evolved much later.

Flower structure has evolved over time

The flowers of the most basal clades of angiosperms have a large and variable number of tepals (or sepals and petals), carpels, and stamens (**Figure 29.11A**). Evolutionary change within the angiosperms has included some striking modifications of this early condition: reductions in the number of each type of floral organ to a fixed number, differentiation of petals from sepals, and changes in symmetry from radial (as in a lily or magnolia) to bilateral (as in a sweet pea or orchid), often accompanied by an extensive fusion of parts (**Figure 29.11B**).

According to one theory, the first carpels to evolve were modified leaves, folded but incompletely closed, and thus differing from the scales of the gymnosperms, which are modified branches. In the groups of angiosperms that evolved later, the carpels fused and became progressively more buried in receptacle tissue (**Figure 29.12A**). In the flowers of the most recent groups, the other floral organs are attached at the very top of the ovary, rather than at the bottom as in Figure 29.9. The stamens of the most ancient flowers may have appeared leaflike (**Figure 29.12B**), little resembling those of the generalized flower in Figure 29.9.

Why do so many flowers have pistils with long styles and anthers with long filaments? Natural selection has favored length in both of these structures, probably because length increases the likelihood of successful pollination. Long filaments may bring the anthers into contact with insect bodies, or they may place the anthers in a better position to catch the wind. Similar arguments apply to long styles.

A perfect flower represents a compromise of sorts. In attracting a pollinating bird or insect, the plant is attending to both its female and male functions with a single flower type, whereas plants with imperfect flowers must create that attraction twice—once for each type of flower. On the other hand, the perfect flower can favor self-pollination, which is usually disadvantageous. Another potential problem is that the female and male functions might interfere with each other—for example, the stigma might be so placed as to

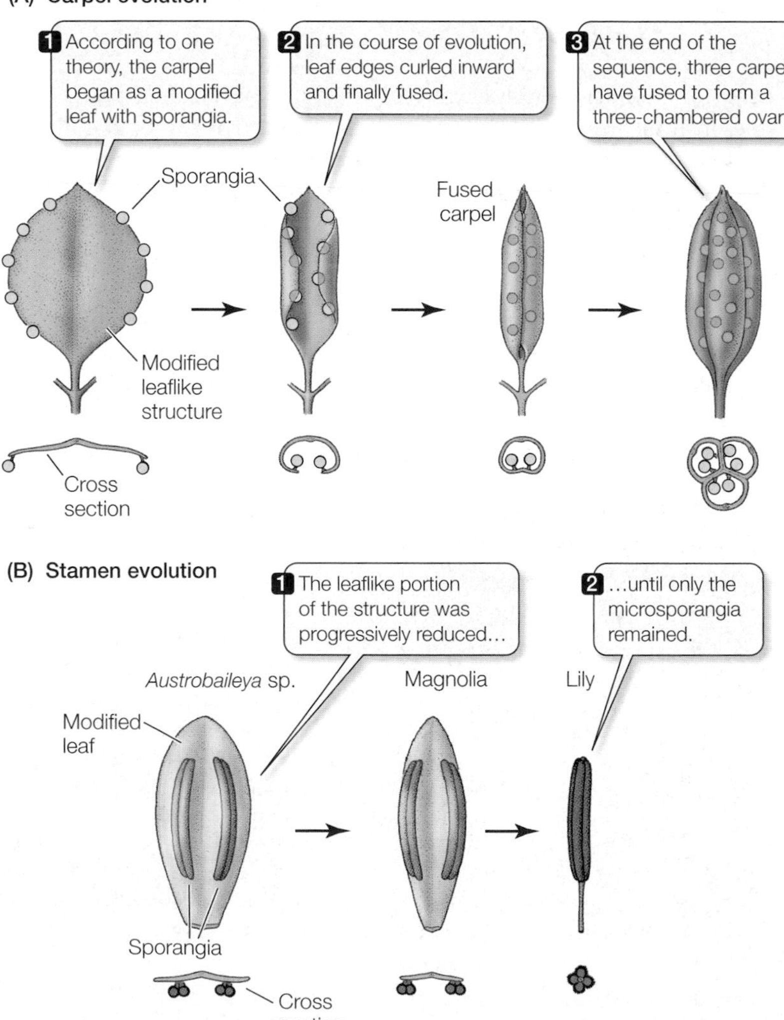

29.12 Carpels and Stamens Evolved from Leaflike Structures (A) Possible stages in the evolution of a carpel from a more leaflike structure. (B) The stamens of three modern plants show the various stages in the evolution of that organ. It is *not* implied that these species evolved one from another; they simply illustrate the structures.

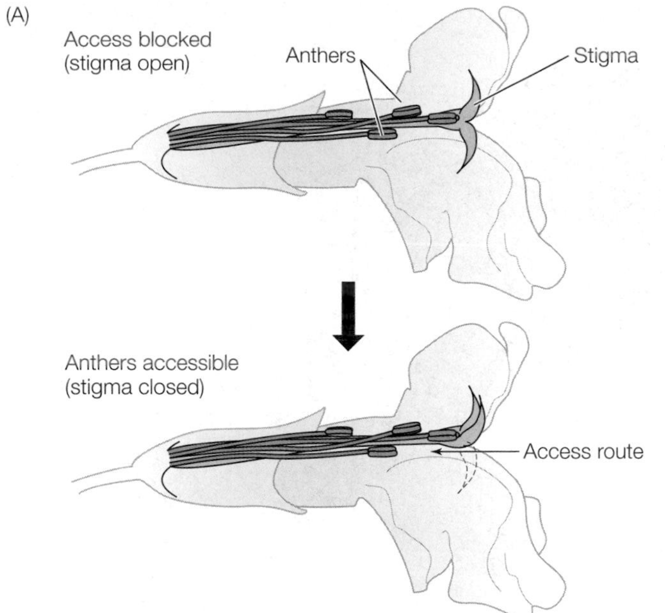

(B)

EXPERIMENT

HYPOTHESIS: Stigma responses in the bush monkeyflower favor the export of pollen.

METHOD

1. Set up experimental arrays of monkeyflowers such that only one flower in each array can donate pollen.
2. Pollen donors in some arrays are normal controls (untouched open stigmas); stigmas of other donors are artificially closed; a third group of donor stigmas are permanently propped open.
3. After hummingbirds visit the arrays, count the grains from each donor on the stigma of the next flower visited.

RESULTS

Almost twice as much pollen was exported from control flowers (i.e., those whose stigmas functioned normally) as from those whose stigmas were experimentally propped open. Experimentally closing the stigmas resulted in even greater pollen dispersal.

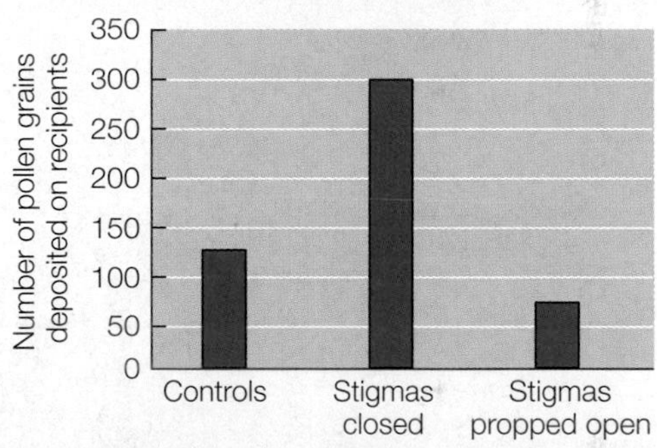

CONCLUSION: Stigma responses enhance the male function of the flower (pollen dispersal) once its female function (pollen deposition) has been performed.

29.13 Stigma Behavior Increases Pollen Export in Monkeyflowers (A) Initially, the stigma of the bush monkeyflower is open, blocking access to the anthers. A hummingbird's touch as it deposits pollen on the stigma causes one lobe of the stigma to retract, creating a path to the anthers. (B) Elizabeth Fetscher explored the reproductive consequences of stigma retraction in this unusual flower. FURTHER RESEARCH: How might you test how this mechanism affects self-pollination of the flower?

make it difficult for pollinators to reach the anthers, thus reducing the export of pollen to other flowers.

Might there be a way around these problems? One solution is seen in the bush monkeyflower (*Mimulus aurantiacus*), which is pollinated by hummingbirds and has a stigma that initially serves as a screen, hiding the anthers. Once a hummingbird touches the stigma, one of the stigma's two lobes folds, providing access by subsequent pollinators to the previously screened anthers (**Figure 29.13A**). The first bird can transfer pollen to the stigma, eventually leading to fertilization. Later visitors can pick up pollen from the anthers, fulfilling the flower's male function. The experiment that revealed the function of this mechanism is described in **Figure 29.13B**.

Angiosperms have coevolved with animals

Whereas many gymnosperms are wind-pollinated, most angiosperms are animal-pollinated. Many flowers entice animals to visit them by providing food rewards. Some flowers produce a sugary fluid called nectar, and the pollen grains themselves may serve as food for animals. In the process of visiting flowers to obtain nectar or pollen, animals often carry pollen from one flower to another or from one plant to another. Thus, in its quest for food, the animal contributes to the genetic diversity of the plant population. Insects, especially bees, are among the most important pollinators; birds and some species of bats also play major roles as pollinators.

For more than 130 million years, angiosperms and their animal pollinators have coevolved in the terrestrial environment. The animals have affected the evolution of the plants, and the plants have affected the evolution of the animals. Flower structure has become incredibly diverse under these selection pressures. Some of the products of coevolution are highly specific; for example, some yucca species are pollinated by only one species of moth. Pollination by just one or a few animal species provides a plant species with a reliable mechanism for transferring pollen from one of its members to another.

Most plant–pollinator interactions are much less specific; that is, many different animal species pollinate the same plant species, and the same animal species pollinate many different plant species. However, even these less specific interactions have developed some specialization. Bird-pollinated flowers are often red and odorless. Many insect-pollinated flowers have characteristic odors, and bee-pollinated flowers may have conspicuous markings, or *nectar guides*, that may be evident only in the ultraviolet region of the spectrum, where bees have better vision than in the red region.

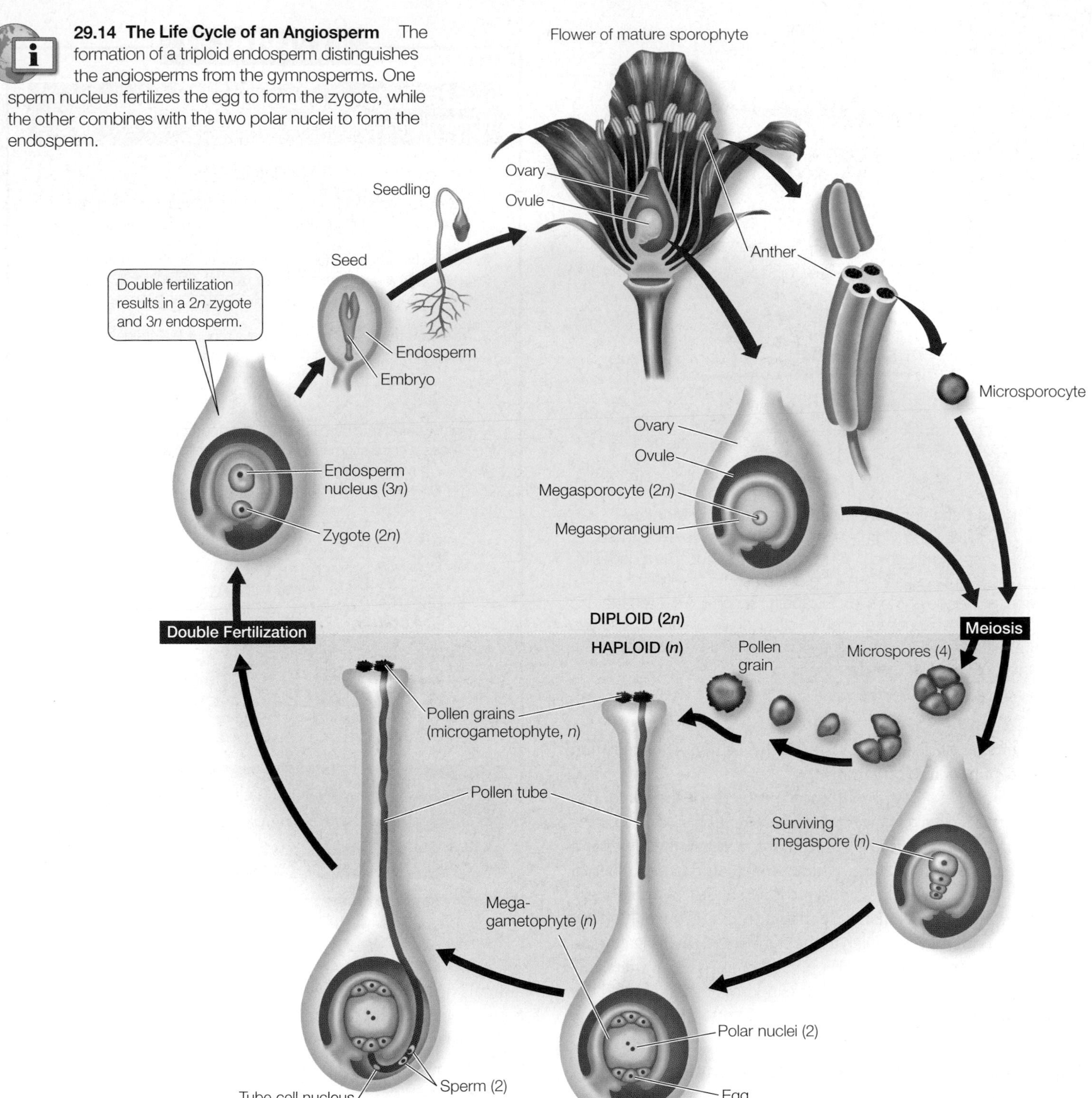

29.14 The Life Cycle of an Angiosperm The formation of a triploid endosperm distinguishes the angiosperms from the gymnosperms. One sperm nucleus fertilizes the egg to form the zygote, while the other combines with the two polar nuclei to form the endosperm.

The angiosperm life cycle features double fertilization

The life cycle of the angiosperms is summarized in **Figure 29.14**. The angiosperm life cycle will be considered in detail in Chapter 38, but let's look at it briefly here and compare it with the conifer life cycle in Figure 29.8.

Like all seed plants, angiosperms are heterosporous. As we have seen, their ovules are contained within carpels, rather than being exposed on the surfaces of scales, as in most gymnosperms. The male gametophytes, as in the gymnosperms, are pollen grains.

The ovule develops into a seed containing the products of the double fertilization that characterizes angiosperms: a diploid zygote and a triploid endosperm. The endosperm serves as storage tissue for starch or lipids, proteins, and other substances that will be needed by the developing embryo.

(A)

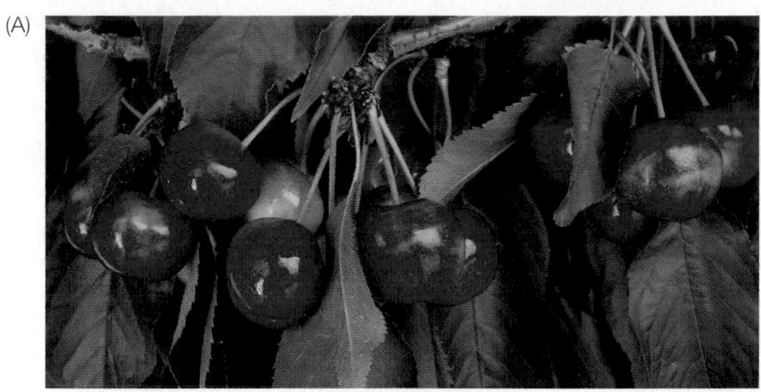

(B)

(C)

(D)

29.15 Fruits Come in Many Forms and Flavors (A) A simple fruit (sour cherry). (B) An aggregate fruit (raspberry). (C) A multiple fruit (pineapple). (D) An accessory fruit (strawberry).

The zygote develops into an embryo, which consists of an embryonic axis (the "backbone" that will become a stem and a root) and one or two **cotyledons**, or seed leaves. The cotyledons have different fates in different plants. In many, they serve as absorptive organs that take up and digest the endosperm. In others, they enlarge and become photosynthetic when the seed germinates. Often they play both roles.

Angiosperms produce fruits

The ovary of a flower (together with the seeds it contains) develops into a fruit after fertilization. The fruit protects the seeds and can also promote seed dispersal by becoming attached to or being eaten by an animal. A fruit may consist only of the mature ovary and its seeds, or it may include other parts of the flower or structures associated with it. A *simple fruit*, such as a cherry (**Figure 29.15A**), is one that develops from a single carpel or several united carpels. A raspberry is an example of an *aggregate fruit* (**Figure 29.15B**)—one that develops from several separate carpels of a single flower. Pineapples and figs are examples of *multiple fruits* (**Figure 29.16C**), formed from a cluster of flowers (an inflorescence). Fruits derived from parts in addition to the carpel and seeds are called *accessory fruits* (**Figure 29.15D**); examples are apples, pears, and strawberries. The development, ripening, and dispersal of fruits will be considered in Chapters 37 and 38.

29.3 RECAP

The synapomorphies of angiosperms include double fertilization, triploid endosperm, flowers, fruit, and distinctive cells in their xylem and phloem.

- Can you distinguish between pollination and fertilization?
- Can you give examples of how animals have affected the evolution of the angiosperms? See p. 641
- Do you understand the roles of the two sperm in double fertilization? See p. 642 and Figure 29.14
- Do you understand the difference between seeds and fruits, and the difference in their roles? See p. 643

We have been considering the shared characteristics of the angiosperms. What are the differences that separate the various angiosperm clades, and where did the angiosperms come from?

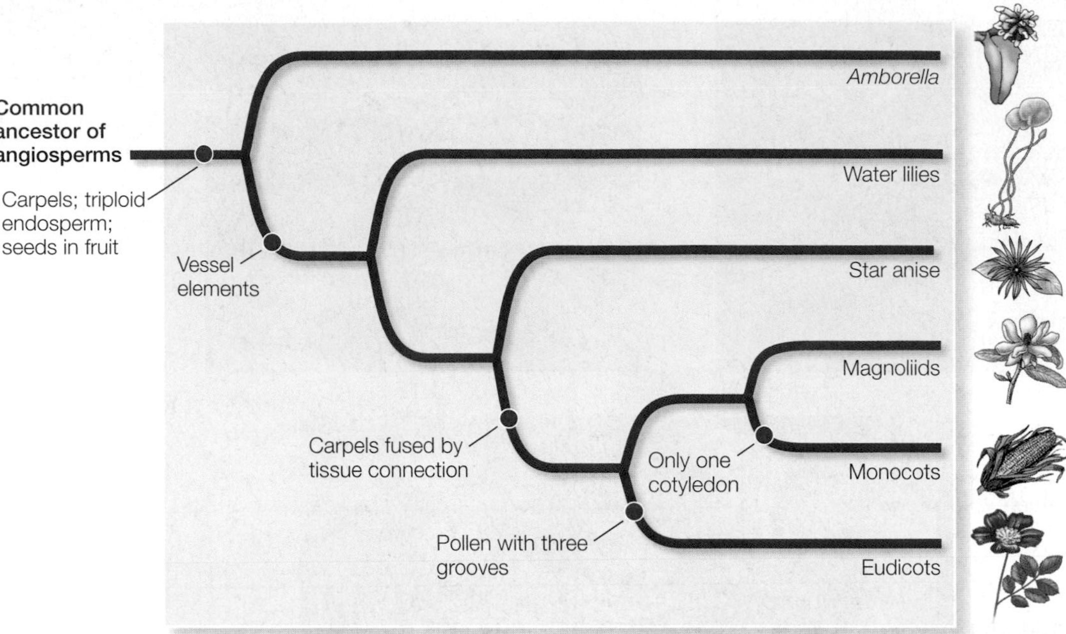

29.16 Evolutionary Relationships among the Angiosperms This diagram is a conservative interpretation of current data on relationships among the clades.

(A) *Amborella trichopoda*

(B) *Nymphaea* sp.

(C) *Illicium floridanum*

(D) *Piper nigrum*

(E) *Aristolochia grandiflora*

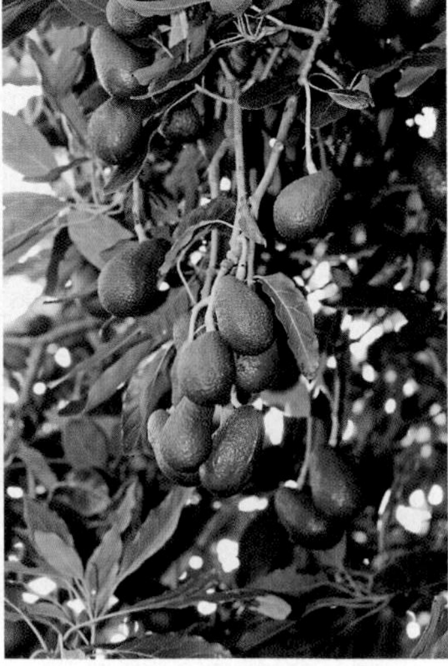
(F) *Persea* sp.

29.17 Monocots and Eudicots Are Not the Only Surviving Angiosperms (A) *Amborella*, a shrub, is the closest living relative of the first angiosperms; its clade is sister to the remaining extant angiosperms. (B) The water lily clade is the next most basal clade after *Amborella*'s. (C) Star anise and its relatives belong to another basal clade. (D–F) The largest clade other than the monocots and eudicots is the magnoliid complex, represented here by (D) a black pepper, (E) Dutchman's pipe, and (F) an avocado tree. The magnolia in Figure 29.11A is another magnoliid.

29.4 How Did the Angiosperms Originate and Diversify?

The relationships among certain angiosperm clades are controversial, but one conservative interpretation of those relationships is shown in **Figure 29.16**. Two large clades include the great majority of angiosperm species: the **monocots** and the **eudicots**. The monocots are so called because they have a single embryonic cotyledon; the eudicots have two. (We will describe other differences between these two groups in Chapter 34).

Some familiar angiosperms belong to clades other than the monocots and eudicots (**Figure 29.17**). These clades include the water lilies, star anise and its relatives, and the magnoliid complex. The magnoliids are less numerous than the monocots and eudicots, but they include many familiar and useful plants such as magnolias, avocados, cinnamon, and black pepper.

Although the structure of grasses is not favorable for fossilization, recently unearthed fossils in India have established that several grass clades were abundant at least 66 million years ago—much earlier than we once thought. These grasses were preserved in the fossilized dung of titanosaurs, among the heaviest of the dinosaurs.

The monocots (**Figure 29.18**) include grasses, cattails, lilies, orchids, and palms. The eudicots (**Figure 29.19**) include the vast majority of familiar seed plants, including most herbs (i.e., nonwoody plants), vines, trees, and shrubs. Among them are such diverse plants as oaks, willows, beans, snapdragons, and sunflowers.

The basal angiosperm clade is a matter of controversy

Which angiosperms constitute the basal clade of flowering plants is a matter of great controversy. Two leading candidates were the magnolia family (see Figures 29.17D–F) and another family, the Chloranthaceae, whose flowers are much simpler than those of the magnolias. At the close of the twentieth century, however, an impressive convergence of evidence led to the conclusion that the most basal living angiosperm belongs to neither of those families, but rather to a clade that today consists of a single species of the genus *Amborella* (see Figure 29.17A). This woody shrub, with cream-colored flowers, lives only on New Caledonia, an island in the South Pacific. Its 5 to 8 carpels are in a spiral arrangement, and it has 30 to 100 stamens. The xylem of *Amborella* lacks vessel elements, which appeared later in angiosperm evolution. The characteristics of *Amborella* give us an idea of what the first angiosperms might have been like. But are there extinct angiosperms that may represent still more ancient lineages?

In 2002, Chinese and American botanists examined fossils of two species of a 125-million-year-old aquatic genus, *Archaefructus* (see Figure 21.18). Their studies established an extinct group, Archaefructaceae, that is posited to be the sister taxon of all other angiosperms. The flower of these plants had its ovules enclosed in carpels, as in all angiosperms. The flower had neither petals nor sepals, however, and its carpels and stamens were arranged spirally around elongated shoots. This arrangement of carpels and stamens is seen today in the magnolias.

(A) *Phoenix dactylifera*

29.18 Monocots (A) Palms are among the few monocot trees. Date palms are a major food source in some areas of the world. (B) Grasses such as this cultivated wheat and the fountain grass in Figure 29.10C are monocots. (C) Monocots also include popular garden flowers such as these lilies. Orchids (Figure 29.11B) are highly sought-after monocot flowers.

(B) *Triticum* sp.

(C) *Lilium* sp.

(A) *Opuntia* sp.

(B) *Cornus florida*

(C) *Rosa rugosa*

29.19 Eudicots (A) The cactus family is a large group of eudicots, with about 1,500 species in the Americas. This cactus bears yellow flowers for a brief period of the year. (B) The flowering dogwood is a small eudicot tree. (C) Climbing Cape Cod roses are members of the eudicot family Rosaceae, as are the familiar roses from your local florist.

The origin of the angiosperms remains a mystery

We have learned a lot about evolution within the angiosperm clade. But how did the angiosperms first arise? Are the angiosperms sister to any single gymnosperm group? Charles Darwin himself brooded long and hard over the origin of the angiosperms. The problem vexed him so deeply that he referred to it as "an abominable mystery." A few years ago, it seemed that we were on the verge of solving that mystery. We do not yet have the answer, however.

Botanists were long intrigued by certain similarities between the gnetophytes and the angiosperms. One similarity is the presence of vessel elements in the xylem of both groups. Another is the occurrence of two fertilization events in both. Other similarities that gnetophytes and flowering plants shared with certain other groups known only as fossils led to the "anthophyte hypothesis," which groups all of these plants in a single clade (called anthophytes) and leaves the remaining gymnosperms as a paraphyletic group. Support for the anthophyte hypothesis comes from both morphological and molecular evidence.

Other molecular evidence, however, places the gnetophytes firmly within the gymnosperms. This alternative hypothesis leaves the extant gymnosperms as a sister clade to the angiosperms. It is possible that the vessel elements and "double fertilization" of the gnetophytes are simply analogous, rather than homologous, to those of the angiosperms. Reexamination of certain fossils has led some botanists to doubt their phylogenetic relationship to the angiosperms, further discrediting the anthophyte hypothesis. Figure 29.2 assumes that the anthophyte hypothesis is invalid, and it accepts the gnetifer hypothesis (see Figure 29.6A). Both hypotheses are still the subject of intense research, however.

How will we resolve these uncertainties? What morphological characters should be selected as important, or should they all be treated as equally important? Are all molecular differences and similarities significant, or are some of them incidental? Which fossils should be chosen for comparisons? Molecular biologists and paleobotanists are focusing their efforts on gathering new data and determining how best to interpret them. We eagerly await their findings.

29.4 RECAP

The largest angiosperm clades are the monocots and the eudicots. We still have questions about some aspects of angiosperm phylogeny.

- Can you describe the relationship between *Amborella* and the other angiosperms? See p. 645 and Figure 29.16
- Why is *Archaefructus* considered to have been an angiosperm? See p. 645

The remarkable diversity of the seed plants has been shaped in part by the different environments in which these and other plants have evolved. In turn, land plants—and seed plants in particular—affect their environments.

29.5 How Do Plants Support Our World?

Plants make profound contributions to **ecosystem services**—processes by which the environment maintains resources that benefit humans. These benefits include the effects of plants on soils, water, the atmosphere, and the climate. As we'll see in Section 36.3, plants play important roles in soil formation and in renewing the fertility of soils. Plant roots help hold soil in place, resisting erosion by wind and water. Plants store water in their bodies. They also moderate local climate in various ways, such as by increasing humidity, providing shade, and blocking wind.

Plants are **primary producers**; that is, their photosynthesis traps energy and carbon, making those resources available not only for their own needs, but also for the herbivores and omnivores that consume them, for the carnivores and omnivores that eat the herbivores, and for the prokaryotes and fungi that complete the food chain.

Seed plants are our primary food source

Twelve seed plant species stand between our species and starvation: rice, coconut, wheat, corn (also called maize), potato, sweet potato, cassava (also called tapioca or manioc), sugarcane, sugar beet, soybean, common bean, and banana. Other seed plants are cultivated for food, but none rank with these twelve in importance.

Indeed, more than half of the world's human population derives the bulk of its food energy from the seeds of a single plant: rice, *Oryza sativa*. Rice is particularly important in the diets of people in the Far East, where it has been cultivated for nearly 5,000 years. People also use rice straw in many ways, such as thatching for roofs, food and bedding for livestock, and clothing. Rice hulls, too, have many uses, ranging from fertilizer to fuel.

Let's look at what another one of these dozen plants provides. In some cultures the coconut palm (*Cocos nucifera*, **Figure 29.20**) is called the Tree of Life because every aboveground part of the plant has value to humans. People use the stem (the trunk) of this tropical coastal lowland monocot tree as lumber. They dry the sap from its trunk for use as a sugar, or they ferment it to drink. They use the leaves to thatch their homes and to make hats and baskets. They eat the apical bud at the top of the trunk in salads.

The coconut fruit serves many purposes. The hard shell can be used as a container or burnt as fuel. The fibrous middle layer, or coir, of the fruit wall can be made into mats and rope. The seed of the coconut palm contains both "milk" and "meat." Because the refreshing and delicious coconut milk contains no bacteria or other pathogens, it is particularly important wherever the water is not fit for drinking. Millions of people get most of their protein from coconut meat. Much coconut meat is dried and marketed as copra, from which coconut oil is pressed. Coconut oil is the most widely used vegetable oil in the world; it is used in the manufacture of a range of products from hydraulic brake fluid to synthetic rubber and, although nutritionally poor, as food. Ground copra serves as fertilizer and as food for livestock.

Seed plants have been sources of medicines since ancient times

One likely candidate for the oldest human profession is that of medicine man or shaman—a person who cures others with medicines derived from seed plants. It is claimed that a legendary Chinese emperor around 2700 B.C.E. knew some 365 medicinal plants. We use many medicines derived from molds, lichens, and actinobacteria as well as synthetic ones. However, we still rely on seed plants for many of our medicines, just a few of which are shown in **Table 29.1**.

How are plant-based medicines discovered? These days many are found by systematic testing of tremendous numbers of plants from all over the world, a process that began in the 1960s. One example is taxol, an important anticancer drug. Among the myriad plant samples that had been tested by 1962, extracts of the bark of Pacific yew (*Taxus brevifolia*) showed antitumor activity in tests against rodent tumors. The active ingredient, taxol, was isolated in 1971 and finally tested against human cancers in 1977. After another 16 years, the U.S. Food and Drug Administration approved it for human use, and taxol is now widely used in treating breast and ovarian cancers as well as several other types of cancers.

This type of widespread screening of plant samples eventually was deemphasized in favor of a purely chemical approach. Using

29.20 The Tree of Life Fruits of the coconut palm await harvest on a plantation in the South Pacific.

TABLE 29.1

Some Medicinal Plants and Their Products

PRODUCT	PLANT SOURCE	MEDICAL APPLICATION
Atropine	Belladonna	Dilating pupils for eye examination
Bromelain	Pineapple stem	Controlling tissue inflammation
Digitalin	Foxglove	Strengthening heart muscle contraction
Ephedrine	*Ephedra*	Easing nasal congestion
Menthol	Japanese mint	Relief of coughing
Morphine	Opium poppy	Relief of pain
Quinine	Cinchona bark	Treatment of malaria
Taxol	Pacific yew	Treatment of ovarian and breast cancers
Tubocurarine	Curare plant	As muscle relaxant in surgery
Vincristine	Periwinkle	Treatment of leukemia and lymphoma

automation and miniaturization, pharmaceutical laboratories generate vast numbers of compounds that are screened just as plant materials were screened in the search for taxol and other plant-based medicines. Now, however, the plant screening is getting renewed interest. Both approaches are based on trial and error.

The other leading source of medicinal plants is work by *ethnobotanists*, who study how and why people use and view plants in their local environments. This work proceeds all over the globe today. An older example is the discovery of quinine as a treatment for malaria. In 1630, Spanish priests in Peru successfully used the bark of the local cinchona tree as a cure for malaria. They were aware that the Peruvians had been using the bark to treat fevers. Word of the successful treatment made its way to Europe, and cinchona bark became the standard treatment for malaria. The active ingredient, quinine, was finally identified in 1820.

CHAPTER SUMMARY

29.1 How did seed plants become today's dominant vegetation?

Only seed plants show **secondary growth**, producing wood.

The surviving groups of seed plants are the **gymnosperms** and **angiosperms**.

All seed plants are heterosporous, and their gametophytes are much smaller than, and dependent on, their sporophytes. Review Figure 29.3

An **ovule** consists of the seed plant megagametophyte and the **integument** which protects it. The ovule develops into a seed.

Pollen grains, the microgametophytes, do not require liquid water to perform their functions. Following **pollination**, a **pollen tube** emerges from the pollen grain and elongates to deliver gametes to the megagametophyte. Pollen grains are enclosed in highly resistant sporopollenin walls.

Seeds are well protected, and they are often capable of long periods of dormancy, germinating when conditions are favorable.

29.2 What are the major groups of gymnosperms?

The extant gymnosperms may be a clade, as are at least three of the four gymnosperm groups. Review Figures 29.2 and 29.6

The gymnosperm clades are the **cycads**, **ginkgos**, and **gnetophytes**. **Conifers** are the most abundant gymnosperms.

The megaspores of pines are produced in **cones**, and microspores are produced in **strobili**. Pollen reaches the megasporangium by way of the **micropyle**, an opening in the integument of the ovule. Review Figures 29.7 and 29.8, See Web/CD Activity 29.1 and Tutorial 29.1

29.3 What features distinguish the angiosperms?

Only angiosperms have **flowers** and **fruits**. Review Figure 29.9, Web/CD Activity 29.2

The ovules and seeds of angiosperms are enclosed in and protected by **carpels**. Review Figure 29.12

Angiosperms have **double fertilization**, resulting in the production of a zygote and a triploid **endosperm**. Review Figure 29.14, Web/CD Tutorial 29.2

The xylem and phloem of angiosperms are more complex and efficient than those of the gymnosperms. **Vessel elements**, **fibers**, and **companion cells** contribute to the efficiency.

The floral organs, from the apex to the base of the flower, are the **pistil**, **stamens**, **petals**, and **sepals**. Stamens bear microsporangia in **anthers**. The pistil (consisting of one or more carpels) includes an **ovary** containing ovules. The **stigma** is the receptive surface of the pistil. The floral organs are born on the **receptacle**.

A flower with both megasporangia and microsporangia is **perfect**; all other flowers are **imperfect**. Flowers may be grouped to form an **inflorescence**. Most flowers are pollinated by animals.

A **monoecious** species has megasporangiate and microsporangiate flowers on the same plant. A **dioecious** species is one in which megasporangiate and microsporangiate flowers never occur on the same plant.

29.4 How did the angiosperms originate and diversify?

The most abundant angiosperm clades are the **monocots** and the **eudicots**. The relationships of these to the other extant angiosperm clades are not fully resolved. Review Figure 29.16

The most basal living angiosperm appears to be a single species in the genus *Amborella*. The most basal angiosperm clade appears to be the extinct Archaefructaceae. We do not know what group is sister to the angiosperms.

29.5 How do plants support our world?

Plants provide **ecosystem services** that affect soils, water, the air, and the climate.

Plants are **primary producers**, providing food for the entire terrestrial food web.

Plants provide many important medicines. Review Table 29.1

SELF-QUIZ

1. Which of the following statements about seed plants is true?
   a. The phylogenetic relationships among the major groups have been well established.
   b. The sporophyte generation is more reduced than in the ferns.
   c. The gametophytes are independent of the sporophytes.
   d. All seed plant species are heterosporous.
   e. The zygote divides repeatedly to form the gametophyte.
2. The gymnosperms
   a. dominate all land masses today.
   b. have never dominated land masses.
   c. have active secondary growth.
   d. all have vessel elements.
   e. lack sporangia.
3. Conifers
   a. produce ovules in strobili and pollen in cones.
   b. depend on liquid water for fertilization.
   c. have triploid endosperm.
   d. have pollen tubes that release two sperm.
   e. have vessel elements.
4. Angiosperms
   a. have ovules and seeds enclosed in a carpel.
   b. produce triploid endosperm by the union of two eggs and one sperm.
   c. lack secondary growth.
   d. bear two kinds of cones.
   e. all have perfect flowers.
5. Which statement about flowers is *not* true?
   a. Pollen is produced in the anthers.
   b. Pollen is received on the stigma.
   c. An inflorescence is a cluster of flowers.
   d. A species having female and male flowers on the same plant is dioecious.
   e. A flower with both megasporangia and microsporangia is said to be perfect.
6. Which statement about fruits is *not* true?
   a. They develop from ovaries.
   b. They may include other parts of the flower.
   c. A multiple fruit develops from several carpels of a single flower.
   d. They are produced only by angiosperms.
   e. A cherry is a simple fruit.
7. Which statement is *not* true of angiosperm pollen?
   a. It is the male gamete.
   b. It is haploid.
   c. It produces a long tube.
   d. It interacts with the carpel.
   e. It is produced in microsporangia.
8. Which statement is *not* true of carpels?
   a. They are thought to have evolved from leaves.
   b. They bear megasporangia.
   c. They may fuse to form a pistil.
   d. They are floral organs.
   e. They were absent in *Archaefructus*.
9. *Amborella*
   a. was the first flowering plant.
   b. belongs to the first gymnosperm clade.
   c. belongs to the oldest angiosperm clade still extant.
   d. is a eudicot.
   e. has vessel elements in its xylem.
10. The eudicots
    a. include many herbs, vines, shrubs, and trees.
    b. and the monocots are the only extant angiosperm clades.
    c. are not a clade.
    d. include the magnolias.
    e. include orchids and palm trees.

FOR DISCUSSION

1. In most seed plant species, only one of the products of meiosis in the megasporangium survives. How might this be advantageous?
2. Suggest an explanation for the great success of the angiosperms in occupying terrestrial habitats.
3. In many locales, large gymnosperms predominate over large angiosperms. Under what conditions might gymnosperms have the advantage, and why?
4. Not all flowers possess all of the following floral organs: sepals, petals, stamens, and carpels. Which floral organ or organs do you think might be found in the flowers that have the smallest number of floral organ types? Discuss the possibilities, both for a single flower and for a species.
5. The problem of the origin of the angiosperms has long been "an abominable mystery," as Charles Darwin once put it. Scientists still do not know the nearest relatives of the angiosperms. It has often been suggested (correctly or incorrectly) that the gnetophytes are sister to the angiosperms. What pieces of evidence suggested this connection?

FOR INVESTIGATION

The flower of a particular species of orchid has a long, spurlike tube into which a pollinating insect can insert its proboscis to suck nectar. The spurs are of different lengths in different habitats, apparently correlated with the lengths of the probosces of the local pollinators. How might you test the hypothesis that this correlation increases reproductive success, in terms of pollen transfer to the flower?

CHAPTER 30 Fungi: Recyclers, Pathogens, Parasites, and Plant Partners

A fungus battles witchweed

About 300 million Africans in 25 countries are suffering because of the invasion of crops by witchweed (*Striga*), a parasitic flowering plant. This parasite has attacked more than two-thirds of the sorghum, corn, and millet crops in sub-Saharan Africa, doing damage estimated at U.S.$7 billion each year.

A team of Canadian scientists set out to find a biological solution to the *Striga* problem. Their strategy was to look for an organism that would destroy witchweed in the fields. They succeeded in isolating a strain of a fungus—the mold *Fusarium oxysporum*—that has several outstanding properties. First, it grows on *Striga*, killing a high percentage of these parasitic plants. Second, it is not toxic to humans, nor does it attack the crop plants on which *Striga* is growing. In subsequent fieldwork in Mali, the scientists established techniques for proper application of *F. oxysporum* to witchweed. Now farmers apply the fungus to their crops and are rewarded by greatly increased crop yields as *Striga* is held in check.

It may be possible to repeat the *Striga* story—the use of a fungus to wipe out a particular type of flowering plant—in a very different context. A different strain of *F. oxysporum* preferentially attacks coca plants (the source of cocaine). A controversial proposal to use *F. oxysporum* to wipe out the coca plantations of Andean South America and in other parts of the world has been proposed. Some naturally occurring strains of *F. oxysporum* attack important crops in various parts of the world. What can we do about that? Recent work in India has identified two *other* species of fungi that inhibit the development of *F. oxysporum* growing on safflower when their spores are sprayed on the plants, allowing the crop to grow better. Thus we are using fungi to battle fungi! Fortunately, these other fungi are not harmful to safflower or any other crop plants that have been tested.

In another promising development, research teams reported in 2005 that two fungi, *Beauveria bassiana* and *Metarhizium anisopliae*, killed malaria-carrying mosquitoes when applied to mosquito netting. Certain fungi are already in use against other insect pests, notably termites and aphids. And in research conducted thus far, insect pests have not developed resistance to fungi as they can to DDT and other pesticides.

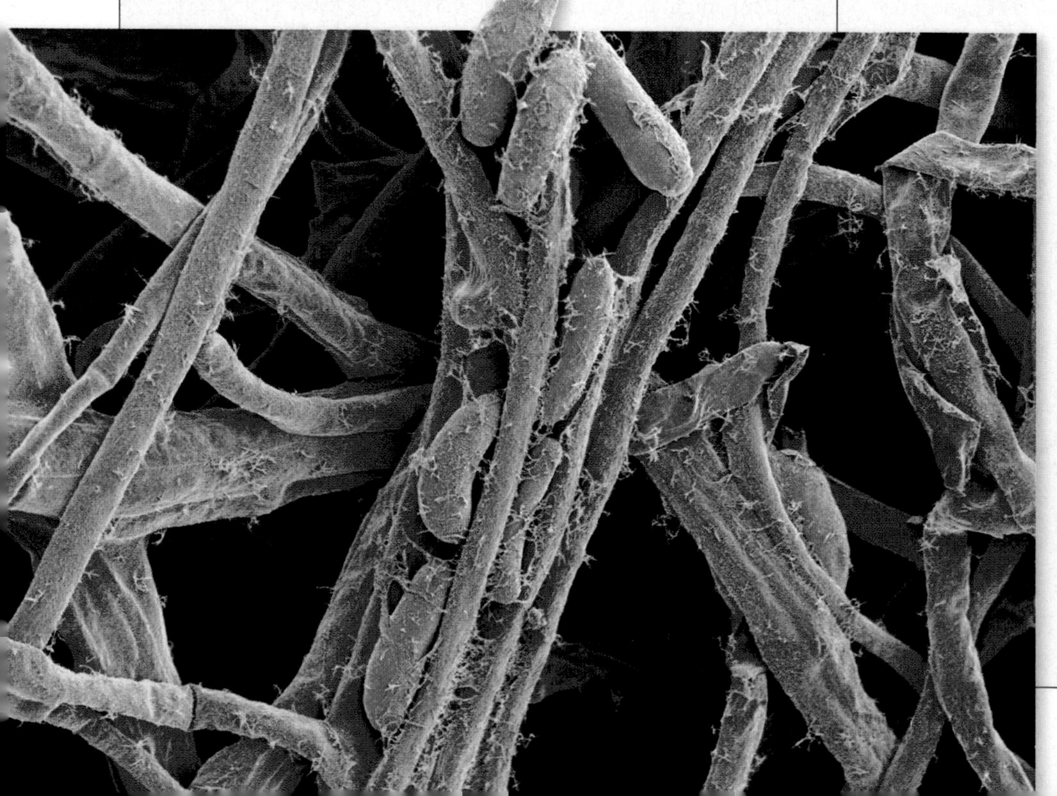

Pathogenic Fungus, Parasitic Plant The fungus *Fusarium oxysporum* is a potent pathogen of witchweed (*Striga*), a parasitic plant that attacks crops. The fungal spores are shown in blue; the fungal filaments are in tan. Both colors were added to enhance this electron micrograph.

An Alien Meal The tropical fungus whose fruiting body is growing on the stalk projecting from this ant's carcass developed internally in the host, from a spore ingested by the ant.

Of course, fungi don't need our help to find organisms to grow on. Fungal spores ingested by animals such as ants can germinate in their new host's gut. They develop into multicellular individuals that absorb nutrients from their unwitting hosts. Eventually they kill their hosts, producing new spores to infect other organisms as they do so.

Fungi interact with other organisms in many different ways, some of which are helpful and some harmful to the other organism. As we begin our study, recall that the fungi and the animals are descended from a common ancestor, and thus molds and mushrooms are more closely related to you than they are to the flowers we admired in Chapter 29.

IN THIS CHAPTER we will see how fungi differ from other eukaryotes in some very interesting ways. We will explore the diversity of body forms, reproductive structures, and life cycles that have evolved among the five groups of fungi, as well as examining the mutually beneficial associations of certain fungi with other organisms.

CHAPTER OUTLINE

30.1 How Do Fungi Thrive in Virtually Every Environment?

Modern fungi are believed to have evolved from a unicellular protistan ancestor that had a flagellum. The probable common ancestor of the animals was also a flagellated microbial eukaryote that may have been similar to the existing choanoflagellates (see Figure 27.30). Current evidence suggests that today's choanoflagellates, fungi, and animals share a single common ancestor, and thus the three lineages are often grouped together as the *opisthokonts*. The synapomorphies that distinguish the fungi among the opisthokonts are absorptive heterotrophy, and the presence of chitin in their cell walls (**Figure 30.1**).

The fungi live by **absorptive nutrition**: they secrete digestive enzymes that break down large food molecules in the environment, then absorb the breakdown products through the plasma membranes of their cells. Absorptive heterotrophy is successful in virtually every conceivable environment. Many fungi are **saprobes** that absorb nutrients from dead organic matter. Others are **parasites** that absorb nutrients from living hosts, such as the ant shown at the opening of this chapter. Still others are **mutualists** that live in intimate associations with other organisms that benefit both partners.

We will discuss five major groups of fungi: chytrids, zygomycetes, ascomycetes, basidiomycetes, and glomeromycetes. The first four of these groups were originally defined by their methods and structures for sexual reproduction and also, to a lesser extent, by other morphological differences. More recently, biologists have turned to evidence from DNA analyses, which established the glomeromycetes as a group independent of the zygomycetes. The chytrids and zygomycetes appear to be paraphyletic, although their phylogenetic relationships are not yet well understood. The other three groups are clades (**Figure 30.2**). The chytrids are aquatic, but the other groups are terrestrial.

The body of a multicellular fungus is composed of hyphae

Most fungi are multicellular, but single-celled species are found in most of the fungal groups. Unicellular members of all but the basal group of fungi (the chytrids) are called **yeasts** (**Figure 30.3**).

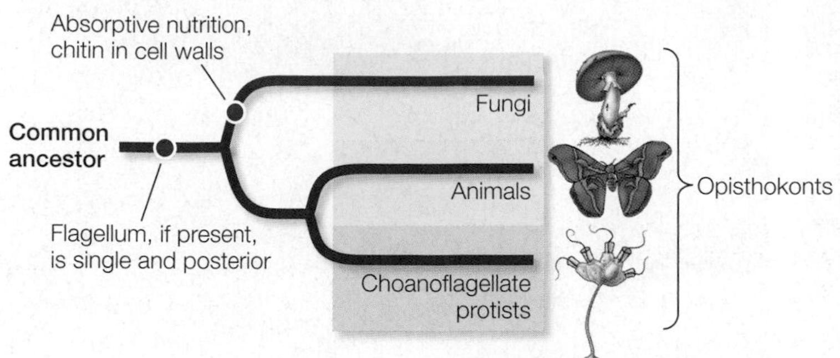

30.1 Fungi in Evolutionary Context Shared derived traits (synapomorphies) distinguish the fungi and other opisthokonts from one another and from the rest of the tree of life.

There are no unicellular glomeromycetes. Yeasts live in liquid or moist environments and absorb nutrients directly across their cell surfaces.

The body of a multicellular fungus is called a **mycelium** (plural, *mycelia*). It is composed of rapidly growing individual tubular filaments called **hyphae** (singular, *hypha*). The cell walls of the hyphae are greatly strengthened by microscopic fibrils of *chitin*, a nitrogen-containing polysaccharide. Hyphae may be subdivided into cell-like units by *incomplete* cross-walls called **septa** (singular, *septum*). Septa do not completely close off compartments in the hyphae; gaps in the septa known as *pores* allow organelles—sometimes even nuclei—to move in a controlled way between cells (**Figure 30.4**). Hyphae that have septa are referred to as **septate**; hyphae that lack septa—but may contain hundreds of nuclei—are referred to as **coenocytic**. The coenocytic condition results from repeated nuclear divisions unaccompanied by cytokinesis.

Certain modified hyphae, called *rhizoids*, anchor some fungi to their substratum (the dead organism or other matter on which they feed). These rhizoids are not homologous to the rhizoids of plants because they are not specialized to absorb nutrients and water. Parasitic fungi may possess modified hyphae that take up nutrients from their host.

The total hyphal growth of a mycelium (not the growth of an individual hypha) may exceed 1 kilometer per day. The hyphae may be widely dispersed to forage for nutrients over a large area, or they may clump together in a cottony mass to exploit a rich nutrient source. In some members of some fungal groups, when sexual spores are produced, the mycelium becomes reorganized into a reproductive *fruiting body* such as a mushroom.

Fungi are in intimate contact with their environment

The filamentous hyphae of a fungus give it a unique relationship with its physical environment. The fungal mycelium has an enormous surface area-to-volume ratio compared with that of most large multicellular organisms. This large ratio is a marvelous adaptation for absorptive nutrition. Throughout the mycelium (except in fruiting structures), all the hyphae are very close to their environmental food source.

There is a downside to the great surface area-to-volume ratio of the mycelium, however: it results in a tendency to lose water rapidly in a dry environment. Thus fungi are most common in moist environments. You have probably observed the tendency of molds, toadstools, and other fungi to appear in damp places.

Another characteristic of some fungi is their tolerance for highly hypertonic environments (those with a solute concentration higher than their own; see Figure 5.9). Many fungi are more resistant than bacteria to damage in hypertonic surroundings. Jelly in the refrigerator, for example, will not become a growth medium for bacteria because it is too hypertonic to those organisms, but it may eventually harbor mold colonies. Their presence in the refrigerator illustrates another trait of many fungi: tolerance of temperature extremes. Many fungi tolerate temperatures as low as –6°C, and some tolerate temperatures over 50°C.

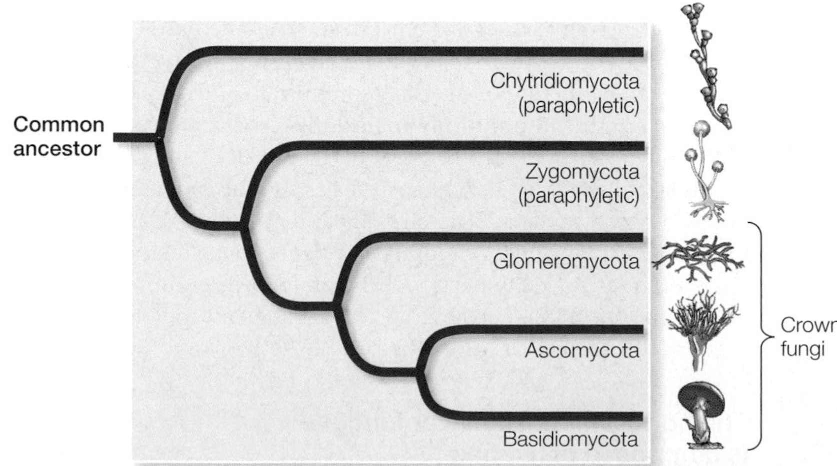

30.2 Phylogeny of the Fungi Five groups are recognized among the fungi. The chytrids and the zygomycetes are paraphyletic. Each of the other groups is a clade, and together the three groups comprise the crown fungi, which is also a clade.

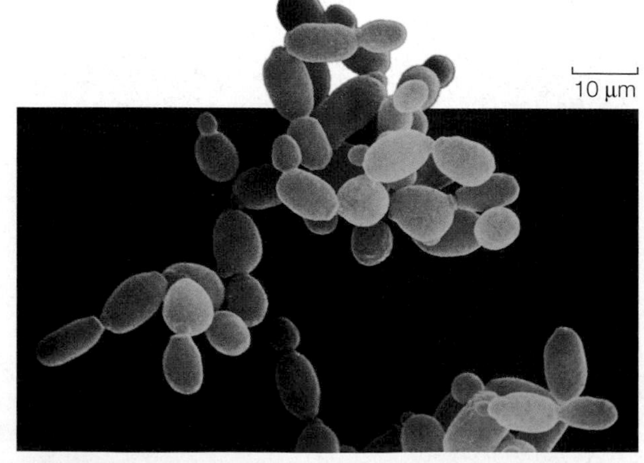

Saccharomyces sp.

30.3 Yeasts Are Unicellular Fungi Unicellular zygomycetes, ascomycetes, and basidiomycetes are known as yeasts. Many yeasts reproduce by budding—mitosis followed by asymmetrical cell division—as those shown here are doing.

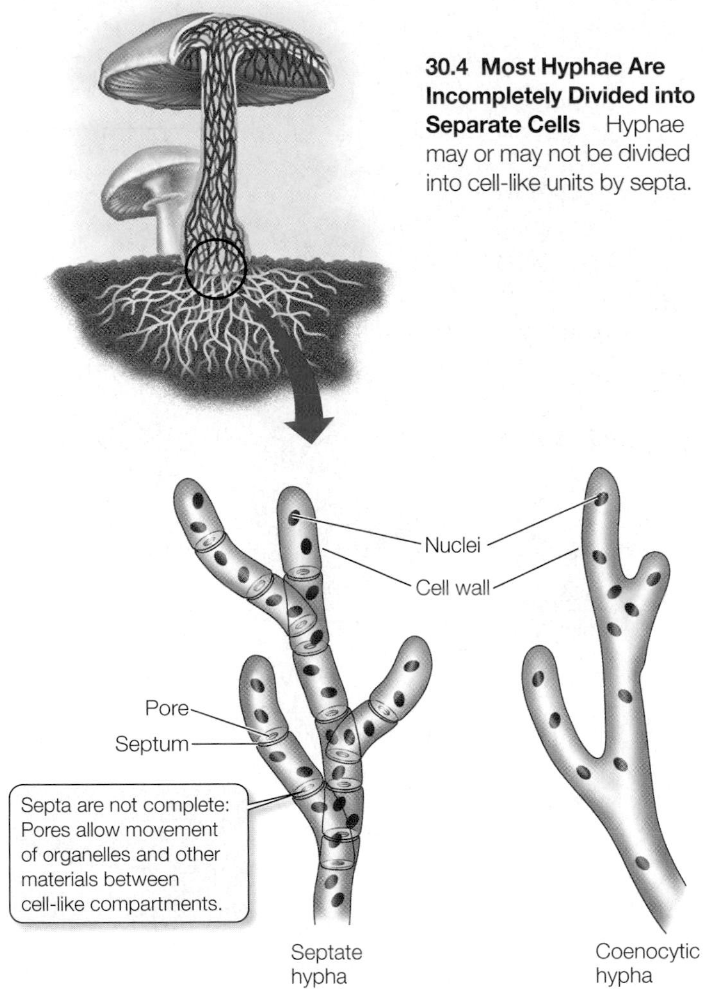

30.4 Most Hyphae Are Incompletely Divided into Separate Cells Hyphae may or may not be divided into cell-like units by septa.

How large can a fungus get? One fungus in Michigan covers 15 hectares (37 acres). At the surface, only isolated clumps of mushrooms are visible. But, growing underground, the vast mycelium of the fungus weighs more than a blue whale. We know from exhaustive analysis of its DNA that this giant fungus is a single individual.

Fungi exploit many nutrient sources

Different types of fungi find different sources of nutrients. While the majority are saprobes, obtaining their energy, carbon, and nitrogen directly from dead organic matter, many form mutualistic associations with other organisms. Others are parasitic—even predatory.

Because many saprobic fungi are able to grow on artificial media, we can perform experiments to determine their exact nutritional requirements. Sugars are their favored source of carbon. Most fungi obtain nitrogen from proteins or the products of protein breakdown. Many fungi can use nitrate (NO_3^-) or ammonium (NH_4^+) ions as their sole source of nitrogen. No known fungus can get its nitrogen directly from nitrogen gas, as can some bacteria and plant–bacteria associations (see Section 36.4). Nutritional studies also reveal that most fungi are unable to synthesize certain vitamins and must absorb them from their environment. On the other hand, fungi can synthesize some vitamins that animals cannot. Like all organisms, fungi also require some mineral elements.

PARASITIC FUNGI Nutrition in the parasitic fungi is particularly interesting to biologists. *Facultative* parasites can attack living organisms but can also be grown by themselves on artificial media. *Obligate* parasites cannot be grown on any available medium; they can grow only on their specific living host, usually a plant species. Because their growth is limited to a living host, such fungi must have specialized nutritional requirements.

The filamentous structure of fungal hyphae is especially well suited to a life of absorbing nutrients from plants. The slender hyphae of a parasitic fungus can invade a plant through stomata, through wounds, or in some cases, by direct penetration of epidermal cells (**Figure 30.5A**). Once inside the plant, the hyphae form a mycelium. Some hyphae produce **haustoria**, branching projections that push into living plant cells, absorbing the nutrients within those cells. The haustoria do not break through the plant cell plasma membranes; they simply press into the cells, with the membrane fitting them like a glove (**Figure 30.5B**). Fruiting structures may form, either within the plant body or on its surface. This type of symbiosis is not usually lethal to the plant.

Some parasitic fungi not only derive nutrition from the bodies of other organisms, but also sicken or kill those hosts. Such a fungus is called a *pathogen*.

SOME PARASITIC FUNGI ARE PATHOGENS Although most human diseases are caused by bacteria or viruses, fungal pathogens are a major cause of death among people with compromised immune systems. Most people with AIDS die of fungal diseases, such as the pneumonia caused by *Pneumocystis carinii* or incurable diarrhea caused by other fungi. *Candida albicans* and certain other yeasts also cause severe diseases, such as esophagitis (which impairs swallowing), in individuals with AIDS and in individuals taking immunosuppressive drugs. Fungal diseases are a growing international health problem, requiring vigorous research. Our limited understanding of the basic biology of these fungi still hampers our ability to treat the diseases they cause. Various fungi cause other, less threatening human diseases, such as ringworm and athlete's foot.

Fungi are by far the most important plant pathogens, causing crop losses amounting to billions of dollars. Bacteria and viruses are less important as plant pathogens. Major fungal diseases of crop plants include black stem rust of wheat and other diseases of wheat, corn, and oats. The agent of black stem rust is *Puccinia graminis*, whose complicated life cycle we will discuss later in the chapter. In an epidemic in 1935, *P. graminis* was responsible for the loss of about one-fourth of the entire wheat crop in Canada and the United States. However, as we saw at the beginning of this chapter, pathogenic fungi that kill certain weed species can be a boon to agriculture.

PREDATORY FUNGI Some fungi have adaptations that enable them to function as active predators, trapping nearby microscopic protists or animals. The most common predatory strategy seen in fungi is to secrete sticky substances from the hyphae so that passing organisms stick tightly to them. The hyphae then quickly invade

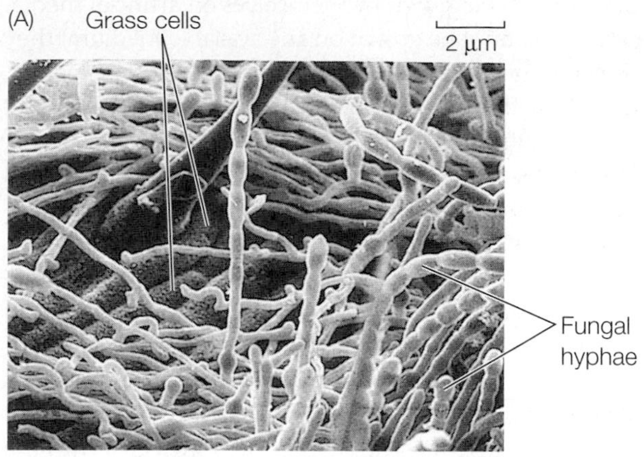

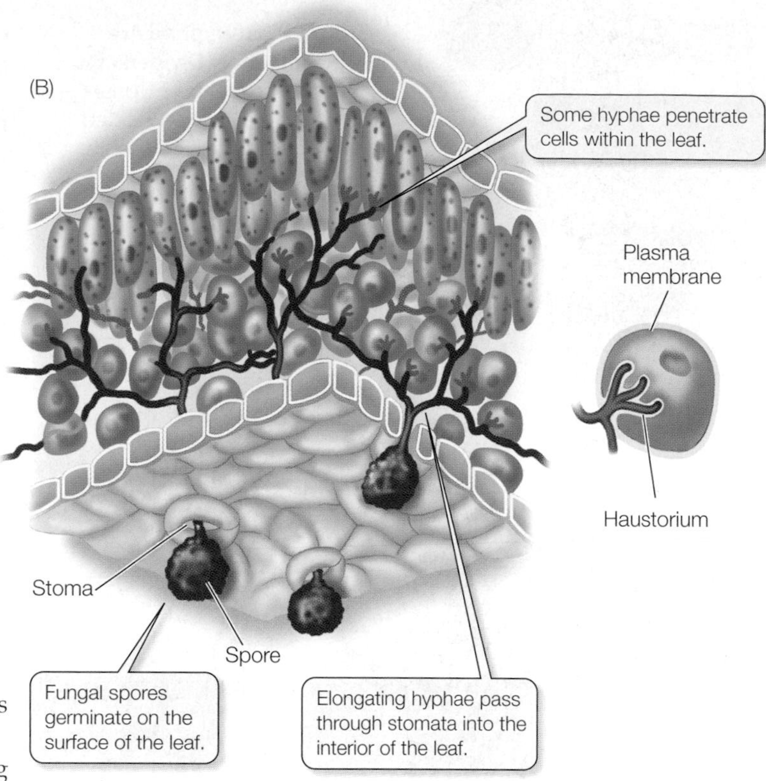

30.5 Attacks on a Leaf (A) The white structures in the micrograph are hyphae of the parasitic fungus *Blumeria graminis* growing on the dark surface of the leaf of a grass. (B) Haustoria are fungal hyphae that push into the living cells of plants, from which they absorb nutrients.

the prey, growing and branching within it, spreading through its body, absorbing nutrients, and eventually killing it.

A more dramatic adaptation for predation is the constricting ring formed by some species of *Arthrobotrys*, *Dactylaria*, and *Dactylella* (**Figure 30.6**). All of these fungi grow in soil. When nematodes (tiny roundworms) are present in the soil, these fungi form three-celled rings with a diameter that just fits a nematode. A nematode crawling through one of these rings stimulates the fungus, causing the cells of the ring to swell and trap the worm. Fungal hyphae quickly invade and digest the unlucky victim.

Fungi balance nutrition and reproduction

What happens if a fungus faces a dwindling food supply? A common strategy is to reproduce rapidly and abundantly. Even when conditions are good, fungi produce great quantities of spores. But the rate of spore production commonly goes up when nutrient supplies go down.

Not only are fungal spores abundant in number, but they are extremely tiny and easily spread by wind or water (**Figure 30.7**). This virtually assures that the individual that produced them will have many progeny, scattered over sometimes great distances. No wonder we find fungi just about everywhere.

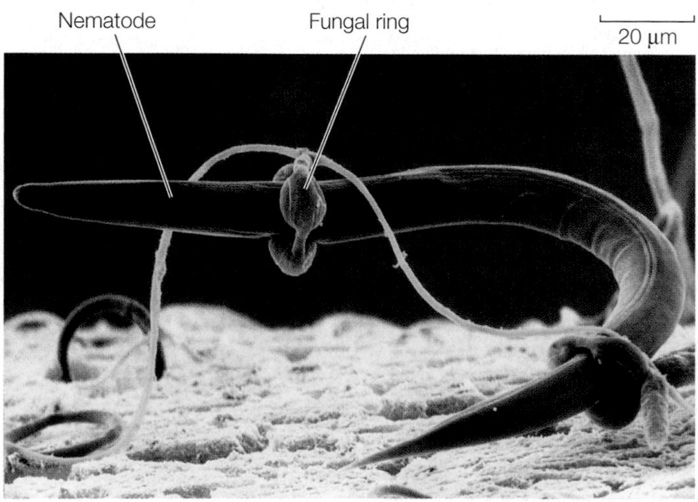

30.6 Some Fungi Are Predators A nematode is trapped in a constricting ring of the soil-dwelling fungus *Arthrobotrys anchonia*.

30.1 RECAP

The rapid growth and large surface area-to-volume ratio of fungal hyphae allow fungi to practice absorptive nutrition efficiently in moist environments.

- Can you explain the relationship between fungal structure and absorptive nutrition? See p. 652
- Can you distinguish among the nutritional modes of saprobes, parasites, and mutualists? See p. 653

The thought of all those fungal spores blowing around can be alarming. But fungi are not always a threat to the organisms around them. There are ways in which they benefit other organisms, and even the biosphere as a whole.

Lycoperdon sp.

30.7 Spores Galore Puffballs disperse trillions of spores in great bursts. Very few of the spores, however, travel very far; some 99 percent of the spores will fall within 100 m of the parent puffball.

30.2 How Are Fungi Beneficial to Other Organisms?

Without the fungi, our planet would be very different. Picture Earth with only a few stunted plants and with watery environments choked with the remains of dead organisms. The colonization of the terrestrial environment was made possible in large part by associations fungi formed with other organisms. Fungi also do much of Earth's garbage disposal. Fungi that absorb nutrients from dead organisms not only help clean up the landscape and form soil, but also play a key role in recycling mineral elements.

Saprobic fungi dispose of Earth's garbage and contribute to the planetary carbon cycle

Saprobic fungi, along with bacteria, are the major decomposers on Earth, contributing to decay and thus to the recycling of the elements used by living things. In forests, for example, the mycelia of fungi absorb nutrients from fallen trees, thus decomposing their wood. Fungi are the principal decomposers of cellulose and lignin, the main components of plant cell walls (most bacteria cannot break down these materials). Other fungi produce enzymes that decompose keratin and thus break down animal structures such as hair and nails.

Were it not for the fungi, great quantities of carbon atoms would have remained trapped on forest floors and elsewhere almost forever—the carbon cycle would have failed. Instead, that carbon is returned to the atmosphere as respiratory CO_2, available for photosynthesis by plants. There was in fact a time when there was a decline in the population of saprobic fungi. During the Carboniferous period, plants in tropical swamps died and began to form peat (see Section 28.4). Peat formation led to acidification of the swamps; in turn, the acidity drastically reduced the fungal population. The result? With the decomposers largely absent, the dead plant material remained on the floor of the swamp and over time was converted to peat and, eventually, coal.

The fungi rose to the occasion during the Permian, a quarter of a billion years ago, when a meteorite crashed to Earth, triggering a planet-wide extinction event (see Section 21.3). The fossil record shows that even though 96 percent of all species became extinct, fungi flourished, demonstrating both their hardiness and their role in recycling the elements in the dead bodies of plants and animals.

As decomposers, the saprobes are essential to life as a whole. There are also many fungi that interact more specifically with other organisms to play key roles as mutualists.

Mutualistic relationships are beneficial to both partners

Certain kinds of relationships between fungi and other organisms have nutritional consequences for the fungal partner. Two of these relationships are highly specific and are **symbiotic** (the partners live in close, permanent contact with one another) as well as **mutualistic** (the relationships benefit both partners).

Lichens are associations of a fungus with a cyanobacterium, a unicellular photosynthetic alga, or both. **Mycorrhizae** (singular *mycorrhiza*) are associations between fungi and the roots of plants. In these associations, the fungus obtains organic compounds from its photosynthetic partner, but provides it with minerals and water in return, so that the partner's nutrition is also promoted. In fact, many plants could not grow at all without their fungal partners.

Lichens can grow where plants cannot

A lichen is not a single organism, but rather a meshwork of two radically different organisms: a fungus and a photosynthetic microorganism. Together the organisms constituting a lichen can survive some of the harshest environments on Earth. The biota of Antarctica, for example, features more than a hundred times as many species of lichens as of plants.

In spite of this hardiness, lichens are very sensitive to air pollution because they are unable to excrete toxic substances that they absorb. Hence they are not common in industrialized cities. Because of their sensitivity, lichens are good biological indicators of air pollution.

The fungal components of most lichens belong to the clade called ascomycetes, or sac fungi, which also includes various cup fungi, yeasts, and molds such as the *Fusarium* mentioned at the beginning of the chapter. The photosynthetic component of a lichen is most often a unicellular green alga, but may be a cyanobacterium, or may include both. Relatively little experimental work has focused on lichens, perhaps because they grow so slowly—typically less than 1 centimeter per year.

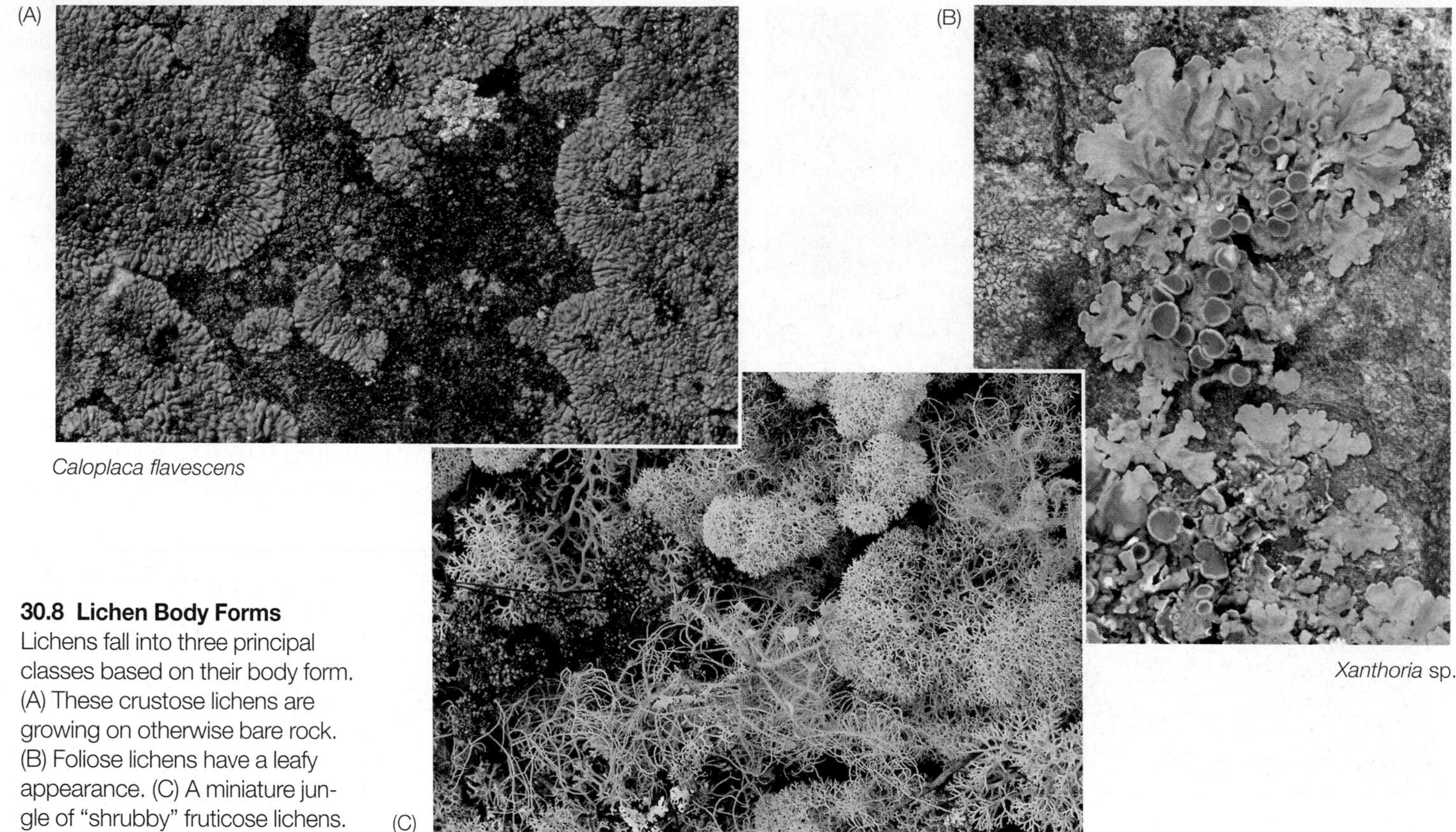

30.8 Lichen Body Forms Lichens fall into three principal classes based on their body form. (A) These crustose lichens are growing on otherwise bare rock. (B) Foliose lichens have a leafy appearance. (C) A miniature jungle of "shrubby" fruticose lichens.

There are more than 15,000 "species" of lichens, each of which is assigned the name of its fungal component. These fungal components may constitute as many as 20 percent of all fungal species. Some of these fungi are able to grow independently without a photosynthetic partner, but others have not been observed in nature other than in a lichen association. Lichens are found in all sorts of exposed habitats: on tree bark, on open soil, and on bare rock. Reindeer moss (actually not a moss at all, but the lichen *Cladonia subtenuis*) covers vast areas in Arctic, sub-Arctic, and boreal regions, where it is an important part of the diets of reindeer and other large mammals. Lichens come in various forms and colors. *Crustose* (crustlike) lichens look like colored powder dusted over their substratum (**Figure 30.8A**); *foliose* (leafy) and *fruticose* (shrubby) lichens may have complex forms (**Figure 30.8B,C**).

The most widely held interpretation of the lichen relationship is that it is a mutually beneficial symbiosis. The hyphae of the fungal mycelium are tightly pressed against the algae or cyanobacteria and sometimes even invade them. The bacterial or algal cells not only survive these indignities, but continue their growth and photosynthesis. In fact, the algal cells in a lichen "leak" photosynthetic products at a greater rate than do similar cells growing on their own. On the other hand, photosynthetic cells from lichens grow more rapidly on their own than when associated with a fungus. On this basis, we could consider lichen fungi to be parasitic on their photosynthetic partners. However, in many places where lichens grow, the photosynthetic cells would not grow at all on their own.

Lichens can reproduce simply by fragmentation of the vegetative body, which is called the *thallus*, or by means of specialized structures called *soredia* (singular *soredium*). Soredia consist of one or a few photosynthetic cells surrounded by fungal hyphae. The soredia become detached from the lichen, are dispersed by air currents, and upon arriving at a favorable location, develop into a new lichen. Alternatively, the fungal partner may go through its sexual cycle, producing haploid spores. When these spores are discharged, however, they disperse alone, unaccompanied by the photosynthetic partner, and thus may not be capable of reestablishing the lichen association, or even of surviving on their own.

Visible in a cross section of a typical foliose lichen are a tight upper region of fungal hyphae, a layer of cyanobacteria or algae, a looser hyphal layer, and finally hyphal rhizoids that attach the whole structure to its substratum (**Figure 30.9**). The meshwork of fungal hyphae takes up some nutrients needed by the photosynthetic cells and provides a suitably moist environment for them by holding water tenaciously. The fungi derive fixed carbon from the photosynthetic products of the algal or cyanobacterial cells.

Lichens are often the first colonists on new areas of bare rock. They get most of the nutrients they need from the air and rainwater, augmented by minerals absorbed from dust. A lichen begins to grow shortly after a rain, as it begins to dry. As it grows, the lichen acidifies its environment slightly, and this acidity contributes to the slow breakdown of rocks, an early step in soil formation. After further drying, the lichen's photosynthesis ceases. The water content of the lichen may drop to less than 10 percent of its dry weight, at which point it becomes highly insensitive to extremes of temperature.

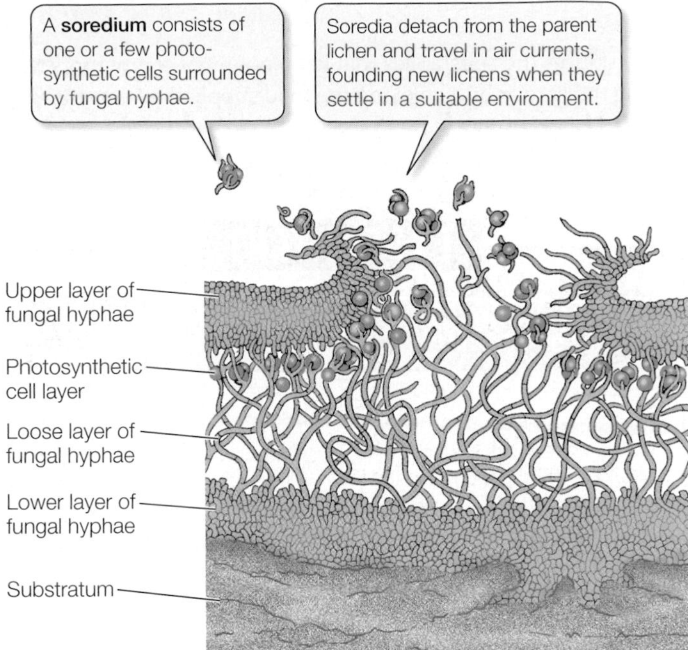
A **soredium** consists of one or a few photosynthetic cells surrounded by fungal hyphae.
Soredia detach from the parent lichen and travel in air currents, founding new lichens when they settle in a suitable environment.
Upper layer of fungal hyphae
Photosynthetic cell layer
Loose layer of fungal hyphae
Lower layer of fungal hyphae
Substratum

30.9 Lichen Anatomy Cross section showing the layers of a foliose lichen and the release of soredia.

Mycorrhizae are essential to most plants

Almost all vascular plants require a symbiotic association with fungi. Unassisted, the root hairs of such plants do not take up enough water or minerals to sustain growth. However, their roots usually do become infected with fungi, forming an association called a mycorrhiza. Mycorrhizae are of two types, based on whether or not the fungal hyphae penetrate the plant cells.

In *ectomycorrhizae*, the fungus wraps around the root, and its mass is often as great as that of the root itself (**Figure 30.10A**). The fungal hyphae wrap around individual cells in the root, but do not penetrate them. An extensive web of hyphae penetrates the soil in the area around the root, so that up to 25 percent of the soil volume near the root may be fungal hyphae. The hyphae attached to the root increase the surface area for the absorption of water and minerals, and the mass of the mycorrhiza in the soil, like a sponge, holds water efficiently in the neighborhood of the root. Infected roots characteristically branch extensively and become swollen and club-shaped, and they lack root hairs.

In *arbuscular mycorrhizae* the fungal hyphae enter the root and penetrate the cell wall of the root cells, forming arbuscular (tree-like) structures inside the cell wall, but outside the plasma membrane. These structures, like the haustoria of parasitic fungi, become the primary site of exchange between plant and fungus (**Figure 30.10B**). As with the ectomycorrhizae, the fungus forms a vast web of hyphae leading from the root surface into the surrounding soil.

The mycorrhizal association is important to both partners. The fungus obtains needed organic compounds, such as sugars and amino acids, from the plant. In return, the fungus, because of its very high surface area-to-volume ratio and its ability to penetrate the fine structure of the soil, greatly increases the plant's ability to absorb water and minerals (especially phosphorus). The fungus may also provide the plant with certain growth hormones and may protect it against attack by disease-causing microorganisms. Plants that have active arbuscular mycorrhizae typically are a deeper green and may resist drought and temperature extremes better than plants of the same species that have little mycorrhizal development. Attempts to introduce some plant species to new areas have failed until a bit of soil from the native area (presumably containing the fungus necessary to establish mycorrhizae) was provided. Trees without ectomycorrhizae will not grow well in the absence of abundant nutrients and water, so the health of our forests depends on the presence of ectomycorrhizal fungi.

The partnership between plant and fungus results in a plant that is better adapted for life on land. It has been suggested that the evolution of mycorrhizae was the single most important step in the colonization of the terrestrial environment by living things. Fossils of mycorrhizal structures 460 million years old have been found. Some liverworts, which are among the most ancient terrestrial plants (see Section 28.4), form mycorrhizae.

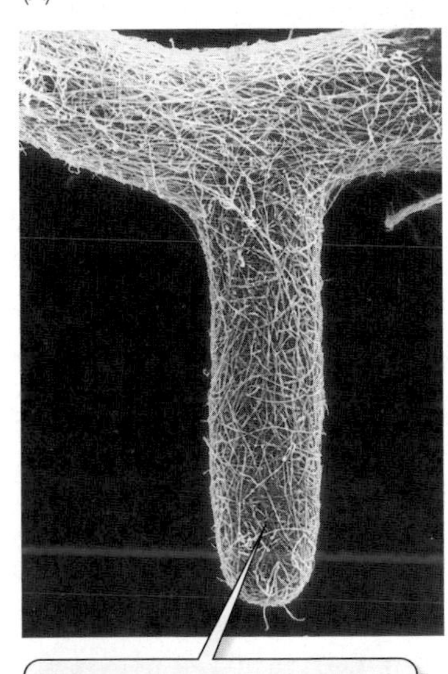
(A)
Hyphae of the fungus *Pisolithus tinctorius* cover a eucalyptus root.

(B)

Hyphae

30.10 Mycorrhizal Associations (A) Ectomycorrhizal fungi wrap themselves around a plant root, increasing the area available for absorption of water and minerals. (B) Hyphae of arbuscular mycorrhizal fungi infect the root internally and penetrate the root cells, branching within the cells and forming tree-like (arbuscular) structures that provide the plant with nutrients. Hyphae fill much of the cell outside the nucleus.

Certain plants that live in nitrogen-poor habitats, such as cranberry bushes and orchids, invariably have mycorrhizae. Orchid seeds will not germinate in nature unless they are already infected by the fungus that will form their mycorrhizae. Plants that lack chlorophyll always have mycorrhizae, which they often share with the roots of green, photosynthetic plants. In effect, these plants without chlorophyll are feeding on nearby green plants, using the fungus as a bridge.

Biologists had long suspected that roots secrete a chemical signal that enables fungi to find and invade them to form arbuscular mycorrhizae. This was proved to be correct in 2005 when researchers succeeded in isolating the signaling compound. Might the compound also be used by parasitic plants to attack their host plants? Indeed it is. *Striga*, discussed at the beginning of this chapter, turns out to be one of the parasitic plants that use exactly this signal. Thus, in attracting its helper fungus, a plant may also attract a dangerous parasite.

Endophytic fungi protect some plants from pathogens, herbivores, and stress

In a tropical jungle, 10,000 or more fungal spores land on a single leaf each day. Some are plant pathogens, some do not attack the plant at all, and some invade the plant in a beneficial way. Fungi that live within aboveground parts of plants are called **endophytic fungi**. Recent research has shown that endophytic fungi are abundant in plants in all terrestrial environments.

Grasses with endophytic fungi are more resistant to pathogens and to insect and mammalian herbivores than are grasses lacking endophytes. The fungi produce alkaloids (nitrogen-containing compounds) that are toxic to animals. The alkaloids do not harm the host plant; in fact, some plants produce alkaloids (such as nicotine) themselves. The fungal alkaloids also increase the ability of host plants to resist stress of various types, including drought (water shortage) and salty soils. Such resistance is useful in agriculture.

The role, if any, of endophytic fungi in most trees is unclear. They may simply occupy space within leaves, without conferring any benefit but also without doing harm.

Some fungi are food for the ants that farm them

In another kind of association, some leaf-cutting ants "farm" fungi, feeding the fungi and later eating the specialized fruiting bodies (called *gongylidia*) the fungi produce. The ants collect leaves and flower petals, chew them into small bits, and "plant" bits of fungal mycelium on their surfaces. The fungi in this "garden" secrete enzymes that digest the plant material, breaking it down into products that the fungi absorb for their own nutrition. The fungi then produce the gongylidia that are harvested and eaten by the ants.

An established garden is not a mixture of fungi; rather, it contains a single clone of fungus. When ants bring other fungi to the garden, the newcomers are killed by substances contained in the ants' feces. Do these substances derive from the ants, or from the fungal garden itself, which thus denies the ants the possible benefits of farming a genetically more diverse garden? A simple experiment performed in 2005 showed that it is indeed the fungus that produces the substances that keep potential competitors from joining the garden and enjoying the food supplied by the ants (**Figure 30.11**).

EXPERIMENT

HYPOTHESIS: Fungi, not ants, produce the substance that keeps "foreign" fungi from growing in the garden.

METHOD

1. Collect samples of distinct fungus gardens and ants from the field and establish colonies in the laboratory.
2. Some ants from each of two colonies (1 and 2) are forced to feed for 10 days on fungi obtained from the other colony. Other ants continue to feed on their "home" gardens.
3. At the beginning and end of the 10-day period, place fecal droplets of ants on fungi from their own garden or from the "alien"garden.
4. Test for incompatibility reactions (avoidance of droplet by fungus, toxicity to fungus).

RESULTS

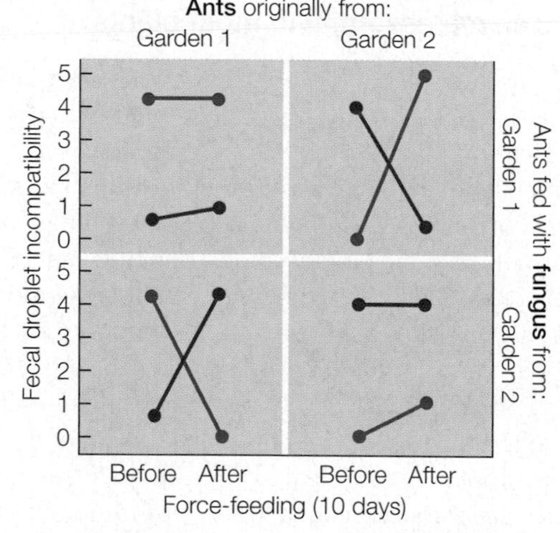

Fungus tested:
- Fungus 1
- Fungus 2

Incompatibility:
- 0 = Complete compatibility
- 5 = Complete incompatibility

CONCLUSION: Foreign fungi are deterred by substances produced by the fungus an ant eats, not by the ants themselves.

30.11 Keeping Fungal Interlopers Away Only a single clone of a single fungal species is maintained in the fungal garden of leaf-cutting ants. What keeps these gardens from becoming contaminated by other fungi? An experiment performed by Michael Poulsen and Jacobus Boomsma explored this question. FURTHER RESEARCH: Some termites raise fungal gardens, but the fungi feed on the insects' fecal matter rather than on plant material. How would you determine whether a similar system to avoid contamination by incompatible fungal species is maintained in the termite-fungus symbiosis?

30.2 RECAP

Fungi form many beneficial associations with other organisms. Mycorrhizal fungi are essential for the survival of most plant species.

- What is the role of fungi in the planetary carbon cycle? See p. 655
- Can you explain the nature and benefits of the lichen association? See pp. 655–656
- Do you understand why plants grow better in the presence of mycorrhizal fungi? See p. 657

Most lichen fungi belong to one particular fungal clade, and arbuscular mycorrhizal fungi are members of another. One of the most important criteria for assigning fungi to taxonomic groups, before molecular techniques made things simpler and more certain, was the nature of their life cycles.

30.3 How Do Fungal Life Cycles Differ from One Another?

Different fungal groups have different life cycles. One group features alternation of generations, a type of life cycle found in all plants and some protists. Other groups have an unusual life cycle that is unique to the fungi, featuring a stage called a *dikaryon*. We will contrast these cycles after surveying the reproductive modes of the various fungal groups. Let's begin by briefly reviewing asexual and sexual reproduction in general terms.

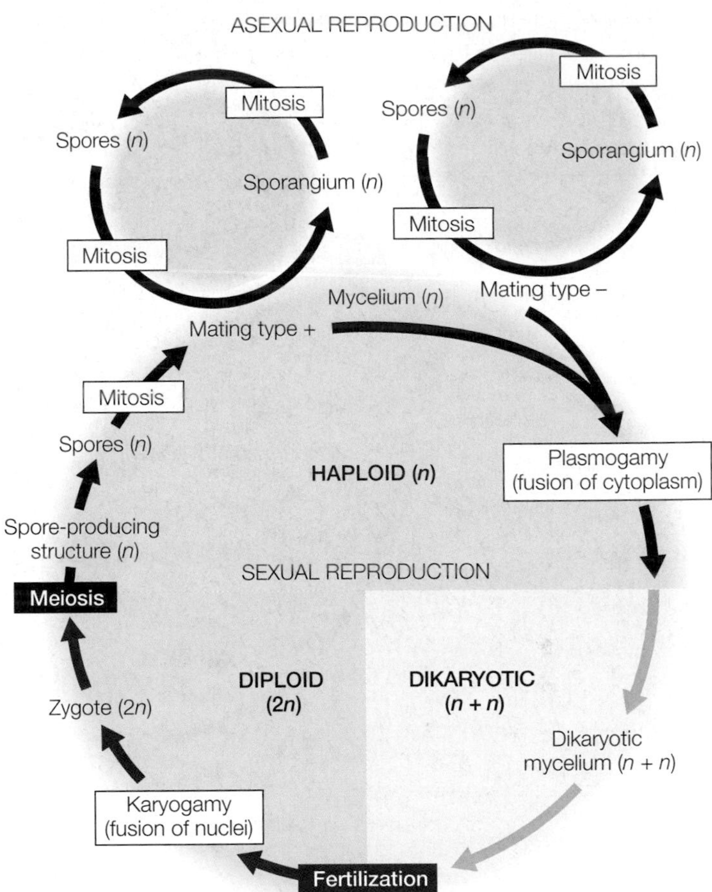

30.12 Asexual and Sexual Reproduction in a Fungal Life Cycle Environmental conditions may determine which mode of reproduction takes place at a given time.

Fungi reproduce both sexually and asexually

Both asexual and sexual reproduction are common among the fungi (**Figure 30.12**). Asexual reproduction takes several forms:

- The production of (usually) haploid spores within structures called *sporangia*
- The production of naked spores (not enclosed in sporangia) at the tips of hyphae; such spores are called *conidia* (from the Greek *konis*, "dust")
- Cell division by unicellular fungi—either a relatively equal division (called *fission*) or an asymmetrical division in which a small daughter cell is produced (called *budding*)
- Simple breakage of the mycelium

Asexual reproduction in fungi can be spectacular in terms of quantity. A 2.5-centimeter colony of *Penicillium*, the mold that produces the antibiotic penicillin, can produce as many as 400 million conidia. The air we breathe contains as many as 10,000 fungal spores per cubic meter.

Sexual reproduction in many fungi features an interesting twist. There is often no morphological distinction between female and male structures, or between female and male individuals. Rather, there is a genetically determined distinction between two *or more* **mating types**. Individuals of the same mating type cannot mate with one another, but they can mate with individuals of another mating type within the same species. This distinction prevents self-fertilization. Individuals of different mating types differ genetically from one another, but are often visually and behaviorally indistinguishable. Many protists also have mating type systems.

Fungi reproduce sexually when hyphae (or, in one group, motile cells) of different mating types meet and fuse (**Figure 30.13**). In many fungi, the zygote nuclei formed by sexual reproduction are the only diploid nuclei in the life cycle. These nuclei undergo meiosis, producing haploid nuclei that become incorporated into spores. Haploid fungal spores, whether produced sexually in this manner or asexually, germinate, and their nuclei divide mitotically to produce hyphae. This type of life cycle, called a *haplontic* life cycle, is also characteristic of many microbial eukaryotes.

The alternation between multicellular haploid (n) and multicellular diploid ($2n$) generations that evolved in plants and certain protist groups (see Section 28.2) is found in the chytrids and certain yeasts (**Figure 30.13A**). Alternation of generations does not appear in the other groups. As one might expect, these basal fungi, which are aquatic, possess flagellated gametes; they also have flagellated spores. Flagella have been lost in the terrestrial fungi.

What are the consequences of alternation of generations in the chytrids? It is possible that the multicellular haploid organisms serve as a "filter" for harmful mutations. A haploid individual with such a mutation would die, and the mutant allele would not be passed to

(A) Chytrids
The life cycle of the aquatic chytrids features alternation of generations. They have no dikaryotic stage.
Sporangium
Haploid zoospores (n)
Multicellular haploid chytrid (n)
~30 µm
Female gametangium
Male gametangium
HAPLOID (n)
DIPLOID (2n)
Female gamete (n)
Male gamete (n)
Meiosis
Multicellular diploid chytrid (2n)
Fertilization
Zygote (2n)

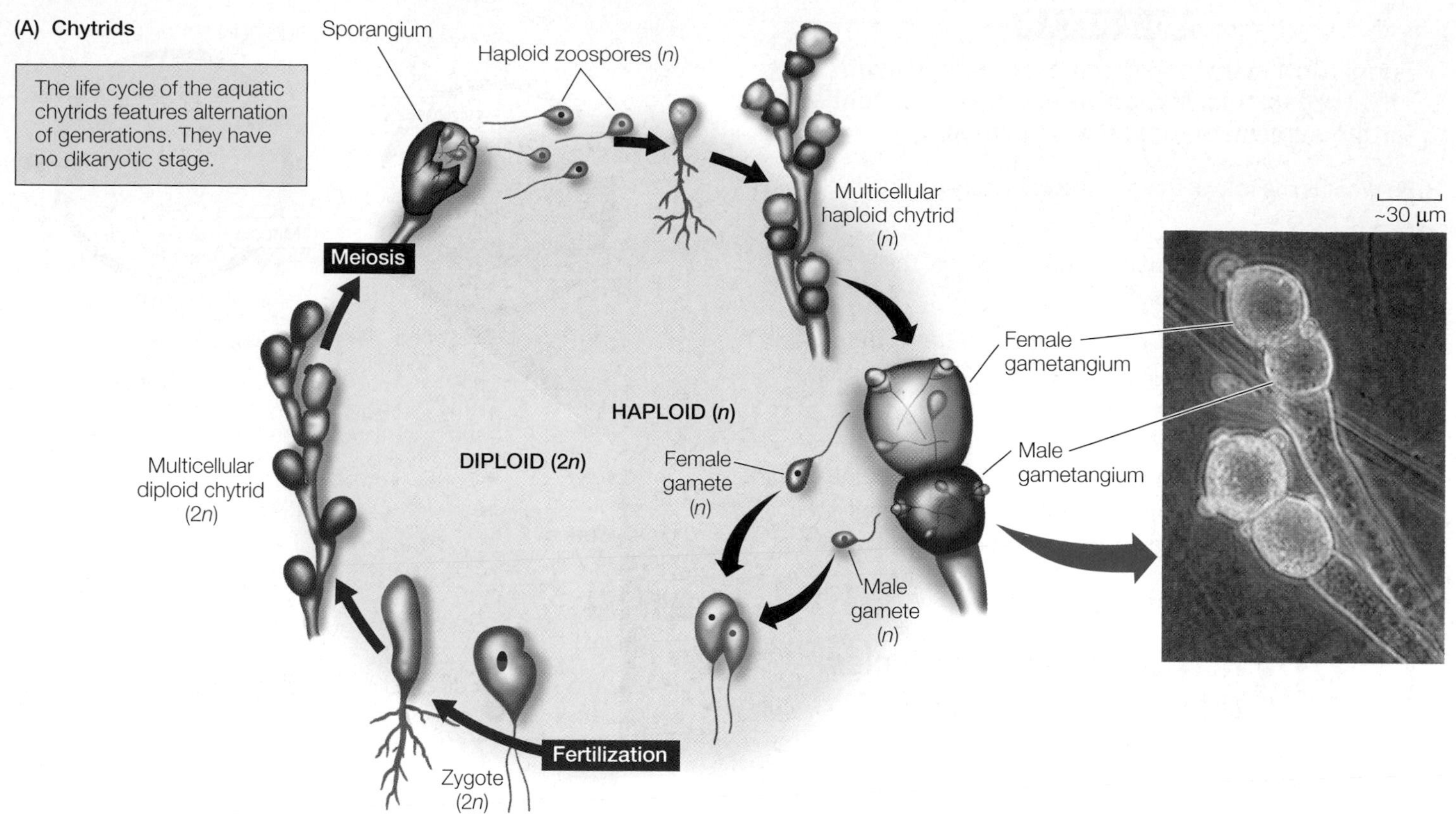

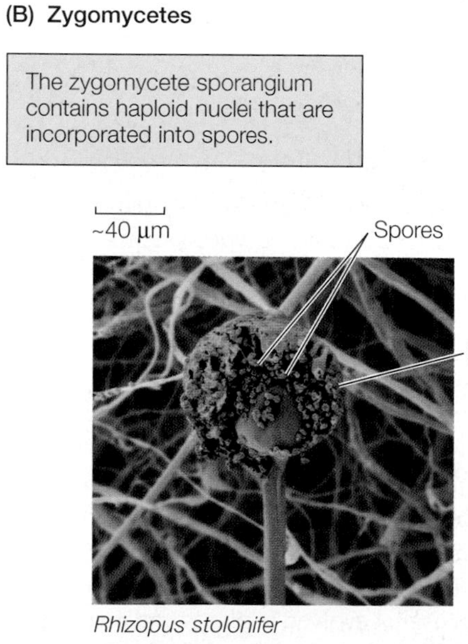

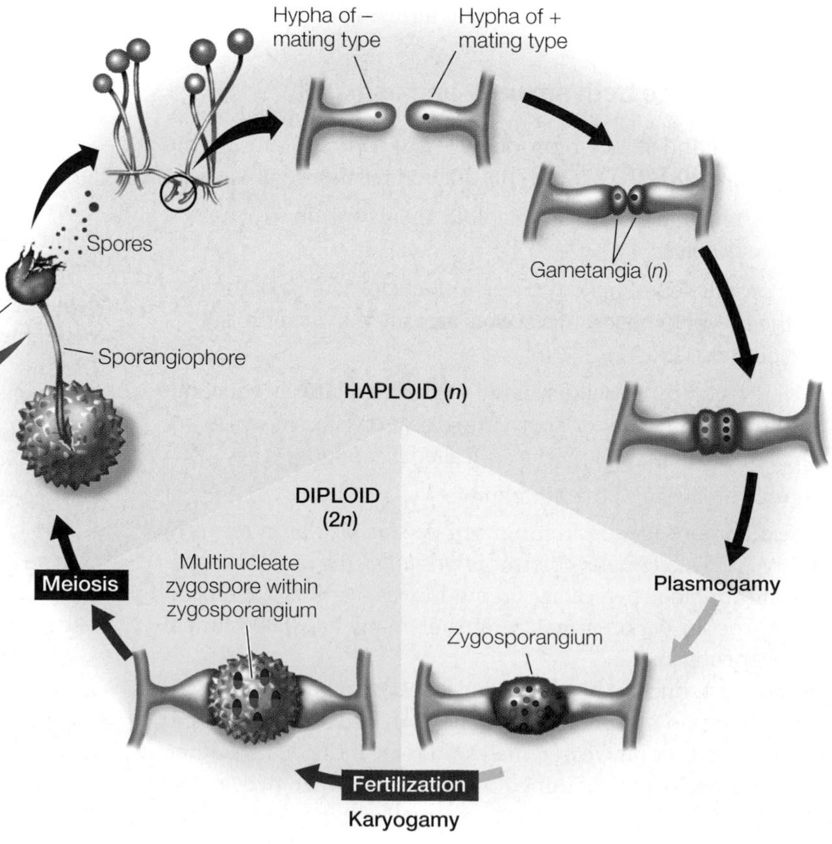

(B) Zygomycetes
The zygomycete sporangium contains haploid nuclei that are incorporated into spores.
~40 µm
Spores
Rhizopus stolonifer

Hypha of – mating type
Hypha of + mating type
Spores
Gametangia (n)
Sporangium
Sporangiophore
HAPLOID (n)
DIPLOID (2n)
Multinucleate zygospore within zygosporangium
Meiosis
Plasmogamy
Zygosporangium
Fertilization
Karyogamy

30.13 Sexual Life Cycles Vary among Different Groups of Fungi (A) The chytrids are the only fungi that posses flagella at any stage of the life cycle. Their flagellated gametes and zoospores link them to the animals. (B) A multinucleate zygospore is unique to the zygomycetes. (C,D) The dikaryon stage is definitive of the ascomycetes (C) and the basidiomycetes (D). Glomeromycetes (the mycorrhizal fungi) reproduce asexually only and are not depicted here.

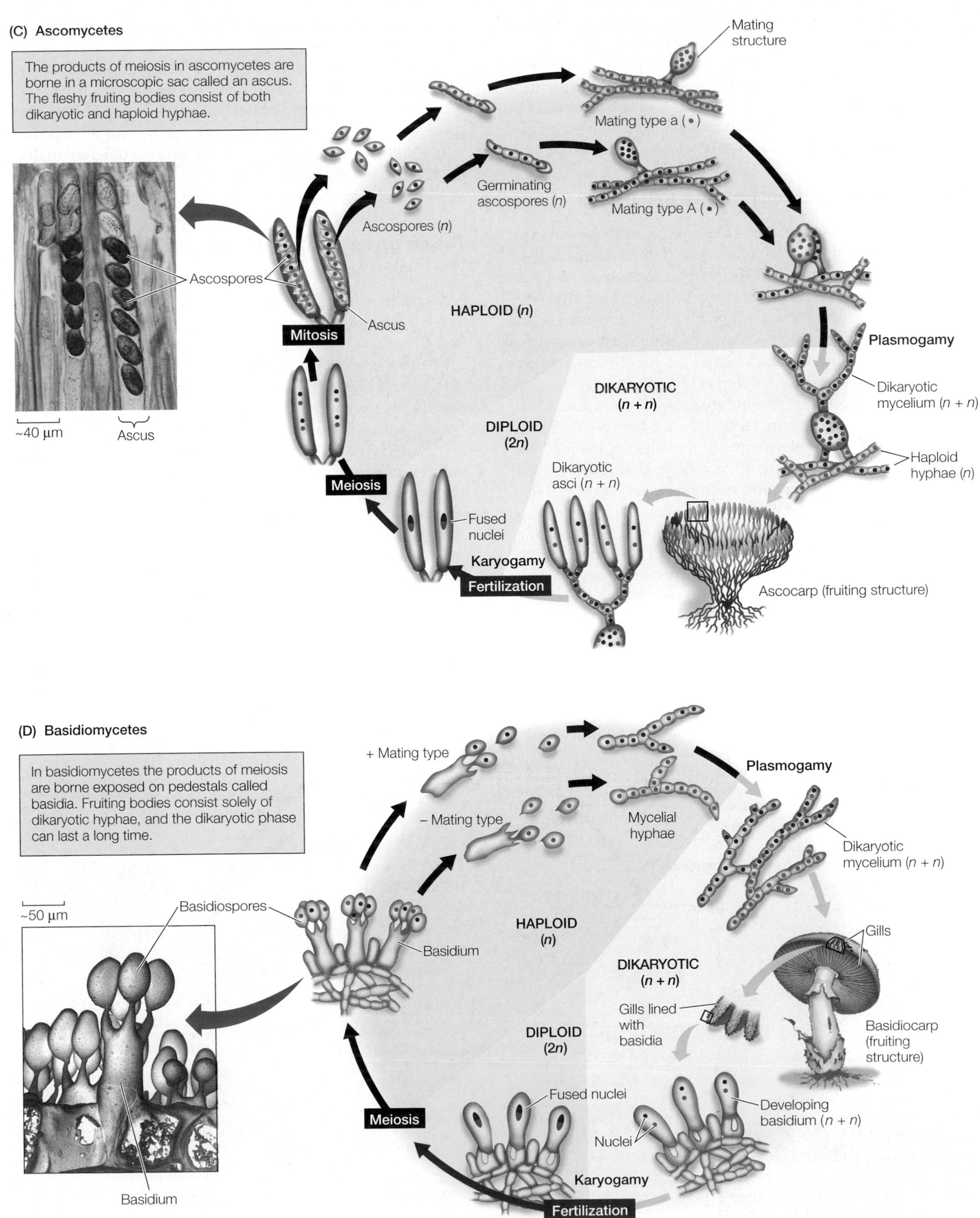
(C) Ascomycetes
The products of meiosis in ascomycetes are borne in a microscopic sac called an ascus. The fleshy fruiting bodies consist of both dikaryotic and haploid hyphae.
Ascospores
~40 µm
Ascus
Mating structure
Mating type a (•)
Germinating ascospores (n)
Mating type A (•)
Ascospores (n)
Ascus
Mitosis
HAPLOID (n)
Plasmogamy
Dikaryotic mycelium (n + n)
DIKARYOTIC (n + n)
DIPLOID (2n)
Haploid hyphae (n)
Dikaryotic asci (n + n)
Meiosis
Fused nuclei
Karyogamy
Fertilization
Ascocarp (fruiting structure)
(D) Basidiomycetes
In basidiomycetes the products of meiosis are borne exposed on pedestals called basidia. Fruiting bodies consist solely of dikaryotic hyphae, and the dikaryotic phase can last a long time.
+ Mating type
– Mating type
Plasmogamy
Mycelial hyphae
Dikaryotic mycelium (n + n)
~50 µm
Basidiospores
Basidium
HAPLOID (n)
Gills
DIKARYOTIC (n + n)
Gills lined with basidia
DIPLOID (2n)
Basidiocarp (fruiting structure)
Basidium
Fused nuclei
Developing basidium (n + n)
Nuclei
Meiosis
Karyogamy
Fertilization

progeny. In these basal fungi, the multicellular diploid stage includes a resistant structure capable of withstanding freezing or drying.

Although the terrestrial fungi grow in moist places, their gamete nuclei are not motile and are not released into the environment (**Figure 30.13B–D**). Instead, the cytoplasms of two individuals of different mating types fuse (**plasmogamy**) before their nuclei fuse (**karyogamy**). Therefore, liquid water is not required for fertilization.

The dikaryotic condition is unique to the fungi

Certain hyphae of some terrestrial ascomycetes and basidiomycetes have a nuclear configuration other than the familiar haploid or diploid states (**Figures 30.13C,D**). In these fungi, sexual reproduction begins in an unusual way: karyogamy occurs long after plasmogamy, so that *two genetically different haploid nuclei coexist and divide within the same hypha*. Such a hypha is called a **dikaryon** ("two nuclei"). Because the two nuclei differ genetically, such a hypha is also called a *heterokaryon* ("different nuclei").

Eventually, specialized fruiting structures form, within which the pairs of genetically dissimilar nuclei—one from each parent—fuse, giving rise to zygotes long after the original "mating." The diploid zygote nucleus undergoes meiosis, producing four haploid nuclei. The mitotic descendants of those nuclei become spores, which give rise to the next generation of hyphae.

This type of life cycle displays several unusual features. First, there are no gamete *cells*, only gamete *nuclei*. Second, there is never any true diploid tissue, although for a long period the genes of both parents are present in the dikaryon and can be expressed. In effect, the hypha is neither diploid ($2n$) nor haploid (n); rather, it is *dikaryotic* ($n + n$). A harmful recessive mutation in one nucleus may be compensated for by a normal allele on the same chromosome in the other nucleus. Dikaryosis is perhaps the most distinctive of the genetic peculiarities of the fungi.

The dikaryotic condition often lasts for months or even years. Basidiomycetes have an elegant mechanism that ensures that the dikaryotic condition is maintained as new cells are formed. One consequence of a sustained dikaryotic condition is an increased opportunity for multiple fusions of hyphae of different mating types before fruiting bodies are formed. This could allow for more genetic recombination than with just two mating types.

The life cycles of some parasitic fungi require two hosts

Some parasitic fungi are very specific about what host organism they use as a source of nutrition, and some even use different hosts for different stages of their life cycle. One of the most striking examples of a fungal life cycle that involves two different hosts is that of *Puccinia graminis*, the agent of black stem rust of wheat mentioned earlier in the chapter (see p. 653).

Let's examine a year in the life cycle of *P. graminis* (**Figure 30.14**; the figure is keyed to the text by parenthetical numbers). During the summer, dikaryotic hyphae of *P. graminis* proliferate in the stem and leaf tissues of wheat plants. These dikaryotic hyphae produce great quantities of dikaryotic summer spores called *uredospores* **(1)**. The one-celled, orange uredospores are scattered by the wind and infect other wheat plants, on which the summer hyphae then proliferate.

Dark brown winter spores—*teliospores*—begin to appear on special hyphae in late summer. Each teliospore consists initially of two dikaryotic cells. The two haploid nuclei in each cell fuse to form a single diploid ($2n$) nucleus. These are the only diploid cells in the entire life cycle **(2)**. Both cells have thick walls and usually survive

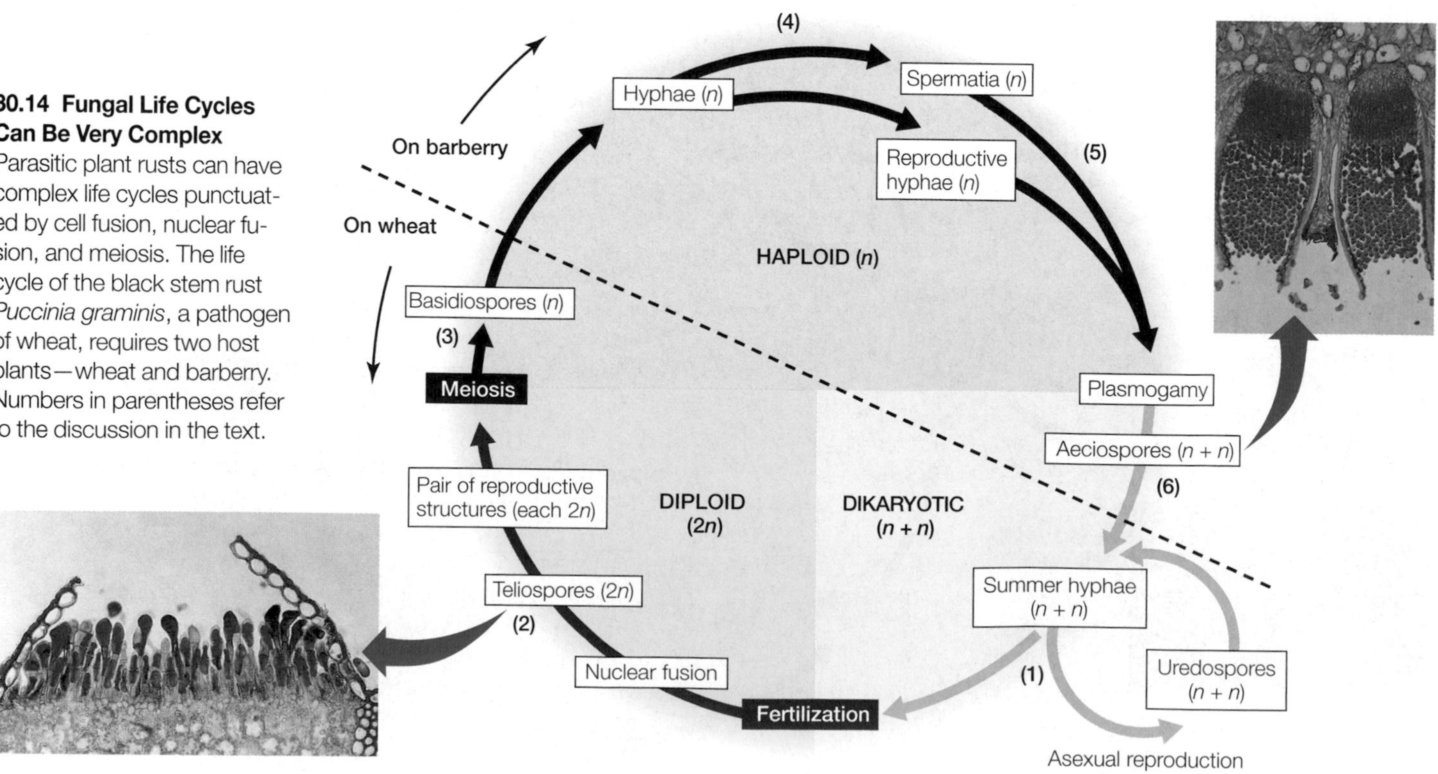

30.14 Fungal Life Cycles Can Be Very Complex Parasitic plant rusts can have complex life cycles punctuated by cell fusion, nuclear fusion, and meiosis. The life cycle of the black stem rust *Puccinia graminis*, a pathogen of wheat, requires two host plants—wheat and barberry. Numbers in parentheses refer to the discussion in the text.

freezing. The teliospores remain dormant until spring, when they germinate. Each of the two cells develops into a reproductive structure, within which the nucleus divides meiotically to produce four haploid *basidiospores* **(3)**.

Basidiospores are one of two mating types: plus (+) or minus (–). They are carried by the wind, and if they land on a leaf of the common barberry plant, they germinate and produce haploid hyphae (+ or –, depending on the mating type of the germinating basidiospore). These hyphae invade the barberry leaf, forming flask-shaped structures on the upper surface of the leaf, within which some of the hyphae pinch off tiny, colorless, haploid *spermatia* from their tips **(4)**.

The hyphae also form another type of structure, called an *aecial primordium*, near the lower surface of the leaf. Thus the leaf contains "flasks" on the upper surface and aecial primordia near the lower one, and these structures are connected by hyphae. Insects attracted to a sweetish liquid produced in the flasks carry spermatia from one flask to another, initiating the next step in the cycle.

A spermatium of one mating type fuses with a receptive hypha of the other type within a flask. The nucleus of the spermatium repeatedly divides mitotically, and the products move through the hyphae into immature aecial primordia, where they produce dikaryotic cells with two haploid nuclei, one of each mating type. These cells develop into *aeciospores*, each containing two unlike nuclei **(5)**. The aeciospores are scattered by the wind, and some land on wheat plants, where they continue the life cycle **(6)**. When the dikaryotic aeciospores germinate on wheat, they produce the summer hyphae with two nuclei in each cell **(1)**.

The different types of spores produced during the life cycle of *P. graminis* play very different roles. Consider first the stages on the wheat host. Windborne uredospores are the primary agents for spreading the rust from wheat plant to wheat plant and to other fields. Resistant teliospores allow the rust to survive the harsh winter, but contribute little to the spreading of the rust. Basidiospores spread the rust from wheat to barberry plants. On the barberry host, spermatia and receptive hyphae initiate the sexual cycle of the rust. Finally, aeciospores spread the rust from barberry to wheat plants. The adaptive value of this division of the life cycle between two host plants is still a matter of speculation. Life cycles including two hosts have evolved many times in many groups of organisms—recall, for example, the parasitic protist that causes malaria (see Figure 27.3).

"Imperfect fungi" lack a sexual stage

As we have just seen, mechanisms of sexual reproduction help distinguish members of the groups of fungi from one another. But many fungi, including both saprobes and parasites, appear to lack sexual stages entirely; presumably these stages have been lost during the evolution of these species or have not yet been observed. Classifying these fungi used to be difficult, but biologists can now assign most of them to one of the five groups on the basis of their DNA sequences.

Fungi that have not yet been placed in any of the existing groups are pooled together in a polyphyletic group called **deuteromycetes**, informally known as "imperfect fungi." Thus the deuteromycete group is a holding area for species whose status is yet to be resolved. At present, about 25,000 species are classified as imperfect fungi.

If sexual structures are found on a fungus classified as a deuteromycete, that fungus is reassigned to the appropriate group. That happened, for example, to a fungus that produces plant growth hormones called gibberellins (see Section 37.1). Originally classified as the deuteromycete *Fusarium moniliforme*, this fungus was later found to produce a reproductive structure characteristic of an ascomycete, whereupon it was renamed and transferred to that group.

30.3 RECAP

Fungi have a variety of life cycles, including one requiring multiple hosts.

- Do you understand the concept of mating types? See p. 659
- Can you explain the phenomenon of dikaryosis in terms of plasmogamy and karyogamy? See p. 662

In examining the most important properties of the fungi as a clade, we have often drawn examples from specific groups. Now let's briefly consider the diversity of the fungi.

30.4 How Do We Tell the Fungal Groups Apart?

In this section we'll examine representative species from each of the five major groups of fungi—chytrids, zygomycetes, glomeromycetes, ascomycetes, and basidiomycetes (**Table 30.1**). The chytrids and the zygomycetes are not monophyletic, but the last three groups are all clades; furthermore, together they form a clade called the **crown fungi**.

Chytrids are the only fungi with flagella

Chytrids
Zygomycetes
Glomeromycetes
Ascomycetes
Basidiomycetes

The **chytrids** are aquatic microorganisms once classified with the protists. However, morphological evidence (cell walls that consist primarily of chitin) and molecular evidence support their classification as basal fungi. In this book we use the term "chytrid" to refer to all basal fungi, but some mycologists reserve the term to apply to a particular clade within that group. There are fewer than 1,000 described species of chytrids.

Like the animals, the chytrids possess flagellated gametes. The retention of this character reflects the aquatic environment in which fungi first evolved. Chytrids are the only fungi that have flagella at any life cycle stage.

Chytrids are either parasitic (on organisms such as algae, mosquito larvae, and nematodes) or saprobic. Chytrids in the compound stomachs of foregut-fermenting animals such as cattle may be an exception, living in a mutualistic association with their hosts. Most chytrids live in freshwater habitats or in moist soil, but some are marine. Some chytrids are unicellular, others have rhizoids and still others have coenocytic hyphae (**Figure 30.15**). Chytrids reproduce both sexually and asexually, but they do not have a dikaryon stage.

Allomyces, a well-studied genus of chytrids, displays alternation of generations. A haploid *zoospore* (a spore with flagella) comes to rest on dead plant or animal material in water and germinates to form a small, multicellular haploid mycelium. That mycelium produces female and male *gametangia* (gamete cases; see Figure 30.13A). *Mitosis* in the gametangia results in the formation of haploid gametes, each with a single nucleus.

TABLE 30.1

A Classification of the Fungi

GROUP	COMMON NAME	FEATURES	EXAMPLES
Chytridiomycetes	Chytrids	Aquatic; zoospores have flagella	*Allomyces*
Zygomycetes	Conjugating fungi	Zygosporangium; no regularly occurring septa; usually no fleshy fruiting body	*Rhizopus*
Glomeromycetes	Mycorrhizal fungi	Form arbuscular mycorrhizae on plant roots	*Glomus*
Ascomycetes	Sac fungi	Ascus; perforated septa	*Neurospora*
Basidiomycetes	Club fungi	Basidium; perforated septa	*Armillariella*

Both female and male gametes have flagella. The motile female gamete produces a *pheromone*, a chemical signal that attracts the swimming male gamete. The two gametes fuse, and then their nuclei fuse to form a diploid zygote. Mitosis and cytokinesis in the zygote gives rise to a small, multicellular, diploid organism, which produces numerous diploid flagellated zoospores. These diploid zoospores disperse and germinate to form more diploid organisms. Eventually, the diploid organism produces thick-walled resting sporangia that can survive unfavorable conditions such as dry weather or freezing. Nuclei in the resting sporangia eventually undergo meiosis, giving rise to haploid zoospores that are released into the water and begin the cycle anew.

Zygomycetes reproduce sexually by fusion of two gametangia

Most **zygomycetes** ("conjugating fungi") are terrestrial, living on soil as saprobes or as parasites on insects, spiders, and other animals. They produce no cells with flagella, and only one diploid cell—the zygote—appears in the entire life cycle. Their hyphae are coenocytic. The mycelium of a zygomycete spreads over its substratum, growing forward by means of vegetative hyphae. Most zygomycetes do not form a fleshy fruiting structure; rather, the hyphae spread in an apparently random fashion, with occasional stalked **sporangiophores** reaching up into the air (**Figure 30.16**). These reproductive structures may bear one or many sporangia.

Chytrids
Zygomycetes
Glomeromycetes
Ascomycetes
Basidiomycetes

More than 700 species of zygomycetes have been described. A zygomycete that you may have seen is *Rhizopus stolonifer*, the black bread mold. *Rhizopus* reproduces asexually by producing many stalked sporangiophores, each bearing a single

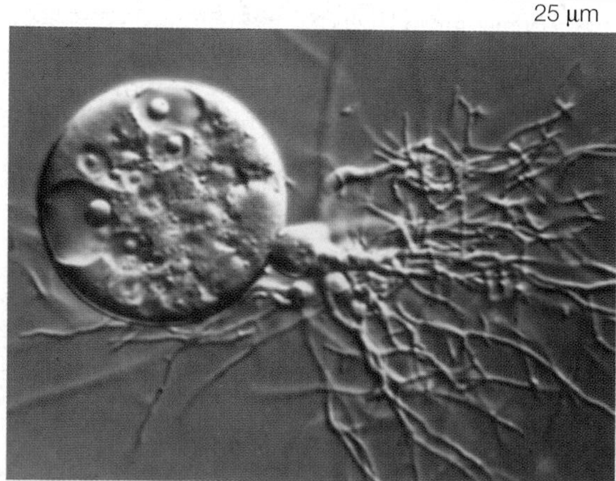

Chytriomyces hyalinus

30.15 A Chytrid Branched rhizoids emerge from the fruiting body of a mature chytrid.

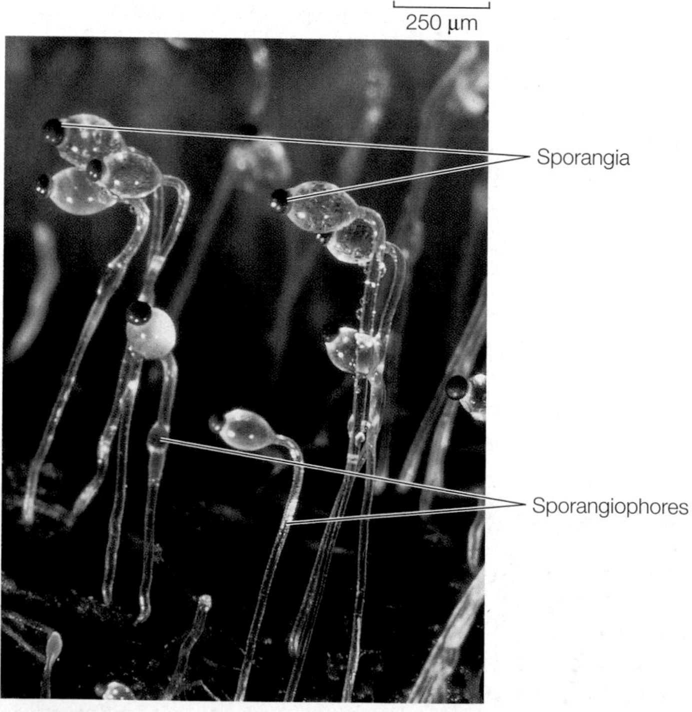

Pilobolus sp.

30.16 Zygomycetes Produce Sporangiophores These transparent structures are sporangiophores (spore-bearing hyphae) growing on decomposing animal dung. Sporangiophores grow toward the light and end in tiny sporangia, which these filamentous structures can eject as far as 2 meters. Animals ingest those sporangia that land on grass and then disseminate the spores in their feces.

sporangium containing hundreds of minute spores (see Figure 30.13B). As in other filamentous fungi, the spore-forming structure is separated from the rest of the hypha by a wall.

The zygomycete *Rhizopus microsporus* has been known as a pathogen of rice. However, it isn't the fungus that kills the rice, but a bacterial symbiont living inside the fungal cells. The bacterium produces rhizoxin, a rice-killing toxin that also happens to be a potent antitumor agent

---

Zygomycetes reproduce sexually when adjacent hyphae of two different mating types release pheromones, which cause them to grow toward each other. These hyphae produce gametangia, which fuse to form a **zygosporangium** (see Figure 30.13B). Sometime later, the gamete nuclei now contained within the zygosporangium fuse to form a single multinucleate *zygospore*. The zygosporangium develops a thick, multilayered wall that protects the zygospore. The highly resistant zygospore may remain dormant for months before its nuclei undergo meiosis and a sporangiophore sprouts. The sporangium contains the products of meiosis: haploid nuclei that are incorporated into spores. These spores disperse and germinate to form a new generation of haploid hyphae.

Glomeromycetes form arbuscular mycorrhizae

Chytrids
Zygomycetes
Glomeromycetes
Ascomycetes
Basidiomycetes

The **glomeromycetes** are strictly terrestrial. They associate with plant roots to form arbuscular mycorrhizae, so they are crucial to the plant world and thus to us (see Figure 30.10B). About half the fungi found in soils are glomeromycetes. Fewer than 200 species have been described, but 80 to 90 percent of all plants have associations with them. These fungi were originally classified among the zygomycetes, but molecular systematic studies have shown that they are a distinct clade, sister to the ascomycetes and basidiomycetes. The glomeromycetes, ascomycetes, and basidiomycetes constitute the clade known as the crown fungi.

The hyphae of glomeromycetes are coenocytic and do not form interconnected mycelia. These fungi use glucose from their plant partners as their primary energy source. They then convert the glucose to other, fungus-specific sugars that cannot return to the plant. Glomeromycetes reproduce asexually; there is no evidence that they reproduce sexually.

The next two fungal clades that we'll discuss are related groups with many similarities, including a dikaryon stage and septate hyphae. A key feature distinguishing between them is whether the sexual spores are borne inside a sac (in the ascomycetes) or on a pedestal (in the basidiomycetes).

The sexual reproductive structure of ascomycetes is the ascus

The **ascomycetes** are a large and diverse group of fungi found in marine, freshwater, and terrestrial habitats. There are approximately 60,000 known species of ascomycetes, about half of which are the fungal partners in lichens. Ascomycete hyphae are segmented by more or less regularly spaced septa. A pore in each septum permits extensive movement of cytoplasm and organelles (including nuclei) from one segment to the next.

Chytrids
Zygomycetes
Glomeromycetes
Ascomycetes
Basidiomycetes

The ascomycetes are distinguished by the production of sacs called **asci** (singular *ascus*), which contain sexually produced *ascospores* (see Figure 30.13C). The ascus is the characteristic sexual reproductive structure of the ascomycetes.

The ascomycetes can be divided into two broad groups, depending on whether the asci are contained within a specialized fruiting structure. Species that have this fruiting structure, the **ascocarp**, are collectively called **euascomycetes** ("true ascomycetes"); those without ascocarps are called **hemiascomycetes** ("half ascomycetes").

Most hemiascomycetes are microscopic, and many species are unicellular. Perhaps the best known are the ascomycete yeasts, especially baker's or brewer's yeast (*Saccharomyces cerevisiae*; see Figure 30.3). These yeasts are among the most important domesticated fungi. *S. cerevisiae* metabolizes glucose obtained from its environment to ethanol and carbon dioxide by fermentation. It forms carbon dioxide bubbles in bread dough and gives baked bread its light texture. Although they are baked away in bread making, the ethanol and carbon dioxide are both retained when yeast ferments grain into beer. Other yeasts live on fruits such as figs and grapes and play an important role in the making of wine.

Hemiascomycete yeasts reproduce asexually either by fission (splitting in half after mitosis) or by budding. Sexual reproduction takes place when two adjacent haploid cells of opposite mating types fuse. In some species, the resulting zygote buds to form a diploid cell population. In others, the zygote nucleus undergoes meiosis immediately; when this happens, the entire cell becomes an ascus. Depending on whether the products of meiosis then undergo mitosis, a yeast ascus contains either eight or four ascospores. The ascospores germinate to become haploid cells. Hemiascomycetes have no dikaryon stage.

The euascomycetes include the cup fungi (**Figure 30.17**). In most of these organisms the ascocarps are cup-shaped and can be as large as several centimeters across. The inner surfaces of the cups, which are covered with a mixture of vegetative hyphae and asci, produce huge numbers of spores. The edible ascocarps of some species, including morels and truffles, are regarded by humans as gourmet delicacies.

Truffles grow underground in a mutualistic association with the roots of oak trees. Europeans traditionally used pigs to find truffles because some truffles secrete a substance that has an odor similar to a pig's sex pheromone.

The euascomycetes also include many of the filamentous fungi known as molds. Many euascomycetes are parasites on flowering plants. Chestnut blight and Dutch elm disease are both caused by euascomycetes. Between its introduction to the United States

(A) *Morchella esculenta*

(B) *Sarcoscypha coccinea*

30.17 Two Cup Fungi (A) Morels, which have a spongelike ascocarp and a subtle flavor, are considered a delicacy by humans. (B) These brilliant red cups are the ascocarps of another cup fungus.

in the 1890s and 1940, the chestnut blight fungus destroyed the American chestnut as a commercial species. Before the blight, this species accounted for over half the trees in the great forests of the eastern United States. Another familiar story is that of the American elm, once considered the ideal street tree. Sometime before 1930, the Dutch elm disease fungus (first discovered in the Netherlands) was introduced into the United States on infected elm logs from Europe. Spreading rapidly—sometimes by way of connected root systems—the fungus destroyed great numbers of American elm trees.

Other euascomycete plant pathogens include the powdery mildews that infect cereal grains, lilacs, and roses, among many other plants. Mildews can be a serious problem to farmers and gardeners, and a great deal of research has focused on ways to control these agricultural pests.

Brown molds of the genus *Aspergillus* are important in some human diets. *A. tamarii* acts on soybeans in the production of soy sauce, and *A. oryzae* is used in brewing the Japanese alcoholic beverage sake. Some species of *Aspergillus* that grow on grains and on nuts such as peanuts and pecans produce extremely carcinogenic (cancer-inducing) compounds called *aflatoxins*. In the United States, moldy grain infected with *Aspergillus* is thrown out; In Africa, where food is scarcer, the grain gets eaten, moldy or not, and causes severe health problems.

The witchcraft trials that took place in Salem, Massachusetts in 1692 have been linked to the mold *Claviceps purpurea* (whose common name is ergot), a parasite of rye. The bizarre behavior of the "possessed" young women who accused their neighbors of witchcraft may have been a case of "ergotism"—the powerful hallucinogenic effects of toxins produced by *C. purpurea* that can result when a person eats bread made with contaminated rye flour.

Penicillium is a genus of green molds, of which some species produce the antibiotic penicillin, presumably for defense against competing bacteria. Two species, *P. camembertii* and *P. roquefortii*, are the organisms responsible for the characteristic strong flavors of Camembert and Roquefort cheeses, respectively.

The euascomycetes reproduce asexually by means of conidia that form at the tips of specialized hyphae (**Figure 30.18**). Small chains of conidia are produced by the millions and can survive for weeks in nature. The conidia are what give molds their characteristic colors. *Fusarium oxysporum*, the plant pathogen mentioned at the beginning of this chapter, is a euascomycete with no known sexual stage. It produces conidia in abundance.

The sexual reproductive cycle of euascomycetes includes the formation of a dikaryon, although this stage is relatively brief. Most euascomycetes form mating structures, some "female" and some "male" (see Figure 30.13C). Nuclei from a male structure on one hypha enter a female mating structure on a hypha of a compatible mating type. Dikaryotic *ascogenous* (ascus-forming) hyphae develop from the now dikaryotic female mating structure. The introduced nuclei divide simultaneously with the host nuclei. Eventually asci form at the tips of the ascogenous hyphae. Only with the formation of asci do the nuclei finally fuse. Both nuclear fusion and the subsequent meiosis of the resulting diploid nucleus take place within individual asci. The meiotic products are incorporated into ascospores that are ultimately shed by the ascus to begin the new haploid generation.

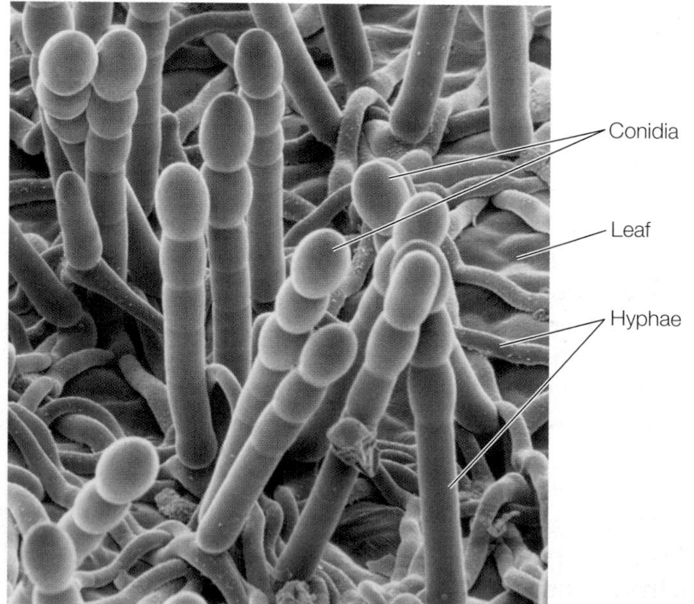

Erysiphe sp.

30.18 Conidia Chains of conidia are developing at the tips of specialized hyphae arising from this powdery mildew growing on a leaf.

(A) *Amanita muscaria*

(B) *Laetiporus sulphureus*

30.19 Basidiomycete Fruiting Structures The basidiocarps of the basidiomycetes are probably the most familiar structures produced by fungi. (A) These mushrooms were produced by a member of a highly poisonous genus, *Amanita*, that forms mycorrhizal relationships with trees. (B) This edible bracket fungus is parasitizing a tree.

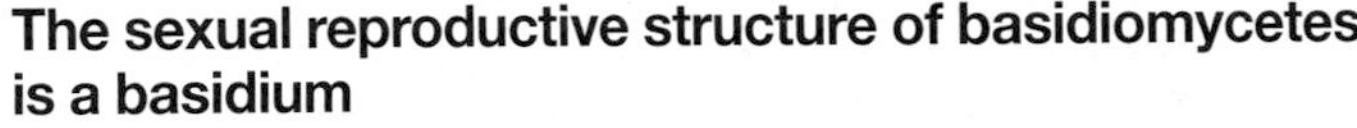

The sexual reproductive structure of basidiomycetes is a basidium

About 25,000 species of **basidiomycetes** (club fungi) have been described. Basidiomycetes produce some of the most spectacular fruiting structures found anywhere among the fungi. These fruiting structures, called **basidiocarps**, include puffballs (see Figure 30.7), some of which may be more than half a meter in diameter, mushrooms of all kinds, and the bracket fungi often encountered on trees and fallen logs in a damp forest. There are more than 3,250 species of mushrooms, including the familiar *Agaricus bisporus* you may enjoy on your pizza as well as poisonous species, such as members of the genus *Amanita* (**Figure 30.19A**). Bracket fungi (**Figure 30.19B**) do great damage to cut lumber and stands of timber. Some of the most damaging plant pathogens are basidiomycetes, including the rust fungi and the smut fungi that parasitize cereal grains. In contrast, other basidiomycetes contribute to the survival of plants as fungal partners in ectomycorrhizae.

Basidiomycete hyphae characteristically have septa with small, distinctive pores. The **basidium** (plural *basidia*), a swollen cell at the tip of a hypha, is the characteristic sexual reproductive structure of the basidiomycetes. It is the site of nuclear fusion and meiosis. Thus the basidium plays the same role in the basidiomycetes as the ascus does in the ascomycetes and the zygosporangium does in the zygomycetes.

As seen in Figure 30.13D, after nuclei fuse in the basidium, the resulting diploid nucleus undergoes meiosis, and the four resulting haploid nuclei are incorporated into haploid *basidiospores*, which form on tiny stalks on the outside of the basidium. A single basidiocarp of the common bracket fungus *Ganoderma applanatum* can produce as many as 4.5 *trillion* basidiospores in one growing season. Basidiospores typically are forcibly discharged from their basidia and then germinate, giving rise to haploid hyphae. As these hyphae grow, haploid hyphae of different mating types meet and fuse, forming dikaryotic hyphae, each cell of which contains two nuclei, one from each parent hypha. The dikaryotic mycelium grows and eventually, when triggered by rain or another environmental cue, produces a basidiocarp. The dikaryon stage may persist for years—some basidiomycetes live for decades or even centuries. This pattern contrasts with the life cycle of the ascomycetes, in which the dikaryon is found only in the stages leading up to formation of the asci.

30.4 RECAP

Five major taxonomic groups of fungi can be distinguished on the basis of differences in their life cycles and reproductive structures.

- What feature of the chytrids suggests an aquatic ancestor to the fungi? See p. 663 and Figure 30.13A
- What happens inside a zygosporangium? See p. 665 and Figure 30.13B
- What distinguishes the fruiting bodies of ascomycetes from those of basidiomycetes? See p. 667 and Figure 30.13C and D

Whether living on their own or in symbiotic associations, fungi have spread successfully over much of Earth since their origin from a protist ancestor. That ancestor also gave rise to the choanoflagellates and the animals, as we will describe in Chapter 31.

CHAPTER SUMMARY

30.1 How do fungi thrive in virtually every environment?

Fungi are heterotrophic organisms with **absorptive nutrition** and with chitin in their cell walls. Fungi have various nutritional modes: some are **saprobes**, others are **parasites**, and some are **mutualists**. **Yeasts** are unicellular fungi. See Web/CD Activity 30.1

The body of a multicellular fungus is a **mycelium**—a meshwork of **hyphae** that may be **septate** (having **septa**) or **coenocytic**. Review Figure 30.4

The cells of fungi live in intimate association with their surroundings, and are well adapted for absorptive nutrition in many environments. Many are parasitic plant pathogens, harvesting nutrients from plant cells by means of **haustoria**. Review Figure 30.5

Fungi are tolerant to hypertonic environments, and many are tolerant to low or high temperatures.

30.2 How are fungi beneficial to other organisms?

Saprobic fungi, as decomposers, make crucial contributions to the recycling of elements. Certain fungi have relationships with other organisms that are both **symbiotic** and **mutualistic**.

Some fungi associate with cyanobacteria or green algae to form **lichens**, which contribute to soil formation. Review Figure 30.9

Mycorrhizae are mutualistic associations of fungi with plant roots. They improve a plant's ability to take up nutrients. Review Figure 30.10

Endophytic fungi protect their plant hosts.

30.3 How do fungal life cycles differ from one another?

See Web/CD Tutorial 30.1

Most fungi reproduce sexually, although several modes of asexual reproduction are used by fungi. In many fungi, reproductive pairing is determined by two or more morphologically indistinguishable **mating types** rather than by two sexes.

Chytrids and some yeasts are the only fungi whose life cycle includes alternation of generations.

In the sexual reproduction of all other fungal groups, hyphae fuse, allowing "gamete" nuclei to be transferred. In the ascomycetes and basidiomycetes, **plasmogamy** precedes **karyogamy**, with the result that a **dikaryon** is formed. This $(n + n)$ dikaryotic condition is unique to the fungi. Review Figures 30.13C and D, Web/CD Activity 30.2

The **deuteromycetes** are a polyphyletic group of fungi for which no sexual stage has been observed and which have not yet been assigned to one of the other five fungal groups.

30.4 How do we tell the fungal groups apart?

There are five major fungal groups, of which the chytrids and the zygomycetes are paraphyletic. The glomeromycetes, ascomycetes, and basidiomycetes are each a clade, and together they form the **crown fungi** clade. Review Figure 30.2 and Table 30.1

The **chytrids**, a paraphyletic group of aquatic fungi, have flagellated gametes. Their life cycle displays alternation of generations. Review Figure 30.13A

The **zygomycetes** are a paraphyletic group with coenocytic hyphae. Gametangia fuse to form a **zygosporangium** within which a zygospore (a single cell containing many diploid nuclei) develops. Review Figure 30.13B

Glomeromycetes form arbuscular mycorrhizae with plant roots. They reproduce only asexually. Their hyphae are coenocytic.

Ascomycetes have septate hyphae; their sexual reproductive structures are **asci**. Many ascomycetes are partners in lichen associations. Most **hemiascomycetes** are yeasts. **Euascomycetes** produce fleshy fruiting bodies called **ascocarps**. The dikaryon stage in the ascomycete life cycle is relatively brief. Review Figure 30.13C

Basidiomycetes have septate hyphae. Their fruiting bodies are called **basidiocarps**, and their sexual reproductive structures are **basidia**. The dikaryon stage may last for years. Review Figure 30.13D

SELF-QUIZ

1. Which statement about fungi is *not* true?
   a. A multicellular fungus has a body called a mycelium.
   b. Hyphae are composed of individual mycelia.
   c. Many fungi tolerate highly hypertonic environments.
   d. Many fungi tolerate low temperatures.
   e. Some fungi are anchored to their substrate by rhizoids.
2. The absorptive nutrition of fungi is aided by
   a. dikaryon formation.
   b. spore formation.
   c. the fact that they are all parasites.
   d. their large surface area-to-volume ratio.
   e. their possession of chloroplasts.
3. Which statement about fungal nutrition is *not* true?
   a. Some fungi are active predators.
   b. Some fungi form mutualistic associations with other organisms.
   c. All fungi require mineral nutrients.
   d. Fungi can make some of the compounds that are vitamins for animals.
   e. Facultative parasites can grow only on their specific hosts.
4. Which statement about dikaryosis is *not* true?
   a. The cytoplasm of two cells fuses before their nuclei fuse.
   b. The two haploid nuclei are genetically different.
   c. The two nuclei are of the same mating type.
   d. The dikaryon stage ends when the two nuclei fuse.
   e. Not all fungi have a dikaryon stage.
5. Reproductive structures consisting of one or more photosynthetic cells surrounded by fungal hyphae are called
   a. ascospores.
   b. basidiospores.
   c. conidia.
   d. soredia.
   e. gametes.
6. The zygomycetes
   a. have hyphae without regularly occurring septa.
   b. produce motile gametes.
   c. form fleshy fruiting bodies.
   d. are haploid throughout their life cycle.
   e. have sexual reproductive structures similar to those of the ascomycetes.

7. Which statement about ascomycetes is *not* true?
   a. They include yeasts.
   b. They form reproductive structures called asci.
   c. Their hyphae are segmented by septa.
   d. Many of their species have a dikaryotic state.
   e. All have fruiting structures called ascocarps.
8. The basidiomycetes
   a. often produce fleshy fruiting structures.
   b. have hyphae without septa.
   c. have no sexual stage.
   d. produce basidia within basidiospores.
   e. form diploid basidiospores.
9. The deuteromycetes
   a. have distinctive sexual stages.
   b. are all parasitic.
   c. have "lost" some members to other fungal groups.
   d. include the ascomycetes.
   e. are never components of lichens.
10. Which statement about lichens is *not* true?
   a. They can reproduce by fragmentation of the vegetative body.
   b. They are often the first colonists in a new area.
   c. They render their environment more basic (alkaline).
   d. They contribute to soil formation.
   e. They may contain less than 10 percent water by weight.

FOR DISCUSSION

1. You are shown an object that looks superficially like a pale green mushroom. Describe at least three criteria (including anatomical and chemical traits) that would enable you to tell whether the object is a piece of a plant or a piece of a fungus.
2. Differentiate among the members of the following pairs of related terms:
   a. hypha/mycelium
   b. euascomycete/hemiascomycete
   c. ascus/basidium
   d. ectomycorrhiza/arbuscular mycorrhiza
3. For each type of organism listed below, give a single characteristic that may be used to differentiate it from the other, related organism(s) in parentheses.
   a. zygomycete (ascomycete)
   b. basidiomycete (deuteromycete)
   c. ascomycete (basidiomycete)
   d. baker's yeast (*Penicillium*)
4. Many fungi are dikaryotic during part of their life cycle. Why are dikaryons described as $n + n$ instead of $2n$?
5. If all the fungi on Earth were suddenly to die, how would the surviving organisms be affected? Be thorough and specific in your answer.
6. How might the first mycorrhizae have arisen?
7. What attributes might account for the ability of lichens to withstand the intensely cold environment of Antarctica? Be specific in your answer.
8. What factors must be taken into account in using fungi to combat agricultural pests?

FOR INVESTIGATION

Recall Poulsen and Boomsma's studies on how the fungi in ant "gardens" prevent the growth of competing fungi (see Figure 30.11). How would you test the hypothesis that a green mold that you found on an orange prevented the growth of other molds?

CHAPTER 31 Animal Origins and the Evolution of Body Plans

Dinosaur embryos illuminate evolution

The images of dinosaurs that are vividly etched in most people's minds include the frightening bipedal carnivores such as *Tyrannosaurus rex* on the one hand, and, on the other, the massive quadrupedal herbivores—the largest land animals ever to walk the planet. In fact, the fossil record indicates that all of the earliest dinosaur species were probably bipedal.

Quadrupedal dinosaurs achieved sizes far too great to be supported by hindlimbs alone, but paleontologists have long believed that these four-legged giants evolved from smaller, bipedal species. The recent discovery of the fossilized eggs of one of the earliest known dinosaur species has shed new light on this hypothesis.

Massospondylus was a bipedal, plant-eating dinosaur that lived in Africa about 190 million years ago. Fully grown adults of this genus measured about 28 meters from head to tail. Within the fossilized eggs of *Massospondylus carinatus*, researchers discovered remarkably well preserved, nearly fully developed embryos about 10 centimeters long.

Embryos reveal the form of an animal when it is born or hatched—and this form can be very different from that of the adult. The *M. carinatus* embryos revealed an animal that was born with a short tail, a neck that it carried horizontally, long forelimbs, a large head, and no teeth. The hatchlings would have been awkward little animals that moved around on all four limbs. To achieve the bipedal adult form, the necks and the hindlimbs of the young animals would need to have grown more rapidly than their front legs and heads.

By measuring *Massospondylus* fossils of various sizes, researchers determined that these dinosaurs walked on all fours while young, but walked and ran on their hindlimbs as adults. These findings suggest that the giant quadrupedal dinosaurs could have evolved from bipedal ancestors by means of rather straightforward changes in their growth pattern: If forelimb growth rate shifted slightly so that the forelimbs continued to grow as rapidly as the hindlimbs, an individual would retain the quadrupedal juvenile body form into adulthood.

The discovery of mature dinosaur embryos is just one example showing how the rapidly expanding fossil record is yielding valuable information about the forms of ancient animals and how they may have changed during their life cy-

Fossilized Embryos Help Explain Evolution This remarkably preserved embryo of *Massospondylus carinatus* revealed to paleontologists that the young of this bipedal dinosaur walked on all fours.

A Giant Quadruped Brachiosaurs lived around 130 million years ago, during the late Jurassic. This individual could have weighed as much as 80 tons. Its front legs were markedly longer than its hind legs—the reverse of the bipedal condition, but one that could have evolved from bipedal ancestors whose hatchlings were quadrupedal.

cle. Mutations in the genes governing development can result in dramatic changes in the forms of adult animals. Developmental and genetic knowledge, combined with fossil evidence, can help explain how a few fundamental body plans could have been modified to yield the remarkable variety of animal forms that we will describe in this and the following two chapters.

IN THIS CHAPTER we first enumerate the evidence that has led biologists to conclude that the animals are monophyletic and present a current phylogenetic tree of animals. We then describe how the diverse animal forms are based on modifications of a few key features within a small array of body plans. We will discuss how animals obtain food and describe their varied life cycles—how they are born, grow, disperse, and reproduce. Finally, we describe the members of several clades of structurally "simpler" animals.

CHAPTER OUTLINE

31.1 What Evidence Indicates the Animals Are Monophyletic?

We have no trouble identifying frogs, lizards, birds, and dogs as animals, but things are not always so simple. For example, many aquatic animals, such as sea anemones, look like plants. Indeed, many of these species were thought to be plants when they were first described. Sponges, for example, were not recognized as animals until 1765.

What traits distinguish the animals from the other groups of organisms?

- In contrast to the Bacteria, Archaea, and most microbial eukaryotes (see Chapters 26 and 27), all animals are *multicellular*. Animal life cycles feature complex patterns of *development* from a single-celled zygote into a multicellular adult.
- In contrast to most plants (see Chapters 28 and 29), all animals are *heterotrophs*. Animals are able to synthesize very few organic molecules from inorganic chemicals, so they must take in nutrients from their environment.
- The fungi (see Chapter 30) are also heterotrophs. In contrast to the fungi, however, animals use *internal* processes to break down materials from their environment into the organic molecules they need most. Most animals *ingest* food into an internal *gut* that is continuous with the outside environment, in which digestion takes place.
- In contrast to plants, most animals can *move*. Animals must move to find food or bring food to them. Animals have specialized *muscle* tissues that allow them to move, and many animal body plans are specialized for movement.

Animal monophyly is supported by gene sequences and morphology

The most convincing evidence that all the organisms considered to be animals share a common ancestor comes from their many shared derived molecular and morphological traits, most of which we have described in earlier chapters:

- Many gene sequences, such as the ribosomal RNA genes (see Section 26.4), support the monophyly of animals.

- Animals display similarities in the organization and function of their Hox genes (see Chapter 20).
- Animals have unique types of junctions between their cells (tight junctions, desmosomes, and gap junctions; see Figure 5.7).
- Animals have a common set of *extracellular matrix* molecules, including collagen and proteoglycans (see Figure 4.25).

Although there are animals in a few clades that lack one or another of these *synapomorphies*, these species apparently once possessed the traits and lost them during their later evolution.

The ancestor of the animal clade was probably a colonial flagellated protist similar to existing colonial choanoflagellates (see Figure 27.30A). The most reasonable current scenario postulates a choanoflagellate lineage in which certain cells within the colony began to be specialized—some for movement, others for nutrition, others for reproduction, and so on. Once this *functional specialization* had begun, cells could have continued to differentiate. Coordination among groups of cells could have improved by means of specific regulatory molecules that guided the differentiation and migration of cells in developing embryos. Such coordinated groups of cells eventually evolved into the larger and more complex organisms that we call animals.

More than a million animal species have been named and described, and there are doubtless millions of living species that have yet to be named. The synapomorphies that indicate animal monophyly cannot be used to infer evolutionary relationships among animals, because nearly all animals have them. Clues to the evolutionary relationships among animal groups thus must be sought in *derived traits* that are found in some groups but not in others. Such characteristics can be found in fossils, in patterns of embryonic development, in the morphology and physiology of living animals, in the structure of animal molecules, and in the genomes of animals (for example, in mitochondrial and ribosomal RNA genes).

Developmental patterns show evolutionary relationships among animals

Differences in patterns of embryonic development traditionally provided some of the most important clues to animal phylogeny, although analyses of gene sequences are now showing that some developmental patterns are more evolutionarily labile than previously thought. We will discuss the details of animal development in Chapter 43; here we will describe the basic developmental patterns that vary among the animal groups shown in **Figure 31.1**.

The first few cell divisions of a zygote are known as *cleavage*. In general, the number of cells in the embryo doubles with each cleavage. A number of different **cleavage patterns** exist among animals.

As will be described in Section 43.1, cleavage patterns are influenced by the configuration of the *yolk*, the nutritive material that

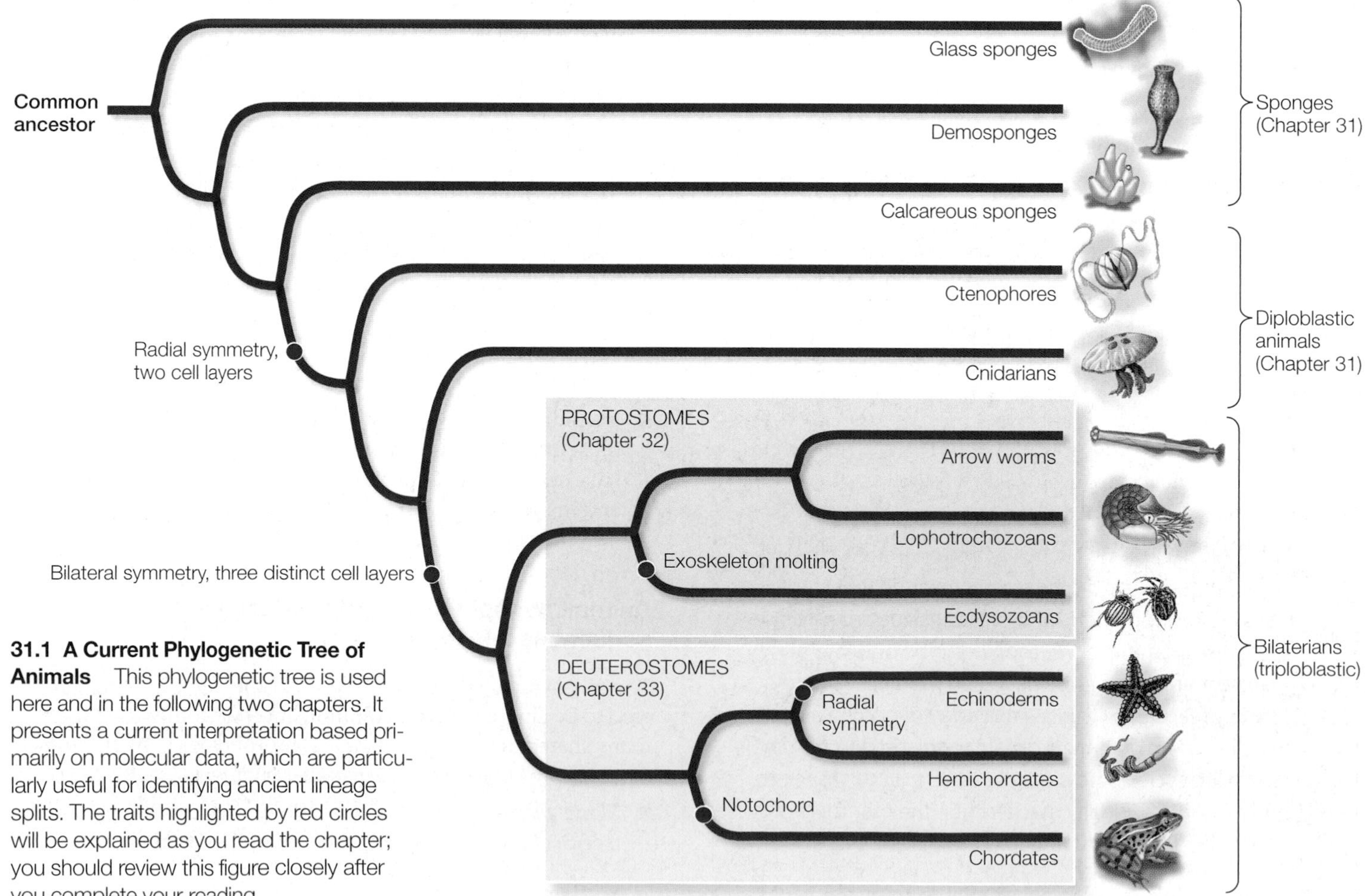

31.1 A Current Phylogenetic Tree of Animals This phylogenetic tree is used here and in the following two chapters. It presents a current interpretation based primarily on molecular data, which are particularly useful for identifying ancient lineage splits. The traits highlighted by red circles will be explained as you read the chapter; you should review this figure closely after you complete your reading.

(A)

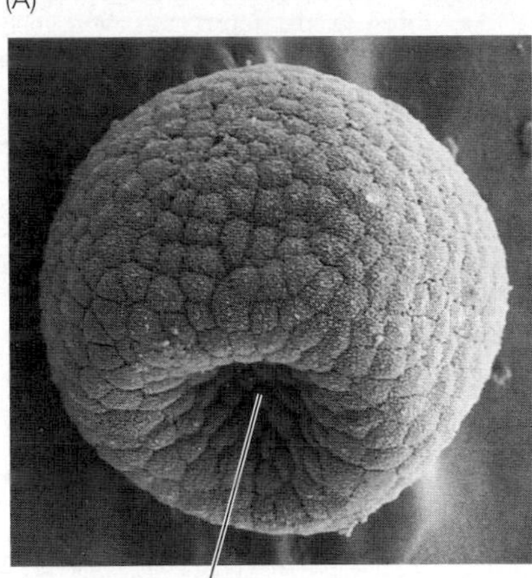

(B)

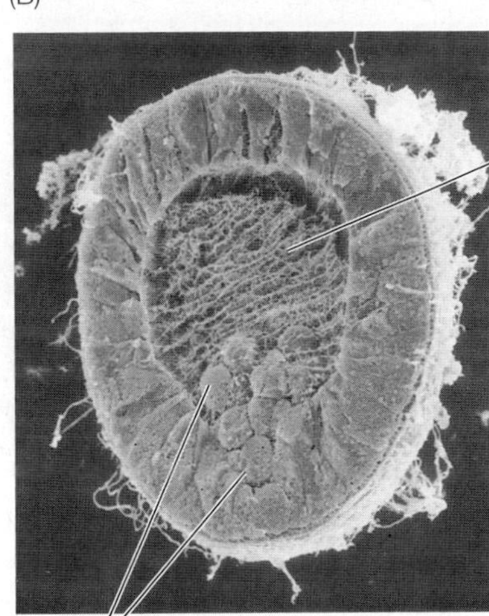

31.2 Gastrulation Illuminates Evolutionary Relationships (A) The blastopore is clear in this scanning electron micrograph of a sea urchin gastrula. Because sea urchins (echinoderms) are deuterostomes, this blastopore will eventually become the anal end of the animal's gut. (B) In this cross section through a later-stage sea urchin gastrula, the cells are beginning to look different from one another. The molecules of the extracellular matrix guide cell movement.

nourishes the growing embryo. In reptiles, for example, the presence of a large body of a cellular yolk within the fertilized egg creates an *incomplete* cleavage pattern in which the dividing cells form an embryo on top of the yolk mass (see Figure 43.3C). In echinoderms such as sea urchins, small yolk particles are evenly distributed throughout the egg cytoplasm, so cleavage is *complete,* with the fertilized egg cell dividing in an even pattern known as *radial cleavage* (see Figure 43.3A). Radial cleavage is the ancestral condition for eumetazoans, so it is found among many protostomes and diploblastic animals as well as deuterostomes. *Spiral cleavage,* a complicated derived permutation of radial cleavage, is found among many lophotrochozoans, such as earthworms and clams. Lophotrochozoans with spiral cleavage are thus sometimes known as *spiralians.* The early branches of the ecdysozoans have radial cleavage, although most ecdysozoans have an idiosyncratic cleavage pattern that is neither radial nor spiral in organization.

During the early development of most animals, distinct layers of cells form. These cell layers differentiate into specific organs and organ systems as development continues. The embryos of **diploblastic** animals have only two of these cell layers: an outer *ectoderm* and an inner *endoderm.* The embryos of **triploblastic** animals have, in addition to ectoderm and endoderm, a third distinct cell layer, the *mesoderm,* which lies between the ectoderm and the endoderm. The existence of three cell layers is a synapomorphy of triploblastic animals, whereas the paraphyletic diploblastic animals (ctenophores and cnidarians) exhibit the ancestral condition.

During early development in many animals, a hollow ball one cell thick indents to form a cup-shaped structure. This process is known as *gastrulation.* The opening of the cavity formed by this indentation is called the *blastopore* (**Figure 31.2**). The pattern of development after formation of the blastopore has been used to divide the triploblastic animals into two major groups. Among members of the first group, the **protostomes** (Greek, "mouth first"), the mouth arises from the blastopore; the anus forms later. This appears to be the derived condition. Among the **deuterostomes** ("mouth second"), the blastopore becomes the anus; the mouth forms later. This is thought to be the ancestral condition. We now know that the developmental patterns of animals are more varied than suggested by this simple dichotomy, but the protostomes and deuterostomes are still recognized as distinct animal clades based upon sequence similarities of their genes.

> The embryologist Lewis Wolpert once stated that "It is not birth, marriage or death, but gastrulation, which is truly the most important time in your life." Aside from its significance to the individual organism, the process of gastrulation is central to our understanding of evolutionary relationships among the animals.

31.1 RECAP

The animals are thought to be monophyletic because they share many derived traits, including multicellularity, mobility, and a heterotrophic lifestyle based on the ingestion of outside nutrients. Evolutionary relationships among animals are inferred from fossils and from molecular and developmental traits that are shared by different groups of animals.

- Do you understand why the traits biologists use to determine evolutionary relationships among animals must differ from those they use to infer that all animals share a common ancestor? See p. 672
- Can you describe the differences between diploblastic and triploblastic embryos, and between protostomes and deuterostomes? See p. 673

We devote Chapter 32 to the protostomes and Chapter 33 to the deuterostomes. Later in this chapter we will describe several groups of animals with relatively simple structural organization. Millions of species of triploblastic animals with complex structures eventually evolved from these ancestral states.

31.2 What Are the Features of Animal Body Plans?

The general structure of an animal, the arrangement of its organ systems, and the integrated functioning of its parts are referred to as its **body plan**. Although animal body plans are extremely varied, they can be seen as variations on four key features:

- The *symmetry* of the body
- The structure of the *body cavity*
- The *segmentation* of the body
- *External appendages* that move the body

All of these features affect the way in which an animal moves and interacts with its environment. These four attributes vary across a number of basic animal body plans.

As we learned in Chapter 20, the *regulatory genes* that govern the development of body symmetry, body cavities, segmentation, and appendages are widely shared among the different animal groups. Thus we might expect animals to share body plans.

Most animals are symmetrical

The overall shape of an animal can be described by its **symmetry**. An animal is said to be *symmetrical* if it can be divided along at least one plane into similar halves. Animals that have no plane of symmetry are said to be *asymmetrical*. Many sponges are asymmetrical, but most other animals have some kind of symmetry, which is governed by the expression of regulatory genes.

The simplest form of symmetry is **spherical symmetry**, in which body parts radiate out from a central point. An infinite number of planes passing through the central point can divide a spherically symmetrical organism into similar halves. Spherical symmetry is widespread among unicellular protists, but most animals possess other forms of symmetry.

In organisms with **radial symmetry**, there is one main axis around which body parts are arranged. Two animal groups—ctenophores and cnidarians—are composed primarily of radially symmetrical animals (**Figure 31.3A**). A perfectly radially symmetrical animal can be divided into similar halves by any plane that contains the main axis. However, most radially symmetrical animals—including the adults of echinoderms such as sea stars and sand dollars—are slightly modified so that fewer planes can divide them into identical halves. Many radially symmetrical animals are sessile (sedentary). Others move slowly, but can move equally well in any direction.

Bilateral symmetry is characteristic of animals that move in one direction. A bilaterally symmetrical animal can be divided into mirror-image (left and right) halves by a single plane that passes through the midline of its body (**Figure 31.3B**). This plane runs from the tip, or **anterior** of the body to its tail, or **posterior**.

A plane at right angles to the midline divides the body into two dissimilar sides. The back of a bilaterally symmetrical animal is its **dorsal** surface; the belly, which contains the mouth, is its **ventral** surface.

Bilateral symmetry is strongly correlated with **cephalization**, which is the concentration of sensory organs and nervous tissues

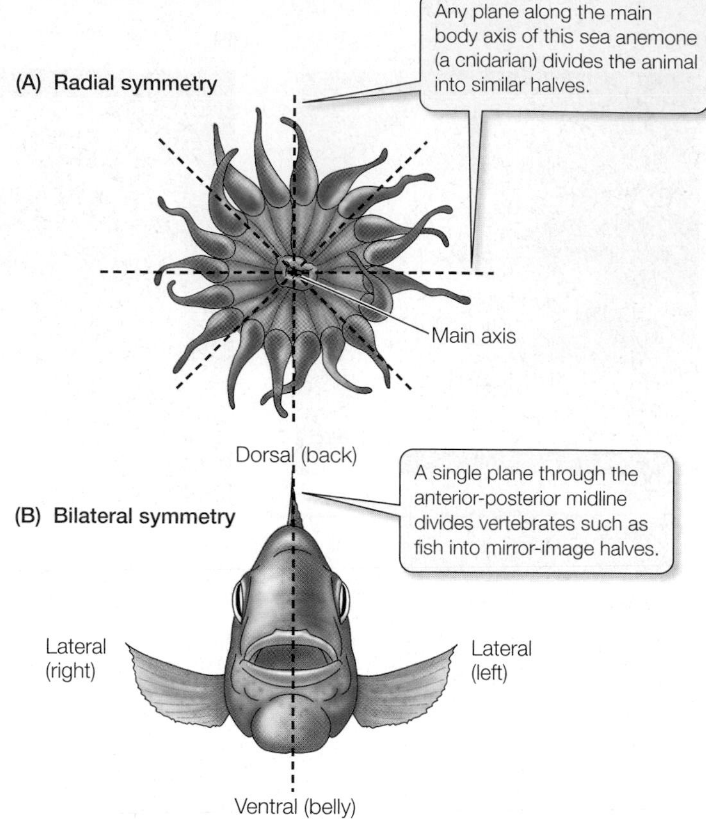

(A) Radial symmetry
Any plane along the main body axis of this sea anemone (a cnidarian) divides the animal into similar halves.
Main axis
(B) Bilateral symmetry
Dorsal (back)
A single plane through the anterior-posterior midline divides vertebrates such as fish into mirror-image halves.
Lateral (right)
Lateral (left)
Ventral (belly)

31.3 Body Symmetry Most animals are either radially or bilaterally symmetrical.

in a head at the anterior end of the animal. Cephalization has been evolutionarily favored because the anterior end of a bilaterally symmetrical animal typically encounters new environments first.

The structure of the body cavity influences movement

Animals can be divided into three types—*acoelomate, pseudocoelomate,* and *coelomate*— based on the presence and structure of an internal, fluid-filled **body cavity**. The structure of an animal's body cavity strongly influences the ways in which it can move.

Acoelomate animals such as flatworms lack an enclosed, fluid-filled body cavity. Instead, the space between the gut (derived from endoderm) and the muscular body wall (derived from mesoderm) is filled with masses of cells called *mesenchyme* (**Figure 31.4A**). These animals typically move by beating cilia.

Body cavities come in two types. Both types lie between the ectoderm and the endoderm; they are differentiated by their relationship to the mesoderm.

- **Pseudocoelomate** animals have a body cavity called a *pseudocoel*, a fluid-filled space in which many of the internal organs are suspended. A pseudocoel is enclosed by muscles (mesoderm) only on its outside; there is no inner layer of mesoderm surrounding the internal organs (**Figure 31.4B**).
- **Coelomate** animals have a *coelom*, a body cavity that develops within the mesoderm. It is lined with a layer of muscular tissue called the *peritoneum*, which also surrounds the internal

(A) Acoelomate (flatworm)

Gut (endoderm)
Muscle layer (mesoderm)
Ectoderm
Mesenchyme
Acoelomates do not have enclosed body cavities.

(B) Pseudocoelomate (roundworm)

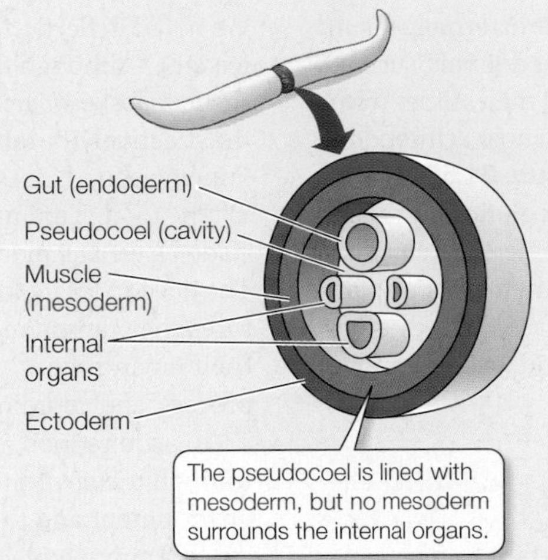

(C) Coelomate (earthworm)

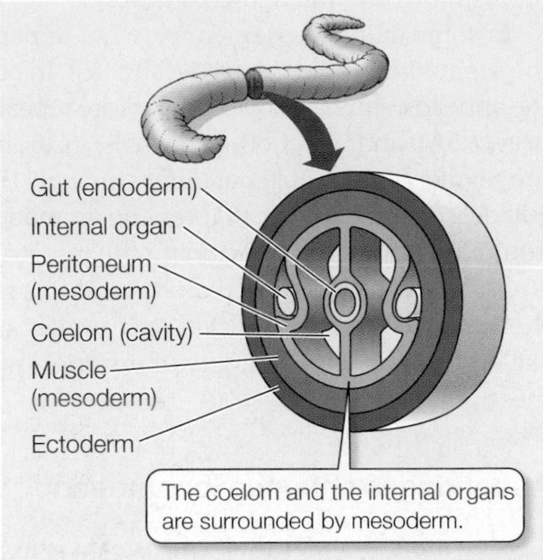

31.4 Animal Body Cavities (A) Acoelomates do not have enclosed body cavities. (B) Pseudocoelomates have a body cavity enclosed by only one layer of mesoderm, which lies outside the cavity. (C) Coelomates have a peritoneum surrounding the internal organs.

organs. The coelom is thus enclosed on both the inside and the outside by mesoderm (**Figure 31.4C**). A coelomate animal has better control over the movement of the fluids in its body cavity than a pseudocoelomate animal does.

The body cavities of many animals function as **hydrostatic skeletons**. Fluids are relatively incompressible, so when the muscles surrounding them contract, they move to another part of the cavity. If the body tissues around the cavity are flexible, fluids squeezed out of one region can cause some other region to expand. The moving fluids can thus move specific body parts. (You can see how a hydrostatic skeleton works by watching a snail emerge from its shell.) An animal with both *circular muscles* (encircling the body cavity) and *longitudinal muscles* (running along the length of the body) has even greater control over its movement.

Although the hydrostatic function of fluid-filled body cavities is important, most animals also have hard skeletons that provide protection and facilitate movement. Muscles are attached to those firm structures, which may be inside the animal or on its outer surface (in the form of a shell or cuticle).

Segmentation improves control of movement

Many animals have bodies that are divided into segments. **Segmentation** facilitates specialization of different body regions. Segmentation also allows an animal to alter the shape of its body in complex ways and to control its movements precisely. If an animal's body is segmented, muscles in each individual segment can change the shape of that segment independently of the others. In only a few segmented animals is the body cavity separated into discrete compartments, but even partly separated compartments allow better control of movement. As we will see, segmen-

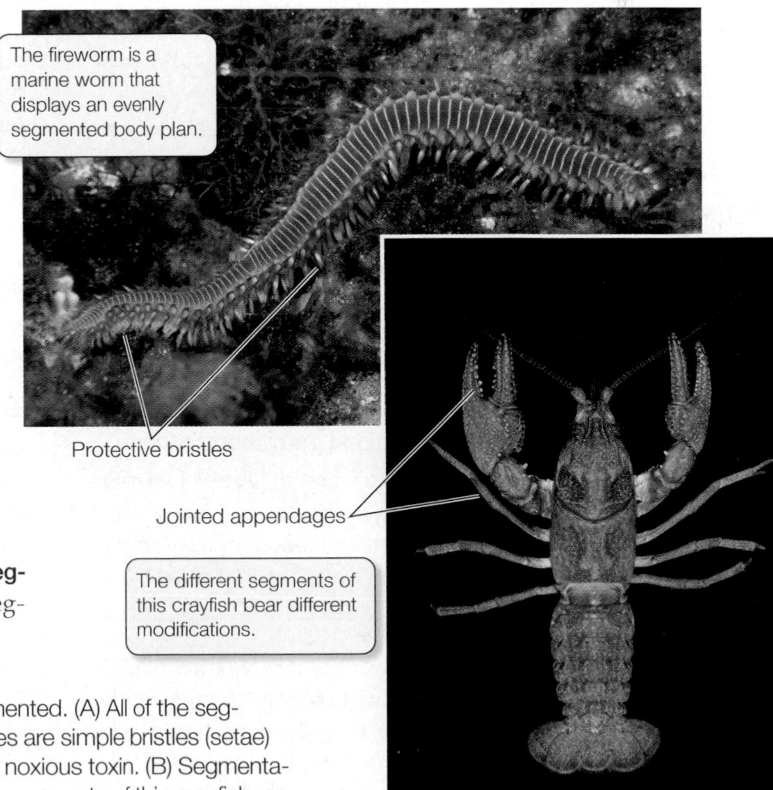

31.5 Segmentation The body cavities of many animals are segmented. (A) All of the segments of this marine fireworm, an annelid, are similar. Its appendages are simple bristles (setae) that, in this animal, serve mainly for protection—the setae contain a noxious toxin. (B) Segmentation allows the evolution of differentiation among the segments. The segments of this crayfish, an arthropod, differ in their form, function, and the appendages they bear.

tation evolved independently several different times in both the protostomes and the deuterostomes.

In some animals, segments are not apparent externally (as in the segmented vertebrae of vertebrates). In other animals, such as the annelids, similar body segments are repeated many times (**Figure 31.5A**); and in yet other animals, including most arthropods, the segments are visible but differ strikingly (**Figure 31.5B**). As we will describe in the next chapter, the dramatic evolutionary radiation of the arthropods (including the insects, spiders, centipedes, and crustaceans) was based on changes in a segmented body plan that features muscles attached to the inner surface of an external skeleton, along with a variety of external appendages that move these animals.

Appendages enhance locomotion

Getting around under their own steam is important to many animals. It allows them to obtain food, to avoid predators, and to find mates. Even some sessile species, such as sea anemones, have larval stages that use cilia to swim, thus increasing the animal's chances of finding a suitable habitat to settle on.

Appendages that project externally from the body greatly enhance an animal's ability to move around. Many echinoderms, including sea urchins and sea stars, have myriad *tube feet* that allow them to move slowly across the substratum. Highly controlled, rapid movement is greatly enhanced in animals whose appendages have become modified into specialized *limbs*. In two animal groups, the arthropods and the vertebrates, the presence of *jointed limbs* has been a prominent factor in their evolutionary success (see Figure 31.5B).

In several independent instances—among the arthropod insects, the pterosaurs, the birds, and the bats—body plans emerged in which limbs were modified into wings, allowing animals to take to the air (see Figure 20.12).

31.2 RECAP

The body plans of animals are all variations on patterns of symmetry, body cavities, segmentation, and appendages. All four can have bearing on movement and locomotion, which are important aspects of the animal way of life.

- Can you describe the main types of symmetry found in animals? Do you see how an animal's symmetry can influence the way it moves? See p. 674 and Figure 31.3
- Can you explain several ways in which body cavities and segmentation improve control over movement? See pp. 674–676

Many of the modifications to their body plans involve ways of finding, capturing, and processing food. Evolutionary changes in the symmetry, body cavities, appendages, and segmentation of animals have played a key role in enabling them to obtain food from their environments, as well as helping them avoid becoming food for other animals.

31.3 How Do Animals Get Their Food?

We noted in Section 31.1 that animals are heterotrophs, or "other-feeders." Although there are many animals that rely on photosynthetic *endosymbionts* for nutrition (see Figure 4.15B), most animals must actively obtain an outside source of nutrition, otherwise known as food.

The food of animals includes most other members of their own clade as well as members of all other groups of living organisms. The need to locate food has favored the evolution of sensory structures that can provide animals with detailed information about their environment, as well as nervous systems that can receive, process, and coordinate that information.

To acquire food, animals must expend energy, either to move through the environment to where food is located or to move the environment and the food it contains to them. Animals that can move from one place to another are **motile**; animals that stay in one place are **sessile**.

The feeding strategies that animals use fall into a few broad categories:

- *Filter feeders* capture small organisms delivered to them by their environment.
- *Herbivores* eat plants or parts of plants.
- *Predators* capture and eat other animals that typically are relatively large in relation to themselves.
- *Parasites* live in or on other organisms from which they obtain energy and nutrients.
- *Detritivores* actively feed on dead organic material.

Each of these modes of feeding can be found in many different animal groups, and none of them is limited to a single group. In addition, individuals of some species may employ more than one feeding strategy, and some animals employ entirely different feeding strategies at different points in their life cycle. The constant and ongoing need to obtain food, the variety of nutrient sources available in any given environment, and the necessity of competing with other animals to get food means that a variety of feeding strategies can be found among all the major animal groups.

Filter feeders capture small prey

Air and water often contain small organisms and organic molecules that are potential food for animals. Moving air and water may carry those items to an animal that positions itself in a good location. These **filter feeders** then use some kind of straining device to filter the food from the environment. Many sessile aquatic animals rely on water currents to bring prey to them (**Figure 31.6A**).

Baleen whales—including the blue whale, the largest animal on Earth—rely on some of the smallest organisms for their nutrition. These whales filter tons of tiny, often microscopic zooplankton out of the ocean water that passes through comblike baleen plates in their upper jaws.

(A)

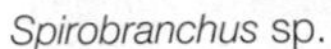

Spirobranchus sp.

(B)

Phoenicopterus ruber

31.6 Filter Feeding Strategies (A) Sessile marine filter feeders such as this "Christmas tree worm," a polychaete, allow the ocean currents to bring their food—plankton—to them. (B) The greater flamingo of South America is a motile filter feeder, using its appendages (legs) to stir up mud as it wades through ocean lagoons and salty lakes. The birds then use their beaks to strain small organisms out of the muddy mixture.

Motile filter feeders bring the nutrient-containing medium to them. The serrated beak of the flamingo, for example, filters small organisms out of the muddy mixture it picks up as it wades through shallow water (**Figure 31.6B**).

Some sessile filter feeders expend energy to move water past their food-capturing devices. Sponges, for example, bring water into their bodies by beating the flagella of their specialized feeding cells, called **choanocytes** (**Figure 31.7**). It is these flagellated feeding cells of sponges that link the animals with choanoflagellate protists and the fungi. The choanoflagellates and the animals are most closely related to the fungi, and these three groups form a clade known as the *opisthokonts* (see p. 605).

Herbivores eat plants

Animals that eat plants are referred to as **herbivores**. An individual plant has many different structures—leaves, wood, sap, flowers, fruits, nectar, and seeds—that animals can consume. Not surprisingly, then, many different kinds of herbivores may feed on a single kind of plant, consuming different parts of the plant or eating the same part in different ways (**Figure 31.8**). An individual animal that is captured by a predator is likely to die, but herbivores often feed on plants without killing them.

Animals do not need to expend energy subduing and killing plants. However, they do need to digest them. This can be difficult for terrestrial herbivores, because the dominant land plants tend to have many different kinds of tissues, many of which are tough or fibrous. Plant tissues may also contain chemicals that must be detoxified before they can be ingested. Herbivorous animals typically have long, complex guts to accomplish the tasks involved in digesting plants (see Section 50.2).

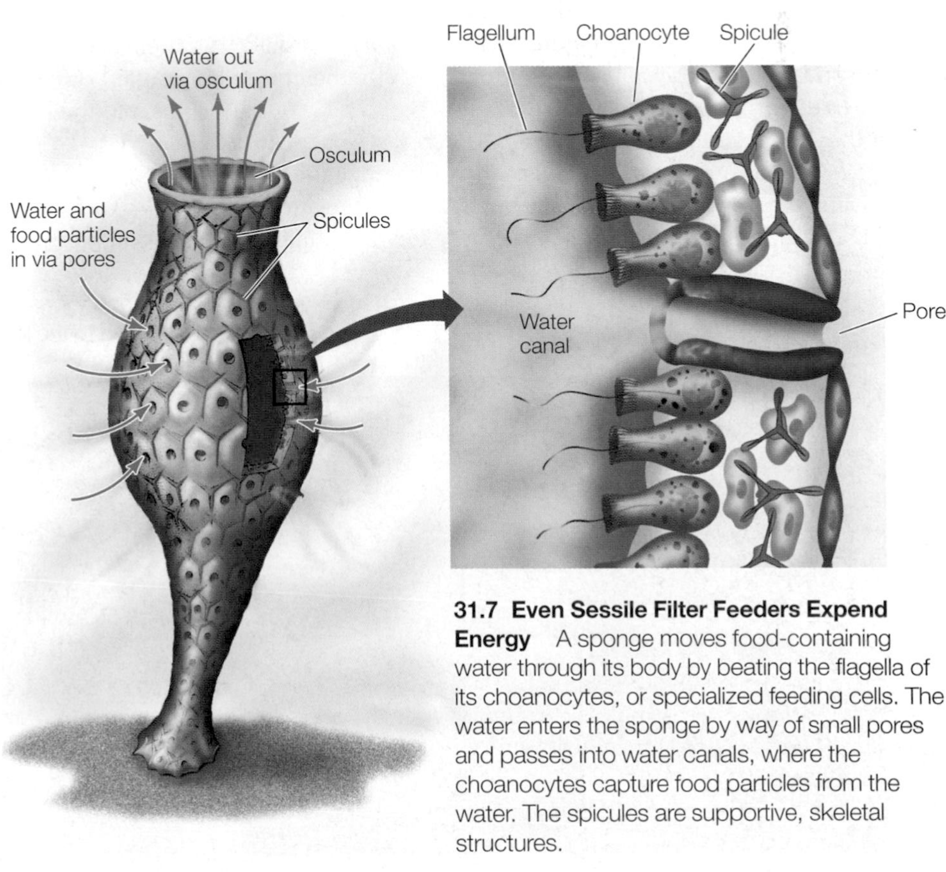

31.7 Even Sessile Filter Feeders Expend Energy A sponge moves food-containing water through its body by beating the flagella of its choanocytes, or specialized feeding cells. The water enters the sponge by way of small pores and passes into water canals, where the choanocytes capture food particles from the water. The spicules are supportive, skeletal structures.

31.8 A Single Plant Can Feed Many Different Herbivores Many species of insects feed on a single species of willow, each species consuming different tissues in different ways.

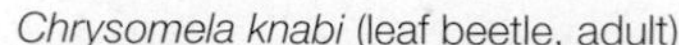

Chrysomela knabi (leaf beetle, adult)

Nematus sp. (willow sawfly, larvae)

Papilio sp. (tiger swallowtail butterfly, larva)

Salix sericea (silky willow)

Plagiodera versicolora (leaf beetle, larva)

Predators capture and subdue large prey

Predators possess features that enable them to capture and subdue relatively large animals (referred to as their **prey**). Many vertebrate predators have sensitive sensory organs that enable them to locate prey as well as sharp teeth or claws that allow them to capture and subdue large prey (**Figure 31.9**). Predators may stalk and pursue their prey, or wait (often camouflaged) for their prey to come to them.

Another weapon of predators (as well as of prey; see page 486) is toxins. We are all aware of the dangers of encountering the toxins of a venomous snake. Cnidarians (jellyfish and their relatives), which are among the simplest of animals, use toxins to capture and subdue prey that are much larger and more complex than themselves. Their tentacles are covered with specialized cells that contain stinging organelles called **nematocysts**, which inject toxins into their prey (**Figure 31.10**).

31.9 Tooth and Claw (A) The teeth of this Kodiak brown bear are adapted to an omnivorous diet that includes fish. (B) The appendages (legs and wings) of the bald eagle, along with its strong beak, are adapted to the life of a predatory hunter.

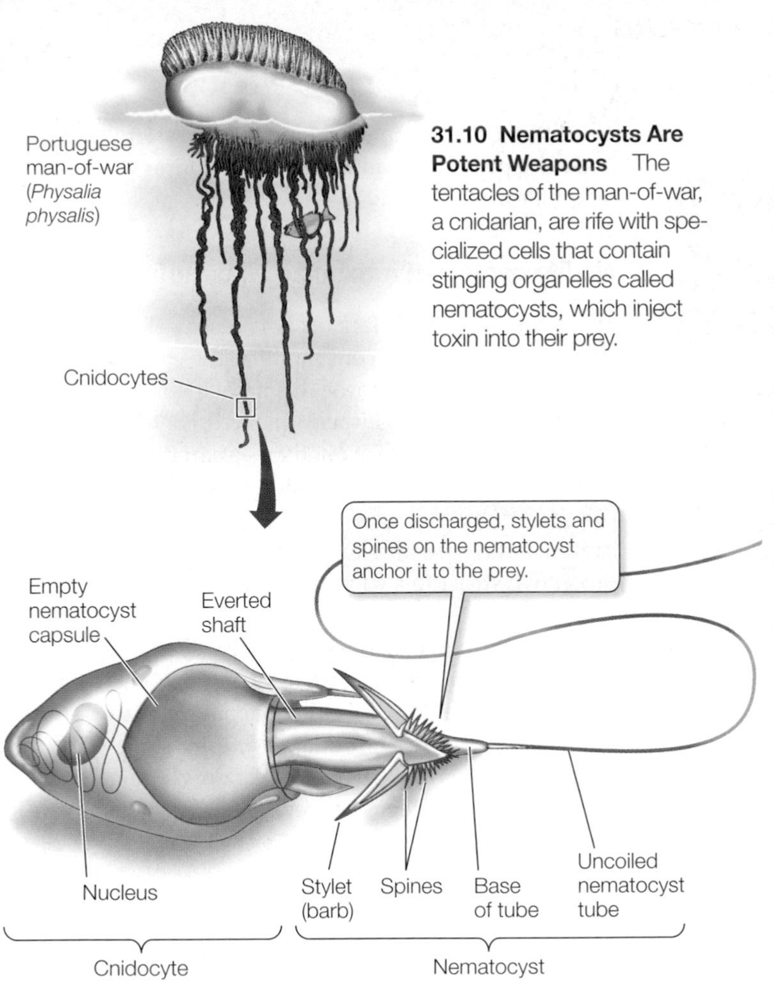

31.10 Nematocysts Are Potent Weapons The tentacles of the man-of-war, a cnidarian, are rife with specialized cells that contain stinging organelles called nematocysts, which inject toxin into their prey.

Cone-shelled snails capture large prey by injecting them with toxins from a highly modified tooth. To date more than 80 U.S. patents have been awarded for various medical uses of cone shell toxins.

Many animals, such as raccoons and humans, eat both plants and other animals. These animals are called **omnivores**. There are also many animals whose diet differs at different life stages, such as the many songbirds that eat fruit or seeds as adults but feed their young on insects.

Parasites live in or on other organisms

Animals that live in or on another organism—called a *host*—and obtain their nutrients by consuming parts of that organism are called **parasites**. Most animal parasites are much smaller than their hosts, and many parasites can consume parts of their host without killing it. However, they first must overcome the host's defenses. Parasites often have complex life cycles, that rely on multiple hosts, as we detail in the next section.

Parasites that live inside their hosts are called *endoparasites* and are often morphologically very simple. They can often function without a digestive system because they absorb food directly from the host's gut or bodily tissues. Many flatworms are endoparasites of humans and other mammals, as will be described in Chapter 32.

Parasites that live outside their hosts—*ectoparasites*—are generally more morphologically complex than endoparasites. They have digestive tracts and mouthparts that enable them to pierce the host's tissues or suck on their body fluids. Fleas and ticks are widely known ectoparasitic arthropods that many humans have unfortunately experienced.

31.3 RECAP

Animals have many ways of acquiring food. Filter feeders strain food particles from the water or air. Herbivores have digestive adaptations that allow them to eat plants, while predators are physically adapted to capture and subdue other animals (prey) and consume them. Parasites obtain their nutrition from a host organism.

- Can you describe the different types of adaptations that are necessary for animals that eat plants as opposed to the adaptations needed for a predatory lifestyle? See pp. 677–678
- Looking at the brief overview of some of the diverse feeding modes of animals presented here, and using whatever you already know about different animals, how useful do you think feeding behavior would be as a criterion for grouping animals into broader categories?

As an animal grows from a single cell into a larger, more complex adult, its body structure, its diet, and the environment in which it lives may all change. In the next section we describe some animal life cycles and discuss why they are so varied.

31.4 How Do Animal Life Cycles Differ?

The **life cycle** of an animal encompasses its embryonic development, birth, growth to maturity, reproduction, and death. During its life an individual animal ingests food, grows, interacts with other individuals of the same and other species, and reproduces.

In some groups of animals, newborns are similar in many ways to adults (a pattern called *direct development*). Newborns of most species, however, differ dramatically from adults. An immature life cycle stage that has a form different from that of the adult is called a **larva** (plural *larvae*). Some of the most striking life cycle changes are found among insects such as beetles, flies, moths, butterflies, and bees, which undergo radical changes (called *metamorphosis*) between their larval and adult stages (**Figure 31.11**). In these animals, one stage may be specialized for feeding and the other for reproduction. Adults of most moth species, for example, do not eat. Alternatively, individuals of all life cycle stages may eat, but what they eat may change. Butterfly larvae, known as *caterpillars*, eat leaves and flowers; most adult butterflies eat only nectar. Having different life cycle stages that are specialized for different activities may increase the efficiency with which the animal performs particular tasks.

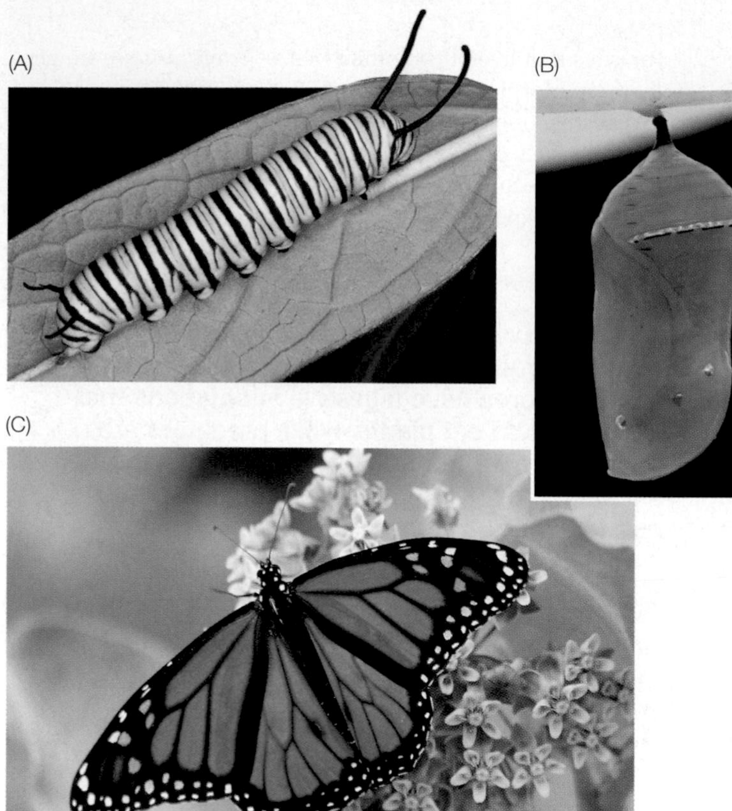

31.11 A Life Cycle with Metamorphosis (A) The larval stage (caterpillar) of the monarch butterfly, *Danaus plexippus*, is specialized for feeding. (B) The pupa is the stage during which the transformation to the adult form occurs. (C) The adult butterfly is specialized for dispersal and reproduction.

All life cycles have at least one dispersal stage

At some time during its life, an animal moves, or is moved, so that it does not die exactly where it was born. Such movement is called **dispersal**.

Animals that are sessile as adults typically disperse as eggs or larvae. This pattern is common among sessile marine animals, most of which discharge their small eggs and sperm into the water, where fertilization takes place. A larva soon hatches and floats freely in the plankton as it filters small prey from the water.

Many animals that live on the seafloor, including polychaete worms and mollusks, have a common larval form, the **trochophore** (**Figure 31.12A**). Some other marine animals, such as crustaceans, have a different, bilaterally symmetrical larval form, called a **nauplius** (**Figure 31.12B**). Both types of larvae feed for some time in the plankton before settling on a substratum and transforming into adults. Larvae that feed in, and are dispersed by the movement of, water probably evolved in these diverse groups of animals because they are all filter feeders on small organisms that are widely dispersed in the water column.

Most animals that are motile as adults disperse when they are mature. A caterpillar, for example, may spend its entire larval stage feeding on a single plant, but after its metamorphosis into a flying adult—a butterfly—it may fly to and lay eggs on other plants located far from the one where it spent its caterpillar days. In some species, individuals disperse during several different life cycle stages.

No life cycle can maximize all benefits

The common saying, "Jack of all trades, master of none," suggests why there are constraints on the evolution of life cycles. The characteristics of an animal in any one life cycle stage may improve its performance in one activity, but reduce its performance in another—a situation known as a **trade-off**. An animal that is good at filtering small food particles from the water, for example, probably cannot capture large prey. Similarly, energy devoted to building protective structures such as shells cannot be used for growth.

Some major trade-offs can be seen in animal reproduction. Some animals produce large numbers of small eggs, each with a small energy store (**Figure 31.13A**). Other animals produce a small number of large eggs, each with a large energy store (**Figure 31.13B**). With a fixed amount of available energy, a female animal can produce many small eggs or a few large eggs, but she cannot do both.

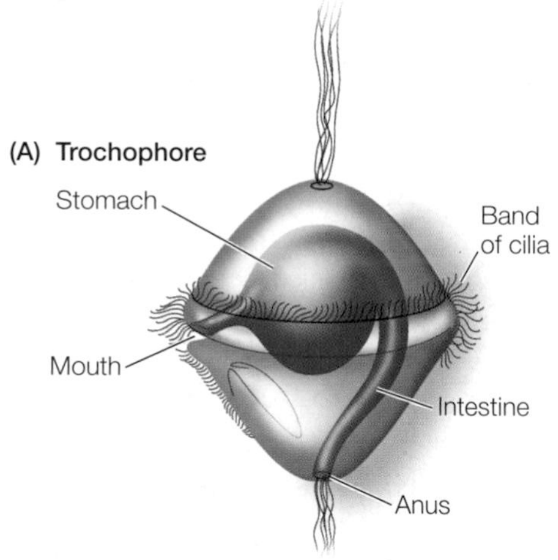

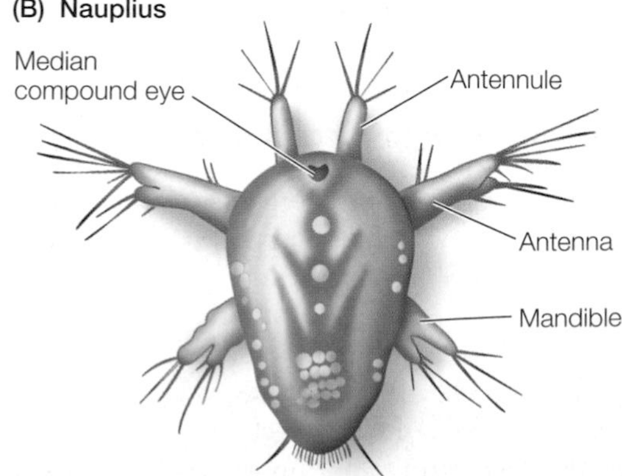

31.12 Planktonic Larval Forms of Marine Animals (A) The trochophore ("wheel-bearer") is a distinctive larval form found in several marine animal clades with spiral cleavage, most notably the polychaete worms and the mollusks. (B) This nauplius larva will mature into a crustacean with a segmented body and jointed appendages.

31.13 Many Small or Few Large Allocation of energy to eggs requires trade-offs. (A) This frog has divided her reproductive energy among a large number of small eggs. (B) This penguin has invested all her reproductive energy in one large egg.

(A) *Rana temporaria* (B) *Pygoscelis antarctica*

The larger the energy store in an egg, the longer an offspring can develop before it must either find its own food or be fed by its parents. Birds of all species lay relatively small numbers of relatively large eggs, but incubation periods vary. In some species, eggs hatch when the young are still helpless (**Figure 31.14A**). Such *altricial* young must be fed and cared for until they can feed themselves; parents can provide for only a small number of altricial offspring. On the other hand, some bird species incubate their eggs longer, and the hatchlings are developed to a point that they are able to forage for themselves almost immediately (**Figure 31.14B**). The young of such species are called *precocial*.

Parasite life cycles evolve to facilitate dispersal and overcome host defenses

Animals that live as internal parasites are bathed in the nutritious tissues of their host or in the digested food that fills their host's digestive tract. Thus they may not need to exert much energy to obtain food, but to survive, they must overcome the host's defenses. Furthermore, either they or their offspring must disperse to new hosts while their host is still living, because they die when their host dies.

31.14 Helpless or Independent (A) The altricial young of the blue tit are essentially helpless when they hatch. Their parents feed and care for them for several weeks. (B) Canada geese hatchlings are precocial. They are ready to swim and feed independently almost immediately after hatching.

(A) *Parus caeruleus* (B) *Branta canadensis*

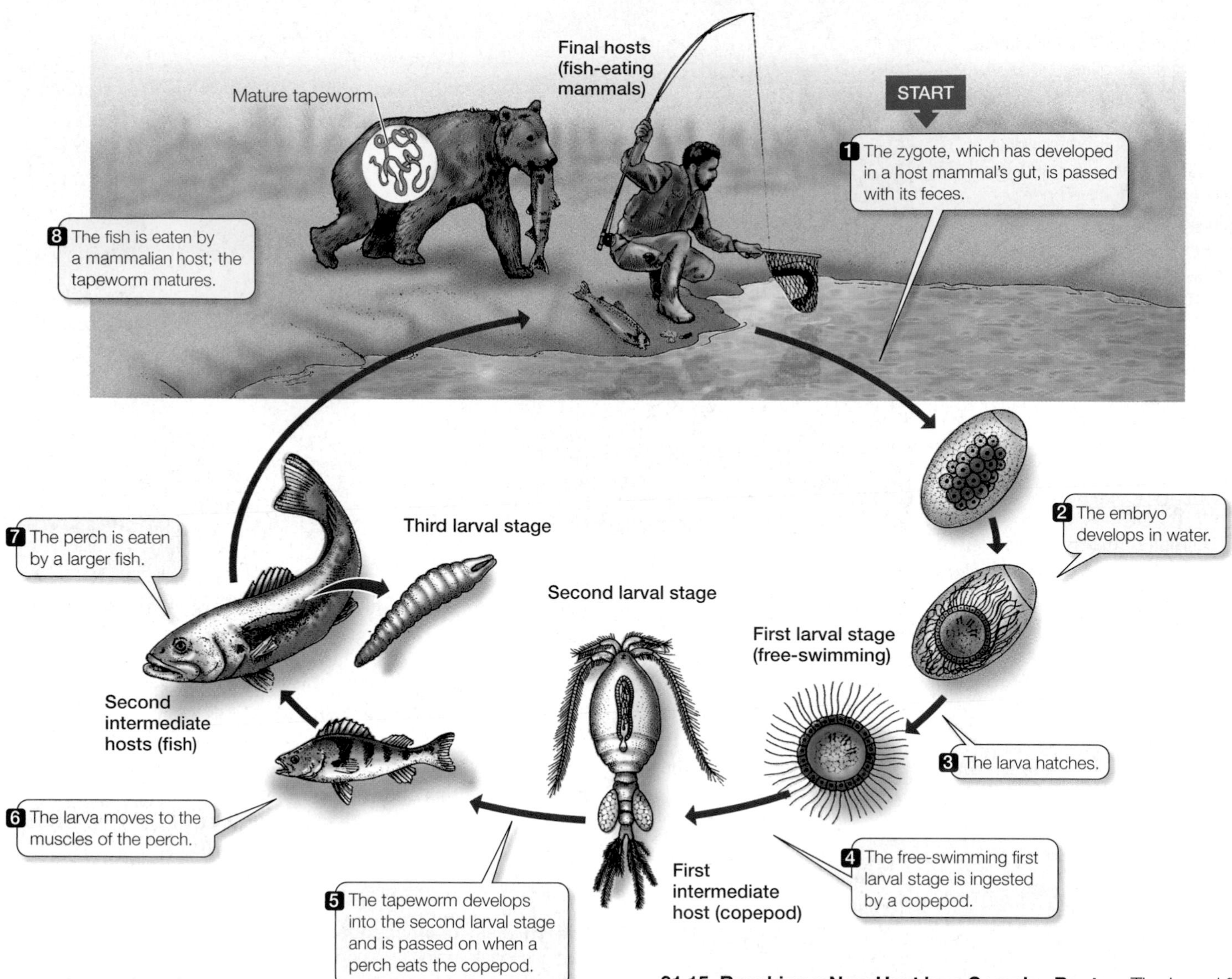

31.15 Reaching a New Host by a Complex Route The broad fish tapeworm, *Diphyllobothrium latum*, must pass through the bodies of a copepod (a type of crustacean) and a fish before it can reinfect its primary host, a mammal. Such complex life cycles assist the parasite's colonization of new host individuals, but they also provide opportunities for humans to break the cycle with hygienic measures.

The fertilized eggs of some parasites are voided with the host's feces and later ingested directly by other host individuals. Most parasite species, however, have complex life cycles involving one or more intermediate hosts and several larval stages (**Figure 31.15**). Some intermediate hosts transport individual parasites directly between other hosts. Others house and support the parasite until another host ingests it. Thus complex life cycles may facilitate the transfer of individual parasites among hosts.

31.4 RECAP

Different life cycle stages of animals may differ in form and be specialized for different activities.

- How do trade-offs constrain the evolution of life cycles? See p. 680
- Can you explain why parasites often have complicated life cycles? See pp. 681–682 and Figure 31.15

31.5 What Are the Major Groups of Animals?

The millions of species of animals display variations in body symmetry and body cavity structure and have a host of different life cycles, patterns of development, and survival strategies. In the remainder of this chapter and in the following two chapters we will become acquainted with the major animal groups and learn how the general characteristics described in this chapter apply to each of them.

Table 31.1 provides a summary of the living members of the major animal groups. The **Bilateria** is a large monophyletic group embracing all animals other than sponges, ctenophores, and cnidarians. The traits that support the monophyly of Bilateria are

TABLE 31.1

Summary of Living Members of the Major Groups of Animals

	APPROXIMATE NUMBER OF LIVING SPECIES DESCRIBED	MAJOR GROUPS
Glass sponges	500	
Demosponges	7,000	
Calcareous sponges	500	
Ctenophores	100	
Cnidarians	11,000	Anthozoans: Corals, sea anemones
		Hydrozoans: Hydras and hydroids
		Scyphozoans: Jellyfishes
PROTOSTOMES		
Arrow worms	100	
Lophotrochozoans		
Ectoprocts	4,500	
Flatworms	25,000	Free-living flatworms; flukes and tapeworms (all parasitic); monogeneans (ectoparasites of fishes)
Rotifers	1,800	
Ribbon worms	1,000	
Phoronids	20	
Brachiopods	335	
Annelids	16,500	Polychaetes (all marine)
		Clitellates: Earthworms, freshwater worms, leeches
Mollusks	95,000	Monoplacophorans
		Chitons
		Bivalves: Clams, oysters, mussels
		Gastropods: Snails, slugs, limpets
		Cephalopods: Squids, octopuses, nautiloids

	APPROXIMATE NUMBER OF LIVING SPECIES DESCRIBED	MAJOR GROUPS
Ecdysozoans		
Kinorhynchs	150	
Loriciferans	100	
Priapulids	16	
Horsehair worms	320	
Nematodes	25,000	
Onychophorans	150	
Tardigrades	800	
Arthropods:		
Crustaceans	50,000	Crabs, shrimps, lobsters, barnacles, copepods
Hexapods	1,000,000	Insects and relatives
Myriapods	14,000	Millipedes, centipedes
Chelicerates	89,000	Horseshoe crabs, arachnids (scorpions, harvestmen, spiders, mites, ticks)
DEUTEROSTOMES		
Echinoderms	7,000	Crinoids (sea lilies and feather stars); brittle stars; sea stars; sea daisies; sea urchins; sea cucumbers
Hemichordates	95	Acorn worms and pterobranchs
Urochordates	3,000	Ascidians (sea squirts)
Cephalochordates	30	Lancelets
Vertebrates	52,000	Hagfish; lampreys
		Cartilaginous fishes
		Ray-finned fishes
		Coelacanths
		Amphibians
		Reptiles (including birds)
		Mammals

bilateral symmetry, three cell layers, and the presence of at least seven Hox genes (see Chapters 19 and 20).

The bilaterian animals comprise the two major categories mentioned earlier in this chapter, and are classified as either **protostomes** or **deuterostomes** (see Figure 31.1). These two groups have been evolving separately for over 500 million years, since the early Cambrian or late Precambrian. We will describe the protostomes in Chapter 32 and the deuterostomes in Chapter 33.

The remainder of this chapter describes those animal groups that are not bilaterians. The simplest animals, the sponges, have no cell layers and no organs. Sponges are not a clade, but the name is used for three groups that exhibit the ancestral body organization of animals. All other animals, including the bilatarians, are known as **eumetazoans**. They have obvious body symmetry, a gut, a nervous system, special types of cell junctions, and well-organized tissues in distinct cell layers (although there have been secondary losses of some of these structures in some eumetazoans). Sponges lack all these features.

Sponges are loosely organized animals

Sponges are the simplest of animals. They have some specialized cells, but no distinct cell layers and no true organs. Early naturalists thought that they were plants because they lacked body symmetry.

Glass sponges
Demosponges
Calcareous sponges
Eumetazoans

Sponges have hard skeletal elements called **spicules**, which may be small and simple or large and complex. Recent analyses of ribosomal RNA genes suggest that there are three major groups of sponges, which are paraphyletic with respect to the remaining animals. Members of two groups (*glass sponges* and *demosponges*) have skeletons composed of silicaceous spicules made of hydrated silicon dioxide (**Figure 31.16A,B**). These spicules are remarkable in having greater flexibility and toughness than synthetic glass rods of similar length. Members of the third group, the *calcareous sponges*, take their name from their calcium carbonate skeletons (**Figure 31.16C**). It is the latter group that is most closely related to the eumetazoans.

The body plan of sponges of all three groups—even large ones, which may reach a meter or more in length—is an aggregation of cells built around a water canal system. Water, along with any food particles it contains, enters the sponge by way of small pores and passes into the water canals, where choanocytes capture food particles (see Figures 31.7 and 31.16A).

Sponges have been collected for thousands of years. The sponge industry peaked in 1938, when more than 3.6 million kilograms were harvested and marketed. Today most commercial "sponges" are made of synthetic materials created in laboratories.

A skeleton of simple or branching spicules, and often a complex network of elastic fibers, supports the bodies of most sponges. Sponges also have an extracellular matrix, composed of collagen, adhesive glycoproteins, and other molecules, that holds the cells together. Most species are filter feeders; a few species are carnivores that trap prey on hook-shaped spicules that protrude from the body surface.

Most of the 8,000 species of sponges are marine animals; only about 50 species live in fresh water. Sponges come in a wide variety of sizes and shapes that are adapted to different movement patterns of water. Sponges living in intertidal or shallow subtidal environments with strong wave action are firmly attached to the substratum. Most sponges that live in slowly flowing water are flattened and are oriented at right angles to the direction of current flow. They intercept water and the prey it contains as it flows past them.

Sponges reproduce both sexually and asexually. In most species, a single individual produces both eggs and sperm, but individuals do not self-fertilize. Water currents carry sperm from one individual to another. Asexual reproduction is by budding and fragmentation.

Ctenophores are radially symmetrical and diploblastic

Glass sponges
Demosponges
Calcareous sponges
Ctenophores
Cnidarians
Bilaterians (protostomes and deuterostomes)

The **ctenophores**, also known as the comb jellies, lack most of the Hox genes possessed by all other eumetazoans. Ctenophores have a radially symmetrical, diploblastic body plan, with the two cell layers separated by a thick, gelatinous **mesoglea**. They have low metabolic rates because the mesoglea is an inert extracellular matrix. Ctenophores have a *complete gut*, with an entrance and an exit. Food enters through a mouth, and wastes are eliminated through two anal pores.

Ctenophores have eight comblike rows of fused plates of cilia, called **ctenes** (**Figure 31.17**). A ctenophore moves through the water by beating these cilia rather than by muscular contractions. Its feeding tentacles are covered with cells that discharge adhesive material when they contact prey. After capturing its prey, a ctenophore retracts its tentacles to bring the food to its mouth. In some species, the entire surface of the body is coated with sticky mucus that captures prey. All of the 100 known species of ctenophores eat small planktonic organisms. They are common in open seas.

(A) *Xestospongia testudinaria*

(B) *Euplectella aspergillum*

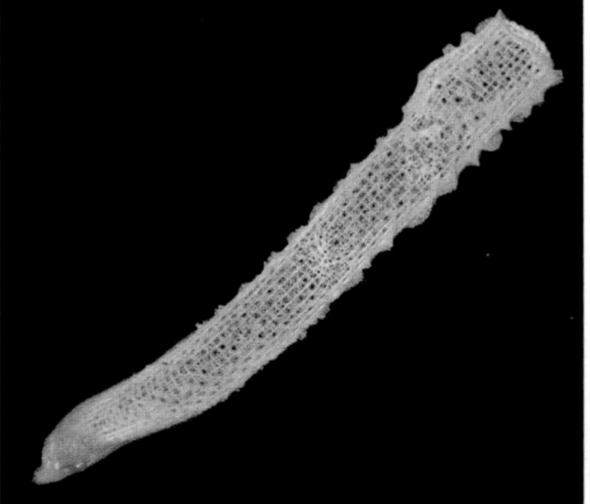

(C) *Leucilla nuttingi*

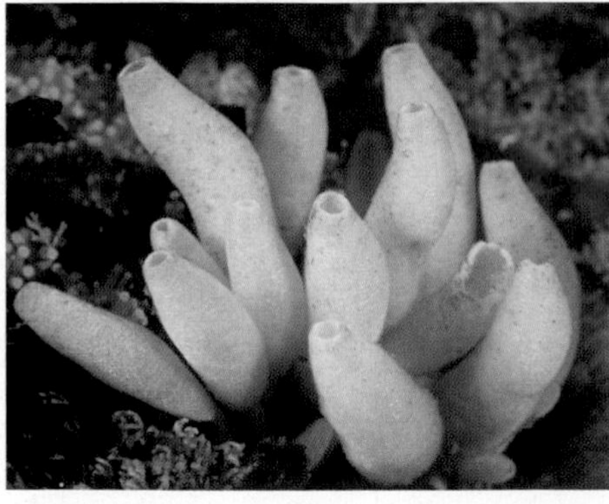

31.16 There Are Three Groups of Sponges (A) The great majority of sponge species are demosponges such as these Pacific barrel sponges. The system of pores and water canals "typical" of the sponge body plan is apparent. (B) The supporting structures of both demosponges and glass sponges are silicaceous spicules, seen here in the skeleton of a glass sponge. (C) The skeletons of calcareous sponges are made of calcium carbonate. Calcareous sponges are more closely related to the eumetazoans than to the other two groups of sponges.

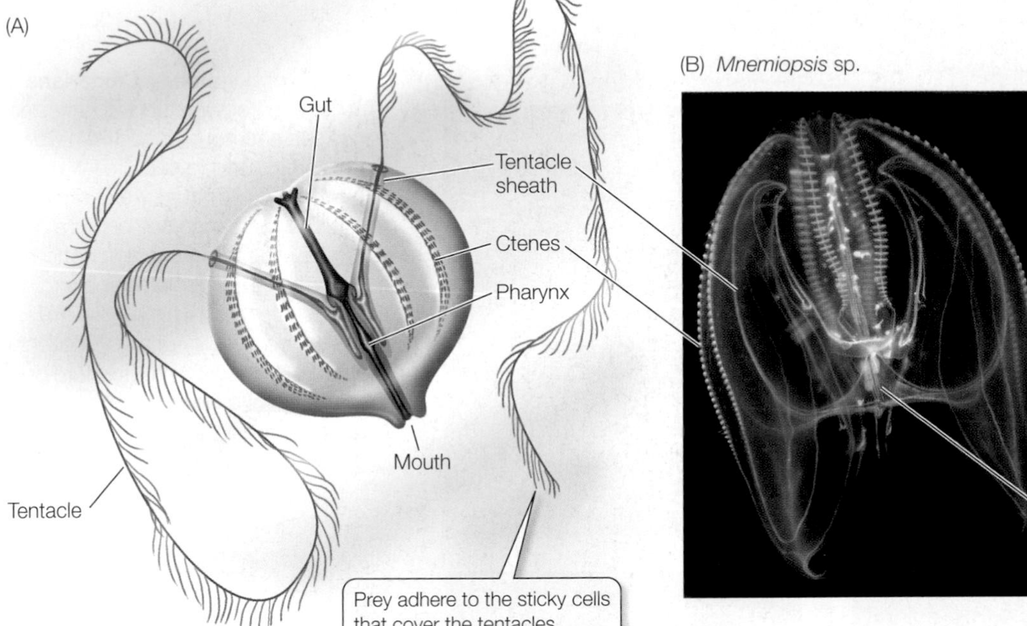

31.17 Comb Jellies Feed with Tentacles (A) The body plan of a typical ctenophore. The long, sticky tentacles sweep through the water, efficiently harvesting small prey. (B) This comb jelly photographed in Sydney Harbor, Australia, has short tentacles.

Ctenophore life cycles are uncomplicated. Gametes are released into the body cavity and then discharged through the mouth or the anal pores. Fertilization takes place in open seawater. In nearly all species, the fertilized egg develops directly into a miniature ctenophore that gradually grows into an adult.

Cnidarians are specialized carnivores

One branch of the next split in the animal lineage led to the **cnidarians** (jellyfishes, sea anemones, corals, and hydrozoans). The mouth of a cnidarian is connected to a blind sac called the **gastrovascular cavity** (thus it does not have a complete gut). The gastrovascular cavity functions in digestion, circulation, and gas exchange, and it also acts as a hydrostatic skeleton. The single opening serves as both mouth and anus.

The life cycle of most cnidarians has two distinct stages, one sessile and the other motile (**Figure 31.18**). In the sessile **polyp** stage, a cylindrical stalk is attached to the substratum. Individual polyps may reproduce asexually by budding, thereby forming a colony. The motile **medusa** (plural *medusae*) is a free-swimming stage shaped like a bell or an umbrella. It typically floats with its mouth and feeding tentacles facing downward. Medusae of many species produce eggs and sperm and release them into the water. A fertilized egg develops into a free-swimming, ciliated larva called a **planula**, which eventually settles to the bottom and develops into a polyp.

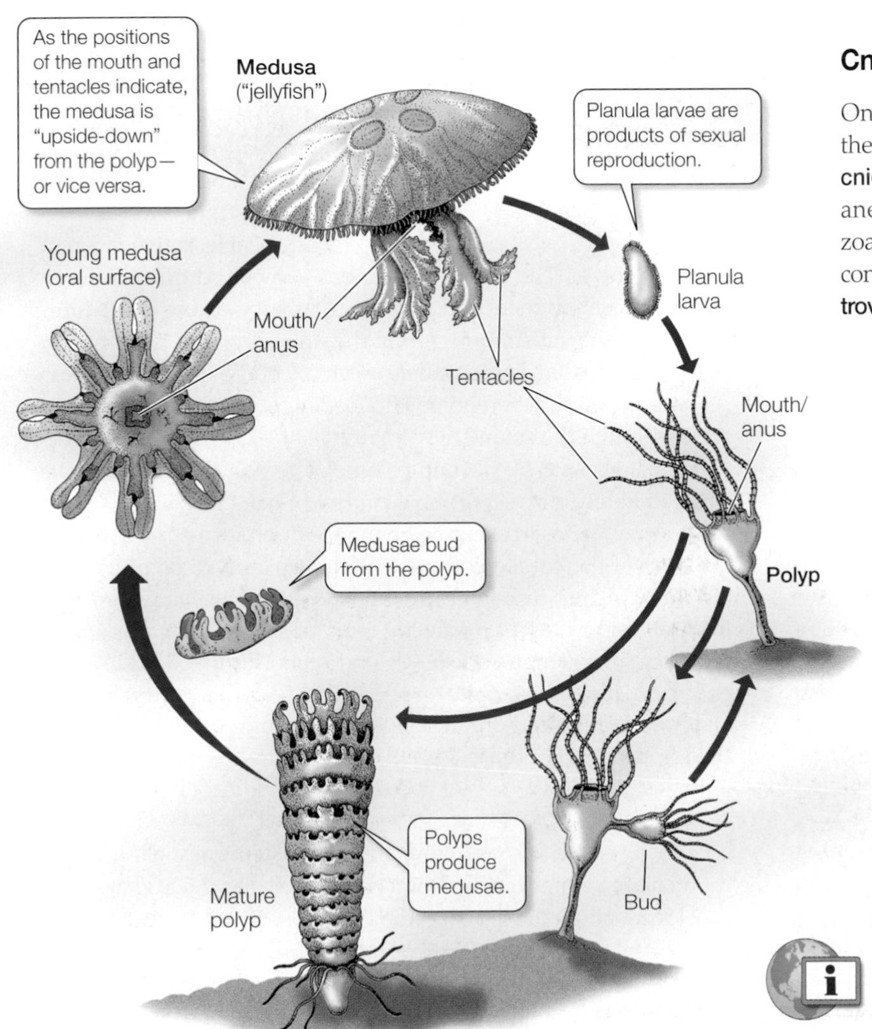

31.18 The Cnidarian Life Cycle Has Two Stages The life cycle of a scyphozoan (jellyfish) exemplifies the typical cnidarian body forms: the sessile, asexual polyp; and the motile, sexual medusa.

(A) *Anthopleura elegantissima*

(B) *Ptilosarcus gurneyi*

(C) *Phyllorhyza punctata*

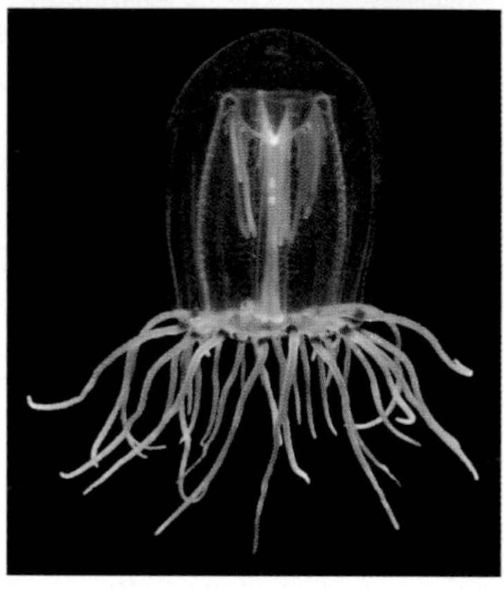

(D) *Polyorchis penicillatus*

31.19 Diversity among Cnidarians (A) The nematocyst-studded tentacles of this sea anemone from British Columbia are poised to capture large prey carried to the animal by water movement. (B) The orange sea pen is a colonial cnidarian that lives in soft bottom sediments and projects polyps above the substratum. (C) This jellyfish illustrates the complexity of a scyphozoan medusa. (D) The internal structure of the medusa of a North Atlantic colonial hydrozoan is visible here.

Cnidarians have epithelial cells with muscle fibers whose contractions enable the animals to move, as well as simple *nerve nets* that integrate their body activities. They also have specialized structural molecules (collagen, actin, and myosin) and Hox genes. They are specialized carnivores, using the toxin in their nematocysts to capture relatively large and complex prey (see Figure 31.10). Some cnidarians, including corals and anemones, gain additional nutrition from photosynthetic protists that live in their tissues. Cnidarians, like ctenophores, are largely made up of inert mesoglea. They have low metabolic rates and can survive in environments where they encounter prey only infrequently.

Of the roughly 11,000 cnidarian species living today, all but a few live in the oceans (**Figure 31.19**). The smallest cnidarians can hardly be seen without a microscope; the largest known jellyfish is 2.5 meters in diameter. We will describe three clades of cnidarians that have many species: the scyphozoans, anthozoans, and hydrozoans.

SCYPHOZOANS The several hundred species of scyphozoans are all marine. The mesoglea of their medusae is thick and firm, giving rise to their common name—jellyfishes or sea jellies. The medusa rather than the polyp dominates the life cycle of scyphozoans. An individual medusa is male or female, releasing eggs or sperm into the open sea. The fertilized egg develops into a small planula larva that quickly settles on a substratum and develops into a small polyp. This polyp feeds and grows and may produce additional polyps by budding. After a period of growth, the polyp begins to bud off small medusae, which feed, grow, and transform themselves into adult medusae (see Figures 31.18 and 31.19C).

ANTHOZOANS Members of the **anthozoan** clade include sea anemones, sea pens, and corals. Sea anemones (**Figure 31.19A**), all of which are solitary, are widespread in both warm and cold ocean waters. Sea pens (**Figure 31.19B**), by contrast, are colonial. Each colony consists of at least two different kinds of polyps. The primary polyp has a lower portion anchored in the bottom sediment and a branched upper portion, which projects above the substratum. Along the upper portion, the primary polyp produces smaller secondary polyps by budding. Some of these secondary polyps differentiate into feeding polyps; others circulate water through the colony.

Corals also are sessile and colonial. The polyps of most corals form a skeleton by secreting a matrix of organic molecules on which they deposit calcium carbonate, which forms the eventual skeleton of the coral colony. As the colony grows, old polyps die but their calcium carbonate skeletons remain. The living members form a layer on top of a growing bank of skeletal remains, eventually forming chains of islands and reefs (**Figure 31.20A**). The common names of coral groups—horn corals, brain corals, staghorn corals, and organ pipe corals, among others—describe their appearance (**Figure 31.20B**).

The Great Barrier Reef along the northeastern coast of Australia is a system of coral formations more than 2,000 km long—about the distance from New York City to St. Louis. A single coral reef in the Red Sea has been calculated to contain more material than all the buildings in the major cities of North America combined.

(A)

(B) *Diploria labyrinthiformis*

31.20 Corals (A) Many different coral species form an Indian Ocean reef in Chumbe Coral Park off the coast of Zanzibar. (B) The very descriptive common name of this Caribbean coral is "brain coral." The Latinate species name reflects its maze-like, or "labyrinthine," appearance.

Corals flourish in clear, nutrient-poor tropical waters. They can grow rapidly in such environments because unicellular photosynthetic protists live endosymbiotically within their cells. These protists provide the corals with products of photosynthesis; the corals, in turn, provide the protists with nutrients and a place to live. This endosymbiotic relationship explains why reef-forming corals are restricted to clear surface waters, where light levels are high enough to support photosynthesis.

Coral reefs throughout the world are threatened both by global warming, which is raising the temperatures of shallow tropical ocean waters (see Figure 57.10), and by polluted runoff from development on adjacent shorelines. An overabundance of nitrogen in the runoff gives an advantage to algae, which overgrow and eventually smother the corals.

HYDROZOANS Hydrozoans have diverse life cycles. The polyp typically dominates the life cycle, but some species have only medusae; others have only polyps. Most hydrozoans are colonial. A single planula larva eventually gives rise to a colony of many polyps, all interconnected and sharing a continuous gastrovascular cavity (**Figure 31.21**). Within such a colony some polyps have tentacles with many nematocysts; they capture prey for the colony. Others lack tentacles and are unable to feed, but are specialized for the production of medusae. Still others are fingerlike and defend the colony with their nematocysts.

31.5 RECAP

The bilaterian animals fall into two major clades, the protostomes and the deuterostomes. The nonbilateran animals—the sponges, ctenophores, and cnidarians—have relatively simple structures and correspondingly simple feeding strategies.

- Do you understand why sponges are considered to be animals, even though they lack the complex body structures found among the other animal groups? See p. 683 and Figure 31.7
- What are some major features of the cnidarian clade? See pp. 685–686 and Figure 31.18

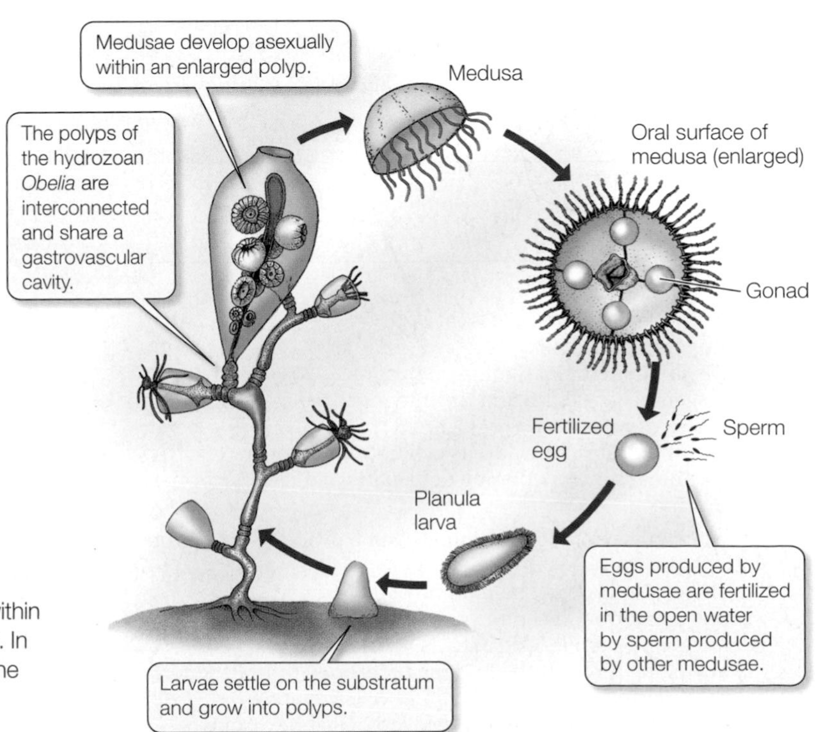

Medusae develop asexually within an enlarged polyp.
Medusa
The polyps of the hydrozoan *Obelia* are interconnected and share a gastrovascular cavity.
Oral surface of medusa (enlarged)
Gonad
Fertilized egg
Sperm
Planula larva
Eggs produced by medusae are fertilized in the open water by sperm produced by other medusae.
Larvae settle on the substratum and grow into polyps.

31.21 Hydrozoans Often Have Colonial Polyps The polyps within a hydrozoan colony may differentiate to perform specialized tasks. In the species whose life cycle is diagrammed here, the medusa is the sexual reproductive stage, producing eggs and sperm in organs called gonads.

CHAPTER SUMMARY

31.1 What evidence indicates the animals are monophyletic?

Most **animals** share a set of derived traits not found in other groups of organisms. These traits include similarities in their ribosomal RNA and Hox genes, cell junctions, and an extracellular matrix.

Patterns of embryonic development provide important clues to the evolutionary relationships among animals. **Diploblastic** animals develop two embryonic cell layers; **triploblastic** animals develop three.

Differences in patterns of early development also characterize two major clades of triploblastic animals, the **protostomes** and **deuterostomes**.

31.2 What are the features of animal body plans?

Animal **body plans** can be described in terms of **symmetry**, **body cavity** structure, **segmentation**, and **appendages**.

A few animals have **spherical** symmetry, but most animals have either **radial** or **bilateral** symmetry. Review Figure 31.3

Most animals with radial symmetry move slowly or not at all, while animals with bilateral symmetry are able to move more rapidly. Many bilaterally symmetrical animals are **cephalized**, with sensory and nervous tissues in an anterior head.

On the basis of their body cavity structure, animals can be described as **acoelomates**, **pseudocoelomates**, or **coelomates**. Review Figure 31.4

Segmentation, which takes many forms, improves control of movement, especially if the animal also has appendages.

31.3 How do animals get their food?

Motile animals can move to find food; **sessile** animals stay in one place and capture food that is brought to them.

Filter feeders filter small organisms and organic molecules from their environment.

Predators have features that enable them to capture and subdue large animal prey.

Herbivores consume plants, usually without killing them.

Parasites live in or on other organisms and obtain nutrition from these host individuals.

31.4 How do animal life cycles differ?

The stages of an animal's **life cycle** may be specialized for different activities.

An immature stage that is dramatically different from the adult stage is called a **larva**.

All life cycles have at least one **dispersal** stage, so that the animal does not die in the same place where it was born. Many sessile marine animals can be grouped by the presence of one of two distinct dispersal stages, the **trochophore** larva and the **nauplius** larva.

Parasites have complex life cycles that may involve one or more hosts and several larval stages. Review Figure 31.15

A characteristic of an animal or a life cycle stage may improve its performance in one activity, but reduce its performance in another, a situation known as a **trade-off**.

31.5 What are the major groups of animals?

All animals other than sponges, ctenophores, and cnidarians belong to a large monophyletic group called the **Bilateria**. The clade **Eumetazoa** embraces all animals other than sponges.

Sponges are simple animals that lack cell layers and true organs. They have skeletons made up of silicaceous or calcareous **spicules**. They create water currents and capture food with flagellated feeding cells called **choanocytes**. Choanocytes are an evolutionary link between the animals and choanoflagellate protists. Review Figure 31.7

The **ctenophores** and **cnidarians** are diploblastic, radially symmetrical animals.

The two cell layers of ctenophores are separated by an inert extracellular matrix called **mesoglea**. They move by beating fused plates of cilia called **ctenes**. Review Figure 31.17

The life cycle of cnidarians has two distinct stages: a sessile **polyp** stage and a motile **medusa**. A fertilized egg develops into a free-swimming **planula** larva, which settles to the bottom and develops into a polyp. Review Figures 31.18 and 31.21, Web/CD Tutorial 31.1

See Web/CD Activities 31.1 and 31.2 for a concept review of this chapter.

SELF-QUIZ

1. The body plan of an animal is
   a. its general structure.
   b. the integrated functioning of its parts.
   c. its general structure and the integrated functioning of its parts.
   d. its general structure and its evolutionary history.
   e. the integrated functioning of its parts and its evolutionary history.
2. A bilaterally symmetrical animal can be divided into mirror images by
   a. any plane through the midline of its body.
   b. any plane from its anterior to its posterior end.
   c. any plane from its dorsal to its ventral surface.
   d. any plane through the midline of its body from its anterior to its posterior end.
   e. a single plane through the midline of its body from its dorsal to its ventral surface.
3. Among protostomes, cleavage of the fertilized egg is
   a. delayed while the egg continues to mature.
   b. always radial.
   c. spiral in some species and radial in others.
   d. triploblastic.
   e. diploblastic.
4. Many parasites evolved complex life cycles because
   a. they are too simple to disperse readily.
   b. they are poor at recognizing new hosts.
   c. they were driven to it by host defenses.
   d. complex life cycles increase the probability of a parasite's transfer to a new host.
   e. their ancestors had complex life cycles and they simply retained them.

5. Animals are believed to be monophyletic because
a. their presumed ancestor has been found in the fossil record.
b. they all have the same number of Hox genes.
c. they all share a set of derived traits not found in other clades.
d. they all lack ribosomal RNAs.
e. they all lack extracellular matrix molecules.

6. In the common ancestor of the protostomes and deuterostomes, the pattern of early cleavage was
a. spiral.
b. radial.
c. biradial.
d. deterministic.
e. haphazard.

7. A fluid-filled body cavity can function as a hydrostatic skeleton because
a. fluids are moderately compressible.
b. fluids are highly compressible.
c. fluids are relatively incompressible.
d. fluids have the same density as body tissues.
e. fluids can be moved by ciliary action.

8. Which of the following is *not* a feature that enables some animals to capture large prey?
a. Sharp teeth
b. Claws
c. Toxins
d. A filtering device
e. Tentacles with stinging cells

9. The sponge body plan is characterized by
a. a mouth and digestive cavity but no muscles or nerves.
b. muscles and nerves but no mouth or digestive cavity.
c. a mouth, digestive cavity, and spicules.
d. muscles and spicules but no digestive cavity or nerves.
e. the lack of a mouth, digestive cavity, muscles, or nerves.

10. Cnidarians have the ability to
a. live in both salt and fresh water.
b. move rapidly in the water column.
c. capture and consume large numbers of small prey.
d. survive where food is scarce because of their low metabolic rate.
e. capture large prey and to move rapidly.

FOR DISCUSSION

1. Differentiate among the members of each of the following sets of related terms:
a. radial symmetry/bilateral symmetry
b. protostome/deuterostome
c. diploblastic/triploblastic
d. coelomate/pseudocoelomate/acoelomate

2. In this chapter we listed some of the traits shared by all animals that convince most biologists that all animals are descendants of a single common ancestral lineage. In your opinion, which of these traits provides the most compelling evidence that animals are monophyletic? If morphological and molecular data do not agree, should one type of evidence be given greater weight? If so, which one?

3. Describe some features that allow animals to capture prey that are larger and more complex than they themselves are.

4. Why is bilateral symmetry strongly associated with cephalization, the concentration of sense organs in an anterior head?

5. How does a slow metabolic rate enable an animal to live in an unproductive environment?

FOR INVESTIGATION

The discovery of fossils of mature dinosaur embryos yielded valuable information about how these animals changed over the course of their lives. What further investigations might be carried out using information from fossils to tell us what modern developmental biology theories must explain?

CHAPTER 32 Protostome Animals

Tiny parasites exert mind control

Most people have never heard of a strepsipteran, and even those who *have* heard of strepsipterans probably have never seen one. They don't know what they are missing! These tiny insects—there are some 600 different species of them—parasitize hundreds of other insect species, including bees, wasps, ants, grasshoppers, and cockroaches. Males and females of most strepsipteran species parasitize the same host species, although in one clade males parasitize ants while females parasitize grasshoppers.

Strepsipteran males and females are often so different that even determining that they are members of the same species requires a DNA analysis, and they have some of the strangest life cycles of any animal. Once grown to maturity within their hosts (whose internal organs they consume), the males of most species emerge looking like a "typical" insect. The females also consume the host from within, but they usually remain inside the host. A mature female extrudes her head and parts of her body from the body of her host. The extruded body parts contain an opening that receives sperm from a male. Much later, this opening becomes an exit for the strepsipteran larvae. The host insects are left dead or severely damaged and produce no offspring of their own.

Strepsipterans dramatically change the behavior of their hosts in ways that help them complete their life cycles at the expense of their hosts' reproduction. For example, when wasps—a typical host—are parasitized by strepsipterans, the parasites generate signals that induce the wasps to leave their nest and form a mating aggregation. This aggregation, however, serves the strepsipterans, not the wasps. Once the wasps aggregate, the male strepsipterans emerge from their hosts to search for and mate with the females whose heads are now poking out of the bodies of other wasps.

Adult male strepsipterans live only a few hours, during which they must find a female and mate. Because the protruding part of a female's body is barely visible, the males have unusually large eyes, and about 75 percent of their brain cells are allocated to vision. This sensory system serves a single purpose: to help the male find a female.

♂

♀

Same Parasite, Different Lifestyles Strepsipterans are parasitic insects that grow to maturity within host insects. Most male strepsipterans look insect-like upon reaching maturity and leave their host to find and mate with a very different-appearing female strepsipteran, who remains inside her host.

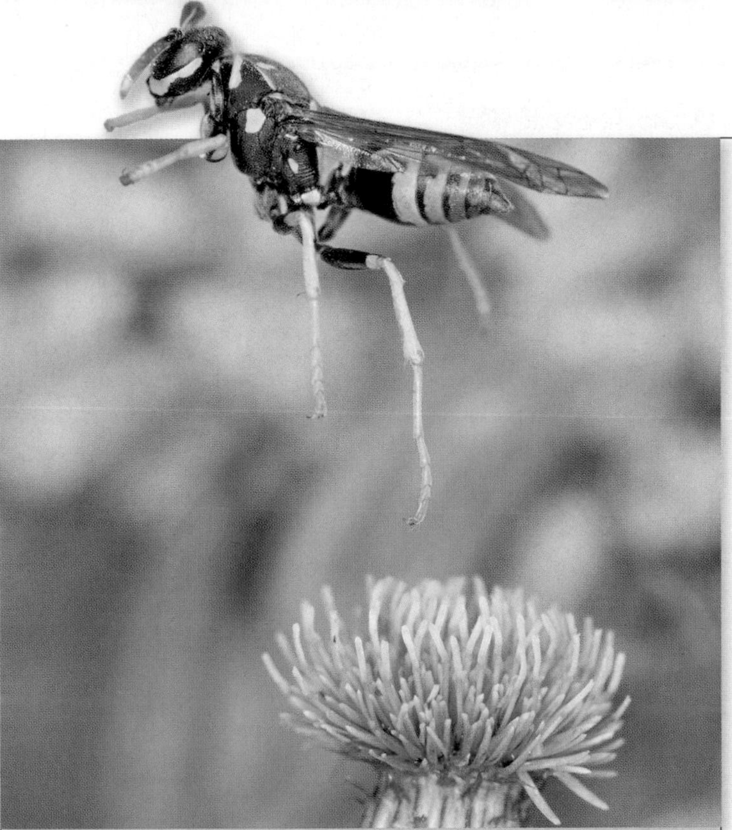

A Host Insect Strepsipterans parasitize many different insect species; wasps of the genus *Polistes* (paper wasps) are common hosts. Molecular signals generated by the parasites can dramatically affect the behavior of the wasps.

Strepsipterans and their hosts are all insects, and insects account for more than half of the described species of protostomes. Other protostome groups, such as mollusks, nematodes, crabs, spiders, and ticks, are also species-rich. Many of the protostome species are parasites. Parasites often live within their hosts and absorb nutrients through their body walls. Some parasites, including the strepsipterans, can more technically be described as *parasitoids*, which consume the host's tissues as they develop from eggs laid on or within the host's body and often grow to be almost as large as their hosts.

IN THIS CHAPTER we will describe the characteristics of protostome animals and describe the members of two major protostome clades, the lophotrochozoans and the ecdysozoans. We will give particular attention to the arthropods, an incredibly species-rich group of ecdysozoans with rigid exoskeletons and jointed appendages.

CHAPTER OUTLINE

32.1 What Is a Protostome?

Some time after the origin of the diplobastic radiate animals, (the cnidarians and ctenophores), a third embryological germ layer—the mesoderm—arose. This feature is found in the two great triploblastic animal clades, the protostomes and the deuterostomes. If we were to judge solely on the basis of numbers, both of species and of individuals, the protostomes would emerge as by far the more successful of the two groups.

As described in Chapter 31, the name protostome means "mouth first," and was applied because in most of these species the embryonic blastopore becomes the mouth; this is in contrast to deuterostome animals, in which the blastopore becomes the anal opening of the digestive tract (see Figure 31.2). This trait is not universally shared, however; for example, no blastopore forms during the early development of insects.

The protostomes are extremely varied, but they are all bilaterally symmetrical animals whose bodies exhibit two major derived traits:

- An anterior *brain* that surrounds the entrance to the digestive tract
- A ventral *nervous system* consisting of paired or fused longitudinal nerve cords

Other aspects of protostome body organization can be quite different from group to group (**Table 32.1**). There are several coelomate groups as well as several groups of pseudocoelomates; one important clade, the flatworms, are acoeolomate (see Figure 31.4). In two of the most prominent clades, the coelom has been secondarily modified:

- The *arthropods* have lost the ancestral coelom over the course of evolution. Their internal body cavity has become a *hemocoel,* or "blood chamber," in which fluid from an open (i.e., no blood vessels) circulatory system bathes the internal organs.
- The *mollusks* are also generally characterized by an open circulatory system and have some of the attributes of the hemocoel, but they retain vestiges of an enclosed coelom around their major organs.

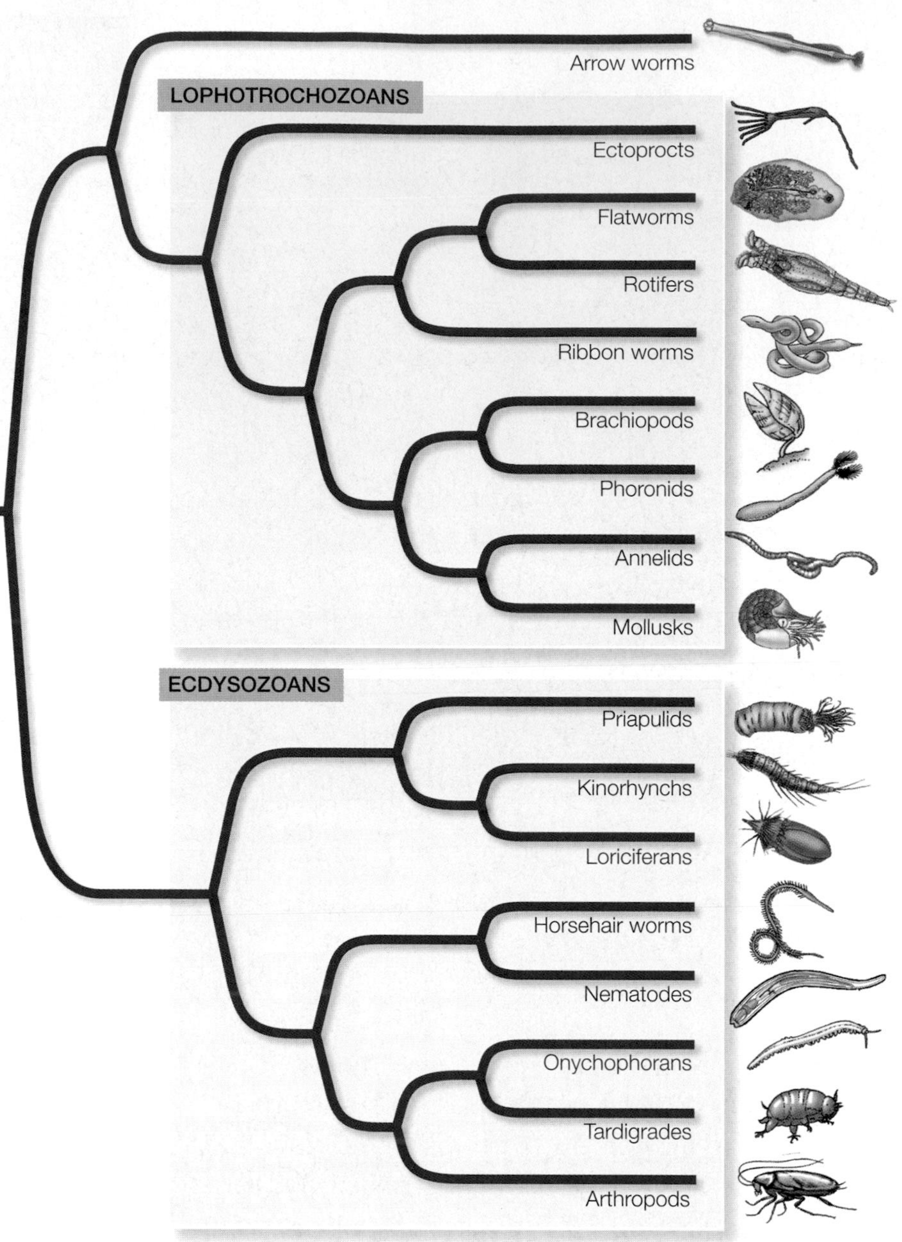

32.1 A Current Phylogenetic Tree of Protostomes Two major lineages, the lophotrochozoans and the ecdysozoans, dominate the tree. Many small clades are not included. The position of the arrow worms is somewhat uncertain; another possibility is that they are the sister group of all other protostomes, not just the lophotrochozoans.

In ancient times, the protostomes split into two major clades—*lophotrochozoans* and *ecdysozoans*—that have been evolving independently ever since (**Figure 32.1**).

Trochophores, lophophores, and spiral cleavage evolved among the lophotrochozoans

The skeletons of most species of **lophotrochozoans** are internal, which means these animals grow by adding to the size of their skeletal elements. Some of them use cilia for locomotion; many groups have a type of free-living larva known as a *trochophore* that moves by beating a band of cilia (see Figure 31.12A).

Several distantly related groups of lophotrochozoans (including ectoprocts, brachiopods, and phoronids) have a **lophophore**, a circular or U-shaped ridge around the mouth that bears one or two rows of ciliated, hollow tentacles (**Figure 32.2**). This complex structure is an organ for both food collection and gas exchange. Biologists once grouped these taxa together as *lophophorates*, but it is now clear that they are not each other's closest relatives. The lophophore appears to have evolved independently at least twice, or else it is an ancestral feature of lophotrochozoans and has been lost in many groups. Nearly all animals with a lophophore are sessile as adults. They use the tentacles and cilia of the lophophore to capture small floating organisms from the water. Other sessile lophotrochozoans use less well developed tentacles for the same purpose.

As discussed in the previous chapter, some lophotrochozoans (including flatworms, ribbon worms, annelids, and mollusks) exhibit a derived form of cleavage in early development known as *spiral cleavage*. Some biologists group these taxa

TABLE 32.1

Anatomical Characteristics of Some Major Protostome Groups[a]

GROUP	BODY CAVITY	DIGESTIVE TRACT	CIRCULATORY SYSTEM
Arrow worms	Coelom	Complete	None
LOPHOTROCHOZOANS			
Flatworms	None	Dead-end sac	None
Rotifers	Pseudocoelom	Complete	None
Ectoprocts	Coelom	Complete	None
Brachiopods	Coelom	Complete in most	Open
Phoronids	Coelom	Complete	Closed
Ribbon worms	Coelom	Complete	Closed
Annelids	Coelom	Complete	Closed or open
Mollusks	Reduced coelom	Complete	Open except in cephalopods
ECDYSOZOANS			
Horsehair worms	Pseudocoelom	Greatly reduced	None
Nematodes	Pseudoceolom	Complete	None
Arthropods	Hemocoel	Complete	Open

[a]Note that all protostomes have bilateral symmetry.

Ectoprocts can oscillate, rotate, and retract their lophophore tentacles.

Lophopus crystallinus

32.2 Ectoprocts Use Their Lophophore to Feed The extended lophophore dominates the anatomy of the colonial ectoprocts. This species inhabits freshwater, although most ectoprocts are marine. Ectoproct colonies can grow to contain over a million individuals, all stemming from the asexual reproduction of the colony's founder.

together as *spiralians,* although phylogenetic analyses of gene sequences do not support spiralians as monophyletic.

Members of several of the groups with spiral cleavage are *wormlike,* which means they are bilaterally symmetrical, legless, soft-bodied, and at least several times longer than they are wide. A wormlike body form enables animals to burrow efficiently through muddy and sandy marine sediment or soil. However, as we will describe later in this chapter, the *mollusks*—the most species-rich of the groups with spiral cleavage—have a very different body plan.

Ecdysozoans must shed their exoskeletons

Ecdysozoans have an external skeleton, or **exoskeleton**, which is a nonliving covering secreted by the underlying *epidermis* (the outermost cell layer). The exoskeleton provides these animals with both protection and support. Once formed, however, an exoskeleton cannot grow. How, then, can ecdysozoans increase in size? They do so by shedding, or **molting**, the exoskeleton and replacing it with a new, larger one. This molting process gives the clade its name (from the Greek word *ecdysis,* "to get out of").

A recently discovered fossil of a Cambrian soft-bodied arthropod, preserved in the process of molting, shows that molting evolved more than 500 million years ago (**Figure 32.3A**). An increasingly rich array of molecular and genetic evidence, including a set of Hox genes shared by all ecdysozoans, suggests that they have a single common ancestor. Thus molting of an exoskeleton is a trait that may have evolved only once during animal evolution.

Before an ecdysozoan molts, a new exoskeleton is already forming underneath the old one. Once the old exoskeleton is shed, the new one expands and hardens. But until it has hardened, the animal is vulnerable to its enemies, both because its outer surface is easy to penetrate and because an animal with a soft exoskeleton can move slowly or not at all (**Figure 32.3B**).

(A)

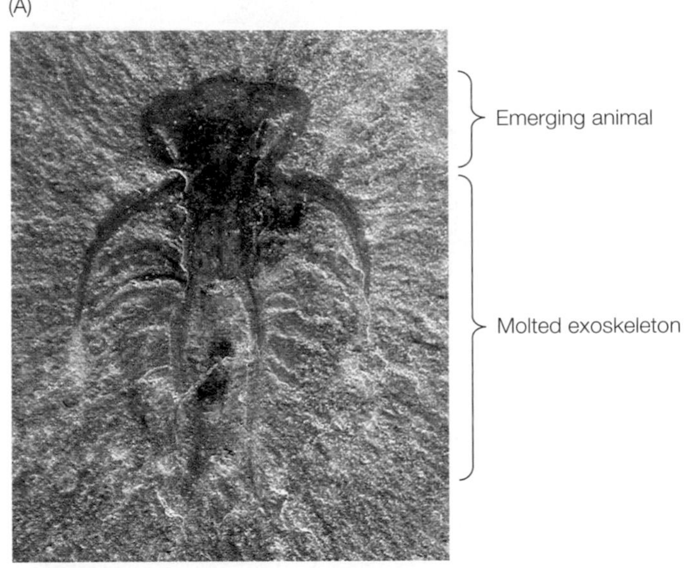

(B) *Phrynus parvulus*

32.3 Molting: Past and Present (A) This 500 million-year-old fossil from the Cambrian captured an individual of a long-extinct arthropod species in the process of molting and shows that the molting process is an evolutionarily ancient trait. (B) This whip scorpion has just emerged from its discarded exoskelton and will be highly vulnerable until its new cuticle has hardened.

Some ecdysozoans have wormlike bodies covered by exoskeletons that are relatively thin and flexible. Such an exoskeleton, called a **cuticle**, offers the animal some protection, but provides only modest body support. A thin cuticle allows the exchange of gases, minerals, and water across the body surface, but restricts the animal to moist habitats. Many species of ecdysozoans with thin cuticles live in marine sediments from which they obtain prey, either by ingesting sediments and extracting organic material from them, or by capturing larger prey using a toothed *pharynx* (a muscular organ at the anterior end of the digestive tract). Some freshwater species absorb nutrients directly through their thin cuticles, as do parasitic species that live within their hosts. Many wormlike ecdysozoans are predators, eating protists and small animals.

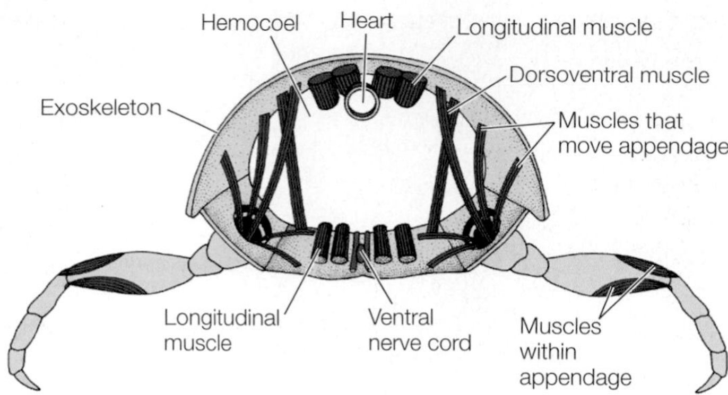

32.4 Arthropod Skeletons Are Rigid and Jointed This cross section through a thoracic segment of a generalized arthropod illustrates the arthropod body plan, which is characterized by a rigid exoskeleton with jointed appendages.

The exoskeletons of other ecdysozoans are thickened by the incorporation of layers of protein and a strong, waterproof polysaccharide called **chitin**. An animal with a rigid, chitin-reinforced body covering can neither move in a wormlike manner nor use cilia for locomotion. A hard exoskeleton also impedes the passage of oxygen and nutrients into the animal, presenting new challenges in other areas besides growth. Therefore, ecdysozoans with hard exoskeletons evolved new mechanisms of locomotion and gas exchange.

To move rapidly, an animal with a rigid exoskeleton must have extensions of the body that can be manipulated by muscles. Such *appendages* evolved in the late Precambrian, leading to the **arthropod** ("jointed foot") clade. Arthropod appendages exist in an amazing variety of forms. They serve many functions, including walking and swimming, gas exchange, food capture and manipulation, copulation, and sensory perception. Arthropods grasp food with their mouths and associated appendages and digest it internally. Their muscles are attached to the inside of the exoskeleton. Each segment has muscles that operate that segment and the appendages attached to it (**Figure 32.4**).

The arthropod exoskeleton has had a profound influence on the evolution of these animals. Encasement within a rigid body covering provides support for walking on dry land, and the waterproofing provided by chitin keeps the animal from dehydrating in dry air. Aquatic arthropods were, in short, excellent candidates to invade terrestrial environments. As we will see, they did so several times.

Arrow worms retain some ancestral developmental features

Nearly all triploblastic animal groups can be readily classified as being either protostomes or deuterostomes, but the evolutionary relationships of one small group, the **arrow worms**, were debated for many years. The early development of arrow worms looks similar to that of deuterostomes, although it is now known that they merely retain developmental features that are ancestral to triploblastic animals in general. Recent studies of gene sequences clearly identify them as protostomes. There is still some question as to whether they are the closest relatives of the lophotrochozoans (as shown in Figure 32.1), or possibly the sister group of all other protostomes.

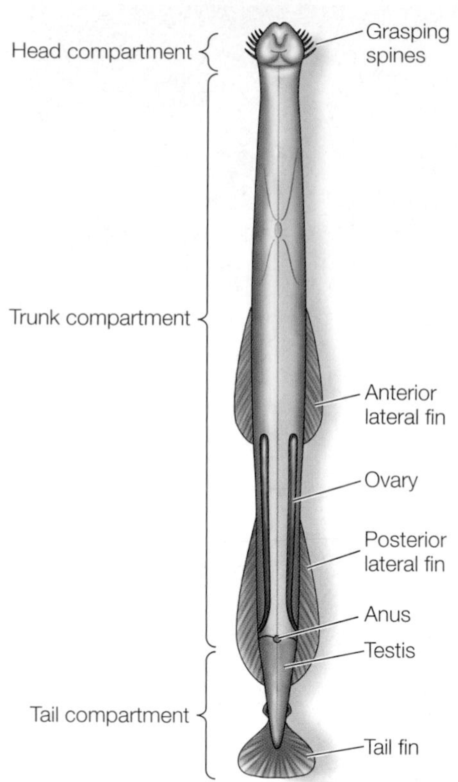

32.5 An Arrow Worm Arrow worms have a three-part body plan. Their fins and grasping spines are adaptations for a predatory lifestyle. Individuals are hermaphroditic, producing both eggs and sperm (ovary and testis).

The arrow worm body plan is based on a coelom divided into three compartments: head, trunk, and tail (**Figure 32.5**). Their bodies are transparent or translucent. Most arrow worms swim in the open sea. A few species live on the seafloor. Their abundance as fossils indicates that they were common more than 500 million years ago. The 100 or so living species of arrow worms are small enough—ranging from 3 mm to less than 12 cm in length—that their gas exchange and waste excretion requirements are met by diffusion through the body surface. They lack a circulatory system; wastes and nutrients are moved around the body in the coelomic fluid, which is propelled by cilia that line the coelom. There is no distinct larval stage. Miniature adults hatch directly from eggs that are fertilized internally following elaborate courtship between two hermaproditic individuals.

Arrow worms are stabilized in the water by means of one or two pairs of lateral fins and a tail fin. They are major predators of planktonic organisms in the open ocean, ranging in size from small protists to young fish as large as the arrow worms themselves. An arrow worm typically lies motionless in the water until water movement signals the approach of prey. The arrow worm then darts forward and grasps the prey with the stiff spines adjacent to its mouth.

32.1 RECAP

The shared derived traits of protostomes include an anterior brain and a ventral nervous system. Many lophotrochozoans grow by adding to the size of their skeletal elements, whereas ecdysozoans, which have an exoskeleton, must molt periodically in order to grow.

- How does an animal's body covering influence the way it breathes, feeds, and moves? See pp. 693–694
- What features made arthropods well adapted for colonizing terrestrial environments? See p. 694

32.2 What Are the Major Groups of Lophotrochozoans?

Lophotrochozoans come in a variety of sizes and shapes, ranging from relatively simple animals with a blind gut (that is, a gut with only one opening) and no internal transport system to animals with a complete gut (having separate entrance and exit openings) and a complex internal transport system. They include some highly species-rich groups, such as flatworms, annelids, and mollusks. A number of these groups exhibit wormlike bodies, but the lophotrochozoans encompass a wide diversity of morphologies, including several groups with external shells. Some lophotrochozoan groups have only recently been discovered by biologists.

Ectoprocts live in colonies

The 4,500 species of **ectoprocts** (also called bryozoans, or "moss animals") are colonial animals that live in a "house" made of material secreted by the external body wall. Almost all ectoprocts are marine, although a few species occur in fresh or brackish water. An ectoproct colony consists of many small (1–2 mm) individuals connected by strands of tissue along which nutrients can be moved (**Figure 32.6**). The colony is created by the asexual reproduction of its founding member, and a single colony may contain as many as 2 million individuals. Rocks in coastal regions in many parts of the world are covered with luxuriant growths of ectoprocts. Some ectoprocts create miniature reefs in shallow waters. In some species, the individual colony members are differentially specialized for feeding, reproduction, defense, or support.

Individual ectoprocts are able to oscillate and rotate their lophophore to increase contact with prey. They can also retract it into their "house" (see Figure 32.2). Ectoprocts also can reproduce sexually by releasing sperm into the water, which carries them to other individuals. Eggs are fertilized internally; developing embryos are brooded before they exit as larvae to seek suitable sites for attachment to the substratum.

Sertella septentrionalis

32.6 An Ectoproct Colony The membranous orange tissue of this ectoproct colony connects and supplies nutrients to thousands of individual animals (see Figure 32.2).

Flatworms, rotifers, and ribbon worms are structurally diverse relatives

The flatworms, rotifers, and ribbon worms are structurally a diverse group, and only recently have their relationships to one another been elucidated. Before the availability of gene sequences for phylogenetic analysis, biologists considered the structure of the body cavity to be an important feature in animal classification. However, this monophyletic group of animals includes subgroups that are acoelomate (flatworms), pseudocoelomate (rotifers), and coelomate (ribbon worms). Thus, systematists today understand that the forms of body cavities have undergone considerable convergence in the course of animal evolution (see Table 32.1).

Flatworms lack specialized organs for transporting oxygen to their internal tissues. Their lack of a gas transport system dictates that each cell must be near a body surface, a requirement met by the dorsoventrally flattened body form. The digestive tract of a flatworm consists of a mouth opening into a blind sac. The sac is often highly branched, however, forming intricate patterns that increase the surface area available for the absorption of nutrients. Some small free-living flatworms are cephalized, with a head bearing chemoreceptor organs, two simple eyes, and a tiny brain composed of anterior thickenings of the longitudinal nerve cords. Free-living flatworms glide over surfaces, powered by broad bands of cilia (**Figure 32.7A**).

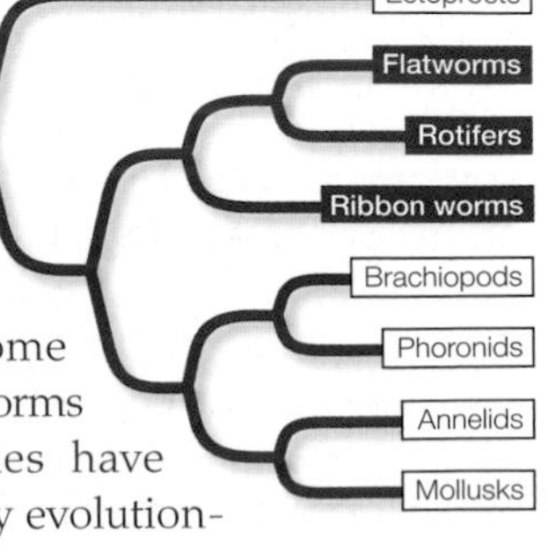

Most flatworms are internal parasites; others feed externally on animal tissues (living or dead), and some graze on plants. Although many flatworms are free-living, many other species have evolved to become parasites. A likely evolution-

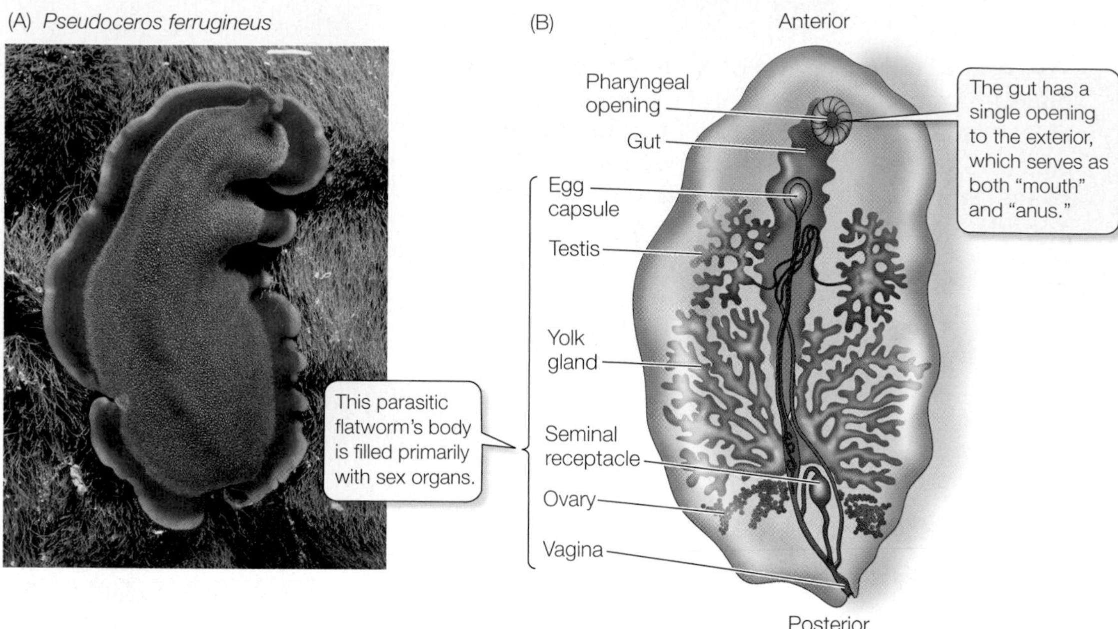

32.7 Flatworms May Live Freely or Parasitically (A) Some flatworm species, such as this Pacific marine flatworm, are free-living. (B) The fluke diagrammed here lives parasitically in the gut of sea urchins and is representative of parasitic flatworms. Because their hosts provide all the nutrition they need, these internal parasites do not require elaborate feeding or digestive organs and can devote most of their bodies to reproduction.

ary transition was from feeding on dead organisms to feeding on the body surfaces of dying hosts to invading and consuming parts of healthy hosts. Most of the 25,000 species of living flatworms are *tapeworms* and *flukes*; members of these two groups are internal parasites, particularly of vertebrates (**Figure 32.7B**). They absorb digested food from the digestive tracts of their hosts; many of them lack digestive tracts of their own. Some cause serious human diseases, such as schistosomiasis. Members of another flatworm group, the *monogeneans*, are external parasites of fishes and other aquatic vertebrates. The *turbellarians* include most of the free-living species.

Most **rotifers** are tiny (50–500 μm long)—smaller than some ciliate protists—but they have specialized internal organs (**Figure 32.8**). A complete gut passes from an anterior mouth to a posterior anus; the body cavity is a pseudocoel that functions as a hydrostatic skeleton. Most rotifers propel themselves through the water by means of rapidly beating cilia rather than by muscular contraction.

The most distinctive organ of rotifers is a conspicuous ciliated organ called the *corona*, which surmounts the head of many species. Coordinated beating of the cilia sweeps particles of organic matter from the water into the animal's mouth and down to a complicated structure called the *mastax*, in which the food is ground into small pieces. By contracting the muscles around the pseudocoel, a few rotifer species that prey on protists and small animals can protrude the mastax through the mouth and seize small objects with it. Males and females are found in some species, but the bdelloid rotifers have only females, which produce eggs that develop without being fertilized by a male. This group is the only known case of a group of animals that appear to have existed for millions of years without the benefits of sexual reproduction.

Most of the 1,800 known species of rotifers live in fresh water. Members of a few species rest on the surfaces of mosses or lichens in a desiccated, inactive state until it rains. When rain falls, they absorb water and become mobile, feeding in the films of water that temporarily cover the plants. Most rotifers live no longer than a week or two.

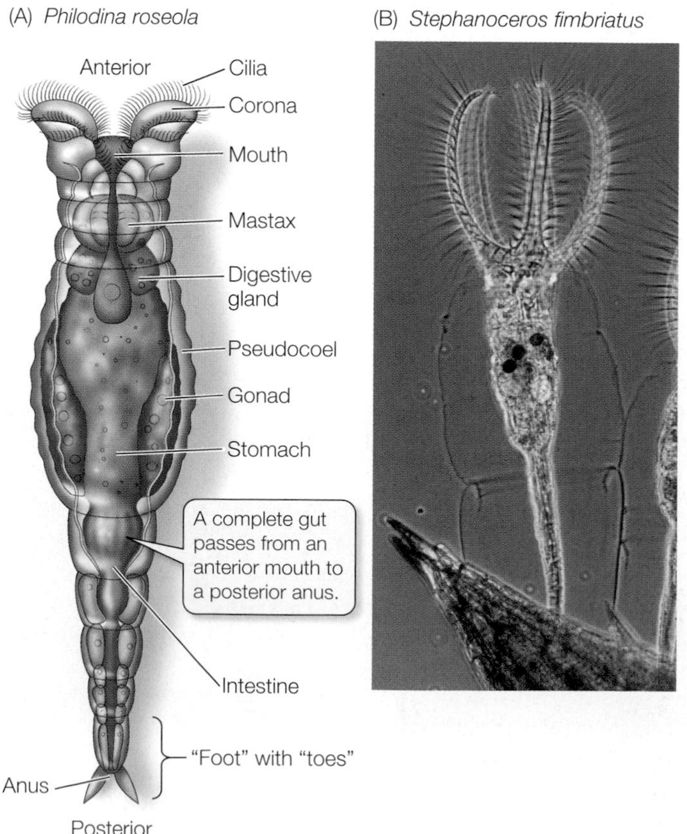

32.8 Rotifers (A) The rotifer diagrammed here reflects the general structure of many rotifers. (B) A micrograph reveals the internal complexity of a rotifer that has ingested photosynthetic protists.

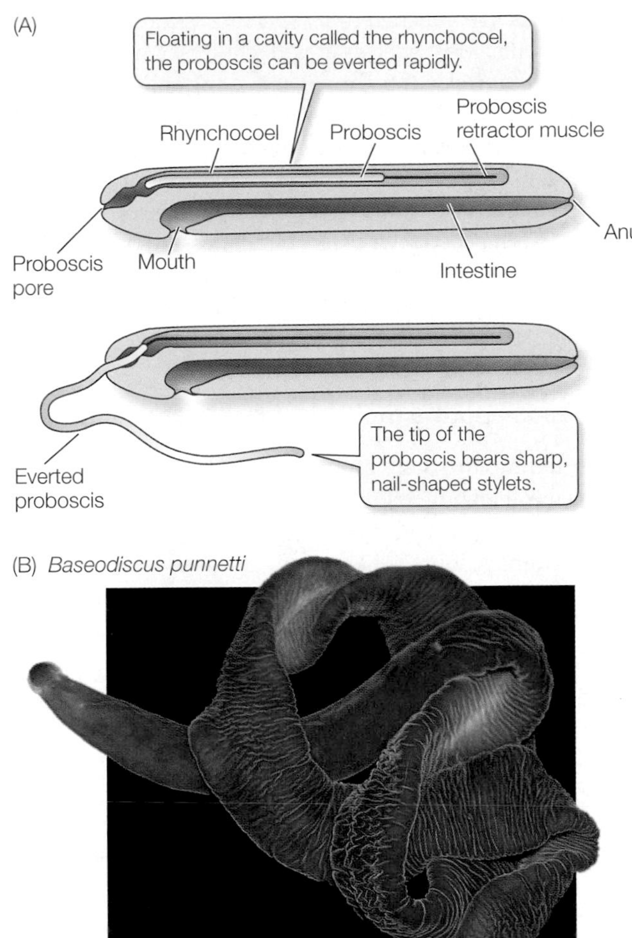

32.9 Ribbon Worms (A) The proboscis is the ribbon worm's feeding organ. (B) This deep-water nemertean is an impressive length despite the fact that its proboscis is not visible.

Ribbon worms (nemerteans) have simple nervous and excretory systems similar to those of flatworms. Unlike flatworms, however, they have a complete digestive tract with a mouth at one end and an anus at the other. Small ribbon worms move slowly by beating their cilia. Larger ones employ waves of muscle contraction to move over the surface of sediments or to burrow into them.

Within the body of nearly all of the 1,000 species of ribbon worms is a fluid-filled cavity called the *rhynchocoel*, within which lies a hollow, muscular *proboscis*. The proboscis, which is the worm's feeding organ, may extend much of the length of the body. Contraction of the muscles surrounding the rhynchocoel causes the proboscis to exit explosively through an anterior pore (**Figure 32.9A**). The proboscis may be armed with sharp stylets that pierce prey and discharge paralysis-causing toxins into the wound.

Ribbon worms are largely marine, although there are species that live in freshwater or on land. Most species are less than 20 cm in length, but individuals of some species reach 20 *meters* or more. Some genera feature species that are conspicuous and brightly colored (**Figure 32.9B**).

Phoronids and brachiopods use lophophores to extract food from the water

Earlier we discussed the ectoprocts, which use a lophophore to feed. Phoronids and brachiopods also feed using a lophophore, but this structure appears to have evolved independently (or else it has been lost in other lophotrochozoan groups). Although neither the phoronids nor the brachiopods are represented by many living species, the brachiopods (which have shells, and so leave an excellent fossil record) are known to have been much more abundant during the Paleozoic and Mesozoic eras.

Ectoprocts
Flatworms
Rotifers
Ribbon worms
Brachiopods
Phoronids
Annelids
Mollusks

The 20 known species of **phoronids** are small (5–25 cm in length), sessile worms that live in muddy or sandy sediments or attached to rocky substrata. Phoronids are found in marine waters, from the intertidal zone to about 400 meters deep. They secrete tubes made of chitin, inside which they live (**Figure 32.10**). Their cilia drive water into the top of the lophophore, and the water exits through the narrow spaces between the tentacles. Suspended food particles are caught and transported to the mouth by ciliary action. In most species, eggs are released into the water, where they are fertilized, but some species produce large eggs that are fertilized internally and retained in the parent's body, where they are brooded until they hatch.

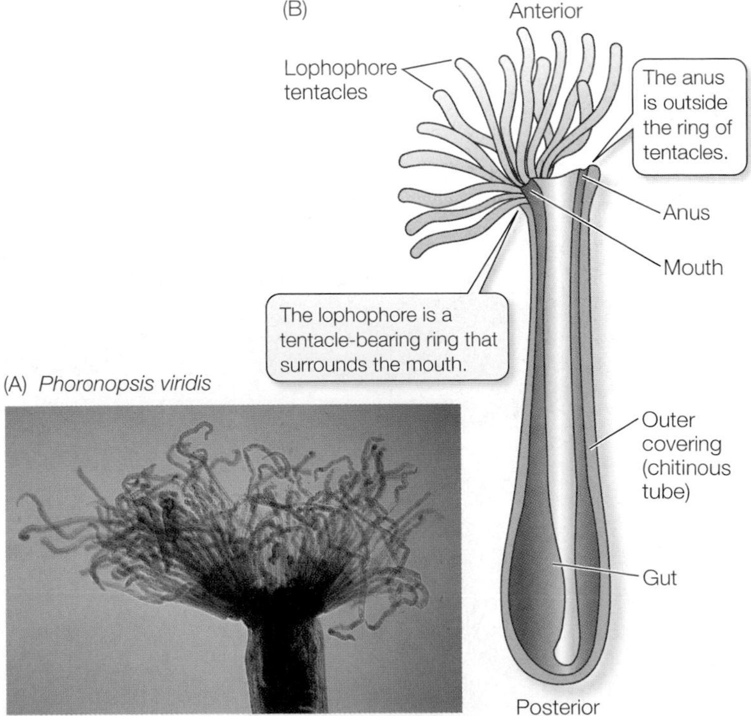

32.10 Phoronids Have Impressive Lophophores (A) The photograph shows the tentacles of a phoronid's lophophore. (B) The phoronid gut is U-shaped, as seen in brown in this generalized diagram.

32.11 A Brachiopod The lophophore of this North Pacific brachiopod can be seen between the valves of its shell.

Brachiopods are solitary marine animals. They have rigid shells that are divided into two parts connected by a ligament (**Figure 32.11**). The two halves can be pulled shut to protect the soft body. Brachiopods superficially resemble bivalve mollusks, but shells have evolved independently in the two groups. The two halves of the brachiopod shell are dorsal and ventral, rather than lateral, as in bivalves. The lophophore is located within the shell. The beating of cilia on the lophophore draws water into the slightly opened shell. Food is trapped in the lophophore and directed to a ridge, along which it is transferred to the mouth. Most brachiopods are 4–6 centimeters long, but they can be as long as 9 centimeters.

Brachiopods live attached to a solid substratum or embedded in soft sediments. Most species are attached by means of a short, flexible stalk that holds the animal above the substratum. Gases are exchanged across body surfaces, especially the tentacles of the lophophore. Most brachiopods release their gametes into the water, where they are fertilized. The larvae remain among the plankton for only a few days before they settle and develop into adults.

Brachiopods reached their peak abundance and diversity in Paleozoic and Mesozoic times. More than 26,000 fossil species have been described. Only about 335 species survive, but they are common in some marine environments.

The annelids and the mollusks are sister groups

Annelids and mollusks are among the most familiar lophotrochozoans. Annelids are another wormlike group of animals, but unlike other worms, they are clearly segmented. The many species of mollusks, in contrast, are not wormlike but display modifications of a unique tripartite body plan.

Ectoprocts
Flatworms
Rotifers
Ribbon worms
Brachiopods
Phoronids
Annelids
Mollusks

Annelids have segmented bodies

The bodies of **annelids** are clearly segmented. Segmentation, as we saw in Section 31.2, allows an animal to move different parts of the body independently of one another, giving it much better control of its movement. The earliest segmented worms, preserved as fossils from the middle Cambrian, were burrowing marine annelids.

In most annelids, the coelom in each segment is isolated from those in other segments (**Figure 32.12**). A separate nerve center called a *ganglion* controls each segment; nerve cords that connect the ganglia coordinate their functioning. Most annelids lack a rigid external protective covering; instead, they have a thin, permeable body wall that serves as a general surface for gas exchange. Thus most annelids are restricted to moist environments because they lose body water rapidly in dry air. The approximately 16,500 described species live in marine, freshwater, and moist terrestrial environments.

POLYCHAETES More than half of all annelid species are *polychaetes* ("many hairs"), most of which are marine animals. Many polychaetes live in burrows in soft sediments. Most of them have one or more pairs of eyes and one or more pairs of tentacles, with which they filter prey from the surrounding water, at the an-

Ganglion in ventral nerve cord
Ring vessels of circulatory system
Brain
Sperm receptacles
Testes and sperm sacs
Ovary
Oviduct
Sperm duct
Setae (bristles)
Segments
Excretory organs
Circular muscle
Longitudinal muscle
Blood vessel
Excretory organ
Coelom
Intestine
Setae (bristles)
Nerve cord
Blood vessels
Cross section

32.12 Annelids Have Many Body Segments The segmented structure of the annelids is apparent both externally and internally. Many organs of this earthworm are repeated serially.

32.13 Diversity among the Annelids
(A) "Cluster dusters" or "social feathers" are sessile marine polychaetes that grow in masses. They filter food from the water with their elegant tentacles. (B) These pogonophorans live around hydrothermal vents deep in the ocean. Their tentacles can be seen protruding from their chitinous tubes. (C) Earthworms are hermaphroditic; when they copulate, each individual donates and receives sperm. (D) The medicinal leech has been a tool of physicians and healers for many centuries. Even today leeches have uses in modern clinical practice.

(A) *Bispira brunnea* (B) *Riftia* sp. (C) *Lumbricus terrestris* (D) *Hirudo medicinalis*

terior end of the body (**Figure 32.13A**; see also Figure 31.6A). In the marine species, the body wall of most segments extends laterally as a series of thin outgrowths called *parapodia*. The parapodia function in gas exchange, and some species use them to move. Stiff bristles called *setae* protrude from each parapodium, forming temporary attachments to the substratum that prevent the animal from slipping backward when its muscles contract. Recent molecular studies suggest that polychaetes are probably paraphyletic with respect to the remaining annelids.

Members of one clade of polychaetes, the *pogonophorans*, lost their digestive tract (they have no mouth or gut). They are burrowing animals with a crown of tentacles through which gases are exchanged. Pogonophorans secrete tubes made of chitin and other substances, in which they live (**Figure 32.13B**). The pogonophoran coelom consists of an anterior compartment, into which the tentacles can be withdrawn, and a long, subdivided cavity that extends much of the length of the body. The posterior end of the body is segmented. Pogonophorans take up dissolved organic matter at high rates from the sediments in which they live or the surrounding water. This uptake is facilitated by endosymbiotic bacteria that live in a specialized organ known as the *trophosome*. Developmentally, the trophosome develops from the embryonic gut, so the organ has retained its nutritional function, although the process of feeding has changed considerably.

Pogonophorans were not discovered until the twentieth century, when deep-sea explorers found them living many thousands of meters below the ocean surface. In these deep oceanic sediments, they may reach densities of many thousands per square meter. About 160 species have been described. The largest and most remarkable pogonophorans are 2 meters or more in length and live near deep-sea *hydrothermal vents*—volcanic openings in the seafloor through which hot, sulfide-rich water pours. The tissues of these species harbor endosymbiotic bacteria that fix carbon using energy obtained from the oxidation of hydrogen sulfide (H_2S).

CLITELLATES Most of the approximately 3,000 described species of the other major annelid group, the *clitellates*, live in freshwater or terrestrial environments. The clitellates are likely phylogenetically nested within the polychaetes, although the exact relationships are not yet clear. There are two major groups of clitellates, the oligochaetes and the leeches.

Oligochaetes ("few hairs") have no parapodia, eyes, or anterior tentacles, and they have relatively few setae. Earthworms—the most familiar oligochaetes—burrow in and ingest soil, from which they extract food particles. All oligochaetes are *hermaphroditic*; that is, each individual is both male and female. Sperm are exchanged simultaneously between two copulating individuals (**Figure 32.13C**). Eggs are laid in a cocoon outside the adult's body. The cocoon is shed, and when development is complete, miniature worms emerge and immediately begin independent life.

Leeches, like oligochaetes, lack parapodia and tentacles. The coelom of leeches is not divided into compartments; the coelomic space is largely filled with undifferentiated tissue. Groups of segments at each end of the body are modified to form suckers, which serve as temporary anchors that aid the leech in movement. With its posterior sucker attached to a substratum, the leech extends its body by contracting its circular muscles. The anterior sucker is then attached, the posterior one detached, and the leech shortens itself by contracting its longitudinal muscles. Leeches live in freshwater or terrestrial habitats.

A leech makes an incision in its host, from which blood flows. It can ingest so much blood in a single feeding that its body may enlarge several-fold. The leech secretes an anticoagulant into the wound that keeps the host's blood flowing. For centuries, medical practitioners have employed leeches to draw blood to treat diseases they believed were caused by an excess of blood or by "bad blood." Although most leeching practices (such as inserting a leech in a person's throat to alleviate swollen tonsils) have been abandoned, *Hirudo medicinalis* (the medicinal leech; **Figure 32.13D**) is used today to reduce fluid pressure and prevent blood clotting in damaged tissues, to eliminate pools of coagulated blood, and to prevent scarring. The anticoagulants of certain other leech species that also contain anesthetics and blood vessel dilators are being studied for possible medical uses.

Mollusks have undergone a dramatic evolutionary radiation

The ancestors of modern **mollusks** were probably unsegmented, wormlike animals. However, the origin of a tripartite body organization in this lineage set the stage for one of the most dramatic of animal evolutionary radiations.

The molluscan "body plan" has three major components: a foot, a visceral mass, and a mantle (**Figure 32.14**).

- The *foot* is a large, muscular structure that originally was both an organ of locomotion and a support for the internal organs. In squids and octopuses, the foot has been modified to form arms and tentacles borne on a head with complex sense organs. In other groups, such as clams, the foot is a burrowing organ. In some groups the foot is greatly reduced.
- The heart and the digestive, excretory, and reproductive organs are concentrated in a centralized, internal *visceral mass*.
- The *mantle* is a fold of tissue that covers the organs of the visceral mass The mantle secretes the hard, calcareous shell that is typical of many mollusks.

In most mollusks, the mantle extends beyond the visceral mass to form a *mantle cavity*. Within this cavity lie *gills* that are used for gas exchange. When cilia on the gills beat, they create a current of water. The tissue of the gills, which is highly *vascularized* (contains many blood vessels), takes up oxygen from the water and releases carbon dioxide. Many mollusk species use their gills as filter-feeding devices, whereas others feed using a rasping structure known as the *radula* to scrape algae from rocks. In some mollusks, such as the marine cone snails, the radula has been modified into a drill or poison dart.

Molluscan blood vessels do not form a closed system. Blood and other fluids empty into a large, fluid-filled *hemocoel*, through which fluids move around the animal and deliver oxygen to the internal organs. Eventually, the fluids re-enter the blood vessels and are moved by a heart.

Monoplacophorans were the most abundant mollusks during the Cambrian period, 500 million years ago, but only a few species

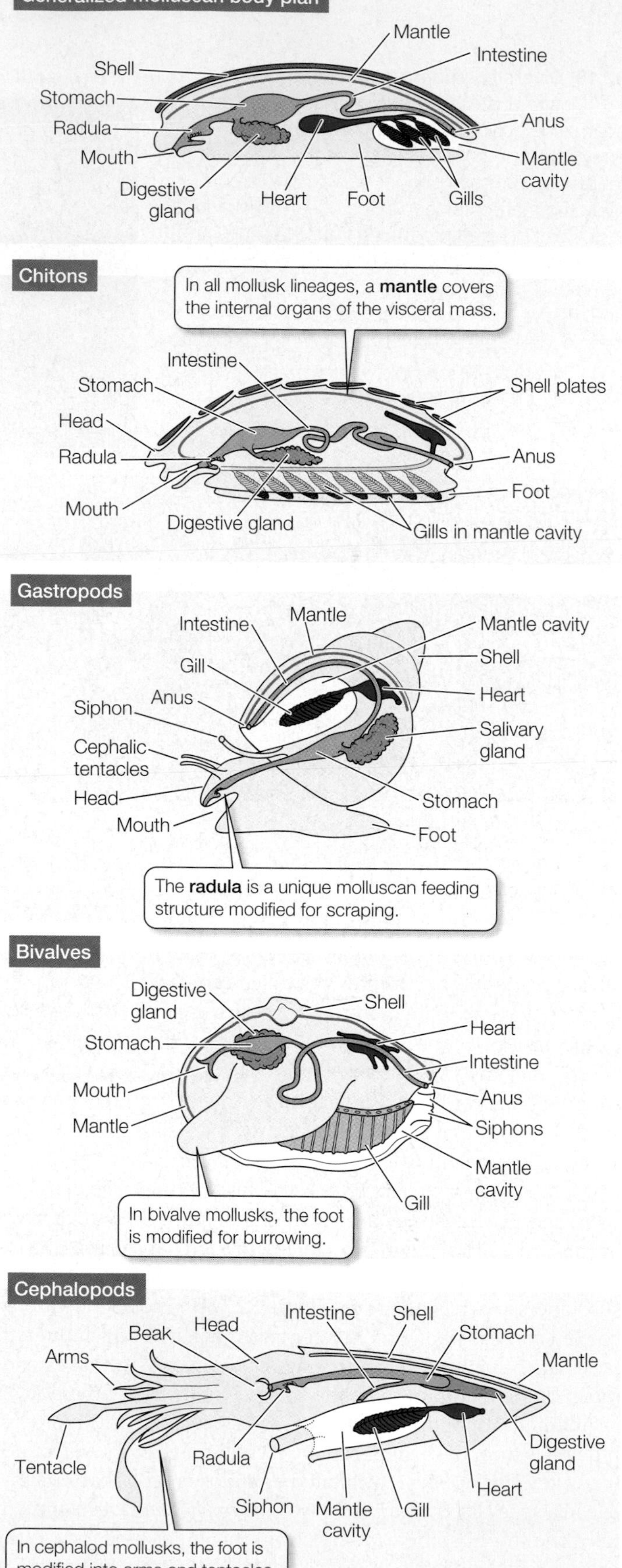

32.14 Molluscan Body Plans The mollusks are all variations on a general body plan that includes three major components: a foot, a visceral mass of internal organs, and a mantle. The mantle may secrete a calcareous shell, as in the gastropods and bivalves.

(A) *Tonicella* sp.

(B) *Tridacna gigas*

(C) *Phidiana hiltoni*

(D) *Helminthoglypta walkeriana*

(E) *Octopus bimaculoides*

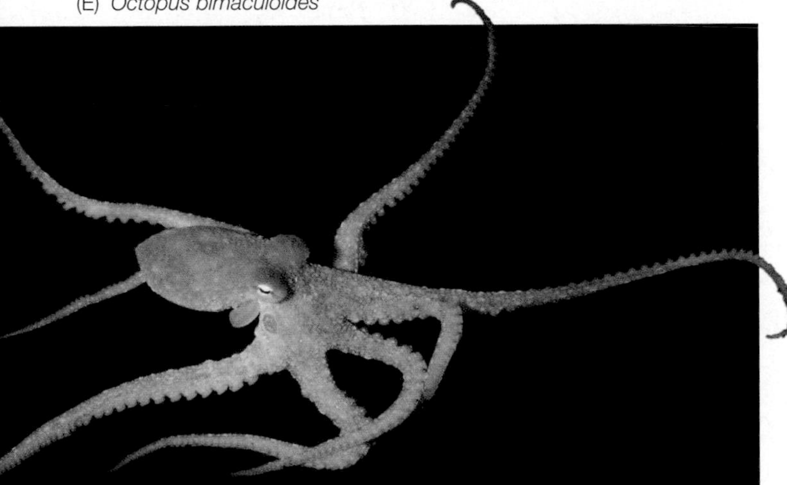

(F) *Nautilus pompilius*

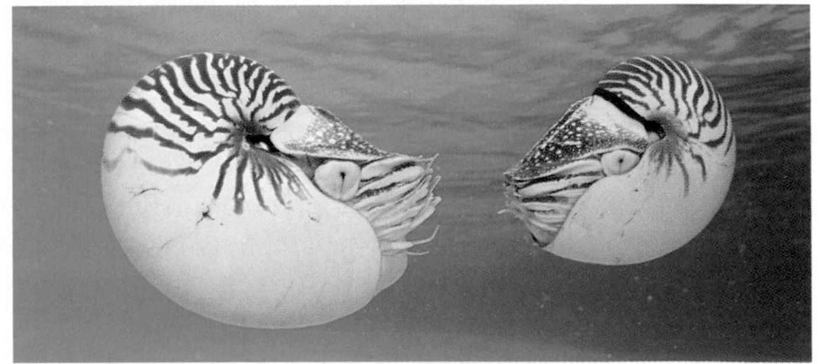

32.15 Diversity among the Mollusks (A) Chitons are common in the intertidal zones of temperate-zone coasts. (B) The giant clam of Indonesia is among the largest of the bivalve mollusks. (C) This shell-less nudibranch ("naked gills"), or sea slug, is colored to signal its toxicity. (D) Land snails are shelled, terrestrial gastropods. (E) Cephalopods such as the octopus are active predators. (F) The boundaries of chambers are clearly visible on the outer surfaces of these shelled cephalopods, known as chambered nautiluses.

survive today. They look something like the gastropod limpets, but in contrast to all other living mollusks, the gas exchange organs, muscles, and excretory pores of monoplacophorans are repeated over the length of the body.

The four major molluscan clades living today—chitons, bivalves, gastropods, and cephalopods—are illustrated in Figure 32.15. The species shown here are a tiny sample of the 95,000 living species of mollusks.

CHITONS Multiple gills and shell plates characterize the **chitons**, but most of their body organs are not iterated (**Figure 32.15A**). The chiton body is bilaterally symmetrical, and the internal organs, particularly the digestive and nervous systems, are relatively simple. Most chitons are marine herbivores that scrape algae from rocks with their sharp radulae. An adult chiton spends most of its life clinging tightly to rock surfaces with its large, muscular, mucus-covered foot. It moves slowly by means of rippling waves of muscular contraction in the foot. Fertilization in most chitons takes place in the water, but in a few species fertilization is internal and embryos are brooded within the body.

BIVALVES Clams, oysters, scallops, and mussels are all familiar **bivalves**. They are found in both marine and freshwater environments. Bivalves have very small heads and hinged, two-part shells that extend over the sides of the body as well as the top (**Figure 32.15B**). Many clams use the foot to burrow into mud and sand. Bivalves feed by taking in water through an opening called an *incurrent siphon* and filtering food from the water with their large gills, which are also the main sites of gas exchange. Water and gametes exit through the *excurrent siphon*. Fertilization takes place in open water in most species.

GASTROPODS **Gastropods** are the most species-rich and widely distributed mollusks. Snails, whelks, limpets, slugs, nudibranchs (sea slugs), and abalones are all gastropods. Most species move by gliding on the muscular foot, but in a few species—the sea butterflies and heteropods—the foot is a swimming organ, with which the animal moves through open ocean waters. The nudibranchs, or sea slugs, have lost their protective shells over the course of evolution; their sometimes brilliant coloration is *aposematic*; that is, bright colors often serve to warn potential predators of toxicity (**Figure 32.15C**). Other nudibranch species exhibit camouflaged coloration.

Shelled gastropods have one-piece shells. The only mollusks that live in terrestrial environments—land snails and slugs—are gastropods (**Figure 32.15D**). In these terrestrial species, the mantle tissue is modified into a highly vascularized lung.

CEPHALOPODS The **cephalopods**—squids, octopuses, and nautiluses—first appeared near the beginning of the Cambrian period. By the Ordovician period a wide variety of types were present.

In the cephalopods, the excurrent siphon is modified to allow the animal to control the water content of the mantle cavity. The modification of the mantle into a device for forcibly ejecting water from the cavity through the siphon enables these animals to move rapidly by "jet propulsion" through the water. With their greatly enhanced mobility, cephalopods became the major predators in the open waters of the Devonian oceans. They remain important marine predators today.

Cephalopods capture and subdue their prey with their tentacles; octopuses also use their tentacles to move over the substratum (**Figure 32.15E**). As is typical of active, rapidly moving predators, cephalopods have a head with complex sensory organs, most notably eyes that are comparable to those of vertebrates in their ability to resolve images. The head is closely associated with a large, branched foot that bears the tentacles and a siphon. The large, muscular mantle provides a solid external supporting structure. The gills hang in the mantle cavity. Many cephalopods have elaborate courtship behavior, which may involve striking color changes.

Many early cephalopods had chambered shells divided by partitions penetrated by tubes through which gases and liquids could be moved to control their buoyancy. Nautiloids (genus *Nautilus*) are the only cephalopods with such external chambered shells that survive today (**Figure 32.15F**).

32.2 RECAP

Lophotrochozoans include animals with diverse body types. Wormlike forms include some flatworms, ribbon worms, phoronids, and annelids. There has been convergent evolution of lophophores (in ectoprocts versus brachiopods and phoronids) and in external two-part shell coverings (in brachiopods versus bivalve mollusks).

- How can flatworms survive without a gas transport system? See p. 695 and Figure 32.7
- Do you understand why most annelids are restricted to moist environments? See p. 698
- Describe how the basic body plan of mollusks has been modified to yield a wide diversity of animals. See p. 700 and Figure 32.14

The second of the two major protostome clades, the ecdysozoans, contains the vast majority of the Earth's animal species. Let's see what aspects of their body plan led to this massive amount of diversity.

32.3 What Are the Major Groups of Ecdysozoans?

Many ecdysozoans are wormlike in form, although the *arthropods* have limblike appendages. In this section we will look at the two clades of wormlike ecdysozoans: the priapulids, kinorhynchs, and loriciferans in one group, and the horsehair worms and nematodes in the other. Section 32.4 will describe the most familiar and diverse ecdysozoans—the arthropods and their relatives—and the many forms their appendages take in the different arthropod groups.

Several marine groups have relatively few species

Members of several species-poor clades of wormlike marine ecdysozoans—priapulids, kinorhynchs, and loriciferans—have relatively thin cuticles that are molted periodically as the animals grow to full size. In 2004, embryos of a fossil species in this group were discovered in sediments laid down in China about 500 million years ago. These remarkable discoveries show that the ancestors of these animals developed directly from an egg to the adult form, as their modern descendants do.

The 16 species of **priapulids** are cylindrical, unsegmented, wormlike animals with a three-part body plan consisting of a proboscis, trunk, and caudal appendage ("tail"). It should be clear from their appearance why they were named after the Greek fertility god Priapus (**Figure 32.16A**). Priapulids range in size from half a millimeter to 20 centimeters in length. They burrow in fine marine sediments and prey on soft-bodied invertebrates such as polychaetes. They capture prey with a toothed, muscular pharynx that they evert through the mouth and then withdraw into the body together with the grasped prey. Fertilization is external, and most species have a larval form that also lives in the mud.

About 150 species of **kinorhynchs** have been described. They live in marine sands and muds and are virtually microscopic; no kinorhynch species is larger than 1 millimeter in length. Their bodies are divided into 13 segments, each with a separate cuticular plate (**Figure 32.16B**). These plates are periodically molted during growth. Kinorhynchs feed by ingesting sediments through their retractable proboscis (*kinorhynch* means "movable snout"). They then digest the organic material found in the sediment, which may include living algae as well as dead matter. Kinorhynchs have no distinct larval stage; fertilized eggs develop directly into juveniles, which emerge from their egg cases with 11 of the 13 body segments already formed.

Loriciferans are also minute animals less than a millimeter in length. They were not discovered until 1983. About 100 living species are known to exist, although many of these are still being described. The body is divided into a head, neck, thorax, and abdomen and is covered by six plates, from which the loriciferans get their name (Latin *lorica*, "corset"). The plates around the base of the neck bear anteriorly directed spines of unknown function (**Figure 32.16C**). Loriciferans live in coarse marine sediments. Little is known about what they eat, but some species apparently eat bacteria.

Nematodes and their relatives are abundant and diverse

About 320 species of the unsegmented **horsehair worms** have been described. As their name implies, these animals are extremely thin in diameter; horsehair worms range from a few millimeters up to a meter in length (**Figure 32.17**). Most adult worms live in fresh water, among leaf litter and algal mats near the edges of streams and ponds. A few species live in damp soil. The larvae are internal parasites of terrestrial and aquatic insects and freshwater crayfish.

An adult horsehair worm has no mouth opening, and its gut is greatly reduced and probably nonfunctional. Some of these worms may feed only as larvae, absorbing nutrients from their hosts across the body wall. Others, however, continue to grow after they have left their hosts, shedding their cuticles, suggesting that adult worms may also absorb nutrients from their environment.

32.16 Wormlike Marine Ecdysozoans (A) A priapulid. Priapulids are marine worms that live, usually in burrows, on the ocean floor. (B) Kinorhynchs are virtually microscopic. The cuticular plates that cover their bodies are periodically molted. (C) Six cuticular plates form a "corset" around the loriciferan body.

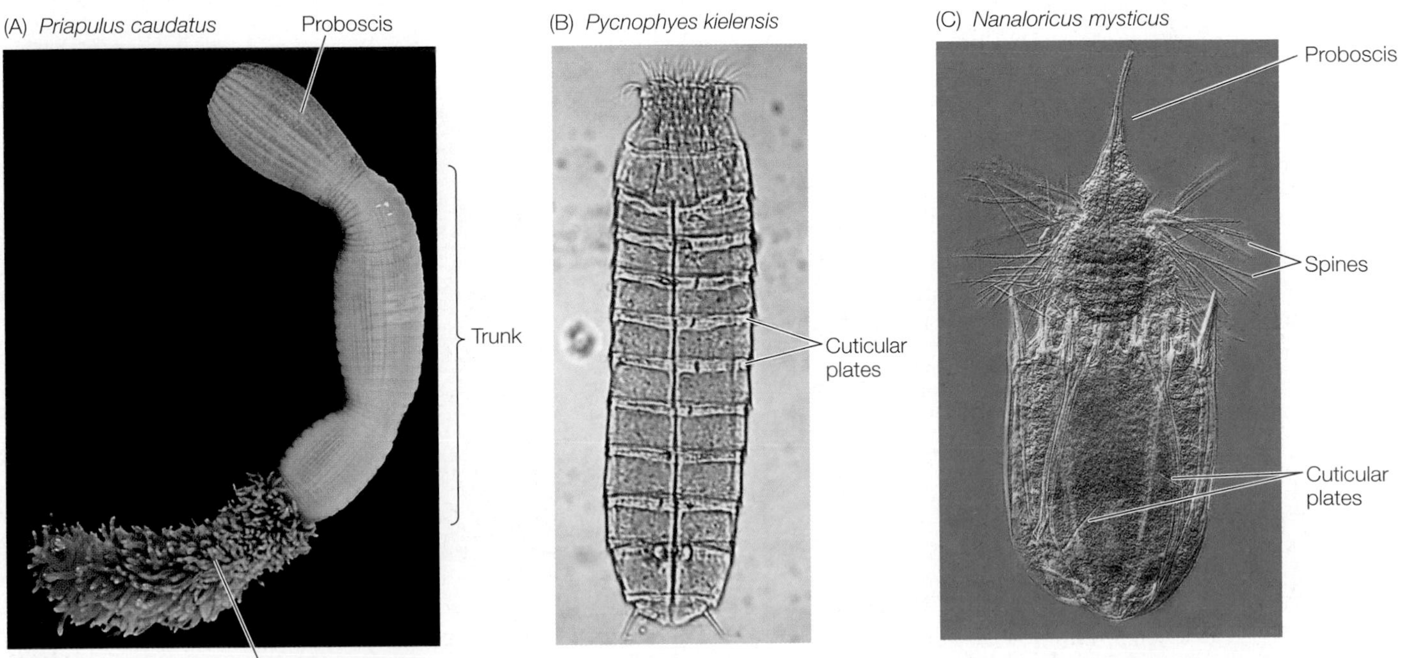

Paragordius sp.

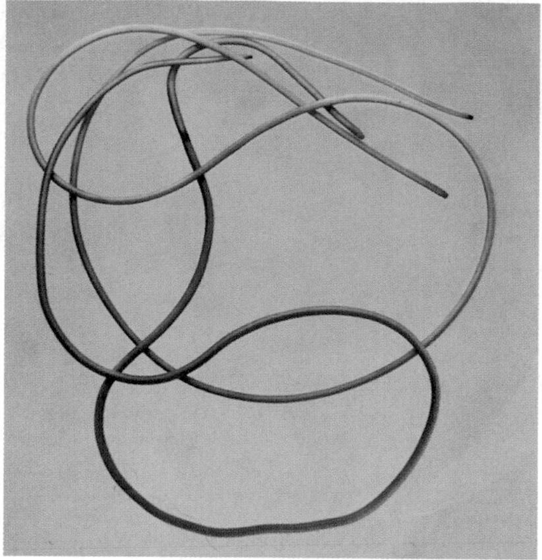

32.17 A Horsehair Worm Horsehair worms get their name from their hairlike or threadlike shape. They can grow to be up to a meter long.

Nematodes (roundworms) have a thick, multilayered cuticle that gives their unsegmented body its shape (**Figure 32.18**). As a nematode grows, it sheds its cuticle four times. Nematodes exchange oxygen and nutrients with their environment through both the cuticle and the gut, which is only one cell layer thick. Materials are moved through the gut by rhythmic contraction of a highly muscular organ, the *pharynx*, at the worm's anterior end. Nematodes move by contracting their longitudinal muscles.

Nematodes are one of the most abundant and universally distributed of all animal groups. Many are microscopic; the largest known nematode, which reaches a length of 9 meters, is a parasite in the placentas of female sperm whales. About 25,000 species have been described, but the actual number of living species may be more than a million. Countless nematodes live as scavengers in the upper layers of the soil, on the bottoms of lakes and streams, and in marine sediments. The topsoil of rich farmland may contain from 3 to 9 billion nematodes per acre. A single rotting apple may contain as many as 90,000 individuals.

> The zoologist Ralph Buchsbaum wrote that "If all the matter in the universe except the nematodes were swept away, our world would still be dimly recognizable ... we should find its mountains, hills, vales, rivers, lakes, and oceans represented by a film of nematodes. ... The location of the various plants and animals would still be decipherable."

One soil-inhabiting nematode, *Caenorhabitis elegans*, serves as a "model organism" in the laboratories of geneticists and developmental biologists. It is ideal for such research because it is easy to cultivate, matures in three days, and has a fixed number of body cells. Its genome has been completely mapped.

Many nematodes are predators, feeding on protists and small animals (including other roundworms). Most significant to humans, however, are the many species that parasitize plants and animals. The nematodes that are parasites of humans (causing serious tropical diseases such as trichinosis, filariasis, and elephantiasis), domestic animals, and economically important plants have been studied intensively in an effort to find ways of controlling them.

The structure of parasitic nematodes is similar to that of free-living species, but the life cycles of many parasitic species have special stages that facilitate the transfer of individuals among hosts. *Trichinella spiralis*, the species that causes the human disease trichi-

(B)

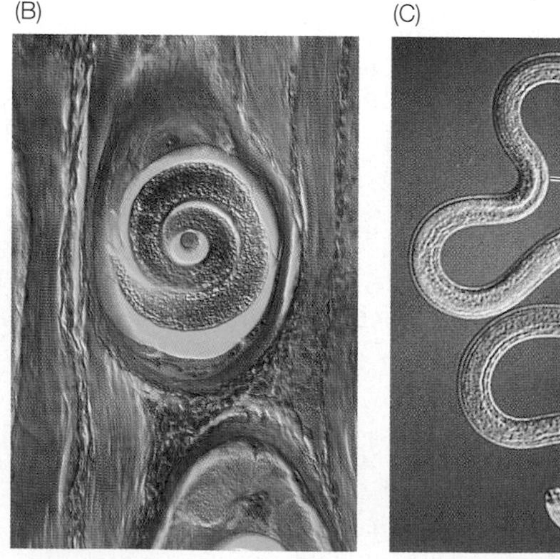

(C)

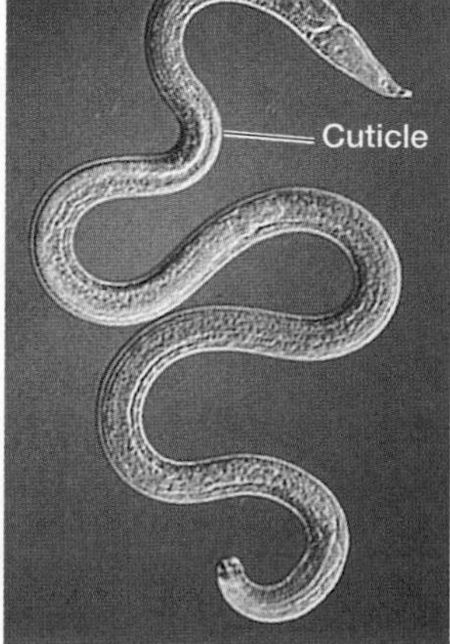

32.18 Nematodes (A) The body plan of *Trichinella spiralis*, a parasitic nematode that causes trichinosis. (B) A cyst of *Trichinella spiralis* in the muscle tissue of a host. (C) This free-living nematode moves through marine sediments.

nosis, has a relatively simple life cycle. A person may become infected by eating the flesh of an animal (usually a pig) containing larvae of *Trichinella* encysted in its muscles (see Figure 32.18B). The larvae are activated in the person's digestive tract, emerge from their cysts, and attach to the intestinal wall, where they feed. Later, they bore through the intestinal wall and are carried in the bloodstream to muscles, where they form new cysts. If present in great numbers, these cysts can cause severe pain or death.

32.3 RECAP

The priapulids, kinorhynchs, and loriciferans are relatively small, poorly known groups of wormlike marine ecdysozoans. The horsehair worms and the nematodes are also wormlike, but the biology of some species has been studied extensively. The nematodes are among the most abundant and species-rich groups on Earth.

- Can you describe at least three different ways in which nematodes have significant impact on humans? See pp. 704–705

We now turn to the animals that not only dominate the ecdysozoan clade, but are also the most numerous animals on Earth.

32.4 Why Do Arthropods Dominate Earth's Fauna?

The arthropods and their relatives are ecdysozoans with limblike appendages. Collectively, arthropods are the dominant animals on Earth, both in numbers of species (over a million described) and numbers of individuals (estimated at some 10^{18} individuals, or a billion billion).

Several key features have contributed to the success of the arthropods. Their bodies are segmented, and their muscles are attached to the inside of their rigid exoskeletons. Each segment has muscles that operate that segment and the jointed appendages attached to it (see Figure 32.4). The jointed appendages permit complex movement patterns, and different appendages are specialized for different functions. Encasement of the body within a rigid exoskeleton provides the animal with support for walking in the water or on dry land and provides some protection against predators. The waterproofing provided by chitin keeps the animal from dehydrating in dry air.

Representatives of the four major arthropod groups living today are all species-rich: the crustaceans (including shrimps, crabs, and barnacles), hexapods (insects and their relatives), myriapods (millipedes and centipedes), and chelicerates (including the arachnids—spiders, scorpions, mites and their relatives). The phylogenetic relationships among arthropod groups are currently being reexamined in the light of a wealth of new information, much of it based on gene sequences (see Chapter 24). As shown in **Figure 32.19,** most of the current uncertainty concerns the relationship of the myriapods to the other three groups. However, it appears clear that the arthropods constitute a monophyletic group, and most analyses support the close relationship between hexapods and crustaceans (the lower two trees in Figure 32.19).

Arthropods evolved from ancestors with simple, unjointed appendages. The exact forms of those ancestors are unknown, but some arthropod relatives with segmented bodies and unjointed appendages survive today. Before we describe the modern arthropods, we will discuss those arthropod relatives as well as an early arthropod clade that went extinct, but left an important fossil record.

Arthropod relatives have fleshy, unjointed appendages

Until fairly recently, biologists debated whether the **onychophorans** (velvet worms) were more closely related to annelids or arthropods, but molecular evidence clearly links them to the latter. In-

32.19 The Placement of Myriapods Among Arthropods Is Under Study Although many aspects of arthropod phylogeny are well understood, the placement of the myriapods is still debated among systematists. These three trees show three possible placements of the myriapods; most analyses favor one of the lower two trees.

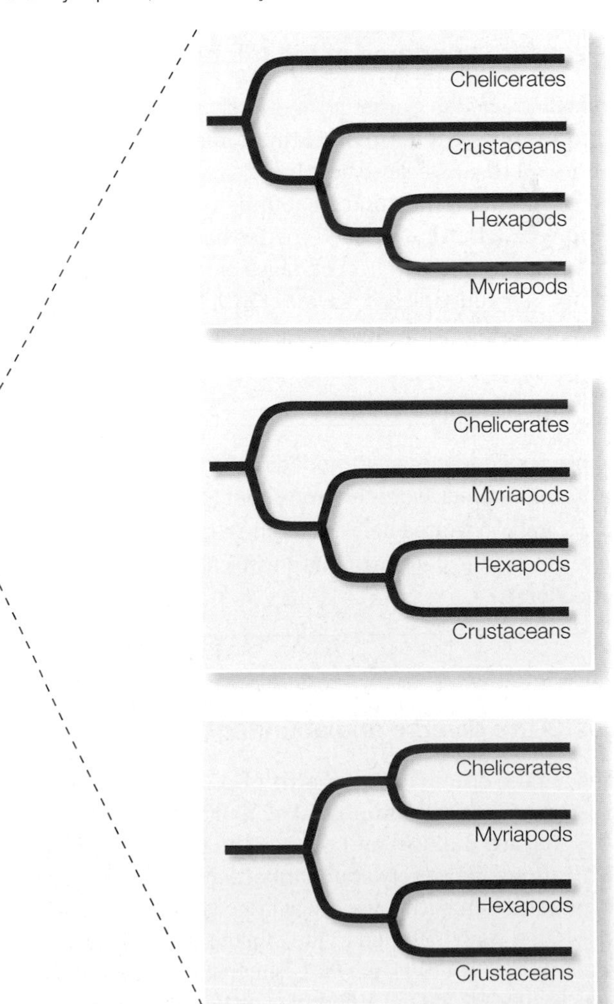

32.20 Arthropod Relatives with Unjointed Appendages
(A) Onychophorans, also called velvet worms, have unjointed legs and use the body cavity as a hydrostatic skeleton. (B) The appendages and general anatomy of tardigrades superficially resemble those of onychophorans.

(A) *Peripatus* sp.

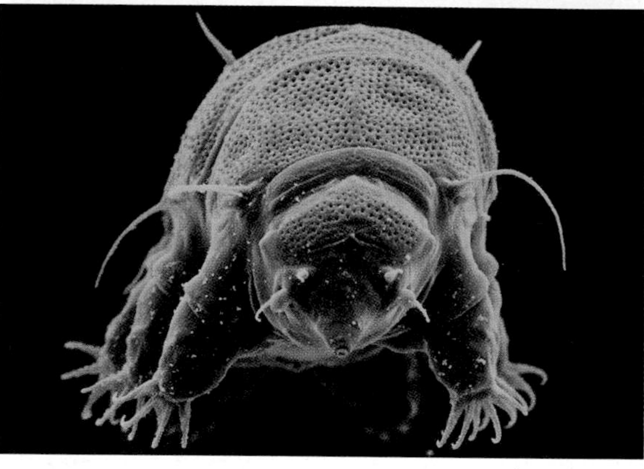

(B) *Echiniscus springer* 50 μm

deed, with their soft, fleshy, unjointed, claw-bearing legs, onychophorans may be similar in appearance to the ancestors of arthropods (**Figure 32.20A**). The 150 species of onychophorans live in leaf litter in humid tropical environments. They have soft, segmented bodies that are covered by a thin, flexible cuticle that contains chitin. They use their fluid-filled body cavities as hydrostatic skeletons. Fertilization is internal, and the large, yolky eggs are brooded within the body of the female.

Tardigrades (water bears) also have fleshy, unjointed legs and use their fluid-filled body cavities as hydrostatic skeletons (**Figure 32.20B**). Tardigrades are extremely small (0.1–0.5 mm in length) and lack both circulatory systems and gas exchange organs. The 800 known extant species live in marine sands and on temporary water films on plants. When these films dry out, the animals also lose water and shrink to small, barrel-shaped objects that can survive for at least a decade in a dormant state. Tardigardes have been found to have densities as high as 2 million per square meter of moss.

Jointed legs first appeared in the trilobites

The **trilobites** flourished in Cambrian and Ordovician seas, but they disappeared in the great Permian extinction at the close of the Paleozoic era (251 mya). Because they had heavy exoskeletons that readily fossilized, they left behind an abundant record of their existence (**Figure 32.21**). About 10,000 species have been described.

The body segmentation and appendages of trilobites followed a relatively simple, repetitive plan, but some of their jointed appendages were modified for different functions. This specialization of appendages is a theme in the continuing evolution of the arthropods.

The jointed appendages of arthropods gave the clade its name, from the Greek words *arthron*, "joint," and *podos*, "foot" or "limb." A related derivation can be discerned that describes the widespread mammalian affliction *arthritis* (inflamed joints).

Crustaceans are diverse and abundant

Crustaceans are the dominant marine arthropods today. The most familiar crustaceans are the shrimps, lobsters, crayfishes, and crabs (all *decapods*; **Figure 32.22A**) and the sow bugs (*isopods*; **Figure 32.22B**). *Krill* are small but very abundant oceanic crustaceans that are important food items for a variety of large vertebrates, including baleen whales. Also included in the crustacean clade are a variety of small *copepod* species, many of which superficially resemble shrimps (**Figure 32.22C**). Copepods are also abundant in the open oceans. Additional species-rich groups of crustaceans include *amphipods* and *ostracods* (both found in freshwater and marine environments).

Barnacles are unusual crustaceans that are sessile as adults (**Figure 32.22D**). Adult barnacles look more like mollusks than like other crustaceans, but, as the zoologist Louis Agassiz remarked more than a century ago, a barnacle is "nothing more than a little shrimp-like animal, standing on its head in a limestone house and kicking food into its mouth."

Odontochile rugosa

32.21 A Trilobite The relatively simple, repetitive segments of the now-extinct trilobites are illustrated by a fossil trilobite from the shallow seas of the Devonian period, some 400 million years ago.

(A) *Grapsus grapsus*

(C) *Cyclops* sp.

(B) *Ligia occidentalis*

(D) *Lepas pectinata*

32.22 Crustacean Diversity (A) This decapod crustacean is one of several crab species referred to as "Sally Lightfoots" on account of their entrancing "tiptoe" walk. Charles Darwin described this genus, which he found on the Galápagos Islands. (B) This isopod, commonly called the "rock louse," is found on rocky beaches of the California coast. (C) This microscopic freshwater copepod is only about 30 μm long. (D) Gooseneck barnacles attach to a substratum by their muscular stalks and feed by protruding and retracting their feeding appendages.

Most of the 50,000 described species of crustaceans have a body that is divided into three regions: *head, thorax,* and *abdomen* (**Figure 32.23**). The segments of the head are fused together, and the head bears five pairs of appendages. Each of the multiple thoracic and abdominal segments usually bears one pair of appendages. The appendages on different parts of the body are specialized for different functions, such as gas exchange, chewing, sensing, walking, and swimming. In some cases, the appendages are branched, with different branches serving different functions. In many species, a fold of the exoskeleton, the *carapace,* extends dorsally and laterally back from the head to cover and protect some of the other segments.

The fertilized eggs of most crustacean species are attached to the outside of the female's body, where they remain during their early development. At hatching, the young of some species are released as larvae; those of other species are released as juveniles that are similar in form to the adults. Still other species release eggs into the water or attach them to an object in the environment. The typical crustacean larva, called a *nauplius,* has three pairs of appendages and one central eye (see Figure 31.12B). In many crustaceans, the nauplius larva develops within the egg before it hatches.

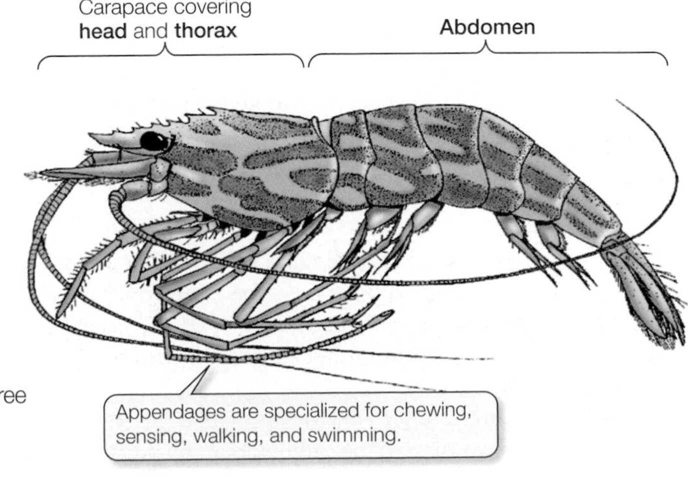

32.23 Crustacean Structure The bodies of crustaceans are divided into three regions—the head, thorax, and abdomen. Each region bears specialized appendages, and a thin, shell-like carapace covers the head and thorax.

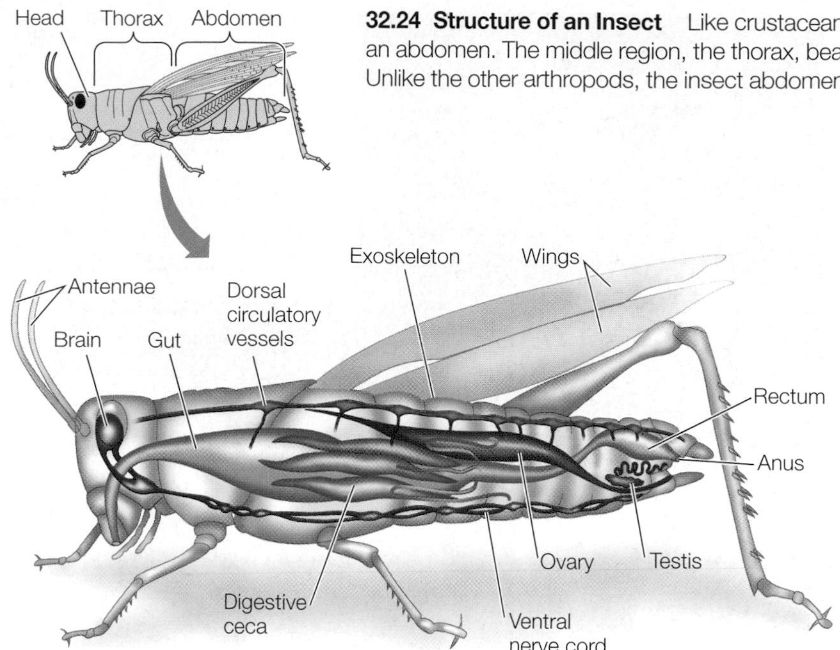

32.24 Structure of an Insect Like crustaceans, the insect's three-part body plan has a head, a thorax, and an abdomen. The middle region, the thorax, bears three pairs of legs and, in most groups, two pairs of wings. Unlike the other arthropods, the insect abdomen bears no appendages (see Figure 20.7).

Insects are the dominant terrestrial arthropods

During the Devonian period, more than 400 million years ago, some arthropods colonized terrestrial environments. Of the several groups that successfully colonized the land, none is more prominent today than the six-legged *hexapods*: the **insects** and their relatives. Insects are abundant and diverse in terrestrial and freshwater environments; only a few live in the oceans.

Insects, like crustaceans, have bodies with three regions—head, thorax, and abdomen. In addition, insects have a unique mechanism for gas exchange in air: a system of air sacs and tubular channels called *tracheae* (singular *trachea*) that extend from external openings called *spiracles* inward to tissues throughout the body. They have a single pair of antennae on the head and three pairs of legs attached to the thorax (**Figure 32.24**). Unlike the other arthropods, insects have no appendages growing from their abdominal segments.

Insects use nearly all species of plants and many species of animals as food. Herbivorous insects can consume massive amounts of plant matter. Gypsy moth caterpillars, for example, have been known to denude entire forests of their hardwood tree leaves, and the cyclical depradations wrought by locust hordes can be of biblical proportions. Many insects are predatory, feeding on small animals or on other insects. *Detrivorous* insects such as dung beetles are important in the recycling of chemicals through ecosystems (see Section 56.1). Some insects are internal parasites of plants or animals; others suck their host's blood or consume surface body tissues.

As recently as 1982, most biologists thought that about half of existing insect species had been described, but today they think that the approximately 1,000,000 species of described insects are a small fraction of the total number of living species. Why did they change their minds?

A simple but important field study provided the stimulus. One entomologist, Terry Erwin of the Smithsonian Institution in Washington, D.C., was aware that the insects of tropical forests (the most species-rich habitat on Earth) were poorly known. To determine how little we knew, he decided to sample the beetles in the canopies of a single species of tropical forest tree, *Luehea seemannii*. Erwin fogged the canopies of several large *L. seemannii* individuals with a pesticide and collected the insects that fell from the tree in collection nets (**Figure 32.25**). His sample contained an amazing number of species, many of them undescribed. He then used a set of assumptions to estimate the total number of species of insects in tropical evergreen forests.

The assumptions Erwin used to make his calculations were uncertain. Had he based his calculations on

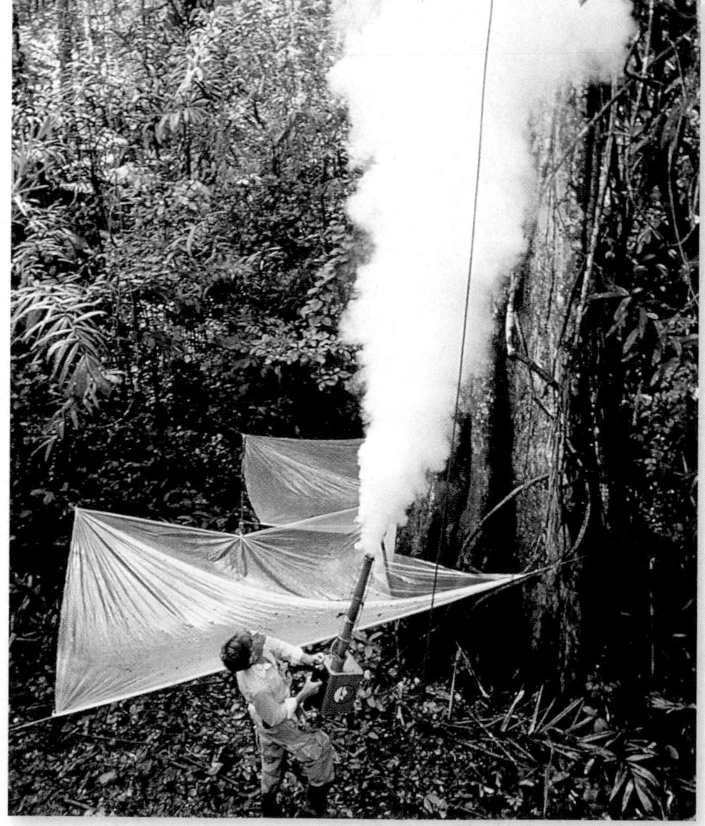

32.25 Erwin's Research Project In an important study, entomologist Terry Erwin fogged tree canopies in the Central American rainforest with insecticide. Using the number of different species represented among the fallen insects as a base, he applied extrapolatory mathematical equations to estimate the total number of insect species on Earth. FURTHER RESEARCH: What kinds of experiments would you perform to refine Erwin's estimates of the number of species of insects on Earth?

different assumptions, his estimates of the number of species of insects could have changed dramatically. For example, if he overestimated the proportion of species that are specialists, his calculations would have overestimated the total number of species; but if he underestimated the proportion of specialists, the reverse would be true. Erwin assumed that he had adequately sampled the insect fauna of *Luehea seemannii* trees, but had he fogged the canopies of more individual trees, especially from different areas, he might have found many more species, which would mean his estimate was too low. But despite its uncertainties, Erwin's pioneering study alerted biologists to the fact that we live on a poorly known planet, most of whose species have yet to be named and described.

Terry Erwin's field within entomology (the study of insects) is *coleoptology*—the study of beetles. This specialty is not as esoteric as it might sound; beetles account for nearly 40 percent of the known insect species. Noted Scottish evolutionary biologist J. B. S. Haldane, when asked in 1951 what his studies revealed about the Creator, is said to have replied, "An inordinate fondness for beetles."

TABLE 32.2

The major insect groups[a]

GROUP	APPROXIMATE NUMBER OF DESCRIBED LIVING SPECIES
Jumping bristletails (*Archaeognatha*)	300
Silverfish (*Thysanura*)	370
PTERYGOTE (WINGED) INSECTS (*PTERYGOTA*)	
Mayflies (*Ephemeroptera*)	2,000
Dragonflies and damselflies (*Odonata*)	5,000
Neopterans (*Neoptera*)[b]	
Ice-crawlers (*Grylloblattodea*)	25
Gladiators (*Mantophasmatodea*)	15
Stoneflies (*Plecoptera*)	1,700
Webspinners (*Embioptera*)	300
Angel insects (*Zoraptera*)	30
Earwigs (*Dermaptera*)	1,800
Grasshoppers and crickets (*Orthoptera*)	20,000
Stick insects (*Phasmida*)	3,000
Cockroaches (*Blattodea*)	3,500
Termites (*Isoptera*)	2,750
Mantids (*Mantodea*)	2,300
Booklice and barklice (*Psocoptera*)	3,000
Thrips (*Thysanoptera*)	5,000
Lice (*Phthiraptera*)	3,100
True bugs, cicadas, aphids, leafhoppers (*Hemiptera*)	80,000
Holometabolous neopterans (*Holometabola*)[c]	
Ants, bees, wasps (*Hymenoptera*)	125,000
Beetles (*Coleoptera*)	375,000
Strepsipterans (*Strepsiptera*)	600
Lacewings, ant lions, dobsonflies (*Neuropterida*)	4,700
Scorpionflies (*Mecoptera*)	600
Fleas (*Siphonaptera*)	2,400
True flies (*Diptera*)	120,000
Caddisflies (*Trichoptera*)	5,000
Butterflies and moths (*Lepidoptera*)	250,000

[a] The hexapod relatives of insects include the springtails (*Collembola*; 3,000 spp.), two-pronged bristletails (*Diplura*; 600 spp.), and proturans (*Protura*; 10 spp.). All are wingless and have internal mouthparts.

[b] Neopteran insects can tuck their wings close to their bodies

[c] Holometabolous insects are neopterans that undergo complete metamorphosis.

Entire textbooks—and hefty ones at that—have been written about the described groups of insects. Here and in **Table 32.2** we present a brief outline of the major groups covered by these detailed entomology texts.

The wingless relatives of the insects—springtails (**Figure 32.26**), two-pronged bristletails, and proturans—are probably the most similar of living forms to insect ancestors. These relatives of insects have a simple life cycle; they hatch from eggs as miniature adults. They differ from insects in having internal mouth parts. Springtails can be extremely abundant (up to 200,000 per square meter) in soil, leaf litter, and on vegetation, and are the most abundant hexapods in the world.

Insects can be distinguished from other hexapods by their external mouthparts and paired antennae that contain a sensory receptor called Johnston's organ. Two groups of insects—the jumping bristletails and silverfish—are wingless and have simple life cycles, like the springtails and other close insect relatives. The remaining insects are the *pterygote insects*. Pterygotes have two pairs of wings, except in some groups where the wings have been secondarily lost.

Sminthurides aquaticus

32.26 Wingless Hexapods The wingless hexapods, such as this springtail, have a simple life cycle. They hatch looking like miniature adults, then grow by successive molts of the cuticle.

Hatchling pterygote insects do not look like adults, and they undergo substantial changes at each molt. The immature stages of insects between molts are called **instars.** As we saw in Section 31.4, a substantial change that occurs between one developmental stage and another is called **metamorphosis**. If the changes between its instars are gradual, an insect is said to have **incomplete metamorphosis.** If the change between at least some instars is dramatic, an insect is said to have **complete metamorphosis**.

The most familiar examples of species with complete metamorphosis are butterflies (see Figure 31.11). The wormlike butterfly larva, called a *caterpillar*, transforms itself during a specialized phase, called the **pupa**, in which many larval tissues are broken down and the adult form develops. In many insects with complete metamorphosis, the different life stages are specialized for living in different environments and using different food sources. In many species, the larvae are adapted for feeding and growing; the adults are specialized for reproduction and dispersal.

Pterygote insects were the first animals in evolutionary history to achieve the ability to fly (although members of several groups of winged insects—including parasitic lice and fleas, as well some beetles and the worker individuals in many ant groups—subsequently lost their wings and their ability to fly). Flight opened up many new lifestyles and feeding opportunities that only the insects could exploit, and it is almost certainly one of the reasons for the remarkable numbers of insect species and individuals, and for their unparalleled evolutionary success.

The adults of most flying insects have two pairs of stiff, membranous wings attached to the thorax. True flies, however, have only one pair of wings. In beetles, one pair of wings—the forewings—forms heavy, hardened wing covers. Flying insects are important pollinators of flowering plants.

Two groups of pterygote insects, the mayflies and dragonflies (**Figure 32.27A**), cannot fold their wings against their bodies. This is the ancestral condition for pterygote insects, and these two groups are not closely related to one another. All members of these groups have predatory aquatic larvae that transform themselves into flying adults after they crawl out of the water. Many of them are excellent flyers. Dragonflies (and their relatives the damselflies) are active predators as adults, but adult mayflies lack functional digestive tracts; adult mayflies live only about a day, just long enough to mate and lay eggs.

All other pterygote insects—the *neopterans*—can tuck their wings out of the way upon landing and crawl into crevices and other tight places. Many neopteran groups have an incomplete metamorphosis, so hatchlings of these insects are sufficiently similar in form to adults to be recognizable. Examples include the grasshoppers (**Figure 32.27B**), roaches, mantids, stick insects, termites, stoneflies, earwigs, thrips, true bugs (**Figure 32.27C**), aphids, cicadas, and leafhoppers. They acquire adult organ systems, such as wings and compound eyes, gradually through several juvenile instars. Remarkably, a new major group, the mantophasmatodeans, was described only recently, in 2002 (**Figure 32.27D**). These small insects are common in the Cape Region of South Africa, an area of exceptional species richness and endemism for many animal and plant groups.

Over 80 percent of all insects are a subgroup of the neopterans called the *holometabolous insects* (see Table 32.2), which have complete metamorphosis. The many species of beetles comprise almost half of this group (**Figure 32.27E**). Also included are lacewings and their relatives; caddisflies; butterflies and moths (**Figure 32.27F**); sawflies; true flies (**Figure 32.27G**); and bees, wasps, and ants, some species of which display unique and highly specialized social behaviors (**Figure 32.27H**).

32.27 The Diversity of Winged Insects (A) Unlike most flying insects, this dragonfly cannot fold its wings over its back. (B) The Mexican bush katydid represents an orthopteran. (C) Dock bugs are "true" bugs (hemipterans) specialized for feeding on a plants of the genus *Rumex*. (D) This mantophasmatodean represents a recently discovered insect group found only in southern Africa. (E) A predatory diving beetle (a coleopteran). (F) The California dogface butterfly is a lepidopteran. (G) Flies such as this green bottle fly are dipterans. (H) Many hymenopteran genera, such as the honeybees seen here on their hive, are social insects.

Molecular data suggest that the lineage leading to the insects separated from the lineage leading to modern crustaceans about 450 million years ago, about the time of the appearance of the first land plants. These ancestral hexapods penetrated a terrestrial environment that lacked any other similar organisms, which in part accounts for their remarkable success. But the success of the insects is also due to their wings. Homologous genes control the development of insect wings and crustacean appendages, suggesting that that the insect wing evolved from a dorsal branch of a crustacean-like limb (**Figure 32.28**). The dorsal limb branch of crustaceans is used for gas exchange. Thus the insect wing probably evolved from a gill-like structure that had a gas exchange function.

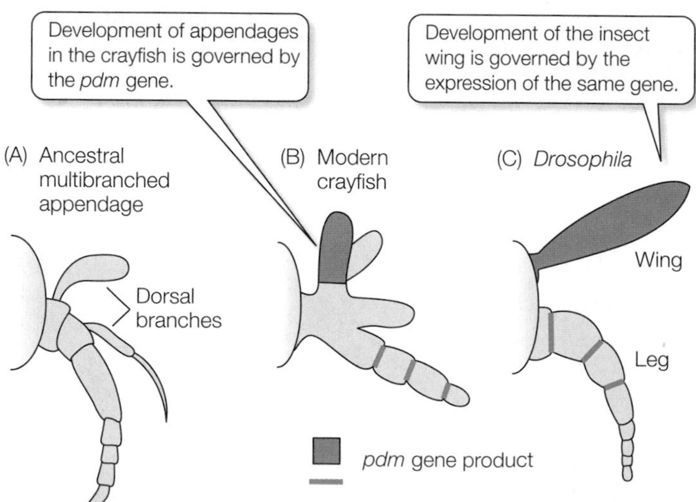
Development of appendages in the crayfish is governed by the *pdm* gene.
Development of the insect wing is governed by the expression of the same gene.
(A) Ancestral multibranched appendage
(B) Modern crayfish
(C) *Drosophila*
Dorsal branches
Wing
Leg
pdm gene product

32.28 The Origin of Insect Wings? The insect wing may have evolved from an ancestral appendage similar to that of modern crustaceans. (A) A diagram of the ancestral, multibranched arthropod limb. (B,C) The *pdm* gene, a Hox gene, is expressed throughout the dorsal limb branch and walking leg of the thoracic limb of a crayfish (B) and in the wings and legs of *Drosophila* (C).

(A) *Libellula luctuosa*

(B) *Scudderia mexicana*

(C) *Coreus marginatus*

(D) *Mantophasma zephyra*

(E) *Dytiscus marginalis*

(F) *Colias eurydice*

(G) *Phaenicia sericata*

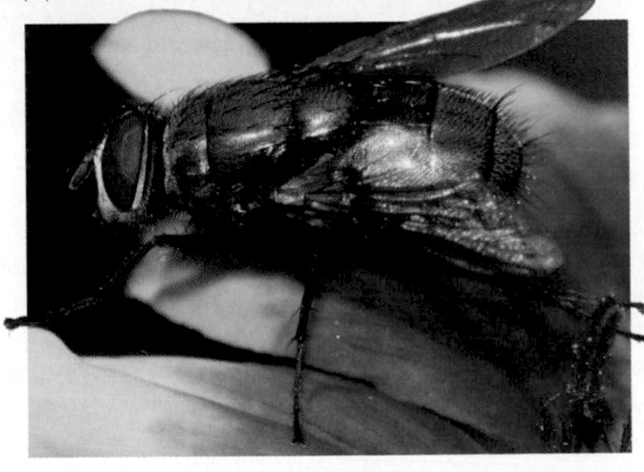

(H) *Apis mellifera*

Myriapods have many legs

As we have seen, insects and most crustaceans have a three-part body plan, with a head, thorax, and abdomen. Members of two other arthropod groups, the myriapods and chelicerates, have a body plan with just two regions: a head and a trunk.

The **myriapods** comprise the centipedes, millipedes, and two other groups. Centipedes and millipedes have a well-formed head and a long, flexible, segmented trunk that bears many pairs of legs (**Figure 32.29**). Centipedes, which have one pair of legs per segment, prey on insects and other small animals. In millipedes, two adjacent segments are fused so that each fused segment has two pairs of legs. Millipedes scavenge and eat plants. More than 3,000 species of centipedes and 11,000 species of millipedes have been described; many more species probably remain unknown. Although most myriapods are less than a few centimeters long, some tropical species are ten times that size.

Most chelicerates have four pairs of legs

In the two-part body plan of **chelicerates**, the head bears two pairs of appendages modified to form mouthparts. In addition, many chelicerates have four pairs of walking legs. The 89,000 described species are usually placed in three clades: pycnogonids, horseshoe crabs, and arachnids.

The *pycnogonids*, or sea spiders, are a poorly known group of about 1,000 marine species (**Figure 32.30A**). Most are small, with leg spans less than 1 centimeter, but some deep-sea species have leg spans up to 60 centimeters. A few pycnogonids eat algae, but most are carnivorous, eating a variety of small invertebrates.

There are four living species of *horseshoe crabs*, but many close relatives are known only from fossils. Horseshoe crabs, which have changed very little morphologically during their long fossil history, have a large horseshoe-shaped covering over most of the body. They are common in shallow waters along the eastern coasts of North America and the southern and eastern coasts of Asia, where

(A) *Scolopendra gigantea*

(B) *Sigmoria trimaculata*

32.29 Myriapods (A) Centipedes have powerful jaws for capturing active prey and one pair of legs per segment. (B) Millipedes, which are scavengers and plant eaters, have smaller jaws and legs. They have two pairs of legs per segment.

(A) *Colossendeis megalonyx*

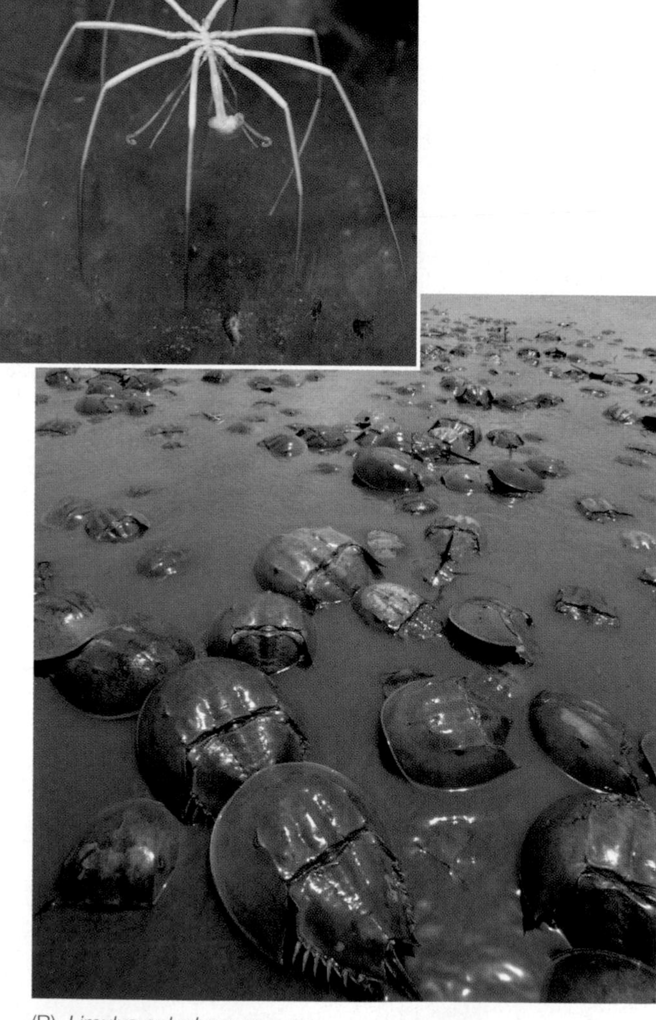

(B) *Limulus polyphemus*

32.30 Some Small Groups of Chelicerates (A) Although they are not spiders, it is easy to see why sea spiders were given their common name. (B) This spawning aggregation of horseshoe crabs was photographed on a sandy beach of the Delaware Bay. Horseshoe crabs are an ancient group that has changed very little in morphology over time; such species are sometimes referred to as "living fossils."

(A) *Poecilotheria metallica*

(C) *Hadrobunus maculosus*

(B) *Pseudouroctonus minimus*

(D) *Brevipalpus phoenicis*

32.31 Arachnid Diversity (A) The name "tarantula" encompasses several hundred species of hairy, ground-dwelling spiders, some of which can grow to the size of a dinner plate. Their venomous bite, although painful, is not usually deadly to humans. (B) Scorpions are nocturnal predators. (C) Harvestmen, also called daddy longlegs, are scavengers. (D) Mites are blood-sucking, external parasites.

they scavenge and prey on bottom-dwelling animals. Periodically they crawl into the intertidal zone in large numbers to mate and lay eggs (**Figure 32.30B**).

Arachnids are abundant in terrestrial environments. Most arachnids have a simple life cycle in which miniature adults hatch from internally fertilized eggs and begin independent lives almost immediately. Some arachnids retain their eggs during development and give birth to live young.

The most species-rich and abundant arachnids are the spiders, scorpions, harvestmen, mites, and ticks (**Figure 32.31**). The 50,000 described species of mites and ticks live in soil, leaf litter, mosses, and lichens, under bark, and as parasites of plants and animals. Mites are vectors for wheat and rye mosaic viruses; they cause mange in domestic animals and skin irritation in humans.

Spiders (of which 38,000 species have been described) are important terrestrial predators. Some have excellent vision that enables them to chase and seize their prey. Others spin elaborate webs made of protein threads in which they snare prey. The threads are produced by modified abdominal appendages connected to internal glands that secrete the proteins, which dry on contact with air. The webs of different groups of spiders are strikingly varied, and this variation enables the spiders to position their snares in many different environments.

32.4 RECAP

All arthropods have segmented bodies. Muscles in each segment operate that segment and the appendages attached to it. Their jointed, specialized appendages permit complex patterns of movement, including the ability of the insects to fly.

- What features have contributed to making arthropods the dominant animals on Earth, both in number of species and number of individuals? Can you think of some possible reasons why there are so many species of insects?
- Do you understand the difference between incomplete and complete metamorphosis? See p. 710

The protostomes encompass the vast majority of the Earth's animal species, so it is not surprising that the different protostome groups display a huge variety of different characteristics.

An Overview of Protostome Evolution

The myriad protostome species encompass a staggering number of different life forms. Consider how the following aspects of protostome evolution have contributed to this enormous diversity:

- The evolution of *segmentation* permitted some protostome groups to move different parts of the body independently of one another. Species in some groups gradually evolved the ability to move rapidly over and through the substratum, through water, and through air.
- *Complex life cycles* with dramatic changes in form between one stage and another allow individuals of different stages to specialize on different resources.
- *Parasitism* has evolved repeatedly, and many protostome groups parasitize multicellular plants and animals.
- The evolution of *diverse feeding structures* allowed protostomes to specialize on many different food sources. Early protostomes were mostly filter-feeders because dissolved organic matter and very small organisms were the most readily available food sources in the oceans. Although filter-feeding remains common, the ability to move rapidly on land and through water and air allowed the evolution of predation and herbivory.
- Predation was a major selection pressure favoring the development of *hard external body coverings* (exoskeletons). Such coverings evolved independently in many lophotrochozoan and ecdysozoan groups. In addition to providing protection, they became key elements in the development of new systems of locomotion.
- *Better locomotion* permitted prey to escape from predators, but also allowed predators to pursue their prey more effectively. Thus the evolution of animals has been, and continues to be, a complex "arms race" among predators and prey.

Many major evolutionary trends among the protostomes are shared by the deuterostomes, which includes the chordates, the group to which humans belong.

CHAPTER SUMMARY

32.1 What is a protostome?

Protostomes ("mouth first") have an anterior brain that surrounds the entrance to the digestive tract and a ventral nervous system. The major protostome clades are grouped into the **lophotrochozoans** and the **ecdysozoans**. Review Figure 32.1

Lophotrochozoans include a wide diversity of animals. Within this group, **lophophores**, free-living **trochophore** larvae, and **spiral cleavage** evolved.

Ecdysozoans have an **exoskeleton**, which they must **molt** in order to grow. Some ecdysozoans have a relatively thin exoskeleton, called a **cuticle**. Others, especially the **arthropods**, have a rigid exoskeleton reinforced with **chitin**. New mechanisms of locomotion and gas exchange evolved in these animals. Review Figure 32.4

32.2 What are the major groups of lophotrochozoans?

Lophotrochozoans range from relatively simple animals having only one entrance to the digestive tract and no oxygen transport system to animals with complete digestive tracts and complex internal transport systems.

Lophophores, wormlike body forms, and external shells are each found in multiple unrelated groups of lophotrochozoans.

The most species-rich groups of lophotrochozoans are the **flatworms**, **annelids**, and **mollusks**.

Segmentation of the body first evolved among the annelids. Review Figure 32.12

Mollusks underwent a dramatic evolutionary radiation based on a body plan consisting of three major components: a **foot**, a **mantle**, and a **visceral mass**. The five major molluscan clades—the **monoplacophorans**, **chitons**, **bivalves**, **gastropods**, and **cephalopods**—demonstrate the diversity that evolved from this three-part body plan. Review Figure 32.14

32.3 What are the major groups of ecdysozoans?

Many groups of ecdysozoans are wormlike in form. Members of several species-poor clades of marine wormlike ecdysozoans (**priapulids**, **kinorhynchs**, and **loriciferans**) have thin cuticles.

Horsehair worms are extremely thin; many are internal parasites as larvae.

Nematodes, or roundworms, have thick, multilayered cuticles. Nematodes are one of the most abundant and universally distributed of all animal groups. Review Figure 32.17

One major ecdysozoan clade, the **arthropods** and their relatives, have evolved limblike appendages. Collectively, the arthropods are the dominant animals on Earth, both in number of species and number of individuals.

32.4 Why do arthropods dominate Earth's fauna?

Encasement within a rigid exoskeleton provides arthropods with support for walking in the water and on dry land as well as some protection from predators. The waterproofing provided by chitin keeps arthropods from dehydrating in dry air.

Jointed appendages permit complex movement patterns. Each arthropod segment has muscles attached to the inside of the exoskeleton that operate that segment and the appendages attached to it.

Two groups of arthropod relatives, the **onychophorans** and the **tardigrades**, have simple, unjointed appendages. The **trilobites** were early marine arthropods that disappeared in the Permian extinction.

Crustaceans are the dominant marine arthropods. Their segmented bodies are divided into three regions (head, thorax, abdomen) with different, specialized appendages in each region. Review Figure 32.23

Hexapods—insects and their relatives—are the dominant terrestrial arthropods. They have the same three body regions as crustaceans, but no appendages form in their abdominal segments. Wings and the ability to fly first evolved among the insects, allowing them to exploit new lifestyles. Review Figure 32.24

The bodies of **myriapods** have only two regions, a head, and a long trunk with many segments, each of which carries appendages. **Chelicerates** also have a two-part body plan; most chelicerates have four pairs of walking legs.

See Web/CD Activities 32.1 and 32.2 for a concept review of this chapter.

SELF-QUIZ

1. Members of which groups have lophophores?
   a. Phoronids, brachiopods, and nematodes
   b. Phoronids, brachiopods, and ectoprocts
   c. Brachiopods, ectoprocts, and flatworms
   d. Phoronids, rotifers, and ectoprocts
   e. Rotifers, ectoprocts, and brachiopods
2. Which of the following is *not* part of the molluscan body plan?
   a. Mantle
   b. Foot
   c. Radula
   d. Visceral mass
   e. Jointed skeleton
3. Nautiluses control their buoyancy by
   a. adjusting salt concentrations in their blood.
   b. forcibly expelling water from the mantle.
   c. pumping water and gases in and out of internal chambers.
   d. using the complex sense organs in their heads.
   e. swimming rapidly.
4. The outer covering of ecdysozoans
   a. is always hard and rigid.
   b. is always thin and flexible.
   c. is present at some stage in the life cycle but not always among adults.
   d. ranges from very thin to hard and rigid.
   e. prevents the animals from changing their shapes.
5. Nematodes are abundant and diverse because
   a. they are both parasitic and free-living and eat a wide variety of foods.
   b. they are able to molt their exoskeletons.
   c. their thick cuticle enables them to move in complex ways.
   d. their body cavity is a pseudocoelom.
   e. their segmented bodies enable them to live in many different places.
6. The arthropod exoskeleton is composed of a
   a. mixture of several kinds of polysaccharides.
   b. mixture of several kinds of proteins.
   c. single complex polysaccharide called chitin.
   d. single complex protein called arthropodin.
   e. mixture of layers of proteins and a polysaccharide called chitin.
7. Which groups are arthropod relatives with unjointed legs?
   a. Trilobites and onychophorans
   b. Onychophorans and tardigrades
   c. Trilobites and tardigrades
   d. Onychophorans and chelicerates
   e. Tardigrades and chelicerates
8. The body plan of insects is composed of which of the three following regions?
   a. Head, abdomen, and trachea
   b. Head, abdomen, and cephalothorax
   c. Cephalothorax, abdomen, and trachea
   d. Head, thorax, and abdomen
   e. Abdomen, trachea, and mantle
9. Insects whose hatchlings are sufficiently similar in form to adults to be recognizable are said to have
   a. instars.
   b. neopterous development.
   c. accelerated development.
   d. incomplete metamorphosis.
   e. complete metamorphosis.
10. Factors that may have contributed to the remarkable evolutionary diversification of insects include
    a. the terrestrial environments penetrated by insects lacked any other similar organisms.
    b. the ability to fly.
    c. complete metamorphosis.
    d. a new mechanism for delivering oxygen to their internal tissues.
    e. all of the above

FOR DISCUSSION

1. Segmentation has arisen several times during animal evolution. What advantages does segmentation provide? Given these advantages, why do so many unsegmented animals survive? Why have some animals lost their segmentation?
2. Major structural novelties have arisen only infrequently during the course of evolution. Which of the features of protostomes do you think are major evolutionary novelties? What criteria do you use to judge whether a feature is a major as opposed to a minor novelty?
3. There are more described and named species of insects than of all other animal groups combined. However, only a very few species of insects live in marine environments, and those species are restricted to the intertidal zone or the ocean surface. What factors may have contributed to this lack of success in the oceans?

FOR INVESTIGATION

If you were given funding to carry out studies to improve estimates of the number of species of insects on Earth, how would you spend it? Would you do the same things if your objective were to determine the number of species of nematodes? Mites?

CHAPTER 33 Deuterostome Animals

Hobbits of Flores Island

In 2004, archeological workers on the Indonesian island of Flores unearthed the skeleton of an adult hominid female who stood less than a meter tall, weighed about 20 kilograms, and had a brain the size of a chimpanzee's. More fossils of these diminutive hominids were subsequently discovered on Flores, and radioactive dating indicated that some of them were shockingly recent—a mere 18,000 years old.

Although not all anthropologists are convinced, many experts think that the remains are those of a new species, dubbed *H. floresiensis,* and that this species is more closely related to *Homo erectus*, an extinct hominid species, than it is to *Homo sapiens*. *H. floresiensis* is thought to have evolved from *H. erectus* ancestors that arrived on Flores at least 840,000 years ago, at a time when the island lacked any other flightless mammals except rodents and pygmy elephants. The fossils tell us that these "hobbits" hunted pygmy elephants, which may have been their major food source.

Assuming that the majority opinion is correct, the discovery of a small hominid species surviving until relatively recent times stimulates several questions. How did their ancestors reach Flores? Why did they evolve to be so small? How did they manage to avoid extermination by the much larger modern humans who spread across Indonesia and reached Australia at least 46,000 years ago?

A probable scenario is that ancestral *H. erectus* colonized Flores during a period of glacial expansion, when sea levels were about 150 meters lower than they are today. At that time, Java and Bali would have been part of the Asian mainland. Even so, to reach Flores, the colonists had to cross three water gaps. These gaps were narrow enough that an island on the other side would have been visible, and reachable on a simple raft or canoe.

That the "hobbits" evolved small size is not surprising; biogeographers have observed that when large mammals colonize small islands, they typically evolve smaller size. Pygmy hippos, buffaloes, ground sloths, elephants, deer, and other mammals have all evolved on islands. In part this may be because islands typically lack both the resources to sustain large animals and the kinds of predators that feed on smaller animals. *Homo*

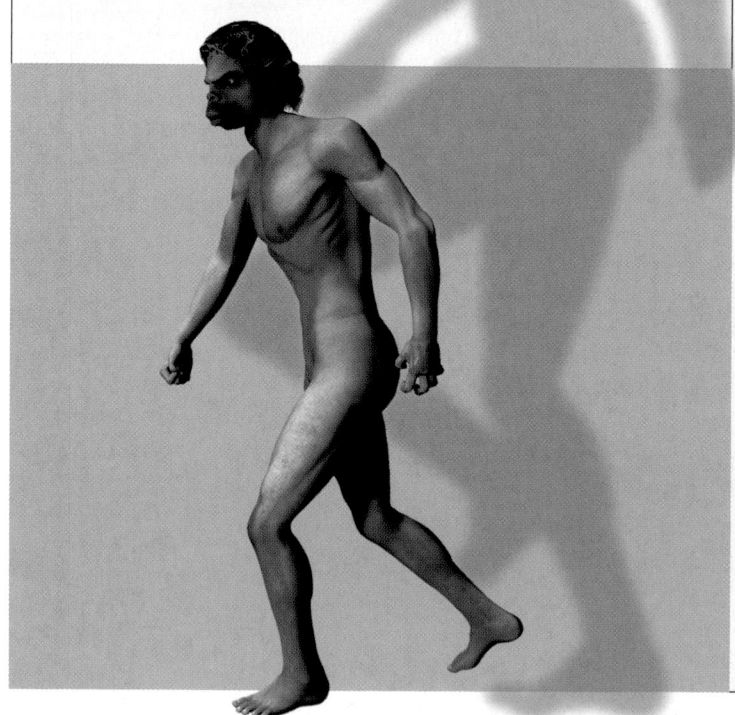

Hominids Are Deuterostomes In this artist's rendition, the extinct hominid *Homo floresiensis*, at less than a meter tall, is dwarfed by the silhouette of a modern human, *Homo sapiens*. Modern and extinct hominid species are vertebrates and thus members of the deuterostome clade.

Different Deuterostomes A sea star rests on a colony of tunicates ("sea squirts"); both of these species are deuterostomes. Although their morphology is not like that of any vertebrate, both animals share several synapomorphies with the vertebrates.

CHAPTER OUTLINE

floresiensis and the pygmy elephants they hunted would be examples of this well-known phenomenon.

The fact that the *H. floresiensis* lineage survived was probably due in large part to the three water gaps, which would have fluctuated in size with glacial fluctuations. Later hominid expansion across Indonesia to New Guinea could have occurred via a more easily traversed northerly route, leaving the "hobbits" isolated and untouched for many millennia.

The study of our own history—the evolution of *Homo sapiens*—has its own field, anthropology. But it is also the legitimate study of biologists, as we will see when we view humans in the light of their position in the realm of the deuterostomes.

IN THIS CHAPTER we introduce the deuterostomes and describe the major deuterostome groups—echinoderms, hemichordates, and chordates. We will then discuss the features that allowed some chordates to colonize dry land and will trace the evolution of those terrestrial vertebrates, taking an especially close look at the primate lineage that includes our own species.

33.1 What is a Deuterostome?

It may surprise you to learn that you and a sea urchin are both deuterostomes. Adult sea stars, sea urchins, and sea cucumbers—the most familiar echinoderms—look so different from adult vertebrates (fish, frogs, lizards, birds, and mammals) that it may be difficult to believe all these animals are closely related to one another. The evidence that deuterostomes all share a common ancestor that is different from the common ancestor of the protostomes is provided by their early developmental patterns and by phylogenetic analysis of gene sequences, neither of which is apparent in the forms of the adult animals.

Three early developmental patterns characterize the deuterostomes:

- Radial cleavage
- Formation of the mouth at the opposite end of the embryo from the blastopore (the pattern that gives the deuterostomes their name; see Figure 31.2)
- Development of a coelom from mesodermal pockets that bud off from the cavity of the gastrula rather than by splitting of the mesoderm, as occurs among protostomes

The first two of these features represent the ancestral condition for bilaterian animals in general, and as we saw in the previous chapter, some protostomes also have radial cleavage and other developmental similarities to deuterostomes. (In fact, some of the groups we now consider protostomes were once thought to be deuterostomes because of their retained ancestral developmental similarities with echinoderms and chordates.) Therefore, although developmental features that were once thought to be uniquely derived in deuterostomes have historically been important for hypotheses about the group's monophyly, these features are now known to be ancestral and therefore not indicative of monophyly. Nonetheless, we still recognize the close evolutionary relationships of echinoderms, hemichordates, and chordates (the groups that now compose the deuterostomes) because phylogenetic analyses of DNA sequences of many slowly evolving genes support their common ancestry.

There are many fewer species of deuterostomes than of protostomes (see Table 31.1), but we have a special interest in deuterostomes, in part because we are members of that clade.

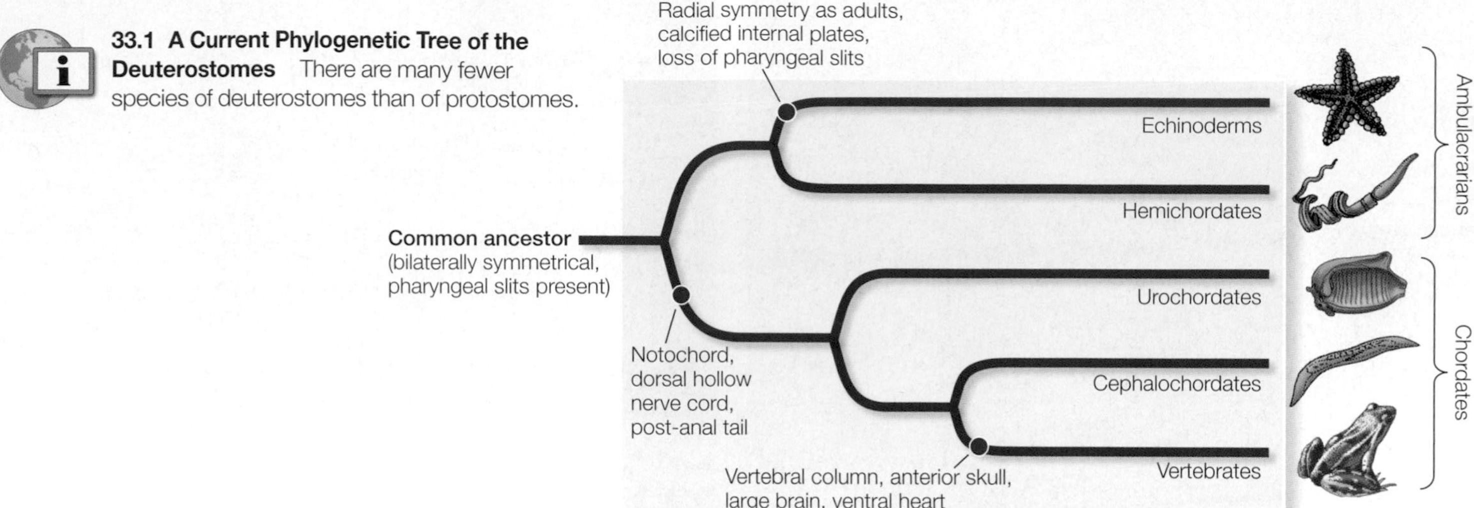

33.1 A Current Phylogenetic Tree of the Deuterostomes There are many fewer species of deuterostomes than of protostomes.

We are also interested in them because they include the largest living animals; these large deuterostomes strongly influence the characteristics of ecosystems.

Living deuterostomes include three major clades (**Figure 33.1**):

- *Echinoderms*: sea stars, sea urchins, and their relatives
- *Hemichordates*: acorn worms and pterobranchs
- *Chordates*: sea squirts, lancelets, and vertebrates

All deuterostomes are triploblastic, coelomate animals (see Figure 31.4) with internal skeletons. Some species have segmented bodies, but the segments are less obvious than those of annelids and arthropods.

Scientists are learning much about the ancestors of modern deuterostomes from recently discovered fossils of several early forms. The most important finds have come from 520-million-year-old fossil beds in China. The *homalozoans* had a skeleton similar to that of a modern echinoderm, but they had pharyngeal slits and bilateral symmetry. The *vetulicosystids*, which were first discovered in 2002, also had pharyngeal slits. Many fossils of a third kind, the *yunnanozoans*, were discovered in China's Yunnan Province. These well-preserved animals had a large mouth, six pairs of external gills, and a segmented posterior body section bearing a light cuticle (**Figure 33.2**).

The features of these fossil animals support the findings from phylogenetic analyses of living species in showing that the earliest deuterostomes were bilaterally symmetrical, segmented animals with a pharynx that had slits through which water flowed (see Figure 33.1). Echinoderms evolved their adult forms with unique pentaradial symmetry much later, whereas other deuterostomes retained the ancestral bilateral symmetry.

Yunnanozoon lividum

33.2 Ancestral Deuterostomes Had External Gills The extinct yunnanozoans may be ancestral to all deuterostomes. This fossil, which dates from the Cambrian, shows the six pairs of external gills and segmented posterior body that characterized these animals.

33.2 What Are the Major Groups of Echinoderms and Hemichordates?

About 13,000 species of echinoderms in 23 major groups have been described from their fossil remains. They are probably only a small fraction of those that actually lived. Only 6 of the 23 groups known from fossils are represented by species that survive today; many clades became extinct during the periodic mass extinctions that have occurred throughout Earth's history. Nearly all of the 7,000 extant species of echinoderms live only in marine environments. There are only 95 known living species of hemichordates.

The echinoderms and hemichordates (together known as *ambulacrarians*) have a bilaterally symmetrical, ciliated larva (**Figure 33.3**). Adult hemichordates also are bilaterally symmetrical. Echinoderms, however, undergo a radical change in form as they develop into adults, changing from a bilaterally symmetrical larva to an adult with **pentaradial symmetry** (symmetry in five or multiples of five). As is typical of animals with radial symmetry, echinoderms have no head, and they move slowly and equally well in

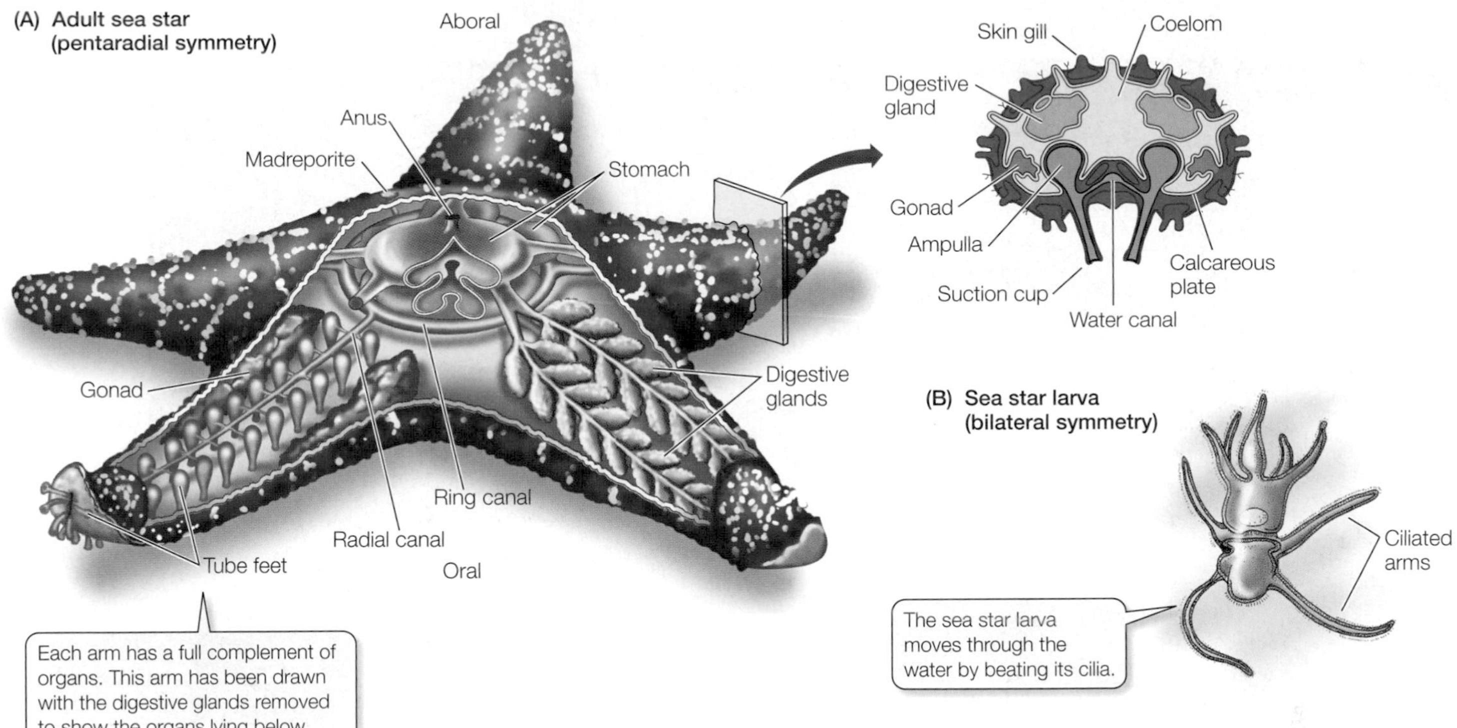

33.3 Evolutionary Innovations of Echinoderms (A) A dorsal view of a sea star displays the canals and tube feet of the echinoderm water vascular system, as well as the calcified internal skeleton. (B) The ciliated larva of a sea star has bilateral symmetry.

many directions. Rather than having an anterior-posterior (head-tail) and dorsal-ventral (back-belly) body organization, echinoderms have an *oral* side (containing the mouth) and an opposite *aboral* side (containing the anus).

Echinoderms have a water vascular system

In addition to having pentaradial symmetry, adult echinoderms have two unique structural features. One is a system of calcified internal plates covered by thin layers of skin and some muscles. The calcified plates of most echinoderms are thick, and they fuse inside the entire body, forming an *internal skeleton*. The other unique feature is a **water vascular system**, a network of water-filled canals leading to extensions called **tube feet**. This system functions in gas exchange, locomotion, and feeding (see Figure 33.3). Seawater enters the system through a perforated structure called a *madreporite*. A calcified canal leads from the madreporite to another canal that rings the *esophagus* (the tube leading from the mouth to the stomach). Other canals radiate from this *ring canal*, extending through the arms (in species that have arms) and connecting with the tube feet. Echinoderms use their tube feet in many different ways to move and to capture prey. These structural innovations have been modified in many ways to result in a striking array of very different animals.

Members of one major extant echinoderm clade, the *crinoids* (sea lilies and feather stars), were more abundant and species-rich 300–500 million years ago than they are today. Most of the 80 living species of sea lilies attach to a substratum by means of a flexible stalk consisting of a stack of calcareous discs. The main body of the animal is a cup-shaped structure that contains a tubular digestive system. Five to several hundred arms, usually in multiples of five, extend outward from the cup. The jointed calcareous plates of the arms enable them to bend (**Figure 33.4A**).

Feather stars are similar to sea lilies, but they grasp the substratum with their flexible appendages (**Figure 33.4B**). They can walk on the tips of their arms or swim by rhythmically beating them. Feather stars feed in much the same manner as sea lilies. About 600 living species have been described.

Most surviving echinoderm species—including sea urchins, sea cucumbers, sea stars, and brittle stars—are not crinoids. Sea urchins are hemispherical in shape and lack arms (**Figure 33.4C**). They are covered with spines that are attached to the underlying skeleton via ball-and-socket joints. These joints enable the spines to be moved so that they can converge toward a point that has been touched. The spines come in varied sizes and shapes; a few produce toxic substances. They provide effective protection for the urchin, as many a scuba diver has found out the hard way. Sand dollars are flattened and disc-shaped relatives of sea urchins.

The sea cucumbers also lack arms, and their bodies are oriented in an atypical manner for an echinoderm (**Figure 33.4D**). The mouth is anterior and the anus is posterior, not oral and aboral as in other echinoderms. Sea cucumbers use most of their tube feet primarily for attaching to the substratum rather than for moving.

Sea stars are the most familiar echinoderms (**Figure 33.4E**). Their gonads and digestive organs are located in the arms, as seen in Figure 33.3. Their tube feet serve as organs of locomotion, gas exchange, and attachment. Each tube foot of a sea star consists of an internal *ampulla* connected by a muscular tube to an external suction cup that can stick to the substratum. The tube foot is

(A) *Nominus novus*

(B) *Oxycomanthus bennetti*

(C) *Strongylocentrotus purpuratus*

(D) *Bohadschia argus*

(E) *Henricia leviuscula*

(F) *Ophiothrix spiculata*

33.4 Diversity among the Echinoderms (A) Sea lilies can have up to several hundred arms, usually in multiples of five. (B) The flexible arms of the golden feather star are clearly visible. (C) Purple sea urchins are important grazers on algae in the intertidal zone of the Pacific Coast of North America. (D) This sea cucumber lives on rocky substrata in the seas around Papua New Guinea. (E) Sea stars are important predators on bivalve mollusks such as mussels and clams. Suction tips on its tube feet allow this animal to grasp both shells of the bivalve and pull them apart. (F) The arms of the brittle star are composed of hard but flexible plates.

moved by expansion and contraction of the circular and longitudinal muscles of the tube. Brittle stars are similar in structure to sea stars, but their flexible arms are composed of jointed hard plates (**Figure 33.4F**).

Sea daisies were discovered only in 1986; little is known about them. They have tiny, disc-shaped bodies with a ring of marginal spines, and two ring canals, but no arms. Sea daisies live on rotting wood in ocean waters. They apparently eat prokaryotes, which they digest outside their bodies and absorb either through a membrane that covers the oral surface or via a shallow, saclike stomach. Recent molecular data suggest that they are greatly modified sea stars.

Echinoderms use their tube feet in a great variety of ways to capture prey. Sea lilies, for example, feed by orienting their arms in passing water currents. Food particles then strike and stick to the tube feet, which are covered with mucus-secreting glands. The tube feet transfer these particles to grooves in the arms, where ciliary action carries the food to the mouth. Sea cucumbers capture food with their anterior tube feet, which are modified into large, feathery, sticky tentacles that can be protruded from the mouth (see Figure 33.4D). Periodically, a sea cucumber withdraws the tentacles, wipes off the material that has adhered to them, and digests it.

Many sea stars use their tube feet to capture large prey such as polychaetes, gastropod and bivalve mollusks, small crustaceans such as crabs, and fish. With hundreds of tube feet acting simultaneously, a sea star can grasp a bivalve in its arms, anchor the arms with its tube feet, and, by steady contraction of the muscles in its arms, gradually exhaust the muscles the bivalve uses to keep its shell closed (see Figure 33.4E). To feed on a bivalve, a sea star can push its stomach out through its mouth and then through the narrow space between the two halves of the bivalve's shell. The sea star's stomach then secretes enzymes that digest the prey.

Members of other echinoderm groups do not use their tube feet to capture food. Most sea urchins eat algae, which they scrape from rocks with a complex rasping structure. Most of the 2,000 species of brittle stars ingest particles from the upper layers of sediments and assimilate the organic material from them, although some species filter suspended food particles from the water and others capture small animals.

Hemichordates have a three-part body plan

Echinoderms
Hemichordates
Chordates

Hemichordates—acorn worms and pterobranchs—have a three-part body plan, consisting of a *proboscis*, a *collar* (which bears the mouth), and a *trunk* (which contains the other body parts). The 75 known species of *acorn worms* range up to 2 meters in length (**Figure 33.5A**). They live in burrows in muddy and sandy marine sediments. The digestive tract of an acorn worm consists of a mouth behind which are a muscular *pharynx* and an *intestine*. The pharynx opens to the outside through a number of *pharyngeal slits* through which water can exit. Highly vascularized tissue surrounding the pharyngeal slits serves as a gas exchange apparatus. Acorn worms breathe by pumping water into the mouth and out through the pharyngeal slits. They capture prey with the large proboscis, which is coated with sticky mucus to which small organisms in the sediment stick. The mucus and its attached prey are conveyed by cilia to the mouth. In the esophagus, the food-laden mucus is compacted into a ropelike mass that is moved through the digestive tract by ciliary action.

The 20 living species of *pterobranchs* are sedentary marine animals up to 12 millimeters in length that live in a tube secreted by the proboscis. Some species are solitary; others form colonies of individuals joined together (**Figure 33.5B**). Behind the proboscis is a collar with 1–9 pairs of arms. The arms bear long tentacles that capture prey and function in gas exchange.

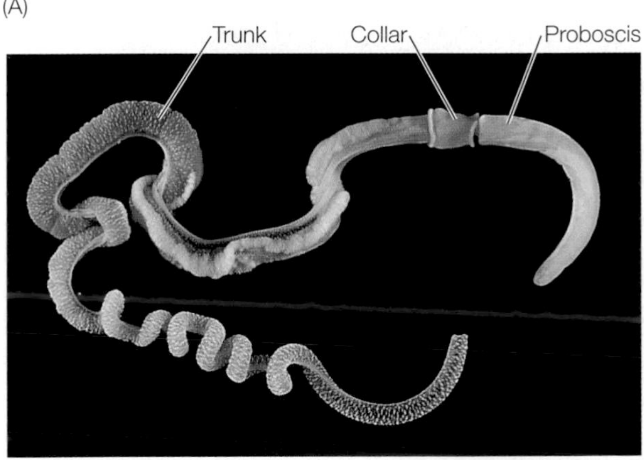

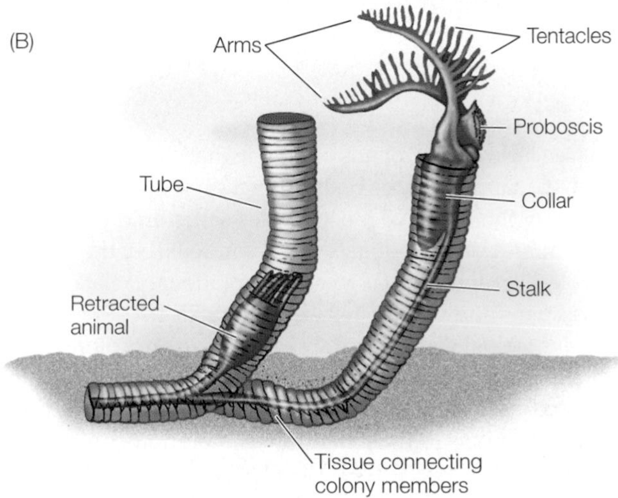

33.5 Hemichordates (A) The proboscis of an acorn worm is modified for burrowing. (B) The structure of a colonial pterobranch.

33.2 RECAP

Echinoderms have an internal skeleton of calcified plates and a unique water vascular system. Hemichordates have a bilaterally symmetrical body divided into three parts: proboscis, collar, and trunk.

- What two unique structural features characterize the echinoderms? See p. 719
- Can you describe some of the ways that echinoderms use their tube feet to obtain food? See p. 721
- How do hemichordates obtain food? See p. 721

We have described the deuterostome groups that are most distantly related to us. Now we turn our attention to the chordates, the clade to which humans belong.

33.3 What New Features Evolved in the Chordates?

It is not obvious from examining adult animals that echinoderms and chordates share a common ancestor. For the same reason, the evolutionary relationships among some chordate groups are not immediately apparent. The features that reveal the evolutionary relationships both among the chordates and between chordates and echinoderms are primarily seen in the larvae—in other words, it is during the early developmental stages that their evolutionary relationships are evident.

There are three chordate clades: the **urochordates**, the **cephalochordates**, and the **vertebrates**. There are about 3,000 species of urochordates and 50,000 species of vertebrates, but only about 30 species of cephalochordates.

Adult chordates vary greatly in form, but all chordates display the following derived structures at some stage in their development (**Figure 33.6**):

- A dorsal, hollow nerve cord
- A tail that extends beyond the anus
- A dorsal supporting rod called the *notochord*

The **notochord** is the most distinctive derived chordate trait. It is composed of a core of large cells with turgid fluid-filled vacuoles, which make it rigid but flexible. In the urochordates the notochord is lost during metamorphosis to the adult stage. In most vertebrate species, it is replaced by skeletal structures that provide support for the body.

The ancestral pharyngeal slits (not a derived feature of this group) are present at some developmental stage of chordates, although they are often lost in adults. The *pharynx*, which develops around the pharyngeal slits, functioned in chordate ancestors as the site for oxygen uptake and the elimination of carbon dioxide and water (as in acorn worms). The pharynx is much enlarged in some chordate species (as in the *pharyngeal basket* of the lancelet in **Figure 33.6**), but has been lost in others.

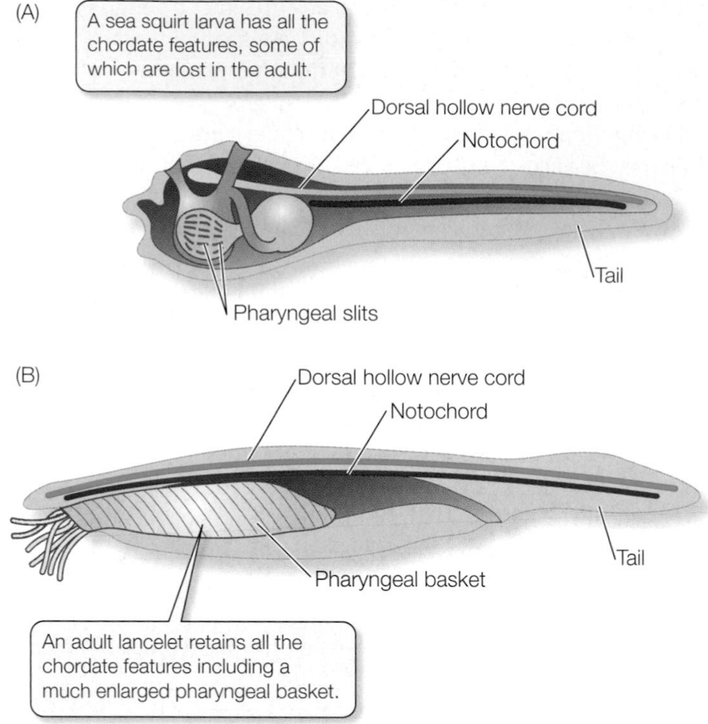

Branchiostoma sp.

33.6 The Key Features of Chordates Are Most Apparent in Early Developmental Stages The pharyngeal slits of both the urochordate sea squirt and the cephalochordate lancelet develop into pharyngeal baskets. (A) The sea squirt larva (but not the adult) has all three chordate features. (B) The pharyngeal basket of the adult lancelet features external gill slits.

Adults of most urochordates and cephalochordates are sessile

All members of the three major urochordate groups—the ascidians, thaliaceans, and larvaceans—are marine animals. More than 90 percent of the known species of urochordates are *ascidians* (sea squirts). Individual ascidians range in size from less than 1 millimeter to 60 centimeters in length, but some form colonies by asexual budding from a single founder. These colonies may measure several meters across. The baglike body of an adult ascidian is enclosed in a tough tunic, leading to its alternate name of "tunicate" (**Figure 33.7A**). The tunic is composed of proteins and a complex polysaccharide secreted by epidermal cells. The ascidian pharynx is enlarged into a *pharyngeal basket* that filters prey from the water passing through it.

Echinoderms
Hemichordates
Urochordates
Cephalochordates
Vertebrates

In addition to its pharyngeal slits, an ascidian larva has a dorsal, hollow nerve cord and a notochord that is restricted to the tail region (see Figure 33.6). Bands of muscle that surround the notochord provide support for the body. After a short time swimming in the plankton, the larvae of most species settle on the seafloor and transform into sessile adults. As Darwin realized, the swimming, tadpole-like larvae suggest a close evolutionary relationship between ascidians and vertebrates (see Figure 25.4).

Thaliaceans (salps) can live singly or in chainlike colonies up to several meters long (**Figure 33.7B**). Salps float in tropical and subtropical oceans at all depths down to 1,500 meters. *Larvaceans* are solitary planktonic animals that retain their notochords and nerve cords throughout their lives. Most larvaceans are less than 5 millimeters long, but some species that live near the bottom of deep ocean waters build delicate casings of slime that may be more than a meter wide. They snare sinking organic particles (their pri-

(A) *Rhopalaea* sp.

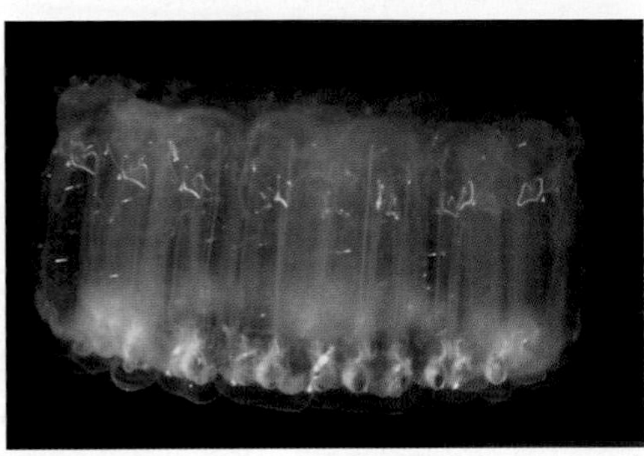

(B) *Pegea socia*

33.7 Adult Urochordates (A) The iridescent tunic is clearly visible in these transparent ascidians ("sea squirts," also known as tunicates). Two different species of the same genus appear in this photograph. (B) A chainlike colony of salps floats in tropical waters.

mary food source) with these slimy "houses." When the old "house" gets clogged, they build new a one.

The 30 species of cephalochordates, or *lancelets*, are small animals that rarely exceed 5 centimeters in length. The notochord, used in burrowing, extends the entire length of the body throughout their lives (see Figure 33.6B). Lancelets are found in shallow marine and brackish waters worldwide. Most of the time they lie covered in sand with their heads protruding above the sediment, but they can swim. They extract prey from the water with their pharyngeal baskets. During the reproductive season, the gonads of the males and females greatly enlarge. At spawning, the walls of the gonads rupture, releasing eggs and sperm into the water column, where fertilization takes place.

A new dorsal supporting structure replaces the notochord in vertebrates

In one chordate group, a new dorsal supporting structure evolved. The **vertebrates** take their name from the jointed, dorsal **vertebral column** that replaces the notochord during early development as their primary supporting structure.

The lineage that led to the vertebrates is thought to have evolved in an estuarine environment (where freshwater meets salt water). Modern vertebrates have since radiated into marine, freshwater, and terrestrial environments worldwide (**Figure 33.8**).

33.8 Vertebrates Have Colonized a Wide Diversity of Environments This current phylogenetic hypothesis shows the distribution of the major vertebrate groups over marine, freshwater, terrestrial, and estuarine (where fresh and salt water meet) environments. Representatives of the amniote vertebrates can be found in all four of these environments, as well as in the air (birds and bats).

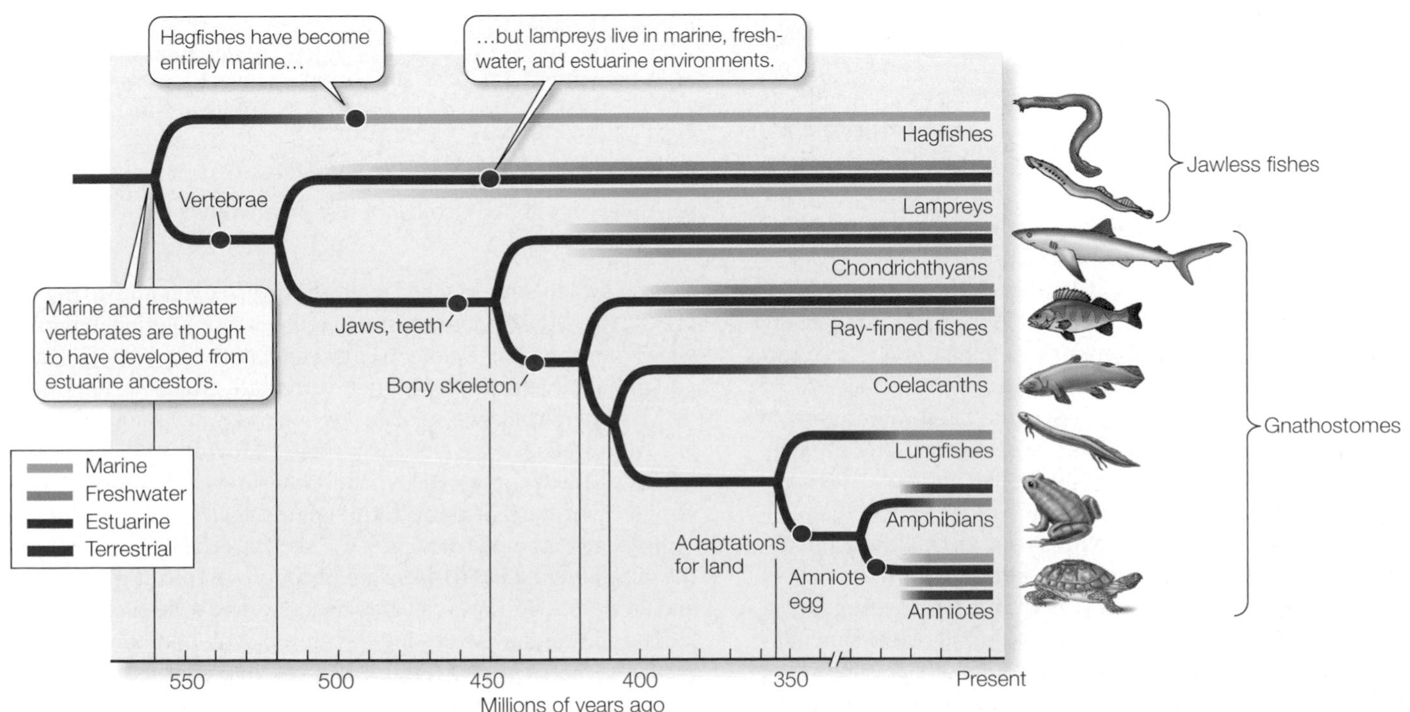

(A) *Eptatretus stouti*

(B) *Petromyzon marinus*

33.9 Modern Jawless Fishes (A) Hagfish burrow in the ocean mud, from which they extract small prey. They are also scavengers on dead or dying fish. Hagfish have degenerate eyes, which has led to their being miscalled "blind eels." (B) Some lampreys can live in either fresh or salt water, although all lampreys breed in freshwater. Many species are ectoparasites that attach to the bodies of living fish and use their large, jawless mouths to suck blood and flesh.

Echinoderms
Hemichordates
Urochordates
Cephalochordates
Vertebrates

The *hagfishes* are thought by many biologists to be the sister group to the remaining vertebrates (as shown in Figure 33.8). Hagfishes (**Figure 33.9A**) have a weak circulatory system, with three small accessory hearts (rather than a single, large heart), a partial *cranium* or skull (containing no *cerebrum* or *cerebellum*, two main regions of the brain of other vertebrates), and no jaws or stomach. They also lack vertebrae, and have a skeleton composed of cartilage. Thus, some biologists do not consider hagfishes vertebrates, but instead call the hagfishes plus the vertebrates the *craniates*. On the other hand, recent analyses of some gene sequences suggest that hagfishes may be more closely related to the vertebrate lampreys (**Figure 33.9B**), in which case the two groups are placed together in the vertebrates as *cyclostomes* ("circle mouths"). If this latter hypothesis is correct, then hagfishes must have secondarily lost many of the major vertebrate morphological features during their evolution.

The 58 known species of hagfishes are unusual marine animals that produce copious quantities of slime as a defense. They are virtually blind, and rely largely on the four pairs of sensory tentacles around their mouths to detect food. Although they have no jaws, they have a tongue-like structure in their mouths that is equipped with tooth-like rasps, which hagfishes can use to tear apart dead organisms and to capture their principal prey (polychaete worms). They have direct development (no larvae), and individuals may actually change sex from year to year (from male to female and vice versa).

The nearly 50 species of lampreys either live in freshwater, or live in coastal salt water and then move into freshwater to breed. Although the lampreys and hagfishes may look superficially similar in body form (with elongate eel-like bodies and no paired fins), they are otherwise very different in their biology.

Lampreys have a complete braincase and true (although rudimentary) vertebrae, all cartilaginous rather than bony. Lampreys undergo a complete metamorphosis from filter-feeding larvae known as *ammocoetes*, which are morphologically quite similar in general structure to adult lancelets. The adults of many species of lampreys are parasitic, although several lineages of lampreys evolved to become non-feeding as adults. These non-feeding adult lampreys survive only a few weeks to breed after metamorphosis. In the species that are parasitic as adults, the round mouth is a rasping and sucking organ (see Figure 33.9B) that is used to attach to their prey and rasp at the flesh. One predatory species, the sea lamprey, began spreading through man-made canals into the Great Lakes in 1835, and contributed to enormous losses of commercial fisheries there until control measures were introduced in the middle of the 1900s.

The vertebrate body plan can support large animals

Vertebrates are characterized by several key features:

- A rigid internal *skeleton* supported by the vertebral column
- An anterior *skull* with a large brain
- Internal organs *suspended in a coelom*
- A well-developed *circulatory system*, driven by contractions of a ventral *heart*

This body plan is exemplified by the bony fish diagrammed in **Figure 33.10**. Fishes underwent many millenia of evolution in Earth's watery environments before the first vertebrate colonization of land and remain the most species-rich vertebrate group.

Many kinds of jawless fishes were found in the seas, estuaries, and fresh waters of the Devonian period. However, hagfishes and lampreys are the only jawless fishes that survived beyond the Devonian. During that period, the *gnathostomes* ("jaw mouths") evolved jaws via modifications of the skeletal arches that supported the gills (**Figure 33.11A**). Jaws greatly improved feeding efficiency, and an animal with jaws can grasp, subdue, and swallow large prey.

The earliest jaws were simple, but the evolution of *teeth* made predators more effective (**Figure 33.11B**). In predators, teeth function crucially both in grasping and in breaking up prey. In both

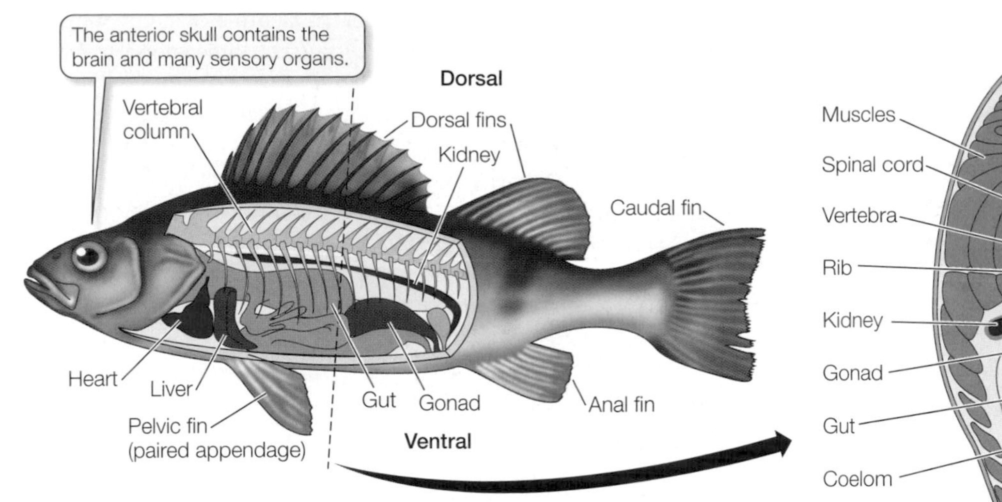
The anterior skull contains the brain and many sensory organs.
Dorsal
Vertebral column
Dorsal fins
Kidney
Caudal fin
Heart
Liver
Gut
Gonad
Anal fin
Pelvic fin (paired appendage)
Ventral

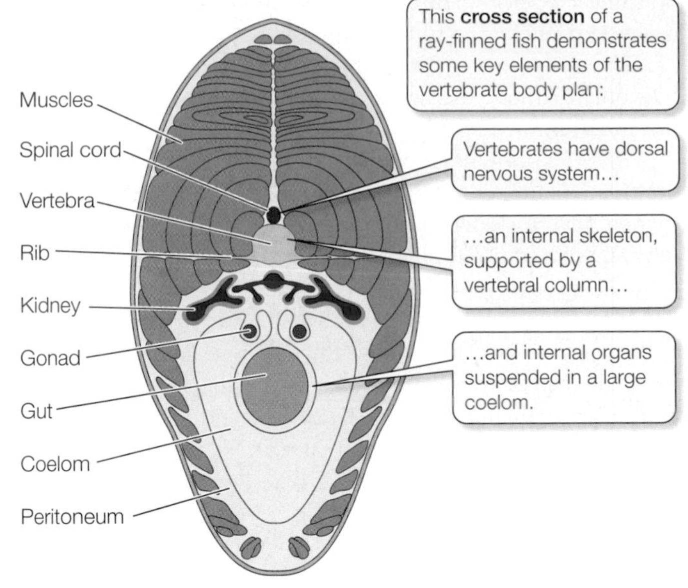
This cross section of a ray-finned fish demonstrates some key elements of the vertebrate body plan:
Muscles
Spinal cord
Vertebra
Rib
Kidney
Gonad
Gut
Coelom
Peritoneum
Vertebrates have dorsal nervous system...
...an internal skeleton, supported by a vertebral column...
...and internal organs suspended in a large coelom.

33.10 The Vertebrate Body Plan A ray-finned fish is used here to illustrate the structural elements common to all vertebrates. In addition to the paired pelvic fins ("hindlimbs"), these fishes have paired pectoral fins ("forelimbs") on the sides of their bodies (not seen in this cutaway view).

predators and herbivores, teeth enable their possessors to chew both soft and hard body parts of their food (see Figure 50.7). Chewing also aids chemical digestion and improves the animal's ability to extract nutrients from its food, as will be described in Chapter 50. Vertebrates are remarkable in the diversity of their jaws and teeth.

Fins and swim bladders improved stability and control over locomotion

Jawed fishes stabilize their position in water, as well as propel themselves through water, using their *fins*. Most fishes have a pair of pectoral fins just behind the gill slits, and a pair of pelvic fins anterior to the anal region (see Figure 33.10). Median dorsal and anal fins stabilize the fish as it moves, or may be used for propulsion

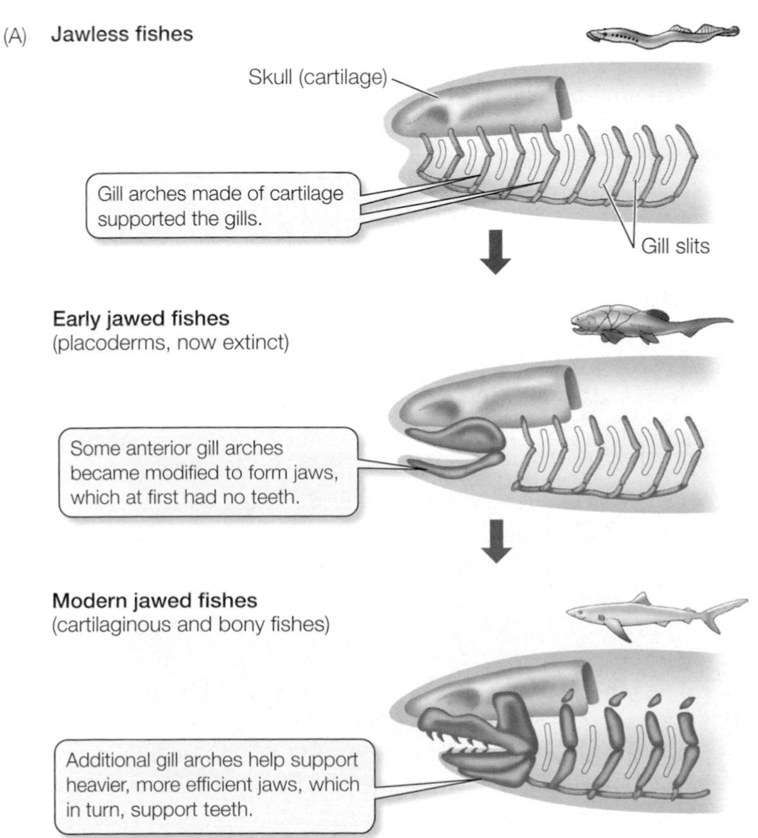
(A) Jawless fishes
Skull (cartilage)
Gill arches made of cartilage supported the gills.
Gill slits
Early jawed fishes (placoderms, now extinct)
Some anterior gill arches became modified to form jaws, which at first had no teeth.
Modern jawed fishes (cartilaginous and bony fishes)
Additional gill arches help support heavier, more efficient jaws, which in turn, support teeth.

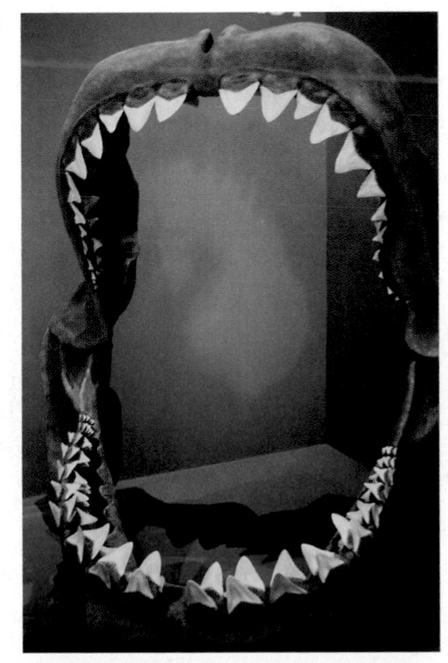
(B)

33.11 Jaws and Teeth Increased Feeding Efficiency (A) This series of diagrams illustrates one probable scenario for the evolution of jaws from the anterior gill arches of fishes. (B) Jaws of the extinct giant shark (*Carcharodon megalodon*) display the teeth that indicate an extreme predatory lifestyle.

in some species. In many fishes, the caudal fins help propel the animal and enable it to turn rapidly.

Several groups of finned fishes became abundant during the Devonian. Among them were **chondrichthyans**—sharks, skates and rays (820 known species), and chimaeras (30 known species). Like hagfishes and lampreys, these fishes have a skeleton composed entirely of a firm but pliable material called *cartilage*. Their skin is flexible and leathery, sometimes bearing scales that give it the consistency of sandpaper. Sharks move forward by means of lateral undulations of their bodies and caudal (tail) fins (**Figure 33.12A**). Skates and rays propel themselves by means of vertical undulating movements of their greatly enlarged pectoral fins (**Figure 33.12B**).

Most sharks are predators, but some feed by straining plankton from the water. Most skates and rays live on the ocean floor, where they feed on mollusks and other animals buried in the sediments. Nearly all cartilaginous fishes live in the oceans, but a few are estuarine or migrate into lakes and rivers. One group of stingrays is found in river systems of South America. The less familiar chimaeras (**Figure 33.12C**) live in deep-sea or cold waters.

In some early fishes, gas-filled sacs supplemented the gas exchange function of the gills by giving the animals access to atmospheric oxygen. These features enabled those fishes to live where oxygen was periodically in short supply, as it often is in freshwater environments. These lunglike sacs evolved into *swim bladders*, which are organs of buoyancy. By adjusting the amount of gas in its swim bladder, a fish can control the depth at which it remains suspended in the water while expending very little energy to maintain its position.

Ray-finned fishes, and most remaining groups of vertebrates, have internal skeletons of calcified, rigid *bone* rather than flexible cartilage. The outer body surface of most species of ray-finned fishes is covered with flat, thin, lightweight scales that provide some protection or enhance movement through the water. The gills of ray-finned fishes open into a single chamber covered by a hard flap, called an *operculum*. Movement of the operculum improves the flow of water over the gills, where gas exchange takes place.

Ray-finned fishes radiated during the Tertiary into about 24,000 species, encompassing a remarkable variety of sizes, shapes, and lifestyles (**Figure 33.13**). The smallest are less than 1 centimeter long as adults; the largest weigh as much as 900 kilograms. Ray-finned fishes exploit nearly all types of aquatic food sources. In the oceans they filter plankton from the water, rasp algae from rocks, eat corals and other soft-bodied colonial animals, dig animals from soft sediments, and prey on virtually all kinds of other fishes. In fresh water they eat plankton, devour insects, eat fruits that fall into the water in flooded forests, and prey on other aquatic vertebrates and, occasionally, terrestrial vertebrates. Many fishes are solitary, but in open water others form large aggregations called *schools*. Many fishes perform complicated behaviors to maintain schools, build nests, court and choose mates, and care for their young.

Although ray-finned fishes can readily control their position in open water using their fins and swim bladders, their eggs tend to sink. Some species produce small eggs that are buoyant enough to complete their development in the open water, but most marine

33.12 Chondrichthyans (A) Most sharks, such as this whitetip reef shark, are active marine predators. (B) Skates and rays, represented here by an eagle ray, feed on the ocean bottom. Their modified pectoral fins are used for propulsion, and their other fins have secondarily become vestigial. (C) A chimaera, or ratfish. These deep-sea fish often possess modified dorsal fins that contain toxins.

(A) *Sphyraena barracuda*

(B) *Chromis punctipinnis*

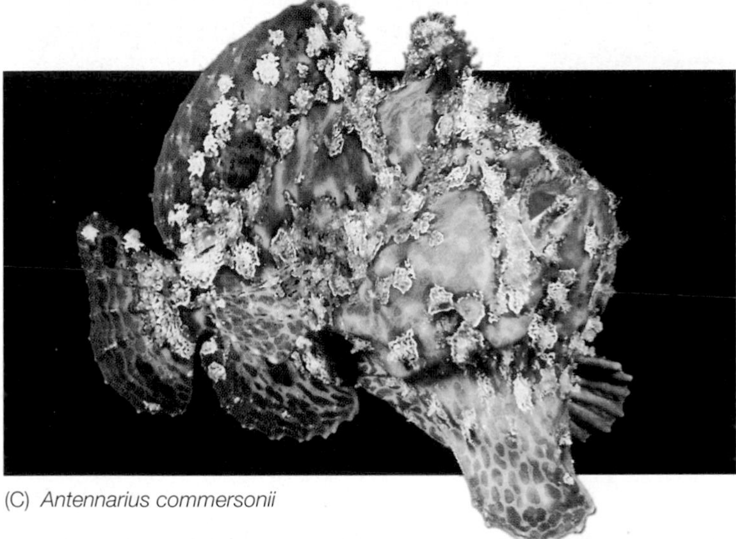
(C) *Antennarius commersonii*

(D) *Phycodurus eques*

33.13 Diverse Ray-Finned Fishes (A) The barracuda has the large teeth and powerful jaws of a predator. (B) The blacksmith fish of North America's Pacific coastal waters displays the "typical" body form of fishes that school in the open ocean. (C) Commerson's frogfish can change its color over a range from pale yellow to orange-brown to deep red, thus enhancing its camouflage abilities. (D) This leafy sea dragon is difficult to see when it hides in vegetation. It is a larger relative of the more familiar seahorse.

fishes move to food-rich shallow waters to lay their eggs. That is why coastal waters and estuaries are so important in the life cycles of many marine species. Some, such as salmon, abandon salt water when they breed, ascending rivers to spawn in freshwater streams and lakes.

How do the larvae of coral reef fish that hatch out in the open ocean find their way back to the reefs? Perhaps their relatives call them home. Biologists built artificial reefs off the coast of Australia and played recordings of calls of reef fishes on half of them. Many more young fish were attracted to the noisy reefs than to the quiet ones.

33.3 RECAP

Chordates are characterized by a dorsal hollow nerve chord, a post-anal tail, and a dorsal supporting rod called a notochord (although all of these features are not found at all life stages). Specialized structures for support (such as vertebrae), locomotion (such as fins), and feeding (such as jaws and teeth) evolved among the vertebrates, which allowed them to colonize and adapt to most of Earth's environments.

- Can you remember the synapomorphies that characterize the chordates and those that characterize the vertebrates? See pp. 722–724 and Figures 33.6 and 33.10
- Can you describe the ways that hagfishes differ from lampreys in their morphology and life history? Do you understand why some biologists do not consider the hagfishes to be vertebrates? See p. 724
- There are more species of fishes than of any other single vertebrate group. Why do you think this is so? Can you describe the adaptations that differentiate the major groups of fishes? See pp. 724–726

In some fishes, the lunglike sacs that gave rise to swim bladders became specialized for another purpose: breathing air. That adaptation set the stage for the vertebrates to move onto the land.

33.4 How Did Vertebrates Colonize the Land?

Coelacanths
Lungfishes
Amphibians
Amniotes

The evolution of lunglike sacs in fishes set the stage for the invasion of the land. Some early ray-finned fishes probably used those structures to supplement their gills when oxygen levels in the water were low, as lungfishes and many groups of ray-finned fishes do today. But with their unjointed fins, those fishes could only flop around on land, as do most modern fishes. Changes in the structure of the fins first allowed some fishes to support themselves better in shallow water and, later, to move better on land.

Jointed fins enhanced support for fishes

Jointed fins evolved in the ancestor of the **sarcopterygians,** which includes coelacanths, lungfishes, and tetrapods. The coelacanths flourished from the Devonian until about 65 million years ago, when they were thought to have become extinct. However, in 1938, a commercial fisherman caught a living coelacanth off South Africa. Since that time, hundreds of individuals of this extraordinary fish, *Latimeria chalumnae,* have been collected. A second species, *L. menadoensis,* was discovered in 1998 off the Indonesian island of Sulawesi. *Latimeria,* a predator of other fish, reaches a length of about 1.8 meters and weighs up to 82 kilograms (**Figure 33.14A**). Its skeleton is mostly composed of cartilage, not bone. A cartilaginous skeleton is a derived feature in this clade because they had bony ancestors.

Lungfishes, which also have jointed fins, were important predators in shallow-water habitats in the Devonian, but most lineages died out. The six surviving species live in stagnant swamps and muddy waters in South America, Africa, and Australia (**Figure 33.14B**). Lungfishes have lungs derived from the lunglike sacs of their ancestors as well as gills. When ponds dry up, individuals of most species can burrow deep into the mud and survive for many months in an inactive state while breathing air.

It is believed that some early aquatic sarcopterygians began to use terrestrial food sources, became more fully adapted to life on land, and eventually evolved to become ancestral **tetrapods** (four-legged vertebrates).

How was the transition from an animal that swam in water to one that walked on land accomplished? Early in 2006, scientists reported the discovery of a Devonian fossil they believe represents an intermediate between the finned appendages of fishes and the limbs of terrestrial tetrapods (**Figure 33.14C**). It appears that limbs able to prop up a large fish with the front-to-rear movement necessary for walking may have evolved while the animals still lived in water.

Amphibians adapted to life on land

During the Devonian, the first tetrapods arose from an aquatic ancestor. In this lineage, stubby, jointed fins evolved into walking legs. The basic elements of those legs have remained throughout the evolution of terrestrial vertebrates, although they have changed considerably in their form.

Most modern **amphibians** are confined to moist environments because they lose water rapidly through the skin when exposed to dry air. In addition, their eggs are enclosed within delicate membranous envelopes that cannot prevent water loss in dry conditions. In many temperate-zone amphibian species, adults live mostly on land, but they return to fresh water to lay their eggs,

(A) *Latimeria chalumnae*

(B) *Neoceratodus forsteri*

Tiktaalik's pectoral fins show some of the skeletal structures of tetrapod limbs.

(C) *Tiktaalik roseae*

33.14 The Closest Relatives of Tetrapods (A) This coelacanth, found in deep waters of the Indian Ocean, represents one of two surviving species of a group that was once thought to be extinct. (B) All surviving lungfish species live in the Southern Hemisphere. (C) A recently discovered fossil provides further insight into the evolution of limbs from pectoral fins.

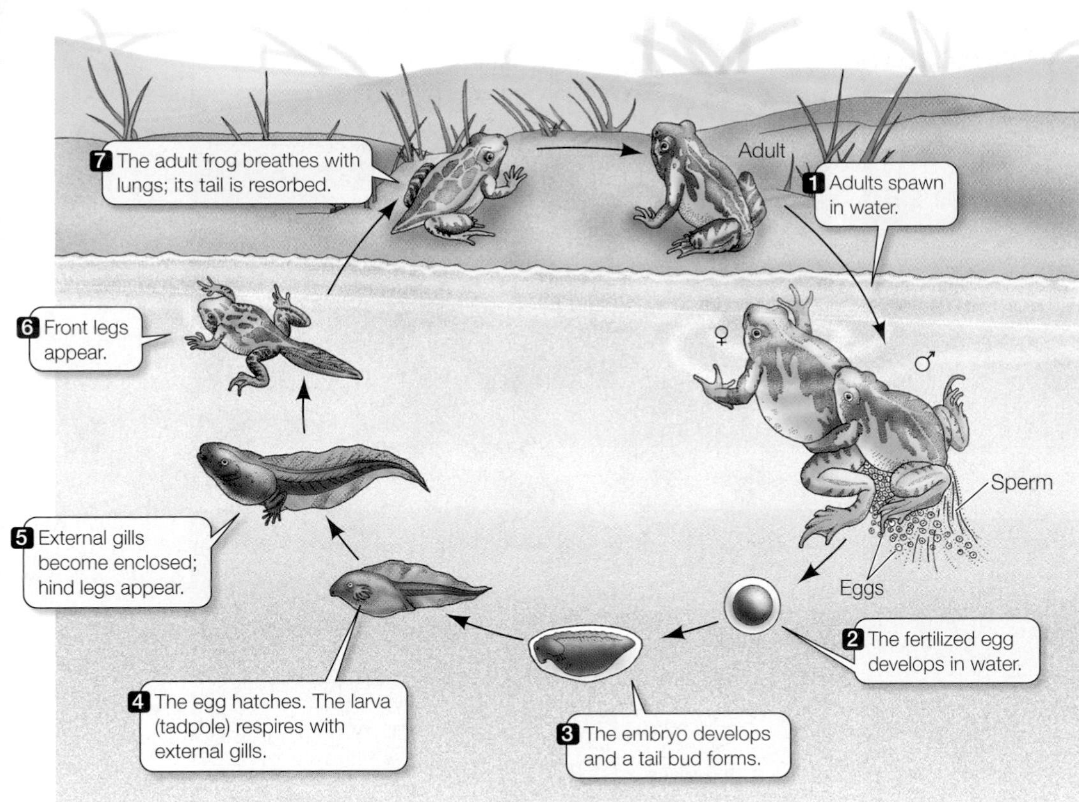
7 The adult frog breathes with lungs; its tail is resorbed.
Adult
1 Adults spawn in water.
6 Front legs appear.
♀
♂
Sperm
5 External gills become enclosed; hind legs appear.
Eggs
2 The fertilized egg develops in water.
4 The egg hatches. The larva (tadpole) respires with external gills.
3 The embryo develops and a tail bud forms.

33.15 In and Out of the Water Most early stages in the life cycle of temperate-zone amphibians take place in water. The aquatic tadpole is transformed into a terrestrial adult through metamorphosis.

which are usually fertilized external to the body (**Figure 33.15**). The fertilized eggs give rise to larvae that live in water until they undergo metamorphosis to become terrestrial adults. However, many amphibians (especially those in tropical and subtropical areas) have evolved a wide diversity of additional reproductive modes and types of parental care. Internal fertilization evolved several times among the amphibians. Many species develop directly into adult-like forms from fertilized eggs laid on land or carried by the parents. Other species of amphibians are entirely aquatic, never leaving the water at any stage of their lives, and many of these species retain a larval-like morphology.

The more than 6,000 known species of amphibians living on Earth today belong to three major groups: the wormlike, limbless, tropical, burrowing or aquatic *caecilians* (**Figure 33.16A**), the tailless frogs and toads (collectively called *anurans;* **Figure 33.16B**), and the tailed *salamanders* (**Figure 33.16C,D**).

Anurans are most diverse in wet tropical and warm temperate regions, although a few are found at very high latitudes. There are far more anurans than any other amphibians, with over 5,300 described species and many more being discovered every year. Some anurans have tough skins and other adaptations that enable them to live for long periods in very dry deserts, whereas others live in moist terrestrial and arboreal environments. Some species are completely aquatic as adults. All anurans have a very short vertebral column, with a strongly modified pelvic region that is modified for leaping, hopping, or propelling their bodies through water by kicking their hind legs.

Salamanders are most diverse in temperate regions of the Northern Hemisphere, but many species are also found in cool, moist environments in mountains of Central America. Many salamanders live in rotting logs or moist soil. One major group has lost lungs, and these species exchange gases entirely through the skin and mouth lining—body parts that all amphibians use in addition to their lungs. Through *paedomorphosis* (retention of the juvenile state; see Chapter 20), a completely aquatic lifestyle has evolved several times among the salamanders (see Figure 33.16D). Most species of salamanders have internal fertilization, which is usually achieved through the transfer of a small jelly-like, sperm-containing capsule (called a *spermatophore*).

Many amphibians have complex social behaviors. Most male anurans utter loud, species-specific calls to attract females of their own species (and sometimes to defend breeding territories), and they compete for access to females that arrive at the breeding sites. Many amphibians lay large numbers of eggs, which they abandon once they are deposited and fertilized. Some species, however, lay only a few eggs, which are fertilized and then guarded in a nest, or carried on the backs, in the vocal pouches, or even in the stomachs of one of the parents. A few species of frogs, salamanders, and caecilians are *viviparous,* meaning that they give birth to well-developed young that have received nutrition from the female during gestation.

Amphibians are the focus of much attention today because populations of many species are declining rapidly, especially in mountainous regions of western North America, Central and South America, and northeastern Australia. Scientists are investigating several hypotheses to account for amphibian population declines, including the adverse effects of habitat alteration by humans, increased solar radiation caused by destruction of the Earth's ozone layer, pollution from urban and industrial areas and airborne agricultural pesticides and herbicides, and the spread of a pathogenic chytrid fungus that attacks amphibians. Scientists have documented the spread of the chytrid fungus through Central Amer-

(A) *Dermophis mexicanus*

(B) *Bufo periglenes*

(C) *Gyrinophilus porphyriticus*

(D) *Necturus* sp.

33.16 Diversity among the Amphibians (A) Burrowing caecilians superficially look more like worms than amphibians. (B) Male golden toads in the Monteverde cloud forest of Costa Rica. This species has recently become extinct, one of many amphibian species to do so in the past few decades. (C) An adult spring salamander. (D) The mudpuppy is a salamander that has developed an entirely aquatic life; it has no terrestrial portion of its life cycle.

ica, where many species of amphibians have become extinct, including Costa Rica's golden toad (see Figure 33.16B).

Amniotes colonized dry environments

Several key innovations contributed to the ability of members of one clade of tetrapods to exploit a wide range of terrestrial habitats. The animals that evolved these water-conserving traits are called **amniotes**.

The **amniote egg** (which gives the group its name) is relatively impermeable to water and allows the embryo to develop in a contained aqueous environment (**Figure 33.17**). The leathery or brittle, calcium-impregnated shell of the amniote egg retards evaporation of the fluids inside, but permits passage of oxygen and carbon dioxide. The egg also stores large quantities of food in the form of *yolk*, permitting the embryo to attain a relatively advanced state of development before it hatches. Within the shell are *extraembryonic membranes* that protect the embryo from desiccation and assist its gas exchange and excretion of waste nitrogen.

In several different groups of amniotes, the amniote egg became modified, allowing the embryo to grow inside (and receive nutrition from) the mother. For instance, the mammalian egg lost its shell and yolk while the functions of the extraembryonic membranes were retained and expanded; we will examine the roles of these membranes in detail in Chapter 43.

Other innovations were found in the organs of terrestrial adults. A tough, impermeable skin, covered with scales or modifications

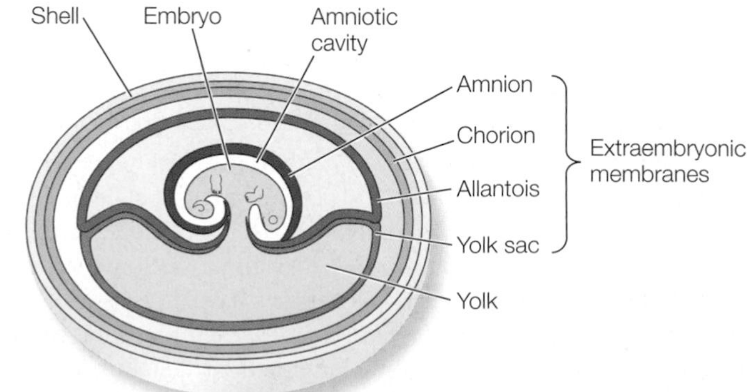

33.17 An Egg for Dry Places The evolution of the amniote egg, with its water-retaining shell, four extraembryonic membranes, and embryo-nourishing yolk, was a major step in the colonization of the terrestrial environment.

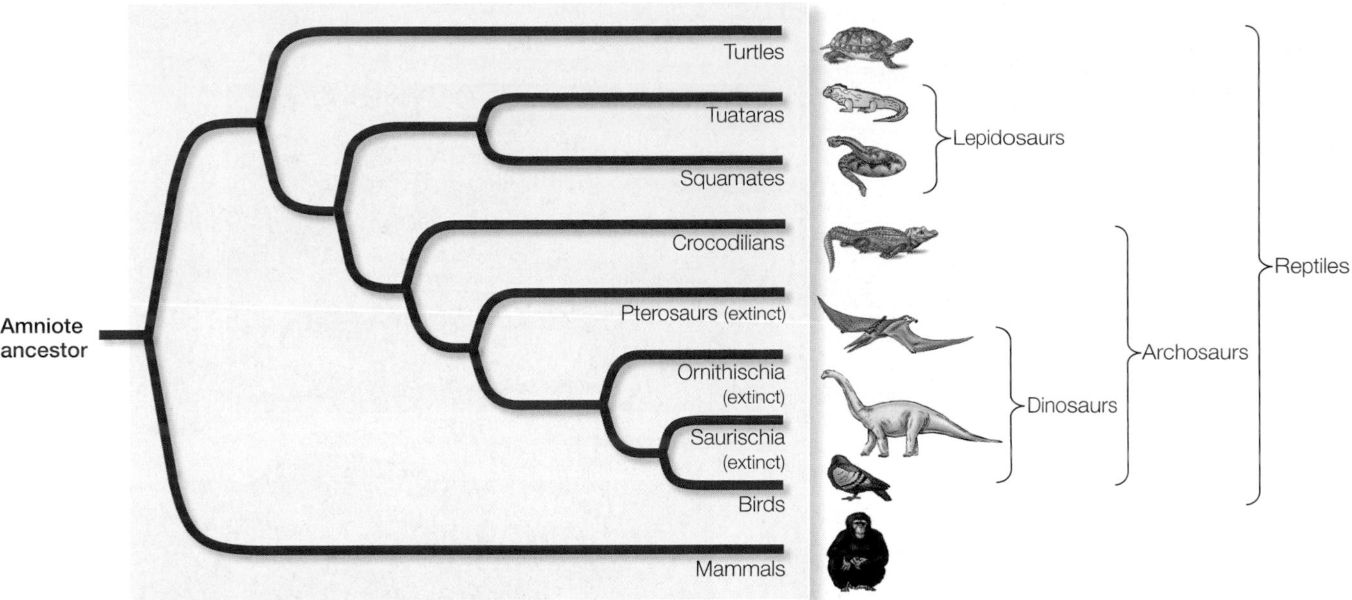

33.18 A Current Phylogenetic Tree of Amniotes This tree of amniote relationships shows the primary split between mammals and reptiles. The reptile portion of the tree shows a lineage that led to the turtles, another to the lepidosaurs (snakes, lizards, and tuataras), and a third branch that includes all the archosaurs (crocodiles, several extinct groups, and the birds).

of scales such as hair and feathers, greatly reduced water loss. And, perhaps most importantly, adaptations to the vertebrate excretory organs, the kidneys, allowed amniotes to excrete concentrated urine, ridding the body of waste nitrogen without losing a large amount of water in the process (see Chapter 51).

During the Carboniferous, amniotes split into two major groups, the mammals and reptiles (**Figure 33.18**). More than 17,200 species of **reptiles** exist today, over half of which are *birds*. Birds are the only living members of the otherwise extinct *dinosaurs*, the dominant terrestrial predators of the Mesozoic.

Reptiles adapted to life in many habitats

The lineage leading to modern reptiles began to diverge from other amniotes about 250 million years ago. One reptilian group that has changed very little over the intervening millennia is the *turtles*. The dorsal and ventral bony plates of turtles form a shell into which the head and limbs can be withdrawn in many species (**Figure 33.19A**). The dorsal shell is an expansion of the ribs, and it is a mystery how the pectoral girdles evolved to be inside the ribs of turtles, unlike any other vertebrates. Most turtles live in aquatic environments, but several groups, such as tortoises and box turtles, are terrestrial. Sea turtles spend their entire lives at sea except when they come ashore to lay eggs. Human exploitation of sea turtles and their eggs has resulted in worldwide declines of these species, all of which are now endangered. A few species of turtles are strict herbivores or carnivores, but most species are omnivores that eat a variety of aquatic and terrestrial plants and animals.

The *lepidosaurs* constitute the second-most species-rich clade of living reptiles. This group is composed of the *squamates* (lizards, snakes, and amphisbaenians—the latter a group of mostly legless, wormlike, burrowing reptiles with greatly reduced eyes) and the *tuataras*, which superficially resemble lizards, but differ from them in tooth attachment and several internal anatomical features. Many species related to the tuataras lived during the Mesozoic era, but today only two species, restricted to a few islands off the coast of New Zealand, survive (**Figure 33.19B**).

The skin of a lepidosaur is covered with horny scales that greatly reduce loss of water from the body surface. These scales, however, make the skin unavailable as an organ of gas exchange. Gases are exchanged almost entirely via the lungs, which are proportionally much larger in surface area than those of amphibians. A lepidosaur forces air into and out of its lungs by bellows-like movements of its ribs. The lepidosaur heart is divided into *chambers* that partially separate oxygenated blood from the lungs from deoxygenated blood returning from the body. With this type of heart, lepidosaurs can generate high blood pressure and can sustain relatively high levels of metabolism.

Most lizards are insectivores, but some are herbivores; a few prey on other vertebrates. The largest lizard, which grows as long as 3 meters, is the predaceous Komodo dragon of the East Indies. Most lizards walk on four limbs (**Figure 33.19C**), although limblessness has evolved repeatedly in the group, especially in burrowing and grassland species. One major group of limbless squamates are known as snakes (**Figure 33.19D**). All snakes are carnivores; many can swallow objects much larger than themselves. Several snake groups evolved venom glands and the ability to inject venom rapidly into their prey.

Crocodilians and birds share their ancestry with the dinosaurs

Another reptilian group, the *archosaurs*, includes the crocodilians, dinosaurs, and birds. *Dinosaurs* rose to prominence about 215 million years ago and dominated terrestrial environments for about 150 million years; only one group of dinosaurs, the *birds*, survived the mass extinction event at the Cretaceous–Tertiary boundary.

(A) *Chelonia mydas*

(B) *Sphenodon punctatus*

(C) *Chlamydosaurus kingii*

(D) *Diadophis punctatus*

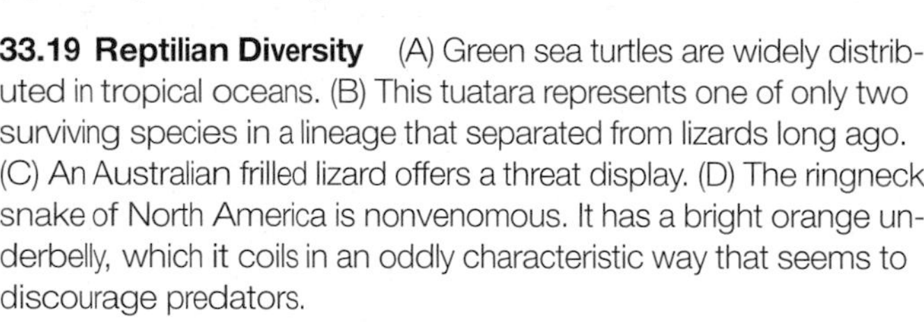

33.19 Reptilian Diversity (A) Green sea turtles are widely distributed in tropical oceans. (B) This tuatara represents one of only two surviving species in a lineage that separated from lizards long ago. (C) An Australian frilled lizard offers a threat display. (D) The ringneck snake of North America is nonvenomous. It has a bright orange underbelly, which it coils in an oddly characteristic way that seems to discourage predators.

During the Mesozoic, most terrestrial animals more than a meter in length were dinosaurs. Many were agile and could run rapidly; they had special muscles that enabled the lungs to be filled and emptied while the limbs moved. We can infer the existence of such muscles in dinosaurs from the structure of the vertebral column in fossils. Some of the largest dinosaurs weighed as much as 80 tons (see the opening of Chapter 31).

Until recently, scientists assumed that the non-avian dinosaurs grew slowly. However, analyses of growth lines in fossil bones indicate that *Tyrannosaurus rex* reached full size in 15 to 18 years—much faster than the 25 to 35 years it takes a smaller African elephant to achieve full size.

Modern *crocodilians*—crocodiles, caimans, gharials, and alligators—are confined to tropical and warm temperate environments (**Figure 33.20A**). Crocodilians spend much of their time in water, but they build nests on land or on floating piles of vegetation. The eggs are warmed by heat generated by decaying organic matter that the female places in the nest. Typically, the female guards the eggs until hatching; and she often facilitates hatching. In some species, the female continues to guard and communicate with her offspring after they hatch. All crocodilians are carnivorous. They eat vertebrates of all kinds, including large mammals.

Biologists have long accepted the phylogenetic position of birds among the reptiles, although birds clearly have many unique, derived morphological features. In addition to the strong morphological evidence for the placement of birds among the reptiles, fossil and molecular data emerging over the last few decades have provided definitive supporting evidence. Birds are thought to have emerged among the *theropods*, a group of predatory dinosaurs that share such traits as bipedal stance, hollow bones, a *furcula* ("wishbone"), elongated metatarsals with three-fingered feet, elongated forelimbs with three-fingered hands, and a pelvis that points backwards.

The living bird species fall into two major groups that diverged during the late Cretaceous, about 80–90 million years ago, from a flying ancestor. The few modern descendants of one lineage include a group of secondarily flightless and weakly flying birds, some of which are very large. This group, called the *palaeognaths*, includes the South and Central American tinamous and several large flightless birds of the southern continents—the rhea, emu, kiwi, cassowary, and the world's largest bird, the ostrich (**Figure 33.20B**). The second lineage (*neognaths*) has left a much larger number of descendants, most of which have retained the ability to fly.

(A) *Crocodylus niloticus*

(B) *Struthio camelus*

33.20 Archosaurs (A) The Nile crocodile. Crocodiles, alligators, and their relatives live in tropical and warm temperate climates. (B) Birds are the other living archosaur group, represented here by the winged but flightless ostrich.

The evolution of feathers allowed birds to fly

Fossil dinosaurs discovered recently in early Cretaceous deposits in Liaoning Province, in northeastern China, show that the scales of some small predatory dinosaurs were highly modified to form *feathers*. The feathers of one of these dinosaurs, *Microraptor gui*, were structurally similar to those of modern birds (**Figure 33.21A**).

During the Mesozoic era, about 175 million years ago, a lineage of theropods gave rise to the birds. The oldest known avian fossil, *Archaeopteryx*, which lived about 150 million years ago, had teeth, but was covered with feathers that are virtually identical to those of modern birds (**Figure 33.21B**). It also had well-developed wings, a long tail, and a furcula, or "wishbone," to which some of the flight muscles were probably attached. *Archaeopteryx* had clawed fingers on its forelimbs, but it also had typical perching bird claws on its hind limbs. It probably lived in trees and shrubs and used the fingers to assist it in clambering over branches.

The evolution of feathers was a major force for diversification. Feathers are lightweight but are strong and structurally complex (**Figure 33.22**). The large quills that support wing feathers arise from the skin of the forelimbs to create the flying surfaces. Other strong feathers sprout like a fan from the shortened tail and serve as stabilizers during flight. The feathers that cover the body, along with an underlying layer of down feathers, provide birds with insulation that helps them to survive in virtually all of Earth's climates.

The bones of theropod dinosaurs, including birds, are hollow with internal struts that increase their strength. Hollow bones would have made early theropods lighter and more mobile; later they facilitated the evolution of flight. The sternum (breastbone) of flying birds forms a large, vertical keel to which the flight muscles are attached.

Flight is metabolically expensive. A flying bird consumes energy at a rate about 15–20 times faster than a running lizard of the same weight. Because birds have such high metabolic rates, they generate large amounts of heat. They control the rate of heat loss using their feathers, which may be held close to the body or elevated to alter the amount of insulation they provide. Another adaptation to the needs of flight is found in the lungs of birds, which allow air to flow through unidirectionally rather than pumping air in and out (see Section 48.2).

There are about 9,600 species of living birds, which range in size from the 150-kilogram ostrich to a tiny hummingbird weighing

(A)

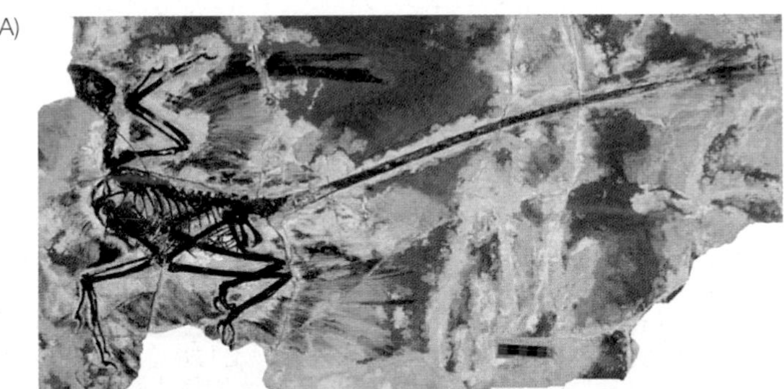

(B)

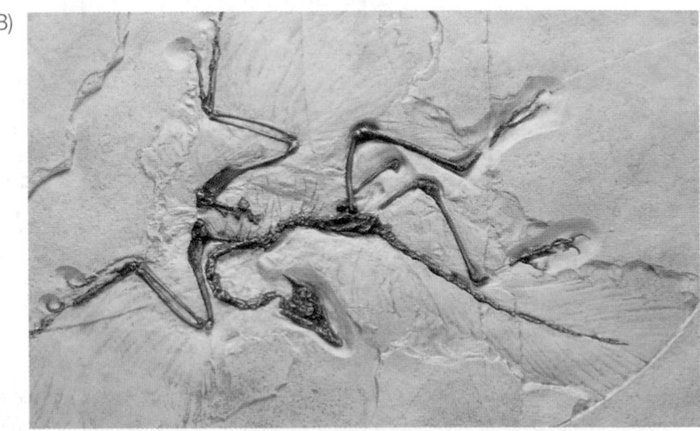

33.21 Mesozoic Bird Fossils Fossil remains demonstrate the evolution of birds from other dinosaurs. (A) *Microraptor gui*, a feathered dinosaur from the early Cretaceous (about 140 mya). (B) *Archaeopteryx* is the oldest known birdlike fossil.

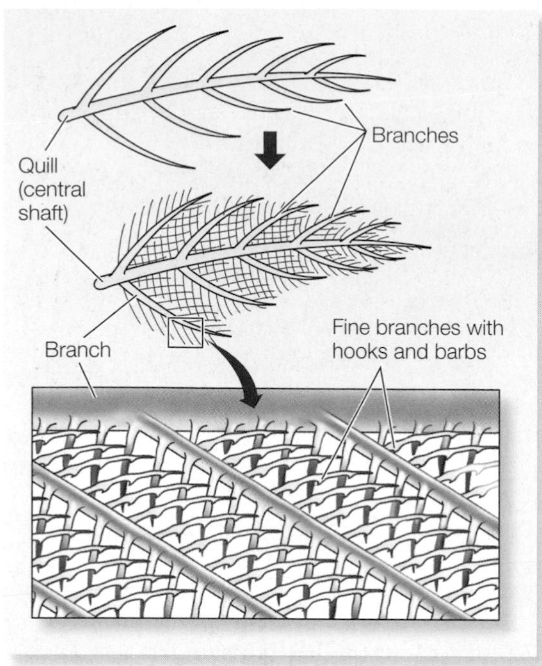

33.22 A Major Evolutionary Breakthrough Early feathers may have been a simple central shaft, or quill, with side branches. A pattern of fine branches with interlocking hooks and barbs creates a strong, lightweight surface that allows flight.

only 2 grams. The teeth so prominent among other dinosaurs were secondarily lost in the ancestral birds, but birds nonetheless eat almost all types of animal and plant material. Insects and fruits are the most important dietary items for terrestrial species. Birds also eat seeds, nectar and pollen, leaves and buds, carrion, and other vertebrates. By eating the fruits and seeds of plants, birds serve as major agents of seed dispersal. Representatives of some of the major groups of birds are shown in **Figure 33.23**.

Mammals radiated after the extinction of dinosaurs

Reptiles
Prototherians
Marsupials
Eutherians

Small and medium-sized **mammals** coexisted with the dinosaurs for millions of years. After the non-avian dinosaurs disappeared during the mass extinction at the close of the Mesozoic era, mammals increased dramatically in numbers, diversity, and size. Today mammals range in size from tiny shrews and bats weighing only about 2 grams to the blue whale, the largest animal on Earth, which measures up to 33

(A) *Aix galericulata*

(B) *Frigata minor*

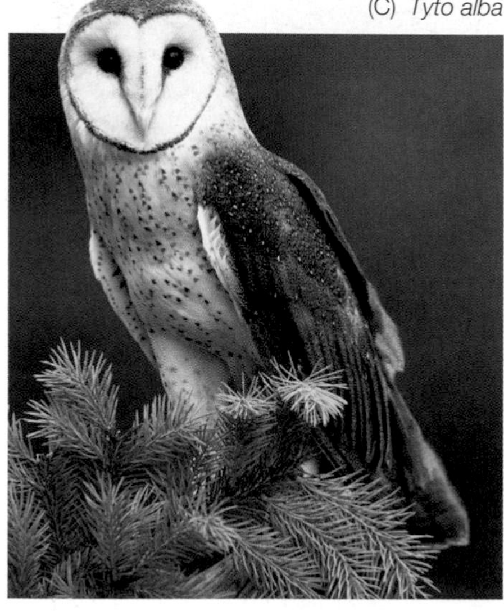
(C) *Tyto alba*

(D) *Cardinalis cardinalis*

33.23 Diversity among the Birds (A) This bright-plumaged male Mandarin duck is a member of the group that includes ducks, geese, and swans. (B) Frigate birds are among the ocean-going birds that are often found miles from shore. This male is in display mode, with an inflated pouch that advertises his strength to females. (C) This barn owl is a nighttime predator that can find prey using its sensitive auditory "sonar" system. (D) Perching, or passeriform, birds such as this male northern cardinal are the most species-rich of all bird group.

meters long and can weigh as much as 160,000 kilograms. Mammals have far fewer, but more highly differentiated, teeth than fishes, amphibians, and reptiles do. Differences among mammals in the number, type, and arrangement of teeth reflect their varied diets (see Figure 50.7).

Four key features distinguish the mammals:

- *Sweat glands,* whose secretions cool the animals as they evaporate.
- *Mammary glands* in female mammals secrete a nutritive fluid (milk) on which newborn individuals feed.
- *Hair,* which gives a protective and insulating covering.
- A *four-chambered heart* that completely separates the oxygenated blood coming from the lungs from the deoxygenated blood returning from the body (this last characteristic is convergent with the archosaurs, including modern birds and crocodiles).

Mammalian eggs are fertilized within the female's body, and the embryos undergo a period of development within the female's body in an organ called the *uterus* prior to being born. Most mammals have a covering of hair (fur), which is luxuriant in some species but has been greatly reduced in others, including the cetaceans (whales and dolphins) and humans. Thick layers of insulating fat (blubber) replace hair as a heat retention mechanism in the cetaceans; humans learned to use clothing for this purpose when they dispersed from warm tropical areas.

The approximately 5,000 species of living mammals are divided into two major groups. The three species of *prototherians* are found only in Australia and New Guinea. These mammals, the duck-billed platypus and two species of echidnas, differ from other mammals in lacking a placenta, laying eggs, and having sprawling legs (**Figure 33.24**). Prototherians supply milk for their young, but they have no nipples on their mammary glands; the milk simply oozes out and is lapped off the fur by the offspring.

(A) *Tachyglossus aculeata*

(B) *Ornithorhynchus anatinus*

33.24 Prototherians (A) The short-beaked echidna is one of the two surviving species of echidnas. (B) The duck-billed platypus is another surviving prototherian species.

Most mammals are therians

Members of the *therian* clade are further subdivided into two more groups, marsupials and eutherians. Females of most species of *marsupials* have a ventral pouch in which they carry and feed their offspring (see Figure 33.25A). Gestation (pregnancy) in marsupials is brief; the young are born tiny but with well-developed forelimbs, with which they climb to the pouch. They attach to a nipple, but cannot suck. The mother ejects milk into the tiny offspring until it grow large enough to suckle. Once her offspring have left the uterus, a female marsupial may become sexually receptive again. She can then carry fertilized eggs capable of initiating development and replacing the offspring in her pouch should something happen to them.

At one time marsupials were found on all continents, but the approximately 330 living species are now restricted to the Australian region (**Figure 33.25A,B**) and the Americas (especially South America; **Figure 33.25C**). One marsupial species, the Virginia opossum, is widely distributed in North America. Marsupials radiated to become herbivores, insectivores, and carnivores, but no marsupials live in the oceans. None can fly, although some *arboreal* (tree-dwelling) marsupials are gliders. The largest living marsupials are the kangaroos of Australia, which can weigh up to 90 kilograms. Much larger marsupials existed in Australia until humans exterminated them soon after reaching that continent about 40,000 years ago (see Figure 57.1).

Most bat species are predators, primarily of insects. A number of bats feed on fruit, however, and some fruit-eating species became specialized fluid feeders, able to pierce the skin of fruit and feed on the juice. It is probably from such juice-drinking ancestors that the three species of notorious "vampire" bats arose, able to pierce the skin and feed on the blood of other mammals.

The largest group of therians are the *eutherians*. Eutherians are sometimes called *placental mammals,* but this name is inappropriate because some marsupials also have placentas. Eutherians are more developed at birth than are marsupials; no external pouch houses them after they are born.

The more than 4,500 species of eutherians belong to one of 20 major groups (**Table 33.1**). The largest group is the rodents, with over 2,000 species. Rodents are traditionally defined by the unique morphology of their teeth, which are adapted for gnawing through substances such as wood. The next largest group comprises the approximately 1,000 bat species—the flying mammals. The bats are followed by the moles and shrews, with more than 400 species. The relationships of the major groups of mammals to

33.25 Marsupials (A) Australia's eastern gray kangaroos are among the largest living marsupials. This female carries her young offspring in the characteristic marsupial pouch. (B) The carnivorous Tasmanian devil is found only in Tasmania, an island off Australia's southern coast. (C) This arboreal opossum is a South American marsupial species.

(A) *Macropus giganteus* (B) *Sarcophilus harrisii* (C) *Caluromys philander*

one another have been difficult to determine, because most of the major groups diverged in a short period of time during an explosive adaptive radiation.

Eutherians are extremely varied in their form and ecology (**Figure 33.26**). The extinction of the non-avian dinosaurs at the end of the Cretaceous may have made it possible for them to diversify and radiate into a large range of ecological *niches.* Many eutherian species grew to become quite large in size, and some assumed the role of dominant terrestrial predators previously occupied by the large dinosaurs. Among these predators, social hunting behavior evolved in a number of species, including members of the canid (wolf/dog), felid (cat), and primate lineages.

Grazing and *browsing* by members of several eutherian groups helped transform the terrestrial landscape. Herds of grazing herbivores feed on open grasslands, while browsers feed on shrubs and trees. The effects of herbivores on plant life favored the evolution of the spines, tough leaves, and difficult-to-eat growth

TABLE 33.1

Major groups of living eutherian mammals

GROUP	APPROXIMATE NUMBER OF LIVING SPECIES	EXAMPLES
Gnawing mammals (*Rodentia*)	>2,000	Rats, mice, squirrels, woodchucks, ground squirrels, beaver, capybara
Flying mammals (*Chiroptera*)	1,000	Bats
Soricomorph insectivores (*Soricomorpha*)	430	Shrews, moles
Even-toed hoofed mammals and cetaceans (*Cetartiodactyla*)	300	Deer, sheep, goats, cattle, antelope, giraffes, camels, swine, hippopotamus, cetaceans (whales, dolphins)
Carnivores (*Carnivora*)	280	Wolves, dogs, bears, cats, weasels, pinnipeds (seals, sea lions, walruses)
Primates (*Primates*)	235	Lemurs, monkeys, apes, humans
Lagomorphs (*Lagomorpha*)	80	Rabbits, hares, pikas
African insectivores (*Afrosoricida*)	30	Tenrecs and golden moles
Spiny insectivores (*Erinaceomorpha*)	24	Hedgehogs
Armored mammals (*Cingulata*)	21	Armadillos
Tree shrews (*Scandentia*)	20	Tree shrews
Odd-toed hoofed mammals (*Perissodactyla*)	20	Horses, zebras, tapirs, rhinoceros
Long-nosed insectivores (*Macroscelidea*)	15	Elephant shrews
Pilosans (*Pilosa*)	10	Anteaters and sloths
Pholidotans (*Pholidota*)	8	Pangolins
Sirenians (*Sirenia*)	5	Manatees, dugongs
Hyracoids (*Hyracoidea*)	4	Hyraxes, dassies
Elephants (*Proboscidea*)	3	African and Indian elephants
Dermopterans (*Dermoptera*)	2	Flying lemurs
Aardvark (*Tubulidentata*)	1	Aardvark

(A) *Xerus inauris*

(B) *Myotis* sp.

(C) *Stenella longirostris*

(D) *Rangifer tarandus*

33.26 Diversity among the Eutherians (A) The Cape ground squirrel of South Africa is one of many species of small, diurnal rodents. (B) Virtually all bat species are nocturnal. Many predatory bat species find their prey via a unique sound wave system similar to sonar. (C) These Hawaiian spinner dolphins are a type of cetacean, a cetartiodactyl group that returned to the marine environment. (D) Large hoofed mammals are important herbivores in terrestrial environments. Although this bull is grazing by himself, caribou are usually found in huge herds.

forms found in many plants. In turn, adaptations to the teeth and digestive systems of many herbivore lineages allowed these species to consume many plants despite such defenses—a striking example of coevolution. A large animal can survive on food of lower quality than a small animal can, and large size evolved in several of the grazing and browsing animals (see Figure 33.26D). The most striking examples of large body size, of course, were found among the giant herbivorous dinosaurs (see pp. 670–671). The evolution of large herbivores, in turn, favored the evolution of large carnivores able to attack and overpower them.

Several lineages of terrestrial eutherians subsequently returned to the aquatic environments their ancestors had left behind. The completely aquatic marine cetaceans—whales and dolphins—evolved from artiodactyl ancestors (whales are closely related to the hippopotamus). The seals, sea lions, and walruses also returned to the marine environment and their limbs became modified into flippers. Weasel-like otters retain their limbs but have also returned to aquatic environments, colonizing both fresh and salt water. The manatees and dugongs colonized estuaries and shallow seas.

33.4 RECAP

The vertebrate colonization of dry land was facilitated by the evolution of an impermeable body covering, efficient kidneys, and the amniote egg—a structure that resists desiccation and provides an aqueous internal environment in which the embryo grows. Major amniote groups include turtles, lepidosaurs, archosaurs (crocodilians and birds), and mammals.

- In the not-too-distant past, the idea that birds were reptiles met with extreme skepticism. Do you understand how fossils, morphology, and molecular evidence support the position of birds vis-à-vis the reptiles? See pp. 732–733
- In reviewing the discussion of the various vertebrate groups, can you identify several reasons why tooth structure is such an important area of study?

The evolutionary history of one eutherian group—the primates—is of special interest to us because it includes the human lineage. The primates have been the subject of extensive research, both physical and molecular. Let's take a closer look at the characteristics and evolutionary history of the primates.

33.5 What Traits Characterize the Primates?

The eutherian **primates** underwent extensive evolutionary radiation from an ancestral small, arboreal, insectivorous mammal. A nearly complete fossil of an early primate, *Carpolestes*, was found in Wyoming and dated at 56 million years ago; it had grasping feet with an opposable big toe that had a nail rather than a claw. Grasp-

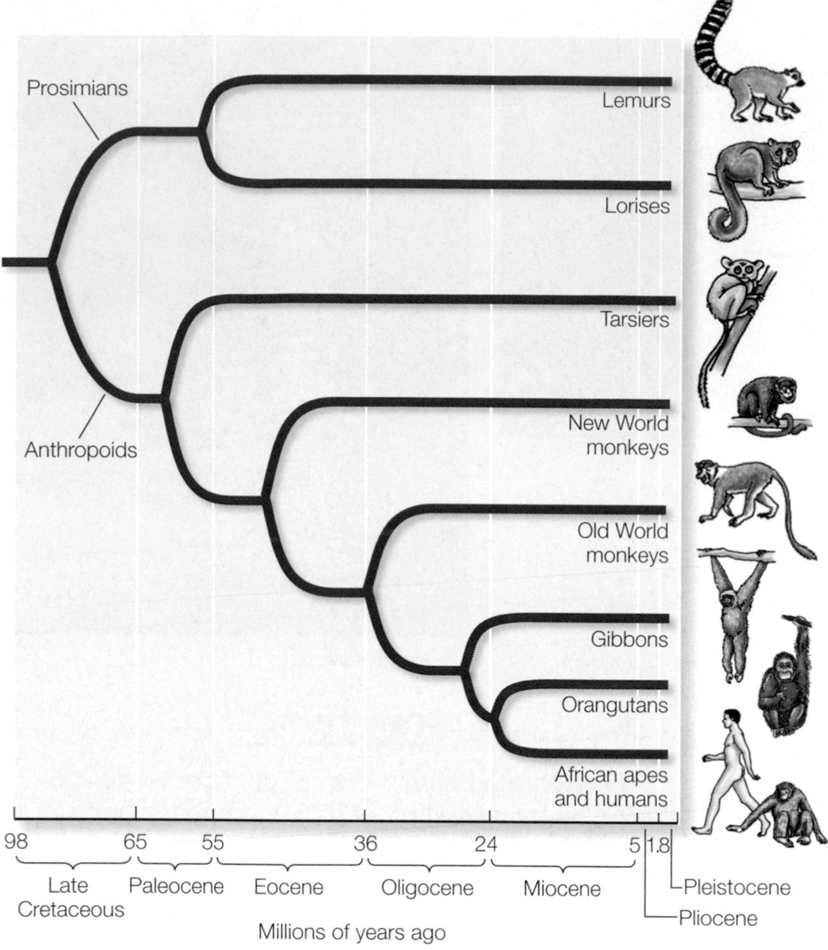

33.27 A Current Phylogenetic Tree of the Primates The phylogeny of primates is among the best studied of any major group of mammals. This tree is based on evidence from many genes, morphology, and fossils.

ing limbs with opposable digits are one of the major adaptations to arboreal life that distinguish primates from other mammals.

Early in their evolutionary history, the primates split into two main clades, the prosimians and the anthropoids (**Figure 33.27**). *Prosimians*—lemurs, pottos, and lorises—once lived on all continents, but today they are restricted to Africa, Madagascar, and tropical Asia. All mainland prosimian species are arboreal and nocturnal (**Figure 33.28**). On the island of Madagascar, however, the site of a remarkable radiation of lemurs, there are also diurnal and terrestrial species.

A second primate lineage, the *anthropoids*—tarsiers, Old World monkeys, New World monkeys, apes, and humans—appeared about 65 million years ago in Africa or Asia. New World monkeys diverged from Old World monkeys at a slightly later date, but early enough that they might have reached South America from Africa when those two continents were still close to each other. All New World monkeys are arboreal (**Figure 33.29A**). Many of them have long, prehensile tails with which they can grasp branches. Many Old World monkeys are arboreal as well, but a number of species are terrestrial (**Figure 33.29B**). No Old World primate has a prehensile tail.

About 35 million years ago, a lineage that led to the modern apes separated from the Old World monkeys. Between 22 and 5.5 million years ago, dozens of species of apes lived in Europe, Asia, and Africa. The Asian apes—gibbons and orangutans (**Figure 33.30A,B**)—descended from two of these ape lineages. Another extinct genus, *Dryopithecus*, is the sister group of the modern African apes—gorillas and chimpanzees (**Figure 33.30C,D**)—and of humans.

Human ancestors evolved bipedal locomotion

About 6 million years ago in Africa, a lineage split occurred that would lead to the chimpanzees on the one hand, and to the *hominid* clade that includes modern humans and their ancestors on the other.

The earliest protohominids, known as *ardipithecines*, had distinct morphological adaptations for *bipedal locomotion* (walking on two legs). Bipedal locomotion frees the forelimbs to manipulate objects and to carry them while walking. It also elevates the eyes, enabling the animal to see over tall vegetation to spot predators and prey. Bipedal locomotion is also energetically more economical than quadrupedal locomotion. All three advantages were probably important for the ardipithecines and their descendants, the australopithecines.

The first *australopithecine* skull was found in South Africa in 1924. Since then australopithecine fossils have been found at many sites in Africa. The most complete fossil skeleton of an australopithecine yet found was discovered

Eulemur coronatus

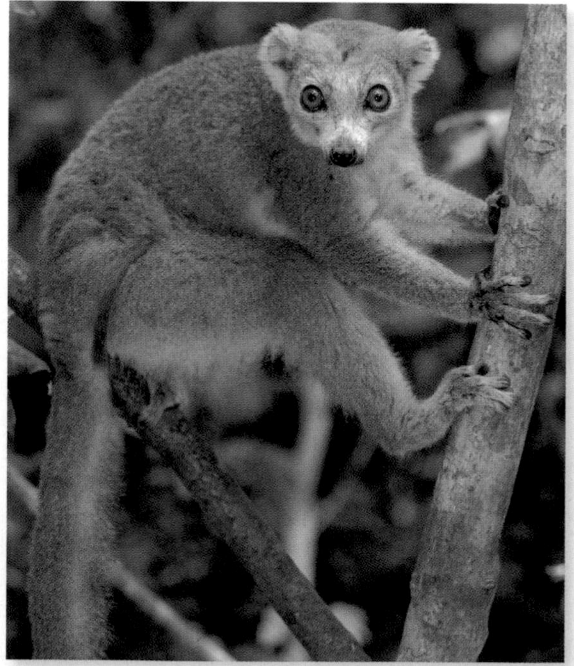

33.28 A Prosimian The crowned lemur is one of the many lemur species found in Madagascar, where they are part of a unique assemblage of endemic plants and animals.

(A) *Ateles geoffroyi*

(B) *Mandrillus leucophaeus*

33.29 Monkeys (A) The spider monkeys of Central America are typical of the New World monkeys, all of which are arboreal. (B) Although many Old World monkey species are arboreal, these drills are among the many terrestrial groups.

in Ethiopia in 1974. The skeleton was approximately 3.5 million years old and was that of a young female who has since become known to the world as "Lucy." Lucy was assigned to the species *Australopithecus afarensis*, and her discovery captured worldwide interest. Fossil remains of more than 100 *A. afarensis* individuals have since been discovered, and there have been recent discoveries of fossils of other australopithecines who lived in Africa 4–5 million years ago.

Experts disagree over how many species are represented by australopithecine fossils, but it is clear that at least two distinct types lived together over much of eastern Africa several million years ago (**Figure 33.31**). The larger type (about 40 kilograms) is represented by at least two species (*Paranthropus robustus* and *P. boisei*), both of which died out about 1.5 million years ago. The smaller type, represented by *A. afarensis*, probably gave rise to the genus *Homo*.

Early members of the genus *Homo* lived contemporaneously with australopithecines in Africa for perhaps half a million years. Some

(A) *Hylobates mulleri*

(B) *Pongo pygmaeus*

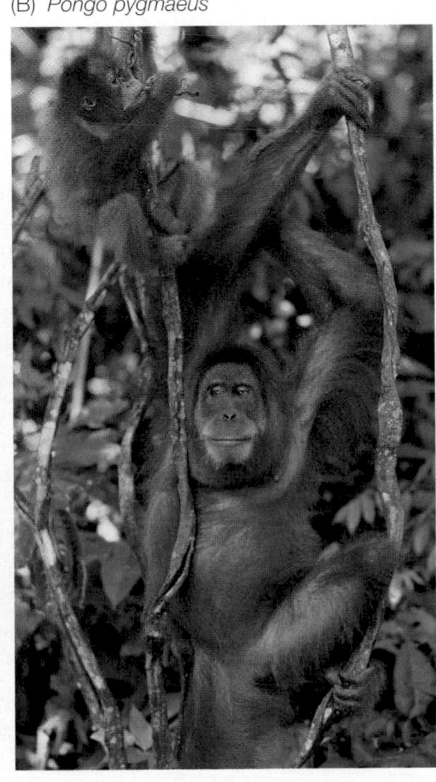

(C) *Gorilla gorilla*

(D) *Pan troglodytes*

33.30 Apes (A) Gibbons are the smallest of the apes. They are found in Asia; Indonesia is home to this Borneo gibbon. (B) Orangutans live in the forests of Sumatra and Borneo. (C) Gorillas, the largest apes, are restricted to humid African forests. This male is a lowland gorilla. (D) Chimpanzees, our closest relatives, are found in forested regions of Africa.

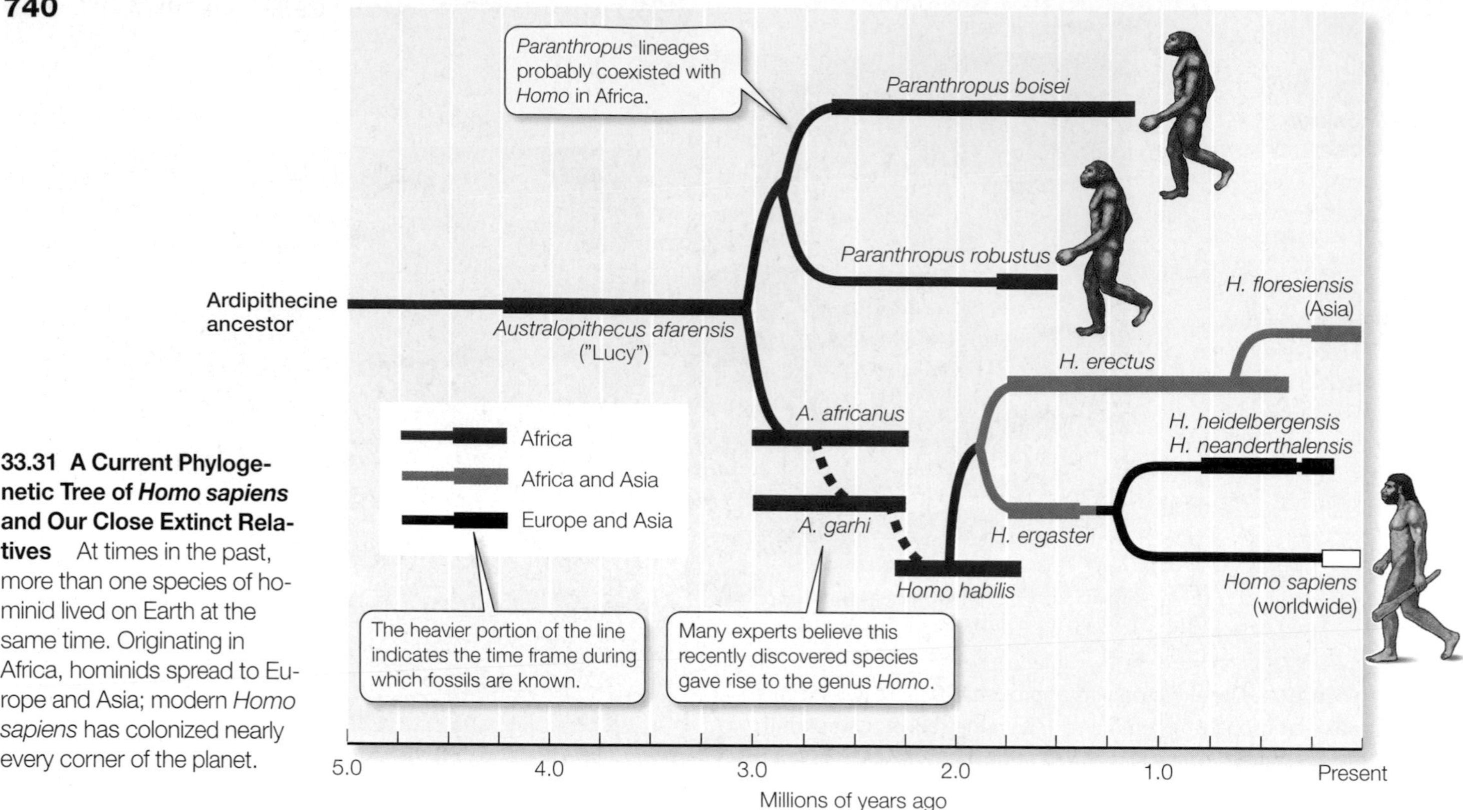

33.31 A Current Phylogenetic Tree of *Homo sapiens* and Our Close Extinct Relatives At times in the past, more than one species of hominid lived on Earth at the same time. Originating in Africa, hominids spread to Europe and Asia; modern *Homo sapiens* has colonized nearly every corner of the planet.

2-million-year-old fossils of an extinct species called *H. habilis* were discovered in the Olduvai Gorge, Tanzania. Other fossils of *H. habilis* have been found in Kenya and Ethiopia. Associated with the fossils are tools that these early hominids used to obtain food.

Another extinct hominid species, *Homo erectus*, evolved in Africa about 1.6 million years ago. Soon thereafter it had spread as far as eastern Asia. Members of *H. erectus* were as large as modern people, but their brains were smaller and they had unusually thick skulls. The cranium, which had thick, bony walls, may have been an adaptation to protect the brain, ears, and eyes from impacts caused by a fall or a blow from a blunt object. What would have been the source of such blows? Fighting with other *H. erectus* males is a highly probable answer.

Homo erectus used fire for cooking and for hunting large animals, and made characteristic stone tools that have been found in many parts of the Old World. Although *H. erectus* survived in Eurasia until about 250,000 years ago, by about 200,000 years ago only *Homo sapiens* was found in most tropical regions. However, as described at the opening of this chapter, a *H. erectus* descendant, *Homo floresiensis*, survived until as recently as 18,000 years ago.

Human brains became larger as jaws became smaller

In the hominid lineage leading to *Homo sapiens*, the brain increased rapidly in size, reaching its modern size by about 160,000 years ago. At the same time, the powerful jaw muscles found in living apes and the australopithecines dramatically decreased in size. These two changes were simultaneous, suggesting that they might have been functionally correlated. A mutation in a regulatory gene that is expressed only in the head may have removed a barrier that had previously prevented this remodeling of the human cranium.

The striking enlargement of brains relative to body size in the hominid lineage was probably favored by an increasingly complex social life. Any features that allowed group members to communicate more effectively with one another would have been valuable in cooperative hunting and gathering and for improving one's status in the complex social interactions that must have characterized early human societies, just as they do in ours today.

Several *Homo* species existed during the mid-Pleistocene epoch, from about 1.5 million to about 300,000 years ago. All were skilled hunters of large mammals, but plants were important components of their diets. During this period another distinctly human trait emerged: rituals and a concept of life after death. Deceased individuals were buried with tools and clothing, supplies for their presumed existence in the next world.

One species, *Homo neanderthalensis*, was widespread in Europe and Asia between about 75,000 and 30,000 years ago. Neanderthals were short, stocky, and powerfully built. Their massive skulls housed brains somewhat larger than our own. They manufactured a variety of tools and hunted large mammals, which they probably ambushed and subdued in close combat. For a short time, their range overlapped that of the *H. sapiens* known as Cro-Magnons, but then the Neanderthals abruptly disappeared. Many scientists believe that they were exterminated by the Cro-Magnons after the latter moved out of Africa into the range of Neanderthals.

Cro-Magnon people made and used a variety of sophisticated tools. They created the remarkable paintings of large mammals, many of them showing scenes of hunting, found in European caves. The animals depicted were characteristic of the cold steppes and grasslands that occupied much of Europe during periods of glacial expansion. Cro-Magnon people spread across Asia, reaching North America perhaps as early as 20,000 years ago, although the date of their arrival in the New World is still uncertain. Within a few thousand years, they had spread southward through North America to the southern tip of South America.

Humans developed complex language and culture

As our ancestors evolved larger brains, their behavioral capabilities increased, especially the capacity for language. Most animal communication consists of a limited number of signals, which refer mostly to immediate circumstances and are associated with charged emotional states induced by those circumstances. Human language is far richer in its symbolic character than other animal vocalizations. Our words can refer to past and future times and to distant places. We are capable of learning thousands of words, many of them referring to abstract concepts. We can rearrange words to form sentences with complex meanings.

The expanded mental abilities of humans enabled the development of a complex **culture**, in which knowledge and traditions are passed along from one generation to the next by teaching and observation. Cultures can change rapidly because genetic changes are not necessary for a cultural trait to spread through a population. On the other hand, cultural norms are not transferred automatically, but must be deliberately taught to each generation.

Cultural transmission greatly facilitated the development and use of domestic plants and animals and the resultant conversion of most human societies from ones in which food was obtained by hunting and gathering to ones in which *pastoralism* (herding large animals) and *agriculture* provided most of the food. The development of agriculture led to an increasingly sedentary life, the growth of cities, greatly expanded food supplies, rapid increases in the human population, and the appearance of occupational specializations, such as artisans, shamans, and teachers.

33.5 RECAP

Grasping limbs with opposable digits distinguish primates from other mammals. Human ancestors developed bipedal locomotion and large brains.

- What are some major trends in primate evolution?
- Can you describe the differences between Old World and New World monkeys? See p. 738
- Do you understand how cultural evolution differs from genetic evolution? See p. 741

CHAPTER SUMMARY

33.1 What is a deuterostome?

Deuterostomes vary greatly in adult form, but because they share distinctive patterns of early development, they are judged to be monophyletic. There are many fewer species of deuterostomes than of protostomes, but many deuterostomes are large and ecologically important. Review Figure 33.1, Web/CD Activity 33.1

33.2 What are the major groups of echinoderms and hemichordates?

Echinoderms and hemichordates both have a bilaterally symmetrical larva. Adult echinoderms, however, have **pentaradial symmetry**. Echinoderms have an internal skeleton of calcified plates and a unique **water vascular system** connected to extensions called **tube feet**. Review Figure 33.3

Hemichordates are bilaterally symmetrical and have a three-part body that is divided into a proboscis, collar, and trunk. They include the acorn worms and the pterobranchs. Review Figure 33.5

33.3 What new features evolved in the chordates?

Chordates fall into three major subgroups: **urochordates**, **cephalochordates**, and **vertebrates**.

At some stage in their development, all chordates have a dorsal, hollow nerve cord, a post-anal tail, and a **notochord**. Review Figure 33.6

Urochordates include the ascidians (sea squirts) and salps. Cephalochordates are the lancelets, which live buried in the sand of shallow marine and brackish waters.

The vertebrate body plan is characterized by a rigid internal skeleton, which is supported by a **vertebral column** that replaces the notochord, and an anterior skull with a large brain. Review Figure 33.10

The evolution of jaws from gill arches enabled individuals to grasp large prey and, together with teeth, cut them into small pieces. Review Figure 33.11

Chondrichthyans have skeletons of cartilage; almost all are marine. The skeletons of ray-finned fishes are made of bone, and they have colonized all aquatic environments.

33.4 How did vertebrates colonize the land?

Lungs and jointed appendages enabled vertebrates to colonize the land. The earliest tetrapod vertebrates were the **amphibians**. Most modern amphibians are confined to moist environments because they and their eggs lose water rapidly. Review Figure 33.16, Web/CD Tutorial 33.1

An impermeable skin, efficient kidneys and an egg that could resist desiccation evolved in the **amniotes** (reptiles and mammals). Review Figure 33.17, Web/CD Activity 33.2

The major reptile groups are the turtles; the lepidosaurs (tuataras, lizards, snakes, and amphisbaenians); and the archosaurs (crocodilians and birds). Review Figure 33.18

Mammals are unique among animals in supplying their young with a nutritive fluid (milk) secreted by **mammary glands**. There are two mammalian clades: the three **protherian** species, and the species-rich **therian** clade, which is further subdivided into the **marsupials** and the **eutherians**. Review Table 33.1

33.5 What traits characterize the primates?

Grasping limbs with opposable digits distinguish **primates** from other mammals. The **prosimian** clade includes the lemurs and lorises; the **anthropoid** clade includes monkeys, apes, and humans. Review Figure 33.27

Hominid ancestors were terrestrial primates who developed efficient bipedal locomotion. In the lineage leading to modern *Homo sapiens*, brains became larger as jaws became smaller; the two events may have been functionally linked. Review Figure 33.31

See Web/CD Activity 33.3 for a concept review of this chapter.

SELF-QUIZ

1. Which of the following deuterostome groups have a three-part body plan?
   a. Acorn worms and urochordates
   b. Acorn worms and pterobranchs
   c. Pterobranchs and urochordates
   d. Pterobranchs and lancelets
   e. Urochordates and lancelets
2. The structure used by adult urochordates to capture food is a
   a. pharyngeal basket.
   b. proboscis.
   c. lophophore.
   d. mucus net.
   e. radula.
3. The pharyngeal gill slits of chordate ancestors functioned as sites for
   a. uptake of oxygen only.
   b. release of carbon dioxide only.
   c. both uptake of oxygen and release of carbon dioxide.
   d. removal of small prey from the water.
   e. forcible expulsion of water to move the animal.
4. The key to the vertebrate body plan is a
   a. rigid internal skeleton supported by a vertebral column.
   b. vertebral column to which internal organs are attached.
   c. vertebral column to which two pairs of appendages are attached.
   d. vertebral column to which a pharyngeal basket is attached.
   e. pharyngeal basket and two pairs of appendages.
5. In most fishes, lunglike sacs evolved into
   a. pharyngeal gill slits.
   b. true lungs.
   c. coelomic cavities.
   d. swim bladders.
   e. none of the above
6. Most temperate-zone amphibians return to water to lay their eggs because
   a. water is isotonic to egg fluids.
   b. adults must be in water while they guard their eggs.
   c. there are fewer predators in water than on land.
   d. amphibians need water to produce their eggs.
   e. amphibian eggs quickly lose water and desiccate if their surroundings are dry.
7. The horny scales that cover the skin of reptiles prevent them from
   a. using their skin as an organ of gas exchange.
   b. sustaining high levels of metabolic activity.
   c. laying their eggs in water.
   d. flying.
   e. crawling into small spaces.
8. Which statement about bird feathers is *not* true?
   a. They are highly modified reptilian scales.
   b. They provide insulation for the body.
   c. They exist in two layers.
   d. They help birds fly.
   e. They are important sites of gas exchange.
9. Prototherians differ from other mammals in that they
   a. do not produce milk.
   b. lack body hair.
   c. lay eggs.
   d. live in Australia.
   e. have a pouch in which the young are raised.
10. Bipedalism is believed to have evolved in the human lineage because bipedal locomotion is
    a. more efficient than quadrupedal locomotion.
    b. more efficient than quadrupedal locomotion, and it frees the forelimbs to manipulate objects.
    c. less efficient than quadrupedal locomotion, but it frees the forelimbs to manipulate objects.
    d. less efficient than quadrupedal locomotion, but bipedal animals can run faster.
    e. less efficient than quadrupedal locomotion, but natural selection does not act to improve efficiency.

FOR DISCUSSION

1. In what animal groups has the ability to fly evolved? How do the structures used for flying differ among these animals?
2. Extracting suspended food from the water is a common mode of feeding among animals. Which groups contain species that extract prey from the air? Why is this mode of obtaining food so much less common than extracting prey from the water?
3. What risks and benefits are posed by large size?
4. Amphibians have survived and prospered for many millions of years, but today many species are disappearing, and populations of others are declining seriously. What features of amphibian life histories might make them especially vulnerable to the kinds of environmental changes now happening on Earth?
5. The body plan of most vertebrates is based on four appendages. What are the varied forms that these appendages take, and how are they used?
6. Compare the ways in which different animal lineages colonized the land. How were those ways influenced by the body plans of animals in the different groups?

FOR INVESTIGATION

A mutation in the gene that encodes the myosin heavy chain (see Chapter 47) decreased the size of the jaw muscles in human ancestors. This mutation may have enabled a restructuring of the cranium and larger brain size. How could we determine when this mutation arose in the phylogenetic history of the primates?

PART SEVEN

Flowering Plants: Form and Function

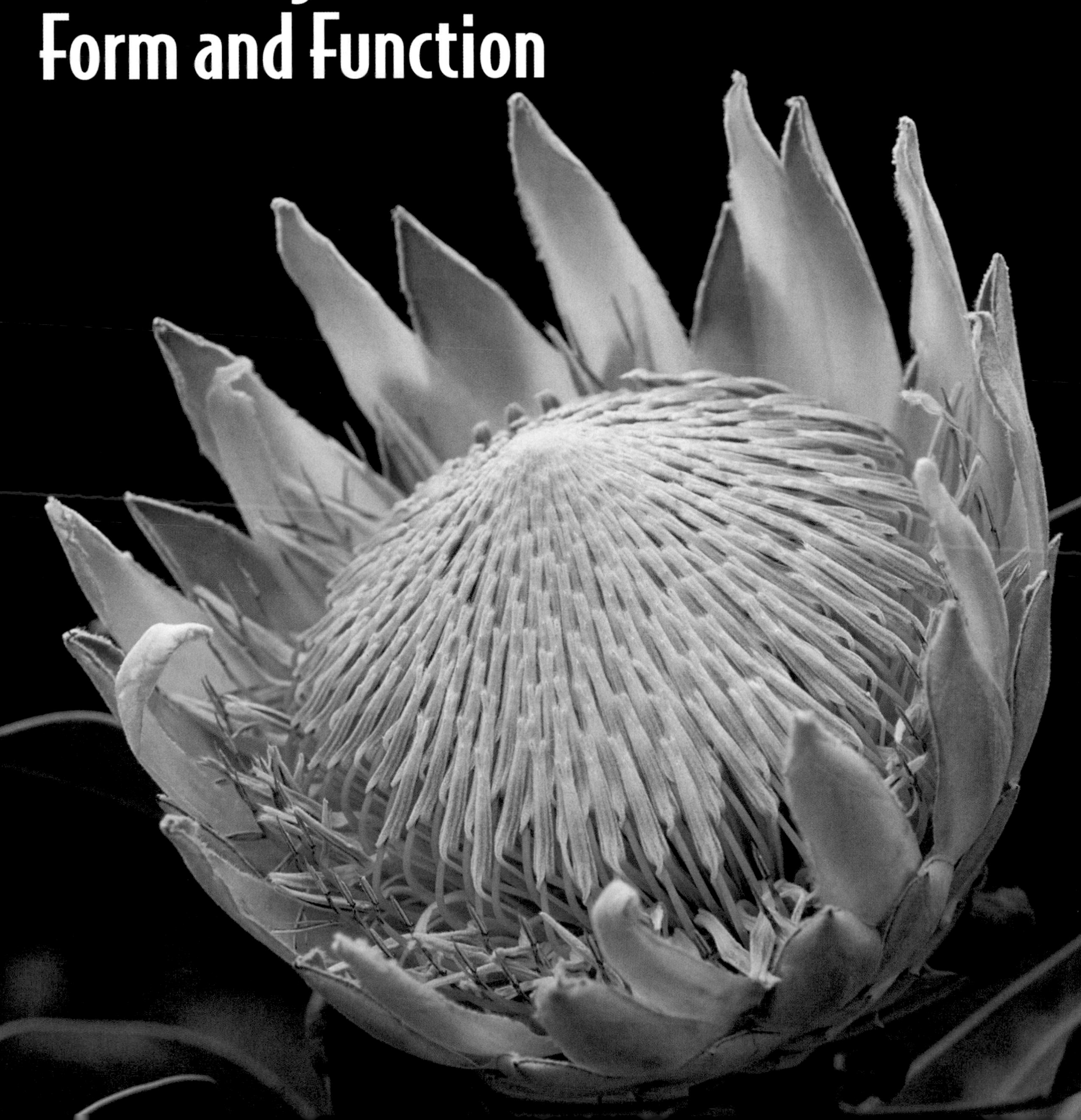

CHAPTER 34 The Plant Body

CSI: Wood anatomy convicts a killer

In the evening of May 21, 1927 Charles Lindbergh landed his plane, *Spirit of St. Louis*, in Paris, becoming the first person to fly nonstop across the Atlantic Ocean. He also became a national hero in the United States, but five years later, "Lucky Lindy" suffered a terrible tragedy when his 20-month-old son was kidnapped. Clues were few: the handwriting on twelve ransom notes, footprints (which were never measured!), a carpenter's chisel, indentations in the ground below the baby's window, and, about 60 feet away in some bushes, a crude wooden ladder in three parts.

WANTED

INFORMATION AS TO THE WHEREABOUTS OF

CHAS. A. LINDBERGH, JR.

OF HOPEWELL, N. J.

SON OF COL. CHAS. A. LINDBERGH

World-Famous Aviator

This child was kidnaped from his home in Hopewell, N. J., between 8 and 10 p. m. on Tuesday, March 1, 1932.

DESCRIPTION:

Age, 20 months — Hair, blond, curly
Weight, 27 to 30 lbs. — Eyes, dark blue
Height, 29 inches — Complexion, light
Deep dimple in center of chin
Dressed in one-piece coverall night suit

ADDRESS ALL COMMUNICATIONS TO
COL. H. N. SCHWARZKOPF, TRENTON, N. J., or
COL. CHAS. A. LINDBERGH, HOPEWELL, N. J.

ALL COMMUNICATIONS WILL BE TREATED IN CONFIDENCE

March 11, 1932

COL. H. NORMAN SCHWARZKOPF
Supt. New Jersey State Police, Trenton, N. J.

The case was assigned to the superintendent of the New Jersey State Police, H. Norman Schwarzkopf (father of General Norman Schwarzkopf, of Gulf War fame), who decided that the ladder was his best clue. He hired Arthur Koehler from the U.S. Forest Products Laboratory in Madison, Wisconsin, to examine the ladder for evidence. What could wood anatomy tell Koehler about the kidnapping?

Microscopic examination of the ladder revealed its cell structure, enabling Koehler to learn that some of the rungs were made from Douglas fir and others from Ponderosa pine, rails were made from Douglas fir and North Carolina pine, and dowels for the railing from birch. Clearly, the ladder was assembled from scraps. Koehler began to extend his examination to other features of the wood.

In the meantime, thousands of leads poured in. The baby's body was discovered, intensifying the search. Gold certificates from the ransom began to appear, eventually leading to the arrest of Bruno Richard Hauptmann. Hauptmann resembled a description provided by the man who had passed the ransom in the dark of night, and his handwriting strongly resembled that on the ransom notes. This and other evidence was circumstantial, however, and prosecutors sought further evidence that would convince a jury that Hauptmann was the kidnapper.

Koehler worked painstakingly to amass evidence based on the wood in the ladder. His most exhausting task was canvassing some 1,600 lumber mills that processed North Carolina pine. He was seeking a particular machine planer that could have left the distinc-

A Desperate Search In the spring of 1932, people across the United States were glued to newspaper and radio updates on the kidnapping of the son of national hero Charles Lindbergh. Posters like this one resulted in thousands of tips from concerned citizens but, sadly, what the police eventually discovered was the child's murdered body.

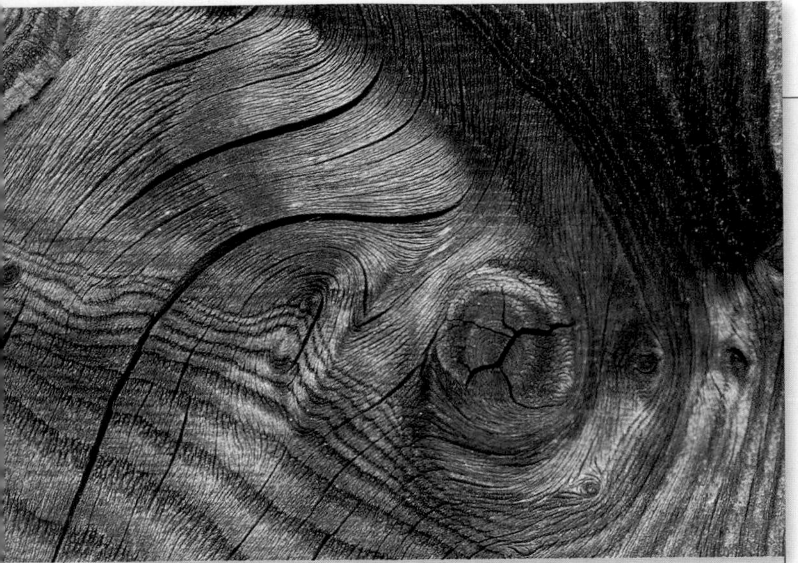

Fingerprints in Wood Wood's unique grain is affected by year-to-year variations in moisture and temperature, by branching (knots), uneven shading of the leaf canopy, whether the wood came from the tree's trunk or a branch, and many other factors. Forensic analysis of wood is one more tool in the arsenal of crime-fighters.

tive pattern of grooves found on the rails. He found it at a mill in South Carolina and traced a shipment from that mill to a lumber company in the Bronx. The Bronx store had employed Hauptmann and sold him lumber.

In court, Koehler used a hand plane from Hauptmann's own toolbox to show that it produced the same pattern of ridges found on one of the rails. He matched three nail holes with protruding nails in a joist in Hauptmann's floor.

Most convincing of all was Koehler's courtroom comparison of the wood grain in a floorboard from Hauptmann's attic—missing a section—and a section of a ladder rail. In spite of a gap in the rail, Koehler showed clearly that the grain of the two pieces matched. That is, the "rings" in the wood of the two sections coincided perfectly, proving that wood from the attic was used to construct the ladder. A jury found Hauptmann guilty, and he was executed on April 3, 1936.

IN THIS CHAPTER we will examine plant structure at the levels of organs, cells, tissues, and tissue systems. We will see how organized groups of dividing cells, called meristems, contribute to the growth of the plant body. The chapter concludes with a consideration of how leaf structure supports photosynthesis.

CHAPTER OUTLINE

34.1 How Is the Plant Body Organized?

The lumber Hauptmann used to build his ladder was the product of the plant body—specifically, the woody stems of trees. All vascular plants have essentially the same structural organization. This chapter describes the basic architecture of the angiosperm body plan.

Most angiosperms (flowering plants) belong to one of two major clades. **Monocots** are generally narrow-leaved flowering plants such as grasses, lilies, orchids, and palms. **Eudicots** are broad-leaved flowering plants such as soybeans, roses, sunflowers, and maples. These two clades, which account for 97 percent of flowering plant species, differ in several important basic characteristics (**Figure 34.1**). Most of the remaining species (including water lilies and magnoliids, discussed in Section 29.4) are structurally similar to the eudicots.

As Chapter 29 described, angiosperms are vascular plants characterized by double fertilization, a triploid endosperm, and seeds enclosed in modified leaves called carpels. These are all reproductive characteristics; flowers, which are the plant's devices for sexual reproduction, consist of modified leaves and stems and will be considered in detail in Chapter 38. But flowering plants also possess three kinds of *vegetative* (nonreproductive) organs: roots, stems, and leaves. In both monocots and eudicots, all the organs are organized in two systems: the **shoot system** and the **root system**. The basic body plans of a generalized monocot and a generalized eudicot are shown in **Figure 34.2**.

- The *shoot system* of a plant consists of the stems, leaves, and flowers. Broadly speaking, the **leaves** are the chief organs of photosynthesis. The **stems** hold and display the leaves to the sun and provide connections for the transport of materials between roots and leaves. The **nodes** are the points of attachment of leaf to stem, and the stem regions between successive nodes are **internodes**.
- The *root system* anchors the plant in place and provides nutrition. The extreme branching of plant roots and their high surface area-to-volume ratio allow them to absorb water and mineral nutrients from the soil.

Each of the vegetative organs can be understood in terms of its structure. By *structure* we mean both its overall form, called its *morphology*, and its component cells and tissues and their

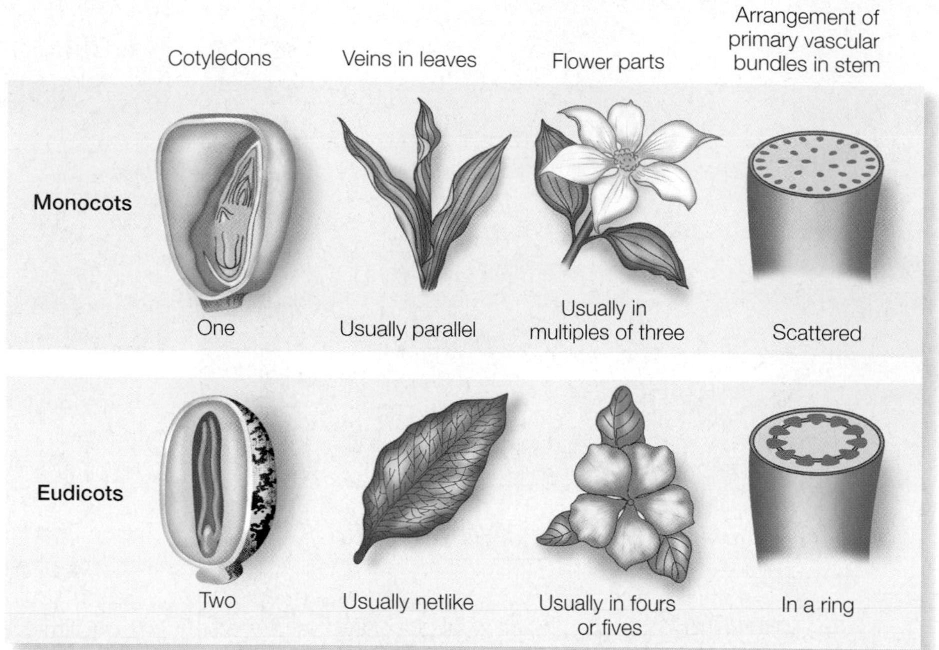

34.1 Monocots versus Eudicots The possession of a single cotyledon clearly distinguishes the monocots from the other angiosperms. Several other anatomical characteristics also differ between the monocots and the eudicots. Most angiosperms that do not belong to either clade resemble eudicots in the characteristics shown here.

arrangement, called its *anatomy*. Let's first consider the overall forms of roots, stems, and leaves.

Roots anchor the plant and take up water and minerals

In most plants, water and minerals enter the plant through the root system, which lies in the soil, where light does not penetrate. Roots typically lack the capacity for photosynthesis even when removed from the soil and placed in light.

There are two principal types of root systems. Many eudicots have a *taproot system*: a single, large, deep-growing primary root accompanied by less prominent lateral roots. The taproot itself often functions as a nutrient storage organ, as in carrots (**Figure 34.3A**).

By contrast, monocots and some eudicots have a *fibrous root system*, which is composed of numerous thin roots that are all roughly equal in diameter (**Figure 34.3B**). Many fibrous root systems have a large surface area for the absorption of water and minerals. A fibrous root system clings to soil very well. The fibrous root systems of grasses, for example, may protect steep hillsides where runoff from rain would otherwise cause erosion.

Some plants have *adventitious roots*. These roots arise above ground from points along the stem; some even arise from the leaves. In many species, adventitious roots can form when a piece of shoot is cut or broken from the plant and placed in water or soil. Adventitious rooting enables the cutting to establish itself in the soil as a new plant. Such a cutting is a form of asexual reproduction, also called *vegetative reproduction* in plants, which we will discuss in Chapter 38. Some plants—corn, banyan trees, and some palms, for example—use adventitious roots as props to help support the shoot.

Stems bear buds, leaves, and flowers

The central function of the **stem** is to elevate and support the reproductive organs (flowers) and the photosynthetic organs (leaves).

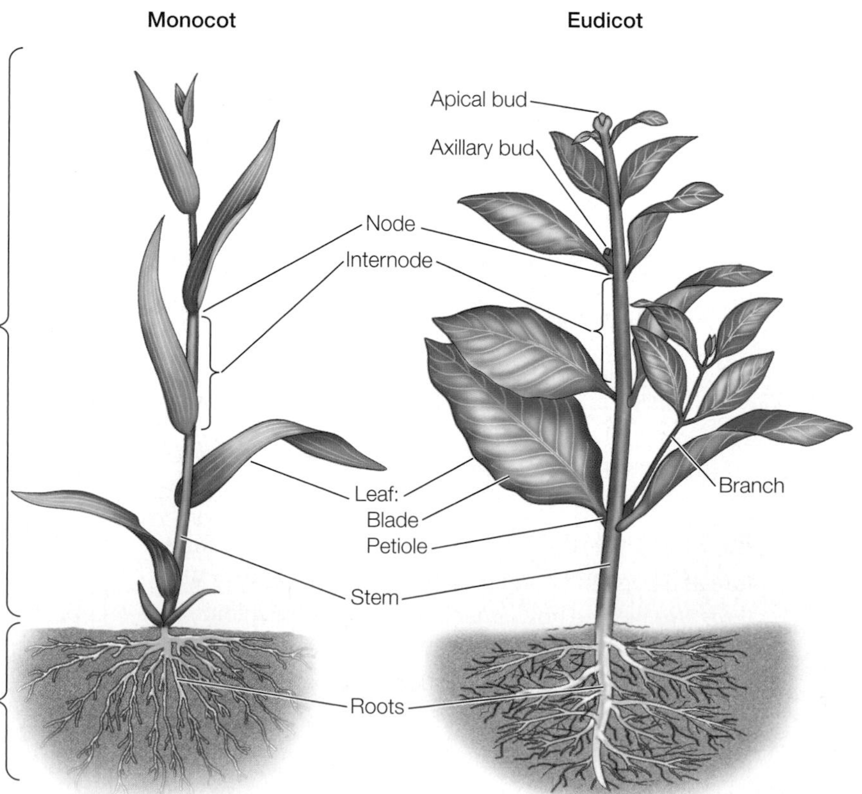

34.2 Vegetative Organs and Systems The basic plant body plan and the principal vegetative organs are similar in monocots and eudicots.

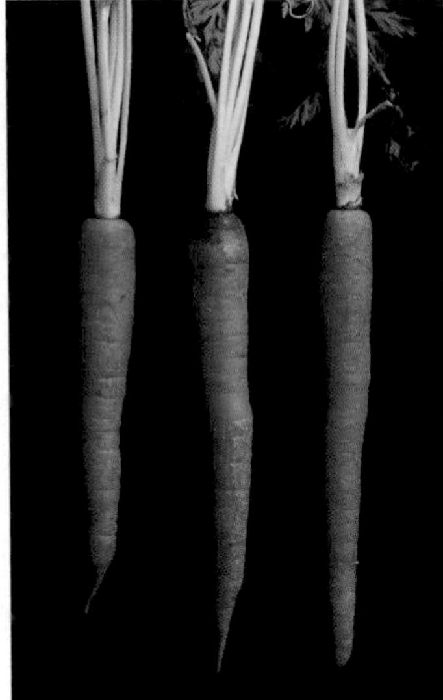

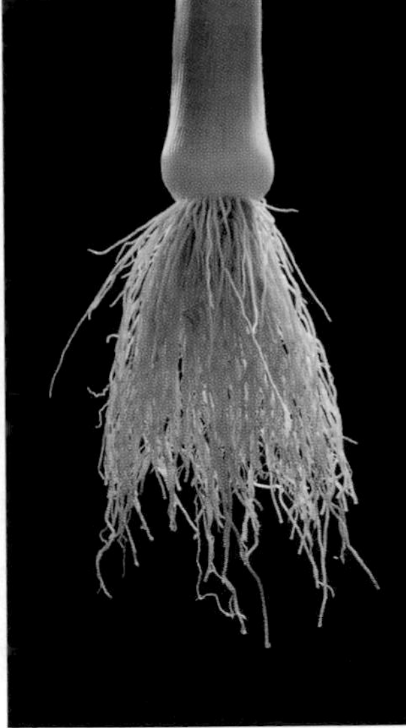

34.3 Root Systems (A) The taproot system of a carrot contrasts with (B) the fibrous root system of a leek.

Unlike roots, stems bear buds of various types. A *bud* is an embryonic shoot. A stem bears leaves at its nodes, and in the angle (axil) where each leaf meets the stem there is an **axillary bud** (see Figure 34.2). If it becomes active, the axillary bud can develop into a new *branch*, or extension of the shoot system. The branching patterns of plants are highly variable, depending on the species, environmental conditions, and a gardener's pruning activities.

At the tip of each stem or branch is an **apical bud**, which produces the cells for the upward and outward growth and development of that shoot. Under appropriate conditions, other buds form that develop into flowers.

Some stems are highly modified. The *tuber* of a potato, for example—the part of the plant eaten by humans—is an underground stem rather than a root. Its "eyes" are depressions containing axillary buds; thus, a sprouting potato is just a branching stem (**Figure 34.4A**). Many desert plants have enlarged, water-retaining stems (**Figure 34.4B**). The *runners* of strawberry plants and Bermuda grass are horizontal stems from which roots grow at frequent intervals (**Figure 34.4C**). If the links between the rooted portions are broken, independent plants can develop on each side of the break—a form of vegetative reproduction.

Although young stems are usually green and capable of photosynthesis, they usually are not the principal sites of photosynthesis. Most photosynthesis takes place in leaves.

Leaves are the primary sites of photosynthesis

In gymnosperms and most flowering plants, the leaves are responsible for most of the plant's photosynthesis, producing energy-rich organic molecules and releasing oxygen gas (see Section 8.1). In certain plants, the leaves are highly modified for more specialized functions, as we will see below.

As photosynthetic organs, leaves are marvelously adapted for gathering light. Typically, the **blade** of a leaf is a thin, flat structure attached to the stem by a stalk called a **petiole** (see Figure 34.2). In many plants, the leaf blade is held by its petiole at an angle almost perpendicular to the rays of the sun. This orientation, with the leaf surface facing the sun, maximizes the amount of light available for photosynthesis. Some leaves track the sun over the course of the day, moving so that they constantly face it.

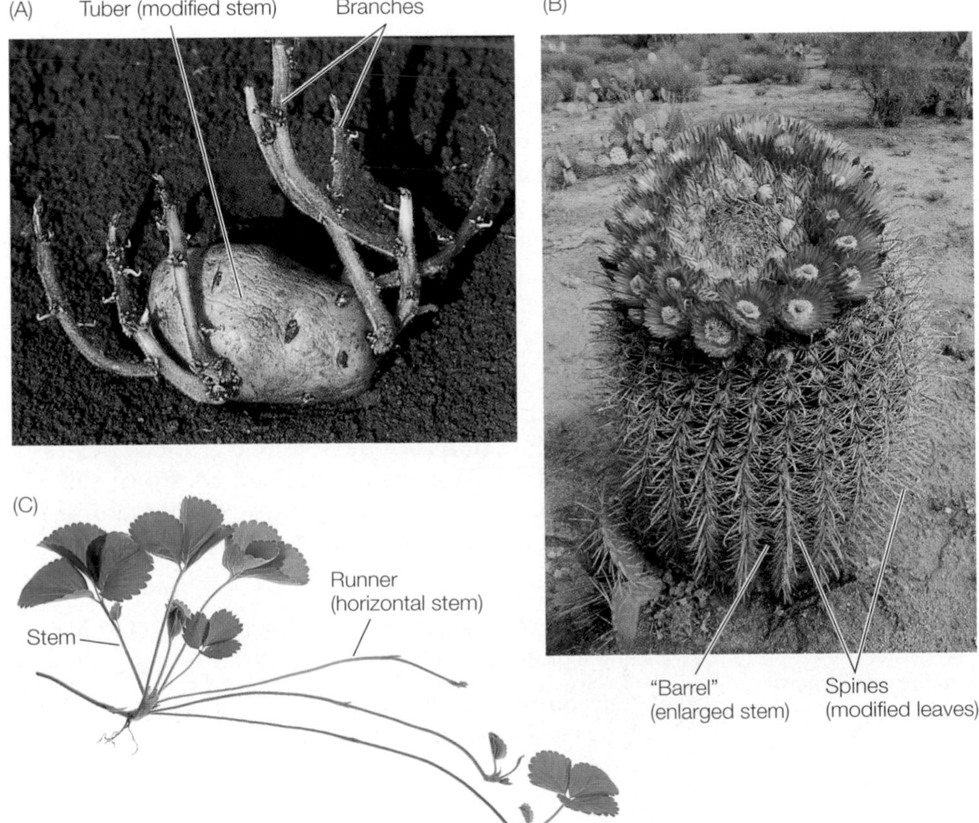

34.4 Modified Stems (A) A potato is a modified stem called a tuber; the sprouts that grow from its eyes are shoots, not roots. (B) The stem of this barrel cactus is enlarged to store water. Its highly modified leaves serve as thorny spines. (C) The runners of beach strawberry are horizontal stems that produce roots at intervals. Runners provide a local water supply and allow rooted portions of the plant to live independently if the runner is cut.

34.5 Simple and Compound Leaves Simple leaves are those with a single blade. Some compound leaves consist of leaflets arranged along a central axis. Further division of the axis results in a doubly compound leaf.

The leaves at different sites on a single plant may have quite different shapes. These shapes result from a combination of genetic, environmental, and developmental influences. Most species, however, bear similar, if not identical, leaves of a particular broadly defined type. A leaf may be *simple,* consisting of a single blade, or *compound,* with multiple blades called *leaflets* arranged along an axis or radiating from a central point (**Figure 34.5**). In a simple leaf, or in a leaflet of a compound leaf, the veins may be parallel to one another, as in most monocots, or in a netlike arrangement, as in eudicots. Differential growth of the leaf veins and the tissue between the veins determines the overall shape of a blade or leaflet.

During development in some plant species, leaves are highly modified for special functions. For example, modified leaves serve as storage depots for energy-rich molecules, as in the bulbs of onions. In other species, the leaves store water, as in succulents. The spines of cacti are modified leaves (see Figure 34.4B). Many plants, such as peas, have modified portions of leaves called *tendrils* that support the plant by wrapping around other structures or plants.

What do all the vegetative organs of flowering plants have in common? Roots, stems, and leaves are composed of three tissue systems. Let's see what they are and what they do.

The tissue systems support the plant's activities

A **tissue** is an organized group of cells that have features in common and that work together as a structural and functional unit. Tissues, in turn, are grouped into **tissue systems.** Three tissue systems extend throughout the body of the vascular plant: *vascular, dermal,* and *ground* tissues in a concentric arrangement (**Figure 34.6**).

The **vascular tissue system** is the plant's plumbing or transport system. Its two constituent tissues, the xylem and phloem, distribute materials throughout the plant. The **xylem** distributes water and mineral ions taken up by the roots to all the cells of the stem and leaves. As a result of its cellular complexity, xylem can perform a variety of functions, including transport, support, and storage. All the living cells of the plant body require a source of energy and chemical building blocks. The **phloem** meets these needs by transporting carbohydrates from sites of production (called *sources,* primarily leaves) to sites of utilization or storage (called *sinks,* such as growing tissue, storage tubers, and developing flowers).

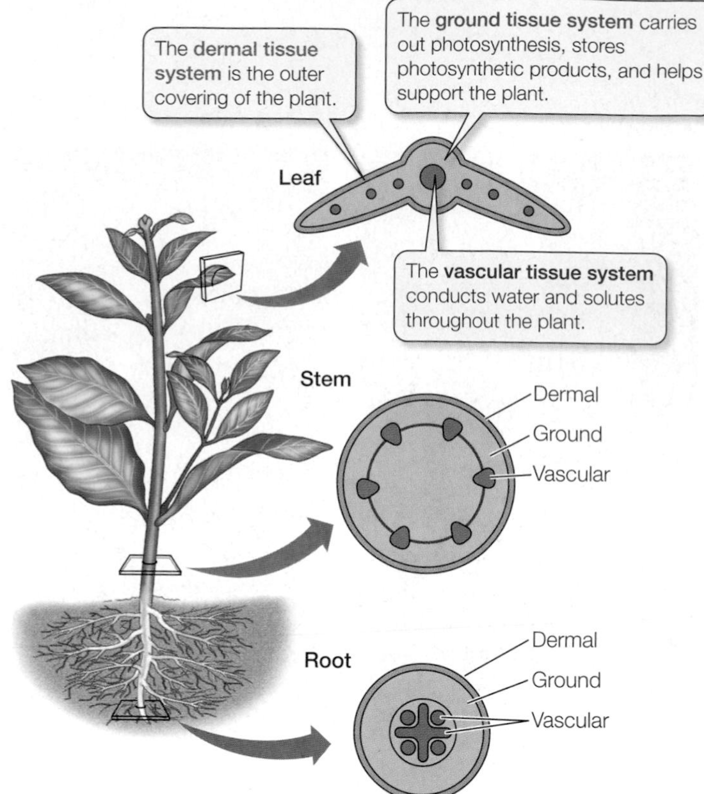

The dermal tissue system is the outer covering of the plant.
The ground tissue system carries out photosynthesis, stores photosynthetic products, and helps support the plant.
Leaf
The vascular tissue system conducts water and solutes throughout the plant.
Stem
Dermal
Ground
Vascular
Root
Dermal
Ground
Vascular

34.6 Three Tissue Systems Extend Throughout the Plant Body The arrangement shown here is typical of eudicots, but the three tissue systems are continuous in the bodies of all vascular plants.

The **dermal tissue system** is the outer covering of the plant. All parts of the young plant body are covered by an **epidermis,** which may be a single layer of cells or several layers. The epidermis is a complex tissue that may include specialized cell types, such as the *guard cells* that form stomata (pores) for gas exchange in leaves. The shoot epidermis secretes a layer of wax-covered *cutin,* the **cuticle,** that helps retard water loss from stems and leaves. The stems and roots of woody plants have a dermal tissue system called the *periderm,* a protective covering that will be discussed later in this chapter.

The **ground tissue system** makes up the rest of the plant. Ground tissue functions primarily in storage, support, photosynthesis, and the production of defensive and attractive substances.

34.1 RECAP

The basic body plan of both monocots and eudicots consists of a root system and a shoot system. Stems and leaves are part of the shoot system. A bud is an embryonic shoot. Plant tissues form three tissue systems: vascular, dermal, and ground.

- How would you distinguish between a piece of stem and a piece of root? See pp. 746–747
- Can you distinguish among the three tissue systems in terms of their location and function? See p. 748 and Figure 34.6

All plant organs are composed of tissues and tissue systems. Let's now consider the basic structural and functional units of plant tissues: their cells.

34.2 How Are Plant Cells Unique?

Plant cells have all the essential organelles common to eukaryotes (see Figure 4.7), but certain additional structures and organelles distinguish them from many other eukaryotes:

- They contain *chloroplasts* or other plastids.
- They contain *vacuoles*.
- They possess cellulose-containing *cell walls*.

Plant cells are alive when they divide and grow, but certain cells function only after their living parts have died and disintegrated. Other plant cells develop specialized metabolic capabilities; for example, some can perform photosynthesis, and others produce and secrete waterproofing materials. A plant has several different types of cells that differ dramatically in the composition and structure of their cell walls. The walls of each cell type have a composition and structure that correspond to its special functions.

Cell walls may be complex in structure

The cytokinesis of a plant cell is completed when the two daughter cells are separated by a cell plate (see Figure 9.12B). The daughter cells then deposit a gluelike substance within the cell plate; this substance constitutes the **middle lamella**. Next, each daughter cell secretes cellulose and other polysaccharides to form a **primary wall**. This deposition and secretion continue as the cell expands to its final size (**Figure 34.7**).

Once cell expansion stops, a plant cell may deposit one or more additional cellulosic layers to form a **secondary wall** internal to the primary wall (see Figure 34.7). Secondary walls are often impregnated with unique substances that give them special properties. Those impregnated with the polymer *lignin* become strong, as in wood cells. Walls to which the complex lipid *suberin* is added become waterproof.

Although it lies outside the plasma membrane, the cell wall is not a chemically inactive region. In addition to cellulose and other polysaccharides, the cell wall contains proteins, some of which are enzymes. Chemical reactions in the wall play important roles in cell expansion and in defense against invading organisms. Cell walls may thicken or be sculpted or perforated as cells differentiate into specialized cell types. The genome of the tiny plant *Arabidopsis thaliana* contains more than a thousand genes related to cell wall biosynthesis and function. Except where the secondary wall is waterproofed, the cell wall is permeable to water and mineral ions and allows molecules other than large macromolecules to reach the plasma membrane.

Localized modifications in the walls of adjacent cells allow water and dissolved materials to move easily from cell to cell. The primary wall usually has regions where it becomes quite thin. In these regions, cytoplasm-filled canals called **plasmodesmata** (singular, *plasmodesma*) pass through the primary wall, allowing direct com-

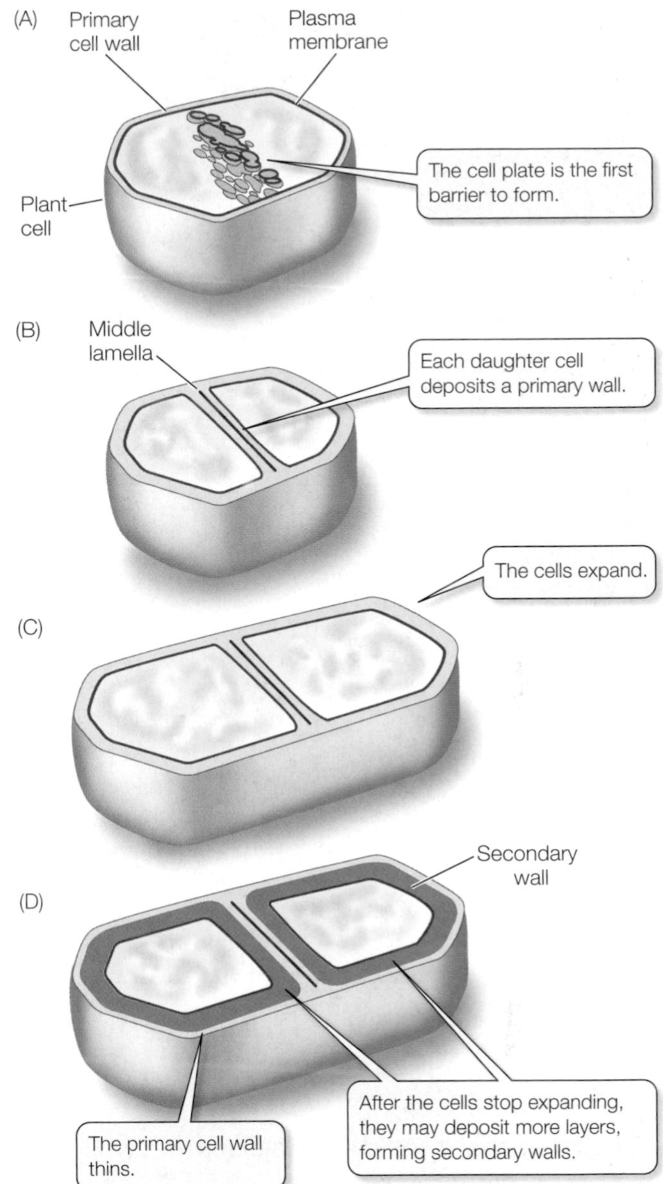
(A)
Primary cell wall
Plasma membrane
Plant cell
The cell plate is the first barrier to form.
(B)
Middle lamella
Each daughter cell deposits a primary wall.
The cells expand.
(C)
(D)
Secondary wall
The primary cell wall thins.
After the cells stop expanding, they may deposit more layers, forming secondary walls.

34.7 Cell Wall Formation Plant cell walls form as the final step in cell division.

munication between plant cells. A plasmodesma is traversed by a strand of endoplasmic reticulum (**Figure 34.8**). Under certain circumstances, a plasmodesma can enlarge dramatically, allowing even macromolecules and viruses to pass directly between cells. Macromolecules passing through enlarged plasmodesmata include transcription factors and RNAs. Substances can move from cell to cell through plasmodesmata without having to cross a plasma membrane.

Parenchyma cells are alive when they perform their functions

The most numerous cell type in young plants is the **parenchyma cell** (**Figure 34.9A**). Parenchyma cells usually have thin walls, con-

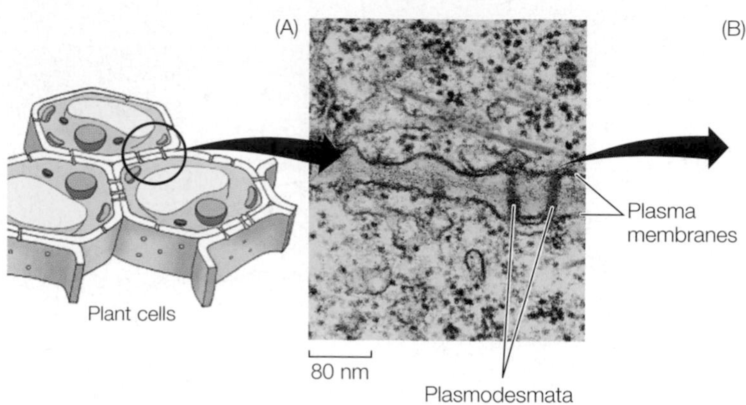

34.8 Plasmodesmata (A) An electron micrograph shows that cell walls are traversed by strandlike structures called plasmodesmata (dark stain). The green objects are cytoskeletal microtubules (see Section 4.3). (B) Plasmodesmata contain strands of endoplasmic reticulum.

sisting only of a primary wall and the shared middle lamella. Many parenchyma cells have shapes with multiple faces. Most have large central vacuoles.

The photosynthetic cells in leaves are parenchyma cells that contain numerous chloroplasts. Some nonphotosynthetic parenchyma cells store substances such as starch or lipids. In the cytoplasm of these cells, starch is often stored in specialized plastids called *leucoplasts* (see Figure 4.16B). Lipids may be stored as oil droplets, also in the cytoplasm. Some parenchyma cells appear to serve as "packing material" and play a vital role in supporting the stem. Many retain the capacity to divide and hence may give rise to new cells, as when a wound results in cell proliferation.

Collenchyma cells provide flexible support while alive

Collenchyma cells are supporting cells. Their primary walls are characteristically thick at the corners of the cells (**Figure 34.9B**). Collenchyma cells are generally elongated. In these cells, the primary wall thickens, but no secondary wall forms. Collenchyma provides support to leaf petioles, nonwoody stems, and growing organs. Tissue made of collenchyma cells is flexible, permitting stems and petioles to sway in the wind without snapping. The familiar "strings" in celery consist primarily of collenchyma cells.

Sclerenchyma cells provide rigid support

In contrast to collenchyma cells, **sclerenchyma** cells have thickened secondary walls that perform their major function: support. Many sclerenchyma cells die after laying down their cell walls and thus perform their supporting function when dead. There are two types of sclerenchyma cells: elongated **fibers** and variously shaped **sclereids**. Fibers provide relatively rigid support in wood and other parts of the plant, where they are often organized into bundles (**Figure 34.9C**). The bark of trees owes much of its mechanical strength to long fibers. Sclereids may pack together densely, as in a nut's shell or in some seed coats (**Figure 34.9D**). Isolated clumps of sclereids, called *stone cells,* in pears and some other fruits give them their characteristic gritty texture.

Fibers from the stems of hemp, *Cannabis sativus*, gave strength to sails, ropes, and rigging of sailing vessels and durability to the Gutenberg Bible, the Declaration of Independence, and the Constitution of the United States. The original Levi's jeans were made of hemp cloth.

Cells of the xylem transport water and minerals from roots to stems and leaves

Xylem contains conducting cells called **tracheary elements**, which undergo programmed cell death (apoptosis; see Section 9.6) before they assume their function of transporting water and dissolved minerals. There are two types of tracheary elements. The evolutionarily more ancient tracheary elements, found in gymnosperms and other vascular plants, are spindle-shaped cells called **tracheids** (**Figure 34.9E**). When the cell contents—nucleus and cytoplasm—disintegrate upon cell death, water and minerals can move with little resistance from one tracheid to its neighbors by way of *pits*, interruptions in the secondary wall that leave the primary wall unobstructed.

Flowering plants evolved a water-conducting system made up of *vessels*. The individual cells that form vessels, called **vessel elements**, must also die and become empty before they can transport water. Vessel elements have pits in their cell walls as do tracheids, but are generally larger in diameter than tracheids. Vessel elements secrete lignin into their secondary cell walls, then partially break down their end walls, and finally die and disintegrate, resulting in a hollow tube. They are laid down end-to-end, so that each ves-

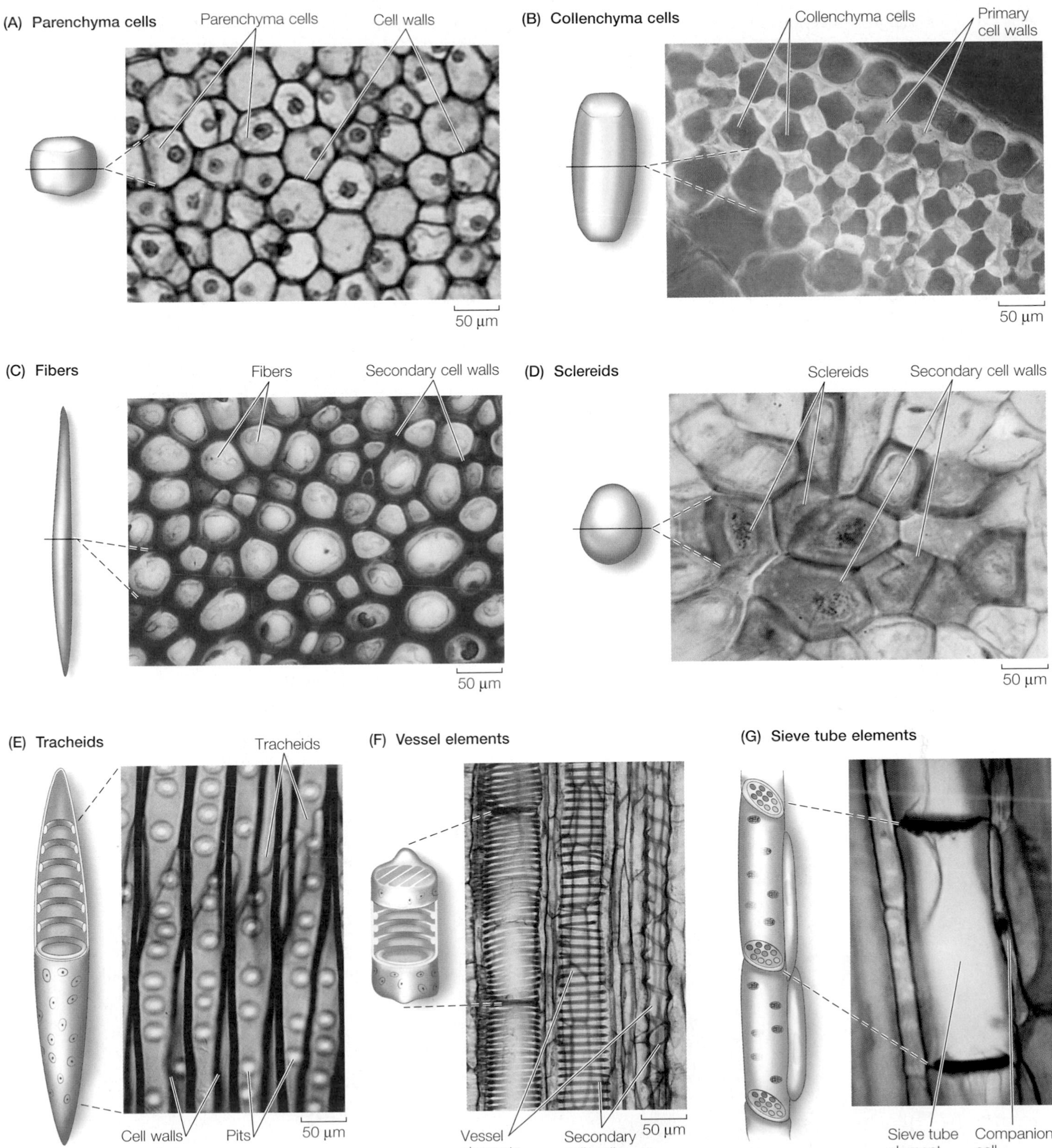

34.9 Plant Cell Types (A) Parenchyma cells in the petiole of *Coleus*. Note the thin, uniform cell walls. (B) Collenchyma cells make up the five outer cell layers of this spinach leaf vein. Their cell walls are thick at the corners of the cells and thin elsewhere. (C) Sclerenchyma: Fibers in a sunflower plant (*Helianthus*). The thick secondary walls are stained red. (D) Sclerenchyma: Sclereids. The extremely thick secondary walls of sclereids are laid down in layers. They provide support and a hard texture to structures such as nuts and seeds. (E) Tracheary elements: Water-conducting tracheids in pine wood. The thick cell walls are stained dark red. (F) Tracheary elements: Vessel elements in the stem of a squash. The secondary walls are stained red; note the different patterns of thickening, including rings and spirals. (G) Food-conducting sieve tube elements and companion cells in the stem of a cucumber.

sel is a continuous hollow tube consisting of many vessel elements, providing an open pipeline for water conduction (**Figure 34.9F**). In the course of angiosperm evolution, vessel elements have become shorter, and their end walls have become less and less obliquely oriented and less obstructed, presumably increasing the efficiency of water transport through them. The xylem of many angiosperms also includes tracheids.

Cells of the phloem translocate carbohydrates and other nutrients

The transport cells of the phloem, unlike those of the mature xylem, are living cells. In flowering plants, the characteristic cells of the phloem are **sieve tube elements** (**Figure 34.9G**). Like vessel elements, these cells meet end-to-end. They form long *sieve tubes,* which transport carbohydrates and many other materials from their sources to tissues that consume or store them. In plants with mature leaves, for example, products of photosynthesis move from leaves to root tissues.

Unlike vessel elements, which break down their end walls as they mature, sieve tube elements contain plasmodesmata in the end walls that enlarge to form pores, enhancing the connection between neighboring cells. The result is end walls that look like sieves, called **sieve plates** (**Figure 34.10**). As the holes in the sieve plates expand, the membrane that encloses the central vacuole, called the *tonoplast,* disappears. The nucleus and some cytoplasmic components also break down, and thus do not clog the pores of the sieve.

At functional maturity, a sieve tube element is filled with **phloem sap**, consisting of water, dissolved sugars, and other solutes. This solution moves from cell to cell along the sieve tube. The moving sap solution is distinct from the layer of cytoplasm at the periphery of a sieve tube element, next to the cell wall. This stationary layer of cytoplasm contains the organelles remaining in the sieve tube element.

Each sieve tube element has one or more **companion cells** (see Figure 34.10), produced as a daughter cell along with the sieve tube element when a parent cell divides. Numerous plasmodesmata link a companion cell with its sieve tube element. Companion cells retain all their organelles and, through the activities of their nuclei, they may be thought of as the "life support systems" of the sieve tube elements.

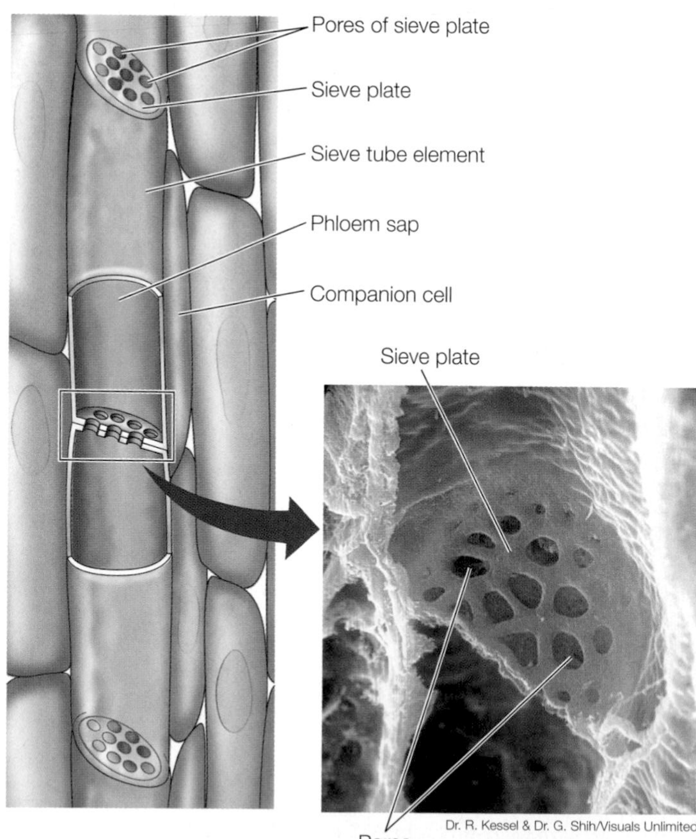

34.10 Sieve Tubes Individual sieve tube elements join together to form long tubes that transport carbohydrates and other nutrient molecules throughout the plant body in the phloem. Sieve plates form at the ends of each sieve tube element, and phloem sap passes through the pores in the sieve plate.

34.2 RECAP

There are several types of plant cells, differing in structure and function. Certain cells specialized for support or transport do not assume their function until they have died, but most plant cells function only when alive.

- Do you understand the structure and role of plasmodesmata? See p. 750 and Figure 34.8
- What structural differences make tissues made of collenchyma cells more flexible than those consisting primarily of sclerenchyma? See p. 750
- How are the transport cells of the phloem different from those of the xylem? What are their respective functions? See pp. 750, 752 and Figure 34.10

In the discussions that follow, we will examine how the cells and tissue systems are organized in the different organs of a flowering plant. Let's begin by seeing how this organization develops as the plant grows.

34.3 How Do Meristems Build the Plant Body?

In its early embryonic stages, a plant establishes the basic body plan for its mature form. Two patterns contribute to the plant body plan:

- The arrangement of cells and tissues along the main axis from root to shoot
- The concentric arrangement of the tissue systems

Both patterns arise through orderly development and are best understood in developmental terms.

Plants and animals grow differently

As the plant body grows, it may lose parts, and it forms new parts that may grow at different rates. The growing stem consists of modules, or units, laid down one after another. Each module consists of a node with its attached leaf or leaves, the internode below that node, and the axillary bud or buds at the base of that internode (see Figure 34.2). New modules are formed as long as the stem continues to grow.

- Each *branch* of a plant may be thought of as a module that is in some ways independent of the other branches. A branch does not bear the same relationship to the remainder of the plant body as a limb does to the remainder of an animal body. Among other things, branches form one after another (unlike limbs, which form simultaneously during embryonic development). Also, branches often differ from one another in their number of leaves and in the degree to which they themselves branch. Branches, like stems, are long-lived, lasting from years to centuries.
- *Leaves* are modules of another sort. They are usually short-lived, lasting weeks to a few years.
- *Root systems* are also branching structures, and lateral roots are semi-independent units. As the root system grows, penetrating and exploring the soil environment, many roots die and are replaced by new ones.

Many seed plants may be thought of as having two units of yet another sort: the primary plant body and the secondary plant body. All seed plants have a **primary plant body**, which consists of *all the non-woody parts* of the plant. Many plants—monocots in particular—consist entirely of primary plant body. Trees and shrubs, however, also have a **secondary plant body** consisting of *wood and bark*. The tissues of the secondary plant body are laid down as the stems and roots thicken; the primary plant body includes leaves, flowers, and all parts of the body that were laid down before thickening began. The secondary plant body continues to grow and thicken throughout the life of the plant. The primary plant body also continues to grow, *lengthening* the shoot and root systems and forming new leaves.

The localized regions of cell division in plants are called **meristems** (**Figure 34.11**). Meristems are forever young, retaining the ability to produce new cells indefinitely. The cells that perpetuate the meristems, called *initials*, are comparable to the stem cells found in animals (discussed in Section 19.2). When an initial divides, one daughter cell develops into another meristem cell the size of its parent, while the other daughter cell differentiates into a more specialized cell.

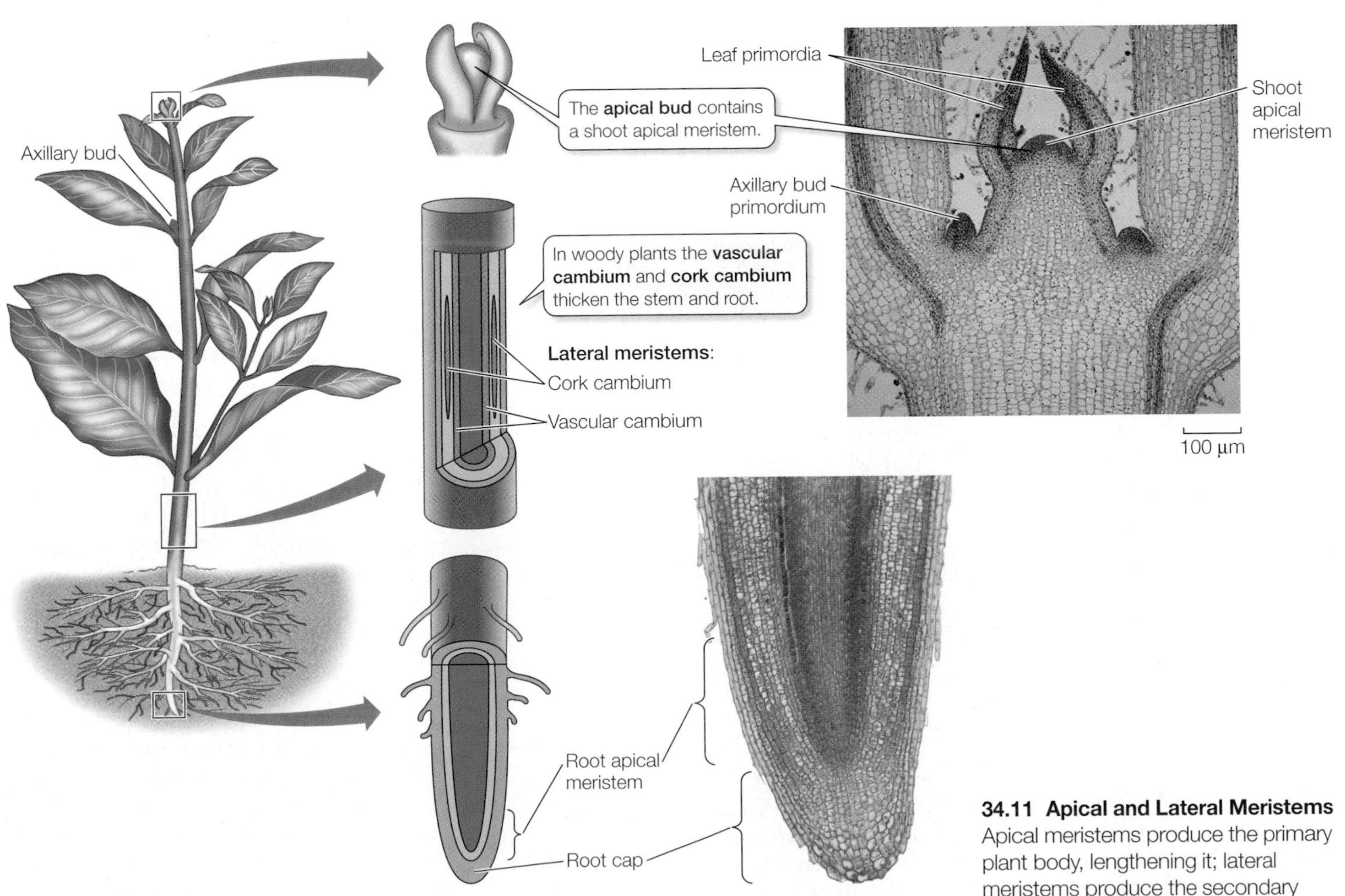

34.11 Apical and Lateral Meristems Apical meristems produce the primary plant body, lengthening it; lateral meristems produce the secondary plant body, thickening it.

Although all parts of the animal body grow as an individual develops from embryo to adult, in most animals, this growth is *determinate*. That is, the growth of the individual and all its parts ceases when the adult state is reached. Determinate growth is also characteristic of some plant parts, such as leaves, flowers, and fruits. The growth of stems and roots, by contrast, is *indeterminate*, and it is generated from specific regions of active cell division and cell expansion.

A hierarchy of meristems generates a plant's body

Two types of meristems contribute to the growth and development of the plant:

- **Apical meristems** give rise to the primary plant body.
- **Lateral meristems** give rise to the secondary plant body.

APICAL MERISTEMS Apical meristems are located at the tips of roots and stems and in buds and are responsible for **primary growth**, which leads to elongation of shoots and roots and formation of organs (see Figure 34.11). All plant organs arise ultimately from cell divisions in the apical meristems, followed by cell expansion and differentiation. Primary growth gives rise to the primary plant body, which is the entire body of many plants.

- *Shoot apical meristems* supply the cells that extend stems and branches, allowing more leaves to form and photosynthesize.
- *Root apical meristems* supply the cells that extend roots, enabling the plant to "forage" for water and minerals.

Apical meristems in both the shoot and the root give rise to a set of cylindrical *primary meristems* that produce the primary tissues of the plant body.

From the outside to the inside of the root or shoot, which are both cylindrical organs, the primary meristems are the **protoderm**, the **ground meristem**, and the **procambium**. These in turn give rise to the three tissue systems:

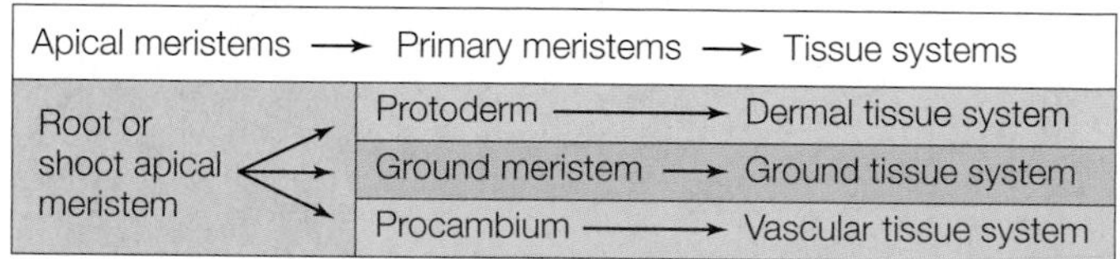

Apical meristems →	Primary meristems →	Tissue systems
Root or shoot apical meristem	Protoderm →	Dermal tissue system
	Ground meristem →	Ground tissue system
	Procambium →	Vascular tissue system

Because meristems can continue to produce new organs throughout the lifetime of the plant, the plant body is much more variable in form than the animal body, whose organs are produced only once.

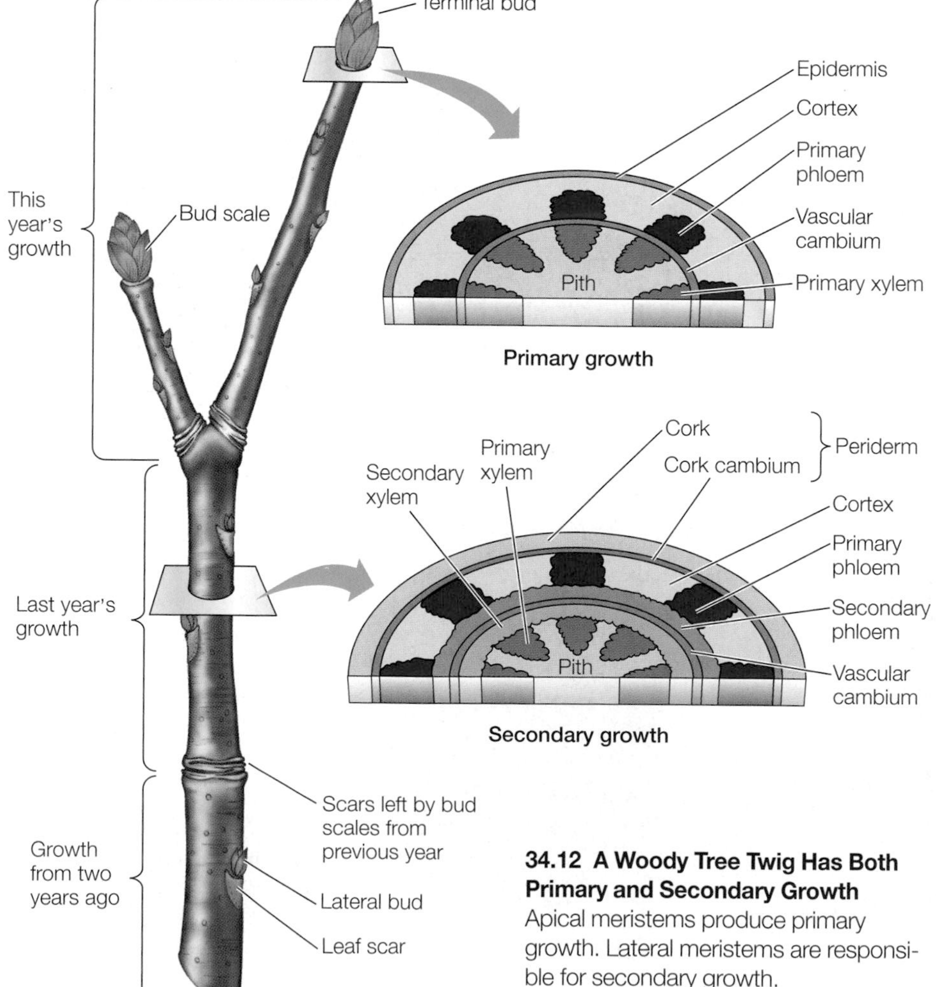

34.12 A Woody Tree Twig Has Both Primary and Secondary Growth Apical meristems produce primary growth. Lateral meristems are responsible for secondary growth.

LATERAL MERISTEMS: VASCULAR AND CORK CAMBIA Some roots and stems develop a secondary plant body, the tissues of which we commonly refer to as *wood* and *bark*. These complex tissues are derived by **secondary growth** from two lateral meristems:

- The **vascular cambium** is a cylindrical tissue consisting predominantly of vertically elongated cells that divide frequently. It supplies the cells of the secondary xylem and phloem, which in trees eventually become wood and bark.
- The **cork cambium** produces mainly waxy-walled *cork cells*. It supplies some of the cells that become bark.

Wood is secondary xylem. **Bark** is everything external to the vascular cambium (periderm plus secondary phloem). Toward the inside of the stem or root, the dividing cells of the vascular cambium form new xylem, the *secondary xylem*, and toward the outside they form new phloem, the *secondary phloem*.

Each year, deciduous trees lose their leaves, leaving bare branches and twigs in winter. These twigs illustrate both primary and secondary growth (**Figure 34.12**). The apical meristems of the twigs and their branches are enclosed in buds protected by bud scales. When the buds begin to grow in the spring, the scales fall away, leaving scars that show us where the bud was and identifying each year's growth. The dormant twig shown in Figure 34.12 is the product of pri-

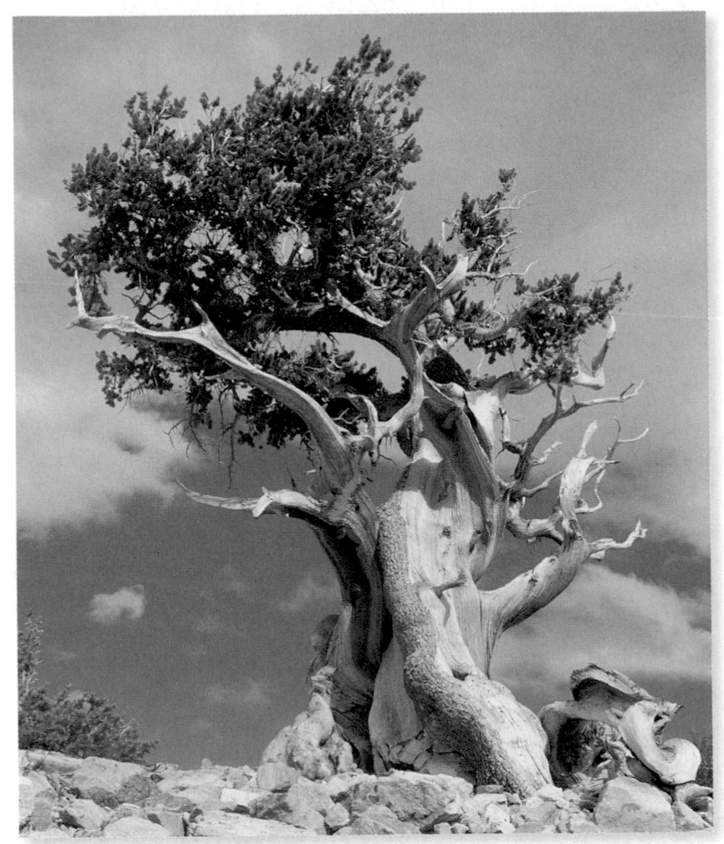

34.13 An Ancient Individual Bristlecone pines (*Pinus longaeva*) can live for centuries. The oldest known living organism is a bristlecone pine that has been alive for almost 5,000 years—almost as long as recorded human history.

mary and secondary growth. Only the buds consist entirely of primary tissues.

As a tree trunk grows in diameter, the outermost layers of the stem, including the epidermis, crack and fall off. Without the activity of the cork cambium, this sloughing off of tissues would expose the tree to potential damage, including excessive water loss or invasion by microorganisms. The cork cambium produces new protective cells, primarily in the outward direction. The walls of these cork cells become impregnated with suberin. The mass of waterproofed cells produced by the cork cambium is called the **periderm**.

In some plants, meristems may remain active for years or even centuries. The oldest known individual plant is a bristlecone pine that has lived for more than 4,900 years—almost 50 centuries (**Figure 34.13**). In contrast, it is doubtful that any animal has ever lived much longer than 2 centuries. Plants such as the bristlecone pine grow in size, or at least in diameter, throughout their lives. In the sections that follow, we will examine how the various meristems give rise to the plant body.

The root apical meristem gives rise to the root cap and the root primary meristems

The root apical meristem produces all the cells that contribute to growth in the length of the root (**Figure 34.14A**). Some of the daughter cells from the apical (tip) end of the root apical meristem contribute to a **root cap**, which protects the delicate growing region of the root as it pushes through the soil. The cells of the root cap are often damaged or scraped away and must therefore be replaced constantly. The root cap is also the structure that detects the pull of gravity and thus controls the downward growth of roots.

In the middle of the root apical meristem is a *quiescent center*, in which cell divisions are rare. The quiescent center can become more active when needed—following injury, for example.

The daughter cells produced above the quiescent center (that is, away from the root cap) elongate and lengthen the root. After they elongate, these cells differentiate, giving rise to the various tissues of the mature root. The growing region farther above the apical meristem comprises the three cylindrical primary meristems: the protoderm, the ground meristem, and the procambium (see page 756). These primary meristems give rise to the three tissue systems of the root.

The apical and primary meristems constitute the **zone of cell division**, the source of all the cells of the root's primary tissues. Just above this zone is the **zone of cell elongation**, where the newly formed cells are elongating and thus pushing the root farther into

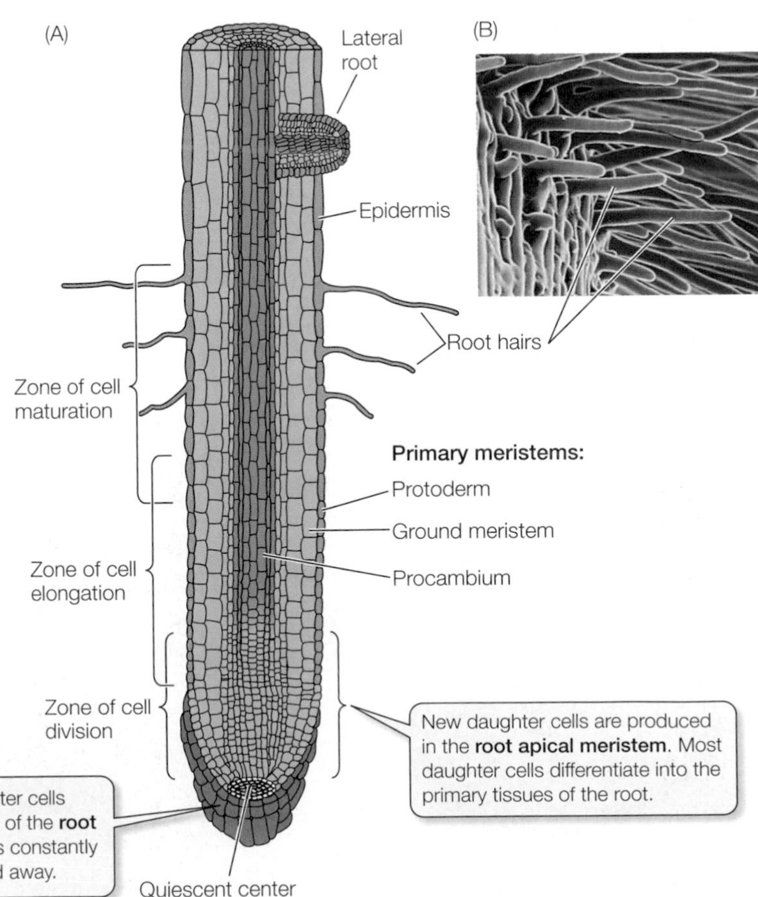

34.14 Tissues and Regions of the Root Tip (A) Extensive cell division creates the complex structure of the root. (B) Root hairs, seen with a scanning electron microscope.

the soil. Above this is the **zone of maturation**, where the cells are differentiating, taking on specialized forms and functions such as water transport or mineral uptake. These three zones grade imperceptibly into one another; there is no abrupt line of demarcation. In the zone of maturation, many of the epidermal cells produce amazingly long, delicate **root hairs**, which vastly increase the surface area of the root (**Figure 34.14B**). Root hairs grow out among the soil particles, probing nooks and crannies and taking up water and minerals.

It has been estimated that the root system of a single mature rye plant has a total absorptive surface of more than 600 square meters (almost half the area of a basketball court).

In the great majority of plants, and especially in trees, a fungus is closely associated with the root tips (see Figure 30.10). Such roots have poorly developed or no root hairs. This association, called a *mycorrhiza*, increases the plant's absorption of minerals and water and in fact these plants cannot survive without the mycorrhizae.

The products of the root's primary meristems become root tissues

The products of the three primary meristems are shown in **Figure 34.15**:

- The **protoderm** gives rise to the outer layer of cells—the **epidermis**—which is adapted for protection of the root and absorption of mineral ions and water.
- Internal to the epidermis, the *ground meristem* gives rise to a region of ground tissue that is many cells thick, called the **cortex**. The innermost layer of the cortex is the **endodermis** of the root.
- Moving inward past the endodermis, we enter the vascular cylinder, or **stele**, produced by the *procambium*.

The cells of the *cortex* are relatively unspecialized and often function in nutrient storage. Unlike those of other cortical cells, the cell walls of the endodermal cells contain suberin. The placement of this waterproofing substance in only certain parts of the cell wall enables the cylindrical ring of endodermal cells to control the access of water and dissolved ions to the vascular tissues.

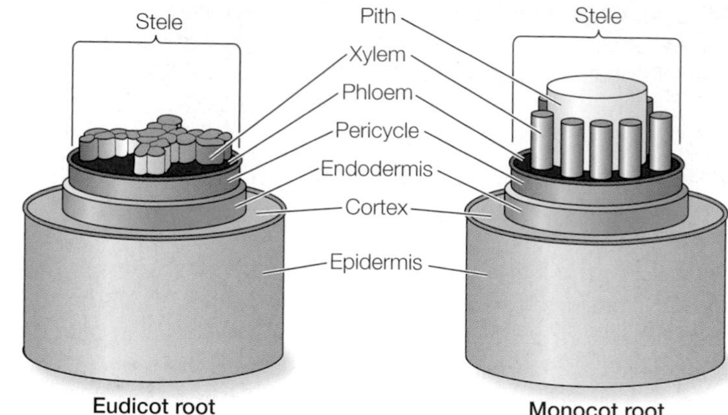

34.15 Products of the Root's Primary Meristems The protoderm gives rise to the outermost layer (epidermis). The ground meristem produces the cortex, the innermost layer of which is the endodermis. The primary vascular tissues of the root are found in the stele, which is the product of the procambium. The arrangement of tissues in the stele differs in the roots of eudicots and monocots.

The *stele* consists of three tissues: pericycle, xylem, and phloem (see Figure 34.15). The **pericycle** consists of one or more layers of relatively undifferentiated cells. It has three important functions:

- It is the tissue within which lateral roots arise (**Figure 34.16A**).
- It can contribute to secondary growth by giving rise to lateral meristems that thicken the root.
- Its cells contain membrane transport proteins that export nutrient ions into the cells of the xylem.

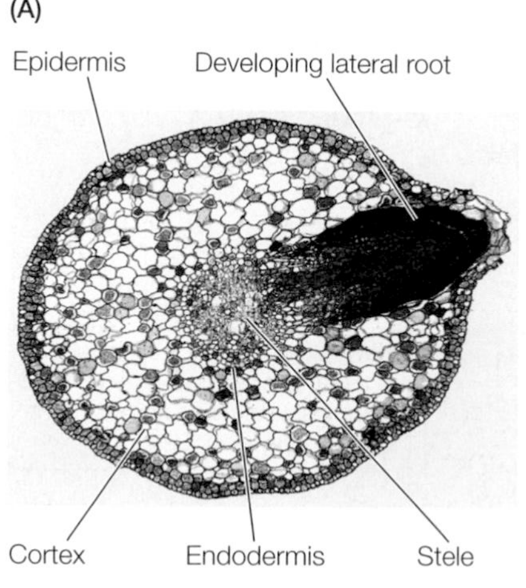

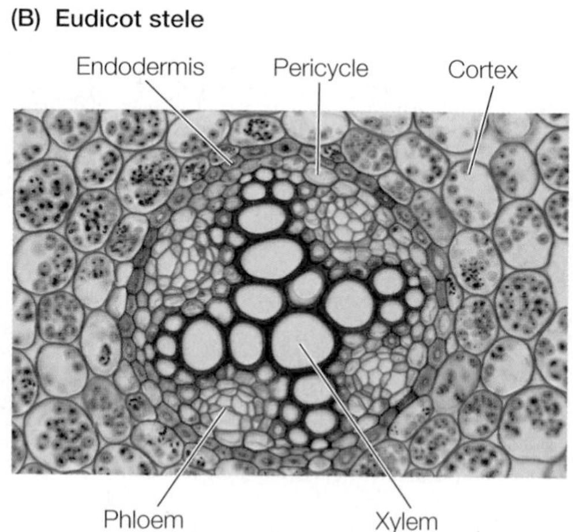

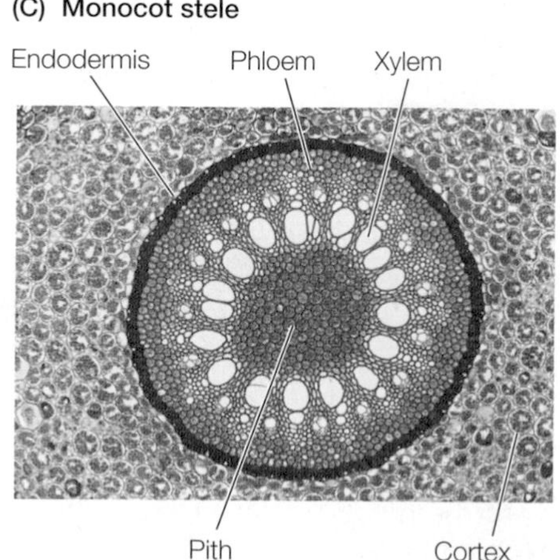

34.16 Root Anatomy (A) Cross section through the tip of a lateral root in a willow tree. Cells in the pericycle divide and the products differentiate, forming the tissues of a lateral root. (B, C) Cross sections showing the primary root tissues of (B) a eudicot (the buttercup, *Ranunculus*) and (C) a monocot (corn, *Zea mays*).

At the very center of the root of a eudicot lies the xylem—seen in cross section in the shape of a star with a variable number of points (**Figure 34.16B**). Between the points are bundles of phloem. In monocots, a region of parenchyma cells, called the **pith**, lies in the center of the root (**Figure 34.16C**). The pith often stores carbohydrate reserves and is also found in the stems of both eudicots and monocots.

The products of the stem's primary meristems become stem tissues

The shoot apical meristem, like the root apical meristem, forms three primary meristems: the protoderm, the ground meristem, and the procambium. These primary meristems, in turn, give rise to the three tissue systems. The shoot apical meristem also repetitively lays down the beginnings of leaves and axillary buds. Leaves arise from bulges called **leaf primordia**, which form as cells divide on the sides of shoot apical meristems (see Figure 34.11). **Bud primordia** form at the bases of the leaf primordia. The growing stem has no protective structure analogous to the root cap, but the leaf primordia can act as a protective covering for the shoot apical meristem.

The plumbing of angiosperm stems differs from that of roots. In a root, the vascular tissue lies deep in the interior, with the xylem at or near the very center (see Figure 34.16B,C). The vascular tissue of a young stem, however, is divided into discrete **vascular bundles** (**Figure 34.17**). Each vascular bundle contains both xylem and phloem. In eudicots, the vascular bundles generally form a cylinder, but in monocots, they are seemingly scattered throughout the stem.

In addition to the vascular tissues, the stem contains other important storage and supportive tissues. Internal to the ring of vascular bundles in eudicots is a storage tissue, the pith, and to the outside lies a similar storage tissue, the cortex. The cortex may contain supportive collenchyma cells with thickened walls. The pith, the cortex, and the regions between the vascular bundles in eudicots—called *pith rays*—constitute the ground tissue system of the stem. The outermost cell layer of the young stem is the epidermis, the primary function of which is to minimize the loss of water from the tissues within.

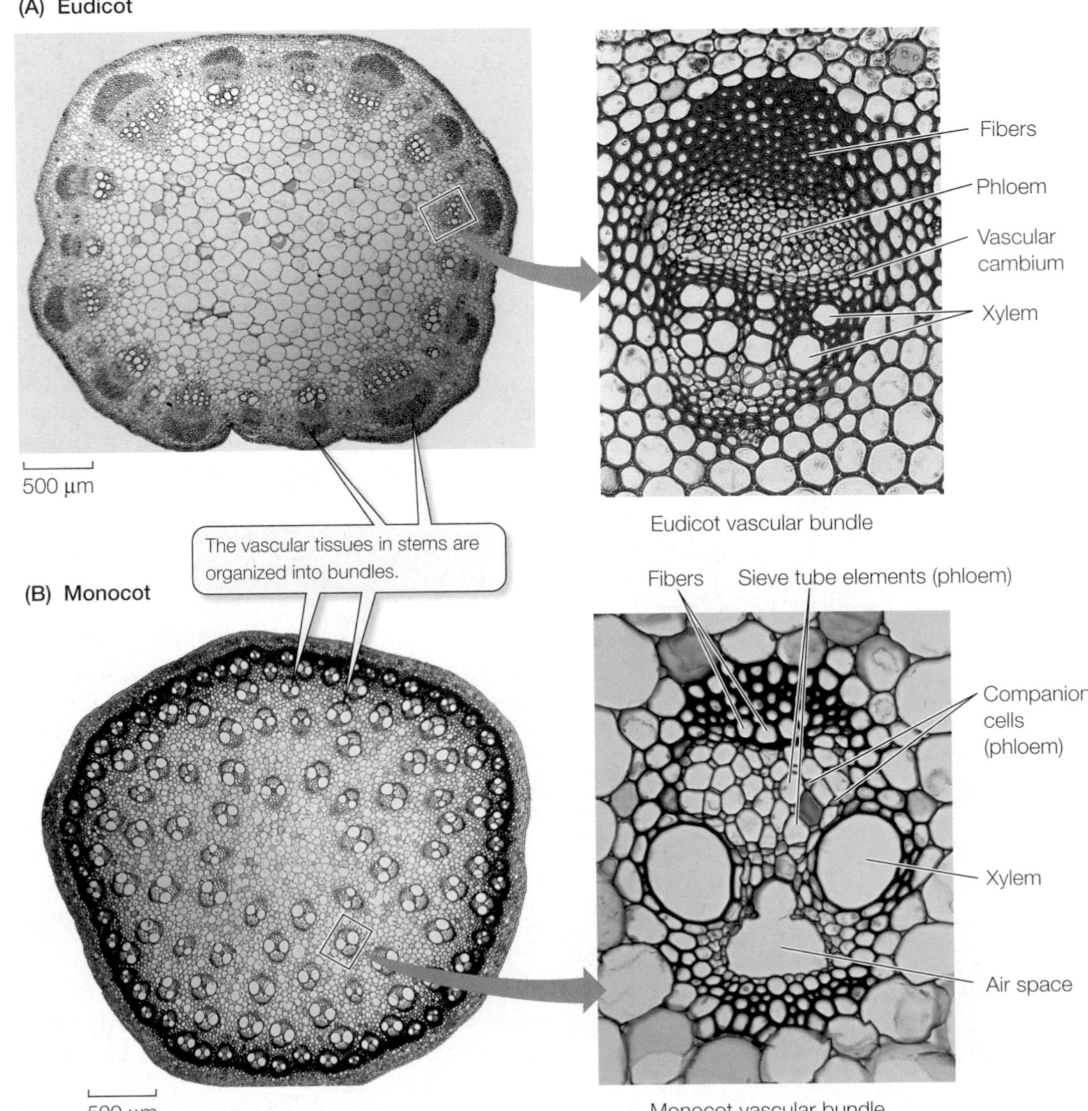

34.17 Vascular Bundles in Stems (A) In eudicot stems, the vascular bundles are arranged in a cylinder, with the pith in the center and the cortex outside the cylinder. (B) A scattered arrangement of vascular bundles is typical of monocot stems.

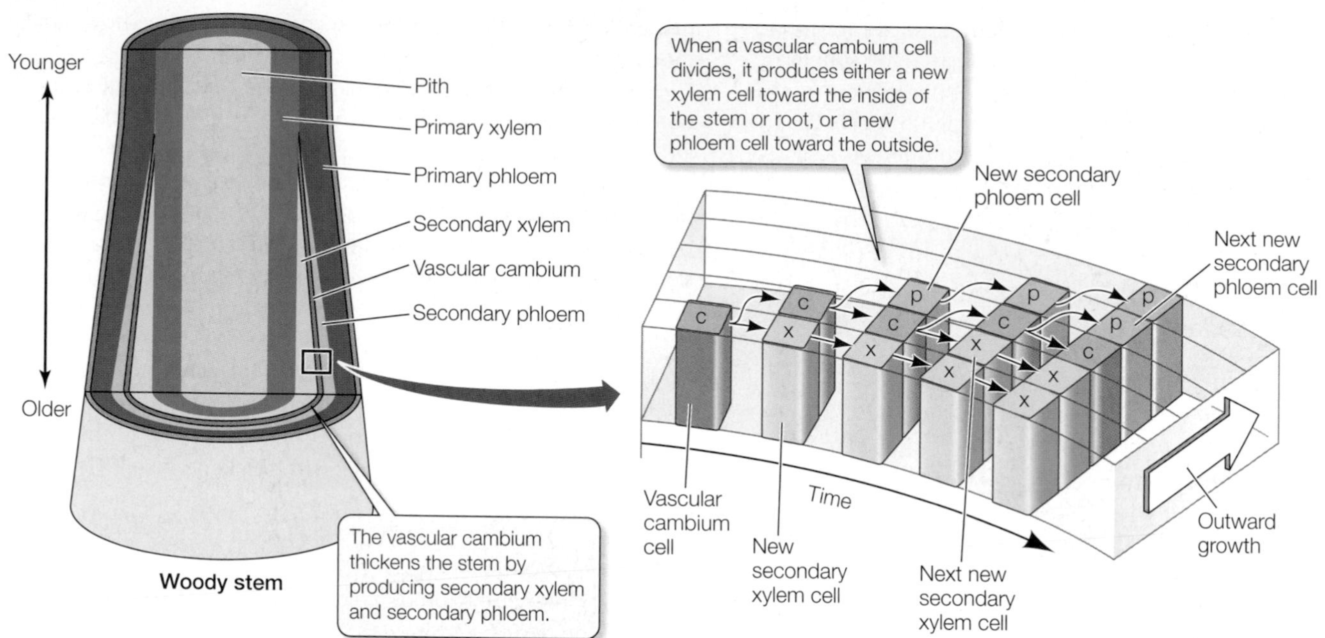

34.18 Vascular Cambium Thickens Stems and Roots Stems and roots grow thicker because a thin layer of cells, the vascular cambium, remains meristematic. These highly diagrammatic images emphasize the pattern of deposition of secondary xylem and phloem by the vascular cambium.

Many eudicot stems and roots undergo secondary growth

Some stems and roots remain slender and show little or no secondary growth. However, in many eudicots, secondary growth thickens stems and roots considerably. This process gives rise to wood and bark, and it makes the support of tall trees possible. As described earlier in the chapter, secondary growth results from the activity of the two lateral meristems, the vascular cambium and cork cambium.

The vascular cambium is initially a single layer of cells lying between the primary xylem and the primary phloem (see Figure 34.17A). The root or stem increases in diameter when the cells of the vascular cambium divide, producing secondary xylem cells toward the inside of the root or stem and producing secondary phloem cells toward the outside (**Figure 34.18**). In the stems of woody plants, cells in the pith rays between the vascular bundles also divide, forming a continuous cylinder of vascular cambium running the length of the stem. This cylinder, in turn, gives rise to complete cylinders of secondary xylem (wood) and secondary phloem, which contributes to the bark.

As the vascular cambium produces secondary xylem and phloem, its principal cell products are vessel elements, supportive fibers, and parenchyma cells in the xylem and sieve tube elements, companion cells, fibers, and parenchyma cells in the phloem. The parenchyma cells in the xylem and phloem store carbohydrate reserves in the stem and root.

Living tissues such as this storage parenchyma must be connected to the sieve tubes of the phloem, or they will starve to death. These connections are provided by **vascular rays**, which are composed of cells derived from the vascular cambium. These rays, laid down progressively as the cambium divides, are rows of living parenchyma cells that run perpendicular to the xylem vessels and phloem sieve tubes (**Figure 34.19**). As the root or stem con-

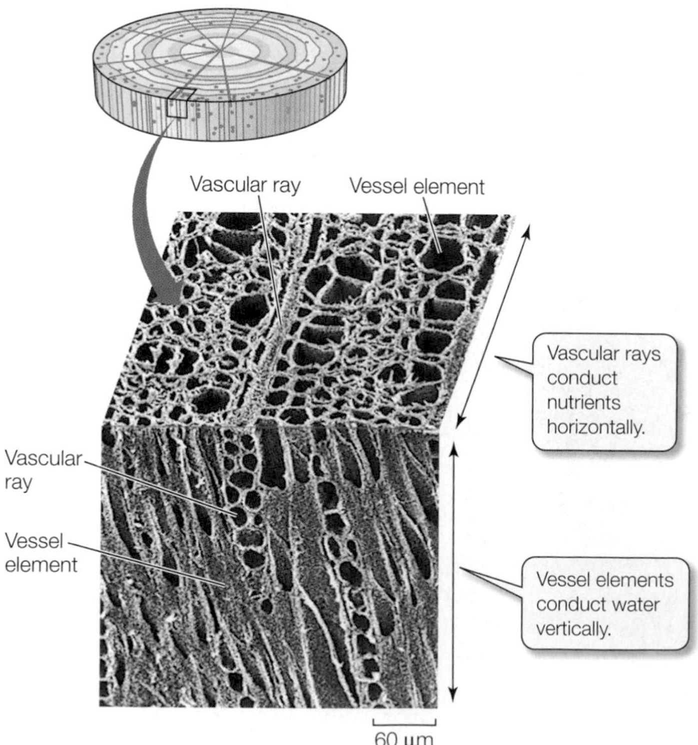

34.19 Vascular Rays and Vessel Elements In this sample of wood from the tulip poplar, the orientation of vascular rays is perpendicular to that of the vessel elements. The vascular rays transport phloem sap horizontally from the phloem to storage parenchyma cells.

tinues to increase in diameter, new vascular rays are initiated so that this storage and transport tissue continues to meet the needs of both the bark and the living cells in the xylem.

The vascular cambium itself increases in circumference with the growth of the root or stem. To do this, some of its cells divide in a plane at right angles to the plane that gives rise to secondary xylem and phloem. The products of each of these divisions lie within the vascular cambium itself and increase its circumference.

Only eudicots and other non-monocot angiosperms have a vascular cambium and a cork cambium and thus undergo secondary growth. The few monocots that form thickened stems—palm trees, for example—do so without using vascular cambium or cork cambium. Palm trees have a very wide apical meristem that produces a wide stem, and dead leaf bases also add to the diameter of the stem. Basically, monocots grow in the same way as do other angiosperms that lack secondary growth.

Wood and bark, consisting of secondary phloem, are unique to plants showing secondary growth. These tissues have their own patterns of organization and development.

WOOD Cross sections of most tree trunks (mature stems) in temperate-zone forests show *annual rings* (**Figure 34.20**), which result from seasonal environmental conditions. In spring, when water is relatively plentiful, the tracheids or vessel elements produced by the vascular cambium tend to be large in diameter and thin-walled. Such wood is well adapted for transporting water and minerals. As water becomes less available during the summer, narrower cells with thicker walls are produced, making this summer wood darker and perhaps more dense than the wood formed in spring. Thus each growing season is usually recorded in a tree trunk by a clearly visible annual ring. Trees in the moist tropics do not undergo seasonal growth, so they do not lay down such obvious regular rings. Variations in temperature or water supply can lead to the formation of more than one "annual" ring in a single year, but commonly each year brings a new annual ring and a new batch of leaves.

How do annual rings and leaves relate to each other? For example, do the needles on a pine tree connect with the current year's annual ring (xylem) or with a previous year's ring (**Figure 34.21**)? As it turns out, the answer varies from species to species.

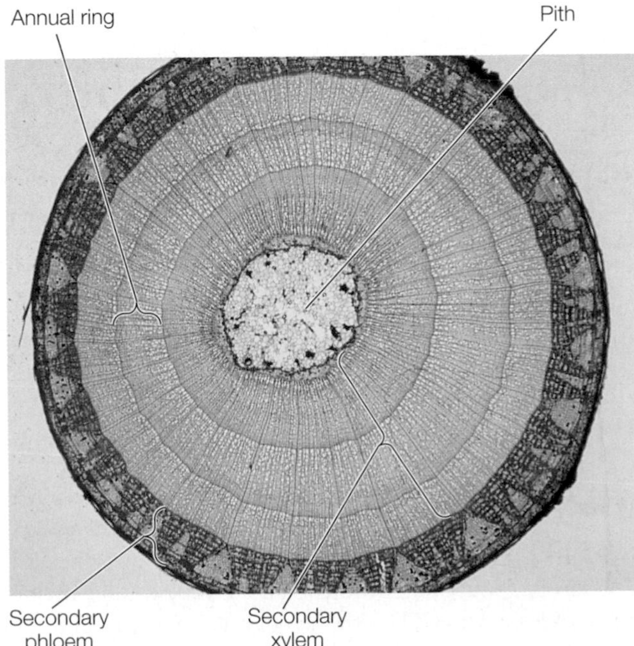

34.20 Annual Rings Rings of secondary xylem are the most noticeable feature of this cross section from a 3-year-old basswood stem.

EXPERIMENT

HYPOTHESIS: A needle of a ponderosa pine tree connects with the xylem laid down in the year the needle was formed.

METHOD

1. Immerse the basal ends of 2-cm-long segments of young branches in a dye solution.
2. Clip the tip of one needle in the segment, and apply vacuum to the cut needle.
3. After 5 minutes of the vacuum treatment, cut the segment several mm above the base. Observe which annual ring(s) contain the dye.

RESULTS

When 1-year-old needles were tested, the dye was always found in the annual ring formed 1 year before the experiment (y–1). When 2-year-old needles were tested, the dye was always found in the annual ring formed 2 years before the experiment (y–2), and sometimes also in the ring formed the following year.

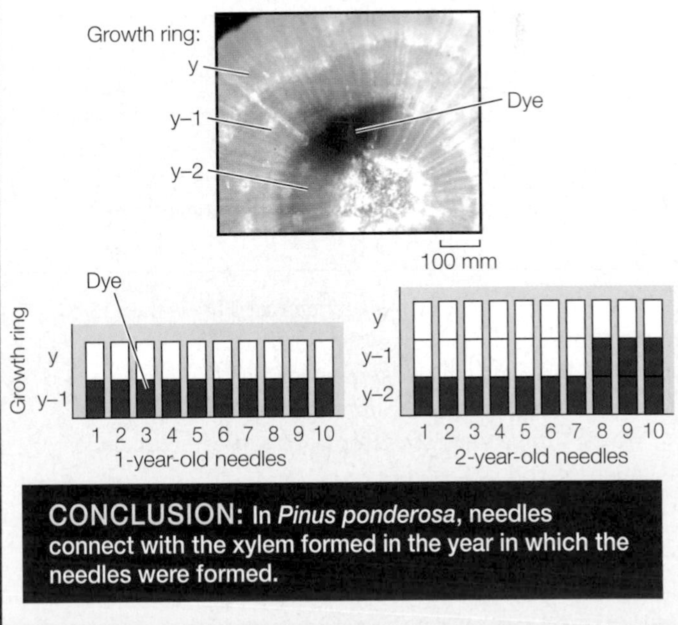

CONCLUSION: In *Pinus ponderosa*, needles connect with the xylem formed in the year in which the needles were formed.

34.21 How Do Leaves Relate to Annual Rings? Clarice Maton and Barbara L. Gartner determined which annual rings supply water to which needles of various gymnosperms. FURTHER RESEARCH: Is this a general phenomenon? That is, does it appear in other woody plants? Suggest how you could perform this experiment on a twig of an angiosperm tree such as a maple.

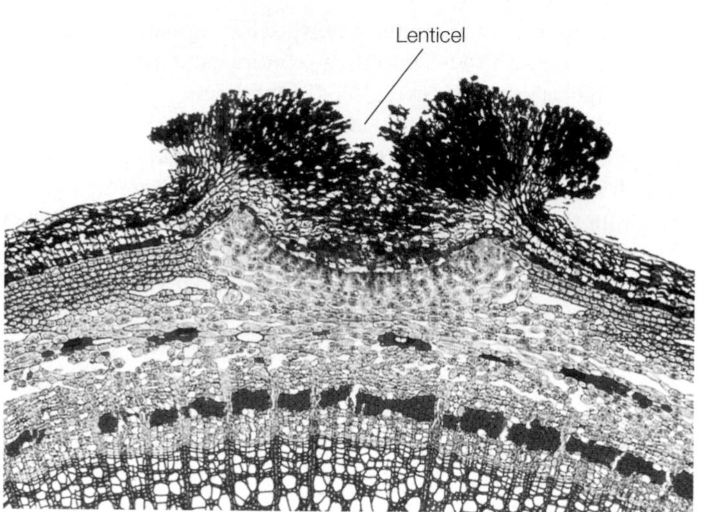

34.22 Lenticels Allow Gas Exchange through the Periderm The region of periderm that appears broken open is a lenticel in a year-old elderberry (*Sambucus*) twig; note the spongy tissue that constitutes the lenticel.

The difference between old and new regions of wood also contributes to its appearance. As a tree grows in diameter, the xylem toward the center becomes clogged with water-insoluble substances and ceases to conduct water and minerals; this *heartwood* appears darker in color. The portion of the xylem that actively conducts water and minerals throughout the tree is called *sapwood* and is lighter in color and more porous than heartwood.

The knots that we find attractive in knotty pine but regard as a defect in structural timbers are cross sections of branches. As a trunk grows, the bases of branches become buried in the trunk's new wood and appear as knots when the trunk is cut lengthwise.

The overall appearance of wood, resulting from annual rings, vascular rays, and unevenness from sawing, knots, and so forth, is referred to as the wood's *figure*. Recall that one of the most convincing pieces of evidence in the "Lindbergh trial" discussed at the beginning of this chapter was a comparison of the figure of two pieces of sawn wood. Recall, too, that wood from no fewer than four tree species was used to construct a single ladder.

BARK As secondary growth of stems or roots continues, the expanding vascular tissue stretches and breaks the epidermis and cortex, which ultimately flake away. Tissue derived from the secondary phloem then becomes the outermost part of the stem. Before the dermal tissues are broken away, cells lying near the surface of the secondary phloem begin to divide and produce layers of **cork**, a tissue composed of cells with thick walls waterproofed with suberin. The cork soon becomes the outermost tissue of the stem or root (see Figure 34.12). The dividing cells, derived from the secondary phloem, form a cork cambium. Sometimes the cork cambium produces cells to the inside as well as to the outside; these cells constitute what is known as the **phelloderm**.

The international wine market once used more than 15 billion cork bottle stoppers every year. For centuries, people around the western Mediterranean have harvested great sheets of this plant tissue from cork oak trees. The use of plastic stoppers and screw top bottles threatens to undermine the cork oak economy.

Cork, cork cambium, and phelloderm make up the periderm of the secondary plant body. As the vascular cambium continues to produce secondary vascular tissue, the corky layers are in turn lost, but the continuous formation of new cork cambia in the underlying phloem gives rise to new corky layers.

When periderm forms on stems and roots, the underlying tissues still need to release carbon dioxide and take up oxygen for cellular respiration. **Lenticels** are spongy regions in the periderm of stems and roots that allow such gas exchange (**Figure 34.22**).

34.3 RECAP

Apical and lateral meristems are responsible for producing the primary and secondary plant bodies. The primary meristems are the protoderm, the ground meristem, and the procambium. The two lateral meristems, from which wood and bark arise, are the vascular cambium and the cork cambium.

- Can you explain the differences among the three types of primary meristems in both roots and shoots? See p. 754
- Can you describe the cells derived from the root apical meristem and the general process of root growth? See pp. 755–756 and Figure 34.14
- Do you understand how the vascular cambium gives rise to thicker stems and roots? See Figure 34.18

Of the three types of vegetative organs, only roots and stems may undergo secondary growth. Leaves do not. The primary function of leaves is one that is essential not only to the life of the plant but to all other life on the planet: photosynthesis.

34.4 How Does Leaf Anatomy Support Photosynthesis?

We can think of roots and stems as important supporting actors that sustain the activities of the real stars of the plant body, the leaves—the organs of photosynthesis. Leaf anatomy is beautifully adapted to carry out photosynthesis and to support it by exchanging the gases O_2 and CO_2 with the environment, limiting evaporative water loss, and exporting the products of photosynthesis to the rest of the plant. **Figure 34.23A** shows a section of a typical eudicot leaf in three dimensions.

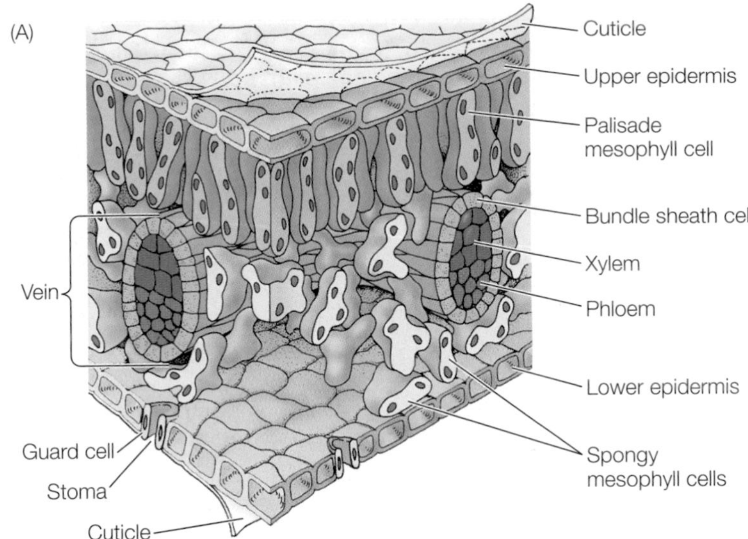

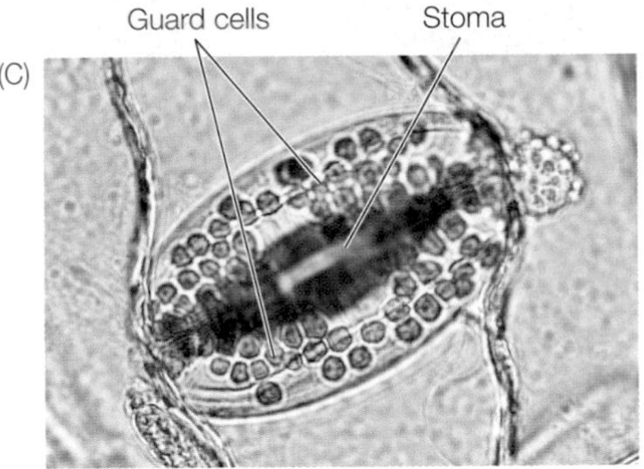

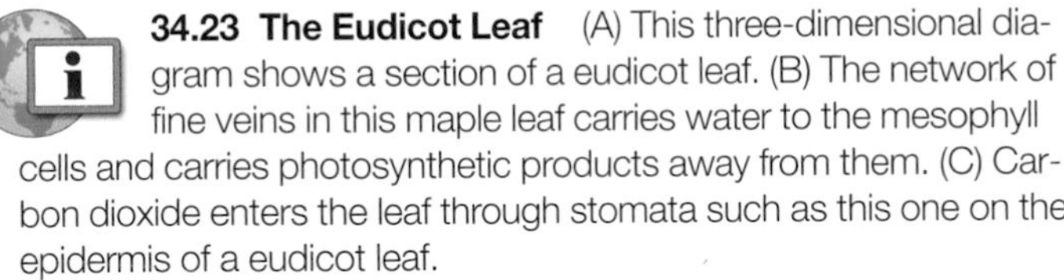

34.23 The Eudicot Leaf (A) This three-dimensional diagram shows a section of a eudicot leaf. (B) The network of fine veins in this maple leaf carries water to the mesophyll cells and carries photosynthetic products away from them. (C) Carbon dioxide enters the leaf through stomata such as this one on the epidermis of a eudicot leaf.

Most eudicot leaves have two zones of photosynthetic parenchyma tissue referred to as **mesophyll**, which means "middle of the leaf." The upper layer or layers of mesophyll consist of elongated cells; this zone is referred to as *palisade mesophyll*. The lower layer or layers consist of irregularly shaped cells; this zone is called *spongy mesophyll*. Within the mesophyll is a great deal of air space through which carbon dioxide can diffuse to and be absorbed by photosynthesizing cells.

Vascular tissue branches extensively throughout the leaf, forming a network of **veins** (**Figure 34.23B**). Veins extend to within a few cell diameters of all the cells of the leaf, ensuring that the mesophyll cells are well supplied with water and minerals. The products of photosynthesis are loaded into the phloem of the veins for export to the rest of the plant.

Covering the entire leaf on both its upper and lower surfaces is a layer of nonphotosynthetic cells, the epidermis. The epidermal cells have an overlying waxy cuticle that is highly impermeable to water. Although this impermeability prevents excessive water loss, it also poses a problem: While keeping water in the leaf, the epidermis also keeps carbon dioxide—the other raw material of photosynthesis—out.

The problem of balancing water retention and carbon dioxide availability is solved by an elegant regulatory system that will be discussed in more detail in the next chapter. *Guard cells* are modified epidermal cells that change their shape, thereby opening or closing pores called *stomata* (singular, *stoma*), which serve as passageways between the environment and the leaf's interior (**Figure 34.23C**). When the stomata are open, carbon dioxide can enter and oxygen can leave, but water vapor can also be lost.

Section 8.4 describes C_4 plants, which can fix carbon dioxide efficiently even when the carbon dioxide supply in the leaf decreases to a level at which the photosynthesis of C_3 plants is inefficient. One adaptation that helps C_4 plants fix carbon dioxide is their modified leaf anatomy (see Figure 8.17). The photosynthetic cells in the C_4 leaf are grouped around the veins in concentric layers, forming an outer mesophyll layer and an inner *bundle sheath*. These layers each contain different types of chloroplasts, leading to the biochemical division of labor illustrated in Figure 8.18.

34.4 RECAP

Leaves are the organs of photosynthesis. They exchange gases with the atmosphere, obtain water and nutrients from the roots, and export the products of photosynthesis by way of the phloem.

- How is a leaf adapted to carry out photosynthesis? See Figure 34.23
- Do you understand how stomata serve the needs of a leaf? See p.19

Leaves receive water and mineral nutrients from the roots by way of the stems. In return, the leaves export products of photosynthesis, providing a supply of chemical energy to the rest of the plant body. And, as we have just seen, leaves exchange gases, including water vapor, with the environment by way of the stomata. All three of these processes will be considered in detail in the next chapter.

CHAPTER SUMMARY

34.1 How is the plant body organized?

Most flowering plants belong to one of two major clades: **monocots** and **eudicots**. Monocots differ from eudicots in a number of structural respects. Review Figure 34.1

The vegetative organs of flowering plants are roots, which form a **root system**, and stems and leaves, which form a **shoot system**.

Stems bear embryonic shoots called buds. **Axillary buds** can develop into branches. **Apical buds** found at the tips of stems and branches produce cells for the elongating shoots. **Leaves** are the primary sites of photosynthesis. The leaf **blade** is attached to the stem by a **petiole**. Review Figure 34.2

Three tissue systems extend throughout the plant body: **vascular tissue**, **dermal tissue**, and **ground tissue**. Review Figure 34.6

The plant vascular tissue system includes the **xylem**, which conducts water and minerals absorbed by the roots, and the **phloem**, which conducts the products of photosynthesis throughout the plant body.

The dermal tissue system protects the plant body surface. In plants without secondary growth, it consists of the **epidermis**.

The ground tissue system produces and stores nutrients and other substances and provides mechanical support.

34.2 How are plant cells unique?

Plant cells are different from other eukaryotic cells in having chloroplasts or other plastids, vacuoles, and cellulose-containing **cell walls**.

The walls of individual cells are separated by a **middle lamella**; each cell also has its own **primary wall**, and some produce a thick **secondary wall**. Review Figure 34.7

Plasmodesmata connect adjacent plant cells. Review Figure 34.8

Many **parenchyma** cells store starch or lipids; some carry out photosynthesis. **Collenchyma** cells provide flexible support. **Sclerenchyma** cells include **fibers** and **sclereids** that provide strength and often do not function until they die. Review Figure 34.9

Tracheary elements include **tracheids** and **vessel elements**, which are conducting cells of the xylem. **Sieve tube elements** are the conducting cells of the phloem; their activities are often controlled by **companion cells**. **Phloem sap** passes from cell to cell through sieve plates. Review Figures 34.9 and 34.10

34.3 How do meristems build the plant body?

All seed plants possess a **primary plant body** consisting of nonwoody tissues. Shrubs and trees also possess a **secondary plant body** consisting of wood and bark.

A hierarchy of **meristems** (localized regions of cell division) generates the plant body. **Apical meristems** at the tips of stems and roots give rise to three primary meristems (**protoderm**, **ground meristem**, and **procambium**) that in turn produce the three tissue systems of those organs. Apical meristems are responsible for **primary growth** (growth in length). Review Figure 34.11

The **root apical meristem** gives rise to the **root cap** and to the three primary meristems. Root tips have overlapping **zones of cell division**, **elongation**, and **maturation**. Review Figure 34.14

The vascular tissue in young roots is contained within the **stele**. Review Figures 34.15 and 34.16, Web/CD Activities 34.1 and 34.2

The **shoot apical meristem** also gives rise to the three primary meristems. **Leaf primordia** on the sides of the apical meristem develop into leaves.

The vascular tissue in young stems is divided into **vascular bundles**, each containing both xylem and phloem. Review Figure 34.17, Web/CD Activities 34.3 and 34.4

Two **lateral meristems**, the **vascular cambium** and **cork cambium**, are responsible for **secondary growth** (growth in width) when it occurs. Review Figure 34.12

In stems and roots with secondary growth, lateral meristems give rise to **wood** (secondary xylem) and **bark** (secondary phloem plus **cork**). Review Figure 34.18, Web/CD Tutorial 34.1

The **periderm** consists of cork, cork cambium, and phelloderm, all pierced at intervals by lenticels that allow gas exchange. Review Figure 34.22

34.4 How does leaf anatomy support photosynthesis?

Veins bring water and minerals to the **mesophyll** (the photosynthetic tissue) and carry the products of photosynthesis to other parts of the plant body. Review Figure 34.23

A waxy cuticle retards water loss from the leaf and is impermeable to carbon dioxide. Guard cells control openings (stomata) in the leaf that allow CO_2 to enter, but also allow some water to escape. Review Figure 34.23, Web/CD Activity 34.5

SELF-QUIZ

1. Which of the following is *not* a difference between monocots and eudicots?
   a. Eudicots more frequently have broad leaves.
   b. Monocots commonly have flower parts in multiples of three.
   c. Monocot stems do not generally undergo secondary thickening.
   d. The vascular bundles of monocot stems are commonly arranged as a cylinder.
   e. Eudicot embryos commonly have two cotyledons.
2. Roots
   a. always form a fibrous root system that holds the soil.
   b. possess a root cap at their tip.
   c. form branches from axillary buds.
   d. are commonly photosynthetic.
   e. do not show secondary growth.
3. The plant cell wall
   a. lies immediately inside the plasma membrane.
   b. is an impermeable barrier between cells.
   c. is always waterproofed with either lignin or suberin.
   d. always consists of a primary wall and a secondary wall, separated by a middle lamella.
   e. contains cellulose and other polysaccharides.
4. Which statement about parenchyma cells is *not* true?
   a. They are alive when they perform their functions.
   b. They typically lack a secondary wall.
   c. They often function as storage depots.
   d. They are the most numerous cells in the young plant body.
   e. They are found only in stems and roots.

5. Tracheids and vessel elements
   a. must die to become functional.
   b. are important constituents of all seed plants.
   c. have walls consisting of middle lamella and a primary wall.
   d. are always accompanied by companion cells.
   e. are found only in the secondary plant body.
6. Which statement about sieve tube elements is *not* true?
   a. Their end walls are called sieve plates.
   b. They must die to become functional.
   c. They link end-to-end, forming sieve tubes.
   d. They form the system for translocation of organic nutrients.
   e. They lose the membrane that surrounds their central vacuole.
7. The pericycle
   a. is the innermost layer of the cortex.
   b. is the tissue within which branch roots arise.
   c. consists of highly differentiated cells.
   d. forms a star-shaped structure at the very center of the root.
   e. is waterproofed by suberin.
8. Secondary growth of stems and roots
   a. is brought about by the apical meristems.
   b. is common in both monocots and eudicots.
   c. is brought about by vascular and cork cambia.
   d. produces only xylem and phloem.
   e. is brought about by vascular rays.
9. Periderm
   a. contains lenticels that allow for gas exchange.
   b. is produced during primary growth.
   c. is permanent; it lasts as long as the plant does.
   d. is the innermost part of the plant.
   e. contains vascular bundles.
10. Which statement about leaf anatomy is *not* true?
   a. Stomata are controlled by paired guard cells.
   b. The cuticle is secreted by the epidermis.
   c. The veins contain xylem and phloem.
   d. The cells of the mesophyll are packed together, minimizing air space.
   e. C_3 and C_4 plants differ in leaf anatomy.

FOR DISCUSSION

1. When a young oak was 5 m tall, a thoughtless person carved his initials in its trunk at a height of 1.5 m above the ground. Today that tree is 10 m tall. How high above the ground are those initials? Explain your answer in terms of the manner of plant growth.
2. Consider a newly formed sieve tube element in the secondary phloem of an oak tree. What kind of cell divided to produce the sieve tube element? What kind of cell divided to produce that parent cell? Keep tracing back until you arrive at a cell in the apical meristem.
3. Distinguish between sclerenchyma cells and collenchyma cells in terms of structure and function.
4. Distinguish between primary and secondary growth. Do all angiosperms undergo secondary growth? Explain.
5. What anatomical features make it possible for a plant to retain water as it grows? Describe the plant tissues and how and when they form.

FOR INVESTIGATION

Maton and Gartner found that a few other species behaved like ponderosa pine, but that different relationships between needles and annual rings characterized other species. For example, in some species the needles, regardless of age, are served by the current year's xylem. How might you try to make sense of such differences?

CHAPTER 35 Transport in Plants

The curious curate

The curate of Teddington had a lively curiosity and a strong practical bent. As a clergyman he believed that God had "observed the most exact proportions of number, weight and measure in the make of all things." He also believed that we should understand nature by observation and experimentation, and thus he spent much of his life observing and experimenting, and then trusting what his investigations told him. He was several things at once: a clergyman, a physiologist, a chemist, and an inventor. He firmly established experimentation with appropriate controls as an indispensable tool.

The Reverend Stephen Hales (1677–1761) is considered the father of the discipline of plant physiology and is one of the great figures in the history of animal physiology. The leading scientists of his day presumed that the sap of trees circulates like the blood in our bodies. They did not question Aristotle's pronouncement in the fourth century B.C.E. that plants can be understood by analogy to animals.

Hales recognized, however, that experimentation was more likely than Aristotelian dogma to reveal how sap moves in plants. He approached the question by measuring both the uptake of water by a plant (the amount he added) and the amount of water lost by the leaves (measured by change in weight). A single sunflower plant, he found, takes up and releases 17 times as much water in a day as does a human, weight for weight. He also showed that the movement of the sap is always upward and never downward—that is, the sap does *not* circulate.

Trying to understand how "nature wonderfully contrived ... most powerfully to raise and keep in motion the sap," he experimented on what we now call root pressure. He determined how root pressure varies by time of day.

Hales also investigated the relationship between leaves and the atmosphere. He was the first to show that plants take up "food" of some sort from the atmosphere, and that this process requires light. His studies of gases, done before anyone had conceived of such things as oxygen or carbon dioxide, were early landmarks in the history of chemistry. He developed methods for collecting gases over water that were later used by Priestley and Lavoisier in the discovery of oxygen and other gases.

He studied the movement of fluid in animals as well as in plants. He was the first to measure blood pressure and the rate of blood flow in the capillaries.

Sunflowers Sweat More Than You Do Stephen Hales measured the amount of water "imbibed" and the amount "perspired" by sunflower plants.

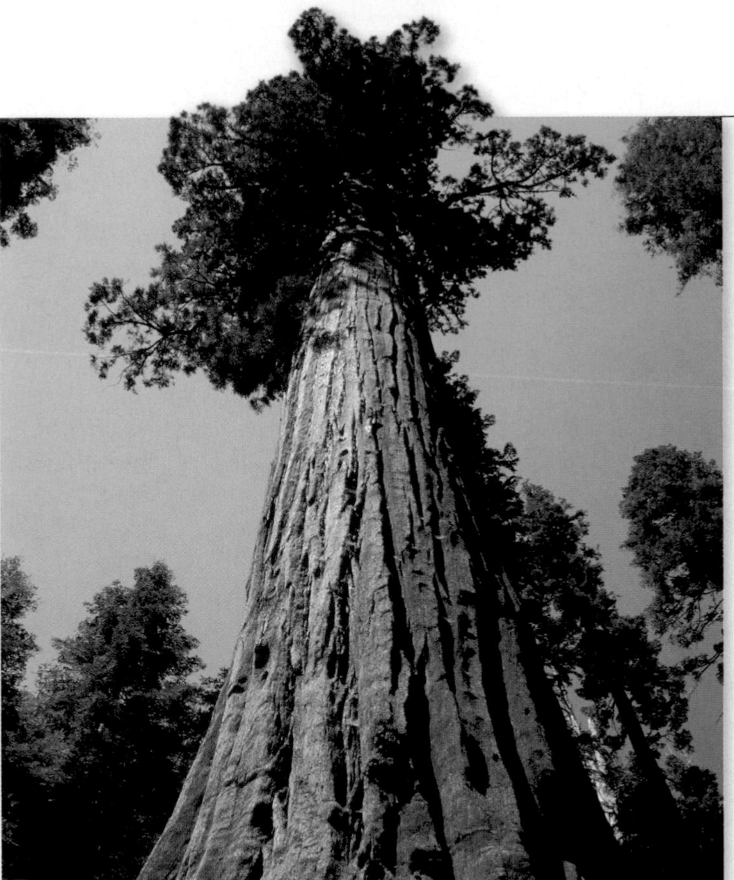

Water Has to Be Transported a Long Way It took more than 200 years for plant biologists to extend Stephen Hales's work and satisfactorily explain transport of water in plants. From sunflowers to the tallest gymnosperms (the coast redwood, *Sequoia sempervirens*), the leaves of plants pull the sap up through the xylem.

Among his many practical inventions was a ventilator, made from the bellows of a church organ, to refresh the air in prisons, hospitals, and the holds of ships.

Hales did not work in obscurity. In his time he received many honors for his contributions. Now every two years the American Society of Plant Biologists gives its prestigious Stephen Hales Prize to a person who has served the science of plant biology in some noteworthy manner. Over the years, the prize has gone to many of the people responsible for the progress described in the chapters of Part Seven.

IN THIS CHAPTER we will consider the uptake of water and minerals from the soil, the transport of these materials up the plant in the xylem, the control of evaporative water loss from leaves, and the translocation (movement from one location to another) of dissolved substances in the phloem.

CHAPTER OUTLINE

35.1 How Do Plants Take Up Water and Solutes?

Terrestrial plants must obtain both water and mineral nutrients from the soil, usually by way of their roots. The roots, in turn, obtain carbohydrates and other important materials from the leaves (**Figure 35.1**). We learned in Section 8.1 that water is one of the ingredients required for carbohydrate production by photosynthesis in the leaves. Water is also essential for transporting solutes both upward and downward, for cooling the plant, and for developing the internal pressure that supports the plant body. As Stephen Hales showed, plants lose large quantities of water to evaporation, and this water must be continually replaced.

How do leaves high in a tree obtain water from the soil? What are the mechanisms by which water and mineral ions enter the plant body through the roots and ascend as sap in the xylem? Because neither water nor minerals can move through the plant into the xylem without crossing at least one plasma membrane, we will first focus on osmosis. Then we will examine the active uptake of mineral ions by the plant and follow the pathway by which both water and minerals move through the root to gain entry to the xylem.

Water moves through a membrane by osmosis

Osmosis, the movement of water through a membrane in accordance with the laws of diffusion, is described in Section 5.3. The **solute potential** (also called the *osmotic potential*) of a solution is a measure of the effect of dissolved solutes on the osmotic behavior of the solution. The following statement presents an opportunity for confusion, so study it carefully:

- The greater the solute concentration of a solution, the more negative its solute potential, and the greater the tendency of water to move into it from another solution of lower solute concentration (and less negative solute potential).

For osmosis to occur, the two solutions must be separated by a *selectively permeable* membrane (i.e., a membrane that is permeable to water but relatively impermeable to the solute). Recall from Figure 5.9 that if pure water (less negative solute potential) is separated from a solution with a high salt concen-

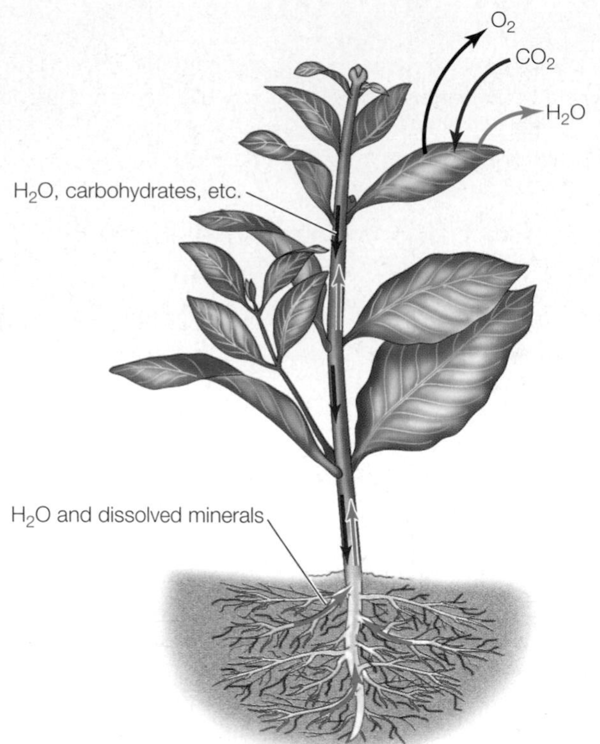

35.1 The Pathways of Water and Solutes in the Plant Water travels from the soil to the atmosphere, and it circulates within the plant, carrying important solutes with it.

tration (more negative solute potential) by a selectively permeable membrane, water molecules will travel across the membrane from the pure water side to the high-salt side. Recall, too, that osmosis is a *passive* process—no direct input of energy is required.

Unlike animal cells, plant cells are surrounded by a relatively rigid cell wall. As water enters a plant cell due to its negative solute potential, the entry of more water is increasingly resisted by an opposing **pressure potential** (called *turgor pressure* in plants). (Pressure potential is a hydraulic pressure analogous to the air pressure in an automobile tire; it is a mechanical pressure that can be measured with a pressure gauge.) As more and more water enters, the pressure potential becomes greater and greater.

Owing to the rigidity of the cell wall, plant cells do not burst the way animals cells do when placed in pure water; instead, water enters plant cells by osmosis until the pressure potential exactly balances the solute potential. At this point, the cell is *turgid;* that is, it has a significant positive pressure potential.

The overall tendency of a solution to take up water from pure water, across a membrane, is called its **water potential** and is represented as ψ, the Greek letter psi (pronounced "sigh") (**Figure 35.2**). The water potential of a solution is simply the sum of its (negative) solute potential (ψ_s) and its (usually positive) pressure potential (ψ_p):

$$\psi = \psi_s + \psi_p$$

For pure water open to the atmosphere and therefore under no applied pressure, all three of these parameters are defined as zero.

Whenever water moves by osmosis, the following important rule applies:

- Water always moves across a selectively permeable membrane toward the region of lower (more negative) water potential.

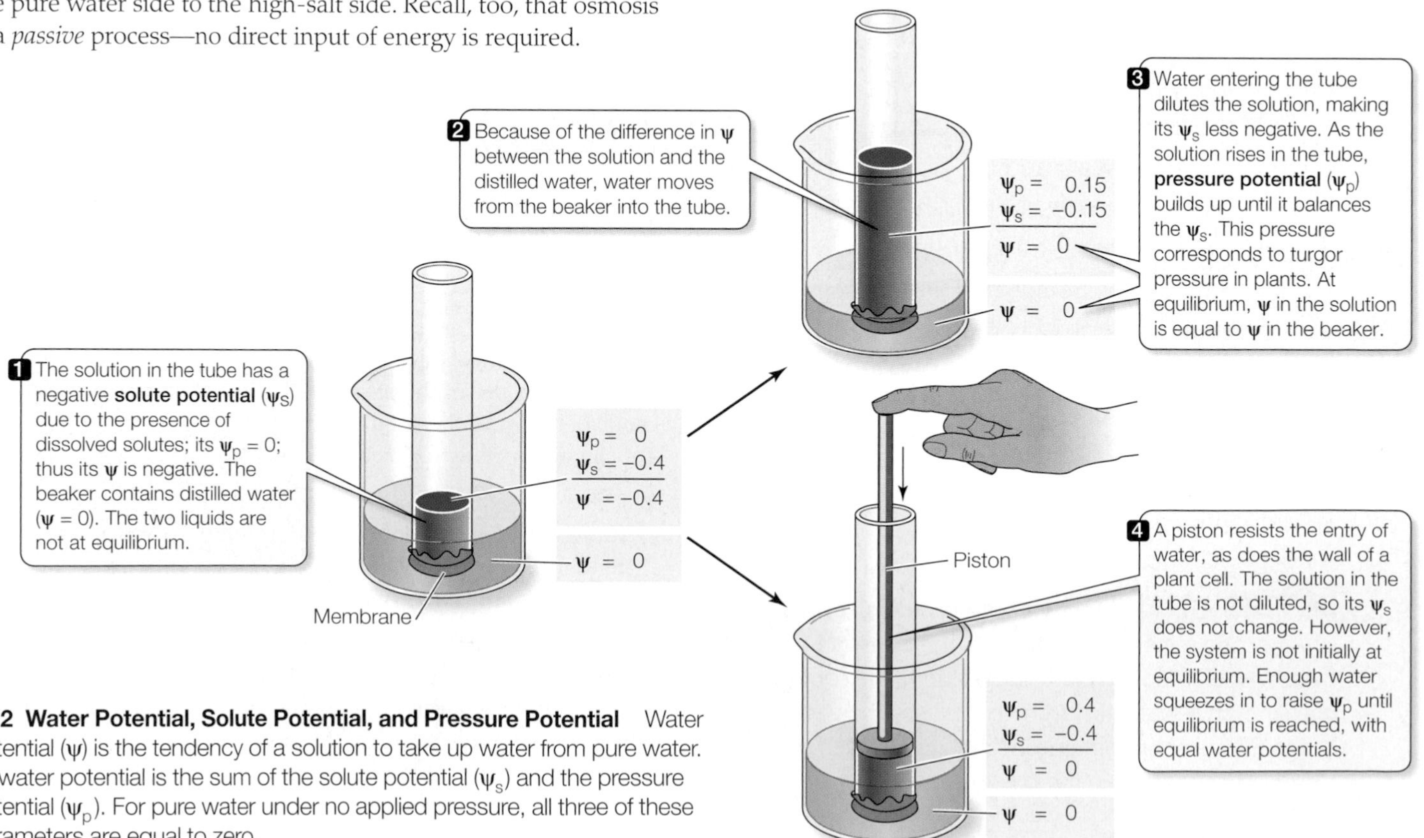

35.2 Water Potential, Solute Potential, and Pressure Potential Water potential (ψ) is the tendency of a solution to take up water from pure water. Its water potential is the sum of the solute potential (ψ_s) and the pressure potential (ψ_p). For pure water under no applied pressure, all three of these parameters are equal to zero.

We can measure solute potential, pressure potential, and water potential in *megapascals* (MPa), a unit of pressure. Atmospheric pressure, "one atmosphere," is about 0.1 MPa, or 14.7 pounds per square inch; typical pressure in an automobile tire is about 0.2 MPa.

Osmosis is of great importance to plants. The physical structure of many plants is maintained by the (positive) pressure potential of their cells; if the pressure potential is lost, the plant *wilts*. Within living tissues, the movement of water from cell to cell follows a gradient of water potential.

Over longer distances, in unobstructed tubes such as xylem vessels and phloem sieve tubes, the flow of water and dissolved solutes is driven by a *gradient of pressure potential*, not a gradient of water potential. The movement of a solution due to a difference in pressure potential between two parts of a plant is called **bulk flow**. We'll see that bulk flow in the xylem is between regions of differing *negative* pressure potential (tension) while bulk flow in the phloem is between regions of differing *positive* pressure potential (turgidity).

Aquaporins facilitate the movement of water across membranes

Water moves readily through biological membranes. How can this be? **Aquaporins** are membrane channel proteins through which water can move without interacting with the hydrophobic environment of the membrane's phospholipid bilayer (see Section 5.3). These proteins, important in both plants and animals, allow water to move rapidly from environment to cell and from cell to cell. Their abundance in the plasma membrane and tonoplast (vacuolar membrane) varies with environmental conditions, depending on a cell's need to obtain and retain water. The permeability of some aquaporins can be regulated, changing the *rate* of osmosis across the membrane. However, water movement through aquaporins is always passive, so the *direction* of water movement is unchanged by alterations in aquaporin permeability.

Uptake of mineral ions requires membrane transport proteins

Mineral ions, which carry electric charges, generally cannot move across a membrane unless they are aided by transport proteins, including ion channels and carrier proteins (see Section 5.3). The ions would otherwise be blocked by the hydrophobic interior of the membrane, and they are too large to pass through aquaporins.

We have just seen that water moves through a water-permeable membrane in response to a gradient of water molecules. Other molecules and ions follow their concentration gradients as permitted by the characteristics of the membrane. When the concentration of these charged ions in the soil is greater than that in the plant, transport proteins can move them into the plant by facilitated diffusion, which is a passive process. The concentrations of most ions in the soil solution, however, are lower than those required inside the plant. Thus the plant must actively take up ions *against* their concentration gradients—a process that requires energy.

Electric charge differences also play a role in the uptake of mineral ions. Movement of a negatively charged ion into a negatively charged region is movement against an electrical gradient and therefore requires energy. The combination of concentration and electrical gradients is called an *electrochemical gradient*. Uptake against an electrochemical gradient involves *active transport*, which is fueled by ATP generated by cellular respiration. Active transport requires specific transport proteins.

Unlike animals, plants do not have a sodium–potassium pump (see Section 5.4) for active transport. Rather, plants have a **proton pump**, which uses energy obtained from ATP to move protons out of the cell against a proton concentration gradient (**Figure 35.3, step 1**). Because protons (H^+) are positively charged, their accumulation outside the cell has two results:

- An electrical gradient is created such that the region outside the cell becomes positively charged with respect to the region inside.
- A proton concentration gradient develops, with more protons outside the cell than inside.

Each of these results has consequences for the movement of other ions. Because the inside of the cell is now more negative than the outside, cations (positively charged ions) such as potassium (K^+) move into the cell by facilitated diffusion through their specific membrane

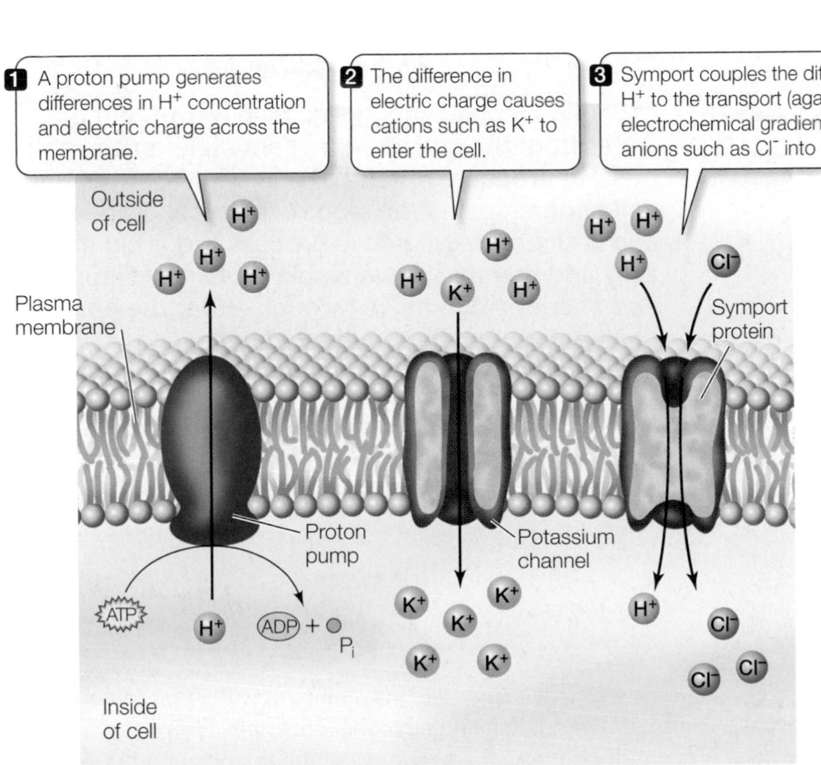

1 A proton pump generates differences in H+ concentration and electric charge across the membrane.
2 The difference in electric charge causes cations such as K+ to enter the cell.
3 Symport couples the diffusion of H+ to the transport (against an electrochemical gradient) of anions such as Cl− into the cell.
Outside of cell
Plasma membrane
Symport protein
Proton pump
Potassium channel
ATP
ADP + Pi
H+
K+
Cl−
Inside of cell

35.3 The Proton Pump in Active Transport of K^+ and Cl^- The buildup of hydrogen ions (H^+) transported outside the cell by the proton pump (1) drives the movement of both cations (2) and anions (3) into the cell.

channels (**Figure 35.3, step 2**). In addition, the proton concentration gradient can be harnessed to drive secondary active transport, in which anions (negatively charged ions) such as chloride (Cl^-) are moved into the cell against an electrochemical gradient by a symport protein that couples their movement with that of H^+ (**Figure 35.3, step 3**). In sum, there is a vigorous traffic of ions across plant cell membranes, involving specific membrane transport proteins and both active and passive processes.

The proton pump and the coordinated activities of other membrane transport proteins cause the interior of a plant cell to be very negatively charged with respect to the exterior; that is, they build up a significant *membrane potential*. Biologists can measure the membrane potential of a plant cell with microelectrodes, just as they can measure similar charge differences in nerve cells and other animal cells (see Section 44.2). Most plant cells maintain a membrane potential of at least –120 millivolts (mV).

Water and ions pass to the xylem by way of the apoplast and symplast

Mineral ions enter and move through plants in various ways. Where water is moving by bulk flow, dissolved minerals are carried along in the stream. Both water and minerals also move by diffusion. At certain sites, where plasma membranes are being crossed, some mineral ions are moved by active transport. One such site is the surface of a root hair, where mineral ions first enter the cells of the plant. Later, within the stele, the ions must cross another plasma membrane before entering the nonliving vessels and tracheids of the xylem.

The movement of ions across membranes can also result in the movement of water. Water moves into a root because the root has a more negative water potential than does the soil solution. Water moves from the cortex of the root into the stele (which is where the vascular tissues are located) because the stele has a more negative water potential than does the cortex.

Water and minerals from the soil can pass through the dermal and ground tissues to the stele via two pathways, the *apoplast* and the *symplast* (**Figure 35.4**):

- The **apoplast** (Greek *apo*, "away from"; *plast*, "living material") consists of the cell walls, which lie outside the plasma membranes, and the intercellular spaces (spaces between cells) that are common to many tissues. The apoplast is a continuous meshwork through which water and dissolved substances can flow or diffuse without ever having to cross a membrane. Movement of materials through the apoplast is thus unregulated—until it reaches the *Casparian strips* of the endodermis.
- The **symplast** (Greek, *sym*, "together with") passes through the continuous cytoplasm of the living cells connected by plasmodesmata. The selectively permeable plasma membranes of the root hair cells control access to the symplast, so movement of water and dissolved substances into the symplast is tightly regulated.

Water and minerals that pass from the soil solution through the apoplast can travel as far as the endodermis, the innermost layer of the root cortex. The endodermis is distinguished from the rest

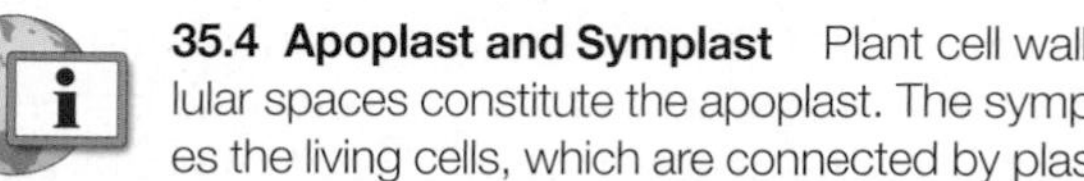

35.4 Apoplast and Symplast Plant cell walls and intercellular spaces constitute the apoplast. The symplast comprises the living cells, which are connected by plasmodesmata. To enter the symplast, water and solutes must pass through a plasma membrane. No such selective barrier limits movement through the apoplast. Casparian strips in the endodermis of the cortex are impregnated with the water-repelling substance suberin and separate apoplast in the cortex from apoplast in the stele.

of the ground tissue by the presence of **Casparian strips**. These waxy, suberin-impregnated regions of the endodermal cell wall form a water-repelling (hydrophobic) belt around each endodermal cell where it is in contact with other endodermal cells. Casparian strips act as a seal that prevents water and ions from moving between the cells (see Figure 35.4).

The Casparian strips of the endodermis completely separate the apoplast of the cortex from the apoplast of the stele. However, they do not obstruct the outer or inner faces of the endodermal cells. Accordingly, water and ions can enter the stele only by way of the symplast—that is, by entering and passing through the cytoplasm of the endodermal cells. Thus transport proteins in the plasma membranes of these cells determine which mineral ions pass into the stele, and at what rates.

Once they have passed the endodermal barrier, water and minerals leave the symplast and enter the apoplast of the stele. Parenchyma cells in the pericycle or xylem can aid this process.

As mineral ions move into the solution in the cell walls, the water potential in the apoplast becomes more negative; thus water moves out of the cells and into the apoplast by osmosis. In other words, active transport of ions moves the ions directly, and water follows passively. The end result is that water and minerals end up in the xylem, where they constitute the *xylem sap*.

35.1 RECAP

Osmotic mechanisms govern the movement of water from the soil into the plant stele; this is a passive process. Uptake of minerals from the soil occurs against an electrochemical gradient and is therefore an active process requiring energy and membrane transport proteins. Water and minerals can move through either the apoplast or the symplast, but must enter and leave the symplast to reach the xylem.

- Can you distinguish among water potential, solute potential, and pressure potential? See pp. 765–766
- Can you explain what happens when you drop a piece of stem into pure water? See p. 766 and Figure 35.2
- Can you distinguish between the apoplast and the symplast? See p. 767 and Figure 35.4

So far we've described the movement of water and minerals into plant roots and their entry into the root xylem. How does the xylem sap move on from the root system into the plant body?

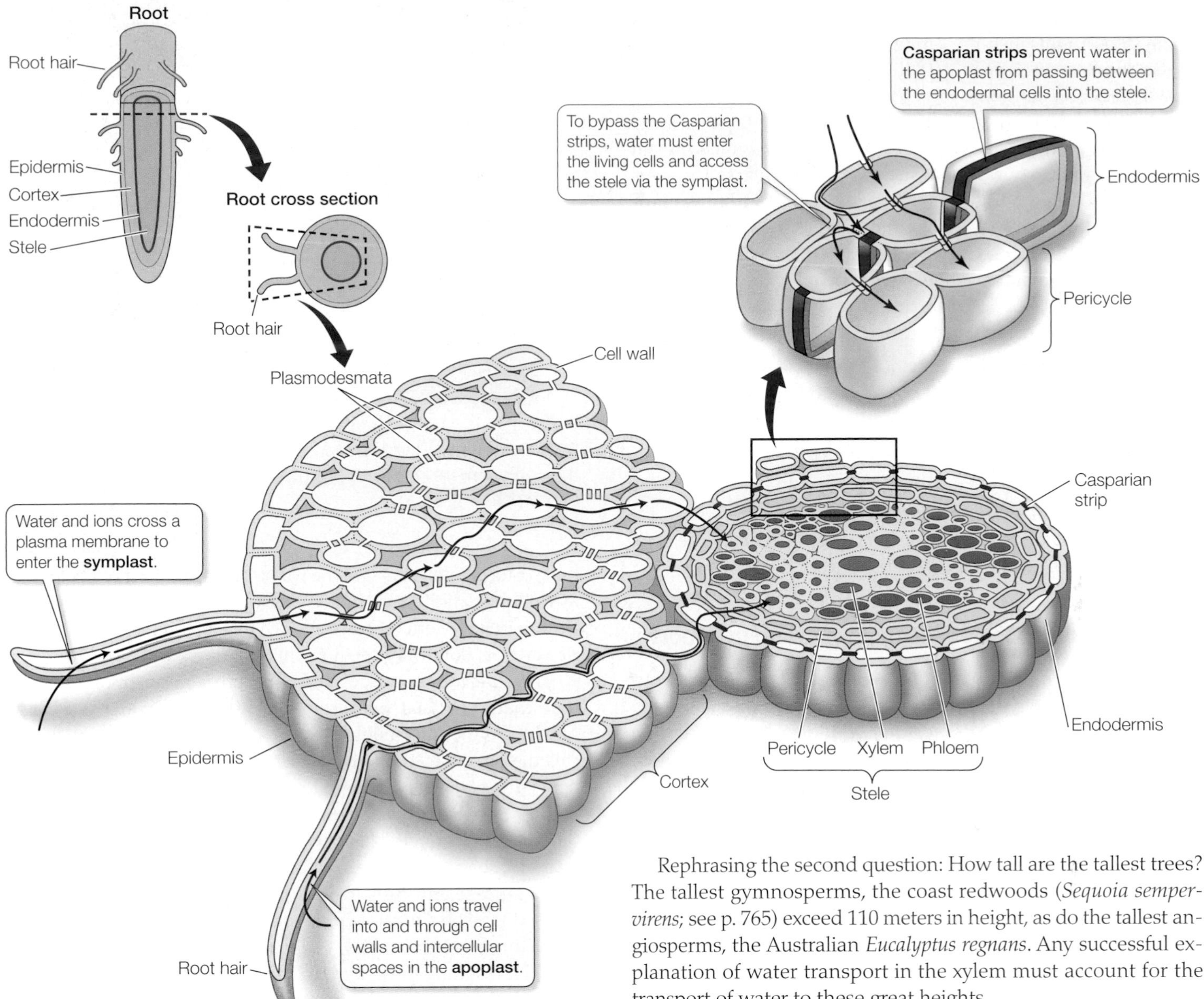
Root
Root hair
Epidermis
Cortex
Endodermis
Stele
Root cross section
Root hair
Plasmodesmata
Cell wall
To bypass the Casparian strips, water must enter the living cells and access the stele via the symplast.
Casparian strips prevent water in the apoplast from passing between the endodermal cells into the stele.
Endodermis
Pericycle
Casparian strip
Water and ions cross a plasma membrane to enter the symplast.
Epidermis
Cortex
Pericycle
Xylem
Phloem
Stele
Endodermis
Root hair
Water and ions travel into and through cell walls and intercellular spaces in the apoplast.

Rephrasing the second question: How tall are the tallest trees? The tallest gymnosperms, the coast redwoods (*Sequoia sempervirens*; see p. 765) exceed 110 meters in height, as do the tallest angiosperms, the Australian *Eucalyptus regnans*. Any successful explanation of water transport in the xylem must account for the transport of water to these great heights.

Movies featuring Tarzan show him swinging through the jungle canopy on lianas—twining jungle vines. And when his exertions leave him thirsty, he just chops open a vine and drinks the water from its stem. Neither activity is particularly realistic, although, like all plants, lianas do move water through the xylem in their stems.

35.2 How Are Water and Minerals Transported in the Xylem?

Before we consider the underlying mechanism of xylem transport, let's pause to reflect on the magnitude of what it accomplishes. Consider two questions: How much water is transported? And how high can water be transported?

The following example illustrates how much water an individual plant can transport. A single maple tree 15 meters tall has been estimated to have some 177,000 leaves, with a total leaf surface area of 675 square meters—half again the area of a basketball court. During a summer day, that tree loses 220 liters of water *per hour* to the atmosphere by evaporation from the leaves. To prevent wilting, the xylem needs to transport 220 liters of water from the roots to the leaves every hour. (By comparison, a 50-gallon drum holds 189 liters.)

Various hypotheses to explain the ascent of xylem sap have been offered over the years. We begin by reviewing some illuminating experiments that ruled out early models, and then turn to evidence in support of the current model.

Experiments ruled out xylem transport by the pumping action of living cells

Some of the earliest attempts to explain the rise of sap in the xylem were based on the hypothesis that a pumping action by living cells in the stem might push the sap upward. However, experiments conducted and published in 1893 by the German botanist Eduard Strasburger definitively ruled out such models.

Strasburger worked with trees about 20 meters tall. He sawed through the trunk of each tree at its base and plunged the cut end into a solution of a poison, such as copper sulfate. The solution rose through the trunk, as was readily evident from the progressive death of the bark higher and higher up. When the solution reached the leaves, the leaves died, too, at which point the movement of the solution stopped (as shown by the liquid level in the bucket, which stopped dropping).

This simple experiment established three important points:

- Living, "pumping" cells were not responsible for the upward movement of the solution, because the solution itself killed all living cells with which it came in contact.
- The leaves played a crucial role in transport. As long as they were alive, the solution continued to move upward; when the leaves died, movement ceased.
- The movement was not caused by the roots, because the trunk had been completely separated from the roots.

Root pressure does not account for xylem transport

In spite of Strasburger's observations, some plant physiologists hypothesized that xylem transport was based on **root pressure**—pressure exerted by the root tissues that would force liquid up the xylem. The basis for root pressure is a higher solute concentration, and accordingly a more negative water potential, in the xylem sap than in the soil solution. This water potential draws water into the stele; once there, the water has nowhere to go but up, so it rises in the vessels and tracheids.

There is good evidence that root pressure exists—for example, the phenomenon of *guttation*, in which liquid water is forced out through openings at the margins of leaves (**Figure 35.5**). Guttation occurs only under conditions of high atmospheric humidity and plentiful water in the soil, which occur most commonly at night. Root pressure is also the source of the sap that oozes from the cut stumps of some plants when their tops are cut off. Stephen Hales was the first to study root pressure quantitatively.

Root pressure, however, cannot account for the ascent of sap in trees. Root pressure seldom exceeds 0.1–0.2 MPa (1–2 atmospheres). If root pressure were driving sap up the xylem, we would observe a positive pressure potential in the xylem at all times. In fact, as we are about to see, the xylem sap in most trees is under *tension*—has a negative pressure potential—when it is ascending. Furthermore, as Strasburger had already shown, materials can be transported upward in the xylem even when the roots have been removed. If the roots are not pushing the xylem sap upward, what causes it to rise?

The transpiration–cohesion–tension mechanism accounts for xylem transport

An alternative to pushing is pulling: The leaves pull the xylem sap upward, which was first demonstrated by Stephen Hales. The evaporative loss of water from the leaves indirectly generates a pulling force—**tension**—on the water in the apoplast of the leaves, as we'll see. Hydrogen bonding between water molecules makes the sap in the xylem cohesive enough to withstand the tension and rise by bulk flow. Let's see how this process works.

35.5 Guttation Root pressure is responsible for forcing water through openings in the margins of this leaf on a lady's mantle plant (*Alchemilla*).

The concentration of water vapor in the atmosphere is lower than that in the leaf. Because of this difference, water vapor diffuses from the intercellular spaces of the leaf, through openings called stomata, to the outside air in the process called **transpiration**. Within the leaf blade, water evaporates from the moist walls of the mesophyll cells and enters the intercellular spaces. As water evaporates from the film coating each cell, the film shrinks back into tiny spaces in the cell walls, increasing the curvature of the water surface and thus increasing its surface tension. This increased tension (negative pressure potential) in the surface film draws more water into the cell walls, replacing that which was lost. The resulting tension in the mesophyll draws water from the xylem of the nearest vein into the apoplast surrounding the mesophyll cells. The removal of water from the veins, in turn, establishes tension on the entire column of water contained within the xylem, so that the column is drawn upward all the way from the roots (**Figure 35.6**).

The ability of water to be pulled upward through tiny tubes results from the remarkable **cohesion** of water—the tendency of water molecules to stick to one another through hydrogen bonding (see Section 2.4). The narrower the tube, the greater the tension the water column can withstand without breaking. The integrity of the column is also maintained by the *adhesion* of water to the xylem walls. The cohesion of water in the xylem is great enough to withstand even the tensions developed there.

In summary, the key elements of water transport in the xylem are

- *Transpiration*, the evaporation of water from the leaves
- *Tension* in the xylem sap resulting from transpiration
- *Cohesion* in the xylem sap from the leaves to the roots

This **transpiration–cohesion–tension mechanism** requires no work (that is, no expenditure of energy) on the part of the plant. At each step between soil and atmosphere, water moves passively toward a region with a more negative water potential. Dry air has the most negative water potential (–95 MPa at 50% relative humidity), and the soil solution has the least negative water potential (between –0.01 and –3 MPa). Xylem sap has a water potential more negative than that of cells in the cortex of the root, but less negative than that of mesophyll cells in the leaf.

In the tallest trees, such as a 110-meter *Sequoia*, the difference in pressure potential between the top and the bottom of the column may be as great as 3 MPa. Recall that the pressure in a typical automobile tire is 0.2 MPa.

Mineral ions contained in the xylem sap rise passively with water as it ascends from root to leaf. In this way the nutritional needs of the shoot are met. Some of the mineral elements brought to the leaves are subsequently redistributed to other parts of the plant by way of the phloem, but the initial delivery from the roots is through the xylem.

In addition to promoting the transport of minerals, transpiration contributes to temperature regulation. As water evaporates from mesophyll cells, heat is taken up from the cells, and the leaf temperature drops. This cooling effect of evaporation (so evident in the cooling of our skin when we sweat) is important in enabling plants to live in hot environments. A farmer can hold a leaf between thumb and forefinger to estimate its temperature; if the leaf doesn't feel cool, that means that transpiration is not occurring, so it must be time to water.

A pressure chamber measures tension in the xylem sap

The transpiration–cohesion–tension model holds true only if the column of sap in the xylem is under tension (has a negative pressure potential). The most elegant demonstrations of this tension, and of its adequacy to account for the ascent of xylem sap in tall trees, were performed by the biologist Per Scholander, who measured tension in stems with an instrument called a **pressure chamber** (**Figure 35.7**).

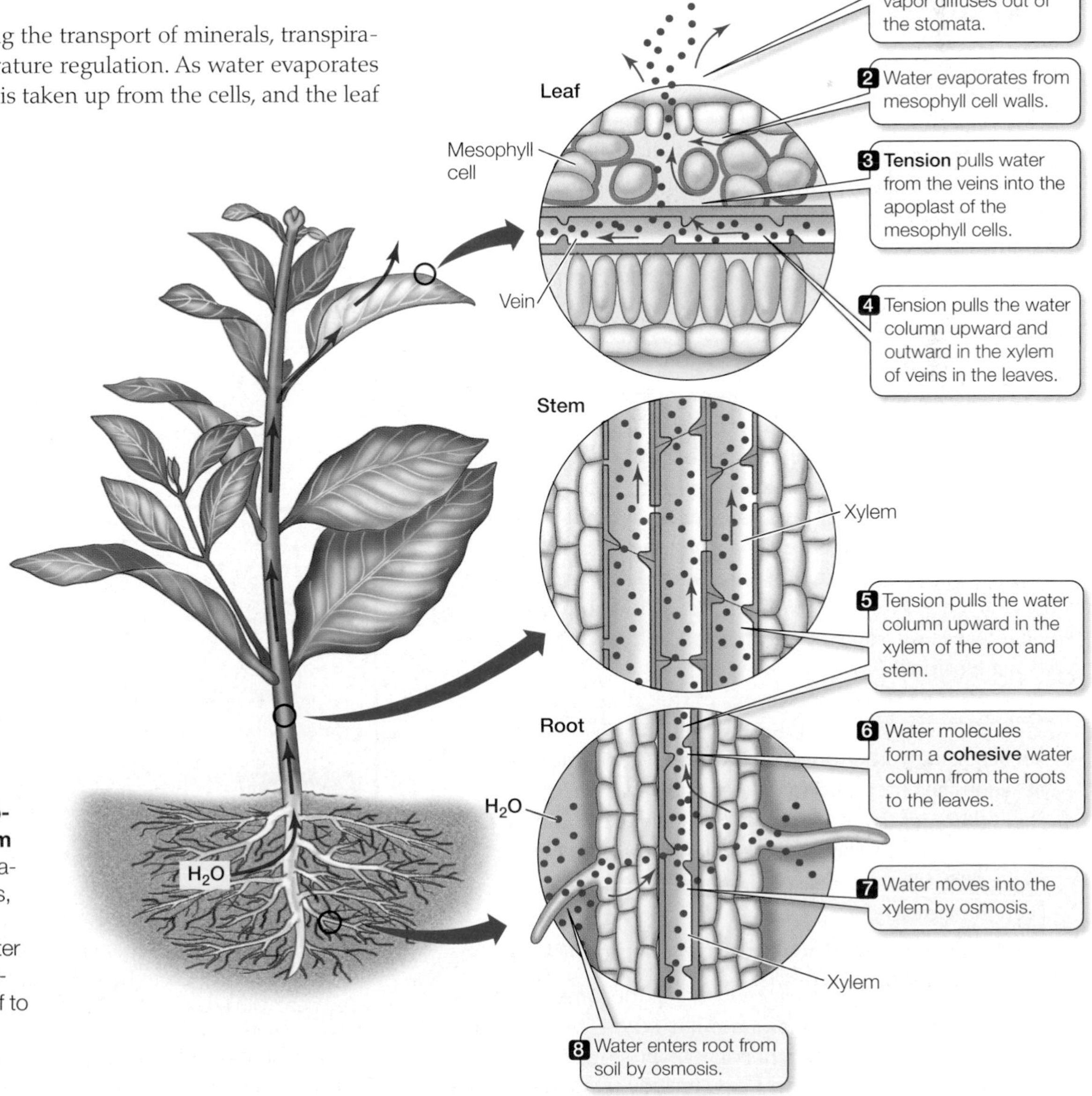
1 Transpiration: water vapor diffuses out of the stomata.
Leaf
2 Water evaporates from mesophyll cell walls.
Mesophyll cell
3 Tension pulls water from the veins into the apoplast of the mesophyll cells.
Vein
4 Tension pulls the water column upward and outward in the xylem of veins in the leaves.
Stem
Xylem
5 Tension pulls the water column upward in the xylem of the root and stem.
Root
6 Water molecules form a cohesive water column from the roots to the leaves.
H_2O
H_2O
7 Water moves into the xylem by osmosis.
Xylem
8 Water enters root from soil by osmosis.

35.6 The Transpiration–Cohesion–Tension Mechanism Transpiration causes evaporation from mesophyll cell walls, generating tension on the xylem. Cohesion among water molecules in the xylem transmits the tension from the leaf to the root, causing water to move from the soil to the atmosphere.

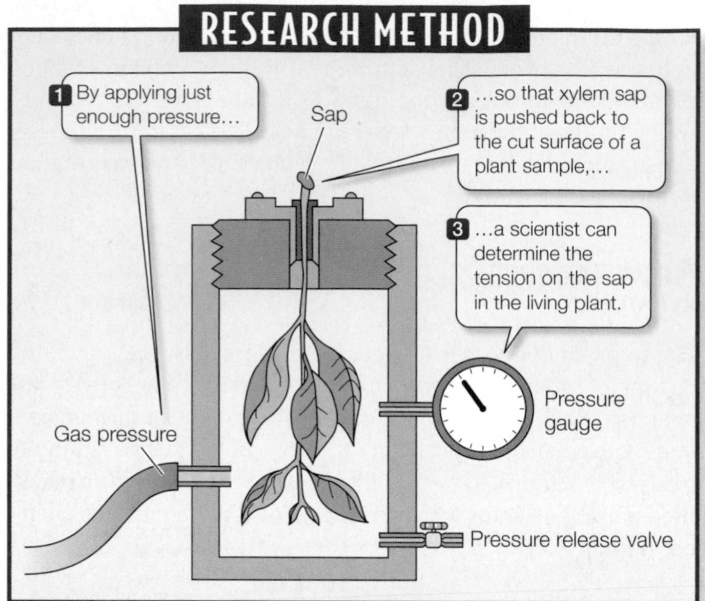

1 By applying just enough pressure...

Pressure release valve

35.7 A Pressure Chamber The amount of tension on the sap in different types of plants can be measured with this device.

Consider a stem in which the xylem sap is under tension. If the stem is cut, the sap pulls away from the cut, into the stem. This behavior indicates that the pressure in the intact xylem is lower than that of the atmosphere. Now the stem is quickly placed in the pressure chamber, in which the pressure may be raised. The cut surface remains outside the chamber. As gas pressure is applied to the plant parts within the chamber, the xylem sap is pushed back to the cut surface. When the sap first becomes visible again at the cut surface, the pressure in the chamber is recorded. This pressure is equal in magnitude but opposite in sign to the tension (negative pressure potential) originally present in the xylem.

Scholander used the pressure chamber to study dozens of plant species, from diverse habitats, growing under a variety of conditions. Whenever xylem sap was ascending, he found that it was under tension. He also noticed that the tension disappeared in some of the plants at night, when transpiration ceased. In developing vines, the xylem sap was under no tension until leaves formed. Once leaves developed, transport in the xylem began and so did the tensions.

Per Scholander used surveying instruments to locate twigs at various heights on tall trees, then had a sharpshooter shoot the twigs down with a rifle. Testing these twigs in the pressure chamber, Scholander confirmed that tension differences at different heights were indeed great enough to keep the sap ascending.

The rate at which xylem sap ascends is not the same at all times. No flow of xylem sap takes place at night, when there is little or no transpiration. By day, when the sap is ascending, the rate of ascent depends on several factors. These include temperature, light intensity, and wind velocity, all of which affect the transpiration rate, and hence the rate of sap flow. Another factor may be the concentration of K^+ in the sap: In greenhouse experiments utilizing tobacco plants, the rate of flow increases as the K^+ concentration increases (**Figure 35.8**). K^+ appears to affect a cell wall component in the membrane of pits between vessel elements, changing the size of the pit. This effect may increase the rate of water flow through a vessel when a neighboring vessel is blocked, as by an air bubble.

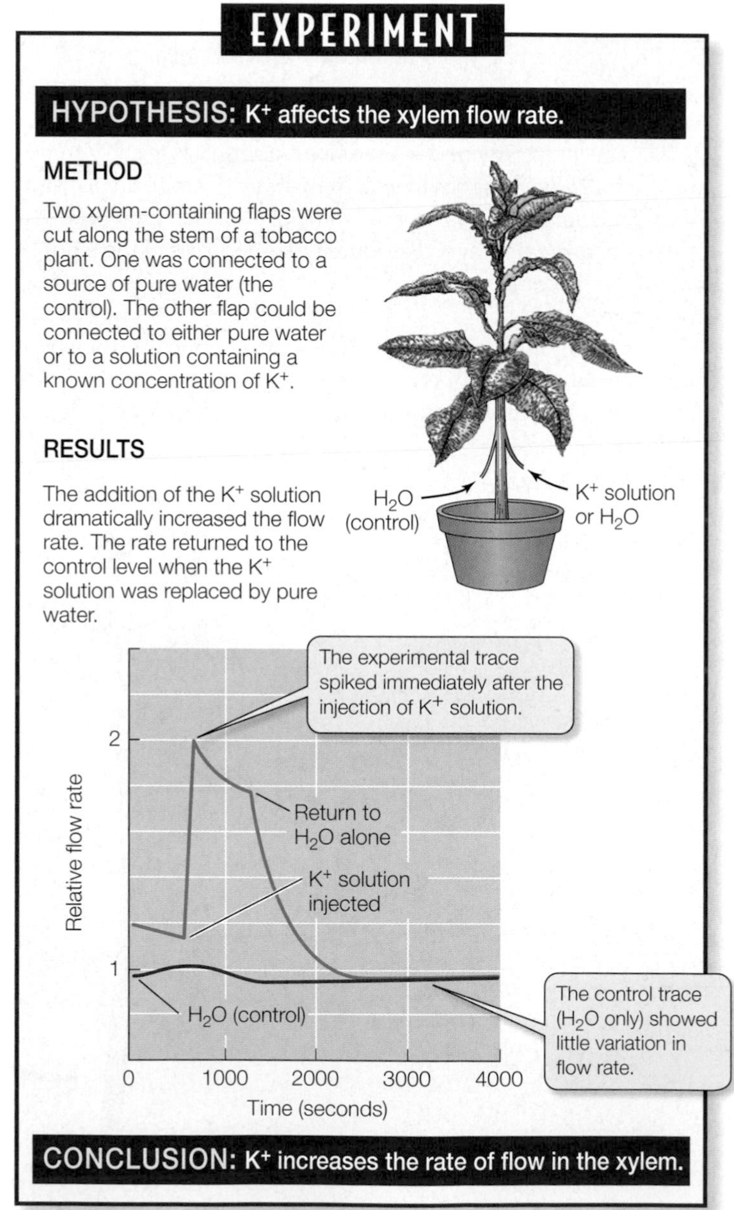

EXPERIMENT
HYPOTHESIS: K+ affects the xylem flow rate.
METHOD
Two xylem-containing flaps were cut along the stem of a tobacco plant. One was connected to a source of pure water (the control). The other flap could be connected to either pure water or to a solution containing a known concentration of K+.
RESULTS
The addition of the K+ solution dramatically increased the flow rate. The rate returned to the control level when the K+ solution was replaced by pure water.
H2O (control)
K+ solution or H2O
The experimental trace spiked immediately after the injection of K+ solution.
Return to H2O alone
K+ solution injected
H2O (control)
The control trace (H2O only) showed little variation in flow rate.
Relative flow rate
1
2
0
1000
2000
3000
4000
Time (seconds)
CONCLUSION: K+ increases the rate of flow in the xylem.

35.8 Potassium Ions Speed Transport in the Xylem This experiment by Maciej Zwieniecki and colleagues showed that the rate of fluid ascending through the xylem spiked when a solution with a known concentration of potassium ions was injected. Repeating the experiment with solutions of different concentrations of K^+ showed that the higher the K^+ concentration, the greater the flow rate. FURTHER RESEARCH: What experiments would you perform to determine whether this effect is specific to K^+?

35.2 RECAP

The transpiration-cohesion-tension mechanism explains the ascent of xylem sap. Transpiration draws water out of leaves, resulting in tension that pulls water from the xylem. Because of cohesion between water molecules, water is pulled passively through the xylem vessels in continuous columns, always toward a region with a more negative water potential.

- Do you understand the roles of transpiration, cohesion, and tension in xylem transport? See pp. 770–771 and Figure 35.6
- What is measured by the pressure chamber technique? See p. 772 and Figure 35.7

Although transpiration provides the impetus for the transport of water and minerals in the xylem, it also results in the loss of tremendous quantities of water from the plant. How plants control this loss is the subject of the next section.

35.3 How Do Stomata Control the Loss of Water and the Uptake of CO_2?

The epidermis of leaves and stems minimizes transpirational water loss by secreting a waxy cuticle, which is impermeable to water. However, the cuticle is also impermeable to carbon dioxide. This poses a problem: How can the leaf balance its need to retain water with its need to obtain CO_2 for photosynthesis?

Plants have evolved an elegant compromise in the form of **stomata** (singular *stoma*), or pores, in the epidermis of their leaves. A pair of specialized epidermal cells, called **guard cells**, controls the opening and closing of each stoma (**Figure 35.9A**). When the stomata are open, CO_2 can enter the leaf by diffusion—but water vapor is lost in the same way. Closed stomata prevent water loss, but also exclude CO_2 from the leaf.

Most plants open their stomata only when the light intensity is sufficient to maintain a moderate rate of photosynthesis. At night, when darkness precludes photosynthesis, their stomata remain closed; no CO_2 is needed at this time, and water is conserved. Even during the day, the stomata close if water is being lost at too great a rate.

The stoma and guard cells seen in Figure 35.9A are typical of eudicots. Monocots typically have specialized epidermal cells associated with their guard cells. The principle of operation, however, is the same for both monocot and eudicot stomata. In what follows, we describe the regulation and mechanism of stomatal opening and the normal cycle of opening and closing.

The guard cells control the size of the stomatal opening

Light causes the stomata of most plants to open, admitting CO_2 for photosynthesis. Another cue for stomatal opening is the level of CO_2 in the intercellular spaces inside the leaf. A low level favors opening of the stomata, thus allowing the uptake of more CO_2.

Water stress is a common problem for plants, especially on hot, sunny, windy days. Plants have a protective response to these conditions, which uses the water potential of the mesophyll cells as a

(A)

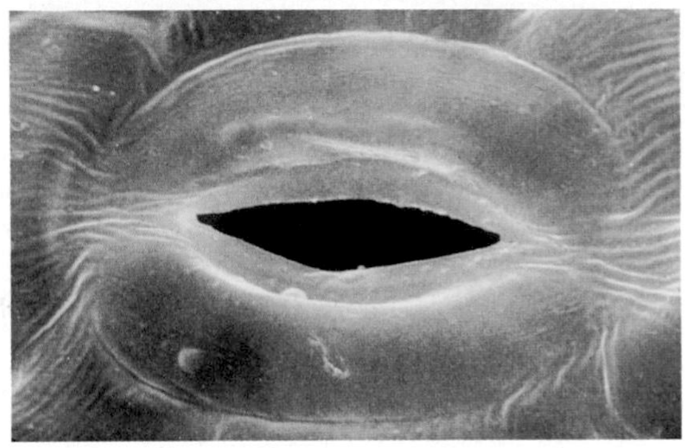

(B)

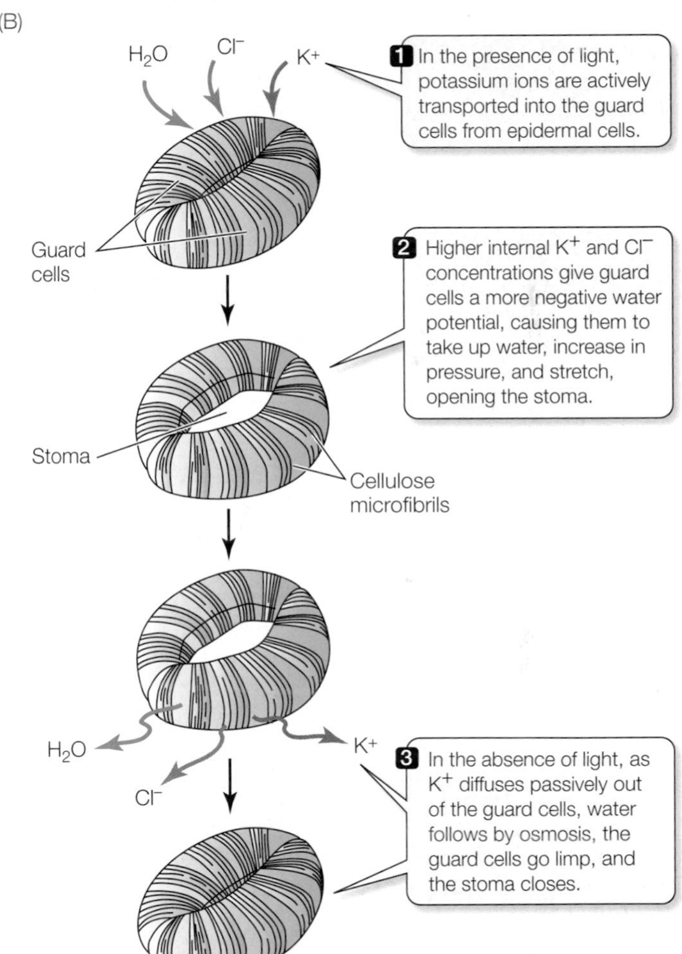

H2O
Cl–
K+
1 In the presence of light, potassium ions are actively transported into the guard cells from epidermal cells.
Guard cells
2 Higher internal K+ and Cl– concentrations give guard cells a more negative water potential, causing them to take up water, increase in pressure, and stretch, opening the stoma.
Stoma
Cellulose microfibrils
H2O
Cl–
K+
3 In the absence of light, as K+ diffuses passively out of the guard cells, water follows by osmosis, the guard cells go limp, and the stoma closes.

35.9 Stomata (A) Scanning electron micrograph of an open stoma formed by two sausage-shaped guard cells. (B) Potassium ion concentrations affect the water potential of the guard cells, controlling the opening and closing of stomata. Negatively charged ions that accompany K^+ maintain electrical balance and contribute to the changes in water potential that open and close the stomata.

cue. Even when the CO_2 level is low and the sun is shining, if the mesophyll is too dehydrated—that is, if the water potential of the mesophyll is too negative—the mesophyll cells release a plant hormone called *abscisic acid*. Abscisic acid acts on the guard cells, causing them to close the stomata and prevent further drying of the leaf. This response reduces the rate of photosynthesis, but it protects the plant.

The opening and closing of stomata is regulated by control of the K^+ concentration in the guard cells (review Figure 35.3). Blue light, absorbed by a pigment in the guard cell plasma membrane, activates a proton pump, which actively transports H^+ out of the guard cells and into the apoplast of the surrounding epidermis. The resulting proton gradient drives K^+ into the guard cell, where it accumulates (**Figure 35.9B**). The increasing internal concentration of K^+ makes the water potential of the guard cells more negative. Water enters the guard cells by osmosis, increasing their pressure potential. The arrangement of the cellulose microfibrils in their cell walls causes the guard cells to respond to this increase by changing their shapes so that a gap—the stoma—appears between them.

The stoma closes by the reverse process when active transport ceases in response to the absence of blue light or the presence of abscisic acid. Potassium ions diffuse passively out of the guard cells, water follows by osmosis, the pressure potential decreases, and the guard cells sag together and seal off the stoma. Negatively charged chloride ions and organic ions also move into and out of the guard cells along with the potassium ions, maintaining electrical balance and contributing to the change in the solute potential of the guard cells.

Showing how much potassium moves into the guard cells to open a stoma was a difficult feat. A typical guard cell has a total volume of less than 0.03 nanoliters when the stoma is closed and almost 0.05 nanoliters when it is open. The scientists who solved the problem used an *electron probe microanalyzer*, an instrument normally used by metallurgists to study the fine structure of alloys (**Figure 35.10**).

EXPERIMENT

HYPOTHESIS: Guard cells of open stomata contain more potassium ions than do those of closed stomata.

METHOD

1. Peel strips of epidermis from leaves of broad beans in the dark (closed stomata) and in the light (open stomata).
2. Examine the strips to locate stomata.
3. Scan across guard cells with the electron probe microanalyzer set to measure K^+ concentration.

RESULTS

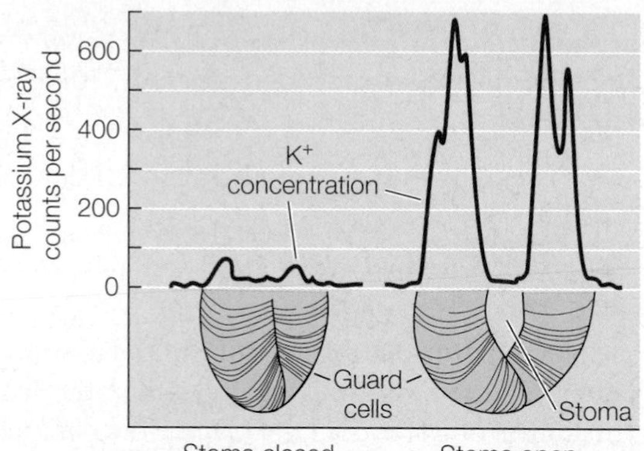

CONCLUSION: K^+ concentration within the guard cells surrounding an open stoma was much greater than that in the guard cells surrounding a closed stoma.

35.10 Measuring Potassium Ion Concentration in Guard Cells G. D. Humble and Klaus Raschke used the electron probe microanalyzer to examine individual stomata of the broad bean. They determined that K^+ concentration in each guard cell increased more than twentyfold as the stomatal aperture increased sixfold, and the guard cell volume nearly doubled. FURTHER RESEARCH: What other ion or ions would you study in order to further explore the mechanism of stomatal opening?

Transpiration from crops can be decreased

Stomata are the "referees" mediating the admission of CO_2 for photosynthesis and the exit of water by transpiration. Farmers would like their crops to transpire less, thus reducing the need for irrigation. Similarly, nurseries and gardeners would like to be able to reduce the amount of water lost by plants that are to be transplanted, because transplanting often damages the roots, causing the plant to wilt or die. What they need is a good **antitranspirant**: a compound that can be applied to plants to reduce water loss from the stomata without excessively limiting CO_2 uptake.

Abscisic acid and its commercial chemical analogs have been found to work as antitranspirants in small-scale tests, but their high cost has precluded commercial use. So research has turned to the question of whether plants can be made more sensitive to their own abscisic acid. The guard cells of transgenic plants with a mutant allele of the *era* gene are highly sensitive to abscisic acid and hence resistant to wilting during drought stress. This recent discovery might lead to an agricultural application.

A totally different type of antitranspirant temporarily seals off the leaves from the atmosphere. Growers use a variety of compounds, most of which form polymer films around leaves, to form a barrier to evaporation by sealing the stomata. These compounds cause undesirable side effects, however, and can be used only for short periods of time. Their most common use is in the transplanting of nursery stock.

35.3 RECAP

Leaf pores called stomata admit the CO_2 needed for photosynthesis but also permit the exit of water by transpiration. Stomata can be opened or closed by guard cells to regulate water loss.

- Can you explain the role of K^+ ions in the functioning of guard cells? See p. 774 and Figure 35.9
- Under what circumstances do the mesophyll cells release abscisic acid, and what is its effect? See p. 774

In the absence of antitranspirants, stomata are normally open during daylight hours, allowing CO_2 to be fixed and converted to the products of photosynthesis. In the next section we'll see how these products are delivered to other parts of the plant, supporting plant growth.

35.4 How Are Substances Translocated in the Phloem?

Photosynthesis takes place in the mesophyll cells and, in C_4 plants, in the bundle sheath cells of the leaf (see Figure 8.18). The products of photosynthesis (primarily carbohydrates) diffuse to the nearest small vein, where they are actively transported into sieve tube elements. The movement of carbohydrates and other solutes through the plant in the phloem is called **translocation.**

Substances in the phloem are translocated from sources to sinks. A **source** is an organ (such as a mature leaf or a storage root) that *produces* (by photosynthesis or by digestion of stored reserves) more sugars than it requires. A **sink** is an organ (such as a root, a flower, a developing fruit or tuber, or an immature leaf) that *consumes* sugars for its own growth and storage needs. Sugars (primarily sucrose), amino acids, some minerals, and a variety of other solutes are translocated between sources and sinks in the phloem.

How do we know that such organic solutes are translocated in the phloem, rather than in the xylem? Just over 300 years ago, the Italian scientist Marcello Malpighi performed a classic experiment in which he removed a ring of bark (containing the phloem) from the trunk of a tree—that is, he *girdled* the tree (**Figure 35.11**). Over time, the bark in the region above the girdle swelled. We now know that the swelling resulted from the accumulation of organic solutes that came from higher up the tree and could no longer continue downward because of the disruption of the phloem. Later, the bark below the girdle died because it no longer received sugars from the leaves. Eventually the roots, and then the entire tree, died.

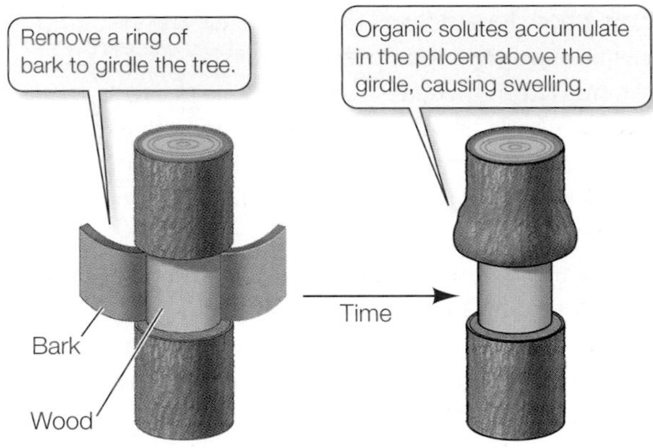

35.11 Girdling Blocks Translocation in the Phloem By girdling—removing a ring of bark containing the phloem—Malpighi blocked the translocation of organic solutes in a tree. Bark below the girdle died because it no longer received nutrients; eventually the entire tree died.

Any explanation of the translocation of organic solutes must account for a few important observations:

- Translocation stops if the phloem tissue is killed by heating or other methods; thus the mechanism must be different from that of transport in the xylem.
- Translocation often proceeds in both directions simultaneously—up the stem and down the stem—depending on the location of sources and sinks.
- Translocation is inhibited by compounds that inhibit respiration and thus limit the ATP supply in the source.

To investigate translocation, plant physiologists needed to obtain samples of pure sieve tube sap from individual sieve tube elements. This difficult task was simplified when it was discovered that a common garden pest, the aphid, feeds on plants by drilling into a sieve tube. An aphid inserts its specialized feeding organ, called a *stylet*, into a stem until the stylet enters a sieve tube. The pressure within the sieve tube is greater than that in the surrounding plant tissues or outside the plant, so the nutritious sieve tube sap is forced through the stylet and into the aphid's digestive tract. So great is the pressure that sugary liquid is forced through the insect's body and out the anus (**Figure 35.12**). This works because the phloem sap is under strongly positive pressure, unlike the negative pressure potential in the xylem.

Plant physiologists use aphids to collect phloem sap. When liquid appears on the aphid's abdomen, indicating that the insect has connected with a sieve tube, the physiologist quickly freezes the aphid and cuts its body away from the stylet, which remains in the sieve tube element. For hours, phloem sap continues to exude from the cut stylet, where it may be collected for analysis. Chemical analysis of phloem sap collected in this manner reveals the contents of a single sieve tube element over time. Physiologists can also infer the rates at which different substances are translocated by

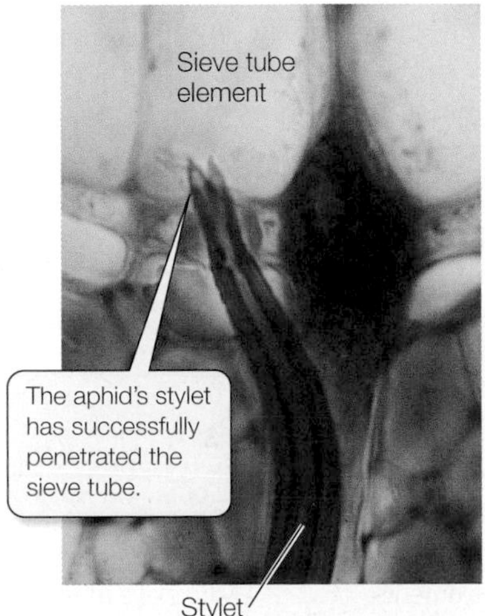

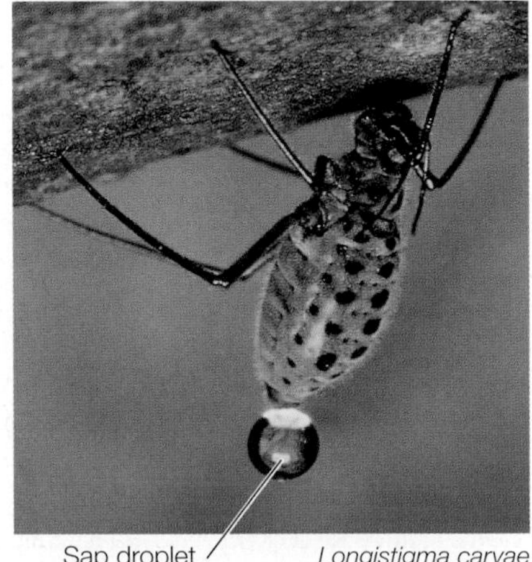

35.12 Aphids Collect Sap Aphids feed on sap drawn from a sieve tube, which they penetrate with a modified feeding organ, the stylet. Pressure inside the sieve tube forces sap through the aphid's digestive tract, from which it can be harvested.

measuring how long it takes for radioactive tracers administered to a leaf to appear at stylets at different distances from the leaf.

These methods have allowed us to understand how, at times, different substances might move in opposite directions in the phloem of a stem. Experiments with aphid stylets have shown that all the contents of any given sieve tube element move in the same direction. Thus, bidirectional translocation can be understood in terms of different sieve tubes conducting sap in opposite directions. These and other experiments led to the general adoption of the *pressure flow model* as an explanation for translocation in the phloem.

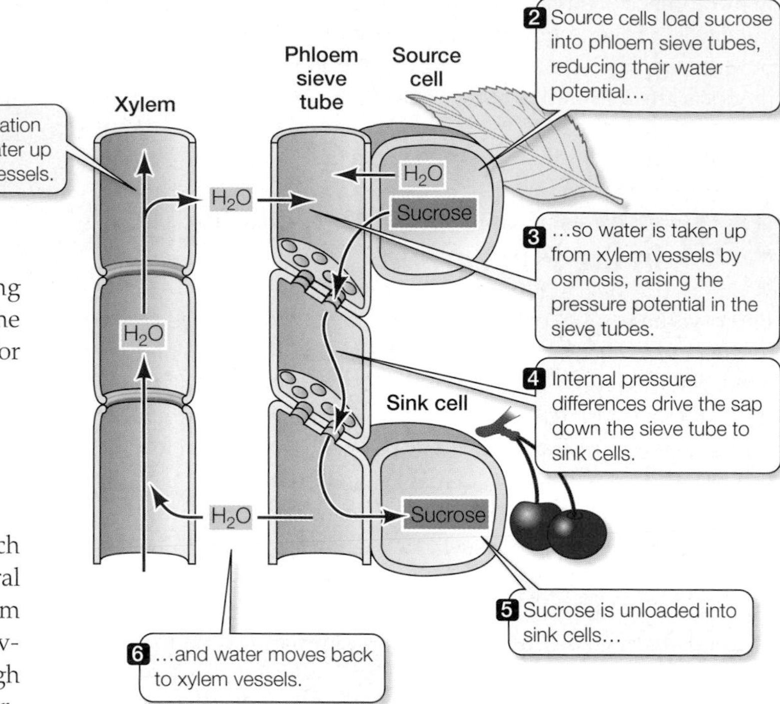

35.13 The Pressure Flow Model Combined pressure potential and water potential differences drive the bulk flow of phloem sap from a source to a sink.

The pressure flow model appears to account for translocation in the phloem

During sieve tube element development, the tonoplast and much of the cytosol breaks down, allowing the contents of the central vacuole to combine with much of the cytosol to form the phloem sap. The sap flows under pressure through the sieve tubes, moving from one sieve tube element to the next by bulk flow through the sieve plates, without crossing a membrane. We need to understand how this pressure is generated in order to understand translocation in the phloem.

Two steps in translocation require metabolic energy:

- Transport of sucrose and other solutes from sources into the sieve tubes, called *loading*
- Removal of the solutes from the sieve tubes into sinks, called *unloading*

According to the **pressure flow model** of translocation in the phloem, sucrose is actively transported into sieve tube elements at a source, giving those cells a greater sucrose concentration than the surrounding cells. Water therefore enters the sieve tube elements by osmosis. The entry of this water causes a greater pressure potential at the source end of the sieve tube, so that the entire fluid content of the sieve tube is pushed toward the sink end of the tube—in other words, the sap moves by bulk flow in response to a pressure gradient (**Figure 35.13**). In the sink, the sucrose is unloaded by active transport, maintaining the gradients of solute potential and water potential needed for movement.

The pressure flow model of translocation in the phloem is contrasted with the transpiration–cohesion–tension model of xylem transport in **Table 35.1**.

The pressure flow model has been experimentally tested

The pressure flow model was first proposed more than half a century ago, but some of its features are still being debated. Other mechanisms have been proposed to account for translocation in sieve tubes, but some have been disproved, and no other theory has as much support as the pressure flow model.

Two essential requirements must be met in order for the pressure flow model to be valid:

- The sieve plates must be unobstructed, so that bulk flow from one sieve tube element to the next is possible.
- There must be an effective method for loading sucrose and other solutes into the phloem in source tissues and removing them in sink tissues.

Let's see whether these requirements are met.

TABLE 35.1

Mechanisms of Sap Flow in Plant Vascular Tissues

	XYLEM	PHLOEM
Driving force for bulk flow	Transpiration from leaves	Active transport of sucrose at source
Site of bulk flow	Non-living vessel elements and tracheids (cohesion)	Living sieve tube elements
Pressure potential in sap	Negative (pull from top; tension)	Positive (push from source; pressure)

ARE THE SIEVE PLATES OPEN? Early electron microscopic studies of phloem samples cut from plants seemed to contradict the pressure flow model. The pores in the sieve plates always appeared to be plugged with masses of a fibrous protein, suggesting that sieve tube sap could not flow freely. The function of that fibrous protein was unclear, however. What purpose would it serve at the sieve plates?

Researchers contemplated the possibility that this protein functions to block the sieve plates when sieve tube elements have been damaged. Perhaps the protein is usually distributed more or less at random throughout the sieve tube elements but damage propels a sudden surge of sap toward the cut surface, carrying the protein into the pores, blocking them and preventing the loss of valuable nutrients. How might this hypothesis be tested? How might phloem for microscopic observation be obtained without causing the sap to surge to the cut surface?

One way to prevent the surge of the sap is to freeze plant tissue rapidly before cutting it. Another way is to let the tissue wilt so that no pressure exists in the phloem before cutting. If these methods are utilized, the sieve plates are not clogged by the protein, indicating that sieve tubes are unobstructed unless damaged. Thus, the first condition of the pressure flow model is met.

HOW ARE SIEVE TUBE ELEMENTS LOADED AND UNLOADED? If the pressure flow model is correct, there must be mechanisms for loading sugars and other solutes into the phloem in source regions and for unloading them in sink regions.

Sugars and other solutes produced in the mesophyll pass from cell to cell in the leaf and eventually enter the sieve tubes of the phloem. In some plants these substances leave the mesophyll cells and enter the apoplast. Then specific sugars and amino acids are actively transported into cells of the phloem, thus reentering the symplast. This passage through the apoplast and back into the symplast allows the selection of substances to be translocated by forcing them to pass through selectively permeable membranes. In many plants, solutes reenter the symplast at the sieve tube elements' companion cells, which then transfer the solutes to adjacent sieve tube elements.

A form of secondary active transport (see Section 5.4) loads sucrose into the companion cells and sieve tubes. Sucrose is carried across the plasma membrane from apoplast to symplast by sucrose–proton symport; thus the entry of sucrose and protons is strictly coupled. For this symport to work, the apoplast must have a high concentration of protons; these protons are supplied by a primary active transport system, the proton pump. The protons then diffuse back into the cell through the symport protein, bringing sucrose with them.

In sink regions, the solutes are actively transported *out* of the sieve tube elements and into the surrounding tissues. This unloading serves two purposes: It helps to maintain the gradient of solute potential and hence of pressure potential in the sieve tubes, and it promotes the buildup of sugars and starch to high concentrations in storage regions, such as developing fruits and seeds. Thus the second requirement of the pressure flow model is met, and the model is supported.

Plasmodesmata allow the transfer of material between cells

Many substances move from cell to cell within the symplast by way of plasmodesmata (see Figures 34.8 and 35.4). Among their other roles, plasmodesmata participate in the loading and unloading of sieve tube elements. The mechanisms vary among plant species, but the story in tobacco plants is a common one. In tobacco, sugars and other solutes in source tissues enter companion cells by active transport from the apoplast and move on to the sieve tube elements through plasmodesmata. In sink tissues, plasmodesmata connect sieve tube elements, companion cells, and the cells that will receive and use the transported compounds.

Plasmodesmata undergo developmental changes as an immature sink leaf becomes a mature source leaf. Plasmodesmata in sink tissues favor rapid unloading: They are more abundant, and they allow the passage of larger molecules. Plasmodesmata in source tissues are fewer in number.

It was long thought that only substances with molecular weights of less than 1 kilodalton could fit through a plasmodesma. Then biologists discovered that cells infected with tobacco mosaic virus (TMV) could allow molecules with molecular weights as great as 20 kDa to exit. We now know that TMV encodes a "movement protein" that produces this change in the permeability of the plasmodesmata—and that the plants themselves normally produce at least one such movement protein. Even large molecules such as proteins and RNAs, with molecular weights up to at least 50,000, can move between living plant cells. We will see some consequences of this movement of macromolecules through plasmodesmata in later chapters. Biologists are exploring possible ways to regulate the permeability, number, and form of plasmodesmata as a means of modifying traffic in the plant. Such modifications might, for example, allow the diversion of more of a grain crop's photosynthetic products into the seeds, increasing the crop yield.

35.4 RECAP

Carbohydrates produced by photosynthesis are translocated from source to sink via the phloem by a pressure flow mechanism.

- Can you distinguish between a source and a sink? See p. 775
- How does loading of sucrose at the source result in bulk flow toward the sink according to the pressure flow model? See p. 776 and Figure 35.13

CHAPTER SUMMARY

35.1 How do plant cells take up water and solutes?

Water moves through biological membranes by osmosis, always moving toward cells with a more negative water potential. The **water potential** (ψ) of a cell or solution is the sum of the **solute potential** and the **pressure potential**. Review Figure 35.2

The movement of a solution due to a difference in pressure potential between two parts of a plant is called **bulk flow**.

Aquaporins allow water molecules to pass through biological membranes without interacting with the hydrophobic interior of the membrane.

Mineral uptake requires transport proteins. Some minerals enter the plant passively by facilitated diffusion; others enter by active transport. A **proton pump** provides energy for the active transport of many mineral ions across membranes in plants. Review Figure 35.3

Water and minerals pass from the soil to the xylem by way of the **apoplast** and **symplast**. In the root, water and minerals can move from the cortex into the stele only by way of the symplast because **Casparian strips** in the endodermis block their movement through the apoplast. Review Figure 35.4, Web/CD Activity 35.1

35.2 How are water and minerals transported in the xylem?

Root pressure is responsible for **guttation** and for the oozing of sap from cut stumps, but it cannot account for the ascent of xylem sap in trees.

Water transport in the xylem results from the combined effects of **transpiration**, **cohesion**, and **tension**. Evaporation from the leaf produces tension in the mesophyll cells, which pulls a column of water—held together by cohesion—up through the xylem from the root. Dissolved minerals are carried passively in the water. Review Figure 35.6

Transport in the xylem is by bulk flow. It does not require the expenditure of energy.

35.3 How do stomata control the loss of water and the uptake of CO_2?

Stomata allow a compromise between water retention and carbon dioxide uptake.

A pair of **guard cells** controls the size of the stomatal opening. A proton pump, activated by blue light, pumps protons out of the guard cells to the walls of surrounding epidermal cells, setting up a proton gradient that drives the active transport of potassium ions into the cells. Water follows osmotically, swelling the cells and opening the stomata. Review Figure 35.9

When threatened by dehydration, mesophyll cells release abscisic acid, which causes guard cells to close the stomata.

Antitranspirants may be used to reduce transpiration rates in some applications but are not used for crop plants.

35.4 How are substances translocated in the phloem?

Products of photosynthesis, as well as some minerals, are translocated through sieve tubes in the phloem by way of living sieve tube elements.

Translocation in the phloem can proceed in both directions in the stem, although in a single sieve tube it goes only one way. Translocation requires a supply of ATP.

Translocation in the phloem is explained by the **pressure flow model**: The difference in solute concentration between **sources** and **sinks** creates a difference in (positive) pressure potential along the sieve tubes, resulting in bulk flow. Review Figure 35.13 and Table 35.1, Web/CD Tutorial 35.1

SELF-QUIZ

1. Osmosis
   a. requires ATP.
   b. results in the bursting of plant cells placed in pure water.
   c. can cause a cell to become turgid.
   d. is independent of solute concentrations.
   e. continues until the pressure potential equals the water potential.
2. Water potential
   a. is the difference between the solute potential and the pressure potential.
   b. is analogous to the air pressure in an automobile tire.
   c. is the movement of water through a membrane.
   d. determines the direction of water movement between cells.
   e. is defined as 1.0 MPa for pure water under no applied pressure.
3. Which statement about aquaporins is *not* true?
   a. They are membrane transport proteins.
   b. Water movement through aquaporins is always active.
   c. The permeability of some aquaporins is subject to regulation.
   d. They are found in both animals and plants.
   e. They enable water to pass through the phospholipid bilayer without encountering a hydrophobic environment.
4. Which statement about proton pumping across the plasma membrane of plants is *not* true?
   a. It requires ATP.
   b. The region inside the membrane becomes positively charged with respect to the region outside.
   c. It enhances the movement of K^+ ions into the cell.
   d. It pushes protons out of the cell against a proton concentration gradient.
   e. It can drive the secondary active transport of negatively charged ions.
5. Which statement is *not* true?
   a. The symplast is a meshwork consisting of the (connected) living cells.
   b. Water can enter the stele without entering the symplast.
   c. The Casparian strips prevent water from moving between endodermal cells.
   d. The endodermis is a cell layer in the cortex.
   e. Water can move freely in the apoplast without entering cells.
6. In the xylem,
   a. the products of photosynthesis travel down the stem.
   b. living, pumping cells push the sap upward.
   c. the driving force is in the roots.
   d. the sap is often under tension.
   e. the sap must pass through sieve plates.

7. Which of the following is *not* part of the transpiration–cohesion–tension mechanism?
   a. Water evaporates from the walls of mesophyll cells.
   b. Removal of water from the xylem exerts a pull on the water column.
   c. Water is remarkably cohesive.
   d. The wider the tube, the greater the tension its water column can withstand.
   e. At each step, water moves to a region with a more strongly negative water potential.
8. Stomata
   a. control the opening of guard cells.
   b. release less water to the environment than do other parts of the epidermis.
   c. are usually most abundant on the upper epidermis of a leaf.
   d. are covered by a waxy cuticle.
   e. close when water is being lost at too great a rate.
9. Which statement about phloem transport is *not* true?
   a. It takes place in sieve tubes.
   b. It depends on mechanisms for loading solutes into the phloem in sources.
   c. It stops if the phloem is killed by heat.
   d. A high pressure potential is maintained in the sieve tubes.
   e. In sinks, solutes are actively transported into sieve tube elements.
10. The fibrous protein in sieve tube elements
   a. may plug leaks when a plant is damaged.
   b. clogs the sieve plates at all times.
   c. never clogs the sieve plates.
   d. serves no known function.
   e. provides the driving force for transport in the phloem.

FOR DISCUSSION

1. Epidermal cells protect against excess water loss. How do they perform this function? What differences might you expect to find in the structure of the epidermis in stems, roots, and leaves?
2. Phloem transports material from sources to sinks. Give examples of each. How might the distribution of sources and sinks change in the course of a year?
3. What is the minimum number of plasma membranes a water molecule would have to cross in order to get from the soil solution to the atmosphere by way of the stele? To get from the soil solution to a mesophyll cell in a leaf?
4. Transpiration exerts a powerful pulling force on the water column in the xylem. When would you expect transpiration to proceed most rapidly? Why? Describe the source of the pulling force.

FOR INVESTIGATION

The experiments by Zwieniecki and colleagues showed that the flow rate in the xylem of the tobacco stem increased with increasing K^+ content. It has been argued that this effect, while interesting, may not be relevant to the xylem of plants in the field. How might you try to resolve this uncertainty?

CHAPTER 36 Plant Nutrition

When the land blew away

The world has known many natural disasters: plagues, famines, wildfires, tsunamis, hurricanes, floods, droughts, and earthquakes. One of the greatest disasters of the twentieth century occurred in North America during the 1930s when a prolonged drought, combined with a culturally modified landscape, turned the central plains of the continent into the Dust Bowl.

The native vegetation in the Plains States in the nineteenth century was grass—long grass in the east and short grass in the west. Cattlemen moved in to where their herds could graze their fill on a seemingly endless supply of food. But in fact, the supply wasn't endless. As one area was overgrazed, the cattle were moved on to new areas, leaving damaged soil in their wake. Settlers followed, "busted the sod" with plows, planted crops, and disrupted vegetation cycles that had gone on for centuries.

In the modern world, events in one location can have consequences far away. In 1914, the first year of World War I, the Turkish navy blocked the supply of Russian wheat to the rest of the world. The resulting shortages prompted farmers in the Plains States to seek a profit in wheat. They plowed both good and marginal soils and planted wheat, and still more wheat. But wheat requires more water than did the native grasses, and rainfall was irregular throughout the 1920s. In 1932 rainfall failed almost completely, not to return in good supply until 1939.

Without water, crops failed—if they even started to grow. The Plains States are windy states, and without plant roots there was nothing to hold the soil in place when the winds blew. The farms literally blew away. Farmers spent their last money to buy seed, hoping for a green year; but dry year followed dry year. The stone-broke farmers migrated westward, along with others whose livelihoods had depended on the farmers. Recall also that these events took place during the period of world economic history known as the Great Depression.

In the meantime, massive dust storms blew the desiccated topsoil eastward. One such storm in May 1934 blew soil all the way to the decks of ships in the Atlantic Ocean, depositing dust on desks in Washington, D.C. and darkening the midday sun

Dreams Disappeared in a Cloud of Dust This photograph of a family displaced by the Dust Bowl was taken by Dorothea Lange in the winter of 1936. The family had traveled to northern California looking for work as migrant farm labor. In Lange's words, "I saw and approached the hungry and desperate mother…I did not ask her name or her history. …She said that they had been living on frozen vegetables from the surrounding fields, and birds that the children killed. She had just sold the tires from her car to buy food."

The Problem Has Not Gone Away This photograph was taken near Fort Benton, Montana, in early May 2002. Drifting topsoil has rendered the plow useless on a field that in previous years would have been ready for spring planting. Unusually dry conditions combined with high, persistent winds turned this agricultural region into a present-day dust bowl.

CHAPTER OUTLINE

in New York City. The worst of the many Dust Bowl storms occurred on "Black Sunday," April 14, 1935, creating a spectacular "black blizzard."

What did these black blizzards mean? They didn't just knock down fences and bury homes in the Dust Bowl while damaging crops to the east. They removed the mineral nutrients that once sustained the grasses of the prairies. The loss of nutrient-containing topsoil in the Dust Bowl left the region scarred for years in terms of its ability to sustain profitable agriculture.

This kind of disaster continues to strike all over the world. Today, parts of sub-Saharan Africa's farmland are losing their topsoil as a result of poor land management, swelling populations, and a challenging climate. Crop failures, starvation, and large-scale human displacements are inevitable consequences.

IN THIS CHAPTER we consider the conditions that foster healthy and sustained plant growth. We identify nutrients that are essential to plants and how they acquire them. Because most plant nutrients come from the soil, we discuss the formation of soils and the effects of plants on soils. We devote a section to nitrogen metabolism in plants, and we conclude with a look at carnivorous and parasitic plants.

36.1 How Do Plants Acquire Nutrients?

Every living thing must obtain raw materials from its environment. These **nutrients** include the major ingredients of macromolecules: carbon, hydrogen, oxygen, and nitrogen. Carbon enters the living world as atmospheric carbon dioxide through the carbon-fixing reactions of photosynthesis. Hydrogen and oxygen enter plants mainly as water, so these elements are in plentiful supply.

Later in this chapter, we will focus on nitrogen, which is in relatively short supply for plants. The movement of nitrogen from the atmosphere into organisms begins with processing by some highly specialized bacteria living in the soil. Some of these bacteria act on nitrogen gas, converting it into a form usable by plants. The plants, in turn, provide organic nitrogen (and carbon) to animals, fungi, and many microorganisms.

Living organisms need certain **mineral nutrients** in addition to nitrogen, of course. The proteins of organisms contain sulfur (S), and their nucleic acids contain phosphorus (P). Chlorophyll contains magnesium (Mg), and many important compounds, such as the cytochromes, contain iron (Fe). Within the soil, these and other minerals dissolve in water, forming a solution—called the **soil solution**—that contacts the roots of plants. Plants take up most of these mineral nutrients from the soil solution in ionic form.

Autotrophs make their own organic compounds

Autotrophs are organisms that make their own *organic* (carbon-containing) compounds from simple inorganic nutrients—carbon dioxide, water, nitrogen-containing ions, and many other soluble mineral nutrients. Plants, some protists, and some bacteria are autotrophs. Plants provide carbon, oxygen, hydrogen, nitrogen, and sulfur to most of the rest of the living world. *Heterotrophs* are organisms, such as animals and fungi, that require pre-formed organic compounds as food. Heterotrophs depend directly or indirectly on autotrophs as their source of nutrition.

Most autotrophs are *photosynthesizers*—that is, they use light as their source of energy for synthesizing organic compounds from inorganic raw materials. Some autotrophs, however, are

chemolithotrophs, deriving their energy not from light but from reduced inorganic substances, such as hydrogen sulfide (H_2S), in their environment. All chemolithotrophs are bacteria. As we'll see below, some chemolithotrophic bacteria in the soil contribute to the nutrition of plants by increasing the availability of nitrogen and sulfur.

How does a stationary organism find nutrients?

Many heterotrophs can move from place to place to find the nutrients they need. An organism that cannot move, termed a *sessile* organism, must obtain nutrients and energy from sources that are somehow brought to it. Most sessile animals, such as clams and barnacles, depend primarily on the movement of water to bring them raw materials and energy in the form of food, but a plant's supply of energy arrives at the speed of light from the sun. However, with the exception of carbon and oxygen in CO_2, a plant's supply of nutrients is strictly local, and the plant may use up the water and mineral nutrients in its local environment as it develops. How does a plant cope with the problem of scarce nutrient supplies?

One way is to extend itself by growing in search of new resources. Growth is a plant's version of movement. Among plant organs, the roots obtain most of the mineral nutrients needed for growth. By growing through the soil, roots mine it for new sources of mineral nutrients and water that help leaves and stems grow. The growth of leaves helps a plant secure light and carbon dioxide, which in turn allows the roots to continue their growth in the soil. A plant may compete with other plants for light by outgrowing and shading them.

As it grows, a plant—or even a single root—must deal with a variable environment. Animal droppings create high local concentrations of nitrogen. A particle of calcium carbonate in the soil may make a tiny area alkaline, while dead organic matter may make a nearby area acidic. Such microenvironments encourage or discourage the proliferation of a root system and help direct its growth.

36.1 RECAP

Plants are autotrophs that obtain carbon by photosynthesis, and mineral nutrients and water from the soil.

- Can you distinguish between autotrophs and heterotrophs? See p. 781
- Can you explain how plants, being sessile, seek out nutrients? See p. 782

We know that plants need nutrients to support their growth. Let's look in more detail at the specific mineral nutrients they need.

36.2 What Mineral Nutrients Do Plants Require?

As roots grow through the soil, what important mineral nutrients do plants take up from their environment, and what are the roles of those nutrients? **Table 36.1** lists the mineral nutrients that have been determined to be essential for plants. Some plants require additional mineral nutrients. Except for nitrogen, they all derive from rock. All of them are usually taken up from the soil solution.

The criterion for calling something an **essential element** is that it be required for the plant to complete its life cycle. An essential element cannot be replaced by another element. In this section,

TABLE 36.1

Mineral Elements Required by Plants

ELEMENT	ABSORBED FORM	MAJOR FUNCTIONS
MACRONUTRIENTS		
Nitrogen (N)	NO_3^- and NH_4^+	In proteins, nucleic acids, etc.
Phosphorus (P)	$H_2PO_4^-$ and HPO_4^{2-}	In nucleic acids, ATP, phospholipids, etc.
Potassium (K)	K^+	Enzyme activation; water balance; ion balance; stomatal opening
Sulfur (S)	SO_4^{2-}	In proteins and coenzymes
Calcium (Ca)	Ca^{2+}	Affects the cytoskeleton, membranes, and many enzymes; second messenger
Magnesium (Mg)	Mg^{2+}	In chlorophyll; required by many enzymes; stabilizes ribosomes
MICRONUTRIENTS		
Iron (Fe)	Fe^{2+} and Fe^{3+}	In active site of many redox enzymes and electron carriers; chlorophyll synthesis
Chlorine (Cl)	Cl^-	Photosynthesis; ion balance
Manganese (Mn)	Mn^{2+}	Activation of many enzymes
Boron (B)	$B(OH)_3$	Possibly carbohydrate transport (poorly understood)
Zinc (Zn)	Zn^{2+}	Enzyme activation; auxin synthesis
Copper (Cu)	Cu^{2+}	In active site of many redox enzymes and electron carriers
Nickel (Ni)	Ni^{2+}	Activation of one enzyme
Molybdenum (Mo)	MoO_4^{2-}	Nitrate reduction

we'll consider the symptoms of particular mineral deficiencies, the roles of some of the mineral nutrients, and the technique by which the essential elements for plants were identified.

Essential elements fall roughly into two categories: *macronutrients* and *micronutrients* (see Table 36.1).

- Plants need **macronutrients** in concentrations of at least 1 gram per kilogram of their dry matter.*
- Plants need **micronutrients** in concentrations of less than 100 milligrams per kilogram of their dry matter.

These two categories differ only with regard to the amounts required by plants. Did you notice the gap between those numbers? The macronutrients are needed in much greater concentrations, but both the macronutrients and the micronutrients are essential for the plant to complete its life cycle from seed to seed. How do we know if a plant is getting enough of a particular nutrient?

Deficiency symptoms reveal inadequate nutrition

Before a plant that is deficient in an essential element dies, it usually displays characteristic **deficiency symptoms**, such as discoloration or deformation of its leaves. **Table 36.2** lists the symptoms of some common mineral deficiencies. Such symptoms help horticulturists diagnose mineral nutrient deficiencies in plants. With proper diagnosis, appropriate treatment can be applied in the form of a **fertilizer** (an added source of mineral nutrients).

Nitrogen deficiency is the most common mineral deficiency in both natural and agricultural environments. Plants in natural environments are almost always limited by nitrogen, but they seldom display deficiency symptoms. Instead, their growth slows to match the available supply of nitrogen. Crop plants, on the other hand, show deficiency symptoms if a formerly abundant supply of nitrogen becomes limiting. The visible symptoms of nitrogen deficiency include uniform yellowing of older leaves (**Figure 36.1, right**).

Dry matter, or *dry weight*, is what remains after all the water has been removed from a plant tissue sample.

TABLE 36.2

Some Mineral Deficiencies in Plants

DEFICIENCY	SYMPTOMS
Calcium	Growing points die back; young leaves are yellow and crinkly
Iron	Young leaves are white or yellow
Magnesium	Older leaves have yellow in stripes between veins
Manganese	Younger leaves are pale with green veins
Nitrogen	Oldest leaves turn yellow and die prematurely; plant is stunted
Phosphorus	Plant is dark green with purple veins and is stunted
Potassium	Older leaves have dead edges
Sulfur	Young leaves are yellow to white with yellow veins
Zinc	Young leaves are abnormally small; older leaves have many dead spots

Chlorophyll, which is responsible for the green color of leaves, contains nitrogen. Without nitrogen there is no chlorophyll, and without chlorophyll, the yellow carotenoid pigments in the leaves become visible.

Nitrogen deficiency is not the only cause of yellowing. Inadequate iron in the soil can also cause yellowing because iron, although it is not contained in the chlorophyll molecule, is required for chlorophyll synthesis. However, iron deficiency commonly causes yellowing of the *youngest* leaves (**Figure 36.1, center**). Nitrogen is readily translocated in the plant and can be redistributed from older tissues to younger tissues to favor their growth. Iron, on the other hand, cannot be readily redistributed. Younger tissues that are actively growing and synthesizing compounds needed for their growth show iron deficiency before older leaves, which have already completed their growth.

36.1 Mineral Deficiency Symptoms The plants on the left were grown with a full complement of essential nutrients. The center plants were deprived of iron, while the plants at the right are deficient in essential nitrogen.

Several essential elements fulfill multiple roles

Essential elements may play several different roles in plant cells—some structural, others catalytic. Magnesium, as we have mentioned, is a constituent of the chlorophyll molecule and hence is essential to photosynthesis. It is also required as a cofactor by numerous enzymes involved in cellular respiration and other metabolic pathways.

Phosphorus, usually in phosphate functional groups, is found in many organic compounds, particularly in nucleic acids and in the intermediates of the energy-harvesting pathways of photosynthesis and glycolysis. The transfer of phosphate groups occurs in many energy-storing and energy-releasing reactions, notably those that use or produce ATP (see Chapter 7). Phosphorylation—the addition or removal of phosphate groups—is also used to activate or inactivate enzymes (see Chapter 15).

Calcium plays many roles in plants. Its function in the processing of hormonal and environmental cues is a subject of great biological interest, as we'll see in the next chapter. Calcium also affects membranes and cytoskeletal activity, participates in spindle formation for mitosis and meiosis, and is a constituent of the middle lamella of cell walls. Other elements, such as iron and potassium, also play multiple roles in plants.

We know that *all* of these elements are essential to the life of all plants; how did biologists discover which elements are essential?

Experiments were designed to identify essential elements

An element is considered essential to plants if a plant fails to complete its life cycle or grows abnormally when that element is absent or insufficient. The essential elements for plants were identified by growing plants **hydroponically**—that is, with their roots suspended in nutrient solutions without soil (**Figure 36.2**). In the first successful experiments of this type, performed a century and a half ago, plants grew seemingly normally in solutions containing only calcium nitrate [$Ca(NO_3)_2$], magnesium sulfate [$MgSO_4$], and potassium phosphate [KH_2PO_4]. Omission of any of these compounds resulted in a solution that was incapable of supporting normal growth. Tests with other compounds including these elements soon established six macronutrients—calcium, nitrogen, magnesium, sulfur, potassium, and phosphorus—as essential elements.

Identifying essential *micronutrients* by this experimental approach proved to be more difficult. Nineteenth-century experiments on plant nutrition sometimes used chemicals so impure that they provided micronutrients that investigators thought they had excluded. Furthermore, some micronutrients are required in such tiny amounts that there may be enough of the substance in a seed to supply the embryo and the resultant second-generation plant throughout its lifetime and leave enough in the next seed to get the third generation well started. Such difficulties make it necessary to perform nutrition experiments in tightly controlled laboratories with special air filters that exclude microscopic salt particles in the air, and to use only chemicals that have been purified to the highest degree attainable by modern chemistry.

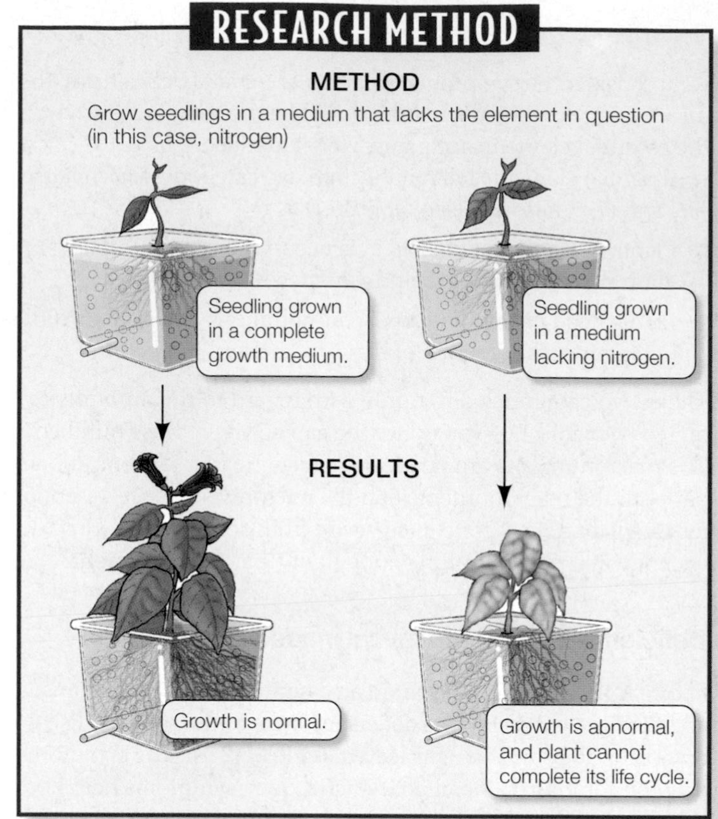

RESEARCH METHOD
METHOD
Grow seedlings in a medium that lacks the element in question (in this case, nitrogen)
Seedling grown in a complete growth medium.
Seedling grown in a medium lacking nitrogen.
RESULTS
Growth is normal.
Growth is abnormal, and plant cannot complete its life cycle.

36.2 Identifying Essential Elements for Plants This figure illustrates the classic procedure for identifying nutrients essential to plants, using nitrogen as an example.

Even simply touching a plant may give it a significant supply of chlorine, in the form of chloride ions from human sweat.

Iron was the first micronutrient to be clearly established as essential, in the 1840s. The last micronutrient to be listed as essential was nickel, in 1983 (the experiment is described in **Figure 36.3**). Only rarely are new essential elements reported now. Either the list is complete or, perhaps, we will need more sophisticated techniques to add to it.

36.2 RECAP

Mineral nutrients required by plants are classified as macronutrients and micronutrients, depending on the amount needed. Micronutrients are often needed in such minute amounts that only sophisticated chemical experiments can determine their essentiality.

- Can you describe some specific deficiency symptoms seen in plants? See p. 783 and Table 36.2
- Can you outline an experimental method for determining whether an element is essential to a plant? See Figures 36.2 and 36.3

EXPERIMENT

HYPOTHESIS: Nickel is an essential element for a plant to complete its life cycle.

METHOD

1. Grow barley plants for 3 generations in nutrient solutions containing 0, 0.6, and 1.0 μM $NiSO_4$.
2. Harvest seeds from 5–6 third-generation plants in each of the groups.
3. Determine the nickel concentration in seeds from each plant.
4. Germinate other seeds from the same plants and plot the success of germination against nickel concentration.

RESULTS

There was a positive correlation between seed germination and seed nickel concentration. There was no germination at the lowest nickel concentration.

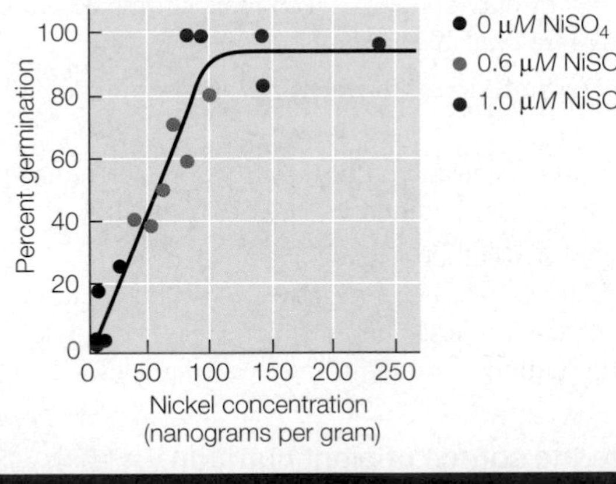

CONCLUSION: Barley seeds from nickel-free plants require nickel in order to germinate, and thereby complete the life cycle.

36.3 Is Nickel an Essential Element for Plant Growth? Using highly purified salts in the growth media, Patrick Brown and his colleagues tested whether barley can complete its life cycle in the absence of nickel. Other investigators showed that no other element could substitute for nickel.

When humans are not adding fertilizer or otherwise "feeding" a plant its mineral nutrients, where does its mineral nutrition come from? The answer lies in the minerals found in the soil, but nutrition is not the only role the soil plays.

36.3 What Are the Roles of Soil?

Most terrestrial plants live their lives anchored to the soil. Soils give growing plants mechanical support, of course, but soils and plants interact in many other noteworthy ways. Plants obtain their mineral nutrients from the soil solution. Water for terrestrial plants also comes from the soil, as does the supply of oxygen for the roots. Soils harbor bacteria, some of which are beneficial to plant life; they may also contain organisms that are harmful to plants, and some contain toxic levels of metal ions such as cadmium, chromium, and lead (see Chapter 39).

In this section, we examine the composition, structure, and formation of soils. We will consider their role in plant nutrition, their care and supplementation in agriculture, and how they are modified in turn by the plants that grow in them.

Soils are complex in structure

Soils are complex systems of living and nonliving components (**Figure 36.4**). The living components include plant roots as well as populations of bacteria, fungi, protists, and animals such as earthworms and insects. The nonliving portion of the soil includes rock fragments ranging in size from large rocks and pebbles through *sand* and *silt* and finally to tiny particles of *clay* that are 2 μm or less in diameter. Soil also contains water and dissolved mineral nutrients, air spaces, and dead organic matter. The air spaces are crucial sources of oxygen (in the form of O_2) for plant roots.

The characteristics of soils are not static. Soils change constantly through the effects of natural phenomena such as rain, extremes of temperature, and the activities of plants and animals, as well as the practices of humans—agriculture in particular.

The structure of many soils changes with depth, revealing a *soil profile*. Although soils differ greatly, almost all soils consist of sev-

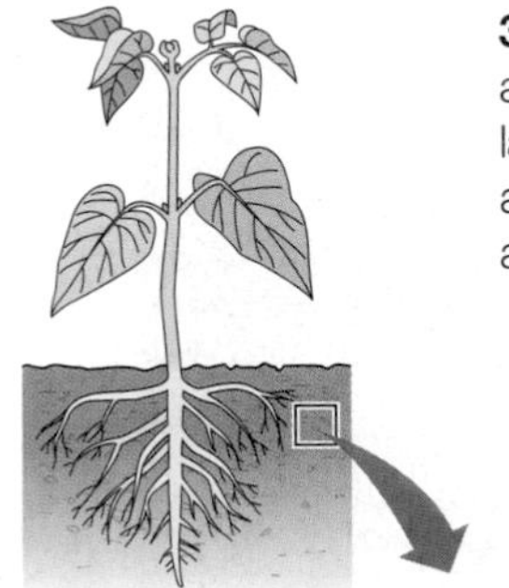

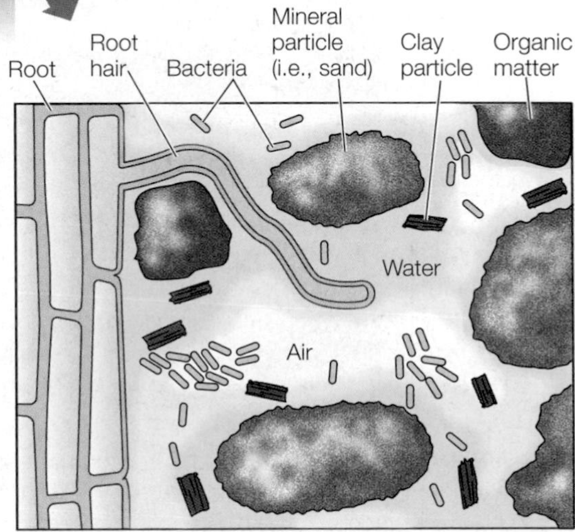

36.4 The Complexity of Soil Soils favorable for plant growth contain both clay and larger mineral particles as well as water, air, and organic matter. Other organisms are also present.

eral recognizable horizontal layers, called **horizons**, lying on top of one another. Mineral nutrients tend to be **leached** from the upper horizons—dissolved in rain or irrigation water and carried to deeper horizons, where they are unavailable to plant roots.

Soil scientists recognize three major *horizons*—termed A, B, and C—in the profile of a typical soil (**Figure 36.5**). **Topsoil** is the **A horizon**, from which mineral nutrients may be depleted by leaching. Most of the soil's dead and decaying organic matter is located in the A horizon, as are most plant roots, earthworms, insects, nematodes, and microorganisms. Successful agriculture depends on the presence of a suitable A horizon; the A horizon is what blew away from the U.S. plains states during the Dust Bowl.

Topsoils are composed of different proportions of sand, silt, and clay. Pure sand contains abundant air spaces between the relatively large particles, but binds little water and does not make mineral nutrients easily available to plant roots. Clay binds more water than sand does, but the tiny clay particles pack tightly together, leaving little space to trap air. As we will see, the charged surfaces of clay particles bind mineral element ions and make them available to plants. A little bit of clay goes a long way in affecting soil properties. A **loam** is a soil that has significant amounts of sand, silt, and clay, and thus has sufficient levels of air, water, and available nutrients for plants. Loams also contain organic matter. Most of the best topsoils for agriculture are loams.

Below the A horizon is the **B horizon**, or **subsoil**, which is the zone of infiltration and accumulation of materials leached from above. Farther down, the **C horizon** is the **parent rock**, also called bedrock, that is breaking down to form soil. Some deep-growing roots extend into the B horizon to obtain water and nutrients, but roots rarely enter the C horizon.

36.5 A Soil Profile The A, B, and C horizons can sometimes be seen in road cuts such as this one in Australia. The dark upper layer (the A horizon) is home to most of the living organisms in the soil.

Soils form through the weathering of rock

The type of soil in a given area depends on many factors, including the type of parent rock from which it formed, the climate, the landscape features, the organisms living there, and the length of time that soil-forming processes have been acting (sometimes millions of years). Both the physical and chemical properties of soils depend to a considerable extent on the amounts and kinds of **clay** particles they contain. These tiny particles, which bind mineral nutrients and aggregate into larger particles (see Figure 36.4), are extremely important to plant growth.

Rocks are broken down into soil particles in part by *mechanical weathering*, which is the physical breakdown—without any accompanying chemical changes—of materials by wetting, drying, and freezing. The most important parts of soil formation, however, include *chemical weathering*, the chemical alteration of at least some of the materials in the rocks. Several types of chemical weathering are required:

- *Oxidation* by atmospheric oxygen makes some essential elements more available to plants.
- *Hydrolysis* (reaction with water) releases some mineral nutrients.
- *Acids* (carbonic acid in particular) free some essential elements from their parent salts.

These chemical weathering reactions leave the surface of clay particles with an abundance of negatively charged chemical groups, to which certain mineral nutrients bind. Let's see how plant roots take up these mineral nutrients from clay particles.

Soils are the source of plant nutrition

The availability of mineral nutrients to plant roots depends on the presence of clay particles in the soil. The negatively charged clay particles bind the positively charged ions (cations) of many minerals that are important for plant nutrition, such as potassium (K^+), magnesium (Mg^{2+}), and calcium (Ca^{2+}). Ammonium ions (NH_4^+), a major form of nitrogen, are also bound by clay. To become available to plants, these cations must be detached from the clay particles.

Plants acquire important cations through reactions generated by protons (hydrogen ions, H^+). These protons are released into the soil by roots, which also release CO_2 through cellular respiration. The CO_2 dissolves in the soil water and reacts with it to form carbonic acid, which then ionizes to form bicarbonate and free protons:

$$CO_2 + H_2O \rightleftharpoons H_2CO_3 \rightleftharpoons H^+ + HCO_3^-$$

These protons bind more strongly to the clay particles than do the mineral cations; in essence, they trade places with the cations in a process called **ion exchange** (**Figure 36.6**). Ion exchange puts important cations back into the soil solution, from which they are taken up by the roots. The capacity of a soil to support plant growth, called *soil fertility*, is determined in part by its ability to provide nutrients in this manner.

Clay particles effectively hold and exchange cations, and cations tend to be retained in the A horizon. However, there is no compa-

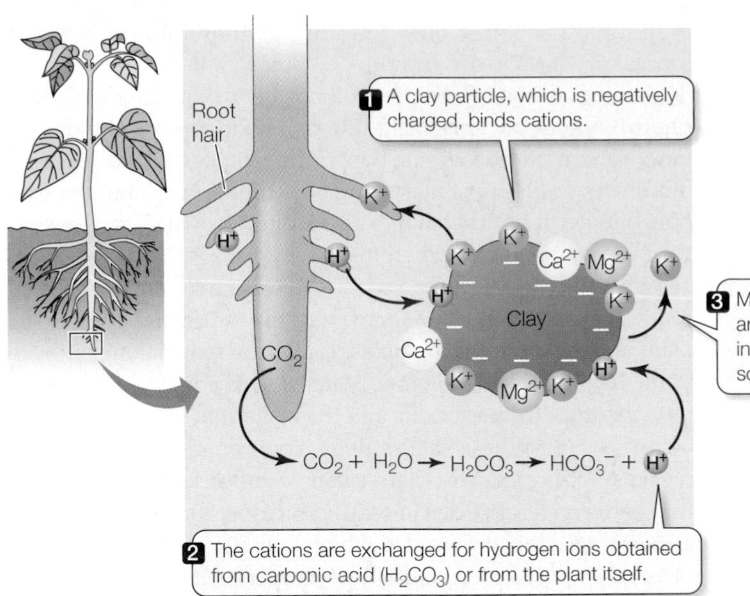
Root hair
1 A clay particle, which is negatively charged, binds cations.
Clay
K^+
H^+
Ca^{2+}
Mg^{2+}
CO_2
3 Mineral cations are released into the soil solution.
$CO_2 + H_2O \rightarrow H_2CO_3 \rightarrow HCO_3^- + H^+$
2 The cations are exchanged for hydrogen ions obtained from carbonic acid (H_2CO_3) or from the plant itself.

36.6 Ion Exchange Plants obtain mineral nutrients from the soil primarily in the form of positive ions; potassium (K^+) is the example shown here.

rable mechanism for exchanging anions, the negatively charged ions. As a result, important anions such as nitrate (NO_3^-) and sulfate (SO_4^{2-})—direct sources of nitrogen and sulfur, respectively—may leach rapidly from the A horizon. As a consequence of such leaching, the primary soil reservoir of nitrogen is not always in the form of nitrate ions. Most of the nitrogen in the A horizon is found in the organic matter in the soil, which slowly decomposes to release nitrogen as ammonium ions, a form that can be absorbed and used by plants.

Fertilizers and lime are used in agriculture

Agricultural soils often require fertilizers because irrigation and rainwater leach mineral nutrients from the soil and because the harvesting of crops removes the nutrients that the plants took up from the soil during their growth. Crop yields decrease if any essential element is depleted. Mineral nutrients may be replaced by organic fertilizers, such as rotted manure, or by inorganic fertilizers of various types.

ORGANIC AND INORGANIC FERTILIZERS The three elements most commonly added to agricultural soils are nitrogen (N), phosphorus (P), and potassium (K). Commercial fertilizers are characterized by their "N-P-K" percentages. A 5-10-10 fertilizer, for example, contains 5 percent nitrogen, 10 percent phosphate (P_2O_5), and 10 percent potash (K_2O) by weight.* Sulfur, in the form of ammonium sulfate, is also occasionally added to soils.

Either organic or inorganic fertilizers can provide the necessary mineral nutrients for plants. Organic fertilizers such as manure or crop residues release nutrients slowly, which results in less leaching than occurs with a one-time application of an inorganic fertilizer. However, the nutrients from organic fertilizers are not immediately available to plants. Organic fertilizers also contain residues of plant or animal materials that improve the structure of the soil, providing spaces for air movement, root growth, and drainage. Inorganic fertilizers, on the other hand, provide a supply of soil nutrients that is almost immediately available for absorption. Furthermore, inorganic fertilizers can be formulated to meet the requirements of a particular soil and a particular crop.

pH EFFECTS ON NUTRIENTS The availability of nutrient ions, whether they are naturally present in the soil or added as fertilizer, is altered by changes in soil pH. The optimal soil pH for most crops is about 6.5, but so-called acid-loving crops such as blueberries prefer a pH closer to 4. Rainfall and the decomposition of organic substances lower the pH of the soil. Such acidification can be reversed by **liming**—the application of compounds commonly known as *lime*, such as calcium carbonate, calcium hydroxide, or magnesium carbonate. The addition of these compounds leads to the removal of H^+ ions from the soil. Liming also increases the availability of calcium to plants.

Sometimes, on the other hand, a soil is not acidic enough for a crop. In this case, sulfur can be added in the form of elemental sulfur, which soil bacteria convert to sulfuric acid. Iron and some other elements are more available to plants at a slightly acidic pH. Soil pH testing is useful for home gardens and lawns as well as for agriculture. The test results indicate what amendments should be made to the soil.

SPRAY APPLICATION OF NUTRIENTS Spraying leaves with a nutrient solution is another effective way to deliver some essential elements to growing plants. Plants take up more copper, iron, and manganese when these elements are applied as foliar (leaf) sprays than when they are added to the soil. Such foliar application of micronutrients is sometimes used in wheat production, but fertilizers are still delivered most commonly by way of the soil.

The relationship between plants and soils is not a one-way affair—soils affect plants, but plants also affect soils.

Plants affect soil fertility and pH

The soil that forms in a particular place depends on the types of plants growing there as well as on mechanical weathering, the underlying parent rock, and other factors. Plant litter, such as dead roots and fallen leaves, is the major source of the carbon-rich materials that break down to form **humus**—dark-colored organic material, each particle of which is too small to be recognizable with the naked eye. Soil bacteria and fungi produce humus by breaking down plant litter, animal feces, dead organisms, and other organic material. Humus is rich in mineral nutrients, especially nitrogen that was excreted by animals. Humus favors plant growth by trap-

*The analysis is by weight of the nutrient-containing compound and not as weights of the elements N, P, and K. A 5-10-10 fertilizer actually does contain 5 percent nitrogen, but only 4.3 percent phosphorus and 8.3 percent potassium on an elemental basis.

ping supplies of water and oxygen for absorption by roots. Looking at the big picture, we see that successful plant growth can help create conditions that support further plant growth.

Plants also affect the pH of the soil in which they grow. Roots maintain a balance of electric charges. If they absorb more cations than anions, they excrete H^+ ions, thus lowering the soil pH. If they absorb more anions than cations, they excrete OH^- ions or HCO_3^- ions, raising the soil pH. Roots can also actively change the pH in their immediate vicinity by exuding organic acids such as citric and malic acids that acidify the soil, making it easier to take up certain ions such as Fe^{3+}.

36.3 RECAP

Land plants live anchored in the soil and obtain water and mineral nutrients from it. Plants and soil interact in many ways, and plants affect the soils in which they grow.

- Can you name types of mechanical and chemical weathering that form soil from rock? See p. 786
- Can you explain how soil fertility is enhanced by the process of ion exchange? See p. 786 and Figure 36.6

The essential mineral nutrient most commonly in short supply, in both natural and agricultural situations, is nitrogen, even though elemental nitrogen makes up almost four-fifths of Earth's atmosphere. Why is nitrogen so scarce in soil, and how do plants acquire it?

36.4 How Does Nitrogen Get from Air to Plant Cells?

The Earth's atmosphere is a vast reservoir of nitrogen in the form of nitrogen gas (N_2). However, plants cannot use N_2 directly as a nutrient. The triple bond linking the two nitrogen atoms is extremely stable, and a great deal of energy is required to break it; thus N_2 is a highly unreactive substance. How, then, do plants obtain usable nitrogen for the synthesis of proteins and nucleic acids?

A few species of bacteria have an enzyme that enables them to convert N_2 into a more reactive and biologically useful form by a process called **nitrogen fixation**. These prokaryotic organisms—*nitrogen fixers*—convert N_2 to ammonia (NH_3). Although there are relatively few species of nitrogen fixers, and their biomass is small compared to that of other organisms that depend on them, these talented prokaryotes are essential to the biosphere as we know it.

Nitrogen fixers make all other life possible

By far the greatest share of total world nitrogen fixation is performed biologically by **nitrogen-fixing bacteria**, which fix approximately 170 million metric tons of nitrogen per year. About 80 million metric tons is fixed industrially by humans. Smaller amounts of nitrogen, about 20 million metric tons per year, are fixed in the atmosphere by nonbiological means such as lightning, volcanic eruptions, and forest fires. Rain brings these atmospherically formed products to the ground.

Several groups of bacteria fix nitrogen. In the oceans, various photosynthetic bacteria, including cyanobacteria, fix nitrogen. In fresh water, cyanobacteria are the principal nitrogen fixers. On land, free-living soil bacteria make some contribution to nitrogen fixation, but they fix only what they need for their own use and release the fixed nitrogen only when they die.

Other nitrogen-fixing bacteria live in close association with plant roots. They release up to 90 percent of the nitrogen they fix to the plant and excrete some amino acids into the soil, making nitrogen immediately available to other organisms. The plant obtains fixed nitrogen from the bacterium, and the bacterium obtains the products of photosynthesis, high-energy compounds, from the plant. Such associations are excellent examples of *mutualism*, an interaction between two species in which both species benefit. They are also examples of *symbiosis*, in which two different species live in physical contact for a significant portion of their life cycles.

Bacteria of the genus ***Rhizobium*** fix nitrogen in close, mutualistic association with the roots of plants in the legume family. The legumes include peas, soybeans, clover, alfalfa, and many tropical shrubs and trees. The bacteria infect the plant's roots, and the roots develop nodules in response to their presence (**Figure 36.7**). (How nodules develop is further described below.) The various species of *Rhizobium* show a high specificity for the species of legume they infect. Farmers and gardeners coat legume seeds with *Rhizobium* to make sure the bacteria are present. Some farmers alternate their crops, planting clover or alfalfa occasionally to increase the available nitrogen content of the soil.

The legume–*Rhizobium* association is not the only bacterial association that fixes nitrogen. Some cyanobacteria fix nitrogen in association with fungi in lichens or with ferns, cycads, or nonvascular plants. Rice farmers can increase crop yields by growing the water fern *Azolla*, with its symbiotic nitrogen-fixing cyanobacterium, in the flooded fields where rice is grown. Another group of bacteria, the filamentous actinobacteria, fix nitrogen in associ-

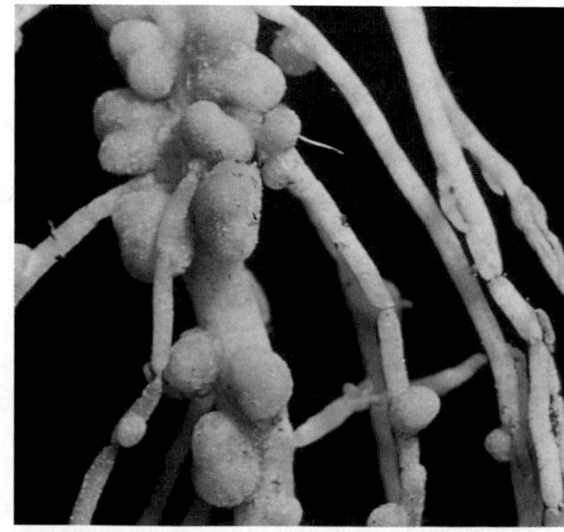

36.7 Root Nodules Large, round nodules are visible in the root system of a pea plant. These nodules house nitrogen-fixing bacteria.

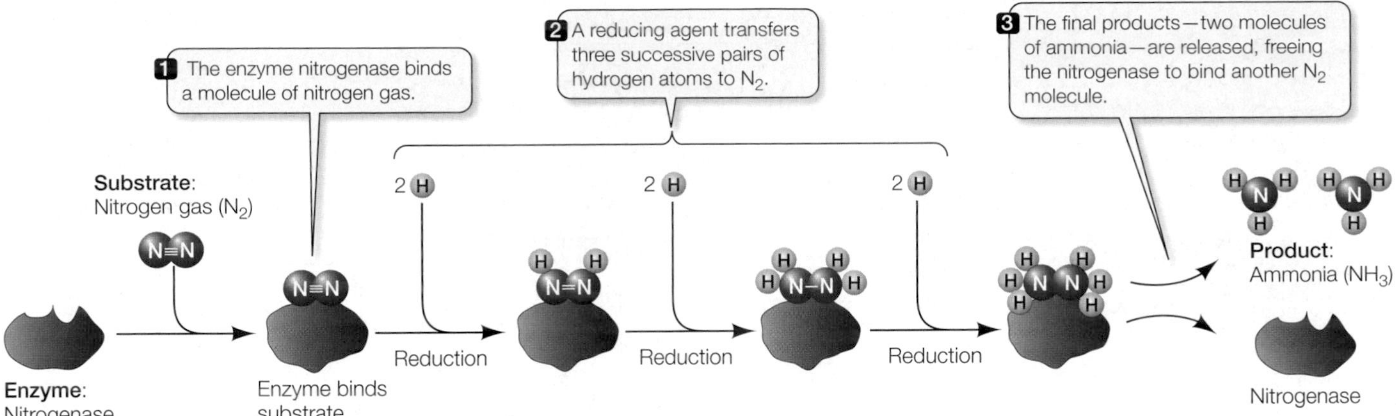
1 The enzyme nitrogenase binds a molecule of nitrogen gas.
2 A reducing agent transfers three successive pairs of hydrogen atoms to N_2.
3 The final products—two molecules of ammonia—are released, freeing the nitrogenase to bind another N_2 molecule.
Substrate: Nitrogen gas (N_2)
Enzyme: Nitrogenase
Enzyme binds substrate
2 H
Reduction
Reduction
Reduction
Product: Ammonia (NH_3)
Nitrogenase

36.8 Nitrogenase Fixes Nitrogen Throughout the chemical reactions of nitrogen fixation, the reactants are bound to the enzyme nitrogenase. A reducing agent transfers hydrogen atoms to nitrogen, and eventually the final product—ammonia—is released.

ation with root nodules on woody species such as alder and mountain lilacs.

How does biological nitrogen fixation work? In the four sections that follow, we'll consider the role of the enzyme *nitrogenase*, the mutualistic collaboration of plant and bacterial cells in **root nodules**, the need to supplement biological nitrogen fixation in agriculture, and the contributions of plants and bacteria to the global nitrogen cycle.

Nitrogenase catalyzes nitrogen fixation

Nitrogen fixation is the *reduction* of nitrogen gas (see Section 7.1). It proceeds by the stepwise addition of three pairs of hydrogen atoms to N_2 (**Figure 36.8**). In addition to N_2, these reactions require three things:

- A strong reducing agent to transfer hydrogen atoms to N_2 and to the intermediate products of the reaction
- A great deal of energy, which is supplied by ATP
- The enzyme **nitrogenase**, which catalyzes the reaction

Depending on the species of nitrogen fixer, either respiration or photosynthesis may provide both the necessary reducing agent and ATP.

Nitrogenase is so strongly inhibited by oxygen that its presence in biochemical extracts was obscured and its discovery delayed because investigators had not thought to seek it under anaerobic conditions. It is therefore not surprising that many nitrogen fixers are anaerobes and live in environments with little or no O_2. Because this crucial enzyme is so inhibited by O_2, it was at first surprising that the reaction occurs in legumes, which respire aerobically, as do *Rhizobium*. Investigation of the root nodules where nitrogenase is found revealed how the enzyme could operate there.

Within a root nodule, O_2 is maintained at a low level sufficient to support respiration, but not so high as to inactivate nitrogenase. The plant makes this possible by producing the protein **leghemoglobin** in the cytoplasm of the nodule cells. Leghemoglobin is a close relative of hemoglobin, the red, oxygen-carrying pigment of animals. Some plant nodules contain enough of it to be bright pink when viewed in cross section. Leghemoglobin, with its iron-containing heme groups, transports enough oxygen to the *bacteroids* to support their respiration.

Some plants and bacteria work together to fix nitrogen

Neither free-living *Rhizobium* species nor uninfected legumes can fix nitrogen. Only when the two are closely associated in root nodules does the reaction take place. The establishment of this symbiosis between *Rhizobium* and a legume requires a complex series of steps, with active contributions by both the bacteria and the plant root (**Figure 36.9**). First the root releases flavonoids and other chemical signals that attract soil-living *Rhizobium* to the vicinity of the root. Flavonoids trigger the transcription of bacterial *nod* genes, which encode Nod (nodulation) factors. These factors, secreted by the bacteria, cause cells in the root cortex to divide, leading to the formation of a primary nodule meristem. This meristem gives rise to the plant tissue that constitutes the nodule.

Among the products of the nodule meristem is a layer of cells that excludes O_2 from the interior of the nodule. The function of leghemoglobin is to carry O_2 across this barrier. Within a nodule, the bacteria take the form of **bacteroids** within membranous vesicles. Bacteroids are swollen, deformed bacteria that can fix nitrogen—in effect, nitrogen-fixing structures.

The partnership between bacterium and plant in nitrogen-fixing nodules is not the only case in which plants depend on other organisms for assistance with their nutrition. Another example is that of **mycorrhizae**, root–fungus associations in which the fungus greatly increases the absorption of water and minerals (especially phosphorus) by the plant (see Figure 30.12). A growing body of evidence suggests that nodule formation depends on some of the same genes and mechanisms that allow mycorrhizae to develop.

Biological nitrogen fixation does not always meet agricultural needs

Bacterial nitrogen fixation is not sufficient to support the needs of agriculture. Traditional farmers used to plant dead fish along with corn; the decaying fish released fixed nitrogen that the developing corn could use. Industrial nitrogen fixation is becoming ever more important to world agriculture because of the need to feed a rapidly expanding population.

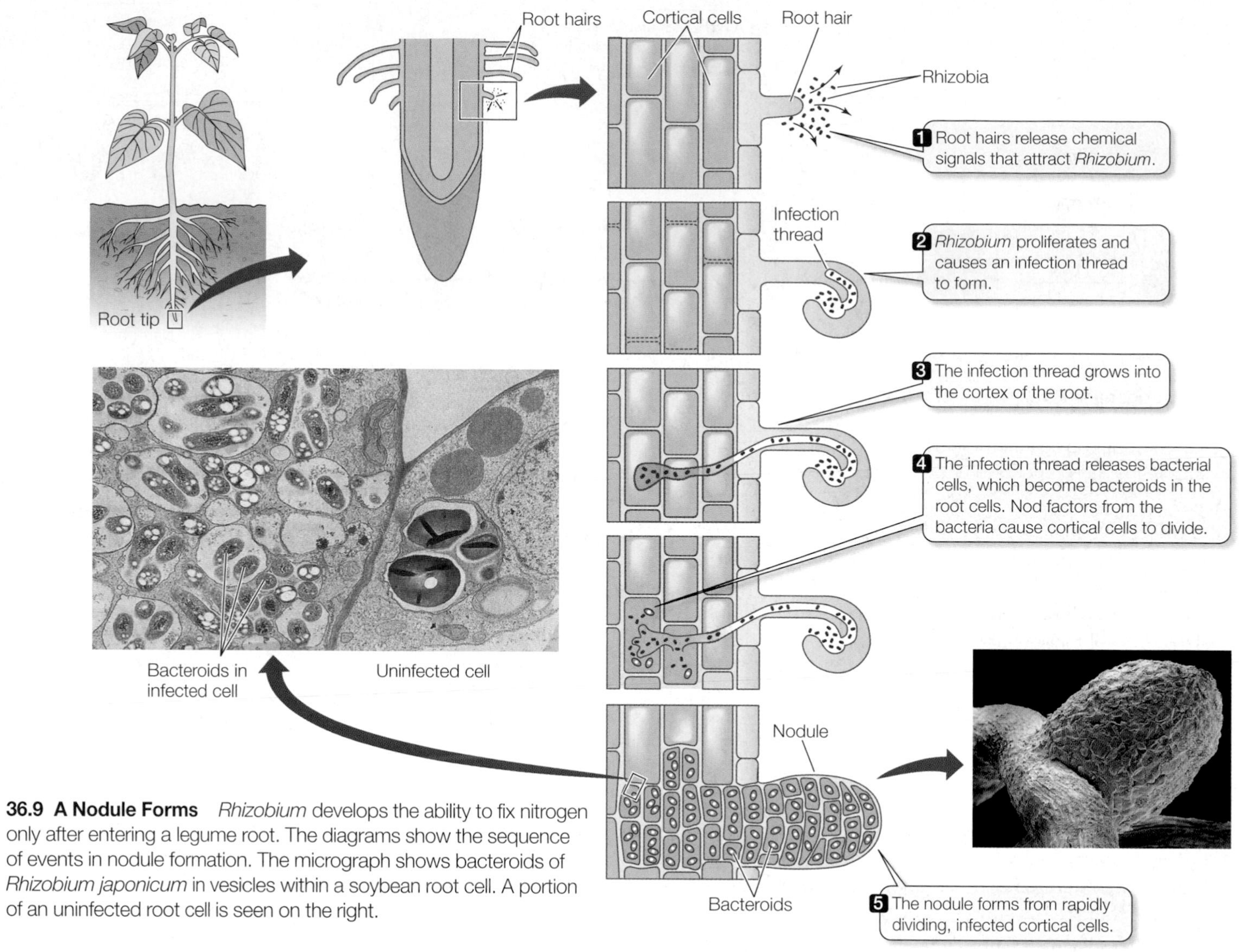

Root hairs
Cortical cells
Root hair
Rhizobia
1 Root hairs release chemical signals that attract *Rhizobium*.
Infection thread
2 *Rhizobium* proliferates and causes an infection thread to form.
3 The infection thread grows into the cortex of the root.
4 The infection thread releases bacterial cells, which become bacteroids in the root cells. Nod factors from the bacteria cause cortical cells to divide.
Root tip
Bacteroids in infected cell
Uninfected cell
Nodule
Bacteroids
5 The nodule forms from rapidly dividing, infected cortical cells.

36.9 A Nodule Forms *Rhizobium* develops the ability to fix nitrogen only after entering a legume root. The diagrams show the sequence of events in nodule formation. The micrograph shows bacteroids of *Rhizobium japonicum* in vesicles within a soybean root cell. A portion of an uninfected root cell is seen on the right.

North Korean agriculture was able to feed that country's population until soon after the 1989 collapse of the Soviet Union, which had provided North Korea with chemicals and petroleum. This loss of support was followed by three years of drought, hailstorms, and floods. Today, North Korea cannot produce the fertilizer it needs and is a starving country with a failed farming system.

Most industrial nitrogen fixation is done by a chemical process called the *Haber process*, which requires a great deal of energy. An alternative is urgently needed because of the rising cost of fossil fuel-generated energy. At present, in the United States, the manufacture of nitrogen-containing fertilizer takes more energy than does any other aspect of crop production. The primary energy sources for industrial production of fertilizer are natural gas and hydroelectric power. In biological systems, nitrogen fixation requires a great deal of ATP—about 16 to 20 ATP per N fixed.

Research on biological nitrogen fixation is being pursued, with commercial applications in mind. One line of investigation centers on recombinant DNA technology as a means of engineering new plant–bacterium associations that produce their own nitrogenase. Currently attempts are underway to transfer genes from *Rhizobium* into bacteria that already live in the roots of important cereal plants such as rice. So far the attempts have been unsuccessful.

Plants and bacteria participate in the global nitrogen cycle

Essential nitrogen cycles through the biosphere in a complex **global nitrogen cycle** that we outline in **Figure 36.10**. The nitrogen released into the soil by nitrogen fixers is primarily in the form of ammonia (NH_3) and ammonium ions (NH_4^+). Although ammonia can be toxic to plants if it accumulates in tissues, ammonium ions can be taken up safely at low concentrations. Soil bacteria called **nitrifiers** oxidize ammonia to nitrate ions (NO_3^-)—another

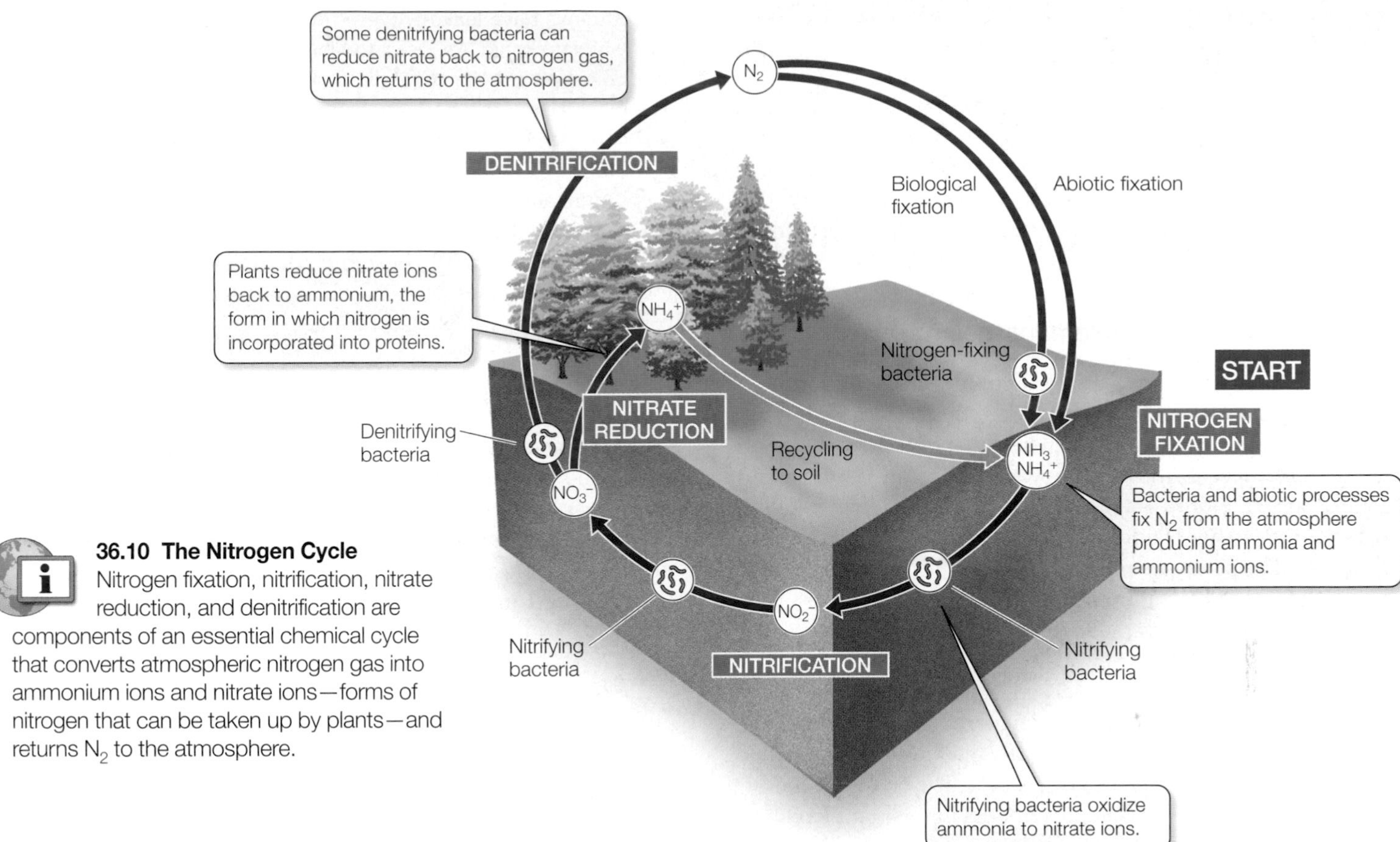

Some denitrifying bacteria can reduce nitrate back to nitrogen gas, which returns to the atmosphere.
N_2
DENITRIFICATION
Biological fixation
Abiotic fixation
Plants reduce nitrate ions back to ammonium, the form in which nitrogen is incorporated into proteins.
NH_4^+
Nitrogen-fixing bacteria
START
NITRATE REDUCTION
Denitrifying bacteria
Recycling to soil
NH_3 NH_4^+
NITROGEN FIXATION
NO_3^-
Bacteria and abiotic processes fix N_2 from the atmosphere producing ammonia and ammonium ions.
NO_2^-
Nitrifying bacteria
NITRIFICATION
Nitrifying bacteria
Nitrifying bacteria oxidize ammonia to nitrate ions.

36.10 The Nitrogen Cycle Nitrogen fixation, nitrification, nitrate reduction, and denitrification are components of an essential chemical cycle that converts atmospheric nitrogen gas into ammonium ions and nitrate ions—forms of nitrogen that can be taken up by plants—and returns N_2 to the atmosphere.

form that plants can take up—by the process of **nitrification**. Soil pH affects the uptake of nitrogen: Nitrate ions are taken up preferentially under more acidic conditions, ammonium ions under more basic ones.

These two initial processes are carried out by bacteria: they *reduce* N_2 to ammonia in nitrogen fixation (see Figure 36.8) and *oxidize* ammonia to nitrate in nitrification. The next steps are carried out by plants, which reduce the nitrate they have taken up all the way back to ammonia. All the reactions of **nitrate reduction** are carried out by the plant's own enzymes. The later steps, from nitrite (NO_2^-) to ammonia, take place in the chloroplasts. The plant uses the ammonia thus formed to manufacture amino acids, from which the plant's proteins and all its other nitrogen-containing compounds are formed. Animals cannot reduce nitrogen, and they depend on plants to supply them with reduced nitrogenous compounds.

Bacteria called **denitrifiers** return nitrogen from soil nitrate to the atmosphere as N_2. This process is called **denitrification**. In combination with leaching and the removal of crops, denitrification keeps the level of available nitrogen in soils low.

Thus, the cycle of nitrogen through the biosphere includes four key steps:

1. *fixation* of atmospheric N_2 to NH_3 and NH_4^+ by bacteria and by abiotic processes;
2. *nitrification* of these molecules to nitrate by bacteria;
3. *nitrate reduction* by plants;
4. *denitrification* of nitrate by bacteria back to N_2, which is then released to the atmosphere to begin another cycle.

The nitrogen cycle is essential for life on Earth: nitrogen-containing compounds constitute 5–30 percent of a plant's total dry weight. The nitrogen content of animals is even higher, and all the nitrogen in the animal world arrives there by way of the plant kingdom. Other elements, sulfur for example, also undergo biogeochemical cycling.

36.4 RECAP

Bacteria in soils and root nodules convert nitrogen from an inert gas into forms that plants can use. Nitrogen-fixing bacteria are essential to the biosphere because all of the nitrogen that animals need comes from plants. Denitrification returns nitrogen from dead organisms and animal waste back to the atmosphere, continuing the global nitrogen cycle.

- To reduce nitrogen gas to a form plants can use requires the enzyme nitrogenase and what else? See p. 789 and Figure 36.8
- Can you explain how a root nodule forms on a legume? See p. 789 and Figure 36.9

Let's conclude by looking at some unusual plant species that have evolved special ways to obtain the nutrients they need.

36.5 Do Soil, Air, and Sunlight Meet the Needs of All Plants?

Most plants obtain their mineral nutrients from the soil solution, which is also their source of water. Atmospheric CO_2 is their source of carbon atoms, and sunlight meets their energy needs. Some plants must look to other sources for some or all of these needed commodities. Carnivorous and parasitic plants are examples of plants with such special needs.

Carnivorous plants supplement their mineral nutrition

Some plants that are found primarily in nitrogen-deficient soils augment their nitrogen and phosphorus supply by capturing and digesting flies and other insects. There are about 450 of these **carnivorous plant** species, the best-known of which are Venus flytraps (genus *Dionaea*; **Figure 36.11A**), sundews (genus *Drosera*; **Figure 36.11B**), and pitcher plants (genus *Sarracenia*).

Carnivorous plants are normally found in boggy regions where the soils are extremely nutrient deficient. The carnivorous plants have evolved adaptations that allow them to augment their supply of nitrogen by capturing animals and digesting their proteins.

The Venus flytraps have specialized leaves with two halves that fold together. When an insect trips trigger hairs on a leaf, its two halves quickly come together, their spiny margins interlocking and imprisoning the insect before it can escape. The closing of the flytrap's leaf is one of the fastest movements in the plant world—it requires only one tenth of a second, during which the leaf halves reverse their curvature. The leaf then secretes enzymes that digest its prey. The leaf absorbs the products of digestion, especially amino acids, and uses them as a nutritional supplement.

Pitcher plants produce pitcher-shaped leaves that collect small amounts of rainwater. Insects are attracted into the pitchers either by bright colors or by scent and are prevented from getting out again by stiff, downward-pointing hairs. The insects eventually die and are digested by a combination of enzymes and bacteria in the water. Even rats have been found in large pitcher plants.

Sundews have leaves covered with hairs that secrete a clear, sticky, sugary liquid. An insect touching one of these hairs becomes stuck, and more hairs curve over the insect and stick to it as well. The plant secretes enzymes to digest the insect and eventually absorbs the carbon- and nitrogen-containing products of digestion.

None of the carnivorous plants *must* feed on insects; they can grow quite adequately without insects, but in their natural habitats they grow faster and are a darker green when they succeed in capturing insects. They use the additional nitrogen from the insects to make more proteins, chlorophyll, and other nitrogen-containing compounds.

Parasitic plants take advantage of other plants

Some **parasitic plants** derive their mineral nutrients from the living bodies of other plants. Most of these parasites are autotrophs, but a few plant species have, in the course of their evolution, lost the ability to sustain themselves by photosynthesis. To meet their needs for energy and carbon, these heterotrophs parasitize other, photosynthesizing plants.

Albino mutant plants cannot produce chlorophyll or photosynthesize, and normally die at an early seedling stage. Some 65 years ago, H. A. Spohr grew albino corn seedlings by "feeding" them sucrose solution. The plants remained white but increased in dry weight, produced the same number of leaves as normal plants, and even flowered—they just couldn't produce their own sugar.

Perhaps the most familiar parasitic plants are the several genera of mistletoes and dodders (**Figure 36.12**). Mistletoes are green and carry on some photosynthesis, but they parasitize other plants for water and mineral nutrients and may derive photosynthetic products from them as well. Mistletoes and dodders extract nutrients from the vascular tissues of their hosts by forming absorptive organs called *haustoria*, which invade the host plant's tissues.

Another parasitic plant, the Indian pipe (*Monotropa uniflora*), once was thought to obtain its nutrients from dead organic matter. It is now known to get its nutrients, with the help of fungi, from nearby actively photosynthesizing plants. Hence it, too, is a parasite.

Dwarf mistletoe (*Arceuthobium americanum*) is a serious parasite in forests of the western United States, destroying more than

(A) *Dionaea muscipula*

(B) *Drosera rotundifolia*

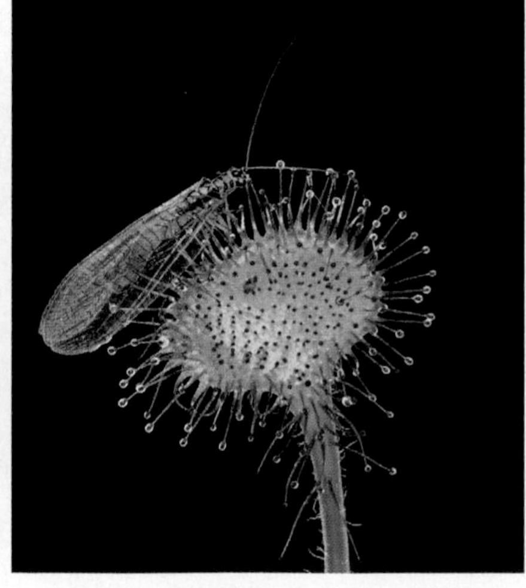

36.11 Carnivorous Plants Some plants have adapted to nitrogen-poor environments by becoming carnivorous. (A) The Venus flytrap obtains nitrogen and phosphorus from the bodies of insects trapped inside the plant when its hinges snap shut. (B) Sundews trap insects on sticky hairs. Secreted enzymes will digest the carcass externally.

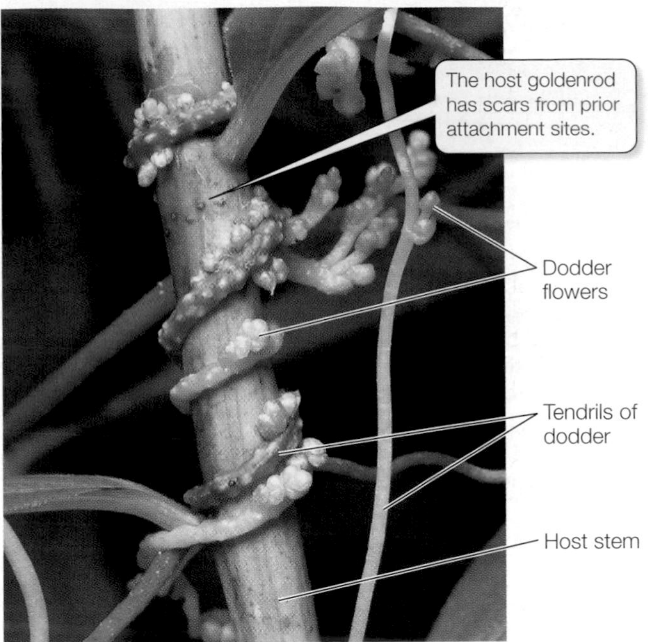

36.12 A Parasitic Plant Tendrils of dodder (genus *Cuscuta*) wrap around a goldenrod (genus *Solidago*). The parasitic dodder obtains water, sugars, and other nutrients through tiny, rootlike protuberances that penetrate the surface of the host plant.

3 billion board feet of lumber per year. Parasitic plants are a much more urgent problem in developing countries. Witchweed (*Striga*) imperils more than 300 million sub-Saharan Africans by attacking their cereal and legume crops. In the Middle East and North Africa, broomrape (*Orobanche*) ravages many crops, especially vegetables and sunflowers.

36.5 RECAP

Carnivorous and parasitic plants supplement their nutrition by extracting materials from animals or other plants.

- How do the needs of heterotrophic parasitic plants differ from those of carnivorous plants? See p. 792

Bad land management led to the Dust Bowl and to current problems in Africa and elsewhere. Invasive parasitic plants can wreak havoc even if the soil contains adequate nutrients. The world's peoples depend on a sufficient nutrient supply to sustain the crops that feed humankind.

CHAPTER SUMMARY

36.1 How do plants acquire nutrients?

Plants are photosynthetic **autotrophs** that can produce all their organic molecules from carbon dioxide, water, and minerals, including a nitrogen source.

Mineral nutrients are obtained from the **soil solution**.

Chemolithotrophic bacteria in the soil increase the availability of nitrogen and sulfur to plants.

Root growth allows plants, which are sessile, to search for mineral resources.

Microenvironments within the soil such as acidic or alkaline areas affect the direction of root growth.

36.2 What mineral nutrients do plants require?

Plants require 14 **essential mineral elements**. Of these, six are **macronutrients** and eight are **micronutrients**. **Deficiency symptoms** suggest what essential element a plant lacks. Review Table 36.1

The requirement for each essential element was discovered by growing plants on **hydroponic solutions** lacking that element. Review Figure 36.2, Web/CD Tutorial 36.1

36.3 What are the roles of soil?

Soils contain water, air, and inorganic and organic substances. Soils have living and nonliving components. Review Figure 36.4

A soil typically consists of two or three horizontal zones called **horizons**. **Topsoil** forms the uppermost or A horizon. Topsoil tends to lose mineral nutrients through **leaching**. **Loams** are excellent agricultural topsoils, with a good balance of sand, silt, clay, and organic matter.

Soils form by mechanical and chemical **weathering** of rock. Chemical weathering imparts mineral nutrients to **clay** particles. Plant litter decomposes to form **humus**. Plants obtain some mineral nutrients through ion exchange between the soil solution and the surface of clay particles. Review Figure 36.6

Farmers use **fertilizers** to make up for deficiencies in soil mineral nutrient content. **Liming** can reverse acidification.

36.4 How does nitrogen get from air to plant cells?

Some **nitrogen-fixing bacteria** live free in the soil; others live symbiotically as **bacteroids** within plant roots. In **nitrogen fixation**, nitrogen gas (N_2) is reduced to ammonia (NH_3) or ammonium ions (NH_4^+) in a reaction catalyzed by **nitrogenase**. Review Figure 36.8

Nitrogenase requires anaerobic conditions, but the bacteroids in **root nodules** require oxygen, which is maintained at the proper level by **leghemoglobin**.

The formation of a root nodule requires interaction between the root system of a legume and ***Rhizobium***. Review Figure 36.9

Mycorrhizae are root-fungus associations that greatly increase a plant's absorption of water and minerals.

Plants and bacteria interact in the **global nitrogen cycle**, which involves series of reductions and oxidations of nitrogen-containing molecules. Review Figure 36.10, Web/CD Activity 36.1

Nitrification by bacteria converts ammonia to nitrate ions in the soil. **Nitrate reduction** is carried out by the plant's own enzymes, enabling plants to form their own nitrogen compounds. **Denitrification** returns nitrogen from animal wastes and dead organisms to the atmosphere.

36.5 Do soil, air, and sunlight meet the needs of all plants?

Carnivorous plants are autotrophs that supplement a low nitrogen supply by feeding on insects. **Parasitic plants** draw on other plants to meet their needs, which may include minerals, water, or the products of photosynthesis.

SELF-QUIZ

1. Macronutrients
   a. are so called because they are more essential than micronutrients.
   b. include manganese, boron, and zinc, among others.
   c. function as catalysts.
   d. are required in concentrations of at least 1 gram per kilogram of plant dry matter.
   e. are obtained by the process of photosynthesis.
2. Which of the following is *not* an essential mineral element for plants?
   a. Potassium
   b. Magnesium
   c. Calcium
   d. Lead
   e. Phosphorus
3. Fertilizers
   a. are often characterized by their N-P-O percentages.
   b. are not required if crops are removed frequently enough.
   c. restore needed mineral nutrients to the soil.
   d. are needed to provide carbon, hydrogen, and oxygen to plants.
   e. are needed to destroy soil pests.
4. In a typical soil,
   a. the topsoil tends to lose mineral nutrients by leaching.
   b. there are four or more horizons.
   c. the C horizon consists primarily of loam.
   d. the dead and decaying organic matter gathers in the B horizon.
   e. more clay means more air space and thus more oxygen for roots.
5. Which of the following is *not* an important step in soil formation?
   a. Removal of bacteria
   b. Mechanical weathering
   c. Chemical weathering
   d. Clay formation
   e. Hydrolysis of soil minerals
6. Nitrogen fixation is
   a. performed only by plants.
   b. the oxidation of nitrogen gas.
   c. catalyzed by the enzyme nitrogenase.
   d. a single-step chemical reaction.
   e. possible because N_2 is a highly reactive substance.
7. Nitrification is
   a. performed only by plants.
   b. the reduction of ammonium ions to nitrate ions.
   c. the reduction of nitrate ions to nitrogen gas.
   d. catalyzed by the enzyme nitrogenase.
   e. performed by certain bacteria in the soil.
8. Nitrate reduction
   a. is performed by plants.
   b. takes place in mitochondria.
   c. is catalyzed by the enzyme nitrogenase.
   d. includes the reduction of nitrite ions to nitrate ions.
   e. is known as the Haber process.
9. Which of the following is a parasite?
   a. Venus flytrap
   b. Pitcher plant
   c. Sundew
   d. Dodder
   e. Tobacco
10. All carnivorous plants
    a. are parasites.
    b. depend on animals as a source of carbon.
    c. are incapable of photosynthesis.
    d. depend on animals as their sole source of phosphorus.
    e. obtain supplemental nitrogen from animals.

FOR DISCUSSION

1. Methods for determining whether a particular element is essential have been known for more than a century. Since these methods are so well established, why was the essentiality of some elements discovered only recently?
2. If a Venus flytrap were deprived of soil sulfates and hence made unable to synthesize the amino acids cysteine and methionine, would it die from lack of protein? Explain.
3. Soils are dynamic systems. What changes might result when land is subjected to heavy irrigation for agriculture after being relatively dry for many years? What changes in the soil might result when a virgin deciduous forest is cut down and replaced by crops that are harvested each year?
4. We mentioned that important positively charged ions are held in the soil by clay particles, but other, equally important, negatively charged ions are leached deeper into the soil's B horizon. Why doesn't leaching cause an electrical imbalance in the soil? (Hint: Think of the ionization of water.)
5. The biosphere of Earth as we know it depends on the existence of a few species of nitrogen-fixing prokaryotes. What do you think might happen if one of these species were to become extinct? If all of them were to disappear?

FOR INVESTIGATION

Brown and coworkers established nickel as an essential element for plant growth, although nickel deficiency has never been observed in nature. What sort of evidence could lead you to suspect that some other element, such as rubidium, was an essential element? (It isn't, by the way.) What sorts of experiments might you perform to test your suspicion?

CHAPTER 37 Regulation of Plant Growth

More rubber, please

Within weeks of attacking Pearl Harbor on December 7, 1941, Japanese forces invaded Malaysia and the Dutch East Indies, thus taking control of most of the world's rubber supply. The governments of Canada and the United States reacted swiftly to what was truly a crisis, since rubber was crucial both to the war effort and to the domestic economy. Rubber tires were the first commodity to be rationed in the United States. The main purpose of gas rationing and a national 35-miles-per-hour speed limit was to conserve tires.

Although rubber trees (*Hevea brasiliensis*) are native to South America, British colonialists established plantations of these magnificent trees in Southeast Asia during the late nineteenth century, and that region continues to be the source of 90 percent of the world supply of natural rubber. A North American shrub, guayule (*Parthenium argentatum*), was used commercially to a limited extent before World War II, but its rubber yield was far from sufficient to ease the shortage.

During the course of the war, scientists developed synthetic rubber, which helped address the problem. Taking a different approach, James Bonner and other plant biologists worked to increase the supply of natural rubber. Bonner and his colleagues launched an urgent study of guayule, but determined that this plant could not become a major source of rubber. Bonner turned his attention back to *H. brasiliensis*, but could do little work with that species until the war ended and access to the great rubber plantations was again possible.

Eventually Bonner became chairman of the Agricultural Science and Biology Subcommittee of the Malaysian Rubber Research and Development Board. Among his many accomplishments was the discovery of a method to speed the collection of rubber-containing latex from *H. brasiliensis*.

To collect latex, workers make a V-shaped cut in the bark of the rubber tree and insert a spout at the bottom of the V. Latex flows from the cut surface into a cup attached to the tree. However, the latex at the cut surfaces coagulates within 1–3 hours and the flow stops, making it necessary to cut frequently. Bonner and his colleagues discovered that coagulation slowed dramatically, and far more latex could be collected, if the region of the cut was exposed to the gas

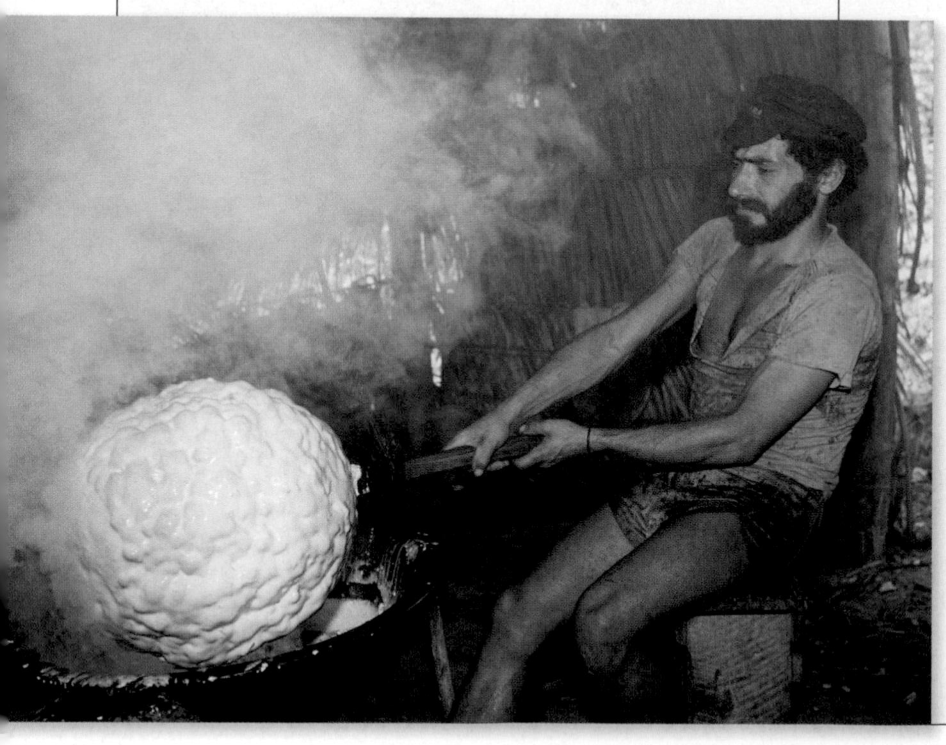

Rubber Balls Make the World Go Around *Hevea brasiliensis* trees are native to Brazil, where it was discovered that boiling and otherwise processing the milky liquid—latex—that flowed from their cut bark resulted in resilient rubber. Today the worldwide demand for rubber is at an all-time high.

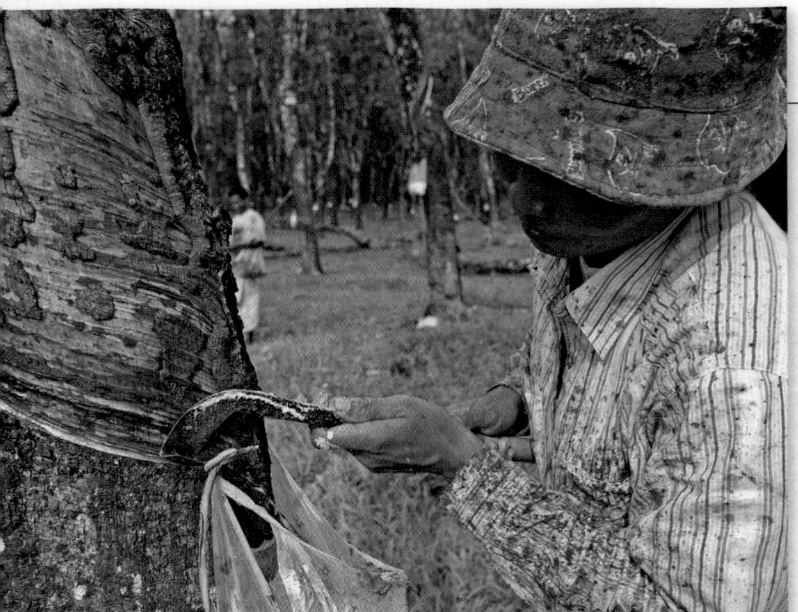

More of a Valuable Substance Latex from *H. brasiliensis* is the source of natural rubber. Workers make V-shaped slashes in the trees from which latex flows; treating the slashes with ethylene keeps latex flowing longer.

ethylene, a natural plant product with dramatic effects on physiology.

Treating cut bark with ethylene, along with improved fertilizers and better control of rubber-tree diseases, ultimately led to a doubling of the world's supply of natural rubber. Even though synthetic substitutes surpassed the production of natural rubber by the late 1950s, consumption of rubber products is so great that today the need for natural rubber is at an all-time high.

In the meantime, biologists have determined that ethylene is an important naturally occurring plant growth hormone. It is but one of the many factors described in this chapter that collaborate to regulate a plant's growth throughout its life cycle, from seed to senescence.

IN THIS CHAPTER we will give a brief overview of the life of a flowering plant and its developmental stages. We will explore the nature of the environmental cues, photoreceptors, and hormones (including ethylene) that regulate plant growth and development. We will also consider the multiple roles and interactions of these different elements.

CHAPTER OUTLINE

37.1 How Does Plant Development Proceed?

The *development* of a plant—the series of progressive changes that take place throughout its life—is regulated in complex ways. Four factors are involved in regulating plant growth:

- *Environmental cues* to which a plant responds
- *Receptors* that allow a plant to sense environmental cues, such as *photoreceptors,* molecules that absorb light
- *Hormones,* chemical signals that mediate the effects of the environmental cues including those sensed by receptors
- The plant's *genome,* which encodes enzymes that catalyze the biochemical reactions of development

Many recent advances in understanding plant growth and development have come from work with *Arabidopsis thaliana,* a weed in the mustard family. This plant is used as a *model* organism by researchers because its body and seeds are tiny, its genome is unusually small for a flowering plant, and it flowers and forms many seeds soon after growth begins. Its genome is fully sequenced, so researchers have an accounting of all genes in the plant. Genes can be inserted or deleted. *Arabidopsis* mutants with altered developmental patterns provide evidence for the existence of hormones and for the mechanisms of hormone and photoreceptor action.

Several hormones and photoreceptors play roles in plant growth regulation

Hormones are regulatory compounds that act at very low concentrations at sites often distant from where they are produced. Unlike animals, which produce each hormone in a specific part of the body, plants produce hormones in many cell types. Each plant hormone plays multiple regulatory roles, affecting several different aspects of plant development (**Table 37.1**). Interactions among these hormones are often complex.

Like hormones, **photoreceptors** are involved in many developmental processes in plants. Unlike plant hormones, which are small molecules, plant photoreceptors are *pigments* (molecules that absorb light) associated with proteins. Light (an en-

TABLE 37.1

Plant Growth Hormones

HORMONE	TYPICAL ACTIVITIES
Abscisic acid	Maintains seed dormancy and winter dormancy; closes stomata
Auxins	Promote stem elongation, adventitious root initiation, and fruit growth; inhibit axillary bud outgrowth and leaf abscission
Brassinosteroids	Promote stem and pollen tube elongation; promote vascular tissue differentiation
Cytokinins	Inhibit leaf senescence; promote cell division and axillary bud outgrowth; affect root growth
Ethylene	Promotes fruit ripening and leaf abscission; inhibits stem elongation and gravitropism
Gibberellins	Promote seed germination, stem growth, and fruit development; break winter dormancy; mobilize nutrient reserves in grass seeds

vironmental cue) acts directly on photoreceptors, which in turn regulate the processes of development, such as the many changes accompanying the growth of a young seedling emerging from the soil and into the light.

No matter what cues regulate development, ultimately the plant's genome determines the limits of plant development. The genome encodes the master plan, but its interpretation depends on conditions in the environment. It is also the target for some hormone actions. For several decades hormones and photoreceptors were the focus of most work on plant development, but recent advances in molecular genetics have allowed us to focus on the underlying processes, such as signal transduction pathways.

Signal transduction pathways are involved in all stages of plant development

Plants, like other organisms, make extensive use of *signal transduction pathways*, sequences of biochemical reactions by which a cell generates a response to a stimulus (Chapter 15). Cell signaling in plant development involves a receptor (for a hormone or for light) and a signal transduction pathway, and concludes with a cellular response. Protein kinase cascades amplify responses to signals in plants, as they do in other organisms (see Figure 15.10).

The details of plant cell signaling are best understood in the context of the general pattern of plant development. The factors just described affect plants through their entire developmental history by acting on three fundamental processes: cell division, cell expansion, and cell differentiation.

The seed germinates and forms a growing seedling

If all developmental activity is suspended in a seed, even when conditions appear to be suitable for its growth, the seed is said to be **dormant**. Cells in dormant seeds do not divide, expand, or differentiate. For the embryo to begin developing, seed dormancy must be broken by one of the mechanisms discussed later in this section.

As the seed begins to **germinate**—to develop into a seedling—it takes up water. The growing embryo then obtains chemical building blocks—monomers—for its development by digesting the polysaccharides, fats, and proteins stored in the seed. The embryos of some plant species secrete hormones that direct the mobilization of these reserves. Germination is completed when the **radicle** (embryonic root) emerges from the seed coat. The plant is then called a **seedling**.

If the seed germinates underground, the new seedling must elongate rapidly (in the right direction!) and cope with a period of life in darkness or dim light. A series of photoreceptors directs this stage of development and prepares the seedling for growth in the light environment.

Early shoot development varies among the flowering plants. **Figure 37.1** shows the shoot development patterns of monocots and eudicots.

Plant growth from seedling to adult is regulated by several hormones. Other hormones are involved in the plant's defenses against herbivores and microorganisms (discussed in Chapter 39).

The plant flowers and sets fruit

Flowering—the formation of reproductive organs—may be initiated when the plant reaches an appropriate age or size. Some plant species, however, flower at particular times of the year, meaning that the plant must be capable of distinguishing different times of the year. In these plants, the leaves measure the length of the night (shorter in summer, longer in winter) with great precision. Light absorption by photoreceptors is the first step in this time-measuring process.

Once the leaves have determined that it is time for the plant to flower, that information must be transmitted as a signal to the places where flowers will form. It was proposed more than 70 years ago that this signal is transmitted from the leaf to the site of flower formation in the form of a "flowering hormone," then termed *florigen*, common to many plants. In spite of intensive research, it was not until 2005 that biologists identified specific compounds that influenced the transition to flowering (see Section 38.2).

After flowers form, hormones play further roles in reproduction. Hormones and other substances control the growth of the pollen tube that brings sperm and egg together (see Figure 38.1). Following fertilization, a fruit develops and ripens under hormonal control.

The plant senesces and dies

Some plants, such as iris and elm, are **perennials**: they continue to grow year after year. **Annuals**, such as petunia and marigold, complete their life cycle in a single year, then **senesce** (deteriorate as a result of aging) and die.

The death of the entire plant, which may be triggered by signals from the environment, follows senescent changes that are controlled by hormones such as ethylene. This life history pattern appears to be an adaptation for producing more offspring by shifting nutrients from nonreproductive tissues into the seeds; in so doing, the parent plant essentially starves itself to death, ensuring that sufficient nutrients are available for seed maturation.

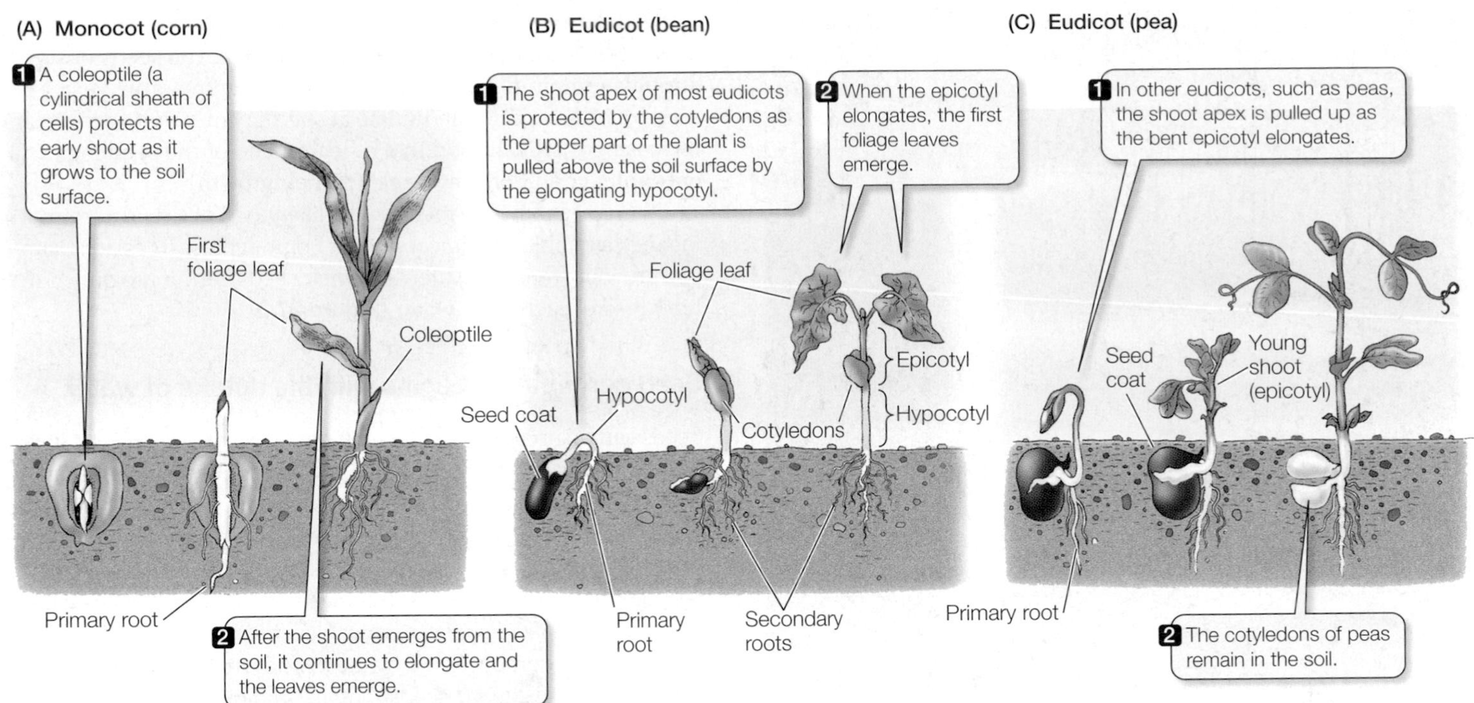

(A) Monocot (corn)
1 A coleoptile (a cylindrical sheath of cells) protects the early shoot as it grows to the soil surface.
First foliage leaf
Coleoptile
Primary root
2 After the shoot emerges from the soil, it continues to elongate and the leaves emerge.
(B) Eudicot (bean)
1 The shoot apex of most eudicots is protected by the cotyledons as the upper part of the plant is pulled above the soil surface by the elongating hypocotyl.
2 When the epicotyl elongates, the first foliage leaves emerge.
Foliage leaf
Seed coat
Hypocotyl
Epicotyl
Hypocotyl
Cotyledons
Primary root
Secondary roots
(C) Eudicot (pea)
1 In other eudicots, such as peas, the shoot apex is pulled up as the bent epicotyl elongates.
Seed coat
Young shoot (epicotyl)
Primary root
2 The cotyledons of peas remain in the soil.

37.1 Patterns of Early Shoot Development (A) In grasses and some other monocots, growing shoots are protected by a coleoptile until they reach the soil surface. (B) In most eudicots, the growing point of the shoot is protected by the cotyledons. (C) In some other eudicots, the cotyledons remain in the soil, and the growing point is protected by the first true leaves.

In many perennials, leaves senesce and fall at the end of the growing season, shortly before the onset of winter. Leaf fall is regulated by an interplay of the hormones ethylene and auxin.

Having described the steps in a plant's life history, let's examine how these various steps are regulated. We'll begin at the start of the life history, with the seed and its germination.

Not all seeds germinate without cues

The seeds of some plant species are capable of germinating as soon as they have matured. All they need for germination is water. But the seeds of many species are dormant at maturity. Seed dormancy may last for weeks, months, years, or even centuries. The mechanisms that maintain seed dormancy are numerous and diverse, but three principal strategies dominate:

- Exclusion of water or oxygen from the embryo by means of an impermeable seed coat
- Mechanical restraint of the embryo by means of a tough seed coat
- Chemical inhibition of embryonic development

Seed dormancy must be broken before germination can begin. The dormancy of seeds with impermeable coats can be broken if the seed coat is abraded as the seed tumbles across the ground or through a creek bed or passes through the digestive tract of an animal. Cycles of freezing and thawing can also aid in making the seed coat permeable. Soil microorganisms probably play a major role in softening seed coats. Fire is a major force for ending seed dormancy (**Figure 37.2**). Fire can melt waterproofing wax in seed coats, allowing water to reach the embryo, and it can release mechanical restraint, cracking the seed coat. Fire can also break down chemical inhibitors of germination. *Leaching*—the dissolving out of water-soluble chemical inhibitors by prolonged exposure to water—is another way in which dormancy can be broken.

Seed dormancy affords adaptive advantages

What are the potential advantages of seed dormancy? For many plant species, dormancy ensures survival through unfavorable conditions and results in germination when conditions are more favorable for growth. To avoid germination in the dry days of late summer, for example, some seeds require exposure to a long cold period before they will germinate. Other seeds will not germinate until a certain amount of time has passed, regardless of how they are treated. This strategy prevents germination while the seeds are still attached to the parent plant. Seeds that must be scorched by fire in order to germinate avoid competition with other plants by germinating only where an area has been cleared by fire. Dormancy also helps seeds to survive long-distance dispersal, allowing plants to colonize new territory.

The dormancy of some seeds is broken by exposure to light. These seeds, which germinate only at or near the surface of the soil, are generally tiny seeds with few food reserves. Such seeds would be incapable of surviving if germination occurred while they were buried deeply. Conversely, the germination of some other seeds is inhibited by light; these seeds germinate only when buried and thus kept in darkness. Light-inhibited seeds are usually large and well stocked with nutrients.

37.2 Fire and Seed Germination This fireweed germinated and flourished after a great fire along the Alaska Highway.

Seed dormancy helps annual plants cope with year-to-year variation in the amount and frequency of rainfall. The seeds of some annuals remain dormant throughout an unfavorable year. The seeds of other species germinate at specific times during the year, increasing the likelihood that at least some of the seedlings will encounter conditions favorable for their growth.

Dormancy may also increase the likelihood of a seed germinating in a favorable ecological setting. Some cypress trees, for example, grow in standing water, and their seeds germinate only if inhibitors are leached by water (**Figure 37.3**).

Seed germination begins with the uptake of water

Seeds can begin to germinate after dormancy is broken and environmental conditions are satisfactory. The first step in germination is the uptake of water, called **imbibition** (from *imbibe,* "to drink"). Typically, only 5 to 15 percent of a seed's weight is water, whereas most other plant parts contain 80 to 95 percent water. A seed's water potential is very negative (see Section 35.1), and water will be taken up if the seed coat is permeable. The magnitude of this water potential is demonstrated by the force exerted by seeds expanding in water. Cocklebur seeds that are imbibing can exert a pressure of up to 1,000 atmospheres (about 100 megapascals) against a restraining force.

As a seed takes up water, it undergoes metabolic changes: Enzymes are activated upon hydration, RNA and then proteins are synthesized, the rate of cellular respiration increases, and other metabolic pathways are activated. In many seeds, no DNA synthesis and no cell division occur during the early stages of germination. Initially, growth results solely from the expansion of small, preformed cells. DNA is synthesized only after the radicle begins to grow and ruptures the seed coat.

The embryo must mobilize its reserves

To fuel these metabolic activities, the embryo must use the reserves of energy and raw materials stored in the seed. Until the young plant is able to photosynthesize, it depends on these reserves, which are stored in the **cotyledons** (see Figure 37.1) or in the **endosperm** (the specialized nutritive tissue) of the seed. The principal reserve of energy and carbon in many seeds is starch. Other seeds store fats or oils. Usually, the endosperm holds amino acid reserves in the form of proteins, rather than as free amino acids.

The giant molecules of starch, lipids, and proteins must be broken down by enzymes into monomers that can enter the cells of the embryo.

37.3 Leaching of Germination Inhibitors The seeds of bald cypress, a tree adapted to moist or wet environments, germinate only after being leached by water, which increases the chances that they will germinate in a location suitable for their growth.

The polymer starch yields glucose for energy metabolism and for the synthesis of cellulose and other cell wall constituents. The digestion of stored proteins provides the amino acids the embryo needs to synthesize its own proteins. Lipids are broken down into glycerol and fatty acids, both of which can be metabolized for energy. Glycerol and fatty acids can also be converted to glucose, which permits fat-storing plants to make all the building blocks they need for growth.

37.1 RECAP

Environmental cues, receptors, hormones, and the genome interact to regulate all stages of plant growth and development. Stages of development include dormancy, germination, seedling growth, flowering, senescence, and death.

- Can you name some circumstances under which seed dormancy is advantageous? See p. 799
- On what can a young plant embryo rely for energy and nutrients before it is able to commence photosynthesis? See pp. 800–801

What signal initiates the breakdown of stored reserves in the seed, releasing energy and building blocks for growth of the new seedling?

37.2 What Do Gibberellins Do?

In germinating barley and other cereal seeds, the embryo secretes **gibberellins**, one of several classes of plant growth hormones. Gibberellins diffuse through the endosperm to a surrounding tissue called the **aleurone layer**, which lies underneath the seed coat. The gibberellins trigger a cascade of events in the aleurone layer, causing it to synthesize and secrete enzymes that digest proteins and starch stored in the endosperm (**Figure 37.4**).

Commercially, gibberellins are used in the brewing industry to enhance the "malting" (germination) of barley and the breakdown of its endosperm, producing sugar that is fermented to alcohol.

Gibberellins play a variety of roles in plant development in addition to triggering the mobilization of seed reserves. We'll begin our discussion of plant growth hormones by describing the discovery of the gibberellins, as well as their many effects.

"Foolish seedling" disease led to the discovery of the gibberellins

The gibberellins are a large family of closely related compounds. Although not steroids themselves, the gibberellins belong to the same broad class of compounds—the *terpenes*—as do rubber, the steroids, and another plant hormone we'll meet later in this chapter. Some gibberellins are found in plants and others in a pathogenic (disease-causing) fungus in which they were first discovered.

In 1809, the study of the gibberellins began indirectly with observations of the *bakanae*, or "foolish seedling," disease of rice. Seedlings affected by this disease grow more rapidly than their healthy neighbors, but this rapid growth gives rise to tall, spindly plants that die before producing seeds (the rice grains used for food). The disease has had a significant impact on rice yields in several parts of the world. It is caused by the ascomycete fungus *Gibberella fujikuroi*.

In 1925, the Japanese biologist Eiichi Kurosawa grew *G. fujikuroi* on a liquid medium and then separated the fungus from the medium by filtration. He heated the filtered medium to kill any re-

37.4 Embryos Mobilize Their Reserves During seed germination in cereal grasses, gibberellins trigger a cascade of events that results in the conversion of starch and protein reserves into monomers that can be used by the developing embryo.

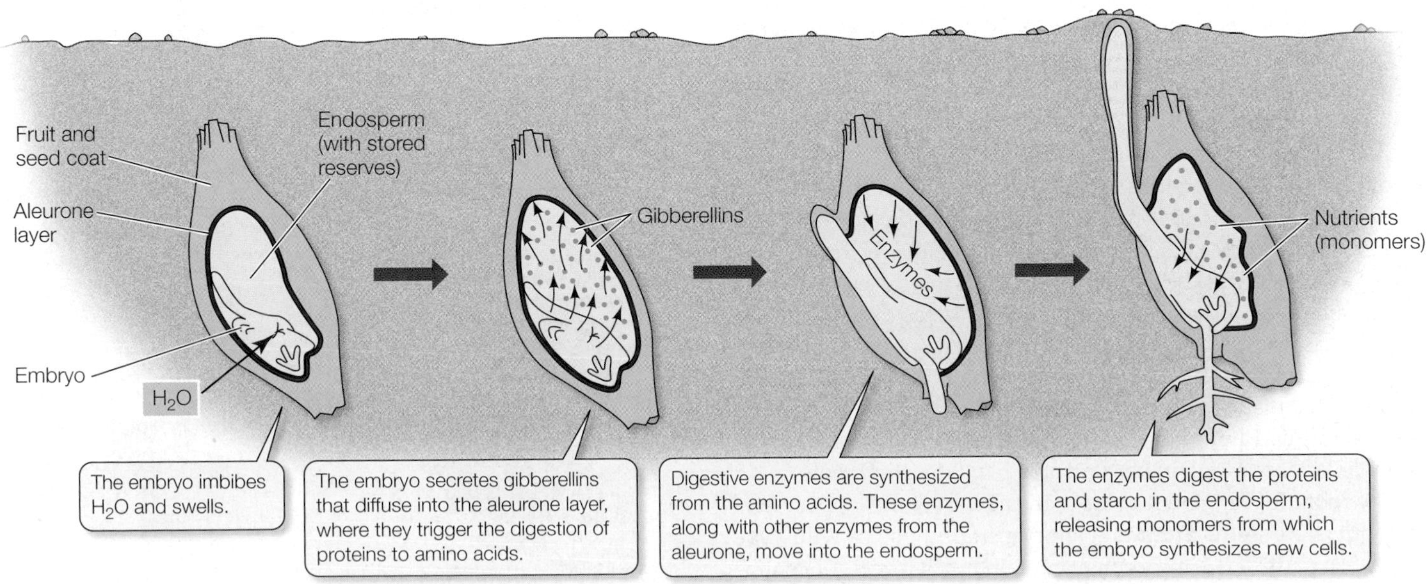

Fruit and seed coat
Endosperm (with stored reserves)
Aleurone layer
Embryo
H_2O
Gibberellins
Enzymes
Nutrients (monomers)
The embryo imbibes H_2O and swells.
The embryo secretes gibberellins that diffuse into the aleurone layer, where they trigger the digestion of proteins to amino acids.
Digestive enzymes are synthesized from the amino acids. These enzymes, along with other enzymes from the aleurone, move into the endosperm.
The enzymes digest the proteins and starch in the endosperm, releasing monomers from which the embryo synthesizes new cells.

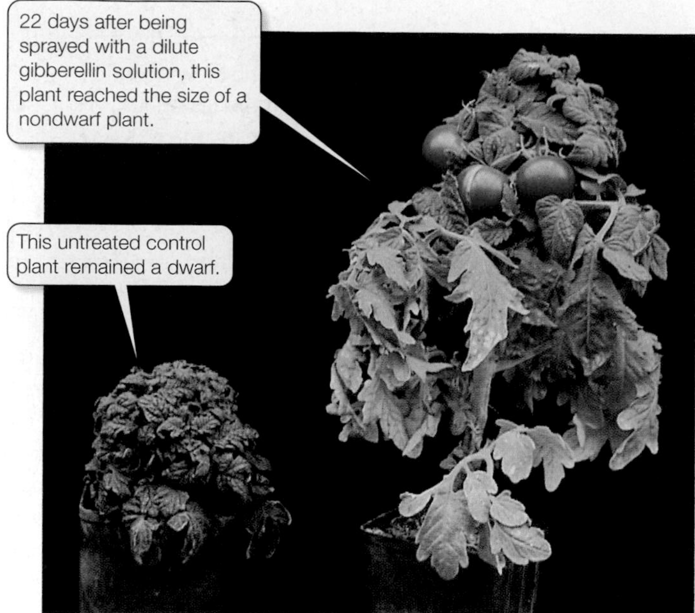

22 days after being sprayed with a dilute gibberellin solution, this plant reached the size of a nondwarf plant.
This untreated control plant remained a dwarf.

37.5 The Effect of Gibberellins on Dwarf Plants Both of the dwarf tomato plants in this photograph were the same size when the one on the right was treated with gibberellins.

maining fungus, but found that the resulting heat-treated filtrate was still capable of inducing rapid growth in rice seedlings. Medium that had never contained the fungus did not have this effect. This experiment established that *G. fujikuroi* produces a growth-promoting chemical substance, which Kurosawa called a gibberellin.

Were the gibberellins simply exotic products of an obscure fungus, or did they play a more general role in plant growth? Bernard O. Phinney of the University of California, Los Angeles, answered this question in part in 1956, when he reported the spectacular growth-promoting effect of gibberellins on dwarf corn seedlings. He used plants that were known to be genetic dwarfs, in which a particular recessive allele (say, *d1*) was present in the homozygous condition (*d1d1*). Gibberellins applied to nondwarf—wild-type—corn seedlings had almost no effect, whereas dwarf seedlings treated with gibberellins grew as tall as their normal relatives. (A comparable effect of gibberellins applied to a dwarf tomato plant is shown in **Figure 37.5**.)

Phinney drew two conclusions from the results of this experiment: first, that gibberellins are normal constituents of corn, and perhaps of all plants, and second, that some dwarf plants are short because they produce insufficient amounts of gibberellins. According to Phinney's hypothesis, nondwarf plants manufacture enough gibberellins to promote their full growth, whereas dwarf plants do not. Extracts from nondwarf plants of numerous species were found to promote growth in dwarf corn. These findings provided direct evidence that plants that are not genetic dwarfs contain gibberellin-like substances. Phinney's work set the stage for today's use of mutant plants to investigate the control of plant development.

The roots, leaves, and flowers of dwarf corn plants appear normal, but their stems are much shorter than those of wild-type plants. All parts of the dwarf plant contain much lower concentrations of gibberellins than do the organs of a wild-type plant. We can infer, then, that normal stem elongation *requires* gibberellins or the products of gibberellin action. We can further infer that gibberellins play a less essential role in the development of roots, leaves, and flowers.

The gibberellins have many effects on plant growth and development

Gibberellins and other hormones regulate the growth of fruits. It has long been known that grapevines that produce seedless grapes develop smaller fruit than varieties that produce seed-bearing grapes. Experimental removal of seeds from immature seeded grapes prevented normal fruit growth, suggesting that the seeds are sources of a growth regulator. It was then shown that spraying young seedless grapes with a gibberellin solution caused them to grow as large as seeded ones. It is now standard commercial procedure to spray seedless grapes with gibberellins. Biochemical studies showed that the developing seeds produce gibberellins, which diffuse out into the immature fruit tissue.

A different type of gibberellin effect is seen in some **biennials**—plants that grow vegetatively in their first year and flower in their second year and then die. Some biennial plants respond dramatically to an increase in the level of gibberellins. In their second year, the apical meristems of these biennials respond to environmen-

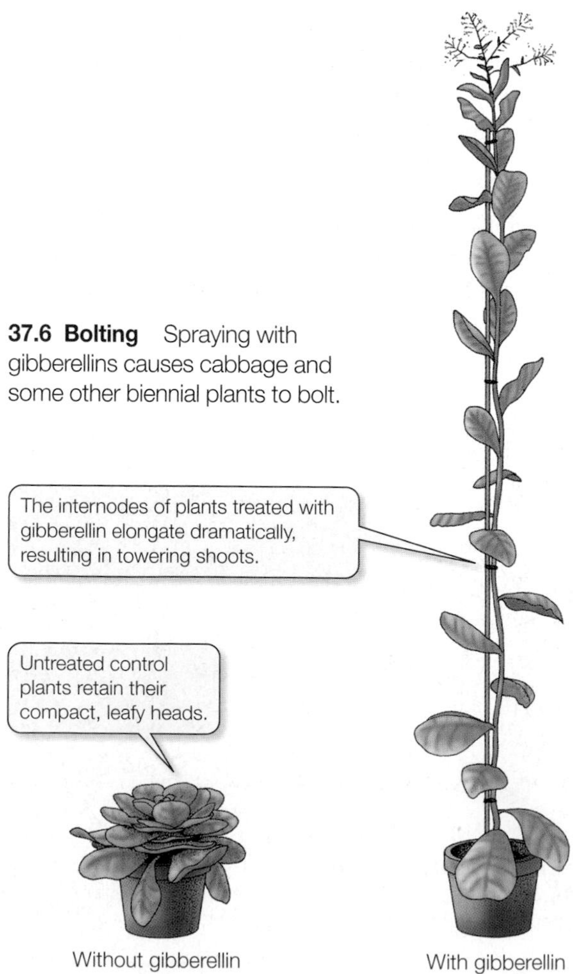

37.6 Bolting Spraying with gibberellins causes cabbage and some other biennial plants to bolt.

tal cues by producing elongated shoots, which eventually bear flowers. This rapid shoot elongation is called **bolting**. When the plant senses the appropriate environmental cue—longer days or a sufficient winter chilling—it produces more gibberellins, raising the gibberellin concentration to a level that causes the shoot to bolt. Some biennial species will bolt when sprayed with a gibberellin solution without exposure to any environmental cue (**Figure 37.6**).

Gibberellins have other important effects. They also cause fruit to grow from unfertilized flowers, promote seed germination, and help bring spring buds out of winter dormancy.

37.2 RECAP

Gibberellins are plant hormones that affect stem growth, fruit size, seed germination, and many other aspects of plant development; the effects vary from species to species.

- Can you explain how gibberellins contribute to the germination of barley seeds? See Figure 37.4
- Why is it believed that gibberellins are more important to the growth of stems than of roots and leaves? See p. 802

Most other hormones, like the gibberellins, have multiple effects within the plant, and they often interact with one another to regulate developmental processes. In controlling stem elongation, for example, gibberellins interact with another hormone, auxin.

37.3 What Does Auxin Do?

If you pinch off the apical bud at the top of a bean plant, inactive axillary buds become active and develop into lateral branches. Similarly, pruning a shrub stimulates the formation of new branches. If you cut off the blade of a leaf but leave its petiole (stalk) attached to the plant, the petiole drops off sooner than it would have if the leaf were intact. If a plant is kept indoors, its shoots will grow toward a window. These diverse responses of shoots are all mediated by plant hormones called **auxins**, of which the most important is *indoleacetic acid* (IAA), a close chemical relative of the amino acid tryptophan.

In this section we will look at the discovery of auxin, its transport within the plant, and its role as a mediator of the effects of light and gravity on plant growth. We'll discover its many effects on vegetative growth and on fruit development. Then we'll examine its mechanism of action.

Phototropism led to the discovery of auxin

The discovery of auxin and its numerous physiological effects on plants can be traced back to work done in the 1880s by Charles Darwin and his son Francis. The Darwins were interested in plant movements. One type of growth movement they studied was **phototropism**, the growth of plant organs toward light (as in most shoots) or away from it (as in roots). They asked, What part of the plant senses the light?

To answer this question, the Darwins worked with canary grass (*Phalaris canariensis*) seedlings grown in the dark. A young grass seedling has a **coleoptile**—a cylindrical sheath a few cells thick that protects the delicate shoot as it pushes through the soil (see Figure 37.1A). When the seedling breaks through the soil surface, the coleoptile soon stops growing, and the shoot emerges unharmed. The coleoptiles of grasses are phototropic—they grow toward the light.

To find the light-receptive region of the coleoptile, the Darwins tried "blindfolding" the coleoptiles of dark-grown canary grass seedlings in various places, then illuminating them from one side (**Figure 37.7**). The coleoptile grew toward the light whenever its tip was exposed. If the top millimeter or more of the coleoptile was covered, however, it showed no phototropic response. Thus, the Darwins were able to conclude that the tip contains the photoreceptor that responds to light. The actual bending toward the light, however, takes place in a growing region a few millimeters below the tip. Therefore, the Darwins reasoned, some type of signal must travel from the tip of the coleoptile to the growing region. Later, others demonstrated that this signal is a chemical substance by showing that it can move through certain permeable materials, such as gelatin, but not through impermeable materials, such as a metal sheet.

Further experiments showed that the tip of the coleoptile produces a hormone that moves down the coleoptile to the growing region, and that this hormone causes cells to grow faster. First, if the tip is removed, the growth of the coleoptile is sharply inhibited. If the tip is carefully replaced, growth resumes—even if the tip and base are separated by a thin layer of gelatin. Furthermore, the hormone moves down from the tip, but it does not move from one side of the coleoptile to the other. If the tip is cut off and moved so that it rests on only one side of the cut end of the coleoptile, the coleoptile curves as the cells on the side below the replaced tip grow more rapidly than those on the other side.

In the 1920s, the Dutch botanist Frits W. Went followed up on the Darwins' experiment. He removed coleoptile tips and placed their cut surfaces on a block of agar. Then he placed pieces of that agar on decapitated coleoptiles—positioned to cover only one side, just as coleoptile tips had been placed in earlier experiments (**Figure 37.8**). As they grew, the coleoptiles curved away from the side with the agar. This curvature demonstrated that a hormone had indeed diffused into the agar block from the isolated coleoptile tips. Went had at last isolated a hormone from a plant. Later chemical analysis showed that this hormone, named auxin, was indoleacetic acid.

Auxin transport is polar and requires carrier proteins

Early experiments showed that the movement of auxin through certain plant tissues is strictly *polar*—that is, it is unidirectional along a line from apex to base. By inverting plants and plant parts, scientists determined that the apex-to-base direction of auxin movement has nothing to do with gravity; the polarity of this movement is a totally biological phenomenon.

Auxin transport is completely or partially polar in many plant parts. In most leaf petioles, for example, auxin moves only from

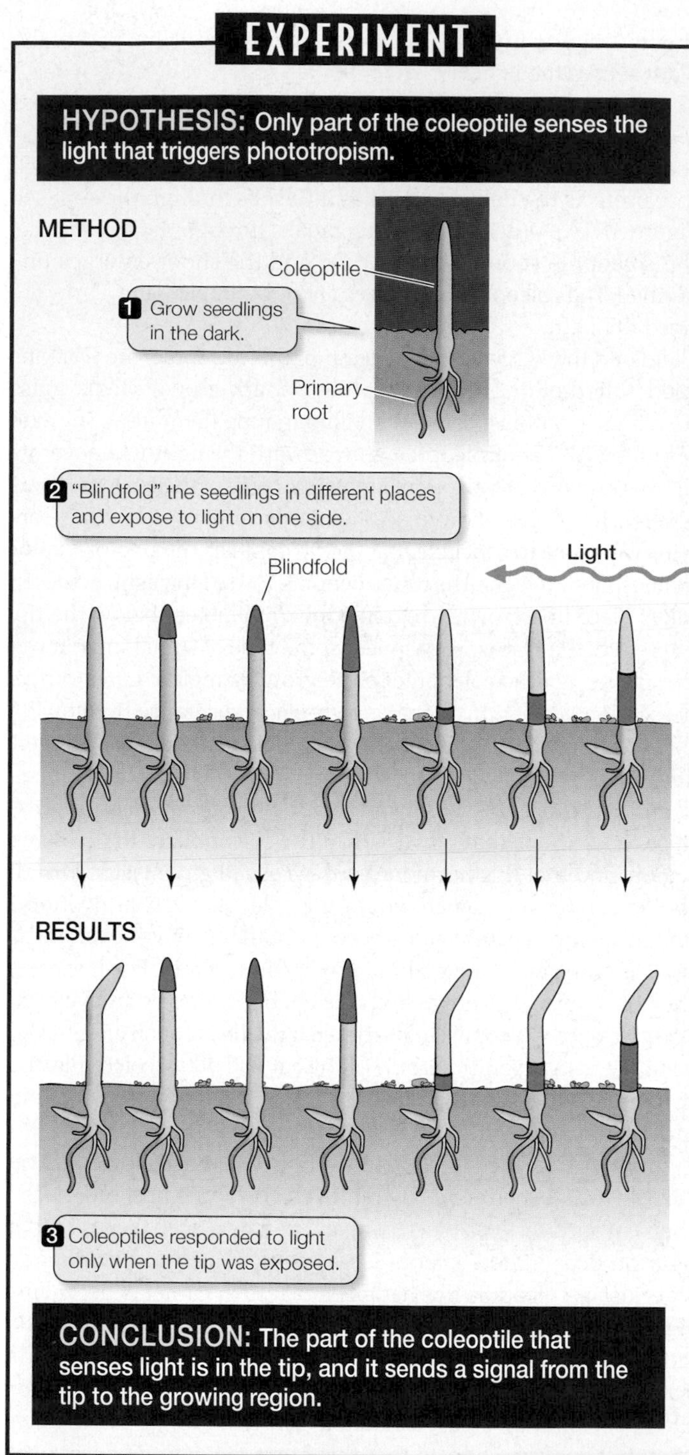

37.7 The Darwins' Phototropism Experiment The series of drawings at the center show some of the ways in which seedlings grown in the dark were "blindfolded"; the drawings below them show the results the Darwins observed in each case. Their observations led them to hypothesize the existence of a growth-promoting signal produced by the coleoptile.

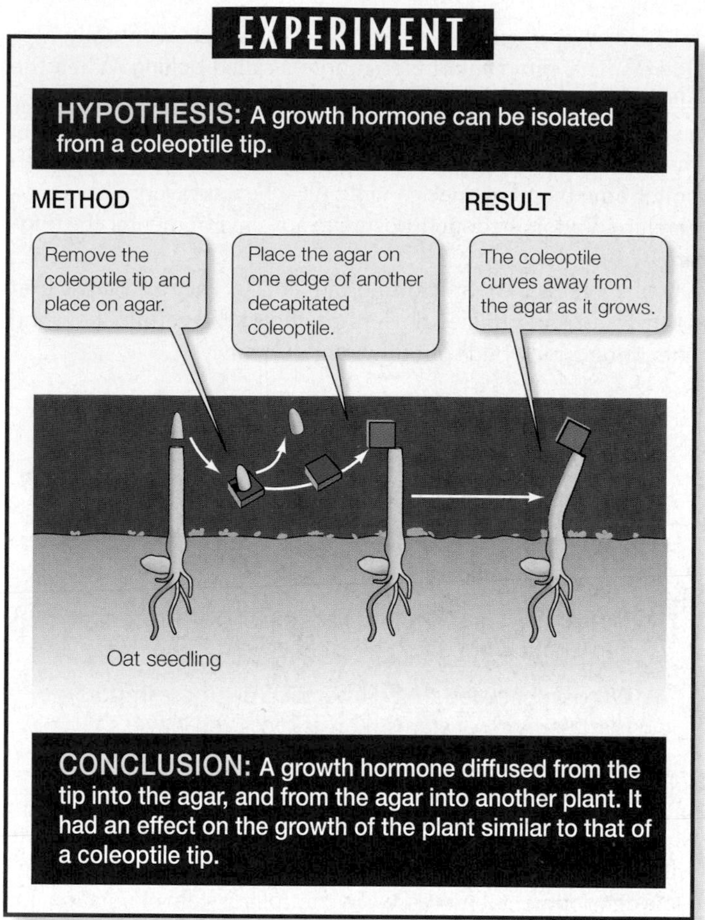

37.8 Went's Experiment By placing coleoptile tips on blocks of agar, Went isolated the growth-promoting hormone whose existence the Darwins had hypothesized.

the blade end toward the stem end. In roots, however, auxin moves toward the root tip, in the phloem. What regulates these movements of auxin?

The polar transport of auxin depends on the location of **auxin anion efflux carriers**, membrane proteins that are confined to the basal ends of cells (the ends closer to the base of the plant than to the shoot apices). Plant cell cytoplasm has a nearly neutral pH, and at this pH auxin exists as an anion—it is negatively charged. Auxin anions can leave the cells only by way of the basally located auxin anion efflux carriers.

Proton pumps in the plasma membrane pump hydrogen ions (H^+) out of the cells, rendering the cell walls acidic. At this lower pH, auxin is present both as an anion and as a free acid. Either form of auxin can enter the cell from any direction. About half the auxin entry into cells is by passive diffusion of the free acid and the other half is by active transport (symport) of the anions along with H^+. However, once auxin molecules enter the cell they become anions, which can depart only from the base (**Figure 37.9**). In this way, auxin anion efflux carriers contribute to the establishment of auxin gradients in the plant. Because it forms a gradient, auxin can act as a *morphogen* (see Section 19.5), instructing cells as to their orientation within the plant and determining how they differentiate.

Other auxin carrier proteins are specific to certain tissues and cells and participate in specific auxin responses. Such auxin carrier proteins are involved in plant responses to light and gravity.

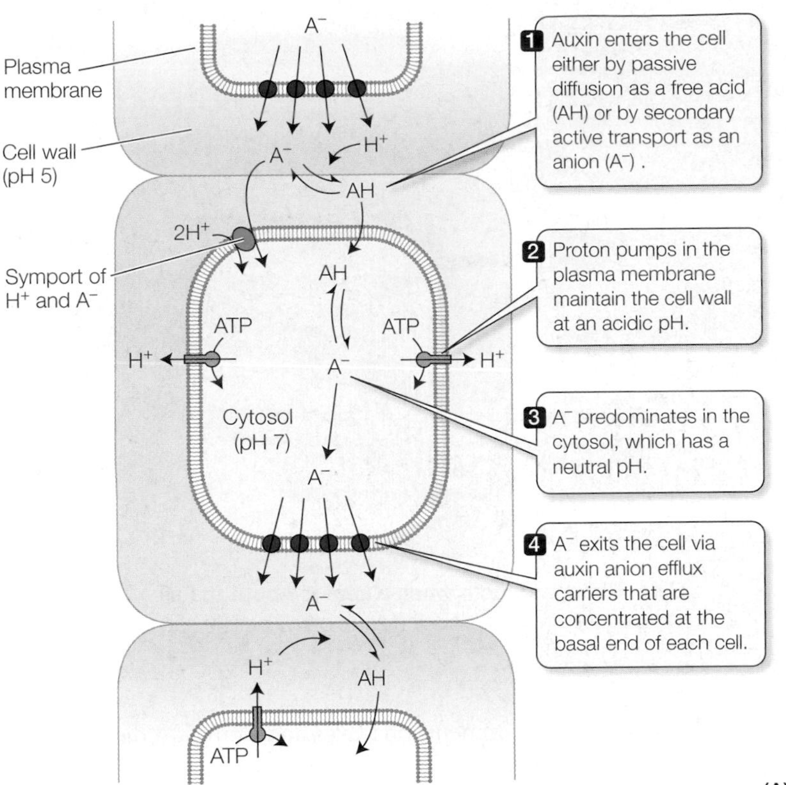

Symport of H+ and A–
Cytosol (pH 7)

3 A– predominates in the cytosol, which has a neutral pH.

37.9 Polar Transport of Auxin Proton pumps and the basally placed auxin anion efflux carriers lead to a net movement of auxin in a basal direction.

Auxin affects plant growth in several ways

Like the gibberellins, auxin has many roles in plant development. It affects the vegetative and reproductive growth of plants in a number of ways.

ROOT INITIATION Cuttings from the shoots of some plants can produce roots and develop into entire new plants. For this to occur, certain undifferentiated cells in the interior of the shoot, originally destined to function only in food storage, must set off on a new mission: They must differentiate and become organized into the apical meristem of a new root. These changes are similar to those in the pericycle of a root when a lateral root forms (see Section 34.3).

Shoot cuttings of many species can be made to develop roots by dipping the cut surfaces into an auxin solution; this observation suggests that the plant's own auxin plays a role in the initiation of lateral roots. Commercial preparations that enhance the rooting of plant cuttings typically contain synthetic auxins.

Light and gravity affect the direction of plant growth

While polar auxin transport establishes the orientation of growth, *lateral* (side-to-side) redistribution of auxin is responsible for plant movements. This redistribution is carried out by an auxin carrier protein that moves to one side of the cell (as opposed to the base) and thus allows auxin to exit the cell only from that side.

When light strikes a grass coleoptile on one side, auxin at the tip moves laterally toward the shaded side. The imbalance thus established is maintained down the coleoptile, so that in the growing region below, the auxin concentration is highest on the shaded side. Cell growth is thus speeded up on that side, causing the coleoptile to bend toward the light (**phototropism**; **Figure 37.10A**). If you have noticed a houseplant bending toward a window, you have observed phototropism.

Even in the dark, auxin moves to the lower side of a shoot that has been tipped over, causing more rapid growth in the lower side and, hence, an upward bending of the shoot. Such growth in a direction determined by gravity is called **gravitropism** (**Figure 37.10B**). The upward gravitropic response of shoots is defined as *negative gravitropism*; that of roots, which bend downward, is *positive gravitropism*.

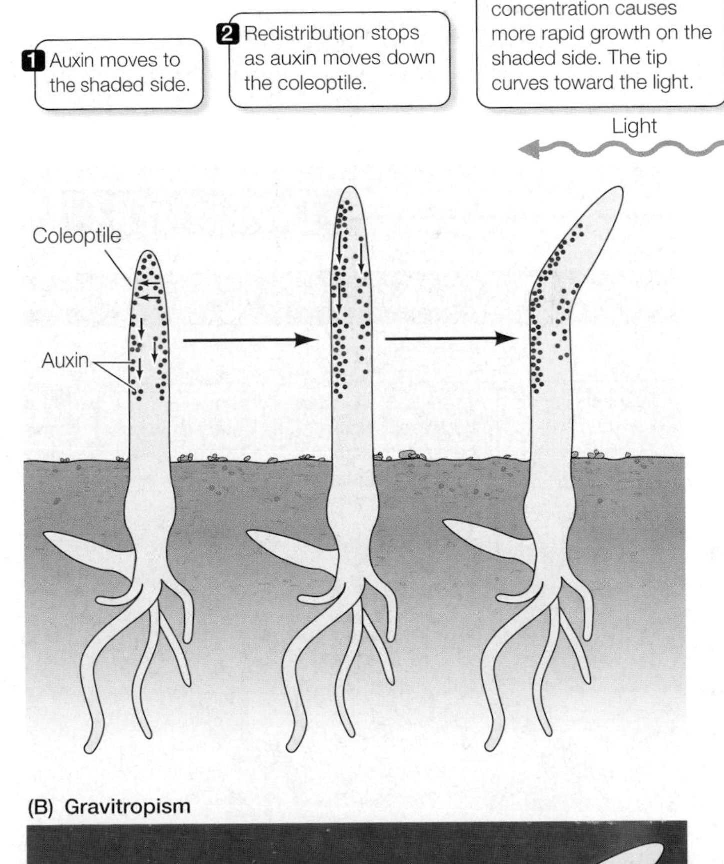

(A) Phototropism
1 Auxin moves to the shaded side.
2 Redistribution stops as auxin moves down the coleoptile.
3 A higher auxin concentration causes more rapid growth on the shaded side. The tip curves toward the light.
Light
Coleoptile
Auxin
(B) Gravitropism
1 Auxin moves downward in response to gravitational stimulus.
2 A higher auxin concentration causes more rapid growth on the lower side. The tip curves upward.

37.10 Plants Respond to Light and Gravity (A) Phototropism and (B) gravitropism occur in shoot apices in response to a redistribution of auxin.

LEAF ABSCISSION In contrast to its stimulatory effect on root initiation, auxin inhibits the detachment of old leaves from stems. This detachment process, called **abscission**, is the cause of autumn leaf fall. Most leaves consist of a blade and a petiole that attaches the blade to the stem. Abscission results from the breakdown of a specific part of the petiole, the *abscission zone* (**Figure 37.11**). If the blade of a leaf is cut off, the petiole falls from the plant more rapidly than if the leaf had remained intact. If the cut surface is treated with an auxin solution, however, the petiole remains attached to the plant, often longer than an intact leaf would have. The timing of leaf abscission in nature appears to be determined in part by a decrease in the movement of auxin, produced in the blade, through the petiole.

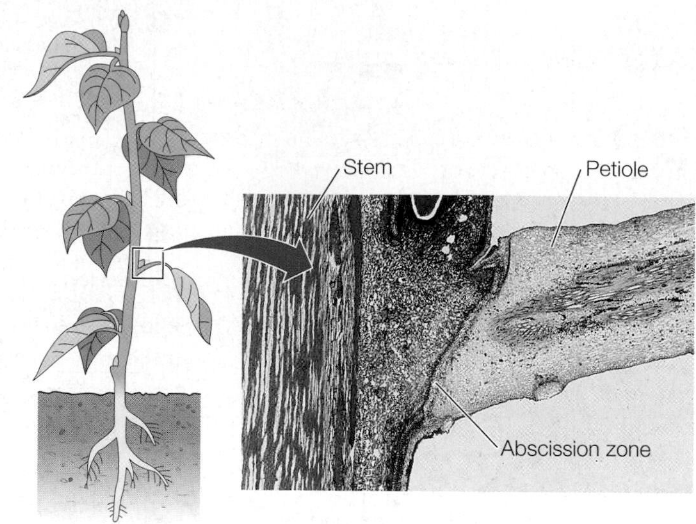

37.11 Changes Occur when a Leaf Is About to Fall The breakdown of cells in the abscission zone of the petiole causes the leaf to fall.

APICAL DOMINANCE Auxin helps maintain **apical dominance**, a phenomenon in which apical buds inhibit the growth of axillary buds, resulting in the growth of a single main stem with minimal branching. This phenomenon can be demonstrated by an experiment with young seedlings. If the plant remains intact, the stem elongates and the axillary buds remain inactive. Removal of the apical bud—the major site of auxin production—results in growth of the axillary buds. If the cut surface of the stem is treated with auxin, however, the axillary buds do not grow (**Figure 37.12**). The apical buds of branches also exert apical dominance: The axillary buds on the branch are inactive unless the apex of the branch is removed. That is why gardeners prune shrubs to encourage branching.

In the two experiments on leaves and stems just discussed, removal of a particular part of the plant elicits a response—abscission or loss of apical dominance—and that response is prevented by treatment with auxin. These results are consistent with other data showing that the excised part of the leaf or stem is an auxin source and that auxin in the intact plant helps maintain apical dominance and delays the abscission of leaves.

STEM AND ROOT ELONGATION Auxin promotes stem elongation but inhibits root elongation. The question of why different organs respond in opposite ways to the same growth hormone remains unanswered, but is a subject of current research.

FRUIT DEVELOPMENT Fruit development normally depends on prior fertilization of the egg, but in many species treatment of an unfertilized ovary with auxin or gibberellins causes **parthenocarpy**—fruit formation without fertilization. Parthenocarpic fruits form spontaneously in some cultivated varieties of plants, including seedless grapes, bananas, and some cucumbers.

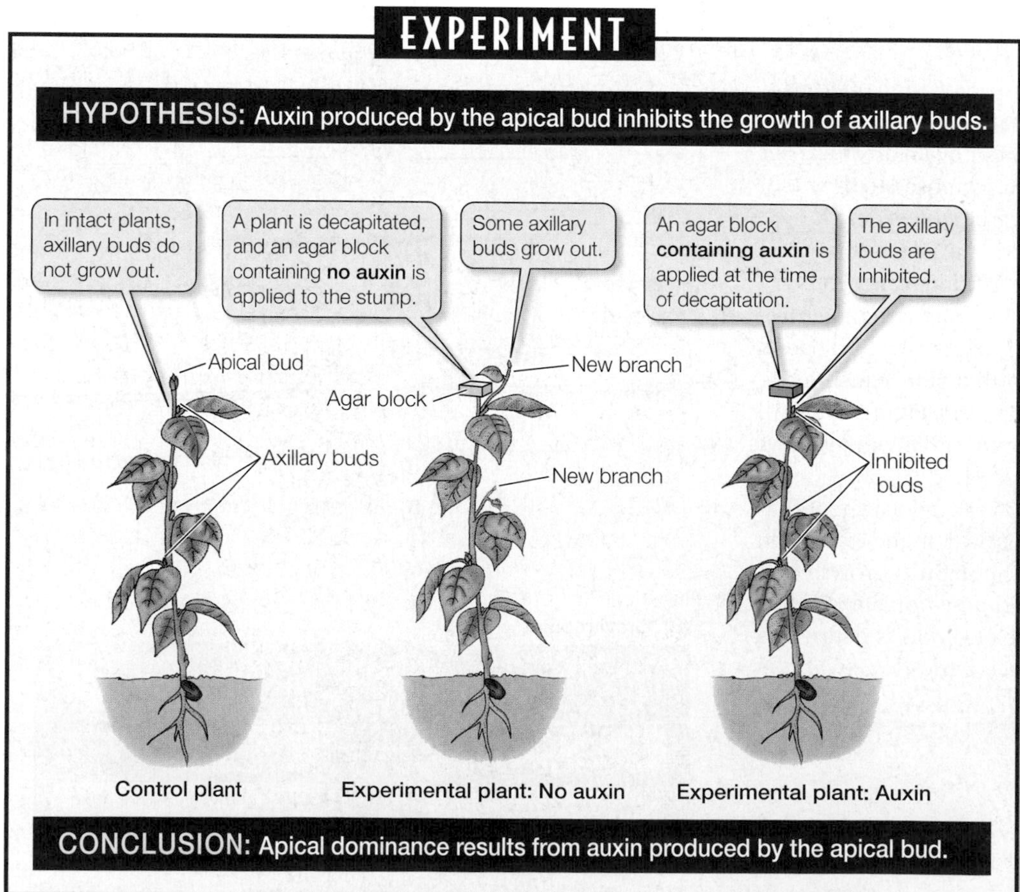

EXPERIMENT
HYPOTHESIS: Auxin produced by the apical bud inhibits the growth of axillary buds.
In intact plants, axillary buds do not grow out.
A plant is decapitated, and an agar block containing no auxin is applied to the stump.
Some axillary buds grow out.
An agar block containing auxin is applied at the time of decapitation.
The axillary buds are inhibited.
Apical bud
Axillary buds
Agar block
New branch
New branch
Inhibited buds
Control plant
Experimental plant: No auxin
Experimental plant: Auxin
CONCLUSION: Apical dominance results from auxin produced by the apical bud.

37.12 Auxin and Apical Dominance Auxin produced by the apical bud maintains apical dominance—the growth of a single main stem with minimal branching.

Auxin analogs as herbicides

Auxin is absolutely essential for plant survival; no mutants lacking auxin have ever been found. Many synthetic auxins—chemical analogs of indoleacetic acid—have been produced and studied. One of them, 2,4-dichlorophenoxyacetic acid (2,4-D), has the striking property of being lethal to eudicots at concentrations that are harmless to monocots. Because 2,4-D cannot be broken down by eudicots, it accumulates and causes the plant to "grow itself to death." This property makes 2,4-D an effective *selective herbicide* that can be sprayed on a lawn or a cereal crop to kill weeds that are eudicots.

All of the activities discussed so far illustrate the great diversity of important roles that auxin plays in plant growth and development. Now let's see *how* auxin plays one of its roles—promoting stem elongation through effects on the cell wall.

Auxin promotes growth by acting on cell walls

The expansion of plant cells is what causes plant growth. Thus the cell wall plays key roles in controlling the rate and direction of growth of a plant cell. Auxin acts on cell walls to regulate this process.

CELL EXPANSION The expansion of a plant cell is driven primarily by the uptake of water, which enters the cytoplasm of the cell and accumulates in its central vacuole (see Section 35.1). As the vacuole expands, the cell grows rapidly, with the vacuole often making up more than 90 percent of the volume of a mature cell. The vacuole presses the cytoplasm against the cell wall as it expands, and the wall resists this force.

The principal strengthening component of the plant cell wall is *cellulose*, a large polymer of glucose. In the wall, string-like cellulose molecules tend to associate in parallel with one another. Bundles of approximately 250 cellulose molecules make up *microfibrils* that are visible under an electron microscope (**Figure 37.13**). What makes the cell wall rigid is a network of cellulose microfibrils connected by bridges of other, smaller polysaccharides. The orientation of the majority of cellulose microfibrils determines the direction of cell expansion (**Figure 37.14**).

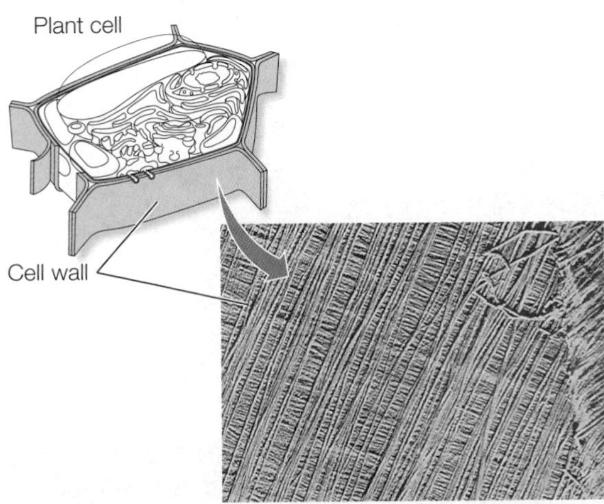

37.13 Cellulose in the Cell Wall The plant cell wall is a network of cellulose microfibrils linked by other polysaccharides. The crisscross pattern results from the deposition of successive layers within each of which the microfibrils are parallel.

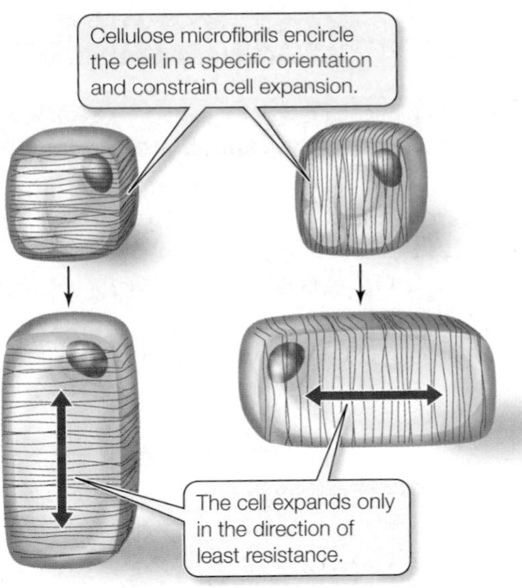

Cellulose microfibrils encircle the cell in a specific orientation and constrain cell expansion.
The cell expands only in the direction of least resistance.

37.14 Plant Cells Expand The orientation of cellulose microfibrils in the plant's cell wall determines the direction of cell expansion.

For the cell to expand, its wall must loosen and be stretched. If the wall were only stretched, however, it would become thinner. Cell expansion involves more than stretching. New polysaccharides are deposited throughout the wall, and new cellulose microfibrils are deposited at the inner surface of the wall, maintaining its thickness. As a consequence of this pattern of cellulose deposition, the microfibrils in the outermost part of the wall are the oldest, and those in the innermost part the youngest. How do these properties of cell walls relate to the action of auxin on plant cell expansion?

CELL WALL LOOSENING Experiments with segments of oat coleoptiles have shown that plant cell walls recover incompletely from being stretched (**Figure 37.15**). Reversible stretching is called *elasticity*, and irreversible stretching is called **plasticity**. Treating the coleoptile segments with auxin before they were stretched significantly increased their plasticity; in other words, it loosened the cell walls. This result suggested that auxin-induced cell expansion might result from just such a loosening effect.

Auxin acts by causing the release of a "wall-loosening factor" from the cytoplasm. Studies in the 1970s indicated that the wall-loosening factor was sometimes simply hydrogen ions (protons, H^+). Acidifying the growth medium (that is, adding H^+) caused segments of stems or coleoptiles to grow as rapidly as segments treated with auxin. Furthermore, treating coleoptile segments with auxin caused acidification of the growth medium. Auxin increases the activity of proton pumps in the plasma membrane, increasing the H^+ concentration in the cell wall. Treatments that block acidification by auxin also block auxin-induced growth. These experimental results led to the hypothesis that hydrogen ions secreted into the cell wall activate one or more cell wall proteins. So, a search began for candidate proteins.

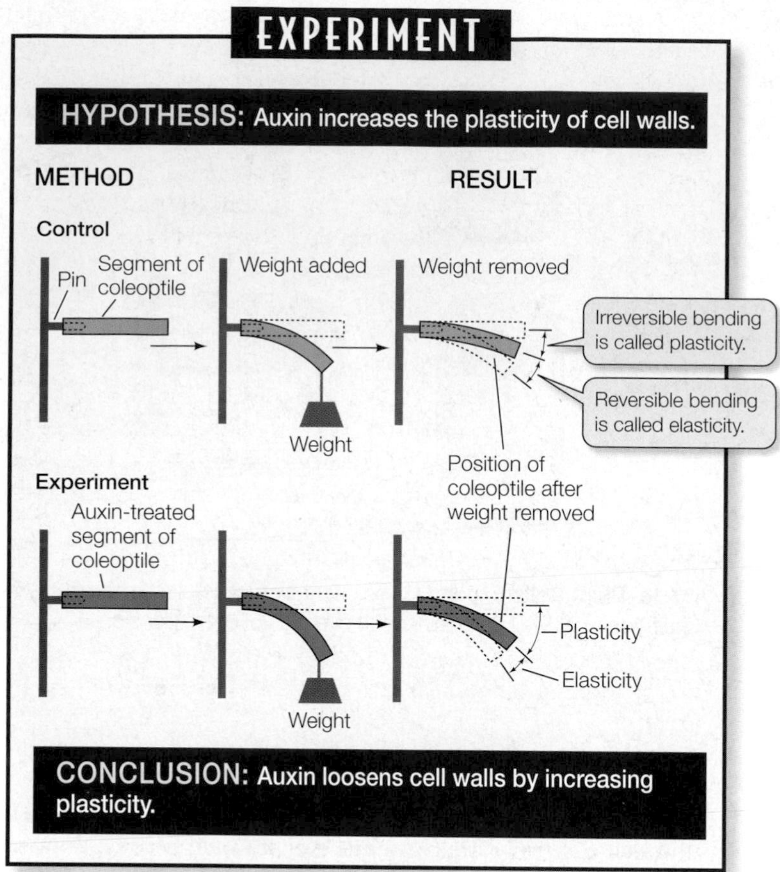
EXPERIMENT
HYPOTHESIS: Auxin increases the plasticity of cell walls.
METHOD
RESULT
Control
Pin
Segment of coleoptile
Weight added
Weight removed
Irreversible bending is called plasticity.
Reversible bending is called elasticity.
Weight
Position of coleoptile after weight removed
Experiment
Auxin-treated segment of coleoptile
Plasticity
Elasticity
Weight
CONCLUSION: Auxin loosens cell walls by increasing plasticity.

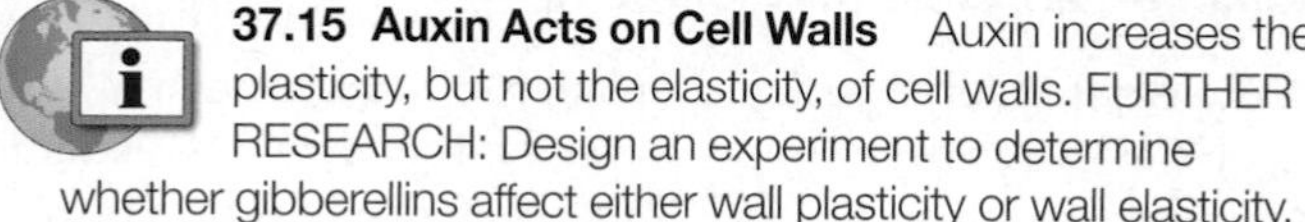

37.15 Auxin Acts on Cell Walls Auxin increases the plasticity, but not the elasticity, of cell walls. FURTHER RESEARCH: Design an experiment to determine whether gibberellins affect either wall plasticity or wall elasticity.

Proteins called *expansins* were isolated and purified from plant cell walls in the 1990s. When the purified proteins were added to isolated cell walls of several plant species, the walls expanded. Expansins are widespread among terrestrial plants. Furthermore, these proteins are activated by hydrogen ions. Expansins modify the pattern of hydrogen bonding between the polysaccharides in the plant cell wall. These modifications may allow polysaccharide macromolecules to slip past each other, so that the wall stretches and the cell expands.

Before auxin can initiate a chain of events such as the one just described, it must first be recognized as a signal by the cell. How does this recognition occur?

Auxin and gibberellins are recognized by similar mechanisms

The initial step in the action of any plant hormone is the binding of that hormone by a specific receptor protein. There are several proteins that can bind various plant hormones, but some of this binding may be nonspecific. To demonstrate that an auxin-binding protein is an auxin receptor, it must be shown that the protein actually mediates one or more of the effects of auxin, whether by promoting the expression of genes, by activating an enzyme, or by activating a proton pump. Different effects of auxin may involve different receptor proteins.

In 2005, more than 20 years after molecular biologists first identified an auxin-binding protein, workers at the University of Indiana and at the University of York (in the United Kingdom) showed conclusively that another protein binds auxin *and* then triggers a normal auxin response. Thus, it is a true auxin receptor. There may be others. When this receptor protein binds an auxin molecule the result is the destruction of proteins that directly inhibit genes involved in certain auxin responses. Thus, auxin indirectly promotes gene expression (**Figure 37.16**).

Almost exactly four months later, workers in Japan reported a strikingly similar finding—a gibberellin receptor protein that binds a gibberellin molecule, resulting in the destruction of proteins that directly inhibit transcription of certain genes. In other words, the mechanism governing the gibberellin receptor protein is exactly

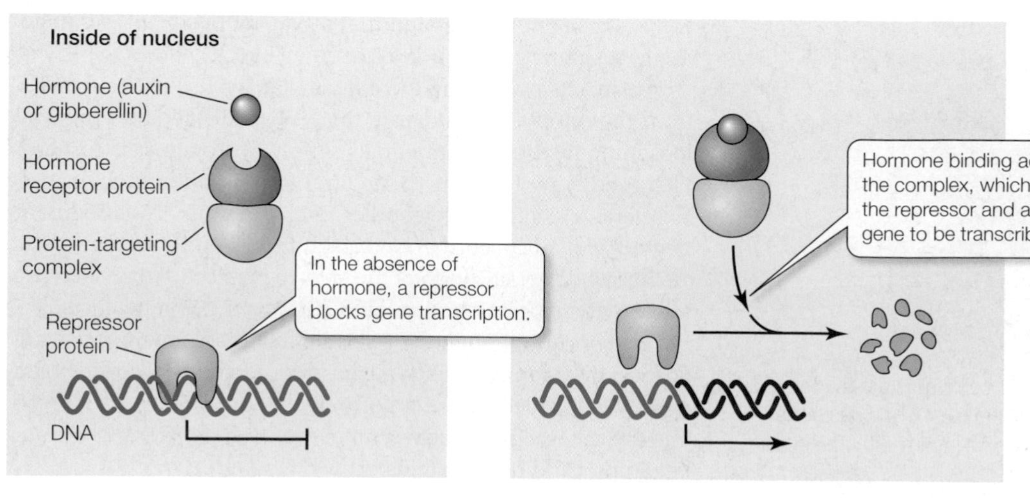
Inside of nucleus
Hormone (auxin or gibberellin)
Hormone receptor protein
Protein-targeting complex
In the absence of hormone, a repressor blocks gene transcription.
Repressor protein
DNA
Hormone binding activates the complex, which degrades the repressor and allows the gene to be transcribed.

37.16 Similar Signal Transduction Pathways Are Used by Auxin and Gibberellins The receptors in some responses to auxin and gibberellin cause repressor inactivation; with these repressors removed from the DNA, genes can be expressed.

analogous to that for the auxin receptor protein. Both receptors reside in the nucleus, but they belong to different families of proteins.

37.3 RECAP

Auxin regulates stem elongation and mediates phototropism and gravitropism; it also plays roles in apical dominance, leaf abscission, root initiation, and other phenomena.

- What are auxin anion efflux carriers and how do they contribute to the polar transport of auxin? See p. 804 and Figure 37.9
- Can you explain why, even though auxin moves *away* from the lighted side of a coleoptile tip, the coleoptile bends *toward* the light? See p. 805 and Figure 37.10
- Can you describe how the signal transduction pathways for auxin and gibberellin are similar? See p. 808 and Figure 37.16

How can a single hormone, such as auxin or a gibberellin, have so many effects? As we have seen, a hormone may have more than one receptor and it may trigger more than one signal transduction pathway, and a single transduction pathway may affect more than one gene or membrane protein. We learn about other important plant hormones in the next section, and they, too, have multiple effects.

37.4 What Do Cytokinins, Ethylene, Abscisic Acid, and Brassinosteroids Do?

What signals tell different types of plant cells and organs to form? That is, what triggers differentiation of plant tissues? Much of the research on this question has been performed using plant tissues that have been removed from the plant body and grown in culture on an artificial medium. One easily grown tissue is pith—the spongy innermost tissue of a stem. Pith cells proliferate rapidly in culture, but show no differentiation. All the cells are similar and unspecialized; they grow into a lump of pith tissue on the surface of the culture medium.

Cutting a notch in the cultured pith tissue and inserting a stem tip into the notch causes the pith cells below the inserted tip to differentiate. Some of them differentiate to form water-conducting xylem cells. Differentiation of pith cells can also be initiated by adding a mixture of auxin and coconut milk (a product of the coconut endosperm and a rich source of plant hormones) to the notch.

It was Dutch plant physiologist Johannes van Overbeek who discovered in 1941 that adding coconut milk to culture medium caused a dramatic increase in the growth of plant embryos. Purification and analysis of coconut milk was instrumental in research that led to the identification of the cytokinins and other plant growth hormones.

A similar effect can be observed in intact plants. If notches are cut in the stems of coleus plants, interrupting some of the strands of vascular tissue, the strands gradually regenerate from the upper side of the cut to the lower side (recall that auxin moves from the apex to the base of a stem). If the leaves above the cut are removed, regeneration is slowed. However, when the missing leaves are replaced with an auxin solution, vascular tissue regenerates normally. These results show that auxin and other plant hormones signal the formation of specific cell types.

Experiments with cultured plant tissues have helped clarify which hormones control organ formation. The eventual fate of an undifferentiated cell can be adjusted depending on hormone signals. Undifferentiated cultures of tobacco pith form roots when treated with an appropriate concentration of auxin. Another group of plant growth hormones—the **cytokinins**—causes buds and then shoots to form in such cultures. The pattern of organ formation depends on the ratio of auxin to cytokinin in the medium. A high proportion of auxin favors roots, and a high proportion of cytokinin favors buds, but both processes are most active when both hormones are present. These results show that other hormones can modify the effects of auxin, reminding us that plant growth is regulated more by hormone interactions than by a single hormone.

Cytokinins are active from seed to senescence

In studies of plant cell division, botanists discovered that the cytokinins powerfully stimulate cell division in tissue cultures. Cytokinins consist of adenine (the nitrogenous base "A" in DNA and RNA) with an attached group. Two closely related cytokinins, called *zeatin* and *isopentenyl adenine,* occur naturally in plants. A third, *kinetin,* may be considered a synthetic cytokinin because it has never been isolated from plant tissue.

No mutants lacking cytokinins have ever been found. Thus, like auxin, cytokinins seem to be required throughout the life of a plant. Cytokinins are synthesized primarily in the roots and move to other parts of the plant. They have a number of different effects:

- Adding an appropriate combination of auxin and cytokinins to a growth medium induces rapid cell proliferation in cultured plant tissues.
- Cytokinins can cause certain light-requiring seeds to germinate even when kept in constant darkness.
- Cytokinins usually inhibit the elongation of stems, but they cause lateral swelling of stems and roots (the fleshy roots of radishes are an extreme example).
- Cytokinins stimulate axillary buds to grow into branches; thus the balance between auxin and cytokinin levels controls the extent of branching (bushiness) of a plant.
- Cytokinins increase the expansion of cut pieces of leaf tissue in culture and may regulate normal leaf expansion.
- Cytokinins delay the senescence of leaves. If leaf blades are detached from a plant and floated on water or a nutrient solution, they quickly turn yellow and show other signs of senescence. If instead they are floated on a solution containing a cytokinin, they remain green and senesce much more slowly.

One effect of cytokinins is to support the production of seeds by rice. Plant biologists in Japan developed a rice variety that combines enhanced seed production and reduced plant height. By reducing the expression of a gene that encodes cytokinin oxidase, an enzyme that degrades cytokinins, they increased cytokinin levels in the developing heads, greatly enhancing grain yield. Why reduce the height? Rice is subject to *lodging*—excessive bending after wind or rain storms—which makes harvesting difficult. Shorter plants bear the increased load and produce a harvestable crop.

Ethylene is a gaseous hormone that hastens leaf senescence and fruit ripening

Whereas the cytokinins delay senescence, another plant hormone promotes it: the gas **ethylene** ($H_2C{=}CH_2$), which is sometimes called the senescence hormone. Ethylene can be produced by all parts of the plant, and like all plant hormones it has several effects.

Back when streets were lit by gas rather than by electricity, leaves on trees near street lamps dropped earlier than those on trees farther from the lamps. We now know that ethylene, a combustion product of the illuminating gas, is what caused the early abscission. Auxin delays leaf abscission, but ethylene strongly promotes it; thus a balance of auxin and ethylene controls abscission.

FRUIT RIPENING By promoting senescence, ethylene also speeds the ripening of fruit. As the fruit ripens, it loses chlorophyll and its cell walls break down; ethylene promotes both of these processes. Ethylene also causes an increase in its own production. Thus, once ripening begins, more and more ethylene forms, and because it is a gas, it diffuses readily throughout the fruit and even to neighboring fruits on the same or other plants.

The old saying "one rotten apple spoils the barrel" is true. That rotten apple is a rich source of ethylene, which speeds the ripening and subsequent rotting of the other fruit in a barrel or other confined space.

Farmers in ancient times slashed developing figs to hasten their ripening. We now know that wounding causes an increase in ethylene production by the fruit and that the raised ethylene level promotes ripening. Today commercial shippers and storers of fruit hasten ripening by adding ethylene to storage chambers. This use of ethylene is the single most important use of a natural plant hormone in agriculture and commerce—even more important than its use in rubber plantations to prevent the latex from coagulating, as described at the beginning of the chapter. Ripening can also be delayed by the use of "scrubbers" and adsorbents that remove ethylene from the atmosphere in fruit storage chambers.

As flowers senesce, their petals may abscise, to the detriment of the cut-flower industry. Growers and florists often immerse the cut stems of ethylene-sensitive flowers in dilute solutions of silver thiosulfate before sale. Silver salts inhibit ethylene action, probably by interacting directly with the ethylene receptor, and thus delay senescence—enabling florists and consumers to keep their cut flowers from dropping their petals prematurely.

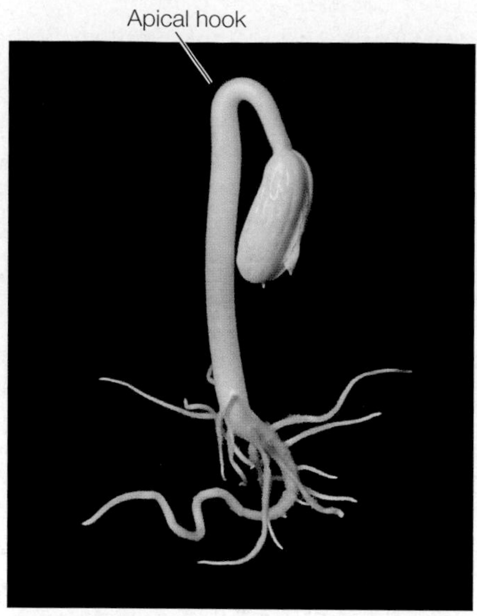

37.17 The Apical Hook of a Eudicot Asymmetrical production of ethylene is responsible for the apical hook of this bean seedling. The ethylene concentration was highest on the right side, so more rapid growth on the left caused and maintained the hook.

STEM GROWTH Although it is associated primarily with senescence, ethylene is active at other stages of plant development as well. The stems of many eudicot seedlings form an **apical hook** that protects the delicate shoot apex while the stem grows through the soil (**Figure 37.17**). The apical hook is maintained through an asymmetrical production of ethylene gas, which inhibits the elongation of cells on the inner surface of the hook. Once the seedling breaks through the soil surface and is exposed to light, ethylene synthesis stops, and the cells of the inner surface are no longer inhibited. These cells now elongate, and the hook unfolds, raising the shoot apex and the expanding leaves into the sun.

Ethylene also inhibits stem elongation in general, promotes lateral swelling of stems (as do the cytokinins), and causes stems to lose their sensitivity to gravitropic stimulation. Together, these three phenomena constitute the *triple response*, a well-characterized stunted growth habit observed when plants are treated with ethylene.

THE ETHYLENE SIGNAL TRANSDUCTION PATHWAY Analysis of *Arabidopsis* mutants has revealed the mechanism of ethylene action. Some of these mutants do not respond to applied ethylene, and others act as if they have been exposed to ethylene even though they have not. Studies of the mutant genes and their protein products, coupled with comparisons of their amino acid sequences with those of other known proteins, have revealed some of the details of the signal transduction pathway through which ethylene produces its effects (**Figure 37.18**). The pathway includes two membrane proteins in the endoplasmic reticulum. The first is an ethylene receptor (labeled A in the figure) and the second is a channel (C). In the absence of ethylene, another protein (B) keeps C inactive. When A binds ethylene it inactivates B. Without B to inactivate it, C acts through a second messenger to activate a transcription factor (D). The transcription factor turns on the genes that produce ethylene's effects in the cell. Ethylene was the first plant hormone to have its mechanism of action elucidated in this way.

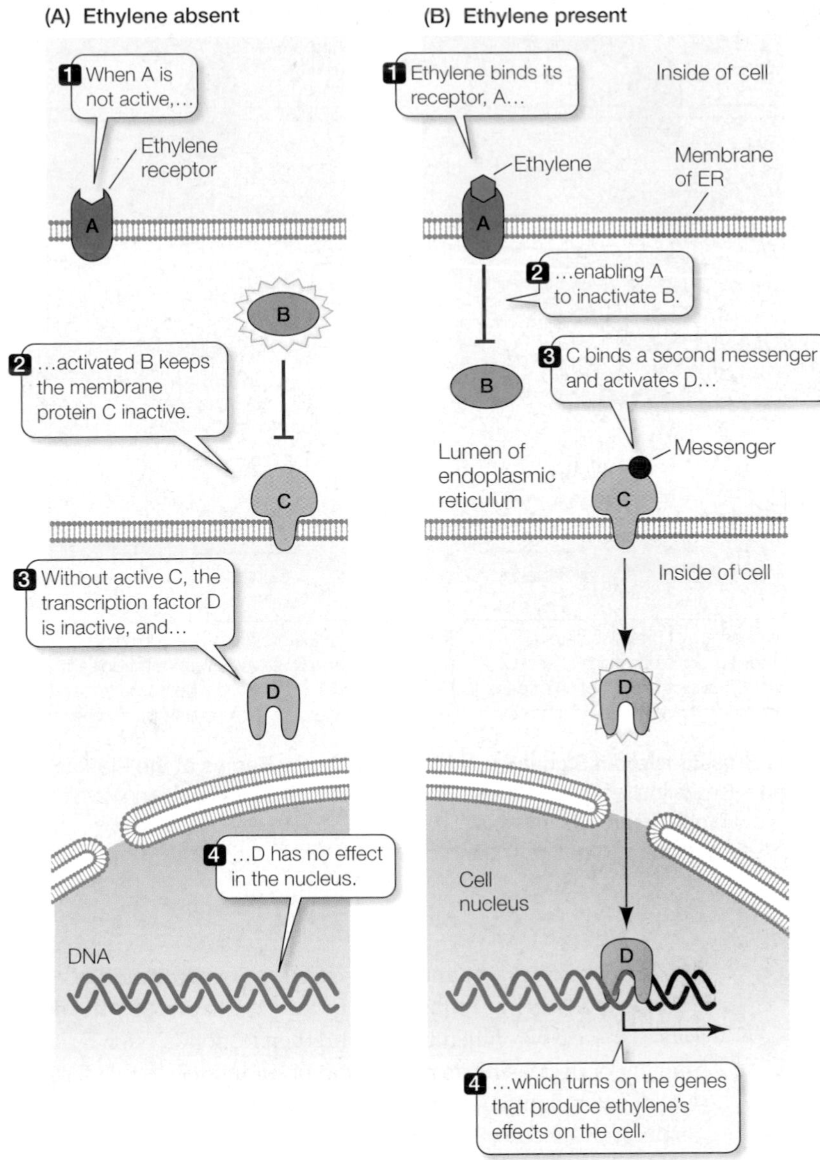
(A) Ethylene absent
1 When A is not active,...
Ethylene receptor
A
B
2 ...activated B keeps the membrane protein C inactive.
C
3 Without active C, the transcription factor D is inactive, and...
D
4 ...D has no effect in the nucleus.
DNA
(B) Ethylene present
1 Ethylene binds its receptor, A...
Inside of cell
Ethylene
Membrane of ER
A
2 ...enabling A to inactivate B.
3 C binds a second messenger and activates D...
B
Lumen of endoplasmic reticulum
Messenger
C
Inside of cell
D
Cell nucleus
D
4 ...which turns on the genes that produce ethylene's effects on the cell.

37.18 The Signal Transduction Pathway for Ethylene This diagram shows the roles of four proteins (A, B, C, and D) in the signal transduction pathway through which ethylene exerts its many effects.

Abscisic acid is the "stress hormone"

Abscisic acid is another hormone with multiple effects in the living plant. During seed formation, abscisic acid promotes the accumulation of storage proteins by allowing the expression of the genes that encode those proteins. It is generally present in high concentrations in dormant buds and some dormant seeds, and it is probably the most common of the chemical inhibitors that initiate and maintain dormancy in mature seeds. Abscisic acid also inhibits stem elongation. It is sometimes referred to as the stress hormone of plants, because it accumulates when plants are deprived of water and it may play a role in maintaining the dormancy of buds in winter. Like the gibberellins, abscisic acid belongs to the class of compounds called terpenes.

Sometimes seed dormancy ends prematurely. Some mutant corn plants, called *vp* mutants, have seeds that germinate while still attached to the cob on the parent plant—a condition called **vivipary** (Latin *vivus*, "alive"; *parere*, "give birth"). Several *vp* mutants are naturally deficient in abscisic acid. Applying abscisic acid to these mutants reduces their tendency to show vivipary. Another type of *vp* mutant fails to respond in any way to applied abscisic acid. These results indicate that abscisic acid is the inhibitor that normally prevents seeds from germinating while still attached to the parent plant. The first type of mutant cannot make enough abscisic acid; the second type of mutant is viviparous because it cannot respond to abscisic acid—its own or any applied to it.

Abscisic acid also regulates gas and water vapor exchange between leaves and the atmosphere through its effects on the guard cells of the leaf stomata. Abscisic acid causes stomata to close, and it also prevents the opening of stomata normally caused by light. Both of these effects involve ion channels in the plasma membrane of the guard cells. The first response of a guard cell to abscisic acid is the opening of calcium channels and the entry of calcium into the cell. This calcium causes the cell's vacuole to release calcium, too. The increased concentration of calcium in the cytoplasm leads to a chain of events that result in the opening of potassium channels, the loss of K^+ and water from the cytoplasm, and the closing of the stoma as the guard cells sag together.

Brassinosteroids are hormones that mediate effects of light

More than 20 years ago, biologists isolated an interesting steroid from the pollen of rape, a member of the Brassicaceae, or mustard family. When applied to various plant tissues, this **brassinosteroid** stimulated cell elongation, pollen tube elongation, and vascular tissue differentiation, but it inhibited root elongation. Since then, dozens of chemically related and growth-affecting brassinosteroids have been found in plants. Treatment with as little as a few nanograms of brassinosteroid per plant is enough to promote growth.

The properties of an *Arabidopsis* mutant called *det2* made it clear that brassinosteroids are naturally occurring plant hormones. When grown in darkness, seedlings homozygous for the *det2* allele differ dramatically from wild-type seedlings grown under the same conditions: In many respects, they look like wild-type seedlings grown in the light. Treatment of dark-grown *det2* mutant seedlings with brassinosteroids causes them to grow normally—that is, like wild-type plants grown in the dark. These results, supported by chemical analysis, showed that *det2* plants are unable to synthesize their own brassinosteroids, and that lack of the hormone results in abnormal growth. The gene product DET2 is a link between hormones and photoreceptors throughout the life cycle of *Arabidopsis*.

The receptor protein and signal transduction pathway for brassinosteroids differ sharply from those for steroid hormones in animals. Receptors for animal steroid hormones exist in the cytoplasm of animal cells and move to the nucleus when bound by steroids. In contrast, the receptor for brassinosteroids is an integral protein in the plasma membrane. Binding of a brassinosteroid by the receptor inactivates the protein that turns certain genes on and others off (**Figure 37.19**).

Some of the effects of light on plant development result from effects on the signal transduction pathway for brassinosteroids. Others may result from alterations in brassinosteroid levels in the plant.

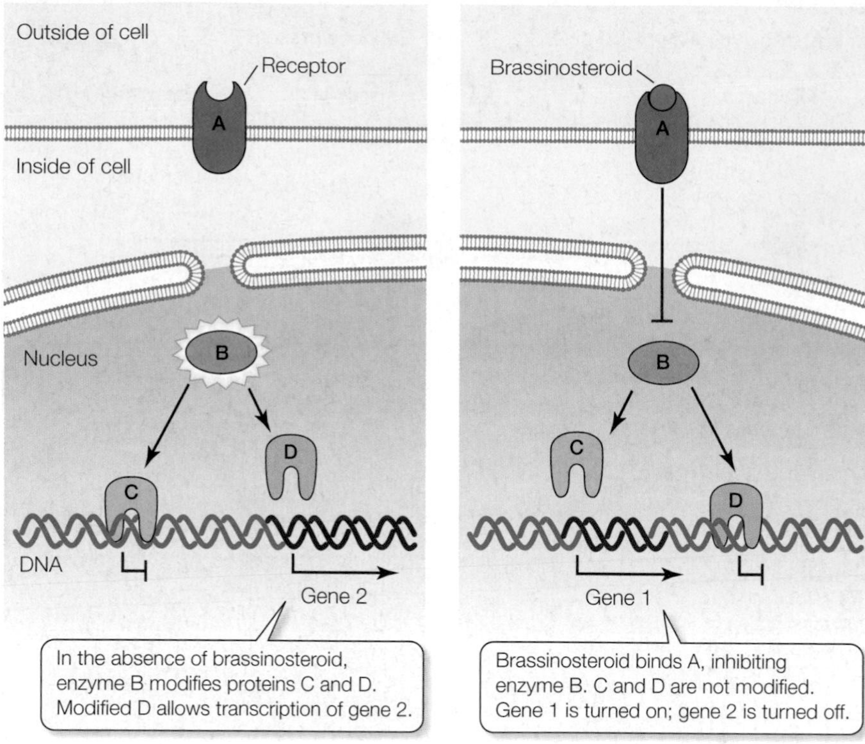
Outside of cell
Receptor
Brassinosteroid
A
Inside of cell
Nucleus
B
C
D
DNA
Gene 2
Gene 1
In the absence of brassinosteroid, enzyme B modifies proteins C and D. Modified D allows transcription of gene 2.
Brassinosteroid binds A, inhibiting enzyme B. C and D are not modified. Gene 1 is turned on; gene 2 is turned off.

37.19 The Brassinosteroid Signal Transduction Pathway Begins at the Plasma Membrane Unlike the receptors for animal steroid hormones, the brassinosteroid receptor is a membrane protein. The signal transduction pathway concludes by activating certain genes and inactivating others.

37.4 RECAP

Cytokinins, abscisic acid, ethylene, and brassinosteroids work in concert with auxin and gibberellins to mediate plant development. Plant growth is regulated more by hormone interactions than by a single hormone.

- How do cytokinins interact with auxin to regulate a plant's development? See p. 809
- Can you explain how the experiments with *vp* mutants revealed the role of abscisic acid in preventing premature seed germination? See p. 811

Plant response to light—the energy source for photosynthesis—is crucial. We saw how their pioneering investigations of phototropism led the Darwins to the discovery of auxin. Let's now look more closely at how plants respond to light.

37.5 How Do Photoreceptors Participate in Plant Growth Regulation?

The length of the night determines the onset of winter dormancy in many plant species. As summer progresses, the days become shorter (that is, the nights become longer). Leaves have a mechanism for measuring the length of the night, as we will see in the next chapter. Measuring night length is an accurate way to determine the season of the year. If a plant determined the season only by the temperature, it might be fooled by a winter warm spell or by unseasonably cold weather in the summer. The length of the night, on the other hand, is determined by Earth's rotation around the sun and, for a given latitude, does not vary randomly. Plants use the environmental cue of night length to time several aspects of their growth and development.

Night length is only one of several environmental cues detected by plants. Environmental conditions are also signaled by light's presence or absence, its intensity, and its spectral properties (specific wavelengths). Even though temperature can vary from its seasonal norms, it nevertheless also provides important environmental cues, both by its value at any particular time and by the distribution of warmer and colder stretches over a period of time. (Temperature cues are considered further in Chapter 38.) The plant senses these environmental cues and then responds, often by stepping up or decreasing its production of hormones. In this section, we focus on light cues: how certain photoreceptors sense light, its duration, and its wavelength distribution.

Light regulates many aspects of plant development in addition to phototropism. Light affects seed germination, shoot elongation, the initiation of flowering, and many other important aspects of plant development. Several photoreceptors take part in these processes. Five **phytochromes** mediate the effects of red and dim blue light. Three or more types of **blue-light receptors**, discovered more recently, mediate the effects of higher-intensity blue light.

Phytochromes mediate the effects of red and far-red light

Some seeds will not germinate in darkness, but do so readily after even a brief exposure to dim light. Blue and red light are highly effective in promoting germination.

Of particular importance to plants is the fact that far-red light *reverses* the effect of a prior exposure to red light. Far-red light is a very deep red, bordering on the limit of human vision and centered on a wavelength of 730 nm; red wavelengths are around 660 nm. If exposed to brief, alternating periods of red and far-red light in close succession, lettuce seeds respond only to the final exposure: If it is red, they germinate; if it is far-red, they remain dor-

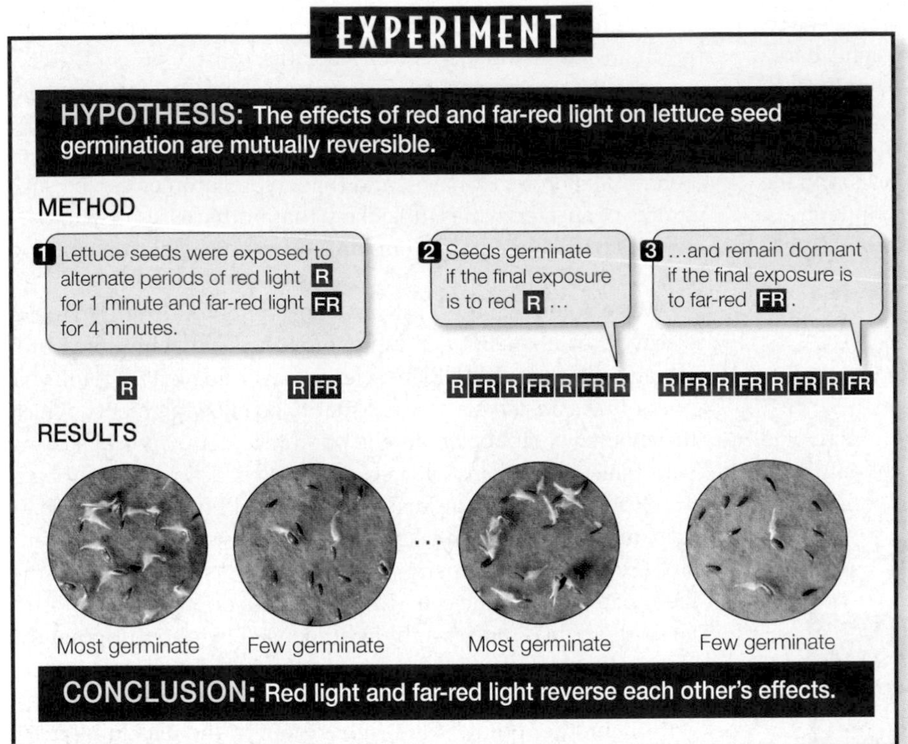

37.20 Sensitivity of Seeds to Red and Far-Red Light In each case, the final exposure determines the seed's germination response.

mant (**Figure 37.20**). This reversibility of the effects of red and far-red light regulates many other aspects of plant development, including flowering and seedling growth.

The basis for these effects of red and far-red light resides in certain bluish photoreceptor proteins called phytochromes. In the cytosol of plants are two interconvertible forms of phytochromes. Light drives the interconversion of the two forms. The form that absorbs principally red light is called $\mathbf{P_r}$. Upon absorption of a photon of red light, a molecule of P_r is converted into $\mathbf{P_{fr}}$. The P_{fr} form absorbs far-red light; when it does so, it is converted to P_r.

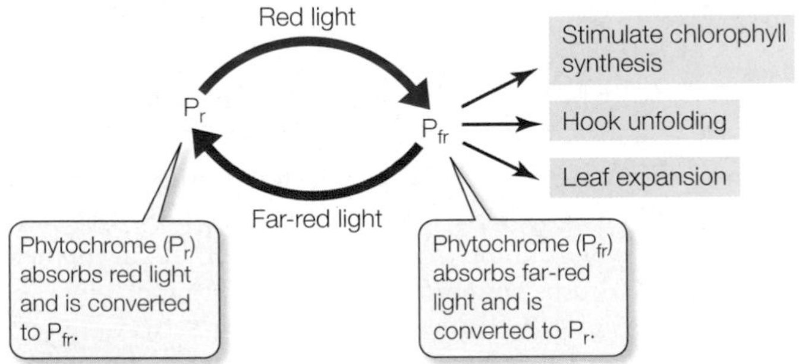

P_{fr} has some important biological effects. As we have just seen, one of them is to initiate germination in certain seeds, such as lettuce.

Phytochromes have many effects on plant growth and development

Phytochromes help to regulate a seedling's early growth. When seeds germinate in the dark below the soil surface, a pale and spindly seedling forms, with undeveloped leaves. Such an **etiolated** seedling cannot carry out photosynthesis. The seedling shoot must reach the soil surface and begin photosynthesis before its nutrient reserves are expended and it starves.

Etiolated plants can be gourmet fare. "Bean sprouts"—a delicacy in China for more than 3,000 years— are mung bean seedlings that are allowed to germinate for several days in either dim light or total darkness. Sprouts grown in total darkness are the whitest, and also the crispest. Similarly, white asparagus spears are grown under a sand cover that keeps out light and allows a distinctive flavor to develop.

Plants have evolved a variety of ways to cope with the problem of germinating underground. Etiolated flowering plants, for example, do not form chlorophyll. Only when they are exposed to light do they synthesize chlorophyll and turn green; this delay conserves the resources needed to make chlorophyll, which would be useless in the dark. An etiolated shoot uses its stored resources to elongate rapidly and hasten its arrival at the soil surface, where photosynthesis quickly begins. To break through the soil yet protect its delicate, underdeveloped leaves, the shoot of an etiolated eudicot seedling forms an apical hook (see Figure 37.17).

All of these etiolation-associated phenomena (lack of chlorophyll, rapid shoot elongation, production of an apical hook, delayed leaf expansion) are regulated by the phytochromes. In a seedling that has never been exposed to light, all the phytochrome is in the P_r (red-absorbing) form. Exposure to light converts P_r to the P_{fr} (far-red-absorbing) form. The P_{fr} initiates a reversal of the etiolation phenomena: Chlorophyll synthesis begins, shoot elongation slows, the apical hook unfolds, and the leaves begin to expand. These changes constitute *photomorphogenesis*.

Multiple phytochromes have different developmental roles

For years, plant biologists had difficulty accounting for some aspects of phytochrome action. A solution to these problems may lie in the discovery of multiple forms of phytochromes and other photoreceptors. *Arabidopsis* has five genes that encode different phytochromes: such diversity has been found throughout the green plants.

The several phytochromes play differing roles during plant development. Some of them even play off each other to fine-tune plant growth during the day. Consider, for example, the light spectrum available to a seedling that is growing in the shade of other plants. Because chlorophyll in the leaves above it absorbs the light first, the shaded seedling receives a spectrum that is relatively rich in far-red wavelengths (and poor in red wavelengths); the ratio of far-red to red is increased as much as 10 to 20 times in the shade. In some shade-intolerant species, the interplay among signal transduction pathways initiated by the different phytochromes leads to an increased rate of stem elongation, moving the shaded leaves to higher light intensities.

Phytochromes act in both the cytoplasm and the nucleus. In the cytoplasm, phytochromes probably function as protein kinases. When they are transferred to the nucleus, phytochromes promote the transcription of many genes.

Cryptochromes, phototropins, and zeaxanthin are blue-light receptors

Cryptochromes are yellow photoreceptor pigments that absorb blue and ultraviolet light. They affect some of the same developmental processes, including seedling development and flowering, that phytochromes do. Unlike phytochromes, cryptochromes play important roles in animals as well as plants.

In contrast to phytochromes, cryptochromes are located primarily in the plant cell nucleus. The exact mechanism of cryptochrome action is not yet known. It may be significant that phytochromes behave like protein kinases, and that cryptochromes can be substrates of such enzymes. It is likely that both classes of photoreceptors participate in protein kinase-based signal transduction pathways (see Section 15.3).

We've noted that the Darwins' study of phototropism led to the discovery of auxin. But a question remained: What photoreceptor initiates phototropism? Plant scientists working with phototropic mutants of *Arabidopsis* showed that two yellow pigments, which they named **phototropins**, are the photoreceptors. Upon absorbing blue light, phototropins initiate a signal transduction pathway leading to phototropic curvature (**Figure 37.21**). Phototropins also participate in the relocation of chloroplasts in response to light intensity. Under low intensity light, chloroplasts gather on the illuminated side of the cell, with their faces oriented perpendicular to the light. Under bright light (sensed by phototropins) the chloroplasts gather on the sides parallel to the light and present their edges, thus absorbing a minimum of light.

Still another type of blue-light receptor, the plastid pigment **zeaxanthin**, appears to participate with phototropin in the light-induced opening of stomata. Zeaxanthin itself is formed in the guard cells in response to light, but it is light absorbed by the zeaxanthin and phototropin that causes the stomata to open.

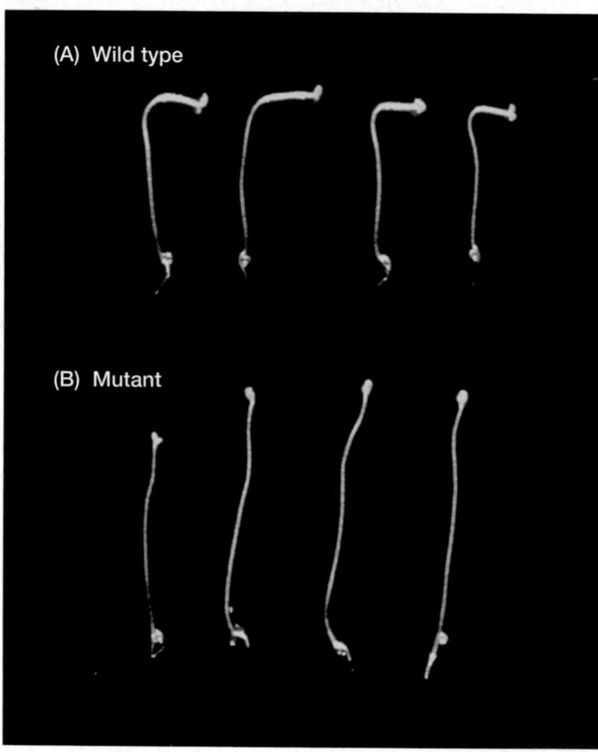

37.21 A Nonphototropic Mutant (A) The four etiolated wild-type *Arabidopsis* seedlings in the top row are demonstrating normal phototropism. (B) These four mutant seedlings cannot produce phototropin. Their lack of phototropic response indicates that phototropin is the photoreceptor that signals the plant to curve toward light.

37.5 RECAP

Phytochromes and blue-light receptors mediate the effects of light on plant growth and development. Several phytochromes exist, each in two forms that are interconvertible by red and far-red light.

- Do you understand why red light affects seed germination differently from far-red light? See p. 813 and Figure 37.20
- Can you explain how the behavior of an etiolated seedling helps it reach the light it needs in order to develop further? See p. 813

Photoreceptors also play a regulatory role in flowering. In addition to light, another environmental cue—temperature—regulates flowering. We will examine these topics and others in the next chapter, which focuses on reproduction in flowering plants.

CHAPTER SUMMARY

37.1 How does plant development proceed?

The environment, photoreceptors, hormones, and the plant's genome all play roles in the regulation of plant development.

Each of several plant hormones plays multiple regulatory roles, affecting several different aspects of development. Interactions among these hormones are often complex. Review Table 37.1

Seed dormancy, which offers adaptive advantages, may be caused by many mechanisms. In nature, it is broken by mechanisms such as abrasion, fire, leaching, and low temperatures. When **dormancy** ends and the seed **imbibes** water, it **germinates** and develops into a **seedling**. Review Figure 37.1, Web/CD Activities 37.1 and 37.2

Hormones and photoreceptors act through signal transduction pathways to regulate seedling development. Before the germinating embryo can begin photosynthesis, it relies on energy reserves in the **cotyledons** and the **endosperm** to grow.

37.2 What do gibberellins do?

The embryos of cereal seeds secrete **gibberellins**, which cause the **aleurone layer** to synthesize and secrete digestive enzymes that break down the large molecules stored in the endosperm. Review Figure 37.4, Web/CD Activity 37.3

Dozens of gibberellins exist. These hormones regulate the growth of stems and of some fruits and cause **bolting** in some **biennial** plants.

37.3 What does auxin do?

See Web/CD Tutorial 37.1

Auxin moves from the tip to the growing region of the **coleoptile**. Review Figures 37.7, 37.8, Web/CD Tutorial 37.2

Auxin transport is polar. **Auxin anion efflux carriers**—membrane proteins confined to the basal ends of cells—cause auxin to move from the tip to the base of the shoot. Review Figure 37.9

Lateral movement of auxin, mediated by auxin carrier proteins, is responsible for **phototropism** and **gravitropism**. Review Figure 37.10

Auxin plays roles in root formation, leaf **abscission**, **apical dominance**, and **parthenocarpic fruit development**. Certain synthetic auxins are used as selective herbicides. Review Figures 37.11 and 37.12

Auxin increases the **plasticity** of the cell wall, promoting cell expansion. It does so by increasing the pumping of protons from the cytoplasm into the cell wall, where the lowered pH activates proteins called expansins. Review Figure 37.15, Web/CD Tutorial 37.3

Receptors for auxin and for gibberellins residing in the cell nucleus initiate similar signal transduction pathways. Review Figure 37.16

37.4 What do cytokinins, ethylene, abscisic acid, and brassinosteroids do?

Cytokinins are adenine derivatives. They promote plant cell division, promote seed germination in some species, inhibit stem elongation, promote lateral swelling of stems and roots, stimulate the growth of axillary buds, promote the expansion of leaf tissue, and delay leaf **senescence**.

A balance between auxin and **ethylene** controls leaf abscission. Ethylene promotes senescence and fruit ripening. It causes the formation of a protective **apical hook** in eudicot seedlings. In stems, it inhibits elongation, promotes lateral swelling, and causes a loss of gravitropic sensitivity. Review Figure 37.18

Abscisic acid maintains winter dormancy in buds, prevents seeds from germinating while still attached to the parent plant, and inhibits stem elongation. It also affects stomatal opening.

Dozens of different **brassinosteroids** affect cell elongation, pollen tube elongation, vascular tissue differentiation, and root elongation. Some effects of light are mediated by changes in the action and levels of brassinosteroids. Review Figure 37.19

37.5 How do photoreceptors participate in plant growth regulation?

Phytochromes are bluish pigments found in the cytosol. They exist in two forms, $\mathbf{P_r}$ and $\mathbf{P_{fr}}$, that are interconvertible by light. They affect seedling growth, flowering, and etiolation. Review Figure 37.20

The five known phytochromes mediate the effects of red, far-red, and dim blue light. They may play different roles in plant development, and their signal transduction pathways may interact to mediate the effects of light environments of differing spectral distribution.

The **blue-light receptors** are yellow pigments that absorb blue and ultraviolet light. **Cryptochromes**, which mediate the effects of high-energy light, interact with phytochromes in controlling seedling development and floral initiation. The signaling pathways for phytochromes and cryptochromes are based on protein kinases.

Other blue-light receptors are **phototropins**, the photoreceptors for phototropism and chloroplast movements, and **zeaxanthin**, which with the phototropins mediates the light-induced opening of stomata.

SELF-QUIZ

1. Which of the following is *not* an advantage of seed dormancy?
   a. It makes the seed more likely to be digested by birds that disperse it.
   b. It counters the effects of year-to-year variations in the environment.
   c. It increases the likelihood that a seed will germinate in the right place.
   d. It favors dispersal of the seed.
   e. It may result in germination at a favorable time of year.
2. Which of the following does/do *not* participate in seed germination?
   a. Imbibition of water
   b. Metabolic changes
   c. Growth of the radicle
   d. Mobilization of nutrient reserves
   e. Extensive mitotic divisions
3. To mobilize its nutrient reserves, a germinating barley seed
   a. becomes dormant.
   b. undergoes senescence.
   c. secretes gibberellins into its endosperm.
   d. converts glycerol and fatty acids into lipids.
   e. takes up proteins from the endosperm.

4. The gibberellins
   a. are responsible for phototropism and gravitropism.
   b. are gases at room temperature.
   c. are produced only by fungi.
   d. cause bolting in some biennial plants.
   e. inhibit the synthesis of digestive enzymes by barley seeds.
5. In coleoptile tissue, auxin
   a. is transported from base to tip.
   b. is transported from tip to base.
   c. can be transported toward either the tip or the base, depending on the orientation of the coleoptile with respect to gravity.
   d. is transported by simple diffusion, with no preferred direction.
   e. is not transported, because auxin is used where it is made.
6. Which process is *not* directly affected by auxin?
   a. Apical dominance
   b. Leaf abscission
   c. Synthesis of digestive enzymes by barley seeds
   d. Root initiation
   e. Parthenocarpic fruit development
7. Plant cell walls
   a. are strengthened primarily by proteins.
   b. often make up more than 90 percent of the total volume of an expanded cell.
   c. can be loosened by an increase in pH.
   d. become thinner and thinner as the cell grows longer and longer.
   e. are made more plastic by treatment with auxin.
8. Which statement about cytokinins is *not* true?
   a. They promote bud formation in tissue cultures.
   b. They delay the senescence of leaves.
   c. They usually promote the elongation of stems.
   d. They cause certain light-requiring seeds to germinate in the dark.
   e. They stimulate the development of branches from axillary buds.
9. Ethylene
   a. is antagonized by silver salts such as silver thiosulfate.
   b. is liquid at room temperature.
   c. delays the ripening of fruits.
   d. generally promotes stem elongation.
   e. inhibits the swelling of stems, in opposition to cytokinins' effects.
10. Phytochrome
    a. is the only photoreceptor pigment in plants.
    b. exists in two forms interconvertible by light.
    c. is a pigment that is colored red or far-red.
    d. is a green-light receptor.
    e. is the photoreceptor for phototropism.

FOR DISCUSSION

1. How may it be advantageous for some species to have seeds whose dormancy is broken by fire?
2. Cocklebur fruits contain two seeds each that are kept dormant by two different mechanisms. How might this use of two mechanisms of dormancy be advantageous to cockleburs?
3. Whereas relatively low concentrations of auxin promote the elongation of segments cut from young plant stems, higher concentrations generally inhibit their growth, as shown:

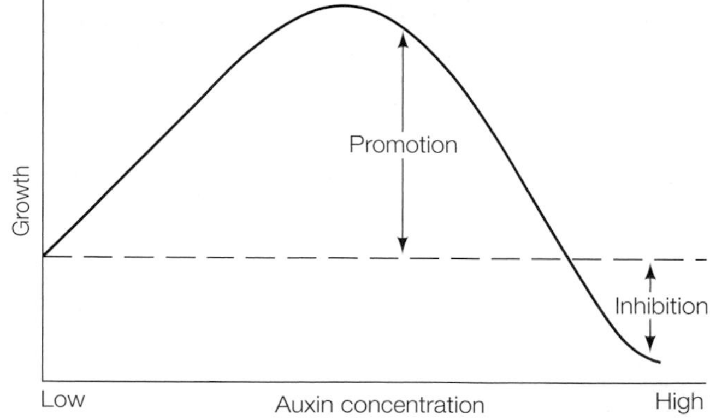

In some plants, the inhibitory effects of high auxin concentrations appear to be secondary: High auxin concentrations cause the synthesis of ethylene, which is what causes the growth inhibition. Silver thiosulfate inhibits ethylene action. How do you think the addition of silver thiosulfate to the solutions in which the stem segments grew would affect the appearance of the graph at the left?

4. Corn stunt virus causes a great reduction in the growth rate of infected corn plants, so the diseased plants take on a dwarfed form. Since their appearance is reminiscent of the genetically dwarfed corn studied by Phinney, you suspect that the virus may inhibit the synthesis of gibberellins by the corn plants. Describe two experiments you might conduct to test this hypothesis, only one of which should require chemical measurement.
5. Some etiolated seedlings develop hairs on their epidermis when exposed to dim light. Describe an experiment to test the hypothesis that a phytochrome is the photoreceptor for this effect.

FOR INVESTIGATION

A. N. J. Heyn investigated the effects of auxin on the plasticity and elasticity of cell walls by bending segments of oat coleoptiles after treating them with or without auxin. Suggest other ways by which one might measure the plasticity and elasticity of cell walls with and without treatment with auxin or exposure to low pH.

CHAPTER 38 Reproduction in Flowering Plants

Big smelly flowers of Sumatra

The British trading colony of Singapore was founded in 1819 by Sir Thomas Stamford Raffles. Today Singapore is a dynamic, independent city-country whose Raffles Hotel is famous as the home of the tropical gin drink known as the Singapore Sling.

Sir Thomas, however, achieved greater things. He abhorred slavery, and during his tenure as the British governor of Sumatra was able to curtail the then-thriving Dutch slave trade. He was also a noteworthy naturalist with a keen interest in the magnificent flora and fauna of the East Indies. The natural history collection he built is the basis of today's Raffles Museum of Biodiversity Research of the National University of Singapore.

In 1818, Sir Thomas traveled into the Sumatran rainforest with an English friend, the botanist Dr. Joseph Arnold. A native guide led the men to an enormous red flower growing in the seeming absence of any leaves or stems. The flower was found to be that of a parasitic plant, with filamentous roots growing inside climbing vines. The parasite gives no visible sign of its presence until a bud appears and grows into a gigantic flower that lasts about a week.

This plant, dubbed *Rafflesia arnoldii*, holds the world record for flower weight—more than 10 kilograms. But it is not the only giant flower found in Sumatra. *Amorphophallus titanum*, also known as titan arum, "voodoo lily," or "corpse flower," is a relative of skunk cabbage and calla lilies. While *Rafflesia* bears a true flower, the "flower" of *A. titanum* is actually an inflorescence, or cluster of many flowers. The towering central structure, called the *spadix*, reaches 2 meters in length; the surrounding petal-like *spathe* can be a meter wide.

Rafflesia, titan arum, and skunk cabbage all smell truly awful ("like a rotting possum under your porch," as one arboretum curator put it). At the time of pollination, these plants all release the protein breakdown products putrescine and cadaverine, which smell just as their names suggest. These malodorous chemicals attract flies and other carrion-feeding insects, which lay eggs and depart, unwittingly carrying pollen. (The insect larvae that hatch, unfortunately for them, find no dead meat to eat, and thus die.)

As in all angiosperms, the flowers of *Rafflesia* and *Amorphophallus* are reproductive equipment. Flowers produce the gametophytes, female and male, that produce the gametes that give rise to the next sporophyte generation. The male gametophytes take the form of pollen grains, and pollination is crucial to angiosperm reproduction.

The World's Largest Flower *Rafflesia arnoldii* produces the world's largest flower; its blossom attains a diameter of nearly a meter and can weigh up to 11 kg. It is rarely seen, as the buds take many months to develop and the blossom lasts for just a few days.

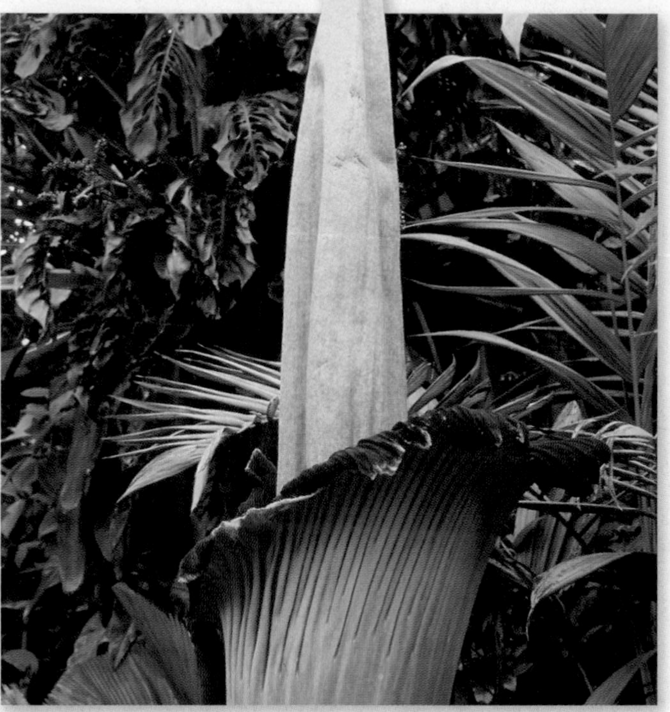

The World's Largest Inflorescence This massive structure is borne by *Amorphophallus titanum*, which came by its name for obvious reasons. Its individual flowers are arranged on a spadix.

Will *Rafflesia's* specialized pollination system be adequate to preserve the species? Its host plant, *Tetrastigma tuberculatum*, grows only in undisturbed rainforests in Southeast Asia, and that habitat is disappearing under human pressures. *Rafflesia* flowers only rarely, and individuals are dioecious. What are the odds of a female *Rafflesia* plant blooming within a very few days of a male plant, and close enough to allow an insect to deliver pollen from the male to the female? The plant does not appear to be common, and its long-term future is doubtful.

IN THIS CHAPTER we contrast sexual and asexual reproduction, focusing on the details of sexual reproduction. We will consider angiosperm gametophytes, pollination, double fertilization, embryo development, and the roles of fruits in seed dispersal. We will examine the transition from the vegetative state to the flowering state, a key event in angiosperm development. We conclude by considering the role of asexual reproduction in nature and in agriculture.

CHAPTER OUTLINE

38.1 How Do Angiosperms Reproduce Sexually?

Angiosperms have many ways of reproducing—and humans have developed even more ways of reproducing them. Flowers contain the sex organs; thus it is no surprise that almost all angiosperms reproduce sexually. But many reproduce asexually as well; some even reproduce asexually most of the time. What are the advantages and disadvantages of these two kinds of reproduction? The answers to this question involve genetic recombination. As we have seen, sexual reproduction produces new combinations of genes and diverse phenotypes. Asexual reproduction, in contrast, produces a clone of genetically identical individuals.

Both sexual and asexual reproduction are important in agriculture. Many important annual crops are grown from seeds, which are the product of sexual reproduction. Seed-grown crops include the great grain crops, all of which are grasses—wheat, rice, corn, sorghum, and millet—as well as plants in other families, such as soybean and safflower. Other crops, such as strawberry, potato, and banana, are usually produced asexually.

Orange trees, which have been under cultivation for centuries, can be grown from seed—except for the navel orange, which has no seeds. Early in the nineteenth century, on a plantation on the Brazilian coast, a single orange seed gave rise to one tree that had aberrant flowers. Parts of the flowers aborted, and seedless fruits formed. Asexual reproduction is the only way of propagating this plant, and every navel orange in the world comes from a tree derived asexually from that original Brazilian navel orange tree.

Unlike navel oranges, strawberries are capable of forming seeds and need not be propagated asexually. Nonetheless, asexual propagation of strawberries is common because individual plants are highly heterozygous and do not breed true to type; thus asexual propagation of a particularly desirable genotype ensures product uniformity and high quality. We will treat asexual reproduction in greater detail at the end of this chapter.

Our concern for now is sexual reproduction. In sexual reproduction, meiosis and subsequent mating between individuals of different genotypes shuffle genes into new combinations, resulting in a diversity of genotypes in each generation, some of which may be superior to those of the parental generation. This genetic diversity may serve the population well if the environment changes or as the population expands into new environments. The adaptability resulting from genetic diversity is the major advantage of sexual reproduction over asexual reproduction, although sexual reproduction can also break up well-adapted combinations of alleles through the same process of recombination.

38.1 Development of Gametophytes and Nuclear Fusion The embryo sac is the female gametophyte; the pollen grain is the male gametophyte. The male and female nuclei meet and fuse within the embryo sac. Most angiosperms have double fertilization, in which a zygote and an endosperm nucleus form from separate fusion events—the zygote from one sperm and the egg, and the endosperm from the other sperm and two polar nuclei.

The flower is an angiosperm's structure for sexual reproduction

A complete flower consists of four groups of organs that are modified leaves: the carpels, stamens, petals, and sepals (see Figure 29.9). The *carpels* and *stamens* are, respectively, the female and male

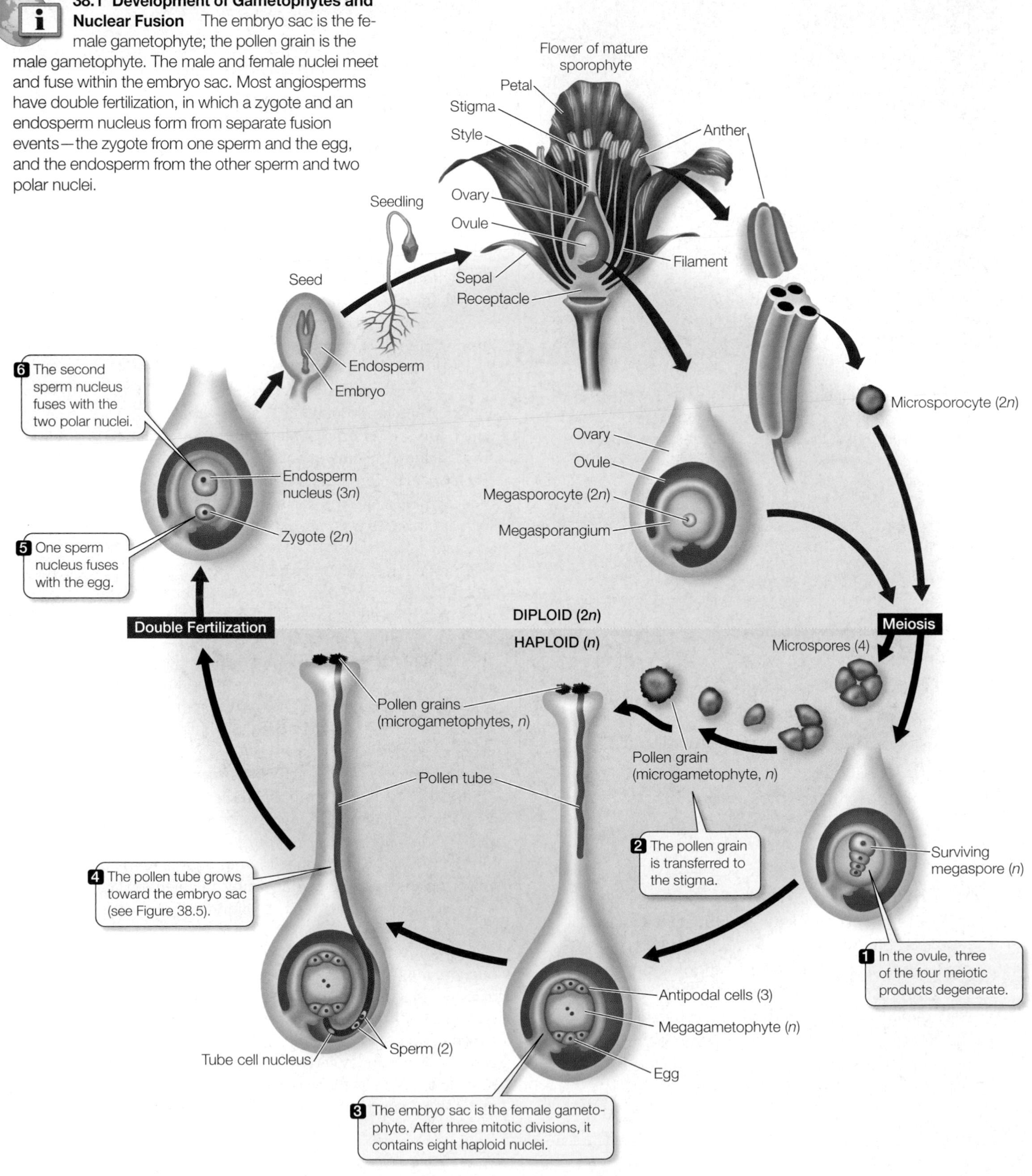

sex organs. A *pistil* is a structure composed of one or more carpels. The base of the pistil, called the *ovary*, contains one or more *ovules*, each of which contains a *megasporangium*, within which a female gametophyte may develop. The stalk of the pistil is the *style*, and the end of that stalk is the *stigma*. Each stamen is composed of a *filament* bearing a two-lobed *anther*, which consists of four *microsporangia* fused together. Male *gametophytes* begin their development within the microsporangia.

The *petals* and *sepals* of many flowers are arranged in whorls (circles) or spirals around the carpels and stamens. Together, the petals constitute the *corolla*. Below them, the sepals constitute the *calyx*. The petals are often colored, attracting pollinating animals; the sepals are often green and photosynthetic. All the parts of the flower are borne on a stem tip, the *receptacle*. Flower parts are very diverse in form, in contrast to the microscopic gametophytes that develop within them.

Flowering plants have microscopic gametophytes

Central to understanding plant reproduction is the concept of *alternation of generations*, in which a multicellular diploid ($2n$) generation alternates with a multicellular haploid (n) generation (see Section 28.2).

In angiosperms, the diploid sporophyte generation is the larger and more conspicuous one. The sporophyte generation produces flowers. The flowers produce spores, which develop into tiny gametophytes that begin and, in the case of the megagametophyte, end their development enclosed by sporophyte tissue.

The haploid gametophytes—the gamete-producing generation—of flowering plants develop from haploid spores in sporangia within the flower (**Figure 38.1**):

- Female gametophytes (megagametophytes), which are called **embryo sacs**, develop in megasporangia.
- Male gametophytes (microgametophytes), which are called **pollen grains**, develop in microsporangia.

Within the ovule, a *megasporocyte*—a cell within the megasporangium—divides meiotically to produce four haploid *megaspores*. In most flowering plants, all but one of these megaspores then degenerate. The surviving megaspore usually undergoes three mitotic divisions, producing eight haploid nuclei, all initially contained within a single cell—three nuclei at one end, three at the other, and two in the middle. Subsequent cell wall formation leads to an elliptical, seven-celled megagametophyte with a total of eight nuclei:

- At one end of the elliptical megagametophyte are three tiny cells: the egg and two cells called *synergids*. The egg is the female gamete, and the synergids participate in fertilization by attracting the pollen tube and receiving the sperm nuclei prior to their movement to the egg and central cell.
- At the opposite end of the megagametophyte are three *antipodal cells*, which eventually degenerate.
- In the large central cell are two **polar nuclei**, which together combine with a sperm nucleus.

The embryo sac (megagametophyte) is the entire seven-cell, eight-nucleus structure. You can review the development of the embryo sac in Figure 38.1.

The pollen grain (microgametophyte) consists of fewer cells and nuclei than the embryo sac (**Figure 38.2**). The development of a pollen grain begins when a *microsporocyte* within the anther divides meiotically. Each resulting haploid *microspore* develops a spore wall, within which it normally undergoes one mitotic division before the anthers open and release these two-celled pollen grains. The two cells are the *tube cell* and the *generative cell*. Further development of the pollen grain, which we will describe shortly, is delayed until the pollen arrives at a stigma. In angiosperms, the transfer of pollen from the anther to the stigma is referred to as **pollination**.

Pollination enables fertilization in the absence of water

Gymnosperms and angiosperms do not require external water as a medium for gamete travel and fertilization—a freedom not shared by other plant groups. The male gametes of gymnosperms and angiosperms are contained within pollen grains. But how do these pollen grains travel from an anther to a stigma?

Many different mechanisms have evolved for pollen transport. In some plants, such as peas and their relatives, self-pollination is accomplished before the flower bud opens. Pollen is transferred by the direct contact of anther and stigma within the same flower, resulting in *self-fertilization*.

Wind is the vehicle for pollen transport in many species. Wind-pollinated flowers have sticky or featherlike stigmas, and they produce pollen grains in great numbers. Some aquatic angiosperms are pollinated by water carrying pollen grains from plant to plant. Animals, including insects, birds, and bats, carry pollen among the

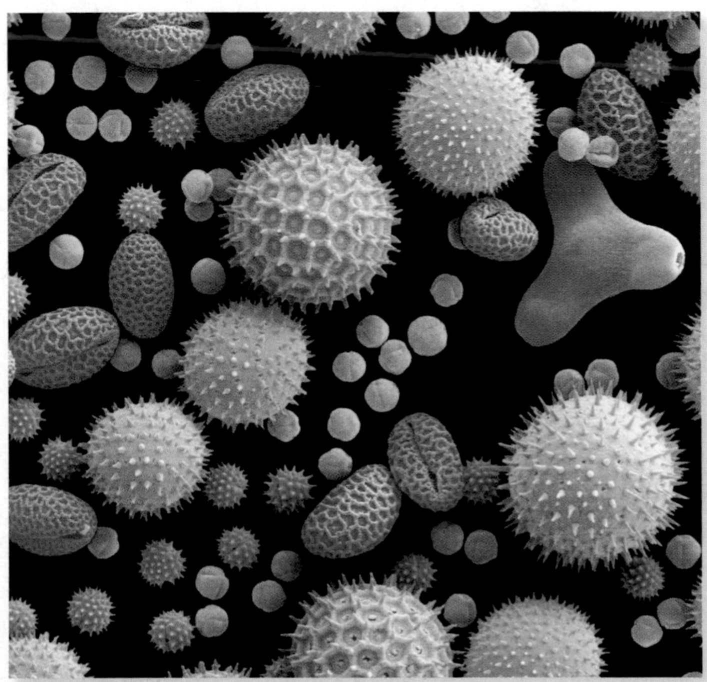

38.2 Pollen Galore Each species' pollen has a characteristic size, shape, and cell wall structure. These grains are the male gametophytes. Color has been artificially added to the micrograph.

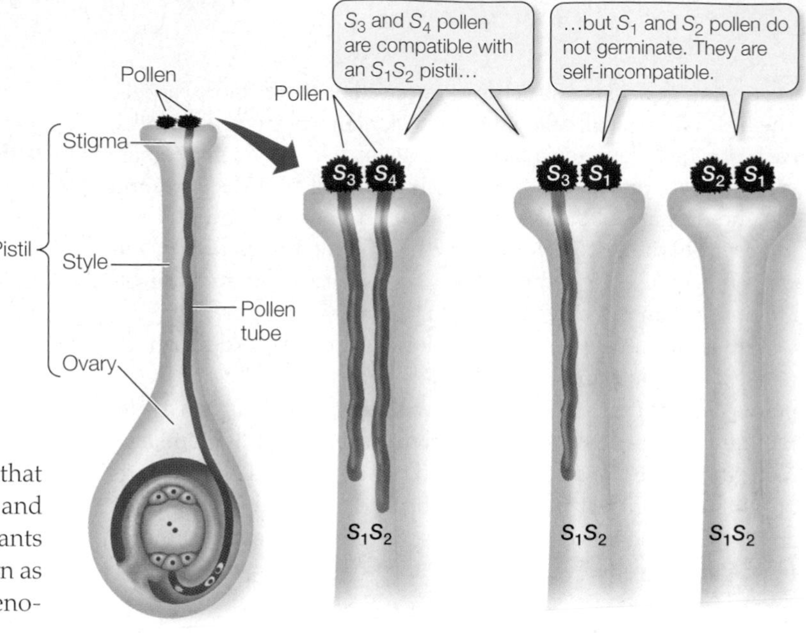

38.3 Self-Incompatibility Pollen grains do not germinate normally if their *S* allele matches one of the *S* alleles of the stigma. Thus, the egg cannot be fertilized by a sperm from the same plant.

flowers of many plants. Pollination among the flowers of different individuals of the same species is called *cross-pollination*.

Some flowering plants practice "mate selection"

In our discussion of Mendel's work (see Section 10.1), we saw that some plants can reproduce sexually by both cross-pollination and self-pollination. But not all plants have this flexibility. Many plants reject pollen from their own flowers. This phenomenon, known as **self-incompatibility**, promotes outcrossing between different genotypes.

A single gene, the *S* gene, is responsible for self-incompatibility in most plants. The *S* gene has dozens of alleles. A pollen grain is haploid and possesses a single *S* allele; the recipient pistil is diploid. In self-incompatible plants, pollen fails to germinate, or the pollen tube fails to traverse the style, if the *S* allele of the pollen matches one of the two *S* alleles in the pistil (**Figure 38.3**).

The stigma plays an important role in "mate selection" by flowering plants. The stigmas of most plants are exposed to the pollen of many other species as well as their own. Pollen from the same species binds strongly to the stigma by cell–cell signaling between the stigma and the cell walls of the pollen grains. In contrast, foreign pollen falls off readily or fails to germinate—or the pollen tubes are incapable of penetrating the stigma.

A pollen tube delivers sperm cells to the embryo sac

When a functional pollen grain lands on the stigma of a compatible pistil, it germinates. Germination for a pollen grain is the development of a **pollen tube** (**Figure 38.4**). The pollen tube either traverses the spongy tissue of the style or, if the style is hollow, grows downward on the inner surface of the style until it reaches an ovule. The pollen tube may grow millimeters or even centimeters in the process.

The downward growth of the pollen tube is guided in part by a chemical signal, a small protein produced by the synergids within the ovule. If one synergid is destroyed, the ovule still attracts pollen tubes, but destruction of both synergids renders the ovule unable to attract pollen tubes, and fertilization does not occur.

Angiosperms perform double fertilization

In most angiosperm species, the mature pollen grain consists of two cells, the tube cell and the generative cell. The larger tube cell encloses the much smaller generative cell. Guided by the tube cell nucleus, the pollen tube eventually grows through the megasporangial tissue and reaches the embryo sac. The generative cell, meanwhile, has undergone one mitotic division and cytokinesis to produce two haploid **sperm cells**.

Both of the sperm cells enter the embryo sac, where they are released into the cytoplasm of one of the synergids. This synergid degenerates, releasing the sperm cells (**Figure 38.5**). Each sperm cell then fuses with a different cell of the embryo sac. One sperm cell fuses with the egg cell, producing the diploid zygote. The other fuses with the two polar nuclei in the central cell, forming a **triploid (3*n*) nucleus**. While the zygote nucleus begins mitotic division to form the new sporophyte embryo, the triploid nucleus undergoes rapid mitosis to form a specialized nutritive tissue, the **endosperm**. The endosperm will later be digested by the developing embryo, as discussed in Chapter 37. The antipodal cells and the remaining synergid eventually degenerate, as does the pollen tube nucleus.

This process is known as **double fertilization** because it involves two nuclear fusion events:

- One sperm cell fuses with the egg cell.
- The other sperm cell fuses with the two polar nuclei.

The fusion of a sperm cell nucleus with the two polar nuclei to form endosperm takes place only in angiosperms. The fusion of these

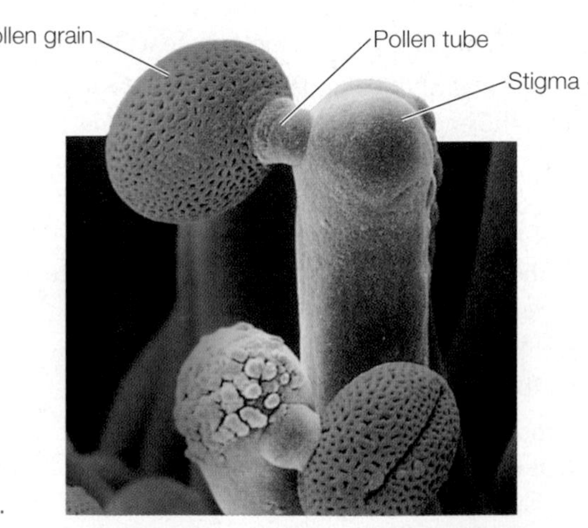

38.4 Pollen Tubes Begin to Grow These pollen grains have landed on fingerlike structures on the stigma of an *Arabidopsis* flower, and pollen tubes have penetrated the stigma.

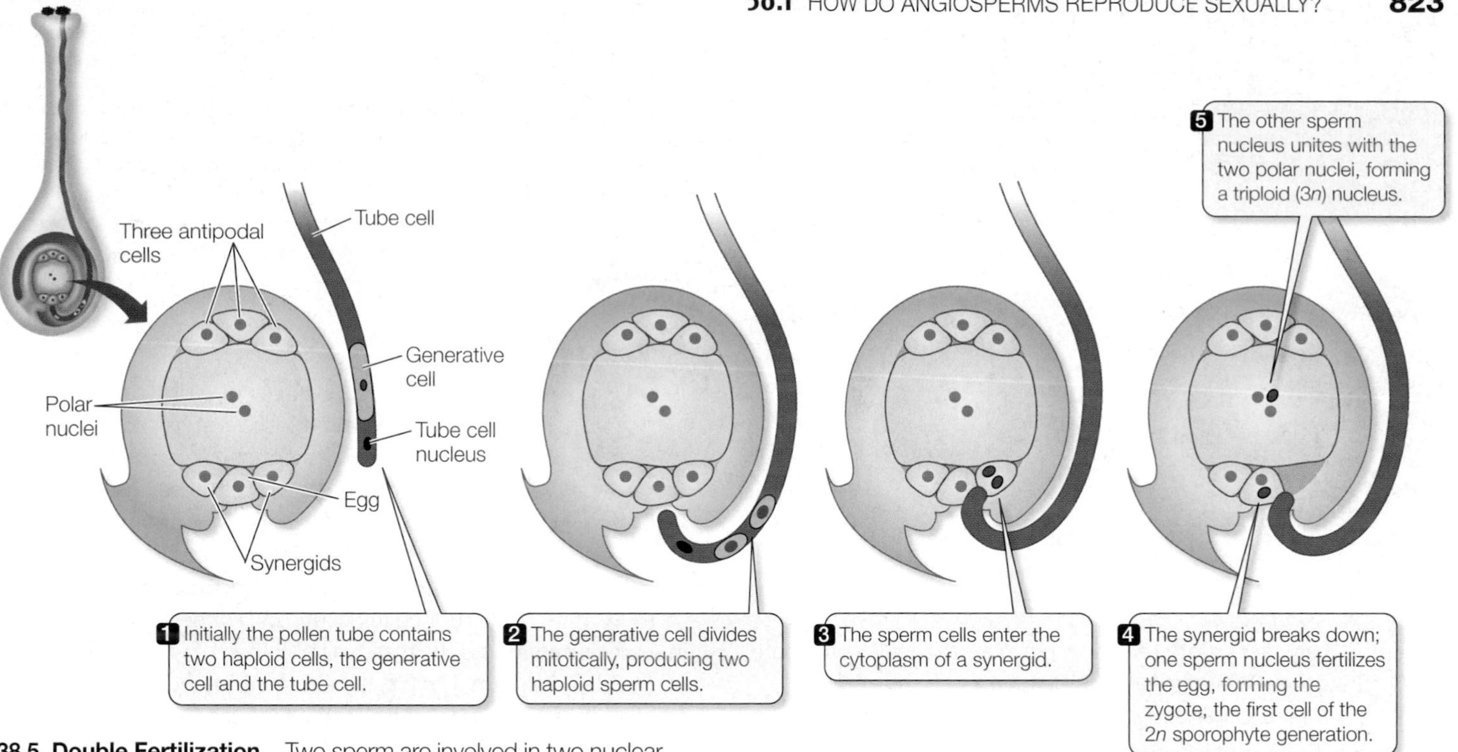

Three antipodal cells
Tube cell
Generative cell
Polar nuclei
Tube cell nucleus
Egg
Synergids
1 Initially the pollen tube contains two haploid cells, the generative cell and the tube cell.
2 The generative cell divides mitotically, producing two haploid sperm cells.
3 The sperm cells enter the cytoplasm of a synergid.
4 The synergid breaks down; one sperm nucleus fertilizes the egg, forming the zygote, the first cell of the 2n sporophyte generation.
5 The other sperm nucleus unites with the two polar nuclei, forming a triploid (3n) nucleus.

38.5 Double Fertilization Two sperm are involved in two nuclear fusion events, hence the term "double fertilization." One sperm is involved in the formation of the diploid zygote and the other results in the formation of the triploid endosperm. Double fertilization is a characteristic feature of angiosperm reproduction.

three nuclei, the possession of flowers, and the formation of fruit are the three most definitive characteristics of angiosperms.

The fusion of a sperm cell with the two polar nuclei produces not a zygote but a nutritionally supportive tissue, the endosperm. You consume endosperm in many foods—including popcorn, which is really popped endosperm.

Embryos develop within seeds

Shortly after fertilization, highly coordinated growth and development of embryo, endosperm, integuments, and carpel ensues. The *integuments*—tissue layers immediately surrounding the megasporangium—develop into the seed coat, and the carpel ultimately becomes the wall of the fruit that encloses the seed.

The first step in the formation of the embryo is a mitotic division of the zygote that gives rise to two daughter cells. These two cells face different fates. An asymmetrical (uneven) distribution of cytoplasm within the zygote causes one daughter cell to produce the embryo proper and the other daughter cell to produce a supporting structure, the **suspensor** (**Figure 38.6**). The suspensor pushes

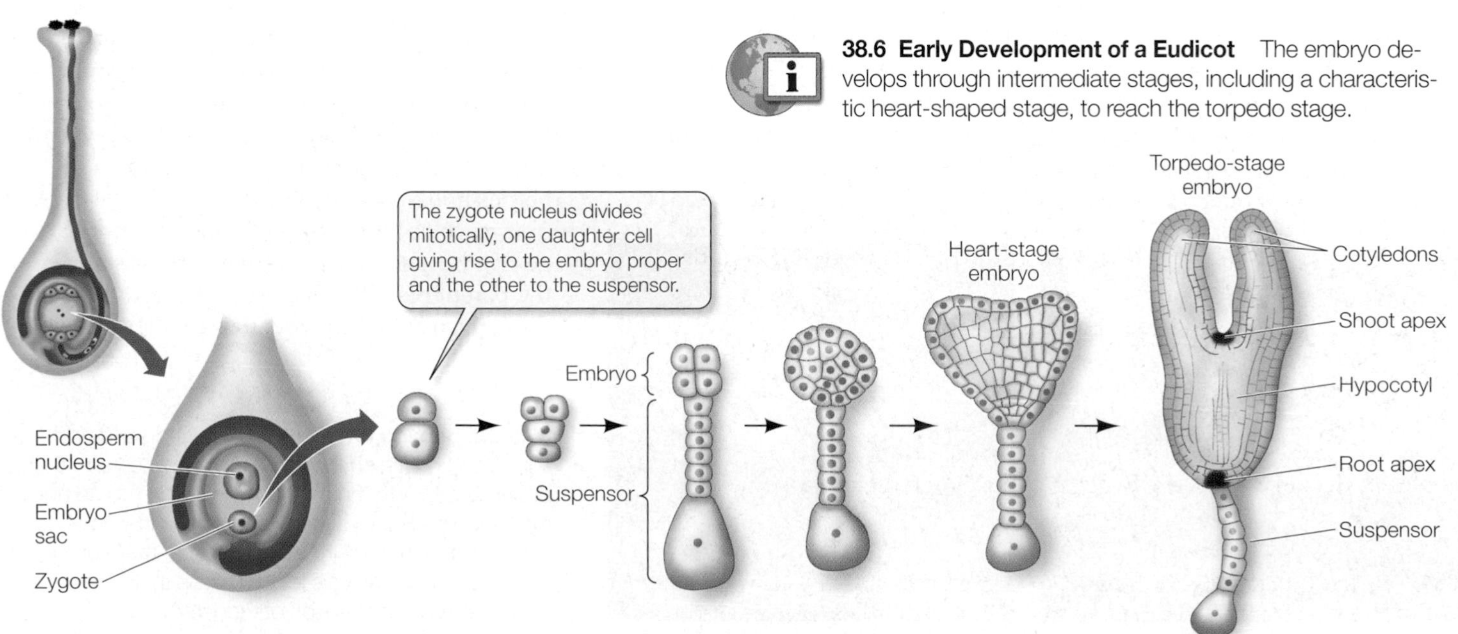

38.6 Early Development of a Eudicot The embryo develops through intermediate stages, including a characteristic heart-shaped stage, to reach the torpedo stage.

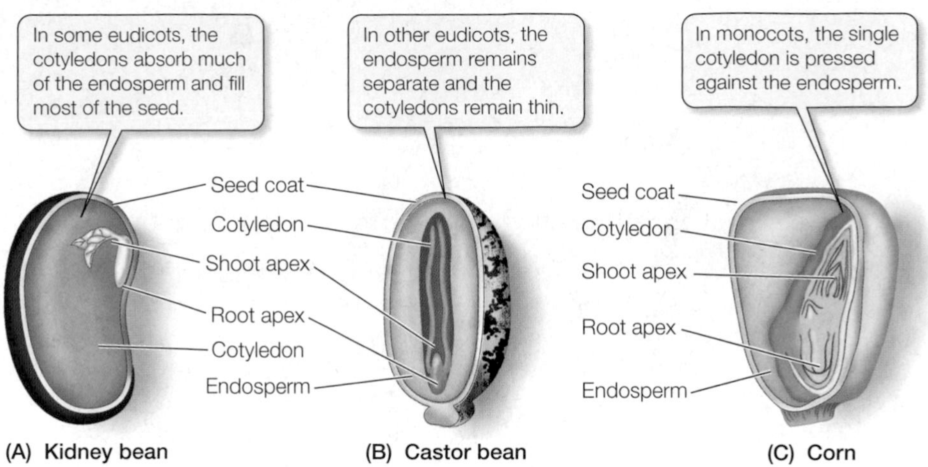
In some eudicots, the cotyledons absorb much of the endosperm and fill most of the seed.
In other eudicots, the endosperm remains separate and the cotyledons remain thin.
In monocots, the single cotyledon is pressed against the endosperm.
Seed coat
Cotyledon
Shoot apex
Root apex
Cotyledon
Endosperm
Seed coat
Cotyledon
Shoot apex
Root apex
Endosperm
(A) Kidney bean
(B) Castor bean
(C) Corn

38.7 Variety in Angiosperm Seeds In some seeds, such as kidney beans (A), the nutrient reserves of the endosperm are absorbed by the cotyledons. In others, such as castor beans (B) and corn (C), the reserves in the endosperm will be drawn upon after germination.

the embryo against or into the endosperm and provides one route by which nutrients pass from the endosperm into the embryo.

The asymmetrical division of the zygote establishes polarity as well as the longitudinal axis of the new plant. A long, thin suspensor and a more spherical or globular embryo are distinguishable after just four mitotic divisions. The suspensor soon ceases to elongate. However, cell divisions continue, the primary meristems form, and the first organs begin to form within the embryo.

In eudicots, the initially globular embryo develops into the characteristic *heart stage* as the cotyledons ("seed leaves") start to grow. Further elongation of the cotyledons and of the main axis of the embryo gives rise to the *torpedo stage*, during which some of the internal tissues begin to differentiate (see Figure 38.6). Between the cotyledons is the *shoot apex*; at the other end of the axis is the *root apex* (**Figure 38.7**). Each of the apical regions contains the meristematic cells that continue to divide to give rise to the organs developing over the life of the plant.

During seed development, large amounts of nutrients are moved in from other parts of the parent plant, and the endosperm accumulates starch, lipids, and proteins. In many species, the cotyledons absorb the nutrient reserves from the surrounding endosperm and grow very large in relation to the rest of the embryo (Figure 38.7A). In others, the cotyledons remain thin (Figure 38.7B) and draw on the reserves in the endosperm as needed when the seed germinates.

In the late stages of embryonic development, the seed loses water—sometimes as much as 95 percent of its original water content. In this desiccated state, the embryo is incapable of further development; it remains quiescent until internal and external conditions are right for germination. (As mentioned in Section 37.1, a necessary early step in seed germination is the massive imbibition of water.) As the embryo and endosperm develop, the structures of the ovule and ovary are also undergoing developmental changes to form a seed and fruit, respectively.

Some fruits assist in seed dispersal

In angiosperms the ovary wall—together with its seeds—develops into a fruit after fertilization has occurred. A **fruit** may consist of only the mature ovary and seeds, or it may include other parts of the flower or structures that are closely related to it. In some species, this process produces fleshy, edible fruits such as peaches and tomatoes, while in other species the fruits are dry or inedible. Some major variations on this theme are illustrated in Figure 29.15, which shows only fleshy, edible fruits. Whatever its form, the fruit serves to promote seed dispersal.

Some fruits help disperse seeds over substantial distances, increasing the probability that at least a few of the many seeds produced by a plant will find suitable conditions for germination and growth to sexual maturity. Various plants, including milkweed and dandelion, produce a fruit with a "parachute" that may be blown some distance from the parent plant by the wind (**Figure 38.8A**). Water disperses some fruits; coconuts have been known to travel thousands of miles between islands. Still other fruits move by hitching rides with animals—either on, as with burrs on your hiking socks (**Figure 38.8B**), or inside them, as with berries in birds. Seeds swallowed whole along with fruits such as berries travel through the animal's digestive tract and are deposited some distance from the parent plant. In some species, seeds must pass through an animal in order to break dormancy.

(A) *Asclepias syriaca*

(B) *Arctium* sp.

38.8 Dispersing Fruit (A) A milkweed seed pod. Silky filaments catch the wind currents and carry the brown seeds with them. (B) Animals who rub up against the "hook and loop" surface of burdock fruit walk away with it attached to their fur, thus making them unwitting agents of dispersal. This feature of the fruit is said to have inspired the invention of modern Velcro™.

38.1 RECAP

Sexual reproduction in angiosperms is accomplished by flowers. After fertilization, the flower develops seed(s) and fruit.

- Explain the relationship between an ovule and an ovary, and between a fruit and a seed. See pp. 821 and 824 and Figure 38.1
- What function does self-incompatibility serve, and how do plants engage in "mate selection"? See p. 822 and Figure 38.3
- Explain the roles of the two sperm nuclei in double fertilization. See p. 822 and Figure 38.5

We have now traced the sexual life cycle of angiosperms from the flower, to the fruit, to the dispersal of seeds. Seed germination and the vegetative development of the seedling are presented in Chapter 37. The next section covers the rest of the angiosperm life cycle—the transition from the vegetative to the flowering state—and how this transition is regulated.

38.2 What Determines the Transition from the Vegetative to the Flowering State?

If we view an angiosperm as something produced by a seed for the function of bearing more seeds, then the act of flowering is one of the supreme events in the plant's life. The transition to the flowering state marks the end of vegetative growth for some plants. In other plants, vegetative growth may accompany flowering or resume after flowering is completed. But whatever the specific pattern, flowering always entails major developmental changes.

Apical meristems can become inflorescence meristems

The first visible sign of a transition to the flowering state may be a change in one or more apical meristems in the shoot system. During vegetative growth, an apical meristem continually produces leaves, axillary buds, and stem (**Figure 38.9A**) in a kind of unrestricted growth called *indeterminate growth* (see Section 34.3).

Flowers may appear singly or in an orderly cluster that constitutes an *inflorescence*. If a vegetative apical meristem becomes an **inflorescence meristem**, it ceases production of leaves and axillary buds and produces other structures: smaller leafy structures called *bracts*, as well as new meristems in the angles between the bracts and the stem (**Figure 38.9B**). These new meristems may also be inflorescence meristems, or they may be **floral meristems**, each of which gives rise to a flower.

Each floral meristem typically produces four consecutive whorls or spirals of organs—the sepals, petals, stamens, and carpels discussed earlier in the chapter—separated by very short internodes, keeping the flower compact (**Figure 38.9C**). In contrast to vegetative apical meristems and some inflorescence meristems, floral meristems are responsible for *determinate* growth—growth of limited duration, like that of leaves.

A cascade of gene expression leads to flowering

How do apical meristems become floral meristems or inflorescence meristems, and how do inflorescence meristems give rise to floral meristems? How does a floral meristem give rise, in short order, to four different floral organs (sepals, petals, stamens, and carpels)? How does each flower come to have the correct number of each of the floral organs? Numerous genes are expressed and interact to produce these results. We'll refer here to some of the genes whose actions have been most thoroughly studied in *Arabidopsis* and snapdragons (*Antirrhinum*).

- Expression of a group of **meristem identity genes** initiates a cascade of further gene expression.
- This cascade begins with **cadastral genes**, which participate in pattern formation—the spatial organization of the whorls of organs.
- Cadastral genes trigger the expression of **floral organ identity genes**, which work in concert to specify the successive whorls (see Figure 19.15).

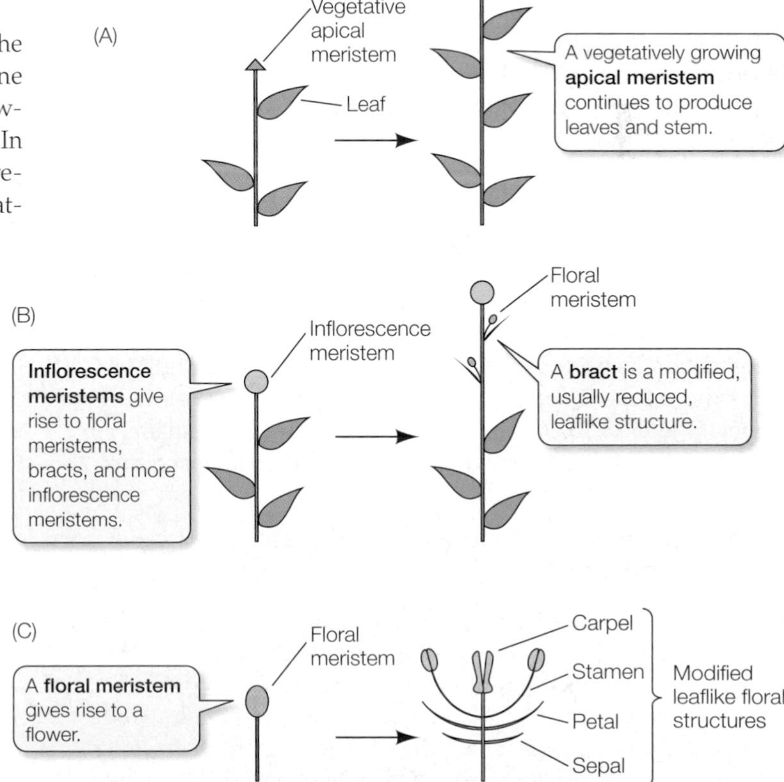

38.9 Flowering and the Apical Meristem A vegetative apical meristem (A) grows without producing flowers. Once the transition to the flowering state is made, inflorescence meristems (B) give rise to bracts and to floral meristems (C), which become the flowers.

Floral organ identity genes are homeotic genes (see Section 19.5), and their products control the transcription of other genes and assign an identity to cells in the floral meristems to control their developmental patterns.

Let's now consider how the transition from the vegetative to the flowering state is initiated.

Photoperiodic cues can initiate flowering

Environmental cues trigger the transition to the flowering state in many cases, depending on the genetic makeup of the species. The life cycles of flowering plants fall into three categories: annual, biennial, and perennial. **Annuals**, such as many food crops, complete their life cycle in one growing season. **Biennials**, such as carrots and sweet William, grow vegetatively for all or part of one growing season, then flower, form seeds, and die in the second growing season. **Perennials**, such as oak trees, live for a few to many growing seasons, during which both vegetative growth and flowering occur once the plants have reached sexual maturity. What control systems give rise to these and other differences in flowering behavior?

In 1920, W. W. Garner and H. A. Allard of the U.S. Department of Agriculture studied the behavior of a newly discovered mutant tobacco plant. The mutant, named 'Maryland Mammoth,' had large leaves and exceptional height. When the other plants in the field flowered, 'Maryland Mammoth' plants continued to grow and remained vegetative. Garner and Allard took cuttings of 'Maryland Mammoth' into their greenhouse, and the plants that grew from those cuttings finally flowered in December.

Garner and Allard guessed that this flowering pattern had something to do with the mutant's response to some environmental cue. They tested several likely environmental variables, such as temperature, but the key variable proved to be day length. By moving plants between light and dark rooms at different times to vary the day length artificially, they were able to establish a direct link between flowering and day length. As we will discuss below, we now know that the key variable is not day length but the length of the *night*; however, Garner and Allard did not make this distinction.

'Maryland Mammoth' plants did not flower if the light period they were exposed to was longer than 14 hours per day, but flowering commenced once the days became shorter than 14 hours. Thus, the **critical day length** for 'Maryland Mammoth' tobacco is 14 hours (**Figure 38.10**). Control of flowering and several other plant responses by the length of day or night is called **photoperiodism**.

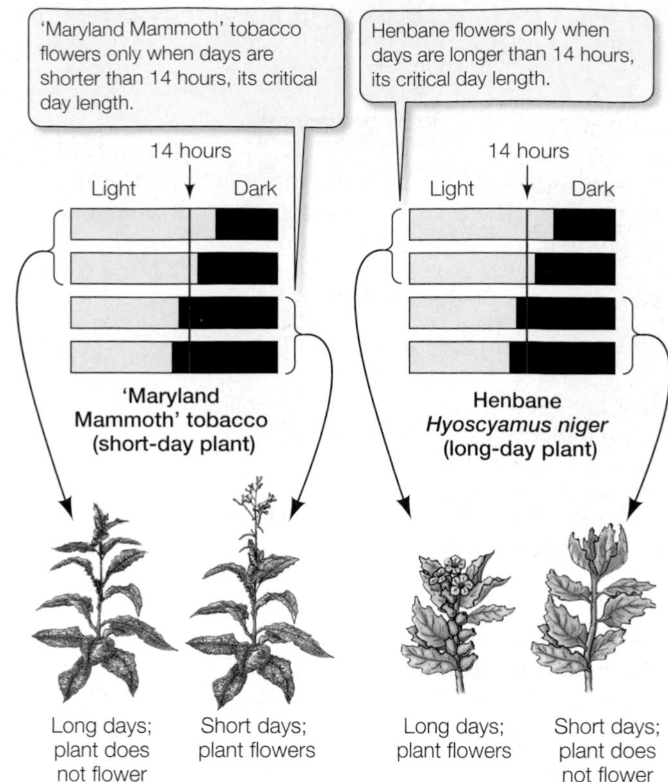
'Maryland Mammoth' tobacco flowers only when days are shorter than 14 hours, its critical day length.
Henbane flowers only when days are longer than 14 hours, its critical day length.
14 hours
Light
Dark
14 hours
Light
Dark
'Maryland Mammoth' tobacco (short-day plant)
Henbane *Hyoscyamus niger* (long-day plant)
Long days; plant does not flower
Short days; plant flowers
Long days; plant flowers
Short days; plant does not flower

38.10 Day Length and Flowering By artificially varying the day length in a 24-hour period, Garner and Allard showed that the flowering of 'Maryland Mammoth' tobacco is initiated when the days become shorter than a critical length. 'Maryland Mammoth' tobacco is thus called a short-day plant. Henbane, a long-day plant, shows an inverse pattern of flowering.

Plants vary in their responses to different photoperiodic cues

Plants that flower in response to photoperiodic stimuli fall into several classes. Poinsettias, chrysanthemums, and 'Maryland Mammoth' tobacco are **short-day plants** (SDPs), which flower only when the day is *shorter* than a critical *maximum*. Thus, for example, we see chrysanthemums in nurseries in the fall, and poinsettias in winter. Spinach and clover are examples of **long-day plants** (LDPs), which flower only when the day is *longer* than a critical *minimum*. Spinach in the garden tends to flower and become bitter in the summer, and is thus normally planted in early spring. Generally, LDPs are triggered to flower in midsummer and SDPs in late summer, fall, or sometimes in the spring. Because short days occur both before and after midsummer, there is a degree of ambiguity in this signal. Could there be a more precise way for plants to regulate flowering?

Some plants require photoperiodic signals that are more complex than just short or long days. One group, the *short-long-day plants*, must experience first short days and then long ones in order to flower. Accordingly, white clover and other short-long-day plants flower during the long days before midsummer. Another group, the *long-short-day plants*, cannot flower until the long days of summer have been followed by shorter ones, so they bloom only in the fall. Kalanchoe (see Figure 38.18B) is a long-short-day plant.

What do these four patterns of photoperiodic response have in common? All of them serve the point of photoperiodism, which is to synchronize the flowering of plants of the same species in a local population to promote cross-pollination and successful reproduction.

Near the Equator, the annual variation in day length is only about 2 minutes. How do angiosperms growing in these regions manage to flower at around the same time? Recent work suggests that their flowering cues on the time of sunrise and/or sunset, which varies predictably by about 30 minutes over the course of a year.

The flowering of some angiosperms, such as corn, roses, and tomatoes, is not photoperiodic. In fact, there are more of these **day-**

neutral plants than there are short-day and long-day plants. Some plants are photoperiodically sensitive only when young and become day-neutral as they grow older. Others require specific combinations of day length and other factors—especially temperature—to flower.

Other processes besides flowering are also under photoperiodic control. We have learned, for example, that short days trigger the onset of winter dormancy in plants. They trigger the formation of tubers in some begonias and storage roots in dahlias. (Animals, too, show a variety of photoperiodic behaviors, as discussed in Chapter 53.)

The length of the night is the key photoperiodic cue determining flowering

The terms "short-day plant" and "long-day plant" became entrenched before scientists determined that photoperiodically sensitive plants actually measure the length of the *night*, or of a period of darkness, rather than the length of the day. This fact was demonstrated by Karl Hamner of the University of California at Los Angeles and James Bonner of the California Institute of Technology (**Figure 38.11**).

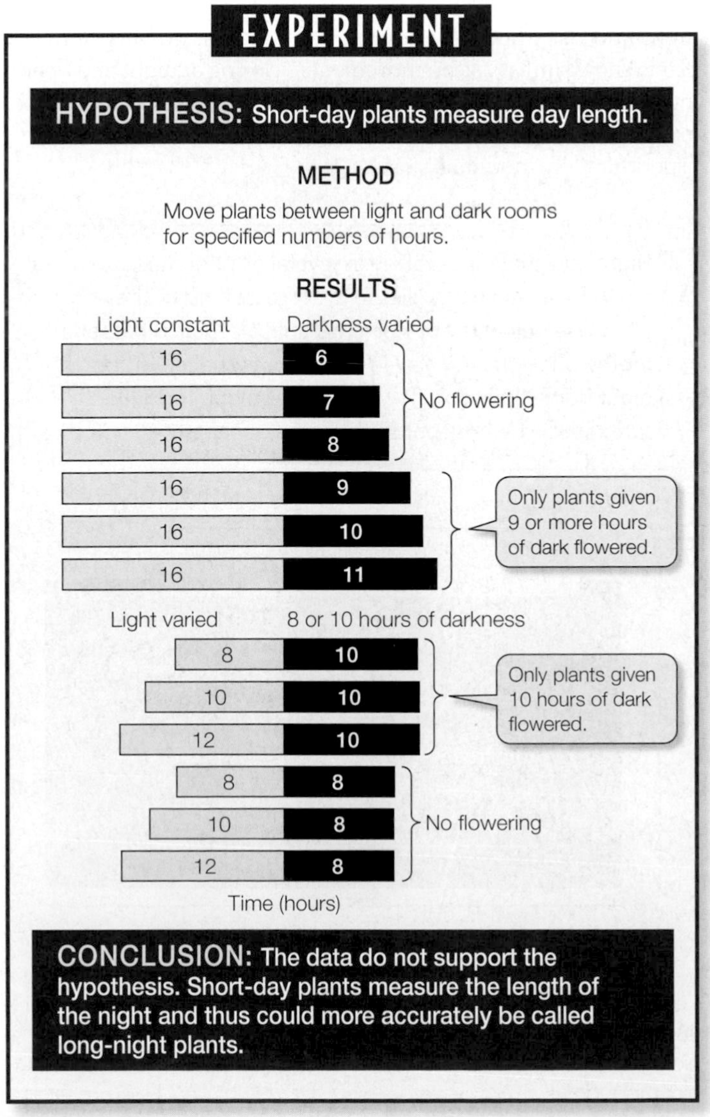

Working with cocklebur, an SDP, Hamner and Bonner ran a series of experiments using two sets of conditions:

- For one group of plants, the light period was kept constant—either shorter or longer than the critical day length—and the dark period was varied.
- For another group of plants, the dark period was kept constant and the light period was varied.

The plants flowered under all treatments in which the dark period exceeded 9 hours, regardless of the length of the light period. Thus, Hamner and Bonner concluded that it is the length of the *night* that matters; for cocklebur, the **critical night length** is about 9 hours. Thus, it would be more accurate to call cocklebur a "long-night plant" than a short-day plant.

In cocklebur, a single long night is sufficient to trigger full flowering some days later, even if the intervening nights are short ones. Most plants are less sensitive than cocklebur and require from two to several nights of appropriate length to induce flowering. For some plants, a single shorter night in a series of long ones, even one day before flowering would have commenced, inhibits flowering.

Through other experiments Hamner and Bonner gained some insight into how plants measure night length. They grew SDPs and LDPs under a variety of light/dark conditions. In some experiments, the dark period was interrupted by a brief exposure to light; in others, the light period was interrupted briefly by darkness. Interruptions of the light period by darkness had no effect on the flowering of either short-day or long-day plants. Even a brief interruption of the dark period by light, however, completely nullified the effect of a long night (experiment A in **Figure 38.12**). An SDP flowered only if the long nights were uninterrupted. An LDP experiencing long nights flowered if those nights were interrupted by exposure to light. Thus, the investigators concluded, these plants must have a timing mechanism that measures the length of a continuous dark period.

The nature of this timing mechanism has been partially revealed, beginning with the determination of the effective wavelengths of light and the identity of the photoreceptors. In the interrupted-night experiments, the most effective wavelengths of light were in the red range (experiment B in **Figure 38.12**), and the effect of a red-light interruption of the night could be fully reversed by a subsequent exposure to far-red light, indicating that a phytochrome is the photoreceptor. Phytochromes and blue-light receptors, which affect several aspects of plant development (Section 37.5), also participate in the photoperiodic timing mechanism. How might such a mechanism operate?

Experiments showed that when a plant is subjected to a dark period several days in duration, the plant's sensitivity to a light flash during the long night varies on a roughly 24-hour cycle. This suggests that phytochrome functions as a photoreceptor only, and that something else plays the timekeeping role. A biological clock is

EXPERIMENT
HYPOTHESIS: Short-day plants measure day length.
METHOD
Move plants between light and dark rooms for specified numbers of hours.
RESULTS
Light constant
Darkness varied
16
6
16
7
16
8
No flowering
16
9
16
10
16
11
Only plants given 9 or more hours of dark flowered.
Light varied
8 or 10 hours of darkness
8
10
10
10
12
10
Only plants given 10 hours of dark flowered.
8
8
10
8
12
8
No flowering
Time (hours)
CONCLUSION: The data do not support the hypothesis. Short-day plants measure the length of the night and thus could more accurately be called long-night plants.

38.11 Night Length and Flowering The length of the dark period, not the length of the light period, determines flowering.

EXPERIMENT A

HYPOTHESIS: Photoperiodically sensitive plants measure the length of the night.

Short-day plants	Light/dark combinations	Long-day plants
No flowering		Flowering
No flowering		Flowering
Flowering		No flowering
No flowering		Flowering

CONCLUSION: Photoperiodic plants measure the length of the night, not the day.

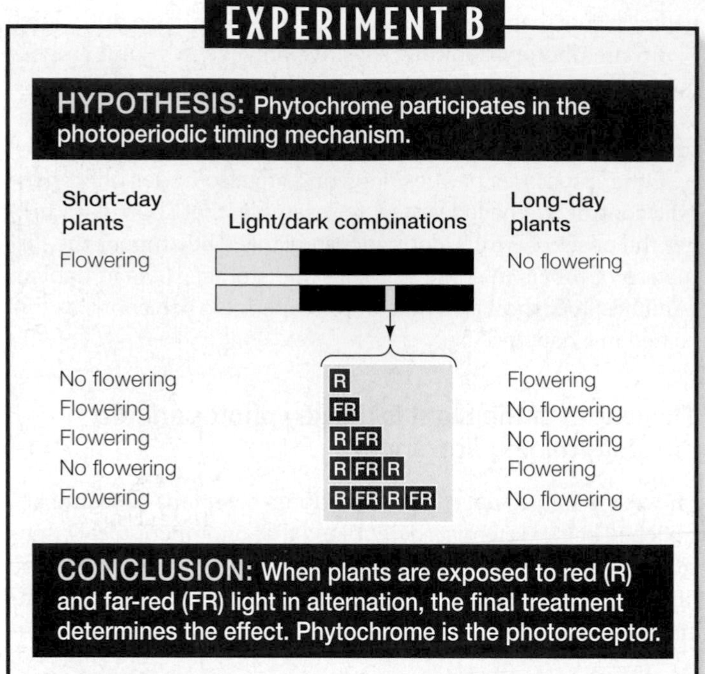

38.12 The Effect of Interrupted Days and Nights (A) Experiments suggest that plants are able to measure the length of a continuous dark period and use this information to trigger flowering. (B) Phytochromes seem to be involved in the photoperiodic timing mechanism.

linked to the phytochrome (which sets the clock) and to the production of flowers.

Circadian rhythms are maintained by a biological clock

It is clear that organisms have some way of measuring time, and that they are well adapted to the 24-hour day–night cycle of our planet. A "biological clock"—an oscillator that "ticks" back and forth between two states at roughly 12-hour intervals—resides within the cells of all eukaryotes and some prokaryotes. The major outward manifestations of this clock are known as **circadian rhythms** (Latin *circa*, "about," and *dies*, "day").

We can characterize circadian rhythms, as well as other regular biological cycles, in two ways: The **period** is the length of one cycle, and the **amplitude** is the magnitude of the change over the course of a cycle (**Figure 38.13**).

The circadian rhythms of cyanobacteria, protists, animals, fungi, and plants share some important characteristics:

- The period is remarkably insensitive to *temperature*, although lowering the temperature may drastically reduce the amplitude of the rhythmic effect.
- Circadian rhythms are *highly persistent;* they may continue for days even in an environment in which there are no environmental cues, such as light–dark periods.
- Circadian rhythms can be *entrained*, within limits, by light–dark cycles that differ from 24 hours. That is, the period an organism expresses can be made to coincide (within limits) with that of the light–dark cycle to which it is exposed.
- A brief exposure to light can shift the peak of the cycle—it can cause a *phase shift.*

Plants provide innumerable examples of circadian rhythms. The leaflets of plants such as clover normally hang down and fold at night and rise and unfold during the day. The flowers of many plants show similar "sleep movements," closing at night and opening during the day. They continue to open and close on an approximately 24-hour cycle even when the light and dark periods are experimentally modified.

Night flowering has evolved in several plant groups, ranging from certain water lilies to many cacti. Most of these plants are pollinated by nocturnal animals such as bats and moths, but some Himalayan bumblebees shelter for the night inside the wooly flowers of night-blooming *Saussurea* species, whose pollen they then transport.

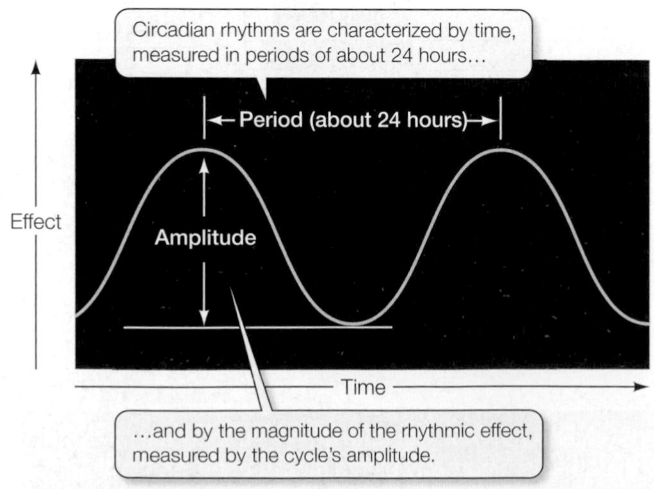

38.13 Features of Circadian Rhythms Circadian rhythms, like all biological rhythms, can be characterized in two ways: by period and by amplitude.

The period of circadian rhythms in nature is approximately 24 hours. If a clover plant, for example, were to be placed in light on a day–night cycle totaling exactly 24 hours, it would express a rhythm with a period of exactly 24 hours. However, if an experimenter used a day–night cycle of, say, 22 hours, then over time the rhythm would change—it would be **entrained** to a 22-hour period.

If an organism is maintained under constant darkness, it will express a circadian rhythm with an approximately 24-hour period. However, a brief exposure to light under these circumstances can cause a **phase shift**—that is, it can make the next peak of activity appear either later or earlier than expected, depending on when the exposure is given. Moreover, the organism does not then return to its old schedule if it remains in darkness. If the first peak is delayed by 6 hours, the subsequent peaks are all 6 hours late. Such phase shifts are permanent—until the organism receives more exposures to light.

EXPERIMENT

HYPOTHESIS: Plants fix more carbon photosynthetically when their circadian clock matches the environment's light-dark cycle.

METHOD

1. Select two mutant strains of *Arabidopsis thaliana*: one (*ztl-1*) with a long-cycle rhythm, and the other (*toc1-1*) with a short-cycle rhythm.
2. Grow plants of both mutant strains under either a 28-hour light-dark cycle or a 20-hour cycle instead of the normal 24-hour cycle.
3. Determine photosynthetic C fixation for each of the four groups.

RESULTS

Long-cycle mutants fixed more carbon per hour when grown on a longer-than-24-hour cycle, while short-cycle mutants performed better on a shorter-than-24-hour cycle.

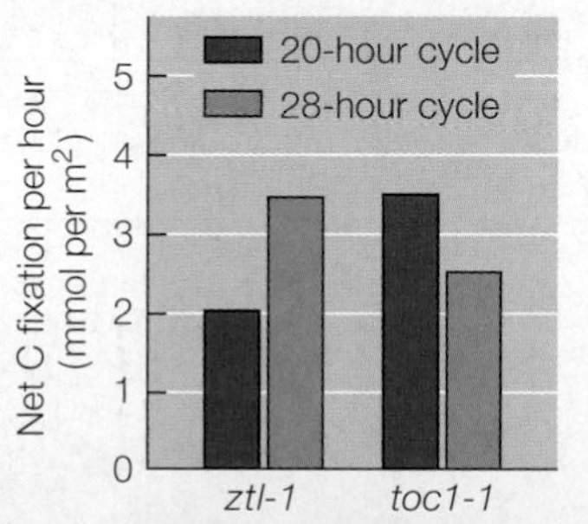

CONCLUSION: Each mutant performed best under the cycle that corresponded to its genetically determined circadian rhythm.

38.14 Does the Circadian Clock Help a Plant Interact with Its Environment? Anthony Dodd and his coworkers studied the relationship between the plants' natural circadian cycles and light-dark cycles by observing the responses of mutant plants to experimentally imposed light-dark cycles. The clock does help the plants perform more efficiently. FURTHER EXPERIMENT: Design experiments to test other measures of plant performance in wild-type plants and the mutants studied by Dodd et al.

The general occurrence of circadian rhythms in plants suggests that they confer a selective advantage, but what might that advantage be? Recent experimental evidence indicates that the advantage consists of a coordination of the expression of certain genes with the phases of the daily light-dark cycles (**Figure 38.14**). The coordination results in increased efficiency—chlorophyll is not produced in the dark, when it is not needed, for example.

Photoreceptors set the biological clock

Phytochromes and blue-light receptors are known to affect the period of the biological clock, with the different pigments "reporting" on different wavelengths and intensities of light. This diversity of photoreceptors could be an adaptation to the changes in the light environment that a plant experiences in the course of a day or a season. How do these photoreceptors interact with a plant's biological clock?

The biological clock of *Arabidopsis* is based on the activities of at least three "clock" genes. The clock genes encode regulatory proteins that interact to produce a circadian oscillation. How does this oscillating clock interact with photoreceptors and the environment?

Arabidopsis is an LDP. Its clock controls the activity of *CONSTANS* (a gene that is *not* part of the clock mechanism) in such a way that the *CONSTANS* product, CO protein, accumulates in one phase of the clock's cycle—the phase in which night falls. Under long nights (short days), CO protein accumulates at night. Under short nights (long days), CO is also relatively abundant at dawn and dusk. When CO protein levels are high, light absorbed by phytochrome A and the blue-light receptor cryptochrome 2 leads to flowering (**Figure 38.15**). Thus *Arabidopsis* flowering results from the coincidence of light (detected by the two photoreceptors) with a sufficient amount of CO protein (which varies according to a circadian rhythm). Where is this coincidence-based photoperiodic mechanism located in relation to where flowering occurs?

The flowering stimulus originates in a leaf

Is the timing mechanism for flowering located in a particular plant tissue or organ, or are all parts able to sense the length of the night? This question was resolved by "blindfold" experiments in which it quickly became apparent that each leaf is capable of timing the night (**Figure 38.16**).

If a cocklebur plant—an SDP that needs long nights to flower—is kept under a regime of short nights and long days, but a leaf is covered so as to give it the needed long nights, the plant will flower (experiment A in Figure 38.16). If one leaf is given a photoperiodic treatment conducive to flowering—called an *inductive* treatment—other leaves kept under noninductive conditions will tend to inhibit flowering. (An inductive treatment works best if only one leaf is left on the plant.)

Although it is the leaves that sense an inductive photoperiod, the flowers form elsewhere on the plant. Thus, some kind of signal must be sent from the leaf to the site of flower formation. Three lines of evidence suggest that this signal is a chemical substance—a flowering hormone.

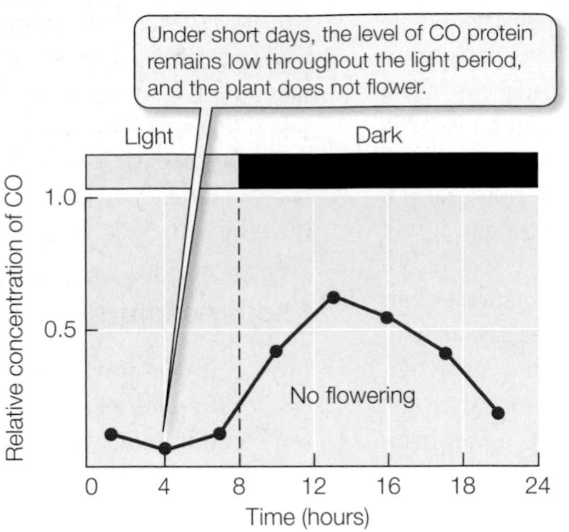
Under short days, the level of CO protein remains low throughout the light period, and the plant does not flower.
Light
Dark
1.0
0.5
Relative concentration of CO
No flowering
0
4
8
12
16
18
24
Time (hours)

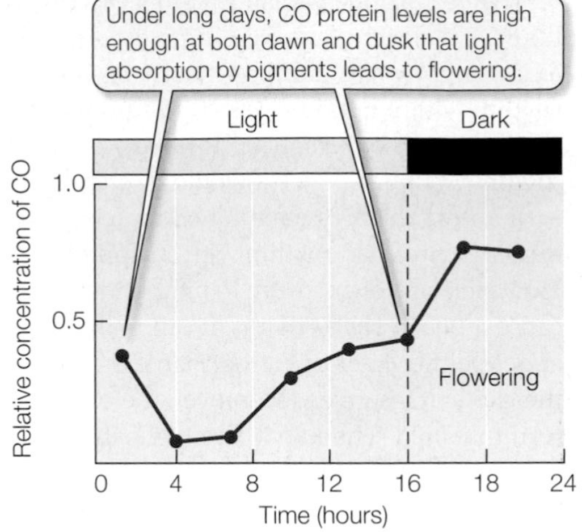
Under long days, CO protein levels are high enough at both dawn and dusk that light absorption by pigments leads to flowering.
Light
Dark
1.0
0.5
Relative concentration of CO
Flowering
0
4
8
12
16
18
24
Time (hours)

38.15 Photoreceptors and the Biological Clock Interact in Photoperiodic Plants One of the genes regulated by the circadian clock in *Arabidopsis* encodes the CO protein. Flowering depends on a combination of CO protein and light: enough CO must be present when photoreceptors have light available to them.

- If a photoperiodically induced leaf is immediately removed from a plant after the inductive dark period, the plant does not flower. If, however, the induced leaf remains attached to the plant for several hours, the plant will flower. This result suggests that something is synthesized in the leaf in response to the inductive dark period, then it moves out of the leaf to induce flowering.
- If two or more cocklebur plants are grafted together, and if one plant is exposed to inductive long nights and its graft partners are exposed to noninductive short nights, all the plants flower (experiment B in Figure 38.16).
- In at least one species, if an induced leaf from one plant is grafted onto another, noninduced plant, the host plant flowers.

38.16 Evidence for a Flowering Hormone If even a single leaf of cocklebur is exposed to inductive conditions, a signal travels to the entire plant (and even to other plants, in grafting experiments), inducing it to flower.

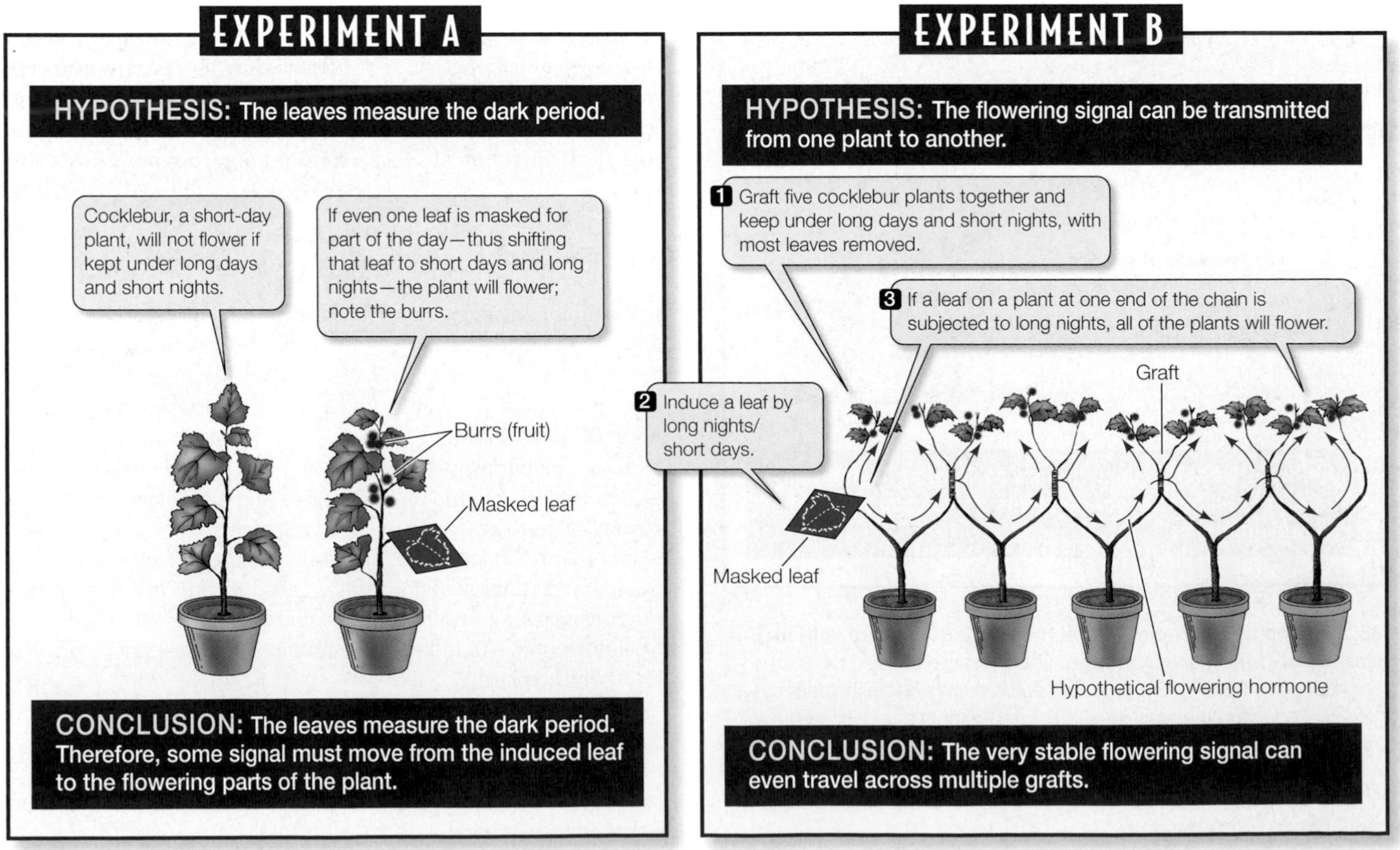

EXPERIMENT A

HYPOTHESIS: The leaves measure the dark period.

Cocklebur, a short-day plant, will not flower if kept under long days and short nights.

If even one leaf is masked for part of the day—thus shifting that leaf to short days and long nights—the plant will flower; note the burrs.

Burrs (fruit)

Masked leaf

CONCLUSION: The leaves measure the dark period. Therefore, some signal must move from the induced leaf to the flowering parts of the plant.

EXPERIMENT B

HYPOTHESIS: The flowering signal can be transmitted from one plant to another.

1 Graft five cocklebur plants together and keep under long days and short nights, with most leaves removed.

2 Induce a leaf by long nights/ short days.

3 If a leaf on a plant at one end of the chain is subjected to long nights, all of the plants will flower.

Graft

Masked leaf

Hypothetical flowering hormone

CONCLUSION: The very stable flowering signal can even travel across multiple grafts.

Some 50 years ago, Jan A. D. Zeevaart, a plant physiologist at Michigan State University, performed this last experiment. He exposed a single leaf of the SDP *Perilla* to a short-day/long-night regime, inducing the plant to flower. Then he detached this leaf and grafted it onto another, noninduced, *Perilla* plant—which responded by flowering. The same leaf grafted onto successive hosts caused each of them to flower in turn. As long as 3 months after the leaf was exposed to the short-day/long-night regime, it could still cause plants to flower.

Experiments such as Zeevaart's led to the conclusion that the photoperiodic induction of a leaf causes a more or less permanent change in the leaf, causing it to start and continue producing a flowering hormone that is transported to other parts of the plant, where the hormone initiates the development of reproductive structures. Biologists named this hypothetical hormone **florigen**, although decades passed without its being isolated or characterized.

An elegant experiment suggested that the florigen of SDPs is identical to that of LDPs, even though SDPs produce it only under long nights and LDPs only under short nights. An SDP and an LDP were grafted together, and both flowered, as long as the photoperiodic conditions were inductive for one of the partners. Either the SDP or the LDP could be the one induced, but both would always flower. These results suggest that a flowering hormone—the elusive florigen—was being transferred from one plant to the other.

The direct demonstration of florigen activity did not occur until 2005. It had long been thought that florigen could be neither a protein nor an RNA because those molecules were too large to pass from one living plant cell to another. However, we now know that such macromolecules can be transferred by way of plasmodesmata. A group in Sweden has shown that exposure of a leaf of *Arabidopsis* (an LDP) to an inductive light regime leads to the production in the leaf of mRNA by the *FLOWERING LOCUS T* (*FT*) gene, and the mRNA travels to the shoot apex. There, other genes are activated, leading to flowering. The *FT* mRNA is, then, at least part of the mobile flower stimulus—florigen.

We have considered the photoperiodic regulation of flowering, from photoreceptors in a leaf to the biological clock to the signal that travels from the induced leaf to the sites of flower formation. Light is not the only environmental variable that affects flowering, however. In some plants, low temperatures are a cue that eventually triggers flowering.

In some plants flowering requires a period of low temperature

Certain cereal grains serve as classic examples of the control of flowering by temperature. In both wheat and rye, we distinguish two categories of flowering behavior: annual and biennial. Spring wheat, for example, is an annual plant: it is sown in the spring and flowers in the same year. Winter wheat is biennial: it must be sown in the fall, and it flowers the following summer (**Figure 38.17**). If winter wheat is not exposed to cold in its first year, it will not flower normally the next year.

The implications of this finding were of great agricultural interest in the newly communist Soviet Union of the 1920s. Winter wheat yields more grain than spring wheat, but it cannot be grown in some parts of Russia because the winters are too cold for its survival. The Soviet agronomist Trofim Lysenko demonstrated that if seeds of winter wheat were premoistened and prechilled, they could be sown in the spring and would develop and flower normally the same year. Thus, high-yielding winter wheat could be grown even in previously hostile regions.*

This induction of flowering by low temperatures is called **vernalization** (Latin *vernum*, "spring"). Vernalization may require as many as 50 days of low temperatures (from about –2° to +12°C). The low-temperature treatment inhibits the expression of a gene whose protein product represses other genes that contribute to flower development. Some plant species require both vernalization and long days to flower. There is a long wait from the cold, short days of winter to the warm, long days of summer, but because the vernalized state easily lasts at least 200 days, these plants do flower when they experience the appropriate night length.

*Lysenko went on to claim that this vernalized state, once imposed by cold treatment, was inherited by the untreated progeny of the treated plants. There was no genetic evidence for his outlandish statement, which repudiated the chromosome model of inheritance and ignored all advances in evolutionary thinking. His argument was well received by the Soviet establishment, however, because it meshed with a Marxist doctrine of inheritance of acquired social and political characteristics. Lysenko was lionized and put in charge of the Soviet Academy of Agricultural Sciences. His reign there, which lasted into the 1960s, destroyed the lives and careers of a generation of Russian geneticists and left Soviet biology in a backward state from which it took years to recover.

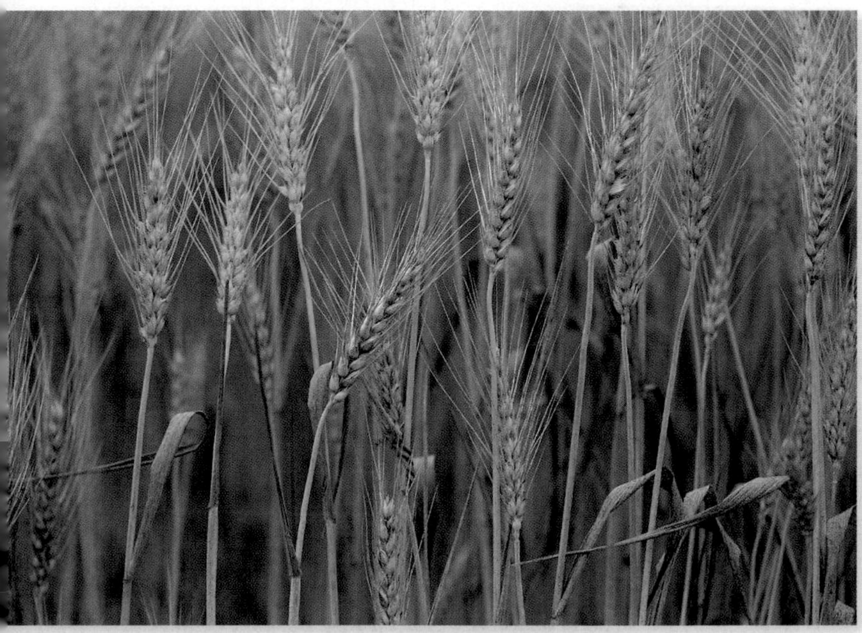

38.17 Vernalization The seeds of winter wheat must be exposed to cold temperatures for some amount of time in order to flower, a phenomenon known as vernalization.

38.2 RECAP

Flowering of some angiosperms is controlled by night length, a phenomenon called photoperiodism. Exposure to low temperatures—vernalization—triggers flowering in others, and some plants require low temperatures followed by an appropriate night length to flower.

- Can you explain the differences between apical meristems, inflorescence meristems, and floral meristems? See p. 825 and Figure 38.9
- Why is "short-day plant" a misleading term? See p. 827
- Do you understand how a biological clock in *Arabidopsis* controls the release of CO protein, and how photoreceptors are responsible for determining when this plant should flower? See p. 829 and Figure 38.15

We have seen how environmental factors interact with genes to control flowering in angiosperms. The entire function of flowers is sexual reproduction, which maintains beneficial genetic variation in a population. Many angiosperms, however, also benefit from being able to reproduce asexually.

38.3 How Do Angiosperms Reproduce Asexually?

Although sexual reproduction takes up most of the space in this chapter, asexual reproduction is responsible for many of the new plant individuals appearing on Earth. This fact suggests that in some circumstances, asexual reproduction must be advantageous.

At the beginning of this chapter, we saw that one of the advantages of sexual reproduction is genetic recombination. Self-fertilization is a form of sexual reproduction, but when a plant self-fertilizes, fewer opportunities for genetic recombination exist than with cross-fertilization. A diploid, self-fertilizing plant that is heterozygous for a certain locus can produce both kinds of homozygotes for that locus plus the heterozygote among its progeny, but it cannot produce any progeny carrying alleles that it does not itself possess. Yet many plant species are capable of self-fertilization and produce viable and vigorous offspring.

Asexual reproduction eliminates genetic recombination altogether. When a plant reproduces asexually, it produces a clone of progeny genetically identical to the parent. If a plant is well adapted to its environment, asexual reproduction may spread its superior genotype throughout that environment. This ability to exploit a particular environment is an advantage of asexual reproduction.

Many forms of asexual reproduction exist

We call stems, leaves, and roots *vegetative organs* to distinguish them from flowers, the reproductive parts of the plant. The modification of a vegetative organ is what makes **vegetative reproduction** in plants—asexual reproduction—possible. In many cases, the stem is the organ that is modified. Strawberries and some grasses, for example, produce horizontal stems, called *stolons* or *runners*, that grow along the soil surface, form roots at intervals, and establish potentially independent plants (see Figure 34.4C). *Tip layers* are upright branches whose tips sag to the ground and develop roots, as in blackberry and forsythia.

Some plants, such as potatoes, form enlarged fleshy tips of underground stems, called *tubers* (see Figure 34.4A). *Rhizomes* are horizontal underground stems that can give rise to new shoots. Bamboo is a striking example of a plant that reproduces vegetatively by means of rhizomes. A single bamboo plant can give rise to a stand—even a forest—of plants constituting a single, physically connected entity.

Whereas stolons and rhizomes are horizontal stems, bulbs and corms are short, vertical, underground stems. Lilies and onions form *bulbs* (**Figure 38.18A**), short stems with many fleshy, highly modified leaves that store nutrients. These storage leaves make up most of the bulb. Bulbs are thus large underground buds. They can give rise to new plants by dividing or by producing new bulbs from axillary buds. Crocuses, gladioli, and many other plants produce *corms*, underground stems that function very much as bulbs do. Corms are disclike and consist primarily of stem tissue; they lack the fleshy modified leaves that are characteristic of bulbs.

Around 1600, the first tulip bulbs were imported into Holland from central Asia. During the 1630s, the flower's popularity blossomed into a bizarre "tulipomania"—at one point a single bulb sold for the equivalent of $76,000! Six weeks later, the price fell to less than a dollar, resulting in a debacle for several Dutch banks.

Not all vegetative organs modified for reproduction are stems. Leaves may also be the source of new plantlets, as in some succulent plants of the genus *Kalanchoe* (**Figure 38.18B**). Many kinds of angiosperms, ranging from grasses to trees such as aspens and poplars, form interconnected, genetically homogeneous populations by means of *suckers*—shoots produced by roots. What appears to be a whole stand of aspen trees, for example, may be a clone derived from a single tree by suckers.

Plants that reproduce vegetatively often grow in physically unstable environments such as eroding hillsides. Plants with stolons or rhizomes, such as beach grasses, rushes, and sand verbena, are common pioneers on coastal sand dunes. Rapid vegetative reproduction enables these plants, once introduced, not only to multiply but also to survive burial by the shifting sand; in addition, the dunes are stabilized by the extensive network of rhizomes or stolons that develops. Vegetative reproduction is also common in some deserts, where the environment is often not suitable for seed germination and the establishment of seedlings.

Dandelions, citrus trees, and some other plants reproduce by the asexual production of seeds, called **apomixis**. As we have seen, meiosis reduces the number of chromosomes in gametes by half, and fertilization restores the sporophytic number of chromosomes in the zygote. Some plants can skip over *both* meiosis and fertilization and still produce seeds. Apomixis produces seeds within

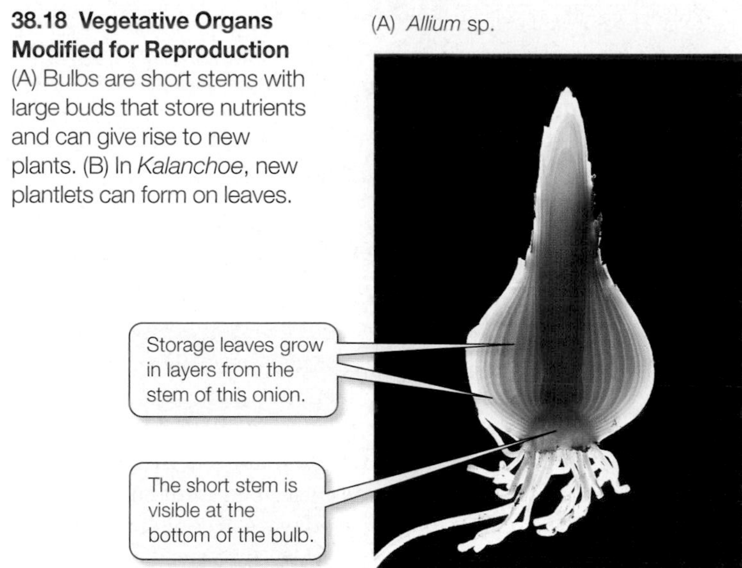

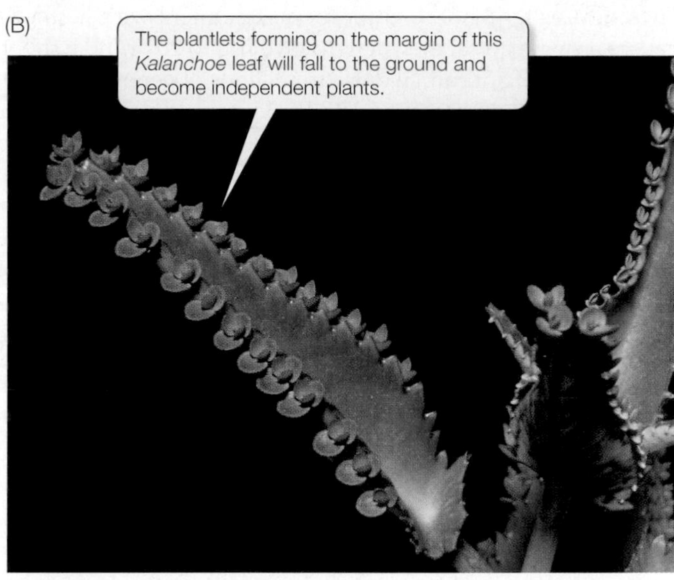

(A) Allium sp.
Storage leaves grow in layers from the stem of this onion.
The short stem is visible at the bottom of the bulb.

(B)
The plantlets forming on the margin of this Kalanchoe leaf will fall to the ground and become independent plants.

38.18 Vegetative Organs Modified for Reproduction (A) Bulbs are short stems with large buds that store nutrients and can give rise to new plants. (B) In *Kalanchoe*, new plantlets can form on leaves.

the ovary without the mingling and segregation of chromosomes and without the union of gametes. Certain cells within the ovule simply develop into seeds, and the ovary wall develops into a fruit. An apomictic embryo has the sporophytic number ($2n$) of chromosomes. The result of apomixis is a fruit with seeds that are genetically identical to the parent plant.

Apomixis sometimes requires pollination. In some apomictic species, a sperm nucleus must combine with the polar nuclei for the endosperm to form. In other apomictic species, the pollen provides the signals for embryo and endosperm formation, although neither sperm nucleus participates in fertilization. This observation emphasizes that pollination and fertilization are not the same thing.

Vegetative reproduction has a disadvantage

Vegetative reproduction is highly efficient in a stable environment. A change in the environment, however, can leave an asexually reproducing species at a disadvantage.

The English elm, *Ulmus procera*, provides a striking example. It was apparently introduced into England as a clone by the ancient Romans. This tree reproduces asexually by suckers and is incapable of sexual reproduction. In 1967, Dutch elm disease first struck the English elms. After 2 millennia of clonal growth the population lacked genetic diversity and was unable to meet the challenge. Today the English elm is all but gone from England.

Vegetative reproduction is important in agriculture

Farmers and gardeners take advantage of some natural forms of vegetative reproduction. They have also developed new types of asexual reproduction by manipulating plants. One of the oldest methods of vegetative reproduction used in agriculture consists of simply making cuttings of stems, inserting them in soil, and waiting for them to form roots and thus become autonomous plants. The cuttings are usually encouraged to root by treatment with a plant hormone, auxin, as described in Section 37.3.

Horticulturists reproduce many woody plants by **grafting**—attaching a bud or a piece of stem from one plant to the root or root-bearing stem of another plant. The part of the resulting plant that comes from the root-bearing "host" is called the **stock**; the part grafted on is the **scion** (**Figure 38.19**).

For a graft to succeed, the vascular cambium of the scion must associate with that of the stock. By cell division, both cambia form masses of wound tissue. If the two masses meet and connect, the resulting continuous cambium can produce xylem and phloem, allowing transport of water and minerals to the scion and of photosynthate to the stock. Grafts are most often successful when the stock and scion belong to the same or closely related species. Much fruit grown for market in the United States is produced on grafted trees.

Scientists in universities and commercial laboratories have been developing new ways to produce useful plants via tissue culture. Because many plant cells are *totipotent*, cultures of undifferentiated tissue can give rise to entire plants, as can small pieces of tissue cut directly from a parent plant (see Figure 19.3). Tissue cultures sometimes are used commercially to produce new plants.

Culturing tiny bits of apical meristem can produce plants free of viruses. Because apical meristems lack developed vascular tis-

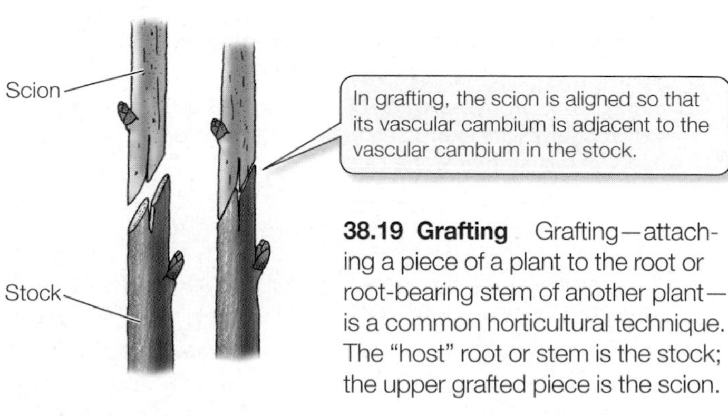

38.19 Grafting Grafting—attaching a piece of a plant to the root or root-bearing stem of another plant—is a common horticultural technique. The "host" root or stem is the stock; the upper grafted piece is the scion.

sues, viruses tend not to enter them. Treatment with hormones causes a single apical meristem to give rise to 20 or more shoots; thus, a single plant can give rise to millions of genetically identical plants within a year by repeated meristem culturing. Using this approach, strawberry and potato producers are able to start each year's crop from virus-free plants.

Recombinant DNA techniques applied to tissue cultures can provide plants with increased resistance to pests or increased nutritive value to humans. There is also interest in making certain valuable, sexually reproducing plants capable of apomixis. By causing cells from sexually incompatible species to fuse, one can obtain plants with exciting new combinations of properties.

38.3 RECAP

Angiosperms may reproduce asexually by means of modified stems, roots, or leaves, or by apomixis.

- Do you see how apomixis differs from sexual reproduction? See pp. 832–833
- Explain how vegetative reproduction of plants can be advantageous to humans. See pp. 833–834

We have seen how angiosperms reproduce sexually and asexually. A disadvantage of asexual reproduction is that its genetic inflexibility may leave a population unable to cope with a changing environment. In the next chapter we see that even unchanging environments, both biological and physical, present challenges to plants—and that plants have evolved ways of coping with them.

CHAPTER SUMMARY

38.1 How do angiosperms reproduce sexually?

Sexual reproduction promotes genetic diversity in a population. The flower is an angiosperm's structure for sexual reproduction.

Flowering plants have microscopic gametophytes. The megagametophyte is the **embryo sac**, which typically contains eight nuclei in a total of seven cells. The microgametophyte is the **pollen grain**, which usually contains two cells. Review Figure 38.1, Web/CD Tutorial 38.1

Following **pollination**, the pollen grain delivers sperm cells to the embryo sac by means of a **pollen tube**. Review Figure 38.3

Most angiosperms exhibit **double fertilization**: One sperm nucleus fertilizes the egg, forming a zygote, and the other sperm nucleus unites with the two **polar nuclei** to form a **triploid endosperm**. Review Figure 38.5

The zygote develops into an embryo (with an attached **suspensor**), which remains quiescent in the seed until conditions are right for germination. Review Figure 38.6, Web/CD Activity 38.1

Ovules develop into seeds, and the ovary wall and the enclosed seeds develop into a **fruit**.

38.2 What determines the transition from the vegetative to the flowering state?

For a vegetatively growing plant to flower, an apical meristem in the shoot system must become an **inflorescence meristem**, which in turn must give rise to one or more **floral meristems**. Review Figure 38.9

Angiosperms are classified as **annuals**, **biennials**, or **perennials**, depending upon the length of their life cycle.

Flowering results from a cascade of gene expression. **Floral organ identity genes** are expressed in floral meristems that give rise to sepals, petals, stamens, and carpels.

Short-day plants flower when the nights are longer than a critical night length specific to each species; **long-day plants** flower when the nights are shorter than a critical night length. Some angiosperms have more complex photoperiodic requirements, but most are **day-neutral**. Review Figure 38.10

Plants exhibit **circadian rhythms**, which are characterized by both their **period** and their **amplitude**. Circadian rhythms can be **entrained** and can be induced to experience a **phase shift**. Review Figure 38.13

The mechanism of photoperiodic control involves phytochromes and a biological clock. Review Figures 38.12 and 38.15, Web/CD Tutorial 38.2

A flowering hormone, called **florigen**, is formed in a photoperiodically induced leaf and is translocated to the sites where flowers will form. Review Figure 38.16

In some angiosperm species, exposure to low temperatures—**vernalization**—is required for flowering.

38.3 How do angiosperms reproduce asexually?

Asexual reproduction allows rapid multiplication of organisms that are well suited to their environment.

Vegetative reproduction involves the modification of a vegetative organ—usually the stem—for reproduction.

Some plant species produce seeds asexually by **apomixis**.

Horticulturists often **graft** different plants together to take advantage of favorable properties of both **stock** and **scion**. Review Figure 38.19

Agriculturalists use natural and artificial techniques of asexual reproduction to reproduce particularly desirable plants.

SELF-QUIZ

1. Sexual reproduction in angiosperms
   a. is by way of apomixis.
   b. requires the presence of petals.
   c. can be accomplished by grafting.
   d. gives rise to genetically diverse offspring.
   e. cannot result from self-pollination.
2. The typical angiosperm female gametophyte
   a. is called a megaspore.
   b. has eight nuclei.
   c. has eight cells.
   d. is called a pollen grain.
   e. is carried to the male gametophyte by wind or animals.
3. Pollination in angiosperms
   a. never requires external water.
   b. never occurs within a single flower.
   c. always requires help by animal pollinators.
   d. is also called fertilization.
   e. makes most angiosperms independent of external water for reproduction.
4. Which statement about double fertilization is *not* true?
   a. It is found in most angiosperms.
   b. It takes place in the microsporangium.
   c. One of its products is a triploid nucleus.
   d. One sperm nucleus fuses with the egg nucleus.
   e. One sperm nucleus fuses with two polar nuclei.
5. The suspensor
   a. gives rise to the embryo.
   b. is heart-shaped in eudicots.
   c. separates the two cotyledons of eudicots.
   d. ceases to elongate early in embryonic development.
   e. is larger than the embryo.
6. Which statement about photoperiodism is *not* true?
   a. It is related to the biological clock.
   b. A phytochrome plays a role in the timing process.
   c. It is based on measurement of the length of the night.
   d. Most plant species are day-neutral.
   e. It is limited to plants.
7. Before florigen was isolated, we thought it exists because
   a. night length is measured in the leaves, but flowering occurs elsewhere.
   b. it is produced in the roots and transported to the shoot system.
   c. it is produced in the coleoptile tip and transported to the base.
   d. we think that gibberellin and florigen are the same compound.
   e. it may be activated by prolonged (more than a month) chilling.
8. Which statement about vernalization is *not* true?
   a. It may require more than a month of low temperatures.
   b. The vernalized state generally lasts for about a week.
   c. Vernalization makes it possible to have a winter wheat crop each year.
   d. It is accomplished by subjecting moistened seeds to chilling.
   e. It was of interest to Russian scientists because of their native climate.
9. Which of the following does *not* participate in asexual reproduction?
   a. Stolon
   b. Rhizome
   c. Zygote
   d. Tuber
   e. Corm
10. Apomixis involves
    a. sexual reproduction.
    b. meiosis.
    c. fertilization.
    d. a diploid embryo.
    e. no production of a seed.

FOR DISCUSSION

1. Which method of reproduction might a farmer prefer for a crop plant that reproduces both sexually and asexually? Why?
2. Thompson Seedless grapes are produced by vines that are triploid. Think about the consequences of this chromosomal condition for meiosis in the flowers. Why are these grapes seedless? Describe the role played by the flower in fruit formation when no seeds are being formed. How do you suppose Thompson Seedless grapes are propagated?
3. Poinsettias are popular ornamental plants that typically bloom just before Christmas. Their flowering is photoperiodically controlled. Are they long-day or short-day plants? Explain.
4. You plan to induce the flowering of a crop of long-day plants in the field by using artificial light. Is it necessary to keep the lights on continuously from sundown until the point at which the critical day length is reached? Why or why not?
5. Dodd and coworkers pointed out the need for crop breeders to be aware if genes for the phase and period of circadian rhythms are closely linked to genes being studied for enhanced crop yields. If such a tight linkage occurs, how might it affect a research program?

FOR INVESTIGATION

The space program may need to deal with methods for crop production on another planet or during transit through space. Outline an experimental program to address this problem in terms of the findings on circadian rhythms by Dodd and coworkers (see Figure 38.14).

CHAPTER 39 Plant Responses to Environmental Challenges

Salt in the delta threatens a hungry nation

Bangladesh is an impoverished country with a high population density and a weak economy. Its 150 million citizens live on an area of about 144,000 square kilometers. (For comparison, imagine half the people in the United States—or more than 4 times the total population of Canada—all living in the single state of Iowa.) Bangladesh is far from being able to feed its own population and must rely heavily on international food aid. It is crucial that the country maximize its ability to feed its citizens, even in the face of ecological challenges.

Aside from the fact of overpopulation, the people of Bangladesh face agricultural problems arising from the country's almost unique geography. Much of its total land area lies at or below sea level in the Ganges River delta. A massive network of 230 rivers threads through the country, carrying runoff from melting Himalayan snow through agricultural land to the Indian Ocean. About one-third of Bangladesh is flooded annually by the torrential rains of the monsoon season (June through October). From November through May, however, a very different problem arises: salinization, or too much salt in the soil.

During the monsoon season, rainfall and rivers provide fresh water that is low in salt content. Things change during the dry season. When the flow of fresh water is sharply reduced, saline (salty) water from the delta's estuaries near the ocean penetrates inland, sometimes as far as 300 kilometers. This results in progressive salinization of the soil. Furthermore, as the soil dries out from November onward, ground water rises and moves laterally, bringing with it dissolved salts from deeper in the soil. With further drying, the soil becomes coated with salt.

Humans have contributed to salinization as well. Some years ago, India diverted part of the flow of the Ganges River toward the city of Calcutta—and away from Bangladesh. The subsequent reduction of the river's flow into the delta meant further intrusion of sea water. Bangladesh has contributed to its own problems by failing to regulate shrimp farming in the delta, which is carried out by allowing brackish water to spread over the soil. Inappropriate agricultural practices continue to favor salinization rather than checking it.

Too much salt in the soil is toxic and can kill a plant outright. A bit less salt can kill a plant osmotically by making it difficult to take up water from the soil. Still lower soil salt concentrations simply reduce

Monsoon Season Each year the monsoon rains of Southeast Asia force many people to retreat from flooded homes.

Salty Soils Much of Bangladesh's agricultural land is in the Ganges River delta, at or below sea level. This cropland suffers from increasing amounts of salt in the soil and diminishing agricultural returns.

plant growth and crop yields. Worldwide, about 20,000 square kilometers of agricultural soil are lost each year to salinization. Even in the United States, almost one-fourth of all irrigated lands experience salinization, resulting in significant crop losses.

Agriculture in many parts of the world is challenged by salty, arid, or waterlogged soils. Some plants have adaptations that allow them to thrive or at least survive in such environments; others simply don't grow in them. In discussing these and other environmental challenges to plant growth, we'll touch on the possibility of engineering plants to become salt-tolerant without becoming unpalatable.

IN THIS CHAPTER we begin by examining interactions between plants and plant pathogens. We then consider interactions between plants and herbivores. Next we will discuss some of the adaptations plants exhibit to cope with temperature extremes, and, finally, we describe how certain plants can survive in physical environments that are either extremely dry or water-saturated, that are dangerously salty, or that contain high concentrations of toxic substances.

CHAPTER OUTLINE

39.1 How Do Plants Deal with Pathogens?

The environment teems with organisms that cause plant diseases. We know of more than a hundred diseases that can kill a tomato plant, each of them caused by a different pathogen, including bacteria, fungi, protists, and viruses. Like animals, plants have a variety of defenses against pathogens. Like the defenses of our own bodies, these mechanisms are not perfect, but they generally keep the plant world in competitive balance with its pathogens.

Plants and pathogens have evolved together in a continuing "arms race." Pathogens have evolved mechanisms with which to attack plants, and plants have evolved mechanisms for defending themselves against pathogens. Each set of mechanisms uses information from the other. For example, the pathogen's enzymes may break down the plant's cell walls, and the breakdown products may signal to the plant that it is under attack. In turn, the plant's defenses alert the pathogen that it is under attack.

What determines the outcome of a battle between a plant and a pathogen? The key to success for the plant is to respond to the information from the pathogen quickly and massively. Plants use both mechanical and chemical defenses in this effort.

Plants seal off infected parts to limit damage

Tissues such as epidermis or cork protect the outer surfaces of plants, and these tissues are generally covered by cutin, suberin, or waxes. This protection is comparable to the nonspecific immune defenses of animals (see Section 18.2). When pathogens pass these barriers, other nonspecific plant defenses are activated.

The defense systems of plants and animals differ. Animals generally repair tissues that have been damaged by pathogens, but plants do not. Instead, they seal off and sacrifice the damaged tissues so that the rest of the plant does not become infected. This approach works because most plants, unlike most animals, are modular and can replace damaged parts by growing new ones.

One of a plant cell's first defensive responses is the rapid deposition of additional polysaccharides on the inside of the cell wall, reinforcing this barrier to invasion by the pathogen (**Figure 39.1**). These polysaccharides block the plasmodesmata, limiting the ability of viral pathogens to move from cell to cell.

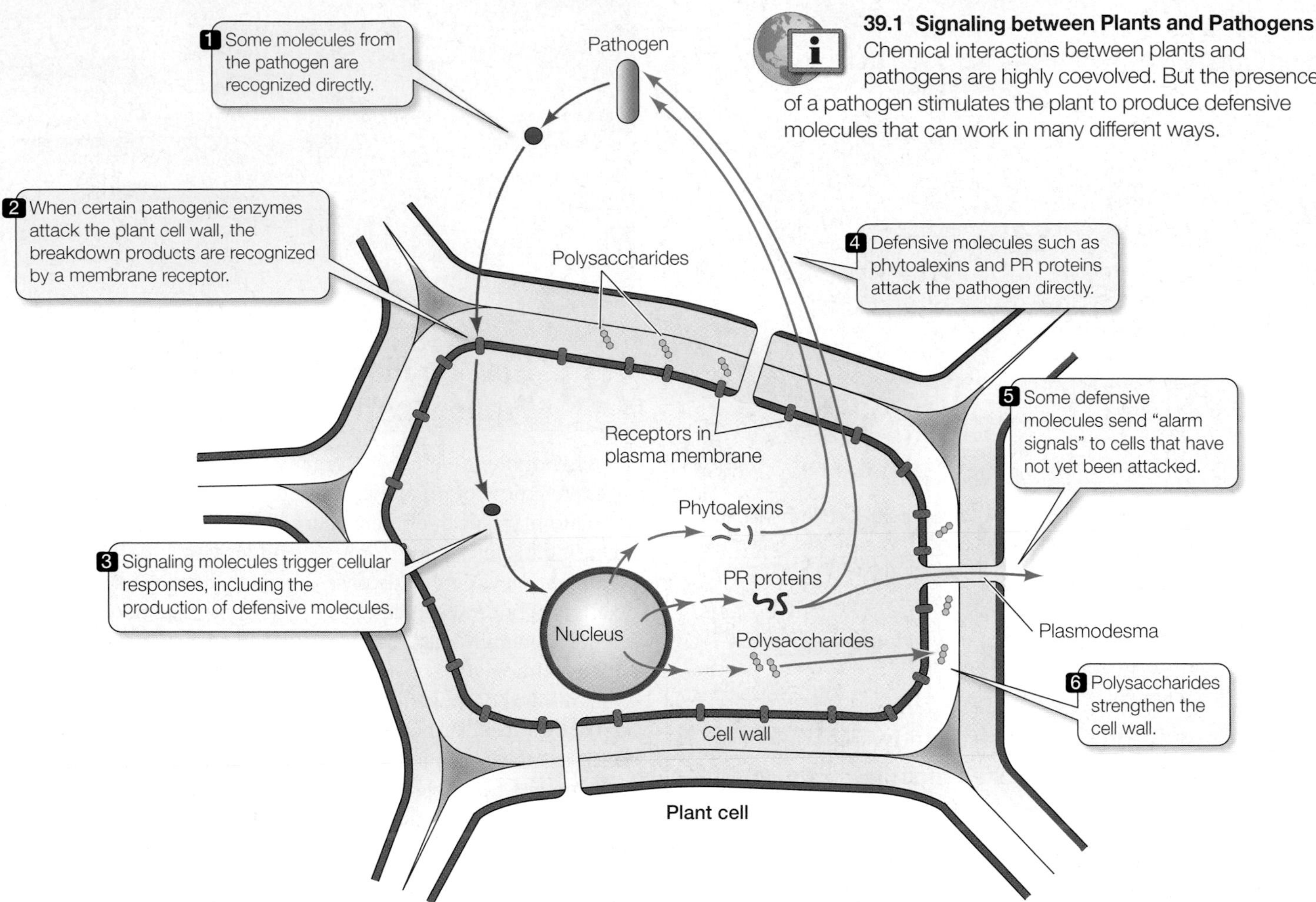

1 Some molecules from the pathogen are recognized directly.
Pathogen
2 When certain pathogenic enzymes attack the plant cell wall, the breakdown products are recognized by a membrane receptor.
Polysaccharides
4 Defensive molecules such as phytoalexins and PR proteins attack the pathogen directly.
5 Some defensive molecules send "alarm signals" to cells that have not yet been attacked.
Receptors in plasma membrane
Phytoalexins
3 Signaling molecules trigger cellular responses, including the production of defensive molecules.
PR proteins
Nucleus
Polysaccharides
Plasmodesma
6 Polysaccharides strengthen the cell wall.
Cell wall
Plant cell

39.1 Signaling between Plants and Pathogens Chemical interactions between plants and pathogens are highly coevolved. But the presence of a pathogen stimulates the plant to produce defensive molecules that can work in many different ways.

They also serve as a base upon which lignin may be laid down. Lignin enhances the mechanical barrier, and the toxicity of lignin precursor chemicals makes the cell inhospitable to some pathogens. These lignin building blocks are only one example of the toxic substances that plants use as chemical defenses.

Some plants have potent chemical defenses against pathogens

When infected by certain fungi and bacteria, plants produce a variety of defensive compounds. Two important kinds of defensive compounds are small molecules called *phytoalexins* and larger proteins called *pathogenesis-related proteins* (see Figure 39.1).

Phytoalexins are toxic to many fungi and bacteria. Most are phenolics or terpenes, compounds that protect plants against herbivores as well as pathogens (**Table 39.1**). They are produced by infected cells and their immediate neighbors within hours of the onset of infection. Enzymes from a pathogenic fungus can cause plant cell walls to release signaling molecules called *oligosaccharins*, which trigger phytoalexin production. Because their antimicrobial activity is nonspecific, phytoalexins can destroy many species of fungi and bacteria in addition to the one that originally triggered their production. Physical injuries, viral infections, and chemical compounds produced in response to damage by herbivores can also induce the production of phytoalexins.

Plants also produce several types of **pathogenesis-related proteins**, or **PR proteins**. Some are enzymes that break down the cell walls of pathogens. These enzymes destroy some of the invading cells, and in some cases the breakdown products of the pathogen's cell walls serve as chemical signals that trigger further defensive responses. Other PR proteins may serve as alarm signals to plant cells that have not yet been attacked. In general, PR proteins appear not to be rapid-response weapons; rather, they act more slowly, perhaps after other mechanisms have blunted the pathogen's attack.

PR proteins and phytoalexins do not act alone. Rather, they are tools used in complex defensive responses, such as the *hypersensitive response* and *systemic acquired resistance*.

The hypersensitive response is a localized containment strategy

Plants that are resistant to fungal, bacterial, or viral diseases generally owe this resistance to the **hypersensitive response**. Cells around the site of infection die, preventing the spread of the pathogen by depriving it of nutrients. Some of the cells produce phytoalexins and other chemicals before they die. The dead tissue, called a *necrotic lesion*, contains and isolates what is left of the microbial invasion (**Figure 39.2**). The rest of the plant remains free of the infecting microbe.

TABLE 39.1

Secondary Plant Metabolites Used in Defense

CLASS	TYPE	ROLE	EXAMPLE
Nitrogen-containing	Alkaloids	Affect herbivore nervous system	Nicotine in tobacco
	Glycosides	Release cyanide or sulfur compounds	Dhurrin in sorghum
	Nonprotein amino acids	Disrupt herbivore protein structure	Canavanine in jack bean
Phenolics	Flavonoids	Phytoalexins	Capsidol in peppers
	Quinones	Inhibit competing plants	Juglone in walnut
	Tannins	Deter herbivores and microbes	Many woods, such as oak
Terpenes	Monoterpenes	Insecticides	Pyrethroids in chrysanthemums
	Sesquiterpenes	Phytoalexins; deter herbivores	Gossypol in cotton
	Steroids	Mimic insect hormones and disrupt insect life cycles	α-Ecdysone in ferns
	Polyterpenes	Feeding deterrent?	Latex in rubber tree

One of the defensive chemicals produced during the hypersensitive response is a close relative of aspirin. Since ancient times, people in Asia, Europe, and the Americas have used willow (*Salix*) leaves and bark to relieve pain and fever. The active ingredient in willow is *salicylic acid*, the substance from which aspirin is derived.

It now appears that all plants contain at least some salicylic acid. This compound often evokes a second complex defensive response, which we will examine next.

Systemic acquired resistance is a form of long-term "immunity"

Systemic acquired resistance is a general increase in the resistance of the entire plant to a wide range of pathogenic species. It is not limited to the pathogen that originally triggered it or to the site of the original infection, and it may have a long-lasting effect.

39.2 The Aftermath of a Hypersensitive Response These necrotic spots on the leaves of a broad bean plant are a response to "chocolate spot" fungus, *Botrytis fabae*.

Systemic acquired resistance is accompanied by the synthesis of PR proteins. Treatment of plants with salicylic acid or aspirin leads to the production of PR proteins and to a resistance to pathogens. Salicylic acid treatment provides substantial protection against tobacco mosaic virus (a well-studied plant pathogen) and some other viruses. In some cases salicylic acid inhibits virus replication and in others it interferes with the movement of viruses out of the infected area.

Salicylic acid also serves as a plant hormone. In some cases, microbial infection in one part of a plant leads to the export of salicylic acid to other parts of the plant, where it causes the production of PR proteins before the infection can spread. The PR proteins then limit the extent of the infection. Infected plant parts also produce the closely related compound *methyl salicylate* (also known as oil of wintergreen). This volatile substance travels to other plant parts through the air, and may trigger the production of PR proteins in neighboring plants that have not yet been infected.

How does a plant know when it should activate the hypersensitive response and systemic acquired resistance? An interaction between plant and pathogen initiates these responses.

Some plant genes match up with pathogen genes

Many plants use the hypersensitive response and systemic acquired resistance as nonspecific defenses against various pathogens. However, the triggering of these responses resides in a highly specific mechanism, called **gene-for-gene resistance**. The ability of a plant to defend itself against a specific strain of a pathogen depends on the presence a particular allele of a gene in the plant that corresponds to a particular allele of a gene in the pathogen (**Figure 39.3**). Let's see how this matching works.

Plants have a large number of ***R* genes** (resistance genes), and many pathogens have sets of ***Avr* genes** (avirulence genes). Dominant *R* alleles favor resistance, and dominant *Avr* alleles make a pathogen less effective. If a particular plant has the dominant allele of one *R* gene and a pathogen strain infecting it has the dominant allele of the corresponding *Avr* gene, the plant will be resistant to that strain. This is true even when none of the other *R-Avr* pairs fea-

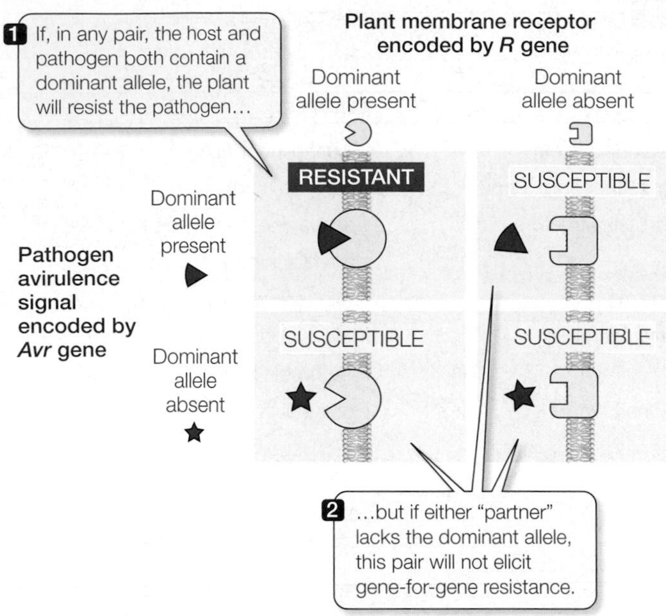
1 If, in any pair, the host and pathogen both contain a dominant allele, the plant will resist the pathogen…
Plant membrane receptor encoded by *R* gene
Dominant allele present
Dominant allele absent
Pathogen avirulence signal encoded by *Avr* gene
Dominant allele present
Dominant allele absent
RESISTANT
SUSCEPTIBLE
SUSCEPTIBLE
SUSCEPTIBLE
2 …but if either "partner" lacks the dominant allele, this pair will not elicit gene-for-gene resistance.

39.3 Gene-for-Gene Resistance A single pair of corresponding dominant alleles promotes resistance even if all the other pairs are mismatches.

tures corresponding dominant alleles. (This effect of one *R-Avr* pair overruling the others is an example of *epistasis*; see Section 10.3.)

The mechanism of gene-for-gene resistance is not completely understood. There are thousands of specific *R* genes among the plants, and their products have different functions. The *Avr* genes in pathogens are simply the genes that cause the pathogen to produce a substance, often toxic, that elicits a defensive response in the plant. Most gene-for-gene interactions trigger the hypersensitive response.

Before we leave the topic of plant defenses against pathogens, let's consider a recently discovered specific defense mechanism directed against RNA viruses (viruses that have RNA instead of DNA as their hereditary material).

Plants develop specific immunity to RNA viruses

Plants respond to attack by RNA viruses by mounting a specific immune response. The plant uses its own enzymes to convert some of the single-stranded RNA of the invading virus into *double-stranded RNA* (dsRNA) and to chop that dsRNA into small pieces called *small interfering RNAs* (siRNAs). Some of the viral RNA is transcribed, forming mRNAs that advance the viral infection. However, the siRNAs interact with another cellular component to degrade those mRNAs, blocking viral replication. This phenomenon is an example of *RNA interference* (*RNAi*), or *posttranscriptional gene silencing* (see p. 365). Molecular biologists are exploring applications of RNAi in plant biotechnology.

The immunity conferred by RNAi spreads quickly throughout the entire plant, by mechanisms not yet fully understood. However, the establishment of immunity depends on the extent of the original infection and the speed of the plant's response. Plant viruses do fight back: most have evolved mechanisms to confound RNA interference. Natural selection favors both improved attack mechanisms for the pathogens and improved defense mechanisms for the plants.

39.1 RECAP

In the hypersensitive response to pathogens, plants seal off infected areas and produce chemical defenses. Systemic acquired resistance, providing a longer-lasting, more general immunity may follow. Gene-for-gene resistance triggers some of these phenomena.

- What are some of the defensive compounds produced by plant cells when infected by bacteria or fungi? See pp. 838–839, Figure 39.1, and Table 39.1
- Can you explain how *R* and *Avr* genes determine which pathogens a plant may be unable to resist? See pp. 839–840 and Figure 39.3

Not all biological threats to plants come from microorganisms and viruses that cause diseases. Another threat comes from the many animals, from inchworms to elephants, that eat plants.

39.2 How Do Plants Deal with Herbivores?

Herbivores—animals that eat plants—depend on plants for energy and nutrients, and they often spread disease. Plants have many defense mechanisms that protect them against herbivores, as we will see. First let's consider how herbivores can have a *positive* effect on some of the plants they eat.

Grazing increases the productivity of some plants

Herbivores are predators that prey on plants, but rarely do they kill their prey. In **grazing**, a herbivore eats part of a plant, such as the leaves, without killing the plant, which then has the potential to grow back.

What are the consequences of grazing? How detrimental is grazing to plants? How well have plants adapted to their place in the food chain? Certain plants and their predators have evolved together, each acting as the agent of natural selection on the other. Coevolution has favored increased photosynthetic production in some grazed-upon plant species.

Removing some leaves from a plant can increase the rate of photosynthesis in the remaining leaves, for several reasons. First, nitrogen obtained from the soil by the roots no longer needs to be divided among as many leaves. Second, the export of sugars and other photosynthetic products from the leaves may be enhanced because the demand for those products in the roots is undiminished, while the sources for those products—leaves—have been decreased. That is, the remaining leaves may compensate by photosynthesizing more rapidly.

A third and particularly significant factor increasing photosynthesis, especially in grasses, is an increase in the availability of light to the younger, more active leaves or leaf parts. The removal of older or dead leaves by a grazer decreases the shading of younger leaves. Unlike most other plants, which grow from their shoot and leaf tips, grasses grow from the base of the shoot and leaf, so their growth is not cut short by grazing.

Mule deer and elk graze many plants, including one called scarlet gilia (*Ipomopsis aggregata*). Although grazing removes about 95 percent of the aboveground plant, the scarlet gilia quickly regrows not one but four replacement stems (**Figure 39.4**). Grazed plants produce three times as many fruits by the end of the growing season as do ungrazed plants.

Some grazed trees and shrubs continue to grow until much later in the season than do ungrazed but otherwise similar plants. The growing season is extended in part by the grazers' removal of apical buds, which stimulates axillary buds to become active and produces a more heavily branched plant. Leaves on ungrazed plants may also die earlier in the growing season than leaves on grazed plants.

A plant also may benefit from moderate herbivory by attracting animals that spread its pollen or that eat its fruit and thus disperse its seeds—in such cases, the benefits to the plant outweigh the costs. Nevertheless, resisting attack by herbivores is often advantageous to a plant.

Some plants produce chemical defenses against herbivores

Although a plant cannot flee its herbivorous enemies, it may be able to defend itself chemically. Many plants attract, resist, and inhibit other organisms by producing special chemicals known as *secondary metabolites*. *Primary metabolites* are substances, such as proteins, nucleic acids, carbohydrates, and lipids, that are produced and used by all living things. **Secondary metabolites** are substances that are not used for basic cellular metabolism. Although all organisms use the same kinds of primary metabolites, plants can differ as radically in their secondary metabolites as they do in their external appearance.

The more than 10,000 known secondary plant metabolites range in molecular weight from about 70 to more than 390,000 daltons, but most have a low molecular weight. Some are produced by only a single species, while others are characteristic of entire genera or even families. The effects of defensive secondary metabolites on animals are diverse. Some secondary metabolites act on the nervous systems of herbivorous insects, mollusks, or mammals. Others mimic the natural hormones of insects, causing some larvae to fail to develop into adults. Still others damage the digestive tracts of herbivores. Some secondary metabolites are toxic to fungal pests. Humans make commercial use of many secondary plant metabolites as fungicides, insecticides, rodenticides, and pharmaceuticals.

The secondary metabolite nicotine was one of the first insecticides to be used by farmers and gardeners. Yet tobacco and related plants that produce nicotine are still attacked by pests such as the tobacco hornworm. The question of whether nicotine deters pests was conclusively investigated by biologists in 2004. The study used tobacco plants in which an enzyme in nicotine biosynthesis had been silenced, lowering the nicotine concentration by more than 95 percent. The low-nicotine plants suffered much more damage than normal plants (**Figure 39.5**).

While many secondary metabolites have protective functions, others are essential as attractants for pollinators and seed dispersers. Table 39.1 lists the major classes of defensive secondary plant metabolites and their biological roles.

Let's look at a specific example of an insecticidal secondary metabolite, canavanine.

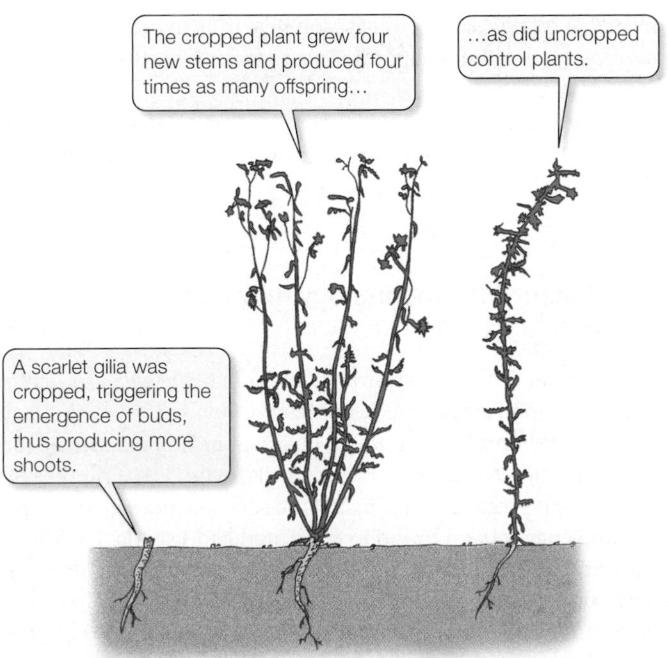

39.4 Overcompensation for Being Eaten Experiments confirm that some plants benefit from grazing.

Some secondary metabolites play multiple roles

Canavanine is an amino acid that is not found in proteins, but is similar to the amino acid arginine, which is found in almost all proteins. Canavanine has two important roles in plants that produce it in significant quantities. The first is as a nitrogen-storing compound in seeds. The second role is defensive, and is based on the similarity of canavanine to arginine:

A seemingly slight chemical difference...

Arginine: NH_2—C(=NH)—N—H — **H—C—H** — H—C—H — H—C—H — H_2N—C—H — C(=O)—OH

Canavanine: NH_2—C(=NH)—N—H — **O** — H—C—H — H—C—H — H_2N—C—H — C(=O)—OH

...produces an inactive protein.

EXPERIMENT

HYPOTHESIS: Nicotine helps protect tobacco plants against insects.

METHOD

1. Create a low-nicotine line of plants by modifying a gene that encodes a key enzyme in the biosynthetic pathway to nicotine.
2. Transplant both low-nicotine and wild-type tobacco plants into a field plantation where they are accessible to insects.
3. Assess the extent of leaf damage by insects at 2-day intervals.

RESULTS

The low-nicotine plants suffered more than twice as much leaf damage as did the wild-type controls.

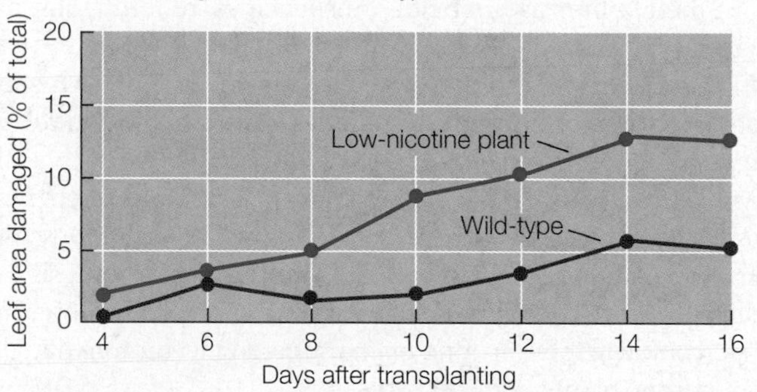

CONCLUSION: Nicotine provides tobacco plants with at least some protection against insects.

39.5 Some Plants Use Nicotine to Reduce Insect Attacks Anke Steppuhn and her coworkers showed that wild-type tobacco plants lost less leaf area to pests than did transgenic plants that contained less nicotine. FURTHER RESEARCH: Treatment of tobacco plants with jasmonate (a hormone) elicits the production of nicotine and other compounds. How would you modify this experiment to determine whether nicotine is the only insecticidal compound produced by tobacco?

When an insect larva consumes canavanine-containing plant tissue, the canavanine is incorporated into the insect's proteins in some of the places where the DNA has coded for arginine, because the enzyme that charges the tRNA specific for arginine fails to discriminate accurately between the two amino acids (see Section 12.4). The structure of canavanine, however, is different enough from that of arginine that some of the resulting proteins end up with a modified tertiary structure and hence reduced biological activity. These defects in protein structure and function lead to developmental abnormalities that kill the insect.

A few insect larvae are able to eat canavanine-containing plant tissue and still develop normally. Why? In these larvae, the enzyme that charges the arginine tRNA discriminates correctly between arginine and canavanine. The canavanine they ingest is thus not incorporated into the proteins they form, and the larvae are not harmed.

In plants that produce it, canavanine is present regardless of whether the plant is under attack. Other chemical defenses come into play only when a predator strikes.

Some plants call for help

If you are attacked, it makes sense to call for help. Some plants do this, too. When caterpillars begin to chew on the leaves of corn, cotton, or some other plant species, the leaves synthesize chemical signals that attract other insects that feed on the caterpillars.

Although most such calls for help are generated in the leaves, in 2005 biologists found that some plant roots attacked by beetle larvae respond by releasing an attractant for a tiny nematode that attacks the larvae (**Figure 39.6**). Perhaps it will soon be possible to enhance the production of such attractants and increase crop yields.

Many defenses depend on extensive signaling

Many plant defenses are activated by a series of signals. Insects feeding on tomato leaves damage the cells, leading to a chain of events that includes the formation of hormones and ends with the production of an insecticide. The signaling steps in the production of one defensive compound, shown in **Figure 39.7**, involve two hormones. **Systemin**, which is formed in response to an insect attack, is a polypeptide hormone—the first polypeptide hormone to be discovered in plants. **Jasmonates**, whose production is initiated by systemin, are formed from the unsaturated fatty acid linolenic acid. The final step in the chain is the synthesis of a protease inhibitor. The inhibitor, once in an insect's gut, interferes with the digestion of proteins and thus stunts the insect's growth.

Jasmonates also take part in the "call for help" caterpillars attack a corn plant. A volatile substance released from the leaves by chewing caterpillars is the first signal, leading to the formation of jasmonates by the plant. The jasmonates, in turn, trigger the formation of the volatile compounds that attract insects that prey on the caterpillars.

Although plants have many effective natural defenses, agricultural researchers are attempting to provide crop plants with even more effective ones, including transfer of natural, effective mechanisms between species.

Recombinant DNA technology may confer resistance to insects

Wild and domesticated common beans (*Phaseolus vulgaris*) differ in their resistance to two species of bean weevils. Some wild bean seeds are highly resistant to these insects, but no cultivated bean seeds show such resistance. Scientists discovered that all weevil-resistant bean seeds contain a specific seed protein, *arcelin*. This protein has never been found in cultivated bean seeds. Therefore, the scientists hypothesized that arcelin is responsible for the resistance of some seeds to predation by the weevils.

To test the hypothesis that arcelin is the compound that confers resistance in wild species, the scientists performed two series of experiments. In one series, they crossed cultivated and wild bean plants. All of the progeny seeds of such crosses that contained

EXPERIMENT

HYPOTHESIS: Corn roots attacked by beetle larvae attract nematodes that will attack the larvae.

METHOD

1. Construct a test system with 6 arms radiating from a central chamber. Add soil, connecting all parts of the system.
2. Three of the control chambers contain only soil. The other three chambers each contain a single corn plant: One healthy plant, one with roots damaged by beetle larvae, and one with roots damaged by stabbing with a metal tool.
3. After three days, add ~2,000 nematodes to the central chamber.
4. After 24 hours count the nematodes in each connecting arm.

RESULTS

Nematodes moved into each of the arms, but by far the most moved into the arm leading to the larvae-damaged plant.

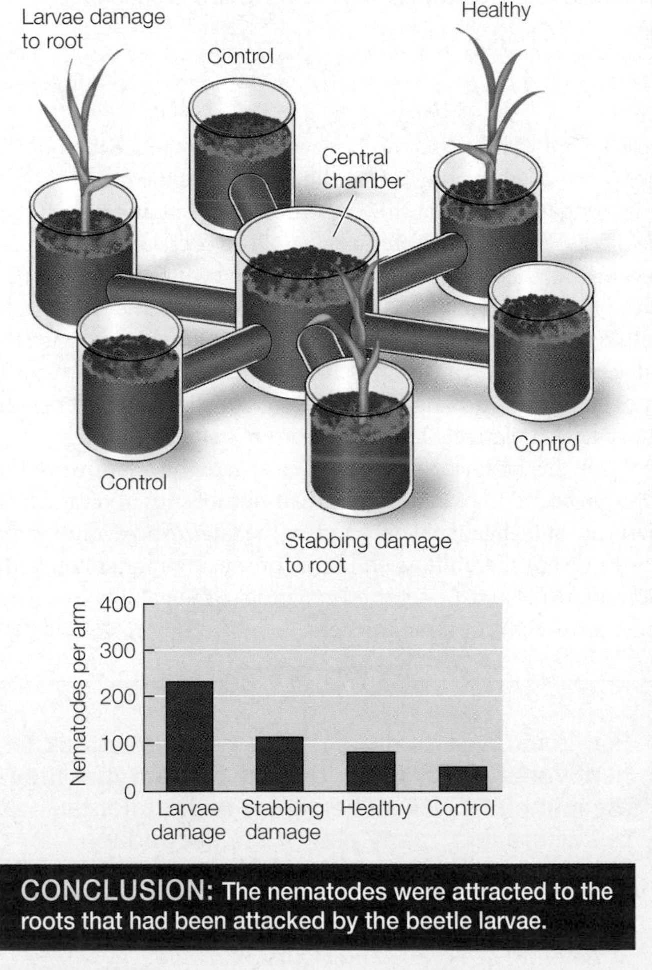

39.6 Roots May Recruit Nematodes as Defenders When attacked by beetle larvae, corn roots generate a chemical signal that attracts nematodes—natural enemies of the larva. FURTHER RESEARCH: The biologists who did this study also showed that roots attacked by the beetle larvae released a secondary metabolite called (E)-β-caryophyllene. How would you test the hypothesis that this compound is the attractant for the nematodes?

arcelin showed resistance to weevils. In the other series of experiments, the scientists removed the seed coats of domesticated beans and ground the remainder of the seeds into flour. They added different concentrations of arcelin to different batches and molded the flour into artificial seeds. They then let weevils attack the artificial seeds. The more arcelin the artificial seeds contained, the more resistant they were to weevils. These data support the hypothesis that arcelin is the active component that confers weevil resistance.

Now that the data indicate that arcelin is the active component in pest resistance, it was of interest to assess its toxicity, if any, in animal systems. Preliminary tests showed that arcelin in cooked beans was not harmful to rats—a first step toward determining whether arcelin is safe in food for humans. Agricultural scientists must sometimes choose between crop protection and appeal to humans. A plant with sturdy chemical defenses may taste bad, make us sick, or even kill us.

Scientists are now seeking to introduce genes for arcelin and other resistance-conferring proteins into agriculturally important crops such as beans. The development of crop plants that produce

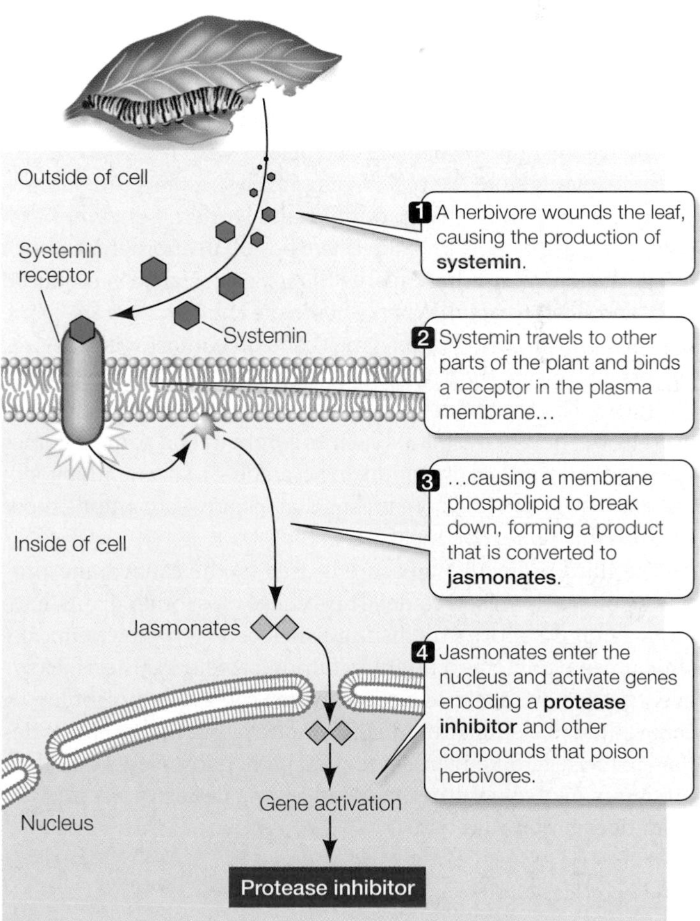

39.7 A Signaling Pathway for Synthesis of a Defensive Secondary Metabolite The chain of events initiated by an insect's attack leads to the production of a defensive chemical and can consist of many steps. These steps may include the synthesis of one or two hormones, binding of receptors, gene activation, and, finally, synthesis of insecticides.

their own pesticides is an active area of research in agricultural biotechnology. One of the most widely applied approaches has been the engineering of several crops, such as tomatoes, corn, and cotton, to express the toxin genes from *Bacillus thuringiensis*. The toxin kills insect pest larvae, enhancing crop yields, as discussed in Section 16.6.

Why don't plants poison themselves?

Why don't the chemicals that are so toxic to herbivores and microbes kill the plants that produce them? Plants that produce toxic secondary metabolites generally use one of the following measures to protect themselves:

- The toxic material is isolated in a special compartment.
- The toxic substance is produced only after the plant's cells have already been damaged.
- The plant uses modified enzymes or modified receptors that do not recognize the toxic substance.

The first method is the most common. Plants using this method store their poisons in vacuoles if they are water-soluble. If they are hydrophobic, the poisons are stored in **laticifers** (tubes containing a white, rubbery latex) or dissolved in waxes on the epidermal surface. This compartmentalized storage keeps the toxic substance away from the mitochondria, chloroplasts, and other parts of the plant's own metabolic machinery. One example of a plant with this mechanism is milkweed, discussed below.

Some plants store the precursors of toxic substances in one area of the plant, such as the epidermis, and store the enzymes that convert those precursors to the active poison in another area, such as the mesophyll. In these plants the toxic substance is produced only after it is damaged. When a herbivore chews part of the plant, the cells rupture, and the enzymes come in contact with the precursors, producing the toxic product. The only part of the plant that is damaged by the toxic substance is that which was already damaged by the herbivore. Plants such as sorghum and some legumes that respond to attack by producing cyanide—a strong inhibitor of cellular respiration in all organisms that respire—are among those that use this protective measure.

The third protective measure is used by the canavanine-producing plants described earlier. They, unlike most other plants, have an enzyme that correctly distinguishes between the chemically similar canavanine and arginine during protein synthesis. However, as we have seen, some herbivores can evade poisoning by canavanine in a similar manner, demonstrating that no plant defense is perfect. Like plants and their pathogens, plants and their predators evolve together in a continuing "arms race," and the plant does not always win.

The plant doesn't always win

Milkweeds such as *Asclepias syriaca* are latex-producing (laticiferous) plants. When damaged, a milkweed releases copious amounts of toxic latex from its laticifers, which run alongside the veins in its leaves. Latex, a milky liquid, has long been suspected to deter insects from eating the plant because some insects that feed on neighboring plants of other species do not attack laticiferous plants. This observed behavior is consistent with, but does not prove, the hypothesis that the latex keeps the insects at bay.

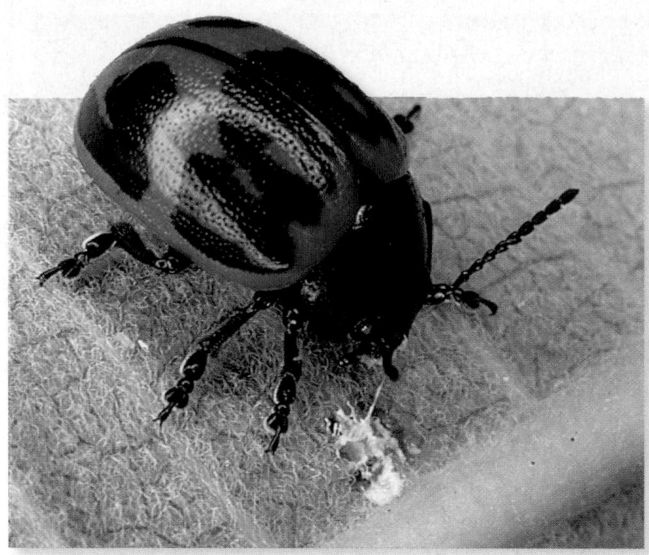

39.8 Disarming a Plant's Defenses This beetle is inactivating a milkweed's defense system by cutting its laticifer supply lines.

Stronger support for this hypothesis was obtained by studying field populations of *Labidomera clivicollis*, a beetle that is one of the few insects that feed on *A. syriaca*. These beetles show a remarkable prefeeding behavior: They cut a few veins in the leaves before settling down to dine (**Figure 39.8**). Cutting the veins, with their adjacent laticifers, causes massive latex leakage and interrupts the latex supply to a downstream portion of the leaf. The beetles then move to the relatively latex-free portion and eat their fill.

Does this behavior of the beetles negate the adaptive value of latex protection? Not entirely. Great numbers of potential insect pests are still effectively deterred by the latex. And evolution proceeds. Over time, milkweed plants producing higher concentrations of toxins may be selected by virtue of their ability to kill even the beetles that cut their laticifers.

39.2 RECAP

Plants must defend themselves against attack by herbivores and disease despite the fact that they are immobile. Not all herbivory is detrimental, but where it is, many plants use secondary metabolites as chemical defenses.

- What roles do systemin and jasmonates play in plant defense? See p. 843 and Figure 39.7
- Why don't a plant's toxic secondary metabolites poison the plant itself? See p. 844

A plant's survival depends not only on successful defense against predators and pathogens. The physical environment can be hostile to a plant as well. Next we consider how plants adapt to climate-imposed threats.

39.9 **Desert Annuals Evade Drought** Seeds of desert plants often lie dormant for long periods awaiting conditions appropriate for germination. When they do germinate, they grow and reproduce rapidly before the short wet season ends. During the long dry spells, only dormant seeds remain alive.

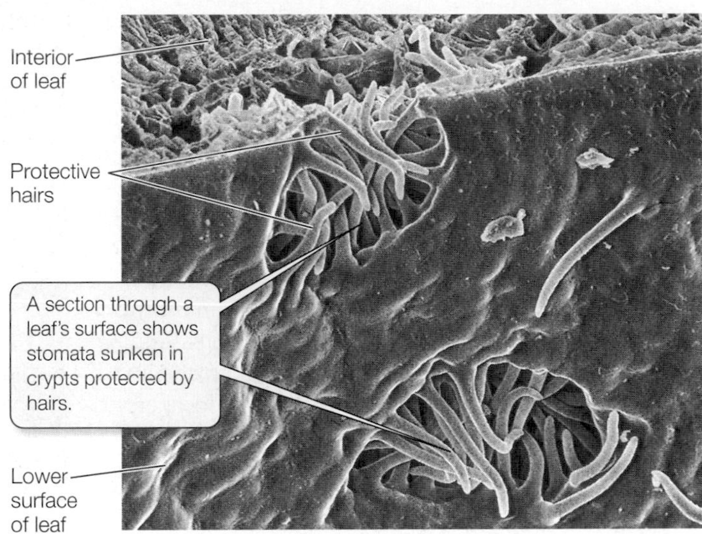

Interior of leaf

A section through a leaf's surface shows stomata sunken in crypts protected by hairs.
Lower surface of leaf

39.10 **Stomatal Crypts** Stomata in the leaves of some xerophytes are located in sunken cavities called stomatal crypts. The hairs covering these crypts trap moist air.

39.3 How Do Plants Deal with Climate Extremes?

Plants are threatened by many aspects of the physical environment, such as excess dryness, waterlogged soils, and extremes of temperature, both high and low. Water is often in short supply in the terrestrial environment. Extreme terrestrial habitats such as deserts intensify this challenge. Some desert plants have no special *structural* adaptations for water conservation other than those found in almost all flowering plants. Instead, they have an alternative *strategy*. These desert annuals simply evade the periods of drought. They carry out their entire life cycle—from seed to seed—during a brief period in which rainfall has made the surrounding desert soil sufficiently moist (**Figure 39.9**).

Many plants that inhabit particularly dry areas, however, have one or more adaptations that allow them to conserve water. Plants adapted to dry environments are called **xerophytes**.

Xeros is the Greek word for "dry." Xerox, one of the most recognized brand names in commercial history, made its reputation (and fortune) from its patent on the first "dry" office copying process, based on powdered toner. Up until that time, office copiers such as mimeograph machines required messy liquids to "duplicate" a master copy.

Some leaves have special adaptations to dry environments

Plants that remain active during dry periods must have structural adaptations that enable them to survive. The secretion of a thick cuticle over the leaf epidermis helps retard water loss in dry environments. An even more common adaptation is a dense covering of epidermal hairs. Some species have stomata only in sunken cavities below the leaf surface, which reduces the drying effects of air currents; often these **stomatal crypts** contain hairs as well (**Figure 39.10**). The hairs slow the air currents around the stomata.

Succulence—the possession of fleshy, water-storing leaves or stems—is an adaptation to dry environments. Ice plants and their relatives have fleshy leaves in which water may be stored. Other xerophytes, such as ocotillo, produce leaves only when water is abundant, shedding them as the soil dries out (**Figure 39.11**). Cacti

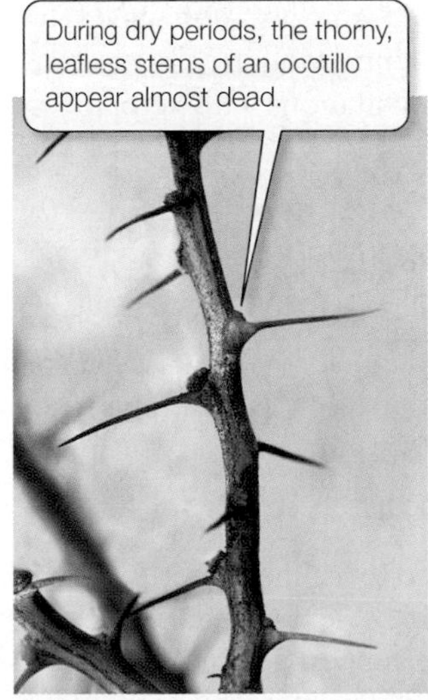

During dry periods, the thorny, leafless stems of an ocotillo appear almost dead.

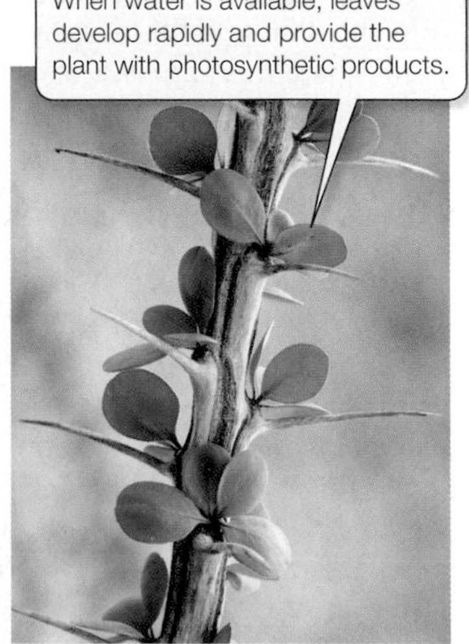

When water is available, leaves develop rapidly and provide the plant with photosynthetic products.

39.11 **Opportune Leaf Production** The ocotillo, a xerophyte that lives in the lower deserts of the southwestern United States and northern Mexico, produces leaves only when there is sufficient water for photosynthesis.

and similar plants have spines rather than typical leaves, and photosynthesis is confined to the fleshy stems. The spines may reflect incident radiation, or they may dissipate heat. Corn and some related grasses have leaves that roll up during dry periods, thus reducing the leaf surface area through which water is lost. Some trees, such as eucalyptus, that grow in arid regions have leaves that hang vertically at all times, thus evading direct rays of the midday sun.

These xerophytic adaptations of leaves minimize water loss by the plant. However, such adaptations simultaneously minimize the uptake of carbon dioxide and thus limit photosynthesis. In consequence, most xerophytes grow slowly, but they utilize water more efficiently than do other plants—that is, they fix more grams of carbon by photosynthesis per gram of water lost to transpiration than other plants do.

39.12 Mining Water with Deep Taproots In Death Valley, California, this mesquite must reach far down into the sand dunes for its water supply.

Plants have other adaptations to a limited water supply

Roots may also be adapted to dry environments. Mesquite trees (genus *Prosopis*; **Figure 39.12**) obtain water through taproots that grow to great depths, reaching water supplies far underground, as well as from condensation on their leaves. The Atacama Desert in northern Chile often goes several years without measurable rainfall. The landscape there is almost barren of plant life, save for many surprisingly large mesquite trees.

A more common adaptation of desert plants is a root system that grows rapidly during rainy seasons but dies back during dry periods. Cacti have shallow but extensive fibrous root systems that effectively intercept water at the surface of the soil following even light rains.

Xerophytes and other plants that receive inadequate water may accumulate the amino acid proline or other solutes to substantial concentrations in their vacuoles. As a consequence, the solute potential and water potential of their cells become more negative; thus these plants tend to extract more water from the soil than do plants that lack this adaptation. Plants living in salty environments share this and several other adaptations with xerophytes, as we will see.

In water-saturated soils, oxygen is scarce

For some plants, the environmental challenge is the opposite of that faced by xerophytes: too much water. Some plants live in environments so wet that the diffusion of oxygen to their roots is severely limited. Since most plant roots require oxygen to support respiration and ATP production, most plants cannot tolerate saturated soil conditions for long.

Some species, however, are adapted to life in a water-saturated habitat. Their roots grow slowly and hence do not penetrate deeply. Because the oxygen level is too low to support aerobic respiration, the roots carry on alcoholic fermentation (an anaerobic process; see Chapter 7), which provides ATP for the activities of the root system. This adaptation explains why their growth is slow—fermentation is much less efficient in producing ATP than aerobic respiration.

The root systems of some plants adapted to swampy environments have **pneumatophores**, which are extensions that grow out of the water and up into the air (**Figure 39.13**). Pneumatophores have lenticels and contain spongy tissues that allow oxygen to diffuse through them, aerating the submerged parts of the root system. Cypresses and some mangroves are examples of plants with pneumatophores.

Submerged or partly submerged aquatic plants often have large air spaces in the leaf parenchyma and in the petioles. Tissue con-

Pneumatophores are root extensions that grow out of the water, under which the rest of the roots are submerged.

39.13 Coming Up for Air The roots of the mangroves in this tidal swamp obtain oxygen through pneumatophores.

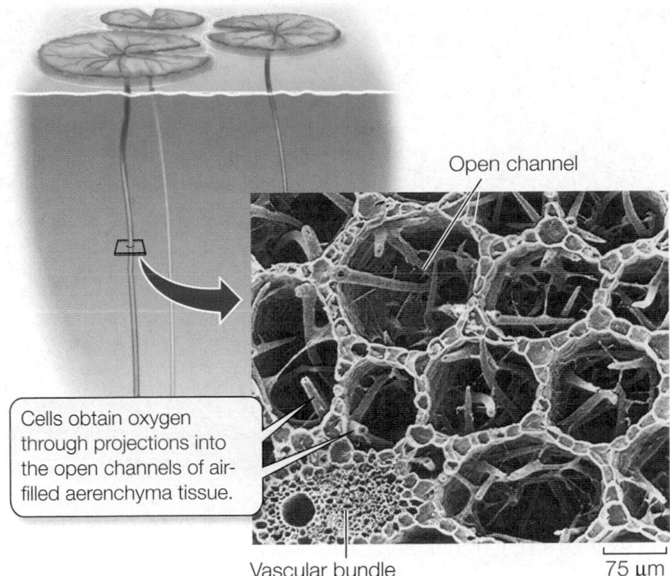

Open channel
Cells obtain oxygen through projections into the open channels of air-filled aerenchyma tissue.
Vascular bundle
75 µm

39.14 Aerenchyma Lets Oxygen Reach Submerged Tissues The scanning electron micrograph, a cross section of a petiole of the yellow water lily, shows the air-filled channels of aerenchyma tissue. The cells that line these channels obtain oxygen by extending projections into these channels.

taining such air spaces is called **aerenchyma** (**Figure 39.14**). Aerenchyma stores oxygen produced by photosynthesis and permits its ready diffusion to parts of the plant where it is needed for cellular respiration. Aerenchyma also imparts buoyancy. Furthermore, because it contains far fewer cells than most other plant tissue, respiratory metabolism in aerenchyma proceeds at a lower rate, and the need for oxygen is much reduced.

We have seen that plants are threatened by shortages or excesses of water in their physical environment. We now examine two more threats in the physical environment: high and low temperatures.

Plants have ways of coping with temperature extremes

Temperatures that are too high or too low can stress plants and even kill them. Plants differ in their sensitivity to heat and cold, but all plants have their limits. Any temperature extreme can damage cellular membranes.

- High temperatures destabilize membranes and denature many proteins, especially some of the enzymes of photosynthesis.
- Low temperatures cause membranes to lose their fluidity and alter their permeabilities to solutes.
- Freezing temperatures may cause ice crystals to form, damaging cellular membranes.

Transpiration (loss of water by plants through evaporation) can cool a plant, but it also increases the plant's need for water. Therefore, it is not surprising that many plants living in hot environments have adaptations similar to those of xerophytes. These adaptations include epidermal hairs and spines that radiate heat, leaf displays that intercept less direct sunlight, and an alternative form of metabolism that allows plants to perform some metabolic processes in the cool of night—crassulacean acid metabolism (CAM; see Section 8.4).

Plants respond within minutes to high temperatures by producing several kinds of **heat shock proteins**. Among these proteins are chaperonins (see Figure 3.12), which help other proteins maintain their structures and avoid denaturation. Threshold temperatures for the production of heat shock proteins vary, but 39°C is sufficient to induce them in most plants. We have much to learn about the dozens of heat shock proteins, but we do know that some other types of stress also induce their formation. Among these stresses are chilling and freezing.

Low temperatures above the freezing point injure many plants, including important crops such as rice, corn, and cotton as well as tropical plants such as bananas. This is referred to as *chilling injury*. Many plant species can be modified to resist the effects of cold spells by a process called **cold-hardening**, which involves repeated exposure to cool, but not injurious, temperatures. The hardening process, however, requires many days. A key change that occurs during the hardening process is an increase in the relative amount of unsaturated fatty acids in membranes. Unsaturated fatty acids solidify at lower temperatures than do saturated ones. Thus, the membranes retain their fluidity and function normally at cooler temperatures.

Low temperatures induce the formation of certain heat shock proteins that protect against chilling injury. There are also cases of "cross-protection" by heat shock proteins that are induced by one type of stress and that protect against other stresses. Tomatoes stressed by 2 days of high temperatures, for example, formed heat shock proteins and became resistant to chilling injury for the next 3 weeks.

If ice crystals form within plant cells, they can kill the cells by puncturing organelles and plasma membranes. More importantly, the growth of ice crystals outside the cells can draw water from the cells and dehydrate them, causing them to plasmolyze. Freezing-tolerant plants have a variety of adaptations to cope with these problems. A common one is the production of **antifreeze proteins** that slow the growth of ice crystals.

39.3 RECAP

Xerophytes are plants that cope with dry environments through structural or behavioral adaptations. Other plants are adapted to life in a water-saturated environment. Many plants are resistant to high and low temperatures. Heat shock proteins play roles in adapting to both extremes.

- Describe the tradeoff between water conservation and photosynthesis in xerophytes. See p. 846
- What conditions induce the formation of heat shock proteins, and what functions do they serve? See p. 847

Just as extremes of climate can limit plant growth, the presence of certain substances can make an environment inhospitable to plant growth. These substances include salt and heavy metals.

39.4 How Do Plants Deal with Salt and Heavy Metals?

Worldwide, no toxic substance restricts angiosperm growth more than salt (sodium chloride) does. *Saline*—salty—habitats support, at best, sparse vegetation. Saline habitats themselves are diverse, ranging from hot, dry, salty deserts to moist, cool, salty marshes. Along the seashore saline environments are created by ocean spray. The ocean itself is a saline environment, as are river estuaries, where fresh and salt water meet and mingle. The salinization of agricultural land is an increasing global problem, not just in Bangladesh, which we described at the beginning of this chapter. Even where crops are irrigated with fresh water, sodium ions from the water accumulate in the soil to ever greater concentrations as the water evaporates.

Saline environments pose an osmotic problem for plants. Because of its high salt concentration, a saline environment has an unusually negative water potential. To obtain water from such an environment, a plant must have an even more negative water potential than that of a plant in a nonsaline environment; otherwise, it would lose water, wilt, and die. A second problem is the potential toxicity of high concentrations of certain ions, notably sodium and chloride.

The **halophytes**—plants adapted to saline habitats—belong to a wide variety of flowering plant groups. How do these plants cope with a saline environment?

Most halophytes accumulate salt

Most halophytes share one adaptation: They accumulate sodium and, usually, chloride ions and transport those ions to the leaves. The accumulated ions are stored in the central vacuoles of leaf cells, away from more sensitive parts of the cells. Nonhalophytes accumulate relatively little sodium, even when placed in a saline environment; of the sodium that is absorbed by their roots, very little is transported to the shoot. The increased salt concentration in the tissues of halophytes makes their water potential more negative, so they can take up water more easily from the saline environment.

Scientists have succeeded in causing the overexpression of a gene that enables tomato plants to take up salt into the vacuoles of cells. This increased sodium uptake converts the tomato to a halophyte, enabling it to grow in saline environments. Crops with such gene overexpression can be watered with diluted seawater.

Some halophytes have other adaptations to life in saline environments. Some, for example, have **salt glands** in their leaves. These glands excrete salt, which collects on the leaf surface until it is removed by rain or wind (**Figure 39.15**). This adaptation, which reduces the danger of poisoning by accumulated salt, is found both in some desert plants, such as tamarisk, and in some mangroves growing in seawater.

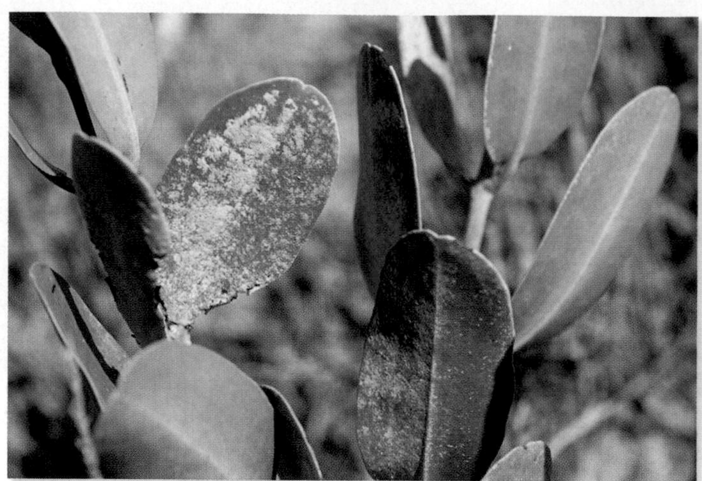

39.15 Excreting Salt This salty mangrove has special salt glands that excrete salt, which appears here as crystals on the leaves.

Salt glands can play multiple roles, as in the desert shrub *Atriplex halimus*. This shrub has glands that secrete salt into small bladders on the leaves, where, by increasing the gradient in water potential, the salt helps the leaves obtain water from the roots. At the same time, by making the water potential of the leaves more negative, the salt reduces the transpirational loss of water to the atmosphere.

The adaptations we have just discussed are specific to halophytes. Several other adaptations are shared by halophytes and xerophytes.

Halophytes and xerophytes have some similar adaptations

Many halophytes, like some xerophytes, accumulate the amino acid proline in their cell vacuoles, making the water potential of their tissues more negative. Unlike sodium, proline is relatively nontoxic.

Succulence is another adaptation that halophytes and xerophytes have in common, as might be expected, since saline environments, like dry ones, make water uptake difficult. Succulence characterizes many halophytes that occupy salt marshes. There the salt concentration in the soil solution may change throughout the day. When the tide is out, evaporation increases the salt concentration. Succulence may offer a reserve of water for the plant during the period of maximum salinity; when the salinity drops as the tide comes in, the leaf's store of water is replenished.

Many succulents—both xerophytes and halophytes—use crassulacean acid metabolism which allows them to take up and store CO_2 as carboxyl groups at night and release the CO_2 for use in photosynthesis during the day. They have reversed stomatal cycles that enable them to conserve water by closing their stomata in the daytime (**Figure 39.16**). Other general adaptations to a saline environment include high root-to-shoot ratios, sunken stomata, reduced leaf areas, and thick cuticles.

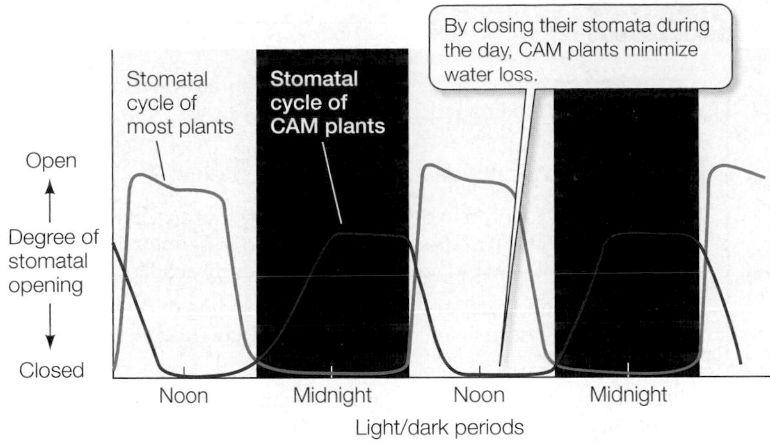

39.16 Stomatal Cycles Most plants open their stomata during the day. CAM plants reverse this stomatal cycle; their stomata open during the night.

Some habitats are laden with heavy metals

Salt is not the only toxic solute found in soils. High concentrations of some heavy metal ions, such as chromium, mercury, lead, and cadmium, poison most plants; these ions often are more toxic than sodium at equivalent concentrations.

Some geographic sites are naturally rich in heavy metals as a result of normal geological processes. In other places, acid rain leads to the release of toxic aluminum ions in the soil. Other human activities, notably the mining of metallic ores, leave localized areas—known as *tailings*—with substantial concentrations of heavy metals and low concentrations of nutrients. Such sites are hostile to most plants, and seeds falling on them generally do not produce adult plants.

39.17 Life after Strip Mining Although high concentrations of heavy metals kill most plants, some plants possess mechanisms to survive the stress and can populate a vacant niche. Here grass colonizes this eroded strip mine in North Park, Colorado.

Mine tailings rich in heavy metals, however, generally are not completely barren (**Figure 39.17**). They may support healthy plant populations that differ genetically from populations of the same species on the surrounding normal soils. How can these plants survive?

Initially, some plants were thought to tolerate heavy metals by excluding them: By not taking up the metal ions, it was believed, the plants avoided being poisoned. However, measurements have shown that tolerant plants growing on mine tailings do take up heavy metals, accumulating them to concentrations that would kill most plants. Thus these plants must have a mechanism for dealing with the heavy metals they take up. Such tolerant plants may be useful agents for *bioremediation*, a decontamination process by which the heavy metal content of some contaminated soils is decreased by living organisms.

We know the mechanism of at least one case of tolerance to a different toxic metal. When the roots of a buckwheat grown in China are exposed to aluminum concentrations high enough to inhibit root growth in other plants, they secrete oxalic acid. Oxalic acid combines with aluminum ions, forming a complex that does not inhibit growth.

From mine to mine, the heavy metals in the soil differ. In Wales and Scotland, bent grass (*Agrostis*) grows near many mines. Samples of bent grass from several such sites were tested for their ability to grow in various solutions, each containing only one heavy metal. In general, the plants tolerated particular heavy metals—the ones most abundant in their habitat—but were sensitive to others. That is, they tolerated only one or two heavy metals, rather than heavy metals as a group.

Tolerant plant populations can evolve and colonize an area surprisingly rapidly. The bent grass population around a particular copper mine in Wales is resistant to copper and is relatively abundant, although the copper-rich soil dates from mining done only a century ago.

39.4 RECAP

Halophytes are adapted to saline habitats by various means, some of which are the same as those employed by xerophytes to cope with dry habitats. A few species are adapted to heavy-metal-rich habitats that are toxic to most other plants.

- What are some roles of salt glands in halophyte leaves? See p. 848
- Can you name some adaptations that are useful to both halophytes and xerophytes? See p. 848 and Figure 39.16
- Can you explain why some heavy metal-tolerant plants may be useful agents for bioremediation? See p. 849

CHAPTER SUMMARY

39.1 How do plants deal with pathogens?

Plants and pathogens evolve together. Review Figure 39.1, Web/CD Tutorial 39.1

Plants can strengthen their cell walls when attacked. Additional lignin and polysaccharides limit the ability of viral pathogens to move from cell to cell.

PR proteins may break down cell walls of pathogens and trigger other defensive responses.

In the **hypersensitive response**, cells produce **phytoalexins** and then die, trapping the pathogens in dead tissue.

The hypersensitive response is often followed by **systemic acquired resistance**, in which salicylic acid activates further synthesis of defensive compounds.

Gene-for-gene resistance matches up alleles in a plant's resistance genes (*R* genes) and a pathogen's **avirulence genes** (***Avr*** **genes**); when these match, the plant mounts a more vigorous defense. Review Figure 39.3

Plants use **RNA silencing** to develop immunity to invading RNA viruses.

39.2 How do plants deal with herbivores?

Grazing by **herbivores** increases the productivity of some plants. Review Figure 39.4

Some plants produce **secondary metabolites** as defenses against herbivores. Review Figures 39.5, 39.6 and Table 39.1

Hormones including **systemin** and **jasmonates** participate in pathways leading to the production of defensive chemicals. Review Figure 39.7

Toxic chemicals produced by plants are often isolated in plant compartments such as vacuoles and **laticifers**. In some plants enzymes that activate precursors of toxic substances come into contact only when a plant is damaged.

39.3 How do plants deal with climate extremes?

Xerophytes are adapted to dry environments. Some adaptations are behavioral: evasion of drought by the timing of germination by desert annuals, and by the shedding of leaves by ocotillo.

Other xerophytic adaptations are structural, including thickened cuticle, epidermal hairs, sunken stomata, **succulence**, and long taproots.

Adaptations to water-saturated habitats include **pneumatophores**, which allow oxygen uptake from the air, and **aerenchyma**, in which oxygen can diffuse and be stored.

Membranes and proteins can be damaged at high or low temperatures. Plants respond to high or low temperatures by producing **heat shock proteins**.

Some plants undergo **cold-hardening**, which includes changes in membrane lipids and production of heat shock proteins.

Some plants resist freezing by producing **antifreeze proteins**.

39.4 How do plants deal with salt and heavy metals?

Most **halophytes** accumulate salt. Some have **salt glands** that excrete the salt to the leaf surface.

Halophytes and xerophytes have some adaptations in common, such as succulence and the ability to make the water potential of their tissues more negative.

Some plants take up and decontaminate heavy metals; others are tolerant to specific heavy metals.

See Web/CD Activity 39.1 for a concept review of this chapter.

SELF-QUIZ

1. Which of the following is *not* a common defense against bacteria, fungi, and viruses?
   a. Lignin formation
   b. Phytoalexins
   c. A waxy covering
   d. The hypersensitive response
   e. Mycorrhizae
2. Plants sometimes protect themselves from their own toxic secondary metabolites by
   a. producing special enzymes that destroy the toxic substances.
   b. storing precursors of the toxic substances in one compartment and the enzymes that convert those precursors to toxic products in another compartment.
   c. storing the toxic substances in mitochondria or chloroplasts.
   d. distributing the toxic substances to all cells of the plant.
   e. performing crassulacean acid metabolism.
3. Herbivory
   a. is predation by plants on animals.
   b. always reduces plant growth.
   c. usually increases the rate of photosynthesis in the remaining leaves.
   d. reduces the rate of transport of photosynthetic products from the remaining leaves.
   e. is always lethal to the grazed plant.
4. Which statement about secondary plant metabolites is *not* true?
   a. Some attract pollinators.
   b. Some are poisonous to herbivores.
   c. Most are proteins or nucleic acids.
   d. Most are stored in vacuoles.
   e. Some mimic the hormones of animals.
5. Which statement about latex is *not* true?
   a. It is sometimes contained in laticifers.
   b. It is typically white.
   c. It is often toxic to insects.
   d. It is a rubbery solid.
   e. Milkweeds produce it.
6. Which of the following is *not* an adaptation to dry environments?
   a. A less negative solute potential in the vacuoles
   b. Hairy leaves
   c. A heavier cuticle over the leaf epidermis
   d. Sunken stomata
   e. A root system that grows each rainy season and dies back when it is dry

7. Some plants adapted to swampy environments meet the oxygen needs of their roots by means of a specialized tissue called
   a. parenchyma.
   b. aerenchyma.
   c. collenchyma.
   d. sclerenchyma.
   e. chlorenchyma.
8. Halophytes
   a. all accumulate proline in their vacuoles.
   b. have solute potentials that are less negative than those of other plants.
   c. are often succulent.
   d. have low root-to-shoot ratios.
   e. rarely accumulate sodium.
9. Which of the following is *not* a commonly toxic heavy metal?
   a. Chromium
   b. Cadmium
   c. Lead
   d. Potassium
   e. Mercury
10. Plants that tolerate heavy metals commonly
    a. differ genetically from other members of their species.
    b. do not take up the heavy metals.
    c. are tolerant to all heavy metals.
    d. are slow to colonize an area rich in heavy metals.
    e. weigh more than plants that are sensitive to heavy metals.

FOR DISCUSSION

1. How might plant adaptations affect the evolution of herbivores? How might the adaptations of herbivores affect plant evolution?
2. The stomata of the common oleander, *Nerium oleander,* are located in sunken crypts in its leaves. Whether or not you know what an oleander is, you should be able to describe an important feature of its natural habitat. What is that feature?
3. Explain why halophytes often use the same mechanisms for coping with their challenging environments as xerophytes do for coping with theirs.
4. In ancient times, people used less sophisticated methods for mining than we use today. Thus ancient mines often yield substantial profits to modern-day miners who find and work them. On the basis of the information in this chapter, how might you try to locate the site of an ancient mine?

FOR INVESTIGATION

The tobacco hornworm *Manduca sexta* is adapted to feeding on nicotine-producing plants. Using the genetically modified tobacco developed by Steppuhn and coworkers (see Figure 39.5), how might you test the hypothesis that dietary nicotine protects the tobacco hornworm against its parasite *Cotesia congregata*?

PART EIGHT

Animals: Form and Function

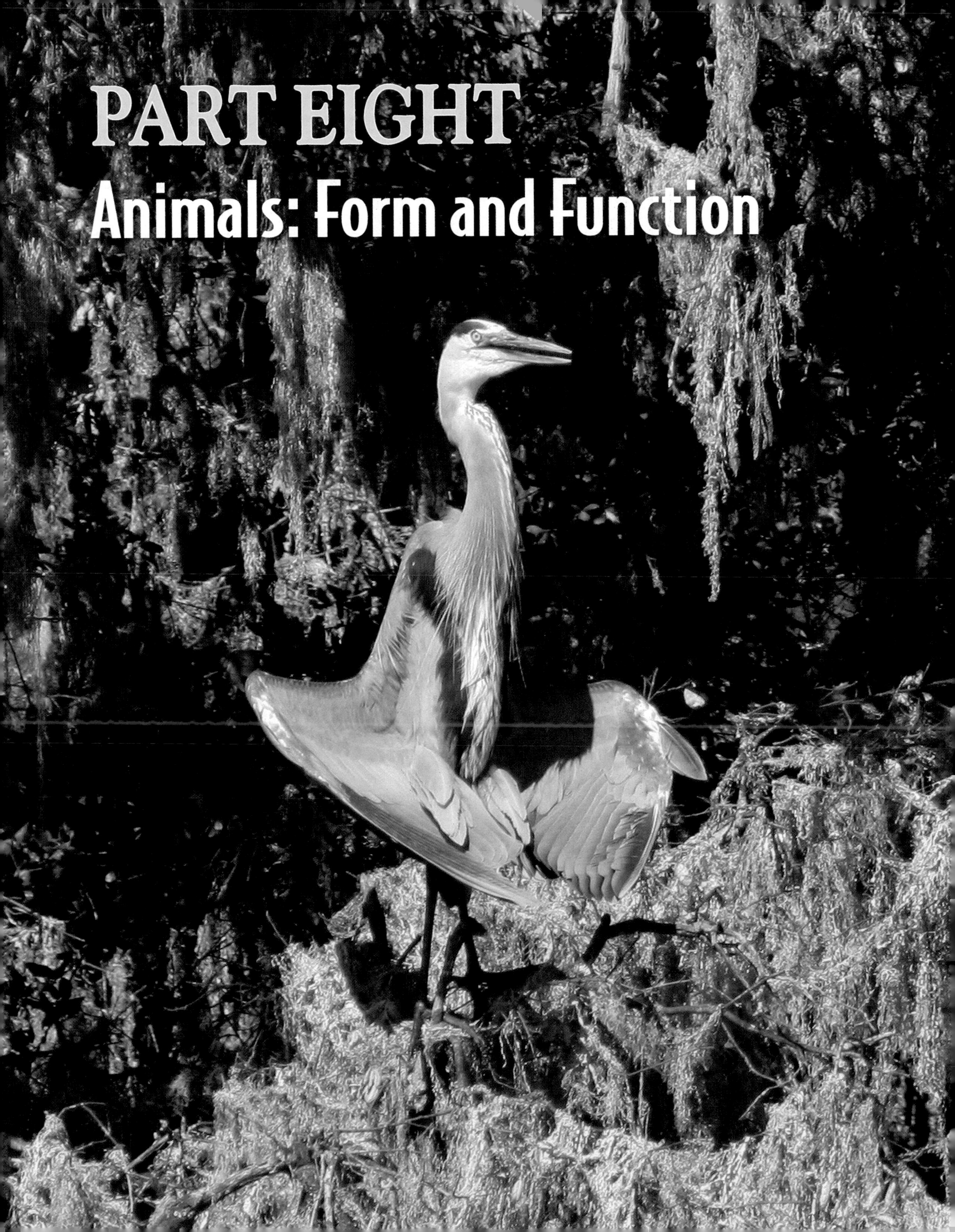

CHAPTER 40 Physiology, Homeostasis, and Temperature Regulation

Cool it!

"A new world record!" These words convey the thrill of world-class athletic competition. But as records are broken by mere centimeters or by fractions of a second, are we reaching absolute limits to human performance? We can assess many physiological limits to extreme performance—maximum breathing rate, for example, or the maximum rate at which the heart can supply blood to the muscles. One less obvious physiological limit is temperature.

The 2004 Olympic women's marathon was held on a hot, humid day in Athens. World record holder Paula Radcliffe was the favorite in the 42-kilometer race. But, overcome by heat stress, Radcliffe collapsed 6 kilometers from the finish line. It could have been much worse. Heat stroke, which can occur when internal body temperature exceeds 41°C, causes failure of major organs such as the heart and brain and results in death in over 20 percent of cases. Soldiers fighting in the desert are at extreme risk of heat-related illnesses, as are firefighters. Agricultural, industrial, and construction workers are also subject to the adverse affects of heat. Biologists at Stanford University developed a technology to cool individuals in such situations, and in the process discovered a way to enhance athletic performance.

Working muscles produce heat. This heat is carried by the blood to skin surfaces, where it is lost to the environment. Not all skin surfaces are equally good at dissipating heat, however. Because fur impedes heat loss, mammals evolved efficient bare skin heat-loss portals such as the nose, tongue, footpads, and parts of the face. These areas have specialized blood vessels that can act like radiators to disperse heat or close down to conserve heat. Although humans are not furred, we retain these specialized blood vessels in our hands, feet, and face (which is why we blush). The Stanford biologists designed a device to amplify heat extraction from these areas.

The heat extractor is a chamber that encases the hand and is sealed at the wrist. The hand is in contact with a cooled surface, but the critical component is a mild vacuum produced inside the chamber. The vacuum pulls more blood into the hand, allowing the cool surface to extract more heat. With this device, an active individual's body temperature rises more slowly, and it cools more rapidly when they rest. An additional unexpected benefit is that cooling reduces fatigue and greatly increases exercise capacity. "Cooled" athletes work out harder and longer, and

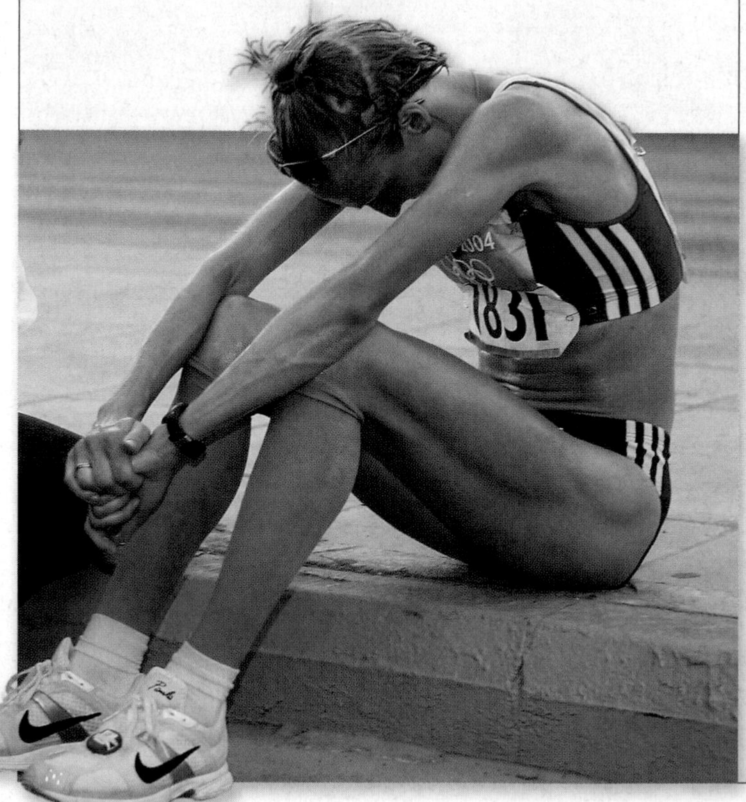

Heat Limits Performance Heat stress forced Paula Radcliffe to drop out of the 2004 Olympic marathon. When the body's internal temperature is stressed by extreme heat, its homeostatic mechanisms may fail.

A Cooling Glove The heat extractor increases heat loss and allows the body to perform at a higher level in severe conditions.

their physical conditioning increases substantially. In one study, college freshmen improved their pushup performance at a rate of 5 pushups a day without cooling, but 9 pushups a day with cooling. Some men and women in the study achieved more than 800 pushups in a workout session.

Human beings can survive in almost any terrestrial environment, from steamy tropical forests to frigid polar regions. Some of the ways we adapt to temperature are behavioral; our ability to control fire undoubtedly affected the course of our evolution. And, like all other mammals, we are adapted to maintain our internal temperature when confronted with heat or cold—an example of the need for living organisms to maintain a "steady state," or *homeostasis*.

IN THIS CHAPTER we explore the internal environment that provides for the needs of all of the body's cells. We survey the cell and tissue types that make up physiological systems and discuss how these systems maintain the internal environment within certain physiological limits, a condition called homeostasis. Homeostasis is explained in detail using one important example, the regulation of body temperature.

CHAPTER OUTLINE

40.1 Why Must Animals Regulate Their Internal Environments?

All animals need nutrients and oxygen and must eliminate carbon dioxide and other waste products of metabolism. Single-celled organisms meet all these needs by direct exchanges with the external environment. Even some simple multicellular animals meet the needs of their cells in this way. Such animals are common in the sea; they tend to be small and flat or, as in sponges, perforated with channels through which seawater can flow. In such an animal, no cell is far from direct contact with seawater, which contains nutrients, absorbs waste, and provides a relatively unchanging physical environment. In larger animals, however, most cells do not have direct contact with the external environment.

An internal environment makes complex multicellular animals possible

The cells of multicellular animals exist within an **internal environment** of extracellular fluid that bathes every cell of the body (**Figure 40.1**). Individual cells get their nutrients from this extracellular fluid and dump their waste products into it. As long as the conditions in the internal environment are held within certain limits, the cells are protected from changes or harsh conditions in the external environment. Thus, a stable internal environment makes it possible for an animal to occupy habitats that would kill its cells if they were exposed to it directly. How is the internal environment kept constant?

As multicellular organisms evolved, cells became specialized for maintaining specific aspects of the internal environment. In turn, the development of an internal environment enabled these specializations, since each cell did not have to be a generalist and provide for all of its own needs. Some cells evolved to be the interface between the internal and the external environments and to provide the necessary transport functions to get nutrients in, move wastes out, and maintain appropriate ion concentrations in the internal environment. Other cells became specialized to provide internal functions such as circulation of the extracellular fluids, energy storage, movement, and information processing. The evolution of physiological systems to maintain different aspects of the internal environment made it possible for multicellular animals to become

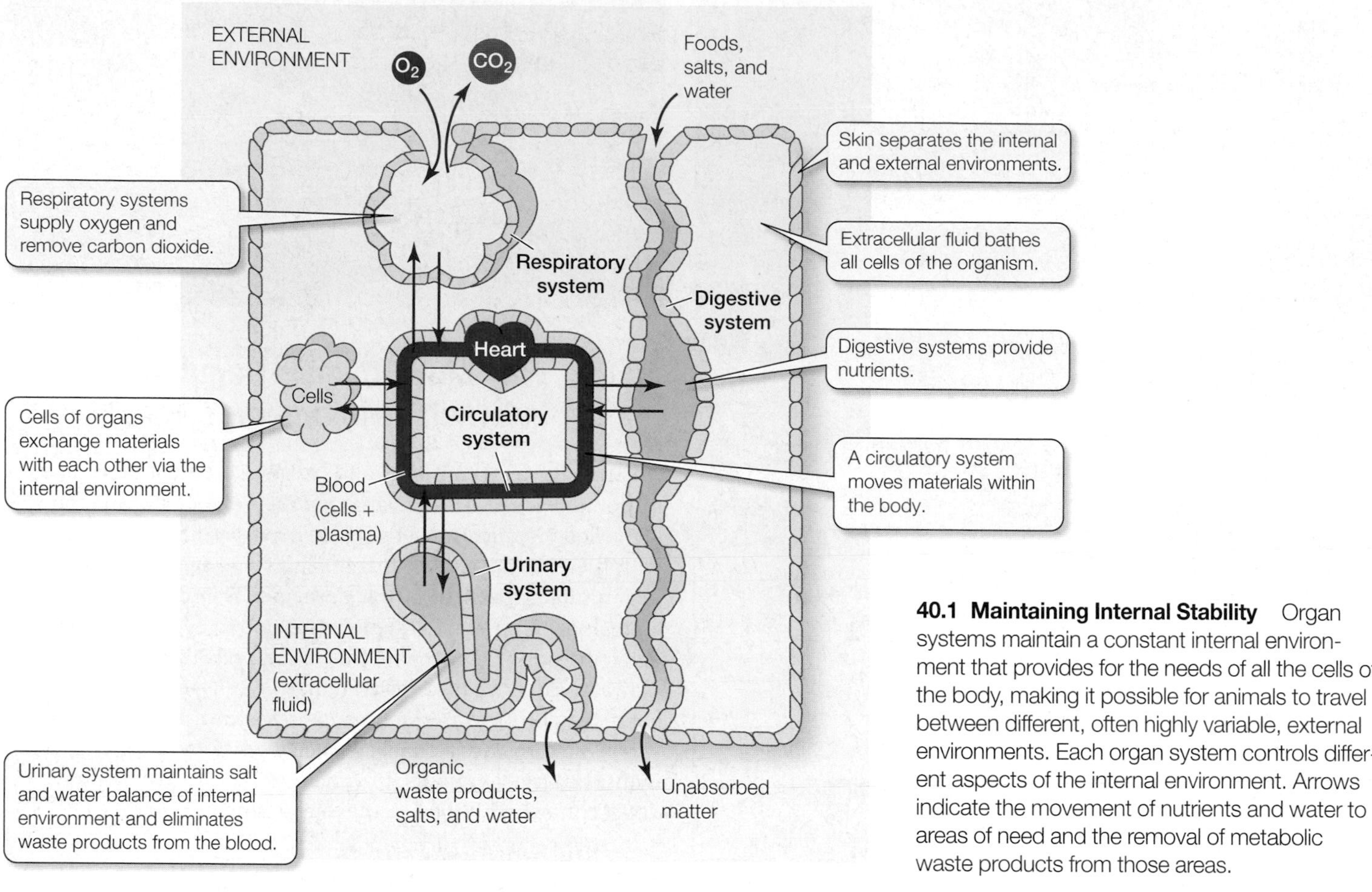

40.1 Maintaining Internal Stability Organ systems maintain a constant internal environment that provides for the needs of all the cells of the body, making it possible for animals to travel between different, often highly variable, external environments. Each organ system controls different aspects of the internal environment. Arrows indicate the movement of nutrients and water to areas of need and the removal of metabolic waste products from those areas.

larger, thicker, more complex, and more adaptable to external environments that are very different from the internal environment.

The composition of the internal environment is constantly being challenged by the external environment and by the metabolic activity of the cells of the body. The maintenance of stable conditions (within a narrow range) in the internal environment is called **homeostasis**. Homeostasis is an essential feature of complex animals. If a physiological system fails to function properly, homeostasis is compromised, and as a result cells are damaged and can die. To avoid loss of homeostasis, physiological systems must be controlled and regulated in response to changes in both the external and internal environments.

You are 60 percent water. One-third of that water is outside of your cells. About 20 percent of that extracellular fluid is circulating in your blood vessels; the rest (some 11 liters) bathes the cells of the body.

Homeostasis requires physiological regulation

The activities of all physiological systems are controlled—speeded up or slowed down—by actions of the nervous and endocrine systems. But to *regulate* the internal environment, information is required.

Think of it this way. You *control* the speed of your car with the accelerator and the brakes, but when you use the accelerator and brakes to *regulate* the speed of your car, you have to know both how fast you are going and how fast you want to go. The desired speed is a **set point**, or reference point, and the reading on your speedometer is **feedback information**. When the set point and the feedback information are compared, any difference between them is an **error signal**. Error signals suggest corrective actions, which you make by using the accelerator or brake (**Figure 40.2**).

Some components of physiological systems are called **effectors** because they *effect* changes in the internal environment. Effectors are **controlled systems** because their activities are controlled by commands from regulatory systems. **Regulatory systems**, in contrast, obtain, process, and integrate information, then issue commands to controlled systems. An important component of any regulatory system is a **sensor**, which provides the feedback information that is compared to the internal set point.

A fundamental way to study a regulatory system is to identify the information it uses. What are its sensors? How is the information from the sensors used? **Negative feedback** is the most common use of sensory information in regulatory systems. The word "negative" indicates that this feedback information causes the effectors to reduce or reverse the process or counteract the influence that created an error signal. In our car analogy, the recognition that you are going too fast is negative feedback if it causes you to slow down. Negative feedback is a stabilizing influence in physiological systems; it tends to return a variable of the internal environment to the set point from which it deviated.

Although not as common as negative feedback, **positive feedback** is also seen in some physiological systems. Rather than returning a system to a set point, positive feedback amplifies a response (i.e., increases the deviation from the set point). Examples of regulatory systems that use positive feedback are the responses that empty body cavities, such as urination, defecation, sneezing, and vomiting. Another example is sexual behavior, in which a little stimulation causes more behavior, which causes more stimulation, and so on.

Feedforward information is another feature of regulatory systems. The function of **feedforward information** is to change the set point. Seeing a deer ahead on the road when you are driving is an example of feedforward information (see Figure 40.2); this information takes precedence over the posted speed limit, and you change your set point to a slower speed. Before the start of a race, the "on your marks, get set" commands are feedforward information that raises your heart rate before you begin to run. Feedforward information anticipates change in the internal environment before that change occurs.

These principles of control and regulation help organize our thinking about physiological systems. Once we understand how a system works, we can then ask how it is regulated. The example we will explore in this chapter is the regulation of body temperature. But before we explore this first example of what will be our recurrent theme—the function, evolution, control, and regulation of each physiological system—we need to get acquainted with the important structural features that all physiological systems have in common.

Physiological systems are made up of cells, tissues, and organs

Each physiological *system* is composed of discrete *organs*, such as the liver, heart, lungs, and kidneys, that serve specific functions in the body. These organs are made up of tissues. A **tissue** is an assemblage of cells and, although there are many, many types of cells, there are only four kinds of tissue: *epithelial*, *connective*, *muscle*, and *nervous*. The word "tissue" is often used in a general way to refer to a piece of an organ, such as "lung tissue" or "kidney tissue"; however, as we will see, a piece of an organ will almost always consist of more than one of the four kinds of tissues.

EPITHELIAL TISSUES **Epithelial tissues** are sheets of densely packed, tightly connected *epithelial cells* that cover inner and outer body surfaces (**Figure 40.3A**). They act as barriers and provide transport across those barriers. Epithelial tissues form the skin and line the hollow organs of the body, such as the gut.

Epithelial cells have many roles in the body. Some secrete hormones, milk, mucus, digestive enzymes, or sweat. Others have cilia that move substances over surfaces or through tubes. Epithelial cells can also provide information to the nervous system. Smell and taste receptors, for example, are epithelial cells that detect specific chemicals. Epithelial cells create boundaries between the inside and the outside of the body and between body compartments; they line the blood vessels and make up various ducts and tubules (**Figure 40.3B**). Filtration and transport are important functions of epithelial cells. They control what molecules and ions can leave the blood to enter the internal environment or the urine. They can selectively transport ions and molecules from one side of an epithelial mem-

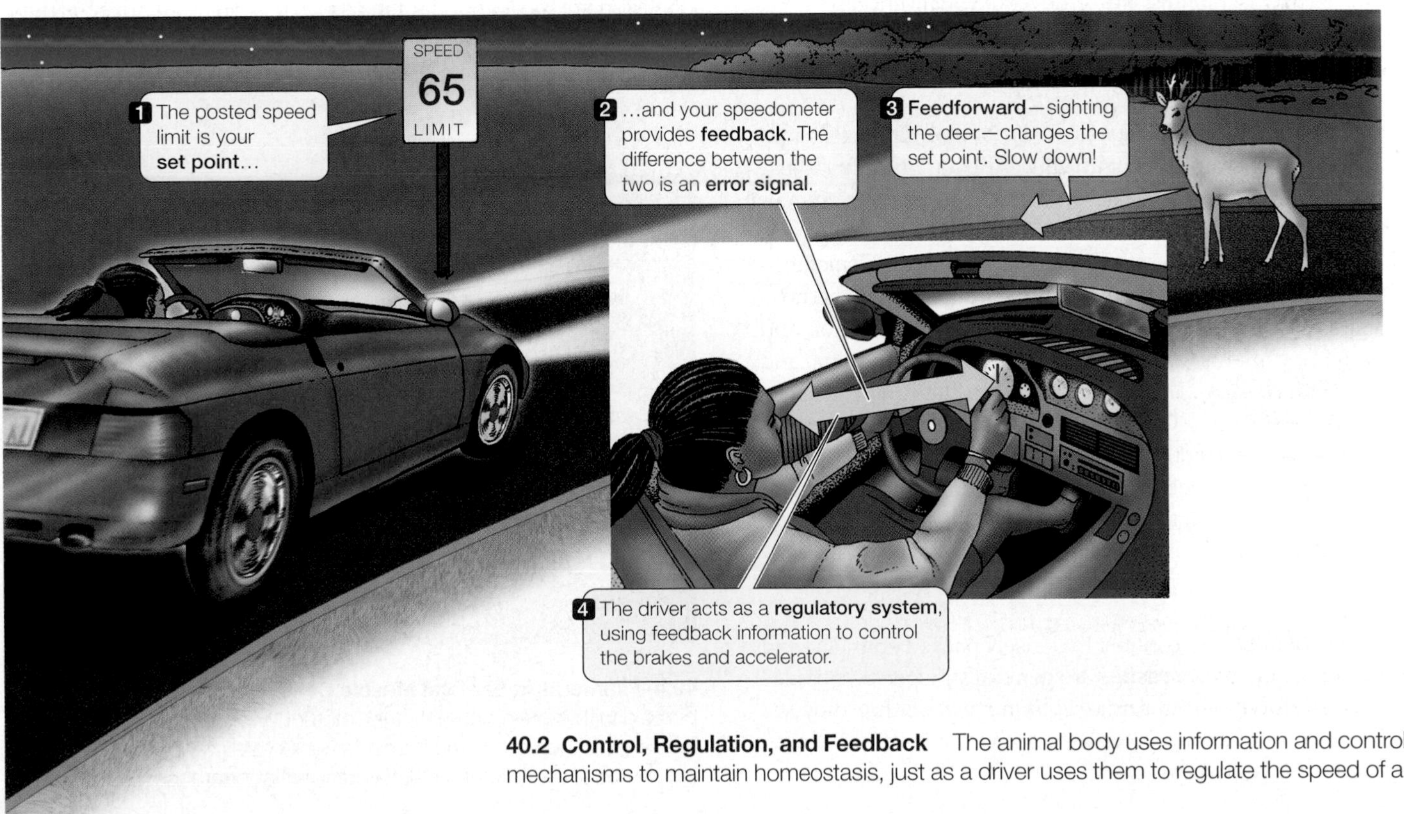

40.2 Control, Regulation, and Feedback The animal body uses information and control mechanisms to maintain homeostasis, just as a driver uses them to regulate the speed of a car.

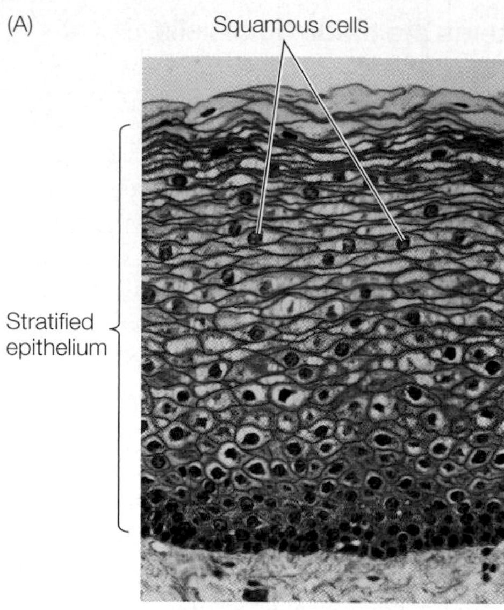

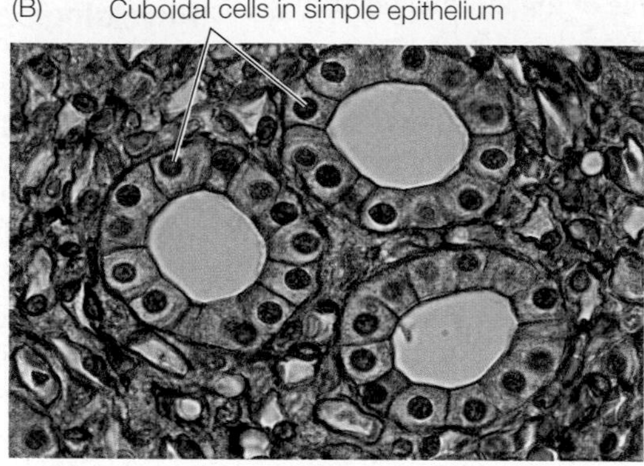

40.3 Epithelial Tissue (A) Epithelial cells make up the outer layers of skin. (B) A single layer of epithelial cells forms a tubule in the kidney. These cells have many transport functions.

brane to the other. Examples are the absorption of nutrient molecules from your gut and the secretion of acid into your stomach.

The skin and the lining of the gut are examples of epithelial tissues that receive much wear and tear. Accordingly, cells in these tissues have a high rate of cell division to replace cells that die and are shed. Dandruff is discarded skin cells.

Parts of you are younger than you think. The epithelium of your digestive tract is renewed about every 5 days, the surface layer of your skin every 2 weeks. Your blood is renewed every 4 months. The turnover of your liver cells is 1–2 years.

MUSCLE TISSUES **Muscle tissues** consist of elongated cells that can contract to generate forces and cause movement. Muscle tissues are the most abundant tissues in the body, and when animals are active, muscles use most of the energy produced in the body. The three types of muscle tissues—**skeletal**, **cardiac**, and **smooth**—are discussed in Section 47.1. Skeletal muscles (so named because they mostly attach to bones) are responsible for locomotion and other body movements such as facial expressions, shivering, and breathing (**Figure 40.4**). These muscles are under both conscious and unconscious control. Cardiac muscle makes up the heart and is responsible for the heart beat and the pumping of blood. Smooth muscle is responsible for movement and generation of forces in many hollow internal organs such as the gut, bladder, and blood vessels. Cardiac and smooth muscle are not under conscious control, but are controlled by physiological regulatory systems.

CONNECTIVE TISSUES In contrast to densely packed epithelial and muscle tissues, **connective tissues** are generally dispersed populations of cells embedded in an *extracellular matrix* that they secrete. The composition and properties of the matrix differ among types of connective tissues.

Protein fibers are an important component of the extracellular matrix of connective tissue cells. The dominant protein in the extracellular matrix is *collagen* (see Figure 4.25). Collagen is the most abundant protein in the human body, representing 25 percent of total body protein. Collagen fibers are strong and resistant to stretch, giving strength to the skin and the connections between bones and between bones and muscles. The fibers provide a netlike framework for organs, giving them shape and structural strength. Connective tissue that fills spaces between organs has a low density of collagen fibers.

Elastin is another type of protein fiber in the extracellular matrix of connective tissues. It is so named because it can be stretched to several times its resting length and then recoil. Fibers composed of elastin are most abundant in tissues that are regularly stretched, such as the walls of the lungs and the large arteries.

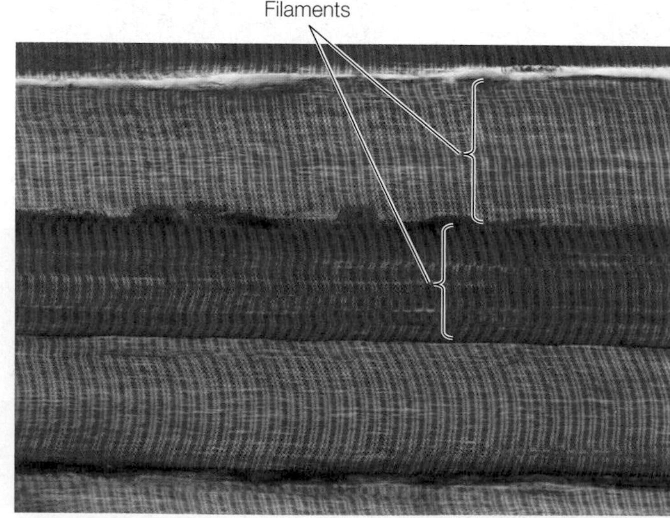

40.4 Filaments in Skeletal Muscle Cells A skeletal muscle cell is packed with protein filaments that interact to cause contraction. The regular arrangement of the filaments, which are made up of two different proteins, results in the striated (striped) appearance.

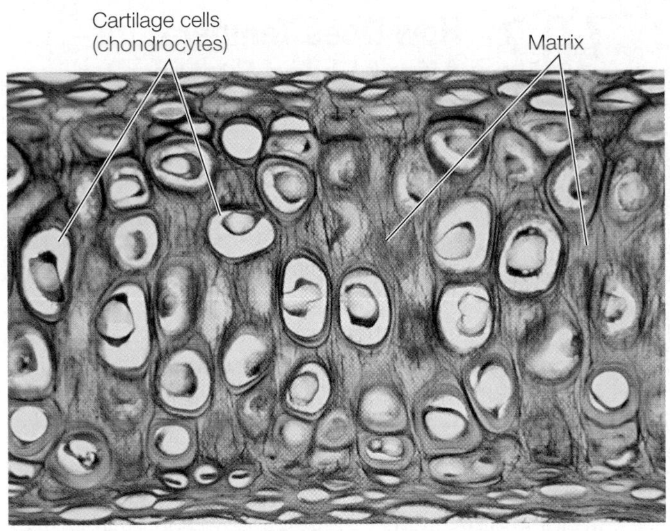

40.5 Cartilage Cartilage makes structures such as the ear stiff but flexible. Cartilage cells, or chondrocytes, secrete an extracellular matrix rich in collagen fibers and elastin fibers. In this micrograph, the elastin fibers are stained dark blue.

Cartilage and bone are connective tissues that provide rigid structural support. In *cartilage*, a network of collagen fibers is embedded in a flexible matrix consisting of a protein–carbohydrate complex, along with a specific type of cell called a *chondrocyte* (**Figure 40.5**). Cartilage, which lines the joints of vertebrates, is resistant to compressive forces. Since it is flexible, it provides structural support for flexible structures such as external ears and noses. The extracellular matrix in *bone* also contains many collagen fibers, but it is hardened by the deposition of the mineral calcium phosphate. Cartilage and bone are discussed in greater detail in Section 47.3.

Adipose tissue is a form of loose connective tissue that includes adipose cells, which form and store droplets of lipids. Adipose tissue, or "fat," is a major source of stored energy. It also serves to cushion organs, and layers of adipose tissue under the skin can provide a barrier to heat loss.

Blood is a connective tissue consisting of cells dispersed in an extensive extracellular matrix, the blood *plasma*. The blood plasma is much more liquid than the extracellular matrices of the other connective tissues, but it too contains an abundance of proteins. Many of the proteins and cellular elements of blood were presented in Section 18.1, and blood will discussed again in Section 49.4.

NERVOUS TISSUES Two basic cell types in **nervous tissues** are *neurons* and *glial cells*. Neurons come in many shapes and sizes, and encode information as electrical signals called nerve impulses. Nerve impulses travel over long extensions of the neurons, called *axons*, to reach other neurons, muscle cells, or secretory cells (**Figure 40.6A**). Where the axon is in close proximity to a target cell, nerve impulses trigger the release of chemical signals that bind to receptors on the target cell and stimulate a response. Neurons are involved in controlling the activities of most organ systems.

Glial cells do not generate or conduct electrochemical signals, but they provide a variety of supporting functions for neurons (**Figure 40.6B**). There are more glial cells than neurons in our nervous system. The properties of nervous tissues are detailed in Chapters 44, 45, and 46.

Organs consist of multiple tissues

Organs include more than one kind of tissue, and most organs include all four. The wall of the stomach is a good example (**Figure 40.7**). The inner surface of the stomach is lined with a sheet of epithelial cells. Different epithelial cells secrete mucus, enzymes, or stomach acid. Beneath the epithelial lining is connective tissue. Within this connective tissue are blood vessels, neurons, and glands (clusters of secretory epithelial cells). Concentric layers of smooth muscle tissue enable the stomach to contract to mix food with digestive juices. A network of neurons between the muscle layers controls these movements and influences the secretions of the stomach. Surrounding the stomach is a sheath of connective tissue.

An individual organ is usually part of an **organ system**—a group of organs that function together to achieve a particular physiological function or set of functions (see Figure 40.1). The stomach is part of the digestive system.

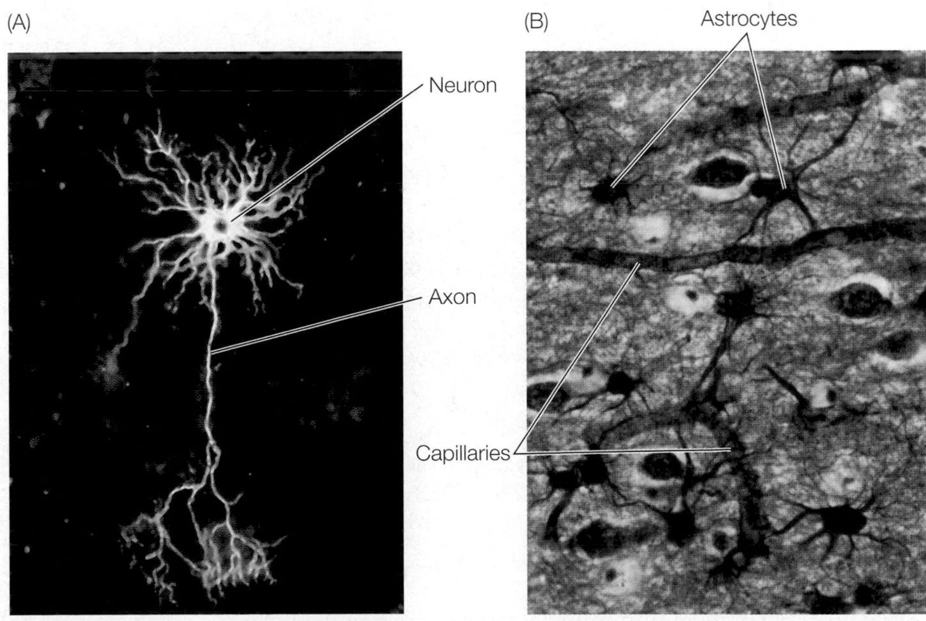

40.6 Nervous Tissue Includes Neurons and Glia (A) This human neuron consists of a cell body, a number of processes that receive input from other neurons, and one long axon that sends information to other cells. (B) A section through human brain tissue shows astrocytes, a type of glia. Glial cells provide support and protection for neurons including creating a barrier that protects the brain from many chemicals circulating in the blood.

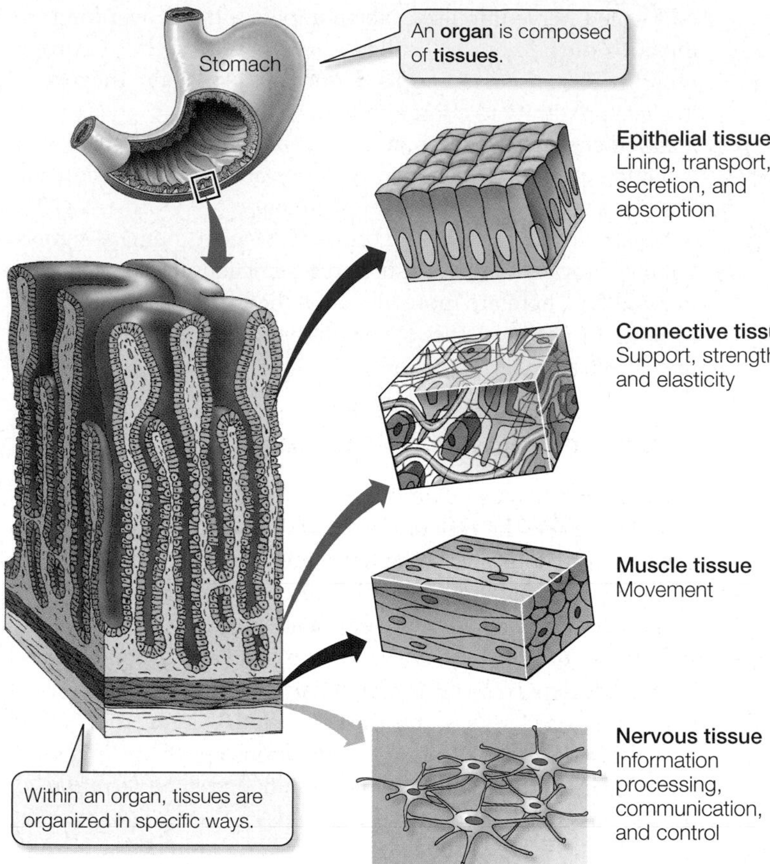

40.7 Tissues Form Organs Most organs contain more than one of the four different kinds of tissue. An organ such as the stomach is made up of all four.

40.1 RECAP

In all but the smallest multicellular animals, the internal environment provides for the needs of all of the cells. Organs and organ systems provide physiological regulation to maintain stability or homeostasis of the internal environment.

- Do you understand how negative feedback regulation tends to return a physiological variable to its set point? See p. 856 and Figure 40.2
- Can you describe a key function of each of the four kinds of tissue found in animals? See pp. 857–859 and Figure 40.7

Subsequent chapters in this unit will describe each of the organ systems mentioned above in much greater detail. The remainder of this chapter will focus on the mechanisms of homeostasis, using one important variable of the internal environment—its temperature—as our example.

40.2 How Does Temperature Affect Living Systems?

Temperatures vary enormously over the face of Earth, from the boiling hot springs of Yellowstone National Park to the interior of Antarctica, where the temperature can fall below –80°C. Living cells, however, can function over only a narrow range of temperatures. If cells cool below 0°C, ice crystals form and damage their structures. Some animals have adaptations, such as antifreeze molecules in their blood, that help them resist freezing; others can survive freezing. Generally, however, cells must remain above 0°C to stay alive.

The upper temperature limit for survival in most cells is about 45°C (although some specialized algae can grow in hot springs at 70°C, and some archaea live at near 100°C). In general, proteins begin to denature and lose their function as temperatures rise above 40°C. Therefore, most cellular functions are limited to the range between 0°C and 40°C, which approximates the thermal limits for life. A particular species, however, usually has much narrower limits.

Q_{10} is a measure of temperature sensitivity

Even between 0°C and 45°C, changes in tissue temperature create problems for animals. Most physiological processes, like the biochemical reactions that constitute them, are temperature-sensitive, going faster at higher temperatures (see Figure 6.22). The temperature sensitivity of a reaction or process can be described in terms of **Q_{10}**, a quotient calculated by dividing the rate of a process or reaction at a certain temperature, R_T, by the rate of that process or reaction at a temperature 10°C lower, R_{T-10}:

$$Q_{10} = \frac{R_T}{R_{T-10}}$$

Q_{10} can be measured for a simple enzymatic reaction or for a complex physiological process, such as rate of oxygen consumption. If a reaction or process is not temperature-sensitive, it has a Q_{10} of 1. Most biological Q_{10} values are between 2 and 3. A Q_{10} of 2 means that the reaction rate doubles as temperature increases by 10°C, and a Q_{10} of 3 indicates a tripling of the rate (**Figure 40.8**).

Changes in tissue temperature can disrupt an animal's physiology because not all of the component reactions that constitute the metabolism of the animal have the same Q_{10}. Biochemical reactions are linked together in complex networks. If different reactions have different Q_{10} values, changes in tissue temperature will shift the rates of some reactions more than others. Thus, a change in tissue temperature can disrupt the balance and integration of the reactions that constitute a physiological process. To maintain homeostasis, organisms must be able to compensate for or prevent changes in body temperature.

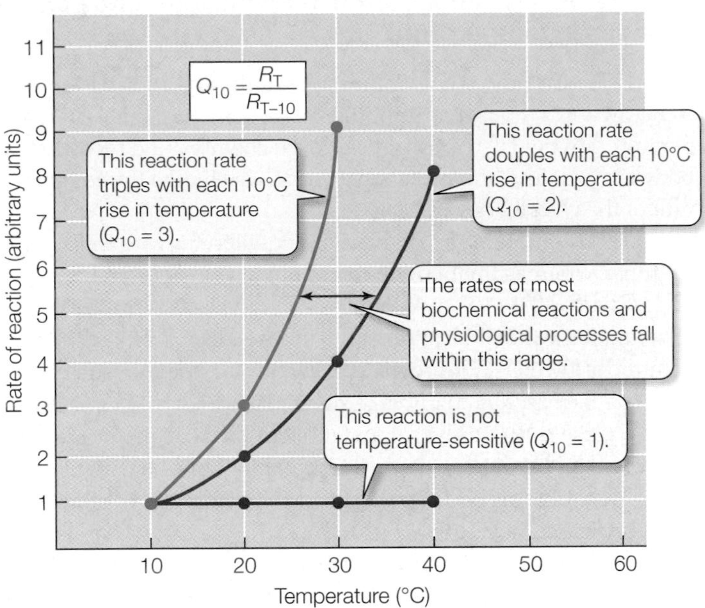

40.8 Q_{10} and Reaction Rate The larger the Q_{10} of a reaction or process, the faster its rate rises in response to an increase in temperature.

Animals can acclimatize to a seasonal temperature change

The body temperature of some animals (especially aquatic animals) is tightly coupled to environmental temperature. The body temperature of a fish in a pond, for example, will always be the same as the water temperature, which might range from 4°C in midwinter to 24°C in midsummer. How do such differences in water temperature affect the fish? We would predict that because of Q_{10} relationships, all of the fish's physiological functions would be much slower in the winter than in the summer. In support of our prediction, we could expose summer fish to a range of temperatures in the laboratory and show that whatever physiological function we measure, it goes slower and slower as we decrease the temperature of the water. However, if we study the same fish in the winter, we see that its physiological functions at low temperatures are not as slow as they were in our experiment with summer fish. The reason is that the fish's physiology has *acclimatized* to the seasonal change in water temperature.

What might account for acclimatization to seasonal changes in temperature in this fish? As discussed in Section 6.5, an organism may call upon one of several different temperature-optimized isozymes to catalyze a given reaction. If the fish can express a number of isozymes that operate at different optimal temperatures, it can catalyze reactions with one set of enzymes in summer and another set in winter. The result is that metabolic functions of the fish are much less sensitive to long-term changes in temperature than they are to short-term changes. Most animals cannot compensate completely for a seasonal change in body temperature. Partial compensation is most commonly observed.

40.2 RECAP

Life can exist in only a narrow range of temperatures, but even changes within that range can be disruptive because different physiological processes have different temperature sensitivities.

- Plot a Q_{10} curve for a physiological process. See p. 860 and Figure 40.8
- Do you understand why a change in body temperature can disrupt physiological processes? See p. 860
- Can you explain how aquatic organisms can acclimatize to seasonal temperatures? See p. 861

We have seen how animals are affected by the temperature of their environment. Now let's take a look at the adaptations of animals that allow them to control and regulate their body temperatures.

40.3 How Do Animals Alter Their Heat Exchange with the Environment?

Many of us learned to think of animals as being either "cold-blooded" or "warm-blooded," which implies a comparison with our own body temperature and sets mammals and birds apart from other animals. This simple classification breaks down when we realize that mammals that hibernate become cold, and that many reptiles and insects can be quite warm when they are active. Physiologists sometimes classify animals according to whether they have a constant body temperature (homeotherms) or a variable body temperature (poikilotherms). But a deep-sea fish has a constant body temperature. Should it be classified with mammals?

A thermal classification system that avoids such irrational results is one based on the source of heat that predominantly determines the temperature of the animal. **Ectotherms** are animals whose body temperatures are determined primarily by external sources of heat. **Endotherms** can regulate their body temperature by producing heat metabolically or by using active mechanisms of heat loss.

Most mammals and birds are endotherms; other animals are ectotherms most of the time. This statement suggests that, like the homeotherm/poikilotherm classification, this scheme is not perfect, and therefore we need a third category. A **heterotherm** is an animal that behaves sometimes as an endotherm and other times as an ectotherm. For example, a mammal that hibernates is a perfect endotherm over the summer, but during the winter it has bouts of hibernation during which its internal heat production falls and it behaves much like an ectotherm.

How do endotherms produce so much heat?

Section 6.1 describes how transfers of energy in biological systems are always inefficient. With every transfer of energy—from food molecules to ATP, from ATP to biological work—some of the energy is lost as heat. This is true for both ectotherms and endotherms, so why do endotherms produce more heat? The answer is that the cells of endotherms are less efficient at using energy than those in ectotherms.

In a resting endotherm, most of the energy expended goes into pumping ions across membranes. K^+ is the dominant positive ion inside cells and Na^+ is the dominant positive ion outside cells. To the extent that cell membranes permit, these ions diffuse down their concentration gradients. To maintain their proper concentrations inside and outside cells, the ions must be transported back "uphill," which requires an expenditure of energy. While this is true for both ectotherms and endotherms, the cells of endotherms tend to be more "leaky" to ions than those of ectotherms. Thus endotherms must expend more energy (and thus release more heat) than do ectotherms to maintain ion concentration gradients. This is akin to running on a treadmill: the faster the treadmill goes (analogous to leaking ions) the faster you have to run (analogous to pumping ions) to remain in the same position.

We can speculate that a mutation allowing seemingly faulty or leaky ion channels may have led to the evolution of endothermy. Such a mutation in a small ectotherm may have promoted sufficient heat production to allow this ectotherm to remain active longer after the sun went down. Thus, for the first endotherms a whole new nocturnal world of ecological opportunities opened, one in which there was less competition from similarly sized ectotherms. Major differences between endotherms and ectotherms are their resting metabolic rates, the sum total of all energy expenditures in their bodies when at rest, and their responses to changes in environmental temperature.

In the early spring, the hooded flower of the skunk cabbage becomes "endothermic" for two weeks and actually melts its way up through overlying ice and snow. During this time the flower's temperature can be 20°C above that of the environment, and its metabolic rate is about the same as a mammal of similar size.

Ectotherms and endotherms respond differently to changes in temperature

Let's compare how two similar-sized animals, a lizard (an ectotherm) and a mouse (an endotherm) respond to changes in temperature. We put each animal in a closed chamber and measure its body temperature and metabolic rate as we change the temperature of the chamber from 37°C to 0°C.

The body temperature of the lizard equilibrates with that of the chamber, whereas the body temperature of the mouse remains at 37°C (**Figure 40.9A**). The metabolic rate of the lizard (already much lower than the metabolic rate of the mouse) decreases as the temperature is lowered (**Figure 40.9B**). In contrast, the mouse's metabolic rate increases as chamber temperature is lowered below 27°C. The increase in the mouse's metabolism produces enough heat to prevent its body temperature from falling. In other words, the mouse regulates its body temperature by increasing its metabolic rate; the lizard does not.

We can test our laboratory observation—that the lizard does not regulate its body temperature—by measuring its temperature after we return it to its desert habitat. In nature, in contrast to in the laboratory, the lizard's body temperature is sometimes considerably different from the environmental temperature, which in the desert can fluctuate by 40°C in a few hours. The lizard achieves this difference by using behavior to alter its heat exchange with the environment (**Figure 40.10A**). Its behavioral strategies include spend-

40.9 Ectotherms and Endotherms React Differently to Environmental Temperatures (A) The body temperatures of a lizard and a mouse of the same body size are different when equilibrated to different environmental temperatures. (B) The metabolic rates of the lizard and mouse react in opposite manners to cooler temperatures. The mouse curve turns upward again at higher temperatures, however, because it takes metabolic energy to sweat or pant to dissipate heat.

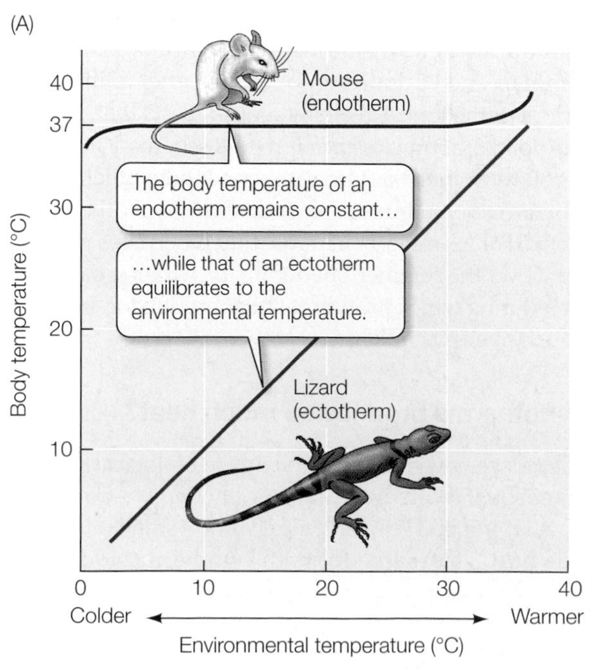

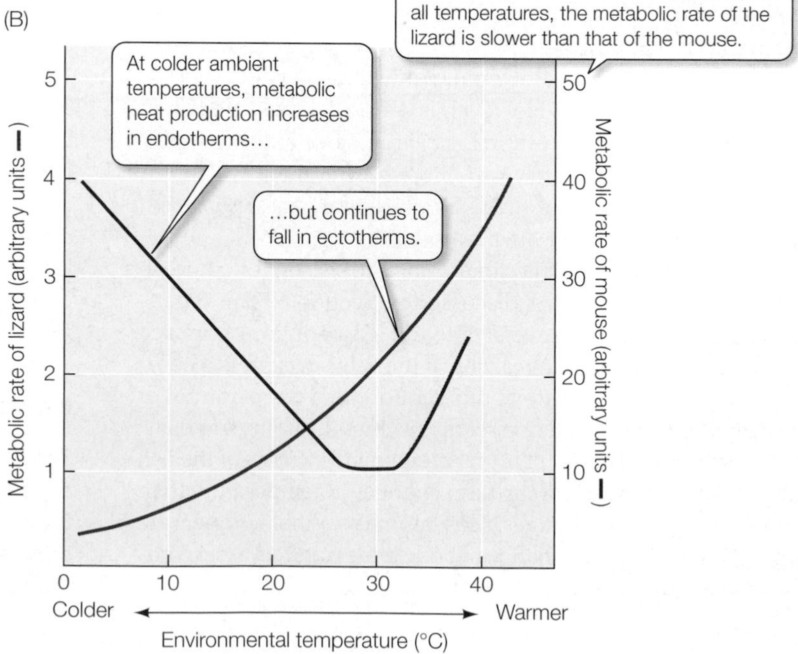

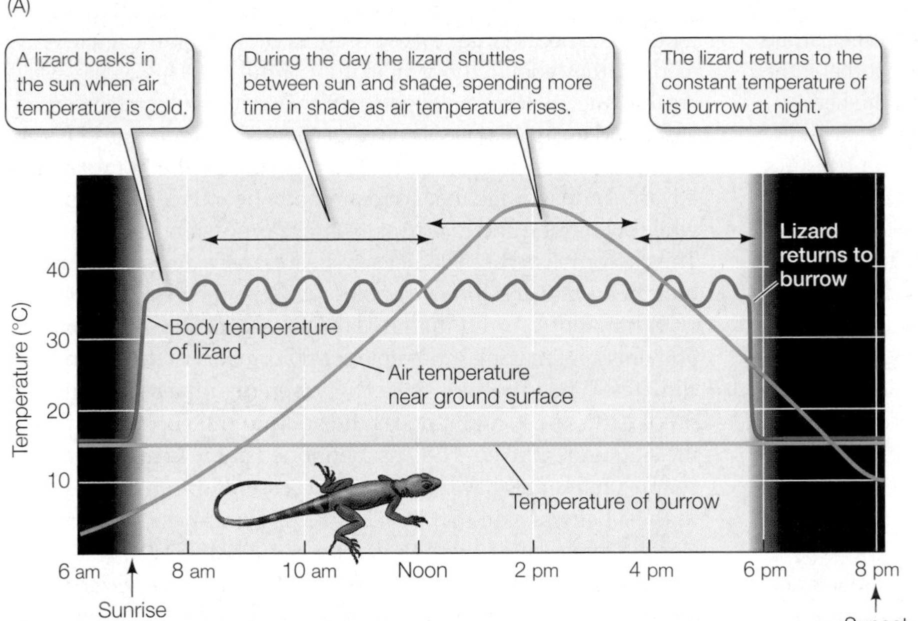

The lizard returns to the constant temperature of its burrow at night.
Lizard returns to burrow

40.10 Ectotherms and Endotherms Use Behavior to Regulate Body Temperature (A) The lizard's body temperature is dependent on environmental heat, but it can regulate its temperature by moving from place to place within its environment. (B) When air temperatures on the African savanna soar, an elephant may thermoregulate by showering itself with water.

ing time in a burrow, basking in the sun, seeking shade, climbing vegetation, and changing its orientation with respect to the sun.

While the lizard can regulate its body temperature quite well, it does so by behavioral mechanisms rather than by altering its internal metabolic heat production. In our laboratory experiment, the lizard could not use its thermoregulatory behavior, but in its natural environment it could move from place to place to alter the heat exchange between its internal and external environments.

Behavioral thermoregulation is not the exclusive domain of ectotherms (**Figure 40.10B**). Endotherms usually select the most comfortable thermal environment possible. They may change their posture, orient to the sun, move between sun and shade, and move between still air and moving air, the same as the ectotherm in our field experiment. Examples of more complex thermoregulatory behavior include nest construction and social behavior such as huddling. Humans put on or remove clothing; we also burn fossil fuels to generate the energy to heat or cool buildings.

Energy budgets reflect adaptations for regulating body temperature

Both ectotherms and endotherms can influence their body temperatures by altering four avenues of heat exchange between their bodies and the environment (**Figure 40.11**):

- **Radiation**: Heat transfers from warmer objects to cooler ones via the exchange of infrared radiation (what you feel when you stand in front of a fire).
- **Conduction**: Heat transfers directly when objects of two different temperatures come into contact (think of putting an icepack on a sprained ankle).
- **Convection**: Heat transfers to a surrounding medium such as air or water as that medium flows over a surface (the wind chill factor).
- **Evaporation**: Heat transfers away from a surface when water evaporates on that surface (the effect of sweating).

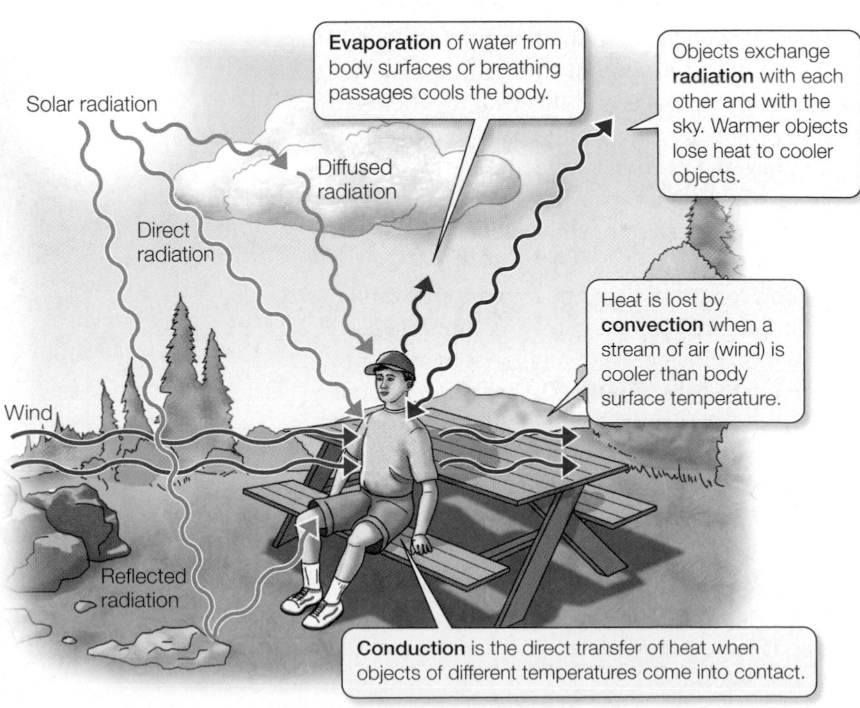

Evaporation of water from body surfaces or breathing passages cools the body.
Objects exchange radiation with each other and with the sky. Warmer objects lose heat to cooler objects.
Solar radiation
Diffused radiation
Direct radiation
Heat is lost by convection when a stream of air (wind) is cooler than body surface temperature.
Wind
Reflected radiation
Conduction is the direct transfer of heat when objects of different temperatures come into contact.

40.11 Animals Exchange Heat with the Environment An animal's body temperature is determined by the balance between internal heat production and four avenues of heat exchange with the environment: radiation, conduction, convection, and evaporation.

The total balance of heat production and heat exchange can be expressed as an **energy budget**, based on the simple fact that if the body temperature of an animal is to remain constant, the heat entering the animal must equal the heat leaving it. The heat coming in usually comes from metabolism and solar radiation (R_{abs}, for radiation absorbed). Heat leaves the body via the four mechanisms listed above: radiation emitted (R_{out}), convection, conduction, and evaporative heat loss. The energy budget takes the following form:

$$\overbrace{\text{metabolism} + R_{abs}}^{\text{heat}_{in}} = \overbrace{R_{out} + \text{convection} + \text{conduction} + \text{evaporation}}^{\text{heat}_{out}}$$

Anyone who has experienced a very hot environment will realize that heat can also *enter* the body through convection (e.g., the hot desert wind) and conduction (e.g., a hot car seat—ouch!). In that case, the values of those factors will become negative in the energy budget equation.

The energy budget is a useful concept because any adaptation that influences the ability of an animal to deal with its thermal environment must affect one or more components of the budget. So the energy budget gives us the ability to quantify and compare the thermal adaptations of animals. One interesting observation is that all of the components on the right side of the energy budget equation—the heat loss side—depend on the surface temperature of the animal. One way surface temperature can be controlled is by altering the flow of blood to the skin.

Both ectotherms and endotherms control blood flow to the skin

Heat exchange between the internal environment and the skin occurs largely through blood flow. As described at the beginning of this chapter, when body temperature rises due to exercise, blood flow to the skin increases, and the skin surface becomes warm. The heat brought from the body core to the skin by the blood is lost to the environment through the four avenues listed above, and this heat loss helps to bring the body temperature back to normal. In contrast, when body temperature is too low or the environment is too cold, the blood vessels supplying the skin constrict, reducing heat loss to the environment.

The control of blood flow to the skin can be an important adaptation for an ectotherm such as the marine iguana (a reptile) of the Galápagos archipelago (**Figure 40.12**). The Galápagos are volcanic islands that lie on the Equator but are bathed by cold ocean currents. The iguanas bask on hot black lava rocks on shore, then enter the cold ocean water to feed on seaweed. When the iguanas are feeding, they cool to the temperature of the sea. This cooling lowers their metabolism, making them slower, more vulnerable to predators, and incapable of efficient digestion. They therefore alternate between feeding in the cold sea and basking on the hot rocks. It is advantageous for iguanas to retain body heat as long as possible while swimming and to warm up as fast as possible when basking. They accomplish these changes in heat transfer rates by changing their heart rate and the rate of blood flow to their skin.

What about furred mammals? Fur acts as *insulation* to keep body heat in, making it possible for mammals to live in very cold climates. When they are active, however, mammals still must get rid of excess heat, and it does little good to transport that heat to the skin under the fur. Thus, as mentioned at the beginning of this chapter, mammals have special blood vessels for transporting heat to their hairless skin surfaces. Heat loss from these areas of skin is tightly controlled by the opening and closing of these special blood vessels. When you are cold, the blood flow to your hands and feet decreases and they can feel very cold, but when you exercise, these same surfaces can get very hot quickly.

Some fish elevate body temperature by conserving metabolic heat

Active fish can produce substantial amounts of metabolic heat, but they have difficulty in retaining any of that heat. Blood pumped from the heart goes directly to the gills, where it comes very close to the surrounding water to exchange respiratory gases. So any

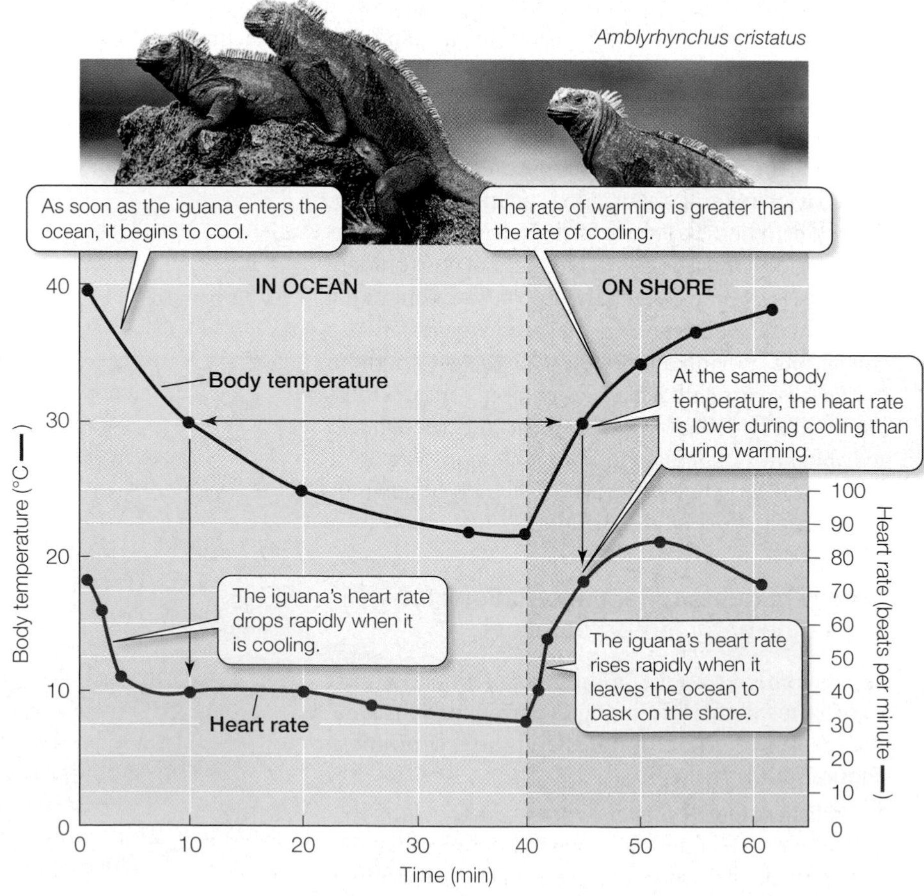

Amblyrhynchus cristatus
As soon as the iguana enters the ocean, it begins to cool.
The rate of warming is greater than the rate of cooling.
IN OCEAN
ON SHORE
Body temperature
At the same body temperature, the heart rate is lower during cooling than during warming.
The iguana's heart rate drops rapidly when it is cooling.
The iguana's heart rate rises rapidly when it leaves the ocean to bask on the shore.
Heart rate
Body temperature (°C —)
Heart rate (beats per minute —)
Time (min)
0
10
20
30
40
50
60
70
80
90
100

40.12 Some Ectotherms Regulate Blood Flow to the Skin Galápagos marine iguanas control blood flow to the skin to alter their heating and cooling rates.

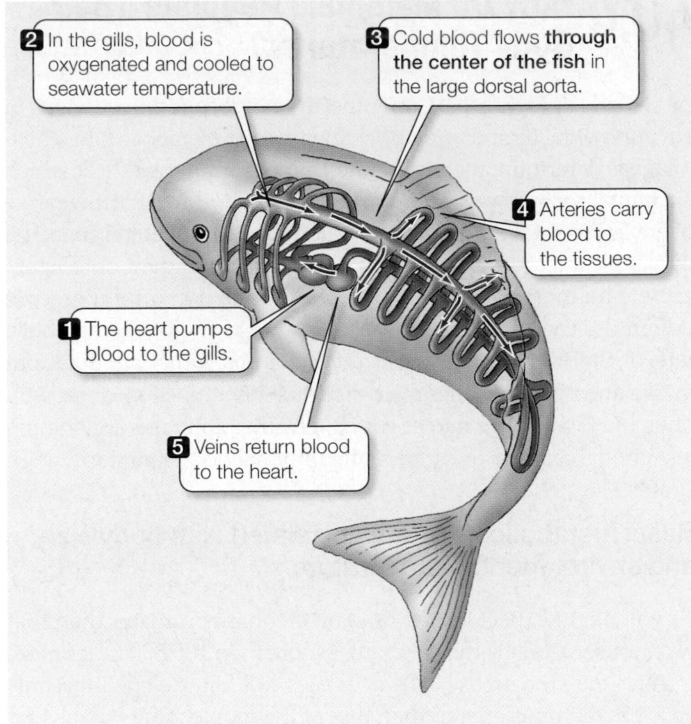
(A) "Cold" fish
2 In the gills, blood is oxygenated and cooled to seawater temperature.
3 Cold blood flows through the center of the fish in the large dorsal aorta.
4 Arteries carry blood to the tissues.
1 The heart pumps blood to the gills.
5 Veins return blood to the heart.

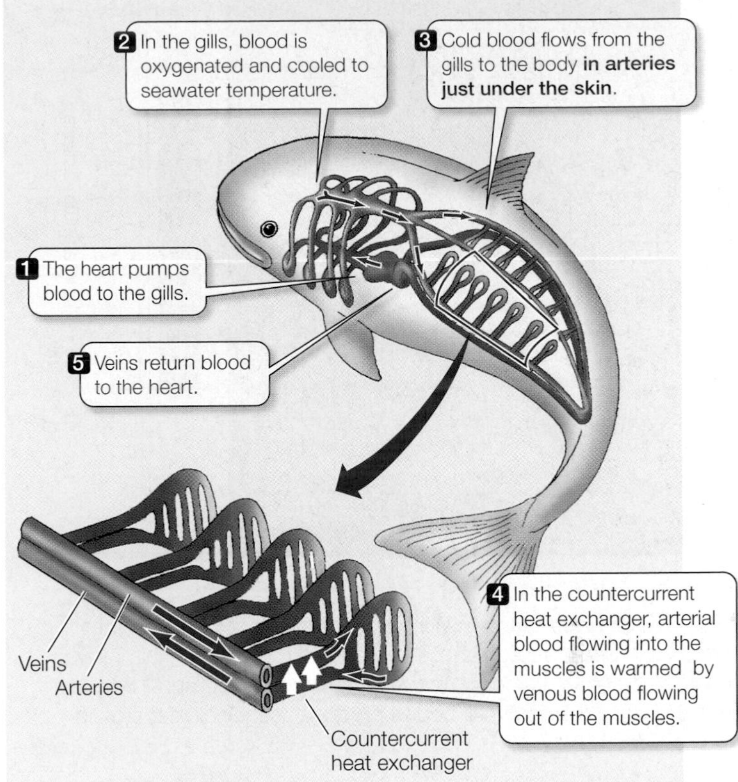
(B) "Hot" fish
2 In the gills, blood is oxygenated and cooled to seawater temperature.
3 Cold blood flows from the gills to the body in arteries just under the skin.
1 The heart pumps blood to the gills.
5 Veins return blood to the heart.
Veins
Arteries
Countercurrent heat exchanger
4 In the countercurrent heat exchanger, arterial blood flowing into the muscles is warmed by venous blood flowing out of the muscles.

40.13 "Cold" and "Hot" Fish (A) Most fish species are "cold" fish. Their circulatory systems conduct cool, oxygenated blood from the gills through a large dorsal aorta to the rest of the body. (B) The anatomy of "hot" fish species includes a mechanism—the countercurrent heat exchanger—that allows heat to pass into cold arterial blood from venous blood that has been warmed by the metabolism of the muscles.

heat that the blood picks up from metabolically active muscles is lost to the surrounding water as it flows through the gills. It is thus surprising that some large, rapidly swimming fishes, such as bluefin tuna and great white sharks, can maintain temperature differences as great as 10°–15°C between their bodies and the surrounding water. The heat comes from their powerful swimming muscles, and the ability of these "hot" fish to conserve that heat is based on the remarkable arrangements of their blood vessels.

In the usual ("cold") fish circulatory system, oxygenated blood from the gills collects in a large dorsal vessel, the aorta, which travels through the center of the fish, distributing blood to all organs and muscles (**Figure 40.13A**). "Hot" fish have a smaller central dorsal aorta, and most of their oxygenated blood is transported in large vessels just under the skin (**Figure 40.13B**). The cold blood from the gills is thus kept close to the surface of the fish. Smaller vessels transporting this cold blood into the muscle mass run parallel to vessels transporting warm blood from the muscle mass back toward the heart. Since the vessels carrying the cold blood into the muscle are in close contact with the vessels carrying warm blood away, heat flows from the warm to the cold blood and is therefore retained in the muscle mass.

Because heat is exchanged between blood vessels carrying blood in opposite directions, this adaptation is called a **countercurrent heat exchanger**. It keeps the heat within the muscle mass, enabling these fish to have an internal body temperature considerably above the water temperature. Why is it advantageous for the fish to be warm? Each 10°C rise in muscle temperature increases the fish's sustainable power output almost threefold!

Some ectotherms regulate heat production

Some ectotherms raise their body temperature by producing heat. For example, the powerful flight muscles of many insects must reach 35°–40°C before the insects can fly, and they must maintain these high temperatures during flight. Such insects produce the required heat by contracting their flight muscles in a manner analogous to shivering in mammals. The heat-producing ability of insects can be quite remarkable. Probably the most impressive case is a species of scarab beetle that lives mostly underground in mountains north of Los Angeles, California. To mate, these beetles come above ground, and males fly in search of females. They undertake this mating ritual at night, in winter, and only during snowstorms.

Honeybees regulate temperature as a group. They live in large colonies consisting mostly of female worker bees that maintain the hive and rear the larval offspring of the single queen bee. During winter, honeybee workers cluster around the brood of larvae. They adjust their individual metabolic heat production and density of clustering so that the brood temperature remains remarkably constant, at about 34°C, even as the outside air temperature drops below freezing (**Figure 40.14**).

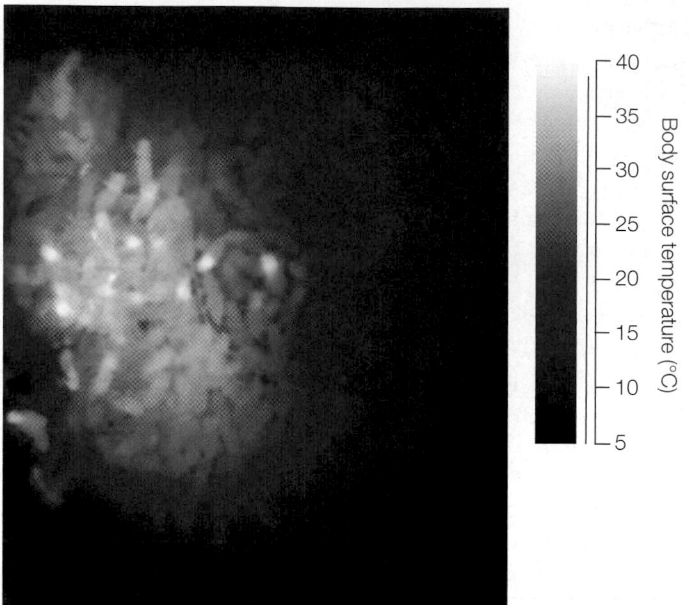

40.14 Bees Keep Warm in Winter Honeybee colonies survive winter cold because workers generate metabolic heat. In this infrared photograph of the center of an overwintering hive, individual bees are discernible by the heat their bodies produce as they cluster around their queen.

40.3 RECAP

Animals that metabolically produce their own heat are called endotherms. Those that depend on environmental sources of heat are called ectotherms. Heat exchange between an animal and its environment occurs via radiation, convection, conduction, and evaporation.

- In terms of the energy budget relationship, why is the control of blood flow to the skin so important for thermoregulation? See p. 864
- Can you explain how countercurrent heat exchange makes it possible for some fish to have a body temperature above that of the surrounding water? See pp. 864–865 and Figure 40.13

Endotherms must keep their body temperatures within a critical physiological range. Let's look more closely at the evolutionary adaptations that enable endothermic mammals to maintain this optimal temperature range.

40.4 How Do Mammals Regulate Their Body Temperatures?

As we saw in Figure 40.9, endotherms can respond to changes in environmental temperature by changing their metabolic rate. Physiologists determine metabolic rate by measuring the rate at which an animal consumes O_2 and produces CO_2. Within a narrow range of environmental temperatures, called the **thermoneutral zone**, the metabolic rate of endotherms is low and independent of temperature. The metabolic rate of a resting animal at a temperature within the thermoneutral zone is known as the **basal metabolic rate**, or **BMR**. It is usually measured in animals that are quiet but awake and not using energy for digestion, reproduction, or growth. Thus the BMR is the rate at which a resting animal is consuming just enough energy to carry out its minimal body functions.

Basal metabolic rates are correlated with body size and environmental temperature

As you might expect, the BMR of an elephant is greater than that of a mouse. After all, the elephant is more than 100,000 times more massive than the mouse. However, the BMR of the elephant is only about 7,000 times greater than that of the mouse. That means that a gram of mouse tissue uses energy at a rate 20 times greater than a gram of elephant tissue (**Figure 40.15**). Across all of the endotherms, BMR per gram of tissue increases as animals get smaller.

Why should this disproportionate difference exist? We don't know for sure. As animals get bigger, they have a smaller ratio of surface area to volume (see Figure 4.2). Since heat production is related to the volume, or mass, of the animal, but its capacity to dissipate heat is related to its surface area, it was once reasoned that larger animals evolved lower metabolic rates to avoid overheating. This explanation is insufficient for several reasons, one being that the relationship between body mass and metabolic rate holds for even very small organisms and for ectotherms, in which overheating is not a problem.

In an endotherm, the basal metabolic rate versus ambient temperature curve represents the integrated response of all of the animal's thermoregulatory adaptations (**Figure 40.16**). The thermoneutral zone is bounded by a *lower* and an *upper critical temper-*

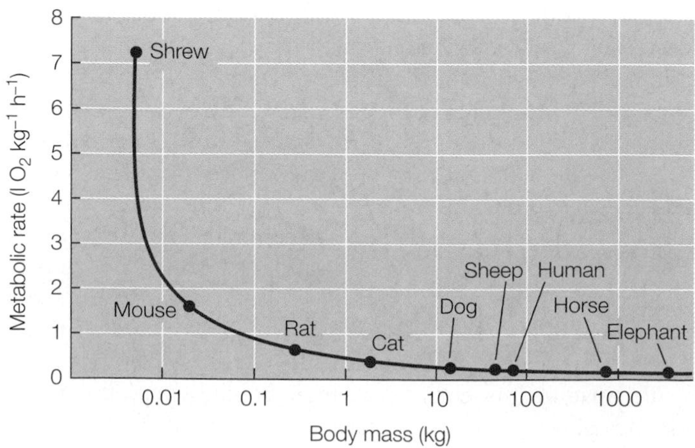

40.15 The Mouse-to-Elephant Curve On a weight-specific basis, the metabolic rate of small endotherms is much greater than that of larger endotherms. This graph plots O_2 consumption per kg of body weight (a measure of the metabolic rate) against a logarithmic plot of body weight.

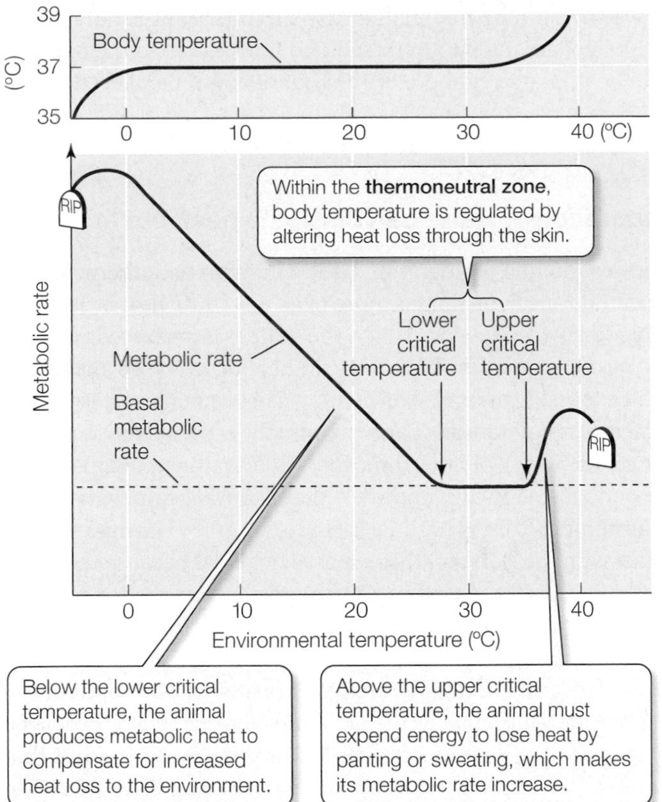

Body temperature
(°C)
39
37
35
0
10
20
30
40 (°C)
Within the **thermoneutral zone**, body temperature is regulated by altering heat loss through the skin.
Lower critical temperature
Upper critical temperature
Metabolic rate
Metabolic rate
Basal metabolic rate
RIP
RIP
Environmental temperature (°C)
Below the lower critical temperature, the animal produces metabolic heat to compensate for increased heat loss to the environment.
Above the upper critical temperature, the animal must expend energy to lose heat by panting or sweating, which makes its metabolic rate increase.

40.16 Environmental Temperature and Mammalian Metabolic Rates Outside the thermoneutral zone, maintaining a constant body temperature requires the expenditure of energy. Outside extreme limits (0°C and 40°C in this instance), the animal cannot maintain its body temperature and dies.

ature. Within its thermoneutral zone, an endotherm's thermoregulatory adaptations do not require much energy and could be considered passive; such adaptations include changing posture, fluffing fur, and controlling blood flow to the skin. Outside its thermoneutral zone, however, thermoregulatory responses are active and require considerable metabolic energy.

One interesting aspect of the thermoneutral zone is that many endotherms can survive at a far greater range of temperatures *below* the lower critical temperature than *above* it (see Figure 40.16). The coldest habitats on Earth are the Arctic, the Antarctic, and the high mountains. Many birds and mammals, but almost no reptiles or amphibians, live in these places. Adaptations to life in the cold have been a big part of the evolutionary success of these animals.

Endotherms respond to cold by producing heat and reducing heat loss

When environmental temperatures fall below the lower critical temperature, endotherms must produce heat to compensate for the heat they lose to the environment. Mammals can accomplish this in two ways: shivering and nonshivering heat production. Birds use only shivering heat production.

Shivering uses the contractile machinery of skeletal muscles to consume ATP without causing any visible behavior. Shivering muscles pull against each other so that little movement other than a tremor results. The energy from the conversion of ATP to ADP in this process is released as heat. Shivering heat production is perhaps too narrow a term, however; increased muscle tone and increased body movements also contribute to increased heat production in cold environments.

Most *nonshivering heat production* occurs in a specialized adipose tissue called **brown fat** (**Figure 40.17**). This tissue looks brown because of its abundant mitochondria and rich blood supply. In brown fat cells, a protein called *thermogenin* uncouples proton movement from ATP production, allowing protons to leak across the inner mitochondrial membrane rather than having to pass through the ATP synthase and generate ATP (review the discussion of the chemiosmotic mechanism in Section 7.4). As a result, metabolic fuels are consumed without producing ATP, but heat is still released. Brown fat is especially abundant in newborn infants of many mammalian species, including humans, in some adult mammals that are small and acclimatized to cold, and in mammals that hibernate.

The most important adaptations of endotherms to cold environments are those that reduce heat loss. Since most heat is lost from the body surface, many cold-climate species have a smaller surface area than their warm-climate cousins, even when their body masses are the same. Rounder body shapes and shorter appendages reduce the surface area-to-volume ratios of some cold-climate species (**Figure 40.18**).

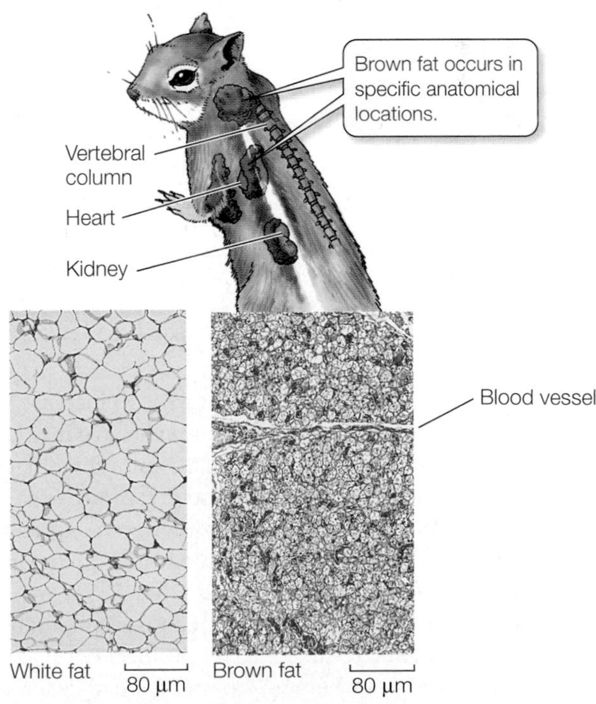

Brown fat occurs in specific anatomical locations.
Vertebral column
Heart
Kidney
Blood vessel
White fat
80 μm
Brown fat
80 μm

40.17 Brown Fat In many mammals, specialized brown fat tissue produces heat. When viewed through a microscope at similar magnifications, we see that white fat cells (left) contain large droplets of lipid but have few organelles and limited blood supply, while brown fat cells (right) are packed with mitochondria and richly supplied with blood.

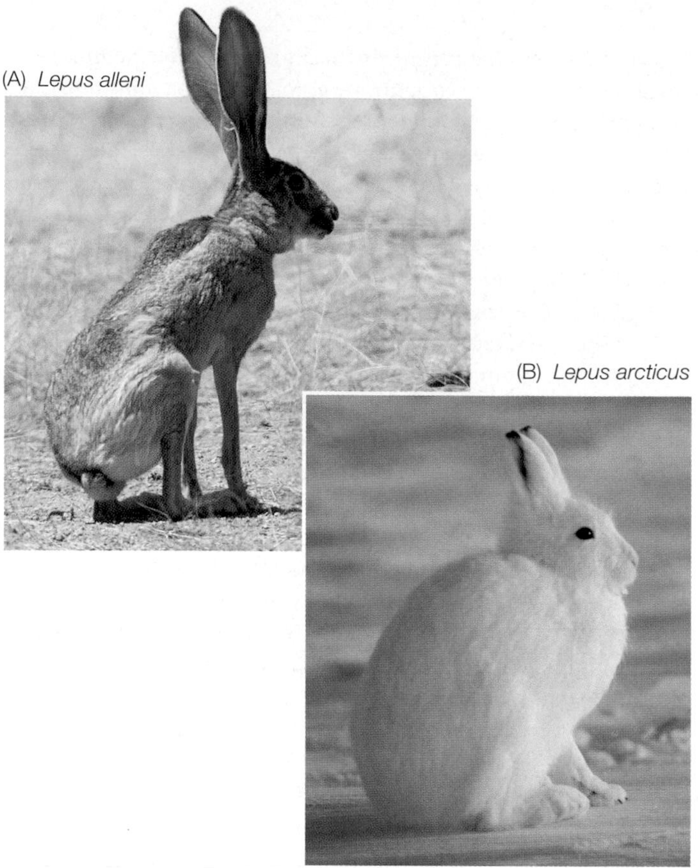

40.18 Adaptating to Hot and Cold Climates (A) The antelope jackrabbit is found in the Sonoran Desert of Arizona. Its large ears serve as heat exchangers, passing heat from the animal's blood to the surrounding air. (B) The thick fur of the arctic hare provides insulation in the frigid winter. Its ears and extremities are relatively smaller than those of the jackrabbit.

Because each strand is hollow, a polar bear's fur is an even better insulator than normal hair. It also acts like optical fibers, transmitting solar radiation to the skin. Beneath its white fur coat the bear's skin is black and readily absorbs the solar radiation that reaches it.

Another means of decreasing heat loss is to increase thermal insulation. Animals adapted to cold climates have much thicker layers of fur, feathers, or fat than do their warm-climate relatives. The fur of an arctic fox or a northern sled dog provides such good thermal insulation that these animals don't even begin to shiver until the air temperature drops considerably below freezing. Fur and feathers are good insulators because they trap a layer of still, warm air close to the skin surface. If that air is displaced by water, insulation is drastically reduced. In many species, oil secretions spread through fur or feathers by grooming are critical for resisting wetting and maintaining a high level of insulation.

Decreasing blood flow to the skin is an important thermoregulatory adaptation in the cold. Constriction of blood vessels in the skin, and especially in the appendages, greatly improves the ability of an animal to conserve heat. Countercurrent heat exchange like we saw in the "hot" fish is also an important adaptation in the appendages of endotherms. Blood flowing out to the paw of a wolf, the hoof of a caribou, or the foot of a bird parallels the flow of the blood returning. Heat is transferred from the outgoing to the returning blood, thus retaining heat in the animal's core.

Evaporation of water can dissipate heat, but at a cost

As environmental temperature rises within an endotherm's thermoneutral zone, it dissipates more of its metabolic heat by increasing blood flow to the skin. When the temperature exceeds the upper critical temperature, however, overheating becomes a problem. For an exercising animal, overheating can occur at even low environmental temperatures. Large mammals, especially those in hot habitats such as elephants, rhinoceroses, and water buffaloes, have little or no insulating fur and seek places to wallow in water when the air temperature is high (see Figure 40.10B). Having water in contact with the skin greatly increases heat loss because the heat-absorbing capacity of water is much greater than that of air.

Evaporation of water from external or internal body surfaces through sweating or panting can also cool an endotherm. A gram of water absorbs about 580 calories of heat when it evaporates. If this evaporation occurs on the skin, most of that heat is absorbed from the skin and the underlying blood. Any sweat or saliva that falls off of the body, however, provides no cooling. Thus when the need for heat loss is greatest, a lot of water from the internal environment can be squandered with no cooling benefit. Since water is heavy, animals do not carry an excess supply of it, and many hot environments are also arid. In habitats that are both hot and dry, sweating and panting are cooling adaptations of last resort.

Sweating and panting are *active* processes that require the expenditure of metabolic energy. That is why the metabolic rate increases when the upper critical temperature is exceeded (the rising curve Figure 40.16). A sweating or panting animal is generating heat in the process of dissipating heat, which can be a losing battle.

The vertebrate thermostat uses feedback information

The thermoregulatory mechanisms and adaptations we have just discussed work through a regulatory system that integrates information from environmental and physiological sources and then issues commands that control body temperature. Such a regulatory system is based on feedback information, and can be thought of as a *thermostat*. We focus now on the vertebrate thermostat, particularly in the mammal.

Where is the vertebrate thermostat? Its major integrative center is at the bottom of the brain in a structure called the **hypothalamus**. If you slide your tongue back as far as possible along the roof of your mouth, it will be just a few centimeters below your hypothalamus. The hypothalamus is a key part of many regulatory systems, so we refer to it again in chapters to come. If the hypothalamus of a mammal's brain is damaged, the animal loses (among other things) its ability to regulate its body temperature, which then rises in warm environments and falls in cold ones.

In many species, the temperature of the hypothalamus itself is the major source of feedback information to the thermostat. Cooling the hypothalamus causes fish and reptiles to seek a warmer

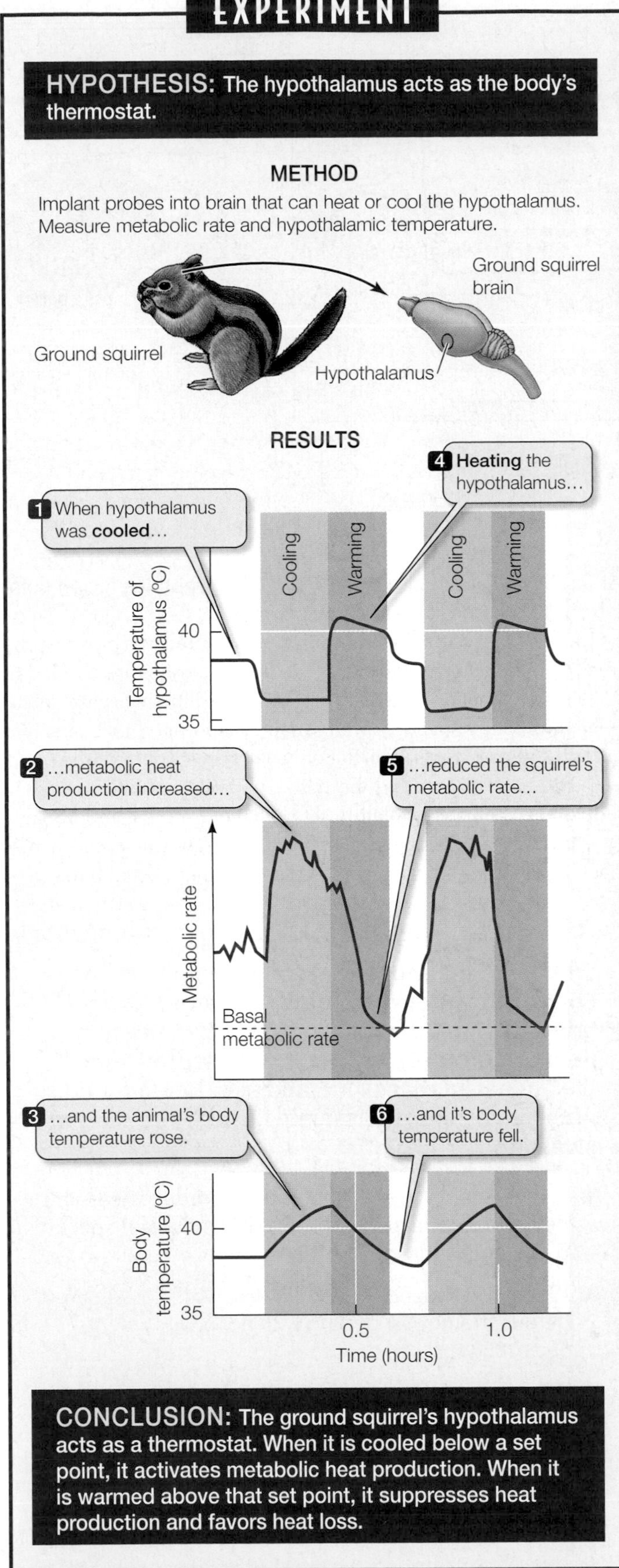

EXPERIMENT
HYPOTHESIS: The hypothalamus acts as the body's thermostat.
METHOD
Implant probes into brain that can heat or cool the hypothalamus. Measure metabolic rate and hypothalamic temperature.
Ground squirrel brain
Ground squirrel
Hypothalamus
RESULTS
1 When hypothalamus was cooled...
4 Heating the hypothalamus...
Cooling
Warming
Cooling
Warming
Temperature of hypothalamus (°C)
40
35
2 ...metabolic heat production increased...
5 ...reduced the squirrel's metabolic rate...
Metabolic rate
Basal metabolic rate
3 ...and the animal's body temperature rose.
6 ...and it's body temperature fell.
Body temperature (°C)
40
35
0.5
1.0
Time (hours)
CONCLUSION: The ground squirrel's hypothalamus acts as a thermostat. When it is cooled below a set point, it activates metabolic heat production. When it is warmed above that set point, it suppresses heat production and favors heat loss.

40.19 The Hypothalamus Regulates Body Temperature The observation that damage to the hypothalamus disrupts thermoregulation led to the finding that the hypothalamus acts as a thermostat in the vertebrate body.

environment, and warming the hypothalamus causes them to seek a cooler environment. In mammals, cooling the hypothalamus can stimulate constriction of the blood vessels supplying the skin and increase metabolic heat production. Because it activates these thermoregulatory responses, cooling the hypothalamus causes the body temperature to rise. Conversely, mild warming of the hypothalamus stimulates dilation of the blood vessels supplying the skin, while stronger hypothalamic heating stimulates sweating or panting. Consequently heating the hypothalamus can actually cause the overall body temperature to fall (**Figure 40.19**).

The hypothalamus generates a set point like a setting on the thermostat of a house. When the temperature of the hypothalamus exceeds or drops below that set point, thermoregulatory responses (the controlled system) are activated to reverse the direction of temperature change. Hence, hypothalamic temperature is a negative feedback signal.

Experimental warming and cooling of the hypothalamus show that mammals have separate set points for activating different thermoregulatory responses. If the hypothalamus of a mammal is cooled, the vessels supplying blood to the skin constrict at a specific hypothalamic temperature. A slightly lower hypothalamic temperature initiates shivering. If the hypothalamic temperature is then raised, shivering ceases; then blood vessels supplying the skin dilate. At still higher hypothalamic temperatures, sweating or panting starts.

The vertebrate thermoregulatory system has adjustable set points and integrates sources of information in addition to hypothalamic temperature. For example, temperature sensors in the skin register environmental temperature; change in skin temperature is feedforward information that shifts the hypothalamic set point for thermoregulatory responses. The set point for metabolic heat production is higher when skin is cold and lower when skin is warm.

Other factors can shift hypothalamic set points for thermoregulatory responses. Set points are higher during wakefulness than during sleep, and they are higher during the active part of the daily cycle than during the inactive part, even if the animal is awake at both times. Even when an endotherm is kept under constant environmental conditions, its body temperature displays a daily cycle of changes in set point. This kind of cycle is controlled by an internal *circadian rhythm*; these endogenous bodily rhythms are discussed further in Chapter 53.

Fever helps the body fight infections

Growing evidence suggests that fever is an adaptive response that helps the body fight pathogens. A fever is a rise in body temperature in response to substances called **pyrogens**. *Exogenous pyrogens* come from foreign substances such as bacteria or viruses that invade the body. *Endogenous pyrogens* are produced by cells of the immune system in response to infection.

The presence of a pyrogen in the body causes a rise in the hypothalamic set point for the metabolic heat production response. As a result, you shiver, put on a sweater, or crawl under a blanket, and your body temperature rises until it matches the new set point. At the higher body temperature you no longer feel cold, and you may not feel hot, but someone touching your forehead will say that you are "burning up." Taking aspirin lowers your set point to normal.

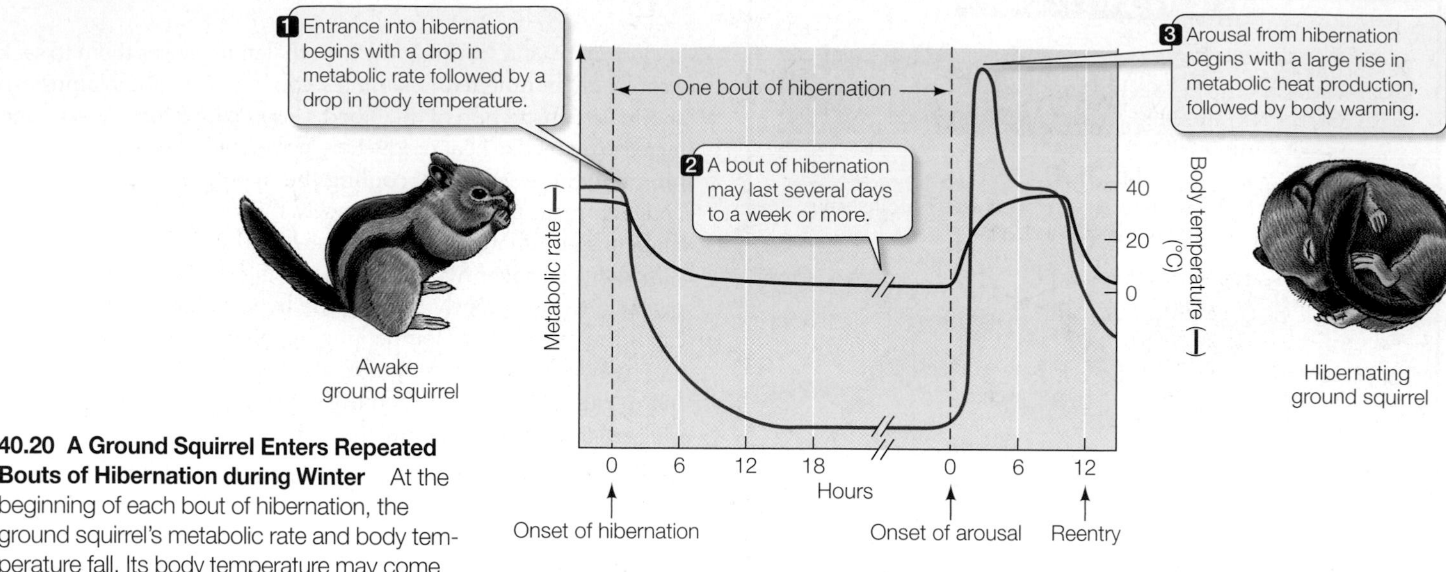

1 Entrance into hibernation begins with a drop in metabolic rate followed by a drop in body temperature.
One bout of hibernation
2 A bout of hibernation may last several days to a week or more.
3 Arousal from hibernation begins with a large rise in metabolic heat production, followed by body warming.
Metabolic rate
Body temperature (°C)
40
20
0
0
6
12
18
0
6
12
Hours
Onset of hibernation
Onset of arousal
Reentry
Awake ground squirrel
Hibernating ground squirrel

40.20 A Ground Squirrel Enters Repeated Bouts of Hibernation during Winter At the beginning of each bout of hibernation, the ground squirrel's metabolic rate and body temperature fall. Its body temperature may come into equilibrium with the temperature of its nest and stay at that level for days. The bout is ended by a rise in metabolic heat production that returns body temperature to a normal level.

Now you feel hot, take off clothes, and even sweat until your elevated body temperature returns to normal. Although modest fevers help the body fight infections, extreme fevers can be dangerous and must be controlled, usually with fever-reducing drugs.

Turning down the thermostat

Hypothermia refers to a state of below-normal body temperature. It can result from a natural turning down of the thermostat or from traumatic events such as starvation (lack of metabolic fuel), exposure to extreme cold, serious illness, or anesthesia. Many species of birds and mammals use *regulated hypothermia* as a means of surviving periods of cold and food scarcity. Some even become hypothermic daily.

Hummingbirds, for example, are very small endotherms with a high metabolic rate. Just getting through a single day without food could exhaust their metabolic reserves. Hummingbirds and other small endotherms can extend the period over which they can survive without food by dropping their body temperature during the portion of day or night when they would normally be inactive. This adaptive hypothermia is called **daily torpor**. Body temperature can drop 10°–20°C during daily torpor—an enormous savings of metabolic energy.

Regulated hypothermia that lasts for days or even weeks, with drops to very low body temperatures, is called **hibernation** (**Figure 40.20**). During the deep sleep of hibernation, the body's thermostat is turned extremely low to maximize energy conservation. Arousal from hibernation occurs when the hypothalamic set point returns to the normal level for a mammal.

Many hibernators maintain body temperatures close to the freezing point during hibernation. The metabolic rate needed to sustain a hibernating animal may be only one-fiftieth its basal metabolic rate. Many species of mammals, including bats, bears, and ground squirrels, hibernate, but only one species of bird (the poorwill) has been shown to hibernate. The ability of hibernators to reduce their thermoregulatory set point so dramatically probably evolved as an extension of the set point decrease that accompanies sleep even in nonhibernating species of mammals and birds.

40.3 RECAP

Within the thermoneutral zone, an endotherm controls its body temperature by altering insulation and blood flow to the skin. When the temperature drops below the thermoneutral zone, the animal increases metabolic heat production. Above this zone, it dissipates heat by panting or sweating.

- Can you describe how endotherms produce heat, and how heat production changes with body size? See p. 866 and Figure 40.15
- Why is dependence on evaporative water loss a dangerous strategy for dealing with hot environments? See p. 868
- What is the nature of negative feedback information and feedforward information used by the mammalian thermostat? See pp. 868–869 and Figure 40.19

CHAPTER SUMMARY

40.1 Why must animals regulate their internal environments?

Multicellular animals provide for the needs of all of their cells by maintaining a stable internal environment, which consists of the **extracellular fluid**.

Each cell, tissue, and organ contributes to **homeostasis** of the internal environment. Review Figure 40.1

The regulation of physiological systems is mostly through **negative feedback regulation. Feedforward information** functions to change **set points**. Review Figure 40.2

Tissues are assemblages of different cells. **Organs** are made up of tissues, and most organs contain all four kinds of tissue. Organs are grouped into **organ systems**. Review Figure 40.7

Epithelial tissues provide barriers and have secretory and transport functions.

Muscle tissues contract. The three types of muscle tissue are skeletal, cardiac, and smooth muscle.

Connective tissues, in which cells are embedded in an extracellular matrix, provide support. Cartilage, bone, adipose tissue, and blood are types of connective tissue.

Nervous tissues process and communicate information. They contain two cell types, neurons and glial cells.

40.2 How does temperature affect living systems?

Life is sustained within a narrow range of environmental temperatures. **Q_{10}** is a measure of the sensitivity of a life process to temperature. A Q_{10} of 2 means that the reaction rate doubles as temperature increases by 10°C. Review Figure 40.8

Metabolic rate is a measure of the energy turnover of an animal's cells and is often measured as the rate of O_2 consumption.

40.3 How do animals alter their heat exchange with the environment?

Endotherms can produce considerable metabolic heat to compensate for heat loss to the environment. **Ectotherms** generally do not. Review Figure 40.9

Energy budgets describe all pathways for heat exchange between an organism and its environment. The four avenues of heat exchange are **radiation**, **conduction**, **convection**, and **evaporation**. Review Figure 40.11

Skin temperature is an important variable and it can be influenced by blood flow. Circulatory system adaptations such as countercurrent heat exchange can conserve metabolic heat. Review Figures 40.12 and 40.13

40.4 How do mammals regulate their body temperatures?

Within the **thermoneutral zone**, mammals have a minimal or **basal metabolic rate.** BMR scales with body size. Review Figures 40.15 and 40.16, Web/CD Activity 40.1

In vertebrates, the control of thermoregulatory effectors relies on commands from a regulatory center in the **hypothalamus.** This thermostat uses its own temperature as a major negative feedback signal and skin temperature as a feedforward signal. Review Figure 40.19, Web/CD Tutorial 40.1

Fever is a regulated increase in body temperature and hibernation is a regulated decrease. Review Figure 40.20

SELF-QUIZ

1. Which of the following characterizes the protein elastin?
   a. It functions predominantly in muscle tissue to resist excess stretching.
   b. It is found predominantly in epithelial tissue.
   c. It is found in the extracellular matrix of connective tissue.
   d. It is the most abundant protein in the body.
   e. It is responsible for the elasticity of the long extensions of neurons.
2. If the Q_{10} of the metabolic rate of an animal is 2, then
   a. the animal is better acclimatized to a cold environment than if its Q_{10} is 3.
   b. the animal is an ectotherm.
   c. the animal consumes half as much oxygen per hour at 20°C as it does at 30°C.
   d. the animal's metabolic rate is not at basal levels.
   e. the animal produces twice as much heat at 20°C as it does at 30°C.
3. Which statement about brown fat is true?
   a. It produces heat without producing ATP.
   b. It insulates animals acclimatized to cold.
   c. It is a major source of heat production for birds.
   d. It is found only in hibernators.
   e. It provides fuel for muscle cells.
4. Which of the following is the most important and most general characteristic of endotherms adapted to cold climates compared to those adapted to warm climates?
   a. Higher basal metabolic rates
   b. Higher Q_{10} values
   c. Brown fat
   d. Greater insulation
   e. Ability to hibernate
5. Which of the following would cause a decrease in the hypothalamic temperature set point for metabolic heat production?
   a. Entering a cold environment
   b. Taking an aspirin when you have a fever
   c. Arousing from hibernation
   d. Getting an infection that causes a fever
   e. Cooling the hypothalamus
6. Mammalian hibernation
   a. occurs when animals run out of metabolic fuel.
   b. is a regulated decrease in body temperature.
   c. is less common than hibernation in birds.
   d. can occur at any time of year.
   e. lasts for several months, during which body temperature remains close to environmental temperature.
7. Which of the following is an important difference between an ectotherm and an endotherm of similar body size?
   a. The ectotherm has higher Q_{10} values.
   b. Only the ectotherm uses behavioral thermoregulation.
   c. Only the endotherm can constrict and dilate the blood vessels to the skin to alter heat flow.
   d. Only the endotherm can have a fever.
   e. At a body temperature of 37°C, the ectotherm has a lower metabolic rate than the endotherm.

8. How would you describe the role of skin temperature in the human thermoregulatory system?
   a. It provides feedforward information.
   b. It acts as a set point for metabolic heat production.
   c. It provides positive feedback information.
   d. It provides an error signal.
   e. It provides negative feedback information.
9. What is the biggest difference between a "cold" fish such as a trout and a "hot" fish such as a tuna?
   a. The temperature of the blood leaving the heart.
   b. The temperature of the blood entering the gills.
   c. The arrangement of blood vessels in the gills.
   d. The temperature of the brain.
   e. The volume of blood flowing in lateral arteries just under the skin.
10. Which of the following statements about the thermoneutral zone is true?
   a. Metabolic heat production is variable.
   b. Skin blood flow is variable.
   c. The environmental temperature equals body temperature.
   d. Its lower boundary (lower critical temperature) is lower for small than for large endotherms.
   e. It is the range of hypothalamic temperatures that do not alter metabolic heat production.

FOR DISCUSSION

1. What is the advantage of feedforward information for homeostasis? Can you suggest what some sources of feedforward information could be for regulation of breathing, blood pressure, secretion of digestive juices, and elimination of wastes?
2. In some epithelial tissues there are "tight junctions" between the individual cells that prevent anything from passing between them (see Figure 5.7), and in other cases the junctions between epithelial cells are quite loose. What are the possible advantages in different organs of loose versus tight junctions between epithelial cells? Give some examples in which these differences would be important.
3. Newton's law of cooling describes how a physical object comes into thermal equilibrium with its environment. The law is expressed

$$HL = K(T_o - T_a)$$

   HL is the rate of heat loss, K is the thermal conductance constant (how easily an object loses heat), T_o is the temperature of the object and T_a is the ambient temperature. Compare this expression with the metabolic rate/temperature curve for endotherms. In Newton's law of cooling, K is a constant reflecting the properties of the object. What would K represent for an endotherm? Using a version of Newton's law that replaces T_o with T_b (body temperature), explain why the metabolic rate curve projects to zero at an ambient temperature that equals body temperature.
4. The range of temperatures compatible with life is about 0°C to the low 40s. Endotherms have regulated body temperatures much closer to the upper limit of this range than to the lower. What are the advantages of living so close to the upper limit?
5. Discuss what it means when we say that the metabolic rate of mammals scales to the 3/4 power of body mass. In contrast, heart size of mammals scales according to the first power of body mass. What does this difference imply for the functions of the hearts of mammals of different sizes?

FOR INVESTIGATION

1. The text described the drop of body temperature of a hibernator as regulated hypothermia—a turning down of the thermostat. Yet we also saw that if we put an ectotherm in a cold environment, its body temperature will also fall. What experiment could you do to prove that the mammalian hibernator did not just simply turn off or inactivate its thermoregulatory system in order to behave like an ectotherm in the cold?
2. The observations on the Galápagos marine iguana showed that its body temperature rose faster in air than it fell in water. The inference was that the iguana was influencing its gain or loss of heat by altering the blood flow to its skin. However, the thermal properties of air and water are different, and in the case described in the text, the animal was breathing when in air but not when diving in the water. What experiment could you do to strengthen the argument that the iguana was actively altering the flow of heat across its skin?

CHAPTER 41 Animal Hormones

Testosterone abuse

The use of performance-enhancing steroid drugs—generally known as *anabolic steroids*—has become a scandal in athletic competition. Olympic champions have lost medals, professional athletes have been suspended, coaches have lost their jobs, suppliers have been arrested, and record-breaking performances have been expunged from the books. The problem may be epitomized by the prestigious and venerable international bicycle race, the Tour de France. In recent years this grueling competition, a month-long trek over flatlands and mountains, has been undermined by accusations and revelations of illegal anabolic steroid use by many top competitors, including the presumed winner of the 2006 race. But just what is the problem?

Shortly before puberty, the male reproductive system increases its production of an important chemical signal—testosterone, the male sex hormone. Testosterone enters cells, where it binds to certain receptors and alters gene expression. The cells that have these receptors are those involved in the development of male secondary sex characteristics such as deep voice, facial and body hair, and increased muscle mass. Supplemental anabolic steroids are used therapeutically to treat conditions such as delayed puberty, some types of impotence, and the loss of muscles that occurs with certain diseases. But hormone signals in the body usually have many different actions, and testosterone is no exception.

When a muscle is exercised, an interaction between the exercise and the sex steroids results in growth of the muscle—something that is very obvious among body builders. Body builders who abuse anabolic steroids typically use them in doses 10–100 times greater than the therapeutic doses. The resulting extreme growth of skeletal muscle mass extends to women as well, because a female's cells have steroid receptors, although females normally have only low levels of testosterone. When women body builders use anabolic steroids they develop male muscle patterns. They also develop deep voices and body and facial hair, and because these steroids generate negative feedback in the control of the female reproductive system, their breast tissue diminishes, they stop menstruating, and they become infertile. Similar negative feedback in males causes their sex organs to shrink and become nonfunctional. Even more serious (for both men and women) is a greatly increased risk of cancer and of heart, liver, and kidney disease.

A Grueling Race The prestigious Tour de France is one example of an international sporting event marred by revelations about the illegal use of anabolic steroids among some competitors.

Anabolic Steroids Build Big Muscles Anabolic steroids greatly enhance the development of skeletal muscle in response to exercise. Steroids have this effect on women as well as men.

CHAPTER OUTLINE

Despite the risks, steroid drugs are consistently taken by those seeking an unfair advantage; athletic governing organizations thus seek to detect their use. Anabolic steroids circulate in the blood and are broken down in the liver; some residual byproducts are excreted in the urine. Illicit drug makers constantly seek to design new anabolic steroids that produce the desired physical results but do not produce detectable byproducts.

IN THIS CHAPTER we will examine how hormones—chemical signals such as the sex steroids—produce and coordinate anatomical, physiological, and behavioral changes in animals. First we examine the hormonal control of invertebrate life cycles. Building on that example, we discuss the general characteristics of hormones and their receptors. We then describe the functions, control, and mechanisms of action of mammalian hormones, paying particular attention to the extensive interactions between the neural and hormonal information systems in mammals.

41.1 What Are Hormones and How Do They Work?

Control and regulation require information. In multicellular animals, most of this information is transmitted as electric signals and as chemical signals. The electric signals are impulses generated by the nervous system, conducted along cell processes of nerve cells to their targets on specific cells. The chemical signals are **hormones**, secreted by cells of the **endocrine system** into the extracellular fluid.

To compare the two informational systems of the body—the nervous and endocrine systems—think of neuronal communication (which will be detailed in Chapters 44–46) as a telephone system that sends specific messages to specific receivers. Hormonal communication, in contrast, is like a radio or TV network that sends out a broadcast message that can be picked up by whoever has an appropriate receiver that is turned on and tuned in.

Hormones can act locally or at a distance

The cells that secrete hormones are called **endocrine cells**, and the cells that have receptors for those hormones are called **target cells**. Hormones secreted into the extracellular fluid can diffuse into the blood, which distributes them throughout the body so they can activate target cells far from the site of release (**Figure 41.1A**). Such hormones are called **circulating hormones**; testosterone is an example of a circulating hormone.

Some hormones are released in such tiny quantities, or are so rapidly inactivated by enzymes, or are taken up so efficiently by local cells that they never diffuse into the blood in sufficient amounts to act on distant cells. Because these hormones affect only target cells near their release site, they are called **paracrine hormones** (**Figure 41.1B**). An example of a paracrine hormone is histamine, one of the mediators of inflammation (see Figure 18.4). The most local action a hormone can have is when there are receptors on the same cell that released it. When a hormone influences the cell that released it, it is said to have an **autocrine** function. Autocrine functions can provide negative feedback to control rates of secretion.

Some endocrine cells exist as single cells within a tissue. Hormones of the digestive tract, for example, are secreted by isolated endocrine cells in the wall of the stomach and small intestine. Many hormones, however, are secreted by aggregations of endocrine cells forming secretory organs called **en-**

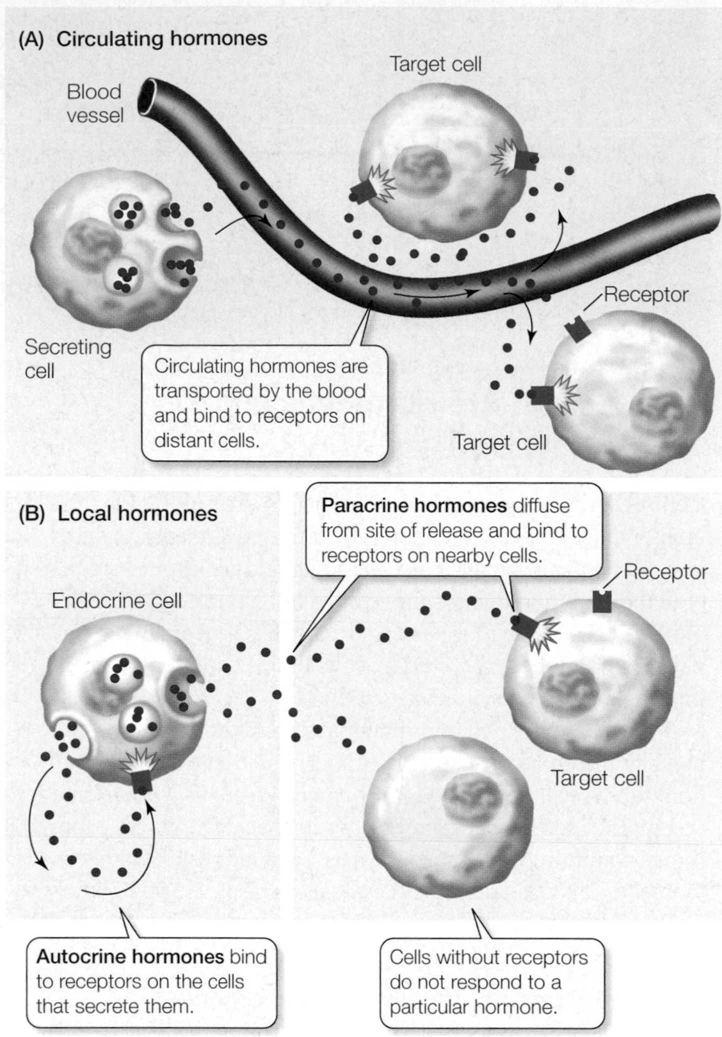

(A) Circulating hormones
Target cell
Blood vessel
Receptor
Secreting cell
Circulating hormones are transported by the blood and bind to receptors on distant cells.
Target cell
(B) Local hormones
Paracrine hormones diffuse from site of release and bind to receptors on nearby cells.
Receptor
Endocrine cell
Target cell
Autocrine hormones bind to receptors on the cells that secrete them.
Cells without receptors do not respond to a particular hormone.

41.1 Chemical Signaling Systems (A) Most hormones are distributed throughout the body by the circulatory system. (B) An autocrine hormone influences the cell that releases it; a paracrine hormone diffuses to nearby cells.

docrine glands. The name "endocrine" reflects the fact that these glands do not have ducts that lead to the outside of the body; rather, they secrete their products directly into the extracellular fluid. In contrast, **exocrine glands** have ducts that carry their products to the surface of the skin (for example, sweat glands) or to the surface of a body passageway that leads to the outside of the body (salivary glands). A single endocrine gland may secrete several different hormones.

Hormonal communication arose early in evolution

Plants do not have nervous systems, but they do have hormones. The most primitive of the multicellular animals, the sponges, also do not have a nervous system, but they do have chemical communication. Even a protist, the social amoeba, which produces multicellular fruiting bodies by the aggregation of individuals, coordinates the aggregation with a chemical signal, cAMP (see Figure 27.32). This chapter provides only a few examples of invertebrate hormone action, but there are many. Here we will explore the hormonal control of molting and metamorphosis—two important events in the lives of arthropods, the most numerous animal group on Earth (see Chapter 32). The hormones involved represent an ancient system of hormonal communication that may be related to the anabolic steroid system discussed in the chapter's opening paragraphs.

Hormones from the head control molting in insects

The largest group of arthropods are the insects, and like all arthropods they have rigid exoskeletons. Therefore, their growth is episodic, punctuated with *molts* (shedding of the exoskeleton). Each growth stage between two molts is called an *instar*.

The British physiologist Sir Vincent Wigglesworth was a pioneer in the study of the hormonal control of growth and development in insects. Wigglesworth conducted experiments on the blood-sucking bug *Rhodnius prolixus*. Upon hatching, *Rhodnius* looks like a miniature version of an adult, lacking some adult features. The juvenile molts five times before developing into a mature adult; a blood meal triggers each episode of molting and growth.

Rhodnius is a hardy experimental animal; it can live a long time even after it is decapitated. If decapitated within an hour after having a blood meal, *Rhodnius* may live up to a year, but it does not molt. If decapitated a week after its blood meal, it does molt (**Figure 41.2, Experiment 1**). These observations led Wigglesworth to the hypothesis that something diffusing slowly from the head controls molting.

Wigglesworth tested his hypothesis with a clever experiment. He decapitated two *Rhodnius*: one shortly after its blood meal and another that had its blood meal a week earlier. The two decapitated bodies were connected with a short piece of glass tubing that allowed body fluid transfer between them. They both molted (**Figure 41.2, Experiment 2**). Thus one or more substances from the bug fed a week earlier must have crossed through the glass tube and stimulated molting in the other bug.

We now know that two hormones working in sequence regulate molting: prothoracicotropic hormone (PTTH) and ecdysone. Cells in the brain produce PTTH, which is why it has also been called brain hormone. PTTH is transported to and stored in a pair of structures called the *corpora cardiaca* attached to the brain. After appropriate stimulation (which for *Rhodnius* is a blood meal), the PTTH is released from these structures and it diffuses in the extracellular fluid to an endocrine gland, the prothoracic gland. PTTH stimulates the prothoracic gland to release the hormone ecdysone. Ecdysone diffuses to target tissues and stimulates molting.

Ecdysone is a lipid-soluble steroid molecule that readily enters its target cells (mostly cells of the epidermis). In the target cells, ecdysone binds to a receptor that is probably ancestral to the vertebrate testosterone receptor. The hormone–receptor complex acts as a transcription factor and induces expression of the genes for enzymes involved in digesting the old cuticle and secreting a new one.

The control of molting by PTTH and ecdysone is a general arthropod hormonal control mechanism and is an example of how a hormonal system works with the nervous system to integrate diverse information and induce a long-term effect. The nervous system receives various types of information (such as day length, temperature, crowding, and nutrition) that help determine the optimal

EXPERIMENT

HYPOTHESIS: Some substance diffusing from the head of *Rhodnius prolixus* (a blood-sucking bug) controls molting.

Experiment 1

METHOD Decapitate juvenile bugs at different times after a blood meal.

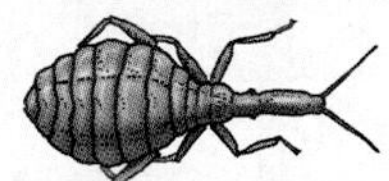

Juvenile bug (third instar)

Decapitation 1 hour after blood meal

Decapitation 1 week after blood meal

RESULTS

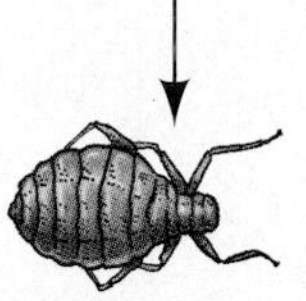

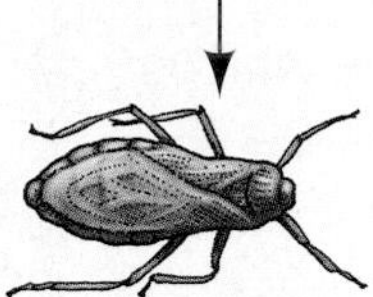

Does not molt (remains a juvenile)

Molts into an adult

CONCLUSION: Whether a decapitated *Rhodnius* will molt depends on the interval between a blood meal and the decapitation, which supports the idea that a substance must pass from head to body.

Experiment 2

METHOD Decapitate juvenile bugs at different times after a blood meal.

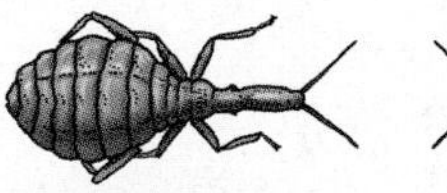

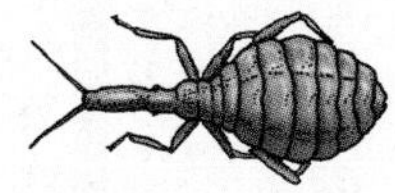

Decapitation 1 hour after blood meal

Decapitation 1 week after blood meal

Join bugs with glass tube

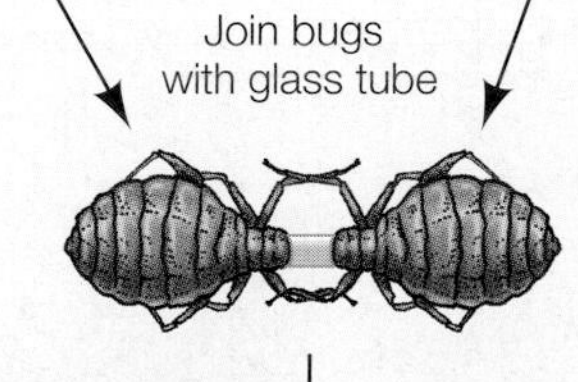

RESULTS

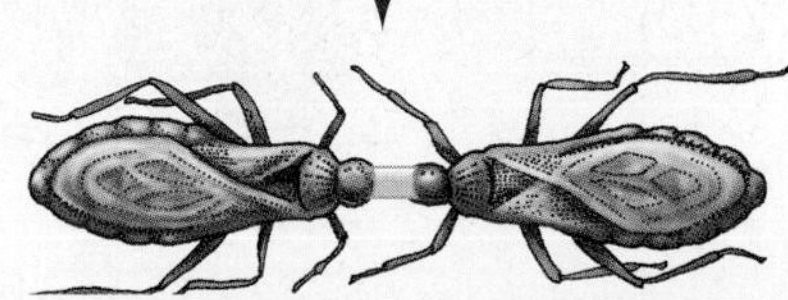

Both bugs molt into adults

CONCLUSION: A diffusible substance is necessary for molting.

41.2 Diffusible Substance Triggers Molting The effect of time since the last blood meal on *Rhodnius* molting (experiment 1) led to the hypothesis that some substance diffusing slowly through the insect's body stimulated molting. Experiment 2 showed that molting is indeed controlled by a substance—a hormone—diffusing from the head.

timing for growth and development. The nervous system (the brain) then controls the endocrine gland (the prothoracic gland) producing the hormone (ecdysone) that orchestrates the physiological processes involved in development and molting. Later in this chapter we will see similar links between the nervous system and endocrine glands in vertebrates.

Juvenile hormone controls development in insects

The *Rhodnius* decapitation experiments yielded a curious result. Regardless of the instar used, the decapitated bug always molted directly into an adult form. Additional experiments by Wigglesworth demonstrated that a hormone other than those responsible for molting determines whether a bug molts into another juvenile instar or into an adult.

Because the head of *Rhodnius* is long, it is possible to remove just the front part of the head, which contains the brain, while leaving the rear part intact. That rear part contains the structures that release the PTTH. When fourth-instar bugs that had been fed a week earlier were partly decapitated, leaving these structures intact, they molted into fifth instars, not into adults.

This experiment was followed by more experiments using glass tubes to connect individual bugs. When an unfed, completely decapitated fifth-instar bug was connected to a fed, partly decapitated fourth-instar bug (with only the front part of its head removed), both bugs molted into juvenile forms. A substance from the rear part of the head of the fourth-instar bug prevented both bugs from molting into adults.

The substance responsible for preventing maturation is **juvenile hormone**, which is released continuously from the same structures that release PTTH in response to feeding. As long as juvenile hormone is present, *Rhodnius* molts into another juvenile instar. Normally *Rhodnius* stops producing juvenile hormone during the fifth instar, and then it molts into an adult.

The control of development by juvenile hormone is more complex in insects, such as butterflies, that undergo complete metamorphosis. These animals undergo dramatic developmental changes in their life cycles. The fertilized egg hatches into a *larva*, which feeds and molts several times, becoming bigger each time. After a fixed number of molts it enters an inactive stage called a *pupa*. The pupa undergoes major body reorganization and finally emerges as an adult.

An excellent example of complete metamorphosis is provided by the silkworm moth, *Hyalophora cecropia* (**Figure 41.3**). As long as juvenile hormone is present in high concentrations, larvae molt into larvae. When the level of juvenile hormone falls, larvae spin cocoons and molt into pupae. Because no juvenile hormone is produced in pupae, they molt into adults.

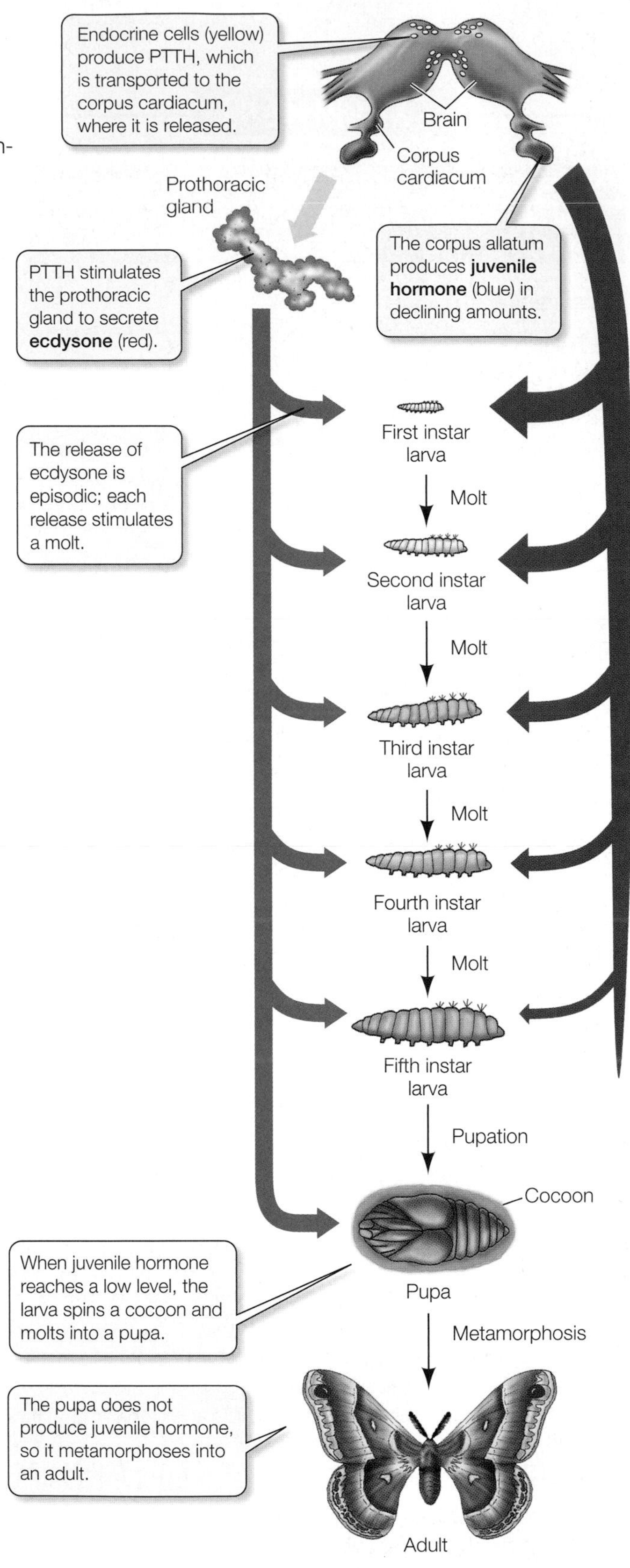

Endocrine cells (yellow) produce PTTH, which is transported to the corpus cardiacum, where it is released.
Brain
Corpus cardiacum
Prothoracic gland
PTTH stimulates the prothoracic gland to secrete **ecdysone** (red).
The corpus allatum produces **juvenile hormone** (blue) in declining amounts.
First instar larva
The release of ecdysone is episodic; each release stimulates a molt.
Molt
Second instar larva
Molt
Third instar larva
Molt
Fourth instar larva
Molt
Fifth instar larva
Pupation
Cocoon
When juvenile hormone reaches a low level, the larva spins a cocoon and molts into a pupa.
Pupa
Metamorphosis
The pupa does not produce juvenile hormone, so it metamorphoses into an adult.
Adult

41.3 Complete Metamorphosis Butterflies and moths undergo complete metamorphosis in which the feeding larvae (caterpillars) bear no resemblance to the reproductive adult. Three hormones control molting and metamorphosis in the silkworm moth *Hyalophora cecropia*.

The existence and function of insect hormones was experimentally demonstrated many years before the hormones were identified chemically. That is not surprising when you consider the tiny amounts of certain hormones that exist in an organism. In one of the earliest studies of ecdysone, biochemists produced only 250 milligrams of pure ecdysone (about one-fourth the weight of an apple seed) from 4 tons of silkworms!

Silk is produced from the fibers of the cocoon that shelters the pupa of the silkworm moth. Silk producers spray the larvae with juvenile hormone to delay pupation. The larvae grow bigger, and bigger larvae make bigger cocoons that yield more silk.

Hormones can be divided into three chemical groups

We have seen examples of the roles hormones can play in long-term physiological and developmental processes in humans and insects. Now we can step back and ask some general questions about hormones. What kinds of hormones exist? What is their chemical nature, and how do they act upon organisms? There is enormous diversity in the chemical structure of hormones, but most of them can be divided into three groups:

- The majority of hormones are *peptides* or *polypeptides* (proteins). These hormones, of which insulin is an example, are water-soluble and are thus easily transported in the blood, but they cannot pass readily through lipid-rich cell membranes. Therefore, peptide and protein hormones are packaged in vesicles in the cells that make them and are released by exocytosis.
- *Steroid hormones* such as testosterone and estrogen are derivatives of the steroid cholesterol. They are lipid-soluble and easily dissolve in and pass through cell membranes. Steroid hormones diffuse out of the cells that make them as they are synthesized. Because steroid hormones are not soluble in blood, however, they must be bound to carrier proteins in order to be transported to their target cells.
- *Amine hormones* are mostly derivatives of the amino acid tyrosine (thyroxine is one example). Some amine hormones are water-soluble and others are lipid-soluble; their modes of release differ accordingly.

Hormone receptors are found on the cell surface or in the cell interior

The chemical structure of hormones is related to the location of their receptors. Lipid-soluble hormones can diffuse through plasma membranes, and therefore their receptors are inside the cell, in either the cytoplasm or the nucleus. In most cases, the complex formed by the lipid-soluble hormone and its receptor acts by altering gene expression in the cell (see Figure 15.8).

Water-soluble hormones cannot readily pass through plasma membranes, so their receptors are on the cell surface. These recep-

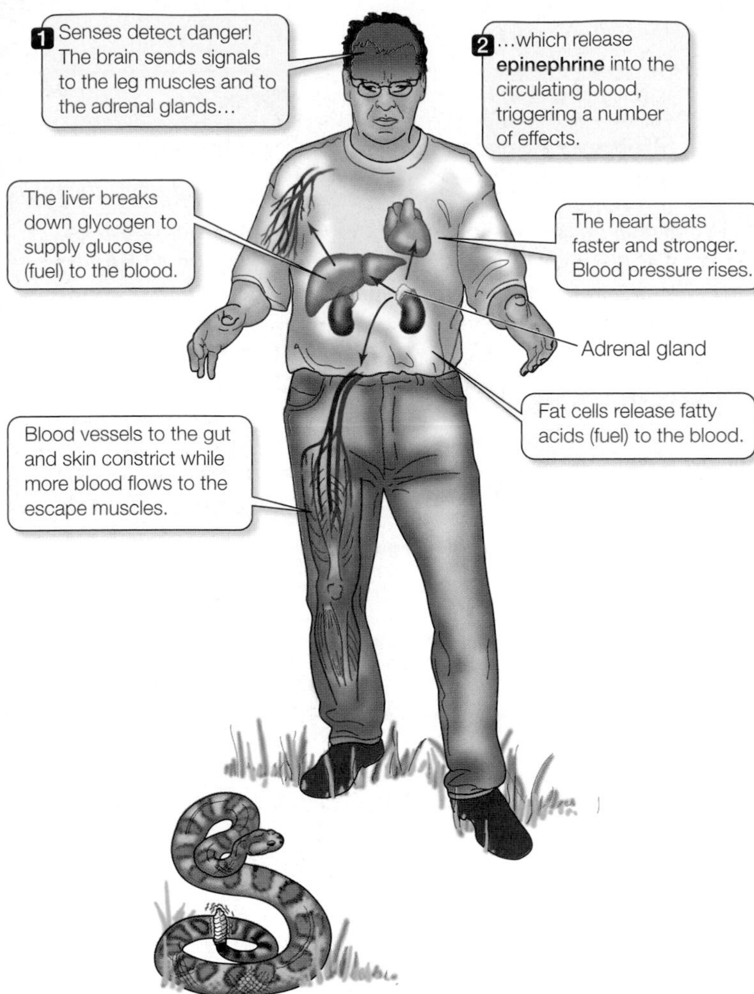

1 Senses detect danger! The brain sends signals to the leg muscles and to the adrenal glands...
2 ...which release epinephrine into the circulating blood, triggering a number of effects.
The liver breaks down glycogen to supply glucose (fuel) to the blood.
The heart beats faster and stronger. Blood pressure rises.
Adrenal gland
Fat cells release fatty acids (fuel) to the blood.
Blood vessels to the gut and skin constrict while more blood flows to the escape muscles.

41.4 Epinephrine Stimulates "Fight or Flight" Responses When surprised by a threat, the brain sends nerve impulses to the adrenal medulla (see Figure 41.11), where epinephrine is released almost instantaneously. Epinephrine circulates around the body and induces different responses in different tissues to help you get out of danger fast.

tors are large glycoprotein complexes with three domains: a **binding domain** that projects outside the plasma membrane, a **transmembrane domain** that anchors the receptor in the membrane, and a **cytoplasmic domain** that extends into the cytoplasm of the cell. The cytoplasmic domain initiates the target cell's response by activating protein kinases or protein phosphatases (see Figures 15.6 and 15.7). In most cases these protein kinases and phosphatases activate or inactivate enzymes in the cytoplasm, which leads to the cell's response, but the signaling cascade initiated by the membrane receptors can also generate chemical signals that enter the nucleus and alter gene expression (see Figure 15.10).

Hormone action depends on the nature of the target cell and its receptors

Most hormones diffuse through the extracellular fluid and are picked up by the blood, which distributes them throughout the body. Wherever such a hormone encounters a cell with a receptor to which it can bind, it triggers a response. The nature of the response depends on the responding cell and its receptors. The same hormone can cause different responses in different types of cells.

Consider the hormone *epinephrine*. Suppose you are walking in the forest and almost trip over a rattlesnake. You jump back. Your heart starts to thump and a whole set of protective actions are set in motion. The jump and the initial heart thumping are driven by your nervous system, which reacts very quickly. Simultaneously with these muscular responses, however, your nervous system stimulates endocrine cells in the adrenal gland just above your kidneys to secrete epinephrine. Within seconds, epinephrine is diffusing into your blood and circulating around your body to activate the many components of the **fight-or-flight response** (**Figure 41.4**).

Epinephrine (an amine) binds to receptors in the heart and blood vessels, sustaining the higher heart rate and causing the heart to beat more strongly. Your heart is now pumping more blood, because your muscles need that blood to fuel your escape. Epinephrine causes more of your circulating blood to flow to the muscles by causing blood vessels in your digestive tract to constrict (digesting lunch can wait!). Similarly, it decreases blood flow to the skin and to the kidneys and suppresses some of the functions of the immune system.

Epinephrine also binds to cells in the liver and to receptors on fat cells. In the liver, epinephrine stimulates the breakdown of glycogen into glucose for a quick energy supply. In fatty tissue, it stimulates the breakdown of fats to yield fatty acids—another source of energy. These are just some of the many actions triggered by one hormone. They all contribute to increasing your chances of escaping from a dangerous situation.

Because it stimulates the heart, epinephrine is a frontline treatment for cardiac arrest (heart attack). Several of its effects, including dilation of the airways, make it a lifesaving drug for people who suffer from asthma or from potentially fatal allergies to things like peanuts and bee stings.

41.1 RECAP

Hormones are chemical signals released by endocrine cells into the extracellular environment, where they diffuse to nearby cells or into the blood. The receptors for water-soluble hormones are on the surface of target cells; receptors for lipid-soluble hormones are in the cytoplasm.

- How does juvenile hormone help effect metamorphosis? See p. 877 and Figures 41.2 and 41.3
- Can you describe the different methods by which water-soluble and lipid-soluble hormones reach their receptors? See p. 879
- Do you understand why a single hormone can have diverse effects in the body?

Since the nervous system and the endocrine system are the two major informational systems of the body, we might expect their activities to be coordinated—and, indeed, they are.

41.2 How Do the Nervous and Endocrine Systems Interact?

The list of hormones known to exist among the vertebrates is long and growing longer. To make the subject manageable, we will focus primarily on the hormones found in mammals (**Figure 41.5**). We will begin our survey by considering the hormones involved in the integration of nervous system and endocrine system functions.

The pituitary connects nervous and endocrine functions

The **pituitary gland** sits in a depression at the bottom of the skull just over the back of the roof of the mouth (**Figure 41.6A**). It is attached by a stalk to the **hypothalamus**, which is involved in many physiological regulatory systems. In turn, the pituitary is involved in the hormonal control of many physiological processes. The nervous system is involved in pituitary functions in two ways. First, two

Hypothalamus (see Figure 41.6)
Release and release-inhibiting hormones control the anterior pituitary
ADH and *oxytocin* are transported to and released from the posterior pituitary

Anterior pituitary (see Figure 41.7)
Thyroid stimulating hormone (TSH): activates the thyroid gland
Follicle stimulating hormone (FSH): in females, stimulates maturation of ovarian follicles; in males, stimulates spermatogenesis
Luteinizing hormone (LH): in females, triggers ovulation and ovarian production of estrogens and progesterone; in males, stimulates production of testosterone
Adrenocorticotropic hormone (ACTH): stimulates adrenal cortex to secrete cortisol
Growth hormone: stimulates protein synthesis and growth
Prolactin: stimulates milk production

Posterior pituitary (see Figure 41.6)
Receives and releases two hypothalamic hormones:
Oxytocin: stimulates contraction of uterus, stimulates flow of milk.
Antidiuretic hormone (ADH): promotes water conservation by kidneys

Thymus gland (disappears in adults)
Thymosin: activates immune system T cells

Pancreas (islets of Langerhans)
Insulin: stimulates cells to take up and use glucose
Glucagon: stimulates liver to release glucose
Somatostatin: slows digestive tract functions including release of insulin and glucagon

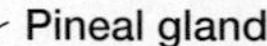

Pineal gland
Melatonin: helps to regulate circadian rhythms

Thyroid gland (see Figures 41.9 and 41.10)
Thyroxine (T_3 and T_4): stimulates cell metabolism
Calcitonin: stimulates incorporation of calcium into bone

Parathyroid glands (on posterior surface of thyroid; see Figure 41.10)
Parathormone (PTH): stimulates release of calcium from bone

Adrenal gland (see Figure 41.11)
Cortex
Cortisol: mediates metabolic responses to stress
Aldosterone: involved in salt and water balance

Medulla
Epinephrine (adrenaline) and *norepinephrine* (noradrenaline): stimulate immediate fight or flight reactions

Gonads (see Chapter 42)
Ovaries (female)
Estrogens: development and maintenance of female sexual characteristics
Progesterone: supports pregnancy

Testes (male)
Testosterone: development and maintenance of male sexual characteristics

Other organs include cells that produce and secrete hormones

Organ	Hormone
Adipose tissue	Leptin
Heart	Atrial natriuretic peptide
Kidney	Erythropoeitin
Stomach	Gastrin
Intestine	Secretin, cholecystokinin
Skin	Vitamin D (cholecalciferol)

41.5 The Endocrine System of Humans The cells that produce and secrete hormones may be organized into discrete endocrine glands, or they may be embedded in the tissues of other organs such as the digestive tract or kidneys. The hypothalamus is part of the brain, but it includes cells that secrete neurohormones into the extracellular fluids.

hormones produced by nerve cells in the hypothalamus are transported to the pituitary via long extensions of those nerve cells and released there. Second, many of the hormones produced by the pituitary are controlled by other hypothalamic hormones that reach the pituitary through the blood.

The two parts of the pituitary gland have different functions and separate developmental origins. The *anterior pituitary* originates as an outpocketing of the roof of the embryonic mouth cavity, and the *posterior pituitary* originates as an outpocketing of the floor of the developing brain. Thus the anterior pituitary originates from gut epithelial tissue and the posterior pituitary originates from neural tissue.

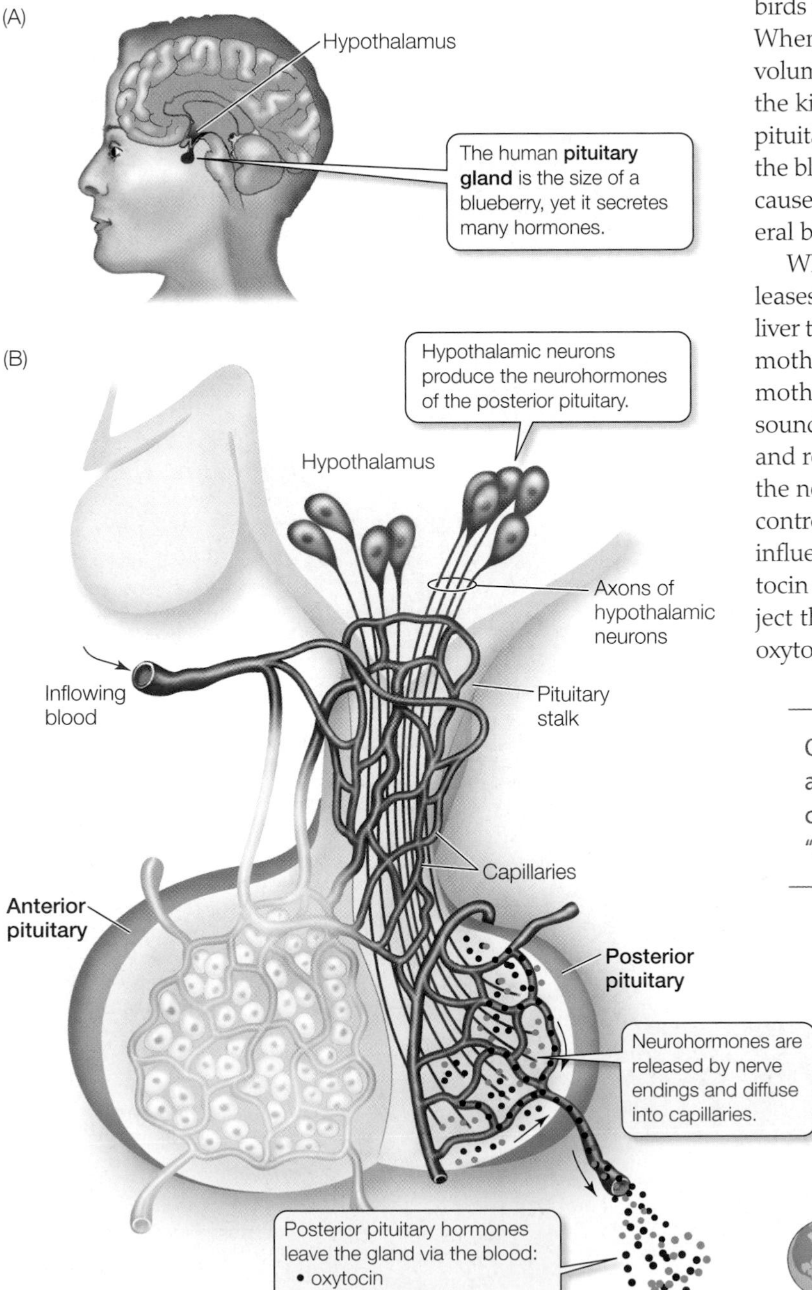
(A)
Hypothalamus
The human **pituitary gland** is the size of a blueberry, yet it secretes many hormones.
(B)
Hypothalamic neurons produce the neurohormones of the posterior pituitary.
Hypothalamus
Axons of hypothalamic neurons
Inflowing blood
Pituitary stalk
Capillaries
Anterior pituitary
Posterior pituitary
Neurohormones are released by nerve endings and diffuse into capillaries.
Posterior pituitary hormones leave the gland via the blood:
• oxytocin
• antidiuretic hormone (ADH)

41.6 The Posterior Pituitary Releases Neurohormones The two hormones stored and released by the posterior pituitary are peptide neurohormones produced in the hypothalamus.

THE POSTERIOR PITUITARY The **posterior pituitary** releases two peptide hormones, antidiuretic hormone (also called vasopressin) and oxytocin. Because these hormones are synthesized in neurons in the hypothalamus, they are called **neurohormones**. The insect hormone PTTH discussed earlier is also a neurohormone (produced in neural tissue). As antidiuretic hormone and oxytocin are produced by cells, they are packaged in vesicles. These vesicles are then transported down long extensions (axons) of the neurons that run from the hypothalamus through the pituitary stalk and terminate in the posterior pituitary. The vesicles are stored until a nerve impulse stimulates their release (**Figure 41.6B**). How do the vesicles move down the axons? Proteins called *kinesins* grab onto the vesicles and, powered by ATP, "walk" step by step down microtubules in the axons.

The main action of **antidiuretic hormone (ADH)** in mammals and birds is to increase the amount of water conserved by the kidneys. When ADH secretion is high, the kidneys produce only a small volume of highly concentrated urine. When ADH secretion is low, the kidneys produce a large volume of dilute urine. The posterior pituitary increases its release of ADH when blood pressure falls or the blood becomes too salty. ADH is also known as *vasopressin* because at high concentrations it causes the constriction of peripheral blood vessels as a means of elevating blood pressure.

When a woman is about to give birth, her posterior pituitary releases **oxytocin**, which stimulates the uterine contractions that deliver the baby. Oxytocin also brings about the flow of milk from the mother's breasts. The baby's suckling stimulates neurons in the mother that cause the secretion of oxytocin. Even the sight and sounds of her baby can cause a nursing mother to secrete oxytocin and release milk from her breasts. This is a good example of how the nervous system integrates information and contributes to the control of hormonally mediated processes. In turn, hormones can influence the nervous system. Oxytocin promotes bonding. If oxytocin release is blocked, new mothers from rats to sheep will reject their newborn offspring, but if a virgin rat is given a dose of oxytocin, she will adopt strange pups as if they were her own.

Oxytocin promotes pair bonding and trust in a variety of animals. In humans, its secretion rises with intimate sexual contact. Not surprisingly, oxytocin has been nicknamed the "cuddle hormone."

THE ANTERIOR PITUITARY Four peptide and protein hormones released by the **anterior pituitary**—*thyrotropin*, *corticotropin* (also known as *adrenocorticotropic hormone*, or *ACTH*), *luteinizing hormone*, and *follicle-stimulating hormone*—are **tropic hormones**, meaning they control the activities of other endocrine glands (**Figure 41.7**).

Each tropic hormone is produced by a different type of pituitary cell. We will say more about the tropic hormones when we describe their target glands (thyroid, adrenal cortex, testes, and ovaries) later in this chapter and in Chapter 42.

Other peptide and protein hormones produced by the anterior pituitary are *growth hormone, prolactin, melanocyte-stimulating hormone, enkephalins,* and *endorphins.*

Growth hormone (GH) acts on a wide variety of tissues to promote growth. One of its important effects is to stimulate cells to take up amino acids. Growth hormone promotes growth also by stimulating the liver to produce chemical messages called *somatomedins* or *insulin-like growth factors (IGFs)*, which stimulate the growth of bone and cartilage. Thus, growth hormone can also be considered in part a tropic hormone because it stimulates liver cells to produce and release hormones.

Overproduction of growth hormone in children causes *gigantism* (individuals may grow to nearly 8 feet tall). Underproduction causes *pituitary dwarfism,* in which individuals fail to reach normal adult height. Beginning in the late 1950s, children with serious growth hormone deficiencies were treated with growth hormone extracted from pituitaries of human cadavers. The treatment was successful in stimulating substantial growth, but a year's supply of the hormone for one individual required up to *50* cadaver pituitaries! In the mid-1980s, scientists using genetic engineering technology isolated the gene for human growth hormone and introduced it into bacteria that could be grown in large quantities, making it possible to purify enough of the hormone to make it widely available.

Prolactin stimulates breast development and the production and secretion of milk in female mammals. In some mammals, prolactin also functions as an important hormone during pregnancy. In human males, prolactin plays a role in controlling the endocrine function of the testes.

Endorphins and **enkephalins** are the body's natural opiates. In the brain, these molecules act as neurotransmitters in pathways that control pain. Their production in the anterior pituitary is normally quite small and probably not significant. They are a byproduct of the production of two other pituitary hormones. One gene encodes a large parent molecule called *pro-opiomelanocortin.* POMC is cleaved to produce several peptides. corticotropin, melanocyte-stimulating hormone, endorphins, and enkephalins all result from the cleavage of POMC.

The anterior pituitary is controlled by hypothalamic hormones

The secretion of hormones by the anterior pituitary is under the control of neurohormones from the hypothalamus. The hypothalamus receives information about conditions in the body and in the external environment through both neuronal signals and hormones that reach it through the circulation. If the connection between the hypothalamus and the pituitary is experimentally cut, pituitary hormones are no longer released in response to changes in the internal or external environment. In experiments in which pituitary cells were maintained in culture, extracts of hypothalamic tissue stimulated some of those cells to release their hormones into the culture medium. Therefore, scientists hypothesized that secretions of the hypothalamic cells control the activities of anterior pituitary cells.

Although hypothalamic neurons do not extend into the anterior pituitary as they do into the posterior pituitary, a special set of **portal blood vessels** connects the hypothalamus and the anterior pituitary (see Figure 41.7). It was thus proposed that secretions from neurons in the hypothalamus enter the blood and are conducted down the portal vessels to the anterior pituitary, where they stimulate the release of anterior pituitary hormones.

In the 1960s, two large teams of scientists, led by Roger Guillemin and Andrew Schally, initiated the search for these hypothalamic secretions. Because the amounts of such neurohormones in any individual mammal would be tiny, massive numbers of hypothalami from pigs and sheep were collected from slaughterhouses and shipped to laboratories in refrigerated trucks. One extraction effort began with the hypothalami from 270,000 sheep

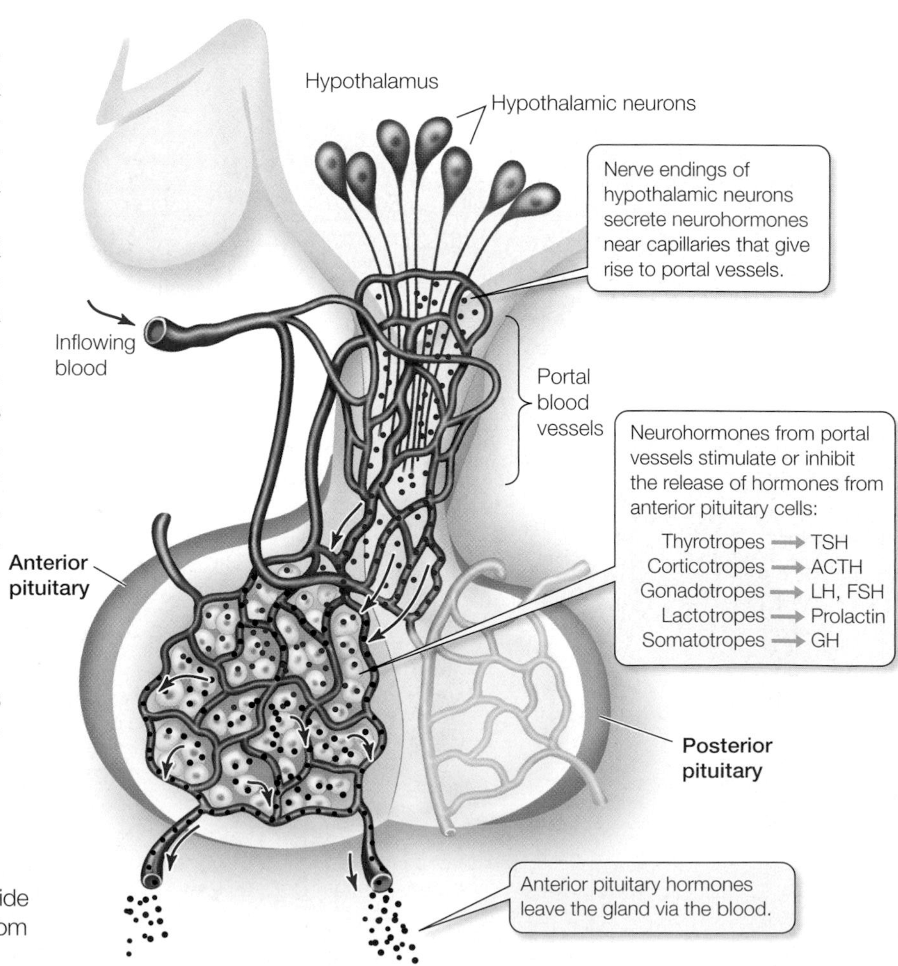
Hypothalamus
Hypothalamic neurons
Nerve endings of hypothalamic neurons secrete neurohormones near capillaries that give rise to portal vessels.
Inflowing blood
Portal blood vessels
Neurohormones from portal vessels stimulate or inhibit the release of hormones from anterior pituitary cells:
Thyrotropes → TSH
Corticotropes → ACTH
Gonadotropes → LH, FSH
Lactotropes → Prolactin
Somatotropes → GH
Anterior pituitary
Posterior pituitary
Anterior pituitary hormones leave the gland via the blood.

41.7 The Anterior Pituitary Produces Many Hormones Cells of the anterior pituitary produce four tropic hormones that control other endocrine glands, and several other peptide hormones. These cells are controlled by neurohormones from the hypothalamus delivered through portal blood vessels.

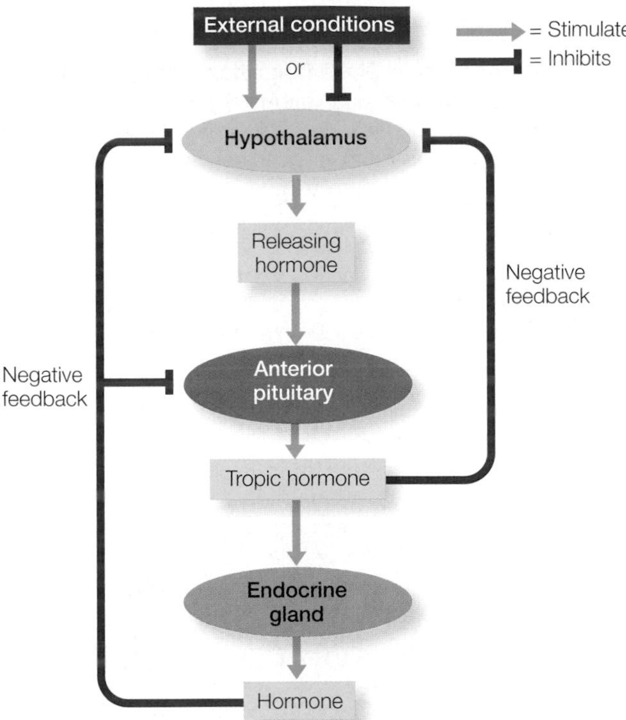

41.8 Multiple Feedback Loops Control Hormone Secretion Multiple negative feedback loops regulate the chain of command from hypothalamus to anterior pituitary to endocrine glands.

and yielded only 1 milligram of purified **thyrotropin-releasing hormone (TRH)**. Biochemical analysis of this pure sample revealed that TRH is a simple tripeptide consisting of glutamine, histidine, and proline. TRH was the first hypothalamic *releasing hormone* (that is, release-stimulating hormone) to be isolated and characterized. It causes certain anterior pituitary cells to release the tropic hormone thyrotropin, which in turn stimulates the activity of the thyroid gland.

Soon after discovering thyrotropin-releasing hormone, Guillemin's and Schally's teams identified **gonadotropin-releasing hormone (GnRH)**, which stimulates certain anterior pituitary cells to release the tropic hormones that control the activity of the gonads (the ovaries and the testes). For these discoveries, Guillemin and Schally received the 1977 Nobel prize in medicine. Many other hypothalamic neurohormones, including both releasing hormones and release-inhibiting hormones, are now known, including:

- Prolactin-releasing and release-inhibiting hormones
- Growth hormone-releasing hormone
- Growth hormone release-inhibiting hormone (somatostatin)
- Adrenocorticotropin-releasing hormone
- Melanocyte-stimulating hormone and release-inhibiting hormone

Negative feedback loops control hormone secretion

As well as being controlled by hypothalamic releasing and release-inhibiting hormones, the endocrine cells of the anterior pituitary are also under direct and indirect negative feedback control by the hormones of the target glands they stimulate (**Figure 41.8**). For example, the hormone cortisol, produced by the adrenal gland in response to corticotropin secreted by the anterior pituitary, returns to the pituitary in the circulating blood and inhibits further release of that tropic hormone. Cortisol also acts as a negative feedback signal to the hypothalamus, inhibiting the release of corticotropin-releasing hormone. In some cases a tropic hormone also exerts negative feedback control on the hypothalamic cells producing the corresponding releasing hormone.

41.2 RECAP

Many interactions between the nervous system and endocrine systems are controlled by the pituitary. The posterior pituitary releases two neurohormones, and the anterior pituitary, under the control of other neurohormones, releases hormones that control other endocrine glands in the body.

- Describe the anatomical and functional relationships between the brain and the two parts of the pituitary. See pp. 881–882 and Figures 41.6 and 41.7
- What are the tropic hormones of the anterior pituitary, and how do they influence endocrine mechanisms? See p. 882 and Figure 41.7

Now that we know some of the mechanisms by which endocrine systems are controlled, we will take a more detailed look at the functions of the major endocrine glands of the body.

41.3 What Are the Major Mammalian Endocrine Glands and Hormones?

Hormones help regulate functions in all physiological systems. In this section we will examine a few major examples of hormonal action in physiological processes, but will add many more in the chapters that follow.

Thyroxine controls cell metabolism

The **thyroid gland** wraps around the front of the windpipe (*trachea*) and expands into a lobe on either side (see Figure 41.5). Two cell types in the thyroid gland produce the hormones thyroxine and calcitonin. Thyroxine is produced, stored, and released by many round structures called *follicles* (**Figure 41.9A,B**). Cells in the spaces between the follicles produce calcitonin.

Thyroxine, which begins as the glycoprotein *thyroglobulin* in the follicle lumen, is built out of two molecules of tyrosine bound to four atoms of iodine.

HO—[ring(I, I)]—O—[ring(I, I)]—CH_2—CH(NH_2)—C(=O)—OH

Thyroxine (T_4)

Thus, thyroxine is called T_4.

(A)

Follicles

Epithelial cells

Calcitonin-producing cells

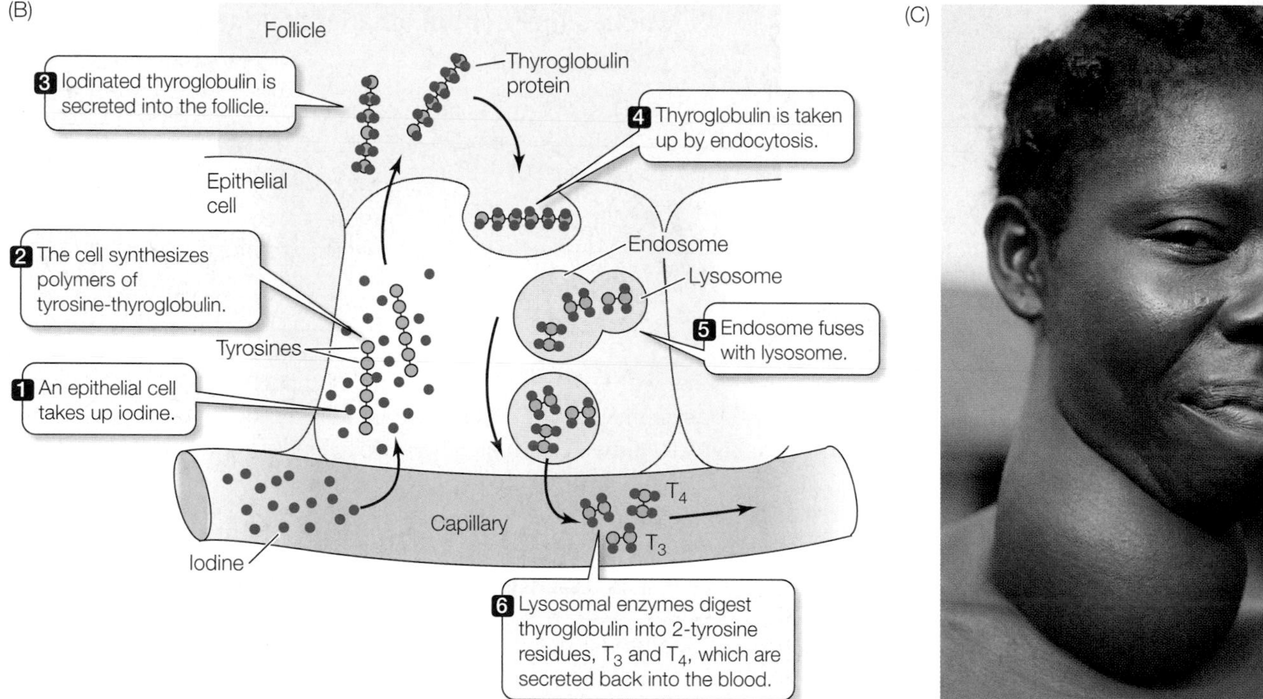
(B)
Follicle
3 Iodinated thyroglobulin is secreted into the follicle.
Thyroglobulin protein
4 Thyroglobulin is taken up by endocytosis.
Epithelial cell
2 The cell synthesizes polymers of tyrosine-thyroglobulin.
Endosome
Lysosome
5 Endosome fuses with lysosome.
Tyrosines
1 An epithelial cell takes up iodine.
Capillary
T4
T3
Iodine
6 Lysosomal enzymes digest thyroglobulin into 2-tyrosine residues, T3 and T4, which are secreted back into the blood.
(C)

41.9 The Thyroid Gland Consists of Many Follicles (A) Cross section through a thyroid gland showing numerous follicles bounded by epithelial cells. Calcitonin-secreting cells are found in the spaces between the follicles. (B) Epithelial cells synthesize and iodinate thyroglobulin. Iodinated thyroglobulin is stored in the follicles until it is processed by epithelial cells to generate T_3 and T_4. (C) Iodine deficiency can result in hypothyroid goiter. In this condition, a lack of functional thyroxine results in oversynthesis of thyroglobulin and subsequent enlarged follicles.

The follicles also produce and release *triiodothyronine*, a version of thyroxine that has only three atoms of iodine and is called T_3:

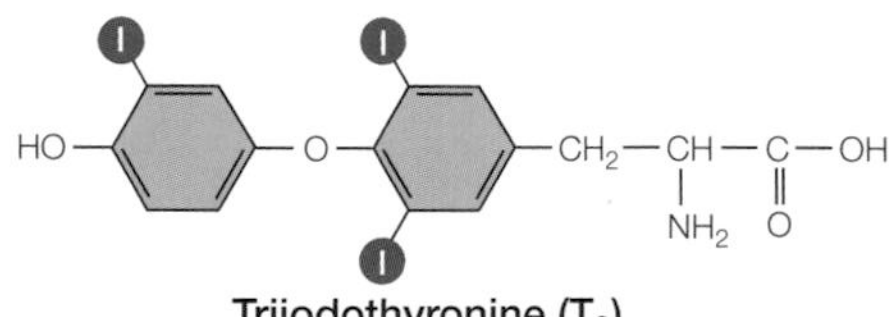
HO
O
CH2—CH—C—OH
NH2
O

Triiodothyronine (T_3)

The thyroid usually releases about four times as much T_4 as T_3. T_3 is the more active hormone in the cells of the body, but when T_4 is in circulation, it can be converted to T_3 by an enzyme within target cells. Each target cell can thus set its own sensitivity to thyroid hormones by controlling the conversion of T_4 to the more active T_3. When you read about thyroxine, keep in mind that the actions discussed are primarily those of T_3.

Thyroxine in mammals plays many roles in regulating cell metabolism. Thyroxine is lipid-soluble, so it enters cells readily where it binds to receptors. The thyroxine–receptor complex acts in the nucleus as a transcription factor that affects the activity of a large number of genes. These genes encode enzymes in energy pathways, transport proteins, and structural proteins. As a result, thyroxine elevates the metabolic rates of most cells and tissues. Exposure to cold for several days leads to an increased release of thyroxine, an increased conversion of T_4 to T_3, and therefore an increased basal metabolic rate (see Section 40.4). Thyroxine is especially crucial during development and growth, as it promotes amino acid uptake and protein synthesis by cells. Insufficient thyroxine in a human fetus or growing child greatly retards physical and mental growth, resulting in a condition known as *cretinism*.

The tropic hormone **thyrotropin**, or **thyroid-stimulating hormone** (**TSH**), produced by the anterior pituitary, activates the thyroxine-producing follicle cells in the thyroid. Thyrotropin-releasing hormone (TRH; discussed earlier in this chapter), produced in the hypothalamus and transported to the anterior pituitary through the portal blood vessels, activates the TSH-producing pituitary cells. The hypothalamus uses environmental information, such as temperature or day length, to determine whether to increase or decrease the secretion of TRH. This sequence of steps is also regulated by a negative feedback loop: circulating thyroxine inhibits the response of pituitary cells to TRH, so less TSH is released when thyroxine levels are high, and more TSH is released when thyroxine levels are low. Circulating thyroxine also exerts negative feedback on the production and release of TRH by the hypothalamus.

Thyroid dysfunction causes goiter

A *goiter* is an enlarged thyroid gland that causes a pronounced bulge on the front and sides of the neck. Goiter results from either **hyperthyroidism** (thyroxine excess) or **hypothyroidism** (thyroxine deficiency). The negative feedback loop whereby thyroxine controls TSH release helps explain how two very different conditions can result in the same symptom, but it is also necessary to understand how the thyroid makes, stores, and releases thyroxine (**Figure 41.9C**).

Each thyroid follicle consists of a layer of epithelial cells surrounding a mass of glycoprotein called **thyroglobulin**. The epithelial cells make the thyroglobulin by creating long polymers of tyrosine residues. As the thyroglobulin is secreted into the lumen of the follicle, iodine is added to the tyrosine residues. When thyroxine is needed, the same epithelial cells that made the thyroglobulin take it back through endocytosis and digest it to the smaller thyroxine molecules.

If there was enough iodine available when the thyroglobulin was made, its digestion releases molecules of T_3 and T_4. If there was not enough iodine available when the thyroglobulin was made, many of the residues released will not be T_3 or T_4 and will not bind to receptors on target cells.

Goiter occurs when the production of thyroglobulin is far above normal and the follicles become greatly enlarged. *Hyperthyroid* goiter results when the negative feedback mechanism fails to turn off the follicle cells even with high blood levels of thyroxine. The most common cause of hyperthyroidism is an autoimmune disease in which an antibody to the TSH receptor is produced. This antibody can bind to the TSH receptor on the follicle cells, causing them to produce and release thyroxine. Even though blood levels of TSH may be quite low because of the negative feedback from high levels of thyroxine, the thyroid remains maximally stimulated and it grows bigger. Hyperthyroid patients have high metabolic rates, are jumpy and nervous, usually feel hot, and may develop a buildup of fat behind the eyeballs, which in turn causes their eyes to bulge.

Hypothyroid goiter results when there is not enough circulating thyroxine to turn off TSH production. The most common cause of this condition is a deficiency of dietary iodide, without which the follicle cells cannot make thyroxine (**Figure 41.9C**). Without sufficient thyroxine, TSH levels remain high, and the thyroid continues to produce large amounts of thyroglobulin. But because insufficient iodide is available, few of the tyrosine residues in the thyroglobulin are iodinated. When this iodine-poor thyroglobulin is digested by the follicle cells, it does not yield functional thyroxine. Without functional thyroxine, the TSH levels remain high and stimulate more and more synthesis of thyroglobulin, and the follicles get bigger. The symptoms of hypothyroidism are low metabolism, intolerance of cold, and general physical and mental sluggishness.

Goiter affects about 5 percent of the world's population. The addition of iodide to table salt has greatly reduced the incidence of hypothyroid goiter in industrialized nations, but the condition is still common in the other parts of the world.

Calcitonin reduces blood calcium

The regulation of calcium levels in the blood is a crucial and difficult task. It is crucial because changes in blood calcium levels can cause serious problems. When blood calcium falls more than 30 percent below normal, the nervous system becomes overly excited, resulting in muscle spasms and even seizures. When blood calcium rises above normal, the nervous system becomes depressed and muscles—including the heart—weaken. Regulation of blood calcium is difficult because only about 0.1 percent of the calcium in the body is located in the extracellular fluids. About 1 percent is within cells, and almost 99 percent is in the bones. Therefore, the body must regulate a tiny pool of calcium in the blood, and that tiny pool can be influenced greatly by relatively small shifts in the much larger pools of calcium in the cells and bones.

There are multiple mechanisms for changing blood calcium levels, including:

- deposition and absorption of bone
- excretion of calcium by the kidneys
- absorption of calcium from the digestive tract

These mechanisms are controlled by the hormones calcitonin, parathyroid hormone, and vitamin D.

Calcitonin lowers the concentration of calcium in the blood (**Figure 41.10**). Bone is continually remodeled through resorption of old bone and synthesis of new bone, as we will see in Section 47.3. Cells called *osteoclasts* break down bone and release calcium; *osteoblasts* take up circulating calcium and deposit new bone. Calcitonin decreases the activity of osteoclasts and thereby shifts the balance of bone turnover to favor removal of calcium from the blood. Because the turnover of bone in adult humans is not very high, calcitonin does not play a major role in calcium homeostasis in adults. It is probably more important in young, growing individuals, but overall calcium levels are more influenced by parathyroid hormone than by calcitonin.

Parathyroid hormone elevates blood calcium

The **parathyroid glands** are four tiny structures embedded in the posterior surface of the thyroid gland. Their single hormone product, **parathyroid hormone** (also called **PTH** or parathormone), is the critical hormone in the regulation of blood calcium levels. Levels of calcium in the blood are sensed by receptors in the plasma membrane of the parathyroid cells. When these receptors are activated, they inhibit the synthesis and release of PTH. A fall in blood calcium removes this inhibition and triggers the synthesis and release of PTH.

PTH raises blood calcium levels in several ways. Its actions provide a good example of the complexity of physiological regulation. One major action of PTH is to stimulate bone turnover and remod-

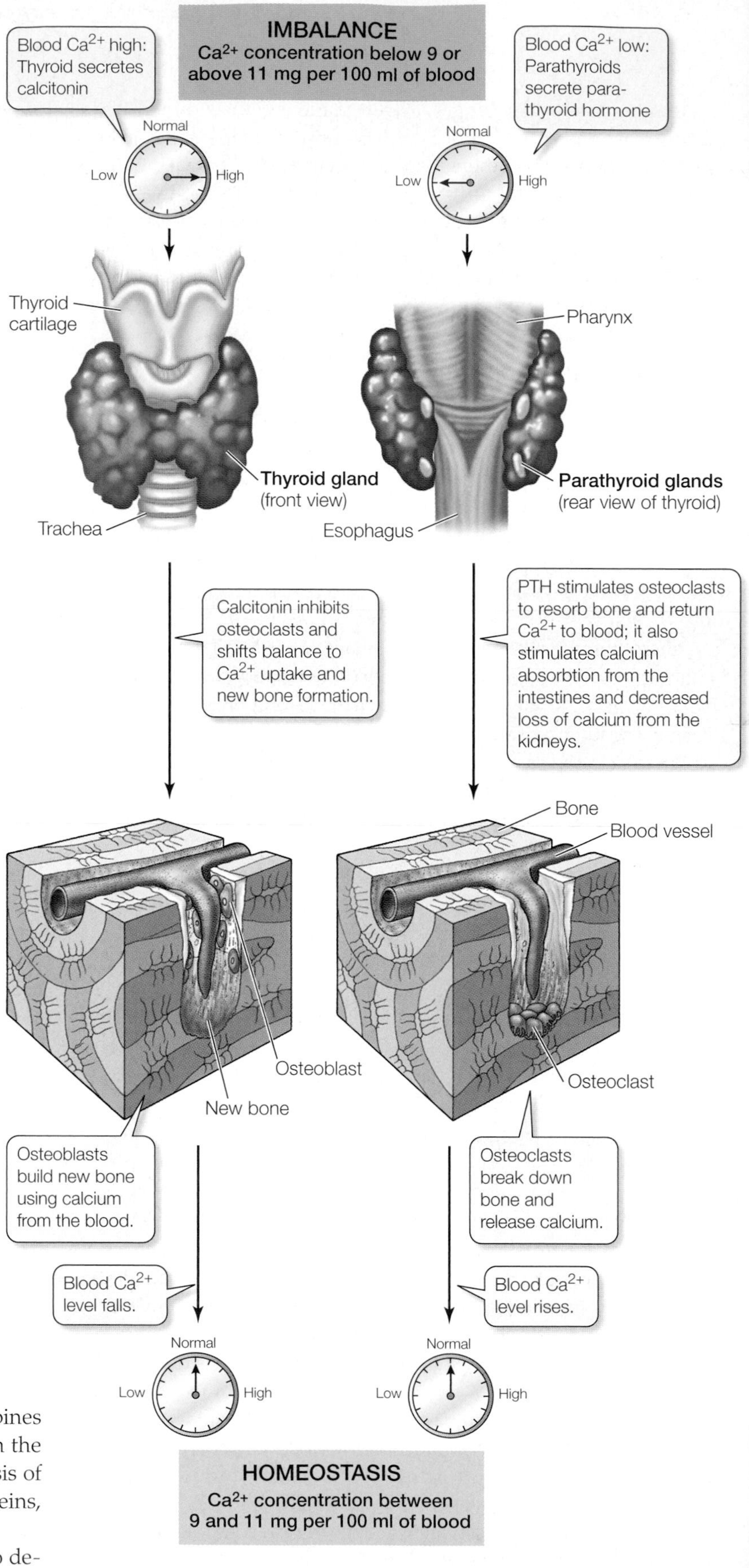

41.10 Hormonal Regulation of Calcium Calcitonin, parathyroid hormone, and vitamin D help regulate calcium levels in the blood.

IMBALANCE
Ca^{2+} concentration below 9 or above 11 mg per 100 ml of blood
Blood Ca^{2+} high: Thyroid secretes calcitonin
Blood Ca^{2+} low: Parathyroids secrete parathyroid hormone
Normal
Low
High
Normal
Low
High
Thyroid cartilage
Pharynx
Thyroid gland (front view)
Parathyroid glands (rear view of thyroid)
Trachea
Esophagus
Calcitonin inhibits osteoclasts and shifts balance to Ca^{2+} uptake and new bone formation.
PTH stimulates osteoclasts to resorb bone and return Ca^{2+} to blood; it also stimulates calcium absorbtion from the intestines and decreased loss of calcium from the kidneys.
Bone
Blood vessel
Osteoblast
New bone
Osteoclast
Osteoblasts build new bone using calcium from the blood.
Osteoclasts break down bone and release calcium.
Blood Ca^{2+} level falls.
Blood Ca^{2+} level rises.
Normal
Low
High
Normal
Low
High
HOMEOSTASIS
Ca^{2+} concentration between 9 and 11 mg per 100 ml of blood

eling. This is a dynamic process that involves both resorption of old bone and laying down of new bone. Osteoblasts have PTH receptors and are stimulated by the hormone to remove calcium from the blood and form new bone. However, stimulated osteoblasts also release paracrine signals—cytokines—that stimulate osteoclasts. This process of bone turnover entails a net loss of calcium from bone. It was previously thought that this indirect path was the only way that PTH influenced bone resorption, but recently investigators have shown that osteoclasts also have PTH receptors and are stimulated by PTH directly (see Figure 41.10). PTH also conserves calcium by stimulating the kidneys to reabsorb it rather than excrete it in the urine. Increased secretion of PTH causes the digestive tract to absorb more calcium from food. This is an indirect effect, however, dependent on vitamin D. PTH activates vitamin D, which in turn causes the digestive tract to enhance absorption of dietary calcium.

Vitamin D is really a hormone

A *vitamin* is a substance that the body needs in small quantities, but cannot synthesize and therefore must obtain from the diet. By this definition, **vitamin D** is not a vitamin, because the body can and does synthesize it.

It had long been known that fragile bones were common among people living at high latitudes where winter days are short and the winter diet often lacks meat, fish, dairy products, and fresh vegetables. Since the condition could be reversed taking cod-liver oil (which, as it turns out, contains large amounts of vitamin D), it was assumed that a dietary vitamin was involved. We now know that vitamin D is synthesized in skin cells, where cholesterol is converted into vitamin D (also called calciferol) by ultraviolet light. Vitamin D circulates in the blood, and acts on distant cells; thus it is actually a hormone.

The vitamin D produced in the skin is not very active, but as it passes through the liver it receives one hydroxyl (—OH) group, and in the kidneys it receives another to become (1,25)-dihydroxyvitamin D, the most active form. PTH stimulates this final step in the kidneys. Active vitamin D is lipid-soluble, so readily enters cells. In cells it combines with a cytoplasmic receptor to form a transcription factor. In the digestive tract, this transcription factor increases the synthesis of calcium pumps, calcium channels, and calcium-binding proteins, all of which promote the uptake of calcium.

In the kidneys, vitamin D acts synergistically with PTH to decrease calcium loss in the urine. In bone, vitamin D, like PTH, stimulates bone turnover and liberates calcium—which seems the opposite of what would be expected. However, through all of its actions, vitamin D raises blood calcium levels, and that is essential to promote bone deposition. Vitamin D also acts on parathyroid cells to inhibit the transcription of the PTH gene, thus forming a negative feedback loop for the regulation of PTH.

PTH lowers blood phosphate levels

Bone minerals are a combination of calcium and phosphate. Thus, when PTH stimulates the release of calcium from bone, it also causes the release of phosphate. Increases in blood levels of both calcium and phosphate can be dangerous. The normal levels of calcium and phosphate in the blood approach the concentration at which they would precipitate out of solution as calcium phosphate salts, leading to maladies such as kidney stones and calcium deposits in the arteries (hardening of the arteries). To reduce this problem, PTH acts on the kidneys to increase the elimination of phosphate via the urine.

Insulin and glucagon regulate blood glucose levels

Before the 1920s, *diabetes mellitus* was a fatal disease, characterized by weakness, lethargy, and a dramatic loss of body mass. The disease was known to be connected somehow with the **pancreas**, a gland located just below the stomach (see Figure 41.5), and with abnormal glucose metabolism, but the link was not clear.

Today we know that diabetes mellitus is caused by a lack of the protein hormone **insulin** (in type I or juvenile-onset diabetes) or by a lack of insulin receptors on the target tissues (in type II or adult-onset diabetes). For patients in which the hormone is lacking, insulin replacement therapy is extremely successful. At present, more than 1.5 million people with diabetes in the United States lead almost normal lives by using manufactured insulin.

Insulin binds to a receptor on the plasma membrane of a target cell, and this insulin–receptor complex allows glucose to enter the cell (see Figure 15.6). In the absence of insulin or insulin receptors, glucose entry into cells is impaired and glucose accumulates in the blood until it is lost in the urine. High levels of blood glucose cause water to move from cells into the blood by osmosis, and the kidneys increase urine output to remove this excess fluid volume from the blood. Because glucose uptake by most cells is impaired without insulin, those cells must use fat and protein for fuel instead of glucose. As a result, the body of the untreated diabetic wastes away, and critical tissues and organs are damaged.

For centuries the prospects for diabetics were bleak. A change came almost overnight in 1921, when the physician Frederick Banting and a medical student, Charles Best of the University of Toronto, discovered that they could reduce the symptoms of diabetes by injecting an extract prepared from pancreatic tissue. The active component of this extract was found to be a small protein hormone—insulin—consisting of just 51 amino acids.

Banting was awarded a Nobel prize for the discovery of insulin, but his co-worker, Charles Best, was not nominated. The rules of the Nobel Committee precluded awards to medical students. Banting protested and shared both the award and the credit with Best.

Insulin is produced in clusters of endocrine cells in the pancreas. These clusters are called **islets of Langerhans** after the German medical student who discovered them. There are three types of cells in the islets:

- Beta (β) cells produce and secrete insulin.
- Alpha (α) cells produce and secrete the hormone glucagon, which has effects opposite from those of insulin.
- Delta (δ) cells produce the hormone somatostatin.

The rest of the pancreas is an exocrine gland producing enzymes and other secretions that travel through ducts to the intestine to participate in digestion.

After a meal, the concentration of glucose in the blood rises as glucose is absorbed from the food in the gut. This increase stimulates the β cells of the pancreas to release insulin. Insulin stimulates cells to use glucose as fuel and to convert it into storage products, such as glycogen and fat. When the gut contains no more food, the glucose concentration in the blood falls, and the pancreas stops releasing insulin. As a result, most cells of the body shift to using glycogen and fat, rather than glucose, for fuel. If the concentration of glucose in the blood falls substantially below normal, the islet α cells release **glucagon**, which stimulates the liver to convert glycogen back to glucose to resupply the blood. These actions will be discussed in greater detail in Section 50.4.

Somatostatin is a hormone of the brain and the gut

Somatostatin is released from the cells of the pancreas in response to rapid rises of glucose and amino acids in the blood. This hormone has paracrine functions within the islets: it inhibits the release of both insulin and glucagon. Its actions outside the pancreas slow the digestive activities of the gut. Pancreatic somatostatin extends the period of time during which nutrients are absorbed from the gut. Somatostatin is also produced in very small amounts by cells in the hypothalamus. Acting as a neurohormone, hypothalamic somatostatin is transported in the portal vessels to the anterior pituitary, where it inhibits the release of growth hormone and thyrotropin.

The adrenal gland is two glands in one

An **adrenal gland** sits above each kidney, just below the middle of your back. Functionally and anatomically, each adrenal gland consists of a gland within a gland (**Figure 41.11**). The core, called the **adrenal medulla**, produces the hormone **epinephrine** (also known as *adrenaline*) and, to a lesser degree, **norepinephrine** (or *noradrenaline*), which also acts as a neurotransmitter in the nervous system. Surrounding the medulla is the **adrenal cortex**, which produces steroid hormones. The medulla develops from nervous tissue and is under the control of the nervous system; the cortex is under hormonal control, largely by **corticotropin** (**ACTH**) from the anterior pituitary.

THE ADRENAL MEDULLA The adrenal medulla produces epinephrine and norepinephrine in response to stressful situations, arousing the body to action. As we saw earlier in this chapter, epinephrine increases heart rate and blood pressure and diverts blood flow to active muscles and away from the gut.

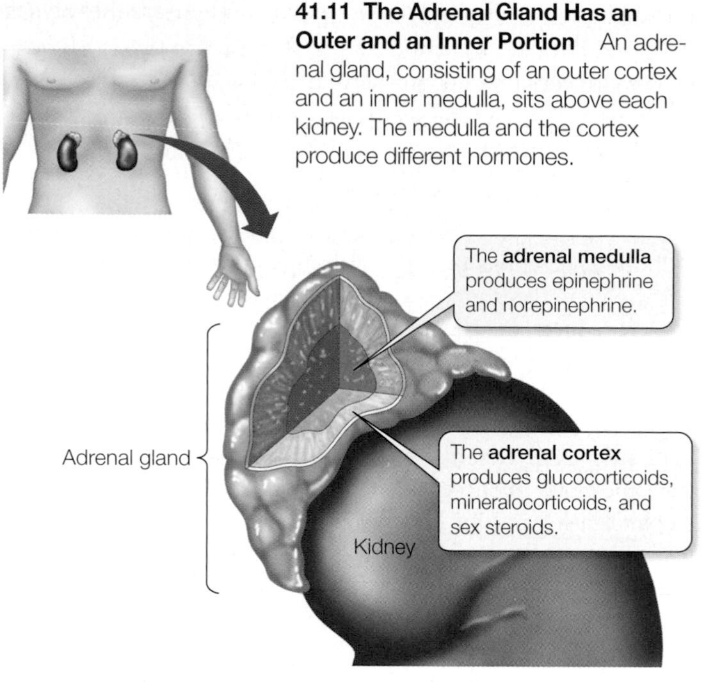

41.11 The Adrenal Gland Has an Outer and an Inner Portion An adrenal gland, consisting of an outer cortex and an inner medulla, sits above each kidney. The medulla and the cortex produce different hormones.

Epinephrine and norepinephrine are both amine hormones—derivatives of the amino acid tyrosine. They are water-soluble, and both bind to the same receptors on the surfaces of target cells. These receptors can be grouped into two general types, *α-adrenergic* and *β-adrenergic receptors*, which stimulate different actions within cells (see Figure 41.17 later in this chapter). Epinephrine acts equally on both types, but norepinephrine acts mostly on α-adrenergic receptors. Therefore, drugs called *beta blockers*, so named because they selectively block β-adrenergic receptors, can reduce the fight-or-flight responses to epinephrine without disrupting the physiological regulatory functions of norepinephrine. Beta blockers are commonly prescribed to reduce the symptoms of anxiety, such as dry mouth and elevated heart rate (heart palpitations).

THE ADRENAL CORTEX The cells of the adrenal cortex use cholesterol to produce three classes of steroid hormones, collectively called **corticosteroids**:

- The *glucocorticoids* influence blood glucose concentrations as well as other aspects of fat, protein, and carbohydrate metabolism.
- The *mineralocorticoids* influence the ionic balance of extracellular fluids.
- The *sex steroids* play roles in sexual development, sex drive, and anabolism.

The adult adrenal cortex secretes sex steroids in only negligible amounts. The major producers of sex steroids are the gonads, as we will see in the following section.

Aldosterone, the main mineralocorticoid (**Figure 41.12A**), stimulates the kidneys to conserve sodium and to excrete potassium. If the adrenal glands are removed from an animal, sodium must be added to its diet, or its sodium will be depleted and it will die.

The main glucocorticoid in humans is **cortisol** (**Figure 41.12B**), which is critical for mediating the body's response to stress. Within minutes of a stressful stimulus (one provoking fear or anger, for example), your blood cortisol level rises. Cells not critical for your action (fight or flight) are stimulated by cortisol to decrease their use of blood glucose and shift instead to utilizing fats and proteins for energy. This is no time to feel sick, have allergic reactions, or heal wounds, so cortisol also blocks immune system reactions. That is why cortisol or drugs that mimic cortisol action are useful for reducing inflammation and allergies.

Cortisol release is controlled by corticotropin from the anterior pituitary, which in turn is controlled by **corticotropin-releasing hormone** from the hypothalamus. Because the cortisol response to a stressor has this chain of steps, each involving secretion, diffusion, circulation, and cell activation, it is much slower than the epinephrine response. Also, many of the actions of cortisol involve changes in gene expression, and that takes time.

Turning off the cortisol response is as important as turning it on. A study of stress in rats showed that old rats could turn on their stress responses as effectively as young rats, but they had lost the ability to turn them off as rapidly. As a result, they suffered from the well-known consequences of stress seen in humans: ulcers, cardiovascular problems, strokes, impaired immune system function, and increased susceptibility to cancers and other diseases. Further research showed that the stress responses are turned off

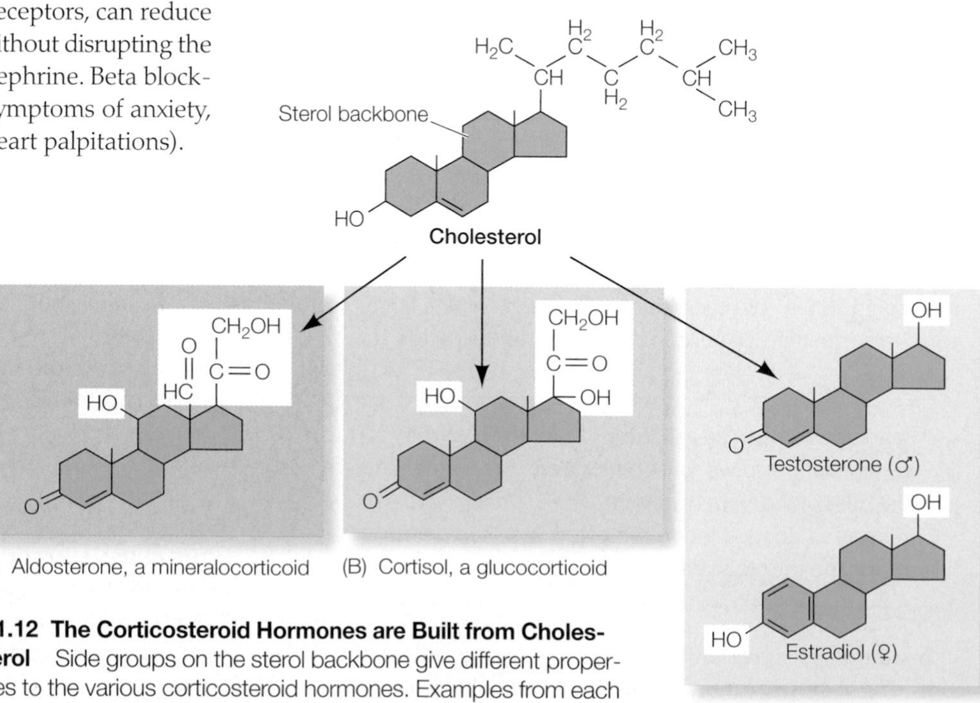

41.12 The Corticosteroid Hormones are Built from Cholesterol Side groups on the sterol backbone give different properties to the various corticosteroid hormones. Examples from each of the three classes of these hormones are shown here.

by the negative feedback action of cortisol on cells in the brain, which causes a decrease in the release of corticotropin-releasing hormone (see Figure 41.8). Repeated activation of this negative feedback mechanism, either through repeated stress or through prolonged medical use of cortisol, leads to a gradual loss of cortisol-sensitive cells in the brain, and therefore to a decreased ability to terminate stress responses.

The sex steroids are produced by the gonads

The **gonads**—the testes of the male and the ovaries of the female—produce hormones as well as sperm and ova. The male steroids are collectively called **androgens**, and the dominant one is *testosterone*. The female steroids are **estrogens** and **progesterone**. The dominant estrogen is *estradiol*, which is made from testosterone. Thus, males and females both synthesize testosterone, but females have an enzyme (aromatase) that converts testosterone to estradiol (**Figure 41.12C**).

The sex steroids have important developmental effects; they determine whether a fetus develops into a female or a male. (A *fetus* is the latter stage of a mammalian embryo; a human embryo is called a fetus from the eighth week of pregnancy to the moment of birth.) After birth, the sex steroids control the maturation of the reproductive organs and the development and maintenance of secondary sexual characteristics, such as breasts and facial hair.

The sex steroids begin to exert their effects in the human embryo in the seventh week of development. Until that time, the embryo has the potential to develop into either sex. In mammals and birds, the instructions for sex determination reside in the genes. In mammals, individuals that receive two X chromosomes normally become females, and individuals that receive an X and a Y chromosome normally become males (see Section 10.4).

These genetic instructions are carried out through the production and action of the sex steroids. In humans, the presence of a Y chromosome normally causes the undifferentiated embryonic gonads to begin producing androgens in the seventh week. In response to the androgens, the reproductive system develops into that of a male. If androgens are not produced at that time, female reproductive structures develop (**Figure 41.13**). In other words, androgens are required to trigger male development in humans, and the default condition is female. The opposite situation exists in birds; male characteristics develop unless estrogens are present to trigger female development.

Changes in control of sex steroid production initiate puberty

Sex steroids have dramatic effects at **puberty**—the time of sexual maturation in humans. Sex steroids are produced at low levels by the juvenile gonads, but their production increases rapidly at the beginning of puberty—around the age of 12 to 13 years. Why does this sudden increase occur?

In the juvenile, as in the adult, the production of sex steroids by the ovaries and testes is controlled by the anterior pituitary tropic hormones **luteinizing hormone** (**LH**) and **follicle-stimulating hormone** (**FSH**), which together are called the **gonadotropins**. The production of these tropic hormones is under the control of the hypothalamic gonadotropin-releasing hormone (GnRH). Before puberty, the gonads can respond to gonadotropins and the pituitary can re-

41.13 The Development of Human Sex Organs The sex organs of early human embryos are similar. Male sex steroids (androgens) promote the development of male sex organs. Without androgen action, female sex organs form, even in genetic males.

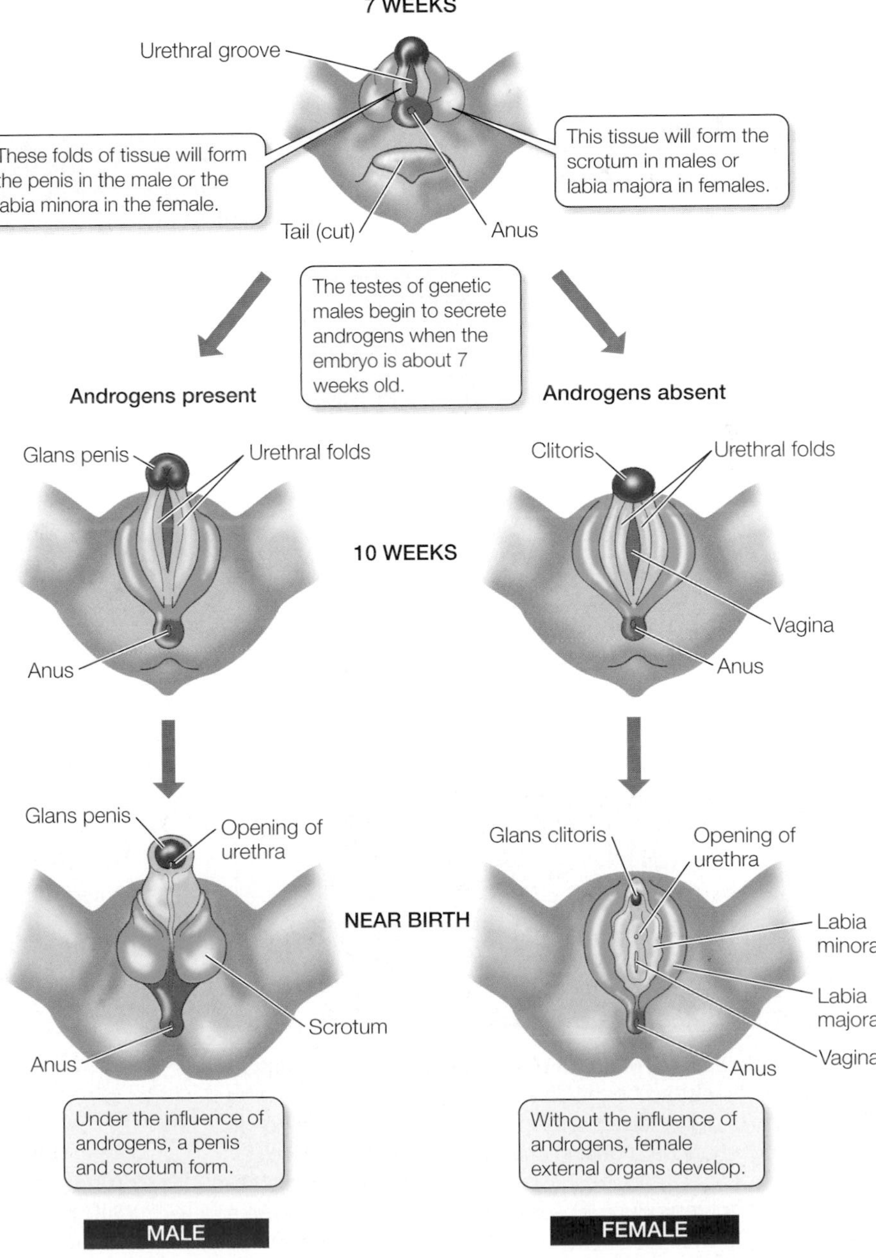
7 WEEKS
Urethral groove
These folds of tissue will form the penis in the male or the labia minora in the female.
This tissue will form the scrotum in males or labia majora in females.
Tail (cut)
Anus
The testes of genetic males begin to secrete androgens when the embryo is about 7 weeks old.
Androgens present
Androgens absent
Glans penis
Urethral folds
Clitoris
Urethral folds
10 WEEKS
Vagina
Anus
Anus
Glans penis
Opening of urethra
Glans clitoris
Opening of urethra
NEAR BIRTH
Labia minora
Labia majora
Vagina
Scrotum
Anus
Anus
Under the influence of androgens, a penis and scrotum form.
Without the influence of androgens, female external organs develop.
MALE
FEMALE

spond to GnRH, but the hypothalamus produces only very low levels of GnRH. Puberty is initiated by a reduction in the sensitivity of hypothalamic GnRH-producing cells to negative feedback from sex steroids and from gonadotropins. As a result, GnRH release increases, stimulating increased production of gonadotropins and hence increased production of sex steroids.

In females, increasing levels of LH and FSH at puberty stimulate the ovaries to begin producing the female sex hormones. The increased circulating levels of these hormones initiate the development of the traits of a sexually mature woman: enlarged breasts, vagina, and uterus; broadened hips; increased subcutaneous fat; growth of pubic hair; and the initiation of the menstrual cycle.

In males, an increasing level of LH stimulates groups of cells in the testes to synthesize testosterone, which in turn initiates the physiological, anatomical, and psychological changes associated with adolescence. The voice deepens, hair begins to grow on the face and body, and the testes and penis grow. As we saw at the beginning of the chapter, androgens also help bones and skeletal muscles grow.

Melatonin is involved in biological rhythms and photoperiodicity

The **pineal gland** is situated between the two hemispheres of the brain and is connected to the brain by a little stalk. It produces the amine hormone **melatonin** from the amino acid tryptophan.

(A)

Plasma melatonin (═)
Light | Dark | Light — Winter (long nights)
Winter
Summer
Light | Dark | Light — Summer (short nights)
Time of day

(B)

41.14 The Release of Melatonin Regulates Seasonal Changes (A) Melatonin is released in the dark and is inhibited by light exposure. The duration of daily melatonin release thus changes as day length (photoperiod) changes, inducing dramatic seasonal physiological changes in some animals. (B) In winter, these Siberian hamsters are white and do not reproduce. In summer, they are mottled brown and breed.

The release of melatonin by the pineal gland occurs in the dark and therefore marks the length of the night. Exposure to light inhibits the release of melatonin.

In vertebrates, melatonin is involved in biological rhythms, including **photoperiodicity**—the phenomenon whereby seasonal changes in day length cause physiological changes in animals. Many species, for example, come into reproductive condition when the days begin to lengthen (**Figure 41.14**). Humans are not strongly photoperiodic, but melatonin in humans may play a role in entraining daily biological rhythms to the daily cycle of light and dark.

The list of hormones is long

We have discussed the major endocrine glands and their hormones in this chapter, but many more hormones exist. As we discuss the organ systems of the body in the chapters that follow, we will frequently describe hormones that their tissues produce as well as hormones that control their functions.

41.3 RECAP

In mammals, the major endocrine glands include the hypothalamus, pituitary, thyroid, parathyroid, pancreas, adrenals, gonads, and pineal gland. Each of these glands secrete, and respond to, hormones that play crucial roles in physiology and development.

- Can you describe how thyroxine is produced and how its production and release are controlled? See p. 884 and Figure 41.9
- How is blood calcium level regulated? See pp. 885–886 and Figure 41.10
- What are the hormonal bases for the two forms of diabetes? See p. 887
- Can you describe the changes in the feedback control of sex steroids that result in puberty? See pp. 889–890

Studies of endocrine systems are not easy. Many hormones are released in very small quantities, and some disappear from the extracellular fluids rapidly. A hormone's receptors may be found on diverse cells around the body, and those different cells can respond in different ways to the same hormone. How have we overcome these difficulties to learning how hormones work?

41.4 How Do We Study Mechanisms of Hormone Action?

We can break the study of hormone actions into different sets of problems. First, we must be able to detect, identify, and measure hormones. Second, we must be able to identify and characterize the receptors for the hormones. Third, we must understand the signal transduction pathways activated by hormones in different tissues.

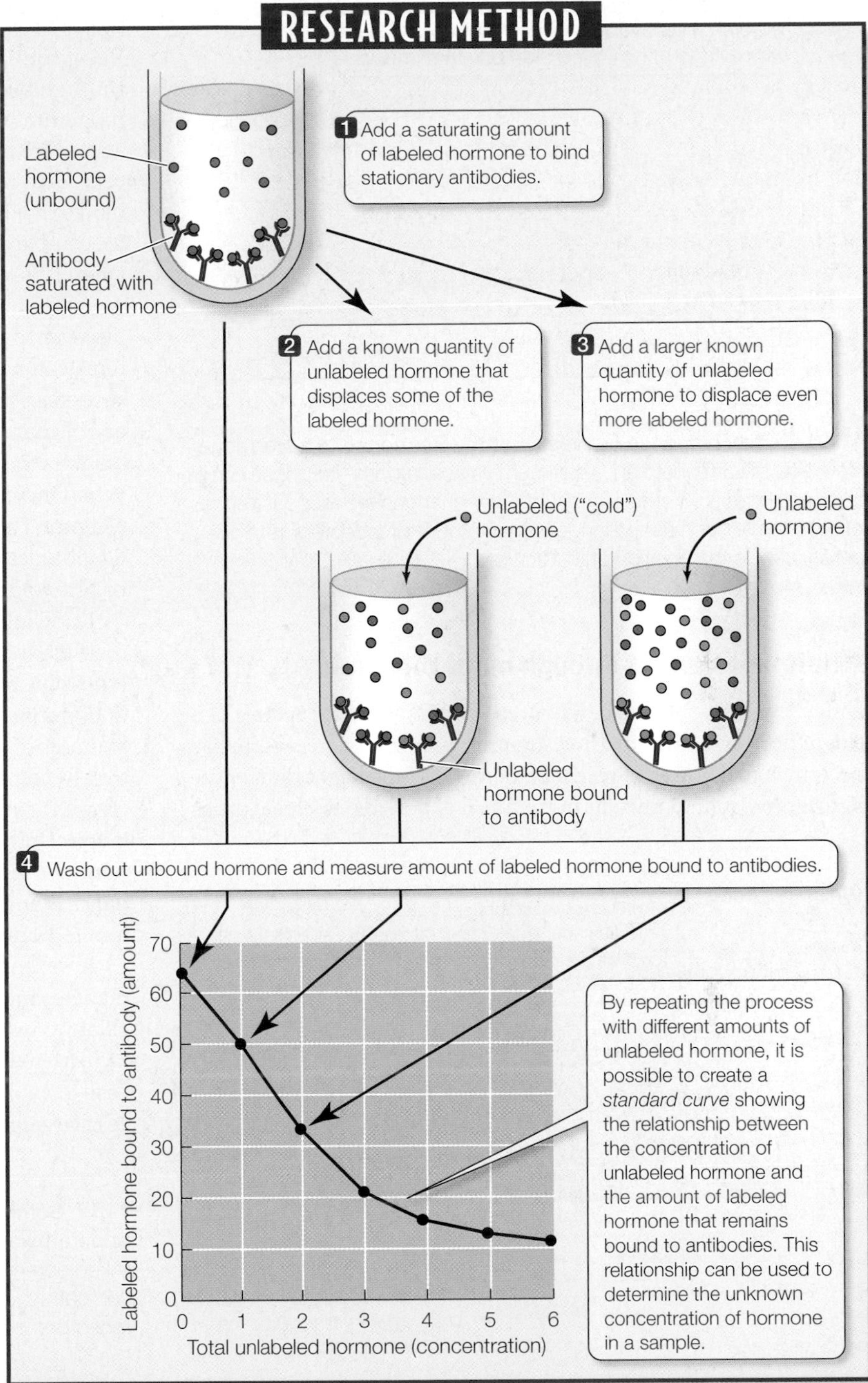
RESEARCH METHOD
Labeled hormone (unbound)
Antibody saturated with labeled hormone
1 Add a saturating amount of labeled hormone to bind stationary antibodies.
2 Add a known quantity of unlabeled hormone that displaces some of the labeled hormone.
3 Add a larger known quantity of unlabeled hormone to displace even more labeled hormone.
Unlabeled ("cold") hormone
Unlabeled hormone
Unlabeled hormone bound to antibody
4 Wash out unbound hormone and measure amount of labeled hormone bound to antibodies.
Labeled hormone bound to antibody (amount)
Total unlabeled hormone (concentration)
0
10
20
30
40
50
60
70
0 1 2 3 4 5 6
By repeating the process with different amounts of unlabeled hormone, it is possible to create a *standard curve* showing the relationship between the concentration of unlabeled hormone and the amount of labeled hormone that remains bound to antibodies. This relationship can be used to determine the unknown concentration of hormone in a sample.

41.15 An Immunoassay Measures Hormone Concentration To develop an immunoassay, the immune response of an animal is used to produce antibodies to the hormone to be measured. A purified sample of the hormone is labeled in some way. A known amount of the antibody is mixed with enough labeled hormone to saturate all of the antibody-binding sites (step 1), and the excess is washed away. When a sample of unlabeled ("cold") hormone is added to the mix, it displaces some of the labeled hormone. The decrease in the amount of label is a measure of the amount of cold hormone added (step 2). By repeating the process using larger known amounts of cold hormone (step 3), a standard curve is created that can be used to determine the amount of hormone in an unknown sample (step 4).

Hormones can be detected and measured with immunoassays

As we have seen, testosterone has many dramatic and diverse effects, yet its concentration in the blood of adult human males is only about 30 to 100×10^{-9} g/ml; that is 30 to 100 billionths of a gram per milliliter. To measure hypothalamic releasing hormones requires calibrations in the range of *trillionths* of a gram per milliliter.

The ability to detect and measure infinitesimal quantities of hormone was an important breakthrough Rosalyn Yalow developed a method called *radioimmunoassay* because it used radioactive labels (she used radioactive isotopes of iodine) to follow interactions between antigens (the hormone to be measured) and antibodies made to that antigen. Today we are more likely to use nonradioactive labels, so we the technique is called simply **immunoassay** (**Figure 41.15**). Being able to measure hormones in the blood made it possible to study many important hormonal mechanisms, and in 1977 Yalow shared the Nobel prize with Guillemin and Schally, who used her technique in their work on hypothalamic neurohormones (see p. 883).

Hormones are not simple on–off switches, and an important characteristic of a hormone is the time course over which it acts. This time course can be measured by the hormone's *half-life* in the blood (defined as the length of time it takes for one-half of the hormone molecules to be depleted. Soon after endocrine cells are stimulated to secrete their hormone, the hormone reaches its maximum concentration in the blood. By taking subsequent blood samples, researchers can determine how long it takes for the circulating hormone to drop to half of that maximum concentration. The fight-or-flight response to epinephrine, for example, is relatively quick in its onset and termination; the half-life of epinephrine in the blood is only 1–3 minutes. The effects of other hormones, such as cortisol and thyroxine, are expressed over much longer periods, and their half-lives are on the order of days or weeks.

Immunoassays have also facilitated the measurement of dose-response curves. Whether one is studying a drug or a natural hormone for therapeutic use, it is critical to know the sensitivity of the body to the drug or hormone. Drugs may stimulate or block release of hormones. Being able to measure the concentrations of the hormones or other molecules in the blood makes it possible to construct dose–response curves that help physicians adjust dosages appropriately. (**Figure 41.16**).

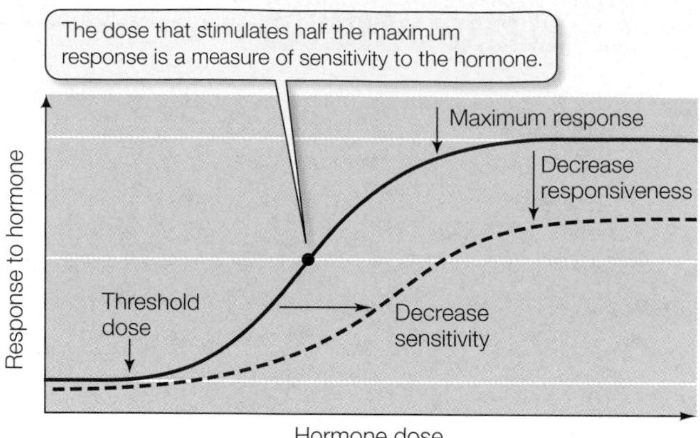

41.16 Dose–Response Curves Quantify Response to a Hormone Between the threshold and maximum values, a dose–response curve frequently has an S shape. Anything that changes the responsiveness of a system—such as a change in the number of receptors in target cells—affects the position of the curve.

A hormone can act through many receptors

Different receptors may be involved in mediating the actions of a single hormone, thus the investigation of hormone receptors has become a major area of research. Because of the slight differences in receptors for a particular hormone, it is possible to create drugs that are very selective in blocking or stimulating specific responses. Receptors have mostly been identified, isolated, and purified through biochemical separation techniques. For example, a hormone can be bound to a substrate such as resin beads packed into a glass column. When an extract of cells suspected of containing receptors to that hormone is run through the column, the receptors bind to the hormone molecules on the beads. The hormone–receptor complexes can be washed off of the beads and the receptors isolated. This technique is called **affinity chromatography**.

As more receptors are isolated and characterized, researchers discover that they frequently exist in families with common structural features. The common features result from common nucleotide sequences in their genes. Genomic analyses have led to the discovery of many receptors. Investigators "scan" the genome for sequences that bear homologies to known receptor gene sequences. When they get a "hit," they have found a candidate gene for a new receptor. They can then identify the molecule to which it binds (its ligand), describe its localization within the body, and characterize its physiological effects.

Knowing the molecular identity of receptors and being able to measure their concentration with immunoassay procedures makes it possible to study their regulation. We saw above that the release of hormones can be under negative feedback control. Similarly, the abundance of receptors for a hormone can be under feedback control. In some cases, continuous high levels of a hormone can decrease the number of its receptors, a process known as **downregulation**. **Upregulation** of receptors can occur when the levels of hormone secretion are suppressed. The regulation of receptor abundance is an important mechanism controlling sensitivity of the system to hormonal signaling.

An example of downregulation of a receptor occurs in type II diabetes mellitus, which is characterized by elevated levels of circulating insulin but a loss of insulin receptors. Although genetic factors are likely involved, a possible immediate cause of the disease is an overstimulation of pancreatic release of insulin by excessive carbohydrate intake, which leads to downregulation of the

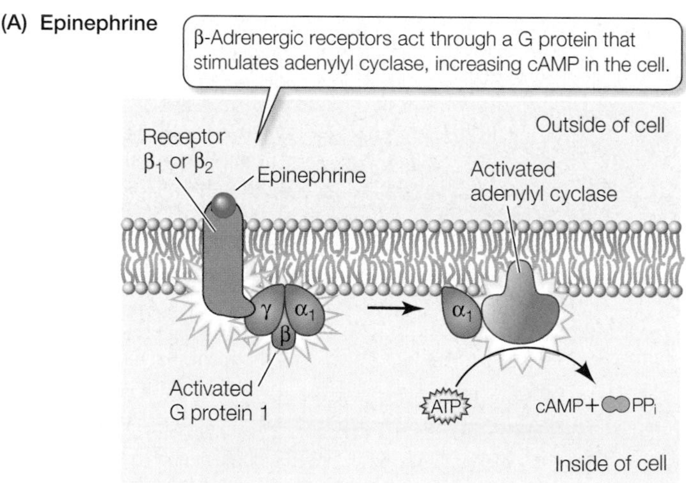

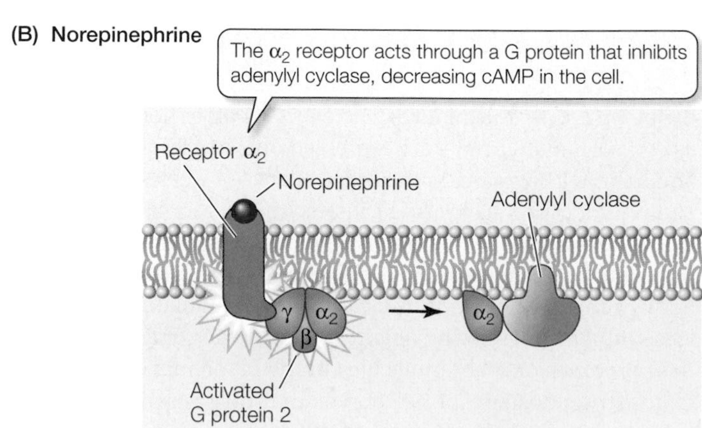

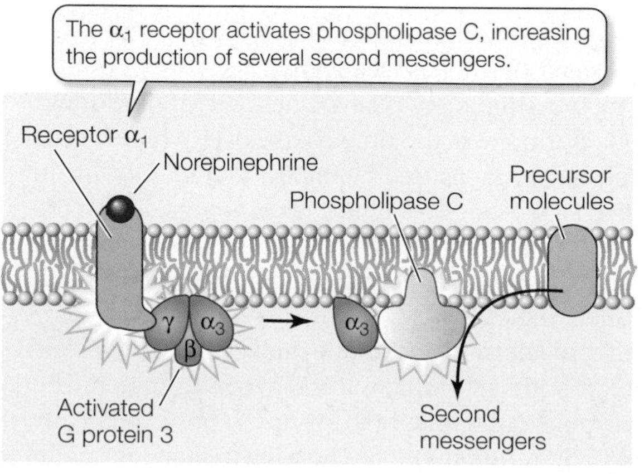

41.17 Some Hormones Can Activate a Variety of Signal Transduction Pathways Epinephrine and norepinephrine bind to G protein-linked adrenergic receptors that act through different signal transduction pathways. Epinephrine acts equally on both α- and β-adrenergic receptors; norepinephrine acts mostly on α-adrenergic receptors

insulin receptors. An example of upregulation occurs when someone has been on a regular dose of beta blockers (see page 888). As the activity of the beta receptors is blocked over time, more receptors are produced. If the person goes off the medication suddenly, the beta receptor effects are amplified, resulting in heightened anxiety. Thus dosage changes in the long-term use of such medications are usually gradual and carefully supervised.

A hormone can act through different signal transduction pathways

A hormone may affect only certain cells, and its effects can differ greatly between different cell types. The cell- or tissue-specific nature of hormone action is due to the fact that only cells with appropriate receptors can respond to the hormone. The diversity of actions of a hormone in different cells, however, arises because receptors can be linked to different signal transduction pathways. For example, the two types of receptors mentioned above for epinephrine and norepinephrine (the α-adrenergic and the β-adrenergic receptors) are both cell-surface G protein-linked receptors, but they connect with different signal transduction pathways within cells (**Figure 41.17**). Therefore, the two hormones can elicit different responses, even in the same cells.

Signal transduction pathways also partly explain why hormones secreted in such tiny quantities can have huge physiological effects. Many of these pathways involve signaling "cascades" in which each step amplifies the signal. For example, the binding of the hormone to its receptor might activate an enzyme (such as adenylyl cyclase), which produces multiple second messenger molecules (cAMP). Each of those may in turn activate another enzyme—a kinase—which can phosphorylate and thereby activate a large number of molecules of another enzyme, each of which in turn will catalyze the production of a final product molecule. Thus, a single event of a hormone binding to its receptor might result in a million or more molecules of a final product. An example is the response of liver cells to epinephrine (see Figure 15.18).

41.4 RECAP

Studies of mechanisms of hormone action involve being able to measure hormone concentrations, identify and characterize hormone receptors, and investigate signal transduction pathways.

- Can you describe what an immunoassay is? See p. 891 and Figure 41.15
- How can you identify receptors to a particular hormone? See p. 892
- Explain how the same hormone can induce different responses in different tissues. See p. 892 and Figure 41.17

CHAPTER SUMMARY

41.1 What are hormones and how do they work?

Endocrine cells secrete chemical messages called **hormones**, which bind to receptors on or in **target cells**. In some cases endocrine cells are aggregated into endocrine glands.

Paracrine hormones diffuse to targets near the site of secretion. **Autocrine** hormones influence the cell that secretes them. Most hormones are delivered to target cells by the circulatory system. Review Figure 41.1

Two diffusible substances, prothoracicotropic hormone and ecdysone, control molting in insects. A third hormone, juvenile hormone, prevents maturation. When an insect stops producing juvenile hormone, it molts into an adult. Review Figures 41.2 and 41.3, Web/CD Tutorial 41.1

Most hormones are either peptides, proteins, steroids, or amines. Peptide and protein hormones and some amines are water-soluble; steroids and some amines are lipid-soluble.

Receptors for water-soluble hormones are on the cell surface. Receptors for lipid-soluble hormones are inside the cell.

Hormones cause different responses in different target cells. Review Figure 41.4

41.2 How do the nervous and endocrine systems interact?

Some vertebrate hormones are released by discrete endocrine glands. Many other hormones are produced and released by endocrine cells incorporated into organs of the body. Review Figure 41.5, Web/CD Activity 41.1

The **pituitary** gland is an interface between the brain and endocrine system. The anterior pituitary develops from embryonic mouth tissue; the posterior pituitary develops from the brain.

The posterior pituitary secretes two **neurohormones**: **antidiuretic hormone** and **oxytocin**. The anterior pituitary secretes **tropic hormones** (**thyrotropin**, **corticotropin**, **luteinizing hormone**, and **follicle-stimulating hormone**) as well as **growth hormone**, **prolactin**, **melanocyte-stimulating hormone**, **endorphins**, and **enkephalins**.

The anterior pituitary is controlled by neurohormones produced by cells in the **hypothalamus** and transported through **portal blood vessels** to the anterior pituitary. Review Figures 41.6 and 41.7, Web/CD Tutorial 41.2

Hormone release in the hypothalamus–pituitary–endocrine gland system is controlled by negative feedback. Review Figure 41.8

41.3 What are the major mammalian endocrine glands and hormones?

The thyroid gland is controlled by **thyrotropin** and secretes **thyroxine**, which controls cell metabolism. Review Figure 41.9

The level of calcium in the blood is regulated by three hormones. **Calcitonin** from the thyroid lowers blood calcium by promoting bone deposition. **Parathyroid hormone** raises blood calcium by promoting bone turnover and decreased calcium excretion. **Vitamin D** promotes calcium absorption from the digestive tract. Review Figure 41.10, Web/CD Tutorial 41.3

CHAPTER SUMMARY

The pancreas secretes three hormones. **Insulin** stimulates glucose uptake by cells and lowers blood glucose, **glucagon** raises blood glucose, and **somatostatin** slows the rate of nutrient processing.

The **adrenal gland** has two portions, one within the other. The hormones of the adrenal medulla, epinephrine and norepinephrine, cause fight-or-flight responses such as stimulating the liver to supply glucose to the blood. Review Figure 41.11

The adrenal cortex produces three classes of corticosteroids: glucocorticoids, mineralocorticoids, and small amounts of sex steroids. Review Figure 41.12

Aldosterone is a mineralocorticoid that stimulates the kidney to conserve sodium and excrete potassium. **Cortisol** is a glucocorticoid that decreases glucose utilization by most cells.

Sex hormones (androgens in males, estrogens and progesterone in females) are produced by the gonads in response to tropic hormones. Sex hormones control sexual development, secondary sexual characteristics, and reproductive functions. Review Figure 41.13

The pineal hormone **melatonin** is involved in controlling biological rhythms and photoperiodism. Review Figure 41.14

41.4 How do we study mechanisms of hormone action?

Immunoassays are used to measure concentrations of hormones and receptors. Review Figure 41.15

The sensitivity of a cell to hormones can be altered by **up- or down-regulation** of the receptors in that cell.

The response of a cell to a hormone depends on its receptors and the signal transduction pathways those receptors activate. Review Figure 41.17

See Web/CD Activity 41.2 for a concept review of this chapter.

SELF-QUIZ

1. Before puberty
   a. the pituitary secretes luteinizing hormone and follicle-stimulating hormone, but the gonads are unresponsive.
   b. the hypothalamus does not secrete much gonadotropin-releasing hormone.
   c. males can stimulate massive muscle development through a vigorous training program.
   d. testosterone plays no role in development of the male sex organs.
   e. genetic females will develop male genitals unless estrogen is present.
2. Both epinephrine and cortisol are secreted in response to stress. Which of the following statements is also true for *both* of these hormones?
   a. They act to increase blood glucose availability.
   b. Their receptors are on the surfaces of target cells.
   c. They are secreted by the adrenal cortex.
   d. Their secretion is stimulated by corticotropin.
   e. They are secreted into the blood within seconds of the onset of stress.
3. Growth hormone
   a. can cause adults to grow taller.
   b. stimulates protein synthesis.
   c. is released by the hypothalamus.
   d. can be obtained only from cadavers.
   e. is a steroid.
4. PTH
   a. stimulates osteoblasts to lay down new bone.
   b. reduces blood calcium levels.
   c. stimulates calcitonin release.
   d. is produced by the thyroid gland.
   e. is released when blood calcium levels fall.
5. Steroid hormones
   a. are produced only by the adrenal cortex.
   b. have only cell surface receptors.
   c. are water-soluble.
   d. act by altering the activity of proteins in the target cell.
   e. act by altering gene expression in the target cell.
6. The hormone ecdysone
   a. is released from the posterior pituitary.
   b. stimulates molting in insects.
   c. maintains an insect in larval stages unless PTTH is present.
   d. stimulates the secretion of juvenile hormone from the prothoracic glands.
   e. keeps the insect exoskeleton flexible to permit growth.
7. The posterior pituitary
   a. synthesizes oxytocin.
   b. is under the control of hypothalamic releasing neurohormones.
   c. secretes tropic hormones.
   d. secretes neurohormones.
   e. is under feedback control by thyroxine.
8. Which of the following contributes to the development of goiter?
   a. Inadequate iodine in the diet
   b. Autoimmune antibodies that stimulate the TSH receptor
   c. Lack of feedback from circulating T_3 and T_4
   d. Overproduction of thyroglobulin
   e. All of the above
9. Which of the following is a likely cause of diabetes?
   a. Overproduction of insulin by beta cells of the pancreas
   b. Loss of alpha cells of the pancreas
   c. Loss of insulin receptors
   d. Overproduction of glucagon
   e. Loss of receptors for somatostatin
10. Which statement is true of all hormones?
    a. They are secreted by glands.
    b. They have receptors on cell surfaces.
    c. They may stimulate different responses in different cells.
    d. They target cells that are distant from their site of release.
    e. When the same hormone occurs in different species, it has the same action.

FOR DISCUSSION

1. Explain how both hyperthyroidism and hypothyroidism can cause goiter. Refer to the roles of the hypothalamus and the pituitary in your answer.
2. There are several apparently enigmatic aspects of the role of PTH in regulating blood calcium. First, PTH raises blood calcium levels, yet osteoclasts may not have PTH receptors. Second, parathyroid cells have calcium-sensing receptors on their plasma membranes, and these receptors are activated by rising levels of blood calcium. Third, bone consists of calcium and phosphate salts, yet PTH causes blood phosphate to decline. Explain these various aspects of PTH actions.
3. Various side effects of anabolic steroid use were mentioned in this chapter. Some of these effects are due to the direct action of the steroid, but others are due to the negative feedback action of the steroid. Discuss an example of each and explain possible mechanisms.
4. Compare the characteristics you would expect of a hormone signaling system that controls a short-term process, such as digestion, with the characteristics you would expect of a hormone signaling system that controls a long-term process, such as development.
5. In the perpetual war between agriculturists and insects, a new weapon has been developed: a chemical similar to juvenile hormone. Explain how this chemical could be used as an insecticide. What factors do you think should be considered before it is released indiscriminately into the environment?

FOR INVESTIGATION

Each spring male deer grow antlers that they use in male–male competition for mates. Each fall they shed those antlers. Although females of most deer species do not grow antlers, the caribou are an exception. Female caribou grow antlers in the spring but do not shed them in the fall. (Over the winter, they use their antlers to defend patches of food from males.) In addition, newborn caribou begin to grow antlers after birth. Assuming that antler growth is controlled by hormones, how would you investigate the control of antler growth in caribou? What hormonal assays might you want to use? What hypotheses would you test with respect to the relationship between time of year and antler growth?

CHAPTER 42 Animal Reproduction

Explosive sex

Producers of valuable honey and pollinators of many crucial plants, the common honeybee, *Apis mellifera*, has been the subject of study and fascination for humans throughout recorded history. The unique sex life of these social insects is among their most intriguing aspects.

Over 99 percent of female honeybees do not reproduce. They exist to help one female—the queen—reproduce. At any given time, there is only one queen in a bee hive. She lays all of the eggs, which is a full-time job. She also determines whether an egg will be fertilized or not. Fertilized eggs develop into females; unfertilized eggs develop into males. Wait, you might think: it's the *male* that fertilizes the egg! Yes, it is the male's sperm that fertilizes the egg, but in this case a female—the queen bee—controls sperm delivery.

Eventually, every hive must have a new queen. Sometimes the old queen dies. Other times, in a phenomenon known as *swarming*, the queen and a retinue of workers leave their current hive to start a new one. Whether the queen dies or leaves, a new queen must be produced. The remaining worker bees enlarge a few cells in the honeycomb that contain fertilized eggs laid by the old queen. The larvae that hatch from those eggs are fed special food that stimulates their growth and development into prospective queens.

The first queen to pupate and emerge from her royal chamber kills any other aspiring queens. She then leaves the hive for her mating flight. Males from all around get the message that a virgin queen is available and congregate around her. While in flight, she will mate with 15–20 males, and each coupling is an event. After a male manages to insert his penis into her vagina, he literally explodes, leaving not only his sperm but also his sex organs (which drop out later) in the female. The process is repeated with other males, all of whom die as a result. The queen returns to the hive with a lifetime supply of sperm and sets about laying eggs. She will live for about 2 years and lay as many as 3000 eggs each day. If she releases sperm, the eggs will be fertilized and develop into females who will devote their lives to feeding her, maintaining the hive, foraging for nectar and pollen, and raising their sisters. An unfertilized egg develops into a male, who will hang around the hive doing nothing useful until he takes off to woo a virgin queen—an unlikely

A Unique Reproductive Strategy Among honeybees and some other insects, the only reproductive female is the queen, seen here in the center as she deposits eggs in the comb. The female workers who attend her are sterile.

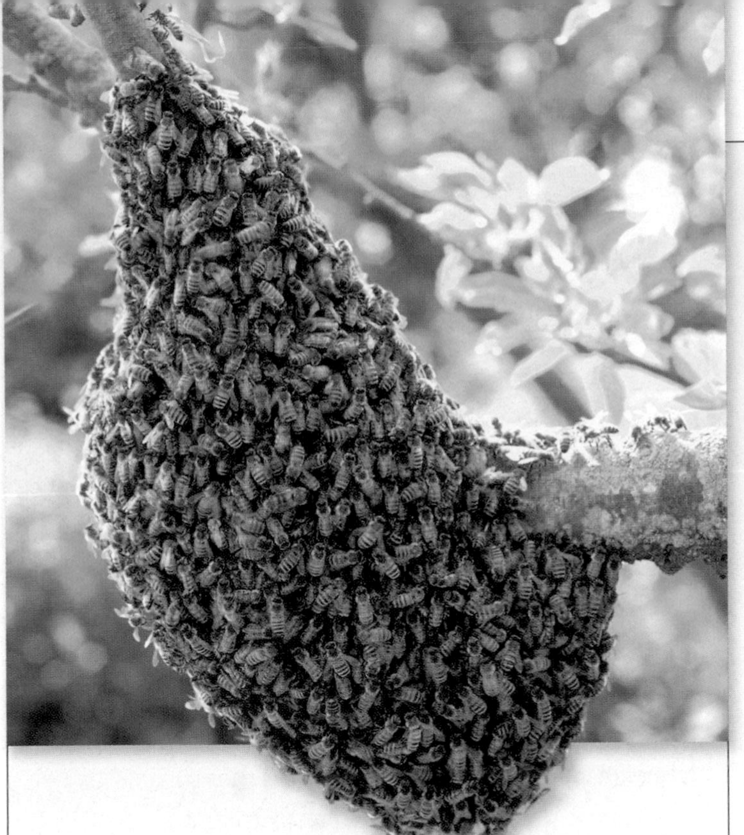

Swarming Will Mean a New Queen Bee When a honeybee colony swarms, as seen here, the queen leaves the hive and takes a retinue of workers with her. A new queen will emerge to take over the old hive and a few males will get to perform their one brief function—fertilizing a virgin queen—before they die.

event. It is thus in the queen's best interest to limit the number of males she produces.

Natural selection has resulted in some amazing adaptations, none more so than those involved in reproduction. Sexual or asexual, bizarre or otherwise, the practices necessary for the continuation of a species must evolve to be advantageous for that species.

IN THIS CHAPTER we examine the diverse ways in which animals produce offspring. We first examine asexual mechanisms of reproduction, in which only a single parent is involved. We then turn to sexual reproduction, in which an egg and a sperm unite to create a new diploid individual. Next we focus on the anatomy, function, and endocrine control of the human reproductive system, as well as the technologies used to limit or to enhance human fertility. We end this chapter with a discussion of human sexual health and sexually transmitted diseases.

CHAPTER OUTLINE

42.1 How Do Animals Reproduce Without Sex?

Sexual reproduction is a nearly universal trait in animals, although many species can also reproduce asexually and some reproduce only asexually. Offspring produced asexually are genetically identical to one another and to their parents. Asexual reproduction is efficient because no mating is required. Furthermore, asexual populations can use resources efficiently because all individuals in the population can convert resources into offspring. However, asexual reproduction does not generate the kind of genetic diversity produced by sexual reproduction, as described in Chapter 9. Genetic diversity is the raw material that enables natural selection to shape adaptations in response to environmental change, and a lack of genetic diversity can be disadvantageous to species in changing environments.

A variety of animals, mostly invertebrates, reproduce asexually. They tend to be species that are sessile and cannot search for mates or that live in sparse populations and rarely encounter potential mates. Asexually reproducing species are likely to be found in relatively constant environments where genetic diversity is less important for species success. Three common modes of asexual reproduction are *budding, regeneration,* and *parthenogenesis.*

Budding and regeneration produce new individuals by mitosis

Many simple multicellular animals produce offspring by **budding**. New individuals form as outgrowths or buds from the bodies of older animals. A bud grows by mitotic cell division, and the cells differentiate before the bud breaks away from the parent (**Figure 42.1A**). The bud is genetically identical to the parent, and it may grow as large as the parent before it becomes independent.

Regeneration is usually thought of as the replacement of damaged tissues or lost limbs, but in some cases pieces of an organism can regenerate complete individuals. Echinoderms, for example, have remarkable abilities to regenerate. If sea stars are cut into pieces, each piece that includes a portion of the central disc grows into a new animal (**Figure 42.1B**). In the early 1900s oyster fishermen in Narragansett Bay tried to eliminate the sea stars (starfish) that were preying on their oysters. Whenever they encountered a sea star, they chopped it up with knives

42.1 Asexual Reproduction in Animals (A) Budding: A new individual forms as an outgrowth from an adult hydra. (B) Regeneration: The single severed arm of a mature sea star is regenerating an entire animal.

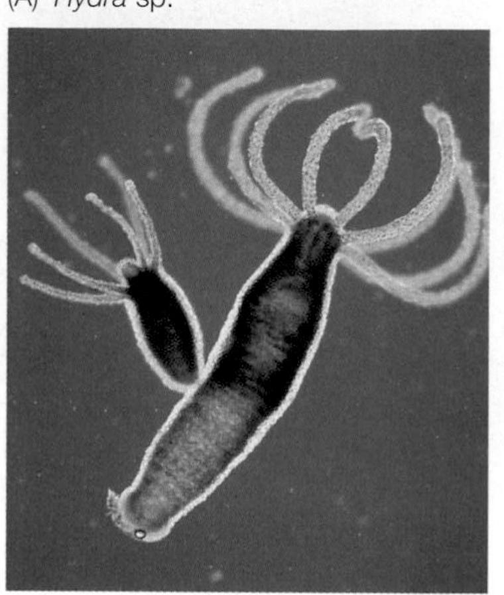

and threw it back into the water. As a result, the sea star population increased explosively.

Regeneration frequently results when an animal is broken by an outside force. A storm, for example, can cause heavy surf that breaks colonial cnidarians such as corals. Pieces broken off the colony can regenerate into new colonies. In some species, breakage occurs in the absence of external forces. Some species of segmented marine worms develop segments with rudimentary heads bearing sensory organs, then break apart. Each fragmented segment forms a new worm.

Parthenogenesis is the development of unfertilized eggs

Not all eggs must be fertilized to develop. A common mode of asexual reproduction in arthropods is the development of offspring from unfertilized eggs. This phenomenon, called **parthenogenesis**, also occurs in some species of fish, amphibians, and reptiles. Most species that reproduce parthenogenetically also engage in sexual reproduction or sexual behavior at other times.

In some species, parthenogenesis is part of the mechanism that determines sex. As we saw at the beginning of this chapter, in honey bees (as well as in most ants and wasps), males develop from unfertilized eggs and are haploid. Females develop from fertilized eggs and are diploid. Most females are sterile workers, but a select few become fertile queens. After a queen mates, she has a supply of sperm that she controls, enabling her to produce either fertilized or unfertilized eggs. Thus the queen determines when and how much of the colony resources are expended on males.

Parthenogenetic reproduction in some species requires sexual behavior even though sperm are not delivered to the female reproductive tract and eggs are not fertilized. One case that has been investigated extensively by David Crews and his students at the University of Texas is parthenogenetic reproduction in a species of whiptail lizard. This species has no males, but females can act as males, engaging in all aspects of courtship display and mating, although no sperm are produced or transferred (**Figure 42.2**). Whether a specific female acts as a female or as a male depends on cyclical hormonal states. When estrogen levels are high, she acts as a female. When her progesterone level peaks, she acts as a male. The stimulation resulting from the sexual activity triggers the release of eggs from the ovary.

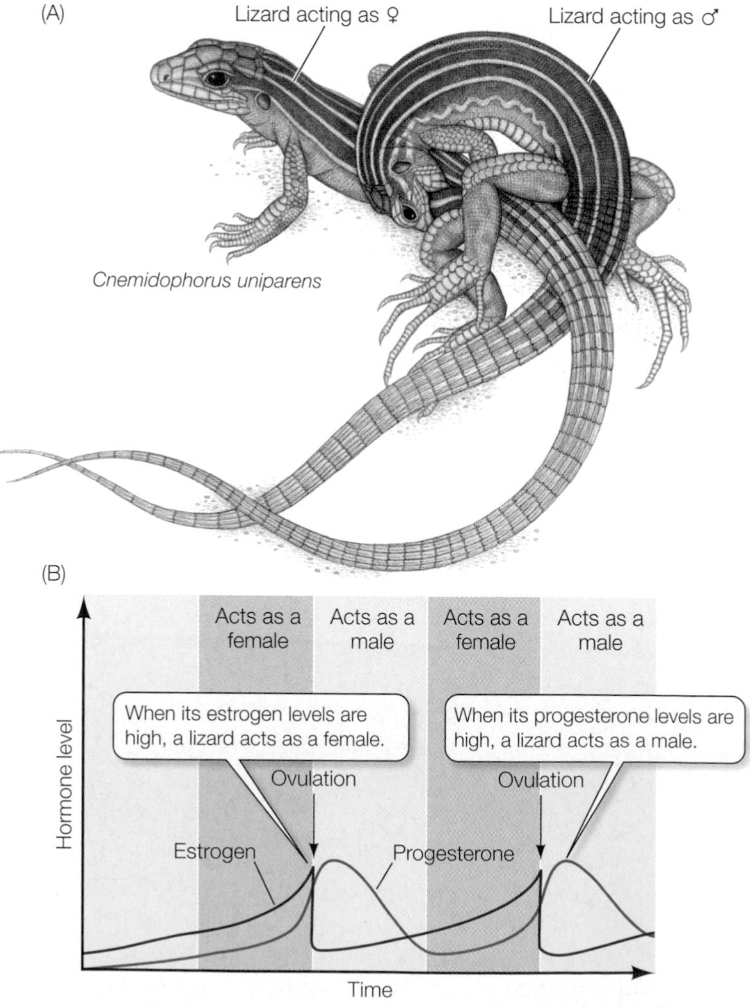

42.2 Sexual Behavior May Be Required for Asexual Reproduction (A) Parthenogenetic whiptail lizards are all female, but take turns acting the male role in reproductive behavior. (B) The stage of the ovarian cycle determines the role an individual whiptail plays.

42.1 RECAP

Most animals reproduce sexually, but many can also or can only reproduce asexually through budding, regeneration, or parthenogenesis.

- Can you explain why asexual reproduction might be disadvantageous for an animal living in a changing environment? See p. 897
- Do you understand how parthenogenesis can be related to sex determination? See p. 898

Asexual reproduction is an efficient way to use resources. However, the fact that sexual reproduction produces genetic diversity must be a tremendous advantage, because most species reproduce sexually.

42.2 How Do Animals Reproduce Sexually?

A large portion of the time and energy budgets of sexually reproducing animals goes into mating, which exposes them to predation, can result in physical damage, and detracts from other useful activities, such as feeding and caring for existing offspring. Furthermore, mating requires that resources be used to maintain a large population of males that do not bear offspring. Despite all these disadvantages, the production of genetic diversity is an overwhelming evolutionary advantage resulting from sexual reproduction.

Sexual reproduction requires the joining of two haploid sex cells to form a diploid individual. These haploid cells, or *gametes*, are produced through **gametogenesis**, a process that involves meiotic cell divisions. Two events in meiosis contribute to genetic diversity: *crossing over* between homologous chromosomes and the *independent assortment* of chromosomes (see Sections 9.6 and 10.1). Sexual reproduction itself also contributes to genetic diversity. The genetic variation among the gametes of a single individual and the genetic variation between any two parents produce an enormous potential for genetic variation between any two offspring of a sexually reproducing pair of individuals.

Sexual reproduction in animals consists of three fundamental steps:

- Gametogenesis (making gametes)
- Mating (getting gametes together)
- Fertilization (fusing gametes)

The process of gametogenesis is very similar across animal species. Processes of fertilization are also rather similar in widely different species. Therefore, while our discussion of gametogenesis will focus generally on mammals, and our discussion of fertilization will feature sea urchins, the facts would not be dramatically different were we to consider different groups of animals. Adaptations for mating, in contrast, show incredible anatomical, physiological, and behavioral diversity across species. Thus, we will take our topics out of order by considering gametogenesis and fertilization first before turning to the more diverse topic of mating and finally discussing human reproduction in detail.

Gametogenesis produces eggs and sperm

Gametogenesis occurs in the **gonads**, which are **testes** (singular *testis*) in males and **ovaries** in females. The tiny gametes of males, called **sperm**, move by beating their flagella. The larger gametes of females, called eggs or **ova** (singular *ovum*), are nonmotile.

Gametes are produced from **germ cells**, which have their origin in the earliest cell divisions of the embryo and remain distinct from the rest of the body. All other cells of the embryo are called *somatic cells*. Germ cells are sequestered in the body of the embryo until its gonads begin to form. The germ cells then migrate to the developing gonads, where they take up residence and proliferate by mitosis, producing **spermatogonia** (singular *spermatogonium*) in males and **oogonia** (singular *oogonium*) in females. Spermatogonia and oogonia, which are diploid, multiply by mitosis, eventually producing **primary spermatocytes** and **primary oocytes**.

Meiosis, the next step in gametogenesis, reduces the chromosomes to the haploid number, and the resulting haploid cells eventually mature into sperm and ova. (You may want to review the discussion of meiosis in Section 9.5 before reading further.) Although the steps of meiosis are similar in males and females, gametogenesis differs between the sexes.

SPERMATOGENESIS The initial proliferation of male germ cells into spermatogonia proceeds by mitosis in the embryo. As illustrated in **Figure 42.3A**, primary spermatocytes then undergo the first meiotic division to form **secondary spermatocytes**. The second meiotic division produces four haploid **spermatids** for each primary spermatocyte that enters meiosis. In mammals, the progeny of primary spermatocytes remain connected by *cytoplasmic bridges* after each division.

One reason that mammalian spermatocytes remain in cytoplasmic contact throughout their development is the asymmetry of sex chromosomes in males. Half the secondary spermatocytes receive an X chromosome, the other half a Y chromosome. The Y chromosome contains fewer genes than the X chromosome, and some of the products of genes found only on the X chromosome are essential for spermatocyte development. By remaining in cytoplasmic contact, all four spermatocytes can share the gene products of the X chromosomes, although only half of them have an X chromosome.

A spermatid bears little resemblance to a mature sperm. Through further differentiation, however, the spermatid becomes compact, streamlined, and motile. We will look at the differentiation of human sperm in more detail below.

OOGENESIS Oogonia, like spermatogonia, proliferate through mitosis (**Figure 42.3B**). The resulting primary oocytes immediately enter prophase of the first meiotic division. In many species, including humans, the oocyte experiences developmental arrest at this point and may remain so for days, months, or years. In the human female, this period of arrest is at least 10 years (i.e., until puberty) and some primary oocytes may remain in prophase I for up to 50 years (i.e., until menopause)! In contrast, male gametogenesis continues steadily to completion once the primary spermatocyte has differentiated.

During this prolonged prophase I, or shortly before it ends, the primary oocyte grows larger through increased production of

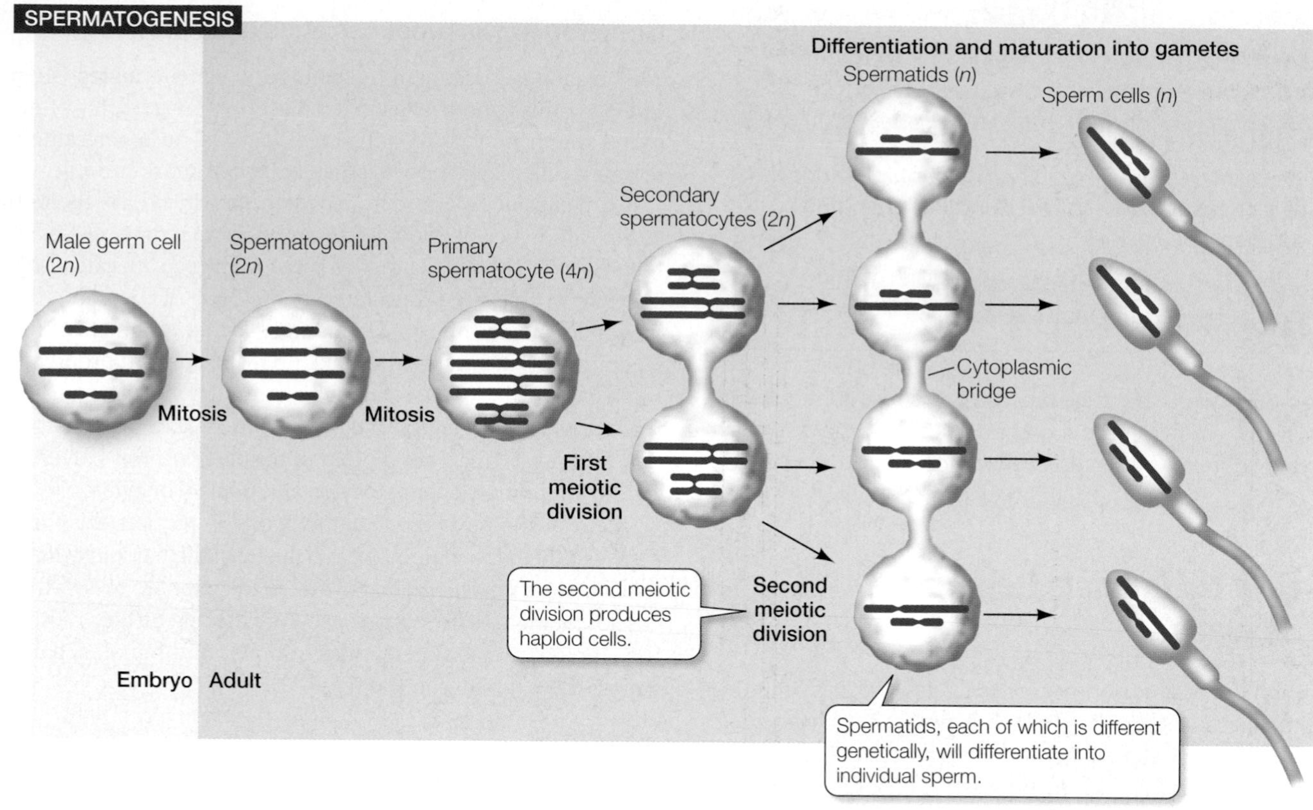

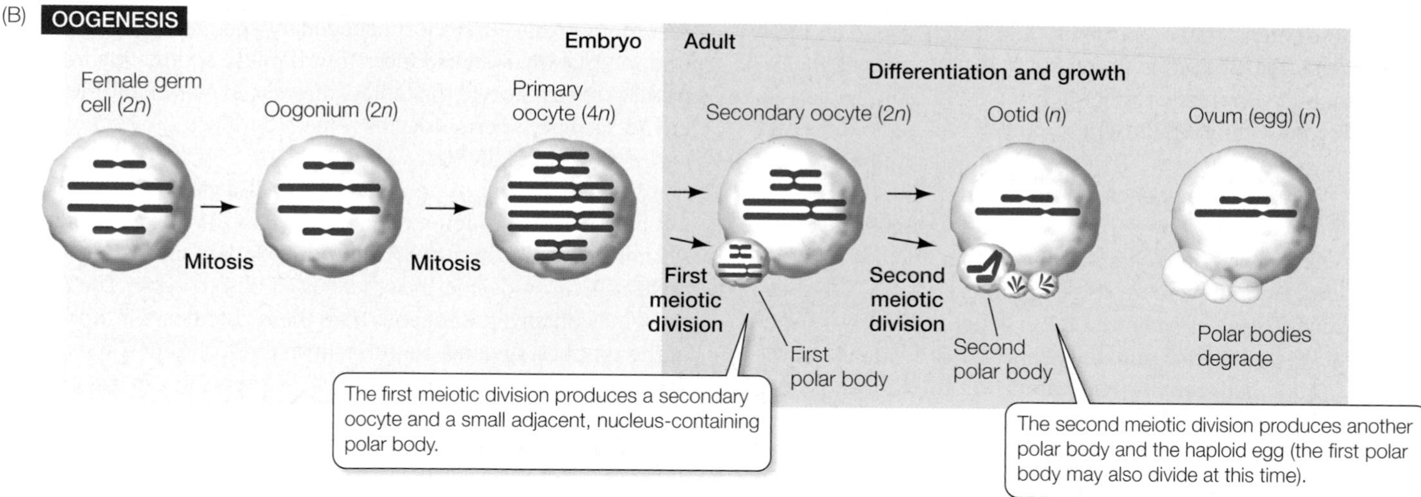

42.3 Gametogenesis Male and female germ cells proliferate by mitosis and produce diploid spermatogonia and oogonia that mature into primary spermatocytes and oocytes before entering meiosis. (A) Spermatogonia continue to divide by mitosis in adults, producing a steady supply of spermatocytes that divide meiotically to produce haploid spermatids, which differentiate into sperm. In many species the progeny of spermatocytes remain in contact through cytoplasmic bridges until the sperm mature. (B) In mammals, oogonia cease division in the embryo, and primary oocytes remain arrested in prophase I of meiosis until they are ovulated and fertilized. Each oocyte will produce one ootid which matures into an ovum.

ribosomes, RNA, cytoplasmic organelles, and energy stores. At this point, the primary oocyte acquires all the energy, raw materials, and RNA that the ovum will need to survive its first cell divisions after fertilization. In fact, the nutrients in the egg must maintain the embryo until it is either nourished by the maternal circulatory system or can feed on its own.

When a primary oocyte resumes meiosis, its nucleus completes the first meiotic division near the surface of the cell. The daughter cells of this division receive grossly unequal shares of cytoplasm. This asymmetry represents another major difference from spermatogenesis, in which cytoplasm is apportioned equally. The daughter cell that receives almost all the cytoplasm becomes the **secondary oocyte**, and the one that receives almost none forms the **first polar body** (see Figure 42.3B).

The second meiotic division of the large secondary oocyte is also accompanied by an asymmetrical division of the cytoplasm. One daughter cell forms the large, haploid **ootid**, which eventually differentiates into a mature ovum, and the other forms the **second polar body**. Polar bodies degenerate, so the end result of oogenesis is only one mature egg for each primary oocyte that entered meiosis. However, that egg is a very large, well-provisioned cell.

A second period of arrested development occurs after the first meiotic division forms the secondary oocyte. The egg may be expelled from the ovary in this condition; however, to simplify discussion, we will call the female gamete an egg once it leaves the ovary. In many species, including humans, the second meiotic division is not completed until the egg is fertilized by a sperm.

Fertilization is the union of sperm and egg

The union of the haploid sperm and the haploid egg in **fertilization** creates a single diploid cell, called a **zygote**, which will develop into an embryo. Fertilization does more, however, than just restore the full genetic complement of the animal. The processes associated with fertilization help eggs and sperm get together, prevent the union of sperm and eggs of different species, and guarantee that only one sperm will enter and activate the egg metabolically. Fertilization involves a complex series of events:

- The sperm and the egg recognize each other.
- The sperm is *activated*, enabling it to gain access to the plasma membrane of the egg.
- The plasma membranes of the sperm and the egg fuse.
- The egg blocks entry of additional sperm.
- The egg is metabolically activated and stimulated to start development.
- The egg and sperm nuclei fuse to create the diploid nucleus of the zygote.

SPECIFICITY IN SPERM–EGG INTERACTIONS Specific recognition molecules mediate interactions between sperm and eggs. These molecules ensure that the activities of sperm are directed toward eggs and not other cells, and they help prevent eggs from being fertilized by sperm from the wrong species. The latter function is particularly important in aquatic species that release eggs and sperm into the surrounding water. The sea urchin is such a species, and its mechanisms of fertilization have been well studied.

The eggs of sea urchins and various other marine invertebrates release chemical attractants that increase the motility of sperm and cause them to swim toward the egg. These chemical attractants are species-specific. For example, eggs of one species of sea urchin release a specific peptide consisting of 14 amino acids. As this peptide diffuses from the egg, it binds to receptors on the sperm of the same species. The sperm respond by increasing their mitochondrial respiration and motility. Before exposure to the peptide, the sperm swim in tight little circles, but after binding the peptide, they swim energetically up the concentration gradient of the peptide until they reach the egg that is releasing it.

When sperm reach an egg, they must get through two protective layers before they can fuse with the egg plasma membrane. The eggs of sea urchins are covered with a **jelly coat**, which surrounds a proteinaceous **vitelline envelope** (**Figure 42.4**). The sperm's

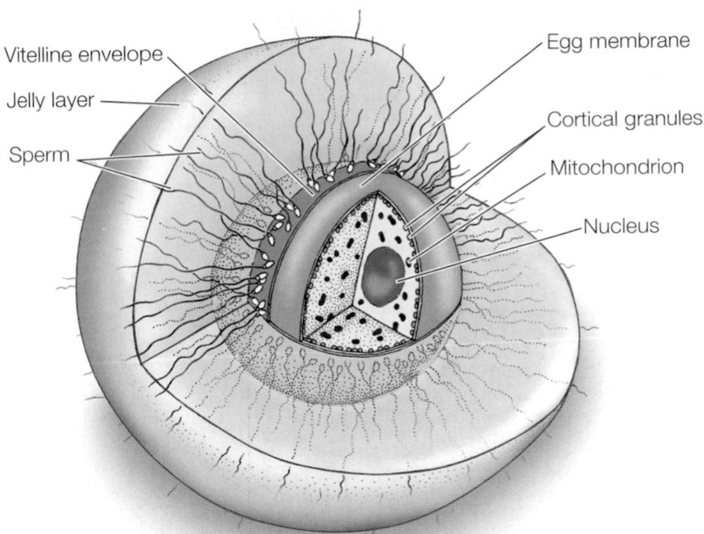

42.4 The Sea Urchin Egg Sea urchin eggs are released into seawater. They are protected by a jelly layer and a proteinaceous vitelline envelope. Sperm must penetrate both to reach the egg plasma membrane. Many sperm attach to the vitelline envelope, but only one will penetrate the egg cell membrane and achieve fertilization.

assault on these protective layers depends on a membrane-enclosed structure called an acrosome.

The **acrosome**, which contains enzymes and other proteins, is located at the front of the sperm head, where it forms a cap over the sperm nucleus. When the sperm makes contact with an egg of its own species, substances in the jelly coat trigger an *acrosomal reaction* which begins with the breakdown of the plasma membrane covering the sperm head and the underlying acrosomal membrane (**Figure 42.5**). The acrosomal enzymes are released, and they digest a hole through the jelly coat.

As a result of the polymerization of actin triggered by the acrosomal reaction, a structure called the *acrosomal process* extends out of the head of the sperm. The acrosomal process is coated with species-specific recognition molecules called *bindin*, and there are bindin receptors on the vitelline envelope of the egg. The interaction of these two molecules enable the sperm to contact the egg plasma membrane. That contact results in fusion of the sperm and egg plasma membranes and the formation of a *fertilization cone* that engulfs the sperm head, bringing it into the egg cytoplasm.

In animals that practice internal fertilization, mating behaviors help guarantee species specificity, but egg–sperm recognition mechanisms still exist. The mammalian egg is surrounded by a thick layer called the **cumulus**, which consists of a loose assemblage of maternal cells in a gelatinous matrix (**Figure 42.6**). Beneath the cumulus is a glycoprotein envelope called the **zona pellucida**, which is functionally similar to the vitelline envelope of sea urchin eggs. When mammalian sperm are deposited in the female reproductive tract, they are metabolically activated and made capable of an acrosomal reaction if they should meet an egg. An activated sperm can penetrate the cumulus and interact with the zona pellucida.

Unlike the jelly coat of sea urchin eggs, the cumulus of mammalian eggs does not trigger the acrosomal reaction. When sperm make contact with the zona pellucida, a species-specific glyco-

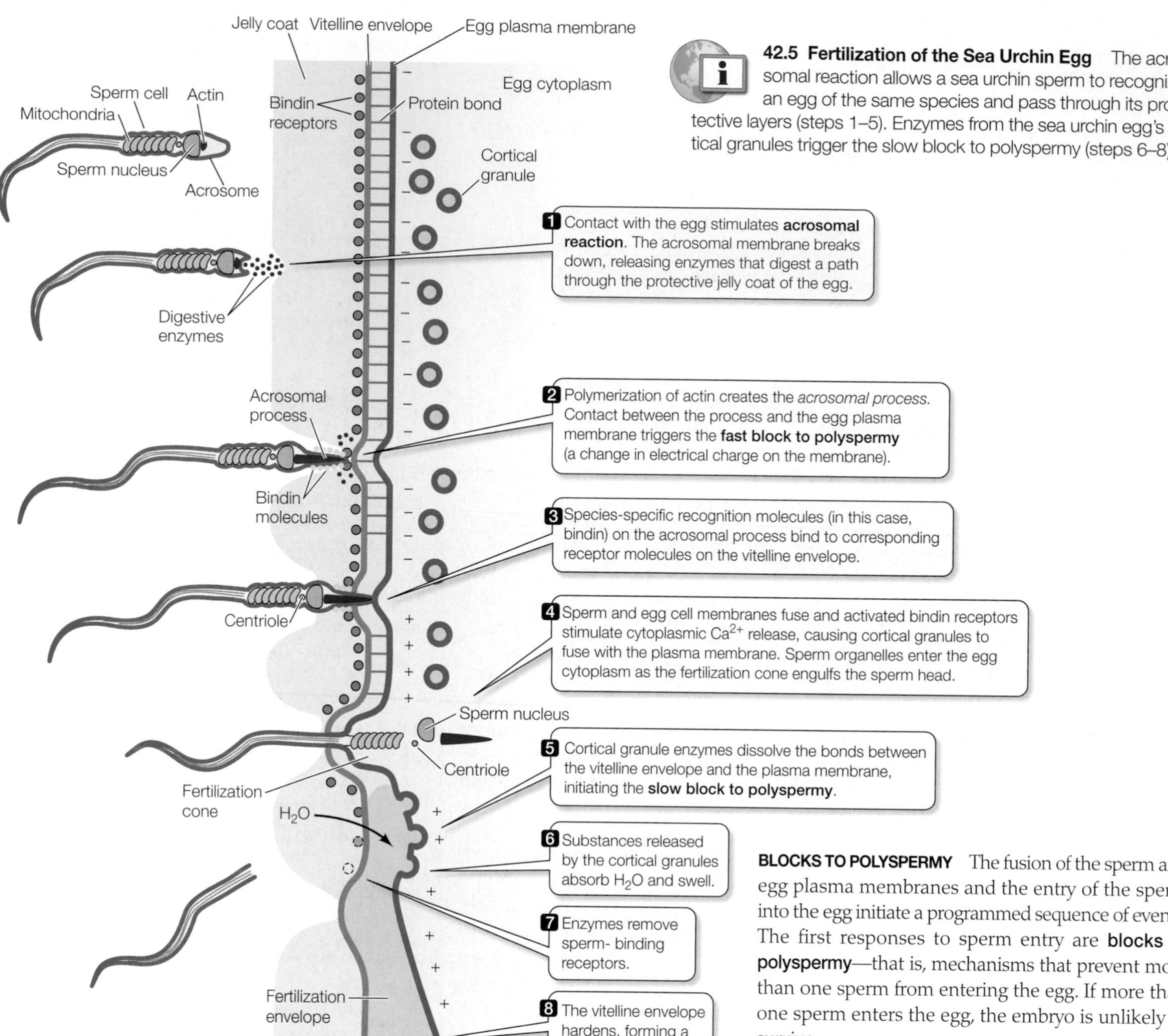
Jelly coat
Vitelline envelope
Egg plasma membrane
Sperm cell
Actin
Mitochondria
Sperm nucleus
Acrosome
Bindin receptors
Protein bond
Egg cytoplasm
Cortical granule
1 Contact with the egg stimulates **acrosomal reaction**. The acrosomal membrane breaks down, releasing enzymes that digest a path through the protective jelly coat of the egg.
Digestive enzymes
Acrosomal process
2 Polymerization of actin creates the *acrosomal process.* Contact between the process and the egg plasma membrane triggers the **fast block to polyspermy** (a change in electrical charge on the membrane).
Bindin molecules
3 Species-specific recognition molecules (in this case, bindin) on the acrosomal process bind to corresponding receptor molecules on the vitelline envelope.
Centriole
4 Sperm and egg cell membranes fuse and activated bindin receptors stimulate cytoplasmic Ca^{2+} release, causing cortical granules to fuse with the plasma membrane. Sperm organelles enter the egg cytoplasm as the fertilization cone engulfs the sperm head.
Sperm nucleus
Centriole
5 Cortical granule enzymes dissolve the bonds between the vitelline envelope and the plasma membrane, initiating the **slow block to polyspermy**.
Fertilization cone
H_2O
6 Substances released by the cortical granules absorb H_2O and swell.
7 Enzymes remove sperm- binding receptors.
Fertilization envelope
8 The vitelline envelope hardens, forming a fertilization envelope.

42.5 Fertilization of the Sea Urchin Egg The acrosomal reaction allows a sea urchin sperm to recognize an egg of the same species and pass through its protective layers (steps 1–5). Enzymes from the sea urchin egg's cortical granules trigger the slow block to polyspermy (steps 6–8).

protein binds to recognition molecules on the head of the sperm. This binding triggers the acrosomal reaction, releasing acrosomal enzymes that digest a path through the zona pellucida. When the sperm head reaches the egg plasma membrane, other proteins facilitate its adhesion to and fusion with the egg plasma membrane.

The importance of the zona pellucida and its sperm-binding molecules as a species-specific recognition mechanism was revealed in experiments on mammalian eggs and sperm in culture dishes. When the zona was stripped from human eggs and they were exposed to hamster sperm, fertilization took place, resulting in a hamster–human hybrid zygote. The hybrid zygote did not survive its first cell division, but the experiment demonstrated that the recognition mechanism in mammalian species resides in the zona pellucida.

BLOCKS TO POLYSPERMY The fusion of the sperm and egg plasma membranes and the entry of the sperm into the egg initiate a programmed sequence of events. The first responses to sperm entry are **blocks to polyspermy**—that is, mechanisms that prevent more than one sperm from entering the egg. If more than one sperm enters the egg, the embryo is unlikely to survive.

Blocks to polyspermy have been studied extensively in sea urchin eggs, which can be fertilized in a dish of seawater. Within seconds after the sperm membrane contacts the egg membrane, an influx of sodium ions occurs, which changes the electric charge difference across the egg's plasma membrane. This *fast block to polyspermy* prevents the fusion of other sperm with the egg plasma membrane, but it is transient. The change in membrane electrical charge lasts only about a minute, but that is enough time to allow a slower block to sperm entry to develop.

Before fertilization, the vitelline envelope is bonded to the egg plasma membrane. Just under the plasma membrane are vesicles called *cortical granules,* which contain enzymes and other proteins that dissolve the bonds between the vitelline envelope and the egg plasma membrane. The sea urchin egg, like all animal cells, contains calcium ions that are sequestered in the endoplasmic reticulum. The *slow block to polyspermy* is initiated by the release of these calcium ions following sperm and egg fusion (see Figure 42.5 and Figure 15.14).

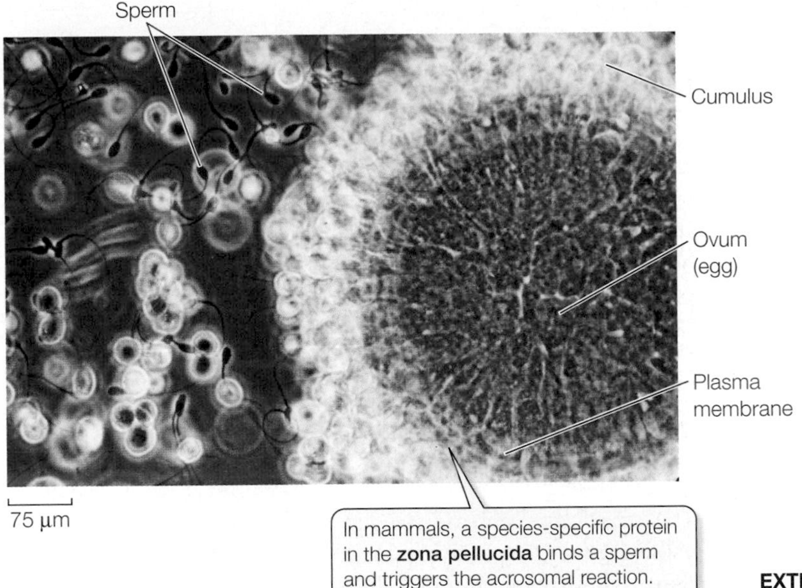
Sperm
Cumulus
Ovum (egg)
Plasma membrane
75 µm
In mammals, a species-specific protein in the **zona pellucida** binds a sperm and triggers the acrosomal reaction.

42.6 A Mammalian Egg Is Surrounded by Barriers to Sperm This human egg is protected by the cumulus and zona pellucida, both of which a sperm must penetrate to fertilize the egg. Only one sperm will penetrate the zona pellucida and fuse with the plasma membrane.

EGG ACTIVATION Sperm entry into the sea urchin egg stimulates the release of calcium from the egg's endoplasmic reticulum. The increase in cytosolic calcium causes the egg's cortical granules to fuse with the plasma membrane and release their contents. The cortical granule enzymes break the bonds between the vitelline envelope and the plasma membrane, and other proteins released from the cortical granules attract water into the space between them. As a result, the vitelline envelope rises to form a *fertilization envelope*. Cortical granule enzymes also degrade sperm-binding molecules on the surface of the fertilization envelope and cause it to harden. The hardened fertilization envelope prevents additional sperm from contacting the egg's plasma membrane.

The mechanism of hardening of the fertilization envelope was discovered only recently. One hallmark of fertilization is a rapid increase in oxygen consumption by the fertilized egg, assumed for almost a hundred years to be due to the activation of the various metabolic processes in the fertilized egg. We now know that the big increase in oxygen consumption is due to the rapid production of hydrogen peroxide (H_2O_2) by an enzyme secreted by the cortical granules. This enzyme, called Udx1 for urchin dual oxidase, has two seemingly opposing functions. First, it catalyzes the production of H_2O_2, which provides the oxidizing capacity for hardening of the fertilization envelope; and second, it catalyzes the breakdown of any unused H_2O_2 so it does not enter the zygote and cause damage. Similar mechanisms of rapid oxidation of the materials surrounding the fertilized egg probably occur in most species, including mammals.

In mammals, sperm entry does not cause a rapid change in membrane potential, but it does trigger a release of calcium from the endoplasmic reticulum. As in the sea urchin, the increased calcium causes the cortical granules to fuse with the egg plasma membrane. A fertilization envelope does not form around the mammalian egg, but the cortical granule enzymes destroy the sperm-binding molecules in the zona pellucida. The rise in cytosolic calcium also activates the egg's metabolism and signals it to complete meiosis. The stage is set for the first cell division.

Mating bring eggs and sperm together

As we have just seen, sexual reproduction requires the production of haploid gametes (gametogenesis) and the joining together of those gametes to form a diploid zygote (fertilization). **Mating**, the step between these two processes, gets eggs and sperm close enough together that fertilization can occur. The simplest distinction in mating systems is whether fertilization occurs externally or internally.

EXTERNAL FERTILIZATION In an aquatic environment, animals can bring their gametes together by simply releasing them into the water. This practice is called **external fertilization**. Many simple aquatic animals are not very mobile, but they produce huge numbers of gametes that can travel far from the point of release. A female oyster, for example, will release millions of eggs when she spawns, and the number of sperm produced by a male oyster is astronomical.

Caviar is fish eggs, most of it coming from sturgeon of the Caspian and Black Seas. Much-prized beluga caviar, which can sell for $4,000 a kilogram, comes from beluga sturgeon (*Husa husa*), an evolutionarily ancient fish that can live for 150 years and be 5 meters long. Today, declining population size due to overfishing and pollution means that the average adult beluga is much younger, and far smaller.

But numbers alone do not guarantee that gametes will meet. The reproductive activities of the males and females of a population must be synchronized, since released gametes have a limited life span. Seasonal breeders may use day length, changes in temperature, or changes in weather to time the production and release of their gametes. Social stimulation is also important. Sexual activity by one member of a population can stimulate others to engage in it.

Behavior can play an important role in bringing gametes together even when fertilization is external. Many species travel great distances to congregate with potential mates and release their gametes at the same time in a suitable environment. Salmon are an extreme example, traveling hundreds of miles to spawn in the stream where they hatched.

INTERNAL FERTILIZATION Terrestrial animals cannot simply release their gametes into the environment. Sperm can move only through liquid, and delicate gametes released into air would dry out and die. Terrestrial animals avoid these problems by **internal fertilization**, the release of sperm directly into the female reproductive tract.

Animals have evolved an astonishing diversity of behavioral and anatomical adaptations for internal fertilization. As we saw above, gametogenesis occurs in the gonads, which are the *primary sex organs*. All additional anatomical components of an animal's reproductive system are called *accessory sex organs*. An obvious accessory sex organ in males of many species is the **penis**, which enables the male to deposit sperm in the female's reproductive tract, called in many species a **vagina**. Accessory sex organs include a variety of glands, tubules, ducts, and other structures.

Copulation is the physical joining of male and female accessory sex organs. Transfer of sperm in internal fertilization can also be indirect. Males of many invertebrate species (for example, mites and scorpions) and a few vertebrates (salamanders) deposit *spermatophores*—packets of sperm—in the environment. When a female mite encounters a spermatophore from a potential mate, she straddles it and opens a pair of plates in her abdomen so that the tip of the spermatophore enters her reproductive tract and allows the sperm to enter.

Male squids and spiders play a more active role in spermatophore transfer. The male spider secretes a drop containing sperm onto a bit of web, then uses a special structure on his foreleg to pick up the sperm-containing web and insert it through the female's genital opening. Male squids use one specialized tentacle to pick up a spermatophore and insert it into the female's genital opening.

Most male insects copulate and transfer sperm to the female's vagina through a penis. The **genitalia**—external sex organs—of insects often have species-specific shapes that match in a lock-and-key fashion. This mechanism ensures a tight, secure fit between the mating pair during the prolonged period of sperm transfer. In some insect species in which females mate with more than one male, the males have elaborate structures on their penises that can scoop sperm deposited by other males out of a female's reproductive tract, replacing it with their own.

A single body can function as both male and female

In most species, gametes are produced by individuals that are either male or female. Species that have separate male and female members are called **dioecious** species (from the Greek for "two houses"). In some species, however, a single individual may produce both sperm and eggs. Such species are called **monoecious** ("one house"), or **hermaphroditic**, species.

Almost all invertebrate groups contain some hermaphroditic species. An earthworm is an example of a *simultaneous hermaphrodite*, meaning that it is both male and female at the same time. When two earthworms mate, they exchange sperm, and as a result, the eggs of each are fertilized (see Figure 32.13C). Some vertebrates, such as the anemone fish or clown fish, which live in small groups within large sea anemones, are *sequential hermaphrodites*, meaning that an individual may function as a male or a female at different times in its life (**Figure 42.7**). All anemone fish are born as males. The largest one in a group becomes a functional female. If she is removed from the group, the largest male becomes a female. Also, the second largest anemone fish in the group is the only male in breeding condition.

What is the selective advantage of hermaphroditism? Some simultaneous hermaphrodites, such as parasitic tapeworms, have a low probability of meeting a potential mate. Even though a tapeworm may be large and troublesome for its host, it may be the only tapeworm in the host. Tapeworms can fertilize their own eggs. Most simultaneous hermaphrodites, however, must mate with another individual; but since each member of the population is both male and female, the probability of encountering a possible mate is double what it would be in strictly monoecious species. In some sequential hermaphrodites, all siblings are either male or female at the same time, thus reducing the incidence of inbreeding.

The evolution of vertebrate reproductive systems parallels the move to land

The earliest vertebrates evolved in aquatic environments. The closest living relatives of those earliest vertebrates are modern-day fishes. They remain exclusively aquatic animals, and most practice external fertilization. The most primitive of the fishes, the lampreys and hagfishes, simply release their gametes into the environment. In most fishes, however, mating behaviors bring females and males into close proximity at the time of gamete release. In some sharks and rays, fins have evolved into claspers that hold the male and female together and enable sperm to be transferred directly into the female reproductive tract.

Amphibians were the first vertebrates to live in terrestrial environments. They dealt with the challenge of a dry environment

42.7 When Size Determines Sex Anemone fish (also known as "clown fish") live in groups of about a dozen centered on a single sea anemone. All anemone fish are born male, and the largest one in the group becomes a female. Thus any one fish may function first as a male and then as a female.

by returning to water to reproduce, as most amphibians still do today.

Reptiles were the first vertebrate group to solve the problem of reproduction in the terrestrial environment. Their solution, the **amniote egg**, is shared with the birds (see Section 33.4). A good example is the chicken egg, which contains a supply of food (yolk) and water for the developing embryo. A hard shell protects the embryo and impedes water loss while allowing the diffusion of oxygen and carbon dioxide (**Figure 42.8A**). The eggshell creates an obvious problem for fertilization, however: Sperm cannot penetrate the shell, so they must reach the egg before the shell forms. Hence internal fertilization and the evolution of accessory sex organs were necessary for the evolution of the amniote egg.

Male snakes and lizards have paired *hemipenes*, which can be filled with blood and thereby extruded from the male's body. Only one hemipenis is inserted into the female's reproductive tract at a time. It is usually rough or spiny at the end to achieve a secure hold while sperm are transferred down a groove on its surface. Retractor muscles pull the hemipenis back into the male's body when mating is completed. Some evolutionarily ancient bird species have erectile penises that channel sperm along a groove into the female's reproductive tract. Birds with more recent evolutionary origins, however, do not have erectile penises; instead, the male and female simply bring their genital openings close together to transfer sperm. Usually this involves the male standing on the female's back (**Figure 42.8B**).

All mammals practice internal fertilization and, except for the prototherian mammals (see Figure 33.25), they have done away with the shelled egg. The developing mammalian embryo is retained in the female reproductive tract, at least through its early stages; mammalian species vary enormously as to the developmental stage of their offspring at the time of birth.

Reproductive systems are distinguished by where the embryo develops

Two patterns of care and nurture of the embryo have evolved in animals: *oviparity* (egg laying) and *viviparity* (live bearing).

- **Oviparous** animals lay eggs in the environment and their embryos develop outside the mother's body. Oviparous terrestrial animals such as insects, reptiles, and birds protect their eggs from desiccation with waterproof membranes or shells.

Oviparity is possible because eggs are stocked with abundant nutrients to supply the needs of the embryo. Some oviparous animals engage in various forms of parental behavior to protect their eggs, but until the eggs hatch, the embryos depend entirely on the nutrients stored in the egg.

- **Viviparous** animals retain the embryo within the mother's body during its early developmental stages. All mammals except the monotremes are viviparous.

Although examples of viviparity exist in all other vertebrate groups except the crocodiles, turtles, and birds (even some sharks retain fertilized eggs in their bodies and give birth to free-living offspring), there is a big difference between viviparity in mammals and in other species. Mammals have a specialized portion of the female reproductive tract, called the **uterus** or *womb*, that holds the embryo and interacts with it to produce a **placenta**, which enables the exchange of nutrients and wastes between the blood of the mother and that of the embryo. Very few nonmammalian species have evolved such a connection between the embryo and the mother.

Most nonmammalian viviparous animals simply retain fertilized eggs in the mother's body until they hatch. These embryos still receive nutrition from stores in the egg, so this reproductive adaptation is called **ovoviviparity**.

42.8 The Shelled Egg The shelled egg was a major evolutionary step that allows reptiles and birds to reproduce in the terrestrial environment. (A) A female green sea turtle deposits her eggs in the sand. (B) The shelled egg requires that sperm meet egg before the shell forms, thus terrestrial animals must practice internal fertilization, as these European bee-eaters are doing.

(A) *Chelonia mydas*

(B) *Merops apiaster*

42.2 RECAP

Sexual reproduction involves gametogenesis, mating, and fertilization. Fertilization can be external or internal and involves mechanisms for ensuring that only one sperm from the right species enters the egg.

- Can you outline the steps by which a sperm penetrates an egg? See Figure 42.5
- Can you explain how polyspermy is prevented, and do you understand why it is crucial to do so? See p. 902 and Figure 42.5
- Can you describe the reproductive adaptations that made life on land possible? See pp. 904–905

Now that we have seen some of the general aspects of animal reproduction in gametogenesis and fertilization, and have been briefly introduced to the great diversity of mating systems, we will consider the human reproductive systems in detail.

42.3 How Do the Human Male and Female Reproductive Systems Work?

In this section we will describe the structures and functions of the male and female sex organs in mammals—specifically, in human beings—and discuss hormonal regulation of both male and female systems. Our discussion will include the primary sex organs (testes in males and ovaries in females) that produce gametes and serve endocrine functions. It will also include the accessory sex organs—the genitalia, the ducts through which the gametes pass, and the various glands that empty into those ducts. We will also refer to *secondary sexual characteristics*, which are not directly involved in reproduction, but comprise the external differences between males and females.

Male sex organs produce and deliver semen

Semen is the product of the male reproductive system. Besides sperm, semen contains a complex mixture of fluids and molecules that support the sperm and facilitate fertilization. Sperm make up less than 5 percent of the volume of the semen.

The male reproductive organs are diagrammed in **Figure 42.9**. Sperm are produced in the testes, the paired male gonads. The testes of most mammals are located outside the body cavity in a pouch of skin called the **scrotum**.

Why should the testes be located outside the body cavity? The optimal temperature for spermatogenesis in most mammals is slightly lower than the normal body temperature. The scrotum keeps the testes at this optimal temperature. Muscles in the scrotum contract in a cold environment, bringing the testes closer to the warmth of the body; in a hot environment they relax, cooling the testes by suspending them farther from the body.

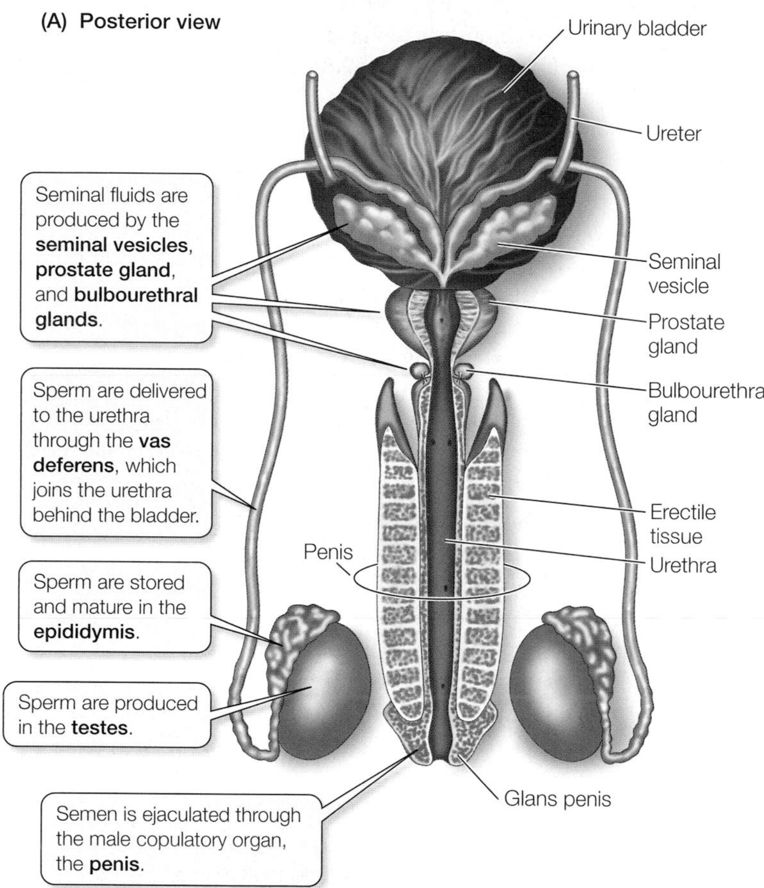

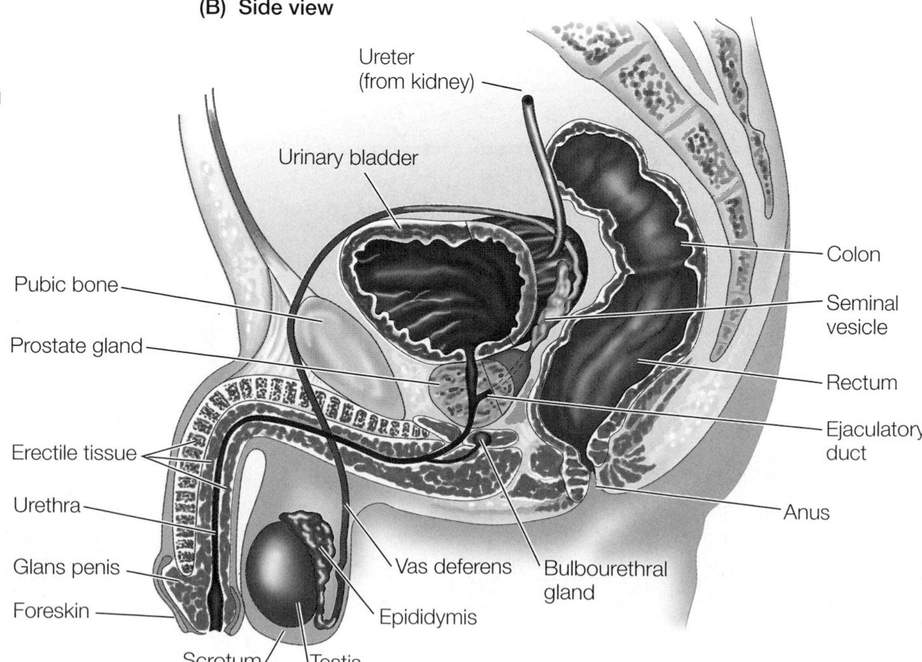

42.9 The Reproductive Tract of the Human Male The male reproductive organs are shown (A) from the rear and (B) in side section.

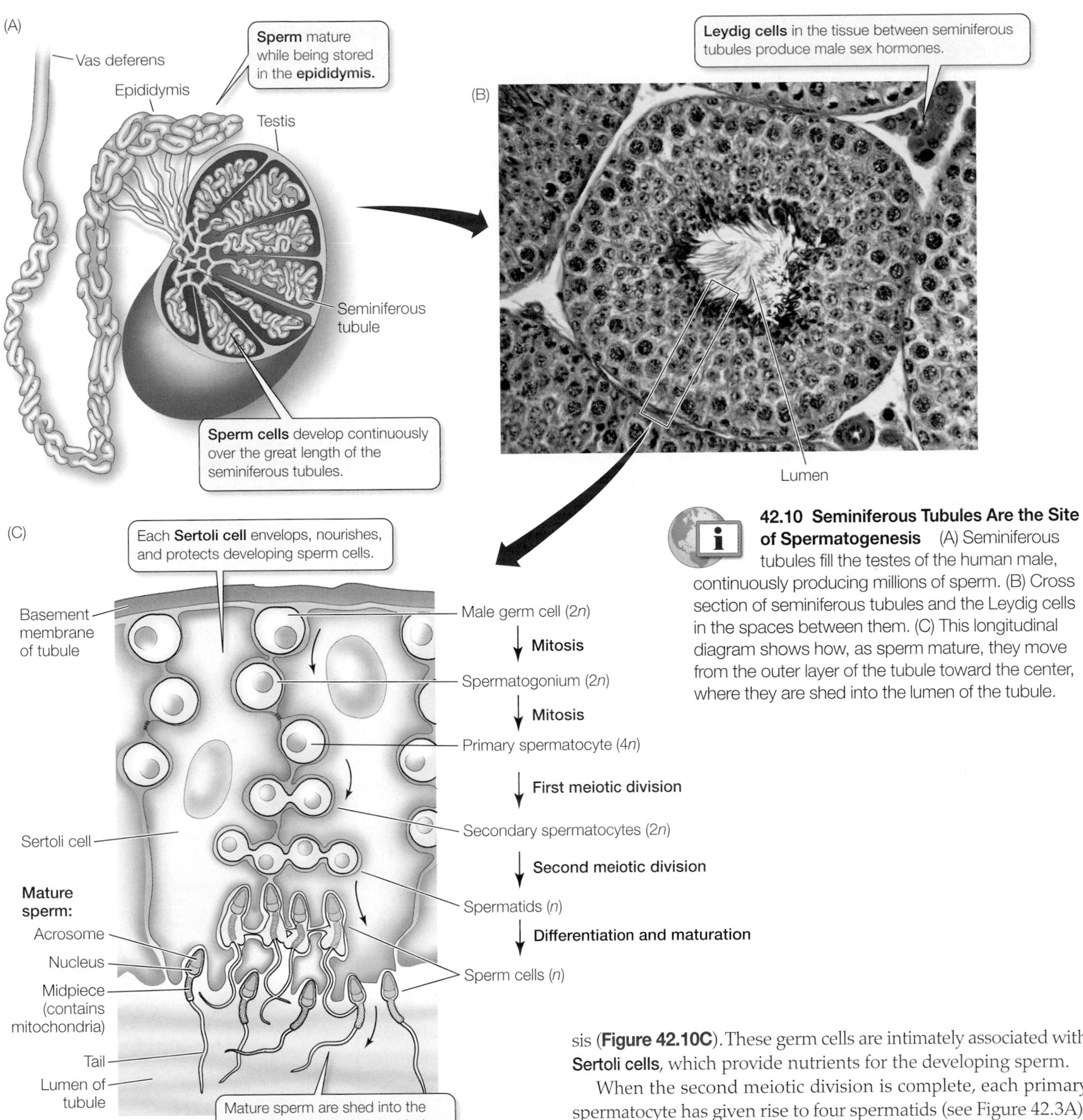

42.10 Seminiferous Tubules Are the Site of Spermatogenesis (A) Seminiferous tubules fill the testes of the human male, continuously producing millions of sperm. (B) Cross section of seminiferous tubules and the Leydig cells in the spaces between them. (C) This longitudinal diagram shows how, as sperm mature, they move from the outer layer of the tubule toward the center, where they are shed into the lumen of the tubule.

Spermatogenesis takes place within the **seminiferous tubules**, which are tightly coiled in each testis (**Figure 42.10A**). Between the seminiferous tubules are clusters of *Leydig cells*, or *interstitial cells*, which produce testosterone (**Figure 4.10B**). Spermatogonia reside in the outer layers of the epithelium that lines the tubules. Moving inward from these outer layers toward the lumen of the tubule, we find germ cells in successive stages of spermatogenesis (**Figure 42.10C**). These germ cells are intimately associated with **Sertoli cells**, which provide nutrients for the developing sperm.

When the second meiotic division is complete, each primary spermatocyte has given rise to four spermatids (see Figure 42.3A), which develop into sperm cells as they continue to migrate toward the lumen of the seminiferous tubule. The nucleus becomes compact and the surrounding cytoplasm is lost. A flagellum, or sperm tail, develops. The mitochondria, which will provide energy for tail motility, become condensed into a midpiece between the head and the tail. An acrosome forms over the nucleus in the head of the sperm. Fully differentiated sperm are shed into the lumen of the seminiferous tubule.

From the seminiferous tubules, sperm move into a storage structure called the **epididymis** (see Figure 42.9), where they mature and become motile. The epididymis connects to the **urethra** via a tube called the **vas deferens** (plural *vasa deferentia*). The urethra originates in the bladder, runs through the penis, and opens to the outside of the body at the tip of the penis. It serves as the common duct for the urinary and reproductive systems. The components of the semen other than sperm come from several accessory glands. About 60 percent of the volume of semen is secreted by the paired **seminal vesicles**, which empty into the vas deferens just before it joins the urethra. Seminal fluid is thick because it contains mucus and fibrinogen, a protein also found in the blood, where it can polymerize to form blood clots. Seminal fluid also contains fructose, an energy source for the sperm.

Men whose work exposes them to high heat (furnace operators and cooks, for example) may experience arrested spermatogenesis and low fertility. This effect may also occur in some men who exercise vigorously or who wear very tight underwear.

The **prostate gland** surrounds the urethra and contributes about 30 percent of the volume of the semen. Prostate fluid is alkaline, so it neutralizes the acidity in the male and female reproductive tracts and makes those environments more hospitable to sperm. The prostate also secretes a clotting enzyme that causes the fibrinogen from the seminal vesicles to convert the semen into a gelatinous mass, facilitating its propulsion into and retention in the upper regions of the female reproductive tract. Another enzyme in the prostate fluid, profibrinolysin, is inactive when secreted, but activated shortly after it enters the female reproductive tract. Active fibrinolysin dissolves the clotted semen and liberates the sperm.

The **bulbourethral glands** produce a small volume of an alkaline, mucoid secretion that helps to neutralize acidity in the urethra and lubricate it to facilitate the passage of semen at the climax of sexual intercourse. Secretions of the bulbourethral glands precede the climax of the sex act and can carry with them residual sperm from prior sexual activity. Therefore, it is possible for pregnancy to occur even if the penis is withdrawn from the female just before climax (a rather ineffective birth control practice known as *coitus interruptus*).

The penis and the scrotum are the male genitalia. The shaft of the penis is covered with normal skin, but the highly sensitive tip, or **glans penis**, is covered with thinner, more sensitive skin that is especially responsive to sexual stimulation. A fold of skin called the *foreskin* covers the glans of the human penis. The procedure known as *circumcision* removes a portion of the foreskin. Considerable controversy exists over this cultural practice. The current position of the American Academy of Pediatrics cites possible medical benefits along with definite associated risks and concludes that scientific evidence does not support circumcision as a routine practice.

Sexual stimulation triggers responses in the nervous system that result in penile **erection.** Nerve endings release a gaseous neurotransmitter, nitric oxide (NO), onto blood vessels leading into the penis. NO stimulates production of the second messenger cGMP (Section 15.3), which causes these vessels to dilate. The increased blood flow that results fills and swells shafts of spongy, erectile tissue located along the length of the penis. The enlargement of these blood-filled cavities compresses the vessels that normally carry blood out of the penis. As a result, the erectile tissue becomes more and more engorged with blood. The penis becomes hard and erect, facilitating its insertion into the female's vagina. Many species of mammals, but not humans, have a bone in the penis, but these species still depend on erectile tissue for copulation.

At the climax of copulation, about 2 to 6 ml of semen is propelled through the vasa deferentia and the urethra in two steps, emission and ejaculation. During **emission** rhythmic contractions of smooth muscles in the vasa deferentia and accessory glands move the semen into the urethra at the base of the penis. **Ejaculation** is caused by contractions of other muscles at the base of the penis surrounding the urethra. The rigidity of the erect penis allows these contractions to force the gelatinous mass of semen through the urethra and out of the penis. The muscle contractions of ejaculation are accompanied by feelings of intense pleasure known as **orgasm.** They are also accompanied by transient increases in heart rate, blood pressure, breathing, and skeletal muscle contractions throughout the body.

After ejaculation, NO release decreases and enzymes break down cGMP, causing the blood vessels flowing into the penis to constrict. The blood pressure in the erectile tissue decreases, relieving the compression of the blood vessels leaving the penis, and the erection declines.

Erectile dysfunction, or *impotence*, is the inability to achieve or sustain an erection. Drugs used to treat erectile dysfunction act by inhibiting the breakdown of cGMP, thus enhancing the effect of NO released in the penis, which improves the ability to achieve and maintain an erection.

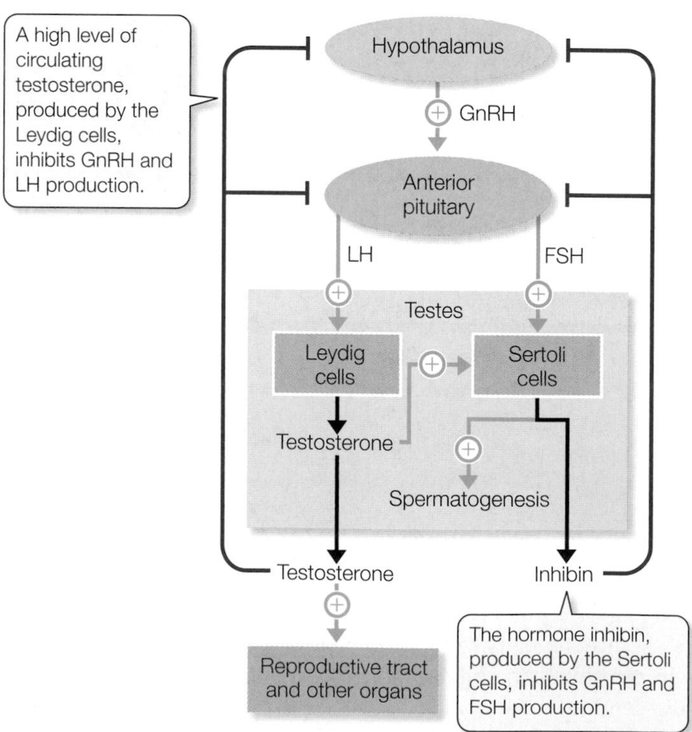

42.11 Hormones Control the Male Reproductive System The male reproductive system is under hormonal control by the hypothalamus and the anterior pituitary.

Male sexual function is controlled by hormones

Spermatogenesis and maintenance of male secondary sexual characteristics depend on testosterone, which is produced by the Leydig cells of the testes. As described in Section 41.3, increased production of testosterone at puberty results from an increased release of gonadotropin-releasing hormone (GnRH) by the hypothalamus, which stimulates anterior pituitary cells to increase their secretion of luteinizing hormone (LH) and follicle-stimulating hormone (FSH) (**Figure 42.11**). Higher levels of LH stimulate the Leydig cells to increase their production and release of testosterone. Testosterone exerts negative feedback on the anterior pituitary and the hypothalamus. At the time of puberty, the sensitivity of the hypothalamus to negative feedback from testosterone declines, and the level of circulating testosterone increases.

Increased testosterone in pubertal males causes the development of secondary sexual characteristics (such as pubic and facial hair, deeper voice, and enlarged genitals) and an increased growth rate. Testosterone also promotes increased muscle mass and maturation of the testes. Continued production of testosterone after puberty is essential for the maintenance of secondary sexual characteristics and the production of sperm.

Spermatogenesis is controlled by the influence of FSH and testosterone on the Sertoli cells in the seminiferous tubules. The Sertoli cells also produce a hormone called *inhibin*, which exerts negative feedback on the anterior pituitary cells that produce and secrete FSH.

Female sex organs produce eggs, receive sperm, and nurture the embryo

When a mammalian egg matures, it is released from the ovary directly into the body cavity. But the egg does not go far. Each ovary is enveloped by the undulating, fringed opening of an **oviduct** (also known as a *Fallopian tube*), which sweeps the egg into that tube (**Figure 42.12**). Fertilization takes place in the oviduct. Whether or not the egg is fertilized, cilia lining the oviduct propel it slowly toward the uterus, a muscular, thick-walled cavity shaped in humans like an upside-down pear. The uterus is where the embryo develops if the egg is fertilized. At the bottom, the uterus narrows into a region called the **cervix**, which leads into the **vagina**.

In humans, two sets of skin folds surround the opening of the vagina and the opening of the urethra, through which urine passes. The inner, more delicate folds are the *labia minora*; the outer, thicker folds are the *labia majora*. At the anterior tip of the labia minora is the *clitoris*, a small bulb of erectile tissue that has the same developmental origins as the penis. The clitoris is highly sensitive and plays an important role in sexual response. The labia minora and the clitoris become engorged with blood in response to sexual stimulation.

The external opening of an infant's vagina is usually, but not always, partly covered by a thin membrane, the *hymen*. Eventually the hymen can be torn by vigorous physical activity or by first sexual intercourse; it can sometimes make first intercourse difficult or painful for the female.

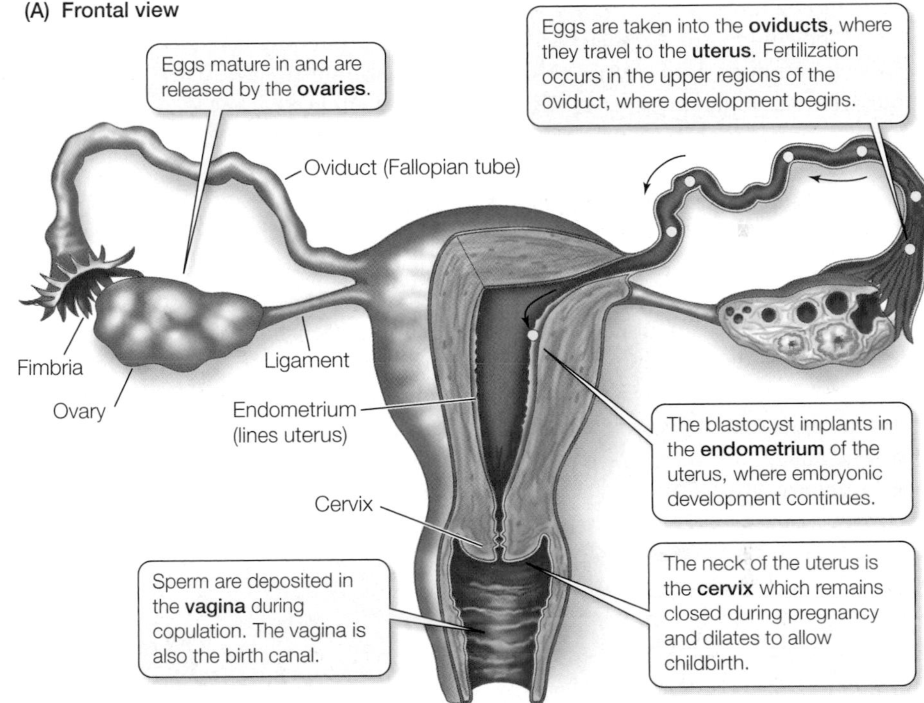

(A) Frontal view
Eggs mature in and are released by the **ovaries**.
Eggs are taken into the **oviducts**, where they travel to the **uterus**. Fertilization occurs in the upper regions of the oviduct, where development begins.
Oviduct (Fallopian tube)
Fimbria
Ovary
Ligament
Endometrium (lines uterus)
The blastocyst implants in the **endometrium** of the uterus, where embryonic development continues.
Cervix
Sperm are deposited in the **vagina** during copulation. The vagina is also the birth canal.
The neck of the uterus is the **cervix** which remains closed during pregnancy and dilates to allow childbirth.

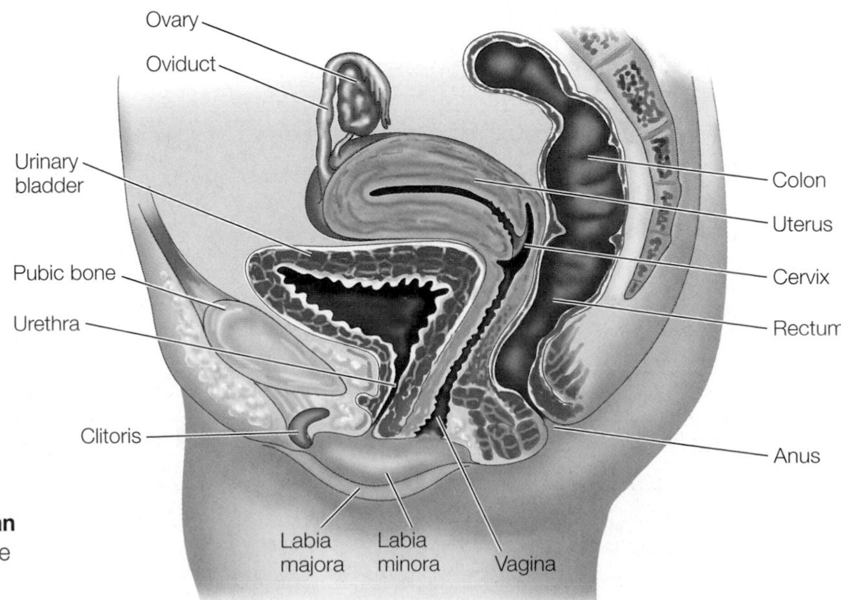

42.12 The Reproductive Tract of the Human Female The female reproductive organs are shown in front and side views.

To fertilize an egg, sperm deposited in the vagina swim and are propelled by contractions of the female reproductive tract through the cervical opening, the uterus, and most of the oviduct. The egg (actually a secondary oocyte) is fertilized in the upper region of the oviduct. Fertilization stimulates the completion of the second meiotic division, after which the haploid nuclei of the sperm and the egg can fuse to produce a diploid zygote nucleus. Still in the oviduct, the zygote undergoes its first few cell divisions to become a **blastocyst**. The blastocyst moves down the oviduct to the uterus, where it attaches itself to the epithelial lining of the uterus, the **endometrium**. Once attached to the endometrium, the blastocyst burrows into it—a process called **implantation**—and interacts with it to form the placenta as we will see in the next chapter. The placenta nurtures the embryo and also produces hormones that help sustain pregnancy.

As the egg matures in the ovary, the endometrium thickens. If a blastocyst does not arrive in the uterus, the endometrium regresses or is sloughed off. Thus the female reproductive cycle actually consists of two linked cycles: an *ovarian cycle* that produces eggs and hormones, and a *uterine cycle* that prepares the endometrium for the arrival of a blastocyst.

The ovarian cycle produces a mature egg

An **ovarian cycle** is about 28 days long in the human female, but it varies considerably among individuals. During the first half of each cycle, at least one primary oocyte matures into a secondary oocyte (egg) and is expelled from the ovary (*ovulation*). During the second half of the cycle, cells in the ovary that were associated with the maturing oocyte develop endocrine functions and then regress if the egg is not fertilized. The progression of these events is shown diagrammatically in **Figure 42.13**.

Some mammals have ovarian cycles longer than 28 days and some shorter. Rats and mice have a 4-day cycle, elephants have a 4-month cycle, and some seasonal breeders have a 1-year cycle. In some species, such as rabbits, ovulation is induced by copulation—a "recipe" for prolific breeding.

A newborn human female has about a million primary oocytes in each ovary. By the time she reaches puberty, she has only about 200,000; the rest have degenerated. During a woman's fertile years, her ovaries will go through about 450 ovarian cycles. During each cycle, a number of oocytes will begin to mature, but usually only one will mature completely and be released; the others degenerate. At around the age of 50, a woman reaches **menopause**—the end of fertility—and may have only a few oocytes left in each ovary. Throughout a woman's life, oocytes are degenerating, and no new ones are produced.

Each primary oocyte in the ovary is surrounded by a layer of ovarian cells. An oocyte and its surrounding cells constitute the functional unit of the ovary, the **follicle**. Between puberty and menopause, six to twelve follicles begin to mature each month. In each follicle, the oocyte enlarges and the surrounding follicular cells proliferate. After about a week, one of these follicles is larger than the rest, and it continues to grow, while the others cease to develop and shrink. In the enlarged follicle, the follicular cells nurture the growing egg, supplying it with nutrients, growth factors, and hormonal stimulation.

In humans, after 2 weeks of follicular growth, **ovulation** occurs: The follicle ruptures and the egg is released. Following ovulation,

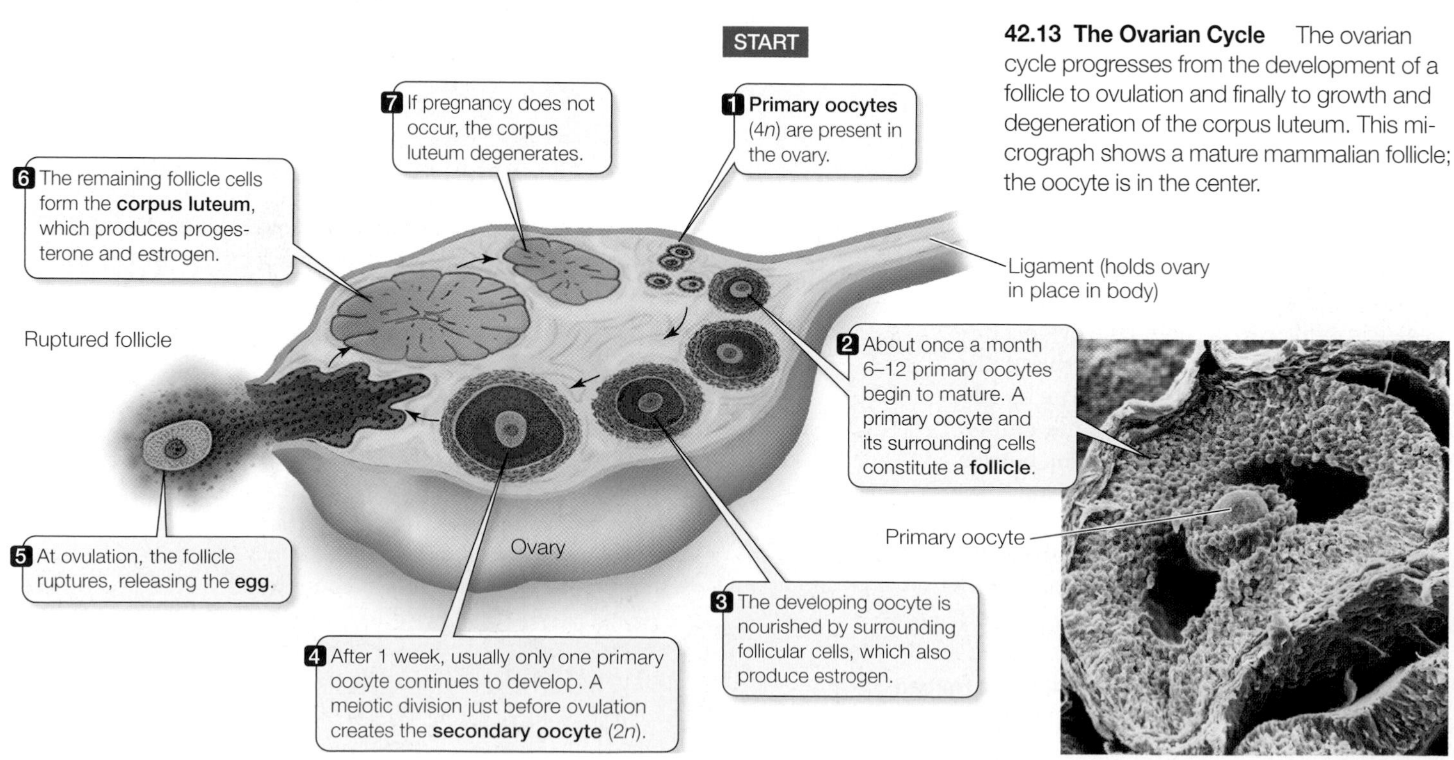

START
1 Primary oocytes (4n) are present in the ovary.
2 About once a month 6–12 primary oocytes begin to mature. A primary oocyte and its surrounding cells constitute a follicle.
3 The developing oocyte is nourished by surrounding follicular cells, which also produce estrogen.
4 After 1 week, usually only one primary oocyte continues to develop. A meiotic division just before ovulation creates the secondary oocyte (2n).
5 At ovulation, the follicle ruptures, releasing the egg.
6 The remaining follicle cells form the corpus luteum, which produces progesterone and estrogen.
7 If pregnancy does not occur, the corpus luteum degenerates.
Ruptured follicle
Ovary
Ligament (holds ovary in place in body)
Primary oocyte

42.13 The Ovarian Cycle The ovarian cycle progresses from the development of a follicle to ovulation and finally to growth and degeneration of the corpus luteum. This micrograph shows a mature mammalian follicle; the oocyte is in the center.

the follicle cells continue to proliferate and form a mass of endocrine tissue about the size of a marble. This structure, which remains in the ovary, is the *corpus luteum* (plural *corpora lutea*). It functions as an endocrine gland, producing estrogen and progesterone for about 2 weeks. It then degenerates unless a blastocyst implants in the endometrium.

The uterine cycle prepares an environment for the fertilized egg

The **uterine cycle** parallels the ovarian cycle and consists of a buildup and then a breakdown of the endometrium (**Figure 42.14**). About 5 days into the ovarian cycle, the endometrium starts to grow in preparation for receiving a blastocyst. The uterus attains its maximum state of preparedness about 5 days after ovulation and remains in that state for another 9 days. If a blastocyst has not arrived by that time, the endometrium begins to break down, and the sloughed-off tissue, including blood, flows from the body through the vagina—the process of **menstruation** (from *menses*, the Latin word for "months").

The uterine cycles of most mammals other than humans do not include menstruation; instead, the uterine lining typically is resorbed. In these species, the most obvious correlate of the ovarian cycle is a state of sexual receptivity called **estrus** around the time of ovulation. You may be aware of the bloody discharge that occurs in dogs at the time of estrus. This discharge is not the same as menstruation—in fact it is exactly the opposite. Bleeding in dogs occurs during the *proliferation* of the uterine lining, which occurs just prior to ovulation. When the female mammal comes into estrus, or "heat," she actively solicits male attention and may be aggressive to other females. Women are unusual among mammals in that they are potentially sexually receptive throughout their ovarian cycles and at all seasons of the year.

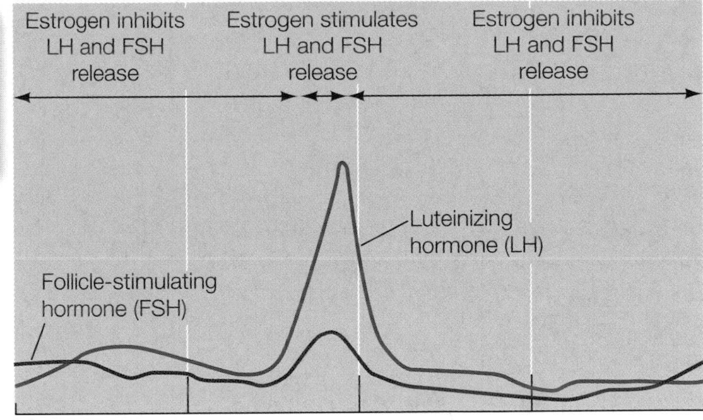

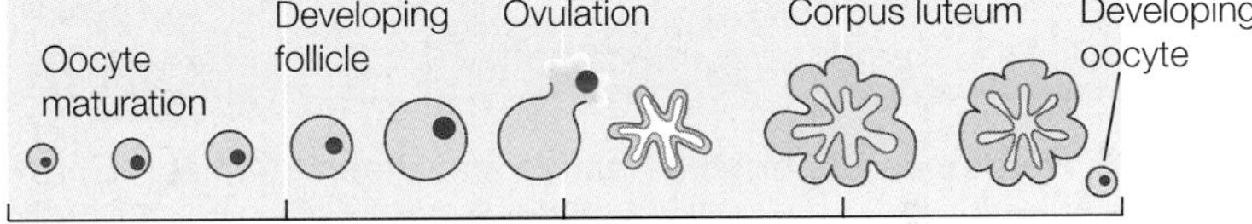

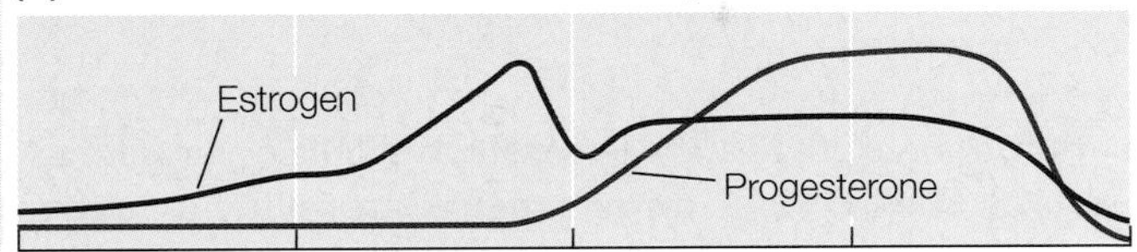

Estrogen inhibits LH and FSH release

Estrogen inhibits LH and FSH release

(C) Ovarian hormones and the uterine cycle
Estrogen and progesterone stimulate the development of the endometrium in preparation for pregnancy.
Estrogen
Progesterone

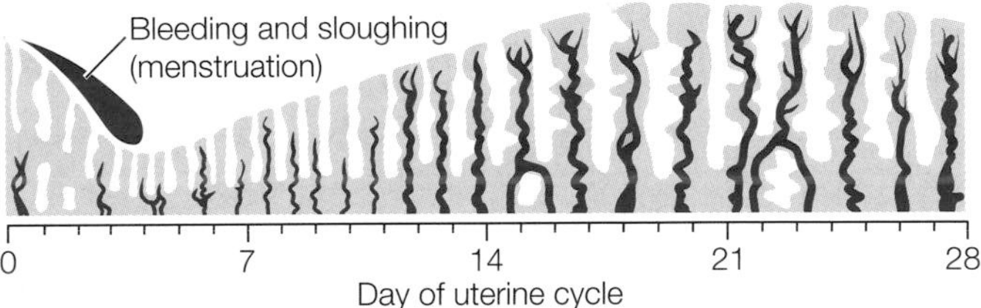

42.14 The Ovarian and Uterine Cycles During a woman's ovarian and uterine cycles, coordinated changes occur in (A) gonadotropin release by the anterior pituitary, (B) the ovary, (C) the release of female sex steroids, and (D) the uterus. The cycles begin with the onset of menstruation; ovulation is at midcycle.

Hormones control and coordinate the ovarian and uterine cycles

The ovarian and uterine cycles are coordinated and timed by the same hormones that initiate sexual maturation. Gonadotropins secreted by the anterior pituitary are the central elements of this control. Before puberty (that is, before about 11 years of age), the secretion of gonadotropins is low and the ovaries are inactive. At puberty, the hypothalamus increases its release of GnRH, stimulating the anterior pituitary to secrete FSH and LH.

In response to FSH and LH, ovarian tissue grows and produces estrogen. The rise in estrogen causes the maturation of the accessory sex organs and the development of female secondary sexual characteristics. Between puberty and menopause, interactions of GnRH, gonadotropins, and sex steroids control the ovarian and uterine cycles.

Menstruation marks the beginning of each uterine and ovarian cycle (see Figure 42.14). A few days before menstruation begins, the anterior pituitary begins to increase its secretion of FSH and LH. In response, some 10 to 20 follicles begin to mature in the ovaries, and these follicles steadily increase their production of estrogen. After about a week, all but one of the follicles wither away.

Estrogen exerts negative feedback control on gonadotropin release by the anterior pituitary during the first 12 days of the ovarian cycle. Then, on about day 12, estrogen exerts positive rather than negative feedback control on the pituitary (**Figure 42.15**).

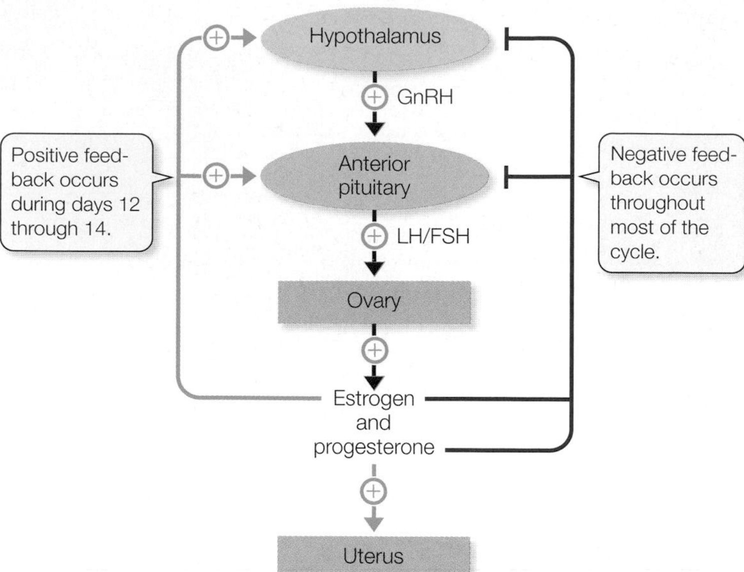

42.15 Hormones Control the Ovarian and Uterine Cycles The ovarian and uterine cycles are under a complex series of positive and negative feedback controls involving several hormones.

As a result, a surge of LH and a lesser surge of FSH occur (see Figure 42.14A). The LH surge triggers the mature follicle to rupture and release its egg, and it stimulates the cells of the ruptured follicle to develop into a corpus luteum.

The corpus luteum becomes an endocrine gland. Estrogen and especially progesterone secreted by the corpus luteum following ovulation are crucial to continued growth and maintenance of the endometrium. In addition, these sex steroids exert negative feedback control on the pituitary, inhibiting gonadotropin release and thus preventing new follicles from beginning to mature.

If the egg is not fertilized, the corpus luteum degenerates on about day 26 of the cycle. Without production of progesterone by the corpus luteum, the endometrium sloughs off, and menstruation occurs. The decrease in circulating steroids also releases the hypothalamus and pituitary from negative feedback control, so GnRH, FSH, and LH all begin to increase. The increase in these hormones induces the next round of follicle development, and the ovarian cycle begins again.

In pregnancy, hormones from the extraembryonic membranes take over

If the egg is fertilized and a blastocyst arrives in the uterus and implants in the endometrium, a new hormone comes into play. A layer of cells covering the blastocyst begins to secrete **human chorionic gonadotropin** (**hCG**). This gonadotropin, a molecule similar to LH, stimulates the corpus luteum to continue to produce estrogen and progesterone to support the growth and maintenance of the endometrium and thereby prevent menstruation. Because hCG is present only in the blood of pregnant women, the presence of this hormone is the basis for pregnancy testing. Modern pregnancy tests make use of an antibody to detect hCG in urine; thus, they take only minutes and can be done at home. These tests are so sensitive that they are 99 percent accurate and in most cases can detect a pregnancy even before the first missed menstrual period.

Tissues derived from the blastocyst also begin to produce estrogen and progesterone, eventually replacing the corpus luteum as the most important source of these sex steroids. Continued high levels of estrogen and progesterone prevent the pituitary from secreting gonadotropins; thus, the ovarian cycle ceases for the duration of pregnancy. The same mechanism is exploited by birth control pills, which contain synthetic hormones resembling estrogen and progesterone that exert negative feedback control on the hypothalamus and pituitary.

Childbirth is triggered by hormonal and mechanical stimuli

Throughout pregnancy, the muscles of the uterine wall periodically undergo slow, weak, rhythmic contractions called *Braxton-Hicks contractions*. These contractions become gradually stronger during the third trimester of pregnancy and are sometimes called *false labor contractions*. True labor contractions usually mark the beginning of childbirth. Both hormonal and mechanical stimuli contribute to the onset of labor.

Progesterone inhibits and estrogen stimulates contractions of uterine muscle. Toward the end of the third trimester, the estrogen–progesterone ratio shifts in favor of estrogen. The onset of labor is marked by increased oxytocin secretion by the pituitaries of both mother and fetus. Oxytocin is a powerful stimulant of uterine muscle contraction.

Mechanical stimuli come from the stretching of the uterus by the fully grown fetus and the pressure of the fetal head on the cervix. These mechanical stimuli increase the release of oxytocin by the posterior pituitary, which in turn increases the activity of uterine muscle, which causes even more pressure on the cervix. This positive feedback loop converts the weak, slow, rhythmic Braxton-Hicks contractions into stronger labor contractions (**Figure 42.16A**).

In the early stage of labor, the contractions of the uterus are 15–20 minutes apart, and each lasts 45–60 seconds. During this time, hormonal changes and pressure created by the contractions cause the cervix to dilate (expand) until it is large enough to allow the baby to pass through. Gradually the contractions become more frequent and more intense. This stage of labor lasts an average of 12 to 15 hours in a first pregnancy; it is usually 8 hours or less in subsequent ones.

The second stage of labor, called *delivery*, begins when the cervix is fully dilated to a diameter of about 10 centimeters (**Figure 42.16B**). The baby's head moves into the vagina; passage of the fetus through the vagina is assisted by the mother's bearing down ("pushing") with her abdominal and other muscles. Once the head and shoulders of the baby clear the cervix, the rest of its body eases out rapidly, but it is still connected to the placenta by the umbilical cord. Delivery may take as little as a minute, or up to half an hour or more in a first pregnancy.

As soon as the baby clears the birth canal, it can start breathing and become independent of its mother's circulation. The umbilical cord may then be clamped and cut. The segment still attached to the baby dries up and sloughs off in a few days, leaving behind its distinctive signature, the belly button—more properly called the

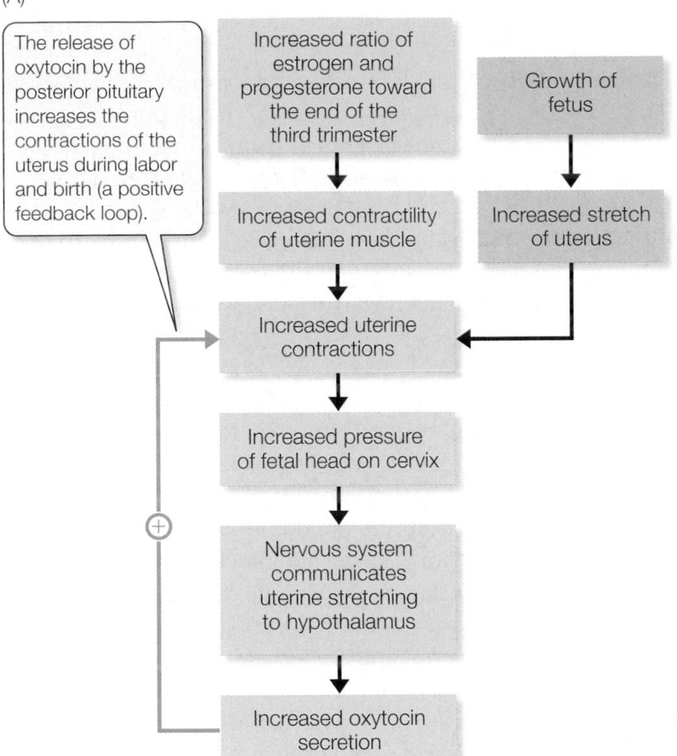

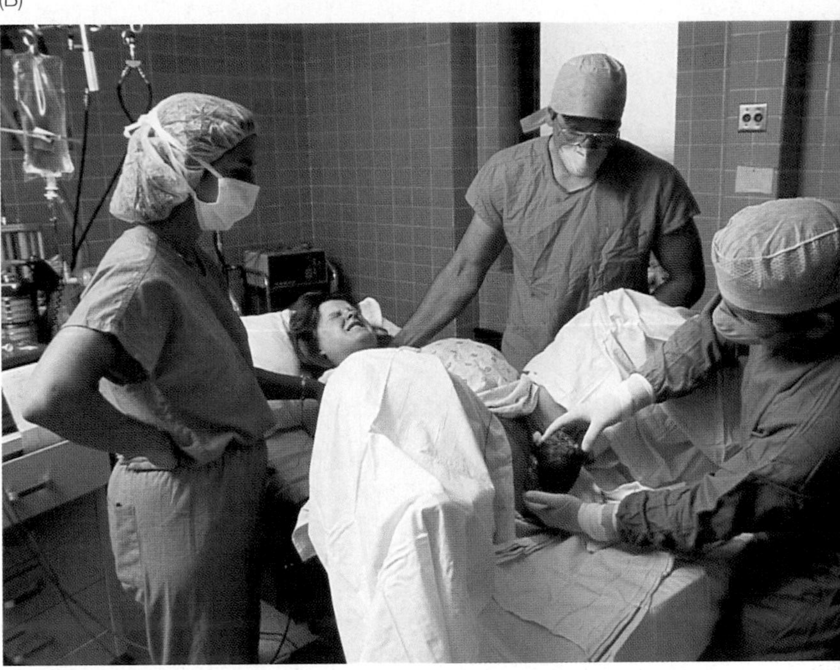

42.16 Control of Uterine Contractions and Childbirth (A) Both mechanical and hormonal signals are involved in stimulating the uterine contractions of labor and delivery. (B) A new person is delivered into the world head first.

umbilicus. The detachment and expulsion of the placenta and fetal membranes takes from a few minutes to an hour, and may be accompanied by uterine contractions. If the baby suckles at the breast immediately following birth, its suckling stimulates additional secretion of oxytocin, which augments uterine contractions that reduce the size of the uterus and help stop bleeding.

42.3 RECAP

The reproductive systems of human males and females produce gametes and hormones, and these functions are controlled by hypothalamic and anterior pituitary hormones. In females the hormonal control of reproductive functions produces linked ovarian and uterine cycles.

- Can you describe the path the human sperm and ovum take in moving from their respective gonads to the point at which fertilization occurs? See Figures 42.9 and 42.12
- Increased production of testosterone at puberty in males stimulates the release of what two hormones of the anterior pituitary? What effect do these hormones have? See p. 909
- Can you explain the events in the ovarian cycle that result in release of a single ovum each month? What events prepare the uterus to receive the egg? See Figures 42.13 and 42.14

Understanding the physiology of human reproduction has led to numerous methods and technologies for controlling it either to prevent unwanted pregnancies or to overcome infertility.

42.4 How Can Fertility Be Controlled and Sexual Health Maintained?

Sexual issues and sexual behavior are dominant aspects of our society, and reproductive technologies have had huge impacts on our sexual and reproductive lives.

Human sexual responses have four phases

The responses of both women and men to sexual stimulation consist of four phases: excitement, plateau, orgasm, and resolution. As sexual *excitement* begins in a woman, her heart rate and blood pressure rise, muscular tension increases, breasts swell, and nipples become erect. Her external genitals, including the sensitive clitoris, swell as they become filled with blood, and the walls of the vagina secrete lubricating fluid that facilitates copulation. In the *plateau* phase, her blood pressure and heart rate rise further, her breathing becomes rapid. The sensitivity once focused in the clitoris spreads over the external genitals, and the clitoris itself becomes even more sensitive. *Orgasm* may last as long as a few minutes, and, unlike men, some women can experience several orgasms in rapid succession. During the *resolution* phase, blood drains from the genitals, and body physiology returns to close to normal.

In the male, the excitement phase is marked by an increase in blood pressure, heart rate, and muscle tension, and penile erection. In the plateau phase, breathing becomes rapid, the diameter of the glans increases, and a clear lubricating fluid from the bulbourethral gland oozes from the penis. Pressure and friction against the nerve endings in the glans and in the skin along the shaft of the penis eventually trigger orgasm. Massive spasms of the muscles in the genital area and contractions in the accessory reproductive organs result in ejaculation.

Within a few minutes after ejaculation, the penis shrinks to its former size, and body physiology returns to resting conditions. The male sexual response includes a *refractory period* immediately after orgasm. During this period, which may last from minutes to hours, a man cannot achieve a full erection or another orgasm, regardless of the intensity of sexual stimulation.

Humans use a variety of methods to control fertility

According to a recent study, almost half of the more than 6 million pregnancies that occur in the United States each year are unintended. The only failure-proof methods of preventing pregnancy are complete abstinence from sexual activity, or the surgical removal of the gonads. Since those approaches are not acceptable to most younger people, they turn to other methods to prevent pregnancy. Many of these methods prevent fertilization or implantation (*conception*) and are therefore referred to as **contraception**.

Some methods of contraception are used by the woman, and others by the man. They vary from blocking gametogenesis to blocking implantation of a blastocyst; they also vary enormously in their effectiveness. **Table 42.1** lists some of the most common methods and their relative failure rates.

For women of college age, a single act of unprotected intercourse in the two days prior to ovulation carries a chance of conception as high as 50 percent.

NONTECHNOLOGICAL APPROACHES An approach to contraception that does not involve physical or pharmacological technologies is periodic abstinence (the "rhythm method"), which attempts to separate sperm and egg in time. The couple avoids sex from day 10 to day 20 of the ovarian cycle, when the woman is most likely to be fertile. The cycle can be tracked by use of a calendar, supplemented by the basal body temperature method, which is based on the observation that a woman's body temperature drops on the day of ovulation and rises sharply on the day after. Changes in the stickiness of the cervical mucus also help identify the day of ovulation.

The "rhythm method" has a high failure rate. Sperm deposited in the female reproductive tract may remain viable for up to 6 days. Similarly, the egg remains viable for 12 to 36 hours after ovulation. These facts, added to individual variation in the timing of ovulation, result in an annual failure rate of between 15 and 35 percent for the rhythm method. In other words, 15 to 35 percent of women using only the rhythm method over the course of 1 year will become pregnant during that time.

Another approach is to try to separate sperm and egg in space through **coitus interruptus**—withdrawal of the penis from the vagina before ejaculation. The annual failure rate of this method (mostly due to lack of willpower) may be as high as 40 percent.

BARRIER METHODS Techniques that place a physical barrier between egg and sperm have been used for centuries. The condom is a sheath made of an impermeable material such as latex that can be fitted over the erect penis. A condom traps semen so that sperm do not enter the vagina. Latex condoms also help prevent the spread of many sexually transmitted diseases. In theory, condom use has a failure rate near zero. However, in practice the annual failure rate is about 15 percent due to leakage because of tearing or poor fit (for example, with the loss of the erection).

Diaphragms and *cervical caps* are dome-shaped pieces of rubber that fit over the woman's cervix to block sperm from entering the uterus. Both the diaphragm and the cervical cap are treated first with jelly or cream containing a *spermicide* (a chemical that kills or incapacitates sperm) and then inserted through the vagina before sexual intercourse. Annual failure rates are about 15 percent, the same as for male condoms. A female condom, which creates an impermeable lining of the vagina, is also available.

Spermicidal foams, jellies, and creams can be used alone by placing them in the vagina with special applicators. Used in this way, they have an annual failure rate of 25 percent or more. *Douching* (flushing the vagina with liquid) after intercourse, despite popular belief, is almost useless as a method of birth control, because sperm can arrive in the oviducts (which cannot be flushed) within 10 minutes after ejaculation.

PREVENTING OVULATION Oral contraceptives, or birth control pills, are the leading method of birth control in the United States and

TABLE 42.1

Methods of Contraception

METHOD	MODE OF ACTION	FAILURE RATE[a]
Unprotected	No form of birth control	85
Douche	Supposedly flushes sperm from vagina	80
Periodic abstinence	Abstinence near time of ovulation	15–35
Coitus interruptus (withdrawal prior to ejaculation)	Prevents sperm from reaching egg	10–40
Vaginal jelly or foam	Kills sperm; blocks sperm movement	3–30
Diaphragm/jelly	Prevents sperm from entering uterus; kills sperm	3–25
Condom	Prevents sperm from entering vagina	3–20
Intrauterine device (IUD)	Prevents implantation of fertilized egg	0.5–6
RU-486	Prevents development of fertilized egg	0–15
Birth control pill	Prevents ovulation	0–3
Vasectomy	Prevents release of sperm	0–0.15
Tubal ligation	Prevents egg from entering uterus	0–0.05

[a]Number of pregnancies per 100 women per year.

are used by almost 12 million women. Used correctly, "the pill" is the most effective method of contraception other than sterilization, with an annual failure rate of less than 1 percent.

Birth control pills work by preventing ovulation. Their mechanisms of action take advantage of the roles of estrogen and progesterone as negative feedback signals to the hypothalamus and the pituitary. The most common pills contain low doses of synthetic estrogens and progesterones (progestins). By keeping the circulating levels of gonadotropins low, these hormones interfere with the maturation of follicles and eggs, suspending the ovarian cycle. The uterine cycle is usually allowed to continue, however, by including in the daily regime pills that contain no hormones every 21 to 23 days.

The negative side effects of oral contraceptives include somewhat increased risk of blood clot formation, heart attack, stroke, and breast cancer. However, these side effects are associated mostly with pills containing higher hormone concentrations than are used in the modern formulations. For birth control pills in use today, the risk is low for non-smokers, especially women under the age of 35. The risk of death from using birth control pills is far less than that associated with a full-term pregnancy.

Long-lasting injectable or implantable steroids are also used to block ovulation through negative feedback effects. Depo-Provera is an injectable progestin that blocks the release of gonadotropins for several months. Another device, called Norplant, consists of thin, flexible tubes filled with progestin. Several of these tubes are inserted under the skin, where they continue to release progestin slowly for years. Another means of controlling blood levels of sex steroids that does not require a daily pill is the transdermal patch.

Another contraceptive device is the plastic *vaginal ring* which is placed around the cervix where it steadily releases estrogen and progestins to prevent ovulation. It is left in place for 3 weeks, removed to allow menstruation, and then a new ring is put in place.

PREVENTING IMPLANTATION A highly effective method of contraception (failure rate 1–7 percent) is the intrauterine device, or IUD, a small piece of plastic or copper that is inserted into the uterus. The IUD works by causing an inflammatory response that includes the release of prostaglandins, which prevent implantation of the fertilized egg.

Another way of interfering with implantation is through the use of "morning-after pills," which deliver high doses of steroids, primarily estrogens. By acting in several ways on the oviducts and the endometrium, this treatment prevents implantation. Morning-after pills can be effective up to several days after sexual intercourse.

The drug RU-486 is not a contraceptive pill, but a *contragestational* pill. It blocks progesterone receptors, thereby interfering with the normal action of progesterone produced by the corpus luteum, which is necessary for the maintenance of the endometrium in early pregnancy. If RU-486 is administered as a morning-after pill, it prevents implantation. However, RU-486 can be effective even if taken at the time of the first missed menstrual period, after implantation has begun. After a few days of treatment with RU-486, the endometrium regresses and sloughs off, along with the embryo, which is in very early stages of development.

STERILIZATION One foolproof method of contraception is sterilization. Male sterilization by **vasectomy** is a simple operation (cutting and tying of the vasa deferentia) that can be performed under a local anesthetic in a doctor's office (**Figure 42.17A**). After this minor surgery, the semen no longer contains sperm. Sperm production continues, but since the sperm cannot move out of the testes, they are destroyed by macrophages. Vasectomy does not affect a man's hormone levels or his sexual responses, and even the amount of semen he ejaculates is essentially unchanged.

In female sterilization, the aim is to prevent the egg from traveling to the uterus and to block sperm from reaching the egg. The most common method is **tubal ligation**: cutting and tying of the oviducts (**Figure 42.17B**). Alternatively, the oviducts may be burned (cauterized) to seal them off. As in the male, these procedures do not alter reproductive hormones or sexual responses. In the United States, female sterilization is second only to use of the pill as a means of limiting fertility.

ABORTION Once a fertilized egg is successfully implanted in the uterus, any termination of the pregnancy is called an **abortion**. A *spontaneous abortion* is the medical term for what most people call a miscarriage. Miscarriages are common early in pregnancy. Most miscarriages are due to a chromosomal abnormality in the fetus or to a breakdown in the process of implantation. Miscarriages often occur before a woman even realizes she is pregnant. Studies indicate that as many as 50 percent of all human conceptions may miscarry before implantation, and of those that implant

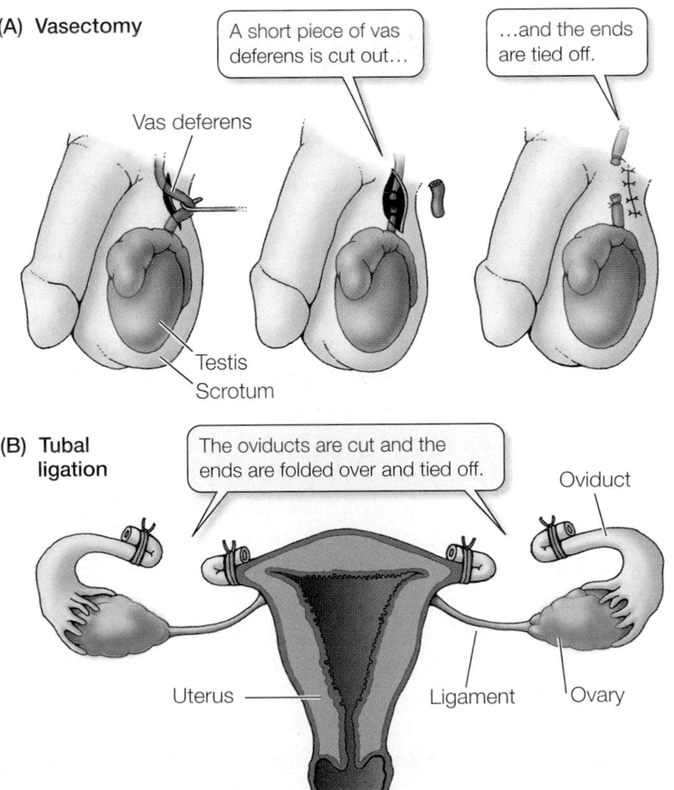
(A) Vasectomy
A short piece of vas deferens is cut out...
...and the ends are tied off.
Vas deferens
Testis
Scrotum
(B) Tubal ligation
The oviducts are cut and the ends are folded over and tied off.
Oviduct
Uterus
Ligament
Ovary

42.17 Sterilization Techniques (A) Vasectomy is the technique for male sterilization. (B) Tubal ligation is the sterilization procedure most commonly performed on human females.

about 25 percent miscarry before the pregnancy can be clinically confirmed. About 10 percent of clinically confirmed pregnancies miscarry.

Abortions that are the result of medical intervention may be performed either for therapeutic purposes or for fertility control. A therapeutic abortion may be necessary to protect the health of the mother, or it may be performed because prenatal testing reveals that the fetus has a severe defect. Of the approximately 3 million unintended pregnancies in the United States each year, almost half are ended by abortion.

In a medical abortion, the cervix is dilated and the endometrium and implanted fetus are removed from the uterus. When performed in the first trimester of a pregnancy, a medical abortion carries less risk of death to the mother than a full-term pregnancy. After the first 12 weeks of pregnancy, the risk rises, but even through the second trimester it is less than that of a full-term pregnancy.

CONTROLLING MALE FERTILITY You may well ask why all the pharmacological approaches to controlling fertility are applied to females. The control of male fertility is a difficult problem. First, spermatogenesis is a continuous rather than a cyclical event, and it is difficult to block a particular step in a continuous process. The ovarian cycle is more vulnerable to manipulation because certain events must happen at certain times and in a certain sequence for ovulation and implantation to occur. Second, the suppression of spermatogenesis must be total to be effective, since technically it takes only a single sperm to fertilize an egg, and normally millions are produced continuously. Such suppression requires powerful and constant chemical intervention, with associated side effects.

Reproductive technologies help solve problems of infertility

There are many reasons why a man and woman may not be able to have children, and a number of technologies have been developed to overcome barriers to conceiving and/or bearing a child.

The simplest treatment available is **artificial insemination**, in which the physician positions sperm in the female's reproductive tract. This technique is useful if the male's sperm count is low, if his sperm lack motility, or if problems in the female's reproductive tract prevent the normal movement of sperm up to and through the oviducts. Artificial insemination is used widely in the production of domesticated animals such as cattle.

More recent advances, called **assisted reproductive technologies**, or **ARTs**, involve procedures that remove unfertilized eggs from the ovary, combine them with sperm outside the body, and then place fertilized eggs or egg–sperm mixtures in the appropriate location in the female's reproductive tract for development to take place.

Approximately 15 percent of all couples in the United States are infertile. The causes are about equally distributed between men and women.

The first successful ART was *in vitro fertilization* (*IVF*). In IVF, the female is treated with hormones that stimulate many follicles in her ovaries to mature. Eggs are collected from these follicles, and sperm are collected from the male. Eggs and sperm are combined in a culture medium outside the body and fertilization takes place. The resulting embryos can be injected into the mother's uterus in the blastocyst stage or kept frozen for implantation later. The first "test-tube baby" resulting from IVF was born in 1978. Since that time, more than 3 million babies have been produced by this ART.

A major cause of failure of IVF is failure of sperm to gain access to the egg plasma membrane (see Figure 42.6). To solve this problem, methods have been developed to inject a sperm cell directly into the cytoplasm of an egg. In *intracytoplasmic sperm injection* (*ICSI*), an egg is held in place by suction applied to a polished glass pipette. A slender, sharp pipette is then used to penetrate the egg and inject a sperm (**Figure 42.18**). This ART was used successfully for the first time in 1992; now thousands of these procedures are performed in U.S. clinics each year, with a success rate of about 25 percent.

IVF, coupled with sensitive techniques of genetic analysis, can eliminate the risk that adults who are carriers of genetic diseases will produce affected children. It is now possible to take a cell from a blastocyst at the 4- or 8-cell stage without damaging its developmental potential. The sampled cell can be subjected to molecular analysis to determine whether it carries the harmful gene. This procedure, called *preimplantation genetic diagnosis* (PGD) makes it possible to determine whether an embryo produced by IVF carries the genetic defect of concern.

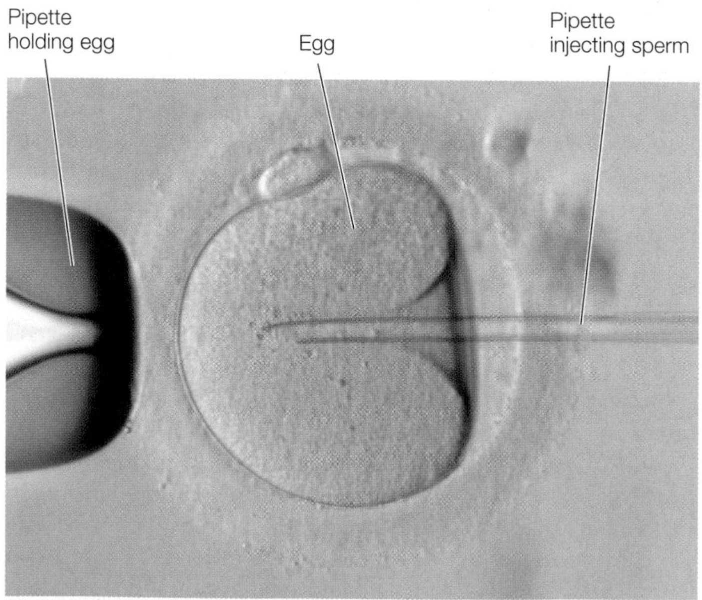

42.18 Intracytoplasmic Sperm Injection In this procedure, sperm are injected directly into a mature egg cell. The fertilized egg is then placed in the female reproductive tract, where it can implant and develop into a fetus.

TABLE 42.2

Some Sexually Transmitted Diseases

DISEASE	INCIDENCE IN UNITED STATES	SYMPTOMS
Syphilis	80,000 new cases/yr	Primary stage (weeks): skin lesion (chancre) at site of infection Secondary stage (months): skin rash and flu-like symptoms; may be followed by a latent period Tertiary stage (years): deterioration of the cardiovascular and central nervous systems; death
Gonorrhea	800,000 new cases/yr	Pus-filled discharge from penis or vagina; burning urination. Infection can also start in throat or rectum
Chlamydia	>4,000,000 new cases/yr	Symptoms similar to gonorrhea, although often there are no obvious symptoms. Can lead to pelvic inflammatory disease in females
Genital herpes	500,000 new cases/yr	Small blisters that can cause itching or burning sensations are accompanied by inflammation and by secondary infections
Genital warts	10% of adults infected	Small growths on genital tissues. Increases risk of cervical cancer in women
Hepatitis B	5–20% of population	Fatigue, fever, nausea, loss of appetite, jaundice, abdominal pain, muscle and joint pain. Can lead to destruction of liver or liver cancer
HIV/AIDS	Approximately 900,000 cases[a]	Failure of the immune system (see Section 18.6)

[a]HIV/AIDS is widespread in other parts of the world, most notably in the southern part of the African continent. The infection is spreading rapidly in Southeast Asia and India. Estimated number of people infected with HIV worldwide in 2006 was 40 million.

Sexual behavior transmits many disease organisms

Disease-causing organisms usually have a very limited ability to survive outside a host organism, which means getting from host to host is a major evolutionary challenge. One of the most intimate types of contact that hosts can have is copulation. It is not surprising, then, that many pathogens have evolved to depend on sexual contact between their hosts as their means of transmission. These organisms are the causes of **sexually transmitted diseases** (commonly referred to as **STDs**), and they include viruses, bacteria, yeasts, and protists. A summary of the most common STDs is presented in **Table 42.2**.

STDs have been with humans since ancient times, and they are one of the most serious public health problems today. Over 10 million new cases of STDs occur each year in the United States, and about two-thirds of these cases occur in people between the ages of 15 and 30. About half of U.S. youth will contract an STD before the age of 25. The only contraceptive device that is effective against the transmission of STDs is the condom.

42.4 RECAP

Controlling fertility and maintaining sexual health are important aspects of human life. Decreasing the probability of pregnancy is achieved through methods that prevent sperm and egg from meeting and from preventing implantation. Pregnancies can be facilitated through medical technology.

- Can you describe the various barrier methods of contraception? See Table 42.1
- Do you understand how implantation of a blastocyst can be prevented? See p. 915

CHAPTER SUMMARY

42.1 How do animals reproduce without sex?

Asexual reproduction produces offspring that are genetically identical to their parent and to one another. A disadvantage of asexual reproduction is that no genetic diversity is produced.

Means of asexual reproduction include **budding**, **regeneration**, and **parthenogenesis**. Review Figures 42.1 and 42.2

42.2 How do animals reproduce sexually?

Sexual reproduction consists of three basic steps: **gametogenesis**, **mating**, and **fertilization**.

Gametogenesis and fertilization are similar in all animals, but mating includes a great variety of anatomical, physiological, and behavioral adaptations.

In sexually reproducing species, genetic diversity is created by crossing over and independent assortment of chromosomes during gametogenesis. Fertilization also contributes to genetic diversity.

Gametogenesis occurs in **testes** and **ovaries**. In **spermatogenesis** (the production of sperm) and **oogenesis** (the production of eggs), the **germ cells** proliferate mitotically, undergo meiosis, and mature into gametes.

Each **primary spermatocyte** can produce four haploid sperm through the two divisions of meiosis. Review Figure 42.3A

Primary oocytes immediately enter prophase of the first meiotic division, and in many species, including humans, their development is arrested at this point. Each **oogonium** produces only one egg. Review Figure 42.3B

Fertilization involves sperm activation, species-specific binding of sperm to egg, the **acrosomal reaction**, digestion of a path through the protective coverings of the egg, and fusion of sperm and egg plasma membranes. Fusion of these two membranes triggers **blocks to polyspermy**, which prevent additional sperm from entering the egg, and in mammals, signal the egg to complete meiosis and begin development. Review Figure 42.5, Web/CD Tutorial 42.1

Fertilization can occur externally, as is common in aquatic species, or internally, as is common in terrestrial species. Internal fertilization usually involves copulation.

Hermaphroditic, or **monoecious**, species have both male and female reproductive systems in the same individual, either sequentially or simultaneously. **Dioecious** species have separate male and female members.

Internal fertilization is necessary for terrestrial species. The shelled egg is an important adaptation to the terrestrial environment.

Animals can be classified as **oviparous** or **viviparous**, depending on whether the early stages of development occur outside or inside the mother's body.

42.3 How do the human male and female reproductive systems work?

Males produce **semen** and deliver it into the female reproductive tract. Semen consists of sperm suspended in seminal fluid, which nourishes them and facilitates fertilization.

Sperm are produced in the **seminiferous tubules** of the testes, mature in the **epididymis**, and are delivered to the **urethra** through the **vas deferens**. Other components of semen are produced in the **seminal vesicles**, **prostate gland**, and **bulbourethral gland**. Review Figures 42.9 and 42.10, Web/CD Activities 42.1 and 42.2

All components of the semen join in the urethra at the base of the penis and are ejaculated through the erect penis by muscle contractions at the culmination of copulation.

Spermatogenesis depends on testosterone secreted by the **Leydig cells** of the testes, which are under the control of LH from the anterior pituitary. Spermatogenesis is also controlled by FSH from the pituitary. Hypothalamic GnRH controls pituitary secretion of LH and FSH. The production of these hormones by the hypothalamus and pituitary is controlled by negative feedback from testosterone and another hormone, inhibin, produced by the **Sertoli cells** of the testes. Review Figure 42.11

Eggs mature in the female's ovaries and are released into the **oviducts**. Sperm deposited in the vagina during copulation move up through the **cervix** and **uterus** into the oviducts. Review Figure 42.12, Web/CD Activity 42.3

Fertilization occurs in the upper regions of the oviducts. The zygote becomes a **blastocyst** as it passes down the oviduct. Upon arrival in the uterus, the blastocyst implants in the **endometrium** and forms a **placenta**.

The maturation and release of eggs constitute an **ovarian cycle**. In humans, this cycle takes about 28 days. Review Figure 42.13

The uterus also undergoes a cycle that prepares it for receipt of a blastocyst. If no blastocyst is implanted, the lining of the uterus sloughs off in the process of **menstruation**. Review Figure 42.14, Web/CD Tutorial 42.2

Both the ovarian and the uterine cycles are under the control of hypothalamic and pituitary hormones, which in turn are under the feedback control of estrogen and progesterone. Review Figure 42.15

Childbirth is initiated by hormonal and mechanical stimuli that increase the contraction of uterine muscle. Review Figure 42.16

42.4 How can fertility be controlled and sexual health maintained?

Human sexual responses consist of four phases: excitement, plateau, orgasm, and resolution. In addition, males have a refractory period during which renewed excitement is not possible.

Methods of contraception include abstention from copulation and the use of technologies that decrease the probability of fertilization. Review Table 42.1

Barrier methods of contraception, such as condoms, diaphragms, and spermicidal substances, kill sperm or block their passage through the female reproductive tract.

Methods to prevent ovulation, such as birth control pills and other hormonal treatments, interfere with the ovarian cycle so that mature, fertile eggs are not produced and released.

Males and females can be sterilized by surgical blockage of the vasa deferentia (**vasectomy**) or oviducts (**tubal ligation**). Review Figure 42.17

Methods to prevent implantation include intrauterine devices, excess doses of steroids, and a progesterone receptor blocker. After implantation, the termination of a pregnancy is called an **abortion**.

Assisted reproductive technologies have been developed to increase fertility.

Many disease-causing organisms are transmitted through sexual behavior. Many **sexually transmitted diseases** are curable if treated early, but can have serious long-term consequences if not treated. Review Table 42.2

SELF-QUIZ

1. A species in which the individual possesses both male and female reproductive systems is termed
   a. dioecious.
   b. parthenogenetic.
   c. hermaphroditic.
   d. monoecious.
   e. ovoviviparous.
2. The major advantage of internal fertilization is that
   a. it ensures paternity.
   b. it permits the fertilization of many gametes.
   c. it reduces the incidence of destructive competitive interactions between the members of a group.
   d. it increases the number of sperm that have access to each egg.
   e. it gives the developing organism a greater degree of protection during the early phases of development.
3. Which statement about oocytes is *true*?
   a. At birth, the human female has produced all the oocytes she will ever produce.
   b. At the onset of puberty, ovarian follicles produce new oocytes in response to hormonal stimulation.
   c. At the onset of menopause, the human female stops producing oocytes.
   d. Oocytes are produced by the human female throughout adolescence.
   e. Oocytes produced by the female are stored in the oviducts.
4. Spermatogenesis and oogenesis differ in that
   a. spermatogenesis produces gametes with greater stores of raw materials than those produced by oogenesis.
   b. spermatocytes remain in prophase of the first meiotic division longer than oocytes.
   c. oogenesis produces four equally functional haploid cells per meiotic event and spermatogenesis does not.
   d. spermatogenesis produces many gametes with meager energy reserves, whereas oogenesis produces relatively few, well-provisioned gametes.
   e. spermatogenesis begins before birth in humans, whereas oogenesis does not start until the onset of puberty.
5. Semen contains all of the following except
   a. fructose.
   b. mucus.
   c. clotting enzymes.
   d. substances to lower the pH of the uterine environment.
   e. substances to increase the contraction of the uterine muscle.
6. During oogenesis in mammals, the second meiotic division occurs
   a. in the formation of the primary oocyte.
   b. in the formation of the secondary oocyte.
   c. before ovulation.
   d. after fertilization.
   e. after implantation.
7. One of the major differences between the sexual responses of human males and females is
   a. the increase in blood pressure in males.
   b. the increase in heart rate in females.
   c. the presence of a refractory period in females.
   d. the presence of a refractory period in males after orgasm.
   e. the increase in muscle tension in males.
8. Which of the following is *true* of sexually transmitted diseases?
   a. They are always caused by viruses or bacteria.
   b. Using contraception will prevent them.
   c. The organisms that cause them have evolved to depend on intimate physical contact between hosts as their means of transmission.
   d. Their transmission has a high probability of failure.
   e. You cannot catch one from someone you love.
9. Contractions of muscles in the uterine wall and in the breasts are stimulated by
   a. progesterone.
   b. estrogen.
   c. prolactin.
   d. oxytocin.
   e. human chorionic gonadotropin.
10. Which method of contraception is *most* likely to fail?
    a. Rhythm method
    b. Birth control pills
    c. Diaphagms
    d. Vasectomy
    e. Condoms

FOR DISCUSSION

1. In the very deep ocean, there are species of fish in which the male is very much smaller than the female and actually lives attached to her body. In terms of the selective pressures that operate on sexual and asexual reproduction and in terms of the deep-sea environment, what factors do you think resulted in the evolution of this extreme sexual dimorphism?
2. What are two main differences between the immediate products of the first and second meiotic divisions in spermatogenesis and oogenesis? Why do these differences exist?
3. At the beginning of each ovarian cycle in humans, six to twelve follicles begin to develop in response to rising levels of FSH, but after a week, only one follicle continues to develop, and the others wither away. Given that follicles produce estrogen, estrogen stimulates follicle cells to produce FSH receptors, and estrogen exerts negative feedback on FSH production in the pituitary, can you explain how one follicle is "selected" to grow?
4. Compare the actions of LH and FSH in the ovaries and testes.
5. Ovarian and uterine events in the month following ovulation differ depending on whether fertilization occurs. Describe the differences and explain their hormonal controls.

FOR INVESTIGATION

No male contraceptive methods exist other than the condom. How would you go about developing a pharmacological method to block sperm production that could lead to a male pill without affecting the maintenance of male secondary sexual characteristics or male sexual behavior?

CHAPTER 43 Animal Development: From Genes to Organisms

Thar she blows!

The whale blows its nose from the top of its head. The spout from the whale's blowhole is exhaled air and water vapor from its nasal passages. It is adaptive for a marine mammal to breathe out of the top of its head because most of its body can remain under water as it breathes. But in most terrestrial mammals, the nose is on the front of the head. How did the whale's nose get to the top of its head? This is an evolutionary question, but the answer is to be found in development—the processes whereby a fertilized egg becomes an adult organism.

The vertebrate body varies enormously among species in form and function, yet its basic structural design does not. For example, the whale flipper, the bat wing, and the human arm all have the same bones. During development, these bones assume different shapes and dimensions to adapt the forelimbs of each species to various functions: swimming, flying, and tool use.

Similarly, all vertebrates have the same bones in their heads, but through development, these bones grow differently, and therefore the skull takes on different shapes in different species. In both whales and humans, the nasal passages are in the nasal bone, which is just above the bones of the upper jaw. In the human, that places the nasal bone just above the jaw on the front of the face. Things are different in the whale. During development, the bones of the whale's upper jaw grow enormously relative to the other bones of its skull. These jaw bones project far forward to form the cavernous mouth. As a result of this differential forward growth of the jaw bone, the nasal bone ends up on the top of the skull, rather than on the front. Thus, the answers to how the whale's nose ends up on the top of its head and how its forelimbs become flippers are found in the processes of development. These processes form and shape the components of the basic vertebrate body plan.

Development begins with the joining of sperm and egg. As described in Section 42.2, in many species the egg will attract a large number of sperm, but only one sperm can supply the genetic material to fertilize the egg. Biochem-

Thar She Blows! The nasal passages of whales such as *Balaenoptera musculus* (the blue whale) are on top of its head because of the extreme growth of its jaw bones during development.

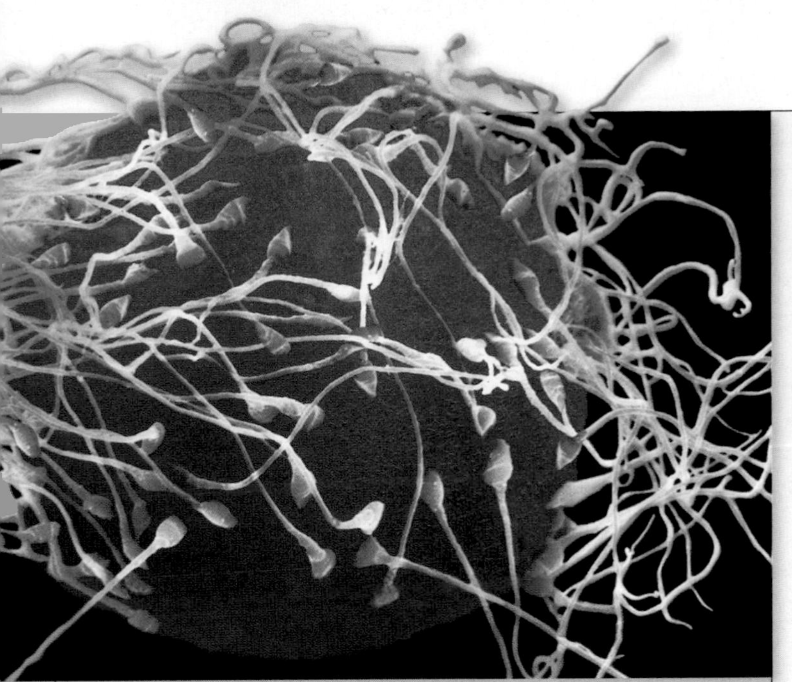

When Sperm Meet Egg Development begins with the fertilization of an egg by a single sperm. Humans are typical of animals in that the egg (artificially colored pink) is much larger than the sperm (gold). The egg cytoplasm is loaded with factors that will direct development and nourish the growing embryo.

ical blocks to polyspermy ensure that the proper genetic complement for the species will be achieved.

The fertilized egg goes through an initial rapid series of cell divisions without growth that subdivides the egg cytoplasm into a mass of smaller undifferentiated cells. Although this mass of cells shows no hints of the eventual body plan, the uneven distribution of molecules in the cytoplasm of the fertilized egg provides positional information that will direct the fates of cells and set up the body plan.

IN THIS CHAPTER we will see how a single cell becomes a multicellular animal through orderly cell movements that create multiple layers and set up cell–cell interactions. The regional and temporal differences in gene expression that control cell differentiation, described in Chapters 19 and 20, lead to the emergence of the body plan of the animal. We will discuss these early developmental steps in four model organisms that have been studied extensively: sea urchins, frogs, chicks, and humans.

CHAPTER OUTLINE

43.1 How Does Fertilization Activate Development?

Two things must be noted at the outset of this discussion. First, in animals that reproduce asexually, development proceeds without fertilization. And second, critical developmental steps have already been taken in the maturation of the egg in all animals. Therefore, the question we really are asking is how the act of fertilization re-starts or activates development in sexually reproducing animals.

In sexually reproducing animals, fertilization does more than just restore a full diploid complement of maternal and paternal genes. The entry of a sperm into an egg sets up blocks to polyspermy, causes ion fluxes, changes pH, stimulates protein synthesis, increases the metabolism of the egg, and initiates the rapid series of cell divisions that produce a multicellular embryo. In many species, the point of entry of the sperm creates an asymmetry in the radially symmetrical egg. This asymmetry enables the emergence of a bilateral body plan from the radial symmetry of the egg. We described the mechanisms of fertilization in Section 42.2. Here we take a closer look at the cellular and molecular interactions of sperm and egg that initiate the first steps of development.

The sperm and the egg make different contributions to the zygote

As shown in the above micrograph, eggs are much larger than sperm. Egg cytoplasm is well stocked with organelles, nutrients, and a variety of molecules including transcription factors and mRNAs. The sperm is little more than a DNA delivery vehicle. Therefore, nearly everything that the embryo needs during its early stages of development comes from the mother. In addition to its haploid nucleus, however, the sperm makes one other important contribution to the zygote in some species: a centriole. The centriole becomes the centrosome of the zygote, which produces the mitotic spindles for subsequent cell divisions (see Figure 9.9).

The entry of the sperm into the egg stimulates rearrangements of the egg cytoplasm that establish the polarity of the embryo. As mentioned in Section 19.4, the nutrients and molecules in the cytoplasm of the zygote are not homogeneously distributed, and therefore will not be divided equally among all daughter cells when cell divisions begin. This unequal distribution of cytoplasmic determinants sets the stage for the signaling cascades that orchestrate the sequential steps of development: determination, differentiation, and morphogenesis. A good example of these events is provided by the frog, an organism in which they have been well studied.

Rearrangements of egg cytoplasm set the stage for determination

The rearrangements of egg cytoplasm in some frog species are easily observed because of pigments in the egg cytoplasm. The nutrient molecules in an unfertilized frog egg are dense, and they are therefore concentrated by gravity in the lower half of the egg, which is called the **vegetal hemisphere**. The haploid nucleus of the egg is located at the opposite end of the egg, in the **animal hemisphere**. The outermost (*cortical*) cytoplasm of the animal hemisphere is heavily pigmented, and the underlying cytoplasm has more diffuse pigmentation. The vegetal hemisphere is not pigmented. How is the cytoplasm rearranged when the egg is fertilized?

The surface of the frog egg has specific sperm-binding sites, but a sperm always enter the egg in the animal hemisphere. When a sperm enters, the cortical cytoplasm rotates toward the site of sperm entry. This rotation reveals a band of diffusely pigmented cytoplasm on the side of the egg opposite the site of sperm entry. This band, called the **gray crescent**, will be the site of important developmental events (**Figure 43.1**).

The cytoplasmic rearrangements that create the gray crescent bring different regions of cytoplasm into contact with each other on opposite sides of the egg. Therefore, bilateral symmetry is imposed on what was a radially symmetrical egg. In addition to the up–down difference of the animal and vegetal hemispheres that defines the anterior–posterior axis of the embryo, the movement of the cytoplasm sets the stage for the creation of the dorsal–ventral (back–belly) body axis. In the frog, the site of sperm entry will become the *ventral* region of the embryo, and the gray crescent will become the *dorsal* region. The region of the gray crescent that borders the animal pole cortical cytoplasm points anteriorly, and the gray crescent region that borders the vegetal pole cytoplasm points posteriorly.

The one non-nuclear organelle that the sperm contributes to the egg—the centriole—initiates cytoplasmic reorganization. The sperm centriole organizes the microtubules in the vegetal hemisphere cytoplasm into a parallel array that guides the movement of the cortical cytoplasm. These microtubules also appear to be directly responsible for movement of specific organelles and proteins, because these organelles and proteins move from the vegetal hemisphere to the gray crescent region even faster than the cortical cytoplasm rotates.

As a result of movement of cytoplasm, proteins, and organelles, the distribution of critical developmental signals changes. A key transcription factor in early development is β-catenin, which is produced from maternal mRNA and is found throughout the cytoplasm of the egg. Also present throughout the egg cytoplasm is a protein kinase called glycogen synthase kinase-3 (GSK-3), which phosphorylates and thereby targets β-catenin for degradation. However, an inhibitor of GSK-3 is segregated in the vegetal cortex of the egg. After sperm entry, this inhibitor is moved along microtubules to the gray crescent, where it prevents the degradation of β-catenin. As a result, the concentration of β-catenin is higher on the dorsal side than on the ventral side of the developing embryo (**Figure 43.2**).

Evidence supports the hypothesis that β-catenin is a key player in the cell–cell signaling cascade that begins the process of cell determination and the formation of the embryo in the region of the gray crescent. But before cell–cell signaling can occur, multiple cells must be in place; let's turn to the early series of cell divisions that transforms the zygote into a multicellular embryo.

Cleavage repackages the cytoplasm

The transformation of the diploid zygote into a mass of cells occurs through a rapid series of cell divisions called **cleavage**. Because the cytoplasm of the zygote is not homogeneous, these first cell divisions result in the differential distribution of nutrients and cytoplasmic determinants among the cells of the early embryo. In most animals, cleavage proceeds with rapid DNA replication and mitosis, but no cell growth and little gene expression. The embryo becomes a solid ball of smaller and smaller cells, called a *morula* (from the Latin word for "mulberry"). Eventually, this ball forms a central fluid-filled cavity called a **blastocoel**, at which point the embryo is called a **blastula**. Its individual cells are called **blastomeres**. The pattern of cleavage in different species influences the form of their blastulas.

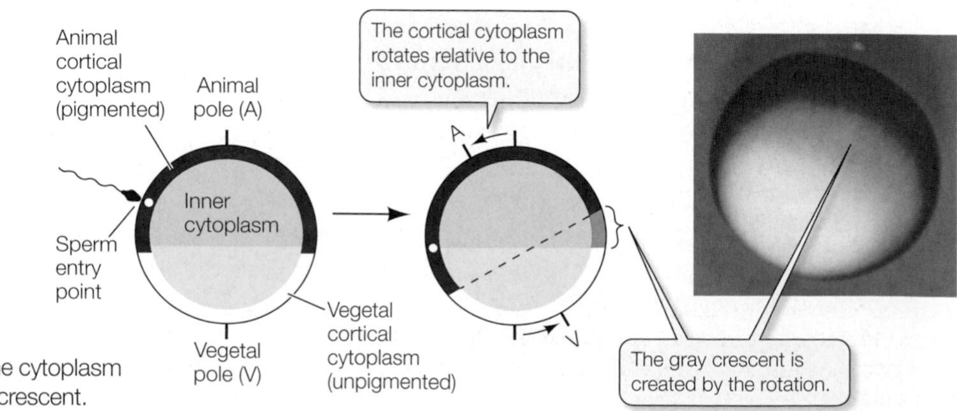

43.1 The Gray Crescent Rearrangement of the cytoplasm of the frog eggs after fertilization creates the gray crescent.

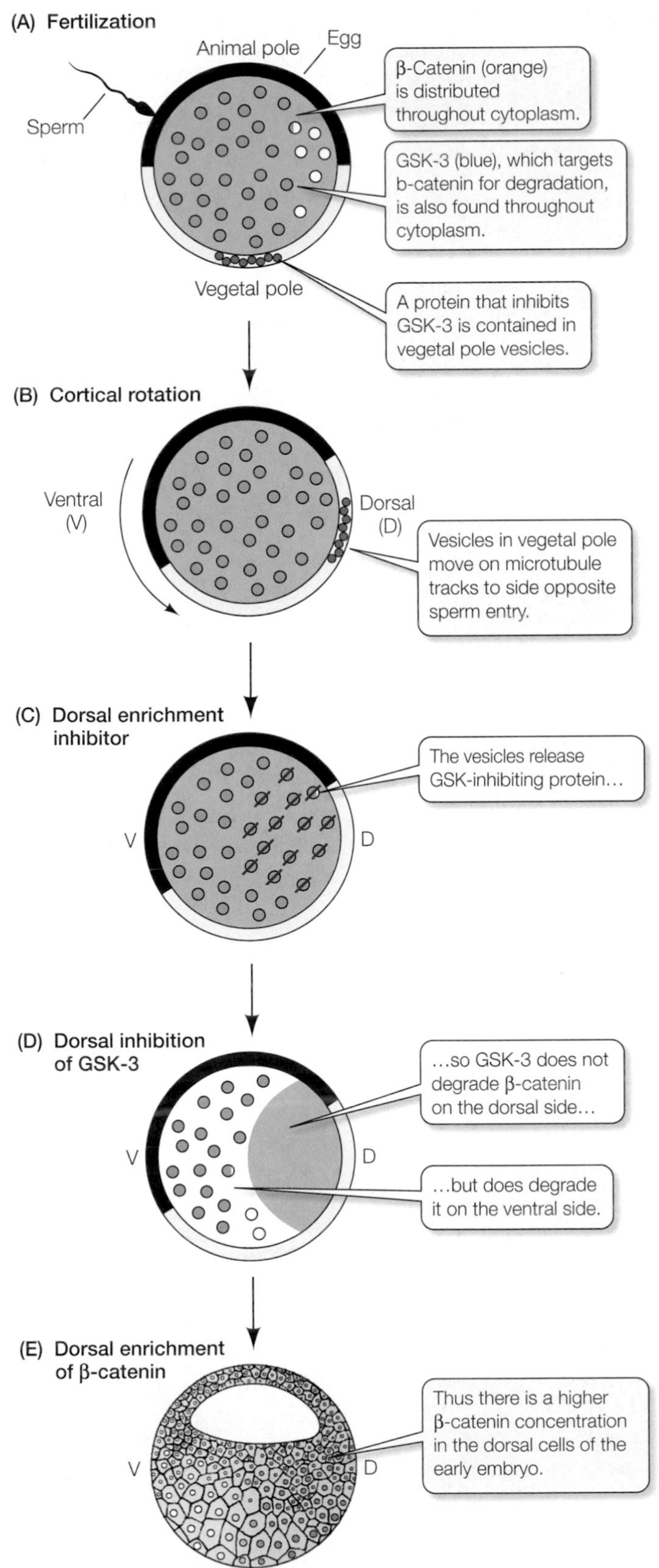

(A) Fertilization
Animal pole
Egg
Sperm
β-Catenin (orange) is distributed throughout cytoplasm.
GSK-3 (blue), which targets b-catenin for degradation, is also found throughout cytoplasm.
Vegetal pole
A protein that inhibits GSK-3 is contained in vegetal pole vesicles.
(B) Cortical rotation
Ventral (V)
Dorsal (D)
Vesicles in vegetal pole move on microtubule tracks to side opposite sperm entry.
(C) Dorsal enrichment inhibitor
V
D
The vesicles release GSK-inhibiting protein...
(D) Dorsal inhibition of GSK-3
V
D
...so GSK-3 does not degrade β-catenin on the dorsal side...
...but does degrade it on the ventral side.
(E) Dorsal enrichment of β-catenin
V
D
Thus there is a higher β-catenin concentration in the dorsal cells of the early embryo.

43.2 Cytoplasmic Factors Set Up Signaling Cascades Cytoplasmic movement changes the distributions of critical developmental signals. In the frog zygote, the interaction of the protein kinase GSK-3, its inhibitor, and the protein β-catenin are crucial in specifying the dorsal–ventral (back–belly) axis of the embryo.

- **Complete cleavage** occurs in most eggs that have little **yolk** (stored nutrients). In this pattern, early cleavage furrows divide the egg completely and the daughter cells are of similar size. The sea urchin egg provides an example (**Figure 43.3A**). The frog egg also undergoes complete cleavage, but because the vegetal pole of the frog egg contains more yolk, the division of the cytoplasm is unequal and the daughter cells in the animal hemisphere are smaller than those in the vegetal hemisphere (**Figure 43.3B**).
- **Incomplete cleavage** occurs in many species in which the egg contains a lot of yolk which the cleavage furrows do not penetrate. **Discoidal cleavage** is a type of incomplete cleavage common in fishes, reptiles, and birds, the eggs of which contain a dense mass of yolk. The embryo forms as a disc of cells, called a **blastodisc**, that sits on top of the yolk mass (**Figure 43.3C**).
- **Superficial cleavage** is a variation of incomplete cleavage that occurs in insects such as the fruit fly (*Drosophila*). Early in development, cycles of mitosis occur without cell division, producing a *syncytium*—a single cell with many nuclei. The nuclei eventually migrate to the periphery of the egg, and after several more mitotic cycles, the plasma membrane of the egg grows inward, partitioning the nuclei into individual cells surrounding a core of yolk (**Figure 43.3D**).

The positions of the mitotic spindles during cleavage are not random but are defined by cytoplasmic factors that were produced from the maternal genome and stored in the egg (see Section 19.4). The orientation of the mitotic spindles can determine the planes of cleavage and, therefore, the arrangement of the daughter cells.

In complete cleavage, if the mitotic spindles of successive cell divisions form parallel or perpendicular to the animal–vegetal axis of the zygote, the cleavage pattern is **radial**, as in the sea urchin and the frog. In these organisms, the first two cell divisions are parallel to the animal–vegetal axis and the third is perpendicular to it (see Figures 43.3A,B). **Spiral cleavage** results when the mitotic spindles are at oblique angles to the animal–vegetal axis. In spiral cleavage, each new cell layer is shifted to the left or right, depending on the orientation of the mitotic spindles. Mollusks have spiral cleavage, which is reflected by the coiling pattern of snail shells. If each new layer of cells shifts to the left, the snail shell will coil to the left (sinistral). If each new layer of cells shifts to the right, the snail shell will coil to the right (dextral). Sinistral and dextral coiling snails are mirror images of each other.

Cleavage in mammals is unique

Several features of cleavage in eutherians—placental mammals—are very different from those seen in other animal groups. First, their pattern of cleavage is *rotational*: the first cell division is parallel to the animal–vegetal axis, yielding two blastomeres. The subsequent cell division of those two blastomeres occurs at right angles to each other: one blastomere divides parallel to the animal–vegetal axis, while the other divides perpendicular to it (**Figure 43.4A**).

Cleavage in mammals is very slow; cell divisions are 12–24 hours apart, compared with tens of minutes to a few hours in nonmammalian species. Also, the cell divisions of mammalian blastomeres are not in synchrony with each other. Because the blas-

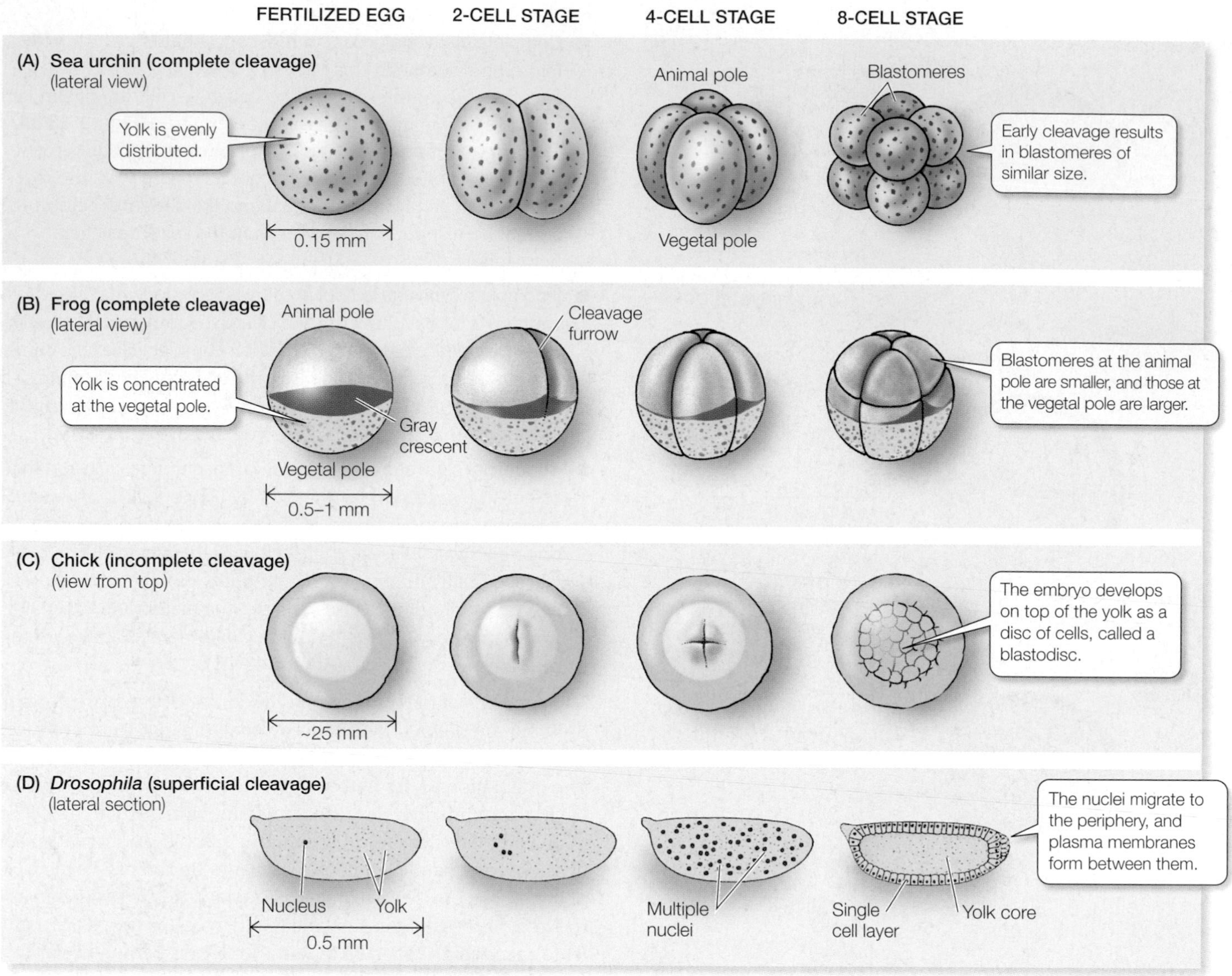

43.3 Patterns of Cleavage in Four Model Organisms Differences in patterns of early embryonic development reflect differences in the way the egg cytoplasm is organized.

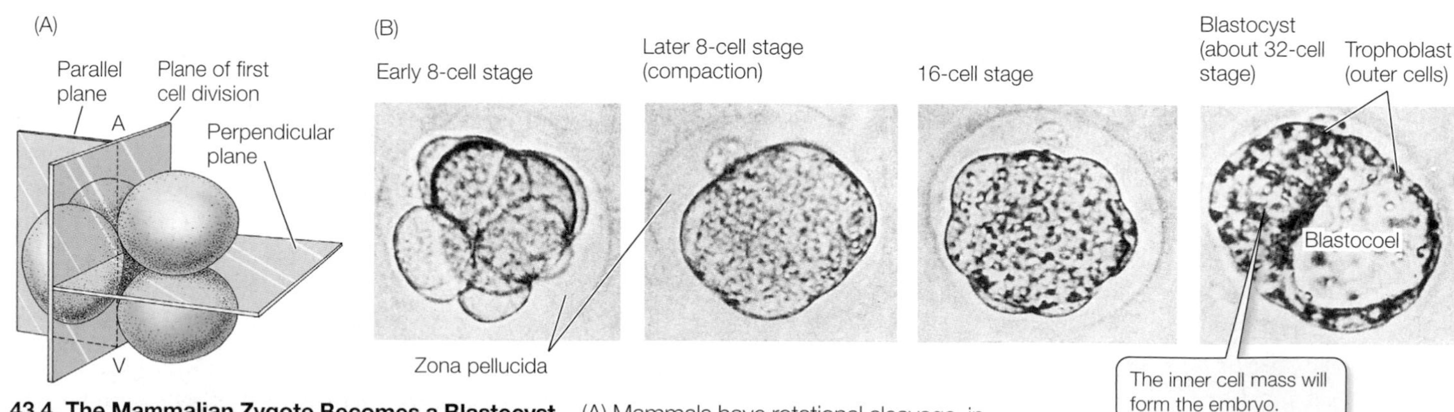

43.4 The Mammalian Zygote Becomes a Blastocyst (A) Mammals have rotational cleavage, in which the plane of the first cleavage is parallel to the animal–vegetal (A, V) axis, but the planes of the second cell division (shown in beige) are at right angles to each other. (B) Starting late in the eight-cell stage, the mammalian embryo undergoes compaction of its cells, resulting in a blastocyst. At the thirty two-cell stage, the blastocyst is a dense mass of cells adjacent to a fluid filled blastocoel and surrounded by trophoblast cells.

tomeres do not undergo mitosis at the same time, the number of cells in the embryo does not increase in the regular (2, 4, 8, 16, 32, etc.) progression typical of other species.

Another unique feature of the slow mammalian cleavage is that the products of genes expressed at this time play roles in cleavage. In animals such as sea urchins and frogs, gene transcription does not occur in the blastomeres, and cleavage is directed exclusively by molecules that were present in the egg before fertilization.

As in other animals that have complete cleavage, the early cell divisions in a mammalian zygote produce a loosely associated ball of cells. However, at about the eight-cell stage, the behavior of the mammalian blastomeres changes. They change shape to maximize their surface contact with one another, form tight junctions, and become a very compact mass of cells (**Figure 43.4B**).

At the transition from the 16- to the 32-cell stage (the fourth division), the cells separate into two groups. The **inner cell mass** will become the embryo, while the surrounding outer cells become an encompassing sac called the **trophoblast**. Trophoblast cells secrete fluid, creating a cavity—the *blastocoel*—with the inner cell mass at one end (final photo in Figure 43.4B). At this stage, the mammalian embryo is called a **blastocyst**, distinguishing it from the blastulas of other animal groups.

Why is mammalian cleavage so different? Remember that placental mammals are *viviparous*: the embryo develops within the uterus of the mother. To support the developing embryo, a connection has to be developed between the circulatory system of the embryo and that of the mother. As we will see later in this chapter, the placenta and the umbilical cord are the structures that provide this connection. Thus, the mammalian blastocyst must produce both the embryo (from the inner cell mass) and its support structures (from the trophoblast).

Fertilization in mammals occurs in the upper reaches of the mother's oviduct, and cleavage occurs as the zygote travels down the oviduct to the uterus. When the blastocyst arrives in the uterus, the trophoblast adheres to the *endometrium* (the lining of the uterus). This event begins the process of **implantation**. In humans, implantation begins on about the sixth day after fertilization. The trophoblast cells of the blastocyst secrete adhesion molecules and enzymes to burrow into the wall of the uterus (**Figure 43.5**). As the blastocyst moves down the oviduct to the uterus, it must not embed itself in the oviduct (Fallopian tube) wall, or the result will be an *ectopic* or *tubal pregnancy*—a very dangerous condition. Early implantation is normally prevented by the zona pellucida, which surrounded the egg and remains around the cleaving ball of cells (see Section 42.2). At about the time the blastocyst reaches the uterus, it hatches from the zona pellucida, and implantation can occur.

Specific blastomeres generate specific tissues and organs

In all animal species, cleavage results in a repackaging of the egg cytoplasm into a large number of small cells surrounding a central cavity. Little cell differentiation and little if any gene expression occur during cleavage. Nevertheless, cells in different regions of the blastula possess different complements of the nutrients and cytoplasmic determinants that were present in the egg.

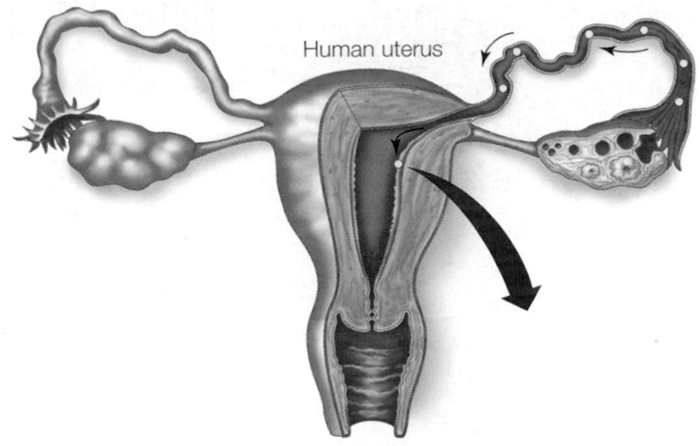

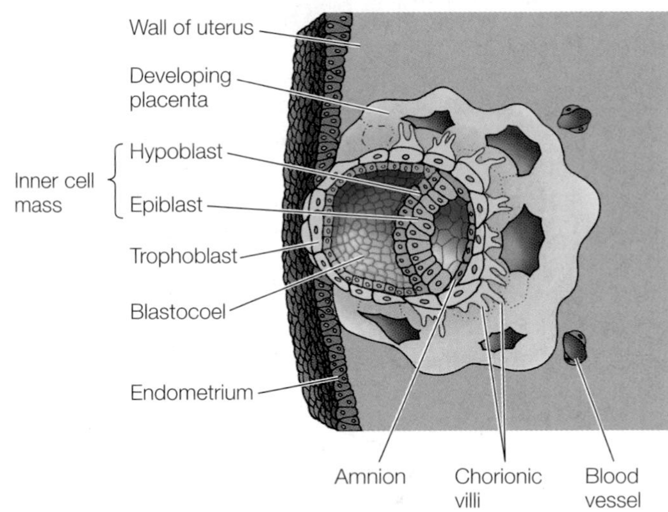

43.5 A Human Blastocyst at Implantation Adhesion molecules and proteolytic enzymes secreted by trophoblast cells allow the blastocyst to burrow into the endometrium. Once implanted within the wall of the uterus, the trophoblast cells send out numerous projections—the chorionic villi—which increase the embryo's area of contact with the mother's bloodstream.

The blastocoel prevents cells from different regions of the blastula from coming into contact and interacting, but that will soon change. During the next stage of development, the cells of the blastula will move around and come into new associations with one another, communicate instructions to one another, and begin to differentiate. In many animals, these movements of the blastomeres are so regular and well orchestrated that it is possible to label a specific blastomere with a dye and identify the tissues and organs that form from its progeny. Such labeling experiments produce **fate maps** of the blastula (**Figure 43.6**).

Blastomeres become **determined**—committed to specific fates—at different times in different species. In some species, such as roundworms, the fates of blastomeres are restricted as early as the two-cell stage. If one of these blastomeres is experimentally removed, a particular portion of the embryo will not form. This type of development has been called **mosaic development** because each blastomere seems to contribute a specific set of "tiles" to the final "mosaic" that is the adult animal. In contrast, in **regulative devel-**

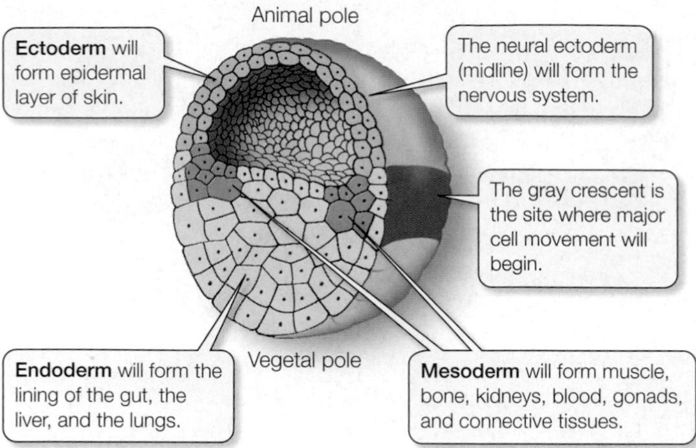

Animal pole
Ectoderm will form epidermal layer of skin.
The neural ectoderm (midline) will form the nervous system.
The gray crescent is the site where major cell movement will begin.
Vegetal pole
Endoderm will form the lining of the gut, the liver, and the lungs.
Mesoderm will form muscle, bone, kidneys, blood, gonads, and connective tissues.

43.6 Fate Map of a Frog Blastula The colors indicate the portions of the blastula that will form the three germ layers and subsequently the frog's tissues and organs.

opment, the loss of some cells during cleavage does not affect the developing embryo, because the remaining cells compensate for the loss. Regulative development is typical of many vertebrate species, including humans. The pluripotent cells of the blastocyst are also known as *embryonic stem cells*, currently the subject of much research, particularly because of their therapeutic potential. (Section 19.2 describes the nature of embryonic stem cells and stem cell research.)

In about 1 out of 50,000 human pregnancies, genetic or environmental factors cause the inner cell mass to split partially. The result is twins that are *conjoined* at some point on their bodies, usually sharing some or most of their organs and/or limbs.

If some blastomeres can change their fate to compensate for the loss of other cells during cleavage and blastula formation, can those cells form an entire embryo? To a certain extent, yes. During cleavage or early blastula formation in mammals, for example, if the blastomeres are physically separated into two groups, both groups can produce complete embryos (**Figure 43.7**). Since the two embryos come from the same zygote, they will be *monozygotic twins*—genetically identical. Nonidentical twins occur when two separate eggs are fertilized by two separate sperm. Thus, while identical twins are always of the same sex, nonidentical twins have a 50 percent chance of being the same sex.

43.1 RECAP

The egg is stocked with nutrients and informational molecules that power and direct the early stages of development. Fertilization activates the egg and stimulates a rearrangement of the cytoplasm, setting up the body axes and positional information that initiate signaling cascades, which control determination and differentiation.

- Can you explain how β-catenin becomes concentrated in only certain blastomeres? See p. 922 and Figure 43.2
- Describe in general terms the difference between complete and incomplete cleavage. See p. 923 and Figure 43.3
- What does a fate map tell us? How are fate maps constructed? See p. 925 and Figure 43.6

We learned in Section 31.1 that the patterns and processes of early development are essential tools in recreating evolutionary history and discerning relationships among the different animal groups, because it is these patterns that result in the diversity of animal body plans. We next describe the events of *gastrulation*, a crucial series of events that initiate the differentiation of embryonic cells and tissues and sets the stage for the emergence of the organs and other structures of the body.

43.2 How Does Gastrulation Generate Multiple Tissue Layers?

The blastula is typically a fluid-filled ball of cells. How does this simple ball of cells become an embryo made up of multiple tissue layers with head and tail ends and dorsal and ventral sides? **Gastrulation** is the process whereby the blastula is transformed by massive movements of cells into an embryo with multiple tissue layers and visible body axes. The resulting spatial relationships between tissues make possible the inductive interactions that trigger differentiation and organ formation (see Figure 19.11).

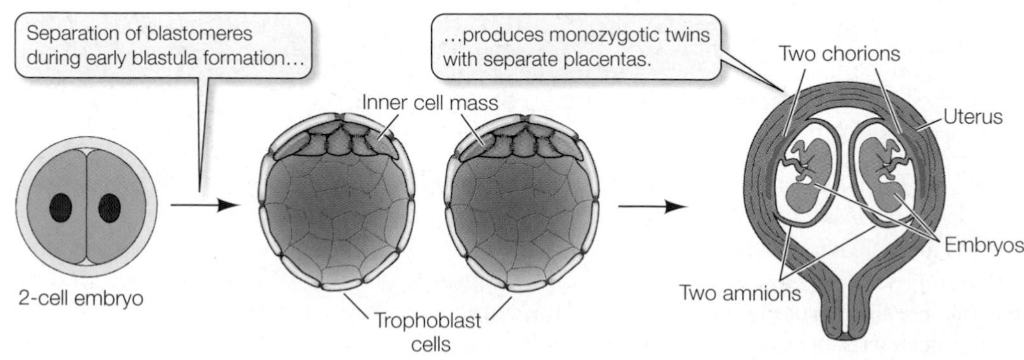

43.7 Twinning in Humans Because humans have regulative development, remaining cells can compensate when cells are lost in early cleavages. Monozygotic (identical) twins can result when cells in the early blastula become physically separated and each group of cells goes on to produce a separate embryo.

During gastrulation, three **germ layers** (also called *cell layers* or *tissue layers*) form (see Figure 43.6):

- The **endoderm** is the innermost germ layer, created as some blastomeres move together as a sheet to the inside of the embryo. The endoderm gives rise to the lining of the digestive tract, respiratory tract, pancreas, and liver.
- The **ectoderm** is the outer germ layer, formed from those cells remaining on the outside of the embryo. The ectoderm gives rise to the nervous system, including the eyes and ears; and to the epidermal layer of the skin and structures derived from skin, such as hair, feathers, nails or claws, sweat glands, and oil glands.
- The **mesoderm**, or middle layer, is made up of cells that migrate between the endoderm and the ectoderm. The mesoderm contributes tissues to many organs, including heart, blood vessels, muscle, and bones.

Some of the most interesting and important challenges in animal development have dealt with two related questions: what directs the cell movements of gastrulation, and what is responsible for the resulting patterns of cell differentiation and organ formation? Scientists have made significant progress in answering both these questions at the molecular level. In the discussion that follows, we will consider the similarities and differences in gastrulation among sea urchins, frogs, reptiles, birds, and mammals. We'll also review some of the exciting discoveries about the mechanisms underlying these phenomena.

Invagination of the vegetal pole characterizes gastrulation in the sea urchin

The sea urchin blastula is a simple, hollow ball of cells that is only one cell thick. The end of the blastula stage is marked by a dramatic slowing of the rate of mitosis, and the beginning of gastrulation is marked by a flattening of the vegetal hemisphere (**Figure 43.8**). Some cells at the vegetal pole bulge into the blastocoel and migrate into the cavity. These cells become *primary mesenchyme* cells—cells of the middle germ layer, the mesoderm. Mesenchymal cells are not organized in tightly packed sheets or tubes like epithelial cells are, and they act more as independent units migrating into and among the other tissue layers.

The flattening at the vegetal pole results from changes in the shape of the individual blastomeres. These cells, which are originally rather cuboidal, become wedge-shaped, with smaller outer edges and larger inner edges. As a result of these shape changes, the vegetal pole bulges inward, or *invaginates*, as if someone were poking a finger into a hollow ball (see Figure 43.8). Some of the cells that invaginate become mesoderm; others become the endoderm and form the primitive gut called the *archenteron*. At the tip of the archenteron more cells enter the blastocoel to form more mesoderm, the *secondary mesenchyme*.

Changes in cell shapes cause the initial invagination of the archenteron but eventually it is pulled by the secondary mesenchyme cells. These cells, attached to the tip of the archenteron, send out extensions called filopodia that adhere to the overlying ectoderm and contract. Where the archenteron eventually makes contact with the ectoderm, the mouth of the animal will form. The opening created by the invagination of the vegetal pole is called the **blastopore**; it will become the anus of the animal.

What mechanisms control the various cell movements of sea urchin gastrulation? The immediate answer is that specific properties of particular blastomeres change. For example, some vegetal cells change shape and bulge into the blastocoel, and these cells become the primary mesenchyme. Once they lose contact with their neighboring cells on the surface of the blastula, they send out filopodia that then move along an extracellular matrix of proteins laid down by the cells lining the blastocoel.

A deeper understanding of gastrulation requires that we discover the molecular mechanisms whereby certain blastomeres develop properties different from those of others. Cleavage systematically divides up the cytoplasm of the egg. The sea urchin blastula at the 64-cell stage is radially symmetrical, but it has *polarity*, as described in Section 19.4. It consists of tiers of cells. As in the frog blastula, the top is the animal pole and the bottom the vegetal pole.

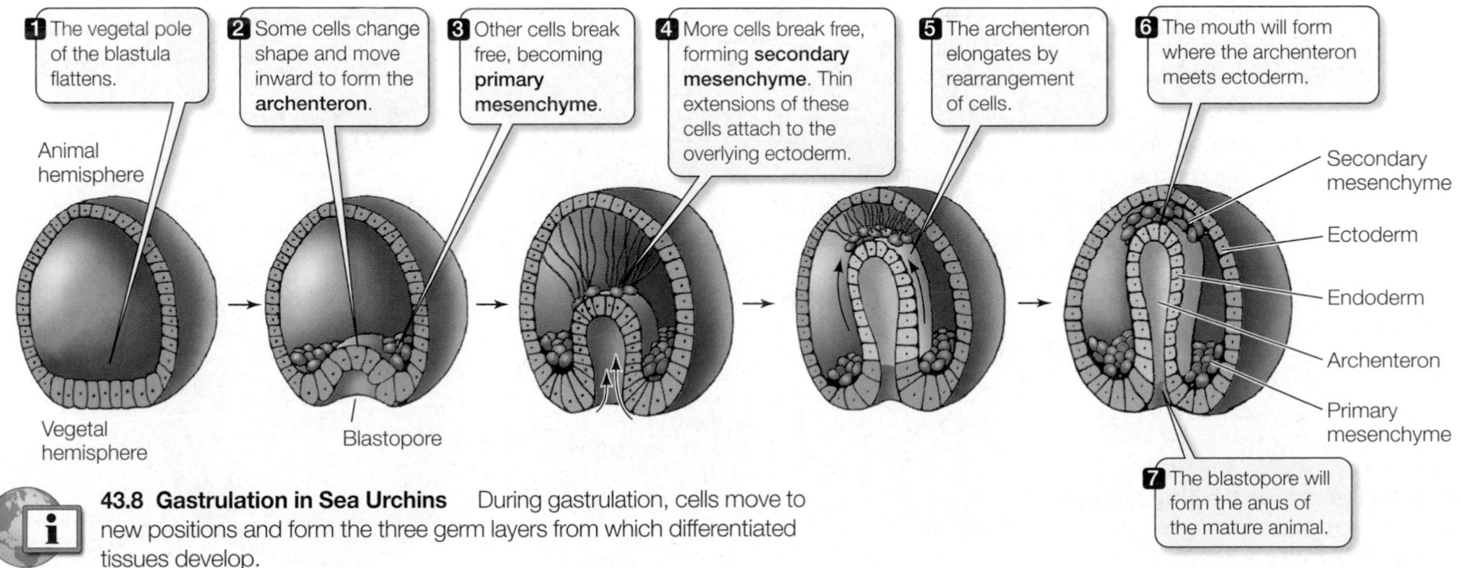

43.8 Gastrulation in Sea Urchins During gastrulation, cells move to new positions and form the three germ layers from which differentiated tissues develop.

If different tiers of blastula cells are separated, they show different developmental potentials; only cells from the vegetal pole are capable of initiating the development of a complete larva (see Figure 19.9). It has been proposed that these differences are due to uneven distribution of various transcriptional regulatory proteins in the egg cytoplasm. As cleavage progresses, these proteins end up in different combinations in different groups of cells. Therefore, specific sets of genes are activated in different cells, determining their different developmental capacities. Let's turn now to gastrulation in the frog, in which a number of key signaling molecules have been identified.

Gastrulation in the frog begins at the gray crescent

Amphibian blastulas have considerable yolk and are more than one cell thick; therefore, gastrulation is more complex in amphibians than in sea urchins. Variation is considerable among different species of amphibians, but in this brief account, we will mix results from studies done on different species to produce a generalized picture of amphibian development.

Amphibian gastrulation begins when certain cells in the gray crescent region change their shapes and cell adhesion properties. These cells bulge inward toward the blastocoel while they remain attached to the outer surface of the blastula by slender necks. Because of their shape, these cells are called *bottle cells*.

The bottle cells mark the spot where the **dorsal lip** of the blastopore will form (**Figure 43.9**). As the bottle cells move inward, the dorsal lip is created, and a sheet of cells moves over it into the blastocoel. This process is called **involution**. One group of involuting cells is the prospective endoderm, and they form the primitive gut, or archenteron. Another group will move between the endoderm and the outermost cells to form the mesoderm. As gastrulation proceeds, cells from the animal hemisphere move toward the site of involution in a process called **epiboly**. The blastopore lip widens and eventually forms a complete circle surrounding a "plug" of yolk-rich cells. As cells continue to move inward through the blastopore, the archenteron grows, gradually displacing the blastocoel.

As gastrulation comes to an end, the amphibian embryo consists of three germ layers: ectoderm on the outside, endoderm on the inside, and mesoderm between. The embryo also has a dorsal–ventral and anterior–posterior organization. Most importantly, however, the fates of specific regions of the endoderm, mesoderm, and ectoderm have been determined. The discovery of the events whereby determination takes place in the amphibian embryo is one of the most exciting stories in animal development.

The dorsal lip of the blastopore organizes embryo formation

In the 1920s, the German biologist Hans Spemann was studying the development of salamander eggs. He was interested in finding out whether the nuclei of blastomeres remain capable of directing the development of complete embryos. With great patience

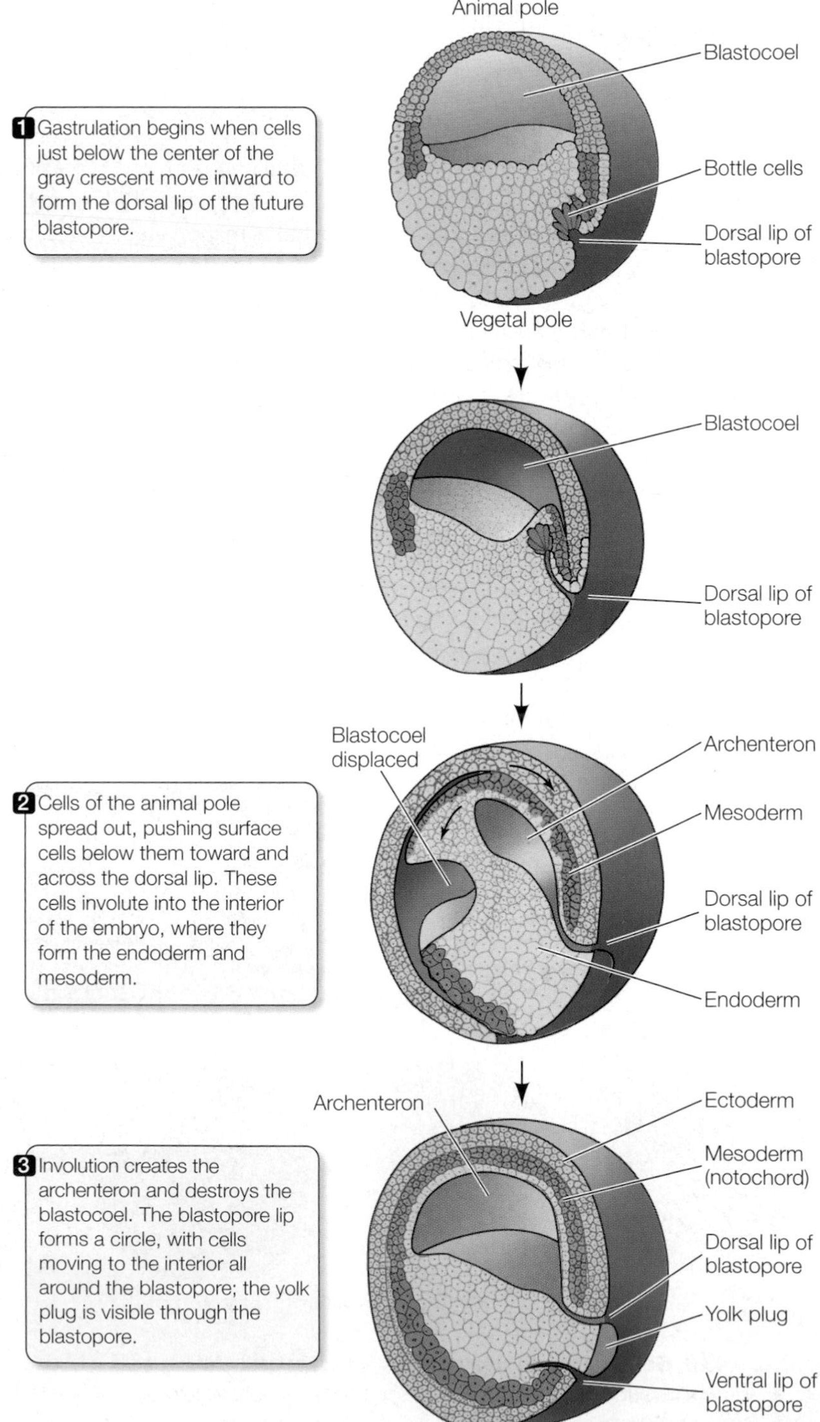
Animal pole
Blastocoel
1 Gastrulation begins when cells just below the center of the gray crescent move inward to form the dorsal lip of the future blastopore.
Bottle cells
Dorsal lip of blastopore
Vegetal pole
Blastocoel
Dorsal lip of blastopore
Blastocoel displaced
Archenteron
2 Cells of the animal pole spread out, pushing surface cells below them toward and across the dorsal lip. These cells involute into the interior of the embryo, where they form the endoderm and mesoderm.
Mesoderm
Dorsal lip of blastopore
Endoderm
Archenteron
Ectoderm
3 Involution creates the archenteron and destroys the blastocoel. The blastopore lip forms a circle, with cells moving to the interior all around the blastopore; the yolk plug is visible through the blastopore.
Mesoderm (notochord)
Dorsal lip of blastopore
Yolk plug
Ventral lip of blastopore

43.9 Gastrulation in the Frog Embryo The colors in this diagram are matched to those in the frog fate map (see Figure 43.6).

and dexterity, he formed loops from single human baby hairs to constrict fertilized eggs, effectively dividing them in half.

When Spemann's loops bisected the gray crescent, both halves of the zygote gastrulated and developed into complete embryos (**Figure 43.10, Experiment 1**). But when the gray crescent was on only one side of the constriction, only that half of the zygote developed into a complete embryo. The half lacking gray crescent material became a clump of undifferentiated cells that Spemann called a "belly piece" (**Figure 43.10, Experiment 2**). Spemann hypothesized that cytoplasmic factors unequally distributed in the fertilized egg were necessary for gastrulation and thus for the development of a normal organism.

To test the hypothesis that cells receiving different complements of cytoplasmic factors had different developmental fates, he transplanted pieces of early gastrulas to various locations on other gastrulas. Guided by fate maps (see Figure 43.6), he was able to take a piece of ectoderm he knew would develop into skin and transplant it to a region that normally becomes part of the nervous system, and vice versa.

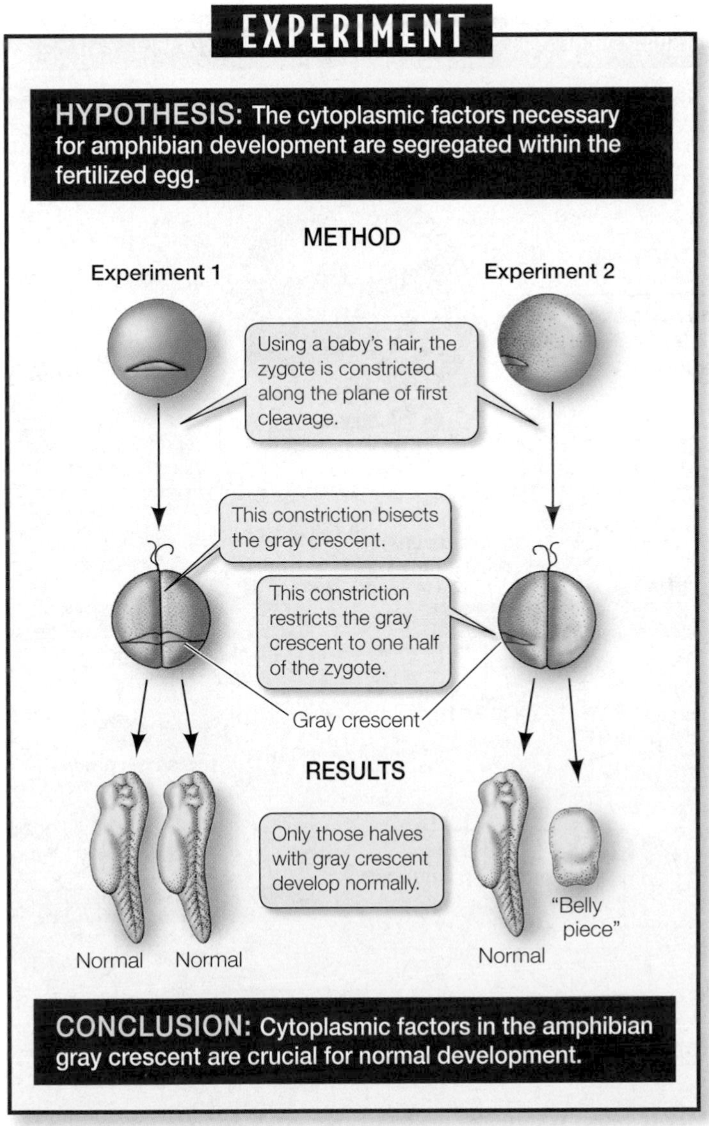

When he performed these transplants in early gastrulas—when the blastopore was just beginning to form—the transplanted pieces always developed into tissues that were appropriate for the location where they were placed. Donor-presumptive epidermis (that is, cells destined to become epidermis in their original location) developed into host neural ectoderm (nervous system tissue), and donor-presumptive neural ectoderm developed into host epidermis. Thus, he learned that the fates of the transplanted cells had not been determined before the transplantation (see Figure 19.2). In late gastrulas, however, the same experiment yielded opposite results. Donor-presumptive epidermis produced patches of skin cells in the host nervous system, and donor-presumptive neural ectoderm produced nervous system tissue in the host skin. At some point during gastrulation, the fates of the embryonic cells had become determined.

Spemann's next experiment, done with his student Hilde Mangold, produced momentous results: they transplanted the dorsal lip of the blastopore (**Figure 43.11**). When this small piece of tissue was transplanted into the presumptive belly area of another gastrula, it stimulated a second site of gastrulation—and a second complete embryo formed belly-to-belly with the original embryo! Because the dorsal lip of the blastopore was apparently capable of inducing the host tissue to form an entire embryo, Spemann and Mangold dubbed the dorsal lip tissue the **primary embryonic organizer**, or simply the **organizer**. For more than 80 years, the organizer has been an active area of developmental biology research.

EXPERIMENT

HYPOTHESIS: The cytoplasmic factors necessary for amphibian development are segregated within the fertilized egg.

METHOD

Experiment 1
Experiment 2
Using a baby's hair, the zygote is constricted along the plane of first cleavage.
This constriction bisects the gray crescent.
This constriction restricts the gray crescent to one half of the zygote.
Gray crescent

RESULTS

Only those halves with gray crescent develop normally.
Normal
Normal
Normal
"Belly piece"

CONCLUSION: Cytoplasmic factors in the amphibian gray crescent are crucial for normal development.

43.10 Spemann's Experiment Spemann's research revealed that gastrulation and subsequent normal development in salamanders depended on cytoplasmic determinants localized in the gray crescent.

The molecular mechanisms of the organizer involve multiple transcription factors

The primary embryonic organizer has been studied intensively to discover the molecular mechanisms involved in its action. The distribution of the transcription factor β-catenin in the late blastula corresponds to the location of the organizer in the early gastrula, so β-catenin is a candidate for the initiator of organizer activity. To prove that a protein is an inductive signal, it has to be shown that it is both *necessary* and *sufficient* for the proposed effect. In other words, the effect should not occur if the candidate protein is not present (necessity), and the candidate protein should be capable of inducing the effect where it would otherwise not occur (sufficiency).

The criteria of necessity and sufficiency have indeed been satisfied for β-catenin. If β-catenin mRNA transcripts are depleted by injections of antisense RNA into the egg (see Section 16.5), gastrulation does not occur. If β-catenin is experimentally overexpressed in another region of the blastula, it can induce a second axis of embryo formation, as the transplanted dorsal lip did in the Spemann–Mangold experiments. Thus, β-catenin appears to be both necessary and sufficient for the formation of the primary embryonic organizer—but it is only one component of a complex signaling process.

How the presence of β-catenin creates the organizer and how the organizer then induces the beginnings of the body plan involves a complex series of interactions between transcription fac-

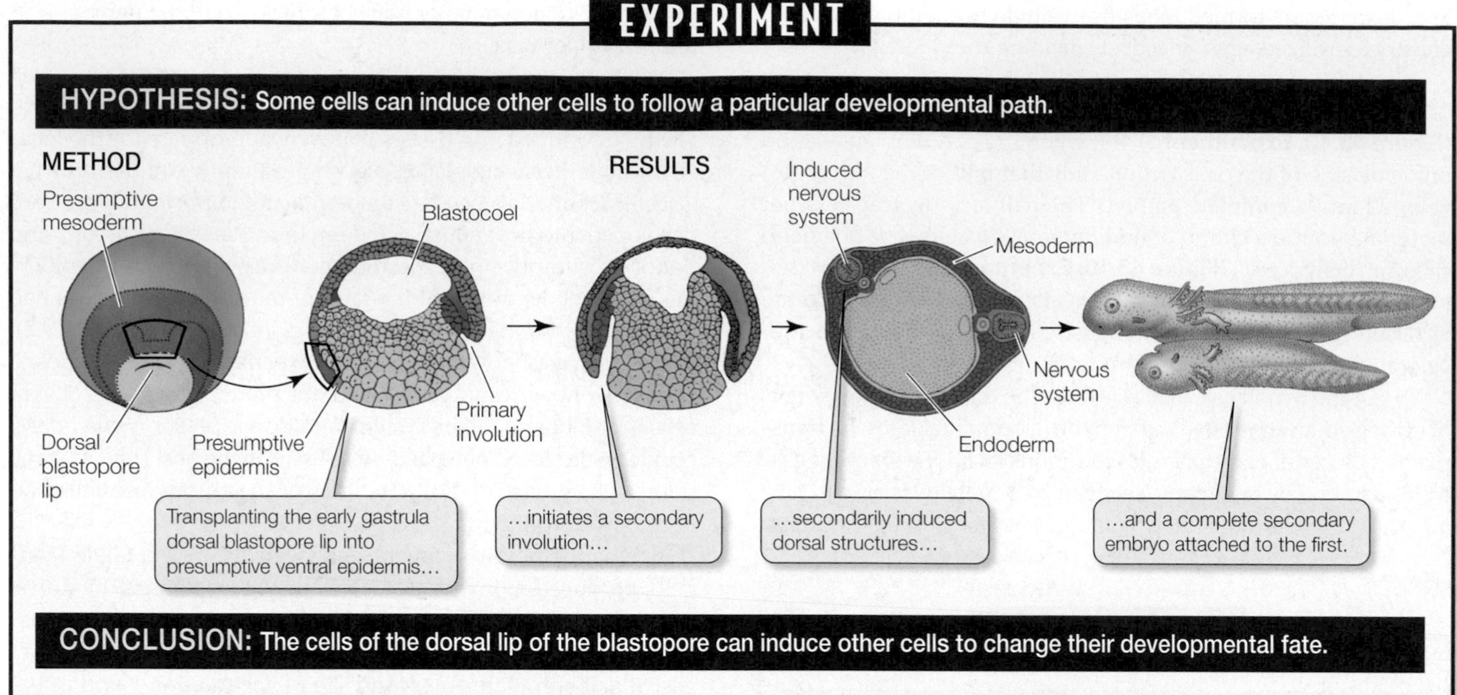

EXPERIMENT
HYPOTHESIS: Some cells can induce other cells to follow a particular developmental path.
METHOD
Presumptive mesoderm
Dorsal blastopore lip
Presumptive epidermis
Blastocoel
RESULTS
Primary involution
Induced nervous system
Mesoderm
Nervous system
Endoderm
Transplanting the early gastrula dorsal blastopore lip into presumptive ventral epidermis...
...initiates a secondary involution...
...secondarily induced dorsal structures...
...and a complete secondary embryo attached to the first.
CONCLUSION: The cells of the dorsal lip of the blastopore can induce other cells to change their developmental fate.

43.11 The Dorsal Lip Induces Embryonic Organization In a famous experiment, Spemann and Mangold transplanted the dorsal lip of the blastopore. The transplanted tissue induced a second site of gastrulation and the formation of a second embryo.

tors and growth factors. What follows is only a portion of this complex and still emerging story. What you should take from this description is not the names of the genes and gene products involved. Rather, we hope you will gain a basic appreciation of how signaling molecules interact to produce different combinations of signals that convey positional and temporal information that guides cells into different paths of determination and differentiation.

Studies of early gastrulas revealed that primary embryonic organizer activity is generated in vegetal cells just below the gray crescent. The concentration of β-catenin is highest here. One critical property of the organizer is expression of the transcription factor Goosecoid. Expression of the *goosecoid* gene depends on two signaling pathways, both of which involve β-catenin.

The first of these pathways involves a *goosecoid*-promoting transcription factor called Siamois. The *siamois* gene is normally repressed by a ubiquitous transcription factor called Tcf-3, but in cells in which β-catenin is present, an interaction between Tcf-3 and β-catenin induces *siamois* expression (**Figure 43.12**). But Siamois protein alone is not sufficient for *goosecoid* expression.

Vegetal cells receive mRNA transcripts from the original egg cytoplasm for proteins in the transforming growth factor-β (TGF-β) superfamily of cell signaling molecules. The signaling pathways activated by one or more proteins from this superfamily interact with the Siamois protein by cooperatively activating the promoter of the *goosecoid* gene and thereby controlling its transcription. Thus it is a particular combination of factors that determine which cells become the primary organizer.

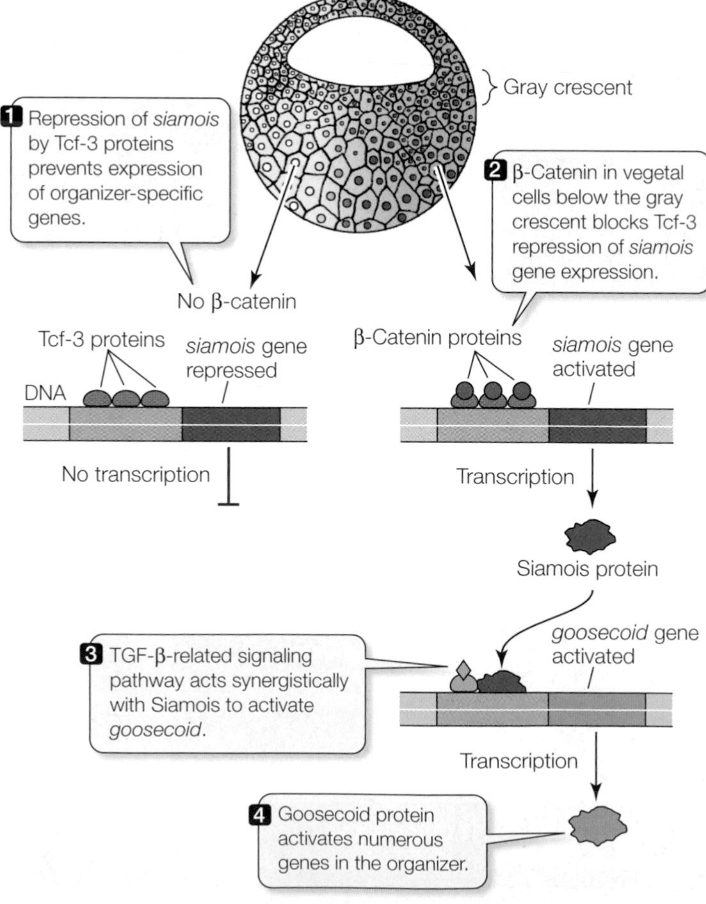

43.12 Molecular Mechanisms of the Primary Embryonic Organizer The organizing potential of the gray crescent depends on the activity of the *goosecoid* gene, which in turn is activated by signaling pathways set up in the vegetal cells below the gray crescent.

The organizer changes its activity as it migrates from the dorsal lip

Organizer cells begin the process of formation of the dorsal lip of the blastopore. Specifically, these cells are at the center of the dorsal lip and involute, moving forward on the midline (i.e., the middle of the anterior–posterior axis) to become mesoderm. Those organizer cells that involute first will move the farthest forward and will induce neighboring cells to participate in making structures of the head. Later organizer cells will induce structures of the trunk, and the last of the organizer cells to move inward from the dorsal lip will induce structures of the tail. How does the capacity of the organizer cells change to enable them to induce head, trunk, or tail structures?

As we learned above, the early organizer cells express the transcription factor Goosecoid, which activates genes encoding soluble signals that influence activity of the cells in contact with the organizer cells in their eventual anterior position. Those neighboring cells produce a number of growth factors. Inhibition of certain of these growth factors is critical for determination of head structures. Under the influence of Goosecoid, the anterior organizer cells produce antagonists to those growth factors. The induction of trunk structures requires inhibition of a different set of growth factors. In organizer cells that involute later than the head organizers, Goosecoid is no longer the dominant transcription factor, and these cells express different growth factor antagonists. The induction of tail structures requires still different activities of the organizer cells that involute last. Thus, the organizer cells express appropriate sets of growth factor antagonists at the right times to achieve different patterns of differentiation on the anterior–posterior axis.

How the activity of the organizer cells changes is a more complex story we will take up later; the main points to grasp are, first, that it is common for mechanisms of development to employ different interactions of transcription factors and growth factors; and, second, the critical interactions are frequently inhibitory.

Reptilian and avian gastrulation is an adaptation to yolky eggs

The eggs of reptiles and birds contain a mass of yolk, and the blastulas of these groups develop as a disc of cells on top of the yolk (see Figure 43.3C). We will use the chicken egg to show how gastrulation proceeds in a flat disc of cells rather than in a ball of cells.

Cleavage in the chick results in a flat, circular layer of cells called a blastodisc (**Figure 43.13**). Between the blastodisc and the yolk mass is a fluid-filled space. Some cells from the blastodisc break free and move into this space. These cells come together to form a continuous layer called the **hypoblast**, which will later contribute to *extraembryonic membranes* that will support and nourish the developing embryo. The overlying cells make up the **epiblast**, from which the embryo proper will form. Thus, the avian blastula is a flattened structure consisting of an upper epiblast and a lower hypoblast, which are joined at the margins of the blastodisc. The blastocoel is the fluid-filled space between the epiblast and hypoblast.

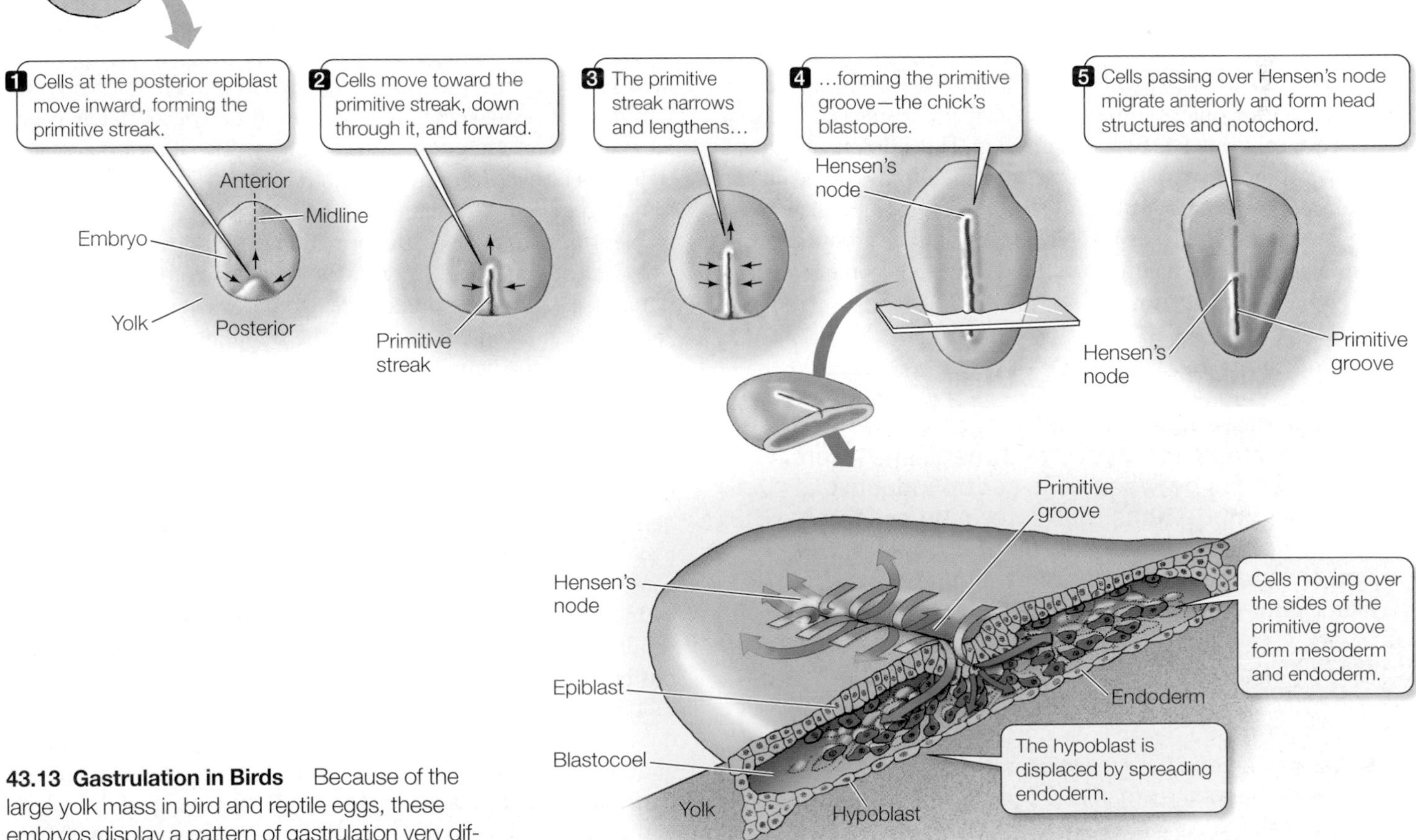
Chick embryo viewed from above
Blastodisc
Yolk
1 Cells at the posterior epiblast move inward, forming the primitive streak.
2 Cells move toward the primitive streak, down through it, and forward.
3 The primitive streak narrows and lengthens...
4 ...forming the primitive groove—the chick's blastopore.
5 Cells passing over Hensen's node migrate anteriorly and form head structures and notochord.
Anterior
Midline
Embryo
Yolk
Posterior
Primitive streak
Hensen's node
Hensen's node
Primitive groove
Primitive groove
Hensen's node
Epiblast
Blastocoel
Yolk
Hypoblast
Endoderm
Cells moving over the sides of the primitive groove form mesoderm and endoderm.
The hypoblast is displaced by spreading endoderm.
Cross section through chick embryo

43.13 Gastrulation in Birds Because of the large yolk mass in bird and reptile eggs, these embryos display a pattern of gastrulation very different from that of sea urchins and amphibians.

Gastrulation begins with a thickening in the posterior region of the epiblast caused by the movement of cells toward the midline and then forward along the midline (see Figure 43.13). The result is a midline ridge called the *primitive streak*. A depression called the *primitive groove* forms along the length of the primitive streak. The primitive groove functions as the blastopore, and cells migrate through it into the blastocoel to become endoderm and mesoderm.

In the chick embryo, no archenteron forms, but the endoderm and mesoderm migrate forward to form the gut and other structures. At the anterior end of the primitive groove is a thickening called **Hensen's node**, which is the equivalent of the dorsal lip of the amphibian blastopore. Many signaling molecules that have been identified in the frog organizer are also expressed in Hensen's node. Cells that pass over Hensen's node become determined by the time they reach their final destination, where they differentiate into certain tissues and structures of the head and dorsal midline.

Placental mammals have no yolk but retain the avian–reptilian gastrulation pattern

Both mammals and birds evolved from reptilian ancestors, so it is not surprising that they share patterns of early development, even though the mammalian eggs have no yolk. Earlier we described the development of the mammalian trophoblast and the inner cell mass, which is the equivalent of the avian epiblast.

As in avian development, the inner cell mass splits into an upper layer called the epiblast and a lower layer called the hypoblast, with a fluid-filled cavity between them. The embryo will form from the epiblast, and the hypoblast will contribute to the extraembryonic membranes that will encase the developing embryo and help form the placenta (see Figure 43.5). The epiblast also contributes to the extraembryonic membranes; specifically, it splits off an upper layer of cells that will form the amnion. The amnion will grow to surround the developing embryo as a membranous sac filled with amniotic fluid. Gastrulation occurs in the mammalian epiblast just as it does in the avian epiblast. A primitive groove forms, and epiblast cells migrate through the groove to become layers of endoderm and mesoderm.

43.2 RECAP

The cell movements of gastrulation convert the blastula into an embryo with three tissue layers. New contacts between cells set up inductive signaling interactions that determine cell fates. Dorsal lip tissue is the source of organizer cells that induce development of preliminary head, trunk, and tail structures.

- Compare the cell movements that occur during gastrulation in a sea urchin, a frog, and a bird. See Figures 43.8, 43.9, and 43.13
- Can you explain the molecular basis for the inductive capabilities of the organizer? See pp. 929–930 and Figure 43.12

We have described how the fertilized egg develops into an embryo with three germ layers and how cellular signals trigger different patterns of differentiation. In the next section we describe how organs and organ systems develop in the early embryo.

43.3 How Do Organs and Organ Systems Develop?

Gastrulation produces an embryo with three germ layers that are positioned to influence one another through inductive interactions. During the next phase of development, called **organogenesis**, many organs and organ systems develop simultaneously and in coordination with one another. An early process of organogenesis in chordates that is directly related to gastrulation is neurulation. **Neurulation** is the initiation of the nervous system. We will examine neurulation in the amphibian embryo, but it occurs in a similar fashion in reptiles, birds, and mammals.

The stage is set by the dorsal lip of the blastopore

As we learned in the previous section, one group of cells that passes over the dorsal lip of the blastopore moves anteriorly and becomes the endodermal lining of the digestive tract. The other group of cells that involutes over the dorsal lip will become mesoderm, and these cells that are closest to midline have organizer functions (see Figure 43.9). This mesoderm is called the *chordamesoderm* because it produces a rod of connective tissue called the **notochord**. The notochord gives structural support to the developing embryo; it is eventually replaced by the vertebral column. After gastrulation, the organizing capacity of the chordamesoderm induces the overlying ectoderm to begin forming the nervous system by expressing signaling molecules (one appropriately called Noggin and another one called Chordin). These molecules neutralize an already-present growth factor that inhibits the determination of neural structures.

Neurulation involves the formation of an internal neural tube from an external sheet of cells. The first signs of neurulation are flattening and thickening of the ectoderm overlying the notochord; this thickened area forms the *neural plate* (**Figure 43.14**). The edges of the neural plate that run in an anterior–posterior direction continue to thicken to form ridges or folds. Between these neural folds, a groove forms and deepens as the folds roll over it to converge on the midline. The folds fuse, forming a cylinder, the **neural tube**, and a continuous overlying layer of epidermal ectoderm. The neural tube develops bulges at the anterior end, which become the major divisions of the brain; the rest of the tube becomes the spinal cord.

In humans, failure of the neural tube to develop normally can result in serious birth defects. If the neural folds fail to fuse in a posterior region, the result is a condition known as *spina bifida*. If they fail to fuse at the anterior end, an infant can develop without a forebrain—a condition called *anencephaly*. Although several genetic factors can cause neural tube defects, other factors are environmental, including diet. The incidence of neural tube defects in the United States in the early 1900s was as high as 1 in 300 live births, but it has declined to less than 1 per 1,000 today. A major

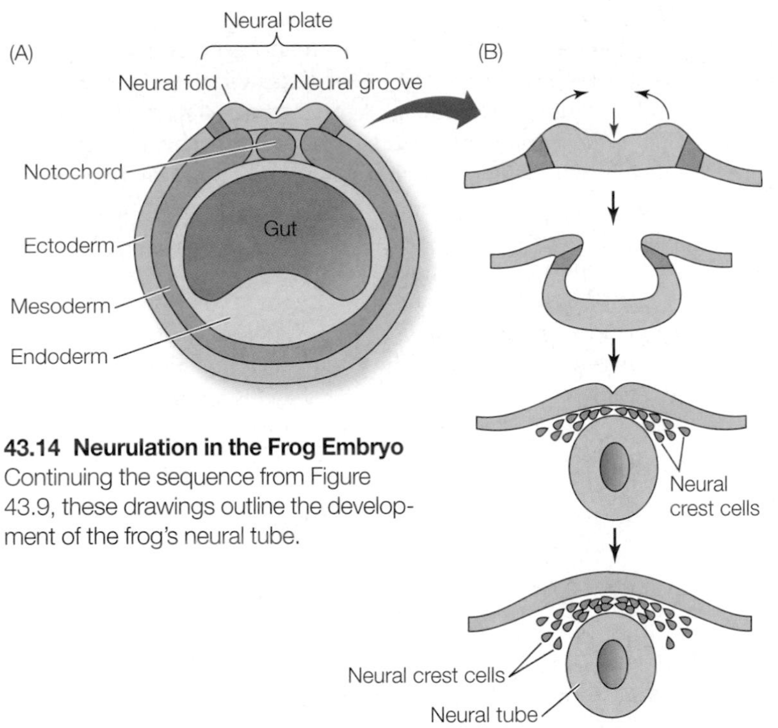

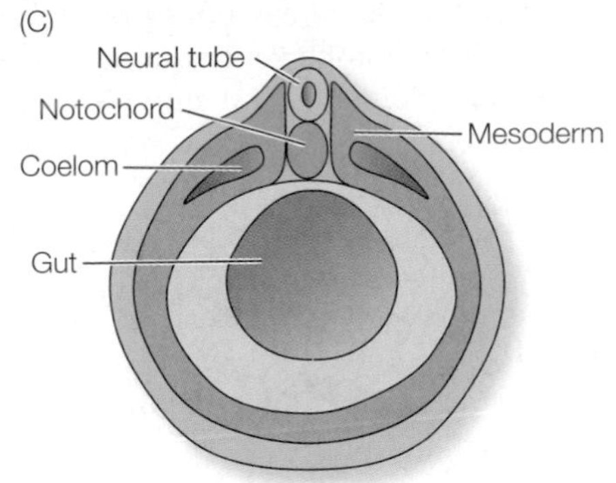

43.14 Neurulation in the Frog Embryo Continuing the sequence from Figure 43.9, these drawings outline the development of the frog's neural tube.

factor in this improvement has been the inclusion of an adequate amount of folic acid (a B vitamin) in the mother's diet.

Body segmentation develops during neurulation

Like the arthropods, vertebrates have a body plan consisting of repeating segments that are modified during development. These segments are most evident as the repeating patterns of vertebrae, ribs, nerves, and muscles along the anterior–posterior axis.

As the neural tube forms, mesodermal tissues gather along the sides of the notochord to form separate, segmented blocks of cells called **somites** (**Figure 43.15**). The somites produce cells that will become the vertebrae, ribs, muscles of the trunk and limbs, and the lower layer of the skin.

The nerves that connect the brain and spinal cord with tissues and organs throughout the body are also arranged segmentally. The somites help guide the organization of these peripheral nerves, but the nerves are not of mesodermal origin. When the neural tube fuses, cells adjacent to the line of closure break loose and migrate inward between the epidermis and the somites and through the somites. These cells, called *neural crest cells*, contribute to a number of structures, including the peripheral nerves, which grow out to the body tissues and back into the spinal cord.

As development progresses, the different segments of the body change. Regions of the spinal cord differ, regions of the vertebral column differ in that some vertebrae grow ribs of various sizes and others do not, forelegs arise in the anterior part of the embryo, and hind legs arise in the posterior region.

Hox genes control development along the anterior–posterior axis

How is mesoderm in the anterior part of a mouse embryo programmed to produce forelegs rather than hind legs? In Section 19.5, we saw how homeotic genes control body segmentation in *Drosophila*. We also learned that all homeotic genes contain a DNA sequence called the *homeobox*. Some of the genes directing gastrulation in the frog are homeobox genes—for example, *goosecoid* and *siamois*. In the mouse, four families of homeotic **Hox genes** control differentiation along the anterior–posterior body axis.

Each mammalian Hox gene family resides on a different chromosome, in clusters of about 10 genes each. Remarkably, the temporal and spatial expression of these genes follows the same pattern as their linear order on their chromosome. That is, the Hox genes closest to the 3′ end of each

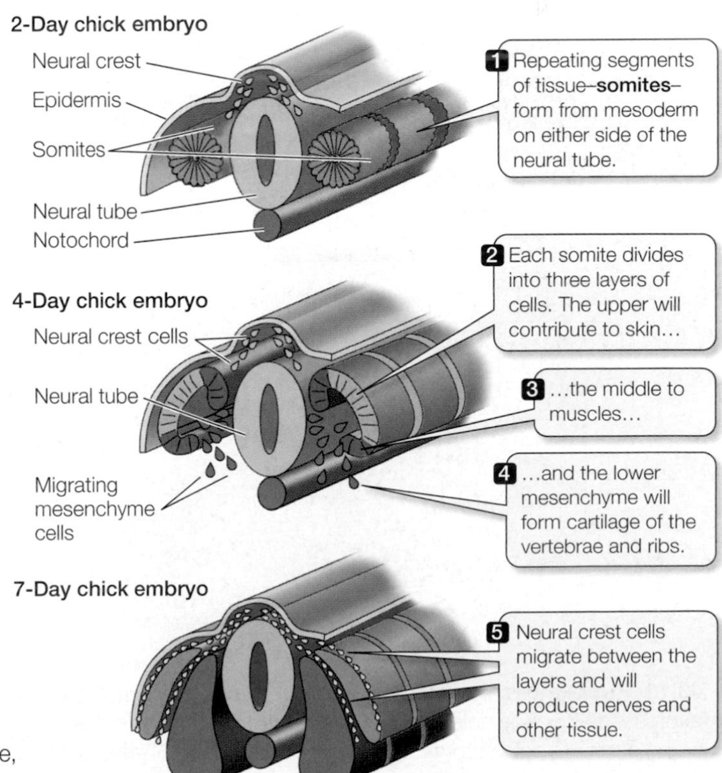

2-Day chick embryo
Neural crest
Epidermis
Somites
Neural tube
Notochord
1 Repeating segments of tissue–somites–form from mesoderm on either side of the neural tube.
4-Day chick embryo
Neural crest cells
Neural tube
Migrating mesenchyme cells
2 Each somite divides into three layers of cells. The upper will contribute to skin...
3 ...the middle to muscles...
4 ...and the lower mesenchyme will form cartilage of the vertebrae and ribs.
7-Day chick embryo
5 Neural crest cells migrate between the layers and will produce nerves and other tissue.

43.15 The Development of Body Segmentation Repeating blocks of tissue called somites form on either side of the neural tube. Muscle, cartilage, bone, and the lower layer of the skin form from the somites.

gene complex are expressed first and in the anterior of the embryo. The Hox genes closer to the 5′ end of the gene complex are expressed later and in a more posterior part of the embryo. As a result, different segments of the embryo receive different combinations of Hox gene products, which serve as transcription factors (**Figure 43.16**; see also Figure 20.1).

Whereas Hox genes give cells information about their position on the anterior–posterior (head–tail) body axis, other genes provide information about their dorsal–ventral (back–belly) position. Tissues in each segment of the body differentiate according to their dorsal–ventral location. In the spinal cord, for example, sensory nerve connections develop in the dorsal region, and motor nerve connections develop in the ventral region. In the somites, dorsal cells develop into skin and muscle and ventral cells develop into cartilage and bone (see Figure 43.15). An example of a gene that provides dorsal–ventral information in vertebrates is *sonic hedgehog*, which is expressed in the mammalian notochord and induces cells in the overlying neural tube to have fates characteristic of ventral spinal cord cells.

The mammalian *sonic hedgehog* gene is homologous to a *Drosophila* gene known simply as *hedgehog*—a gene whose name comes from the distinctive spiny appearance of the back in mutant flies. The mammalian homolog was given the name of the video game character to distinguish it from its insect counterpart.

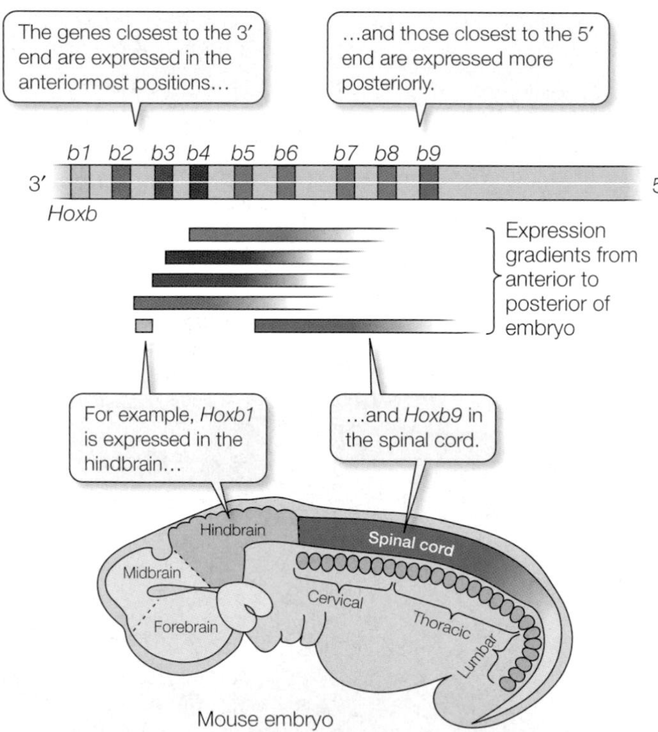
The genes closest to the 3′ end are expressed in the anteriormost positions...
...and those closest to the 5′ end are expressed more posteriorly.
b1 b2 b3 b4 b5 b6 b7 b8 b9
3′
5′
Hoxb
Expression gradients from anterior to posterior of embryo
For example, Hoxb1 is expressed in the hindbrain...
...and Hoxb9 in the spinal cord.
Hindbrain
Spinal cord
Midbrain
Forebrain
Cervical
Thoracic
Lumbar
Mouse embryo

43.16 Hox Genes Control Body Segmentation Hox genes are expressed along the anterior–posterior axis of the embryo in the same order as their arrangement between the 3′ and 5′ ends of the gene complex.

One family of homeobox genes, the *Pax* genes, plays many roles in nervous system and somite development. The role of *Pax6* in eye development was described at the start of Chapter 20. Another gene in the same family, *Pax3*, is expressed in those neural tube cells that will develop into dorsal spinal cord structures. Sonic hedgehog protein represses the expression of the *Pax3* gene, and their interaction is one source of dorsal–ventral information for the differentiation of the spinal cord.

After the development of body segmentation, the formation of organs and organ systems progresses rapidly. The development of an organ involves extensive inductive interactions of the kind we saw in Section 19.4 in the example of the vertebrate eye. These inductive interactions are a current focus of study for developmental biologists.

43.3 RECAP

Gastrulation sets up tissue interactions that initiate organogenesis. Neurulation is initiated by organizer mesoderm that forms the notochord.

- Describe the formation of the neural tube in vertebrates. See p. 932 and Figure 43.14
- Can you explain how somites relate to segmentation of the body axis? See p. 933 and Figure 43.15
- Using information from this chapter and from Chapters 19 and 20, can you explain what Hox genes are and how they instruct patterns of differentiation along the body axis? See Figures 19.19, 20.1, and 43.16

You may be aware that in mammals the circulatory systems of the fetus and mother are separate and that nourishment reaches the fetus through the placenta and the umbilical cord. In the next section we will examine the developmental events that result in the creation of the placenta.

43.4 What Is the Origin of the Placenta?

There is more to a developing reptile, bird, or mammal than the embryo itself. As mentioned earlier, the embryos of these vertebrates are surrounded by several **extraembryonic membranes**, which originate from the embryo but are not part of it. The extraembryonic membranes function in nutrition, gas exchange, and waste removal. In mammals, they interact with tissues of the mother to form the placenta.

Extraembryonic membranes form with contributions from all germ layers

We will use the chick to demonstrate how the extraembryonic membranes form from the germ layers created during gastrulation. In the chick, four membranes form—the *yolk sac*, the *allantoic membrane*, the *amnion*, and the *chorion*. The **yolk sac** is the first to form, and it does so by extension of the endodermal tissue of the hypoblast layer along with some adjacent mesoderm. The yolk sac grows to enclose

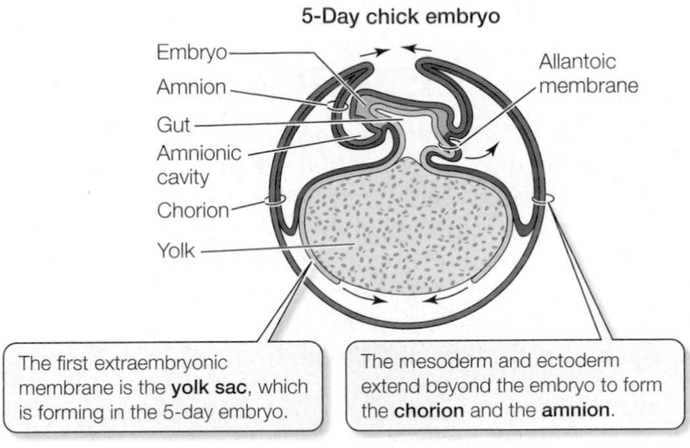

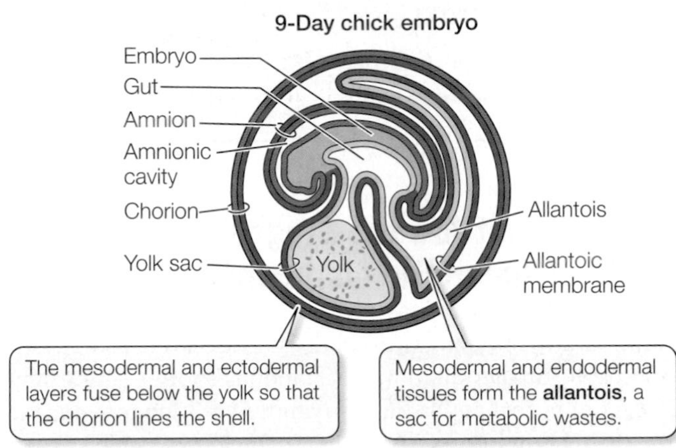

43.17 The Extraembryonic Membranes In birds, reptiles, and mammals, the embryo constructs four extraembryonic membranes. The yolk sac encloses the yolk, and the amnion and chorion enclose the embryo. Fluids secreted by the amnion fill the amniotic cavity, providing an aqueous environment for the embryo. The chorion, along with the allantois, mediates gas exchange between the embryo and its environment. The allantois stores the embryo's waste products.

the entire body of yolk in the egg (**Figure 43.17**). It constricts at the top to create a tube that is continuous with the gut of the embryo. However, yolk does not pass through this tube. Yolk is digested by the cells of the yolk sac, and the nutrients are transported to the embryo through blood vessels that form from mesoderm and line the outer surface of the yolk sac. The **allantoic membrane** is also an outgrowth of the extraembryonic endoderm plus adjacent mesoderm. It forms the *allantois*, a sac for storage of metabolic wastes.

Just as the endoderm and mesoderm of the hypoblast grow out from the embryo to form the yolk sac and the allantoic membrane, ectoderm and mesoderm combine and extend beyond the limits of the embryo to form the other extraembryonic membranes. Two layers of cells extend all along the inside of the eggshell, both over the embryo and below the yolk sac. Where they meet, they fuse, forming two membranes, the inner **amnion** and the outer **chorion**. The amnion surrounds the embryo, forming the amniotic cavity. The amnion secretes fluid into the cavity, providing a protective environment for the embryo. The outer membrane, the chorion, forms a continuous membrane just under the eggshell (see Figure 43.17). It limits water loss from the egg and also works with the enlarged allantoic membrane to exchange respiratory gases between the embryo and the outside world.

Extraembryonic membranes in mammals form the placenta

In mammals, the first extraembryonic membrane to form is the trophoblast, which is already apparent by the fifth cell division (see Figure 43.4). When the blastocyst reaches the uterus and hatches from its encapsulating zona pellucida, the trophoblast cells interact directly with the endometrium. Adhesion molecules expressed on the surfaces of these cells attach them to the uterine wall. By secreting proteolytic enzymes, the trophoblast burrows into the endometrium, beginning the process of implantation (see Figure 43.5). Eventually, the entire trophoblast is within the wall of the uterus. The trophoblast cells then send out numerous projections, or villi, to increase the surface area of contact with maternal blood.

Meanwhile, the hypoblast cells proliferate to form what in the bird would be the yolk sac. But there is no yolk in eggs of placental mammals, so the yolk sac contributes mesodermal tissues that interact with trophoblast tissues to form the chorion. The chorion, along with tissues of the uterine wall, produces the **placenta**, the organ that exchanges nutrients, respiratory gases, and metabolic wastes between the mother and the embryo (**Figure 43.18**).

At the same time the yolk sac is forming from the hypoblast, the epiblast produces the amnion, which grows to enclose the entire embryo in a fluid-filled amniotic cavity. The rupturing of the

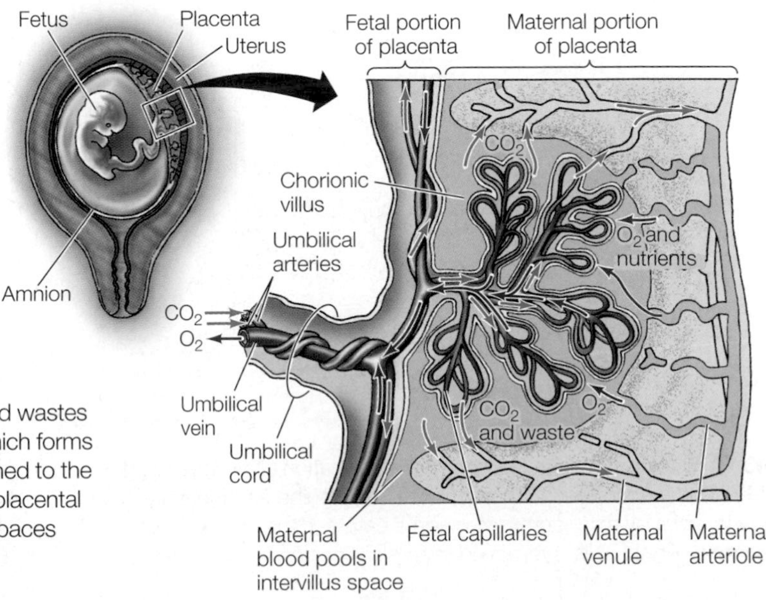

43.18 The Mammalian Placenta In most mammals, nutrients and wastes are exchanged between maternal and fetal blood in the placenta, which forms from the chorion and tissues of the uterine wall. The embryo is attached to the placenta by the umbilical cord. Embryonic blood vessels invade the placental tissue to form fingerlike chorionic villi. Maternal blood flows into the spaces surrounding the villi.

amnion and chorion and the loss of the amniotic fluid (a process called "water breaking") herald the onset of labor in humans.

An allantois also develops in mammals, but its importance depends on how well nitrogenous wastes can be transferred across the placenta. In humans the allantois is minor; in pigs it is important. In humans and other mammals, allantoic tissues contribute to the formation of the umbilical cord, by which the embryo is attached to the chorionic placenta. It is through the blood vessels of the umbilical cord that nutrients and oxygen from the mother reach the developing fetus, and wastes, including carbon dioxide and urea, are removed (see Figure 43.18).

The extraembryonic membranes provide means of detecting genetic diseases

Cells slough off of the developing human embryo and float in the amniotic fluid that bathes it. Later in development, a small volume of the amniotic fluid may be extracted with a needle as the first step of a process called **amniocentesis**. Cells from the fluid can be cultured and used for biochemical and genetic analyses that can reveal the sex of the fetus, as well as genetic markers for diseases such as cystic fibrosis, Tay-Sachs disease, and Down syndrome.

If amniocentesis is performed, it is usually not until after the fourteenth week of pregnancy, and the tests require two weeks to complete. If abnormalities in the fetus are detected, termination of the pregnancy at that stage would put the mother's health at greater risk than would an earlier abortion. Therefore, a newer technique, called **chorionic villus sampling**, is now in common use. In this test, a small sample of the tissue from the surface of the chorion is taken (**Figure 43.19**). This test can be done as early as the eighth week of pregnancy, and the results are available in several days.

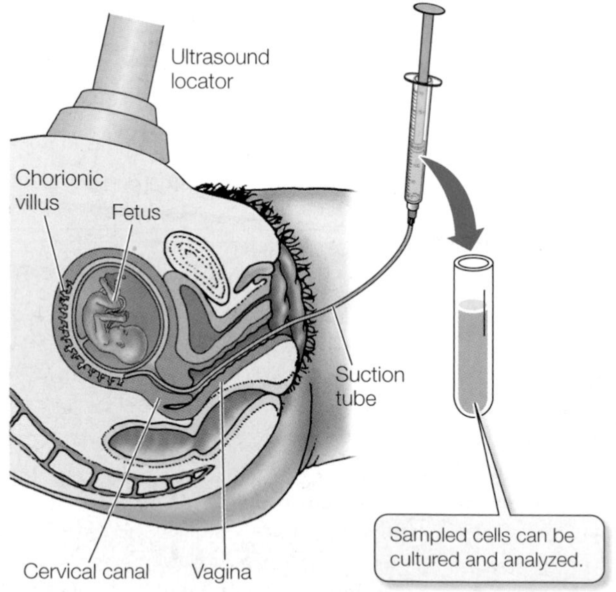

43.19 Chorionic Villus Sampling Information about genetic defects can be obtained from chorionic tissues. The fetus and placenta are imaged by a sonogram to guide a catheter, through which a sample of fetal cells is removed from a chorionic villus for testing.

43.4 RECAP

The extraembryonic membranes of reptiles, birds, and mammals sustain the growing embryo. In reptiles and birds, these membranes surround the embryo within the shelled egg. In mammals the extraembryonic membranes form the placenta, an organ that exchanges nutrients, respiratory gases, and metabolic wastes between the mother and the embryo.

- Can you describe each of the four extraembryonic membranes and their functions in the developing chick egg? See pp. 934–935 and Figure 43.17
- Can you explain the role of the trophoblast in the early development of a mammalian embryo? See p. 935

43.5 What Are the Stages of Human Development?

In humans, **gestation**, or pregnancy, lasts about 266 days, or 9 months. In smaller mammals gestation is shorter—for example, 21 days in mice—and in larger mammals it is longer—for example, 330 days in horses and 600 days in elephants. The events of human gestation can be divided into three periods of roughly 3 months each, called *trimesters*.

The embryo becomes a fetus in the first trimester

Implantation of the human blastocyst begins on about the sixth day after fertilization. After implantation, gastrulation occurs, tissues differentiate, the placenta forms, and organs begin to develop. The heart begins to beat during week 4, and limbs form by week 8 (**Figure 43.20A,B**). By the end of the first trimester, most organs have started to form. The embryo is about 8 centimeters long and weighs about 40 grams (less than 2 ounces); it would fit neatly in a teaspoon. At about this point in time, the human embryo is medically and legally referred to as a **fetus**. (This distinction is not made for other mammals; developing mice, for example, remain embryos until they are born.)

An embryo can be damaged before the mother even knows she is pregnant. A classic and tragic case is that of thalidomide, a drug widely prescribed in Europe in the late 1950s to treat nausea. Women who took this drug in the fourth and fifth week of pregnancy, when the embryo's limbs are beginning to form, gave birth to children with missing or severely malformed arms and legs.

The first trimester is a time of rapid cell division and tissue differentiation. Signal transduction cascades and the resulting branching sequences of developmental processes are in their early stages. Therefore, the first trimester is the period during which the em-

(A) 4 weeks

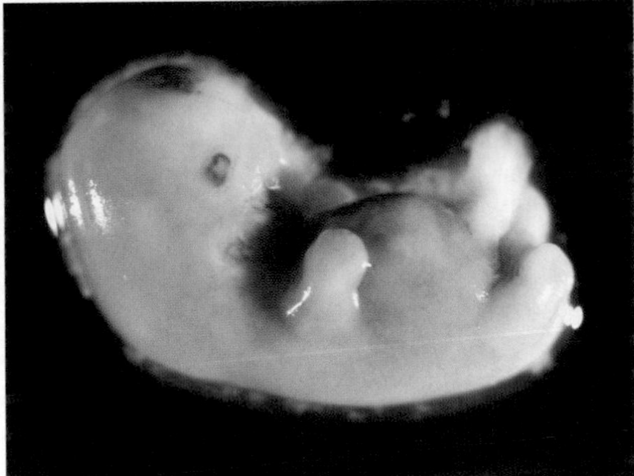

(B) 8 weeks

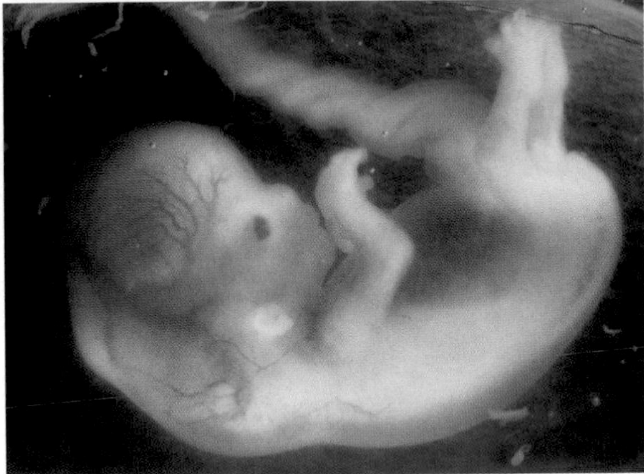

(C) 4 months

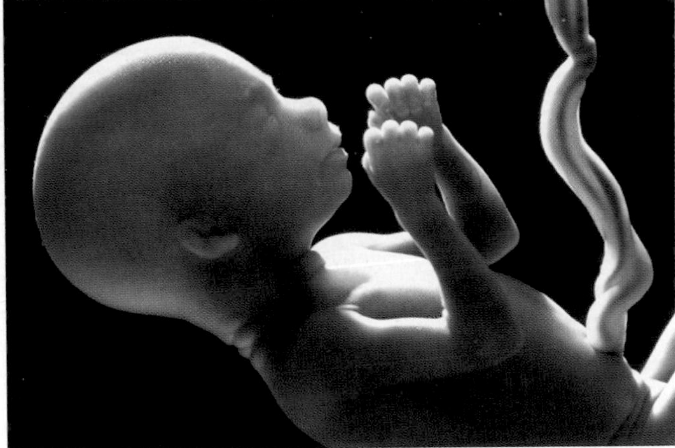

(D) 9 months

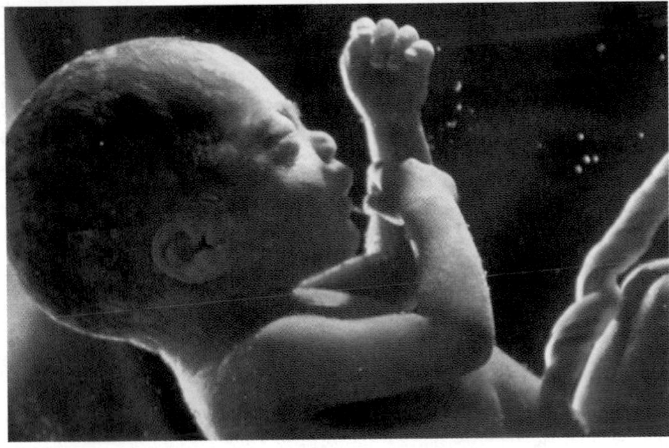

43.20 Stages of Human Development (A) At 4 weeks of gestation, most of the human embryo's organ systems have been formed and the heart is beating. (B) The body structures of this 8-week-old embryo are forming rapidly, and it is visibly a male. The umbilical cord attaches the embryo to the placenta (upper left). (C) At 4 months, the fetus has fully formed limbs with fingers and toes, and moves freely within the amniotic cavity. (D) This fetus is well along in its ninth month. Soon its lungs will be mature enough to trigger the onset of contractions and birth.

bryo is most sensitive to damage from radiation, drugs, chemicals, and pathogens that can cause birth defects.

Hormonal changes cause major and noticeable responses in the mother during the first trimester. Soon after the blastocyst implants itself, it begins to secrete the hormone human chorionic gonadotropin (hCG). Because of this early release, detection of hCG is used as an early test for pregnancy. The presence of hCG stimulates the mother's ovaries to continue producing the hormones estrogen and progesterone, which help maintain pregnancy. These hormonal changes cause pregnancy's well-known symptoms, including morning sickness, mood swings, changes in the senses of taste and smell, and swelling of the breasts.

The fetus grows and matures during the second and third trimesters

During the second trimester the fetus grows rapidly to a weight of about 600 g, and the mother's abdomen enlarges considerably. The limbs of the fetus elongate, and the fingers, toes, and facial features become well formed (**Figure 43.20C**). Eyebrows and fingernails grow. Fetal movements are first felt by the mother early in the second trimester, and they become progressively stronger and more coordinated. By the end of the second trimester, the fetus may suck its thumb. The nervous system undergoes rapid development.

The fetus grows rapidly during the third trimester (**Figure 43.20D**). As the third trimester approaches its end, internal organs mature. The digestive system begins to function, the liver stores glycogen, the kidneys produce urine, and the brain undergoes cycles of sleep and waking. A human infant is born as soon as the last of its critical organs—the lungs—mature. If development were to continue for longer inside the mother's body, the baby's head would grow larger than the birth canal. But if the fetus is born before its lungs mature, the baby cannot breathe on its own.

Although the first-trimester embryo is the most susceptible to adverse effects of drugs, chemicals, and diseases, the potential for serious effects from exposure to environmental factors continues throughout pregnancy. Severe protein malnutrition, alcohol consumption, and cigarette smoking are examples of factors that can result in low birth weight, mental retardation, and other developmental complications.

Developmental changes continue throughout life

Development does not end with birth. Obviously, growth continues until adult size is reached, and even when growth stops, organs of the body continue to repair and renew themselves through cycles of cell replacement by the progeny of undifferentiated stem cells. In humans especially, enormous developmental changes occur in the brain in the years between birth and adolescence. Especially in the early years, there is a great deal of plasticity in the organization of the nervous system as the connections between neurons develop.

For example, if a child is born with its eyes misaligned, a condition known as *strabismus*, he or she will use mostly one eye. The connections to the brain from that eye will become strong, and connections from the other eye will become weak. The child will develop with reduced visual acuity and depth perception. If the eye alignment is corrected in the first 3 years of life, however, the connections between the eyes and the brain will improve, and the child is likely to develop normal vision. If the eye alignment is corrected after 3 years of age, the correct connections between the eyes and the brain are less likely to improve, and visual impairments may persist. Thus, plasticity in the development of the visual system in humans continues for several years after birth, but is gradually lost.

A very exciting area of current research is the role of learning in stimulating the production and differentiation of new neurons in the brains of young and even adult animals (see Section 46.3). The first observations of new neurons were made on song birds, which relearn their songs each year prior to mating. At that time, the regions of their brains involved in vocalization grow and new neurons appear. Subsequently it was shown that mice and rats, when exposed to complex environments offering ample opportunity for exploration and activity, also acquired new neurons in certain parts of their brains. Scientists believe that acquisition of new neurons also occurs in adult humans.

43.5 RECAP

Gestation in humans lasts 9 months and can be divided into three trimesters. By the end of the first trimester, the fetus is very small but most of its organs have begun to form. In the second trimester, limbs elongate and the fetus moves. By the end of the third trimester, most organs have begun to function.

- Do you understand why the first trimester is a time of particular sensitivity for the embryo with respect to environmental risks? See p. 936

CHAPTER SUMMARY

43.1 How does fertilization activate development?

The sperm and the egg contribute differentially to the zygote. The sperm contributes a haploid nucleus and, in some species, a centriole. The egg contributes a haploid nucleus, nutrients, ribosomes, mitochondria, mRNAs, and proteins.

The cytoplasmic contents of the egg are not distributed homogeneously, and they are rearranged after fertilization to set up the major axes of the future embryo. The nutrient molecules are generally found in the **vegetal hemisphere**, whereas the nucleus is found in the **animal hemisphere**. Review Figures 43.1 and 43.2

Cleavage is a period of rapid cell division without cell expansion or gene expression. Cleavage can be complete or incomplete, and the pattern of cell divisions depends on the orientation of the mitotic spindles. The result of cleavage is a ball or mass of cells called a **blastula**. Review Figure 43.3

Cleavage in mammals is unique in that cell divisions are very slow and genes are expressed early in the process. Cleavage results in an inner cell mass that becomes the embryo and an outer cell mass that becomes the **trophoblast**. The mammalian embryo at this stage is called a **blastocyst**. At the time of implantation, the trophoblast secretes molecules that help the blastocyst attach to and penetrate the uterine wall. Review Figures 43.4 and 43.5

A **fate map** can be created by labeling specific blastomeres and observing what tissues and organs are formed by their progeny. Review Figure 43.6

Some species undergo mosaic development, in which the fate of each cell is determined during early divisions. Other species, including vertebrates, undergo regulative development, in which remaining cells can compensate for cells lost in early cleavages.

43.2 How does gastrulation generate multiple tissue layers?

Gastrulation involves massive cell movements that produce three **germ layers** and place cells from various regions of the blastula into new associations with one another. Review Figure 43.8, Web/CD Tutorial 43.1

The initial step of sea urchin and amphibian gastrulation is inward movement of certain blastomeres. The site of inward movement becomes the **blastopore**. Cells that move into the blastula become the **endoderm** and **mesoderm**; cells remaining on the outside become the **ectoderm**. Cytoplasmic factors in the vegetal pole cells are essential to initiate development. Review Figures 43.8 and 43.9

The **dorsal lip** of the amphibian blastopore is a critical site for cell determination. It has been called the primary embryonic **organizer** because it induces determination in cells that pass over it during gastrulation. Review Figures 43.9, 43.10, and 43.11, Web/CD Tutorial 43.2

The protein β-catenin activates a signaling cascade that induces the primary embryonic organizer and sets up the anterior–posterior body axis. Review Figures 43.2 and 43.12

Gastrulation in reptiles and birds differs from that in sea urchins and frogs because the large amount of yolk in their eggs causes the blastula to form a flattened disc of cells. Review Figure 43.13

Mammals have a pattern of gastrulation similar to that of reptiles and birds, although their eggs have no yolk.

CHAPTER SUMMARY

43.3 How do organs and organ systems develop?

Gastrulation is followed by **organogenesis**, the process whereby tissues interact to form organs and organ systems.

In the formation of the vertebrate nervous system, one group of cells that migrates over the blastopore lip is determined to become the **notochord**. The notochord induces the overlying ectoderm to thicken, form parallel ridges, and fold in on itself to form a **neural tube** below the epidermal ectoderm. The nervous system develops from this neural tube. Review Figure 43.14

The notochord and neural crest cells participate in the segmental organization of mesoderm into structures called **somites** along the body axis. Rudimentary organs and organ systems form during these stages. Review Figure 43.15

Four families of **Hox genes** determine the pattern of anterior–posterior differentiation along the body axis in mammals. Other genes, such as *sonic hedgehog*, contribute to dorsal–ventral differentiation. Review Figure 43.16

43.4 What is the origin of the placenta?

The embryos of reptiles, birds, and mammals are protected and nurtured by four **extraembryonic membranes**. In birds and reptiles, the **yolk sac** surrounds the yolk and provides nutrients to the embryo, the **chorion** lines the eggshell and participates in gas exchange, the **amnion** surrounds the embryo and encloses it in an aqueous environment, and the **allantois** stores metabolic wastes. Review Figure 43.17, Web/CD Activity 43.1

In mammals, the chorion and the trophoblast cells interact with the maternal uterus to form a **placenta**, which provides the embryo with nutrients and gas exchange. The amnion encloses the embryo in an aqueous environment. Review Figure 43.18

Samples of amniotic fluid or pieces of the chorion can be analyzed either by **amniocentesis** or **chorionic villus sampling** for genetic analysis that can reveal the presence of genes that can cause birth defects or disease. Review Figure 43.19

43.5 What are the stages of human development?

Human pregnancy, or **gestation**, can be divided into three trimesters. The embryo forms in the first trimester; during this time, it is most vulnerable to environmental factors that can lead to birth defects. During the second and third trimesters the **fetus** grows, the limbs elongate, and the organ systems mature.

Development continues throughout childhood and throughout life.

SELF-QUIZ

1. Fertilization involves all of the following *except*
   a. joining of most cell organelles from sperm and egg.
   b. joining of sperm and egg haploid nuclei.
   c. induction of rearrangements of the egg cytoplasm.
   d. sperm binding to specific sites on the egg surface.
   e. metabolic activation of the egg.
2. Which of the following does *not* occur during cleavage in frogs?
   a. A high rate of mitosis
   b. Reduction in the size of cells
   c. Expression of genes critical for blastula formation
   d. Orientation of cleavage planes at right angles
   e. Unequal division of cytoplasmic determinants
3. How does cleavage in mammals differ from cleavage in frogs?
   a. Slower rate of cell division
   b. Formation of tight junctions
   c. Expression of the embryo's genome
   d. Early separation of cells that will not contribute to the embryo
   e. All of the above
4. Which statement about gastrulation is *true*?
   a. In frogs, gastrulation begins in the vegetal hemisphere.
   b. In sea urchins, gastrulation produces the notochord.
   c. In birds, cells from the surface of the blastodisc move down through the primitive groove to form the hypoblast.
   d. In mammals, gastrulation occurs in the hypoblast.
   e. In sea urchins, gastrulation produces only two germ layers.
5. Which of the following was a conclusion from the experiments of Spemann and Mangold?
   a. Cytoplasmic determinants of development are homogeneously distributed in the amphibian zygote.
   b. In the late blastula, certain regions of cells are determined to form skin or nervous tissue.
   c. The dorsal lip of the blastopore can be isolated and will form a complete embryo.
   d. The dorsal lip of the blastopore can initiate gastrulation.
   e. The dorsal lip of the blastopore gives rise to the neural tube.
6. Which of the following is true of human development?
   a. Most organs begin to form during the second trimester.
   b. Gastrulation takes place in the oviducts.
   c. Genetic diseases can be detected by sampling cells from the chorion.
   d. Implantation occurs through interactions of the zona pellucida with the uterine lining.
   e. Exposure to drugs and chemicals is most likely to cause birth defects when it occurs in the third trimester.
7. Which of the following characterizes neurulation?
   a. The notochord forms a neural tube.
   b. The neural tube is formed from ectoderm.
   c. A neural tube forms around the notochord.
   d. The neural tube forms somites.
   e. In birds, the neural tube forms from the primitive groove.
8. Which statement about trophoblast cells is *true*?
   a. They are capable of producing monozygotic twins.
   b. They are derived from the hypoblast of the blastocyst.
   c. They are endodermal cells.
   d. They secrete proteolytic enzymes.
   e. They prevent the zona pellucida from attaching to the oviduct.
9. Which membrane is part of the embryonic contribution to placenta formation?
   a. Amnion
   b. Chorion
   c. Epiblast
   d. Allantois
   e. Zona pellucida

10. A major factor in the determination and differentiation of tissues along the anterior–posterior axis of the mouse is the
    a. differential expression of Hox genes.
    b. concentration gradient of β-catenin.
    c. differential expression of the *sonic hedgehog* gene.
    d. distance of the tissue from the gray crescent.
    e. distribution of GSK-3, which degrades β-catenin.

FOR DISCUSSION

1. If you found a protein that was localized to a small group of cells in the frog blastula, how would you determine whether that protein played a role in development? Address the issues of sufficiency and necessity.
2. During gastrulation in birds, the *sonic hedgehog* gene is expressed only on the left side of Hensen's node. What might be the significance of this expression pattern?
3. Much of the early work of describing animal development was done on sea urchins, amphibians, and chicks. Most recent work on the molecular mechanisms of animal development has been done on nematodes, fruit flies, zebrafish, and mice. Why do you think there has been a shift in the animal models used by developmental biologists?
4. If all the mitochondria and mitochondrial DNA in the embryo come from the egg, what implications does this have for using mitochondrial DNA for molecular evolutionary studies?
5. There is currently much controversy over therapeutic cloning as a way of obtaining embryonic stem cells to treat diseases. Given that human development is regulative—in other words, twinning can occur if an early blastocyst is divided into two cell masses—can you think of a way to guarantee a source of isogenic (i.e., identically matching a person's own body) stem cells for an individual without resorting to therapeutic cloning? Assume isolated cells can be preserved indefinitely in a frozen state.

FOR INVESTIGATION

The early gastrula is bilaterally symmetrical. However, the fetus is not bilaterally symmetrical. For example, the top of your heart tilts to the right side of your body and the aorta comes off of the left side of the heart. Your spleen is on the left side of your body. Your large intestine goes from right to left. These asymmetries are set up during gastrulation. How would you investigate the mechanisms involved?

CHAPTER 44 Neurons and Nervous Systems

Fear and survival in the brain

Charles Whitman was a normal and responsible child. He became the youngest Eagle Scout in the country. He was a fine son and husband, and received commendations as a U.S. Marine. But while he was in the service, he began having unexplained fits of anger, among other personality disorders. He was discharged from the Marines and entered the University of Texas. Several times he visited campus doctors and complained about having violent thoughts. Then, on August 1, 1966, after killing his wife and mother, he gathered several high-powered rifles, went to the top of the tall clock tower on the University campus, and barricaded himself inside. From this vantage point, he killed 14 people and wounded 38 others before being shot and killed by Austin police. His autopsy revealed a tumor pressing on his amygdala.

The *amygdala* (Latin for "almond," which describes this structure's shape) is the brain's center for the emotion and memory of fear. When the cells of this structure are activated, your heart beats faster, your breathing becomes rapid and shallow, and your hands get cold and clammy. If you watch a horror movie, your amygdala is activated. If you encounter a threatening face, your amygdala is activated. If you are alone at night and hear an unusual noise, your amygdala is activated. What would life be like without an amygdala? You wouldn't get scared—and *not* being scared could be hazardous to your health.

A rare case of brain damage left a woman without a functional amygdala. When shown pictures of faces registering different emotions, she could not pick out the ones that were threatening or scary. She could not recall every having a frightening experience. In tests where she was administered mild electrical shocks, she developed no anticipatory fear; even though she knew that seeing a red card meant she was about to receive a shock, she never reacted to the red card. Thus, in real life, she would not have the reflex to pull away from a threat.

People with damage to the amygdala frequently have trouble engaging in normal social relationships. They cannot "read" the nature, mood, or intentions of other people by looking at their faces. The presence of pressure on Charles Whitman's amygdala may have been a factor in the emotions that drove him to mass

Fear Factor The fear response—nerves tense, heart racing, cold sweat—kicks in when we encounter a scary animal, person, or situation. Even a scary movie can trigger this primitive and protective reaction.

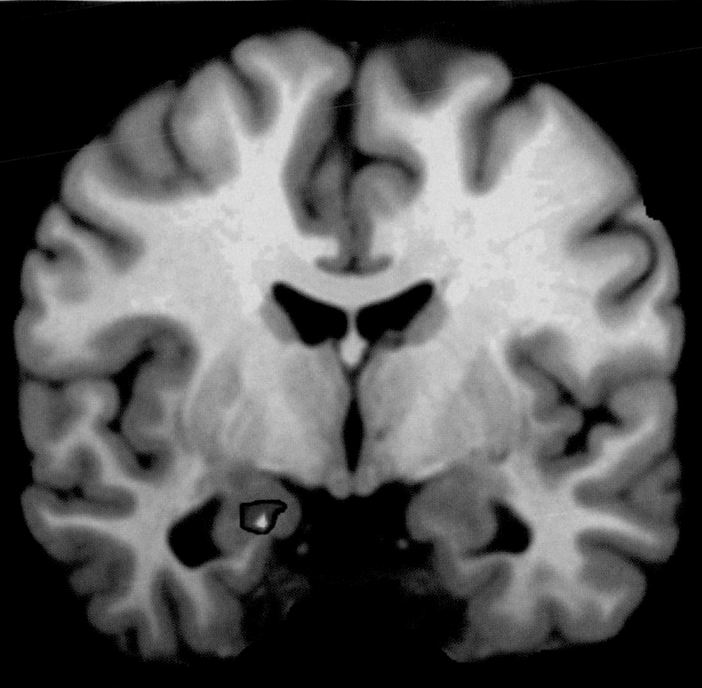

Source of the Fear Response Frightening situations—or even memories of such a situation—activate a group of cells in the amygdala, a structure deep in the brain.

murder; this diagnosis remains a matter for medical speculation.

Our nervous system enables us to experience the world around us and to react to it. But in between sensing and reacting, there is much interpretation based on memory, learning, emotions, and beliefs—all of which are based on the activities of cells in the nervous system. To understand how the eyes see, how the fingers play the piano, or how emotions affect our behavior, we have to understand how cells in different parts of our brains work and interact.

IN THIS CHAPTER we will discuss the general properties of nervous systems. We will begin with a look at the special cells that constitute nervous systems. Next we will see how some of these cells, the neurons, transmit information by generating electrical signals and conducting them from place to place in the body. Finally we will examine the electrical and biochemical mechanisms by which neurons communicate these signals to each other and to other cells in the body.

CHAPTER OUTLINE

44.1 What Cells Are Unique to the Nervous System?

Nervous systems are composed of two unique categories of cells: *nerve cells* or **neurons** and *glial cells* or **glia** (see Figure 40.6). Neurons are *excitable*: they can generate and propagate electrical signals, which are known as nerve impulses, or **action potentials**. Most neurons have long extensions called **axons** that enable them to conduct action potentials over long distances. Glial cells do not conduct action potentials; rather, they support neurons physically, immunologically, and metabolically. A **nerve** (as opposed to a neuron) is a bundle of axons that come from many different neurons.

Nervous systems can process information because their neurons are organized into networks. These networks include three functional categories of cells, which can be thought of as being involved with input, integration, and output. In the first category, **afferent neurons** carry sensory information into the nervous system. That information comes from specialized **sensory neurons** that transduce (convert) various kinds of sensory input into action potentials. Next, **efferent neurons** carry commands to physiological and behavioral *effectors* such as muscles and glands. The third category of cells, called **interneurons**, integrate and store information and facilitate communication between sensors and effectors.

Neuronal networks range in complexity

Simple animals such as cnidarians (e.g., sea anemones) can process information with simple networks of neurons that do little more than provide direct lines of communication from sensory cells to effectors (**Figure 44.1A**). The cnidarian's *nerve net* is most developed around the tentacles and the oral opening, where it facilitates detection of food or danger and causes tentacles to extend or retract. Animals that are more complex and move around the environment to search for food and mates need to process and integrate larger amounts of information. Even animals such as earthworms fit this description, and their increased need for information processing is met by higher numbers of neurons organized into clusters called **ganglia**. Ganglia serving different functions may be distributed around the body, as in the earthworm or the squid (**Figure 41.1B,C**). In animals that are bilaterally symmetrical, ganglia frequently come in pairs, one on each side of the body. Also, as animals increase in complexity, generally one pair of ganglia is larger than the others, and is therefore given the designation of **brain**.

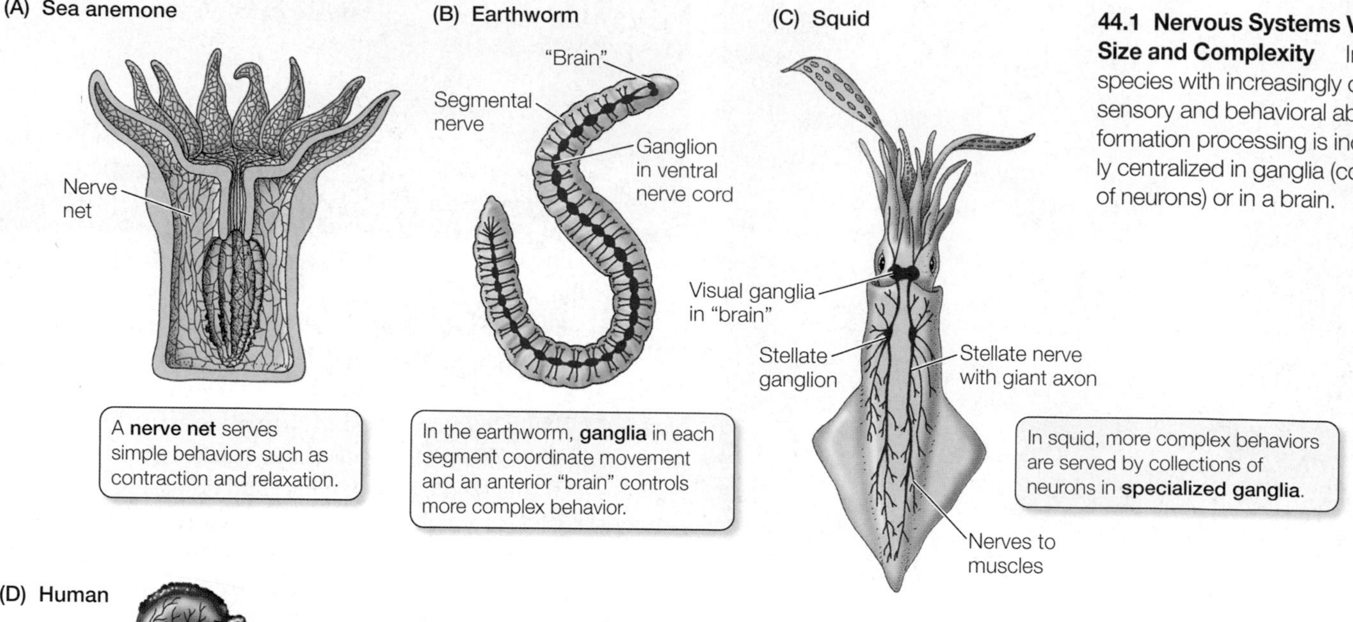

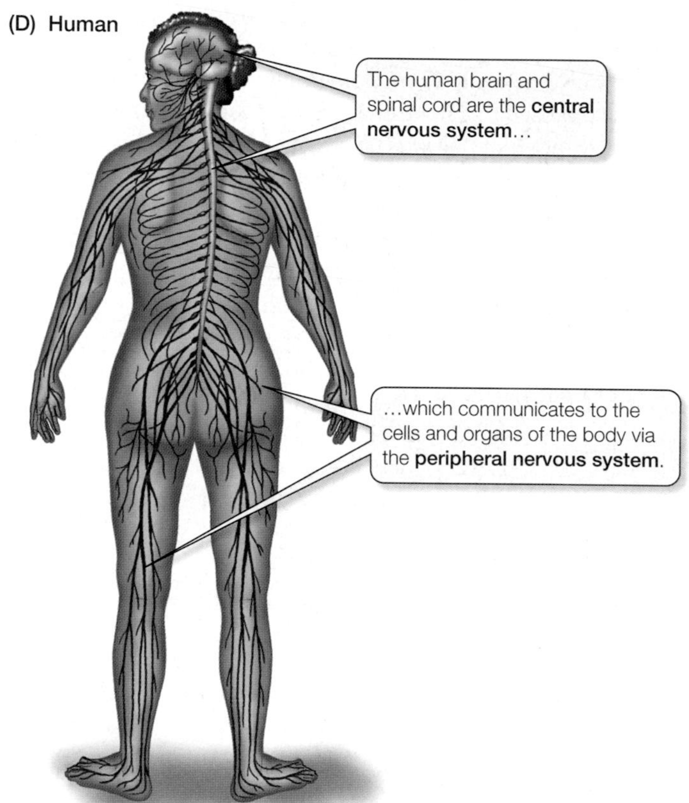

(D) Human
The human brain and spinal cord are the central nervous system...
...which communicates to the cells and organs of the body via the peripheral nervous system.

44.1 Nervous Systems Vary in Size and Complexity In animal species with increasingly complex sensory and behavioral abilities, information processing is increasingly centralized in ganglia (collections of neurons) or in a brain.

The small nervous systems of invertebrates can be remarkably complex. Consider the nervous systems of spiders, which have programmed within them the thousands of precise movements necessary to construct a beautiful web without prior experience or opportunities to learn the specific web architecture of their species.

In vertebrates, most cells of the nervous system are found in the brain and the **spinal cord**, the sites of most information processing, storage, and retrieval (**Figure 44.1D**). Therefore, the brain and spinal cord are called the **central nervous system** (**CNS**). Information is transmitted from sensory cells to the CNS and from the CNS to effectors via neurons that extend or reside outside of the brain and the spinal cord; these neurons and their supporting cells are called the **peripheral nervous system** (**PNS**). Vertebrates differ greatly in their behavioral complexity and in their physiological specializations, and their nervous systems reflect this diversity. **Figure 44.2** shows the brains of four vertebrate species of similar body mass drawn to the same scale.

The human nervous system contains an estimated 10^{11} neurons. Information is passed from one neuron to another where they come into close proximity at structures called **synapses**. The cell that sends the message is the **presynaptic neuron**, and the cell that receives it is the **postsynaptic neuron**. A given neuron in the brain can receive information from a thousand or more synapses. Thus the human brain may contain up to 10^{14} synapses, which can be highly plastic—strengthening with use and weakening with disuse. Therein lies the incredible ability of the human brain to process information, to learn, to do complex tasks, to remember, and even to have emotions. This astronomical number of neurons and synapses is divided into thousands of distinct but interacting networks that function in parallel. Before we can understand how even one of these circuits works, we must understand the properties of individual neurons.

Neurons are the functional units of nervous systems

Although nervous systems of different species vary enormously in structure and function, neurons behave similarly in animals as different as squids and humans. Their plasma membranes generate action potentials and conduct these signals from one location on a neuron to the most distant reaches of that cell—a distance that can be more than a meter in a human and many meters in a whale. Moreover, this transmission of action potentials can be rapid—up to 100 meters per second or more—making it possible to sense, process, and act on information very quickly.

Most neurons have four regions—a *cell body*, *dendrites*, an *axon*, and *axon terminals* (**Figure 44.3A**)—but the variation among different types of neurons is considerable (**Figure 44.3B**). The *cell body* contains the nucleus and most of the cell's organelles. Many pro-

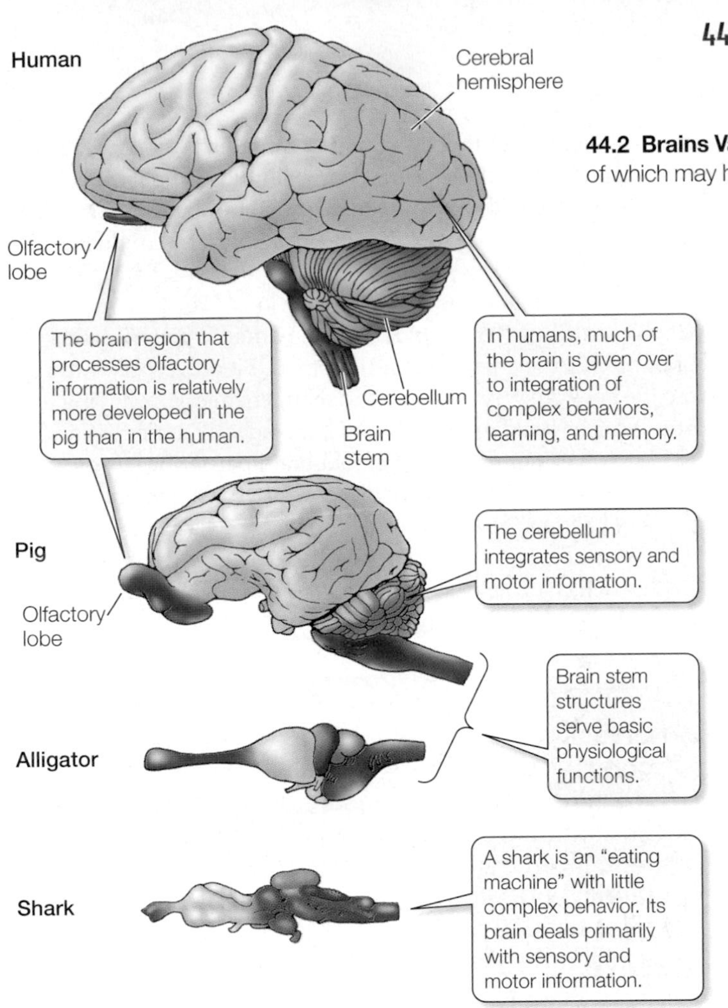
Human
Cerebral hemisphere
Olfactory lobe
The brain region that processes olfactory information is relatively more developed in the pig than in the human.
Cerebellum
Brain stem
In humans, much of the brain is given over to integration of complex behaviors, learning, and memory.
Pig
The cerebellum integrates sensory and motor information.
Olfactory lobe
Brain stem structures serve basic physiological functions.
Alligator
Shark
A shark is an "eating machine" with little complex behavior. Its brain deals primarily with sensory and motor information.

44.2 Brains Vary in Size and Complexity The brains of four vertebrate species—all of which may have a similar body mass—show immense differences.

jections may sprout from the cell body. Most of these projections are shrublike **dendrites** (from the Greek *dendron,* "tree"), which bring information from other neurons or sensory cells to the cell body. The degree of branching of the dendrites differs among different types of neurons. In most neurons, one projection, the axon, is much longer than the others. Axons usually carry information away from the cell body. The length of the axon also differs among different types of neurons—some axons are remarkably long, such as those that run from the spinal cord to the toes.

Axons are the "telephone lines" of the nervous system. Information received by the dendrites can cause the cell body to generate an action potential, which is then conducted along the axon to the cell that is its target. At the target cell the axon divides into a spray of fine nerve endings. At the tip of each of these tiny nerve endings is a swelling, called an **axon terminal**, that comes very close to the target cell to form a synapse.

Most synapses are chemical synapses. A space only about 25 nanometers wide separates the *presynaptic* and *postsynaptic membranes,* and an action potential arriving at an axon terminal causes the release of chemical messenger molecules called **neurotransmit-**

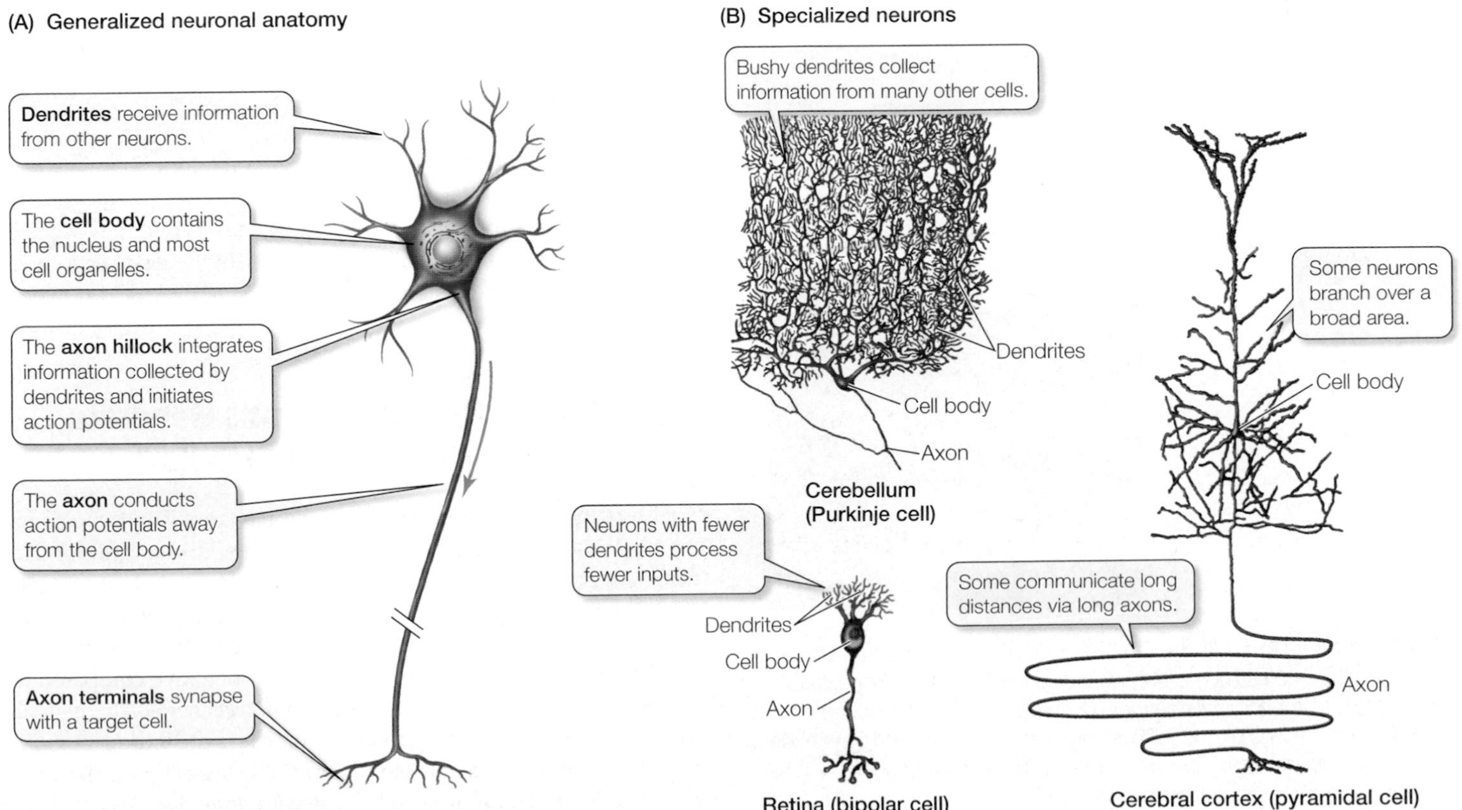
(A) Generalized neuronal anatomy
Dendrites receive information from other neurons.
The cell body contains the nucleus and most cell organelles.
The axon hillock integrates information collected by dendrites and initiates action potentials.
The axon conducts action potentials away from the cell body.
Axon terminals synapse with a target cell.
(B) Specialized neurons
Bushy dendrites collect information from many other cells.
Dendrites
Cell body
Axon
Cerebellum (Purkinje cell)
Neurons with fewer dendrites process fewer inputs.
Dendrites
Cell body
Axon
Retina (bipolar cell)
Some neurons branch over a broad area.
Cell body
Some communicate long distances via long axons.
Axon
Cerebral cortex (pyramidal cell)

44.3 Neurons (A) A generalized diagram of a neuron. (B) Neurons from different parts of the mammalian nervous system are specifically adapted to their functions.

ters from the axon terminal. The released neurotransmitters diffuse across the space and bind to receptors on the plasma membrane of the postsynaptic or target cell. (We will discuss this process of synaptic transmission in more detail later in the chapter.) Integration of information in the nervous system is possible because a neuron can receive information (synaptic inputs) from many sources before producing action potentials that travel down its single axon to target cells. There are also electrical synapses that transmit the action potential from one neuron to the next; we will discuss electrical synapses later.

Glial cells are also important components of nervous systems

Glial cells, or simply *glia*, are another class of nervous system cells. There are many more glial cells than neurons in the human brain. Like neurons, glia come in several forms and have a diversity of functions. They are not excitable and do not transmit electrical signals. Some glia physically support and orient the neurons and help them make the right contacts during embryonic development. Others supply neurons with nutrients, maintain the extracellular environment, consume foreign particles and cellular debris, or insulate axons.

In the CNS some glia called **oligodendrocytes** wrap around the axons of neurons, covering them with concentric layers of insulating plasma membrane. In the PNS, glia called **Schwann cells** perform this function (**Figure 44.4**). **Myelin** is the covering produced by oligodendrocytes and Schwann cells, and it gives many parts of the nervous system a glistening white appearance. Not all axons are myelinated, but those that are can conduct action potentials more rapidly than those axons that are not myelinated. Later in this chapter we will see how the electrical insulation provided by myelin increases the speed with which axons can conduct action potentials.

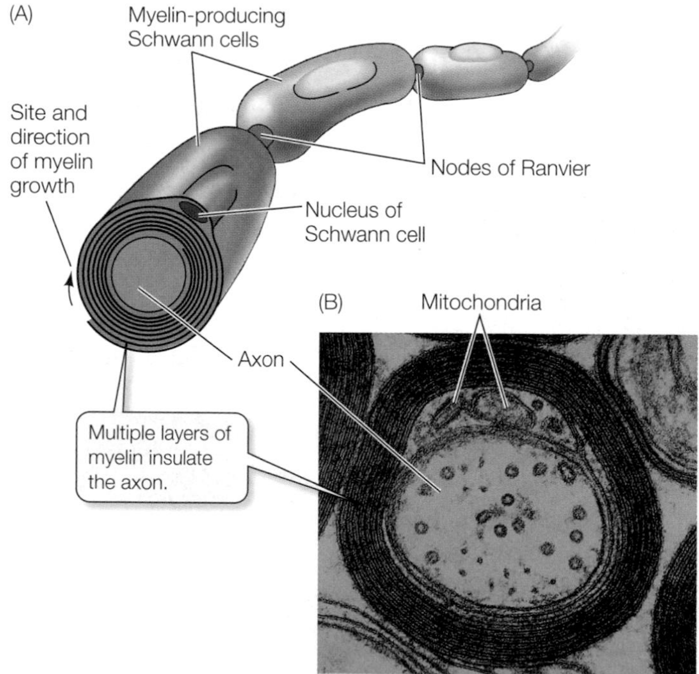

44.4 Wrapping Up an Axon (A) Schwann cells produce layers of myelin, a type of plasma membrane that provides electrical insulation to the axon. At the intervals between Schwann cells—the nodes of Ranvier—the axon is exposed. Action potentials travel along the axon by "jumping" from node to node. (B) A myelinated axon, seen in cross section through an electron microscope.

Glia called **astrocytes** (because they look like stars; see Figure 40.6B) contribute to the **blood–brain barrier**, which protects the brain from toxic chemicals in the blood. Blood vessels throughout the body are very permeable to many chemicals, including toxic ones, which would reach the brain if this special barrier did not exist. Astrocytes help form the blood–brain barrier by surrounding the smallest, most permeable blood vessels in the brain. The barrier is not perfect, however. Since it consists of plasma membranes, it is permeable to fat-soluble substances such as anesthetics and alcohol, which explains why these substances have such rapid and marked effects on the nervous system.

44.1 RECAP

Nervous systems are made up of two unique categories of cells, neurons and glia. Neurons are arranged in circuits. Sensory neurons transduce information from the external or internal environment; interneurons integrate and process information; and efferent neurons carry commands to target tissues of effector organs. Glial cells have a wide variety of supporting roles.

- Comparing nervous systems of anemones, worms, fish, and humans, what trends seem to reflect increasing complexity? See pp. 943–944 and Figures 44.1 and 44.2
- Describe the different parts of neurons and their functions. See pp. 944–945 and Figure 44.3
- What are some types of glial cells and what are their functions? See p. 946

The one feature common to all nervous systems is that they process information in the form of action potentials. In the next section we will focus on how these signals are produced and used by nervous systems.

44.2 How Do Neurons Generate and Conduct Signals?

Action potentials are generated when ion channels in the plasma membranes of neurons open for a short time and permit ions to move across the membrane. The movement of these charged molecules is driven by differences in their concentration gradients and by electrical charge differences on the two sides of the membrane. At rest, the inside of a neuron is electrically negative compared to the outside. Any difference in electric potential across the plasma membrane is a **membrane potential**, which is measured in *millivolts*. When the neuron is resting and not firing action potentials, the membrane potential is called a **resting potential**.

Simple electrical concepts underlie neuronal function

Voltage (electric potential difference) is a force that causes electrically charged particles to move between two points. Voltage is to the flow of electrically charged particles as pressure is to the flow of water. If the negative and the positive poles of a battery are connected by a wire, an electric current will flow through the wire because there is a voltage difference between the two poles of the battery. This flow of electric current can be used to do work, just as a current of water can be used to do work.

In wires, electric current is carried by electrons, but in solutions and across cell membranes, electric current is carried by ions. The major ions that carry electric charges across the plasma membranes of neurons are sodium (Na^+), potassium (K^+), calcium (Ca^{2+}), and chloride (Cl^-). Recall that ions with opposite charges attract one another, and those with like charges repel one another. How do these basic principles of bioelectricity establish the resting potential of the neuronal plasma membrane? And how is the flow of ions through membrane channels turned on and off to generate action potentials? We address these questions next.

Membrane potentials can be measured with electrodes

An *electrode* can be made from a glass pipette with a very sharp tip filled with a solution that conducts electric charges. Using electrodes, we can record electrical events in a cell. If one electrode is placed inside the plasma membrane of an axon, and another electrode is placed just outside of the axon, the difference in voltage can be measured. The typical resting potential of an axon thus measured is a voltage difference usually between –60 and –70 millivolts (mV) (**Figure 44.5**).

The resting potential provides a means for neurons to respond to a stimulus. Because of the voltage difference across the membrane, ions would cross the membrane if they could. Since the inside of the resting cell is negative, for example, positively charged ions such as sodium (Na^+) would enter if they could. Therefore, any chemical or physical stimulus that changes the permeability of the plasma membrane to ions will produce a change in the cell's membrane potential. The most extreme change in membrane potential is the action potential, a sudden and rapid reversal in the voltage across a portion of the plasma membrane. For 1 or 2 milliseconds, positively charged ions flow into the cell, making the inside of the cell *more positive* than the outside. Action potentials are conducted along axons, usually from the cell body to the terminal endings of its axon.

Ion pumps and channels generate membrane potentials

The plasma membranes of neurons, like those of all other cells, are lipid bilayers that are impermeable to ions, but contain many protein molecules that serve as ion channels and ion pumps (see Section 5.3). Ion pumps and channels are responsible for the distribution of charges across the membrane that create resting and action potentials.

Ion pumps require energy to move ions or other molecules against their concentration or electrical gradients. A major ion pump in the plasma membranes of neurons (and all other cells) is the **sodium–potassium pump**, so called because it actively expels Na^+ from inside the cell, exchanging it for K^+ from outside the cell (**Figure 44.6A**). The Na^+–K^+ pump is also known as **sodium–potassium ATPase**, a term emphasizing it as an enzyme complex requir-

44.5 Measuring the Resting Potential If the difference in electric charge across the plasma membrane of an unstimulated neuron, is measured, it is constant (about –60 mV), and is known as the resting potential.

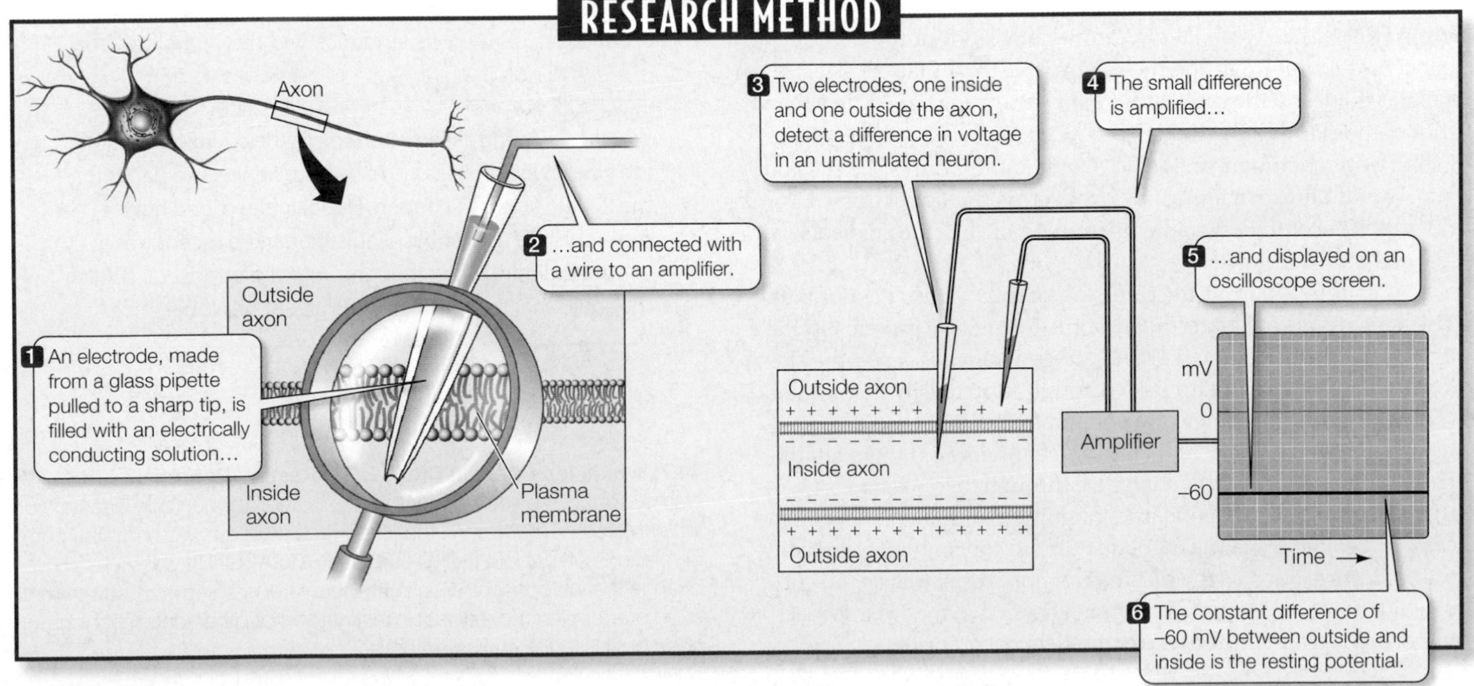

RESEARCH METHOD
Axon
1 An electrode, made from a glass pipette pulled to a sharp tip, is filled with an electrically conducting solution...
2 ...and connected with a wire to an amplifier.
3 Two electrodes, one inside and one outside the axon, detect a difference in voltage in an unstimulated neuron.
4 The small difference is amplified...
5 ...and displayed on an oscilloscope screen.
6 The constant difference of –60 mV between outside and inside is the resting potential.
Outside axon
Inside axon
Plasma membrane
Outside axon
Inside axon
Outside axon
Amplifier
mV
0
–60
Time

(A) $Na^+ - K^+$ pump (ATPase)

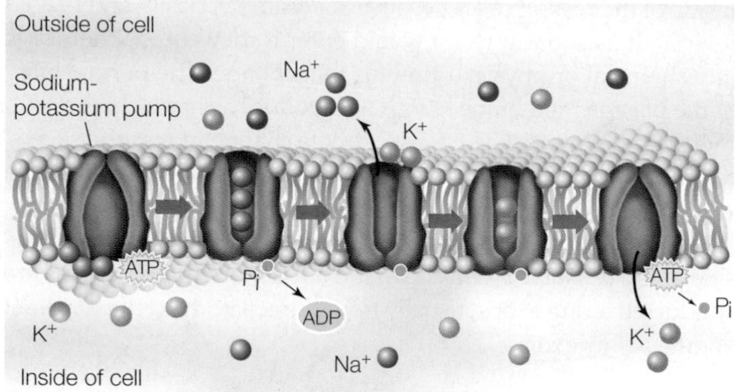

(B) $Na^+ - K^+$ channels

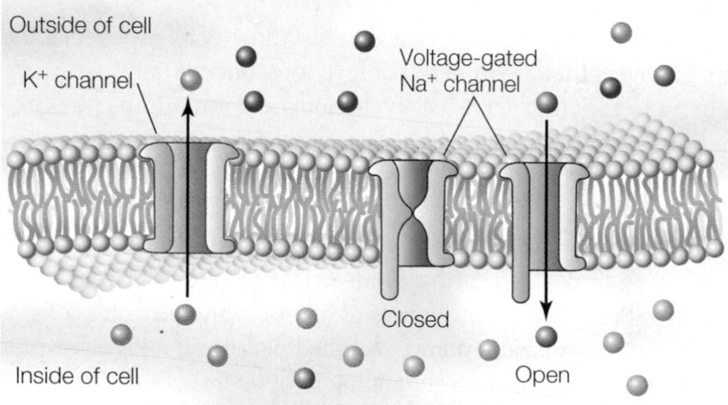

44.6 Ion Pumps and Channels (A) The sodium–potassium pump actively moves K^+ to the inside of a neuron and Na^+ to the outside. (B) Ion channels allow specific ions to diffuse down their concentration gradients; K^+ tends to leave neurons when potassium channels are open, and Na^+ tends to enter neurons when sodium channels are open.

ing ATP to do its work. The sodium–potassium pump keeps the concentration of K^+ inside the cell greater than that of the extracellular fluid, and the concentration of Na^+ inside the cell less than that of the extracellular fluid. The concentration differences established by the pump mean that K^+ would diffuse out of the cell and Na^+ would diffuse in if the ions could cross the lipid bilayer. How do these concentration gradients relate to the electric gradients we learned about above?

Ion channels permit the diffusion of ions across membranes. These channels are water-filled pores formed by proteins in the lipid bilayer (see Section 5.1) and are generally *selective*—they allow some types of ions to pass through more easily than others (**Figure 44.6B**). Thus, there are potassium channels, sodium channels, chloride channels, and calcium channels, and there are different kinds of each. Ions can diffuse through these channels in either direction. The direction and magnitude of the net movement of ions through a channel depends on the concentration gradient of that ion type across the plasma membrane, as well as the voltage difference across that membrane. These two motive forces acting on an ion are termed its **electrochemical gradient**.

Potassium channels are the most common open channels in the plasma membranes of resting (nonstimulated) neurons. As a consequence, resting neurons are more permeable to K^+ than to any other ion. Thus, open potassium channels are largely responsible for the membrane potential. Because the potassium channels make the plasma membrane permeable to K^+, and because the sodium–potassium pump keeps the concentration of K^+ inside the cell much higher than that outside the cell, K^+ tends to diffuse down its electrochemical gradient, out of the cell, through the channels. As these positively charged potassium ions diffuse out of the cell, they leave behind unbalanced negative charges, generating an electric potential across the membrane that tends to pull K^+ back into the cell.

The membrane potential at which the net diffusion of K^+ out of the cell ceases (that is, the point at which K^+ diffusion out due to the concentration gradient is balanced by its movement in due to the negative electric potential) is called the **potassium equilibrium potential**. The value of the potassium equilibrium potential can be calculated from the concentrations of K^+ on the two sides of the membrane using the **Nernst equation** (**Figure 44.7**). This equation, developed in the late 1800s, illustrates that the nature of ion channels in neuronal membranes was hypothesized long before their specific structures and properties were described.

In the late 1940s, A. L. Hodgkin and A. F. Huxley at the University of Cambridge set out to study the electrical properties of axonal membranes. With the techniques available at that time, the necessary measurements could be made only if you had a very large axon to work with. Such an axon exists in nature, in the giant neuron that controls the escape response of squid. Hodgkin and Huxley used electrodes to measure the voltage across the plasma membrane of this large axon, as seen in Figure 44.7, and to pass electric current into it to change its resting potential. They also changed the concentrations of Na^+ and K^+ both inside and outside the squid axon and measured the resulting changes in membrane potential.

On the basis of their many careful experiments, Hodgkin and Huxley developed virtually all of our basic concepts about the electrical properties of neurons, and received the Nobel prize in 1963.

When J. Z. Young called attention to the squid giant axon in 1936, it was a milestone for neuroscience. In a paper for the Royal Society of London, Hodgkin described how "a distinguished neurophysiologist remarked recently at a congress dinner (not, I thought, with the utmost tact), 'It's the squid that really ought to be given the Nobel prize.'"

44.7 Which Ion Channel Creates the Resting Potential? The Nernst equation calculates membrane potential when only one type of ion can cross a membrane that separates solutions with different concentrations of that ion. FURTHER INVESTIGATION: If you were to change the ion concentration in the seawater bathing your squid giant axon preparation, how would investigate what other ion(s) might be contributing to the resting potential?

EXPERIMENT

HYPOTHESIS: The resting potential of neurons is due to permeability of the membrane to potassium ions.

METHOD

1. Measure concentrations of ions inside and outside of a neuron.

To measure the concentration of ions in a neuron, the neuron (and its axon) must be big. Squid have giant neurons that control their escape response. It is possible to sample the cytoplasm of these axons, which are about 1 mm in diameter.

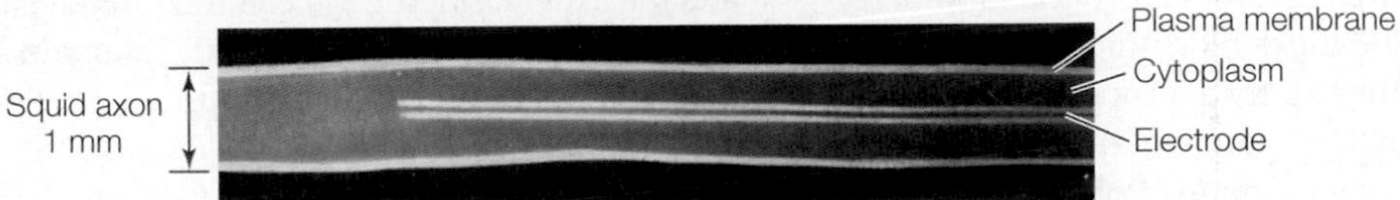

2. Use the Nernst equation to calculate what the membrane potential would be if it were permeable to each of these ions: Na^+, K^+, Ca^{2+}, and Cl^-.

Recall the Nernst equation from Section 5.3. It predicts the membrane potential resulting from membrane permeability to a single type of ion that differs in concentration on the two sides of the membrane. The equation is written

$$E_{ion} = 2.3\frac{RT}{zF} \log \frac{[\text{ion}]_o}{[\text{ion}]_i}$$

where E is the equilibrium (resting) membrane potential (the voltage across the membrane in mV), R is the universal gas constant, T is the absolute temperature, z is the charge on the ion (+1, +1, +2, or –1, respectively, for the ions used here), and F is the Faraday constant. The subscripts o and i indicate the ion concentrations outside and inside the cell, respectively.

At this point you could just "plug and play," but do you understand this equation?

A concentration difference of ions across a membrane creates a *chemical* force that pushes the ions across the membrane; however, the resulting unbalanced *electrical charges* will pull the ions back the other way. At *equilibrium*, the work done moving ions in each direction will be the same.

The *chemical* work pushing the ions will equal 2.3 RT log $[\text{ion}]_o/[\text{ion}]_i$
The *electrical* work pulling the ions will equal zEF. So, at equilibrium:

$$zEF = 2.3\, RT \log \frac{[\text{ion}]_o}{[\text{ion}]_i}$$

Rearranging the equation to solve for E, we get the Nernst equation:

$$E_{ion} = 2.3\frac{RT}{zF} \log \frac{[\text{ion}]_o}{[\text{ion}]_i}$$

We can simplify the equation by picking a temperature—let's use "room temperature," or 20°C—and solving for 2.3 RT/F. At 20°C, 2.3 RT/F equals 58. Thus:

$$E_{ion} = 58/z \log \frac{[\text{ion}]_o}{[\text{ion}]_i}$$

3. Measure the membrane potential across the squid giant axon and compare with calculated values for each ion.

RESULTS

1. Measuring ion concentrations in squid giant axon cytoplasm and in seawater, then solving the Nernst equation for each ion, we find:

Ion	Ion concentration (n*M*) in squid axon	Ion concentration (n*M*) in seawater	Predicted membrane potential (mV)
K^+	400	20	–75
Na^+	50	460	+56
Ca^{2+}	0.5	10	+38
Cl^-	50	560	–60

2. Using the method shown in Figure 44.5, the actual resting membrane potential of a squid giant axon is recorded to be –66 mV.

CONCLUSION: The resting potential of the squid giant axon can be due to permeability to K^+, but there is probably some permeability to another ion as well.

We now know that, in general, the resting potential is less negative than the Nernst equation predicts because resting neurons are also slightly permeable to other ions, such as Na^+ and Cl^-. Another equation, called the Goldman equation, takes all of the ions that can cross the membrane into account and therefore can calculate the membrane potential accurately.

Ion channels and their properties can now be studied directly

Hodgkin and Huxley were working long before the laboratory techniques emerged that enabled the demonstration of ion channels, and so could only hypothesize their properties. With the advent of a technique called **patch clamping**, developed in the 1980s by B. Sakmann and E. Neher, neurobiologists can now record currents caused by the openings and closings of single ion channels.

The patch clamp electrode is a polished glass micropipette filled with an electrically conductive solution that has the same composition as extracellular fluids. The tip of this electrode is then positioned right up against the membrane of a cell. When slight suction is applied to the electrode, it forms a seal with the membrane. Once the seal is formed, any exchanges of ions across the patch of membrane bounded by the seal are exchanges with the fluid in the micropipette, and they can be recorded as electric currents. If a single ion channel or a few ion channels happen to be in that patch of membrane, then the openings and closings of individual channels will be recorded by the pipette-electrode. If the pipette is retracted, it can tear the patched membrane away from the cell, and the activities of the ion channels in the patch can continue to be recorded (**Figure 44.8**). In 1991, Sakmann and Neher received the Nobel prize for their work with the patch clamp technique.

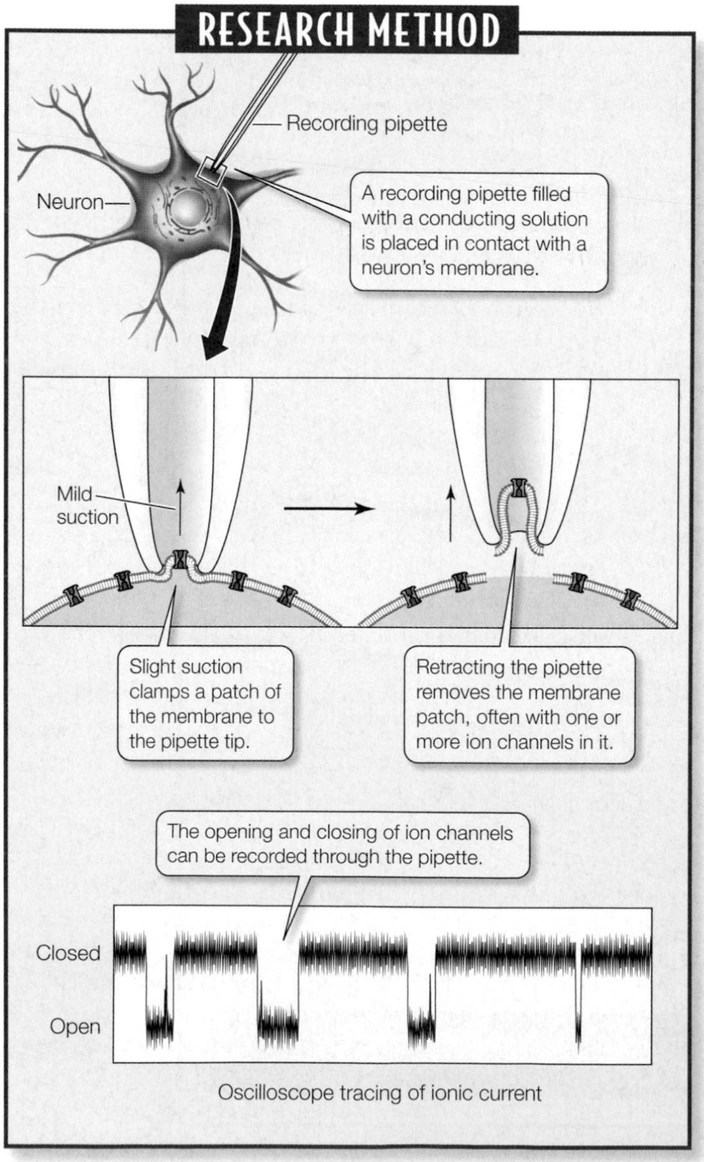

RESEARCH METHOD

Recording pipette

Neuron

A recording pipette filled with a conducting solution is placed in contact with a neuron's membrane.

Mild suction

Slight suction clamps a patch of the membrane to the pipette tip.

Retracting the pipette removes the membrane patch, often with one or more ion channels in it.

The opening and closing of ion channels can be recorded through the pipette.

Closed

Open

Oscilloscope tracing of ionic current

44.8 Patch Clamping The patch clamping technique can record the opening and closing of a single ion channel.

Gated ion channels alter membrane potential

Many ion channels in the plasma membranes of neurons behave as if they contain "gates" that are open under some conditions and closed under other conditions. **Voltage-gated channels** open or close in response to a change in the voltage across the plasma membrane. **Chemically gated channels** open or close depending on the presence or absence of a specific molecule that binds to the channel protein, or to a separate receptor that in turn alters the channel protein. **Mechanically gated channels** open or close in response to mechanical force applied to the plasma membrane. Gated channels play important roles in neuronal function.

Openings and closings of gated channels perturb the resting potential. Imagine what happens, for example, if sodium channels in the plasma membrane open. Na^+ diffuses into the neuron down its electrochemical gradient. As a result of the entry of Na^+, the inside of the cell becomes less negative. When the inside of a neuron becomes less negative (or more positive) in comparison to its resting condition, its plasma membrane is said to be **depolarized** (**Figure 44.9**).

An opposite change in the resting potential occurs if gated K^+ channels open. When K^+ efflux from the neuron increases, the membrane potential becomes even more negative, and the plasma membrane is said to be **hyperpolarized**.

The opening and closing of ion channels, which result in changes in the voltage across the plasma membrane, are the basic mechanisms by which neurons respond to stimuli whether they be electrical, chemical, or mechanical. How does a neuron use a change in its resting membrane potential to process and transmit information?

A change in membrane potential may result from the activity at a synapse. When an action potential traveling along an axon reaches the axon terminal, it causes the release of a chemical neurotransmitter. Neurotransmitter molecules diffuse across the small gap between the presynaptic and postsynaptic membranes and bind to receptors on the postsynaptic membrane. Those receptors may be ion channels or they may control ion channels indirectly. In either case, the neurotransmitter causes the channel to open, ions flow down their electrochemical gradients, and the

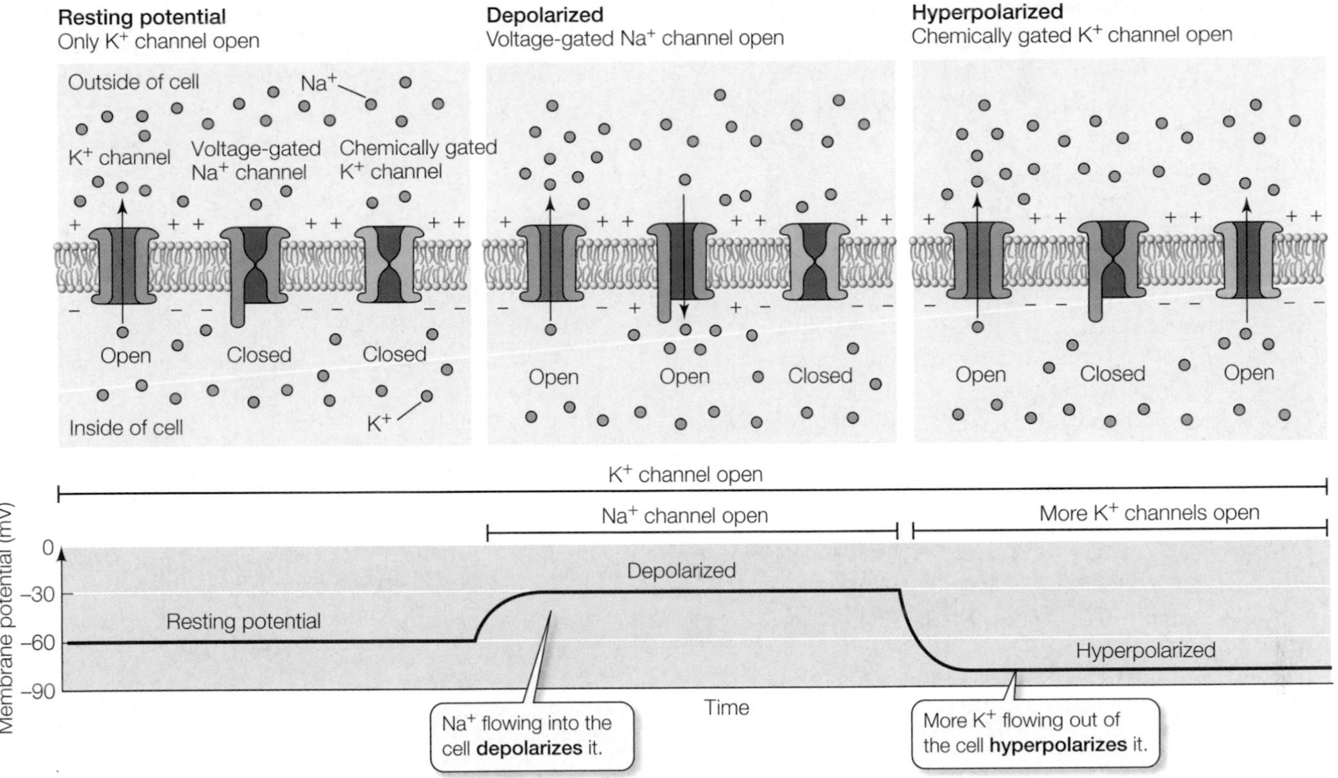

44.9 Membranes Can Be Depolarized or Hyperpolarized The resting potential is produced by open K^+ channels. A shift from the resting potential to a less negative membrane potential, as occurs when Na^+ enters the cell through a gated sodium channel, is called depolarization. Hyperpolarization occurs when the membrane potential becomes more negative, as when additional K^+ leaves the cell through gated K^+ channels.

membrane potential of the postsynaptic neuron changes. How is this local change in membrane potential communicated to other parts of the cell?

A local change in the membrane potential of the postsynaptic neuron causes a flow of ions that spreads the change in membrane potential to adjacent regions of the membrane. For example, when Na^+ enters the postsynaptic neuron through open sodium channels at one location, those positively charged ions are attracted to adjacent areas on the inside of the membrane that are more negative, and thus there is a rapid flow of electric current from the site of the open Na^+ channels. However, this local flow of electric current decays as it spreads and therefore does not spread very far.

Electric currents do not spread far in cells because cell membranes are not completely impermeable to ions. An electric current traveling along a membrane is like water flowing through a leaky hose. The flow of electric current along plasma membranes is useful for transmitting signals over only very short distances. Therefore, axons do not transmit information as a continuous flow of electric current (as telephone wires do). However, the local flow of electric current is an important part of the mechanism by which neurons generate action potentials, the truly long-distance signals.

Sudden changes in Na^+ and K^+ channels generate action potentials

Action potentials are sudden, transient, large changes in membrane potential that are conducted along unmyelinated axons at speeds of up to 2 meters per second, but in myelinated axons the conduction velocity can be 100 meters per second. Think of running the 100-meter dash—the world record is slightly under 10 seconds.

If we place the tips of a pair of electrodes on either side of the plasma membrane of a resting axon and measure the voltage difference, the reading is about –60 mV, as we saw in Figure 44.5. If these electrodes are in place when an action potential travels down the axon, they register a rapid change in membrane potential, from –60 mV to about +50 mV. The membrane potential then rapidly returns to its resting level of –60 mV as the action potential passes (**Figure 44.10**).

Voltage-gated Na^+ and K^+ channels in the plasma membrane of the axon are responsible for action potentials. At the resting potential, most of these channels are closed. Depolarization of the membrane causes them to open. For example, if synaptic input to a neuron is sufficiently strong to cause the plasma membrane of its cell body to depolarize, that depolarization can spread by local current flow to the axon hillock at the base of the axon (see Figure 44.3), where voltage-gated Na^+ channels are concentrated. When the plasma membrane in this area depolarizes, some of these voltage-gated channels open briefly—for less than a millisecond. The Na^+ concentration is much higher outside the axon than inside, and the inside is negatively charged, so when these channels open, Na^+ rushes into the axon. The entering Na^+ depolarizes the membrane even more, causing more Na^+ channels to open—a positive

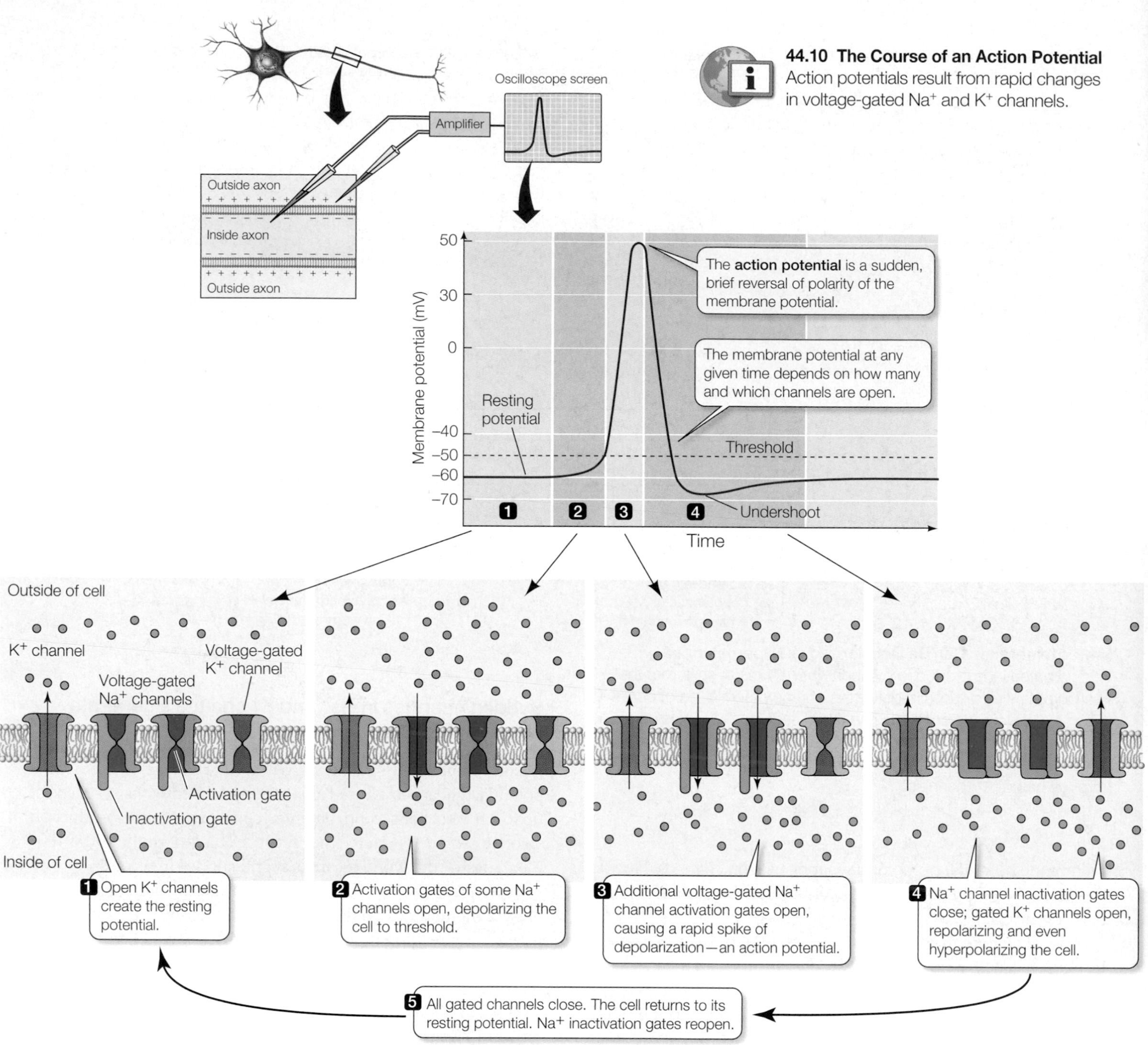

Oscilloscope screen
Amplifier
Outside axon
Inside axon
Outside axon
Membrane potential (mV)
50
30
0
–40
–50
–60
–70
The **action potential** is a sudden, brief reversal of polarity of the membrane potential.
The membrane potential at any given time depends on how many and which channels are open.
Resting potential
Threshold
Undershoot
Time
1
2
3
4
Outside of cell
K^+ channel
Voltage-gated K^+ channel
Voltage-gated Na^+ channels
Activation gate
Inactivation gate
Inside of cell
1 Open K^+ channels create the resting potential.
2 Activation gates of some Na^+ channels open, depolarizing the cell to threshold.
3 Additional voltage-gated Na^+ channel activation gates open, causing a rapid spike of depolarization—an action potential.
4 Na^+ channel inactivation gates close; gated K^+ channels open, repolarizing and even hyperpolarizing the cell.
5 All gated channels close. The cell returns to its resting potential. Na^+ inactivation gates reopen.

44.10 The Course of an Action Potential Action potentials result from rapid changes in voltage-gated Na^+ and K^+ channels.

feedback effect. When the membrane is depolarized about 5 to 10 mV above the resting potential, a **threshold** is reached at which the depolarizing influx of Na^+ can no longer be offset by the efflux of K^+. At this point, a large number of sodium channels open, and the membrane potential becomes positive—an action potential. The rising phase of the action potential halts abruptly in 1 to 2 milliseconds, and the membrane potential rapidly becomes negative once again.

What causes the axon to return to resting potential? There are two contributing factors: The voltage-gated Na^+ channels close, and voltage-gated K^+ channels may open. The voltage-gated K^+ channels open more slowly than the Na^+ channels and stay open longer, allowing K^+ to carry excess positive charges out of the axon. As a result, the membrane potential returns to a negative value and usually becomes even more negative than the resting potential until all of the voltage-gated K^+ channels close.

Another feature of the voltage-gated Na^+ channels is that once they open and close, they have a **refractory period** of 1 to 2 milliseconds during which they cannot open again. This property can be explained by the channels having two gates, an **activation gate** and an **inactivation gate** (see Figure 44.10). Under resting conditions, the activation gate is closed and the inactivation gate is open. Depolarization of the membrane to the threshold level causes both gates to change state, but the activation gate responds faster. As a result, the channel is open for a brief time between the opening of the activation gate and the closing of the inactivation gate. Inactivation gates remain closed for 1 to 2 milliseconds before they spontaneously open again, thus explaining why the membrane has a refractory period before it can fire another action potential. When the inactivation gate finally opens, the activation gate is closed, and the membrane is poised to respond once again to a depolarizing stimulus by firing another action potential. Another contribution to the refractory period is the duration of the opening of the voltage-gated K^+ channels, as we saw above. The dip in the membrane potential following an action potential is called the *after-hyperpolarization* or *undershoot*.

The difference in the concentration of Na^+ across the plasma membrane and the negative resting potential constitute the "battery" that drives action potentials. How rapidly does the battery run down? It might seem that a substantial number of ions would have to cross the membrane for the membrane potential to change from –60 mV to +50 mV and back to –60 mV again. In fact, only a vanishingly small percentage of the Na^+ concentrated outside the plasma membrane moves through the channels during the passage of an action potential. Thus the effect of a single action potential on the concentration gradients of Na^+ and K^+ is very small, and it is possible in most cases for the sodium–potassium pump to keep the "battery" charged, even when the neuron is generating many action potentials every second.

Action potentials are conducted along axons without loss of signal

Action potentials can travel over long distances with no loss of signal. If we place two pairs of electrodes at two different locations along an axon, we can record an action potential at those two locations as it travels along the axon (**Figure 44.11A**). The magnitude of the action potential does not change between the two recording sites. This constancy is possible because an action potential is an all-or-none, self-regenerating event.

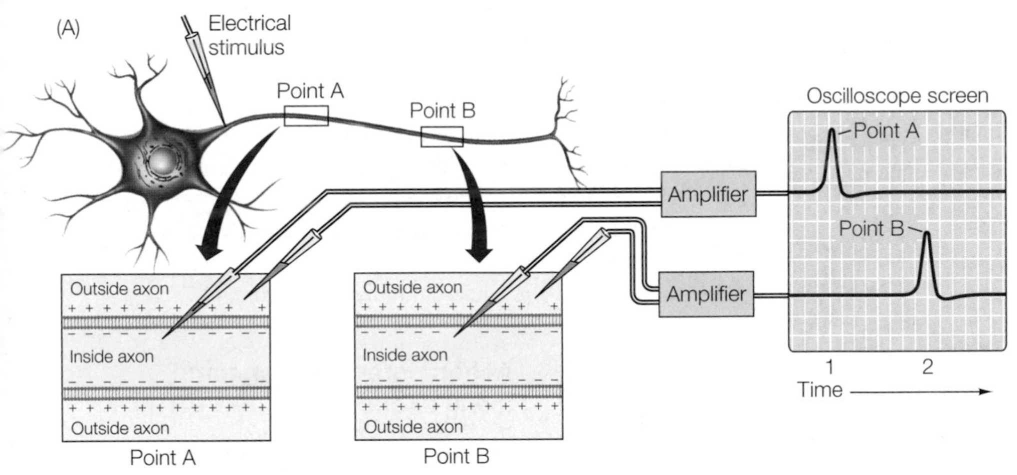

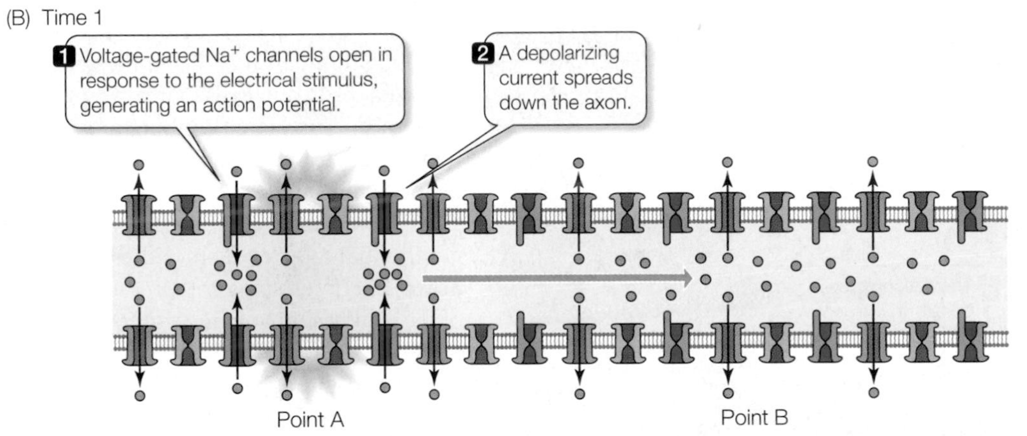

(B) Time 1
1 Voltage-gated Na+ channels open in response to the electrical stimulus, generating an action potential.
2 A depolarizing current spreads down the axon.
Point A
Point B

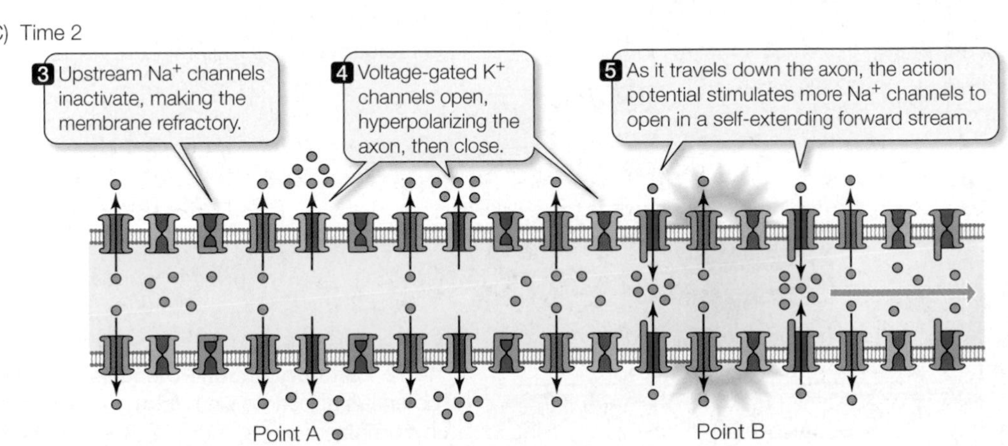

(C) Time 2
3 Upstream Na+ channels inactivate, making the membrane refractory.
4 Voltage-gated K+ channels open, hyperpolarizing the axon, then close.
5 As it travels down the axon, the action potential stimulates more Na+ channels to open in a self-extending forward stream.
Point A
Point B

44.11 Action Potentials Travel along Axons (A) There is no loss of signal as an action potential travels along an axon. (B) When an action potential is stimulated in one region of membrane, electric current flows to adjacent areas of membrane and depolarizes them. (C) The advancing wave of depolarization causes more Na^+ channels to open, and the action potential is generated anew in the next section of membrane. Meanwhile, in the region where the action potential has just fired, the Na^+ channels are inactivated and the voltage-gated K^+ channels are still open, rendering this section of the axon refractory. Hence the action potential cannot "back up," but moves continuously forward along the axon, regenerating itself as it goes.

- An action potential is *all-or-none* because of the interaction between the voltage-gated Na^+ channels and the membrane potential. If the membrane is depolarized slightly, some voltage-gated Na^+ channels open. Some sodium ions cross the plasma membrane and depolarize it even more, opening more voltage-gated Na^+ channels, and so on, until the membrane reaches threshold and generates an action potential. This positive feedback mechanism ensures that action potentials always rise to their maximum value.
- An action potential is *self-regenerating* because it spreads by local current flow to adjacent regions of the plasma membrane. The resulting depolarization brings those neighboring areas of membrane to threshold. So when an action potential occurs at one location on an axon, it stimulates the adjacent region of axon to generate an action potential, and so on down the length of the axon.

We can initiate an action potential by using a stimulating electrode to deliver an electric current that depolarizes the membrane enough to reach threshold. We can then observe the changes in membrane potential associated with the passage of that action potential past the recording electrodes (**Figure 44.11B**). The action potential normally is propagated in only one direction, away from the cell body. It cannot reverse itself because the region of membrane it came from is in its refractory period.

Have you ever touched something very hot and felt the pain before you felt the heat? The neurons that carry the sharp pain sensations are myelinated and transmit action potentials about 10 times faster than the unmyelinated neurons that transmit the sensations of hot or cold.

Action potentials do not travel along all axons at the same speed. They travel faster in myelinated than in nonmyelinated axons, and they travel faster in large-diameter axons than in small-diameter axons. Invertebrates do not have myelin, and therefore the rate of conduction in their axons depends mostly on axon diameter. Invertebrate axons that transmit messages involved in escape behavior are very large, like the squid giant axon in Figure 44.7.

Action potentials can jump along axons

In vertebrate nervous systems, increasing the speed of action potentials by increasing the diameter of axons is not feasible because of the huge number of axons in these organisms. Each of our eyes, for example, has about a million axons connecting it to the brain. These axons conduct action potentials at about the same speed as does the squid giant axon—about 20 meters per second—yet the diameter of each of these axons is about 200 times smaller than the diameter of the squid axon. Imagine your optic nerves being 200 times larger than they are. Each would be over half a meter in diameter! Vertebrates have found a different way of increasing conduction velocity of axons, and that adaptation is myelination.

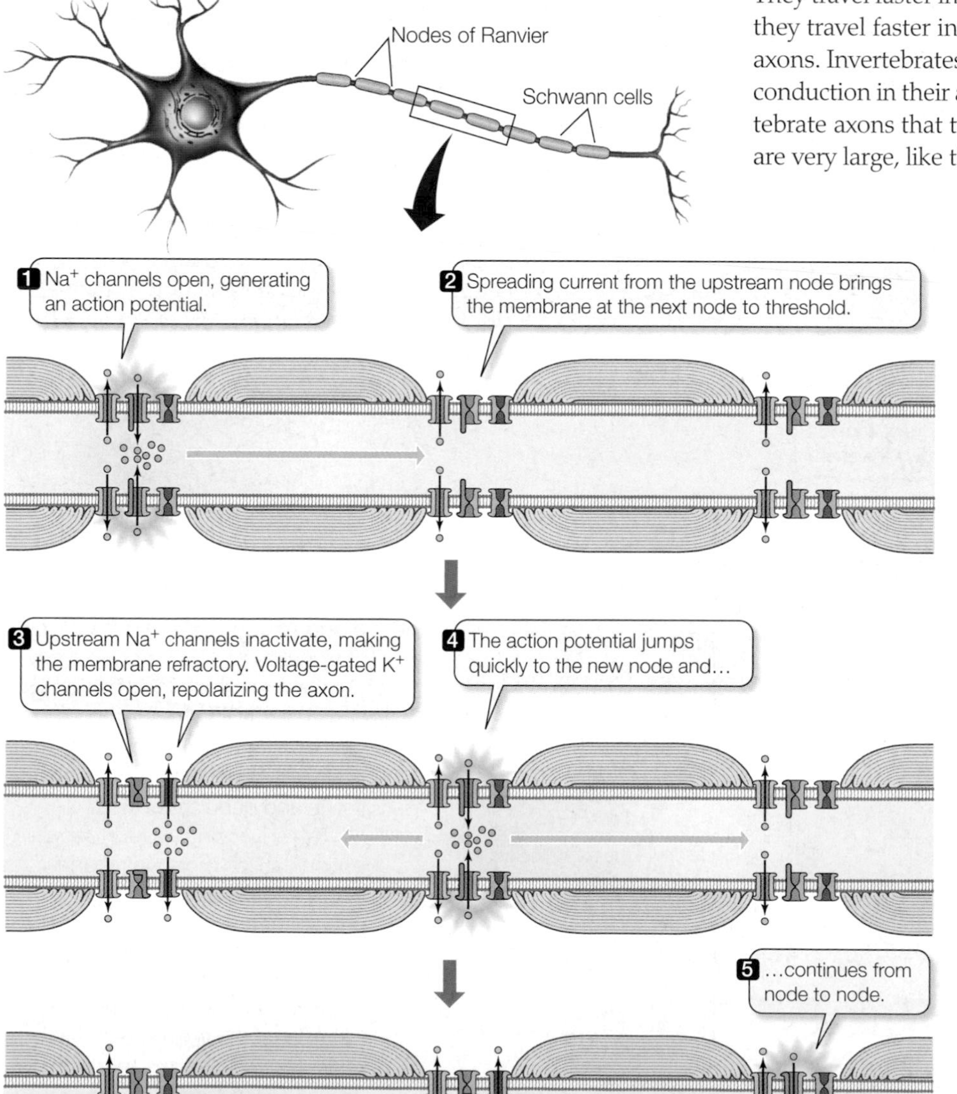

Nodes of Ranvier
Schwann cells
1 Na+ channels open, generating an action potential.
2 Spreading current from the upstream node brings the membrane at the next node to threshold.
3 Upstream Na+ channels inactivate, making the membrane refractory. Voltage-gated K+ channels open, repolarizing the axon.
4 The action potential jumps quickly to the new node and...
5 ...continues from node to node.

44.12 Saltatory Action Potentials Action potentials appear to jump from node to node in myelinated axons.

When glial cells wrap themselves around axons, covering them with concentric layers of myelin (see Figure 44.4), they leave regularly spaced gaps, called **nodes of Ranvier**, where the axon is not covered (**Figure 44.12**). The leakage of ions across the regions of the plasma membrane that are wrapped in myelin is reduced, so electric current can spread farther along the inside of a myelinated axon than it can along a nonmyelinated axon. Additionally, voltage-gated ion channels are clustered at the nodes of Ranvier. Thus an axon can fire action potentials only at nodes, and those action potentials cannot be propagated through the adjacent patch of membrane covered with myelin. The positive charges that flow into the axon at the node do, however, flow down the inside of the axon in the form of electric current. When the current reaches the next node, the plasma membrane at that node is depolarized to threshold and fires another an action potential. Action potentials therefore appear to jump from node to node along the axon.

The speed of conduction is increased in these myelin-wrapped axons because electric current flows much faster through the cytoplasm than ion channels can open and close. This form of rapid impulse propagation is called **saltatory conduction** (Latin *saltare*, "to jump").

44.2 RECAP

Neurons generate action potentials—rapid, all-or-none changes in membrane potential that are conducted along axons from the neuron body to the axonal terminals. At axon teriminals, the information is communicated across synapses to target cells.

- How are membrane resting potentials generated? See pp. 947–948 and Figures 44.6 and 44.9
- How are action potentials generated? See pp. 951–953 and Figure 44.10
- How are action potentials propagated along axons? See pp. 953–955 and Figures 44.11 and 44.12

Once we understand how action potentials are generated in and conducted along axons, the question remains as to exactly what happens at the synapse when an action potential gets to the axon terminal. How do synapses convey information from one cell to another?

44.3 How Do Neurons Communicate with Other Cells?

Neurons communicate with each other and with target cells at synapses. The most common type of synapse in the nervous system is the **chemical synapse**—one in which chemical messages from a presynaptic cell induce changes in a postsynaptic cell. In **electrical synapses** the action potential spreads directly from presynaptic to postsynaptic cell.

The neuromuscular junction is a model chemical synapse

Neuromuscular junctions are synapses between motor neurons and the skeletal muscle cells they innervate. They are excellent models for how chemical synaptic transmission works. Like other neurons, a motor neuron has only one axon, but that axon can have many branches, each with axon terminals that form neuromuscular junctions with a muscle cell. At each axon terminal an enlarged knob or button-like structure contains many vesicles filled with the chemical messenger—neurotransmitter molecules. The neurotransmitter used by all vertebrate motor neurons is **acetylcholine (ACh)**. ACh is released by exocytosis when the membrane of a vesicle fuses with the presynaptic membrane.

Where does the neurotransmitter come from? Some neurotransmitters, such as ACh, are synthesized in the axon terminal and packaged in vesicles. The enzymes required for ACh biosynthesis, however, are produced in the cell body of the motor neuron and are transported along microtubules down the axon in membrane-bound packages to the terminals. Other kinds of neurotransmitters, such as peptide neurotransmitters, are produced in the cell body and transported in membrane-bound packages down the axon to the terminals.

The postsynaptic membrane of the neuromuscular junction is a modified part of the muscle cell plasma membrane called a **motor end plate** (**Figure 44.13**). It appears as a depression in the muscle cell membrane, and the terminals of the motor neuron sit in the depression. The space between the presynaptic membrane and the postsynaptic membrane is the **synaptic cleft**, which in chemical synapses is about 20 to 40 nanometers wide. Neurotransmitter released into the cleft by the presynaptic cell diffuses across to the postsynaptic membrane. The membrane of the motor end plate is highly folded. ACh receptors are on the crests of the folds, and voltage-gated cation channels are at the bottoms of the folds and in the surrounding muscle cell membrane.

The arrival of an action potential causes the release of neurotransmitter

Neurotransmitter is released when an action potential arrives at the axon terminal and causes the opening of voltage-gated Ca^{2+} channels in the presynaptic membrane. Because the Ca^{2+} concentration is greater outside the cell than inside, Ca^{2+} enters the axon terminal near the sites of vesicle exocytosis. The increase in Ca^{2+} inside the axon terminal causes the vesicles containing ACh to fuse with the presynaptic membrane and empty their contents into the synaptic cleft.

Two well-known deadly toxins, botulinum toxin and tetanus toxin, act by interfering with the proteins responsible for the fusion of neurotransmitter vesicles and the presynaptic membrane.

In neuromuscular synapses, vesicle fusion and emptying is all-or-none. The vesicle membrane is incorporated into the presy-

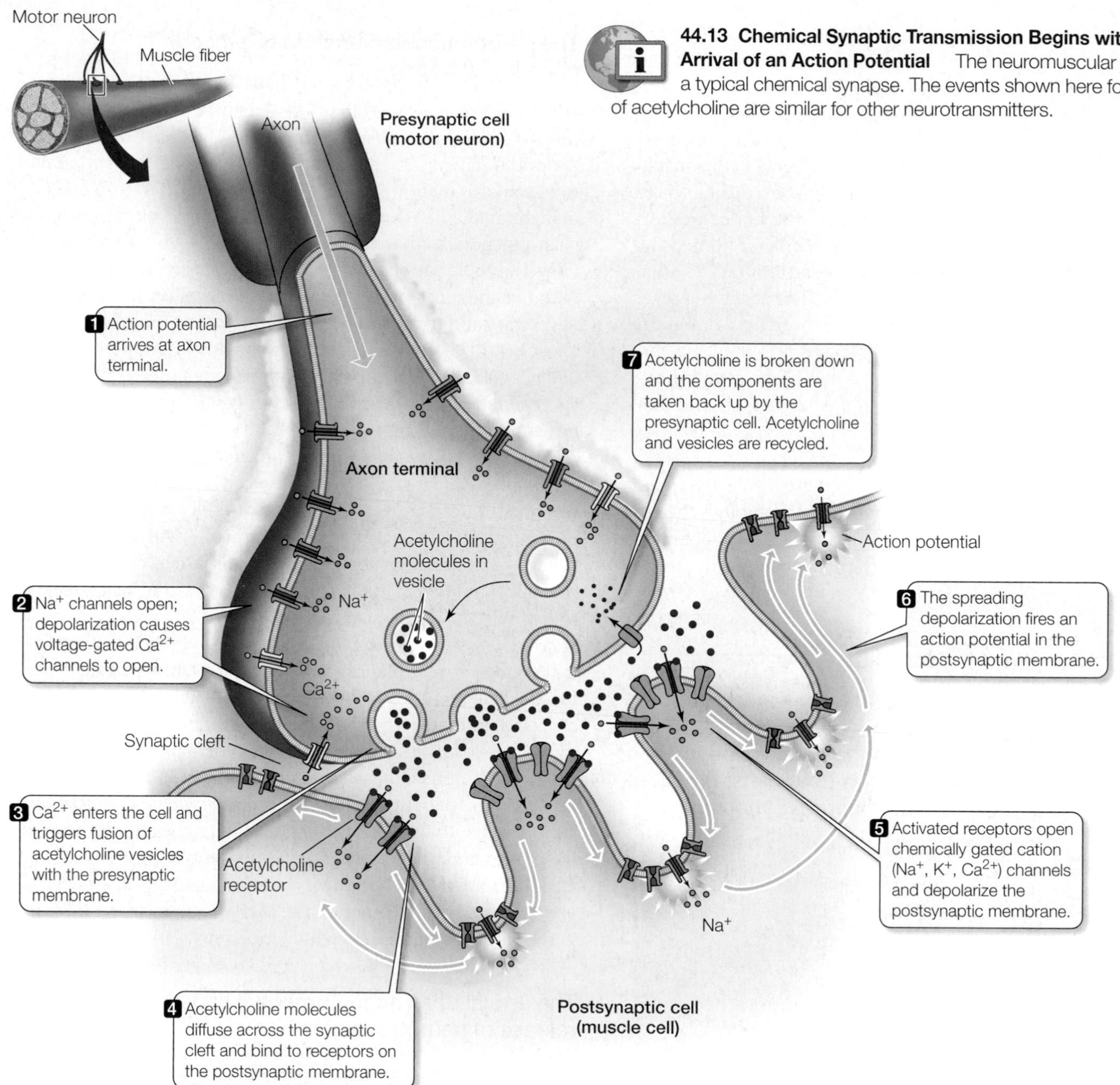

Motor neuron
Muscle fiber
Axon
Presynaptic cell (motor neuron)
1 Action potential arrives at axon terminal.
Axon terminal
7 Acetylcholine is broken down and the components are taken back up by the presynaptic cell. Acetylcholine and vesicles are recycled.
Action potential
Acetylcholine molecules in vesicle
2 Na+ channels open; depolarization causes voltage-gated Ca2+ channels to open.
Na+
Ca2+
6 The spreading depolarization fires an action potential in the postsynaptic membrane.
Synaptic cleft
3 Ca2+ enters the cell and triggers fusion of acetylcholine vesicles with the presynaptic membrane.
Acetylcholine receptor
5 Activated receptors open chemically gated cation (Na+, K+, Ca2+) channels and depolarize the postsynaptic membrane.
Na+
4 Acetylcholine molecules diffuse across the synaptic cleft and bind to receptors on the postsynaptic membrane.
Postsynaptic cell (muscle cell)

44.13 Chemical Synaptic Transmission Begins with the Arrival of an Action Potential The neuromuscular junction is a typical chemical synapse. The events shown here for release of acetylcholine are similar for other neurotransmitters.

naptic membrane, which actually gets larger as a result—at least until the extra membrane is recycled through endocytosis. The recycled membrane is processed through endosomes to become new vesicles that are then refilled with neurotransmitter.

The postsynaptic membrane responds to neurotransmitter

When ACh is released at a synapse, some of it diffuses across the synaptic cleft and binds to ACh receptors on the postsynaptic membrane (**Figure 44.14**). The ACh receptors are channels that allow both Na^+ and K^+ to flow through, but since the electrochemical gradients favor a net influx of Na^+, the response of the motor end plate to ACh is to depolarize. That depolarization spreads to the depths of the folds of the motor end plate membrane and to surrounding muscle cell membrane, which contain voltage-gated Na^+ channels.

If the axon terminal of a motor neuron releases sufficient amounts of ACh to adequately depolarize a motor end plate, that spreading depolarization will activate the voltage-gated Na^+ channels causing the firing of an action potential. This action potential is then conducted throughout the muscle cell's system of membranes, causing the cell to contract. (We'll learn more about muscle membrane action potentials and the contraction of muscle cells in Section 47.1.)

How much neurotransmitter is enough? Neither a single ACh molecule nor the contents of an entire vesicle (about 10,000 ACh molecules) will bring the plasma membrane of a muscle cell to threshold. However, a single action potential in an axon terminal releases the contents of about 100 vesicles, which is more than enough to fire an action potential in the muscle cell and cause it to contract.

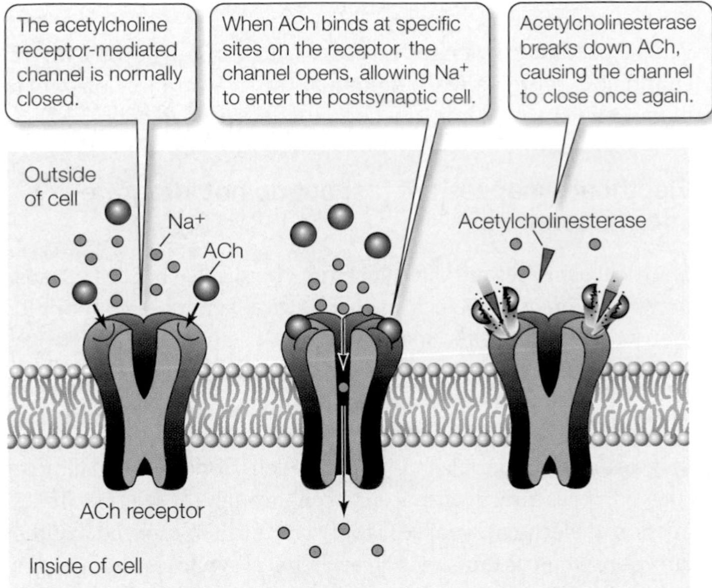

The acetylcholine receptor-mediated channel is normally closed.
When ACh binds at specific sites on the receptor, the channel opens, allowing Na+ to enter the postsynaptic cell.
Acetylcholinesterase breaks down ACh, causing the channel to close once again.
Outside of cell
Na+
ACh
Acetylcholinesterase
ACh receptor
Inside of cell

44.14 The Acetylcholine Receptor Is a Chemically Gated Channel The motor end plate contains acetylcholine receptors, which are chemically gated ion channels. When one of these receptors binds ACh, its channel pore opens, and Na^+ ions move into the postsynaptic cell, depolarizing its plasma membrane. Acetylcholinesterase breaks down ACh in the synapse, closing the channel; the breakdown products are then taken up by the presynaptic membrane and resynthesized into more ACh.

Synapses between neurons can be excitatory or inhibitory

In vertebrates, the synapses between motor neurons and muscle cells are always **excitatory**; that is, motor end plates always respond to ACh by depolarizing the postsynaptic membrane. Synapses between neurons, however, are not always excitatory.

Recall that a neuron may have many dendrites. Axon terminals from many other neurons may form synapses with those dendrites and with the cell body. The axon terminals of different presynaptic neurons may store and release different neurotransmitters, and the plasma membrane of the dendrites and cell body of a postsynaptic neuron may have receptors for a variety of neurotransmitters. Thus, at any one time, a postsynaptic neuron may receive a several different chemical messages. If the postsynaptic neuron's response to a neurotransmitter is depolarization, as at the neuromuscular junction, the synapse is excitatory; if its response is hyperpolarization, the synapse is **inhibitory**.

The postsynaptic cell sums excitatory and inhibitory input

What determines when an individual neuron will fire an action potential? Neurons sum excitatory and inhibitory postsynaptic potentials, and when the sum of the potentials surpasses the threshold, the neuron generates an action potential. This summation ability is the major mechanism by which the nervous system integrates information. Each neuron may receive a thousand or more synaptic inputs, but it has only one output: an action potential in a single axon. All of the inputs a neuron receives are reduced to the rate at which that neuron generates action potentials in its axon.

For most neurons, summation takes place in the **axon hillock**, the region of the cell body at the base of the axon (see Figure 44.3). The plasma membrane of the axon hillock is not insulated by glial cells and has many voltage-gated Na^+ channels. Excitatory and inhibitory postsynaptic potentials from synapses anywhere on the dendrites or the cell body may spread to the axon hillock by local current flow. If the resulting combined potential depolarizes the axon hillock to threshold, it fires an action potential. Because postsynaptic potentials decrease in strength as they spread from the site of the synapse, a synapse at the tip of a dendrite has less influence than a synapse on the cell body, near the axon hillock.

Excitatory and inhibitory postsynaptic potentials are summed over space or over time. **Spatial summation** adds up the simultaneous influences of synapses at different sites on the postsynaptic cell (**Figure 44.15A**). **Temporal summation** adds up postsynaptic potentials generated at the same site in a rapid sequence (**Figure 44.15B**).

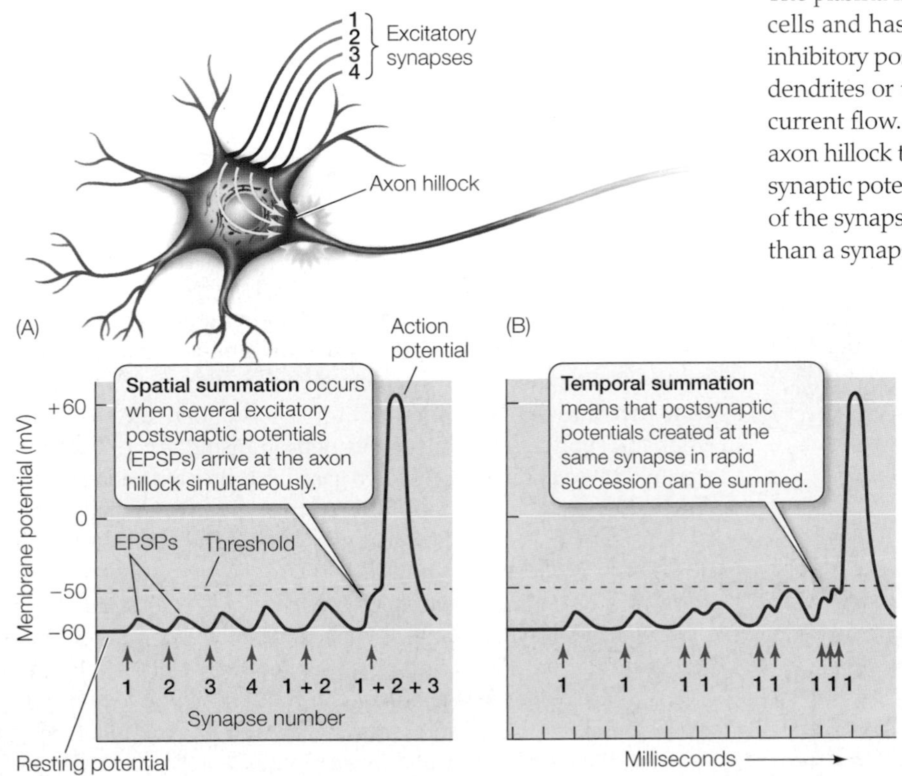

44.15 The Postsynaptic Neuron Sums Information Individual neurons sum excitatory and inhibitory postsynaptic potentials over space (A) and time (B). When the sum of the potentials depolarizes the axon hillock to threshold, the neuron generates an action potential.

Synapses can be fast or slow

Most neurotransmitter receptors induce changes in postsynaptic cells by opening or closing ion channels. How they do so is the basis for grouping receptors into two general categories:

- **Ionotropic receptors** are ion channels themslves. Neurotransmitter binding to an ionotropic receptor causes a direct change in ion movement across the plasma membrane of the postsynaptic cell. These proteins enable fast, short-lived responses.
- **Metabotropic receptors** are not ion channels, but they induce signaling cascades in the postsynaptic cell that secondarily lead to changes in ion channels. Postsynaptic cell responses mediated by metabotropic receptors are generally slower and longer-lived than those induced by ionotropic receptors.

The ACh receptor of the motor end plate is an example of an ionotropic receptor. It consists of five subunits, each of which extends through the plasma membrane. When assembled, the subunits create a central pore that allows ions to pass through (see Figure 44.14). Of several different kinds of subunits, only one kind has the ability to bind ACh. Each functional receptor has two of the ACh-binding subunits and three other subunits.

Metabotropic receptors are also transmembrane proteins, but instead of acting as ion channels, they initiate an intracellular signaling process that can result in the opening or closing of an ion channel. These receptors have seven transmembrane domains, and they are linked to G proteins (**Figure 44.16**). When a neurotransmitter binds to the extracellular domain of a metabotropic receptor, the intracellular domain activates a G protein. In its inactive state, the G protein has three subunits, one of which (the α subunit) is bound to a molecule of GDP. When the receptor binds its neurotransmitter, the GDP is replaced by a GTP molecule, and the α subunit separates from the other two subunits (called β and γ). The subunits move laterally in the membrane until they bind to and open an ion channel or bind to an effector protein that activates a second messenger cascade, which in turn opens an ion channel. G proteins working in just this way are key players in many cellular signal transduction processes (see Section 15.2).

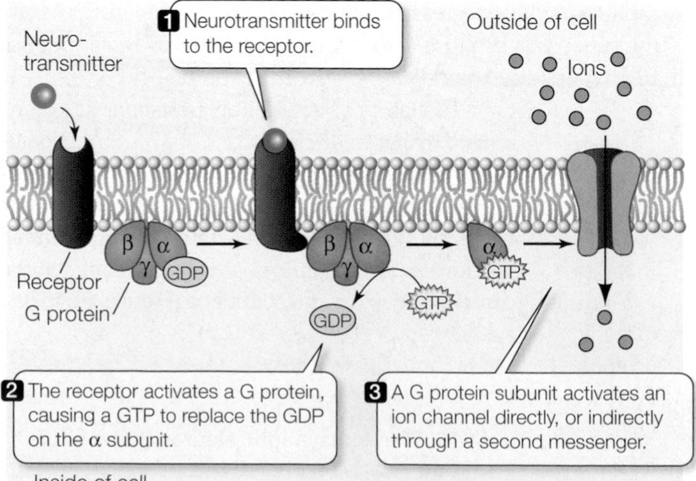

44.16 Metabotropic Receptors Act through G Proteins
Metabotropic receptors activate G proteins, which can influence ion channels directly or through second messengers.

Electrical synapses are fast but do not integrate information well

Electrical synapses are different from chemical synapses because they couple neurons electrically. Electrical synapses contain numerous *gap junctions* (see Figures 5.7C and 15.19) At these synapses, the presynaptic and postsynaptic cell membranes are separated by a space of only 2 to 3 nanometers, and membrane proteins called *connexons* link the two neurons by forming pores that connect the cytoplasm of the two cells. Ions and small molecules can pass directly from cell to cell through these pores. Transmission at electrical synapses is very fast and can proceed in either direction, whereas transmission at chemical synapses is slower and unidirectional.

Electrical synapses are less common in the nervous systems of vertebrates than are chemical synapses for several reasons. First, electrical continuity between neurons does not allow temporal summation of synaptic inputs. Second, an effective electrical synapse requires a large area of contact between the presynaptic and postsynaptic cells. This condition rules out the possibility of thousands of synaptic inputs to a single neuron—which is the norm in complex nervous systems. Third, electrical synapses cannot be inhibitory. Thus, electrical synapses are useful for rapid communication, but they are less useful for processes of integration and learning.

The action of a neurotransmitter depends on the receptor to which it binds

More than 50 neurotransmitters are now recognized, and more will surely be discovered. **Table 44.1** describes some of the best-known neurotransmitters. ACh, as we have seen, is an important neurotransmitter because it is how the nervous system commands muscles to contract. ACh also plays roles in certain synapses between neurons in the CNS, but it accounts for only a small percentage of the total neurotransmitter content of the CNS. The workhorse neurotransmitters of the CNS are simple amino acids: glutamate (excitatory) and glycine and γ-aminobutyrate (GABA, a derivative of glutamate) (inhibitory). Another important group of neurotransmitters in the CNS is the monoamines, which are derivatives of amino acids. They include dopamine and norepinephrine (derivatives of tyrosine) and serotonin (a derivative of tryptophan). Peptides also function as neurotransmitters. An exciting recent discovery revealed that two gases, carbon monoxide and nitric oxide, are used by neurons as intercellular messengers, although the neurons do not have associated receptors (see Figure 15.15).

Neurotransmission is complex in part because each neurotransmitter has multiple receptor types. Acetylcholine, for example, has two receptor types: *nicotinic receptors*, which are ionotropic, and *muscarinic receptors*, which are metabotropic. Both types of acetylcholine receptors are found in the CNS, where nicotinic receptors tend to be excitatory and muscarinic receptors tend to be in-

TABLE 44.1

Some Well-Known Neurotransmitters

NEUROTRANSMITTER	ACTIONS	COMMENTS
Acetylcholine	The neurotransmitter of vertebrate motor neurons and of some neural pathways in the brain	Broken down in the synapse by acetylcholinesterase; blockers of this enzyme are powerful poisons
MONOAMINES		
Norepinephrine	Used in certain neural pathways in the brain. Also found in the peripheral nervous system, where it causes gut muscles to relax and the heart to beat faster	Related to epinephrine and acts at some of the same receptors
Dopamine	A neurotransmitter of the central nervous system	Involved in schizophrenia. Loss of dopamine neurons is the cause of Parkinson's disease
Histamine	A minor neurotransmitter in the brain	Involved in maintaining wakefulness
Serotonin	A neurotransmitter of the central nervous system that is involved in many systems, including pain control, sleep/wake control, and mood	Certain medications that elevate mood and counter anxiety act by inhibiting the reuptake of serotonin (so it remains active longer)
PURINES		
ATP	Co-released with many neurotransmitters	Large family of receptors may shape postsynaptic responses to classical neurotransmitters
Adenosine	Transported across cell membranes; not synaptically released	Not released synaptically; mainly has inhibitory effects on neighboring cells
AMINO ACIDS		
Glutamate	The most common excitatory neurotransmitter in the central nervous system	Some people have reactions to the food additive monosodium glutamate because it can affect the nervous system
Glycine Gamma-aminobutyric acid (GABA)	Common inhibitory neurotransmitters	Drugs called benzodiazepines, used to reduce anxiety and produce sedation, mimic the actions of GABA
PEPTIDES		
Endorphins Enkephalins	Modulation of pain pathways	Receptors are activated by narcotic drugs: opium, morphine, heroin, codeine
Substance P	Used by certain sensory nerves, especially in pain pathways	Released by neurons sensitive to heat and pain
GAS		
Nitric oxide	Widely distributed in the nervous system	Not a classic neurotransmitter, it diffuses across membranes rather than being released synaptically. A means whereby a postsynaptic cell can influence a presynaptic cell

hibitory. Acetylcholine actions can differ outside of the CNS as well. Acetylcholine acting through nicotinic receptors causes the smooth muscle of the gut to depolarize and therefore increases its motility, but acetylcholine acting through muscarinic receptors causes cardiac muscle to hyperpolarize and therefore decreases the contractility of the heart.

We could give many more examples of neurotransmitters that have different effects in different tissues, but the important thing to remember is that the action of a neurotransmitter depends on the receptor to which it binds.

Glutamate receptors may be involved in learning and memory

Glutamate is a neurotransmitter that can bind to a variety of receptors, including both metabotropic and ionotropic receptors. The glutamate receptors are divided into several classes because they can be differentially activated by other chemicals that mimic the action of glutamate. One class of ionotropic glutamate receptors is the *NMDA receptors*, which can be activated by the chemical *N*-methyl-D-aspartate. Another class of ionotropic glutamate receptors is activated by a different chemical, abbreviated as *AMPA* (which stands for α-amino-3-hydroxy-5-methyl-4-isoxazole propionate).

Glutamate is an excitatory neurotransmitter, so activation of ionotropic glutamate receptors always results in Na^+ entry into the neuron and depolarization. But the *timing* of the response to activation by these different types of receptors differs significantly. The AMPA receptors allow a rapid influx of Na^+ into the postsynaptic cell, while the NMDA receptors allow a slower and longer-lasting influx of Na^+. The NMDA receptors also require that the cell be somewhat depolarized through the action of other receptors before their pores will open and permit Na^+ influx. When they do open, these receptors also allow Ca^{2+} to enter the cell. Calcium ions

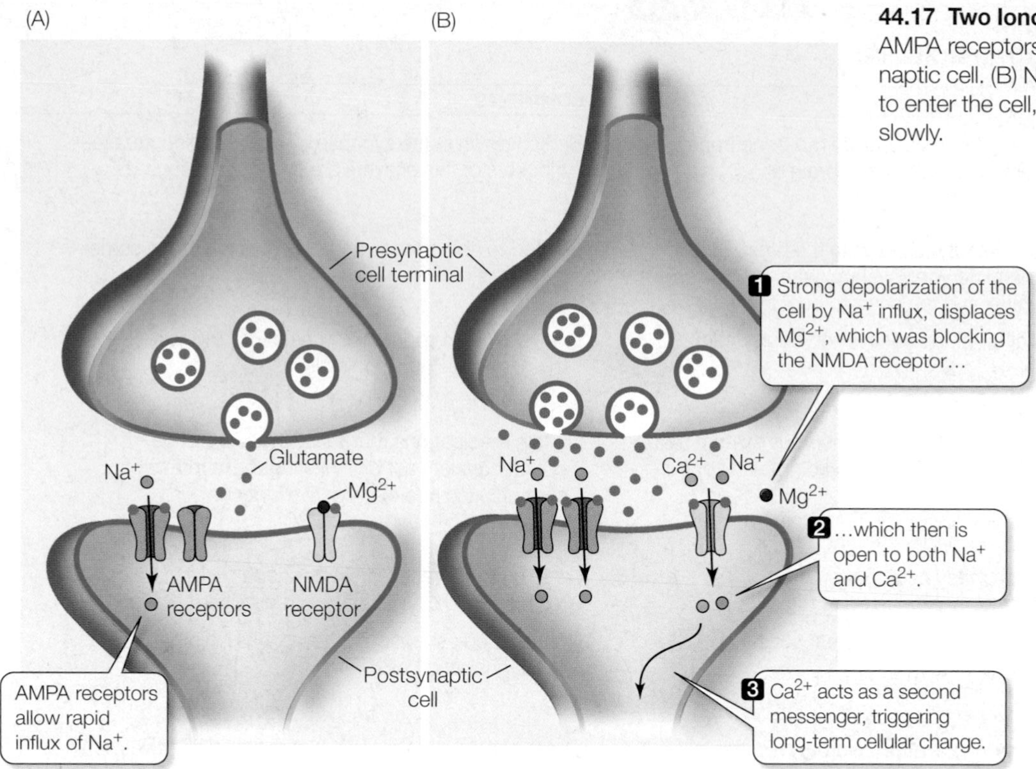
(A)
(B)
Presynaptic cell terminal
Na+
Glutamate
Mg2+
AMPA receptors
NMDA receptor
AMPA receptors allow rapid influx of Na+.
Postsynaptic cell
Na+
Ca2+
Na+
Mg2+
1 Strong depolarization of the cell by Na+ influx, displaces Mg2+, which was blocking the NMDA receptor…
2 …which then is open to both Na+ and Ca2+.
3 Ca2+ acts as a second messenger, triggering long-term cellular change.

44.17 Two Ionotropic Glutamate Receptors (A) AMPA receptors allow rapid influx of Na^+ into the postsynaptic cell. (B) NMDA receptors allow both Na^+ and Ca^{2+} to enter the cell, but respond to synaptic input more slowly.

act as second messengers in the cell and can trigger a variety of long-term cellular changes.

Figure 44.17 shows how the AMPA and NMDA receptors can work in concert. At resting potential, the NMDA receptor is blocked by a magnesium ion (Mg^{2+}). Strong depolarization of the neuron due to other inputs—such as the activation of AMPA receptors—displaces Mg^{2+} from the NMDA receptors and allows Na^+ and Ca^{2+} to pass through them when they are activated by glutamate. These special properties of the NMDA receptor are probably involved in learning and memory.

Most of the synaptic events we have studied so far happen very quickly. It is therefore a special challenge to understand how the messages carried by action potentials can result in long-term events such as learning and memory. Our understanding of these processes has been greatly affected by study of a phenomenon called **long-term potentiation** (**LTP**), which is studied by neurobiologists working with slices of brain kept alive in dishes of culture medium. Using these brain slice preparations, it is possible to stimulate and record from specific brain regions, and even specific neurons.

In the studies leading to the discovery of LTP, experimenters repeatedly stimulated synaptic inputs to a particular neuron and observed the usual action potential response. When the neuron was stimulated many times in rapid succession, however, they found that the properties of the neuron changed. The magnitude of the postsynaptic response was enhanced, or *potentiated*, and this change lasted for days or weeks.

How does potentiation of a synapse occur? The answer, at least for some areas of the brain, now seems clear. With low levels of stimulation, the glutamate released by presynaptic cells activates only AMPA receptors, and the postsynaptic membrane simply responds with action potentials. With higher levels of stimulation, however, NMDA receptors are also activated, allowing both Na^+ and Ca^{2+} ions to enter the postsynaptic neuron. The Ca^{2+} ions induce long-term changes in the postsynaptic membrane that make it more sensitive to synaptic input (**Figure 44.18**).

Exploiting the LTP system, Joe Tsien and his students and collaborators at Princeton University genetically engineered mice so that their NMDA receptors had a slightly altered structure and were activated for a longer time whenever they bound a molecule of glutamate. These mice learned tasks better, ran mazes faster, and remembered the mazes longer than normal mice. These exciting experiments show that we are on the right track to understanding how the brain achieves learning and memory.

To turn off responses, synapses must be cleared of neurotransmitter

Turning off the action of neurotransmitters is as important as turning it on. If released neurotransmitter molecules simply remained in the synaptic cleft, the postsynaptic membrane would become saturated with neurotransmitter, and receptors would be constantly activated. As a result, the postsynaptic cell would remain hyperpolarized or depolarized and would be unresponsive to short-term changes in the presynaptic cell. The more discrete each separate neuronal signal is, the more information can be processed in a given time. Thus neurotransmitter must be cleared from the synaptic cleft shortly after it is released by the axon terminal.

Neurotransmitter action may be terminated in several ways. First, enzymes may destroy the neurotransmitter. ACh, for example, is rapidly destroyed by the enzyme acetylcholinesterase, which is present in the synaptic cleft in close association with the ACh

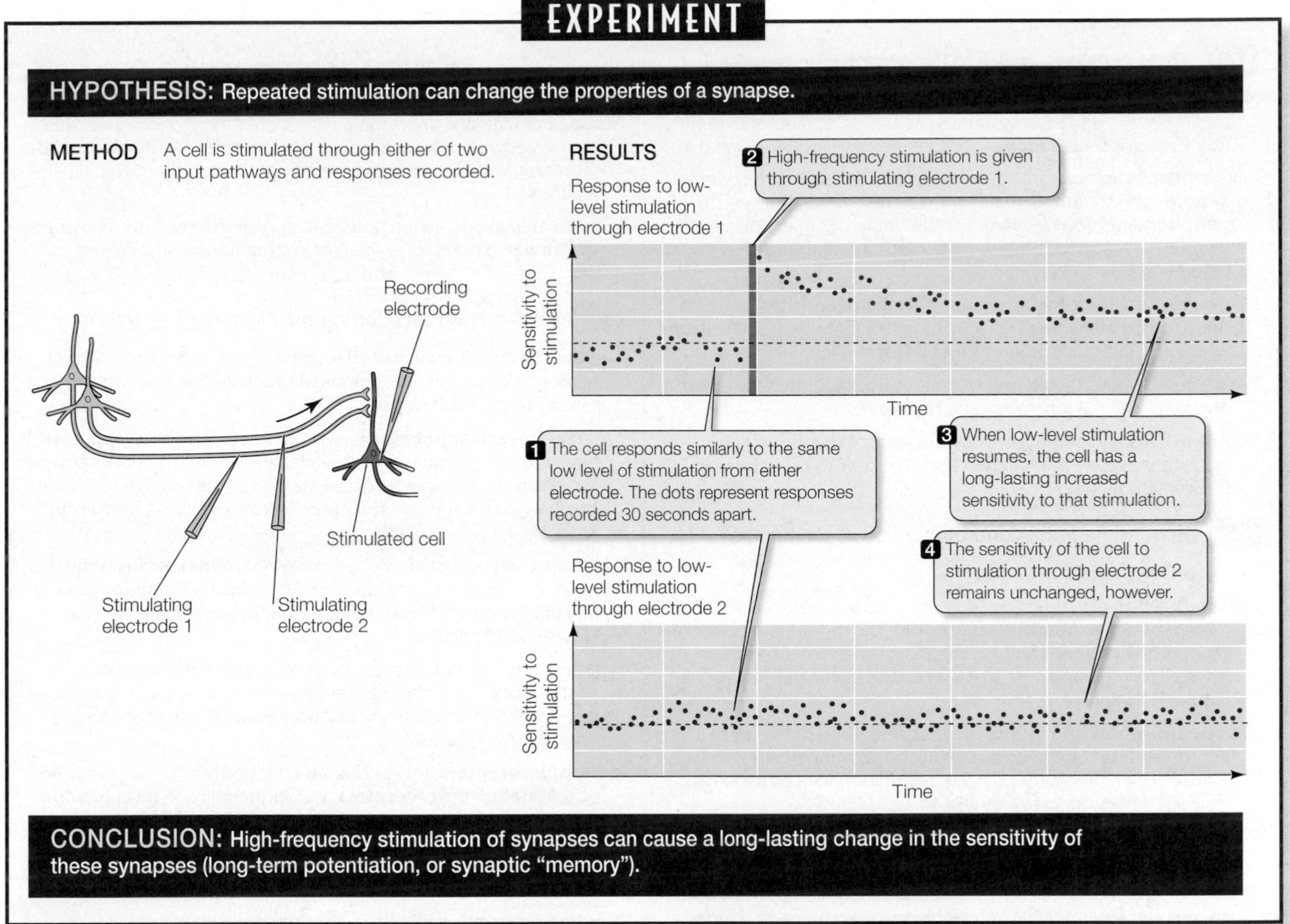

EXPERIMENT
HYPOTHESIS: Repeated stimulation can change the properties of a synapse.
METHOD A cell is stimulated through either of two input pathways and responses recorded.
Recording electrode
Stimulated cell
Stimulating electrode 1
Stimulating electrode 2
RESULTS
Response to low-level stimulation through electrode 1
Sensitivity to stimulation
Time
2 High-frequency stimulation is given through stimulating electrode 1.
1 The cell responds similarly to the same low level of stimulation from either electrode. The dots represent responses recorded 30 seconds apart.
3 When low-level stimulation resumes, the cell has a long-lasting increased sensitivity to that stimulation.
4 The sensitivity of the cell to stimulation through electrode 2 remains unchanged, however.
Response to low-level stimulation through electrode 2
Sensitivity to stimulation
Time
CONCLUSION: High-frequency stimulation of synapses can cause a long-lasting change in the sensitivity of these synapses (long-term potentiation, or synaptic "memory").

44.18 Repeated Stimulation Can Cause Long-Term Potentiation When a cell receives low-frequency synaptic input, the resulting postsynaptic response remains constant. If, however, that same synaptic pathway is stimulated briefly at a high frequency, the subsequent sensitivity of the postsynaptic cell to the original level of synaptic input is potentiated for a longer period.

receptors on the postsynaptic membrane (see Figure 44.14). Some of the most deadly nerve gases developed for chemical warfare work by inhibiting acetylcholinesterase. As a result, ACh lingers in the synaptic cleft, causing the victim to die of spastic (contracted) muscle paralysis. Some agricultural insecticides, such as malathion, also inhibit acetylcholinesterase and can poison farm workers if used without safety precautions.

Neurotransmitter also may simply diffuse away from the cleft, or be taken up via active transport by nearby cell membranes. Prozac, a drug commonly prescribed to treat depression, slows the reuptake of the neurotransmitter serotonin, thus enhancing its activity at the synapse.

44.3 RECAP

Chemical synapses involve the release of neurotransmitter molecules stored in vesicles in the presynaptic terminal. Action potentials reaching that terminal cause the fusion of vesicles with the presynaptic membrane releasing neurotransmitter that can then bind to receptors on the postsynaptic membrane.

- Can you explain the role of Ca^{2+} channels in synaptic events? See pp. 955–956 and Figure 44.13
- How can some synapses be excitatory and others inhibitory? See p. 957
- How do neurons integrate the input from various synapses? See pp. 957–958 and Figure 44.15

CHAPTER SUMMARY

44.1 What cells are unique to the nervous systems?

Nervous systems include **neurons** and **glial cells**. Neurons are organized in circuits with sensory inputs, integration, and outputs to effectors. Glial cells serve support functions. Review Figures 44.1 and 44.3

In vertebrates, the brain and spinal cord form the **central nervous system**, which communicates with the rest of the body via the **peripheral nervous system**. The CNS increases in complexity from invertebrates to vertebrates and from fish to mammals. Review Figures 44.1 and 44.2

Neurons generally receive information via their **dendrites**, of which there can be many, and transmit information via their single **axons**, which end in axon terminals. Review Figure 44.3

Where neurons and their target cells meet, information is transmitted across specialized junctions called **synapses**.

Schwann cells and **oligodendrocytes** produce **myelin**, which insulates neurons. **Astrocytes** create the **blood–brain barrier**. Review Figure 44.4

44.2 How do neurons generate and conduct signals?

See Web/CD Tutorial 44.1

Neurons have an electric charge difference across their plasma membranes, the **membrane potential**. The membrane potential is created by ion pumps and ion channels. When a neuron is not active, its membrane potential is a **resting potential**. Review Figures 44.5 and 44.6

The **sodium–potassium pump** concentrates K^+ on the inside of a neuron and Na^+ on the outside. Potassium channels allow K^+ to diffuse out of the neuron, leaving behind unbalanced negative charges. Review Figures 44.6 and 44.7

Patch clamping allows the study of single ion channels. Review Figure 44.8

The resting potential is perturbed when ion channels open or close, changing the permeability of the plasma membrane to charged ions. Through this mechanism, the plasma membrane can become **depolarized** or **hyperpolarized**. Review Figure 44.9

An **action potential** is a rapid reversal in charge across a portion of the plasma membrane resulting from the sequential opening and closing of **voltage-gated Na^+ and K^+ channels**. These changes in voltage-gated channels occur when the plasma membrane depolarizes to a **threshold level**. Review Figure 44.10, Web/CD Tutorial 44.2

Action potentials are all-or-none, self-regenerating events. They are conducted down axons because local current flow depolarizes adjacent regions of membrane and brings them to threshold. Review Figure 44.11

In myelinated axons, action potentials appear to jump between **nodes of Ranvier**, patches of axonal plasma membrane that are not covered by myelin. Review Figure 44.12

44.3 How do neurons communicate with other cells?

Neurons communicate with each other and with other cells by transmitting information over **chemical synapses** (with neurotransmitters) or **electrical synapses**.

The **neuromuscular junction** is a well-studied chemical synapse between a motor neuron and a skeletal muscle cell. Its neurotransmitter is **ACh**, which causes a depolarization of the postsynaptic membrane when it binds to its receptor. Review Figure 44.13, Web/CD Tutorial 44.3

When an action potential reaches an **axon terminal**, it causes the release of neurotransmitters, which diffuse across the **synaptic cleft** and bind to receptors on the postsynaptic membrane. Review Figures 44.13 and 44.14

Synapses between neurons can be either excitatory or inhibitory. A postsynaptic neuron integrates information by **summing** excitatory and inhibitory postsynaptic potentials in both space and time. Review Figure 44.15

Ionotropic receptors are ion channels or directly influence ion channels. **Metabotropic receptors** are G protein-linked receptors that influence the postsynaptic cell through various signal transduction pathways and result in the opening or closing of ion channels. The actions of ionotropic synapses are generally faster than those of metabotropic synapses. Review Figure 44.16

There are many different neurotransmitters and even more types of receptors. The action of a neurotransmitter depends on the receptor to which it binds. Review Table 44.1, Web/CD Activity 44.1

With repeated stimulation, a neuron can become more sensitive to its inputs through **LTP**. The properties of the NMDA glutamate receptor appear to explain LTP. Review Figures 44.17 and 44.18

SELF-QUIZ

1. The rising phase of an action potential is due to the
   a. closing of K^+ channels.
   b. opening of chemically gated Na^+ channels.
   c. closing of voltage-gated Ca^{2+} channels.
   d. opening of voltage-gated Na^+ channels.
   e. spread of positive current along the plasma membrane.
2. The resting potential of a neuron is due mostly to
   a. local current spread.
   b. open Na^+ channels.
   c. synaptic summation.
   d. open K^+ channels.
   e. open Cl^- channels.
3. Which statement about synaptic transmission is *not* true?
   a. The synapses between neurons and muscle cells use ACh as their neurotransmitter.
   b. A single vesicle of neurotransmitter cannot cause a muscle cell to contract.
   c. The release of neurotransmitter at the neuromuscular junction causes the motor end plate to fire action potentials.
   d. In vertebrates, the synapses between motor neurons and muscle fibers are always excitatory.
   e. Inhibitory synapses cause the resting potential of the postsynaptic membrane to become more negative.
4. Which statement accurately describes an action potential?
   a. Its magnitude increases along the axon.
   b. Its magnitude decreases along the axon.
   c. All action potentials in a single neuron are of the same magnitude.
   d. During an action potential the membrane potential of a neuron remains constant.
   e. An action potential permanently shifts a neuron's membrane potential away from its resting value.

5. A neuron that has just fired an action potential cannot be immediately restimulated to fire a second action potential. The short interval of time during which restimulation is not possible is called
   a. hyperpolarization.
   b. the resting potential.
   c. depolarization.
   d. repolarization.
   e. the refractory period.
6. The rate of propagation of an action potential depends on
   a. whether or not the axon is myelinated.
   b. the axon's diameter.
   c. whether or not the axon is insulated by glial cells.
   d. the cross-sectional area of the axon.
   e. all of the above.
7. The binding of an inhibitory neurotransmitter to the postsynaptic receptors results in
   a. depolarization of the membrane.
   b. generation of an action potential.
   c. hyperpolarization of the membrane.
   d. increased permeability of the membrane to sodium ions.
   e. increased permeability of the membrane to calcium ions.
8. The difference between slow and fast synapses is
   a. the width of the synaptic cleft.
   b. the size of the synapse.
   c. whether or not the neurotransmitter acts directly on ion channels.
   d. the density of receptors on the postsynaptic membrane.
   e. whether or not the presynaptic vesicles kiss and run.
9. Whether a synapse is excitatory or inhibitory depends on the
   a. type of neurotransmitter.
   b. presynaptic axon terminal.
   c. size of the synapse.
   d. nature of the postsynaptic receptors.
   e. concentration of neurotransmitter in the synaptic space.
10. Which of the following is a likely mechanism for long-term potentiation?
   a. When glutamate binds to postsynaptic AMPA receptors, it activates G proteins that trigger intracellular changes.
   b. When glutamate binds to NMDA receptors, it allows magnesium ions to enter the cell, which initiate intracellular changes.
   c. When sufficient glutamate is released by the presynaptic neuron, it causes an increase in the number of AMPA receptors on the postsynaptic cell.
   d. When sufficient glutamate is released, both AMPA and NMDA receptors are activated, and NMDA receptors allow Ca^{2+} as well as Na^{+} to enter the cell, thus initiating intracellular changes.
   e. When both glutamate and ACh are released together, they create a long-lasting depolarization of the postsynaptic cell.

FOR DISCUSSION

1. The language of the nervous system consists of one "word," the action potential. How can this single message convey a diversity of information, how can that information be quantitative, and how can it be integrated?
2. If you stimulate an axon in the middle, action potentials are conducted in both directions. Yet when an action potential is generated at the axon hillock, it goes only toward the axon terminals and does not backtrack. Explain why action potentials are bidirectional in the first example and unidirectional in the second.
3. The nature of synapses presents various opportunities for plasticity in the nervous system. Discuss at least four synaptic mechanisms that could be altered to change the response of a neuron to a specific input.
4. Benzodiazepines are drugs that act through GABA receptors and open Cl^{-} channels. What effects would you expect these drugs to have?

FOR INVESTIGATION

The patch clamping technique shown in Figure 44.8 produces an "inside-out" patch because the cytoplasmic side of the membrane faces away from the pipette tip and the surface side of the membrane is exposed to the solution in the pipette. Another patch preparation is the "outside-out patch" in which the cytoplasmic side of the membrane is exposed to the solution in the pipette. Describe what kind of experiments you could conduct with the "inside-out" patch, and what kind of experiments you could conduct with the "outside-out" patch.

CHAPTER 45 Sensory Systems

Out of range

A rattlesnake can see to strike and kill a running rodent in complete darkness. How can this be, when "seeing" means using the eyes to detect light, and "complete darkness" means the absence of any light? It is possible because these definitions are based on human capabilities. What we call "light" is actually only a small portion (red, orange, yellow, green, blue, indigo, and violet) of the spectrum of electromagnetic radiation. Other animals see wavelengths we cannot. For example, insects and birds see patterns on flowers that reflect ultraviolet wavelengths invisible to us. Similarly, rattlesnakes can "see" infrared wavelengths that are invisible to us (although at high enough levels of intensity, humans feel these wavelengths as heat).

It is not the snake's eyes that perceive infrared wavelengths. Rattlesnakes and their relatives have *pit organs* containing high densities of infrared-sensitive neurons. The two pits, located between the nostrils and the eyes on each side of the skull, are positioned in such a way that sensory receptor cells in the pits receive directional information. The fields of "view" of the bilateral pits are overlapping and thus convey a three-dimensional perspective. Information from the pit organs goes to the same region of the brain as information from the eyes, so rattlesnakes actually do "see" the world in a range of electromagnetic radiation that is different from the human visual spectrum.

Our definition of silence is as arbitrary as our definition of darkness. "Sound" is actually pressure waves in the environment, and many animals are sensitive to pressure waves with frequencies we cannot hear. Elephants communicate in sound waves that are below human hearing range; such long waves travel great distances, an advantage to large animals that roam over extensive areas. Bats emit incredibly loud, short wavelength sound pulses that are far above our range of hearing, and a flying bat hears echoes of these pulses bouncing off objects in the environment. The pulses are so loud and the echoes are so weak that it is rather like a construction worker trying to overhear a whispered conversation while using a pneumatic drill. Why don't the loud pulses "drown out" the weak echoes for the bat? Small muscles in the bat's ears contract to dampen their hearing sensitivity while the sounds are being emitted, but

Sensing Infrared Radiation The hole to the left of this Aruba rattlesnake's eye is one of its bilateral pit organs. These organs detect infrared radiation from their preferred prey—small rodents—with unerring precision, even in total darkness. The forked tongue also provides positional information, picking up molecular signals which are transmitted to the brain by a specialized organ in the roof of the snake's mouth.

Echolocating around an Obstacle Course A bat's ability to echolocate using sound waves is so precise that, in a totally dark room strung with fine wires, bats can capture small insects while avoiding the wires.

relax in time for the bat to hear the echo—a truly remarkable ability, since the pulses are emitted at rates of 20–80 per second.

Our senses are our windows on the world. "Reality" is what our eyes see, our ears hear, our noses smell, what we touch and taste. But human beings sense only a limited range of the information available. Animals with different ranges of sensitivity process different sources of information and may perceive "reality" quite differently.

IN THIS CHAPTER we will examine the general properties of sensory receptor cells and see how they convert environmental stimuli into the electrochemical signals of nervous systems. We will examine in detail the diversity of cells responsible for our senses and see how they are incorporated into sensory systems that provide the central nervous system with information about the world around and within us. In the course of our study of sensory systems, we will learn about the unusual sensory abilities of many other animals.

CHAPTER OUTLINE

45.1 How Do Sensory Cells Convert Stimuli into Action Potentials?

Sensory cells convert, or *transduce*, physical and chemical stimuli such as light waves, sound waves, touch, pain, and odorant and taste molecules into neuronal signals. These signals are then transmitted to the central nervous system for processing and interpretation.

Sensory receptor proteins act on ion channels

As we learned in Section 44.2, cells use energy to create large gradients of charged ions across their plasma membranes. In this way, cells are like batteries: batteries store potential energy by separating electrical charges between their poles, and cells store potential energy in the ionic gradients across their plasma membranes. If the two poles of a battery are connected by a wire and a switch, current flows through the wire when the switch is closed. Ion channels are like switches. When they open, charged ions can flow down their electrochemical gradient. When the ion channel is selective for one type of ion, the flow of that charged ion creates a change in the electrical potential across the membrane. **Sensory transduction** begins with a membrane receptor protein on a sensory receptor cell that can detect a specific stimulus such as heat, light, chemicals, mechanical force, or electrical fields. The **receptor protein** then directly or indirectly opens or closes ion channels, affecting the resting potential of the cell. This change in resting potential then either causes the sensory cell to fire an action potential or, through release of neurotransmitter, leads an associated neuron one or even two synapses away to fire action potentials.

In Section 44.3 we learned that synaptic receptor proteins are either *ionotropic* or *metabotropic*; the same distinction can be applied to sensory receptor proteins. Ionotropic sensory receptor proteins are either ion channels themselves, or they directly affect the opening of an ion channel. Examples are receptors that respond to physical force (called *mechanoreceptors*), to temperature (*thermoreceptors*), and probably to salty taste. Although *electrosensors* most likely have no receptor protein at all, they are grouped with the ionotropic receptors; the membrane of the sensory cell is sensitive to the voltage across it and releases neurotransmitter in response to slight changes in membrane potential. Metabotropic sensory receptor proteins influence ion channels indirectly, through G proteins and sec-

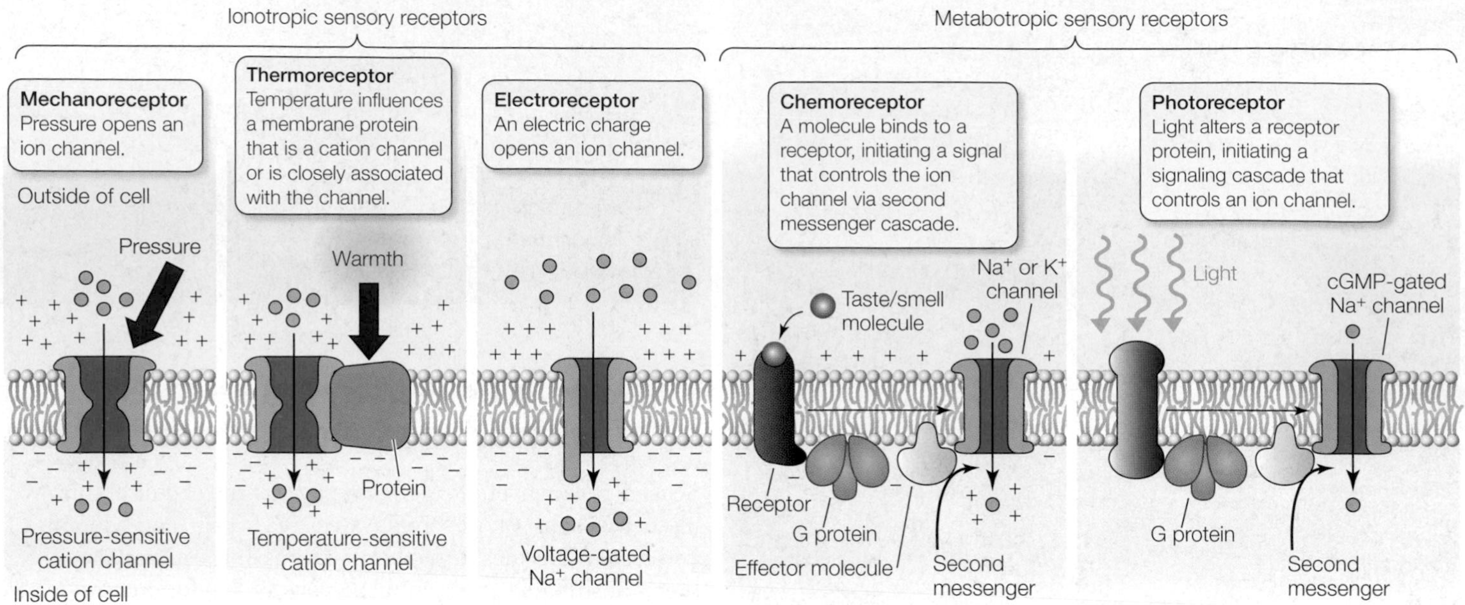

45.1 Sensory Cell Membrane Receptor Proteins Respond to Stimuli The receptor proteins in mechanoreceptors are ion channels. The activated receptor proteins of metabotropic chemoreceptors and photoreceptors initiate signal transduction cascades that eventually open or close ion channels.

ond messengers. Examples are *chemoreceptors* and *photoreceptors* (**Figure 45.1**).

Sensory transduction involves changes in membrane potentials

In this chapter we will examine several sensory systems. In each case, we can ask the same general question: How do **sensory receptor cells**, also known simply as *receptors* or *sensory cells*, transduce energy from a stimulus into a change in membrane potential? The details differ for different receptors, but those details all fit into one general pattern.

A change in the resting membrane potential of a sensory receptor cell in response to a stimulus is called a **receptor potential**. Receptor potentials can spread by local current flow over short distances, but to travel long distances in the nervous system, they must be converted into action potentials. Receptor potentials produce action potentials in two ways: by generating action potentials within the receptor cell itself, or by causing the release of a neurotransmitter that induces an associated neuron to generate action potentials.

A good example of a sensory receptor cell that can generate action potentials is the stretch receptor of a crayfish (**Figure 45.2**).

By placing an electrode in the dendrites or cell body of a crayfish stretch receptor cell, we can record the receptor potentials that result from stretching of the muscle to which the dendrites of the cell are attached. These receptor potentials spread to the base of the cell's axon (the axon hillock), which contains voltage-gated sodium channels. Action potentials generated here travel down

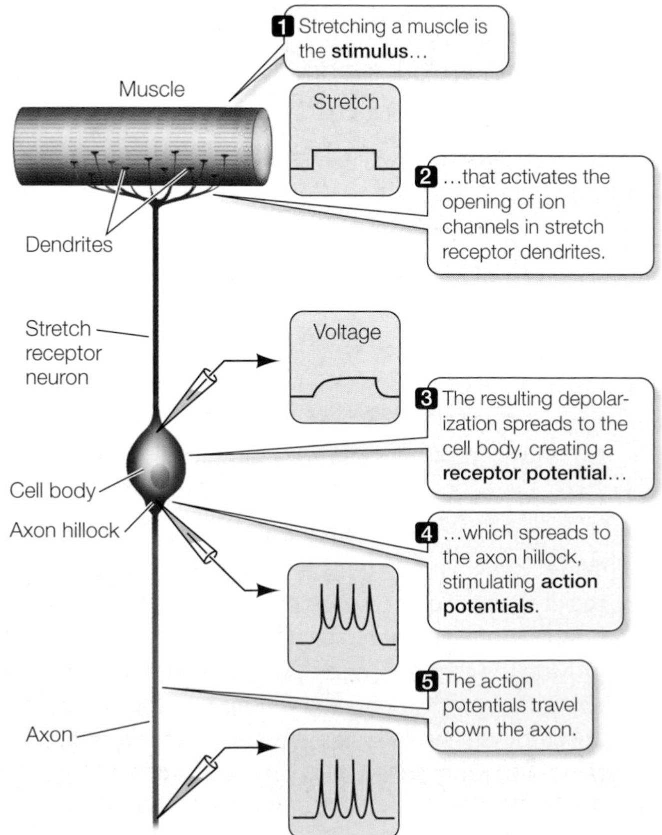

1 Stretching a muscle is the stimulus...
Muscle
Stretch
2 ...that activates the opening of ion channels in stretch receptor dendrites.
Dendrites
Stretch receptor neuron
Voltage
3 The resulting depolarization spreads to the cell body, creating a receptor potential...
Cell body
Axon hillock
4 ...which spreads to the axon hillock, stimulating action potentials.
5 The action potentials travel down the axon.
Axon

45.2 Stimulating a Sensory Cell Produces a Receptor Potential Signal transduction in the stretch receptor of a crayfish can be investigated by measuring the membrane potential at different places on the stretch receptor neuron while stretching the muscle innervated by that sensory neuron.

the axon to the CNS. The rate at which action potentials are fired by the axon depends on the magnitude of the receptor potential; that, in turn, depends on how much the muscle is stretched.

In a receptor cell that does not fire action potentials, the spreading receptor potential reaches a presynaptic patch of plasma membrane and induces the release of a neurotransmitter. The intensity of the stimulus influences how much neurotransmitter is released. That neurotransmitter binds to receptor proteins on an associated sensory neuron, altering its membrane potential and perhaps causing it to increase or decrease its rate of firing action potentials. In a few cases, this second cell also responds only by changing the rate at which it releases neurotransmitter onto another neuron. Eventually, however, the stimulation of the sensory cell will be coded as a change in firing of action potentials in a sensory circuit.

Sensation depends on which neurons receive action potentials from sensory cells

All the sensory systems process information in the form of action potentials. So how do we perceive different sensations? Sensations such as heat, pressure, pain, light, smell, and sound differ because the messages from different kinds of sensory cells arrive at different places in the CNS. Action potentials arriving in the visual cortex of the brain are interpreted as light, in the auditory cortex as sound, in the olfactory bulb as smell, and so forth.

A small patch of skin on your arm contains some sensory receptor cells that increase their firing rates when the skin is warmed and others that increase their activity when the skin is cooled. Other types of sensory cells in the same patch of skin respond to touch, movement of hairs, irritants such as mosquito bites, and the pain from cuts or burns. These receptor cells transmit their messages through axons that enter the CNS at the spinal cord. The synapses made by those axons in the spinal cord and the subsequent pathways of transmission determine whether the stimulation of the patch of skin on your arm is perceived as warmth, cold, touch, tickle, itch, or pain. So, even though the action potentials carried by all of these sensory axons look the same, the connectivity of each axon is specific for a given sensory modality.

How is the intensity of the stimulus encoded if, as Section 44.2 describes, each action potential is an all-or-none event? Intensity of sensation is coded as the frequency of the action potentials.

Some sensory cells transmit information about internal conditions in the body, but we may not be consciously aware of that information. The brain continuously receives information about body temperature, blood carbon dioxide and oxygen concentrations, arterial pressure, muscle tension, and the positions of the limbs—all of which are important for the maintenance of homeostasis. All sensory cells produce information that the nervous system can use, but that information does not always result in conscious sensation.

Some sensory receptor cells are assembled with other types of cells into *sensory organs*, such as eyes, ears, and noses, that enhance the ability of the sensory cells to collect, filter, and amplify stimuli. We therefore refer to **sensory systems**, which include the sensory cells, the associated structures, and the neuronal networks that process the information.

Many receptors adapt to repeated stimulation

Some sensory cells give gradually diminishing responses to maintained or repeated stimulation. This phenomenon is known as **adaptation**, and it enables an animal to ignore background or unchanging conditions while remaining sensitive to changes or to new information. (Note that this use of the term "adaptation" is different from its application in an evolutionary context.) When you dress, you feel each item of clothing touch your skin, but the sensation of clothes touching your skin is not constantly on your mind throughout the day. You are immediately aware, however, when a seam rips, your shoe comes untied, or someone touches your back.

Animals can discriminate between continuous and changing stimuli partly because some sensory cells adapt; it is also a result of information processing by the CNS. Some sensory cells adapt very little or very slowly; examples are some types of pain receptors and the mechanoreceptors for balance.

In the rest of this chapter we will learn how sensory systems gather and filter stimuli, transduce specific stimuli into action potentials, and transmit action potentials to the CNS.

45.1 RECAP

Sensory receptor cells have receptor proteins that respond to specific stimuli from the external or internal environment by opening or closing ion channels, which results in the generation of action potentials in sensory neurons.

- Can you explain the difference between ionotropic and metabotropic sensory receptor proteins? See p. 965 and Figure 45.1
- How are we able to perceive action potentials—which are all essentially the same—as different sensations? See p. 967

Now that we have a general view of how sensory systems code and process information, we will look in more depth at specific sensory modalities, beginning with chemosensation—the basis of smell and taste.

45.2 How Do Sensory Systems Detect Chemical Stimuli?

A colony of corals responds to a small amount of meat extract in the seawater bathing it by extending bodies and tentacles and searching for food. A solution of a single amino acid can stimulate this response. Conversely, a small amount of sea water from a container in which corals were crushed will stimulate a defensive retraction of the coral polyps. Humans also react strongly to certain chemical stimuli. When we smell freshly baked bread, we salivate and feel hungry, but when we smell diamines from rotting meat we gag and feel nauseated. All animals receive information about chemical stimuli through **chemoreceptors**, which are receptor proteins that bind various ligands and are responsible for smell and taste. Chemoreceptors are also responsible for monitoring aspects of the internal environment such as the level of carbon dioxide in the blood. Information from chemoreceptors can cause powerful physiological behavioral and responses.

Arthropods provide good examples for studying chemoreception

Arthropods use chemical signals to attract mates. These signals, called **pheromones**, demonstrate the sensitivity of chemosensory systems. One of the best-studied examples of this phenomenon is the silkworm moth.

> Flies have chemoreceptors on their feet that respond to sugars, amino acids, salts, and even distilled water. A fly tastes a potential food by stepping in it.

To attract a mate, the female silkworm moth (*Bombyx mori*) releases a pheromone called *bombykol* from a gland at the tip of her abdomen. The male silkworm moth has receptors for this molecule on his antennae (**Figure 45.3**). Each feathery antenna carries about 10,000 bombykol-sensitive hairs. A single molecule of bombykol may be sufficient to generate action potentials in the antennal nerve that transmits the signal to the CNS. Because of the male's great sensitivity, the sexual message of a female moth is likely to reach any male within a downwind area stretching over several kilometers. When approximately 200 hairs per second are activated, the male flies upwind in search of the female. Because the rate of firing in the male's sensory nerves is proportional to the bombykol concentration in the air, he can follow the airborne concentration gradient and home in on the signaling female.

Olfaction is the sense of smell

The sense of smell, **olfaction**, depends on chemoreceptors. In vertebrates, the olfactory sensors are neurons embedded in a layer of epithelial tissue at the top of the nasal cavity. These neurons project their axons to the olfactory bulb of the brain, while their dendrites end in olfactory hairs on the surface of the nasal epithelium. A protective layer of mucus covers the epithelium. Molecules from the environment must diffuse through this mucus to reach the receptor proteins on the olfactory hairs. When you have a cold, the amount of mucus in your nose increases, and the epithelium swells. With this in mind, study **Figure 45.4**, and you will easily understand why respiratory infections can cause you to lose your sense of smell. Humans have a fairly sensitive olfactory system, but we are unusual among mammals in that we depend more on vision than on olfaction; a dog has up to 40 million nerve endings per square centimeter of nasal epithelium, many more than we do.

An **odorant** is a molecule that activates an olfactory receptor protein. Odorants bind to receptor proteins on the olfactory cilia of the olfactory neurons. Olfactory receptor proteins are specific for particular odorant molecules. When an odorant molecule binds to its receptor on an olfactory neuron, it activates a G protein. The G protein in turn activates an enzyme that causes an increase of a second messenger (cAMP in vertebrates) in the cytoplasm (see Figures 15.17 and 44.16). The second messenger binds to cation channels in the sensory cell's plasma membrane and opens them, causing an influx of Na^+. The sensory neuron depolarizes to threshold and fires action potentials.

The olfactory world has an enormous number of odors, and accordingly, there are a large number of olfactory receptor proteins. In the 1990s, Linda Buck and Richard Axel discovered in mice a family of about a thousand genes (about 3 percent of the genome) that code for olfactory receptor proteins. Humans have about one-third that number of functional olfactory receptor genes. Each receptor protein that is expressed is found in a limited number of sensory receptor cells in the olfactory epithelium, and each cell expresses just one receptor. Using a combination of patch clamping and molecular techniques, the investigators were able to match specific gene products with the odorants they detect.

45.3 Some Scents Travel Great Distances Mating in silkworm moths of the genus *Bombyx* is coordinated by a pheromone called bombykol.

(A)
The female moth releases a pheromone from a gland at the tip of her abdomen. The pheromone can travel thousands of meters downwind.

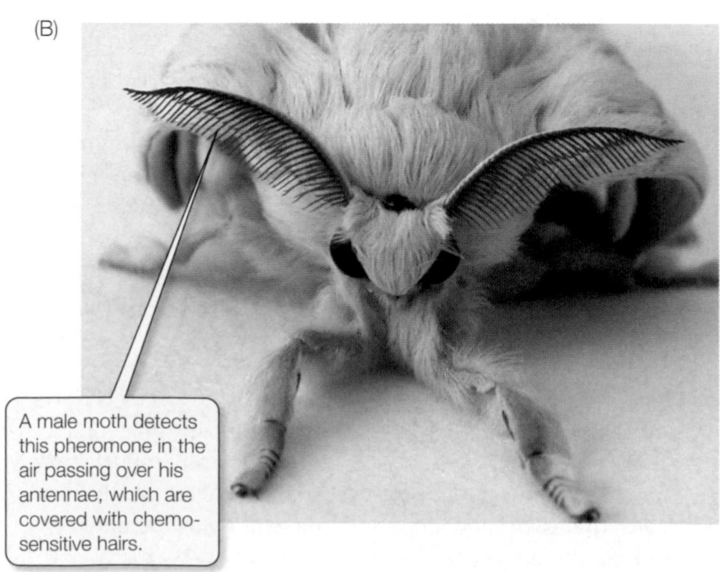

(B)
A male moth detects this pheromone in the air passing over his antennae, which are covered with chemo-sensitive hairs.

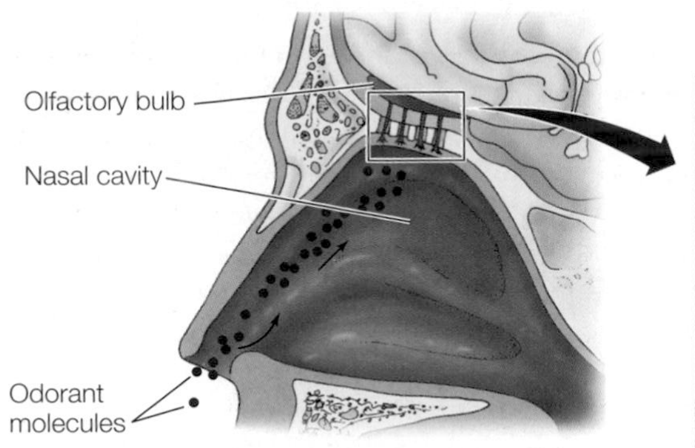

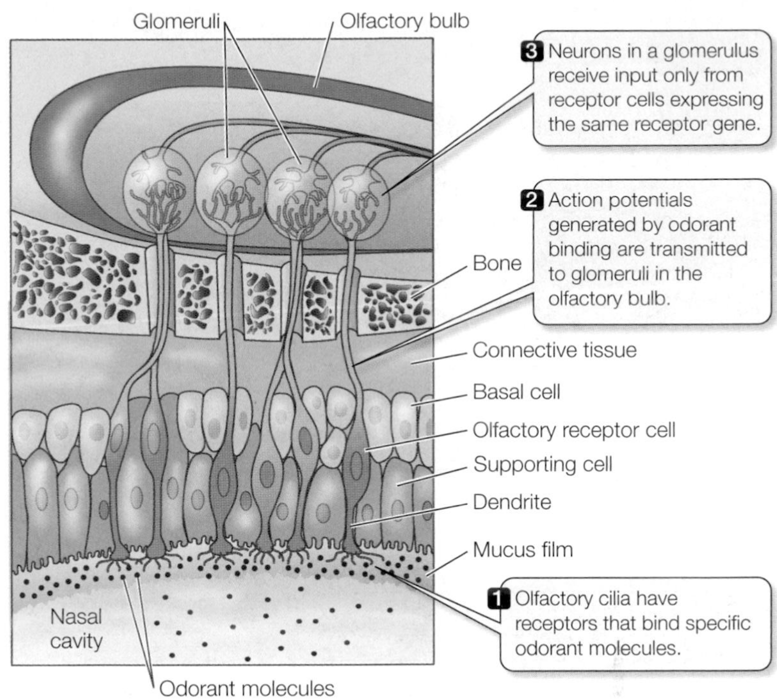

45.4 Olfactory Receptors Communicate Directly with the Brain The receptor cells of the human olfactory system are embedded in epithelial tissues lining the nasal cavity and send their axons to the olfactory bulb of the brain.

Olfactory sensitivity, however, enables discrimination of many more odorants than there are olfactory receptors. An odorant molecule can be quite complex, and different regions of that molecule may bind to different receptor proteins. The next stage of processing of olfactory information is in the olfactory bulb of the brain. In the olfactory bulb, axons from neurons expressing the same receptor protein converge on *glomeruli* (see Figure 45.4), which are clusters of olfactory bulb neurons. Therefore, a complex odorant molecule can activate a unique combination of glomeruli in the olfactory bulb, so an olfactory system with hundreds of different receptor proteins can discriminate an astronomically large number of smells. For their discoveries of the molecular nature of the olfactory system, Buck and Axel received the Nobel prize in 2004.

How does the olfactory receptor cell signal the intensity of a smell? The more odorant molecules that bind to receptors, the greater the frequency of action potentials and the greater the intensity of the perceived smell.

The vomeronasal organ senses pheromones

The **vomeronasal organ** (**VNO**) is a small, paired tubular structure embedded in the nasal epithelium of amphibians, reptiles, and many mammals (maybe even humans). In mammals, the VNO is located on the septum dividing the two nostrils. The VNO has a pore that opens into the nasal cavity. When the animal sniffs, the VNO pulsates and draws a sample of nasal fluid over the chemoreceptors embedded in its walls. The information from these chemoreceptors goes to an accessory olfactory bulb in the brain, and information from there goes to brain regions involved in sexual and other instinctive behaviors.

In 2002, Lawrence Katz and his colleagues at Duke University recorded the neuronal activity in the accessory olfactory bulbs of conscious, behaving mice. To test the hypothesis that the VNO detects pheromones, they placed a mouse attached to recording electrodes with other mice of same or different gender and same or different strain. These neurons were activated when the mouse with the implanted electrodes sniffed another mouse that was placed in the same cage. However, the neurons fired differentially depending on the gender and strain of the strange mouse. These observations support the hypothesis that the VNO is a specialized olfactory organ for pheromones.

In snakes, the VNO opens into the roof of the mouth cavity. Each time the snake's forked tongue darts in and out, the forks fit into the VNO openings and present to the chemoreceptors a sample of molecules from the surrounding air (see the chapter-opening photo). Thus the snake uses its tongue to smell its environment, not to taste it. Why doesn't the snake simply use the flow of air to and from its lungs, as we do, to smell the environment? In reptiles, air flows to and from the lungs slowly (and can even stop entirely for long periods of time), but the tongue can dart in and out many times in a second. It is a quick source of olfactory information.

Gustation is the sense of taste

The sense of taste, or **gustation**, in humans and other vertebrates depends on clusters of chemoreceptor cells called **taste buds**. The taste buds of terrestrial vertebrates are confined to the mouth cavity, but some fishes have taste buds in the skin that enhance their ability to sense their environment. Some fishes living in murky water are very sensitive to small amounts of amino acids in the water around them and can find food without the use of vision. The duck-billed platypus, a prototherian mammal (see Figure 33.24), has similar talents as a result of taste buds on the sensitive skin of its bill.

A human tongue has approximately 10,000 taste buds. The taste buds are embedded in the epithelium of the tongue, and most are found on the papillae of the tongue. (Look at your tongue in a mirror—the papillae make it look fuzzy.) Each papilla has many taste buds—mostly on the sides. The outer surface of a taste bud has a pore that exposes the tips of the sensory receptor cells. Microvilli

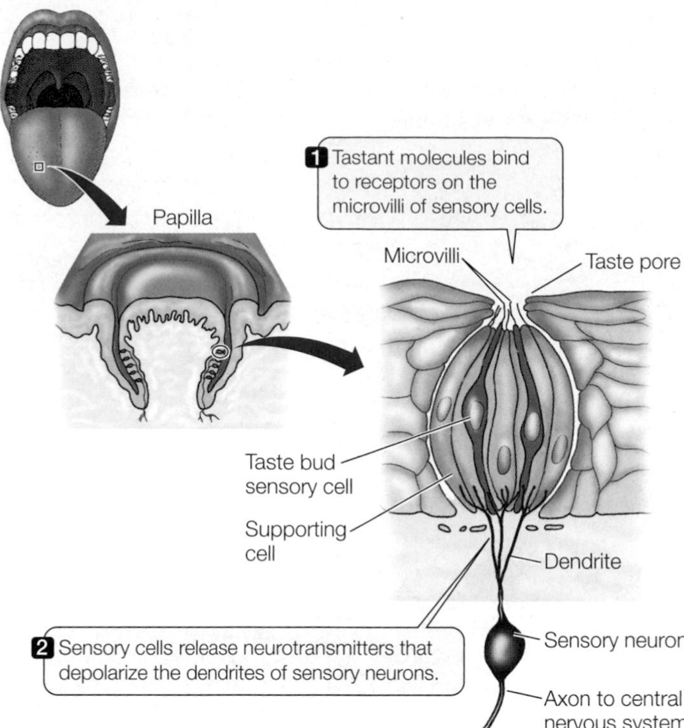

1 Tastant molecules bind to receptors on the microvilli of sensory cells.
Papilla
Microvilli
Taste pore
Taste bud sensory cell
Supporting cell
Dendrite
2 Sensory cells release neurotransmitters that depolarize the dendrites of sensory neurons.
Sensory neuron
Axon to central nervous system

45.5 Taste Buds Are Clusters of Sensory Cells Each taste bud contains a number of sensory cells.

(tiny hairlike projections) increase the surface area of these cells where their tips converge at the pore (**Figure 45.5**). The cells do not generate action potentials, but changes in their membrane potential influence the release of neurotransmitter at their base where they form synapses with sensory neurons.

The tongue does a lot of hard work, so its epithelium, along with cells of its taste buds, are shed and replaced at a rapid rate. Individual taste bud cells last about 10 days before they are replaced, but the sensory neurons associated with them live on, always forming new synapses as new taste buds form.

You may have heard that humans can perceive only four tastes: sweet, salty, sour, and bitter. However, taste buds can distinguish among a variety of sweet-tasting molecules and a variety of bitter-tasting molecules. Recently small families of receptor protein genes for sweet and bitter tastes have been discovered. In addition, we now recognize a fifth taste, called **umami**. Umami is a savory, meaty taste that originates from receptors for amino acids, including monosodium glutamate (MSG). The full complexity of the chemosensitivity that enables us to enjoy the subtle flavors of food comes from the combined activation of gustatory and olfactory receptors, which is why you lose much of your sense of taste when you have a cold.

Gustation begins with receptor proteins in the membranes of the microvilli. The nature of these proteins and the mechanisms by which they depolarize the sensory receptor cell differ for the different basic tastes. Saltiness receptors are ionotropic and simply respond to Na^+ diffusing through open Na^+ channels depolarizing the sensory cell. Sourness receptors are also probably ionotropic. Depolarization of these receptors is similarly due to a direct effect of H^+ ions on Na^+ channels. In contrast, sweetness and bitterness involve families of receptor proteins similar to those involved in olfaction, and they are metabotropic. The bitter taste probably evolved as a protective mechanism enabling animals to detect avoid toxic plant compounds such as quinine, caffeine, and nicotine. Since plants have evolved many such molecules to repel herbivorous predators, a variety of receptors is essential. Similarly a large number of molecules in food could indicate nutritional value, so a variety of receptors is of value. The diversity of sweet receptors helps explain why it has been possible to invent many different artificial sweeteners.

In all cases of taste sensation, changes in the membrane potential of sensory receptor cells cause the cells to release neurotransmitters onto the dendrites of the sensory neurons. Sensory neurons fire action potentials that are conducted to the CNS, where the information is interpreted as specific taste sensations.

45.2 RECAP

Chemoreceptors are the basis of the sensations of olfaction and gustation and the reception of pheromones.

- Why are we able to distinguish so many different smells? Can you see why some people experience more or different odors than others? See pp. 968–969 and Figure 45.4
- Can you describe how different substances in food are transduced into action potentials in taste buds? See pp. 969–970 and Figure 45.5

Chemoreceptors have diverse structures that bind to a tremendous variety of stimulus molecules. Mechanoreceptors, however, need respond only to physical forces. Nevertheless, a considerable diversity of mechanosensory cells and mechanisms has evolved.

45.3 How Do Sensory Systems Detect Mechanical Forces?

Mechanoreceptors are sensory cells that respond to mechanical forces. Physical distortion of a mechanoreceptor's plasma membrane causes ion channels to open, altering the membrane potential of the cell, which in turn leads to the generation of action potentials. The rate of action potentials tells the CNS the strength of the stimulus to the mechanoreceptor. Mechanoreceptor cells are involved in many sensory systems, ranging from interpreting skin sensations to sensing blood pressure.

Many different cells respond to touch and pressure

Objects touching the skin generate varied sensations because skin is packed with diverse mechanoreceptor cells (**Figure 45.6**). The most important tactile receptors found in both hairy and nonhairy skin are *Merkel's discs*, which adapt rather slowly and provide continuous information about things touching the skin. Other mechanoreceptors, called *Meissner's corpuscles*, found primarily in nonhairy skin, are very sensitive, but adapt rapidly, so they provide information

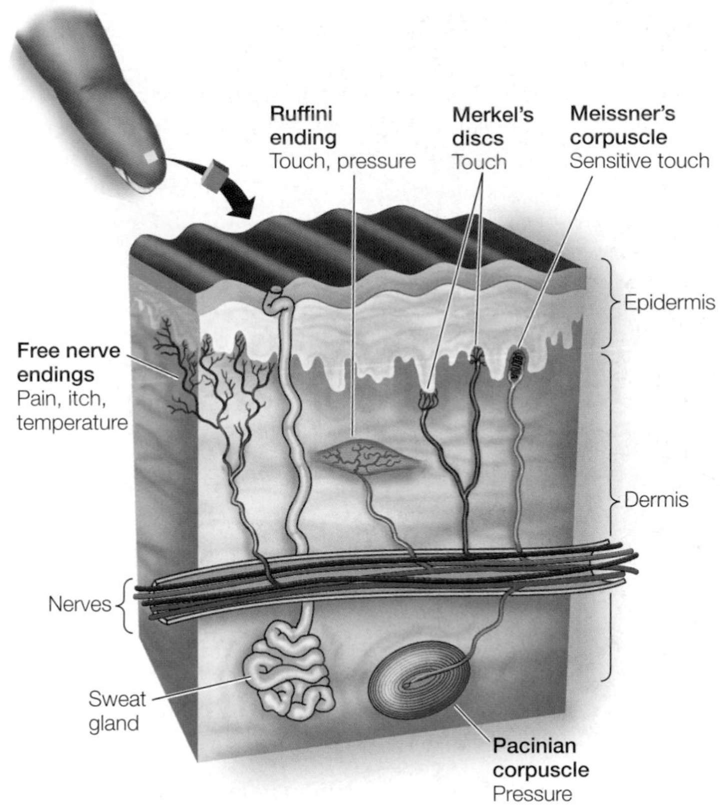

45.6 The Skin Feels Many Sensations Even a very small patch of skin contains a variety of sensory cells.

about changes in things touching the skin. The rapid adaptation of these tactile sensors is why you roll a small object between your fingers, rather than holding it still, to discern its shape and texture. As you roll it, you continue to stimulate Meissner's corpuscles.

Two other kinds of mechanoreceptor cells are found deeper in the skin. *Ruffini endings,* which adapt slowly, are good at providing information about vibrating stimuli of low frequencies. *Pacinian corpuscles,* which adapt rapidly, are good at providing information about vibrating stimuli of higher frequencies. Even deeper in the skin, dendrites of sensory neurons wrap around hair follicles. When the hairs are displaced, those neurons are stimulated.

The density of tactile mechanoreceptor cells varies across the surface of the body. A two-point spatial discrimination test demonstrates this fact. If you lightly touch someone's skin with two toothpicks simultaneously, you can determine how far apart the two stimuli have to be before the person can tell whether he or she is being touched by one or by two toothpicks. On the back, the stimuli have to be rather far apart before they are perceived as two discrete stimuli. The same test applied to the person's lips or fingertips reveals finer spatial discrimination; that is, the person can identify as separate two stimuli that are close together.

Mechanoreceptors are found in muscles, tendons, and ligaments

An animal receives information from mechanoreceptor cells about the position of its limbs and the stresses on its muscles and joints. These mechanoreceptors supply information continuously to the CNS, and this information is essential for postural control and the coordination of movements.

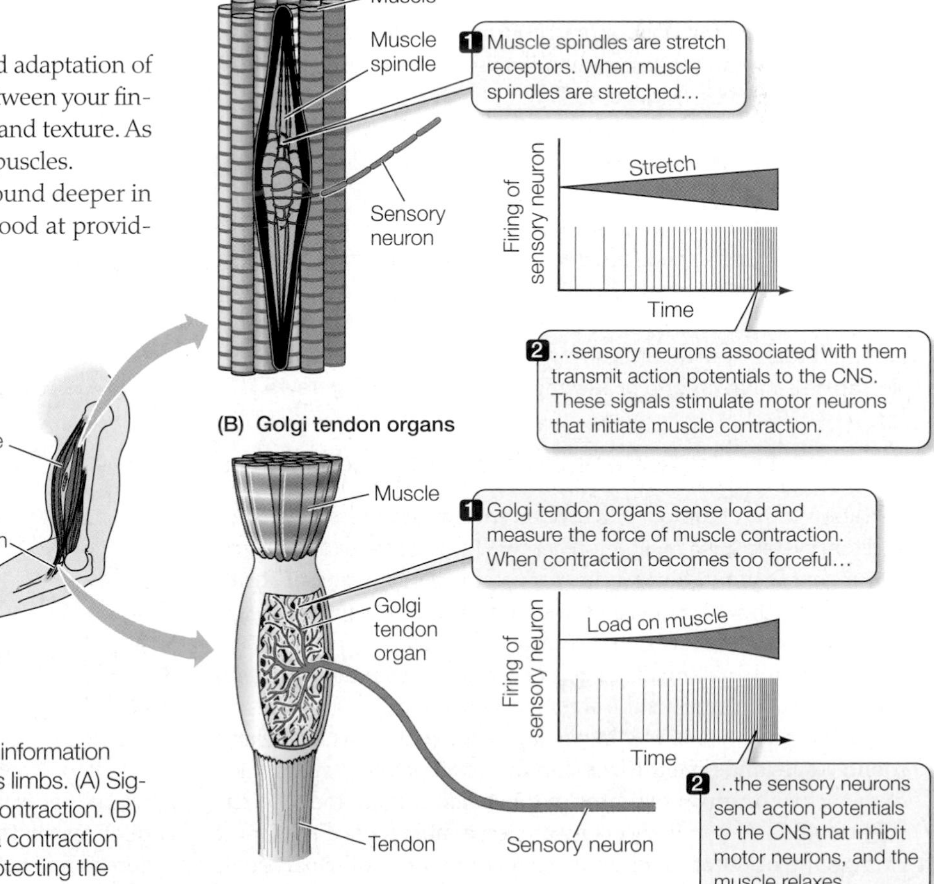

45.7 Stretch Receptors Stretch receptors provide information about the stresses on muscles and joints in an animal's limbs. (A) Signals from muscle spindles to the CNS initiate muscle contraction. (B) Golgi tendon organs in tendons and ligaments inhibit a contraction that becomes too forceful, triggering relaxation and protecting the muscle from tearing.

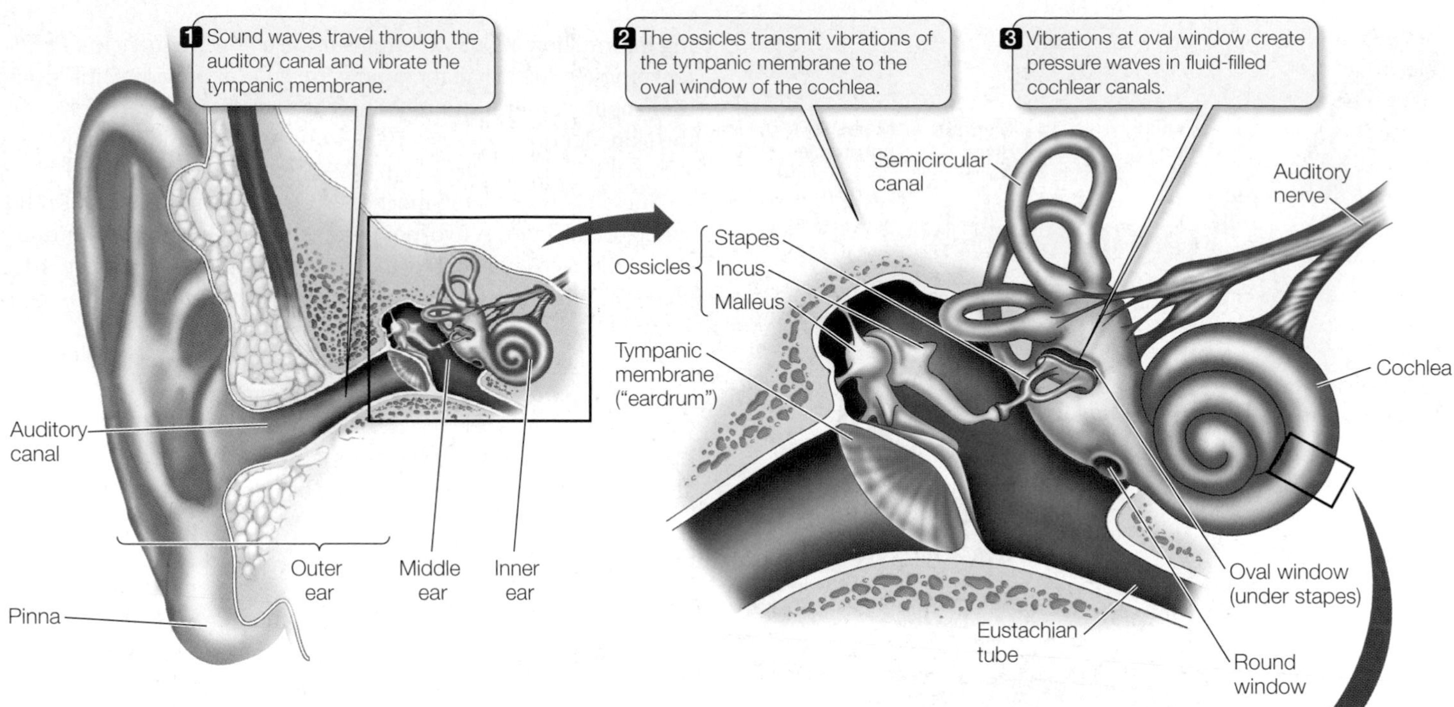

1 Sound waves travel through the auditory canal and vibrate the tympanic membrane.
2 The ossicles transmit vibrations of the tympanic membrane to the oval window of the cochlea.
3 Vibrations at oval window create pressure waves in fluid-filled cochlear canals.
Semicircular canal
Auditory nerve
Ossicles
Stapes
Incus
Malleus
Tympanic membrane ("eardrum")
Cochlea
Auditory canal
Outer ear
Middle ear
Inner ear
Pinna
Oval window (under stapes)
Eustachian tube
Round window

The mechanoreceptor cells found in skeletal muscle are called *muscle spindles*. These are **stretch receptors**, modified muscle cells that are embedded in connective tissue within muscles and innervated by sensory neurons. (Earlier in this chapter, we saw how crayfish stretch receptors transduce physical force into action potentials; see Figure 45.2.) The actions of muscle spindles are similar. Whenever the muscle is stretched, muscle spindles are also stretched, and the neurons transmit action potentials to the central nervous system (**Figure 45.7A**). The CNS uses this information to adjust the strength of contraction of the muscle to match the load put on the muscle. Thus a bartender can hold a beer mug in the same position as he fills it from the tap.

Another type of mechanoreceptor cell, the *Golgi tendon organ*, is found in tendons and ligaments and provides information about the force generated by a contracting muscle. When a contraction becomes too forceful, action potentials from the Golgi tendon organ inhibit the spinal cord motor neurons innervating that muscle, causing it to relax and protecting it from tearing (**Figure 45.7B**).

Auditory systems use hair cells to sense sound waves

The stimuli that animals perceive as sounds are pressure waves. **Auditory systems** use mechanoreceptors to convert pressure waves into receptor potentials. Auditory systems include special structures that gather sound waves, direct them to the sensory organ, and amplify their effect on the mechanoreceptors.

Human hearing provides a good example of an auditory system. The organs of hearing are the ears. The two prominent structures on the sides of our heads are the *pinnae*. The pinna of an ear collects sound waves and directs them into the *auditory canal*, which leads to the actual hearing apparatus in the *middle ear* and the *inner ear* (**Figure 45.8**). If you have ever watched a rabbit, a horse, or a cat change the orientation of its ear pinnae to focus on a particular sound, then you have witnessed the role of pinnae in hearing.

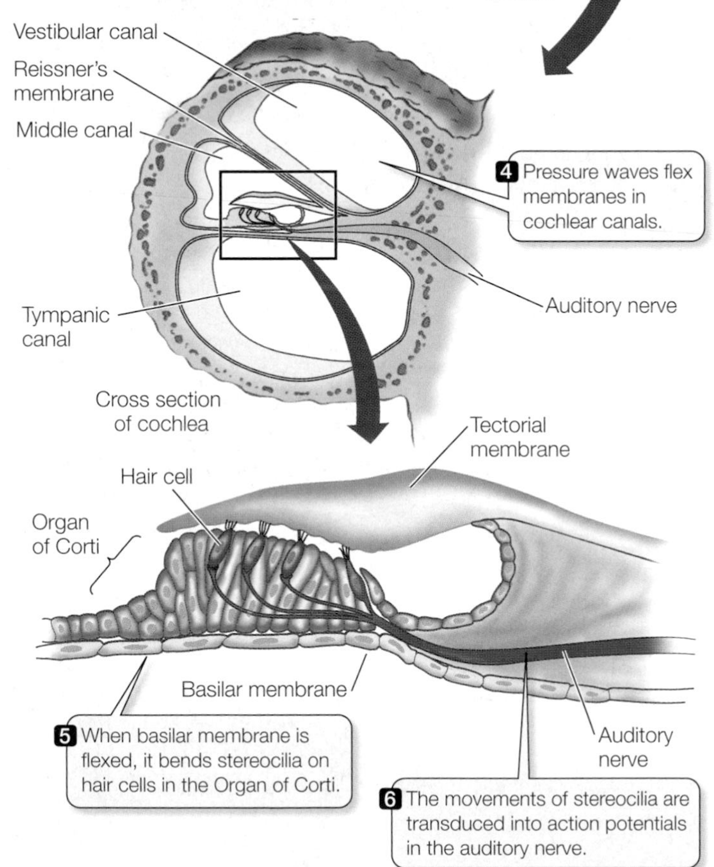

Vestibular canal
Reissner's membrane
Middle canal
4 Pressure waves flex membranes in cochlear canals.
Auditory nerve
Tympanic canal
Cross section of cochlea
Tectorial membrane
Hair cell
Organ of Corti
Basilar membrane
5 When basilar membrane is flexed, it bends stereocilia on hair cells in the Organ of Corti.
Auditory nerve
6 The movements of stereocilia are transduced into action potentials in the auditory nerve.

45.8 Structures of the Human Ear The human ear uses hair cells to transduce sound waves into action potentials.

The eardrum, or **tympanic membrane**, covers the end of the auditory canal. The tympanic membrane vibrates in response to pressure waves traveling down the auditory canal. The middle ear, an air-filled cavity, lies on the other side of the tympanic membrane.

The middle ear is open to the throat at the back of the mouth through the *eustachian tube*. Because the eustachian tube is also filled with air, pressure equilibrates between the middle ear and the outside world. When you have a cold or allergy, the tube can become blocked by mucus or by tissue swelling, so you have difficulty "clearing your ears," or equilibrating the pressure in the middle ear with the outside air pressure, which you have to do when ascending or descending in a plane or in a dive using SCUBA gear.

The middle ear contains three delicate bones called the **ossicles**, individually named the *malleus* (Latin, "hammer"), *incus* ("anvil"), and *stapes* ("stirrup"). The ossicles transmit the vibrations of the tympanic membrane to another flexible membrane called the **oval window**. The ossicles act as a lever—like a crow bar—translating a large movement of the tympanic membrane into a smaller movement of the oval window, but a movement of greater force. Also, because the oval window is much smaller than the tympanic membrane, the pressure the stapes transmits to the oval window is more than 20 times greater than the pressure exerted by the sound wave on the tympanic membrane. Behind the oval window lies the fluid-filled inner ear. Movements of the oval window result in pressure changes in the inner ear. These pressure waves in the inner ear are transduced into action potentials.

The inner ear is a long, tapered, coiled chamber called the **cochlea** (from Latin and Greek words for "snail" or "spiral shell"). A cross section of this chamber reveals that it is composed of three parallel canals separated by two membranes: **Reissner's membrane** and the **basilar membrane** (see Figure 45.8). Sitting on the basilar membrane is the **organ of Corti**, the apparatus that transduces pressure waves into action potentials. The organ of Corti contains *hair cells* with *stereocilia* that are in contact with an overhanging, rigid shelf called the *tectorial membrane*. Hair cells do not fire action potentials. However, they form synapses with associated sensory neurons whose axons make up the auditory nerve. When the basilar membrane flexes, the tectorial membrane bends the hair cell stereocilia. When the stereocilia bend, they alter the membrane potential of the hair cell and the rate at which it releases neurotransmitter onto its sensory neuron. As a result, the rates of action potentials traveling to the brain in the auditory nerve change.

What causes the basilar membrane to flex, and how does this mechanism distinguish sounds of different frequencies? In **Figure 45.9**, the cochlea is shown uncoiled to make it easier to understand its structure and function. The upper and lower canals separated by the basilar membrane are joined at the distal end of the cochlea (the end farthest from the oval window), making one continuous canal that turns back on itself. Just as the oval window is a flexible membrane at the beginning of the cochlea, the **round window** is a flexible membrane at the end of the long cochlear canal.

Air is highly compressible, but fluids are not. Therefore, a pressure wave can travel through air without much displacement of the air, but a pressure wave in fluid causes displacement of the fluid. When the stapes pushes on the oval window, the fluid in the upper canal of the cochlea is displaced. If this movement of the oval window occurs slowly, the cochlear fluid pressure wave travels down the upper canal, around the bend, and back through the lower canal. At the end of the lower canal, the displacement pressure is dissipated by the outward bulging of the round window.

If the oval window vibrates in and out rapidly, however, the waves of fluid pressure do not travel all the way to the end of the upper canal and back through the lower canal. Instead, they take a shortcut by crossing the basilar membrane, producing a traveling wave that flexes the basilar membrane. The more rapid the vibration, the greater is the amplitude of the flexion of the basilar membrane closer to the oval and round windows. Slower vibrations cause the largest amplitude flexions of the basilar membrane farther from the oval and round windows. Thus, different pitches of sound flex the basilar membrane at different locations and activate different sets of hair cells. Action potentials stimulated by the mechanoreceptors at different positions along the organ of Corti travel to the brain stem along the auditory nerve.

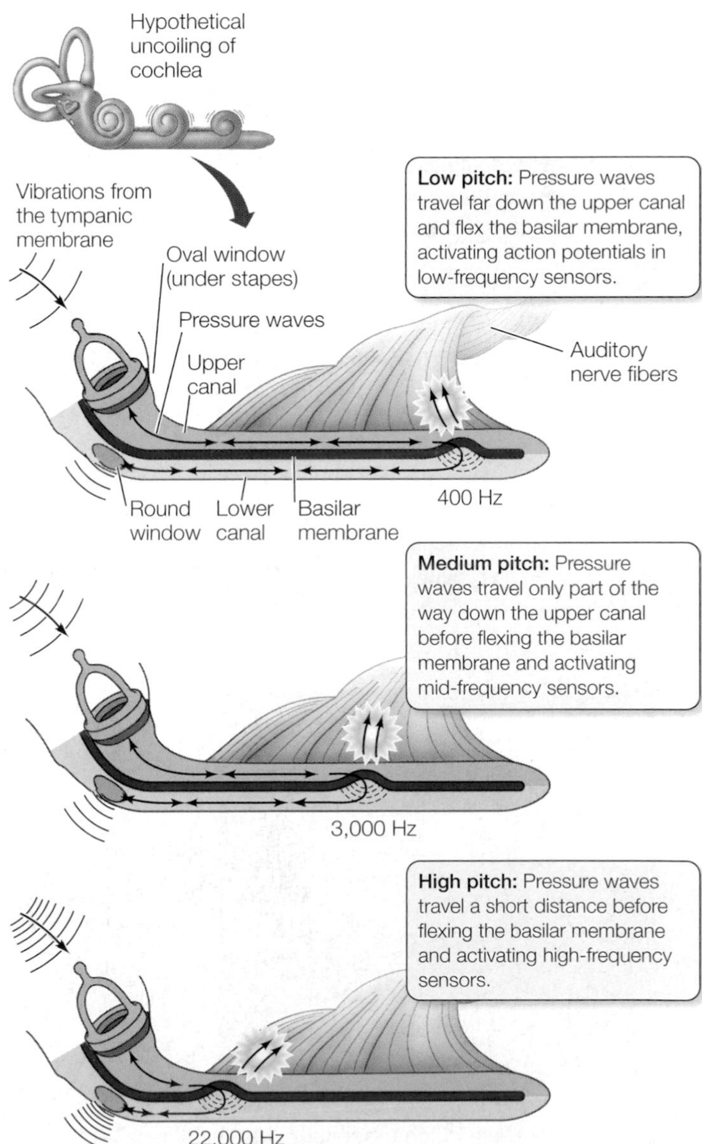

45.9 Sensing Pressure Waves in the Inner Ear Pressure waves of different frequencies flex the basilar membrane at different locations. Information about sound frequency is specified by which hair cells are activated. For simplicity, this representation illustrates the cochlea as uncoiled, and leaves out the middle canal.

Deafness, the loss of the sense of hearing, has two general causes. *Conduction deafness* is caused by the loss of function of the tympanic membrane and/or the ossicles of the middle ear. Repeated infections of the middle ear can cause scarring of the tympanic membrane and stiffening of the connections between the ossicles. The consequence is less efficient conduction of sound waves from the tympanic membrane to the oval window. With increasing age, the ossicles progressively stiffen, resulting in a gradual loss of the ability to hear high-frequency sounds. *Nerve deafness* is caused by damage to the inner ear or the auditory pathways. A common cause of nerve deafness is damage to the hair cells of the delicate organ of Corti by exposure to loud sounds such as jet engines, pneumatic drills, or highly amplified music. This damage is cumulative and irreversible.

Earphones can put you at risk for hearing loss. Although they are small, they can generate high-pressure sound waves close to your tympanic membrane. Consistent exposure to sounds above 85 decibels can damage hearing. Personal stereo earphones can reach 120 decibels, and people commonly use them at 100 decibels (equivalent to being at a rock concert).

Hair cells provide information about displacement

Hair cells are the mechanoreceptors found in organs of hearing and, as we will see, equilibrium. Projecting from the surface of each hair cell is a set of stereocilia, which looks like a set of organ pipes (**Figure 45.10A**). When these stereocilia (which are really microvilli) are bent, they alter ion channels in the hair cell's plasma membrane. When the stereocilia are bent in one direction, open channels close, and the membrane is hyperpolarized (the potential becomes more negative); when they are bent in the opposite direction, closed channels open, and the membrane is depolarized (more positive). When the membrane is adequately depolarized, the hair cell releases a neurotransmitter to the sensory neuron associated with it, and the sensory neuron sends action potentials to the CNS.

How does the bending of the stereocilia open ion channels? The ion channels that are opened are at the ends of the stereocilia. This was discovered by exploring the areas around the stereocilia with microelectrodes and seeing that local currents were created near the tips of the stereocilia when they were bent. Then, careful electron microscopic work revealed minute filaments that connected the tip of each stereocilium to its taller neighbor. It is hypothesized that these filaments are tethers on mechanoreceptor ion channels in the stereocilium plasma membrane, and they act like springs that open the channels. If the taller neighboring stereocilium is bent away, the spring tightens, and the ion channel is opened. If the taller neighbor bends toward its shorter neighbor, the spring is loosened and the channel closes (see **Figure 45.10B**).

Vertebrate organs of equilibrium use hair cells to detect the position of the body with respect to gravity. Within the mammalian inner ear, three **semicircular canals** at angles to one another sense the position and orientation of the head (see Figure 45.8). The **vestibular apparatus** has two chambers that sense the static position of the head as well as linear acceleration produced by movement. The structure and function of these organs are described in **Figure 45.11**.

A very simple use of hair cells to measure displacement can be seen in the **lateral line** sensory system of fishes. The lateral line is a canal just under the surface of the skin that runs down each side of the fish (**Figure 45.12**). Hair cells line the canal and their stereocilia, protected by capsules of gelatinous material called *cupulae* (singular *cupula*), project into the stream of water that flows

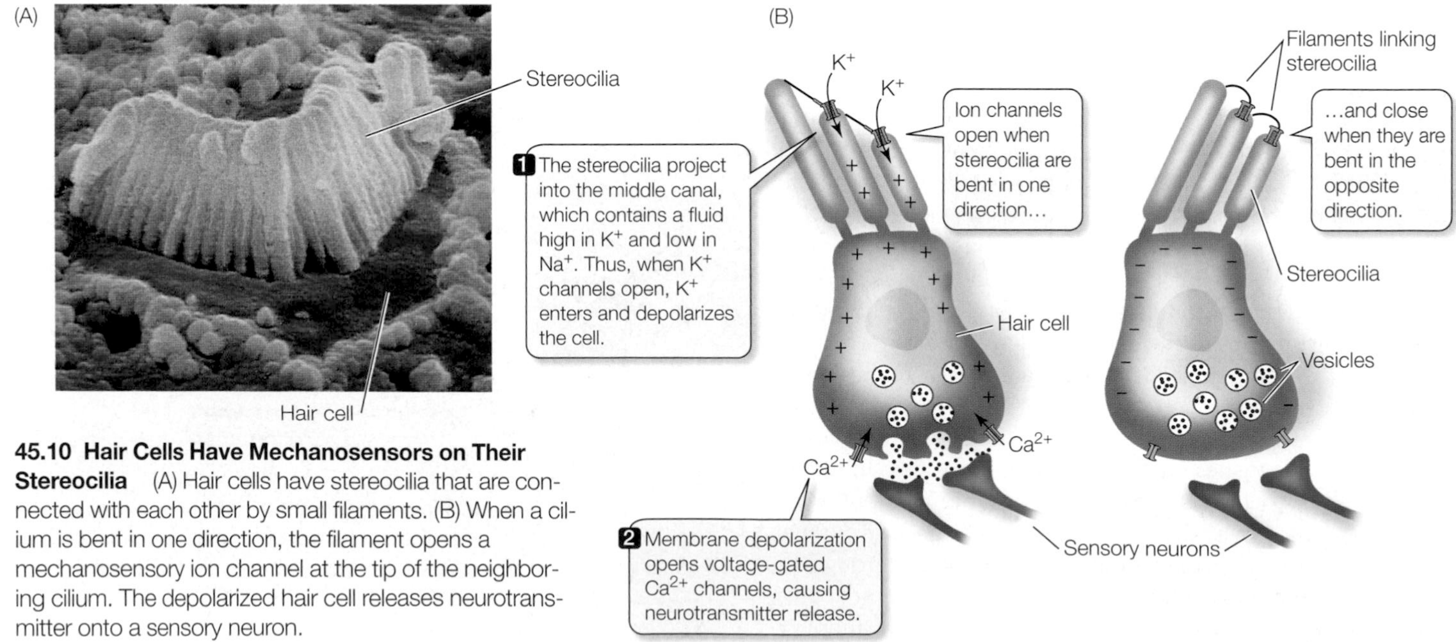

45.10 Hair Cells Have Mechanosensors on Their Stereocilia (A) Hair cells have stereocilia that are connected with each other by small filaments. (B) When a cilium is bent in one direction, the filament opens a mechanosensory ion channel at the tip of the neighboring cilium. The depolarized hair cell releases neurotransmitter onto a sensory neuron.

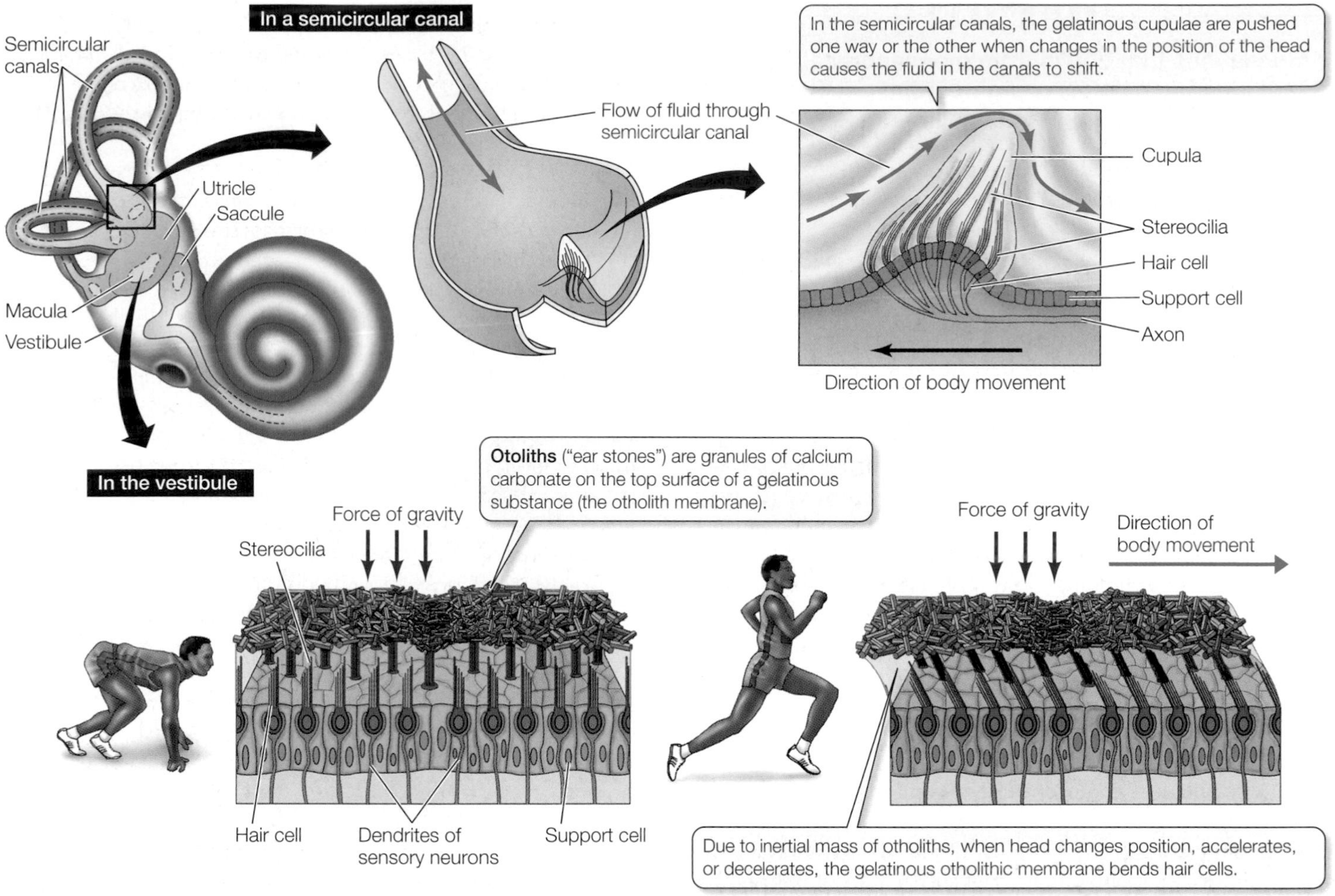

Semicircular canals
Utricle
Saccule
Macula
Vestibule
In a semicircular canal
Flow of fluid through semicircular canal
In the semicircular canals, the gelatinous cupulae are pushed one way or the other when changes in the position of the head causes the fluid in the canals to shift.
Cupula
Stereocilia
Hair cell
Support cell
Axon
Direction of body movement
In the vestibule
Otoliths ("ear stones") are granules of calcium carbonate on the top surface of a gelatinous substance (the otholith membrane).
Force of gravity
Stereocilia
Hair cell
Dendrites of sensory neurons
Support cell
Force of gravity
Direction of body movement
Due to inertial mass of otoliths, when head changes position, accelerates, or decelerates, the gelatinous otholithic membrane bends hair cells.

45.11 Organs of Equilibrium The bony inner ear of a human includes organs of equilibrium—three semicircular canals and two vestibular organs—as well as the cochlea, the auditory organ described in Figure 45.9.

through the lateral line canal when the fish moves through the water. Forward movement of the fish puts pressure on the cupulae, causing the stereocilia to bend in the direction that depolarizes the hair cells. Since water is incompressible, disturbances in the water around the fish are translated into pressure waves that can be picked up by the lateral line stereocilia. Thus, the lateral line system provides information about the fish's movements as well as about other moving objects, such as predators or prey. Sound in water also consists of pressure waves, so fish "hear" through their lateral lines.

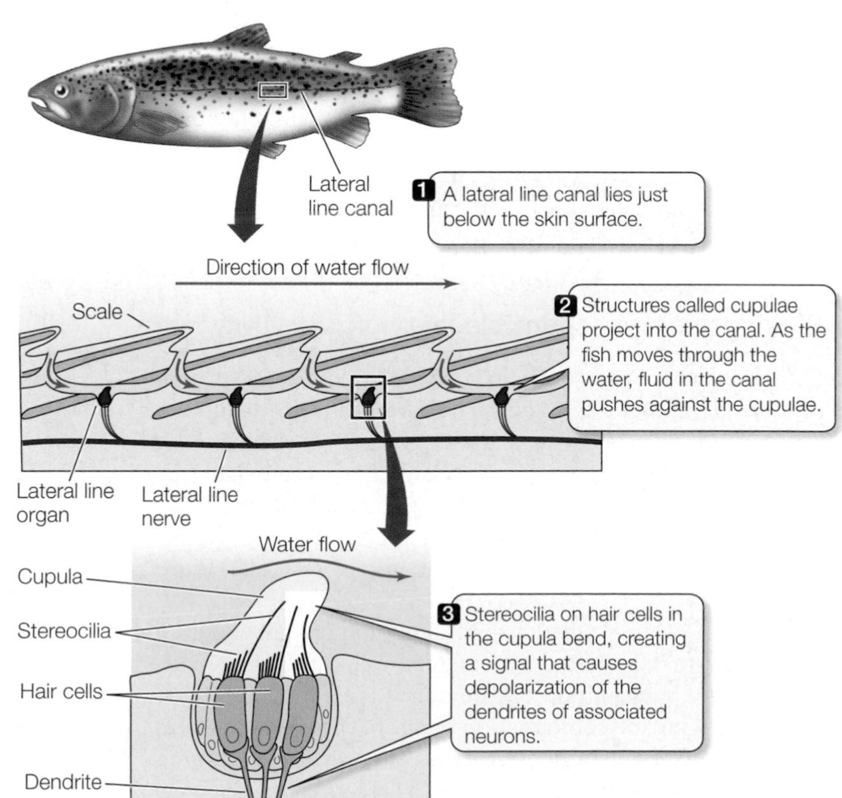

Lateral line canal
1 A lateral line canal lies just below the skin surface.
Direction of water flow
Scale
2 Structures called cupulae project into the canal. As the fish moves through the water, fluid in the canal pushes against the cupulae.
Lateral line organ
Lateral line nerve
Water flow
Cupula
Stereocilia
Hair cells
Dendrite
3 Stereocilia on hair cells in the cupula bend, creating a signal that causes depolarization of the dendrites of associated neurons.

45.12 The Lateral Line Acoustic System Contains Mechanosensors Hair cells in the lateral line of a fish detect movement of the water around the animal, giving the fish information about its own movements and the movements of objects nearby.

45.3 RECAP

Sensations that derive from mechanoreceptors include touch, tickle, pressure, joint position, muscle load, hearing, and equilibrium.

- Describe some of the different mechanoreceptor cells in the skin and their properties. See p. 971 and Figure 45.6
- How do different frequencies of sound result in action potentials being fired in different acoustic neurons? See pp. 972–973 and Figures 45.8 and 45.9
- Can you explain how hair cells transduce force into action potentials? See p. 974 and Figure 45.10

Chemoreception gave us good examples of metabotropic sensory receptors, and mechanoreception has given us good examples of ionotropic sensory receptors. Now we will turn to another example of metabotropic sensory reception, but one in which light energy capture is the stimulus. We will see how light enegy is converted into action potentials.

45.4 How Do Sensory Systems Detect Light?

Sensitivity to light—**photosensitivity**—confers on the simplest animals the ability to orient to the sun and sky and gives more complex animals rapid and extremely detailed information about objects in their environment. It is not surprising that both simple and complex animals can sense and respond to light. What is remarkable is that across the entire range of animal species, evolution has conserved the same basis for photosensitivity: a family of pigments called **rhodopsins**.

In this section we will learn how rhodopsin molecules respond when stimulated by light energy and how that response is transduced into neuronal signals. We will also examine the structures of eyes, the organs that gather light energy and focus it onto **photoreceptor cells**, the metabotropic sensory receptors that transform light energy into action potentials.

Rhodopsins are responsible for photosensitivity

Photosensitivity depends on the ability of rhodopsins to absorb photons of light and to undergo a change in conformation. A rhodopsin molecule consists of a protein, **opsin** (which alone is not photosensitive) and an associated nonprotein light-absorbing group, **11-*cis*-retinal**, cradled in the center of the opsin and bound covalently to it. The entire rhodopsin molecule sits within the plasma membrane of a photoreceptor cell (**Figure 45.13**).

When the 11-*cis*-retinal absorbs a photon of light energy, it changes into a different isomer of retinal, called all-*trans*-retinal. This change puts a strain on the bonds between retinal and opsin, changing the conformation of opsin. This change signals the detection of light. In vertebrate eyes, the retinal and the opsin eventually separate from each other—a process called *bleaching*, which causes the molecule to lose its photosensitivity. A series of enzymatic reactions is then required to return the all-*trans*-retinal to the 11-*cis* isomer, which then recombines with opsin so that it once again becomes the photosensitive pigment rhodopsin.

> The amino acid sequence of opsin is similar to that of olfactory receptors, and the retinal binding site is similar to the odorant binding site. Did photoreception evolve as an ability to "smell" light?

How does the conformational change of rhodopsin transduce light into a cellular response? After retinal is converted from the 11-*cis* into the all-*trans* form, its interactions with opsin pass through several unstable intermediate stages. One of these stages triggers a cascade of reactions involving a G protein signaling mechanism that results in the alteration of membrane potential that is the photoreceptor cell's response to light (see Figure 45.13).

To get a better idea of how rhodopsin alters the membrane potential of a photoreceptor cell and how that photoreceptor cell

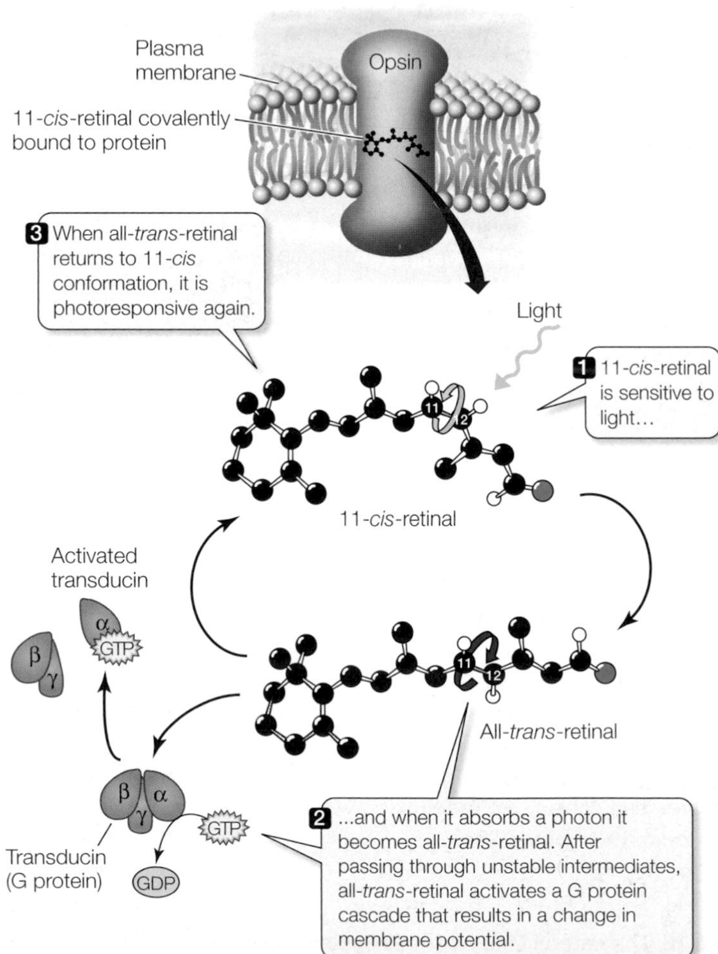
Plasma membrane
Opsin
11-*cis*-retinal covalently bound to protein
3 When all-*trans*-retinal returns to 11-*cis* conformation, it is photoresponsive again.
Light
1 11-*cis*-retinal is sensitive to light...
11
12
11-*cis*-retinal
Activated transducin
β
γ
α
GTP
All-*trans*-retinal
2 ...and when it absorbs a photon it becomes all-*trans*-retinal. After passing through unstable intermediates, all-*trans*-retinal activates a G protein cascade that results in a change in membrane potential.
Transducin (G protein)
GDP

45.13 Light Changes the Conformation of Rhodopsin The light-absorbing molecule 11-*cis*-retinal bonds with the protein opsin to form the pigment rhodopsin, the molecular agent of photosensitivity.

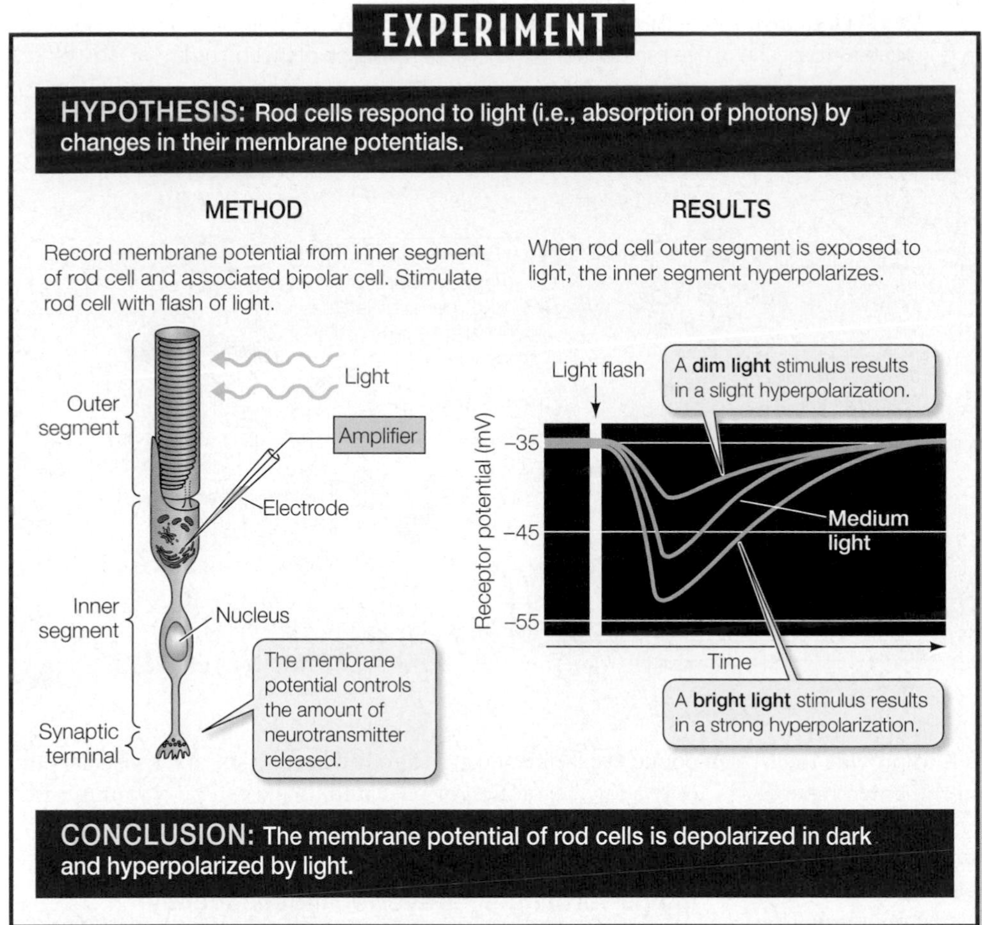

EXPERIMENT
HYPOTHESIS: Rod cells respond to light (i.e., absorption of photons) by changes in their membrane potentials.
METHOD
Record membrane potential from inner segment of rod cell and associated bipolar cell. Stimulate rod cell with flash of light.
Light
Outer segment
Amplifier
Electrode
Inner segment
Nucleus
Synaptic terminal
The membrane potential controls the amount of neurotransmitter released.
RESULTS
When rod cell outer segment is exposed to light, the inner segment hyperpolarizes.
Light flash
A dim light stimulus results in a slight hyperpolarization.
Receptor potential (mV)
-35
-45
-55
Medium light
Time
A bright light stimulus results in a strong hyperpolarization.
CONCLUSION: The membrane potential of rod cells is depolarized in dark and hyperpolarized by light.

45.14 A Rod Cell Responds to Light The vertebrate rod cell is a neuron modified for photosensitivity. The membranes of a rod cell's discs are densely packed with rhodopsin. The plasma membrane of a rod cell hyperpolarizes—becomes more negative—in response to a flash of light. Rod cells do not fire action potentials. FURTHER RESEARCH: How would you investigate the effect of background illumination on the rod cell's response to light?

signals that it has been stimulated by light, let's look at one type of vertebrate photoreceptor cell, the **rod cell**. The rod cell, named for its shape, is a modified neuron that does not produce action potentials. Rod cells release neurotransmitter from their bases where they form synapses with the next neurons in the visual pathway (**Figure 45.14**). Each rod cell has an outer segment, an inner segment, and a synaptic terminal. The outer segment is highly specialized and contains a stack of disks of plasma membrane densely packed with rhodopsin. The function of the disks is to capture photons of light passing through the rod cell. The inner segment contains the cell nucleus and abundant mitochondria. The synaptic terminal is where the rod cell communicates with other neurons.

To see how a rod cell responds to light, we can penetrate a single rod cell with an electrode and record its membrane potential in the dark and in the light. From what we have learned about other types of sensory receptors, we might expect stimulation of the rod cell by light would make its membrane potential less negative. But the opposite is true. When a rod cell is kept in the dark, it has a relatively depolarized resting potential in comparison with other neurons. In fact, the plasma membrane of the rod cell is almost as permeable to Na^+ as to K^+. In the dark, Na^+ continually enters the outer segment of the cell—the dark current.

When light is flashed on the dark-adapted rod cell, its membrane potential becomes more negative—it hyperpolarizes (see Figure 45.14). The rate of neurotransmitter release changes as membrane potential changes. As the rod cell hyperpolarizes, its release of neurotransmitter decreases.

How does the absorption of light by rhodopsin hyperpolarize the rod cell? When rhodopsin is excited by light, it initiates a cascade of events (**Figure 45.15**). The photoexcited rhodopsin combines with and activates a G protein called *transducin*. Activated transducin in turn activates a phosphodiesterase (PDE), which converts cyclic GMP (cGMP) to GMP. This reaction plays a central role in phototransduction. In the dark, the cGMP in the outer segment binds to cation channels, keeping them open and allowing Na^+ to enter the outer segment. As cGMP is converted to GMP, the channels close, and the cell hyperpolarizes.

This mechanism may seem like a roundabout way of doing business, but its advantage is its enormous amplification ability. Each molecule of photoexcited rhodopsin can activate several hundred transducin molecules, thus activating a large number of PDE molecules. The catalytic capacity of a molecule of PDE is great: It can hydrolyze more than 4,000 molecules of cGMP per second. The bottom line is that a single photon of light can cause a huge number of sodium channels to close.

The dark current in photoreceptors makes evolutionary sense. The ability to detect shadows is a significant selective advantage in avoiding predators.

Invertebrates have a variety of visual systems

Photoreceptors are incorporated into a variety of visual systems, from simple to complex. Flatworms obtain directional information about light from photoreceptor cells that are organized into **eye cups**. The eye cups are paired bilateral structures, each partly shielded from light by a layer of pigmented cells lining the cup. The photoreceptors on the two sides of the animal are unequally stimulated unless the animal is facing directly toward or away from a light source. The flatworm generally uses directional information from the eye cups to move away from light.

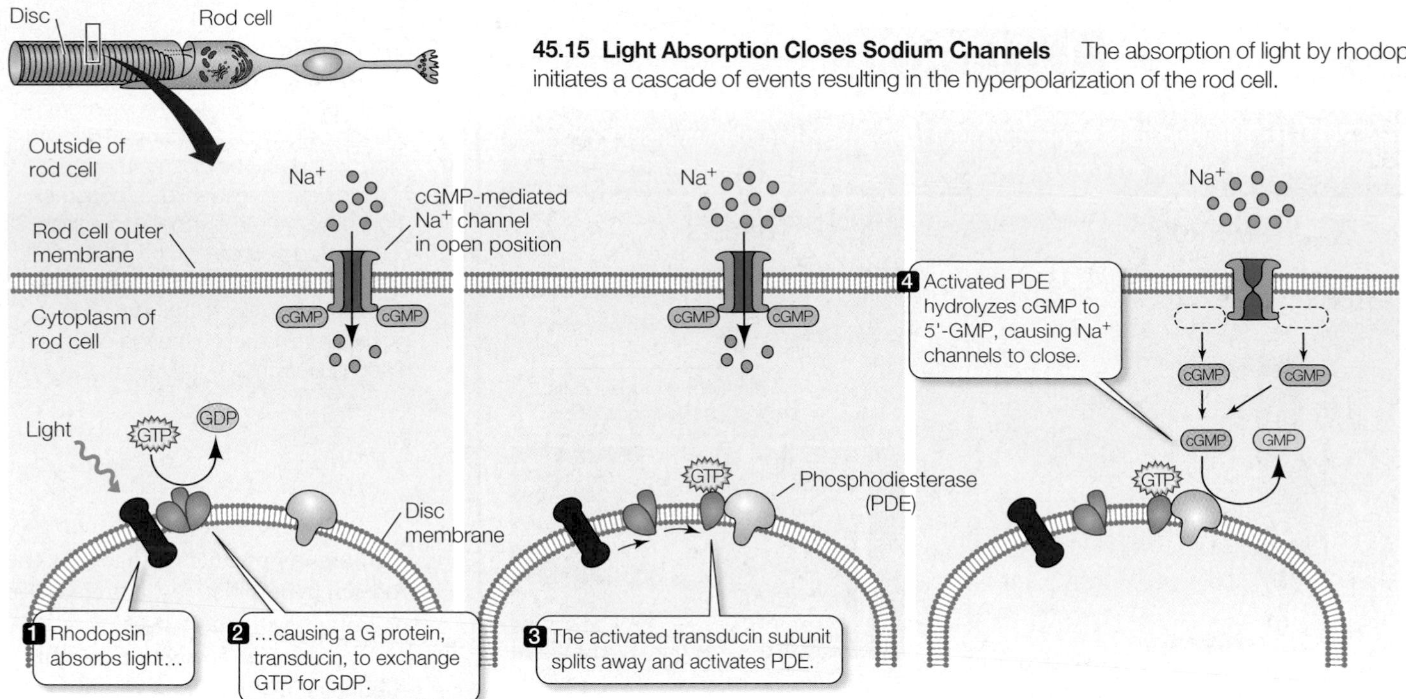

45.15 Light Absorption Closes Sodium Channels The absorption of light by rhodopsin initiates a cascade of events resulting in the hyperpolarization of the rod cell.

Arthropods have evolved **compound eyes** that provide them with information about patterns or images in the environment. These eyes are called compound because each eye consists of many optical units called **ommatidia** (singular *ommatidium*), each with its own narrow-angle lens (**Figure 45.16**). In contrast, a vertebrate eye consists of just one optical unit with a wide-angle lens. The number of ommatidia in a compound eye varies from only a few in some ants, to 800 in fruit flies, to 30,000 in some dragonflies.

Each ommatidium has a lens structure that directs light onto photoreceptor cells. Flies, for example, have eight elongated photoreceptors in each ommatidium. The inner borders of the photoreceptors are covered with microvilli that contain rhodopsin and trap light. Axons from the photoreceptors send the light information to the nervous system. Since each ommatidium of a compound eye is directed at a slightly different part of the visual world, only a low-resolution or pixillated image can be communicated from the compound eye to the CNS.

Image-forming eyes evolved independently in vertebrates and cephalopods

Both vertebrates and cephalopod mollusks have evolved **image-forming eyes**—eyes with exceptional abilities to form detailed images of the visual world. Like cameras, these eyes focus images on an internal surface that is sensitive to light. Considering that they evolved independently of each other, their high degree of similarity is remarkable (**Figure 45.17**).

The vertebrate eye is a spherical, fluid-filled structure bounded by a tough connective tissue layer called the *sclera*. At the front of the eye, the sclera forms the transparent **cornea**, through which light passes to enter the eye. Just inside the cornea is the pigmented **iris**, which gives the eye its color. The function of the iris is to con-

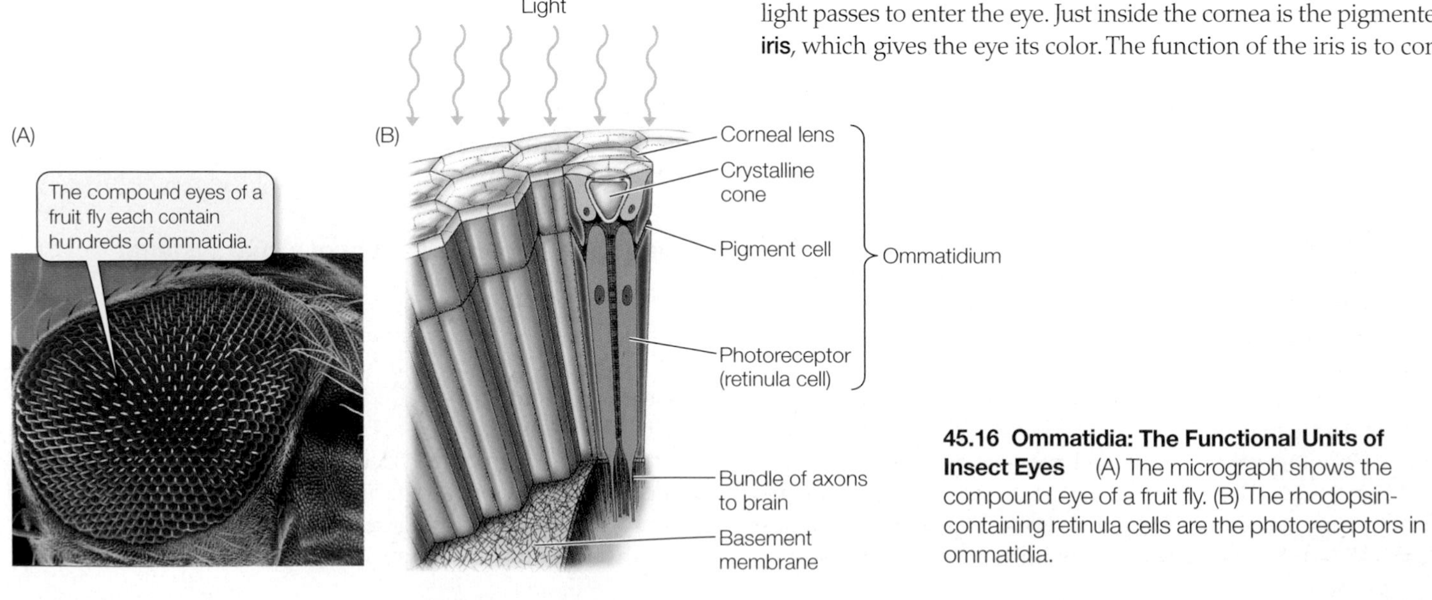

45.16 Ommatidia: The Functional Units of Insect Eyes (A) The micrograph shows the compound eye of a fruit fly. (B) The rhodopsin-containing retinula cells are the photoreceptors in ommatidia.

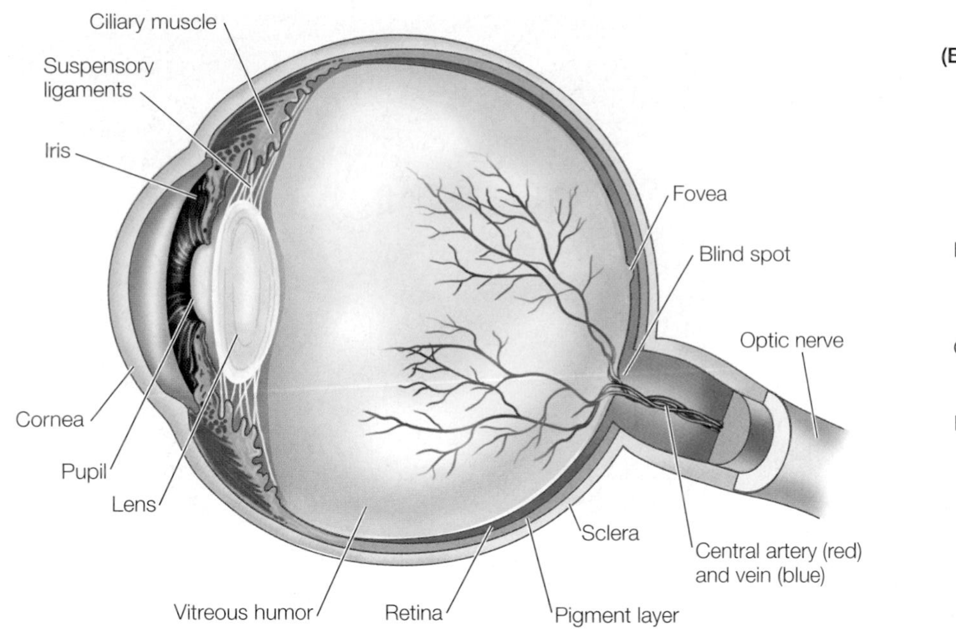

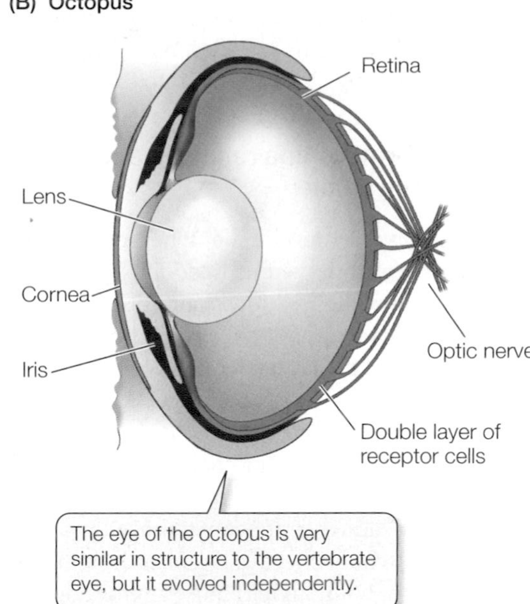

45.17 Eyes Like Cameras The lenses of vertebrate (A) and cephalopod (B) eyes focus images on layers of photoreceptor cells.

trol the amount of light that reaches the photoreceptor cells at the back of the eye, just as the diaphragm of a camera controls the amount of light reaching the film. The central opening of the iris is the **pupil**. The iris is under neuronal control. In bright light, the iris constricts, and the pupil is very small. As light levels fall, the iris relaxes, and the pupil enlarges.

> Lenses become less elastic with age, so we lose the ability to focus on objects close at hand without the help of corrective lenses. As a consequence, most people over the age of 45 need the assistance of bifocal lenses or reading glasses.

Behind the iris is the crystalline protein **lens**, which makes fine adjustments in the focus of images falling on the photosensitive layer, the **retina**, at the back of the eye. The cornea and the fluids within the eye are mostly responsible for focusing light on the retina, but the lens allows the eye to *accommodate*—that is, to focus on objects at various locations in the near visual field (**Figure 45.18**). To focus a camera on objects close at hand, you adjust the distance between the lens and the film. Fishes, amphibians, and reptiles accommodate in a similar manner, moving the lenses of their eyes closer to or farther from their retinas. Mammals and birds use a different method: They alter the shape of the lens.

The lens is contained in a connective tissue sheath that tends to keep it in a spherical shape, but it is attached to suspensory ligaments that pull it into a flatter shape. Circular muscles called the *ciliary muscles* counteract the pull of the suspensory ligaments and permit the lens to round up. When the ciliary muscles are at rest, the flatter lens has the correct optical properties to focus dis-

Fovea
A camera's lens focuses an inverted image on the film in the same way the eye's lens focuses an image on the retina.
For near vision, ciliary muscles contract, causing the lens to round up.
For distant vision, ciliary muscles relax and suspensory ligaments pull the lens into a flatter shape.

45.18 Staying in Focus Mammals and birds focus their eyes by changing the shape of the lens.

tant images on the retina, but not close images. Contracting the ciliary muscles rounds up the lens, changing its light-bending properties to bring close images into focus.

The vertebrate retina receives and processes visual information

During embryonic development, neuronal tissue grows out from the brain to form the retina. In addition to a layer of photoreceptor cells, the retina includes four additional layers of cells that process visual information from the photoreceptors. Light must pass through all the layers of retinal cells before being captured by rhodopsin. Light that is not captured by rhodopsin is absorbed by a black pigment layer behind the retina. In contrast, nocturnal animals such as deer and raccoons have a white reflective layer behind the retina to maximize the capture of photons. Therefore, the deer in the headlights appears to have bright white eyes. We do not have the white reflective layer in our retinas, but photographic flashes are bright enough to cause a reflection that appears red on photos because of the abundance of blood vessels in the retina.

The pigmented epithelium also plays a role in the renewal of the photoreceptors. New disks are continuously being generated by the inner segments and at the distal ends of the outer segments disks are being shed. The pigmented epithelial cells phagocytose the shed disks. Each outer segment is totally renewed about every two weeks.

THE PHOTORECEPTORS OF THE RETINA Until now we have referred to only one kind of photoreceptor, the rod cell. But there is another major kind of vertebrate photoreceptor, also named for its shape: the **cone cell** (**Figure 45.19**). A human retina has about 5 million cones and about 100 million rods. The density of rods and cones is not the same across the entire retina. In humans, light coming from the center of the visual field falls on the **fovea**, where the density of cone cells is highest. The human fovea has about 160,000 cones per square millimeter. The fovea of a hawk has almost twice the number of photoreceptors per square millimeter, making its vision sharper than ours. In addition, the hawk has two foveas in each eye: one receives light from straight ahead, while the other receives light from below. Thus, while the hawk is flying, it sees both its projected flight path and the ground below (where it might detect a mouse scurrying in the grass).

Because cones have low sensitivity to light, they contribute little to night vision. Night vision depends mostly on rod cells and therefore vision in dim light is mostly in shades of gray and acuity is low. You may have trouble seeing a small object such as a keyhole at night when you are looking straight at it—that is, when its image is falling on your fovea. If you look a little to the side, so that the image falls on a rod-rich area of your retina, you can see the object better. Astronomers looking for faint objects in the sky learned this trick a long time ago.

The human retina has three kinds of cone cells, each containing slightly different opsin molecules, which differ in the wavelengths of light they absorb best. Although the same 11-*cis*-retinal group is the light absorber in all three kinds of cones (see Figure 45.14), its molecular interactions with opsin determine the spectral sensitivity of the rhodopsin molecule as a whole (**Figure 45.20**).

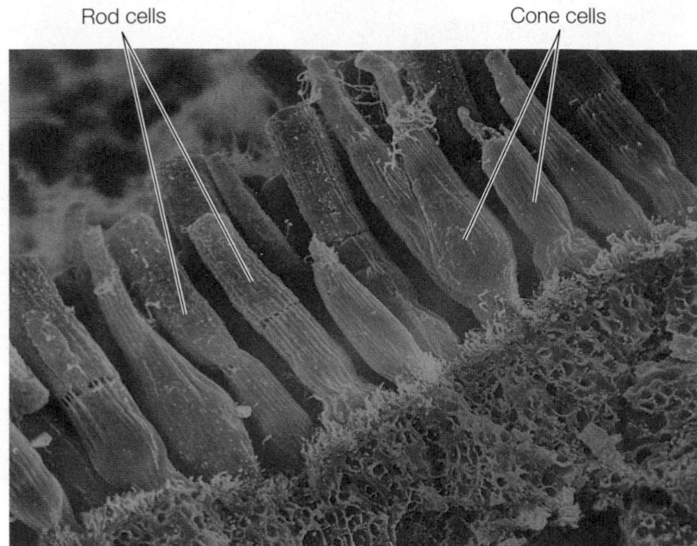

45.19 Rods and Cones This scanning electron micrograph of photoreceptors in the retina of a mud puppy (an amphibian) shows cylindrical rods and tapered cones.

Animals that are nocturnal (such as flying squirrels) have retinas containing a high percentage of rods and may have poor color vision. By contrast, some animals that are active only during the day (such as chipmunks) have mostly cones in their retinas.

Where blood vessels and the optic nerve pass through the back of the eye, there are no photoreceptors, resulting in a blind spot on the retina (see Figure 45.17). You are normally not aware of your blind spot, but you can find it. Stare straight ahead, holding a pencil in your outstretched hand so that the eraser is in the center of your field of vision. While continuing to stare straight ahead, slowly move the pencil to the side until the eraser disappears. When this happens, the light from the eraser is focused directly on your blind spot.

INFORMATION FLOW IN THE RETINA The human retina is organized into five layers of neurons (including the photoreceptor cells) that receive visual information and process it before sending it to the brain (**Figure 45.21**). A first step in understanding how the retina tells the brain what it is seeing is to study how these layers are interconnected and how they influence one another. From our discussion of rod cells above, we know that the photoreceptor cells at the back of the retina hyperpolarize in response to light and do not generate action potentials. The cells at the front of the retina (the cells closest to the lens) are **ganglion cells**. They do fire action potentials, and their axons form the optic nerve that travels to the brain. The layers of cells between the photoreceptors and the ganglion cells process information about the visual field.

The photoreceptors and ganglion cells are connected by *bipolar cells*. Changes in the membrane potential of rods and cones in response to light alter the rates at which the rods and cones release neurotransmitter at their synapses with the bipolar cells. In response to this neurotransmitter, the membrane potentials of the

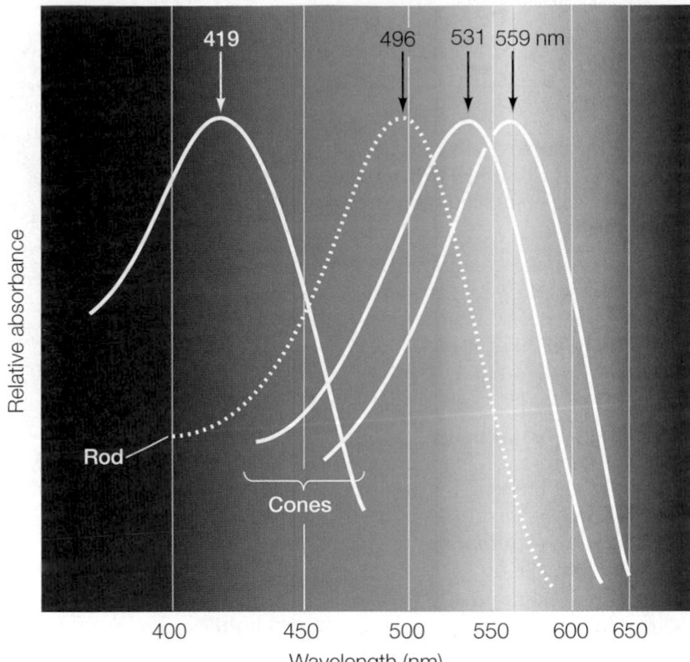

45.20 Absorption Spectra of Cone Cells The three kinds of cone cells contain slightly different opsin molecules, which absorb different wavelengths of light.

bipolar cells change, altering the rate at which they release neurotransmitter onto ganglion cells. The rate of neurotransmitter release from the bipolar cells determines the rate at which the ganglion cells fire action potentials. Thus, the direct flow of information in the retina is from photoreceptor to bipolar cell to ganglion cell. The ganglion cells send the information to the brain.

The other two cell layers, the horizontal cells and the amacrine cells, communicate laterally across the retina. *Horizontal cells* form synapses with neighboring photoreceptors. Thus, light falling on one photoreceptor can influence the sensitivity of its neighbors to light. This lateral flow of information enables the retina to sharpen the perception of contrast between light and dark patterns.

Amacrine cells form local interconnections between bipolar cells and ganglion cells. Some amacrine cell types are highly sensitive to changing illumination or to motion. Others assist in adjusting the sensitivity of the eyes according to the overall level of light falling on the retina. When background light levels change, amacrine cell connections to the ganglion cells help the ganglion cells remain sensitive to temporal changes in stimulation. Thus, even with large changes in background illumination, the eyes are sensitive to smaller, more rapid changes in the pattern of light falling on the retina.

Knowing the path of information in the retina still does not tell us how that information is processed by the brain. What does the eye tell the brain in response to a pattern of light falling on the retina? In Chapter 46 we will learn how the brain reassembles that information into our view of the world.

45.4 RECAP

A family of photopigments called rhodopsins are responsible for light sensitivity in all animals. Receptor cells, including rod and cone cells in humans, transduce the photosensitivity of rhodopsins to light and use it to form images of the environment.

- How does a photon of light change the membrane potential in a rod cell? See p. 977 and Figures 45.14 and 45.15
- What is the mechanism of color vision? See p. 980 and Figure 45.20
- Describe the flow of signals that occurs in the eye in response to light. See pp. 980–981 and Figure 45.21

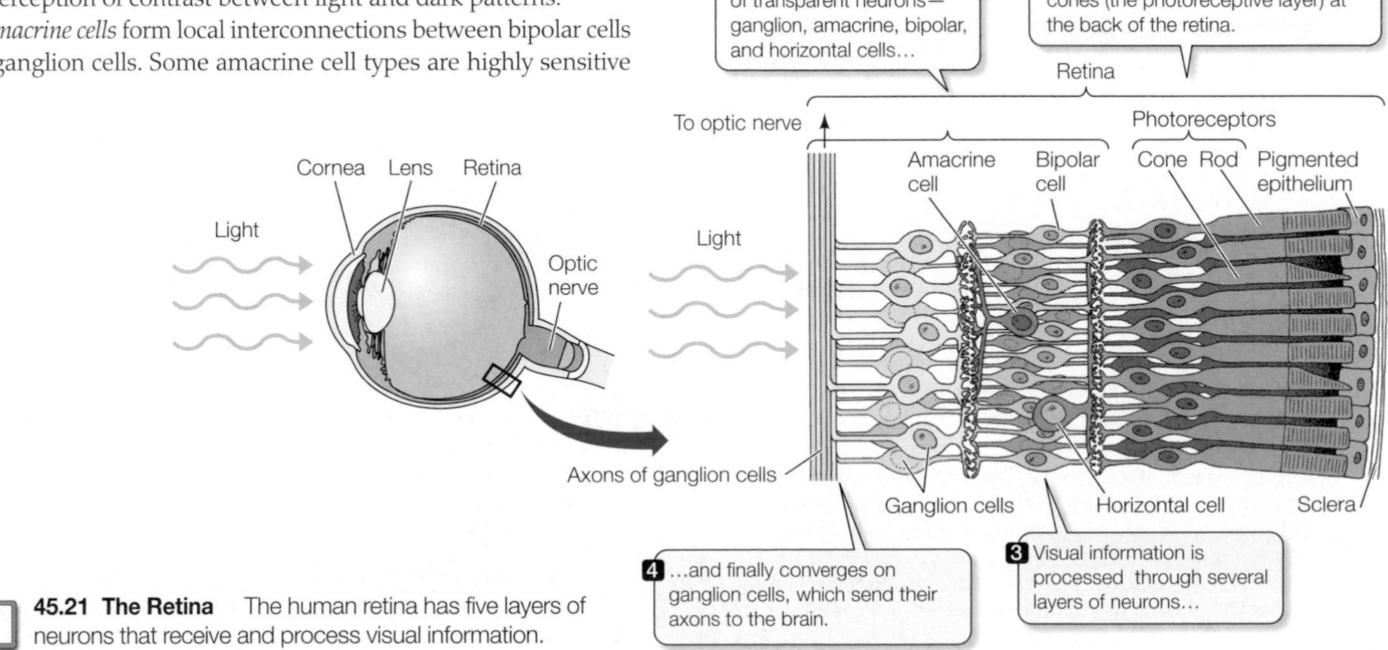

45.21 The Retina The human retina has five layers of neurons that receive and process visual information.

CHAPTER SUMMARY

45.1 How do sensory cells convert stimuli into action potentials?

Sensory receptor cells, also known as sensory cells or simply receptors, transduce information about an animal's external and internal environment into action potentials. Those sensory receptor cells that can fire action potentials are called sensory neurons.

The interpretation of action potentials as particular sensations depends on which neurons in the CNS receive them.

Sensory receptor cells have membrane **receptor proteins** that cause ion channels to open or close, affecting the resting potential of the cell. Metabotropic receptors act through signal transduction pathways to generate **receptor potentials**. Mechanoreceptors are ionotropic sensory receptors that open ion channels physically through forces such as pressure or stretch. Review Figure 45.1

Receptor potentials initiated by a sensory cell can spread to regions of the cell's plasma membrane that generate action potentials, or they can release neurotransmitter in response to changes in membrane potential. Review Figure 45.2

Adaptation enables the nervous system to ignore irrelevant or continuous stimuli while remaining responsive to relevant or new stimuli.

45.2 How do sensory systems detect chemical stimuli?

Chemoreceptors are responsible for smell, taste, and the sensing of **pheromones**.

Mammalian olfactory sensors project directly to the **olfactory bulb** of the brain. Sensors for the same **odorant** project to the same area of the olfactory bulb.

Each olfactory receptor cell expresses one receptor protein that can bind a specific molecule or ion. Binding causes a second messenger to open ion channels, which creates an action potential. Review Figure 45.4

Taste buds in the mouth cavities of vertebrates are responsible for the sense of **gustation**. The five basic tastes are sweet, salt, sour, bitter, and umami. Review Figure 45.5

45.3 How do sensory systems detect mechanical forces?

See Web/CD Tutorial 45.1

The skin contains a variety of ionotropic **mechanoreceptors** that respond to touch and pressure. The density of mechanoreceptors in any skin area determines the sensitivity of that area. Review Figure 45.6

Stretch receptors in muscle spindles and in the Golgi tendon organ found in tendons and ligaments inform the CNS of the positions of and loads on parts of the body. Review Figure 45.7

In mammalian auditory systems, ear pinnae collect and direct sound waves to the **tympanic membrane**, which vibrates in response to sound waves. The movements of the tympanic membrane are amplified through a chain of **ossicles** that conduct the vibrations to the **oval window**. Movements of the oval window create pressure waves in the fluid-filled **cochlea**. Review Figure 45.8, Web/CD Activity 45.1

The **basilar membrane** running down the center of the cochlea is distorted by sound waves at specific locations that depend on their frequency. These distortions cause the bending of hair cells in the **organ of Corti**. Receptor potentials in hair cells cause them to release neurotransmitter, which creates action potentials in the **auditory nerve**. Review Figure 45.9

Hair cells are also mechanoreceptors. The bending of their stereocilia alters receptor proteins and therefore their membrane potentials. Hair cells are found in the auditory organs and organs of equilibrium such as the **lateral line** system of fishes and the **semicircular canals** and **vestibular apparatus** of mammals. Review Figures 45.10, 45.11, and 45.12

45.4 How do sensory systems detect light?

Photosensitivity depends on the absorption of photons of light by **rhodopsin**, a **photoreceptor** molecule that consists of a protein called **opsin** and a light-absorbing prosthetic group called **retinal**. Absorption of light by retinal is the first step in a cascade of intracellular events leading to a change in the membrane potential of the photoreceptor cell. Review Figure 45.13

When excited by light, vertebrate photoreceptor cells hyperpolarize and release less neurotransmitter onto the neurons with which they form synapses. They do not fire action potentials. Review Figures 45.14 and 45.15

Visual systems range from the simple **eye cups** of flatworms, which sense the direction of a light source, to the **compound eyes** of arthropods, which detect shapes and patterns, to the **image-forming eyes** of vertebrates and cephalopods. Review Figures 45.16 and 45.17, Web/CD Activity 45.2

Vertebrate and cephalopod eyes focus detailed images of the visual field onto dense arrays of photoreceptors that transduce the visual image into neuronal signals. Review Figure 45.18

Vertebrates have two types of photoreceptors, **rod cells** and **cone cells**. In humans, the **fovea** contains almost exclusively cone cells, which are responsible for color vision but are not very sensitive in dim light. Color vision arises from three types of cone cells with different spectral absorption properties. Review Figures 45.19 and 45.20

The vertebrate **retina** consists of five layers of neurons lining the back of the eye. The light-absorbing photoreceptor cells are at the back of the retina. Review Figure 45.21, Web/CD Activity 45.3

The axons of the **ganglion cells**, in the innermost layer of the retina, are bundled together in the optic nerve. Between the photoreceptors and the ganglion cells are neurons that process information from the photoreceptors.

SELF-QUIZ

1. Which statement about sensory systems is *not* true?
   a. Sensory transduction involves the conversion (direct or indirect) of a physical or chemical stimulus into changes in membrane potentials.
   b. In general, a stimulus causes a change in the flow of ions across the plasma membrane of a sensory receptor cell.
   c. The term "adaptation" refers to the process by which a sensory system becomes insensitive to a continuing source of stimulation.
   d. The more intense a stimulus, the greater the magnitude of each action potential fired by a sensory neuron.
   e. Sensory adaptation plays a role in the ability of organisms to discriminate between important and unimportant information.

2. The female silkworm moth releases a chemical called bombykol from a gland at the tip of her abdomen. Bombykol is
   a. a sex hormone.
   b. detected by the male only when present in large quantities.
   c. not species-specific.
   d. detected by hairs on the antennae of male silkworm moths.
   e. a chemical basic to the taste process in arthropods.
3. Which statement about olfaction is *not* true?
   a. In general, mammals depend more on vision than on olfaction as their dominant sensory modality.
   b. Olfactory stimuli are recognized by the interaction between odorant molecules and receptor proteins on olfactory hairs.
   c. The more odorant molecules that bind to receptors, the more action potentials are generated.
   d. The greater the number of action potentials generated by an olfactory receptor, the greater the intensity of the perceived smell.
   e. The perception of different smells results from the activation of different combinations of olfactory receptors.
4. The touch receptors located very close to the skin surface
   a. are relatively insensitive to light touch.
   b. adapt very quickly to stimuli.
   c. are uniformly distributed throughout the surface of the body.
   d. are called Pacinian corpuscles.
   e. adapt slowly and only partially to stimuli.
5. The membrane that is most directly responsible for the ability to discriminate different pitches of sound is the
   a. round window.
   b. oval window.
   c. tympanic membrane.
   d. tectorial membrane.
   e. basilar membrane.
6. Which statement is *not* true?
   a. The transmembrane potential of a rod cell becomes more negative when the rod cell is exposed to light.
   b. A photoreceptor releases the most neurotransmitter when in total darkness.
   c. Whereas in vision the intensity of a stimulus is encoded by the degree of hyperpolarization of photoreceptors, in hearing the intensity of a stimulus is encoded by changes in firing rates of sensory neurons.
   d. Stiffening of the ossicles in the middle ear can lead to deafness.
   e. The interaction among hammer (malleus), anvil (incus), and stirrup (stapes) conducts sound waves across the fluid-filled middle ear.
7. In humans, the region of the retina where the central part of the visual field falls is the
   a. central ganglion cell.
   b. fovea.
   c. optic nerve.
   d. cornea.
   e. pupil.
8. The region of the vertebrate eye where the optic nerve passes out of the retina is the
   a. fovea.
   b. iris.
   c. blind spot.
   d. pupil.
   e. visual cortex.
9. Which statement about the cone cells in a human eye is *not* true?
   a. They are responsible for our sharpest vision.
   b. They are responsible for color vision.
   c. They are more sensitive to light than rods are.
   d. They are fewer in number than rods.
   e. They exist in high numbers at the fovea.
10. The color in color vision results from the
    a. ability of each cone cell to absorb all wavelengths of light equally.
    b. lens of the eye acting like a prism and separating the different wavelengths of light.
    c. differential absorption of wavelengths of light by different kinds of rod cells.
    d. three different isomers of opsin in cone cells.
    e. absorption of different wavelengths of light by amacrine and horizontal cells.

FOR DISCUSSION

1. Compare and contrast the functioning of olfactory receptors and photoreceptors. How do these sensory cells enable the CNS to discriminate between an apple and an orange?
2. Amplification of signal is an important feature of sensory systems. Compare mechanisms of amplification in olfactory, visual, and auditory systems.
3. If you were blindfolded and placed in a wheelchair, how would you know if you were being pushed forward or backward?
4. Animals can use visual, olfactory, tactile, and auditory signals to communicate. From what you know about these sensory systems, discuss the relative advantages and disadvantages of these systems for communication.

FOR INVESTIGATION

A certain region of the brain contains neurons that are sensitive to the osmolarity of the blood. You could imagine that these cells are sensitive to the concentration of a particular solute such as NaCl or to tension in their plasma membranes if they take up or lose water osmotically. Using a slice of this brain region in a culture dish, and patch pipettes, how could you investigate the mechanism of signal transduction in these neurons?

CHAPTER 46 The Mammalian Nervous System: Structure and Higher Function

Can our brains be full?

A famous "Far Side" cartoon by Gary Larson shows a classroom in which one student with a noticeably small head has his hand raised, saying, "Mr. Anderson, may I be excused? My brain is full." This lighthearted scene actually suggests a deep question: What is the capacity of the human brain? Is it limited by the organ's size? By the number of synapses? By the number of neurons?

Scientists long believed that we are born with a certain number of neurons, that we steadily lose neurons throughout life, and that we do not grow any new ones. The evidence seemed clear: in the adult brain, we don't see neurons with mitotic structures that would indicate cell division, nor do we see neurons at different stages of maturation. Also, it is hard to imagine how a new neuron, with its many complex connections, could be inserted into an adult brain. Our lifelong ability to learn has been explained solely by the formation and strengthening of synapses.

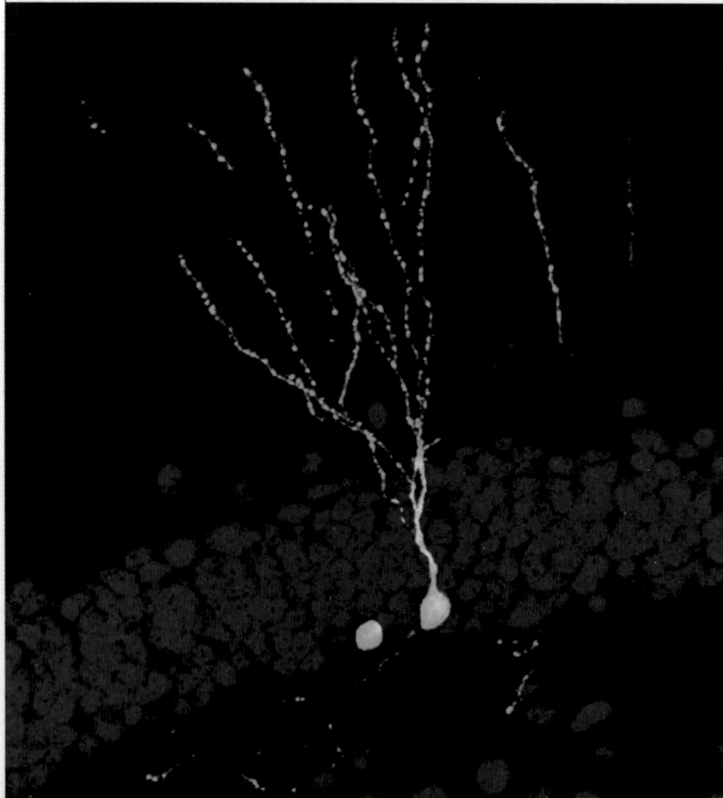

But one of the most exciting recent discoveries in neurobiology is that new neurons *are* formed in adult avian and mammalian brains, and that the formation of these new neurons seems to be stimulated by experience and learning. The birth of new neurons was first seen when adult rats were injected with radioactively labeled thymidine, which is incorporated into new DNA when cells divide. It came as a huge surprise when thymidine-labeled neurons were discovered in the brains of these animals.

Many reasons were advanced suggesting why these labeled cells could not really be new neurons, and few scientists were ready to give up the old dogma. The debate picked up, however, when Fernando Nottebohm and his colleagues at Rockefeller University showed that new neurons are formed in parts of the bird brain responsible for song at the time of the year when birds are ready to mate and males begin to sing. The researchers further showed that sex hormones and hearing song stimulated the birth of new neurons.

The discovery of neuron generation in bird brains gave new impetus to mammalian studies, and these studies revealed that two structures of the adult mammalian brain can acquire new neurons. One is the olfactory bulb (not surprising, because olfactory bulb neurons extend into the nasal epithelium, which sheds regularly). The other structure is the *hippocampus*, a brain region involved in forming long-term memories. As in birds, experience and learning stimulate neurogenesis in the mammalian hippocampus. Fred Gage and his colleagues at the Salk Institute have recorded action potentials from new hippocampal

New Neurons in an Adult Mouse An adult mouse was injected with a vector carrying a gene for green fluorescent protein (GFP). The gene was taken up and expressed only by dividing cells. Neurons labeled with GFP were later found in the mouse's brain—and these neurons must have arisen *after* injection of the labeled gene.

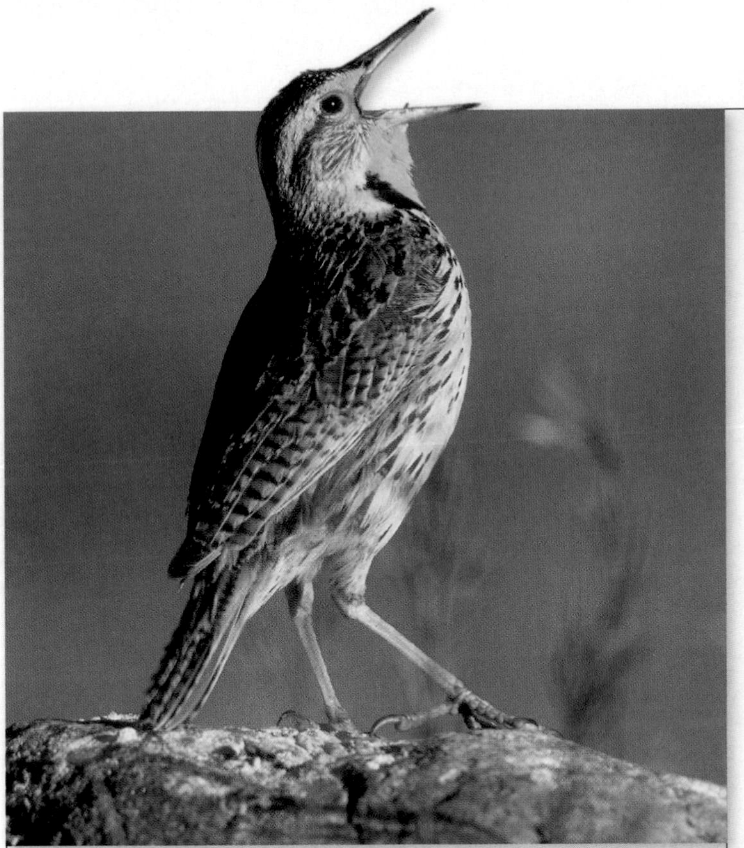

Birdsong Requires New Neurons A male meadowlark (*Sturnella neglecta*) sings a specific song in order to attract a mate. New neurons are formed in the parts of brain responsible for producing the song, stimulated by sex hormones and hearing the songs of other male meadowlarks.

neurons in mice, showing that they mature into functional neurons with properties identical to those of older neighboring cells.

Many questions arise from these studies. Could neurogenesis be stimulated to repair damage and to counter the effects of aging? Are neurons regularly lost and replaced? Do new neurons facilitate learning? Will you generate new neurons by reading this chapter?

IN THIS CHAPTER we explore the cellular basis of several important subsystems of the human nervous system. The human brain has about 100 billion neurons and probably a thousand times as many synapses, which account for its ability to handle vast amounts of information. But the ability of the brain to serve specific functions—evaluating sensory input, controlling emotions, generating motor output, learning and remembering—arises from the organization of the nervous system into functional subsystems.

CHAPTER OUTLINE

46.1 How Is the Mammalian Nervous System Organized?

We can describe the organization of the mammalian nervous system anatomically or functionally. In anatomical terms, all vertebrate nervous systems consist of three parts: a brain, a spinal cord, and a set of peripheral nerves that reach to all parts of the body. As we learned at the start of Chapter 44, the brain and spinal cord are referred to as the **central nervous system**, or **CNS**, and the cranial and spinal nerves that connect the CNS to all of the tissues of the body are referred to as the **peripheral nervous system** (**PNS**). An additional division of the nervous system exists in the gut—the enteric nervous system, which we will discuss in Chapter 50.

Recall from Section 44.1 that a *neuron* is an electrically excitable cell that communicates via an axon, and a *nerve* is a bundle of axons that carries information about many things simultaneously. Some axons in a nerve may be carrying information to the CNS, while other axons in the same nerve are carrying information from the CNS to the organs of the body. As we will see in this chapter, we can further divide the anatomy of the brain, spinal cord, and PNS into smaller units.

A functional organization of the nervous system is based on flow and type of information

The major avenues of information flow through the nervous system are illustrated in **Figure 46.1**. The **afferent** portion of the peripheral nervous system carries sensory information to the central nervous system. We are conscious of much of this information (for example, light, sound, skin temperature, pain, the position of limbs). We are usually not conscious, however, of the afferent information involved in physiological regulation (for example, blood pressure, deep body temperature, and blood oxygen supply).

The **efferent** portion of the PNS carries information from the CNS to the muscles and glands of the body. Efferent pathways can be divided into a *voluntary* division, which executes our conscious movements, and an *involuntary*, or **autonomic**, division, which controls physiological functions.

In addition to the neuronal information it receives from the PNS, the CNS receives chemical information from hormones circulating in the blood. **Neurohormones** released by neurons into the extracellular fluids of the brain can send chemical information to other neurons in the brain or can leave the brain and enter the circulation. In Chapter 42 we discussed the

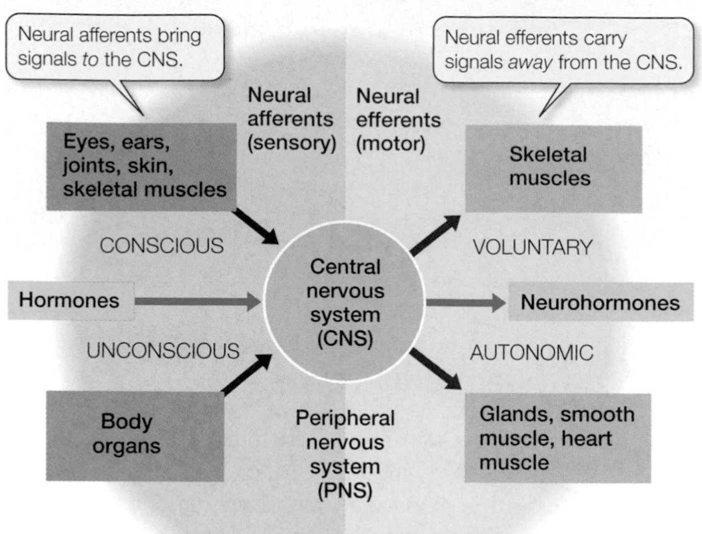

46.1 Organization of the Nervous System The peripheral nervous system (pink, blue) carries information both to and from the central nervous system (gold). The CNS also receives hormonal inputs and produces hormonal outputs (green).

important role of neurohormones (such as GnRH) in the control of the anterior pituitary, and we described how two other neurohormones, oxytocin and ADH, are released into the general circulation from the posterior pituitary.

The vertebrate CNS develops from the embryonic neural tube

Early in the development of a vertebrate embryo, a hollow tube of neural tissue forms (see Section 43.3). This **neural tube** runs the length of the embryo on its dorsal side. At the anterior end of the embryo, the neural tube forms three swellings that become the **hindbrain**, the **midbrain**, and the **forebrain**. The rest of the neural tube becomes the spinal cord (**Figure 46.2**). The cranial and spinal nerves sprout from the neural tube. From these early developmental stages we see the linear axis of information flow in the nervous system. Although the developing brain will fold and become a complex structure, the information flow in the adult nervous system still follows the paths that emerge from the simple linear neural tube.

Each of the three regions of the embryonic brain develops into several structures in the adult brain. From the hindbrain come the **medulla**, the **pons**, and the **cerebellum**. The medulla is continuous with the spinal cord. The pons is in front of the medulla, and the cerebellum is a dorsal outgrowth of the pons. The medulla and pons contain distinct groups of neurons that are involved in the control of physiological functions such as breathing and circulation or basic motor patterns such as swallowing and vomiting. All information traveling between the spinal cord and higher brain areas must pass through the pons and the medulla.

The cerebellum is like the conductor of an orchestra; it receives "copies" of the commands going to the muscles from higher brain areas, and it receives information coming up the spinal cord from

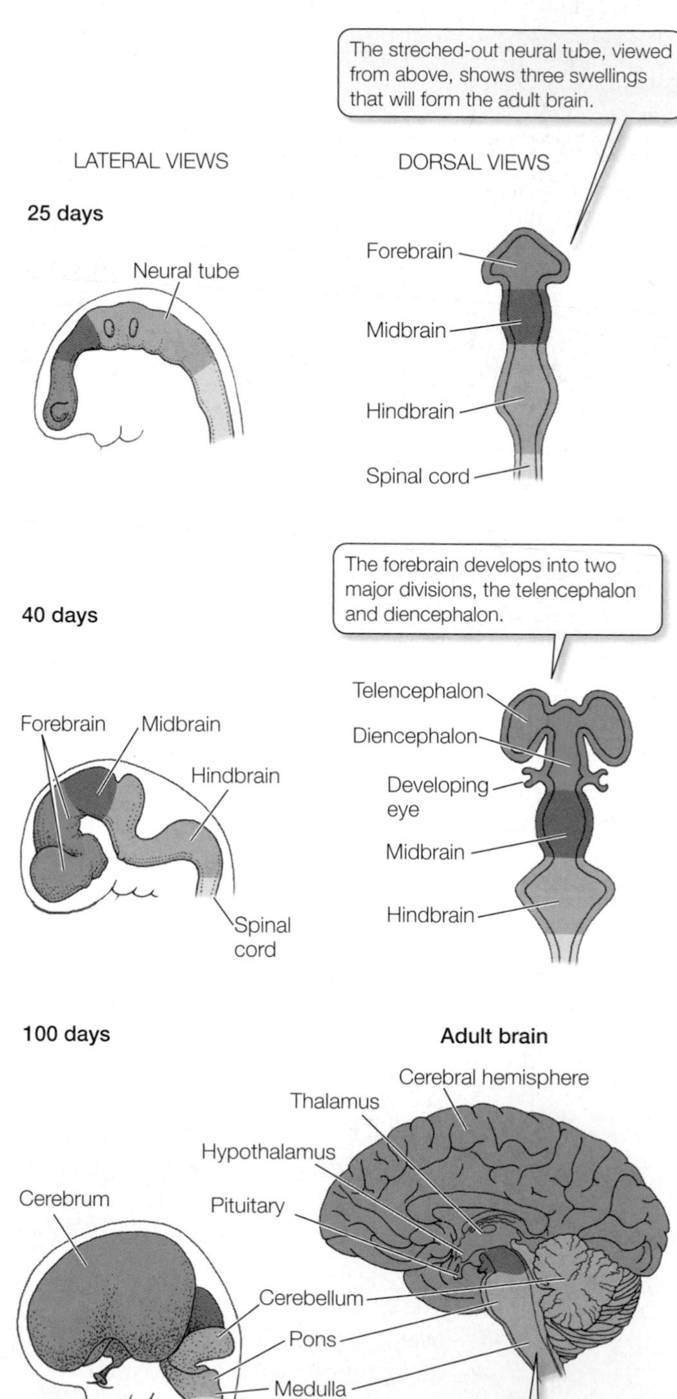

46.2 Development of the Human Nervous System Three swellings at the anterior end of the hollow neural tube in the early vertebrate embryo develop into the regions of the adult brain. The final panel shows an adult human brain cut in half through the midline, a view known as a midsagittal section.

the joints and muscles. It compares the motor "score" with the actual behavior of the muscles and thus refines the motor commands.

From the embryonic midbrain come structures that process aspects of visual and auditory information. In addition, all information traveling between higher brain areas and the spinal cord must pass through the midbrain. The hindbrain and the midbrain are collectively known as the **brain stem**.

The embryonic forebrain develops a central region called the **diencephalon** and a surrounding structure called the **telencephalon**. The diencephalon is the core of the forebrain and consists of an upper structure called the **thalamus** and a lower structure called the **hypothalamus**. The thalamus is the final relay station for sensory information going to the telencephalon, and the hypothalamus regulates many physiological functions (see Section 41.2) and biological drives such as hunger and thirst. The hypothalamus receives a lot of physiological information of which we are not conscious.

The telencephalon consists of two **cerebral hemispheres**, left and right (and is also referred to as the **cerebrum**). In humans, the telencephalon is by far the largest part of the brain and plays major roles in sensory perception, learning, memory, and conscious behavior.

If we compare the classes of vertebrates from fish through amphibians, reptiles, birds, and mammals, the telencephalon increases in size, complexity, and importance—an evolutionary trend called *telencephalization* (see Figure 44.2). The forebrain dominates the nervous systems of mammals, and damage to this region results in severe impairment of sensory, motor, or cognitive functions, and even coma. In contrast, a shark with its telencephalon removed can swim almost normally.

The spinal cord transmits and processes information

The spinal cord conducts information in both directions between the brain and the organs of the body. It also integrates much of the information coming from the PNS, and it responds to that information by issuing motor commands.

A cross section of the spinal cord reveals a central area of gray matter in the shape of a butterfly, surrounded by an area of white matter (**Figure 46.3**). In the nervous system, **gray matter** is rich in neuronal cell bodies, and **white matter** contains axons. The gray matter of the spinal cord contains the cell bodies of the spinal neurons; the white matter contains the axons that conduct information up and down the spinal cord. The white appearance is due to the myelin that wraps most of the axons. Spinal nerves extend from the spinal cord at regular intervals on each side. Each spinal nerve has two roots, one connecting with the *dorsal horn* of the gray matter, and the other connecting with the *ventral horn*. The afferent (sensory) axons in a spinal nerve enter the spinal cord through the *dorsal root*, and the efferent (motor) axons in a spinal nerve leave the spinal cord through the *ventral root*.

The conversion of afferent to efferent information in the spinal cord without participation of the brain is called a **spinal reflex**. The simplest type of spinal reflex involves only two neurons and one synapse and is therefore called a **monosynaptic reflex**. An example is the knee-jerk reflex, which your physician checks with a mallet tap just below your knee. We can diagram the wiring of a monosynaptic reflex by following the flow of information through the spinal cord, as shown in Figure 46.3.

In the case of the knee-jerk reflex, sensory information comes from stretch receptors in the leg muscle that is suddenly stretched when the mallet strikes the tendon that runs over the knee. Each stretch receptor initiates action potentials that are conducted by the axon of a sensory neuron through the dorsal horn of the spinal cord and all the way to the ventral horn. In the ventral horn, the

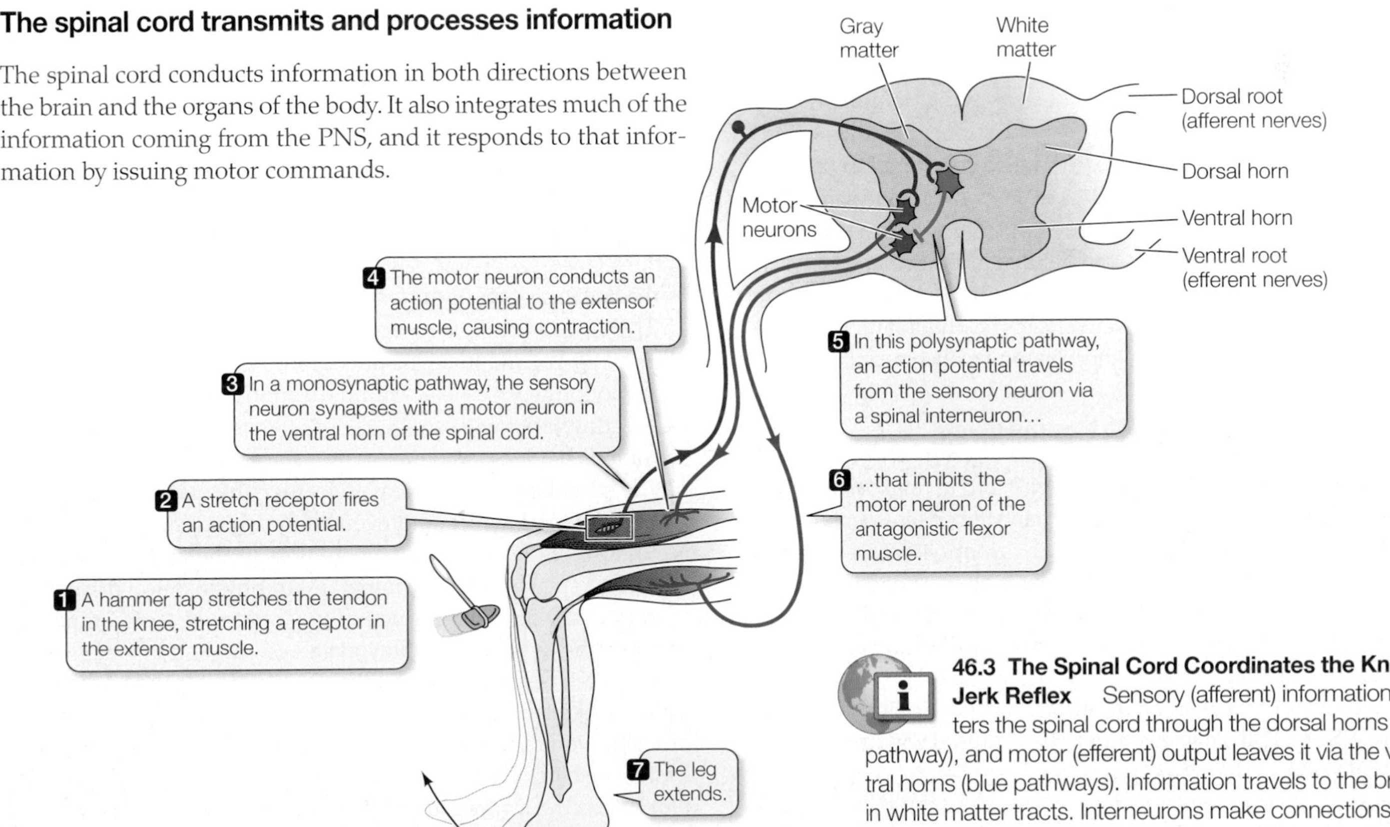

Gray matter
White matter
Dorsal root (afferent nerves)
Dorsal horn
Ventral horn
Ventral root (efferent nerves)
Motor neurons
1 A hammer tap stretches the tendon in the knee, stretching a receptor in the extensor muscle.
2 A stretch receptor fires an action potential.
3 In a monosynaptic pathway, the sensory neuron synapses with a motor neuron in the ventral horn of the spinal cord.
4 The motor neuron conducts an action potential to the extensor muscle, causing contraction.
5 In this polysynaptic pathway, an action potential travels from the sensory neuron via a spinal interneuron...
6 ...that inhibits the motor neuron of the antagonistic flexor muscle.
7 The leg extends.

46.3 The Spinal Cord Coordinates the Knee-Jerk Reflex Sensory (afferent) information enters the spinal cord through the dorsal horns (red pathway), and motor (efferent) output leaves it via the ventral horns (blue pathways). Information travels to the brain in white matter tracts. Interneurons make connections within the spinal cord that result in a complex, coordinated behavior pattern.

sensory neuron synapses with motor neurons, causing them to fire action potentials that are then conducted back to the leg extensor muscle, causing it to contract. The function of this simple circuit is to sense an increased load on the limb and to increase the strength of muscle contraction to compensate for the added load and thereby keep muscle length constant.

Most spinal circuits are more complex than this monosynaptic reflex, as we can demonstrate by building on the circuit we have just traced. Limb movement is controlled by *antagonistic* sets of muscles—muscles that work against each other. When one member of an antagonistic set of muscles contracts, it bends, or flexes, the limb; it is therefore called a *flexor*. The antagonist to this muscle straightens, or extends, the limb, and is called an *extensor*. For a limb to move, one muscle of the pair must relax while the other contracts. Thus, sensory input that activates the motor neuron of one muscle also inhibits its antagonist. This coordination is achieved by an **interneuron**, which makes an inhibitory synapse onto the motor neuron of the antagonistic muscle (see Figure 46.3). Thus the reciprocal inhibition of antagonistic muscles involves an interneuron between the sensory cell and the motor neuron of the inhibited muscle, and therefore at least two synapses.

The withdrawal reflex is an example of a polysynaptic spinal reflex that involves many interneurons. When you step on a tack, you immediately pull back the injured foot: the tack stimulates pain receptors in the foot, and the sensory neurons transmit action potentials into the dorsal horn of the spinal cord on the same side of the body. In the dorsal horn, these neurons synapse with interneurons that send information through their axons to the brain resulting in the conscious sensation of pain. But even before the brain is aware of the pain, synapses of the sensory neurons with other interneurons stimulate and inhibit directly a variety of different motor neurons in the spinal cord. Interneurons on the same side of the spinal cord coordinate the activity of the muscles that withdraw the foot and leg. To pull away, however, the other leg has to extend and balance must be shifted. The coordination of these activities involves interneurons that make connections across the spinal cord to motor neurons on the opposite side. Thus a rather complex suite of movements is coordinated in the spinal cord without the brain's participation. Spinal circuits can even generate repetitive motor patterns (such as a shark's swimming movements) even when the individual's telencephalon has been removed.

The reticular system alerts the forebrain

Sensory information ascending the spinal cord to final destinations in the forebrain passes through the brain stem. Many sensory axons give off collateral branches that form synapses with a network of brain stem neurons called the **reticular system**. The reticular system is a highly complex network of axons and dendrites. Within the reticular system are many discrete groups of neurons that share a common characteristic such as the neurotransmitter they produce and release. Such an anatomically distinct group of neurons in the CNS is called a **nucleus** (not to be confused with the nucleus of a single cell).

As axons carrying sensory information ascend through the reticular formation, they make connections with nuclei that are involved in controlling many functions of the body. Information from joints and muscles, for example, is directed to nuclei in the pons and cerebellum that are involved in balance and coordination. Sensory information also goes to reticular formation nuclei that control sleep and wakefulness. High reticular formation activity produces waking; in the absence of such stimulation, sleep occurs. Because of the alerting function of the reticular core of the brain stem, it is called the *reticular activating system*.

If the brain is damaged at midbrain or higher levels, the alerting action of the reticular system on the forebrain can be lost and a person loses the ability to be conscious—they enter a coma. Damage to the brain stem or the spinal cord below the reticular system may cause paralysis but leaves a person with normal patterns of sleep and waking.

The core of the forebrain controls physiological drives, instincts, and emotions

As mentioned earlier, the midbrain connects to the forebrain through the diencephalon, which includes two important structures, the thalamus and the hypothalamus. The thalamus communicates sensory information to the cerebral cortex, and the hypothalamus receives information about physiological conditions in the body and regulates many homeostatic functions. Section 40.4 describes how the hypothalamus is involved in the regulation of body temperature, and Section 41.2 discusses the intimate association between the hypothalamus and the pituitary gland and its control of many homeostatic functions. Axons from neurons in the cerebral cortex that travel down through the brainstem to the spinal cord make up white matter tracts that go around the diencephalon.

The telencephalon is large in birds and mammals in comparison to the other vertebrates, but the regions of the telencephalon that border the diencephalon are the more primitive areas found in all vertebrates. Structures in this phylogenetically older region of the telencephalon form a group of structures collectively known as the **limbic system** (**Figure 46.4**).

The limbic system is responsible for basic physiological drives such as hunger and thirst, instincts, long-term memory formation, and emotions such as fear. Within the limbic system are areas that, when stimulated with small electric currents, can cause intense sensations of pleasure, pain, or rage. If a rat is given the opportunity to stimulate its own pleasure centers by pressing a switch, it will ignore food, water, and even sex, pushing the switch until it is exhausted. Pleasure and pain centers in the limbic system are believed to play roles in learning and in physiological drives.

As we saw in the introduction to Chapter 44, one component of the limbic system—the **amygdala**—is involved in fear and fear memory. If a certain portion of the amygdala is damaged or chemically blocked, an animal cannot learn to be afraid of a stimulus or a situation that would normally induce a strong fear reaction. Moreover, blocking protein synthesis in this part of the limbic system blocks the formation of fear memory.

Another part of the limbic system, the **hippocampus**, is necessary in humans for the transfer of short-term memory to long-term

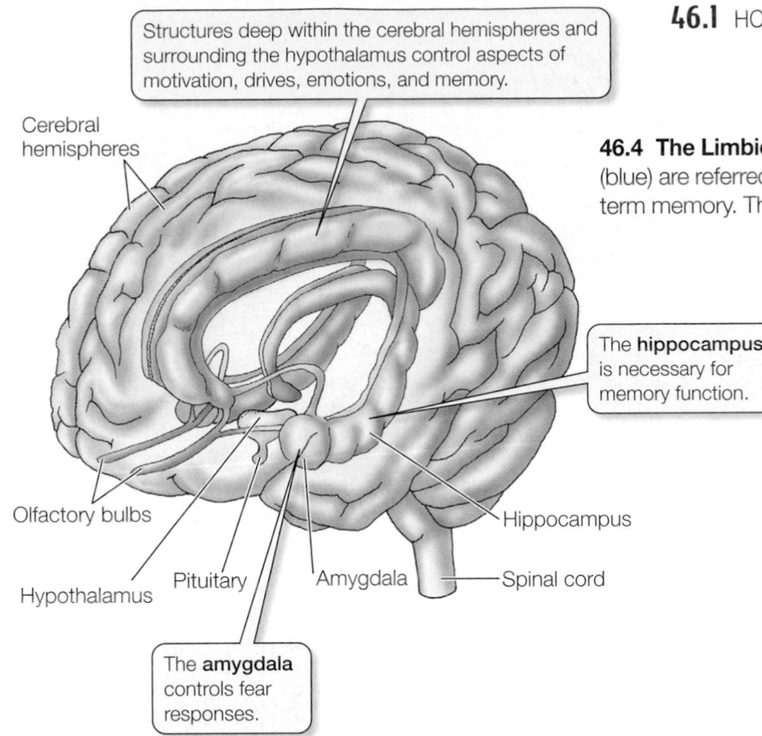

Structures deep within the cerebral hemispheres and surrounding the hypothalamus control aspects of motivation, drives, emotions, and memory.
Cerebral hemispheres
The **hippocampus** is necessary for memory function.
Olfactory bulbs
Hippocampus
Hypothalamus
Pituitary
Amygdala
Spinal cord
The **amygdala** controls fear responses.

46.4 The Limbic System The evolutionarily primitive parts of the telencephalon (blue) are referred to as the limbic system. The hippocampus is involved in forming long term memory. The amygdala is responsible for fear emotions and fear memories.

memory. If you are told a new telephone number, you may be able to hold it in short-term memory for a few minutes, but within half an hour it is forgotten unless you make a real effort to remember it. The phenomenon of remembering something for more than a few minutes requires its transfer to long-term memory.

Regions of the telencephalon interact to produce consciousness and control behavior

The cerebral hemispheres are the dominant structures in the mammalian brain. In humans, they are so large that they cover all other parts of the brain except the cerebellum (**Figure 46.5A**). A sheet of gray matter called the **cerebral cortex** covers each cerebral hemisphere. It is about 4 mm thick and covers a total surface area over both hemispheres of 1 square meter. The cerebral cortex is convoluted, or folded, into ridges called *gyri* (singular *gyrus*) and valleys called *sulci* (singular *sulcus*). These convolutions allow it to fit into the skull. Under the cerebral cortex is white matter, made up of the axons that connect the cell bodies in the cortex with one another and with other areas of the brain.

A curious feature of our nervous system is that the left side of the body is served (in both sensory and motor aspects) mostly by the right side of the brain, and the right side of the body is served mostly by the left side of the brain. Thus, sensory input from the right hand goes to the left cerebral hemisphere, and sensory input from the left hand goes to the right cerebral hemisphere. The exception is the head region where left side of head is controlled by the left cerebral hemisphere and right side is controlled by the right cerebral hemisphere. The two hemispheres, are not symmetrical with respect to all functions. Language abilities, for example, reside predominantly in the left hemisphere, as we will see below.

If we compare the ratio of brain size to body size for all mammals, humans and dolphins top the list. Humans, however, have the largest amount of association cortex relative to body mass of any mammal (although no measure of size of any area of the human brain correlates with human intelligence).

46.5 The Human Cerebrum (A) Each cerebral hemisphere is divided into four lobes. (B) Different functions are localized in particular areas of the cerebral lobes.

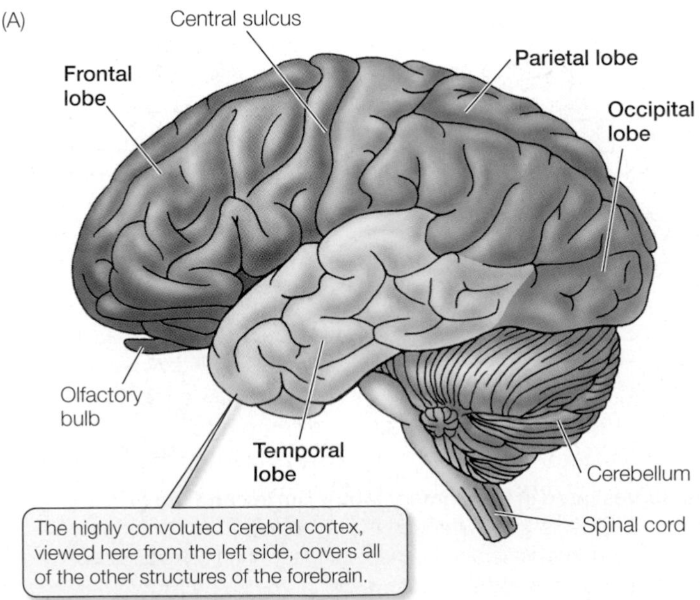

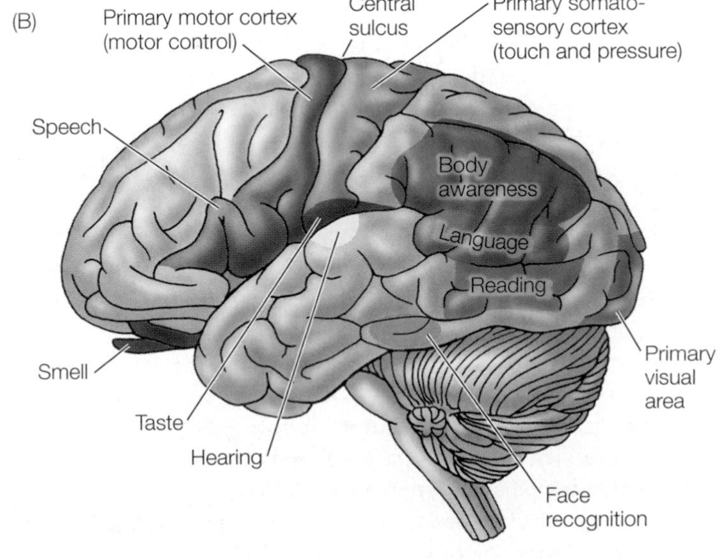

Different regions of the cerebral cortex have specific functions (**Figure 46.5B**). Some of those functions are easily defined, such as receiving and processing sensory information, but most of the cortex is involved in higher-order information processing that is less easy to define. These latter areas are given the general name of **association cortex**, so named because they integrate, or *associate*, information from different sensory modalities and from memory.

To understand the cerebral cortex, it helps to have an anatomical road map. As viewed from the left side, a left cerebral hemisphere looks like a boxing glove for the right hand with the fingers pointing forward, the thumb pointing out, and the wrist at the rear (see Figure 46.5A). The "thumb" area is the **temporal lobe**, the fingers the **frontal lobe**, the back of the hand the **parietal lobe**, and the wrist the **occipital lobe**. A mirror image of this arrangement characterizes the right cerebral hemisphere. Let's look at each lobe of the cerebral cortex separately.

As we explore the functions of the regions of the cerebral cortex and other parts of the brain, you will note frequent mention of persons with damage to their brains. Until recently, the study of such individuals has been the main source of functional information about the human brain, but new imaging technologies such as PET (positron emission tomography) and MRI (magnetic resonance imaging) are providing a wealth of new information and opportunities to study the functioning of the human brain in real time.

THE TEMPORAL LOBE The upper region of the temporal lobe receives and processes auditory information. The association areas of the temporal lobe are involved in the recognition, identification, and naming of objects. Damage to the temporal lobe results in disorders called *agnosias*, in which the individual is aware of a stimulus but cannot identify it.

Damage to a certain area of the temporal lobe results in the inability to recognize faces. Even old acquaintances cannot be identified by facial features, although they may be identified by other attributes such as voice, body features, and characteristic style of walking. Using monkeys, it has been possible to record the activity of neurons in this region that respond selectively to faces in general (**Figure 46.6**). These neurons do not respond to other stimuli in the visual field, and their responsiveness decreases if some of the features of the face are missing or appear in inappropriate locations. Damage to other association areas of the temporal lobe causes deficits in understanding spoken language, although speaking, reading, and writing abilities may be intact.

THE FRONTAL LOBE The frontal and parietal lobes are separated by a deep valley called the *central sulcus*. A strip of the frontal lobe cortex just in front of the central sulcus is called the **primary motor cortex** (see Figure 46.5B). The neurons in this region control muscles in specific parts of the body. The parts of the body have been mapped onto the primary motor cortex largely during neurosurgical procedures. As part of these procedures, electrodes were used to stimulate small areas of cortex. In the area just anterior to the central sulcus, stimulation causes specific muscles to contract. Parts of the body with fine motor control, such as the face and hands, have disproportionate representation (**Figure 46.7**). Stimulation of neurons in the primary motor cortex causes twitches of muscles, not coordinated movements.

The association functions of the frontal lobe are diverse. They are best described as having to do with planning, and they con-

46.6 Neurons in One Region of the Temporal Lobe Respond to Faces The traces represent the firing rate of a neuron in the temporal lobe of a monkey in response to the pictures shown below them.

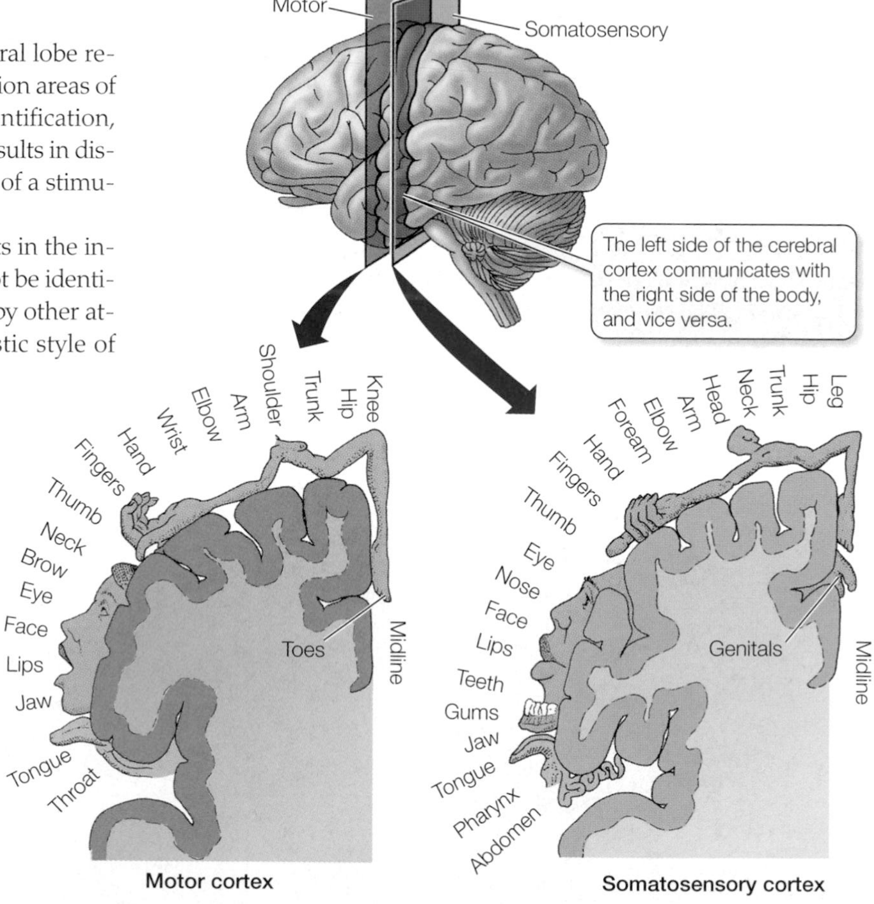

46.7 The Body Is Represented in the Primary Motor Cortex and the Primary Somatosensory Cortex Cross sections through the primary motor and primary somatosensory cortexes can be represented as maps of the human body. Body parts are shown in proportion to the brain area devoted to them.

tribute very significantly to personality. People with frontal lobe damage have drastic alterations of personality because they are unable to perceive themselves in the context of the world around them and cannot plan for future events. A dramatic case of frontal lobe damage is the story of Phineas Gage, who was an industrious, responsible, considerate young railroad construction foreman in 1848. Then a blasting accident shot a meter-long, 3-cm-wide iron tamping rod through his brain. The tamping iron entered Gage's head below his left eye, passed through his frontal lobe, and exited the top of his head (**Figure 46.8**).

Remarkably, Gage survived this terrible accident, but he was a completely different person. He was quarrelsome, bad-tempered, lazy, and irresponsible. He was impatient and obstinate, and he used profane language, which he had never done before. He lost his railroad job and spent the rest of his days as a drifter, earning money by telling his story and exhibiting his scars (and the tamping iron). He died of a seizure disorder in 1860, at the age of 38. If you are in Boston, Massachusetts, you can pay him a visit—his skull, death mask, and the tamping iron are on display in the Warren Anatomical Museum of Harvard Medical School.

THE PARIETAL LOBE The strip of parietal lobe cortex just behind the central sulcus is the **primary somatosensory cortex** (see Figure 46.5B). This area receives touch and pressure information relayed from the body through the thalamus.

The whole body surface can be mapped onto the primary somatosensory cortex (see Figure 46.7). Areas of the body that have a high density of tactile mechanoreceptors and are capable of making fine discriminations in touch (such as the lips and the fingers) have disproportionately large representation. If a very small area of the primary somatosensory cortex is stimulated electrically, the subject reports feeling specific sensations, such as touch, in a localized part of the body.

A major association function of the parietal lobe is attending to complex stimuli. Damage to the right parietal lobe causes a condition called *contralateral neglect syndrome,* in which the individual tends to ignore stimuli from the left side of the body or the left visual field. Such individuals have difficulty performing complex tasks, such as dressing the left side of the body; an afflicted man may not be able to shave the left side of his face. When asked to copy simple drawings, a person who exhibits this syndrome can do well with the right side of the drawing but not the left (**Figure 46.9**). The parietal cortex is not symmetrical with respect to its role in attention, however. Damage to the left parietal cortex does not cause the same degree of neglect of the right side of the body. We will see similar asymmetries in cortical function when we discuss language.

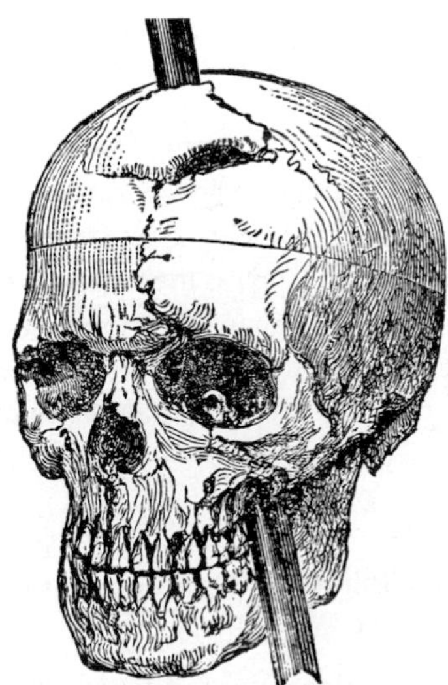

46.8 A Mind-Altering Experience In a nineteenth-century railroad construction accident, an explosion blew a tamping iron through the brain of Phineas Gage. Unbelievably, Gage survived, but his personality was radically changed.

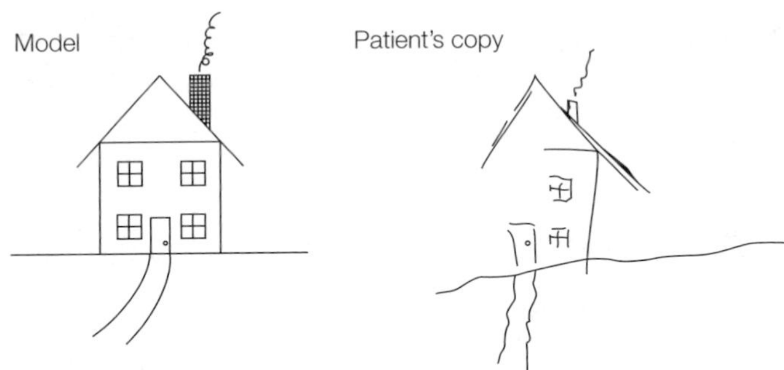

46.9 Contralateral Neglect Syndrome A person with damage to the right parietal association cortex will neglect the left side of a drawing when asked to copy a model.

THE OCCIPITAL LOBE The occipital lobe receives and processes visual information; we'll learn more about that process later in this chapter. The association areas of the occipital cortex are essential for making sense of the visual world and translating visual experience into language. Some deficits resulting from damage to these areas are specific. In one case, a woman with limited damage was unable to see motion. Her vision was intact, but she could see a waterfall only as a still image, and a car approaching only as a series of scenes of a stationary object at different distances.

46.1 RECAP

The central nervous system communicates with the rest of the body through the peripheral nervous system. We are conscious of some sensory information coming into the CNS, but are not conscious of other afferent information used in physiological regulation. Different regions of the brain have specific functions.

- Can you relate the major functional divisions of the nervous system to their origins in the embryonic neural tube? See p. 986–987 and Figure 46.2
- Can you trace the information flow that is involved in a spinal reflex? See p. 987–988 and Figure 46.3
- Can you describe the spatial relations and the functions of the major divisions of the telencephalon? See p. 989–991 and Figure 46.5

Now that we have knowledge of the structure and functions of different regions of the nervous system, we can explore some examples of how information is processed in the neuronal circuitry in some specific regions.

46.2 How Is Information Processed by Neuronal Networks?

We have seen that specific functions are localized in specific parts of the CNS. Therefore, those functions must depend on the neuronal circuits, or networks, in those CNS structures. A major focus of modern neuroscience is to understand how the various functions of the nervous system, ranging from simple reflexes to complex learning and memory, are accomplished by the interactions of neurons in circuits. We now consider two examples of how neuronal networks process information: the autonomic nervous system, which constitutes the efferent pathways of autonomic networks, and the visual system, which processes the information coming from the eyes.

The autonomic nervous system controls involuntary physiological functions

The **autonomic nervous system**, or **ANS**, comprises the output pathways of the CNS that control involuntary functions. The ANS has two divisions, **sympathetic** and **parasympathetic**, that work in opposition to each other in their effects on most organs: one division will cause an increase in an activity and the other a decrease. The sympathetic and parasympathetic divisions of the ANS are easily distinguished by their anatomy, their neurotransmitters, and their actions (**Figure 46.10**).

The best known functions of the ANS are those of the sympathetic division that produce the "fight-or-flight" response: increasing heart rate, blood pressure, and cardiac output and preparing the body for emergencies (see Figure 41.4). In contrast, the parasympathetic division slows the heart and lowers blood pressure, and its actions have been characterized as "rest and digest." It is tempting to think of the sympathetic division as the one that speeds things up and the parasympathetic division as the one that slows things down, but it is not that simple. For example, the sympathetic division slows the digestive system, and the parasympathetic division accelerates it.

Whether sympathetic or parasympathetic, every autonomic efferent pathway begins with a *cholinergic neuron* (one that uses acetylcholine as its neurotransmitter) that has its cell body in the brain stem or spinal cord. These cells are called *preganglionic neurons* because the second neuron in the pathway with which they synapse resides in a collection of neurons outside of the CNS called a *ganglion*. The second neuron is called a *postganglionic neuron* because its axon extends out from the ganglion. The axon of the postganglionic neuron synapses with cells in the target organs.

The postganglionic neurons of the sympathetic division are called *noradrenergic* because they use norepinephrine (also known as noradrenaline) as their neurotransmitter. In contrast, the postganglionic neurons of the parasympathetic division are mostly cholinergic. In organs that receive both sympathetic and parasympathetic input, the target cells respond in opposite ways to norepinephrine and to acetylcholine. A region of the heart called the *pacemaker*, which generates the heartbeat, is an example. Stimulating the sympathetic nerve to the heart or dripping norepinephrine onto the pacemaker region depolarizes the pacemaker cells, increases their firing rate, and causes the heart to beat faster. Stimulating the parasympathetic nerve to the heart or dripping acetylcholine onto the pacemaker region hyperpolarizes the pacemaker cells, decreases their firing rate, and causes the heart to beat more slowly. In contrast, in the digestive tract, norepinephrine hyperpolarizes muscle cells, which slows digestion, and acetylcholine depolarizes muscle cells, which accelerates digestion.

The sympathetic and parasympathetic divisions of the ANS can also be distinguished by anatomy (see Figure 46.10). The preganglionic neurons of the parasympathetic division come from the brain stem and the last segment of the spinal cord (the *sacral* region). The preganglionic neurons of the sympathetic division come from the upper regions of the spinal cord below the neck (the *thoracic* and *lumbar* regions). Most of the ganglia of the sympathetic division are lined up in two chains, one on either side of the spinal cord. The parasympathetic ganglia are close to and sometimes located in the walls of the target organs.

A specialization of the sympathetic division is its innervation of the adrenal gland, which is critical for the "fight or flight" response. The preganglionic sympathetic neuron sends its axon all the way out to the adrenal gland. The medulla (core) of the gland is composed of hormone-secreting cells that are really modified postganglionic sympathetic neurons that have lost their axons, and they secrete their "neurotransmitters" (epinephrine and norepinephrine) into the extracellular fluid (see Section 41.1).

The ANS is an important link between the CNS and many physiological functions. Its control of diverse organs and tissues is crucial to homeostasis. Despite its complexity, work by neurobiologists and physiologists over many decades has made it possible to understand its functions in terms of neuronal properties and circuits. For example, information from pressure receptors in the blood vessels is transmitted to the CNS, where it produces autonomic signals that control the rate of the heartbeat (see Section 49.5).

The first isolation of acetylcholine receptors was possible because of their extremely high concentration in the electric organs of some fish, such as the torpedo ray. Acetylcholine receptors are ion channels, and when many open in a small volume of muscle tissue, they can generate an electric pulse powerful enough to stun a human.

Patterns of light falling on the retina are integrated by the visual cortex

In Section 45.4 we learned how light falling on the retina produces signals that are transmitted through the cellular circuits of the retina resulting in action potentials in the optic nerve. How do these action potentials result in the reconstruction of the visual world in the brain?

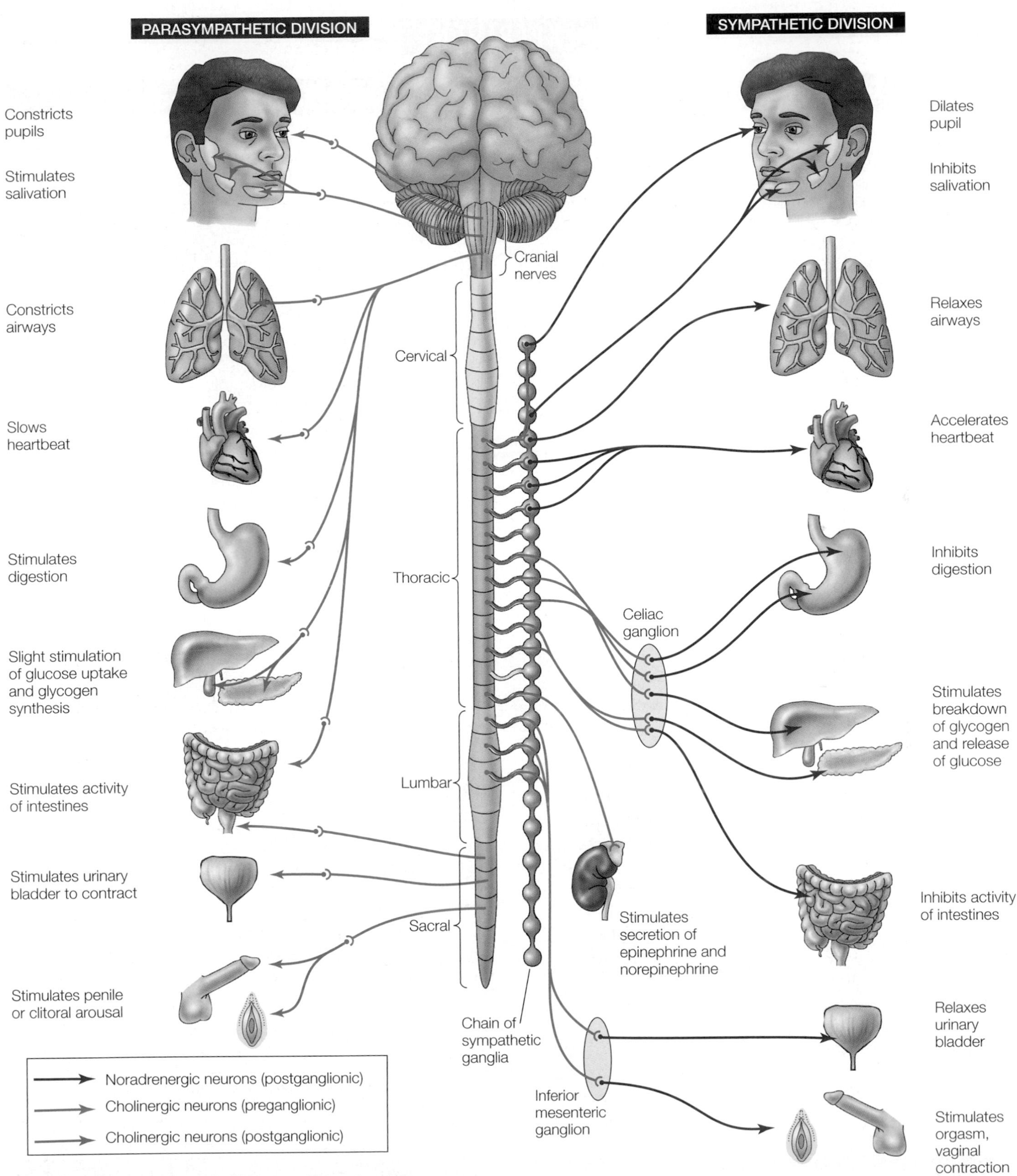

46.10 The Autonomic Nervous System The autonomic nervous system is divided into the sympathetic and parasympathetic divisions. The two divisions which work in opposition to each other in their effects on most organs, with one causing an increase and the other a decrease in activity.

HYPOTHESIS: Retinal ganglion cells are excited or inhibited by light and dark stimuli falling on local areas of the retina.

METHOD

Place electrodes in the optic nerve to record from the axons of ganglion cells while stimulating the retina with different combinations of light and dark stimuli. The stimuli are moved around the retina to find the area of sensitivity for a particular ganglion cell.

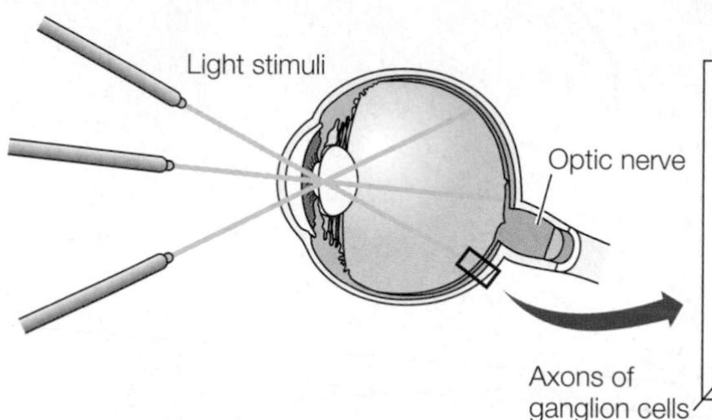

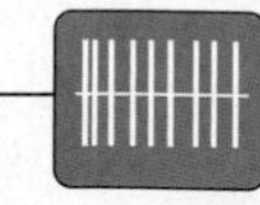

RESULTS

Some ganglion cells are maximally stimulated by light falling on the center of their receptive fields. Others are maximally stimulated by light falling on the surround of their receptive fields.

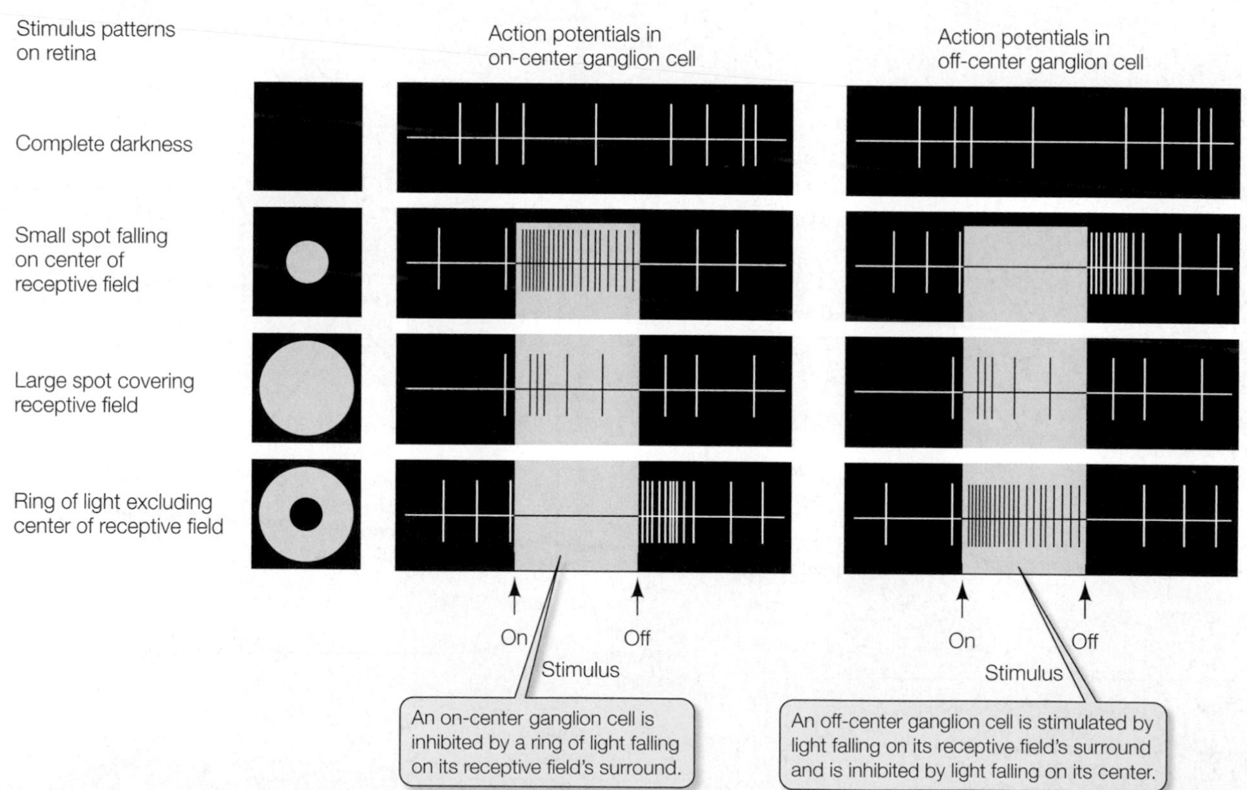

CONCLUSION: Ganglion cells use a center-surround dichotomy to encode patterns of contrast between light and dark.

46.11 What Does the Eye Tell the Brain? When the retina is stimulated with dots and rings of light, individual ganglion cells show different responses.

RETINAL RECEPTIVE FIELDS One aspect of the processing of visual information through the retina is the *convergence of information*. Each human retina contains over 100 million photoreceptors, but only about 1 million *ganglion cells*. It is the axons of the ganglion cells that communicate information from the eyes to the brain. How is the information from so many photoreceptors integrated by the many fewer ganglion cells?

This question was addressed in classic experiments by Stephen Kuffler at Johns Hopkins University in 1953, in which he used electrodes to record the activity in the axons of single ganglion cells while the retina of a cat was stimulated with spots of light (**Figure 46.11**). These experiments were the starting point for our understanding of how the brain assembles information from single cells to create visual images—in other words, of how the brain sees.

Kuffler's studies revealed that each ganglion cell has a well-defined **receptive field** composed of a group of photoreceptor cells that receive light from a small area of the whole visual field. Stimulating these photoreceptors with light activates the ganglion cell, which sends action potentials to the thalamus and on to the visual cortex (the area of the occipital lobe where visual information is processed; see Figure 46.5B). Information from many photoreceptors is therefore communicated to the brain as a single message. However, individual photoreceptors may contribute to the receptive fields of multiple ganglion cells.

The receptive fields of most ganglion cells are circular, but whether a spot of light falling on a receptive field excites or inhibits that ganglion cell depends both on the nature of the receptive field and on where the spot of light falls on the receptive field. Receptive fields have two concentric regions—a *center* and a *surround*—and the field can be either *on-center* or *off-center*. Light falling on an on-center receptive field excites the ganglion cell, and light falling on an off-center receptive field inhibits the ganglion cell. The surround area has the opposite effect: the surround for an on-center receptive field inhibits the ganglion cell and the surround for an off-center field is excitatory. Thus the activity of the ganglion cell reflects how much of the light stimulus is on the center and how much is on the surround of its receptive field (see Figure 46.11).

Center effects are always stronger than surround effects, however. Thus a small dot of light directly on the center of a receptive field has the maximal effect, and a larger light stimulus illuminating the center and parts of the surround has less of an effect. A uniform patch of light falling equally on center and surround has very little effect on the firing rate of the ganglion cell for that receptive field.

How are cells in the retina connected to each other to create receptive fields? The photoreceptors in a receptive field's center are connected to the associated ganglion cell by *bipolar cells*. The photoreceptors in the surround modify the communication between the center photoreceptors and their bipolar cells through the lateral connections of *horizontal cells* (see Figure 45.21). Thus, the receptive field of a ganglion cell results from a pattern of synapses between photoreceptors, horizontal cells, amacrine, and bipolar cells.

Now things can get a bit confusing. Remember that vertebrate photoreceptors are depolarized in the dark and hyperpolarized by light. Thus these cells are releasing neurotransmitter in the dark, whereas light reduces their rate of neurotransmitter release (see Section 45.4). The neurotransmitter of photoreceptors is glutamate which is usually an excitatory neurotransmitter (see Section 44.3). For bipolar cells, however, glutamate can be excitatory or inhibitory depending on the receptors they have. Thus, for bipolar cells of on-center receptive fields, glutamate is inhibitory. In the dark, when the central photoreceptors are releasing the most glutamate, the bipolar cells of on-center receptive fields are hyperpolarized. When light shines on those photoreceptors, they decrease their release of glutamate causing their bipolar cells to depolarize. As a result, these bipolar cells release more neurotransmitter onto their ganglion cells increasing their firing rates. The opposite is true for the bipolar cells of off-center receptive fields as they have glutamate receptors that are excitatory. The general lesson to learn from this seemingly confusing chain of events is that inhibition can be as important as excitation in neural circuits, and sometimes excitation results from inhibition of an inhibitory synapse.

In summary, the neuronal circuitry of the retina results in the generation of signals in the axons of the optic nerve that communicate simple information about the patterns of light and dark falling on different parts of the retina.

RECEPTIVE FIELDS OF CELLS IN THE VISUAL CORTEX We learned how photoreceptors in the retina transduce light into neural signals and how the neural circuits in the retina organize those signals into information about patterns of light and dark that is transmitted to the brain. But once the action potentials in the optic nerve reach their destinations, how does the brain integrate them to construct visual image of the outside world?

The axons of the optic nerves terminate in a region of the thalamus which is a relay station that integrates information from the right and left eyes. From the thalamus the information encoded in the activity of axons in the optic nerves is relayed to the visual cortex in the occipital lobes at the back of the brain. In the 1960s, David Hubel and Torsten Wiesel of Harvard University studied the activity of neurons in the visual cortex by shining spots and bars of light on retinas while recording the activities of single cells in the cortex. They found that neurons in the visual cortex, like retinal ganglion cells, have receptive fields. For their pioneering work in visual neurophysiology, Hubel and Wiesel received the Nobel prize in 1981.

Receptive fields of neurons in the visual cortex, however, differ from the circular receptive fields of retinal ganglion cells. Cortical neurons labeled *simple cells* are maximally stimulated by bars of light with a particular orientation and falling on a small region of the retina. These simple cells probably receive input from several ganglion cells whose circular retinal receptive fields are lined up in a row, creating a bar of sensitivity.

Another class of cortical cells is *complex cells*. These cells responded maximally to bars of light with particular orientations, but the bar of light could fall anywhere on a large area of retina. Thus the receptive field of a complex cell appears to be due to that cell receiving inputs from a number of simple cells that share a certain stimulus orientation but have receptive fields in different locations on the retina (**Figure 46.12**). Some complex cells respond most strongly when the bar of light moves in a particular direction.

The concept that emerges from these experiments is that the brain assembles a mental image of the visual world by analyzing edges in patterns of light falling on the retina. This analysis is conducted in a massively parallel fashion. Each retina sends a million axons to the

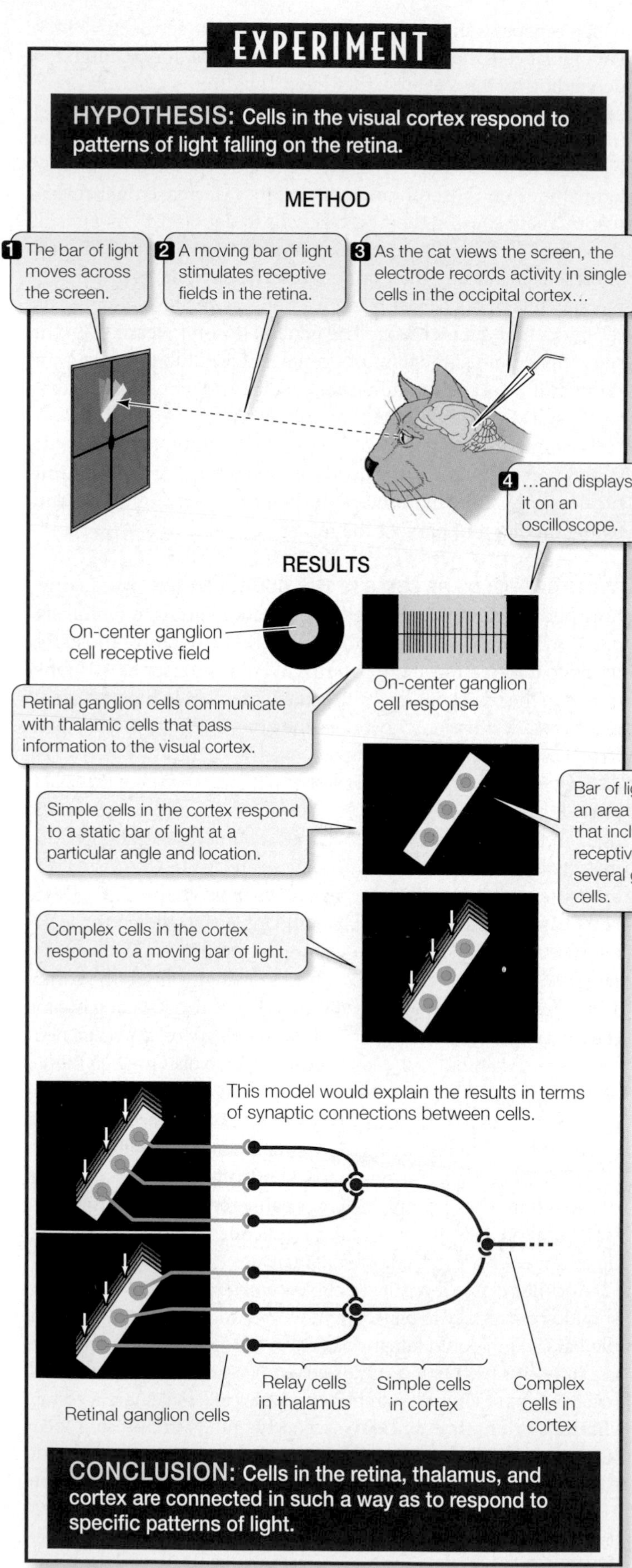

EXPERIMENT
HYPOTHESIS: Cells in the visual cortex respond to patterns of light falling on the retina.
METHOD
1 The bar of light moves across the screen.
2 A moving bar of light stimulates receptive fields in the retina.
3 As the cat views the screen, the electrode records activity in single cells in the occipital cortex...
4 ...and displays it on an oscilloscope.
RESULTS
On-center ganglion cell receptive field
On-center ganglion cell response
Retinal ganglion cells communicate with thalamic cells that pass information to the visual cortex.
Simple cells in the cortex respond to a static bar of light at a particular angle and location.
Bar of light covers an area of retina that includes receptive fields of several ganglion cells.
Complex cells in the cortex respond to a moving bar of light.
This model would explain the results in terms of synaptic connections between cells.
Retinal ganglion cells
Relay cells in thalamus
Simple cells in cortex
Complex cells in cortex
CONCLUSION: Cells in the retina, thalamus, and cortex are connected in such a way as to respond to specific patterns of light.

46.12 Receptive Fields of Cells in the Visual Cortex Cells in the visual cortex respond to specific patterns of light falling on the retina. Ganglion cells that transmit information about circular receptive fields converge on the cortex's simple cells, which have linear receptive fields. Simple cells transmit information to complex cells, which respond to linear stimuli falling on different areas of the retina.

brain, but there are *hundreds of millions* of neurons in the visual cortex. The action potentials from one retinal ganglion cell are received by hundreds of cortical neurons, each responsive to a different combination of orientation, position, color, and movement of contrasting lines in the patterns of light and dark falling on the retina.

Cortical cells receive input from both eyes

How do we see objects in three dimensions? The quick answer is that a person's two eyes, located at the front of the head, see overlapping, yet slightly different, visual fields—that is, humans have **binocular vision**. A person who is blind in one eye has great difficulty discriminating distances. Animals whose eyes are on the sides of the head have minimal overlap in their fields of vision and, as a result, poor depth vision; however, they can see predators creeping up from all sides.

The story of how the brain integrates information from two eyes begins with the paths of the optic nerves. From the underside of the brain, the optic nerves from the two eyes appear to join together just under the hypothalamus and then separate again (**Figure 46.13**). The place where they join is called the **optic chiasm**. Axons from the half of each retina closest to your nose cross in the optic chiasm and go to the opposite side of your brain. The axons from the other half of each retina go to the same side of the brain. This means that visual information from the left visual field (everything left of straight ahead) goes to the right side of the brain, as shown in red in Figure 46.13, while visual information from the right visual field goes to the left side of the brain (shown in green). Both eyes transmit information about a specific spot in your right visual field, for example, to the same place in the left visual cortex.

Cells in the visual cortex are organized in stripes and columns. Stripes refer to the organization across the surface of the cortex, and columns refer to the organization through the depth of the cortex. Both stripes and columns alternate according to the source of their input: left eye, right eye, left eye, right eye, and so on. Cells closest to the border between two stripes or columns receive input from both eyes and are therefore called *binocular cells*. Binocular cells interpret distance by measuring the *disparity* between the points at which the same stimulus falls on the two retinas.

What is disparity? Hold your finger out in front of you and look at it, closing one eye and then the other. Your finger appears to jump back and forth, because its image falls on a different position on each retina. Repeat the exercise with an object at a distance. It doesn't appear to jump back and forth as much, because there is less disparity in the positions of the image on the two retinas. Certain binocular cells respond optimally to a stimulus falling on both retinas with a particular disparity. Which set of binocular cells is stimulated depends on how far away the stimulus is.

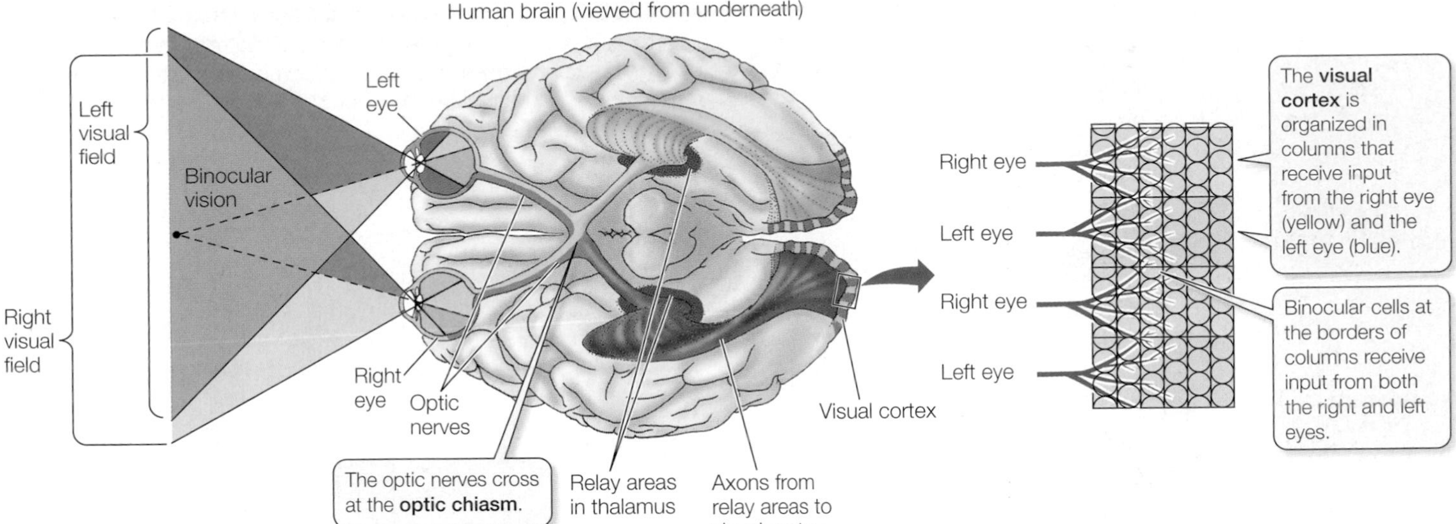

46.13 The Anatomy of Binocular Vision Each eye transmits information to both sides of the brain; however, the right side of the brain processes all information from the left visual field, and the left side of the brain processes all information from the right visual field. The visual cortex sorts visual field information according to whether it comes from the right eye or the left eye.

When we look at something, we can detect its shape, color, depth, and movement. Where does all this information come together? Is there a single cell that fires only when a red sports car drives by? The answer is no. Specific visual experience comes from simultaneous activity in a large collection of cells. In addition, most visual experiences are enhanced by information from the other senses and from memory, which helps explain why about 75 percent of the cerebral cortex is association cortex.

46.2 RECAP

Information in the nervous system is processed by cellular interactions in neuronal networks. The opposing actions of the sympathetic and parasympathetic divisions of the ANS can be understood in terms of neural pathways consisting of just two neurons. Vision involves a more complex interaction of neurons to accomplish the processing of important sensory information.

- Can you describe the anatomical and functional differences between the sympathetic and parasympathetic divisions? See p. 992 and Figure 46.10
- Do you understand the cellular basis for the receptive fields of retinal ganglion cells? See p. 995 and Figure 46.11
- How do cells in the visual cortex get information about the distance of an object? See p. 996 and Figure 46.13

By studying the neural circuitry of the visual system and the autonomic nervous system, you have gained some understanding of how information reaches the central nervous system and how the CNS controls various functions of the body. But what about the higher functions of the mammalian CNS—the complex functions between input and output, such as language, learning, memory, and dreams?

46.3 Can Higher Functions Be Understood in Cellular Terms?

The higher brain functions discussed in the remaining pages of this chapter are undeniably complex. Nevertheless, neuroscientists, using a wide range of techniques, are making considerable progress in understanding some of the cellular and molecular mechanisms involved in those processes. The following discussion presents several complex aspects of brain and behavior that present challenges to neuroscientists: sleep and dreaming, learning and memory, language use, and consciousness.

Sleep and dreaming are reflected in electrical patterns in the cerebral cortex

A dominant feature of behavior is the daily cycle of sleep and waking. All birds and mammals, probably all other vertebrates, and also many invertebrates, sleep. We humans spend one-third of our lives sleeping, yet we do not know why or how. We do know, however, that we need to sleep. Loss of sleep impairs alertness and performance. Many people in our society—certainly most college students—are chronically sleep-deprived. Every day, accidents and serious mistakes that endanger lives can be attributed to impaired alertness due to lack of sleep. Insomnia (difficulty in falling or staying asleep) is one of the most common medical complaints.

(A)

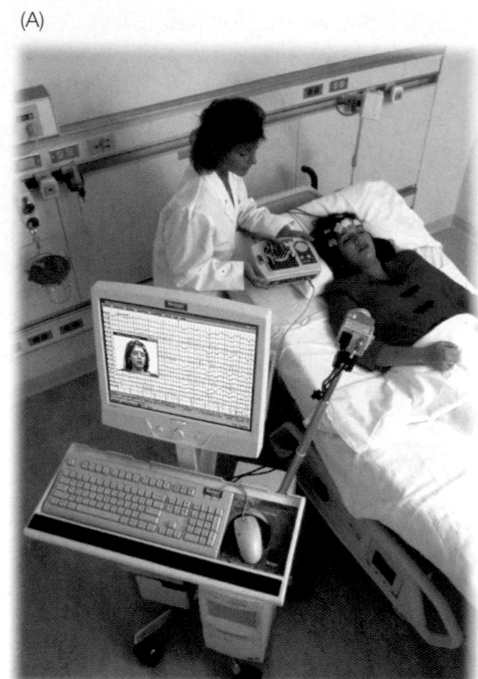

46.14 Patterns of Electrical Activity in the Cerebral Cortex Characterize Stages of Sleep (A) Electrical activity in the cerebral cortex is detected by electrodes placed on the scalp and recorded as changes in voltage between the electrodes through time. (B) The resulting record is an electroencephalogram (EEG). (C) Humans cycle through different stages of sleep throughout the night.

(B)

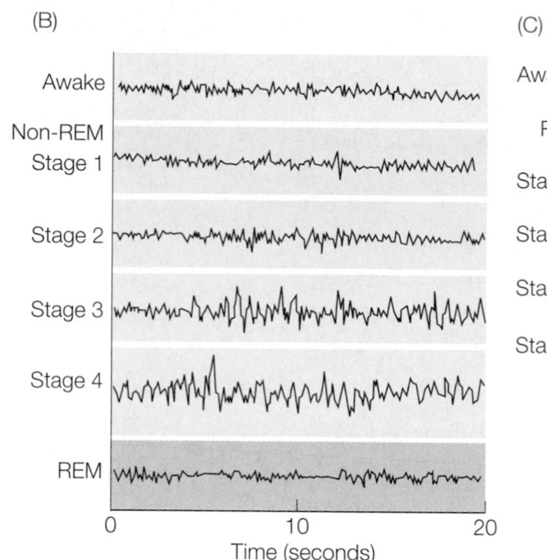

(C)

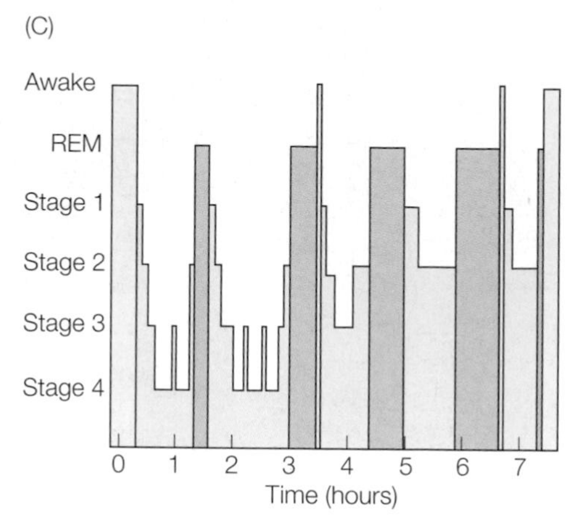

THE ELECTROENCEPHALOGRAM A common tool of sleep researchers is the **electroencephalogram**, or **EEG** (**Figure 46.14A**). Rather than recording the activity of single neurons, the EEG characterizes activity in huge numbers of neurons. EEG electrodes are much larger than the very fine electrodes used to detect single cell activity. EEG electrodes are placed at different locations on the scalp, and changes in the electric potential differences between electrodes are recorded over time. These electric potential differences reflect the electrical activity of the neurons in the brain regions under the electrodes, primarily regions of the cerebral cortex. Usually, the electrical activity of one or more skeletal muscles is also recorded on the chart; this record is called an *electromyogram (EMG)*. Movements of the eyes are recorded as an *electrooculogram (EOG)*.

EEG, EMG, and EOG patterns reveal the transition from being awake to being asleep. They also reveal that there are different states of sleep. In mammals other than humans, two major sleep states are easily distinguished: They are **slow-wave sleep** and **rapid-eye-movement (REM) sleep**. In humans, sleep states are characterized as **non-REM sleep** and **REM sleep**. Of the four stages of human non-REM sleep, only the two deepest stages are considered true slow-wave sleep.

When you fall asleep at night, the first sleep state entered is non-REM sleep, which progresses from stage 1 to stage 4 (**Figure 46.14B**). Stages 3 and 4 are deep, restorative, slow-wave sleep. This first full episode of non-REM sleep is followed by an episode of REM sleep. Throughout the night, you experience four or five cycles of non-REM and REM sleep (**Figure 46.14C**). About 80 percent of your sleep is non-REM sleep.

Vivid dreams and nightmares occur during the 20 percent of sleep that is REM sleep, which gets its name from the jerky movements of the eyeballs that occur during this state. The most remarkable feature of REM sleep is that inhibitory commands from the brain almost completely paralyze the skeletal muscles. Occasional muscle twitches break through the paralysis, as can be seen in a dog that appears to be trying to run in its sleep. The function of muscle paralysis during REM sleep may be to prevent the acting out of dreams. Sleepwalking occurs during non-REM sleep.

CELLULAR CHANGES DURING SLEEP Striking neurophysiological differences distinguish non-REM from REM sleep. Non-REM sleep is characterized by a decrease in the responsiveness of neurons in the thalamus and cerebral cortex. Remember that neurons have a negative resting membrane potential and a threshold for firing action potentials (see Section 44.2). Usually the resting potential is near or slightly below the threshold potential, so the neuron is firing at a low rate. When synaptic input causes the membrane potential to become less negative (depolarized), the cell can exceed threshold and fire action potentials at a higher rate.

During waking, several nuclei in the brain stem reticular formation are continuously active. Axons from neurons in these nuclei extend to the thalamus and throughout the cerebral cortex where they release depolarizing neurotransmitters (acetylcholine, norepinephrine, and serotonin). These broadly distributed neurotransmitters keep the resting potential of the neurons of the thalamus and cortex close to threshold and sensitive to synaptic inputs, thereby maintaining waking. As mentioned earlier, brain damage that cuts off the influence of these brainstem nuclei on the forebrain results in coma.

With the onset of sleep, activity in these brain stem nuclei decreases, and their axon terminals release less neurotransmitter. With the withdrawal of the depolarizing neurotransmitters, the resting potentials of the cells of the thalamus and cortex become more negative (hyperpolarized), and the cells are less sensitive to excitatory synaptic input. Their processing of information is inhibited, and consciousness is lost.

An interesting neuronal event happens as a result of this hyperpolarization: The cells begin to fire action potentials in bursts. The synchronization of these bursts over broad areas of cerebral cortex results in the EEG slow-wave pattern that characterizes non-REM sleep. Studies of neurons of the thalamus and the cortex have shown that their hyperpolarization during non-REM sleep is due to increased opening of K^+ channels, and the bursting is due to Ca^{2+} channels whose inactivation gates close rapidly and require hyperpolarization to be reopened. We can therefore explain the EEG pattern of non-REM sleep in terms of the properties of neurons and ion channels.

At the transition from non-REM to REM sleep, dramatic changes occur. Some of the brain stem nuclei that were inactive during non-REM sleep become active again, causing a general depolarization of cortical neurons. Thus the synchronized bursts of firing cease, and the EEG resembles that of the waking brain. Because the resting potentials of the neurons return to near threshold levels, the cortex can process information, and vivid dreams can occur. However, during non-REM sleep the brain inhibits both afferent (sensory) and efferent (motor) pathways; therefore, the activity in the cortex is unconstrained by its usual sources of information, which is why events in dreams are often quite bizarre.

Knowing the cellular mechanisms of sleep has not led to an understanding of its function, and many questions remain. Why do we have two sleep states with very different neurophysiological characteristics? Why does non-REM sleep always occur first? Why do the two states oscillate on a regular basis through the rest period? We know sleep is essential for life, but we don't know why. One prominent hypothesis is that sleep is necessary for the maintenance and repair of neuronal connections and for the neuronal changes involved in learning and memory—and possibly forgetting. Evidence for such functions is still meager, however.

Some learning and memory can be localized to specific brain areas

Learning is the modification of behavior by experience. *Memory* is the ability of the nervous system to retain what is learned and what is experienced. Even very simple animals can learn and remember, but these two abilities are most highly developed in humans. Consider the amount of information associated with learning a language. The capacity of memory and the rate at which memories can be retrieved are remarkable features of the human nervous system.

LEARNING Learning that leads to long-term memory and modification of behavior must involve long-lasting synaptic changes. An experimental model of how long-term synaptic changes might arise is called **long-term potentiation**, or **LTP** (see Figure 44.18). LTP involves the high-frequency electrical stimulation of certain identifiable circuits that makes these circuits more sensitive to subsequent stimulation. In contrast, continuous, repetitive, low-level stimulation of these hippocampal circuits reduces their responsiveness, a phenomenon that has been called **long-term depression** (**LTD**). LTP and LTD may be fundamental cellular or molecular mechanisms involved in learning and memory.

Several kinds of learning exist. A form of learning that is widespread among animal species is **associative learning**, in which two unrelated stimuli become linked to the same response. The simplest example of associative learning is the **conditioned reflex**, discovered by the Russian physiologist Ivan Pavlov. Pavlov was studying the control of digestive functions in dogs and observed that a dog salivates at the sight or smell of food—a simple autonomic reflex. He discovered that if he rang a bell just before food was presented to the dog, after a few trials the dog would salivate at the sound of the bell, even if no food followed. The salivation reflex was conditioned to be associated with the sound of a bell, a stimulus that normally is unrelated to feeding and digestion.

This simple form of learning has been studied extensively in efforts to understand its underlying neuronal mechanisms. In a series of studies led by Richard Thompson, the eye-blink reflex of a rabbit in response to a puff of air directed at its eyes was conditioned to be associated with a tone stimulus. After conditioning, the rabbit blinked when it heard the tone. A small and specific area of the cerebellum was discovered to be necessary for this conditioned reflex. Thus, it was possible to localize one form of learning to an identifiable set of synapses in the mammalian brain.

MEMORY Attempts to treat human neurological diseases have led to the identification of areas of the brain involved in the formation and recall of memories. Epilepsy is a disorder characterized by uncontrollable increases in neuronal activity in specific parts of the brain. The resulting *seizures*, or "epileptic fits," can endanger the afflicted individual. In the past, serious cases of epilepsy were sometimes treated by destroying the part of the brain from which the surge of activity originated.

To find the right area, the surgery was done under local anesthesia, and different regions of the brain were electrically stimulated with electrodes while the patient reported on the resulting sensations. Stimulation of some regions of the association cortex elicited recall of vivid memories. Such observations were the first evidence that specific areas in the brain are associated with specific memories and that memory can be attributed to properties of neurons and networks of neurons. The destruction of a small area of the brain does not completely erase a memory, however, so it is postulated that memory is a function distributed over many brain regions and can be stimulated via many different routes.

You may recognize several forms of memory from your own experience. There is **immediate memory** for events that are happening now. Immediate memory is almost perfectly photographic, but it lasts only seconds. **Short-term memory** contains less information, but it lasts longer—on the order of 10 to 15 minutes. If you are introduced to a group of new people, you may remember most of their names for 5 or 10 minutes, but you will have forgotten them in an hour or so if you have not repeated them, written them down, or used them in a conversation. Repetition, use, or reinforcement by something that gets your attention (such as the title President) facilitates the transfer of short-term memory to **long-term memory**, which can last for days, months, years, or a lifetime.

Knowledge about neuronal mechanisms for the transfer of short-term memory to long-term memory has come from observations of persons who have lost parts of the limbic system, notably the hippocampus. A famous case is that of a man identified as Henry M., whose hippocampus on both sides of the brain was removed in an effort to control severe epilepsy. After that surgery,

H.M. was not able to transfer information to long-term memory. If someone was introduced to him, had a conversation with him, and then left the room for several minutes, when that person returned, H.M did not know him—it was as if the conversation had never taken place. H.M. retained memories of events that happened before his surgery, but could not remember postsurgery events for more than 10 or 15 minutes.

Memory of people, places, events, and things is called **declarative memory** because you can consciously recall and describe them. Another type of memory, called **procedural memory**, cannot be consciously recalled and described: It is the memory of how to perform a motor task. When you learn to ride a bicycle, ski, or use a computer keyboard, you form procedural memories. Although H.M. was incapable of forming declarative memories, he could form procedural memories. When taught a motor task day after day, he could not recall the lessons of the previous day, yet his performance steadily improved. Thus procedural learning and memory must involve mechanisms different from those used in declarative learning and memory.

Memories can have considerable emotional content. As mentioned earlier, the limbic system plays a major role in controlling emotions. The amygdala is necessary for the emotion of fear and the formation of fear memories. Patients with damage to the amygdala can form declarative memories but do not associate fear reactions with those memories. Normal subjects can be conditioned easily to show anticipatory fear. For example, subjects may be shown cards of different colors, one of which is accompanied by an unpleasant electrical shock, which causes the sympathetic nervous system response of increasing heart rate. Once trained, the heart rate of subjects will increase upon seeing the card even if the shock is not delivered. Patients with damage to the amygdala, however, show the sympathetic response to the shock, but not in response to seeing the appropriate card. They will report that they expect to receive a shock, but do not exhibit a fear response.

Language abilities are localized in the left cerebral hemisphere

No aspect of brain function is as integrally related to human consciousness and intellect as is language. Therefore, studies of the brain mechanisms that underlie the acquisition and use of language are extremely interesting to neuroscientists. A curious observation about language abilities is that they are usually located in only one cerebral hemisphere—which in 97 percent of people is the left hemisphere. This phenomenon is referred to as the *lateralization* of language functions.

Some of the most fascinating research on this subject was conducted by Roger Sperry and his colleagues at the California Institute of Technology; Sperry received a Nobel prize for this work in 1981. The two cerebral hemispheres are connected by a tract of white matter called the *corpus callosum*. In one severe form of epilepsy, bursts of action potentials travel from hemisphere to hemisphere across the corpus callosum. Cutting the tract eliminates the problem, and patients function nearly normally following the surgery. But the "split-brain" subjects of this surgery displayed interesting deficits in language ability. Lacking the connecting tissue between the two hemispheres, knowledge or experience of the right hemisphere could no longer be expressed in language—a function localized in the left hemisphere. Nor could language be used to communicate with the right hemisphere. A patient could respond to verbal instructions to use his right hand (controlled by the left cerebral cortex), but not his left hand (controlled by the right cerebral cortex).

The mechanisms of language in the left hemisphere have been the focus of much research. The experimental subjects are persons who have suffered damage to the left hemisphere and are left with one of many forms of **aphasia**, a deficit in the ability to use or understand words. These studies have identified several language areas in the left hemisphere.

Broca's area, located in the frontal lobe just in front of the primary motor cortex, is essential for speech. Damage to Broca's area results in halting, slow, poorly articulated speech or even complete loss of speech, but the patient can still read and understand language. In the temporal lobe, close to its border with the occipital lobe, is *Wernicke's area*, which is more involved with sensory than with motor aspects of language. Damage to Wernicke's area can cause a person to lose the ability to speak sensibly while retaining the abilities to form the sounds of normal speech and to imitate its cadence. Moreover, such a patient cannot understand spoken or written language. Near Wernicke's area is the *angular gyrus*, which is believed to be essential for integrating spoken and written language.

Normal language ability depends on the flow of information among various areas of the left cerebral cortex. Input from spoken language travels from the auditory cortex to Wernicke's area (**Fig-**

46.15 Language Areas of the Cortex Different regions of the left cerebral cortex participate in the processes of (A) repeating a word that is heard and (B) speaking a written word.

(A) Repeating a heard word

Motor
Broca's area
Speech
Hearing
Wernicke's area
Vision

(B) Speaking a written word

Angular gyrus

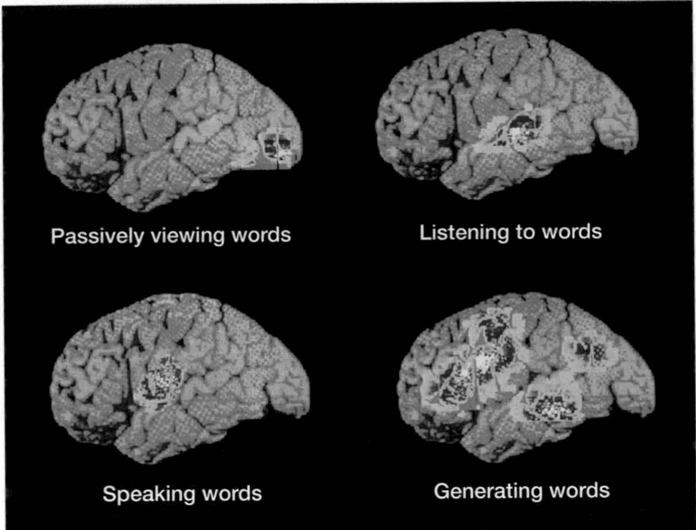

46.16 Imaging Techniques Reveal Active Parts of the Brain Positron emission tomography (PET) scanning reveals the brain regions that are activated by different aspects of language use. Radioactively labeled glucose is given to the subject. Brain areas take up radioactivity in proportion to their metabolic use of glucose. The PET scan visualizes levels of radioactivity in specific brain regions when a particular activity is performed. The red and white areas are the most active.

ure 46.15A). Input from written language, which is happening right now as you read this sentence, travels from the visual cortex to the angular gyrus to Wernicke's area (**Figure 46.15B**). Commands to speak are formulated in Wernicke's area and travel to Broca's area and from there to the primary motor cortex. Damage to any one of those areas or the pathways between them can result in aphasia.

Using modern methods of functional brain imaging, it is possible to see the metabolic activity in different brain areas when the brain is using language (**Figure 46.16**). As noted earlier, most knowledge of regional functions of the human brain have come from studies of patients who have sustained damage—usually injuries from war or accidents. Modern methods of imaging can be used on persons with either damaged or normal brains to study activity in real time; thus, these methods have greatly increased our ability to investigate human brain functions.

What is consciousness?

This chapter has only scratched the surface of our knowledge about the organization and functions of the human brain. Even the most sophisticated new research tools may not allow us to answer the overriding question of "What is consciousness?"

If you see a black dog running across a field, you are conscious that it is a dog, it is black, and it is a Labrador retriever. You may also remember that the dog's name is Sarina, that she belongs to your friend Meera, and that Sarina is 6 years old. From what you have learned in this chapter, imagine how many neurons would be active during this experience: neurons in the visual system, the language areas, and in different regions of association cortex. But is being *conscious* of the dog simply a result of all of these neurons firing at the same time? Your brain is simultaneously processing many other sensory inputs, but you are not necessarily conscious of those inputs. What makes you conscious of the black dog and associated memories and not of other information the brain is processing at the same time?

If we could describe all the neurons and all the synapses involved in the conscious experience of seeing and naming a black dog, and then build a computer with devices that modeled all these neurons and connections, would that computer be conscious? It has been said that the question of consciousness resolves into two types of problems: "easy" and "hard." The easy problems deal with all the cells and circuits that process the information that is involved in conscious experience. The implication of "easy" is that we seem to have the tools to solve these kinds of problems, as complex as they may be. The hard problems involve explaining how properties of cells and networks result in consciousness, and we seem to lack the proper tools or concepts even to begin to solve these problems.

46.3 RECAP

Even complex functions of the nervous system are beginning to be understood in terms of the properties of neurons and neuronal networks.

- Can you describe the differences between slow-wave sleep and REM sleep? See p. 998 and Figure 46.14
- Can you describe the role of the hippocampus in memory and the different types of memories it is involved in? See pp. 999–1000

CHAPTER SUMMARY

46.1 How is the mammalian nervous system organized?

The brain and spinal cord make up the **central nervous system** (**CNS**); the cranial and spinal nerves make up the **peripheral nervous system** (**PNS**).

The nervous system can be modeled conceptually in terms of the direction of information flow and whether we are conscious of the information. The **afferent** component carries information from the PNS to the CNS, and the **efferent** component directs information from the CNS to the peripheral parts of the body. Review Figure 46.1

The vertebrate nervous system develops from a hollow dorsal **neural tube**. The brain forms from three swellings at the anterior end of the neural tube, which become the **hindbrain**, the **midbrain**, and the **forebrain**. Review Figure 46.2

The forebrain develops into the **cerebral hemispheres** (the **telencephalon**) and the underlying **thalamus** and **hypothalamus** (the **diencephalon**). The midbrain and hindbrain develop into the **brain stem** and the **cerebellum**. Review Figure 46.2, Web/CD Tutorial 46.1

The spinal cord communicates information between the brain and the rest of the body. It can issue some commands to the body without input from the brain. Review Figure 46.3

The **reticular system** is a complex network that directs incoming information to appropriate brain stem **nuclei** that control **autonomic functions**, and transmits the information to the forebrain that results in conscious sensation. The reticular system controls the level of arousal of the nervous system, including sleep and wakefulness.

The **limbic system** is an evolutionarily primitive part of the telencephalon that is involved in emotions, physiological drives (such as hunger or thirst), instincts, and memory. Review Figure 46.4

The cerebral hemispheres are the dominant structures of the human brain. Their surfaces are layers of neurons called the **cerebral cortex**. The cerebral hemispheres can be divided into **temporal**, **frontal**, **parietal**, and **occipital lobes**. Many motor functions are localized in parts of the frontal lobe. Information from many sensory receptors projects to a region of the parietal lobe. Visual information projects to the occipital lobe, and auditory information projects to a region of the temporal lobe. Review Figures 46.5, 46.6, and 46.7, Web/CD Activity 46.1

46.2 How is information processed by neuronal networks?

The **autonomic nervous system** (**ANS**) consists of efferent pathways that control the physiological function of organs and organ systems. Its **sympathetic** and **parasympathetic** divisions are characterized by their anatomy, neurotransmitters, and effects on target tissues. Review Figure 46.10

The neuronal network of vision involves patterns of light falling on **receptive fields** in the retina. Receptive fields have a center and a surround, which have opposing effects on ganglion cell firing. Review Figure 46.11, Web/CD Tutorial 46.2

Information from retinal ganglion cells is communicated via the optic nerve to the thalamus and then to the visual cortex. The visual cortex seems to assemble an image of the visual world by analyzing edges of patterns of light. Review Figure 46.12

Binocular vision is possible because information from both eyes is communicated to binocular cells in the visual cortex. These cells interpret distance by measuring the disparity between where the same stimulus falls on the two retinas. Review Figure 46.13

46.3 Can higher functions be understood in cellular terms?

Humans have a daily cycle of sleep and waking. Sleep can be divided into **rapid-eye-movement** (**REM**) **sleep** and **slow-wave** (**non-REM**) **sleep**. Review Figure 46.14

Some learning and memory processes have been localized to specific brain areas. Long-lasting changes in synaptic properties referred to as **long-term potentiation** (**LTP**) and **long-term depression** (**LDP**) may be involved in learning and memory.

Complex memories can be elicited by stimulating small regions of association cortex. Damage to the hippocampus can destroy the ability to form long-term **declarative memories**, but not **procedural memories**.

Language abilities are localized mostly in the left cerebral hemisphere, a phenomenon known as lateralization. Different areas of the left hemisphere—including **Broca's area**, **Wernicke's area**, and the **angular gyrus**—are responsible for different aspects of language. Review Figure 46.15, Web/CD Activity 46.2

See Web/CD Activity 46.3 for a concept review of this chapter.

SELF-QUIZ

1. Which of the following describes the route of sensory information from the foot to the brain?
   a. Ventral horn, spinal cord, medulla, cerebellum, midbrain, thalamus, parietal cortex
   b. Dorsal horn, spinal cord, medulla, pons, midbrain, hypothalamus, frontal cortex
   c. Dorsal horn, spinal cord, medulla, pons, midbrain, thalamus, parietal cortex
   d. Ventral horn, spinal cord, pons, cerebellum, midbrain, thalamus, parietal cortex
   e. Dorsal horn, spinal cord, medulla, pons, midbrain, thalamus, frontal cortex
2. Which statement about the reticular system is *not* true?
   a. Increased activity in the reticular system induces sleep.
   b. The reticular system is located in the brain stem.
   c. Damage to the reticular system in the midbrain can result in coma.
   d. Information from the spinal cord is routed to different nuclei in the reticular system and to the forebrain.
   e. There are groups of neurons called nuclei in the reticular system.
3. Which statement about afferent and efferent pathways is *not* true?
   a. Sympathetic and parasympathetic pathways carry only efferent information.
   b. Visceral afferents carry information about physiological functions of which we are not consciously aware.
   c. The voluntary division of the efferent portion of the peripheral nervous system executes conscious movements.
   d. The cranial nerves and spinal nerves are part of the peripheral nervous system.
   e. Afferent and efferent axons never travel in the same nerve.

4. Which statement about the limbic system is *not* true?
   a. Damage to one structure in the limbic system makes it impossible to form a fear memory.
   b. The limbic system is involved in basic physiological drives, instincts, and emotions.
   c. The limbic system consists of primitive forebrain structures.
   d. In humans, the limbic system is the largest part of the brain.
   e. In humans, a part of the limbic system is necessary for the transfer of short-term memory to long-term memory.
5. Which of the following represents the largest portion of the human cerebral cortex?
   a. The frontal lobes
   b. The primary somatosensory cortex
   c. The temporal cortex
   d. The association cortex
   e. The occipital cortex
6. Which statement about the autonomic nervous system is true?
   a. The sympathetic division is afferent, and the parasympathetic division is efferent.
   b. The transmitter norepinephrine is always excitatory, and acetylcholine is always inhibitory.
   c. Each pathway in the autonomic nervous system includes two neurons, and the neurotransmitter of the first neuron is acetylcholine.
   d. The cell bodies of many sympathetic preganglionic neurons are in the brain stem.
   e. The cell bodies of most parasympathetic postganglionic neurons are in or near the thoracic and lumbar spinal cord.
7. Which statement about cells in the visual cortex is *not* true?
   a. Many cortical cells receive inputs directly from single retinal ganglion cells.
   b. Many cortical cells respond most strongly to bars of light falling at specific locations on the retina.
   c. Some cortical cells respond most strongly to bars of light falling anywhere over large areas of the retina.
   d. Some cortical cells receive inputs from both eyes.
   e. Some cortical cells respond most strongly to an object when it is a certain distance from the eyes.
8. Which of the following characterizes non-REM sleep?
   a. Dreaming
   b. Paralysis of skeletal muscles
   c. EEG slow waves
   d. Rapid and jerky eye movements
   e. It makes up about 20 percent of total sleep time
9. Which conclusion was supported by experiments on split-brain patients?
   a. Language abilities are localized mostly in the left cerebral hemisphere.
   b. Language abilities require both Wernicke's area and Broca's area.
   c. The ability to speak depends on Broca's area.
   d. The ability to read depends on Wernicke's area.
   e. The left hand is served by the left cerebral hemisphere.
10. In the withdrawal reflex,
   a. Action potentials in a pain sensory neuron enter the spinal cord through the ventral root on the same side of the body.
   b. The axon from a pain sensory neuron makes inhibitory synapses with motor neurons on the same side of the spinal cord and inhibitory synapses with motor neurons on the other side of the spinal cord.
   c. Coordinated escape reactions can be initiated before the brain registers the painful sensation.
   d. Axons from the pain sensory neuron make synapses with sympathetic preganglionic neurons in the spinal cord.
   e. Axons from motor neurons on the same side of the spinal cord that receives the input from the pain receptors cross the spinal cord to stimulate muscles on the other side of the body.

FOR DISCUSSION

1. A person receives a stab wound to the left side of his neck. Miraculously, blood vessels are spared. However, following this trauma, his left pupil remains more constricted than his right pupil, and he drools out of the left side of his mouth. How can you explain these symptoms?
2. The stretch receptors in muscles are modified muscle cells, and they have their own motor neurons. What is the function of those motor neurons? To think about this question, remember that the function of the monosynaptic reflex is to adjust muscle tension to a change in load so that the position of the limb does not change.
3. A patient is unable to speak coherently. He can read and write, and he has no obvious loss of muscle function. Where would you expect to find an abnormality if you did brain scans of this patient?
4. How can you investigate how visual images falling on the retina are communicated to the brain? Start with the retina, and imagine how you would continue your investigation in visual areas of the brain.
5. We described the organization of the visual cortex as columns of cells that alternately receive input from the left eye and the right eye. If a young kitten is allowed to see light out of only one eye for a day, more synapses are maintained in the cortical columns receiving input from that eye, while synapses decrease in the intervening columns. This redistribution of synapses does not occur, however, if the kitten is not allowed to sleep. What hypotheses could you propose on the basis of these results?

FOR INVESTIGATION

Figure 46.16 describes technology that enables us to image regional activity in the brain. How would you use this imaging methodology to investigate a higher brain function that is of interest to you? First, formulate your hypothesis. Then design your investigation, keeping in mind that your subject has to lie motionless in a large machine during the imaging. You will have to plan how you will deliver the appropriate stimulus to elicit the response or activity you are interested in. Also, describe what controls you will use to draw clear conclusions about the stimulus-response relationships you expect to observe.

CHAPTER 47 Effectors: How Animals Get Things Done

Champion jumpers

The Olympic record for the woman's long jump is 7.4 meters, set in 1988 by Jackie Joyner-Kersee. Another world record long jump that still stands was set two years earlier by Rosie the Ribeter, who jumped 6.5 meters. Rosie was a frog competing in the Calaveras County Jumping Frog Contest. In some ways, Rosie's jump is more impressive—while Jackie's jump was about 5 times her body length (ie., her height), Rosie's was about *20 times* her body length.

Jackie's jump and Rosie's jump were both powered by skeletal muscle. Muscle tissue is an *effector* that responds to commands from the nervous system. The molecular and cellular mechanisms of muscle contraction are essentially the same in the frog and the human, so why is the frog's jump so much more impressive? The answer involves the concept of *leverage*.

Both frog and human jumping muscles pull on bones that are connected at joints to make levers. A lever makes it possible for the same force to move a large mass a small distance or a small mass a large distance. The ratio of a frog's leg length to its body mass is simply greater than that of a human. Thus the frog's legs are better at moving a small mass a long distance than are the human's legs.

Let's add a flea to our interspecies Olympic competition. The flea can jump over 200 times its body length. This incredible performance is not due to feats of leverage, because no muscle can contract fast enough to explain the take-off velocity of the flea. A different effector mechanism evolved in the flea—a kind of slingshot action. At the base of the flea's jumping legs is an elastic material that is compressed by muscles while the flea is resting. When a trigger mechanism is released, the elastic material recoils and "fires" the flea into the air.

In a contest of jumping endurance, the uncontested champion would be the kangaroo. As a human runs faster, the number of strides and the energy expended per minute increase rapidly. Neither is true for the kangaroo. When moving at speeds from about 5 to 25 kilometers per hour, the kangaroo takes the same number of strides per minute and its metabolic rate does not increase. Why is this so?

A Champion Jumper Jackie Joyner-Kersee set an Olympic record for the women's long jump at the Seoul Olympics in 1988.

Champion Jumpers Relative to their size, many animals have more impressive jumping skills than humans. This Pacific chorus frog (*Pseudacris gegilla*) can leap distances up to 20 times its body length.

CHAPTER OUTLINE

In kangaroos, as in frogs and humans, the muscles used to jump are attached to bones by tendons. Like the material at the base of the flea's legs, tendons can be elastic. The kangaroo's tendons stretch when it lands, and their recoil helps power the next jump—similar to the action of a pogo stick. In order to move faster, the kangaroo simply increases the length of its stride, thereby increasing the stretch on its tendons each time it lands and the magnitude of the recoil at the initiation of each jump.

As we saw in Chapter 31, the ability to move is one of the things that distinguishes animals from the other multicellular organisms (i.e., plants and fungi). Our muscles and skeletons—the *musculoskeletal system*—are the *effectors* that produce movement.

IN THIS CHAPTER we will begin by describing muscle structure and the molecular mechanisms of contraction. We then discuss properties of muscles that adapt different types of muscles to different kinds of tasks. To generate specific kinds of movement, muscles must have something to pull on, so we describe the roles of skeletal systems in generating movement. This chapter ends with brief considerations of effector mechanisms other than musculoskeletal ones: glands that secrete, organs that emit light or sound, and structures that produce strong electrical pulses.

47.1 How Do Muscles Contract?

Most behavior and many physiological actions, such as beating of the heart and moving of food through the digestive tract, depend on muscle contraction. Wherever tissues contract, muscle cells are responsible. There are three types of vertebrate muscle:

- **Skeletal muscle** is responsible for all voluntary movements, such as running or playing a piano. It also generates the unconscious movements of breathing.
- **Cardiac muscle** is responsible for the beating action of the heart.
- **Smooth muscle** creates the movement in many hollow internal organs, such as the gut, bladder, and blood vessels, and it is under the control of the autonomic (involuntary) nervous system.

All three types of muscle use the same contractile mechanism, and we begin our study of effectors by describing the molecular mechanisms underlying this ability to contract. We will use vertebrate skeletal muscle as our primary example here, and discuss its structure in detail. Later we will learn about the differences in cardiac and smooth muscle that adapt them to their particular functions.

Sliding filaments cause skeletal muscle to contract

Skeletal muscle is also called *striated muscle* because of its striped appearance (see Figure 47.7A). Skeletal muscle cells, called **muscle fibers**, are large and have many nuclei. These multinucleate cells form in development through the fusion of many individual embryonic muscle cells called *myoblasts*. A specific muscle such as your biceps (which bends your arm) is composed of hundreds or thousands of muscle fibers bundled together by connective tissue (**Figure 47.1**).

Muscle contraction is due to the interaction between the contractile proteins **actin** and **myosin**. Within muscle cells, actin and myosin molecules are organized into filaments consisting of many molecules. Actin filaments are also called *thin filaments*, and myosin filaments are *thick filaments*. The two kinds of filaments lie parallel to each other. When muscle contraction is triggered, the actin and myosin filaments slide past each other in a telescoping fashion.

What is the relation between a skeletal muscle fiber and the actin and myosin filaments responsible for its contraction? Each

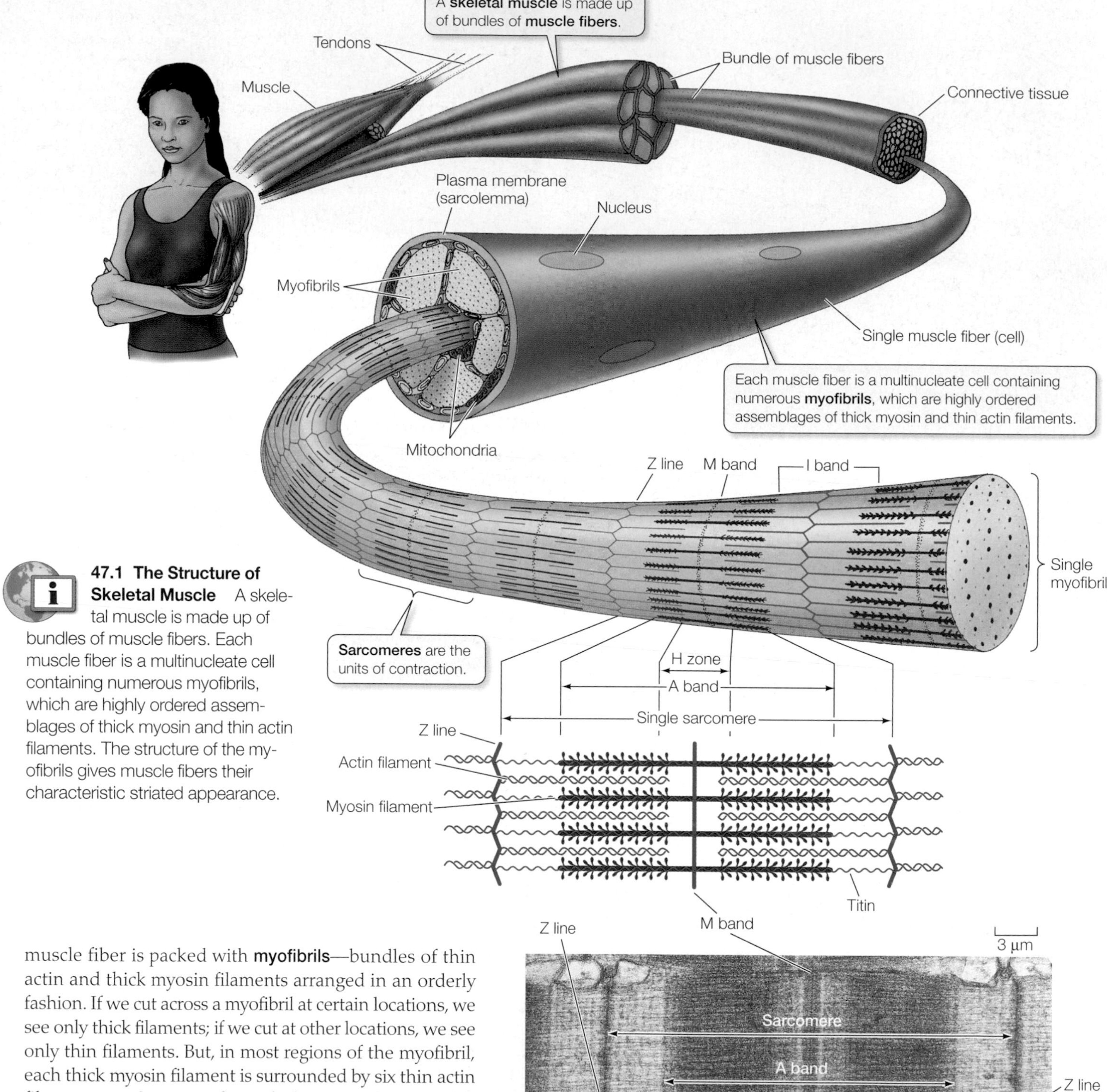

47.1 The Structure of Skeletal Muscle A skeletal muscle is made up of bundles of muscle fibers. Each muscle fiber is a multinucleate cell containing numerous myofibrils, which are highly ordered assemblages of thick myosin and thin actin filaments. The structure of the myofibrils gives muscle fibers their characteristic striated appearance.

muscle fiber is packed with **myofibrils**—bundles of thin actin and thick myosin filaments arranged in an orderly fashion. If we cut across a myofibril at certain locations, we see only thick filaments; if we cut at other locations, we see only thin filaments. But, in most regions of the myofibril, each thick myosin filament is surrounded by six thin actin filaments, and conversely, each thin actin filament sits within a triangle of three thick myosin filaments.

A longitudinal view of a myofibril reveals why skeletal muscle appears striated. The myofibril consists of repeating units called **sarcomeres**. Each sarcomere is made of overlapping filaments of actin and myosin, which create a distinct banding pattern (see Figure 47.2). The bundles of myosin filaments are held in a centered position within the sarcomere by a protein called **titin**. Titin is the largest protein in the body; it runs the full length of the sarcomere from Z line to Z line. Each titin molecule runs right through a myosin bundle. Between the ends of the myosin bundles and the Z lines, titin molecules are very stretchable, like bungee cords. In a relaxed skeletal muscle, resistance to stretch is mostly due to the elasticity of the titin molecules.

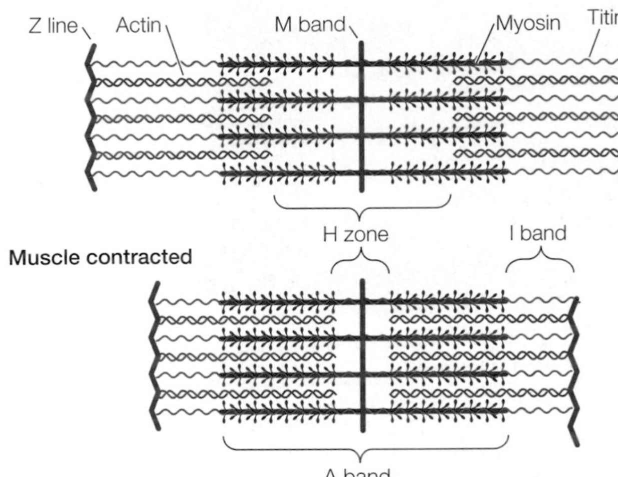

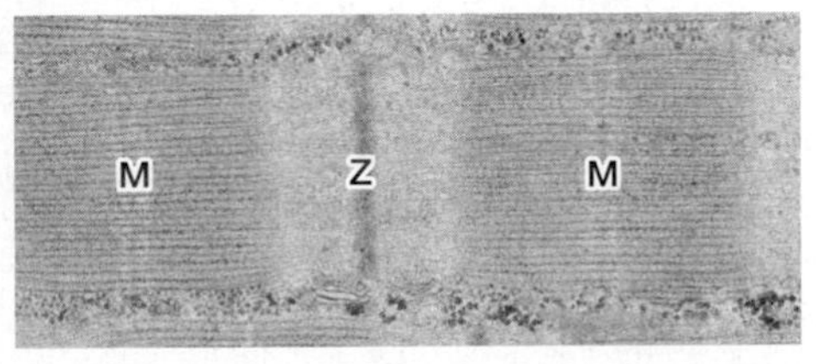

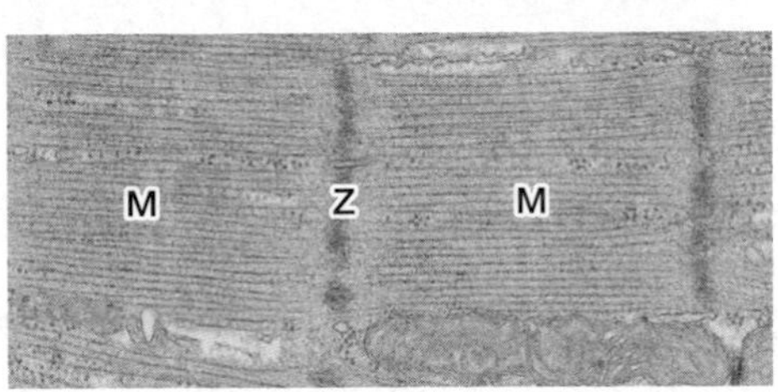

47.2 Sliding Filaments The banding pattern of the sarcomere changes as it shortens. Observations of electron micrographs such as those on the right led to the sliding filament hypothesis of muscle contraction.

Before the molecular nature of the muscle banding pattern was known, the bands were given names that are still used today. Each sarcomere is bounded by *Z lines*, which anchor the thin actin filaments. Centered in the sarcomere is the *A band*, which contains all the myosin filaments. The *H zone* and the *I band*, which appear light, are regions where actin and myosin filaments do not overlap in the relaxed muscle. The dark stripe within the H zone is called the *M band*; it contains proteins that help hold the myosin filaments in their regular arrangement.

As the muscle contracts, the sarcomeres shorten, and the band pattern changes. The H zone and the I band become much narrower, and the Z lines move toward the A band as if the actin filaments were sliding into the region occupied by the myosin filaments (**Figure 47.2**). In the mid 1950s, this observation independently led two teams of British biologists to independently propose the **sliding filament theory** of muscle contraction. It is not uncommon in science for critical breakthroughs to be made simultaneously in different laboratories, but the arm of coincidence was long indeed in this case. The leaders of the two teams were named Hugh Huxley and Andrew Huxley—but they were not related. Working in separate Cambridge University labs, the two groups proposed the sliding filament model at the same time, and both papers were published in the same issue of the journal *Nature.*

Actin–myosin interactions cause filaments to slide

To understand how the sliding filament model explains muscle contraction, we must first examine the structures of actin and myosin (**Figure 47.3**). A myosin molecule consists of two long polypeptide chains coiled together, each ending in a large globular head. A myosin filament is made up of many myosin molecules arranged in parallel, with their heads projecting laterally from one or the other end of the filament. An actin filament consists of two chains of actin monomers twisted together like two strands of pearls in a helix. Twisting around the actin chains is another protein, *tropomyosin,* and attached to the tropomyosin at intervals are molecules of *troponin*. We'll discuss the roles of these last two proteins in the following section.

The myosin heads can bind specific sites on actin and thereby form cross-bridges between the myosin and the actin filaments. Moreover, when a myosin head binds to an actin filament, its conformation changes, and it bends and exerts a force that causes the actin filament to slide 5–10 nanometers relative to the myosin filament. The myosin heads also have ATPase activity; when they are bound to actin, they can bind and hydrolyze ATP. The energy released when this happens changes the conformation of the myosin head, causing it to release the actin and return to its extended position, from which it again can bind to actin. Together, these details explain the cycle of events that cause the actin and myosin filaments to slide past each other and shorten the sarcomere.

47.3 Actin and Myosin Filaments Overlap to Form Myofibrils Myosin filaments are bundles of molecules with globular heads and polypeptide tails; the protein titin holds these filaments centered within the sarcomere. Actin filaments consist of two chains of actin monomers twisted together. They are wrapped by chains of the polypeptide tropomyosin and studded at intervals with another protein, troponin.

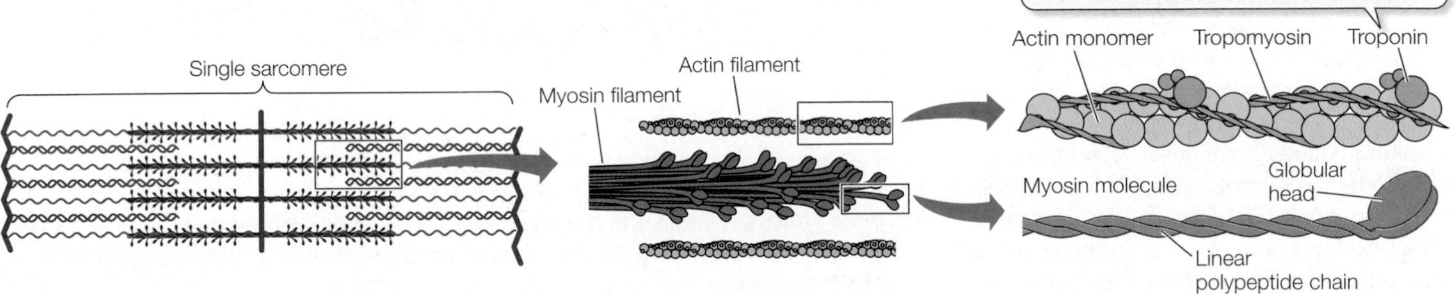

ATP is needed to break the actin–myosin bonds, explaining why muscles stiffen soon after death—*rigor mortis*. ATP production ceases with death, so the actin–myosin bonds cannot be broken, and the muscles stiffen. Eventually, however, the proteins begin to lose their integrity, and the muscles soften. The timing of these events help a coroner estimate the time of death.

As an analogy, think of a person firing an old-fashioned pistol. A finger pulls the pistol hammer back against the force of a spring, cocking the hammer into a high-energy, unstable position. Similarly, ATP hydrolysis converts the potential chemical energy of a phosphate bond into the kinetic energy that "cocks" the myosin head into its unstable, extended position. Pressure on the trigger of a pistol can release the hammer from the cocked position, allowing the stretched spring attached to the hammer to slam it back into its original position. Similarly, the binding of actin destabilizes the myosin head from its extended position, and the spring-like myosin molecule snaps back to its bent configuration, dragging the bound actin filament along with it.

We have been discussing the cycle of contraction in terms of a single myosin head. Remember that each myosin filament has many myosin heads at both ends and is surrounded by six actin filaments; thus the contraction of the sarcomere involves a great many cycles of interaction between actin and myosin molecules. That is why when a single myosin head breaks its contact with actin, the actin filaments do not slip backward.

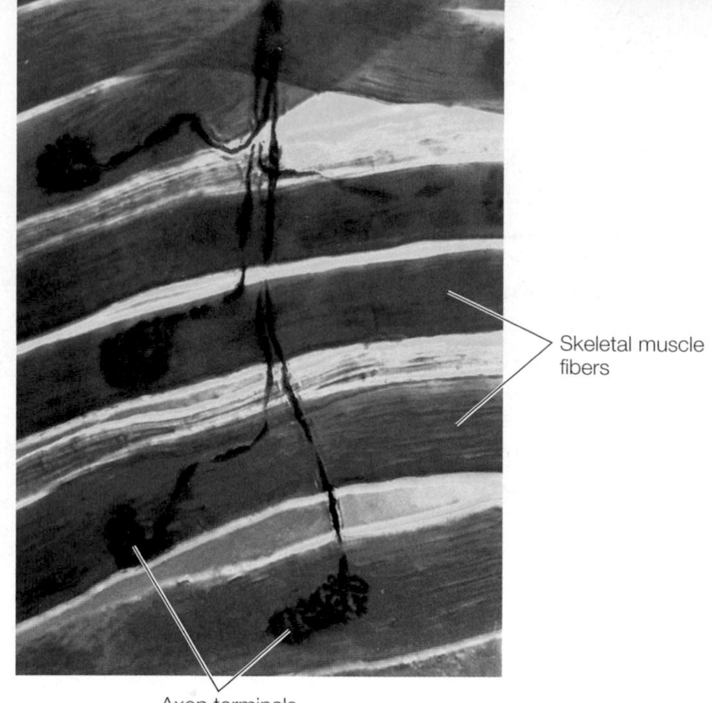

47.4 The Neuromuscular Junction Axon terminals from a single motor neuron innervate multiple skeletal muscle fibers.

Actin–myosin interactions are controlled by calcium ions

Muscle contractions are initiated by action potentials from motor neurons arriving at the *neuromuscular junction* (**Figure 47.4**; also see Figure 44.13). The axon terminals of motor neurons are generally highly branched and can form synapses with hundreds of muscle fibers. All the fibers activated by a single motor neuron constitute a **motor unit** and contract simultaneously in response to action potentials fired by that motor neuron. A muscle can consist of many motor units. Thus, there are two ways to increase the strength of contraction of a muscle—increase the rate of firing in an individual motor neuron or recruit more motor neurons to fire.

Like neurons, muscle cells are *excitable*; that is, their plasma membranes can generate and conduct action potentials. In skeletal muscle fibers (but not smooth or cardiac muscle fibers), all action potentials are initiated by motor neurons. When an action potential arrives at the neuromuscular junction, the neurotransmitter acetylcholine is released from the motor neuron terminals, diffuses across the synaptic cleft, binds to receptors in the postsynaptic membrane, and causes ion channels in the motor end plate to open. Most of the ions that flow through these channels are Na^+, and therefore the motor end plate is depolarized. The depolarization spreads to the surrounding plasma membrane of the muscle fiber, which contains voltage-

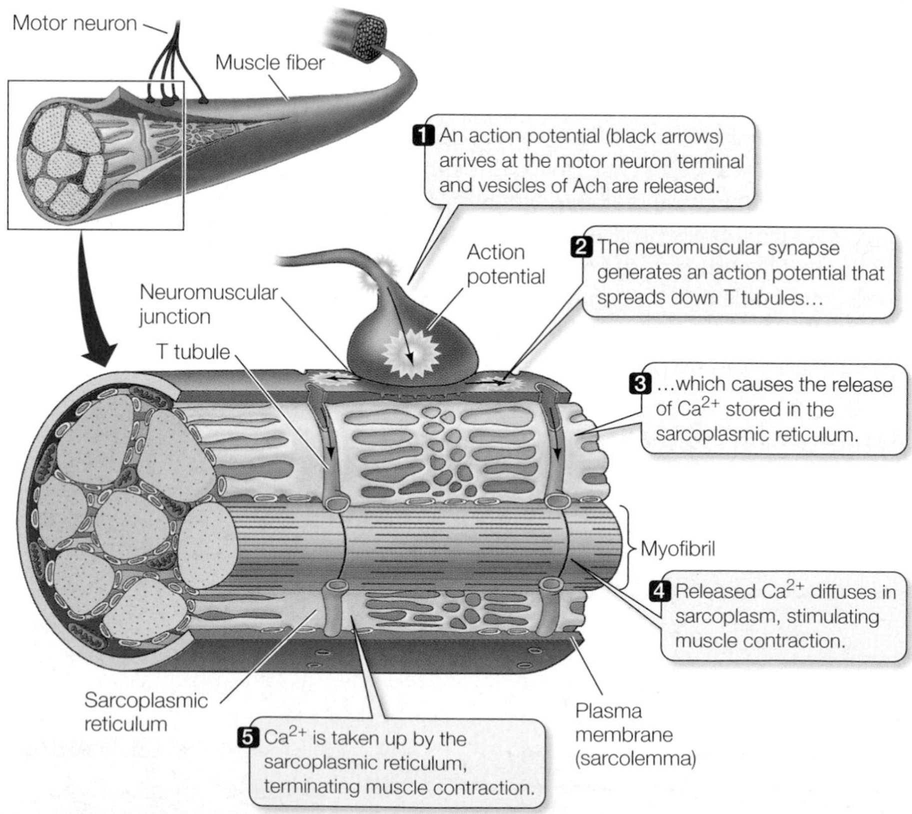

Motor neuron
Muscle fiber
1 An action potential (black arrows) arrives at the motor neuron terminal and vesicles of Ach are released.
Action potential
2 The neuromuscular synapse generates an action potential that spreads down T tubules...
Neuromuscular junction
T tubule
3 ...which causes the release of Ca2+ stored in the sarcoplasmic reticulum.
Myofibril
4 Released Ca2+ diffuses in sarcoplasm, stimulating muscle contraction.
Sarcoplasmic reticulum
5 Ca2+ is taken up by the sarcoplasmic reticulum, terminating muscle contraction.
Plasma membrane (sarcolemma)

47.5 T Tubules in Action An action potential at the neuromuscular junction spreads throughout the muscle fiber via a network of T tubules, triggering the release of Ca^{2+} from the sarcoplasmic reticulum.

gated sodium channels. When threshold is reached, the plasma membrane fires an action potential that is conducted rapidly to all points on the surface of the muscle fiber.

An action potential in a muscle fiber also travels deep within the cell. The plasma membrane is continuous with a distribution system of tubules that descend into the muscle fiber cytoplasm (also called the **sarcoplasm**) (**Figure 47.5**). The action potential that spreads over the plasma membrane also spreads through this system of transverse tubules, or **T tubules**.

The T tubules come very close to the endoplasmic reticulum of the muscle cell. In muscle cells the ER is called the **sarcoplasmic reticulum**, and it is a closed compartment surrounding every myofibril. Calcium pumps in the sarcoplasmic reticulum cause it to take up Ca^{2+} ions from the sarcoplasm. Therefore, when the muscle fiber is at rest, there is a higher concentration of Ca^{2+} in the sarcoplasmic reticulum and a lower concentration of Ca^{2+} in the sarcoplasm.

Spanning the space between the membranes of the T tubules and the membranes of the sarcoplasmic reticulum are two proteins. One protein, the *dihydropyridine* (DHP) *receptor*, is located in the T tubule membrane; it is voltage-sensitive and changes its conformation when an action potential reaches it. The other protein, the *ryanodine receptor*, is located in the sarcoplasmic reticulum membrane; it is a Ca^{2+} channel. These two proteins are physically connected. When the DHP receptor is activated by an action potential, it changes conformation, causing the ryanodine receptor to allow Ca^{2+} to leave the sarcoplasmic reticulum. Ca^{2+} ions diffuse into the sarcoplasm surrounding the actin and myosin filaments and trigger the interaction of actin and myosin and the sliding of the filaments. How do the Ca^{2+} ions do this?

An actin filament, as we have seen, is a helical arrangement of two strands of actin monomers. Lying in the grooves between the two actin strands is the two-stranded protein **tropomyosin** (**Figure 47.6**; also see Figure 47.3). At regular intervals, the filament also includes a globular protein, **troponin**. The troponin molecule has three subunits: One binds actin, one binds tropomyosin, and one binds Ca^{2+}.

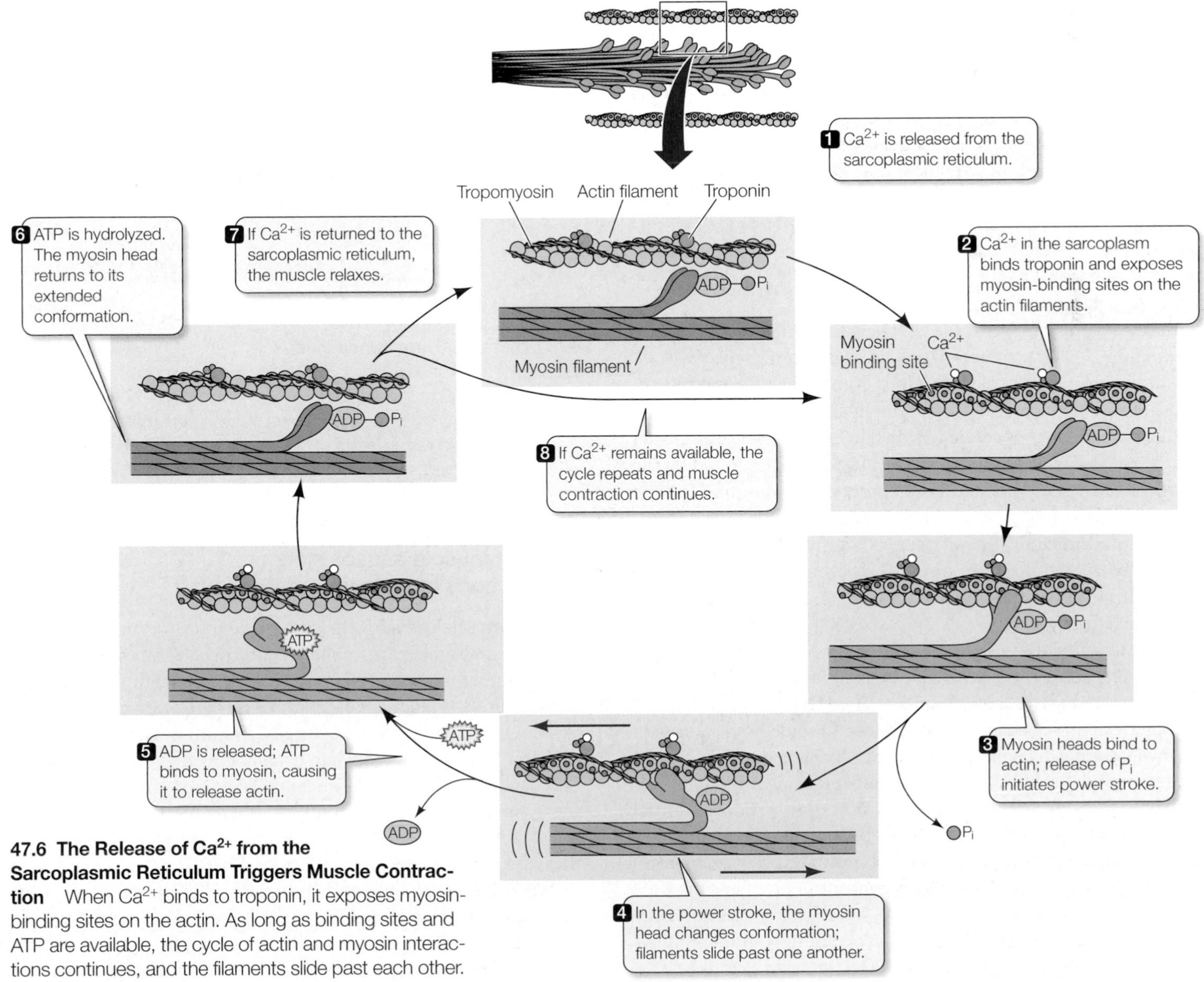

1 Ca2+ is released from the sarcoplasmic reticulum.
Tropomyosin
Actin filament
Troponin
ADP
Pi
Myosin filament
2 Ca2+ in the sarcoplasm binds troponin and exposes myosin-binding sites on the actin filaments.
Myosin binding site
Ca2+
3 Myosin heads bind to actin; release of Pi initiates power stroke.
4 In the power stroke, the myosin head changes conformation; filaments slide past one another.
5 ADP is released; ATP binds to myosin, causing it to release actin.
ATP
6 ATP is hydrolyzed. The myosin head returns to its extended conformation.
7 If Ca2+ is returned to the sarcoplasmic reticulum, the muscle relaxes.
8 If Ca2+ remains available, the cycle repeats and muscle contraction continues.

47.6 The Release of Ca^{2+} from the Sarcoplasmic Reticulum Triggers Muscle Contraction When Ca^{2+} binds to troponin, it exposes myosin-binding sites on the actin. As long as binding sites and ATP are available, the cycle of actin and myosin interactions continues, and the filaments slide past each other.

When the muscle is at rest, the tropomyosin strands are positioned so that they block the sites on the actin filament where myosin heads can bind. When Ca^{2+} is released into the sarcoplasm, it binds to troponin, changing its conformation. Because the troponin is bound to the tropomyosin, this conformational change twists the tropomyosin enough to expose the actin–myosin binding sites. Thus the cycle of making and breaking actin–myosin bonds is initiated, the filaments are pulled past each other, and the muscle fiber contracts. When the calcium pumps remove the Ca^{2+} ions from the sarcoplasm, the tropomyosin returns to the position in which it blocks the binding of myosin heads to actin, and the muscle fiber returns to its resting condition. Figure 47.6 summarizes this cycle.

Cardiac muscle causes the heart to beat

Cardiac muscle appears striated as does skeletal muscle because of the regular arrangement of actin and myosin filaments into sarcomeres (**Figure 47.7A,B**). The difference between cardiac and skeletal muscle is that cardiac muscle cells are much smaller and have only one nucleus each. Cardiac muscle cells branch, and the branches of adjoining cells interdigitate into a meshwork that is resistant to tearing. As a result, the heart walls can withstand high pressures while pumping blood without the danger of developing leaks. Adding to the strength of cardiac muscle are *intercalated discs* that provide strong mechanical adhesions between adjacent cells. Gap junctions—protein structures that allow cytoplasmic continuity between cells—in the intercalated discs offer low-resistance pathways for ionic currents to flow between cells. Therefore, cardiac muscle cells are electrically coupled. An action potential initiated at one point in the heart spreads rapidly through a large mass of cardiac muscle.

Certain cardiac muscle cells are specialized for generating and conducting electrical signals. These *pacemaker* and *conducting cells* do not have contractile elements, but they do initiate and coordinate the rhythmic contractions of the heart. (The molecular basis for this pacemaking function is covered in Section 49.3.) Pacemaker cells make the vertebrate heartbeat *myogenic*—generated by the heart muscle itself. The autonomic nervous system modifies the rate of the pacemaker cells, but is not essential for their continued rhythmic function. A heart removed from a vertebrate can continue to beat with no input from the nervous system. The myogenic nature of the heartbeat is a major factor in making heart transplants possible.

The mechanism of excitation–contraction coupling in cardiac muscle cells is different from that in skeletal muscle cells. The T tubules are larger and the voltage-sensitive DHP proteins in the T tubules are actual calcium channels. These T tubule proteins are not physically connected with the ryanodine receptors in the sarcoplasmic reticulum. Instead, the ryanodine receptors are ion-gated Ca^{2+} channels that are sensitive to Ca^{2+}. When an action potential spreads down the T tubules, it causes the voltage-gated channels to open, allowing extracellular Ca^{2+} to flow into the sarcoplasm. This slight rise in sarcoplasmic Ca^{2+} concentration opens the Ca^{2+} channels in the sarcoplasmic reticulum, which in turn causes a huge rise in sarcoplasmic calcium ion concentration, resulting in fiber contraction. This mechanism is called *Ca^{2+}-induced Ca^{2+} release.*

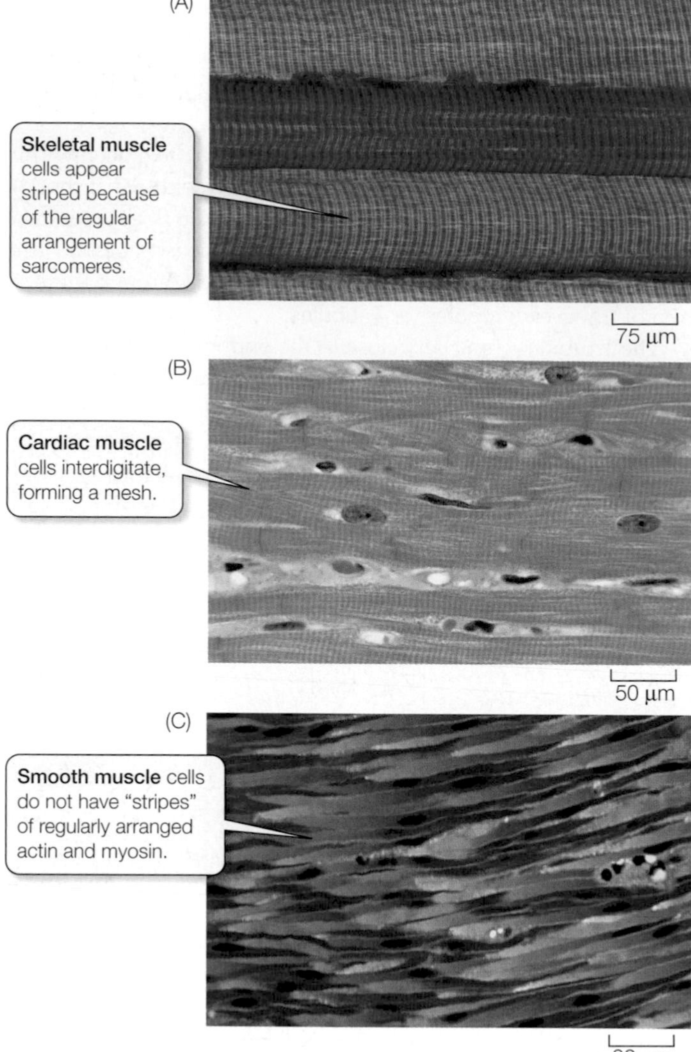
(A)
Skeletal muscle cells appear striped because of the regular arrangement of sarcomeres.
75 μm
(B)
Cardiac muscle cells interdigitate, forming a mesh.
50 μm
(C)
Smooth muscle cells do not have "stripes" of regularly arranged actin and myosin.
30 μm

47.7 There are Three Kinds of Muscle All muscle cells contain filaments of actin and myosin proteins. (A) Skeletal muscle cells are very large and form fibers. Each cell contains multiple nuclei. (B) Cardiac muscle cells and (C) smooth muscle cells are smaller, have individual nuclei, and form sheets of contractile tissue in which cells are electrically coupled.

Smooth muscle causes slow contractions of many internal organs

Smooth muscle provides the contractile force for most of our internal organs, which are under the control of the autonomic nervous system. Smooth muscle moves food through the digestive tract, controls the flow of blood through blood vessels, and empties the urinary bladder. Structurally, smooth muscle cells are the simplest muscle cells. They are usually long and spindle-shaped, and each cell has a single nucleus. They are "smooth" because the actin and myosin filaments are not as regularly arranged as they are in skeletal and cardiac muscle, and therefore do not produce the striated appearance (**Figure 47.7C**).

Smooth muscle tissue, such as that from the wall of the digestive tract, has interesting properties. The cells are arranged in sheets, and individual cells in a sheet are in electrical contact with one another through gap junctions, as they are in cardiac muscle. As a result, an action potential generated in the membrane of one smooth muscle cell can spread to all the cells in the sheet of tissue. Thus the cells in the sheet can contract in a coordinated fashion.

The plasma membranes of smooth muscle cells are sensitive to stretch, with important consequences. If the wall of the digestive tract is stretched in one location (as by a mouthful of food passing down the esophagus to the stomach), the membranes of the stretched cells depolarize, reach threshold, and fire action potentials, which cause the cells to contract. Thus, smooth muscle contracts after being stretched, and the harder it is stretched, the stronger it contracts.

The neurotransmitters of the autonomic nervous system can alter the membrane potential of smooth muscle cells as well (**Figure 47.8**). For example, acetylcholine causes smooth muscle cells

EXPERIMENT

HYPOTHESIS: Stretch and parasympathetic stimulation induce contraction in gut smooth muscle.

METHOD

Incubate a strip of smooth (intestinal) muscle in a saline bath. Measure action potentials and force of contraction.

Experiment 1 Stretch intestinal muscle and analyze response.

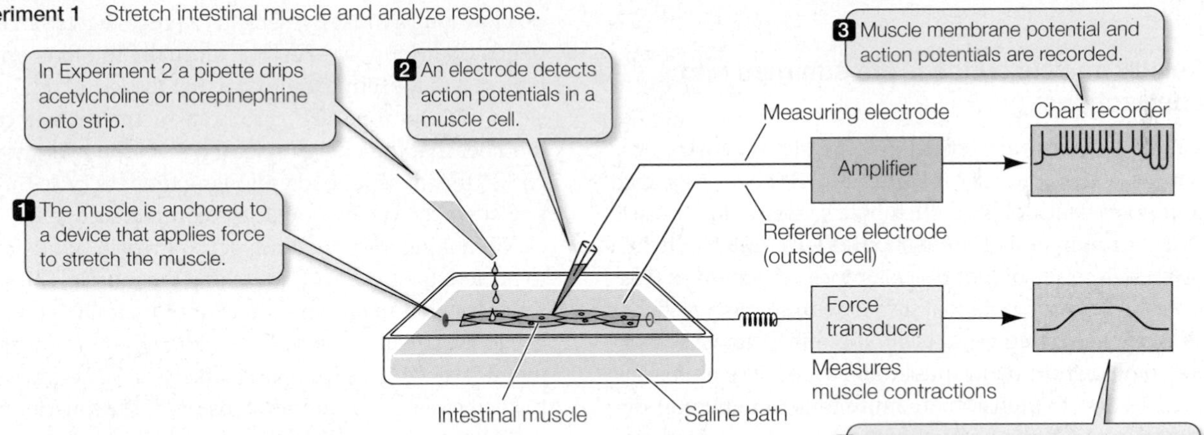

In Experiment 2 a pipette drips acetylcholine or norepinephrine onto strip.

2 An electrode detects action potentials in a muscle cell.

3 Muscle membrane potential and action potentials are recorded.

Measuring electrode

Chart recorder

Amplifier

1 The muscle is anchored to a device that applies force to stretch the muscle.

Reference electrode (outside cell)

Force transducer

Measures muscle contractions

Intestinal muscle

Saline bath

4 The force of contraction of the muscle is measured by a force transducer.

RESULTS Stretching depolarizes the smooth muscle membrane. The depolarization causes action potentials that activate the contractile mechanism.

Experiment 2 Response of muscle strip to neurotransmitters of the autonomic nervous system.

When acetylcholine is dripped onto the muscle, the cells depolarize, fire action potentials more rapidly, and increase their force of contraction.

Norepinephrine, on the other hand, causes the cells to hyperpolarize, decreasing their rate of firing, and decreasing their force of contraction.

Apply acetylcholine

Wash out acetylcholine

Apply norepinephrine

Wash out norepinephrine

Membrane potential (mV)

+25

0

–25

–50

Force

Muscle contracts

Muscle relaxes

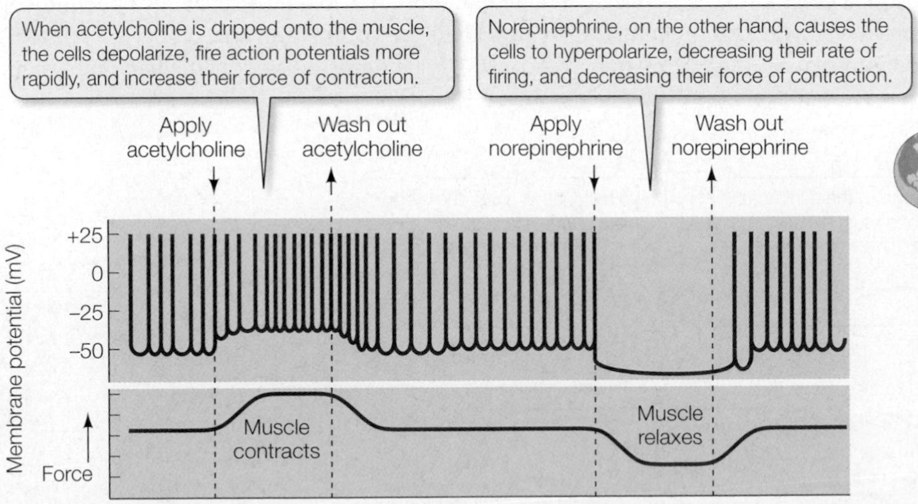

RESULTS Autonomic neurotransmitters alter membrane resting potential and thereby determine the rate that smooth muscle cells fire action potentials.

CONCLUSION: Gut smooth muscle contraction is stimulated by stretch and by the neurotransmitter acetylcholine.

47.8 Mechanisms of Smooth Muscle Activation This experiment showed that stretching depolarizes the membrane of smooth muscle cells, and this depolarization causes action potentials that activate the contractile mechanism. The neurotransmitters acetylcholine and norepinephrine alter the membrane potential of smooth muscle, making it more or less likely to contract.

of the digestive tract to depolarize; they are thus more likely to fire action potentials and contract. Antagonistically, norepinephrine causes these muscle cells to hyperpolarize, and they are therefore less likely to fire action potentials and contract.

Although smooth muscle cell contraction is not controlled by the troponin–tropomyosin mechanism, calcium still plays a critical role. A Ca^{2+} influx into the sarcoplasm of a smooth muscle cell can be stimulated by action potentials, hormones, or stretching. The Ca^{2+} that enters the sarcoplasm combines with a protein called *calmodulin*. The calmodulin–Ca^{2+} complex activates an enzyme called myosin kinase, which can phosphorylate myosin heads. When the myosin heads in smooth muscle are phosphorylated, they can undergo cycles of binding and releasing actin, causing muscle contraction. As Ca^{2+} is removed from the sarcoplasm, it dissociates from calmodulin, and the activity of myosin kinase falls. An additional enzyme, myosin phosphatase, dephosphorylates the myosin to help stop actin–myosin interactions.

Single skeletal muscle twitches are summed into graded contractions

In skeletal muscle, the arrival of an action potential at a neuromuscular junction causes an action potential in a muscle fiber. The spread of that action potential through the T tubule system of the muscle fiber causes a minimum unit of contraction, called a **twitch**. A twitch can be measured in terms of the *tension*, or force, it generates (**Figure 47.9A**). A single action potential stimulates a single twitch, but the ultimate force generated by an action potential can vary enormously depending on how many muscle fibers are in the motor unit it innervates. The level of tension an entire muscle generates depends on two factors: (1) the number of motor units activated, and (2) the frequency at which the motor units are firing. In muscles responsible for fine movements, such as those of the fingers, a motor neuron may innervate only one or a few muscle fibers, but in a muscle that produces large forces, such as the biceps, a motor neuron innervates a large number of muscle fibers.

At the level of the single muscle fiber, a single action potential stimulates a single twitch. If action potentials reaching the muscle fiber are adequately separated in time, each twitch is a discrete, all-or-none phenomenon. If action potentials are fired more rapidly, however, new twitches are triggered before the myofibrils have had a chance to return to their resting condition. As a result, the twitches sum, and the tension generated by the fiber increases and becomes more sustained. Thus an individual muscle fiber can show a graded response to increased levels of stimulation by its motor neuron.

Twitches sum at high levels of stimulation because the calcium pumps in the sarcoplasmic reticulum are not able to clear the Ca^{2+} ions from the sarcoplasm between action potentials. Eventually a stimulation frequency can be reached that results in continuous presence of Ca^{2+} in the sarcoplasm at high enough levels to cause continuous activation of the contractile machinery—a condition known as **tetanus** (**Figure 47.9B**). (Do not confuse this condition with the disease *tetanus*, which is caused by a bacterial toxin and is characterized by spastic contractions of skeletal muscles.)

How long a muscle fiber can maintain a tetanic contraction depends on its supply of ATP. Eventually the fiber will become fatigued. It may seem paradoxical that the *lack* of ATP causes fatigue, since the action of ATP is to break actin–myosin bonds. But remember that the energy released from the hydrolysis of ATP "recocks" the myosin heads, allowing them to cycle through another power stroke. When a muscle is contracting against a load, the cycle of making and breaking actin–myosin bonds must continue to prevent the load from stretching the muscle. The situation is like rowing a boat upstream: You cannot maintain your position relative to the stream bank by just holding the oars out against the current; you have to keep rowing. Likewise, actin–myosin bonds have to keep cycling to maintain tension in the muscle.

Many muscles of the body maintain a low level of tension even when the body is at rest. For example, the muscles of the neck, trunk, and limbs that maintain our posture against the pull of gravity are always working, even when we are standing or sitting still. **Muscle tone** comes from the activity of a small but changing number of motor units in a muscle; at any one time, some of the muscle's fibers are contracting and others are relaxed. Muscle tone is constantly being readjusted by the nervous system.

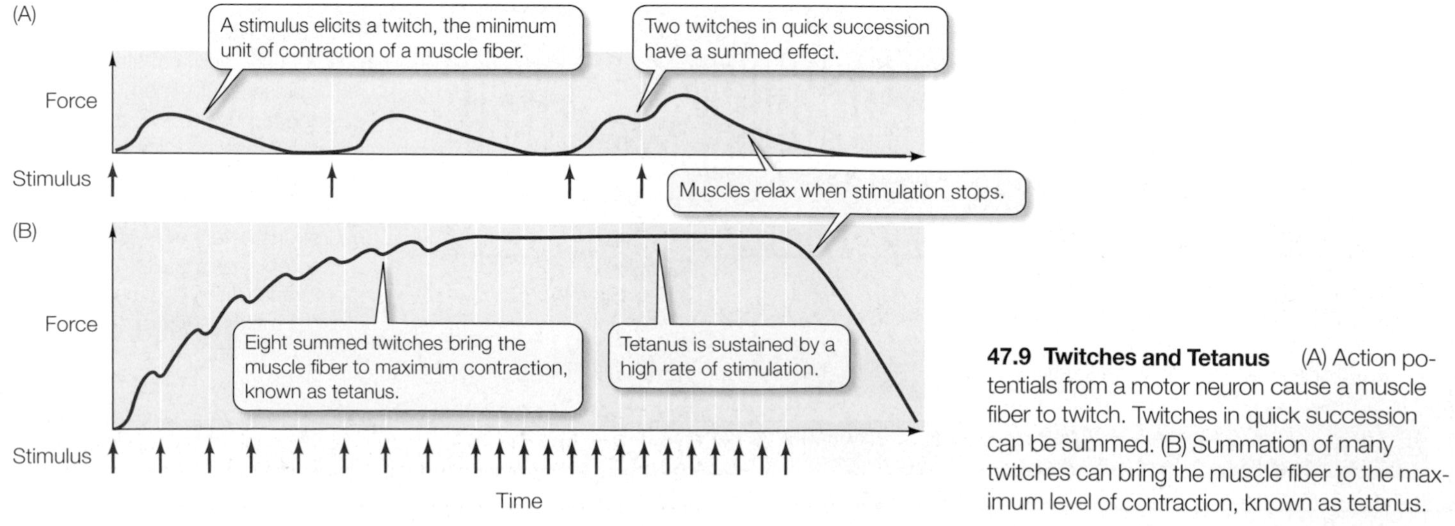
(A)
A stimulus elicits a twitch, the minimum unit of contraction of a muscle fiber.
Two twitches in quick succession have a summed effect.
Force
Stimulus
Muscles relax when stimulation stops.
(B)
Force
Eight summed twitches bring the muscle fiber to maximum contraction, known as tetanus.
Tetanus is sustained by a high rate of stimulation.
Stimulus
Time

47.9 Twitches and Tetanus (A) Action potentials from a motor neuron cause a muscle fiber to twitch. Twitches in quick succession can be summed. (B) Summation of many twitches can bring the muscle fiber to the maximum level of contraction, known as tetanus.

47.1 RECAP

The most ubiquitous effector is muscle. The contractile ability of muscle derives from interactions between actin and myosin filaments.

- Can you describe how the cellular and subcellular structure of skeletal muscle relate to the sliding filament theory of muscle contraction? See pp. 1007–1008 and Figures 47.1, 47.2, and 47.3
- What is the role of Ca^{2+} in the contractile mechanism of skeletal, cardiac, and smooth muscle? See pp. 1008–1010 and Figures 47.5 and 47.6
- Do you understand the role ATP plays in the actin and myosin interactions that produce contraction? See Figure 47.6

Now that we understand how muscles generate force, we can ask why different muscles have different characteristics, and how individual muscles can change their characteristics with regular use and conditioning.

47.2 What Determines Muscle Strength and Endurance?

The functions that different muscles perform place different demands on them. Some muscles, such as postural muscles, must sustain a load continuously over long periods of time. Other muscles, such as those that control your fingers, generally do not have to sustain long contractions, but they must be able to contract quickly. What differences adapt muscles to specific functions? Can those differences be amplified through conditioning to improve different kinds of human physical performance? A sprinter wants muscles that generate maximum force rapidly, but they do not need to sustain a particular load for a long time. On the other hand, a marathon runner wants muscles with maximum endurance. What properties of muscles determine these functional characteristics?

Muscle fiber types determine endurance and strength

Not all skeletal muscle fibers are alike, and a single muscle contains more than one type of fiber. The two major types of skeletal muscle fibers express different genes for their myosin molecules, and these myosin variants have different rates of ATPase activity. Those with high ATPase activity can recycle their actin–myosin cross-bridges rapidly and are therefore called fast-twitch fibers. Slow-twitch fibers have lower ATPase activity, so they develop tension more slowly but can maintain it longer.

Slow-twitch fibers are also called *oxidative* or *red muscle* because they contain the oxygen-binding protein *myoglobin*, they have many mitochondria, and they are well supplied with blood vessels. These characteristics both increase their capacity for oxidative metabolism and result in their red appearance. The maximum tension a slow-twitch fiber produces is low and develops slowly but is highly resistant to fatigue. Slow-twitch fibers have substantial reserves of fuel (glycogen and fat), so they can maintain steady, prolonged production of ATP as long as oxygen is available. Muscles with high proportions of slow-twitch fibers are good for long-term *aerobic* work (that is, work that requires oxygen). Long-distance runners, swimmers, cyclists, and other athletes whose activities require endurance have leg and arm muscles consisting mostly of slow-twitch fibers (**Figure 47.10**).

Some **fast-twitch fibers** are also called *glycolytic* or *white muscle* because, in comparison to slow-twitch fibers, they have few mitochondria, little or no myoglobin, and fewer blood vessels; thus they look pale. Fast-twitch, glycolytic fibers can develop maximum tension more rapidly than slow-twitch fibers can, and that maximum tension is greater. However, fast-twitch fibers fatigue rapidly. The myosin of these fibers puts the energy of ATP to work very rapidly, but the fibers cannot replenish ATP quickly enough to sustain contraction for a long time. Fast-twitch fibers are especially good for short-term work that requires maximum strength. Weight lifters and sprinters have leg and arm muscles with high proportions of fast-twitch fibers.

There are also fast-twitch fibers that are somewhat oxidative, and therefore intermediate in their properties between slow-twitch and fast glycolytic fibers. These intermediate fibers can become more oxidative with endurance training.

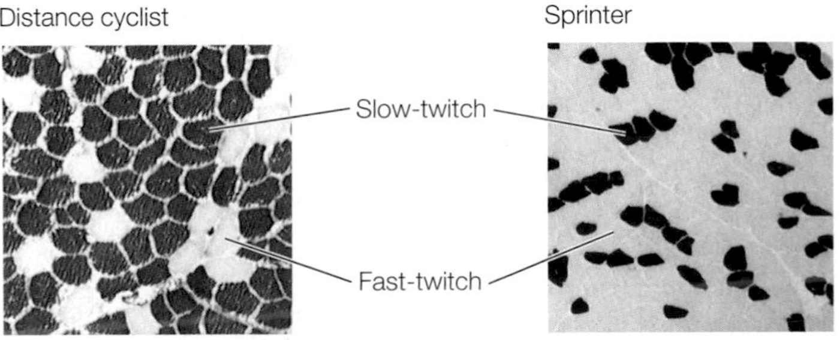

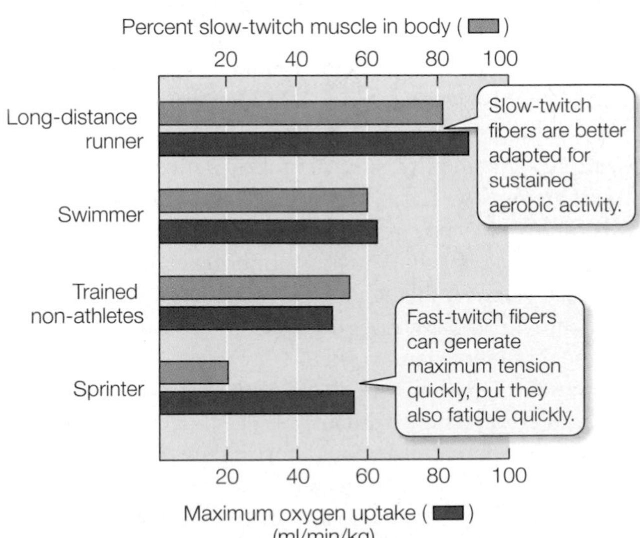

47.10 Slow-and Fast-Twitch Muscle Fibers The skeletal muscles in the micrographs were stained with a reagent that shows slow-twitch fibers as dark; fast-twitch muscle shows up as a light area. Athletes in different sports have different distributions of muscle fiber types.

What determines the proportion of fast- and slow-twitch fibers in your skeletal muscles? The most important factor is genetic heritage, so there is some truth to the statement that champions are born, not made. To a certain extent, you can alter the properties of your muscle fibers through training. But a person born with a high proportion of fast-twitch fibers will never become a champion marathon runner, and one born with a high proportion of slow-twitch fibers will never become a champion sprinter.

A muscle has an optimal length for generating maximum tension

Have you ever done a pull-up? If you have, you know that two parts of this exercise are especially difficult. When you are hanging from the bar with your arms fully extended, it is hard to get the pull-up started; and when your chin is just about to the bar, pulling yourself up the last small distance is difficult. These experiences are explained by the structure of the sarcomere.

When a muscle is stretched and the sarcomeres are lengthened, there is less overlap between the actin and myosin filaments; therefore, fewer cross-bridges can form, and less force can be produced. In fact, if the sarcomeres are stretched too much, actin and myosin do not overlap and no force can be produced. How would a muscle recover from such a difficult situation? The bungee cord-like titin molecules create enough elastic recoil to pull the actin and myosin fibrils back into an overlapping arrangement.

When the muscle is fully contracted, the actin and myosin filaments overlap so much that the myosin bundles are pressed up against the Z lines. Because they have no place to go, additional shortening is difficult. You can see the relationship between the length of a muscle fiber and its ability to develop tension in **Figure 47.11**.

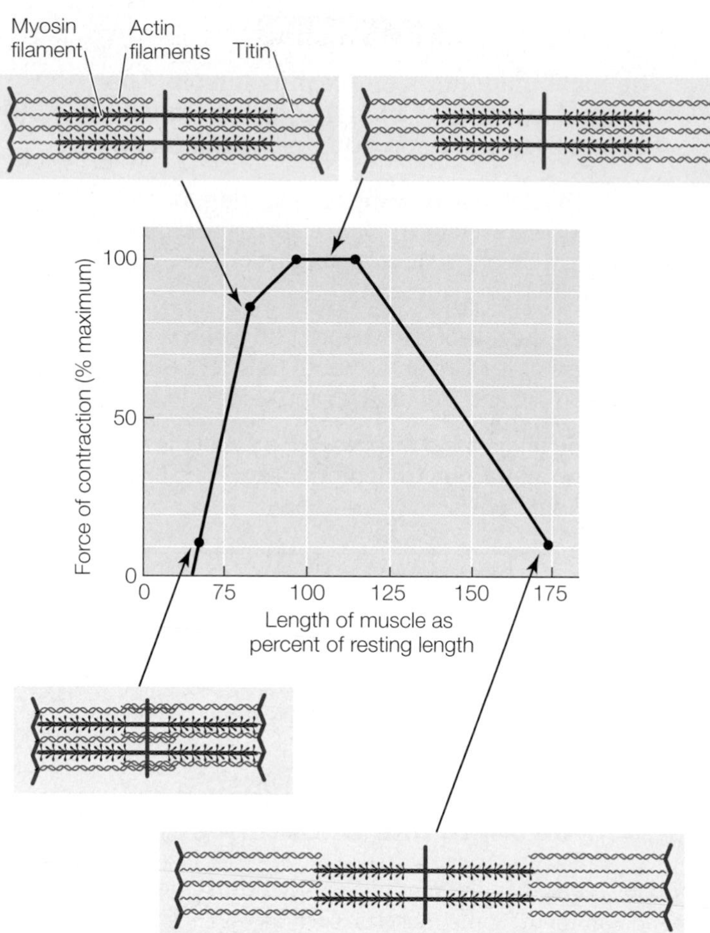

47.11 Strength and Length The amount of force a sarcomere can generate depends on its resting length. When a muscle is stretched, the sarcomeres lengthen, there is less overlap between the actin and myosin filaments, and less force is produced. Overstretched sarcomeres produce no force because there is no overlap between the actin and myosin.

Exercise increases muscle strength and endurance

Different types of exercise produce different physical conditioning responses. In general, anaerobic activities, such as weight lifting, increase strength, and aerobic activities, such as jogging, increase endurance. What is the physiological basis for these differences? Strength is simply a function of the cross-sectional area of muscles: the more actin and myosin filaments in a muscle or a muscle fiber, the more tension it can produce. When athletes undertake strength training, they use weights or exercises such as pull-ups to repeatedly contract specific muscles under heavy loads. Repetitions are usually done until the muscle is completely fatigued. Such stress on a muscle probably does minor tissue damage—hence the soreness the day after a hard workout—but it also induces the formation of new actin and myosin filaments in existing muscle fibers. The muscle fibers, and hence the muscles, get bigger and stronger. In extreme cases, and after serious muscle damage, new muscle fibers can also be produced from stem cells called *satellite cells* in the muscle. In general, however, the major effect of strength training is to produce bigger, rather than more, muscle fibers.

Aerobic exercise has a completely different effect on muscles: it enhances their oxidative capacity. This effect comes from increases in the number of mitochondria, in enzymes involved in energy utilization, and in the density of capillaries that deliver oxygen to the muscle. *Myoglobin* also increases in skeletal muscle cells. **Myoglobin** is an oxygen-binding protein similar to hemoglobin in red blood cells. However, myoglobin has a higher affinity for oxygen than does hemoglobin. Therefore, myoglobin accepts oxygen from the blood, facilitates the diffusion of oxygen throughout the muscle, and provides a store of oxygen for use when oxygen delivery by the blood is insufficient. By increasing the capacity of muscle to use oxygen to produce ATP, aerobic training increases the work load that muscles can sustain through time.

Muscle ATP supply limits performance

Muscles have three systems for obtaining the ATP they need for contraction:

- The *immediate system* uses preformed ATP and creatine phosphate.
- The *glycolytic system* metabolizes carbohydrates to lactate and pyruvate.

- The *oxidative system* metabolizes carbohydrates or fats all the way to H_2O and CO_2.

The capacity of these three systems and the rates at which they can produce ATP determine both work capacity and endurance (**Figure 47.12**).

ATP is stored in muscles in very small amounts. However, muscle fibers also contain a storage compound called *creatine phosphate* (CP). This molecule stores energy in a phosphate bond, which it can transfer to ADP. The total energy available in all the muscles of your body in the form of ATP and CP—the immediate energy system—is only about 10 kilocalories. (A kilocalorie is the amount of energy necessary to raise the temperature of 1 liter of water 1°C). However, the energy from ATP and CP is available immediately, and it enables fast-twitch fibers to generate a lot of force quickly. The immediate system is exhausted in only a few seconds.

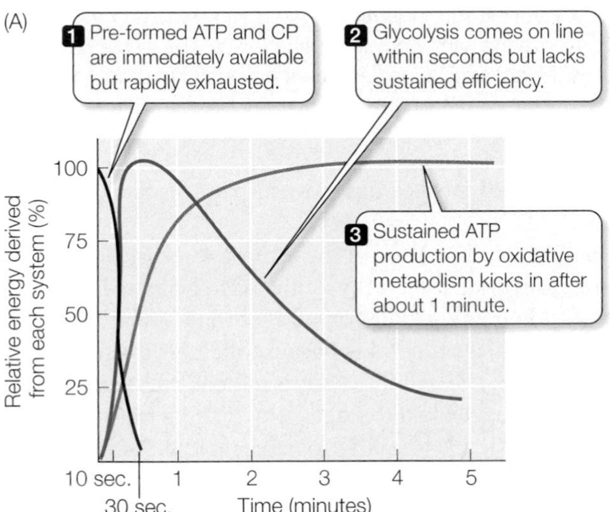

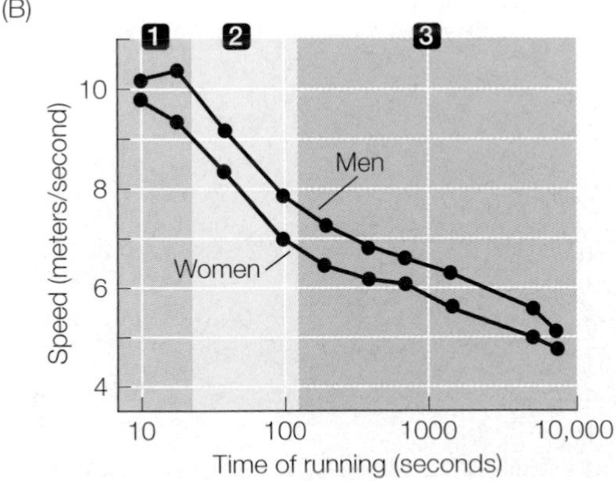

47.12 Supplying Fuel for High Performance (A) Muscles have three systems for obtaining the ATP they need for contraction during exertion such as running. (B) Plotting the time course of world records for running events of different durations, you can see that the performance of world-class athletes corresponds to the time courses of the three energy systems.

The glycolytic system activates within a few seconds to replace the ATP depleted at the onset of muscle activity. The glycolytic enzymes are located in the cytoplasm of the muscle fiber, and therefore the ATP they generate is rapidly available to the myosin filaments. However, as we saw in Chapter 7, glycolysis alone is an inefficient way to produce ATP, and it leads to the accumulation of lactic acid, which slows the process. Thus, the glycolytic system and the immediate system together can provide most of the energy for active muscles for less than a minute (see Figure 47.12).

Oxidative metabolism becomes fully active in about a minute, producing relatively huge amounts of ATP because it can completely metabolize carbohydrates and fats. However, it requires many reactions (see Chapter 7), and it takes place in the mitochondria, so O_2 and substrate must diffuse into the mitochondria, and the formed ATP must diffuse from the mitochondria to the myosin filaments in the muscle. These processes are not instantaneous, so the rate at which oxidative metabolism can make ATP available to do work is slower than the rate at which the other two systems can supply ATP.

The fuel supply available to the muscles influences how long someone can sustain a high level of aerobic exercise. From the circulating blood, muscle receives glucose and free fatty acids, which it can metabolize to generate ATP. At high levels of aerobic exercise, however, most of the fuel used by muscles to produce ATP comes from the reserve of glycogen stored in the muscle itself. Depletion of muscle glycogen results in fatigue. The rate at which muscle glycogen is replenished depends on diet: it is high with a high-carbohydrate diet, low with a high-fat diet, and intermediate with a mixed diet.

After muscle glycogen is depleted by exercise, it is replenished more rapidly by a high-carbohydrate diet than by a diet of fats and protein. This fact is the basis for a practice called "carbo-loading." For 3–5 days, the athlete exercises at a level that depletes muscle glycogen. Then, 2 or 3 days before the event, she tapers down the level of training and eats a diet rich in complex carbohydrates. The result can be glycogen supercompensation, in which the restoration of muscle glycogen stores "overshoots" and reaches above-normal levels.

47.2 RECAP

Depending on the function a muscle serves, it may need to generate maximum force rapidly or sustain activity for a long period. Properties of muscles can adapt them to either of these types of functions.

- Can you describe the differences between slow-twitch and fast-twitch fibers? See p. 1013 and Figure 47.10
- How does exercise influence muscle strength and endurance? See p. 1014
- How do the different sources of ATP influence performance in different types of events? See pp. 1014–1015 and Figure 47.12

47.3 What Roles Do Skeletal Systems Play in Movement?

Muscles can only contract and relax. To create significant movement, they must have something to pull on. In some cases, muscles pull on each other—consider the trunk of the elephant or the arms of an octopus. In most cases, however, **skeletal systems** provide rigid supports against which muscles can pull, creating directed movements. In this section, we'll examine the three types of skeletal systems found in animals: hydrostatic skeletons, exoskeletons, and endoskeletons.

A hydrostatic skeleton consists of fluid in a muscular cavity

The simplest type of skeleton is the hydrostatic skeleton of cnidarians, annelids, and many other soft-bodied invertebrates. As Section 31.2 discusses, a **hydrostatic skeleton** consists of a volume of fluid enclosed in a body cavity surrounded by muscle. When muscles oriented in a certain direction contract, the fluid-filled body cavity bulges out in the opposite direction.

An earthworm uses its hydrostatic skeleton to crawl. The earthworm's body cavity is divided into many separate segments, each of which contains a compartment filled with extracellular fluid. The body wall surrounding each segment has two muscle layers: a circular layer and a longitudinal layer. If the circular muscles in a segment contract, the compartment in that segment narrows and elongates. If the longitudinal muscles in a segment contract, the compartment shortens and bulges outward. Alternating contractions of the earthworm's circular and longitudinal muscles create waves of narrowing and widening, lengthening and shortening, that travel down the body. Bulging, shortened segments serve as anchors as long, narrow segments project forward and longitudinal contractions pull other segments forward. Bristles help the widest parts of the body to hold firm against the substratum (**Figure 47.13**).

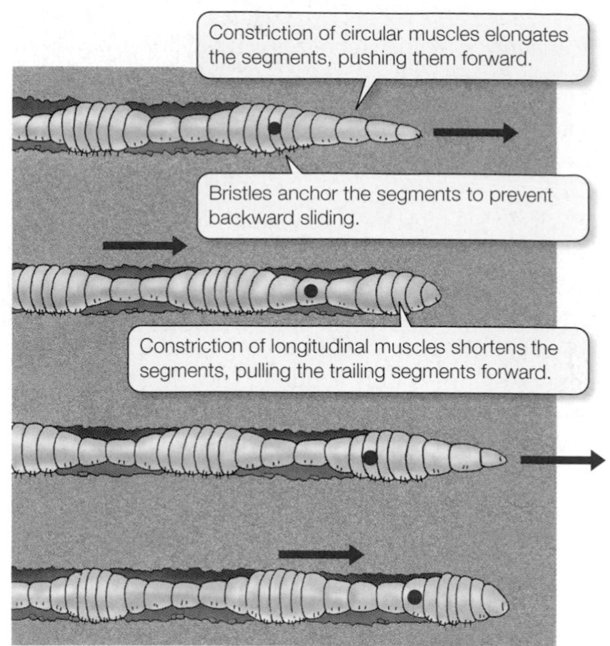

Constriction of circular muscles elongates the segments, pushing them forward.
Bristles anchor the segments to prevent backward sliding.
Constriction of longitudinal muscles shortens the segments, pulling the trailing segments forward.

47.13 A Hydrostatic Skeleton Alternating waves of muscle contraction move the earthworm through the soil. The red dot enables you to follow the changes in one segment as the worm moves forward.

Exoskeletons are rigid outer structures

An **exoskeleton** is a hardened outer surface to which muscles can be attached. Contractions of the muscles cause jointed segments of the exoskeleton to move relative to each other. The simplest example of an exoskeleton is the shell of a mollusk. Some marine mollusks, such as clams and snails, have shells composed of protein strengthened by crystals of calcium carbonate (a rock-hard material). These shells can be massive, affording significant protection against predators. The shells of land snails generally lack the hard mineral component and are much lighter. Molluscan shells can grow as the animal grows, and growth rings are usually apparent on the shells (**Figure 47.14**).

The most complex exoskeletons are found among the arthropods. An exoskeleton, or **cuticle**, covers all the outer surfaces of the arthropod's body and all its appendages. It is made up of plates secreted by a layer of cells just below the exoskeleton. The cuticle contains stiffening materials everywhere except at the joints, where flexibility must be retained. Muscles attached to the inner surfaces of the arthropod exoskeleton move its parts around the joints (see Figure 32.4).

The greatest drawback of the arthropod exoskeleton is that it cannot grow. Therefore, if the animal is to become larger, it must *molt*, shedding its exoskeleton and forming a new, larger one. A molting animal is vulnerable because the new exoskeleton takes time to harden. The animal's body is temporarily unprotected, and without a firm exoskeleton against which its muscles can exert maximum tension, it is unable to move rapidly. Soft-shelled crabs, a gourmet delicacy, are crabs caught when they are molting.

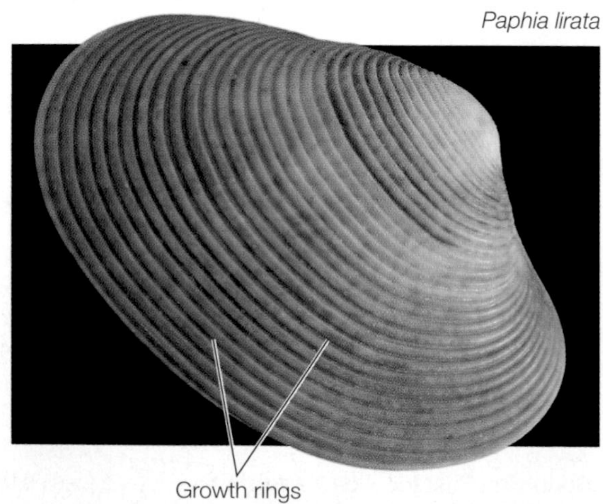

47.14 The Clam Shell Is an Exoskeleton The clam is a bivalve, meaning its exoskeleton consists of an upper and lower shell that are hinged. As the clam grows, the shells grow bigger and are marked by growth rings.

Spiders have an unique mechanism of movement. By pumping extracellular fluid into their hollow legs, they cause those legs to extend. Decreasing fluid pressure allows the legs to flex. This is why the legs of dead spiders curl up underneath them.

Vertebrate endoskeletons provide supports for muscles

The **endoskeleton** of vertebrates is an internal scaffolding. Muscles are attached to it and pull against it. Endoskeletons are composed of rodlike, platelike, and tubelike bones connected to one another at a variety of joints that allow a wide range of movements. An advantage of endoskeletons over the exoskeletons of arthropods is that bones within the body can enlarge without the animal shedding its skeleton.

The human skeleton consists of 206 bones, some of which are shown in **Figure 47.15**. It can be divided into an *axial skeleton*, which includes the skull, vertebral column, and ribs, and an *appendicular skeleton*, which includes the pectoral girdle, the pelvic girdle, and the bones of the arms, legs, hands, and feet.

The vertebrate endoskeleton consists of two kinds of connective tissue, *cartilage* and *bone*, which are produced by two kinds of connective tissue cells. *Cartilage cells* produce an extracellular matrix that is a tough, rubbery mixture of polysaccharides and proteins—mainly fibrous collagen. Collagen fibers run in all directions like reinforcing cords through the gel-like matrix and give it the well-known strength and resiliency of "gristle." This matrix, called **cartilage**, is found in parts of the endoskeleton where both stiffness and resiliency are required, such as on the surfaces of joints where bones move against one another. Cartilage is also the supportive tissue in stiff but flexible structures such as the larynx (voice box), the nose, and the ear pinnae. Sharks and rays are called *cartilaginous fishes* because their skeletons are composed entirely of cartilage. In all other vertebrates, cartilage is the principal component of the embryonic skeleton, but during development most of it is gradually replaced by bone.

Bone also contains collagen fibers, but it gets its rigidity and hardness from an extracellular matrix of insoluble calcium phosphate crystals. Bone serves as a reservoir of calcium for the rest of the body and is in dynamic equilibrium with soluble calcium in the extracellular fluids of the body. This equilibrium is under the control of calcitonin and parathyroid hormone (see Figure 41.10). If too much calcium is taken from the skeleton, the bones are seriously weakened. The living cells of bone—called *osteoblasts*, *osteocytes*, and *osteoclasts*—are responsible for the constant dynamic remodeling of bone (**Figure 47.16**). **Osteoblasts** lay down new matrix material on bone surfaces. These cells gradually become surrounded by matrix and eventually become enclosed within the bone, at which point they cease laying down matrix, but continue to exist within small lacunae (cavities) in the bone. In this state they are called **osteocytes**. Despite the vast amounts of matrix between them, osteocytes remain in contact with one another through long cellular extensions that run through tiny channels in the bone. Communication between osteocytes is important in controlling the activities of the cells that are laying down or removing bone.

The cells that resorb bone are the **osteoclasts**. They are derived from the same cell lineage that produces the white blood cells. Osteoclasts erode bone, forming cavities and tunnels. Osteoblasts follow osteoclasts, depositing new bone. Thus the interplay of osteoblasts and osteoclasts constantly replaces and remodels the bones.

How the activities of the bone cells are coordinated is not understood, but stress placed on bones somehow provides them with information. A remarkable finding in studies of astronauts who spent

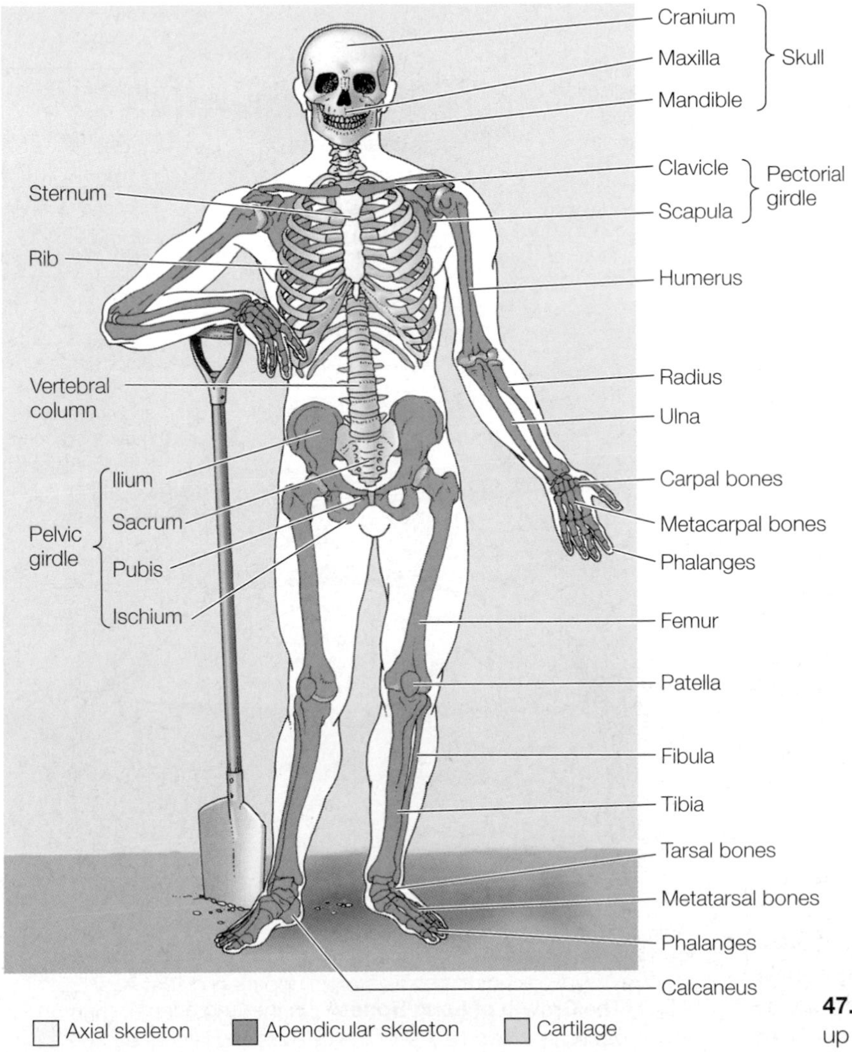

47.15 The Human Endoskeleton Cartilage and bone make up the internal skeleton of a human being.

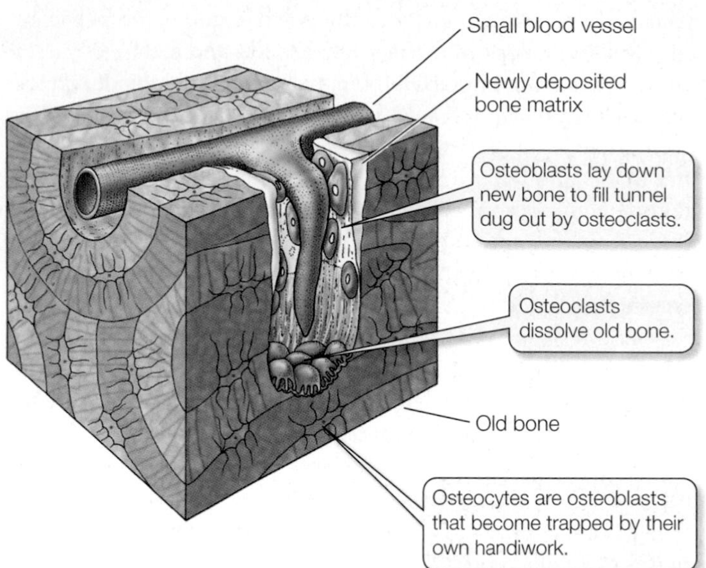
Small blood vessel
Newly deposited bone matrix
Osteoblasts lay down new bone to fill tunnel dug out by osteoclasts.
Osteoclasts dissolve old bone.
Old bone
Osteocytes are osteoblasts that become trapped by their own handiwork.

47.16 Renovating Bone Bones are constantly being remodeled by osteoblasts, which lay down bone, and osteoclasts, which resorb bone.

long periods in zero gravity was that their bones decalcified. Conversely, certain bones of athletes thicken during training. Both thickening and thinning of bones are experienced by anyone who has had a leg in a cast for a long time. The bones of the uninjured leg carry the person's weight and thicken, while the bones of the inactive leg in the cast thin.

Because of the positive effects of physical stress on bone deposition, weight-bearing exercise is effective in preventing and treating the loss of bone density (and hence strength) known as *osteoporosis.* Over 25 million people in the United States suffer from this debilitating condition. Although osteoporosis is most commonly a problem for postmenopausal women, it can occur in younger people as a result of malnutrition and sedentary lifestyle.

Bones develop from connective tissues

Bones are divided into two types on the basis of how they develop. **Membranous bone** forms on a scaffold of connective tissue membrane. **Cartilage bone** forms first as a cartilaginous structure resembling the future mature bone, then gradually hardens or *ossifies* to become bone. The outer bones of the skull are membranous bones; the bones of the limbs are cartilage bones.

Cartilage bones can grow throughout the ossification process. The long bones of the legs and arms, for example, ossify first at the centers and later at each end (**Figure 47.17**). Growth can continue until these areas of ossification join. The membranous bones forming the skull cap grow until their edges meet. The soft spot on the top of a baby's head (the fontanelle) is the point at which the skull bones have not yet joined.

The structure of bone may be **compact** (solid and hard) or **cancellous** (having numerous internal cavities that make it appear spongy, although it is rigid). The architecture of a specific bone depends on its position and function, but most bones have both compact and cancellous regions. The shafts of the long bones of the limbs, for example, are cylinders of compact bone surrounding central cavities that contain the bone marrow, where the cellular elements of the blood are made. The ends of the long bones are cancellous (see Figure 47.17). Cancellous bone is lightweight because of its numerous cavities, but it is also strong because its internal meshwork constitutes a support system. It can withstand considerable forces of compression. The rigid, tubelike shaft of compact bone can withstand compression and bending forces. Architects and nature alike use hollow tubes as lightweight structural elements.

Most of the compact bone in mammals is called *Haversian bone* because it is composed of structural units called **Haversian systems** (**Figure 47.18**). Each Haversian system is a set of thin, concentric bony cylinders, between which are the osteocytes in their lacunae. Through the center of each Haversian system runs a narrow canal containing blood vessels and nerves. Adjacent Haversian systems are separated by boundaries called *glue lines*. Haversian bone is resistant to fracturing because cracks tend to stop at glue lines.

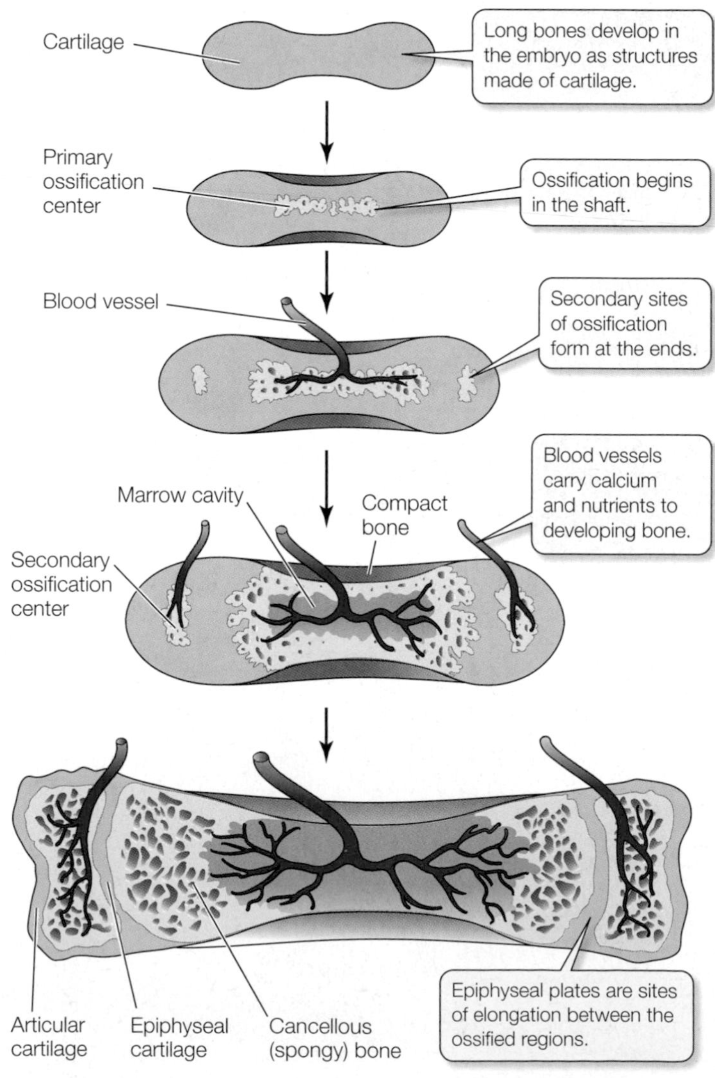
Cartilage
Long bones develop in the embryo as structures made of cartilage.
Primary ossification center
Ossification begins in the shaft.
Blood vessel
Secondary sites of ossification form at the ends.
Marrow cavity
Compact bone
Blood vessels carry calcium and nutrients to developing bone.
Secondary ossification center
Articular cartilage
Epiphyseal cartilage
Cancellous (spongy) bone
Epiphyseal plates are sites of elongation between the ossified regions.

47.17 The Growth of Long Bones In the long bones of human limbs, ossification occurs first at the centers and later at each end.

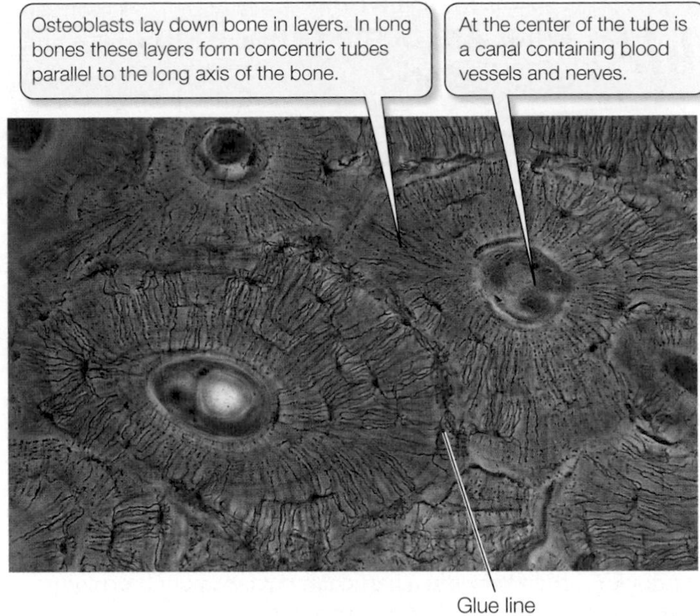

Osteoblasts lay down bone in layers. In long bones these layers form concentric tubes parallel to the long axis of the bone.
At the center of the tube is a canal containing blood vessels and nerves.
Glue line

47.18 Most Compact Bone Is Composed of Haversian Systems A micrograph of a section of a long bone shows Haversian systems with their central canals. Glue lines separate Haversian systems.

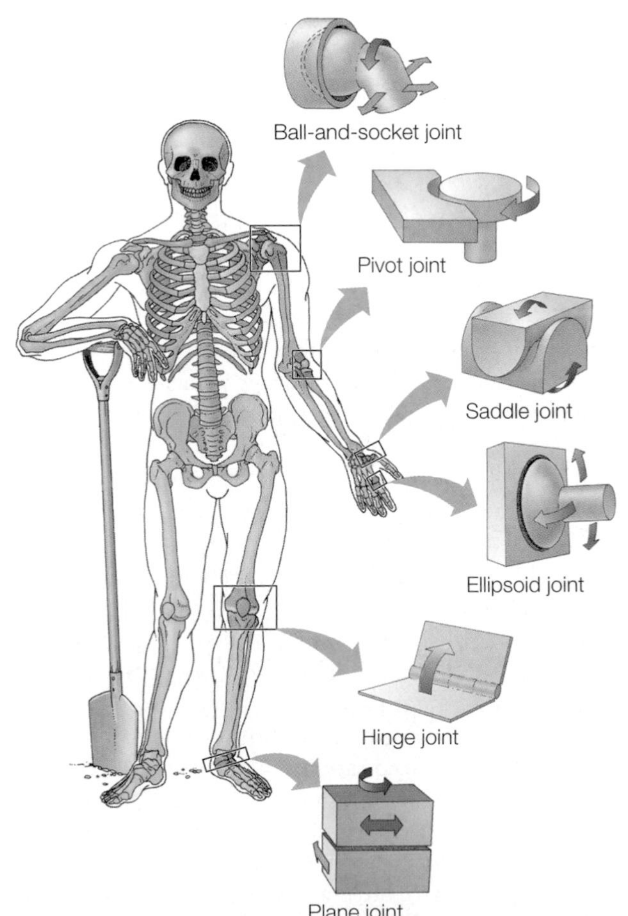

47.19 Types of Joints The designs of joints are similar to mechanical counterparts and enable a variety of movements.

Bones that have a common joint can work as a lever

Muscles and bones work together around **joints**, where two or more bones come together. Different kinds of joints allow motion in different directions (**Figure 47.19**), but muscles can exert force in only one direction. Therefore, muscles create movement around joints by working in antagonistic pairs: When one contracts, the other relaxes. When both contract, the joint becomes rigid. With respect to a particular joint, such as the knee, we can refer to the muscle that bends, or flexes, the joint as the **flexor**, and the muscle that straightens, or extends, the joint as the **extensor**. The bones that meet at the joint are held together by **ligaments**, which are flexible bands of connective tissue. Other straps of connective tissue, called **tendons**, attach the muscles to the bones (**Figure 47.20**). In many kinds of joints, only the tendon spans the joint, sometimes moving over the surfaces of the bones like a rope over a pulley. The tendon of the quadriceps muscle traveling over the knee joint is what is tapped to elicit the knee-jerk reflex (see Figure 46.3).

Bones constitute a system of levers that are moved around joints by the muscles. A lever has a *power arm* and a *load arm* that work around a *fulcrum* (pivot). The length ratio of the two arms determines whether a particular lever can exert a lot of force over a short distance or is better at translating force into large or fast movements. Compare the jaw joint and the knee joint, for example (**Figure 47.21**). The power arm of the jaw is long relative to the load arm, allowing the jaw to apply great force over a small distance. Think of the powerful jaws of carnivores that can easily crack bones. The power arm of the lower leg, on the other hand, is short relative to the load arm, so you can run fast, jump high, and deliver swift kicks.

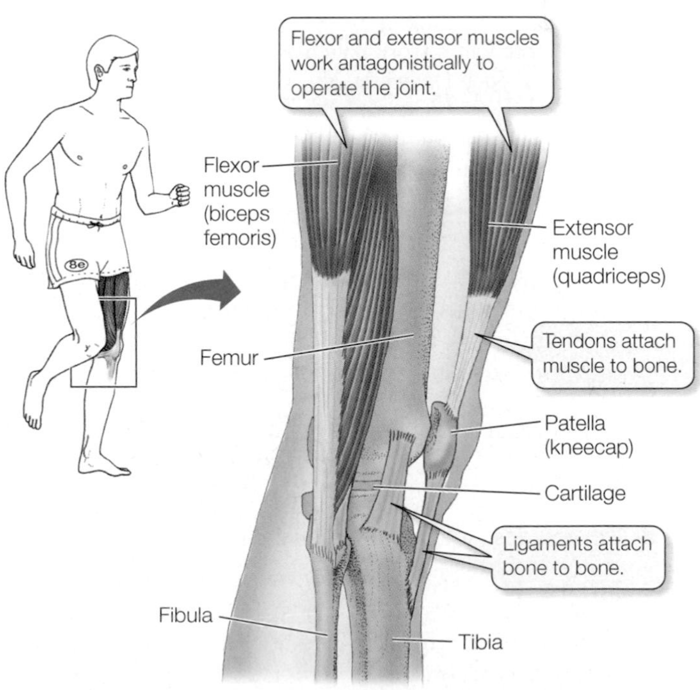

47.20 Joints, Ligaments, and Tendons A side view of the knee shows the interactions of muscle, bone, cartilage, ligaments, and tendons at this crucial and vulnerable human joint.

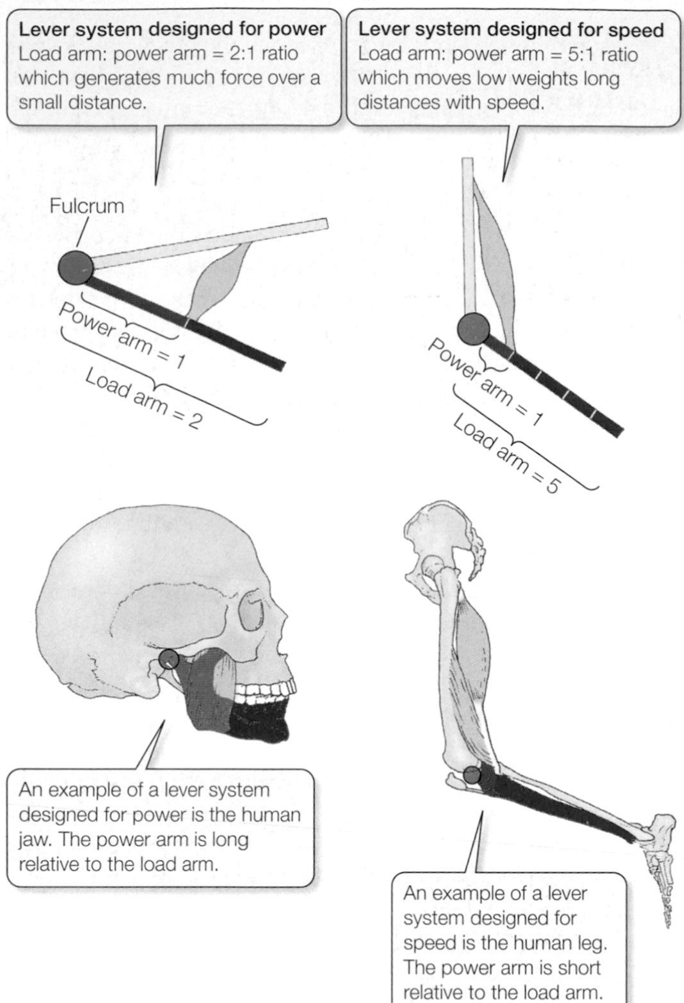

Lever system designed for power
Load arm: power arm = 2:1 ratio which generates much force over a small distance.
Lever system designed for speed
Load arm: power arm = 5:1 ratio which moves low weights long distances with speed.
Fulcrum
Power arm = 1
Load arm = 2
Power arm = 1
Load arm = 5
An example of a lever system designed for power is the human jaw. The power arm is long relative to the load arm.
An example of a lever system designed for speed is the human leg. The power arm is short relative to the load arm.

47.21 Bones and Joints Work Like Systems of Levers A lever system can be designed for either power or speed.

47.3 RECAP

Muscles can only contract and relax; to achieve organized movement they have to pull against rigid structures—other muscles, hydrostatic skeletons, exoskeletons, or endoskeletons.

- Can you describe how muscles and a fluid-filled body cavity interact to enable an earthworm to crawl? See p. 1016 and Figure 47.13
- Can you describe the difference between membranous and cartilaginous bone and between compact and cancellous bone? See p. 1018 and Figure 47.17
- In terms of levers can you explain how specific joints can produce maximum force versus maximum speed? See p. 1019 and Figure 47.21

47.4 What Are Some Other Kinds of Effectors?

Musculoskeletal systems, whether they be hydrostatic skeletons, exoskeletons, or endoskeletons, are just one of many effector systems that allow different species of animals to accomplish defense, communication, feeding, and reproduction. For example, in Chapter 31 we learned how cnidarians such as jellyfish use *nematocysts,* an effector that fires miniature"harpoons"to capture prey and repel predators (see Figure 31.10). Several other effector systems are described next, revealing the evolutionary diversity of the mechanisms animals use to get things done.

Chromatophores allow an animal to change its color or pattern

A change in body color is a response that some animals use to camouflage themselves in a particular environment or to communicate with other animals. **Chromatophores** are pigment-containing cells in the skin that can change the color and pattern of the animal. Chromatophores are under neuronal or hormonal control, or both; in most cases, they can effect a change within minutes or even seconds.

Chromatophores enable squids, sole, and flounder, all of which spend much time on the seafloor, as well as the famous chameleons (a group of African lizards; see Figure 52.10) and a few other animals, to blend in with the background on which they are resting and thus escape discovery by predators. Chromatophores with different pigments enable animals to assume different hues or to become mottled to match the background more precisely. In other mollusks, fishes, and lizards, a color change sends a signal to potential mates and territorial rivals of the same species.

There are three principal types of chromatophore cells. The most common type has fixed cell boundaries, within which pigmented granules may be moved about by microfilaments. When the pigment is concentrated in the center of each chromatophore, the animal is pale; the animal turns darker when the pigment is dispersed throughout the cell. Another type of chromatophore is capable of amoeboid motion. These cells can mold themselves into shapes with a minimal surface area, leaving the tissue relatively pale, or they can flatten out to make the tissue appear darker. The third type of chromatophore changes shape as a result of the action of muscle fibers radiating outward from the cell (**Figure 47.22A**). When the muscle fibers are relaxed, the chromatophores are small and compact, and the animal is pale. To darken the animal, the muscle fibers contract and spread the chromatophores over more of the body surface. These chromatophores can change so rapidly that they are used in some species for communication during courtship and aggressive interactions. For example, the cuttlefish, a cephalopod, can signal courtship intentions to a potential mate on one side of its body while signaling aggressive threats to a rival on the other side (**Figure 47.22B**).

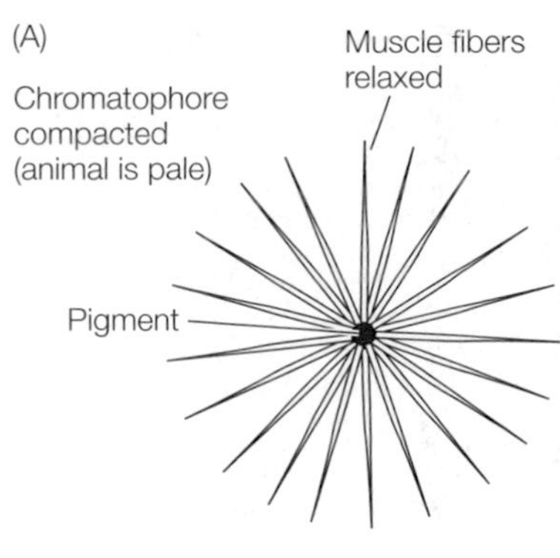

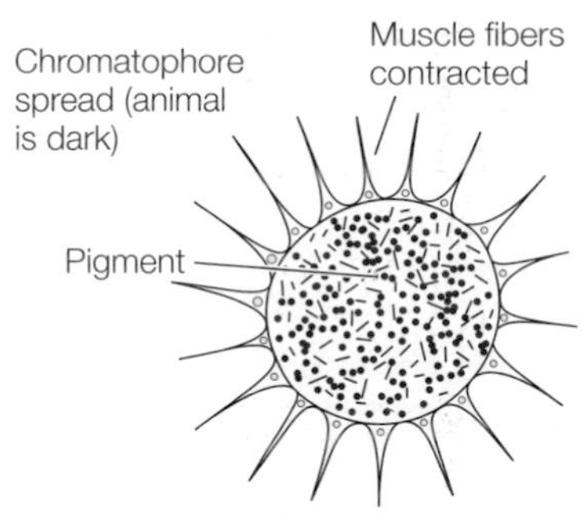

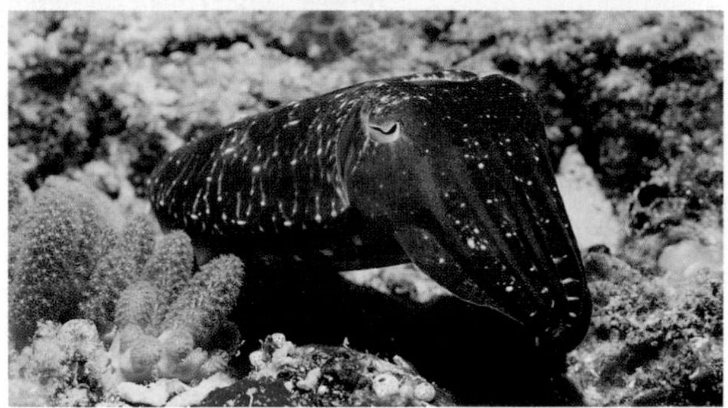

47.22 Chromatophores Help Animals Camouflage Themselves or Communicate (A) Muscle fibers around chromatophores cause the chromatophores to contract. (B) Cuttlefish are cephalopod mollusks that can change color patterns so fast that these changes can be used for rapid communication.

Glands secrete chemicals for defense, communication, or predation

Glands are effector organs that produce and release chemicals. As we learned in Chapter 41, endocrine glands produce hormones for internal signaling and exocrine glands secrete substances into the gut or onto the body surface. In certain animals, some of these secretions are used defensively or to capture prey. Others are *pheromones*, chemical signals released into the environment for communication with other individuals of the same species.

Not all defensive secretions are poisonous; a well-known example is mercaptan, the odoriferous chemical sprayed by skunks as an effective deterrent to most predators.

Certain reptiles, amphibians, mollusks, and fishes have poison glands that are used in capturing prey or defending against predators. Many of the poisons produced by these glands are extremely specific in their modes of action. For example, the poison dendrotoxin, which certain tribes of the Amazonian rainforest use on the tips of their arrows for hunting, comes from the skin of a frog and blocks certain potassium channels. The snake venom α-bungarotoxin inactivates the acetylcholine receptors at the neuromuscular junction. Tetrodotoxin, or TTX (see the opening pages of Chapter 22) blocks voltage-gated sodium channels. Conotoxin, the poison produced by cone snails, blocks calcium channels.

Electric organs generate electricity used for sensing, communication, defense, or attack

Various fishes can generate electricity. These species include the electric eel, the knife fish, the torpedo (a type of ray), and the electric catfish. The electric fields they generate help them sense the environment, communicate, and stun potential predators or prey. The electric organs of these animals evolved from muscles, and they produce electric potentials in the same general way as nerves and muscles do.

Electric organs consist of very large, disc-shaped cells arranged in long rows like stacks of batteries. When these cells discharge simultaneously, the electric organ can generate far more voltage and current than can nerve or muscle tissue. Electric eels, for example, can produce up to 600 volts with an output of approximately 100 watts—enough to light a row of light bulbs or to temporarily stun a person.

Light-emitting organs use enzymes to produce light

Diverse organisms can emit light—**bioluminescence**. Some, such as fireflies, can produce their own light, and others, such as some deep sea fish, use symbiotic luminescent bacteria sequestered in special organs as their light source. The ability to produce light depends on enzymatic oxidation of a substrate. The enzyme is called *luciferase* and the substrate *luciferin*. Bioluminescence serves various functions, including social signaling, attraction of prey, or even avoidance of prey.

47.4 RECAP

Although muscles and the movements they create are ubiquitous effectors in animals, a host of other types of effectors exist, many of which are highly specialized and found in only a limited number of species.

- Can you explain how chromatophores work? See p. 1020 and Figure 47.22
- Can you describe two examples of how effectors are used both for defense and for communication? See p. 1021

CHAPTER SUMMARY

47.1 How do muscles contract?

See Web/CD Tutorial 47.1

Skeletal muscle consists of bundles of **muscle fibers**. Each skeletal muscle fiber is a huge cell containing multiple nuclei.

Skeletal muscles contain numerous **myofibrils**, which are bundles of **actin** and **myosin** filaments. The regular, overlapping arrangement of the actin and myosin filaments into **sarcomeres** gives skeletal muscle its striated appearance. Review Figure 47.1, Web/CD Activity 47.1

The changes in the banding patterns of sarcomeres led to the sliding filament model of muscle contraction. Review Figure 47.2

The molecular mechanism of muscle contraction involves the binding of the globular heads of myosin molecules to actin. Upon binding, the myosin head changes its conformation, causing the two filaments to slide past each other (the **sliding filament** mechanism). Release of the myosin heads from actin and their return to their original conformation requires ATP. Review Figures 47.3 and 47.6

All the fibers activated by a single motor neuron constitute a **motor unit**. When an action potential spreads across the plasma membrane and through the **T tubules**, Ca^{2+} is released from the **sarcoplasmic reticulum**. Review Figure 47.5, Web/CD Activity 47.2

The Ca^{2+} binds to **troponin** and changes its conformation, pulling the **tropomyosin** strands away from the myosin-binding sites on the actin filament. The muscle fiber contracts until the Ca^{2+} is returned to the sarcoplasmic reticulum. Review Figure 47.6

Cardiac muscle cells are striated, uninucleate, branching, and electrically connected by gap junctions, so that action potentials spread rapidly throughout sheets of cardiac muscle and cause coordinated contractions. Some cardiac muscle cells are pacemaker cells that generate and conduct electrical signals.

Smooth muscle provides contractile force for internal organs. Smooth muscle cells respond to stretch and to neurotransmitters from the autonomic nervous system. See Web/CD Tutorial 47.2

In skeletal muscle, a single action potential causes a minimum unit of contraction called a **twitch**. Twitches occurring in rapid succession can be summed, thus increasing the strength of contraction. Maximum sustained tension is known as **tetanus**. Review Figure 47.9

47.2 What determines muscle strength and endurance?

Slow-twitch muscle fibers are adapted for extended, aerobic work; **fast-twitch** fibers are adapted for generating maximum forces for short periods of time. The ratio of slow-twitch to fast-twitch fibers in the muscles of an individual is mostly genetically determined. Review Figure 47.10

The force that a muscle fiber can produce depends on its initial state of extension or contraction. Review Figure 47.11

Anaerobic exercise stimulates the enlargement of muscle fibers through production of new microfilaments. Through aerobic conditioning, muscle fibers can acquire greater oxidative capacity.

Muscle performance depends on ATP supply. Available ATP and creatine phosphate can fuel maximum tension immediately, but are exhausted in seconds. Glycolysis can regenerate ATP rapidly, but is rapidly slowed by accumulation of lactic acid. Oxidative metabolism delivers ATP more slowly, but can continue to do so for a long time. Review Figure 47.12

47.3 What Roles Do Skeletal Systems Play in Movement?

Skeletal systems provide supports against which muscles can pull.

Hydrostatic skeletons are fluid-filled body cavities that can be squeezed by muscles. Review Figure 47.13

Exoskeletons are hardened outer surfaces to which internal muscles are attached.

Endoskeletons are internal systems of rigid rodlike, platelike, and tubelike supports, consisting of **bone** and **cartilage**, to which muscles are attached. Review Figure 47.15

Bone is continually remodeled by **osteoblasts**, which lay down new bone, and **osteoclasts**, which erode bone. Review Figure 47.16

Bones develop from connective tissue membranes (**membranous bone**) or from cartilage (**cartilage bone**) through ossification. Cartilage bone can grow until centers of ossification meet. Review Figure 47.17

Bone can be solid and hard (**compact bone**), or it can contain numerous internal spaces (**cancellous bone**). Most of the compact bone of mammals is composed of **Haversian systems**. Review Figure 47.18

Joints enable muscles to power movements in different directions. Review Figure 47.19, Web/CD Activity 47.3

Tendons connect muscles to bones; **ligaments** connect bones to one another. Review Figure 47.20

Muscles and bones work together around **joints** as systems of levers. Review Figures 47.21

47.4 What Are Some Other Kinds of Effectors?

Effector organs other than muscles include nematocysts, chromatophores, glands, electric organs, and light-emitting organs.

SELF-QUIZ

1. Smooth muscle differs from both cardiac and skeletal muscle in that
   a. it can act as a pacemaker for rhythmic contractions.
   b. contractions of smooth muscle are not due to interactions between neighboring microfilaments.
   c. neighboring cells are electrically connected by gap junctions.
   d. neighboring cells are tightly coupled by intercalated discs.
   e. the membranes of smooth muscle cells are depolarized by stretching.
2. Fast-twitch fibers differ from slow-twitch fibers in that
   a. they are more common in the leg muscles of champion sprinters than marathon runners.
   b. they have more mitochondria.
   c. they fatigue less rapidly.
   d. their abundance is more a product of training than of genetics.
   e. they are more common in postural muscles than in finger muscles.

3. The role of Ca^{2+} in the control of muscle contraction is to
   a. cause depolarization of the T tubule system.
   b. change the conformation of troponin, thus exposing myosin-binding sites.
   c. change the conformation of myosin heads, thus causing microfilaments to slide past each other.
   d. bind to tropomyosin and break actin–myosin cross-bridges.
   e. block the ATP-binding site on myosin heads, enabling muscles to relax.
4. Fifteen minutes into a 10-km run, what is the major energy source of the leg muscles?
   a. Preformed ATP
   b. Glycolysis
   c. Oxidative metabolism
   d. Pyruvate and lactate
   e. High-protein drink consumed right before the race
5. Which statement about skeletal muscle contraction is *not* true?
   a. A single action potential at the neuromuscular junction is sufficient to cause a muscle to twitch.
   b. Once maximum muscle tension is achieved, no ATP is required to maintain that level of tension.
   c. An action potential in the muscle cell activates contraction by releasing Ca^{2+} into the sarcoplasm.
   d. Summation of twitches leads to a graded increase in the tension that can be generated by a single muscle fiber.
   e. The tension generated by a muscle can be varied by controlling how many of its motor units are active.
6. Which statement about the structure of skeletal muscle is *true*?
   a. The light bands of the sarcomere are the regions where actin and myosin filaments overlap.
   b. When a muscle contracts, the A bands of the sarcomere lengthen.
   c. The myosin filaments are anchored in the Z lines.
   d. When a muscle contracts, the H zone of the sarcomere shortens.
   e. The sarcoplasm of the muscle cell is contained within the sarcoplasmic reticulum.
7. The long bones of our arms and legs are strong and can resist both compressional and bending forces because
   a. they are solid rods of compact bone.
   b. their extracellular matrix contains crystals of calcium carbonate.
   c. their extracellular matrix consists mostly of collagen and polysaccharides.
   d. they have a very high density of osteoclasts.
   e. they consist of lightweight cancellous bone with an internal meshwork of supporting elements.
8. If we compare the jaw joint with the knee joint as lever systems,
   a. the jaw joint can apply greater compressional forces.
   b. their ratios of power arm to load arm are about the same.
   c. the knee joint has greater rotational abilities.
   d. the knee joint has a greater ratio of power arm to load arm.
   e. only the jaw is a hinged joint.
9. Which statement about skeletons is *true*?
   a. They can consist of mostly cartilage.
   b. Hydrostatic skeletons cannot be used for locomotion.
   c. An advantage of exoskeletons is that they can continue to grow throughout the life of the animal.
   d. External skeletons must remain flexible, so they never include calcium carbonate crystals, as bones do.
   e. Internal skeletons consist of four different types of bone: compact, cancellous, membranous, and Haversian.
10. Which of the following effectors is *not* used both for avoiding predators and for communication?
   a. Chromatophores
   b. Electric organs
   c. Skeletal muscle
   d. Glandular secretions
   e. All of the above

FOR DISCUSSION

1. You can see from the structure of a sarcomere that it can shorten only by a certain percentage of its resting length. Yet muscles can cause a wide variety of ranges of movement—compare the range of movement of a toe and a leg. What are two adaptive design features of muscles and skeletons that can maximize the ability of a muscle to cause a greater range of movement of an appendage?
2. If athletes train up until a day before competition and then take a day of rest, performance in which types of events will be most affected by what they eat during the rest day?
3. Wombats are powerful digging animals, and kangaroos are powerful jumping animals. How do you think the structures of their legs would compare in terms of their designs as lever systems?
4. Why are ducks better long-distance fliers than chickens?
5. If an adolescent breaks a leg bone close to the ankle joint, after the break heals, that leg may not grow as long as the other one. Why?

FOR INVESTIGATION

For aerobic exercise endurance, the most important limiting factor is muscle glycogen reserve, but the muscle also uses glucose from the blood. How would you design an experiment to test whether or not intake of carbohydrate during exercise will improve endurance, and if it does, what is its effect on the sparing of muscle glycogen?

CHAPTER 48 Gas Exchange in Animals

High fliers

"And so back up the gradual slopes, the wind behind me. A much greater effort this, stopping every few yards with a slight anxiety lest I should not make the distance. As I approached the tents, I was astonished to see a bird, a chough, strutting about on the stones near me. … During this day, too, Charles Evans saw what must have been a migration of small grey birds. … Neither of us had thought to find any signs of life as high as this."

Sir John Hunt related the above incident in his 1953 book *The Ascent of Everest*, the story of the first expedition to reach the summit of Mount Everest—at 8,850 meters, the highest point on Earth. Hunt's encounter with the chough occurred at the last campsite before the summit attempt, some 8,000 meters up. At that altitude, humans are incapacitated if they do not breathe supplemental oxygen from pressurized bottles. Just before the moment described, Hunt had gone a short distance downhill from his tent without supplemental oxygen.

Humans have a limited capacity to exist when oxygen is in short supply and, to make it worse, we may not even realize when we are in trouble. One example was the infamous flight of the hot air balloon Zenith in 1875, in which three French scientists decided to study the effect of high altitude on humans—themselves. Intending to take physiological measurements on each other during the ascent, the three men loaded the *Zenith* with scientific apparatus and took off. As they reached higher and higher altitudes, the writing in their notebooks became less legible, and finally non-sensical. At no point did they register concern or alarm as they continued to cut away the ballast bags and ascend even higher. Finally all three men went unconscious, and when the balloon finally descended on its own, only one regained consciousness; the other two were dead. What limits our survival at high altitudes, and why didn't these men realize they had reached these limits?

Unlike humans, some birds do indeed function at extreme altitudes. Barhead geese and other birds migrate over the Himalayas at elevations above 9,000 meters. The highest recorded altitude for a bird is for a vulture that collided with an airliner at 11,278 meters. These numbers are even more impressive when you consider that birds in flight consume oxygen at a rate that a well-conditioned human athlete can sustain for only minutes. Birds sustain high rates of

"I Was Astonished To See a Bird" Bar-headed geese (*Anser indicus*) are among those birds that can sustain the high metabolic costs of flight even at the high altitudes of the Himalayas, where oxygen is scarce.

Aquatic Feats Many open-ocean fish such as these barracuda (*Sphyraena genie*) can swim great distances at high speeds, even though the oxygen content of water is only about 5 percent that of the air humans breathe.

CHAPTER OUTLINE

oxygen consumption during very long flights and at extremely high altitudes.

Birds are not the only animals that travel where humans cannot. Fish can swim much faster, farther, and longer than the best human swimmers. Yet the fish are breathing water that has less than 5 percent of the oxygen content of the air humans breathe. The ability of birds and fish to maintain high rates of metabolism even when oxygen is severely limited is explained by adaptations of their respiratory gas exchange systems.

IN THIS CHAPTER we explore adaptations of the respiratory systems of both water and air breathers for exchanging oxygen and carbon dioxide with the environment. We first discuss the physical factors that limit respiratory gas exchange and identify those factors that natural selection has optimized. We then examine the respiratory gas exchange organs of a variety of species and describe the adaptations of the blood for transporting respiratory gases. Finally, we will see how respiratory gas exchange systems are controlled and regulated.

48.1 What Physical Factors Govern Respiratory Gas Exchange?

The **respiratory gases** that animals must exchange are oxygen (O_2) and carbon dioxide (CO_2). Cells need to obtain O_2 from the environment to produce an adequate supply of ATP by cellular respiration (see Chapter 7). CO_2 is an end product of cellular respiration, and it must be removed from the body to prevent toxic effects.

Diffusion is the only means by which respiratory gases are exchanged between the internal body fluids of an animal and the outside medium (air or water). There are no active transport mechanisms to move respiratory gases across biological membranes. Because diffusion is a physical process, knowing the physical factors that influence rates of diffusion helps us understand the diverse adaptations of gas exchange systems. (You might want to review what you learned about the physical nature of diffusion in Section 5.3.) Here we will discuss environmental factors that influence diffusion rates, then consider respiratory system adaptations that facilitate the diffusion of respiratory gases.

Diffusion is driven by concentration differences

Because diffusion results from the random motion of molecules, the net movement of a molecule is always down its concentration gradient. One way biologists express the concentrations of different gases in a mixture is by the *partial pressures* of those gases. First, we have to know what the total pressure is, and we measure that with an instrument called a barometer. There are many types of barometers, but the classical one is a glass tube closed at one end and filled with mercury. This is then inverted over a pool of mercury with the open end of the tube under the surface of the mercury. At sea level, the pressure exerted by the atmosphere will support, and therefore be equal to, a column of mercury in the tube that is about 760 mm high (depending on the weather). Therefore, **barometric pressure** (atmospheric pressure) at sea level is 760 millimeters of mercury (mm Hg).* Because dry air is 20.9 percent O_2, the **partial pressure of oxygen** (P_{O_2}) at sea level is 20.9 percent of 760 mm Hg,

*In S.I. units of pressure, this is 10.1 kilopascals (kPa).

or about 159 mm Hg. That is, the contribution of O_2 to the total air pressure is about 159 mm Hg.

Describing the concentration of respiratory gases in a liquid such as water is a little more complicated because another factor is involved—the solubility of the gas in the liquid. Thus, the actual amount of a gas in a liquid depends on the partial pressure of that gas in the gas phase in contact with the liquid as well as on the solubility of that gas in that liquid. However, the diffusion of the gas between the gaseous phase and the liquid still depends on the partial pressures of the gas in the two phases.

Fick's law applies to all systems of gas exchange

Diffusion is a physical phenomenon that can be described quantitatively with a simple equation called **Fick's law of diffusion**. All environmental variables that limit respiratory gas exchange and all adaptations that maximize respiratory gas exchange are reflected in one or more components of this equation. Fick's law is written as

$$Q = DA\frac{P_1 - P_2}{L}$$

where

- Q is the rate at which a gas such as O_2 diffuses between two locations.
- D is the *diffusion coefficient*, which is a characteristic of the diffusing substance, the medium, and the temperature. (For example, perfume has a higher D than motor oil vapor, and all substances diffuse faster at higher temperatures, as well as diffusing faster in air than in water.)
- A is the cross-sectional area through which the gas is diffusing.
- P_1 and P_2 are the partial pressures of the gas at the two locations.
- L is the path length, or distance, between the two locations.

Therefore, $(P_1 - P_2)/L$ is a *partial pressure gradient*.

Animals can maximize D for respiratory gases by using air rather than water as their gas exchange medium whenever possible; doing so greatly increases Q. All other adaptations for maximizing respiratory gas exchange must influence the surface area (A) for gas exchange or the partial pressure gradient $[(P_1 - P_2)/L]$ across that surface area.

Air is a better respiratory medium than water

Oxygen can be obtained more easily from air than from water for several reasons:

- The O_2 content of air is much higher than the O_2 content of an equal volume of water. The maximum O_2 content of a bubbling stream in equilibrium with air is less than 10 ml of O_2 per liter of water. The O_2 content of the air over the stream is about 200 ml of O_2 per liter of air.
- Oxygen diffuses about 8,000 times more rapidly in air than in water. That is why the O_2 content of a stagnant pond can be zero only a few millimeters below the surface.
- When an animal breathes, it does work to move water or air over its specialized gas exchange surfaces. More energy is required to move water than to move air because water is 800 times more dense than air and about 50 times more viscous.

The slow diffusion of O_2 molecules in water affects air-breathing animals as well as water-breathing ones. Eukaryotic cells carry out cellular respiration in their mitochondria, which are located in the cytoplasm—an aqueous medium. Cells are bathed in extracellular fluid—also an aqueous medium. In addition, all respiratory surfaces must be protected from desiccation by a thin film of fluid through which O_2 must diffuse. The slow rate of O_2 diffusion in water limits the efficiency of O_2 distribution from gas exchange surfaces to the sites of cellular respiration even in air-breathing animals.

Diffusion of O_2 in water is so slow that even animal cells with low rates of metabolism can be no more than a couple of millimeters away from a good source of environmental O_2. Therefore, there are severe size and shape limits on the many species of invertebrates that lack internal systems for distributing O_2. Most of these species are very small, but some have grown larger by evolving a flat, thin body with a large external surface area (**Figure 48.1A**). Still others have very thin bodies that are built around a central cavity through which water circulates (**Figure 48.1B**). A critical factor enabling larger, more complex animal bodies has been the evolution of specialized respiratory systems with large surface areas for enhancing respiratory gas exchange (**Figure 48.1C**).

High temperatures create respiratory problems for aquatic animals

Animals that breathe water are in a double bind when environmental temperatures rise. Most water breathers are *ectotherms*—their body temperatures are closely tied to the temperature of the water around them. As the temperature of the water gets warmer, the ectotherm's body temperature and metabolic rate rise (see Figure 40.9). Thus, water breathers need more O_2 as the water gets warmer. But warm water holds less dissolved gas than cold water does (just think of the gases that escape when you open a warm bottle of soda). In addition, if the animal performs work to move water across its gas exchange surfaces (as fish do), the energy the animal must expend to breathe increases as water temperature rises. Therefore, as water temperature goes up, the water breather must extract more and more O_2 from an environment that is increasingly O_2 deficient, and a lower percentage of that O_2 is available to support activities other than breathing (**Figure 48.2**).

O_2 availability decreases with altitude

Just as a rise in temperature reduces the supply of O_2 available to water breathers, an increase in altitude reduces the O_2 supply for air breathers. At all altitudes, O_2 makes up 20.9 percent of the dry air; however, as you go up in altitude, the total amount of gas per unit volume decreases, as reflected in the barometric pressure. For example, at an altitude of 5,800 m, barometric pressure is only half what it is at sea level, so the P_{O_2} at that altitude is only about 80 mm Hg. At the summit of Mount Everest (8,850 m), P_{O_2} is only about 50 mm Hg—roughly one-third what it is at sea level. Since the movement of O_2 across respiratory gas exchange surfaces and into the body depends on diffusion, its rate of movement depends on the P_{O_2} difference between the air and the body fluids. Therefore, the drastically reduced P_{O_2} in the air at high altitudes constrains O_2 uptake. Because of these constraints, mountain climbers who ven-

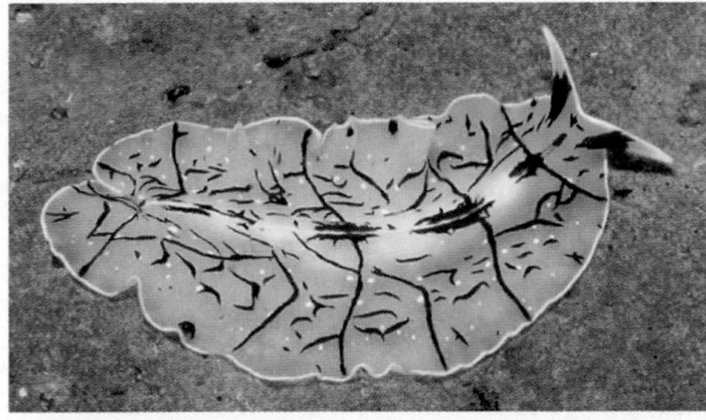

48.1 Keeping in Touch with the Medium (A) No cell in the leaflike body of this marine flatworm is more than a millimeter away from seawater. (B) Sponges have body walls perforated by many channels, which allow water to flow between the outside world and a central cavity. No cell in the sponge is more than a millimeter away from seawater. (C) A feathery fringe of gills on this larval salamander provides a large surface area for gas exchange. Blood circulating through the gills comes into close contact with the respiratory medium.

ture to the heights of Mount Everest usually breathe O_2 from pressurized bottles.

CO_2 is lost by diffusion

Respiratory gas exchange is a two-way process: CO_2 diffuses out of the body as O_2 diffuses in. The direction and rate of diffusion of the respiratory gases across the respiratory exchange surfaces depend on the partial pressure gradients of the gases. The partial pressure gradients of O_2 and CO_2 across these gas exchange surfaces are quite different. The amount of CO_2 in the atmosphere is extremely low (0.03 percent), so for air-breathing animals there is always a large concentration gradient for diffusion of CO_2 from the body to the environment. Whereas the partial pressure gradient for O_2 decreases with altitude, the partial pressure gradient for CO_2 does not. The partial pressure of CO_2 in the atmosphere is close to zero both at sea level and on top of Mount Everest.

In general, getting rid of CO_2 is not a problem for water-breathing animals because CO_2 is much more soluble in water than is O_2. Even in stagnant water, where the P_{CO_2} is higher than in fresh water, the lack of O_2 becomes a problem for the animal long before CO_2 exchange difficulties arise.

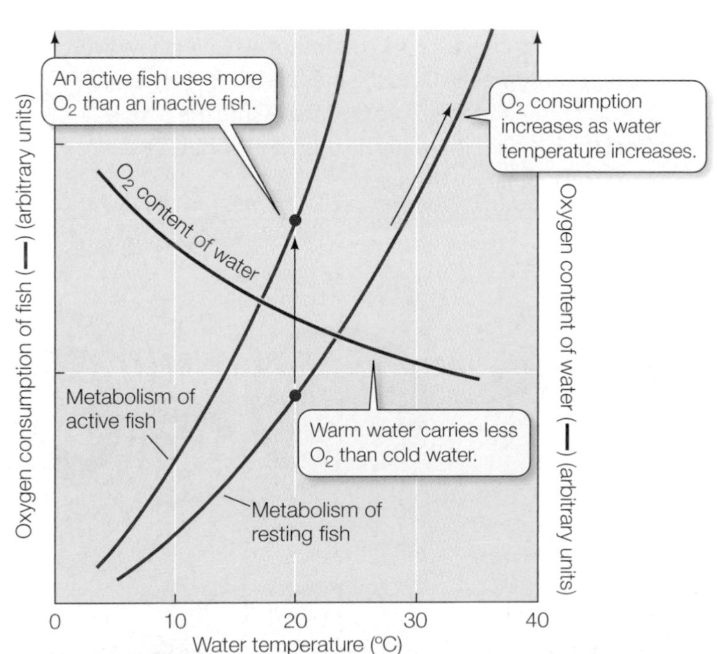

48.2 The Double Bind of Water Breathers Fish need more O_2 when the water is warmer, but warm water carries less O_2 than cold water.

48.1 RECAP

Respiratory gases are exchanged by diffusion only. Air is a better respiratory medium than water because there is more O_2 in a given volume of air than in the same volume of water, O_2 diffuses faster in air than in water, and it requires less work to move air over respiratory exchange surfaces than water.

- Can you explain the concept of partial pressures of gases and how they relate to rates of diffusion of O_2 and CO_2 at different altitudes? See pp. 1025–1026
- Why does a rise in water temperature create a double-bind situation for aquatic animals? See p. 1026 and Figure 48.2
- Can you describe how the variables in Fick's law of diffusion relate to respiratory systems?

Now that we know the physical factors that influence the rates of diffusion of respiratory gases between an animal and its environment, let's take a look at some of the adaptations animals have evolved for maximizing their respiratory gas exchange.

48.2 What Adaptations Maximize Respiratory Gas Exchange?

Some common ways the respiratory systems of different organisms maximize the exchange of O_2 and CO_2 with the environment include adaptations for increasing the surface area over which diffusion of gases can occur; for maximizing partial pressure gradients; and for minimizing the diffusion path length through an aqueous medium.

Respiratory organs have large surface areas

Many anatomical adaptations maximize the specialized body surface area (*A*) over which respiratory gases can diffuse. **External gills** are highly branched and folded extensions of the body surface that provide a large surface area for gas exchange with water (**Figure 48.3A**). External gills are found in larval amphibians and in the larvae of many insect species. Because they consist of thin, delicate tissues, they minimize the length of the path (*L*) traversed by diffusing molecules of O_2 and CO_2. External gills are vulnerable to damage, however, and are tempting morsels for predators, so in many animals protective body cavities for gills have evolved. Such **internal gills** are found in many mollusks and arthropods, and in all fishes (**Figure 48.3B**).

Air-breathing vertebrates also have surprisingly large surface areas for gas exchange. **Lungs** are internal cavities for respiratory gas exchange with air. Their structure is quite different from that of gills (**Figure 48.3C**). Lungs have a large surface area because they are highly divided, and they are elastic so that they can be inflated with air and deflated.

The most abundant air-breathing invertebrates are insects, which have a respiratory gas exchange system consisting of a network of air-filled tubes called **tracheae** that branch through all tissues of the insect's body (**Figure 48.3D**). The terminal branches of these tubes are so numerous that they have an enormous surface area compared to the external surface area of the insect's body.

Transporting gases to and from the exchange surfaces optimizes partial pressure gradients

Fick's law of diffusion includes variables (other than surface area) that can affect the rate of gas exchange. Partial pressure gradients $[(P_1 - P_2)/L]$ drive diffusion across gas exchange surfaces. These gradients can be maximized in several ways:

- *Minimization of path length*: Very thin tissues in gills and lungs reduce the diffusion path length (*L*).
- *Ventilation*: Actively moving the respiratory medium over the gas exchange surfaces (breathing) exposes those surfaces regularly to fresh respiratory medium containing maximum O_2 and minimum CO_2 concentrations. Thus, the concentration gradient is maximized.
- *Perfusion*: Circulating blood over the internal side of the exchange surfaces transports CO_2 to those surfaces and O_2 away from those surfaces, thus maximizing the concentration gradients driving diffusion.

An animal's **gas exchange system** is made up of its gas exchange surfaces and the mechanisms it uses to ventilate and perfuse those surfaces. This chapter describes four gas exchange systems. First we'll look at the gas exchange system of insects. Then we will describe fish gills and bird lungs—two remarkably efficient systems. The next section details the structure and functioning of human lungs.

Insects have airways throughout their bodies

The tracheal system that enables insects to exchange respiratory gases extends to all tissues in the insect body. Thus, respiratory gases diffuse through air most of the way to and from every cell. The insect respiratory system communicates with the outside environment through gated openings called *spiracles* in the sides of the abdomen (**Figure 48.4A,B**). The spiracles can open to allow gas exchange, and then close to decrease water loss. Spiracles open into tubes called *tracheae* that branch into even finer tubes, or *tracheoles*, until they end in tiny *air capillaries* (**Figure 48.4C**). In the insect's flight muscles and other highly active tissues, every mitochondrion is close to an air capillary.

Some aquatic insects carry a bubble of air when they dive. As they consume the O_2 in the bubble, the P_{O_2} in the bubble decreases and therefore O_2 diffuses into the bubble from the surrounding water. In this way the bubble serves as a "scuba tank" for the insect.

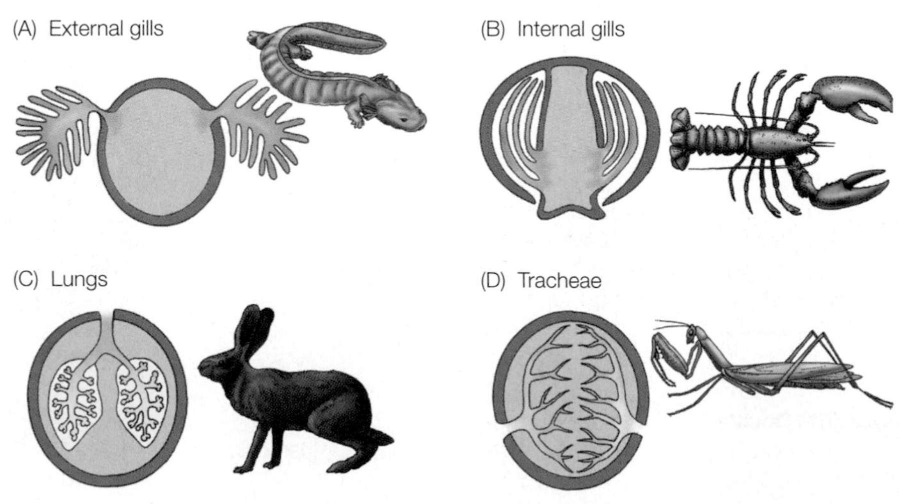

48.3 Gas Exchange Systems Large surface areas (blue in these diagrams) for the diffusion of respiratory gases are common features of animals. Both external (A) and internal (B) gills are adaptations for gas exchange with water. Lungs (C) and tracheae (D) are organs for gas exchange with air.

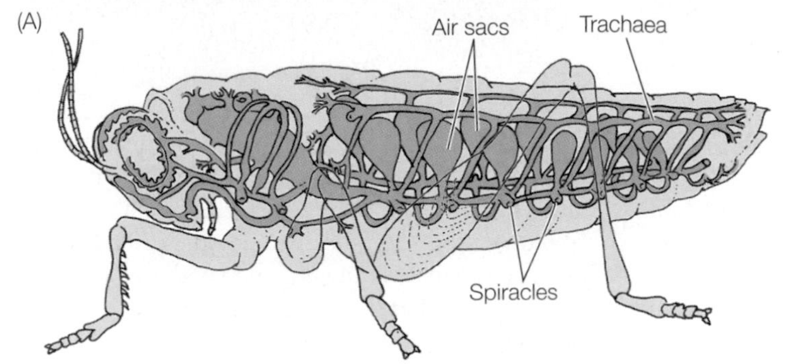

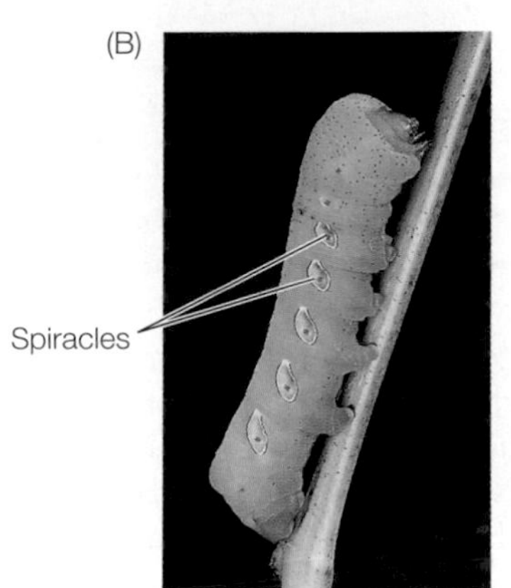

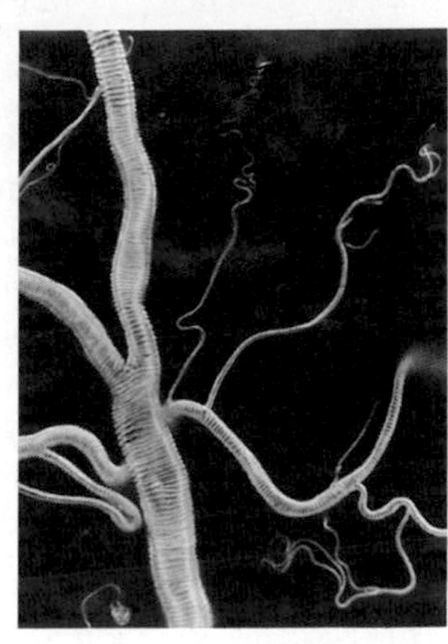

48.4 The Tracheal Gas Exchange System of Insects (A) In insects, respiratory gases diffuse through a system of air tubes (tracheae) that open to the external environment through holes called spiracles. (B) The spiracles of a sphinx moth larva run down its sides. (C) A scanning electron micrograph shows an insect trachea dividing into smaller tracheoles and still finer air capillaries.

Fish gills use countercurrent flow to maximize gas exchange

The internal gills of fish are supported by *gill arches* that lie between the mouth cavity and the protective *opercular flaps* on the sides of the fish just behind the eyes (**Figure 48.5A**). Water flows *unidirectionally* into the fish's mouth, over the gills, and out from under the opercular flaps. Thus, the gills are continuously bathed with fresh water. This constant, one-way flow of water moving over the gills maximizes the P_{O_2} on the external gill surfaces. On the internal side of the gill membranes, the circulation of blood minimizes the P_{O_2} by sweeping the O_2 away as rapidly as it diffuses across.

Gills have an enormous surface area for gas exchange because they are so highly divided. Each gill consists of hundreds of leaf-

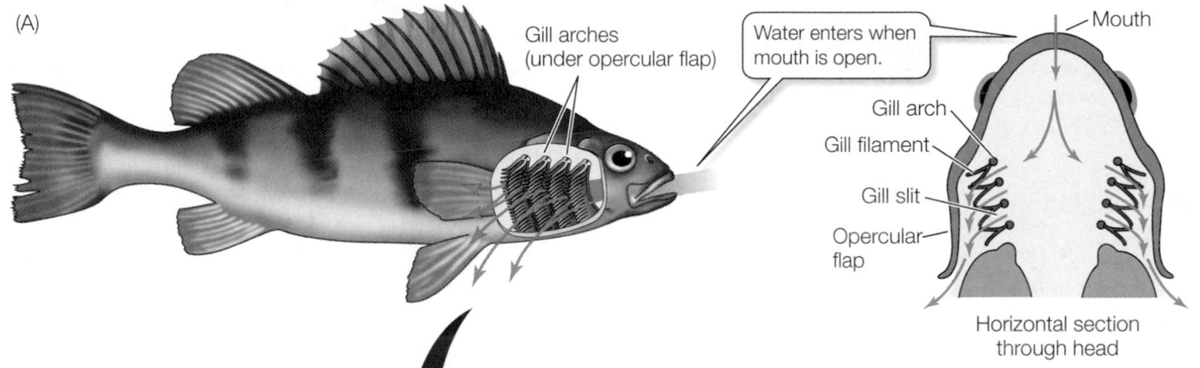

(A)
Gill arches (under opercular flap)
Water enters when mouth is open.
Mouth
Gill arch
Gill filament
Gill slit
Opercular flap
Horizontal section through head

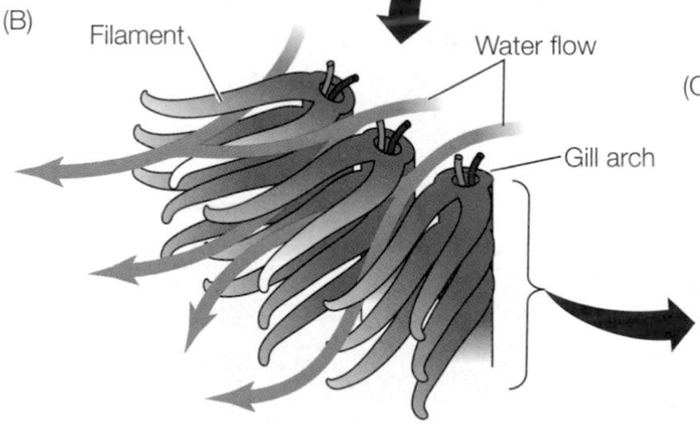

(B)
Filament
Water flow
Gill arch

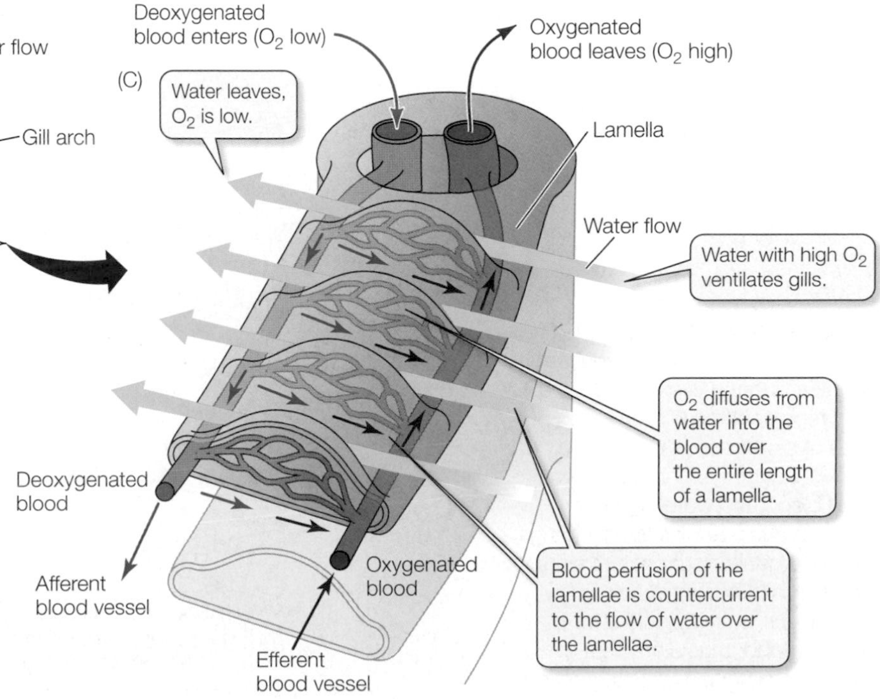

(C)
Deoxygenated blood enters (O_2 low)
Oxygenated blood leaves (O_2 high)
Water leaves, O_2 is low.
Lamella
Water flow
Water with high O_2 ventilates gills.
O_2 diffuses from water into the blood over the entire length of a lamella.
Deoxygenated blood
Afferent blood vessel
Oxygenated blood
Efferent blood vessel
Blood perfusion of the lamellae is countercurrent to the flow of water over the lamellae.

48.5 Fish Gills (A) Water flows unidirectionally over the gills of a fish. (B) Gill filaments have a large surface area and thin tissues. (C) Blood flows through the lamellae in the direction opposite (left to right, in this depiction) to the flow of water (right to left) over the lamellae.

shaped *gill filaments* (**Figure 48.5B**). The upper and lower flat surfaces of each gill filament are covered with rows of evenly spaced folds, or *lamellae*. The lamellae are the actual gas exchange surfaces. Their delicate structure minimizes the path length (L) for diffusion of gases between blood and water. The surfaces of the lamellae consist of highly flattened epithelial cells, so the water and the fish's red blood cells are separated by little more than 1 or 2 μm.

The flow of blood perfusing the inner surfaces of the lamellae, like the flow of water over the gills, is unidirectional. *Afferent* blood vessels bring blood to the gills, while *efferent* blood vessels take blood away from the gills (**Figure 48.5C**). Blood flows through the lamellae in the direction opposite to the flow of water over the lamellae. This **countercurrent flow** optimizes the P_{O_2} gradient between water and blood, making gas exchange more efficient than it would be in a system using concurrent (parallel) flow (**Figure 48.6**).

Some fish, including anchovies, tuna, and certain species of sharks, ventilate their gills by swimming almost constantly with their mouths open. Most fish, however, ventilate their gills by means of a two-pump mechanism. The closing and contracting of the mouth cavity pushes water over the gills, and the expansion of the opercular cavity prior to opening of the opercular flaps pulls water over the gills.

These adaptations allow fish to extract an adequate supply of O_2 from meager environmental sources by maximizing the surface area (A) for diffusion, minimizing the path length (L) for diffusion, and maximizing the P_{O_2} gradient by means of constant, unidirectional, countercurrent flow of blood and water over the opposite sides of their gas exchange surfaces.

Birds use unidirectional ventilation to maximize gas exchange

As we saw at the beginning of this chapter, birds can sustain high levels of activity much longer than mammals can—even at high altitudes where mammals cannot even survive. Yet the lungs of a bird are smaller than the lungs of a similar-sized mammal, and bird lungs expand and contract less during a breathing cycle than do mammalian lungs. Even more puzzling, bird lungs contract during inhalation and expand during exhalation!

The structure of bird lungs allows air to flow unidirectionally through the lungs, rather than bidirectionally through all of the same airways, as it does in mammals. Because mammalian lungs are never completely emptied of air during exhalation, there is always some lung volume that is not ventilated with fresh air. This nonventilated volume is called dead space. Because of continuous/unidirectional airflow, bird lungs have very little dead space, and the fresh incoming air is not mixed with stale air. In this way, a high P_{O_2} gradient is maintained.

In addition to lungs, birds have **air sacs** at several locations in their bodies. The air sacs are interconnected with each other, with the lungs, and with air spaces in some of the bones (**Figure 48.7A**). The air sacs receive inhaled air, but they are not gas exchange surfaces. As in other air-breathing vertebrates, air enters and leaves a bird's gas exchange system through a **trachea** (commonly known as the *windpipe*, and not to be confused with the air-conducting tracheae of insects), which divides into smaller airways called **bronchi** (singular bronchus).

The bronchi divide into tiny tubelike **parabronchi** that run parallel to one another through the lungs (**Figure 48.7B**). Branching off the parabronchi are numerous tiny *air capillaries*. Air flows through the lungs in the parabronchi and diffuses into the air capillaries, which are the gas exchange surfaces. They are so numerous that they provide an enormous surface area for gas exchange. The parabronchi coalesce into larger bronchi that take the air out of the lungs and back to the trachea. Thus the anatomy of the airways of birds allow air to flow unidirectionally and continuously through the lungs.

The puzzle of how birds breathe was solved by an experiment in which small O_2 sensors were placed at different locations in the air sacs and airways of birds. The bird was then exposed to pure O_2 for just a single breath, which made it possible to track that particular inhalation. The experiment demonstrated that a single breath remains in the bird's gas exchange system for two cycles of inhalation and exhalation, and that the air sacs work like bellows; inhalation expands the sacs, and exhalation compresses them to maintain a continuous and unidirectional flow of fresh air through the lungs (**Figure 48.8**).

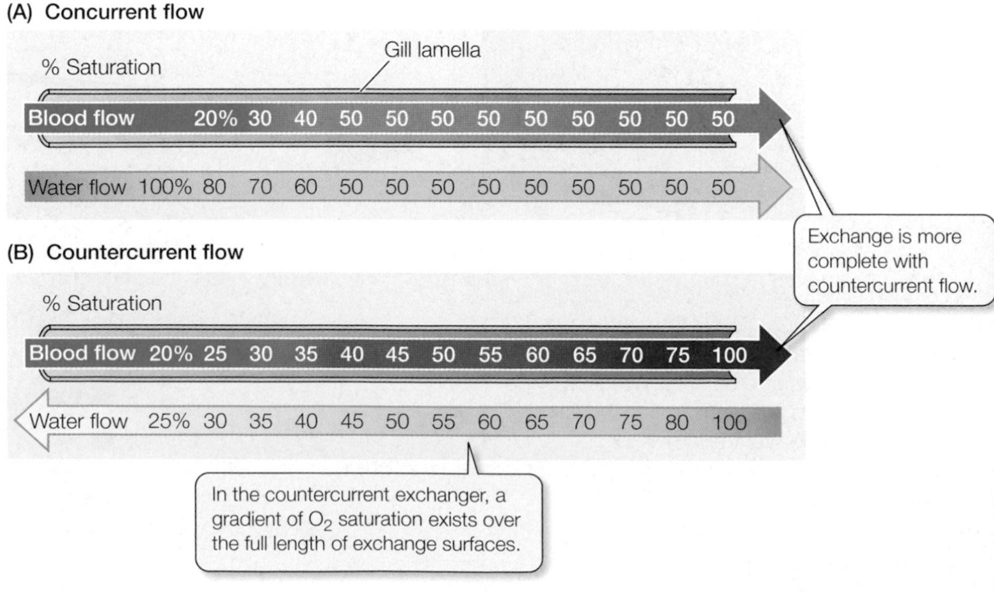

48.6 Countercurrent Exchange Is More Efficient In these models of concurrent and countercurrent gas exchange, the numbers represent the O_2 saturation percentages of blood and water. (A) In a concurrent exchanger, the percentages of saturation of blood and water reaches equilibrium halfway across the exchange surface. (B) A countercurrent exchanger allows more complete gas exchange because the water is always more O_2-saturated than the blood; thus a gradient of O_2 saturation is maintained.

(A) Avian air sacs and lungs

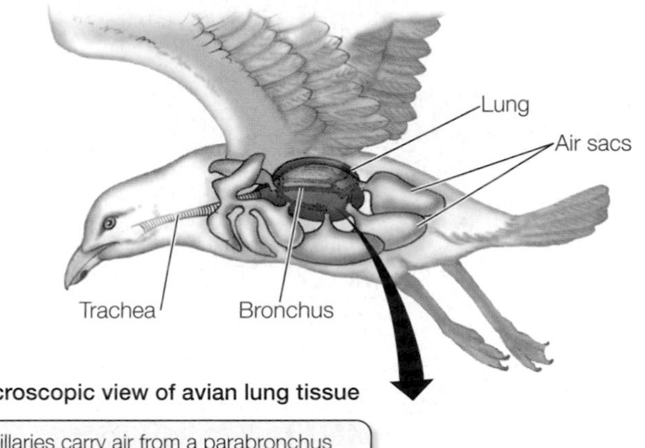

(B) Microscopic view of avian lung tissue

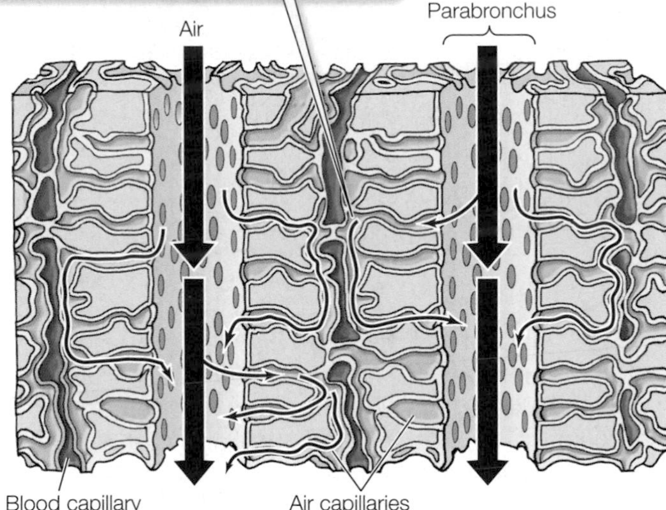

48.7 The Respiratory System of a Bird (A) The air sacs and air spaces in the bones are unique to birds. (B) Air flows through bird lungs unidirectionally in parabronchi. Air capillaries, the site of gas exchange, branch off the parabronchi.

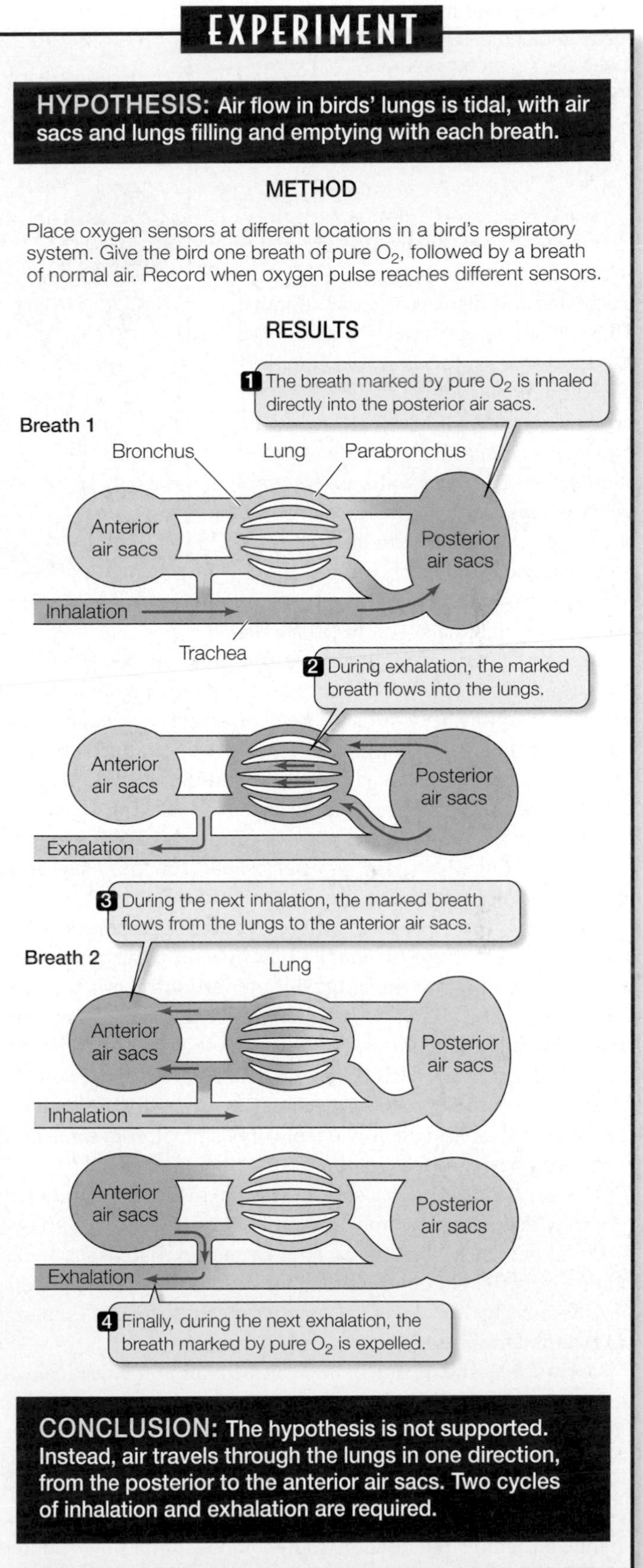

EXPERIMENT
HYPOTHESIS: Air flow in birds' lungs is tidal, with air sacs and lungs filling and emptying with each breath.
METHOD
Place oxygen sensors at different locations in a bird's respiratory system. Give the bird one breath of pure O_2, followed by a breath of normal air. Record when oxygen pulse reaches different sensors.
RESULTS
1 The breath marked by pure O_2 is inhaled directly into the posterior air sacs.
Breath 1
Bronchus
Lung
Parabronchus
Anterior air sacs
Posterior air sacs
Inhalation
Trachea
2 During exhalation, the marked breath flows into the lungs.
Anterior air sacs
Posterior air sacs
Exhalation
3 During the next inhalation, the marked breath flows from the lungs to the anterior air sacs.
Breath 2
Lung
Anterior air sacs
Posterior air sacs
Inhalation
Anterior air sacs
Posterior air sacs
Exhalation
4 Finally, during the next exhalation, the breath marked by pure O_2 is expelled.
CONCLUSION: The hypothesis is not supported. Instead, air travels through the lungs in one direction, from the posterior to the anterior air sacs. Two cycles of inhalation and exhalation are required.

48.8 The Path of Air Flow through Bird Lungs The air a bird takes in with one breath (blue) travels through the lungs in one direction, from the posterior to the anterior air sacs.

The advantages of the bird gas exchange system are similar to those of fish gills. The air sacs keep fresh air flowing unidirectionally and continuously over the gas exchange surfaces. Thus, the bird can supply its gas exchange surfaces with a continuous flow of fresh air that has a P_{O_2} close to that of the ambient air. Even when the P_{O_2} of the ambient air is only slightly above that of the blood, O_2 can diffuse from air to blood.

Tidal ventilation produces dead space that limits gas exchange efficiency

Lungs evolved in the first "air-gulping" vertebrates as outpocketings of the digestive tract. Although their structure has evolved considerably, lungs remain dead-end sacs in all air-breathing vertebrates except birds. Because lungs are dead-end sacs, ventilation cannot be constant and unidirectional, but must be **tidal**: air flows in and exhaled gases flow out by the same route. Since the lungs can never be completely emptied of air, the residual air in the lungs after exhalation represents dead space. We can measure how large that dead space is.

A *spirometer* is a device that measures the volumes of air that a person breathes in or breathes out. A person breathes through a

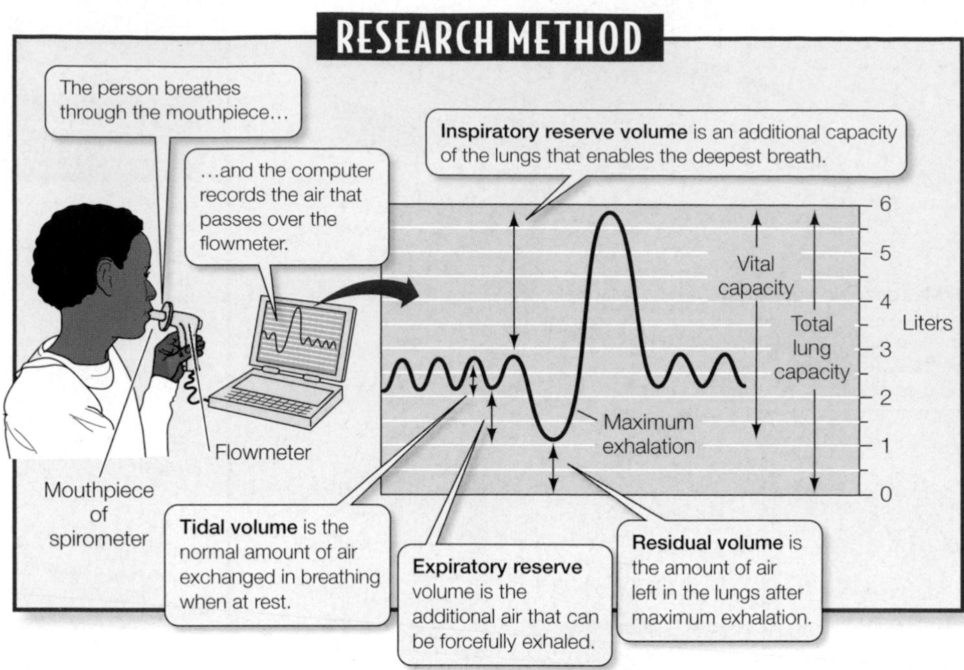

RESEARCH METHOD
The person breathes through the mouthpiece…
…and the computer records the air that passes over the flowmeter.
Inspiratory reserve volume is an additional capacity of the lungs that enables the deepest breath.
Vital capacity
Total lung capacity
Liters
6
5
4
3
2
1
0
Maximum exhalation
Flowmeter
Mouthpiece of spirometer
Tidal volume is the normal amount of air exchanged in breathing when at rest.
Expiratory reserve volume is the additional air that can be forcefully exhaled.
Residual volume is the amount of air left in the lungs after maximum exhalation.

48.9 Measuring Lung Ventilation A spirometer measures the volume of air that a person breathes through a mouthpiece. The combined tidal volume, inspiratory reserve volume, and expiratory reserve volume is the vital capacity.

mouthpiece, an electronic flowmeter measures air volumes, and a computer calculates and displays the data (**Figure 48.9**). When we are at rest, the amount of air that moves in and out per breath is called the **tidal volume** (about 500 ml for an average human adult). We can breathe much more deeply and inhale more air than our resting tidal volume; the additional volume of air we can take in above normal tidal volume is our **inspiratory reserve volume.** Conversely, we can forcefully exhale more air than we normally do during a resting exhalation. This additional amount of air is the **expiratory reserve volume.** The combined tidal volume, inspiratory reserve volume, and expiratory reserve volume is the **vital capacity.** The vital capacity of an athlete is generally greater than that of a non-athlete, and vital capacity decreases with age.

Vital capacity is not the entire lung volume. Even after the most extreme exhalation possible, some air remains in the dead space of the lungs. The lungs and airways cannot be collapsed completely; they always contain a *residual volume.* The **total lung capacity** is the sum of the residual volume and the vital capacity.

Tidal breathing severely limits the partial pressure gradient available to drive the diffusion of O_2 from air into the blood. Fresh air is not moving into the lungs during part of the breathing cycle; therefore, the average P_{O_2} of air in the lungs is considerably less than it is outside the lungs. Furthermore, the incoming fresh air mixes with the stale air that was not expelled by the previous exhalation. The volume of this stale air is the sum of the residual volume and, depending on how deeply one is breathing, some or all of the expiratory reserve volume.

The scale in Figure 48.9 shows a typical resting breathing pattern in which a tidal volume of 500 ml of fresh air mixes with over 2,000 ml of stale air before reaching the gas exchange surfaces in our lungs. In this situation, although the P_{O_2} in the ambient air may be 159 mm Hg, the P_{O_2} of the air that reaches the gas exchange surfaces is only about 100 mm Hg.

Considering the mixing of fresh air with air in the residual volume, you can understand why reductions in tidal volume can be a problem—and therefore why patients recovering from surgery are encouraged to breathe deeply, even if it hurts. In addition to reducing the P_{O_2} gradient, tidal breathing makes countercurrent gas exchange impossible. Because air enters and leaves the gas exchange structures by the same route, blood cannot flow countercurrent to the air flow. Offsetting the inefficiencies of tidal breathing, mammalian lungs have some design features to maximize the rate of gas exchange: an enormous surface area and a very short path length for diffusion.

48.2 RECAP

The major adaptations of animals that increase their efficiency of respiratory gas exchange are a large surface area for exchange and maximized concentration gradient across that surface.

- Describe three different ways that the concentration gradient for gas exchange is maximized across fish gills. See pp. 1029–1030 and Figures 48.5 and 48.6
- What respiratory adaptations enable birds to fly at extremely high altitudes? See pp. 1030–1031 and Figures 48.7 and 48.8
- Can you explain why residual volume limits the efficiency of tidal breathing? See pp. 1031–1032

Despite their limitations, mammalian lungs serve the respiratory needs of mammals well. We will use as our example the human respiratory system.

48.3 How Do Human Lungs Work?

In humans, air enters the lungs through the oral cavity or nasal passage, which join together in the *pharynx* (**Figure 48.10A**). Below the pharynx, the esophagus conducts food to the stomach, and the trachea leads to the lungs. At the beginning of this airway is the *larynx,* or voice box, which houses the vocal cords. The larynx is the "Adam's apple" that you can see or feel on the front of your neck. The trachea is about 2 cm in diameter. Its thin walls are prevented from collapsing by C-shaped bands of cartilage as air pressure changes during the breathing cycle. If you run your fingers down the front of your neck just below your larynx, you can feel a couple of these bands of cartilage.

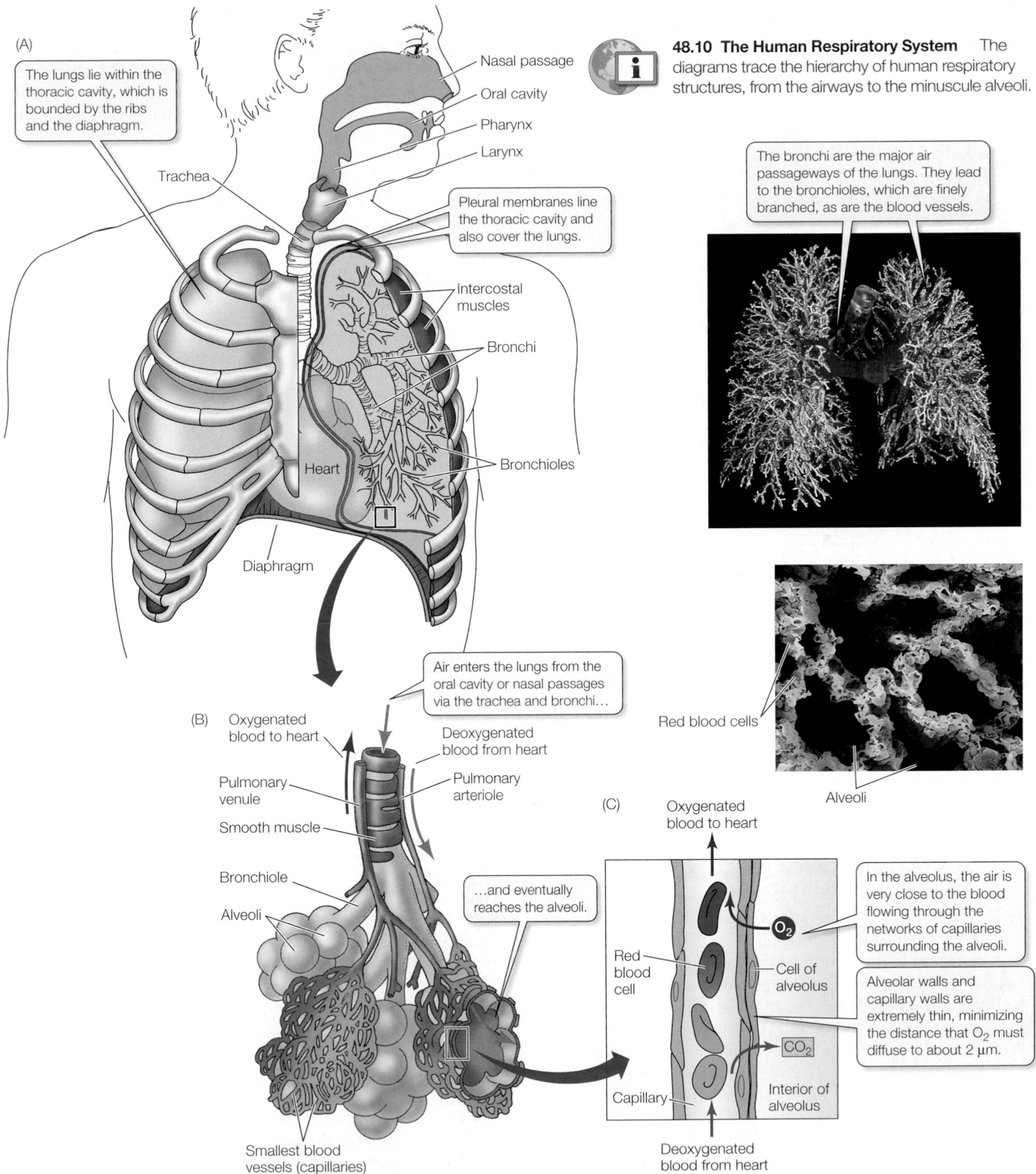
(A)
The lungs lie within the thoracic cavity, which is bounded by the ribs and the diaphragm.
Nasal passage
Oral cavity
Pharynx
Larynx
Trachea
Pleural membranes line the thoracic cavity and also cover the lungs.
Intercostal muscles
Bronchi
Heart
Bronchioles
Diaphragm
The bronchi are the major air passageways of the lungs. They lead to the bronchioles, which are finely branched, as are the blood vessels.
Red blood cells
Alveoli
Air enters the lungs from the oral cavity or nasal passages via the trachea and bronchi...
(B)
Oxygenated blood to heart
Deoxygenated blood from heart
Pulmonary venule
Pulmonary arteriole
Smooth muscle
Bronchiole
Alveoli
...and eventually reaches the alveoli.
Smallest blood vessels (capillaries)
(C)
Oxygenated blood to heart
In the alveolus, the air is very close to the blood flowing through the networks of capillaries surrounding the alveoli.
O_2
Red blood cell
Cell of alveolus
Alveolar walls and capillary walls are extremely thin, minimizing the distance that O_2 must diffuse to about 2 μm.
CO_2
Capillary
Interior of alveolus
Deoxygenated blood from heart

48.10 The Human Respiratory System The diagrams trace the hierarchy of human respiratory structures, from the airways to the minuscule alveoli.

The trachea branches into two **bronchi**, one leading to each lung. The bronchi branch repeatedly to generate a treelike structure of progressively smaller airways extending to all regions of the lungs. After four branchings, the cartilage supports disappear, marking the transition to **bronchioles**. After about 16 branchings, the bronchioles are less than a millimeter in diameter, and tiny, thin-walled air sacs called **alveoli** begin to appear. Alveoli are the sites of gas exchange. Thus, the bronchioles before the appearance of alveoli are called the conducting bronchioles, and those that sprout alveoli are called respiratory bronchioles. After alveoli begin to appear, about six more branchings of the airways occur that then end in clusters of alveoli (**Figure 48.10B**). Because the airways conduct air only to and from the alveoli and do not themselves participate in gas exchange, their volume is dead space.

Human lungs have about 300 million alveoli. Even though each alveolus is very small, their combined surface area for diffusion of respiratory gases is about 70 m^2—about one-fourth the size of a basketball court. Each alveolus is made of very thin cells. Between and surrounding the alveoli are networks of capillaries whose walls are also made up of exceedingly thin cells. Where capillary meets alveolus, very little tissue separates them (**Figure 48.10C**), so the length of the diffusion path between air and blood is less than 2 μm.

Emphysema is a lethal condition in which, over time, inflammation destroys the walls of the alveoli. It is the fourth largest cause of death in the United States. Although genetic factors can influence an individual's susceptibility to emphysema, the principal cause of the disease is smoking.

Respiratory tract secretions aid ventilation

Mammalian lungs produce two secretions that do not directly influence their gas exchange but do affect the process of ventilation: mucus and surfactant.

Many cells lining the airways produce sticky mucus that captures bits of dirt and microorganisms that are inhaled. Other cells lining the airways have cilia whose beating continually sweeps the mucus, with its trapped debris, up toward the pharynx, where it can be swallowed or spit out. This phenomenon, called the *mucus escalator*, can be adversely affected by inhaled pollutants. Smoking one cigarette can immobilize the cilia of the airways for hours. A smoker's cough results from the need to clear the obstructing mucus from the airways when the mucus escalator is out of order. The genetic disease *cystic fibrosis* causes respiratory problems by affecting the respiratory mucus (see Figure 17.3B).

A **surfactant** is a substance that reduces the surface tension of a liquid. **Surface tension** gives the surface of a liquid such as water the properties of an elastic membrane, and it is why certain insects, such as water-striders, can walk on water. As discussed in Section 2.4, surface tension is the result of chemical forces of attraction between water molecules. The attractive forces working on the water molecules at the surface are pulling from below and from the sides, but not from above. This imbalance of forces creates surface tension. The thin film of fluid covering the air-facing surfaces of the alveoli has surface tension that contributes to the elasticity of the lungs. To inflate the lungs, enough force has to be generated to overcome both the elasticity of the lung tissue and the surface tension in the alveoli.

Lung surfactant is a fatty substance that is critical for reducing the work necessary to inflate the lungs. Certain cells in the alveoli release surfactant molecules when they are stretched. If a baby is born more than a month prematurely, however, these cells may not have developed the ability to produce surfactant. Therefore, the baby has great difficulty breathing because an enormous effort is required to stretch the alveoli. A baby with this condition, known as *respiratory distress syndrome*, may die from exhaustion and lack of O_2. Common treatments for premature babies have been to put them on respirators to assist their breathing and to give them hormones to speed lung development. A new approach, however, is to apply surfactant to the lungs via an aerosol.

Lungs are ventilated by pressure changes in the thoracic cavity

Human lungs are suspended in a right and a left **thoracic cavity**. Each thoracic cavity is a closed compartment bounded on the bottom by the domed sheet of muscle called the **diaphragm** (Figure 48.10A). Lining each thoracic cavity and covering the lung in that cavity is a continuous sheet of tissue called the **pleural membrane**. Because the pleural membrane is continuous, it encloses a **pleural space**. There is no real space between the pleural membranes, but there is a thin film of fluid. This fluid lubricates the inner surfaces of the pleural membranes so they can slip and slide against each other during breathing movements. Just as we mentioned above in the explanation of surface tension, there are forces of attraction between the molecules of the fluid in the pleural space. As a result, it is difficult to pull the pleural membranes apart. Think of two wet panes of glass, or two wet microscope slides. You can easily slide them past each other, but it is difficult to separate them. While the inner surfaces of the pleural membranes are "stuck" to each other by surface tension, the outer surfaces of the pleural membranes are attached to the wall of the thoracic cavity and to the surface of the lungs.

Breathing involves changes in the volume of the thoracic cavity (**Figure 48.11**). Because the pleural membranes are attached to the walls of the thoracic cavity, and because the pleural space is a closed compartment, any attempt to increase the volume of the thoracic cavity creates a subatmospheric pressure (which we will refer to as a *negative pressure*) inside the pleural space. Even between breaths, there is normally a slight negative pressure in the pleural space because the rib cage is pulling outward and the elasticity of the lung tissue is pulling inward. This slight negative pressure keeps the alveoli partly inflated even at the end of an exhalation. If the thoracic cavity is punctured—by a knife wound, for example—air leaks into the pleural space, the negative pressure in that closed space is lost, and the lung collapses. If the wound is not sealed, breathing movements pull air into the pleural space rather than into the lung, and there is no ventilation of the alveoli in that lung.

Inhalation is initiated by contraction of the muscular diaphragm (see Figure 48.11). As the domed diaphragm contracts, it pulls down, expanding the thoracic cavity and pulling on the pleural membranes. This pulling on the pleural membranes makes the pressure in the pleural space more negative. The closed pleural space cannot expand, so the pleural membranes pull on the lungs. The lungs are not a closed cavity; they have an airway to the atmosphere, and they can expand in volume. When the diaphragm contracts, air rushes in through the trachea from the outside and the lungs expand. Exhalation begins when the contraction of the

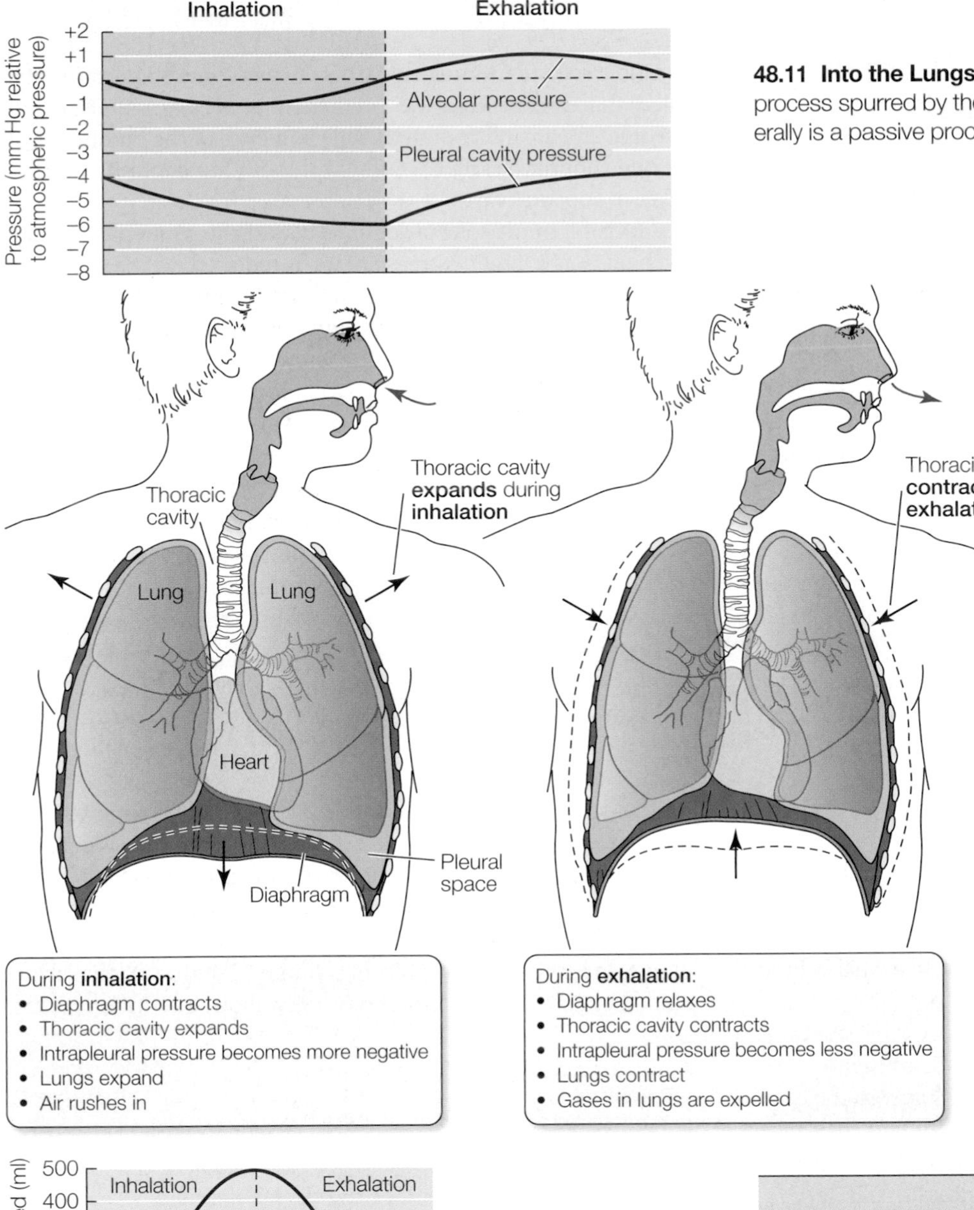

48.11 Into the Lungs and Out Again Inhalation is an active process spurred by the contraction of the diaphragm. Exhalation generally is a passive process as the diaphragm relaxes.

diaphragm ceases. As the diaphragm relaxes, the elastic recoil of the lung tissues pulls the diaphragm up and pushes air out through the airways. When a person is at rest, inhalation is an active process and exhalation is a passive process.

The diaphragm is not the only muscle that can change the volume of the thoracic cavity. Between the ribs are two sets of **intercostal muscles**. The *external intercostal muscles* expand the thoracic cavity by lifting the ribs up and outward. The *internal intercostal muscles* decrease the volume of the thoracic cavity by pulling the ribs down and inward. During strenuous exercise, the external intercostal muscles increase the volume of air inhaled, making use of the inspiratory reserve volume, and the internal intercostal muscles increase the amount of air exhaled, making use of the expiratory reserve volume. The abdominal muscles can also aid in breathing. When they contract, they cause the abdominal contents to push up on the diaphragm and thereby contribute to the expiratory reserve volume.

Remember that ventilation and perfusion work together to maximize the partial pressure gradients across the gas exchange surface. Ventilation delivers O_2 to the environmental side of the exchange surface, where some diffuses into the body and is swept away by perfusion. The reverse is true for the exchange of CO_2. Perfusion delivers CO_2 to the exchange surface, where it diffuses out and is swept away by ventilation.

48.3 RECAP

The mammalian respiratory system consists of a highly branching system of airways that lead to blind end sacs called alveoli which are the gas exchange surfaces. CO_2 and O_2 are exchanged across thin capillary and alveoli walls by diffusion.

- Can you describe the path that a breath of air takes from the nose to the gas exchange surfaces? See p. 1033 and Figure 48.10
- What roles do mucus and surfactant play in maintaining the function of the mammalian respiratory system? See p. 1034
- Can you explain the anatomical and functional relationships between the thoracic cavity, the pleural space, and the lungs? See p. 1034 and Figure 48.11

Now that we have seen how respiratory gases get to and from the environmental side of the gas exchange membranes through ventilation, we can look at how these gases get to and from the internal side of the gas exchange membranes through perfusion.

48.4 How Does Blood Transport Respiratory Gases?

Perfusion of the lungs is one of the functions of the circulatory system. The circulatory system uses a pump (the heart) and a network of vessels to transport extracellular fluids and associated cells (blood) around the body. Circulatory systems are the subject of the next chapter, so here we will discuss only one aspect of perfusion: how the respiratory gases are transported in the blood.

The liquid part of the blood, the *blood plasma*, carries some O_2 in solution, but its ability to transport O_2 is quite limited. The blood plasma of a human can contain in solution only about 0.3 ml of O_2 per 100 ml of plasma, which is inadequate to support even basal metabolism. To increase its O_2 transport capacity, the blood of most animals, vertebrate and invertebrate, also contains molecules that can bind reversibly to O_2 depending on its partial pressure. These molecules pick up or bind O_2 where its partial pressure is high and release it where its partial pressure is lower. There are many O_2 transport molecules in the animal kingdom, but in vertebrates, this role is played by hemoglobin contained in red blood cells. Hemoglobin increases the capacity of blood to transport O_2 by about sixtyfold.

Hemoglobin combines reversibly with oxygen

Red blood cells contain enormous numbers of hemoglobin molecules. **Hemoglobin** is a protein consisting of four polypeptide subunits (see Figure 3.9). Each of these polypeptides surrounds a *heme group*—an iron-containing ring structure that can reversibly bind a molecule of O_2. Thus, each molecule of hemoglobin can bind up to four molecules of O_2.

As O_2 diffuses into the red blood cells, it binds to hemoglobin. Once O_2 is bound, it cannot diffuse back across the red cell plasma membrane. By binding O_2 molecules as they enter the red blood cells, hemoglobin maximizes the partial pressure gradient driving the diffusion of O_2 into the cells. In addition, it enables the red blood cells to carry a large amount of O_2 to the tissues of the body.

The ability of hemoglobin to pick up or release O_2 depends on the P_{O_2} of its environment. When the P_{O_2} of the blood plasma is high, as it usually is in the lung capillaries, each molecule of hemoglobin can carry its maximum load of four molecules of O_2. As the blood circulates through the rest of the body, it encounters lower P_{O_2} values. At these lower P_{O_2} values, the hemoglobin releases some of the O_2 it is carrying (**Figure 48.12**).

The relation between P_{O_2} and the amount of O_2 bound to hemoglobin is not linear, but S-shaped (sigmoidal). The sigmoidal hemoglobin–oxygen binding curve in Figure 48.12 reflects interactions between the four subunits of the hemoglobin molecule. At low P_{O_2} values, only one subunit will bind an O_2 molecule. When it does so, the shape of that subunit changes, causing an alteration in the quaternary structure of the whole hemoglobin molecule. That structural change makes it easier for the other subunits to bind a molecule of O_2; that is, their O_2 *affinity* is increased. Therefore, a smaller increase in P_{O_2} is necessary to get the hemoglobin molecules to bind a second O_2 molecule (that is, to become 50 percent saturated) than was necessary to get them to bind one O_2 molecule (to become 25 percent saturated). This influence of the binding of O_2 by one subunit on the O_2 affinity of the other subunits is called **positive cooperativity**.

Once the third molecule of O_2 is bound, the relationship seems to change, as a larger increase in P_{O_2} is required for the hemoglobin to reach 100 percent saturation. This upper bend of the sigmoid curve is due to a probability phenomenon. The closer we get to having all subunits occupied, the less likely it is that any particular O_2 molecule will find a place to bind. Therefore, it takes a relatively greater P_{O_2} to achieve 100 percent saturation.

Carbon monoxide (CO) from faulty home heating systems can cause unconsciousness and death because CO binds to hemoglobin with a 240-fold higher affinity than O_2. This tightly bound CO prevents hemoglobin from transporting O_2 to the tissues of the body, and the brain is most vulnerable to lack of O_2. Annually more than 5,000 people in the United States die from carbon monoxide poisoning.

The oxygen-binding/dissociation properties of hemoglobin help get O_2 to the tissues that need it most. In the lungs, where the P_{O_2} is about 100 mm Hg, hemoglobin is 100 percent saturated. The P_{O_2} in blood returning to the heart from the body is usually about 40 mm Hg. You can see from Figure 48.12 that at this P_{O_2}, the hemoglobin is still about 75 percent saturated. This means that as the blood circulates around the body, only about one in four of the

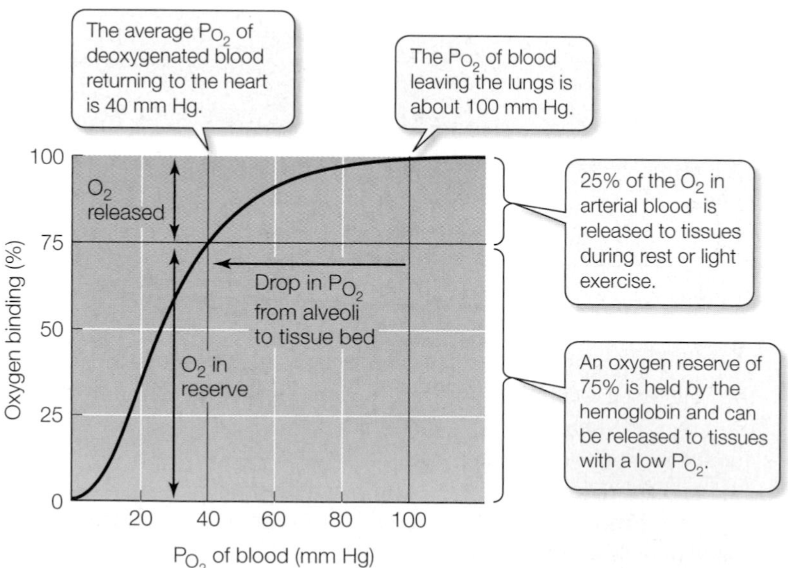

48.12 The Binding of O_2 to Hemoglobin Depends on P_{O_2} Hemoglobin in blood leaving the lungs is 100 percent saturated (four molecules of O_2 are bound to each hemoglobin molecule). Most hemoglobin molecules will drop only one of their four O_2 molecules as they circulate through the body, and are still 75 percent saturated when the blood returns to the lungs. The steep portion of this oxygen binding curve comes into play when tissue P_{O_2} falls below the normal 40 mm Hg, at which point the hemoglobin will "unload" its O_2 reserves.

O_2 molecules it carries is released to the tissues. This system seems inefficient, but it is really quite adaptive, because the hemoglobin keeps 75 percent of its O_2 in reserve to meet peak demands.

When a tissue becomes starved of O_2 and its local P_{O_2} falls below 40 mm Hg, the hemoglobin flowing through that tissue is on the steep portion of its sigmoid binding/dissociation curve. That means that relatively small decreases in P_{O_2} below 40 mm Hg will result in the release of lots of O_2 to the tissue. Thus hemoglobin is very effective in making O_2 available to the tissues precisely when and where it is needed most.

Myoglobin holds an oxygen reserve

Muscle cells have their own oxygen-binding molecule, **myoglobin**. Myoglobin consists of just one polypeptide chain associated with an iron-containing ring structure that can bind one molecule of O_2. Myoglobin has a higher affinity for O_2 than hemoglobin does, so it picks up and holds O_2 at P_{O_2} values at which hemoglobin is releasing its bound O_2 (**Figure 48.13**).

Myoglobin facilitates the diffusion of O_2 in muscle cells and provides a reserve of O_2 for times when metabolic demands are high and blood flow is interrupted. Interruption of blood flow in muscles is common because contracting muscles squeeze blood vessels. When tissue P_{O_2} values are low and hemoglobin can no longer supply more O_2, myoglobin releases its bound O_2. Diving mammals such as seals have high concentrations of myoglobin in their muscles, which is one reason they can stay under water for so long. (We will learn more about adaptations for diving in the next chapter.) Even in nondiving animals, muscles called on for extended periods of work frequently have more myoglobin than muscles that are used for short, intermittent periods, as we saw in the previous chapter.

The affinity of hemoglobin for oxygen is variable

Various factors influence the oxygen-binding/dissociation properties of hemoglobin, thereby influencing O_2 delivery to tissues. In this section we examine three of those factors: the chemical composition of the hemoglobin, pH, and the presence of 2,3-bisphosphoglyceric acid (BPG).

HEMOGLOBIN COMPOSITION There is more than one type of hemoglobin, because the chemical composition of the polypeptide chains that form the hemoglobin molecule varies. The normal hemoglobin of adult humans has two each of two kinds of polypeptide chains—two α-globin chains and two β-globin chains—and the oxygen-binding characteristics shown in Figure 48.12.

Before birth, the human fetus has a different form of hemoglobin, consisting of two α-globin and two γ-globin chains. The functional difference between fetal and adult hemoglobin is that the fetal hemoglobin has a higher affinity for O_2. Therefore, the hemoglobin–oxygen binding/dissociation curve of fetal hemoglobin is shifted to the left in comparison to the curve for adult hemoglobin (see Figure 48.13). You can see from these curves that if both types of hemoglobin are at the same P_{O_2}, the fetal hemoglobin will pick up O_2 more easily than can the maternal hemoglobin. This difference in O_2 affinities facilitates the transfer of O_2 from the mother's blood to the blood of the fetus in the placenta.

Llamas and vicuñas are native to the Andes Mountains of South America. In the natural habitat of these mammals, more than 5,000 m above sea level, the P_{O_2} is below 85 mm Hg, and the P_{O_2} in their lungs is about 50 mm Hg. Thus, the hemoglobins of these animals must pick up O_2 in an environment that has a low P_{O_2}. The hemoglobins of llamas and vicuñas have oxygen binding/dissociation curves to the left of the curves of hemoglobins of most other mammals—in other words, their hemoglobin can become saturated with O_2 at lower P_{O_2} values than those of other mammals.

HEMOGLOBIN AND pH The oxygen-binding properties of hemoglobin are also influenced by physiological conditions. The influence of pH on the function of hemoglobin is known as the **Bohr effect**. As blood passes through metabolically active tissue such as exercising muscle, it picks up acidic metabolites such as lactic acid, fatty acids, and CO_2. As a result, blood pH falls. The excess H^+ binds preferentially to deoxygenated hemoglobin and decreases the affinity of that hemoglobin for O_2. As a result, the oxygen binding/dissociation curve of hemoglobin shifts to the right (see Figure 48.13). This shift means that the hemoglobin will release more O_2 in tissues where pH is low—another way that O_2 is supplied where and when it is most needed.

2,3-BISPHOSPHOGLYCERIC ACID BPG is a metabolite of glycolysis (see Figure 7.5). Mammalian red blood cells have a high concentration of BPG, which serves as an important regulator of hemoglobin function. BPG, like excess H^+, reversibly combines with deoxygenated hemoglobin and lowers its affinity for

Myoglobin
Llama hemoglobin
Human fetal hemoglobin
Human maternal hemoglobin (pH 7.4)
Human hemoglobin (pH 7.2)
Oxygen binding (%)
100
80
60
40
20
20
40
60
80
100
P_{O_2} (mm Hg)

Llama guanicoe

48.13 Oxygen-Binding Adaptations The different hemoglobins and myoglobin have different oxygen-binding properties adapted to different circumstances. The hemoglobin of llamas, for example, is adapted for binding O_2 at high altitudes, where P_{O_2} is low.

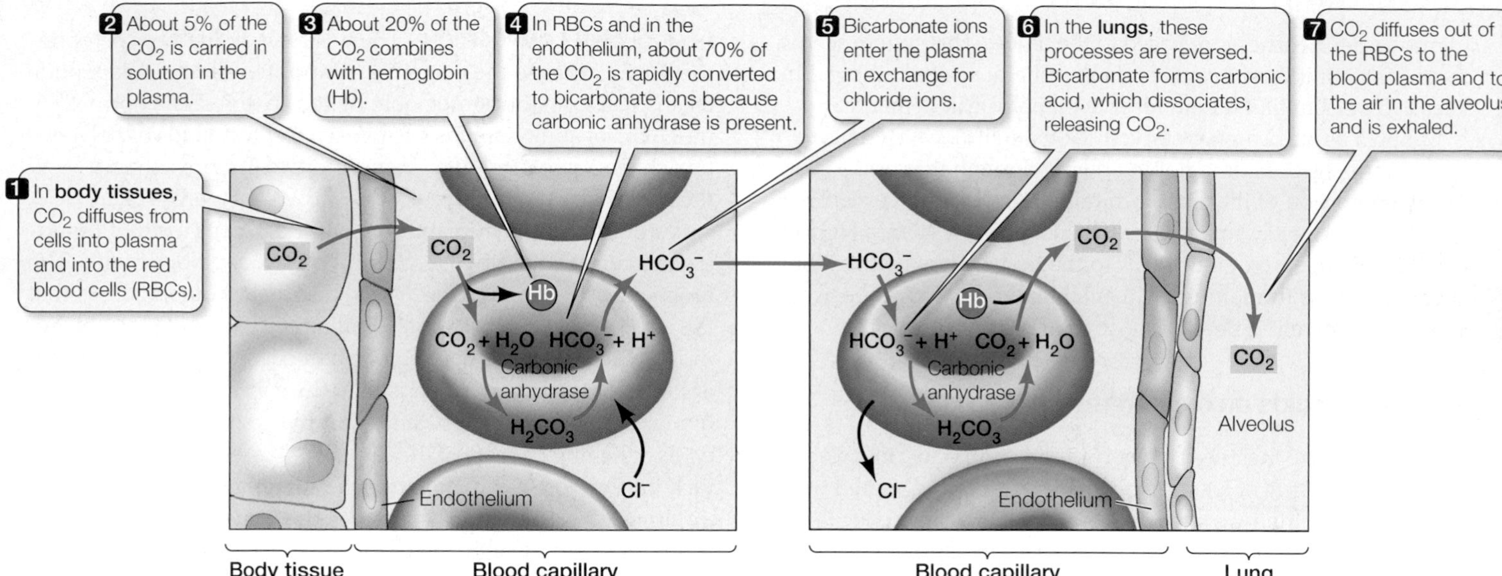

2 About 5% of the CO_2 is carried in solution in the plasma.
3 About 20% of the CO_2 combines with hemoglobin (Hb).
4 In RBCs and in the endothelium, about 70% of the CO_2 is rapidly converted to bicarbonate ions because carbonic anhydrase is present.
5 Bicarbonate ions enter the plasma in exchange for chloride ions.
6 In the lungs, these processes are reversed. Bicarbonate forms carbonic acid, which dissociates, releasing CO_2.
7 CO_2 diffuses out of the RBCs to the blood plasma and to the air in the alveolus and is exhaled.
1 In body tissues, CO_2 diffuses from cells into plasma and into the red blood cells (RBCs).
CO_2
CO_2
Hb
HCO_3^-
$CO_2 + H_2O$ $HCO_3^- + H^+$
Carbonic anhydrase
H_2CO_3
Cl^-
Endothelium
HCO_3^-
Hb
CO_2
$HCO_3^- + H^+$ $CO_2 + H_2O$
Carbonic anhydrase
H_2CO_3
Cl^-
Endothelium
CO_2
Alveolus
Body tissue
Blood capillary
Blood capillary
Lung

48.14 Carbon Dioxide Is Transported as Bicarbonate Ions Carbonic anhydrase in capillary endothelial cells and in red blood cells facilitates conversion of CO_2 produced by tissues into bicarbonate ions carried by the plasma. In lungs, the process is reversed as CO_2 is exhaled.

O_2. The result is that at any P_{O_2}, hemoglobin releases more of its bound O_2 than it otherwise would. In other words, BPG shifts the oxygen binding/dissociation curve of mammalian hemoglobin to the right. When humans go to high altitudes, or when they cease being sedentary and begin to exercise, the level of BPG in their red blood cells goes up and makes it easier for the hemoglobin to deliver more O_2 to the tissues. The reason that fetal hemoglobin has a left-shifted hemoglobin–oxygen binding/dissociation curve is that the γ-globin chains of fetal hemoglobin have a lower affinity for BPG than do the β-globin chains of adult hemoglobin.

Carbon dioxide is transported as bicarbonate ions in the blood

Delivering O_2 to the tissues is only half of the respiratory function of the blood. The blood also must take CO_2, a metabolic waste product, away from the tissues (**Figure 48.14**). CO_2 is highly soluble and readily diffuses through cell membranes, moving from its site of production in the tissues into the blood, where the partial pressure of carbon dioxide (P_{CO_2}) is lower. However, very little dissolved CO_2 is transported by the blood. Most CO_2 produced by the tissues is transported to the lungs in the form of **bicarbonate ions**, HCO_3^-. CO_2 is converted to HCO_3^-, transported to the lungs, and then converted back to CO_2 in several steps.

When CO_2 dissolves in water, some of it slowly reacts with the water molecules to form carbonic acid (H_2CO_3), some of which then dissociates into a proton (H^+) and a bicarbonate ion (HCO_3^-). This reversible reaction is expressed as follows:

$$CO_2 + H_2O \rightleftharpoons H_2CO_3 \rightleftharpoons H^+ + HCO_3^-$$

In the blood plasma, the reaction between CO_2 and H_2O proceeds slowly. But it is a different story in the endothelial cells of the capillaries and in the red blood cells, where the enzyme *carbonic anhydrase* speeds up the conversion of CO_2 to H_2CO_3. The newly formed carbonic acid dissociates, and the resulting bicarbonate ions enter the plasma in exchange for Cl^- (see Figure 48.14). By converting CO_2 to H_2CO_3, carbonic anhydrase reduces the P_{CO_2} in these cells and in the plasma, facilitating the diffusion of CO_2 from tissue cells to endothelial cells, plasma, and red blood cells. Some CO_2 is also carried in chemical combination with hemoglobin.

In the lungs, the reactions involving CO_2 and bicarbonate ions are reversed. Remember that an enzyme such as carbonic anhydrase only speeds up a reversible reaction; it does not determine its direction. The direction is determined by concentrations of reactants and products (see Section 6.1). Ventilation keeps the CO_2 concentration in the alveoli low, so CO_2 diffuses from the blood plasma into the alveoli, lowering the CO_2 concentration in the blood, which favors the conversion of HCO_3^- into CO_2.

48.4 RECAP

O_2 is transported from the lungs to the tissues of the body in reversible combination with hemoglobin. Each molecule of hemoglobin can reversibly combine with four molecules of O_2; the saturation of the binding sites is a function of the P_{O_2} in the environment of the hemoglobin.

- Can you explain the advantage of having hemoglobin hold on to three molecules of O_2 at the usual P_{O_2} of mixed venous blood? See pp. 1036–1037 and Figure 48.12
- How is the oxygen binding/dissociation curve of hemoglobin influenced by pH? By BPG? By development from fetus to newborn infant? See pp. 1037–1038 and Figure 48.13
- How is CO_2 transported in the blood? See p. 1038 and Figure 48.14

We must breathe every minute of our lives, but we don't usually worry about it, or even think about it very often. In the next section we will examine how the regular breathing cycle is generated and controlled by the central nervous system.

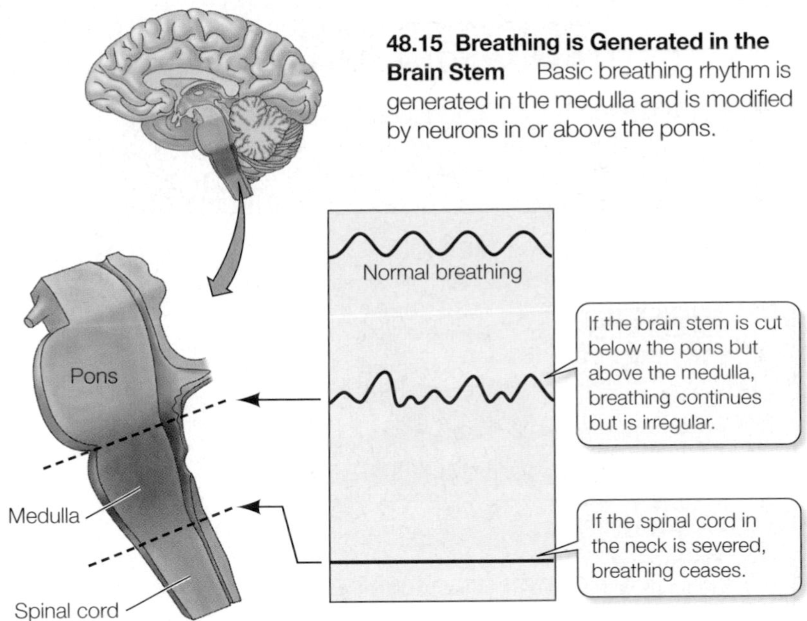

48.15 Breathing is Generated in the Brain Stem Basic breathing rhythm is generated in the medulla and is modified by neurons in or above the pons.

48.5 How is Breathing Regulated?

Breathing is an autonomic function of the central nervous system. The breathing pattern easily adjusts itself around other activities (such as speech and eating), and breathing rates change to match the metabolic demands of our bodies. How is this accomplished?

Breathing is controlled in the brain stem

The autonomic nervous system maintains breathing and modifies its depth and frequency to meet the demands of the body for O_2 supply and CO_2 elimination. Breathing ceases if the spinal cord is severed in the neck region, showing that the breathing pattern is generated in the brain. If the brain stem is cut just above the medulla, the segment of the brain stem just above the spinal cord, an irregular breathing pattern remains (**Figure 48.15**).

Groups of respiratory motor neurons within the medulla increase their firing rates just before an inhalation begins. As more and more of these neurons fire—and fire faster and faster—the diaphragm contracts. Suddenly the neurons stop firing, the diaphragm relaxes, and exhalation begins. Exhalation is usually a passive process that depends on the elastic recoil of the lung tissues. When breathing demand is high, however, as during strenuous exercise, motor neurons for the intercostal muscles are recruited, which increases both the inhalation and the exhalation volumes. Brain areas above the medulla modify breathing to accommodate speech, ingestion of food, coughing, and emotional states.

Regulating breathing requires feedback information

When breathing or metabolism changes, it alters the P_{O_2} and the P_{CO_2} in the blood. We should therefore expect the blood levels of one or both of these gases to provide feedback information to the breathing rhythm generator in the medulla. Experiments in which subjects breathe air with different P_{O_2} and P_{CO_2} concentrations lead us to conclude that humans (and other mammals) are remarkably insensitive to falling blood levels of O_2, but very sensitive to increases in the P_{CO_2} of the blood (**Figure 48.16**). This relationship is reversed for water-breathing animals, in which O_2 is the primary feedback stimulus for breathing. Remember that CO_2 is readily lost across gill membranes, so its concentrations in the blood would not be a good index of metabolic needs.

We might ask whether it is an increase in the P_{CO_2} of the blood that stimulates increased breathing when we exercise. To answer this question, researchers in the lab of C. R. Bainton observed dogs running on treadmills at different speeds. As the speed of the treadmill increased, the respiratory gas exchange rate of the dogs increased, but the P_{CO_2} of their blood remained constant. Before concluding that their hypothesis was not supported, and that blood P_{CO_2} does not control breathing rate, the researchers changed their experiment. Instead of increasing the speed of the treadmill, they gradually increased its slope so that the dogs were running at the same speed, but were working harder because they were running uphill (**Figure 48.17**).

In this second experiment, the P_{CO_2} of the blood increased as the slope of the treadmill increased and as the respiratory gas exchange rate increased. The researchers concluded that the P_{CO_2} of the blood is the primary metabolic feedback information for breathing. However, when an animal starts to run or changes its running speed, additional feedback information from receptors in muscles and joints changes the sensitivity of the CO_2 sensors—

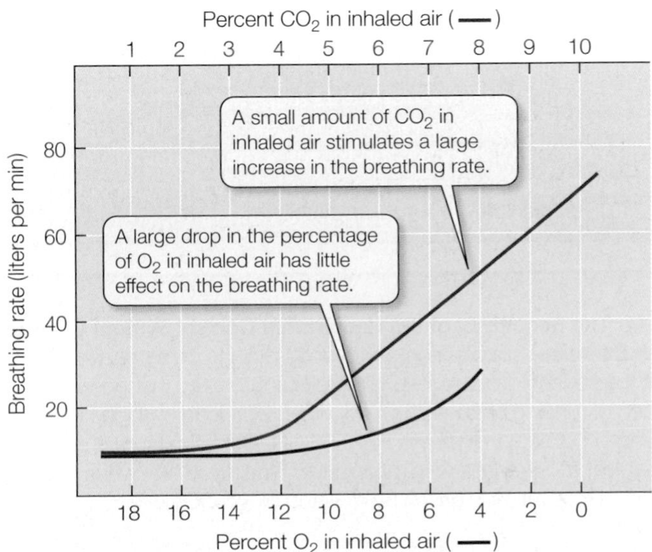

48.16 Carbon Dioxide Affects Breathing Rate Breathing is more sensitive to increased CO_2 content in inhaled air than to its decreased O_2 content.

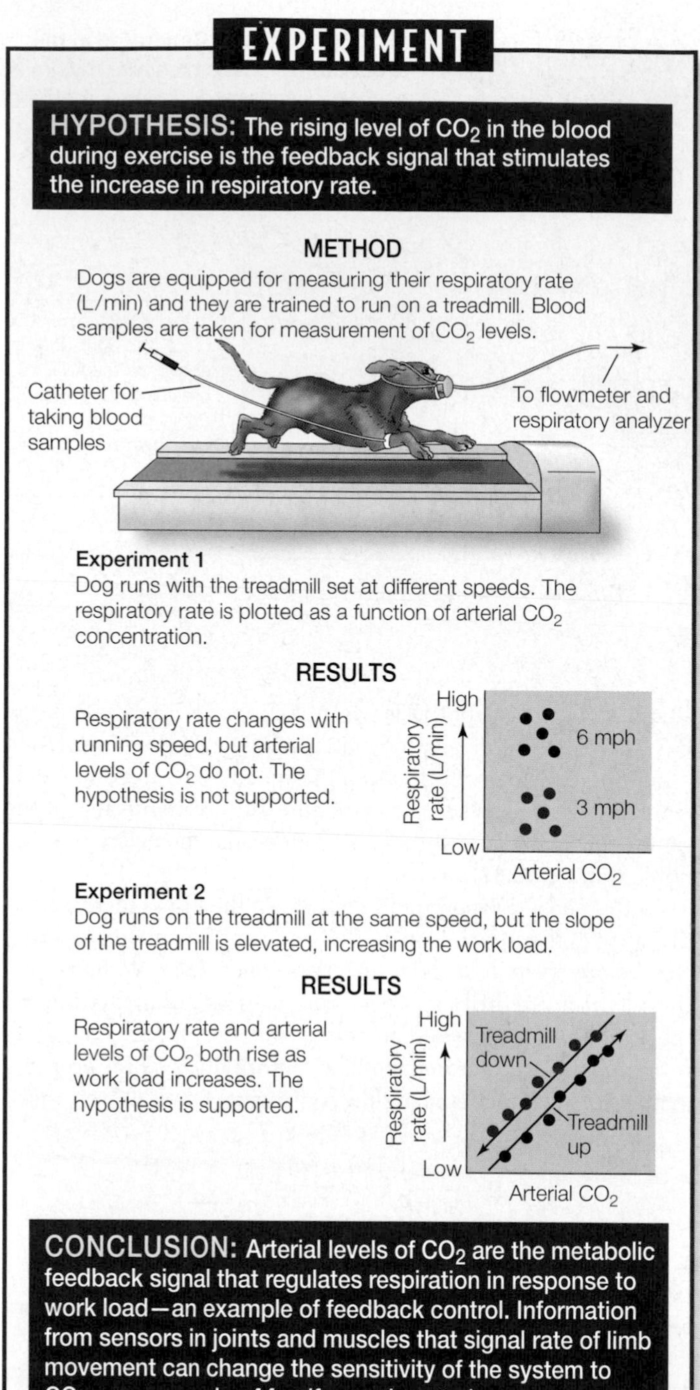

EXPERIMENT
HYPOTHESIS: The rising level of CO_2 in the blood during exercise is the feedback signal that stimulates the increase in respiratory rate.
METHOD
Dogs are equipped for measuring their respiratory rate (L/min) and they are trained to run on a treadmill. Blood samples are taken for measurement of CO_2 levels.
Catheter for taking blood samples
To flowmeter and respiratory analyzer
Experiment 1
Dog runs with the treadmill set at different speeds. The respiratory rate is plotted as a function of arterial CO_2 concentration.
RESULTS
Respiratory rate changes with running speed, but arterial levels of CO_2 do not. The hypothesis is not supported.
High
Low
Respiratory rate (L/min)
6 mph
3 mph
Arterial CO_2
Experiment 2
Dog runs on the treadmill at the same speed, but the slope of the treadmill is elevated, increasing the work load.
RESULTS
Respiratory rate and arterial levels of CO_2 both rise as work load increases. The hypothesis is supported.
High
Low
Respiratory rate (L/min)
Treadmill down
Treadmill up
Arterial CO_2
CONCLUSION: Arterial levels of CO_2 are the metabolic feedback signal that regulates respiration in response to work load—an example of feedback control. Information from sensors in joints and muscles that signal rate of limb movement can change the sensitivity of the system to CO_2—an example of feedforward control.

48.17 The Sensitivity of the Respiratory Control System Changes with Exercise Experiments with dogs running on a treadmill show that the sensitivity of the respiratory system to CO_2 changes when speed of running changes, but not when workload changes without a change in speed. FURTHER RESEARCH: Cold exposure is another variable that increases respiration. How would you test whether cold-induced respiration is driven by elevated blood CO_2 levels?

another example of feedforward information. Remember from Section 40.1 that feedforward information can change the sensitivity or the set point of a regulatory system.

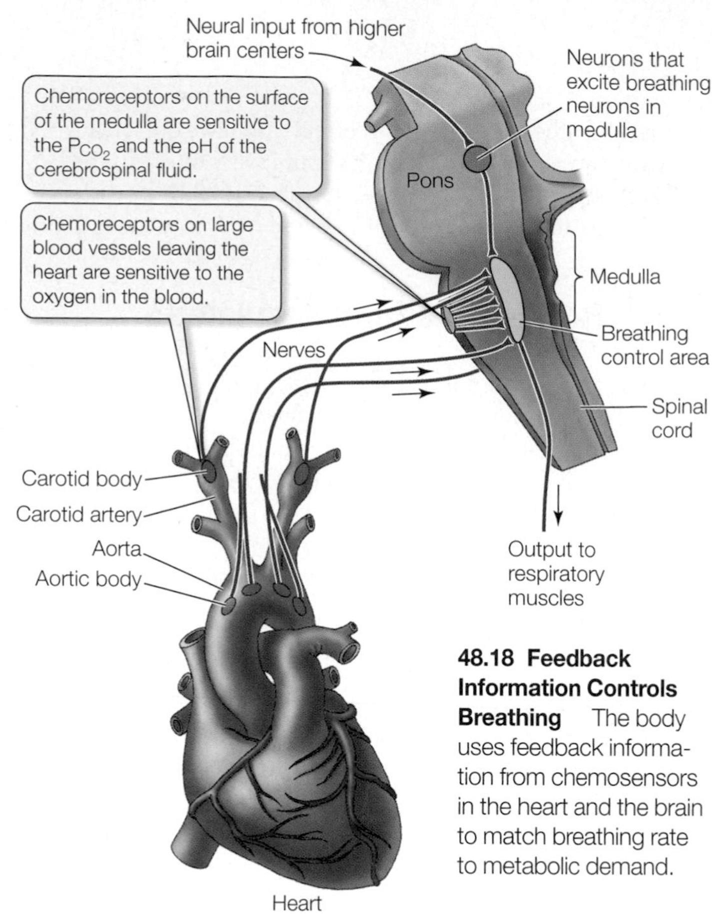

48.18 Feedback Information Controls Breathing The body uses feedback information from chemosensors in the heart and the brain to match breathing rate to metabolic demand.

Where are partial pressures of gases in the blood sensed? The major site of CO_2 sensitivity is an area on the ventral surface of the medulla, not far from the groups of neurons that generate the breathing rhythm. Primary sensitivity to P_{O_2} in the blood resides in nodes of tissue on the large blood vessels leaving the heart: the aorta and the carotid arteries (**Figure 48.18**). These **carotid** and **aortic bodies** receive enormous supplies of blood relative to their small size, and they contain chemoreceptors. If the blood supply to these structures decreases, or if the P_{O_2} of the blood falls dramatically, the chemoreceptors are activated and send nerve impulses to the breathing control center. Although we are not very sensitive to changes in blood P_{O_2}, the carotid and aortic bodies can stimulate increases in breathing during exposure to very high altitudes or when blood volume or blood pressure is very low.

48.5 RECAP

The rhythmic contractions of the respiratory muscles that drive breathing are generated by neurons in the brainstem.

- What is the primary chemical stimulus for controlling the respiratory rate and where is it sensed? See p. 1039 and Figures 48.16 and 48.18
- Do you understand what feedforward information is? Can you give an example of feedforward information in the respiratory control system? See Figure 48.17
- What are the functions of the carotid and aortic bodies? See p. 1040 and Figure 48.18

CHAPTER SUMMARY

48.1 What physical factors govern respiratory gas exchange?

Most cells require a constant supply of oxygen (O_2) and continuous removal of carbon dioxide (CO_2). These **respiratory gases** are exchanged between the body fluids of an animal and its environment by diffusion.

Fick's law of diffusion shows how various physical factors influence the rate of diffusion of gases. Adaptations to maximize respiratory gas exchange influence one or more variables of Fick's law.

In aquatic animals, gas exchange is limited by the low diffusion rate and low amount of O_2 in water. Aquatic animals face a double bind in that the amount of O_2 in water decreases, but their metabolism and the amount of work required to move water over their **gas exchange surfaces** increase, as water temperature rises. Review Figure 48.2

In air, the **partial pressure of oxygen** decreases with altitude.

48.2 What adaptations maximize respiratory gas exchange?

Adaptations to maximize gas exchange include increasing the surface areas for gas exchange, maximizing partial pressure gradients across those exchange surfaces, **ventilating** the outer surface with the respiratory medium, and **perfusing** the inner surface with blood.

Insects distribute air throughout their bodies in a system of **tracheae**, tracheoles, and air capillaries. Review Figure 48.4

The **gills** of fish have large gas exchange surface areas that are ventilated continuously and unidirectionally with water. **Countercurrent blood flow** helps increase the efficiency of gas exchange. Review Figures 48.5 and 48.6

The gas exchange system of birds includes **air sacs** that communicate with the lungs, but are not used for gas exchange. Air flows unidirectionally through bird lungs; gases are exchanged in air capillaries that run between **parabronchi**. Review Figure 48.7

Each breath of air remains in the bird respiratory system for two breathing cycles. The air sacs work as bellows to supply the air capillaries with a continuous unidirectional flow of fresh air. Review Figure 48.8, Web/CD Tutorial 48.1

Breathing in vertebrates other than birds is **tidal** and is therefore less efficient than gas exchange in fish or birds. Although the volume of air exchanged with each breath can vary considerably, the inhaled air is always mixed with stale air. Review Figure 48.9

48.3 How do human lungs work?

See Web/CD Tutorial 48.2

In mammalian lungs, the gas exchange surface area provided by the millions of **alveoli** is enormous, and the diffusion path length between the air and perfusing blood is short. Surface tension in the alveoli would make inflation of the lungs difficult if the alveoli did not produce **surfactant**. Review Figure 48.10, Web/CD Activity 48.1

Inhalation occurs when contractions of the **diaphragm** create subatmospheric pressure in the **thoracic cavities**. Relaxation of the diaphragm increases pressure in the thoracic cavities and causes **exhalation**. Review Figure 48.11

During periods of heavy metabolic demands such as strenuous exercise, the **intercostal muscles**, located between the ribs, increase the volume of air inhaled and exhaled.

48.4 How does blood transport respiratory gases?

See Web/CD Activity 48.2

O_2 is reversibly bound to **hemoglobin** in red blood cells. Each molecule of hemoglobin can carry a maximum of four molecules of O_2. Because of **positive cooperativity**, the affinity of hemoglobin for O_2 depends on the P_{O_2} to which the hemoglobin is exposed. Therefore, hemoglobin picks up O_2 as it flows through respiratory exchange structures and gives up O_2 in metabolically active tissues. Review Figure 48.12

Myoglobin serves as an O_2 reserve in muscle.

There is more than one type of hemoglobin. Fetal hemoglobin has a higher affinity for O_2 than does maternal hemoglobin, allowing fetal blood to pick up O_2 from the maternal blood in the placenta. Review Figure 48.13

Carbon dioxide is transported in the blood principally as bicarbonate ions (HCO_3^-). Review Figure 48.14

48.5 How is breathing regulated?

The breathing rhythm is an autonomic function generated by neurons in the medulla and modulated by higher brain centers. The most important feedback stimulus for breathing is the level of CO_2 in the blood. Review Figures 48.16 and 14.17

The breathing rhythm is sensitive to feedback from chemoreceptors on the ventral surface of the medulla and in the carotid and aortic bodies on the large vessels leaving the heart. Review Figure 48.18

See Web/CD Activity 48.3 for a concept review of this chapter.

SELF-QUIZ

1. Which of the following statements is *not* true?
   a. Respiratory gases are exchanged only by diffusion.
   b. O_2 has a lower rate of diffusion in water than in air.
   c. The O_2 content of water falls as the temperature of water rises.
   d. The amount of O_2 in the atmosphere decreases with increasing altitude.
   e. Birds have evolved active transport mechanisms to augment their respiratory gas exchange.
2. Which statement about the gas exchange system of birds is *not* true?
   a. Respiratory gases are not exchanged in the air sacs.
   b. A bird can achieve more complete exchange of O_2 from air to blood than can the human gas exchange system.
   c. Air passes through birds' lungs in only one direction.
   d. The gas exchange surfaces in bird lungs are the alveoli.
   e. A breath of air remains in the system for two breathing cycles.
3. Which statement about gas exchange in fish is true?
   a. Blood flows over the gas exchange surfaces in a direction opposite to the flow of water.
   b. Gases are exchanged across the gill filaments.
   c. Ventilation of the gills is tidal in fast-swimming fishes.
   d. Less work is needed to ventilate gills in warm water than in cold water.
   e. The path length for diffusion of respiratory gases is determined by the length of the gill filaments.

4. In the human gas exchange system,
   a. the lungs and airways are completely collapsed after a forceful exhalation.
   b. the average P_{O_2} concentration of air inside the lungs is always lower than that in the air outside the lungs.
   c. the P_{O_2} of the blood leaving the lungs is greater than that of the exhaled air.
   d. the amount of air that is moved per breath during normal, at-rest breathing is termed the total lung capacity.
   e. O_2 and CO_2 are actively transported across the alveolar and capillary membranes.
5. Which statement about the human gas exchange system is *not* true?
   a. During inhalation, a subatmospheric pressure exists in the space between the lung and the thoracic wall.
   b. Smoking one cigarette can immobilize the cilia lining the airways for hours.
   c. The respiratory control center in the medulla responds more strongly to changes in arterial O_2 concentration than to changes in arterial CO_2 concentration.
   d. Without surfactant, the work of breathing is greatly increased.
   e. The diaphragm contracts during inhalation and relaxes during exhalation.
6. The hemoglobin of a human fetus
   a. is the same as that of an adult.
   b. has a higher affinity for O_2 than adult hemoglobin has.
   c. has only two protein subunits instead of four.
   d. is supplied by the mother's red blood cells.
   e. has a lower affinity for O_2 than adult hemoglobin has.
7. The amount of O_2 carried by hemoglobin depends on the P_{O_2} in the blood. Hemoglobin in active muscles
   a. becomes saturated with O_2.
   b. takes up only a small amount of O_2.
   c. readily unloads O_2.
   d. tends to decrease the P_{O_2} in the muscle tissues.
   e. is denatured.
8. Most CO_2 in the blood is carried
   a. in the cytoplasm of red blood cells.
   b. dissolved in the plasma.
   c. in the plasma as bicarbonate ions.
   d. bound to plasma proteins.
   e. in red blood cells bound to hemoglobin.
9. Myoglobin
   a. binds O_2 at P_{O_2} values at which hemoglobin is releasing its bound O_2.
   b. has a lower affinity for O_2 than hemoglobin does.
   c. consists of four polypeptide chains, just as hemoglobin does.
   d. provides an immediate source of O_2 for muscle cells at the onset of activity.
   e. can bind four O_2 molecules at once.
10. When the level of CO_2 in the bloodstream *increases,*
    a. the rate of respiration decreases.
    b. the pH of the blood rises.
    c. the respiratory centers become dormant.
    d. the rate of respiration increases.
    e. the blood becomes more alkaline.

FOR DISCUSSION

1. The blood of a certain species of fish that lives in Antarctica has no hemoglobin. What anatomical and behavioral characteristics would you expect to find in this fish, and why is its distribution limited to the waters of Antarctica?
2. Blood banks store whole blood for a much shorter period than they store blood plasma because when blood that has been stored for too long is infused into a patient, it can actually decrease the O_2 availability to the patient's tissues. Why is this so? Explain in terms of the different physiological functions of BPG.
3. From your knowledge of respiratory gas exchange and regulation of respiration, how would you explain the tragic episode of the scientists who ascended in the hot-air balloon *Zenith,* described on the opening page of this chapter? Why do you think the three men continued tossing out ballast to ascend higher and higher, without realizing they were in mortal danger of asphyxiation?
4. In the disease emphysema, the fine structures of alveoli break down, resulting in the formation of larger air cavities in the lungs. Also, the tissue of the lungs becomes less elastic. Explain at least two reasons why patients with emphysema have a low tolerance for exercise.
5. A condition called "the bends" affects scuba divers who surface too quickly after spending an extended period in deep water, where they have been breathing pressurized air. The cause of the bends is tiny bubbles of nitrogen coming out of solution in the blood plasma. Seals spend much more time under water and at deeper depths than scuba divers, yet they do not suffer the bends. Why?

FOR INVESTIGATION

When you suddenly travel to a location at high altitude, you will notice an unusual breathing pattern when you are resting. For a while you stop breathing completely; then suddenly you start breathing rapidly for a short time; then you stop breathing again. This can go on and on in a cyclical pattern called Cheyne-Stokes breathing. It generally goes away after spending a few days at altitude. Can you hypothesize what causes Cheyne-Stokes breathing and design an experiment to test your hypothesis?

CHAPTER 49 Circulatory Systems

You gotta have heart

On April 29, 1993, the Boston Celtics met the Charlotte Hornets in a National Basketball Association playoff game at the Boston Garden. The Celtics captain and star player, 27-year-old Reggie Lewis, had just scored 10 points in 3 minutes when he went up for a rebound and suddenly slumped forward, falling to the floor "as if he'd been shot in the back." He was examined by the team doctor (an orthopedic specialist), who allowed him to return to the game in the second half. But Lewis's legs were "wobbly" and he played only briefly.

Lewis had been experiencing dizzy spells for about a month prior to his collapse. After the playoff incident, he underwent rigorous testing by cardiologists, who diagnosed Lewis as having a dangerous arrhythmia (irregular heartbeat) caused by cardiomyopathy (diseased cardiac muscle). Accepting this diagnosis would mean the end of his professional athletic career, and Lewis sought a second opinion. After another battery of tests, a second medical team felt that Lewis had undergone a transient irregular heartbeat attributable to normal athletic stresses. The condition was deemed treatable, and to have been in part the result of Lewis's enlarged heart—a condition sometimes seen in high-performing athletes. In July of 1993, after an hour spent shooting baskets in a pick-up game, Reggie Lewis collapsed and died of heart failure.

Your heart is a muscular pump that, at rest, beats an average of 72 times per minute. With each beat it circulates about 70 milliliters of blood through the body. Without taking work or exercise into account, that is 300 liters per hour, 7,200 liters per day, 2.6 million liters per year—no time outs. Heart failure accounts for about one-third of the deaths (about 900,000 deaths) each year in the United States, making it the country's leading cause of death. Heart failure is most commonly the result of blockage of the vessels that supply the heart muscle with blood. The risk of such heart failure tends to increase with age. But heart failure is also the leading cause of death among young athletes.

Why does the heart of a young, fit athlete fail? It is usually not due to vessel blockage, but to a mutation that affects the contractile proteins of heart muscle. The heart is good at compensating for these mutant

An Athlete's Heart Boston Celtics basketball star Reggie Lewis died of heart failure at the age of 27. The exact medical situation underlying his heart problems remains clouded in controversy.

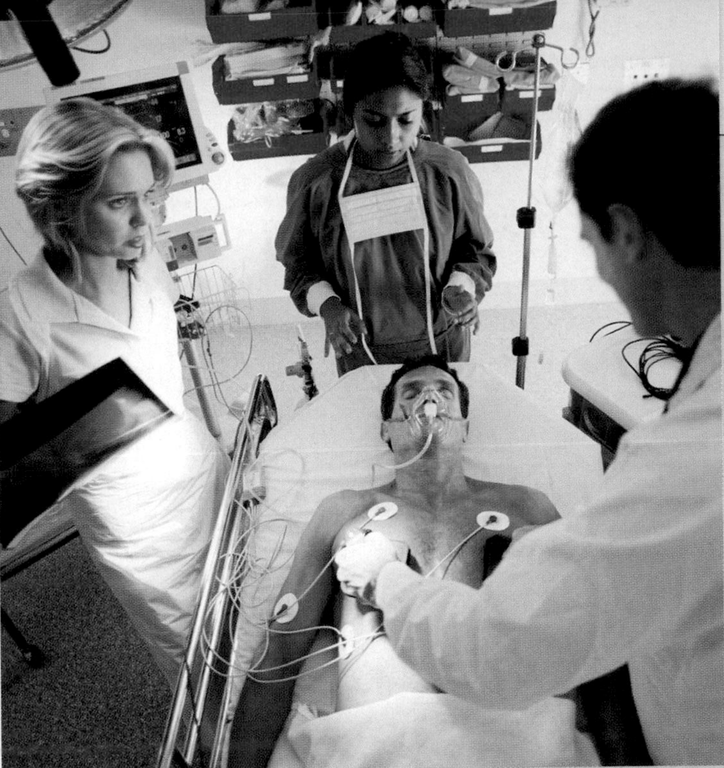

Clear! Hospital emergency rooms deal with heart attacks on a daily basis. Advances in heart surgery and emergency resuscitation techniques have saved many lives, but heart failure remains the number one cause of death in the United States and Europe.

proteins; most people can live their entire lives with the condition and never have a symptom. But with continued heavy exercise, the heart compensates by getting larger. Eventually, the increase in heart size can disrupt the electrical impulses that coordinate the contractions of the heart muscle. When heavy demand is placed on such an enlarged heart, muscle fiber contractions can suddenly become uncoordinated and the heart cannot pump blood. The lack of any prior symptoms is why the condition usually goes undiagnosed in athletes.

IN THIS CHAPTER we will learn about the adaptations of circulatory systems that enable them to match blood supply with demand in a wide variety of species. Taking the human circulatory system as a model, we will explore the mechanics of the beating heart and the characteristics of the vascular system: the arteries, capillaries, and veins. Another component of the circulatory system is the blood, and we will describe the features of this fluid tissue. The chapter ends with a discussion of the hormonal and neuronal regulation of the mammalian circulatory system.

CHAPTER OUTLINE

49.1 Why Do Animals Need a Circulatory System?

A **circulatory system** consists of a muscular pump (the **heart**), a fluid (**blood**), and a series of conduits (**blood vessels**) through which the fluid can be pumped around the body. Heart, blood, and vessels are also known collectively as a **cardiovascular system** (from the Greek *kardia*, "heart," and the Latin *vasculum*, "small vessel"). The function of circulatory systems is to transport things around the body. In preceding chapters we have learned that circulatory systems transport: heat, hormones, respiratory gases, blood cells, platelets, and elements of the immune system. In coming chapters we will add nutrients and waste products to that list. These all seem like important tasks, so why do so many species not have circulatory systems? In this section we will answer that question and examine the general types of circulatory systems found in animals.

Some animals do not have circulatory systems

Single-celled organisms serve all of their needs through direct exchanges with the environment. Such organisms are mostly found in aquatic environments or very moist terrestrial environments. Similarly, for multicellular organisms a circulatory system is unnecessary if all of their cells are close enough to the external environment that nutrients, respiratory gases, and wastes can diffuse between the cells and the environment. Small aquatic invertebrates have structures and body shapes that permit direct exchanges between cells and environment. Many of these animals have flattened, thin body shapes that maximize the amount of surface area that is in contact with the external environment and minimize the diffusion path length for exchanges between the cells and the external environment. The cells of some other aquatic invertebrates are served by highly branched central cavities called *gastrovascular systems* that essentially bring the external environment into the animal. All the cells of a sponge, for example, are in contact with, or very close to, the water that surrounds the animal and circulates through its central cavity. Very small animals without circulatory systems can maintain high levels of metabolism and activity, but bigger animals without circulatory systems such as sponges, coelenterates, and flatworms tend to be inactive, slow, or even sedentary. Larger, active animals need circulatory systems.

Larger and more active animals must support the metabolism of their cells by delivering nutrients to them and taking wastes away from them with circulatory systems. The critical environment for each cell is the extracellular fluid that surrounds it. All of the cell's nutrients—oxygen, fuel, essential molecules—comes from that fluid, and the waste products of the cell go into that fluid. In many animals, the extracellular fluid is continuous with the fluid in the circulatory system. The vessels of these animals empty their fluid directly into the tissues. At other locations the extracellular fluid flows back into the circulatory system to be pumped back out again. These circulatory systems are called *open*. In contrast, *closed* circulatory systems completely contain the circulating fluid, blood, in a continuous system of vessels. However, even in closed circulatory systems, liquid and low-molecular-weight solutes are exchanged between the blood and the extracellular fluids surrounding the cells of the body. In fact, the term *extracellular fluid* refers to both the fluid in the circulatory system and the fluid between the cells of the body. To distinguish the two, we refer to the fluid in the circulatory system as the *blood plasma* and the fluid around the cells as *interstitial fluid*. A normal 70 kilogram person has a total extracellular fluid volume of about 14 liters. A little more than a quarter of it, about 3 liters, is the blood plasma; the rest is interstitial fluid.

Circulatory systems carry materials to and from all regions of the body to maintain the optimum composition of the interstitial fluids, which in turn serve the needs of the cells. Importantly, circulatory systems speed up the delivery of nutrients and deliver them preferentially to the most active tissues.

Open circulatory systems move extracellular fluid

In **open circulatory systems** extracellular fluid squeezes through intercellular spaces as the animal moves. A muscular pump usually assists the distribution of the fluid in these systems. The contractions of this simple heart propel the extracellular fluid through vessels leading to different regions of the body, but the fluid leaves those vessels to trickle through the tissues and eventually return to the heart. Open circulatory systems are found in arthropods, mollusks, and some other invertebrate groups. In the generalized arthropod shown in **Figure 49.1A**, the fluid returns to the heart through valved openings called *ostia*. In mollusks (**Figure 49.1B**), open vessels aid in the return of extracellular fluid to the heart.

Closed circulatory systems circulate blood through a system of blood vessels

In a **closed circulatory system**, a system of vessels keeps circulating blood separate from the interstitial fluid. Blood is pumped through this *vascular system* by one or more muscular hearts, and some components of the blood never leave the vessels. Closed circulatory systems characterize vertebrates, annelids, and some other invertebrate groups.

A simple example of a closed circulatory system is that of the earthworm (**Figure 49.1C**). One large ventral blood vessel carries blood from its anterior end to its posterior end. Smaller vessels branch off and transport the blood to even smaller vessels serving the tissues in each segment of the worm's body. In the smallest vessels, respiratory gases, nutrients, and metabolic wastes diffuse between the blood and the interstitial fluid. The blood then flows from these vessels into larger vessels that lead into one large dorsal vessel, which carries the blood from the posterior to the an-

(A) Arthropod

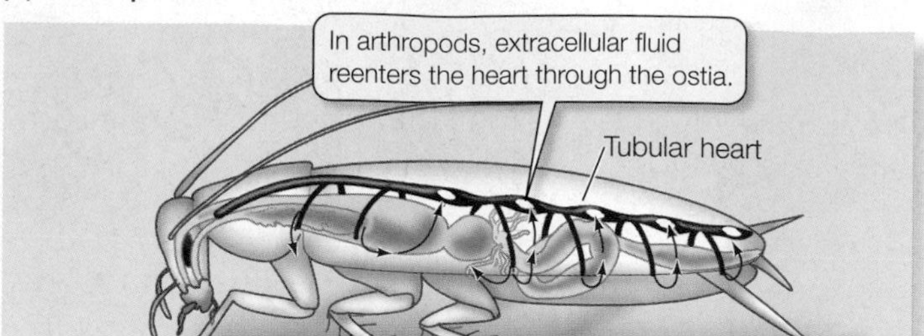
In arthropods, extracellular fluid reenters the heart through the ostia.
Tubular heart

(B) Mollusk

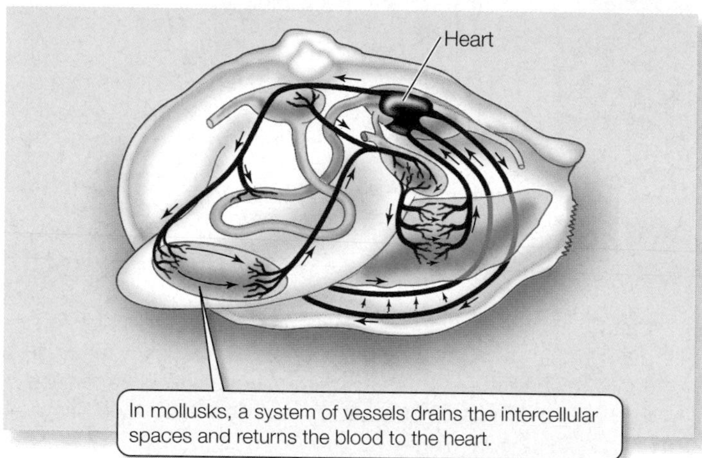
Heart
In mollusks, a system of vessels drains the intercellular spaces and returns the blood to the heart.

(C) Annelid worm

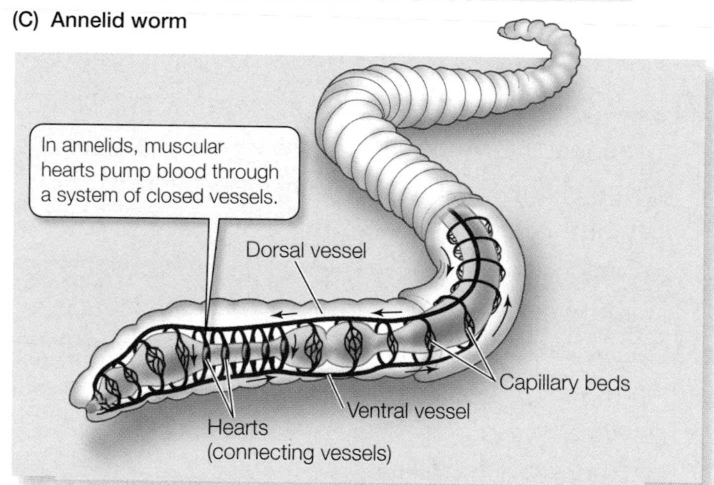
In annelids, muscular hearts pump blood through a system of closed vessels.
Dorsal vessel
Capillary beds
Ventral vessel
Hearts
(connecting vessels)

49.1 Circulatory Systems Arthropods, illustrated here by an insect (A) and mollusks such as clams (B) have open circulatory systems. Blood is pumped by a tubular heart and directed to different regions of the body through vessels that open into intercellular spaces. (C) Annelids such as earthworms have closed circulatory systems, in which the cellular and macromolecular elements of the blood are confined in a system of vessels and the blood is pumped through those vessels by one or more muscular hearts.

terior end of the body. Five pairs of muscular vessels connect the large dorsal and ventral vessels in the anterior end, thus completing the circuit. The dorsal vessel and the five connecting vessels serve as hearts for the earthworm; their contractions keep the blood circulating. The direction of circulation is determined by one-way valves in the dorsal and connecting vessels.

Closed circulatory systems have several advantages over open systems:

- Fluid can flow more rapidly through vessels than through intercellular spaces, and can therefore transport nutrients and wastes to and from tissues more rapidly.
- By changing resistance in the vessels, closed systems can be selective in directing blood to specific tissues.
- Specialized cells and large molecules that aid in the transport of hormones and nutrients can be kept within the vessels, but can drop their cargo in the tissues where it is needed.

It seems logical to accept the premise that in all but very small animals, closed circulatory systems can support higher levels of metabolic activity than open systems can. But we then have to ask: How do highly active insect species achieve high levels of metabolic output with their open circulatory systems? The reason is that insects do not depend on their circulatory systems for respiratory gas exchange. Recall from the last chapter that respiratory gas exchange in insects is through a system of air-filled tubes (see Figure 48.4).

49.1 RECAP

Circulatory systems consist of a pump and an open or closed set of vessels through which is pumped a fluid that transports oxygen, nutrients, wastes, and a variety of other substances.

- Do you understand the difference between extracellular fluid and interstitial fluid? See p. 1046
- Why are most animals with open or with no circulatory systems rather inactive, and why does that not pertain to insects? See p. 1046
- What are some advantages of a closed circulatory system? See pp. 1046–1047

This brief overview of the open and closed systems of several invertebrates introduced some basic concepts about circulatory systems. Now we turn to a more detailed description of the more complex circulatory systems of vertebrates.

49.2 How Have Vertebrate Circulatory Systems Evolved?

Vertebrates have closed circulatory systems and hearts with two or more chambers. When a heart chamber contracts, it squeezes the blood, putting it under pressure. Blood then flows out of the heart and into vessels where pressure is lower. Valves prevent the backflow of blood as the heart cycles between contraction and relaxation.

As we explore the features of the circulatory systems of different classes of vertebrates, a general evolutionary theme will emerge: as circulatory systems become more complex, the blood that flows to the gas exchange organs becomes more completely separated from the blood that flows to the rest of the body. In fish, the phylogenetically oldest vertebrates, blood is pumped from the heart to the gills and then to the tissues of the body and back to the heart. In birds and mammals, the phylogenetically youngest vertebrates, blood is pumped from the heart to the lungs and back to the heart in a **pulmonary circuit**, and then from the heart to the rest of the body and back to the heart in a **systemic circuit**. We will trace the evolution of the separation of the circulation into two circuits.

The closed vascular system of vertebrates begins with vessels called **arteries** that carry blood away from the heart. Arteries give rise to smaller vessels called **arterioles**, which feed blood into **capillary beds. Capillaries** are the tiny, thin-walled vessels where materials are exchanged between the blood and the tissue fluid. Small vessels called **venules** drain capillary beds. The venules join together to form larger vessels called **veins**, which deliver blood back to the heart.

We can trace the evolutionary history of vertebrate circulatory systems by comparing the circulatory systems of fish, lungfish, amphibians, reptiles and crocodilians, and mammals.

Fish have two-chambered hearts

The fish heart has two chambers. An **atrium** (plural *atria*) receives blood from the body and pumps it into a more muscular chamber, the **ventricle**. The ventricle pumps the blood to the gills, where gases are exchanged. Blood leaving the gills collects in a large dorsal artery, the **aorta**, which distributes blood to smaller arteries and arterioles leading to all the organs and tissues of the body. In the tissues, blood flows through beds of tiny capillaries, collects in venules and veins, and eventually returns to the atrium of the heart.

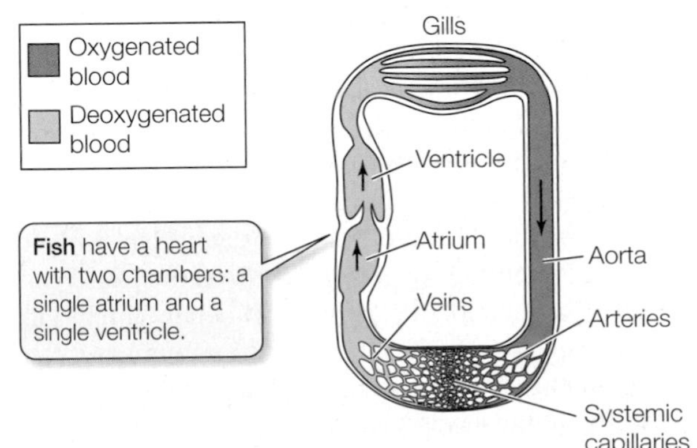

Most of the pressure imparted to the blood by the contraction of the ventricle is dissipated by the high resistance of the narrow spaces in the gill lamellae through which blood flows. As a result, blood leaving the gills and entering the aorta is under low pressure, limiting the maximum capacity of the fish circulatory system to supply the tissues with oxygen and nutrients. Yet this limitation on arterial blood pressure does not seem to hamper the performance of many rapidly swimming species, such as tuna and marlin.

The evolutionary transition from breathing water to breathing air had important consequences for the vertebrate circulatory system. An example of how the system changed to serve a primitive lung can be seen the African lungfish.

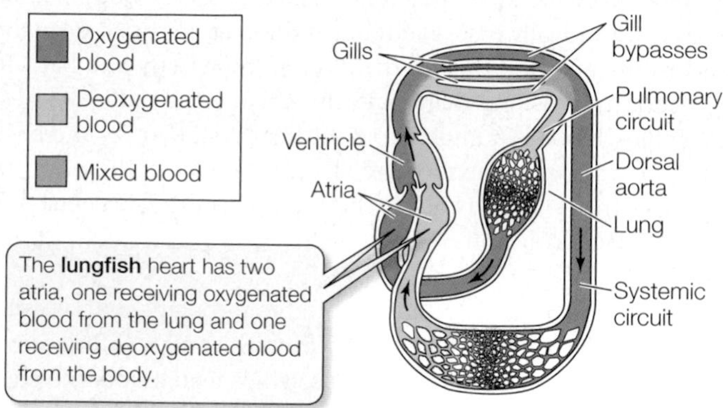

The **lungfish** heart has two atria, one receiving oxygenated blood from the lung and one receiving deoxygenated blood from the body.

Lungfish are periodically exposed to water with low oxygen content or to situations in which their aquatic environment dries up. Their adaptation for dealing with these conditions is an outpocketing of the gut that serves as a lung. The lung contains many thin-walled blood vessels, so blood flowing through those vessels can pick up oxygen from air gulped into the lung.

How does the lungfish circulatory system take advantage of this new organ? The posterior pair of gill arteries has been modified to carry blood to the lung, and a new vessel carries oxygenated blood from the lung back to the heart. In addition, two anterior gill arches have lost their gills, and their blood vessels deliver blood from the heart directly to the dorsal aorta. Because a few of the gill arches retain gills, the African lungfish can breathe either air or water.

The lungfish heart has adaptations that partially separate the flow of its blood into pulmonary and systemic circuits. Unlike other fish, the lungfish has a partly divided atrium; the left side receives oxygenated blood from the lungs, and the right side receives deoxygenated blood from the other tissues. These two bloodstreams stay mostly separate as they flow through the ventricle and the large vessel leading to the gill arches. As a result, oxygenated blood mostly goes to the anterior gill arteries leading to the dorsal aorta, and the deoxygenated blood mostly goes to the other gill arches that have functional gills as well as to the gill arteries that serve the lung.

We can conclude that the lung of the lungfish evolved as a means of supplementing oxygen uptake from the gills. When the water is fully oxygenated, the lungfish can depend on its gills; but in oxygen-depleted water, the lungfish can augment its oxygen intake by gulping air. This trait in turn resulted in associated modifications of the lungfish vascular system that set the stage for the evolution of separate pulmonary and systemic circulations.

Amphibians have three-chambered hearts

Pulmonary and systemic circulation are partly separated in adult amphibians. A single ventricle pumps blood to the lungs and to the rest of the body. Two atria receive blood returning to the heart. One receives oxygenated blood from the lungs and the other receives deoxygenated blood from the body.

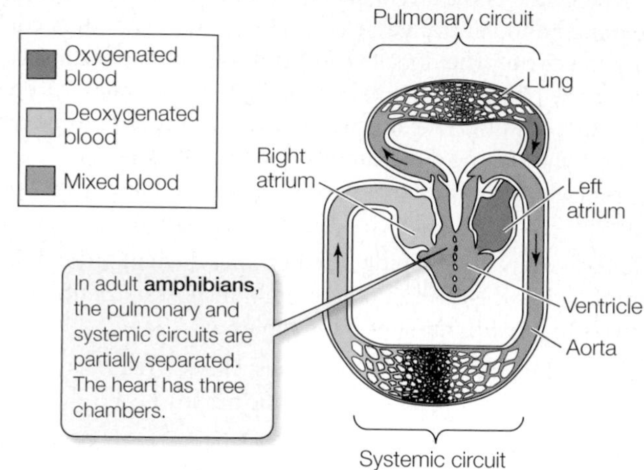

In adult **amphibians**, the pulmonary and systemic circuits are partially separated. The heart has three chambers.

Because both atria deliver blood to the same ventricle, the oxygenated and deoxygenated blood could mix, so that blood going to the tissues would not carry a full load of oxygen. Mixing is limited, however, because anatomical features of the ventricle direct the flow of deoxygenated blood from the right atrium to the pulmonary circuit and the flow of oxygenated blood from the left atrium to the aorta.

Partial separation of pulmonary and systemic circulation has the advantage of allowing blood destined for the tissues to sidestep the high flow resistance of the gas exchange organ. Blood leaving the amphibian heart for the tissues moves directly to the aorta, and hence to the body, at a higher pressure than if it had first flowed through the lungs.

Reptiles have exquisite control of pulmonary and systemic circulation

Turtles, snakes, and lizards are commonly said to have three-chambered hearts because their ventricles are not completely separated into left and right chambers. Crocodilians (crocodiles and alligators), however, have two completely separated ventricles, creating a four-chambered heart. This anatomical description, however, does not reveal the amazing ability of these animals to control the proportion of the blood that is flowing through the lungs and the proportion that is flowing through the rest of the body.

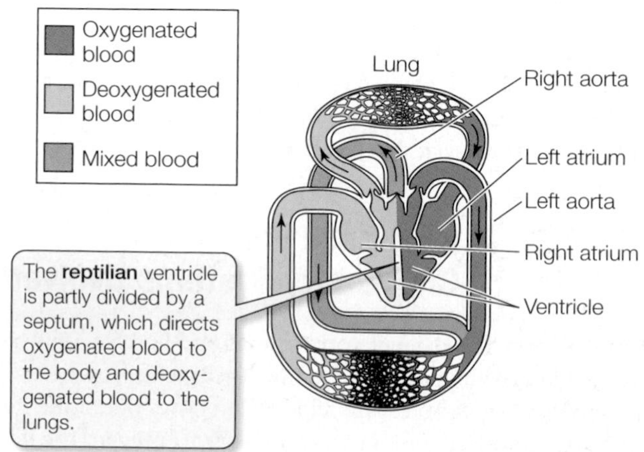

The **reptilian** ventricle is partly divided by a septum, which directs oxygenated blood to the body and deoxygenated blood to the lungs.

Consider the behavior, ecology, and physiology of reptiles. Many reptiles and crocodilians are active, powerful, fast animals. But their activity is in bursts that are interspersed with long periods of inactivity, during which their metabolic rates are much lower than the resting metabolic rates of birds and mammals. So enormous is the range of metabolic demand in these animals that they do not have to breathe continuously. Some species are accomplished divers and spend long periods under water, where they cannot breathe. When these animals are not breathing, it is a waste of energy for them to pump blood through their lungs. Thus, they have evolved the capability of sending blood to lungs and to the rest of the body when they are breathing, but when they are not breathing, they can bypass the pulmonary circuit and pump all the blood to the body. How do they do this?

Reptiles have two aortas—a left and a right. The left aorta receives oxygenated blood from the left side of the ventricle and carries it to the tissues of the body. The right aorta, however, can receive blood from either the right side or the left side of the ventricle. The two sides of the ventricle are partially divided by a *septum*. When the animal is breathing air, the resistance in the pulmonary circuit is lower than the resistance in the systemic circuit, so blood from the right side of the ventricle tends to flow into the pulmonary artery rather than the right aorta. When the animal is not breathing, pulmonary vessels constrict, resistance in the pulmonary circuit goes up, and therefore blood from the right side of the ventricle tends to flow into the right aorta. As a result, blood from both sides of the ventricle flows through both aortas to the systemic circuit.

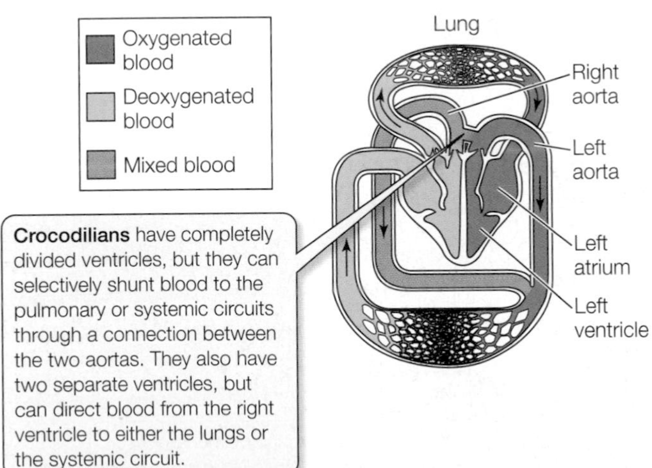
Oxygenated blood
Deoxygenated blood
Mixed blood
Lung
Right aorta
Left aorta
Left atrium
Left ventricle
Crocodilians have completely divided ventricles, but they can selectively shunt blood to the pulmonary or systemic circuits through a connection between the two aortas. They also have two separate ventricles, but can direct blood from the right ventricle to either the lungs or the systemic circuit.

The crocodilians have completely separated ventricles with one aorta originating in each. They can alter the proportion of blood going to their pulmonary and systemic circuits, however, because of a connection between the two aortas just as they leave the heart. When the crocodile or alligator is breathing and resistance in the pulmonary circuit is low, backpressure from the stronger left ventricle prevents blood from the right ventricle from entering the right aorta and it flows into the pulmonary circuit instead. When the animal stops breathing, pulmonary vessels constrict, resistance in the pulmonary circuit rises, and blood from the right ventricle flows into the right aorta.

Birds and mammals have fully separated pulmonary and systemic circuits

The four-chambered hearts of birds and mammals have completely separate pulmonary and systemic circuits. Separate circuits have several advantages for active animals with continuously high metabolic rates:

- Oxygenated and deoxygenated blood cannot mix; therefore, the systemic circuit always receives blood with the highest oxygen content.
- Respiratory gas exchange is maximized because the blood with the lowest oxygen content and highest CO_2 content is sent to the lungs.
- Separate systemic and pulmonary circuits can operate at different pressures.

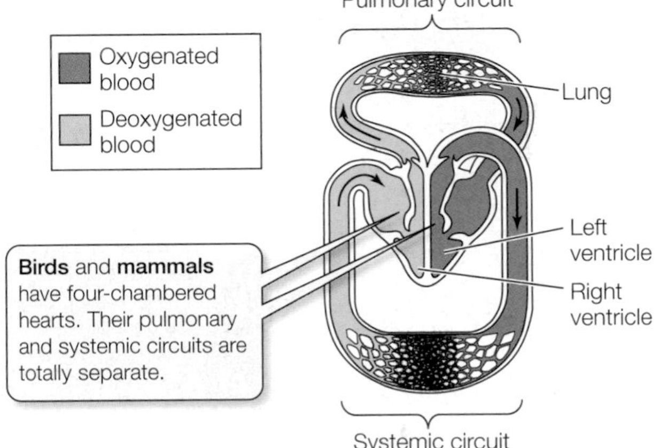
Pulmonary circuit
Oxygenated blood
Deoxygenated blood
Lung
Left ventricle
Right ventricle
Birds and **mammals** have four-chambered hearts. Their pulmonary and systemic circuits are totally separate.
Systemic circuit

The tissues of birds and mammals have high nutrient demands and thus a very high density of the smallest vessels, the capillaries. Many small vessels present high resistance to the flow of blood. Therefore, high pressure is required in the systemic circuits of birds and mammals. Their pulmonary circuits have fewer capillaries and lower resistance than their systemic circuits, so the pulmonary circuits of birds and mammals can function at lower pressures.

49.2 RECAP

The closed circulatory system of vertebrates has evolved from a two-chambered heart in fish to a four-chambered heart in mammals, birds, and crocodilians.

- Can you explain why fish cannot supply blood to their tissues at high pressure? See p. 1047
- By comparing lungfish and amphibian circulatory systems, can you describe how a three-chambered heart could have evolved? See p. 1048
- Can you name some advantages of a four-chambered heart for mammals and birds? See p. 1049

We now take a closer look at the structure and function of the mammalian heart, using the human heart as our example.

49.3 How Does the Mammalian Heart Function?

The human heart is typical of all mammalian hearts; it has four chambers: two atria and two ventricles (**Figure 49.2**). The atrium and ventricle on the right side of your body are called the right heart. The atrium and ventricle on the left are the left heart. The right heart pumps blood through the pulmonary circuit, and the left heart pumps blood through the systemic circuit.

Valves between the atria and ventricles, the **atrioventricular valves**, prevent backflow of blood into the atria when the ventricles contract. The **pulmonary valve** and the **aortic valve**, positioned between the ventricles and the major arteries, prevent the backflow of blood into the ventricles when they relax.

In this section, we first focus on the flow of blood through the heart and through the body. Next we examine the unique electrical properties of cardiac muscle that result in the heartbeat, and see how the heart's electrical activity can be recorded in an ECG (electrocardiogram).

Blood flows from right heart to lungs to left heart to body

The heart's right atrium receives deoxygenated blood from the **superior** (upper) **vena cava** and the **inferior** (lower) **vena cava** (see Figure 49.2), large veins that collect blood returning to the heart from the upper and lower body, respectively. The veins of the heart itself also drain into the right atrium. From the right atrium, the blood flows through an atrioventricular (AV) valve into the right ventricle. Most of the filling of the ventricle results from passive flow while the heart is relaxed between beats. Just at the end of this period of passive ventricular filling, the atrium contracts and adds a little more blood to the ventricular volume. The right ventricle then contracts, causing the AV valve to close and pumping the blood into the **pulmonary artery** to the lungs.

After gas exchange occurs in the lungs, **pulmonary veins** return oxygenated blood from the lungs to the left atrium, from which

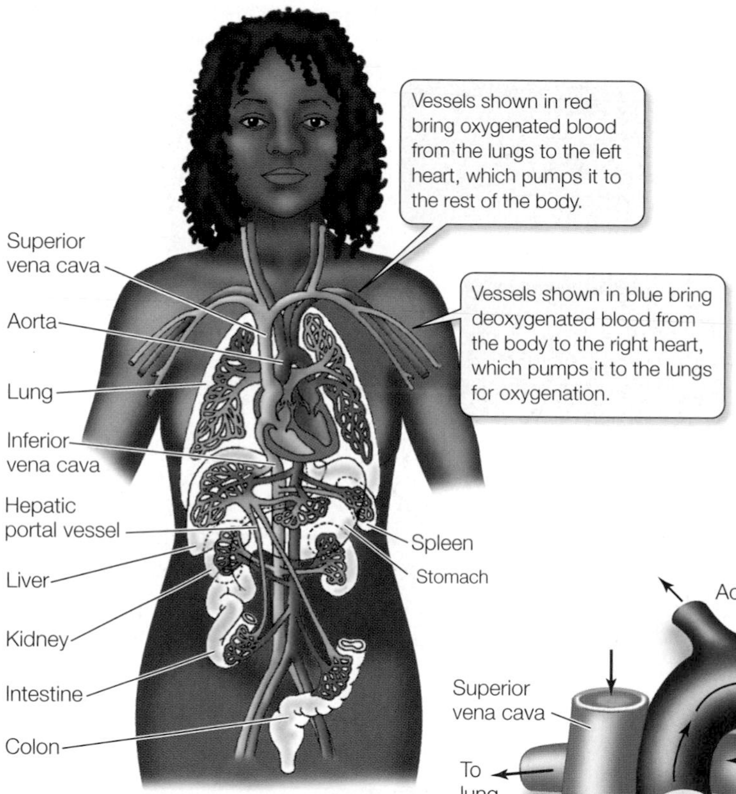

49.2 The Human Heart and Circulation In the human heart, blood flows from right heart to lungs to left heart to body. The atrioventricular valves prevent blood from flowing back into the atria when the ventricles contract. The pulmonary and aortic valves prevent blood from flowing back into ventricles from the arteries when the ventricles relax.

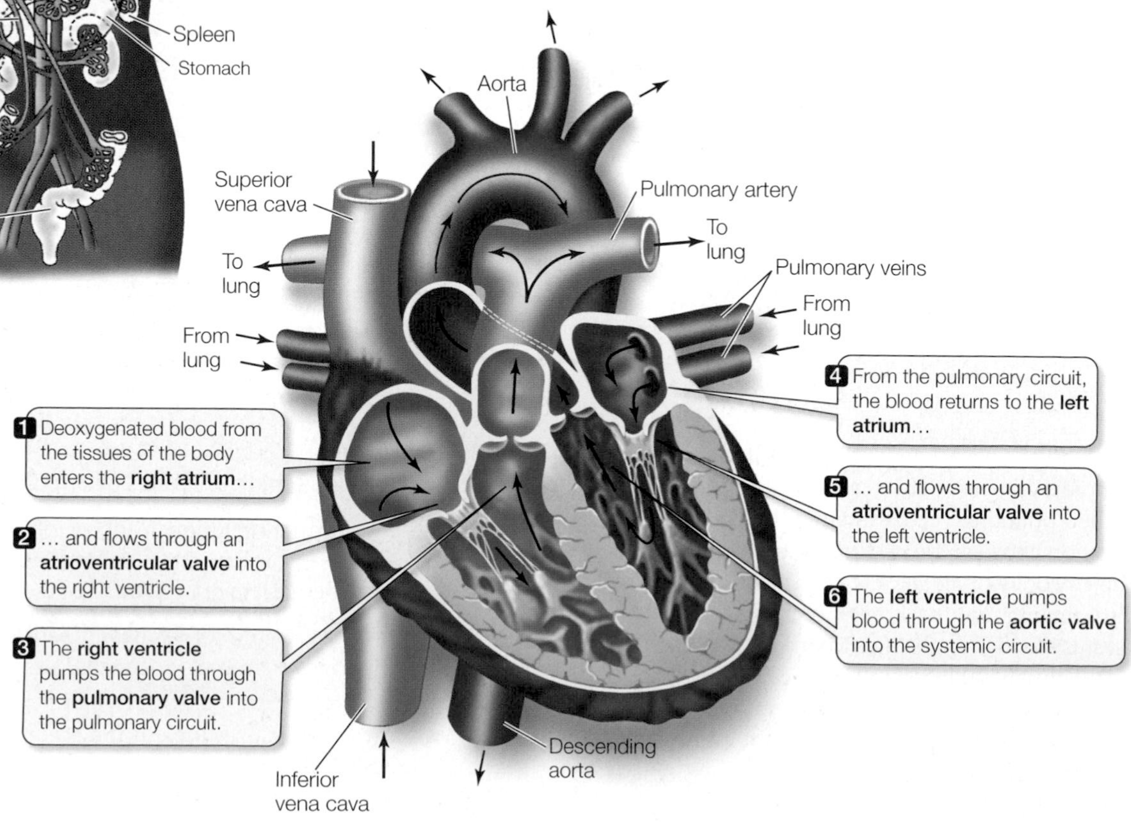

the blood enters the left ventricle through another AV valve. As with the right side of the heart, most left ventricular filling is passive, but the ventricle is topped off by contraction of the atrium just at the end of the period of passive filling.

The walls of the left ventricle are powerful muscles that contract around the blood with a wringing motion starting from the bottom. When pressure in the left ventricle is high enough to push open the aortic valve, the blood rushes into the aorta to begin its circulation throughout the body. In Figure 49.2, observe that the walls at the left ventricle are thicker than those of the right ventricle. The left ventricle has to propel the blood through many more kilometers of blood vessels than does the right ventricle, and must therefore push against more resistance, even though both ventricles pump the same volume of blood.

Both sides of the heart contract at the same time. The contraction of the two atria, followed by the contraction of the two ventricles and then relaxation, is the **cardiac cycle**. The cardiac cycle is divided into two phases: **systole**, when the ventricles contract, and **diastole**, when the ventricles relax (**Figure 49.3**). Just at the end of diastole, the atria contract and top off the volume of blood in the ventricles. The sounds of the cardiac cycle, the "lub-dup" heard through a stethoscope placed on the chest, are created by the heart valves slamming shut. The closing and opening of these valves are simple mechanical events resulting from pressure differences on the two sides of the valves. As the ventricles begin to contract, the pressure in the ventricles rises above the pressure in the atria, so blood starts flowing back into the atria, and the atrioventricular valves close ("lub"). When the ventricles begin to relax, the high pressure in the aorta and pulmonary artery causes blood to start to flow back into the ventricles, and this flow of blood closes the aortic and pulmonary valves ("dup"). Defective valves produce turbulent blood flow and produce the sounds known as *heart murmurs*. For example, if an atrioventricular valve is defective and does not close completely, blood will flow back into the atrium with a "whoosh" sound following the "lub."

The rhythm of the cardiac cycle can be felt in the pulsation of arteries such as the one that supplies blood to your hand. You can

49.3 The Cardiac Cycle The rhythmic contraction (systole) and relaxation (diastole) of the ventricles is called the cardiac cycle. The graphical representation below shows pressure and volume changes during the cardiac cycle for the left ventricle only.

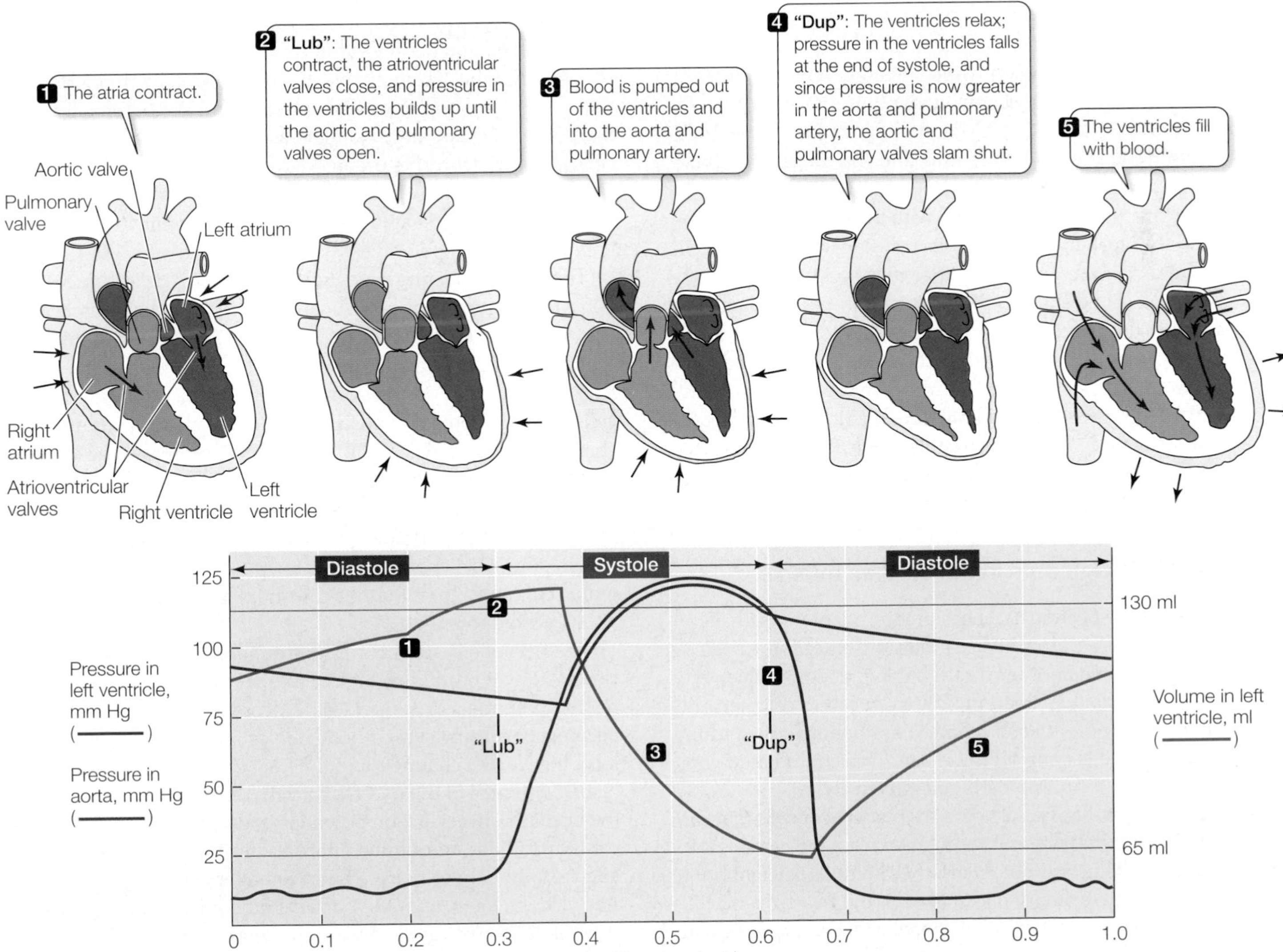

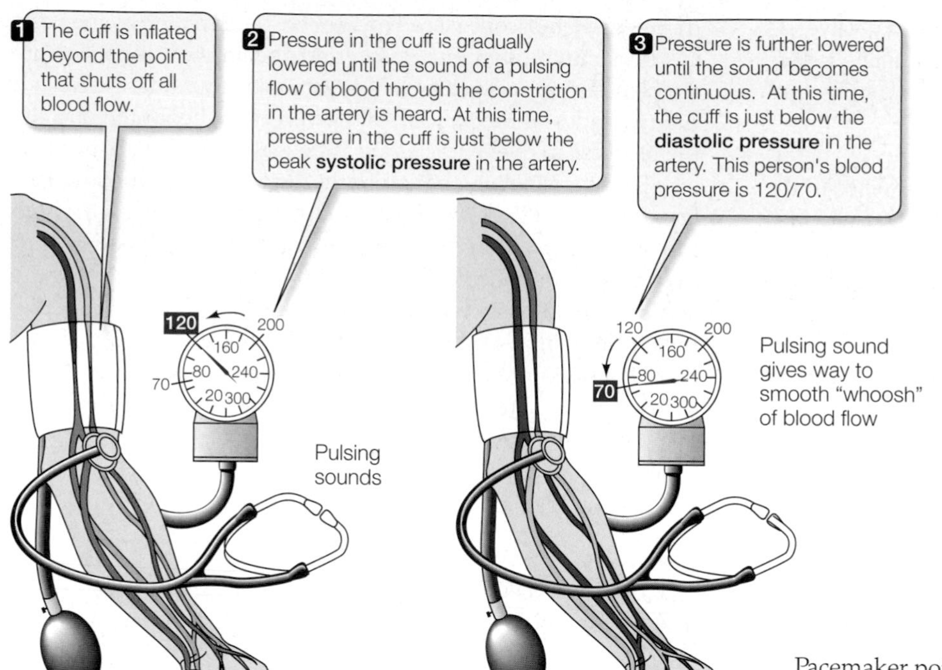

1 The cuff is inflated beyond the point that shuts off all blood flow.
2 Pressure in the cuff is gradually lowered until the sound of a pulsing flow of blood through the constriction in the artery is heard. At this time, pressure in the cuff is just below the peak **systolic pressure** in the artery.
3 Pressure is further lowered until the sound becomes continuous. At this time, the cuff is just below the **diastolic pressure** in the artery. This person's blood pressure is 120/70.
120
200
160
80
240
70
20
300
Pulsing sounds
Pulsing sound gives way to smooth "whoosh" of blood flow

49.4 Measuring Blood Pressure Blood pressure in the major artery of the arm can be measured with a device called a sphygmomanometer, which combines an inflatable cuff and a pressure gauge. A stethoscope is also used to detect sounds created by the blood vessels in response to changes in pressure during the cardiac cycle.

feel your pulse by placing two fingers from one hand lightly over the wrist of the other hand just below the thumb. During systole, the pressure wave created by the contraction of the left ventricle surges through the arteries of your arm, and you can feel this pressure wave as a pulsing of the artery in your wrist.

Blood pressure changes associated with the cardiac cycle can be measured in the large artery in your arm by using an inflatable pressure cuff and a pressure gauge, together called a *sphygmomanometer,* and a stethoscope (**Figure 49.4**). This method measures the minimum pressure necessary to compress an artery so that blood does not flow through it at all (the systolic value) and the minimum pressure that causes intermittent flow through the artery (the diastolic value). A conventional blood pressure reading is expressed as the systolic value placed over the diastolic value. Healthy values for a young adult might be 120 millimeters of mercury (mm Hg) during systole and 70 mm Hg during diastole, or 120/70.

The heartbeat originates in the cardiac muscle

Cardiac muscle (see Section 47.1) has unique adaptations that enable it to function as a pump. First, cardiac muscle cells are in electrical contact with one another through gap junctions, which enable action potentials to spread rapidly from cell to cell. Because a spreading action potential stimulates contraction, large groups of cardiac muscle cells contract in unison. This coordinated contraction is essential for pumping blood effectively.

Second, some cardiac muscle cells are **pacemaker cells** that can initiate action potentials without stimulation from the nervous system. When they fire action potentials, they stimulate neighboring cells to contract. The primary pacemaker of the heart is a nodule of modified cardiac muscle cells, the **sinoatrial node**, located at the junction of the superior vena cava and right atrium. The resting membrane potentials of these cells are less negative than other cardiac muscle cells and they are not stable, but gradually become even less negative until they reach threshold for initiating an action potential. The action potentials of pacemaker cells are very different from those of neurons and other muscle cells. They are slower to rise; they are broader; and they are slower to return to resting potential. These properties of pacemaker cells are due to the ion channels in their membranes (**Figure 49.5**).

Pacemaker potentials involve Na^+, Ca^{2+}, and K^+ channels. Remember from Section 44.2 that when Na^+ or Ca^{2+} channels open, positive charges flow into the cell and the membrane potential becomes less negative. When K^+ channels open, positive charges flow out of the cell and the membrane potential becomes more negative. Because the Na^+ channels of pacemaker cells are more open than are those of other cardiac muscle cells, the pacemaker resting potential is less negative. The action potential of pacemaker cells is not due to voltage-gated Na^+ channels, rather to voltage-gated Ca^{2+} channels. These ion channels open and close more slowly than voltage-gated Na^+ channels, explaining the shape of pacemaker action potentials.

The unstable resting potential of pacemaker cells is due to the behavior of cation channels. As in neurons and other muscle cells, the opening of K^+ channels helps return the cell to its resting potential following an action potential. The resting potential is due to the balance of K^+ ions leaving the cell and Na^+ ions entering the cell, and in pacemaker cells this balance is controlled by cation channels through which both K^+ and Na^+ can pass. These channels open when the cell returns to resting potential following an action potential, and since they are more permeable to Na^+ than to K^+, the resting potential gradually becomes less negative. At a certain point Ca^{2+} channels begin to open, and the membrane potential continues to rise until a threshold is reached that results in a Ca^{2+} action potential.

The nervous system controls the heartbeat (speeds it up or slows it down) by influencing the rate at which the pacemaker cell resting potentials drift upwards. Norepinephrine released onto pacemaker cells by sympathetic nerves increases the permeability of the Na^+/K^+ channels and the Ca^{2+} channels. The result is the resting potential of the pacemaker cells drifts up more rapidly and the interval between action potentials is decreased. Conversely, the parasympathetic neurotransmitter, acetylcholine, has the opposite effect on the rate of rise of the pacemaker cell resting potential. Acetylcholine increases the permeability of K^+ channels and decreases the permeability of Ca^{2+} channels. The result is the interval between pacemaker action potentials shortens.

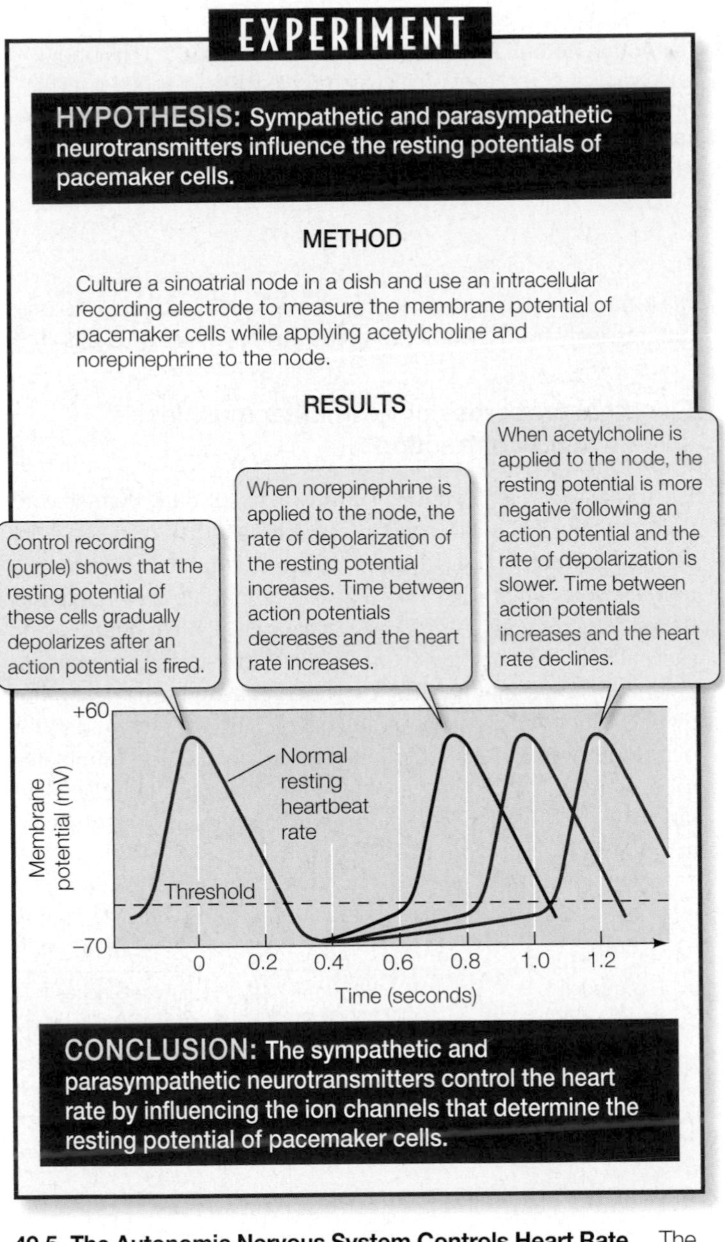

EXPERIMENT
HYPOTHESIS: Sympathetic and parasympathetic neurotransmitters influence the resting potentials of pacemaker cells.
METHOD
Culture a sinoatrial node in a dish and use an intracellular recording electrode to measure the membrane potential of pacemaker cells while applying acetylcholine and norepinephrine to the node.
RESULTS
Control recording (purple) shows that the resting potential of these cells gradually depolarizes after an action potential is fired.
When norepinephrine is applied to the node, the rate of depolarization of the resting potential increases. Time between action potentials decreases and the heart rate increases.
When acetylcholine is applied to the node, the resting potential is more negative following an action potential and the rate of depolarization is slower. Time between action potentials increases and the heart rate declines.
+60
–70
Membrane potential (mV)
Normal resting heartbeat rate
Threshold
0
0.2
0.4
0.6
0.8
1.0
1.2
Time (seconds)
CONCLUSION: The sympathetic and parasympathetic neurotransmitters control the heart rate by influencing the ion channels that determine the resting potential of pacemaker cells.

49.5 The Autonomic Nervous System Controls Heart Rate The plasma membranes of pacemaker cells spontaneously depolarize to threshold and fire action potentials. Signals from the two divisions of the autonomic nervous system raise and lower the heart rate by altering the rate at which the cells depolarize. Their effects can be seen in the dynamics of the resting membrane potential.

A conduction system coordinates the contraction of heart muscle

A normal heartbeat begins with an action potential in the sinoatrial node (**Figure 49.6**). This action potential spreads rapidly throughout the electrically coupled cells of the atria, causing them to contract in unison. Since there are no gap junctions between the cells of the atria and those of the ventricles, the action potential does not spread directly to the ventricles. Therefore, the ventricles do not contract in unison with the atria.

The action potential initiated in the atria is regenerated and conducted to and through the ventricular muscle mass by a system of modified, noncontractile cardiac muscle cells. Situated at the junction of the atria and the ventricles is a nodule of modified cardiac muscle cells—the **atrioventricular node**—which is stimulated by the depolarization of the atria. With a slight delay, it generates action potentials that are conducted to the ventricles via the **bundle of His**, which consists of modified cardiac muscle fibers that do not contract. These fibers divide into right and left *bundle branches* that run to the tips of the ventricles and then spread throughout

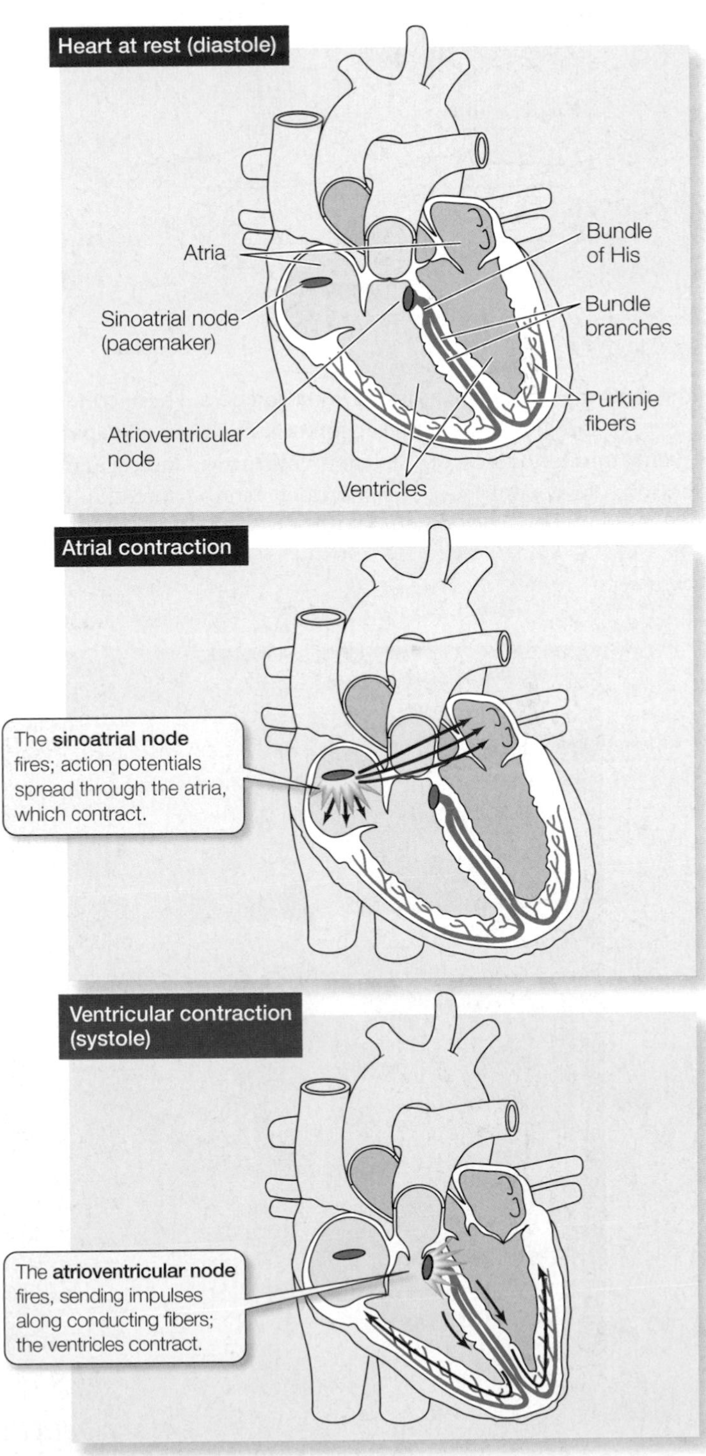

49.6 The Heartbeat Pacemaker cells in the sinoatrial node initiate the heartbeat by firing action potentials.

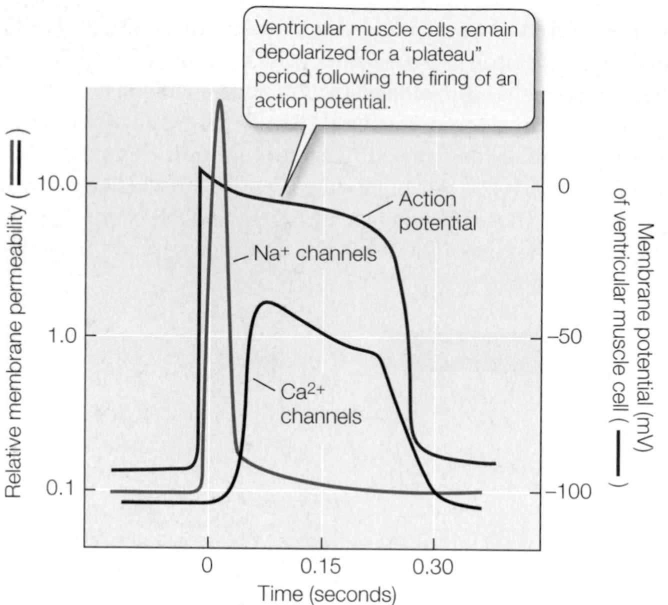

49.7 The Action Potential of Ventricular Muscle Fibers The rising phase of the action potential of ventricular muscle fibers (graphed in red) is due to the opening of voltage-gated Na^+ channels (bluegreen). However, the membrane remains in a depolarized state for a prolonged time because of the opening of voltage-gated Ca^{2+} channels (black).

the ventricular muscle mass as **Purkinje fibers**. These conducting fibers ensure that the cardiac action potential spreads rapidly and evenly throughout the ventricular muscle mass, starting at the very bottom of the ventricles. The short delay in the spread of the action potential imposed by the atrioventricular node ensures that the atria contract before the ventricles do, so that the blood passes progressively from the atria to the ventricles to the arteries.

Electrical properties of ventricular muscles sustain heart contraction

Electrical properties of ventricular muscle fibers allow them to contract for about 300 milliseconds—much longer than those of skeletal muscle fibers. Like neuronal and skeletal muscle action potentials, ventricular muscle cell action potentials are initiated by the opening of voltage-gated Na^+ channels. Unlike neurons and skeletal muscle fibers, however, ventricular muscle cells remain depolarized for a long time. The extended plateau of the action potential is due to sustained opening of voltage-gated calcium channels (**Figure 49.7**). Like other muscle, cardiac muscle is stimulated to contract when Ca^{2+} is available to bind with troponin (see Figure 47.6). As long as their calcium channels remain open, the ventricular muscle cells continue to contract.

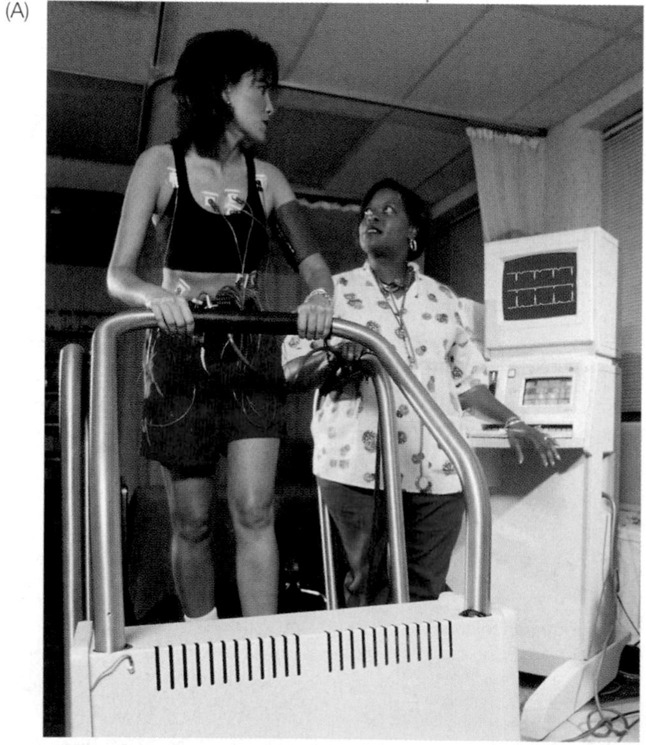

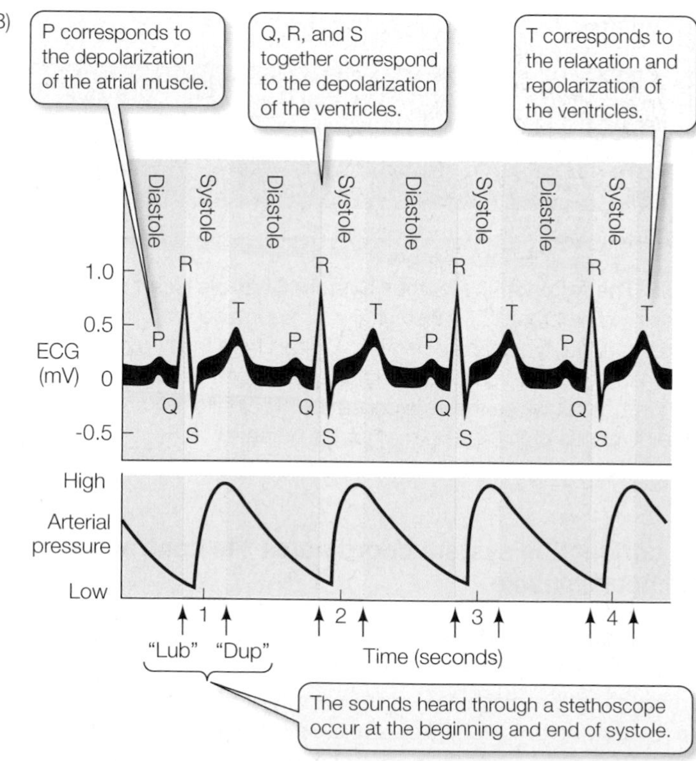

49.8 The Electrocardiogram (A) An electrocardiogram (abbreviated variously as ECG or EKG) is used to monitor heart function. Electrodes attached to the person on the treadmill record an ECG that is amplified and displayed on a monitor. (B) Variations from the normal pattern shown here can be used to diagnose heart problems.

In 1775 a British physician learned that a gypsy woman was successfully treating people for heart problems using an herbal remedy. He discovered that the remedy contained an extract of the purple foxglove plant, *Digitalis purpurea*. This drug, called digitalis and still in use today, increases Ca^{2+} concentration in cardiac muscle cells, thereby strengthening and prolonging systole. The physician, incidentally, became rich; the gypsy woman did not.

The ECG records the electrical activity of the heart

Electrical events in the cardiac muscle during the cardiac cycle can be recorded by electrodes placed on the surface of the body. Such a recording is an **electrocardiogram**, or **ECG** (EKG is also used because German physicians who invented the method used the Greek word for heart and called it the *electrokardiogramm*). The ECG is an important tool for diagnosing heart problems (**Figure 49.8A**).

The action potentials that sweep through the muscles of the atria and the ventricles before they contract are such massive, localized electrical events that they cause electric currents to flow outward from the heart to all parts of the body. Surface electrodes placed at different locations on the body detect those electric currents at different times, and therefore register a voltage difference. The appearance of the ECG depends on the placement of the electrodes. Electrodes placed on the right wrist and left ankle produced the normal ECG shown in **Figure 49.8B**. The wave patterns of the ECG are designated P, Q, R, S, and T, each letter representing a particular event in the cardiac muscle, as shown in the figure.

49.3 RECAP

The mammalian heart is organized into two atria and two ventricles. Modified cardiac muscle tissue in the right atrium functions to spontaneously generate pacemaker action potentials. Other modified cardiac muscle tissue between the atria and ventricles and throughout the ventricles conducts that signal and coordinates the heart contraction.

- Trace the path of blood through both sides of the heart, naming the major blood vessels and heart valves. See p. 1050 and Figure 49.2
- Can you differentiate systole and diastole and describe the events of the cardiac cycle? See p. 1051 and Figure 49.3
- Can you explain how cells of the sinoatrial node generate the heart beat? See pp. 1053–1054 and Figures 49.6 and 49.7

We next consider the composition of the blood and the characteristics of the vessels through which blood circulate around the body, illustrating once again how structure serves function. We will also consider the role of the lymphatic vessels that return interstitial fluid to the blood.

49.4 What Are the Properties of Blood and Blood Vessels?

Blood is a connective tissue. It consists of cells suspended in an extracellular matrix of complex, yet specific, composition. The unusual feature of blood is that the extracellular matrix is a liquid, so blood is a fluid tissue.

The cells of the blood can be separated from the fluid matrix, called **plasma**, by centrifugation (**Figure 49.9**). If a sample of blood is spun in a centrifuge, all the cells move to the bottom of the tube, leaving the clear, straw-colored plasma on top. The *packed-cell volume*, or *hematocrit*, is the percentage of the blood volume made up by cells. Normal hematocrit is about 42 percent for women and 46 percent for men, but these values can vary considerably. They are usually higher, for example, in people who live and work at high altitudes, because the low oxygen concentrations at high altitudes stimulate the production of more red blood cells.

Here we will consider two elements in blood: the *red blood cells* and the *platelets*, which are pinched-off fragments of cells. Another important class of blood cells—*white blood cells*, or leukocytes—is discussed in Chapter 18.

Red blood cells transport respiratory gases

Most of the cells in the blood are **erythrocytes**, or red blood cells. Mature red blood cells are biconcave, flexible discs packed with hemoglobin. Their function is to transport respiratory gases. Their shape gives them a large surface area for gas exchange, and their flexibility enables them to squeeze through narrow capillaries. There are 5 to 6 million red blood cells per microliter of blood.

Red blood cells, as well as all the other cellular components of blood, are generated by stem cells in the bone marrow, particularly in the ribs, breastbone, pelvis, and vertebrae. Red blood cell production is controlled by a hormone, **erythropoietin**, which is released by cells in the kidney in response to insufficient oxygen—**hypoxia**. Many tissues respond to hypoxia by expressing a transcription factor called *hypoxia-inducible factor 1* (*HIF-1*). When the kidney becomes hypoxic and expresses HIF-1, one of the actions of the transcription factor is to activate the gene encoding erythropoietin.

Under normal conditions, your bone marrow produces about 2 million red blood cells every second. Developing, immature red blood cells divide many times while still in the bone marrow, and during this time they produce hemoglobin. When the hemoglobin content of a red blood cell approaches about 30 percent, the nucleus, endoplasmic reticulum, Golgi apparatus, and mitochondria of the cell begin to break down. This process is almost complete when the new red blood cell squeezes between the endothelial cells of blood vessels in the bone marrow and enters the circulation. Loss of organelles is seen only in mammalian red blood cells. The red blood cells of other vertebrates are nucleated.

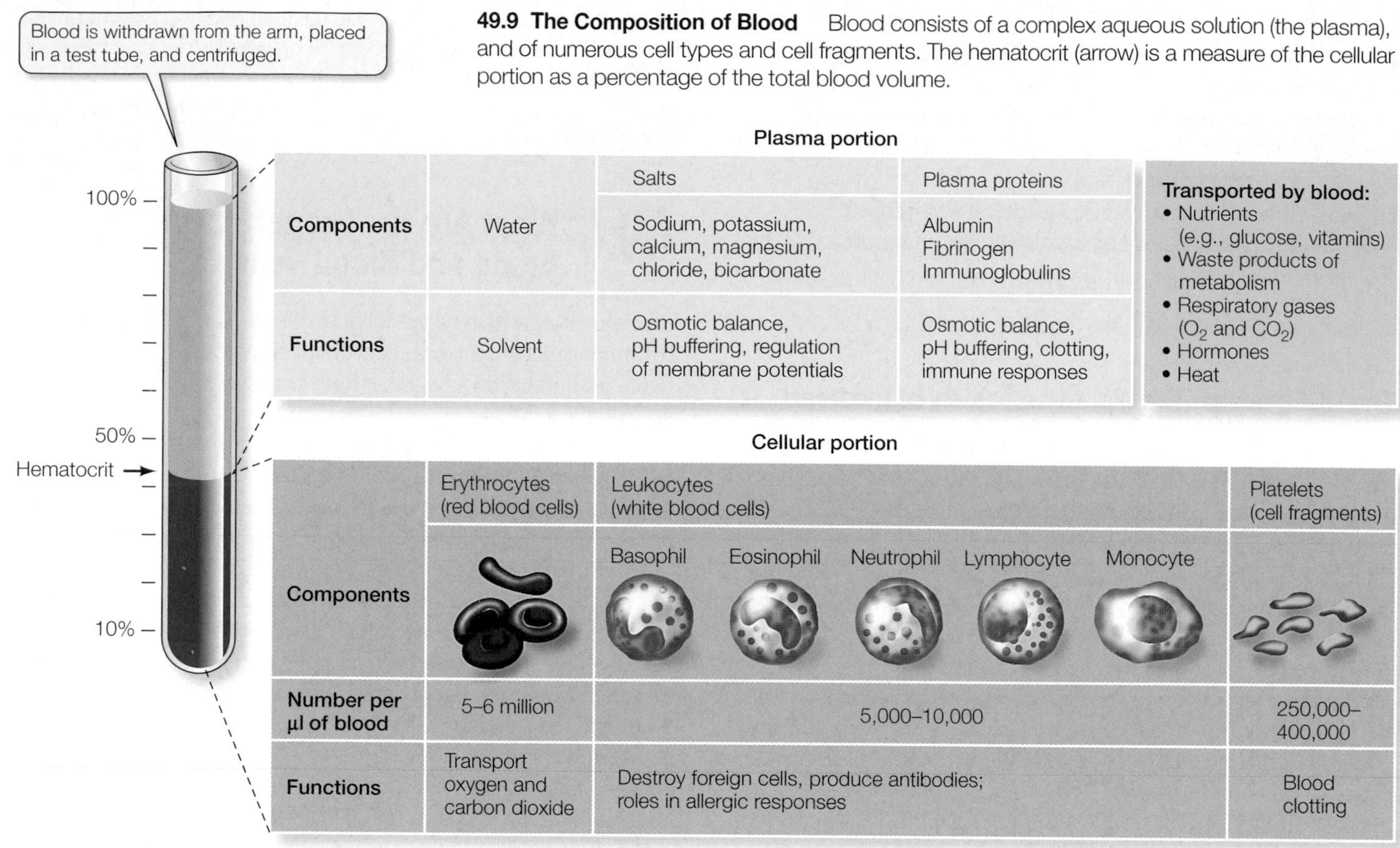

49.9 The Composition of Blood Blood consists of a complex aqueous solution (the plasma), and of numerous cell types and cell fragments. The hematocrit (arrow) is a measure of the cellular portion as a percentage of the total blood volume.

Each red blood cell circulates for about 120 days. As it gets older, its membrane becomes less flexible and more fragile. Therefore, old red blood cells can rupture as they bend to fit through narrow capillaries. One place where they are really squeezed is in the **spleen**, an organ that sits near the stomach in the upper left side of the abdominal cavity. The spleen has many venous cavities, or sinuses, that serve as a reservoir for red blood cells, but to get into the sinuses, the red blood cells must squeeze between spleen cells. When old red blood cells are ruptured by this squeezing, their remnants are taken up and degraded by macrophages.

Platelets are essential for blood clotting

Besides producing erythrocytes and leukocytes, the bone marrow stem cells described in Section 18.1 also produce cells called *megakaryocytes*. Megakaryocytes are large cells that remain in the bone marrow and continually break off cell fragments called **platelets**. A platelet is just a tiny fragment of a cell without cell organelles, but it is packed with enzymes and chemicals necessary for its function: sealing leaks in blood vessels and initiating **blood clotting** (**Figure 49.10**).

Damage to a blood vessel exposes collagen fibers. An encounter with collagen fibers activates a platelet. The platelet swells, becomes irregularly shaped and sticky, and releases chemicals called **clotting factors**, which activate other platelets and initiate the clotting of blood. The sticky platelets also form a plug at the damaged site.

The clotting of blood requires many steps and many clotting factors. The absence of any one of these factors can impair clotting and cause excessive bleeding. Because the liver produces most of the clotting factors, liver diseases such as hepatitis and cirrhosis can result in excessive bleeding. People with hemophilia experience uncontrolled bleeding due to an inherited inability to produce one of the clotting factors.

Blood clotting factors participate in a cascade of chemical reactions that activate other substances circulating in the blood. The cascade begins with cell damage and platelet activation and leads to the conversion of an inactive circulating enzyme, **prothrombin**, to its active form, **thrombin**. Thrombin cleaves molecules of **fibrinogen**, a plasma protein, forming insoluble threads of **fibrin**. The fibrin threads form the meshwork that clots the blood, seals the vessel, and provides a scaffold for the formation of scar tissue (see Figure 49.10).

Plasma is a complex solution

Plasma, the clear, straw-colored liquid portion of the blood, contains dissolved gases, ions, nutrient molecules, proteins, and other molecules, such as hormones and vitamins (see Figure 49.9). Most of the ions are Na^+ and Cl^- (hence the salty taste of blood), but many other ions are also present. The nutrient molecules in plasma include glucose, amino acids, lipids, cholesterol, and lactic acid. The circulating proteins in plasma include the blood clotting factors that we have just mentioned as well as many other proteins.

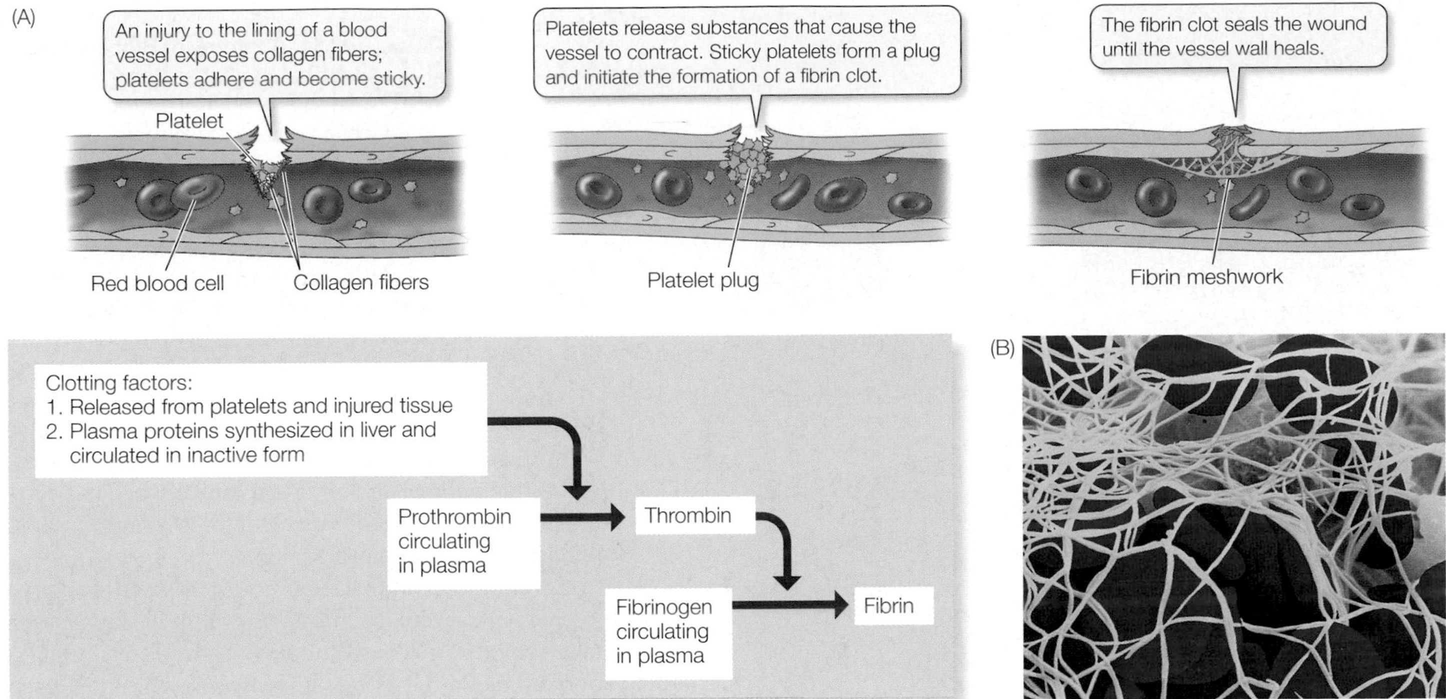

(A)
An injury to the lining of a blood vessel exposes collagen fibers; platelets adhere and become sticky.
Platelet
Red blood cell
Collagen fibers
Platelets release substances that cause the vessel to contract. Sticky platelets form a plug and initiate the formation of a fibrin clot.
Platelet plug
The fibrin clot seals the wound until the vessel wall heals.
Fibrin meshwork
Clotting factors:
1. Released from platelets and injured tissue
2. Plasma proteins synthesized in liver and circulated in inactive form
Prothrombin circulating in plasma
Thrombin
Fibrinogen circulating in plasma
Fibrin
(B)

49.10 Blood Clotting (A) Damage to a blood vessel initiates a cascade of events that produces a fibrin meshwork. (B) As the meshwork forms, red blood cells are enmeshed in the fibrin threads, forming a clot, as shown in this color-enhanced electron micrograph.

Plasma is very similar to interstitial fluid in composition, and most of its components move readily between these two fluid compartments of the body. The main difference between the two fluid compartments is the higher concentration of proteins in the plasma.

Blood circulates throughout the body in a system of blood vessels

As we mentioned in Section 49.1, blood circulates through the vertebrate body in a system of closed vessels. Blood leaves the heart in arteries and is distributed throughout the tissues of the body in arterioles, which feed capillary beds. Exchanges of nutrients, wastes, respiratory gases, and hormones occur in the capillaries. Blood leaving capillary beds collects in venules, which empty into veins that conduct the blood back to the heart.

The walls of the large arteries have many extracellular collagen and elastin fibers, which enable them to withstand the high pressures of blood flowing rapidly from the heart (**Figure 49.11A**). These elastic tissues have another important function: they are stretched during systole, and thereby store some of the energy imparted to the blood by the heart. Elastic recoil during diastole returns this energy to the blood by squeezing it and pushing it forward. As a result, even though pressure in the arteries pulsates with the beating of the heart, the flow of blood is smoother than it would be through a system of rigid pipes.

Smooth muscle cells in the walls of the arteries and arterioles allow those vessels to constrict or dilate. When the diameter of the vessels changes, their resistance to blood flow changes as well, and the amount of blood flowing through them changes as a result. Neuronal and hormonal mechanisms control the resistance of the vessels—and therefore the distribution of blood to the different tissues of the body— by influencing the contraction and relaxation of the smooth muscle in the vessel walls. The arteries and arterioles are referred to as the *resistance vessels* because their resistance can vary.

Materials are exchanged in capillary beds by filtration, osmosis, and diffusion

Beds of capillaries lie between arterioles and venules (**Figure 49.11B**). Few cells of the body are more than a couple of cell diameters away from a capillary (notable exceptions include developing oocytes and the cells of the lens and cornea). The cells' needs are served by the exchange of materials between blood and interstitial fluid across the capillary walls. Capillaries have thin, permeable walls, and blood flows through these vessels very slowly, facilitating this exchange.

To anyone who has played with a garden hose, it may seem strange that blood flows through the large arteries rapidly at high pressures, but when it reaches the small capillaries, the pressure and rate of flow decrease (**Figure 49.11C**). When you restrict the diameter of a garden hose by placing your thumb over the opening, the pressure in the hose increases, which in turn increases the velocity of the water spraying out of the hose. This puzzle is solved by two more pieces of information. First, arterioles are highly branched. When flow through one branch is restricted, blood flows into other branches, so pressure does not build up quickly. Second, each arteriole gives rise to a large number of capillaries. Even though each capillary has a diameter so small that red blood cells must pass through in single file, there are so many capillaries that their total cross-sectional area is much greater than that of any

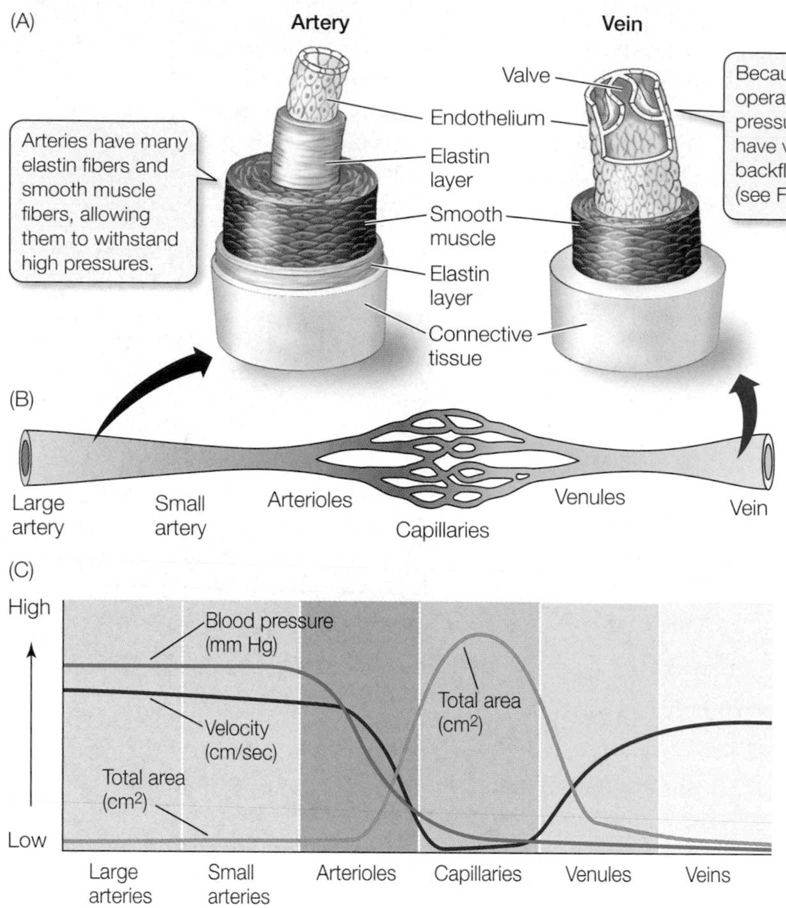

49.11 Anatomy of Blood Vessels (A) The different anatomical characteristics of arteries and veins match their functions. (B) Blood from the arterial system feeds into capillary beds, which exchange nutrients and wastes with the body's cells. The venous system returns the blood to the heart. (C) The area encompassed by each vessel type is graphed along with the pressure and velocity of the blood within them.

other class of vessels. As a result, all of the capillaries together have a much greater capacity for blood than do the arterioles. Returning to our garden hose analogy, if we connected the hose to many junctions leading to small irrigation tubes, the pressure and the flow in each of the irrigation tubes would be quite low.

Capillary walls consist of a single layer of thin endothelial cells (**Figure 49.12**). In most tissues of the body other than the brain, capillaries have tiny holes, or *fenestrations* (Latin *fenestra*, "windows"). Capillaries are permeable to water, some ions, and some small molecules, but not to large molecules such as proteins. Blood pressure therefore squeezes water and some small solutes out of the capillaries and into the surrounding intercellular spaces. Why don't water and small-molecular-weight solutes collect in the intercellular spaces? How is the blood volume maintained if fluid is continuously leaking out of the capillaries?

An answer to this question was put forth more than a hundred years ago by the physiologist E. H. Starling. Starling suggested that water balance in capillary beds is a result of two opposing forces, which have come to be known as **Starling's forces**. One force is blood pressure, which squeezes water and small solutes out of the capillaries, and the other is osmotic pressure created by the large protein molecules that cannot leave the capillaries. Starling called this second force *colloidal osmotic pressure*. He hypothesized that blood pressure is high at the arterial end of a capillary bed and drops steadily as blood flows to the venous end (**Figure 49.13**). The colloidal osmotic pressure, however, is constant along the capillary. As long as the blood pressure is above the osmotic pressure, fluid leaves the capillary, but when blood pressure falls below the osmotic pressure, fluid returns to the capillary. The actual numbers for a normal capillary bed in a resting person suggest that there would be a *slight* net loss of fluid to the intercellular spaces.

Several observations supported Starling's model. In people with severe liver disease or protein starvation, a fall in blood protein concentration leads to an accumulation of fluid in the extracellular spaces, which results in tissue swelling, or **edema**. Edema is also characteristic of the inflammation accompanying tissue damage or allergic responses (see Figure 18.4). *Histamine*, a mediator of inflammation released by certain white blood cells, increases capillary permeability and relaxes the smooth muscles of the arterioles, raising blood pressure in the capillaries, leading to fluid leakage into tissues. Study the trends in Figure 49.13 and you will see that edema should occur in all of these cases.

A few situations are not explained by Starling's hypothesis. During strenuous exercise, the blood pressure in the arterioles serving the muscles rises substantially but does not result in edema. In birds, the blood pressure in arterioles is much higher than in mammals, and the colloidal osmotic pressure is lower. If edema is not a chronic problem in exercising muscles and in birds, what is missing from Starling's model?

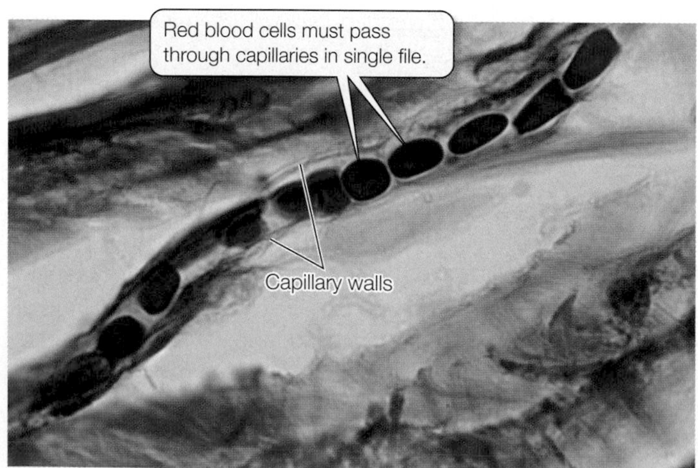

49.12 A Narrow Lane Capillaries have a very small diameter, and blood flows through them slowly.

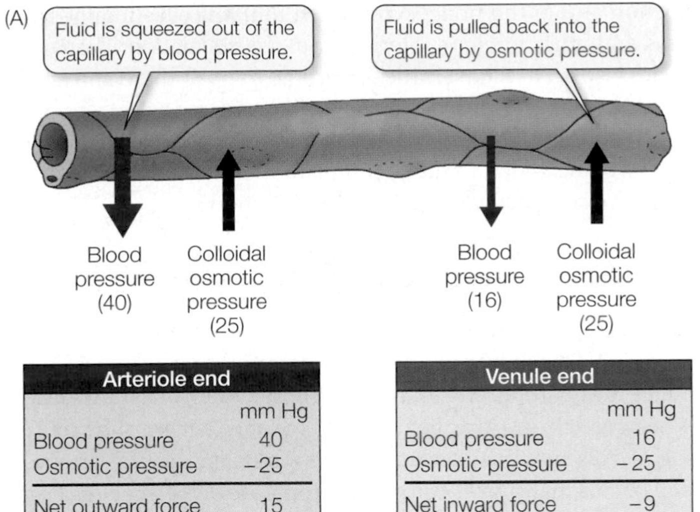

(A)
Fluid is squeezed out of the capillary by blood pressure.
Fluid is pulled back into the capillary by osmotic pressure.
Blood pressure (40)
Colloidal osmotic pressure (25)
Blood pressure (16)
Colloidal osmotic pressure (25)
Arteriole end
mm Hg
Blood pressure 40
Osmotic pressure −25
Net outward force 15
Venule end
mm Hg
Blood pressure 16
Osmotic pressure −25
Net inward force −9

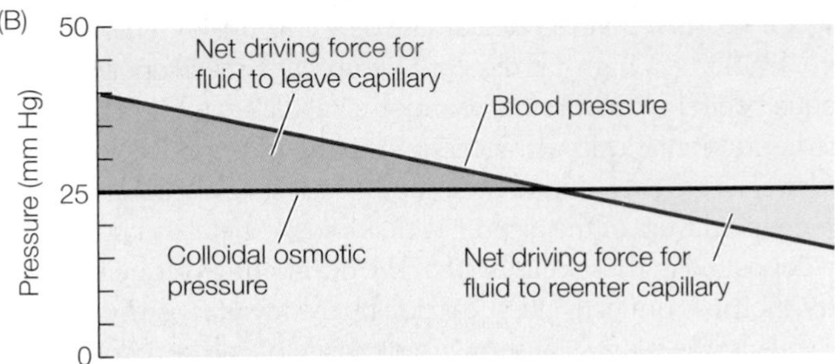

49.13 Starling's Forces Starling's model explains how blood volume is maintained in the capillary beds. (A) When blood pressure is greater than the colloidal osmotic pressure, fluid leaves the capillary; when blood pressure falls below this osmotic pressure, fluid returns to the capillary. (B) The balance of these two forces changes over the capillary bed as blood pressure falls.

Recent research suggests that bicarbonate ions (HCO_3^-) in the blood plasma contribute significantly to the osmotic attraction that draws water back into the capillary. The CO_2 produced by cellular metabolism diffuses into the endothelial cells of the capillaries where it is converted into HCO_3^- and released into the plasma. When the subject is at rest, the increasing HCO_3^- concentration can cause the osmotic pressure of the blood at the venous end to be 30 mm Hg higher than at the arterial end, and during strenuous exercise this difference can be much higher. Thus it appears that CO_2 and HCO_3^- are major factors that pull water back into the capillaries.

Lipid-soluble substances and many small solute molecules can easily pass through capillary walls from an area of higher concentration to one of lower concentration. The capillaries in different tissues, however, are differentially selective as to the sizes of molecules that can pass through them. You can imagine that this is an important issue in the design and delivery of drugs. All capillaries are permeable to O_2, CO_2, glucose, lactate, and small ions such as Na^+ and Cl^-. The capillaries of the brain do not have fenestrations, and therefore not much else can pass through them other than lipid-soluble substances, such as alcohol and anesthetics. This high selectivity of brain capillaries is known as the **blood–brain barrier**. Much less selective capillaries are found in the digestive tract, where nutrients are absorbed, and in the kidneys, where wastes are filtered. Even in the brain there are specific regions where the capillaries are more permeable, enabling the brain to detect non-lipid soluble hormones.

Blood flows back to the heart through veins

The pressure of the blood flowing from capillaries to venules is extremely low, and is insufficient to propel blood back to the heart. The walls of veins are more expandable than the walls of arteries, and blood tends to accumulate in veins. As much as 60 percent of the total blood volume may be in the veins of a resting individual. Because of their high capacity to stretch and store blood, veins are called *capacitance vessels*.

Blood flow through veins that are above the level of the heart is assisted by gravity. Below the level of the heart, however, venous return is against gravity. The most important force propelling blood from these regions is the squeezing of the veins by the contractions of surrounding skeletal muscles. As muscles contract, the vessels are compressed, and blood is squeezed through them. Blood flow may be temporarily obstructed during a prolonged muscle contraction, but when muscles relax, blood is free to move again. One-way valves within the veins of the extremities prevent backflow of blood. Thus, whenever a vein is squeezed, blood is propelled forward toward the heart (**Figure 49.14**).

In a resting person, gravity causes blood accumulation in the veins of the lower body and exerts back pressure on the capillary beds. This back pressure shifts the balance between blood pressure and osmotic pressure, causing increased loss of fluid to the intercellular spaces. That is why your feet swell during a long airline flight.

Varicose veins develop when the veins (especially those of the legs) become so stretched out that the valves can't prevent backflow. People with varicose veins may wear support hose or periodically elevate their legs above heart level to aid venous return.

Because of the one-way valves in the veins of the legs, the contractions of leg muscles act as auxiliary vascular pumps when an animal walks or runs and facilitate the return of blood to the heart from the veins of the lower body. As a greater volume of blood is returned to the heart, the heart contracts more forcefully and its pumping action is enhanced. The heartbeat gets stronger due to a property of cardiac muscle cells described by the **Frank–Starling law**: If the cardiac muscle cells are stretched, as they are when the volume of returning blood increases, they contract more forcefully. The actions of breathing also help return venous blood to the heart. The muscles involved in inhalation create negative pressure that pulls air into the lungs (see Figure 48.11), and this negative pressure also pulls blood toward the chest, increasing venous return to the right atrium. In addition, some of the largest veins closest to the heart contain smooth muscle that contracts at the onset of exercise. Contraction of veins can rapidly increase venous return and

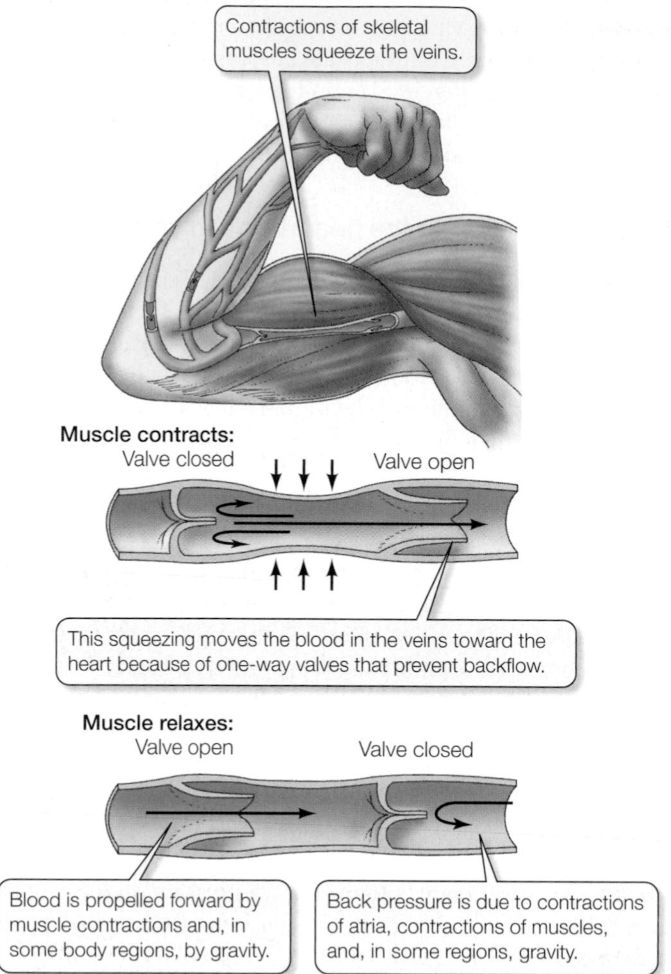

Contractions of skeletal muscles squeeze the veins.
Muscle contracts:
Valve closed
Valve open
This squeezing moves the blood in the veins toward the heart because of one-way valves that prevent backflow.
Muscle relaxes:
Valve open
Valve closed
Blood is propelled forward by muscle contractions and, in some body regions, by gravity.
Back pressure is due to contractions of atria, contractions of muscles, and, in some regions, gravity.

49.14 One-Way Flow Veins have valves that prevent blood from flowing backward, and contractions of skeletal muscle help to move blood toward the heart.

stimulate the heart in accord with the Frank–Starling law, increasing cardiac output.

Lymphatic vessels return interstitial fluid to the blood

The interstitial fluid contains water and small molecules, but no red blood cells, and less protein than found in blood. A separate system of vessels—the **lymphatic system**—returns interstitial fluid to the blood.

Once it enters the lymphatic vessels, the interstitial fluid is called **lymph**. Fine lymphatic capillaries merge into progressively larger vessels and ultimately into two lymphatic vessels—the **thoracic ducts**—that empty into large veins at the base of the neck (see Figure 18.1). The left thoracic duct carries most of the lymph from the lower part of the body and is much larger than the right thoracic duct. Lymph, like blood, is propelled toward the heart by skeletal muscle contractions and breathing movements, and lymphatic vessels, like veins, have one-way valves that keep the lymph flowing toward the thoracic duct.

Mammals and birds have **lymph nodes** along the major lymphatic vessels. Lymph nodes are a major site of lymphocyte production and of the phagocytic action that removes microorganisms and other foreign materials from the circulation. The lymph nodes also act as filters. Particles become trapped there and are digested by phagocytes in the nodes. When you get an infection, the lymph nodes closest to the infection may become swollen and sore. The swelling is due to the accumulation of immune system cells that have been activated to fight the infection.

Vascular disease is a killer

As mentioned at the start of this chapter, cardiovascular disease is responsible for about one-third of all deaths each year in the United States and Europe. The immediate cause of most of these deaths is heart attack or stroke, but those events are frequently the end result of a disease called **atherosclerosis** ("hardening of the arteries") that begins many years before symptoms are detected. Hence atherosclerosis is one of the "silent killer" diseases.

Healthy arteries have a smooth internal lining of endothelial cells (**Figure 49.15A**) that can be damaged by chronic high blood pressure, smoking, a high-fat diet, or microorganisms. Deposits called **plaque** begin to form at sites of endothelial damage. First, the damaged endothelial cells attract certain white blood cells to the site. These cells are then joined by smooth muscle cells migrating from the deeper layers of the arterial wall. Lipids, especially cholesterol, are deposited in these cells, so that the developing plaque becomes fatty. Fibrous connective tissue made by the invading smooth muscle cells in the plaque, along with deposits of calcium, makes the artery wall less elastic—hence, "hardening of the arteries." The growing plaque deposit narrows the artery and causes turbulence in the blood flow. Blood platelets stick to the plaque (see Figure 49.10) and initiate the formation of an intravascular blood clot, a **thrombus**, which can quickly block the artery (**Figure 49.15B**).

The blood supply to the heart muscle flows through the **coronary arteries** which are highly susceptible to atherosclerosis. As these arteries narrow, blood flow to the heart muscle decreases. Chest pain and shortness of breath during mild exertion are symptoms of this condition. A person with atherosclerosis is at high risk of forming a thrombus in a coronary artery. This condition, called **coronary thrombosis**, can totally block the vessel, causing a *heart attack*, or **myocardial infarction**.

A piece of a thrombus that breaks loose, called an **embolus**, is likely to travel to and become lodged in a vessel of smaller diameter, blocking its flow (an **embolism**). Arteries already narrowed by plaque formation are likely places for an embolism. An embolism in an artery in the brain causes the cells fed by that artery to die. This event is a **stroke**. The specific damage resulting from a stroke, such as memory loss, speech impairment, or paralysis, depends on the location of the blocked artery.

Probably the most important determinants of whether you will get atherosclerosis are your genetic predisposition and your age. Environmental risk factors play a large role, however. If you do have a genetic predisposition to atherosclerosis, it is even more important to minimize environmental risk factors. These factors include high-fat and high-cholesterol diets, smoking, and a sedentary lifestyle. Certain untreated medical conditions such as *hypertension* (high blood pressure), obesity, and diabetes are also risk factors for atherosclerosis. Changes in diet and behavior and treatment of

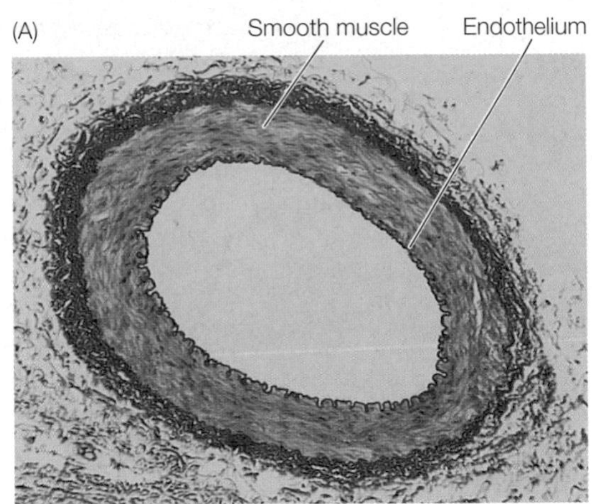

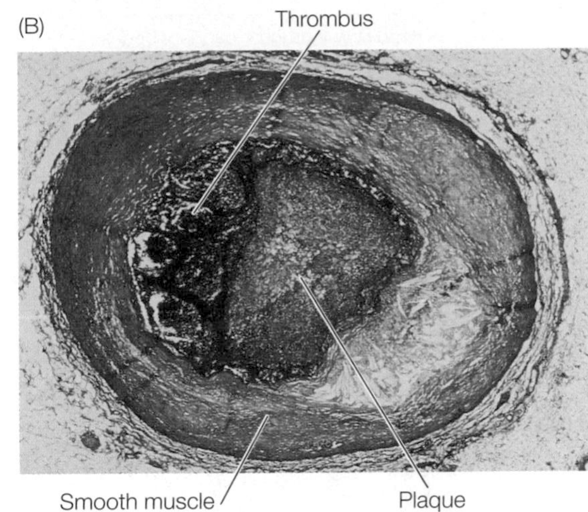

49.15 Atherosclerotic Plaque (A) A healthy, clear artery. (B) An atherosclerotic artery, clogged with plaque and a thrombus.

predisposing medical conditions can prevent and reverse early atherosclerosis and help fend off this silent killer.

49.4 RECAP

Blood is a fluid tissue with cellular components that play roles in transport of respiratory gases, immune system function, and blood clotting. Blood is distributed throughout the body in a system of vessels, and exchanges between the blood and the interstitial fluids occur in the smallest of those vessels, the capillaries.

- Why are arteries called resistance vessels and veins called capacitance vessels? See pp. 1057 and 1059
- What factors control the movement of fluids between the vascular and extravascular spaces? See pp. 1057–1059 and Figure 49.13
- What propels blood from the lower part of the body back to the heart? See pp. 1059–1060 and Figure 49.14

Every tissue in the body requires an adequate supply of oxygen-saturated blood that carries essential nutrients and is relatively free of waste products. But the nervous system cannot monitor and control every capillary bed in the body. We will now discuss how is blood flow is regulated.

49.5 How Is the Circulatory System Controlled and Regulated?

The circulatory system is controlled and regulated by neuronal and hormonal mechanisms at both the local and systemic levels that ensure each tissue is appropriately nourished and maintained. Regulation begins at the local level with each tissue controlling its own blood flow through **autoregulatory mechanisms** that cause the arterioles supplying the tissue to constrict or dilate.

The collective autoregulatory actions of every capillary bed in every tissue in the body influence the pressure and composition of the arterial blood. If many arterioles suddenly dilate, for example, allowing blood to flow through many more capillary beds, arterial blood pressure falls. If all these newly filled capillary beds contribute CO_2 to the blood at one time, the concentration of CO_2 in the blood returning to the heart increases. The nervous and endocrine systems respond to such changes by changing breathing rate, heart rate, and blood distribution to match the metabolic needs of the body.

Autoregulation matches local blood flow to local need

The autoregulatory mechanisms that adjust the flow of blood to a tissue are local but can be influenced by the nervous system and by certain hormones.

The amount of blood that flows through a capillary bed is controlled by the smooth muscle of the arteries and arterioles feeding that bed. The flow of blood in a typical capillary bed is diagrammed in **Figure 49.16**. Blood flows into the bed from an arteriole. Smooth muscle "cuffs," or **precapillary sphincters**, on the arteriole can shut off the supply of blood to the capillary bed. When the precapillary sphincters are relaxed and the arteriole is open, the arterial blood pressure pushes blood into the capillaries.

Autoregulation depends on the sensitivity of the smooth muscle to its local chemical environment. Low O_2 concentrations and high CO_2 concentrations cause the smooth muscle to relax, thus increasing the supply of blood, which brings in more O_2 and carries away CO_2—a response known as **hyperemia**, which means "excess blood." Increases in other by-products of metabolism, such as lactic acid, hydrogen ions, potassium, and adenosine (all of which increase in exercising muscle), promote hyperemia through the same mechanism. Hence, activities that increase the metabolism of a tissue also induce hyperemia in that tissue.

Arterial pressure is controlled and regulated by hormonal and neuronal mechanisms

Control and regulation of the cardiovascular system begins with the local autoregulatory mechanisms we have just described. As more blood flows into the tissues, the central blood pressure falls, and the composition of the blood returning to the heart reflects the exchanges that are occurring in the tissues. Changes in central blood

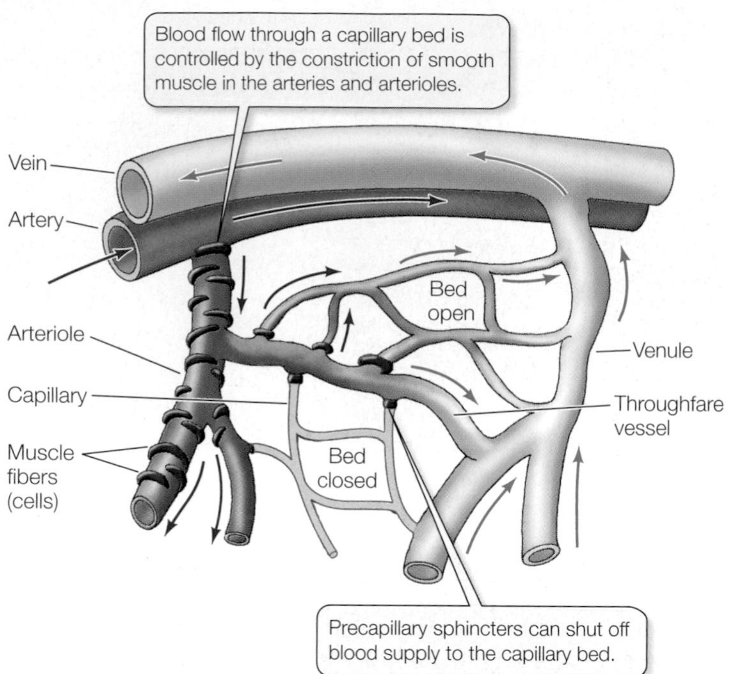

Blood flow through a capillary bed is controlled by the constriction of smooth muscle in the arteries and arterioles.
Vein
Artery
Arteriole
Capillary
Muscle fibers (cells)
Bed open
Bed closed
Venule
Throughfare vessel
Precapillary sphincters can shut off blood supply to the capillary bed.

49.16 Local Control of Blood Flow Low O_2 concentrations or high levels of metabolic by-products cause the smooth muscle of the arteries and arterioles to relax, thus increasing the supply of blood to the capillary bed.

pressure and blood composition are sensed, and both endocrine and central nervous system responses are activated to return blood pressure and composition to normal. Thus circulatory functions are matched to the regional and overall needs of the body.

Arteries and arterioles are innervated by the autonomic nervous system, particularly the sympathetic division. The sympathetic postganglionic neurotransmitter norepinephrine causes the smooth muscle cells of arterioles to contract, thus constricting the vessels and reducing blood flow. An exception is found in skeletal muscle, in which specialized sympathetic neurons release acetylcholine, causing the smooth muscle of the arterioles to relax and the vessels to dilate, increasing blood to flow to the muscle.

Hormones also play a role in the regulation of arterial pressure. *Epinephrine*, which has actions similar to those of norepinephrine, is released from the adrenal medulla during massive sympathetic activation stimulated by a fall in arterial pressure or by the activation of the fight-or-flight response to a dangerous threat. *Angiotensin* is produced when blood pressure to the kidneys falls (**Figure 49.17**). These hormones influence arterioles located for the most part in peripheral tissues (extremities) or in tissues whose functions need not be maintained continuously (such as the gut). By reducing blood flow in those arterioles, these hormones increase central blood pressure and blood flow to essential organs such as the heart, brain, and kidneys.

The autonomic nervous system activity that controls heart rate and constriction of blood vessels originates in a cardiovascular control center in the medulla. Many inputs converge on this central integrative network and influence the commands it issues via parasympathetic and sympathetic nerves (**Figure 49.18**). Of special importance is incoming information about changes in blood pressure and composition from stretch receptors and **chemoreceptors** in the walls of the large arteries leading to the brain—the aorta and the carotid arteries. Stretch receptors are called **baroreceptors** because they are sensitive to the blood pressure in the aorta and carotid arteries.

Increased activity in the stretch receptors of the large arteries signals rising blood pressure and inhibits sympathetic nervous system signaling to arteries and arterioles while increasing parasympathetic signaling to the heart's pacemaker. As a result, the heart slows and arterioles in peripheral tissues dilate. If pressure in the large arteries falls, the activity of the stretch receptors decreases, stimulating sympathetic output to the arteries and arterioles while reducing parasympathetic output to the heart's pacemaker. As a result, the heart beats faster, and the arterioles in peripheral tissues constrict. Another hormone that helps to stabilize blood pressure is antidiuretic hormone (ADH, also called vasopressin), which is secreted by the posterior pituitary in response to a fall in the activity of the baroreceptors, signaling a fall in arterial pressure. ADH causes the kidneys to resorb more water and thereby maintain blood volume. Increased activity of the

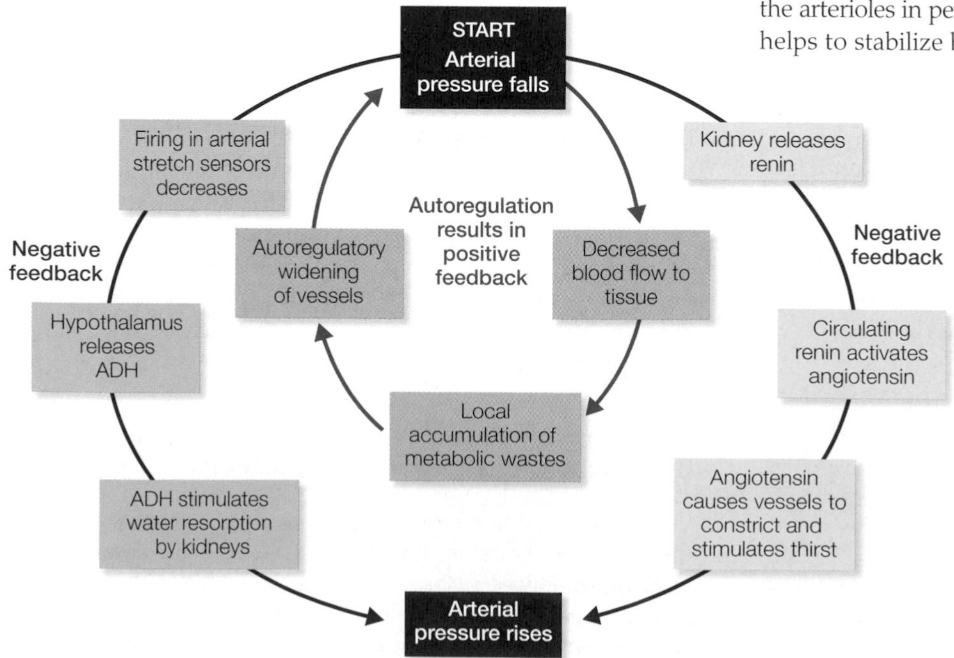

START
Arterial pressure falls
Firing in arterial stretch sensors decreases
Kidney releases renin
Autoregulation results in positive feedback
Negative feedback
Autoregulatory widening of vessels
Decreased blood flow to tissue
Negative feedback
Hypothalamus releases ADH
Circulating renin activates angiotensin
Local accumulation of metabolic wastes
ADH stimulates water resorption by kidneys
Angiotensin causes vessels to constrict and stimulates thirst
Arterial pressure rises

49.17 Control of Blood Pressure through Vascular Resistance A drop in arterial pressure reduces blood flow to tissues, resulting in local accumulation of metabolic wastes. This change in the extracellular environment stimulates autoregulatory opening of the arteries and would lead to a further decrease in central blood pressure if this were not prevented by negative feedback mechanisms, which work by promoting the constriction of arteries in less essential tissues and by maintaining blood volume.

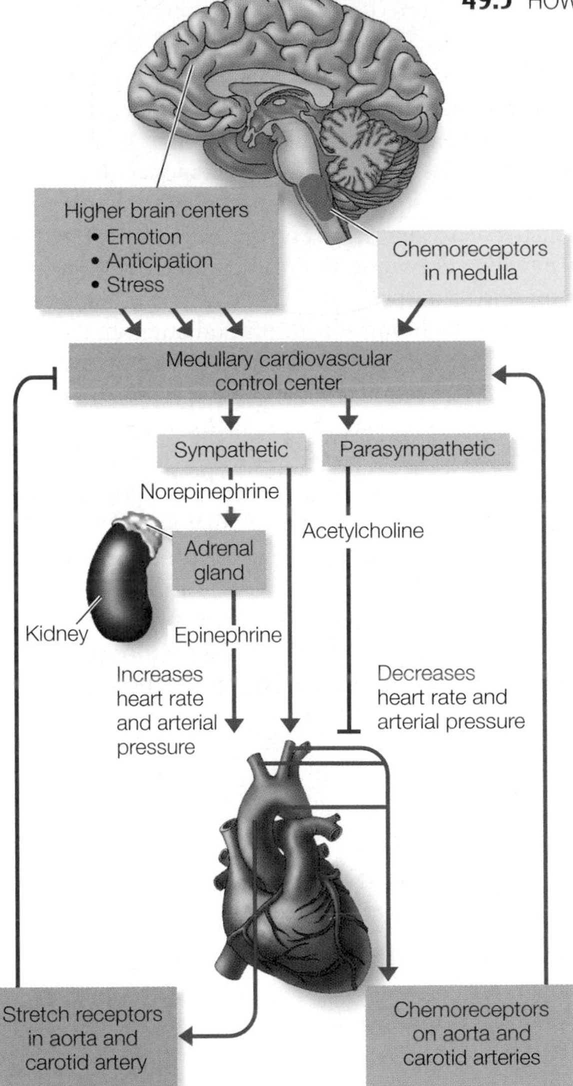

49.18 Regulating Blood Pressure The autonomic nervous system controls heart rate in response to information about blood pressure and blood composition that is integrated by regulatory centers in the medulla.

baroreceptors inhibits the release of ADH, and as a result the kidney excretes more water reducing blood volume and contributing to a fall in arterial pressure (see Figure 49.17).

Other information that causes the cardiovascular control center to increase heart rate and blood pressure comes from chemoreceptors in the medulla, aorta, and the carotid arteries. The medullary chemosensors are activated by increases in arterial CO_2 levels, and the carotid and aortic bodies are activated by falls in arterial O_2 levels. Chemosensors send signals to the regulatory center.

Cardiovascular control in diving mammals conserves oxygen

This chapter and Chapter 48 explained how animals, especially highly active birds and mammals, obtain needed oxygen from the environment and deliver it to their tissues. What happens when environmental oxygen is not available—as when an air-breathing mammal is under water?

EXPERIMENT

HYPOTHESIS: Northern elephant seals spend extended periods of time underwater. A slowed heart rate is part of the physiological reflex that makes extended dives possible.

Mirounga angustirostris

METHOD

While seals are on land for the breeding season, attach electronic tracking and recording devices to individual animals to obtain time and location data, along with readings of heart rate, water pressure (indicates depth of dive), and breathing pattern.

RESULTS

Heart rate (beats/min): 0, 25, 50, 75, 100, 125
Dive depth (m): 0, 200, 400, 600
Time of day: 22:00, 23:00, 24:00, 01:00, 02:00

During many days at sea, seals spent more than 80 percent of their time under water, making repeated dives to depths of 300–600 meters. Each dive lasted an average 20 minutes; surface time averaged less than 3 minutes per dive. Heart rate averaged 110 beats per minute at the surface but only 30 bpm while diving, with heart rate dropping as low as 3–4 bpm during parts of the dive.

CONCLUSION: Northern elephant seals spend more than 80 percent of their time at sea underwater. Reduced heart rate is part of the diving reflex that makes this possible.

49.19 Elephant Seal Diving Ability Northern elephant seals spend months at sea. New field recording technologies reveal the remarkable diving activities of these free-ranging marine mammals.

The diving capabilities of marine mammals are quite remarkable. Several adaptations enable diving seals to spend most of their time underwater, either by increasing the seal's oxygen stores or by limiting its use of oxygen.

Seals have greater oxygen stores in their bodies relative to other mammals because of a greater blood volume, a greater concentration of hemoglobin in that blood, and a greater concentration of myoglobin in their muscles. Yet the body stores of oxygen a seal takes into a dive are only about twice what a human carries into a dive.

The adaptations most critical for sustaining a seal's dive are cardiovascular mechanisms that limit blood flow to non-critical organs. This suite of adaptations is known as the **diving reflex**, which slows the heart (**Figure 49.19**) and constricts the major blood vessels going to all tissues except the critical tissues of the nervous system, heart, and eyes. The seal's central blood pressure remains high, but blood flow to its tissues decreases. This reduced blood flow has two effects: tissues switch to glycolytic (anaerobic) metabolism, and their metabolism drops considerably. While diving, the seal accumulates lactic acid in its muscles, which constitutes an "oxygen debt" to be paid back through elevated metabolism after the dive ends. But the total metabolic "debt" is much less than the metabolism that would have occurred over the same period of time had the seal not dived. The diving reflex causes the seal to be **hypometabolic** (to have a metabolic rate below its basal rate) during the dive. Hypometabolism, increased oxygen stores, and a high capacity for anaerobic metabolism make it possible for the seal to perform its amazing diving feats.

The seal's diving reflex may seem like a unique adaptation, but it provides yet another example of how natural selection shapes biological traits that are widely shared among related species. Humans also have a diving reflex. It is controlled by the vagus nerve and the parasympathetic nervous system. When our faces are submerged, we experience a mild slowing of our heart rate. This reflex probably serves as a protective response during the birth process, when pressure on the umbilical cord can deprive the fetus of maternal oxygen before breathing can begin. There are many cases, in which drowning victims have been submerged in cold water for rather long periods of time, yet have survived with no brain damage. The human diving reflex, along with the rapid cooling of the brain as body heat is lost to the water, is the probable explanation for these remarkable cases of survival.

49.5 RECAP

The delivery of blood to tissues is controlled locally by autoregulatory mechanisms that cause dilation or constriction of arterioles. These local actions are translated into alterations in central blood pressure and composition that are detected by neuronal and hormonal mechanisms, which then mediate corrective cardiovascular adjustments.

- What are the roles of hormones in the regulation of blood pressure? See pp. 1061–1062 and Figure 49.17
- Can you describe the role of stretch receptors and chemoreceptors in regulating blood pressure? See pp. 1062–1063 and Figure 49.18
- What are the major adaptations that enable marine mammals to sustain underwater activity for long periods without breathing? See pp. 1063–1064 and Figure 49.19

CHAPTER SUMMARY

49.1 Why do animals need a circulatory system?

The metabolic needs of the cells of many small animals are met by direct exchange of materials with the external medium. The metabolic needs of the cells of larger animals are met by a circulatory system that transports nutrients, respiratory gases, and metabolic wastes throughout the body.

In **open circulatory systems**, extracellular fluid leaves vessels and percolates through tissues. In **closed circulatory systems**, the blood is contained in a system of vessels. Closed circulatory systems have several advantages, including the ability to selectively direct blood, hormones, and nutrients to specific tissues. Review Figure 49.1

49.2 How has the vertebrate circulatory systems evolved?

See Web/CD Activity 49.1

The circulatory systems of vertebrates consist of a **heart** and a closed system of **vessels** containing blood that is separate from the interstitial fluid. **Arteries** and **arterioles** carry blood from the heart; **capillaries** are the site of exchange between blood and interstitial fluid; **venules** and **veins** carry blood back to the heart.

The vertebrate heart evolved from two chambers in fish to three in amphibians and reptiles and four in crocodilians, mammals, and birds. This evolutionary progression has led to an increasing separation of blood that flows to the gas exchange organs and blood that flows to the rest of the body. Review pages 1047–1049.

The simplest (two-chambered heart) has an **atrium** that receives blood from the body, and a **ventricle** that pumps blood out of the heart. An **aorta** distributes blood to arteries.

In birds and mammals, blood circulates through two circuits: the **pulmonary** which transports blood between the heart and lungs, and the **systemic**, which transports oxygen-rich blood between the heart and tissues.

49.3 How does the mammalian heart function?

The human heart has four chambers. **Valves** in the heart prevent the backflow of blood. Review Figure 49.2, Web/CD Activity 49.2

The **cardiac cycle** has two phases: **systole**, in which the ventricles contract; and **diastole**, in which the ventricles relax. The sequential heart sounds ("lub-dup") are made by the closing of the heart valves. Review Figure 49.3, Web/CD Tutorial 49.1

CHAPTER SUMMARY

Blood pressure can be measured using a sphygmomanometer and a stethoscope. Review Figure 49.4

The autonomic nervous system controls heart rate: Sympathetic activity increases heart rate, and parasympathetic activity decreases it. Norepinephrine increases and acetylcholine decreases the rate of depolarization of the plasma membranes of **pacemaker cells**, affecting heart rate. Review Figure 49.5

The **sinoatrial node** controls the cardiac cycle by initiating a wave of depolarization in the atria, which is conducted to the ventricles through a system consisting of the **atrioventricular node**, the **bundle of His**, and the **Purkinje fibers**. Review Figure 49.6

The sustained contraction of ventricular muscle cells is due to long duration action potentials that are generated by voltage-gated Na^+ and Ca^{2+} channels. Review Figure 49.7

An **electrocardiogram** (**ECG**) records electrical events associated with the contraction and relaxation of the cardiac muscles. Review Figure 49.8

49.4 What are the properties of blood and blood vessels?

Blood can be divided into a **plasma** portion (water, salts, and proteins) and a cellular portion (**erythrocytes** or red blood cells, **platelets**, and **white blood cells**). All of the cellular components are produced from stem cells in the bone marrow. Review Figure 49.9

Erythrocytes transport oxygen. Their production in the bone marrow is stimulated by **erythropoietin**, which is produced in response to **hypoxia** (low oxygen levels) in the tissues.

Platelets, along with circulating proteins, are involved in **blood clotting**, which results in a meshwork of **fibrin threads** that help seal vessels. Review Figure 49.10

Abundant smooth muscle cells allow vessels to change their diameter, altering their resistance and thus blood flow. Arteries and arterioles have many elastic fibers that enable them to withstand high pressures. Review Figure 49.11, Web/CD Activity 49.3

Capillary beds are the site of exchange of materials between blood and tissue fluid.

Starling's forces suggest that blood volume is maintained in the capillary beds by an exchange of fluids driven by both blood pressure and osmotic pressure. Review Figure 49.13

An accumulation of fluid in the extracellular spaces leads to **edema**. Bicarbonate ions in the blood plasma contribute to the osmotic forces that draw water back into capillaries.

The ability of a specific molecule to cross a capillary wall depends on the architecture of the capillary, the type of substance, and the concentration gradient between the blood and the tissue fluid.

Veins have a high capacity for storing blood. Aided by gravity, by contractions of skeletal muscle, and by the actions of breathing, they return blood to the heart. Review Figure 49.14

The **Frank–Starling law** describes forces that increase cardiac output, such as stretch of the cardiac muscles cells due to increased venous return.

The lymphatic system returns the interstitial fluid to the blood.

49.5 How is the circulatory system controlled and regulated?

Blood flow through capillary beds is controlled by local autoregulatory mechanisms, hormones, and the autonomic nervous system. Review Figure 49.16

Blood pressure is controlled in part by the hormones vasopressin and angiotensin, which stimulate contraction of blood vessels. Review Figure 49.17

Heart rate is controlled by the autonomic nervous system, which responds to information about blood pressure and blood composition that is integrated by regulatory centers in the medulla. Review Figure 49.18

Diving mammals conserve blood oxygen stores by slowing the heart rate during dives. Review Figure 49.19

SELF-QUIZ

1. An open circulatory system is characterized by
   a. the absence of a heart.
   b. the absence of blood vessels.
   c. blood with a composition different from that of tissue fluid.
   d. the absence of capillaries.
   e. a higher-pressure circuit through gills than to other organs.
2. Which statement about vertebrate circulatory systems is *not* true?
   a. In fish, oxygenated blood from the gills returns to the heart through the left atrium.
   b. In mammals, deoxygenated blood leaves the heart through the pulmonary artery.
   c. In amphibians, deoxygenated blood enters the heart through the right atrium.
   d. In reptiles, the blood in the pulmonary artery has a lower oxygen content than the blood in the aorta.
   e. In birds, the pressure in the aorta is higher than the pressure in the pulmonary artery.
3. Which statement about the human heart is true?
   a. The walls of the right ventricle are thicker than the walls of the left ventricle.
   b. Blood flowing through atrioventricular valves is always deoxygenated blood.
   c. The second heart sound is due to the closing of the aortic valve.
   d. Blood returns to the heart from the lungs in the vena cava.
   e. During systole, the aortic valve is open and the pulmonary valve is closed.
4. The pacemaker action potentials in the heart
   a. are due to opposing actions of norepinephrine and acetylcholine.
   b. are generated by the bundle of His.
   c. depend on the gap junctions between the cells that make up the atria and those that make up the ventricles.
   d. are due to spontaneous depolarization of the plasma membranes of modified cardiac muscle cells.
   e. result from hyperpolarization of cells in the sinoatrial node.
5. Blood velocity through capillaries is slow because
   a. much blood volume is lost from the capillaries.
   b. the pressure in venules is high.
   c. the total cross-sectional area of capillaries is larger than that of arterioles.
   d. the osmotic pressure in capillaries is very high.
   e. erythrocytes must pass through in single file.

6. How are lymphatic vessels like veins?
   a. Both have nodes where they join together into larger common vessels.
   b. Both carry blood under low pressure.
   c. Both are capacitance vessels.
   d. Both have valves.
   e. Both carry fluids rich in plasma proteins.
7. The production of erythrocytes
   a. ceases if the hematocrit falls below normal.
   b. is stimulated by erythropoietin.
   c. is about equal to the production of white blood cells.
   d. is inhibited by prothrombin.
   e. occurs in bone marrow before birth and in lymph nodes after birth.
8. Which of the following does *not* increase blood flow through a capillary bed?
   a. High concentration of CO_2
   b. High concentration of lactate and hydrogen ions
   c. Histamine
   d. ADH
   e. Increase in arterial pressure
9. Blood clotting
   a. is impaired in patients with hemophilia because they do not produce platelets.
   b. is initiated when platelets release fibrinogen.
   c. involves a cascade of factors produced in the liver.
   d. is initiated by leukocytes forming a meshwork.
   e. requires production of angiotensin.
10. Autoregulation of blood flow to a tissue is due to
   a. sympathetic innervation.
   b. the release of ADH by the hypothalamus.
   c. increased activity of stretch receptors.
   d. chemoreceptors in the aorta and the carotid arteries.
   e. the effect of the local chemical environment on arterioles.

FOR DISCUSSION

1. At the beginning of a race, cardiac output increases immediately before there is any change in blood O_2 or CO_2 concentrations. Explain two factors that contribute to this effect. Include the Frank–Starling law in your answer.
2. Explain how the hearts of crocodilians have the advantages of mammalian hearts during exercise but the efficiency of reptilian hearts during rest.
3. A sudden and massive loss of blood results in a decrease in blood pressure. Describe several mechanisms that help return blood pressure to normal.
4. You can describe the cycle of events in a ventricle of the heart by a graph that plots the pressure in the ventricle on the *y* axis and the volume of blood in the ventricle on the *x* axis. What would such a graph look like? Where would the heart sounds be on this graph? How would the graph differ for the left and the right ventricles?
5. If the major arteries become clogged with plaque and become less elastic because of atherosclerosis, the left ventricle must work harder and harder to pump an adequate supply of blood to the body. As a result, the left ventricle can become weakened and begin to fail, even though the right ventricle is healthy. A heart attack primarily affecting the left ventricle can have the same effect. This condition is known as congestive heart failure, and it commonly leads to fatal pulmonary edema. Explain how left ventricular failure can result in pulmonary edema, and why is it said that this condition creates a vicious circle that makes itself worse rapidly.

FOR INVESTIGATION

1. There are strains of rats that are spontaneously hypertensive—they naturally have high blood pressure. Can you formulate a hypothesis about the basis for such inherited hypertension and explain experiments you would do to test your hypothesis?
2. If you were in training to compete in the Tour de France, a grueling 3-week-long bicycle race, a trainer might tell you to "train high, compete low." What do you think is the physiological basis for such advice, and how would you test whether or not that reasoning was correct?

CHAPTER 50 Nutrition, Digestion, and Absorption

An obesity epidemic

For thousands of years, the Pima of southwestern North America were hunters and gatherers who supplemented their diet with subsistence agriculture. Their environment was arid, so they developed sophisticated irrigation systems; even so, they frequently encountered drought and subsequent starvation. Today most individuals of the ethnic Pima population in North America are clinically obese. In fact, as a population, they are one of the heaviest in the world. With obesity come related health problems such as diabetes, high blood pressure, and heart disease. The incidence of diabetes among these native American people rose from an extremely high level of 45 percent of adults in 1965 to a staggering 80 percent in 1999. Moreover, diabetes is occurring in younger individuals than ever before. What could have caused such a radical health change in an entire population? At least two interacting factors are involved: genetics and lifestyle.

Geneticists hypothesize that recurring episodes of starvation produce strong selective pressure for "thrifty genes"—particular alleles of the genes involved in digestion, absorption, and energy storage that result in greater-than-average efficiency in converting food into energy and into energy reserves, such as fat. Thrifty genes would give individuals a strong selective advantage when food is scarce. An example of a "thrifty" phenotype is seen among the Pima. They have a very low resting metabolic rate and convert food into fat readily. As we will see later in this chapter, the hormone insulin facilitates the conversion of dietary sugar into fat tissue. For many Pima, consuming a standard amount of glucose causes their insulin levels to rise three times higher than it does in Americans of European ancestry.

The other factor in the Pima obesity epidemic is an abrupt change in their traditional lifestyle. When food is plentiful and has high caloric content, thrifty genes contribute to obesity by maximizing fat storage. Today the Pima eat a modern Western diet that includes high-fat, high-calorie fast foods. In general, they also engage in less physical activity than their ancestors did.

A comparative study supports the hypotheses that have been put forward to explain the obesity epidemic in the Pima. Another population of Pima live in the Sierra Madre mountains of northern Mexico.

Efficiency Genes The Pima are an example of a population that has probably undergone selection in the past for genes that improve the efficiency of managing the energy obtained from food. With modern diets and lifestyles, these "efficiency genes" can contribute to obesity.

The Great American Lunch High-fat fast foods have become prevalent in much of the developed world. A steady diet of such foods will mean weight gain for most people, and is a major reason for the "obesity epidemic" in the United States.

Genetically they are the same as the Arizona population. However, they live a traditional lifestyle and eat traditional foods. Whereas the Arizona Pima engage in an average of only 2 hours of physical work per week, the Mexico Pima average 23 hours per week. Obesity and diabetes are not prevalent among the Mexican Pima.

High-calorie diet and sedentary lifestyle affect not just the Pima, but contribute to the overall increase in obesity throughout the U.S. population. Researchers are studying the Pima to learn more about the genetics of obesity and related diseases.

IN THIS CHAPTER we will review the nutrients required by organisms for energy, for molecular building blocks, and for specific biochemical functions. We examine diverse adaptations for acquiring, ingesting, and digesting food and absorbing nutrients. Then we will learn how the body regulates its traffic in metabolic fuels, and we return to the question of control of body mass. Lastly, we will explore the issue of natural and artificial toxins in food.

CHAPTER OUTLINE

50.1 What Do Animals Require from Food?

Animals are **heterotrophs**—they derive their nutrition from eating other organisms. In contrast, **autotrophs** (most plants, some bacteria, some archaea, and some protists) can use solar energy or inorganic chemical energy to synthesize all of their components. Directly and indirectly, heterotrophs take advantage of—indeed, depend on—the organic synthesis carried out by autotrophs, and have evolved an enormous diversity of adaptations to exploit this resource (**Figure 50.1**). In this section we will discover how animals use food, be it plants or other animals, to obtain energy and building blocks of complex molecules.

Energy can be measured in calories

Chapters 6 and 7 described how energy in the chemical bonds of food molecules is transferred to the high-energy phosphate bonds of adenosine triphosphate (ATP). ATP provides animals with energy for cellular work; however, each conversion of energy from food molecules to ATP and from ATP to cellular work is inefficient. In fact, most of the energy that was in the food is lost as heat. Even the energy the animal uses is eventually reduced to heat, as molecules that were synthesized are broken down and the energy of movement is dissipated by friction. Therefore, we can talk about the energy requirements of animals and the energy content of food in terms of a measure of heat energy: the calorie.

A **calorie** is the amount of heat necessary to raise the temperature of 1 gram of water 1°C. Since this is a tiny amount of energy compared with the energy requirements of many animals, physiologists commonly use the **kilocalorie** (**kcal**) as a unit of measure (1 kcal = 1,000 calories). Nutritionists also use the kilocalorie as a standard unit of energy, but they traditionally refer to it as the **Calorie** (**Cal**), which is always capitalized to distinguish it from the single calorie. (Scientists are abandoning the calorie as an energy unit as they switch to the International System of Units. In this system, the basic unit of energy is the joule: 1 joule = 0.239 calories.)

The *metabolic rate* of an animal (see Section 40.4) is a measure of the overall energy needs that must be met by the animal's ingestion and digestion of food. The basal metabolic rate

(A) *Ailuropoda melanoleuca*

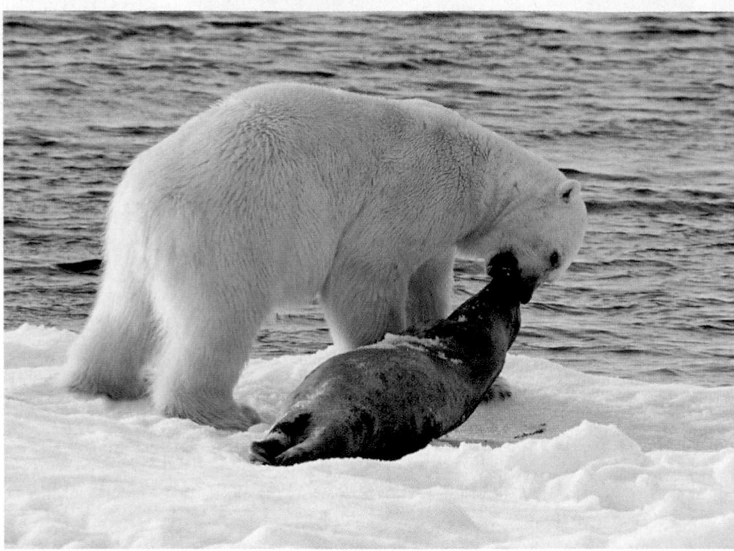

(B) *Ursus maritimus*

50.1 Heterotrophs Get Energy from Autotrophs (A) Herbivores get their energy directly from autotrophs. Large herbivores such as the giant panda must consume huge amounts of plant matter (this particular animal eats only bamboo plants) to fulfill their nutritional needs. (B) Polar bears are carnivores and ferocious predators. A carnivore's energy is indirectly obtained from autotrophs, since the energy stored in a prey animal was originally obtained from autotrophs.

of a human is about 1,300–1,500 kcal/day for an adult female and 1,600–1,800 kcal/day for an adult male. Physical activity adds to this basal energy requirement. For a person doing sedentary work, about 30 percent of total energy consumption is due to skeletal muscle activity, and for a person doing heavy physical labor, over 95 percent of total caloric expenditure is due to skeletal muscle activity. The components of food that provide energy are fats, carbohydrates, and proteins. Fats yield 9.5 kcal/gram, carbohydrates 4.2 kcal/gram, and proteins about 4.1 kcal/gram. Some equivalencies of food, energy, and energy consumption are shown in **Figure 50.2**.

Energy budgets reveal how animals use their resources

It is possible to quantify the caloric value of any food an animal eats. It is also possible to quantify the caloric cost of anything an animal does. By comparing calories consumed with calories expended, we can construct **energy budgets** that allow ecologists and evolutionists to apply a cost–benefit analysis to behavior.

Consider, for example, territorial aggression: Under what circumstances does it "pay" to fight over a food resource? Such a study was led by Larry Wolf of Syracuse University on African sunbirds, which feed on the nectar of a particular flower. These birds defend feeding territories in some habitats but not others. The in-

50.2 Food Energy and How We Use It The energy contained in several common food items is shown at the left. The graphs indicate about how long it would take a person with a basal metabolic rate of about 1,800 kcal/day to utilize the equivalent amount of energy while involved in various activities.

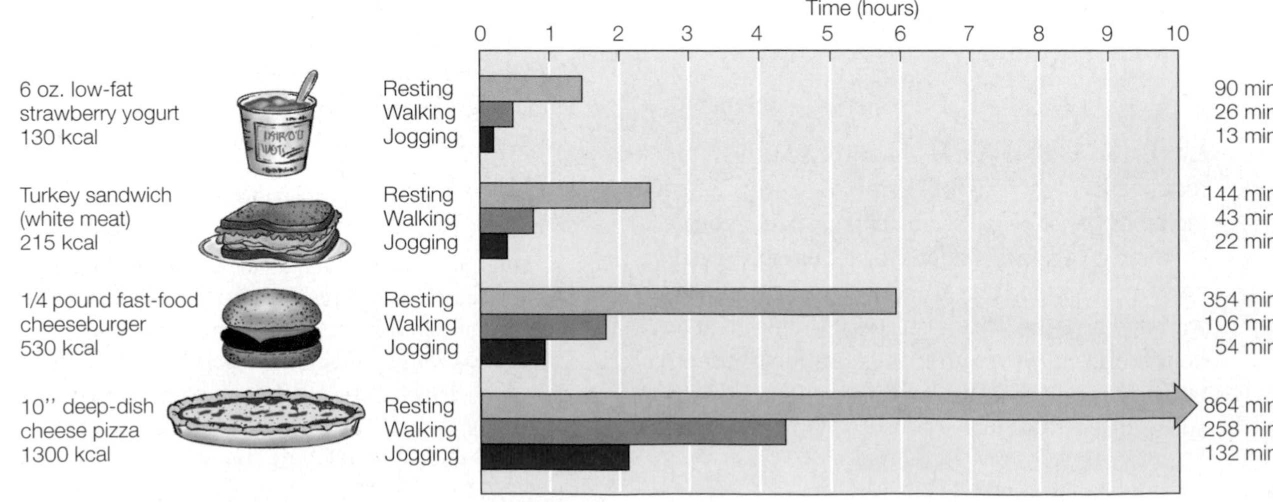

vestigators hypothesized that the birds could "afford" to be territorial only if the food resource was rich enough to support the metabolic cost of aggressive behavior. Accordingly, they determined how much nectar a flower produces and how this amount differs with habitat type. They then determined how many calories the birds spent when they were resting, when they were foraging, and when they were aggressively defending a patch of flowers.

According to the investigators' calculations, if a bird had a choice between a patch of flowers that produced nectar at a rate of 1 μl/day per blossom or a patch that produced nectar at 2/μl day per blossom, by selecting the richer patch it could meet its daily caloric requirement with 4 hours of foraging instead of 8 hours of foraging. Since caloric expenditure during foraging is 600 calories/hour greater than when resting, the bird could save 2,400 calories (4 hours × 600 calories/hour) by feeding on the more productive flowers. However, territorial defense costs 2,000 calories more per hour than foraging. Therefore, if a bird has to spend much more than an hour a day chasing intruders away from its rich flower patch, it is better off being nonaggressive and feeding on the less productive flowers. Direct observation and measurement have shown that the actual behavior of the birds tends to agree with predictions based on energy budget calculations.

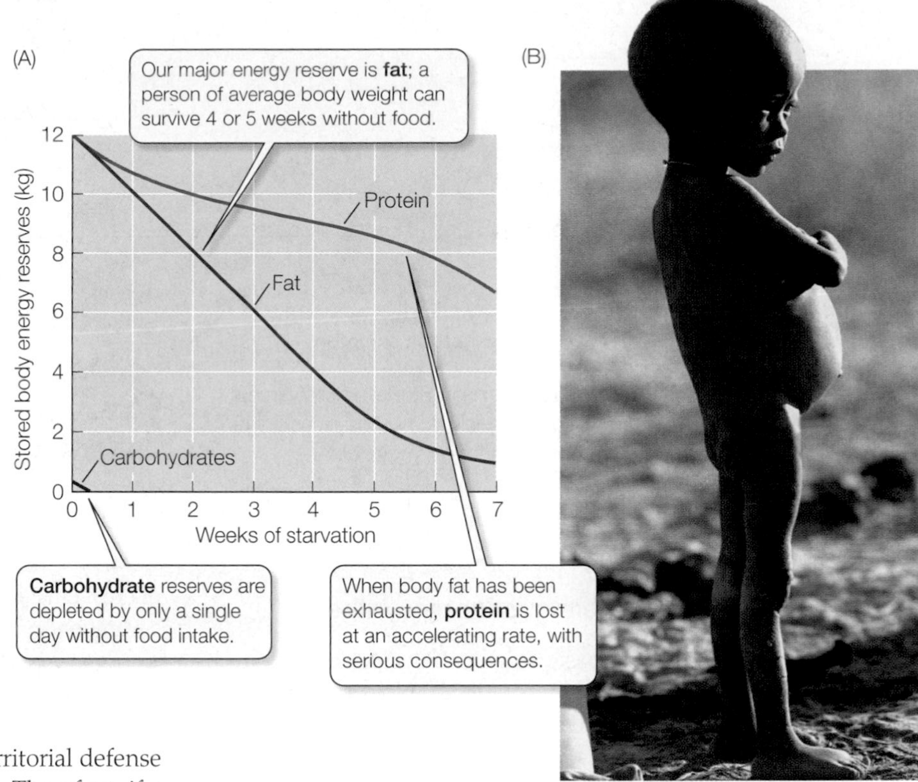

50.3 The Course of Starvation (A) In a person subjected to undernutrition, the energy reserves of the body are eventually depleted. (B) The swollen (due to edema) abdomen, face, hands, and feet of this young Somali girl, as well as her spindly limbs, are symptoms of kwashiorkor. This syndrome results from the body breaking down blood proteins and muscle tissue to fuel metabolism.

Sources of energy can be stored in the body

Although the cells of the body use energy continuously, most animals do not eat continuously. Therefore, animals must store fuel molecules that can be released as needed between meals. Carbohydrates are stored in liver and muscle cells as glycogen, but the total glycogen store represents only about a day's basal energy requirements (1,500–2,000 kcal). Fat is the most important form of stored energy in the bodies of animals. Not only does fat have more energy per gram than glycogen, but it can be stored with little associated water, making it more compact. Migrating birds store energy as fat to fuel their long flights; if they had to store the same amount of energy as glycogen, they would be too heavy to fly! Proteins are not used as energy storage compounds, although body protein can be metabolized as an energy source of last resort.

If an animal takes in too little food to meet its energy requirements, it is **undernourished** and must start metabolizing some of the molecules of its own body. This "self-consumption" begins with the energy storage compounds glycogen and fat. Protein loss is rapid at first, as readily available amino acids are consumed by ongoing protein synthesis. The rate of protein loss lessens as the body maximizes fat utilization and conserves protein (**Figure 50.3A**). Decreased protein synthesis in the liver, however, causes a decrease in blood proteins, resulting in increased loss of fluid from the blood to the interstitial spaces (*edema*; see Section 49.4). Accumulation of fluid in the extremities and abdomen are the classic signs of *kwashiorkor*, a disease caused by chronic protein deficiency (**Figure 50.3B**). When fat reserves are almost gone, protein loss accelerates to the point of damaging the organs of the body, leading eventually to death.

When an animal consistently takes in *more* food than it needs to meet its energy requirements, it is **overnourished**, and the excess nutrients are stored as increased body mass. First, glycogen reserves build up; then additional dietary carbohydrates, fats, and proteins are converted to body fat. In some species, such as hibernators, seasonal overnutrition is an important adaptation for surviving periods when food is not available. In humans, however, overnutrition can be a serious health hazard, increasing the risk of high blood pressure, heart attack, diabetes, and other disorders.

Ironically, at a time when a billion people—one-sixth of the world's population—do not have adequate food, the leading cause of life-threatening undernourishment in some Western nations is a self-imposed starvation syndrome called *anorexia nervosa*. This condition is found mainly among adolescent girls and young women who develop a pathological aversion to real or imagined body fat.

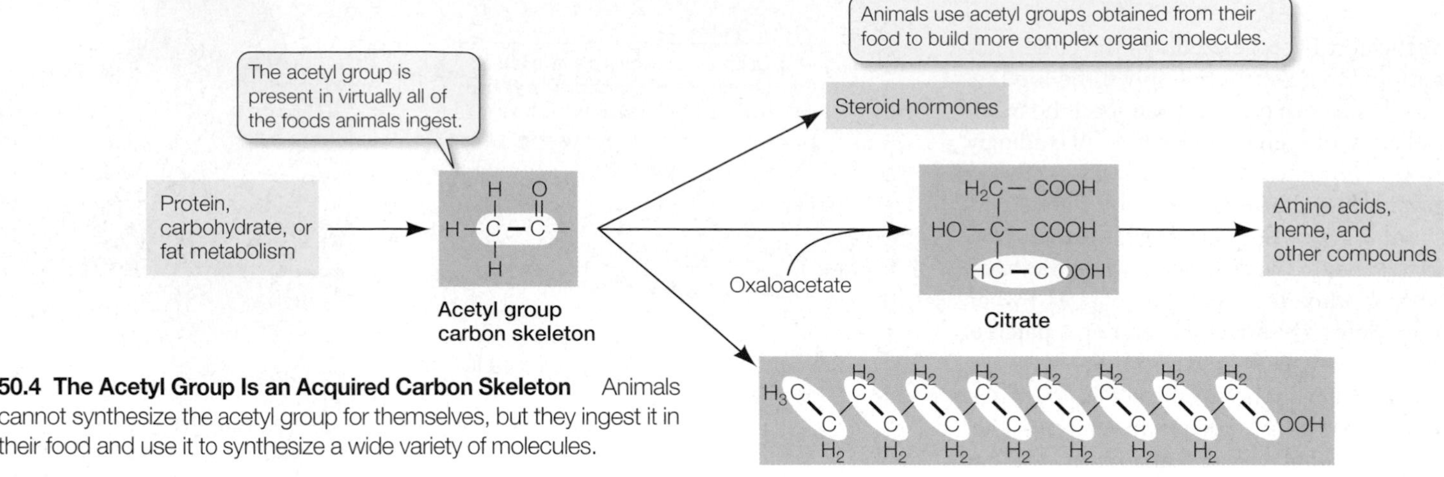

50.4 The Acetyl Group Is an Acquired Carbon Skeleton Animals cannot synthesize the acetyl group for themselves, but they ingest it in their food and use it to synthesize a wide variety of molecules.

Food provides carbon skeletons for biosynthesis

Every animal requires certain basic organic molecules that it cannot synthesize for itself, but needs as building blocks for its own complex organic molecules. The acetyl group (CH_3CO—) is one such required building block supplying the **carbon skeleton** of larger organic molecules (**Figure 50.4**). Animals cannot make acetyl groups from carbon, oxygen, and hydrogen molecules; they must obtain acetyl groups from food. Acetyl groups can be derived from the metabolism of almost any food; they are unlikely ever to be in short supply for an adequately nourished animal. However, some groups supplying carbon skeletons can be deficient in an animal's diet even if its caloric intake is adequate.

Amino acids, the building blocks of proteins, are a good example of carbon skeletons that can be in short supply. Animals can synthesize some of their own amino acids by utilizing carbon skeletons from acetyl or other groups and transferring to them amino groups (—NH_2) derived from other amino acids. Most animals cannot synthesize all the amino acids they need. Each species must obtain certain **essential amino acids** from food. Essential amino acids vary by species. In general, herbivores have fewer essential amino acids than carnivores. If an animal does not take in enough of even one of its essential amino acids, its protein synthesis is impaired.

Humans must obtain eight essential amino acids from their food: isoleucine, leucine, lysine, methionine, phenylalanine, threonine, tryptophan, and valine. All eight are available in milk, eggs, meat, and soybean products, but most plant foods do not contain adequate quantities of all eight. A strict vegetarian diet, therefore, is subject to a risk of protein malnutrition. A complementary dietary mixture of plant foods, however, supplies all eight essential amino acids (**Figure 50.5**). In general, grains (such as rice, wheat, and corn) are complemented by legumes (such as beans and peas). Long before the chemical basis for this complementarity was understood, societies with little access to meat developed complementary diets. Many Central and South American peoples traditionally eat beans with corn, and the native peoples of North America complemented their beans with squash.

Why are dietary proteins completely digested to their constituent amino acids before being used by the body? Wouldn't it be more energy-efficient to reuse some dietary proteins directly? There are several reasons why ingested proteins are not used "as is":

- Macromolecules such as proteins are not readily absorbed by the cells of the gut, but their constituent monomers (such as amino acids) are readily absorbed.
- Protein structure and function are highly species-specific. A protein that functions optimally in one species might not function well in another.
- Foreign proteins entering the body directly from the gut would be recognized as invaders and would be attacked by the immune system.

Humans can synthesize almost all the lipids required by the body using acetyl groups obtained from food (see Figure 50.4), but we must have a dietary source of certain **essential fatty acids**—notably, linoleic acid—that we cannot synthesize. Linoleic acid is an unsaturated fatty acid needed by mammals to synthesize other unsaturated fatty acids, such as arachidonic acid, which is a component of several signaling molecules, including prostaglandins. A deficiency of linoleic acid can lead to problems such as infertility and impaired lactation. Essential fatty acids are also necessary components of membrane phospholipids.

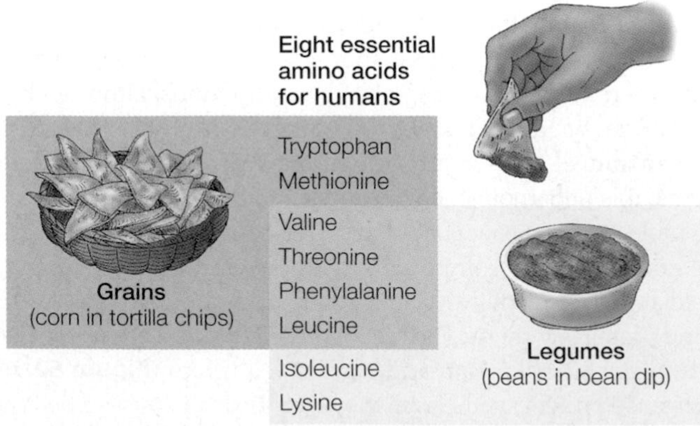

50.5 A Strategy for Vegetarians By combining cereal grains with legumes, a vegetarian can obtain all eight essential amino acids.

Animals need mineral elements for a variety of functions

The principal mineral elements that animals require are listed in **Table 50.1**. Elements required in large amounts are called **macronutrients**; elements required in only tiny amounts are called **micronutrients**. Some micronutrients are required in such minute amounts that deficiencies are never observed, but they are nevertheless essential elements.

Calcium is an example of a macronutrient. It is the fifth most abundant element in the body; a 70-kg person contains about 1.2 kg of calcium. Calcium phosphate is the principal structural material in bones and teeth. Muscle contraction, neuronal function, and many other intracellular functions in animals require calcium ions (Ca^{2+}). The turnover of calcium in the extracellular fluid is high, as bones are constantly being remodeled and calcium is constantly entering and leaving cells. Calcium is lost from the body in urine, sweat, and feces, so it must be replaced regularly. Humans require about 800–1,000 mg of calcium per day in the diet.

Iron is an example of a micronutrient. Iron is found everywhere in the body because it is the oxygen-binding atom in hemoglobin and myoglobin and is a component of enzymes in the electron transport chain. Nevertheless, the total amount of iron in a 70-kg person is only about 4 grams, and since iron is recycled so efficiently in the body and is not lost in the urine, we require only about 15 mg per day in our food. Despite the small amount required, insufficient iron is the most common mineral nutrient deficiency in the world today.

Animals must obtain vitamins from food

Vitamins are another group of essential nutrients. Like essential amino acids and fatty acids, vitamins are carbon compounds that an animal requires for growth and metabolism but cannot synthesize for itself. Most vitamins function as coenzymes or parts of coenzymes and are required in very small amounts compared to the essential amino acids and fatty acids, which have structural roles.

TABLE 50.1

Mineral Elements Required by Animals

ELEMENT	SOURCE IN HUMAN DIET	MAJOR FUNCTIONS
MACRONUTRIENTS		
Calcium (Ca)	Dairy foods, eggs, green leafy vegetables, whole grains, legumes, nuts, meat	Found in bones and teeth; blood clotting; nerve and muscle action; enzyme activation
Chlorine (Cl)	Table salt (NaCl), meat, eggs, vegetables, dairy foods	Water balance; digestion (as HCl); principal negative ion in extracellular fluid
Magnesium (Mg)	Green vegetables, meat, whole grains, nuts, milk, legumes	Required by many enzymes; found in bones and teeth
Phosphorus (P)	Dairy, eggs, meat, whole grains, legumes, nuts	Found in nucleic acids, ATP, and phospholipids; bone formation; buffers; metabolism of sugars
Potassium (K)	Meat, whole grains, fruits, vegetables	Nerve and muscle action; protein synthesis; principal positive ion in cells
Sodium (Na)	Table salt, dairy foods, meat, eggs	Nerve and muscle action; water balance; principal positive ion in extracellular fluid
Sulfur (S)	Meat, eggs, dairy foods, nuts, legumes	Found in proteins and coenzymes; detoxification of harmful substances
MICRONUTRIENTS		
Chromium (Cr)	Meat, dairy, whole grains, legumes, yeast	Glucose metabolism
Cobalt (Co)	Meat, tap water	Found in vitamin B_{12}; formation of red blood cells
Copper (Cu)	Liver, meat, fish, shellfish, legumes, whole grains, nuts	Found in active site of many redox enzymes and electron carriers; production of hemoglobin; bone formation
Fluorine (F)	Most water supplies	Found in teeth; helps prevent decay
Iodine (I)	Fish, shellfish, iodized salt	Found in thyroid hormones
Iron (Fe)	Liver, meat, green vegetables, eggs, whole grains, legumes, nuts	Found in active sites of many redox enzymes and electron carriers, hemoglobin, and myoglobin
Manganese (Mn)	Organ meats, whole grains, legumes, nuts, tea, coffee	Activates many enzymes
Molybdenum (Mo)	Organ meats, dairy, whole grains, green vegetables, legumes	Found in some enzymes
Selenium (Se)	Meat, seafood, whole grains, eggs, milk, garlic	Fat metabolism
Zinc (Zn)	Liver, fish, shellfish, and many other foods	Found in some enzymes and some transcription factors; insulin physiology

he list of vitamins varies from species to species. Most mammals, for example, can make their own ascorbic acid (vitamin C). Primates (including humans) cannot, so for primates, ascorbic acid is a vitamin. If we do not get vitamin C in our food, we develop a disease known as scurvy, characterized by bleeding gums, loss of teeth, subcutaneous hemorrhages, and slow wound healing. Scurvy was a serious and frequently fatal problem for sailors on long voyages until a Scottish physician, James Lind, discovered that the disease could be prevented if the sailors ate fresh greens or fresh fruit. Eventually the British Admiralty made limes standard provisions for its ships, and ever since, British sailors have been called "limeys." The active ingredient in limes was named ascorbic ("without scurvy") acid.

Humans require 13 vitamins, which are listed in **Table 50.2**. They are divided into two groups: water-soluble and fat-soluble. When water-soluble vitamins are ingested in excess of bodily needs, they are simply eliminated in the urine. (This is the fate of much of the vitamin C that people take in excessive doses.) The fat-soluble vitamins, however, can accumulate in body fat, and may build up to toxic levels in the liver if taken in excess.

The fat-soluble vitamin D (cholecalciferol), which is essential for the absorption and metabolism of calcium, is a special case because the body can make it. (Recall from Section 41.3 that vitamin D is by definition a hormone.) Certain lipids present in the human body can be converted into vitamin D by the action of ultraviolet light on the skin. Thus vitamin D must be obtained in the diet only by individuals with inadequate exposure to the sun, such as people who live in cold climates where clothing usually covers most of the body and where the sun may not shine for long periods of time.

The need for vitamin D may have been an important factor in the evolution of skin color. Human races that are adapted to equatorial and low latitudes have dark skin pigmentation as a protection against the damaging effects of ultraviolet radiation. These peoples generally expose extensive areas of skin to the sun on a regular basis, so their skin synthesizes adequate amounts of vitamin D. Most human races that became adapted to higher latitudes lost this dark skin pigmentation. Lighter skin facilitates vitamin D production in the relatively small areas of skin exposed to sunlight during the short days of winter. The Inuit peoples of the Arctic represent an exception to the correlation between latitude and skin pigmentation. These dark-skinned people obtain plenty of vitamin D from the large amounts of meat and fish oils in their diet; they don't require exposure to sunlight to obtain this vitamin.

Nutrient deficiencies result in diseases

The lack of any essential nutrient in the diet produces a state of deficiency called **malnutrition**, and chronic malnutrition leads to a characteristic **deficiency disease**. We have already discussed kwashiorkor (protein deficiency) and scurvy (vitamin C deficiency). A shortage of any of the vitamins results in specific deficiency symptoms (see Table 50.2). Another deficiency disease, *beriberi*, was directly involved in the discovery of vitamins. Beriberi means "extreme weakness." It became prevalent in Asia in the nineteenth century after it became standard practice to mill rice to a high,

TABLE 50.2

Vitamins in the Human Diet

VITAMIN	SOURCE	FUNCTION	DEFICIENCY SYMPTOMS
WATER-SOLUBLE			
B_1 (thiamin)	Liver, legumes, whole grains	Coenzyme in cellular respiration	Beriberi, loss of appetite, fatigue
B_2 (riboflavin)	Dairy, meat, eggs, green leafy vegetables	Coenzyme in FAD	Lesions in corners of mouth, eye irritation, skin disorders
Niacin	Meat, fowl, liver, yeast	Coenzyme in NAD and NADP	Pellagra, skin disorders, diarrhea, mental disorders
B_6 (pyridoxine)	Liver, whole grains, dairy foods	Coenzyme in amino acid metabolism	Anemia, slow growth, skin problems, convulsions
Pantothenic acid	Liver, eggs, yeast	Found in acetyl CoA	Adrenal problems, reproductive problems
Biotin	Liver, yeast, bacteria in gut	Found in coenzymes	Skin problems, loss of hair
B_{12} (cobalamin)	Liver, meat, dairy foods, eggs	Formation of nucleic acids, proteins, and red blood cells	Pernicious anemia
Folic acid	Vegetables, eggs, liver, whole grains	Coenzyme in formation of heme and nucleotides	Anemia
C (ascorbic acid)	Citrus fruits, tomatoes, potatoes	Formation of connective tissues; antioxidant	Scurvy, slow healing, poor bone growth
FAT-SOLUBLE			
A (retinol)	Fruits, vegetables, liver, dairy	Found in visual pigments	Night blindness
D (cholecalciferol)	Fortified milk, fish oils, sunshine	Absorption of calcium and phosphate	Rickets
E (tocopherol)	Meat, dairy foods, whole grains	Muscle maintenance, antioxidant	Anemia
K (menadione)	Intestinal bacteria, liver	Blood clotting	Blood-clotting problems

white polish and discard the hulls present in brown rice. A critical observation was that birds—chickens and pigeons—developed beriberi-like symptoms when fed only polished rice. In 1912, Casimir Funk cured pigeons of beriberi by feeding them the discarded hulls.

At the time of Funk's discovery, all diseases were thought to be either caused by microorganisms or inherited. Funk suggested the radical idea that beriberi and some other diseases are dietary in origin and result from deficiencies in specific substances. Funk coined the term "vitamines" because he mistakenly thought that all these substances vital for life were compounds with amino groups. In 1926, thiamin (vitamin B_1)—the substance lost in the rice milling process—was the first vitamin to be isolated in pure form.

Deficiency diseases can also result from an inability to absorb or process an essential nutrient even if it is present in the diet. Vitamin B_{12} (cobalamin), for example, is present in all foods of animal origin. Since plants neither use nor produce vitamin B_{12}, a strictly vegetarian diet (not supplemented with dairy products or vitamin pills) can lead to a B_{12} deficiency disease called pernicious anemia, characterized by a failure of red blood cells to mature. The most common cause of pernicious anemia, however, is not a lack of vitamin B_{12} in the diet, but an inability to absorb it. Normally, cells in the stomach lining secrete a peptide called intrinsic factor, which binds to vitamin B_{12} and makes it possible for it to be absorbed in the small intestine. Conditions that damage the stomach lining, such as alcoholism or gastritis, can thus lead to pernicious anemia.

Inadequate mineral nutrition can also lead to deficiency diseases. Examples are hypothyroidism and goiter resulting from iodine deficiency (see Section 41.3), and anemia resulting from iron deficiency.

50.1 RECAP

As heterotrophs, animals must obtain the energy and molecular building blocks for biosynthesis from their food. Energy can come from the metabolism of carbohydrates, fats, and proteins. Molecular building blocks include carbon skeletons, vitamins, and minerals.

- Do you understand how metabolic energy measurement can be used to construct energy budgets? See pp. 1070–1071
- Can you give an example of an essential carbon skeleton? A micronutrient? A macronutrient? See pp. 1072–1073, Figure 50.4, and Table 50.1
- Can you explain why fat-soluble vitamins should not be taken in excess? See p. 1074

We have surveyed the essential elements of nutrition in animals. Next, we will look at various methods and adaptations by which animals obtain the food they need, and mechanisms by which the food is processed in the animal's body to extract nutrients.

50.2 How Do Animals Ingest and Digest Food?

Heterotrophic organisms can be classified by how they acquire their nutrition. **Saprobes** (also called *saprotrophs* or *decomposers*) are organisms—mostly protists and fungi—that absorb nutrients from dead organic matter. **Detritivores**, such as earthworms and crabs, actively feed on dead organic material. Animals that feed on living organisms are **predators: Herbivores** prey on plants, **carnivores** prey on animals, and **omnivores** prey on both. **Filter feeders**, such as clams and blue whales, prey on small organisms by filtering them from the aquatic environment. **Fluid feeders** include mosquitoes, aphids, leeches, and hummingbirds. The anatomical adaptations that enable a species to exploit a particular source of nutrition are usually quite obvious, but physiological and biochemical adaptations are also important.

The food of herbivores is often low in energy and hard to digest

Most vegetation is coarse and difficult to break down physically, but herbivores must process large amounts of it because its energy content is low. Therefore, herbivores spend a great deal of time feeding. Many have striking adaptations for feeding, such as the trunk (a flexible, gripping nose) of the elephant or the long neck of the giraffe. Many types of grinding, rasping, cutting, and shredding mouthparts have evolved in invertebrates for ingesting plant material, and the teeth of herbivorous vertebrates have been shaped by selection to tear, crush, and grind coarse plant matter. The digestive processes of herbivores can also be quite specialized.

Carnivores must detect, capture, and kill prey

The predatory behaviors of many carnivores are legendary. One need only call to mind the hunting skills of hawks, wolves, or tigers. Carnivores have evolved stealth, speed, power, large jaws, sharp teeth, and strong gripping appendages. Carnivores also have evolved remarkable means of detecting prey. Bats use echolocation, pit vipers sense infrared radiation from the warm bodies of their prey, and certain fishes detect electric fields created in the water by their prey.

Adaptations for killing and ingesting prey are diverse and can be highly specialized. These adaptations are especially important when the prey can inflict damage on the predator. A snake may strike with poisonous fangs, using its venom to immobilize its prey, which may include animals that are very active and have dangerous teeth or claws. To swallow large prey, a snake disengages its lower jaw from its joint with the skull (**Figure 50.6**). The tentacles of jellyfishes, the long, sticky tongues of chameleons, and the webs of spiders are other fascinating examples of adaptations for capturing and immobilizing prey. Some predators digest their prey externally. For example, a spider injects its insect prey with digestive enzymes and then sucks out the liquefied contents, leaving behind the empty exoskeletons frequently seen in old spider webs.

50.6 An Adaptation for Carnivory Snakes such as this corn snake (*Elaphe guttata*) can ingest large prey (in this case, a mouse) by dislocating their jaws.

Vertebrate species have distinctive teeth

Teeth are adapted for the acquisition and initial processing of specific types of foods. Because they are among the hardest structures of the body, an animal's teeth remain in the environment long after it dies. Paleontologists use teeth to identify animals that lived in the distant past and to deduce their feeding behavior.

All mammalian teeth have the same general structure, consisting of three layers (**Figure 50.7A**). An extremely hard material called **enamel**, composed principally of calcium phosphate, covers the crown of the tooth. Both the crown and the root contain a layer of bony material called **dentine**, inside of which is a **pulp cavity** containing blood vessels, nerves, and the cells that produce the dentine.

Like you, elephants have wisdom teeth (third molars). They also have two more molars and three premolars, but at any one time only one of these huge teeth serves as the main chewing surface. As it wears out, it is replaced by the next one moving forward. The elephant's equivalent of your wisdom teeth do not emerge until the elephant is about 45 years old.

The shapes and organization of mammalian teeth, however, can be very different, since they are adapted to different diets (**Figure 50.7B**). In general, *incisors* are used for cutting, chopping, or gnawing; *canines* are used for stabbing, gripping, and ripping; and *molars* and *premolars* (the cheek teeth) are used for shearing, crushing, and grinding. The highly varied diet of humans is reflected by our multipurpose set of teeth, as is common among omnivores.

Animals digest their food extracellularly

Animals take food into a body cavity that is continuous with the outside environment. They secrete digestive enzymes into that cavity, and the enzymes break down the food into nutrient molecules that can be absorbed by the cells lining the cavity.

The simplest digestive systems are **gastrovascular cavities**, which connect to the outside world through a single opening. Cnidari-

50.7 Mammalian Teeth (A) A mammalian tooth has three layers: enamel, dentine, and a pulp cavity. (B) The teeth of different mammalian species are specialized for different diets. This illustration depicts the teeth of the lower jaw, viewed from above.

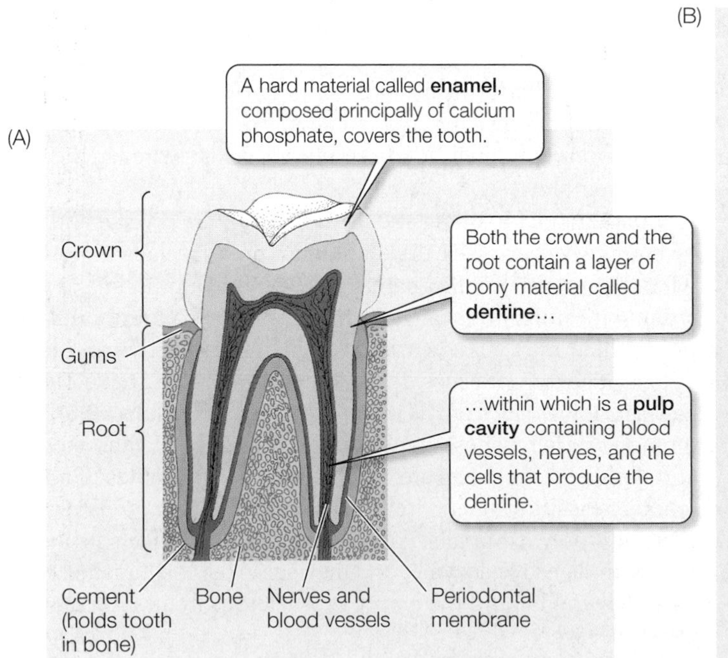

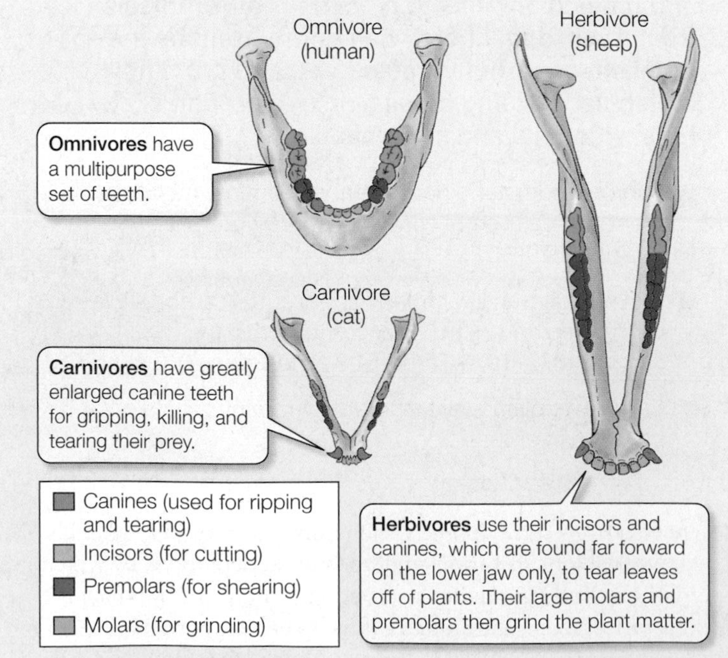

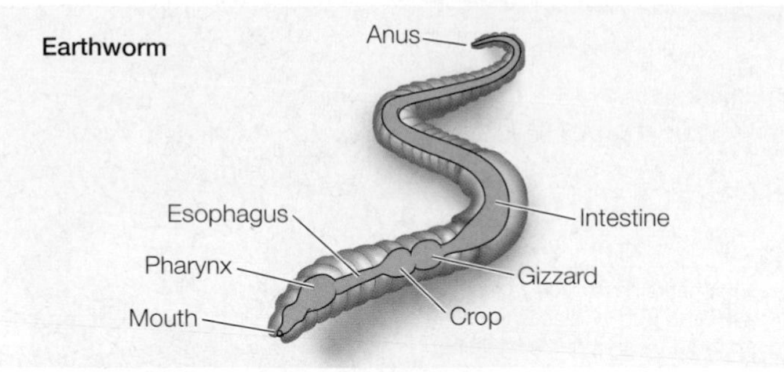

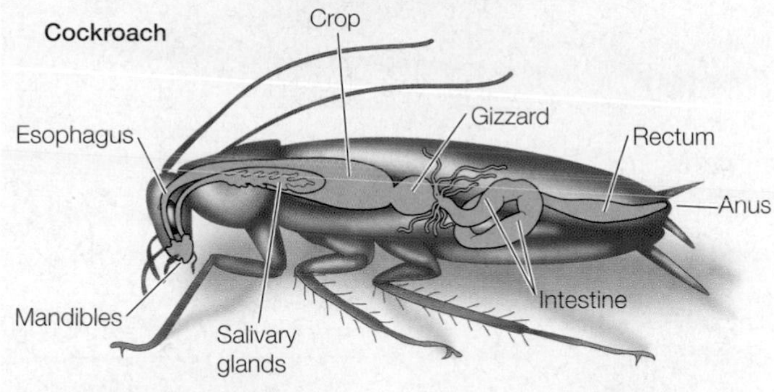

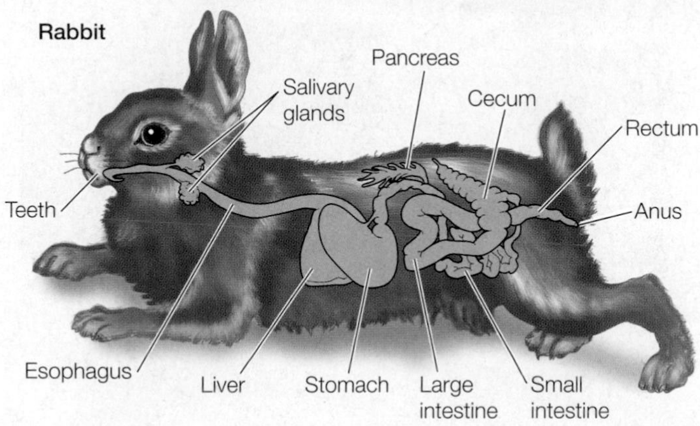

50.8 Compartments for Digestion and Absorption Most animals have tubular guts that begin with a mouth, which takes in food, and end in an anus, which eliminates wastes. Between these two structures are specialized regions for digestion and nutrient absorption; the structures in these regions vary from species to species.

ans, for example, capture prey using their stinging nematocysts (see Figure 31.10) and cram them into their gastrovascular cavities with their tentacles. Enzymes in the gastrovascular cavity partly digest the prey. Cells lining the cavity take in small food particles by endocytosis. The vesicles created by endocytosis then fuse with lysosomes containing digestive enzymes, and intracellular digestion completes the breakdown of the food. Nutrients are released to the cytoplasm as the vesicles break down.

Tubular guts have an opening at each end

The guts of most animals are tubular: A **mouth** takes in food; molecules are digested and absorbed throughout the length of the gut; and solid digestive wastes are eliminated through an **anus**. Different regions in the tubular gut are specialized for particular functions (**Figure 50.8**). These functions must be coordinated so that they occur in the proper sequence and at rates that maximize the efficiency of digestion and absorption of nutrients.

At the anterior end of the gut is the mouth cavity. Food may be physically broken up in the mouth cavity, for example by teeth (in many vertebrates), by the **radula** (in snails), or by **mandibles** (in many arthropods). In most birds food is ground by small stones in an early, muscular, portion of the gut called the **gizzard**. Some animals, such as snakes, simply ingest large chunks of food, with little or no fragmentation. **Stomachs** and **crops** are storage chambers that enable animals to ingest relatively large amounts of food when it is available, then digest it gradually. In these storage chambers, food may be further fragmented and mixed, but digestion may or may not occur there, depending on the species. In any case, food delivered into the next section of the gut, the **midgut** or **intestine**, is in small particles and well mixed.

Most nutrients, water, and ions are absorbed in the midgut. To digest food materials, specialized glands secrete some digestive enzymes into the midgut, and the gut wall itself secretes other enzymes. The **hindgut** recovers water and ions and stores undigested wastes, or **feces**, so that they can be released to the environment at an appropriate time or place. A muscular **rectum** near the anus assists in the expulsion of feces.

Within the hindguts of many species are colonies of endosymbiotic bacteria. These bacteria obtain their nutrition from the food passing through the host's gut while contributing to the digestive processes of the host. Members of the leech genus *Hirudo*, for example, produce no enzymes that can digest the proteins in the blood they suck from vertebrates; instead, a colony of gut bacteria performs this service. The resulting amino acids are subsequently used by both the leech and the bacteria. In many animals, the parts of the gut that absorb nutrients have greater surface areas than would be expected of a simple tube (**Figure 50.9A,B**). In vertebrates, the wall of the gut is richly folded, with the individual folds bearing legions of tiny fingerlike projections called **villi** (**Figure 50.9C**). The cells that line the surfaces of the villi, in turn, have microscopic projections called **microvilli**. The microvilli give the gut an enormous internal surface area for the absorption of nutrients.

Digestive enzymes break down complex food molecules

Protein, carbohydrate, and fat macromolecules are broken down into their simplest monomeric units by hydrolytic enzymes secreted at different locations in the digestive tract. All of these enzymes cleave the chemical bonds of macromolecules through hydrolysis, a reaction that adds a water molecule (see Figure 3.4B). Digestive enzymes are classified according to the substances they hydrolyze: *proteases* break the bonds between adjacent amino acids in proteins; *carbohydrases* hydrolyze carbohydrates; *peptidases*, peptides; *lipases*, fats; and *nucleases*, nucleic acids.

How can an organism produce enzymes that hydrolyze biological macromolecules without digesting itself? Most digestive enzymes are produced in an inactive form, known as a **zymogen**, so that they cannot act on the cells that produce them. When secreted into the gut, a zymogen is generally activated by another enzyme. The cells lining the gut are not digested because they are protected by a covering of mucus.

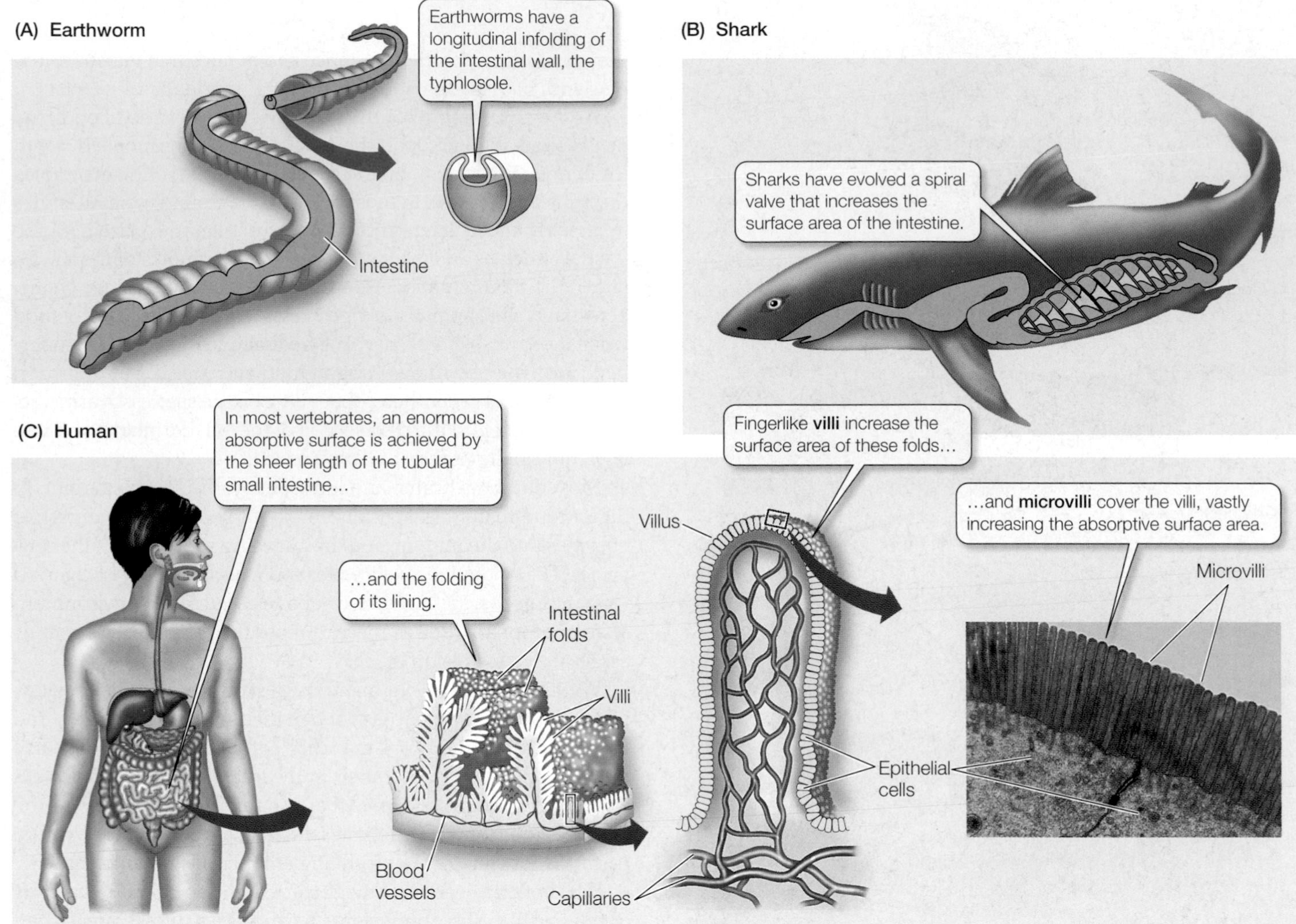

(A) Earthworm
Earthworms have a longitudinal infolding of the intestinal wall, the typhlosole.
Intestine
(B) Shark
Sharks have evolved a spiral valve that increases the surface area of the intestine.
(C) Human
In most vertebrates, an enormous absorptive surface is achieved by the sheer length of the tubular small intestine...
...and the folding of its lining.
Intestinal folds
Villi
Blood vessels
Fingerlike villi increase the surface area of these folds...
Villus
Capillaries
Epithelial cells
...and microvilli cover the villi, vastly increasing the absorptive surface area.
Microvilli

50.9 Greater Intestinal Surface Area Means More Nutrient Absorption The guts of most animals have evolved to maximize their surface area.

50.2 RECAP

Heterotrophs have diverse adaptations for acquiring food. Once captured and/or ingested, food is digested extracellularly by secreted enzymes to release nutrients, which are absorbed into the animal's body, usually via a tubular gut.

- Why do herbivores typically spend a great deal of their time feeding? See p. 1075
- What is the primary purpose of the intestinal microvilli? See p. 1077 and Figure 50.9
- What is a zymogen and why is it important? See p. 1077

Animals have evolved many ways of obtaining and ingesting their food. We have learned that, once ingested, the food may be fragmented and moved into the gut for digestion by hydrolytic enzymes, which release the various nutrients needed by the animal. Next we focus in detail on how those processes occur in vertebrates.

50.3 How Does the Vertebrate Gastrointestinal System Function?

Digestion in vertebrates occurs in the gastrointestinal system, which includes a tubular gut running from mouth to anus and several accessory structures that produce secretions that play important roles in digestion (**Figure 50.10**).

The vertebrate gut consists of concentric tissue layers

The cellular architecture of the vertebrate gut follows a common layered plan throughout its length (**Figure 50.11**). Starting in the cavity, or **lumen**, of the gut, the first layer of cells is the **mucosal epithelium**. These cells have secretory and absorptive functions. Some secrete mucus, which lubricates and protects the walls of the gut; others secrete digestive enzymes or hormones. Mucosal epithelial cells in the stomach secrete hydrochloric acid (HCl). In some regions of the gut, nutrients are absorbed by mucosal epithelial cells. The plasma membranes of these absorptive cells have *microvilli* that increase their surface area (see Figure 50.9C)

At the base of the mucosa are some smooth muscle cells, and just outside the mucosa is the submucosal tissue layer. Here we

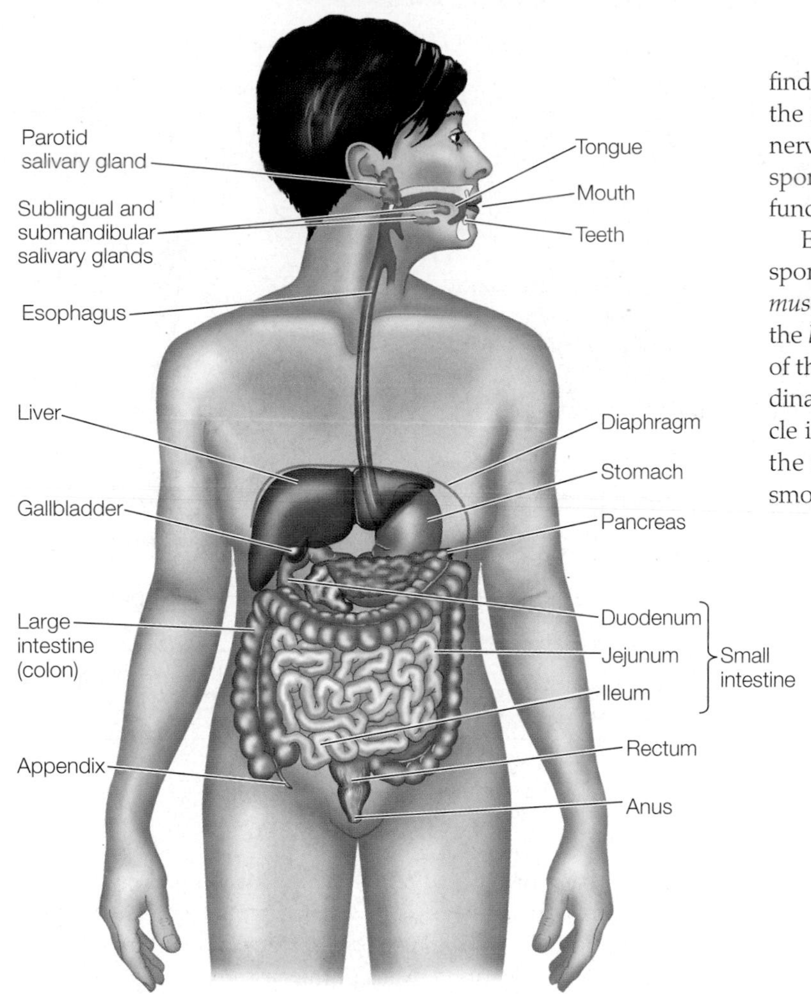

50.10 The Human Digestive System Different compartments within the long tubular gut specialize in digesting food, absorbing nutrients, and storing and expelling wastes. Accessory organs contribute secretions containing enzymes and other molecules.

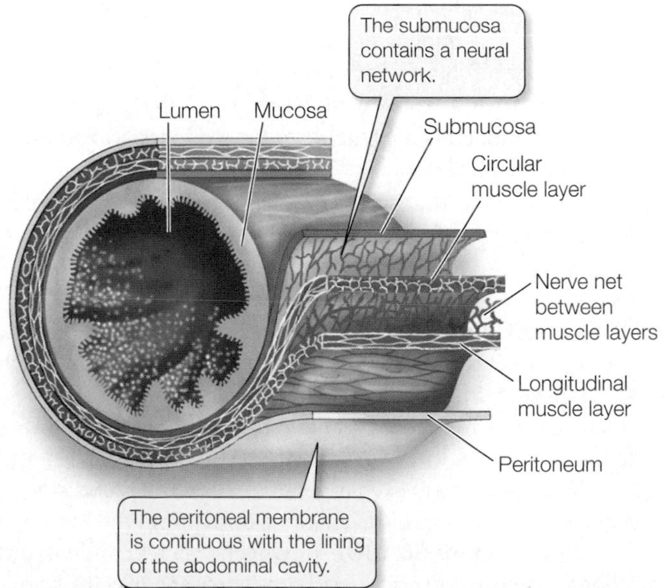

50.11 Tissue Layers of the Vertebrate Gut In all compartments of the gut, the organization of the tissue layers is the same, but specialized adaptations of specific tissues characterize different regions.

find the blood and lymph vessels that carry absorbed nutrients to the rest of the body. The **submucosa** also contains a network of nerves; the neurons in this network have sensory functions (responsible for stomach aches), and they control various secretory functions of the gut.

External to the submucosa are two layers of smooth muscle responsible for the movements of the gut. Innermost is the *circular muscle layer*, with its cells oriented around the gut. Outermost is the *longitudinal muscle layer*, with its cells oriented along the length of the gut. The circular muscles constrict the gut, and the longitudinal muscles shorten it. Between the two layers of smooth muscle is another network of nerves, which controls and coordinates the movements of the gut. The coordinated activity of the two smooth muscle layers moves the gut contents continuously toward the rectum.

A membrane called the **peritoneum** surrounds the gut as it does all of the organs of the abdominal cavity as well as lining the wall of the cavity. The peritoneum includes connective and epithelial tissue, which secrete a fluid that lubricates the organs so they can easily move against each other.

Mechanical activity moves food through the gut and aids digestion

In the mouth cavity, food is chewed and mixed with saliva. Periodically the *tongue* pushes a *bolus* of the chewed food toward the throat. By making contact with the *soft palate* at the back of the mouth cavity, the bolus of food initiates *swallowing*, which is a complex series of reflexes. Swallowing propels the food through the pharynx (where the mouth cavity and the nasal passages join) and into the **esophagus** (the food tube). To prevent the food from entering the trachea (windpipe), the *larynx* (voice box) closes, and a flap of tissue called the *epiglottis* covers the entrance to the larynx (**Figure 50.12**).

Once a bolus of food enters the esophagus, it is moved toward the stomach by waves of muscle contraction called **peristalsis**. The muscle of the upper third of the esophagus is striated (i.e., skeletal muscle) and is controlled by the somatic nervous system; the muscles of the lower two-thirds is smooth muscle, controlled by the autonomic nervous system.

The smooth muscles of the gut contract in response to being stretched. When a bolus of food reaches the smooth-muscle region of the esophagus and stretches it, the muscle responds by contracting, thus pushing the food toward the stomach. Why doesn't the contraction of the esophageal smooth muscle push the food back toward the mouth? The nerve net between the two smooth muscle layers coordinates the muscles so that contraction is always preceded by an *anticipatory wave of relaxation*: When a region of the gut smooth muscle contracts, the smooth muscle just beyond it relaxes, and food is squeezed into that area. The resulting stretch causes that region to constrict and simultaneously the region just beyond it relaxes. In this way peristalsis moves down the

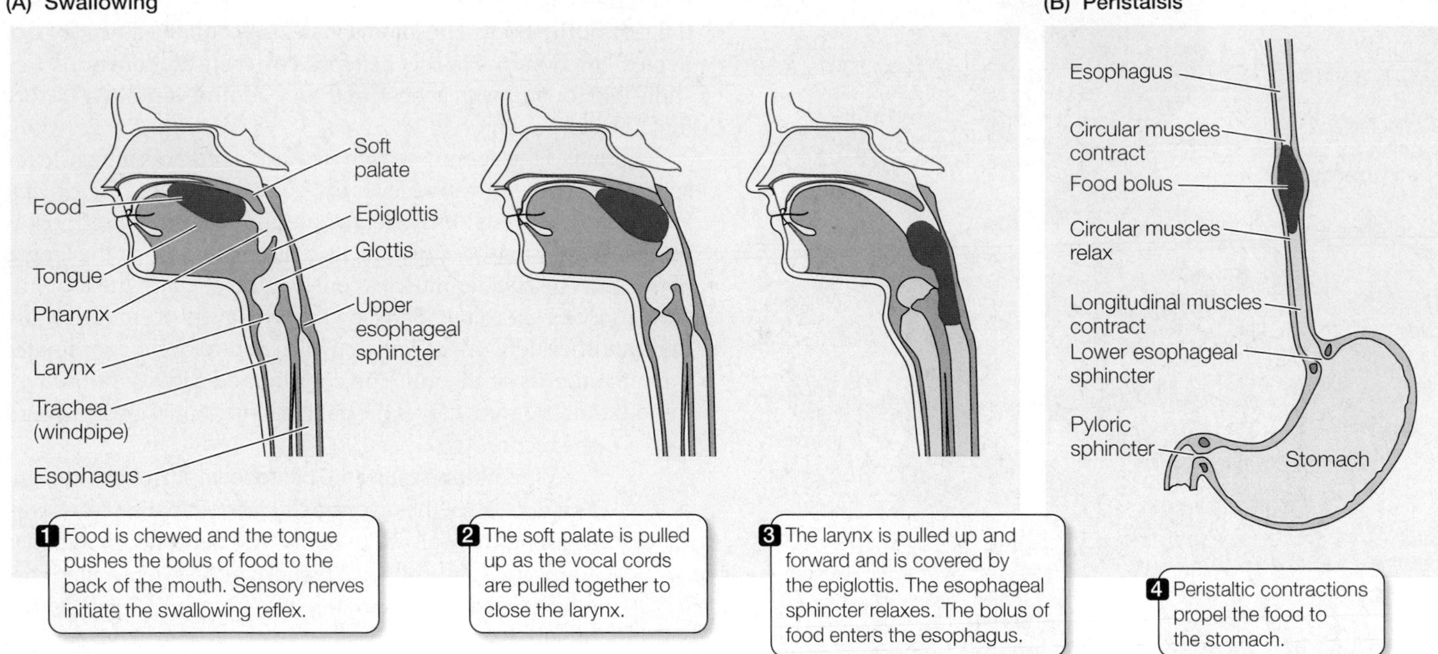
(A) Swallowing
Soft palate
Food
Epiglottis
Glottis
Tongue
Pharynx
Upper esophageal sphincter
Larynx
Trachea (windpipe)
Esophagus
1 Food is chewed and the tongue pushes the bolus of food to the back of the mouth. Sensory nerves initiate the swallowing reflex.
2 The soft palate is pulled up as the vocal cords are pulled together to close the larynx.
3 The larynx is pulled up and forward and is covered by the epiglottis. The esophageal sphincter relaxes. The bolus of food enters the esophagus.
(B) Peristalsis
Esophagus
Circular muscles contract
Food bolus
Circular muscles relax
Longitudinal muscles contract
Lower esophageal sphincter
Pyloric sphincter
Stomach
4 Peristaltic contractions propel the food to the stomach.

50.12 Swallowing and Peristalsis (A) Food pushed to the back of the mouth triggers the swallowing reflex. (B) Once the food bolus enters the esophagus, peristalsis propels it through the gut.

gut from mouth to anus pushing the contents of the gut in that direction.

The backward movement of food from the stomach into the esophagus is normally prevented by the *lower esophageal sphincter,* a thick ring of circular smooth muscle at the junction of the esophagus and the stomach. This sphincter is normally constricted, but waves of peristalsis cause it to relax enough to let food pass from the esophagus into the stomach. Sphincter muscles are found elsewhere in the digestive tract as well. The *pyloric sphincter* governs the passage of stomach contents into the intestine. Another important sphincter surrounds the anus.

Chemical digestion begins in the mouth and the stomach

The enzyme **amylase** is secreted by the salivary glands and mixed with food as it is chewed. Amylase hydrolyzes the bonds between the glucose monomers that make up starch molecules. The action of amylase is what makes a chewed piece of bread or cracker taste slightly sweet if you hold it in your mouth long enough.

The main role of the stomach is to store food so that digestion can occur slower than ingestion. Therefore, the stomach is very expandable, up to about 1.5 liters after a large meal. To make this expansion possible, the smooth muscle of the stomach is less sensitive to stretching than is the smooth muscle of the esophagus.

The stomach also has secretory functions. Deep infoldings in the walls of the stomach called **gastric pits** are lined with three types of secretory cells (**Figure 50.13A**). One type secretes a proteolytic enzyme that begins the digestion of protein. Another type secretes hydrochloric acid (HCl), which kills most ingested microorganisms. This nasty mix of substances could damage the walls of the stomach, but a third cell type secretes mucus that provides a protective coating of the walls of the gastric pits and the stomach. The mucus contains buffers that maintain the pH at the surface of the gastric mucosa near neutrality, and protease inhibitors that reduce damage from stomach enzymes.

The gastric pit cells that secrete the proteolytic enzyme are called *chief cells*. The enzyme is **pepsin**, and it is secreted as an inactive zymogen called **pepsinogen**. The extremely low pH of the stomach juices activates the conversion of pepsinogen to pepsin by cleaving away a sequence of amino acids that masks the active site of the enzyme. Newly activated pepsin can act on other pepsinogen molecules to activate them. Thus, the activation of pepsin in the stomach is a positive feedback process called **autocatalysis** (**Figure 50.13B**).

The gastric pit cells that secrete the HCl are called *parietal cells*. They can secrete so much HCl—about 2 liters per day—that they can bring the pH of the stomach contents below 1, which is the same as battery acid and 10 times more acidic than pure lemon juice. This means that across their plasma membranes, gastric pits can create a H^+ concentration difference of 3 million-fold. Such a feat of transport is not seen anywhere else in the body. How do they do it? The enzyme carbonic anhydrase in these cells catalyzes the hydration of CO_2 to H_2CO_3, which dissociates into H^+ and HCO_3^-. HCO_3^- is actively exchanged for Cl^- on the blood side of the gastric pits . H^+ is actively exchanged for K^+ from the lumen of the gastric pits (**Figure 50.13C**). However, this K^+ can leak out again down its concentration gradient. Thus, the inward transport of K^+ acts like an endless conveyer belt pushing H^+ out into the stomach lumen. Cl^- also passively leaks out of the gastric lumen side of the parietal cells to maintain electrical neutrality.

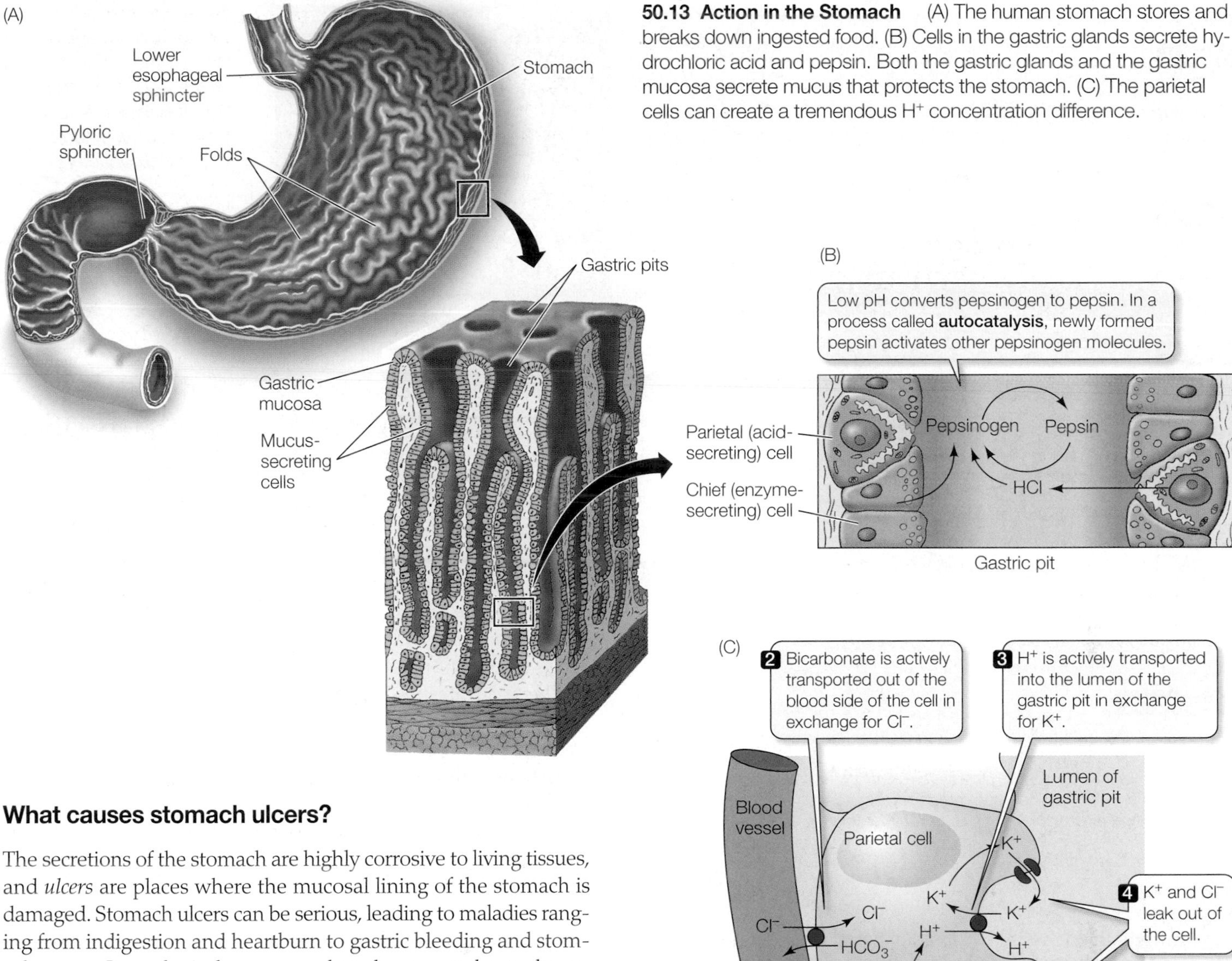

(A)
Lower esophageal sphincter
Stomach
Pyloric sphincter
Folds
Gastric pits
Gastric mucosa
Mucus-secreting cells
(B)
Low pH converts pepsinogen to pepsin. In a process called **autocatalysis**, newly formed pepsin activates other pepsinogen molecules.
Parietal (acid-secreting) cell
Chief (enzyme-secreting) cell
Pepsinogen
Pepsin
HCl
Gastric pit
(C)
2 Bicarbonate is actively transported out of the blood side of the cell in exchange for Cl^-.
3 H^+ is actively transported into the lumen of the gastric pit in exchange for K^+.
Lumen of gastric pit
Blood vessel
Parietal cell
K^+
Cl^-
HCO_3^-
H^+
$H_2O + CO_2$
4 K^+ and Cl^- leak out of the cell.
1 Carbonic anhydrase catalyzes formation of carbonic acid, which dissociates into H^+ and HCO_3^-.

50.13 Action in the Stomach (A) The human stomach stores and breaks down ingested food. (B) Cells in the gastric glands secrete hydrochloric acid and pepsin. Both the gastric glands and the gastric mucosa secrete mucus that protects the stomach. (C) The parietal cells can create a tremendous H^+ concentration difference.

What causes stomach ulcers?

The secretions of the stomach are highly corrosive to living tissues, and *ulcers* are places where the mucosal lining of the stomach is damaged. Stomach ulcers can be serious, leading to maladies ranging from indigestion and heartburn to gastric bleeding and stomach cancer. It was logical to assume that ulcers were due to the actions of the stomach secretions on the stomach mucosa, and therefore that stress and lifestyle issues leading to excess stomach secretions (especially HCl) were the major cause of ulcers. This view led the pharmaceutical industry to develop a plethora of drugs to decrease stomach acid production, which became "billion dollar drugs" because they were prescribed so widely. What seemed like the simplest fact in gastrointestinal medicine, however, was turned on its head by the work of two Australian researchers. In retrospect, their work is a perfect example of the application of Koch's postulates (see Section 26.6) for the proof that a microorganism causes a disease (**Figure 50.14**).

In 1982, Robin Warren, a pathologist, observed an unknown bacterium in biopsies from the stomachs of patients with ulcers. In a study of 100 patients, he and Barry Marshall of the University of Western Australia found that the bacterium was always present in the patients with ulcers. They isolated the bacterium, which they named *Helicobacter pylori*, and grew it in culture. Having thus satisfied Koch's first two postulates, they turned to the last two, as described in Figure 50.14. Their research showed not only that *H. pylori* causes stomach ulcers, but that ulcer patients can be cured with antibiotics.

The medical profession was so certain that no microorganisms could live in the stomach and that stomach acid was the cause of ulcers that at first Warren and Marshall's findings were resisted and even ridiculed. But in 2005 they received the Nobel Prize in Medicine for their important discovery, and antibiotic therapy is now the primary treatment for stomach ulcers worldwide.

Drugs that decrease stomach acid production are still important ulcer medications for several reasons. Once a person has an ulcer, stomach acid exacerbates it. In addition, in many individuals, the lower esophageal sphincter muscle is inadequate to prevent acid from entering the esophagus and causing irritations and

Marshall and Warren set out to satisfy Koch's postulates:

Test 1

The microorganism must be present in every case of the disease.

Results: Biopsies from the stomachs of many patients revealed that the bacterium was always present if the stomach was inflamed or ulcerated.

Test 2

The microorganism must be cultured from a sick host.

Results: The bacterium was isolated from biopsy material and eventually grown in culture media in the laboratory.

Test 3

The isolated and cultured bacteria must be able to induce the disease.

Results: Marshall was examined and found to be free of bacteria and inflammation in his stomach. After drinking a pure culture of the bacterium, he developed stomach inflammation (gastritis).

Test 4

The bacteria must be recoverable from the infected volunteers.

Results: Biopsy of Marshall's stomach 2 weeks after he ingested the bacteria revealed the presence of the bacterium, now christened *Helicobacter pylori*, in the inflamed tissue.

Conclusion

Antibiotic treatment eliminated the bacteria and the inflammation in Marshall. The experiment was repeated on healthy volunteers, and many patients with gastric ulcers were cured with antibiotics. Thus Marshall and Warren demonstrated that the stomach inflammation leading to ulcers is caused by *H. pylori* infections in the stomach.

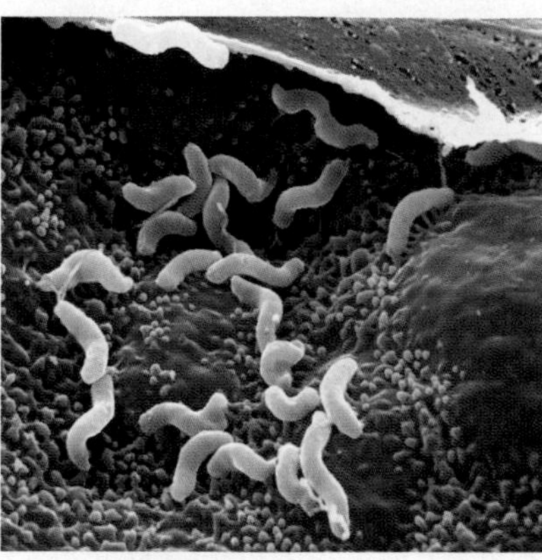

Helicobacter pylori

50.14 Satisfying Koch's Postulates Marshall and Warren showed that ulcers are caused not by the action of stomach acid, but by infection with the bacterium *Helicobacter pylori*.

lesions there. How does *H. pylori* survive in the hostile environment of the stomach? The bacteria live within the mucus layer lining the stomach, and by producing the enzyme urease they convert urea to bicarbonate and ammonium ions, which neutralize the HCl in their local environment.

The stomach gradually releases its contents to the small intestine

Contractions of the smooth muscles in the walls of the stomach churn its contents, thoroughly mixing them with the stomach secretions. The acidic, fluid mixture of gastric juice and partly digested food in the stomach is called **chyme**. A few substances can be absorbed across the stomach wall, including alcohol (hence its rapid effects), aspirin, and caffeine, but even these substances are absorbed in rather small quantities from the stomach.

Contractions of the stomach walls push the chyme toward the bottom end of the stomach. These waves of contractions cause the pyloric sphincter to relax briefly so that little squirts of the chyme can enter the small intestine. In this manner, the human stomach empties itself gradually over a period of approximately 4 hours. This slow introduction of food into the small intestine enables it to work on a little material at a time.

Most chemical digestion occurs in the small intestine

In the **small intestine**, the digestion of carbohydrates and proteins continues, and the digestion of fats and the absorption of nutrients begin. The small intestine takes its name from its diameter; it is in fact a very large organ, about 6 meters long in an adult. Given its length, and because of the folds, villi, and microvilli of its lining, its inner surface area is roughly the size of a tennis court. Across this surface, the small intestine absorbs all the nutrient molecules derived from food.

The small intestine has three sections. The initial section (about 25 cm long) is called the **duodenum** and is the site of most digestion; the **jejunum** and the **ileum** (together about 600 cm) carry out 90 percent of the absorption of nutrients (see Figure 50.10).

Digestion in the small intestine requires many specialized enzymes, as well as several other secretions. Two accessory organs that are not part of the digestive tract—the liver and the pancreas—provide many of these secretions and enzymes.

LIVER The liver synthesizes **bile** from cholesterol. Bile from the liver flows through the *hepatic duct* to the duodenum and through a side branch of the hepatic duct to the **gallbladder** (**Figure 50.15**), where it is stored until it is needed. When fat enters the duodenum, a hormonal signal causes the walls of the gallbladder to contract rhythmically, squeezing bile out of the gallbladder and into the hepatic duct. Below the branch to the gallbladder, the hepatic duct is called the *common bile duct*. Bile flows down the common bile duct to the duodenum.

To understand the role of bile in fat digestion, think of an oil and vinegar salad dressing. The oil, which is hydrophobic, tends to aggregate in large globules. For that reason, many salad dressings include an *emulsifier*—something that prevents oil droplets from aggregating. Mayonnaise, for example, is oil and vinegar with egg yolk added as an emulsifier. Bile contains salts that emulsify fats in the chyme, and thereby greatly enlarge the surface area of the fats exposed to the **lipases**—the enzymes that digest fats. One end of each bile salt molecule is soluble in fat (it is lipophilic and hydrophobic); the other end is soluble in water (it is hydrophilic and lipophobic). Bile molecules bury their lipophilic ends in fat droplets, leaving their hydrophilic ends sticking out. As a result, bile salts prevent the fat droplets from sticking together. The very small fat particles that result are called **micelles** (**Figure 50.16A**).

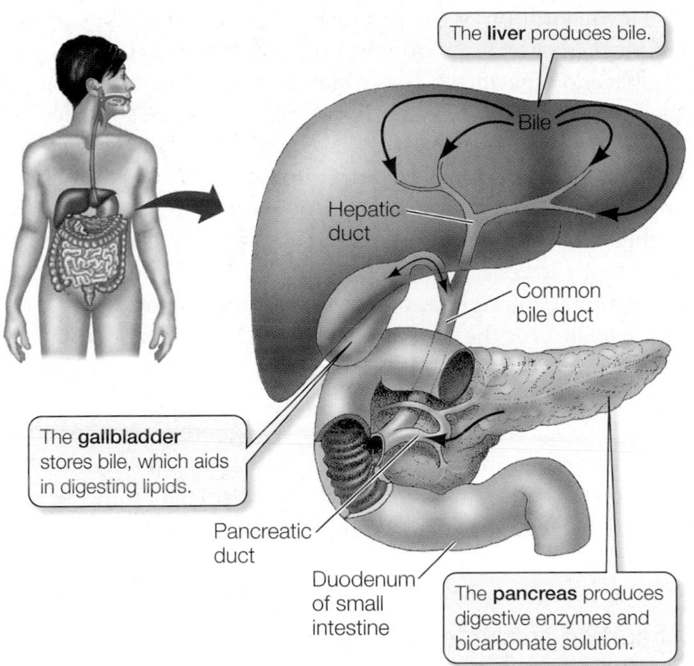

50.15 Ducts of the Gallbladder and Pancreas Bile produced in the liver leaves the liver via the hepatic duct. Branching off this duct is the gallbladder, which stores bile. Below the gallbladder, the hepatic duct is called the common bile duct and is joined by the pancreatic duct before entering the duodenum.

PANCREAS The **pancreas** is a large gland that lies just behind and below the stomach (see Figures 50.10 and 50.15). It is both an endocrine gland (secreting hormones into the tissue fluid; see Section 41.1) and an exocrine gland (secreting other substances through the pancreatic duct to the gut lumen, which is continuous with the outside of the body). The pancreatic duct joins the common bile duct before emptying into the duodenum.

The exocrine tissues of the pancreas produce a host of digestive enzymes, including lipases, amylases, proteases, and nucleases (**Table 50.3**). As in the stomach, some of these enzymes—most notably the proteases—are released as zymogens; otherwise, they would digest the pancreas and its ducts before they ever reached the duodenum. Once in the duodenum, one of these zymogens, **trypsinogen**, is activated by an enzyme called *enterokinase*, which is produced by cells lining the duodenum. This process is similar to the activation of pepsinogen by low pH in the stomach. Active **trypsin** can cleave other trypsinogen molecules to release even more active trypsin. Similarly, trypsin activates other zymogens secreted by the pancreas.

The mixture of zymogens produced by the pancreas can be dangerous if the pancreatic duct is blocked or if the pancreas is injured by infection or physical trauma. A few trypsinogen molecules spontaneously converting to trypsin can initiate a chain reaction of enzyme activity that digests the pancreas in a short time, destroying both its endocrine and exocrine functions.

The pancreas also produces a secretion rich in bicarbonate ions (HCO_3^-). Bicarbonate ions are alkaline (basic) and neutralize the acidic pH of the chyme that enters the duodenum from the stomach. Intestinal enzymes function best at a neutral or slightly alkaline pH.

Nutrients are absorbed in the small intestine

The final step in digesting proteins and carbohydrates and absorbing their components occurs among the microvilli. Mucosal epithelial cells secrete peptidases that cleave polypeptides into absorbable tripeptides, dipeptides, and individual amino acids. These epithelial cells also produce the enzymes maltase, lactase, and sucrase, which cleave the common disaccharides into absorbable monosaccharides—glucose, galactose, and fructose.

Many humans stop producing the enzyme lactase in childhood and thereafter have difficulty digesting lactose (the sugar in milk). Lactose is a disaccharide and cannot be absorbed without being cleaved into its constituents, glucose and galactose. Unabsorbed lactose is metabolized by bacteria in the large intestine, causing intense abdominal cramps, gas, and diarrhea.

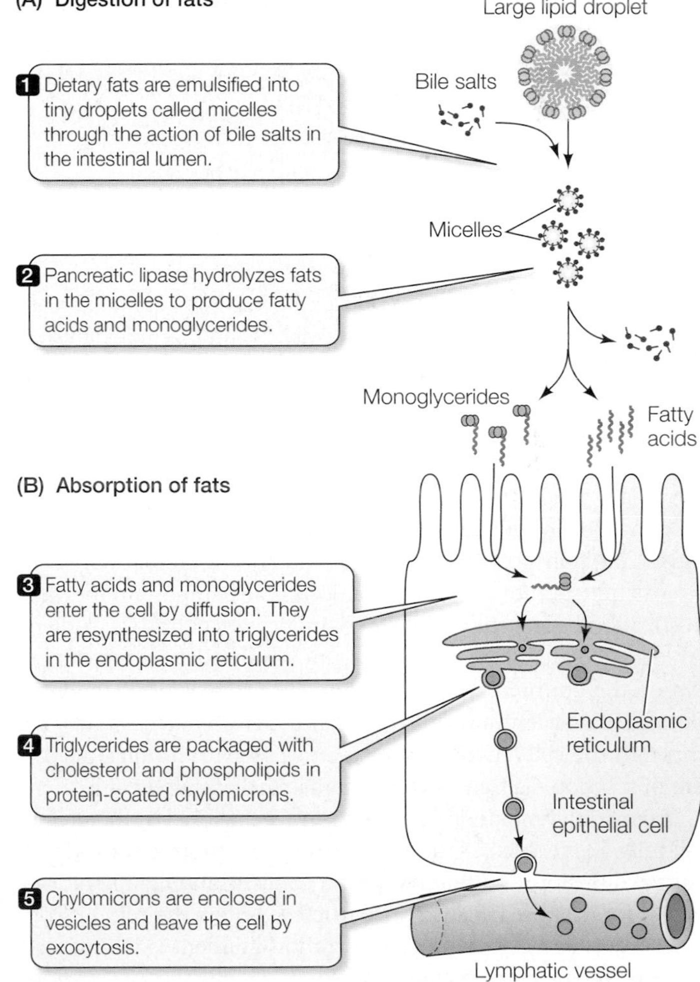

50.16 Digestion and Absorption of Fats (A) Dietary fats are broken up by bile into small micelles that present a large surface area to lipases. (B) The products of fat digestion are absorbed by intestinal mucosal cells, where they are resynthesized into triglycerides and exported to lymphatic vessels.

TABLE 50.3

Major Digestive Enzymes of Humans

SOURCE/ENZYME	ACTION
SALIVARY GLANDS	
Salivary amylase	Starch → Maltose
STOMACH	
Pepsin	Proteins → Peptides; autocatalysis
PANCREAS	
Pancreatic amylase	Starch → Maltose
Lipase	Fats → Fatty acids and glycerol
Nuclease	Nucleic acids → Nucleotides
Trypsin	Proteins → Peptides; zymogen activation
Chymotrypsin	Proteins → Peptides
Carboxypeptidase	Peptides → Shorter peptides and amino acids
SMALL INTESTINE	
Aminopeptidase	Peptides → Shorter peptides and amino acids
Dipeptidase	Dipeptides → Amino acids
Enterokinase	Trypsinogen → Trypsin
Nuclease	Nucleic acids → Nucleotides
Maltase	Maltose → Glucose
Lactase	Lactose → Galactose and glucose
Sucrase	Sucrose → Fructose and glucose

The mechanisms by which cells of the intestinal epithelium absorb nutrients and inorganic ions are diverse and include diffusion, facilitated diffusion, osmosis, active transport, and co-transport. Many inorganic ions such as sodium, calcium, and iron are actively transported by these cells. For example, active Na^+ transporters exist on the basal and lateral sides of the epithelial cells. They maintain a low concentration of Na^+ in those cells so that Na^+ can diffuse in from the chyme in the intestinal lumen. About 30 grams of Na^+ are transported this way every day, and Cl^- follows.

The transport of Na^+ and other ions is also important for water absorption because it creates an osmotic concentration gradient. At least 7–8 liters of water per day moves through the spaces between the epithelial cells in response to this osmotic gradient. Because the water moves through *spaces* between the cells and not through the cells themselves, it can carry with it nutrients that are in solution—a transport mechanism called *solvent drag*.

Many different kinds of transport proteins exist in the epithelial cells. Some, such as the transport protein for fructose, only facilitate diffusion, and that requires a concentration gradient. That works for fructose because once fructose enters the cell it is converted to glucose and the concentration gradient is maintained. A class of transporter proteins known as *sodium co-transporters* exploit the active transport of Na^+, which maintains a low Na^+ concentration in the cells. Co-transporter proteins combine with Na^+ and another molecule, such as glucose, galactose, or an amino acid. As Na^+ is pulled down its concentration gradient into the cell, the "hitchhiking" molecules are carried along with it.

The absorption of the products of fat digestion is relatively simple. Diglycerides, monoglycerides, and fatty acids are lipid-soluble and thus able to pass through the plasma membranes of the microvilli. In the intestinal epithelial cells, these molecules are resynthesized into triglycerides, combined with cholesterol and phospholipids, and coated with protein to form water-soluble **chylomicrons**, which are little particles of fat (**Figure 50.16B**). Rather than entering the blood directly, chylomicrons pass into the lymphatic vessels in the submucosa. They then flow through the lymphatic system and enter the bloodstream through the thoracic ducts at the base of the neck. After a meal rich in fats, chylomicrons can be so abundant in the blood that they give it a milky appearance.

The bile salts that emulsify fats are not absorbed along with the monoglycerides and the fatty acids, but shuttled back and forth between the gut contents and the microvilli. In the ileum, bile salts are actively reabsorbed and returned to the liver via the bloodstream.

Absorbed nutrients go to the liver

All of the blood leaving the digestive tract flows to the liver in the *hepatic portal vein*. This large vein delivers the blood to small spaces called sinusoids between groups of liver cells. These cells absorb the nutrients coming from the digestive tract and either store them or convert them to molecules the body needs. Glucose, sucrose, and fructose are used to synthesize glycogen. Amino acids are used to build proteins. Lipids from the chylomicrons are either stored as triglycerides or used to make lipoproteins.

Water and ions are absorbed in the large intestine

The motility of the small intestine gradually pushes its contents into the *large intestine*, or **colon**. Most of the available nutrients have been removed from the chyme that enters the colon, but it contains a lot of water and inorganic ions.

The colon absorbs water and ions, producing semisolid feces from the chyme it receives from the small intestine. Feces are stored in the rectum of the colon until they are eliminated. Absorption of too much water from the colon can cause *constipation*. The opposite condition, *diarrhea*, results if too little water is absorbed; in this case, water in the colon is excreted with the feces. The excessive diarrhea caused by diseases such as cholera can produce such rapid loss of water and electrolytes that death can occur in hours.

The problem with cellulose

Cellulose is the principal component of the food of herbivores. Most herbivores, however, cannot produce *cellulases*—enzymes that hydrolyze cellulose. Exceptions include silverfish (insects that eat books and stored papers), earthworms, and shipworms. From termites to cattle, herbivores rely on microorganisms in their digestive tracts to digest cellulose.

The stomachs of **ruminants** (cud chewers) such as cattle are large, four-chambered organs that take advantage of their endosymbiotic microorganisms (**Figure 50.17**). The first two chambers, the **rumen** and the **reticulum**, are packed with microorganisms

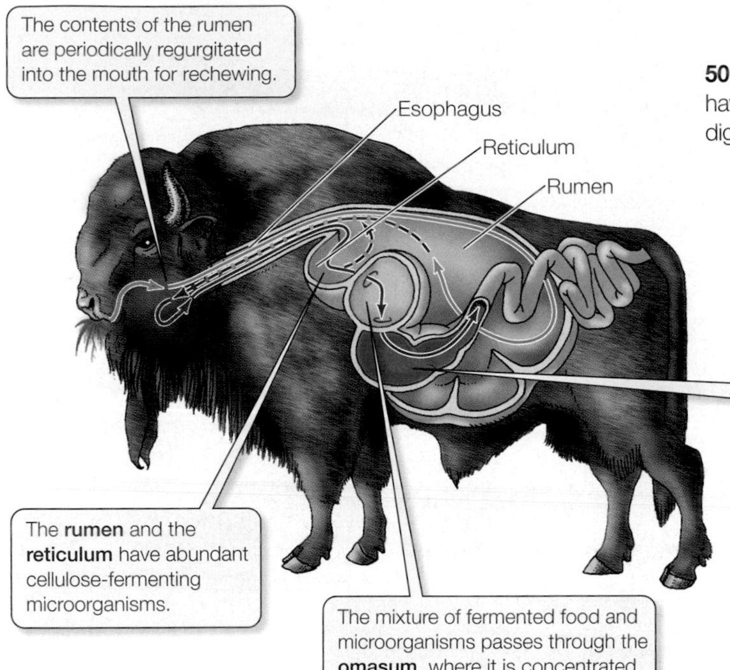
The contents of the rumen are periodically regurgitated into the mouth for rechewing.
Esophagus
Reticulum
Rumen
The abomasum is the "true" stomach, secreting HCl and proteases. The microorganisms are killed by the HCl, digested by the proteases, and passed on to the small intestine for further digestion.
The rumen and the reticulum have abundant cellulose-fermenting microorganisms.
The mixture of fermented food and microorganisms passes through the omasum, where it is concentrated by water absorption.

50.17 A Ruminant's Stomach Bison, like their relatives domestic cattle, have a specialized stomach with four compartments that enables them to digest and obtain energy from coarse plant material.

that break down cellulose by fermentation. The ruminant periodically regurgitates the contents of the rumen (the *cud*) into the mouth for re-chewing. When swallowed again, the vegetal fibers present more surface area to the microorganisms. The microorganisms metabolize cellulose and other nutrients to simple fatty acids, which become nutrients for their host.

Enormous numbers of microorganisms leave the rumen along with the partially digested food. This mass is concentrated by water absorption in the **omasum** before it then enters the true stomach, the **abomasum**, where the microorganisms are killed by secreted hydrochloric acid, digested by proteases, and passed on to the small intestine for further digestion and absorption. A cow derives more than 100 grams of protein per day from digestion of its endosymbiotic microorganisms. The rate of multiplication of microorganisms in the rumen offsets their loss, so a well-balanced, mutually beneficial relationship is maintained.

Intestinal bacteria produce gases such as methane and hydrogen sulfide as by-products of their anaerobic metabolism. Beans contain some carbohydrates that humans cannot digest but their intestinal bacteria can. A meal of beans increases gut bacterial metabolic activity and results in the well-earned reputation of beans.

Some mammalian herbivores have a microbial fermentation chamber called a **cecum** extending from the large intestine. An example is the rabbit (see Figure 50.8). Since the cecum empties into the large intestine, absorption of the nutrients produced by the microorganisms is inefficient. Such species frequently produce two kinds of feces – ones that are pure waste and ones that contain cecal material. In a behavior known as **coprophagy**, they re-ingest the cecal feces directly from the anus so they can digest and absorb the nutrients that would otherwise be lost. In humans, the cecum has become the vestigial **appendix**, which serves no digestive function.

50.3 RECAP

The vertebrate gastrointestinal system is a tubular gut that is adapted to ingest food, fragment it, digest it, and absorb nutrients. Peristalsis moves food through the gut. Absorption of nutrients occurs mostly in the small intestine; water and ions are absorbed in the large intestine.

- What digestive functions occur in the mouth and in the stomach? See p. 1080 and Figure 50.13
- How do bile salts assist in the digestion of fats? See pp. 1082–1083 and Figure 50.16
- Can you describe how sodium co-transport drives the absorption of nutrients? See p. 1084

The steps included in ingestion and digestion of food—from fragmentation in the mouth to the digestive processes in the gastrointestinal tract—serve one purpose: to make nutrients in the food available for absorption and ultimately for metabolism. Let's look at how the processes of digestion are controlled and how nutrients are handled by the body once food has been digested.

50.4 How Is the Flow of Nutrients Controlled and Regulated?

The vertebrate gut is an assembly line in reverse—a *dis*assembly line. As with a standard assembly line, the control and coordination of the sequential processes of digestion is critical. Both neuronal and hormonal controls govern these processes. Once the products of digestion are absorbed, their availability to the cells of the body must also be controlled.

You have certainly experienced salivation at the sight or smell of food. That response is an *unconcious reflex,* as is swallowing. Many such autonomic reflexes coordinate activity in different regions of the digestive tract. For example, the introduction of food into the stomach stimulates increased activity in the colon, which can lead to defecation.

The digestive tract is unusual in that it has an intrinsic (its own independent) nervous system. Neuronal messages can travel from one region of the digestive tract to another without being processed by the central nervous system (CNS). One function of the gut's nervous system is the coordination of motility. Of course, this intrinsic nervous system can communicate information to the CNS and receive input from the CNS, but its most important role is to coordinate actions throughout the digestive tract.

Hormones control many digestive functions

Several hormones control the activities of the digestive tract and its accessory organs (**Figure 50.18**). The first hormone discovered came from the duodenum; it was called **secretin** because it causes the pancreas to secrete digestive juices. We now know that secretin is only one of several hormones that control pancreatic secretion; specifically, secretin stimulates the pancreas to secrete a solution rich in bicarbonate ions.

In response to the presence of fats and proteins in the chyme, the mucosa of the small intestine secretes **cholecystokinin**, a hormone that stimulates the gallbladder to release bile and the pancreas to release digestive enzymes. Cholecystokinin and secretin also slow the movements of the stomach, which slows the delivery of chyme into the small intestine.

The stomach secretes a hormone called **gastrin** into the blood. Cells in the lower region of the stomach release gastrin when they are stimulated by the presence of food. Gastrin circulates in the blood until it reaches cells in the upper areas of the stomach wall, where it stimulates the secretions and movements of the stomach. Gastrin release begins to be inhibited when the pH of the stomach contents falls below 3—an example of negative feedback.

Most animals do not eat continuously. When they do eat, food is present in the gut, and nutrients are being absorbed, for a period of time after a meal, called the **absorptive period**. Once the stomach and small intestine are empty, nutrients are no longer being absorbed. During this **postabsorptive period**, the processes of energy metabolism and biosynthesis must run on internal reserves. Nutrient traffic must be controlled so that reserves accumulate during the absorptive period and are used appropriately during the postabsorptive period.

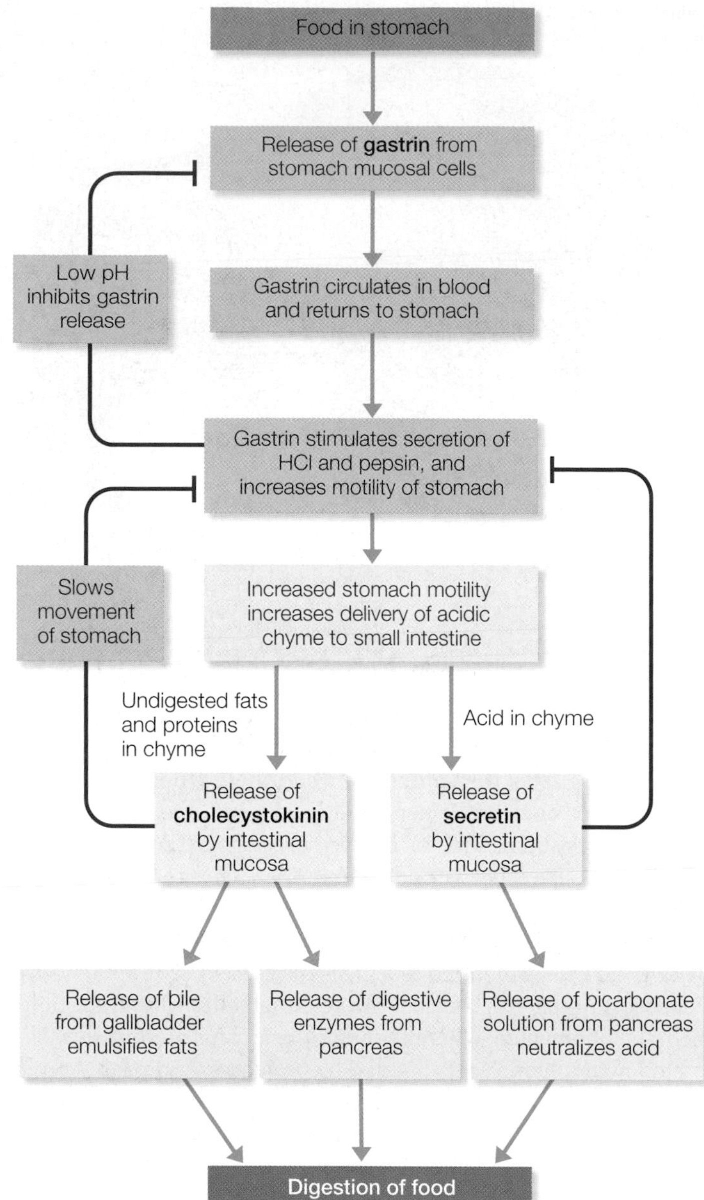

50.18 Hormones Control Digestion The hormones gastrin, cholecystokinin, and secretin are involved in feedback loops that control the sequential processing of food in the digestive tract.

The liver directs the traffic of the molecules that fuel metabolism

When fuel molecules are abundant in the blood, the liver stores them in the form of glycogen and fats. The liver also synthesizes blood plasma proteins from circulating amino acids. When fuel molecule levels in the blood decline, the liver delivers them back into the blood.

The liver has an enormous capacity to interconvert fuel molecules. Liver cells can convert monosaccharides into either glycogen or fats, and vice versa. The liver can also convert certain amino acids and some other molecules, such as pyruvate and lactate, into glucose—this process is termed *gluconeogenesis*. The liver is also the major controller of fat metabolism through its production of *lipoproteins*.

LIPOPROTEINS: THE GOOD, THE BAD, AND THE UGLY In the intestine, bile solves the problem of processing hydrophobic fats in an aqueous medium. The transport of fats in the circulatory system presents the same problem, and lipoproteins are the solution. A **lipoprotein** is a particle made up of a core of hydrophobic fat and cholesterol with a covering of hydrophilic protein that allows it to be suspended in water. The largest lipoprotein particles in the bloodstream are the **chylomicrons** produced by the mucosal cells of the intestine. As the circulation carries chylomicrons through the liver and adipose (fat) tissue, receptors on the capillary walls recognize and bind to the chylomicron protein coats. Lipases begin to hydrolyze the fats, which are then absorbed into liver or fat cells. Thus, the protein coat of the lipoprotein serves as an "address" that directs it to a specific tissue.

Lipoproteins other than chylomicrons are synthesized in the liver. These lipoproteins can be classified according to their density. Fat has a low density (it floats in water) and protein has a high density, so the greater the fat-to-protein ratio in the lipoprotein, the lower its density.

- **High-density lipoproteins (HDLs)** remove cholesterol from tissues and carry it to the liver, where it can be used to synthesize bile. HDL consists of about 50% protein, 35% lipids, and 15% cholesterol.
- **Low-density lipoproteins (LDLs)** transport cholesterol around the body for use in biosynthesis and for storage. LDL consists of about 25% protein, 25% lipids, and 50% cholesterol.
- **Very-low-density lipoproteins (VLDLs)** contain mostly triglyceride fats, which they transport to fat cells in adipose tissues around the body. VLDL consists of about 2% protein, 94% lipids, and 3% cholesterol.

Because of their functions in cholesterol regulation—LDL "adds" it and HDL "removes" it—LDL is sometimes called "bad cholesterol" and HDL "good cholesterol." A high ratio of LDL to HDL in a person's blood is a risk factor for atherosclerotic heart disease. Cigarette smoking lowers HDL levels. Regular exercise increases them.

INSULIN AND GLUCAGON CONTROL FUEL METABOLISM During the absorptive period, blood glucose levels rise as carbohydrates are digested and absorbed. During this time, β cells of the pancreas release insulin, which plays a major role in directing glucose to where it will be used or stored. The actions of insulin vary in different tissues. Glucose enters cells by diffusion facilitated by glucose transporters. However, the glucose transporters in resting skeletal muscle and adipose tissues are normally sequestered in cytoplasmic vesicles until insulin combines with its receptors on the cell surface and triggers the insertion of transporters into the cell membrane. In adipose cells, insulin inhibits lipase and promotes fat synthesis from glucose. In the liver, insulin activates an enzyme that phosphorylates glucose as it enters the cell so it cannot diffuse back out again, enhancing the overall diffusion of glucose into the cells. Insulin also activates the enzymes in liver cells that catalyze the synthesis of glycogen. The many actions of insulin promote the uptake and utilization or storage of glucose by cells all around the body.

During the postabsorptive period, a fall in blood glucose decreases the release of insulin, and the uptake of glucose by most cells is curtailed (**Figure 50.19**). To maintain blood glucose levels, liver cells break down glycogen, which releases glucose into the blood. The liver and the adipose tissues supply fatty acids to the blood, and most cells preferentially use fatty acids as their metabolic fuel. One tissue that does not switch fuel sources during the postabsorptive period, however, is the nervous system.

The cells of the nervous system require a constant supply of glucose, and they can use other fuels to a very limited extent. Most neurons do not require insulin to absorb glucose from the blood, but they do need an adequate glucose concentration gradient to drive the facilitated diffusion of glucose across their plasma membranes. Therefore it is critical that blood glucose levels are maintained during the postabsorptive period. The overall dependence of neuronal tissues on glucose, and their requirement for constant blood glucose levels, are the reasons it is so important for other cells of the body to shift to fat metabolism during the postabsorptive period.

The metabolism of fuel molecules during the postabsorptive period is mostly controlled by the lack of insulin, but if blood glucose falls below a certain level, another pancreatic hormone, **glucagon**, is called into play. Glucagon's effect is opposite that of insulin: it stimulates liver cells to break down glycogen and to carry out gluconeogenesis. Thus, under the influence of glucagon, the liver produces glucose and releases it into the blood.

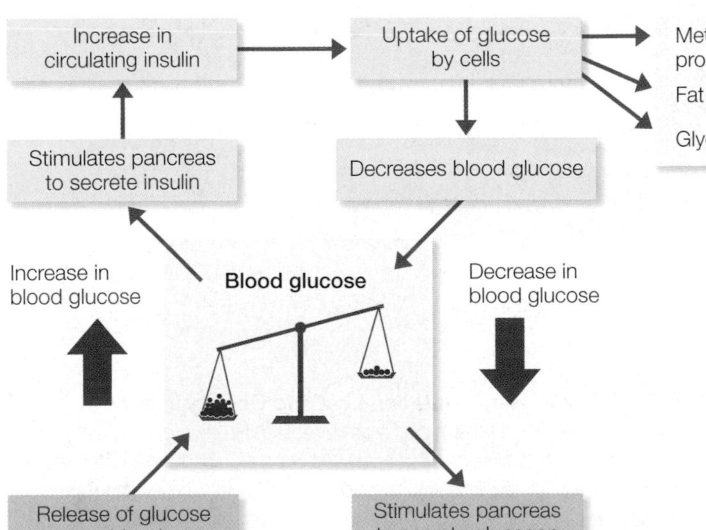

50.19 Regulating Glucose Levels in the Blood Insulin (blue) and glucagon (pink) interactions maintain the homeostasis of circulating glucose.

Regulating food intake is important

Obesity is a major health issue in the United States. People spend billions of dollars on schemes to lose weight, but the problem seems to get worse every year. A simple rule—take in fewer calories than your body burns while eating a balanced diet—should solve the problem, but it doesn't. Why? As we noted at the beginning of this chapter, lifestyle plays a major role in obesity, and genetic and regulatory factors "weigh in" as well.

The amount of food an animal eats is governed by its sensations of hunger and satiety. These sensations are influenced by the hypothalamus (see Section 41.2). If a region in the middle of the hypothalamus of rats, called the ventromedial hypothalamus, is damaged, the animals will increase their food intake and become obese. If a different region, called the lateral hypothalamus, is damaged, rats will decrease their food intake and become thin. In both cases, the rats eventually reach a new equilibrium body weight, which they maintain. Thus, regulation is maintained, but the set point has been changed. Other brain regions have also been implicated in control of hunger and satiety.

In Section 40.1 we learned that regulation involves feedback information and a means of comparing that information with a set point. Some evidence suggests that cells in the hypothalamus and liver are sensitive to glucose and insulin in the blood, with high levels stimulating satiety and low levels stimulating hunger. Even stronger evidence, however, indicates that signals from fat metabolism influence hunger and satiety.

A single-gene mutation in mice, when present in the homozygous condition, results in mice that eat enormous amounts of food and become obese. Geneticists call these mice *ob/ob*, given their double dose of the recessive "obese" allele. Experiments revealed that the wild-type *Ob* allele codes for a protein, which was named **leptin** (Greek *leptos*, "thin"). Leptin is a hormone produced by fat cells and circulates in the blood. Receptors for leptin are found in the regions of the hypothalamus that are involved in control of hunger and satiety. It seems that leptin provides feedback information about the status of the body fat reserves to the brain. When leptin was injected into *ob/ob* mice they ate less and lost body fat. However, a different strain of obese mice, *db/db*, did not respond to leptin injections. Further experiments revealed that *db/db* mice lacked the leptin receptor (**Figure 50.20**).

Could leptin be used to reduce human obesity? In the very few obese people who do not produce the hormone, leptin injections have curbed appetites and facilitated weight loss. Most obese people, however, have higher than normal circulating levels of leptin, so it is likely that their leptin receptors are not completely functional.

Additional feedback signals are most certainly involved in the regulation of food intake. Recently a hormone called **ghrelin** was discovered that is produced and secreted by cells in the stomach. Normally, ghrelin levels rise before a meal and fall after a meal. Fasting causes an increase in ghrelin levels. Ghrelin binding to its receptors in the hypothalamus stimulates appetite. Ghrelin also stimulates cells in the pituitary gland to release growth hormone.

EXPERIMENT

HYPOTHESIS: Strains of mice that become obese lack a satiety factor (or its receptor) in their brains.

METHOD

Using mice from two strains that are genetically obese, surgically join the circulatory systems of normal or wild-type mice with an obese-strain partner so that they share circulating blood (i.e., create parabiotic pairs).

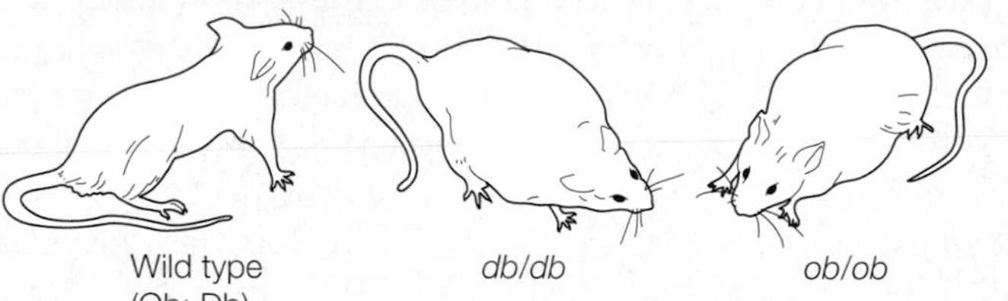

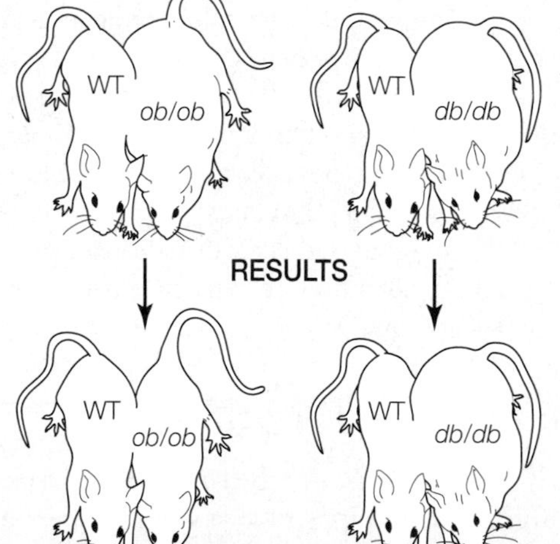

Parabiotic *ob/ob* mice lose fat, but parabiotic *db/db* mice do not. The *ob/ob* mice obtain the satiety factor from the wild-type mouse. *db/db* mice obtain the satiety factor, but are still fat because they lack the receptor.

CONCLUSION: The wild-type *Ob* gene codes for a satiety hormone. This hormone is given the name leptin. The *Db* gene encodes the leptin receptor.

50.20 A Single-Gene Mutation Leads to Obesity in Mice A study of obese and wild-type mice revealed the existence of a satiety factor, leptin, and its receptor. The wild-type *Ob* gene encodes leptin. The recessive allele *ob* is a loss-of-function allele, so *ob/ob* mice do not have leptin; they do not experience satiety and eat so much that they become obese. The wild-type *Db* gene encodes the leptin receptor, so mice homozygous for the loss-of-function allele (*db/db*), even if supplied leptin, cannot use it. The *ob/ob* mice do have the wild-type *Db* gene, so if supplied with leptin they experience satiety and lose fat.

50.4 RECAP

The major controlling factors of gut function are an intrinsic nervous system and the hormones gastrin, secretin, and cholecystokinin. Insulin is the major hormonal controller of fuel metabolism. The hypothalamus controls food intake by generating sensations of hunger and satiety influenced by feedback from blood glucose and the hormones leptin and ghrelin.

- What are the roles of the three different classes of lipoproteins? See pp. 1086–1087
- By what actions does insulin promote uptake and storage of energy during the absorptive period? See p. 1087 and Figure 50.19
- Can you describe the evidence that leptin influences satiety? See p. 1088 and Figure 50.20

The goal of digestion and absorption of food for any animal is nourishment. Not everything an animal ingests, however, is safe; many foods also contain toxins.

50.5 How Do Animals Deal with Ingested Toxins?

Some plant and animal tissues contain more than just nutrients; some can also contain toxic compounds. Plants produce toxic secondary metabolites as defenses against herbivores. One example is the nicotine in tobacco. Animals may use toxins for capturing prey as well as for self-defense, so ingesting certain plant and animal tissues can be dangerous. In addition, human activities add millions of tons of synthetic toxic compounds to the environment every year, and many of these compounds enter the air we breathe, the water we drink, and the food we eat. A new field called **environmental toxicology** addresses the problems of poisons in the environment.

The body cannot metabolize many synthetic toxins

How does the animal body handle synthetic toxins? In many cases, the systems that metabolize natural chemicals can also metabolize synthetic toxins, breaking them apart and eliminating them through the urine. As we saw earlier, everything absorbed from the small intestine is transported first to the liver; liver enzymes called *cytochrome P450*s are largely responsible for the detoxification of absorbed chemicals.

P450s are less specific in their abilities to bind substrates than are most enzymes; thus, each P450 can catalyze reactions with a wide range of compounds, and many P450s exist. Few natural compounds can escape the P450s, even when the body encounters them for the first time.

Some synthetic chemicals, however, fall outside the range of structures that P450s and other enzymes can metabolize. If a synthetic chemical that cannot be metabolized is structurally similar to a hormone, that synthetic chemical may activate the hormone's signaling pathway within target cells. Whereas the natural hormonal signal can be turned off, the synthetic signal cannot be, and control of function is lost.

Some toxins are retained and concentrated

The physical and chemical properties of a toxic compound affect its retention within a biological system. If a compound can dissolve in water, it may be quickly metabolized (and thus detoxified) because it is accessible to the wide variety of enzymes that can break down complex molecules in food. In addition to being degraded or metabolized, many water-soluble compounds can be filtered out of the blood by the kidneys, and therefore do not accumulate in the body. However, some dangerous water-soluble compounds can be incorporated into the body and disrupt normal functions. An example is lead, which can replace iron in blood and calcium in bone.

Lipid-soluble compounds are usually metabolized more slowly than water-soluble compounds, and they are often stored in the body for a long time because they dissolve in adipose tissues. Accumulated lipophilic compounds can reach very high concentrations. Some lipid-soluble toxins, including many pesticides, can **bioaccumulate** in the environment; that is, they can become concentrated in predators that eat contaminated prey. The pesticide load is passed up the food chain from prey to predator, growing increasingly concentrated in the tissues of each consumer. In the top predator, the pesticide may be concentrated thousands or millions of times. Long-lived predators, such as eagles and bears, are particularly at risk for heavy pesticide burdens because they have many years to accumulate them. Bioaccumulated toxins may be responsible for the high rates of cancer and infertility found in some wildlife populations.

One class of toxic synthetic chemicals that has bioaccumulated in animals, including humans, are polychlorinated biphenyls (PCBs). PCBs were produced extensively for use as an insulating fluid in electrical transformers from the 1930s until recently. They are chemically stable, lipophilic, and are now found throughout the environment. They have been shown to bioaccumulate, reaching dangerously high levels in fish from contaminated waters such as the Great Lakes. In communities around the Great Lakes, studies have indicated cognitive impairment in children of mothers with a high body burden of PCBs, probably from eating fish caught in the Great Lakes.

The risks of PCBs are now clear, but it is usually difficult to make a causal connection between a toxin in the environment and specific health effects in a population. Environmental toxicologists must be able to study large populations, use powerful statistical analyses, and do controlled laboratory studies to obtain evidence that will support policy changes to stop and reverse the effects of synthetic environmental toxins.

CHAPTER SUMMARY

50.1 What do animals require from food?

Animals are **heterotrophs** that derive their energy and molecular building blocks, directly or indirectly, from **autotrophs**.

Carbohydrates, fats, and proteins in food supply animals with metabolic energy. A measure of the energy content of food is the **kilocalorie**. Excess caloric intake is stored as glycogen and fat. Review Figure 50.2

For many animals, food provides essential **carbon skeletons** that they cannot synthesize themselves. Review Figure 50.4

Humans require eight **essential amino acids** in the diet. Different animals need mineral elements in different amounts. **Macronutrients** are needed in large quantities. **Micronutrients** are needed in small amounts. Review Figure 50.5 and Table 50.1, Web/CD Activity 50.1

Vitamins are organic molecules that must be obtained in food. Review Table 50.2, Web/CD Activity 50.2

Malnutrition results when any essential nutrient is lacking from the diet. A chronic state of malnutrition causes a **deficiency disease**.

50.2 How do animals ingest and digest food?

Animals can be characterized by how they acquire nutrients: **Saprobes** and **detritivores** depend on dead organic matter, **filter feeders** strain the aquatic environment for small food items, **herbivores** eat plants, and **carnivores** eat animals. Behavioral and anatomical adaptations reflect these feeding strategies. See Web/CD Activity 50.3

Digestion involves the breakdown of complex food molecules into monomers that can be absorbed and utilized by cells. In most animals, digestion takes place in a tubular gut. Review Figure 50.8

Absorptive areas of the gut are characterized by a large surface area produced by extensive folding and numerous **villi** and **microvilli**. Review Figure 50.9

Hydrolytic enzymes break down proteins, carbohydrates, and fats into their monomeric units. To prevent the organism itself from being digested, many of these enzymes are released as inactive **zymogens**, which become activated when secreted into the gut.

50.3 How does the vertebrate gastrointestinal system function?

The vertebrate gut can be divided into several compartments with different functions. Review Figure 50.10, Web/CD Activity 50.4

The cells and tissues of the vertebrate gut are organized in the same way throughout its length. The innermost tissue layer, the **mucosa**, is the secretory and absorptive surface. The **submucosa** contains blood and lymph vessels, and a nerve plexus. External to the submucosa are two **smooth muscle** layers. Between the two muscle layers is another nerve plexus that controls the movements of the gut. Review Figure 50.11

Swallowing is a reflex that pushes the bolus of food into the **esophagus**. **Peristalsis** and other movements of the gut move the bolus down the esophagus and through the entire length of the gut. **Sphincters** block the gut at certain locations, but they relax as a wave of peristalsis approaches. Review Figure 50.12

Digestion begins in the mouth, where **amylase** is secreted with the saliva. Digestion of protein begins in the stomach, where parietal cells secrete HCl and chief cells secrete **pepsinogen**, a zymogen that becomes **pepsin** when activated by low pH and **autocatalysis**. The mucosa also secretes **mucus**, which protects the tissues of the gut. Review Figure 50.13

In the **duodenum**, pancreatic enzymes carry out most of the digestion of food. **Bile** from the **liver** and **gallbladder** emulsify fats into **micelles**. Bicarbonate ions from the pancreas neutralize the pH of the **chyme** entering from the stomach to produce an environment conducive to the actions of pancreatic enzymes such as **trypsin**. Review Figure 50.15 and Table 50.3

Final enzymatic cleavage of polypeptides and disaccharides occurs among the microvilli of the intestinal mucosa. Amino acids, monosaccharides, and inorganic ions are absorbed by the microvilli. Specific transporter proteins are sometimes involved. Sodium co-transport often powers the active transport of nutrients.

Fats broken down by **lipases** are absorbed mostly as monoglycerides and fatty acids and are resynthesized into triglycerides within cells. The triglycerides are combined with cholesterol and phospholipids and coated with protein to form **chylomicrons**, which pass out of the mucosal cells and into lymphatic vessels in the submucosa. Review Figure 50.16, Web/CD Tutorial 50.1

Water and ions are absorbed in the **large intestine** as waste matter is consolidated into **feces**, which is periodically eliminated.

Microorganisms in some compartments of the gut digest materials that their host cannot. Review Figure 50.17

50.4 How is the flow of nutrients controlled and regulated?

Autonomic reflexes coordinate activity of the digestive tract, which has an intrinsic nervous system that can act independently of the CNS.

The actions of the stomach and small intestine are largely controlled by the hormones **gastrin**, **secretin**, and **cholecystokinin**. Review Figure 50.18

The liver plays a central role in directing the traffic of fuel molecules. During the **absorptive period**, the liver takes up and stores fats and carbohydrates, converting monosaccharides to glycogen or fats. The liver also takes up amino acids and uses them to produce blood plasma proteins, and can engage in gluconeogenesis.

Fat and cholesterol are shipped out of the liver as **low-density lipoproteins**. **High-density lipoproteins** act as acceptors of cholesterol and are believed to bring fat and cholesterol back to the liver.

Insulin largely controls fuel metabolism during the absorptive period and promotes glucose uptake as well as glycogen and fat synthesis. During the **postabsorptive** period, lack of insulin blocks the uptake and utilization of glucose by most cells of the body except neurons. If blood glucose levels fall, **glucagon** secretion increases, stimulating the liver to break down glycogen and release glucose to the blood. Review Figure 50.19, Web/CD Tutorial 50.2

Food intake is governed by sensations of hunger and satiety, which are determined by brain mechanisms. Review Figure 50.20

50.5 How do animals deal with ingested toxins?

Toxins in food may come from natural sources, but many come from human activities such as the use of pesticides and the release of pollutants into the environment.

Toxins such as PCBs that accumulate in the bodies of prey are transferred to and further concentrated in the bodies of their predators. This **bioaccumulation** produces high concentrations of toxins in animals high up the food chain.

SELF-QUIZ

1. Most of the metabolic energy needed by a bird for a long-distance migratory flight is stored as
   a. glycogen.
   b. fat.
   c. protein.
   d. carbohydrates.
   e. ATP.
2. Which statement about essential amino acids is true?
   a. They are not found in vegetarian diets.
   b. They are stored by the body until they are needed.
   c. Without them, one is undernourished.
   d. All animals require the same ones.
   e. Humans can acquire all of theirs by eating milk, eggs, and meat.
3. Which statement about vitamins is true?
   a. They are essential inorganic nutrients.
   b. They are required in larger amounts than are essential amino acids.
   c. Many serve as coenzymes.
   d. Vitamin D can be acquired only by eating meat or dairy foods.
   e. When vitamin C is eaten in large quantities, the excess is stored in fat for later use.
4. The digestive enzymes of the small intestine
   a. do not function best at a low pH.
   b. are produced and released in response to circulating secretin.
   c. are produced and released under neuronal control.
   d. are all secreted by the pancreas.
   e. are all activated by an acidic environment.
5. Which statement about nutrient absorption by the intestinal mucosal cells is true?
   a. Carbohydrates are absorbed as disaccharides.
   b. Fats are absorbed as fatty acids and monoglycerides.
   c. Amino acids move across the plasma membrane only by diffusion.
   d. Bile transports fats across the plasma membrane.
   e. Most nutrients are absorbed in the duodenum.
6. Chylomicrons are like the tiny micelles of dietary fat in the lumen of the small intestine in that both
   a. are coated with bile.
   b. are lipid soluble.
   c. travel through the lymphatic system.
   d. contain triglycerides.
   e. are coated with lipoproteins.
7. Microbial fermentation in the gut of a cow
   a. produces fatty acids as a major nutrient for the cow.
   b. occurs in specialized regions of the small intestine.
   c. occurs in the cecum, from which food is regurgitated, chewed again, and swallowed into the true stomach.
   d. produces methane as a major nutrient.
   e. is possible because the stomach wall does not secrete hydrochloric acid.
8. Which of the following is stimulated by cholecystokinin?
   a. Stomach motility
   b. Release of bile
   c. Secretion of hydrochloric acid
   d. Secretion of bicarbonate ions
   e. Secretion of mucus
9. During the absorptive period,
   a. breakdown of glycogen supplies glucose to the blood.
   b. glucagon secretion is high.
   c. the number of circulating lipoproteins is low.
   d. glucose is the major metabolic fuel.
   e. the synthesis of fats and glycogen in muscle is inhibited.
10. During the postabsorptive period,
   a. glucose is the major metabolic fuel.
   b. glucagon stimulates the liver to produce glycogen.
   c. insulin facilitates the uptake of glucose by brain cells.
   d. fatty acids constitute the major metabolic fuel.
   e. liver functions slow down because of low insulin levels.

FOR DISCUSSION

1. Several currently popular diet books recommend high fat and protein intake and low carbohydrate intake as a means of losing body mass. What could the rationale of a high-fat and high-protein diet be, and what health issues should be considered when someone considers going on such a diet?
2. Carnivores generally have more dietary vitamin requirements than herbivores do. Why?
3. It is said that the most important hormonal control of fuel metabolism in the postabsorptive period is the lack of insulin. Explain.
4. Why is obstruction of the common bile duct so serious? Consider in your answer the multiple functions of the pancreas and the way in which digestive enzymes are processed.
5. Trace the history of a fatty acid molecule from a slice of cheese pizza to a plaque on a coronary artery. Into what possible forms and structures might it have been converted as it passed through in the body? Describe a direct and an indirect route it could have taken.

FOR INVESTIGATION

Cystic fibrosis is a genetic disease due to a mutation in the gene for a protein that transports chloride ions out of cells. Heterozygous individuals are normal, but are carriers of the disease. Homozygous individuals have trouble clearing their airways and are subject to respiratory infections. Before the advent of effective respiratory therapy, such individuals succumbed to the disease early in life. How do you think such a gene could have become established in the human population? (A clue might come from our knowledge of the mode of action of the cholera toxin, which overly stimulates chloride transporters in the membranes of intestinal cells, as described on pp. 76–77 of Chapter 5). Could cystic fibrosis, like sickle-cell disease, be a case of heterozygote advantage? How would you investigate that hypothesis?

CHAPTER 51 Salt and Water Balance and Nitrogen Excretion

Blood, sweat, and tears

Blood, sweat, and tears taste salty because they have similar ionic concentrations as the interstitial fluids that bathe the cells of the body. The volume and the composition of the interstitial fluids must remain within certain limits and kept relatively free of wastes. Maintaining homeostasis of the interstitial fluids is the job of the excretory system, and it can be challenging. The nature of the challenge depends on the environment of the animal and its lifestyle. Some animals such as desert insects and small mammals may never experience free water in their environments. They must be able to live their entire lives without drinking. All animals derive water from the metabolism of food, but to make that amount of water do, desert animals must conserve it. Accordingly, they excrete wastes that are extremely concentrated. Insects excrete semisolid wastes and desert rodents excrete urine that is so concentrated it contains crystals of solute.

Animals that live in fresh water have the opposite problem; water is continuously entering their bodies by osmosis and with the food they eat, so they must constantly bail themselves out by producing copious amounts of dilute urine while they conserve the solutes their bodies need. We will see the adaptations such animals have to maintain salt and water balance, but to show how flexible physiological adaptations to the environment can be, we will discuss an animal that experiences both extreme conditions, and within minutes of each other. We will consider the problems of vampires—not the horror film kind, but the bat kind.

Vampire bats are small tropical mammals that feed on the blood of other mammals, such as cattle. The bat lands on an unsuspecting (usually sleeping) victim, bites into a vein, and drinks blood—a high-protein, liquid food. The bat has only a short time to feed before the victim wakes. To maximize the volume of blood it can ingest, it eliminates water from its food as fast as it can by producing a lot of very dilute urine. The warm trickle down the neck of the victim is not blood!

Once feeding ends, this high rate of water loss must stop. Now the vampire bat is digesting protein and must excrete large amounts of nitrogenous breakdown products while conserving its body water. Within minutes, the excretory system of the vampire bat switches from producing abundant, very dilute urine to producing a tiny

Blood as a Fast Food The vampire bat *Desmodus rotundus* is able to adjust its excretory physiology from water-excreting to water-conserving, depending on whether it is ingesting or digesting its blood meal.

Living without Water Kangaroo rats such as *Dipodomys spectabilis*, a denizen of the Arizona desert, may never see free water during their lifetime. Their dry-adapted excretory systems allow these rodents to derive enough water to survive on from their food.

CHAPTER OUTLINE

amount of highly concentrated urine. The vampire bat rapidly switches from an excretory physiology typical of a mammal living in an environment with abundant fresh water (copious amounts of dilute urine) to an excretory physiology typical of a mammal living in an arid desert (small amounts of concentrated urine).

IN THIS CHAPTER we will discover how animals accomplish the various feats of salt and water balance and waste excretion that adapt them to different environments. We will begin by discussing the challenges presented by those different environments. Then we will explore some invertebrates that illustrate basic mechanisms common to all animal excretory systems. Turning to vertebrates, we will learn about the basic anatomical unit of the excretory system—the nephron—and how it evolved. Finally, we will cover the mechanisms that control and regulate salt and water balance in mammals, giving the vampire bat and other species their remarkable abilities to exploit unusual diets and extreme environments.

51.1 What Roles Do Excretory Organs Play in Maintaining Homeostasis?

Excretory organs control the volume, concentration, and composition of the extracellular fluids of animals. Life evolved in the seas, and seawater is the extracellular environment for the cells of the simplest marine animals. The cells of many small marine animals derive their oxygen directly from the seawater and expel carbon dioxide and nitrogenous wastes into the seawater. Larger and more complex marine animals cannot serve the needs of all of their cells by direct exchanges with seawater, so they maintain an internal environment of extracellular fluids. The salt concentration and composition of the extracellular fluids of most marine invertebrates are similar to that of seawater. The characteristics of extracellular fluids of marine vertebrates and all freshwater and terrestrial animals, in contrast, are considerably different from those of seawater. The concentration of solutes in an animal's extracellular fluid determines the water balance of its cells, and the composition of the extracellular fluid influences the health and functions of the cells it bathes. Recall, for example, the importance of ion concentration gradients between the extracellular fluid and the cytoplasm of nerve and muscle cells (see Sections 44.2 and 47.1).

Water enters or leaves cells by osmosis

The volume of cells depends on whether they take up water from or lose water to the extracellular fluids. The movement of water across cell plasma membranes depends on differences in solute concentration. If the solute concentrations are different on two sides of a membrane permeable to water but not the solutes, the water will flow from the side with the lower concentration to the side with the higher concentration of solute (see Sections 5.3 and 35.1 for more about osmosis). Likewise, if the solute concentration of the extracellular fluid surrounding animal cells is less than that of the cytoplasm, water moves into the cells, causing them to swell and possibly burst. If the solute concentration of the extracellular fluid is greater than that of the cytoplasm, the cells lose water and shrink. Thus the solute concentration of the extracellular fluid affects both the volume and the solute concentration of the cells.

Animal physiologists use the term **osmolarity** in discussing osmosis. The osmolarity of a solution is the number of moles of osmotically active solutes per liter of solvent. Thus, a 1 molar solution of glucose is also a 1 osmolar (1 osmole per liter) solution, but a 1 molar solution of sodium chloride (NaCl) is a 2 osmolar solution, because each NaCl molecule dissociates into two osmotically active ions.

To achieve cellular water balance, animals must maintain the osmolarity of their extracellular fluid within an appropriate range. In addition, they must maintain an appropriate solute composition by saving some substances and eliminating others. To accommodate these needs, most animals have excretory organs.

Excretory organs control extracellular fluid osmolarity by filtration, secretion, and reabsorption

Excretory organs control the osmolarity and the volume of the extracellular fluids (blood and interstitial fluid) by excreting solutes that are present in excess (such as NaCl when we eat lots of salty food) and conserving solutes that are valuable or in short supply (such as glucose and amino acids). In terrestrial organisms, these excretory organs also eliminate the waste products of nitrogen metabolism. The output of the excretory organs is called **urine**.

Certain basic principles apply to all animal excretory organs. In some way, the excretory organ filters extracellular fluid to produce a filtrate that contains no cells or large molecules, such as proteins. The composition of this filtrate is then modified to produce urine. In animals with closed circulatory systems, the blood plasma is filtered across the walls of capillaries. The filtration is driven by blood pressure. Water and small molecular weight solutes cross the capillary wall, but large molecular weight solutes and cells remain in the blood. The filtrate (water and small molecules) then flows through tubules. The cells of the tubules change the composition of the filtrate by active secretion and reabsorption of specific solute molecules. We will see that these three mechanisms—filtration, secretion, and reabsorption—are used in the excretory systems of a wide variety of animals. In all excretory systems *there is no active transport of water*. Water must be moved either by a pressure difference or by a difference in osmolarity.

Animals can be osmoconformers or osmoregulators

Animals that live in marine, freshwater, or terrestrial environments face different salt and water balance problems. In the terrestrial environment salts and water can be scarce and usually must be conserved by excretory systems. In the freshwater environment water is plentiful, but salts are scarce. Freshwater animals have to conserve salts and excrete the water that continuously invades their bodies through osmosis.

The osmolarity of ocean water is high—over 1,000 milliosmoles/liter. Most marine invertebrates equilibrate their extracellular fluid osmolarity with the ocean water and are therefore called **osmoconformers**. Other marine animals maintain extracellular fluid osmolarities much lower than seawater and are therefore called **osmoregulators**. All marine vertebrates except for sharks and rays are osmoregulators and maintain their extracellular fluids at about 300 mosm/l like humans do.

There are limits to osmoconformity. Even animals that can osmoconform over a wide range of osmolarities must osmoregulate in extreme environments. For example, no animal could survive if its extracellular fluid had the osmolarity of fresh water, because that would mean there were too few solutes in the extracellular fluid, including nutrients and ions necessary for cell functions. Nor could animals survive with internal osmolarities as high as those that may be reached in an evaporating tide pool. High solute concentrations can cause proteins to denature. A case in point is the brine shrimp *Artemia* (**Figure 51.1**), which can live in environments of almost any osmolarity. *Artemia* are found in huge numbers in the most salty environments known, such as Utah's Great Salt Lake, and in coastal evaporation ponds where salt is concentrated for commercial purposes (see Figure 26.24). The osmolarity of such water reaches 2,500 mosm/l. At these high environmental osmolarities, *Artemia* maintains its tissue fluid osmolarity considerably below that of the environment. Its mechanism of osmoregulation is the active transport of Cl^- from its extracellular fluid out across its gill membranes to the environment. Na^+ ions follow.

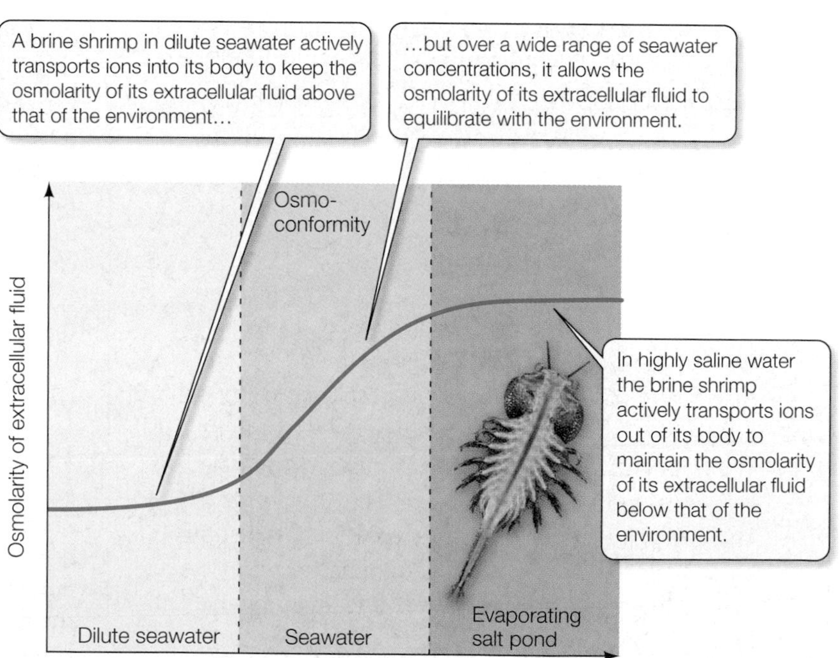
A brine shrimp in dilute seawater actively transports ions into its body to keep the osmolarity of its extracellular fluid above that of the environment...
...but over a wide range of seawater concentrations, it allows the osmolarity of its extracellular fluid to equilibrate with the environment.
Osmo-conformity
In highly saline water the brine shrimp actively transports ions out of its body to maintain the osmolarity of its extracellular fluid below that of the environment.
Osmolarity of extracellular fluid
Dilute seawater
Seawater
Evaporating salt pond
Osmolarity of environment

51.1 Environments Can Vary Greatly in Salt Concentration Animals such as the brine shrimp that live at the extremes of environmental osmolarities display flexible osmoregulatory abilities. When they encounter dilute seawater they can regulate their extracellular fluid osmolarity above that of the environment, and when they encounter extreme salt concentrations they can regulate their extracellular fluid osmolarity below that of the environment.

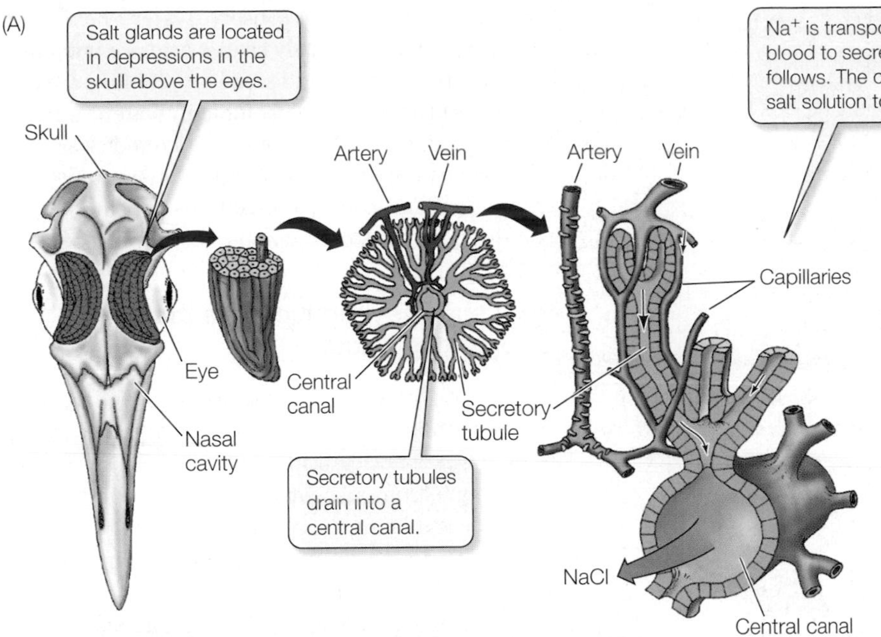

Salt glands are located in depressions in the skull above the eyes.
Skull

Central canal

(B) *Macronectes giganteus*

Note the secretion at the tip of the bird's beak.

51.2 Nasal Salt Glands Excrete Excess Salt (A) Marine birds have nasal salt glands adapted to excrete the excess salt from the seawater they consume with their food. (B) This giant petrel has returned from a feeding trip at sea and is secreting salt through its nasal salt gland.

Artemia cannot survive in fresh water, but it can live in dilute seawater, in which it maintains the osmolarity of its extracellular fluid above that of the environment. Under these conditions, *Artemia* reverses the direction of Cl^- transport across its gill membranes.

Animals can be ionic conformers or ionic regulators

Osmoconformers can also be ionic conformers, allowing the ionic composition, as well as the osmolarity, of their extracellular fluid to match that of the environment. Most osmoconformers, however, are ionic regulators to some degree; they employ active transport mechanisms to excrete some ions and to maintain other ions in their extracellular fluid at optimal concentrations.

Terrestrial animals obtain their salts from food and regulate the ionic composition of their extracellular fluids by conserving some ions and excreting others. For example, herbivores have to conserve Na^+ because the plants they eat have low concentrations of Na^+. Some terrestrial herbivores travel long distances to naturally occurring salt licks. By contrast, birds that feed on marine animals must excrete the excess of sodium they ingest with their food. Their *nasal salt glands* excrete a concentrated solution of NaCl via a duct that empties into the nasal cavity. Birds, such as penguins and seagulls, that have nasal salt glands can be seen frequently sneezing or shaking their heads to get rid of the very salty droplets excreted from their nasal salt glands (**Figure 51.2**).

51.1 RECAP

Excretory organs control water and salt balance and the excretion of nitrogenous waste products through three mechanisms: filtration of body fluids to form urine, active secretion of substances into the urine, and active reabsorption of substances from the urine.

- Can you explain the two mechanisms used to move water across membranes? See pp. 1093–1094
- Can you describe different salt and water balance problems that animals might encounter in marine, freshwater, and terrestrial environments and some ways they meet those challenges? See p. 1094 and Figures 51.1 and 51.2

51.2 How Do Animals Excrete Toxic Wastes from Nitrogen Metabolism?

In addition to maintaining salt and water balance, animals must eliminate the waste products of metabolism from their extracellular fluids. The end products of the metabolism of carbohydrates and fats are water and carbon dioxide, which are not difficult to eliminate. Proteins and nucleic acids, however, contain nitrogen, so their metabolism produces *nitrogenous wastes* in addition to water and carbon dioxide.

Animals excrete nitrogen in a number of forms

Artery

Central canal
Secretory tubules drain into a central canal.
Secretory tubule
Artery

Na⁺ is transported from the blood to secretory tubules. Cl⁻ follows. The central canal carries salt solution to the nasal cavity.
Capillaries

The most common nitrogenous waste is **ammonia** (NH_3). Because it is highly toxic, ammonia must either be excreted continuously to prevent its accumulation, or it must be detoxified by conversion into other molecules (**Figure 51.3**).

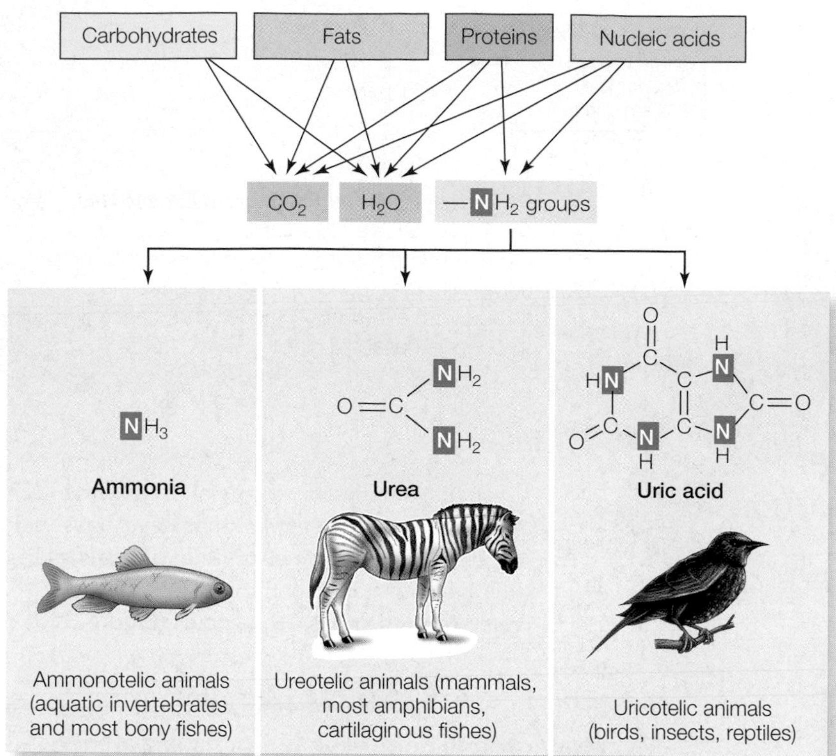

51.3 Waste Products of Metabolism The metabolism of proteins and nucleic acids produces nitrogenous wastes. Most aquatic animals, including most fishes, excrete nitrogenous wastes as ammonia. Most terrestrial animals excrete either urea or uric acid. Urea is more soluble in water and is the major nitrogenous excretory product for mammals, amphibians, and some fishes. Uric acid is not very soluble in water and is the major nitrogenous excretory product for birds, reptiles, and insects.

AMMONIA Ammonia is highly soluble in water and diffuses rapidly, so the continuous excretion of ammonia is relatively simple for aquatic animals. Animals that breathe water continuously lose ammonia from their blood to the environment by diffusion across their gill membranes. Animals that excrete ammonia, such as aquatic invertebrates and bony fishes, are **ammonotelic.**

If ammonia builds up in the extracellular fluids, it becomes toxic at rather low levels and is a dangerous metabolite for terrestrial animals, which cannot continuously excrete ammonia the way aquatic animals do. Therefore, terrestrial (and some aquatic) animals convert ammonia into either **urea** or **uric acid**.

UREA **Ureotelic** animals, such as mammals, amphibians, and cartilaginous fishes (sharks and rays), excrete urea as their principal nitrogenous waste product. Urea is quite soluble in water, but excretion of urea solutions can result in a large loss of water that many terrestrial animals can ill afford. As we will see later in this chapter, mammals have evolved excretory systems that conserve water by producing concentrated urea solutions. The sharks and rays are another story. These marine species keep their extracellular fluids almost isosmotic (same osmotic concentration) to the marine environment by retaining high concentrations of urea.

URIC ACID Animals that conserve water by excreting nitrogenous wastes mostly as uric acid are **uricotelic.** Insects, reptiles, birds, and some amphibians are uricotelic. Uric acid is very insoluble in water, so it can precipitate out of the urine and be excreted as a semisolid (for example, it is the whitish material in bird droppings). A uricotelic animal loses very little water as it disposes of its nitrogenous wastes.

Most species produce more than one nitrogenous waste

Humans are ureotelic, but we also excrete uric acid. The uric acid in human urine comes largely from the metabolism of nucleic acids and caffeine. Humans can also excrete ammonia, which is an important mechanism for regulating the pH of the extracellular fluids. Excreted ammonia buffers the urine and enables the secretion of more hydrogen ions.

Species that live in different habitats at different developmental stages may utilize more than one mechanism of nitrogen excretion. The tadpoles of frogs and toads, for example, excrete ammonia across their gill membranes, but when they develop into adult frogs or toads, they generally excrete urea. Some adult amphibians that live in arid habitats excrete uric acid.

If you wake up at night with a sharp pain in your big toe, you might have the malady of kings—gout. Known for thousands of years, gout was most commonly seen in the overindulgent wealthy. A protein-rich diet coupled with too much alcohol, especially beer, raises blood uric acid levels and causes uric acid crystals to form in joints. (The big toe is especially vulnerable.)

51.2 RECAP

Ammonia is a common toxic waste product of nitrogen metabolism. Aquatic animals excrete ammonia by diffusion into the water. Terrestrial animals detoxify it by conversion to urea or uric acid.

- Why might you expect a species from an arid habitat to use uric acid as its primary nitrogenous waste product? See p. 1096

The challenges animals face with respect to homeostasis of salt and water balance differ with the environment in which they live. Therefore, we can expect to find in animals a variety of adaptations by which they regulate this aspect of their internal environments. However, as we will see in the next section of this chapter, these adaptations are based on a limited number of mechanisms—namely, tubular processing of the extracellular fluids to conserve some solutes and excrete others.

51.3 How Do Invertebrate Excretory Systems Work?

Freshwater and terrestrial invertebrates have a wide variety of adaptations for maintaining salt and water balance and excreting nitrogen. In this section, we will explore three examples of invertebrate excretory systems.

The protonephridia of flatworms excrete water and conserve salts

Many flatworms, such as *Planaria,* live in fresh water. These animals excrete water through an elaborate network of tubules running throughout their bodies. The tubules end in *flame cells,* so called because each cell has a tuft of cilia projecting into the tubule. The beating of the cilia gives the appearance of a flickering flame (**Figure 51.4**). A flame cell and a tubule together form a **protonephridium** (plural protonephridia; from the Greek *proto,* "before," and *nephros,* "kidney").

Extracellular fluid enters the tubules by filtration. The beating of the cilia causes a slight negative pressure within the tubule, and in response to this pressure difference, extracellular fluid flows into the tubule and flows toward the animal's excretory pore. As it flows, the cells of the tubules modify the composition of the fluid. As the modified tubule fluid (urine) leaves the flatworm's body, it is less concentrated than the animal's extracellular fluid, so ions are conserved and water is excreted by the protonephridium.

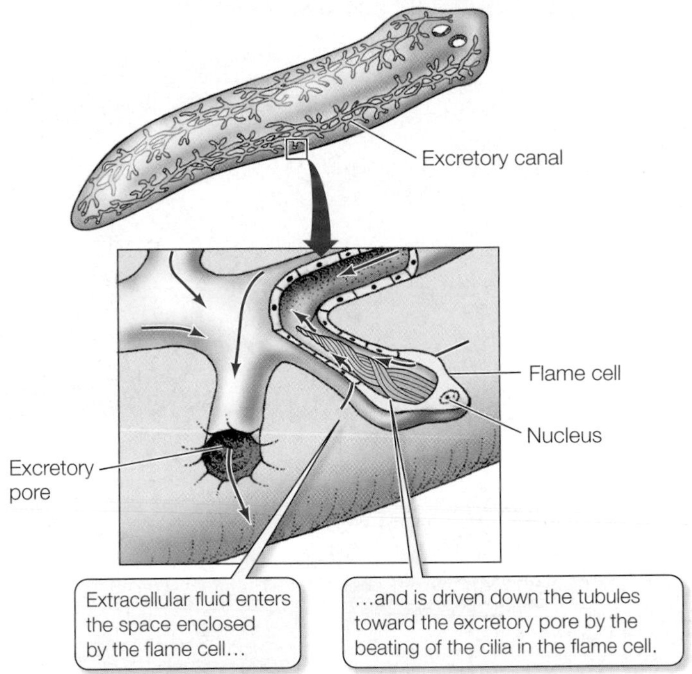

51.4 Protonephridia in Flatworms The protonephridia of the flatworm *Planaria* consist of tubules ending in flame cells. The tubule cells modify the composition of the fluid passing through them.

The metanephridia of annelids process coelomic fluid

Filtration of body fluids and modification of urine by tubules are highly developed processes in annelid worms, such as the earthworm. Recall that annelids are segmented and have a fluid-filled body cavity, called a *coelom,* in each segment (see Figure 32.12). Annelids have a closed circulatory system through which blood is pumped under pressure. The pressure causes the blood to be filtered across the thin, permeable capillary walls into the coelom. The cells and large protein molecules of the blood stay behind in the capillaries, while water and small molecules leave them and enter the coelom. In addition, some waste products, such as ammonia, diffuse directly from the tissues into the coelom. But where does this coelomic fluid go?

Each segment of the earthworm contains a pair of **metanephridia** (singular metanephridium; from the Greek *meta,* "akin to," and *nephros,* "kidney"). Each metanephridium begins in one segment as a ciliated, funnel-like opening in the coelom, called a *nephrostome,* which leads into a tubule in the next segment. The tubule ends in a pore, called the *nephridiopore,* that opens to the outside of the animal (**Figure 51.5**). Coelomic fluid enters the metanephridia through the nephrostomes. As the fluid passes through the tubules, their cells actively reabsorb certain molecules from it and actively secrete other molecules into it. What leaves the animal through the nephridiopores is a dilute urine containing nitrogenous wastes and other solutes.

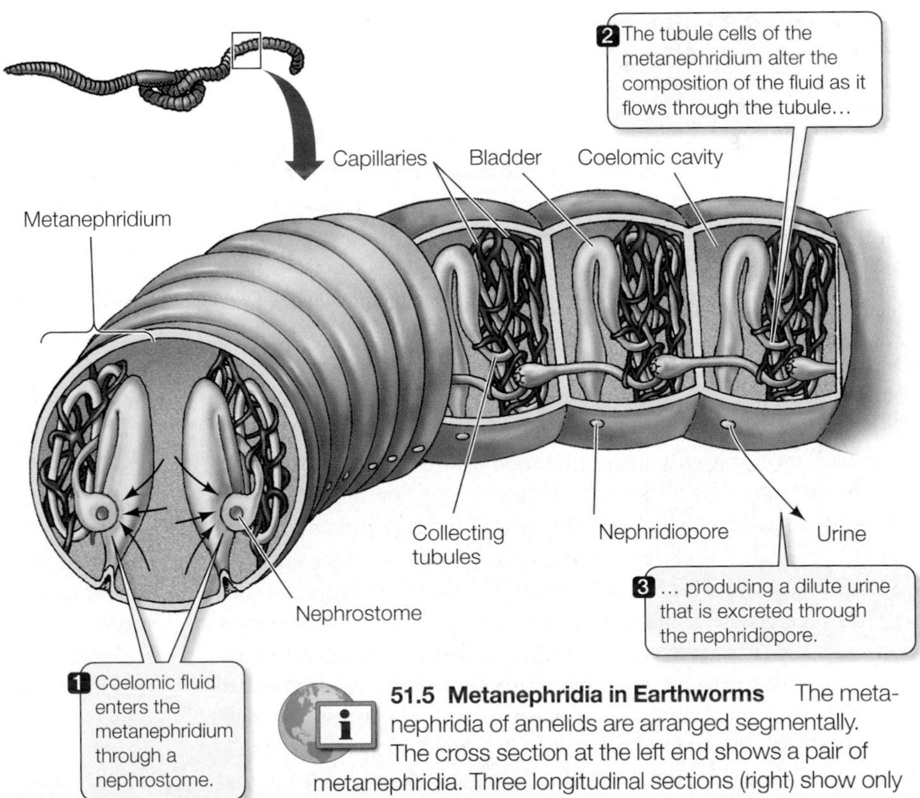

2 The tubule cells of the metanephridium alter the composition of the fluid as it flows through the tubule...
Capillaries
Bladder
Coelomic cavity
Metanephridium
Collecting tubules
Nephridiopore
Urine
3 ... producing a dilute urine that is excreted through the nephridiopore.
Nephrostome
1 Coelomic fluid enters the metanephridium through a nephrostome.

51.5 Metanephridia in Earthworms The metanephridia of annelids are arranged segmentally. The cross section at the left end shows a pair of metanephridia. Three longitudinal sections (right) show only one metanephridium of the two in each segment.

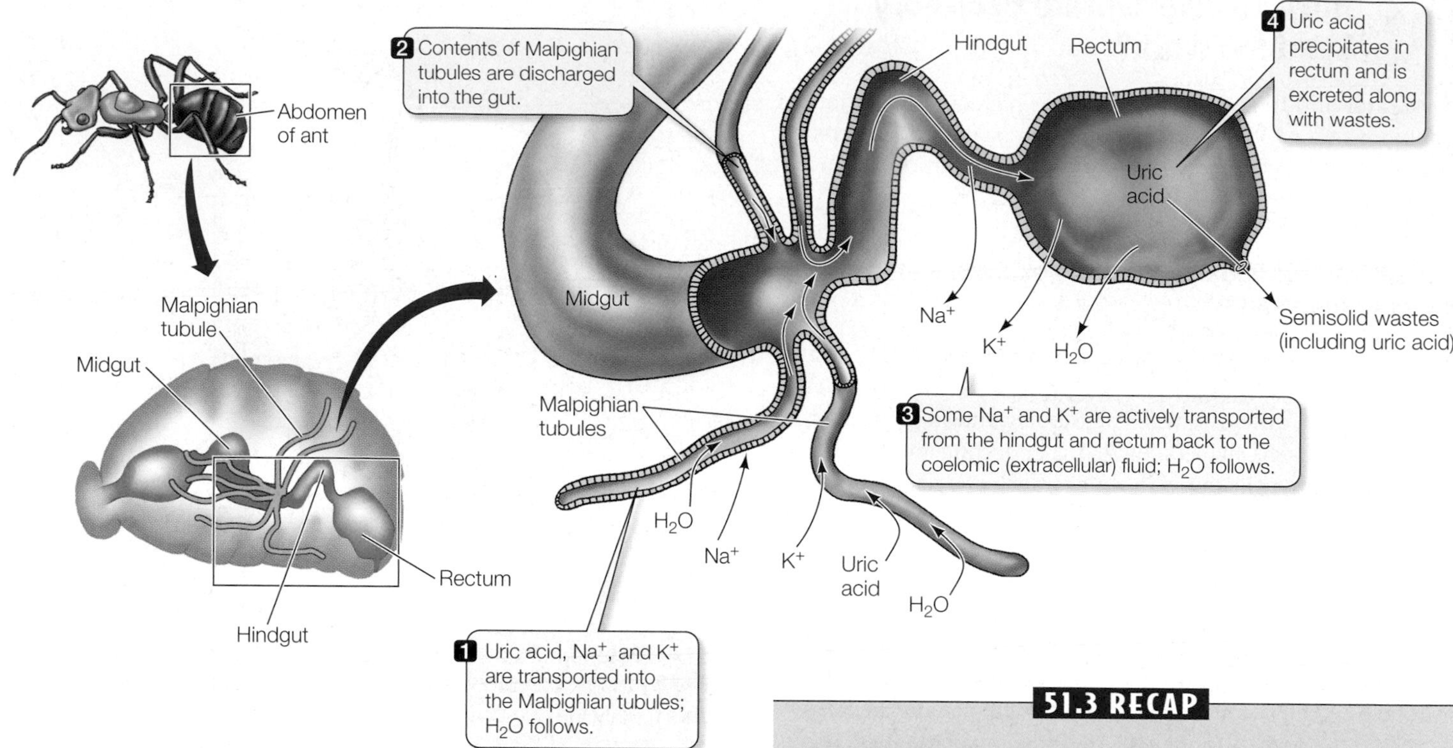

Abdomen of ant
Malpighian tubule
Midgut
Rectum
Hindgut
2 Contents of Malpighian tubules are discharged into the gut.
Midgut
Malpighian tubules
Hindgut
Rectum
Uric acid
4 Uric acid precipitates in rectum and is excreted along with wastes.
Semisolid wastes (including uric acid)
Na+
K+
H2O
3 Some Na+ and K+ are actively transported from the hindgut and rectum back to the coelomic (extracellular) fluid; H2O follows.
H2O
Na+
K+
Uric acid
H2O
1 Uric acid, Na+, and K+ are transported into the Malpighian tubules; H2O follows.

51.6 Malpighian Tubules in Insects The blind, thin-walled Malpighian tubules are attached to the junction of the insect's midgut and hindgut and project into the spaces containing extracellular fluid.

The Malpighian tubules of insects depend on active transport

Insects can excrete nitrogenous wastes with very little loss of water and can therefore live in the driest habitats on Earth. The insect excretory system consists of **Malpighian tubules**. An individual insect has from two to more than a hundred of these blind-ended tubules that open into the gut between the midgut and hindgut (**Figure 51.6**).

Insects have open circulatory systems and therefore cannot use a pressure difference to filter extracellular fluids into the Malpighian tubules. Instead, the cells of the tubules actively transport uric acid, potassium ions, and sodium ions from the extracellular fluid into the tubules. As these solutes are secreted into the tubules, water follows osmotically and flushes the tubule contents toward the gut.

The composition of the tubule fluid changes while it is in the hindgut and rectum. The epithelial cells of the hindgut and rectum actively transport sodium and potassium ions from the gut contents back into the extracellular fluid. This local transport of salts creates an osmotic gradient that pulls water out of the rectal contents. As the uric acid concentration increases, it precipitates and frees more water to be reabsorbed. Remaining in the rectum are crystals of uric acid mixed with other wastes; this semi-dry matter is what the insect eliminates. The Malpighian tubule system is a highly effective mechanism for excreting nitrogenous wastes and some salts without giving up much water.

51.3 RECAP

Protonephridia and metanephridia work by creating a filtrate of the body fluids that is modified by the secretion and reabsorption of specific substances before being excreted. Insect Malpighian tubules actively secrete uric acid and other solutes into closed tubules.

- Do you understand how Malpighian tubules function, and how this system makes it possible for some insect species to survive in the desert? See p. 1098 and Figure 51.6

The survival of animals living in environments with different salt and water combinations depends on their ability to maintain a relatively constant internal environment. Next we will consider how the basic unit of the vertebrate excretory system has evolved to be able to respond to a variety of salt and water balance challenges.

51.4 How Do Vertebrates Maintain Salt and Water Balance?

The main excretory organ of vertebrates is the **kidney**, and the functional unit of the kidney is the **nephron**. The nephron is a structure made up of a long tubule and associated blood vessels that filters the blood and modifies the composition of the filtrate to produce urine. Nephrons can filter large volumes of blood and achieve bulk reabsorption of salts and other valuable molecules such as glucose, making the vertebrate kidney well adapted for the excretion of excess water. If the ancestors of vertebrates evolved in fresh water, as paleontologists propose, the excretory systems of vertebrates would have evolved to excrete excess water. How then have vertebrates adapted to environments where water must be conserved and salts excreted? The answer to this question differs among vertebrate groups. Even among the marine fishes, the excretory adaptations of the bony fishes are different from those of the cartilaginous fishes. The ultimate kidney adaptations for water conservation, however, are found only in birds and mammals.

Marine fishes must conserve water

Marine bony fishes osmoregulate their extracellular fluids to maintain them at one-third to one-half the osmolarity of seawater. Since their only source of water is the sea around them, they must therefore conserve water and excrete excess salts. Marine bony fish cannot produce urine that is more concentrated than their extracellular fluid, so they minimize water loss by producing very little urine. In contrast, fresh water fish produce lots of dilute urine.

If marine bony fish cannot excrete excess salt in their urine, how do they deal with the large salt loads that they ingest with food? Marine bony fish do not absorb from their guts some of the ions they take in, especially divalent ions such as Mg^{2+} or SO_4^{2-}. NaCl, the major salt ingested, is excreted across the gill membranes.

Cartilaginous fishes (sharks and rays) are osmoconformers, but not ionic conformers. They convert nitrogenous wastes to urea and another compound called trimethylamine oxide, and they retain large amounts of these compounds in their extracellular fluids. As a result, their extracellular fluids have an osmolarity close to that of seawater, so they do not lose body water to the environment by osmosis, and may actually gain water. These species have adapted to a concentration of urea in the body fluids that would be toxic to other vertebrates. Sharks and rays still have the problem of excreting the large amounts of salts they take in with their food. They solve this problem by having a gland in the rectum that actively secretes NaCl by a mechanism similar to that of the nasal salt gland of sea birds.

Terrestrial amphibians and reptiles must avoid desiccation

Most amphibians live in or near fresh water, and they stay in humid habitats when they do venture from the water. Like freshwater fishes, most amphibian species produce large amounts of dilute urine and conserve salts. Some amphibians, however, have adapted to habitats that require water conservation.

Amphibians living in very dry terrestrial environments have reduced the permeability of their skin to water. Some secrete a waxy substance over the skin to waterproof it. Several species of frogs that live in arid regions of Australia burrow deep into the ground and remain there during long dry periods. They enter *estivation*, a state of very low metabolic activity and therefore low water turnover. When it rains, the frogs come out of estivation, feed, and reproduce. Their most interesting adaptation is an enormous urinary bladder. Before entering estivation, they fill the bladder with dilute urine, which can amount to one-third of their body weight. This dilute urine serves as a water reservoir that is gradually reabsorbed into the blood during the long period of estivation.

Reptiles occupy habitats ranging from aquatic to extremely hot and dry. In fact, snakes and lizards are among the most prominent members of many desert faunas (including an extensive list of venomous species that inhabit the great deserts of Western Australia).

When Australian aboriginal people need an emergency source of drinking water, they dig up estivating frogs and use the urine in their bladders .

Three major adaptations have freed reptiles from the close association with water that is necessary for most amphibians (see Section 33.4). First, reptiles are *amniotes* that do not need fresh water to reproduce because they employ internal fertilization and lay eggs with shells that retard evaporative water loss. Second, they have a dry, scale-covered epidermis (skin) that retards evaporative water loss. Third, they excrete nitrogenous wastes as uric acid semisolids, losing little water in the process.

Birds and mammals can produce highly concentrated urine

Birds and mammals occupy widely diverse habitats, many of which present special excretory system challenges. The most challenging environments are those in which water is severely limited. Birds and mammals both have adaptations seen in other vertebrate groups that enable them to conserve water. They have skins and surface coverings that impede water loss. Like reptiles, they are amniotes, and thus free from dependence on an aqueous environment for reproduction. Birds have shelled eggs, as reptiles do, whereas mammalian embryos develop internally in the mother (see Section 43.4). Like reptiles, birds produce uric acid as their nitrogenous waste product.

Both mammals and, to a lesser extent birds have a unique ability: they can produce urine that is more concentrated than their blood. They can concentrate their urine because of adaptations of their kidneys that enable a solute concentration gradient to develop that can be used to osmotically reabsorb water from the urine. We will focus on mammals to describe how this important adaptation works, but to do so we first have to understand the structure and function of the vertebrate nephron.

The nephron is the functional unit of the vertebrate kidney

Urine formation in vertebrate nephrons involves three main processes (**Figure 51.7**):

- *Filtration:* The blood is filtered in a dense ball of capillaries called the **glomerulus** (plural glomeruli), which is highly permeable to selected substances but impermeable to large molecules.
- *Tubular reabsorption:* The glomerular filtrate flows into the **renal tubule**. Cells lining the renal tubule modify the glomerular filtrate by reabsorption of specific ions, nutrients, and water, returning these to the blood and leaving behind and concentrating excess ions and waste products such as urea.
- *Tubular secretion:* The glomerular filtrate in the renal tubule is further modified by tubule cells transporting substances into the tubular contents. These are substances that the body needs to excrete.

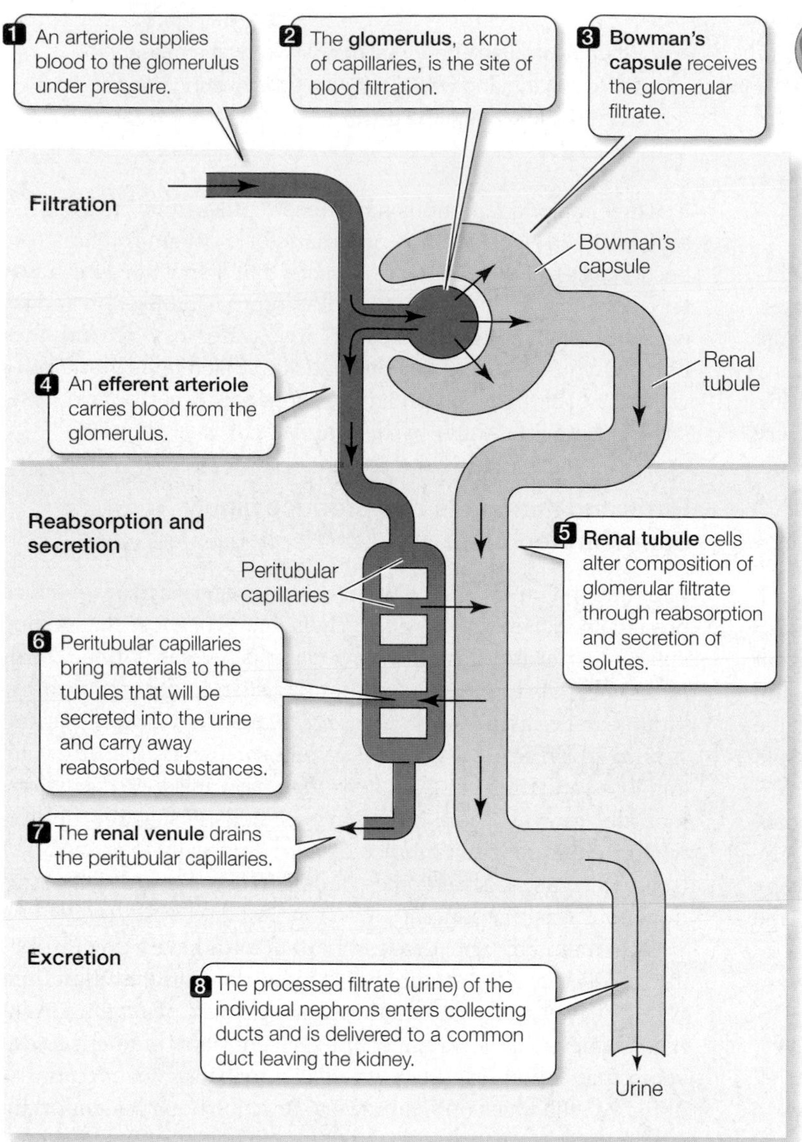

1 An arteriole supplies blood to the glomerulus under pressure.
2 The **glomerulus**, a knot of capillaries, is the site of blood filtration.
3 **Bowman's capsule** receives the glomerular filtrate.
Filtration
Bowman's capsule
Renal tubule
4 An **efferent arteriole** carries blood from the glomerulus.
Reabsorption and secretion
Peritubular capillaries
5 **Renal tubule** cells alter composition of glomerular filtrate through reabsorption and secretion of solutes.
6 Peritubular capillaries bring materials to the tubules that will be secreted into the urine and carry away reabsorbed substances.
7 The **renal venule** drains the peritubular capillaries.
Excretion
8 The processed filtrate (urine) of the individual nephrons enters collecting ducts and is delivered to a common duct leaving the kidney.
Urine

51.7 The Vertebrate Nephron The vertebrate nephron consists of a renal tubule closely associated with two capillary beds, the glomerulus and the peritubular capillaries.

The **peritubular capillaries** are vascular structures closely associated with the renal tubules. These vessels transport substances to and from the renal tubules. By carrying away absorbed substances and delivering substances to be secreted into the urine, the peritubular capillaries work in tandem with the glomerular capillaries.

Blood is filtered in the glomerulus

As seen in Figure 51.7, the vascular arrangement of the nephron is unusual in that the two capillary beds—the glomerulus and the peritubular capillaries—lie in series between the arteriole that supplies them and the venule that drains them. Blood enters the glomerulus through an **afferent arteriole** and exits through an **efferent arteriole** (**Figure 51.8A**). The efferent arteriole gives rise to the peritubular capillaries, which surround the tubule component of the nephron and serve as sites of exchange between the filtrate in the renal tubules and the extracellular fluid.

The tubule component of the nephron—the renal tubule—begins with **Bowman's capsule**, which encloses

51.8 A Tour of the Nephron Scanning electron micrographs illustrate the anatomical basis for blood filtration by the kidneys. (A) In a preparation showing only the blood vessels (tubular tissue has been digested away), the glomeruli appear as balls of capillaries surrounded by the more diffuse peritubular capillaries. (B) Higher magnification of a glomerulus with the tubule cells intact shows the podocytes that wrap around the glomerular capillaries. (C) This cross section of an intact glomerulus shows the tubule cells that form Bowman's capsule.

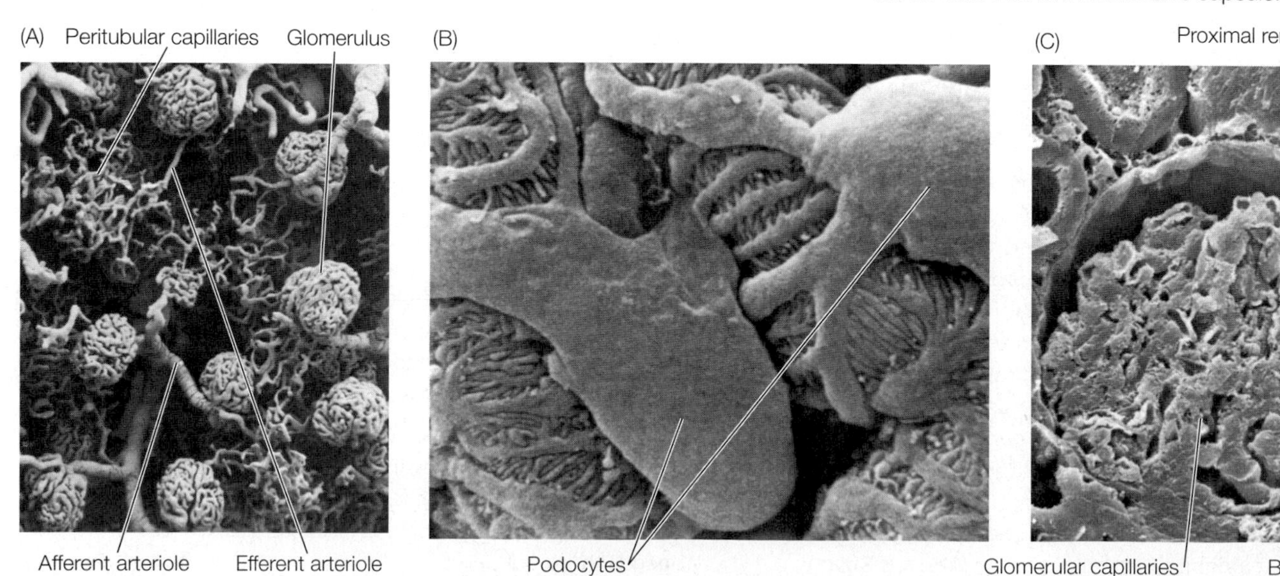

the glomerulus. The glomerulus appears to be pushed into Bowman's capsule much like a fist pushed into an inflated balloon. The cells of the capsule that come into direct contact with the glomerular capillaries are called **podocytes** (**Figure 51.8B**). These highly specialized cells have numerous arm-like extensions, each with hundreds of fine, finger-like processes. The podocytes wrap around the capillaries so that their processes interdigitate and cover the capillaries extensively.

The glomerulus filters the blood to produce a fluid (the renal filtrate) that lacks cells and large molecules. The walls of the capillaries, the basal lamina of the capillary endothelium, and the podocytes of Bowman's capsule all participate in filtration. *Fenestrations* in the walls of the capillaries (see Section 49.4) allow water and many solute molecules, but not red blood cells, to pass through. The meshwork of the basal lamina and the spaces between the processes of the podocytes are even finer and prevent large molecules from leaving the capillaries (**Figure 51.8C**). The arterial pressure of the blood entering the permeable capillaries causes the filtration of water and small molecules in the glomerulus. The glomerular filtration rate is high because blood pressure in the glomerular capillaries is unusually high, and because the capillaries of the glomerulus, along with their covering of podocytes, are more permeable than other capillary beds in the body.

The renal tubules convert glomerular filtrate to urine

The composition of the filtrate that first enters the nephron is similar to that of the blood plasma, with the exception of high-molecular-weight solutes such as proteins. Reabsorption and secretion cause the composition of this fluid to change as it passes down the renal tubule. Cells of the tubule actively reabsorb certain molecules from the tubule fluid (which are returned to the blood flowing through the peritubular capillaries) and actively secrete into the tubule fluid substances delivered by the peritubular capillaries. The urine that eventually exits the body is very different from the original filtrate due to the actions of the renal tubule cells.

51.4 RECAP

The kidney is the major excretory organ of vertebrates. Its functional unit is the nephron, which includes a glomerulus that filters blood and a renal tubule that secretes and reabsorbs solutes, modifying the filtrate to produce urine. The nephron is a mechanism for excreting excess water while conserving valuable solutes.

- Can you explain the difference in osmoregulatory adaptations between marine bony and cartilaginous fishes? See p. 1099
- Can you describe the functional relationships between the glomerular and peritubular capillaries? See pp. 1099–1100 and Figure 51.7
- Can you describe how the blood is filtered in the glomerulus? See p. 1100

The adaptations of mammals and birds that allow them to produce urine that is more concentrated than their extracellular fluids constitute an important step in vertebrate evolution. We now consider the mammalian kidney in more detail.

51.5 How Does the Mammalian Kidney Produce Concentrated Urine?

Mammals and birds have high body temperatures and high metabolic rates, and therefore have the potential for a high rate of water loss. Having an excretory system that minimizes water loss made it possible for these highly active species to occupy arid habitats.

Kidneys produce urine and the bladder stores it

We will use the excretory system of humans as our example of the mammalian excretory system. Humans have two kidneys at the back of the upper region of the abdominal cavity (**Figure 51.9A**). Each kidney filters blood, processes the filtrate into urine, and releases that urine into a duct called the **ureter**. The ureter of each kidney leads to the **urinary bladder**, where the urine is stored until it is excreted through the **urethra**, a short tube that opens to the outside of the body.

Two sphincter muscles surrounding the base of the urethra control the timing of urination. One of these sphincters is a smooth muscle and is controlled by the autonomic nervous system. As the bladder fills, stretch receptors in the walls of the bladder trigger a spinal reflex that relaxes this sphincter. This reflex is the only control of urination in infants. The other sphincter is a skeletal muscle and is controlled by the voluntary nervous system. When the bladder is *very* full, only deliberate conscious effort prevents urination. Infant toilet training involves their learning to control this sphincter.

Nephrons have a regular arrangement in the kidney

The kidney is shaped like a kidney bean; when sliced along its long axis, its key anatomical features are revealed (**Figure 51.9B**). The ureter and the **renal artery** and **renal vein** enter the kidney on its concave (punched-in) side. The ureter extends into the kidney in several branches, the ends of which envelop kidney tissues called **renal pyramids**. The renal pyramids make up the internal core, or **medulla**, of the kidney. The medulla is covered by an outer layer of tissue with a granular appearance, called the **cortex**. In the region between the cortex and the medulla, the renal artery divides into the many arterioles that serve the nephrons. In this same region the renal vein collects the blood from the many venules that drain the peritubular capillaries.

The organization of the nephrons within the kidney is very regular. All of the glomeruli are located in the cortex. The initial segment of a renal tubule is called the **proximal convoluted tubule**—"proximal" because it is the first segment of renal tubule to receive the filtrate from the glomerulus, and "convoluted" because it is twisted (**Figure 51.9C**). All the proximal convoluted tubules are also located in the cortex.

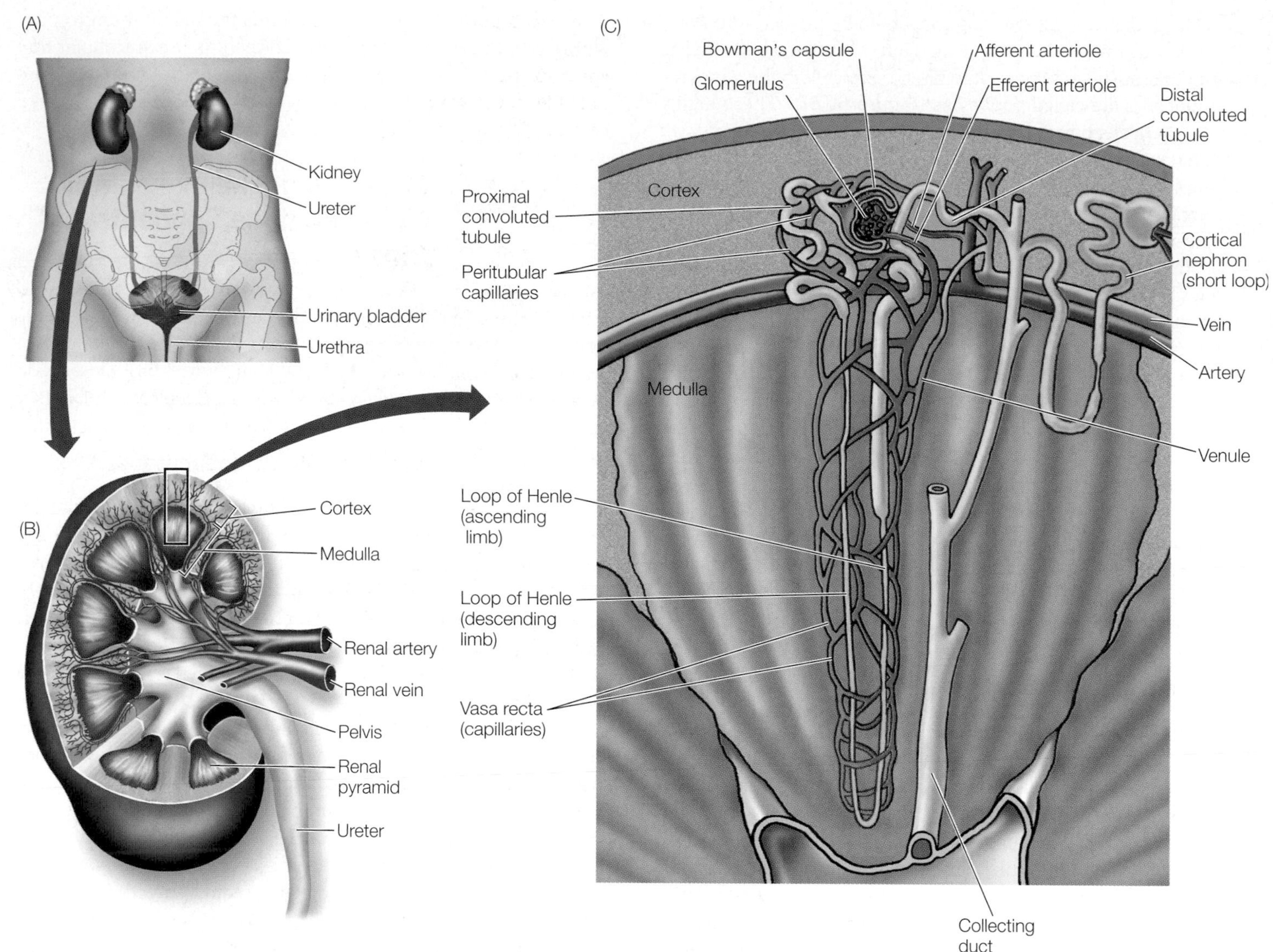

51.9 The Human Excretory System (A) The human kidneys are in the upper dorsal region of the abdominal cavity. (B) The human kidney has a highly organized internal structure that is the basis for its function. (C) The glomeruli and the proximal and distal convoluted tubules are located in the cortex of the kidney, but the loops of Henle run in parallel as straight sections down into the renal medulla and back up to the cortex. Collecting ducts run from the cortex to the inner surface of the medulla, where they open into the ureter.

At the point at which the renal tubule descends into the medulla, it becomes much less convoluted and descends directly down into the medulla. In the medulla the tubule makes a hairpin turn and ascends back to the cortex, forming what is called the **loop of Henle**. Some nephrons have longer loops of Henle than others. Some 20–30 percent of human nephrons that have glomeruli deep in the cortex (i.e., near the border with the medulla) have long loops of Henle that go deep into the medulla. Nephrons that have glomeruli farther up in the cortex generally have short loops of Henle that only descend a short distance into the medulla. As we will see, the long loops are critical to the formation of concentrated urine. The *ascending limb* of the loop of Henle becomes the **distal convoluted tubule** when it reaches the cortex. This section of the renal tubule is called distal because it is farther from the glomerulus. The distal convoluted tubules of many nephrons join a common **collecting duct** in the cortex. The collecting ducts descend back down through the renal pyramid, parallel to and past the tips of the loops of Henle, emptying into the funnel-shaped *pelvis.* Divisions of the pelvis that surround each renal pyramid join together to leave the kidney as the ureter (see Figure 51.9B).

The organization of the blood vessels of the kidney closely parallels the organization of the nephrons (see Figure 51.9C). Smaller and smaller arteries branch from the renal artery and radiate into the cortex. Finally an afferent arteriole carries blood to each glomerulus. Draining each glomerulus is an efferent arteriole that gives rise to the peritubular capillaries, most of which surround the cortical portions of the tubules.

A few peritubular capillaries run into the medulla in parallel with the loops of Henle and the collecting ducts, forming a vascular network called the **vasa recta**. All the peritubular capillaries from a nephron join back together into a venule that joins with venules from other nephrons and eventually leads to the renal vein.

Most of the glomerular filtrate is reabsorbed by the proximal convoluted tubule

Most of the water and solutes filtered out of the glomerulus are reabsorbed and do not appear in the urine. We can reach this conclusion by comparing the rate of filtration by the glomeruli with the rate of urine production. The kidneys receive about 1 liter of blood per minute, or about 1,500 liters of blood per day. How much of this huge volume is filtered out of the glomeruli? The answer is about 12 percent. This is still a large volume—180 liters per day! Since we normally urinate 2–3 liters per day, about 98 percent of the fluid volume that is filtered out of the glomerulus is returned to the blood. Where and how is this enormous fluid volume reabsorbed?

The proximal convoluted tubule (PCT) is responsible for most of the reabsorption of water and solutes from the glomerular filtrate. The cells of this section of the renal tubule have many microvilli that increase their surface area for reabsorption, and they have many mitochondria—an indication that they are metabolically active. PCT cells actively transport Na^+ (with Cl^- following passively) and other solutes, such as glucose and amino acids, out of the tubule fluid. Almost all glucose and amino acid molecules that are filtered from the blood are actively reabsorbed by PCT cells and transported into the extracellular fluid. The active transport of solutes from the tubule into the interstitial fluid causes water to follow osmotically. The water and solutes moved into the interstitial fluid are taken up by the peritubular capillaries and returned to the venous blood.

Despite the large volume of water and solutes reabsorbed by the PCT, the overall osmolarity of the renal filtrate does not change. The fluid that enters the loop of Henle has the same osmolarity as the blood plasma, although its composition is different. How, then, does the kidney produce urine that is more concentrated than the blood plasma?

The loop of Henle creates a concentration gradient in the surrounding tissue

Humans can produce urine that is four times more concentrated than their blood plasma. The vampire bat we encountered at the beginning of this chapter can produce urine that is twenty times more concentrated than its blood plasma. The concentrating ability of the mammalian kidney arises from a **countercurrent multiplier** mechanism made possible by the anatomical arrangement of the loops of Henle. The term "countercurrent" refers to the opposing directions in which the tubule fluid in the descending and ascending limbs flows. The term "multiplier" refers to the ability of this system to create a solute concentration gradient in the renal medulla. The loops of Henle do not themselves produce concentrated urine; rather, they increase the osmolarity of the interstitial fluid in the medulla in a graduated way. The osmolarity of the interstitial fluid of the medulla increases from the top to the bottom of the medulla. More specifically, the interstitial fluid concentration of the medulla increases from about 300 mosm/l (the concentration of blood plasma) at the cortex to about 1,200 mosm/l near the hairpin turn of the loop of Henle (**Figure 51.10**). How do the loops produce this effect?

The segments of the loop of Henle differ anatomically and functionally. Cells of the descending limb and the initial cells of the ascending limb are thin, with no microvilli and few mitochondria. They are not specialized for transport. Partway up the ascending limb, the cells become specialized for active transport. These cells are thick and have many mitochondria. Accordingly, the segments of the loop of Henle are named the *thin descending limb*, the *thin ascending limb*, and the *thick ascending limb*.

The countercurrent multiplier mechanism may be more easily understood by first considering events occurring in the thick ascending limb and their affect on the fluid in the descending limb (see Figure 51.10). The thick ascending limb, with its cells specialized for active transport, actively reabsorbs Na^+ (with Cl^- following passively) from the tubule fluid and moves it into the interstitial fluid. (In the following discussion, it is important to distinguish between the two components of extracellular fluid—the blood plasma and the interstitial fluid.) The thick ascending limb is not permeable to water, so the reabsorption of Na^+ and Cl^- raises the concentration of those solutes in the surrounding interstitial fluid and decreases the concentration of the tubular fluid entering the distal convoluted tubule.

The thin descending limb, in contrast, is highly permeable to water, but not very permeable to Na^+ and Cl^-. Since the local interstitial fluid has been made more concentrated by the Na^+ and Cl^- reabsorbed from the neighboring thick ascending limb, water is withdrawn osmotically from the tubule fluid in the descending limb. Therefore, the fluid in the descending limb becomes more concentrated as it flows toward the hairpin turn at the bottom of the renal medulla.

The thin ascending limb, like the thick ascending limb, is not permeable to water. It is, however, permeable to Na^+ and Cl^-. As the concentrated tubule fluid flows up the thin ascending limb, it is more concentrated than the surrounding interstitial fluid, so Na^+ and Cl^- diffuse out. When the tubule fluid reaches the thick ascending limb, active transport continues to move Na^+ and Cl^- from the tubule fluid to the interstitial fluid.

As a result of the processes described above, the tubule fluid reaching the distal convoluted tubule is less concentrated than the blood plasma, and the solutes that have been left behind in the renal medulla have created a concentration gradient in the interstitial fluid of the medulla. As a result of this countercurrent multiplier mechanism, the interstitial fluid of the renal medulla becomes increasingly concentrated between its border with the cortex and the tips of the renal pyramids.

You may wonder why the blood flow through the medulla does not wash out the concentration gradient established by the loops of Henle. The parallel arrangement of the descending and ascending peritubular capillaries—the vasa recta—in the medulla helps preserve the concentration gradient in the medulla. These capillaries are permeable to salt and water. Therefore, as blood flows down the descending limb of the vasa recta into the increasingly concentrated interstitial fluid of the medulla, it loses water and gains solutes. As the blood flows up from the bottom of the medulla in the ascending limb of the vasa recta, the opposite happens because now the blood is more concentrated than the surrounding interstitial fluid. The dynamics of this countercurrent exchange of salts and water between the blood in the vasa recta and the interstitial fluids result in little change in the composition of the tissue fluid in the medulla, but the excess water that is reabsorbed from the collecting ducts gets returned to the blood.

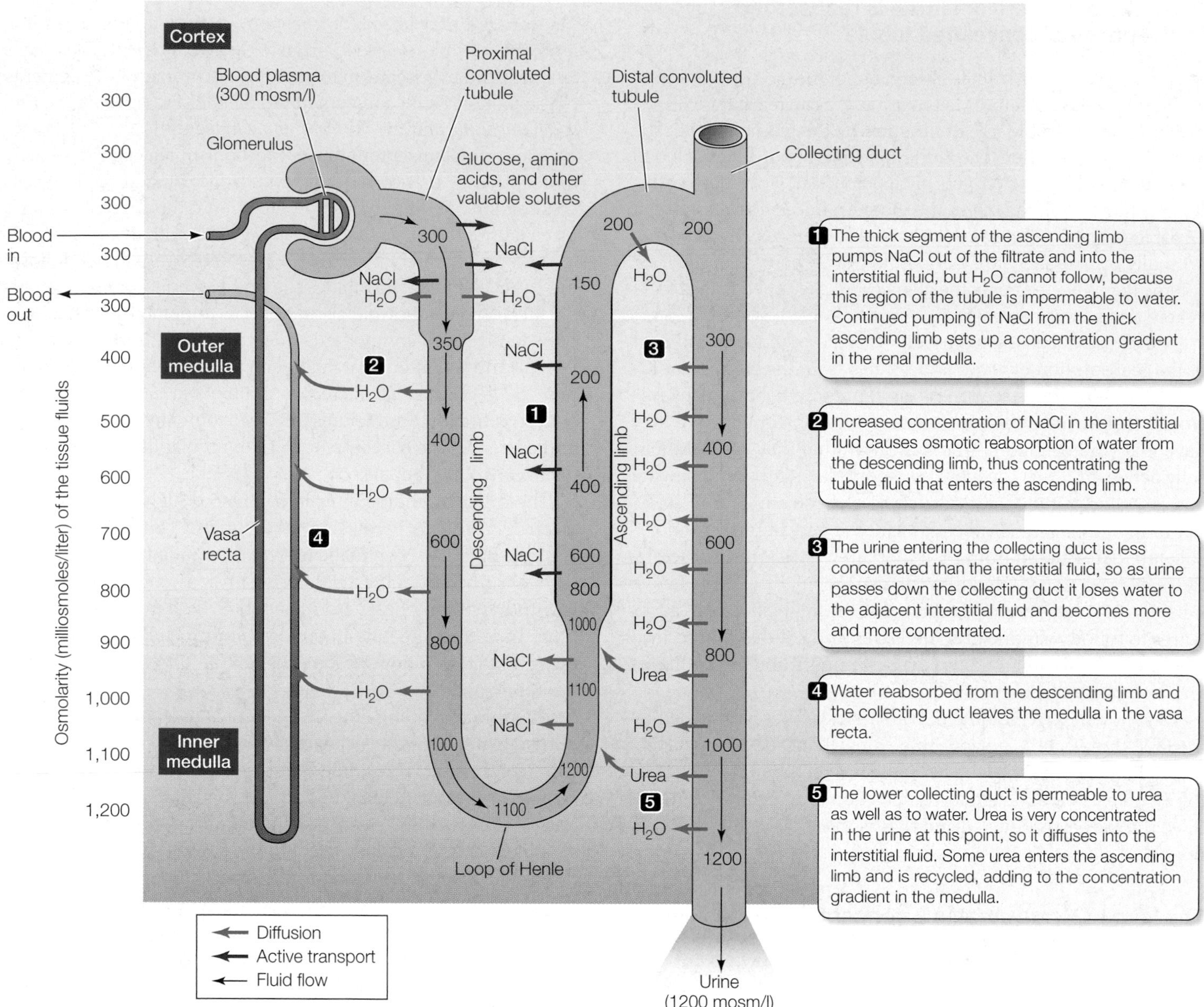

Cortex
Outer medulla
Inner medulla
Osmolarity (milliosmoles/liter) of the tissue fluids
300
300
300
300
300
400
500
600
700
800
900
1,000
1,100
1,200
Blood plasma (300 mosm/l)
Glomerulus
Blood in
Blood out
Proximal convoluted tubule
Glucose, amino acids, and other valuable solutes
Distal convoluted tubule
Collecting duct
Vasa recta
Descending limb
Ascending limb
Loop of Henle
NaCl
H_2O
Urea
Urine (1200 mosm/l)
1 The thick segment of the ascending limb pumps NaCl out of the filtrate and into the interstitial fluid, but H_2O cannot follow, because this region of the tubule is impermeable to water. Continued pumping of NaCl from the thick ascending limb sets up a concentration gradient in the renal medulla.
2 Increased concentration of NaCl in the interstitial fluid causes osmotic reabsorption of water from the descending limb, thus concentrating the tubule fluid that enters the ascending limb.
3 The urine entering the collecting duct is less concentrated than the interstitial fluid, so as urine passes down the collecting duct it loses water to the adjacent interstitial fluid and becomes more and more concentrated.
4 Water reabsorbed from the descending limb and the collecting duct leaves the medulla in the vasa recta.
5 The lower collecting duct is permeable to urea as well as to water. Urea is very concentrated in the urine at this point, so it diffuses into the interstitial fluid. Some urea enters the ascending limb and is recycled, adding to the concentration gradient in the medulla.
Diffusion
Active transport
Fluid flow

51.10 Concentrating the Urine A countercurrent multiplier mechanism enables the kidney to produce urine that is far more concentrated than mammalian blood plasma.

Water permeability of kidney tubules depends on water channels

We have noted that some tubule regions, such as the PCT, are highly permeable to water while others, such as the thick ascending limb of the loop of Henle, are impermeable to water. How do differences in water permeability in different regions of the nephron arise? Regions of the nephron that are highly permeable to water have greater numbers of *aquaporins*, a class of membrane proteins that form water channels (see Section 5.3). Aquaporins are abundant in kidney PCT cells and in descending limbs of the loops of Henle, but not in the ascending limbs of the loop of Henle. The discovery of aquaporins resulted in Peter Agre of Johns Hopkins University receiving the Nobel Prize in Chemistry in 2003.

Water reabsorption begins in the distal convoluted tubule

The first portion of the distal convoluted tubule has properties similar to the thick ascending limb of the loop of Henle. Na^+ and Cl^- are transported out of the tubule fluid, and water cannot follow. As a result, the tubule fluid becomes even more dilute. The later sections of the distal convoluted tubule, however, can be permeable to water, and water can be osmotically drawn from the tubule into the interstitial fluid. As the tubule fluid flows from the distal tubule to the collecting duct, it can be below or equal to the osmolarity of the blood plasma.

Urine is concentrated in the collecting duct

The tubule fluid entering the collecting duct is at the same solute *concentration* as the blood plasma, but its *composition* is considerably different from that of the plasma. The major solute in the tubular fluid is now urea, since salts were reabsorbed earlier in the nephron. As the tubule fluid flows down the collecting duct, it loses water osmotically to the interstitial fluid, and that water returns to the circulatory system via the vasa recta.

The concentration gradient established in the renal medulla by the loops of Henle creates the osmotic potential that withdraws water from the collecting ducts. The collecting ducts begin in the renal cortex and run through the renal medulla before emptying into the ureter at the tips of the renal pyramids. As the solute concentration of the surrounding interstitial fluid increases, more and more water can be absorbed from the urine in the collecting duct. By the time it reaches the ureter, the urine can become greatly concentrated, with urea as a major solute.

As water is withdrawn from the collecting duct, some urea also leaks out into the medullary interstitial fluid, adding to its osmotic potential. This urea diffuses back into the loop of Henle and is returned to the collecting duct. The recycling of urea in the renal medulla contributes significantly to the ability of the kidney to concentrate the urine in the collecting duct.

Overall, the ability of a mammal to concentrate its urine is determined by the maximum concentration gradient it can establish in its renal medulla. An important adaptation for increasing the concentration gradient is to increase the lengths of the loops of Henle. The desert gerbil, for example, has such extremely long loops of Henle that its renal pyramid (each of its kidneys has only one, in contrast to ours) extends far out of the concave surface of the kidney and into the ureter (**Figure 51.11**). These animals are so effective in conserving water that they can survive just on the water released by the metabolism of their food.

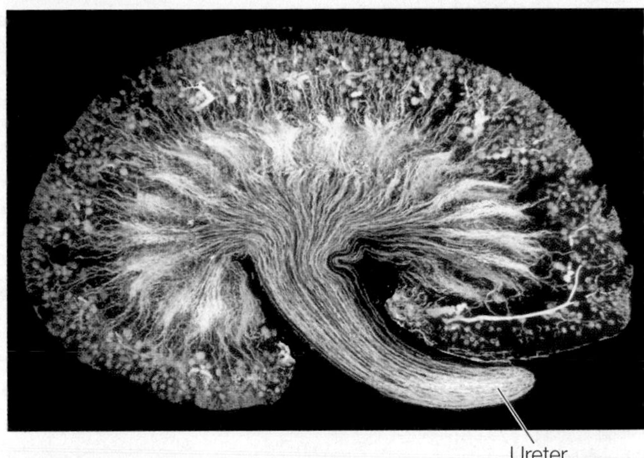

51.11 The Ability to Concentrate The ability of the mammalian kidney to concentrate urine depends on the lengths of its loops of Henle relative to the overall size of the kidney. The kidney of a desert gerbil has a single renal pyramid (the yellow-green area) with loops of Henle so long that the pyramid extends far into the ureter.

Ever wonder how drugs get their names? The prescription drug Lasix® increases water loss by inhibiting sodium reabsorption in the kidney tubules. It is widely used to treat patients with congestive heart failure (a condition that causes them to retain water). The effects of the drug *last* about *six* hours—hence, *Lasix.*

The kidneys help regulate acid–base balance

Besides regulating salt and water balance and excreting nitrogenous wastes, the kidneys have another important role: They regulate the hydrogen ion concentration (the pH) of the blood. Blood pH is a critical variable because it influences the structure, and therefore the function, of proteins.

One way to minimize pH changes in a chemical solution is to add a *buffer*—a substance that can either absorb or release hydrogen ions (Section 2.4). The major buffers in the blood are bicarbonate ions (HCO_3^-; see Figure 48.14) that are formed from the dissociation of carbonic acid, which in turn is formed by the hydration of CO_2 according to the following equilibrium reaction:

$$CO_2 + H_2O \rightleftharpoons H_2CO_3 \rightleftharpoons H^+ + HCO_3^-$$

You can see that if excess hydrogen ions are added to this reaction mixture, the reaction will move to the left and absorb the excess H^+. If hydrogen ions are removed from the reaction mixture, however, the reaction will move to the right and supply more H^+.

The HCO_3^- buffer system is important for controlling the pH of the blood because the reaction can be pushed to the right and pulled to the left physiologically. The lungs control the levels of CO_2 in the blood, thus altering the acid portion of the reaction. CO_2 is considered the acid portion of the reaction because if you add additional CO_2, the reaction shifts to the right, producing more H^+ ions. The kidneys control the base portion of the reaction by removing H^+ from the blood and returning HCO_3^- to the blood. The renal tubules secrete H^+ into the tubule fluid and reabsorb HCO_3^- (**Figure 51.12**). The kidney has other secretory and reabsorption systems that enhances its ability to control blood pH.

Kidney failure is treated with dialysis

Loss of kidney function, or *renal failure,* results in the retention of salts and water (hence high blood pressure), retention of urea (uremic poisoning), and metabolic acids (acidosis). A person who suffers complete renal failure will die within 2 weeks if not treated. A drastic but highly successful treatment is kidney transplant, but it is usually necessary to sustain a patient for considerable time while waiting for a suitable organ to become available. Therefore, artificial kidneys, or renal dialysis machines, are essential modes of treatment.

In a dialysis machine, the patient's blood flows through many small channels made of semipermeable membranes (**Figure 51.13**). A dialysis solution flows on the other side of these membranes, and small molecules can diffuse through these membranes. Molecules and ions diffuse from an area of high concentration to an area of lower concentration, so the composition of the dialysis fluid is crucial. The concentrations of the molecules or ions that need to be conserved must be at the same concentration in the dialysis fluid as they are in the blood. The concentrations of mol-

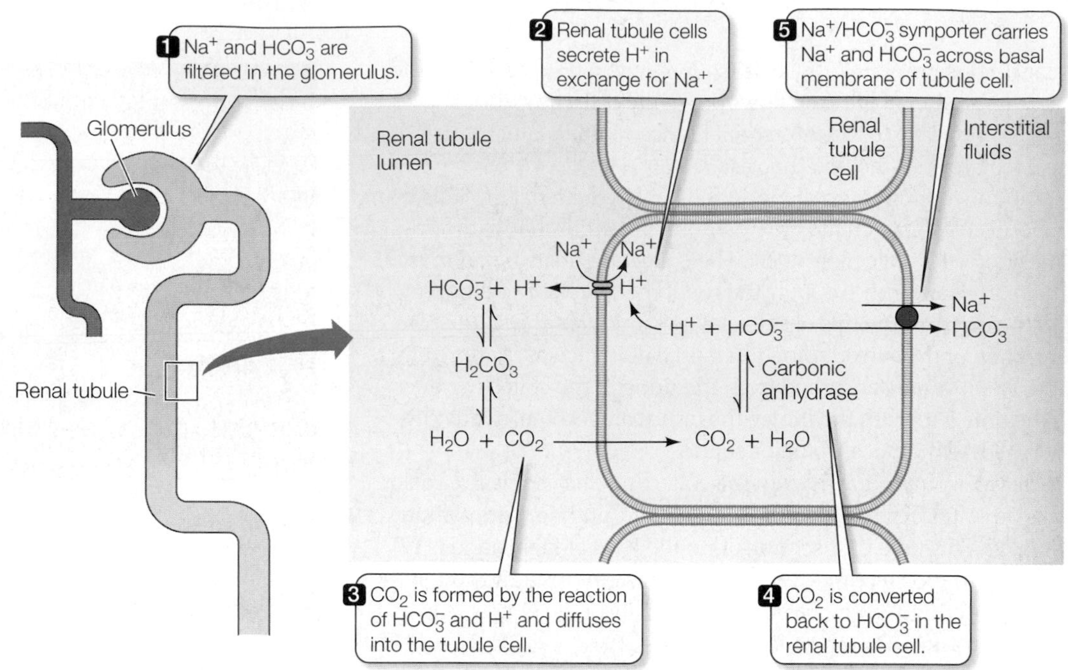

1 Na^+ and HCO_3^- are filtered in the glomerulus.
2 Renal tubule cells secrete H^+ in exchange for Na^+.
5 Na^+/HCO_3^- symporter carries Na^+ and HCO_3^- across basal membrane of tubule cell.
Glomerulus
Renal tubule lumen
Renal tubule cell
Interstitial fluids
Renal tubule
Na^+
Na^+
$HCO_3^- + H^+$
H^+
$H^+ + HCO_3^-$
Na^+
HCO_3^-
H_2CO_3
Carbonic anhydrase
$H_2O + CO_2$
$CO_2 + H_2O$
3 CO_2 is formed by the reaction of HCO_3^- and H^+ and diffuses into the tubule cell.
4 CO_2 is converted back to HCO_3^- in the renal tubule cell.

51.12 The Kidney Excretes Acids and Conserves Bases Bicarbonate ions are filtered out at the glomerulus, and renal tubule cells secrete hydrogen ions into the tubule fluid. In the renal tubule, the filtered bicarbonate buffers the secreted hydrogen ions and keeps the urine from becoming too acidic. The CO_2 formed by the reaction of bicarbonate and hydrogen ions is converted back to bicarbonate by the renal tubule cells and transported back into the interstitial fluid.

ecules and ions that need to be removed from the blood are zero in the dialysis fluid. The total osmotic potential of the dialysis fluid must equal that of the plasma.

About 500 ml of the patient's blood is in the dialysis machine at any one time, and the unit processes several hundred milliliters of blood per minute. A patient with no kidney function must be on the dialysis machine for 4–6 hours at least 3 times a week.

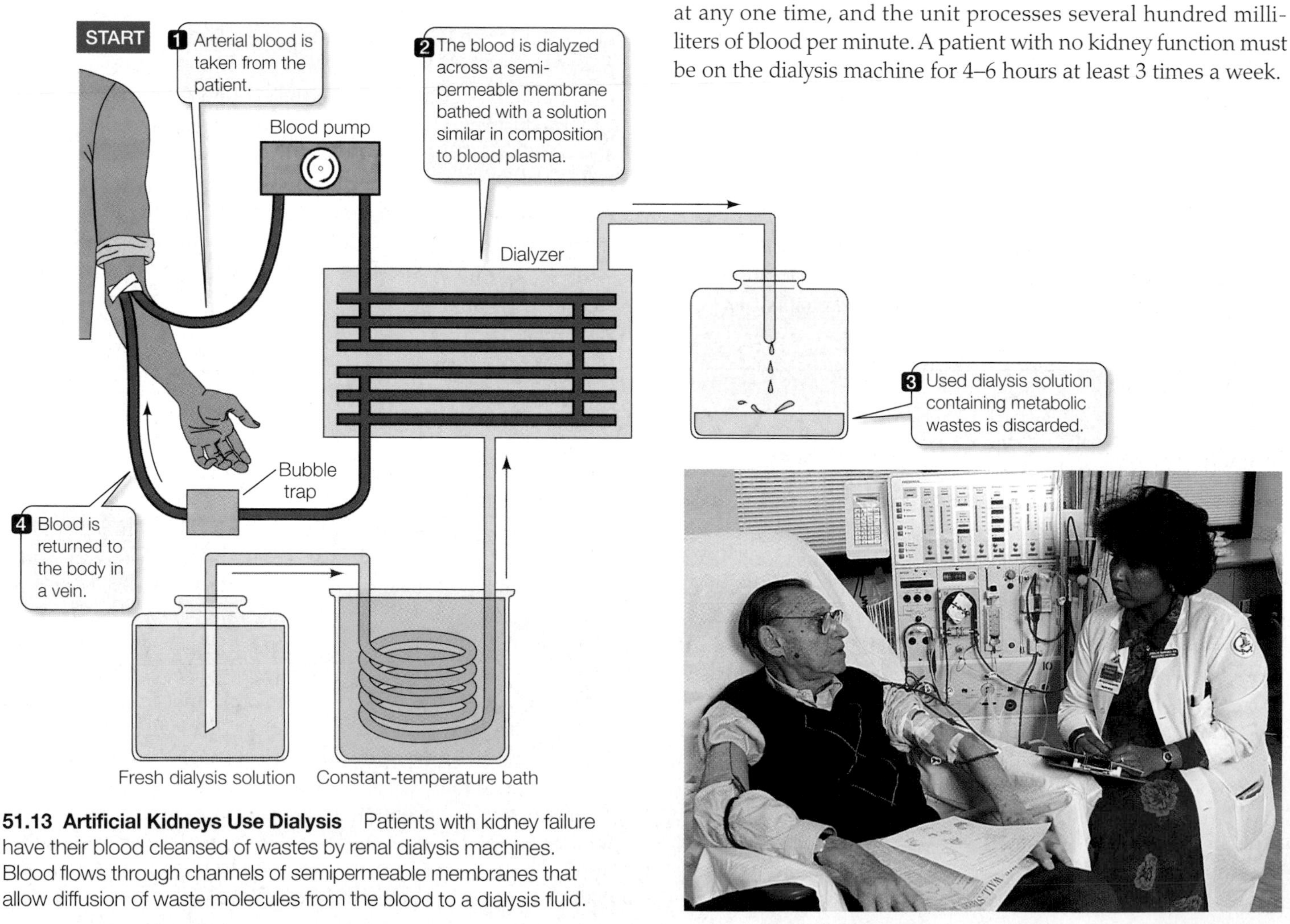

START
1 Arterial blood is taken from the patient.
2 The blood is dialyzed across a semipermeable membrane bathed with a solution similar in composition to blood plasma.
Blood pump
Dialyzer
3 Used dialysis solution containing metabolic wastes is discarded.
Bubble trap
4 Blood is returned to the body in a vein.
Fresh dialysis solution
Constant-temperature bath

51.13 Artificial Kidneys Use Dialysis Patients with kidney failure have their blood cleansed of wastes by renal dialysis machines. Blood flows through channels of semipermeable membranes that allow diffusion of waste molecules from the blood to a dialysis fluid.

51.5 RECAP

The anatomical organization of the nephrons in the mammalian kidney makes it possible for the kidney to produce a urine more concentrated than the blood. Bulk reabsorption of salts, other valuable solutes, and water takes place in the proximal convoluted tubule. The thick ascending limb of the loop of Henle reabsorbs salt, increasing the osmolarity of the interstitial fluids in the renal medulla. Collecting ducts run through the renal medulla and lose water osmotically to the surrounding interstitial fluids, concentrating the urine.

- What is a countercurrent multiplier mechanism, and how does it work in nephrons to produce concentrated urine? See p. 1103 and Figure 51.10
- How does the kidney contribute to acid–base balance? See p. 1105 and Figure 51.12

The kidney contributes to homeostasis in a number of ways, including maintaining the osmotic concentration and ionic composition of the extracellular fluid and by regulating pH. The kidneys also play a major role in regulating blood pressure.

51.6 What Mechanisms Regulate Kidney Function?

Several regulatory mechanisms act on the kidneys to maintain blood osmolarity and blood pressure. We will discuss these mechanisms separately, but keep in mind that they are always working together.

The kidneys maintain the glomerular filtration rate

If the kidneys stop filtering blood, they cannot accomplish any of their functions. The maintenance of a constant **glomerular filtration rate** (**GFR**) depends on an adequate blood supply to the kidneys at an adequate blood pressure. **Autoregulatory mechanisms** ensure adequate blood supply and blood pressure for kidney function regardless of what is happening elsewhere in the body. The kidney's autoregulatory adjustments compensate for decreases in cardiac output or decreases in blood pressure so that the GFR remains constant.

One autoregulatory mechanism is the dilation (expansion) of the afferent renal arterioles when blood pressure falls. This dilation decreases the resistance in the arterioles and helps maintain blood pressure in the glomerulus. If arteriole dilation does not keep the GFR from falling, the kidney releases an enzyme, **renin**, into the blood. Renin acts on a circulating protein to convert it into an active hormone called **angiotensin**.

Angiotensin has several effects that help restore the GFR to normal. First, angiotensin preferentially constricts the efferent renal arterioles, which elevates blood pressure in the glomerular capillaries. Second, it constricts peripheral blood vessels all over the body—an action that elevates central blood pressure. Third, it stimulates the adrenal cortex to release the hormone **aldosterone**. Aldosterone and angiotensin both stimulate sodium reabsorption by the kidney, thereby making its reabsorption of water more effective. Enhanced water reabsorption helps maintain blood volume and therefore central blood pressure. Finally, angiotensin acts on the brain to stimulate thirst. Increased water intake in response to thirst increases blood volume and blood pressure.

Blood osmolarity and blood pressure are regulated by ADH

Cells in the hypothalamus can stimulate the release from the posterior pituitary of a hormone called **antidiuretic hormone** (**ADH**, also called *vasopressin*) that can act upon water channels to increase the permeability of membranes to water. The higher the circulating levels of ADH, the greater the number of active water channels. Various factors can stimulate or inhibit the release of ADH. Of key importance to kidney function are osmoreceptors that monitor blood osmolarity and stretch receptors that monitor blood pressure (**Figure 51.14**).

Osmoreceptors that detect a rise in blood osmolarity will stimulate the release of ADH. ADH helps regulate blood osmolarity by

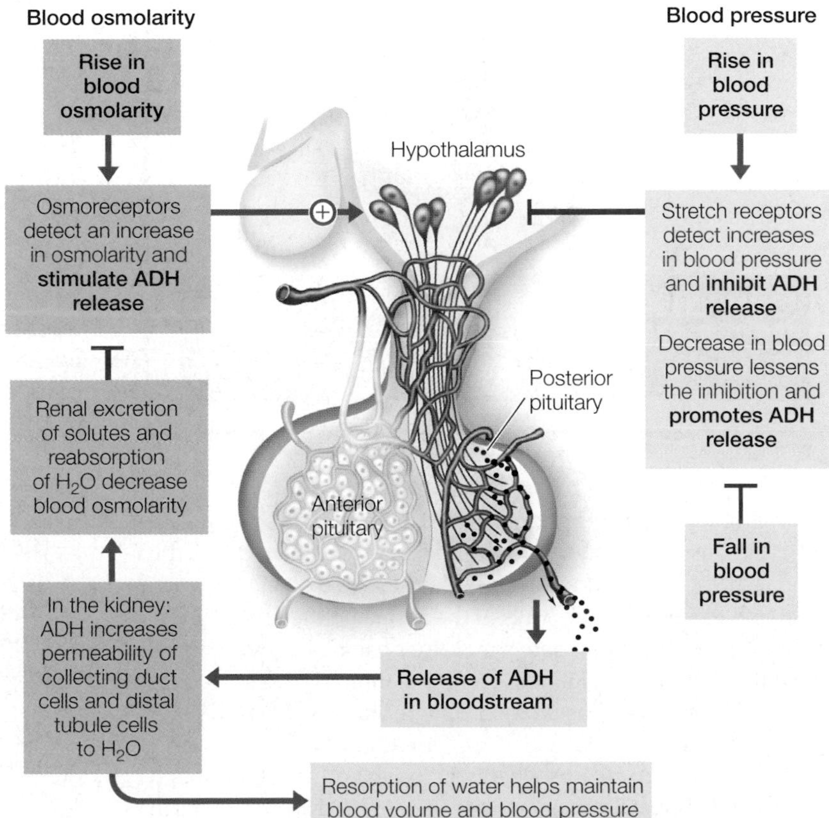

51.14 Antidiuretic Hormone Increases Blood Pressure and Promotes Water Reabsorption ADH is produced by neurons in the hypothalamus and released in the posterior pituitary. The release of ADH is stimulated by hypothalamic osmoreceptors and inhibited by stretch receptors in the great arteries.

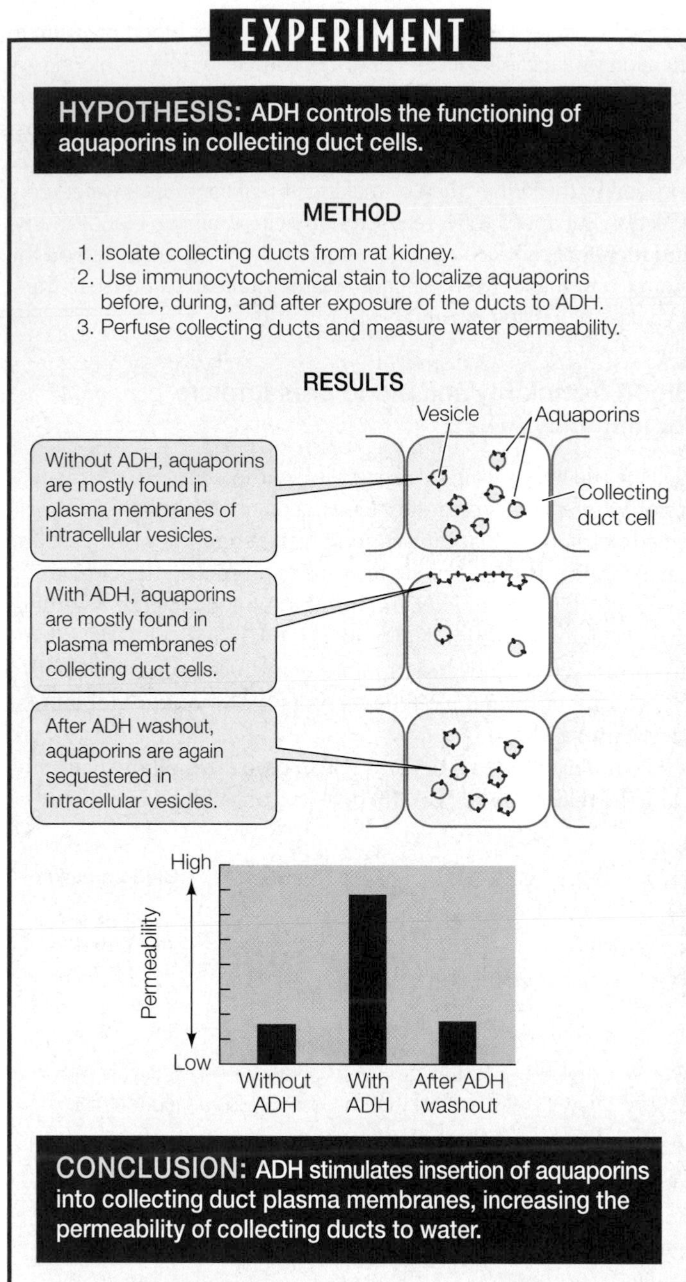

EXPERIMENT
HYPOTHESIS: ADH controls the functioning of aquaporins in collecting duct cells.
METHOD
1. Isolate collecting ducts from rat kidney.
2. Use immunocytochemical stain to localize aquaporins before, during, and after exposure of the ducts to ADH.
3. Perfuse collecting ducts and measure water permeability.
RESULTS
Vesicle
Aquaporins
Collecting duct cell
Without ADH, aquaporins are mostly found in plasma membranes of intracellular vesicles.
With ADH, aquaporins are mostly found in plasma membranes of collecting duct cells.
After ADH washout, aquaporins are again sequestered in intracellular vesicles.
High
Low
Permeability
Without ADH
With ADH
After ADH washout
CONCLUSION: ADH stimulates insertion of aquaporins into collecting duct plasma membranes, increasing the permeability of collecting ducts to water.

51.15 ADH Induces Insertion of Aquaporins into Collecting Duct Cell Plasma Membranes Aquaporin proteins are ordinarily found in the membranes of intracellular vesicles of collecting duct cells and do not contribute to water permeability. In the presence of ADH, however, these vesicles fuse with the plasma membrane, and the cells become more permeable to water.

controlling water reabsorption. The osmoreceptors also stimulate thirst. The resulting water retention and water intake dilute the blood as they expand blood volume.

Stretch receptors in the walls of the aorta and the carotid arteries (see Figure 49.17) that detect an increase in blood pressure will *inhibit* the release of ADH. With less circulating ADH, less water is reabsorbed, which decreases blood volume and hence acts to lower blood pressure.

If blood pressure falls, as when you lose blood volume through hemorrhage or excessive evaporative water loss, activity of the stretch receptors in the aorta and carotid arteries decreases. Input to the hypothalamus from these receptors inhibits the release of ADH, so when the firing rates of these stretch receptors fall, ADH release increases. More ADH results in more efficient water reabsorption and therefore a protection of blood volume and blood pressure.

Alcohol inhibits ADH release, explaining why excessive beer drinking leads to even more excessive urination and dehydration, which contributes to the symptoms of a hangover.

Given the importance of the collecting ducts in controlling the concentration of the urine and the importance of ADH in regulating collecting duct permeability to water, it was surprising that when the first aquaporins were discovered, they were not found in the collecting ducts of the kidney. It was hypothesized that a different aquaporin was expressed in the epithelial cells of the collecting duct. And, indeed, one was found and was simply named AQP-2. This aquaporin is normally sequestered in the membranes of intracellular vesicles until the collecting duct cells are stimulated by ADH. The immediate effect of circulating ADH is the insertion of AQP-2 into the plasma membranes of the cells of the collecting duct and, consequently, an increase in their permeability to water (**Figure 51.15**). When ADH decreases in the blood, the AQP-2 proteins are again internalized in vesicles and the water permeability of the cells drops.

The heart produces a hormone that influences kidney function

When systemic venous return to the heart increases and the atria of the heart become more stretched, the atrial muscle fibers release a peptide hormone called **atrial natriuretic peptide** (**ANP**). This peptide hormone enters the circulation, and when it reaches the kidney, it decreases the reabsorption of sodium. The result is an increased loss of sodium and water, which has the effect of lowering blood volume and blood pressure.

51.6 RECAP

Glomerular filtration is essential for kidney function, and it is sustained by autoregulatory mechanisms. Sensors that monitor blood pressure and blood osmolarity may stimulate or inhibit the release of hormones that regulate kidney function.

- Can you explain how falling GFR results in an increase in circulating angiotensin and how angiotensin restores GFR? See p. 1107
- Can you explain how falling blood pressure or increasing blood osmolarity results in changes in permeability of the collecting ducts? See Figure 51.14

CHAPTER SUMMARY

51.1 What roles do excretory organs play in maintaining homeostasis?

Excretory organs maintain the osmolarity and the volume of the extracellular fluids and eliminate the waste products of nitrogen metabolism through the processes of filtration, reabsorption, and secretion. **Urine** is the output of excretory organs.

There is no active transport of water, so water must be moved across membranes by a difference in either osmolarities or pressures.

Water enters or leaves cells by osmosis. To achieve cellular water balance, animals must maintain the **osmolarity** of their extracellular fluids within an acceptable range.

Marine animals can be **osmoconformers** or **osmoregulators**. Freshwater animals must be osmoregulators and must continually excrete water and conserve salts. Most animals are ionic regulators to some degree.

Apart from regulating osmolarity of cells and extracellular fluids, most animals must also regulate their ionic composition by conserving some ions and secreting others. Salt glands are adaptations for secretion of NaCl. Review Figure 51.2

51.2 How do animals excrete toxic wastes from nitrogen metabolism?

Aquatic animals can eliminate nitrogenous wastes such as ammonia by diffusion across their gill membranes. Terrestrial animals must detoxify ammonia by converting it to **urea** or **uric acid** before excretion. Review Figure 51.3

Depending on the form in which they excrete their nitrogenous wastes, animals are classified as **ammonotelic**, **ureotelic**, or **uricotelic**.

51.3 How do invertebrate excretory systems work?

The **protonephridia** of flatworms consist of flame cells and excretory tubules. Extracellular fluid is filtered into the tubules, which process the filtrate to produce a dilute urine. Review Figure 51.4

In annelid worms, blood pressure causes filtration of the blood across capillary walls. The filtrate enters the coelomic cavity, where it is taken up by **metanephridia**, which alters the composition of the filtrate by active transport mechanisms. Review Figure 51.5, Web/CD Activity 51.1

The **Malpighian tubules** of insects receive ions and nitrogenous wastes by active transport across the tubule cells. Water follows by osmosis. Ions and water are reabsorbed from the rectum, so the insect excretes semisolid wastes. Review Figure 51.6

51.4 How do vertebrates maintain salt and water balance?

Marine and terrestrial animals conserve water in various ways. Marine bony fishes produce little urine. Cartilaginous fishes retain urea so that the osmolarity of their body fluids remains close to that of seawater. Amphibians remain close to water. Reptiles have scaly skin with low water permeability, lay shelled eggs, and excrete nitrogenous wastes as uric acid.

Only birds and mammals can produce urine more concentrated than their extracellular fluids.

The **nephron**, the functional unit of the vertebrate **kidney**, consists of a **glomerulus**, in which blood is filtered, a **renal tubule**, which processes the filtrate into urine to be excreted, and a system of **peritubular capillaries**, which surround the tubule and serve as site of secretion and reabsorption. Review Figure 51.7, Web/CD Activity 51.2

51.5 How does the mammalian kidney produce concentrated urine?

See Web/CD Tutorial 51.1

The concentrating ability of the mammalian kidney is a function of its anatomy, which is ideal for **countercurrent exchange**. Review Figure 51.9, Web/CD Activity 51.3

The glomeruli and the **proximal** and **distal convoluted tubules** are located in the **cortex** of the kidney. Certain molecules are actively reabsorbed from the glomerular filtrate by the tubule cells, and other molecules are actively secreted. Straight sections of renal tubules called **loops of Henle** and **collecting ducts** are arranged in parallel in the **medulla** of the kidney.

Salts and water are reabsorbed in the proximal convoluted tubule without the renal filtrate becoming more concentrated, although its composition changes.

The loops of Henle create a concentration gradient in the **interstitial** fluid of the renal medulla by a **countercurrent multiplier mechanism**. Urine flowing down the collecting ducts to the ureter is concentrated by the osmotic reabsorption of water caused by the concentration gradient in the surrounding **interstitial fluid**. Review Figure 51.10

Hydrogen ions secreted by the renal tubules are buffered in the urine by bicarbonate and other chemical buffering systems. Review Figure 51.12

Artificial kidneys use dialysis to remove wastes from the blood. Review Figure 51.13

51.6 What mechanisms regulate kidney function?

Kidney function in mammals is controlled by **autoregulatory mechanisms** that maintain a constant high **glomerular filtration rate** even if blood pressure varies.

An important autoregulatory mechanism is the release of **renin** by the kidney when blood pressure falls. Renin activates **angiotensin**, which causes the constriction of efferent glomerular arterioles and peripheral blood vessels, causes the release of **aldosterone** (which enhances water reabsorption), and stimulates thirst.

Changes in blood pressure and osmolarity influence the release of **antidiuretic hormone** (**ADH**), which controls the permeability of the collecting duct to water and therefore the amount of water that is reabsorbed from the urine. ADH stimulates the expression of and controls the intracellular location of aquaporins, which serve as water channels in the membranes of collecting duct cells. Review Figures 51.14 and 51.15

When the volume of blood returning to the heart increases and stretches the atrial walls, **atrial natriuretic peptide** (**ANP**) is released, which causes increased excretion of salt and water.

See Web/CD Activity 51.4 for a review of the major human organ systems.

SELF-QUIZ

1. Which statement is true?
   a. Most marine invertebrates are osmoregulators.
   b. All freshwater invertebrates are osmoconformers.
   c. Marine bony and cartilaginous fishes have similar interstitial osmolarities.
   d. Freshwater fish are ionic regulators.
   e. Marine mammals gain water osmotically.
2. The excretion of nitrogenous wastes
   a. by humans can be in the form of urea and uric acid.
   b. by mammals is never in the form of uric acid.
   c. by marine fishes is mostly in the form of urea.
   d. does not contribute to the osmolarity of the urine.
   e. requires more water if the waste product is the rather insoluble uric acid.
3. How are earthworm metanephridia like mammalian nephrons?
   a. Both process coelomic fluid.
   b. Both take in fluid through a ciliated opening.
   c. Both produce urine more concentrated than the blood.
   d. Both employ tubular secretion and reabsorption to control urine composition.
   e. Both involve a countercurrent multiplier effect.
4. What is the role of renal podocytes?
   a. They prevent red blood cells and large molecules from entering the renal tubules.
   b. They reabsorb most of the glucose that is filtered from the plasma.
   c. They control the glomerular filtration rate by changing the resistance of renal arterioles.
   d. They provide a large surface area for tubular secretion and reabsorption.
   e. They release renin when the glomerular filtration rate falls.
5. Which of the following are *not* found in a renal pyramid?
   a. Collecting ducts
   b. Vasa recta
   c. Peritubular capillaries
   d. Convoluted tubules
   e. Loops of Henle
6. Which part of the nephron is responsible for most of the difference in mammals between the glomerular filtration rate and the urine production rate?
   a. The glomerulus
   b. The proximal convoluted tubule
   c. The loops of Henle
   d. The distal convoluted tubule
   e. The collecting duct
7. For mammals of the same size, what feature of their excretory systems would give them the greatest ability to concentrate their urine?
   a. Higher glomerular filtration rate
   b. Longer convoluted tubules
   c. Increased number of nephrons
   d. More permeable collecting ducts
   e. Longer loops of Henle
8. Which of the following would *not* be a response stimulated by a large drop in blood pressure?
   a. Constriction of afferent renal arterioles
   b. Increased release of renin
   c. Increased release of antidiuretic hormone
   d. Increased thirst
   e. Constriction of efferent renal arterioles
9. Which statement about angiotensin is true?
   a. It is secreted by the kidney when the glomerular filtration rate falls.
   b. It is released by the posterior pituitary when blood pressure falls.
   c. It stimulates thirst.
   d. It increases the permeability of the collecting ducts to water.
   e. It decreases glomerular filtration rate when blood pressure rises.
10. Birds that feed on marine animals ingest a lot of salt, but they excrete most of it by means of
    a. Malpighian tubules.
    b. rectal salt glands.
    c. gill membranes.
    d. concentrated urine.
    e. nasal salt glands.

FOR DISCUSSION

1. Why is it said that the oceans are a physiological desert? For what animals would this apply?
2. Persons with uncontrolled diabetes mellitus can have very high levels of glucose in their blood. Why do such individuals have a high level of urine production?
3. Inulin is a molecule that is filtered out of the glomerulus, but is not secreted or reabsorbed by the renal tubules. If you injected inulin into an animal and after a brief time measured the concentration of inulin in its blood and urine, how could you determine the animal's glomerular filtration rate? Assume that the rate of urine production is 1 milliliter per minute.
4. After you did the inulin experiment to measure glomerular filtration rate, how could you use that information to determine whether another substance is secreted or reabsorbed by the renal tubules? Assume you can measure the concentration of that substance in the blood and in the urine. Urine production is still 1 milliliter per minute.
5. Explain what happens with respect to control and regulation of your salt and water balance when you go to a movie and eat a lot of very salty popcorn.

FOR INVESTIGATION

We mentioned that the GFR is autoregulated, and this regulation is achieved by control over constriction/dilation of the afferent and the efferent renal arterioles. What information could be used in this process? A clue comes from the anatomy of the nephron. When the ascending limb of Henle reaches the cortex, it makes direct contact with the afferent and efferent arterioles of the glomerulus that produced the filtrate flowing in that particular nephron. What aspect of the tubular fluid in the early distal tubule would reflect changes in the volume of filtrate entering the nephron? How might the cells of the early distal tubule communicate with the smooth muscle cells of the arterioles?

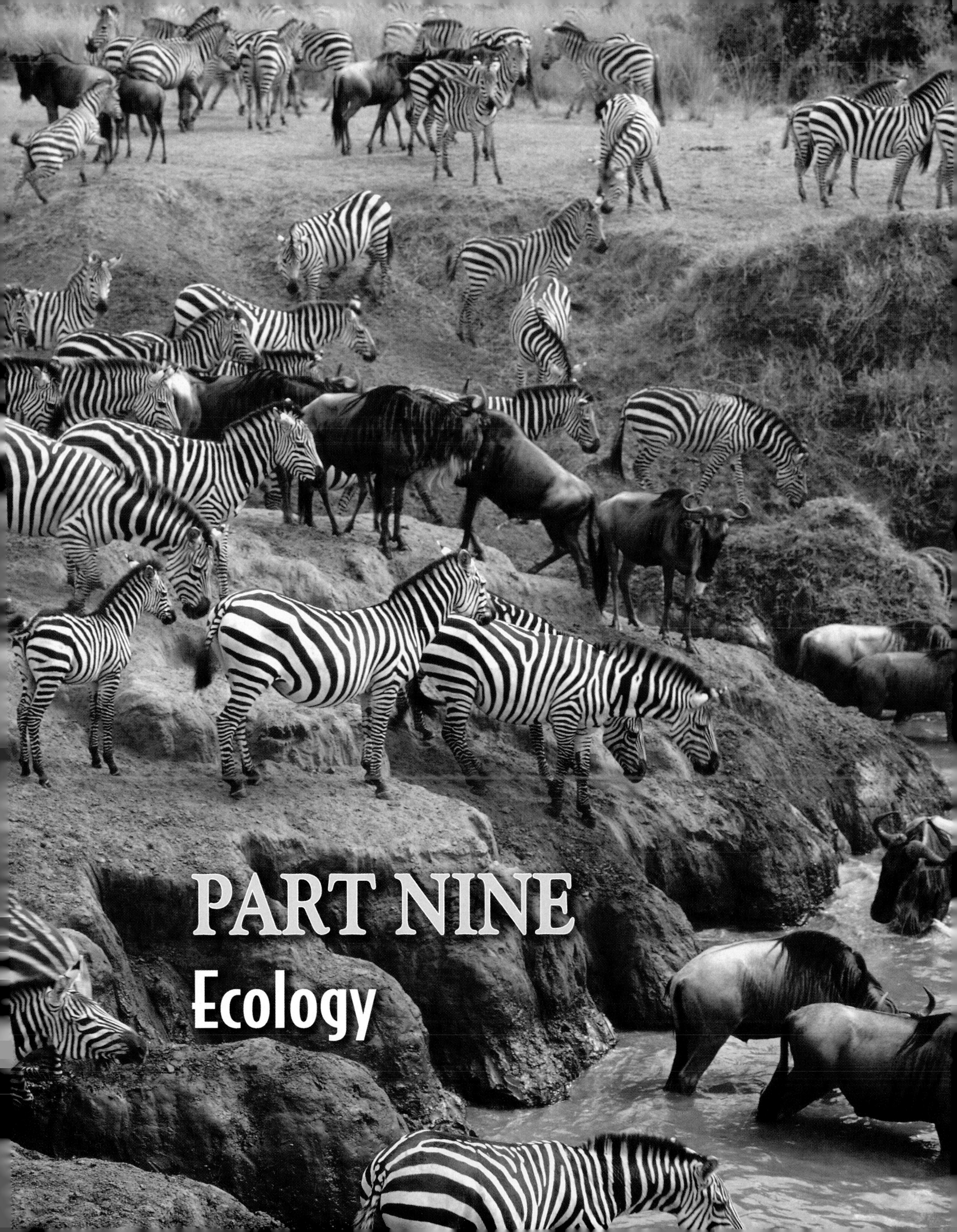

PART NINE
Ecology

CHAPTER 52 Ecology and the Distribution of Life

Shelter from the storm

The city of New Orleans lies along the Mississippi River at the point where the river spreads into a magnificent, low-lying delta of swamps and wetlands. More than half the city is below sea level. An elaborate flood control system of dams and levees constructed by the Army Corps of Engineers holds back both the river and Lake Pontchartrain, a large, brackish body of water to the north of the city.

On August 29, 2005, Hurricane Katrina made landfall just to the east of New Orleans. At first it seemed that the storm, once a terrifying category 5 hurricane, had dissipated enough to spare the city the worst. But over a 24-hour period, the combination of strong winds, heavy rain, and storm surges breached some of the levees, with catastrophic effects. More than 1,800 people died. Many thousands lost their homes. The 2000 U.S. census figures put the population of New Orleans at 485,000; figures collected in June of 2006—nine months after Katrina—estimated the city's population at about 225,000.

Katrina's devastating impact was due in part to a situation that had been developing for decades. The upstream dams that protect New Orleans also prevent the Mississippi River from depositing the sediments that sustained the surrounding wetlands for centuries. The oil and natural gas industry has cut thousands of small canals through the delta wetlands in order to lay pipelines and drilling rigs, and the extraction of oil and gas from beneath the land has caused it to sink. Increased dredging of shipping lanes and rising sea levels due to global warming contribute to a rise in salinity, killing off many of the great cypress tree swamps. It all adds up to the loss of over 80 percent (more than 1.2 *million* acres) of protective wetlands between 1930 and 2005. At the time Katrina made landfall, the greatly reduced coastal wetlands could not protect New Orleans. Storm surges raced along the paths carved by canals and shipping lanes to breach the levees, inundating about 80 percent of the city.

Coastal wetlands provide many services besides protection. Louisiana's wetlands provide winter habitat for about 70 percent of the migrating birds in the huge Mississippi Valley. They are also the spawning grounds for marine organisms, some of which are commercially

Approach of a Mighty Storm A satellite weather photo shows Katrina at full strength, as a category 5 hurricane approaching the U.S. Gulf Coast. Although the storm weakened somewhat before making landfall, the city of New Orleans and other coastal towns suffered devastating damage and destruction.

The Sheltering Bayou Bald cypress (*Taxodium distichum*) growing in freshwater swamps are a distinctive and vital part of the Louisiana bayou. Their destruction by encroaching salt water from the Gulf of Mexico is endangering the region.

CHAPTER OUTLINE

valuable. The shrimping industry of that region has long been renowned, and the Gulf Coast contributes about 30 percent by weight of the total commercial fish harvest in the continental U.S.

Around the world, valuable, indeed, indispensable, coastal wetlands and other ecosystems are being destroyed or altered at a frightening rate. Some of the services provided by these systems can be replaced—usually at tremendous expense—by human technology, but some of them are irreplaceable. Ecologists work to find ways to manage Earth's ecosystems and to guarantee the sustained flow of their benefits to future generations.

IN THIS CHAPTER we define the field of ecology and the kinds of questions ecologists try to answer. Then we describe the distribution of Earth's climates and the forces that create them. We explain some of the ways in which climates and other features of the physical environment influence where different organisms are found. We conclude by describing the major biomes and biogeographic regions that reflect the distribution of life on Earth.

52.1 What Is Ecology?

Ecology is the scientific study of the rich and varied interactions between organisms and their environment. Ecologists study these interactions at many levels. As we will see in the next chapter, behavioral interactions among individuals of the same species may give rise to elaborate social systems. Organisms also interact with individuals of different species (for example, as predators, prey, mutualists, and competitors) and with their physical environment. These interactions, in turn, influence the structure of **communities** (systems embracing all the organisms living together in the same area), **ecosystems** (systems embracing all organisms in an area plus their physical environment), and the **biosphere** (the system that embraces all regions of the planet where organisms live).

The term **environment**, as used by ecologists, encompasses both **abiotic** (physical and chemical) factors, such as water, mineral nutrients, light, temperature, and wind, and **biotic** factors (living organisms). Interactions between organisms and their environment are two-way processes: organisms both influence and are influenced by their environment. Indeed, dealing with environmental changes caused by our own species is one of the major challenges facing humanity today. For this reason, ecologists are often asked to help analyze the causes of environmental problems and to assist in finding solutions.

There are many reasons to care about ecology. Our lives are enriched by the fascinating interactions among species that we encounter whenever we venture outside. We watch a butterfly pollinating flowers (**Figure 52.1**) and want to know more about how species interact with one another. We are curious about the consequences of those interactions.

Beyond simple curiosity, however, information and insights from ecological science are needed to solve many practical problems. Humans depend on vibrant, functioning ecosystems to provide many goods and services—including the essential services of clean air and water. An understanding of ecology allows us to manage ecosystems so as to maintain the availability of those goods and services. An understanding of ecology allows us to grow food, control pests and diseases, and deal with natural disasters such as floods without causing a cascade of other unanticipated consequences.

52.1 An Ecological Interaction The pollination of flowers by butterflies is mutualism that benefits both participants.

Ecological information can help us solve practical problems only if we know how and why those problems arise. Thus ecologists must become familiar with various environments and understand how organisms adapt to them. This chapter will focus on the big picture: Earth's climates and general patterns in the distribution of life. In subsequent chapters we will look in greater detail at the ways in which individual organisms interact, the dynamics of their populations, and the ecological communities in which they live.

52.1 RECAP

Ecology is the scientific investigation of the interactions among organisms and between organisms and their physical environment.

- What are the components of the environment as defined by ecologists? See p. 1113
- At what levels do ecologists study ecological systems? See p. 1113
- Do you understand why understanding ecology can be considered essential to human survival? See p. 1113

Climate is one of the abiotic factors that determines what kinds of organisms can survive and reproduce in a particular place. We begin our study of ecology with an overview of Earth's climates.

52.2 How Are Climates Distributed on Earth?

The **climate** of a region is the average of the atmospheric conditions (temperature, precipitation, and wind direction and velocity) found there over the long term. *Weather* is the short-term state of those conditions. In other words, climate is what you expect; weather is what you get! Climates vary greatly from place to place on Earth, primarily because different places receive different amounts of solar energy. In this section we will examine how these differences in solar energy input determine atmospheric and oceanic circulation patterns—the factors that most strongly influence climates.

Solar energy drives global climates

The differences in air temperature among different places on Earth are largely determined by differences in solar energy input. Every place on Earth receives the same total number of hours of sunlight each year—an average of 12 hours per day—but not the same amount of solar energy. The rate at which solar energy arrives on Earth per unit of Earth's surface depends primarily on the angle of sunlight. If the sun is low in the sky, a given amount of solar energy is spread over a larger area (and is thus less intense) than if the sun is directly overhead. In addition, when the sun is low in the sky, its light must pass through more of Earth's atmosphere, so more of its energy is absorbed and reflected before it reaches the ground. Thus higher latitudes (closer to the poles) receive less solar energy than latitudes closer to the equator. On average, mean annual air temperature decreases about 0.4°C for every degree of latitude (about 110 kilometers) at sea level. In addition, higher latitudes experience greater variation in both day length and the angle of arriving solar energy over the course of a year, leading to more seasonal variation in temperature.

Air temperature also decreases with elevation. As a parcel of air rises, it expands (its molecules move farther apart), its pressure and temperature drop, and it releases moisture. When a parcel of air descends, it is compressed, its pressure rises, its temperature increases, and it takes up moisture.

Global air circulation patterns result from the global variation in solar energy input that we have just described and from the spinning of Earth on its axis (**Figure 52.2**). Air rises when it is heated by the sun, so warm air rises in the tropics, which receive the greatest solar energy input. This rising air is replaced by air that flows in toward the equator from the north and south. The coming together of these air masses produces the **intertropical convergence zone**. Cool air cannot hold as much moisture as warm air, so heavy rains fall in the intertropical convergence zone as the rising air cools and releases its moisture. The intertropical convergence zone shifts latitudinally with the seasons, following the shift in the zone of greatest solar energy input. This shift results in predictable rainy and dry seasons in tropical and subtropical regions.

The air that moves into the intertropical convergence zone to replace the rising air is replaced, in turn, by air from aloft that descends at roughly 30° north and south latitudes after having traveled away from the equator high in the atmosphere. This air cooled and lost its moisture while it rose at the equator. It now descends, warms, and takes up, rather than releases, moisture. Many of Earth's deserts, such as the Sahara and the Australian deserts, are located at these latitudes where dry air descends.

At about 60° north and south latitudes, air rises again and moves either toward or away from the equator. At the poles, where there is little solar energy input, air descends. These movements of air masses are largely responsible for global wind patterns.

The spinning of Earth on its axis also influences surface winds because Earth's velocity is rapid at the equator, where its diameter is greatest, but relatively slow close to the poles. A stationary air

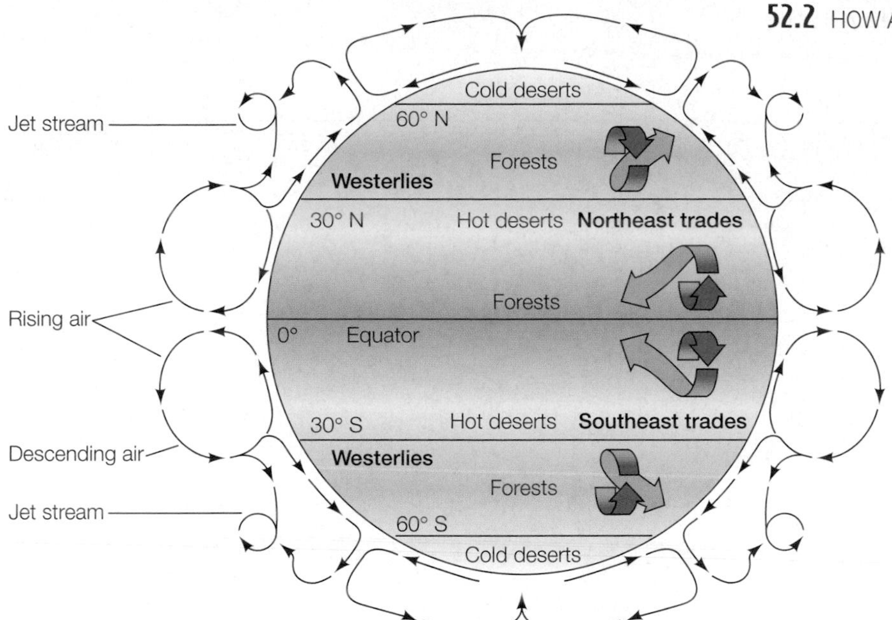

52.2 The Circulation of Earth's Atmosphere If we could stand outside Earth and observe air movements in its atmosphere, we would see vertical air circulation patterns similar to those indicated by the black and red arrows and surface winds similar to those shown by the blue arrows.

mass has the same velocity as Earth at the same latitude. As an air mass moves toward the equator, it confronts an increasingly faster spin, and its rotational movement is slower than that of Earth beneath it. Conversely, as an air mass moves poleward, it confronts an increasingly slower spin, and it speeds up relative to Earth beneath it. Therefore, air masses moving latitudinally are deflected to the right in the Northern Hemisphere and to the left in the Southern Hemisphere. The air masses moving toward the equator from the north and south veer to become the northeast and southeast *trade winds*, respectively. Air masses moving away from the equator also veer and become the *westerly* winds that prevail at mid-latitudes.

When these prevailing winds bring air masses into contact with a mountain range, the air rises to pass over the mountains, cooling as it does so. Thus clouds frequently form on the windward side of mountains (the side facing into the winds) and release moisture as rain or snow. On the leeward side (opposite from the direction of the winds) of mountains, the now-dry air descends, warms, and once again picks up moisture. This pattern often results in a dry area called a **rain shadow** on the leeward sides of mountain ranges (**Figure 52.3**).

Global oceanic circulation is driven by wind patterns

The global air circulation patterns that we have just described drive the circulation patterns of surface ocean waters, known as *currents* (**Figure 52.4**). The trade winds blowing toward the equator from the northeast and southeast cause water to converge at the equator and move westward until it encounters a continental land mass. At that point the water splits, with some of it moving north and some of it moving south along continental shores. This poleward movement of ocean water that has been warmed in the tropics transfers large amounts of heat to high latitudes. As these currents move toward the poles, the water, driven by the winds, veers right in the Northern Hemisphere and left in the Southern Hemisphere. Thus water flowing toward the poles turns eastward until it encounters another continent and is deflected laterally along its shores. In both hemispheres, water flows toward the equator along the western sides of continents, continuing to veer right or left until it meets at the equator and flows westward again.

Organisms must adapt to changes in their environment

Some kinds of changes in an organism's environment, such as the approach of a fire, storm, or predator, require immediate responses; other changes allow time for a more gradual response. Many plants adapt to hot conditions by reducing water loss and avoid overheating by shifting the position of their leaves during the day so that they intercept sunlight early and late in the day, but do not overheat at midday. Lizards bask in sunshine in the morning to raise their body temperature, but move into the shade when it gets too hot.

In addition to short-term changes in behavior, the evolution of many morphological and physiological features has enabled organisms to function in a variable physical environment. Many such features are described in Parts Seven and Eight of this book.

Few individuals die exactly where they were born; at some time during their lives, most organisms move, or are moved, to a new place, a phenomenon known as **dispersal**. Individuals may leave the site of their birth to find a better place to reproduce. Others may seek new places to live when local conditions deteriorate.

If repeated seasonal changes alter an environment in predictable ways, organisms may evolve life cycles that appear to anticipate those changes. **Migration** is one response to such cycli-

2 On the windward side of the mountain, air rises and cools, releasing moisture in the form of rain or snow.
1 Prevailing winds pick up moisture over water bodies.
3 On the leeward side of the mountain, air descends, warms, and picks up moisture, which results in little rain.

52.3 A Rain Shadow Average annual rainfall tends to be lower on the leeward side of a mountain range than on the windward side.

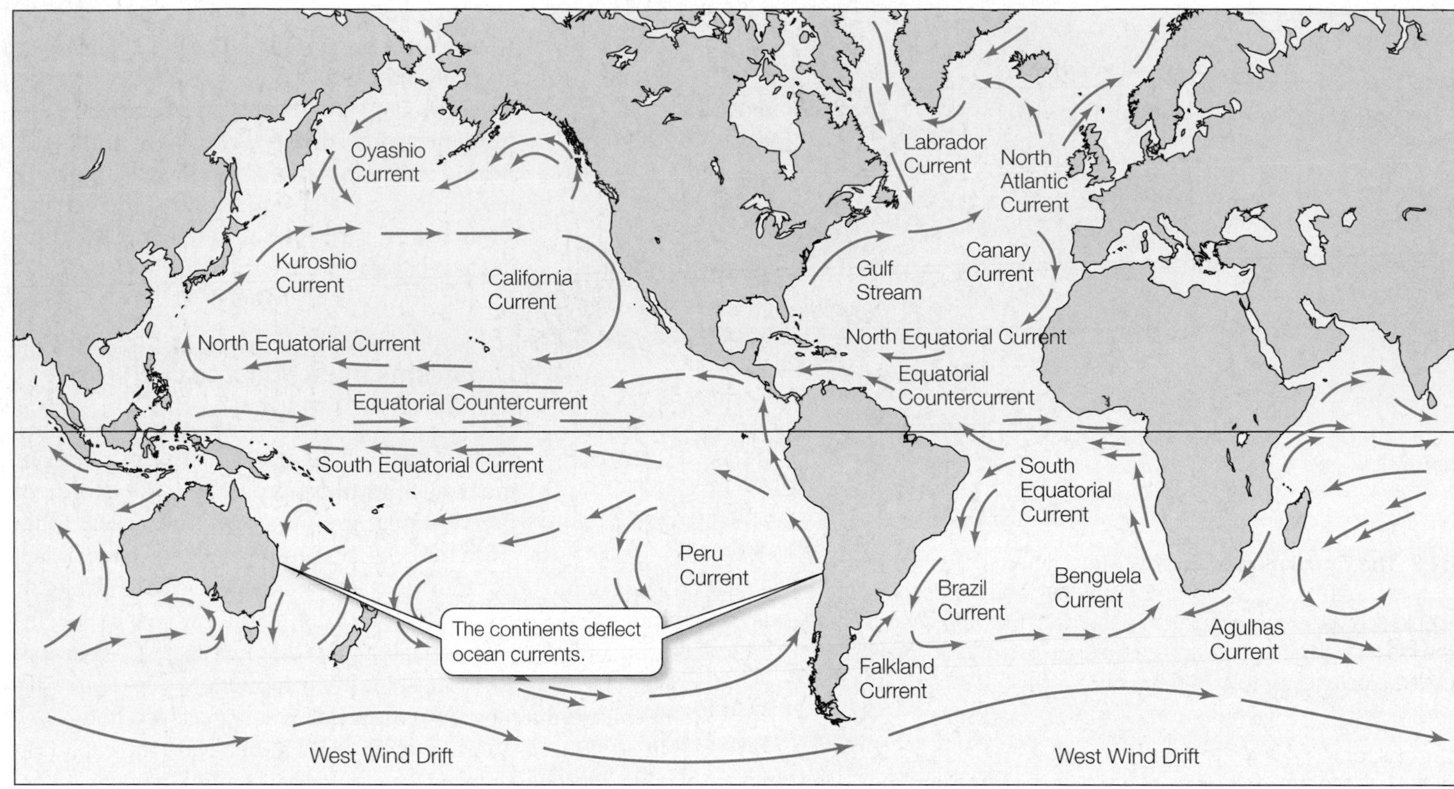

52.4 Global Oceanic Circulation To see that surface ocean currents are driven primarily by winds, compare the currents shown here with the prevailing winds shown in Figure 52.2.

cal environmental changes. Other animals enter a resting state (estivation, hibernation, or diapause) before adverse conditions materialize. They remain in that state until environmental signals indicate that conditions have improved.

Humans have clearly influenced Earth's climate, but such effects were thought to be only recent. However, climate models built by scientists at the University of Colorado suggest that if the early human occupants of Australia had not extensively burned that continent's forests, Australia's climate would be much wetter today than it actually is.

Most changes in the physical environment, both short-term and long-term, happen independently of anything that organisms do. Glaciers advance and retreat, storms brew, heat waves and cold fronts come and go, uninfluenced by the organisms that must deal with them. Other environmental changes, however, have been, and continue to be, influenced by the activities of organisms. For example, as Chapter 21 describes, organisms generated Earth's oxygenated atmosphere and produced Earth's soils. As we will see in the following chapters, many significant changes in the environment to which organisms must adapt are caused by other organisms, and many characteristics of organisms have been determined to a large degree by the history of such interactions.

52.2 RECAP

Differences in solar energy input create patterns of atmospheric circulation, and these prevailing winds affect oceanic circulation patterns. Organisms adapt to both short- and long-term climatic changes in their environment.

- Why does mean annual air temperature decrease with both latitude and altitude? See p. 1114
- Do you understand how variations in solar energy drive global air circulation patterns? See p. 1115 and Figure 52.2
- How do global air circulation patterns drive ocean currents? See p. 1115 and Figure 52.4

The great variability of the effects of Earth's climate has given rise to many different assemblages of organisms. Ecologists have found it useful to classify these assemblages into different ecosystem types. Depending on what parts of the system they intend to study, ecologists may classify ecosystems into biomes or biogeographic regions.

52.3 What Is a Biome?

When ecologists identify ecosystems, how do they decide where to locate the boundaries between different ecosystem types? The growth forms of the dominant plants of a particular environment, by determining the structure of the vegetation and by modifying the climate near the ground, strongly influence the lives of the other organisms that live there. A **biome** is a terrestrial environment defined by the growth forms of its plants. Common biomes include forests, grassland, desert, and tundra (**Figure 52.5**).

The distributions of plants are strongly influenced by annual patterns of temperature and rainfall. In some biomes, such as temperate deciduous forest, precipitation is relatively constant throughout the year, but temperature varies strikingly between summer and winter. In other biomes, both temperature and precipitation change seasonally. In still other biomes, temperatures are nearly constant, but rainfall varies seasonally. In the tropics, where seasonal temperature fluctuations are small, annual cycles are dominated by wet and dry seasons. Tropical biome types are determined primarily by the length of the dry season.

It is easiest to grasp the similarities and differences between biomes by means of a combination of photographs and graphs of temperature, precipitation, and biological activity, supplemented by a few words that describe other attributes of those biomes. In the following pages, each biome is represented by a map showing its locations and two photographs that illustrate either the biome at different times of year or representatives of the biome in different places on Earth. One set of graphs plots seasonal patterns of temperature and precipitation at a site in the biome. Other graphs show activity patterns of different kinds of organisms during the year. (For high-latitude biomes, patterns in the Southern Hemisphere are 6 months out of phase with those shown, which represent the Northern Hemisphere.) Levels of biological activity, shown by the width of the horizontal bars, change either because resident organisms become more or less active (produce leaves, come out of hibernation, hatch, or reproduce) or because organisms migrate into and out of the biome at different times of the year. A small box describes the growth forms of the plants that dominate the vegetation in the biome and its pattern of **species richness** (the number of species present in its communities).

These descriptions of biomes are very general and cannot begin to describe the variation that exists within each biome. For example, the temperate deciduous forest biome contains lakes, rivers, low-growing vegetation on cliffs, and grassy areas recovering from fire or other types of disturbance, as well as forests. In addition, the boundaries between biomes are somewhat arbitrary. Although sometimes an abrupt change can be seen in a landscape, more often one biome gradually merges into another. For example, the boundary between a forest and a grassland may not be distinct; instead, the spaces between trees may gradually increase, allowing more grasses to grow among them. Nevertheless, it is useful to recognize the major biomes of the world.

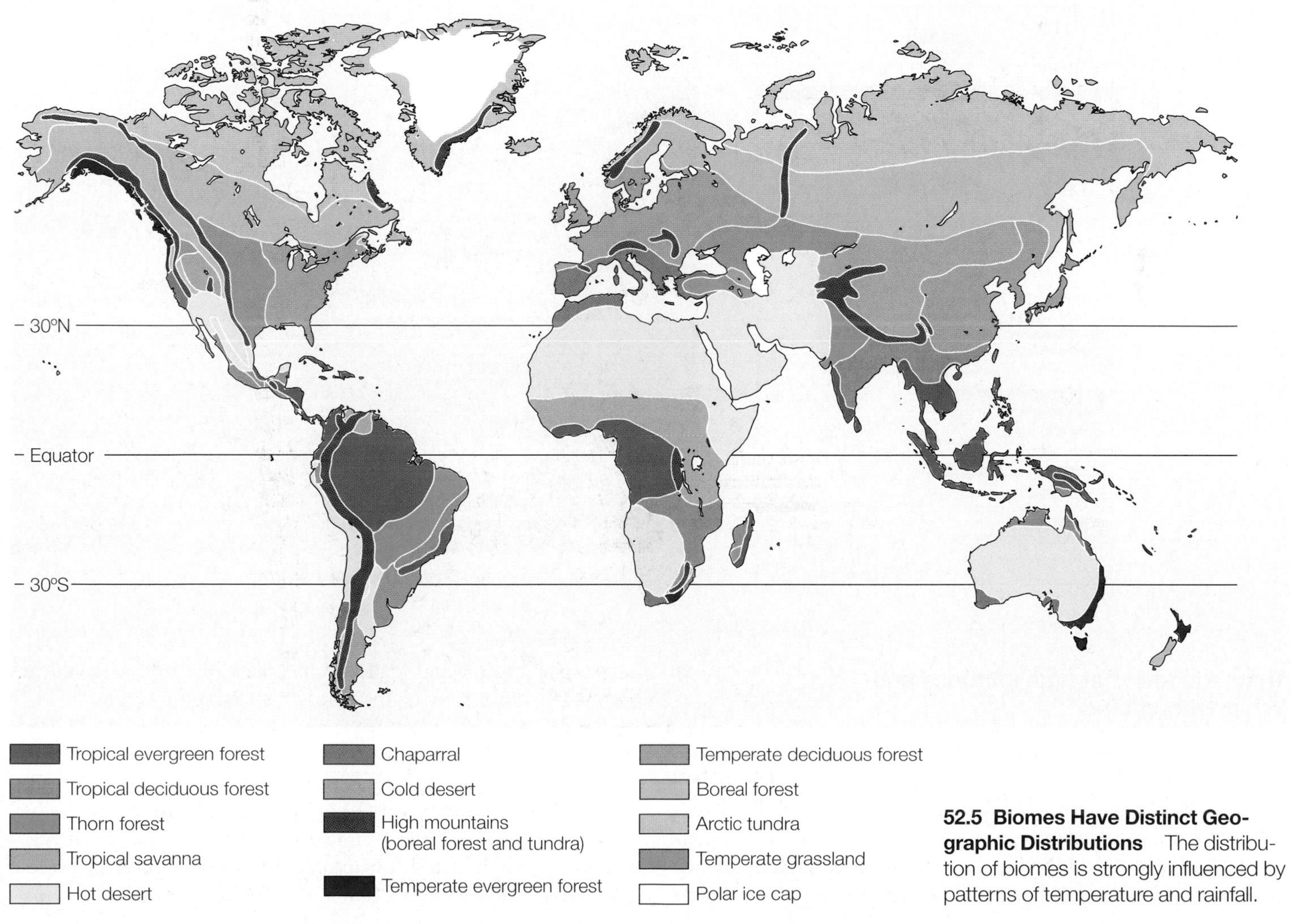

52.5 Biomes Have Distinct Geographic Distributions The distribution of biomes is strongly influenced by patterns of temperature and rainfall.

TUNDRA

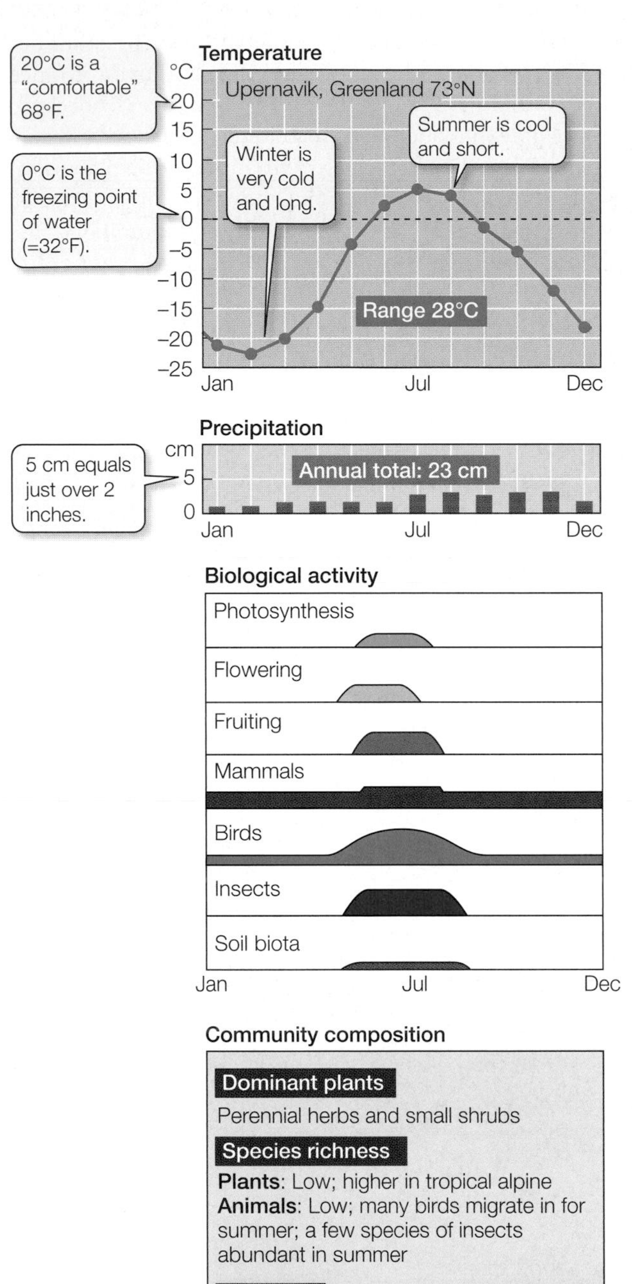

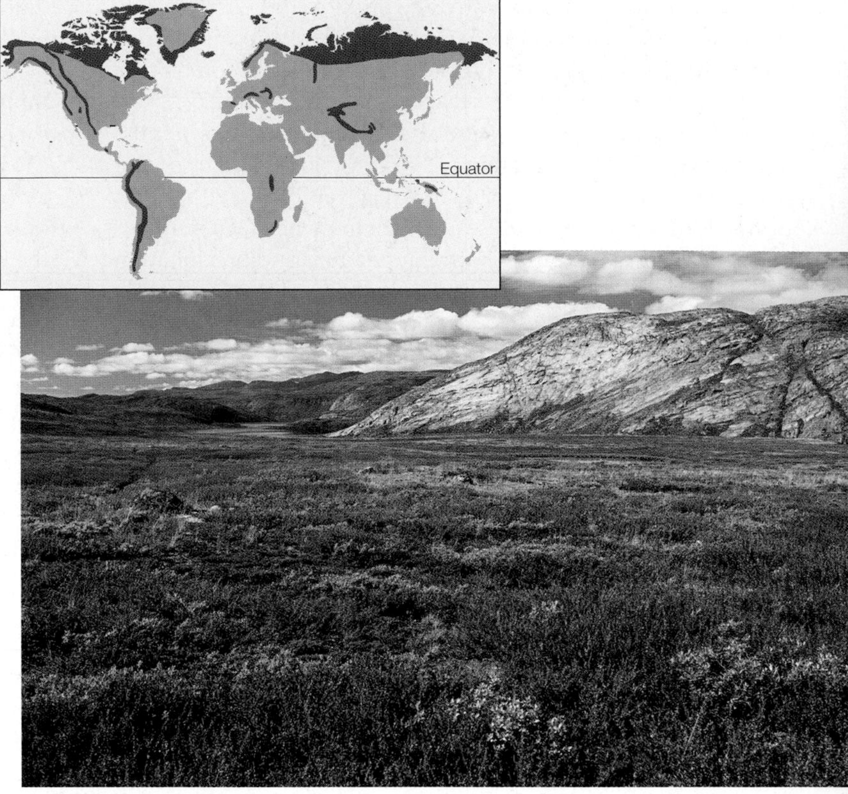

Arctic tundra, Greenland

Tropical alpine tundra, Teleki Valley, Mt. Kenya, Kenya

Tundra is found at high latitudes and in high mountains

The **tundra** biome is found in the Arctic and at high elevations in mountains at all latitudes. In *Arctic tundra*, the vegetation, which consists of low-growing perennial plants, is underlain by *permafrost*—soil whose water is permanently frozen. The top few centimeters of soil thaw during the short summers, when the sun may be above the horizon 24 hours a day. Even though there is little precipitation, lowland Arctic tundra is very wet because water cannot drain down through the permafrost. Plants grow for only a few months each year. Most Arctic tundra animals either migrate into the area only for the summer or are dormant for most of the year.

Tropical alpine tundra is not underlain by permafrost, so photosynthesis and most other biological activities continue (albeit slowly) throughout the entire year. As the photo of alpine vegetation on Mt. Kenya shows, more plant growth forms are present in tropical alpine tundra than in Arctic tundra vegetation.

BOREAL FOREST and TEMPERATE EVERGREEN FOREST

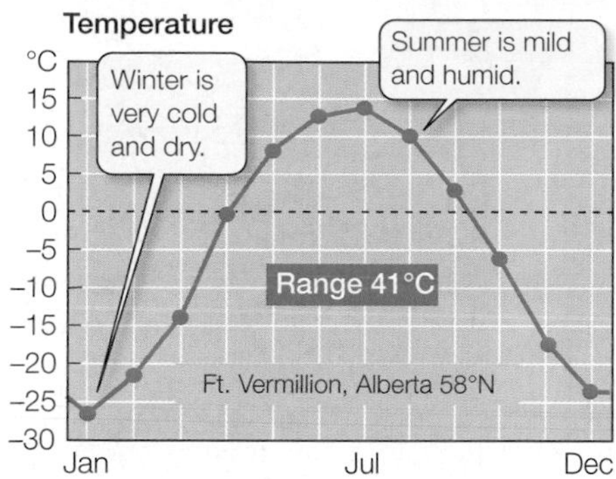

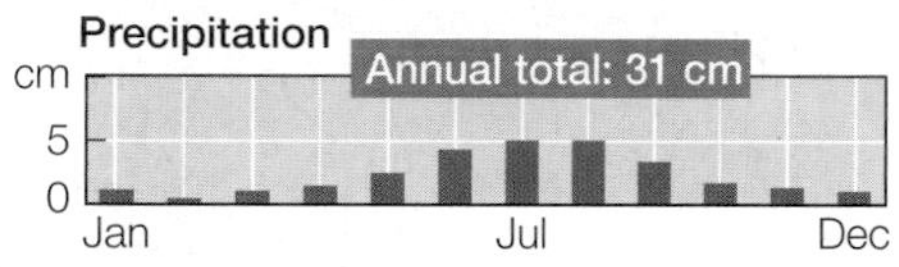

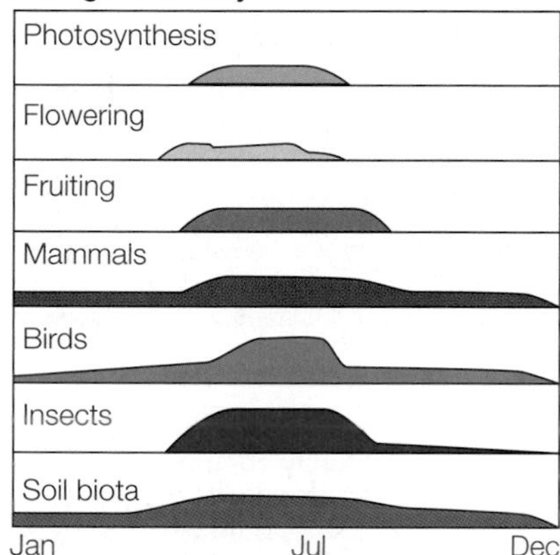

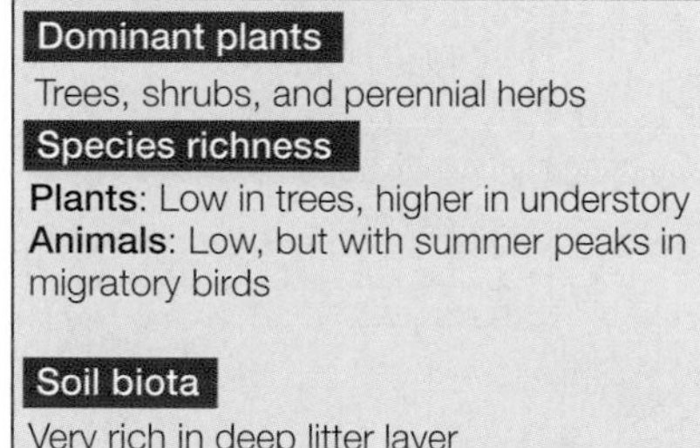

Community composition
Dominant plants
Trees, shrubs, and perennial herbs
Species richness
Plants: Low in trees, higher in understory
Animals: Low, but with summer peaks in migratory birds
Soil biota
Very rich in deep litter layer

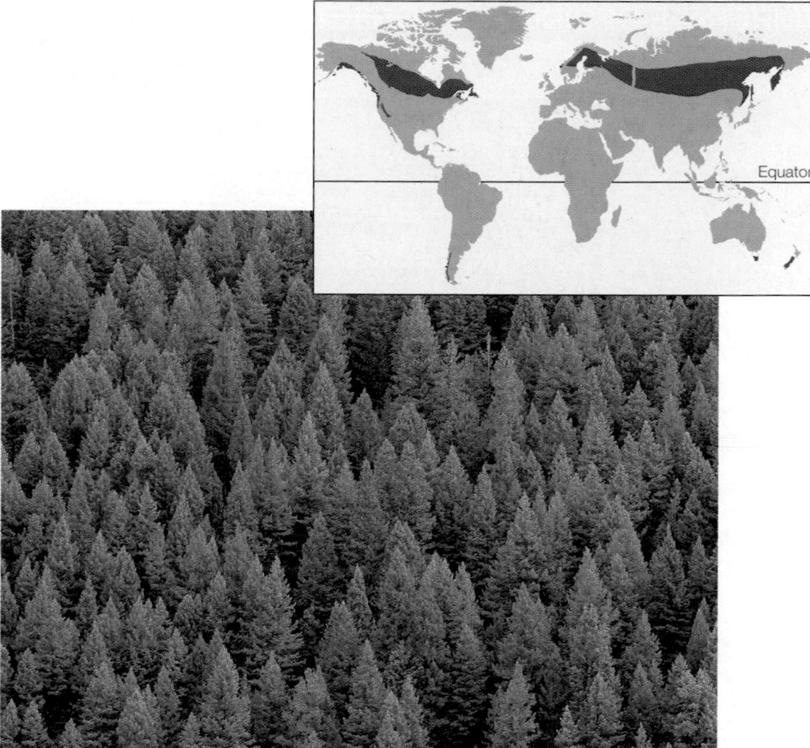

Northern boreal forest, Gunnison National Forest, Colorado

Southern boreal forest, Fiordland National Park, New Zealand

Evergreen trees dominate most boreal forests

The **boreal forest** biome is found toward the equator from Arctic tundra and at lower elevations on temperate-zone mountains. Boreal forest winters are long and very cold; summers are short (although often warm). The shortness of the summer favors trees with evergreen leaves because these trees are ready to photosynthesize as soon as temperatures warm in spring.

The boreal forests of the Northern Hemisphere are dominated by evergreen coniferous gymnosperms. In the Southern Hemisphere the dominant trees are southern beeches (*Nothofagus*), some of which are evergreen. **Temperate evergreen forests** also grow along the western coasts of continents at middle to high latitudes in both hemispheres, where winters are mild but very wet and summers are cool and dry. These forests are home to Earth's tallest trees.

Boreal forests have only a few tree species. The dominant mammals, such as moose and hares, eat leaves. The seeds in the cones of conifers support a fauna of rodents, birds, and insects.

TEMPERATE DECIDUOUS FOREST

Equator

A Rhode Island forest in summer and…

…in winter

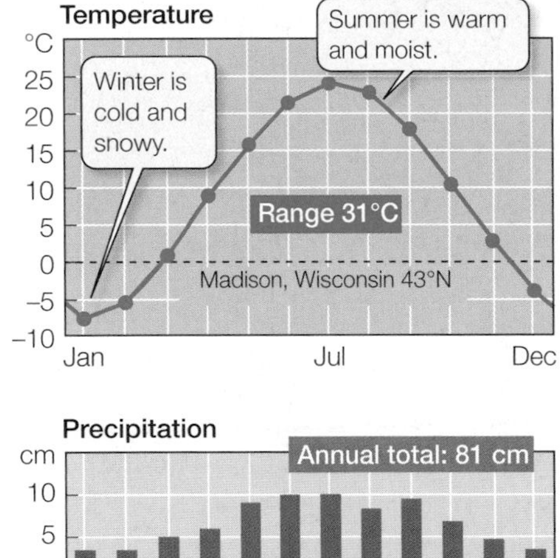

Summer is warm and moist.
Winter is cold and snowy.
Range 31°C

Annual total: 81 cm

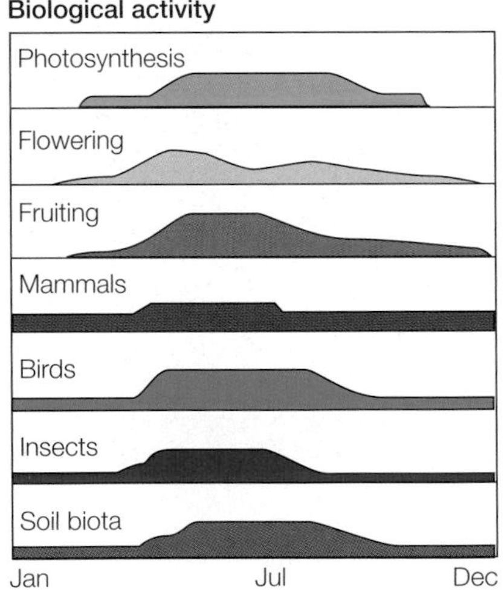

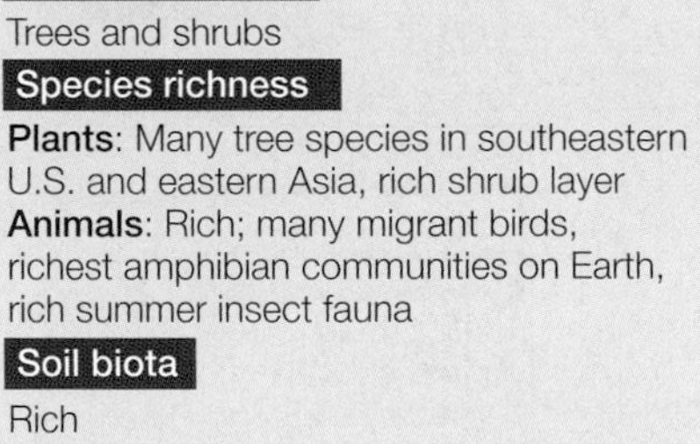

Community composition
Dominant plants
Trees and shrubs
Species richness
Plants: Many tree species in southeastern U.S. and eastern Asia, rich shrub layer
Animals: Rich; many migrant birds, richest amphibian communities on Earth, rich summer insect fauna
Soil biota
Rich

Temperate deciduous forests change with the seasons

The **temperate deciduous forest** biome is found in eastern North America, eastern Asia, and Europe. Temperatures in these regions fluctuate dramatically between summer and winter. Precipitation is relatively evenly distributed throughout the year.

Deciduous trees, which dominate these forests, lose their leaves during the cold winters and produce leaves that photosynthesize rapidly during the warm, moist summers. Many more tree species live here than in boreal forests. The temperate forests richest in species are in the southern Appalachian Mountains of the United States and in eastern China and Japan—areas that were not covered by glaciers during the Pleistocene. Many genera of plants and animals are shared among the three geographically separate regions with a deciduous forest biome.

TEMPERATE GRASSLANDS

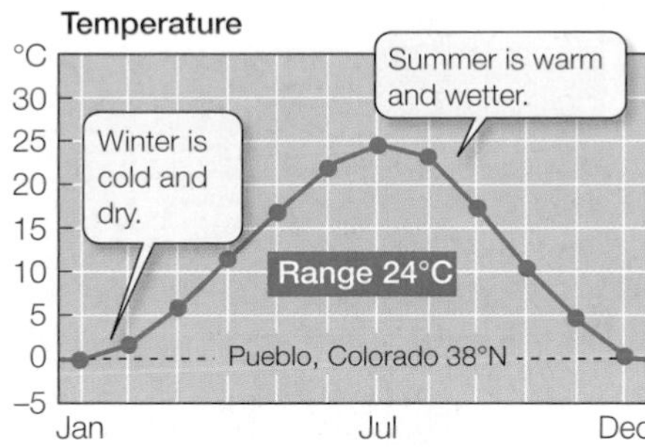

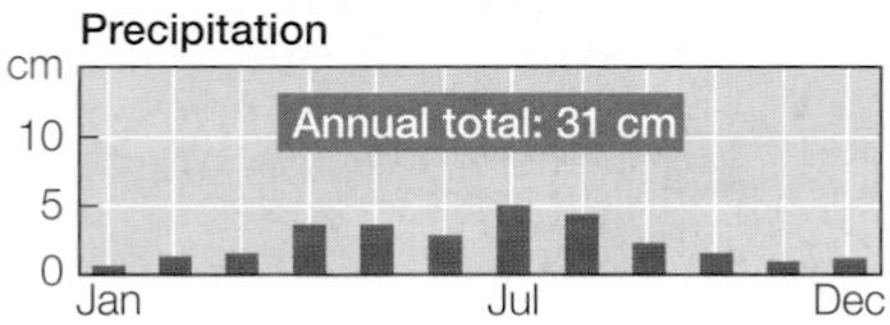

Biological activity

Photosynthesis

Flowering

Fruiting

Mammals

Birds

Insects

Soil biota

Jan Jul Dec

Community composition

Dominant plants

Perennial grasses and forbs

Species richness

Plants: Fairly high

Animals: Relatively few birds because of simple structure; mammals fairly rich

Soil biota

Rich

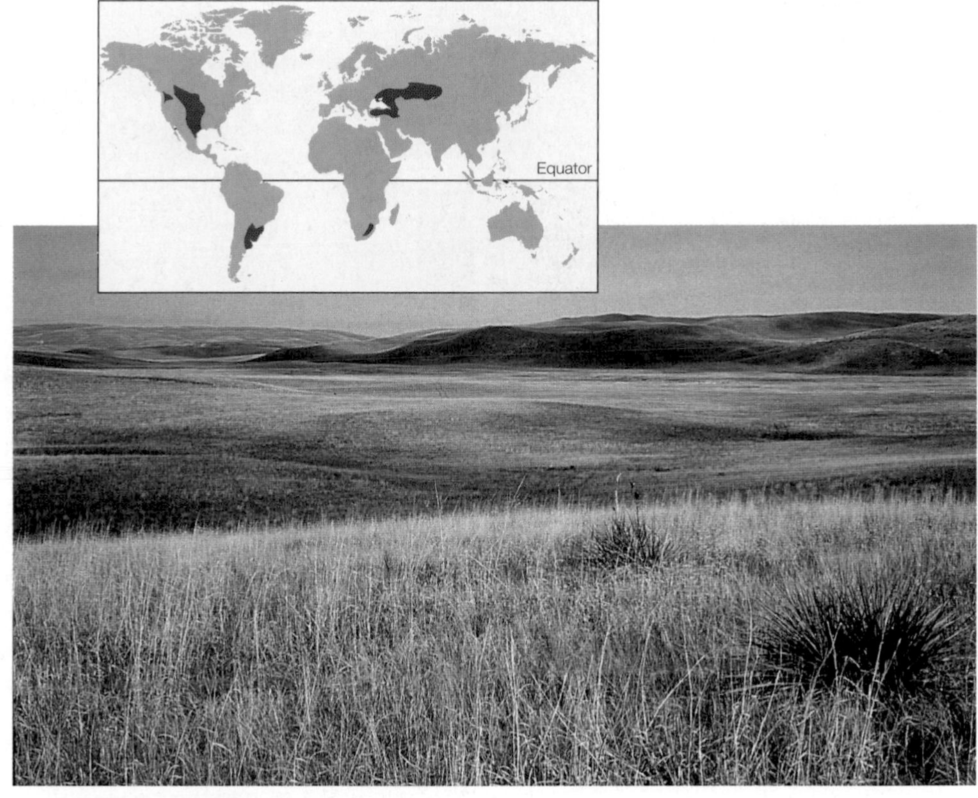

Nebraska prairie in spring

The Veldt, Natal, South Africa

Temperate grasslands are widespread

The **temperate grassland** biome is found in many parts of the world, all of which are relatively dry for much of the year. Most grasslands, such as the pampas of Argentina, the veldt of South Africa, and the Great Plains of North America, have hot summers and relatively cold winters. Most of this biome has been converted to agriculture. In some grasslands, most of the precipitation falls in winter (California grasslands); in others, the majority falls in summer (Great Plains, Russian steppe).

Grassland vegetation is structurally simple, but it is rich in species of perennial grasses, sedges, and *forbs* (herbaceous plants other than grasses). Grasslands are often a riot of color when forbs are in bloom. Grassland plants are adapted to grazing and fire. They store much of their energy underground and quickly resprout after they are burned or grazed.

COLD DESERT

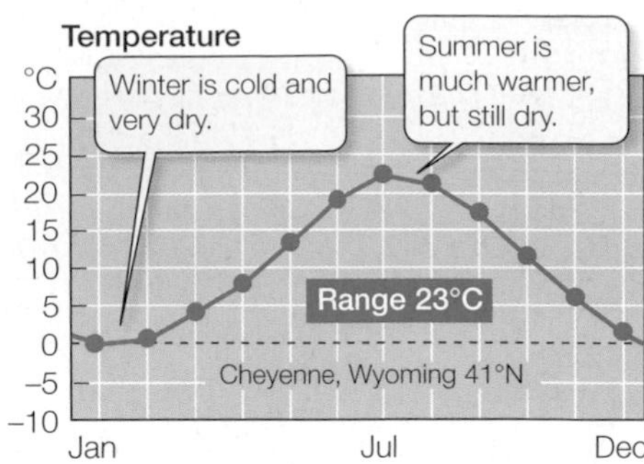

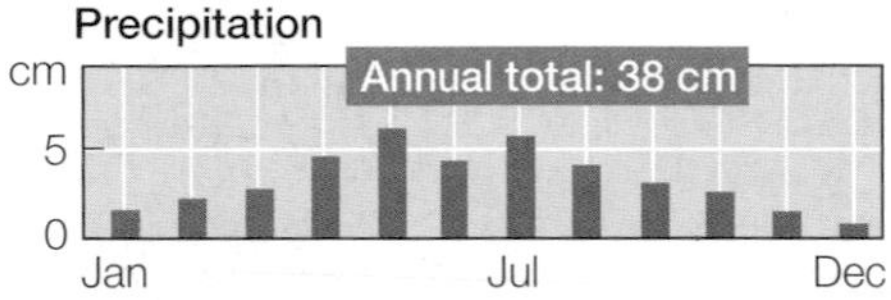

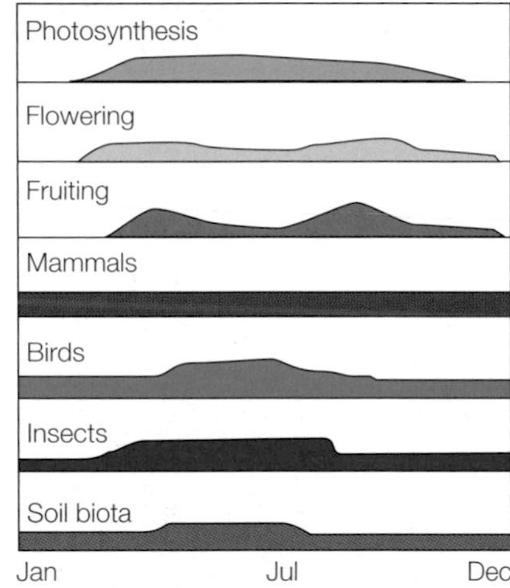

Community composition

Dominant plants

Low-growing shrubs and herbaceous plants

Species richness

Plants: Few species
Animals: Rich in seed-eating birds, ants, and rodents; low in all other taxa

Soil biota

Poor in species

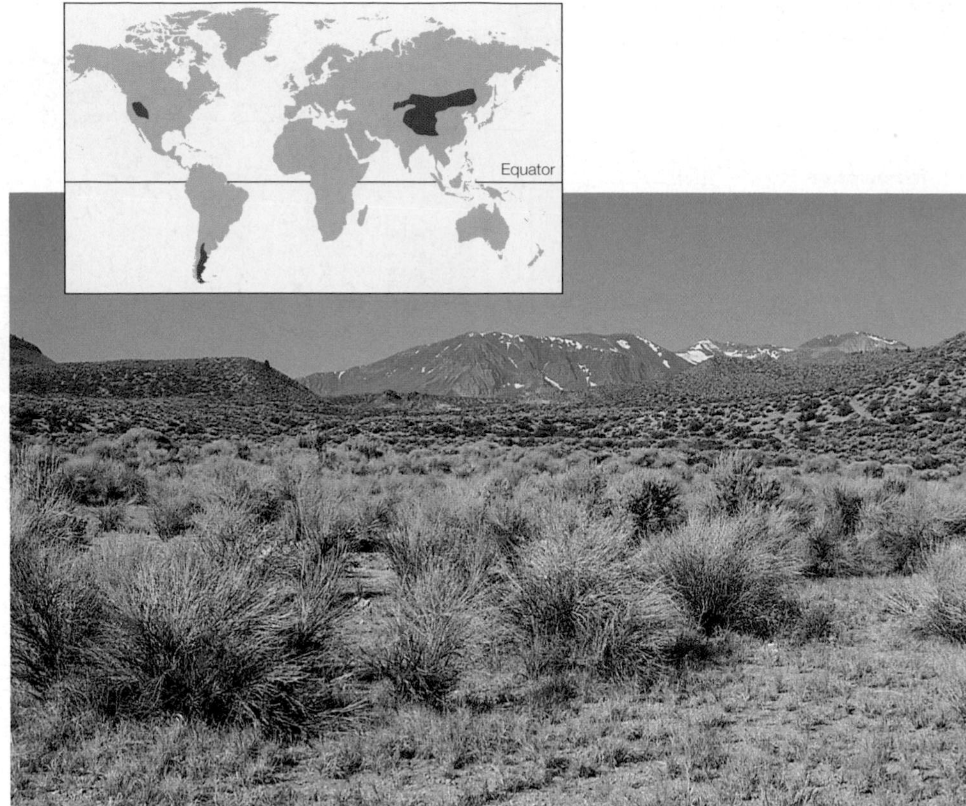

Sagebrush steppe near Mono Lake, California

Patagonia, Argentina

Cold deserts are high and dry

The **cold desert** biome is found in dry regions at middle to high latitudes, especially in the interiors of large continents in the rain shadows of mountain ranges. Seasonal changes in temperature are great.

Cold deserts are dominated by a few species of low-growing shrubs. The surface layers of the soil are recharged with moisture in winter, and plant growth is concentrated in spring. Because soils dry rapidly in spring, annual productivity is low. Cold deserts are relatively poor in species of most taxonomic groups, but the plants of this biome tend to produce large numbers of seeds, supporting many species of seed-eating birds, ants, and rodents.

HOT DESERT

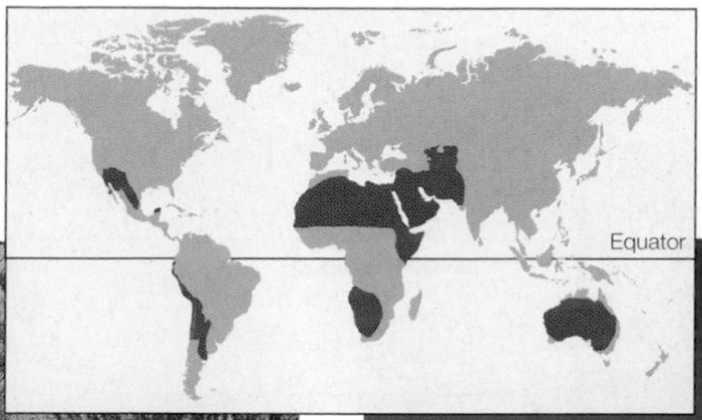

Equator

Anza Borrego Desert, California

Simpson Desert, Australia, following rain

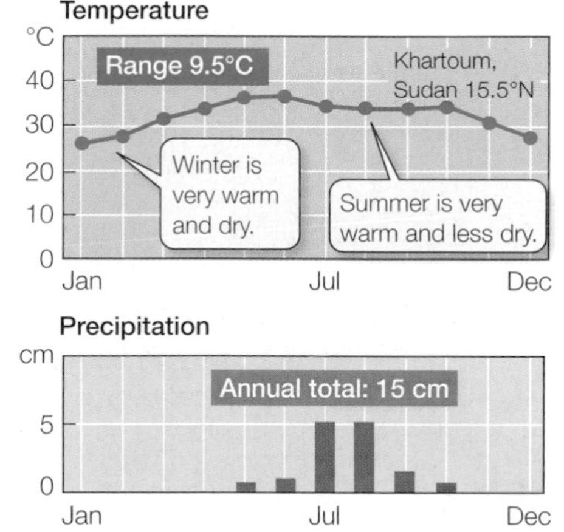

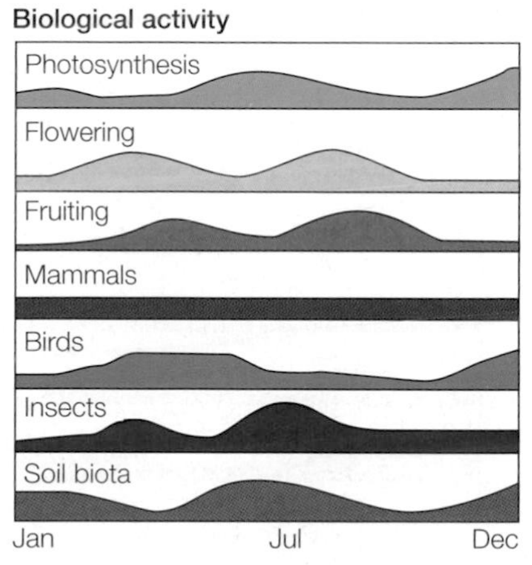

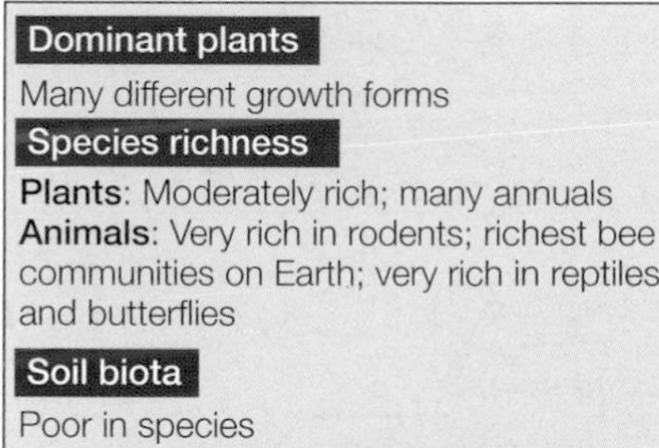

Community composition
Dominant plants
Many different growth forms
Species richness
Plants: Moderately rich; many annuals
Animals: Very rich in rodents; richest bee communities on Earth; very rich in reptiles and butterflies
Soil biota
Poor in species

Hot deserts form around 30° latitude

The **hot desert** biome is found in two belts, centered around 30° north and 30° south latitudes, where air descends, warms, and picks up moisture. Hot deserts receive most of their scarce rainfall in summer, but they also receive winter rains from storms that form over the mid-latitude oceans. The driest large regions, where summer and winter rains rarely penetrate, are in the center of Australia and the middle of the Sahara Desert of Africa.

Except in the driest regions, hot deserts have richer and structurally more diverse vegetation than cold deserts. Succulent plants (such as cacti) that store water in their expandable stems are conspicuous in some hot deserts. When rain falls, annual plants germinate and grow in abundance. Pollination and dispersal of fruits by animals are common. A rich fauna of rodents, termites, ants, lizards, and snakes is found in hot deserts.

CHAPARRAL

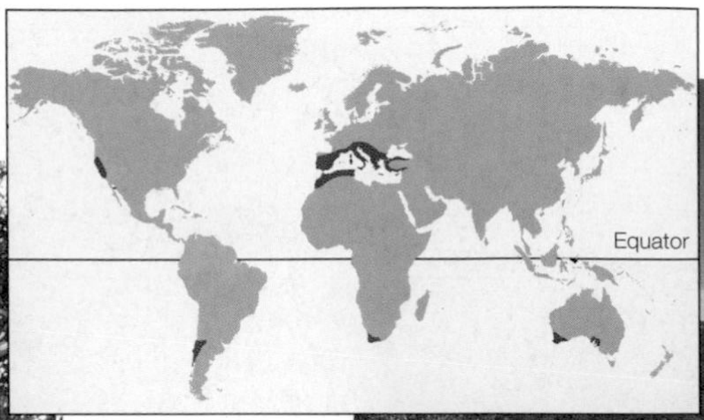

Equator

Southwest Australia

Santa Barbara County, California

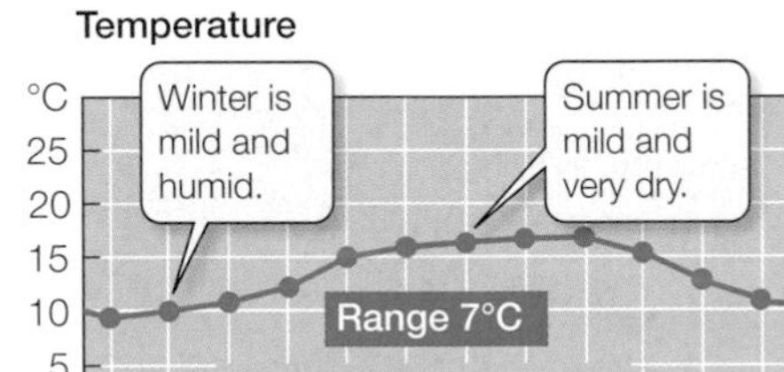

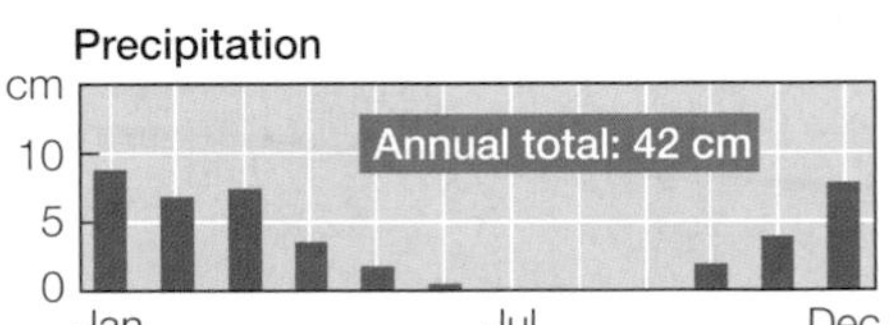

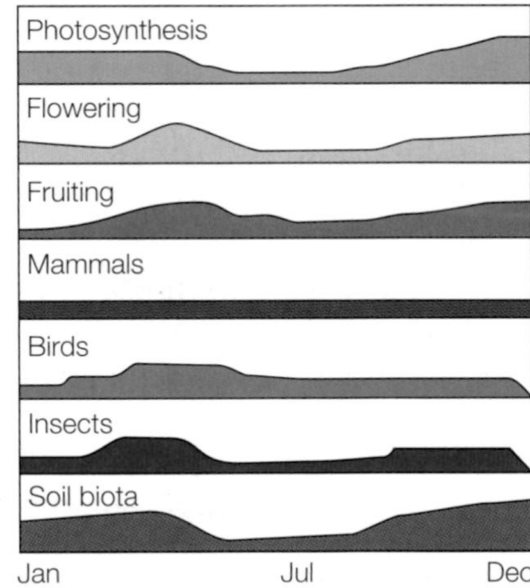

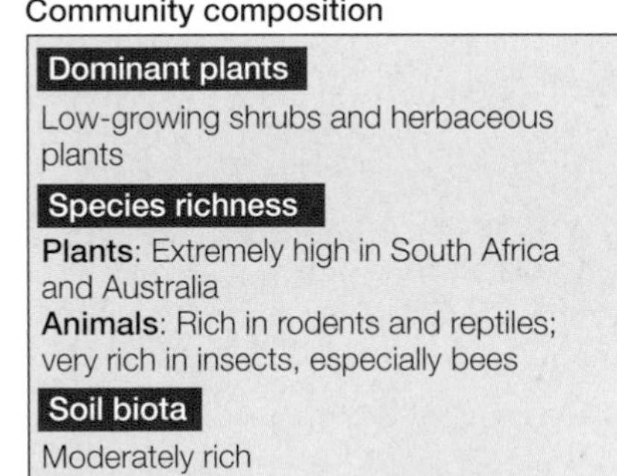

Community composition
Dominant plants
Low-growing shrubs and herbaceous plants
Species richness
Plants: Extremely high in South Africa and Australia
Animals: Rich in rodents and reptiles; very rich in insects, especially bees
Soil biota
Moderately rich

The chaparral climate is dry and pleasant

The **chaparral** biome is found on the western sides of continents at mid-latitudes (around 30°), where cool ocean currents flow offshore. Winters in this biome are cool and wet; summers are warm and dry. Such climates are found in the Mediterranean region of Europe, coastal California, central Chile, extreme southern Africa, and southwestern Australia.

The dominant plants of chaparral vegetation are low-growing shrubs and trees with tough, evergreen leaves. The shrubs carry out most of their growth and photosynthesis in early spring, when insects are active and birds breed. Annual plants are abundant and produce copious seeds that fall onto the soil. This biome thus supports large populations of small rodents, most of which store seeds in underground burrows. Chaparral vegetation is naturally adapted to survive periodic fires. Many shrubs of Northern Hemisphere chaparral produce bird-dispersed fruits that ripen in the late fall, when large numbers of migrant birds arrive from the north.

THORN FOREST and TROPICAL SAVANNA

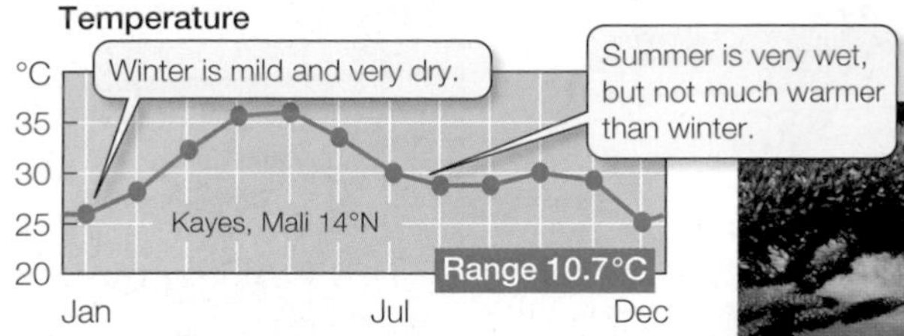

Temperature

°C

Winter is mild and very dry.

Summer is very wet, but not much warmer than winter.

35
30
25
20

Kayes, Mali 14°N

Range 10.7°C

Jan Jul Dec

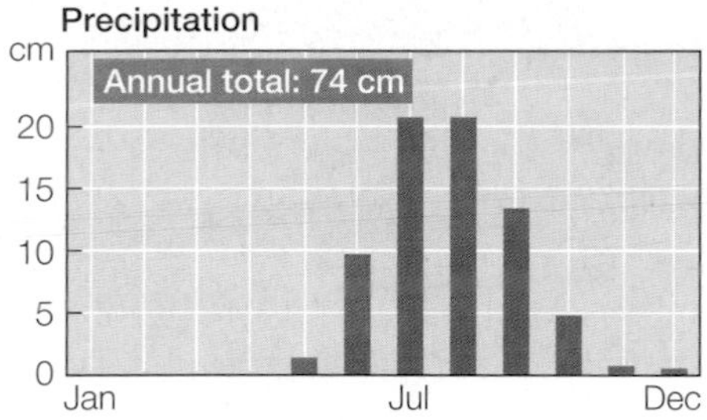

Precipitation

cm

Annual total: 74 cm

20
15
10
5
0

Jan Jul Dec

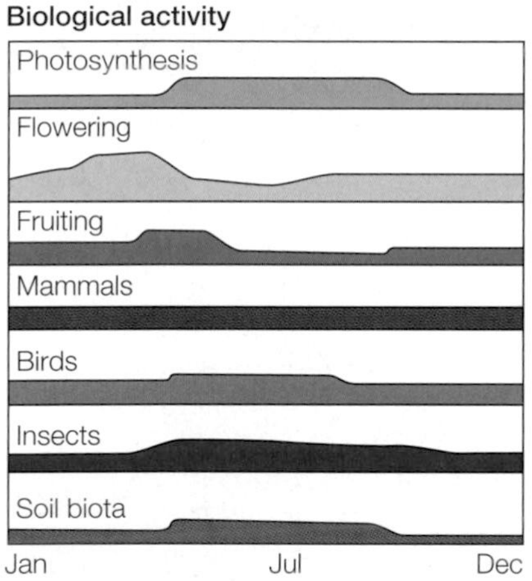

Biological activity

Photosynthesis

Flowering

Fruiting

Mammals

Birds

Insects

Soil biota

Jan Jul Dec

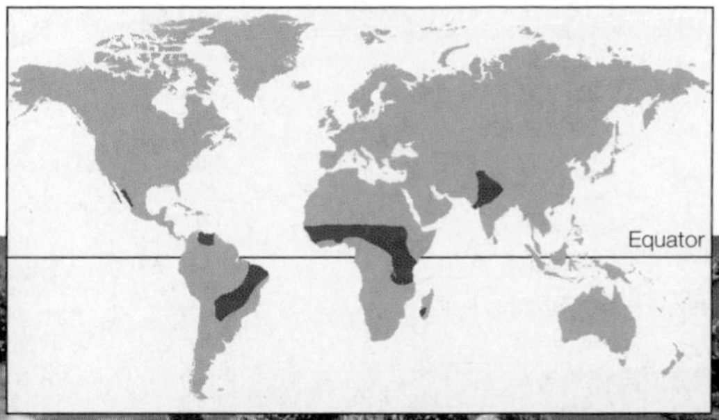

Community composition

Dominant plants

Shrubs and small trees; grasses

Species richness

Plants: Moderate in thorn forest; low in savanna

Animals: Rich mammal faunas; moderately rich in birds, reptiles, and insects

Soil biota

Rich

Equator

Thorn forest in Madagascar

KwaZulu-Natal, South Africa

Thorn forests and tropical savannas have similar climates

The **thorn forest** biome is found on the equatorial sides of hot deserts. The climate is semiarid; little or no rain falls during winter, but rainfall may be heavy during summer. Thorn forests contain many plants similar to those found in hot deserts. The dominant plants are spiny shrubs and small trees, many of which drop their leaves during the long, dry winter. Members of the genus *Acacia* are common in thorn forests worldwide.

The dry tropical and subtropical regions of Africa, South America, and Australia have extensive areas of the **tropical savanna** biome—expanses of grasses and grasslike plants with scattered trees. The largest tropical savannas are found in central and eastern Africa, where the biome supports huge numbers of grazing and browsing mammals and many large carnivores that prey on them. The grazers and browsers maintain the savannas. If savanna vegetation is not grazed, browsed, or burned, it typically reverts to dense thorn forest.

TROPICAL DECIDUOUS FOREST

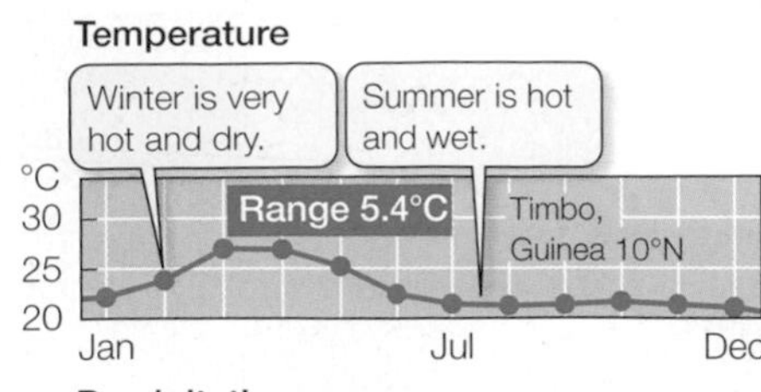

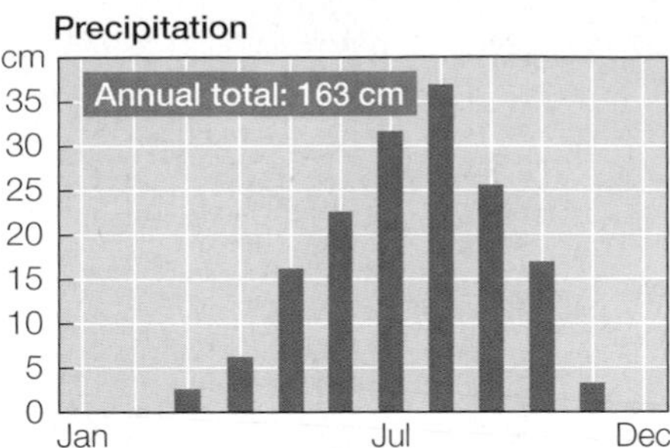

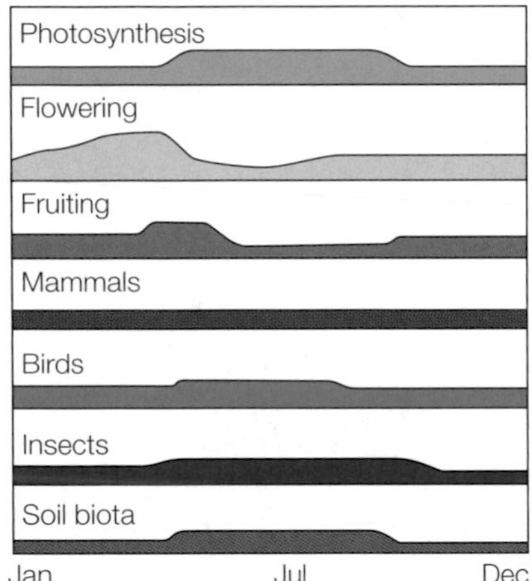

Community composition

Dominant plants
Deciduous trees

Species richness
Plants: Moderately rich in tree species
Animals: Rich mammal, bird, reptile, and amphibian communities; rich in insects

Soil biota
Rich, but poorly known

Palo Verde National Park, Costa Rica, in the rainy season...

...and in the dry season

Tropical deciduous forests occur in hot lowlands

As the length of the rainy season increases toward the equator, the **tropical deciduous forest** biome replaces thorn forests. Tropical deciduous forests have taller trees and fewer succulent plants than thorn forests, and they are much richer in plant and animal species. Most of the trees, except for those growing along rivers, lose their leaves during the long, hot dry season. Many of them flower while they are leafless, and most species are pollinated by animals. During the hot rainy season, biological activity is intense.

The soils of the tropical deciduous forest biome are some of the best soils in the tropics for agriculture because they contain more nutrients than the soils of wetter areas. As a result, most tropical deciduous forests worldwide have been cleared for agriculture and cattle grazing. Restoration efforts are under way on several continents.

TROPICAL EVERGREEN FOREST

The exterior of lowland wet forest…

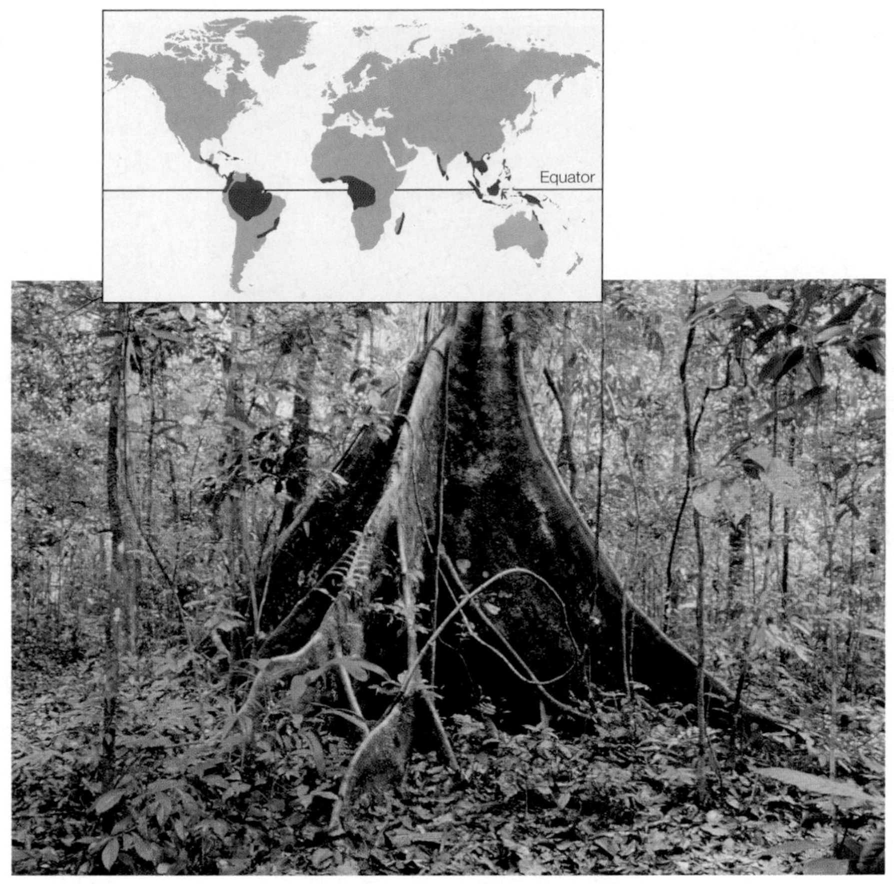

…and its interior, Cocha Cashu, Peru

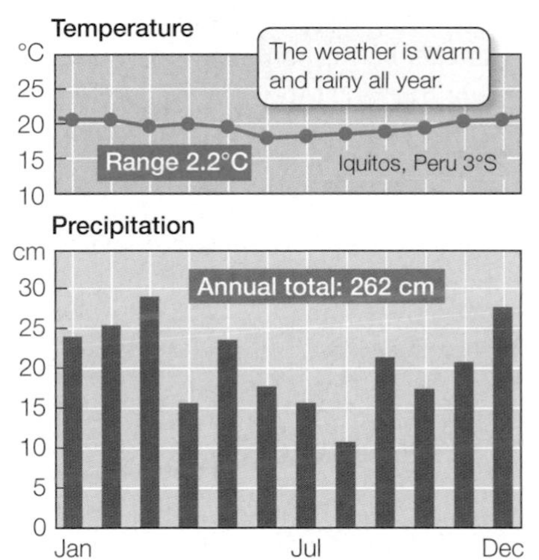

The weather is warm and rainy all year.
Range 2.2°C
Iquitos, Peru 3°S

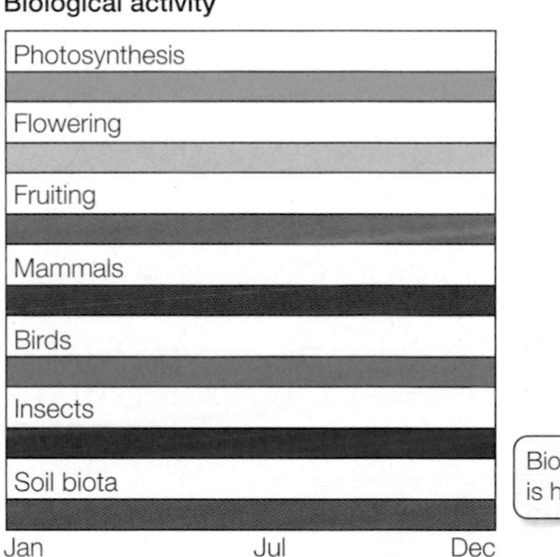

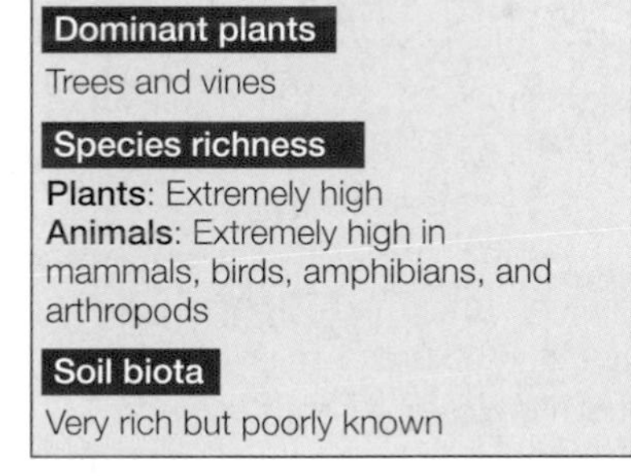

Community composition
Dominant plants
Trees and vines
Species richness
Plants: Extremely high
Animals: Extremely high in mammals, birds, amphibians, and arthropods
Soil biota
Very rich but poorly known

Tropical evergreen forests are species-rich

The **tropical evergreen forest** biome is found in equatorial regions where total rainfall exceeds 250 centimeters annually and the dry season lasts no longer than 2 or 3 months. It is the richest of all biomes in numbers of species of both plants and animals, with up to 500 species of trees per square kilometer. Along with their immense species richness, tropical evergreen forests have the highest overall productivity of all ecological communities. However, most mineral nutrients are tied up in the vegetation. The soils usually cannot support agriculture without massive applications of fertilizers.

On the slopes of tropical mountains, trees are shorter than lowland tropical trees. Their leaves are smaller, and there are more *epiphytes* (plants that grow on other plants, deriving their nutrients and moisture from air and water rather than soil).

52.6 Australian Deserts Differ from Those on Other Continents Extensive areas in Australia are dominated by a tough, fibrous grass, called spinifex (*Triodia*) that burns readily when green.

The distribution of biomes is not determined solely by climate

From the previous discussion, you might conclude that only climate determines the distribution of biomes and the characteristics of the plants that dominate them. Climate is important, but other factors, particularly soil fertility and fire, also exert strong influences on the structure of the vegetation in an area. For example, the vegetation of Australian deserts grows on extremely nutrient-poor soils. The leaves of those plants are so difficult to digest that herbivores consume little of them. Thus flammable litter accumulates rapidly under the trees, and intense fires periodically sweep across the landscape. As a result, succulent plants, which are easily killed by fires, are not found in Australia, although they are common in deserts on other continents (see p. 1123). Australian deserts have distinctive vegetation (**Figure 52.6**).

52.3 RECAP

Ecologists recognize a number of large ecological units called biomes, which are based on the growth forms of the dominant vegetation. The distribution of terrestrial biomes is determined primarily by temperature and precipitation, but is also influenced by soil fertility and fire.

Climate explains why chaparral is confined to the western coasts of continents at mid-latitudes, but it cannot explain why the chaparral in California and the chaparral in South Africa do not share any species in common. To explain the distribution of species on Earth, we need to study past as well as current physical environments.

52.4 What Is a Biogeographic Region?

Until European naturalists traveled the globe, they had no way of knowing what organisms were found elsewhere. Alfred Russel Wallace, who along with Charles Darwin put forth the idea that natural selection could account for the evolution of life on Earth (see Section 22.1), was one of those global travelers. He returned to England in the spring of 1862 after having spent seven years traveling in the Malay Archipelago. During his travels he noticed some remarkable patterns in the distributions of organisms across the archipelago. For example, he described the dramatically different birds that inhabited Bali and Lombok, two adjacent islands:

> In Bali we have barbets, fruit-thrushes and woodpeckers; on passing over to Lombock these are seen no more, but we have an abundance of cockatoos, honeysuckers, and brush-turkeys, which are equally unknown in Bali, or any island further west. The strait here is fifteen miles wide, so that we may pass in two hours from one great division of the earth to another, differing as essentially in their animal life as Europe does from America.

Wallace pointed out that these striking biological differences could not be explained by climate or geology, because in those respects, Bali and Lombok were essentially identical.

Wallace saw that he could draw a line through the Malay Archipelago that would divide it into two distinct halves based on the distributions of plant and animal species. He correctly deduced that these dramatic differences in flora and fauna were related to the depth of the channel separating Bali and Lombok. This channel is so deep that it remained a barrier to the movement of terrestrial animals even during the Pleistocene glaciations, when sea levels dropped more than 100 meters and Java and other islands to the west were connected to the Asian mainland (**Figure 52.7**).

With these insights, Wallace established the conceptual foundations of **biogeography**, the scientific study of the patterns of distribution of populations, species, and ecological communities across Earth. He pointed out that the geological history of regions profoundly influences the kinds of organisms found in them. Today we call the line he drew through the Malay Archipelago "Wallace's line."

The flora, fauna, and microorganisms of different parts of the world—that is, their *biota*—differ enough to allow us to divide Earth into several major **biogeographic regions** (**Figure 52.8**). Biogeographic regions are based on the taxonomic composition of the organisms living in them. Their boundaries are set where species compositions change dramatically over short distances. Wallace's line, for example, separates the Malay Archipelago into two distinct biogeographic regions. The biotas of biogeographic regions differ because oceans, mountains, deserts, and other barriers restrict the dispersal of organisms between them. Although organisms do disperse between adjacent biogeographic regions, such interchanges have not been frequent enough or massive enough to eliminate the striking differences that have resulted from speciation and extinction within each region. Most species are confined to a single biogeographic region.

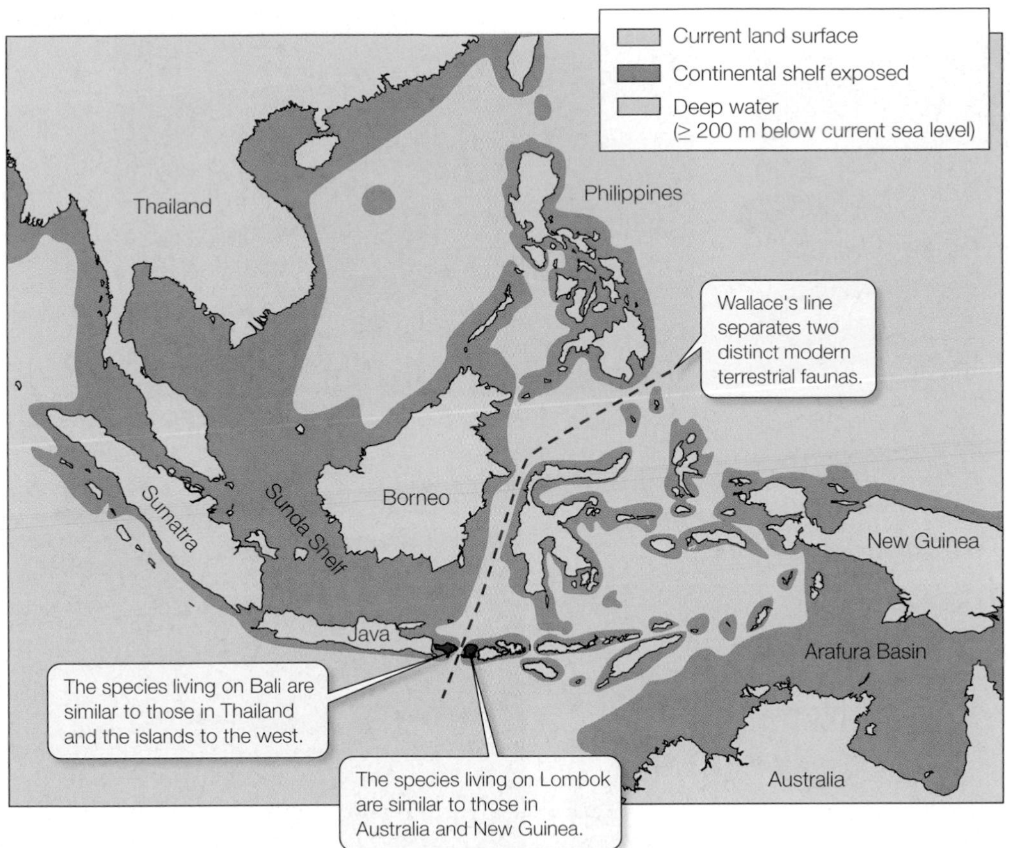

52.7 The Malay Archipelago during the Most Recent Glacial Maximum Wallace's line, which divides two distinct geographic regions, corresponds to a deep-water channel between the islands of Bali and Lombok. This channel is deep enough to have blocked the movement of terrestrial organisms even during the glaciations of the Pleistocene era when the sea level was 100 meters lower than it is today.

Today the most widespread terrestrial animal species is *Homo sapiens*, but there are also a few bird species—for example, the great egret, osprey, peregrine falcon, and barn owl—that are found on all continents except Antarctica.

A species found only within a certain region is said to be **endemic** to that region. Remote islands typically have distinctive endemic biotas because water barriers greatly restrict immigration. For example, nearly all the species of vascular plants and vertebrates of Madagascar, a large island off the eastern coast of Africa, are endemic to that island (**Figure 52.9**). Madagascar by itself could be considered a biogeographic region, but because dozens of islands would qualify as biogeographic regions on the basis of the distinctiveness of their biotas, islands are not called biogeographic regions.

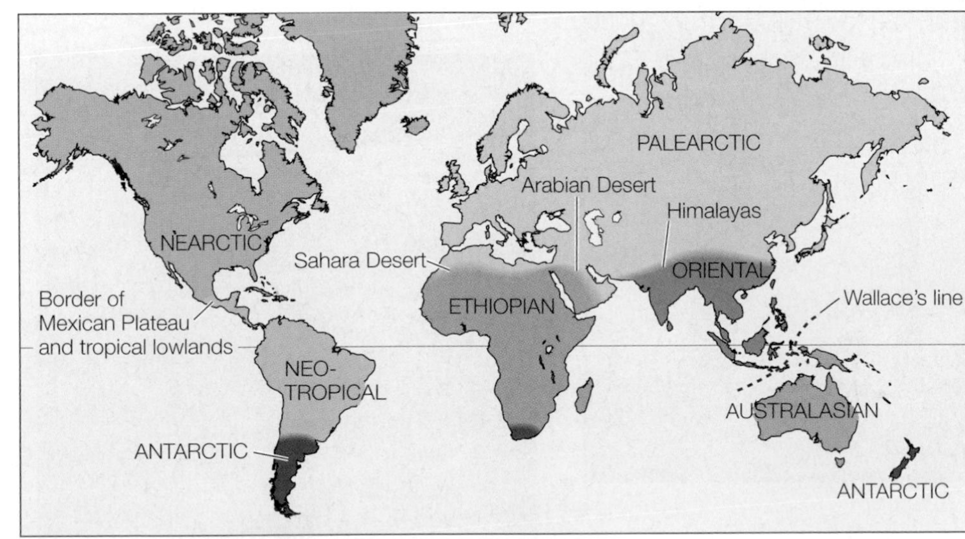

52.8 Major Biogeographic Regions The biotas of Earth's major biogeographic regions differ strikingly from one another. These regions are separated by topographic, climatic, or aquatic barriers to dispersal that cause their biotas to differ from one another.

Three scientific advances changed the field of biogeography

For many decades after biogeographers determined that the biotas of the major biogeographic regions are strikingly different from one another, they devoted their efforts to assembling information on the distributions of organisms. They speculated about the causes of the distributional patterns they found, but the field remained primarily descriptive. Biogeographers did gain insights from the study of fossils, which can show how long a taxon has been present in an area and whether its members formerly lived in areas where they are no longer found. But it took three scientific ad-

Chamaeleo parsonii

Hemicentetes semispinosus (yellow-streaked tenrec)

Cryptoprocta ferox (fossa)

Lemur catta (ring-tailed lemur) climbing *Alluaudia procera* (Madagascar ocotillo)

Adansonia grandidieri (giant baobab tree)

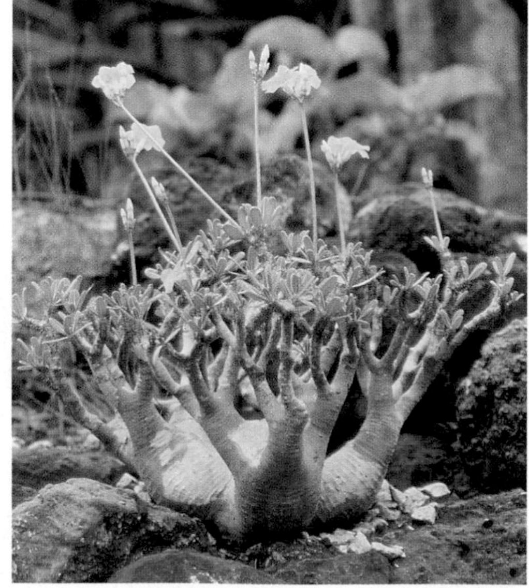

Pachypodium rosulatum (elephant's foot)

52.9 Madagascar Abounds with Endemic Species The majority of the plant and vertebrate species found on the island of Madagascar are found nowhere else on Earth.

vances in the latter part of the twentieth century to change biogeography into the dynamic multidisciplinary field it is today:

- The acceptance of the theory of *continental drift*
- The development of *phylogenetic taxonomy*
- The development of the theory of *island biogeography*

Let's see how each of these advances contributed to the development of biogeography.

CONTINENTAL DRIFT Prior to the mid-nineteenth century, most people believed that Earth was young and had changed little over time. Carolus Linnaeus, the naturalist who initiated the classification of species some 250 years ago, believed that all organisms had been created in one place (which he called Paradise), from which they later dispersed. Indeed, because most people believed that the continents were fixed in their positions, the only way to account for the current distributions of organisms was to invoke massive dispersal.

The notion that the continents might have moved was not seriously considered until 1912, when Alfred Wegener put forward the concept of continental draft, proposing that the continents had changed position over time. When Wegener proposed this idea, few scientists took him seriously, but as Section 21.2 described, geological evidence and plausible mechanisms that could move continents were eventually discovered.

About 280 million years ago, the continents were united to form a single land mass, Pangaea (see Figure 21.15). The continents then began to separate from one another, but by the Triassic period, when they were still very close to one another (about 245 mya), many groups of terrestrial and freshwater organisms, such as insects, freshwater fishes, frogs, and vascular plants, had already evolved. The ancestors of some organisms that live on widely separated continents today were probably present on those land masses when they were part of Pangaea.

PHYLOGENETIC TAXONOMY As discussed in Chapter 25, taxonomists have developed powerful methods of reconstructing phylogenetic relationships among organisms. Biogeographers have adapted these methods to help them answer biogeographic questions. Biogeographers can transform phylogenetic trees into **area phylogenies** by replacing the names of the taxa on a tree with the names of the places where those taxa live or lived. For example, we know from the fossil record that the earliest ancestors of horses evolved in North America, but an area phylogeny of horses makes it possible to follow their evolution and dispersal much further. It suggests that the ancestors of today's horses dispersed from North America to Asia, and then from Asia to Africa, and further, that the speciation of zebras took place entirely in Africa (**Figure 52.10**).

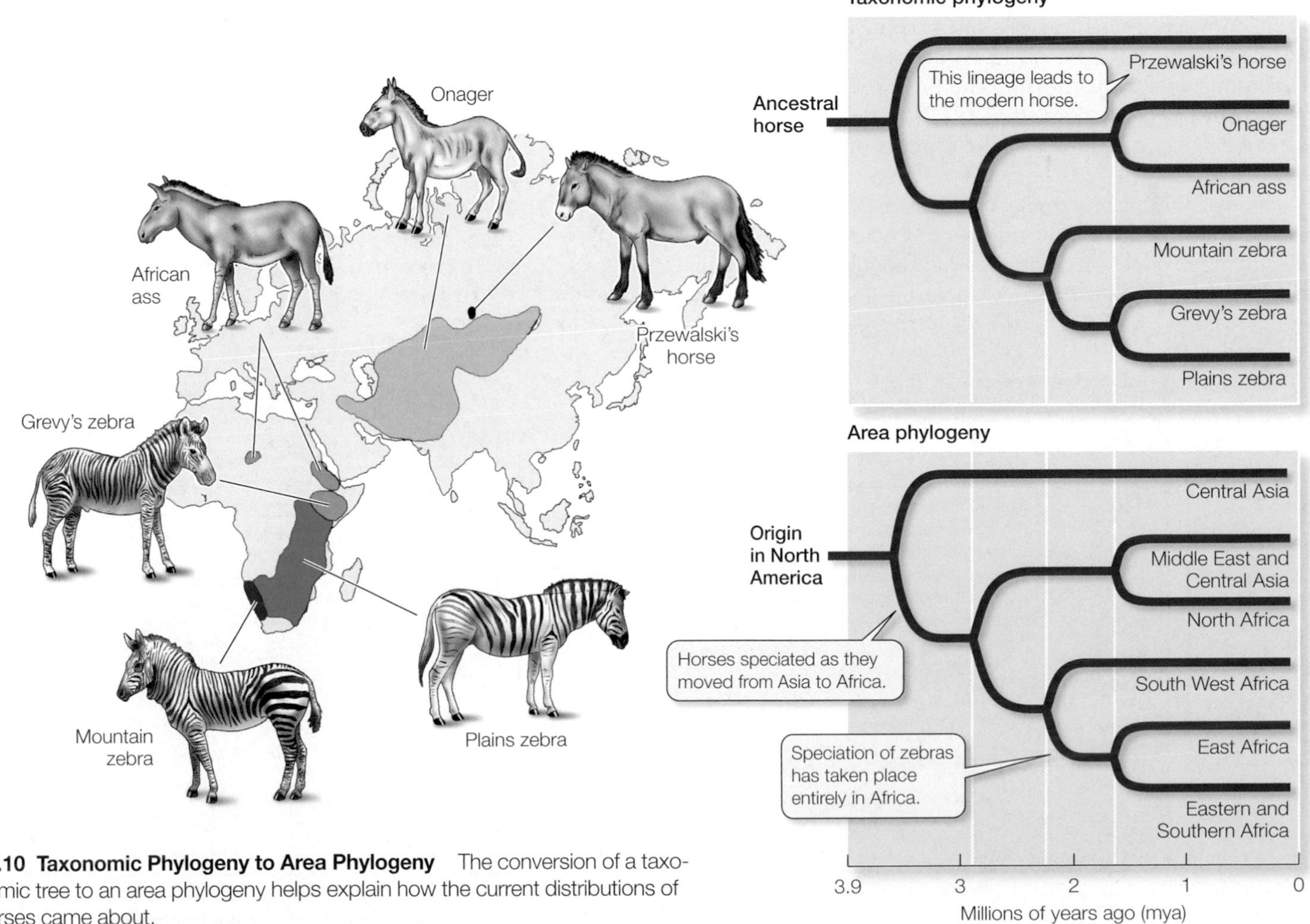

52.10 Taxonomic Phylogeny to Area Phylogeny The conversion of a taxonomic tree to an area phylogeny helps explain how the current distributions of horses came about.

ISLAND BIOGEOGRAPHY Robert MacArthur and Edward O. Wilson developed the theory of **island biogeography** to help explain why oceanic islands always have fewer species than a mainland area of equivalent size. They based their theory on just two processes: the immigration of new species to an island and the extinction of species already present on that island. They modeled the unfolding of these two processes over periods of only a few hundred years or less, during which they assumed that no speciation took place.

Imagine a newly formed oceanic island that receives colonists from a mainland area. The list of species on the mainland that might possibly colonize the island constitutes the **species pool**. The first colonists to arrive on the island are all "new" species because no species live there initially. As the number of species on the island increases, a larger fraction of colonizing individuals will be members of species already present. Therefore, even if the same number of species arrive as before, the rate of arrival of new species decreases, until it reaches zero when the island has all the species in the species pool.

Now consider extinction rates. At first there will be only a few species on the island, and their populations may grow large. As more species arrive and their populations increase, the resources of the island will be divided among more species. Therefore, the average population size of each species will become smaller as the number of species increases. The smaller a population, the more likely it is to become extinct (see Section 54.4). In addition, the number of species that can possibly become extinct increases as species accumulate on the island. Furthermore, new arrivals on the island may include pathogens and predators that increase the probability of extinction for other species. For all these reasons, the rate of extinction increases as the number of species on the island increases.

Because the rate of arrival of new species decreases and the extinction rate increases as the number of species increases, eventually the number of species on the island should reach an equilibrium at which the rates of arrival and extinction are equal (**Figure 52.11A**). If there are more species than the equilibrium number, extinctions should exceed arrivals, and species richness should decline. If there are fewer species than the equilibrium number, arrivals should exceed extinctions, and species richness should increase. The equilibrium is dynamic because if either rate fluctuates, as they generally do, the equilibrium number of species shifts up or down.

MacArthur and Wilson's model can also be used to predict how species richness should differ among islands of different sizes and at different distances from the mainland. We expect extinction rates to be higher on small islands than on large islands because species' populations are, on average, smaller there. Similarly, we expect fewer immigrants to reach distant islands. **Figure 52.11B** gives hypothetical relative species numbers for islands of different sizes and distances from the mainland. As you can see, the number of species should be highest for islands that are relatively large and relatively close to the mainland (such as Madagascar). These principles can be applied not only to oceanic islands, but also to "habitat islands" in terrestrial landscapes, as we will see in Section 57.2.

The most important contribution of the theory of island biogeography was to demonstrate that biogeography could actually have an experimental component. Immigration and extinction rates can be measured and, sometimes, manipulated to test hypotheses. In addition, major disturbances, which can serve as "natural experiments," sometimes permit scientists to estimate colonization and extinction rates. In August 1883, Krakatau, an island in the Sunda Strait between Sumatra and Java, was devastated by a series of volcanic eruptions that destroyed all life on the island's surface. After the lava cooled, plants and animals from Sumatra to

52.11 The Theory of Island Biogeography (A) The rate of arrival of new species and the rate of extinction of species already present determine the equilibrium number of species on an island. (B) These rates are affected by the size of the island and its distance from the mainland.

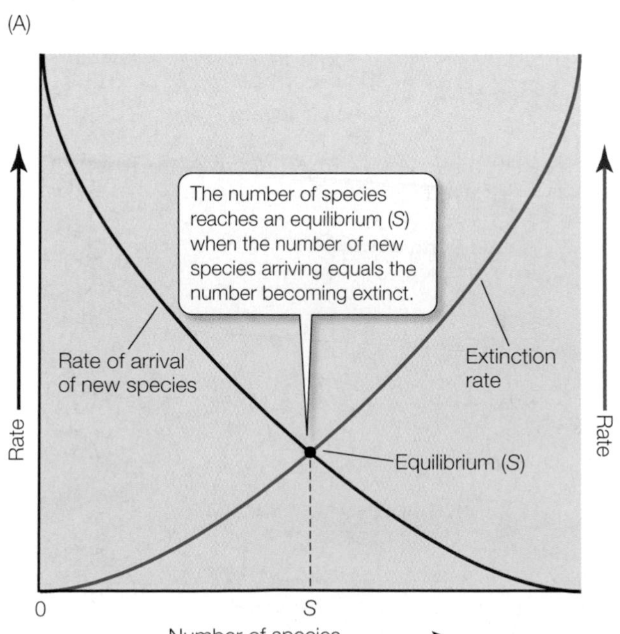

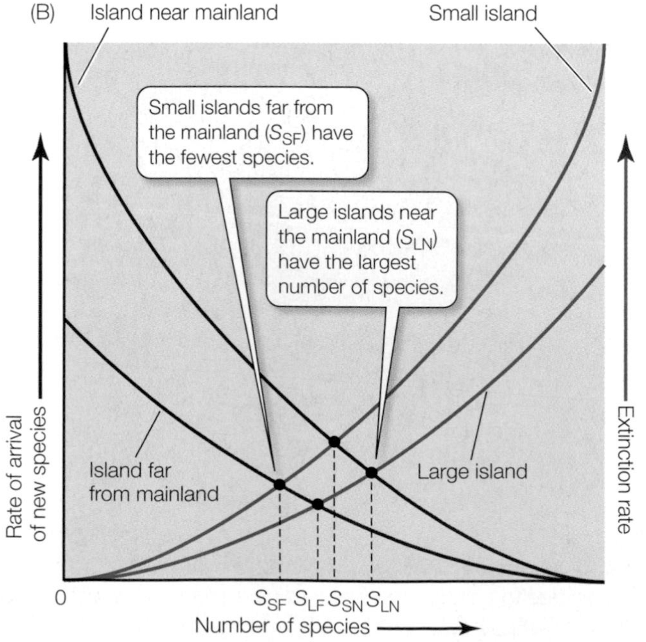

TABLE 52.1

Number of Species of Resident Land Birds on Krakatau

PERIOD	NUMBER OF SPECIES	EXTINCTIONS	COLONIZATIONS
1908	13		
1908–1919		2	17
1919–1921	28		
1921–1933		3	4
1933–1934	29		
1934–1951		3	7
1951	33		
1952–1984		4	7
1984–1996	36		

the west and Java to the east rapidly colonized Krakatau. By 1933, the island was again covered with a tropical evergreen forest, and 271 species of plants and 27 species of resident land birds were found there.

During the 1920s, as a forest canopy was developing on Krakatau, both birds and plants colonized the island at a high rate (**Table 52.1**). Birds probably brought the seeds of many plants because, between 1908 and 1934, both the percentage (from 20% to 25%) and the absolute number (from 21 to 54) of plant species with bird-dispersed seeds increased. Today the numbers of plant and bird species are not increasing as fast as they did during the 1920s, but colonizations and extinctions continue, as predicted by the theory of island biogeography.

Experiments can also be conducted to test components of the theory. To test the hypothesis that there is an equilibrium number of species on islands, Daniel Simberloff and Edward O. Wilson of Harvard University defaunated (eliminated all animals from) tiny islets of red mangroves in the Florida Keys and monitored their recolonization (**Figure 52.12**). The islands were rapidly recolonized, and all of them eventually supported roughly the same number of species that they had before defaunation. In addition, the rate of recolonization was slowest on the most remote island.

The role of experimentation in biogeography is necessarily limited, however, because the processes that drive biogeographic patterns are fully expressed only after long time spans, usually extending over millions of years. The striking patterns that Alfred Russel Wallace observed had their origins deep in time. Let's now consider some of the long-term processes that result in recognizably different biogeographic regions.

EXPERIMENT

HYPOTHESIS: Defaunated islands will be rapidly recolonized, eventually achieving about the same number of species that they had prior to defaunation.

METHOD

Erect scaffolding and tent to enclose islets. Fumigate small islets with a chemical (methyl bromide) that kills arthropods but does not harm plants. Periodically monitor recolonizations and extinctions of arthropods on the islands.

RESULTS

Recolonization was rapid, turnover rates were high, and the rate of recolonization was slowest on the most remote island.

CONCLUSION: An island can support a certain equilibrium number of species.

52.12 Experimental Island Defaunation Simberloff and Wilson surrounded several small islands with scaffolding so that they could be covered and fumigated to remove all arthropods. The researchers then counted and monitored arthropods as they recolonized the islands. FURTHER RESEARCH: Although the results of Wilson and Simberloff's defaunation experiment were dramatic, the experiment lasted only a short time, and only arthropods were counted. What experiments could you conduct to assess some of the longer-term predictions of the theory of island biogeography?

A single barrier may split the ranges of many species

Two major processes—vicariance and dispersal—generate biogeographic patterns. The appearance of a physical barrier that splits the range of a species is called a **vicariant event**. A vicariant event divides the population of a species into two or more discontinuous populations, even though no individuals have dispersed to new areas. If, however, members of a species cross an already existing barrier and establish a new population, the species' discontinuous range is considered the result of dispersal.

By studying a clade, a biogeographer may discover evidence suggesting that the distribution of an ancestral species was influenced by a vicariant event, such as a change in sea level or mountain building (see Figure 23.3). If that inference is correct, it is reasonable to assume that other ancestral species would have been affected by the same event and that similar distribution patterns would be seen in other clades. Differences in distribution patterns among clades may indicate that they responded differently to the same vicariant events, that they diverged at different times, or that the clades have had very different dispersal histories. By analyzing such similarities and differences, biogeographers seek to discover the relative roles of vicariant events and dispersal in determining today's distribution patterns.

Biotic interchange follows fusion of land masses

When formerly separated land masses fuse, as may happen through sea level changes or continental drift, two different biotas can merge. Many species of both biotas are likely to disperse into the region they had not previously inhabited. This phenomenon is called **biotic interchange**. Such a massive colonization of new areas by mammals occurred when the Central American land bridge formed about 4 million years ago. This land bridge connected North and South America for the first time in about 65 million years. While the two continents were separated, their mammals evolved independently of one another because terrestrial mammals (with the exception of bats) are poor dispersers across water barriers. South America evolved a distinctive mammalian fauna dominated by marsupials, primates, edentates (armadillos, sloths, ground sloths, and anteaters), and caviomorph rodents (porcupines, capybaras, pacas, agoutis, guinea pigs, and chinchillas).

Many species of mammals dispersed across the newly established land bridge. Only a few South American species—the porcupine, nine-banded armadillo, and Virginia opossum—became established north of the tropical forests of Mexico (**Figure 52.13A**). However, many North America mammals—rabbits, mice, foxes, bears, raccoons, weasels, cats, tapirs, peccaries, camels, and deer—successfully colonized South America. The North American invasion apparently caused the extinction of several kinds of large marsupial carnivores and the large herbivores that were their prey. Subsequently, the northern invaders gave rise to new species that today exist only in South America (**Figure 52.13B**).

The exchange of mammals between North and South America did not eliminate the differences between the mammalian faunas of the two continents. North American visitors to South America today still enjoy seeing a rich array of mammals that are unfamiliar to them, most of which are endemic to South America or penetrate only a short distance into Central America. But without knowing the long-term history of the region, we would not be aware of the many recently extinct South American species, nor could we know which of the species there today are recent arrivals.

(A)

Dasypus novemcinctus

Erethizon dorsatum

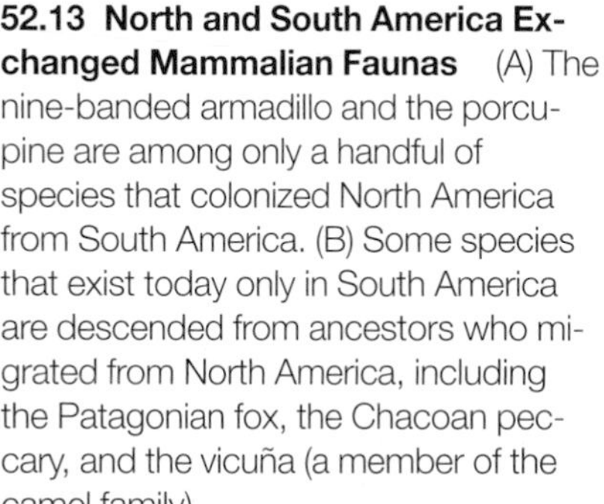

52.13 North and South America Exchanged Mammalian Faunas (A) The nine-banded armadillo and the porcupine are among only a handful of species that colonized North America from South America. (B) Some species that exist today only in South America are descended from ancestors who migrated from North America, including the Patagonian fox, the Chacoan peccary, and the vicuña (a member of the camel family).

(B)

Pseudalopex griseus

Catagonus wagneri

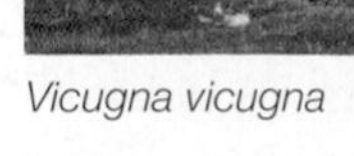

Vicugna vicugna

Both vicariance and dispersal influence most biogeographic patterns

If both vicariance and dispersal influence distribution patterns, how can we determine their relative importance in particular cases? When several hypotheses can explain a pattern, scientists typically prefer the most parsimonious one—the one that requires the smallest number of unobserved events to account for it. We saw how the parsimony principle is used in the reconstruction of phylogenies in Section 25.2. To see how it is applied to biogeography, consider the distribution of the New Zealand flightless weevil *Lyperobius huttoni*, a species that is found in the mountains of South Island and on sea cliffs at the extreme southwestern corner of North Island (**Figure 52.14**). If you knew only its current distribution and the current positions of the two islands, you might surmise that, even though this weevil cannot fly, it had somehow managed to cross Cook Strait, the 25-kilometer body of water that separates the two islands.

More than 60 other animal and plant species, however, including other species of flightless insects, live on both sides of Cook Strait. Although organisms do cross physical barriers, it is unlikely that all of these species made the same ocean crossing. In fact, we do not need to make that assumption. Geological evidence indicates that the present-day southwestern tip of North Island was formerly united with South Island. Therefore, none of the 60 species need have made a water crossing. A single vicariant event—the separation of the northern tip of South Island from the remainder of the island by the newly formed Cook Strait—could have split all of the distributions.

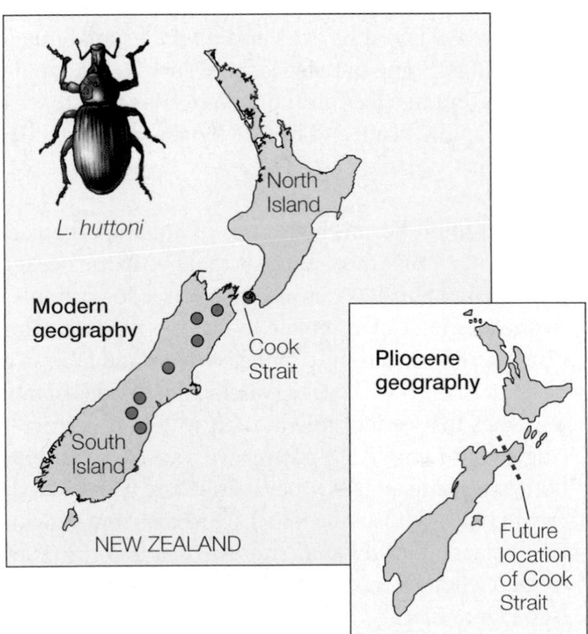

52.14 A Vicariant Distribution Explained Blue circles indicate the current distribution of the weevil *Lyperobius huttoni*. A comparison of the present New Zealand geography with that of the Pliocene, when the southern part of today's North island was part of South Island, suggests that a vicariant event—a physical split separating populations—explains this distribution.

52.4 RECAP

Biogeographic regions are based on the taxonomic similarities of the organisms living in them. Biotas of biogeographic regions differ because barriers restrict the dispersal of organisms between them.

- What determines the boundaries of Earth's major biogeographic regions? See p. 1128 and Figure 52.8
- Do you understand how the concepts of continental drift, phylogenetic taxonomy, and island biogeography have transformed and energized the field of biogeography? See pp. 1131–1133
- How do vicariance and dispersal interact to generate major biogeographic patterns? See pp. 1133–1135

Now that we have seen how species are distributed over Earth's land masses, let's take a look at the other three-quarters of the planet—the blue world of water.

52.5 How Is Life Distributed in Aquatic Environments?

Most of Earth's aquatic environment is saltwater oceans and seas. However, the small percentage of the watery world that constitutes freshwater lakes, rivers, ponds, and streams hosts a significant number of ecosystems and species.

Currents create biogeographic regions in the oceans

Earth's oceans form one large, interconnected water mass interrupted only partly by continents. Therefore, we might not expect to find biogeographic regions in the oceans, but we would be wrong. Currents generate striking physical and biological discontinuities, dividing the oceans into distinct regions.

Figure 52.4 depicted the great circular patterns of the world's ocean currents. Even organisms with limited swimming abilities can travel long distances simply by floating with the ocean currents. Nevertheless, most marine organisms have restricted ranges. Why is this true?

The oceans may be connected, but water temperatures, salinities, and food supplies all change spatially. Surviving under these varied conditions requires different physiological tolerances and morphological attributes. Changes in water temperatures, for example, can be barriers to dispersal because many marine organisms function well in only a relatively narrow range of temperatures. The main biogeographic divisions of the ocean coincide with regions where the surface water temperatures and salinities change relatively abruptly as a result of horizontal and vertical currents (**Figure 52.15**).

Primary food production by photosynthesis also varies across the oceans. Temperature changes, in combination with seasonal changes in the amount of sunlight, determine the seasons of maximum photosynthesis. Different species of marine algae photosyn-

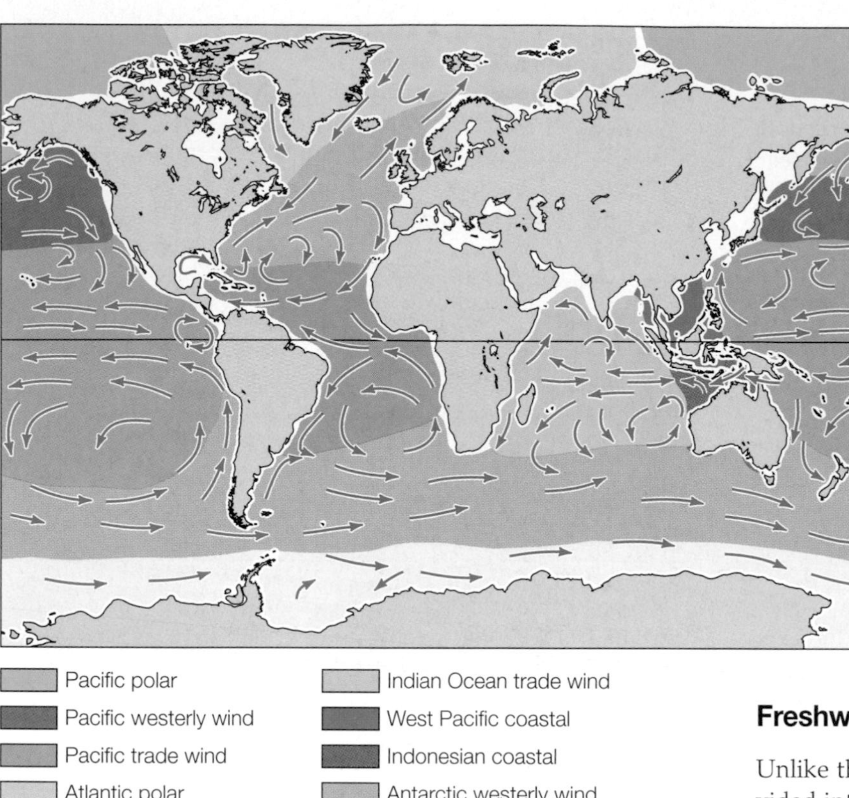

52.15 Oceanic Biogeographic Regions Are Determined by Ocean Currents The arrows represent ocean currents. Different biogeographic regions, in which photosynthesis is maximized at different seasons, are indicated by different colors.

thesize either in summer or in winter, but not during both seasons. Mangrove forests grow in estuaries in tropical regions, corals generate complex structures around tropical islands, and large algae form beds along many coasts. But over most of the oceans, where the dominant photosynthesizers are unicellular organisms, water temperature, salinity, and currents determine the physical environment to which organisms must adjust.

Deep ocean waters are barriers to the dispersal of marine organisms that live only in shallow water. The distance that ocean currents carry eggs and larvae of many marine organisms is determined in large part by the time it takes for larvae to metamorphose into sedentary adults. Relatively few species have eggs and larvae that survive long enough to disperse across wide areas of deep water. As a result, the richness of shallow-water species in the intertidal and subtidal zones of isolated islands in the Pacific Ocean decreases with distance from New Guinea (**Figure 52.16**).

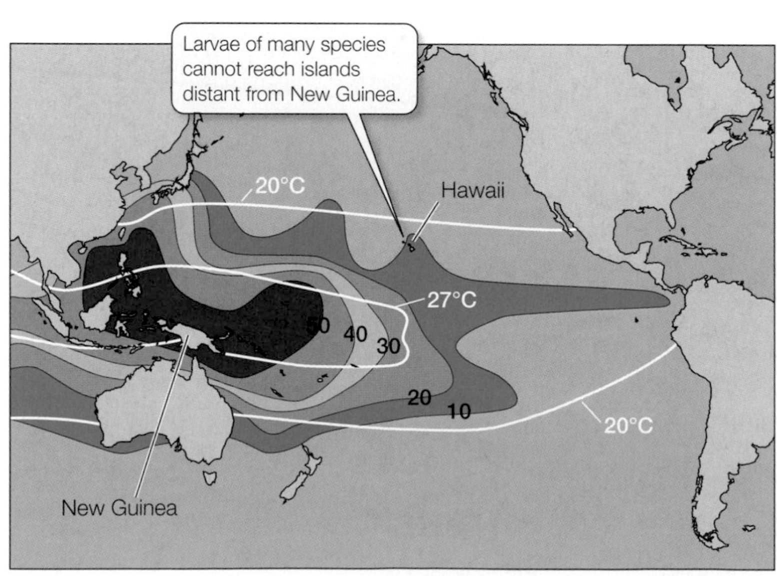

52.16 Generic Richness of Reef-Building Corals Declines with Distance from New Guinea The colored zones represent areas with equal numbers of coral genera. The 20°C and 27°C mean annual temperature isotherms are also shown.

Freshwater environments may be rich in species

Unlike the oceans, the world's freshwater environments are divided into river basins and into thousands of relatively isolated lakes. Although only about 2.5 percent of Earth's water is found in ponds, lakes, and streams, about 10 percent of all aquatic species live in these freshwater ecosystems. Prominent among freshwater taxa are the more than 25,000 species of insects that have at least one aquatic stage in their life cycle. Many freshwater insects are capable of dispersing across terrestrial barriers and are found over several continents. Typically, eggs and juveniles are aquatic; the adults have wings. Some of these insects, such as dragonflies, are powerful flyers, but mayflies and some other species are weak flyers, desiccate rapidly in air, and live no longer than a few days. As you would expect, oceanic islands have few, if any, species of these weak flyers.

Discontinuities in the ranges of aquatic organisms are also seen because most animals that live in the oceans cannot survive in fresh water, and vice versa. For example, most freshwater fishes are unable to live in salt water. They can disperse only within the connected rivers and lakes of a river basin or in brackish coastal waters. Most clades of freshwater fishes that cannot tolerate salt water are restricted to a single continent. Those groups with species distributed on both sides of major saltwater barriers are believed to be ancient clades whose ancestors were distributed widely in Laurasia or Gondwana, the two supercontinents that formed when Pangaea broke up some 250 million years ago (see Figure 21.17).

52.5 RECAP

Despite being connected, the oceans are divided into distinct regions by currents that generate striking physical and biological discontinuities. Fresh waters, in contrast, are divided into river basins and thousands of relatively isolated lakes.

- Why do the main biogeographic divisions of the ocean coincide with regions where the surface water temperatures and salinities change relatively abruptly? See p. 1135 and Figure 52.15
- How do organisms that live in fresh water disperse among watersheds? See p. 1136

Now that we have described how the physical environment influences the distribution of species and ecosystems on Earth, we are ready to delve into the details of the interactions among organisms and their environments. In Chapter 53 we will consider the behavior of individual organisms and the choices they make that enable them to survive and reproduce in their environments. In Chapter 54 we will consider the dynamics of populations. In Chapter 55 we will discuss how populations of different species interact and how these interactions influence ecological communities. The final two chapters in the book recall global ecological patterns and emphasize the importance of ecology both to humanity and to our planet as a whole.

CHAPTER SUMMARY

52.1 What is ecology?

Ecology is the field of science that investigates interactions among organisms and between organisms and the physical environment.

An organism's **environment** encompasses both **abiotic** (physical and chemical) and **biotic** factors (other living organisms).

Ecologists study interactions on many levels, from the level of an individual organism's interactions with members of its own species to the functioning of **communities**, **ecosystems**, and the entire **biosphere**.

52.2 How are climates distributed on Earth?

See Web/CD Tutorial 52.1

The **climate** of a region is the average of the atmospheric conditions (temperature, precipitation, and wind direction and velocity) found there over time.

The rate at which solar energy arrives on Earth per unit of Earth's surface depends primarily on the angle at which sunlight arrives.

Mean annual air temperature decreases with both latitude and altitude.

Global atmospheric circulation patterns drive ocean currents and strongly influence global climates. Review Figures 52.2 and 52.4

Organisms respond behaviorally to short-term environmental changes.

Morphological and physiological features have evolved in organisms in response to long-term changes in the physical environment.

52.3 What is a biome?

See Web/CD Tutorial 52.2

Biomes are large ecological units that are classified by the structure of the dominant vegetation. Review Figure 52.5

Many biomes are recognized, including Arctic and alpine **tundra**, **boreal forest**, **temperate evergreen forest**, **temperate deciduous forest**, **temperate grassland**, **hot** and **cold deserts**, **chaparral**, **thorn forest** and **tropical savanna**, **tropical deciduous forest**, and **tropical evergreen forest**.

The distribution of biomes is determined primarily by climate, but other factors, such as soil fertiltity and fire, also influence patterns of vegetation.

52.4 What is a biogeographic region?

Biogeography is the scientific study of the distributional patterns of populations, species, and ecological communities. **Biogeographic regions** are based on the taxonomic similarities of the organisms living within them. Review Figure 52.8, Web/CD Activity 52.1

A species that is found only within a certain region is said to be **endemic** to that region.

Three scientific advances have played a role in explaining the distributions of species: the understanding of continental drift; the development of phylogenetic taxonomy; and the theory of island biogeography.

Area phylogenies are phylogenetic trees that show where species originated and where they now live. Review Figure 52.10

The theory of **island biogeography** predits an equlibrium number of species on an island based on rates of immigration of new species and rates of extinction of species already present. Review Figure 52.11, Web/CD Tutorial 52.3

Both vicariant events and dispersal generate biogeographic patterns. When formerly separated land masses come together, many species from both biotas may disperse into the other region, a phenomenon known as **biotic interchange**.

52.5 How is life distributed in aquatic environments?

The oceans are divided into distinct biogeographic regions by currents that generate striking physical and biological discontinuities. The main biogeographic divisions of the oceans coincide with regions where the surface water temperatures and salinities change relatively abruptly. Review Figure 52.15

Fresh waters are divided into river basins and thousands of relatively isolated lakes. Most organisms that live in fresh water cannot survive in the oceans, and vice versa.

SELF-QUIZ

1. Ecosystems are composed of
   a. all the organisms that live in an area.
   b. the organisms that most strongly influence the environment of other organisms in an area.
   c. all the organisms that live in an area plus the physical environment.
   d. the geology, soils, weather, and climate of an area.
   e. Different ecosystems have different components.
2. A biome is
   a. a large ecological unit based on the growth forms of the dominant plants.
   b. a large ecological unit based on the life forms of the dominant animals.
   c. a large ecological unit based on the regional climate.
   d. a large ecological unit based on both topography and climate.
   e. a large ecological unit based on biogeochemical cycles.
3. The rate at which solar energy arrives per unit of Earth's surface depends primarily on
   a. the angle of the sun's rays.
   b. the moisture content of the air.
   c. the amount of cloud cover.
   d. the strength of the winds.
   e. day length.
4. When a region lies under the intertropical convergence zone,
   a. it experiences unusually strong winds.
   b. it is in the middle of the dry season
   c. it is in the middle of the wet season.
   d. the winds are not unusually strong but they come from multiple directions.
   e. day lengths change rapidly.
5. Biogeography as a science began when
   a. nineteenth-century naturalists first noted intercontinental differences in the distributions of organisms.
   b. Europeans went to the Middle East during the Crusades.
   c. phylogenetic methods were developed.
   d. the fact of continental drift was accepted.
   e. Charles Darwin proposed the theory of natural selection.
6. Vicariant events
   a. are infrequent in nature.
   b. were common in the past but are rare today.
   c. separate species ranges in the absence of dispersal.
   d. were rare in the past but are common today.
   e. caused most of today's discontinuous distributions.
7. Marine biogeographic regions exist even though the oceans are all connected because
   a. the rate of photosynthesis is low in the oceans.
   b. ocean currents keeps organisms close to where they were born.
   c. most taxa of marine organisms evolved before the oceans were separated by continental drift.
   d. water temperatures and salinities often change abruptly where ocean currents meet.
   e. oceanic circulation is too slow to carry marine organisms from one ocean to another.
8. A parsimonious interpretation of a distribution pattern is one that
   a. requires the smallest number of undocumented vicariant events.
   b. requires the smallest number of undocumented dispersal events.
   c. requires the smallest total number of undocumented vicariant plus dispersal events.
   d. accords with the phylogeny of a lineage.
   e. accounts for centers of endemism.
9. The only major biogeographic region that today is completely isolated by water from other biogeographic regions is
   a. Greenland.
   b. Africa.
   c. South America.
   d. Australasia.
   e. North America.
10. According to the theory of island biogeography, equilibrium species richness is reached when
    a. immigration rates of new species and extinction rates of species are equal.
    b. immigration rates of all species and extinction rates of species are equal.
    c. the rate of vicariant events equals the rate of dispersal.
    d. the rate of island formation equals the rate of island loss.
    e. No equilibrium number of species exists according to the theory.

FOR DISCUSSION

1. Horses evolved in North America, but subsequently became extinct there. They survived to modern times only in Africa and Asia. In the absence of a fossil record, we would probably infer that horses originated in the Old World. Today, the Hawaiian Islands have by far the greatest number of species of fruit flies (*Drosophila*). Would you conclude that the genus *Drosophila* originally evolved in Hawaii and spread to other regions? Under what circumstances do you think it is safe to conclude that a group of organisms evolved close to where the greatest number of species live today?
2. Processes in nature do not always conform to the parsimony principle. Why, then, do biogeographers often use the parsimony principle to infer geographic histories of species and lineages?
3. A well-known legend states that Saint Patrick drove the snakes out of Ireland. Give some alternative explanations, based on sound biogeographic principles, for the absence of indigenous snakes in that country.
4. Most of the world's flightless birds are either nocturnal and secretive (such as the kiwi of New Zealand) or large, swift, and powerful (such as the ostrich of Africa). The exceptions are found primarily on islands. Many of these island species have become extinct with the arrival of humans and their domestic animals. What special conditions on islands might permit the survival of flightless birds? Why has human colonization so often resulted in the extinction of such birds? The power of flight has been lost secondarily in representatives of many groups of birds and insects; what are some possible evolutionary advantages of flightlessness that might offset its obvious disadvantages?
5. MacArthur and Wilson's theory of island biogeography incorporates almost nothing about the biology of species. What traits of species should be incorporated into more realistic models of rates of colonization and extinction of species on islands?
6. A legislator introduces a controversial bill into the U.S. Congress that would ban all introductions of exotic species to the Hawaiian Islands. Would you vote in favor of this bill if you were in Congress? Why or why not?

FOR INVESTIGATION

Alfred Russel Wallace observed that dramatically different birds inhabited Bali and Lombok, even though the strait that separates them is only 15 miles wide. But most birds can easily fly 15 miles, and can probably see the other island across the water. What kinds of data, in addition to those gathered by Wallace, could be obtained to determine why birds do not fly across the strait or, if indeed they do cross the water, why they fail to colonize the island on the other side?

CHAPTER 53 Behavior and Behavioral Ecology

Monkey see, monkey do

Scientists studying a troop of Japanese macaques often fed the monkeys by throwing pieces of sweet potato onto the beach from a passing boat. The monkeys tried to brush the sand off the potato pieces, but they were still gritty. One day a young female macaque took her sweet potatoes to the water and began washing them. Soon her siblings and other juveniles in her playgroup were imitating her new behavior. Next their mothers began washing their potatoes. None of the adult males imitated this behavior, but young males learned the behavior from their mothers and their siblings.

The scientists were fascinated by the way in which the creative, insightful behavior of one juvenile female spread throughout the population, so they presented the monkeys with a new challenge: they threw grains of wheat onto the beach. Picking the wheat out of the sand was tedious and difficult. The same juvenile female came up with a solution: she carried handfuls of sand and wheat to the water and threw them in. The sand sank, but the wheat floated, enabling her to skim the grain off the surface and eat it. This behavior spread throughout the population just as washing sweet potatoes had—first to other juveniles, then to mothers, and then from mothers to their male and female offspring, but not to adult males.

The macaques of that troop now routinely wash their food. They play in the water, which they never did before, and they have added some marine food items to their diet. Clearly, this population of monkeys has acquired a *culture*: a set of behaviors shared by members of the population and transmitted through learning.

Animals also display many elaborate behaviors that they do not have to learn. Web spinning by spiders, for example, requires no prior experience. In many species of spiders, when juvenile spiders hatch, their mother is already dead (remember *Charlotte's Web* by E. B. White?). They have no experience with their mother's web. Yet when they construct their own web, they do it perfectly, without the benefit of experience or a model to copy. When generations do not overlap, we can rule out parental guidance as a factor in the acquisition of a behavior.

Web spinning requires thousands of movements performed in just the right sequence. For any given spider species, most of that sequence is *stereotypic*—

Shared Learned Behaviors Become a Culture In the space of a single generation, a population of Japanese macaques (*Macaca fuscata*) learned and transmitted a set of behaviors that included washing food, playing in the water, and eating marine food items—a new "culture" of water-related behaviors.

Spiders Are Born Web Designers   Each spider performs a stereotypic sequence of movements that results in a species-specific web design. Spiders such as this orb weaver are born with this ability, with no need to learn from experience or to model their movements after those of a parent.

CHAPTER OUTLINE

performed almost exactly the same way every time. Different spider species spin webs of different designs, using different sequences of movements. Yet every spider knows how to spin its web at birth; the behavior is part of its genetic inheritance.

The relationship between learned and inherited behavior has been a fascinating one for psychologists, sociologists, and philosophers as well as for biologists. In addition, the behavioral responses of individuals, both learned and inherited, are the foundation of much of ecology. The interactions, densities, and distributions of populations, and the effects of these population interactions on ecosystems, may all change as a result of the behavioral responses of individual animals.

IN THIS CHAPTER we will see how biologists identify the hereditary and experiential underpinnings of behavior. We will consider how genes and environment interact to shape the development of both the behavior of individuals and the long-term evolution of behavior. We will discuss several types of animal behaviors: how animals respond to changes in the environment, decide where to carry out their activities, select the resources they need (food, water, shelter, nest sites), respond to predators and competitors, and associate with other members of their own species.

53.1 What Questions Do Biologists Ask about Behavior?

Niko Tinbergen, one of the founders of **ethology**—the study of animal behavior from an evolutionary perspective—pointed out that to gain a complete understanding of any form of behavior, we need to ask several questions about it. Some questions focus on describing behavior patterns and how and when individuals perform them. They also address the *proximate* mechanisms that underlie behavior—the neuronal, hormonal, and anatomical mechanisms that we described in Part Eight of this book. Other questions concern the origins and means of acquisition of specific behaviors, especially on the relative roles of genes and experience. Most behaviors result from complex interactions between inherited anatomical and physiological mechanisms and the ability to modify behavior as a result of experience.

Other questions concern the *ultimate* causes of behavior—the selection pressures that shaped its evolution. In what ways does the behavior contribute to the fitness of the organism that possesses it (see Section 22.3)?

The behavior of the macaques described at the beginning of this chapter stimulates many questions. How did one young female macaque first think of washing sweet potatoes? Did the innovative young female differ genetically from the other monkeys, and if so, how was she different? Why were females more likely than males to learn the behavior? Why didn't any adult males learn the new behavior? The observations also stimulate questions about how changes in behavior influence the functioning of populations. How do animals incorporate new foods into their diets? How do individuals benefit from being members of social groups? How do differences in the benefits individuals receive from group membership influence how groups form and why they remain cohesive?

For many animals, much of their behavior is like the web-spinning behavior of spiders—unlearned and highly stereotypic (i.e., always exactly the same). Stereotypic behavior is often *species-specific,* in that most individuals of a given species perform the behavior in the same way. We can identify different species of spiders by the patterns of their webs. Why are the patterns so different? We know that a spider's genes code for the proteins of the silk from which its web is spun; but how do genes encode the limb movements necessary to produce the web? The questions raised by animal behavior are endless and fascinating.

53.1 RECAP

The study of animal behavior focuses on descriptive, mechanistic, and ultimate questions about behavior.

- Do you understand the difference between proximate and ultimate causes of a behavior? See p. 1141
- Can you explain what we mean when we say that a behavior is stereotypic and species-specific? See p. 1141

Distinguishing the various genetic and environmental influences on behavior is not always straightforward. The next section introduces some experimental techniques that can help us make this distinction. Then we'll investigate how inheritance and learning interact to shape certain behaviors.

53.2 How Do Genes and Environment Interact to Shape Behavior?

The observation that a given behavior is stereotypic and requires little or no learning tells biologists little about the relative roles of genes and experience in the development of behavior. An animal may fail to perform even a "genetically controlled behavior" if the environmental conditions needed to stimulate it are absent. Conversely, all individuals may behave in the same way not because of their genes but because they all imitated the same teacher.

Genes do not encode behaviors. Rather, gene products such as enzymes can affect behavior by setting in motion a series of gene-environment interactions that underlie the development of proximate mechanisms that enable individuals to make certain behavioral responses. We next describe some of the methods biologists use to determine how genes affect behavior and how genes and experience interact to influence how behaviors develop and when they are performed.

Experiments can distinguish between genetic and environmental influences on behavior

Experiments that either carefully control the environment to eliminate opportunities for learning or that modify the organism's genome can help us distinguish between genetic and environmental influences on the development of behavior. Two experimental approaches are especially useful to biologists in evaluating how genes and experience interact to shape behaviors:

- In a *deprivation experiment*, investigators rear a young animal so that it is deprived of all experience relevant to the behavior under study. If it still exhibits the behavior, we may assume that the behavior can develop without opportunities to learn it.
- In *genetic experiments*, investigators alter the genomes of organisms by interbreeding closely related species, by comparing individuals that differ in only one or a few genes, or by knocking out or inserting specific genes to determine how these manipulations affect their behavior.

DEPRIVATION EXPERIMENTS A simple deprivation experiment can yield useful information. A newborn tree squirrel was reared in isolation, on a liquid diet, and in a cage without soil or other particulate matter. When the juvenile squirrel was given a nut, it put the nut in its mouth and ran around the cage. Eventually it made stereotypic digging movements in the corner of its cage, placed the nut in the imaginary hole, went through the motions of refilling the hole, and ended by tamping the nonexistent soil with its nose. The squirrel had never handled a food object and had never experienced soil, yet it fully expressed the stereotypic behavior of a squirrel burying a nut. This experiment showed that heredity underlies the food-storing behavior of this tree squirrel species, but the behavior was expressed only when the environment provided conditions that stimulated the behavior (the presence of a nut).

SELECTIVE BREEDING *Selective breeding* is a means of genetic manipulation that has been in use since plants and animals were first domesticated. Selective breeding has been used extensively to select for both anatomical traits and behavior, as can be seen in many breeds of dogs, such as retrievers, pointers, shepherds, and bloodhounds. Thus, although its approach is applied and not based in theoretical science, it does provide insights about the effect genetic constitution has on behavior.

53.1 The Mallard Courtship Display The courtship display of the male mallard duck contains ten elements. The displays of closely related duck species contain some of the same ten elements, but have other elements not displayed by mallards. The elements of the courtship display and their sequence are species-specific and act to discourage mating between species.

1. Tail shake
2. Head flick
3. Tail shake
4. Bill shake
5. Grunt whistle
6. Tail shake

(A)

(B)

EXPERIMENT

HYPOTHESIS: Absence of the *fosB* gene changes the maternal behavior of female mice.

METHOD

1. Inactivate ("knock out") the *fosB* gene in a strain of female mice. When mature, allow them to mate and give birth at the same time wild-type females of the same strain do so.
2. Immediately upon birth, separate mouse pups from mothers. Place three newborn pups in opposite corners of a cage with their mother.
3. Count number of pups each mother retrieves within 20 minutes.

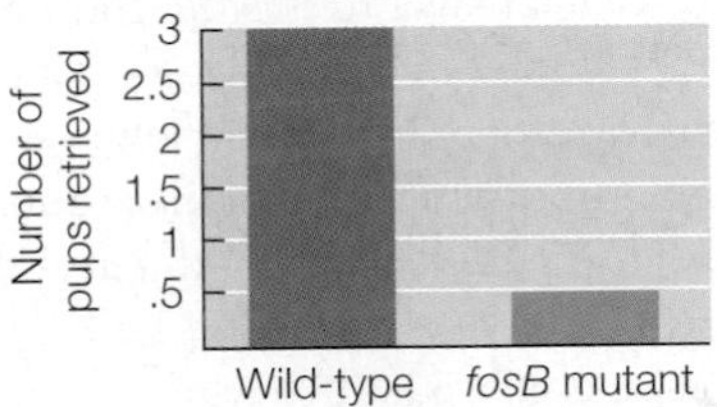

RESULTS

Normal females retrieved all of their pups within 20 minutes, whereas *fosB* mutant females retrieved an average of only 0.5 pups.

CONCLUSION: The protein product of the *fosB* gene is needed in order for female mice to display normal maternal behavior.

53.2 A Single Gene Affects Maternal Behavior in Mice (A) A wild-type female mouse gathers her pups together and crouches over them (top), but a female lacking the *fosB* gene product (bottom) does not exhibit these behaviors; her pups can be seen scattered in the foreground of the photograph. (B) An experiment confirms that *fosB* knockout mice do not exhibit a basic maternal behavior.

INTERBREEDING Konrad Lorenz, one of the pioneers in ethology, used interbreeding (hybridization) of duck species to investigate the hereditary basis of their elaborate courtship displays. Some duck species, such as mallards, teals, pintails, and gadwalls, are closely related to one another and can interbreed, but they rarely do so in nature. Each male duck performs a carefully choreographed water ballet that is typical of his species (**Figure 53.1**). A female is not likely to accept his advances unless the entire display is successfully and correctly completed, and she uses his skill in performing the display to judge his quality.

When Lorenz crossbred these duck species, the hybrid offspring expressed some elements of each parent's courtship display, but in new combinations. Furthermore, Lorenz observed that the hybrids sometimes exhibited display elements that were not in the repertoire of either parent species, but were characteristic of other species. Lorenz's hybridization studies clearly demonstrated that the stereotypic motor patterns of the courtship displays are inherited. The observation that females are not interested in males performing hybrid displays is evidence that *sexual selection* (see Section 22.3) has shaped these genetically determined behaviors.

GENE KNOCKOUT EXPERIMENTS Gene mutations that affect behavior have been studied in fruit flies and rodents using gene knockout and gene silencing techniques such as those described in Section 16.5. One striking example of the influence of a single gene mutation on a complex behavior is provided by experiments with house mice.

Female mice in which the *fosB* gene is active gather their pups together, keep them warm, and nurse them. Females carrying a mutation that inactivates ("knocks out") the *fosB* gene appear normal, but they differ from other mice in the way they treat their newborn pups. After giving birth, they inspect their pups but then ignore them.

Investigators displaced pups from the litters of both normal and *fosB* mutant females and recorded how many pups each mother retrieved. Normal females retrieved all of their pups within 20 minutes, whereas mutant females almost never retrieved their pups (**Figure 53.2**).

How does the *fosB* gene influence maternal behavior? The answer is not definite, but it would seem that the protein encoded by *fosB* is involved in stimulating neural changes in the hypothalamus of the mother's brain, possibly in response to olfactory (odor) molecules she encounters upon her initial inspection of the pups. These altered neural connections apparently play a part in motivating the mother to retrieve and care for her pups; the neural changes do not occur if the *fosB* gene is inactivated.

Genetic control of behavior is adaptive under many conditions

As we saw at the opening of this chapter, the ability to learn and to modify behavior as a result of experience can be highly adaptive. So why are behavior patterns in many species so strongly influenced by genes? One answer to this question has already been pointed out. Without role models and opportunities for learning—as in species with nonoverlapping generations—individuals might fail to acquire the appropriate behavior, or acquire inappropriate behavior, if genes did not exert strong influences on the development of the behavior.

Inherited behavior is also adaptive when mistakes are costly or dangerous. Mating with a member of the wrong species is a costly mistake, and in an environment in which incorrect as well as correct models exist, learning the wrong pattern of courtship behavior would be possible (see the story of whooping cranes and sandhill cranes at the start of Chapter 57).

Inheritance of behavior patterns used to avoid predators or capture dangerous prey is obviously adaptive. These situations allow no room for mistakes: if the behavior is not performed promptly and accurately the first time, the animal will probably not get a second chance.

Kangaroo rats are small, nocturnal desert rodents with powerful jumping legs. They can avoid rattlesnakes in total darkness because as soon as a rattlesnake begins to strike, they hear it moving through the air and jump out of the way. This jumping behavior does not need to be learned.

Even if it is not expressed during a deprivation experiment or in genetically modified individuals, a behavior may nonetheless be genetically influenced. Some behaviors are expressed only under certain conditions. The deprived squirrel described earlier, for example, did not exhibit digging and burying behaviors until it had been given a nut; that is, the nut triggered the behavior. The nut acts as a **releaser**—an object, event, or condition required to elicit a behavior. The sight of red feathers on the breasts of adult male European robins, for example, may release aggressive behavior in other males. During the breeding season, when males are defending breeding territories, the sight of an adult male robin stimulates another male robin to sing, perform aggressive displays, and attack the intruder. An immature male robin, whose feathers are brown, does not elicit this aggressive behavior. A tuft of red feathers on a stick, however, is a sufficient releaser for male aggressive behavior, even though it looks nothing like a real robin. The response to the releaser may depend on the motivational state of the animal; a male robin does not respond to red feathers when he is not on his territory and in breeding condition.

Even though genetic control of behavior is adaptive under many conditions, complete stereotypy may not be. Thus spiders adjust some details of their webs to accommodate the geometry of the structures to which they anchor their webs. As discussed in Section 20.4, the genes that govern development often allow organ-

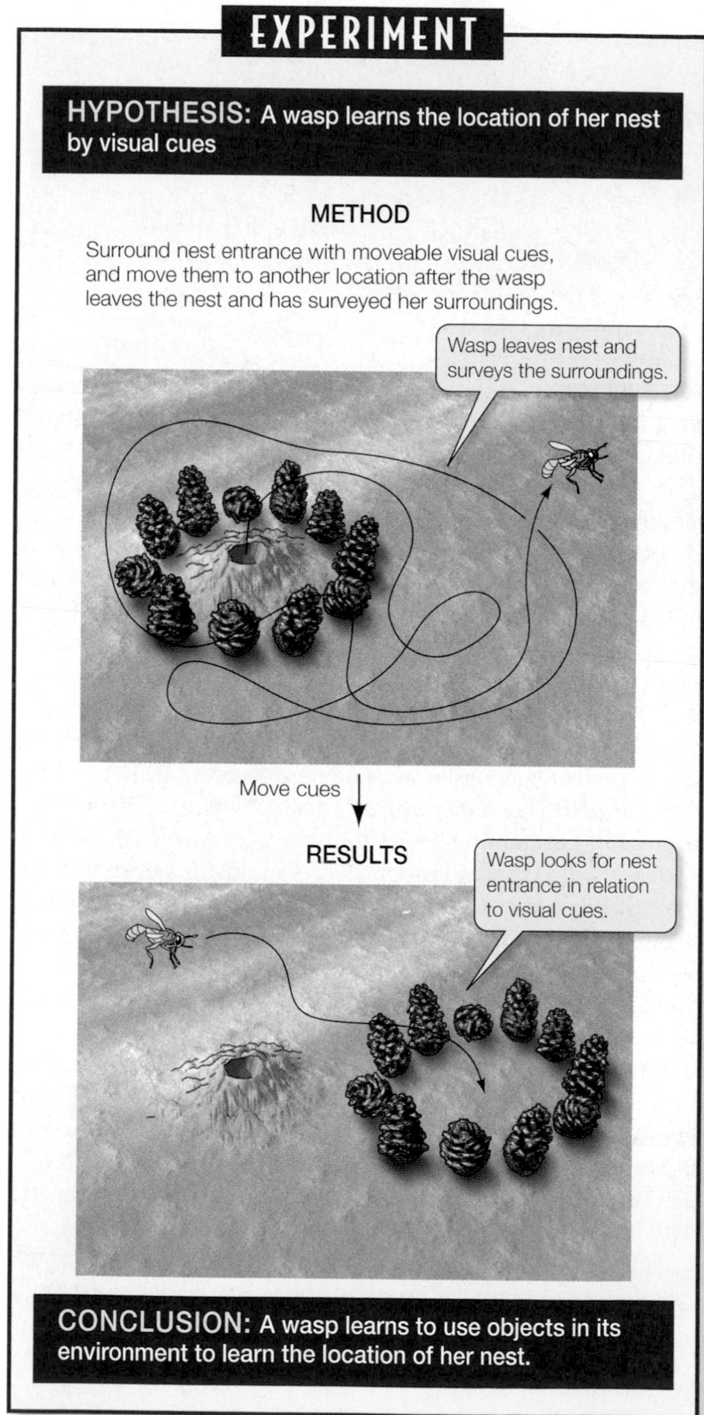

EXPERIMENT
HYPOTHESIS: A wasp learns the location of her nest by visual cues
METHOD
Surround nest entrance with moveable visual cues, and move them to another location after the wasp leaves the nest and has surveyed her surroundings.
Wasp leaves nest and surveys the surroundings.
Move cues
RESULTS
Wasp looks for nest entrance in relation to visual cues.
CONCLUSION: A wasp learns to use objects in its environment to learn the location of her nest.

53.3 Spatial Learning by a Wasp Tinbergen's classic experiment showed that a female digger wasp learns the positions of objects in her environment.

isms to adjust their forms and behavior to the particular environment in which they develop.

For example, many animals select nesting sites with specific features, but no two sites are identical. To relocate its nest, an individual must learn and remember the particular features around it. By placing pine cones near the entrance of a female digger wasp's nest, Niko Tinbergen demonstrated that the wasp used features in her environment to learn the location of her nest. After the wasp left her nest, he moved the cones a short distance away. Upon returning, the wasp oriented to the displaced cones and could not find her nest entrance (**Figure 53.3**).

Imprinting takes place at a specific point in development

Some types of learning take place only at a specific time in the animal's development, called a **critical period**. In a type of learning called **imprinting**, an animal learns a set of stimuli during a limited critical period. The classic example of a behavior learned by imprinting is the recognition of offspring by their parents and of parents by their offspring, which is essential if parents and their young are to find one another in a crowded situation such as a colony or a herd. Individual recognition must often be learned quickly because the opportunity to do so may arise only once. Konrad Lorenz demonstrated that young goslings imprint on their parents at birth by making sure that he was the first thing they saw upon hatching. They imprinted on him and followed him around thereafter as if he were their parent goose (**Figure 53.4A**).

Imprinting requires only a brief exposure, but its effects are strong and can last a long time. Emperor penguins reproduce during the coldest, darkest time of the year in Antarctica. The parents walk up to 150 kilometers inland to form a dense colony where a female lays her egg. Her mate incubates the egg while the mother walks back to the ocean to feed. By the time she returns, the chick has hatched. She takes over its care and feeding, and the father walks back to the ocean to feed. Before he gets back, the mother must leave to avoid starvation. Thus, after being away for weeks, the father has to find his chick in a crowded, milling colony of chicks, all calling for their parents (**Figure 53.4B**). Yet he finds it by recognizing its call, which he learned before he left to feed.

The critical period for imprinting may be determined by a brief developmental or hormonal state. For example, if a mother goat does not nuzzle and lick her newborn within 10 minutes after its birth, she will not recognize it as her own offspring later. The high levels of the hormone oxytocin in the mother's circulatory system at the time she gives birth and olfactory cues from the newborn kid determine the critical period.

Some behaviors result from intricate interactions between inheritance and learning

Male songbirds use a species-specific song in territorial displays and courtship. For many species, such as the white-crowned sparrow, learning is an essential step in the acquisition of this song. If eggs of white-crowned sparrows are hatched in an incubator and the young male birds are reared in isolation, they will not sing the typical species-specific song when they mature. White-crowned sparrows cannot produce their species-specific song unless they hear it as nestlings (**Figure 53.5**). But even though a male white-crowned sparrow must hear the song of his own species as a nestling to sing it as an adult, he does not sing for almost a year. Instead, the sounds he heard as a nestling form a song memory in his nervous system.

(A)

(B) *Aptenodytes forsteri*

53.4 Imprinting Helps Parents and Offspring Recognize One Another (A) Graylag geese that imprinted on Konrad Lorenz as hatchlings followed him everywhere he went. (B) A male emperor penguin must quickly find his own chick among many others.

As the young male sparrow approaches sexual maturity the following spring, he begins to sing. Eventually he matches his stored song memory through trial and error. If a bird that has heard his species-specific song as a nestling is deafened before he begins to sing, he will not develop his species-specific song (Experiment 2 in Figure 53.5). He must be able to hear himself to match his song memory. If he is deafened *after* he sings his correct species-specific song, however, he will continue to sing like a normal bird. Thus there are two critical periods for song learning: the first in the nestling stage, the second as the bird approaches sexual maturity.

Although a nestling male white-crowned sparrow in the wild is likely to hear the songs of his father and neighboring male white-crowned sparrows, he is also likely to hear the songs of many other species of birds. Deprivation experiments have demonstrated that young male sparrows do not learn the songs of other

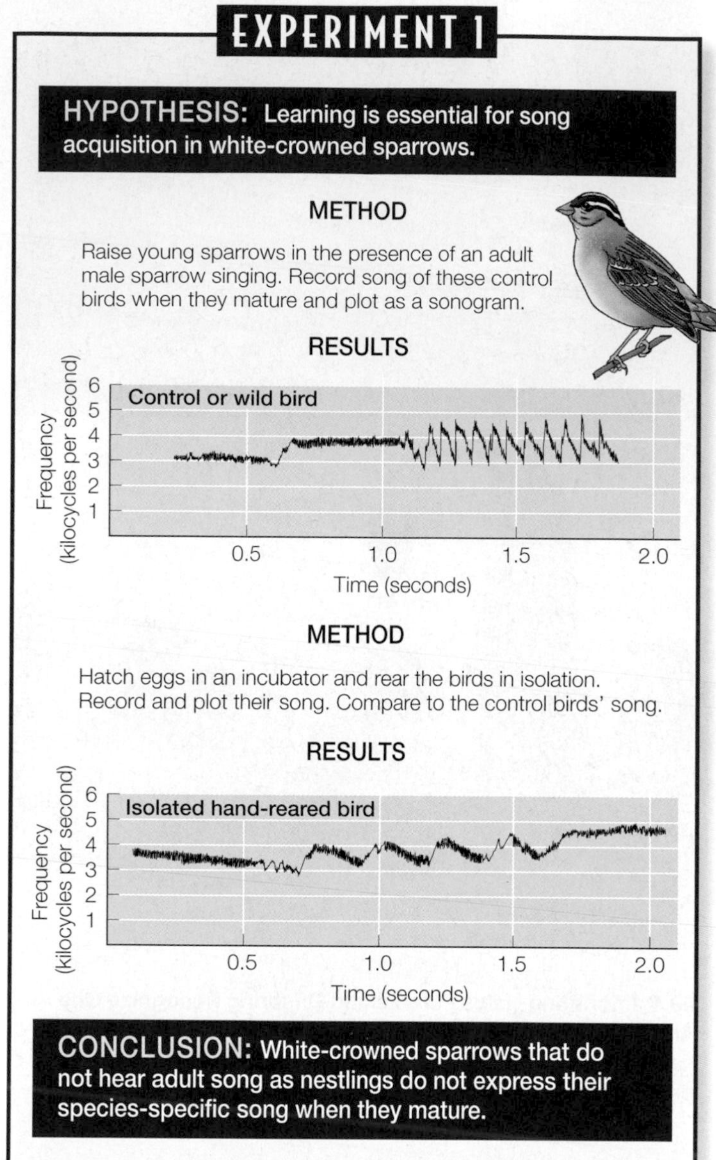

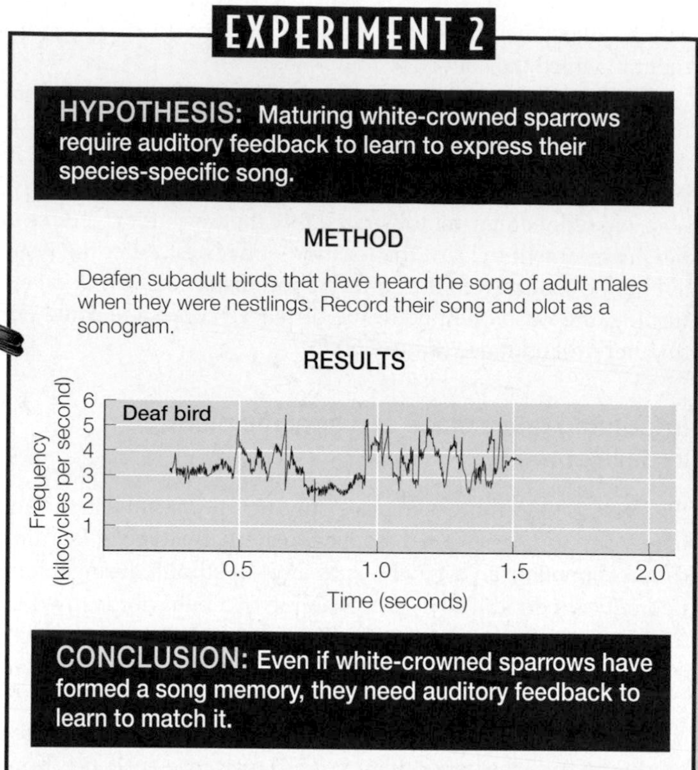

53.5 Two Critical Periods for Song Learning To sing his species-specific song as an adult, a male white-crowned sparrow must acquire a song memory by hearing the song as a nestling, and must be able to hear himself as he attempts to match his singing to that song memory.

species, even if they hear them many times, but exposure to just a few songs of their own species is sufficient for imprinting. Thus, although the sparrows must learn their songs, their genes make it very difficult for them to learn the songs of other species.

Hormones influence behavior at genetically determined times

In multicellular organisms, all behavior depends on the nervous system for initiation, coordination, and execution. Frequently, however, the hormones of the endocrine system, through controlling influences on development and on the physiological state of the animal (see Chapter 41), determine when a particular behavior is performed, as well as when certain behaviors can be learned.

We have already seen that learning is essential for the acquisition of song by some birds. Both male and female songbirds hear their species-specific song as nestlings, but only the males of most migratory species sing as adults. Male birds use song to claim and advertise a breeding territory, compete with other males, and declare dominance. They also use song to attract females, which suggests that the females recognize the song of their species even if they do not sing it. What is responsible for this difference in the learning and expression of song in male and female songbirds?

After leaving the nest where they heard their fathers' or neighbors' songs, young songbirds from temperate and Arctic habitats migrate and associate with other species in mixed-species flocks. During this time they do not sing, and they do not hear their species-specific song again until the following spring. As spring approaches and the days become longer, the young male's testes begin to grow and mature. As his testosterone level rises, he begins to sing, matching his own vocalizations to his imprinted memory of his father's song.

Why don't the females of these species sing? Can't they learn the vocalization patterns of their species-specific song? Do they lack the muscular or nervous system capabilities necessary to sing? Or do they simply lack the hormonal stimulus for developing the behavior? To answer these questions, investigators injected female songbirds with testosterone in the spring. In response to these injections, the females developed their species-specific song and sang just as the males did. Apparently females form a memory of their species-specific song when they are nestlings and have the capability to express it, but they normally lack the hormonal stimulation.

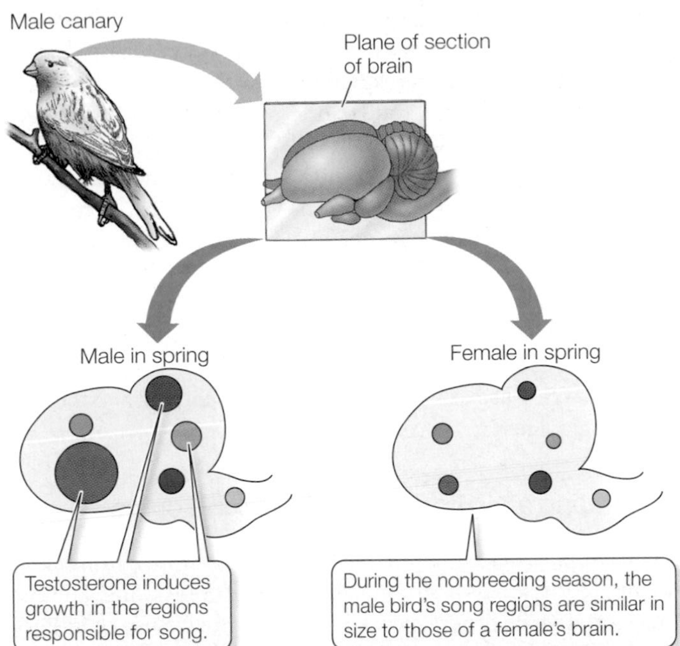
Male canary
Plane of section of brain
Male in spring
Female in spring
Testosterone induces growth in the regions responsible for song.
During the nonbreeding season, the male bird's song regions are similar in size to those of a female's brain.

53.6 Effects of Testosterone on Bird Brains In spring, rising testosterone levels in the male cause the song regions of the brain to develop. The size of each circle is proportional to the volume of the brain occupied by that region.

How does testosterone cause a male songbird to sing? A remarkable study revealed that each spring, an increase in testosterone causes certain parts of the male's brain necessary for learning and developing song to grow larger (**Figure 53.6**). Individual neurons in those regions of the brain increase in size and grow longer extensions, and the number of neurons in those regions increases. Such research on the neurobiology of bird song has revealed that, contrary to the widely held belief that new neurons are not produced in the brains of adult vertebrates, hormones can control behavior by changing brain structure as well as brain function, both developmentally and seasonally.

53.2 RECAP

Some behaviors are strongly influenced by genes, resulting in highly stereotypic behavior patterns. Genes also play a role in shaping what individuals learn. Much adaptive behavior results from interactions between heredity and learning, often modified by hormones.

- What kinds of experiments can show whether a behavior is strongly influenced by genes? See pp. 1142–1143 and Figure 53.2
- What adaptive advantages does imprinting offer? See p. 1145
- Can you explain how inheritance and learning interact when songbirds learn their species-specific songs? See pp. 1145–1146 and Figure 53.5

Let's look now at the many decisions about how and when to behave in a certain way that animals must make during their lives, and that influence their fitness.

53.3 How Do Behavioral Responses to the Environment Influence Fitness?

Any animal that reaches old age has made decisions throughout its life as it grew to maturity, reproduced, and experienced the effects of aging. It is likely to have chosen where to settle, how long to stay there, and when, if ever, to leave. It will also have found its way around the environment and selected places for specific activities, such as resting and nesting. It will have decided which things to eat from among the rich array of potential food sources in its immediate environment. Most animals also choose with whom to associate and for what purposes.

All of these decisions are made in an environment that varies dramatically in space and in time. For example, in the Sierra Nevada of California, you can drive from a hot, dry sagebrush desert to alpine snowfields in a half hour. In the woodlands of Pennsylvania it can be hot and humid in the summer and below freezing in the winter. In the marine environment, only a few kilometers may separate the rocky intertidal zone from the dark, deep ocean floor. Given this great variation, how do animals find a good place to live and decide where to go if the local environment deteriorates?

Choosing where to live influences survival and reproductive success

Selecting a place in which to live is one of the most important decisions an individual makes. The environment in which an organism lives is called its **habitat**. Once a habitat is chosen, an animal seeks its food, resting places, nest sites, and escape routes within that habitat. A challenge to behavioral ecologists is to discover what information animals use to make their habitat choices, and how that information relates to environmental qualities that influence their fitness. Some of the ways in which organisms make their choices are surprisingly simple. The cues most organisms use to select suitable habitats have a common feature: *they are good predictors of conditions suitable for future survival and reproduction*.

The red abalone, a marine mollusk, sticks to rocks and does not move very far as an adult. Abalones, however, shed their eggs and sperm into the water. The fertilized eggs and the larvae that hatch from them drift and swim in the open ocean for about 8 days before they drop to the ocean floor, choose places to settle, and develop into adults. The best place for a red abalone is on coralline algae, its major source of food. Red abalone larvae recognize coralline algae by a chemical the algae produce. In the laboratory, abalone larvae will settle on any surface on which these molecules have been placed, but in nature, only coralline algae produce them. By using this simple chemosensory cue, the larvae always settle on a surface that has the potential of supplying the food necessary for their future survival and reproductive success.

Visual information can also provide useful cues about the quality of a habitat. Many animals use the presence of already settled individuals as an indication that the habitat may be good. While collared flycatchers are on their breeding grounds in spring, they regularly peer into the nests of other individuals. Observing this behavior, the Swedish and French biologists Blandine Doligez, Etienne Danchin, and Jean Clobert hypothesized that the flycatchers were assessing the quality of the habitat by seeing how well their neighbors were doing. To test this hypothesis, they created some areas with super-sized broods—normally an indication of abundant food—by taking young birds from some nests and adding them to nests in another area. The next year, flycatchers preferentially settled in the areas where broods had been artificially enlarged (**Figure 53.7**).

Some highly social animals actually "vote" on the quality of habitats. For example, honeybees, live in large colonial nests that consist of vertical sheets of hexagonal cells that they build out of wax. Some of the cells are used to raise larvae, some are used to store honey, and some are used to store pollen, a source of protein. When a bee colony outgrows its hive space, the queen leaves it, along with a huge retinue of worker bees—the whole group is called a *swarm*. The queen is not a very good flyer, so the swarm does not go far before coming to rest as a huge mass of bees tightly clustered around the queen. From this cluster, individual worker bees fly out to search for a cavity that would be a good site for a new colony. When a worker (a scout) finds a cavity, she explores it by walking around inside. She then returns to the swarm and dances on its surface. Her dance conveys information to other bees in the swarm about the distance and direction to the potential nest site she has discovered (we will see how she does so in Section 53.4). The vigor of her dance reflects the quality of the nest site. In response to her dance, other worker bees fly to the site and assess it. When they return to the swarm, they also dance to communicate information about the site and recruit more workers to visit it. Different workers typically provide information to the swarm about several potential sites at the same time. Eventually the site that excites the most workers is chosen, and the swarm takes flight again to move to the new location.

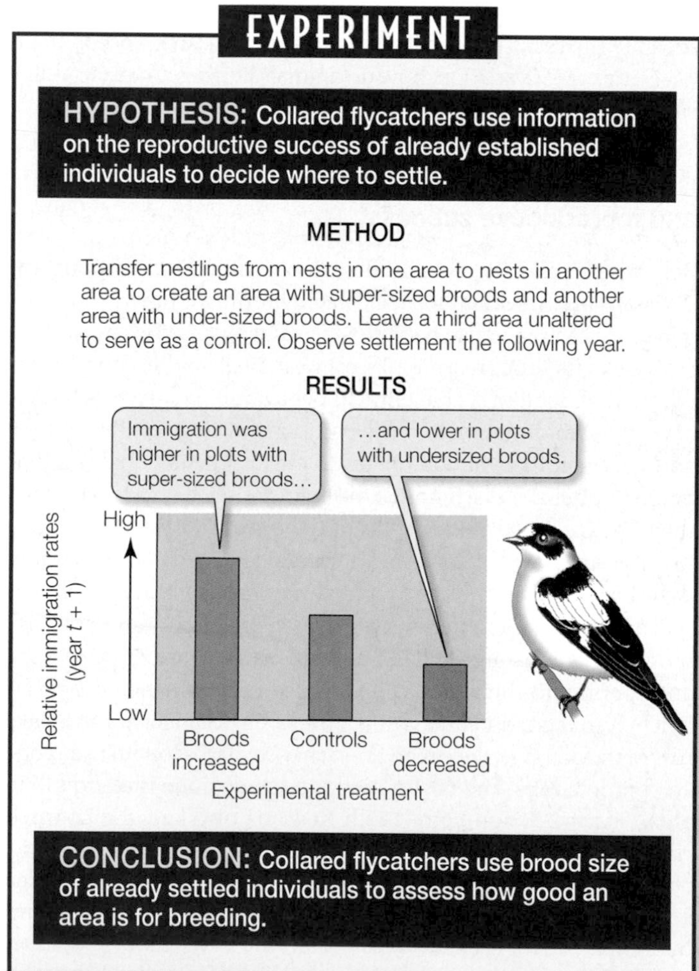

EXPERIMENT
HYPOTHESIS: Collared flycatchers use information on the reproductive success of already established individuals to decide where to settle.
METHOD
Transfer nestlings from nests in one area to nests in another area to create an area with super-sized broods and another area with under-sized broods. Leave a third area unaltered to serve as a control. Observe settlement the following year.
RESULTS
Immigration was higher in plots with super-sized broods...
...and lower in plots with undersized broods.
Relative immigration rates (year t + 1)
High
Low
Broods increased
Controls
Broods decreased
Experimental treatment
CONCLUSION: Collared flycatchers use brood size of already settled individuals to assess how good an area is for breeding.

53.7 Flycatchers Use Neighbors' Success to Assess Habitat Quality Collared flycatchers (*Ficedula albicollis*) settled at higher densities in areas where experimenters had artificially enlarged the broods of other flycatchers.

Defending a territory has benefits and costs

When high-quality habitats are in limited supply, animals may compete for access to them. Under these conditions, an animal may improve its fitness by establishing exclusive use of its chosen habitat. The most common way of doing so is to establish a **territory** from which it excludes *conspecifics* (other individuals of the same species)—and sometimes individuals of other species as well—by advertising that it owns the area and, if necessary, chasing others away. But advertising and chasing take time and energy that could have been used for other beneficial purposes, such as finding food and watching out for predators.

To understand the evolution of behaviors that involve these kinds of trade-offs, ecologists often use a cost–benefit approach. A **cost–benefit approach** assumes that an animal has only a limited amount of time and energy to devote to its activities. Animals cannot long perform behaviors whose total costs are greater than the sum of their benefits. A cost–benefit approach provides a framework that behavioral ecologists can use to make predictions, design experiments, and make observations that can explain why behavior patterns evolve as they do.

The benefits of a behavior are the improvements in survival and reproductive success that the animal achieves by performing the behavior. The total cost of any particular behavior typically has three components:

- **Energetic cost** is the difference between the energy the animal would have expended had it rested and the energy expended in performing the behavior.
- **Risk cost** is the increased chance of being injured or killed as a result of performing the behavior, compared with resting.
- **Opportunity cost** is the sum of the benefits the animal forfeits by not being able to perform other behaviors during the same time interval. An animal that devotes all of its time to foraging, for example, cannot achieve high reproductive success!

Michael Moore and Catherine Marler at Arizona State University performed an experiment to estimate the costs incurred by male lizards when defending a territory. Male Yarrow's spiny lizards defend territories that include the habitats of several females, from which they exclude conspecific males. They normally do so most vigorously during September and October, when females are most receptive to mating. To assess the costs of territorial behavior, the researchers inserted small capsules containing testosterone, a hormone that the lizards normally produce in the fall and that induces aggressive territorial behavior, beneath the skin of some males. They performed the experiment in June and July, a time of year when the lizards are normally only weakly territorial. They also captured and released control males, but the controls received no testosterone implants.

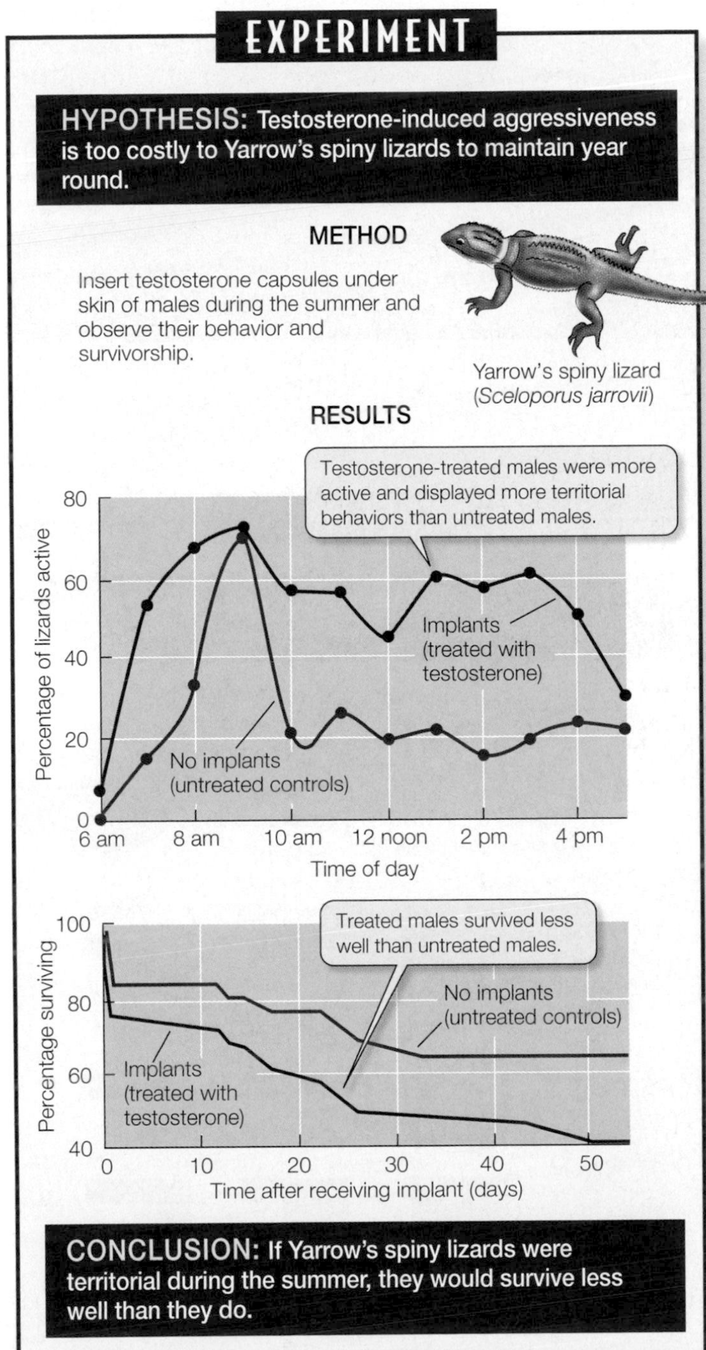

EXPERIMENT
HYPOTHESIS: Testosterone-induced aggressiveness is too costly to Yarrow's spiny lizards to maintain year round.
METHOD
Insert testosterone capsules under skin of males during the summer and observe their behavior and survivorship.
Yarrow's spiny lizard (*Sceloporus jarrovii*)
RESULTS
Testosterone-treated males were more active and displayed more territorial behaviors than untreated males.
Percentage of lizards active
0
20
40
60
80
Implants (treated with testosterone)
No implants (untreated controls)
6 am
8 am
10 am
12 noon
2 pm
4 pm
Time of day
Treated males survived less well than untreated males.
Percentage surviving
40
60
80
100
No implants (untreated controls)
Implants (treated with testosterone)
0
10
20
30
40
50
Time after receiving implant (days)
CONCLUSION: If Yarrow's spiny lizards were territorial during the summer, they would survive less well than they do.

53.8 The Costs of Defending a Territory By using testosterone implants to increase territorial behavior, Moore and Marler measured the costs to male lizards of defending a territory during the summer months.

Males with implanted testosterone capsules spent more time patrolling their territories, performed more advertising displays, and expended about one-third more energy (an energetic cost) than control males. As a result, they had less time to feed (an opportunity cost), captured fewer insects, stored less energy, and died at a higher rate (a risk cost) (**Figure 53.8**). In June and July, when females are less receptive, these high costs of vigorous territorial defense probably outweigh the reproductive benefits of territoriality. Probably that is why the lizards normally wait to perform these behaviors until later in the summer.

Some animals defend all-purpose territories that include all of the resources the animal needs for nesting, mating, and finding food. Tigers defend territories that are many square kilometers in extent by leaving scent marks on their boundaries (**Figure 53.9A**). Many songbirds defend all-purpose territories by singing and threat postures.

The food supplies of some animals cannot be defended because they are widely distributed in space or are subject to fluctuations in availability. For example, the open ocean where seabirds feed cannot be defended. However, most seabirds nest on islands or on rocky cliffs, which offer protection from predators. Such nesting sites can be in very short supply and are vigorously defended. Some of these territories are no larger than the distance the bird can reach while sitting on its nest (**Figure 53.9B**).

Some animals defend territories that are used only for mating. When the breeding season arrives, males of some grouse species congregate on display grounds, where they defend very small areas (**Figure 53.9C**). Their vigorous displays show other males how strong they are and show the females their quality as mates. A female visits a display ground and selects the male with whom she will mate. She then leaves and raises her brood alone. The males often expend so much energy defending their little display areas that less exhausted males may eventually evict them.

Animals choose what foods to eat

Cost–benefit analyses have also been used to investigate the food choices animals make. When an animal *forages* (looks for food), how much time should it spend searching an area before moving to another site? When many different types of food are available, which ones should it eat, and which ones should it ignore? By applying cost–benefit approaches to feeding behavior, scientists have produced a body of knowledge known as **foraging theory**, which helps us understand the survival (ultimate) value of feeding choices. The primary benefits of foraging are the nutritional value of the food obtained—its energy, minerals, and vitamins. The costs of foraging are similar to those for territorial defense: energy expended, time lost for other activities that could enhance fitness, and the risk of increased exposure to predators.

As an example, let's consider some tests of the hypothesis that animals make choices among available prey in such a way as to maximize the rate at which they obtain energy. This is a plausible hypothesis because the more rapidly an animal captures food, the more time and energy it will have for other activities, such as reproduction or avoiding predators. To determine how an animal should choose food items to maximize its energy intake rate, behavioral ecologists can characterize each type of available food item in two

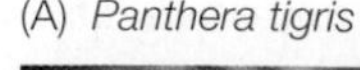

(A) *Panthera tigris*

(B) *Morus serrator*

(C) *Lyrurus tetrix*

53.9 Animals Defend Territories of Different Sizes (A) To mark the boundaries of her extensive territory, this female tiger rubs the pheromone-producing scent glands on her face against a tree. Other tigers passing the spot will know that the area is "claimed," as well as knowing something about the individual who claimed it. (B) The nesting territories of some colonial birds are no larger than the distance the incubating bird can reach. (C) Some territorial displays are related to mating. Here two male black grouse defend a small territory in snowy Sweden; the winner will mate with nearby females.

ways: by the time it takes the animal to pursue, capture, and consume an individual item, and by the amount of energy an individual item contains. They can then rank the food types according to the amount of energy the animal gets relative to the time it spends pursuing, capturing, and handling the item. The most valuable food type is the one that yields the most energy per unit of time expended.

With this information, behavioral ecologists can determine the rate at which an animal would obtain energy given a particular foraging strategy. They can then compare alternative foraging strategies and determine the one that yields the highest rate of energy intake. Such calculations show that, if the most valuable food type is abundant enough, an animal gains the most energy per unit of time spent foraging by taking only the most valuable type and ignoring all others. However, as the abundance of the most valuable type decreases, an energy-maximizing animal adds less valuable types to its diet in order of the energy per unit of time that they yield.

Earl Werner and Donald Hall of Michigan State University performed laboratory experiments with bluegill sunfish to test the energy maximization hypothesis (**Figure 53.10**). In preparation for their experiments, they measured the energy content of water fleas of different sizes (the different food types), how much time bluegill sunfish (the foragers) needed to capture and eat these different food types, the energy they spent pursuing and capturing the different food types, and the rates at which they encountered the different food types under different food densities. Werner and Hall then stocked experimental environments with different densities and proportions of large, medium, and small water fleas. They made two predictions from the energy maximization hypothesis: first, that in an environment with abundant large water fleas, the fish would ignore smaller water fleas, and second, that in an environment stocked with low densities of all three sizes of water fleas, the fish

HYPOTHESIS: Bluegills select prey so as to maximize their energy intake.

METHOD

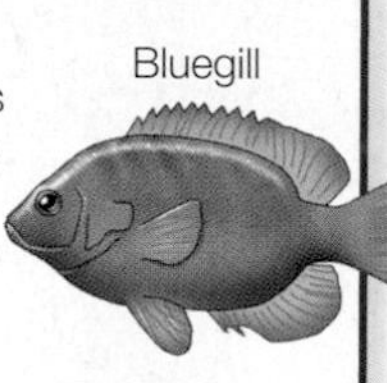

Provide bluegills with varying proportions of *Daphnia* (water fleas) of different sizes and in differing densities. Compare prey actually eaten with the predictions of the energy maximization hypothesis.

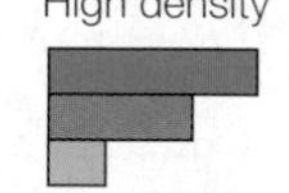

RESULTS

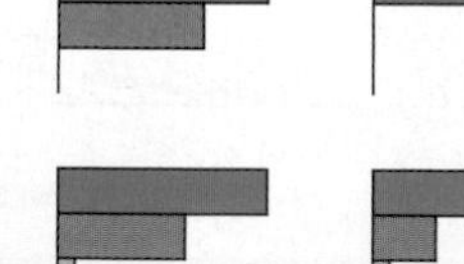

CONCLUSION: Bluegills select prey to maximize their rate of energy intake.

53.10 Bluegills Are Energy Maximizers The prey choices of bluegill sunfish (*Lepomis macrochirus*) were very similar to those predicted by the energy maximization hypothesis.

53.11 Mineral Seekers Red-and-green macaws obtain mineral nutrients from a clay lick in the Amazon jungle of Peru.

would eat every water flea they encountered. The proportions of large, medium, and small water fleas taken by the fish under different conditions were close to those predicted by the hypothesis.

In the bluegill experiment, only the energy content of water fleas mattered, but minerals are also important to some foragers. Some animals incur large energetic costs and the risks of traveling great distances to obtain essential minerals. Many species of mammals and birds, particularly herbivores and seed eaters, get mineral nutrients by eating soil at particular sites where mineral-rich soil is exposed (**Figure 53.11**). Moose wade into streams and ponds to eat plants that contain more sodium than the terrestrial plants from which they get most of their energy.

Animals may ingest some foods for reasons other than the energy or nutrients they provide. For example, some species of frogs have poisons in their skins. Like many other animals that defend themselves from predators with toxic compounds, these frogs have bright, conspicuous coloration that warns potential predators of the dangers of eating them. Where do these frogs get their poisons? Some species get them by eating certain species of ants that have evolved poisons as their own predator defense mechanisms. The frogs can eat these ants because they are immune to the ants' poisons.

A striking example of the ingestion of non-nutritive foods is the use of spices by humans in food preparation. Spices must provide some benefit to humans because we value them so highly. Spices contain chemicals known to protect the plants that produce them against bacteria and fungi, so perhaps the use of spices in food preparation protects people from contaminated food. Zhi-He Liu and Hiroyuki Nakano of Hiroshima University in Japan tested this antimicrobial hypothesis by exposing food-borne bacteria to chemicals found in spices. Most of the commonly used spices inhibited the growth of more than one kind of food-borne bacteria (**Figure 53.12**).

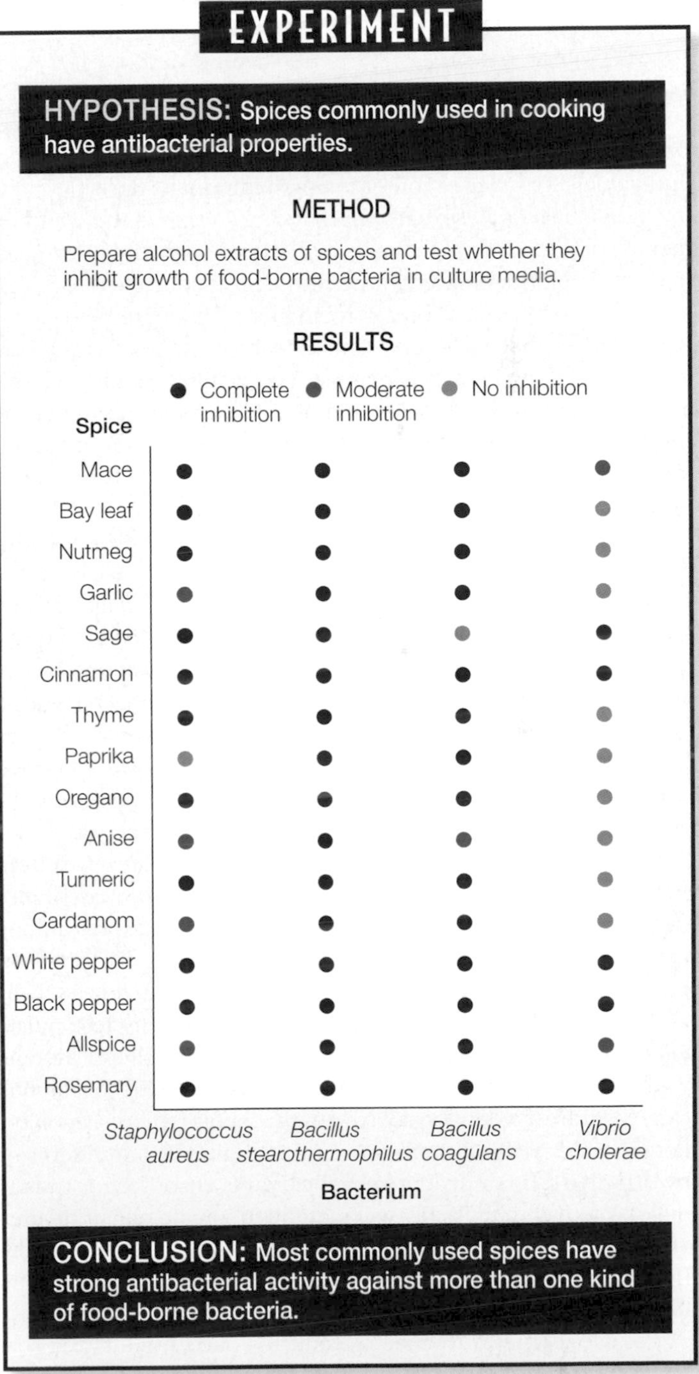

53.12 Most Spices Have Antimicrobial Activity Laboratory tests have shown that most commonly used spices can inhibit the growth of bacteria.

Although these tests support the antimicrobial hypothesis, they do not exclude the possibility that an alternative hypothesis might also explain the human fondness for spices. One alternative hypothesis is that spices disguise the smell and taste of spoiled foods. This hypothesis cannot be tested easily, but it predicts that spoiled foods could be safely consumed if aversions to their smell and taste could be overcome. We know, however, that toxins from food-borne bacteria kill thousands of people every year and debilitate millions more. Therefore, eating spoiled food by covering up its bad flavors would be a dangerous thing to do, even for a starving person. Natural selection is not likely to have favored people who ate rancid food, no matter how tasty.

An animal's choice of associates influences its fitness

Most animals do not lead solitary lives. They associate with other individuals for a variety of reasons. An important decision made by individuals of all sexually reproducing species is the choice of mating partners. These choices may be based on the inherent qualities of a potential mate, on the resources it controls (food, nest sites, escape places), or on a combination of the two.

The ways that males and females choose mating partners are often very different. Males of most species initiate courtship, seldom reject receptive females, and often fight for opportunities to mate with females. Females of most species seldom fight over males, and they often reject courting males. Why are these sexual roles so different?

The answer lies in part in the costs of producing sperm and eggs. Because sperm are small and cheap to produce, one male produces enough to father a very large number of offspring—usually many more than the number of eggs a female can produce or the number of young she can nourish. Therefore, males of many species can increase their reproductive success by mating with many females.

Eggs, on the other hand, are typically much larger than sperm and are expensive to produce. Consequently, a female usually cannot increase her reproductive output very much by increasing the number of males she mates with. The reproductive success of a female may depend on the quality of the genes she receives from her mate, the resources he controls, or the amount of assistance he provides in the care of their offspring. By their choices among males, females may cause the evolution of traits that reliably signal male quality, a component of sexual selection.

Males employ a variety of tactics to induce females to copulate with them. A male may use courtship behavior that signals in some way that he is in good health, that he is a good provider of parental care, that he can control resources, or that he has a good genotype. For example, males of some species of hangingflies court females by offering them a dead insect, a valuable resource for the female. A female hangingfly will mate with a male only if he provides her with food in this manner. The bigger the food item, the longer she copulates with him, and the more of her eggs he fertilizes (**Figure 53.13**).

Females can improve their reproductive success if they can correctly assess the genetic quality and health of potential mates, the quantity of parental care they may provide, and the quality of the resources they control. But how can females make such assessments when all males would benefit by attempting to signal that they excel in the favored traits? The answer is that by paying particular attention to those signals that males cannot fake, females have favored the evolution of "reliable" signals. Possession of a large dead insect reliably indicates that a male hangingfly is a good forager.

53.13 A Male Wins His Mate The male hangingfly on the right has just presented a moth to his mate, thus demonstrating his foraging skills. He also provides a material reward to his partner, who eats it while they copulate. The bigger the moth, the longer they copulate, and the more eggs he fertilizes.

Responses to the environment must be timed appropriately

Most environments are not constant, and therefore behaviors that are adaptive at one time may not be adaptive at another time. Earth turns on its axis once every 24 hours, generating daily cycles of light and dark, temperature, humidity, and tides. In addition, Earth is tilted on its axis, so the light–dark cycle, and hence the seasons, change as it revolves around the sun. These environmental changes have been discussed in detail as they pertain to plants (see especially Section 38.2) but such cycles profoundly influence the physiology and behavior of animals as well.

CIRCADIAN RHYTHMS CONTROL THE DAILY CYCLE OF BEHAVIOR If animals are kept in constant darkness, at a constant temperature, with food and water available all the time, they will still demonstrate daily cycles of activities such as sleeping, eating, drinking, and just about anything else that can be measured. The persistence of these daily cycles in the absence of environmental time cues suggests that animals have an endogenous (internal) clock. Although these daily cycles seldom are exactly 24 hours long, they are known as **circadian rhythms** (Latin *circa*, "about," and *dies*, "day").

As we learned in Chapter 38, any biological rhythm can be thought of as a series of cycles, and the length of one of those cycles is the *period* of the rhythm (see Figure 38.13). Any point on the cycle is a *phase* of that cycle: Hence, when two rhythms completely match, they are *in phase*, and if a rhythm is shifted (as in the resetting of a clock), it is *phase-advanced* or *phase-delayed*. Since the period of a circadian rhythm is not exactly 24 hours, it must be phase-advanced or phase-delayed each day to remain in phase with the daily cycle of the environment. In other words, the rhythm has to be **entrained** to the cycle of light and dark in the animal's environment.

> When you fly across time zones, your circadian rhythm gets out of phase, resulting in "jet lag." Because your biorhythms can only be shifted by 30–60 minutes each day, it may take several days to re-entrain your internal clock to reflect the environmental time in your new location.

An animal kept in constant conditions will not be entrained to the cycle of the environment, and its circadian clock will run according to its natural period—it will be *free-running*. If its period is less than 24 hours, the animal will begin its activity a little earlier each day (**Figure 53.14**). The free-running circadian rhythm is under genetic control. Different species may have different average periods, and within a species, mutations can lead to different period lengths.

Under natural conditions, environmental time cues, such as the onset of light or dark, entrain the free-running rhythm to the light-dark cycle of the environment. In the laboratory, it is possible to entrain the circadian rhythms of free-running animals with short pulses of light or dark administered every 24 hours (bottom panel of Figure 53.14).

A variety of different sensory capabilities have evolved in response to daily cycles of light and dark. Animals that are active at night are *nocturnal* and have different sensory capabilities than those that are *diurnal*, or active during the day. Diurnal animals (including humans and most birds) tend to be highly visual, whereas nocturnal animals depend more on their abilities to hear and smell and use tactile information. Among animals that rely mostly on vision, adaptations have evolved to support diurnal versus nocturnal habits. Ground squirrels are diurnal and have retinas made up entirely of cone cells (visual receptors that can process light wavelengths into the perception of color but are not sensitive to low light levels). In contrast, flying squirrels are nocturnal and their retinas are composed entirely of rods (visual receptors that are highly sensitive to light stimuli but do not provide information about color; see the discussion of photoreceptors in Section 45.4).

PHOTOPERIOD AND CIRCANNUAL RHYTHMS CONTROL SEASONAL BEHAVIORS Seasonal changes in the environment present challenges to many species. Most animals reproduce most successfully if they time their reproductive behavior to coincide with the most favorable time of year for the survival of their offspring. It is advantageous for animals to be able to anticipate when to shift their reproductive systems into breeding mode, because it may take considerable time to prepare physiologically for reproduction.

For many species, a change in day length—the *photoperiod*—is a reliable indicator of seasonal changes to come. For some animals, however, change in day length is not a reliable seasonal cue. Hi-

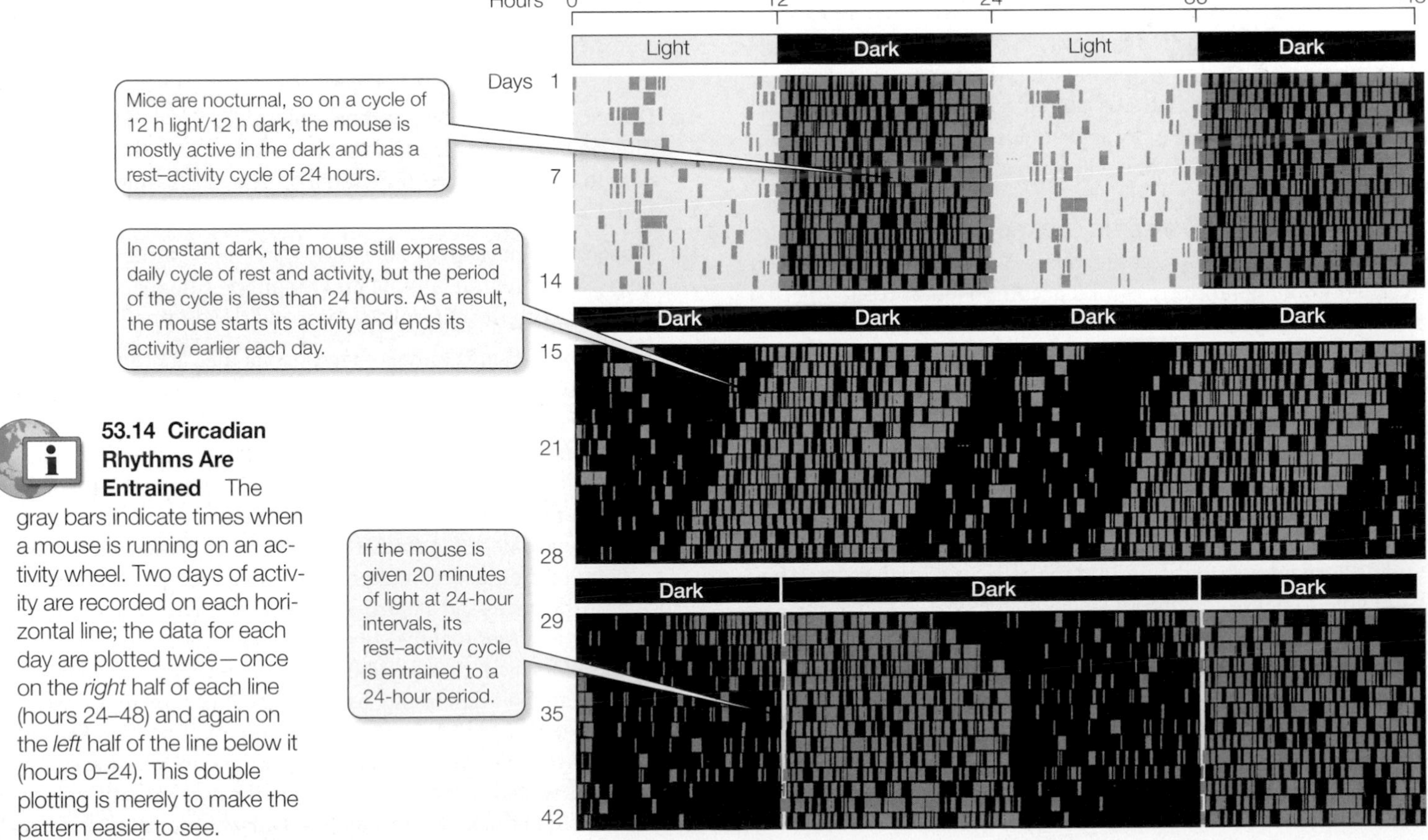

Hours
0
12
24
36
48
Light
Dark
Light
Dark
Days
1
7
14
15
21
28
29
35
42
Mice are nocturnal, so on a cycle of 12 h light/12 h dark, the mouse is mostly active in the dark and has a rest–activity cycle of 24 hours.
In constant dark, the mouse still expresses a daily cycle of rest and activity, but the period of the cycle is less than 24 hours. As a result, the mouse starts its activity and ends its activity earlier each day.
Dark
Dark
Dark
Dark
If the mouse is given 20 minutes of light at 24-hour intervals, its rest–activity cycle is entrained to a 24-hour period.
Dark
Dark
Dark

53.14 Circadian Rhythms Are Entrained The gray bars indicate times when a mouse is running on an activity wheel. Two days of activity are recorded on each horizontal line; the data for each day are plotted twice—once on the *right* half of each line (hours 24–48) and again on the *left* half of the line below it (hours 0–24). This double plotting is merely to make the pattern easier to see.

bernators, for example, spend long months in dark burrows underground, but have to be physiologically prepared to breed almost as soon as they emerge in the spring. A bird overwintering near the equator cannot use changes in photoperiod as a cue to time its migration north to the breeding grounds. Hibernators and equatorial migrants have *circannual rhythms*, built-in neural calendars that keep track of the time of year.

Arctic ground squirrels have a very short season for breeding and raising young. Females are ready to mate almost as soon as they end hibernation and may remain in estrus for only a day. Males must be prepared! They stop hibernating a month earlier, but remain in their dens, burning precious metabolic fuel. Why? Because cold testes do not produce sperm.

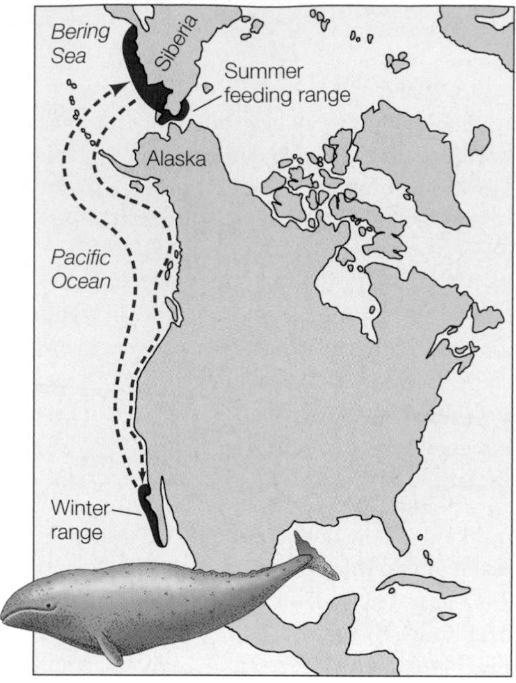

53.15 Piloting Gray whales migrate south in winter from the Bering Sea to the coast of Baja California, in part by following the western coast of North America.

Animals must find their way around their environment

To locate good habitats, find food and mates, and escape from predators and bad weather, an animal needs to be able to find its way around its environment. Within its local habitat, an animal can organize its behavior spatially by orienting to landmarks, as digger wasps do (see Figure 53.3). But what if its destination is a considerable distance away?

PILOTING: ORIENTATION BY LANDMARKS Most animals find their way by knowing and remembering the structure of their environment in a process called **piloting**. Gray whales, for example, migrate seasonally between the Bering Sea and the coastal lagoons of Mexico (**Figure 53.15**). They find their way in part by following the west coast of North America. Coastlines, mountain chains, rivers, water currents, and wind patterns can all serve as piloting cues. But some remarkable cases of long-distance orientation and movement cannot be explained by piloting.

HOMING: RETURN TO A SPECIFIC LOCATION The ability to return over long distances to a nest site, burrow, or other specific location is called **homing**. Homing can be accomplished by piloting in a known environment, but some animals perform much more sophisticated homing.

Pigeons, for example, can return home after being transported to remote sites where they have never been. They do not search around until they encounter familiar territory, but fly fairly directly home from the point of release. In one series of experiments, pigeons were fitted with frosted contact lenses so that they could see nothing but the degree of light and dark. These pigeons still homed and fluttered down to the ground in the vicinity of their loft. They were able to navigate without visual images of the landscape.

MIGRATION: TRAVELING GREAT DISTANCES For as long as humans have inhabited temperate latitudes, they must have been aware that whole populations of animals, especially birds, disappear and reappear seasonally—that is, they migrate. Not until the early nineteenth century, however, were patterns of migration established by marking individual birds with identification bands around their legs. Only when individuals could be unmistakably identified was it possible to show that the same birds and their offspring often returned to the same breeding grounds year after year, and that these same birds could be found during the nonbreeding season at locations hundreds or even thousands of kilometers from their breeding grounds.

Because many homing and migrating species are able to take direct routes to their destinations through environments they have never experienced, they must have mechanisms of navigation other than piloting. There are two systems of navigation:

- **Distance-and-direction navigation** requires knowing in what direction and how far away the destination is. With a compass to determine direction and a means of measuring distance, humans can navigate.
- **Bicoordinate navigation**, also known as *true navigation*, requires knowing the latitude and longitude (the map coordinates) of both the current position and the destination.

The behavior of many animals suggests that they are capable of bicoordinate navigation. Gray-headed albatrosses, for example, breed on oceanic islands. When a young gray-headed albatross first leaves its parents' nest, it flies widely over the southern oceans for 8 or 9 years before it reaches reproductive maturity (**Figure 53.16A**). At that time, it flies back to the island where it was raised to select a mate and build a nest (**Figure 53.16B**). How can it find a tiny island in such an enormous ocean after years of wandering? A circadian clock probably gives the albatross enough information about time of day and the position of the sun to determine their coordinates—much as sailors did in the days before global positioning satellites.

How do animals determine distance and direction? Two obvious means of determining direction are the sun and the stars. During the day, the sun can serve as a compass, as long as the time of day is known. As we have seen, animals can tell the time of day by means of their circadian clocks. Experiments in which the internal clocks of

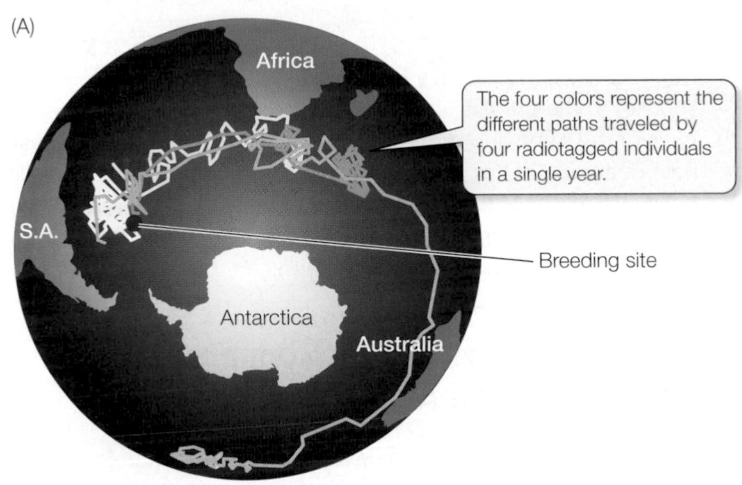

The four colors represent the different paths traveled by four radiotagged individuals in a single year.

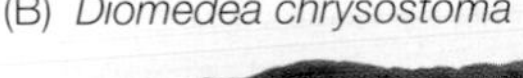

(B) *Diomedea chrysostoma*

53.16 Coming Home (A) Gray-headed albatrosses are born on islands in the subantarctic oceans. Young birds roam widely over the southern oceans for 8 or 9 years. (B) Once reaching maturity, the birds return to the island on which they were born, where they mate and raise their own young. A courting couple is shown here.

animals are shifted have demonstrated that animals can use their circadian clocks to determine direction using the position of the sun.

Researchers placed pigeons in a circular cage that enabled them to see the sun and sky, but no other visual cues. Food bins were arranged around the sides of the cage, and the birds were trained to expect food in the bin at one particular direction—south, for example. After training, no matter what time they were fed, and even if the cage was rotated between feedings, the birds always went to the bin at the southern end of the cage for food, even if that bin contained no food (**Figure 53.17**).

Next the birds were placed in a room with a controlled light cycle, and their circadian rhythms were shifted by turning the lights on at midnight and off at noon. After about 2 weeks, the birds' circadian clocks had been advanced by 6 hours. Then the birds were returned to the circular cage under natural light conditions, with sunrise at 6:00 A.M. Because of the shift in their circadian rhythms, their endogenous clocks anticipated that the sun would come up at noon. Assuming that birds use the sun as a compass, if the birds expected to find food in the south bin, then at sunrise, they should have looked for food 90 degrees to the right of the direction of the sun. But because their circadian clocks were telling them it was noon, they looked for food in the direction of the sun—in the east bin. The 6-hour shift in their circadian clocks resulted in a 90-

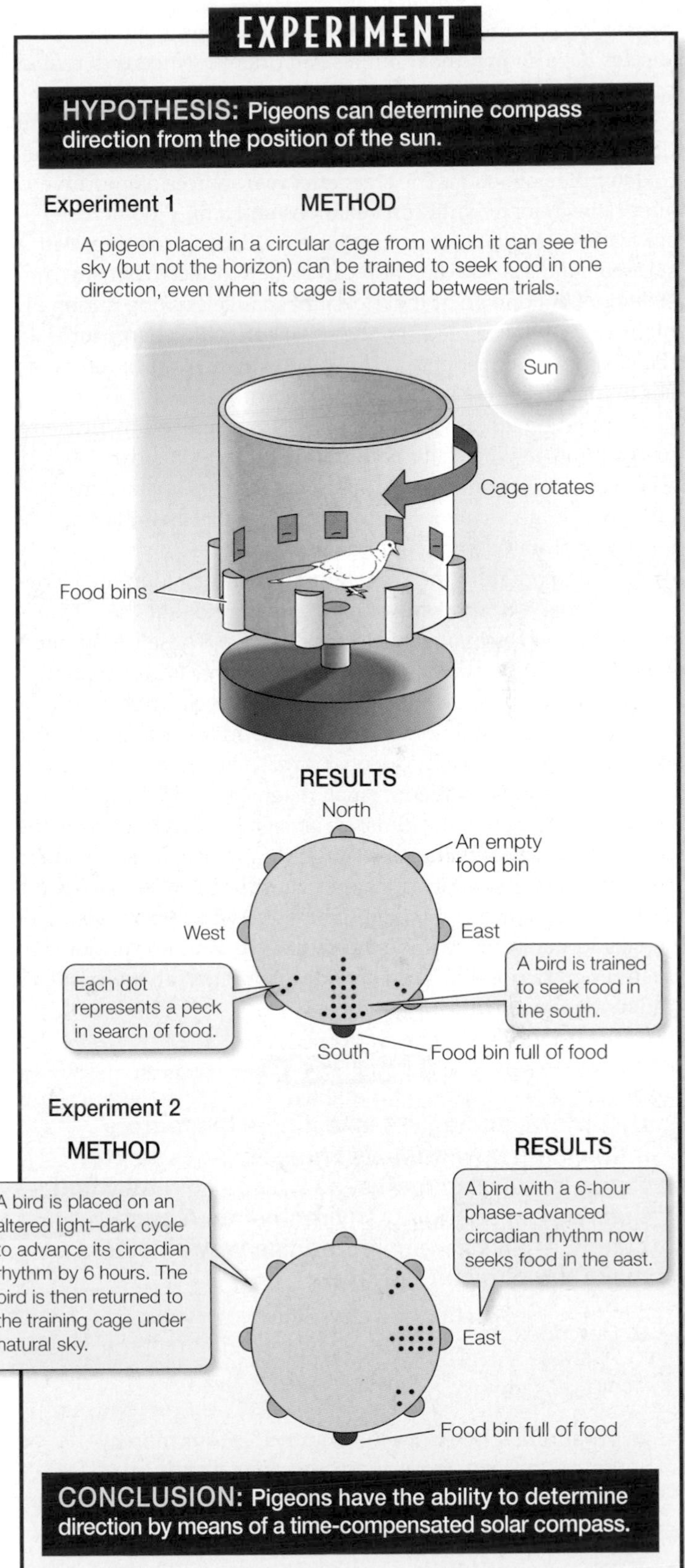

EXPERIMENT
HYPOTHESIS: Pigeons can determine compass direction from the position of the sun.
Experiment 1
METHOD
A pigeon placed in a circular cage from which it can see the sky (but not the horizon) can be trained to seek food in one direction, even when its cage is rotated between trials.
Sun
Cage rotates
Food bins
RESULTS
North
An empty food bin
West
East
Each dot represents a peck in search of food.
A bird is trained to seek food in the south.
South
Food bin full of food
Experiment 2
METHOD
RESULTS
A bird is placed on an altered light–dark cycle to advance its circadian rhythm by 6 hours. The bird is then returned to the training cage under natural sky.
A bird with a 6-hour phase-advanced circadian rhythm now seeks food in the east.
East
Food bin full of food
CONCLUSION: Pigeons have the ability to determine direction by means of a time-compensated solar compass.

53.17 The Time-Compensated Solar Compass Pigeons whose circadian rhythms were phase shifted forward by 6 hours oriented as though the dawn sun was at its noon position. These results show that birds are capable of using their circadian clocks to determine direction from the position of the sun.

degree error in their orientation. Similar experiments on many species have shown that animals can orient by means of a *time-compensated solar compass*.

Many animals are normally active at night; in addition, many diurnal bird species migrate at night and thus cannot use the sun to determine direction. The stars offer two sources of information about direction: moving constellations and a fixed point. The positions of constellations change because Earth is rotating. With a star map and a clock, direction can be determined from any constellation. But one point that does not change position during the night is the point directly over the axis on which Earth turns. In the Northern Hemisphere, a star called Polaris, or the North Star, lies in that position and always indicates north.

Stephen Emlen at Cornell University raised young birds in a planetarium, in which star patterns are projected on the ceiling of a large, domed room. The star patterns in the planetarium could be slowly rotated to simulate the rotation of Earth. If the star patterns were rotated each night as the young birds matured, they were able to orient in the planetarium. Birds reared in the planetarium under a nonmoving sky moved around, but they did not orient themselves in any particular direction. These experiments showed that birds can learn to use star patterns for orientation.

Animals cannot use sun and star compasses when the sky is overcast, yet they still home and migrate under such conditions. Pigeons home as well on overcast days as on clear days, but this ability is severely impaired if small magnets are attached to their heads—evidence that the birds use a magnetic sense, although the neurophysiology of this sense is largely unknown. Although there is as yet no experimental evidence for these possibilities, birds could also use the plane of polarization of light, which can give directional information even under heavy cloud cover, and very low frequencies of sound, which can provide information about coastlines and mountain chains.

53.3 RECAP

Behavioral ecologists investigate the factors influencing how animals choose places to live, foods to eat, and mates, as well as how they find their way in changing environments. A cost–benefit analysis helps determine why animals make the choices they do.

- Describe three kinds of costs an animal may incur by defending a territory. See pp. 1148–1149 and Figure 53.8
- What prediction does the energy maximization hypothesis make about a foraging animal's choice of food types? See pp. 1149–1151 and Figure 53.10
- Can you explain why the ways that males and females choose mating partners are often very different? See p. 1152
- Describe some ways in which circadian and circannual rhythms improve an animal's fitness. See pp. 1152–1154

Many of the behaviors we have discussed in this section are responses to the physical environment, but responses to other animals of the same and different species are also an important aspect of animal behavior. Let's consider how individual animals behave in response to interactions with other animals of their own and different species and how those interactions have influenced the evolution of animal communication.

53.4 How Do Animals Communicate with One Another?

As individuals interact, they convey information; therefore, their behavior can evolve into systems of information exchange, or **communication**. The behaviors of individuals may become elaborated as **displays** or **signals**, however, only if the transmission of information benefits both the sender and the receiver. To understand why these conditions must be met, consider male courtship displays (see Figure 53.13). A courtship display may benefit a male if it improves his attractiveness to females, but a female will not benefit from responding to such displays unless they allow her to assess whether the male is of the right species, and whether he is strong, healthy, and has other attributes that will make him a good mate or father for her offspring. Female red deer, for example, pay attention to the loud vocalizations of competing stags (**Figure 53.18**) because only large and healthy males can sustain them. Weaker males lose these roaring contests.

Animals communicate in various sensory modes. These modes differ not only in how the communication is performed and in the properties of the signals, but also in what can be communicated. Let's look at some examples of different sensory modes that animals use to communicate with one another.

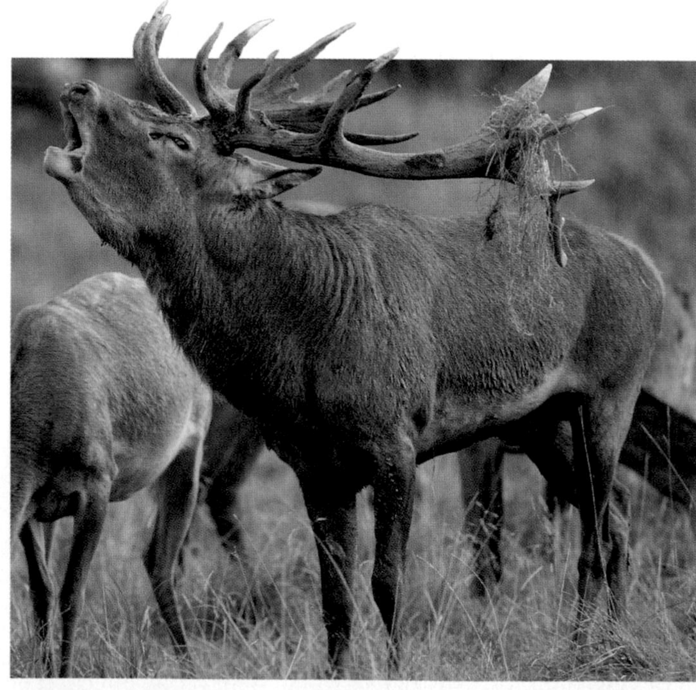

53.18 Strong Males Win Roaring Contests Weak male red deer cannot sustain long roaring bouts, so females can use the results of these contests to judge the quality of potential mates.

Visual signals are rapid and versatile

Visual signals are easy to produce, come in an endless variety, can be changed rapidly, and clearly indicate the position of the signaler. Most animals are sensitive to light and can therefore receive visual signals. However, the extreme directionality of visual signals means that the receiver must be looking directly at the signaler.

Visual communication is not useful at night or in environments that lack light, such as caves and the ocean depths. Some species have surmounted this constraint by evolving their own light-emitting mechanisms. Fireflies, for example, use an enzymatic mechanism to create flashes of light. By emitting flashes in species-specific patterns, fireflies can advertise for mates at night.

Fireflies also illustrate how some species can exploit the communication systems of other species. There are predatory species of fireflies that mimic the mating flashes of females of other species. When an eager suitor approaches the signaling "female," he is eaten. Thus deception can be part of animal communication systems, just as it is part of human use of language.

Chemical signals are durable

Molecules used for chemical communication between individuals of the same species are called **pheromones**. Because of the diversity of their molecular structures, pheromones can communicate very specific, information-rich messages. The mate attraction pheromone of the female Japanese silkworm moth, for example, enables male moths as far as several kilometers downwind to determine that a female of their species is sexually receptive (see Figure 45.3). By orienting to the wind direction and following the concentration gradient of the pheromone, they can find her.

Pheromone messages left in the environment by mammals who mark their territories with urine or other secretions can reveal a great deal of information about the signaler: species, individual identity, reproductive status, size (indicated by the height of the message), and how recently the animal has been in the area (indicated by the strength of the scent).

Pheromones remain in the environment for some time after they are released. In contrast, vocal and visual displays disappear as soon as the animal stops signaling or displaying. The durability of pheromones makes them useful for marking trails (as ants do), marking territories (as many mammals do), or indicating location (as with the moth sex attractant). Their durability, however, makes pheromones unsuitable for a rapid exchange of information.

Auditory signals communicate well over a distance

Compared with visual communication, auditory communication has both advantages and disadvantages. As is implied by the expression "A picture is worth a thousand words," auditory signals cannot convey complex information as rapidly as visual signals can. But auditory signals can be used at night and in dark environments. They can go around objects that would interfere with visual signals, so they can be transmitted in complex environments such as forests. They are often better than visual signals at getting the attention of a receiver because the receiver does not have to be focused on the signaler for the message to be received. Sounds are also useful for communicating over long distances. Even though the intensity of a sound decreases with distance from the source, loud sounds can be used to communicate over distances much greater than those possible with visual signals.

The complex songs of humpback whales, when produced at a depth of about 1,000 meters, can be heard hundreds of kilometers away. In this way, humpback whales locate one another across vast expanses of ocean.

Tactile signals can communicate complex messages

Communication by touch is common, although not always obvious. Animals in close contact use tactile interactions extensively, especially under conditions in which visual communication is difficult. The best-studied use of tactile communication is the dance of honeybees, first described by Karl von Frisch. When a forager bee finds food, she returns to the hive and communicates her discovery to her hivemates by dancing in the dark on the vertical surface of the honeycomb. Other bees follow and touch the dancer to interpret the message.

If the food source the forager has found is more than about 80 meters away from the hive, she performs a *waggle dance* (**Figure 53.19**), which conveys information about both the distance and the direction of the food source. The forager bee repeatedly traces out a figure-eight pattern as she runs on the honeycomb. She alternates half-circles to the left and right with vigorous wagging of her abdomen in the short, straight run between turns. The angle of the straight run indicates the direction of the food source relative to the direction of the sun. The speed of the dancing indicates the distance to the food source: the farther away it is, the slower the waggle run.

If the food she has found is less than 80 meters from the hive, the forager performs a *round dance*, running rapidly in a circle and reversing her direction after each circumference. The odor on her body and the round dance combine tactile and chemical cues: the odor indicates the flower to be looked for, and the dance communicates the fact that the food source is within 80 meters of the hive.

Electric signals can convey messages in murky water

Some species of fish generate electric fields in the water around them by emitting a series of electric pulses. These pulse trains can be used both for sensing objects in the immediate surroundings and for communication.

For example, each individual glass knife fish emits pulses at a different frequency, and the frequency each fish uses relates to its status in the population. Males emit lower frequencies than females; the most dominant male has the lowest frequency, and the most dominant female has the highest frequency. When a new individual is introduced into the group, the other individuals adjust their frequencies so that they do not overlap with the newcomer's, and the newcomer's signal indicates its position in the hierarchy. In their natural environment—the murky waters of tropical evergreen forests—these fish can determine the identity, sex, and social position of another fish by its electric signals. They can also locate prey by the distortions in the electric field they cause.

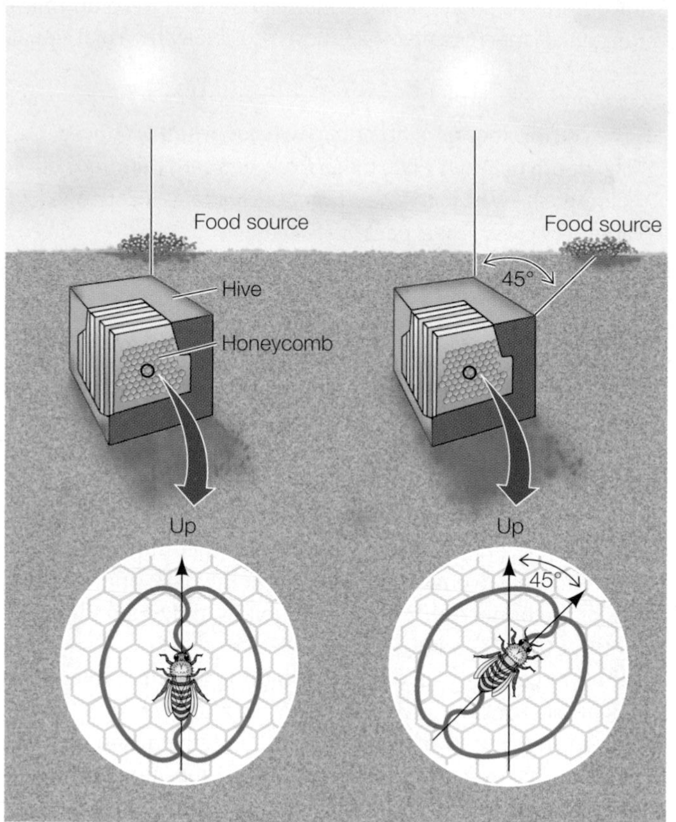

53.19 The Waggle Dance of the Honeybee (A) A honeybee runs straight up on the vertical surface of the honeycomb in the dark hive while wagging her abdomen to tell her hivemates that there is a food source in the direction of the sun. (B) When her waggle runs are at an angle from the vertical, the other bees know that the same angle separates the direction of the food source from the direction of the sun.

53.4 RECAP

A communication system can evolve only if both the sender and receiver of a signal benefit from the information provided by a signal.

- What are the advantages and disadvantages of chemical and visual signals with respect to information transfer? See p. 1157
- Do you understand how honeybee "dances" communicate the location of a good food source? See p. 1157 and Figure 53.19

The use of communication signals allows animals to function not only individually but as part of a pair or larger group. Many animals must associate with each other for mating, but some form groups that remain together for long time periods, eventually evolving into complex societies. Why do animals form such associations?

53.5 Why Do Animal Societies Evolve?

Social behavior evolves when, by cooperating, conspecific individuals can achieve, on average, higher rates of survival and reproduction than they would if they lived alone. The elaborate colonies of ants, bees, and wasps and the social groups of lions and primates are examples of complex social systems. How did these complex animal societies evolve? Because behavior leaves few traces in the fossil record, biologists must infer possible routes for the evolution of social systems by studying current patterns of social organization. Fortunately, many degrees of social system complexity exist among living species; the simpler systems suggest stages through which the more complex ones may have passed.

As we describe a few animal social systems, keep in mind that social systems are dynamic; individuals repeatedly communicate with one another and adjust their relationships. Their relationships change regularly because the costs and benefits experienced by individuals in a social system change with their age, sex, physiological condition, and status.

Also keep in mind that social systems are best understood not by asking how they benefit the species as a whole, but by asking how the individuals that join together benefit from doing so. All individuals must benefit to some degree from being a member of the group; otherwise they would leave it. The option of solitary living may confer few benefits, but it is likely to be better than living in a social group that offers no benefits or imposes high costs.

Group living confers benefits but also imposes costs

Living in groups may confer many types of benefits. It may improve hunting success or expand the range of food that can be captured. For example, by hunting in groups, our ancestors were able to kill large mammals they could not have subdued as individual hunters. These social humans could also defend their prey and themselves from other carnivores, and they could tell one another about the locations of food and enemies.

Many small birds forage in flocks. To test whether flocking provides protection against predators, an investigator released a trained goshawk near wood pigeons in England. The hawk was most successful when it attacked solitary pigeons. Its success in capturing a pigeon decreased as the number of pigeons in the flock increased (**Figure 53.20**). The larger the flock of pigeons, the sooner some individual in the flock spotted the hawk and flew away. This escape behavior stimulated other individuals in the flock to take flight as well.

Social behavior, however, has many costs as well as benefits. The pigeons in a flock interfere with one another's ability to find seeds. Social individuals may inhibit one another's attempts to reproduce or injure one another's offspring. An almost universal cost associated with group living is higher exposure to diseases and parasites. Long before the causes of human diseases were known, people knew that association with sick persons increased their chances of getting sick. Quarantine has been used to combat the spread of disease throughout recorded history. The diseases of wild animals are not well understood, but many of them are probably spread by close contact as well.

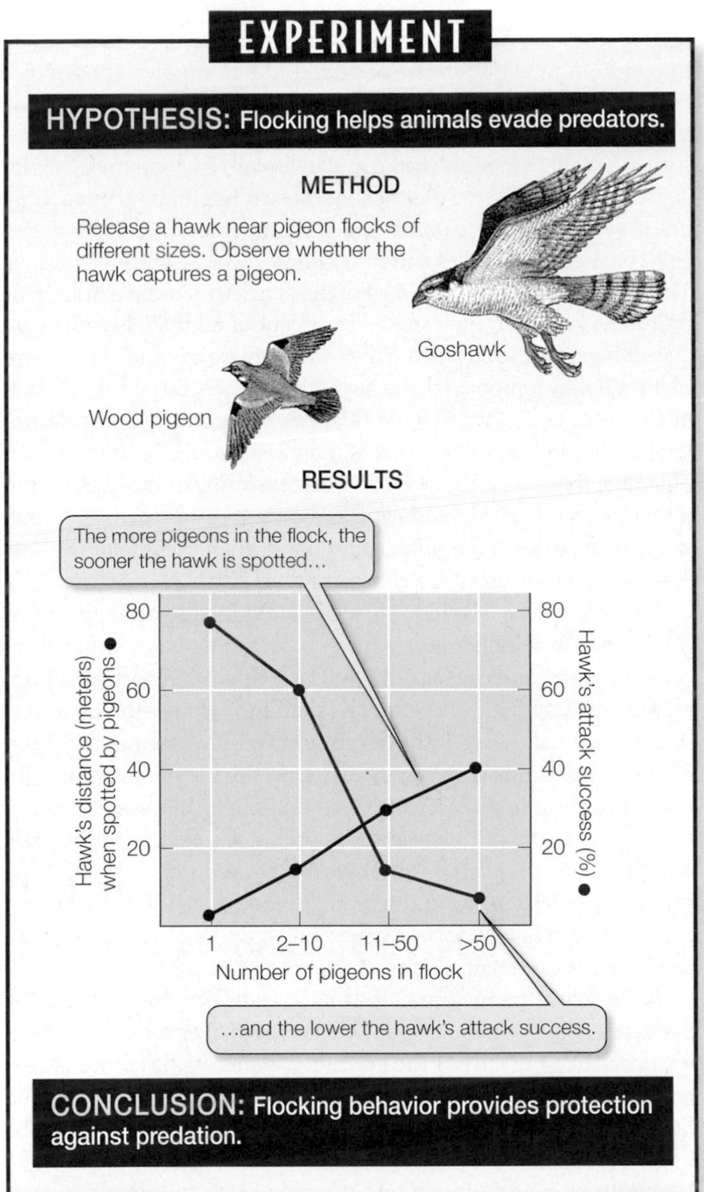

53.20 Groups Provide Protection from Predators The larger a flock of pigeons, the greater the distance at which they detect an approaching hawk, and the less likely the hawk is to succeed in capturing a pigeon.

Parental care can evolve into more complex social systems

The most widespread form of social system is an association of one or two parents with their immature, dependent offspring. Some family units include more adult individuals, typically older offspring that remain with their parents and help care for their younger siblings. Florida scrub jays, for example, live year-round on territories, each of which contains a breeding pair and up to six helpers that bring food to the nest. Nearly all helpers are offspring from the previous breeding season that remain with their parents.

Many mammalian species have also evolved social systems based on extended families. In simple mammalian social systems, solitary females or male–female pairs care for their young. As the period of parental care increases, older offspring may still be present when the next generation is born. They often help rear their younger siblings. In most social mammal species, female offspring remain in the group in which they were born, but males tend to leave, or are driven out. Therefore, among mammals, most helpers are females. In some cases, these related females remain a unit even after they begin having their own offspring. For example, the females in a pride of lions are related, and they care for each other's offspring. Males from the outside compete to take over prides and mate with the females.

Caring for young involves tremendous costs for parents and helpers. Animals that provide food for their young may sacrifice food for themselves, and protecting its young may put an animal in danger. It is easy to understand how natural selection favors parental care in spite of such costs when the benefit is the survival of the animal's own offspring. But natural selection can favor **altruistic** acts—behaviors that reduce the helper's reproductive chances, but increase the fitness of the helped individual. Why?

Altruism can evolve by contributing to an animal's inclusive fitness

Helping behaviors exhibited by parents toward their offspring are easily understood in terms of the close genetic relatedness between them. An animal's offspring contribute to its **individual fitness**. Genetic relatedness extends beyond the parent–offspring relationship, however. In diploid organisms, two offspring of the same parents share, on average, 50 percent of the same alleles, and an individual is likely to share 25 percent of its alleles with its sibling's offspring. Therefore, by helping its relatives, an individual can increase the representation of some of its own alleles in the population. Together, individual fitness plus the fitness gained by increasing the reproductive success of nondescendant kin determine the **inclusive fitness** of an individual. Occasional altruistic acts may eventually evolve into altruistic behavior patterns if their benefits (in terms of inclusive fitness) exceed their costs (in terms of decreases in the altruist's own reproductive success).

Many social groups consist of some individuals that are close relatives and others that are unrelated or distantly related. Individuals of some species recognize their relatives and adjust their behavior accordingly. White-fronted bee-eaters are African birds that nest colonially. Most breeding pairs are assisted by nonbreeding adults that help incubate eggs and feed nestlings. Nearly all of these helpers assist close relatives (**Figure 53.21**). When helpers have a choice of two nests at which to help, about 95 percent of the time they choose the nest with the young more closely related to them. Individual bee-eaters improve their inclusive fitness by helping because nests with helpers produce more fledglings than do nests without helpers. Helpers also increase their chances of inheriting the nesting territory when the owners die.

Species whose social groups include sterile individuals are said to be **eusocial**. This extreme form of social behavior has evolved in termites and many hymenopterans (ants, bees, and wasps). In these species, most females are nonreproductive *workers* that forage for the colony or defend it against predators. Workers may include *soldiers* with large, specialized defensive weapons (**Fig-**

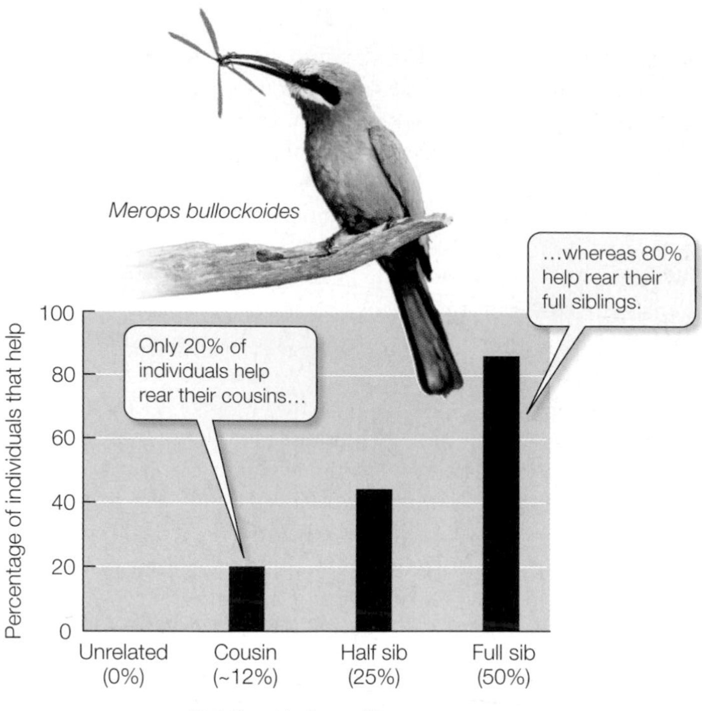

53.21 White-Fronted Bee-Eaters Are Discriminating Altruists Bee-eaters preferentially help close relatives rear their nestlings. The relationship axis gives the approximate percentage of alleles that would be identical between two relatives. Because some individuals help out more than one nest, the percentages add to more than 100 percent.

ure 53.22). They may be killed while defending the colony. Only a few females, known as *queens*, are fertile; they produce all the offspring of the colony.

Both genetic and environmental factors may facilitate the evolution of eusociality. The British evolutionary biologist W. D. Hamilton first suggested that eusociality evolved among the Hymenoptera because its members have an unusual sex determination system in which males are haploid but females are diploid. Among the Hymenoptera, a fertilized (diploid) egg hatches into a female; an unfertilized (haploid) egg hatches into a male.

If a female hymenopteran copulates with only one male, all the sperm she receives are identical because a haploid male has only one set of chromosomes, all of which are transmitted to every sperm cell. Therefore, a female's daughters share all of their father's genes. They also share, on average, half of the genes they receive from their mother. As a result, they share 75 percent of their alleles, on average, rather than the 50 percent they would share if both parents were diploid. If they reproduced, they would share only 50 percent of their alleles with their own female offspring. Since workers are more genetically similar to their sisters than they would be to their own offspring, they can potentially increase their fitness more by caring for their sisters than by producing and caring for their own offspring. If, however, a queen copulates with more than one male, workers may not be so closely related to all of their sisters.

Eusociality may also be favored if it is difficult or dangerous to establish new colonies. Nearly all eusocial animals construct elaborate nests or burrow systems within which their offspring are reared (see Figure 53.24 below). Naked mole-rats—the most eusocial mammals—live in underground colonies containing 70 to 80 individuals. Sterile workers maintain the colony's tunnel systems. Breeding is restricted to a single female and several males that live in a nest chamber in the center of the colony. Individuals attempting to found new colonies are at high risk of being captured by predators, and most founding events fail. These circumstances, which favor cooperation among founding individuals, may facilitate the evolution of eusociality.

Inbreeding—the mating of individuals who are closely related—increases genetic relatedness within a group. Even if two parents are unrelated, but each is the product of generations of intense inbreeding, all of their offspring may be genetically nearly identical. Such offspring would increase their fitness by helping to rear siblings. Genetic similarity generated by inbreeding could explain the evolution of eusociality among the many hymenopteran species in which queens mate with many males, and among termites and naked mole-rats, in which both sexes are diploid.

Eciton hamatum

53.22 Sterile Workers are Extreme Altruists Eusocial insect species contain classes of sterile worker individuals. The army ant soldier from Costa Rica protects her colonymates with her large, powerful jaws.

53.5 RECAP

Social behavior evolves when the benefits of group living outweigh its costs. Most complex social systems evolved from extended families.

- How does genetic relatedness among group members influence the evolution of altruistic behavior? See p. 1159 and Figure 53.21
- Why is eusociality particularly common in hymenopterans? See p. 1160

Animals with complex social systems often achieve remarkably dense populations. Why? In the next section we'll see how the ways in which animals interact with members of their own and other species influence what habitats and foods a species uses in nature, which species live together, and how abundant they are.

53.6 How Does Behavior Influence Populations and Communities?

Individuals use the behaviors we have described in this chapter to perform the activities they need to grow, reproduce, and avoid becoming food for other organisms. These behaviors have evolved primarily because they have provided benefits to the individuals that performed them. In addition, however, the ways in which animals make decisions about habitats, food, and associates influence the structure and functioning of populations, communities, and ecosystems.

Habitat and food selection influence the distributions of organisms

Although animals settle preferentially in good habitats, individuals that are already there may prevent them from using those habitats. If population densities are high, many individuals of a species may be forced to settle in poor habitats. And if prior residents do not exclude newcomers, their activities may lower the quality of the best habitats.

Foraging behavior can influence population dynamics in many ways. Some animals buffer fluctuations in food availability by storing food. The energy resources gained from food can be stored within an organism's body; animals typically store them as fat, and plants typically store them as starch. In cold and arid environments, where food does not decompose quickly, animals may store large amounts of food outside their bodies (**Figure 53.23**). Some food types, notably seeds, spores, wood, fungi, leaves, and nectar, are more readily stored than others, either because they already have a low moisture content or because they can be dried prior to storage. Storing food enables animals to live in environments where food would otherwise be unavailable seasonally.

How animals choose their food may influence the species composition and abundance of their prey. For example, fish influence the composition of communities of small planktonic animals in lakes by selectively capturing and eating the largest of these animals (as we saw in Figure 53.10). If predatory fish are absent, large plankton outcompete smaller plankton; consequently, large planktonic species dominate fishless lakes, but small planktonic species dominate lakes with plankton-eating fish.

Territoriality influences community structure

Individuals of most territorial species defend their territories only against other individuals of the same species, but some animals benefit by defending their territories against individuals of other species as well. Typically the winner of interspecific territorial contests is the species with the larger body size. Because individuals of larger species need to harvest energy at a higher rate than those of smaller species, the individuals of the dominant species may be able to survive and reproduce only in high-productivity habitats. If the dominant species prevents subordinate species from living in those habitats, it follows that subordinate species would occupy those habitats if the dominant individuals were removed. Removal of the subordinate species, however, should have little effect on the distribution of the dominant species.

Melanerpes formicivorus

53.23 Storing Energy for a Poor Season Acorn woodpeckers drill holes in dead trees in which they hoard acorns to sustain them over long dry or cold seasons. The birds may form social nesting groups, and members of the group will defend their cache tree against potential "robbers."

Stuart Pimm and Michael Rosenzweig tested this hypothesis with hummingbirds in southern Arizona. They created artificial "flower patches" by setting up an array of feeders. Some feeders contained artificial nectar rich in sucrose. Others contained a more dilute sucrose solution. Male blue-throated hummingbirds, which weigh an average of 8.3 grams, dominated male black-chinned hummingbirds, which weigh only 3.2 grams. When male blue-throats were absent, black-chin males fed almost exclusively at the rich feeders, but when male blue-throats were present, male black-chins fed extensively at the poorer-quality feeders. They were able to maintain their weight on the poorer-quality feeders, whereas male blue-throats could not.

Social animals may achieve great population densities

The population densities achieved by some social animals are impressive. For example, up to 94 percent of the individuals and 86 percent of the biomass of arthropods in the canopies of tropical evergreen forests are eusocial ants. Termites (also eusocial insects) are the primary consumers of plant tissues in the savannas of Africa and Australia. They live in and build large mounds, within which many other species of animals live (**Figure 53.24**). In parts of Australia, termite density may reach 1,000 colonies per hectare.

Ants and termites have achieved these remarkable abundances in part because their social organization allows them to exploit the services of other organisms in harvesting vital resources. The most abundant and productive ants and termites actively cultivate fungi that break down difficult-to-digest plant tissues, including wood. Some ants protect aphids and other insects that tap phloem fluids

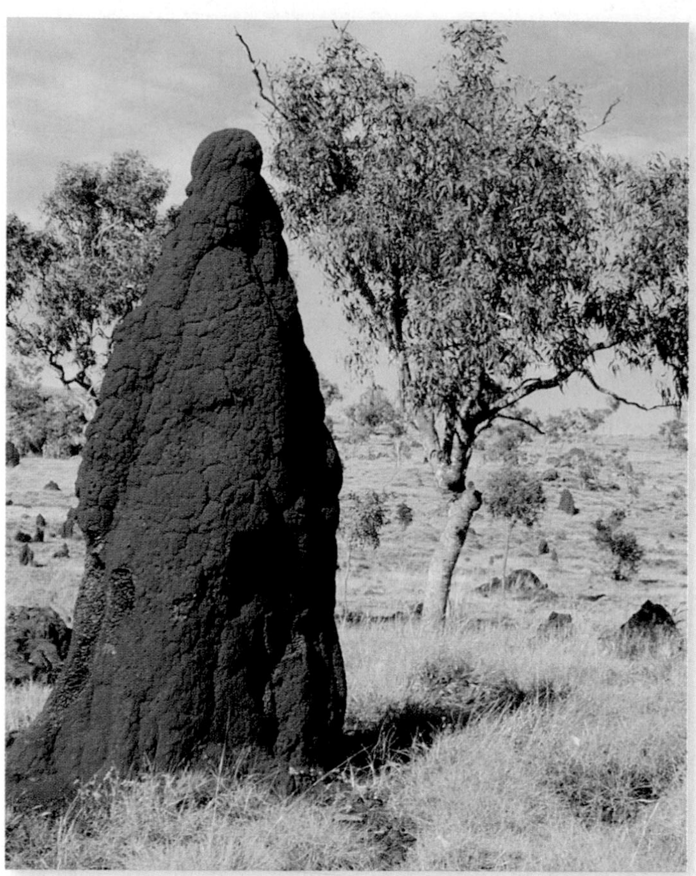

53.24 Termite Mounds Are Large and Complex These immense Australian termite mounds are constructed over many years by millions of worker termites. Elaborate nests or burrows, which are very costly to construct and maintain, characterize nearly all eusocial animals.

of plants. Phloem is rich in carbohydrates but poor in proteins, so phloem suckers ingest more carbohydrates than they can use. They eject the excess in the form of sugar-rich anal drops (see Figure 35.13), which the ants eat. Because the ants can easily obtain enough carbohydrates from the anal drops, they need to get only proteins in other ways, such as by eating other insects. Moreover, with their high metabolic rates fueled by ready access to carbohydrates, the ants can expend the energy needed to drive other predatory insects away from their food sources. Ants strongly influence the community of insects in tropical forest canopies.

The population densities achieved by our own species serve as striking reminders of the consequences of social behavior (**Figure 53.25**). Social living enabled members of human groups to specialize in different activities. Among the benefits of specialization were the domestication of plants and animals and the cultivation of land. These innovations enabled our ancestors to increase the resources at their disposal dramatically. Those increases, in turn, stimulated rapid population growth up to the limit determined by the agricultural productivity that was possible with human- and animal-powered tools. Agricultural machines and artificial fertilizers, made possible by the tapping of fossil fuels, greatly increased agricultural productivity and removed that earlier limit. In addition, the development of modern medicine reduced mortality rates in human populations. Medicine and better hygiene have also allowed people to live in large numbers in areas where diseases formerly kept numbers very low. However, these successes are accompanied by many social and environmental problems, some of which we will discuss in subsequent chapters.

53.6 RECAP

The ways in which individuals choose habitats and foods influence interactions both within and among species. Social animals can achieve great population densities and can therefore be a major influence on communities and ecosystems.

- How does foraging behavior enable animals to live in environments with fluctuating food supplies? See p. 1161
- Can you explain why highly social animals often achieve high population densities? See pp. 1161–1162

53.25 Social Organization Allows Humans to Live at High Densities Densely populated human cities, such as Benidorm on Spain's Costa Blanca, serve as examples of how social organization allows our species to achieve and sustain extreme population densities.

Observations of the behavior of animals show us that the ways in which animals choose what to eat, where to seek food, and with whom to associate influence the sizes and distributions of populations of many species and how they interact in nature. These aspects of populations will form the focus of the next chapter.

CHAPTER SUMMARY

53.1 What questions do biologists ask about behavior?

Scientists who study behavior describe what they observe and then try to answer either proximate or ultimate questions about the behavior.

53.2 How do genes and environment interact to shape behavior?

Animals perform many stereotypic and species-specific behaviors without prior experience.

In a deprivation experiment, an animal is deprived of all experience relevant to the behavior under study so that its genetic component can be assessed.

In a genetic experiment, investigators are able to compare the behaviors of individuals that differ in only one or a few known genes. Review Figure 53.2

Inherited behaviors are often triggered by simple stimuli called **releasers**. Such behavior is adaptive when opportunities to learn are lacking and when mistakes are costly or even lethal.

Imprinting is a type of learning that takes place during a **critical period** in an animal's development.

Learning abilities require proximate mechanisms whose construction requires genetic information. Review Figure 53.5

Hormones can influence the development and expression of behavior patterns. Review Figure 53.6

53.3 How do behavioral responses to the environment influence fitness?

The cues most organisms use to select suitable **habitats** are good predictors for their future survival in those habitats.

Animals may establish and defend a **territory**, giving themselves exclusive use of that space.

A **cost–benefit approach** analyzes the total cost of any particular behavior in terms of **energetic**, **risk**, and **opportunity costs**. Review Figure 53.8, Web/CD Tutorial 53.1

Foraging theory helps behavioral ecologists understand the survival value of animals' food choices. When choosing food, animals often function as energy maximizers. Review Figure 53.10, Web/CD Tutorial 53.2

Individuals of all sexually reproducing species choose mating partners based on the inherent qualities of a potential mate, on the resources it controls, or on a combination of the two.

Circadian rhythms enable animals to anticipate daily changes in the environment. Photoperiods and circannual rhythms enable them to anticipate seasonal changes. Review Figure 53.14, Web/CD Tutorial 53.3

Piloting involves navigating by landmarks; **homing** is the ability to return home regardless of distance. **Migration** often requires long-distance navigation.

Humans can use both **distance-and-direction** and **bicoordinate navigation**. It is not known whether animals can use bicoordinate navigational techniques, but they can use the sun and stars to determine direction. Review Figure 53.17, Web/CD Tutorial 53.4

53.4 How do animals communicate with one another?

Interactions between individuals can evolve into **communication systems** if the transmission of information by **displays** or **signals** benefits both signaler and receiver.

Animals communicate in different sensory modes, which differ in their properties and in what can be communicated by using them.

Visual signals can convey complex information rapidly. Chemical signals such as **pheromones** can convey complex information and last a long time. Auditory signals can convey complex information over long distances in complex habitats. Tactile and electric signals can convey complex information rapidly where other kinds of signals are not possible. See Web/CD Activity 53.1

53.5 Why do animal societies evolve?

Group living inevitably imposes both costs and benefits. Social behavior evolves when cooperation among conspecifics results in higher rates of survival and reproduction than solitary individuals can achieve. Review Figure 53.20

The simplest social systems consist of one or two parents with their immature, dependent offspring. Other social systems include members of extended families.

An animal's own reproductive success contributes to its **individual fitness**. The reproductive success of its close relatives, with whom it shares alleles, contributes to its **inclusive fitness**. An animal that performs altruistic behavior reduces its own reproductive success but may gain inclusive fitness by increasing the fitness of its relatives. Review Figure 53.21

Eusocial species live in groups that include sterile individuals that help close relatives reproduce.

53.6 How does behavior influence populations and communities?

The ways in which animals select habitats and food and interact with individuals of their own and other species can influence the range of habitats and foods a species uses in nature, which species live together, and how abundant they are.

Animals with complex social systems often achieve remarkably high population densities.

See Web/CD Activity 53.2 for a concept review of this chapter.

SELF-QUIZ

1. In a deprivation experiment, young animals are reared
   a. with only young, naive individuals of their own species.
   b. with conspecifics of all ages, but in a limited environment.
   c. so that they have no experience relevant to the behavior being studied.
   d. in an environment that lacks the conditions that would stimulate them to perform the behavior.
   e. under conditions that deprive them of food.
2. Which of the following is *not* a component of the cost of performing a behavior?
   a. Its energetic cost
   b. The risk of being injured
   c. Its opportunity cost
   d. The risk of being attacked by a predator
   e. Its information cost
3. Birds that migrate at night
   a. inherit a star map.
   b. determine direction by knowing the time and the position in the sky of a constellation.
   c. orient to a particular point in the sky.
   d. imprint on one or more key constellations.
   e. determine distance, but not direction, from the stars.
4. If a bird is trained to seek food on the western side of a cage open to the sky, its circadian rhythm is then delayed by 6 hours, and it is returned to the open cage at noon in real time, it will seek food in the
   a. north.
   b. south.
   c. east.
   d. west.
5. To be able to pilot, an animal must
   a. have a time-compensated solar compass.
   b. orient to a fixed point in the night sky.
   c. know the distance between two points.
   d. know landmarks.
   e. know its longitude and latitude.
6. Which of the following statements about communication is true?
   a. Complex information can be conveyed most rapidly by pheromones.
   b. Visual signaling is advantageous in complex environments.
   c. Auditory communication always reveals the location of the signaler.
   d. An advantage of pheromones is that the message can persist through time.
   e. The dance of bees is an example of visual signaling.
7. One cost commonly associated with group living is
   a. increased risk of predation.
   b. interference with foraging.
   c. higher exposure to diseases and parasites.
   d. poorer access to mates.
   e. All of the above
8. The choice of a mating partner may be based on
   a. the inherent qualities of a potential mate.
   b. the resources held by a potential mate.
   c. both the inherent qualities of a potential mate and the resources it holds.
   d. the courtship display of a potential mate.
   e. All of the above
9. Altruistic behavior
   a. confers a benefit on the performer by inflicting some cost on some other individual.
   b. confers a benefit both on the performer and on some other individual.
   c. inflicts a cost both on the performer and on some other individual.
   d. confers a benefit on an individual other than the performer's offspring at some cost to the performer.
   e. inflicts a cost on the performer without benefiting any other individual.
10. An animal is said to be *eusocial* if
   a. group members interact very intensively.
   b. some group members produce many more offspring than others do.
   c. a dominance hierarchy exists among group members.
   d. young individuals remain in the group to help their parents rear other offspring.
   e. the social group contains sterile individuals.

FOR DISCUSSION

1. Cowbirds are nest parasites. A female cowbird lays her eggs in the nest of another bird species, which then incubates the eggs and raises the young. What do you think would characterize the acquisition of song in cowbirds? In a given area, cowbirds tend to parasitize the nests of particular bird species. How do you think female cowbirds learn this behavior? How would you test your hypothesis?
2. Male dogs lift a hind leg when they urinate; female dogs squat. If a male puppy receives an injection of estrogen when it is a newborn, it will never lift its leg to urinate for the rest of its life; it will always squat. How might this result be explained?
3. The short-tailed shearwater is a bird that winters in Antarctica and summers in the Arctic. What problems would this species have in using either the sun or the stars for navigation? What is the most likely means it uses to find its way to its summer and its winter feeding grounds?
4. Most hawks are solitary hunters. Swallows often hunt in groups. What are some plausible explanations for this difference? How could you test your ideas?
5. Among birds, males of species that mate with many females and perform communal courtship displays are usually much larger and more brightly colored than females, whereas among species that form monogamous pairs, males are usually similar to females in size, whether or not they are more brightly colored. What hypotheses can be advanced to explain this difference?
6. Many animals defend territories, but the sizes of the territories they defend and the resources those areas provide vary enormously. Why don't all animals defend the same type and size of territory?
7. Among vertebrates, helpers are individuals capable of reproducing, and most of them later breed on their own. Among eusocial insects, sterile castes have evolved repeatedly. What differences between vertebrates and insects might explain the failure of sterile castes to evolve in the former?

FOR INVESTIGATION

Experiments on animal migration show that animals can use a time-compensated solar compass or identification of a fixed point in the night sky as a basis for distance and direction navigation. However, the observations on the homing ability of seabirds such as the albatross indicate that animals might be capable of true or bicoordinate navigation. What experiments could you do to prove that animals have such abilities, and what experiments could you do to test hypotheses about the mechanisms they might use? At least two hypotheses might involve using the elevation of the sun or the angles of Earth's magnetic lines of force as a means of determining latitude.

CHAPTER 54 Population Ecology

Exotic moth controls exotic cactus

In the early 1830s, someone thought the South American prickly-pear cactus *Opuntia stricta* would make a nice ornamental ground cover in the semiarid climate of Australia. The cactus reacted well to its new home—by 1925 it had covered more than 25 million hectares of eastern Australia's valuable sheep-grazing land.

In 1926, another *exotic* (non-native) species was introduced in Australia, this time to control the burgeoning *Opuntia* population. The caterpillar larvae of the moth *Cactoblastis cactorum* feed on the cactus. About 2 million eggs of this moth, whose native origins are also in South America, were imported and dispersed among the Australian *Opuntia.* The strategy was extremely successful, and today the total numbers of both the cactus and the moth in Australia are constant and fairly low, although there are many local population *oscillations.*

A female *Cactoblastis* moth lays her eggs anchored to a cactus spine. Hundreds of caterpillars hatch and feed communally inside the leathery cactus pads, which the caterpillars reduce to a gooey green mess. *Cactoblastis* caterpillars can completely destroy *Opuntia* patches, but new patches arise in other places from seeds dispersed by birds. The new patches flourish until a female *Cactoblastis* finds them and lays large numbers of eggs on them. In Australia, both *Opuntia* and *Cactoblastis* are now distributed as scattered subpopulations among which individuals occasionally disperse.

Various *Opuntia* speces native to North and South America have been introduced to Europe and Africa, as well as to Hawaii and many Caribbean islands. In some of these places, opuntias have become invasive and *C. cactorum* has been introduced to control them. The moth was introduced onto the Caribbean island of Nevis, where it successfully controlled an invasive exotic *Opuntia* species, but also destroyed some native species of *Opuntia*. The moth reached Florida in 1989. Today it is also found in the southwestern United States, where it has become a pest on the native opuntias, many of which are threatened species. In Mexico, where opuntias are economically important producers of fruit, fodder, dyes, and medicinals, the moth is a serious pest.

Why has *Cactoblastis* been so successful in Australia, yet caused so many problems in North America? The answer is probably that

A Pest in Australia Humans introduced *Opuntia* into Australia, where there are no native cactus species. The exotic cactus quickly overran valuable grazing land, but was brought under control by importing eggs of the moth *Cactoblastis cactorum*, whose caterpillars feed voraciously on the fleshy cactus pads.

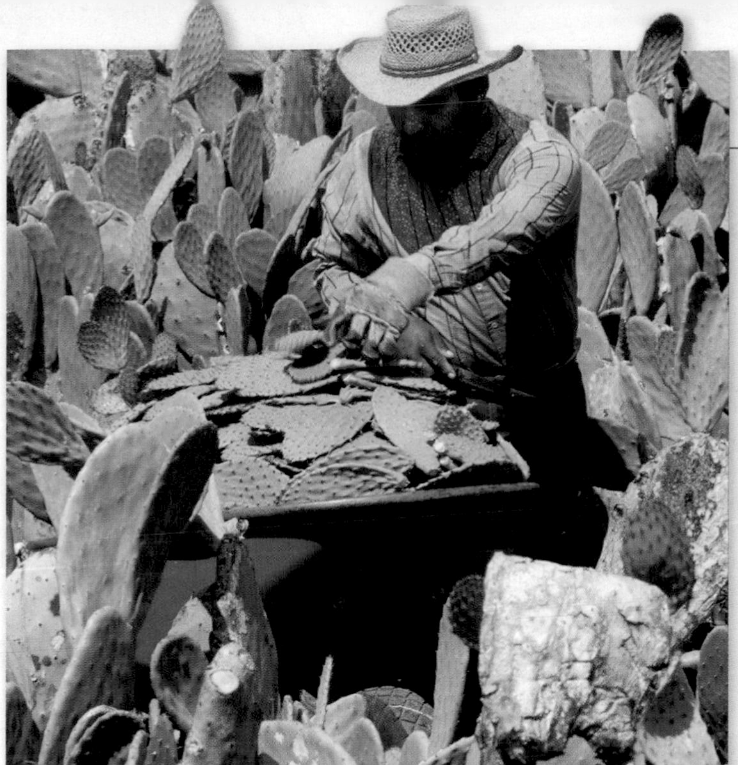

A Valuable Crop in Mexico In Mexico, many native species of *Opuntia* are economically valuable farm crops. The presence of the accidentally introduced *Cactoblastis* moth threatens some of the native cacti.

Australia has no native cactus species, and so far the moth has not attacked any other type of plant. In areas where there are other native species of *Opuntia*, the exotic moth is destructive.

Species introduced to regions beyond their original ranges often achieve much greater population densities than they had in their native ranges. This may happen because in the new region, no predators exist to attack them—which is why we often introduce predators and parasites in an attempt to control an invasive species. As we will see in this chapter, surprises often accompany our efforts to manage populations.

IN THIS CHAPTER we first explore how ecologists study populations and then describe some of the relationships between the life history traits of organisms and the dynamics of their populations. Next we will discuss the factors that influence population densities and explore how environmental variation over space affects population dynamics. Finally, we will show how ecological knowledge is used to manage populations that are of special interest or importance to humans.

CHAPTER OUTLINE

54.1 How Do Ecologists Study Populations?

Before we consider how ecologists study populations, we need to know how they define populations. A **population** consists of the individuals of a species within a given area. At any given moment, an individual organism occupies only one spot in space, and it is of a particular age and size. The members of a population, however, are distributed over space, and they differ in age and size. The age distribution of individuals in a population and the way those individuals are spread over the environment describe its **population structure**. Ecologists study population structure because the spatial distributions of individuals and their ages influence the stability of a population and affect how that population interacts with other species.

The number of individuals of a population per unit of area (or volume) is its **population density**. Population densities exert strong influences on the ways in which members of a population interact with one another and with populations of other species. Scientists working in agriculture, conservation, and medicine typically try to maintain or increase population densities of some species (crop plants, game animals, aesthetically attractive species, threatened or endangered species) and reduce the densities of others (agricultural pests, pathogens). To manage populations, we need to know what factors cause populations to grow or decrease in density and how those factors work.

The structure of a population changes continually because **demographic events**—births, deaths, immigration (movement of individuals into the area), and emigration (movement of individuals out of the area)—are common occurrences. Knowledge of when individuals are born and when they die provides a surprising amount of information about a population. The study of the birth, death, and movement rates that create *population dynamics* (changes in population structure and density) is known as *demography*.

Ecologists use a variety of devices to track individuals

In order to study populations, ecologists determine how many individuals are found in an area and where they are located. They also measure the rates at which individual are born, die, and move into and out of an area. How do they make those measurements?

Chapter 53 describes various behaviors that affect animal population dynamics. For example, sometimes individual animals increase their inclusive fitness by helping to care for close relatives. Individuals also change their locations by migrating or dispersing. To recognize and study such events, investigators need to be able to recognize and track individual animals. Although an experienced investigator may be able to distinguish individuals by their appearance—elephants by their ears or orcas by differences in the pattern of white blotches on their bodies (**Figure 54.1A**)—differences among individuals are often too minor to detect in the field. Most field studies of animal populations require tagging or marking individuals in some way.

No single form of marking can be used on all species. Birds are typically marked by colored bands on their legs (**Figure 54.1B**), butterflies by placing colored spots on their wings, bees by placing numbered tags on their bodies, and mammals by tags or dyeing their fur. Plants are marked by tags tied to their branches or, because they do not move, by stakes in the ground nearby.

Until recently, most field techniques for tracking animals depended on the ability of researchers to see them. Today's tracking devices are so sophisticated that they can record and transmit information not only about an animal's location, but also about its physiology (by measuring its heartbeat), feeding behavior (its stomach temperature), and social behavior (its vocalizations) many times per second. Microchips and other forms of electronic tagging are used on organisms of all sizes (**Figure 54.1C**). The latest devices can also record information about the animal's environment.

Molecular markers can be used to determine the movement of individuals over long distances. American redstarts (*Setophaga ruticilla*) breed in the deciduous forests of eastern North America. Following the energetically demanding breeding season, these small songbirds migrate south to the Caribbean and Central America for the winter. It is easy to determine where redstarts breed, but how might we determine where they molt? One way is to analyze the chemical composition of the feathers that the birds molt as they migrate south. Scientists can determine where they molted by evaluating hydrogen isotopes in feathers because there is a strong latitudinal gradient in stable hydrogen isotopes in precipitation. These hydrogen isotopes are incorporated into the tissues of plants, the animals that eat the plants (such as herbivorous insects), and eventually, the animals, such as redstarts, that eat the herbivores. Because feathers are metabolically inert after they have formed, the hydrogen isotopes in redstart feathers indicate the latitude at which the individuals grew those feathers (**Figure 54.2**). Most individuals molted their feathers close to the breeding ground.

Population densities can be estimated from samples

Because organisms and their environments differ, population densities must be measured in more than one way. Ecologists usually measure the densities of organisms in terrestrial environments as the number of individuals per unit of area. The number per unit of volume is generally a more useful measure for aquatic organisms. For species whose members differ markedly in size, as do most plants and some animals (such as mollusks, fishes, and non-avian reptiles), the percentage of ground covered or the total mass of individuals may be more useful measures of density than the number of individuals.

The most accurate way to determine the density and structure of a population is to count every individual and note its location. Ecologists studying populations of woody plants can often do this. In most field studies of animal populations, however, counting every individual would be impossible. Fortunately,

(A) *Orcinus orca*

(B) *Calidris alba*

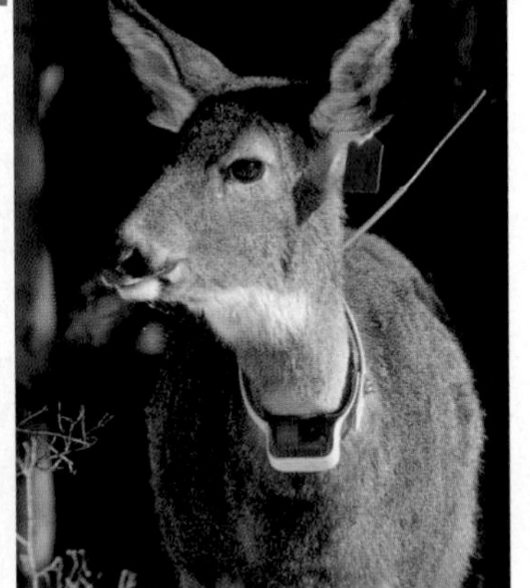
(C) *Odocoileus virginianus*

54.1 By Their Marks You May Know Them
(A) Many individual animals are distinguishable from each other by subtle differences in their size and markings. The slight differences in the white markings of these orcas are discernible when they are side by side. (B) Colored bands on the legs of this sanderling (a small shorebird) allow researchers to identify this individual. (C) A female white-tailed deer has an individually identifying ear tag. In addition, researchers have attached a collar that emits a radio signal so that the deer's movements can be monitored even when she cannot be seen.

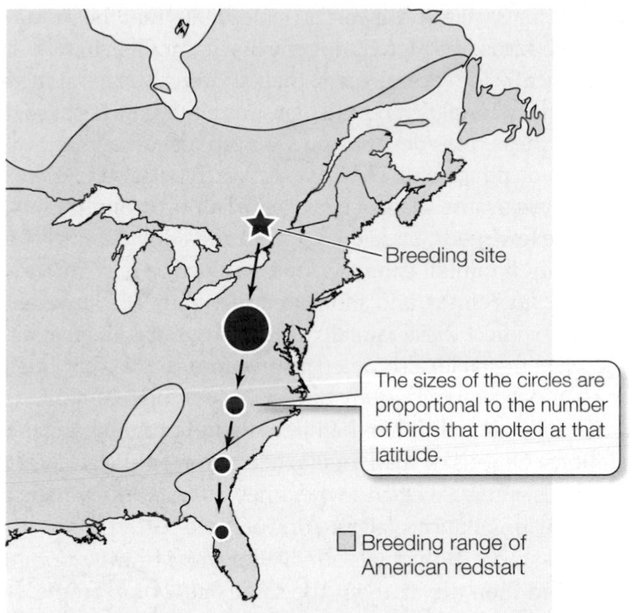

54.2 Hydrogen Isotopes Tell Where Migratory American Redstarts Molted Their Feathers Redstarts that breed in southern Ontario migrate to the Caribbean for the winter. The values of a stable hydrogen isotope in their tail feathers indicate where the feathers were molted.

statistical methods allow us to reliably estimate population densities from representative samples without locating and counting every individual.

Estimating population densities is easiest for sedentary organisms. Investigators need only count the number of individuals in a sample of representative habitats and extrapolate the counts to the entire ecosystem. Individuals may be counted on sampling plots of various shapes (square, rectangular, circular) or along linear transects across the habitat. By making repeated censuses, investigators can also determine trends in the size and distribution of the population.

Counting mobile organisms is much more difficult because individuals move into and out of census areas. Estimating the number of individuals in a population often involves capturing, marking, and then releasing a number of individuals. Later, after the marked individuals have had time to mix with the unmarked individuals in the population, another sample of individuals is taken. The proportion of individuals in the new sample that is marked can be used to estimate the size of the population using the formula

$$\frac{m_2}{n_2} = \frac{n_1}{N}$$

where

n_1 = the number of marked individuals in the first sample (all get marked)

n_2 = the total number of individuals in the second sample

m_2 = the number of marked individuals in the second sample

N = the estimated size of the total population

You may already have guessed that an estimate of total population size computed from such a capture–mark–recapture effort would be accurate only if the marked individuals had randomly mixed with the unmarked individuals and if marked and unmarked individuals were equally likely to survive and be captured. Estimates of total population size will be inaccurate if marked individuals either learn to avoid traps or become "trap-happy" (because, say, they get a quick meal), or if, as a result of being captured, individuals leave the study area. Ecologists have developed statistical techniques that can correct for these errors and improve the accuracy of population estimates.

Birth and death rates can be estimated from population density data

Ecologists use estimates of population densities to estimate the *rate* (number per unit of time) at which births, deaths, and movements take place in a population, and they study how these rates are influenced by environmental factors, life histories, and population densities.

The number of individuals in a population at any given time is equal to the number present at some time in the past, plus the number born between then and now, minus the number that died, plus the number that immigrated into the population, minus the number that emigrated from the population. That is, the number of individuals at a given time, N_1, is given by the equation

$$N_1 = N_0 + B - D + I - E$$

where

N_1 = the number of individuals at time 1

N_0 = the number of individuals at time 0

B = the number of individuals born between time 0 and time 1

D = the number that died between time 0 and time 1

I = the number that immigrated between time 0 and time 1

E = the number that emigrated between time 0 and time 1

If we make these counts over many time intervals, we can determine how a population's density changes over time.

A useful way to display information about birth and death rates in a population is to construct a **life table**. We can construct a life table by tracking a group of individuals born at the same time (called a **cohort**) and determining the number that are still alive at later dates (**survivorship**). Some life tables also include the number of offspring produced by the cohort (*fecundity*) during each time interval.

Life tables were first developed by the Romans nearly 2,000 years ago to determine how much money needed to be set aside to compensate families of soldiers who might be killed in battle. Today, life insurance companies use life tables ("actuarial tables") to determine how much to charge people of different ages for insurance policies.

TABLE 54.1

Life Table of the 1978 Cohort of the Cactus Finch (*Geospiza scandens*) on Isla Daphne

AGE IN YEARS (x)	NUMBER ALIVE	SURVIVORSHIP[a]	SURVIVAL RATE[b]	MORTALITY RATE[c]
0	210	1.000	0.434	0.566
1	91	0.434	0.857	0.143
2	78	0.371	0.898	0.102
3	70	0.333	0.928	0.072
4	65	0.309	0.955	0.045
5	62	0.295	0.678	0.322
6	42	0.200	0.548	0.452
7	23	0.109	0.652	0.348
8	15	0.071	0.933	0.067
9	14	0.067	0.786	0.214
10	11	0.052	0.909	0.091
11	10	0.048	0.400	0.600
12	4	0.019	0.750	0.250
13	3	0.014	0.996	

[a]Survivorship = the proportion of newborns who survive to age *x*.
[b]Survival rate = the proportion of individuals of age *x* who survive to age *x* + 1.
[c]Mortality rate = the proportion of individuals of age *x* who die before the age of *x* + 1.

Biologists can use life tables to predict future trends in populations. A life table based on an intensive study of the seed-eating cactus finch, carried out on Isla Daphne in the Galápagos Archipelago, is shown in **Table 54.1**. The data come from 210 birds that hatched in 1978 and were followed until 1991, at which time only 3 individuals were still alive. The table shows that the mortality rate for these birds was high during the first year of life. It then dropped dramatically for several years, then showed a general increase in later years. Mortality rates fluctuated among years because the survival of these birds depends on seed production, which is strongly correlated with rainfall. The Galápagos archipelago experiences both drought years and years of heavy rain. During drought years, plants produce few seeds, birds do not nest, and adult survival is poor. In years when rainfall is heavy, seed production is high, most birds breed several times, and adult survival is high. The survival rates in the table reflect these rainfall fluctuations. Reproductive rates (not shown in the table) also varied greatly from year to year, but females of all ages bred successfully during years of high rainfall.

Graphs are helpful for highlighting important changes in populations. Graphs of survivorship in relation to age show when individuals survive well and when they do not. Survivorship curves for many populations fall into one of three patterns. In some populations, most individuals survive for most of their potential life span and then die at about the same age. For example, because of intensive parental care and the availability of medical services, the survivorship of humans in the United States is high for many decades, but then declines rapidly in older individuals (**Figure 54.3A**). In a second pattern, which is characteristic of many songbirds, the probability of surviving to the next year is about the same over most of the life span once individuals are a few months old (**Figure 54.3B**). A third widespread survivorship pattern is found among organisms that produce a large number of offspring, each of which receives little investment of energy or parental care. In these species, high death rates of young individuals are followed by high survival rates during the middle part of the life span. *Spergula vernalis*, an annual plant that grows on sand dunes in Poland, illustrates this pattern (**Figure 54.3C**).

The age distribution of individuals in a population reveals much about the recent history of births and deaths in the population. The timing of births and deaths can influence age distributions for many years in populations of long-lived species. The human population of the United States is a good example. Between 1947 and 1964, the United States experienced what is known as the post-World War II baby boom. During these years, average family size grew from 2.5 to 3.8 children; an unprecedented 4.3 million babies were born in

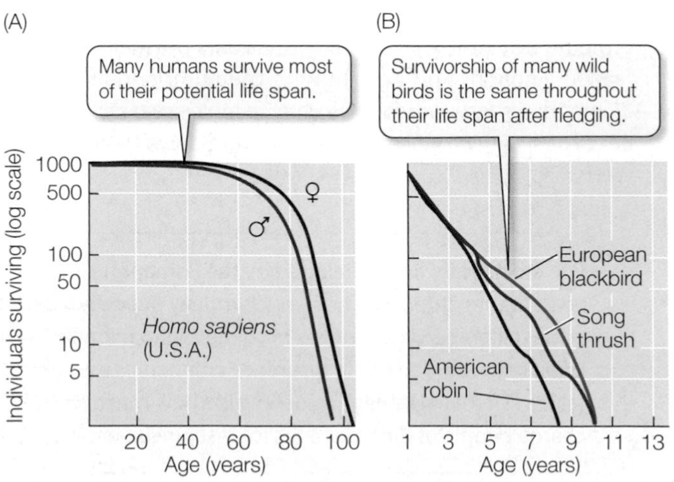

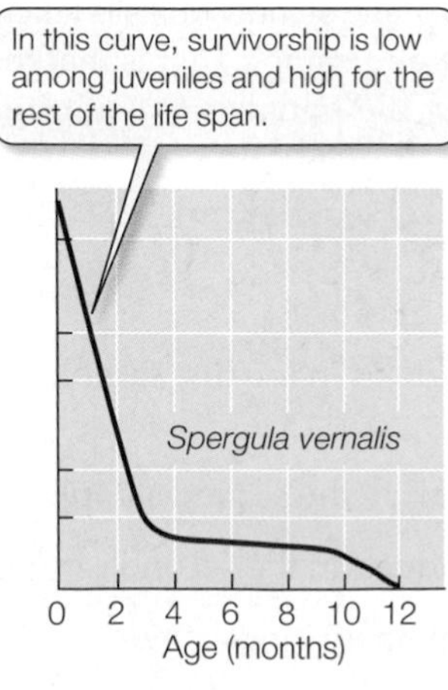

54.3 Survivorship Curves Three common survivorship curves. (A) For some species, mortality is highest at advanced ages. (B) For other species, the probability of survivorship is similar throughout much of the life span. (C) Still other species have high mortality rates at early ages, but survive well once past a critical point.

1957. Birth rates declined during the 1960s, but Americans born during the baby boom still constitute the dominant age class in the first part of the twenty-first century (**Figure 54.4**). "Baby boomers" became parents in the 1980s, producing another bulge in the age distribution—a "baby boom echo"—but they had, on average, fewer children than their parents did, so the bulge is not as large.

By summarizing information on when individuals are born and die, life tables help us understand why population densities change over time. Life table data can also be used to determine how heavily a population can be harvested and which age groups should be the focus of our efforts to save rare species.

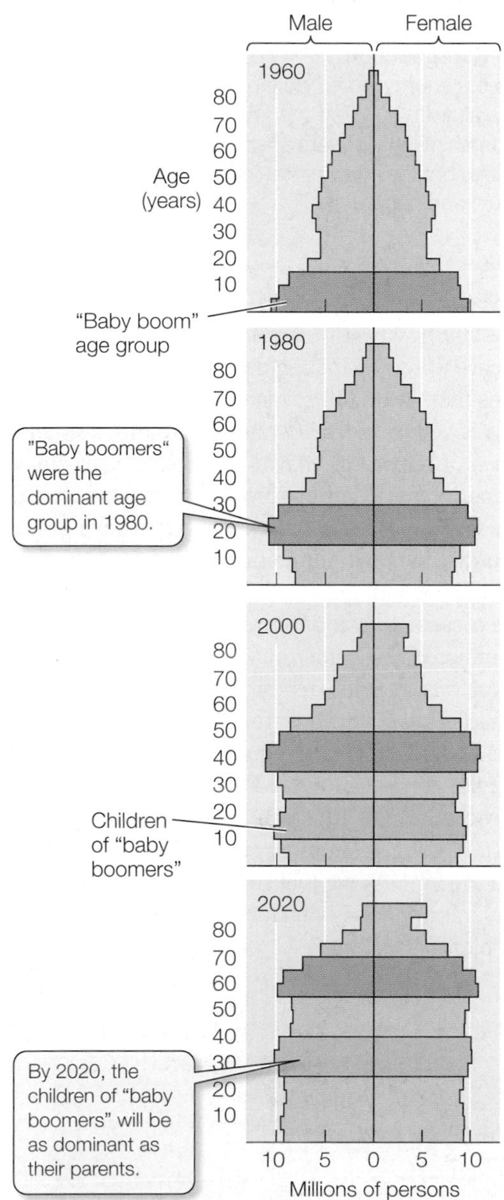

54.4 Age Distributions Change over Time The graphs shows age distributions for the human population of the United States from 1960 to 2020. The high birth rate during the "baby boom" has influenced the structure of the population over many decades.

54.1 RECAP

A population consists of the individuals of a species within a particular area. Population density is a measure of the number of individuals per unit of area or volume. Birth rates, death rates, and other aspects of population dynamics can be estimated from population data collected over time.

- Describe some ways in which population density can be measured. See pp. 1168–1169
- What kinds of information can life tables provide about a population? See pp. 1169–1170 and Table 54.1

During its life, an individual organism ingests nutrients, grows, interacts with other individuals of the same and other species, and usually moves or is moved so that it does not die exactly where it was born. In the next section we discuss the ecological significance of the different ways in which organisms allocate their time and energy to these activities.

54.2 How Do Ecological Conditions Affect Life Histories?

An organism's **life history** describes how it allocates its time and energy among the various activities that occupy its life. The life histories of different organisms vary dramatically. Some plants grow rapidly, produce large numbers of small seeds, and die soon thereafter. Other plants grow slowly, do not reproduce until they are many years old, produce only a few large seeds, and continue to reproduce for decades. Some animals, such as elephants and humans, usually give birth to a single offspring in each reproductive episode; others, such as oysters, produce thousands of eggs in one batch. Some organisms, such as agaves and salmon, usually reproduce only once and then die (**Figure 54.5**). Ecologists study life histories because they influence how populations grow and are distributed.

The life history of the black rockfish (*Sebastes melanops*), which lives off the Pacific coast of North America, offers an example of how life history traits influence the growth of a population that people would like to manage. Female rockfish continue to grow throughout their lives. Larger females are much more productive than smaller females because the number of eggs a female produces is proportional to her size. In addition, older, larger females produce eggs containing larger oil droplets. These droplets provide energy to newly hatched fish, giving them a head start in their independent lives (**Figure 54.6**). Larvae that hatch from eggs with large oil droplets grow faster and survive better than do larvae that hatch from eggs with small oil droplets. These facts have important implications for the harvesting of rockfish populations. Intensive bottom fishing off the Oregon coast from 1996 to 1999 reduced the average age of female rockfish from 9.5 to 6.5 years, so reproductive females were much smaller in 1999 than they were in 1996. This age reduction decreased the number of eggs produced by females in the population and reduced the average growth rates

(A) *Agave americana*

(B) *Oncorhyncus* sp.

54.5 Big Bang Reproduction (A) Agaves, also known as century plants, mobilize the energy stored during their long lives to produce a large flowering stalk with hundreds of flowers, literally reproducing themselves to death. (B) After years spent feeding and growing in the open sea, salmon fight their way up the river in which they hatched. There they spawn and soon die.

of larvae by about 50 percent. Thus maintaining productive populations of rockfish may require setting aside no-fishing zones where some females can grow to very large sizes without being harvested.

Ecological interactions influence the evolution of life histories. The influence of predation on the evolution of life history traits was tested in experiments on guppies (*Poecilia reticulata*). In Trinidad, guppies live in streams where they are preyed on by larger fish. But some streams have waterfalls that predatory fishes are unable to surmount. Guppies that live in the predator-free areas upstream from those waterfalls have low mortality rates; those that live below the falls have high mortality rates. David Reznick and his colleagues reared in the laboratory 240 guppies from high-predation and low-predation sites. They supplied some guppies from each group with plentiful food and others with limited food to match the variation that the fish would encounter in nature. In the laboratory, where no predators were present, they found that guppies from high-predation sites matured earlier, reproduced more frequently, and produced more offspring in each brood than guppies from low-predation sites, no matter how much food they received. Thus predation favored early and frequent reproduction, leading to change in the guppy genotype.

As these examples illustrate, a study of the population dynamics of an organism must take its life history characteristics into account. As we will see at the end of this chapter, life history information is vital for designing population management plans as well.

(A) *Sebastes melanops*

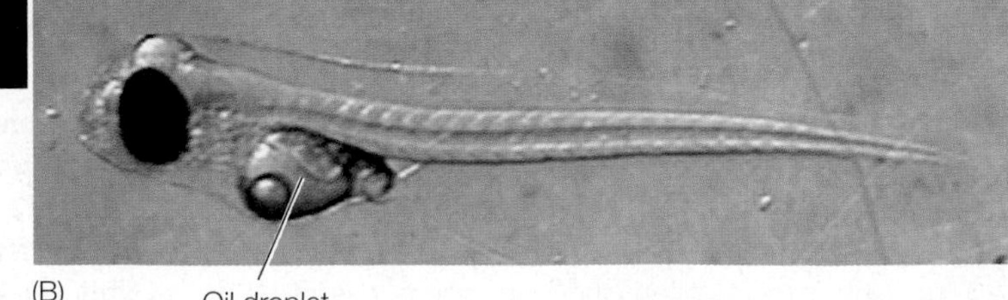

(B)

54.6 An Oil Droplet Is an Energy Kick Start (A) Among black rockfish, the older, larger females are more reproductively successful, producing both more eggs and eggs with larger stores of nutritive oil. (B) The oil droplet beneath this rockfish larva provides it with nutrition to fuel its growth until it can feed on its own.

54.2 RECAP

The life history of an organism describes how it allocates its time and energy among growth, reproduction, and other activities.

- How can age-related differences in fecundity influence population dynamics? See pp. 1171–1172
- How do mortality patterns influence the evolution of life history traits? See p. 1172

A locally rare species may be abundant somewhere else; but some species exist at low population densities everywhere they are found. A species that is rare at a given time may be abundant at some later time, or vice versa. What factors determine population densities, and why do they vary so much?

54.3 What Factors Influence Population Densities?

Although many populations fluctuate markedly in density, none fluctuate as dramatically as is theoretically possible. Consider, for example, a single bacterium selected at random from the surface of this book. If all its descendants were able to grow and reproduce in an environment with unlimited resources, the population would grow explosively. In a month, this bacterial colony would weigh more than the visible universe and would be expanding outward at the speed of light. Similarly, a single pair of Atlantic cod and their descendants, reproducing at the maximum rate of which they are capable, would fill the Atlantic Ocean basin in 6 years if none of them died. What prevents such dramatic population growth from happening in nature?

All populations have the potential for exponential growth

All populations have the potential for explosive growth. As the number of individuals in a population increases, the number of new individuals added per unit of time accelerates, even if the rate of increase expressed on a per individual basis—called the *per capita growth rate*—remains constant. If births and deaths occur continuously and at constant rates, a graph of the population size over time forms a continuous upward curve (see Figure 54.7): this pattern, known as **exponential growth**, can be expressed as

$$\text{Rate of increase in number of individuals} = \begin{pmatrix}\text{Average per capita birth rate}\\ -\text{Average per capita death rate}\end{pmatrix} \times \text{Number of individuals}$$

or, more concisely,

$$r = \frac{\Delta N}{\Delta t} = (b - d)N$$

where

r = the net reproductive rate

ΔN = the change in number of individuals

Δt = change in time

b = the average population per capita birth rate

d = the average population per capita death rate

The term $\Delta N/\Delta t$ thus conveys the rate of change in the size of the population over time. In other words, the difference between the average per capita birth rate in a population (b) and its average per capita death rate (d) is the *net reproductive rate* (r). (In these equations, b includes both births and immigrations, and d includes both deaths and emigrations.)

The highest possible value for the net reproductive rate—the rate at which the population would grow under optimal conditions—is called r_{max}, or the **intrinsic rate of increase**; r_{max} has a characteristic value for each species. The intrinsic rate of increase can be expressed as

$$\frac{\Delta N}{\Delta t} = r_{max} N$$

For very short periods, some populations may grow at rates close to the intrinsic rate of increase. For example, northern elephant seals were hunted nearly to extinction in the late nineteenth century. In 1890, only about 20 animals remained, confined to Isla Guadalupe off the northwestern coast of Mexico. Ample elephant seal habitat remained available, however, so once the hunting was stopped, the population began to increase rapidly (**Figure 54.7**). Elephant seals recolonized Año Nuevo Island near Santa Cruz, California, in 1960. During the 20 years following colonization, the population breeding on the island expanded exponentially.

Population growth is limited by resources and biotic interactions

No real population can maintain exponential growth for very long. As a population increases in density, environmental limits cause birth rates to drop and death rates to rise. The simplest way to picture the limits imposed by the environment is to assume that an environment can support no more than a certain number of individuals of any particular species per unit of area (or volume). This number, called the **environmental carrying capacity** (***K***), is determined by the availability of resources—such as food, nest sites, or shelter—as well as by disease, predators, and, in some cases, social interactions.

The growth of a population typically slows down as its density approaches the environmental carrying capacity because resource limitations and the activities of predators and pathogens lower birth rates and increase death rates. A graph of population size over time typically forms an S-shaped curve; this pattern is known as **logistic growth** (**Figure 54.8**). The simplest way to generate an S-shaped growth curve is to add to the equation for exponential growth a term that slows the population's growth as it approaches the carrying capacity. This term, expressed as $(K - N)/K$, implies

Number of pups born
Year
Theoretically, a population in an environment with unlimited resources could grow like this indefinitely.

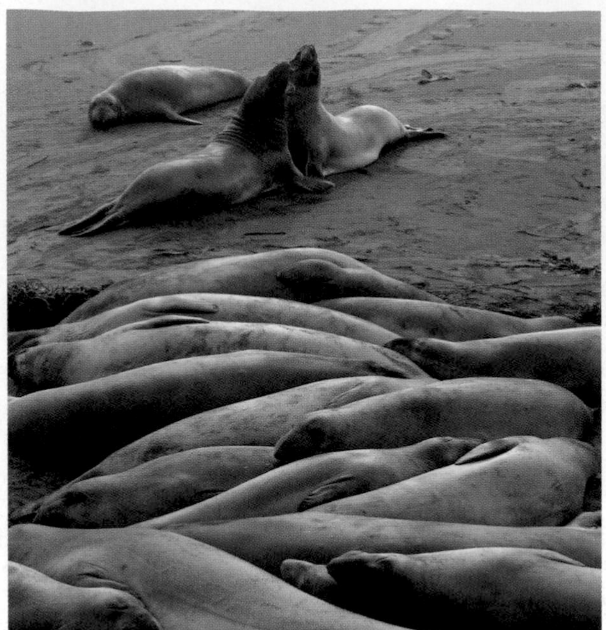

Mirounga angustirostris

54.7 Exponential Population Growth The elephant seal population on Año Nuevo Island, California grew exponentially between 1960 and 1980. Since no resource on Earth is unlimited, this pattern cannot continue for very long. The growth of the seal population on Año Nuevo has indeed slowed down.

that each individual added to the population depresses population growth by an equal amount:

$$\frac{\Delta N}{\Delta t} = r\left(\frac{K - N}{K}\right)N$$

Population growth stops when $N = K$ because then $(K - N) = 0$, so $(K - N)/K = 0$, and thus $\Delta N/\Delta t = 0$.

Population densities influence birth and death rates

Because each additional individual typically makes things worse for other members of the population in an environment with limited resources, per capita birth and death rates usually change together with changes in population density; that is, they are **density-dependent**. Birth and death rates may be density-dependent for several reasons:

- As a species increases in abundance, it may deplete its food supply, reducing the amount of food available to each individual. Poorer nutrition may then increase death rates and decrease birth rates.
- Predators may be attracted to areas with high densities of their prey. If predators capture a larger proportion of the prey than they did when the prey were scarce, the per capita death rate of the prey rises.
- Diseases can spread more easily in dense populations than in sparse populations.

Not all factors affecting population size act in a density-dependent way. A cold spell in winter or a hurricane that blows down most of the trees in its path may kill a large proportion of the individuals in a population regardless of its density. Factors that change per capita birth and death rates in a population independently of its density are said to be **density-independent**.

Fluctuations in the density of a population are determined by all of the density-dependent and density-independent factors acting on it. The combined action of these factors can be seen in a study of the dynamics of a population of song sparrows (*Melospiza melodia*) on Mandarte Island, off the coast of British Columbia, Canada. Over a period of 12 years, the number of song sparrows fluctuated between 4 and 72 breeding females and between 9 and 100 territorial males. Death rates were high during particularly cold, snowy winters, regardless of the density of the population. Several density-dependent factors also contributed to fluctuations in the density of the population. The number of breeding males, for example, was limited by territorial behavior: the larger the number of males, the larger the number that failed to gain a territory and lived as "floaters" with little chance of reproducing (**Figure 54.9A**). In addition, the larger the number of breeding females, the fewer offspring each female fledged (**Figure 54.9B**). And the more birds alive

Environmental carrying capacity (K)
Population size
Time
1 The rate accelerates.
2 A maximum growth rate is reached.
3 The rate slows down.

54.8 Logistic Population Growth Typically, a population in an environment with limited resources stops growing exponentially long before it reaches the environmental carrying capacity.

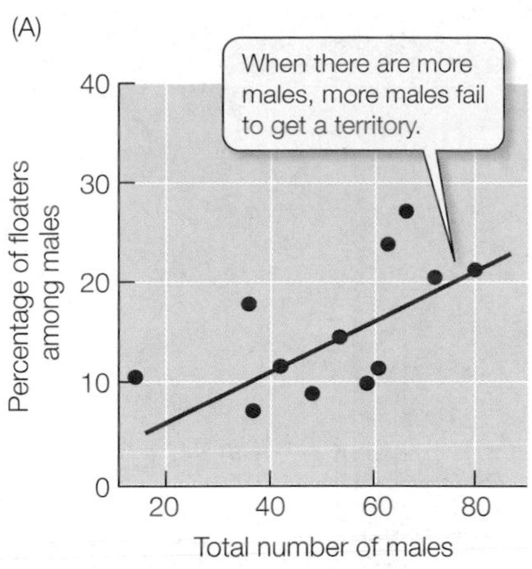

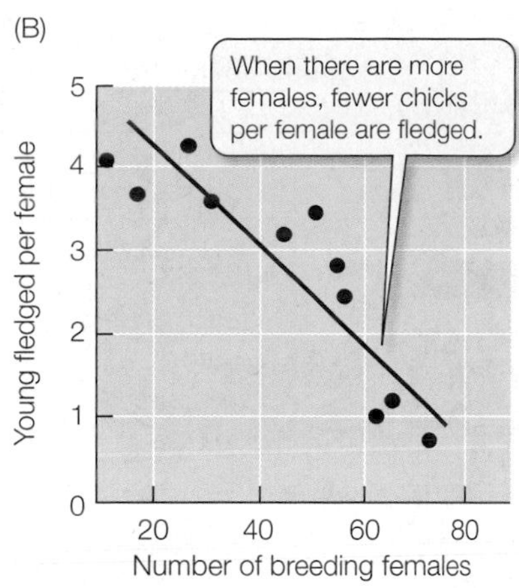

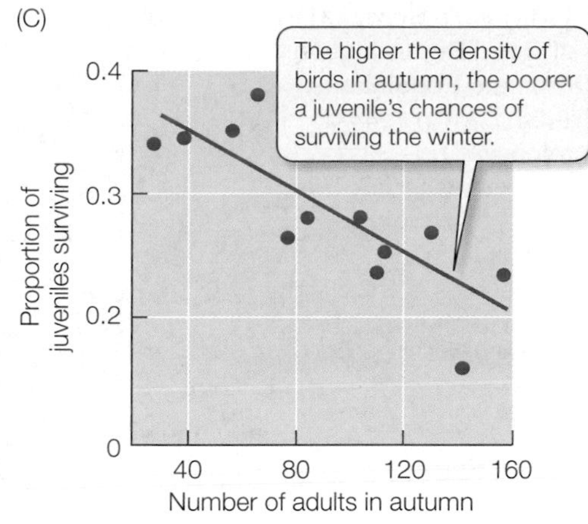

54.9 Regulation of an Island Population of Song Sparrows The size of the population of song sparrows (*Melospiza melodia*) on Mandarte Island, British Columbia, is determined in part by the severity of winter weather. In addition, the population is regulated by density-dependent factors, including (A) the territorial behavior of males, (B) the reproductive success of females, and (C) the survival of juveniles over the winter.

in autumn, the poorer were the chances that juveniles born in that year would survive the winter (**Figure 54.9C**). Thus the number of males and females breeding each year was influenced by both density-independent and density-dependent factors.

All populations fluctuate less than the theoretical maximum, but the sizes of some populations fluctuate remarkably little. In general, more stable populations are seen in species with long-lived individuals that have low reproductive rates than in short-lived species that have high reproductive rates. Small, short-lived individuals are generally more vulnerable to environmental changes than long-lived individuals. Insect population densities tend to fluctuate much more than those of birds and mammals, and population densities of annual plants fluctuate much more than those of trees.

Most fluctuations in the densities of populations are driven by changes in the biotic and abiotic environment that change the environmental carrying capacity for the species. Let's consider two examples.

EPISODIC REPRODUCTION GENERATES POPULATION FLUCTUATIONS

For most species, some years are better than others for reproductive success. In Lake Erie, 1944 was such an excellent year for whitefish reproduction that individuals born that year dominated whitefish catches in the lake for several years thereafter (**Figure 54.10A**). Similarly, most of the individuals found in a population of black cherry trees in a Wisconsin forest in 1971 had become established between 1931 and 1941 (**Figure 54.10B**). Population densities increase following years of good reproductive success, but they decrease following years of poor reproduction.

RESOURCE FLUCTUATIONS GENERATE CONSUMER FLUCTUATIONS

Densities of populations of species that depend on a single or just a few resources are likely to fluctuate more than those of species that use a greater variety of resources. Several species of birds and mammals that live in boreal forests feed on the seeds found in conifer cones. Most trees in these forests reproduce *synchronously* (all at the same time) and *episodically* (on an irregular basis). Over large areas, there are years of massive seed production and years of little or no seed production. Some birds (such as crossbills) wander over large areas, looking for places where cones have been produced. Other birds (such as jays and nutcrackers) and some mammals (squirrels) store seeds during years of high production, but they often suffer high mortality rates during years when the trees in their area produce few or no seeds.

Several factors explain why some species are more common than others

The processes that we have just discussed enable us to understand how populations grow, why they fluctuate in size, and why fluctuations in their densities are much less than would be theoretically possible. But they do not explain why some species are common whereas others are rare. Many factors determine why typical population densities vary so greatly among species, but four of them—resource abundance, the size of individuals, the length of time a species has lived in an area, and social organization—exert especially strong influences.

- *Species that use abundant resources generally reach higher population densities than species that use scarce resources.* Thus, on average, animals that eat plants are typically more common than animals that eat other animals.
- *Species with small body sizes generally reach higher population densities than species with large body sizes.* In general, population density decreases as body size increases, because small individuals require less energy to survive than large individuals.

The relationship between body size and population density is illustrated by a logarithmic plot of population density against body size for mammal species worldwide (**Figure 54.11**). Although the relationship is strong, the great scatter of the points on the graph shows that some small species use scarce resources and some large species use abundant resources.

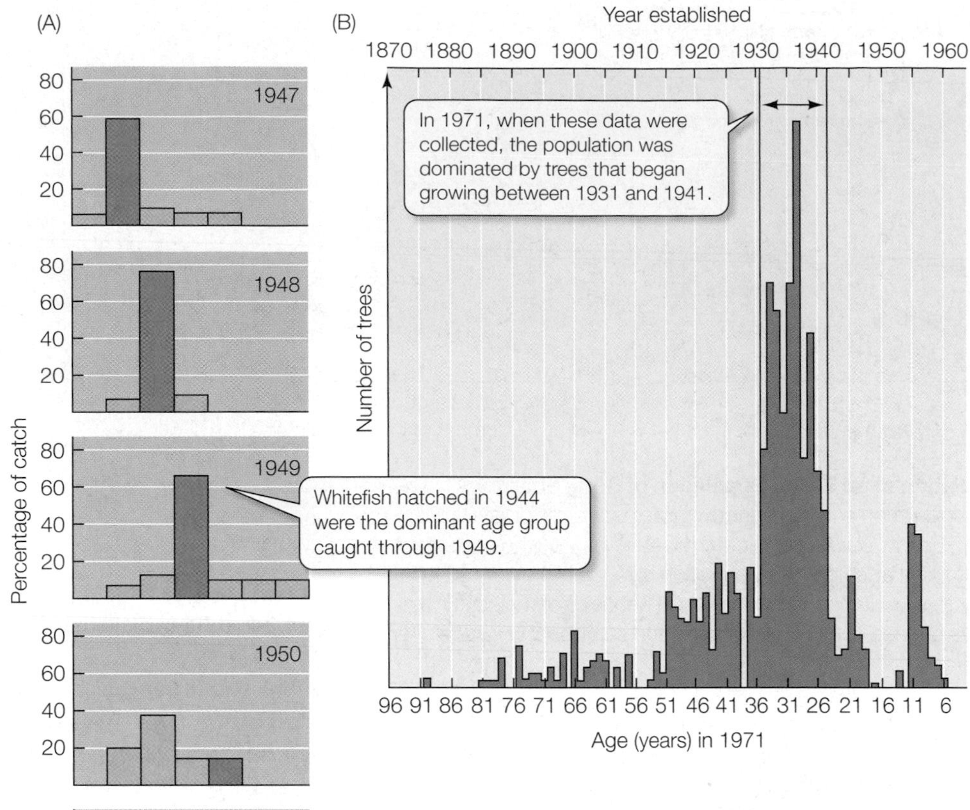

54.10 Individuals Born during Years of Good Reproduction May Dominate Populations (A) Whitefish born in 1944 dominated catches in Lake Erie for many years thereafter. (B) The population of black cherry trees (*Prunus serotina*) in a Wisconsin forest in 1971 was dominated by trees that became established between 1931 and 1941.

- *Some newly introduced species reach high population densities.* Species that have recently escaped control by the factors that normally prevent them from becoming more abundant may achieve temporarily high population densities. Species that are introduced into a new region, where their normal predators and pathogens are absent, sometime reach population densities much higher than ever found in their native ranges.

The zebra mussel (*Dreissena polymorpha*), whose larvae were carried from Europe in the ballast water of commercial cargo ships, became established in the Great Lakes in about 1985. Zebra mussels spread rapidly, and today they occupy much of the Great Lakes and the Mississippi River drainage (**Figure 54.12**). In some places these mussels have reached densities as high as 400,000 individuals per square meter; such densities are never found in Europe. Densities of zebra mussels in North America may decrease in the future if local predators and pathogens begin to attack them.

- *Complex social organization may facilitate high densities.* As we saw in Section 53.6, highly social species, including ants, termites, and humans, can achieve remarkably high population densities.

Important though these four factors are, they cannot explain many differences in abundances of species. For example, both Douglas firs and giant sequoias are large trees that use the same source

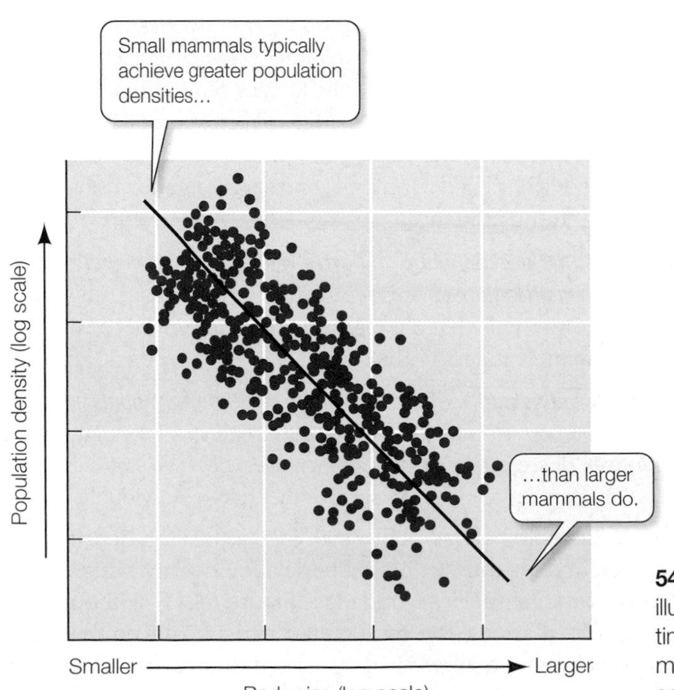

54.11 Population Density Decreases as Body Size Increases This trend is illustrated by a logarithmic plot (that is, each tick mark represents a number 10 times greater than the one before it) of population density against body size for mammals of different sizes. Each dot represents a different mammalian species, and the resulting slope (straight line) is determined algebraically.

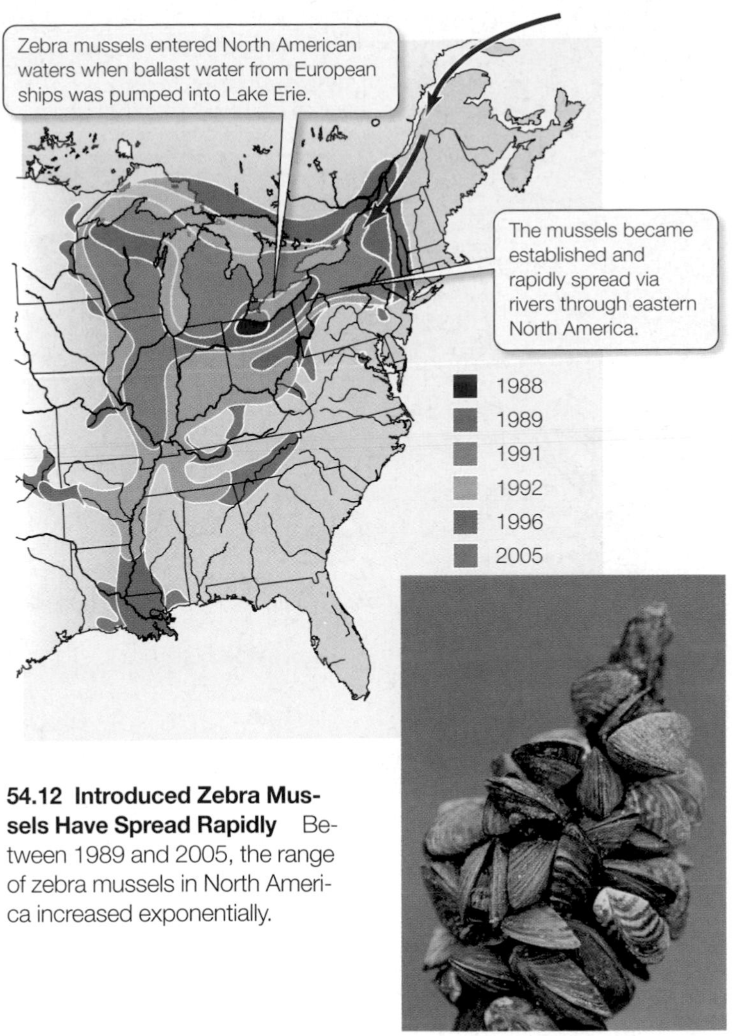

Zebra mussels entered North American waters when ballast water from European ships was pumped into Lake Erie.
The mussels became established and rapidly spread via rivers through eastern North America.
1988
1989
1991
1992
1996
2005

Dreissena polymorpha

54.12 Introduced Zebra Mussels Have Spread Rapidly Between 1989 and 2005, the range of zebra mussels in North America increased exponentially.

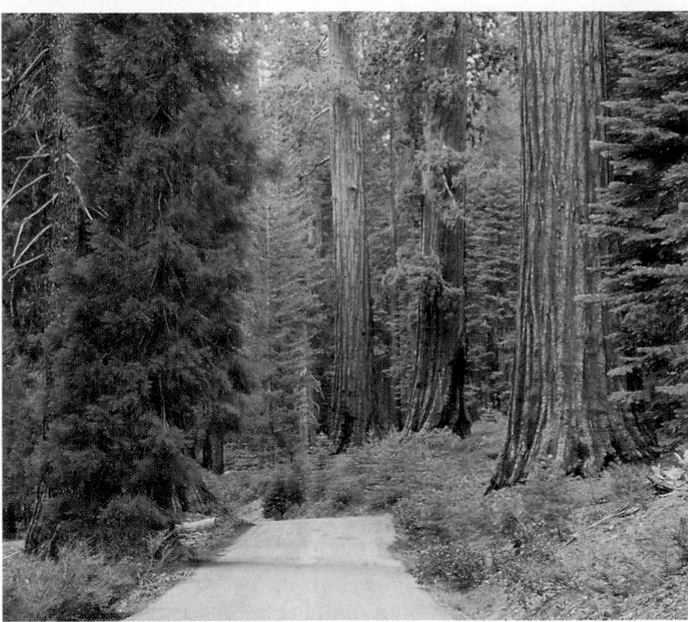

Sequoiadendron giganteum

54.13 The Last Refuge The range of the giant sequoia has progressively shrunk over thousands of years, primarily because of changes in climate. Only a few groves of trees remain, scattered in the southern Sierra Nevada of California.

of energy (sunlight) and need the same nutrients. Yet Douglas firs are widespread and abundant in western North America, whereas giant sequoias are restricted to a few groves in the southern Sierra Nevada of California (**Figure 54.13**). Similarly, several species of desert pupfish are each restricted to a single spring in Death Valley, California, whereas smallmouth bass live in most of the rivers and lakes in eastern North America. To explain these differences, we also need to know about the origins and long-term histories of species.

As Chapter 23 describes, a new species can originate in several ways. A species that arises by polyploidy inevitably begins with a very small, local population. Many polyploid plant species that have formed only recently have not spread much beyond the site of their origin. They have small ranges, although their local population densities may be high. Similarly, species that arise through founder events typically begin their history with only a few individuals. In contrast, most species that arise from a vicariant event (see Section 52.4) begin with large populations and ranges. Finally, as a species declines toward extinction, as may be happening to giant sequoias, its range shrinks until it vanishes when the last individual dies.

Interactions with other species may also limit both the densities and ranges of species. We will explore the consequences of those interactions in Chapter 55.

54.3 RECAP

Many populations fluctuate markedly in density, but even the most dramatic fluctuations are much less than those that are theoretically possible. Population sizes are limited by the environmental carrying capacity, which is determined by biotic interactions and by the availability of resources.

- Why can populations grow exponentially only for short periods? See pp. 1173–1174
- How do density-dependent and density-independent factors interact to determine population densities? See pp. 1174-1175 and Figure 54.9
- How do resource availability, the size of individuals, the length of time a species has lived in an area, and social organization influence population densities? See pp. 1175–1177

All species, no matter how abundant, are found in only those habitats in which they can survive and reproduce well enough to persist over time. Yet a species is rarely found in all of the habitats that seem suitable for it. The next section explores why this is the case.

54.4 How Do Spatially Variable Environments Influence Population Dynamics?

Most natural history field guides display maps that show the geographic range over which a species is found. But you know that you will not find individuals of a species everywhere within the area indicated on the map. No species, not even the most abundant, is found everywhere within its mapped range. To know where to look for the species, you consult the text that describes the habitat in which the species lives.

Many populations live in separated habitat patches

Most populations are divided into separated, discrete *subpopulations* that live in distinct **habitat patches**—areas of a particular kind of environment that are surrounded by other kinds. The larger population to which such subpopulations belong is referred to as a **metapopulation**. Each subpopulation has a probability of "birth" (colonization of that habitat patch) and "death" (extinction in that patch). Growth occurs in each subpopulation in the ways we have just described, but because the subpopulations are much smaller than the metapopulation, local disturbances and random fluctuations in numbers of individuals are more likely to cause the extinction of a subpopulation than the extinction of an entire metapopulation. However, if individuals move frequently between subpopulations, immigrants may prevent declining subpopulations from becoming extinct, a process called the **rescue effect**.

The bay checkerspot butterfly (*Euphydryas editha bayensis*) provides a good illustration of the dynamics of metapopulations. The caterpillars (larvae) of this butterfly feed on only a few species of annual plants, which are restricted to outcrops of serpentine rock on hills south of San Francisco, California. The bay checkerspot has been studied for many years by Stanford University biologists. During drought years, most host plants die early in spring, before the caterpillars have developed far enough to be able to enter their summer resting stage. At least three butterfly subpopulations became extinct during a severe drought in 1975–1977. The largest patch of suitable butterfly habitat, Morgan Hill, typically supported thousands of butterflies (**Figure 54.14**). Until the population on Morgan Hill became extinct recently, it probably served as a source of individuals that dispersed to and recolonized small patches where the butterflies had become extinct.

In another study, ecologists manipulated the habitat of tiny arthropods (springtails—tiny hexapods without wings—and mites) to investigate the metapopulation dynamics of these animals. In one experiment, they created isolated patches of their habitat—mosses growing on rocks—by clearing moss from parts of the rock surface (**Figure 54.15, Experiment 1**). The number of species present in these patches declined about 40 percent within a year, with more rare species than common species disappearing from the patches. The experiment demonstrated that small populations were more likely to become extinct than large populations.

In a second experiment, the investigators created similar patches, but these patches were connected by narrow corridors of moss that were either intact or disrupted by a barrier only 10 millimeters wide

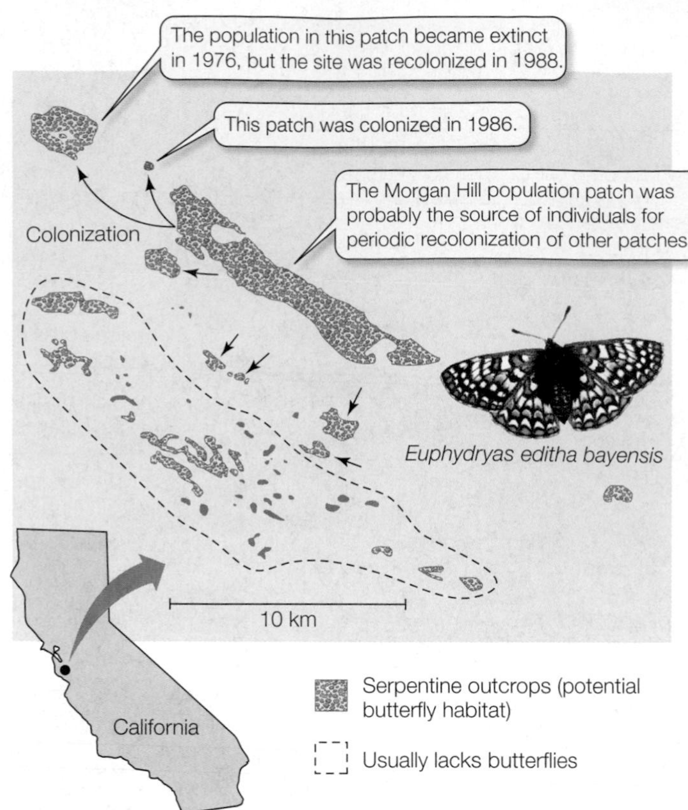
The population in this patch became extinct in 1976, but the site was recolonized in 1988.
This patch was colonized in 1986.
The Morgan Hill population patch was probably the source of individuals for periodic recolonization of other patches.
Colonization
Euphydryas editha bayensis
10 km
California
Serpentine outcrops (potential butterfly habitat)
Usually lacks butterflies

54.14 Metapopulation Dynamics The bay checkerspot butterfly population is divided into a number of subpopulations confined to patches of habitat (serpentine rock outcrops) that contain the food plants of its larvae. Extinction of these subpopulations is common. Indeed, no butterflies survive today in any of the habitat patches.

(**Figure 54.15, Experiment 2**). Moss patches connected by unbroken corridors contained more species of arthropods 6 months later than patches connected by discontinuous "pseudocorridors." Thus a gap between patches of only 10 millimeters was sufficient to reduce the rescue effect for these tiny organisms.

Distant events may influence local population densities

Field guides to birds have many maps that show both breeding and winter ranges of migratory species. In some cases the winter range is on another continent. Populations of migratory species may be influenced by events on both the breeding and wintering grounds, as well as in the places where the individuals stop to rest and feed during migration. To understand fluctuations in populations of migratory species, ecologists may need to study distant events.

A census of breeding birds conducted in Eastern Wood, in southeastern England, illustrates this point. Between 1950 and 1980, populations of some species increased while others decreased (**Figure 54.16**). The population of wood pigeons more than doubled, but the population of garden warblers decreased to zero in 1971; no more than two pairs have bred in the wood since then. The population of blue tits increased from just a few pairs to an average of more than 15 pairs. Why did these populations show such different dynamics?

No matter how intensively ecologists might have studied the birds of Eastern Wood, they could not have answered that question, because populations of two of these three species were

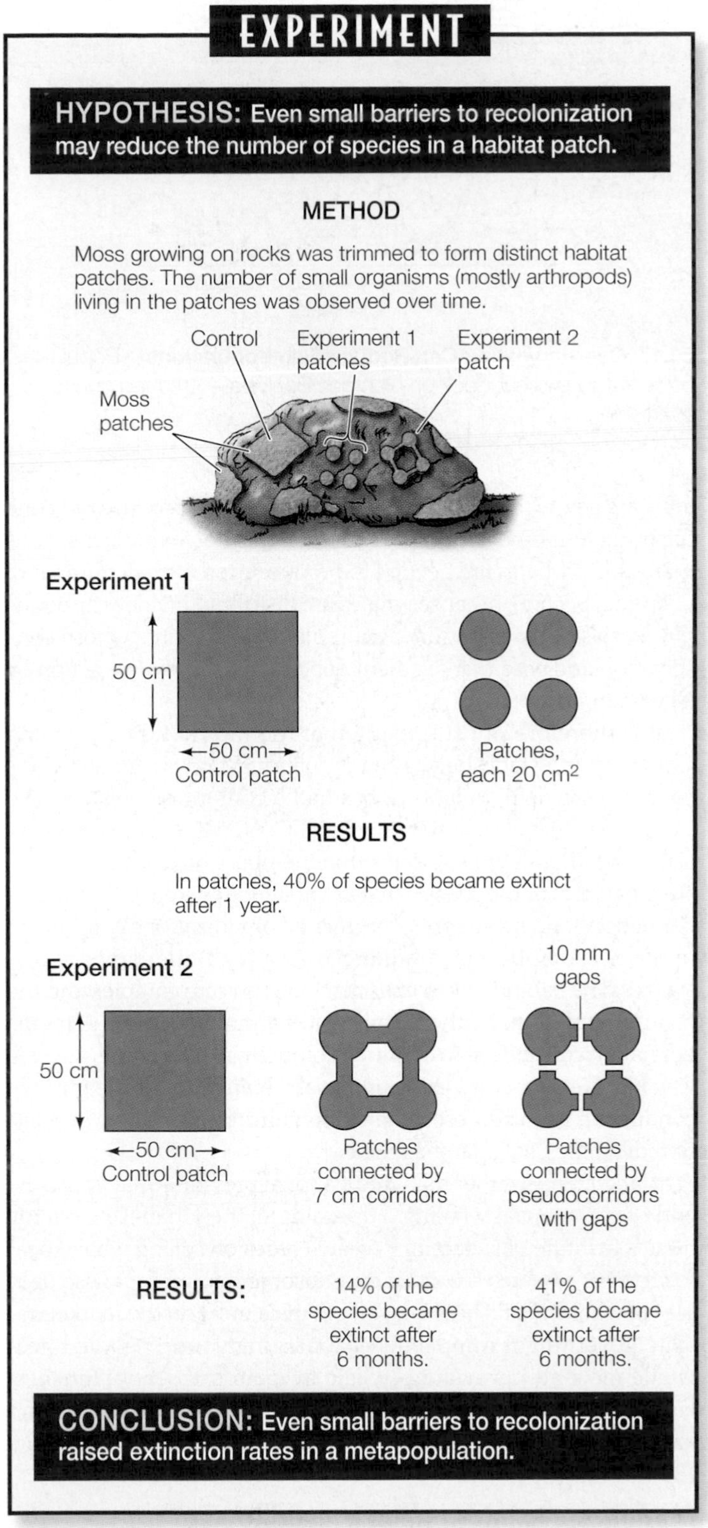

EXPERIMENT

HYPOTHESIS: Even small barriers to recolonization may reduce the number of species in a habitat patch.

METHOD

Moss growing on rocks was trimmed to form distinct habitat patches. The number of small organisms (mostly arthropods) living in the patches was observed over time.

Control
Experiment 1 patches
Experiment 2 patch
Moss patches

Experiment 1

50 cm
50 cm
Control patch
Patches, each 20 cm^2

RESULTS

In patches, 40% of species became extinct after 1 year.

Experiment 2

10 mm gaps
50 cm
50 cm
Control patch
Patches connected by 7 cm corridors
Patches connected by pseudocorridors with gaps

RESULTS: 14% of the species became extinct after 6 months. 41% of the species became extinct after 6 months.

CONCLUSION: Even small barriers to recolonization raised extinction rates in a metapopulation.

54.15 Narrow Barriers Suffice to Separate Arthropod Subpopulations Many species of small arthropods went extinct in isolated habitat patches. Barriers to dispersal as small as 10 millimeters prevented recolonization of patches.
FURTHER RESEARCH: These experiments investigated the dispersal of very small organisms over very short distances over very short time spans. The number of species in the isolated patches was unlikely to have reached equilibrium during that time. How could the longer-term effects of imposing barriers to dispersal be investigated?

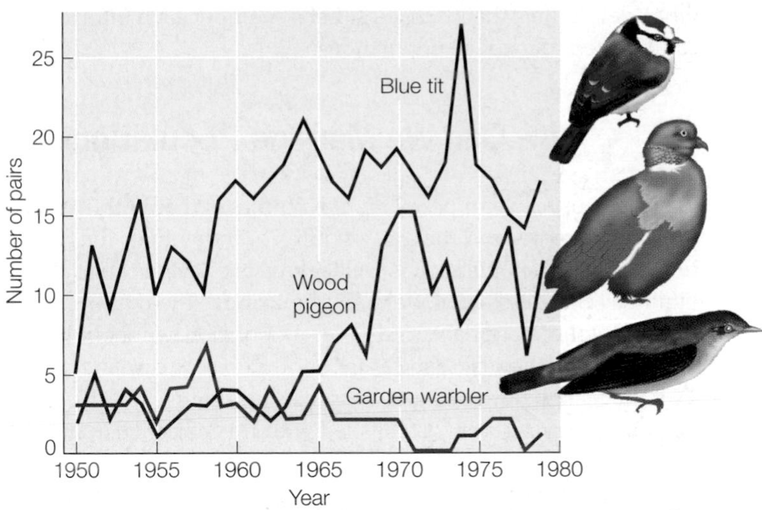

54.16 Populations May Be Influenced by Remote Events Populations of some birds increased in Eastern Wood, England, but others decreased. The wood pigeon (*Columba palumbus*) and garden warbler (*Sylvia borin*) population shifts were strongly influenced by different events that took place far from Eastern Wood. Only the blue tit (*Parus caeruleus*) population was affected most strongly by events within Eastern Wood itself.

strongly influenced by events outside Eastern Wood. Wood pigeons increased greatly over most of southern England during this 30-year period because of the widespread adoption of oilseed rape (the source of canola oil) as an agricultural crop. Rape fields provide wood pigeons with abundant winter food. Garden warblers decreased because their overwinter survival was poor due to a severe drought on their wintering grounds in West Africa.

The population of blue tits was influenced primarily by changes within Eastern Wood itself. Until the early 1950s, trees in Eastern Wood were periodically felled and sold for timber. After the cutting stopped, more holes, in which blue tits nest, became available in mature and dead trees.

54.4 RECAP

Most populations are divide into subpopulations that inhabit patches of suitable habitat.

- Why are many populations divided into subpopulations? See p. 1178 and Figure 54.14
- How can barriers to dispersal reduce species richness in a habitat patch? See p. 1178 and Figure 54.15
- How can distant events influence local population densities? See p. 1179

For many centuries, people have tried to reduce populations of species they consider undesirable and maintain or increase populations of desirable or useful species. Efforts to control and manage populations of organisms are more likely to be successful if they are based on knowledge of how those populations grow and what determines their densities. Let's see how such information can be used to manage populations.

54.5 How Can We Manage Populations?

A general principle of population dynamics is that both the total number of births and the growth rates of individuals tend to be highest when a population is well below its carrying capacity (see Figure 54.8). Therefore, if we wish to maximize the number of individuals that can be harvested from a population, we should manage the population so that it is far enough below carrying capacity to have high birth and growth rates. Hunting seasons for game birds and mammals are established with this objective in mind.

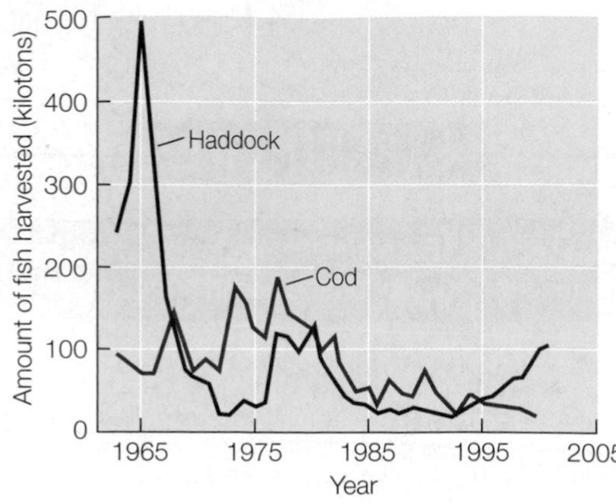

54.17 Overharvesting Can Reduce Fish Populations Populations of cod and haddock on Georges Bank have crashed due to overfishing.

Demographic traits determine sustainable harvest levels

Populations that have high reproductive capacities can persist even if harvest rates are high. In such populations (which include many species of fish), each female may produce thousands or millions of eggs. In these fast-reproducing populations, growth rates of individuals are often density-dependent. Therefore, if prereproductive individuals are harvested at a high rate, the remaining individuals may grow faster. Some fish populations can be harvested heavily on a sustained basis because a modest number of females can produce sufficient eggs to maintain the population.

Fish can, of course, be overharvested, as illustrated by the story of the black rockfish in Section 54.2. Many fish populations have been greatly reduced because so many individuals were harvested that the few surviving reproductive adults could not maintain the population. Georges Bank off the coast of New England—a source of cod, haddock, and other prime food fishes—was exploited so heavily during the twentieth century that many fish stocks were reduced to levels insufficient to support a commercial fishery (**Figure 54.17**). The haddock fishery has rebounded enough to support a valuable fishery today because commercial fishing ceased and was restarted only after the population had recovered. In contrast, managers reduced fishing pressure on cod only slowly; the cod population has failed to increase.

Mid-nineteenth-century New England fishing logs contain geographically specific catch records. Using these documents, scientists estimated that in 1852, the biomass of cod on the rich fishing bank south of Nova Scotia was about 1,260,000 metric tons. Today the cod biomass is less than 50,000 metric tons.

The whaling industry has also engaged in overharvesting. Twentieth-century whalers hunted the blue whale, Earth's largest animal, nearly to extinction. They then turned to smaller species of whales that were still numerous enough to support commercially viable whaling operations. Most whale populations have failed to recover.

Management of whale populations is difficult for two reasons. First, unlike most fish, whales reproduce at very low rates. They live for years before becoming reproductively mature, produce only one offspring at a time, and have long intervals between births. Thus many adult whales are needed to produce even a small number of offspring. Second, because whales are distributed widely throughout Earth's oceans, they are an international resource whose conservation and wise management depend on cooperative action by all whaling nations.

It is therefore not surprising that the International Whaling Commission, established as an international body to guide the recovery of whale populations, has had a contentious history. The Commission's member nations initially voted to ban all commercial harvesting of whales, but some members have lobbied to restore harvests of the species that are not endangered. Other member nations and most nongovernmental organizations continue to oppose resumption of commercial whaling under any circumstances. The behind-the-scenes machinations are complex and the recommendations of the Commission are not binding. A member nation can choose to continue commercial harvesting, as Norway has done, or conduct commercial whaling under the guise of conducting scientific research to better understand whale population dynamics, as Japan has done.

Market forces rather than international pressures may cause the end of commercial whaling. For example, the Japanese Government's Institute of Cetacean Research relies on selling whale meat to fund its "research" program, but fewer and fewer Japanese people eat whale meat. There is a glut of whale meat on the market despite government campaigns encouraging young people to eat whale meat and providing "whale burgers" for school lunches. Faced with a declining market, the Japanese government may no longer have reason to push for resumption of commercial whaling.

Demographic information is used to control populations

The same management principles apply if we wish to reduce the size of populations of undesirable species and keep them at low densities. At densities well below carrying capacity, populations typically have high birth rates and can therefore withstand higher death rates than they can when they are closer to carrying capacity. When population dynamics are influenced primarily by factors that operate in a density-dependent manner, killing part of a population typically reduces it to a density at which it reproduces at a

Bufo marinus

54.18 Biological Control Gone Awry The toad *Bufo marinus* not only failed to control destructive cane beetles in Australia, but increased dramatically in abundance and now threatens many species of Australian animals.

higher rate. A more effective approach to reducing such a population is to remove its resources, thereby lowering the carrying capacity of its environment. For example, we can rid our cities of rats more easily by making garbage unavailable (reducing the carrying capacity of the rats' environment) than by poisoning rats (which only increases their reproductive rate). However, this option is not useful as a means to control agricultural pests because a high density of the crop is the management objective.

As we saw at the opening of this chapter, humans may attempt to control populations of an undesirable species by introducing one of its predators; this worked for *Opuntia* and *Cactoblastis* in Australia. Sometimes such efforts are successful, but just as often they are not.

Sometimes an introduced predator or parasite simply fails to control its host; more serious consequences occur when a species introduced to control an exotic pest not only attacks the pest species but also destroys other species that are considered valuable. This happened with *Cactoblastis* in Mexico and the southwestern U.S. It is also what happened when the toad *Bufo marinus* was introduced into Australia from Central America (**Figure 54.18**), only to become one of Australia's worst ecological disasters.

The idea was for the toads to control the cane beetles that were attacking sugar cane fields in northern Australia. The toads were believed to have controlled cane beetles in Hawaii, so a number of them were released in Australian fields. Unfortunately, the toads could not reach the beetles, which stayed high on the upper stalks of the sugar cane plants, so they had no effect on the beetle population. But they had massive effects on other species.

All stages of the *Bufo marinus* life cycle are poisonous, and Australian snakes and mammals that eat them usually die. The toads grow fast and outcompete native amphibian species for resources. They have spread from northern Australia down the east coast, where they are threatening the native frog species. Currently the Australian government is spending millions of dollars in an attempt to genetically manipulate a pathogen that will kill *B. marinus* exclusively.

Can we manage our own population?

Managing our own population has become a matter of great concern because the size of the human population contributes to most of the environmental problems we are facing today, from pollution to extinctions of other species. For thousands of years, Earth's carrying capacity for humans was set at a low level by food and water supplies and disease. We saw in Section 53.6 how human social behavior and specialization have allowed us to develop technologies for increasing our resources and combating diseases. Our social behavior, the domestication of plants and animals, improved crops and farm yields, mining and use of fossil fuels, and the development of modern medicine have all contributed to a staggering increase in Earth's human population.

It took more than 10,000 years for Earth's population of *Homo sapiens* to reach 1 billion, an event that occurred in the late nineteenth century. In the subsequent 125 years, the human population has increased to 6.5 billion. The growth rate has slowed somewhat in the last two decades; the World Resources Institute estimates the current worldwide rate of population increase to be about 1.1 percent per year. But with a base of 6.5 billion, even a minimal growth rate means millions more individuals.

If half the people alive today were to mysteriously vanish overnight, there would still be more than twice as many people on Earth as there were in 1900.

Earth's present carrying capacity for humans is set in part by the biosphere's ability to absorb the by-products—especially carbon dioxide—of our enormous consumption of fossil fuel energy; by water availability (in many areas); and by whether we are willing to cause the extinction of millions of other species to accommodate our increasing use of Earth's resources. We will explore the global cycles of these resources in Chapter 56, and we will enumerate some of the consequences of high human population densities and high per capita use of resources for the survival of other species in Chapter 57. Chapter 57 will also discuss the actions being taken to try and maintain Earth's biological diversity.

54.5 RECAP

Efforts to control and manage populations are more likely to be successful if they are based on knowledge of how those populations grow and what determines their densities. At densities well below carrying capacity, populations typically have high birth rates and can therefore withstand higher death rates than they can when they are closer to carrying capacity.

- What general ecological principles guide human efforts to manage other species?

CHAPTER SUMMARY

54.1 How do ecologists study populations?

A **population** consists of the individuals of a species within a particular area.

The distribution of the ages and locations of individuals in a population describes the **population structure**.

Population density is the number of individuals per unit of area or volume.

Demographic events such as births, deaths, immigration, and emigration affect the structure of a population.

Ecologists often mark and track individual animals in population studies.

Population sizes can be estimated from representative samples.

Life tables provide summaries of births, deaths, and other demographic events in a population. Review Table 54.1

A life table tracks a **cohort** of individuals born at the same time and records the **survivorship** of those individuals over time.

54.2 How do ecological conditions affect life histories?

The **life history** of an organism describes how it allocates its time and energy among growth, reproduction, and other activities.

Mortality rates can influence the evolution of life history characteristics.

54.3 What factors influence population densities?

Many populations fluctuate markedly in density, but even the most dramatic fluctuations are much less than those that are theoretically possible.

Populations can exhibit **exponential growth** for short periods, but eventually the **environmental carrying capacity** is approached, causing birth rates to drop and death rates to rise. Review Figure 54.7, Web/CD Activity 54.1 and Tutorial 54.1

Logistic growth is the pattern seen when the growth of a population slows as its density approaches the environmental carrying capacity. Review Figure 54.8, Web/CD Tutorial 54.2

Population densities are determined by the combined influences of **density-dependent** and **density-independent** factors. Review Figure 54.9

When and how species form influences their range and local population density.

Several factors—including resource abundance, the size of individuals in a population, the length of time a species has occupied an area, and social organization—exert strong influences on the densities achieved by populations of different species.

54.4 How do spatially variable environments influence population dynamics?

No species is found everywhere within its mapped range. Members of most species live as separate subpopulations within suitable **habitat patches**. See Web/CD Tutorial 54.3

A **metapopulation** consists of separate subpopulations among which some individuals move frequently.

Extinction of a subpopulation may be prevented by immigration of individuals from another subpopulation; this process is known as the **rescue effect**.

Distant events may influence local population densities.

54.5 How can we manage populations?

To maximize the number of individuals that can be harvested from a population, the population should be kept well below carrying capacity.

Species that have high reproductive capacities can persist even if harvested at high rates.

Reducing the carrying capacity of the environment is a more effective way to reduce an unwanted population than killing its members.

Predators may be introduced to control populations of introduced species, but they may cause other problems.

Earth's carrying capacity for humans depends on our use of resources and the effects of our activities on other species.

SELF-QUIZ

1. The distribution of the ages of individuals in a population and the way those individuals are spread over the environment describes
   a. population dynamics.
   b. population regulation.
   c. population structure.
   d. subpopulation structure.
   e. biomass distribution.
2. The age distribution of a population is determined by
   a. the timing of births.
   b. the timing of deaths.
   c. the timing of both births and deaths.
   d. the rate at which the population is growing.
   e. all of the above
3. Which of the following is *not* a demographic event?
   a. Growth
   b. Birth
   c. Death
   d. Immigration
   e. Emigration
4. A group of individuals born at the same time is known as a
   a. deme.
   b. subpopulation.
   c. Mendelian population.
   d. cohort.
   e. taxon.
5. A population grows at a rate closest to its intrinsic rate of increase when
   a. its birth rates are the highest.
   b. its death rates are the lowest.
   c. environmental conditions are optimal.
   d. it is close to the environmental carrying capacity.
   e. it is well below the environmental carrying capacity.
6. The process by which immigrants prevent a subpopulation from becoming extinct is called the
   a. colonization effect.
   b. rescue effect.
   c. metapopulation effect.
   d. genetic drift effect.
   e. salvage effect.

7. Density-dependent factors have the greatest effect on population densities when
   a. only birth rates change in response to density.
   b. only death rates change in response to density.
   c. diseases spread in populations at all densities.
   d. both birth and death rates change in response to density.
   e. population densities fluctuate very little.
8. A metapopulation is
   a. an unusually large population.
   b. a population that is spread out over a very large area.
   c. a group of subpopulations among which some individuals move.
   d. a group of subpopulations that are isolated from one another.
   e. a group of subpopulations among which individuals move frequently.
9. The best way to reduce the population of an undesirable species in the long term is to
   a. reduce the carrying capacity of the environment for the species.
   b. selectively kill reproducing adults.
   c. selectively kill prereproductive individuals.
   d. attempt to kill individuals of all ages.
   e. sterilize individuals.
10. Populations that are most readily overharvested are characterized by having
   a. very long-lived adults.
   b. short prereproductive periods and many offspring.
   c. short prereproductive periods and few offspring.
   d. long prereproductive periods and few offspring.
   e. long prereproductive periods and many offspring.

FOR DISCUSSION

1. Most organisms whose populations we wish to manage for higher densities are long-lived and have low reproductive rates, whereas most organisms whose populations we attempt to reduce are short-lived, but have high reproductive rates. What is the significance of these differences for management strategies and the effectiveness of management practices?
2. In the mid-nineteenth century, the human population of Ireland was largely dependent on a single food crop, the potato. When a disease caused the potato crop to fail, the Irish population declined drastically for three reasons: (1) a large percentage of the population emigrated to the United States and other countries; (2) the average age of a woman at marriage increased from about 20 to about 30 years; and (3) many families starved to death rather than accept food from Britain. None of these social changes was planned at the national level, yet all contributed to adjusting the population size to the new carrying capacity. Discuss the ecological principles involved, using examples from other species. What would you have done had you been in charge of the national population policy for Ireland at that time?
3. Because some species introduced to control a pest have become pests themselves, some scientists argue that species introductions should not be used under any circumstances to control pests. Others argue that, provided they are properly researched and controlled, we should continue to use introductions as part of our set of tools for managing pest populations. Which view do you support, and why?

FOR INVESTIGATION

The species whose metapopulation dynamics were studied in the experiment described in Figure 54.15 were all tiny animals with limited dispersal abilities. How could experiments be designed to test the role of barriers to recolonization on the metapopulation dynamics of species such as birds, lizards, and mammals, which readily disperse across large areas? Would you expect the results of such experiments to be similar to those using small arthropods on rocks? Why or why not?

CHAPTER 55 Community Ecology

Host sweet host

Many plant species produce sweet nectar in their flowers. This floral nectar attracts pollinators—animals the plant relies on in order to reproduce. But plants in at least several hundred genera also produce nectar on their nonreproductive (vegetative) parts; this *extrafloral* nectar attracts ants. The plant provides the ants with the nectar as well as with other food rewards and, in some cases, nesting sites. For their part the ants patrol the plants and attack herbivores, pathogens, and competing plants.

Some of these ant-hosting plants are vitally dependent on the insects, which in turn live on and depend on a single host plant species. The phenomenon has been extensively studied among Central American thorn trees of the genus *Acacia* and ants of the genus *Pseudomyrmex*. Because both partners in this *mutualism* reproduce independently, the association must be reestablished in each subsequent generation. Individuals of competing species of ants that consume nectar but do not defend the plant may arrive on a young plant before its mutualistic ants have colonized that plant and can drive them away. How can a plant attract the ants that will help it while discouraging ants that will eat its nectar and then move on?

One way in which plants can attract the right insects is by controlling the composition of their nectars. The nectar produced by most plants contains sucrose and varying amounts of glucose and fructose. Sucrose is a particularly important food for most ant species, which produce an enzyme called invertase that cleaves sucrose into monomers that are easily transported through cell membranes. A surprising finding is that, although the nectars of acacia species that do not have intimate associations with ants contain sucrose, the nectar produced by several species of acacias that are defended by specialist ants does not contain sucrose.

Nectar that lacks sucrose is not attractive to generalist ants because they cannot digest its sugars. The specialist ants, on the other hand, readily eat and efficiently digest the acacia's sucrose-free nectar. Why? Because the nectar of the host plant has a trait that makes digestion of its nectar by specialist ants possible: it contains an enzyme that stimulates specific enzymatic activity in the guts of the specialist ants that enables them to digest the sugars found in the nectar.

Thus, the nectar produced by those acacia species that have intimate associations with ants differs chemically from the nectar produced by other acacias, even closely related species. In addition, the digestive enzymes of specialist ants differ from those of generalist

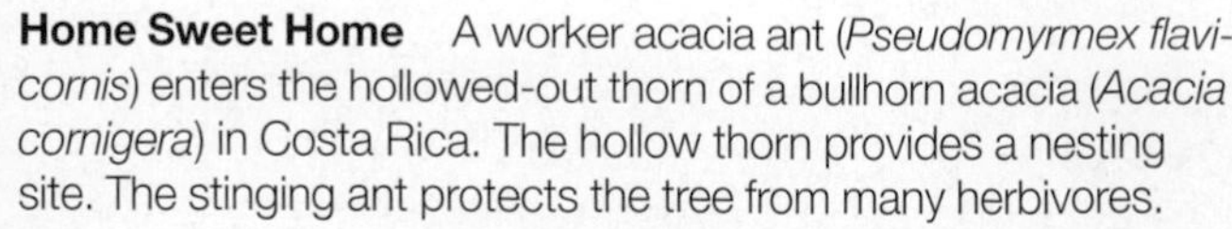

Home Sweet Home A worker acacia ant (*Pseudomyrmex flavicornis*) enters the hollowed-out thorn of a bullhorn acacia (*Acacia cornigera*) in Costa Rica. The hollow thorn provides a nesting site. The stinging ant protects the tree from many herbivores.

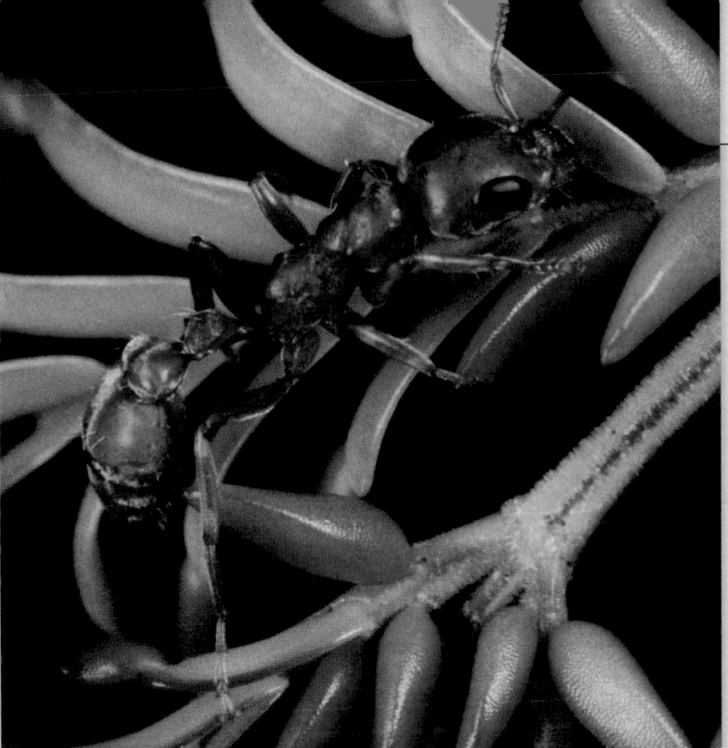

All You Can Eat The acacia ant *Pseudomyrmex ferrugineus* gathers carrot-like growths known as beltian bodies from *Acacia drepanolobium*. Beltian bodies have no known function beyond serving as food for the ant larvae; the tree gains no benefit beyond the presence of the protective ants.

ants. These differences imply that the mutually beneficial association between the acacias and the ants has existed for a very long time. During that time, traits evolved in both partners that benefit the specialist ants but exclude other species that would not help the plant.

All species interact with other species in various ways. Most of those associations are not as specialized as the ones between individual ant and acacia species, but they nevertheless influence the structure and dynamics of ecological communities. How they do so is the subject of this chapter.

IN THIS CHAPTER we will describe ecological communities, discuss the processes that determine community structure, and show how those processes interact in nature. We will also consider how disturbances affect communities and how ecological communities are assembled over time. We will conclude by considering factors that determine how many species can live together in ecological communities.

CHAPTER OUTLINE

55.1 What Are Ecological Communities?

Charles Darwin is remembered mostly for his contributions to evolutionary theory, but as the following quote from *The Origin of Species* shows, he was also a pioneering ecologist who understood the nature and complexity of the interactions among the species of organisms that live in a particular place.

> It is interesting to contemplate an entangled bank, clothed with many plants of many kinds, with birds singing on the bushes, with various insects flitting about, and with worms crawling through the damp earth, and to reflect that these elaborately constructed forms, so different from each other, and dependent on each other in so complex a manner, have all been produced by the laws acting around us.

The species that live and interact in an area constitute an ecological **community**. The "entangled bank" near Darwin's home was an ecological community that had obvious boundaries defined by adjacent crops, pastures, and gardens. But the organisms living in the bank were not confined within those boundaries. Some of the seeds that landed in the bank and grew into trees and shrubs came from parent plants living far away. The insects and birds Darwin observed must have flown into and out of the bank from a large area. To understand which species live in the bank and how they interact, he would have needed to know about such movements. Knowledge of how the bank had changed over time would also be relevant. Glaciers had covered the area 10,000 years earlier. The plant species Darwin observed had colonized Britain at different times over the thousands of years since the glaciers melted.

Communities are loose assemblages of species

Early in the twentieth century, two leading North American plant ecologists debated the nature of communities. Henry Gleason argued in 1926 that plant communities were loose assemblages of species, each of which was individually distributed according to its unique interactions with the physical environment. In contrast, in a paper published in 1936, Frederick Clements argued that plant communities were tightly inte-

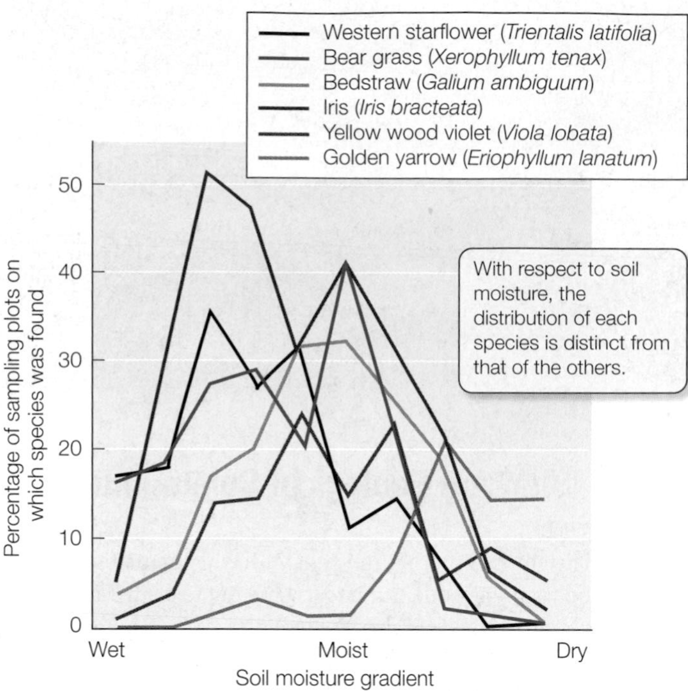

55.1 Plant Distributions along an Environmental Gradient The abundances of different plant species change gradually and individually along a soil moisture gradient in Oregon's Siskiyou Mountains.

grated "superorganisms," and that communities in similar environments would have the same species composition unless they had been recently disturbed.

The debate was resolved by detailed studies of the distributions of plants. Especially influential were analyses of the vegetation of the Siskiyou Mountains of Oregon carried out by Robert Whittaker, who showed that different combinations of plant species are found at different locations. Species enter and drop out of communities independently over environmental gradients (**Figure 55.1**). These and other results generally supported Gleason's view of the nature of communities. However, where environmental conditions change abruptly, as they do at the edges of lakes and streams, the ranges of many species may terminate at the same place.

Thus ecological communities are not assemblages of organisms that move together as units when environmental conditions change. Rather, each species has unique interactions with its biotic and abiotic environments. But if ecological communities are just loose assemblages of species, why do we care about them? We care about ecological communities in part because we wish to know how these assemblages of species, however loose they might be, function. But we also care about them because we too are members of those communities. We interact with many other species, and as we will see, those interactions affect human welfare in many ways.

The organisms in a community use diverse sources of energy

Most ecological communities contain hundreds of species that interact with one another and the physical environment in multitudinous ways, so trying to understand how communities function might seem to be an impossible task. Fortunately, we do not need to know all of the details to make considerable progress. We can understand a great deal by simply finding out who eats whom. The organisms in a community can be divided into trophic levels based on the source of their energy (**Table 55.1**). A **trophic level** consists of the organisms whose energy source has passed through the same number of steps to reach them. Plants and other photosynthetic organisms (*autotrophs*) get their energy directly from sunlight. Collectively, they constitute a trophic level called *photosynthesizers*, or **primary producers**. They produce the energy-rich organic molecules that nearly all other organisms consume.

In most ecological communities, all nonphotosynthetic organisms (*heterotrophs*) consume, either directly or indirectly, the energy-rich organic molecules produced by primary producers. Organisms that eat plants constitute a trophic level called *herbivores* or **primary consumers**. Organisms that eat herbivores are called **secondary consumers**. Those that eat secondary consumers are called *tertiary consumers*, and so on. Organisms that eat the dead bodies of organisms or their waste products are called *detritivores* or **decomposers**. Organisms that obtain their food from more than one trophic level are called *omnivores*. Because many species are omnivores, trophic levels are often not clearly distinct, but if we remember that boundaries between trophic levels are fuzzy, the con-

TABLE 55.1

The Major Trophic Levels

TROPHIC LEVEL	SOURCE OF ENERGY	EXAMPLES
Photosynthesizers (primary producers)	Solar energy	Green plants, photosynthetic bacteria and protists
Herbivores (primary consumers)	Tissues of primary producers	Termites, grasshoppers, gypsy moth larvae, anchovies, deer, geese, white-footed mice
Primary carnivores (secondary consumers)	Herbivores	Spiders, warblers, wolves, copepods
Secondary carnivores (tertiary consumers)	Primary carnivores	Tuna, falcons, killer whales
Omnivores	Several trophic levels	Humans, opossums, crabs, robins
Detritivores (decomposers)	Dead bodies and waste products of other organisms	Fungi, many bacteria, vultures, earthworms

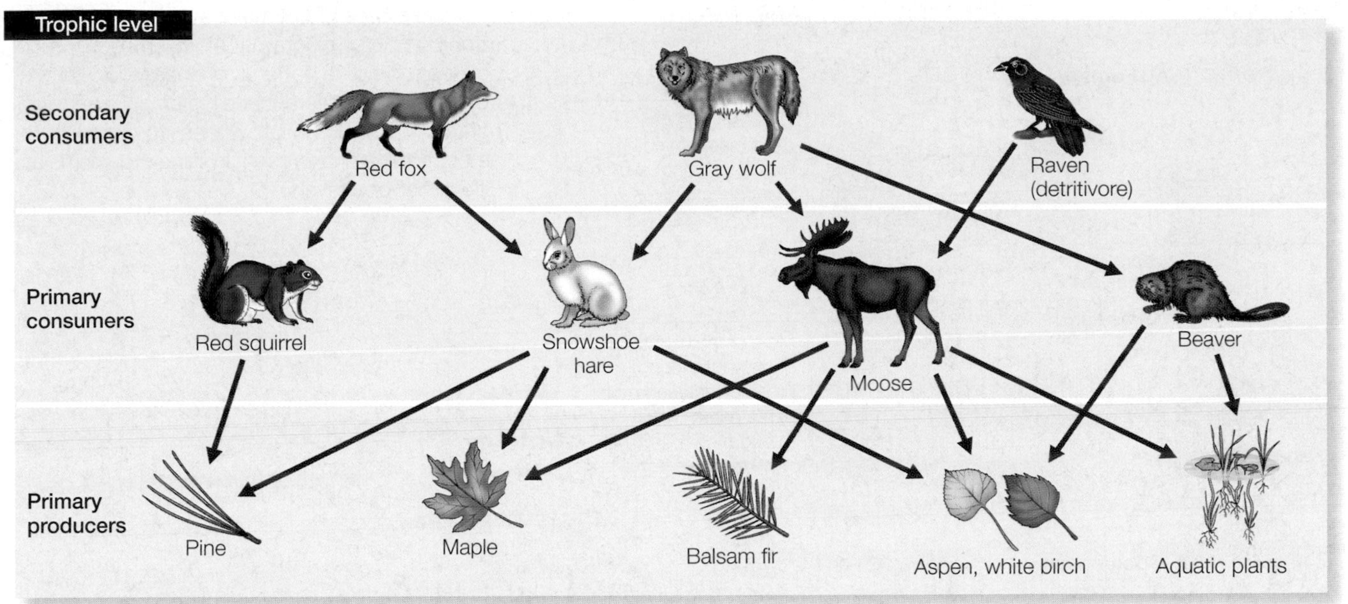

55.2 Food Webs Show Trophic Interactions in a Community This food web for Isle Royale National Park, located on a large island in Lake Superior, includes only large vertebrates and the plants on which they depend. Even with these restrictions, the web is complex. The arrows show who eats whom.

cept still provides a useful way of thinking about energy flow through communities.

A sequence of interactions in which a plant is eaten by an herbivore, which is in turn eaten by a secondary consumer, and so on, can be diagrammed as a **food chain**. Food chains are usually interconnected to make a **food web** because most species in a community eat and are eaten by more than one other species (**Figure 55.2**). Ecological communities contain so many species that it is impossible to show all of them in a food web. Even so, simplified diagrams of food webs can help us understand the trophic interactions among organisms in an ecosystem.

Despite their considerable differences, most communities have only three to five trophic levels. Why are there so few levels? Loss of energy between trophic levels is partly responsible. To show how energy decreases at each step as it flows from lower to higher trophic levels, ecologists construct diagrams that show the distribution of energy or **biomass** (the weight of living matter) at each trophic level in a community. Diagram such as those in **Figure 55.3** show the amount of energy or biomass that is available at a given time for organisms at the next trophic level.

Distributions of energy and biomass for a particular ecosystem usually have similar shapes. Variations in their dimensions depend on the nature of the dominant organisms at each trophic level and how they allocate their energy. In most terrestrial ecosystems, photosynthetic plants dominate, both in terms of the energy they represent and the biomass they contain. They store energy for long periods, some of it in difficult-to-digest forms (such as cellulose and lignin). In forests, the biomass at the primary producer level is mostly wood, which is rarely eaten unless the plant is diseased or otherwise weakened. In contrast, grassland plants produce few hard-to-digest woody tissues. Mammals may consume 30–40 percent of the annual aboveground grassland plant biomass; insects may consume an additional 5–15 percent. Soil organisms, primarily nematodes, may consume 6–40 percent of the belowground biomass in grasslands. Thus, relative to the biomass of plants, the biomass of herbivores is larger in grasslands than in forests (**Figure 55.3A,B**).

In most aquatic ecosystems, the dominant photosynthesizers are bacteria and protists. Those unicellular organisms have such high rates of cell division that a small biomass of photosynthesizers can feed a much larger biomass of herbivores, which grow and reproduce much more slowly. This pattern can result in an inverted distribution of biomass, even though the energy distribution for the same ecosystem has the typical shape (**Figure 55.3C**).

Much of the energy ingested by organisms is converted to biomass that is eventually consumed by decomposers, members of a trophic level not shown in Figure 55.3. **Detritivores** such as bacteria, fungi, worms, mites, and many insects, transform *detritus* (the dead remains and waste products of organisms) into free mineral nutrients that can again be taken up by plants. If there were no detritivores, most nutrients would eventually be tied up in dead bodies, where they would be unavailable to plants. Continued ecosystem productivity depends on the rapid decomposition of detritus.

Dense populations of detrivorous crabs live in the nutrient-poor hydrothermal vent communities of certain shallow ocean waters. During slack tide, when water currents cease, the sulfurous vent plumes asphyxiate vast numbers of copepods, which sink to the ocean floor. The crabs swarm out of cracks in the surrounding rocks and feed on the dead copepods.

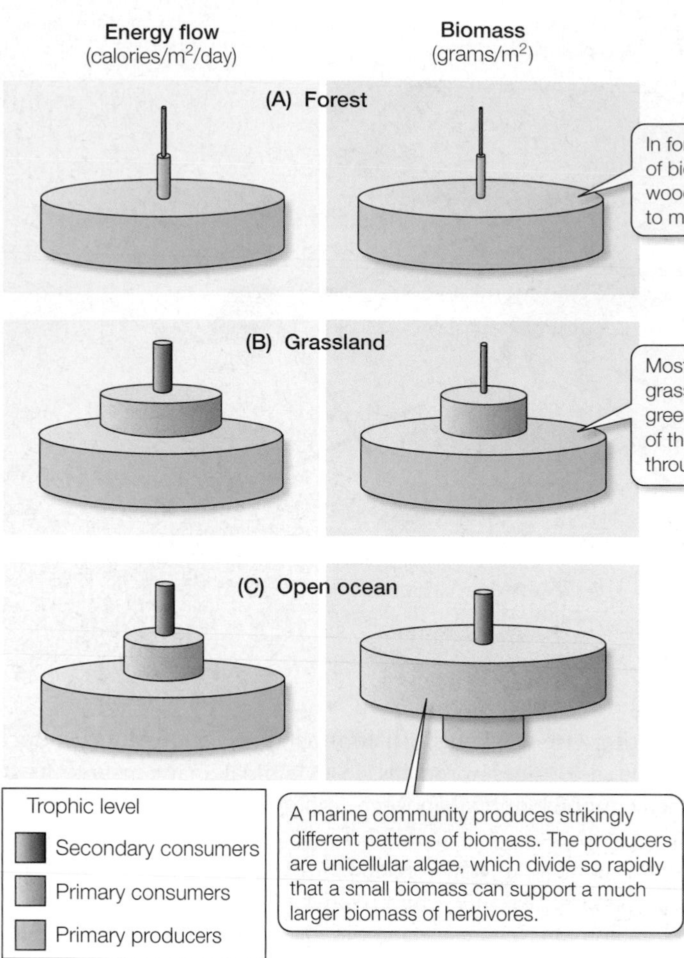
Energy flow
(calories/m²/day)
Biomass
(grams/m²)
(A) Forest
In forests, the majority of biomass is tied up in wood and is not available to most herbivores.
(B) Grassland
Most of the biomass in a grassland is found in the green plants, and most of the energy flows through them.
(C) Open ocean
A marine community produces strikingly different patterns of biomass. The producers are unicellular algae, which divide so rapidly that a small biomass can support a much larger biomass of herbivores.
Trophic level
Secondary consumers
Primary consumers
Primary producers

55.3 Diagrams of Biomass and Energy Distributions Energy diagrams (left column) allow ecologists to compare patterns of energy flow through trophic levels in different ecosystems. Biomass diagrams (right column) allow them to compare the amount of material present in living organisms at different trophic levels.

55.1 RECAP

The species that live and interact in an area constitute an ecological community. Although each species has unique interactions within the ecological community, the community as a whole can be studied on the basis of the distribution of energy and biomass within it.

- Do you understand how the concept of trophic levels is useful when describing ecological communities? See p. 1186 and Table 55.1
- What are food webs and what do they describe? See p. 1187 and Figure 55.2
- What is the primary cause of the differences between patterns of distribution of energy and biomass in communities? See p. 1187 and Figure 55.3

Energy and biomass distributions reveal some important things about ecological communities, but they do not tell us which processes most strongly influence the structure and dynamics of a community. In the next section we explore different types of interactions between species and how they influence the properties of communities as a whole.

55.2 What Processes Influence Community Structure?

The properties of ecological communities are influenced not only by who eats whom, but also by how organisms affect one another as they seek food. Fortunately, the many ways in which organisms interact with one another also fall into a small number of categories (**Table 55.2**):

- **Predation** or **parasitism**: interactions in which one participant is harmed, but the other benefits (+/– interactions)
- **Competition**: interactions in which two organisms use the same resources and those resources are insufficient to supply their combined needs (–/– interactions)
- **Mutualism**: interactions in which both participants benefit (+/+ interactions)
- **Commensalism**: interactions in which one participant benefits but the other is unaffected (+/0 interactions)
- **Amensalism**: interactions in which one participant is harmed but the other is unaffected (0/– interactions)

All five types of interactions, combined with the effects of the physical environment, influence the population densities of species. They may also restrict the range of environmental conditions under which species can persist. If there were no competitors, predators, or pathogens in their environment, most species would be able to persist under a broader array of abiotic conditions than they do in the presence of other species. On the other hand, the presence of mutualists may increase the range of physical conditions under which a species can persist.

Predation and competition might, at first glance, seem to be distinct processes, but they interact strongly because most organisms are preyed on by more than one other species, and because most predators include many other species in their diets. Competitors are often predators that share diets. Let's look more closely at all of these types of interactions to see under what conditions they operate and how they influence community dynamics.

TABLE 55.2

Types of Ecological Interactions

		EFFECT ON ORGANISM 2		
		HARM	BENEFIT	NO EFFECT
EFFECT OF ORGANISM 1	HARM	Competition (–/–)	Predation or parasitism (–/+)	Amensalism (–/0)
	BENEFIT	Predation or parasitism (+/–)	Mutualism (+/+)	Commensalism (+/0)
	NO EFFECT	Amensalism (0/–)	Commensalism (0/+)	—

Predation and parasitism are universal

Predation and parasitism are universal processes. Every species is eaten by at least one other species, and no species is entirely free of parasites and pathogens. *Parasites* are typically smaller than their hosts. They may live inside or outside the bodies of their hosts. Parasites often feed on their hosts without killing them. Some parasites are only slightly smaller than their hosts, but *microparasites,* such as pathogenic viruses, bacteria, and protists, are much smaller than their hosts. Multiple generations of microparasites may reside within a single host individual, and a host may harbor thousands or millions of them. As a result, parasite–host interactions differ in interesting ways from predator–prey interactions.

Predators are typically larger than and live outside the bodies of their prey. Herbivores are predators of plants; they may feed on many individual plants without killing them, whereas predators of animals typically kill their prey.

PREDATOR AND PREY POPULATIONS OFTEN OSCILLATE You might think that predators would simply reduce the sizes of their prey populations, but the consequences of predator–prey interactions are much more complex. Predators often do depress populations of their prey, but they also cause fluctuations in prey population *densities*. In part because of this aspect of predation, predator–prey interactions may actually result in increases in the prey populations.

Growth of a predator population nearly always lags behind growth in its prey populations. As a predator population grows, it may eat most of its prey populations. The predator population, which no longer has enough food, then crashes. Population oscillations among small mammals and their predators living at high latitudes, where there are only a few species of predators and prey, are the best-known examples of such population density fluctuations driven by predator–prey interactions. Populations of Arctic lemmings and their chief predators—snowy owls, jaegers, and Arctic foxes—oscillate with a 3- to 4-year periodicity. Populations of Canadian lynx and their principal prey, snowshoe hares, oscillate on a 9- to 11-year cycle (**Figure 55.4**).

For many years, ecologists thought that hare–lynx population oscillations were caused only by interactions between hares and lynx. Recently, Charles Krebs and his associates at the University of British Columbia performed experiments in Yukon Territory, Canada, to test the hypothesis that the lynx–hare oscillations are caused by fluctuations in the hares' food supply as well as by lynx predation. They enclosed some areas with fences through which hares, but not lynx, could pass, and they provided food in some of the enclosures. The results of the experiments showed that the oscillations are driven both by lynx predation and by interactions between hares and their food supply (**Figure 55.5**).

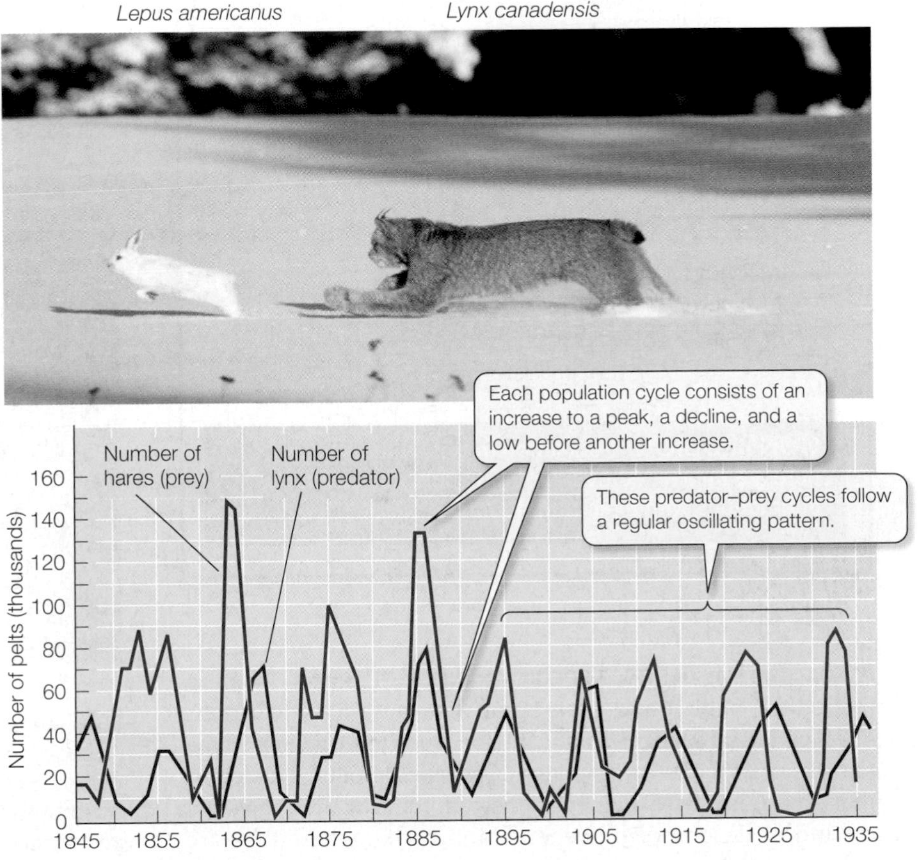

55.4 Hare and Lynx Populations Cycle in Nature The 9- to 11-year population cycle of the snowshoe hare and its major predator, the Canadian lynx, was revealed in records of the number of pelts that were sold to the Hudson's Bay Company by fur trappers.

EXPERIMENT

HYPOTHESIS: Population cycles of hares are influenced by both food supply and predators.

METHOD

1. Select 9 1-km^2 blocks of undisturbed coniferous forest.
2. In two of the blocks give the hares supplemental food year-round.
3. Erect an electric fence around two other blocks, with mesh large enough to allow hares, but not lynxes, to pass through.
4. Provide extra food in one of these enclosed blocks.
5. In two other blocks add fertilizer to increase food quality.
6. Use three other blocks as unmanipulated controls.

RESULTS

Food added

Adding food tripled hare density.

Predators excluded

Excluding predators doubled hare density.

Fertilizer added

Fertilizing vegetation to increase its food quality had no significant effect.

Food added and predators excluded

Both adding food and excluding predators increased hare density dramatically.

Ratio of hare density to controls

Control

One hare population cycle (11 years)

CONCLUSION: Population cycles of the snowshoe hare are influenced by their food supply as well as by interactions with their predators.

55.5 Prey Population Cycles May Have Multiple Causes Experiments showed that both food supply (but not food quality) and predation by lynx affected the population densities of snowshoe hares.

PREDATORS MAY RESTRICT SPECIES' RANGES Predators may also restrict the habitat and geographic distribution of their prey. The Australasian biogeographic region (see Figure 52.8) is home to a group of birds—megapodes, also called mound-builders—that do not incubate their eggs. Instead, they lay their eggs in a mound of decomposing vegetable material, where they are warmed by the heat of decomposition (**Figure 55.6**). The parent megapodes regularly visit the nest mound and, if necessary, add or remove decaying material to maintain the eggs at an appropriate temperature.

Megapodes are excellent dispersers. They have colonized many remote oceanic islands and are found on many islands west of Wallace's line (see Figure 52.7), but they are absent from all islands with Asian mammalian predators. Unattended eggs in large, conspicuous mounds of decomposing vegetation would be obvious to egg-eating mammals. Megapodes have survived only in regions where the predators are all marsupials, few of which eat eggs.

MIMICRY EVOLVES IN RESPONSE TO PREDATION Predators do not capture prey individuals randomly. Prey individuals vary in ways that make them more or less susceptible to being captured. Therefore, prey species have evolved a rich variety of adaptations that make them more difficult to capture, subdue, and eat. Among these adaptations are toxic hairs and bristles, tough spines, noxious chemicals, camouflage, and mimicry of inedible objects or of larger and more dangerous organisms. Predators, in turn, have evolved to be more effective at overcoming such prey defenses.

Mimicry is the best-studied adaptation of prey to predation. A palatable species may mimic an unpalatable or noxious one—a process called **Batesian mimicry**—or two or more unpalatable or noxious species may converge to resemble one another—a process called **Müllerian mimicry**. Batesian mimicry works because a predator that captures an individual of an unpalatable or noxious species learns to avoid other prey individuals of similar appearance. However, if a predator captures a palatable mimic, it is re-

Leipoa ocellata

55.6 Megapode Distributions are Limited by Mainland Predators A megapode at its nest mound. Egg-eating mammalian predators quickly destroy the easily accessible eggs; thus megapodes are limited to areas where the only mammalian predators are marsupials, among whom egg-eating is rare.

warded with food. It learns to associate palatability with the appearance of that prey. As a result, individuals of unpalatable species are attacked more often than they would be if they had no Batesian mimics. Unpalatable individuals that differ from their mimics more than the average are less likely to be attacked by predators that have eaten a mimic. Therefore, directional selection (see Section 22.3) causes unpalatable species to evolve away from their mimics. Batesian mimicry systems can evolve and be maintained only if the mimic evolves toward an unpalatable species faster than the unpalatable species evolves away from it. Generally, this happens only if the mimic is less common than the unpalatable species.

All species in a Müllerian mimicry system benefit when inexperienced predators eat individuals of any of the species because the predators learn that that all species of similar appearance are unpalatable. Some of the most spectacular tropical butterflies are members of Müllerian mimicry systems (**Figure 55.7**), as are many kinds of bees and wasps.

HOSTS RESIST INFECTION BY MICROPARASITES For a microparasite population to persist in a host population, at least one new host individual, on average, must become infected with the microparasite before each infected host dies. Members of host populations involved in a microparasite–host interaction fall into three distinct classes: susceptible, infected, or recovered (and thus immune; see Chapter 18). Changes in the numbers of individuals in each class depend on births, deaths, infections, and development and loss of immunity.

A microparasite can readily invade a host population dominated by susceptible individuals, but as the infection spreads, fewer and fewer susceptible individuals remain. Eventually a point is reached at which infected individuals do not, on average, transmit the infection to at least one other individual. Then the infection dies out. As a result, rates of microparasite infection typically rise, then fall, and do not rise again until a sufficiently dense population of susceptible host individuals has reappeared.

A microparasite can be transferred from one host individual to another by direct body contact; via the breath, or body fluids, or waste products of infected individuals; by water; or by an animal vector. A single infected host can most readily infect a large number of other individuals if it continues to infect others for a long time, even after it has died. An infected host individual can most easily spread an infection to many other hosts if the microparasite is dispersed by water. Cholera, one of the most deadly human diseases, is caused by the bacterium *Vibrio cholerae*, which usually lives in the ocean where it infects copepods and other small planktonic animals, but also lives in fresh water. People ingest *V. cholerae* by drinking contaminated water. The bacterium produces a toxin that damages the salt balance mechanisms of the cells that line the small intestine, and infected persons defecate large quantities of the bacteria (see the opening of Chapter 5). A single infected person can release thousands of pathogenic bacteria into local waters and stimulate a new infection. In Bangladesh, where most people get their drinking water directly from rivers, filtering river water through folded cloth can cut infection rates by 50 percent (**Figure 55.8**).

Researchers recently found six viruses of primate origin in the blood of 930 people in Cameroon who had eaten freshly killed primates, or "bush meat." Logging of tropical forests, which increases hunters' access to primates and generates local markets for bush meat, is thus implicated in the transfer of potentially dangerous viruses to humans.

55.7 Batesian and Müllerian Mimicry Systems By converging in appearance, the unpalatable Müllerian mimics among these different species of Costa Rican butterflies and moths reinforce one another's ability to deter predators. The palatable Batesian mimics benefit because predators learn to associate these color patterns with unpalatability.

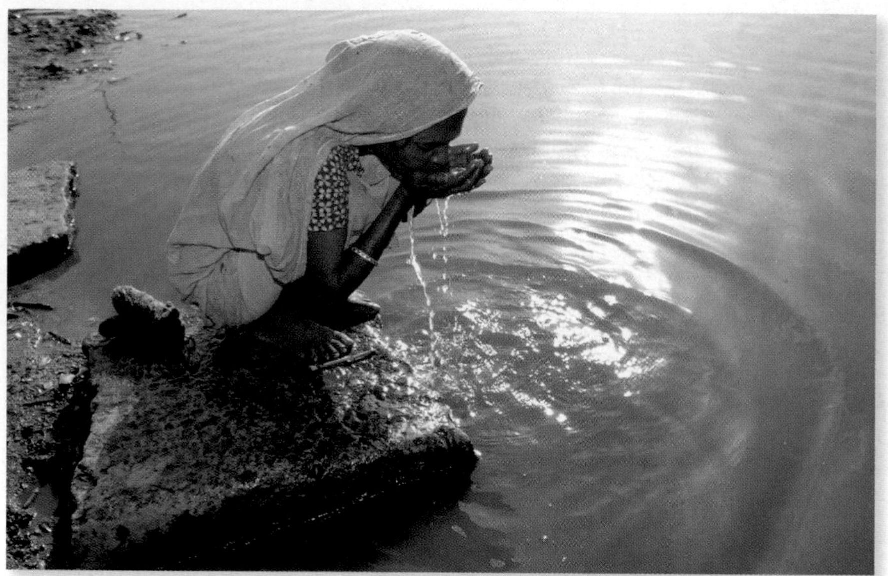

55.8 Filtering Water Can Help Combat Cholera The *Vibrio cholerae* bacterium is spread in contaminated water supplies. When drinking water is obtained from open supplies such as this one, even the simple act of filtering water through a cloth can lessen the rate of infection.

Competition is widespread because all species share resources

Nearly all species share at least part of their diet and use of other resources with other species, but resource sharing influences the abundances and distributions of species only if individuals reduce the ability of others to access resources, either by interfering with their activities—**interference competition**—or by reducing the available resources—**exploitation competition**.

Competition may occur among individuals of a single species or among individuals of different species. *Intraspecific* competition—competition among individuals of the same species—may result in reduced growth and reproductive rates for some individuals, may exclude some individuals from better habitats, and may cause the deaths of others. Intraspecific competition is a primary cause of the density-dependent birth and death rates that we discussed in Section 54.3. *Interspecific* competition—competition among individuals of different species—affects individuals in the same way, but in addition, a superior competitor can prevent all members of another species from using a habitat, a phenomenon called **competitive exclusion**.

COMPETITION MAY RESTRICT SPECIES' HABITAT USE Photosynthetic plants typically compete for space, which is why gardeners and farmers weed their crops. Occupation of space gives a plant access to sunlight (for which shoots compete) and to water and mineral nutrients (for which roots compete).

Competition among sessile animals may also restrict their habitat distribution. For example, two species of barnacles, *Balanus balanoides* and *Chthamalus stellatus*, compete for space in intertidal zones (**Figure 55.9**). The planktonic larvae of both species settle between high and low tide levels on the rocky shorelines of the North Atlantic Ocean and metamorphose into sessile adults. Curiously, however, the two populations end up occupying distinct settlement areas. Adult *Chthamalus* generally live higher in the intertidal zone than do adult *Balanus*, and there is little overlap be-

55.9 Competition Restricts the Intertidal Ranges of Barnacles Interspecific competition between *Balanus* and *Chthamalus* makes the zone each species occupies smaller than the zone it could occupy in the absence of the other species. The widths of the red and blue bars are proportional to the densities of the populations.

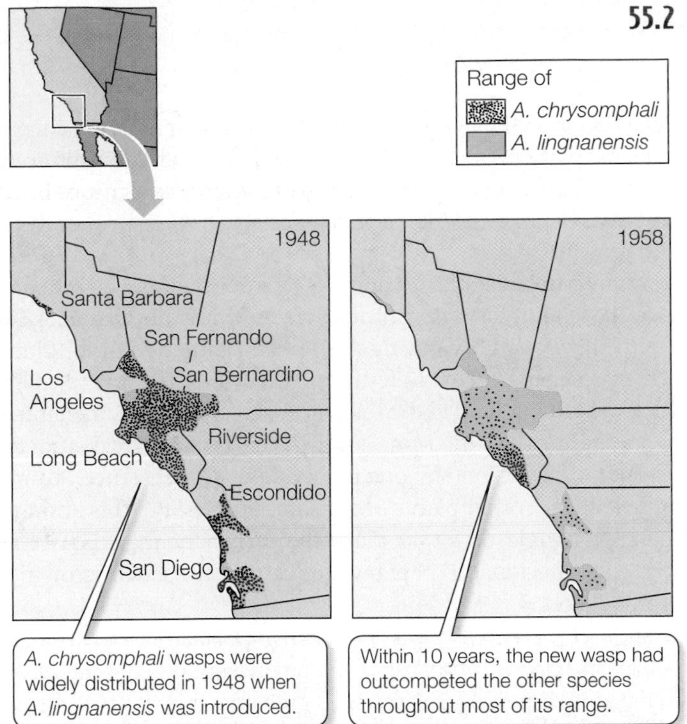

55.10 A Parasitoid Wasp Outcompetes Its Close Relative *Aphytis lingnanensis* displaced *A. chrysomphali* over most of its range within a decade of its introduction.

tween areas occupied by the two species. What explains their distinct distributions in the intertidal zone?

By experimentally removing one or the other species, Joseph Connell showed that the vertical range of adults of each species is greater in the absence of the other species. *Chthamalus* larvae normally settle in large numbers in the *Balanus* zone. If *Balanus* are absent, young *Chthamalus* survive and grow well in the *Balanus* zone, but if *Balanus* are present, they smother, crush, or undercut the *Chthamalus*. *Balanus* larvae also settle in the *Chthamalus* zone, but the young *Balanus* grow slowly there because they lose water rapidly when exposed to air, so *Chthamalus* outcompete *Balanus* in that zone. The result of the competitive interaction between the two species is a pattern of *intertidal zonation*, with *Chthamalus* growing above *Balanus*.

COMPETITION MAY RESTRICT SPECIES' RANGES A species can restrict the range of another species by reducing populations of shared prey to such low levels that the other species cannot persist. Consider the distributions of two species of parasitoid wasps that parasitize scale insects. The wasps, whose larvae burrow into, eat, and kill scale insects, were introduced to southern California to control outbreaks of scale insects that were damaging citrus orchards. The Mediterranean wasp *Aphytis chrysomphali* was introduced around 1900, but it failed to control the scale insects. Therefore, a close relative from China, *A. lingnanensis*, was introduced in 1948. *A. lingnanensis*, which has a higher reproductive rate, increased rapidly. Within a decade it had greatly reduced population densities of the scale insects and had displaced *A. chrysomphali* from most of its range in California (**Figure 55.10**).

Commensal and amensal interactions are widespread

Amensalisms are widespread and inevitable interactions. For example, herds of mammals drinking at a water hole may trample and kill many plants. There is no benefit to the mammals in trampling the plants; the destruction of the plants is unintentional but inevitable. Trees drop dead leaves and branches, often harming smaller plants and unsuspecting animals beneath them. Commensal and mutualistic relationships are also everywhere.

Consider a large herbivorous mammal such as a rhinoceros grazing on the African plains (**Figure 55.11**). As it forages, the mammal disturbs multitudes of insects in the grass, crushing some and inadvertently consuming others in an amensal relationship.

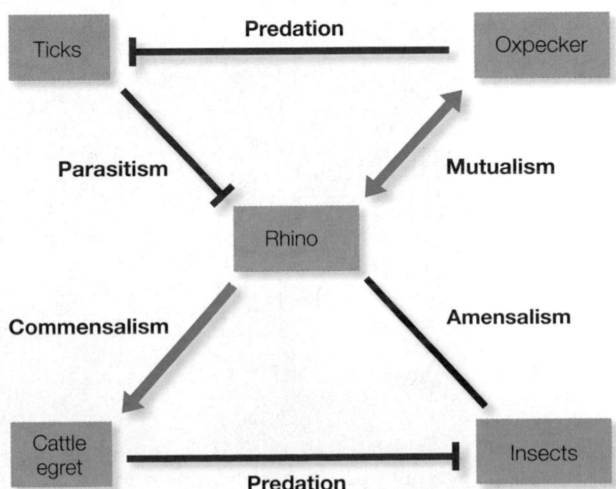

55.11 A Single Small Community Demonstrates Many Interactions On the African plain, large herbivores such as the black rhinoceros disturb insect communities. The cattle egret pounces on the displaced insects; neither the amensal displacement of the insects nor the commensal activity of the birds has any effect on the rhino. Oxpeckers (the small, dark bird on the rhino's back) remove ticks from the skin of many large African mammals. In this mutualism, the birds get food and the mammals get relief from parasites. The diagram at right reveals the ecological complexity of the simple scene.

In addition, a commensal relationship has evolved between large herbivores and some bird species that are predators of insects. Birds such as cattle egrets typically forage on the ground near the mammals' heads and feet, where they capture insects the insects flushed by their hooves and mouths Cattle egrets foraging close to grazing mammals capture more food for less effort than egrets foraging away from grazing mammals. The benefit to the egrets is clear; the mammal neither gains nor loses. In another aspect of the community shown in Figure 55.11, a *mutualistic* relationship has evolved between the rhino and another bird species; birds known as oxpeckers pluck blood-sucking ticks from the skin of the grazing mammal. The bird gains a meal, and the mammal gains some protection from a parasite. Such mutualisms exist between many species and have been the subject of wide study.

Most organisms participate in mutualistic interactions

Mutualistic interactions exist between plants and microorganisms, between protists and fungi, among plants and insects, among animals, and among plants. Most plants have critical and beneficial associations with soil-inhabiting fungi called mycorrhizae, which enhance the plant's ability to extract minerals from the soil (see Figure 30.10). The mutualistic relationship between plants and nitrogen-fixing bacteria of the genus *Rhizobium* is the basis of much of life as we know it (see Section 36.4).

Examples of mutualisms between animals and protists abound. Corals and some other marine organisms gain most of their energy from photosynthetic protists living within their tissues. In exchange, they provide the protists with nutrients from the small planktonic animals they capture. Termites have protists in their guts that help them digest the cellulose in the wood they eat. The termites provide the protists with a suitable environment in which to live and an abundant supply of cellulose.

Terrestrial plants have many mutualistic associations with animals. As we saw at the opening of this chapter, many plants produce nectar on their vegetative parts that attracts ants, which provide the plants with protection from their predators and competitors. Experiments with Central American acacias have shown that trees deprived of their ants are heavily attacked by herbivores and grow poorly.

Many plants depend on animals to move their pollen and provide those animals with nutrient-rich rewards (**Figure 55.12A**). The plants benefit by having their pollen carried to other plants and by receiving pollen to fertilize their ovules. The animals benefit by obtaining food in the form of nectar and pollen. Movement to another plant of the same species is encouraged by the limited amount of nectar on any one plant and by the existence of similar rewards on other plants of the same species. But this arrangement has a trade-off for the plant: the energy and materials it uses to produce nectar and other rewards cannot be used for growth or reproduction.

Interactions between plants and their pollinators and seed dispersers are clearly mutualistic, but they are not purely mutualistic. Many seed dispersers are also seed predators that destroy some of the seeds they remove from plants. Some animals that visit flowers cut holes in the petals and gain access to the nectar without transferring any pollen. On the other hand, some plants exploit their pollinators. The flowers of certain orchids, for example, mimic female insects, enticing male insects to copulate with them (**Figure 55.12B**). The male insects neither sire any offspring nor obtain any reward, but they transfer pollen between flowers, benefiting the orchids.

55.12 Plant–Animal Mutualisms Are Important in Pollination
(A) A bat of the genus *Brachyphylla* obtains nectar from an orchid plant in the West Indies. Some yellow pollen usually adheres to the bat's mouthparts and head and is then spread to other flowers. (B) In a twist, pollination of the *Ophrys scolopax* orchid is no longer a mutualism. The male bee (*Eucera longicornis*) is deceived by its odor and appearance into attempting copulation. The orchid will be pollinated, but the bee will receive no reward and has wasted valuable energy.

(A)

(B)

55.2 RECAP

Species interactions can be grouped into five categories: predator–prey (or parasite–host) interactions; competition; commensalisms; amensalisms; and mutualisms.

- Can you explain why predator and prey population densities often undergo oscillations? See p. 1189 and Figure 55.5
- Can you name two different ways in which organisms compete? See p. 1192
- Can you name at least two widespread kinds of mutualisms that are vital to sustain the plants that support all terrestrial life? See p. 1194

As illustrated in Figure 55.11, many interactions are going on at the same time in any given community. How do these multiple interactions influence the properties of ecological communities?

55.3 How Do Species Interactions Cause Trophic Cascades?

The interactions of a single predator species in a community can cause a progression of indirect effects across successively lower trophic levels. Such a pattern is called a **trophic cascade** and can be illustrated by the effects of the wolf population in the Lamar Valley of Yellowstone National Park.

One predator can affect many different species

The food web in Yellowstone National Park is extremely complex. Wolves in the park feed on pronghorn, mule deer, elk, bison, and bighorn sheep in addition to moose. They share these prey with coyotes, mountain lions, and grizzly and black bears. Despite the complexity of interactions in Yellowstone, wolves exert particularly strong effects on the Park's community structure and dynamics.

America's first national park, Yellowstone was established in 1872, but unrestricted hunting of mammals continued within and around the park for years after its establishment. In 1886, the U.S. Army assumed responsibility for protecting wildlife resources in the park, but the soldiers continued to kill mammalian predators (except for bears) in order to increase the populations of the large herbivores visitors wanted to see. Killing of predators continued when the National Park Service took over management of the park in 1918. By 1926, wolves had been extirpated from the park (**Figure 55.13A**).

Annual censuses of elk were initiated in Yellowstone in 1920. To prevent elk from exceeding the park's carrying capacity, the park service culled elk herds until 1968, when, in response to public pressure, elk kills were stopped. When the culling ended, the elk population rapidly increased (**Figure 55.13B**). When wolves were absent, the elk population browsed aspen trees so intensely that no young trees were recruited to the population after 1920 (**Figure 55.13C,D**). The elk also severely browsed streamside willows,

(A)

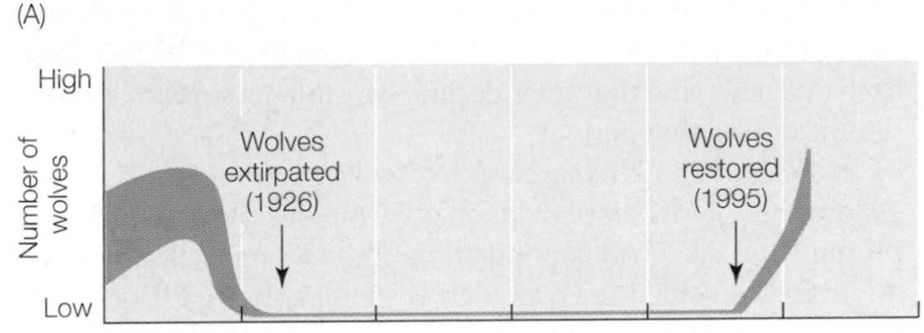

(B)

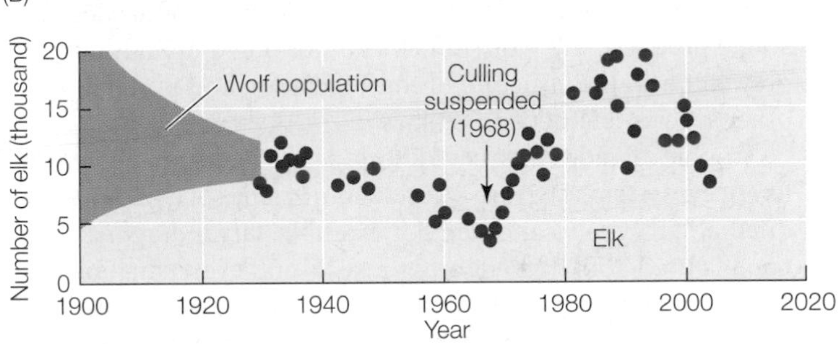

(C)

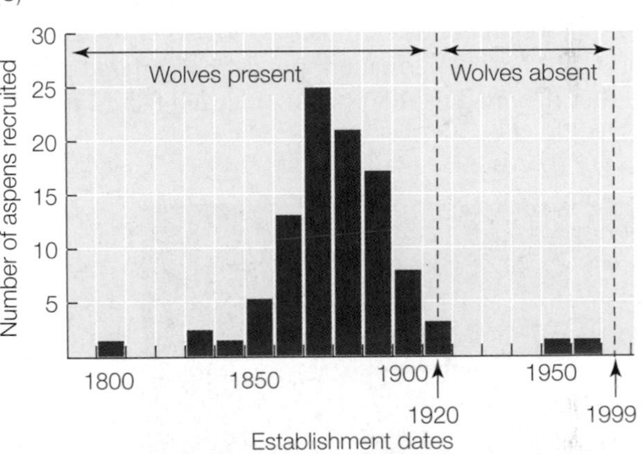

(D)

Populus tremuloides

55.13 Wolves Initiated a Trophic Cascade (A) Wolves were eliminated from Yellowstone National Park in 1926 and reintroduced in 1995. (B) In the absence of wolves, culling controlled elk populations until 1968. When culling of elk ended in 1968, elk populations grew rapidly. (C) In the absence of wolves, the elk prevented recruitment of aspens. No young individuals are recruiting under these old trees (D).

with the result that beavers in Lamar Valley were nearly exterminated. (Aspen and willows grew well in areas from which elk were excluded, showing that their decline was not due to climate conditions during that period).

In 1995, after a 70-year absence, wolves were reintroduced to Yellowstone, and their population grew rapidly. The wolves preyed primarily on elk. The elk population of the Lamar Valley dropped and the elk avoided the aspen groves where they were very vulnerable to wolves. Young aspen began to grow, willows regrew along streams, and the number of beaver colonies increased from one in 1996 to seven in 2003. Thus the presence or absence of a single predator, the wolf, influenced not only populations of its prey, but also the structure of the vegetation and populations of other species that depend on it.

Trophic cascades may have effects across multiple and very different ecosystems because individuals of many species move from one habitat type to another. For example, larval dragonflies are aquatic predators that eat other insects and even small fish; the larvae are eaten in turn by larger fish. The surviving larvae metamorphose into adult dragonflies that eat flying insects.

Dragonfly larvae are abundant in ponds without fish; they are much less common in ponds with fish. Ecologists at the University of Florida studied 12 permanent ponds that differed primarily in whether they had fish (8 ponds) or lacked fish (4 ponds).

55.15 Some Sea Stars are Keystone Species Ochre sea stars (*Pisaster ochraceus*) have harvested all the mussels from the lower parts of these rocks on the Olympic Peninsula of Washington. By consuming mussels, *Pisaster* creates bare spaces on the rocks that are in turn occupied by a variety of other species.

(A)

(B)

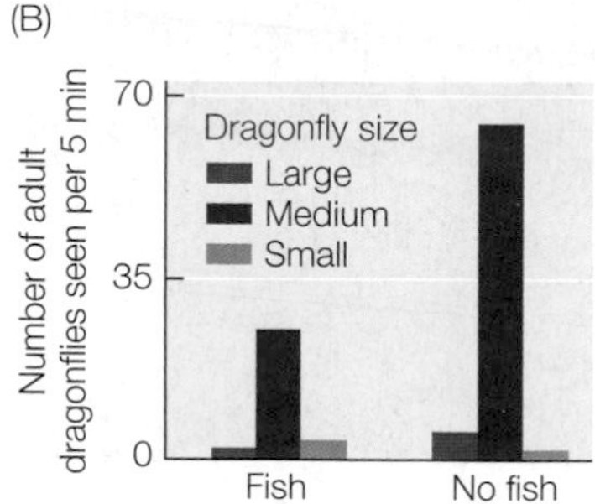

(C)

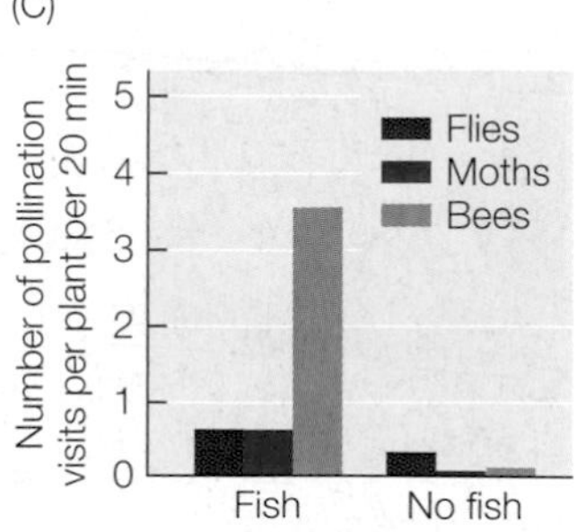

55.14 Trophic Cascades May Cross Habitats (A) Dragonflies affect the composition of communities in several habitats. (B) Adult dragonfly populations are greater near ponds without fish than near ponds with fish. (C) More pollinating insects are found around ponds with fish than around ponds without fish because fish reduce populations of dragonflies, which are major predators on the pollinators.

They found that adult dragonflies (**Figure 55.14A**) were much more abundant around fish-free ponds than around ponds with fish (**Figure 55.14B**). Insect pollinators that were preyed on by the dragonflies were much *less* common around fish-free ponds. Flowers of Saint-John's-wort, the most common plant near the ponds, were visited much less frequently near fish-free ponds than near ponds with fish (**Figure 55.14C**). The plants near fish-free ponds also produced fewer seeds; by artificially pollinating some plants, the ecologists showed that they produced fewer seeds because they did not receive enough pollen. Thus the effects of fish predation in one habitat reached into another habitat where fish do not live.

Beavers cause trophic cascades both through what they eat and through what they construct. By preferentially cutting down some species of trees, they alter the composition of the vegetation. By building dams, they create meadows and ponds that are habitats for other species that would otherwise not be able to live in the area. Organisms that create structures are called **ecosystem engineers**.

Keystone species have wide-ranging effects

A species that exerts an influence out of proportion to its abundance is called a **keystone species**. Keystone species may influence both the species richness of communities and the flow of energy and materials through ecosystems. The sea star *Pisaster ochraceus*, which lives in rocky intertidal ecosystems on the Pacific coast of North America, is an example of a keystone species. Its preferred prey is the mussel *Mytilus californianus*. In the absence of sea stars, these mussels crowd out other competitors in a broad belt of the intertidal zone. By consuming mussels, *P. ochraceus* creates bare spaces that are taken over by a variety of other species (**Figure 55.15**).

Robert Paine of the University of Washington demonstrated the influence of *Pisaster* on species richness by removing sea stars from selected parts of the intertidal zone repeatedly over a 5-year period. Two major changes occurred in the areas from which he removed sea stars. First, the lower edge of the mussel bed extended farther down into the intertidal zone, showing that sea stars are able to eliminate mussels completely where they are covered with water most of the time. Second, and more dramatically, 28 species of animals and algae disappeared from the sea star removal zone. Eventually only *Mytilus*, the dominant competitor, occupied the entire substratum. Through its effect on competitive relationships, predation by *Pisaster* largely determines which species live in these rocky intertidal ecosystems.

Keystone species are not necessarily predators. A plant species that serves as food for many different animals can also be a keystone species, as has been suggested for fig tree species in tropical forests. Figs ripen their fruits during times of the year when fruit is otherwise scarce. The populations of dozens of fruit-eating species depend on figs when no other fruits are present in the forests.

55.3 RECAP

Some interspecific interactions result in a trophic cascade of indirect effects on species at lower trophic levels. Keystone species have an especially strong influence on the species richness of communities and on the flow of energy and materials through ecosystems.

- Describe an example of how a species causes indirect effects across many trophic levels. See pp. 1195–1196 and Figure 55.13
- What are keystone species, and how do they affect the environment of other species? See pp. 1196–1197

Now that we've seen some of the ways in which species interact among themselves to influence the structure of ecological communities, let's look at how abiotic physical disturbances affect what species live in a particular community, and how the assemblage of species in a community can change over time after a disturbance.

55.4 How Do Disturbances Affect Ecological Communities?

A **disturbance** is an event that changes the survival rate of one or more species in an ecological community. Disturbances may remove some species from a community, but may open up space and resources for other species. Keystone species generate disturbances, as we saw in the case of sea stars, but so do physical events. Logs carried by waves may crush algae and animals attached to rocks in an intertidal community. A windstorm may blow down trees, crushing shrubs and herbs. The effects of such disturbances are typically limited to small areas. Other kinds of disturbances, such as hurricanes and volcanic eruptions, may affect much larger areas. Small disturbances are much more common than large disturbances, but a few large events may cause most of the changes in a community. One hurricane, for example, may fell more trees than years of "normal" storms. The effects of disturbances also depend on how often they occur. If strong windstorms are frequent, for example, trees may never have the opportunity to grow tall.

A particular type of disturbance can have a variety of effects. For example, in 1988, massive fires burned one-third of Yellowstone National Park, but they created a mosaic that included unburned patches, areas where only herbs and shrubs were burned, and areas where all trees were consumed (**Figure 55.16**). How does such an ecological community recover after a disturbance?

Succession is a change in a community after a disturbance

A change in the composition of an ecological community following a disturbance is called **succession**. Ecologists divide succession into two major types: primary succession and secondary succession. **Primary succession** begins on sites that lack living organisms. **Secondary succession** begins on sites where some organisms have survived the most recent disturbance. The patterns and causes of ecological succession are varied, but the species that colonize a site soon after the disturbance often alter environmental conditions for other species that come after them.

A good example of primary succession is seen in the changes in plant communities that have followed the retreat of a glacier in Glacier Bay, Alaska, over the last 200 years. The melting and retreating glacier left a series of *moraines*—gravel deposits formed where the

55.16 Fires Create Mosaics of Burned and Unburned Patches This view of Yellowstone National Park was taken a year after the massive forest fires of 1988.

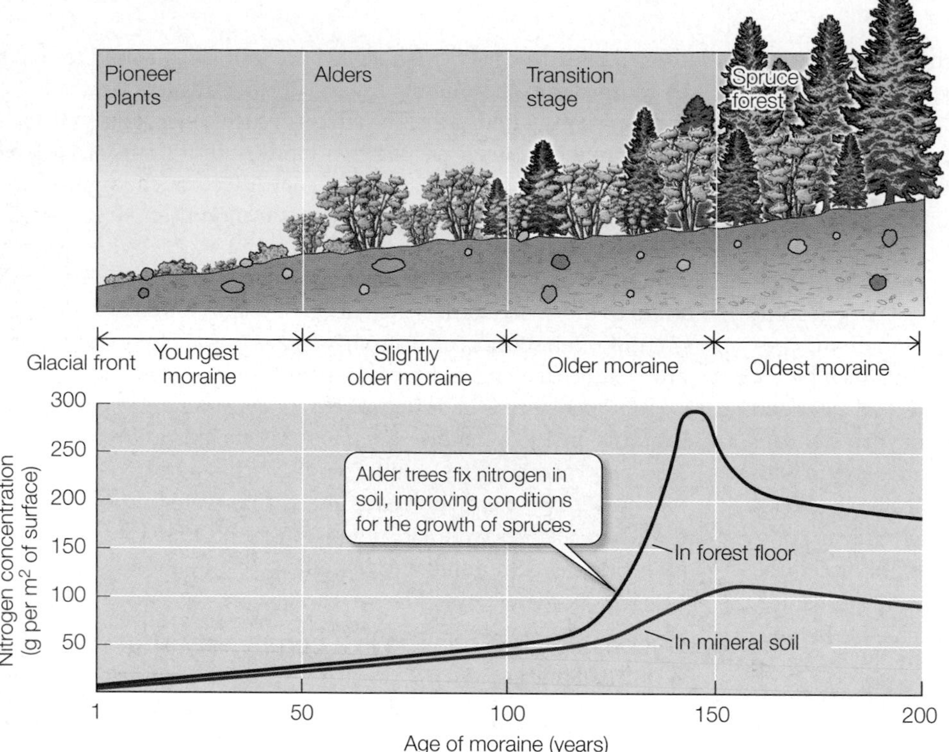

55.17 Primary Succession on a Glacial Moraine As the plant community occupying a glacial moraine at Glacier Bay, Alaska, changes from an assemblage of pioneer plants such as *Dryas* to a spruce forest, nitrogen accumulates in the mineral soil.

glacial front was stationary for a number of years. No human observer was present to measure changes over the entire 200-year period, but ecologists have inferred the temporal pattern of succession by studying plant communities on moraines of different ages. The youngest moraines, closest to the current glacial front, are populated with bacteria, fungi, and photosynthetic microorganisms. Slightly older moraines farther from the glacial front have lichens, mosses, and a few species of shallow-rooted herbs. Still farther from the glacial front, successively older moraines have shrubby willows, alders, and spruces.

By comparing moraines of different ages, ecologists deduced the pattern of plant succession and changes in soil nitrogen content on a glacial moraine (**Figure 55.17**). Succession is caused in part by changes in the soil brought about by the plants themselves. Nitrogen is virtually absent from glacial moraines, so the plants that grow best on recently formed moraines at Glacier Bay are the herbaceous plant *Dryas* and alder trees (*Alnus*), both of which have nitrogen-fixing bacteria in nodules on their roots (see Figure 36.7). Nitrogen fixation by *Dryas* and alders improves the soil so that spruces can grow. Spruces then outcompete and displace the alders and *Dryas*. If the local climate does not change dramatically, a forest community dominated by spruces is likely to persist for many centuries on old moraines at Glacier Bay.

Secondary succession may begin with the dead parts of organisms. The succession of fungal species that decompose pine needles in litter beneath Scots pines (*Pinus sylvestris*) is shown in **Figure 55.18**. New needles continuously fall from the pines, so the surface layer of litter is the youngest and deeper layers are progressively older. Decomposition begins when the first group of fungi starts attacking the needles soon after they fall. Each group of fungi derives its energy by decomposing certain compounds, convert-

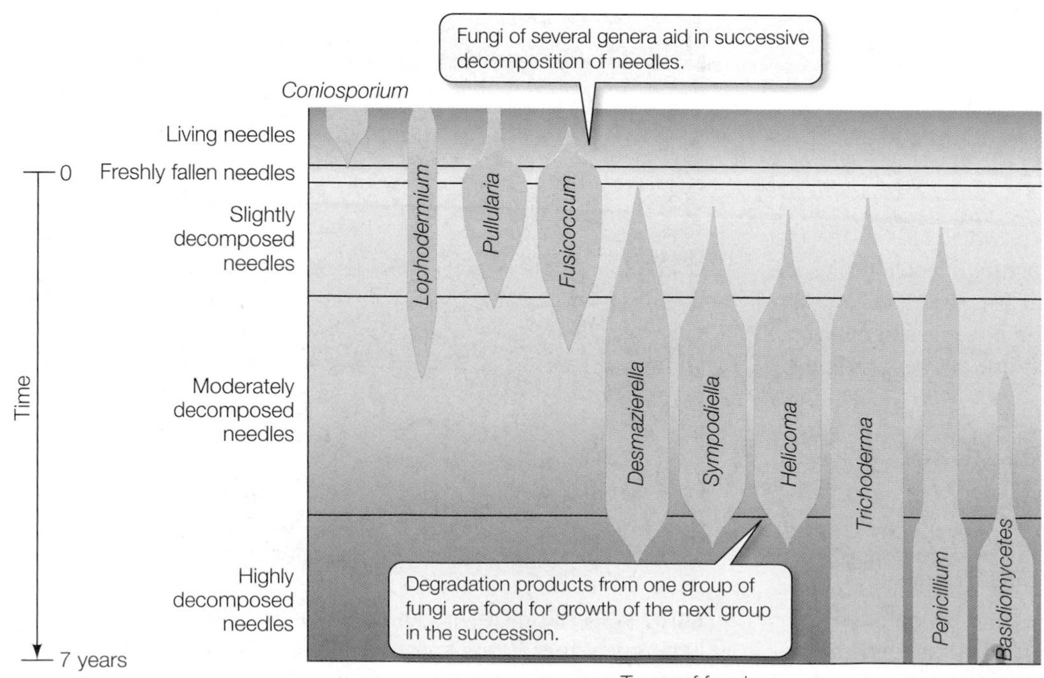

55.18 Secondary Succession on Pine Needles As indicated by the widths of the gray bars, the abundances of ten types of fungi in pine needle litter change over time as the needles are decomposed.

ing them to other compounds that are used by the next group of species. This process continues over about 7 years, by which time the last group of fungi—basidiomycetes—has decomposed the last remaining compounds.

Species richness is greatest at intermediate levels of disturbance

Although the consequences of various kinds of disturbances are highly variable, their results conform to a general pattern: communities with very high levels of disturbance and those with very low levels of disturbance have fewer species than communities subjected to intermediate levels of disturbance. The discovery of this general pattern generated the **intermediate disturbance hypothesis,** which explains the low species richness in areas with high disturbance levels by suggesting that only species with great dispersal abilities and high reproductive rates can persist in such areas. Conversely, the hypothesis explains the decline in species richness where disturbance levels are low by suggesting that competitively dominant species displace other species, as mussels did when sea stars were removed.

The intermediate disturbance hypothesis was tested using boulders in the intertidal zone on beaches in California. Wayne Sousa observed that boulders of intermediate size had more species of algae and barnacles attached to them than smaller boulders did. He hypothesized that this pattern existed because waves move small boulders more easily, more often, and farther than large boulders, not because small boulders are intrinsically unsuitable habitats for many species. When a wave moves a boulder, its motion crushes organisms living on the boulder's surface. To test his hypothesis, Sousa altered disturbance levels by gluing some small boulders to the substratum. Within 6 months, these boulders had more species than the unsecured small boulders (**Figure 55.19**). The results of the experiment supported the intermediate disturbance hypothesis and also refuted the hypothesis that species not normally found on small boulders were absent because small boulders were not suitable habitat for them.

The boulders in Sousa's experiments accumulated species rapidly. Most ecological succession progresses much more slowly, as happened at Glacier Bay.

EXPERIMENT

HYPOTHESIS: Small boulders have fewer species growing on them than larger boulders because they are subjected to high levels of disturbance.

METHOD

Sterilize a number of small boulders. Secure some of them to the natural substratum with glue. Leave other small boulders unsecured to serve as controls. Observe accumulation of species on the boulders over time.

RESULTS

Secured small boulders accumulated many more species than unsecured small boulders.

Small boulders secured with glue
Medium boulders
Large boulders
Small boulders free to move
Number of species per boulder
Experiment month

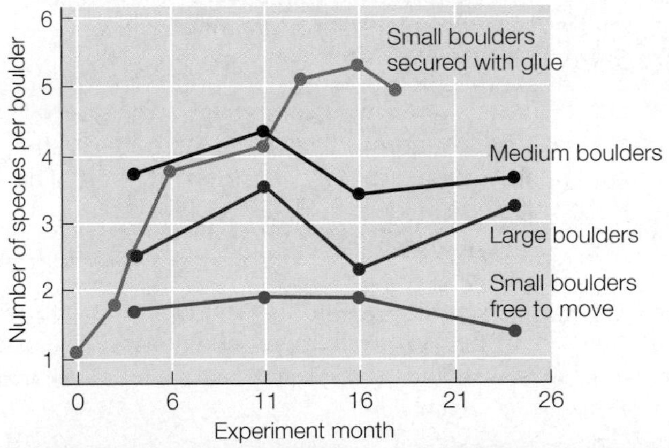

CONCLUSION: Small boulders have fewer species because the higher rates at which they are moved by waves prevent many species from surviving on them, not because they are unsuitable habitat for local species.

55.19 Species Richness Depends on Level of Disturbance By gluing boulders to the substratum, an ecologist showed that small boulders can support more species if they are not disturbed at high rates. Why they accumulated more species than larger boulders is unknown.

Both facilitation and inhibition influence succession

Succession at Glacier Bay illustrates how early colonizers of a changed environment can further change it in ways that *facilitate* the establishment of other species. Is this always the case, or might established species in some cases *inhibit* colonization by other species?

Wayne Sousa again turned to intertidal boulders to test the possible influence of inhibition on succession. He removed all organisms from some boulders and placed them in the intertidal zone. The first species to colonize these boulders were the green algae *Ulva* and *Enteromorpha*. These species were replaced much later by large brown algae, and finally by the red alga *Gigartina canaliculata*. To assess the role of inhibition during this succession, Sousa removed *Ulva* from some of the boulders. The boulders from which *Ulva* had been removed soon had a much greater population density of *Gigartina* than the others.

Inhibition clearly influenced the pattern of succession on intertidal boulders, but its effects lasted only a few months. Can inhibition have an effect lasting centuries or even millennia? Most ecological communities will reach an equilibrium number of species within a few decades to a few centuries after a disturbance. But the species living in some communities may have been assembled over periods of up to many millennia (as Darwin noted in describing an entangled bank at the start of this chapter).

As an example, we learned in Section 52.5 that South American ecological communities include many mammalian species whose ancestors arrived in South America from North America over the past 4 million years. North American communities, on the other hand, contain few mammalian species whose ancestors recently came to North America from South America (see Figure 52.13).

Such a pattern suggests that the effects of inhibition may persist for millions of years, but experiments cannot evaluate processes that take place over such long time spans. To evaluate such processes, scientists use the comparative method and examine patterns of species richness in ecological communities worldwide.

55.4 RECAP

Disturbances are events that change the survival rate of one or more species in an ecological community. Ecological succession—a process of change in community structure—typically follows a disturbance.

- Describe an example of primary succession. See pp. 1197–1198 and Figure 55.17
- Name some ways in which already-established species facilitate colonization by other species. See p. 1198
- Do you understand why communities with intermediate levels of disturbance often have more species than communities with either very low or high levels of disturbance? See p. 1199 and Figure 55.19

Some ecological communities are home to many species; other communities have only a few. In the next section we will discuss patterns of species richness and the factors that influence them.

55.5 What Determines Species Richness in Ecological Communities?

The number of species living in a community is its **species richness**. Biologists have been aware of a geographic pattern in species richness for many years: more species are found in low-latitude than in high-latitude regions. **Figure 55.20** shows this latitudinal gradient in species richness for mammals in North and Central America. Similar patterns exist for birds, frogs, trees, and many groups of marine organisms. The figure also shows that more species are found in mountainous regions than in relatively flat areas because more vegetation types and climates exist within these topographically complex areas. Species richness on islands and peninsulas is always less than that in an equivalent area on the nearest mainland. Patterns of species richness on islands can be explained in large part by differences in immigration and extinction rates (see Section 52.5). What other processes determine species richness in mainland ecological communities?

Species richness is influenced by productivity

The species richness of ecological communities is correlated with ecosystem productivity, but the relationship between these two factors is complex. Ecologists first observed that species richness often increases with productivity up to a point, but then decreases (**Figure 55.21**). The increase occurs because the number of individuals an area can support increases with productivity, and with larger population sizes, species extinction rates are lower. But why should species richness decrease when productivity is still higher?

One hypothesis postulates that interspecific competition becomes more intense when productivity is very high, resulting in competitive exclusion of some species. This hypothesis is supported by the results of a long-term experiment at the Rothamstead Experiment Station in England begun in 1855. Fertilizer has been added regularly to selected plots of land to increase their productivity, and fertilized and unfertilized plots have been monitored continuously. Over this period, the number of plant species in the unfertilized plots has remained roughly constant, whereas species richness has declined in the fertilized plots.

Species richness and productivity influence ecosystem stability

We have seen that, up to a point, higher ecosystem productivity favors increased species richness. How might species richness in turn influence ecosystem productivity? Ecologists hypothesized that species richness might enhance productivity because no two

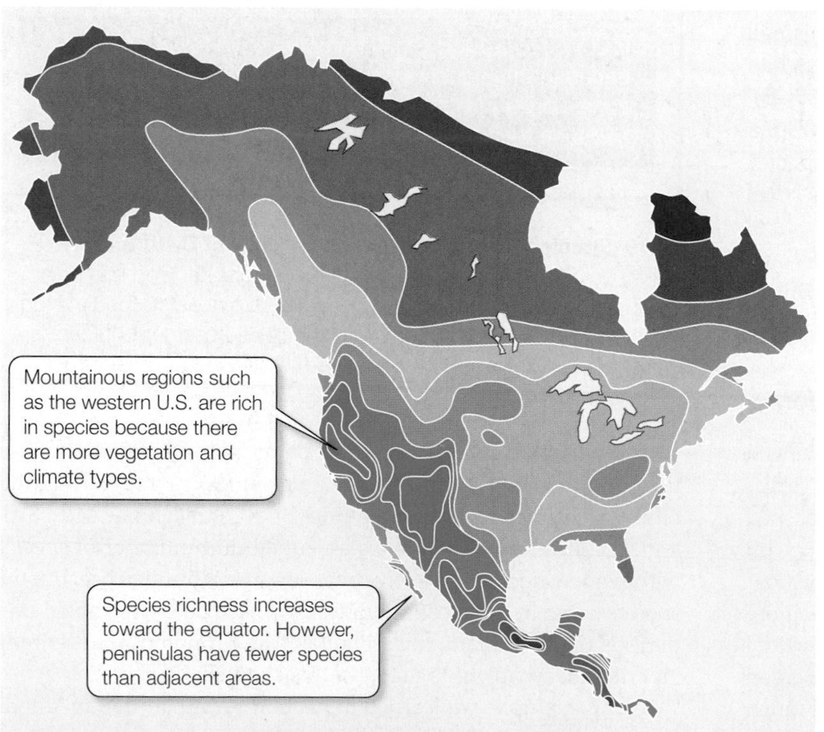

Mountainous regions such as the western U.S. are rich in species because there are more vegetation and climate types.
Species richness increases toward the equator. However, peninsulas have fewer species than adjacent areas.

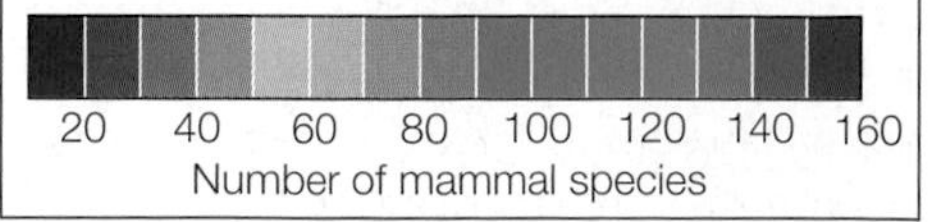

55.20 The Latitudinal Gradient of Species Richness of North American Mammals The colored zones represent regions with equal numbers of species. A similar pattern is seen in the Southern Hemisphere.

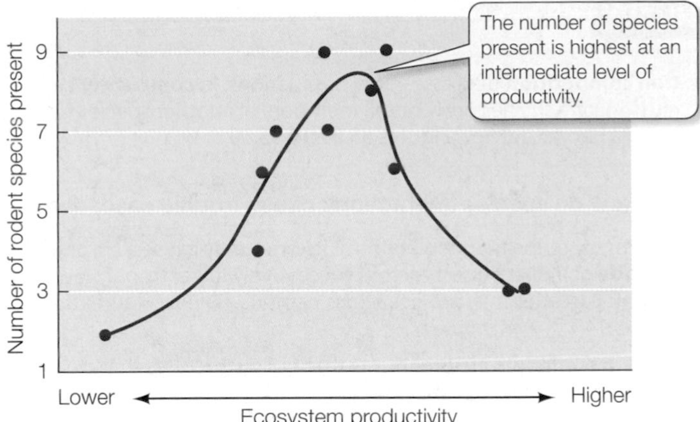

55.21 Species Richness Peaks at Intermediate Productivity
The number of species of rodents that live on gravel and rocky plains in the Gobi Desert peaks at intermediate levels of productivity.

species in a community have the same relationship with the environment. Therefore, a richer mixture of species should result in a more complete use of the available resources. In addition, if the environment changes, a species-rich ecosystem is more likely to contain species that are already adapted to the new conditions than is a species-poor ecosystem. If that is the case, then a species-rich ecosystem should also be more *stable*; that is, it should change less over time in both productivity and species composition than a species-poor ecosystem.

To test this hypothesis, David Tilman and his colleagues at the University of Minnesota cleared several outdoor plots. In them they planted grasses in mixtures that ranged from a few to 25 species. At the end of each growing season, they measured total grass biomass and the population densities of all the grasses in each plot. Over a period of 11 years, which included a serious drought, the plots with more species were more productive, and their productivity varied less from one year to another, supporting the first part of the hypothesis (**Figure 55.22**). However, the population densities of individual species in the plots were not stable over the years (regardless of a plot's species richness), because different species performed better during drought years and wet years.

EXPERIMENT

HYPOTHESIS: Communities with many species should have higher productivity and stability than communities with few species.

METHOD

Clear and plant plots with different numbers and mixtures of grass species. Measure productivity and species composition of the plots over 11 years.

RESULTS

(A) Productivity increases with species richness

Total plant cover (%)
60
55
50
45
40
35
30
25
0
5
10
15
20
25
Number of species

(B) Variation in productivity decreases with species richness

High
Variation in primary productivity (year-to-year)
Low
0
2
4
6
8
10
12
Average number of species

CONCLUSION: Plots with more species were more productive and varied less in productivity.

55.22 Species Richness Enhances Community Productivity
Net primary productivity was greater, and variation in net primary productivity from one year to the next was less, in species-rich than in species-poor plots of grasses.

55.5 RECAP

More species are found in low-latitude than in high-latitude regions. Species richness correlates positively with ecosystem productivity up to a point, then declines at very high productivity levels.

- Why does species richness decline at very high levels of productivity? See p. 1200 and Figure 55.21
- Why are communities with many species more productive and more stable than communities with fewer species? See p. 1201 and Figure 55.22

CHAPTER SUMMARY

55.1 What are ecological communities?

The species that live and interact in an area constitute an ecological **community**.

Ecological communities are loose assemblages of organisms. Review Figure 55.1

The organisms in a community can be divided into **trophic levels** based on the source of their energy. **Primary producers** get their energy from sunlight. Herbivores that get their energy by eating primary producers are **primary consumers**; organisms that get their energy by eating herbivores are **secondary consumers**; and so on. Review Table 55.1, Web/CD Activity 55.1

A **food chain** diagrams who eats whom. A **food web** shows how food chains are interconnected in an ecological community. Review Figure 55.2

Other kinds of diagrams show how energy decreases as it flows from lower to higher trophic levels and the **biomass** of organisms present at each trophic level. Review Figure 55.3

Most of the energy ingested by organisms that is converted to biomass is eventually consumed by **decomposers**.

55.2 What processes influence community structure?

Species interactions fall into five categories: **predation** or parasitism benefits the predator while harming the prey; **competition** is harmful to all participants; **mutualisms** are associations that benefit all participants; **commensal** interactions benefit one party while neither harming nor benefiting the other; **amensal** interactions have no effect on one party but harm another. Review Table 55.2, Web/CD Activity 55.2

Predator–prey interactions typically undergo oscillations. As predator populations grow, they may eat most of their prey; the predator population then crashes. Review Figure 55.4, Web/CD Tutorial 55.1

Predators may restrict the ways in which prey species make use of a habitat and may restrict the geographic distributions of their prey. Review Figure 55.6

Mimicry is an adaptation by prey to predation. In **Batesian mimicry**, a palatable species mimics an unpalatable species. In **Müllerian mimicry**, two or more unpalatable species converge to resemble one another. Review Figure 55.7

Microparasite populations can persist only if, on average, each infected host individual transmits the infection to one other individual.

Competition may restrict the abundances and ranges of species. **Interference competition** makes foraging more difficult; **exploitation competition** depresses prey populations. In **competitive exclusion**, one species prevents all members of another species from using a habitat. Review Figures 55.9 and 55.10

55.3 How do species interactions cause trophic cascades?

A predator, by reducing populations of its prey, may cause a **trophic cascade** of indirect effects across successively lower trophic levels. Trophic cascades may project across habitats. Review Figures 55.13 and 55.14

Ecosystem engineers are organisms that build structures that create environments for other species.

A **keystone species** affects the entire community out of proportion to its abundance.

55.4 How do disturbances affect ecological communities?

A **disturbance** is an event that changes the survival rate of one or more species in a community.

Ecological **succession** is a change in community composition following a disturbance.

Primary succession begins on sites that lack living organisms. **Secondary succession** begins on sites where some organisms have survived the most recent disturbance. Review Figures 55.17 and 55.18, Web/CD Tutorial 55.2

The **intermediate disturbance hypothesis** explains why communities with intermediate levels of disturbance often have more species than communities with very low or high levels of disturbance. Review Figure 55.19

Species that have already become established may facilitate or inhibit colonization by other species.

55.5 What determines species richness in ecological communities?

The number of species living in a community is its **species richness**. More species of most clades are found in low-latitude than in high-latitude regions. Review Figure 55.20

Species richness often increases with productivity, but only up to a point. Review Figure 55.21

Species-rich ecosystems tend to vary less in both productivity and species composition than species-poor ecosystems. Review Figure 55.22

SELF-QUIZ

1. An ecological community is
   a. all the species of organisms that live and interact with one another in an area.
   b. all the species that live and interact with one another in an area together with the abiotic environment.
   c. all the species in an area that belong to a particular trophic level.
   d. all the species that are members of a local food web.
   e. all of the above
2. A trophic level consists of the organisms
   a. whose energy source has passed through the same number of steps to reach them.
   b. that use similar foraging methods to obtain food.
   c. that are eaten by a similar set of predators.
   d. that eat both plants and other animals.
   e. that compete with one another for food.
3. Two organisms that use the same resources when those resources are in short supply are said to be
   a. predators.
   b. competitors.
   c. mutualists.
   d. commensalists.
   e. amensalists.

4. Damage caused to shrubs by branches falling from overhead trees is an example of
   a. interference competition.
   b. partial predation.
   c. amensalism.
   d. commensalism.
   e. diffuse coevolution.
5. The diagrams of energy and biomass distribution for forests and grasslands differ because
   a. forests are more productive than grasslands.
   b. forests are less productive than grasslands.
   c. large mammals avoid living in forests.
   d. trees store much of their energy in difficult-to-digest wood, whereas grassland plants produce few difficult-to-digest tissues.
   e. grasses grow faster than trees.
6. Keystone species
   a. influence the communities in which they live more than expected on the basis of their abundance.
   b. may influence the species richness of communities.
   c. may influence the flow of energy and nutrients through ecosystems.
   d. are not necessarily predators.
   e. all of the above
7. What is the general relationship between species richness and disturbance?
   a. Species richness peaks at low levels of disturbance.
   b. Species richness peaks at high levels of disturbance.
   c. Species richness peaks at intermediate levels of disturbance.
   d. Species richness is less at intermediate levels of disturbance.
   e. There is no general relationship between species richness and level of disturbance.
8. Ecological succession is
   a. the changes in species over time.
   b. the changes in community composition after a disturbance.
   c. the changes in a forest as the trees grow larger.
   d. the process by which a species becomes abundant.
   e. the buildup of soil nutrients.
9. Primary succession begins
   a. soon after a disturbance ends.
   b. at varying times after a disturbance ends.
   c. at sites where some organisms survived the disturbance.
   d. at sites were no organisms survived the disturbance.
   e. at sites where only primary producers survived the disturbance.
10. A latitudinal gradient in species richness
   a. is found in North America but not in South America.
   b. exists for birds, frogs, and mammals but not for plants.
   c. exists because tropical regions are more mountainous than high-latitude regions.
   d. exists because there are fewer peninsulas in the tropics.
   e. exists on land, but not in the oceans.

FOR DISCUSSION

1. Some evidence suggests that interspecific competition is responsible for the decrease in species richness at high levels of productivity. What other hypotheses might explain this puzzling relationship? How would you test them?
2. The increased productivity and stability of species-rich communities could be explained by ecological differences among the species or by the fact that the more species in a community, the greater the chance that it will contain an unusually productive species. How could you distinguish between these competing hypotheses?
3. If species-rich communities are more productive than species-poor communities, how can modern agriculture, which is based almost entirely on cultivating a single species on a plot, be so productive?
4. Figures 55.17 and 55.18 illustrated succession with two examples from forests. How might ecological succession differ in grasslands? In deserts? In the rocky intertidal zone?
5. Many conservationists believe that our greatest efforts should be expended to save undisturbed ecosystems. Many users of natural resources, on the other hand, argue that disturbing ecosystems to extract resources will actually improve species richness. Is the latter view an appropriate invocation of the intermediate disturbance hypothesis?

FOR INVESTIGATION

The experiments illustrated in Figure 55.5, even though they were done on enclosed plots 1 km^2 in size, were small in scale compared with the scale over which the natural hare–lynx cycle is synchronized in the boreal forest. How could scientists investigate the roles of food supply, predation, and other factors in synchronizing the cycle over large areas?

CHAPTER 56 Ecosystems and Global Ecology

Ecologists swing in the canopy to measure carbon's fate

Photosynthesis by terrestrial plants transforms (*fixes*) the carbon atoms in atmospheric carbon dioxide into the carbohydrate sugars that fuel the fungal and animal worlds. Concentrations of CO_2 in the atmosphere have been rising steadily over the last century, largely as a result of human activities, and some scientists have hypothesized that we can expect rates of photosynthesis—and thus CO_2 fixation and carbon storage in ecosystems—to increase. But ecologists are discovering that this "silver lining" scenario is probably not going to be the case.

Forest studies are important for determining the responses of vegetation to atmospheric carbon enrichment. Most photosynthesis in any forest takes place high up in tree canopies where sunlight is most intense. Until recently, making measurements in forest canopies required climbing or use of balloons. Then, in 1992, Alan Smith at the Smithsonian Tropical Research Institute in Panama realized that he could use a construction crane to gain access to the forest canopy. He recognized that by swinging the crane's boom in a circle, shuttling the gondola along its length, and lowering the cage to different heights, researchers could reach and measure biological processes over a large volume of the forest canopy. Smith's idea caught on quickly, and today ecologists on four continents use cranes to study forest canopies.

When they monitored growth, reproduction, and carbon storage in large trees, Swiss ecologists found that, as expected, canopy trees increased their growth rates in a carbon-enriched atmosphere; but unlike young, fast-growing saplings, they did not store large amounts of carbon in their trunks and branches. Instead, they transported more carbon first to the soil, and then to the atmosphere (via their roots). The leaf litter they produced was richer in starch and sugar, but poorer in lignin; the litter decomposed faster as soil microbes rapidly respired the carbon back to the atmosphere. These and other experiments suggest that young forests are likely to respond to greater atmospheric CO_2 concentrations by growing faster. However, mature forests, along with other types of vegetation such as grasslands, shrublands, and tundra, are not likely to store significantly more carbon in response to CO_2 enrichment.

Studying the Canopy A research crane 42 meters high allows scientists to study the forest canopy at the Smithsonian Tropical Research Institute in Parque Metropolitana, Panama. One aspect of these studies involves taking measurements that reveal the amount of carbon being stored in forest trees.

Enriching the Canopy Researchers from a team of Swiss ecologists install a mechanism of fine tubes that administers controlled amounts of CO_2 to enrich the trees' carbon supply. These experiments mimic what will happen as atmospheric CO_2 levels continue to increase.

As Section 52.3 described, the global ecosystem is composed of many different types of vegetation (biomes), many of which have few woody plants. To understand how ecosystems function at large scales, scientists must make observations and conduct experiments on many vegetation types. Moreover, because the responses of natural ecosystems to carbon enrichment will depend on many other factors—such as temperature, water, wind, and soil fertility—ecologists must study the interactions among these factors to understand how they influence ecosystem processes.

IN THIS CHAPTER we will begin describing the compartments of the global ecosystem. We then trace the flow of energy through the ecosystem, following key chemical elements such as carbon as they cycle through its compartments. We will highlight the many ways in which human activities have altered these cycles, and conclude by considering the many goods and services that ecosystems provide to human society.

CHAPTER OUTLINE

56.1 What Are the Compartments of the Global Ecosystem?

Earth is an essentially closed system with respect to atomic matter, but it is an open system with respect to energy. The sun delivers a nearly constant amount of energy to Earth every day and has done so for billions of years. Energy from the sun, combined with energy from the radioactive decay that melts the magma in Earth's interior, drives the processes that move materials around the planet. Many of these processes are cyclic (**Figure 56.1**). Almost all of the rocks that compose the continents have been processed at least once through a complex chemical and physical cycle involving weathering, formation of sediments, and movement into Earth's interior by means of the processes that result in continental drift (see Figure 21.2). Deep within the Earth, these rocks are subjected to great heat and pressure to form new rocks. The water in the oceans has evaporated, condensed in the atmosphere, precipitated, and returned to the oceans via rivers and groundwater flow millions of times. The carbon (C), nitrogen (N), oxygen (O), and sulfur (S) in atmospheric gases have been cycled repeatedly through living organisms.

We take for granted many features of Earth, but Earth is actually a very unusual planet. Its unusual properties include the presence of life, oceans, a moderate surface temperature, continental drift, and a large moon. The moon has had a profound effect on Earth's history. It stabilizes the tilt of Earth on its axis. The degree of tilt strongly influences climate. If Earth's tilt were 50° or more, for example, equatorial regions would receive less solar energy over the year than would polar regions! The moon also plays a major role in producing ocean tides, and it slows Earth's rotation.

The atmospheres of planets without life are in chemical equilibrium, but Earth's atmosphere is not. Oxygen gas (O_2), nitrogen gas (N_2), and water vapor (H_2O) are the main molecules of Earth's atmosphere, but without living organisms these gases would combine to produce nitric acid (HNO_3), which would dissolve in the oceans and remain there. Living organisms, by continually generating O_2, prevent this from happening.

All organisms depend on inputs of energy (in the form of sunlight or high-energy molecules), water, and nutrients for their metabolism and growth. As Section 55.1 explained, energy flows through ecosystems from producers to consumers.

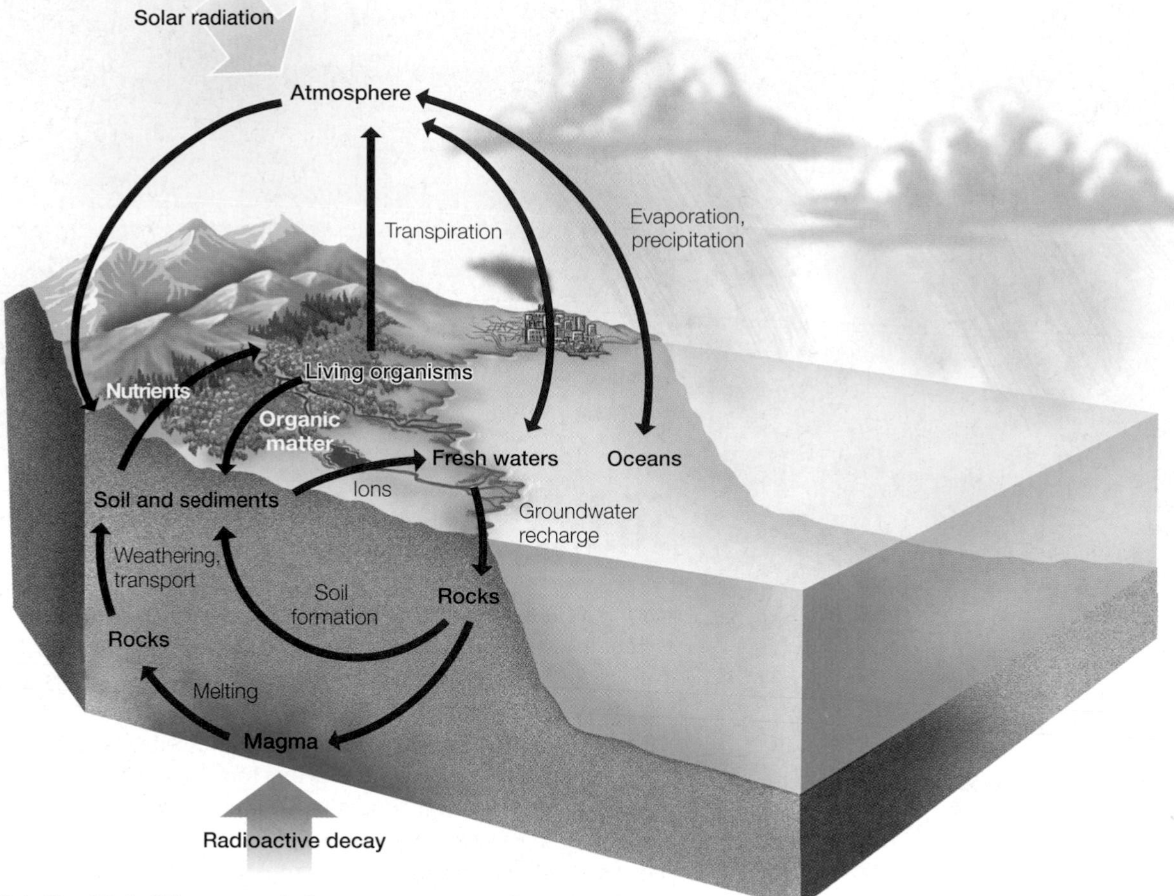

56.1 The Global Ecosystem's Compartments are Connected by the Flow of Elements The arrows indicate flow of materials between the compartments. The atmosphere exchanges some components rapidly. Exchange of materials between rocks, soils, and the oceans is generally much slower.

At each transformation, much of this energy is used to power metabolism and is dissipated as heat. Other organisms cannot use that heat to power their metabolism. Chemical elements, on the other hand, are not altered when they are transferred among organisms. The availability of the chemical elements of which living organisms are composed—carbon, nitrogen, phosphorus, calcium, sodium, sulfur, hydrogen, oxygen, and a few others—is strongly influenced by how organisms get them, how long they retain them, and what they do with them while they have them.

To understand the movement of energy and materials on Earth, it is convenient to divide the physical environment into four interacting compartments: oceans, fresh waters, atmosphere, and land. These four compartments, and the types of organisms living in them, are very different. Therefore, the quantities of different elements in each compartment, what happens to them, and the rates at which they enter and leave the compartments differ strikingly. After we have described the four compartments, we will discuss how energy flows and elements cycle among them.

Oceans receive materials from the land and atmosphere

Over time scales of hundreds to thousands of years, most materials that cycle through the four compartments end up in the oceans. The oceans are enormous, but they exchange materials with the atmosphere only at their surface, so they respond very slowly to inputs from other compartments. They receive materials from land primarily in runoff from rivers.

Except on continental shelves—the shallow ocean waters surrounding large land masses—ocean waters mix slowly. Most materials that enter the oceans from other compartments gradually sink to the seafloor. They may remain there for millions of years until the seafloor sediments are uplifted above sea level by mountain building. Therefore, concentrations of mineral nutrients are very low in most ocean waters. Some elements are brought back to the surface near the coasts of continents, where offshore winds push surface waters away from shore, causing cold bottom water to rise to the surface (**Figure 56.2**). The nutrient-rich waters in these **upwelling zones** support high rates of photosynthesis and dense animal populations. Most of the world's great fisheries are concentrated there.

Water moves rapidly through lakes and rivers

The freshwater compartment of the global ecosystem is comprised of rivers, lakes, and **groundwater** (water in soil and rocks). Only a small fraction of Earth's water resides at any one time in lakes and rivers, but water moves rapidly through them. Some mineral nutrients enter the freshwater compartment in rainfall, but most are released by the weathering of rocks. They are carried to lakes and rivers by surface flow or by movement of groundwater.

After entering rivers, mineral nutrients are usually carried rapidly to lakes or to the oceans. In lakes, they are taken up by organisms and incorporated into their cells. Those organisms eventually

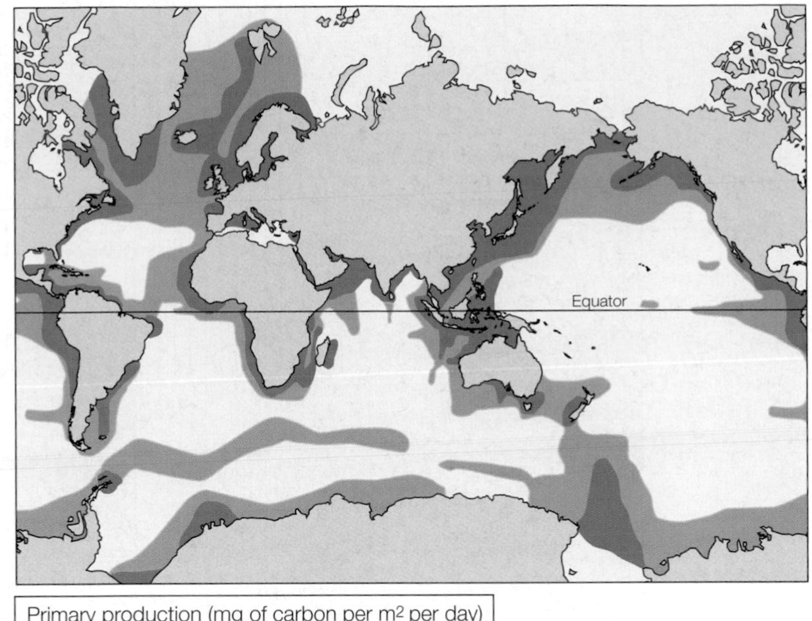

56.2 Zones of Upwelling are Productive Biological productivity is greatest near continents where surface waters, driven by prevailing winds, move offshore and are replaced by cool, nutrient-rich water upwelling from below.

die and sink to the bottom, taking the nutrients with them. Decomposition of their tissues by detritivores consumes the O_2 in the bottom water. The surface waters of lakes thus quickly become depleted of nutrients, while deeper waters become depleted of O_2. This process is countered, however, by vertical movements of water called **turnover**, which bring nutrients and dissolved CO_2 to the surface and O_2 to deeper water.

Wind is an important agent of turnover in shallow lakes, but in deeper lakes it usually mixes only surface waters. Deep lakes in temperate climates have an annual turnover cycle that is driven by temperature (**Figure 56.3**). These lakes turn over because water is most dense at 4°C; above and below that temperature, it expands. Thus in winter, the coldest water in a lake floats at its surface, often covered by a layer of ice. The waters below remain at 4°C. When the spring sun warms the surface of the lake, it brings the

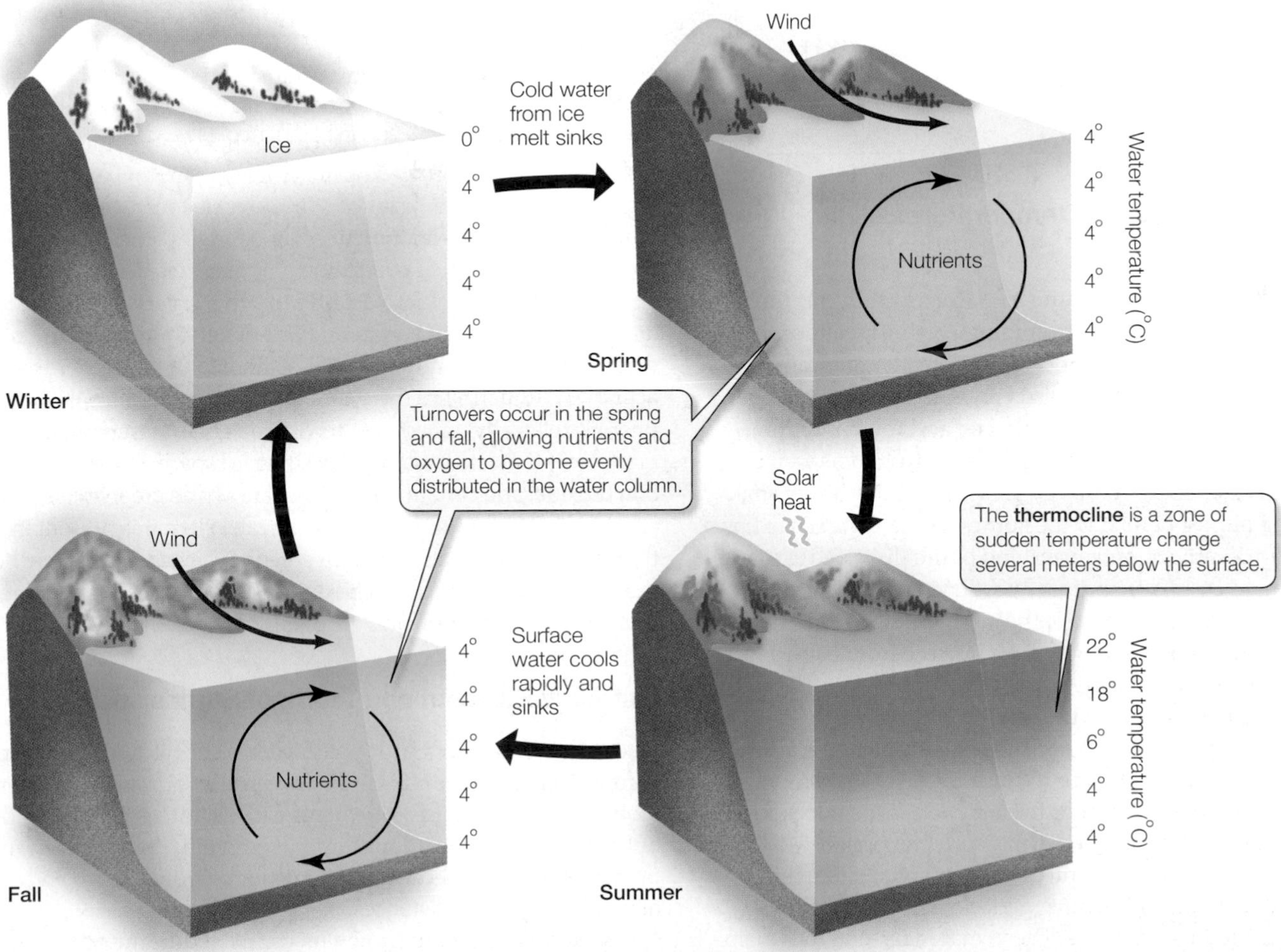

56.3 The Turnover Cycle in a Temperate Lake Turnovers that occur in spring and fall allow nutrients and oxygen to become evenly distributed in the water column. The vertical temperature profiles shown here are typical of temperate-zone lakes that freeze in the winter.

temperature of the surface water up to 4°C. At this point, the density of the water is uniform throughout the lake, and even modest winds readily mix the entire water column.

As spring and summer progress, the surface water becomes warmer, and the depth of the warm layer gradually increases. There is still a well-defined *thermocline*, however, where the temperature drops abruptly. Only if the lake is shallow enough to warm right to the bottom does the temperature of the deepest water rise above 4°C. Another turnover occurs in autumn as the surface of the lake cools, and the cooler surface water—which is now denser than the warmer water below it—sinks and is replaced by warmer water from below.

Arctic lakes turn over only once each year. Deep tropical and subtropical lakes may have permanent thermoclines because they never become cool enough to have uniformly dense water. Their bottom waters lack oxygen because decomposition quickly depletes any oxygen that reaches them. However, many tropical lakes are turned over at least periodically by strong winds, so that their deeper waters are occasionally oxygenated. If turnover does not happen, lake productivity can be drastically affected.

Lake Tanganyika is an extremely deep tropical lake in East Africa. Fishermen there have seen their catches decline by as much as 50 percent over the past 30 years, in part because warmer water temperatures and a decade of slack wind conditions have prevented the lake water from turning over.

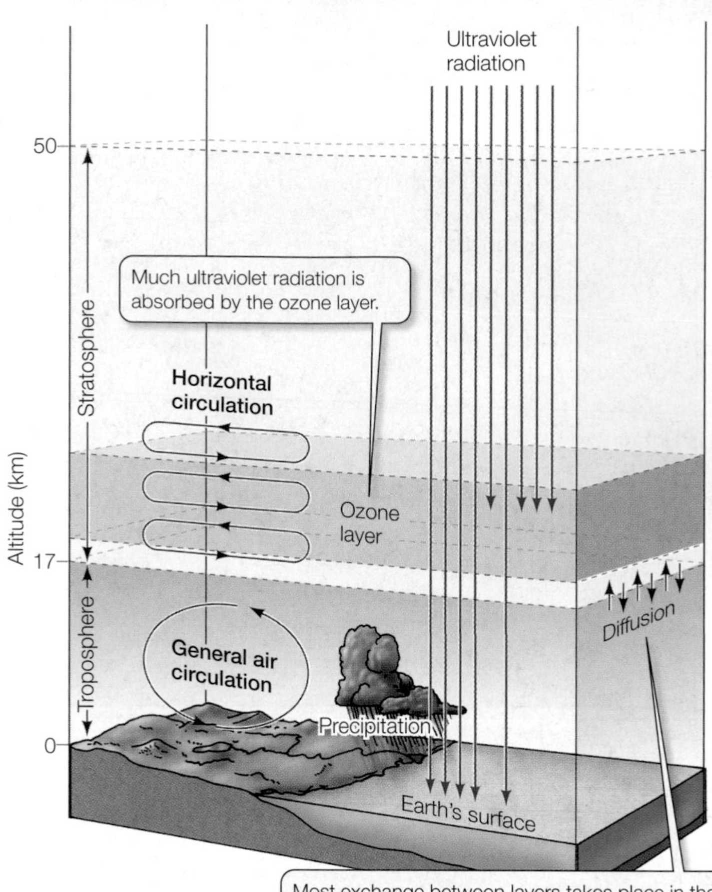

56.4 The Two Layers of Earth's Atmosphere The troposphere and the stratosphere differ in their circulation patterns, the amount of water vapor they contain, and the amount of ultraviolet radiation they receive.

The atmosphere regulates temperatures close to Earth's surface

The third major compartment of the global ecosystem, the atmosphere, is a thin layer of gases surrounding Earth. The atmosphere is 78.08 percent N_2, 20.95 percent O_2, 0.93 percent argon, and 0.03 percent carbon dioxide (CO_2). It also contains traces of hydrogen gas, neon, helium, krypton, xenon, ozone, and methane. The atmosphere contains Earth's biggest pool of nitrogen as well as a large proportion of its O_2. Although CO_2 constitutes a very small fraction of the atmosphere, it is the source of the carbon used by terrestrial photosynthetic organisms and of the dissolved carbonate in water used by marine primary producers.

About 80 percent of the mass of the atmosphere lies in its lowest layer, the **troposphere**, which extends upward from Earth's surface about 17 kilometers in the tropics and subtropics, but only about 10 kilometers at higher latitudes. Most global air circulation takes place within the troposphere, and virtually all of the atmospheric water vapor is found there (**Figure 56.4**).

The **stratosphere**, which extends from the top of the troposphere up to about 50 kilometers above Earth's surface, contains very little water vapor. Most materials enter the stratosphere from the troposphere in the intertropical convergence zone, where air heated by the sun rises to high altitudes. These materials tend to remain in the stratosphere for a relatively long time. A layer of ozone (O_3) in the stratosphere absorbs most of the biologically damaging short-wavelength ultraviolet radiation that enters the atmosphere, which is why the development of an ozone hole in the Southern Hemisphere is of great concern.

The atmosphere plays a decisive role in regulating temperatures at and close to Earth's surface. If Earth had no atmosphere, its average surface temperature would be about –18°C, rather than its actual +17°C. Earth has this warm temperature because the atmosphere is relatively transparent to visible light, but traps a large part of the heat that Earth radiates back toward space. Water vapor, carbon dioxide, and certain other gases, known as **greenhouse gases**, are especially important trappers of heat. When it is present in the troposphere, ozone also acts as a greenhouse gas. We will see how human-caused increases in greenhouse gases are contributing to global climate warming in Section 56.3.

Land covers about one-fourth of Earth's surface

About one-fourth of Earth's surface, most of it in the Northern Hemisphere, is currently land above sea level. Even though the global supply of chemical elements is constant, regional and local deficiencies of particular elements strongly affect ecosystem processes on land, because elements move slowly there and usually over only short distances.

The terrestrial compartment is connected to the atmospheric compartment by organisms that take chemical elements from and, some time later, release them back to the air. Chemical elements in soils are carried in solution into groundwater and eventually into the oceans, where they are unavailable to terrestrial organisms unless an episode of uplifting raises marine sediments and a new cycle of weathering begins. The type of soil that exists in an area depends on the underlying rock from which it forms, as well as on climate, topography, the organisms living there, and the length of time that soil-forming processes have been acting. Very old soils are much less fertile than most young soils because nutrients leach out of them over time.

Although land covers only a small proportion of Earth's surface, human life depends intimately on soil fertility and the productivity of terrestrial and freshwater ecosystems. The land and fresh waters are the global ecosystem compartments most strongly affected by human activities.

56.1 RECAP

Earth is a closed system with respect to atomic matter, but an open system with respect to energy. Earth's physical environment can be divided into four compartments—oceans, fresh waters, atmosphere, and land—through which energy flows and elements cycle.

- What keeps Earth's atmosphere from reaching chemical equilibrium? See pp. 1206–1206
- Can you describe the process of turnover in a temperate-zone lake in autumn? See p. 1207 and Figure 56.3
- How does the atmosphere keep temperatures at and close to Earth's surface warmer than they would be in its absence? See p. 1208 and Figure 56.4

Elements cycle through the global ecosystem, but energy flows through it unidirectionally. In the next section we introduce some concepts that help us track global energy flows.

56.2 How Does Energy Flow Through the Global Ecosystem?

We begin our discussion by explaining how solar energy drives ecosystem processes and describing the geographical distributions of the amount of energy assimilated by primary producers. We then show how human activities are modifying energy flow in the global ecosystem.

Solar energy drives ecosystem processes

With the exception of a few ecosystems in which solar energy is not the main energy source (some caves, deep-sea hydrothermal vent ecosystems), all energy utilized by organisms comes (or once came) from the sun. Even the fossil fuels—coal, oil, and natural gas—on which the economy of modern human civilization is based are reserves of captured solar energy locked up in the remains of organisms that lived millions of years ago.

Solar energy enters ecosystems by way of plants and other photosynthetic organisms. Only about 5 percent of the solar energy that arrives on Earth is captured by photosynthesis. The remaining energy is either radiated back into the atmosphere as heat or taken up by the evaporation of water from plants and other surfaces. **Gross primary productivity** (**GPP**) is the rate at which energy is incorporated into the bodies of photosynthetic organisms. The accumulated energy is called **gross primary production**. In other words, productivity is a rate; production is its product. Primary producers use some of this accumulated energy for their own metabolism; the rest is stored in their bodies or used for their growth and reproduction. The energy available to organisms that eat primary producers, called **net primary production** (**NPP**), is gross primary production minus the energy expended by the primary producers during their metabolism. Only the energy of an organism's net production—its growth plus reproduction—is potentially available to other organisms that consume it (**Figure 56.5**).

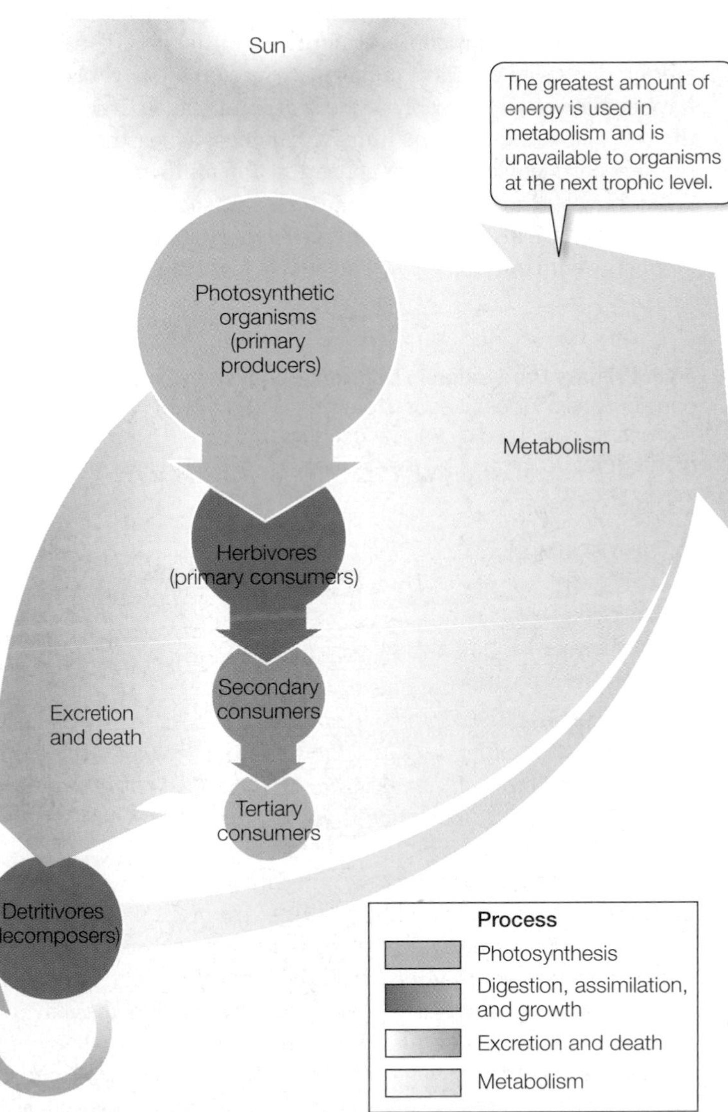

56.5 Energy Flow through an Ecosystem Only the energy of an organism's net production—its growth plus reproduction—is potentially available to other organisms that consume it. In this diagram, the width of each arrow is roughly proportional to the amount of energy flowing through that channel. Arrows indicate directions of energy flow.

The geographic distribution of the energy assimilated by primary producers reflects the distribution of land masses, temperature, and moisture on Earth (**Figure 56.6**). Close to the equator at sea level, temperatures are high throughout the year, and the water supply typically is adequate for plant growth much of the time. In these climates, productive forests thrive. In lower-latitude and mid-latitude deserts, where plant growth is limited by lack of moisture, primary production is low. At still higher latitudes, even though moisture is generally available, primary production is also low because it is cold much of the year (**Figure 56.7**). Production in aquatic ecosystems is limited by light, which decreases rapidly with depth; by nutrients, which sink and must be replaced by upwelling of water; and by temperature.

Human activities modify flows of energy

Human activities modify the ways in which energy flows between the global ecosystem compartments and how it is distributed among them. The effects of human activities on the flow of energy have accelerated remarkably in the last century and particularly in the last five decades. Some human activities decrease net global primary productivity (such as converting forests to grasslands and urban developments) and some increase it (such as intensifying agriculture). We also influence energy flows by supplementing solar energy with our ever-increasing use of fossil fuels. Satellite and climate data indicate that humans are appropriating about 20 percent of the average annual net primary production on Earth. The percentage varies strikingly among regions: urban areas consume as much as 300 times the NPP they generate, but people in sparsely inhabited parts of the Amazon Basin appropriate almost nothing.

56.2 RECAP

Nearly all energy utilized by living organisms comes from the sun. Human activities influence energy flows and appropriate a great deal of Earth's net terrestrial primary productivity.

- What is the difference between gross primary production and net primary production? See p. 1209
- What percentage of Earth's average annual net primary production is appropriated by humans, and how does that percentage vary regionally? See p. 1210

Understanding how human appropriation of net primary production affects quantities and patterns of energy flow requires knowing how the compounds of carbon, in which most energy is stored, are formed and move among Earth's physical compartments. In the next section we will consider how carbon, as well as the other chemical elements living organisms need, cycle through the compartments of the global ecosystem.

56.6 Primary Production in Different Ecosystem Types The contributions of different ecosystem types to Earth's primary production can be measured by (A) their geographic extent; (B) their net primary production; and (C) their percentage of Earth's total primary production.

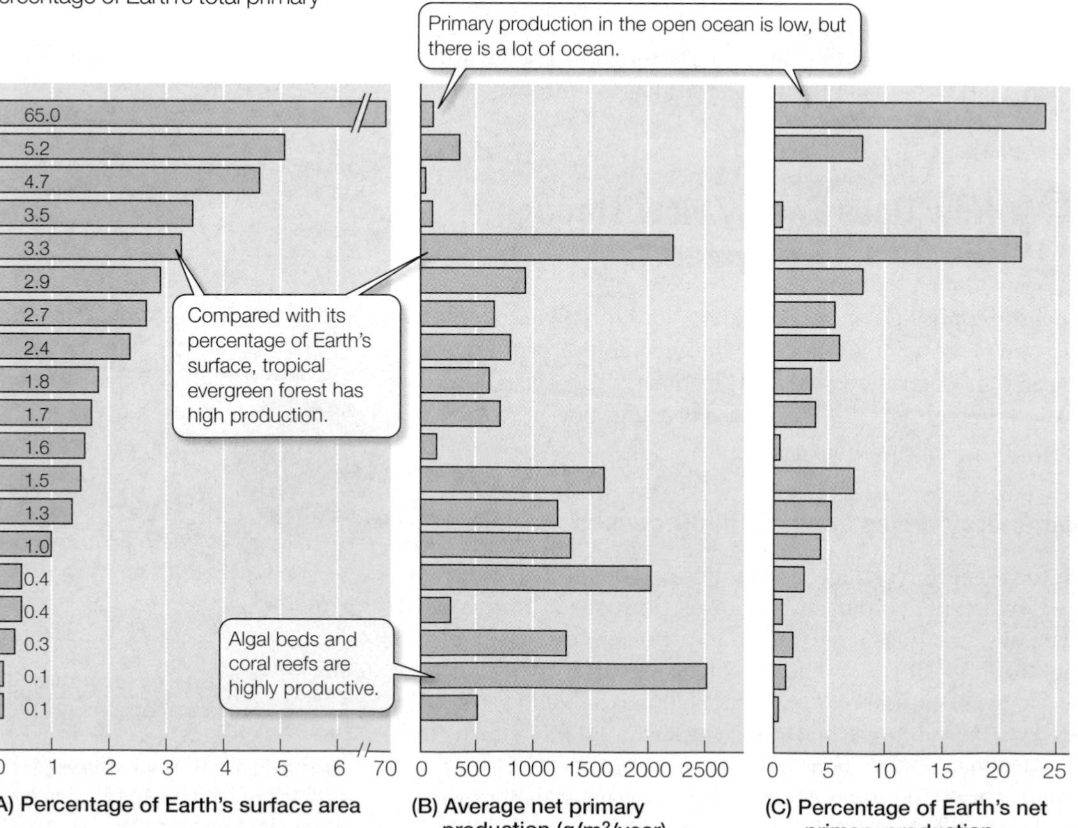

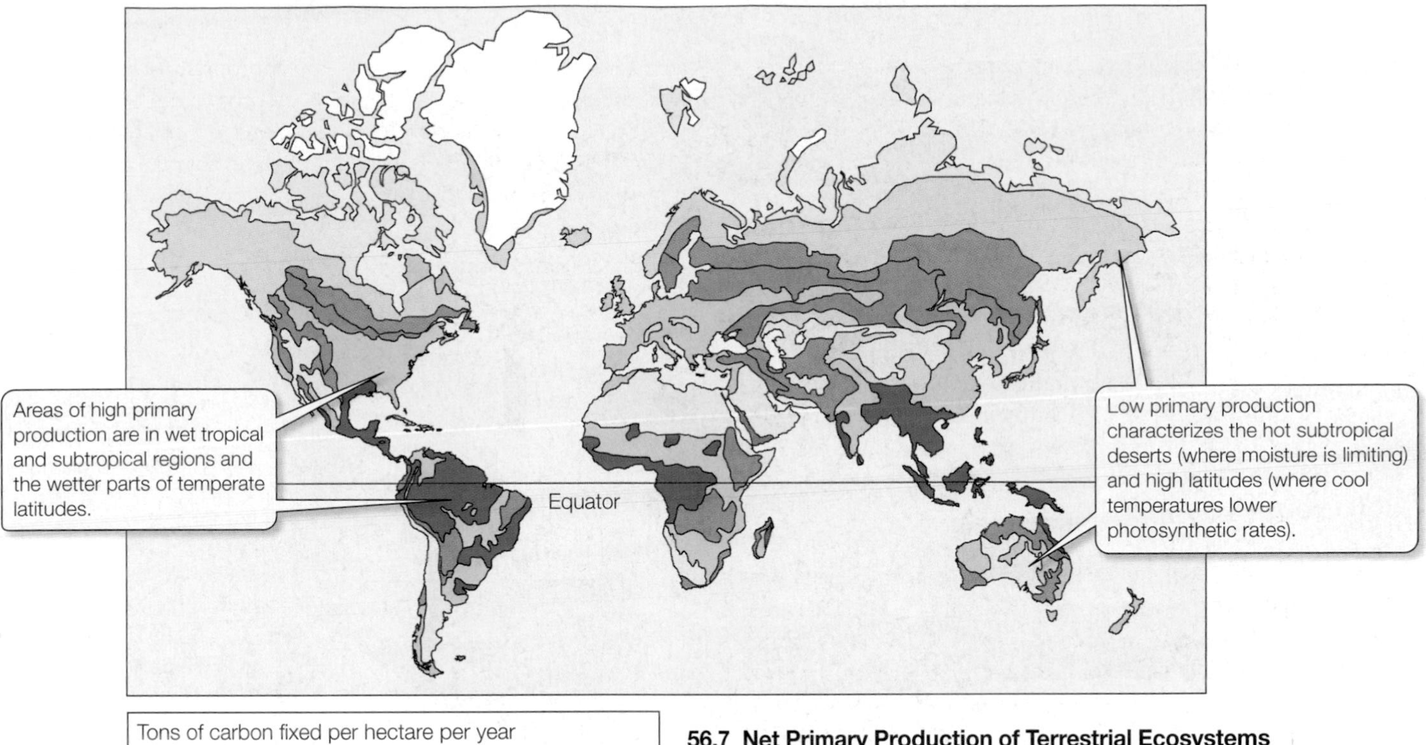

Areas of high primary production are in wet tropical and subtropical regions and the wetter parts of temperate latitudes.
Equator
Low primary production characterizes the hot subtropical deserts (where moisture is limiting) and high latitudes (where cool temperatures lower photosynthetic rates).
Tons of carbon fixed per hectare per year
Ice caps 0
0–2.5
2.5–6.0
6.0–8.0
8.0–10.0
10.0–30.0
>30.0

56.7 Net Primary Production of Terrestrial Ecosystems Variations in temperature and water availability over Earth's land surface influence net primary production.

56.3 How Do Materials Cycle Through the Global Ecosystem?

The chemical elements organisms need in large quantities—carbon, hydrogen, oxygen, nitrogen, phosphorus, and sulfur—cycle through organisms to the physical environment and back again. The pattern of movement of a chemical element through organisms and the other compartments of the global ecosystem is called its **biogeochemical cycle**.

Each element has a distinctive biogeochemical cycle whose properties depend on the physical and chemical nature of the element and the ways in which organisms use it. All chemical elements cycle quickly through organisms because no individual, even of the longest-lived species, lives very long in geological terms.

Before we describe the biogeochemical cycles of individual elements, however, we must discuss the roles of water and fire.

- The movement of *water* transfers many chemical elements between the atmosphere, land, fresh waters, and oceans.
- *Fire* is a powerful agent that speeds the cycling of chemical elements.

> An atom of phosphorus "X" had marked time in the limestone ledge since the Paleozoic seas covered the land. Time, to an atom locked in a rock, does not pass. The break came when a bur-oak root nosed down a crack and began prying and sucking. In the flash of a century the rock decayed, and X was pulled out and up into the world of living things. He helped build a flower, which became an acorn, which fattened a deer, which fed an Indian all in a single year.
>
> Aldo Leopold, *A Sand County Almanac*

Water transfers materials from one compartment to another

The cycling of water through the oceans, atmosphere, fresh waters, and land is known as the **hydrological cycle**. The hydrological cycle operates because more water is evaporated from the surface of the oceans than is returned to them as *precipitation* (rain or snow). The excess evaporated water is carried by winds over the land, where it falls as precipitation. Water also evaporates from soils, from lakes and rivers, and from the leaves of plants (transpiration), but the total amount evaporated from those surfaces is less than the amount that falls on them as precipitation. Excess terrestrial precipitation eventually returns to the oceans via rivers, coastal runoff, and groundwater flows (**Figure 56.8**). Earth's 16 largest rivers account for more than one-third of total water runoff from the land, more than half of which comes from the three largest rivers (the Amazon, Congo, and Yangtze). Despite their relatively small volume, rivers play a disproportionate role in the hydrological cycle because the average residence time of a water molecule in lakes and rivers is only 4.3 years, compared with 2,640 years in the oceans. The average residence time of water in living things is particularly short—about 5.6 days.

By building dams, canals, and reservoirs and by diverting huge quantities of water to irrigated fields, humans have had major effects on the temporal and spatial distribution of fresh water on Earth. The most important consequence of human activities is that

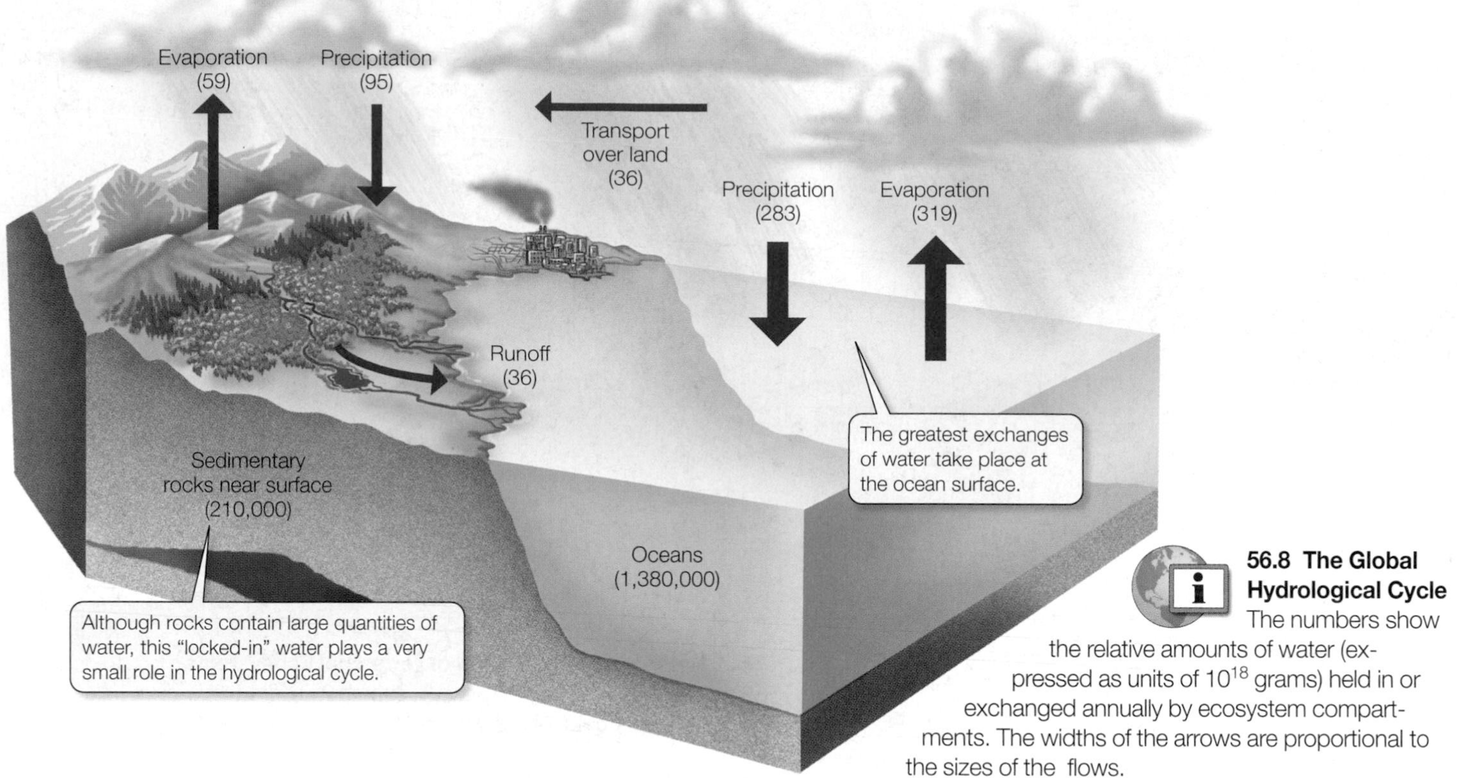

56.8 The Global Hydrological Cycle The numbers show the relative amounts of water (expressed as units of 10^{18} grams) held in or exchanged annually by ecosystem compartments. The widths of the arrows are proportional to the sizes of the flows.

more water now evaporates from land than before the Industrial Revolution. In addition, freshwater flow patterns are being seriously altered. For example, dams on the Columbia River in Washington State are managed to reduce the variation in water flow rates during the year (**Figure 56.9**). An unintended result of this practice is reduced autumn and winter ocean surface salinity from the mouth of the river north along the coast to the Aleutian Islands, with negative consequences for the growth and survival of salmon in that part of the Pacific Ocean.

Although large amounts of groundwater are present in underground pools called **aquifers**, this water has a long residence time underground and plays only a small role in the hydrological cycle. In some places, however, groundwater is being seriously depleted because humans are using it more rapidly than it can be replaced, primarily by pumping it for irrigation. In some areas of the High Plains of the central United States, more than half of the groundwater has been removed. On the North China Plain, depletion of shallow aquifers is forcing people to sink wells more than 1,000 meters deep to reach water.

Much of the groundwater used today in the Northern Hemisphere was deposited during the last Ice Age, when regional precipitation was much greater than it is now. Using this groundwater for irrigation and other purposes has increased flows of water to the oceans and has contributed to the sea level rise of the past century.

If current water consumption patterns continue, by 2025 at least 3.5 billion people (48 percent of the world's population) will live in areas with inadequate water supplies. The population in Asia will probably be most severely affected. Fortunately, current water consumption patterns need not continue. Per capita water consumption in the United States and Europe is dropping because of increasing use of water-efficient home appliances, increasing prices for water, and development of regulations that restrict water use. If such programs to improve water use efficiency were implemented vigorously, global water use in 2025 could be less than it is today, despite continued population growth.

Fire is a major mover of elements

Every year, 200–400 million hectares* of savannas, 5–15 million hectares of boreal forests, and lesser amounts of other biomes burn. Lightning ignites some fires, but humans start most fires to manage vegetation (as when they cut down and burn forests to clear land for growing crops). Fires consume the energy of, and release chemical elements from, the vegetation they burn. Some nutrients, such as nitrogen, sulfur, and selenium, are easily vaporized by fire. They are discharged to the atmosphere in smoke or carried into groundwater by rain falling on burned ground.

Fires also release large amounts of carbon into the atmosphere. The global annual flow of carbon to the atmosphere from savanna and forest fires is estimated to be in the range of 1.7 to 4.1 petagrams.* Biomass burning is responsible for about 40 percent of Earth's annual production of CO_2. It is also a significant contributor to the production of other greenhouse gases, such as carbon monoxide (CO), methane (CH_4), and tropospheric ozone (O_3).

*1 hectare = 10^4 square meters, or 2.5 acres; 1 petagram = 10^{15} grams

(A)

(B)

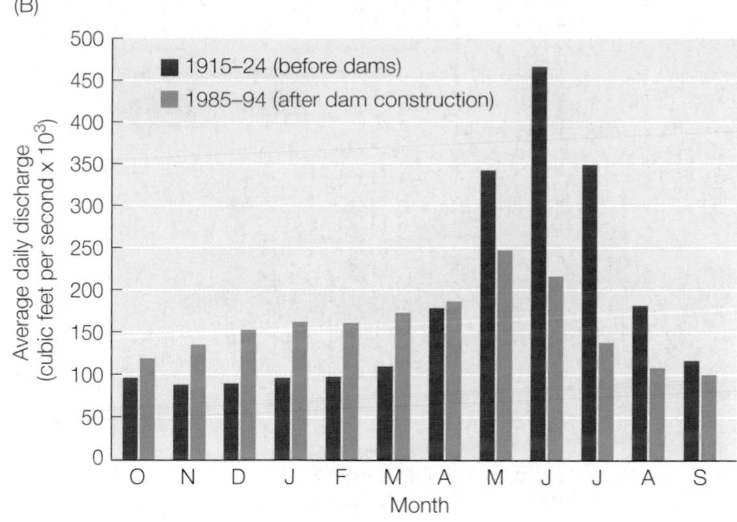

56.9 Columbia River Flows Have Been Massively Altered
(A) The Bonneville Dam, near Portland, Oregon, is just one of dozens of dams that have been built along the Columbia River and its tributaries to regulate the river's flow and generate hydroelectric power. (B) Before the construction of dams, rates of freshwater flow from the Columbia River to the Pacific Ocean varied greatly with the seasons. Human intervention has substantially reduced this variation.

Today's oceans absorb 20–25 million tons of CO_2 every day—a faster rate than at any time during the past 20 million years. As a result, water near the ocean surface is becoming more acidic. How this increasing acidity will affect marine organisms is unknown.

Because humans deliberately start most fires, we can take steps to reduce their undesirable consequences. Periodic burning of fire-prone vegetation can prevent the buildup of large fuel loads and reduce the frequency of high-intensity canopy fires, which discharge great quantities of materials to the atmosphere. Improvements in the productivity of tropical agriculture can reduce the rate at which forests are cleared and burned to create more agricultural land. Expanded use of simple solar cookers can reduce the need to harvest firewood for cooking meals.

Our discussion of water and fire shows that the biogeochemical cycles of different chemical elements are connected to one another. However, it is easier to understand the interactions of these cycles if we first discuss each one separately. Let's begin by tracing the biogeochemical cycle of carbon.

The carbon cycle has been altered by industrial activities

As we saw in Part Two of this book, all of the important macromolecules that compose living things contain carbon, and much of the energy that organisms need to fuel their metabolic activities comes from carbon-containing (organic) compounds. Carbon is incorporated into organic molecules by photosynthesis in the cells of autotrophs. All heterotrophic organisms get their carbon by consuming autotrophs or other heterotrophs, their remains, or their waste products.

Most of Earth's carbon is stored in rocks and marine sediments, and dissolved in ocean waters. In the terrestrial compartment, most carbon that is available to organisms is stored in soils. Biological processes move carbon between the atmospheric and terrestrial compartments, removing it from the atmosphere during photosynthesis and returning it to the atmosphere during metabolism (**Figure 56.10**).

At times in the remote past, great quantities of carbon were removed from the global carbon cycle when organisms died in large numbers and were buried in sediments lacking O_2. In such anaerobic environments, detritivores do not reduce organic carbon to CO_2. Instead, organic molecules accumulate and eventually are transformed into deposits of oil, natural gas, coal, or peat. Humans have discovered and used these deposits, known as **fossil fuels**, at ever-increasing rates during the past 150 years. As a result, CO_2, one of the final products of the burning of fossil fuels, is being released into the atmosphere faster today than it is dissolving in the surface waters of the oceans or being incorporated into terrestrial biomass. Based on a variety of calculations, atmospheric scientists believe that before the Industrial Revolution, the concentration of atmospheric CO_2 was probably about 265 parts per million. Today, it is 380 parts per million (**Figure 56.11**), representing a rate of increase more than ten times faster than at any other time for millions of years.

WHERE HAS ALL THE CARBON GONE? Less than half of the vast amount of CO_2 released to the atmosphere by human activities remains in the atmosphere. Where is the rest? Much of the remainder is dissolved in the oceans. Over decades to centuries, the oceans, which contain 50 times the amount of dissolved inorganic carbon as the atmosphere, determine atmospheric CO_2 concentrations. The rate at which CO_2 moves from the atmosphere to the oceans depends, in part, on photosynthesis by phytoplankton in

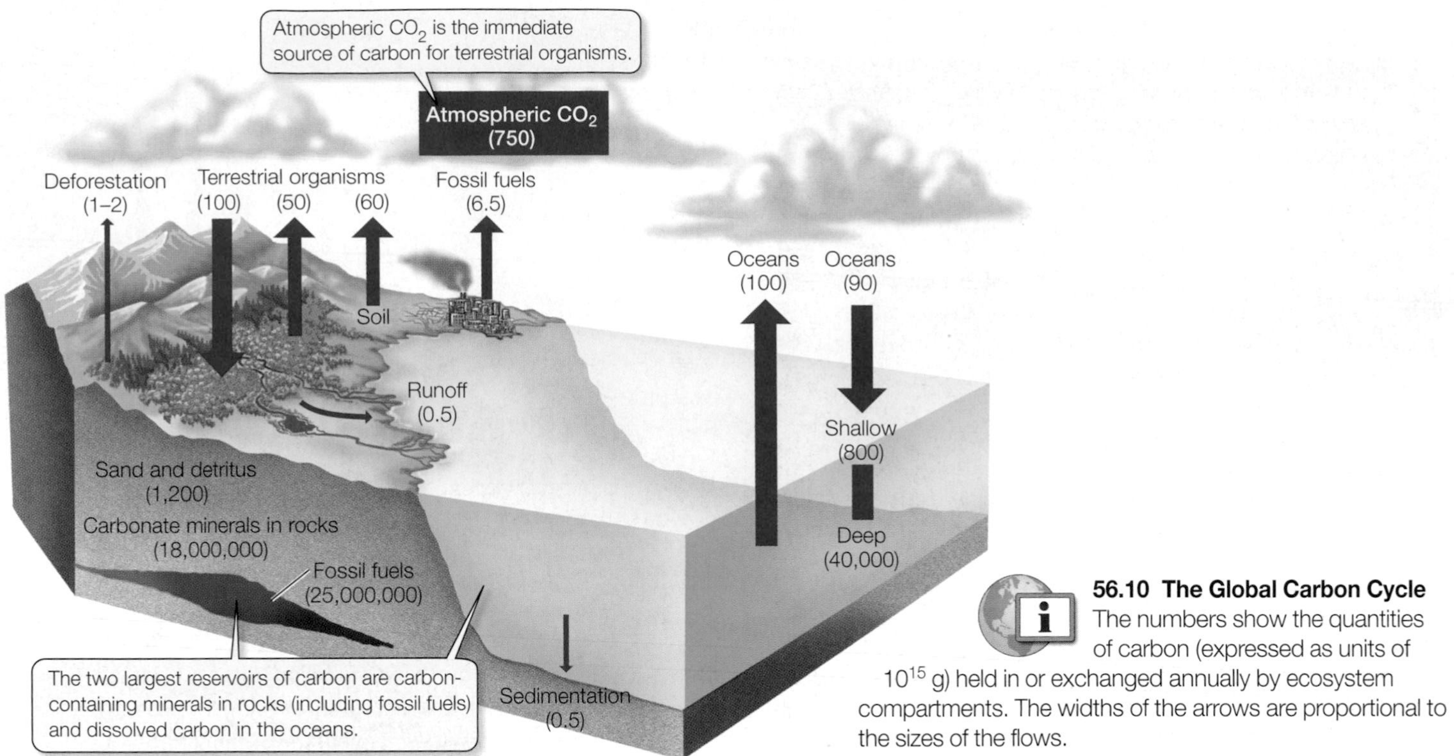

56.10 The Global Carbon Cycle The numbers show the quantities of carbon (expressed as units of 10^{15} g) held in or exchanged annually by ecosystem compartments. The widths of the arrows are proportional to the sizes of the flows.

the surface waters. These organisms remove carbon from the water, thereby increasing its absorption of carbon from the atmosphere. In addition, many marine organisms form calcium carbonate ($CaCO_3$) shells, which eventually sink to the ocean floor. Thus, by removing carbon from surface waters, marine organisms increase absorption of carbon from the atmosphere.

Between 1800 and 1994, the oceans absorbed a total of about 118 petagrams of carbon, accounting for nearly half of the carbon in the total CO_2 emissions from human activities, which would otherwise have remained in the atmosphere. This human-generated carbon is not evenly distributed in the oceans, however. The North Atlantic Ocean covers only 15 percent of the global ocean area, but it stores 23 percent of the human-generated carbon. By contrast, the much larger Southern Ocean, south of 50°S latitude, contains only 9 percent of the total carbon. The North Atlantic is an especially important carbon sink (a *sink* is a location in which an element is taken out of circulation) because of a circulation pattern called the *ocean conveyor belt*, which is driven by the sinking of dense, saline water between Greenland and Europe. By removing carbon-rich waters from the ocean surface, the ocean conveyor belt increases the transfer rate of carbon from the atmosphere to the oceans.

Some human-generated carbon is stored on land. Photosynthesis by terrestrial vegetation, principally in forests and savannas, typically absorbs about the same amount of carbon that is released by metabolism, about half by plants and half by microbes decomposing organic matter produced by the plants. Currently, however, the photosynthetic consumption of CO_2 exceeds the metabolic production of CO_2. Thus Earth's terrestrial vegetation is storing carbon that would otherwise be increasing atmospheric CO_2 concentrations.

Can terrestrial ecosystems help absorb the additional CO_2 that human activities are releasing into the atmosphere? As we saw at the opening of this chapter, recent investigations of forest trees suggest that we cannot count on terrestrial vegetation to store much more carbon for us. Christian Körner and his colleagues at the University of Basel in Switzerland monitored tree growth, reproduction, and carbon storage at these high CO_2 concentrations over a 4-year period. They found that the trees, which had reached only about two-thirds of their maximum size at the time of the experiment, did increase their growth rates, but that various species responded differently. For example, beech trees kept their leaves longer, but oaks held their leaves for less time.

56.11 Atmospheric Carbon Dioxide Concentrations Are Increasing These carbon dioxide concentrations, expressed as parts per million by volume of dry air, were recorded on top of Mauna Loa, Hawaii, far from most sources of human-generated CO_2 emissions.

The results of these and other experiments suggest that although *young* forests are likely to store more carbon in a carbon-enriched world, terrestrial vegetation, taken overall, may actually *lose* carbon rather than storing it under these conditions. For example, a recent intensive survey, based on almost 6,000 sites representing all types of land uses in England and Wales, showed that about 13 million tons of carbon had been lost from the soils of those countries during the past 25 years. Soil organic carbon was lost from all ecosystem types, including natural forests, but losses were greater from agricultural soils. Why the losses occurred is not clear, but climate warming (which increases plant metabolism) is the most likely factor causing losses of carbon from natural forests.

ATMOSPHERIC CO_2 AND GLOBAL WARMING How will global climates and ecosystems change in response to the rapid carbon dioxide enrichment caused by human activities? The buildup of atmospheric CO_2 that has resulted from the burning of fossil fuels is warming Earth. The concentration of CO_2 in air trapped in the Antarctic and Greenland ice caps during the last Ice Age—between 15,000 and 30,000 years ago—was as low as 200 parts per million. During a warm interval 5,000 years ago, atmospheric CO_2 may have been slightly higher than it is today. This long-term record shows that Earth has been warmer when CO_2 levels were higher and cooler when they were lower (**Figure 56.12**). The atmosphere would have warmed even more than it has if the oceans had not absorbed vast amounts of heat. The upper 100 to 200 meters of all the oceans have warmed dramatically, and warming has penetrated as deep as 700 meters in both the North Atlantic and South Atlantic oceans (**Figure 56.13**).

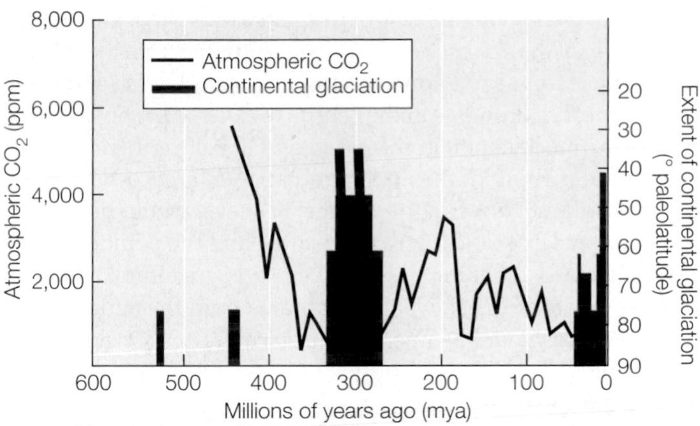

56.12 Higher Atmospheric CO_2 Concentrations Correlate with Warmer Temperatures The geological record for variations in carbon dioxide over the last 500 million years (black trace) shows that CO_2 levels have been lowest during times of extensive glaciation (red bars), and thus presumably lower temperatures.

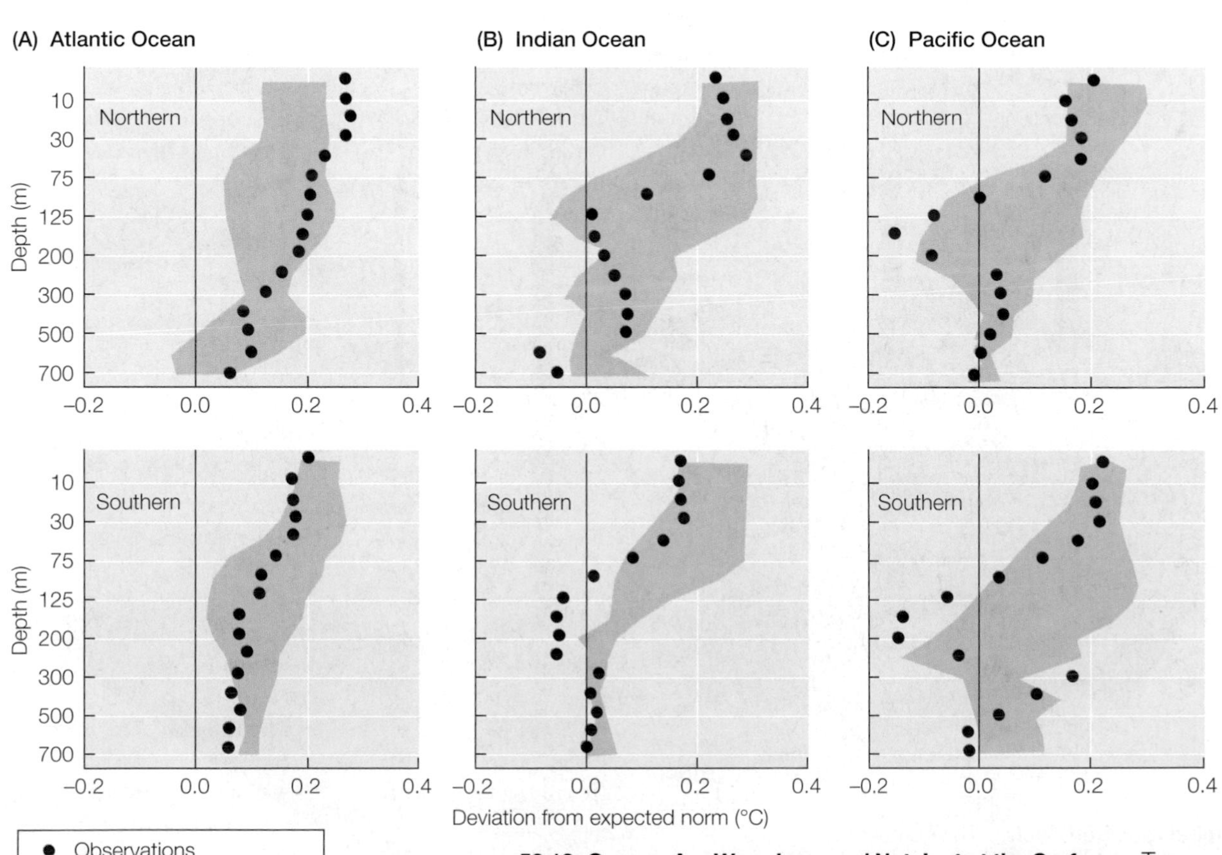

56.13 Oceans Are Warming—and Not Just at the Surface Temperatures are plotted as deviations from the value expected (0.0) at the depth of the observations. Temperature readings from the northern portions of the oceans are shown above; southern readings are below. Both sections of the Atlantic Ocean are abnormally warm to a depth of 700 meters.

Complex computer models of the global ecosystem indicate that a doubling of the atmospheric CO_2 concentration from today's level would increase mean annual temperatures worldwide and would probably cause droughts in the central regions of continents while increasing precipitation in coastal areas. Global warming could result in the melting of the Greenland and Antarctic ice caps, and would continue to warm the oceans. Sea level would rise (due to both thermal expansion of ocean waters and the addition of glacial meltwater), flooding coastal cities and agricultural lands. Because the rate at which water evaporates from the surface of the oceans increases with temperature, tropical storms are likely to become more intense. Indeed, during the past two decades, the number of hurricanes generated per year worldwide has not increased, but the number of severe (category 4 and 5) storms has.

Another result of global warming is increasing outbreaks of many diseases. Winter cold typically kills many pathogens, sometimes eliminating as much as 99 percent of a pathogen population. If climate warming reduces this population bottleneck, some diseases will become more common in temperate regions. For example, dengue fever and its mosquito vector are now spreading to higher latitudes where they were formerly absent. Several plant diseases are more severe after milder winters or during periods of warmer temperatures. For example, during a 39-year study in Maine, beech bark canker, caused by a fungus (*Nectria* spp.), was found to be more severe after mild winters or dry autumns. These weather conditions favor the survival and spread of the beech scale insect, which weakens the trees and predisposes them to the fungal infection. The spread of this pathogen poses a serious problem for the timber industry.

The 1990 Assessment Report of the Intergovernmental Panel on Climate Change (IPCC) paid little attention to risks to human health. In contrast, IPCC's 1996 Assessment Report gave detailed consideration to the potential effects of climate change on human health. Our increasingly interconnected global society enables disease organisms to travel rapidly around the world. Infected people carried the SARS virus, for example, from China to Canada within a few days. The combination of human mobility and climate warming poses serious challenges to human health throughout the world.

Recent perturbations of the nitrogen cycle have had adverse effects on ecosystems

Nitrogen gas (N_2) makes up 78 percent of Earth's atmosphere, but most organisms cannot use nitrogen in its gaseous form. Only a few species of microorganisms can convert atmospheric N_2 into forms that are usable by plants, a process called *nitrogen fixation*. Other microorganisms carry out *denitrification*, the principal process that removes nitrogen from the biosphere and returns it to the atmosphere (see Section 36.4 and Figure 36.10). These movements of nitrogen account for about 95 percent of all natural nitrogen flows on Earth (**Figure 56.14**).

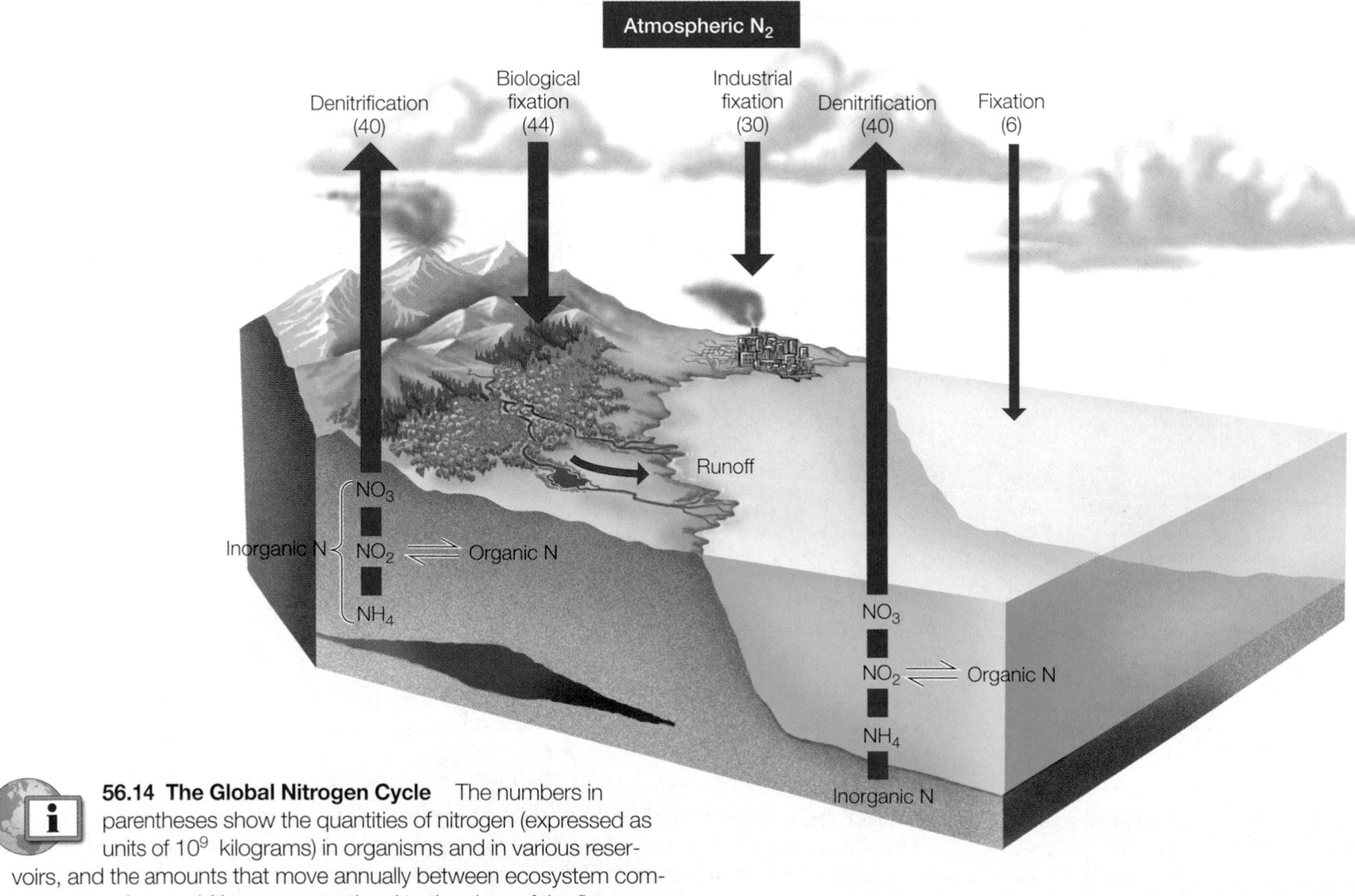

56.14 The Global Nitrogen Cycle The numbers in parentheses show the quantities of nitrogen (expressed as units of 10^9 kilograms) in organisms and in various reservoirs, and the amounts that move annually between ecosystem compartments. Arrow widths are proportional to the sizes of the flows.

Biologically usable nitrogen is often in short supply in ecosystems. That is why nearly all fertilizers contain nitrogen. Populations of nitrogen-fixing organisms rarely increase to such an extent that nitrogen is no longer limiting because nitrogen tends to be lost rapidly from ecosystems by leaching, vaporization of ammonia, and denitrification. In addition, fixing nitrogen requires large amounts of energy. Nitrogen-fixing organisms often lose out in competition with non-fixers when nitrogen becomes more readily available.

As a result of the extensive use of fertilizers on agricultural crops and the burning of fossil fuels (which generates nitric oxide), the total nitrogen fixation by humans today is about equal to global natural nitrogen fixation (**Figure 56.15**). This human-generated nitrogen flow is expected to continue to increase during the coming decades. A variety of adverse effects are associated with these large perturbations of the nitrogen cycle. They include the contamination of groundwater by nitrate (NO_3^-) from agricultural runoff, increases in atmospheric nitrous oxide (N_2O) and tropospheric O_3 (both greenhouse gases), and smog production.

When more nitrogen is applied to croplands than is taken up by plants, the excess nitrogen moves downward into groundwater or as runoff, into rivers, lakes, and oceans. The addition of nutrients to these bodies of water, known as **eutrophication**, can have a number of negative effects on aquatic ecosystems, as we'll see in our discussion of the phosphorus cycle. The "dead zone" in the Gulf of Mexico that has formed near the mouth of the Mississippi River is a result of flows of nitrogen-enriched water from agricultural fields in the Upper Midwest (**Figure 56.16**). Outbreaks of the toxic dinoflagellate *Pfiesteria* in estuaries on the Atlantic coast of North America are also triggered by nitrogen enrichment.

Some of the nitrogen that enters the atmosphere falls back to land in precipitation or as dry particles. This deposition of nitrogen affects the composition of terrestrial vegetation by favoring species that are better adapted to high nutrient levels. Atmospheric deposition of nitrogen has increased dramatically during recent decades due to human activity. Spatial variation in deposition rates allowed ecologists to determine that plant species richness is inversely correlated with the rate of nitrogen deposition in grasslands in the United Kingdom. Rates of nitrogen deposition are high enough over much of Europe and eastern North America to cause substantial reductions in species richness in grasslands on both continents.

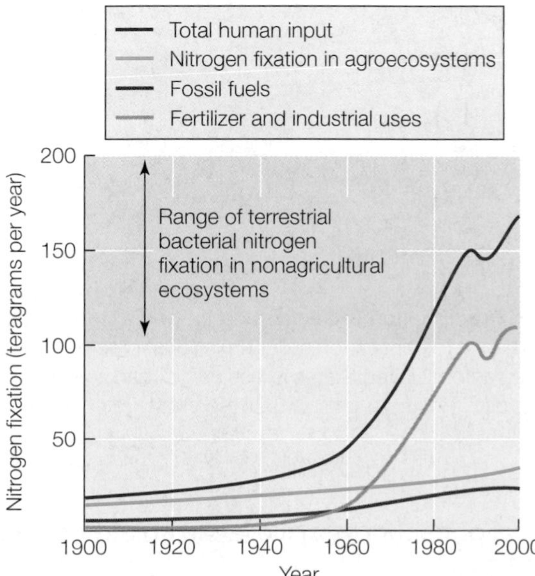

56.15 Human Activities Have Increased Nitrogen Fixation Most of the biologically reactive nitrogen produced by humans is manufactured for use in agriculture (synthetic fertilizers) and industry. Some nitrogen is also a by-product of burning fossil fuels.

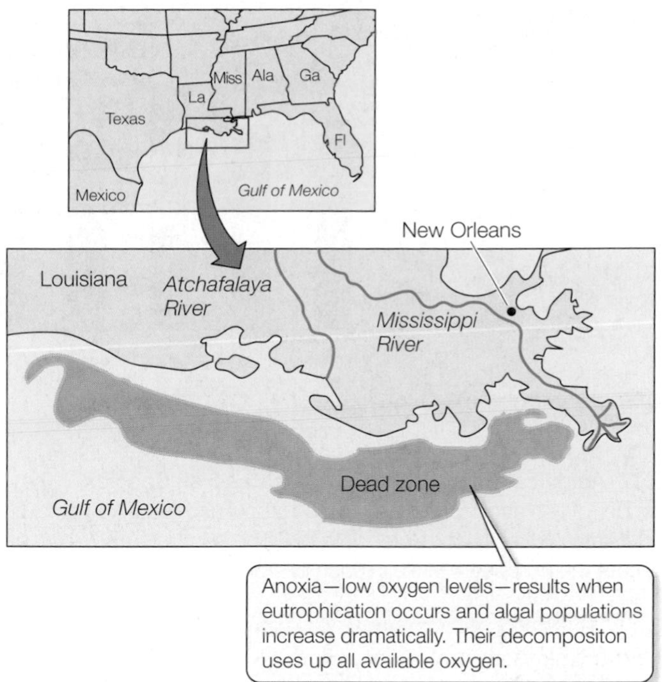

56.16 A "Dead Zone" Near the Mouth of the Mississippi River Rising amounts of nitrogen and phosphorus in the runoff from agricultural lands in the midwestern United States are carried by the Mississippi River to the Gulf of Mexico. This nutrient enrichment causes excessive algal growth, depleting the oxygen in the water and creating a deoxygenated "dead zone" in which most aquatic organisms cannot survive.

The burning of fossil fuels affects the sulfur cycle

Emissions of the gases sulfur dioxide (SO_2) and hydrogen sulfide (H_2S) from volcanoes and fumaroles (vents for hot gases) are the only significant natural nonbiological flows of sulfur. These sources release, on average, between 10 and 20 percent of the total natural flow of sulfur to the atmosphere, but they vary greatly in time and space. Large volcanic eruptions spread great quantities of sulfur over broad areas, but they are rare events. Terrestrial and marine organisms also emit compounds containing sulfur. Certain marine algae produce large amounts of dimethyl sulfide (CH_3SCH_3), which accounts for about half of the biotic component of the sulfur cycle (hence the smell of rotting seaweeds). Sulfur is apparently always abundant enough to meet the needs of living organisms.

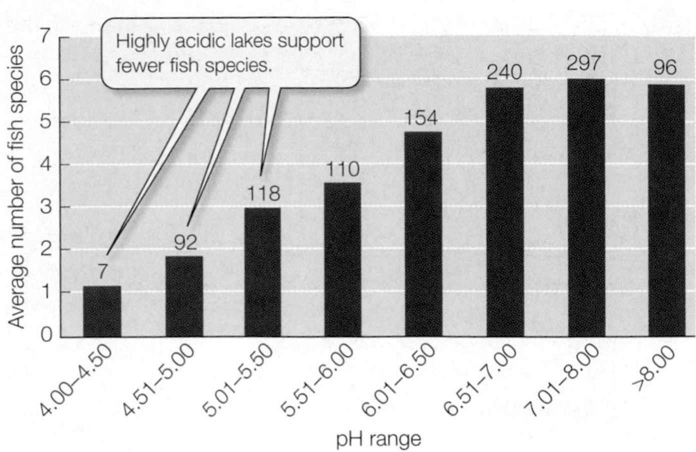

56.17 Acidification of Lakes Exterminates Fish Species The number of fish species found in lakes in the Adirondack region of New York State is directly correlated with pH. The numbers above the bars indicate the number of lakes in each pH range.

Sulfur plays an important role in global climate. Even if air is moist, clouds do not form readily unless there are small particles in the atmosphere around which water can condense. Dimethyl sulfide is the major component of such particles. Therefore, increases or decreases in atmospheric sulfur levels can change cloud cover and hence climate.

Humans have altered the sulfur cycle, as well as the nitrogen cycle, by the burning of fossil fuels. An important regional effect of these alterations is **acid precipitation**: rain or snow whose pH is lowered by the presence of sulfuric acid (H_2SO_4) and nitric acid (HNO_3). These acids may travel hundreds of kilometers in the atmosphere before they settle to Earth in precipitation or as dry particles.

Acid precipitation now falls in all major industrialized countries and is particularly widespread in eastern North America and Europe. The normal pH of precipitation in New England was about 5.6, but precipitation there now averages about pH 4.4, and there have been occasional rain- or snowfalls with a pH as low as 3.0. Precipitation with a pH of about 3.5 or lower causes direct damage to the leaves of plants and reduces rates of photosynthesis. Acidification of lakes in the Adirondack region of New York State has reduced fish species richness by causing the extinction of acid-sensitive species (**Figure 56.17**). Fortunately, as a result of the establishment of a flexible regulatory system under the 1990 Clean Air Act amendments, precipitation in much of the eastern United States is less acidic today than it was 20 years ago, primarily because of reduced sulfur emissions (**Figure 56.18**).

David Schindler at the University of Alberta studied the effects of acid precipitation on small Canadian lakes by adding enough H_2SO_4 to two lakes to reduce their pH from about 6.6 to 5.2. In both lakes, nitrifying bacteria failed to survive under these moderately acidic conditions. As a result, the nitrogen cycle was blocked, and ammonium ions accumulated in the water. When Schindler stopped adding acid to one of the lakes, its pH increased to 5.4, and nitrification resumed. After about a year, the pH of the lake returned to its original value. These experiments show that lakes are very sensitive to acidification, but that pH can return rapidly to normal values because water in lakes is exchanged rapidly.

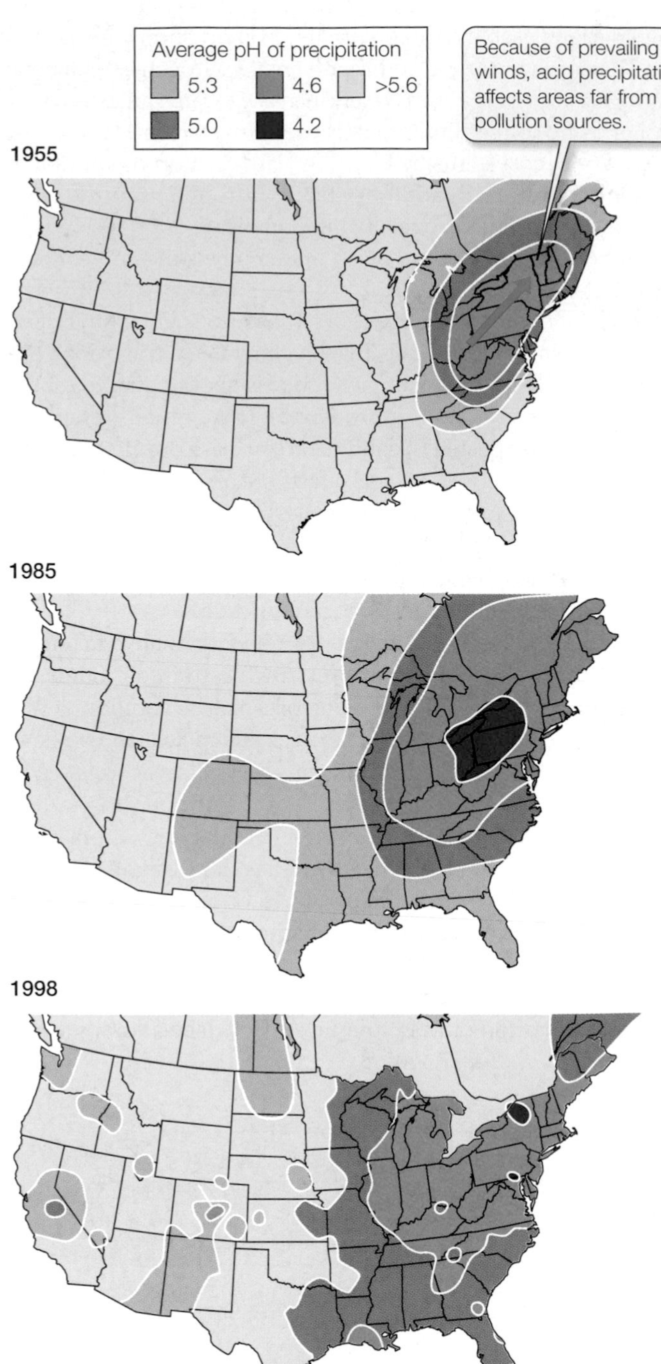

56.18 Acid Precipitation Is Decreasing in the Eastern United States Thanks to controls on sulfur emissions, precipitation in many parts of the eastern United States is less acidic than it was a decade ago, but acidity is increasing in parts of the West.

The global phosphorus cycle lacks an atmospheric component

Phosphorus accounts for only about 0.1 percent of Earth's crust, but it is an essential nutrient for all life forms. It is a key component of DNA, RNA, and ATP. Unlike the other elements discussed in this section, phosphorus lacks a gaseous phase. Some phospho-

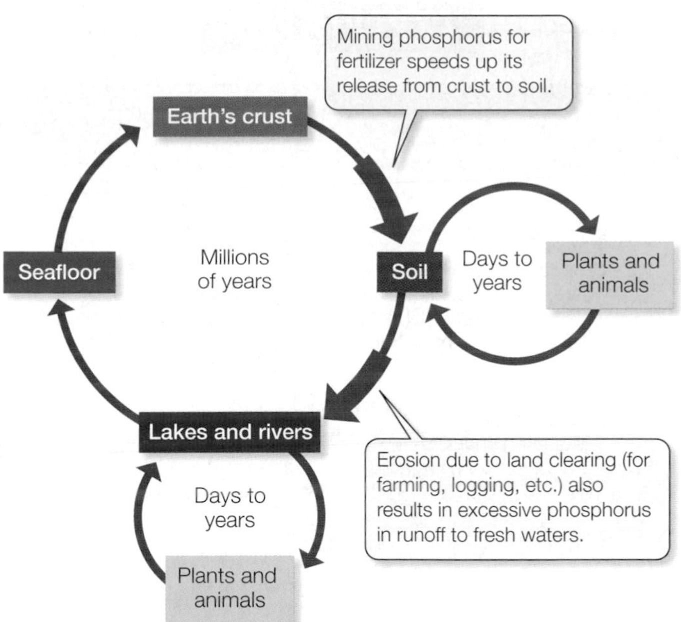

56.19 The Phosphorus Cycle The widths of the arrows are proportional to the flow rates. Two great increases in phosphorus flows (wider arrows) are due to human activities.

rus is transported on dust particles, but very little of the phosphorus cycle takes place in the atmospheric compartment. The global phosphorus cycle takes millions of years to complete because the processes of sedimentary rock formation on the ocean bottom, subsequent uplifting, and the weathering of rock into soil all act slowly (**Figure 56.19**). Phosphorus often cycles rapidly among organisms, however.

Human activity has radically accelerated some parts of the phosphorus cycle. About 90 percent of the phosphorus that is mined is used to produce fertilizers and animal feeds. One consequence of the massive use of phosphorus for fertilizer is that between 10.5 and 15.5 teragrams (1 teragram = 10^{12} grams) of phosphorus are accumulating in soils each year, primarily on agricultural lands. Increasing concentrations of phosphorus in agricultural soils are certain to lead to increased runoff of phosphorus into streams and lakes. Soil erosion due to the clearing of land for purposes such as agriculture and logging also increases the amount of phosphorus and other nutrients in runoff.

Phosphorus is often a limiting nutrient in soils and fresh waters, which is why adding phosphorus to agricultural soils and lakes increases rates of photosynthesis and biological productivity. Most extra phosphorus enters lakes as phosphates derived from fertilizers and household detergents. The resulting eutrophication allows algae and bacteria to multiply, forming blooms that turn the water green. When these organisms die, their decomposition consumes all of the O_2 in the lake, and anaerobic organisms come to dominate the bottom sediments. These anaerobic organisms do not break down carbon compounds all the way to CO_2. Their metabolic end products, many of which have unpleasant odors, build up.

Two hundred years ago, Lake Erie, one of the Great Lakes on the border between the United States and Canada, had only moderate rates of photosynthesis and clear, oxygenated water. Today, however, the more than 15 million people that live in the Lake Erie drainage basin pour more than 250 billion liters of sewage and industrial wastes into the lake annually. The entire basin is intensely farmed and heavily fertilized. In the early part of the twentieth century, when industrialization began, nutrient concentrations in the lake increased greatly, and algae proliferated. At the water filtration plant in Cleveland, Ohio, algae increased from 81 per milliliter in 1929 to 2,423 per milliliter in 1962. Populations of bacteria also increased: the numbers of the bacterium *Escherichia coli* increased enough to cause the closing of many of the lake's beaches as health hazards.

Since 1972, the United States and Canada have invested more than U.S. $9 billion to improve municipal waste treatment facilities and reduce discharges of phosphorus into Lake Erie. As a result, the amount of phosphate added to Lake Erie has decreased more than 80 percent from its highest level, and phosphorus concentrations in the lake have declined substantially. The deeper waters of Lake Erie still become poor in O_2 during the summer months, but the rate of O_2 depletion is declining.

Fortunately, the potential to recover and recycle phosphorus is high. The phosphorus contained in sewage and animal wastes could supply much of the needs of the detergent and fertilizer industries. Careful application of fertilizers on agricultural lands could reduce the rate of phosphorus accumulation in soils without reducing crop yields. However, reduction of phosphorus in soils takes many decades after remedial actions are initiated. During that time, increased eutrophication of lakes and streams is certain to happen.

Other biogeochemical cycles are also important

In addition to the common elements involved in biogeochemical cycles that we have just discussed, some rare elements are also very important ecologically because, in addition to macronutrients, there is an array of essential nutrients that organisms need only in very small amounts.

One important such nutrient is *iron* (Fe), which is required for photosynthesis. Iron is readily available on land, from which it is transported to coastal waters by rivers and to the open ocean by atmospheric dust. Because iron is insoluble in oxygenated water, it rapidly sinks to the ocean floor. Therefore, over much of the ocean, the rate of photosynthesis is limited by iron.

Three micronutrients that occur at particularly low concentrations, iodine (I), cobalt (Co), and selenium (Se), are of particular importance. Land plants require these micronutrients in minimal concentrations, but animals, particularly endothermic vertebrates, require I, Co, and Se in concentrations that exceed the supply in many environments. Large herbivores may need to obtain I, Co, and Se by drinking water at mineralized springs or by eating enriched earth.

- *Iodine* is an essential component of the hormone thyroxine, which governs many metabolic processes. The ocean is the main reservoir of iodine; fresh water is poor in iodine. Iodine, which is supplied to land primarily by marine algae that release it to the air, is therefore often deficient in continental interiors.

- *Cobalt* is an essential component of vitamin B_{12} and is necessary for the synthesis of proteins. Bacteria that fix atmospheric nitrogen require cobalamin and cobamide coenzymes. Herbivores cannot synthesize cobalamin, so microbes in their guts are essential not only for fixation of atmospheric nitrogen and digestion of plant fibers, but also for synthesis of cobalamin.
- *Selenium* is a component of several important enzymes, such as the antioxidant glutathione peroxidase, which protects tissues from oxidative damage.

Iodine, together with Co and Se, regulates the production and functioning of many metabolic agents, including hormones, vitamins, and enzymes that contain zinc and copper. Although I, Co, and Se are needed by animals in very small amounts, they are potentially deficient in their plant food, which has even lower concentrations than those found in soils and rocks. Deficiencies of I, Co, and Se are not frequent everywhere, but Australia, which is characterized by nutrient-poor soils, contains very few places where herbivores can obtained the necessary supplies of I, Co, and Se. Cobalt deficiency is prevalent among domestic ruminants in Western Australia.

Biogeochemical cycles interact

Because biogeochemical cycles interact in significant ways, alterations in any one cycle affect the others. We have already seen how human alterations of several biogeochemical cycles are increasing the concentrations of greenhouse gases in the atmosphere and warming Earth's climate. As we saw earlier, acid precipitation is a result of the combined influences of sulfuric and nitric acids released into the atmosphere by human activities. Nitrate released by humans is a powerful oxidant; by oxidizing iron, it increases the movement of both iron and arsenic in lakes. Arsenic is a toxic element of considerable importance in the United States and south central Asia. Iron interacts with silicon to influence phytoplankton productivity in some areas of the oceans. Every year, scientists are discovering new interactions of which they were previously unaware, and experiments exploring biogeochemical interactions are increasing in number.

The interaction between elevated concentrations of atmospheric CO_2 and nitrogen fixation was studied by a group of ecologists in scrub oak vegetation on sandy, acidic soils in central Florida. They artificially increased CO_2 concentrations surrounding a nitrogen-fixing vine, *Galactia elliottii*, and measured its response. Higher CO_2 levels increased nitrogen fixation during the first year of the experiment, but, surprisingly, the positive effect disappeared by the third year. Elevated CO_2 levels actually depressed nitrogen fixation during the fifth, sixth, and seventh years of the experiment (**Figure 56.20**). The experimenters suspected that shortages of other micronutrients, such as iron and molybdenum, caused the depression of nitrogen fixation, so they measured concentrations of those nutrients in the leaves of *G. elliottii*. Concentrations of molybdenum (Mo) were particularly low in the leaves because accumulation of soil organic matter and lowering of soil pH—both of which strengthen Mo attachment to soil particles—reduced the availability of Mo to the plants.

EXPERIMENT

HYPOTHESIS: Higher CO_2 levels enhance nitrogen fixation.

METHOD

Artificially increase concentration of atmospheric CO_2 around plants. Measure rates of nitrogen fixation over 7 years. Determine concentration of iron and molybdenum in leaves of experimental plants. Maintain normal CO_2 levels around control plants.

RESULTS

Although higher CO_2 levels initially increased the rate of nitrogen fixation, the effect was quickly reversed. By year 4, nitrogen fixation rates were lower than those in the control plants.

Elevated CO_2 results in a doubling of rate of N-fixation the first year...

...but N fixation rates are depressed in later years.

CO_2 effect on N-fixation: 100, 80, 60, 40, 20, 0, −20, −40

Year: 1996, 1998, 2000, 2002

60, 40, 20

Control area

Experimental area

Iron (mg/g) Molybdenum (ng/g)

CONCLUSION: To predict the likely effects of CO_2 enrichment we need to investigate the interactions among many factors that might alter the processes we are studying.

56.20 High Atmospheric CO_2 Concentrations Negatively Affect Nitrogen Fixation Although the investigators' original expectation was that CO_2 enrichment would enhance nitrogen fixation, changes in soil pH made molybdenum (a key constituent of nitrogenase) less available to plants, thereby depressing nitrogen fixation.

56.3 RECAP

The pattern of movement of a chemical element through organisms and the physical environment is its biogeochemical cycle. Human activities have affected many biogeochemical cycles, especially those of water, carbon, nitrogen, sulfur, and phosphorus. Increasing concentrations of carbon dioxide and other greenhouse gases in the atmosphere are implicated in global warming.

- What drives the hydrological cycle? See pp. 1211–1212 and Figure 56.8
- Can you describe how biological processes move carbon from the atmospheric compartment to the terrestrial compartment and then return it to the atmosphere? See pp. 1213–1215 and Figure 56.10
- What are the results of human-induced alterations of the sulfur and nitrogen cycles? See pp. 1216–1218

The biogeochemical cycles of chemical elements are intimately connected with the functioning of ecosystems. Just as human alterations of these cycles are having many effects on ecosystems worldwide, the resulting changes in those ecosystems are having profound effects on human lives.

56.4 What Services Do Ecosystems Provide?

Ecosystems provide people with a variety of goods and services: food, clean water, clean air, flood control, soil stabilization, pollination, climate regulation, spiritual fulfillment, and aesthetic enjoyment, to name just a few. Most of these benefits either are irreplaceable or the technology necessary to replace them is prohibitively expensive. For example, potable fresh water can be provided by desalinating seawater, but only at great cost.

The rapidly expanding human population has greatly modified Earth's ecosystems to increase their ability to provide some of the goods and services it needs, particularly food, fresh water, timber, fiber, and fuel. These modifications have contributed substantially to human well-being and economic development. However, the benefits have not been equally distributed, and some people have actually been harmed by these changes. Moreover, short-term increases in some ecosystem goods and services have come at the cost of the long-term degradation of others. For example, efforts to increase the production of food and fiber have decreased the ability of some ecosystems to provide clean water, regulate flooding, and support biodiversity.

Increasing awareness of the complex linkages and trade-offs among the goods and services provided by ecosystems led to the establishment in 2001 of the Millennium Ecosystem Assessment (MA), a project that involved more than a thousand scientists and managers worldwide (**Figure 56.21**). The goals of the MA were to provide a global assessment of Earth's ecosystems, determine trends in the services they provide, and assess their importance for

Different strategies and interventions can be applied at many points in this framework to enhance human well-being and conserve ecosystems.
Human well-being and poverty reduction
• Basic material for a good life
• Health
• Good social relations
• Security
• Freedom of choice and action
Indirect drivers of change
• Demographic (population size, age structure, etc.)
• Economic (e.g., globalization, trade, market and policy framework)
• Sociopolitical (e.g., governance, institutional and legal framework)
• Science and technology
• Cultural and religious (e.g., beliefs, consumption choices)
Ecosystem services
• Provisioning (e.g., food, water, fiber, and fuel)
• Regulating (e.g., climate regulation, water, and disease)
• Cultural (e.g., spiritual, aesthetic, recreation, and education)
• Supporting (e.g., primary production, and soil formation)
Direct drivers of change
• Changes in local land use and cover
• Species introduction or removal
• Technology adaptation and use
• External inputs (e.g., fertilizer use, pest control, and irrigation)
• Harvest and resource consumption
• Climate change
• Natural, physical, and biological drivers (e.g., evolution, volcanoes)

56.21 The Millennium Ecosystem Assessment The chart shows the conceptual framework of the Millennium Assessment including the four categories of ecosystem services, the connections of these services with human well-being, the drivers of changes in ecosystems that can affect their ability to provide these services, and several points where interventions (dark green, curved arrows) could prevent or mitigate undesirable changes.

human well-being. To achieve these goals, the MA divided ecosystem services into four categories: provisioning services, regulating services, cultural services, and supporting services. The MA determined the drivers of change in the ecosystems that provide these services, and it identified strong links among these services and components of human well-being. It also sought to identify options for mitigating undesirable changes in the availability of ecosystem services.

The most important driver of alterations in ecosystems and the services they provide has been changes in land use as natural ecosystems have been converted to other, more intensive uses. More land was converted to cropland between 1950 and 1980 than in the 150 years between 1700 and 1850. Conversions recently have been particularly rapid in tropical and subtropical terrestrial ecosystem types (**Figure 56.22**). Aquatic ecosystems have suffered losses as well: between 1970 and 2000 about 20 percent of the world's coral reefs were lost; an additional 20 percent were degraded. So much water is impounded behind dams that about six times as much water is held in reservoirs today as in natural rivers. The amount of water withdrawn from rivers has doubled since 1960, mostly for agriculture. More than half of all the synthetic nitrogen fertilizer ever used on Earth has been used since 1985. Changes in these components of the global ecosystem continue rapidly today.

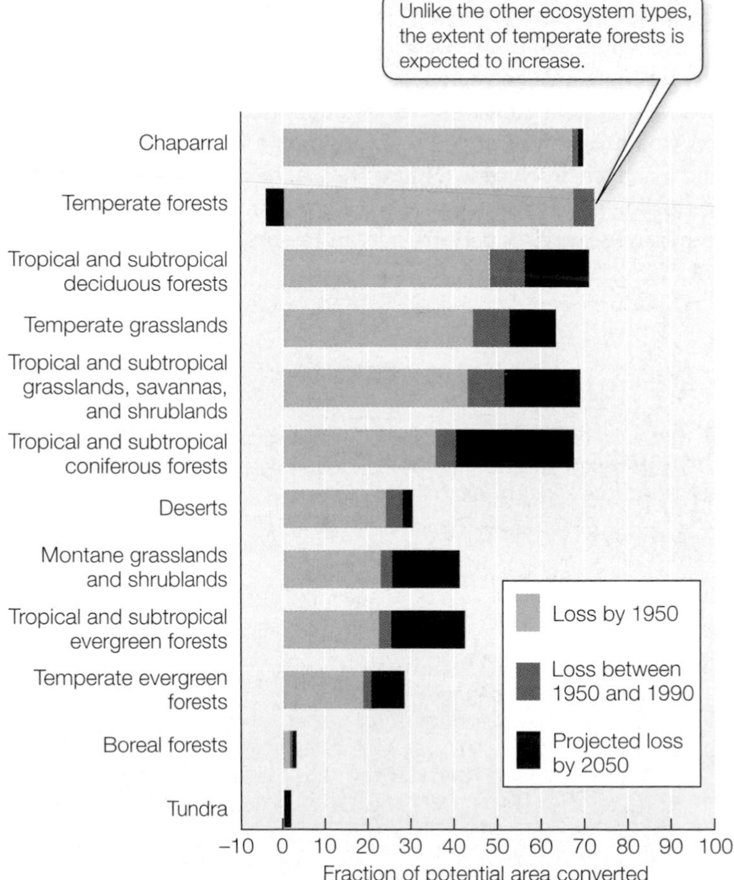

56.22 Some Ecosystem Types Have Suffered Extensive Losses Changes in land use have been the most important driver of changes in ecosystems, but some terrestrial ecosystem types have been converted to other uses more rapidly and extensively than others.

These altered ecosystems are delivering services that have contributed to many components of a good life—health, good social relations, and security. Food production contributes by far the most to economic activity and employment. The global market value of food production was $981 billion in 2000. About 22 percent of the world's population is employed in agriculture. More than 40 percent of people live in agriculturally based households, but the percentage is much lower in industrialized countries. Timber, marine fisheries, marine aquaculture, and recreational hunting and fishing are activities that use ecosystem services to make important contributions to human well-being.

Unfortunately, the modification of ecosystems to benefit human beings in one way has often resulted in the degradation of other services. For example, the spread of agriculture into marginal lands has increased soil degradation and reduced the ability of many ecosystems to provide clean water. Extensive use of pesticides controls pest insects, but also greatly reduces populations of pollinators and the services they provide to both crops and native plants. As described at the start of Chapter 52, the loss of wetlands and other natural buffers has reduced the ability of ecosystems to regulate natural hazards. The damage from the tsunami that hit Indonesia and other southeast Asian countries in December of 2004 was greater in many places than it would have been if the mangrove forests that protect the coast had not been cut down and converted to agriculture. Hurricane Katrina, which struck the U.S. Gulf Coast less than a year later, would not have caused as much flooding in New Orleans if the wetlands surrounding the city had been left intact (see p. 1112).

56.4 RECAP

Ecosystems provide human society with indispensable goods and services. Human alterations in these ecosystems may alter their ability to provide these goods and services.

- Can you describe some of the essential goods and services that ecosystems provide to humans? See pp. 1221–1222
- Do you understand how human efforts to increase the provision of some ecosystem services can cause degradation of other ecosystem services?

Human society faces the challenge of determining how we can obtain the goods and services we desire from ecosystems without compromising their ability to provide those goods and services over the long term. Fortunately, many options exist for such sustainable management of ecosystems.

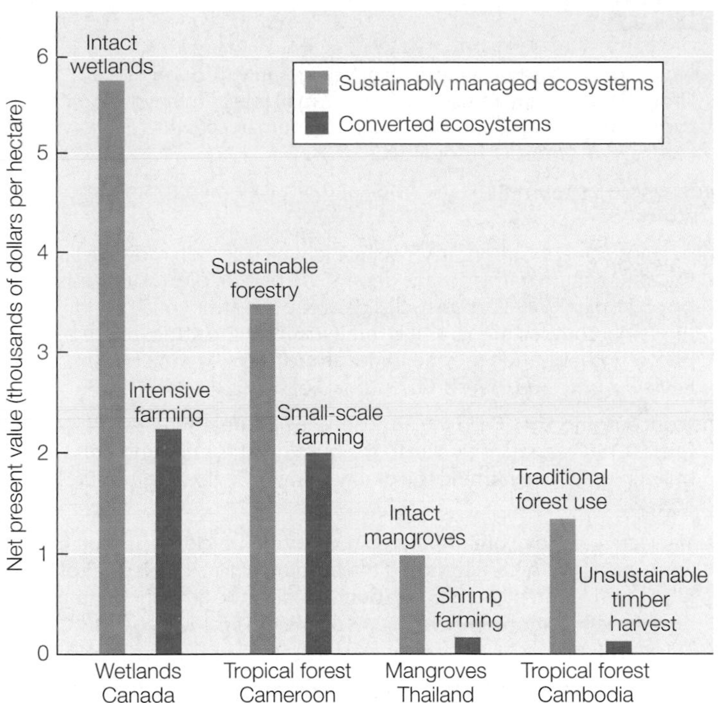

56.23 Economic Values of Sustainably Managed and Converted Ecosystems The Millenium Assessment estimates that sustainably managed ecosystems often provide more goods and services than those that have been converted to intensive uses.

56.5 What Options Exist to Manage Ecosystems Sustainably?

Despite the degradation of some important ecosystem services, there are many ways to conserve or enhance specific ecosystem services without compromising others. Often the total economic value of a sustainably managed ecosystem is higher than that of an intensively exploited ecosystem, such as one subjected to clear-cut logging or intensive agriculture (**Figure 56.23**). The long-term economic benefits of preventing overexploitation of marine fish populations, for example, are enormous; the collapse of the Newfoundland cod fishery in the 1990s due to overfishing resulted in the loss of tens of thousands of jobs and cost at least $2 billion in income support and retraining. As we saw in Section 54.5, it can take a long time for such fisheries to recover.

A major barrier to achieving these greater long-term benefits is that many ecosystem services are considered "public goods" that have no market value. Nobody has an incentive to pay for those services, whereas the individuals that own converted ecosystems can capture the great private economic benefits yielded by the conversions. It may take government action to create incentives encouraging sustainable ecosystem management. Elimination of subsidies that promote excessive exploitation of ecosystems is one such action governments might take. For example, between 2001 and 2003, governments of developed countries subsidized domestic agriculture each year by more than $324 billion. These subsidies led to greater food production in those countries than the global market warranted. They also promoted excessive use of fertilizers and reduced the profitability of agriculture in developing countries.

More sustainable use of fresh water can be achieved by charging users the full cost of providing the water, by developing methods to use water more efficiently in agriculture (the activity that consumes much more water than any other), and by altering the allocation of water rights so that the incentives favor conservation.

Sustainable use of marine fisheries can be achieved by establishing more marine protected reserves and flexible no-take zones where fish can readily grow to reproductive age. For some fisheries, establishing and implementing a limit to the number of fishing permits can increase the total economic value of the fishery and provide better fish to consumers. Such improvements followed the establishment of permits that could be bought and sold for the North Pacific halibut fishery.

Although many options exist to manage ecosystems in ways that will increase the total value of ecosystem goods and services and achieve a more equitable distribution of their benefits among people, implementing them will not be easy. Most people are not aware of how much human activities affect the functioning of ecosystems or of the great long-term value of ecosystem services, so active public education programs are needed. Maintaining and enhancing ecosystem services that have no established market value will be especially difficult. Probably the most difficult service to maintain in the face of increasingly intensive human use of global ecosystems is biological diversity. We devote the final chapter of this book to this important topic.

CHAPTER SUMMARY

56.1 What are the compartments of the global ecosystem?

Earth is a closed system with respect to atomic matter, but an open system with respect to energy.

Earth's physical environment can be divided into four compartments: oceans, fresh waters, atmosphere, and land. Review Figure 56.1

Most materials that cycle through the four compartments end up in the oceans. Cold, nutrient-rich water rises to the ocean surface in coastal **upwelling zones**.

Only a small fraction of Earth's water resides in the freshwater compartment, but water moves rapidly through it. Nutrients reach lakes and rivers by surface flow or by the movement of **groundwater**.

Vertical **turnover** of the water brings nutrients and dissolved CO_2 to the surface and O_2 to the deeper waters of freshwater lakes. Turnover occurs in temperate-zone lakes in spring and in autumn. Review Figure 56.3

The atmosphere plays a decisive role in regulating temperatures at and close to Earth's surface. Most global air circulation takes place in its lowest layer, the **troposphere**. A layer of ozone in the **stratosphere** absorbs ultraviolet radiation. Review Figure 56.4

Water vapor, carbon dioxide, and other **greenhouse gases** in the atmosphere are transparent to sunlight, but trap heat, warming Earth's surface.

Land covers only one-fourth of Earth's surface, but human life strongly depends on the productivity of that compartment.

56.2 How does energy flow through the global ecosystem?

See Web/CD Activity 56.1

Gross primary productivity is the rate at which energy is incorporated into the bodies of primary producers (which obtain their energy directly from the sun). The accumulated energy is **gross primary production** (**GPP**).

The energy available to organisms that eat primary producers, or **net primary production** (**NPP**), is gross primary production minus the energy expended by primary producers on their own metabolism.

The geographic distribution of primary production reflects the distribution of land masses, temperature, and moisture. See Figure 56.7

Humans appropriate about 20 percent of Earth's average annual net primary production, although the percentage varies regionally.

56.3 How do materials cycle through the global ecosystem?

See Web/CD Tutorial 56.1

The pattern of movement of a chemical element through organisms and the compartments of the global ecosystem is its **biogeochemical cycle**.

The **hydrological cycle** is driven by evaporation of water from the surface of the oceans. The excess terrestrial precipitation that falls on land eventually returns to the oceans, primarily through rivers. Groundwater in **aquifers** plays only a small role in the hydrological cycle, but is being seriously depleted by human activities. Review Figure 56.8, Web/CD Tutorial 56.2

Fires release large amounts of carbon and other elements into the atmosphere.

The oceans act as a carbon sink. The rate at which carbon moves from the atmosphere to the oceans depends in part on photosynthesis by phytoplankton. Carbon cycles between the atmospheric and terrestrial compartments as it is removed from the atmosphere by photosynthesis and returned to the atmosphere by metabolism. Review Figure 56.10, Web/CD Tutorial 56.3

The concentration of CO_2 in the atmosphere has greatly increased in the last 150 years, largely due to the burning of **fossil fuels**. This buildup of CO_2 is warming the global climate. Review Figures 56.11 and 56.12

Human activities have greatly altered the nitrogen cycle. As a result of agricultural use of fertilizers and the burning of fossil fuels, the total nitrogen fixation by humans is about equal to the amount of natural nitrogen fixation. Review Figures 56.14 and 56.15, Web/CD Tutorial 56.4

Human alterations of the nitrogen cycle have resulted in excess nitrogen compounds and other nutrients in bodies of water, leading to **eutrophication**. Review Figure 56.16

The burning of fossil fuels increases the amount of nitrogen and sulfur in the atmosphere, leading to **acid precipitation**.

Human activities such as agricultural use of fertilizers and clearing of land have dramatically increased the input of phosphorus into soils and fresh waters. Review Figure 56.19

56.4 What services do ecosystems provide?

Ecosystem services include food, clean water, soil stabilization by plants, pollination, climate regulation, spiritual fulfillment, and aesthetic enjoyment. Most ecosystem services either are irreplaceable or the technology necessary to replace them is prohibitively expensive.

Efforts to enhance the capacity of an ecosystem to provide some goods and services often come at the cost of its ability to provide others.

56.5 What options exist to manage ecosystems sustainably?

The economic value of a sustainably managed ecosystem is often higher than that of a converted or intensively exploited ecosystem. Review Figure 56.23

Recognition of the value of a sustainably managed ecosystem for "public goods" may induce organizations and government entities to protect them. Public education is needed to make people aware of how much human activities affect ecosystem sustainability.

SELF-QUIZ

1. Earth is not in chemical equilibrium because
   a. Earth has a moon.
   b. organisms dissipate energy as heat.
   c. most continents are in the Northern Hemisphere.
   d. Earth has living organisms that generate O_2.
   e. Earth is tilted on its axis.

2. Marine upwelling zones are important because
   a. they help scientists measure the chemistry of deep ocean water.
   b. they bring to the surface organisms that are difficult to observe elsewhere.
   c. ships can sail faster in these zones.

d. they increase marine productivity by bringing nutrients back to surface ocean waters.
e. they bring oxygenated water to the surface.

3. Which of the following is *not* true of the troposphere?
a. It contains nearly all of the atmospheric water vapor.
b. Materials enter it primarily at the intertropical convergence zone.
c. It is about 17 kilometers thick in the tropics and subtropics.
d. Most global air circulation takes place there.
e. It contains about 80 percent of the mass of the atmosphere.

4. The hydrological cycle is driven by
a. the flow of water into the oceans via rivers.
b. evaporation (transpiration) of water from the leaves of plants.
c. evaporation of water from the surface of the oceans.
d. precipitation falling on the land.
e. the fact that more water falls on the ocean as precipitation than evaporates from its surface.

5. Carbon dioxide is called a greenhouse gas because
a. it is used in greenhouses to increase plant growth.
b. it is transparent to heat, but traps sunlight.
c. it is transparent to sunlight, but traps heat.
d. it is transparent to both sunlight and heat.
e. it traps both sunlight and heat.

6. Micronutrients whose cycles are particularly important for animals include
a. iodine, cobalt, and molybdenum.
b. selenium, iodine, and cobalt.
c. molybdenum, iodine, and iron.
d. iodine, zinc, and selenium.
e. cobalt, selenium, and molybdenum .

7. The cycle of phosphorus differs from the cycles of carbon and nitrogen in that
a. phosphorus lacks an atmospheric component.
b. phosphorus lacks a liquid phase.
c. only phosphorus is cycled through marine organisms.
d. living organisms do not need phosphorus.
e. The phosphorus cycle does not differ importantly from the carbon and nitrogen cycles.

8. The sulfur cycle influences the global climate because
a. sulfur compounds are important greenhouse gases.
b. sulfur compounds help transfer carbon from the atmosphere to the oceans.
c. sulfur compounds in the atmosphere are components of particles around which water condenses to form clouds.
d. sulfur compounds contribute to acid precipitation.
e. The sulfur cycle does not influence the global climate.

9. Acid precipitation results from human modifications of
a. the carbon and nitrogen cycles.
b. the carbon and sulfur cycles.
c. the carbon and phosphorus cycles.
d. the nitrogen and sulfur cycles.
e. the nitrogen and phosphorus cycles.

10. Maintaining the capacity of ecosystems to provide goods and services to humanity is important because
a. most ecosystem services cannot be replicated by any other means.
b. most ecosystem services can be replaced by technology, but only at great cost.
c. technological substitutes take up valuable land.
d. governments cannot function without taxing ecosystem services.
e. It is not important; humans could survive quite well even if ecosystem services declined greatly.

FOR DISCUSSION

1. A string of powerful hurricanes struck the east coast of the United States over the course of a single year's hurricane season. Some people claim that this disaster was due to warming of the oceans caused by greenhouse gases in the atmosphere. Others assert that global warming is not responsible because hurricanes have occurred for many centuries. How would you evaluate these conflicting claims?
2. The waters of Lake Washington, adjacent to the city of Seattle, rapidly returned to their pre-industrial condition when sewage was diverted from the lake to Puget Sound, an arm of the Pacific Ocean. Would all lakes being polluted with sewage clean themselves up as quickly as Lake Washington if pollutant inputs were stopped? What characteristics of a lake are most important to its rate of recovery following reduction of pollutant inputs?
3. Tropical forests are being cut at a very rapid rate. Does this necessarily mean that deforestation is a major source of carbon dioxide inputs to the atmosphere? If not, why not?
4. What types of experiments would you conduct to assess the likely consequences of fertilization of the oceans with iron to increase rates of photosynthesis? At what spatial and temporal scales should they be conducted?
5. A government official authorizes the construction of a large coal-burning power plant in a former wilderness area. Its smokestacks discharge great quantities of combustion wastes. List and describe all likely effects on ecosystems at local, regional, and global levels. If the wastes were thoroughly scrubbed from the stack gases, which of the effects you have just outlined would still happen?
6. Many nations have signed the Kyoto Accord, which commits them to reducing their emissions of carbon dioxide to the atmosphere. The United States has so far refused to sign the treaty, claiming that it is against the country's interests. The United States has been severely criticized for this action. Is such criticism warranted? Justify your answer.

FOR INVESTIGATION

The experiment described in Figure 56.20 showed that interactions between elevated atmospheric carbon dioxide and nitrogen fixation can reduce the availability of other essential nutrients. The investigators suggested an expansion in the suite of elements that should be studied to determine whether elevated levels of carbon dioxide are likely to result in a large carbon sink in terrestrial ecosystems. What other elements should be considered, and how might their influence be investigated?

CHAPTER 57 Conservation Biology

Ultralight and Ultrahelpful

On October 10, 2004, a flock of young whooping cranes took to the air above Neceda, Wisconsin along with three ultralight aircraft operated by pilots from an organization called Operation Migration. The captive-reared birds had been trained from a young age to follow a slow-moving ultralight aircraft on the ground. Now the birds were ready to fly with the ultralights, which would guide them across seven states to their traditional winter home near the Gulf of Mexico. Two months and 1,200 miles later, the ultralights and their avian entourage landed safely in a national wildlife refuge in Florida. Remarkably, the following spring, most of the cranes accomplished the return trip to Wisconsin on their own—they had learned the migration route during their southward journey.

When the first Europeans arrived in North America, large whooping crane populations bred across the Gulf Coast, in the Upper Midwest, and in Canada. By the middle of the twentieth century, hunting and habitat loss had reduced them to a tiny group of fewer than 20 individuals. These few remaining birds spent the winter in the Aransas National Wildlife Refuge on the Texas coast and bred in Wood Buffalo National Park in Canada's Northwest Territories.

The recovery of whooping cranes involves cooperation among the U.S. Fish and Wildlife Service, Operation Migration, the Whooping Crane Eastern Partnership, and the International Crane Foundation (ICF). The ICF raises young cranes in captivity in ways that ensure that the birds never see a human being. They are fed by people wearing crane costumes. These methods are vital to the recovery effort because young whooping cranes learn to know who they are by *imprinting* on their caretakers (see Section 53.2).

These were not the first efforts to save the endangered crane species. The organizations' early efforts to establish a whooping crane breeding population in Montana failed. Whooping crane eggs were placed in the nests of (non-endangered) sandhill cranes. The sandhill "foster parents" raised whooping crane chicks successfully, but there was no way the young whooping cranes could learn the specific courtship and mating behaviors of their own species—they simply behaved like sandhills, and thus were unable to mate successfully. A few of these adult whooping cranes still migrate with the sandhills to the Bosque del Apache National Wildlife Refuge in New

Operation Migration An ultralight aircraft piloted by a human leads young whooping cranes on their first long migration to their overwintering grounds. After being guided once in this fashion, the birds can find their own way along the migration route.

Saving a Species Young cranes imprint on the aircraft and on the human in crane costume who feeds them. They thus grow up primed to follow the aircraft as their "leader."

Mexico, but they are a genetic dead end that will leave no offspring.

Today there is a nonmigratory population of 50 whooping cranes in Florida, 64 birds in the migratory flock that summers in Wisconsin, and 214 birds that winter in Texas. Because so much of its habitat has been lost, the whooping crane will never be as common as it was when Europeans arrived in North America, but the species has been moved back from the brink of extinction. Its future is now secure, as long as we continue to protect its breeding habitat.

IN THIS CHAPTER we will describe the rapidly expanding field of conservation biology, which is devoted to preserving the diversity of life. We will show how conservation biologists predict species extinctions, and we will see how some human activities are causing those extinctions. Finally, we will describe the strategies conservation biologists use to reduce extinction rates, prevent species from becoming threatened with extinction, and help populations of endangered species recover.

CHAPTER OUTLINE

57.1 What Is Conservation Biology?

Conservation biology is an applied scientific discipline devoted to preserving the diversity of life on Earth. People have been practicing conservation biology for centuries, but conservation biology today differs from past efforts in two important ways. First, modern conservation biology is supported by, and integrated with, other scientific disciplines. Conservation biologists draw heavily on concepts and knowledge from population genetics, evolution, ecology, biogeography, wildlife management, economics, and sociology. In turn, the needs and goals of conservation biology are stimulating new research in those disciplines.

Second, until recently, humans directed conservation efforts primarily toward species from which they gained direct economic benefits. Management actions were designed to maximize yields of useful products from populations of wild plants and animals. Today, conservation biologists study the full array of goods and services that humans derive from species and ecosystems, including aesthetic and spiritual benefits. Understanding the global ecosystem and the effects of human activities on that system is now understood to be essential to our long-term well-being.

Nearly every natural ecosystem on the planet has been altered by the activities of the 6.5 billion humans that live on Earth today. Many habitats have disappeared completely, and many others have been greatly modified. Even Earth's climate and its great biogeochemical cycles have been altered. One consequence of these changes is a rapid increase in the rate at which species are becoming extinct.

Conservation biology is a normative scientific discipline

Conservation biology is a **normative** discipline—that is, it embraces certain values and applies scientific methods to the goal of achieving these values. Conservation biologists are not neutral with respect to the preservation of biodiversity. They are motivated by the belief that preservation of biodiversity is good and that its loss is bad. Some people have criticized the discipline on the grounds that science is "supposed" to be neutral, but in fact most applied sciences are normative. Medical science, for example, is motivated by a desire to improve human health—health is good, illness is bad. Good science is as easily carried out in normative as in value-free scientific disciplines, provided the practitioners carry out their investigations using the standards and practices of established scientific methods.

Conservation biology is guided by three basic principles:

- *Evolution is the process that unites all of biology.* To be effective in preserving biodiversity, we need to know how evolutionary processes generate and maintain that diversity.
- *The ecological world is dynamic.* There is no static "balance of nature" that can serve as a goal of conservation activities.
- *Humans are a part of ecosystems.* Human activities and needs must be incorporated into conservation goals and practices.

Conservation biology aims to prevent species extinctions

Why should we worry about species extinctions today? After all, many natural events have led to species extinctions—some of them massive, as we saw in Section 21.2. Organisms have always altered Earth's ecosystems. The very first organisms probably reduced the supply of energetically and structurally useful compounds (at least in their immediate vicinity), replacing these compounds with their waste products. Early photosynthetic prokaryotes and eukaryotes generated oxygen, making Earth's atmosphere unsuitable for anaerobic organisms. When plants colonized the land, they accelerated the weathering of rocks, thereby gaining access to rock-bound nutrients. The weathering of phosphorus increased global productivity, further contributing to the rise of atmospheric oxygen concentrations. The rise of vascular plants further increased atmospheric oxygen concentrations while decreasing carbon dioxide concentrations. All of these changes, and many others we have not mentioned, created conditions that favored some existing organisms but negatively affected others.

Human beings have likewise caused extinctions of other species for thousands of years. When people first arrived in North America about 20,000 years ago, they encountered a rich fauna of large mammals, including bison, camels, horses, mammoths, and giant ground sloths. Most of those species were exterminated—probably by overhunting—within a few thousand years. A similar extermination of large animals followed the human colonization of Australia, about 40,000 years ago. At that time, Australia had 13 genera of marsupials larger than 50 kilograms (**Figure 57.1**), a genus of gigantic lizards, and a genus of heavy, flightless birds. All of the species in 13 of those 15 genera had gone extinct by 18,000 years ago. When Polynesian people settled in Hawaii about 2,000 years ago, they exterminated, probably by overhunting, at least 39 **endemic** species of land birds—species that are found nowhere else.

If organisms have continually changed the nature of Earth's ecosystems, and if humans have been exterminating species for thousands of years, why should we worry about the changes humans are causing today? Life survived those changes. Indeed, both the productivity and richness of Earth's biota has increased during the long course of life's evolution. But, although a long-term historical perceptive might lead us to be complacent about today's human-caused changes, the current situation is unique. For the first time in history, all the major environmental changes are being caused by a single species. Unlike all previous ecosystem engineers, we are aware of what we are doing. If we deliberately or inadvertently exterminate a species, we have irreversibly destroyed a potentially valuable resource.

57.1 Extinct Australian Megafauna A number of large marsupial mammals, including *Diprotodon* (background), *Palorchestes* (center) and *Zygomaturus* (foreground), were found in Australia during the Pleistocene. These and other large-bodied species became extinct after the arrival of humans on the continent.

People are concerned about maintaining Earth's biodiversity because they recognize its value. People value biodiversity for many reasons:

- Humans depend on other species for food, fiber, and medicine. More than half of all medical prescriptions written in the United States contain or are based on a natural plant or animal product.
- Species are necessary for the functioning of ecosystems and the many benefits and services those ecosystems provide to humanity (such as clean water and purification of the air; see Section 56.4).
- Humans derive enormous aesthetic pleasure from interacting with other organisms. Many people would consider a world with far fewer species to be a less desirable place in which to live.
- Extinctions deprive us of opportunities to study and understand ecological relationships among organisms. The more species that are lost, the more difficult it will be to understand the structure and functioning of ecological communities and ecosystems.

- Living in ways that cause the extinction of other species raises serious ethical issues that increasingly concern philosophers, ethicists, and religious leaders because those species are judged to have intrinsic value.

Later in this chapter we will see how conservation biologists incorporate these values into strategies for preserving biodiversity.

57.1 RECAP

Conservation biology is an applied, normative scientific discipline devoted to preserving biodiversity.

- What is a normative discipline, and how does this influence its practices and its goals? See pp. 1227–1228
- If ecosystem change and species extinctions have always been part of Earth's evolutionary history, why are biologists today so concerned about preserving biodiversity? See p. 1228

To preserve biodiversity, conservation biologists must understand the biodiversity that exists today and how it is changing. We next consider how conservation biologists predict rates of extinctions and infer their causes.

57.2 How Do Biologists Predict Changes in Biodiversity?

As described Chapter 23, the rich array of species on Earth has been generated by speciation processes that have functioned for several billion years. Even if humans did not exist, many species would become extinct during the next 100 years. But human activities are causing more extinctions than would occur in our absence. To preserve Earth's biodiversity, we need to both maintain the processes that generate new species and provide conditions that will keep extinction rates no higher than typical levels.

For four reasons, scientists cannot accurately predict the number of extinctions that will occur during the coming century:

- *We do not know how many species live on Earth.* Many of the species that may become extinct during the next 50 years have not even been named and described!
- *We do not know where species live.* The ranges of most described species are poorly known.
- *It is difficult to determine when a species actually becomes extinct.*
- *We do not know what will happen in the future.* The number of extinctions that actually occur will depend both on what humans do and on unexpected natural events.

We live on a poorly explored planet, and even in the case of conspicuous species in biologically well-studied areas, our knowledge of biodiversity is incomplete. For example, North America's largest woodpecker, the ivory-billed woodpecker (**Figure 57.2A**), was believed to have been glimpsed in Arkansas late in 2004, after 60 years without any confirmed sightings. Most ornithologists had assumed that the species was extinct; a field guide to the birds of North America published in 2000 did not picture it. Although there have now been several reputably documented sightings of the woodpecker (sightings that require birdwatchers to sit motionless for long hours in swampy terrain; **Figure 57.2B**), a clear photograph of the living bird has so far eluded the watchers.

When the Laotian rock rat (*Laonastes aenigmamus*) was discovered and described in 2005, it was so distinct from all other living rodents that it was placed in a lineage all its own. In 2006, further research indicated that this animal actually belongs among the diatomyids, a rodent lineage known from the fossil record but believed to have been extinct for 11 million years.

(A) *Campephilus principalis*

(B)

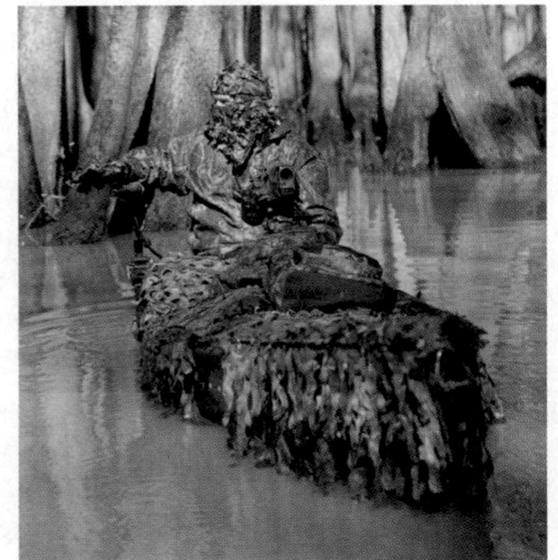

57.2 Back from Extinction? (A) The ivory-billed woodpecker, shown here in a nineteenth century Audubon print, was presumed to be extinct until 2004, when sightings were reported in a wetland forest. (B) Ornithologists and amateur birdwatchers alike have spent long, silent hours searching for signs of the majestic bird in the Cache River National Wildlife Refuge. This naturalist camouflaged himself and his canoe for the wait.

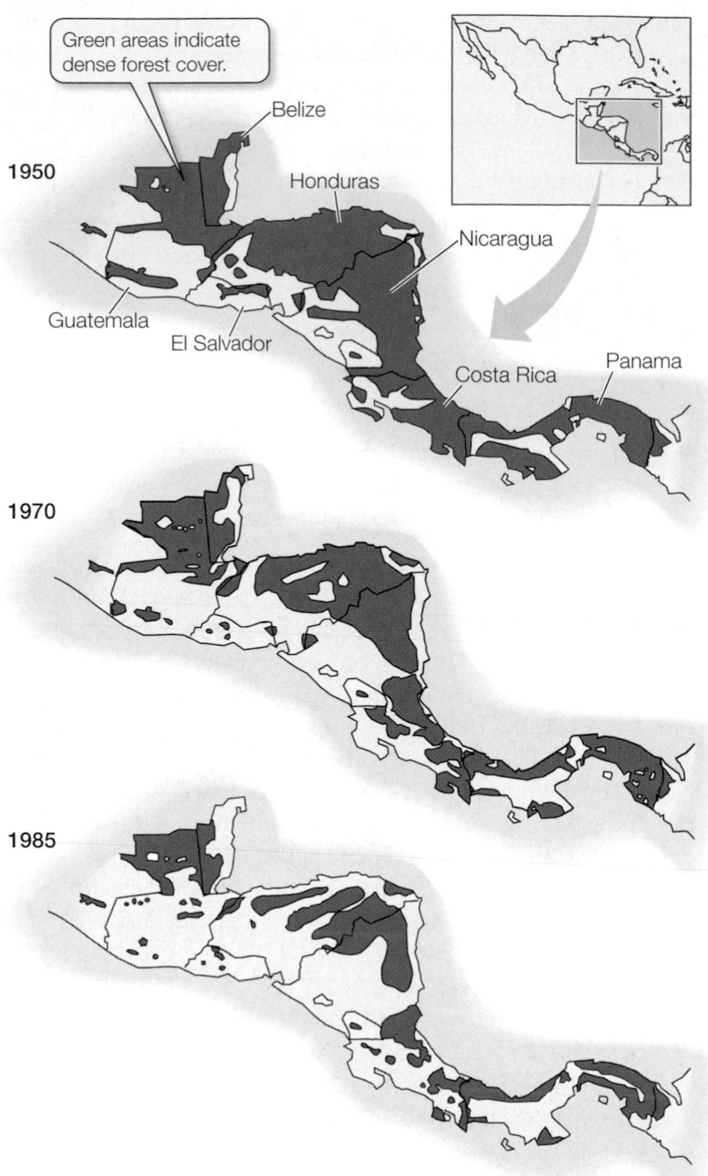

57.3 Deforestation Rates are High in Tropical Forests Less than half the tropical forest that existed in Central America in 1950 remained as of 1985. Much of the remaining forest is in small patches.

Despite these difficulties, methods exist for estimating probable rates of extinction resulting from human actions such as habitat destruction. One tool often used in making such estimations is the **species–area relationship**, a well-established mathematical relationship between the size of an area and the number of species that area contains. Conservation biologists have measured the rate at which species richness decreases with decreasing habitat patch size. Their findings suggest that, on average, a 90 percent loss of habitat will result in the loss of half of the species that live in and depend on that habitat.

Similar calculations can be made for the total area of a habitat type left on Earth. The current rate of loss of tropical evergreen forests—Earth's most species-rich biome—is about 2 percent of the remaining forest each year, due to the increasing demand for forest resources by a rapidly expanding human population (**Figure 57.3**). If this rate of loss continues, at least 1 million species that live in tropical evergreen forests could become extinct during this century.

To estimate the risk that a population will become extinct, conservation biologists develop statistical models that incorporate information about a population's size, its genetic variation, and the morphology, physiology, and behavior of its members.

Species in imminent danger of extinction in all or a significant part of their range are labeled *endangered species*. *Threatened species* are those that are likely to become endangered in the near future. Rarity in and of itself is not always a cause for concern, because some species that live in highly specialized habitats have probably always been rare and are well adapted to those conditions. However, species whose populations suddenly are shrinking rapidly—the "newly rare"—usually are at high risk. Species with special habitat or dietary requirements are more likely to become extinct than species with more generalized requirements.

Populations with only a few individuals confined to a small range can easily be eliminated by local disturbances such as fires, unusual weather, climate change, disease, habitat destruction, or predators. For example, the golden toad (*Bufo periglenes*), which lived only in the Monteverde Cloud Forest Preserve in Costa Rica, disappeared during the 1980s (see Figure 33.17B). At that time, the altitude at which clouds formed moved higher up the mountains, most likely as a result of climate warming. The forests became too dry for the toad, whose range was already near the summit of the mountains. Similarly, a group of scientists carrying out a rapid inventory of species in the western Andes of Ecuador found 90 endemic plant species on a single ridge. Shortly after they completed their survey, the entire ridge was cleared for agriculture. The populations of all 90 species were destroyed. Future surveys may reveal that some of the species exist elsewhere, but many of them are probably extinct.

57.2 RECAP

The species–area relationship can be used to predict rates of extinction in areas that are subject to habitat loss. Models that take into account variables such as population size, dietary and habitat requirements, and genetic variation help predict a species' risk of extinction.

- Why are species with only a few individuals confined to a small range especially vulnerable to extinction? See p. 1230
- Do you understand how a species can be rare but not necessarily in danger of extinction? See p 1230

Now that we've looked at some of the factors that place species at risk of extinction, let's see how human activities contribute to those risks.

57.3 What Factors Threaten Species Survival?

The human activities that threaten the survival of species include habitat destruction and modification, the introduction of exotic species, overexploitation, and climate change. Conservation biologists determine how these activities are affecting species and use that information to devise strategies to preserve species that are endangered or threatened.

Species are endangered by habitat loss, degradation, and fragmentation

Habitat loss is the most important cause of species endangerment in the United States, especially for species that live in fresh waters, the habitat that has been most extensively degraded (**Figure 57.4**). Obviously, when habitats are destroyed or made uninhabitable, the organisms living there are lost. But habitat loss affects even those remaining habitats that are not destroyed. As habitats are progressively lost to human activities, the remaining habitat patches become smaller and more isolated. In other words, the habitat becomes increasingly **fragmented**.

Small habitat patches are qualitatively different from larger patches of the same habitat in ways that affect the survival of species. Small patches cannot maintain populations of species that require large areas, and they can support only small populations of many of the species that can survive in small patches. In addition, the fraction of a patch that is influenced by factors originating outside it increases rapidly as patch size decreases (**Figure 57.5**). Close to the edges of forest patches, for example, winds are stronger, temperatures are higher, humidity is lower, and light levels are higher than they are farther inside the forest. Species from surrounding habitats often colonize the edges of patches to compete with or prey on the species living there. The effects of such factors are known as **edge effects**.

Often we do not know which organisms lived in an area before the habitat became fragmented. Timely surveys can provide some of this information. For example, a major research project in a tropical evergreen forest near Manaus, Brazil, was launched before logging took place. Landowners agreed to preserve forest patches of certain sizes and configurations (**Figure 57.6**). Biologists counted the species in those future "patches" while they were still part of the continuous forest. Soon after the surrounding forest was cut and converted to pasture, species began to disappear from isolated patches. The first species to be eliminated were monkeys that travel over large areas. Army ants and the birds that follow army ant swarms also soon disappeared.

Species that are lost from small habitat fragments are unlikely to become reestablished there because dispersing individuals are unlikely to find isolated fragments. As Section 54.4 points out, however, a species may persist in a small patch if it is connected to other patches by *corridors* of habitat through which individuals can disperse. Some of the pastures that surrounded the experimental forest fragments in Brazil have been

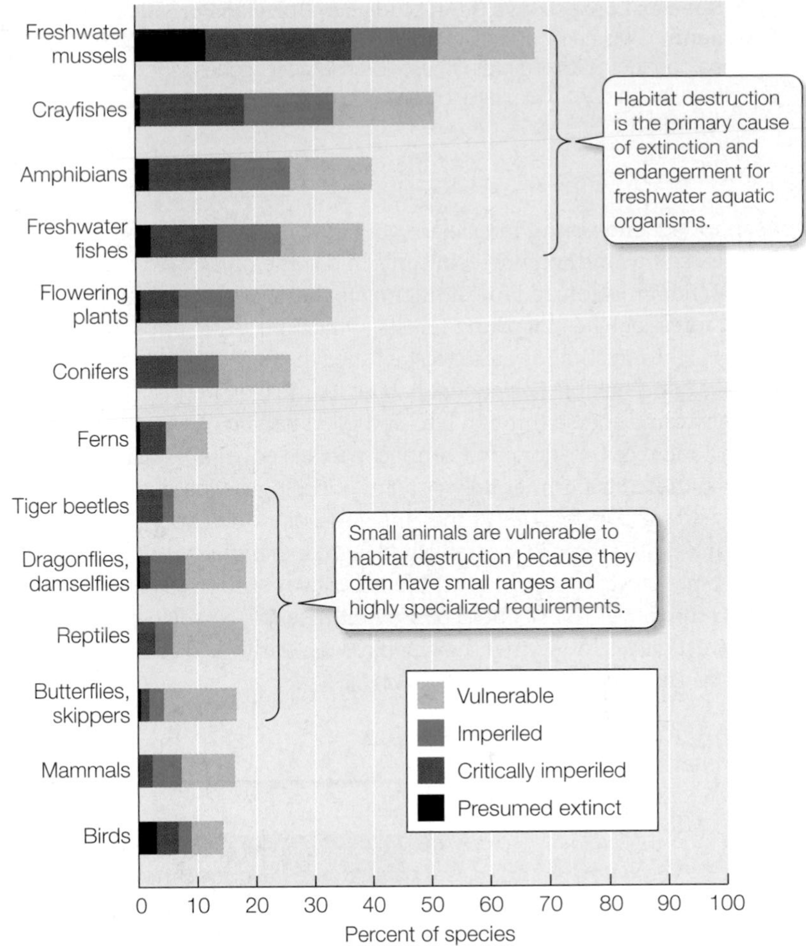

57.4 Proportions of U.S. Species Extinct or Threatened The groups of species that are most threatened with extinction—mussels, crayfishes, amphibians, and fishes—live in freshwater habitats, which have been extensively destroyed, degraded, and polluted.

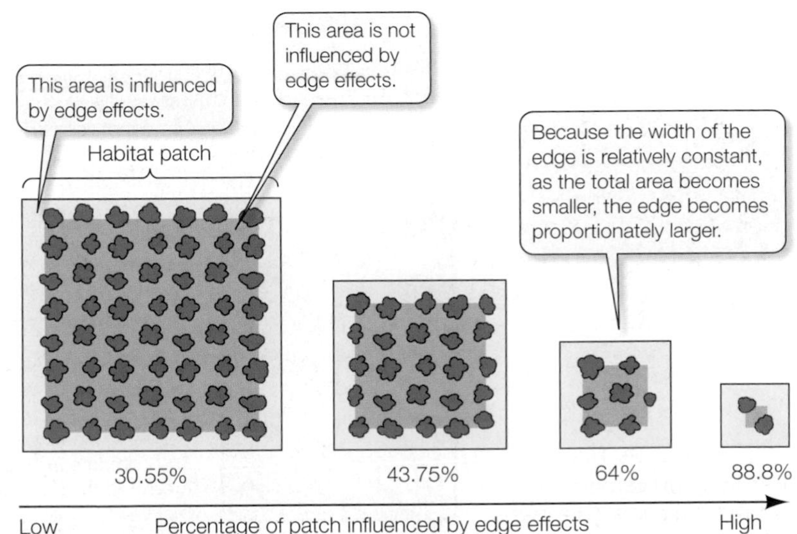

57.5 Edge Effects The smaller a patch of habitat, the greater the proportion of that patch that is influenced by conditions in the surrounding environment.

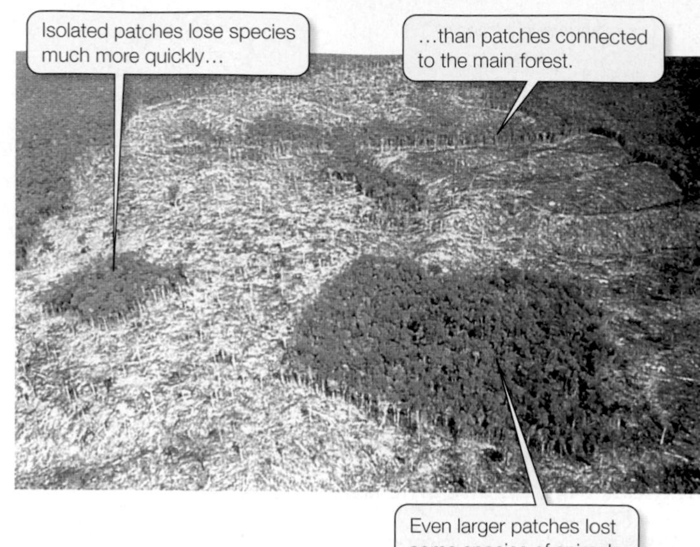

57.6 Species Losses Have Been Studied in Brazilian Forest Fragments Isolated patches lost species much more quickly than patches connected to the main forest. Even larger patches, such as the one in the foreground, were too small to maintain populations of some species.

abandoned, and young forests are now growing in them. Within 7–9 years of abandonment, army ants and some of the birds that follow them recolonized forest fragments that were connected to larger forest patches by young forests. Other species of birds that forage in the forest canopy also reestablished themselves. The young forest is not a suitable permanent habitat for most of these species, but they can disperse through it to find good habitat.

The value of corridors that connect patches of suitable habitat can be studied experimentally. A team of ecologists led by Douglas Levey and colleagues at the University of Florida and North Carolina State University established eight experimental sites in open pine forests to test whether bluebirds use corridors. By monitoring the movements of both birds and the seeds they dispersed, the investigators found that the bluebirds regularly used corridors to move among patches (**Figure 57.7**).

Overexploitation has driven many species to extinction

Humans have caused extinctions for thousands of years, but until recently, they did so primarily by overhunting. Some species are still threatened by overexploitation today. Elephants and rhinoceroses are threatened in much of Africa and Asia because poachers kill

EXPERIMENT

HYPOTHESIS: Bluebirds use corridors to move between open patches within a forest.

METHOD

1. Create suitable patches of bluebird habitat in pine forests by cutting down trees to create the open conditions favored by bluebirds. Connect two patches by habitat corridors.
2. In one of the connected patches, plant wax myrtle bushes with fluorescently tagged fruit. Because birds that eat the fruit carry seeds in their guts and later defecate them, fluorescently tagged seeds serve as a tracking device.
3. Observe birds as they fly from wax myrtle bushes until they fly out of sight. Record the direction and distance each bird travels. Check surrounding patches for fluorescent seeds defecated by the birds. Develop a computer model that predicts the movement of seeds based on observations of the birds.

Rectangular patch
Seed traps
Edges
Tagged fruit
Connected patches
Forested
"Winged" patches
150 m

RESULTS

Bluebirds usually remained within a habitat patch when they encountered an edge, but they did move through corridors. Many more fluorescent seeds were moved between patches connected by corridors than between unconnected patches. The computer model correctly predicted the number of seeds that were moved.

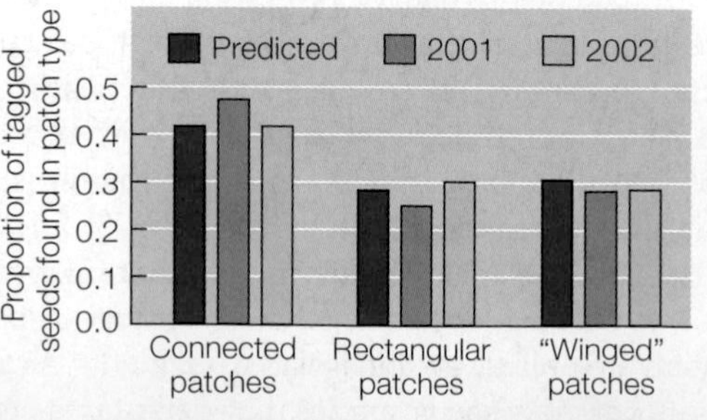

CONCLUSION: Bluebirds use corridors to move between patches of suitable habitat.

57.7 Habitat Corridors Facilitate Movement Bluebirds moved more frequently between habitat patches connected by corridors than between unconnected patches.

(A) *Dicerorhinus sumatrensis*

(B) *Saiga tatarica*

57.8 Endangered by Medical Practices (A) Endangered rhinoceros are killed for their horns, which are widely used in Oriental folk medicine. (B) An attempt to encourage the use of Mongolion saiga antelope horns in lieu of rhino horns succeeded too well—the antelope is now a threatened species.

them for their tusks and horns, which are used for ornaments and knife handles; in addition, some men believe that powdered rhinoceros horn enhances their sexual potency, and rhinoceros horn is used extensively in Chinese traditional medicine (**Figure 57.8**). The use of animal parts in traditional medical practices is a threat to some species. Massive international trade in pets, ornamental plants, and tropical forest hardwoods has decimated many species of tropical fishes, corals, parrots, reptiles, orchids, and trees.

Invasive predators, competitors, and pathogens threaten many species

People have moved many species to regions outside their original range, deliberately or accidentally. Some of these exotic (foreign) species become **invasive**—that is, they spread widely and become unduly abundant, often at a cost to the native species of the region. Weed seeds have been carried around the world accidentally in sacks of crop seeds, and marine organisms have been spread throughout the oceans by ballast water from ships. Europeans deliberately introduced rabbits and foxes to Australia for sport hunting. Nearly half of the small to medium-sized marsupials and rodents of Australia have been exterminated during the last 100 years by a combination of competition with introduced rabbits and predation by introduced cats, dogs, and foxes.

The brown tree snake, *Boiga irregularis* (**Figure 57.9**), arrived in air cargo on Guam sometime in the 1940s. Until then, the only snake on Guam was a tiny insectivorous species. For unknown reasons, brown tree snakes remained rare until the 1960s, when they began to appear in large numbers. Today they can be found at densities as high as 5,000 individuals per square kilometer. The snake has exterminated 15 species of birds, 3 of which were found only on Guam.

Invasive plants can have many negative effects on ecosystems. Most introduced species are imported without their natural enemies, while native plants must devote considerable energy to defending themselves against the native herbivores. The invasive plants typically have high rates of growth and reproduction, in part because they invest less energy in producing defensive compounds. The invaders thus tend to take advantage of soil nutrients in ways that the native plants cannot.

Disease-causing organisms may also proliferate quickly following their introduction to new continents. Introduced pathogens have destroyed whole populations of several eastern North American forest trees. The chestnut blight (*Cryphonectria parasitica*), caused by an introduced fungus, reduced the American chestnut (*Castanea dentata*), formerly an abundant tree in Appalachian Mountain forests, to an understory shrub. Dutch elm disease, caused by the fungus *Ophiostoma ulmi*, introduced in North America in 1930, has killed nearly all American elms (*Ulmus americana*) over large areas of the East and Midwest. Ecologists suspect that intercontinental movement of disease organisms caused extinctions in the past, but disease outbreaks usually leave no traces in the fossil record.

In the Hawaiian Islands, nearly all endemic bird species living below 1,500 meters elevation have been eliminated by avian malaria, which was introduced to the islands with exotic birds. The native birds, never having been exposed to malaria, were highly susceptible to the disease. Species that live above the current range

Boiga irregularis

57.9 Agent of Extinction Since its accidental introduction on Guam, the brown tree snake has eaten 15 species of land birds to extinction.

of the mosquitoes that transmit the disease have fared better, but the insects' range may be expanding upward as the climate warms.

Rapid climate change can cause species extinctions

Scientists predict that, as a result of human activites, average temperatures in North America will increase 2°C–5°C by the end of the twenty-first century. If the climate warms by only 1°C, the average temperature currently found at any particular location in North America today will be found 150 kilometers to the north. If the climate warms 2°C–5°C, some species will need to shift their ranges by as much as 500 to 800 kilometers within a single century. Some habitats, such as alpine tundra, could be eliminated as forests expand up mountain slopes.

Conservation biologists cannot alter rates of global warming, but their research can help us predict how the resulting climate changes will affect organisms and find ways to mitigate those effects. Their research activities include analyses of past climatic events and studies of sites currently undergoing rapid climate change. It is helpful to know, for example, how rapidly species ranges shifted during the last 10,000 years of postglacial warming. Which species were and were not able to keep pace with climate change by shifting their ranges? How much and in what ways did past ecological communities differ from those of today as a result of differences in the rates at which species changed their ranges?

Organisms that are able to disperse easily, such as most birds, may be able to shift their ranges as rapidly as the climate changes, provided that appropriate habitats exist in the new areas. However, the ranges of species with sedentary habits are likely to shift slowly. As the glaciers retreated in North America about 8,000 years ago, for example, the ranges of some coniferous trees expanded northward, so that today they grow as far north as the current climate permits (see Figure 21.22). On the other hand, some earthworm species spread only very slowly into the areas that had been covered by ice.

If Earth's surface warms as predicted, entirely new climates will develop, and some existing climates will disappear. New climates are certain to develop at low elevations in the tropics because a warming of even 2°C would result in climates near sea level that are warmer than those found anywhere in the humid tropics today. Adaptation to those climates may prove difficult even for many tropical organisms. Although there has been little recent climate warming in tropical regions, nights are now slightly warmer than they were only a few decades ago. Since the mid-1980s, the average minimum nightly temperature at the La Selva Biological Station, in the Caribbean lowlands of Costa Rica, has increased from about 20°C to 22°C. During these warmer nights, trees use more of their energy reserves. The result has been a reduction of about 20 percent in the average growth rates of six different tree species.

Staghorn coral, *Acropora cervicornis*, was recently discovered near Fort Lauderdale, Florida. Elkhorn coral, *A. palmata*, was discovered in the northern Gulf of Mexico. Both sites are 50 kilometers north of the previous range limits of these species, providing the first evidence of range expansions of Caribbean corals in response to climate warming.

In 1998, the highest sea surface temperatures ever recorded caused corals to lose their endosymbiotic dinoflagellates (a phenomenon called *bleaching*) and increased their mortality worldwide (**Figure 57.10**). If warming of the oceans continues as predicted, about 40 percent of coral reefs worldwide are likely to be killed by 2010. To identify possible ways to help preserve coral reefs, conservation biologists are measuring conditions in places where corals have escaped bleaching. They have found that reefs adjacent to cool, upwelling waters and reefs in cloudy waters, both of which have relatively low temperatures, are generally healthy. These reefs are receiving special protection because corals are likely to continue to survive well there. Corals from those reefs could be used as colonists for reestablishing bleached reefs if cooler ocean temperatures return in the future.

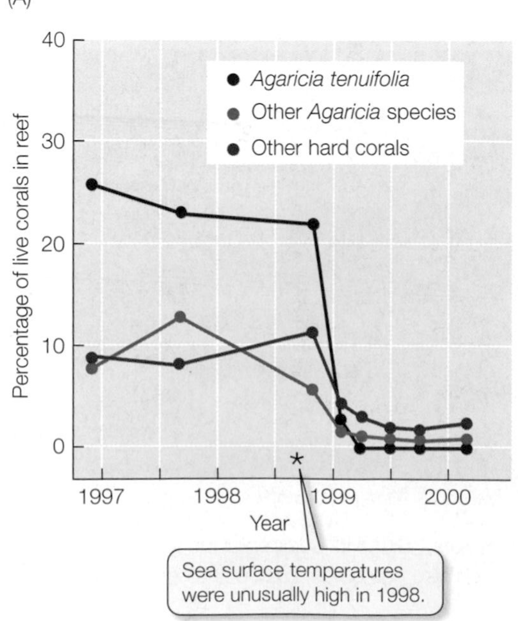

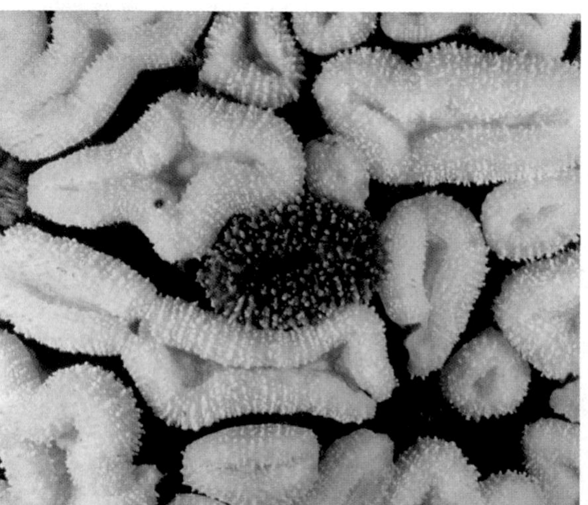

57.10 Global Warming Threatens Corals
(A) Unusually high sea surface temperatures in 1998 caused massive bleaching and death of corals on a reef in Belize. (B) A single photosynthesizing polyp (center) remains alive among the bleached remnants of this Indonesian coral colony.

57.3 RECAP

A number of human activities threaten the survival of species, including habitat destruction and fragmentation, the introduction of invasive species, overexploitation, and rapid climate change.

- Can you explain why extinction rates are high in small habitat patches? See pp. 1231–1232 and Figure 57.5
- Do you understand the particular worries conservation biologists have about climate change? See p. 1234

Acquiring data to demonstrate that species or communities are endangered is an empty exercise if we cannot implement a plan of action to save them. In the next section we consider some of the positive steps that can be taken to preserve biodiversity.

57.4 What Strategies Do Conservation Biologists Use?

Conservation biologists use data, concepts, and tools from a variety of disciplines to help preserve endangered and threatened species and communities. They determine what factors, including human activities, are affecting species health and numbers, and they use that information to devise a plan of action. What information is relevant depends on which of the factors that threaten biodiversity are most important in particular cases. Reducing threats to species often requires changes in national legislation or international rules and treaties. Therefore, biologists regularly join sociologists, political scientists, and environmental activists to influence legislation. Let's look at some of the actions that conservation biologists take to maintain biological diversity.

Protected areas preserve habitat and prevent overexploitation

Establishing *protected areas* is an important component of efforts to preserve biological diversity. Protected areas that preserve habitat while preventing the human exploitation of the species living there may serve as nurseries from which individuals disperse into exploited areas, replenishing populations that might otherwise become extinct. The importance of protected areas was highlighted in the United Nations Convention on Biological Diversity, a document generated by the Rio de Janeiro Earth Summit in 1992: "The fundamental requirement for the conservation of biological diversity is the *in situ* conservation of ecosystems and natural habitats and the maintenance and recovery of viable populations of species in their natural surroundings."

But how should we select the areas to be protected? Two obvious criteria are the number of species living in an area—its *species richness*—and the number of endemic species. Using these criteria, Norman Myers identified a number of biodiversity "hotspots" of unusual richness and endemism (**Figure 57.11**). These hotspots occupy only 15.7 percent of Earth's land surface, but they are home to 77 percent of Earth's terrestrial vertebrate species. Most of these hotspots are also regions of high human population density where habitat destruction is a major problem. Consequently, much effort is being devoted to saving particular habitats in those regions.

The hotspot concept successfully directs attention to places harboring unusual species richness, but hotspots do not represent all of Earth's biodiversity. Many areas with lower species richness and less endemism are nonetheless very important biologically. By using taxonomic uniqueness, unusual ecological or evolutionary phenomena, and global rarity, as well as species richness and endemism, scientists at the World Wildlife Fund identified 200 ecoregions of great conservation importance (**Figure 57.12**). Some of these "Global 200" ecoregions are marine areas. The list also includes tundra, boreal forests, and deserts, ecosystems that are missed by the hotspot approach.

Identifying focal areas for preservation is only the first step in a conservation program, however. Developing a conservation strategy for an ecoregion requires both a detailed analysis of the distributions of species and the locations of special resources, such as caves, springs, and migratory stopover areas for birds, and an

Centers of bird species richness

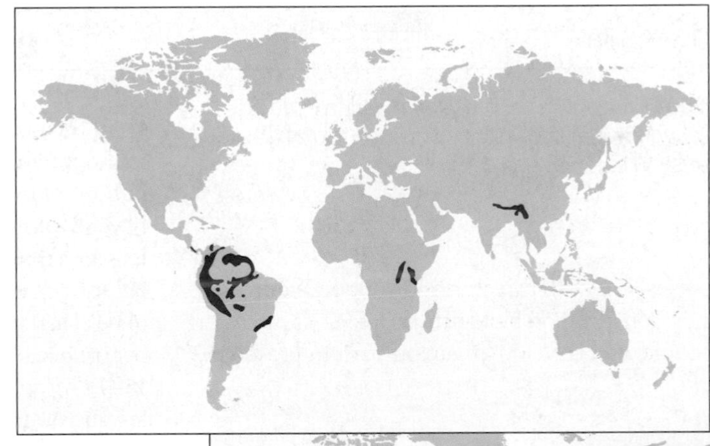

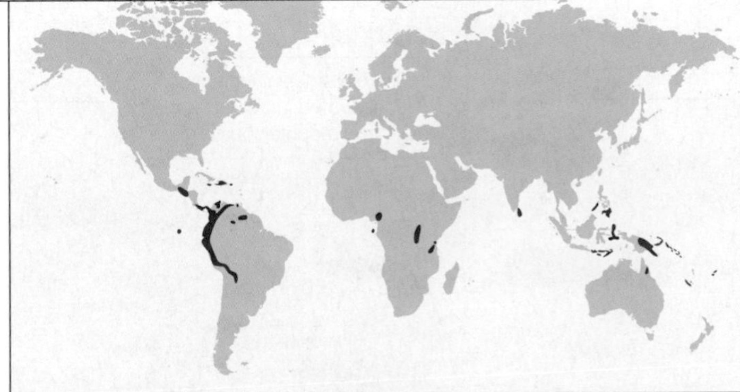

Centers of endemic bird species

57.11 Hotspots of Avian Biodiversity The regions marked in red contain rich avian biodiversity in terms of either total number of species or number of endemic species (species found nowhere else in the world).

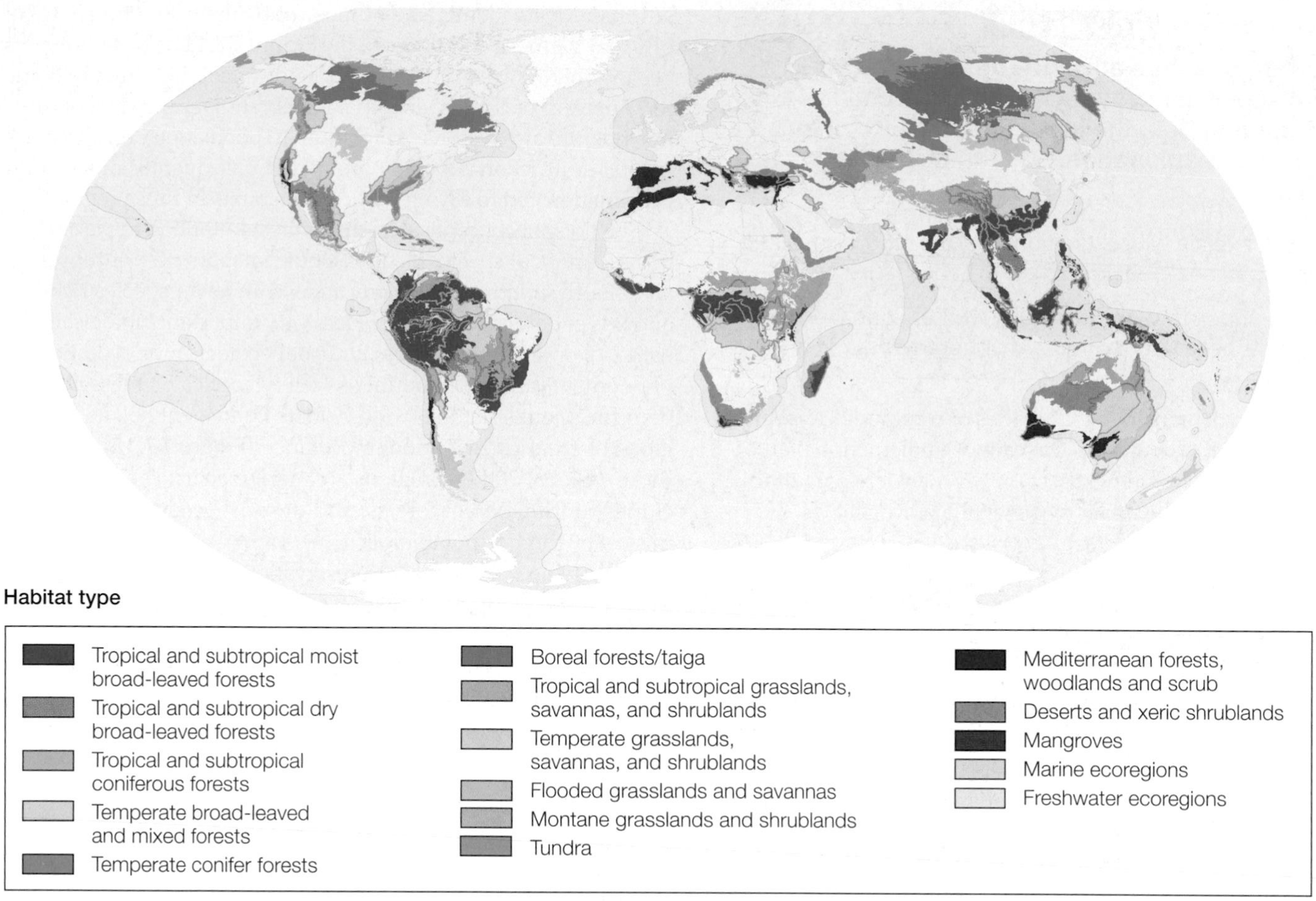

Habitat type
Tropical and subtropical moist broad-leaved forests
Tropical and subtropical dry broad-leaved forests
Tropical and subtropical coniferous forests
Temperate broad-leaved and mixed forests
Temperate conifer forests
Boreal forests/taiga
Tropical and subtropical grasslands, savannas, and shrublands
Temperate grasslands, savannas, and shrublands
Flooded grasslands and savannas
Montane grasslands and shrublands
Tundra
Mediterranean forests, woodlands and scrub
Deserts and xeric shrublands
Mangroves
Marine ecoregions
Freshwater ecoregions

57.12 The "Global 200" Ecoregions The World Wildlife Fund has designated 200 areas worldwide (shown in color) as particularly important for the preservation of biodiversity.

analysis of the processes that both threaten and support biodiversity in the region. Then conservation biologists, together with other experts and local people, can develop an action plan to preserve the ecoregion.

In the effort to identify sites with threatened species that are found nowhere else, conservation biologists have analyzed distributions of mammals, birds, reptiles, amphibians, and conifers and have identified 595 "centers of imminent extinction." These centers are concentrated in tropical forests, on islands, and in mountainous regions (**Figure 57.13**). The sites harbor 794 species judged to be at serious risk of extinction—more than three times the number of species in those groups known to have become extinct since 1500. Only one-third of the sites are legally protected. Most of them are surrounded by rapid human development. Urgent action is needed in these areas if species extinctions are to be avoided.

Degraded ecosystems can be restored

If the cause of a species' endangerment is modification, rather than loss, of its habitat, preserving the species may require that the habitat be restored to its natural state. Practitioners of **restoration ecology** are developing methods that attempt to restore natural habitats. Such interventions are often needed because many degraded ecosystems will not recover, or will do so only very slowly, without human assistance.

"Centers of imminent extinction" are concentrated in tropical forests, islands, and mountainous regions.

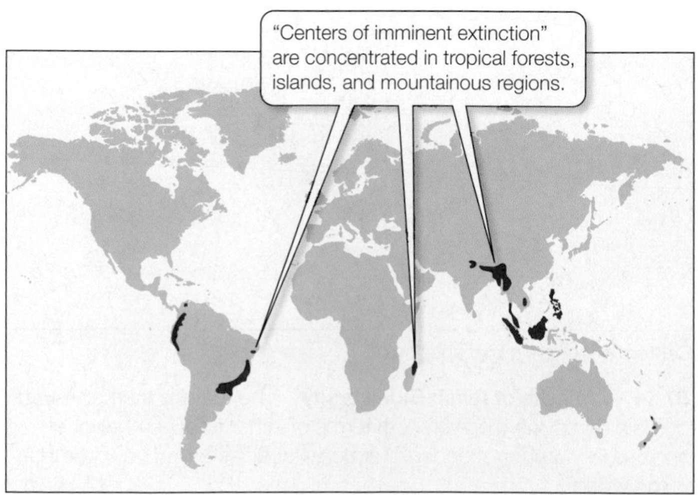

57.13 Centers of Imminent Extinction The areas shown in red include 595 "centers of imminent extinction." These regions are home to 794 species that are at serious risk of extinction.

Many grassland habitats grow on rich soils and humans have eagerly converted them to agricultural use. By the middle of the twentieth century, for example, most North American prairies had been converted to cropland or were heavily grazed by domesticated livestock. The populations of large mammals that roamed the prairies when Europeans arrived on the continent are now reduced to tiny remnants confined to small areas. Most of these remaining populations are too small to maintain their genetic diversity or their original ecological roles.

The species *have* survived, however, so opportunities exist to reintroduce them if the prairie habitat can be restored. A major prairie restoration project is under way in northeastern Montana. When Lewis and Clark mapped this region 200 years ago, it supported large herds of bison, elk, deer, and pronghorn, as well as their predators. The goal of the restoration project, run by the World Wildlife Fund and the American Prairie Foundation in cooperation with public land managers, is to restore the native prairie and its fauna in a 15,000-square-kilometer area near the Missouri River (**Figure 57.14A**).

This ambitious, multi-decade project is feasible for three reasons. First, the private land in the area is owned by a small number of ranchers. Each ranch owns extensive grazing leases on nearby public lands administered by the Bureau of Land Management, the Fish and Wildlife Service, or the state of Montana. Second, most of the land has never been plowed, so the native vegetation is likely to recover rapidly when grazing pressures are reduced. Third, the area is steadily losing its human population. In 1920 there were 9,300 people in Phillips County, which encompasses a large portion of the project area. Only 4,200 remained there in 2000. The ranchers are aging. Their children are choosing the excitement of cities and a regular paycheck rather than the hard labor and uncertain profits of ranching. The ranchers need to sell their land to fund their retirement; their grazing leases transfer automatically to the new owner.

The American Prairie Foundation is buying ranches from willing sellers with the objective of reintroducing bison, greatly expanding populations of prairie dogs, and restoring a viable population of black-footed ferrets, the most endangered mammal in North America. The prairie dogs, by digging extensive burrows and manipulating vegetation, in turn support dozens of species of birds, mammals, reptiles, and invertebrates (**Figure 57.14B**). In November 2005, the first 16 bison, obtained from Wind Cave National Park, were reintroduced to a ranch purchased by the Foundation (**Figure 57.14C**). Once a free-ranging herd of several thousand bison and large numbers of elk and their predators (wolves) have been established, nature-minded tourists should flock to the area to view the restored wildlife spectacle. The ecotourists will, in turn, spend money in the region, providing Phillips County with a new economic base. Over the long term, the restored ecosystem should deliver major economic benefits to the region.

Disturbance patterns sometimes need to be restored

Many species depend on particular patterns of disturbance on the landscape, such as fires, windstorms, and grazing. Conservation biologists work to assess whether reestablishment of historic disturbance patterns can help preserve biodiversity. Humans often try to reduce the frequency and intensity of such disturbances. For example, although many plant species require periodic fires for successful establishment and survival, for many years the official policy in the United States, symbolized by Smokey Bear, was to suppress all forest fires. Today, however, controlled burning is a common forest management tool, particularly in western North America. But to determine how to do this, we need to know the historical pattern of fires in an area.

Scars in the annual growth rings of trees preserve evidence of past fires that did not kill them. Tree ring researchers can determine when fires occurred, how severe they were, and when fire patterns changed. Annual growth rings on ponderosa pines show that low-intensity ground fires were common near Los Alamos, New Mexico, until about 1900 (**Figure 57.15A**). After that time, cattle and sheep grazing in pine forests and fire suppression greatly reduced the frequency of low-intensity fires. Without these fires,

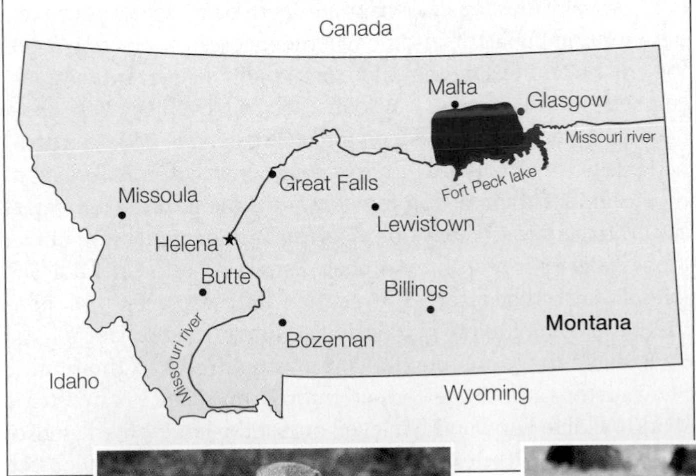

57.14 An American Prairie Is Being Restored (A) A major prairie restoration project is underway north of the Missouri River in the state of Montana. (B) Burrowing prairie dogs manipulate vegetation and are crucial in sculpting the natural ecosystem. (C) Bison have been reintroduced.

(A)

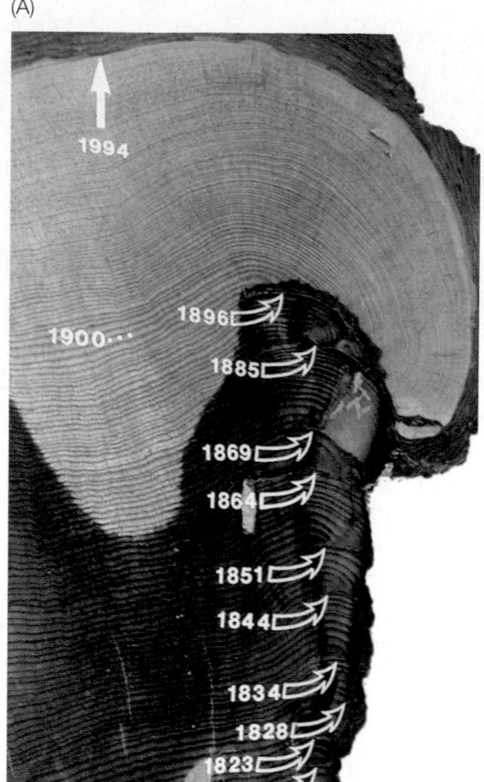

(B)

57.15 The Frequency and Intensity of Fires Affect Ecosystems (A) As revealed by scars (arrows) in the growth rings of this Ponderosa Pine tree, low-intensity ground fires were frequent in the pine forests of the southwestern United States prior to fire suppression. (B) Fire suppression results in the buildup of large quantities of fuel, so that subsequent fires are likely to spread to the canopy and kill most trees.

dead branches and needles accumulated in the forest. When fires inevitably did occur, the buildup of fuel made them more likely to become intense, tree-killing canopy fires (**Figure 57.15B**). Today, ground fires are deliberately started in many areas to reduce fuel loads and prevent destructive fires.

New habitats can be created

In the United States, the belief that humans know how to create functioning ecosystems has resulted in policies that make it easy to get permits for developments that destroy habitats. Developers need only state that they will create new habitats to substitute for the ones they are destroying. Of special concern is destruction of wetlands, which is permitted because regulators believe that alternative wetlands can be created. Creating new wetlands that can support all the species that live in those being destroyed is difficult, however, and requires detailed ecological knowledge.

In southern California, where 90 percent of the coastal wetlands have been destroyed, wetland restoration is a high priority. Because species have been lost from degraded coastal wetlands, restoration requires species introductions, but deciding which species should be introduced is not easy. Early attempts at restoration, in which one or two common, easily grown wetland species were planted, did not succeed; other wetland-associated species failed to re-colonize the "rehabilitated" wetlands. To understand why, conservation biologists established a large field experiment at the Tijuana Estuary to examine the effects of plant species richness on the success of wetland restoration. They found that experimental plots planted with species-rich mixtures developed a complex vegetation structure, which is important to insects and birds. The species-rich plots also accumulated nitrogen faster than species-poor plots (**Figure 57.16**).

We use markets to influence exploitation of species

Many people would like to consume only natural products that have been harvested in ways that protect biodiversity and ecosystem productivity. To enable consumers of forest products to exercise that choice, a consortium of environmental organizations and members of the forest products industry launched the Forest Stewardship Council (FSC) in 1993. FSC establishes criteria that a forest products company must meet for its products to be certified. Special certification companies determine whether a forestry operation meets the criteria and ensure that there is a chain of custody that tracks certified products on their way to market. By November 2005, about 34 million hectares of managed forests worldwide had been certified by FSC in 66 countries on five continents. FSC initially had its most significant impact on forest management in the temperate zone, but certification of tropical forests is growing rapidly. In October 2005, for example, more than 2 million hectares of Bolivian forests were certified by FSC. More than 400 companies in 18 countries have committed to purchasing certified wood products.

To serve the same function for marine products, the Marine Stewardship Council was formed through an alliance between the World Wildlife Fund and Unilever, one of the largest marketers of frozen seafood. Its first certified product, Australian rock lobster, came to market in 2000. Alaskan salmon has also been certified; other major fisheries are in the process of becoming certified. This action, combined with the elimination of government subsidies, can help reduce the current overexploitation of many marine fish stocks.

Ending trade is crucial to saving some species

Most endangered species cannot cope with any further reductions in their breeding populations. The legal mechanism for prohibiting

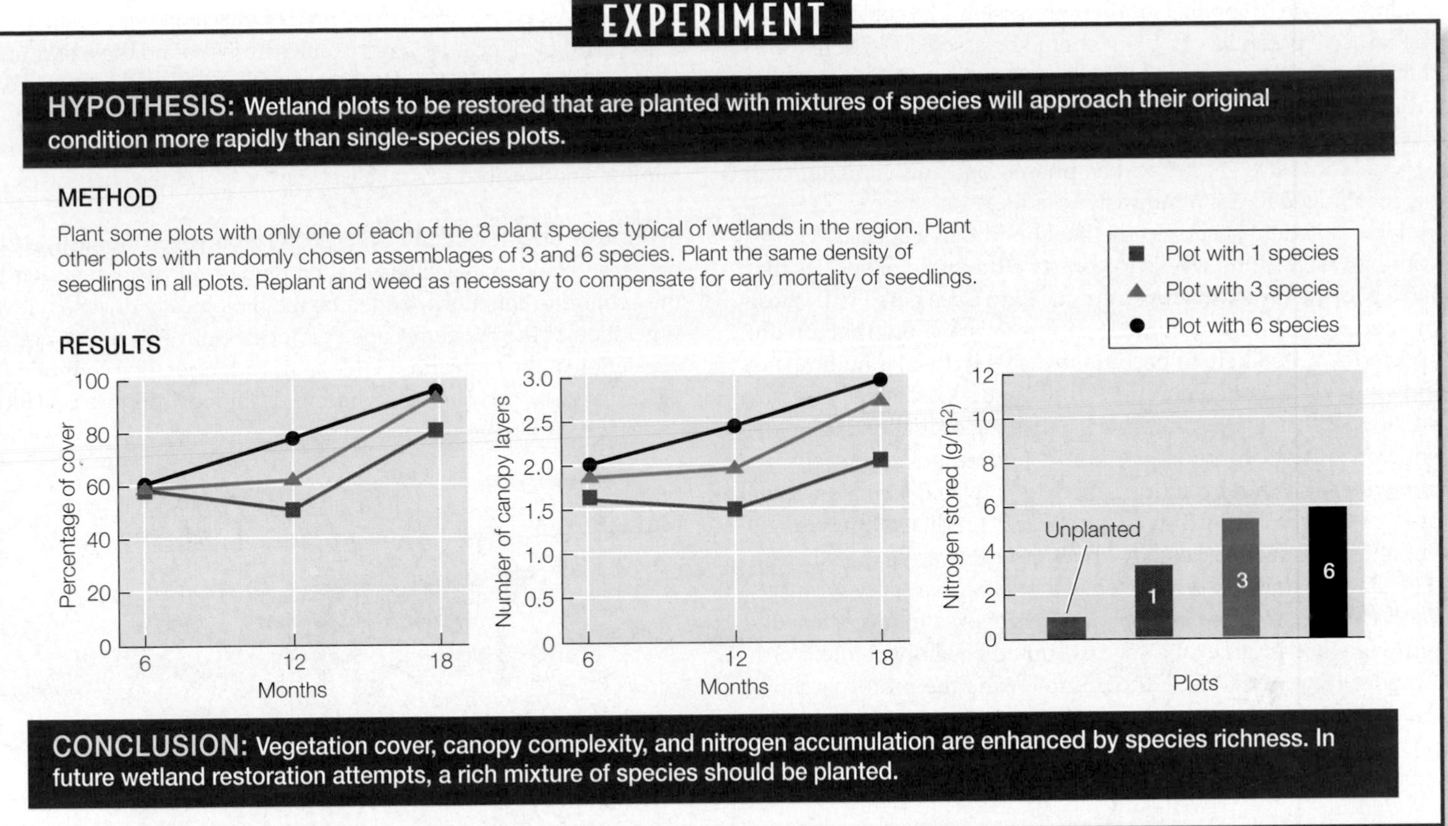

EXPERIMENT
HYPOTHESIS: Wetland plots to be restored that are planted with mixtures of species will approach their original condition more rapidly than single-species plots.
METHOD
Plant some plots with only one of each of the 8 plant species typical of wetlands in the region. Plant other plots with randomly chosen assemblages of 3 and 6 species. Plant the same density of seedlings in all plots. Replant and weed as necessary to compensate for early mortality of seedlings.
Plot with 1 species
Plot with 3 species
Plot with 6 species
RESULTS
Percentage of cover
Months
Number of canopy layers
Months
Nitrogen stored (g/m²)
Unplanted
Plots
CONCLUSION: Vegetation cover, canopy complexity, and nitrogen accumulation are enhanced by species richness. In future wetland restoration attempts, a rich mixture of species should be planted.

57.16 Species Richness Enhances Wetlands Restoration Both vegetation complexity and nitrogen accumulation are greater in species-rich than in species-poor experimental plots that had been degraded and lacked wetland plants.

exploitation of these species is an international agreement called the Convention on International Trade in Endangered Species (CITES). Most nations of the world are members of CITES. National representatives meet every two years to review the status of species on its protected species lists, to determine which species may no longer need protection, and to add new species. CITES rules currently prohibit international trade in items such as whale meat, rhinoceros horn, and many species of parrots and orchids.

CITES instituted a ban on international trade in elephant ivory in 1989, but demand for ivory remains strong, especially in Japan and China. As a result, poaching of elephants continues in the forests of central Africa and in East Africa. Southern African countries (Botswana, Namibia, South Africa, Zambia, and Zimbabwe), however, have so many elephants that government officials must kill many of them to control populations in the limited areas where they are allowed to roam. These countries would like to sell the ivory from the culled elephants to fund their conservation efforts. Other countries are worried that if trade restrictions are relaxed, poaching will escalate over the entire continent. Control and regulation of ivory trade might be possible if scientists could determine where the ivory comes from.

Samuel Wasser and his colleagues at the University of Washington have identified 16 microsatellite DNA markers that can be extracted from elephant feces. Park rangers in Malawi and Zambia were able to sample the elephant populations in their countries in just two weeks by collecting fresh scat while they were on routine patrols. The source of an elephant tusk can now be determined by matching the DNA extracted from the ivory to the geographically based frequencies of the 16 microsatellites. In June 2002, 6.5 tons of illegal ivory were seized in Singapore. DNA analyses showed that the elephants were killed in Zambia, providing an early warning system for directing anti-poaching efforts.

Controlling invasions of exotic species is important

Because some species are endangered by invasive exotic species, controlling these invasives is an important component of conservation biology. The best way to reduce the damage caused by invasive species, of course, is to prevent their introduction in the first place. Given the immense amount of global traffic between continents, it might seem impossible to curtail the spread of exotic species. Some promising options, however, do exist. For example, transoceanic transport of invasive species in ballast water could be largely eliminated by the simple procedure of deoxygenating ballast water before it is pumped out. This practice kills most organisms in the water. It also extends the life of ballast tanks, providing an economic benefit to shippers.

Regulating the importation and sale of exotic species can reduce deliberate introductions. In 2003, the Connecticut General Assembly established a penalty of $100 per plant for the sale of any of 81 plant species judged to be invasive in the state. In 2002, some members of the American horticultural industry crafted a voluntary code of conduct for their profession. The code states that the invasive potential of a plant should be assessed prior to introducing and marketing it. Horticulturists work with conservation biologists to determine which species are currently invasive, or likely to become so, and to identify suitable alternative species. Stocks of invasive species will be phased out, and gardeners will be encouraged to use noninvasive plants.

How do scientists assess the likelihood that a species will become invasive? One way is to compare the traits of species that have become invasive when introduced to a new area with those of species that have not. Such comparisons show that a plant species is more likely to become invasive if it has a high rate of growth, a short generation time and small seeds, is dispersed by vertebrates, has a large range in its native region, depends on nonspecific mutualists (root symbionts, pollinators, and seed dispersers), and is not evolutionarily closely related to plants in the area to which it is introduced. The best predictor, however, is whether the species is already known to be invasive elsewhere.

Using the traits that characterize most invasive species, conservation biologists have developed a decision tree to help them determine whether an exotic species should be allowed into North America (**Figure 57.17**). Although following the protocols stipulated by this decision tree cannot eliminate the introduction of all potentially invasive species, if it is used conscientiously, its application can greatly reduce the risk.

Biodiversity can be profitable

Much of the value of ecosystems to humans depends on their biodiversity. But it has been difficult to assess the value of biodiversity in monetary terms. When an ecosystem is perceived to have economic value, industries and government agencies are given a greater incentive to preserve it than if no monetary values are evident. Fortunately, our ability to assess the potential economic value of ecosystem services is improving rapidly.

Enough is known about the value of biological diversity and its role in the functioning of natural ecosystems to establish markets for ecosystem services. In 2005, Ecosystem Marketplace, the first global clearinghouse for information on this emerging trade, was launched. The clearinghouse is sponsored not only by environmental organizations, such as The Nature Conservancy, but also by large corporations, such as Citigroup and the reinsurance company Swiss Re. Its Web site contains information on the basic work of forests, including water filtration, soil quality maintenance, habitats, and carbon sequestration. The Web site encourages trade by providing details of transactions to potential buyers and sellers who are considering entering into what may seem to be a risky market. It includes news features from around the world, and a "Market Watch" page that tracks money flows into ecosystem services. Ecosystem services may soon become big business!

There are many instances demonstrating the profit value of sustaining biodiversity. Let's take a closer look at three such cases.

FYNBOS OF SOUTH AFRICA Studies by a group of economists, ecologists, and land managers have attempted to calculate the value of the economic benefits provided by the biologically diverse native vegetation of the Western Cape Province, South Africa. The native vegetation of the highlands of this area is a species-rich community of shrubs known as *fynbos* (pronounced "fainbos"; **Figure 57.18A**).

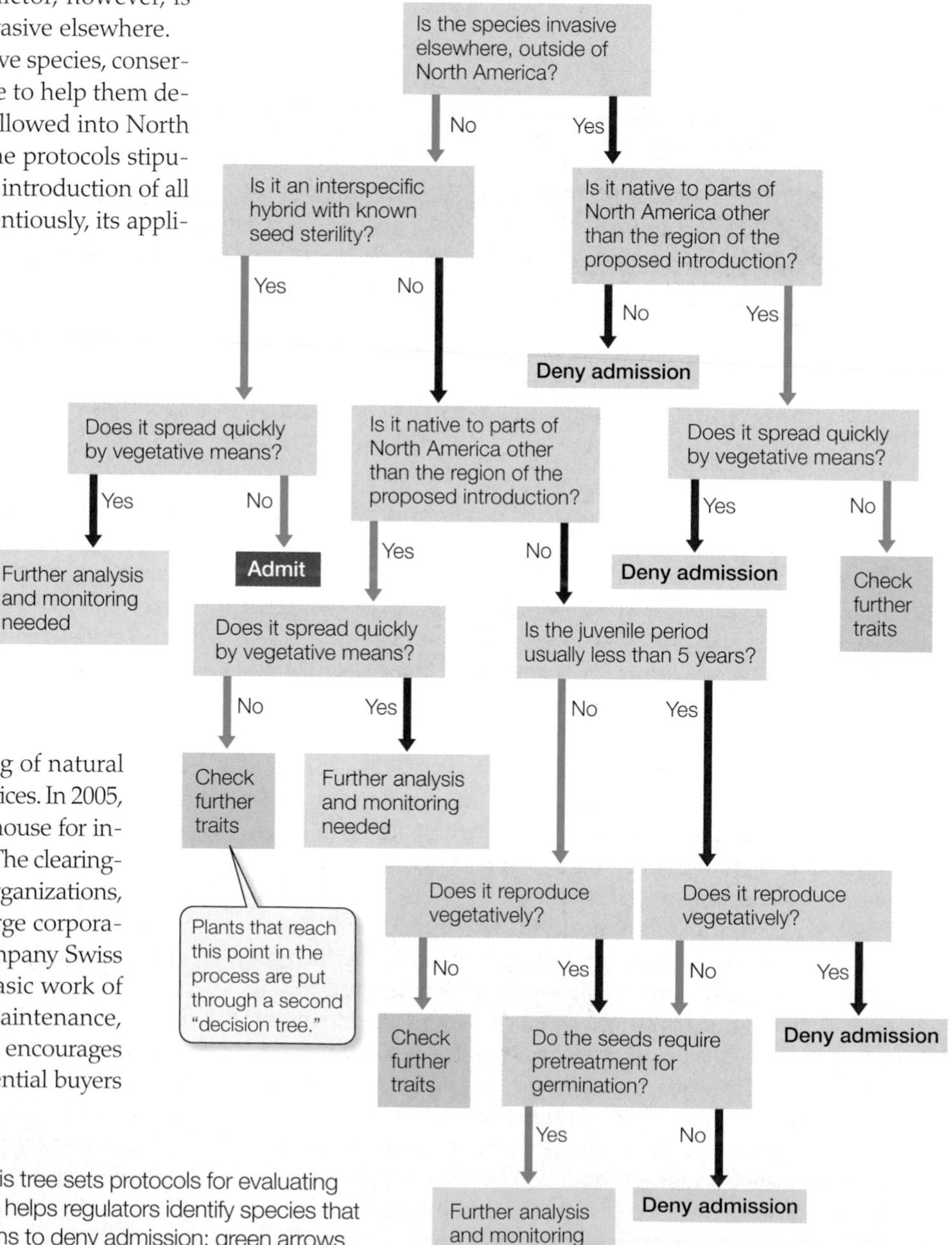

57.17 A "Decision Tree" for Exotic Species This tree sets protocols for evaluating the proposed introduction of an exotic species and helps regulators identify species that may become invasive. Red arrows indicate decisions to deny admission; green arrows can result in a decision to admit the species.

(A)

(B)

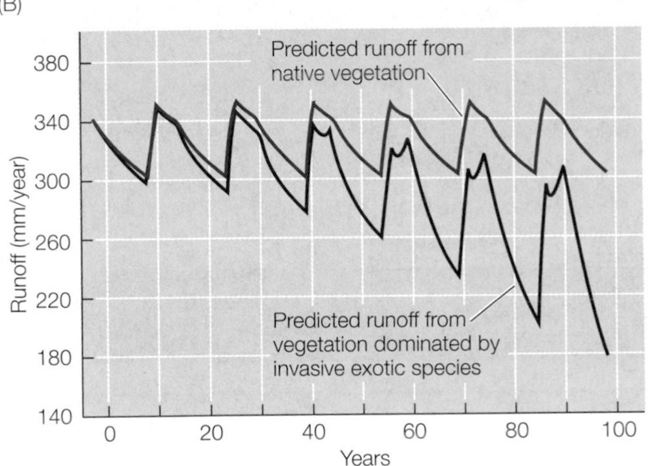

57.18 Invasive Species Disrupt Ecosystem Functioning (A) The unique fynbos ecosystem of the Western Cape Province of South Africa provides much of the area's water. (B) A computer simulation of stream flows from watersheds that have and have not been invaded by exotic trees.

These shrubs survive regular summer droughts, nutrient-poor soils, and the fires that periodically sweep through the highlands.

The fynbos-clad highlands provide a crucial economic service—about two-thirds of the Western Cape's water supply comes from them. In addition, some of the endemic plants are harvested for cut and dried flowers and thatching grass. The combined value of these harvests in 1993 was about $19 million. Some of the income from tourism in the region comes from people who want to see the fynbos. About 400,000 people visit the Cape of Good Hope Nature Reserve each year, primarily to see the endemic plants.

During recent decades, a number of plants introduced into South Africa from other continents have invaded the fynbos. Because they are taller and grow faster than the native plants, these exotics increase the intensity and severity of fires. By transpiring larger quantities of water, they decrease stream flows to less than half the amount flowing from mountains covered with native plants, reducing the water supply (**Figure 57.18B**). Removing the exotic plants by felling and digging out invasive trees and shrubs and managing fire costs between $140 and $830 per hectare, depending on the densities of invasive plants. Annual follow-up operations cost about $8 per hectare.

The services provided by fynbos vegetation could be replaced, but only at a much higher cost. A sewage purification plant that would deliver the same volume of water to the Western Cape Province as a well-managed watershed of 10,000 hectares would cost $135 million to build and $2.6 million per year to operate. Desalination of seawater would cost four times as much. Thus the available alternatives would deliver water at a cost between 1.8 and 6.7 times more than the cost of maintaining natural vegetation in the watershed. The technologically sophisticated methods that could substitute for the services provided by the biodiversity of the fynbos are more expensive than labor-intensive, employment-generating methods of maintaining those services.

WILD DOGS AND ECOTOURISM *Ecotourism* exists because of biodiversity and is a major source of income for many countries. For example, because diseases have caused populations of African wild dogs (*Lycaon pictus*) to plummet throughout Africa, tourists are now increasingly interested in seeing wild dogs when they sign up for a safari. South Africa has about 400 of Africa's remaining 5,700 dogs, most in Kruger National Park. A survey of South African tourists found that nearly three-fourths of them were willing to pay an extra U.S. $12 to see the dogs. In other words, a pack of 10 dogs would generate an added annual income of about U.S. $90,000. The investigators are now working with lodge owners and ranchers elsewhere in South Africa and in Kenya to engage them in efforts to reestablish wild dogs in areas from which they have disappeared.

COFFEE AND THE BEE Taylor Ricketts and colleagues at Stanford University assessed the economic value of the pollination services provided by the bees that live in, and depend on, tropical forest patches adjacent to a coffee plantation in Costa Rica. In a landscape where coffee plants are intermixed with forest patches, they found that coffee production was highest at the sites that were closest to forest patches (**Figure 57.19**). They also hand-pollinated some coffee plants to show that the difference in production was a result of pollination services rather than some other environmental condition. The investigators calculated that the value of pollination services to the plantation on which the experiments were carried out was about $60,000 per year, more than the current conservation incentive payments offered to landowners to preserve forest patches.

Living lightly helps preserve biodiversity

Protected areas, as we have seen, are an essential component of efforts to maintain biodiversity. More of them need to be established, but by themselves, protected areas cannot do the job. The extensive landscapes in which people live and extract resources must also play important roles in biodiversity conservation. The good news is that, carefully used, these lands can contribute much more to conservation than they currently do. The practice of using exploited lands in ways that sustain biodiversity is becoming known as **reconciliation ecology**.

57.19 Establishing the Economic Value of Forest Patches Coffee plants in plantations located close to forest patches, where they can benefit from the services of native pollinators (bees) living in the forests, produce more coffee beans (the seeds of coffee plants) than plantations farther from forest patches.

EXPERIMENT

HYPOTHESIS: Native pollinators (bees) of coffee flowers that live in forests increase both the quantity and quality of coffee seeds produced on adjacent plantations.

METHOD

Establish 12 sites that are near (50 m), intermediate (800 m), and far (1600 m) from forest patches. Measure the quality of seeds produced per plant and the frequency of small, misshapen seeds. To test whether the observed differences are due to the frequency of pollinator visits, hand-pollinate flowers.

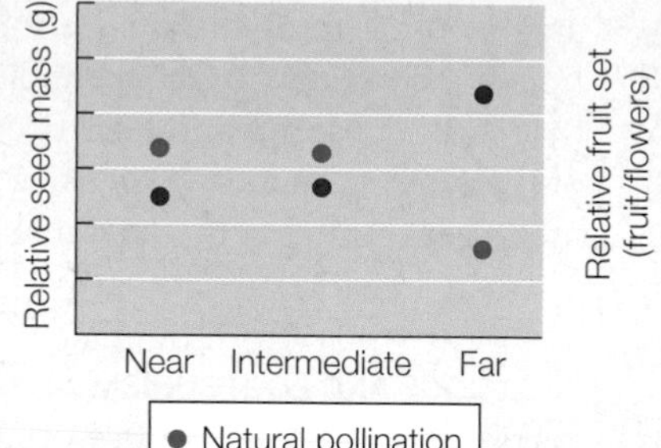

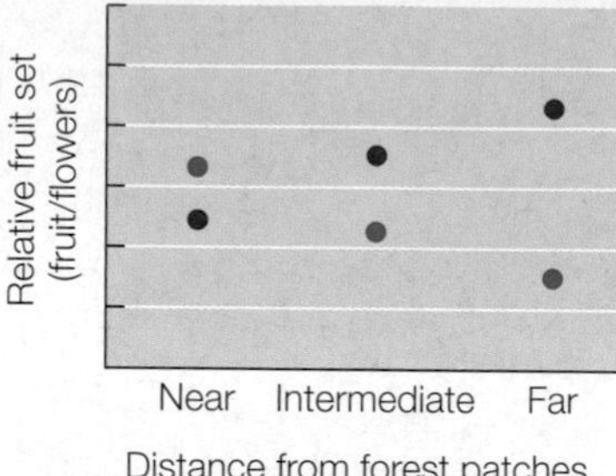

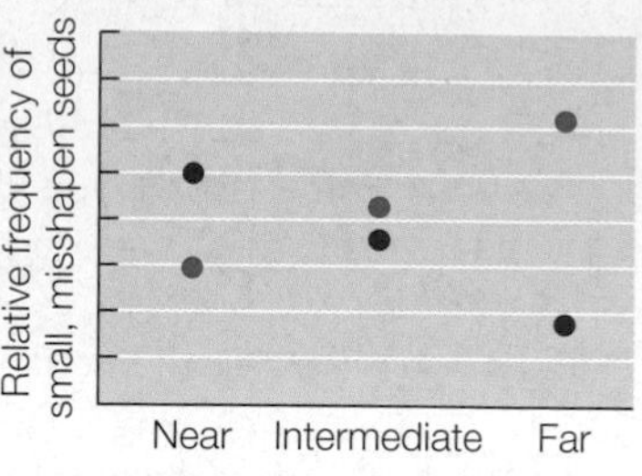

Distance from forest patches

- Natural pollination
- Hand pollination

RESULTS

Forest-dwelling bees increased coffee yields by 20% at sites within 1 kilometer of a forest patch. There were sufficient visits from bees to support full production at the near and intermediate sites. At those sites, hand pollination did not increase seed production. However, at far sites, hand pollination increased the mass of harvested seeds and fruit set. Hand pollination also decreased the proportion of small, misshapen seeds.

CONCLUSION: Coffee plants close to forest patches produced more seeds and had fewer misshapen seeds than coffee plants farther from the forest patches.

Most ecosystem services are provided locally. Fortunately, it is easier to motivate people to work to protect their local interests than it is to stimulate them to work on national or global issues. The National Wildlife Federation has established a very successful program in which people petition to have their backyards certified as wildlife-friendly. Criteria for certification include planting shrubs that provide food for birds and not using pesticides on lawns. The city of Tucson, Arizona, has launched a major project to make the city into an important habitat for many species of birds, not just the typical urban birds that live in most cities in North America.

Sometimes the alterations that make a big difference in biodiversity are remarkably simple. During the warmer part of the year, about 1.5 million Mexican free-tailed bats roost under the Congress Avenue Bridge in Austin, Texas. This is the largest population of bats in North America, much larger than the famous colony that lives in Carlsbad Caverns National Park. The nightly spectacle of their departure from the roost to feed is a major tourist attraction. Twenty years ago, however, only a few thousand bats roosted under the bridge. When the bridge was reconstructed a few years ago, engineers, with no intention of helping bats, added expansion joints that are about an inch wide and 16 inches deep. These crevices are perfect roosting sites for the bats, which quickly moved in by the tens of thousands.

The Turkey Point power plant in southern Florida consists of two fossil fuel generating units and two units powered by nuclear fuel. These four units generate lots of hot water. To cool the discharged water, the Florida Power & Light Company dug a system of 38 canals that covers 6,000 acres. The cooling canals are separated by low-lying berms that support a variety of native and exotic plants. Red mangroves grow along the edges of the canals. Today, the canals support a thriving population of American crocodiles, a highly endangered species. Crocodiles living in the canals yield about 10 percent of all young crocodiles in the United States. Having discovered the biodiversity value of its cooling system, the company employs biologists to monitor the crocodiles and does what it can to ensure their continued reproductive success.

The Elephant Pepper Development Trust of Africa promotes using chili peppers as a buffer crop around fields of maize, sorghum, and millet to deter elephants, buffalos, and other large mammals—who are repelled by the taste of the hot peppers—from invading the fields and destroying crops. Alternatively, farmers spray a mixture of capsaicin (the active ingredient in chilies) and axle grease on the string fences surrounding their fields.

Captive breeding programs can maintain a few species

A few of the world's many endangered species can be maintained in captivity while the external threats to their existence are reduced or removed. However, captive propagation is only a temporary measure that buys time to deal with those threats. Existing zoos, aquariums, and botanical gardens do not have enough space to maintain adequate populations of more than a small fraction of Earth's rare and endangered species. Nonetheless, captive propagation can play an important role by maintaining species during critical periods and by providing a source of individuals for reintroduction into the wild. Captive propagation projects in zoos also raise public awareness of threatened and endangered species.

As we saw at the beginning of this chapter, captive propagation helped save the whooping crane. The California condor, North America's largest bird, survives today only because of captive propagation (**Figure 57.20**). Two hundred years ago, condors ranged from southern British Columbia to northern Mexico, but by 1978, the wild population was plunging toward extinction—only 25 to 30 birds remained in southern California. Many birds died from ingesting carcasses containing lead shot.

To save the condor from certain extinction, biologists initiated a captive breeding program in 1983. The first chick conceived in captivity hatched in 1988. By 1993, nine captive pairs were producing chicks, and the captive population had increased to more than 60 birds. Six captive-bred birds were released in the mountains north of Los Angeles in 1992. These birds are provided with lead-free food in remote areas. They are using the same roosting sites, bathing pools, and mountain ridges as their predecessors did. Captive-reared birds have also been released in northern Arizona and Baja California. As of October 2005, there were 121 wild condors in California, Arizona, and Baja California. Three wild-born chicks are now flying free for the first time in 20 years in Arizona and California. Lead poisoning is still a problem, but an effort to encourage hunters to use non-lead ammunition is under way.

Gymnogyps californianus

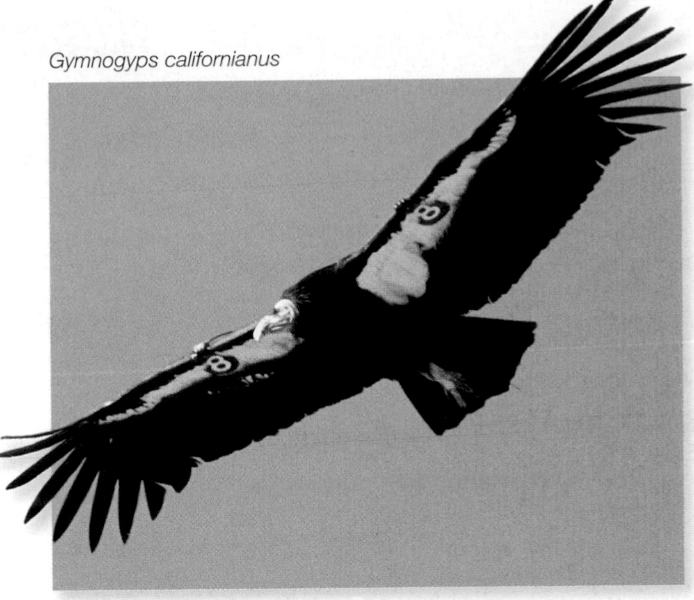

57.20 A California Condor Soars California condors raised in captivity have successfully survived after being released into the wild. Numbered wing tags allow conservation biologists to identify the individual bird and track its movements. The survival of North America's largest bird species depends on this captive propagation project.

The legacy of Samuel Plimsoll

During the nineteenth century, many British merchant ships sailed Earth's oceans. At that time, there were no undersea telegraph cables or shipboard radios. Once a ship left a harbor, it was out of contact with the rest of the world; in the case of a shipwreck, rescue was impossible. Owners could maximize their profits by overloading their ships, even though this caused some of them to be unseaworthy and sink. Samuel Plimsoll, a member of England's Parliament, became concerned about the rate of loss of British vessels and sailors. He convinced Parliament to require that a "load line" be painted on the hull of every large oceangoing vessel. The position of the line was calculated using factors such as the structural strength of the vessel and the shape of its hull. If the load line was under water, the ship was not permitted to leave the harbor. The "Plimsoll line," as it has come to be known, dramatically reduced the rate of loss of British ships and sailors at sea.

The increasing loss of Earth's species suggests that the load of human activities has pushed the hull of Noah's Ark below the Plimsoll line. But where and how should society draw that line? The decision should be based on scientific information, but just as in Samuel Plimsoll's time, science cannot determine an "acceptable rate of loss." Ethical considerations will figure prominently in the decisions societies make concerning how much to change the way we use ecosystems so that other species can survive on Earth with us.

57.4 RECAP

To preserve biodiversity, we must be willing to set aside protected areas, restore habitat, develop programs to increase populations of endangered species, enforce laws that restrict transport of invasive species, and otherwise recognize the benefits of maintaining well-functioning ecosystems.

- Can you enumerate the priorities conservation biologists consider when establishing protected areas?
- Why is restoration of degraded ecosystems often necessary? See pp. 1236–1237
- Explain why controlling the importation of exotic species is an important component of conservation biology. See pp. 1239–1240 and Figure 57.17

CHAPTER SUMMARY

57.1 What is conservation biology?

Conservation biology is an applied scientific discipline devoted to preserving biodiversity.

Conservation biology is a normative discipline, but its practitioners adhere to scientific standards and practices.

Conservation biologists recognize that Earth's ecosystems are dynamic and that people are integral components of ecosystems.

There are many compelling reasons for preserving biodiversity, including maintaining well-functioning ecosystems that provide humans with many goods and services.

57.2 How do biologists predict changes in biodiversity?

The **species–area relationship** estimates the number of species that can be supported by an area of a given size.

To estimate probable rates of extinction, statistical models take into account data on population sizes, genetic variation, physiology, morphology, and behavior.

Rarity is not always a cause for concern, but species whose populations are shrinking rapidly are usually at risk.

Populations with only a few individuals confined to a small range may be eliminated by local disturbances such as fires, unusual weather, disease, habitat destruction, and predators.

57.3 What factors threaten species survival?

Habitat loss is the most important cause of endangerment of species worldwide. As habitats become increasingly **fragmented**, more species are lost from those habitats. Small habitat patches can support only small populations and are adversely influenced by **edge effects.** Review Figure 57.5, Web/CD Tutorial 57.1

Species introduced to regions outside their original range often become **invasive**, causing extinctions of native species by competing with them, eating them, or transmitting diseases to them.

Overexploitation has historically been the most important cause of species extinctions, and overexploitation continues today.

Climate change is likely to become an increasingly important cause of extinctions for those species that cannot shift their ranges as rapidly as the climate warms.

57.4 What strategies do conservation biologists use?

Establishing protected areas is crucial to preserving biodiversity. Protected areas are selected by taking into account species richness, **endemism**, imminence of threats, and the need to protect representative ecosystems.

Restoration ecology is an important conservation strategy because many degraded ecosystems will not recover, or will do so only very slowly, without human assistance. Review Figure 57.16

International trade in endangered species is controlled by regulations that most countries endorse.

Determining which species are likely to become invasive and preventing their introduction to new areas is an important component of conservation biology.

Calculating the economic value of ecosystem services is helping establish incentives to preserve them.

Even within the extensive landscapes where people live and extract resources, steps may be taken to preserve biodiversity. This approach is becoming known as **reconciliation ecology**.

Captive breeding programs can maintain selected endangered species while threats to their existence are reduced or removed.

See Web/CD Activity 57.1 for a concept review of this chapter.

SELF-QUIZ

1. Which of the following is *not* currently a major cause of species extinctions?
   a. Habitat destruction
   b. Rising sea levels
   c. Overexploitation
   d. Introduction of predators
   e. Introduction of diseases
2. The most important cause of endangerment of species in the United States currently is
   a. pollution.
   b. invasive species.
   c. overexploitation.
   d. habitat destruction.
   e. loss of mutualists.
3. People care about species extinctions because
   a. more than half of the medical prescriptions written in the United States contain a natural plant or animal product.
   b. people derive aesthetic pleasure from interacting with other organisms.
   c. causing species extinctions raises serious ethical issues.
   d. biodiversity helps maintain valuable ecosystem services.
   e. all of the above
4. As a habitat patch gets smaller, it
   a. cannot support populations of species that require large areas.
   b. supports only small populations of many species.
   c. is influenced to an increasing degree by edge effects.
   d. is invaded by species from surrounding habitats.
   e. all of the above
5. A plant species is most likely to become invasive when introduced to a new area if it
   a. grows tall.
   b. has become invasive in other places where it has been introduced.
   c. is closely related to species living in the area where it has been introduced.
   d. has specialized disseminators of its seeds.
   e. has a long life span.
6. Conservation biologists are concerned about global warming because
   a. the rate of change in climate is projected to be faster than the rate at which many species can shift their ranges.
   b. it is already too hot in the tropics.
   c. climates have been so stable for thousands of years that many species lack the ability to tolerate variable temperatures.
   d. climate change will be especially harmful to rare species.
   e. none of the above

7. Scientists can determine the historical frequency of fires in an area by
   a. examining charcoal in sites of ancient villages.
   b. measuring carbon in soils.
   c. radioactively dating fallen tree trunks.
   d. examining fire scars in growth rings of living trees.
   e. determining the age structure of forests.
8. Captive propagation is a useful conservation tool, provided that
   a. there is space in zoos, aquariums, and botanical gardens for breeding a few individuals.
   b. the areas of origin of all individuals are known.
   c. the threats that endangered the species are being alleviated so that captive-reared individuals can later be released back into the wild.
   d. there are sufficient caretakers.
   e. none of the above. Captive propagation should never be used because it directs attention away from the need to protect the species in their natural habitats.
9. Restoration ecology is an important field because
   a. many areas have been highly degraded.
   b. many areas are vulnerable to global climate change.
   c. many species suffer from demographic stochasticity.
   d. many species are genetically impoverished.
   e. fire is a threat to many areas.
10. The new discipline of reconciliation ecology has developed because
   a. all other methods of preserving biodiversity have failed.
   b. protected areas should be able to maintain biodiversity.
   c. protected areas alone are not sufficient to maintain biodiversity.
   d. scientists are unable to control diseases today.
   e. we are not reconciled with other species.

FOR DISCUSSION

1. Most species driven to extinction by humans in the past were large vertebrates. Do you expect this pattern to persist into the future? If not, why not?
2. Conservation biologists have debated extensively which is better: many small protected areas—which may contain more species—or a few large protected areas—which may be the only ones that can support populations of species that require large areas. What ecological processes should be evaluated in making judgments about the sizes and locations of protected areas?
3. During World War I, doctors adopted a "triage" system for dealing with wounded soldiers. The wounded were divided into three categories: those almost certain to die no matter what was done to help them, those likely to recover even if not assisted, and those whose probability of survival was greatly increased if they were given immediate medical attention. Limited medical resources were directed primarily at the third category. What are some implications of adopting a similar attitude toward species preservation?
4. Utilitarian arguments dominate discussions about the importance of preserving the biological richness of the planet. In your opinion, what role should ethical and moral arguments play?
5. The desert bighorn sheep of the southwestern United States is endangered. Its major predator, the puma, is also threatened in the region. Under what conditions, if any, would it be appropriate to suppress the population of one rare species to assist another rare species?

FOR INVESTIGATION

Forest-dwelling bees increased seed production in coffee plants growing within a kilometer of a forest patch in Costa Rica (see Figure 57.19). What forest organisms are likely to provide benefits, such as control of insect pests, to surrounding agricultural crops in addition to pollination? Over what distances are those effects likely to be felt? What experiments could be designed to test the importance of those effects and how they vary with distance from forest patches?

Appendix A: The Tree of Life

Phylogeny is the organizing principle of modern biological taxonomy, and a guiding principle of modern phylogeny is *monophyly*: a monophyletic group is considered to be one that contains an ancestral lineage and *all* of its descendants. Any such a group can be extracted from a phylogenetic tree with a single cut. The tree shown here provides a guide to the relationships among the major groups of the extant (living) organisms in the Tree of Life as we have presented them throughout this book. We do include three groups that are not believed to be monophyletic; these are designated with quotation marks.

The position of the branching "splits" indicates the relative branching order of the lineages of life, but the timing of splits in different groups is not drawn on a comparable time scale. In addition, the groups appearing at the branch tips do not necessarily carry equal phylogenetic "weight." For example, the ginkgo [55] is indeed at the apex of its lineage; this gymnosperm group consists of a single living species. In contrast, a phylogeny of the angiosperms [52] would continue on from this point to fill many more trees the size of this one.

The glossary entries that follow are informal descriptions of some major features of the organisms described in Part Six of this book. Each entry gives the group's common name, followed by the formal scientific name of the group (in parentheses). Numbers in square brackets reference the location of the respective groups on the tree.

It is sometimes convenient to use an informal name to refer to a collection of organisms that are not monophyletic but nonetheless all share (or all lack) some common attribute. We call these "convenience terms"; such groups are indicated in these entries by quotation marks, and we do not give them formal scientific names. Examples include "prokaryotes," "protists," and "algae." Note that these groups cannot be removed with a single cut; they represent a collection of distantly related groups that appear in different parts of the tree.

– A –

acorn worms (*Enteropneusta*) Benthic marine hemichordates [109] with an acorn-shaped proboscis, a short collar (neck), and a long trunk.

"algae" A convenience term encompassing various distantly related groups of aquatic, photosynthetic chromalveolates [5] and certain members of the Plantae [8].

alveolates (*Alveolata*) [7] Unicellular eukaryotes with a layer of flattened vesicles (alveoli) supporting the plasma membrane. Major alveolate groups include the dinoflagellates [49], apicomplexans [50], and ciliates [51].

ambulacrarians (*Ambulacraria*) [27] The echinoderms [108] and hemichordates [109].

amniotes (*Amniota*) [34] Mammals, reptiles, and their extinct close relatives. Characterized by many adaptations to terrestrial life, including an amniotic egg (with a unique set of membranes—the amnion, chorion, and allantois), a water-repellant epidermis (with epidermal scales, hair, or feathers), and, in males, a penis that allows internal fertilization.

amoebozoans (*Amoebozoa*) [76] A group of eukaryotes [4] that use lobe-shaped pseudopods for locomotion and to engulf food. Major amoebozoan groups include the loboseans, plasmodial slime molds, and cellular slime molds.

amphibians (*Amphibia*) [118] Tetrapods [33] with glandular skin that lacks epidermal scales, feathers, or hair. Many amphibian species undergo a complete metamorphosis from an aquatic larval form to a terrestrial adult form, although direct development is also common. Major amphibian groups include frogs and toads (anurans), salamanders, and caecilians.

amphipods (*Amphipoda*) Small crustaceans [106] that are abundant in many marine and freshwater habitats. They are important herbivores, scavengers, and micropredators, and are an important food source for many aquatic organisms.

angiosperms (*Anthophyta* or *Magnoliophyta*) [52] The flowering plants. Major angiosperm groups include the monocots, eudicots, and magnoliids.

animals (*Animalia* or *Metazoa*) [19] Multicellular heterotrophic eukaryotes. The majority of animals are bilaterians [21]. Other major groups are the cnidarians [87], ctenophores [86], calcareous sponges [85], demosponges [84], and glass sponges [83]. The closest living relatives of the animals are the choanoflagellates [82].

annelids (*Annelida*) [94] Segmented worms, including earthworms, leeches, and polychaetes. One of the major groups of lophotrochozoans [23].

anthozoans (*Anthozoa*) One of the major groups of cnidarians [87]. Includes the sea anemones, sea pens, and corals.

anurans (*Anura*) Comprising the frogs and toads, this is the largest group of living amphibians [118]. They are tailless, with a shortened vertebral column and elongate hind legs modified for jumping. Many species have an aquatic larval form known as a tadpole.

apicomplexans (*Apicomplexa*) [50] Parasitic alveolates [7] characterized by the possession of an apical complex at some stage in the life cycle.

arachnids (*Arachnida*) Chelicerates [104] with a body divided into two parts: a cephalothorax that bears six pairs of appendages (four pairs of which are usually used as legs) and an abdomen that bears the genital opening. Familiar arachnids include spiders, scorpions, mites and ticks, and harvestmen.

archaeans (*Archaea*) [3] Unicellular organisms lacking a nucleus and lacking peptidoglycan in the cell wall. Once grouped with the bacteria, archaeans possess distinctive membrane lipids.

archosaurs (*Archosauria*) [36] A group of reptiles [35] that includes dinosaurs and crocodilians [123]. Most dinosaur groups became extinct at the end of the Cretaceous; birds [122] are the only surviving dinosaurs.

arrow worms (*Chaetognatha*) [96] Small planktonic or benthic predatory marine worms with fins and a pair of hooked, prey-grasping spines on each side of the head.

arthropods (*Arthropoda*) The largest group of ecdysozoans [24]. Arthropods are characterized by a stiff exoskeleton, segmented bodies, and jointed appendages. Includes the chelicerates [104], myriapods [105], crustaceans [106], and hexapods (insects and their relatives) [107].

ascidians (*Ascidiacea*) "Sea squirts"; the largest group of urochordates [110]. Also known as tunicates, they are sessile (as adults), marine, sac-like filter feeders.

ascomycetes (*Ascomycota*) [78] Fungi that bear the products of meiosis within sacs (asci) if the organism is multicellular. Some are unicellular.

– B –

bacteria (*Eubacteria*) [2] Unicellular organisms lacking a nucleus, possessing distinctive ribosomes and initiator tRNA, and generally containing peptidoglycan in the cell wall. Different bacterial groups are distinguished primarily on nucleotide sequence data.

barnacles (*Cirripedia*) Crustaceans [106] that undergo two metamorphoses—first from a feeding planktonic larva to a nonfeeding swimming larva, and then to a sessile adult that forms a "shell" composed of four to eight plates cemented to a hard substrate.

basidiomycetes (*Basidiomycota*) [77] Fungi [17] that, if multicellular, bear the products of meiosis on club-shaped basidia and possess a long-lasting dikaryotic stage. Some are unicellular.

bilaterians (*Bilateria*) [21] Those animal groups characterized by bilateral symmetry and three distinct tissue types (endoderm, ectoderm, and mesoderm). Includes the protostomes [22] and deuterostomes [26].

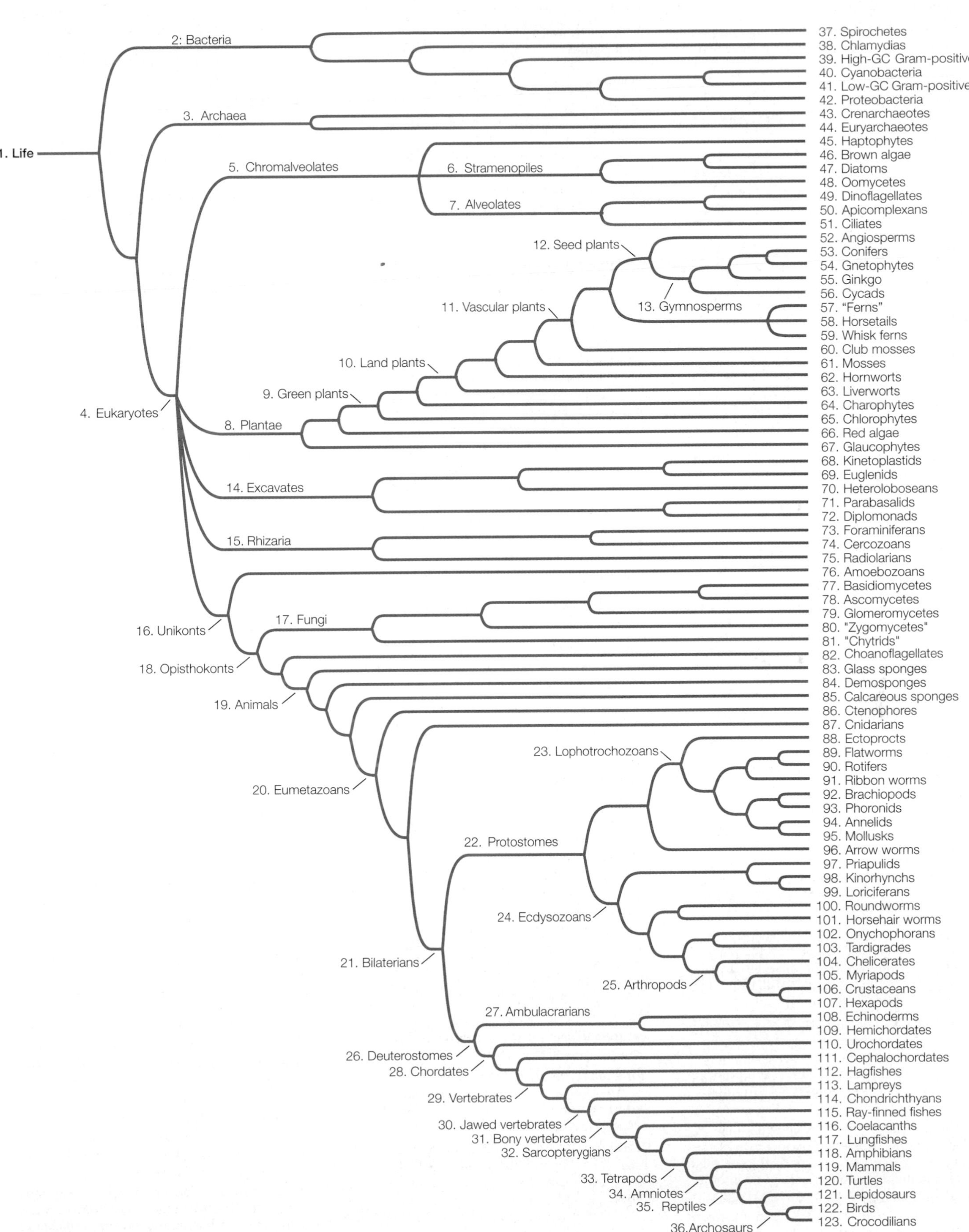
1. Life
2: Bacteria
3. Archaea
4. Eukaryotes
5. Chromalveolates
6. Stramenopiles
7. Alveolates
8. Plantae
9. Green plants
10. Land plants
11. Vascular plants
12. Seed plants
13. Gymnosperms
14. Excavates
15. Rhizaria
16. Unikonts
17. Fungi
18. Opisthokonts
19. Animals
20. Eumetazoans
21. Bilaterians
22. Protostomes
23. Lophotrochozoans
24. Ecdysozoans
25. Arthropods
26. Deuterostomes
27. Ambulacrarians
28. Chordates
29. Vertebrates
30. Jawed vertebrates
31. Bony vertebrates
32. Sarcopterygians
33. Tetrapods
34. Amniotes
35. Reptiles
36.Archosaurs
37. Spirochetes
38. Chlamydias
39. High-GC Gram-positives
40. Cyanobacteria
41. Low-GC Gram-positives
42. Proteobacteria
43. Crenarchaeotes
44. Euryarchaeotes
45. Haptophytes
46. Brown algae
47. Diatoms
48. Oomycetes
49. Dinoflagellates
50. Apicomplexans
51. Ciliates
52. Angiosperms
53. Conifers
54. Gnetophytes
55. Ginkgo
56. Cycads
57. "Ferns"
58. Horsetails
59. Whisk ferns
60. Club mosses
61. Mosses
62. Hornworts
63. Liverworts
64. Charophytes
65. Chlorophytes
66. Red algae
67. Glaucophytes
68. Kinetoplastids
69. Euglenids
70. Heteroloboseans
71. Parabasalids
72. Diplomonads
73. Foraminiferans
74. Cercozoans
75. Radiolarians
76. Amoebozoans
77. Basidiomycetes
78. Ascomycetes
79. Glomeromycetes
80. "Zygomycetes"
81. "Chytrids"
82. Choanoflagellates
83. Glass sponges
84. Demosponges
85. Calcareous sponges
86. Ctenophores
87. Cnidarians
88. Ectoprocts
89. Flatworms
90. Rotifers
91. Ribbon worms
92. Brachiopods
93. Phoronids
94. Annelids
95. Mollusks
96. Arrow worms
97. Priapulids
98. Kinorhynchs
99. Loriciferans
100. Roundworms
101. Horsehair worms
102. Onychophorans
103. Tardigrades
104. Chelicerates
105. Myriapods
106. Crustaceans
107. Hexapods
108. Echinoderms
109. Hemichordates
110. Urochordates
111. Cephalochordates
112. Hagfishes
113. Lampreys
114. Chondrichthyans
115. Ray-finned fishes
116. Coelacanths
117. Lungfishes
118. Amphibians
119. Mammals
120. Turtles
121. Lepidosaurs
122. Birds
123. Crocodilians

birds (*Aves*) [122] Feathered, flying (or secondarily flightless) tetrapods [33].

bivalves (*Bivalvia*) Major mollusk [95]group; clams and mussels. Bivalves typically have two similar hinged shells that are each asymmetrical across the midline.

bony vertebrates (*Osteichthyes*) [31] Vertebrates [29] in which the skeleton is usually ossified to form bone. Includes the ray-finned fishes [115], coelacanths [116], lungfishes [117], and tetrapods [33].

brachiopods (*Brachiopoda*) [92] Lophotrochozoans [23] with two similar hinged shells that are each symmetrical across the midline. Superficially resemble bivalve mollusks, except for the shell symmetry.

brittle stars (*Ophiuroidea*) Echinoderms [108] with five long, whip-like arms radiating from a distinct central disk that contains the reproductive and digestive organs.

brown algae (*Phaeophyta*) [46] Multicellular, almost exclusively marine stramenopiles [6] generally containing the pigment fucoxanthin as well as chlorophylls *a* and *c* in their chloroplasts.

– C –

caecilians (*Gymnophiona*) A group of burrowing or aquatic amphibians [118]. They are elongate, legless, with a short tail (or none at all), reduced eyes covered with skin or bone, and a pair of sensory tentacles on the head.

calcareous sponges (*Calcarea*) [85] Filter-feeding marine sponges with spicules composed of calcium carbonate.

cellular slime molds (*Dictyostelida*) Amoebozoans [76] in which individual amoebas aggregate under stress to form a multicellular pseudoplasmodium.

cephalochordates (*Cephalochordata*) [111] A group of weakly swimming, eel-like benthic marine chordates [28]; also called lancelets. Thought to be the closest living relatives of the vertebrates [29].

cephalopods (*Cephalopoda*) Active, predatory mollusks [95] in which the molluscan foot has been modified into muscular hydrostatic arms or tentacles. Includes octopuses, squids, and nautiluses.

cercozoans (*Cercozoa*) [74] Unicellular eukaryotes [4] that feed by means of threadlike pseudopods. Together with foraminiferans [73] and radiolarians [75], the cercozoans comprise the group *Rhizaria* [15].

charophytes (*Charales*) [64] Multicellular green algae with branching, apical growth and plasmodesmata between adjacent cells. The closest living relatives of the land plants [10], they retain the egg in the parent organism.

chelicerates (*Chelicerata*) [104] A major group of arthropods [25] with pointed appendages (chelicerae) used to grasp food (as opposed to the chewing mandibles of most other arthropods). Includes the arachnids, horseshoe crabs, pycnogonids, and extinct sea scorpions.

chimaeras (*Holocephali*) A group of bottom-dwelling, marine, scaleless chondrichthyan fishes [114] with large, permanent, grinding tooth plates (rather than the replaceable teeth found in other chondrichthyans).

chitons (*Polyplacophora*) Flattened, slow-moving mollusks [95] with a dorsal protective calcareous covering made up of eight articulating plates.

chlamydias (*Chlamydiae*) [38] A group of very small Gram-negative bacteria; they live as intracellular parasites of other organisms.

chlorophytes (*Chlorophyta*) [65] The most abundant and diverse group of green algae, including freshwater, marine, and terrestrial forms; some are unicellular, others colonial, and still others multicellular. Chlorophytes use chlorophylls *a* and *c* in their photosynthesis.

choanoflagellates (*Choanozoa*) [82] Unicellular eukaryotes [4] with a single flagellum surrounded by a collar. Most are sessile, some are colonial. The closest living relatives of the animals [19]

chondrichthyans (*Chondrichthyes*) [114] One of the two main groups of jawed vertebrates [30]; includes sharks, rays, and chimaeras. They have cartilaginous skeletons and paired fins.

chordates (*Chordata*) [28] One of the two major groups of deuterostomes [26], characterized by the presence (at some point in development) of a notochord, a hollow dorsal nerve cord, and a post-anal tail. Includes the urochordates [110], cephalochordates [111], and vertebrates [29].

chromalveolates (*Chromalveolata*) [5] A contested group, said to have arisen from a common ancestor with chloroplasts derived from a red alga and supported by molecular evidence. Major chromalveolate groups include the alveolates [7] and stramenopiles [6].

"chytrids" [81] A convenience term used for a paraphyletic group of mostly aquatic, microscopic fungi [17] with flagellated gametes. Some exhibit alternation of generations.

ciliates (*Ciliophora*) [51] Alveolates [7] with numerous cilia and two types of nuclei (micronuclei and macronuclei).

clitellates (*Clitellata*) Annelids [94] with gonads contained in a swelling (called a clitellum) toward the head of the animal. Includes earthworms (oligochaetes) and leeches.

club mosses (*Lycophyta*) [60] Vascular plants [11] characterized by microphylls.

cnidarians (*Cnidaria*) [87] Aquatic, mostly marine eumetazoans [20] with specialized stinging organelles (nematocysts) used for prey capture and defense, and a blind gastrovascular cavity. The closest living relatives of the bilaterians [21].

coelacanths (*Actinista*) [116] A group of marine sarcopterygians [32] that was diverse from the Middle Devonian to the Cretaceous, but is now known from just two living species. The pectoral and anal fins are on fleshy stalks supported by skeletal elements, so they are also called lobe-finned fishes.

conifers (*Pinophyta* or *Coniferophyta*) [53] Cone-bearing, woody seed plants.

copepods (*Copepoda*) Small, abundant crustaceans [106] found in marine, freshwater, or wet terrestrial habitats. They have a single eye, long antennae, and a body shaped like a teardrop.

craniates (*Craniata*) Some biologist exclude the hagfishes [112] from the vertebrates [29], and use the term craniates to refer to the two groups combined.

crenarchaeotes (*Crenarchaeota*) [43] A major and diverse group of archaeans [3], defined on the basis of rRNA base sequences. Many are extremophiles (inhabit extreme environments), but the group may also be the most abundant archaeans in the marine environment.

crinoids (*Crinoidea*) Echinoderms [108] with a mouth surrounded by feeding arms, and a U-shaped gut with the mouth next to the anus. They attach to the substratum by a stalk or are free-swimming. Crinoids were abundant in the middle and late Paleozoic, but only a few hundred species have survived to the present. Includes the sea lilies and feather stars.

crocodilians (*Crocodylia*) [123] A group of large, predatory, aquatic archosaurs [36]. The closest living relatives of birds [122]. Includes alligators, caimans, crocodiles, and gharials.

crustaceans (*Crustacea*) [106] Major group of marine, freshwater, and terrestrial arthropods [25] with a head, thorax, and abdomen (although the head and thorax may be fused), covered with a thick exoskeleton, and with two-part appendages. Crustaceans undergo metamorphosis from a nauplius larva. Includes decapods, isopods, krill, barnacles, amphipods, copepods, and ostracods.

ctenophores (*Ctenophora*) [86] Radially symmetrical, diploblastic marine animals [19], with a complete gut and eight rows of fused plates of cilia (called ctenes).

cyanobacteria (*Cyanobacteria*) [40] A group of unicellular, colonial, or filamentous bacteria that conduct photosynthesis using chlorophyll *a*.

cycads (*Cycadophyta*) [56] Palmlike gymnosperms with large, compound leaves.

cyclostomes (*Cyclostomata*) This term refers to the possibly monophyletic group of lampreys [113] and hagfishes [112]. Molecular data support this group, but morphological data suggest that lampreys are more closely related to jawed vertebrates [30] than to hagfishes.

– D –

decapods (*Decapoda*) A group of marine, freshwater, and semiterrestrial crustaceans [106] in which five of the eight pairs of thoracic appendages function as legs (the other three pairs, called maxillipeds, function as mouthparts). Includes crabs, lobsters, crayfishes, and shrimps.

demosponges (*Demospongiae*) [84] The largest of the three groups of sponges, accounting for 90 percent of all sponge species. Demosponges have spicules made of silica, spongin fiber (a protein), or both.

deuterostomes (*Deuterostomia*) [26] One of the two major groups of bilaterians [21], in which the mouth forms at the opposite end of the embryo from the blastopore in early development (contrast with protostomes). Includes the ambulacrarians [27] and chordates [28].

diatoms (*Bacillariophyta*) [47] Unicellular, photosynthetic stramenopiles [6] with glassy cell walls in two parts.

dinoflagellates (*Dinoflagellata*) [49] A group of alveolates [7] usually possessing two flagella, one in an equatorial groove and the other in a longitudinal groove; many are photosynthetic.

diplomonads (*Diplomonadida*) [72] A group of eukaryotes [4] lacking mitochondria; most have two nuclei, each with four associated flagella.

– E –

ecdysozoans (*Ecdysozoa*) [24] One of the two major groups of protostomes [22], characterized by periodic molting of their exoskeletons. Roundworms [100] and arthropods [25] are the largest ecdysozoan groups.

echinoderms (*Echinodermata*) [108] A major group of marine deuterostomes [26] with fivefold radial symmetry (at some stage of life) and an endoskeleton made of calcified plates and spines. Includes sea stars, crinoids, sea urchins, sea cucumbers, and brittle stars.

ectoprocts (*Ectoprocta*) [88] A group of marine and freshwater lophotrochozoans [23] that live in colonies attached to substrata. Also known as bryozoans or moss animals.

elasmobranchs (*Elasmobranchii*) The largest group of chondrichthyan fishes [114]. Includes sharks, skates, and rays. In contrast to the other group of living chondrichthyans (the chimaeras), they have replaceable teeth.

eudicots (*Eudicotyledones*) A group of angiosperms [52] with pollen grains possessing three openings. Typically with two cotyledons, net-veined leaves, taproots, and floral organs typically in multiples of four or five.

euglenids (*Euglenida*) [69] Flagellate excavates characterized by a pellicle composed of spiraling strips of protein under the plasma membrane; the mitochondria have disk-shaped cristae. Some are photosynthetic.

eukaryotes (*Eukarya*) [4] Organisms made up of one or more complex cells in which the genetic material is contained in nuclei. Contrast with archaeans [3] and bacteria [2].

eumetazoans (*Eumetazoa*) [20] Those animals [19] characterized by body symmetry, a gut, a nervous system, specialized types of cell junctions, and well-organized tissues in distinct cell layers (although there have been secondary losses of some of these characteristics in some eumetazoans).

euphyllophytes (*Euphyllophyta*) This clade is sister to the club mosses [60] and includes all plants with megaphylls.

euryarchaeotes (*Euryachaeota*) [44] A major group of archaeans [3], diagnosed on the basis of rRNA sequences. Includes many methanogens, extreme halophiles, and thermophiles.

eutherians (*Eutheria*) A group of viviparous mammals [119], eutherians are well developed at birth (contrast to prototherians and marsupials, the other two groups of mammals). Most familiar mammals outside the Australian and South American regions are eutherians (see Table 33.1).

excavates (*Excavata*) [14] Diverse group of unicellular, flagellate eukaryotes, many of which possess a feeding groove; some lack mitochondria.

– F –

"ferns" [57] Vascular plants [11] usually possessing large, frond-like leaves that unfold from a "fiddlehead." Not a monophyletic group, although most fern species are encompassed in a monophyletic clade, the leptosporangiate ferns.

flatworms (*Platyhelminthes*) [89] A group of dorsoventrally flattened and generally elongate soft-bodied lophotrochozoans [23]. May be free-living or parasitic, found in marine, freshwater, or damp terrestrial environments. Major flatworm groups include the tapeworms, flukes, monogeneans, and turbellarians.

flowering plants *See* angiosperms.

flukes (*Trematoda*) A group of wormlike parasitic flatworms [89] with complex life cycles that involve several different host species. May be paraphyletic with respect to tapeworms.

foraminiferans (*Foraminifera*) [73] Amoeboid organisms with fine, branched pseudopods that form a food-trapping net. Most produce external shells of calcium carbonate.

fungi (*Fungi*) [17] Eukaryotic heterotrophs with absorptive nutrition based on extracellular digestion; cell walls contain chitin. Major fungal groups include the "chytrids" [81], "zygomycetes" [80], glomeromycetes [79], ascomycetes [78], and basidiomycetes [77].

– G –

gastropods (*Gastropoda*) The largest group of mollusks [95]. Gastropods possess a well-defined head with two or four sensory tentacles (often terminating in eyes) and a ventral foot. Most species have a single coiled or spiraled shell. Common in marine, freshwater, and terrestrial environments.

ginkgo (*Ginkgophyta*) [55] A gymnosperm [13] group with only one living species. The ginkgo seed is surrounded by a fleshy tissue not derived from an ovary wall and hence not a fruit.

glass sponges (*Hexactinellida*) [83] Sponges with a skeleton composed of four- and/or six-pointed spicules made of silica.

glaucophytes (*Glaucophyta*) [67] Unicellular freshwater algae with chloroplasts containing traces of peptidoglycan, the characteristic cell wall material of bacteria.

glomeromycetes (*Glomeromycota*) [79] A group of fungi [17] that form arbuscular mycorrhizae.

gnathostomes (*Gnathostomata*) *See* jawed vertebrates.

gnetophytes (*Gnetophyta*) [54] A gymnosperm [13] group with three very different lineages; all have wood with vessels, unlike other gymnosperms.

green plants (*Viridiplantae*) [9] Organisms with chlorophylls *a* and *b*, cellulose-containing cell walls, starch as a carbohydrate storage product, and chloroplasts surrounded by two membranes.

gymnosperms (*Gymnospermae*) [13] Seed plants [12] with seeds "naked" (i.e., not enclosed in carpels). Probably monophyletic, but status still in doubt. Includes the conifers [53], gnetophytes [54], ginkgo [55], and cycads [56].

– H –

hagfishes (*Myxini*) [112] Elongate, slimy-skinned vertebrates [29] with three small accessory hearts, a partial cranium, and no stomach or paired fins. *See also* craniata; cyclostomes.

haptophytes (*Haptophyta*) [45] Unicellular, photosynthetic stramenopiles [6] with two slightly unequal, smooth flagella. Abundant as phytoplankton, some form marine algal blooms.

hemichordates (*Hemichordata*) [109] One of the two primary groups of ambulacrarians [27]; marine wormlike organisms with a three-part body plan.

heteroloboseans (*Heterolobosea*) [70] Colorless excavates [14] that can transform among amoeboid, flagellate, and encysted stages.

hexapods (*Hexapoda*) [107] Major group of arthropods [25] characterized by a reduction (from the ancestral arthropod condition) to six walking appendages, and the consolidation of three body segments to form a thorax. Includes insects and their relatives (see Table 32.2).

high-GC Gram-positives (*Actinobacteria*) [39] Gram-positive bacteria with a relatively high G+C/A+T ratio of their DNA, with a filamentous growth habit.

hornworts (*Anthocerophyta*) [62] Nonvascular plants with sporophytes that grow from the base. Cells contain a single large, platelike chloroplast.

horsehair worms (*Nematomorpha*) [101] A group of very thin, elongate, wormlike freshwater ecdysozoans [24]. Largely nonfeeding as adults, they are parasites of insects and crayfish as larvae.

horseshoe crabs (*Xiphosura*) Marine chelicerates [104] with a large outer shell in three parts: a carapace, an abdomen, and a tail-like telson. Only five living species remain, but many additional species are known from fossils.

horsetails (*Sphenophyta* or *Equisetophyta*) [58] Vascular plants [11] with reduced megaphylls in whorls.

hydrozoans (*Hydrozoa*) A group of cnidarians [87]. Most species go through both polyp and mesuda stages, although one stage or the other is eliminated in some species.

– I –

insects (*Insecta*) The largest group within the hexapods [107]. Insects are characterized by exposed mouthparts and one pair of antennae containing a sensory receptor called a Johnston's organ. Most have two pairs of wings as adults. There are more described species of insects than all other groups of life [1] combined, and many species remain to be discovered. The major insect groups are described in Table 32.2.

isopods (*Isopoda*) Crustaceans [106] characterized by a compact head, unstalked compound eyes, and mouthparts consisting of four pairs of appendages. Isopods are abundant and widespread in salt, fresh, and brackish water, although some species (the sow bugs) are terrestrial.

– J –

jawed vertebrates (*Gnathostomata*) [30] A major group of vertebrates [29] with jawed mouths. Includes chondrichthyans [114], ray-finned fishes [115], and sarcopterygians [32].

– K –

kinetoplastids (*Kinetoplastida*) [68] Unicellular, flagellate organisms characterized by the presence in their single mitochondrion of a kinetoplast (a structure containing multiple, circular DNA molecules).

kinorhynchs (*Kinorhyncha*) [98] Small (<1 mm) marine ecdysozoans [24] with bodies in 13 segments and a retractable proboscis.

korarchaeotes (*Korarchaeota*) A group of archaeans [3] known only by evidence from nucleic acids derived from hot springs. Its phylogenetic relationships within the Archaea are unknown.

krill (*Euphausiacea*) A group of shrimplike marine crustaceans [106] that are important components of the zooplankton.

– L –

lampreys (*Petromyzontiformes*) [113] Elongate, eel-like vertebrates [29] that often have rasping and sucking disks for mouths.

lancelets (*Cephalochordata*) *See* cephalochordates.

land plants (*Embryophyta*) [10] Plants with embryos that develop within protective structures; sporophytes and gametophytes are multicellular. Land plants possess a cuticle. Major groups are the liverworts [63], hornworts [62], mosses [61], and vascular plants [11].

larvaceans (*Larvacea*) Solitary, planktonic urochordates [110] that retain both notochords and nerve cords throughout their lives.

lepidosaurs (*Lepidosauria*) [121] Reptiles [35] with overlapping scales. Includes tuataras and squamates (lizards, snakes, and amphisbaenians).

life (*Life*) [1] The monophyletic group that includes all known living organisms. Characterized by a nucleic-acid based genetic system (DNA or RNA), metabolism, and cellular structure. Some parasitic forms, such as viruses, have secondarily lost some of these features and rely on the cellular environment of their host.

liverworts (*Hepatophyta*) [63] Nonvascular plants lacking stomata; stalk of sporophyte elongates along its entire length.

loboseans (*Lobosea*) A group of unicellular amoebozoans [76]; includes the most familiar amoebas (e.g., *Amoeba proteus*).

"lophophorates" Not a monophyletic group. A convenience term used to describe several groups of lophotrochozoans [23] that have a feeding structure called a lophophore (a circular or U-shaped ridge around the mouth that bears one or two rows of ciliated, hollow tentacles).

lophotrochozoans (*Lophotrochozoa*) [23] One of the two main groups of protostomes [22]. This group is morphologically diverse, and is supported primarily on information from gene sequences. Includes ectoprocts [88], flatworms [89], rotifers [90], ribbon worms [91], brachiopods [92], phoronids [93], annelids [94], and mollusks [95].

loriciferans (*Loricifera*) [99] Small (< 1 mm) ecdysozoans [24] with bodies in four parts, covered with six plates.

low-GC Gram-positives (*Firmicutes*) [41] A diverse group of bacteria [2] with a relatively low G+C/A+T ratio of their DNA, often but not always Gram-positive, some producing endospores.

lungfishes (*Dipnoi*) [117] A group of aquatic sarcopterygians [32] that are the closest living relatives of the tetrapods [33]. They have a modified swim bladder used to absorb oxygen from air, so some species can survive the temporary drying of their habitat.

– M –

magnoliids Major group of angiosperms [52] possessing two cotyledons and pollen grains with a single opening. The group is defined primarily by nucleotide sequence data; it is more closely related to the eudicots and monocots than to three other small angiosperm groups.

mammals (*Mammalia*) [119] A group of tetrapods [33] with hair covering all or part of their skin; females produce milk to feed their developing young. Includes the prototherians, marsupials, and eutherians.

marsupials (*Marsupialia*) Mammals [119] in which the female typically has a marsupium (a pouch for rearing young, which are born at an extremely early stage in development). Includes such familiar mammals as opossums, koalas, and kangaroos.

metazoans (*Metazoa*) *See* animals.

mollusks (*Mollusca*) [95] One of the major groups of lophotrochozoans [23], mollusks have bodies composed of a foot, a mantle (which often secretes a hard, calcareous shell), and a visceral mass. Includes monoplacophorans, chitons, bivalves, gastropods, and cephalopods.

monocots (*Monocotyledones*) Angiosperms [52] characterized by possession of a single cotyledon, usually parallel leaf veins, a fibrous root system, pollen grains with a single opening, and floral organs usually in multiples of three.

monogeneans (*Monogenea*) A group of ectoparasitic flatworms [89].

monoplacophorans (*Monoplacophora*) Mollusks [95] with segmented body parts and a single, thin, flat, rounded, bilateral shell.

mosses (*Bryophyta*) [61] Nonvascular plants with true stomata and erect, "leafy" gametophytes; sporophytes elongate by apical cell division.

multicellular eukaryotes *See* "protists."

myriapods (*Myriapoda*) [105] Arthropods [25] characterized by an elongate, segmented trunk with many legs. Includes centipedes and millipedes.

– N –

nanoarchaeotes (*Nanoarchaeota*) A hypothetical group of extremely small, thermophilic archaeans [3] with a much-reduced genome. The only described example can survive only when attached to a host organism.

nematodes (*Nematoda*) [100] A very large group of elongate, unsegmented ecdysozoans [24] with thick, multilayer cuticles. They are among the most abundant and diverse animals, although most species have not yet been described. Include free-living predators and scavengers, as well as parasites of most species of land plants [10] and animals [19].

neognaths (*Neognathae*) The main group of birds [122], including all living species except the ratites (ostrich, emu, rheas, kiwis, cassowaries) and tinamous (*see* palaeognaths).

– O –

oligochaetes (*Oligochaeta*) An annelid [94] group whose members lack parapodia, eyes, and anterior tentacles, and have few setae. Earthworms are the most familiar oligochaetes.

onychophorans (*Onychophora*) [102] Elongate, segmented ecdysozoans [24] with many pairs of soft, unjointed, claw-bearing legs. Also known as velvet worms.

oomycetes (*Oomycota*) [48] Water molds and relatives; absorptive heterotrophs with nutrient-absorbing, filamentous hyphae.

opisthokonts (*Opisthokonta*) [18] A group of unikonts [16] in which the flagellum on motile cells, if present, is posterior. The opisthokonts include the fungi [17], animals [19], and choanoflagellates [82].

ostracods (*Ostracoda*) Marine and freshwater crustaceans [106] that are laterally compressed and protected by two clam-like calcareous or chitinous shells.

– P –

palaeognaths (*Palaeognathae*) A group of secondarily flightless or weakly flying birds [122]. Includes the flightless ratites (ostrich, emu, rheas, kiwis, cassowaries) and the weakly flying tinamous.

parabasalids (*Parabasalia*) [71] A group of unicellular eukaryotes [4] that lack mitochondria; they possess flagella in clusters near the anterior of the cell.

phoronids (*Phoronida*) [93] A small group of sessile, wormlike marine lophotrochozoans [23] that secrete chitinous tubes and feed using a lophophore.

placoderms (*Placodermi*) An extinct group of jawed vertebrates [30] that lacked teeth. Placoderms were the dominant predators in Devonian oceans.

Plantae [8] The most broadly defined plant group. In most parts of this book, we use the word "plant" as synonymous with "land plant" [10], a more restrictive definition.

plasmodial slime molds (*Myxogastrida*) Amoebozoans [76] that in their feeding stage consist of a coenocyte called a plasmodium.

pogonophorans (*Pogonophora*) Deep-sea annelids [94] that lack a mouth or digestive tract; they feed by taking up dissolved organic matter, facilitated by endosymbiotic bacteria in a specialized organ (the trophosome).

polychaetes (*Polychaeta*) A group of mostly marine annelids [94] with one or more pairs of eyes and one or more pairs of feeding tentacles; parapodia and setae extend from most body segments. May be paraphyletic with respect to the clitellates.

priapulids (*Priapulida*) [97] A small group of cylindrical, unsegmented, wormlike marine ecdysozoans [24] that takes its name from its phallic appearance.

"progymnosperms" Paraphyletic group of extinct vascular plants [11] that flourished from the Devonian through the Mississippian periods. The first truly woody plants, and the first with vascular cambium that produced both secondary xylem and secondary phloem, they reproduced by spores rather than by seeds.

"prokaryotes" Not a monophyletic group; as commonly used, includes the bacteria [2] and archaeans [3]. A term of convenience encompassing all cellular organisms that are not eukaryotes.

proteobacteria (*Proteobacteria*) [42] A large and extremely diverse group of Gram-negative bacteria that includes many pathogens, nitrogen fixers, and photosynthesizers. Includes the alpha, beta, gamma, delta, and epsilon proteobacteria.

"protists" This term of convenience does not describe a monophyletic group but is used to encompass a large number of distinct and distantly related groups of eukaryotes, many but far from all of which are microbial and unicellular. Essentially a "catch-all" term for any eukaryote group not contained within the land plants [10], fungi [17], or animals [19].

protostomes (*Protostomia*) [22] One of the two major groups of bilaterians [21]. In protostomes, the mouth typically forms from the blastopore (if present) in early development (contrast with deuterostomes). The major protostome groups are the lophotrochozoans [23] and ecdysozoans [24].

prototherians (*Prototheria*) A mostly extinct group of mammals [119], common during the Cretaceous and early Cenozoic. The three living species—the echidnas and the duck-billed platypus—are the only extant egg-laying mammals.

pterobranchs (*Pterobranchia*) A small group of sedentary marine hemichordates [109] that live in tubes secreted by the proboscis. They have one to nine pairs of arms, each bearing long tentacles that capture prey and function in gas exchange.

pteridophytes (*Pteridophyta*) A group of vascular plants [11], sister to the seed plants [12], characterized by overtopping and possession of megaphylls. The pteridophytes include the horsetails [58], whisk ferns [59], and "ferns" [57].

pycnogonids (*Pycnogonida*) Treated in this book as a group of chelicerates [104], but sometimes considered an independent group of arthropods [25]. Pycnogonids have reduced bodies and very long, slender legs. Also called sea spiders.

– R –

radiolarians (*Radiolaria*) [75] Amoeboid organisms with needle-like pseudopods supported by microtubules. Most have glassy internal skeletons.

ray-finned fishes (*Actinopterygii*) [115] A highly diverse group of freshwater and marine bony vertebrates [31]. They have reduced swim bladders that often function as hydrostatic organs and fins supported by soft rays (lepidotrichia). Includes most familiar fishes.

red algae (*Rhodophyta*) [66] Mostly multicellular, marine algae characterized by the presence of phycoerythrin in their chloroplasts.

reptiles (*Reptilia*) [35] One of the two major groups of extant amniotes [34], supported on the basis of similar skull structure and gene sequences. The term "reptiles" traditionally excluded the birds [122], but the resulting group is then clearly paraphyletic. As used in this book, the reptiles include turtles [120], lepidosaurs [121], birds [122], and crocodilians [123].

rhizaria (*Rhizaria*) [15] Mostly amoeboid unicellular eukaryotes with pseudopods, many with external or internal shells. Includes the foraminiferans [73], cercozoans [74], and radiolarians [75].

rhyniophytes (*Rhyniophyta*) A group of early vascular plants [11] that appeared in the Silurian and became extinct in the Middle Devonian. Possessed dichotomously branching stems with terminal sporangia but no true leaves or roots.

ribbon worms (*Nemertea*) [91] A group of unsegmented lophotrochozoans [23] with an eversible proboscis used to capture prey. Mostly marine, but some species live in fresh water or on land.

rotifers (*Rotifera*) [90] Tiny (< 0.5 mm) lophotrochozoans [23] with a pseudocoelomic body cavity that functions as a hydrostatic organ and a ciliated feeding organ called the corona that surrounds the head. They live in freshwater and wet terrestrial habitats.

roundworms (*Nematoda*) [100] *See* nematodes.

– S –

salamanders (*Caudata*) A group of amphibians [118] with distinct tails in both larvae and adults and limbs set at right angles to the body.

salps *See* thaliaceans

sarcopterygians (*Sarcopterygii*) [32] One of the two major groups of bony vertebrates [31], characterized by jointed appendages (paired fins or limbs).

scyphozoans (*Scyphozoa*) Marine cnidarians [87] in which the medusa stage dominates the life cycle. Commonly known as jellyfish.

sea cucumbers (*Holothuroidea*) Echinoderms [108] with an elongate, cucumber-shaped body and leathery skin. They are scavengers on the ocean floor.

sea spiders *See* pycnogonids.

sea squirts *See* ascidians.

sea stars (*Asteroidea*) Echinoderms [108] with five (or more) fleshy "arms" radiating from an indistinct central disk. Also called starfishes.

sea urchins (*Echinoidea*) Echinoderms [108] with a test (shell) that is covered in spines. Most are globular in shape, although some groups (such as the sand dollars) are flattened.

"seed ferns" A paraphyletic group of loosely related, extinct seed plants that flourished in the Devonian and Carboniferous. Characterized by large, frond-like leaves that bore seeds.

seed plants (*Spermatophyta*) [12] Heterosporous vascular plants [11] that produce seeds; most produce wood; branching is axillary (not dichotomous). The major seed plant groups are gymnosperms [13] and angiosperms [52].

sow bugs *See* isopods.

spirochetes (*Spirochaetes*) [37] Motile, Gram-negative bacteria with a helically coiled structure and characterized by axial filaments.

"sponges" A term of convenience used for a paraphyletic group of relatively asymmetric, filter-feeding animals that lack a gut or nervous system and generally lack differentiated tissues. (*See* glass sponges [83], demosponges [84], and calcareous sponges [85].)

springtails (*Collembola*) Wingless hexapods [107] with springing structures on the third and fourth segments of their bodies. Springtails are extremely abundant in some environments (especially in soil, leaf litter, and vegetation).

squamates (*Squamata*) The major group of lepidosaurs [121], characterized by the possession of movable quadrate bones (which allow the upper jaw to move independently of the rest of the skull) and hemipenes (a paired set of eversible penises, or penes) in males. Includes the lizards (a paraphyletic group), snakes, and amphisbaenians.

starfish (*Asteroidea*) *See* sea stars

stramenopiles (*Heterokonta* or *Stramenopila*) [6] Organisms having, at some stage in their life cycle, two unequal flagella, the longer possessing rows of tubular hairs. Chloroplasts, when present, surrounded by four membranes. Major stramenopile groups include the brown algae [46], diatoms [47], and oomycetes [48].

– T –

tapeworms (*Cestoda*) Parasitic flatworms [89] that live in the digestive tracts of vertebrates as adults, and usually in various other species of animals as juveniles.

tardigrades (*Tardigrada*) [103] Small (< 0.5 mm) ecdysozoans [24] with fleshy, unjointed legs and no circulatory or gas exchange organs. They live in marine sands, in temporary freshwater pools, and on the water films of plants. Also called water bears.

tetrapods (*Tetrapoda*) [33] The major group of sarcopterygians [32]; includes the amphibians [118] and the amniotes [34]. Named for the presence of four jointed limbs (although limbs have been secondarily reduced or lost completely in several tetrapod groups).

thaliaceans (*Thaliacea*) A group of solitary or colonial planktonic marine urochordates [110]. Also called salps.

therians (*Theria*) Mammals [119] characterized by viviparity (live birth). Includes eutherians and marsupials.

theropods (*Theropoda*) Archosaurs [36] with bipedal stance, hollow bones, a furcula ("wishbone"), elongated metatarsals with three-fingered feet, and a pelvis that points backwards. Includes many well-known extinct dinosaurs (such as *Tyrannosaurus rex*), as well as the living birds [122].

trilobites (*Trilobita*) An extinct group of arthropods [25] related to the chelicerates [104]. Trilobites flourished from the Cambrian through the Permian.

tuataras (*Rhyncocephalia*) A group of lepidosaurs [121] known mostly from fossils; there are just two living tuatara species. The quadrate bone of the upper jaw is fixed firmly to the skull. Sister group of the squamates.

tunicates *See* ascidians.

turbellarians (*Turbellaria*) A group of free-living, generally carnivorous flatworms [89]. Their monophyly is questionable.

turtles (*Testudines*) [120] A group of reptiles [35] with a bony carapace (upper shell) and plastron (lower shell) that encase the body.

– U –

unikonts (*Unikonta*) [16] A group of eukaryotes [4] whose motile cells possess a single flagellum. Major unikont groups include the amoebozoans [76], fungi [17], and animals [19].

urochordates (*Urochordata*) [110] A group of chordates [28] that are mostly saclike filter feeders as adults, with motile larvae stages that resemble a tadpole.

– V –

vascular plants (*Tracheophyta*) [11] Plants with xylem and phloem. Major groups include the club mosses [60] and euphyllophytes.

vertebrates (*Vertebrata*) [29] The largest group of chordates [28], characterized by a rigid endoskeleton supported by the vertebral column and an anterior skull encasing a brain. Includes hagfishes [112], lampreys [113], and the jawed vertebrates [30], although some biologists exclude the hagfishes from this group (see craniates).

– W –

water bears *See* tardigrades.

whisk ferns (*Psilotophyta*) [59] Vascular plants [11] lacking leaves and roots.

– Y –

"yeasts" A convenience term for several distantly related groups of unicellular fungi [17].

– Z –

"zygomycetes" [80] A convenience term for a paraphyletic group of fungi [17] in which hyphae of differing mating types conjugate to form a zygosporangium.

Appendix B: Some Measurements Used in Biology

MEASURES OF	UNIT	EQUIVALENTS	METRIC → ENGLISH CONVERSION
Length	meter (m)	base unit	1 m = 39.37 inches = 3.28 feet
	kilometer (km)	1 km = 1000 (10^3) m	1 km = 0.62 miles
	centimeter (cm)	1 cm = 0.01 (10^{-2}) m	1 cm = 0.39 inches
	millimeter (mm)	1 mm = 0.1 cm = 10^{-3} m	1 mm = 0.039 inches
	micrometer (μm)	1 μm = 0.001 mm = 10^{-6} m	
	nanometer (nm)	1 nm = 0.001 μm = 10^{-9} m	
Area	square meter (m^2)	base unit	1 m^2 = 1.196 square yards
	hectare (ha)	1 ha = 10,000 m^2	1 ha = 2.47 acres
Volume	liter (L)	base unit	1 L = 1.06 quarts
	milliliter (mL)	1 mL = 0.001 L = 10^{-3} L	1 mL = 0.034 fluid ounces
	microliter (μL)	1 μL = 0.001 mL = 10^{-6} L	
Mass	gram (g)	base unit	1 g = 0.035 ounces
	kilogram (kg)	1 kg = 1000 g	1 kg = 2.20 pounds
	metric ton (mt)	1 mt = 1000 kg	1 mt = 2,200 pounds = 1.10 ton
	milligram (mg)	1 mg = 0.001 g = 10^{-3} g	
	microgram (μg)	1 μg = 0.001 mg = 10^{-6} g	
Temperature	degree Celsius (°C)	base unit	°C = (°F – 32)/1.8
			0°C = 32°F (water freezes)
			100°C = 212°F (water boils)
			20°C = 68°F ("room temperature")
			37°C = 98.6°F (human internal body temperature)
	Kelvin (K)*	°C + 273	0 K = –460°F
Energy	joule (J)		1 J ≈ 0.24 calorie = 0.00024 kilocalorie†

*0 K (–273°C) is "absolute zero," a temperature at which molecular oscillations approach 0—that is, the point at which motion all but stops.

†A *calorie* is the amount of heat necessary to raise the temperature of 1 gram of water 1°C. The *kilocalorie*, or nutritionist's calorie, is what we commonly think of as a calorie in terms of food.

Glossary

- A -

abdomen (ab′ duh mun) [L. *abdomin*: belly] In arthropods, the posterior segments of the body; in mammals, the part of the body containing the intestines and most other internal organs, posterior to the thorax.

abiotic (a′ bye ah tick) [Gk. *a*: not + *bios*: life] Nonliving. (Contrast with biotic.)

abscisic acid (ab sighs′ ik) A plant growth substance having growth-inhibiting action. Causes stomata to close.

abscission (ab sizh′ un) [L. *abscissio*: break off] The process by which leaves, petals, and fruits separate from a plant.

absorption (1) Of light: complete retention, without reflection or transmission. (2) Of liquids: soaking up (taking in through pores or cracks).

absorption spectrum A graph of light absorption versus wavelength of light; shows how much light is absorbed at each wavelength.

absorptive period When there is food in the gut and nutrients are being absorbed.

abyssal zone (uh biss′ ul) [Gk. *abyssos*: bottomless] The deep ocean, below the point that light can penetrate.

accessory pigments Pigments that absorb light and transfer energy to chlorophylls for photosynthesis.

acetylcholine A neurotransmitter substance that carries information across vertebrate neuromuscular junctions and some other synapses.

acetylcholinesterase An enzyme that breaks down acetylcholine.

acetyl coenzyme A (acetyl CoA) Compound that reacts with oxaloacetate to produce citrate at the beginning of the citric acid cycle; a key metabolic intermediate in the formation of many compounds.

acid [L. *acidus*: sharp, sour] A substance that can release a proton in solution. (Contrast with base.)

acid precipitation Precipitation that has a lower pH than normal as a result of acid-forming precursors introduced into the atmosphere by human activities.

acidic Having a pH of less than 7.0 (a hydrogen ion concentration greater than 10^{-7} molar).

acoelomate Lacking a coelom.

Acquired Immune Deficiency Syndrome *See* AIDS.

acrosome (a′ krow soam) [Gk. *akros*: highest + *soma*: body] The structure at the forward tip of an animal sperm which is the first to fuse with the egg membrane and enter the egg cell.

ACTH (adrenocorticotropin) A pituitary hormone that stimulates the adrenal cortex.

actin [Gk. *aktis*: ray] One of the two major proteins of muscle; it makes up the thin filaments. Forms the microfilaments found in most eukaryotic cells.

action potential An impulse in a neuron taking the form of a wave of depolarization or hyperpolarization imposed on a polarized cell surface.

action spectrum A graph of a biological process versus light wavelength; shows which wavelengths are involved in the process.

activating enzymes Enzymes that catalyze the addition of amino acids to their appropriate tRNAs. Also called aminoacyl-tRNA synthetases.

activation energy (E_a) The energy barrier that blocks the tendency for a set of chemical substances to react.

active site The region on the surface of an enzyme where the substrate binds, and where catalysis occurs.

active transport The energy-dependent transport of a substance across a biological membrane against a concentration gradient—that is, from a region of low concentration (of that substance) to a region of high concentration. (See also primary active transport, secondary active transport; contrast with facilitated diffusion.)

adaptation (a dap tay′ shun) (1) In evolutionary biology, a particular structure, physiological process, or behavior that makes an organism better able to survive and reproduce. Also, the evolutionary process that leads to the development or persistence of such a trait. (2) In sensory neurophysiology, a sensory cell's loss of sensitivity as a result of repeated stimulation.

adaptive radiation The proliferation of members of a single clade into a variety of different adaptive forms.

adenine (A) (a′ den een) A nitrogen-containing base found in nucleic acids, ATP, NAD, and other compounds.

adenosine triphosphate *See* ATP.

adenylate cyclase Enzyme catalyzing the formation of cyclic AMP (cAMP) from ATP.

adrenal (a dree′ nal) [L. *ad*: toward + *renes*: kidneys] An endocrine gland located near the kidneys of vertebrates, consisting of two glandular parts, the cortex and medulla.

adrenaline *See* epinephrine.

adrenocorticotropin *See* ACTH.

adsorption Binding of a gas or a solute to the surface of a solid.

aerobic (air oh′ bic) [Gk. *aer*: air + *bios*: life] In the presence of oxygen; requiring oxygen.

afferent (af′ ur unt) [L. *ad*: toward + *ferre*: to carry] Carrying to, as in a neuron that carries impulses to the central nervous system, or a blood vessel that carries blood to a structure. (Contrast with efferent.)

AIDS (acquired immune deficiency syndrome) Condition caused by a virus (HIV) in which the body's helper T lymphocytes are reduced, leaving the victim subject to opportunistic diseases.

alcoholic fermentation Breakdown of glucose in cells under anaerobic conditions to produce alcohol.

aldehyde (al′ duh hide) A compound with a —CHO functional group. Many sugars are aldehydes. (Contrast with ketone.)

aldosterone (al dohs′ ter own) A steroid hormone produced in the adrenal cortex of mammals. Promotes secretion of potassium and reabsorption of sodium in the kidney.

allantois (al lan′ to is) A sac-like extraembryonic membrane that contains nitrogen waste from embryo.

allele (a leel′) [Gk. *allos*: other] The alternate forms of a genetic character found at a given locus on a chromosome.

allele frequency The relative proportion of a particular allele in a specific population.

allergy [Ger. *allergie*: altered reaction] An overreaction to amounts of an antigen that do not affect most people; often involves IgE antibodies.

allometric growth A pattern of growth in which some parts of the body of an organism grow faster than others, resulting in a change in body proportions as the organism grows.

allopatric speciation (al′ lo pat′ rick) [Gk. *allos*: other + *patria*: homeland] The formation of two species from one when reproductive isolation occurs because of the interposition of (or crossing of) a physical geographic barrier such as a river. Also called geographic speciation. (Contrast with parapatric speciation, sympatric speciation.)

allopolyploidy The possession of more than two entire chromosomes sets that are derived from a single species.

allostery (al′ lo steer y) [Gk. *allos*: other + *stereos*: structure] Regulation of the activity of a protein by the binding of an effector molecule at a site other than the active site.

alpha (α) helix A prevalent type of secondary protein structure; a right-handed spiral.

alternation of generations The succession of multicellular haploid and diploid phases in some sexually reproducing organisms, notably plants.

alternative splicing A process for generating different mature mRNAs from a single gene by splicing together different sets of exons during RNA processing.

altruism Behavior that harms the individual who performs it but benefits other individuals.

alveolus (al ve′ o lus) (plural: alveoli) [L. *alveus*: cavity] A small, baglike cavity, especially the blind sacs of the lung.

amensalism (a men′ sul ism) Interaction in which one animal is harmed and the other is unaffected. (Contrast with commensalism, mutualism.)

amine An organic compound with an amino group. (Compare with amino acid.)

amino acid Organic compounds containing both NH_2 and COOH groups. Proteins are polymers of amino acids.

amino acid replacement A change in a protein sequence in which one amino acid is replaced by another.

ammonotelic (am moan′ o teel′ ic) [Gk. *telos*: end] Describes an organism in which the final product of breakdown of nitrogen-containing compounds (primarily proteins) is ammonia. (Contrast with ureotelic, uricotelic.)

amnion (am′ nee on) The fluid-filled sac in which the embryos of reptiles, birds, and mammals develop.

amniote egg A shelled egg surrounding four extraembryonic membranes and embryo-nourishing yolk. This adaptation allowed animals to colonize the terrestrial environment.

amphipathic (am′ fi path′ ic) [Gk. *amphi*: both + *pathos*: emotion] Of a molecule, having both hydrophilic and hydrophobic regions.

amylase (am′ ill ase) Any of a group of enzymes that digest starch.

anabolic reaction A single reaction that participates in anabolism.

anabolism (an ab′ uh liz′ em) [Gk. *ana*: upward + *ballein*: to throw] Synthetic reactions of metabolism, in which complex molecules are formed from simpler ones. (Contrast with catabolism.)

anaerobic (an ur row′ bic) [Gk. *an*: not + *aer*: air + *bios*: life] Occurring without the use of molecular oxygen, O_2.

anagenesis Evolutionary change in a single lineage over time.

analogy (a nal′ o jee) [Gk. *analogia*: resembling] A resemblance in function, and often appearance as well, between two structures that is due to convergent evolution rather than to common ancestry. (Contrast with homology.)

anaphase (an′ a phase) [Gk. *ana*: upward progress] The stage in nuclear division at which the first separation of sister chromatids (or, in the first meiotic division, of paired homologs) occurs.

anaphylactic shock A precipitous drop in blood pressure caused by loss of fluid from capillaries because of an increase in their permeability stimulated by an allergic reaction.

ancestral trait The trait originally present in the ancestor of a given group; may be retained or changed in the descendants of that ancestor.

androgens (an′ dro jens) The male sex steroids.

aneuploidy (an′ you ploy dee) A condition in which one or more chromosomes or pieces of chromosomes are either lacking or present in excess.

angiotensin (an′ jee oh ten′ sin) A peptide hormone that raises blood pressure by causing peripheral vessels to constrict. Also maintains glomerular filtration by constricting efferent vessels and stimulates thirst and the release of aldosterone.

animal hemisphere The metabolically active upper portion of some animal eggs, zygotes, and embryos; does not contain the dense nutrient yolk. (Contrast with vegetal hemisphere.)

anion (an′ eye on) [Gk. *ana*: upward progress] A negatively charged ion. (Contrast with cation.)

anisogamy (an eye sog′ a mee) [Gk. *aniso*: unequal + *gamos*: marriage] The existence of two dissimilar gametes (egg and sperm).

annual Referring to a plant whose life cycle is completed in one growing season. (Contrast with biennial, perennial.)

antenna system In photosynthesis, a group of different molecules that cooperate to absorb light energy and transfer it to a reaction center.

anterior pituitary The portion of the vertebrate pituitary gland that derives from gut epithelium and produces tropic hormones.

anther (an′ thur) [Gk. *anthos*: flower] A pollen-bearing portion of the stamen of a flower.

antheridium (an′ thur id′ ee um) [Gk. *antheros*: blooming] The multicellular structure that produces the sperm in nonvascular plants and ferns.

antibody One of the millions of proteins produced by the immune system that specifically binds to a foreign substance and initiates its removal from the body.

anticodon The three nucleotides in transfer RNA that pair with a complementary triplet (a codon) in messenger RNA.

antidiuretic hormone (ADH) A hormone that promotes water reabsorption by the kidney. ADH is produced by neurons in the hypothalamus and released from nerve terminals in the posterior pituitary. Also called vasopressin.

antigen (an′ ti jun) Any substance that stimulates the production of an antibody or antibodies in the body of a vertebrate.

antigenic determinant A specific region of an antigen, which is recognized by and binds to a specific antibody.

antiparallel Pertaining to molecular orientation in which a molecule or parts of a molecule have opposing directions.

antipodal cell At one end of the megagametophyte, one of the three cells which eventually degenerate.

antiport A membrane transport process that carries one substance in one direction and another in the opposite direction. (Contrast with symport.)

antisense nucleic acid A single-stranded RNA or DNA complementary to and thus targeted against the mRNA transcribed from a harmful gene such as an oncogene.

anus (a′ nus) Opening through which digestive wastes are expelled, located at the posterior end of the gut.

aorta (a or′ tah) [Gk. *aorte*: aorta] The main trunk of the arteries leading to the systemic (as opposed to the pulmonary) circulation.

aortic body A nodule of modified muscle tissue on the aorta that is sensitive to the oxygen supply provided by arterial blood.

apex (a′ pecks) The tip or highest point of a structure, as the apex of a growing stem or root.

apical (a′ pi kul) Pertaining to the apex, or tip, usually in reference to plants.

apical dominance Inhibition by the apical bud of the growth of axillary buds.

apical meristem The meristem at the tip of a shoot or root; responsible for the plant's primary growth.

apomixis (ap oh mix′ is) [Gk. *apo*: away from + *mixis*: sexual intercourse] The asexual production of seeds.

apoplast (ap′ oh plast) in plants, the continuous meshwork of cell walls and extracellular spaces through which material can pass without crossing a plasma membrane. (Contrast with symplast.)

apoptosis (ap uh toh′ sis) A series of genetically programmed events leading to cell death.

aquaporin A transport protein in plant and animal cells through which water passes in osmosis.

aquatic (a kwa′ tic) [L. *aqua*: water] Living in water. (Compare with marine, terrestrial.)

aqueous (a′ kwee us) Pertaining to water or a watery solution.

archegonium (ar′ ke go′ nee um) [Gk. *archegonos*: first, foremost] The multicellular structure that produces eggs in nonvascular plants, ferns, and gymnosperms.

archenteron (ark en′ ter on) [Gk. *archos*: first + *enteron*: bowel] The earliest primordial animal digestive tract.

area phylogenies Phylogenies in which the names of the taxa are replaced with the names of the places where those taxa live or lived.

arteriosclerosis *See* atherosclerosis.

artery A muscular blood vessel carrying oxygenated blood away from the heart to other parts of the body. (Contrast with vein.)

artficial selection The selection by plant and animal breeders of individuals with certain desirable traits.

ascus (ass′ cuss) [Gk. *askos*: bladder] In ascomycete fungi (sac fungi), the club-shaped sporangium within which spores (ascospores) are produced by meiosis.

asexual Without sex.

assortative mating A breeding system in which mates are selected on the basis of a particular trait or group of traits.

atherosclerosis (ath′ er oh sklair oh′ sis) [Gk. *athero*: gruel, porridge + *skleros*: hard] A disease of the lining of the arteries characterized by fatty, cholesterol-rich deposits in the walls of the arteries. When fibroblasts infiltrate these deposits and calcium precipitates in them, the disease become arteriosclerosis, or"hardening of the arteries."

atmosphere The gaseous mass surrounding our planet. Also a unit of pressure, equal to the normal pressure of air at sea level.

atom [Gk. *atomos*: indivisible] The smallest unit of a chemical element. Consists of a nucleus and one or more electrons.

atomic mass The average mass of an atom of an element; the average depends on the relative amounts of ifferent isotopes of the element on Earth. Also called atomic weight.

atomic number The number of protons in the nucleus of an atom; also equals the number of electrons around the neutral atom. Determines the chemical properties of the atom.

ATP (adenosine triphosphate) An energy-storage compound containing adenine, ribose, and three phosphate groups. When it is formed from ADP, useful energy is stored; when it is broken down (to ADP or AMP), energy is released to drive endergonic reactions.

ATP synthase An integral membrane protein that couples the transport of proteins with the formation of ATP.

atrioventricular node A modified node of cardiac muscle that organizes the action potentials that control contraction of the ventricles.

atrium (a' tree um) [L. *atrium*: central hall] An internal chamber. In the hearts of vertebrates, the thin-walled chamber(s) entered by blood on its way to the ventricle(s). Also, the outer ear.

autocrine Referring to a cell signaling mechanism in which the signal binds to and affects the cell that makes it. An autocrine hormone, for example, is one that influences the cell that releases it. (Compare with endocrine gland;, paracrine)

autoimmune disease A disorder in which the immune system attacks the animal's own antigens.

autonomic nervous system The system that controls such involuntary functions as those of guts and glands. (Compare with central nervous system.)

autopolyploidy The possession of more than two chromosome sets that are derived from more than one species.

autosome Any chromosome (in a eukaryote) other than a sex chromosome.

autotroph (au' tow trow' fik) [Gk. *autos:* self + *trophe*: food] An organism that is capable of living exclusively on inorganic materials, water, and some energy source such as sunlight or chemically reduced matter. (Contrast with heterotroph.)

auxin (awk' sin) [Gk. *auxein*: to grow] In plants, a substance (the most common being indoleacetic acid) that regulates growth and various aspects of development.

auxotroph (awks' o trofe) [Gk. *auxein*: to grow + *trophe*: food] A mutant form of an organism that requires a nutrient or nutrients not usually required by the wild type. (Contrast with prototroph.)

Avogadro's number The number of atoms or molecules in a mole (weighed out in grams) of a substance, calculated to be 6.022×10^{23}.

***Avr* genes (avirulence genes)** Genes in a pathogen that may trigger defenses in plants. *See* gene-for-gene resistance.

axillary bud A bud occurring in the upper angle (axil) between a leaf and stem.

axon [Gk. *axon*: axle] The part of a neuron that conducts action potentials away from the cell body.

axon hillock The junction between an axon and its cell body, where action potentials are generated.

axon terminals The endings of an axon; they form synapses and release neurotransmitter.

axoneme (ax' oh neem) The complex of microtubules and their crossbridges that forms the motile apparatus of a cilium.

- B -

bacillus (bah sil' us) [L: little rod] Any of various rod-shaped bacteria.

bacterial artificial chromosome (BAC) A DNA cloning vector used in bacteria that can carry up to 150,000 base pairs of foreign DNA.

bacteriophage (bak teer' ee o fayj) [Gk. *bakterion*: little rod + *phagein*: to eat] One of a group of viruses that infect bacteria and ultimately cause their disintegration.

bacteroids Nitrogen-fixing organelles that develop from endosymbiotic bacteria.

balanced polymorphism [Gk. *polymorphos*: many forms] The maintenance of more than one form, or the maintenance at a given locus of more than one allele, at frequencies of greater than 1 percent in a population. Often results when heterozygotes are more fit than either homozygote.

bark All tissues outside the vascular cambium of a plant.

baroreceptor [Gk. *baros*: weight] A pressure-sensing cell or organ.

Barr body In female mammals, an inactivated X chromosome.

basal Pertaining to one end—the base—of an axis.

basal body Centriole found at the base of a eukaryotic flagellum or cilium.

basal metabolic rate (BMR) The minimum rate of energy turnover in an awake (but resting) bird or mammal that is not expending energy for thermoregulation.

base (1) A substance tha can accept a hydrogen ion in solution. (Contrast with acid.) (2) In nucleic acids, the purine or pyrimidine that is attached to each sugar in the backbone.

base pairing *See* complementary base pairing.

basic Having a pH greater than 7.0 (i.e., having a hydrogen ion concentration lower than 10^{-7} molar).

basidium (bass id' ee yum) In basidiomycete fungi, the characteristic sporangium in which four spores are formed by meiosis and then borne externally before being shed.

basophils One type of phagocytic white blood cell that releases histamine and may promote T cell development.

Batesian mimicry The convergence in appearance of an edible species (mimic) with an unpalatable species (model).

B cell A type of lymphocyte involved in the humoral immune response of vertebrates. Upon recognizing an antigenic determinant, a B cell develops into a plasma cell, which secretes an antibody. (Contrast with T cell.)

benefit An improvement in survival and reproductive success resulting from performing a behavior or having a trait. (Contrast with cost.)

benign (be nine') Referring to a tumor that grows to a certain size and then stops, uaually with a fibrous capsule surrounding the mass of cells. Benign tumors do not spread (metastasize) to other organs. (Contrast with malignant.)

benthic zone [Gk. *benthos*: bottom] The bottom of the ocean.

beta (β) pleated sheet Type of protein secondary structure; results from hydrogen bonding between polypeptide regions running antiparallel to each other.

biennial Referring to a plant whose life cycle includes vegetative growth in the first year and flowering and senescence in the second year. (Contrast with annual, perennial.)

bilateral symmetry The condition in which only the right and left sides of an organism, divided exactly down the back, are mirror images of each other. (Contrast with biradial symmetry.)

bilayer In membranes, a structure that is two lipid layers in thickness.

bile A secretion of the liver delivered to the small intestine via the common bile duct. In the intestine, bile emulsifies fats.

binary fission Reproduction by cell division of a single-celled organism.

binding domain The region of a receptor molecule where its ligand attaches.

binocular cells Neurons in the visual cortex that respond to input from both retinas; involved in depth perception.

binomial (bye nome' ee al) Consisting of two; for example, the binomial nomenclature of biology in which each species has two names (the genus name followed by the species name).

biodiversity crisis The current high rate of loss of species, caused primarily by human activities.

biofilm A community of microorganisms embedded in a polysaccharide matrix, forming a highly resistant coating on almost any moist surface.

biogeochemical cycle Movement of elements through living organisms and the physical environment.

biogeographic region One of several defined, continental-scale regions of Earth, each of which has a biota distinct from that of the others. (Contrast with biome.)

bioinformatics The use of computers and/or mathematics to analyze complex biological information, such as DNA sequences.

bioluminescence The production of light by biochemical processes in an organism.

biomass The total weight of all the living organisms, or some designated group of living organisms, in a given area.

biome (bye' ome) A major division of the ecological communities of Earth, characterized primarily by distinctive vegetation. A given biogeographic region contains many different biomes.

biosphere (bye' oh sphere) All regions of Earth (terrestrial and aquatic) and Earth's atmosphere in which organisms can live.

biota (bye oh' tah) All of the organisms—animals, plants, fungi, and microorganisms—found in a given area. (Contrast with flora, fauna.)

biotic (bye ah' tick) [Gk. *bios*: life] Alive. (Contrast with abiotic.)

biradial symmetry Radial symmetry modified so that only two planes can divide the animal into similar halves.

blastocoel (blass' toe seal) [Gk. *blastos*: sprout + *koilos*: hollow] The central, hollow cavity of a blastula.

blastocyst (blass' toe cist) An early embryo formed by the first divisions of the fertilized egg (zygote). In mammals, a hollow ball of cells.

blastodisc (blass' toe disk) A disk of cells forming on the surface of a large yolk mass, comparable to a blastula, but occurring in animals such as birds and reptiles, in which the massive yolk restricts cleavage to one side of the egg only.

blastomere A cell produced by the division of a fertilized egg.

blastopore The opening from the archenteron to the exterior of a gastrula.

blastula (blass' chu luh) An early stage in animal embryology; in many species, a hollow sphere of cells surrounding a central cavity, the blastocoel. (Contrast with blastodisc.)

blood–brain barrier A property of the blood vessels of the brain that prevents most chemicals from diffusing from the blood into the brain.

blood plasma (plaz' muh) [Gk. *plassein*: to mold] The liquid portion of blood, in which blood cells and other particulates are suspended.

blue light receptors Pigments in plants that absorbs blue light (400–500 nm). These pigments mediate many plant responses including phototropism, stomatal movements, and expression of some genes.

body cavity Membrane-lined, fluid-filled compartment that lies between the cell layers of many animals.

Bohr effect The fact that low pH decreases the affinity of hemoglobin for oxygen.

bond *See* chemical bond.

bottleneck A stressful period during which only a few individuals of a once large population survive.

Bowman's capsule An elaboration of kidney tubule cells that surrounds a know of capillaries (the glomerulus). Blood is filtered across the walls of these capillaries and the filtrate is collected into Bowman's capsule.

brain stem The portion of the vertebrate brain between the spinal cord and the forebrain.

brassinosteroids Plant steroid hormones that mediate light effects promoting the elongation of stems and pollen tubes.

bronchioles The smallest airways in a vertebrate lung, branching off the bronchi.

bronchus (plural: bronchi) The major airway(s) branching off the trachea into the vertebrate lung.

brown fat Fat tissue in mammals that is specialized to produce heat. It has many mitochondria and capillaries, and a protein that uncouples oxidative phosphorylation.

browser An animal that feeds on the tissues of woody plants.

budding Asexual reproduction in which a more or less complete new organism simply grows from the body of the parent organism and eventually detaches itself.

buffer A substance that can transiently accept or release hydrogen ions and thereby resist changes in pH.

bundle of His Fibers of modified cardiac muscle that conduct action potentials from the atria to the ventricular muscle mass.

bundle sheath cell Part of a tissue that surrounds the veins of plants; contains chloroplasts in C_4 plants.

- C -

C_3 photosynthesis Form of photosynthesis in which 3-phosphoglycerate is the first stable product, and ribulose bisphosphate is the CO_2 receptor.

C_4 photosynthesis Form of photosynthesis in which oxaloacetate is the first stable product, and phosphoenolpyruvate is the CO_2 acceptor. C_4 plants also perform the reactions of C_3 photosynthesis.

calcitonin A hormone produced by the thyroid gland; it lowers blood calcium and promotes bone formation. (Compare with parathyroid hormone.)

calmodulin (cal mod' joo lin) A calcium-binding protein found in all animal and plant cells; mediates many calcium-regulated processes.

calorie [L. *calor*: heat] The amount of heat required to raise the temperature of one gram of water by one degree Celsius (1°C) from 14.5°C to 15.5°C. Calorie spelled with a capital C refers to the kilocalorie (1 kcal = 1,000 cal).

Calvin cycle The stage of photosynthesis in which CO_2 reacts with RuBP to form 3PG, 3PG is reduced to a sugar, and RuBP is regenerated, while other products are released to the rest of the plant. Also known as the Calvin–Benson cycle.

calyx (kay' licks) [Gk. *kalyx*: cup] All of the sepals of a flower, collectively.

CAM *See* crassulacean acid metabolism.

cambium (kam' bee um) [L. *cambiare*: to exchange] A meristem that gives rise to radial rows of cells in stem and root, increasing them in girth; commonly applied to the vascular cambium which produces wood and phloem, and the cork cambium, which produces bark.

Cambrian explosion The rapid diversification of life that took place during the Cambrian period.

cAMP (cyclic AMP) A compound formed from ATP that mediates the effects of numerous animal hormones.

canopy The leaf-bearing part of a tree. Collectively, the aggregate of the leaves and branches of the larger woody plants of an ecological community.

capillaries [L. *capillaris*: hair] Very small tubes, especially the smallest blood-carrying vessels of animals between the termination of the arteries and the beginnings of the veins. Capillaries are the site of exchange of materials between the blood and the interstitial fluid.

capsid The outer shell of a virus that encloses its nucleic acid.

carbohydrates Organic compounds containing carbon, hydrogen, and oxygen in the ratio 1:2:1 (i.e., with the general formula $C_nH_{2n}O_n$). Common examples are sugars, starch, and cellulose.

carboxylase An enzyme that catalyzes the addition of carboxyl groups to a substrate.

carboxylic acid (kar box sill' ik) An organic acid containing the carboxyl group, —COOH, which dissociates to the carboxylate ion, $—COO^-$.

carcinogen (car sin' oh jen) A substance that causes cancer.

cardiac (kar' dee ak) [Gk. *kardia*: heart] Pertaining to the heart and its functions.

cardiac muscle One of the three types of muscle tissue, it makes up the vertebrate heart. Characterized by branching cells with single nuclei and striated (striped) appearance. (Contrast with smooth muscle, striated muscle.)

carnivore [L. *carn*: flesh + *vovare*: to devour] An organism that eats animal tissues. (Contrast with detritivore, herbivore, omnivore.)

carotenoid (ka rah' tuh noid) A yellow, orange, or red lipid pigment commonly found as an accessory pigment in photosynthesis; also found in fungi.

carpel (kar' pel) [Gk. *karpos*: fruit] The organ of the flower that contains one or more ovules.

carrier (1) In facilitated diffusion, a membrane protein that binds a specific molecule and transports it through the membrane. (2) In respiratory and photosynthetic electron transport, a participating substance such as NAD that exists in both oxidized and reduced forms. (3) In genetics, a person heterozygous for a recessive trait.

carrier protein *See* Carrier (1).

cartilage In vertebrates, a tough connective tissue found in joints, the outer ear, and elsewhere. Forms the entire skeleton in some animal groups.

Casparian strip A band of cell wall containing suberin and lignin, found in the endodermis. Restricts the movement of water across the endodermis.

caspase A member of a group of proteases that catalyze cleavage of target proteins and are active in apoptosis.

catabolic reaction A single reaction that participates in catabolism.

catabolism [Gk. *kata*: to break down + *ballein*: to throw] Degradational reactions of metabolism, in which complex molecules are broken down. (Contrast with anabolism.)

catabolite repression In the presence of abundant glucose, the diminished synthesis of catabolic enzymes for other energy sources.

catalyst (cat' a list) [Gk. *kata*: to break down] A chemical substance that accelerates a reaction without itself being consumed in the overall course of the reaction. Catalysts lower the activation energy of a reaction. Enzymes are biological catalysts.

catalytic subunit The polypeptide in an enzyme protein with quaternary structure that contains the active site for the enzyme. (Contrast regulatory subunit.)

cation (cat' eye on) An ion with one or more positive charges. (Contrast with anion.)

caudal [L. *cauda*: tail] Pertaining to the tail, or to the posterior part of the body.

cDNA *See* complementary DNA.

cecum (see′ cum) [L. *caecus*: blind] A blind branch off the large intestine. In many nonruminant mammals, the cecum contains a colony of microorganisms that contribute to the digestion of food.

cell adhesion molecules Molecules on animal cell surfaces that affect the selective association of cells during development of the embryo.

cell cycle The stages through which a cell passes between one division and the next. Includes all stages of interphase and mitosis.

cell division The reproduction of a cell to produce two new cells. In eukaryotes, this process involves nuclear division (mitosis) and cytoplasmic division (cytokinesis).

cell junctions Specialized structures associated with the plasma membranes of epithelial cells. Some contribute to cell adhesion, others to intercellular communication.

cell plate A structure that forms at the equator of the spindle in dividing plant cells.

cell recognition Binding of cells to one another mediated by membrane proteins or carbohydrates.

cell wall A relatively rigid structure that encloses cells of plants, fungi, many protists, and most prokaryotes. Gives these cells their shape and limits their expansion in hypotonic media.

cellular immune response Action of the immune system based on the activities of T cells. Directed against parasites, fungi, intracellular viruses, and foreign tissues (grafts). (Contrast with humoral immune system.)

cellular respiration *See* respiration.

cellulose (sell′ you lowss) A straight-chain polymer of glucose molecules, used by plants as a structural supporting material.

centimorgan In genetic mapping, a recombinant frequency of 0.01; a map unit.

central dogma The statement that information flows from DNA to RNA to polypeptide (in retroviruses, there is also information flow from RNA to cDNA).

central nervous system That part of the nervous system which is condensed and centrally located, e.g., the brain and spinal cord of vertebrates; the chain of cerebral, thoracic and abdominal ganglia of arthropods. (Compare with autonomic nervous system.)

centrifuge [L. *centrum*: center + *fugere*: to flee] A laboratory device in which a sample is spun around a central axis at high speed. Used to separate suspended materials of different densities.

centriole (sen′ tree ole) A paired organelle that helps organize the microtubules in animal and protist cells during nuclear division.

centromere (sen′ tro meer) [Gk. *centron*: center + *meros*: part] The region where sister chromatids join.

centrosome (sen′ tro soam) The major microtubule organizing center of an animal cell.

cephalization (sef ah luh zay′ shun) [Gk. *kephale*: head] The evolutionary trend toward increasing concentration of brain and sensory organs at the anterior end of the animal.

cerebellum (sair uh bell′ um) [L.: diminutive of *cerebrum*, brain] The brain region that controls muscular coordination; located at the anterior end of the hindbrain.

cerebral cortex The thin layer of gray matter (neuronal cell bodies) that overlays the cerebrum.

cerebrum (su ree′ brum) [L. *cerebrum*: brain] The dorsal anterior portion of the forebrain, making up the largest part of the brain of mammals. In mammals, the chief coordination center of the nervous system; consists of two cerebral hemispheres.

cervix (sir′ vix) [L. *cervix*: neck] The opening of the uterus into the vagina.

cGMP (cyclic guanosine monophosphate) An intracellular messenger that is part of signal transmission pathways involving G proteins. (*See* G protein.)

channel protein A membrane protein that forms an aqueous passageway though which specific solutes may pass.

chemical bond An attractive force stably linking two atoms.

chemical equilibrium A state reached in a reversible chemical reaction when the forward and reverse reactions balance each other and there is no net change.

chemical evolution The theory that life originated through the chemical transformation of inanimate substances.

chemical reaction The change in the composition or distribution of atoms of a substance with consequent alterations in properties.

chemiosmosis The formation of ATP in mitochondria and chloroplasts, resulting from a pumping of protons across a membrane (against a gradient of electrical charge and of pH), followed by the return of the protons through a protein channel with ATPase activity.

chemoautotroph *See* chemolithotroph.

chemoheterotroph An organism that must obtain both carbon and energy from organic substances. (Contrast with chemolithotroph, photoautotroph, photoheterotroph.)

chemolithotroph [Gk. *lithos*: stone, rock] An organism that uses carbon dioxide as a carbon source and obtains energy by oxidizing inorganic substances from its environment. (Contrast with chemoheterotroph, photoautotroph, photoheterotroph.)

chemoreceptor A sensory receptor cell or tissue that senses specific substances in its environment.

chemosynthesis Synthesis of food substances, using the oxidation of reduced materials from the environment as a source of energy.

chiasma (kie az′ muh) (plural: chiasmata) [Gk. *chiasmata*: cross] An X-shaped connection between paired homologous chromosomes in prophase I of meiosis. A chiasma is the visible manifestation of crossing over between homologous chromosomes.

chitin (kye′ tin) [Gk. *kiton*: tunic] The characteristic tough but flexible organic component of the exoskeleton of arthropods, consisting of a complex, nitrogen-containing polysaccharide. Also found in cell walls of fungi.

chlorophyll (klor′ o fill) [Gk. *kloros*: green + *phyllon*: leaf] Any of a few green pigments associated with chloroplasts or with certain bacterial membranes; responsible for trapping light energy for photosynthesis.

chloroplast [Gk. *kloros*: green + *plast*: a particle] An organelle bounded by a double membrane containing the enzymes and pigments that perform photosynthesis. Chloroplasts occur only in eukaryotes.

choanocyte (ko′ an uh site) The collared, flagellated feeding cells of sponges.

cholecystokinin (ko′ luh sis tuh kai′ nin) A hormone produced and released by the lining of the duodenum when it is stimulated by undigested fats and proteins. It stimulates the gallbladder to release bile and slows stomach activity.

chorion (kor′ ee on) [Gk. *khorion*: afterbirth] The outermost of the membranes protecting mammal, bird, and reptile embryos; in mammals it forms part of the placenta.

chromatid (kro′ ma tid) Each of a pair of new sister chromosomes from the time at which the molecular duplication occurs until the time at which the centromeres separate at the anaphase of nuclear division.

chromatin The nucleic acid–protein complex found in eukaryotic chromosomes.

chromatin remodeling Changes in chromatin structure to allow transcription, translation, and chromosome condensation.

chromatophore (krow mat′ o for) [Gk. *kroma*: color + *phoreus*: carrier] A pigment-bearing cell that expands or contracts to change the color of the organism.

chromosomal mutation Loss of or changes in position/direction of a DNA segment on a chromosome.

chromosome (krome′ o sowm) [Gk. *kroma*: color + *soma*: body] In bacteria and viruses, the DNA molecule that contains most or all of the genetic information of the cell or virus. In eukaryotes, a structure composed of DNA and proteins that bears part of the genetic information of the cell.

chylomicron (ky low my′ cron) Particles of lipid coated with protein, produced in the gut from dietary fats and secreted into the extracellular fluids.

chyme (kime) [Gk. kymus, juice] Created in the stomach; a mixture of ingested food with the digestive juices secreted by the salivary glands and the stomach lining.

cilium (sil′ ee um) (plural: cilia) [L.: eyelash] Hairlike organelle used for locomotion by many unicellular organisms and for moving water and mucus by many multicellular organisms. Generally shorter than a flagellum.

circadian rhythm (sir kade′ ee an) [L. *circa*: approximately + *dies*: day] A rhythm in behavior, growth, or some other activity that recurs about every 24 hours under constant conditions.

circannual rhythm [L. *circa*: approximately + *annus*: year) A rhythm of behavior, growth, or some other activity that recurs on a yearly basis.

citric acid cycle A set of chemical reactions in cellular respiration, in which acetyl CoA is oxidized to carbon dioxide, and hydrogen atoms are stored as NADH and $FADH_2$. Also called the Krebs cycle.

clade [Gk. *klados*: branch] A monophyletic group made up of an ancestor and all of its descendants.

class I MHC molecules These cell surface proteins participate in the cellular immune response directed against virus-infected cells.

class II MHC molecules These cell surface proteins participate in the cell-cell interactions (of helper T cells, macrophages, and B cells) of the humoral immune response.

class switching The process whereby a plasma cell changes the class of immunoglobulin that it synthesizes by changing the DNA region coding for the C segment.

clathrin A fibrous protein on the inner surfaces of animal cell membranes that strengthens coated vesicles and thus participates in receptor-mediated endocytosis.

cleavage First divisions of the fertilized egg of an animal.

climate The average of the atmospheric conditions (temperature, precipitation, wind direction and velocity) found in a region over time.

cline A gradual change in the traits of a species over a geographical gradient.

cloaca (klo ay' kuh) [L. *cloaca*: sewer] In some invertebrates, the posterior part of the gut; in many vertebrates, a cavity receiving material from the digestive, reproductive, and excretory systems.

clonal anergy Prevention of the synthesis of antibodies against the body's own antigens. When a T cell binds to a self-antigen, it does not receive signals from an antigen-presenting cell; thus the T cell dies (becomes anergic) rather than yielding a clone of active cells.

clonal deletion The inactivation or destruction of lymphocyte clones that would produce immune reactions against the animal's own body.

clonal selection The mechanism by which exposure to antigen results in the activation of selected T- or B-cell clones, resulting in an immune response.

clone [Gk. *klon*: twig, shoot] Genetically identical cells or organisms produced from a common ancestor by asexual means.

cnidocytes (nye' duh sites) The feeding cells of cnidarians, within which nematocysts are housed.

coacervate (ko as' er vate) [L. *coacervare*: to heap up] An aggregate of colloidal particles in suspension.

coated vesicle Cytoplasmic vesicle containing distinctive proteins, including clathrin.

coccus (kock' us) [Gk. *kokkos*: berry, pit] Any of various spherical or spheroidal bacteria.

cochlea (kock' lee uh) [Gk. *kokhlos* snail] A spiral tube in the inner ear of vertebrates; it contains the sensory cells involved in hearing.

codominance A condition in which two alleles at a locus produce different phenotypic effects and both effects appear in heterozygotes.

codon Three nucleotides in messenger RNA that direct the placement of a particular amino acid into a polypeptide chain. (Contrast with anticodon.)

coelom (see' lum) [Gk. *koiloma*: cavity] The body cavity of certain animals; the coelom is lined with cells of mesodermal origin.

coelomate Having a coelom.

coenocyte (seen' a sight) [Gk. *koinos*: common + *kytos*: container] A "cell" enclosed by a single plasma membrane but containing many nuclei.

coenzyme A nonprotein organic molecule that plays a role in catalysis by an enzyme.

cofactor An inorganic ion that is weakly bound to an enzyme and required for its activity.

cohesin Proteins involved in binding chromatids together.

cohesion The tendency of molecules (or any substances) to stick together.

cohort (co' hort) [L. *cohors*: company of soldiers] A group of similar-aged organisms, considered as it passes through time.

coleoptile A sheath that surrounds and protects the shoot apical meristem and young primary leaves of a grass seedling as they move through the soil.

collagen [Gk. *kolla*: glue] A fibrous protein found extensively in bone and connective tissue.

collecting duct In vertebrates, a tubule that receives urine produced in the nephrons of the kidney and delivers that fluid to the ureter for excretion.

collenchyma (cull eng' kyma) [Gk. *kolla*: glue + *enchyma*: infusion] A type of plant cell, living at functional maturity, which lends flexible support by virtue of primary cell walls thickened at the corners. (Contrast with parenchyma, sclerenchyma.)

colon [Gk. *kolon*] The large intestine.

common bile duct A single duct that delivers bile from the gallbladder and secretions from the pancreas into the small intestine.

communication A signal from one organism (or cell) that alters the functioning or behavior of another organism (or cell).

community Any ecologically integrated group of species of microorganisms, plants, and animals inhabiting a given area.

companion cell A specialized cell found adjacent to a sieve tube element in flowering plants.

comparative genomics Computer-aided comparison of DNA sequences between different organisms to reveal genes with related functions.

comparative method An experimental design in which two samples or populations exposed to different conditions or treatments are compared to each other.

compensation point The light intensity at which the rates of photosynthesis and of cellular respiration are equal.

competition In ecology, use of the same resource by two or more species, when the resource is present in insufficient supply for the combined needs of the species.

competitive exclusion A result of competition between species for a limiting resource in which one species completely eliminates the other.

competitive inhibitor A nonsubstrate that binds to the active site of an enzyme and thereby inhibits binding of substrate and reaction from part of the environment.

complement system A group of eleven proteins that play a role in some reactions of the immune system. The complement proteins are not immunoglobulins.

complementary base pairing The AT (or AU), TA (or UA), CG, and GC pairing of bases in double-stranded DNA, in transcription, and between tRNA and mRNA.

complementary DNA (cDNA) DNA formed by reverse transcriptase acting with an RNA template; essential intermediate in the reproduction of retroviruses; used as a tool in recombinant DNA technology; lacks introns.

complete metamorphosis A change of state during the life cycle of an organism in which the body is almost completely rebuilt to produce an individual with a very different body form. Characteristic of insects such as butterflies, moths, beetles, ants, wasps, and flies.

compound (1) A substance made up of atoms of more than one element. (2) A structure made up of many units, as the compound eyes of arthropods.

concerted evolution The common evolution of a family of repeated genes, such that changes in one copy of the gene family are replicated in other copies of the gene family.

condensation reaction A reaction in which two molecules become connected by a covalent bond and a molecule of water is released. ($AH + BOH \rightarrow AB + H_2O$.)

condensing A protein complex involved in chromosome condensation during mitosis and meiosis.

conditional mutations Mutations that show characteristic phenotype only under certain environmental conditions such as temperature.

conduction The transfer of heat from one object to another through direct contact. In neurophysiology, the progression of an action potential along an axon.

cone cells (1) In the vertebrate retina, photoreceptor cells that are responsible for color vision. (2) In gymnosperms, reproductive structures consisting of spore-bearing scales inserted on a short axis; the scales are modified branches. (Contrast with strobilus.)

conformation The three-dimensional shape of a protein or other macromolecule.

conidium (ko nid' ee um) [Gk. *konis*: dust] An asexual fungus spore borne singly or in chains either apically or laterally on a hypha.

conifer (kahn' e fer) [Gr. *konos*: cone + *phero*: carry] One of the cone-bearing gymnosperms, mostly trees, such as pines and firs.

conjugation (kon ju gay' shun) [L. *conjugare*: yoke together] The close approximation of two cells during which they exchange genetic material, as in *Paramecium* and other ciliates, or during which DNA passes from one to the other, as in bacteria.

conjugation tube Cytoplasmic connection between two bacterial cells through which DNA passes during conjugation.

connective tissue An animal tissue that connects or surrounds other tissues; its cells are embedded in a collagen-containing matrix.

connexon In a gap junction, a protein channel linking adjacent animal cells.

consensus sequences Short stretches of DNA that appear, with little variation, in many different genes.

conservation biology An applied, normative science that carries out investigations designed to help preserve the diversity of life on Earth.

constant region For a particular class of immunoglobulin molecules, the region with identical amino acid composition.

constitutive enzyme An enzyme that is present in approximately constant amounts in a system, whether its substrates are present or absent. (Contrast with inducible enzyme.)

consumer An organism that eats the tissues of some other organism.

continental drift The gradual movements of the world's continents that have occurred over billions of years.

contraception The act of preventing union of sperm and egg.

controlled experiment An experimental design in which a sample or population is divided into two groups; one group is exposed to a manipulated variable while the other group serves as a nontreated control. The two groups are compared to see if there are changes in a "dependent" variable as a result of the experimental manipulation.

convection The transfer of heat to or from a surface via a moving stream of air or fluid.

convergent evolution The independent evolution of similar features from different ancestral traits.

copulation Reproductive behavior that results in a male depositing sperm in the reproductive tract of a female.

coprophagy Ingesting ones own feces.

corepressor A low-molecular-weight compound that unites with a protein (the repressor) to prevent transcription in a repressible operon.

cork A waterproofing tissue in plants, with suberin-containing cell walls. Produced by a cork cambium.

cornea The clear, transparent tissue that covers the eye and allows light to pass through to the retina.

corolla (ko role' lah) [L. *corolla*: a small crown] All of the petals of a flower, collectively.

coronary (kor' oh nair ee) [L. *corona*: crown] Referring to the blood vessels of the heart.

coronary thrombosis A fibrous clot that blocks a coronary vessel.

corpus luteum (kor' pus loo' tee um) [L.: yellow body] A structure formed from a follicle after ovulation; it produces hormones important to the maintenance of pregnancy.

cortex [L. *cortex*: covering, rind] (1) In plants, the tissue between the epidermis and the vascular tissue of a stem or root. (2) In animals, the outer tissue of certain organs, such as the adrenal cortex and cerebral cortex.

corticosteroids Steroid hormones produced and released by the cortex of the adrenal gland.

cortisol A corticosteroid that mediates stress responses.

cost–benefit analysis An approach for explaining why the traits of organisms evolve as they do; assumes that an organism has a limited amount of time and energy to devote to its activities, and that each activity has costs (e.g., risks, expenditure of energy) as well as benefits (e.g., obtaining food, mating and successful reproduction). (*See also* trade-off.)

cotyledon (kot' ul lee' dun) [Gk. *kotyledon*: hollow space] A "seed leaf." An embryonic organ that stores and digests reserve materials; may expand when seed germinates.

countercurrent exchange An adaptation that promotes maximum exchange of heat or any diffusible substance between two fluids by the fluids flow in opposite directions through parallel tubes close together

countercurrent multiplier The mechanism of increasing the concentration of the interstitial fluid in the mammalian kidney as a result of countercurrent exchange in the loops of Henle and selective permeability and active transport of ions by segments of the loops of Henle.

covalent bond A chemical bond that arises from the sharing of electrons between two atoms. Usually a strong bond.

crassulacean acid metabolism (CAM) A metabolic pathway enabling the plants that possess it to store carbon dioxide at night and then perform photosynthesis during the day with stomata closed.

crista (plural: cristae) A small, shelflike projection of the inner membrane of a mitochondrion; the site of oxidative phosphorylation.

critical night length In the photoperiodic flowering response of short-day plants, the length of night above which flowering occurs and below which the plant remains vegetative. (The reverse applies in the case of long-day plants.)

critical period The age during which some particular type of learning must take place or during which it occurs much more easily than at other times. Typical of song learning among birds.

cross section A section taken perpendicular to the longest axis of a structure. Also called a transverse section.

crossing over The mechanism by which linked markers undergo recombination. In general, the term refers to the reciprocal exchange of corresponding segments between two homologous chromatids.

cryptic [Gk. *kryptos*: hidden] The resemblance of an animal to some part of its environment, which helps it to escape detection by predators.

cryptochromes [Gk. *kryptos*: hidden + *kroma*: color] Photoreceptors mediating some blue-light effects in plants and animals.

culture (1) A laboratory association of organisms under controlled conditions. (2) The collection of knowledge, tools, values, and rules that characterize a human society.

cuticle A waxy layer on the outer surface of a plant or an insect, tending to retard water loss.

cyanobacteria (sigh an' o bacteria) [Gr. *kuanos*: blue] A group of photosynthetic bacteria, formerly referred to as blue-green algae; they use chlorophyll *a* in photosynthesis.

cyclic AMP *See* cAMP.

cyclic electron transport In photosynthetic light reactions, the flow of electrons that produces ATP but no NADPH or O_2.

cyclins Proteins that activate cyclin-dependent kinases, bringing about transitions in the cell cycle.

cyclin-dependent kinase (Cdk) A kinase whose target proteins are involved in transitions in the cell cycle and which is active only when complexed with additional protein subunits, called cyclins.

cytochromes (sy' toe chromes) [Gk. *kytos*: container + *chroma*: color] Iron-containing red proteins, components of the electron-transfer chains in photophosphorylation and respiration.

cytokine A regulatory protein made by immune system cells that affects other target cells in the immune system.

cytokinesis (sy' toe kine ee' sis) [Gk. *kytos*: container + *kinein*: to move] The division of the cytoplasm of a dividing cell. (Compare with mitosis.)

cytokinin (sy' toe kine' in) A member of a class of plant growth substances playing roles in senescence, cell division, and other phenomena.

cytoplasm The contents of the cell, excluding the nucleus.

cytoplasmic determinants In animal development, gene products whose spatial distribution may determine such things as embryonic axes.

cytoplasmic domain The portion of a membrane bound receptor molecule that projects into the cytoplasm.

cytoplasmic segregation The asymmetrical distribution of cytoplasmic determinants in a developing animal embryo.

cytosine (C) (site' oh seen) A nitrogen-containing base found in DNA and RNA.

cytoskeleton The network of microtubules and microfilaments that gives a eukaryotic cell its shape and its capacity to arrange its organelles and to move.

cytosol The fluid portion of the cytoplasm, excluding organelles and other solids.

cytotoxic T cells (T_C) Cells of the cellular immune system that recognize and directly eliminate virus-infected cells. (Compare with helper T cells.)

- D -

DAG *See* diacylglycerol.

daughter chromosomes During mitosis, the separated chromatids from the beginning of anaphase onward.

deciduous [L. *deciduus*: falling off] Refers to a woody plant that sheds it leaves but does not die.

decomposers Organisms that metabolize organic compounds in debris and dead organisms, releasing inorganic material; found among the bacteria, protists, and fungi. *See also* detritivore.

defensin A type of protein made by phagocytes that kills bacteria and enveloped viruses by insertion into their plasma membranes.

degeneracy The situation in which a single amino acid may be represented by any of two or more different codons in messenger RNA. Most of the amino acids can be represented by more than one codon.

delayed hypersensitivity An increased immune reaction against an antigen that does not appear for 1–2 days after exposure. (Contrast with immediate hypersensitivity.)

deletion A mutation resulting from the loss of a continuous segment of a gene or chromosome. Such mutations never revert to wild type. (Contrast with duplication, point mutation.)

deme (deem) [Gk. *demos:* the populace] Any local population of individuals belonging to the same species that interbreed with one another.

demographic processes Events (births, deaths, immigration, and emigration) that change the number of individuals and the age structure of a population.

demographic stochasticity Random variations in the factors influencing the size, density, and distribution of a population.

denaturation Loss of activity of an enzyme or nucleic acid molecule as a result of structural changes induced by heat or other means.

dendrite [Gk. *dendron*: tree] A fiber of a neuron which often cannot carry action potentials. Usually much branched and relatively short compared with the axon, and commonly carries information to the cell body of the neuron.

denitrification Metabolic activity by which nitrate and nitrite ions are reduced to form nitrogen gas; carried on by certain soil bacteria.

density dependence The state in which changes in the severity of action of agents affecting birth and death rates within populations are directly or inversely related to population density.

density independence The state in which the severity of action of agents affecting birth and death rates within a population does not change with the density of the population.

deoxyribonucleic acid *See* DNA.

deoxyribose A five-carbon sugar found in nucleotides and DNA.

depolarization A change in the electric potential across a membrane from a condition in which the inside of the cell is more negative than the outside to a condition in which the inside is less negative, or even positive, with reference to the outside of the cell. (Contrast with hyperpolarization.)

derived trait A trait that differs from the ancestral trait. (Compare with shared derived trait.)

dermal tissue system The outer covering of a plant, consisting of epidermis in the young plant and periderm in a plant with extensive secondary growth. (Contrast with ground tissue system and vascular tissue system.)

desmosome (dez' mo sowm) [Gk. *desmos*: bond + *soma*: body] An adhering junction between animal cells.

desmotubule A membrane extension connecting the endoplasmic retituclum of two plant cells that traverses the plasmodesma.

determination Process whereby an embryonic cell or group of cells becomes fixed into a predictable developmental pathway.

determined The state of a cell prior to its differentiation but after its developmental fate has been channeled.

detritivore (di try' ti vore) [L. *detritus*: worn away + *vorare*: to devour] An organism that obtains its energy from the dead bodies and/or waste products of other organisms.

development Progressive change, as in structure or metabolism; in most kinds of organisms, development continues throughout the life of the organism.

developmental plasticity The capacity of an organism to alter its pattern of development in response to environmental conditions.

diacylglycerol (DAG) In hormone action, the second messenger produced by hydrolytic removal of the head group of certain phospholipids.

diaphragm (dye' uh fram) [Gk. *diaphrassein*: barricade] (1) A sheet of muscle that separates the thoracic and abdominal cavities in mammals; responsible for breathing. (2) A method of birth control in which a sheet of rubber is fitted over the woman's cervix, blocking the entry of sperm.

diastole (dye ass' toll ee) [Gk. dilation] The portion of the cardiac cycle when the heart muscle relaxes. (Contrast with systole.)

diencephalons The portion of the vertebrate forebrain that becomes the thalamus and hypothalamus.

differential gene expression The hypothesis that, given that all cells contain all genes, what makes one cell type different from another is the difference in transcription and translation of those genes.

differentiation Process whereby originally similar cells follow different developmental pathways. The actual expression of determination.

diffusion Random movement of molecules or other particles, resulting in even distribution of the particles when no barriers are present.

digestion Enzyme-catalyzed process by which large, usually insoluble, molecules (foods) are hydrolyzed to form smaller molecules of soluble substances.

dihybrid cross A mating in which the parents differ with respect to the alleles of two loci of interest.

dikaryon (di care' ee ahn) [Gk. *di*: two + *karyon*: kernel] A cell or organism carrying two genetically distinguishable nuclei. Common in fungi.

dioecious (die eesh' us) [Gk.: *di*: two + *oikos*: house] Refers to organisms in which the two sexes are"housed"in two different individuals, so that eggs and sperm are not produced in the same individuals. Examples: humans, fruit flies, date palms. (Contrast with monoecious.)

diploblastic Having two cell layers. (Contrast with triploblastic.)

diploid (dip' loid) [Gk. *diplos*: double] Having a chromosome complement consisting of two copies (homologs) of each chromosome. Designated 2*n*.

diplontic A type of life cycle in which gametes are the only haploid cells and mitosis occurs only in diploid cells. (Contrast with haplontic.)

direct transduction A cell signaling mechanism in which the receptor acts as the effector in the cellular response. (Contrast with indirect transduction.)

directional selection Selection in which phenotypes at one extreme of the population distribution are favored. (Contrast with disruptive selection, stabilizing selection.)

disaccharide A carbohydrate made up of two monosaccharides (simple sugars).

displacement activity Apparently irrelevant behavior performed by an animal under conflict situations, especially when tendencies to attack and escape are closely balanced.

display A behavior that has evolved to influence the actions of other individuals.

disruptive selection Selection in which phenotypes at both extremes of the population distribution are favored. (Contrast with directional selection; stabilizing selection.)

distal Away from the point of attachment or other reference point. (Contrast with proximal.)

disturbance A short-term event that disrupts populations, communities, or ecosystems by changing the environment.

disulfide bridge The covalent bond between twosulfur atoms (–S—S–) linking to molecules or remote parts of the same molecule.

diverticulum (di ver tik' u lum) [L. *divertere*: turn away] A small cavity or tube that connects to a major cavity or tube.

division A term used by some microbiologists and formerly by botanists, corresponding to the term phylum.

DNA (deoxyribonucleic acid) The fundamental hereditary material of all living organisms. In eukaryotes, stored primarily in the cell nucleus. A nucleic acid using deoxyribose rather than ribose.

DNA chip A small glass or plastic square onto which thousands of single-stranded DNA sequences are fixed. Hybridization of cell-derived RNA or DNA to the target sequences can be performed.

DNA fingerprint An individual's unique pattern of DNA fragments produced by action of restriction endonucleases and separated by electrophoresis.

DNA helicase An enzyme that functions during DNA replication to unwind the double helix.

DNA ligase Enzyme that unites Okazaki fragments of the lagging strand during DNA replication; also mends breaks in DNA strands. It connects pieces of a DNA strand and is used in recombinant DNA technology.

DNA methylation The addition of methyl groups to DNA. Methylation plays a role in regulation of gene expression. Among bacteria, it protects DNA against restriction endonucleases.

DNA polymerase Any of a group of enzymes that catalyze the formation of DNA strands from a DNA template.

DNA sequencing Determining the precise sequence of nucleotides, and thus the sequence of amino acids encoded in a segment DNA.

DNA topoisomerase Enzymes that introduce positive or negative supercoils into the double-stranded DNA of continuous (circular) chromosomes.

docking protein A receptor protein that binds (docks) a ribosome to the membrane of the endoplasmic reticulum by binding the signal sequence attached to a new protein being made at the ribosome.

domain (1) Independent structural elements within proteins that affect the protein's function. Encoded by recognizable nucleotide sequences, a domain often folds separately from the rest of the protein. Similar domains can appear in a variety of different proteins across phylogenetic groups (e.g., "homeobox domain"; "calcium-binding domain.") (2) In phylogenetics, the three monophyletic branches of Life. Members of the three domains (Bacteria, Archaea, and Eukarya) are believed to have been evolving independently of each other for at least a billion years.

dominance In genetics, the ability of one allelic form of a gene to determine the phenotype of a heterozygous individual in which the homologous chromosomes carry both it and a different (recessive) allele. (Contrast with recessive.)

dormancy A condition in which normal activity is suspended, as in some seeds and buds.

dorsal [L. *dorsum*: back] Pertaining to the back or upper surface. (Contrast with ventral.)

dorsal lip In amphibian embryos, the dorsal segment of the blastopore. Also called the "organizer," this region directs the development of nearby embryonic regions.

double fertilization Virtually unique to angiosperms, a process in which the nuclei of two sperm fertilize one egg. One sperm's nucleus combines with the egg nucleus to produce a zygote, while the other combines with the same egg's two polar nuclei to produce the first cell of the triploid endosperm (the tissue that will nourish the growing plant embryo).

double helix In DNA, the natural, right-handed coil configuration of two complementary, antiparallel strands.

duodenum (do' uh dee' num) The beginning portion of the vertebrate small intestine. (Compare with ileum, jejunum.)

duplication A mutation in which a segment of a chromosome is duplicated, often by the attachment of a segment lost from its homolog. (Contrast with deletion.)

dynein [Gk. *dynamis*: power] A protein that plays a part in the movement of eukaryotic flagella and cilia by means of conformational changes.

- E -

ecdysone (eck die' sone) [Gk. *ek*: out of + *dyo*: to clothe] In insects, a hormone that induces molting.

ecological community The species living together at a particular site.

ecological niche (nitch) [L. *nidus*: nest] The functioning of a species in relation to other species and its physical environment.

ecological succession The sequential replacement of one assemblage of populations by another in a habitat following some disturbance.

ecology [Gk. *oikos*: house + *logos*: study] The scientific study of the interaction of organisms with their living (biotic) and nonliving (abiotic) environment.

ecosystem (eek' oh sis tum) The organisms of a particular habitat, such as a pond or forest, together with the physical environment in which they live.

ectoderm [Gk. *ektos*: outside + *derma*: skin] The outermost of the three embryonic tissue layers first delineated during gastrulation. Gives rise to the skin, sense organs, nervous system, etc.

ectotherm [Gk. *ektos*: outside + *thermos*: heat] An animal unable to control its body temperature. (Contrast with endotherm.)

edema (i dee' mah) [Gk. *oidema*: swelling] Tissue swelling caused by the accumulation of fluid.

edge effect The changes in ecological processes in a community caused by physical and biological factors originating in an adjacent community.

effector Any organ, cell, or organelle that moves the organism through the environment or else alters the environment; for example, muscle, exocrine glands, chromatophores.

effector cell A cell responsible for the effector phase of the immune response.

effector phase Stage of the immune response, when cytotoxic T cells attack virus-infected cells, and helper T cells assist B cells to differentiate into plasma cells.

effector protein In cell signaling, a protein responsible for the cellular reponse to a signal transduction pathway.

efferent [L. *ex*: out + *ferre*: to bear] In physiology, conducting outward or away from an organ or structure. (Contrast with afferent.)

egg In all sexually reproducing organisms, the female gamete; in birds, reptiles, and some other vertebrates, a structure within which early embryonic development occurs. (Compare with amniote egg.)

elasticity The property of returning quickly to a former state after a disturbance.

electrocardiogram (ECG or EKG) A graphic recording of electrical potentials from the heart.

electroencephalogram (EEG) A graphic recording of electrical potentials from the brain.

electromagnetic radiation A self-propagating wave that travels though space and has both electrical and magnetic properties.

electromyogram (EMG) A graphic recording of electrical potentials from muscle.

electron A subatomic particle outside the nucleus carrying a negative charge and very little mass.

electron shell The region surrounding the atomic nucleus at a fixed energy level in which electrons orbit.

electron transport The passage of electrons through a series of proteins with a release of energy which may be captured in a concentration gradient or chemical form such as NADH or ATP.

electronegativity The tendency of an atom to attract electrons when it occurs as part of a compound.

electrostatic Pertaining to the attraction and repulsion of negative and positive charges on atoms due to the number and distribution of electrons.

electrophoresis (e lek' tro fo ree' sis) [L. *electrum*: amber + Gk. *phorein*: to bear] A separation technique in which substances are separated from one another on the basis of their electric charges and molecular weights.

element A substance that cannot be converted to simpler substances by ordinary chemical means.

elongation (1) In molecular biology, the addition of monomers to make a longer RNA or protein during synthesis. (2) Growth of a plant axis or cell primarily in the longitudinal direction.

embolus (em' buh lus) [Gk. *embolos*: inserted object; stopper] A circulating blood clot. Blockage of a blood vessel by an embolus or by a bubble of gas is referred to as an embolism. (Contrast with thrombus.)

embryo [Gk. *en*: within + *bryein*: to grow] A young animal, or young plant sporophyte, while it is still contained within a protective structure such as a seed, egg, or uterus.

embryophyte A photosynthetic organism that develops from a protected embryo; synonymous with "land plant." In this book, "plant," when unmodified, is synonymous with "embryophyte."

embryo sac In angiosperms, the female gametophyte. Found within the ovule, it consists of eight or fewer cells, membrane bounded, but without cellulose walls between them.

emergent property A property of a complex system that is not exhibited by its individual component parts.

emigration The deliberate and usually oriented departure of an organism from the habitat in which it has been living.

3′ end (3 prime) The end of a DNA or RNA strand that has a free hydroxyl group at the 3′ carbon of the sugar (deoxyribose or ribose).

5′ end (5 prime) The end of a DNA or RNA strand that has a free phosphate group at the 5′ carbon of the sugar (deoxyribose or ribose).

embolism A blockage of a coronary vessel resulting from a circulating blood clot.

endemic (en dem' ik) [Gk. *endemos*: native, dwelling in] Confined to a particular region, thus often having a comparatively restricted distribution.

endergonic reaction A chemical reaction that requires the input of energy in order to proceed. (Contrast with exergonic reaction.)

endocrine gland (en' doh krin) [Gk. *endo*: within + *krinein*: to separate] Any gland, such as the adrenal or pituitary gland of vertebrates, that secretes certain substances, especially hormones, into the body through the blood. The *endocrine system* consists of all cells and glands in the body that produce and release hormones.

endocytosis A process by which liquids or solid particles are taken up by a cell through invagination of the plasma membrane. (Contrast with exocytosis.)

endoderm [Gk. *endo*: within + *derma*: skin] The innermost of the three embryonic tissue layers delineated during gastrulation. Gives rise to the digestive and respiratory tracts and structures associated with them.

endodermis In plants, a specialized cell layer marking the inside of the cortex in roots and some stems. Frequently a barrier to free diffusion of solutes.

endomembrane system Endoplasmic reticulum plus Golgi apparatus; also lysosomes, when present. A system of membranes that exchange material with one another.

endoplasmic reticulum (ER) [Gk. *endo*: within + L. *plasma*: form + L. *reticulum*: net] A system of membranous tubes and flattened sacs found in the cytoplasm of eukaryotes. Exists in two forms: rough ER, studded with ribosomes; and smooth ER, lacking ribosomes.

endorphins, enkephalins Naturally occurring, opiate-like substances in the mammalian brain.

endoskeleton [Gk. *endo*: within + *skleros*: hard] An internal skeleton covered by other, soft body tissues. (Contrast with exoskeleton.)

endosperm [Gk. *endo*: within + *sperma*: seed] A specialized triploid seed tissue found only in angiosperms; contains stored nutrients for the developing embryo.

endosymbiosis [Gk. *endo*: within + *sym*: together + *bios*: life] Two species living together, with one living inside the body (or even the cells) of the other.

endosymbiotic theory The theory that the eukaryotic cell evolved via the engulfing of one prokaryotic cell by another.

endotherm [Gk. *endo*: within + *thermos*: heat] An animal that can control its body temperature by the expenditure of its own metabolic energy. (Contrast with ectotherm.)

end product inhibition A control capacity of some metabolic pathways in which the final product produced inhibits an early enzyme in the pathway.

energetic cost The difference between the energy an animal expends in performing a behavior and the energy it would have expended had it rested.

energy The capacity to do work or move matter against an opposing force. The capacity to accomplish change.

energy budget A quantitative description of all paths of energy exchange between an animal and its environment.

enhancer In eukaryotes, a DNA sequence, lying on either side of the gene it regulates, that stimulates a specific promoter.

enterocoelous development A pattern of development in which the coelum is formed by an outpocketing of the embryonic gut (enteron).

enterokinase (ent uh row kine′ ase) An enzyme secreted by the mucosa of the duodenum. It activates the zymogen trypsinogen to create the active digestive enzyme trypsin.

enthalpy The sum of the internal energy of a system; the product of the volume multiplied by the pressure.

entrainment With respect to circadian rhythms, the process whereby the period is adjusted to match the 24-hour environmental cycle.

entropy (en′ tro pee) [Gk. *tropein*: to change] A measure of the degree of disorder in any system. Spontaneous reactions in a closed system are always accompanied by an increase in disorder and entropy.

environment Whatever surrounds and interacts with a population, organism, or cell. May be external or internal.

environmental carrying capacity (*K*) The number of individuals in a population that the resources of a habitat can support.

enzyme (en′ zime) [Gk. *zyme*: to leaven (as in yeast bread)] A protein, on the surface of which are chemical groups so arranged as to make the enzyme a catalyst for a chemical reaction.

enzyme–substrate complex The complex that forms when an enzyme binds to its substrate(s).

eosinophils Phagocytic white blood cells that attack multicellular parasites once they have been coated with antibodies.

epi- [Gk.: upon, over] A prefix used to designate a structure located on top of another; for example: epidermis, epiphyte.

epiblast The upper or overlying portion of the avian blastula which is joined to the hypoblast at the margins of the blastodisc.

epiboly The movement of cells over the surface of the blastula toward the forming blastopore.

epicotyl (epp′ i kot′ il) [Gk. *epi*: over + *kotyle*: something hollow] That part of a plant embryo or seedling that is above the cotyledons.

epidermis [Gk. *epi*: over + *derma*: skin] In plants and animals, the outermost cell layers. (Only one cell layer thick in plants.)

epididymis (epuh did′ uh mus) [Gk. *epi*: over + *didymos*: testicle] Coiled tubules in the testes that store sperm and conduct sperm from the seiminiferous tubules to the vas deferens.

epinephrine (ep i nef′ rin) [Gk. *epi*: over + *nephros*: kidney] The "fight or flight" hormone produced by the medulla of the adrenal gland; it also functions as a neurotransmitter. (Also known as adrenaline.)

epiphyte (ep′ e fyte) [Gk. *epi*: over + *phyton*: plant] A specialized plant that grows on the surface of other plants but does not parasitize them.

epistasis Interaction between genes in which the presence of a particular allele of one gene determines whether another gene will be expressed.

epithelium Sheets of cells that line or cover organs, make up tubules, and cover the surface of the body.

equatorial plate In a cell undergoing mitosis, the region in the middle of a cell where the centromeres will align during metaphase.

equilibrium Any state of balanced opposing forces and no net change.

ER *See* endoplasmic reticulum.

erythrocyte (ur rith′ row site) [Gk. *erythros*: red + *kytos*: container] A red blood cell.

erythropoietin A hormone produced by the kidney in response to lack of oxygen. It stimulates the production of red blood cells.

esophagus (i soff′ i gus) [Gk. *oisophagos*: gullet] That part of the gut between the pharynx and the stomach.

essential element A mineral nutrient element required in order for a seed to develop and complete the plant's life cycle, producing viable new seeds.

essential acids Amino acids or fatty acids that an animal cannot synthesize for itself.

ester linkage A condensation (water-releasing) reaction in which the carboxyl group of a fatty acid reacts with the hydroxyl group of an alcohol. Lipids are formed in this way.

estivation (ess tuh vay′ shun) [L. *aestivalis*: summer] A state of dormancy and hypometabolism that occurs during the summer; usually a means of surviving drought and/or intense heat. (Contrast with hibernation.)

estrogen Any of several steroid sex hormones; produced chiefly by the ovaries in mammals.

estrus (es′ truss) [L. *oestrus*: frenzy] The period of heat, or maximum sexual receptivity, in some female mammals. Ordinarily, the estrus is also the time of release of eggs in the female.

ethology An approach to the study of animal behavior developed in Europe. Emphasizes the causes of the evolution of behavior.

ethylene One of the plant growth hormones, the gas $H_2C{=}CH_2$. Involved in fruit ripening and other growth and developmental responses.

euchromatin Chromatin that is diffuse and non-staining during interphase; may be transcribed. (Contrast with heterochromatin.)

eukaryotes (yew car′ ree oats) [Gk. *eu*: true + *karyon*: kernel or nucleus] Organisms whose cells contain their genetic material inside a nucleus. Includes all life other than the viruses, archaea, and bacteria.

eutrophication (yoo trofe′ ik ay′ shun) [Gk. *eu*: truly + *trephein*: to flourish] The addition of nutrient materials to a body of water, resulting in changes in ecological processes and species composition therein.

evaporation The transition of water from the liquid to the gaseous phase.

evolution Any gradual change. Organic or Darwinian evolution, often referred to as evolution, is any genetic and resulting phenotypic change in organisms from generation to generation. (*See* macroevolution, microevolution; compare with speciation.)

evolutionary agent Any factor that influences the direction and rate of evolutionary change.

evolutionarily conserved Refers to traits that have evolved very slowly and are similar or even identical in individuals of highly divergent groups.

evolutionary radiation The proliferation of species within a single evolutionary lineage.

evolutionary reversal The reappearance of an ancestral trait in a group that had previously acquired a derived trait.

excision repair The removal and damaged DNA and its replacement by the appropriate nucleotides.

excited state The state of an atom or molecule when, after absorbing energy, it has more energy than in its normal, ground state. (Compare with ground state.)

excretion Release of metabolic wastes by an organism.

exergonic reaction A reaction in which free energy is released. (Contrast with endergonic reaction.)

exocrine gland (eks' oh krin) [Gk. *exo*: outside + *krinein*: to separate] Any gland, such as a salivary gland, that secretes to the outside of the body or into the gut. (Contrast with endocrine gland.)

exocytosis A process by which a vesicle within a cell fuses with the plasma membrane and releases its contents to the outside. (Contrast with endocytosis.)

exon A portion of a DNA molecule, in eukaryotes, that codes for part of a polypeptide. (Contrast with intron.)

exoskeleton (eks' oh skel' e ton) [Gk. *exos*: outside + *skleros*: hard] A hard covering on the outside of the body to which muscles are attached. (Contrast with endoskeleton.)

exotoxins Highly toxic proteins released by living, multiplying bacteria.

expanding triplet repeat A three-base-pair sequence in a human gene that is unstable and can be repeated a few to hundreds of times. Often, the more the repeats, the less the activity of the gene involved. Expanding triplet repeats occur in some human diseases such as Huntington's disease and fragile-X syndrome.

experiment A testing process to support or disprove hypotheses and to answer questions. The basis of the scientific method. *See* comparative experiment, controlled experiment.

expiratory reserve volume The amount of air that can be forcefully exhaled beyond the normal tidal expiration. (Compares with inspiratory reserve volume, tidal volume, vital capacity.)

exploitation competition Competition in which individuals reduce the quantities of their shared resources. (Compare with interference competition.)

exponential growth Growth, especially in the number of organisms in a population, which is a geometric function of the size of the growing entity: the larger the entity, the faster it grows. (Contrast with logistic growth.)

expression vector A DNA vector, such as a plasmid, that carries a DNA sequence that includes the adjacent sequences for its expression into mRNA and protein in a host cell.

expressivity The degree to which a genotype is expressed in the phenotype; may be affected by the environment.

extensor A muscle the extends an appendage.

extinction The termination of a lineage of organisms.

extrinsic protein A membrane protein found only on the surface of the membrane. (Contrast with intrinsic protein.)

extracellular matrix In animal tissues, a material of heterogeneous composition surrounding cells and performing many functions including adhesion of cells.

extraembryonic membranes Four membranes that support but are not part of the developing embryos of reptiles, birds, and mammals, defining these groups phylogenetically as amniotes. (*See* amnion, allantois, chorion, and yolk sac.)

- F -

F_1 (first filial generation) The immediate progeny of a parental (P) mating.

F_2 (second filial generation) The immediate progeny of a mating between members of the F_1 generation.

facilitated diffusion Passive movement through a membrane involving a specific carrier protein; does not proceed against a concentration gradient. (Contrast with active transport, diffusion.)

facultative anaerobes (alternatively, facultative aerobes) Prokaryotes that can shift their metabolism between anaerobic and aerobic operations depending on the presence or absence of O_2.

FAD *See* flavin adenine dinucleotide.

fast-twitch fibers Skeletal muscle fibers that can generate high tension rapidly, but fatigue rapidly ("sprinter" fibers). Characterized by an abundance of enzymes of glycolysis.

fat A triglyceride that is solid at room temperature. (Contrast with oil.)

fate In an embryo, the type of cell that a particular cell will become in the adult.

fate map A diagram of the blastula showing which cells (blastomeres) are "fated" to contribute to specific tissues and organs in the mature body.

fatty acid A molecule with a long hydrocarbon tail and a carboxyl group at the other end. Found in many lipids.

fauna (faw' nah) All of the animals found in a given area. (Contrast with flora.)

feces [L. *faeces*: dregs] Waste excreted from the digestive system.

feedback information Information relevant to the rate of a process that can be used by a control system to regulate that process at a particular level.

feedforward information Information that can be used to alter the setpoint of a regulatory process.

fermentation (fur men tay' shun) [L. *fermentum*: yeast] The anaerobic degradation of a substance such as glucose to smaller molecules with the extraction of energy.

ferredoxin A protein containing iron that mediates the transfer of electrons in a number of pathways, including the light reactions of photosynthesis.

fertility factor (F factor) A plasmid that confers on a bacterium the ability to act as a DNA donor during conjugation.

fertilization Union of gametes. Also known as syngamy.

fertilization membrane A membrane surrounding an animal egg which becomes rapidly raised above the egg surface within seconds after fertilization, serving to prevent entry of a second sperm.

fetus Medical and legal term for the latter stages of a developing human embryo, from about the eighth week of pregnancy (the point at which all major organ systems have formed) to the moment of birth.

fiber An elongated, tapering cell of flowering plants, usually with a thick cell wall. Serves a support function.

fibrin A protein that polymerizes to form long threads that provide structure to a blood clot.

fibrinogen A circulating protein that can be stimulated to fall out of solution and provide the structure for a blood clot.

Fick's law of diffusion An equation that describes the factors that determine the rate of diffusion of a molecule from an area of higher concentration to an area of lower concentration.

filter feeder An organism that feeds upon much smaller organisms, that are suspended in water or air, by means of a straining device.

filtration In the excretory physiology of some animals, the process by which the initial urine is formed; water and most solutes are transferred into the excretory tract, while proteins are retained in the blood or hemolymph.

first law of thermodynamics Energy can be neither created nor destroyed.

fission Reproduction of a prokaryote by division of a cell into two comparable progeny cells.

fitness The contribution of a genotype or phenotype to the genetic composition of subsequent generations, relative to the contribution of other genotypes or phenotypes. (*See* also inclusive fitness.)

flagellum (fla jell' um) (plural: flagella) [L. *flagellum*: whip] Long, whiplike appendage that propels cells. Prokaryotic flagella differ sharply from those found in eukaryotes.

flavin adenine dinucleotide (FAD) A coenzyme involved in redox reactions and containing the vitamin riboflavin (B_2).

flexor A muscle that flexes an appendage.

flora (flore' ah) All of the plants found in a given area. (Contrast with fauna.)

floral meristem Meristem that forms the sexual parts of flowering plants (sepals, petals, stamens, and carpels).

florigen A plant hormone involved in the conversion of a vegetative shoot apex to a flower.

flower The total reproductive structure of an angiosperm; its basic parts include the calyx, corolla, stamens, and carpels.

fluid mosaic model A molecular model for the structure of biological membranes consisting of a fluid phospholipid bilayer in which suspended proteins are free to move in the plane of the bilayer.

fluorescence The emission of a photon of visible light by an excited atom or molecule.

follicle [L. *folliculus*: little bag] In female mammals, an immature egg surrounded by nutritive cells.

follicle-stimulating hormone A gonadotropic hormone produced by the anterior pituitary.

food chain A portion of a food web, most commonly a simple sequence of prey species and the predators that consume them.

food vacuole Membrane enclosed structure formed by phagocytosis in which engulfed food particles are digested by the action of lysosomal enzymes.

food web The complete set of food links between species in a community; a diagram indicating which ones are the eaters and which are eaten.

forb Any broad-leaved herbaceous plant. Especially applied to such plants growing in grasslands.

fossil Any recognizable structure originating from an organism, or any impression from such a structure, that has been preserved over geological time.

founder effect Random changes in allele frequencies resulting from establishment of a population by a very small number of individuals.

fovea [L. *fovea;* a small pit] In the vertebrate retina, the area of most distinct vision.

frame-shift mutation A mutation resulting from the addition or deletion of one or two consecutive base pairs in the DNA sequence of a gene, resulting in misreading mRNA during translation and production of a nonfunctional protein. (Compare with missense substitution, nonsense substitution, synonymous substitution.)

Frank–Starling law The stroke volume of the heart increases with increased return of blood to the heart.

free energy That energy which is available for doing useful work, after allowance has been made for the increase or decrease of disorder.

freeze-fracturing Method of tissue preparation for transmission and scanning electron microscopy in which a tissue is frozen and a knife is then used to crack open the tissue; the fracture often occurs in the path of least resistance, within a membrane.

frequency In population genetics, the proportion of all alleles or genotypes in a population composed of a particular allele or genotype.

frequency-dependent selection Selection that changes in intensity with the proportion of individuals in a population having the trait.

fruit In angiosperms, a ripened and mature ovary (or group of ovaries) containing the seeds. Sometimes applied to reproductive structures of other groups of plants.

fruiting body Any structure that bears spores.

functional genomics The assignment of functional roles to genes first identified by sequencing entire genomes.

functional group A characteristic combination of atoms that contribute specific properties when attached to larger molecules.

functional mRNA Eukaryotic mRNA that has been modified after transcription by the removal of introns and the addition of a 5′ cap and a 3′ poly(A) tail.

- G -

G cap A chemically modified GTP added to the 5′ end of mRNA; facilitates binding of mRNA to ribosome and prevents mRNA breakdown.

G_1 phase In the cell cycle, the gap between the end of mitosis and the onset of the S phase.

G_2 phase In the cell cycle, the gap between the S (synthesis) phase and the onset of mitosis.

G protein A membrane protein involved in signal transduction; characterized by binding GDP or GTP.

gametangium (gam uh tan′ gee um) [Gk. *gamos*: marriage + *angeion*: vessel] Any plant or fungal structure within which a gamete is formed.

gamete (gam′ eet) [Gk. *gamete/gametes*: wife, husband] The mature sexual reproductive cell: the egg or the sperm.

gametogenesis (ga meet′ oh jen′ e sis) The specialized series of cellular divisions that leads to the production of sex cells (gametes). (Contrast with oogenesis and spermatogenesis.)

gametophyte (ga meet′ oh fyte) In plants and photosynthetic protists with alternation of generations, the multicellular haploid phase that produces the gametes. (Contrast with sporophyte.)

ganglion (gang′ glee un) [Gk.: tumor] A cluster of neurons that have similar characteristics or function.

gap genes During insect development, the first step of segmentation genes to act organizing the anterior-posterior axis.

gap junction A 2.7-nanometer gap between plasma membranes of two animal cells, spanned by protein channels. Gap junctions allow chemical substances or electrical signals to pass from cell to cell.

gas exchange In animals, the process of taking up oxygen from the environment and releasing carbon dioxide to the environment.

gastrovascular cavity Serving for both digestion (gastro) and circulation (vascular); in particular, the central cavity of the body of jellyfish and other cnidarians.

gastrula (gas′ true luh) [Gk. *gaster*: stomach] An embryo forming the characteristic three cell layers (ectoderm, endoderm, and mesoderm) that give rise to all of the major tissue systems of the adult animal.

gastrulation Development of a blastula into a gastrula.

gated channel A membrane protein that opens and closes in response to binding of specific molecules or to changes in membrane potential. When open, it allows specific ions to move across the membrane.

gene [Gk. *genes*: to produce] A unit of heredity. Used here as the unit of genetic function which carries the information for a single polypeptide or RNA.

gene amplification Creation of multiple copies of a particular gene, allowing the production of large amounts of the RNA transcript (as in rRNA synthesis in oocytes).

gene cloning Formation of a clone of bacteria or yeast cells containing a particular foreign gene.

gene family A set of identical (or once-identical) genes derived from a single parent gene; need not be on the same chromosomes. The vertebrate globin genes constitute a classic example of a gene family.

gene flow Exchange of genes between different species (an extreme case referred to as hybridization) or between different populations of the same species caused by migration followed by breeding.

gene-for-gene resistance A mechanism for resistance to pathogens, in which resistance is triggered by the specific interaction of the products of pathogens′ *Avr* genes and plants′ *R* genes.

gene frequency *See* allele frequency.

gene library All of the cloned DNA fragments generated by action of a restriction endonuclease on a genome or chromosome.

gene pool All of the different alleles of all of the genes existing in all individual of a population.

gene therapy Treatment of a genetic disease by providing patients with cells containing functioning alleles of the genes that are nonfunctional in their bodies.

generative cell In a pollen tube, a haploid nucleus that undergoes mitosis to produce the two sperm nuclei that participate in double fertilization. (Contrast with tube cell.)

genetic code The set of instructions, in the form of nucleotide triplets, that translate a linear sequence of nucleotides in mRNA into a linear sequence of amino acids in a protein.

genetic drift Changes in gene frequencies from generation to generation as a result of random (chance) processes.

genetic map The positions of genes along a chromosome as revealed by recombination frequencies.

genetic screening The application of medical tests to determine whether an individual carries a specific allele.

genetic stochasticity Random variation in the frequencies of alleles and genotypes in a population over time. (Compare with demographic stochasticity.)

genetics The study of the structure, functioning, and inheritance of genes, the units of hereditary information.

genome (jee′ nome) All the genes in a complete haploid set of chromosomes. (Compare with proteome.)

genomics The study of entire sets of genes and their interactions.

genomic imprinting When a given gene′s phenotype is determined by whether that gene is inherited from the male or the female parent.

genotype (jean′ oh type) [Gk. *gen*: to produce + *typos*: impression] An exact description of the genetic constitution of an individual, either with respect to a single trait or with respect to a larger set of traits. (Contrast with phenotype.)

genus (jean′ us) (plural: genera) [Gk. *genos*: stock, kind] A group of related, similar species recognized by taxonomists with a distinct name used in binomial nomenclature.

geotropism *See* gravitropism.

germ cell [L. *germen*: to beget] A reproductive cell or gamete of a multicellular organism. (Contrast with somatic cell.)

germ line mutation A change in the genetic material in a germ cell; such mutations are heritable. (Contrast with somatic mutation.)

germ layers The three embryonic tissue layers formed during gastrulation (ectoderm, mesoderm, endoderm).

germination Sprouting of a seed or spore.

gestation (jes tay′ shun) [L. *gestare*: to bear] The period during which the embryo of a mammal develops within the uterus. Also known as pregnancy.

gibberellin (jib er el′ lin) A class of plant growth substances playing roles in stem elongation, seed germination, flowering of certain plants, etc. Named for the fungus *Gibberella*.

gill An organ for gas exchange in aquatic organisms.

gill arch A skeletal structure that supports gill filaments and the blood vessels that supply them.

gizzard (giz′ erd) [L. *gigeria*: cooked chicken parts] A muscular port of the stomach of birds that grinds up food, sometimes with the aid of fragments of stone.

gland An organ or group of cells that produces and secretes one or more substances.

glans penis Sexually sensitive tissue at the tip of the penis.

glia (glee′ uh) [Gk. *glia*: glue] Cells, found only in the nervous system, that do not conduct action potentials.

glomerular filtration rate (GFR) The rate at which the blood is filtered in the glomeruli of the kidney.

glomerulus (glo mare′ yew lus) [L. *glomus*: ball] Sites in the kidney where blood filtration takes place. Each glomerulus consists of a knot of capillaries served by afferent and efferent arterioles.

glucagon Hormone produced by alpha cells of the pancreatic islets of Langerhans. Glucagon stimulates the liver to break down glycogen and release glucose into the circulation.

glucocorticoids Steroid hormones produced by the adrenal cortex. Secreted in response to ACTH, they inhibit glucose uptake by many tissues in addition to mediating other stress responses.

gluconeogenesis The biochemical synthesis of glucose from other substances, such as amino acids, lactate, and glycerol.

glucose[Gk. *gleukos*: sugar, sweet] The most common monosaccharide; the monomer of the polysaccharides starch, glycogen, and cellulose.

glycerol (gliss′ er ole) A three-carbon alcohol with three hydroxyl groups; a component of phospholipids and triglycerides.

glycogen (gly′ ko jen) An energy storage polysaccharide found in animals and fungi; a branched-chain polymer of glucose, similar to starch.

glycolipid A lipid to which sugars are attached.

glycolysis (gly kol′ li sis) [Gk. *gleukos*: sugar + *lysis*: break apart] The enzymatic breakdown of glucose to pyruvic acid. One of the evolutionarily oldest of the cellular energy-yielding mechanisms.

glycoprotein A protein to which sugars are attached.

glycosidic linkage Bond between carbohydrate (sugar) molecules through an intervening oxygen atom (–O–).

glycosylation The addition of carbohydrates to another type of molecule, such as a protein.

glyoxysome (gly ox′ ee soam) An organelle found in plants, in which stored lipids are converted to carbohydrates.

Golgi apparatus (goal′ jee) A system of concentrically folded membranes found in the cytoplasm of eukaryotic cells; functions in secretion from cell by exocytosis.

gonad (go′ nad) [Gk. *gone*: seed] An organ that produces sex cells in animals: either an ovary (female gonad) or testis (male gonad).

gonadotropin A hormone that stimulates the gonads.

gonadotropin-releasing hormone (GnRH) Hypothalamic hormone that stimulates the anterior pituitary to secrete growth hormone.

Gondwana The large southern land mass that existed from the Cambrian (540 mya) to the Jurassic (138 mya). Present-day remnants are South America, Africa, India, Australia, and Antarctica.

graft Tissue artificially and viably transplanted from one organism to another. In agriculture, refers to the transfer of bud or stem segment from one plant onto another plant as a form of asexual reproduction.

Gram stain A differential purple stain useful in characterizing bacteria. The peptidoglycan-rich cell walls of Gram-positive bacteria stain purple; cell walls of Gram-negative bacteria generally stain orange.

granum (plural: grana) Within a chloroplast, a stack of thylakoids.

gravitropism A directed plant growth response to gravity.

grazer An animal that eats the vegetative tissues of herbaceous plants.

green gland An excretory organ of crustaceans.

greenhouse effect The heating of Earth's atmosphere by gases such as water vapor, carbon dioxide, and methane; such greenhouse gases are transparent to sunlight but opaque to heat; thus sunlight-engendered heat builds up at Earth's surface and cannot be dissipated into the atmosphere.

gross primary production The total energy captured by plants growing in a particular area.

gross primary productivity (GPP) The rate of assimilation of energy by plants growing in a particular area.

ground meristem That part of an apical meristem that gives rise to the ground tissue system of the primary plant body.

ground state The lowest energy state of an atom of molecule. (Compare with excited state.)

ground tissue system Those parts of the plant body not included in the dermal or vascular tissue systems. Ground tissues function in storage, photosynthesis, and support.

group transfer The exchange of atoms between molecules.

growth Irreversible increase in volume (an accurate definition, but at best a dangerous oversimplification).

growth hormone A peptide hormone of the anterior pituitary that stimulates many anabolic processes.

guanine (G) (gwan′ een) A nitrogen-containing base found in DNA, RNA, and GTP.

guard cells In plants, specialized, paired epidermal cells that surround and control the opening of a stoma (pore). *See* stoma.

gut An animal's digestive tract.

guttation The extrusion of liquid water through openings in leaves, caused by root pressure.

gymnosperm (jim′ no sperm) [Gk. *gymnos*: naked + *sperma*: seed] A plant, such as a pine or other conifer, whose seeds do not develop within an ovary (hence, the seeds are"naked").

gyrus The raised or ridged portion of the convoluted surface of the brain. (Contrast with sulcus.)

- H -

habitat The environment in which an organism lives.

habituation (ha bich′ oo ay shun) The simplest form of learning, in which an animal presented with a stimulus without reward or punishment eventually ceases to respond.

hair cell A type of mechanoreceptor in animals. Detects sound waves and other forms of motion in air or water.

half-life The time required for half of a sample of a radioactive isotope to decay to its stable, nonradioactive form, or for a drug or other substance to reach half its initial dosage.

halophyte (hal′ oh fyte) [Gk. *halos*: salt + *phyton*: plant] A plant that grows in a saline (salty) environment.

haploid (hap′ loid) [Gk. *haploeides*: single] Having a chromosome complement consisting of just one copy of each chromosome; designated 1*n* or *n*. (Contrast with diploid.)

haplontic A type of life cycle in which the zygote is the only diploid cell and mitosis occurs only in haploid cells. (Contrast with diplontic.)

Hardy–Weinberg equililbrium The allele frequency at a given locus in a sexually reproducing population that is not being acted on by agents of evolution; the conditions that would result in no evolution in a population.

haustorium (haw stor′ ee um) [L. *haustus*: draw up] A specialized hypha or other structure by which fungi and some parasitic plants draw nutrients from a host plant.

Haversian systems Units of organization in compact bone that reflect the action of intercommunicating osteoblasts.

heat of vaporization The energy that must be supplied to convert a molecule from a liquid to a gas at its boiling point.

heat-shock proteins Chaperone proteins expressed in cells exposed to high or low temperatures or other forms of environmental stress.

helical Shaped like a screw or spring; this shape occurs in DNA and proteins.

helper T cells (T_H) T cells that participate in the activation of B cells and of other T cells; targets of the HIV-I virus, the agent of AIDS. (Contrast with cytotoxic T cells.)

hematocrit (heme at′ o krit) [Gk. *heaema*: blood + *krites*: judge] The proportion of 100 cc of blood that consists of red blood cells.

hemizygous (hem′ ee zie′ gus) [Gk. *hemi*: half + *zygotos*: joined] In a diploid organism, having only one allele for a given trait, typically the case for X-linked genes in male mammals and Z-linked genes in female birds. (Contrast with homozygous, heterozygous.)

hemoglobin (hee′ mo glow bin) [Gk. *heaema*: blood + L. *globus*: globe] Oxygen-transporting protein found in the red blood cells of vertebrates (and found in some invertebrates).

Hensen's node In avian embryos, a structure at the anterior end of the primitive groove; determines the fates of cells passing over it during gastrulation.

hepatic (heh pat′ ik) [Gk. *hepar*: liver] Pertaining to the liver.

hepatic duct Conveys bile from the liver to the gallbladder.

herbivore (ur′ bi vore) [L. *herba*: plant + *vorare*: to devour] An animal that eats plant tissues. (Contrast with carnivore, detritivore, omnivore.)

heritable Able to be inherited; in biology, refers to genetically influenced traits.

hermaphroditism (her maf′ row dite ism) [Gk. Hermes (messenger god) + Aphrodite (goddess of love)] The coexistence of both female and male sex organs in the same organism.

hertz (abbreviated Hz) Cycles per second.

hetero- [Gk.: *heteros*: other, different] A prefix specifying that two or more different conditions are involved; for example, heterotroph, heterozygous.

heterochromatin Chromatin that retains its coiling during interphase; generally not transcribed. (Contrast with euchromatin.)

heterochrony Comparing different species, an alternation in the timing of developmental events, leading to different results in the adult.

heterocyst A large, thick-walled cell in the filaments of certain cyanobacteria; performs nitrogen fixation.

heterogeneous nuclear RNA (hnRNA) The product of transcription of a eukaryotic gene, including transcripts of introns.

heterokaryon In fungi, hypha containing two genetically different nuclei.

heteromorphic (het′ er oh more′ fik) [Gk. *heteros*: different + *morphe*: form] Having a different form or appearance, as two heteromorphic life stages of a plant. (Contrast with isomorphic.)

heterosporous (het′ er os′ por us) Producing two types of spores, one of which gives rise to a female megaspore and the other to a male microspore. (Contrast with homosporous.)

heterosis Situation in which heterozygous genotypes are superior to homozygous genotypes with respect to growth, survival, or fertility. Also called hybrid vigor.

heterotherm An animal that regulates its body temperature at a constant level at some times but not others, such as a hibernator.

heterotroph (het′ er oh trof) [Gk. *heteros*: different + *trophe*: food] An organism that requires preformed organic molecules as food. (Contrast with autotroph.)

heterotypic Referring to adhesion of cells of different types. (Contrast with homotypic.)

heterozygous (het′ er oh zie′ gus) [Gk. *heteros*: different + *zygotos*: joined] Of a diploid organism having different alleles of a given gene on the pair of homologs carrying that gene. (Contrast with homozygous.)

hexose [Gk. *hex*: six] A sugar containing six carbon atoms.

hibernation [L. *hibernum*: winter] The state of inactivity of some animals during winter; marked by a drop in body temperature and metabolic rate.

hierarchical sequencing An approach to DNA sequencing in which markers are mapped and DNA sequences are aligned by matching overlapping sites of known sequence.

highly repetitive DNA Short DNA sequences present in millions of copies in the genome, next to each other (in tandem). In reassociation experiments, denatured highly repetitive DNA reanneals very quickly.

hindbrain The region of the developing vertebrate brain that gives rise to the medulla, pons, and cerebellum.

hippocampus [Gr.: sea horse] A part of the forebrain that takes part in long-term memory formation.

histamine (hiss′ tah meen) A substance released by damaged tissue, or by mast cells in response to allergens. Histamine increases vascular permeability, leading to edema (swelling).

histology [Gk. *histos*: weaving] The study of tissues.

histone Any one of a group of basic proteins forming the core of a nucleosome, the structural unit of a eukaryotic chromosome. (Compare with nucleosome.)

hierarchical sequencing An approach to DNA sequencing in which markers are mapped and DNA sequences are aligned by matching overlapping sites of known sequence.

hnRNA *See* heterogeneous nuclear RNA.

homeobox A 180-base-pair segment of DNA found in certain homeotic genes; regulates the expression of other genes and thus controls large-scale developmental processes.

homeostasis (home′ ee o sta′ sis) [Gk. *homos*: same + *stasis*: position] The maintenance of a steady state, such as a constant temperature or a stable social structure, by means of physiological or behavioral feedback responses.

homeotherm (home′ ee o therm) [Gk. *homos*: same + *thermos*: heat] An animal that maintains a constant body temperature by its own internal heating and cooling mechanisms. (Contrast with heterotherm, poikilotherm.)

homeotic genes (home ee ot′ ic) Genes that determine the developmental fate of entire segments of an animal.

homeotic mutations Mutations in homeotic genes that drastically alter the characteristics of a particular body segment, giving it the characteristics of other segments (as when wings grow from a *Drosophila* thoracic segment that should have produced legs).

homing The ability to return over long distances to a specific site.

homo- [Gk. *homos*: same] Prefix indicating two or more similar conditions, structures, or processes. (Contrast with hetero-.)

homolog (home′ o log′) [Gk. *homos*: same + *logos*: word] In cytogenetics, one of a pair (or larger set) of chromosomes having the same overall genetic composition and sequence. In diploid organisms, each chromosome inherited from one parent is matched by an identical (except for mutational changes) chromosome—its homolog—from the other parent.

homology (ho mol′ o jee) [Gk. *homologia*: of one mind; agreement] A similarity between two or more structures that is due to inheritance from a common ancestor. The structures are said to be homologous, and each is a homolog of the others. (Contrast with analogy.)

homoplasy (home′ uh play zee) [Gk. *homos*: same + *plastikos*: shape, mold] The presence in multiple groups of a trait that is not inherited from the common ancestor of those groups. Can result from convergent evolution, evolutionary reversal, or parallel evolution.

homosporous Producing a single type of spore that gives rise to a single type of gametophyte, bearing both female and male reproductive organs. (Contrast with heterosporous.)

homotypic Referring to adhesion of cells of the same type. (Contrast with heterotypic.)

homozygous (home′ oh zie′ gus) [Gk. *homos*: same + *zygotos*: joined] In a diploid organism, having identical alleles of a given gene on both homologous chromosomes. An individual may be a homozygote with respect to one gene and a heterozygote with respect to another. (Contrast with heterozygous.)

hormone (hore′ mone) [Gk. *hormon*: to excite, stimulate] A substance produced in minute amount at one site in a multicellular organism and transported to another site where it acts on target cells.

host An organism that harbors a parasite or symbiont and provides it with nourishment.

Hox genes Conserved homeotic genes found in vertebrates, *Drosophila*, and other animal groups. Hox genes contain the homeobox domain and specify pattern and axis formation in these animals.

human chorionic gonadotropin (hCG) A hormone secreted by the placenta which sustains the corpus luteum and helps maintain pregnancy.

Human Genome Project An effort to determine the DNA sequence of the entire human genome, understand the structures and functions of the genes, make comparisons with other organisms, and understand the social implications of this information.

humoral immune response The part of the immune system mediated by B cells that produce circulating antibodies active against extracellular bacterial and viral infections.

humus (hew′ muss) The partly decomposed remains of plants and animals on the surface of a soil.

hyaluronidase (high′ uh loo ron′ uh dase) An enzyme that digests proteoglycans. In sperm cells, it digests the coatings surrounding an egg so the sperm can enter.

hybrid (high′ brid) [L. *hybrida*: mongrel] (1) The offspring of genetically dissimilar parents. (2) In molecular biology, a double helix formed of nucleic acids from different sources.

hybridize To combine the genetic material of two distinct species or of two distinguishable populations within a species.

hybrid vigor *See* heterosis.

hybridoma A cell produced by the fusion of an antibody-producing cell with a myeloma cell; it produces monoclonal antibodies.

hybrid zone A narrow zone where two populations interbreed, producing hybrid individuals.

hydrocarbon A compound containing only carbon and hydrogen atoms.

hydrogen bond A weak electrostatic bond which arises from the attraction between the slight positive charge on a hydrogen atom and a slight negative charge on a nearby oxygen or nitrogen atom.

hydrological cycle The movement of water from the oceans to the atmosphere, to the soil, and back to the oceans.

hydrolysis (high drol' uh sis) [Gk. *hydro*: water + *lysis*: break apart] A chemical reaction that breaks a bond by inserting the components of water: AB + H_2O → AH + BOH.

hydrophilic (high dro fill' ik) [Gk. *hydro*: water + *philia*: love] Having an affinity for water. (Contrast with hydrophobic.)

hydrophobic (high dro foe' bik) [Gk. *hydro*: water + *phobia*: fear] Having no affinity for water. Uncharged and nonpolar groups of atoms are hydrophobic; for example, fats and the side chain of the amino acid phenylalanine. (Contrast with hydrophilic.)

hydrostatic pressure Pressure generated by compression of liquid in a confined space. Generated in plants, fungi, and some protists with cell walls by the osmotic uptake of water. Generated in animals with closed circulatory systems by the beating of a heart.

hydrostatic skeleton The incompressible internal liquids of some animals that transfer forces from one part of the body to another when acted upon by the surrounding muscles.

hydroxyl group The —OH group found on alcohols and sugars.

hyper- [Gk. *hyper*: above, over] Prefix indicating above, higher, more.

hyperpolarization A change in the resting potential of a membrane so the inside of a cell becomes more electronegative. (Contrast with depolarization.)

hypersensitive response A defensive response of plants to microbial infection; it results in a "dead spot."

hypertension High blood pressure.

hypertonic Having a greater solute concentration. Said of one solution compared to another. (Contrast with hypotonic, isotonic.)

hypha (high' fuh) (plural: hyphae) [Gk. *hyphe*: web] In the fungi and oomycetes, any single filament.

hypo- [Gk. *hypo*: beneath, under] Prefix indicating underneath, below, less.

hypoblast The lower tissue portion of the avian blastula which is joined to the epiblast at the margins of the blastodisc.

hypocotyl [Gk. *hypo*: beneath + *kotyledon*: hollow space] That part of the embryonic or seedling plant shoot that is below the cotyledons.

hypothalamus The part of the brain lying below the thalamus; it coordinates water balance, reproduction, temperature regulation, and metabolism.

hypothesis A tentative answer to a question, from which testable predictions can be generated. (Contrast with theory.)

hypotonic Having a lesser solute concentration. Said of one solution in comparing it to another. (Contrast with hypertonic, isotonic.)

- I -

ileum The final segment of the small intestine.

imaginal disc [L. *imagos*: image, form] In insect larvae, groups of cells that develop into specific adult organs.

imbibition Water uptake by a seed; first step in germination.

immediate hypersensitivity A rapid, extensive immune reaction against an allergen involving IgE and histamine release. (Contrast with delayed hypersensitivity.)

immune system [L. *immunis*: exempt from] A system in vertebrates that recognizes and attempts to eliminate or neutralize foreign substances (e.g., bacteria, viruses, pollutants).

immunization The deliberate introduction of antigen to bring about an immune response.

immunoassay The use of labeled antibodies to measure the concentration of an antigen in a sample.

immunoglobulins A class of proteins, with a characteristic structure, active as receptors and effectors in the immune system.

immunological memory The capacity to more rapidly and massively respond to a second exposure to an antigen than occurred on first exposure.

immunological tolerance A mechanism by which an animal does not mount an immune response to the antigenic determinants of its own macromolecules.

implantation The process by which the early mammalian embryo becomes attached to and embedded in the lining of the uterus.

imprinting (1) In genetics, the differential modification of a gene depending on whether it is present in a male or a female. (2) In animal behavior, a rapid form of learning in which an animal comes to make a particular response, which is maintained for life, to some object or other organism.

inbreeding Breeding among close relatives.

inclusive fitness The sum of an individual's genetic contribution to subsequent generations both via production of its own offspring and via its influence on the survival of relatives who are not direct descendants.

incomplete dominance Condition in which the heterozygous phenotype is intermediate between the two homozygous phenotypes.

incomplete metamorphosis Insect development in which changes between instars are gradual.

incus (in' kus) [L. *incus*: anvil] The middle of the three bones that conduct movements of the eardrum to the oval window of the inner ear. (*See* malleus, stapes.)

independent assortment During meiosis, the random separation of genes carried on nonhomologous chromosomes. Articulated by Mendel as his second law.

indirect transduction A cell signaling mechanism in which a second messenger mediates the interaction between receptor binding and cellular response. (Contrast with direct transduction.)

individual fitness That component of inclusive fitness resulting from an organism producing its own offspring. (Contrast with kin selection.)

indoleacetic acid *See* auxin.

induced fit A change in enzyme conformation upon binding to substrate with an increase in the rate of catalysis.

induced mutation A mutation resulting from treatment with a chemical or other agent.

inducer (1) In enzyme systems, a small molecule which, when added to a growth medium, causes a large increase in the level of some enzyme. (2) In embryology, a substance that causes a group of target cells to differentiate in a particular way.

inducible enzyme An enzyme that is present in much larger amounts when a particular compound (the inducer) has been added to the system. (Contrast with constitutive enzyme.)

inflammation A nonspecific defense against pathogens; characterized by redness, swelling, pain, and increased temperature.

inflorescence A structure composed of several to many flowers.

inflorescence meristem A meristem that produces floral meristems as well as other small leafy structures (bracts).

inhibitor A substance that binds to the surface of an enzyme and interferes with its action on its substrates.

initial cells In plant meristems, undifferentiated cells that retain the capacity to divide producing both undifferentiated cells (initials) and cells committed to differentiation. (Compare with stem cells.)

initiation In molecular biology, the beginning of transcription or translation.

initiation complex Combination of a ribosomal light subunit, an mRNA molecule, and the tRNA charged with the first amino acid coded for by the mRNA; formed at the onset of translation.

initiation factors Proteins that assist in forming the translation initiation complex at the ribosome.

inner cell mass Derived from the mammalian blastula (bastocyst), the inner cell mass will give rise to the yolk sac (via hypoblast) and embryo (via epiblast).

inositol triphosphate (IP_3) An intracellular second messenger derived from membrane phospholipids.

inspiratory reserve volume The amount of air that can be inhaled above the normal tidal inspiration. (Compares with expiratory reserve volume, tidal volume, vital capacity.)

instar (in' star) An immature stage of an insect between molts.

insulin (in' su lin) [L. *insula*: island] A hormone synthesized in islet cells of the pancreas that promotes the conversion of glucose into the storage material, glycogen.

integral membrane protein A membrane protein embedded in the bilayer of the membrane. (Contrast with peripheral membrane protein.)

integrase An enzyme that integrates retroviral cDNA into the genome of the host cell.

integrated pest management Control of pests by the use of natural predators and parasites in conjunction with sparing use of chemicals; an attempt to limit environmental damage.

integument [L. *integumentum*: covering] A protective surface structure. In gymnosperms and angiosperms, a layer of tissue around the ovule which will become the seed coat.

intercalary meristem A meristematic region in plants which occurs not apically, but between two regions of mature tissue. Intercalary meristems occur in the nodes of grass stems, for example.

intercostal muscles Muscles between the ribs that can augment breathing movements by elevating and suppressing the rib cage.

interference competition Competition in which individuals actively interfere with one another's access to resources. (Compare with exploitation competition.)

interference RNA (RNAi) A mechanism for reducing mRNA translation whereby a double-stranded RNA, made by the cell or synthetically, is processed to a small, single-stranded RNA, and binding of this RNA to a target mRNA results in the latter's breakdown.

interferon A glycoprotein produced by virus-infected animal cells; increases the resistance of neighboring cells to the virus.

interkinesis The period between meiosis I and meiosis II.

interleukins Regulatory proteins, produced by macrophages and lymphocytes, that act upon other lymphocytes and direct their development.

intermediate disturbance hypothesis The hypothesis that explains why species richness is lower in areas with both high and low rates of disturbance than in areas with intermediate rates of disturbance.

intermediate filaments Cytoskeletal component with diameters between the larger microtubules and smaller microfilaments.

internal environment The physical and chemical characteristics of the extracellular fluids of the body.

interneuron A neuron that communicates information between two other neurons.

ionotropic receptors A receptor that that directly alters membrane permeability to a type of ion when it combines with its ligand.

internode The region between two nodes of a plant stem.

interphase The period between successive nuclear divisions during which the chromosomes are diffuse and the nuclear envelope is intact. It is during this period that the cell is most active in transcribing and translating genetic information.

interspecific competition Competition between members of two or more species. (Contrast with intraspecific competition.)

intertropical convergence zone The tropical region where the air rises most strongly; moves north and south with the passage of the sun overhead.

intraspecific competition Competition among members of the same species. (Contrast with interspecific competition.)

intrinsic protein A membrane protein that is embedded in the phospholipid bilayer of the membrane. (Contrast with extrinsic protein.)

intrinsic rate of increase The rate at which a population can grow when its density is low and environmental conditions are highly favorable.

intron A portion of a DNA molecule that, because of RNA splicing, is not involved in coding for part of a polypeptide molecule. (Contrast with exon.)

invagination An infolding of cells during animal embryonic development.

inversion A rare 180° reversal of the order of genes within a segment of a chromosome.

invertebrate A "convenience term" that encompasses any animal that is not a vertebrate—that is, whose nerve cord is not enclosed in a backbone of bony segments.

in vitro [L.: in glass] In a test tube, rather than in a living organism. (Contrast with in vivo.)

in vitro evolution Evolution in the laboratory, as used to produce compounds for industrial and pharmaceutical purposes.

in vivo [L.: in the living state] In a living organism. Many processes that occur in vivo can be reproduced in vitro with the right selection of cellular components. (Contrast with in vitro.)

ion (eye' on) [Gk.: *ion*: wanderer] An atom or group of atoms with electrons added or removed, giving it a negative or positive electrical charge.

ion channel A membrane protein that can let ions diffuse across the membrane. The channel can be ion-selective, and it can be voltage-gated or ligand-gated.

ionic bond An electrostatic attraction between positively and negatively charged ions. Usually a strong bond.

iris (eye' ris) [Gk. *iris*: rainbow] The round, pigmented membrane that surrounds the pupil of the eye and adjusts its aperture to regulate the amount of light entering the eye.

irruption A rapid increase in the density of a population. Often followed by massive emigration.

islets of Langerhans Clusters of hormone-producing cells in the pancreas.

iso- [Gk. *iso*: equal] Prefix used two separate entities that share some element of identity.

isogamous Describes male and female gametes that are morphologically identical.

isolating mechanism Geographical, physiological, ecological, or behavioral mechanisms that lead to a reduction in the frequency of successful matings between individuals in separate populations of a species. Can lead to the eventual evolution of separate species.

isomers Molecules consisting of the same numbers and kinds of atoms, but differing in the bonding patterns by which the atoms are held together.

isomorphic (eye so more' fik) [Gk. *isos*: equal + *morphe*: form] Having the same form or appearance, as when the haploid and diploid life stages of an organism appear identical. (Contrast with heteromorphic.)

isotonic Having the same solute concentration; said of two solutions. (Contrast with hypertonic, hypotonic.)

isotope (eye' so tope) [Gk. *isos*: equal + *topos*: place] Isotopes of a given chemical element have the same number of protons in their nuclei (and thus are in the same position on the periodic table), but differ in the number of neutrons

isozymes Enzymes that have somewhat different amino acid sequences but catalyze the same reaction.

- J -

jasmonates Plant hormones that trigger defenses against pathogens and herbivores.

jejunum (jih jew' num) The middle division of the small intestine, where most absorption of nutrients occurs. (*See* duodenum, ileum.)

joule (jool, or jowl) A unit of energy, equal to 0.24 calories.

juvenile hormone In insects, a hormone maintaining larval growth and preventing maturation or pupation.

- K -

karyogamy The fusion of nuclei of two cells. (Contrast with plasmogamy.)

karyotype The number, forms, and types of chromosomes in a cell.

keratin (ker' a tin) [Gk. keras: horn] A protein which contains sulfur and is part of such hard tissues as horn, nail, and the outermost cells of the skin.

ketone (key' tone) A compound with a C=O group attached to two other groups, neither of which is an H atom. Many sugars are ketones. (Contrast with aldehyde.)

keystone species Species that have a dominant influence on the composition of a community.

kidneys A pair of excretory organs in vertebrates.

kin selection That component of inclusive fitness resulting from helping the survival of relatives containing the same alleles by descent from a common ancestor. (Contrast with individual fitness.)

kinase (kye' nase) An enzyme that transfers a phosphate group from ATP to another molecule. Protein kinases transfer phosphate from ATP to specific proteins, playing important roles in cell regulation.

kinesin Motor protein having the capacity to attach to organelles or vesicles and move them along microtubules of the cytoskeleton.

kinetic energy The energy associated with movement.

kinetochore (kin net' oh core) [Gk. *kinetos*: moving] Specialized structure on a centromere to which microtubules attach.

knockout A molecular genetic method in which a single gene of an organism is permanently inactivated.

Koch's posulates A set of rules for establishing that a particular microorganism causes a particular disease.

Krebs cycle *See* citric acid cycle.

- L -

lactic acid fermentation Fermentation whose end product is lactic acid (lactate).

lagging strand In DNA replication, the daughter strand that is synthesized in discontinuous stretches. (*See* Okazaki fragments.)

lamella (la mell' ah) [L. *lamina*: thin sheet] Layer.

larva (plural: larvae) [L. *lares*: guiding spirits] An immature stage of any invertebrate animal that differs dramatically in appearance from the adult.

larynx (lar' inks) [Gk. *larynx*: voice box] A structure between the pharynx and the trachea that includes the vocal cords.

lateral Pertaining to the side.

lateral gene transfer The transfer of genes from one species to another, common among bacteria and archaea.

lateral meristems The vascular cambium and cork cambium, which give rise to secondary tissue in plants.

laticifers (luh tiss' uh furs) In some plants, elongated cells containing secondary plant products such as latex.

leader sequence A sequence of amino acids at the amino-terminal end of a newly synthesized protein; determines where the protein will be placed in the cell.

leading strand In DNA replication, the daughter strand that is synthesized continuously. (Contrast with lagging strand.)

leaf primordium An outgrowth on the side of the shoot apical meristem that will eventually develop into a leaf.

lenticel (len' ti sill) Spongy region in a plant's periderm, allowing gas exchange.

leukocyte (loo' ko sight) [Gk. *leukos*: clear + *kytos*: container] A white blood cell.

lichen (lie' kun) An organism resulting from the symbiotic association of a true fungus and either a cyanobacterium or a unicellular alga.

life cycle The entire span of the life of an organism from the moment of fertilization (or asexual generation) to the time it reproduces in turn.

life history The stages an individual goes through during its life.

life table A table showing, for a group of equal-aged individuals, the proportion still alive at different times in the future and the number of offspring they produce during each time interval.

ligament A band of connective tissue linking two bones in a joint.

ligand (lig' and) Any molecule that binds to a receptor site of another (usually larger) molecule.

light reactions The initial phase of photosynthesis, in which light energy is converted into chemical energy.

light-independent reactions The phase of photosynthesis in which chemical energy captured in the light reactions is used to drive the reduction of CO_2 to form carbohydrates.

lignin The principal noncarbohydrate component of wood, a polymer that binds together cellulose fibrils in some plant cell walls.

limbic system A group of primitive vertebrate forebrain nuclei that form a network and are involved in emotions, drives, instinctive behaviors, learning, and memory.

limiting resource The required resource whose supply most strongly influences the size of a population.

linkage Association between genetic markers on the same chromosome such that they do not show random assortment and seldom recombine; the closer the markers, the lower the frequency of recombination.

lipase (lip' ase; lye' pase) An enzyme that digests fats.

lipids (lip' ids) [Gk. *lipos*: fat] Substances in a cell which are easily extracted by organic solvents; fats, oils, waxes, steroids, and other large organic molecules, including those which, with proteins, make up the cell membranes. (Compare with phospholipids.)

littoral zone The coastal zone from the upper limits of tidal action down to the depths where the water is thoroughly stirred by wave action.

liver A large digestive gland. In vertebrates, it secretes bile and is involved in the formation of blood.

lobes Regions of the human cerebral hemispheres; includes the temporal, frontal, parietal, and occipital lobes.

locus In genetics, a specific location on a chromosome. May be considered to be synonymous with *gene*.

logistic growth Growth, especially in the size of an organism or in the number of organisms in a population, that slows steadily as the entity approaches its maximum size. (Contrast with exponential growth.)

long-day plant (LDP) A plant that requires long days (actually, short nights) in order to flower.

long-term potentiation (LTP) A long-lasting increase in the sensitivity of a neuron resulting from a period of intense stimulation.

loop of Henle (hen' lee) Long, hairpin loop of the mammalian renal tubule that runs from the cortex down into the medulla, and back to the cortex. Creates a concentration gradient in the interstitial fluids in the medulla.

lophophore A U-shaped fold of the body wall with hollow, ciliated tentacles that encircles the mouth of animals in several different groups. Used for filtering prey from the surrounding water.

lumen (loo' men) [L. *lumen*: light] The open cavity inside any tubular organ or structure, such as the gut or a kidney tubule.

luteinizing hormone A gonadotropin produced by the anterior pituitary. It stimulates the gonads to produce sex hormones.

lymph [L. *lympha*: liquid] A clear, watery fluid that is formed as a filtrate of blood; it contains white blood cells; it collects in a series of special vessels and is returned to the bloodstream.

lymph nodes Specialized tissue regions that act as filters for cells, bacteria and foreign matter.

lymphocyte A major class of white blood cells. Includes T cells, B cells, and other cell types important in the immune response.

lymphoid tissue Tissues of the immune defense system dispersed throughout the body and consisting of: thymus, spleen, bone marrow, lymph nodes, blood, and lymph.

lysis (lie' sis) [Gk. *lysis*: break apart] Bursting of a cell.

lysogenic bacteria Bacteria that harbor a viral chromosome capable of the lysogenic cycle.

lysogenic cycle A form of viral replication in which the virus becomes incorporated into the bacterial chromosome and the host cell is not killed. (Contrast with lytic cycle.)

lysosome (lie' so soam) [Gk. *lysis*: break away + *soma*: body] A membrane-enclosed organelle found in eukaryotic cells (other than plants). Lysosomes contain a mixture of enzymes that can digest most of the macromolecules found in the rest of the cell.

lysozyme (lie' so zyme) An enzyme in saliva, tears, and nasal secretions that attacks bacterial cell walls, as one of the body's nonspecific defense mechanisms.

lytic cycle A form of viral reproduction that lyses the host bacterium releasing the new viruses. (Contrast with lysogenic cycle.)

- M -

M phase The portion of the cell cycle in which mitosis takes place.

macroevolution [Gk. *makros*: large, long] Evolutionary changes occurring over long time spans and usually involving changes in many traits. (Contrast with microevolution.)

macromolecule A giant polymeric molecule. The macromolecules are proteins, polysaccharides, and nucleic acids.

macronutrient A mineral element required by plant tissues in concentrations of at least 1 milligram per gram of their dry matter.

macrophage (mac' roh faj) A type of white blood cell that endocytoses bacteria and other cells.

MADS box A DNA-binding domain in many plant transcription factors that is active in development.

major histocompatibility complex (MHC) A complex of linked genes, with multiple alleles, that control a number of cell surface antigens that identify self and can lead to graft rejection.

malignant Referring to a tumor that can grow indefinitely and/or spread from the original site of growth to other locations in the body. (Contrast with benign.)

malleus (mal' ee us) [L. *malleus*: hammer] The first of the three bones that conduct movements of the eardrum to the oval window of the inner ear. (*See* incus, stapes.)

Malpighian tubule (mal pee' gy un) A type of protonephridium found in insects.

mantle A sheet of specialized tissues that covers most of the viscera of mollusks; provides protection to internal organs and secretes the shell.

mapping In genetics, determining the order of genes on a chromosome and the distances between them.

map unit The distance between two genes, a recombinant frequency of 0.01.

marine [L. *mare*: sea, ocean] Pertaining to or living in the ocean. (Contrast with aquatic, terrestrial.)

marker A gene of identifiable phenotype that indicates the presence on another gene, DNA segment, or chromosome fragment.

mass extinctions Periods of evolutionary history during which rates of extinction were much higher than during intervening times.

mass number The sum of the number of protons and neutrons in an atom's nucleus.

mast cells Typically found in connective tissue, mast cells can be provoked by antigens or inflammation to release histamine.

maternal effect genes These genes code for morphogens that determine the polarity of the egg and larva in the fruit fly *Drosophila melanogaster*.

maternal inheritance Inheritance in which the mother's phenotype is exclusively expressed. Mitochondria and chloroplasts are maternally inherited via egg cytoplasm. Also known as cytoplasmic inheritance.

mating type A particular strain of a species that is incapable of sexual reproduction with another member of the same strain but capable of sexual reproduction with members of other strains of the same species.

maximum likelihood A statistical method of determining which of two or more hypotheses (such as phylogenetic trees) best fit the observed data, given an explicit model of how the data were generated.

mechanoreceptor A cell that is sensitive to physical movement and generates action potentials in response.

medulla (meh dull' luh) (1) The inner, core region of an organ, as in the adrenal medulla (adrenal gland) or the renal medulla (kidneys). (2) The portion of the brain stem that connects to the spinal cord.

megaphyll The generally large leaf of a fern, horsetail, or seed plant, with several to many veins. (Contrast with microphyll.)

megaspore [Gk. *megas*: large + *spora*: to sow] In plants, a haploid spore that produces a female gametophyte.

meiosis (my oh' sis) [Gk. *meiosis*: diminution] Division of a diploid nucleus to produce four haploid daughter cells. The process consists of two successive nuclear divisions with only one cycle of chromosome replication. In *meiosis I*, homologous chromosomes separate but retain their chromatids. The second division *meiosis II*, is similar to mitosis, in which chromatids separate.

melatonin A hormone released by the pineal gland that is involved in photoperiodism and circadian rhythms.

membrane potential The difference in electrical charge between the inside and the outside of a cell, caused by a difference in the distribution of ions.

memory cells Long-lived lymphocytes produced by exposure to antigen. They persist in the body and are able to mount a rapid response to subsequent exposures to the antigen.

Mendelian population A local population of individuals belonging to the same species and exchanging genes with one another.

Mendel's first law *See* segregation.

Mendel's second law *See* independent assortment.

menstrual cycle The monthly sloughing off of the uterine lining if fertilization does not occur in the female. Occurs between puberty and menopause.

meristem [Gk. *meristos*: divided] Plant tissue made up of undifferentiated actively dividing cells.

mesenchyme (mez' en kyme) [Gk. *mesos*: middle + *enchyma*: infusion] Embryonic or unspecialized cells derived from the mesoderm.

mesoderm [Gk. *mesos*: middle + *derma*: skin] The middle of the three embryonic tissue layers first delineated during gastrulation. Gives rise to skeleton, circulatory system, muscles, excretory system, and most of the reproductive system.

mesophyll (mez' uh fill) [Gk. *mesos*: middle + *phyllon*: leaf] Chloroplast-containing, photosynthetic cells in the interior of leaves.

mesosome (mez' uh soam') [Gk. *mesos*: middle + *soma*: body] A localized infolding of the plasma membrane of a bacterium.

messenger RNA (mRNA) A transcript of one of the strands of DNA; carries information (as a sequence of codons) for the synthesis of one or more proteins.

meta- [Gk.: between, along with, beyond] A prefix used in biology to denote a change or a shift to a new form or level; for example, as used in metamorphosis.

metabolic compensation Changes in metabolic properties of an organism that render it less sensitive to temperature changes.

metabolic factor In bacteria, a plasmid that carries genes determining unusual metabolic functions, such as the breakdown of hydrocarbons in oil.

metabolic pathway A series of enzyme-catalyzed reactions so arranged that the product of one reaction is the substrate of the next.

metabolism (meh tab' a lizm) [Gk. *metabole*: to change] The sum total of the chemical reactions that occur in an organism, or some subset of that total (as in respiratory metabolism).

metabotropic receptor A receptor that that indirectly alters membrane permeability to a type of ion when it combines with its ligand.

metamorphosis (met' a mor' fo sis) [Gk. *meta*: between + *morphe*: form, shape] A change occurring between one developmental stage and another, as for example from a tadpole to a frog. (*See* complete metamorphosis, incomplete metamorphosis.)

metaphase (met' a phase) The stage in nuclear division at which the centromeres of the highly supercoiled chromosomes are all lying on a plane (the metaphase plane or plate) perpendicular to a line connecting the division poles.

metapopulation A population divided into subpopulations, among which there are occasional exchanges of individuals.

metastasis (meh tass' tuh sis) The spread of cancer cells from their original site to other parts of the body.

methylation The addition of a methyl group (—CH_3) to a molecule. Extensive methylation of cytosine in DNA is correlated with reduced transcription.

MHC *See* major histocompatibility complex.

micelle a particle of lipid covered with bile salts that is produced in the duodenum and facilitates digestion and absorption of lipids.

microbiology [Gk. *mikros*: small + *bios*: life + *logos*: discourse] The scientific study of microscopic organisms, particularly bacteria, protists, and viruses.

microevolution The small evolutionary changes typically occurring over short time spans; generally involving a small number of traits and minor genetic changes. (Contrast with macroevolution.)

microfilament Minute fibrous structure generally composed of actin found in the cytoplasm of eukaryotic cells. They play a role in the motion of cells.

micronutrient A mineral element required by plant tissues in concentrations of less than 100 micrograms per gram of their dry matter.

microphyll A small leaf with a single vein, found in club mosses and their relatives. (Contrast with megaphyll.)

micropyle (mike' roh pile) [Gk. *mikros*: small + *pylon*: gate] Opening in the integument(s) of a seed plant ovule through which pollen grows to reach the female gametophyte within.

micro RNA A small RNA, typically about 21 bases long, that binds to mRNA to reduce its translation.

microspore [Gk. *mikros*: small + *spora*: to sow] In plants, a haploid spore that produces a male gametophyte.

microtubules Minute tubular structures found in centrioles, spindle apparatus, cilia, flagella, and cytoskeleton of eukaryotic cells. These tubules play roles in the motion and maintenance of shape of eukaryotic cells.

microvilli (singular: microvillus) The projections of epithelial cells, such as the cells lining the small intestine, that increase their surface area.

middle lamella A layer of polysaccharides that separates plant cells; a shared middle lamella lies outside the primary walls of the two cells.

migration The regular, seasonal movements of animals.

mineral An inorganic substance other than water.

mineral nutrients Inorganic ions required by organisms for normal growth and reproduction.

mismatch repair When a single base in DNA is changed into a different base, or the wrong base inserted during DNA replication, there is a mismatch in base pairing with the base on the opposite strand. A repair system removes the incorrect base and inserts the proper one for pairing with the opposite strand.

missense substitution A change in a gene from one nucleotide to another that also results in a change in the amino acid specified by the corresponding codon. (Compare with frame-shift mutation, nonsense substitution, synonymous substitution.)

mitochondrial matrix The fluid interior of the mitochondrion, enclosed by the inner mitochondrial membrane.

mitochondrion (my' toe kon' dree un) [Gk. *mitos*: thread + *chondros*: grain] An organelle in eukaryotic cells that contains the enzymes of the citric acid cycle, the respiratory chain, and oxidative phosphorylation.

mitosis (my toe' sis) [Gk. *mitos*: thread] Nuclear division in eukaryotes leading to the formation of two daughter nuclei, each with a chromosome complement identical to that of the original nucleus.

mitotic center Cellular region that organizes the microtubules for mitosis. In animals a centrosome serves as the mitotic center.

moderately repetitive DNA DNA sequences that appear hundreds to thousands of times in the genome. They include the DNA sequences coding for rRNAs and tRNAs, as well as the DNA at telomeres.

modular organism An organism which grows by producing additional units of body construction (modules) that are very similar to the units of which it is already composed.

mole A quantity of a compound whose weight in grams is numerically equal to its molecular weight expressed in atomic mass units. Avogadro's number of molecules: 6.023×10^{23} molecules.

molecular clock The theory that macromolecules diverge from one another over evolutionary time at a constant rate; this rate may provide insight into the phylogenetic relationships among organisms.

molecular tool kit A set of developmental genes and proteins that is common to most animals and is hypothesized to be responsible for the evoltuion of their differing developmental pathways.

molecular weight The sum of the atomic weights of the atoms in a molecule.

molecule A particle made up of two or more atoms joined by covalent bonds or ionic attractions.

molting The process of shedding part or all of an outer covering, as the shedding of feathers by birds or of the entire exoskeleton by arthropods.

monoclonal antibody Antibody produced in the laboratory from a clone of hybridoma cells, each of which produces the same specific antibody.

monocytes White blood cells that produce macrophages.

monoecious (mo nee' shus) [Gk. *mono*: one + *oikos*: house] Describes organisms in which both sexes are "housed" in a single individual that produces both eggs and sperm. (In some plants, these are found in different flowers within the same plant.) Examples include corn, peas, earthworms, hydras. (Contrast with dioecious.)

monohybrid cross A mating in which the parents differ with respect to the alleles of only one locus of interest.

monomer [Gk. *mono*: one + *meros*: unit] A small molecule, two or more of which can be combined to form oligomers (consisting of a few monomers) or polymers (consisting of many monomers).

monophyletic (mon' oh fih leht' ik) [Gk. *mono*: one + *phylon*: tribe] Referring to a group that consists of an ancestor and all of its descendants. (Compare with paraphyletic, polyphyletic.)

monosaccharide A simple sugar. Oligosaccharides and polysaccharides are made up of monosaccharides.

monosomic Referring to an organism with one less than the normal diploid number of chromosomes.

monosynaptic reflex A neural reflex that begins in a sensory neuron and makes a single synapse before activating a motor neuron.

morphogen A diffusible substances whose concentration gradients determine patterns of development in animals and plants.

morphogenesis (more' fo jen' e sis) [Gk. *morphe*: form + *genesis*: origin] The development of form; the overall consequence of determination, differentiation, and growth.

morphology (more fol' o jee) [Gk. *morphe*: form + *logos*: study, discourse] The scientific study of organic form, including both its development and function.

mosaic development Pattern of animal embryonic development in which each blastomere contributes a specific part of the adult body. (Contrast with regulative development.)

motor end plate The modified area on a muscle cell membrane where a synapse is formed with a motor neuron.

motor neuron A neuron carrying information from the central nervous system to an effector such as a muscle fiber.

motor proteins Specialized proteins that use energy to change shape and move cells or structures within cells. *See* dynein, kinesin.

motor unit A motor neuron and the set of muscle fibers it controls.

mRNA *See* messenger RNA.

mucosa (mew koh' sah) An epithelial membrane containing cells that secrete mucus. The inner cell layers of the digestive and respiratory tracts.

Müllerian mimicry The convergence in appearance of two or more unpalatable species.

multifactorial In medicine, referring to a disease with many interacting causes, both genetic and environmental.

muscle fiber A single muscle cell. In the case of skeletal (striated) muscle, a syncitial, multinucleate cell.

muscle tissue Excitable tissue that can contract due to interactions of actin and myosin. Three types are striated, smooth, and cardiac.

muscle tone The degree of contraction of a muscle.

muscle spindle Modified muscle fibers encased in a connective sheat and functioning as stretch receptors.

mutagen (mute' ah jen) [L. *mutare*: change + Gk. *genesis*: source] Any agent (e.g., chemicals, radiation) that increases the mutation rate.

mutation A detectable, heritable change in the genetic material not caused by recombination.

mutation pressure Evolution (change in gene proportions) by different mutation rates alone (i.e., without the influence of natural selection).

mutualism The type of symbiosis, such as that exhibited by fungi and algae or cyanobacteria in forming lichens, in which both species profit from the association.

mycelium (my seel' ee yum) [Gk. *mykes*: fungus] In the fungi, a mass of hyphae.

mycorrhiza (my' ko rye' za) [Gk. *mykes*: fungus + *rhiza*: root] An association of the root of a plant with the mycelium of a fungus.

myelin (my' a lin) A material forming a sheath around some axons. Formed by Schwann cells that wrap themselves about the axon, myelin insulates the axon electrically and increases the rate of transmission of a nervous impulse.

myocardial infarction A blockage of an artery that carries blood to the heart muscle.

myofibril (my' oh fy' bril) [Gk. *mys*: muscle + L. *fibrilla*: small fiber] A polymeric unit of actin or myosin in a muscle.

myogenic (my oh jen' ik) [Gk. *mys*: muscle + *genesis*: source] Originating in muscle.

myoglobin (my' oh globe' in) [Gk. *mys*: muscle + L. *globus*: sphere] An oxygen-binding molecule found in muscle. Consists of a heme unit and a single globiin chain, and carrys less oxygen than hemoglobin.

myosin One of the two major proteins of muscle, it makes up the thick filaments. (*See* actin.)

- N -

NAD (nicotinamide adenine dinucleotide) A compound found in all living cells, existing in two interconvertible forms: the oxidizing agent NAD^+ and the reducing agent $NADH + H^+$.

NADP (nicotinamide adenine dinucleotide phosphate) A compound similar to NAD, but possessing another phosphate group; plays similar roles but is used by different enzymes.

natural killer cells A nonspecific defensive cell (lymphocyte) that attacks tumor cells and virus infected cells.

natural selection The differential contribution of offspring to the next generation by various genetic types belonging to the same population. The mechanism of evolution proposed by Charles Darwin.

necrosis (nec roh' sis) [Gk. *nekros*: death] Tissue damage resulting from cell death.

negative control The situation in which a regulatory macromolecule (generally a repressor) functions to turn off transcription. In the absence of a regulatory macromolecule, the structural genes are turned on.

negative feedback Information relevant to the rate of a process that can be used by a control system to return the outcome of that process to an optimal level.

nematocyst (ne mat' o sist) [Gk. *nema*: thread + *kystis*: cell] An elaborate, threadlike structure produced by cells of jellyfish and other cnidarians, used chiefly to paralyze and capture prey.

nephridium (nef rid′ ee um) [Gk. *nephros*: kidney] An organ which is involved in excretion, and often in water balance, involving a tube that opens to the exterior at one end.

nephron (nef′ ron) [Gk. *nephros*: kidney] The functional unit of the kidney, consisting of a structure for receiving a filtrate of blood, and a tubule that absorbs selected parts of the filtrate back into the bloodstream.

nephrostome (nef′ ro stome) [Gk. *nephros*: kidney + *stoma*: opening] An opening in a nephridium through which body fluids can enter.

Nernst equation A mathematical statement; calculates the potential across a membrane permeable to a single type of ion that differs in concentration on the two sides of the membrane.

nerve A structure consisting of many neuronal axons and connective tissue.

net primary production Total photosynthesis minus respiration by plants.

neural plate A thickened strip of ectoderm along the dorsal side of the early vertebrate embryo; gives rise to the central nervous system.

neural tube An early stage in the development of the vertebrate nervous system consisting of a hollow tube created by two opposing folds of the dorsal ectoderm along the anterior–posterior body axis.

neurohormone A chemical signal produced and released by neurons; the signal then acts as a hormone.

neuromuscular junction The region where a motor neuron contacts a muscle fiber, creating a synapse.

neuron (noor′ on) [Gk. *neuron*: nerve] A nervous system cell that can generate and conduct action potentials along an axon to a synapse with another cell.

neurotransmitter A substance produced in and released by one a neuron (the presynaptic cell) that diffuses across a synapse and excites or inhibits another cell (the postsynaptic cell).

neurula (nure′ you la) Embryonic stage during the dorsal nerve cord forms from two ectodermal ridges.

neurulation A stage in vertebrate development during which the nervous system begins to form.

neutral allele An allele that does not alter the functioning of the proteins for which it codes.

neutral theory A view of molecular evolution that postulates that most mutations do not affect the amino acid being coded for, and that such mutations accumulate in a population at rates driven by genetic drift and mutation rates.

neutron (new′ tron) One of the three most fundamental particles of matter, with mass approximately 1 amu and no electrical charge.

neutrophils Abundant, short-lived phagocytic leukocytes that attack antibody-coated antigens.

nitrate reduction The process by which nitrate (NO_3^-) is reduced to ammonia (NH_3).

nitric oxide (NO) An unstable molecule (a gas) that serves as a second messenger causing smooth muscle to relax. In the nervous system it operates as a neurotransmitter.

nitrification The oxidation of ammonia to nitrite and nitrate ions, performed by certain soil bacteria.

nitrogenase In nitrogen-fixing organisms, an enzyme complex that mediates the stepwise reduction of atmospheric N_2 to ammonia.

nitrogen fixation Conversion of nitrogen gas to ammonia, which makes nitrogen available to living things. Carried out by certain prokaryotes, some of them free-living and others living within plant roots.

node [L. *nodus*: knob, knot] In plants, a (sometimes enlarged) point on a stem where a leaf is or was attached.

node of Ranvier A gap in the myelin sheath covering an axon; the point where the axonal membrane can fire action potentials.

noncompetitive inhibitor An inhibitor that binds the enzyme at a site other than the active site. (Contrast with competitive inhibitor.)

noncyclic electron transport In photosynthesis, the flow of electrons that forms ATP, NADPH, and O_2.

nondisjunction Failure of sister chromatids to separate in meiosis II or mitosis, or failure of homologous chromosomes to separate in meiosis I. Results in aneuploidy.

nonpolar molecule A molecule whose electric charge is evenly balanced from one end of the molecule to the other.

nonrandom mating The selection by individuals of other individuals of particular genotypes as mates.

non-REM sleep A state of sleep characterized by low muscle tone, but not atonia, quiescence,

nonsense substitution A change in a gene from one nucleotide to another that prematurely terminates a polypeptide. Termination occurs when a codon that specifies an amino acid is changed to one of the codons (UAG, UAA, or UGA) that signal termination of translation. (Compare with frame-shift mutation, missense substitution, synonymous substitution.)

nonspecific defenses Immunologic responses directed against any invading agent without reacting to apecific antigens.

nonsynonymous substitution A change in a gene from one nucleotide to another that changes the amino acid specified by the corresponding codon (i.e., AGC $\rightarrow$ AGA, or serine $\rightarrow$ arginine). (Contrast with synonymous substitution.)

nonvascular plants Those plants lacking well-developed vascular tissue; the liverworts, hornworts, and mosses. (Contrast with vascular plants.)

norepinephrine A neurotransmitter found in the central nervous system and also at the postganglionic nerve endings of the sympathetic nervous system. Also called noradrenaline.

notochord (no′ tow kord) [Gk. *notos*: back + *chorde*: string] A flexible rod of gelatinous material serving as a support in the embryos of all chordates and in the adults of tunicates and lancelets.

nuclear envelope The surface, consisting of two layers of membrane, that encloses the nucleus of eukaryotic cells.

nuclear lamina A meshwork of fibers on the inner surface of the nuclear envelope.

nuclear pore complex A protein structure situated in nuclear pores through which RNA and proteins enter and leave the nucleus.

nucleic acid (new klay′ ik) A long-chain alternating polymer of deoxyribose or ribose and phosphate groups, with nitrogenous bases—adenine, thymine, uracil, guanine, or cytosine (A, T, U, G, or C)—as side chains. DNA and RNA are nucleic acids.

nucleic acid hybridization A technique in which a single-stranded nucleic acid probe in made that is complementary to, and binds to, a target sequence, either DNA or RNA. The resulting double-stranded molecule is a hybrid.

nucleoid (new′ klee oid) The region that harbors the chromosomes of a prokaryotic cell. Unlike the eukaryotic nucleus, it is not bounded by a membrane.

nucleolar organizer (new klee′ o lar) A region on a chromosome that is associated with the formation of a new nucleolus following nuclear division. The site of the genes that code for ribosomal RNA.

nucleolus (new klee′ oh lus) A small, generally spherical body found within the nucleus of eukaryotic cells. The site of synthesis of ribosomal RNA.

nucleoplasm (new′ klee o plazm) The fluid material within the nuclear envelope of a cell, as opposed to the chromosomes, nucleoli, and other particulate constituents.

nucleoside A nucleotide without the phosphate group.

nucleosome A portion of a eukaryotic chromosome, consisting of part of the DNA molecule wrapped around a group of histone molecules, and held together by another type of histone molecule. The chromosome is made up of many nucleosomes.

nucleotide The basic chemical unit in a nucleic acid. A nucleotide in RNA consists of one of four nitrogenous bases linked to ribose, which in turn is linked to phosphate. In DNA, deoxyribose is present instead of ribose.

nucleotide substitution A change of one base pair to another in a DNA sequence.

nucleus (new′ klee us) [L. *nux*: kernel or nut] (1) In cells, the centrally located compartment of eukaryotic cells that is bounded by a double membrane and contains the chromosomes. (2) In the brain, an identifiable group of neurons that share common characteristics or functions.

null hypothesis The assertion that an effect proposed by its companion hypothesis does not in fact exist.

nutrient A food substance; or, in the case of mineral nutrients, an inorganic element required for completion of the life cycle of an organism.

- 0 -

obligate anaerobe An anaerobic prokaryote that cannot survive exposure to O_2.

odorant A molecule that can bind to an olfactory receptor.

oil A triglyceride that is liquid at room temperature. (Contrast with fat.)

Okazaki fragments Newly formed DNA making up the lagging strand in DNA replication. DNA ligase links Okazaki fragments together to give a continuous strand.

olfactory [L. *olfacere*: to smell] Having to do with the sense of smell.

oligomer [Gk.: *oligo*: a few + *meros*: units] A compound molecule of intermediate size, made up of two to a few monomers. (Contrast with monomer, polymer.)

oligosaccharide A polymer containing a small number of monosaccharides.

oligosaccharins Plant hormones, derived from the plant cell wall, that trigger defenses against pathogens.

ommatidium [Gk. *omma*: eye] One of the units which, collected into groups of up to 20,000, make up the compound eye of arthropods.

omnivore [L. *omnis*: everything + *vorare*: to devour] An organism that eats both animal and plant material. (Contrast with carnivore, detritivore, herbivore.)

oncogene [Gk. *onkos*: mass, tumor + *genes*: born] Genes that greatly stimulate cell division, giving rise to tumors.

one-gene, one-polypeptide The principle that each gene codes for a single polypeptide.

oocyte (oh′ eh site) [Gk. *oon*: egg + *kytos*: container] The cell that gives rise to eggs in animals.

oogenesis (oh′ eh jen e sis) [Gk. *oon*: egg + *genesis*: source] Female gametogenesis, leading to production of the egg.

oogonium (oh′ eh go′ nee um) In some algae and fungi, a cell in which an egg is produced.

operator The region of an operon that acts as the binding site for the repressor.

operon A genetic unit of transcription, typically consisting of several structural genes that are transcribed together; the operon contains at least two control regions: the promoter and the operator.

opportunity cost The sum of the benefits an animal forfeits by not being able to perform some other behavior during the time when it is performing a given behavior.

opsin (op′ sin) [Gk. *opsis*: sight] The protein portion of the visual pigment rhodopsin. (*See* rhodopsin.)

optic chiasm [Gk. *chiasma*: cross] Structure on the lower surface of the vertebrate brain where the two optic nerves come together.

optical isomers Two isomers that are mirror images of each other.

orbital A region in space surrounding the atomic nucleus in which an electron is most likely to be found.

organ [Gk. *organon*: tool] A body part, such as the heart, liver, brain, root, or leaf. Organs are composed of different tissues integrated to perform a distinct function. Organs are in turn often integrated into systems, such as the digestive or reproductive system.

organ identity genes Plant genes that specify the various parts of the flower. *See* homeotic genes.

organ of Corti Structure in the inner ear that transforms mechanical forces produced from pressure waves ("sound waves") into action potentials that are sensed as sound.

organ system An interrelated and integrated group of tissues and organs that work together in a physiological function.

organelles (or gan els′) Organized structures found in or on eukaryotic cells. Examples include ribosomes, nuclei, mitochrondria, chloroplasts, cilia, and contractile vacuoles.

organic Pertaining to any aspect of living matter, e.g., to its evolution, structure, or chemistry. The term is also applied to any chemical compound that contains carbon.

organism Any living entity.

organizer Region of an early embryo that directs the development of nearby regions. In amphibian early gastrulas, the dorsal lip of the blastopore is the organizer.

organogenesis The formation of organs and organ systems during development.

origin of replication DNA sequence at which helicase unwinds the DNA double helix and DNA polymerase binds to initiate DNA replication.

orthology (or thol′ o jee) A type of homology applied to genes in which the divergence of homologous genes can be traced to speciation events. The genes are said to be *orthologous*, and each is an *ortholog* of the others. (Compare with paralogy)

osmoconformer An aquatic animal that maintains an osmotic concentration of its extracellular fluid that is thet same as that of the external environment.

osmolarity The concentration of osmotically active particles in a solution.

osmoregulation Regulation of the chemical composition of the body fluids of an organism.

osmoreceptor Neuron that converts changes in the solute potential of interstial fluids into action potentials.

osmosis (oz mo′ sis) [Gk. *osmos*: to push] The movement of water across a differentially permeable membrane, from one region to another region where the water potential is more negative.

ossicle (oss′ ick ul) [L. *os*: bone] The calcified construction unit of echinoderm skeletons.

osteoblasts (oss′ tee oh blast) [Gk. *osteon*: bone + *blastos*: sprout] Cells that lay down the protein matrix of bone.

osteoclasts (oss′ tee oh clast) [Gk. *osteon*: bone + *klastos*: broken] Cells that dissolve bone.

otolith (oh′ tuh lith) [Gk. *otikos*: ear + *lithos*: stone[Structures in the vertebrate vestibular apparatus that mechanically stimulate hair cells when the head moves or changes position.

oval window The flexible membrane that, when moved by the bones of the middle ear, produces pressure waves in the inner ear

ovary (oh′ var ee) [L. *ovum*: egg] Any female organ, in plants or animals, that produces an egg.

oviduct [L. *ovum*: egg + *ducere*: to lead] In mammals, the tube serving to transport eggs to the uterus or to outside of the body.

oviparity Reproduction in which eggs are released by the female and development is external to the mother's body. (Contrast with viviparous.)

ovoviviparity Reproduction in which fertilized eggs develop and hatch within the body of the mother but are not attached to the mother by means of a placenta.

ovulation The release of an egg from an ovary.

ovule (oh′ vule) In plants, a structure that contains a gametophyte and, within the gametophyte, an egg; when it matures, an ovule becomes a seed.

ovum (oh′ vum) [L. *ovum*: egg] The egg; the female sex cell.

oxidation (ox i day′ shun) Relative loss of electrons in a chemical reaction; either outright removal to form an ion, or the sharing of electrons with substances having a greater affinity for them, such as oxygen. Most oxidation, including biological ones, are associated with the liberation of energy. (Contrast with reduction.)

oxidative phosphorylation ATP formation in the mitochondrion, associated with flow of electrons through the respiratory chain.

oxidizing agent A substance that can accept electrons from another. The oxidizing agent becomes reduced; its partner becomes oxidized.

oxygenase An enzyme that catalyzes the addition of oxygen to a substrate from O_2.

- P -

P generation Parental generation. The individuals that mate in a genetic cross. Their immediate offspring are the F_1 generation.

pacemaker That part of the heart which undergoes most rapid spontaneous contraction, thus setting the pace for the beat of the entire heart. In mammals, the sinoatrial (SA) node. Also, an artificial device, implanted in the heart, that initiates rhythmic contraction of the organ.

Pacinian corpuscle A modified nerve ending that senses touch and vibration.

pair rule genes Segmentation genes that divide the *Drosophila* larva into two segments each.

pancreas (pan′ cree us) A gland located near the stomach of vertebrates that secretes digestive enzymes into the small intestine and releases insulin into the bloodstream.

Pangaea (pan jee′ uh) [Gk. *pan*: all, every] The single land mass formed when all the continents came together in the Permian period.

para- [Gk. *para*: akin to, beside] Prefix indicating association in being along side or accessory to.

parabronchi Passages in the lungs of birds through which air flows.

paracrine A substance, such as a hormone, that acts locally, near the site of its secretion. (Compare with autocrine, endocrine gland.)

parallel evolution Repeated evolutionary patterns of change that occur independently in multiple lineages.

paralogy (par al′ o jee) A type of homology applied to genes in which the divergence of homologous genes can be traced to gene duplication events. The genes are said to be paralogous, and each is an paralog of the others. (Compare with orthology.)

parapatric speciation [Gk. *para*: along side + *patria*: homeland] Reproductive isolation between subpopulations arising from some non-geographic but physical condition, such as soil nutrient content. (Contrast with allopatric speciation, sympatric speciation.)

paraphyletic (par′ a fih leht′ ik) [Gk. *para*: beside + *phylon*: tribe] Referring to a group that consists of an ancestor and some (but not all) of its descendants. (Compare with monophyletic, polyphyletic.)

parasite An organism that attacks and consumes parts of an organism much larger than itself. Parasites sometimes, but not always, kill their host.

parasympathetic nervous system A portion of the autonomic (involuntary) nervous system. (Contrast with sympathetic nervous system.)

parathyroids Four glands on the posterior surface of the thyroid that produce and release parathormone.

parathyroid hormone Hormone secreted by the parathyroid glands. Stimulates osteoclast activity and raises blood calcium levels.

parenchyma (pair eng' kyma) A plant tissue composed of relatively unspecialized cells without secondary walls.

parsimony The principle of preferring the simplest among a set of plausible explanations of any phenomenon.

parthenocarpy Formation of fruit from a flower without fertilization.

parthenogenesis (par' then oh jen' e sis) [Gk. parthenos: virgin + genesis: source] The production of an organism from an unfertilized egg.

partial pressure The portion of the barometric pressure of a mixture of gases that is due to one component of that mixture. For example, the partial pressure of oxygen at sea level is 20.9% of barometric pressure.

particulate theory In genetics, the theory that genes are physical entities that retain their identities after fertilization.

passive transport Diffusion across a membrane; may or may not require a channel or carrier protein. (Contrast with active transport.)

patch clamping A technique for isolating a tiny patch of membrane to allow the study of ion movement through a particular channel.

pathogen (path' o jen) [Gk. *pathos*: suffering + *genesis:* source] An organism that causes disease.

pattern formation In animal embryonic development, the organization of differentiated tissues into specific structures such as wings.

pedigree The pattern of transmission of a genetic trait within a family.

penetrance Of a genotype, the proportion of individuals with that genotype who show the expected phenotype.

pentose [Gk. *penta:* five] A sugar containing five carbon atoms.

PEP carboxylase The enzyme that combines carbon dioxide with PEP to form a 4-carbon dicarboxylic acid at the start of C_4 photosynthesis or of crassulacean acid metabolism (CAM).

pepsin [Gk. *pepsis*: digestion] An enzyme in gastric juice that digests protein.

pepsinogen Inactive secretory product that is converted into pepsin by low pH or by enzymatic action.

peptide linkage The bond between amino acids in a protein. Formed between a carboxyl group and amino group (CO—NH⁻) with the loss of water molecules.

peptidoglycan The cell wall material of many bacteria, consisting of a single enormous molecule that surrounds the entire cell.

perennial (per ren' ee al) [L. *per*: throughout + *annus*: year] Refers to a plant that survives from year to year. (Contrast with annual, biennial.)

perfect flower A flower with both stamens and carpels, therefore hermaphroditic.

pericycle [Gk. *peri*: around + *kyklos*: ring or circle] In plant roots, tissue just within the endodermis, but outside of the root vascular tissue. Meristematic activity of pericycle cells produces lateral root primordia.

periderm The outer tissue of the secondary plant body, consisting primarily of cork.

period (1) A category in the geological time scale. (2) The duration of a single cycle in a cyclical event, such as a circadian rhythm.

peripheral membrane protein Membrane protein not embedded in the bilayer. (Contrast with integral membrane protein.)

peripheral nervous system Neurons that transmit information to and from the central nervous system and whose cell bodies reside outside the brain or spinal cord.

peristalsis (pair' i stall' sis) [Gk. *peri*: around + *stellein*: place] Wavelike muscular contractions proceeding along a tubular organ, propelling the contents along the tube.

peritoneum The mesodermal lining of the body cavity among coelomate animals.

permease A membrane protein that specifically transports a compound or family of compounds across the membrane.

peroxisome An organelle that houses reactions in which toxic peroxides are formed. The peroxisome isolates these peroxides from the rest of the cell.

petal [Gk. *petalon*: spread out] In an angiosperm flower, a sterile modified leaf, nonphotosynthetic, frequently brightly colored, and often serving to attract pollinating insects.

petiole (pet' ee ole) [L. *petiolus*: small foot] The stalk of a leaf.

pH The negative logarithm of the hydrogen ion concentration; a measure of the acidity of a solution. A solution with pH = 7 is said to be neutral; pH values higher than 7 characterize basic solutions, while acidic solutions have pH values less than 7.

phage (fayj) Short for bacteriophage. A virus that infects bacteria.

phagocyte [Gk. *phagein*: to eat + *kystos*: sac] A white blood cell that ingests microorganisms by endocytosis.

phagocytosis Endocytosis by a cell of another cell or large particle.

pharmacogenomics The relaionship between an individual's genetic makeup and response to drugs.

pharming The use of genetically modified animals to produce medically useful products in their milk.

pharynx [Gk. *pharynx*: throat] The part of the gut between the mouth and the esophagus.

phenotype (fee' no type) [Gk. *phanein*: to show] The observable properties of an individual resulting from both genetic and environmental factors. (Contrast with genotype.)

phenotypic plasticity Refers to the fact that the phenotype of a developing organism is determined by a complex series of processes that are affected by both its genotype and its environment.

pheromone (feer' o mone) [Gk. *pheros*: carry + *hormon*: excite, arouse] A chemical substance used in communication between organisms of the same species.

phloem (flo' um) [Gk. *phloos*: bark] In vascular plants, the tissue that transports sugars and other solutes from sources to sinks. It consists of sieve cells or sieve tubes, fibers, and other specialized cells.

phosphate group The functional group — OPO_3H_2. The transfer of energy from one compound to another is often accomplished by the transfer of a phosphate group.

phosphodiester linkage The connection in a nucleic acid strand, formed by linking two nucleotides.

phospholipids Lipids containing a phosphate group; important constituents of cellular membranes. (*See* lipids.)

phosphorylation The addition of a phosphate group.

photoautotroph An organism that obtains energy from light and carbon from carbon dioxide. (Contrast with chemolithotroph, chemoheterotroph, photoheterotroph.)

photoheterotroph An organism that obtains energy from light but must obtain its carbon from organic compounds. (Contrast with chemolithotroph, chemoheterotroph, photoautotroph.)

photon (foe' ton) [Gk. *photos*: light] A quantum of visible radiation; a "packet" of light energy.

photoperiod (foe' tow peer' ee ud) The duration of a period of light, such as the length of time in a 24-hour cycle in which daylight is present.

photoperiodicity A condition in which physiological and behavioral changes are induced by changes in day length.

photoreceptor (1) A pigment that triggers a physiological response when it absorbs a photon. (2) A sensory receptor cell that senses and responds to light energy.

photorespiration Light-driven uptake of oxygen and release of carbon dioxide, the carbon being derived from the early reactions of photosynthesis.

photosynthesis (foe tow sin' the sis) [literally, "synthesis from light"] Metabolic processes, carried out by green plants, by which visible light is trapped and the energy used to synthesize compounds such as ATP and glucose.

photosystem [Gk. *phos:* light + *systema:* assembly] A light-harvesting complex in the chloroplast thylakoid composed of pigments and proteins.

photosystem I In photosynthesis, the reactions that absorb light at 700 nm, passing electrons to ferrodoxin and thence to NADPH. Rich in chlorophyll *a*.

photosystem II In photosynthesis, the reactions that absorb light at 660 nm, passing electrons to the electron transport chain in the chloroplast. Rich I chlorphyll *b*.

phototropins A class of blue light receptors that mediate phototropism and other plant responses.

phototropism [Gk. *photos*: light + *trope*: turning] A directed plant growth response to light.

phycobilin Photosynthetic pigment that absorbs red, yellow, orange, and green light and is found in cyanobacteria and some red algae.

phylogenetic tree A graphic representation of lines of descent among organisms or their genes.

phylogeny (fy loj' e nee) [Gk. *phylon*: tribe, race + *genesis*: source] The evolutionary history of a particular group of organisms or their genes.

physiology (fiz' ee ol' o jee) [Gk. *physis*: natural form + *logos*: discourse, study] The scientific study of the functions of living organisms and the individual organs, tissues, and cells of which they are composed.

phytoalexins Substances toxic to pathogens, produced by plants in response to fungal or bacterial infection.

phytochrome (fy' tow krome) [Gk. *phyton*: plant + *chroma*: color] A plant pigment regulating a large number of developmental and other phenomena in plants.

pigment A substance that absorbs visible light.

pineal gland A gland located between the cerebral hemispheres that secretes melatonin.

pinocytosis Endocytosis by a cell of liquid containing dissolved substances.

pistil [L. *pistillum*: pestle] The structure of an angiosperm flower within which the ovules are borne. May consist of a single carpel, or of several carpels fused into a single structure. Usually differentiated into ovary, style, and stigma.

pith In plants, relatively unspecialized tissue found within a cylinder of vascular tissue.

pituitary A small gland attached to the base of the brain in vertebrates. Its hormones control the activities of other glands. Also known as the hypophysis.

pits Recessed cavities in the cell walls of a plant vascular element where only the primary wall is present. facilitating the movement of sap between cells.

placenta (pla sen' ta) [Gk. *plax*: flat surface] The organ found in most mammals that provides for the nourishment of the fetus and elimination of the fetal waste products.

placental (pla sen' tal) Pertaining to mammals of the subclass Eutheria, a group characterized by the presence of a placenta; contains the majority of living species of mammals.

plankton Free-floating small organisms inhabiting the surface waters of lakes and oceans. Photosynthetic members of the plankton are referred to as phytoplankton.

plant *See* embryophyte.

planula (plan' yew la) [L. *planum*: flat] The free-swimming, ciliated larva of the cnidarians.

plaque (plack) [Fr.: a metal plate or coin] (1) A circular clearing in a layer (lawn) of bacteria growing on the surface of a nutrient agar gel. (2) An accumulation of prokaryotic organisms on tooth enamel. Acids produced by these microorganisms can cause tooth decay. (3) A region of arterial wall invaded by fibroblasts and fatty deposits (*see* atherosclerosis).

plasma *See* blood plasma.

plasma cell An antibody-secreting cell that developed from a B cell. The effector cell of the humoral immune system.

plasma membrane The membrane that surrounds the cell, regulating the entry and exit of molecules and ions. Every cell has a plasma membrane.

plasmid A DNA molecule distinct from the chromosome(s); that is, an extrachromosomal element. May replicate independently of the chromosome.

plasmodesma (plural: plasmodesmata) [Gk. *plassein*: to mold + *desmos*: band] A cytoplasmic strand connecting two adjacent plant cells.

plasmogamy The fusion of the cytoplasm of two cells. (Contrast with karyogamy.)

plasmolysis (plaz mol' i sis) Shrinking of the cytoplasm and plasma membrane away from the cell wall, resulting from the osmotic outflow of water. Occurs only in cells with rigid cell walls.

plastid Organelle in plants that serves for food manufacture (by photosynthesis) or food storage; bounded by a double membrane.

plastoquinone A mobile electron carrier within the thylakoid membrane of the chloroplast linking photosystems I and II of photosynthesis.

platelet A membrane-bounded body without a nucleus, arising as a fragment of a cell in the bone marrow of mammals. Important to blood-clotting action.

pleiotropy (plee' a tro pee) [Gk. *pleion*: more] The determination of more than one character by a single gene.

pleural membrane [Gk. *pleuras*: rib, side] The membrane lining the outside of the lungs and the walls of the thoracic cavity. Inflammation of these membranes is a condition known as pleurisy.

pluripotent Of a stem cell, having the ability to differentiate into any of a limted number of cell types. (Compare with totipotent.)

podocytes Cells of Bowman's capsule of the nephron that cover the capillaries of the glomerulus, forming filtration slits.

poikilotherm (poy' kill o therm) [Gk. *poikilos*: varied + *thermos*: heat] An animal whose body temperature tends to vary with the surrounding environment. (Contrast with homeotherm, heterotherm.)

point mutation A mutation that results from a small, localized alteration in the chemical structure of a gene; can revert to wild type. (Contrast with deletion.)

polar body A nonfunctional nucleus produced by meiosis, accompanied by very little cytoplasm. The meiosis which produces the mammalian egg produces in addition three polar bodies.

polar molecule A molecule in which the electric charge is not distributed evenly in the covalent bonds.

polar nuclei In flowering plants, the two nuclei in the central cell of the megagametophyte; following fertilization they give rise to the endosperm.

polarity In development, the difference between one end and the other. In chemistry, the property that makes a polar molecule.

pollen [L. *pollin*: fine, powdery flour] In seed plants, the microscopic grains containing the male gametophyte (microgametophyte) and gamete (microspore).

pollination The process of transferring pollen from an anther to the stigma of a pistil in an angiosperm or from a strobilus to an ovule in a gymnosperm.

poly- [Gk. *poly*: many] A prefix denoting multiple entities.

poly(A) tail A long sequence of adenine nucleotides (50–250) added after transcription to the 3′ end of most eukaryotic mRNAs.

polygenes Multiple loci whose alleles increase or decrease a continuously variable phenotypic trait.

polymer [Gk. *poly*: many + *meros*: unit] A large molecule made up of similar or identical subunits called monomers. (Contrast with monomer, oligomer.)

polymerase chain reaction (PCR) An enzymatic technique for the rapid production of millions of copies of a particular stretch of DNA.

polymerization reactions Chemical reactions that generate polymers by linking monomers.

polymorphic Referring to a gene whose most frequent allele in a population is present less than 99% of the time.

polymorphism (pol' lee mor' fiz um) [Gk. *poly*: many + *morphe*: form, shape] In genetics, the coexistence in the same population of two distinct hereditary types based on different alleles.

polyp The sessile asexual stage in the life cycle of most cnidarians.

polypeptide A large molecule made up of many amino acids joined by peptide linkages. Large polypeptides are called proteins.

polyphyletic (pol' lee fih leht' ik) [Gk. *poly*: many + *phylon*: tribe] Referring to a group that consists of multiple distantly related organisms, and does not include the common ancestor of the group. (Compare with monophyletic, paraphyletic.)

polyploidy (pol' lee ploid ee) The possession of more than two entire sets of chromosomes.

polysaccharide A macromolecule composed of many monosaccharides (simple sugars). Common examples are cellulose and starch.

polysome (polyribosome) A complex consisting of a threadlike molecule of messenger RNA and several (or many) ribosomes. The ribosomes move along the mRNA, synthesizing polypeptide chains as they proceed.

polytene (pol' lee teen) [Gk. *poly*: many + *taenia*: ribbon] An adjective describing giant interphase chromosomes, such as those found in the salivary glands of fly larvae. The characteristic pattern of bands and bulges seen on these chromosomes provided a method for preparing detailed chromosome maps of several organisms.

pons [L. *pons*: bridge] Region of the brain stem anterior to the medulla.

population Any group of organisms coexisting at the same time and in the same place and capable of interbreeding with one another.

population bottleneck *See* bottleneck.

population density The number of individuals (or modules) of a population in a unit of area or volume.

population genetics The study of genetic variation and its causes within populations.

population structure The proportions of individuals in a population belonging to different age classes (age structure). Also, the distribution of the population in space.

portal blood vessels Blood vessels that begin and end in capillary beds.

positional cloning A technique for isolating a gene associated with a disease on the basis of its approximate chromosomal location.

positional information Signals by which genes regulate cell functions to locate cells in a tissue during development.

positive control The situation in which a regulatory macromolecule is needed to turn transcription of structural genes on. In its absence, transcription will not occur.

positive cooperativity Occurs when a molecule can bind several ligands and each one that binds alters the conformation of the molecule so that it can bind the next ligand more easily. The binding of four molecules of O_2 by hemoglobin is an example of positive cooperativity.

post [L. *postere*: behind, following after] Prefix denoting something that comes after.

postabsorptive period When there is no food in the gut and no nutrients are being absorbed.

posterior pituitary The portion of the pituitary gland that is derived from neural tissue.

postsynaptic cell The cell whose membranes receive neurotransmitter after its release by another cell (the presynaptic cell) at a synapse.

postzygotic reproductive barriers Barriers to the reproductive process that occur after the union of the nuclei of two gametes. (Contrast with prezygotic reproductive barriers.)

potential energy "Stored" energy not doing work, such as the energy in chemical bonds.

precapillary sphincter A cuff of smooth muscle that can shut off the blood flow to a capillary bed.

pre-mRNA (precursor mRNA) Initial gene transcript before it is modified to produce functional mRNA. Also known as the primary transcript.

predator An organism that kills and eats other organisms.

pressure flow model An effective model for phloem transport in angiosperms. It holds that sieve element transport is driven by an osmotically driven pressure gradient between source and sink.

pressure potential The hydrostatic pressure of an enclosed solution in excess of the surrounding atmospheric pressure. (Contrast with solute potential, water potential.)

presynaptic excitation/inhibition Occurs when a neuron modifies activity at a synapse by releasing a neurotransmitter onto the presynaptic nerve terminal.

prey [L. *praeda*: booty] An organism consumed as an energy source.

prezygotic reproductive barriers Barriers to the reproductive process that occur before the union of the nuclei of two gametes (Contrast with postzygotic reproductive barriers.)

primary active transport Form of active transport in which ATP is hydrolyzed, yielding the energy required to transport ions against their concentration gradients. (Contrast with secondary active transport.)

primary consumer An herbivore; an organism that eats plant tissues.

primary embryonic organizer *See* organizer.

primary growth In plants, growth produced by the apical meristems. (Contrast with secondary growth.)

primary immune response The first response of the immune system to an antigen, involving recognition by lymphocytes and the production of effector cells and memory cells. (Contrast with secondary immune response.)

primary motor cortex The region of the cerebral cortex that contains motor neurons that directly stimulate specific muscle fibers to contract.

primary producer A photosynthetic or chemosynthetic organism that synthesizes complex organic molecules from simple inorganic ones.

primary sex determination Genetic determination of gametic sex, male or female. (Contrast with secondary sex determination.)

primary somatosensory cortex The region of the cerebral cortex that receives input from mechanosensors distributed throughout the body.

primary succession Succession that begins in an area initially devoid of life, such as on recently exposed glacial till or lava flows. (Contrast with secondary succession.)

primary structure The specific sequence of amino acids in a protein.

primary wall Cellulose-rich cell wall layers laid down by a growing plant cell.

primase An enzyme that catalyzes the synthesis of a primer for DNA replication.

primer A short, single-stranded segment of DNA that is the necessary starting material for the synthesis of a new DNA strand, which is synthesized from the 3′ end of the primer.

primitive streak A line running axially along the blastodisc, the site of inward cell migration during formation of the three-layered embryo. Formed in the embryos of birds and fish.

primordium [L. *primordium*: origin] The most rudimentary stage of an organ or other part.

prion An infectious protein that can proliferate by converting other proteins.

pro- [L.: first, before, favoring] A prefix often used in biology to denote a developmental stage that comes first or an evolutionary form that appeared earlier than another. For example, prokaryote, prophase.

probe A segment of single stranded nucleic acid used to identify DNA molecules containing the complementary sequence.

procambium Primary meristem that produces the vascular tissue.

processive Referring to an enzyme that catalyzes many reactions each time it binds to a substrate, as DNA polymerase does during DNA replication.

progesterone [L. *pro*: favoring + *gestare*: to bear] A vertebrate female sex hormone that maintains pregnancy.

prokaryotes (pro kar′ ry otes) [L. *pro*: before + Gk. *karyon*: kernel, nucleus] Organisms whose genetic material is not contained within a nucleus: the bacteria and archaea. Considered an earlier stage in the evolution of life than the eukaryotes.

prometaphase The phase of nuclear division that begins with the disintegration of the nuclear envelope.

promoter The region of an operon that acts as the initial binding site for RNA polymerase.

proofreading The correction of an error in DNA replication just after an incorrectly paired base is added to the growing polynucleotide chain.

prophage (pro′ fayj) The noninfectious units that are linked with the chromosomes of the host bacteria and multiply with them but do not cause dissolution of the cell. Prophage can later enter into the lytic phase to complete the virus life cycle.

prophase (pro′ phase) The first stage of nuclear division, during which chromosomes condense from diffuse, threadlike material to discrete, compact bodies.

prostaglandin Any one of a group of specialized lipids with hormone-like functions. It is not clear that they act at any considerable distance from the site of their production.

prostate gland Glandular tissue that surrounds the male urethra at its junction with the vas deferens; contributes an alkaline fluid to the semen.

prosthetic group Any nonprotein portion of an enzyme.

proteasome In the eukaryotic cytoplasm, a huge protein structure that binds to and digests cellular proteins that have been tagged by ubiquitin.

protein (pro′ teen) [Gk. *protos*: first] One of the most fundamental building substances of living organisms. A long-chain polymer of amino acids with twenty different common side chains. Occurs with its polymer chain extended in fibrous proteins, or coiled into a compact macromolecule in enzymes and other globular proteins.

protein domain *See* domain (1)

protein kinase An enzyme that catalyzes the addition of a phosphate group from ATP to a target protein.

protein kinase cascade aA series of reactions in response to a molecular signal, in which a series of protein kinases activates one another in sequence, amplifying the signal at each step.

proteoglycan A glycoprotein containing a protein core with attached long, linear carbohydrate chains.

proteolysis [protein + Gk. *lysis*: break apart] An enzymatic digestion of a protein or polypeptide.

proteome The total of the different proteins that can be made by an organism. Because of alternate splicing of pre-mRNA, the number of proteins that can be made is usually much larger than the number of protein-coding genes present in the organism's genome.

protobiont [Gk. *protos*: first, before + *bios*: life] Aggregates of abiotically produced molecules that cannot reproduce but do maintain internal chemical environments that differ from their surroundings.

protoderm Primary meristem that gives rise to the plant epidermis.

proton (pro' ton) [Gk. *protos*: first, before] (1) A subatomic particle with a single positive charge. The number of protons in the nucleus of an atom determine its element. (2) A hydrogen ion, H^+.

proton pump An active transport system that uses ATP energy to move hydrogen ions across a membrane generating an electric potential (voltage).

proton motive force A force generated across a membrane expressed in millivolts having two components: a chemical potential (difference in proton concentration) plus an electrical potential due to the electrostatic charge on the proton.

proto-oncogenes The normal alleles of genes possessing oncogenes (cancer-causing genes) as mutant alleles. Proto-oncogenes encode growth factors and receptor proteins.

prototroph (pro' tow trofe') [Gk. *protos*: first + *trophein*: to nourish] The nutritional wild type, or reference form, of an organism. Any deviant form that requires growth nutrients not required by the prototrophic form is said to be a nutritional mutant, or auxotroph.

proximal Near the point of attachment or other reference point. (Contrast with distal.)

pseudocoelom [Gk. *pseudes*: false] A body cavity not surrounded by a peritoneum. Characteristic of nematodes and rotifers.

pseudogene [Gk. *pseudes*: false] A DNA segment that is homologous to a functional gene but is not expressed because of changes to its sequence or changes to its location in the genome.

pseudopod (soo' do pod) [Gk. *pseudes*: false + *podos*: foot] A temporary, soft extension of the cell body that is used in location, attachment to surfaces, or engulfing particles.

pulmonary [L. *pulmo*: lung] Pertaining to the lungs.

punctuated equilibrium An evolutionary pattern in which periods of rapid change are separated by longer periods of little or no change.

Punnett square A method of predicting the results of a genetic cross by arranging the gametes of each parent at the edges of a square.

pupa (pew' pa) [L. *pupa*: doll, puppet] In certain insects (the Holometabola), the encased developmental stage between the larva and the adult.

pupil The opening in the vertebrate eye through which light passes.

purine (pure' een) One of the types of nitrogenous bases. The purines adenine and guanine are found in nucleic acids. (Contrast with pyrimidine.)

Purkinje fibers Specialized heart muscle cells that conduct excitation throughout the ventricular muscle.

pyrimidine (per im' a deen) A type of nitrogenous base. The pyrimidines cytosine, thymine, and uracil are found in nucleic acids.

pyruvate A three-carbon acid; the end product of glycolysis and the raw material for the citric acid cycle.

pyruvate oxidation Conversion of pyruvate to acetyl CoA and CO_2 that occurs in the mitochondrial matrix in the presence of O_2.

- Q -

Q_{10} A value that compares the rate of a biochemical process or reaction over a 10°C range of temperature. A process that is not temperature-sensitive has a Q_{10} of 1; values of 2 or 3 mean the reaction speeds up as temperature increases.

quantum (kwon' tum) [L. *quantus*: how great] An indivisible unit of energy.

quaternary structure The specific three dimensional arrangement of protein subunits.

quiescent center In root meristem, central region where cells do not divide or divide very slowly.

- R -

R factor (resistance factor) A plasmid that contains one or more genes that encode resistance to antibiotics.

***R* genes** Resistance genes that function in plant defenses against bacteria, fungi, and nematodes. *See* gene-for-gene resistance.

R group The distinguishing group of atoms of a particular amino acid.

radial symmetry The condition in which two halves of a body are mirror images of each other regardless of the angle of the cut, providing the cut is made along the center line. Thus, a cylinder cut lengthwise down its center displays this form of symmetry. (Contrast with biradial symmetry.)

radioisotope A radioactive isotope of an element. Examples are carbon-14 (^{14}C) and hydrogen-3, or tritium (^3H).

radiometry The use of the regular, known rates of decay of radioisotopes of elements to determine dates of events in the distant past.

rain shadow The relatively dry area on the leeward side of a mountain range.

rapid-eye-movement *See* REM sleep

reactant A chemical substance that enters into a chemical reaction with another substance.

reaction A chemical change in which changes take place in the kind, number, or position of atoms making up a substance.

reaction center A group of electron transfer proteins that receive energy from light-absorbing pigments and convert it to chemical energy by redox reactions.

receptacle The end of a plant stem to which the parts of the flower are attached.

receptive field The area of visual space that activates a particular cell in the visual system.

receptor A site or protein on the outer surface of the plasma membrane or in the cytoplasm to which a specific ligand from another cell binds.

receptor-mediated endocytosis Endocytosis initiated by macromolecular binding to a specific membrane receptor.

receptor potential The change in the resting potential of a sensory cell when it is stimulated.

recessive In genetics, an allele that does not determine phenotype in the presence of a dominant allele. (Contrast with dominance.)

reciprocal crosses A pair of matings in one of which a female of genotype A mates with a male of genotype B and in the other of which a female of genotype B mates with a male of genotype A.

recognition site *See* restriction site.

recombinant An individual, meiotic product, or single chromosome in which genetic materials originally present in two individuals end up in the same haploid complement of genes. The reshuffling of genes can be either by independent segregation, or by crossing over between homologous chromosomes.

recombinant DNA DNA generated in vitro, from more than one source.

recombinant DNA technology The application of restriction endonucleases, plasmids, and transformation to alter and assemble recombinant DNA, with the goal of producing specific proteins.

recombinant frequency The proportion of offspring of a genetic cross that have phenotypes different from the parental phenotypes due to crossing over between linked genes during gamete formation.

reconciliation ecology The practice of making exploited lands more biodiversity-friendly.

rectum The terminal portion of the gut, ending at the anus.

redox reaction A chemical reaction in which one reactant becomes oxidized and the other becomes reduced.

reducing agent A substance that can donate electrons to another substance. The reducing agent becomes oxidized, and its partner becomes reduced.

reduction Gain of electrons by a chemical reactant; any reduction is accompanied by an oxidation. (Contrast with oxidation.)

reflex An automatic action, involving only a few neurons (in vertebrates, often in the spinal cord), in which a motor response swiftly follows a sensory stimulus.

refractory period Of a neuron, the time interval after an action potential, during which another action potential cannot be elicited.

regulative development A pattern of animal embryonic development in which the fates of the first blastomeres are not absolutely fixed. (Contrast with mosaic development.)

regulator sequence A DNA sequence to which the protein product of a regulatory gene binds.

regulatory gene A gene that codes for a protein that controls the transcription of another gene(s).

regulatory subunit The polypeptide in an enzyme protein with quaternary structure that does not contain the active site, but instead binds non-substrate molecules and changes its structure, in turn changing the structure and function of the active site. (Contrast catalytic subunit.)

regulatory system A system that uses feedback information to maintain a physiological function or parameter at an optimal level.

reinforcement The evolution of enhanced reproductive isolation between populations due to natural selection for greater isolation.

releaser A sensory stimulus that triggers the performance of a stereotyped behavior pattern.

releasing hormone One of several hypothalamic hormones that stimulates the secretion of anterior pituitary hormone.

REM sleep A sleep state characterized by vivid dreams, skeletal muscle relaxation, and rapid eye movements.

renal [L. *renes*: kidneys] Relating to the kidneys.

replication Pertaining to the duplication of genetic material.

replication complex The close association of several proteins operating in the replication of DNA.

replication fork A point at which a DNA molecule is replicating. The fork forms by the unwinding of the parent molecule.

replicon A region of DNA controlled by a single origin of replication.

reporter gene A marker gene included in recombinant DNA to indicate the presence of the recombinant DNA in a host cell.

repressible enzyme An enzyme whose synthesis can be decreased or prevented by the presence of a particular compound. A repressible operon often controls the synthesis of such an enzyme.

repressor A protein coded by the regulatory gene. The repressor can bind to a specific operator and prevent transcription of the operon.

reproductive isolating mechanism Any trait that prevents individuals from two different populations from producing fertile hybrids.

reproductive isolation The condition in which a population is not exchanging genes with other populations of the same species.

rescue effect The process by which a few individuals moving among declining subpopulations of a species and reproducing may prevent their extinction.

resolution Of an optical device such as a microscope, the smallest distance between two lines that allows the lines to be seen as separate from one another.

resource Something in the environment required by an organism for its maintenance and growth that is consumed in the process of being used.

respiration (res pi ra' shun) [L. *spirare*: to breathe] (1) Cellular respiration; the catabolic pathways by which electrons are removed from various molecules and passed through intermediate electron carriers to O_2, generating H_2O and releasing energy. (2) Breathing.

respiratory chain The terminal reactions of cellular respiration, in which electrons are passed from NAD or FAD, through a series of intermediate carriers, to molecular oxygen, with the concomitant production of ATP.

resting potential The membrane potential of a living cell at rest. In cells at rest, the interior is negative to the exterior. (Contrast with action potential, electrotonic potential.)

restoration ecology The science and practice of restoring damaged or degraded ecosystems.

restriction digestion Use of restriction enzymes to cleave DNA into fragments in a test tube.

restriction endonuclease *See* restriction enzyme.

restriction enzyme Any one of several enzymes, produced by bacteria, that break foreign DNA molecules at specific sites. Some produce "sticky ends." Extensively used in recombinant DNA technology.

restriction fragment length polymorphism *See* RFLP.

restriction point The specific time during G1 of the cell cycle at which the cell becomes committed to undergo the rest of the cell cycle.

restriction site A specific DNA base sequence recognized and acted on by a restriction endonuclease cutting the DNA.

reticular system A central region of the vertebrate brain stem that includes complex fiber tracts conveying neural signals between the forebrain and the spinal cord, with collateral fibers to a variety of nuclei that are involved in autonomic functions, including arousal from sleep.

retina (rett' in uh) [L. *rete*: net] The light-sensitive layer of cells in the vertebrate or cephalopod eye.

retinal The light-absorbing portion of visual pigment molecules. Derived from β-carotene.

retinoblastoma protein A protein that inhibits an animal cell from passing through the restriction point; inactivation of this protein is necessary for the cell cycle to proceed.

retrovirus An RNA virus that contains reverse transcriptase. Its RNA serves as a template for cDNA production, and the cDNA is integrated into a chromosome of the mammalian host cell.

reversal *See* evolutionary reversal.

reverse transcriptase An enzyme that catalyzes the production of DNA (cDNA), using RNA as a template; essential to the reproduction of retroviruses.

reversible reaction A chemical change that can occur in both the forward and reverse directions.

RFLP (restriction fragment length polymorphism) Coexistence of two or more patterns of restriction fragments (patterns produced by restriction enzymes), as revealed by a probe. The polymorphism reflects a difference in DNA sequence on homologous chromosomes.

rhizoids (rye' zoids) [Gk. *rhiza*: root] Hairlike extensions of cells in mosses, liverworts, and a few vascular plants that serve the same function as roots and root hairs in vascular plants. The term is also applied to branched, rootlike extensions of some fungi and algae.

rhizome (rye' zome) A special underground stem (as opposed to root) that runs horizontally beneath the ground.

rhodopsin A photopigment used in the visual process of transducing photons of light into changes in the membrane potential of photoreceptor cells.

ribonucleic acid *See* RNA.

ribose A five-carbon sugar in nucleotides and RNA.

ribosomal RNA (rRNA) Several species of RNA that are incorporated into the ribosome. Involved in peptide bond formation.

ribosome A small organelle that is the site of protein synthesis.

ribozyme An RNA molecule with catalytic activity.

risk cost The increased chance of being injured or killed as a result of performing a behavior, compared to resting.

RNA (ribonucleic acid) An often single stranded nucleic acid whose nucleotides use ribose rather than deoxyribose and in which the base uracil replaces thymine found in DNA. Serves as genome from some viruses. (*See* rRNA, tRNA, mRMA, and ribozyme.)

RNA editing The alteration of bases on mRNA prior to its translation.

RNA polymerase An enzyme that catalyzes the formation of RNA from a DNA template.

RNA primase A replication complex enzyme that makes the primer strand of DNA needed to initiate DNA replication.

RNA splicing The last stage of RNA processing in eukaryotes, in which the transcripts of introns are excised through the action of small nuclear ribonucleoprotein particles (snRNP).

rod cells Light-sensitive cell in the vertebrate retina; these sensory receptor cells are sensitive in extremely dim light and are responsible for dim light, black and white vision.

root The organ responsible for anchoring the plant in the soil, absorbing water and minerals, and producing certain hormones. Some roots are storage organs.

root cap A thimble-shaped mass of cells, produced by the root apical meristem, that protects the meristem; the organ that perceives the gravitational stimulus in root gravitropism.

root hair A long, thin process from a root epidermal cell that absorbs water and minerals from the soil solution.

rough ER That portion of the endoplasmic reticulum whose outer surface has attached ribosomes. (Contrast with smooth ER.)

rRNA *See* ribosomal RNA.

RT-PCR A technique in which RNA is first converted to cDNA by the use of the enzyme reverse transcriptase, then the cDNA is amplified by the polymerase chain reaction.

rubisco (ribulose bisphosphate carboxylase/ oxygenase) Acronym for the enzyme that combines carbon dioxide or oxygen with ribulose bisphosphate to catalyze the first step of the Calvin-Benson cycle.

rumen (rew' mun) The first division of the ruminant stomach. It stores and initiates bacterial fermentation of food. Food is regurgitated from the rumen for further chewing.

ruminant Herbivorous, cud-chewing mammals such as cows or sheep. The ruminant stomach consists of four compartments.

- S -

S phase In the cell cycle, the stage of interphase during which DNA is replicated. (Contrast with G_1 phase, G_2 phase, M phase.)

saprobe [Gk. *sapros*: rotten] An organism (usually a bacterium or fungus) that obtains its carbon and energy directly from dead organic matter.

sarcomere (sark′ o meer) [Gk. *sark*: flesh + *meros*: unit] The contractile unit of a skeletal muscle.

sarcoplasm The cytoplasm of a muscle cell.

sarcoplasmic reticulum The endoplasmic reticulum of a muscle cell.

saturated fatty acid A fatty acid usually containing from 12 to 18 carbon atoms and no double bonds.

scientific method A means of gaining knowledge about the natural world by making observations, posing hypotheses, and conducting experiments to test those hypotheses.

schizocoelous development [Gk. *schizo*: split + *koiloma*: cavity] Formation of a coelom during embryological development by a splitting of mesodermal masses.

Schwann cell A glial cell that wraps around part of the axon of a peripheral neuron, creating a myelin sheath.

sclereid [Gk. *skleros*: hard] A type of sclerenchyma cell, commonly found in nutshells, that is not elongated.

sclerenchyma (skler eng′ kyma) [Gk. *skleros*: hard + *kymus*: juice] A plant tissue composed of cells with heavily thickened cell walls, dead at functional maturity. The principal types of sclerenchyma cells are fibers and sclereids.

scrotum A sac of skin that encloses the testicles in male mammals.

second law of thermodynamics States that in any real (irreversible) process, there is a decrease in free energy and an increase in entropy.

second messenger A compound, such as cAMP, that is released within a target cell after a hormone (first messenger) has bound to a surface receptor on a cell; the second messenger triggers further reactions within the cell.

secondary active transport Form of active transport which does not use ATP as an energy source; rather, transport is coupled to ion diffusion down a concentration gradient established by primary active transport.**secondary consumer** An organism that eat primary consumers.

secondary growth In plants, growth produced by vascular and cork cambia, contributing to an increase in girth. (Contrast with primary growth.)

secondary immune response A rapid and intense response to a second or subsequent exposure to an antigen, initiated by memory cells. (Contrast with primary immune response.)

secondary metabolite A compound synthesized by a plant that is not needed for basic cellular metabolism. Typically has an antiherbivore or antiparasite function.

secondary sex determination Formation of nongametic features of sex, such as external organs and body hair. (Contrast with primary sex determination.)

secondary structure Of a protein, localized regularities of structure, such as the α helix and the β pleated sheet.

secondary succession Ecological succession after a disturbance that did not eliminate all the organisms originally living on the site. (Contrast with primary succession.)

secondary wall Wall layers laid down by a plant cell that has ceased growing; often impregnated with lignin or suberin.

secretin (si kreet′ in) A peptide hormone secreted by the upper region of the small intestine when acidic chyme is present. Stimulates the pancreatic duct to secrete bicarbonate ions.

section A thin slice, usually for microscopy, as a tangential section or a transverse section.

seed A fertilized, ripened ovule of a gymnosperm or angiosperm. Consists of the embryo, nutritive tissue, and a seed coat.

seed plant Plants in which the embryo is protected and nourished within a seed; the gymnosperms and angiosperms.

seedling A young plant that has grown from a seed (rather than by grafting or by other means.)

segmentation genes In insect larvae, genes that determine the number and polarity of larval segments.

segment polarity genes Genes that determine the boundaries and front-to-back organization of the segments in the *Drosophila* larva.

segregation In genetics, the separation of alleles, or of homologous chromosomes, from each other during meiosis so that each of the haploid daughter nuclei produced by meiosis contains one or the other member of the pair found in the diploid parent cell, but never both. This principle was articulated by Mendel as his first law.

selective permeability Allowing certain substances to pass through while other substances are excluded; a characteristic of membranes.

self incompatability In plants, the rejection of their own pollen; promotes genetic variation and limits inbreeding.

selfish act A behavioral act that benefits its performer but harms the recipients.

semen (see′ men) [L. *semin*: seed] The thick, whitish liquid produced by the male reproductive organ in mammals, containing the sperm.

semiconservative replication The way in which DNA is synthesized. Each of the two partner strands in a double helix acts as a template for a new partner strand. Hence, after replication, each double helix consists of one old and one new strand.

seminiferous tubules The tubules within the testes within which sperm production occurs.

senescence [L. *senescere*: to grow old] Aging; deteriorative changes with aging; the increased probability of dying with increasing age.

sensory receptor cells Cells that are responsive to a particular type of physical or chemical stimulation.

sensory transduction The transformation of environmental stimuli or information into neural signals.

sepal (see′ pul) [L. *sepalum*: covering] One of the outermost structures of the flower, usually protective in function and enclosing the rest of the flower in the bud stage.

septum [L. *saeptum*: wall, fence] (1) A partition or cross-wall appearing in the hyphae of some fungi. (2) The bony structure dividing the nasal passages.

Sertoli cells Cells in the seminiferous tubules that nuture the developing sperm.

sessile (sess′ ul) [L. *sedere*: to sit] Permanently attached; not moving.

set point In a regulatory system, the threshold sensitivity to the feedback stimulus.

sex chromosome In organisms with a chromosomal mechanism of sex determination, one of the chromosomes involved in sex determination.

sex linkage The pattern of inheritance characteristic of genes located on the sex chromosomes of organisms having a chromosomal mechanism for sex determination.

sex pilus (pill′ us) [L. *pilus*: hair] A structure on the cell wall that allows one bacterium to adhere to another prior to conjugation.

sexual reproduction Reproduction involving union of gametes.

sexual selection Selection by one sex of characteristics in individuals of the opposite sex. Also, the favoring of characteristics in one sex as a result of competition among individuals of that sex for mates.

shared derived trait A trait that arose in the ancestor of a phylogenetic group and is present (sometimes in modified form) in all of its members, thus helping define that group. Also called a synapomorphy.

shoot system The aerial parts of a vascular plant, consisting of the leaves, stem(s), and flowers.

short-day plant (SDP) A plant that requires short days (or long nights) in order to flower.

short tandem repeat (STR) An inherited, short (2–5 base pairs), moderately repetitive sequence of DNA.

shotgun sequencing A relatively rapid method of analyzing DNA sequences in which a large DNA molecule is broken up into overlapping fragments, each fragment is sequenced, and computers are used to analyze and realign the fragments.

sieve tube A column of specialized cells found in the phloem, specialized to conduct organic matter from sources (such as photosynthesizing leaves) to sinks (such as roots). Found principally in flowering plants.

sieve tube element A single cell of a sieve tube, containing cytoplasm but relatively few organelles, with highly specialized perforated end walls leading to elements above and below.

signal A chemical (neurotransmitter or hormone) or light message emitted from a cell or cells or organism(s) and received by others to cause some change in function or behavior.

signal recognition particle (SRP) A complex of RNA and protein that recognizes both the signal sequence on a growing polypeptide and receptor protein on the surface of the ER.

signal sequence The sequence of a protein that directs the protein through a particular cellular membrane.

signal transduction pathway The series of biochemical steps whereby a stimulus to a cell (such as a hormone or neurotransmitter binding to a receptor) is translated into a response of the cell.

silencer sequence A sequence of eukaryotic DNA that binds proteins that inhibit the transcription of an associated gene.

silent substitution A change in gene sequence that, due to the redundancy of the genetic code, has no effect on the amino acid produced, and thus no effect on the protein phenotype. Also called a synonymous substitution.

similarity matrix A matrix used to compare the degree of divergence among pairs of objects. For molecular sequences, constructed by summing the number or percentage of nucleotidies or amino acids that are identical in each pair of sequences.

single nucleotide polymorphisms (SNPs) Inherited variations in a single nucleotide base in DNA.

single-strand binding protein In DNA replication, a protein that binds to single strands of DNA after they have been separated from each other, keeping the two strands separate for replication.

sinoatrial node (sigh′ no ay′ tree al) [L. *sinus*: curve + *atrium*: hall, chamber] The pacemaker of the mammalian heart.

sink In plants, any organ that imports the products of photosynthesis, such as roots, developing fruits, immature leaves. (Contrast with source.)

sinus (sigh′ nus) [L. *sinus*: curve, hollow] A cavity in a bone, a tissue space, or an enlargement in a blood vessel.

sister chromatid In the eukaryotic cell, a chromatid resulting from chromosome replication during interphase.

sister groups Two phylogenetic groups that are each other's closest relatives.

skeletal muscle *See* striated muscle.

sliding DNA clamp A protein complex that keeps DNA polymerase bound to DNA during replication.

sliding filament theory A proposed mechanism of muscle contraction based on formation and breaking of crossbridges between actin and myosin filaments, causing them to slide together.

slow-twitch fibers Skeletal muscle fibers that generate tension slowly, but are resistant to fatigue ("marathon" fibers). They have abundant mitochondria, enzymes of aerobic metabolism, myoglobin, and blood supply.

slow-wave sleep A state of mammalian and avian sleep characterized by high amplitude slow waves in the EEG.

small intestine The portion of the gut between the stomach and the colon; consists of the duodenum, the jejunum, and the ileum.

small nuclear ribonucleoprotein particle (snRNP) A complex of an enzyme and a small nuclear RNA molecule, functioning in RNA splicing.

smooth muscle One of three types of muscle tissue. Usually consists of sheets of mononucleated cells innervated by the autonomic nervous system. (Compare with cardiac muscle, striated muscle.)

sodium–potassium pump (Na–K pump) The complex protein in plasma membranes that is responsible for primary active transport; it pumps sodium ions out of the cell and potassium ions into the cell, both against their concentration gradients.

solute A substance that is dissolved in a liquid (solvent) to form a solution.

solute potential A property of any solution, resulting from its solute contents; it may be zero or have a negative value. The more negative the solute potential, the greater the tendency of the solution to take up water through a differentially permeable membrane. (Contrast with pressure potential, water potential.)

solution A liquid (the solvent) and its dissolved solutes.

somatic [Gk. *soma*: body] Pertaining to the body. Somatic cells are cells of the body (as opposed to germ cells).

somatic mutation Permament genetic change in a somatic cell. These mutations affect the individual only; they are not passed on to offspring. (Contrast with germ line mutation)

somite (so′ might) One of the segments into which an embryo becomes divided longitudinally, leading to the eventual segmentation of the animal as illustrated by the spinal column, ribs, and associated muscles.

source In plants, an organ exporting photosynthetic products in excess of its own needs. For example, a mature leaf or storage organ. (Contrast with sink.)

spatial summation In the production or inhibition of action potentials in a postsynaptic neuron, the interaction of depolarizations and hyperpolarizations produced by several terminal boutons.

spawning The direct release of sex cells into the water.

speciation (spee′ shee ay′ shun) The process of splitting one population into two populations that are reproductively isolated from one another.

species (spee′ shees) [L. *species*: kind] The basic lower unit of classification, consisting of an ancestor–descendant lineage of populations of closely related and similar organisms. The more narrowly defined "biological species" consists of individuals capable of interbreeding freely with each other but not with members of other species.

species–area relationship The relationship between the sizes of areas and the numbers of species they support.

species diversity A weighted representation of the species of organisms living in a region; large and common species are given greater weight than are small and rare ones. (Contrast with species richness.)

species richness The total number of species living in a region. (Contrast with species diversity.)

specific defenses Defensive reactions of the immune system that are based on antibody reaction with a specific antigen.

specific heat The amount of energy that must be absorbed by a gram of a substance to raise its temperature by one degree centigrade. By convention, water is assigned a specific heat of one.

sperm [Gk. *sperma*: seed] A male gamete (reproductive cell).

spermatogenesis (spur mat′ oh jen′ e sis) [Gk. *sperma*: seed + *genesis*: source] Male gametogenesis, leading to the production of sperm.

sphincter (sfink′ ter) [Gk. *sphinkter*: something that binds tightly] A ring of muscle that can close an orifice, for example at the anus.

spindle apparatus An array of microtubules stretching from pole to pole of a dividing nucleus and playing a role in the movement of chromosomes at nuclear division. Named for its shape.

spiracle (spy′ rih kel) [L. *spirare*: to breathe] An opening of the treacheal respiratory system of terrestrial arthorpods.

spleen An organ that serves as a reservoir for venous blood and eliminates old, damaged red blood cells from the circulation.

spliceosome An RNA–protein complex that splices out introns from eukaryotic pre-mRNAs.

splicing Removal of introns and connecting of exons in eukaryotic pre-mRNAs.

spontaneous mutation A genetic change caused by internal cellular mechanisms, such as an error in DNA replication. (Contrast with induced mutation.)

spontaneous reaction A chemical reaction that will proceed on its own, without any outside influence. A spontaneous reaction need not be rapid.

sporangium (spor an′ gee um) [Gk. *spora*: seed + *angeion*: vessel or reservoir] In plants and fungi, any specialized stucture within which one or more spores are formed.

spore [Gk. *spora*: seed] Any asexual reproductive cell capable of developing into an adult organism without gametic fusion. In plants, haploid spores develop into gametophytes, diploid spores into sporophytes. In prokaryotes, a resistant cell capable of surviving unfavorable periods.

sporocyte Specialized cells of the diploid sporophyte that will divide by meiosis to produce four haploid spores. Germination of these spores produces the haploid gametophyte.

sporophyte (spor′ o fyte) [Gk. *spora*: seed + *phyton*: plant] In plants and protists with alternation of generations, the diploid phase that produces the spores. (Contrast with gametophyte.)

stabilizing selection Selection against the extreme phenotypes in a population, so that the intermediate types are favored. (Contrast with disruptive selection.)

stamen (stay′ men) [L. *stamen*: thread] A male (pollen-producing) unit of a flower, usually composed of an anther, which bears the pollen, and a filament, which is a stalk supporting the anther.

starch [O.E. *stearc*: stiff] A polymer of glucose; used by plants to store energy.

start codon The mRNA triplet (AUG) that acts as a signal for the beginning of translation at the ribosome. (Contrast with stop codon.)

stasis [Gk. *stasis*: to stop, stand still] Period during which little or no evolutionary change takes place within a lineage or groups of lineages.

statocyst (stat′ oh sist) [Gk. *statos*: stationary + *kystos*: cell] An organ of equilibrium in some invertebrates.

statolith (stat′ oh lith) [Gk. *statos*: stationary + *lithos*: stone] A solid object that responds to gravity or movement and stimulates the mechanoreceptors of a statocyst.

stele (steel) [Gk. *stylos*: pillar] The central cylinder of vascular tissue in a plant stem.

stem Plant structure that holds leaves and/or flowers; it is the site for transporting and distributing material throughout the plant.

stem cells In animals, undifferentiated cells that are capable of extensive proliferation. A stem cell generates more stem cells and a large clone of differentiated progeny cells.

steroid Any of numerous lipids based on a 17-carbon atom ring system.

sticky ends On a piece of two-stranded DNA, short, complementary, one-stranded regions produced by the action of a restriction endonuclease. Sticky ends allow the joining of segments of DNA from different sources.

stigma [L. *stigma*: mark, brand] The part of the pistil at the apex of the style that is receptive to pollen, and on which pollen germinates.

stimulus [L. *stimulare*: to goad] Something causing a response; something in the environment detected by a receptor.

stolon [L. *stolon*: branch, sucker] A horizontal stem that forms roots at intervals.

stoma (plural: stomata) [Gk. *stoma*: mouth, opening] Small opening in the plant epidermis that permits gas exchange; bounded by a pair of guard cells whose osmotic status regulates the size of the opening.

stop codon Any of the thre mRNA codons that signal the end of protein translation at the ribosome: UAG, UGA, UAA.

stratosphere The upper part of Earth's atmosphere, above the troposphere; extends from approximately 18 kilometers upward to approximately 50 kilometers above the surface.

stratum (plural strata) [L. *stratos*: layer] A layer of sedimentary rock laid down at a particular time in the past.

striated muscle Contractile tissue characterized by multinucleated cells containing highly ordered arrangements of actin and myosin microfilaments. Also known as skeletal muscle. (Compare with cardiac muscle, smooth muscle.)

strobilus A conelike structure consisting of spore-bearing scales (modified leaves) inserted on an axis. (Contrast with cone.)

stroma The fluid contents of an organelle, such as a chloroplast.

stromatolites Composite, flat-to-domed structures composed of successive mineral layers produced by the action of cyanobacteria in water; ancient ones provide evidence for early life on the earth.

structural gene A gene that encodes the primary structure of a protein.

structural isomers Molecules made up of the same kinds and numbers of atoms, in which the atoms are bonded differently.

style [Gk. *stylos*: pillar or column] In flowering plants, a column of tissue extending from the tip of the ovary, and bearing the stigma or receptive surface for pollen at its apex.

sub- [L. *sub*: under] A prefix often used to designate a structure that lies beneath another or is less than another. For example, subcutaneous (beneath the skin); subspecies.

suberin A waxlike lipid that acts as a barrier to water and solute movement across the Casparian strip of the endodermis. Suberin is the waterproofing element in the cell walls of cork.

submucosa (sub mew koe' sah) The tissue layer just under the epithelial lining of the lumen of the digestive tract.

substrate (sub' strayte) The molecule or molecules on which an enzyme exerts catalytic action.

substrate-level phosphorylation Reaction in which ATP is formed from ADP by the addition of a phosphate group directly from a reactant.

substratum The base material on which a sessile organism lives.

succession In ecology, the gradual, sequential series of changes in species composition of a community following a disturbance.

sulcus [L. *sulcare*: to plow] The valleys or creases between the raised portions of the convoluted surface of the brain. (Contrast with gyrus.)

summation The ability of a neuron to fire action potentials in response to numerous subthreshold postsynaptic potentials arriving simultaneously at differentiated places on the cell, or arriving at the same site in rapid succession.

surface area-to-volume ratio For any cell, organism, or geometrical solid, the ratio of surface area to volume; this is an important factor in setting an upper limit on the size a cell or organism can attain.

surface tension The attractive intermolecular forces at the surface of liquid; especially important in water.

surfactant A substance that decreases the surface tension of a liquid. Lung surfactant, secreted by cells of the alveoli, is mostly phospholipid and decreases the amount of work necessary to inflate the lungs.

survivorship The proportion of individuals in a cohort that is alive at some time in the future.

suspensor In the embryos of seed plants, the stalk of cells that pushes the embryo into the endosperm and is a source of nutrient transport to the embryo.

symbiosis (sim' bee oh' sis) [Gk. *sym*: together + *bios*: living] The living together of two or more species in a prolonged and intimate ecological relationship. (Compare with parasitism, mutualism.)

symmetry Describes an attribute of an animal body in which at least one plane can divide the body into similar, mirror-image halves. (*See* bilateral symmetry, biradial symmetry, radial symmetry.)

sympathetic nervous system A division of the autonomic (involuntary) nervous system. (Contrast with parasympathetic nervous system.)

sympatric speciation (sim pat' rik) [Gk. *sym*: same + *patria*: homeland] Speciation due to reproductive isolation without any physical separation of the subpopulation. (Contrast with allopatric speciation, parapatric speciation.)

symplast The continuous meshwork of the interiors of living cells in the plant body, resulting from the presence of plasmodesmata. (Contrast with apoplast.)

symport A membrane transport process that carries two substances in the same direction across the membrane. (Contrast with antiport.)

synapomorphy *See* shared derived trait.

synapse (sin' aps) [Gk. *syn*: together + *haptein*: to fasten] The narrow gap between the terminal bouton of one neutron and the dendrite or cell body of another.

synapsis (sin ap' sis) The highly specific parallel alignment (pairing) of homologous chromosomes during the first division of meiosis.

synaptic vesicle A membrane-bounded vesicle containing neurotransmitter; the neurotransmitter is produced in and discharged by a presynaptic neuron.

synergids [Gk. *syn*: together + *ergos*: performing work] In flowering plants, the two cells accompanying the egg cell at one end of the megmagametophyte.

syngamy (sing' guh mee) [Gk. *syn*: together + *gamos*: marriage] Union of gametes. Also known as fertilization.

synonymous substitution A change of one nucleotide in a sequence to another when that change does not affect the amino acid specified (i.e., UUA → UUG, both specifying leucine). (Compare with nonsynonymous substitution, missense substitution, nonsense substitution.)

systematics The scientific study of the diversity of organisms, and of their relationships.

systemic circulation The part of the circulatory system serving those parts of the body other than the lungs or gills (which are served by the pulmonary circulation.)

systemic acquired resistance A general resistance to many plant pathogens following infection by a single agent.

systemin The only polypeptide plant hormone; participates in response to tissue damage.

systems biology The study of an organism as an integrated and interacting system of genes, proteins, and biochemical reactions.

systole (sis' tuh lee) [Gk. *systole*: contraction] Contraction of a chamber of the heart, driving blood forward in the circulatory system.

- T -

T cell A type of lymphocyte, involved in the cellular immune response. The final stages of its development occur in the thymus gland. (Contrast with B cell; see also cytotoxic T cell, helper T cell, suppressor T cell.)

T cell receptor A protein on the surface of a T cell that recognizes the antigenic determinant for which the cell is specific.

target cell A cell with the appropriate receptors to bind and respond to a particular hormone or other chemical mediator.

taste bud A structure in the epithelium of the tongue that includes a cluster of chemoreceptors innervated by sensory neurons.

TATA box An eight-base-pair sequence, found about 25 base pairs before the starting point for transcription in many eukaryotic promoters, that binds a transcription factor and thus helps initiate transcription.

taxis (tak' sis) [Gk. *taxis*: arrange, put in order] The movement of an organism or its part directly toward or away from the stimulus. For example, positive phototaxis is movement toward a light source, negative geotaxis is movement away from gravity).

taxon A biological group (typically a species or a clade) that is given a name.

telencephalon The frontmost division of the vertebrate brain; becomes the cerebrum.

telomerase An enzyme that catalyzes the addition of telomeric sequences lost from chromosomes during DNA replication.

telomeres (tee' lo merz) [Gk. *telos*: end + *meros*: units, segments] Repeated DNA sequences at the ends of eukaryotic chromosomes.

telophase (tee' lo phase) [Gk. *telos*: end] The final phase of mitosis or meiosis during which chromosomes became diffuse, nuclear envelopes reform, and nucleoli begin to reappear in the daughter nuclei.

template (1) In biochemistry, a molecule or surface upon which another molecule is synthesized in complementary fashion, as in the replication of DNA. (2) In the brain, a pattern that responds to a normal input but not to incorrect inputs.

template strand In double-stranded DNA, the strand that is transcribed to create an RNA transcript that will be processed into a protein. Also refers to a strand of RNA that is used to create a complementary RNA.

temporal summation [L. *tempus*: time; *summus*: highest amount] In the production or inhibition of action potentials in a postsynaptic neuron, the interaction of depolarizations or hyperpolarizations produced by rapidly repeated stimulation of a single point.

tendon A collagen-containing band of tissue that connects a muscle with a bone.

termination The end of protein synthesis triggered by a stop codon which binds release factor that causes the polypeptide to release from the ribosome.

terminator A sequence at the 3′ end of mRNA that causes the RNA strand to be released from the transcription complex.

terrestrial (ter res' tree al) [L. *terra*: earth] Pertaining to the land. (Contrast with aquatic, marine.)

territory A fixed area from which an animal or group of animals excludes other members of the same (and sometimes other) species by aggressive behavior or displays.

tertiary structure In reference to a protein, the relative locations in three-dimensional space of all the atoms in the molecule. The overall shape of a protein. (Contrast with primary, secondary, and quaternary structures.)

test cross Mating of a dominant-phenotype individual (who may be either heterozygous or homozygous) with a homozygous-recessive individual.

testis (tes' tis) (plural: testes) [L. *testis*: witness] The male gonad; the organ that produces the male sex cells.

testosterone (tes toss' tuhr own) A male sex steroid hormone.

tetanus [Gk. *tetanos*: stretched] (1) A state of sustained maximal muscular contraction caused by rapidly repeated stimulation. (2) In medicine, an often fatal disease ("lockjaw") caused by the bacterium *Clostridium tetani*.

tetrad [Gk. *tettares*: four] During prophase I of meiosis, the association of a pair of homologous chromosomes or four chromatids.

thalamus [Gk. *thalamos*: chamber] A region of the vertebrate forebrain; involved in integration of sensory input.

thallus (thal' us) [Gk. *thallos*: sprout] Any algal body which is not differentiated into root, stem, and leaf.

therapeutic cloning The use of cloning by nuclear transfer to produce an embryo that will provide embryonic stem cells to be used in therapy.

theory [Gk. *theoria*: analysis of facts] A far-reaching explanation of observed facts that is supported by such a wide body of evidence, with no significant contradictory evidence, that it is scientifically accepted as a factual framework. Examples are Newton's theory of gravity and Darwin's theory of evolution. (Contrast with hypothesis.)

thermoneutral zone [Gk. *thermos*: temperature] The range of temperatures over which an endotherm does not have to expend extra energy to thermoregulate.

thermoreceptor A cell or structure that responds to changes in temperature.

thoracic cavity [Gk. *thorax*: breastplate] The portion of the mammalian body cavity bounded by the ribs, shoulders, and diaphragm. Contains the heart and the lungs.

thoracic duct The connection between the lymphatic system and the circulatory system.

thorax [Gk. *thorax*: breastplate] In an insect, the middle region of the body, between the head and abdomen. In mammals, the part of the body between the neck and the diaphragm.

thrombin An enzyme that converts fibrinogen to fibrin, thus triggering the formation of blood clots.

thrombus (throm' bus) [Gk. *thrombos*: clot] A blood clot that forms within a blood vessel and remains attached to the wall of the vessel. (Contrast with embolus.)

thylakoid (thigh la koid) [Gk. *thylakos*: sack or pouch] A flattened sac within a chloroplast. Thylakoid membranes contain all of the chlorophyll in a plant, in addition to the electron carriers of photophosphorylation. Thylakoids stack to form grana.

thymine (T) A nitrogen-containing base found in DNA.

thymus [Gk. *thymos*: warty] A ductless, glandular portion of the lymphoid system, involved in development of the immune system of vertebrates. In humans, the thymus degenerates during puberty.

thyroid [Gk. *thyreos*: door-shaped] A two-lobed gland in vertebrates. Produces the hormone thyroxin.

thyrotropin (thyroid-stimulating hormone, TSH) A hormone of the anterior pituitary that stimulates the thyroid gland to produce and release thyroxin.

thyrotropin-releasing hormone (TRH) A hypothalamic hormone that stimulates anterior pituitary cells to release TSH.

thyroxine The hormone produced by the thyroid gland that controls many metabolic processes.

tidal volume The amount of air that is exchanged during each breath when a person is at rest. (Compare with vital capacity.)

tight junction A junction between epithelial cells, in which there is no gap whatever between the adjacent cells. Materials may pass through a tight junction only by entering the epithelial cells themselves.

tissue A group of similar cells organized into a functional unit; usually integrated with other tissues to form part of an organ.

toll A member of a receptor family that responds to the binding of a molecule from a pathogen by initiating a protein kinase cascade, resulting in the synthesis of defensive proteins.

tonus (toe' nuss) [L. *tonus*: tension] A low level of muscular tension that is maintained even when the body is at rest.

topsoil The uppermost soil layer; contains most of the organic matter of soil, but may be depleted of most mineral nutrients.

totipotent [L. *toto*: whole, entire + *potens*: powerful] Of a cell, possessing all the genetic information and other capacities necessary to form an entire individual. (Compare with pluripotent.)

toxic [L. *toxicum*: poison] Injurious to the tissues of the host organism.

trachea (tray' kee ah) [Gk. *trakhoia*: tube] A tube that carries air to the bronchi of the lungs of vertebrates. When plural (*tracheae*), refers to the major airways of insects.

tracheary element Refers to either or both types of conductive xylem cells: tracheids and vessel elements.

tracheid (tray' kee id) A distinctive conducting and supporting cell found in the xylem of nearly all vascular plants, characterized by tapering ends and walls that are pitted but not perforated. (Contrast with vessel element.)

tracheophytes [Gk. *trakhoia*: tube + *phyton*: plant] *See* vascular plants.

trade-off The relationship between the costs of performing a behavior or other trait and the benefits the individual gains from the trait or behavior. (*See also* cost–benefit analysis.)

trait In genetics, one form of a character: eye color is a character; brown eyes and blue eyes are traits. (Compare with character.)

transcription The synthesis of RNA using one strand of DNA as the template.

transcription factors Proteins that assemble on a eukaryotic chromosome, allowing RNA polymerase II to perform transcription.

transduction (1) Transfer of genes from one bacterium to another, with a bacterial virus acting as the carrier of the genes. (2) In sensory cells, the transformation of a stimulus (e.g., light energy, sound pressure waves, chemical or electrical stimulants) into action potentials.

transfection Uptake, incorporation, and expression of recombinant DNA.

transfer cell A modified parenchyma cell that transports solutes from its cytoplasm into its cell wall, thus moving the solutes from the symplast into the apoplast.

transfer RNA (tRNA) A family of double-stranded RNA molecules. Each tRNA carries a specific amino acid and anticodon that will pair with the complementary codon in mRNA during translation.

transformation Mechanism for transfer of genetic information in bacteria in which pure DNA extracted from bacteria of one genotype is taken in through the cell surface of bacteria of a different genotype and incorporated into the chromosome of the recipient cell.

transforming principle An early term for the as yet unidentified chemical substance responsible for bacterial tranformation.

transgenic organism An organism containing recombinant DNA incorporated into its genetic material.

transition-state species A short-lived, unstable intermediate with high potential energy in a chemical reaction.

translation The synthesis of a protein (polypeptide). Takes place on ribosomes, using the information encoded in messenger RNA.

transmembrane domain The portion of a protein that lies inside the membrane bilayer.

transmembrane protein An integral membrane protein that spans the lipid bilayer.

translocation (1) In genetics, a rare mutational event that moves a portion of a chromosome to a new location, generally on a nonhomologous chromosome. (2) In vascular plants, movement of solutes in the phloem.

transpiration [L. *spirare*: to breathe] The evaporation of water from plant leaves and stem, driven by heat from the sun, and providing the motive force to raise water (plus mineral nutrients) from the roots.

transposable element A segment of DNA that can move to, or give rise to copies at, another locus on the same or a different chromosome.

transposon Mobile DNA segment that can insert into a chromosome and cause genetic change.

triglyceride A simple lipid in which three fatty acids are combined with one molecule of glycerol.

triplet *See* codon.

triplet repeat The occurrence of repeated triplet of bases in a gene, often leading to genetic disease, as does excessive repetition of CGG in the gene responsible for fragile-X syndrome.

triploblastic Having three cell layers. (Contrast with diploblastic.)

trisomic Containing three rather than two members of a chromosome pair.

tRNA *See* transfer RNA.

trophic cascade The progression over successively lower trophic levels of the indirect effects of a predator.

trophic level [Gk *trophes*: nourishment] A group of organisms united by obtaining their energy from the same part of the food web of a biological community.

trophoblast [Gk *trophes*: nourishment + *blastos*: sprout] At the 32-cell stage of mammalian development, the outer group of cells that will become part of the placenta and thus nourish the growing embryo. (Comtrast with inner cell mass.)

trochophore (troke' o fore) [Gk. *trochos*: wheel + *phoreus*: bearer] The free-swimming larva of some annelids and mollusks, distinguished by a wheel-like band of cilia around the middle.

tropic hormones Hormones of the anterior pituitary that control the secretion of hormones by other endocrine glands.

tropism [Gk. *tropos*: to turn] In plants, growth toward or away from a stimulus such as light (phototropism) or gravity (gravitropism).

tropomyosin [troe poe my' oh sin] A protein that, along with actin, constitutes the thin filaments of myofibrils. It controls the interactions of actin and myosin necessary for muscle contraction.

troposphere The lowest atmospheric zone, reaching upward from the Earth's surface approximately 17 km in the tropics and subtropics but only to about 10 km at higher latitudes. The zone in which virtually all the water vapor in the atmosphere is located.

true-breeding A genetic cross in which the same result occurs every time with respect to the trait(s) under consideration, due to homozygous parents.

trypsin A protein-digesting enzyme. Secreted by the pancreas in its inactive form (trypsinogen), it becomes active in the duodenum of the small intestine.

T tubules A system of tubules that runs throughout the cytoplasm of muscle fibers, through which action potentials spread.

tube cell The larger of the two cells in a pollen grain; responsible for growth of the pollen tube. *See* generative cell.

tubulin A protein that polymerizes to form microtubules.

tumor [L. *tumor*: a swollen mass] A disorganized mass of cells, often growing out of control. Malignant tumors spread to other parts of the body.

tumor suppressor genes Genes which, when homozygous mutant, result in cancer. Such genes code for protein products that inhibit cell proliferation.

turgor pressure [L. *turgidus*: swollen] *See* pressure potential.

tympanic membrane [Gk. *tympanum*: drum] The eardrum.

- U -

ubiquinone (yoo bic' kwi known) [L. *ubique*: everywhere] A mobile electron carrier of the mitochondrial respiratory chain. Similar to plastoquinone found in chloroplasts.

ubiquitin A small protein that is covalently linked to other cellular proteins identified for breakdown by the proteosome.

umbilical cord Tissue made up of embryonic membranes and blood vessels that connects the embryo to the placenta in eutherian mammals.

understory The aggregate of smaller plants growing beneath the canopy of dominant plants in a forest.

unicellular (yoon' e sell' yer ler) [L. *unus*: one + *cella*: chamber] Consisting of a single cell, as in a unicellular organism. (Contrast with multicellular.)

uniport [L. *unus*: one + portal: doorway] A membrane transport process that carries a single substance. (Contrast with antiport, symport.)

unsaturated hydrocarbon A compound containing only carbon and hydrogen atoms, with one or more pairs of carbon atoms that are connected by double bonds.

upwelling The surfaceward movement of nutrient-rich, cooler water from deeper layers of the ocean.

uracil (U) A pyrimidine base found in nucleotides of RNA.

urea A compound serving as the main excreted form of nitrogen by many animals, including mammals.

ureotelic Describes an organism in which the final product of the breakdown of nitrogen-containing compounds (primarily proteins) is urea. (Contrast with ammonotelic, uricotelic.)

ureter (your' uh tur) A long duct leading from the vertebrate kidney to the urinary bladder or the cloaca.

urethra (you ree' thra) In most mammals, the canal through which urine is discharged from the bladder and which serves as the genital duct in males.

uric acid A compound that serves as the main excreted form of nitrogen in some animals, particularly those which must conserve water, such as birds, insects, and reptiles.

uricotelic Describes an organism in which the final product of the breakdown of nitrogen-containing compounds (primarily proteins) is uric acid. (Contrast with ammonotelic, ureotelic.)

urine (you' rin) In vertebrates, the fluid waste product containing the toxic nitrogenous by-products of protein and amino acid metabolism.

uterus (yoo' ter us) [L. *utero*: womb] The uterus or womb is a specialized portion of the female reproductive tract in certain mammals. It receives the fertilized egg and nurtures the embryo in its early development.

- V -

vaccination Injection of virus or bacteria or their proteins into the body, to induce immunization. The injected material is usually attenuated (weakened) before injection.

vacuole (vac' yew ole) [Fr.: small vacuum] A liquid-filled, membrane-enclosed compartment in cytoplasm; may function as digestive chambers, storage chambers, waste bins.

vagina (vuh jine' uh) [L.: sheath] In female mammals, the passage leading from the external genital orifice to the uterus; receives the copulatory organ of the male in mating.

van der Waals forces Weak attractions between atoms resulting from the interaction of the electrons of one atom with the nucleus of another. This type of attraction is about one-fourth as strong as a hydrogen bond.

variable regions The part of an immunoglobulin molecule or T-cell receptor that includes the antigen-binding site.

vasa recta Blood vessels that parallel the loops of Henle and the collecting ducts in the renal medulla of the kidney.

vascular (vas' kew lar) [L. *vasculum*: a small vessel] Pertaining to organs and tissues that conduct fluid, such as blood vessels in animals and phloem and xylem in plants.

vascular bundle In vascular plants, a strand of vascular tissue, including conducting cells of xylem and phloem as well as thick-walled fibers.

vascular cambium A lateral meristem giving rise to secondary xylem and phloem.

vascular rays In vascular plants, radially oriented sheets of cells produced by the vascular cambium, carrying materials laterally between the wood and the phloem.

vascular system The conductive system of the plant, consisting primarily of xylem and phloem.

vas deferens The duct that transfers sperm from the epididymis to the urethra.

vasopressin *See* antidiuretic hormone.

vector (1) An agent, such as an insect, that carries a pathogen affecting another species. (2) A plasmid or virus that carries an inserted piece of DNA into a bacterium for cloning purposes in recombinant DNA technology.

vegetal hemisphere The lower portion of some animal eggs, zygotes, and embryos, in which the dense nutrient yolk settles. The vegetal pole refers to the very bottom of the egg or embyro. (Contrast with animal hemisphere.)

vegetative Nonreproductive, nonflowering, or asexual.

vegetative reproduction Asexual reproduction.

vein [L. *vena*: channel] A blood vessel that returns blood to the heart. (Contrast with artery.)

ventral [L. *venter*: belly, womb] Toward or pertaining to the belly or lower side. (Contrast with dorsal.)

ventricle A muscular heart chamber that pumps blood through the lungs or through the body.

vernalization [L. *vernalis*: spring] Events occurring during a required chilling period, leading eventually to flowering.

vertebral column [L. *vertere*: to turn] The jointed, dorsal column that is the primary support structure of vertebrates.

vesicle A membrane enclosed compartment within the cytoplasm.

vessel element In plants, a nonliving water-conducting cell with perforated end walls. (Contrast with tracheid.)

vestibular apparatus (ves tib' yew lar) [L. *vestibulum*: an enclosed passage] Structures associated with the vertebrate ear; these structures sense changes in position or momentum of the head, affecing balance and motor skills.

vestigial (ves tij' ee al) [L. *vestigium*: footprint, track] The remains of body structures that are no longer of adaptive value to the organism and therefore are not maintained by selection.

vicariant distribution A population distribution resulting from the disruption of a formerly continuous range by a vicariant event.

vicariant event (vye care' ee unce) [L. *vicus*: change] The splitting of the range of a taxon by the imposition of some barrier to interchange among its members.

villus (vil' lus) (plural: villi) [L. *villus*: shaggy hair or beard] A hairlike projection from a membrane; for example, from many gut walls.

virion (veer' e on) The virus particle, the minimum unit capable of infecting a cell.

viroid (vye' roid) An infectious agent consisting of a single-stranded RNA molecule with no protein coat; produces diseases in plants.

virulent [L. *virus*: poison, slimy liquid] Causing or capable of causing disease and death.

virus Any of a group of ultramicroscopic particles constructed of nucleic acid and protein (and, sometimes, lipid) that require living cells in order to reproduce. Viruses probably evolved from eukaryotic cells, secondarily losing some cellular attributes.

vital capacity The sum total of the tidal volume and the inspiratory and expiratory reserve volumes.

vitamins [L. *vita*: life] Organic compounds that an organism cannot synthesize, but nevertheless requires in small quantity for normal growth and metabolism.

viviparous (vye vip' uh rus) [L. *vivus*: alive] Reproduction in which fertilization of the egg and development of the embryo occur inside the mother's body. (Contrast with oviparous.)

VNTRs (variable number of tandem repeats) In the human genome, short DNA sequences that are repeated a characteristic number of times in related individuals. Can be used to make a DNA fingerprint.

voltage-gated channels Ion channels in membranes that change conformation and therefore ion conductance when a certain potential difference exists across the membrane in which they are inserted.

- W -

waggle dance The running movement of a working honey bee on the hive, during which the worker traces out a repeated figure eight. The dance contains elements that transmit to other bees the location of the food.

water potential In osmosis, the tendency for a system (a cell or solution) to take up water from pure water through a differentially permeable membrane. Water flows toward the system with a more negative water potential. (Contrast with solute potential, pressure potential.)

wavelength The distance between successive peaks of a wave train, such as electromagnetic radiation.

wild-type Geneticists' term for standard or reference type. Deviants from this standard, even if the deviants are found in the wild, are usually referred to as mutant. (Note that this terminology is not usually applied to human genes.)

wood Secondary xylem tissue.

- X -

xanthophyll (zan' tho fill) [Gk. *xanthos*: yellowish-brown + *phyllon*: leaf] A yellow or orange pigment commonly found as an accessory pigment in photosynthesis, but found elsewhere as well. An oxygen-containing carotenoid.

X-linked A character that is coded for by a gene on the X chromosome; a sex-linked trait.

xerophyte (zee' row fyte) [Gk. *xerox*: dry + *phyton*: plant] A plant adapted to an environment with a limited water supply.

xylem (zy' lum) [Gk. *xylon*: wood] In vascular plants, the tissue that conducts water and minerals; xylem consists, in various plants, of tracheids, vessel elements, fibers, and other highly specialized cells.

- Y -

yolk [M.E. *yolke*: yellow] The stored food material in animal eggs, usually rich in protein and lipid.

yolk sac In reptiles, birds, and mammals, the extraembryonic membrane that forms from the endoderm of the hypoblast; it encloses and digests the yolk.

- Z -

Z-DNA A form of DNA in which the molecule spirals to the left rather than to the right.

zeaxanthin A blue-light receptor involved in the opening of plant stomata.

zona pellucida A jellylike substance that surrounds the mammalian ovum when it is released from the ovary.

zoospore (zoe' o spore) [Gk. *zoon*: animal + *spora*: seed] In algae and fungi, any swimming spore. May be diploid or haploid.

zygote (zye' gote) [Gk. *zygotos*: yoked] The cell created by the union of two gametes, in which the gamete nuclei are also fused. The earliest stage of the diploid generation.

zymogen An inactive precursor of a digestive enzyme secreted into the lumen of the gut, where a protease cleaves it to form the active enzyme.

Answers to Self-Quizzes

Chapter 2
1.	b	6.	a
2.	d	7.	d
3.	c	8.	a
4.	c	9.	c
5.	d	10.	b

Chapter 3
1.	e	6.	a
2.	e	7.	c
3.	c	8.	e
4.	d	9.	a
5.	b	10.	d

Chapter 4
1.	b	6.	e
2.	d	7.	a
3.	c	8.	d
4.	e	9.	b
5.	a	10.	d

Chapter 5
1.	e	6.	c
2.	c	7.	c
3.	a	8.	b
4.	d	9.	e
5.	c	10.	c

Chapter 6
1.	c	6.	e
2.	e	7.	d
3.	b	8.	b
4.	c	9.	d
5.	c	10.	e

Chapter 7
1.	d	6.	d
2.	d	7.	a
3.	e	8.	b
4.	e	9.	a
5.	c	10.	e

Chapter 8
1.	c	6.	c
2.	b	7.	c
3.	d	8.	d
4.	b	9.	d
5.	e	10.	b

Chapter 9
1.	d	6.	d
2.	c	7.	e
3.	b	8.	d
4.	d	9.	c
5.	c	10.	c

Chapter 10*
1.	e	6.	d
2.	a	7.	b
3.	d	8.	b
4.	d	9.	b
5.	d	10.	b

Chapter 11
1.	c	6.	b
2.	a	7.	d
3.	c	8.	d
4.	b	9.	c
5.	e	10.	c

Chapter 12
1.	c	6.	d
2.	d	7.	b
3.	e	8.	d
4.	b	9.	d
5.	a	10.	a

Chapter 13
1.	b	6.	d
2.	e	7.	d
3.	a	8.	c
4.	c	9.	b
5.	c	10.	d

Chapter 14
1.	c	6.	c
2.	c	7.	c
3.	a	8.	b
4.	a	9.	e
5.	c	10.	d

Chapter 15
1.	d	6.	a
2.	d	7.	e
3.	c	8.	b
4.	c	9.	c
5.	d	10.	a

Chapter 16
1.	b	6.	b
2.	a	7.	c
3.	a	8.	a
4.	c	9.	e
5.	e	10.	e

Chapter 17
1.	a	6.	b
2.	c	7.	e
3.	b	8.	d
4.	b	9.	c
5.	d	10.	b

Chapter 18
1.	a	6.	a
2.	b	7.	d
3.	e	8.	d
4.	e	9.	a
5.	c	10.	d

Chapter 19
1.	c	6.	c
2.	a	7.	d
3.	a	8.	b
4.	b	9.	a
5.	b	10.	b

Chapter 20
1.	d	6.	b
2.	e	7.	a
3.	c	8.	c
4.	b	9.	b
5.	c	10.	b

Chapter 21
1.	d	6.	a
2.	b	7.	c
3.	e	8.	b
4.	c	9.	c
5.	a	10.	e

*Answers to Chapter 10 Genetics Problems

1. Each of the eight boxes in the Punnett squares should contain the genotype *Tt*, regardless of which parent was tall and which dwarf.

2. See Figure 10.4, page 212.

3. The trait is autosomal. Mother *dp dp*, father *Dp dp*. If the trait were sex-linked, all daughters would be wild-type and sons would be *dumpy*.

4. Yellow parent = $s^Y s^b$; offspring 3 yellow (s^Y–): 1 black ($s^b s^b$). Black parent = $s^b s^b$; offspring all black ($s^b s^b$). Orange parent = $s^O s^b$; offspring 3 orange (s^O–): 1 black ($s^b s^b$). Both s^O and s^Y are dominant to s^b.

5. All females wild-type; all males spotted.

6. F_1 all wild-type, *PpSwsw*; F_2 9:3:3:1 in phenotypes. See Figure 10.7, page 214, for analogous genotypes.

7*a*. Ratio of phenotypes in F_2 is 3:1 (double dominant to double recessive).

7*b*. The F_1 are $Pby\ pB^Y$; they produce just two kinds of gametes (*Pby* and *pBy*). Combine them carefully and see the 1:2:1 phenotypic ratio fall out in the F_2.

7*c*. Pink-blistery.

7*d*. See Figures 9.16 and 9.18 (pages 196–198). Crossing over took place in the F_1 generation.

8. The genotypes are:

PpSwsw
Ppswsw
ppSwsw
ppswsw

Ratio: 1:1:1:1

The phenotypes are:

wild eye, long wing	pink eye, long wing
wild eye, short wing	pink eye, short wing

Ratio: 1:1:1:1

9*a*. 1 black:2 blue:1 splashed white

9*b*. Always cross black with splashed white.

10*a*. $w^+ > w^e > w$

10*b*. Parents $w^e w$ and $w^+ Y$. Progeny $w{+}w^e$, $w{+}w$, $w^e Y$, and wY.

11. All will have normal vision because they inherit dad's wild-type X chromosome, but half of them will be carriers.

12. Agouti parent *AaBb*. Albino offspring *aaBb* and *aabb*; black offspring *Aabb*; agouti offspring *AaBb*.

13. Because the gene is carried on mitochondrial DNA, it is passed through the mother only. Thus if the woman does not have the disease but her husband does, their child will not be affected. On the other hand, if the woman has the disease but her husband does not, their child *will* have the disease.

Chapter 22
1. d 6. d
2. e 7. b
3. d 8. e
4. c 9. b
5. d 10. c

Chapter 23
1. c 6. d
2. e 7. a
3. d 8. a
4. c 9. c
5. a 10. e

Chapter 24
1. a 6. e
2. a 7. e
3. d 8. e
4. a 9. b
5. b 10. e

Chapter 25
1. b 6. b
2. e 7. e
3. a 8. a
4. b 9. e
5. e 10. d

Chapter 26
1. e 6. b
2. e 7. d
3. b 8. d
4. c 9. c
5. e 10. d

Chapter 27
1. a 6. b
2. e 7. d
3. c 8. b
4. d 9. a
5. c 10. d

Chapter 28
1. d 6. e
2. c 7. c
3. e 8. b
4. b 9. b
5. b 10. d

Chapter 29
1. d 6. c
2. c 7. a
3. d 8. e
4. a 9. c
5. d 10. a

Chapter 30
1. b 6. a
2. d 7. e
3. e 8. a
4. c 9. c
5. d 10. c

Chapter 31
1. c 6. b
2. d 7. c
3. c 8. d
4. d 9. e
5. c 10. d

Chapter 32
1. b 6. e
2. e 7. b
3. c 8. d
4. d 9. d
5. a 10. e

Chapter 33
1. b 6. e
2. a 7. a
3. c 8. e
4. c 9. c
5. d 10. b

Chapter 34
1. d 6. b
2. b 7. b
3. e 8. c
4. e 9. a
5. a 10. d

Chapter 35
1. c 6. d
2. d 7. d
3. b 8. e
4. b 9. e
5. b 10. a

Chapter 36
1. d 6. c
2. d 7. e
3. c 8. a
4. a 9. d
5. a 10. e

Chapter 37
1. a 6. c
2. e 7. e
3. c 8. c
4. d 9. a
5. b 10. b

Chapter 38
1. d 6. e
2. b 7. a
3. e 8. b
4. b 9. c
5. d 10. d

Chapter 39
1. e 6. a
2. b 7. b
3. c 8. c
4. c 9. d
5. d 10. a

Chapter 40
1. c 6. b
2. c 7. e
3. a 8. a
4. d 9. e
5. b 10. b

Chapter 41
1. b 6. b
2. a 7. d
3. b 8. e
4. e 9. c
5. e 10. c

Chapter 42
1. c 6. d
2. e 7. d
3. a 8. c
4. d 9. d
5. d 10. a

Chapter 43
1. a 6. c
2. c 7. b
3. e 8. b
4. a 9. b
5. d 10. a

Chapter 44
1. d 6. e
2. d 7. c
3. c 8. c
4. c 9. d
5. e 10. d

Chapter 45
1. d 6. e
2. d 7. b
3. a 8. c
4. b 9. c
5. e 10. d

Chapter 46
1. c 6. c
2. a 7. a
3. e 8. c
4. d 9. a
5. d 10. c

Chapter 47
1. e 6. d
2. a 7. e
3. b 8. a
4. c 9. a
5. b 10. e

Chapter 48
1. e 6. b
2. d 7. c
3. a 8. c
4. b 9. a
5. c 10. d

Chapter 49
1. d 6. d
2. a 7. b
3. c 8. d
4. d 9. c
5. c 10. e

Chapter 50
1. b 6. d
2. e 7. a
3. c 8. b
4. a 9. d
5. b 10. d

Chapter 51
1. d 6. b
2. a 7. e
3. d 8. a
4. a 9. c
5. d 10. e

Chapter 52
1. c 6. c
2. a 7. d
3. a 8. c
4. c 9. d
5. a 10. a

Chapter 53
1. c 6. d
2. e 7. c
3. c 8. c
4. a 9. d
5. d 10. e

Chapter 54
1. c 6. b
2. c 7. d
3. a 8. c
4. d 9. a
5. c 10. d

Chapter 55
1. a 6. e
2. a 7. c
3. b 8. a
4. c 9. d
5. d 10. e

Chapter 56
1. d 6. b
2. d 7. a
3. e 8. c
4. e 9. d
5. c 10. b

Chapter 57
1. b 6. a
2. d 7. d
3. e 8. c
4. e 9. a
5. b 10. c

Illustration Credits

Cover *Amor de Madre* © Max Billder.

Frontispiece © Lynn M. Stone/Naturepl.com.

Part Openers and Back Cover:

Part 1 *Canopy researcher*: © Mark Moffett/ Minden Pictures.

Part 2 *Mitochondrion*: © Bill Loncore/SPL/Photo Researchers, Inc.

Part 3 *T4 bacteriophage*: © Dept. of Microbiology, Biozentrum/SPL/Photo Researchers, Inc.

Part 4 *Dividing cells*: © Steve Gschmeissner/SPL/ Photo Researchers, Inc.

Part 5 *Fossil mesosaur*: © John Cancalosi/AGE Fotostock.

Part 6 *Tiger moth*: © Piotr Naskrecki/Minden Pictures.

Part 7 *Protea flower*: © Gerry Ellis/Minden Pictures.

Part 8 *Great blue heron*: © Larry Jon Friesen.

Part 9 *Zebras and wildebeest*: © Gerry Ellis/ Minden Pictures.

Table of Contents:

p. xix *Leaf*: David McIntyre.

p. xxix *Passionflower*: © Larry Jon Friesen.

p. xxxv *Feather stars*: © André Seale/AGE Fotostock.

Chapter 1 *Frogs*: © Frans Lanting/Minden Pictures. *Pieter Johnson*: © Steven Holt/ stockpix.com. 1.1A: © Eye of Science/SPL/Photo Researchers, Inc. 1.1B: © Dennis Kunkel Microscopy, Inc. 1.1C: © Markus Geisen/NHMPL. 1.1D: © Frans Lanting/Minden Pictures. 1.1E: © Glen Threlfo/Auscape/Minden Pictures. 1.1F: © Piotr Naskrecki/Minden Pictures. 1.1G: © Tui De Roy/Minden Pictures. 1.2A: © Dr. Cecil H. Fox/ Photo Researchers, Inc. 1.2B: From R. Hooke, 1664. *Micrographia*. 1.2C: © Astrid & Hanns-Frieder Michler/Photo Researchers, Inc. 1.2D: © John Durham/SPL/Photo Researchers, Inc. 1.3 *Maple*: © Simon Colmer & Abby Rex/Alamy. 1.3 *Spruce*: David McIntyre. 1.3 *Lily pad*: © Pete Oxford/ Naturepl.com. 1.3 *Cucumber*: David McIntyre. 1.3 *Pitcher plant*: © Nick Garbutt/Naturepl.com. 1.5A: © Bob Bennett/Jupiter Images. 1.5B: © Frans Lanting/Minden Pictures. 1.6 *Organism*: © Sami Sarkis/Painet Inc. 1.6 *Population*: © LogicStock/ Painet Inc. 1.6 *Community*: © Georgie Holland/ AGE Fotostock. 1.6*Biosphere*: Courtesy of NASA. 1.7A: © Frans Lanting/Minden Pictures. 1.7B: © Marguerite Smits Van Oyen/Naturepl.com. 1.8: © Chris Howes/Wild Places Photography/Alamy. 1.10: © Dennis Kunkel Microscopy, Inc. 1.12: Courtesy of Wayne Whippen. 1.13 *Frog (inset)*: © Frans Lanting/Minden Pictures. 1.13, 1.14: After P. T. Johnson et al., 1999. *Science* 284: 802.

Chapter 2 *Enceladus*: Courtesy of the NASA Jet Propulsion Laboratory/Space Science Institute. *Rover*: Courtesy of the NASA Jet Propulsion Laboratory. 2.4: From N. D. Volkow et al., 2001. *Am. J. Psychiatry* 158: 377. 2.14: © Mitsuaki Iwago/ Minden Pictures. 2.15: David McIntyre.

Chapter 3 *Cookies*: David McIntyre. 3.8: Data from PDB 1IVM. T. Obita, T. Ueda, & T. Imoto, 2003. *Cell. Mol. Life Sci.* 60: 176. 3.9A: Data from PDB 2HHB. G. Fermi et al., 1984. *J. Mol. Biol.* 175: 159. 3.16C *left*: © Biophoto Associates/Photo Researchers, Inc. 3.16C *middle*: © Ken Wagner/ Visuals Unlimited. 3.16C *right*: © CNRI/SPL/Photo Researchers, Inc. 3.17 *Cartilage*: © Robert Brons/ Biological Photo Service. 3.17 *Beetle*: © Scott Bauer/ USDA. 3.26: Data from S. Arnott & D. W. Hukins, 1972. *Biochem. Biophys. Res. Commun.* 47(6): 1504. 3.27: Courtesy of NASA.

Chapter 4 *Fossil*: © Stanley M. Awramik/ Biological Photo Service. *Cyanobacterium*: © Dennis Kunkel Microscopy, Inc. *Protobionts*: © Sidney Fox/Science VU/Visuals Unlimited. 4.1: After N. Campbell, 1990. *Biology*, 2nd Ed., Benjamin Cummings. 4.3 *upper row, left*: Courtesy of the IST Cell Bank, Genoa. 4.3 *upper row, center and right, middle row, left*: © Michael W. Davidson, Florida State U. 4.3 *middle row, center*: © Dr. Gopal Murti/SPL/Photo Researchers, Inc. 4.3 *middle row, right*: © Richard J. Green/SPL/Photo Researchers, Inc. 4.3 *bottom row, left*: © Dr. Gopal Murti/Visuals Unlimited. 4.3 *bottom row, center*: © K. R. Porter/ SPL/Photo Researchers, Inc. 4.3 *bottom row, right*: © D. W. Fawcett/Photo Researchers, Inc. 4.4: © J. J. Cardamone Jr. & B. K. Pugashetti/Biological Photo Service. 4.5A: © Dennis Kunkel Microscopy, Inc. 4.5B: Courtesy of David DeRosier, Brandeis U. 4.7 *Mitochondrion*: © K. Porter, D. Fawcett/Visuals Unlimited. 4.7 *Cytoskeleton*: © Don Fawcett, John Heuser/Photo Researchers, Inc. 4.7 *Nucleolus*: © Richard Rodewald/Biological Photo Service. 4.7 *Peroxisome*: © E. H. Newcomb & S. E. Frederick/ Biological Photo Service. 4.7 *Cell wall*: © Biophoto Associates/Photo Researchers, Inc. 4.7 *Ribosome*: From M. Boublik et al., 1990. *The Ribosome*, p. 177. Courtesy of American Society for Microbiology. 4.7 *Centrioles*: © Barry F. King/Biological Photo Service. 4.7 *Plasma membrane*: Courtesy of J. David Robertson, Duke U. Medical Center. 4.7 *Rough ER*: © Don Fawcett/Science Source/Photo Researchers, Inc. 4.7 *Smooth ER*: © Don Fawcett, D. Friend/Science Source/Photo Researchers, Inc. 4.7 *Chloroplast*: © W. P. Wergin, E. H. Newcomb/ Biological Photo Service. 4.7 *Golgi apparatus*: Courtesy of L. Andrew Staehelin, U. Colorado. 4.8 *left*: From U. Aebi et al., 1986. *Nature* 323: 560. © Macmillan Publishers Ltd. 4.8 *upper right*: © D. W. Fawcett/Photo Researchers, Inc. 4.8 *lower right*: Courtesy of Dr. Ron Milligan, Scripps Research Institute. 4.9A: © Barry King, U. California, Davis/Biological Photo Service. 4.9B: © Biophoto Associates/Science Source/Photo Researchers, Inc. 4.10: © Don Fawcett/Visuals Unlimited. 4.11: © B. Bowers/Photo Researchers, Inc. 4.12: © Sanders/Biological Photo Service. 4.13: © K. Porter, D. Fawcett/Visuals Unlimited. 4.14: © W. P. Wergin, E. H. Newcomb/Biological Photo Service. 4.14 *inset*: © W. P. Wergin/Biological Photo Service. 4.15A: © John Durham/SPL/Photo Researchers, Inc. 4.15B: © Paul W. Johnson/ Biological Photo Service. 4.15C: © Gerald & Buff Corsi/Visuals Unlimited. 4.16A: David McIntyre. 4.16A *inset*: © Richard Green/Photo Researchers, Inc. 4.16B: David McIntyre. 4.16B *inset*: Courtesy of R. R. Dute. 4.17: © E. H. Newcomb & S. E. Frederick/Biological Photo Service. 4.18: Courtesy of M. C. Ledbetter, Brookhaven National Laboratory. 4.19: © Michael Abbey/Visuals Unlimited. 4.20: Courtesy of Vic Small, Austrian Academy of Sciences, Salzburg, Austria. 4.21: Courtesy of N. Hirokawa. 4.22A *top*: © SPL/Photo Researchers, Inc. 4.22A *bottom*, 4.22B, C: © W. L. Dentler/Biological Photo Service. 4.23: From N. Pollack et al., 1999. *J. Cell Biol.* 147: 493. Courtesy of R. D. Vale. 4.24: © Biophoto Associates/Photo Researchers, Inc. 4.25 *left*: Courtesy of David Sadava. 4.25 *upper right*: From J. A. Buckwalter & L. Rosenberg, 1983. *Coll. Rel. Res.* 3: 489. Courtesy of L. Rosenberg. 4.25 *lower right*: © J. Gross, Biozentrum/SPL/Photo Researchers, Inc.

Chapter 5 *Vibrio*: © Dennis Kunkel Microscopy, Inc. *Cholera treatment*: Courtesy of the WHO Global Consultation on Child and Adolescent Health and Development. 5.2: After L. Stryer, 1981. *Biochemistry*, 2nd Ed., W. H. Freeman. 5.3: Courtesy of L. Andrew Staehelin, U. Colorado. 5.7A: Courtesy of D. S. Friend, U. California, San Francisco. 5.7B: Courtesy of Darcy E. Kelly, U. Washington. 5.7C: Courtesy of C. Peracchia. 5.17: From M. M. Perry, 1979. *J. Cell Sci.* 39: 26.

Chapter 6 *Enzyme*: Data from PDB 1CW3. L. Ni et al., 1999. *Protein Sci.* 8: 2784. *Champagne*: © Purestock/Alamy. 6.1: Violet Bedell-McIntyre. 6.5B: © Darwin Dale/Photo Researchers, Inc. 6.11A: Data from PDB 1AL6. B. Schwartz et al., 1997. 6.11B: Data from PDB 1BB6. V. B. Vollan et al., 1999. *Acta Crystallogr. D. Biol. Crystallogr.* 55: 60. 6.11C: Data from PDB 1AB9. N. H. Yennawar, H. P. Yennawar, & G. K. Farber, 1994. *Biochemistry* 33: 7326. 6.13: Data from PDB 1A7K. H. Kim & W. G. Hol, 1998. *J Mol. Biol.* 278: 5.

Chapter 7 *Marathoners*: © Chuck Franklin/ Alamy. *Mouse*: © Royalty-Free/Corbis. 7.13: Courtesy of Ephraim Racker.

Chapter 8 *Rainforest*: © Jon Arnold Images/ photolibrary.com. *Meteor*: © Don Davis. 8.1: © Andrew Syred/SPL/Photo Researchers, Inc. 8.12: Courtesy of Lawrence Berkeley National Laboratory. 8.16, 8.18: © E. H. Newcomb & S. E. Frederick/Biological Photo Service. 8.20: © Aflo Foto Agency/Alamy.

Chapter 9 *Cells*: © Dr. Torsten Wittmann/Photo Researchers, Inc. *Henrietta Lacks*: © Obstetrics and Gynaecology/Photo Researchers, Inc. 9.1A: © David M. Phillips/Visuals Unlimited. 9.1B, C: © John D. Cunningham/Visuals Unlimited. 9.2B: © John J. Cardamone Jr./Biological Photo Service. 9.7: Courtesy of G. F. Bahr. 9.8 *inset*: © Biophoto Associates/Science Source/Photo Researchers, Inc. 9.9B: © Conly L. Rieder/Biological Photo Service. 9.10: © Nasser Rusan. 9.12A: © T. E. Schroeder/ Biological Photo Service. 9.12B: © B. A. Palevitz, E. H. Newcomb/Biological Photo Service. 9.13A: © Dr. John Cunningham/Visuals Unlimited. 9.13B:

© Garry T. Cole/Biological Photo Service. 9.14 *left*: © Andrew Syred/SPL/Photo Researchers, Inc. 9.14 *center*: David McIntyre. 9.14 *right*: © Gerry Ellis, DigitalVision/PictureQuest. 9.15: Courtesy of Dr. Thomas Ried and Dr. Evelin Schröck, NIH. 9.16: © C. A. Hasenkampf/Biological Photo Service. 9.17: Courtesy of J. Kezer. 9.21A: © Gopal Murti/ Photo Researchers, Inc.

Chapter 10 *Bris*: © David H. Wells/CORBIS. *Eye test*: © Brand X Pictures/Alamy. 10.1: © the Mendelianum. 10.11: Courtesy the American Netherland Dwarf Rabbit Club. 10.14: Courtesy of Madison, Hannah, and Walnut. 10.15: Courtesy of Pioneer Hi-Bred International, Inc. 10.16: © Carolyn A. McKeone/Photo Researchers, Inc. 10.17: © Peter Morenus/U. of Connecticut. *Bay scallops*: © Barbara J. Miller/Biological Photo Service.

Chapter 11 *T. rex*: © The Natural History Museum, London. *Earrings*: David McIntyre/ Model: Ashley Ying. 11.3: © Biozentrum, U. Basel/ SPL/Photo Researchers, Inc. 11.6 *X-ray crystallograph*: Courtesy of Prof. M. H. F. Wilkins, Dept. of Biophysics, King's College, U. London. 11.6 *Franklin*: © CSHL Archives/Peter Arnold, Inc. 11.8A: © A. Barrington Brown/Photo Researchers, Inc. 11.8B: Data from S. Arnott & D. W. Hukins, 1972. *Biochem. Biophys. Res. Commun.* 47(6): 1504. 11.21C: © Dr. Peter Lansdorp/Visuals Unlimited.

Chapter 12 *Ricinus*: © Alan L. Detrick/Photo Researchers, Inc. *Ribosome*: Data from PDB 1GIX and 1G1Y. M. M. Yusupov et al., 2001. *Science* 292: 883. 12.4: Data from PDB 1I3Q. P. Cramer et al., 2001 *Science* 292: 1863. 12.8: Data from PDB 1EHZ. H. Shi & P. B. Moore, 2000. *RNA* 6: 1091. 12.9B: Data from PDB 1EUQ. L. D. Sherlin et al., 2000. *J. Mol. Biol.* 299: 431. 12.10: Data from PDB 1GIX and 1G1Y. M. M. Yusupov et al., 2001. *Science* 292: 883. 12.14: Courtesy of J. E. Edström and *EMBO J.* 12.18: © Stanley Flegler/Visuals Unlimited.

Chapter 13 *Market*: © Derek Brown/Alamy. *Scientist*: © Tek Image/SPL/Photo Researchers, Inc. 13.1A: © Dept. of Microbiology, Biozentrum/SPL/ Photo Researchers, Inc. 13.1B: © Dennis Kunkel Microscopy, Inc. 13.2A: © Dr. Harold Fisher/ Visuals Unlimited. 13.2B: © E.O.S./Gelderblom/ Photo Researchers, Inc. 13.2C: © BSIP Agency/ Index Stock Imagery/Jupiter Images. 13.7: Courtesy of Roy French. 13.11: Courtesy of L. Caro and R. Curtiss. 13.22: Based on an illustration by Anthony R. Kerlavage, Institute for Genomic Research. *Science* 269: 449 (1995). *For Investigation*: Adapted from J. J. Perry, J. T. Staley, & S. Lory, 2002. *Microbial Life*. Sinauer Associates.

Chapter 14 *Cheetah*: © John Giustina/ ImageState/Jupiter Images. *Panther*: © Lynn M. Stone/Naturepl.com. 14.7: From D. C. Tiemeier et al., 1978. *Cell* 14: 237. 14.18: Courtesy of Murray L. Barr, U. Western Ontario. 14.20: Courtesy of O. L. Miller, Jr.

Chapter 15 *Student*: © Ryan McVay, Photodisc Green/Getty Images. *Ethiopian*: © JTB Photo/ photolibrary.com. 15.2: © Biophoto Associates/ Photo Researchers, Inc. 15.3: From A. M. de Vos, M. Ultsch, & A. A. Kossiakoff, 1992. *Science* 255: 306. 15.14: © Stephen A. Stricker, courtesy of Molecular Probes, Inc.

Chapter 16 *Destruction*: © Suzanne Plunkett/AP Images. *Baby 81*: © Gemunu Amarasinghe/AP Images. 16.2: © Philippe Plailly/Photo Researchers, Inc. 16.5: Bettmann/CORBIS. 16.6: Courtesy of Keith Weller, De Wood, Chris Pooley, & Scott Bauer/USDA ARS. 16.19A: Courtesy of Ingo Potrykus, Swiss Federal Institute of Technology. 16.19B: Joan Gemme. 16.20: Courtesy of Eduardo Blumwald.

Chapter 17 *Champlain*: © North Wind Picture Archives/Alamy. *Family*: © David Sanger Photography/Alamy. 17.6: From C. Harrison et al., 1983. *J. Med. Genet.* 20: 280. 17.11: Courtesy of Harvey Levy and Cecelia Walraven, New England Newborn Screening Program. 17.14: © Dennis Kunkel Microscopy, Inc. 17.19B: © David M. Martin, M.D./SPL/Photo Researchers, Inc. 17.23: From P. H. O'Farrell, 1975. High resolution two-dimensional electrophoresis of proteins. *J. Biol. Chem.* 250: 4007-21. Courtesy of Patrick H. O'Farrell. 17.24: After N. M. Morel et al., 2004. *Mayo Clin. Proc.* 79: 651.

Chapter 18 *Vaccination*: Bettmann/CORBIS. *T cell*: © Dr. Andrejs Liepins/SPL/Photo Researchers, Inc. 18.3: © Dennis Kunkel Microscopy, Inc. 18.8: © Dr. Gopal Murti/SPL/Photo Researchers, Inc. 18.14: © David Phillips/Science Source/Photo Researchers, Inc.

Chapter 19 *Embryo*: © Dr. Yorgas Nikas/Photo Researchers, Inc. *Centrifuge*: Courtesy of Cytori Therapeutics. 19.4: © Roddy Field, the Roslin Institute. 19.5: Courtesy of T. Wakayama and R. Yanagimachi. 19.12: From J. E. Sulston & H. R. Horvitz, 1977. *Dev. Bio.* 56: 100. 19.15B, 19.16 *left*: Courtesy of J. Bowman. 19.16 *right*: Courtesy of Detlef Weigel. 19.17: Courtesy of W. Driever and C. Nüsslein-Vollhard. 19.18B: Courtesy of C. Rushlow and M. Levine. 19.18C: Courtesy of T. Karr. 19.18D: Courtesy of S. Carroll and S. Paddock. 19.20: Courtesy of F. R. Turner, Indiana U.

Chapter 20 *Fly head*: © David Scharf/Photo Researchers, Inc. *Mutant leg*: From G. Halder et al., 1995. *Science* 267: 1788. Courtesy of W. J. Gehring and G. Halder. 20.2A: Courtesy of E. B. Lewis. 20.2B: From H. Le Mouellic et al., 1992. *Cell* 69: 251. Courtesy of H. Le Mouellic, Y. Lallemand, and P. Brûlet. 20.5: Courtesy of J. Hurle and E. Laufer. 20.6: Courtesy of J. Hurle. 20.7 *Cladogram*: After R. Galant & S. Carroll, 2002. *Nature* 415: 910. 20.7 *Beetle*: © Stockbyte/PictureQuest. 20.7 *Centipede*: © Burke/Triolo/Brand X Pictures/PictureQuest. 20.8: Courtesy of S. Carroll and P. Brakefield. 20.9: © Erick Greene. 20.10: Photograph by C. Laforsch & R. Tollrian, courtesy of A. A. Agrawal. 20.11: © Nigel Cattlin, Holt Studios International/Photo Researchers, Inc. 20.13: © Simon D. Pollard/Photo Researchers, Inc. 20.14: Courtesy of Mike Shapiro and David Kingsley.

Chapter 21 *Capybara*: © Frans Lanting/Minden Pictures. *Grand Canyon*: © Robert Fried/Tom Stack & Assoc. 21.4A: David McIntyre. 21.4B: © Robin Smith/photolibrary.com. 21.7: © François Gohier/ The National Audubon Society Collection/Photo Researchers, Inc. 21.8: © Jeff J. Daly/Visuals Unlimited. 21.9 *left*: © Ken Lucas/Visuals Unlimited. 21.9 *right*: © The Natural History Museum, London. 21.10: © Chip Clark. 21.11: © Hans Steur/Visuals Unlimited. 21.12: © Tom McHugh/Field Museum, Chicago/Photo Researchers, Inc. 21.13: Courtesy of Conrad C. Labandeira, Department of Paleobiology, National Museum of Natural History, Smithsonian Institution. 21.14: © Chase Studios, Cedarcreek, MO. 21.16: © The Natural History Museum, London. 21.18: © K. Simons and David Dilcher. 21.20: David McIntyre. 21.22: © John Worrall. 21.23: © Calvin Larsen/Photo Researchers, Inc.

Chapter 22 *Snake*: © Joseph T. Collins/Photo Researchers, Inc. *Newt*: © Robert Clay/Visuals Unlimited. *Pufferfish*: © Georgette Douwma/ Naturepl.com. 22.1A, B: © SPL/Photo Researchers, Inc. 22.2: From W. Levi, 1965. *Encyclopedia of Pigeon Breeds*. T. F. H. Publications, Jersey City, NJ. (A, B: photos by R. L. Kienlen, courtesy of Ralston Purina Company; C, D: photos by Stauber). 22.9A: © S. Maslowski/Visuals Unlimited. 22.9B: © Anthony Cooper/SPL/Photo Researchers, Inc. 22.11, 22.17A: David McIntyre. 22.21A: © Marilyn Kazmers/Dembinsky Photo Assoc. 22.21B: © Paul Osmond/Painet Inc. 22.22 *upper*: © Jeff Foott/ Naturepl.com. 22.22 *lower*: © franzfoto.com/Alamy.

Chapter 23 *Hummingbird*: © Tui De Roy/Minden Pictures. *Swift*: © Kim Taylor/Naturepl.com. 23.1A *left*: © Gary Meszaros/Dembinsky Photo Assoc. 23.1A *right*: © Lior Rubin/Peter Arnold, Inc. 23.1B: © Fi Rust/Painet Inc. 23.8: © Jan Vermeer/Foto Natura/Minden Pictures. 23.9A: © Virginia P. Weinland/Photo Researchers, Inc. 23.9B: © Pablo Galán Cela/AGE Fotostock. 23.10A: © J. S. Sira/ photolibrary.com. 23.10B: © Daniel L. Geiger/ SNAP/Alamy. 23.12: © Boris I. Timofeev/Pensoft. 23.14A: © Tony Tilford/photolibrary.com. 23.14B: © W. Peckover/VIREO. 23.15 *Madia*: © Peter K. Ziminsky/Visuals Unlimited. 23.15 *Argyroxiphium*: © Elizabeth N. Orians. 23.15 *Wilkesia*: © Gerald D. Carr. 23.15 *Dubautia*: © Noble Proctor/The National Audubon Society Collection/Photo Researchers, Inc.

Chapter 24 *Vaccination*: Courtesy of Chris Zahniser, B.S.N., R.N., M.P.H./CDC. *Virus*: © Dr. Tim Baker/Visuals Unlimited. 24.3 *Rice*: data from pdb 1CCR. *Tuna*: data from pdb 5CYT. 24.4: From P. B. Rainey & M. Travisano, 1998. *Nature* 394: 69. © Macmillan Publishers Ltd. 24.7A: © Barrie Britton/Naturepl.com. 24.7B: © M. Graybill/ J. Hodder/Biological Photo Service.

Chapter 25 *HIV*: © James Cavallini/Photo Researchers, Inc. *Chimpanzees*: © John Cancalosi/ Naturepl.com. 25.6A: © Mark Smith/Photo Researchers, Inc. 25.6B: After E. Verheyen et al., 2003. *Science* 300: 325. 25.7: © Larry Jon Friesen. 25.9: © Alexandra Basolo. 25.10: David M. Hillis and Matthew Brauer. 25.11A: © Helen Carr/ Biological Photo Service. 25.11B: © Michael Giannechini/Photo Researchers, Inc. 25.11C: © Skip Moody/Dembinsky Photo Assoc.

Chapter 26 *River*: © Felipe Rodriguez/Alamy. *Salmonella* and *Methanospirillum*: © Kari Lounatmaa/Photo Researchers, Inc. 26.2A: © David Phillips/Photo Researchers, Inc. 26.2B: © R. Kessel & G. Shih/Visuals Unlimited. 26.2C: Courtesy of Janice Carr/NCID/CDC. 26.4: From F. Balagaddé et al., 2005. *Science* 309: 137. Courtesy of Frederick Balagaddé. 26.5A *left*: © David M. Phillips/Visuals Unlimited. 26.5A *right*: Courtesy of Peter Hirsch and Stuart Pankratz. 26.5B *left*: Courtesy of the CDC. 26.5B *right*: Courtesy of Peter Hirsch and Stuart Pankratz. 26.6A: © J. A. Breznak & H. S. Pankratz/Biological Photo Service. 26.6B: © J. Robert Waaland/Biological Photo Service. 26.7: © USDA/Visuals Unlimited. 26.8: © Steven Haddock and Steven Miller. 26.12: Courtesy of David Cox/CDC. 26.13: Courtesy of Randall C. Cutlip. 26.14: © David Phillips/Visuals Unlimited. 26.15A: © Paul W. Johnson/Biological Photo Service. 26.15B: © H. S. Pankratz/Biological Photo Service. 26.15C: © Bill Kamin/Visuals

Unlimited. 26.16: © Dr Kari Lounatmaa/Photo Researchers, Inc. 26.17: © Dr. Gary Gaugler/Visuals Unlimited. 26.18: © Michael Gabridge/Visuals Unlimited. 26.20: © Phil Gates/Biological Photo Service. 26.21: From K. Kashefi & D. R. Lovley, 2003. *Science* 301: 934. Courtesy of Kazem Kashefi. 26.23: © Krafft/Hoa-qui/Photo Researchers, Inc. 26.24: © David Sanger Photography/Alamy. 26.25: From H. Huber et al., 2002. *Nature* 417: 63. © Macmillan Publishers Ltd. Courtesy of Karl O. Stetter.

Chapter 27 *Leishmania*: © Dennis Kunkel Microscopy, Inc. *Macrocystis*: © Karen Gowlett-Holmes/photolibrary.com. 27.1: © Steve Gschmeissner/Photo Researchers, Inc. 27.2: © David Patterson, Linda Amaral Zettler, Mike Peglar, & Tom Nerad/micro*scope. 27.3: © London School of Hygiene/SPL/Photo Researchers, Inc. 27.4A: © Bill Bachman/Photo Researchers, Inc. 27.4B: © Markus Geisen/NHMPL. 27.5: © Astrid & Hanns-Frieder Michler/SPL/Photo Researchers, Inc. 27.9: © Mike Abbey/Visuals Unlimited. 27.12A: © Wim van Egmond/Visuals Unlimited. 27.12B: © Biophoto Associates/Photo Researchers, Inc. 27.18: © Dennis Kunkel Microscopy, Inc. 27.19A: © Mike Abbey/Visuals Unlimited. 27.19B: © Dennis Kunkel Microscopy, Inc. 27.19C: © Paul W. Johnson/Biological Photo Service. 27.19D: © M. Abbey/Photo Researchers, Inc. 27.21: © Manfred Kage/Peter Arnold, Inc. 27.22: After A. Ianora et al., 2005. *Nature* 429: 403. 27.23A: © Duncan McEwan/Naturepl.com. 27.23B, C: © Larry Jon Friesen. 27.23D: © J. N. A. Lott/Biological Photo Service. 27.24: © James W. Richardson/Visuals Unlimited. 27.25A: © Larry Jon Friesen. 27.25B: © Milton Rand/Tom Stack & Assoc. 27.26A: © Carolina Biological/Visuals Unlimited. 27.26B: © Larry Jon Friesen. 27.27A: © J. Paulin/Visuals Unlimited. 27.27B: © Dr. David M. Phillips/Visuals Unlimited. 27.29: © Andrew Syred/SPL/Photo Researchers, Inc. 27.30A: © William Bourland/micro*scope. 27.30B: © David Patterson & Aimlee Laderman/micro*scope. 27.31: © Larry Jon Friesen. 27.32: Courtesy of R. Blanton and M. Grimson.

Chapter 28 *Fossils*: © Sinclair Stammers/SPL/Photo Researchers, Inc. *Rainforest*: © Photo Resource Hawaii/Alamy. 28.2A: © Ronald Dengler/Visuals Unlimited. 28.2B: © Larry Mellichamp/Visuals Unlimited. 28.3: © Brian Enting/Photo Researchers, Inc. 28.5: © J. Robert Waaland/Biological Photo Service. 28.6: After L. E. Graham et al., 2004. *PNAS* 101: 11025. Micrograph courtesy of Patricia Gensel and Linda Graham. 28.8: © U. Michigan Exhibit Museum. 28.10: Courtesy of the Biology Department Greenhouses, U. Massachusetts, Amherst. 28.11: After C. P. Osborne et al., 2004. *PNAS* 101: 10360. 28.13A, B: David McIntyre. 28.13C: © Harold Taylor/photolibrary.com. 28.14: © Daniel Vega/AGE Fotostock. 28.15: © Danilo Donadoni/AGE Fotostock. 28.16A: © Ed Reschke/Peter Arnold, Inc. 28.16B: © Carolina Biological/Visuals Unlimited. 28.17A: © J. N. A. Lott/Biological Photo Service. 28.17B: © David Sieren/Visuals Unlimited. 28.18: Courtesy of the Biology Department Greenhouses, U. Massachusetts, Amherst. 28.19 A: © Rod Planck/Dembinsky Photo Assoc. 28.19B: © Nuridsany et Perennou/Photo Researchers, Inc. 28.19C: Courtesy of the Talcott Greenhouse, Mount Holyoke College. 28.20 *inset*: David McIntyre.

Chapter 29 *Masada*: © Eddie Gerald/Alamy. *Coconut*: © Ben Osborne/Naturepl.com. 29.1 *Seed fern*: David McIntyre. 29.1 *Cycad*: © Patricio Robles Gil/Naturepl.com. 29.1 *Magnolia*: © Plantography/Alamy. 29.4: © Natural Visions/Alamy. 29.5A: © Patricio Robles Gil/Naturepl.com. 29.5B: © Dave Watts/Naturepl.com. 29.5C: © M. Graybill/J. Hodder/Biological Photo Service. 29.5D: © Frans Lanting/Minden Pictures. 29.7A *left*: David McIntyre. 29.7A *right*: © Stan W. Elems/Visuals Unlimited. 29.7B *left*: David McIntyre. 29.7B *right*: © Dr. John D. Cunningham/Visuals Unlimited. 29.10A: David McIntyre. 29.10B: © Aflo Foto Agency/Naturepl.com. 29.10C: © Richard Shiell/Dembinsky Photo Assoc. 29.11A: © Plantography/Alamy. 29.11B: © Thomas Photography LLC/Alamy. 29.15A: © Inga Spence/Tom Stack & Assoc. 29.15B: © Holt Studios/Photo Researchers, Inc. 29.15C: © Catherine M. Pringle/Biological Photo Service. 29.15D: © blickwinkel/Alamy. 29.17A: Courtesy of Stephen McCabe, U. California, Santa Cruz, and UCSC Arboretum. 29.17B: David McIntyre. 29.17C: © Rob & Ann Simpson/Visuals Unlimited. 29.17D: © R. C. Carpenter/Photo Researchers, Inc. 29.17E: © Geoff Bryant/Photo Researchers, Inc. 29.17F: © José Antonio Jiménez/AGE Fotostock. 29.18A: © Andrew E. Kalnik/Photo Researchers, Inc. 29.18B: © Ed Reschke/Peter Arnold, Inc. 29.18C: © Adam Jones/Dembinsky Photo Assoc. 29.19A: © Willard Clay/photolibrary.com. 29.19B: © Adam Jones/Dembinsky Photo Assoc. 29.19C: © Alan & Linda Detrick/The National Audubon Society Collection/Photo Researchers, Inc. 29.20: © Diaphor La Phototheque/photolibrary.com.

Chapter 30 *Fusarium*: © Dr. Gary Gaugler/Visuals Unlimited. *Ant*: © L. E. Gilbert/Biological Photo Service. 30.3: © David M. Phillips/Visuals Unlimited. 30.5A: © G. T. Cole/Biological Photo Service. 30.6: © N. Allin & G. L. Barron/Biological Photo Service. 30.7: © Richard Packwood/photolibrary.com. 30.8A: © Amy Wynn/Naturepl.com. 30.8B: © Geoff Simpson/Naturepl.com. 30.8C: © Gary Meszaros/Dembinsky Photo Assoc. 30.10A: © R. L. Peterson/Biological Photo Service. 30.10B: © Ken Wagner/Visuals Unlimited. 30.13A: © J. Robert Waaland/Biological Photo Service. 30.13B: © M. F. Brown/Visuals Unlimited. 30.13C: © Dr. John D. Cunningham/Visuals Unlimited. 30.13D: © Biophoto Associates/Photo Researchers, Inc. 30.14 *left*: © William E. Schadel/Biological Photo Service. 30.14 *right*: © Dr. John D. Cunningham/Visuals Unlimited. 30.15: © John Taylor/Visuals Unlimited. 30.16: © G. L. Barron/Biological Photo Service. 30.17A: © Richard Shiell/Dembinsky Photo Assoc. 30.17B: © Matt Meadows/Peter Arnold, Inc. 30.18: © Andrew Syred/SPL/Photo Researchers, Inc. 30.19A: © Manfred Danegger/Photo Researchers, Inc. 30.19B: © Botanica/photolibrary.com.

Chapter 31 *Fossil*: Photograph by Diane Scott. *Dinosaur*: © Joe Tucciarone/Photo Researchers, Inc. 31.2A: Courtesy of J. B. Morrill. 31.2B: From G. N. Cherr et al., 1992. *Microsc. Res. Tech.* 22: 11. Courtesy of J. B. Morrill. 31.5A: © Jurgen Freund/Naturepl.com. 31.5B: © Henry W. Robison/Visuals Unlimited. 31.6A: © Brian Parker/Tom Stack & Assoc. 31.6B: © Tui De Roy/Minden Pictures. 31.8: © Colin M. Orians. 31.9A: © T. Kitchin & V. Hurst/Photo Researchers, Inc. 31.9B: © Stockbyte/PictureQuest. 31.10: Adapted from F. M. Bayerand & H. B. Owre, 1968. *The Free-Living Lower Invertebrates*, Macmillan Publishing Co. 31.11A: © Scott Camazine/Alamy. 31.11B, C: David McIntyre. 31.13A: © Stephen Dalton/Minden Pictures. 31.13B: © Gerald & Buff Corsi/Visuals Unlimited. 31.14A: © Dave Watts/Naturepl.com. 31.14B: © Larry Jon Friesen. 31.16A: © Jurgen Freund/Naturepl.com. 31.16B: © Larry Jon Friesen. 31.16C: © David Wrobel/Visuals Unlimited. 31.17B: © Larry Jon Friesen. 31.18: Adapted from F. M. Bayerand & H. B. Owre, 1968. *The Free-Living Lower Invertebrates*, Macmillan Publishing Co. 31.19A: © Larry Jon Friesen. 31.19B: © Michael Patrick O'Neill/Alamy. 31.19C, D: © Larry Jon Friesen. 31.20A: From J. E. N. Veron, 2000. *Corals of the World*. © J. E. N. Veron. 31.20B: © Jurgen Freund/Naturepl.com. 31.21: Adapted from F. M. Bayerand & H. B. Owre, 1968. *The Free-Living Lower Invertebrates*, Macmillan Publishing Co.

Chapter 32 *Strepsipterans*: Courtesy of Dr. Hans Pohl. *Wasp*: © Paulo de Oliveira/OSF/Jupiter Images. 32.2: © Robert Brons/Biological Photo Service. 32.3A: From D. C. García-Bellido & D. H. Collins, 2004. *Nature* 429: 40. Courtesy of Diego García-Bellido Capdevila. 32.3B: © Piotr Naskrecki/Minden Pictures. 32.6: © Alexis Rosenfeld/Photo Researchers, Inc. 32.7A: © Ed Robinson/photolibrary.com. 32.8B: © Robert Brons/Biological Photo Service. 32.9B: © Larry Jon Friesen. 32.10A: © Stan Elems/Visuals Unlimited. 32.11: © David J. Wrobel/Visuals Unlimited. 32.13A: © Jurgen Freund/Naturepl.com. 32.13B: Courtesy of R. R. Hessler, Scripps Institute of Oceanography. 32.13C: © Roger K. Burnard/Biological Photo Service. 32.13D: © Larry Jon Friesen. 32.15A: © Larry Jon Friesen. 32.15B: © Dave Fleetham/photolibrary.com. 32.15C, D, E: © Larry Jon Friesen. 32.15F: © Orion Press/Jupiter Images. 32.16A: Courtesy of Jen Grenier and Sean Carroll, U. Wisconsin. 32.16B: Courtesy of Graham Budd. 32.16C: Courtesy of Reinhardt Møbjerg Kristensen. 32.17: © R. Calentine/Visuals Unlimited. 32.18B, C: © James Solliday/Biological Photo Service. 32.20A: © Michael Fogden/photolibrary.com. 32.20B: © Diane R. Nelson/Visuals Unlimited. 32.21: © Ken Lucas/Visuals Unlimited. 32.22A: © Frans Lanting/Minden Pictures. 32.22B: © Larry Jon Friesen. 32.22C: © Tom Branch/Photo Researchers, Inc. 32.22D: © Norbert Wu/Minden Pictures. 32.25: © Mark Moffett/Minden Pictures. 32.26: © Oxford Scientific Films/Jupiter Images. 32.27A: © Larry Jon Friesen. 32.27B: © Larry Jon Friesen. 32.27C: © Meul/ARCO/Naturepl.com. 32.27D: © Piotr Naskrecki/Minden Pictures. 32.27E: © Colin Milkins/Oxford Scientific Films/photolibrary.com. 32.27F: © Peter J. Bryant/Biological Photo Service. 32.27G: © Larry Jon Friesen. 32.27H: David McIntyre. 32.29A: © John Mitchell/Jupiter Images. 32.29B: © John R. MacGregor/Peter Arnold, Inc. 32.30A: © Norbert Wu/Minden Pictures. 32.30B: © Frans Lanting/Minden Pictures. 32.31A: © Kelly Swift, www.swiftinverts.com. 32.31B: © Larry Jon Friesen. 32.31C: © W. M. Beatty/Visuals Unlimited. 32.31D: Photo by Eric Erbe; colorization by Chris Pooley/USDA ARS.

Chapter 33 *Homo floresiensis*: © Christian Darkin/Alamy. *Tunicates*: © Larry Jon Friesen. 33.2: From S. Bengtson, 2000. Teasing fossils out of shales with cameras and computers. *Palaeontologia Electronica* 3(1). 33.4A: © Larry Jon Friesen. 33.4B: © Hal Beral/Visuals Unlimited. 33.4C: © Randy Morse/Tom Stack & Assoc. 33.4D: © Mark J. Thomas/Dembinsky Photo Assoc. 33.4E: © Peter Scoones/Photo Researchers, Inc. 33.4F: © Larry Jon Friesen. 33.5A: © C. R. Wyttenbach/Biological Photo Service. 33.6: © Robert Brons/Biological Photo Service. 33.7A: © Jurgen Freund/Naturepl.com. 33.7B: © David Wrobel/Visuals Unlimited. 33.9A: © Brian Parker/Tom Stack &

Assoc. 33.9B: © Reijo Juurinen/Naturbild/Naturepl.com. 33.11B: © Roger Klocek/Visuals Unlimited. 33.12A: © Dave Fleetham/Tom Stack & Assoc. 33.12B: © Kelvin Aitken/AGE Fotostock. 33.12C: © Dave Fleetham/Tom Stack & Assoc. 33.13A: © Tobias Bernhard/Oxford Scientific Films/photolibrary.com. 33.13B: © Larry Jon Friesen. 33.13C: © Dave Fleetham/Visuals Unlimited. 33.13D: © Larry Jon Friesen. 33.14A, B: © Tom McHugh, Steinhart Aquarium/The National Audubon Society Collection/Photo Researchers, Inc. 33.14C: © Ted Daeschler/Academy of Natural Sciences/VIREO. 33.16A: © Ken Lucas/Biological Photo Service. 33.16B: © Michael & Patricia Fogden/Minden Pictures. 33.16C: © Gary Meszaros/Dembinsky Photo Assoc. 33.16D: © Larry Jon Friesen. 33.19A: © Dave B. Fleetham/Tom Stack & Assoc. 33.19B: © C. Alan Morgan/Peter Arnold, Inc. 33.19C: © Dave Watts/Alamy. 33.19D: © Larry Jon Friesen. 33.20A: © Frans Lanting/Minden Pictures. 33.20B: © Gerry Ellis, DigitalVision/PictureQuest. 33.21A: From X. Xu et al., 2003. *Nature* 421: 335. © Macmillan Publishers Ltd. 33.21B: © Tom & Therisa Stack/Painet, Inc. 33.23A: © Roger Wilmhurst/Foto Natura/Minden Pictures. 33.23B: © Andrew D. Sinauer. 33.23C: © Tom Vezo/Minden Pictures. 33.23D: © Skip Moody/Dembinsky Photo Assoc. 33.24A: © Ed Kanze/Dembinsky Photo Assoc. 33.24B: © Dave Watts/Tom Stack & Assoc. 33.25A: © Ingo Arndt/Naturepl.com. 33.25B: © JTB Photo Communications/photolibrary.com. 33.25C: © Jany Sauvanet/Photo Researchers, Inc. 33.26A: © Tim Jackson/OSF/Jupiter Images. 33.26B: © Claude Steelman/OSF/Jupiter Images. 33.26C: © Michael S. Nolan/Tom Stack & Assoc. 33.26D: © Erwin & Peggy Bauer/Tom Stack & Assoc. 33.28: © Pete Oxford/Minden Pictures. 33.29A: © mike lane/Alamy. 33.29B: © Cyril Ruoso/JH Editorial/Minden Pictures. 33.30A: © Frans Lanting/Minden Pictures. 33.30B: © Anup Shah/Dembinsky Photo Assoc. 33.30C: © Stan Osolinsky/Dembinsky Photo Assoc. 33.30D: © Anup Shah/Dembinsky Photo Assoc.

Chapter 34 *Poster*: Courtesy of NJSP Museum. *Wood*: © Werner H. Muller/Peter Arnold, Inc. 34.3: David McIntyre. 34.4A: © Joyce Photographics/Photo Researchers, Inc. 34.4B: © C. K. Lorenz/The National Audubon Society Collection/Photo Researchers, Inc. 34.4C: © Renee Lynn/Photo Researchers, Inc. 34.8: © Biophoto Associates/Photo Researchers, Inc. 34.9A: © Biodisc/Visuals Unlimited. 34.9B: © P. Gates/Biological Photo Service. 34.9C: © Biophoto Associates/Photo Researchers, Inc. 34.9D: © Jack M. Bostrack/Visuals Unlimited. 34.9E: © John D. Cunningham/Visuals Unlimited. 34.9F: © J. Robert Waaland/Biological Photo Service. 34.9G: © Randy Moore/Visuals Unlimited. 34.10: © R. Kessel & G. Shih/Visuals Unlimited. 34.11 *upper*: © Larry Jon Friesen. 34.11 *lower*: © Biodisc/Visuals Unlimited. 34.13: © D. Cavagnaro/Visuals Unlimited. 34.14B: © Microfield Scientific LTD/Photo Researchers, Inc. 34.16A: © Larry Jon Friesen. 34.16B: © Ed Reschke/Peter Arnold, Inc. 34.16C: © Dr. James W. Richardson/Visuals Unlimited. 34.17A *left*: © Ed Reschke/Peter Arnold, Inc. 34.17A *right*: © Biodisc/Visuals Unlimited. 34.17B: © Ed Reschke/Peter Arnold, Inc. 34.19: © John N. A. Lott/Biological Photo Service. 34.20: © Jim Solliday/Biological Photo Service. 34.21: From C. Maton & B. L. Gartner, 2005. *Am. J. Botany* 92: 123. Courtesy of Barbara L. Gartner. 34.22: © Biodisc/Visuals Unlimited. 34.23B: Courtesy of Thomas Eisner, Cornell U. 34.23C: © Larry Jon Friesen.

Chapter 35 *Sunflower*: © Aflo/Naturepl.com. *Sequoia*: © Jeff Foott/Naturepl.com. 35.5: David McIntyre. 35.8: After M. A. Zwieniecki, P. J. Melcher, & N. M. Holbrook, 2001. *Science* 291: 1059. 35.9A: © David M. Phillips/Visuals Unlimited. 35.10: After G. D. Humble & K. Raschke, 1971. *Plant Physiology* 48: 447. 35.12: © M. H. Zimmermann.

Chapter 36 *Family*: Dorothea Lange/Library of Congress, Prints & Photographs Division, FSA/OWI Collection. *Plow*: © Liz Hahn. 36.1: David McIntyre. 36.5: © Kathleen Blanchard/Visuals Unlimited. 36.7: © Hugh Spencer/Photo Researchers, Inc. 36.9 *left*: © E. H. Newcomb & S. R. Tandon/Biological Photo Service. 36.9 *right*: © Dr. Jeremy Burgess/SPL/Photo Researchers, Inc. 36.11A: © J. H. Robinson/The National Audubon Society Collection/Photo Researchers, Inc. 36.11B: © Kim Taylor/Naturepl.com. 36.12: Courtesy of Susan and Edwin McGlew.

Chapter 37 *Rubber ball*: © Luiz Claudio Marigo/Naturepl.com. *Tapping tree*: © J. D. Dallet/AGE Fotostock. 37.2: © Tom J. Ulrich/Visuals Unlimited. 37.3: © Joe Sohm/Pan America/Jupiter Images. 37.5: Courtesy of J. A. D. Zeevaart, Michigan State U. 37.11: © Ed Reschke/Peter Arnold, Inc. 37.13: © Biophoto Associates/Photo Researchers, Inc. 37.17, 37.20: David McIntyre. 37.21: Courtesy of Dr. Eva Huala, Carnegie Institution of Washington.

Chapter 38 *Rafflesia*: © Ingo Arndt/Minden Pictures. *Amorphophallus*: © Deni Bown/OSF/Jupiter Images. 38.2: © RMF/Visuals Unlimited. 38.4: From J. Bowman (ed.), 1994. *Arabiopsis: An Atlas of Morphology and Development.* Springer-Verlag, New York. Photo by S. Craig & A. Chaudhury, Plate 6.2. 38.8: David McIntyre. 38.14: After A. N. Dodd et al., 2005. *Science* 309: 630. 38.15: After M. J. Yanovsky & S. A. Kay, 2002. *Nature* 419: 308. 38.17: © Gerry Ellis/Minden Pictures. 38.18A: © Nigel Cattlin, Holt Studios International/Photo Researchers, Inc. 38.18B: © Maurice Nimmo/SPL/Photo Researchers, Inc.

Chapter 39 *Flood*: © Abir Abdullah/Peter Arnold, Inc. *Salty soil*: © STRDEL/AFP/Getty Images. 39.2: © Holt Studios International Ltd/Alamy. 39.5: After A. Steppuhn et al., 2004. *PLoS Biology* 2: 1074. 39.6: After S. Rasmann et al., 2005. *Nature* 434: 732. 39.8: Courtesy of Thomas Eisner, Cornell U. 39.9: © Adam Jones/Dembinsky Photo Assoc. 39.10: © John N. A. Lott/Biological Photo Service. 39.11 *left*: David McIntyre. 39.11 *right*: © Frans Lanting/Minden Pictures. 39.12: © Simon Fraser/SPL/Photo Researchers, Inc. 39.13: © imagebroker/Alamy. 39.14: © John N. A. Lott/Biological Photo Service. 39.15: © Jurgen Freund/Naturepl.com. 39.17: © Budd Titlow/Visuals Unlimited.

Chapter 40 *Radcliffe*: © Nick Laham/Getty Images. *Soldier*: Courtesy of Major Bryan "Scott" Robison. 40.3A: © Gladden Willis/Visuals Unlimited. 40.3B: © Ed Reschke/Peter Arnold, Inc. 40.4: © Manfred Kage/Peter Arnold, Inc. 40.5: © Ed Reschke/Peter Arnold, Inc. 40.6A: © James Cavallini/Photo Researchers, Inc. 40.6B: © Innerspace Imaging/SPL/Photo Researchers, Inc. 40.10B: © Frans Lanting/Minden Pictures. 40.12: © Gerry Ellis/PictureQuest/Jupiter Images. 40.14: Courtesy of Anton Stabentheiner. 40.17: © Gladden Willis/Visuals Unlimited. 40.18A: © Robert Shantz/Alamy. 40.18B: © Jim Brandenburg/Minden Pictures.

Chapter 41 *Bicyclists*: © Peter Dejong/AP Images. *Body builder*: © Tiziana Fabi/AFP/Getty Images. 41.9A: © Ed Reschke/Peter Arnold, Inc. 41.9C: © Scott Camazine/Photo Researchers, Inc. 41.14: Courtesy of Gerhard Heldmaier, Philipps U.

Chapter 42 *Queen bee*: David McIntyre. *Swarm*: © Stephen Dalton/Minden Pictures. 42.1A: © Biophoto Associates/Photo Researchers, Inc. 42.1B: © Constantinos Petrinos/Naturepl.com. 42.2A: © Patricia J. Wynne. 42.6: © SIU/Peter Arnold, Inc. 42.7: © ullstein - Ibis Bildagentur / Peter Arnold, Inc. 42.8A: © Mitsuaki Iwago/Minden Pictures. 42.8B: © Dave Watts/Naturepl.com. 42.10B: © Ed Reschke/Peter Arnold, Inc. 42.13: © P. Bagavandoss/Photo Researchers, Inc. 42.16B: © S. I. U. School of Med./Photo Researchers, Inc. 42.18: Courtesy of The Institute for Reproductive Medicine and Science of Saint Barnabas, New Jersey.

Chapter 43 *Whale*: © François Gohier/Photo Researchers, Inc. *Egg and sperm*: © Dr. David M. Phillips/Visuals Unlimited. 43.1: Courtesy of Richard Elinson, U. Toronto. 43.4: From J. G. Mulnard, 1967. *Arch. Biol. (Liege)* 78: 107. Courtesy of J. G. Mulnard. 43.20A: © CNRI/SPL/Photo Researchers, Inc. 43.20B: © Dr. G. Moscoso/SPL/Photo Researchers, Inc. 43.20C: © Tissuepix/SPL/Photo Researchers, Inc. 43.20D: © Petit Format/Photo Researchers, Inc.

Chapter 44 *Movie still*: © New Line Cinema/Courtesy of the Everett Collection. *Brain scan*: Courtesy of Dr. Kevin LaBar, Duke U. 44.4B: © C. Raines/Visuals Unlimited. 44.7: From A. L. Hodgkin & R. D. Keynes, 1956. *J. Physiol.* 148: 127.

Chapter 45 *Snake*: © James Gerholdt/Peter Arnold, Inc. *Bat*: © Stephen Dalton/Photo Researchers, Inc. 45.3A: © Hans Pfletschinger/Peter Arnold, Inc. 45.3B: David McIntyre. 45.10A: © P. Motta/Photo Researchers, Inc. 45.16A: © Dennis Kunkel Microscopy, Inc. 45.19: © Omikron/Science Source/Photo Researchers, Inc.

Chapter 46 *Neuron*: From H. van Praag et al., 2002. *Nature* 415: 1030. © Macmillan Publishers Ltd. *Meadowlark*: © Werner Bollmann/AGE Fotostock. 46.8: From J. M. Harlow, 1869. *Recovery from the passage of an iron bar through the head.* Boston: David Clapp & Son. 46.14A: © David Joel Photography, Inc. 46.16: © Wellcome Dept. of Cognitive Neurology/SPL/Photo Researchers, Inc.

Chapter 47 *Joyner-Kersee*: © AFP/Getty Images. *Frog*: © Michael Durham/Minden Pictures. 47.1: © Frank A. Pepe/Biological Photo Service. 47.2: © Tom Deerinck/Visuals Unlimited. 47.4: © Kent Wood/Peter Arnold, Inc. 47.7A: © Manfred Kage/Peter Arnold, Inc. 47.7B: © Innerspace Imaging/SPL/Photo Researchers, Inc. 47.7C: © SPL/Photo Researchers, Inc. 47.10: Courtesy of Jesper L. Andersen. 47.14: From the collection of Andrew Sinauer/photo by David McIntyre. 47.18: © Robert Brons/Biological Photo Service. 47.22B *upper*: © Ken Lucas/Visuals Unlimited. 47.22B *lower*: © Fred McConnaughey/The National Audubon Society Collection/Photo Researchers, Inc.

Chapter 48 *Geese*: © Konrad Wothe/Minden Pictures (geese) and John Warden/PictureQuest (mountains). *Barracudas*: © Fred Bavendam/Minden Pictures. 48.1A: © Larry Jon Friesen. 48.1B: © Norbert Wu/Minden Pictures. 48.1C: © Tom McHugh/Photo Researchers, Inc. 48.4B: © Skip Moody/Dembinsky Photo Assoc. 48.4C:

Courtesy of Thomas Eisner, Cornell U. 48.10A: © SPL/Photo Researchers, Inc. 48.10C: © P. Motta/Photo Researchers, Inc. 48.13: © Berndt Fischer/Jupiter Images. 48.17: After C. R. Bainton, 1972. *J. Appl. Physiol.* 33: 775.

Chapter 49 *Lewis*: © NBAE/Getty Images. *Emergency room*: © BananaStock/Alamy. 49.8A: © Brand X Pictures/Alamy. 49.9: After N. Campbell, 1990. *Biology*, 2nd Ed., Benjamin Cummings. 49.10B: © CNRI/Photo Researchers, Inc. 49.12: © Ed Reschke/Peter Arnold, Inc. 49.15A: © Chuck Brown/Science Source/Photo Researchers, Inc. 49.15B: © Biophoto Associates/Science Source/Photo Researchers, Inc. 49.19: © Doc White/Nature Picture Library.

Chapter 50 *Pima*: © Marilyn "Angel" Wynn/Nativestock.com. *Fast food*: © Matt Bowman/FoodPix/Jupiter Images. 50.1A: © Gerry Ellis, DigitalVision/PictureQuest. 50.1B: © Rinie Van Meurs/Foto Natura/Minden Pictures. 50.3: © AP/Wide World Photos. 50.6: © Ace Stock Limited/Alamy. 50.9: © Dennis Kunkel Microscopy, Inc. 50.14: © Eye of Science/SPL/Photo Researchers, Inc. 50.20: © Science VU/Jackson/Visuals Unlimited.

Chapter 51 *Bat*: © Michael & Patricia Fogden/Minden Pictures. *Kangaroo rat*: © Mary McDonald/Naturepl.com. 51.1 *inset*: © Kim Taylor/Naturepl.com. 51.2B: © Rod Planck/Photo Researchers, Inc. 51.8: From R. G. Kessel & R. H. Kardon, 1979. *Tissues and Organs.* W. H. Freeman, San Francisco. 51.11: From L. Bankir & C. de Rouffignac, 1985. *Am. J. Physiol.* 249: R643-R666. Courtesy of Lise Bankir, INSERM Unit, Hôpital Necker, Paris. 51.13: © Hank Morgan/Photo Researchers, Inc.

Chapter 52 *Katrina*: Courtesy of the NOAA Satellite and Information Service. *Swamp*: © Tim Fitzharris/Minden Pictures. 52.1: © Tim Fitzharris/Minden Pictures. *Tundra, upper*: © Tim Acker/Auscape/Minden Pictures. *Tundra, lower*: © Elizabeth N. Orians. *Boreal, upper*: © Carr Clifton/Minden Pictures. *Boreal, lower*: © Robert Harding Picture Library Ltd/Alamy. *Temperate deciduous*: © Paul W. Johnson/Biological Photo Service. *Temperate grasslands, upper*: © Robert & Jean Pollock/Biological Photo Service. *Temperate grasslands, lower*: © Elizabeth N. Orians. *Cold desert, upper*: © Edward Ely/Biological Photo Service. *Cold desert, lower*: © Robert Harding Picture Library Ltd./photolibrary.com. *Hot desert, left*: © Terry Donnelly/Tom Stack & Assoc. *Hot desert, right*: © Dave Watts/Tom Stack & Assoc. *Chaparral, left*: © Elizabeth N. Orians. *Chaparral, right*: © Larry Jon Friesen. *Thorn forest*: © Frans Lanting/Minden Pictures. *Savanna*: © Nigel Dennis/AGE Fotostock. *Tropical deciduous*: Courtesy of Donald L. Stone. *Tropical evergreen*: © Elizabeth N. Orians. 52.6: © Elizabeth N. Orians. 52.9 *Chameleon*: © Pete Oxford/Naturepl.com. 52.9 *Tenrec*: © Nigel J. Dennis/Photo Researchers, Inc. 52.9 *Fossa*: © Frans Lanting/Minden Pictures. 52.9 *Lemur*: © Wendy Dennis/Dembinsky Photo Associates. 52.9 *Baobab*: © Pete Oxford/Naturepl.com. 52.9 *Pachypodium*: © Michael Leach/Oxford Scientific Films/photolibrary.com. 52.12: Courtesy of E. O. Wilson. 52.13A *left*: © Bill Lea/Dembinsky Photo Associates. 52.13A *right*: © Stephen J. Krasemann/Photo Researchers, Inc. 52.13B *left*: © Frans Lanting/Minden Pictures. 52.13B *center*: © Kenneth W. Fink/Photo Researchers, Inc. 52.13B *right*: © Chris Gomersall/Naturepl.com.

Chapter 53 *Macaques*: © Frans de Waal, Emory U. *Spider*: © Mark Gibson/Jupiter Images. 53.2A: From J. R. Brown et al., 1996. *Cell* 86: 297. Courtesy of Michael Greenberg. 53.4A: © Nina Leen/Time Life Pictures/Getty Images. 53.4B: © Frans Lanting/Minden Pictures. 53.9A: © Anup Shah/Naturepl.com. 53.9B: © Mitsuaki Iwago/Minden Pictures. 53.9C: © Konrad Wothe/Minden Pictures. 53.11: © Frans Lanting/Minden Pictures. 53.13: Courtesy of John Alcock, Arizona State U. 53.16: © Tui De Roy/Minden Pictures. 53.18: © Cyril Ruoso/JH Editorial/Minden Pictures. 53.21: © Nigel Dennis/AGE Fotostock. 53.22: © Piotr Naskrecki/Minden Pictures. 53.23: © Steve & Dave Maslowski/Photo Researchers, Inc. 53.24: Courtesy of John Alcock, Arizona State U. 53.25: © José Fuste Raga/AGE Fotostock.

Chapter 54 *Caterpillars*: Courtesy of John R. Hosking, NSW Agriculture, Australia. *Farmer*: © Thomas Shjarback/Alamy. 54.1A: © Flip Nicklin/Minden Pictures. 54.1B: © PhotoStockFile/Alamy. 54.1C: © Robert McGouey/Alamy. 54.5A: © Frans Lanting/Minden Pictures. 54.5B: © David Nunuk/SPL/Photo Researchers, Inc. 54.6A: © Michael Durham/Minden Pictures. 54.6B: Courtesy of Colin Chapman. 54.7: © Larry Jon Friesen. 54.11: After P. A. Marquet, 2000. *Science* 289: 1487. 54.12: © Ed Reschke/Peter Arnold, Inc. 54.13: © Adam Jones/Photo Researchers, Inc. 54.14: © T. W. Davies/California Academy of Sciences. 54.18: © Kathie Atkinson/OSF/Jupiter Images.

Chapter 55 *Ant on spine*: © Piotr Naskrecki/Minden Pictures. *Ant with Beltian bodies*: © Mark Moffett/Minden Pictures. 55.1: After R. H. Whittaker, 1960. *Ecological Monographs* 30: 279. 55.4: © Alan & Sandy Carey/Photo Researchers, Inc. 55.6: © Dave Watts/Naturepl.com. 55.7: © Lawrence E. Gilbert/Biological Photo Service. 55.8: © Shehzad Noorani/Peter Arnold, Inc. 55.11: © Mitsuaki Iwago/Minden Pictures. 55.12A: © Kim Taylor/Naturepl.com. 55.12B: © Perennou Nuridsany/Photo Researchers, Inc. 55.13D: Courtesy of William W. Dunmire/National Park Service. 55.14A: David McIntyre. 55.15: © Jim Zipp/Photo Researchers, Inc. 55.16: Courtesy of Jim Peaco/National Park Service. 55.18: After M. Begon, J. Harper, & C. Townsend, 1986. *Ecology*. Blackwell Scientific Publications.

Chapter 56 *Crane*: © Mark Moffett/Minden Pictures. *Researchers*: Courtesy of Christian Koerner, U. Basel. 56.1: After M. C. Jacobson et al., 2000. *Introduction to Earth System Science.* Academic Press. 56.9A: © Corbis Images/PictureQuest.

Chapter 57 *Flight*: © Tom Hugh-Jones/Naturepl.com. *Crane suit*: © Mark Payne-Gill/Naturepl.com. 57.1: © Michael Long/NHMPL. 57.2B: © Mark Godfrey/The Nature Conservancy. 57.6: Richard Bierregaard, Courtesy of the Smithsonian Institution, Office of Environmental Awareness. 57.7: © Mr_Jamsey/iStockphoto.com. 57.8A: © Terry Whittaker/Alamy. 57.8B: © Paul Johnson/Naturepl.com. 57.9: © Michael & Patricia Fogden/Minden Pictures. 57.10A: After R. B. Aronson et al., 2000. *Nature* 405: 36. 57.10B: © Fred Bavendam/Minden Pictures. 57.12: Courtesy of WWF-US, GIS Map provided by J. Morrison. 57.14B: © Jim Brandenburg/Minden Pictures. 57.14C: Courtesy of Jesse Achtenberg/U.S. Fish and Wildlife Service. 57.15A: Courtesy of Christopher Baisan and the Laboratory of Tree-Ring Research, U. Arizona, Tucson. 57.15B: © Karen Wattenmaker/Painet, Inc. 57.17: After S. H. Reichard & C. W. Hamilton, 1997. *Conservation Biology* 11: 193. 57.18A: © Elizabeth N. Orians. 57.20: © Tom Vezo/Nature Picture Library.

Index

Numbers in **boldface** indicate a definition of the term. Numbers in *italic* indicate the information will be found in a figure, caption, or table.

- A -

- B -

- D -

- E -

- F -

- G -

- I -

- J -

- K -

- L -

- N -

- Q -

- R -

- U -

- X -

- Y -

- Z -

Experiment and Research Method figures in LIFE

Experiment and Research Method figures in LIFE